STRUCTURAL GEOLOGY
OF ROCKS AND REGIONS

STRUCTURAL GEOLOGY
OF ROCKS AND REGIONS

GEORGE H. DAVIS
The University of Arizona, Tucson

STEPHEN J. REYNOLDS
Arizona State University, Tempe

CHARLES F. KLUTH
Colorado School of Mines Geology and
Geological Engineering, Colorado

JOHN WILEY & SONS, INC.

VP & EXECUTIVE PUBLISHER Jay O'Callaghan
EXECUTIVE EDITOR Ryan Flahive
EDITORIAL ASSISTANT Darnell Sessoms
MARKETING MANAGER Margaret Barrett
PRODUCTION MANAGER Janis Soo
SENIOR PRODUCTION EDITOR Joyce Poh

Cover art by David A. Fischer
Cover designed by George H. Davis

This book was set in 10/12 points Times Roman by MPS Limited, a Macmillan Company, Chennai and printed and bound by Courier Westford. The cover was printed by Courier Westford.

This book is printed on acid free paper.

Founded in 1807, John Wiley & Sons, Inc. has been a valued source of knowledge and understanding for more than 200 years, helping people around the world meet their needs and fulfill their aspirations. Our company is built on a foundation of principles that include responsibility to the communities we serve and where we live and work. In 2008, we launched a Corporate Citizenship Initiative, a global effort to address the environmental, social, economic, and ethical challenges we face in our business. Among the issues we are addressing are carbon impact, paper specifications and procurement, ethical conduct within our business and among our vendors, and community and charitable support. For more information, please visit our website: www.wiley.com/go/citizenship.

Library of Congress Cataloging-in-Publication Data:

Davis, George H. (George Herbert), 1942-
Structural geology of rocks and regions / George H. Davis, Stephen J. Reynolds, Charles F. Kluth. — 3rd ed.
p. cm.
Includes bibliographical references and index.
ISBN 978-0-471-15231-6 (hardback : acid free paper)
1. Geology, Structural. I. Reynolds, Stephen J. II. Kluth, Charles F. III. Title.
QE601.D3 2011
551.8—dc23 2011025716

Printed in the United States of America

10 9 8 7 6 5 4 3 2

Preface

High desert near Sheets Gulch, Capitol Reef National Park, Utah. Geologists Steve Ahlgren, Karen Swanberg, Pilar García, and Alex Bump are taking stock of how the geologic structures mapped on individual sheets are matching up and fitting together. The carryall at left shields them and their aerial-photo collage from the intensity of the late afternoon July sun. The field area is amongst the peaks of Navajo Sandstone (Jurassic) in the background. Fundamental to structural geology is mapping and fieldwork. More fundamental yet is thinking hard about what we observe. Unlike what we see at this particular moment in time, we never want to put hard boundaries on the areas and problems we are addressing. [Photograph by G. H. Davis.]

Excellence and learning are not commodities to be bought at the corner store. Rather they dwell among rocks hardly accessible, and we must almost wear our hearts out in search of them.

[From *College Talks* by H. F. Lowry, edited by J. R. Blackwood, p. 116. Published with permission of Oxford University Press, New York, copyright © 1969.]

INTENTIONS

Our purpose in writing this textbook is to communicate the physical and geometric elegance of geologic structures within the Earth's crust and to describe the ways in which they reflect the nature and origin of crustal deformation through time. Geologic structures provide part of the basis for recognizing and reconstructing the profound changes that have marked the physical evolution of the Earth's outermost layers, as observed from the scale of the plates down to the scale of the microscopic. Understanding the nature and extensiveness of geologic structures of deformational origin in the Earth's crust has both scientific and practical benefit. But, there is a philosophical value as well. Our perceptions of who we are and where we are in time and space are shaped by facts and interpretations regarding the historical development of the crust of the planet on which we live. Knowing fully the extent to which our planet is dynamic, not static, is a reminder of the lively and special environment we inhabit.

For example, we have come to understand that our earthly foundations are not fixed. Instead, we live on continent-size plates that are in a continual state of slow motion. The interaction of these plates has played a dominant role in both the formation and deformation of rock bodies in the Earth's crust. Knowledge of present-day plate tectonic processes aids us in interpreting past structural movements. Furthermore, evaluating and interpreting past dynamics may help us predict what the present actions hold for the future.

Structural geology is an essential tool in unraveling the geologic history of any given area or region within the Earth, especially in mountain belts where deformational movements profoundly modify original arrangements of rocks and geologic contacts. The 17th century anatomist, Nicolaus Steno, discerned that the structure of each layer of the earth told us something important about the history of the earth. Steno's fascination with the earth, and his creation of fundamental principles of structural geology, began with a life-changing trip across the Alps and Apennines, where he "*had the chance to see with his own eyes layers of rock packed with fossilized shells, [and] strata raised and contorted into mountains, . . . He would become an anatomist of the world*" (Cutler, 2003, p. 44). In fact, "*Steno was the first to assert* [that] *history might be recoverable from the rocks and* [took] *it upon himself to unravel that history. And he did so . . . based on elementary geometry. . . . above versus below, continuous versus discontinuous, tilted versus horizontal, enclosed versus enclosing*" (Cutler 2003, p. 106).

When viewed at geological scales of time and spatial dimension, rocks must be regarded as materials almost without strength, capable of being deformed continuously even under the slightest of pressures. The geologic mapping of mountain belts has shown this to be true time and time again. The understanding of this paradox, which we try to capture in the expression "*soft as a rock*," yields a transformed view of the strength and behavior of rocks and regions.

Practical applications of structural geology are broad ranging and powerful, traditionally in the fields of petroleum geology, exploration geology, mining engineering, and civil engineering; and increasingly in the fields of environmental geology, hydrology, earthquake engineering, and geothermal energy development. The practical value of structural geology derives largely from the fact that movement and trapping of fluids in the earth are strongly influenced by fracturing, faulting, and folding. Thus, a firm understanding of principles of structural geology is beneficial practically.

This 3rd Edition of *Structural Geology of Rocks and Regions* benefits from a 3rd author, Charles F. Kluth, former coordinator and principle lecturer of the Chevron and Chevron-Texaco Structural Geology Schools. Chuck's range of experience brings to this textbook a useful scale-dimension of thinking about vast tracts of structural geology in the subsurface and in mountain belts, such as the Canadian Rockies. To be sure, we want this book to be helpful to those whose careers involve exploration for and recovery of oil and gas, investigating for the sources of contaminants in a part of the groundwater system, evaluating the energy potential of a natural geothermal system, appraising daring but dangerous proposals for the subterranean disposal of radioactive waste, examining basin sediments as potential groundwater reservoirs, assessing the fundamental stability of steep slopes underlain by rain-infiltrated fractures in the bedrock, or planning the excavation of mines or tunnels. Because most ore deposits owe their existence to the movement of mineral-bearing hydrothermal solutions through fractured bedrock, structural geology is a basic tool in the exploration for metals, and in the development of an ore body. Finally, the understanding of active tectonic events like earthquakes, volcanic eruptions, and massive landsliding, are intellectually demanding activities requiring knowledge derived from intimacy with the principals of fracturing and faulting.

ARRANGEMENT

We have arranged this text in three parts. *Fundamentals* (*Part I*) provides essential background for analyzing *Structures* (*Part II*). *Descriptive Analysis* (*Part III*) describes how to function in the field, and what to do with the data that are collected.

Nature of Structural Geology (*Chapter 1*) introduces the beauty, the challenging geometry, and the practical value of structural geology. It also presents the basic approach to structural geology as used throughout the book. No matter what the structure, we like to start with comprehensive descriptive treatment, i.e., describing the physical and geometric characteristics of the geologic structures and deformed rocks. Then we evaluate strain, and displacement, describing deformation and interpreting the deformation paths by which changes in location, orientation, shape, and size of rocks and regions were achieved. Next is dynamic, i.e., interpreting the forces and stresses that caused deformation, appraising the mechanical strength and behavior of the rock at the time of deformation, and applying (plate) tectonic concepts to deduce the narrative of origin. This approach, which we sometimes refer to as **detailed structural analysis**, provides the leverage for unraveling and interpreting structures and systems of structures at any scale, from rocks to regions.

Displacement and Strain (*Chapter 2*) presents the means for describing deformation, documenting changes in location, orientation, shape, and size of rocks or regions of rocks during the course of deformation, and interpreting deformation paths. *The very presence of structures and systems of structures in rocks in the Earth's crust reflects translation (change in the location), and/or rotation (change in orientation), and/or distortion (change in shape), and/or dilation (change in size) of the rocks in which they are found. If no such changes were required, the structures would have no reason to exist.* Evaluation of displacement and strain is applied at all scales, from movements within the lattices of an individual crystal to the movements of plates.

Stress and Strength (Chapter 3) probes the origin of deformation in terms of force, traction, stress, rock strength, and mechanics. Down-to-earth examples give a 'feel' for force, traction, and stress and the equations that are used to describe these fundamental concepts. Conventional laboratory testing of rock strength is employed as a helpful context to think about the mechanics of origin of deformation and to apply the basic math. Furthermore, laboratory evaluation of the response of rocks to force and stress permits us to evaluate how strength and resistance to deformation are influenced by such factors as rock type, temperature, confining pressure, fluid pressure, and strain rate.

Deformation Mechanisms and Microstructures (Chapter 4) explores what actually takes place at the microscopic and submicroscopic scales that enables 'hard' rock to change shape and size as if it were soft. In this chapter we see that temperature, pressure, mineralogy, strain rate, and the presence (or absence) of fluids determine which mechanisms will operate during deformation. The processes that permit rocks to deform on the grain and subgrain scales range from brittle microfracturing and grinding to plastic deformation involving the subtle 'creep' of crystals. The ease (or difficulty) with which microscopic deformational mechanisms are triggered when rock is stressed controls the level of stress that rock can support.

Part II, Structures, presents the chief classes of structures. Structures are described and analyzed according to the methods and principles presented in Part I, *Fundamentals*. In addition, the mechanics of formation of each class of structure is examined. The geometry and formation of the structures is viewed from the microscopic to the regional, although the greatest descriptive attention is at the outcrop and map scale.

Within *Part II* reside the following chapters: *Joints* (*Chapter 5*), *Faults* (*Chapter 6*), *Folds (Chapter 7)*, *Fault–Fold Interactions* (*Chapter 8*), *Foliation and Lineation* (*Chapter 9*), and *Shear Zones* (*Chapter 10*). These chapters present the 'bread and butter' of structural geology, namely the fundamental categories of ancient or active map-scale structures, which record deformation of the earth's crust. *Part II* concludes with *Active Tectonics* (*Chapter 11*). Within *Active Tectonics* we tour the western United States, distilling an astonishing body of knowledge regarding how, why, and at what rate faults and folds are forming today. This examination of deformation related to Pacific/North American plate motion informs us about the mechanics of formation of geologic structures, and provides a basis for pulling together what is presented throughout the book. *Chapter 11* makes the point that structures do not exist in isolation, but are integrated within networks of interconnecting structures, at all scales, to accomplish just the right amount of deformational response to the prevailing tectonic stresses.

Part III, Descriptive Analysis, is subdivided into sections *A* to *S*. It is intended to be used as preparation for fieldwork, and as a guide to reducing data collected in the field. The sections in *Part III* are entitled: (*A*) Nature of Descriptive Analysis, (*B*) Geologic Mapping, (*C*) Mapping Contact Relations, (*D*) Identifying Primary Structures, (*E*) Measuring The Orientations of Structures, (*F*) Preparing Geologic Cross Sections, (*G*) Preparing Subsurface Contour Maps, (*H*) Using Orthographic Projection, (*I*) Carrying Out Stereographic Projection, (*J*) Evaluating Rotation Using Stereographic Projection, (*K*) Determining Slip on Faults Through Orthographic and Stereographic Projection, (*L*) Carrying Out Strain Analysis, (*M*) Determining the Relationship of Faults To Principal Stress Directions, (*N*) Carrying Out Joint Analysis, (*O*) Engaging in Fault Analysis, (*P*) Carrying Out Fold Analysis, (*Q*) Deciphering Structure in Boreholes, (*R*) Studying Shear Zones in the Field, and (*S*) Determining Focal Mechanisms for Earthquakes.

We have written this book in a way that proceeds from basic physical concepts and methods of analysis to the description and interpretation of structures and systems of structures. We have made each chapter as self-contained as possible, knowing that some will want to cover the material in a different order. Throughout, techniques, methods, experiments, and calculations are described in detail, with the aim of engaging active participation through laboratory and field work.

Each of us likes to tell stories to illustrate and bring to life the material we present. As a story begins, we parenthetically insert who is doing the 'telling:' (GHD vs. SJR vs. CFK). In this way you will know who is telling which personal story within a book that "*WE*" have so enjoyed writing together.

George H. Davis, Stephen J. Reynolds,
& Charles F. Kluth

We dedicate this book to Peter J. Coney. He grasped the significance of things with profound originality, and kept his eye always *on what mattered most. In so doing, he impacted tectonics and structural geology mightily!*

Acknowledgments

Many people helped enormously in the preparation of this book. At the very top of our list we wish to acknowledge the skill, drive, efficiency, and talent of Susie Gillatt, who generated most of the line drawings and art work from materials provided, who sized and enhanced all of the photographs (some directly from digital files, others produced from scans), and who managed the enormous enterprise of tracking, coding, archiving, backing, and uploading illustration files. It was a formidable task simply to keep figure numbers straight, as changes and shifts were made from draft to draft. To appreciate the load of work that Susie lifted, it is only necessary to 'fan' through the text and catch a glimpse of the large number of figures, including line drawings, maps, cross sections, and photographs. What made it all possible is Susie's attention to detail, and her clear grasp of a central principle for this textbook in structural geology, i.e., the need to illustrate in ways that enhance visualization. At times of overload Susie was supported by Cynthia Shaw, another very experienced and artistic scientific illustrator, who prepared some of the more complex and daunting figures. Chuck Carter, a well-known 3D artist, also helped with a number of figures, including some requiring 3D modeling. Julia K. Johnson helped us tremendously in our move to full digital artwork by scanning and fixing hundreds of ancient line-art illustrations and photographs.

The other key member of the scientific illustration team was artist David Fischer, whose talents have been focused mainly, but not exclusively, on some of the special dimensions of the book. David has been a part of *Structural Geology of Rocks and Regions* from the beginning. He is the person to whom we turn to do the book cover (for each of the three editions) and the cartoons. We feel that the cartoons 'teach' in a very special way, and always David takes the concepts, preliminary drawings, or write-ups for new cartoons to a more innovative level through subtle creative nuances.

Among the new additions to this 3rd edition are chapter-opening photographs. We asked Peter Kresan, geologist and professional photographer in Tucson, Arizona, to take the lead on selections and enhancements using largely his own photographs but including contributions from others. It was a pleasure to work with Pete on this part of the project. We thank him for streamlining the process so effectively, and for his artistry.

Yet another photographer we wish to thank is Gary Mackender, Supervisor of the Virtual Reality Annex at The University of Arizona. In a recent project unrelated to this textbook, David carried out all of the streamline digital photography of George Davis' structure collection, and these now are an online resource, designed by Maritza Wright. The collection can be easily accessed as a supplemental resource in the teaching of structural geology (see "*Welcome, Geologic Structures*" http://gdavis.web.arizona.edu/). Then, in illustrating this 3rd edition, we called upon Gary Mackender to prepare macrophotos of a subset of structures in the collection. We thank Gary for his skill and the instantaneous 'turn around' he provided in response to our requests.

The writing of this third edition was preceded by preparation of outlines and objectives. The concepts and materials we produced were sent out for peer review by John Wiley & Sons. Peer input was extremely beneficial, especially as it pertained to strategic and tactical questions regarding what to include and how to internally organize afresh, based upon reflection on past editions and on future directions of structural geology. Accordingly we want to thank the following colleagues who constructively critically reviewed the outlines and organizational plans: Sue Cashman, Humboldt State University; Mihai Ducea, University of Arizona; Suasanne Janecke, Utah State University; Carl Mueller, University of Colorado; Terry Naumann, University of Alaska-Anchorage; and Bob Varga, The College of Wooster.

Also, before initiating the writing of this 3rd edition, and at our suggestion, John Wiley & Sons solicited from Donal Ragan a thorough review and critique of the strain and stress chapters in the 2nd edition of *Structural Geology of Rocks and Regions*, knowing that there was a need to improve these. Donal undertook this with his customary thoroughness and deep grasp of and passion for the subjects. His incisive recommendations and observations were immensely helpful, especially in relation to precision in relation to terms, conventions, and equations. As a result of changes made in response to Donal's input, we believe that Chapters 2 and 3 is this edition are improved and more effective than they would have been otherwise. Our goal was to make the material accessible and yet rigorous, even though not bringing to bear calculus or tensor analysis. We also consulted with Win Means on several matters related to presenting stress and strain, and he too was thoughtful and helpful as we made decisions on specific points.

John Wiley & Sons selected and sent several drafted chapters out for peer review in the early going. Again, input and observations were quite helpful, and influenced positively the final outcomes. Thus we wish to thank the following reviewers: Hassan Babaie, Georgia State University; Rachael Beane, Bowdoin College; Jason Chaytor, Oregon State University; John Dembosky, Methodist University; Beth Lincoln, Albion College; Ryan Mathur, Juniata College; and Edwin Romanowicz, SUNY Plattsburgh.

I (GHD) have experienced a sea-change in textbook preparation *circa* 1980–1984 versus *circa* 1992–1996 versus *circa* 2007–2011, and of course a sea-change in the technologies brought to bear in carrying out research in structural geology and tectonics. The technologies that are advancing the science of structural geology, e.g., geodesy, seismology, paleoseismology, remote sensing, GIS-based geological mapping, are clearly expressed in the sophisticated, detailed, multifaceted nature of maps, cross sections, and reconstructions, which have become the norm in the geological journals. From an authorship point of view, no longer is it even possible to consider 're-drafting' such illustrations, for the time and expense would be overwhelming. Consequently, to achieve the illustration quality of what we expect for this book, we depended on the good will, support, and generosity of colleagues, internationally, from whom we requested digital files. The cooperation we received in this enterprise was remarkable, and humbling. The turn-around time in many cases was within hours or just a couple of days. With digital files in hand, we could make adjustments (with approval in advance) in font and size, achieving the kind of clarity and uniformity that is required of a textbook. In some cases, where the fundamental digital files of mapped information were dense and especially sophisticated, colleagues themselves carried out custom adjustments and sent us finished versions. Consequently, you will understand how important it is for us to list those who were helpful in these ways, including making special digital-photograph files available. We present the following long list of acknowledgment at some risk, for we are aware that we are likely, inadvertently, to leave out a few colleagues, for our records are not perfect. For example, I (GHD) lost an important email pertaining to Figure 9.70*D*, the world's largest pencil structure, . . . so large that the 'pencils' are used as roof beams. I wish I could summon the name of the record holder! Thus we apologize to those who we may have inadvertently failed to list as we thank the following people: Andrew Alden, Mark Allen, Rick Allmendinger, Donald Argus, Jim Aronson, Dov Bahat, Rick Bennett, Steve Boyer, Tom Brocher, Ron Bruhn, Sue Cashman, Wu-Long Chang, Brent Couzens-Schultz, Ray Coveney, Julie Crider, Tony Crone, Greg Davis, Tom Davis, Alexander Davydov, Pete DeCelles, Roy Dokka, Jim Dolan, Fred Donath, Ted Doughty, Jim Ellis, Terry Engelder, Eric Erslev, Jim Faulds, David Ferrill, Lynn Ficter, Gary Fuis, Matt Golombek, Rick Groshong, Michael Gross, Barry Hankley, Tekla Harms, Steve Hickman, Brian Horton, Gene Humphreys, James Jackson, Martin Jackson, Arvid Johnson, Samuel Johnson, Randy Keller, Harvey Kelsey, Young-Seog Kim, Eric Kirby, Bob Krantz, Corné Kreemer, Peter Kukla, Jeff Lee, John Lorenz, Bill Lund, Mike Machette, Jill McCarthy, Vali Memeti, Shankar Mitra, Jay Namson, Alan Nelson, Nick Nickelsen, Tina Niemi, John Oldow, Tom Parsons, Scott Paterson, Nicolas Pinter, Carol Prentice, Ray Price, Donal Ragan, Spencer Reber, Randy Richardson, Michael Rymer, Rick Sibson, Mats Schöpfer, Brian Sherrod, Roger Soliva, Joshuya Spinler, Deborah Spratt, Deepak Srivastava, Fabirzio Stortie, John Suppe, Ben Surpless, David Susong, Barry Sutton, Art Sylvester, Wanda Taylor, Sarah Tindall, Stuart Thomson, Bruce Trudgill, Jeff Unruh, Janos Urai, Bruno Vendeville, Lisa Wald, John Waldron, John Walsh, Ray Weldon, Steve Wesnousky, David Wilson, Nigel Woodcock, Lauren Wright, Tom Yancey, Bob Yeats, Adolph Yonkee, and Mark Zoback.

We want to make a special 'shout out' to several colleagues who prepared original figures to help provide a big picture view of active tectonic kinematics in the western United States, namely Don Argus, Rick Bennett, Arvid Johnson, and John Oldow.

We appreciate and wish to acknowledge the high quality of support rendered by John Wiley & Sons, Inc., in transforming manuscript and figures to printed page. We extend our deepest thanks and heartfelt praise for the chapter-by-chapter steps from copy-editing through production to final printing. The geology team at John Wiley & Sons has given us solid support, including the support received from Joyce Poh, Senior Production Editor; Deepa Chungi, Assistant Editor; Laura Spence Kelleher, Assistant Editor; Denise Powell, Assistant Editor; Veronica Armour, Associate Editor; Darnell Sessoms, Editorial Assistant; and Ryan Flahive, Executive Editor.

One of the challenging completion steps of a book so heavily illustrated is securing copyright permissions, and we have Kathryn McBride for carrying out this process over a period of many months. This critical process requires attention to detail and tenaciousness. We thank Kathryn for her thoroughness.

In spite of all of this support, we recognize that we undoubtedly have introduced errors and missed typos, and have failed to include certain things we probably should have included. We take full responsibility. Hopefully, with feedback, we will be able to address oversights and errors in future printings. In fact, this version of the 3rd edition is the 2nd printing, made clearer because of especially thorough an helpful edits by Kelly Davis.

In the meantime, we hope that students and practioneers of structural geology will find the totality of this effort worthwhile.

Table of Contents: Short Version

Detailed Table of Contents

Part I

Fundamentals

Queen Elizabeth Range, Jasper National Park, Canada. Norm Meader is encountering the structural geology of rocks and regions. He stands on outcrop-scale structures and fabrics, yet views macroscopic properties of geological structures as far as the eye can see. The superposition of glacial flow is a reminder that structures and fabrics at any scale become appreciated, and sometimes understood, in the context of rheology, mechanics of behavior, and the unyielding influence of force and stress. [Photograph and copyright © by Peter Kresan.]

Chapter 1 Nature of Structural Geology

San Miguelito anticline, the westward extension of the Ventura Avenue anticline, Ventura, California. This fold is forming today, and it is producing oil today (note pumpjack). The fold is responding to a relatively constant rate of tectonic shortening of ~9 mm/yr. During the last 100,000 years the core of this anticline has grown at a rate of ~2–5 mm/yr, even while oil and gas are being removed there from the subsurface. The San Miguelito anticline epitomizes the diverse ways in which structures are studied (geodesy, seismology, structural geology, basin analysis, neotectonics) as well as the applications of structural geology to society (e.g., natural resources, seismic hazards). [Photograph courtesy of Art Sylvester.]

Excellence and learning are not commodities to be bought at the corner store.
Rather they dwell among rocks hardly accessible.
And we must almost wear our hearts out in search of them.

[From *College Talks* by H. F. Lowry, edited by J. R. Blackwood, p. 116. Published with permission of Oxford University Press, New York, copyright @ 1969.]

MOTIVATION AND CONTEXT

Our purpose in writing this textbook, now updated to a third-edition view of the world, is to communicate the physical and geometric elegance of geologic structures within the Earth's crust, and to describe the ways in which they reflect the nature and origin of crustal deformation through time (Figure 1.1). Geologic structures provide part of the basis for recognizing and reconstructing the profound changes that have marked the physical evolution of the Earth's outermost layers, as observed from the scale of the plates down to the microscopic. Understanding the nature and extensiveness of geologic structures of deformational origin in the Earth's crust has both scientific and practical benefit. But there is a philosophical value as well. Our perceptions of who we are and where we are in time and space are shaped by facts and

Figure 1.1 Photograph of folding within the Thick White Limestone Beds (Cretaceous/Paleocene) at Mt. Lykaion in the Peloponnesos, Greece. Geologist is Tom Fenn, who is using his hand-held Global Positioning System (GPS) instrument to 'catch' satellite-transmitted radio signals and determine exact location in terms of latitude and longitude. The limestone was subjected to end-on compression during the formation of the Pindos fold and thrust belt in the early Tertiary. [Photograph by G. H. Davis.]

interpretations regarding the historical development of the crust of the lively, dynamic planet on which we live.

Plate tectonics provides a backdrop for understanding the origin and significance of geologic structures, especially regional structures. Plate tectonic analysis is the essential basis for interpreting the dynamic circumstances that give rise to deformational movements. We have come to understand that our earthly foundations are not fixed. Instead, we live on tectonic plates that are in a continual state of slow motion. According to plate tectonic theory, the Earth can be subdivided into discrete fundamental "*rigid*" plates that move in relation to one another (Figure 1.2). The interaction of these plates has played

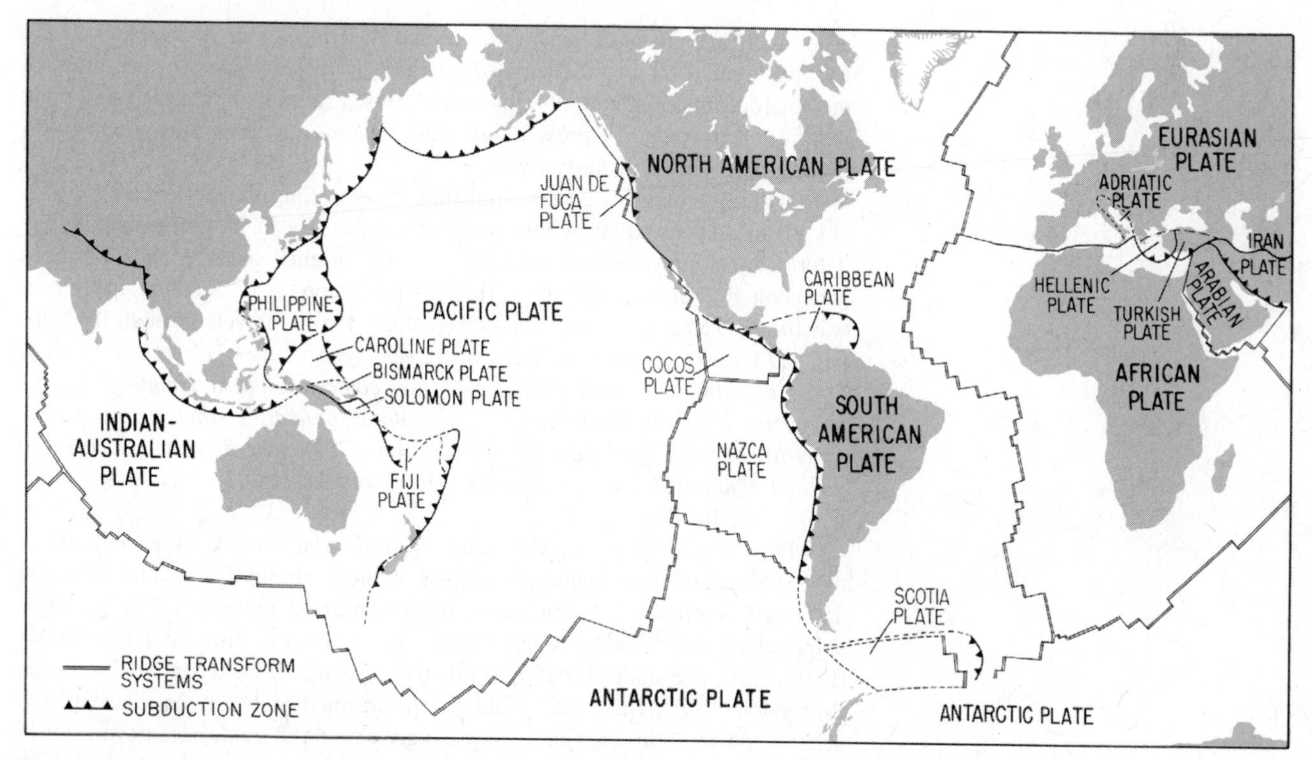

Figure 1.2 Map of the world's plates. [From "*Plate Tectonics*" by J. F. Dewey, copyright © 1972 by *Scientific American, Inc.* All rights reserved.]

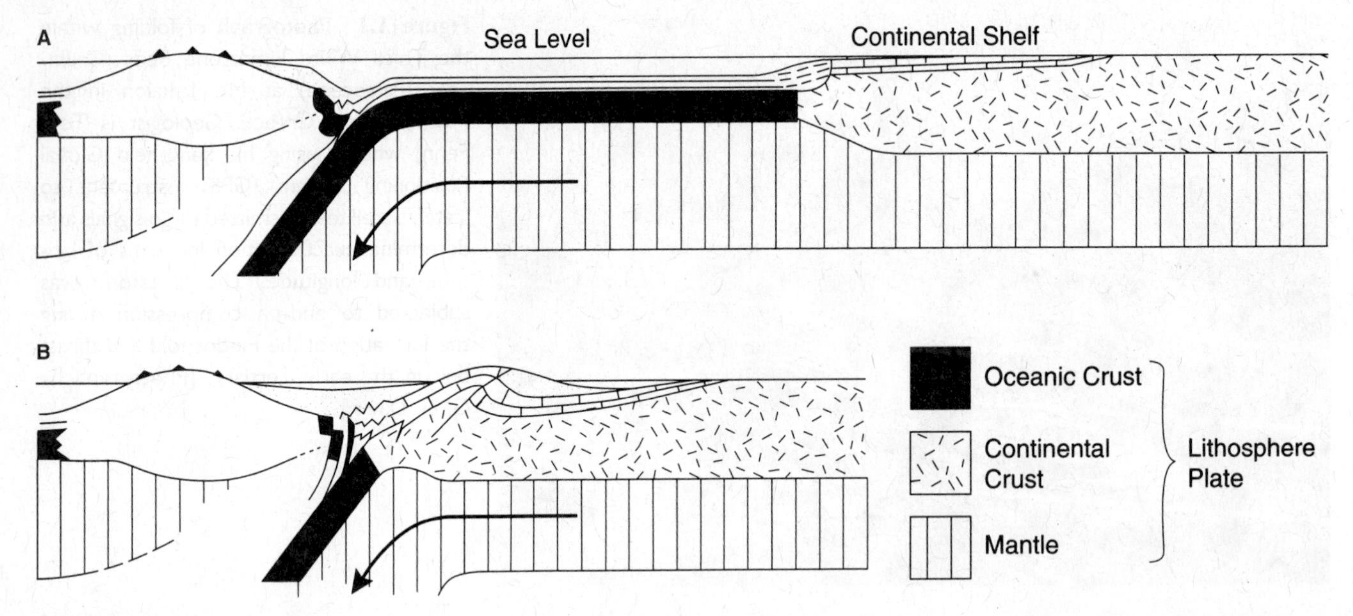

Figure 1.3 Schematic rendition of tectonic convergence between two plates. (*A*) Oceanic lithosphere is being subducted from right to left beneath volcanic arc of basalt and andesite. (*B*) Continental margin makes contact with the volcanic arc. Profound deformation takes place at location of collision. [Adapted from Dewey and Bird, Mountain belts and the new global tectonics, *Journal of Geophysical Research*, v. 75, p. 2625–2647, copyright © 1970 by American Geophysical Union.]

a dominant role in both the formation and deformation of rock bodies in the Earth's crust. The boundaries and margins of plates typically are sites of tectonic deformation. Notably, mountains '*build*' at plate boundaries and plate margins. Knowledge of present-day plate tectonic processes aids us in interpreting past structural movements.

Structural geology is an essential tool in unraveling the geologic history of any given area or region within the Earth, especially in mountain belts where deformational movements profoundly modify original arrangements of rocks and geologic contacts (Figure 1.3). Nicolaus Steno, a 17th-century anatomist, was the first practitioner of structural geology. On his travels through the Alps he "*had the chance to see with his own eyes layers of rock [it reminded him of layers of skin] . . ., [and] strata raised and contorted into mountains.*" . . . "*A mountain cannot be opened up with a scalpel; the earth's strata cannot be cut away to see what lies below the surface. But in Florence, Steno would learn to do that figuratively, if not literally. He would become an anatomist of the world*" (Cutler, 2003, p. 44).

When viewed at geological scales of time and spatial dimension, rocks must be regarded as materials almost without strength, capable of being deformed continuously even under the slightest of pressures. The geologic mapping of mountain belts has shown this to be true time and time again (Figure 1.4). The understanding of this paradox, which we try to capture in the expression "*soft as a rock,*" yields a transformed view of the strength and behavior of rocks and regions. Surprisingly, structural geologists have a lot in common with ocean scientists and atmospheric scientists: we all deal with deformable "*fluids.*"

Figure 1.4 Photograph of folding of marble layers (white) and impure marble and quartzite (dark) along the Snake Range décollement in the Snake Range, western Utah. The variations in layer thickness indicate that the rocks flowed during deformation. Height of cliff face is ~40 m. [Photograph by G. H. Davis.]

PRACTICAL APPLICATIONS

Practical applications of structural geology are broad ranging and powerful. The value of structural geology in petroleum exploration, ore deposits geology, and hydrogeology relates to an understanding of fracturing and faulting in the subsurface, and how fracture permeability created by networks and systems of such brittle structures influences fluid transport. The common occurrence of calcite veins in limestone or marble is an example of this. The ancient marble columns pictured in Figure 1.5 are deeply weathered and as a consequence calcite veining stands out in stark relief. The myriad calcite veinlets disclose the remarkable degree of fracturing that the marble once experienced. $CaCo_3$-bearing solutions clearly permeated the rock at one time.

Most ore deposits owe their existence to the movement of hydrothermal solutions through fractured bedrock, where precipitation of precious metals (gold, silver) and base metals (copper, lead, zinc) creates structurally controlled veins (Figure 1.6). Petroleum geologists are experts on the migration of hydrocarbons traveling through permeable rocks (including fractured rocks), with trapping of hydrocarbons not uncommonly structurally controlled (e.g., in an anticline, or along a fault). Hydrogeologists study fracture-induced permeability and overall shapes, sizes, and orientations of basins of sedimentary rocks in order to model the flow of groundwater, to estimate water reserves, and to site wells.

Opportunities to apply structural geology have expanded to include particularly difficult issues affecting the safety and well-being of people. The challenges include investigating sources of contaminants in groundwater systems; evaluating the potential energy of natural geothermal systems; appraising daring but dangerous proposals for the subterranean disposal of radioactive waste; assessing fundamental stability of steep slopes underlain by rain-infiltrated fractures in bedrock; and mitigating natural disasters related to earthquake, landslide, volcanic, and tsunami events. Assessing such issues are intellectually demanding activities requiring knowledge derived from intimacy with the principals of fracturing and faulting, and leveraging understanding through geophysics and engineering mechanics.

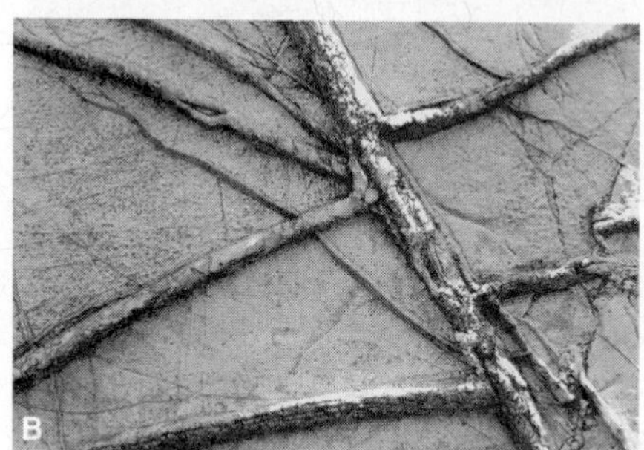

Figure 1.5 *(A)* Weathered ancient marble columns exposed at site on the outskirts of Megalopolis, Greece. *(B)* At close range it can be seen that calcite veins in the marble form networks that stand out in relief from the marble matrix (grey and dull). The field of view for this macrophotograph is ~12 cm. Notice that different generations of calcite veining can be identified on the basis cf cross-cutting relationships. Note also that in addition to the conspicuous veins, there are hundreds of microveins throughout the limestone. $CaCo_3$-bearing waters infiltrated hairline cracks and deposited calcite, forming this meshwork of microveins. [Photographs by G. H. Davis.]

Figure 1.6 Vein of stilpnomelane (black) and quartz (white). Stilpnomelane (from the Greek, "*black*" and "*shiny*") is a mica mineral commonly associated with iron deposits. This vein invaded thin-bedded wall rock in the North Hillcrest mine area in the Iron Range of Minnesota. The very center of the vein bears a faint line, perhaps a vestige of the former fracture trace that guided the hydrothermal solutions that gave rise to the vein. Spreading apart of the walls was directed at right angles to the centerline of the vein and to the contact of the vein with the wall rock. Note that the conspicuous parting (p) in wall rock on the left wall of the vein is offset in a way that perfectly matches the opening direction of the vein. [Photograph by R. G. Schmidt. Courtesy of the United States Geological Survey.]

Studies of natural hazards and geological disasters in relation to society have emerged as very important fields of endeavor. The phenomena are startling! For example, Figure 1.7 shows a massive rockslide that occurred near Mt. Aconcagua, Argentina. Individual blocks at the base are gigantic. The new church just beyond the toe of the slide replaces its predecessor, which lies buried beneath the slide. Structural geologic factors preordained this rock fall. The dip direction of the rock layering favored a sheet of the rock failing, sliding, and breaking up under the influence of gravity, especially given the fact that runoff undeniably penetrated bedrock along layer boundaries, creating a kind of lubrication that reduced friction and resistance to sliding. The "*breakaway*" for this rockslide can be seen in the upper reaches of the ridge, very close to the trace of the white sill.

Figure 1.7 Photograph of rockslide near Sierra Aconcagua, Argentina. It would seem that everything in the Andes is on a gigantic scale, and the sizes of the blocks in this slide are no exception. The church (for scale) is the successor to its predecessor, now buried. The breakaway for this rockslide is quite apparent. See text for details. [Photograph by G. H. Davis.]

The energy bound-up in rockslides pales in comparison to the energy harnessed by earthquakes. Disasters related to major earthquakes are now being fully covered by media, drawing broader and more informed attention to active tectonics and the dynamic earth, and to relief efforts. As reported by the U.S. Geological Survey Earthquake Center, there is a long, disturbing list of relatively recent (post-1995) killer earthquakes. The reporting of fatalities is undoubtedly conservative:

Place of occurrence	*Year*	*Magnitude*	*Fatalities*
Kobe, Japan	1995	M 6.9	5,502
Sakhalin Island	1995	M 7.1	1,989
Northern Iran	1997	M 7.3	1,567
Afghanistan-Tajikistan	1998	M 5.9, M 6.6	6,323
Papua New Guinea	1998	M 7.0	2,183
Columbia	1999	M 6.1	1,185
Izmit, Turkey	1999	M 7.6	17,118
Taiwan	1999	M 7.6	2,400
Gujarat, India	2001	M 7.6	20,023
Hindu Kush, Afghanistan	2002	M 6.1	1,000
Northern Algeria	2003	M 6.8	2,266
Southeastern Iran	2003	M 6.6	31,000
Sumatra-Andaman Islands	2004	M 9.1	227,898
Northern Sumatra	2005	M 8.6	1,313
Pakistan	2005	M 7.6	86,000
Java	2006	M 6.3	5,749
Eastern Sichuan, China	2008	M 7.9	87,587
Southern Sumatra	2009	M 7.5	1,117
Haiti	2010	M 7.0	222,570
Honshu, Japan	2011	M 9.0	20,000

FIELD WORK

Field relations are the primary sources for structural geologists. Dr. Howard Lowry, a brilliant scholar in the field of English literature, captured the significance of primary sources in ways that may have special meaning for geologists:

> *By [primary sources] we mean the first-hand things, the authentic ground of facts and ideas, the original wells and springs out of which all the rest either is drawn or flows. . . . Regard for the primary sources makes one forever the enemy of preconceptions, of manipulated data, raw opinion, and guesswork of all the sleek shortcuts to wisdom in ten easy lessons. . . . Exclusive reliance on second-hand things makes second-hand men and women. It deludes us into thinking we are wiser than we are. . . . Breadth of knowledge, even knowing a little about a lot, has its obvious value. But breadth that perpetually sends down no clean, strong roots in the primary sources—into the deep earth and 'the hidden rivers murmuring in the dark' of the rocks—such breadth clarifies very little. It merely puts our bewilderment on a broader basis. It leads us into incredible naiveté and gullibility. It makes us too quick to believe all we read.* [From *College Talks* by H. F. Lowry, edited by J. R. Blackwood, p. 86–87. Published with permission of Oxford University Press, New York, copyright @ 1969.]

To be sure, this text on structural geology is based on and directed to the primary sources!

"*DEFORMATION*" IS THE HEART OF THE MATTER

The General Concept of "*Deformation*"

Our journey will explore the nature of the architecture of the crust of the planet on which we live. We will be concerned primarily with geologic structures that have developed through deformation as a response to forces and stresses associated with plate tectonic movements. Our goals include learning how to describe the structures of rocks and regions; to understand the movements (both small and large) that accompanied development of the structures; and to interpret structures and systems of structures in relation to mechanics, tectonics, and overall geologic history.

Understanding **deformation** is indeed the heart of the matter. "*Deformation*" is a term that is used in a variety of ways. We use it in reference to geologic structures, such as faults or folds, for instance, "*Look at the deformation! Check out those folds!*" We use it in reference to processes, for instance, fracturing, faulting, or folding, which produce geologic structures: for instance, "*Deformation is taking place today!*" "*Folding is occurring earthquake-by-earthquake in the Los Angeles basin!*" And we use it in a formal, rigorous manner, emphasizing the measurable changes that have occurred in a particular volume of rock (no matter how large or small), such as the volume *V* shown in Figure 1.8*A*. Describing "*deformation*" is geometric analysis. It requires (1) comparing the final location of a volume of rock to where it started out (Figure 1.8); (2) comparing the present orientation of a volume of rock to its original orientation; (3) comparing the present shape of a volume of rock to its original shape; and (4) comparing the present size of a volume of rock to its original size. In fact, this kind of geometric analysis strives to connect *each* material point within the deformed volume of rock to the very same material point in the undeformed volume of rock. For the example shown in Figure 1.8, the material points of interest in the undeformed volume of rock (Figure 1.8*A*) are A, B, C, and D. The location of each of these material points in the deformed volume of rock are A′, B′, C′, and D′ (Figure 1.8*B*). The connecting links (**AA′**, **BB′**, **CC′**, and **DD′**) are **displacement vectors**, which like all vectors have magnitude and direction. The whole family of displacement vectors (go ahead and lightly draw them in for all of the material points in Figure 1.8) is the **displacement field**. The displacement field is in fact the deformation (Hobbs, Means, and Williams, 1976, p. 23).

Applying this kind of thinking to the rockslide shown in Figure 1.7, our immense structural geologic challenge would be to analyze the volume of broken rock spilled along the slope, and to connect the final position of each block to its original position. Doing this for the blocks, and for material points in the block, would result in a displacement field that, in fact, represents the deformation. We will see in *Chapter* 2 (*Displacement and Strain*) that establishing the displacement field sets the stage for interpreting the motions themselves, including the **deformation paths** for each block (and each material point) and the velocity field for the movement. This type of analysis is **kinematic analysis**. It goes well beyond geometric analysis, but does not venture into causation (i.e., forces, stresses, conditions), which is reserved for **dynamic analysis**.

Kinematic analysis of the rockslide pictured in Figure 1.7 would be aided by a video of the rockslide in action, but as it happens we seldom have videos for sudden contemporary tectonic events, and never for ancient deformation. 'Doing without' is the mother of invention, and in this case structural geology is the discipline that has borrowed best practices from continuum mechanics

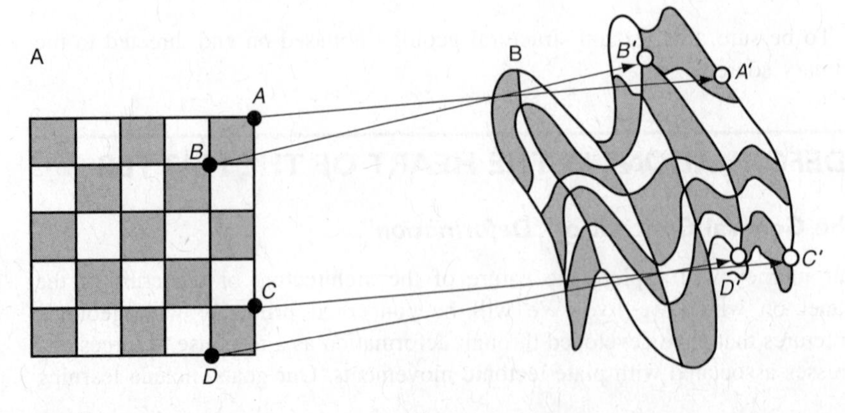

Figure 1.8 Comparison of the initial state of a volume of rock to the final deformed state. See text for explanation. [Modified from Hobbs, Means, and Williams, 1976, Figure 1.11, p. 23.]

and adapted them to analyze deformation of volumes of rocks deformed in the deep past. The concepts and methods are made to work even when exposures are limited. 3D visualization plays an important role in all of this, filling in what is missing. Attention to rates and ages is essential to get to 4D solutions (e.g., deformation histories).

For now, in this introduction, let us simply move from one analogy to another, in this case from rockslides to pizza.

Analyzing Deformation of a Pepperoni Pizza

Several decades ago a situation presented itself to me (GHD) that proved to be a good example of deformation, and a rare kitchen-table opportunity for analysis of deformation. Going to the freezer for a midnight snack, I pulled out a pizza whose form is portrayed in the geologic map I immediately prepared (Figure 1.9*A*). During mapping, I arbitrarily oriented the pizza with respect to north (see Figure 1.9*A*).

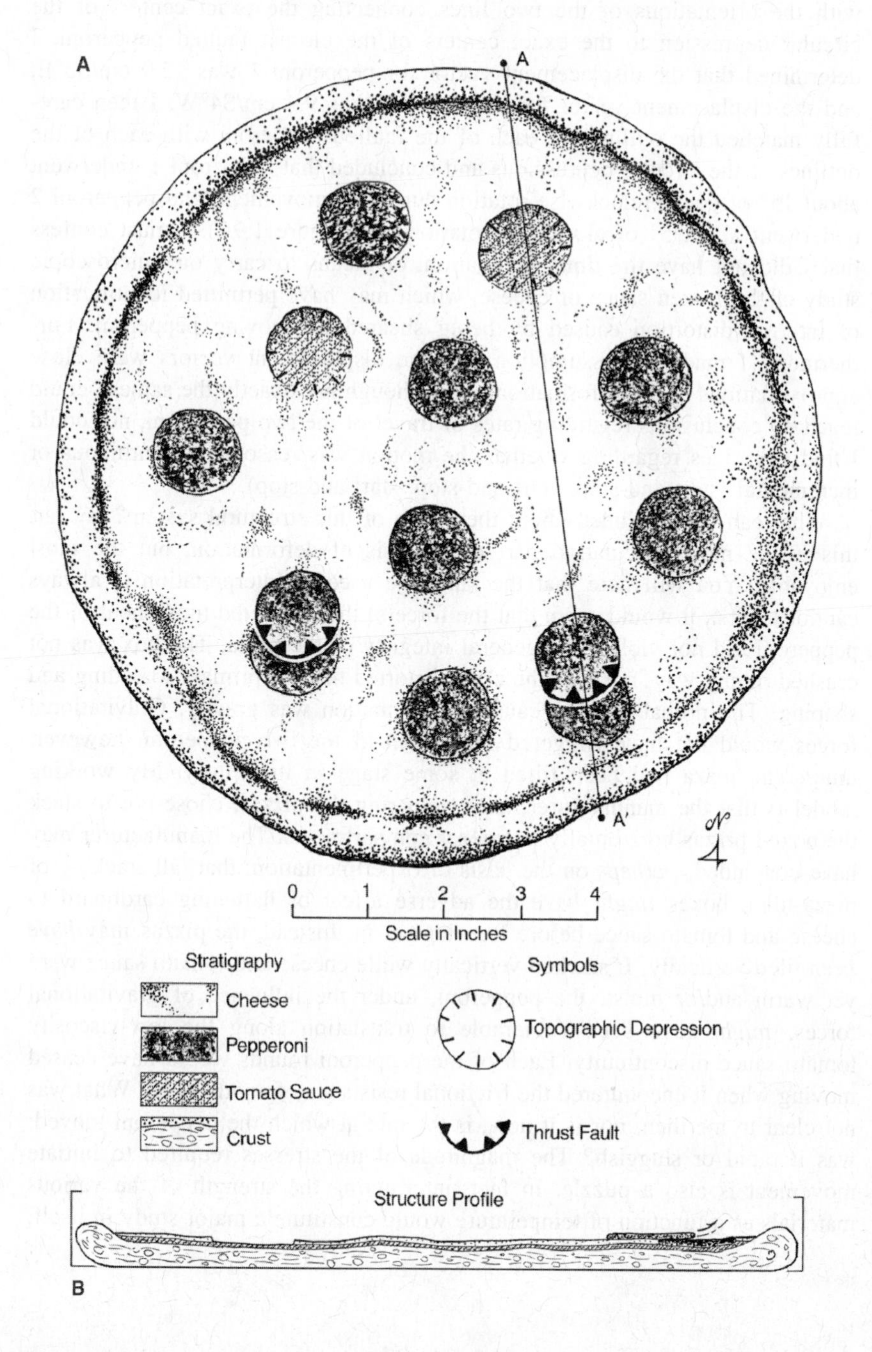

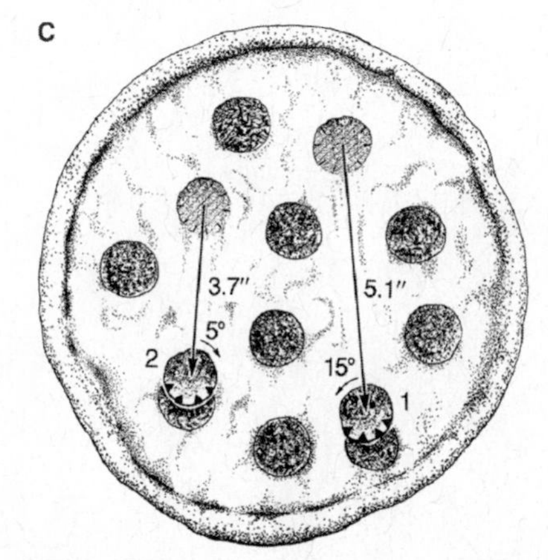

Figure 1.9 (*A*) Geologic map and structure profile of a medium-sized pepperoni pizza. (*B*) Kinematic model of the translation and rotation of the pepperoni. (*C*) Detail of displacement vectors.

In most respects this was a normal pizza. A thin stratum of cheese rested atop tomato sauce and crust. And pepperoni, lightly dusted with cheese, was distributed across the face of the pizza. The diameter of the pizza was 23 cm. Topographic relief of the pizza, as revealed in cross-sectional profile was merely 13 mm (see Figure 1.9*B*). My delight as a structural geologist came upon observing that two pepperoni-sized circular depressions existed near the edge of the pizza in the northeast and northwest quadrants. There was not even a trace of cheese in these depressions, let alone pepperoni. Furthermore, opposite these depressions, in the southeast and southwest quadrants, two pieces of pepperoni overlapped in low-angle (overthrust) fault contact. The cross-section of Figure 1.9*B* clarifies this relationship.

The physical and geometric properties of the pizza demanded the following interpretation of the deformation (Figure 1.9*C*): somehow the pepperoni that had once occupied the circular depressions now devoid of cheese had moved 12.9 cm and 9.4 cm, respectively, to their present locations. I determined the displacement vector for each pepperoni by combining these measurements with the orientations of the two lines connecting the exact centers of the circular depression to the exact centers of the closest faulted pepperoni. I determined that the displacement vector for pepperoni 1 was 12.9 cm/S5°E, and the displacement vector for pepperoni 2 was 9.4 cm/S4°W. I then carefully matched the outlines of each of the faulted pepperoni with each of the outlines of the circular depressions and concluded that pepperoni 1 underwent about 15° of counterclockwise rotation during its movement, and pepperoni 2 underwent about 5° of clockwise rotation (see Figure 1.9*C*). I must confess that I did not have the time, inclination, or means to carry out microscopic study of the frozen sauce or cheese, which may have permitted identification of internal distortion caused by being sheared by moving pepperoni. Furthermore, I made the assumption that the displacement vectors were close approximations to the deformation paths, though not exactly the same. I could reach no conclusions regarding rates of travel of the two pepperoni, nor could I find any clues regarding whether the motion was reasonably continuous or incremental (start-and-stop, start-and-stop, start-and-stop).

What can be concluded about the origin of this structural system? By far, this is the most speculative part of analysis of deformation, but the most enjoyable. You will note that the language used in interpretation is always cautious. First, it would *seem* that the force(s) that triggered movement of the pepperoni did not violate the general integrity of the pizza. The box was not crushed in any way, nor was the crust distorted beyond primary kneading and shaping. The probable force causing deformation was gravity. Gravitational forces would not have triggered movement of the two pepperoni, however, unless the pizza had been tilted at some stage in its history. My working model is that the manufacturer, after preparing the pizzas, chose not to stack the boxed pizzas horizontally in freezer compartments. The manufacturer may have concluded, *perhaps* on the basis of experimentation, that tall stacking of pizza-filled boxes *might have* the adverse affect of flattening cardboard to cheese and tomato sauce before freezing set in. Instead, the pizzas *may have been* filed vertically. *If* stacked vertically while cheese and tomato sauce were yet warm and/or moist, the pepperoni, under the influence of gravitational forces, *might have been* vulnerable to translation along the low-viscosity tomato sauce discontinuity. Each of the pepperoni rounds would have ceased moving when it encountered the frictional resistance of another one. What was not clear to me then, nor is it now, is the rate at which the pepperoni moved: was it rapid or sluggish? The magnitude of the stresses required to initiate movement is also a puzzle. In fact, interpreting the strength of the various materials as a function of temperature would constitute a major study in itself.

My working model is, of course, only one interpretation. *Maybe* the structural event that dislodged the pepperoni was of an entirely different nature. *Maybe* my interpretation is correct except for timing: after all, the pizza, stacked vertically, *may have been* undeformed and solidly frozen . . . until the power failure. Interpreting the timing of structural events is often very difficult.

Deformation, An Outcome of Stress vs. Strength

The strength of any body depends upon its physical makeup and the conditions of the environment it occupies, such as temperature and pressure conditions. Any body of rock, no matter how hard, will deform if the conditions are right. This concept is illustrated in historic photographs of a fence line located at the site of the Hebgen Lake earthquake (M_W 7.5), a destructive event that wracked southwesternmost Montana in 1959 (Witkind and Stickney, 1987). Shifts in the ground surface caused fence lines of certain orientations to shorten. Shifts caused bending where shortening was modest (see Figure 1.10*A*). But where shortening exceeded the bending limit of the wooden slats, the fence fractured and splintered (see Figure 1.10*B*).

Generally speaking, deformation of rock bodies results from the **loading** by gravitational, tectonic, thermal, and/or impact forces, which generate stresses

Figure 1.10 (*A*) Buckled fence and (*B*) broken fence, Hebgen Lake earthquake area, Montana. Fences, like rocks, respond in different ways to shortening. [Photograph by J. R. Stacy. Courtesy of United States Geological Survey.]

Figure 1.11 Photographs underscoring the impossibility of permanently resisting gravitational loading at 2000-m depth in the Mponeng gold mines in South Africa. Depth where photos were taken: 2900 m. Temperature at face: 56°C. Open stopes excavated with about 1–2 m heights completely close up after 2–3 months, leaving nothing but a thin line of carbon where the eucalyptus timbers had been. (*A*) Closely spaced steel struts are used here to support the roof during mining. (*B*) Using everything in site to try to resist gravitational loading. (*C*) Inevitable squeezing down of roof to floor. [Photographs courtesy of Ray Coveney.]

that may exceed rock strength. When rock strength is exceeded, the rock body will behave in a **brittle** manner (by fracturing) or a **ductile** manner (by flowing in the solid state), depending on how the conditions of the physical environment have affected the rock's capacity to resist stress.

Loading can be created in a number of ways, including gravitational loading, tectonic loading, thermal loading, and impact loading. As an example of **gravitational loading**, the weight of thousands of meters of sediments within a depositional basin produces stresses that generally result in the thinning and compaction of the sediments as they are buried deeper and deeper. Platinum-mining operations in the Bushveld complex of South Africa deal with stresses derived from gravitational loading on a day-to-day basis. By mining standards (but not by tectonic standards), the workings in the Bushveld Igneous Complex are *very* deep: up to 2000 m! The gravitational loading by overlying rock is so intense that the intricate timber-and-steel roof support systems simply delay the inevitable; that is, the squeezing down of the openings (Figure 1.11).

Another example of gravitational loading is the gravitational collapse of volcanoes above evacuated magma chambers. Enormous craterlike **calderas**, such as Crater Lake in Oregon, can be produced as a result (Figure 1.12). The weight of prehistoric volcano Mount Mazama combined with partial evacuation of the magma chamber below caused collapse of the volcanic edifice. Steep faults, creating a circular system in map view, accommodated the down-dropping.

An example of **tectonic loading** is seen today in the slow, steady, north-south head-on convergence of continental lithosphere of the Indian-Australian plate against the continental lithosphere of the Eurasian plate (Figure 1.13*A*). The rate of **plate convergence** is ~50 mm/yr, with ~40% taken up in **subduction** of the Indian-Australian plate beneath the Eurasian plate, and ~60% responsible for internal deformation by thrust faulting in the Himalayas and strike-slip faulting in China (see Figure 1.13*A*). This **plate collision**-induced tectonic loading has been elegantly simulated through experiments featuring a rigid indenter acting upon a plasticene Eurasian plate (see Figure 1.13*B*).

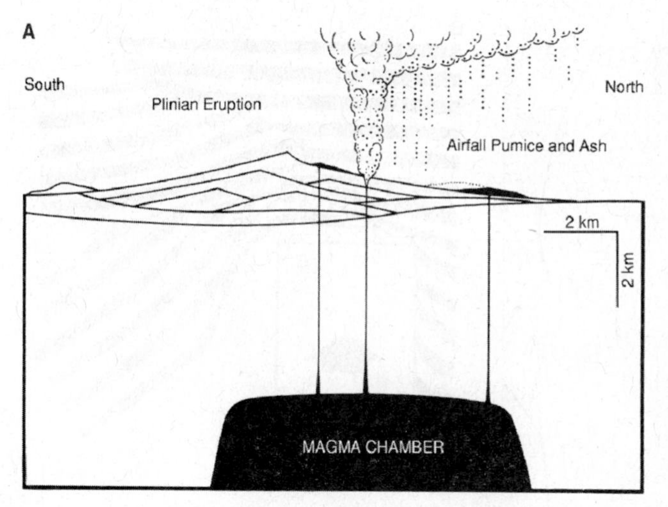

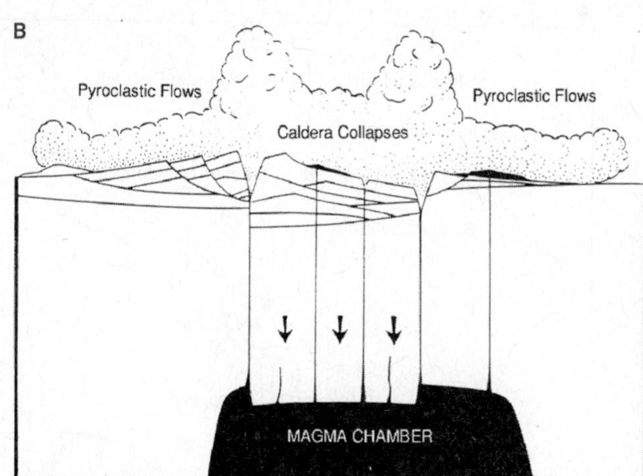

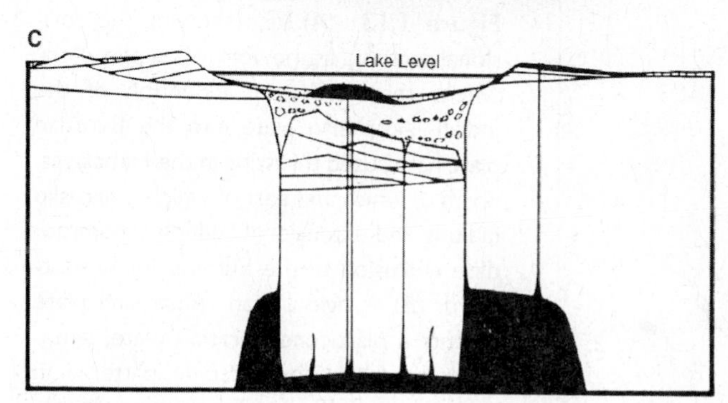

Figure 1.12 (*A*) Eruptive formation of prehistoric volcano, Mount Mazama. (*B*) Caldera collapse of the volcanic edifice into the emptied magma chamber. (*C*) Lake forms within the caldera structure. At the center of the lake minor eruptions build a small cinder cone, part of which (not shown) is Wizard Island. [Drawing by Charles R. Bacon. Courtesy of United States Geological Survey (1988).]

Deformation that accompanies the cooling of certain basalt flows yields spectacular examples of **thermal loading**. The cooling causes uniform shrinkage and contraction expressed in **columnar jointing** of hexagonal character (Figure 1.14*A*). The Giant's Causeway in Northern Ireland is a playground of ~40,000 individual columns (largest is 12 m) created by thermal loading some 50 to 60 million years ago (Figure 1.14*B*). Thermal loading is one interpretation of the formation of columnar basalts at Giant's Causeway. Another is that the Irish giant, Finn McCool, built the road by hand so that he could walk to Scotland and fight his nemesis, Benandonner. What happened on that walk is worth looking into (see Wikipedia, February 9, 2010, http://en.wikipedia.org/wiki/Giant's_Causeway).

Impact loading also causes stress and deformation. We normally think of loading causing stresses to build slowly but persistently, but in some situations incredibly high stresses just show up, through impact loading. As an example of impact loading we have in mind Meteor Crater, located in northern Arizona, where asteroid impact (loading) created a bull's-eye of deformational destruction (Figure 1.15*A*) The now-upturned, pervasively fractured and distorted sedimentary rocks never knew what hit them (Figure 1.15*B*), as revealed by the telltale presence of a peculiar microscopic mineral texture aptly named "*shocked*" quartz (Figure 1.15*C*). Shocked quartz includes coesite

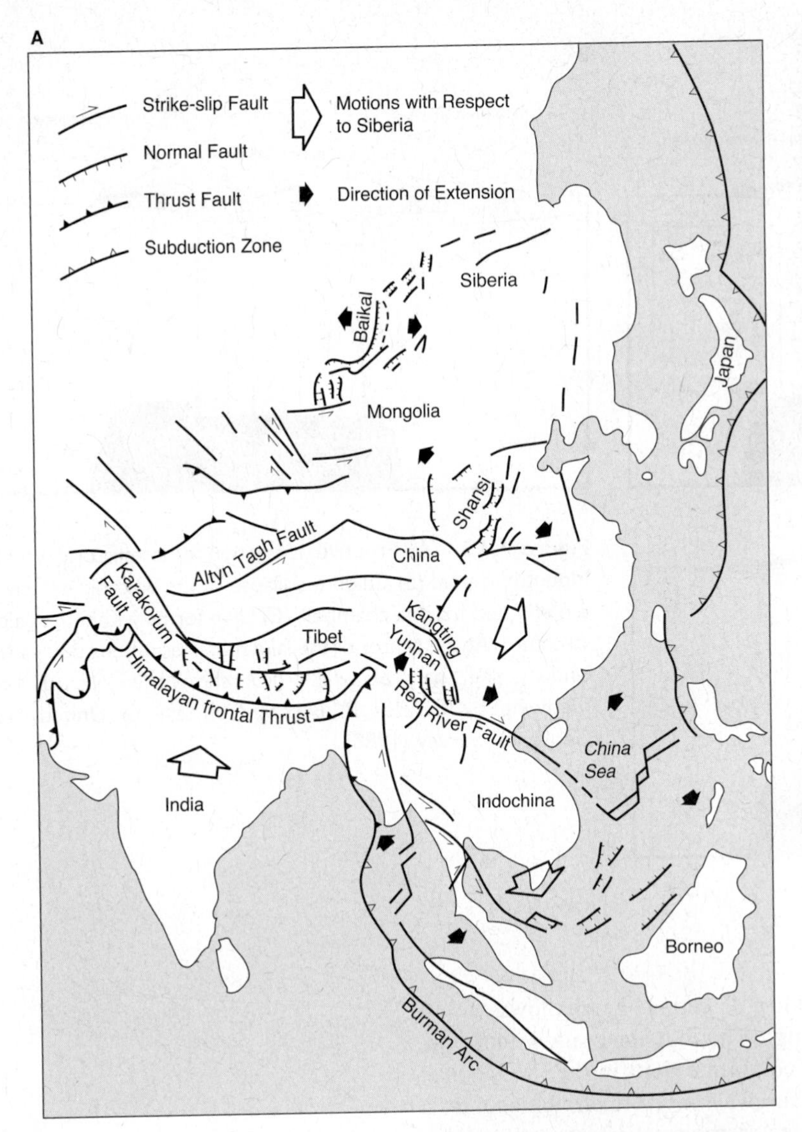

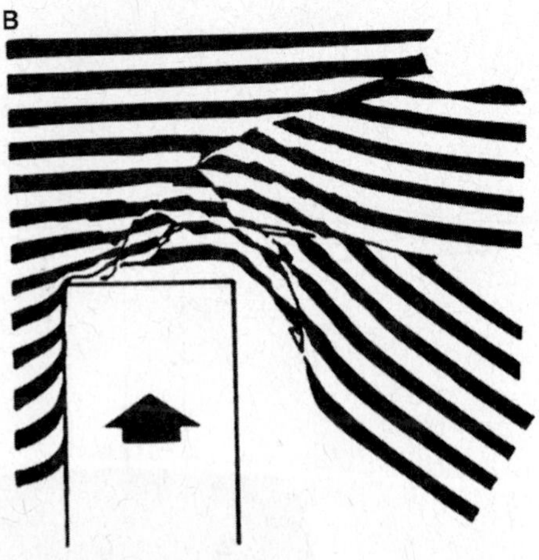

Figure 1.13 (*A*) Map showing the collisional tectonic framework of eastern Asia, where the northward indention of the Indian–Australian plate into the Eurasian plate has caused thrusting in the Himalayas, north of which and east of which strike-slip faulting and extensional faulting accommodate extrusion of the Eurasian plate eastward. (*B*) A rigid Indian–Australian plate indents a plasticene Eurasian plate, simulating the extrusion tectonic patterns in eastern Asia. [From Tapponnier et al., 1982. Published with permission of Geological Society of America.]

Figure 1.14 Photographs of columnar jointing in basalt, Giants Causeway, Northern Ireland. (*A*) Top view of typical polygonal pattern of such jointing in basalt. Columnar jointing results from regular shrinkage and contraction during cooling of the lava flow. (*B*) View of the columns themselves with Matt Davis tucked inside. [Photographs by G. H. Davis.]

Figure 1.15 (*A*) Oblique aerial photo of Meteor Crater, Arizona. [Photograph and copyright © by Peter Kresan.] (*B*) Diagrammatic representation of the formation of Meteor Crater, and the accompanying upturning of formerly horizontal strata. [After Shoemaker, 1979, Fig. 4, p.11.] (*C*) Shocked quartz collected from Meteor Crater. The rock, derived from the Coconino Sandstone, is 75% quartz, 20% coesite, and 5% glass. The light-colored quartz grains are deformed to fit into a mosaic. Individual quartz grains are close to 0.1 mm in diameter. The black opaque areas and the medium gray areas are the main regions of coesite and stishovite [Photomicrograph by Susan Kieffer.]

and stishovite, which are very high-pressure transformations of quartz brought about by phenomena such as asteroid impacts or atomic bomb blasts.

Other telltale signs of impact loading are **shatter cones**, which are linearly etched fractures with distinctive conical shapes (Figure 1.16). Most are associated with craters interpreted to have formed by high-velocity impacts of meteorites, asteroids, and other planetary debris (Dietz, 1968; Grieve and Pesonen, 1992). Debris associated with shatter cones can include glass formed as a result of impact-induced shock melting. Maps of orientations of such cones demonstrate

Figure 1.16 Photograph of shatter cones from the Wells Creek impact site in Tennessee. Rock is Knox Dolomite. It shows the telltale pattern of the diverging linear pattern so typical of shatter cones. [Photograph by John F. McHone.]

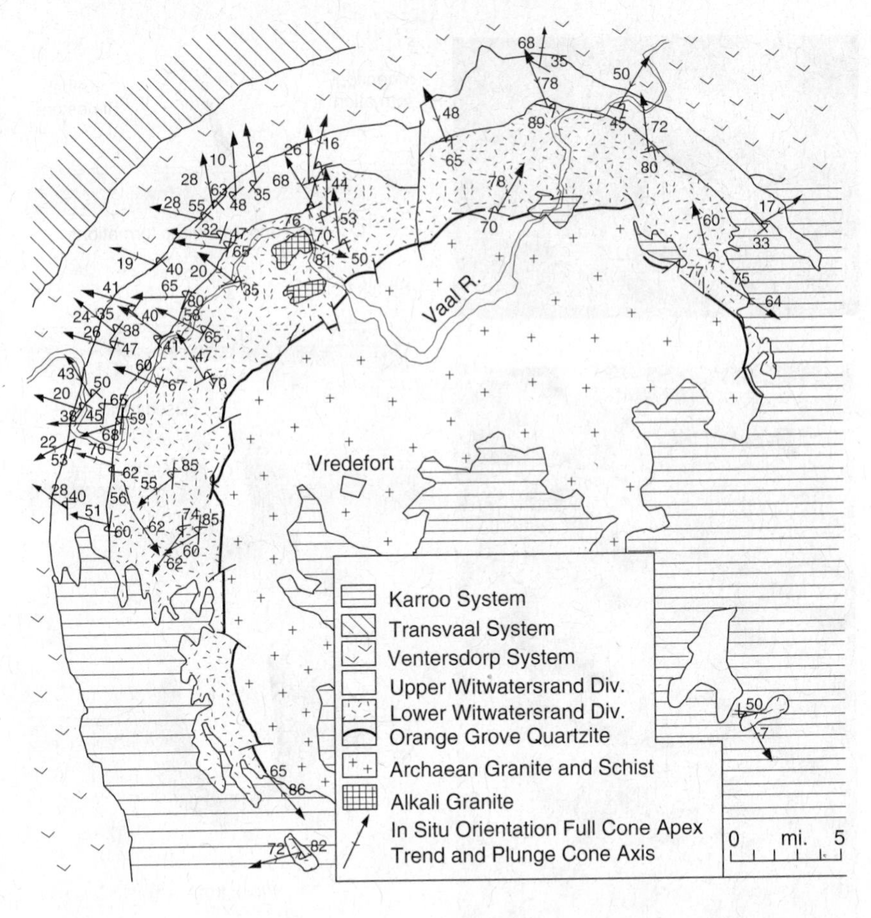

Figure 1.17 Geologic map of the Vredefort dome, South Africa, showing the mapped orientation of shatter cones. Note that these shatter cones occur in bedding that has been overturned. When bedding is restored to its original orientation, the apices point to a common locus of impact in the center of the impact ring. [From Manton. The orientation and origin of shattercones in the Vredefort Ring, 1965. Permission by the *Annals of the New York Academy of Sciences*.]

that they point toward the impact site (Figure 1.17). Thus the orientations of shatter cones can be used to locate buried craters.

ARCHITECTURE AND STRUCTURAL GEOLOGY

Many of us are attracted to the field of structural geology because of the systematic, elegant geometries and patterns of structures; they play on our curiosity. Jacob Bronowski (1973), in his superb set of essays entitled *The Ascent of Man*, suggests that our conception of science today is a description and exploration of underlying structures of nature, and he points out that words like *structure, pattern, plan, arrangement,* and *architecture* constantly occur in every description that we try to make. He believes:

> *The notion of discovering an underlying order in matter is man's basic concept for exploring nature. The architecture of things reveals a structure below the surface, a hidden grain, which when it is laid bare, makes it possible to take natural formations apart.* [From *The Ascent of Man* by J. Bronowski, p. 95. Published with permission of Little, Brown and Company, Boston, copyright © 1973.]

Bronowski's remarks apply beautifully to **structural geology**, which can be most succinctly defined as the "*study of the architecture of the Earth's crust, insofar as it has resulted from deformation*" (Billings, 1972, p. 2). The expression "*architecture of the Earth*" is very appropriate because structural

Figure 1.18 Geologist confronting the structure of nature, in this case exfoliation sheeting (jointing) in granite near Shuteye Peak in the Sierra Nevada. [Photograph by N. K. Huber. Courtesy of United States Geological Survey.]

geology addresses the form, symmetry, geometry, and certainly the elegance and artistic rendering of the components of the Earth's crust on all scales. At the same time, structural geology focuses on the strength and mechanical properties of crustal materials, both at the time of their deformation and now.

Although architecture and structural geology have much in common, the challenges of the architect and the structural geologist are quite different. The architect designs a structure, perhaps a building or a bridge, giving due attention to function, appearance, geometry, material, size, strength, cost, and other such factors. Then the architect supervises the process of construction daily, or perhaps weekly, making changes where necessary. In the end, the architect may be the only person who is aware of discrepancies between the original plan and the final product.

In contrast, the structural geologist is greeted in nature by what looks like a finished product, like the structural product (**exfoliation jointing** in granite) shown in Figure 1.18, and is challenged to ask a number of questions. What is the structure? What starting materials were used? What is the geometry of the structure? How did the materials change shape during deformation? What was the source of the stress that caused deformation? What was the sequence of steps in construction? Attempts to answer these questions generate even more questions. When was the job done? How long did it take? What were the temperature and pressure conditions? How strong were the materials? What "*on earth*" caused it to form in the first place?

The complexity of interpreting natural systems hit home to me (GHD) in the reflections of a small pool within which the surface waters were marked by foam patterns (Figure 1.19). Delicately fashioned, these patterns resembled the layering of metamorphic rocks deformed under hot, deep conditions. Patterns of movement were both complex and ever-changing. Seeing the patterns come and go, my mind shifted to what would happen to these

Figure 1.19 Deformed foam layers in a Rocky Mountain pool. Paper cup (upper right corner) is being blown by the wind from right to left, while the water is being pulled by gravity toward the lower right. [Photograph by G. H. Davis.]

structures when winter set in. Some single pattern would be frozen; one of an infinite number of patterns would be preserved; and yet that pattern might or might not be representative of the kinds of motion I had watched. I began to realize more fully that every geologic record we examine is but one out of millions of possible frozen records, stop-action points, tiny scenarios from a much longer and more complex history that we never will know in full.

PLATE TECTONICS AND STRUCTURAL GEOLOGY

Basics

Plate tectonics provides an essential backdrop for understanding the significance of structures, especially regional structures. It is the basis for understanding the dynamic circumstances that give rise to deformational movements. Plate interactions create rock-forming environments, which in turn give rise to the fundamental, original properties of regional rock assemblages. Furthermore, plate motions, both during and after the construction of regional rock assemblages, generate the stresses that impart to rocks their chief deformational characteristics.

Plate motions in the past have been responsible for shaping mountain belts, high plateaus, and other deformed regions of the earth. Some of the most spectacular expressions of deformation are mountain belts, which are long, broad, and generally linear to curved belts in the Earth's crust where extreme mechanical deformation and/or thermal activity are concentrated. The Appalachians, Alps, Andes, and Himalayas are examples. The major regional structures found in mountain belts reflect systematic distortion of the crust in which the structures are found.

Lithosphere and Asthenosphere, Crust and Mantle

Plates are composed of **lithosphere**, that is, the earth's outermost rind made of material that is rigid enough to withstand very low levels of differential stress indefinitely without flowing (Figure 1.20). Lithosphere consists of the crust (both continental and oceanic) and upper mantle. **Oceanic crust** is relatively thin, ranging from about 4 to 9 km in thickness. It is composed predominantly of rocks of basaltic composition that are relatively high in density (average $\rho = 2.9$). **Continental crust** is relatively thick, ranging from approximately 25 to 70 km, and composed of rocks of granitic composition having relatively low density (average $\rho = 2.7$). The **Moho**, which marks the base of the crust and the top of the mantle, lies *within* lithospheric plates (see Figure 1.20). In fact, it normally lies at a high structural level within lithosphere. The position of the Moho can be identified by a seismic velocity discontinuity. The Moho is thought to mark a lithological transition to underlying ultramafic rocks. It is a *compositional* boundary.

Lithosphere probably is thickest under continents (see Figure 1.20), but this is still somewhat controversial. The relatively thick, convex-downward bulges of lithosphere under continental terranes may serve as "*viscous anchors*" that may halt, or slow down, plate motions (Chapman and Pollack, 1977; Jordan, 1988). Even so, the thickest lithosphere, when viewed at the scale of the Earth as a whole, is mighty thin. Lithospheric plates are like contact lenses on the surface of an eye: exceedingly thin and gently curved. Like contact lenses, the plates move and slip readily.

The lithosphere rides on **asthenosphere** (see Figure 1.20), upper mantle material that is capable of flowing continuously, even under the lowest levels of differential stress (Karato, 1993). The ability of the asthenosphere to flow

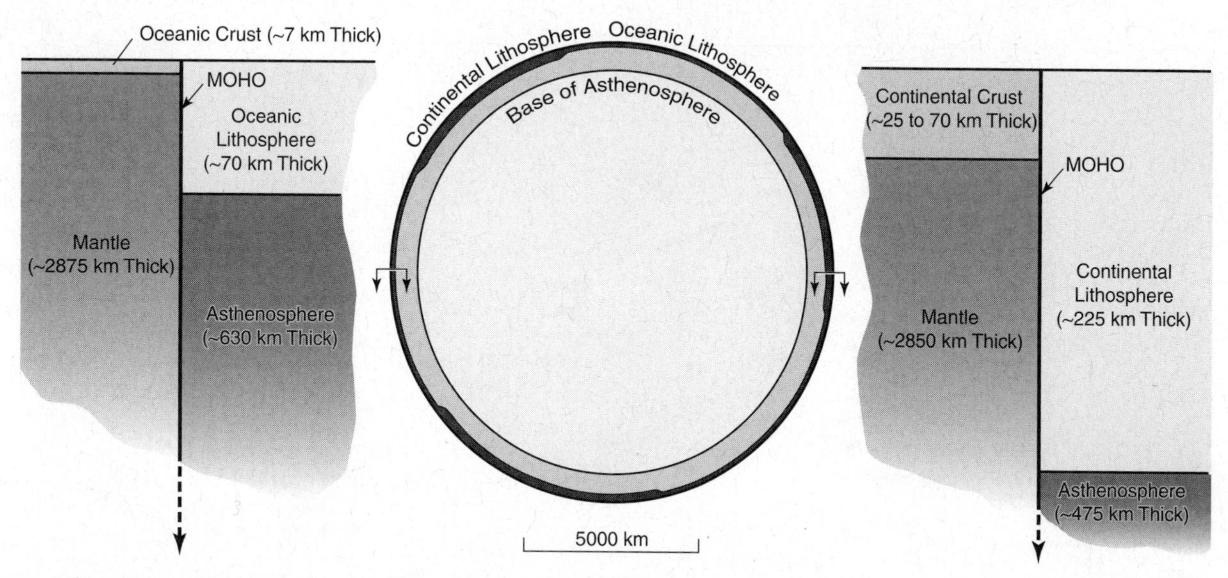

Figure 1.20 Physical/mechanical components of the crust and mantle.

continuously permits movement of the overlying plates of lithosphere. The boundary between lithosphere and asthenosphere is one of major and significant ductility contrast. It is a mechanical boundary, not a compositional one. The asthenosphere may contain a small fraction of partial melt, thus enhancing its capacity to flow in the solid state. The **viscosity** of the lower lithosphere is approximately 10^{23} **poise**, whereas the viscosity of the asthenosphere is approximately 10^{21} poise, two orders of magnitude lower!

The depth of the structural transition between lithosphere and asthenosphere is vague and for now is impossible to define precisely. Under the ocean basins, the depth to the top of the asthenosphere is ~75 km. Under continental crust, the top of the asthenosphere lies at an average depth of ~225 km, but its position is truly variable (Jordan, 1988) (see Figure 1.20). The base of the asthenosphere may lie at a depth of ~700 km. The 700 km depth to the base of the asthenosphere coincides with the depth level of the very deepest earthquakes.

Plate Motions and Plate Boundaries

The lithosphere is not a flawless outer shell. Rather, it is fragmented into many discrete plates that move relative to one another (Figure 1.21). Plate boundaries are the edges of plates, the contacts between adjacent plates. Plate boundaries in modern settings are marked by abundant earthquake activity. In fact, the historic record of earthquake activity around the globe is nearly a tracing of present-day plate boundaries. Depth levels of earthquakes at or near plate boundaries vary from very shallow to very deep, depending on the nature of the plate boundary and the mechanical conditions of the plates brought into contact.

Plate motions can be pictured in terms of convergence, divergence, and transform (Figure 1.22). Relative velocities between plates range from 1 to 22 cm/yr. Plate movements may combine in many ways, depending on the overall plate interactions.

Convergence is marked by a relative movement that brings adjacent plates toward one another (see Figure 1.22). Plates in convergence are in constant competition for space. A common response to the space problem is the structural descent of one plate beneath the other. In effect, rock is "*swallowed*" to greater depth (Bally and Snelson, 1980) through a tectonic process known as

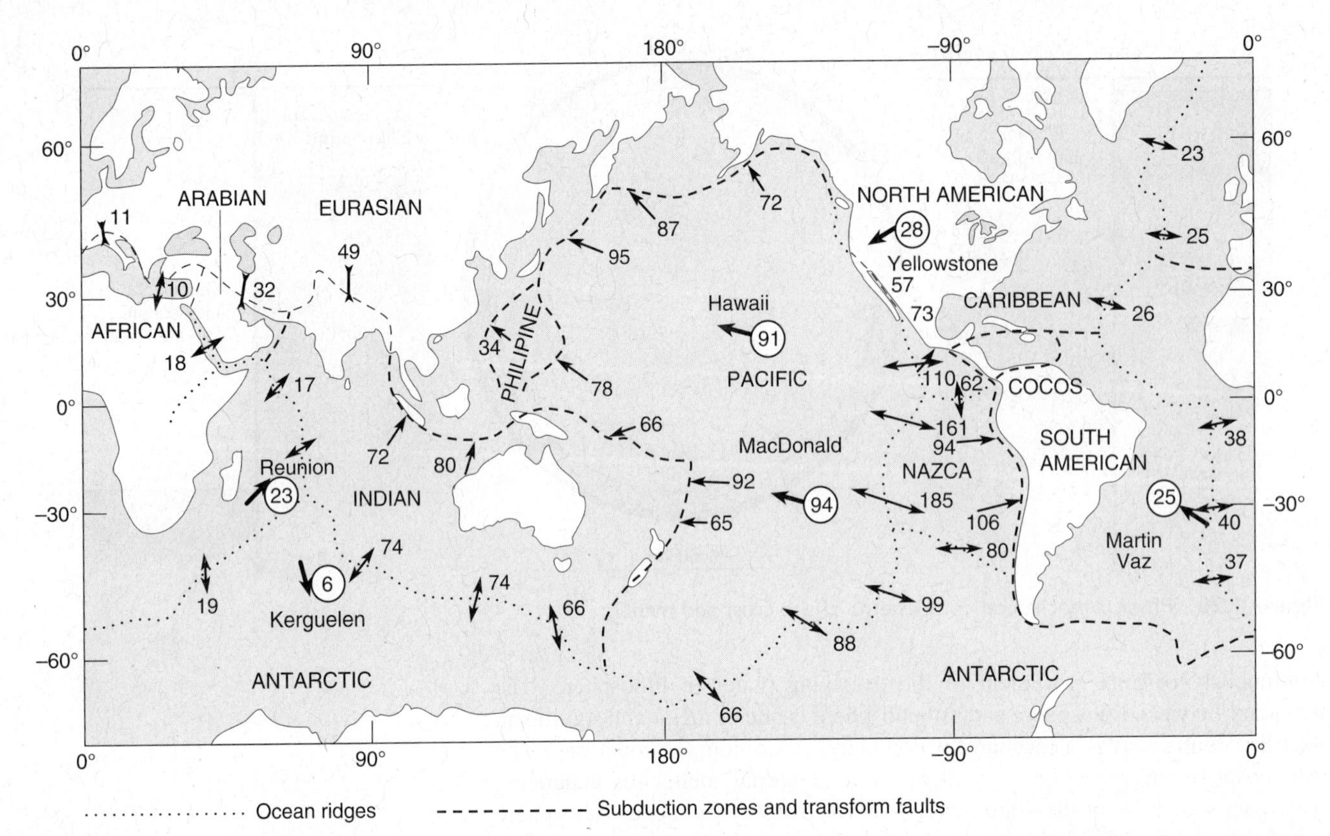

Figure 1.21 The directions of movement and the rates of movement of the relative motions of the plates of the Earth. Velocities are given in mm/yr. Circled numbers (also in mm/yr) are the absolute motions of plates relative to hotspots. [Reprinted from Earth and Planetary Sciences Letters, v. 37, C. G. Chase, Plate kinematics: the Americas, east Africa, and the rest of the world, pg. 355–368, copyright © 1968 by American Geophysical Union.]

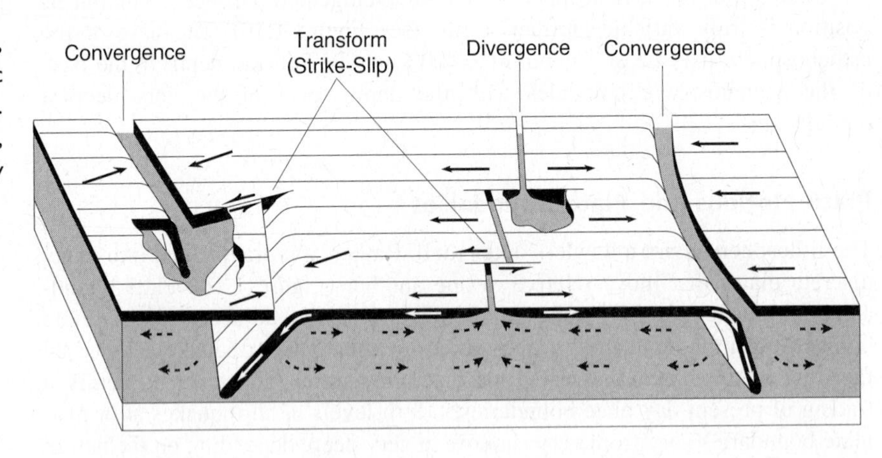

Figure 1.22 Convergence, divergence, and transform (strike-slip) of lithospheric plates. [From B. Isacks, J. Oliver, and L. R. Sykes, *Journal of Geophysical Research*, v. 73, Figure 1, p. 5857, copyright © 1968 by American Geophysical Union.]

subduction. An alternative response of plates to the space problem of convergence is **collision**. Plate collision is like a slow motion, head-on collision on a slippery highway between two cars whose brakes have locked. Converging plates in collision can be thought of as equally buoyant. Still, two plates cannot occupy the same space, and so regional-scale shortening must occur.

Divergence is marked by the movement of plates away from one another (see Figure 1.22). The actual relative movement may be perpendicular or oblique to the boundary between adjacent plates. Without compensation, some void or opening would surely develop between diverging plates. But upwelling igneous intrusions continuously fill in "*would-be*" voids. When solidified, the intrusions and freshly made volcanic and sedimentary accumulations constitute new additions to the lithosphere.

Transform boundaries are where one plate scrapes past another horizontally (Figures 1.22 and 1.23). Steeply dipping fault zones and shear zones absorb the mechanical effects of stresses generated during the frictional movements. Material may be accreted from one plate to another as the plate interaction takes place. Alternatively, material may be sliced off a plate boundary region and removed.

Some plate interactions are accurately described as some combination of convergence, divergence, and transform. The notion is quite similar to describing oblique-slip motions on faults. An oblique *convergence* of plates produces **transpressive deformation** (Figure 1.24). An oblique *divergence* produces **transtensive deformation** (see Figure 1.24).

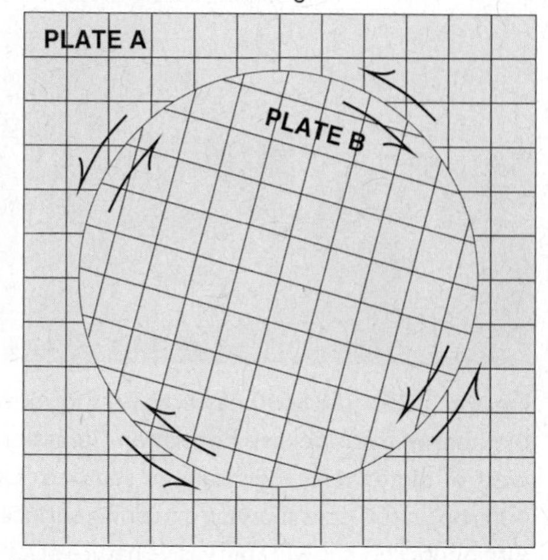

Figure 1.23 Though this is just a cartoon, it underscores that transform faulting is described by motion about a pole of rotation. [Adapted from Cox and Hart, 1986, Box 1.2, p. 13.]

THE FUNDAMENTAL STRUCTURES

To whet our appetites for studying structural geology, we want to introduce the family of fundamental structures. There are three categories of geological structures: contacts, primary structures, and secondary structures.

Contacts, Primary Structures, and Secondary Structures

Contacts are the boundaries that separate one rock body from another. They include normal depositional contacts, unconformities, intrusive contacts, fault contacts, and shear-zone contacts (see *Part III-C, Mapping Contact Relations*). **Primary structures** are typically outcrop-scale features that develop during the formation of a rock body, for example, in sediment before the sediment become sedimentary rock, or in lava or magma before it becomes volcanic or intrusive igneous rock (see *Part III-D, Identifying Primary Structures*). Examples include cross bedding or ripple marks in sandstone, gas vesicles or ropy texture in basalt, and slump structures in clay. Primary structures generally reflect the local conditions of the environment within which the rock forms.

Secondary structures are the principal focus of this text, and, whether big or small, most bear a direct relationship to tectonics and regional deformation. Classically, secondary structures form in sedimentary or igneous rocks after lithification, and in metamorphic rocks during or after metamorphism. The distinction between primary and secondary structures is easily blurred in regions of contemporary regional deformation, where tectonic loading is disturbing yet-unconsolidated sediments that have not had sufficient time to become lithified.

The fundamental (secondary) geologic structures in nature are joints, faults, folds, metamorphic fabrics, and shear zones. We will introduce these here in a general way, but then discuss them in depth in *Part II, Structures*. Keep in mind that each of these categories of fundamental structures is represented by quite a variety of specific forms, and sizes (ranging from hand specimens to outcrop scale to regional).

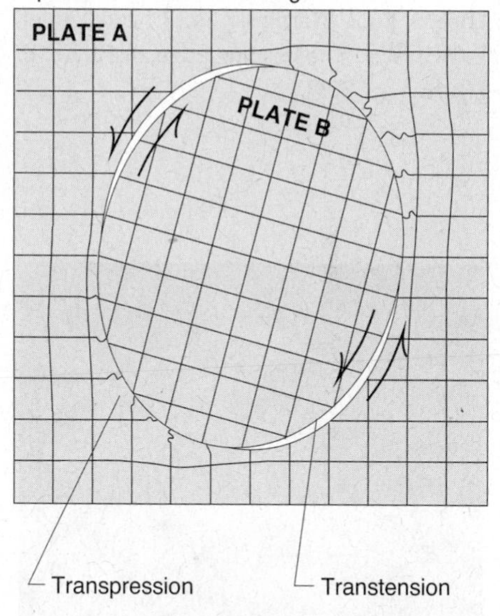

Figure 1.24 Though this is just a cartoon, it underscores that movement along transform faults may not always be perfectly strike-slip in nature. Where transform faulting has a component of compression, the environment produces transpressive deformation. Where transform faulting has a component of extension, the environment produces transtensive deformation.

Joints

Joints are smooth, planar fractures that cut through rock bodies and rock layers, and along which there has been *almost* imperceptible movement (Figure 1.25). **Joints** form perpendicular to the direction in which the rock body or layer is being stretched (i.e., pulled apart). The pulling apart is ever so slight. Joints are often

Figure 1.25 (*A*) Southeasterly view of near-vertical joint faces catching morning sunlight and cutting the Springdale Sandstone Member of the Moenave Formation (Jurassic) on the eastern limb of the Virgin anticline. Location is within Silver Reef mining district, west of the town of Hurricane in southern Utah. Bedding dips approximately 30° southeastward. (*B*) Northwesterly view of same outcrop, in this case showing a bedding surface and the evenly spaced nature of the jointing. (*C*) Northwesterly view of eastern limb of Virgin anticline, showing pervasive nature of jointing in stiff competent beds in the Chinle Formation (Triassic). The traces of the joints are the faint parallel lines on the light-colored bedding surfaces. [Photographs by G. H. Davis.]

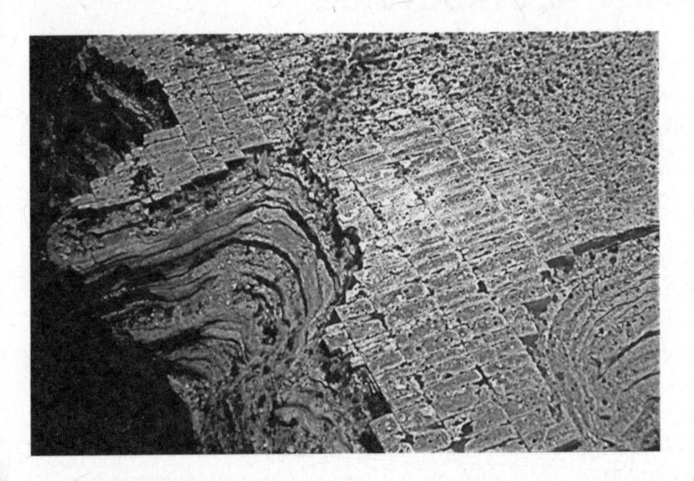

Figure 1.26 Oblique aerial photograph of two sets of regularly spaced, regional fractures in the Cedar Mesa Sandstone, southeastern Utah. Photograph taken by John Lorenz in 1981 while hanging out of a small plane with a good grip on his 35 mm camera. [From Lorenz, J. C., Warpinski, N. R., and Teufel, L. W., 1996, The Leading Edge, v. 15, no. 8, Figure 1, p. 909].

Figure 1.27 Photograph of nearly complete joint surface cutting hard, black, contact-metamorphosed fine-grained sedimentary rock of the Amole Arkose Formation (Cretaceous), collected in the Tucson Mountains, southern Arizona. "*Plume*" pattern on joint face is a signature of pull-apart fracture propagation. The plumes converge to a concave center area, which is the point of initiation for the fracturing. [Collection of G. H. Davis. Photography and editing by Gary Mackender. vr.arizona.edu, glm@email.arizona.edu, © 2009–2010 Arizona Board of Regents.]

remarkable in their consistency of spacing and orientation (see Figure 1.25). Although individual joints commonly are only meters or tens of meters in length and in spacing, we will learn in *Chapter* 5 (*Joints*) that they form in sets that can have regional continuity (Figure 1.26).

The pull-apart nature of joints is revealed in distinctive decorative textural ornamentations (Figure 1.27), which appear on the joint surface itself, and which can be reproduced in laboratory experiments under conditions of extension (stretching). Certain characteristic textures form as the rock rips apart (think of pulling apart Velcro fasteners on a jacket) as the tensional crack propagates at velocities sometimes exceeding half the speed of sound. Elevated fluid pressures in rocks can have a profound influence on the ease of formation, and reactivation, of joints. Movement of fluid along fractures is commonly recorded in the presence of **veins** (Figure 1.28), composed of minerals that precipitated from solution under favorable conditions of temperature and pressure. Where the veins contain precious metals, like gold, things become especially interesting.

Some *joints* prove to be **shear fractures**, which form by an *ever-so-slight* sliding or shearing movement parallel (not perpendicular!) to the plane of the fracture. Movement is so small that even when a shear fracture cuts a distinctive marker, such as a pebble in a conglomerate or a fossil in a limestone, the offset is rarely discernible to the naked eye. Figure 1.29 is an example of shear fracturing. The rock object is a stiff, silicified concretion cut by a network of fractures. These fractures are not "*opening*" fractures (i.e., joints), but

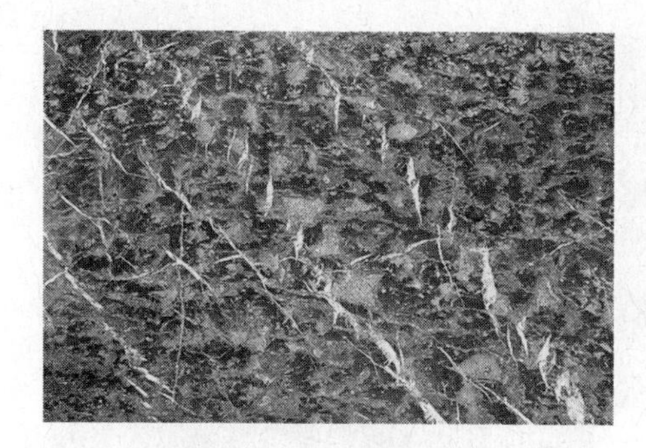

Figure 1.28 Short, stubby, lens-shaped calcite veins (white) in pavement block at the Panhellenic site of Delphi, Greece. These veins are referred to as "*gash veins*," and they exemplify the stretching and pulling apart of the rock. As the gashes opened, they were being simultaneously filled ("*healed*") by precipitations of calcite from circulating fluids. [Photograph by G. H. Davis.]

Figure 1.29 Close-up view of concretion cut by shear fractures. The shear fractures are microfaults along which slip has taken place parallel to the fracture surfaces. As a result of the slip displacements, the original shape of the concretion has become slightly modified. Object was found in highly deformed metasedimentary rocks in the Rincon Mountains, southern Arizona. [Collection of G. H. Davis. Photography and editing by Gary Mackender, vr.arizona.edu, glm@email.arizona.edu, © 2009–2010 Arizona Board of Regents.]

instead they are fractures that accommodated slip parallel to the fracture surfaces. Microfaulting resulted (see Figure 1.29). Imagine if fractures of such small (or smaller) offset cut through rock that had no distinctive internal markers (e.g., pebbles, concretions, fossils). Under such circumstances it might be difficult to determine whether the fracturing was due to jointing or shear fracturing.

It is useful to take just a moment here to think about the term "*shearing*". If Nicholaus Steno, anatomist and structural geologist, were alive today, he would appreciate Dr. R. K. Mishra's (M.D.) (2009) description of how shearing works. (Keep in mind that the original name for "*scissors*" was "*shears*"). There are five steps in the shearing of tissue (which we can think of a rock layer with a bit of elasticity): (1) The blades of the scissors engage the piece of tissue to be cut. (2) When force is applied to close the handles of the scissors, the tissue deforms elastically, in a way that the deformation of the tissue is recoverable. (3) When more force is applied, the tissue suffers irreversible deformation; even if you stop "*cutting*," the tissue will bear the permanent imprint of the deformation. (4) Applying yet more force the tissue will "*fracture*" (Dr. Mishra's word for it!). (5) Finally the tissue separates along the line of the blade, and the line of cutting will continue on the engaged tissue. Rock, like tissue, can be "*engaged*" by shearing, when the boundary conditions of closely adjoined blocks and plates demand it.

Faults

Faults are discrete fracture surfaces, or discrete narrow-to-broad zones, along which rocks have been offset by slip or shearing movements parallel to the fault surface(s) or fault zone(s). In sedimentary sequences, such as that shown in Figure 1.30, the presence of faults can be quite obvious because of abrupt truncation and offset of bedding. The offset is produced earthquake

Figure 1.30 (*A*) Fault cutting Ordovician and Silurian strata in road cut at Canberra traffic circle. Marnie Forster is geologist. (*B*) Closer photograph of same sequence at same general location, showing two faults that abruptly truncate and offset the strata. (*C*) Closer still, this time showing bending (i.e., drag) of strata immediately adjacent to fault. [Photographs by G. H. Davis, while geologist Gordon Lister circled the roundabout 14 times in late afternoon traffic.]

Figure 1.31 (*A*) Photograph of fault surface at the Panhellenic site of Delphi, Greece. Fault surface is polished ("*slickensided*") and marked by slickenlines (i.e., tool marks produced during frictional sliding). The slickenlines plunge directly down the dip of the fault surface. Geologist Phil Nickerson (on the left), and Father Anthony Moschonas (his first slickenlines) closer to camera. (*B*) Close-up view of the fault surface, revealing not only the slickenlines but also the broken, fractured character of the bedrock along the fault zone. [Photographs by G. H. Davis.]

by earthquake over time, and net accumulation of displacement on faults can range from centimeters to kilometers. The direction of fault slip is commonly preserved on fault surfaces in the form of **slickenlines** (Figure 1.31), which for now can simply be thought of as scratches "*tooled*" into the surfaces during the faulting. Fault movement can result in a brittle breakup of rock on either side into various fault rocks and textures, which reveal fault damage.

Certain faults (**normal faults**) permit the crust to be extended and stretched (Figure 1.32*A*). Other faults (**thrust faults**) permit the crust to be shortened (Figure 1.32*B*). Still other faults (**strike-slip faults**) permit parts of the crust to shift horizontally (Figure 1.32*C*). The type of faulting that triggers a given earthquake can be determined through analyzing the pattern of seismic radiation that emanates from slip on a fault lane (see *Part III-S, Determining Focal Mechanisms for Earthquakes*).

Mountain belts are marked by systems of faults, which can completely disrupt and disturb the original stratigraphic relationships. Major faults exert their influence in broad-ranging ways, from creating sites of catastrophic earthquakes, to predetermining locations of "*boomtown*" gold/silver camps where ore deposits are fault controlled. Because of increasing concern about disasters associated with catastrophic earthquakes, the study of contemporary faulting has emerged as a strategically important arena for structural geologists.

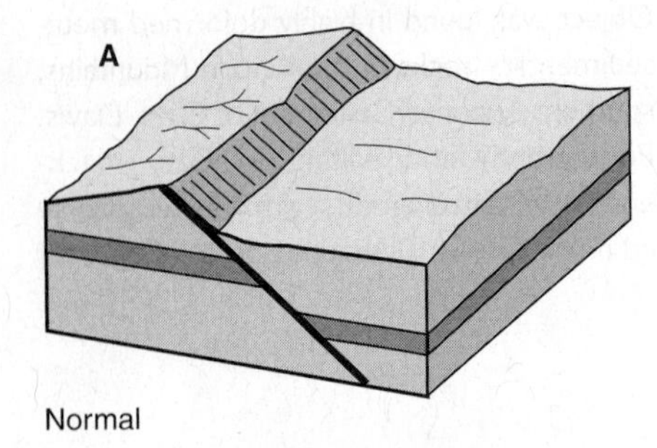

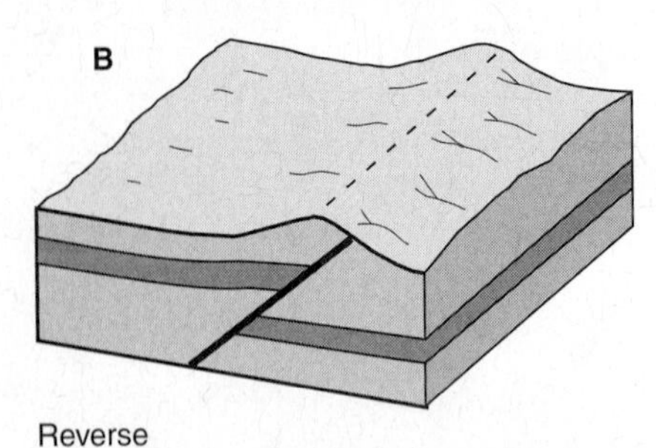

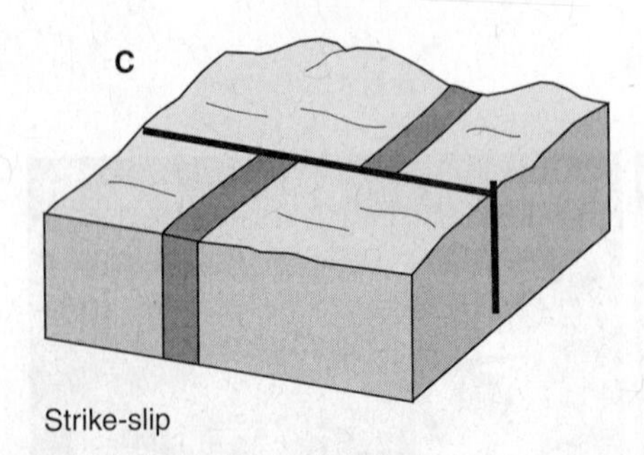

Figure 1.32 Diagrams of three fundamental classes of faulting. (*A*) Normal faulting, which accommodates stretching. (*B*) Thrust faulting (reverse faulting when steeper than 45°), which accommodates shortening. (*C*) Strike-slip faulting, which accommodates horizontal shifting.

Folds

Folds are structures that form when beds and layers are transformed into curved, bent, and crumpled shapes (Figure 1.33). They come in all sizes and shapes (Figure 1.34), revealing in their internal form something of the conditions under which they developed. Folds can form in many ways. They commonly appear to reflect end-on, viselike buckling and shortening of originally horizontal layers (Figure 1.35). Most folds are intimately associated with faults and shear zones, and the geometry of folding has a close dependency on the shape of the master fault beneath. Under high-temperature conditions, plastically deforming layers can produce cascades of folds that pile up like taffy candy (Figure 1.36). In addition, folds may form in unconsolidated sediments that slump and flow, as well as in viscous lava flows or dikes

Figure 1.33 East-directed view of the Raplee anticline, which formed near the eastern margin of the Monument uplift in the vicinity of Mexican Hat, southern Utah. This fold is a "*monocline,*" which is fault-like in that it forms a step in otherwise flat-lying strata (see horizontal beds in both the low foreground and high background). [Photograph by S. J. Reynolds.]

Figure 1.34 Tiny fold (anticline) in siltstone from Cretaceous strata in the Adobe Canyon region of the Santa Rita Mountains, southern Arizona. Next to it is a 1970s-vintage (computer) punch card deck that was placed in a steel pressure vessel and subjected to end-on shortening, causing "*chevron*" folds to form. The deformation experiment was carried out in Berkeley's geology department. [Collection of G. H. Davis. Photography and editing by Gary Mackender, vr.arizona.edu, glm@email.arizona.edu, ©2009–2010 Arizona Board of Regents.]

as these hot bodies move under the influence of gravity. Folds, however formed, offer a wonderful geometric challenge in three-dimensional visualization and analysis. Details of the folded form of a given layer can commonly be related to contrasting stiffness and strength from layer to layer within the sequence of rocks at the time of folding.

When we study large folds that are expressions of regional deformation, we will see that folding and faulting are intimately related. In fact almost always the forms and orientations of big folds reflect the forms and orientations of big faults with which the folds are associated. For this reason we introduce a new chapter in this edition, *Chapter* 8, *Fault–Fold Interactions.*

Figure 1.35 Folded dolomite and limestone near Danby, Vermont. [Photograph by A. Keith. Courtesy of United States Geological Survey.]

Figure 1.36 Cascade of folds in banded gneiss, Rincon Mountains, southern Arizona. [Photograph by G. H. Davis.]

Foliation and Lineation

There is a whole set of challenging small-scale structures that commonly form in metamorphic rocks and igneous rocks. We are familiar with metamorphic rocks such as gneiss, schist, and slate. Each of these rock types is distinguished by the "*three P's*": penetrative parallel planar elements, where "*elements*" covers a wide range of permissible players (2 more "*P's*"). In gneiss, the distinguishing characteristic is **gneissic structure**, marked by penetrative parallel planar layering, compositional bands, and mineral laminae such as ribbons of quartz and aligned augen ("*eyes*") (Figure 1.37*A*). In schist, it is **schistosity**, marked by penetrative parallel planar mica and lath-shaped minerals (Figure 1.37*B*). In slate, the distinguishing characteristic is **slaty cleavage**, marked by penetrative parallel planar elements that, are impossible to discern with the naked eye; yet slate has an undeniable splitting property related to internal alignments of micas and mineral laminae at the finest scale imaginable (Figure 1.37*C*). One aspect of cleavage that structural geologists love is its expression in folds. Cleavage is aligned symmetrically with respect

Figure 1.37 (*A*) Outcrop of banded granite gneiss at Blue Lake spillway at Quandary Mountain, near Breckenridge, Colorado. Geologist is Eliza Davis. Golden Retriever is Lillie. [Photograph by G. H. Davis.] (*B*) Photograph of Otago Schist. Site is the historic Arrowtown Chinese Camp (South Island, New Zealand), where "*In February, 2002, the Government apologized formally to the Chinese community for legal discrimination against New Zealand's early Chinese settlers*," who were invited there in the 19th century to mine placer gold. [Photograph by G. H. Davis.] (*C*) Cleavage in slate near Walland, Tennessee. Note how the orientation of the cleavage (i.e., the splitting property) cross-cuts bedding in this folded formation. [Photograph by A. Keith. Courtesy of United States Geological Survey.]

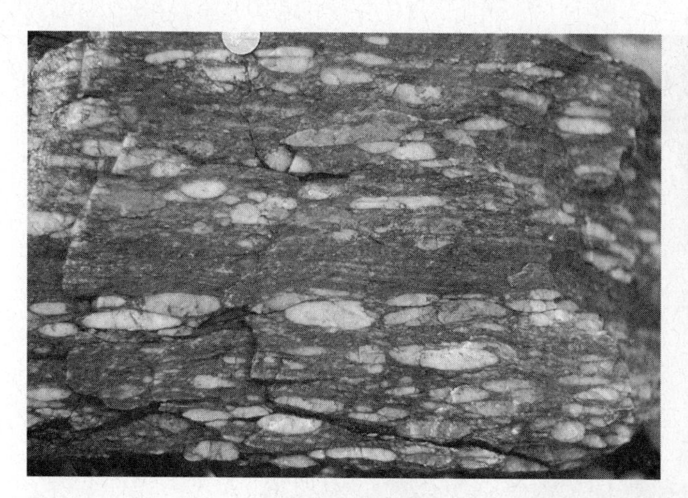

Figure 1.38 Foliation defined by alignment of deformed cobbles in a Triassic conglomerate, Little Piute Mountains, southeastern California. Penny for scale. [Photograph by S. J. Reynolds.]

Figure 1.39 Photograph of alignment of dark "*enclaves*" in outcrop of Catalina Granite (27 Ma) near Tucson, Arizona. These volumes of dark rock were produced at an early stage of intrusion, probably due to magma mixing (mafic magma and felsic magma). Slow flow in the magma caused streamlining and flattening of the globs when magma temperature was quite high. The alignment constitutes foliation, in this case a **primary foliation** produced during the very formation of the igneous rock, as opposed to **secondary foliation** produced by deformation of an already-formed rock body. [Photograph by G. H. Davis.]

to the shape of a given fold; that is, oriented in a way that suggests that the cleavage formed approximately perpendicular to the shortening direction (see Figure 1.37*C*). Cleavage cuts across bedding as if it did not exist!

Thus, gneiss, schist, and slate (and other rocks with the three *P's*) have in common the presence of foliation. **Foliation** is the name we give to the presence of penetrative parallel planar arrangement of elements. Foliation is an umbrella term that subsumes gneissic structure, schistosity, cleavage, and other such fabrics. Even the penetrative parallel alignment of flattened pebbles and cobbles, or deformed worm burrows, in metasedimentary rocks is considered foliation (Figure 1.38).

Foliation is most commonly associated with metamorphic rocks, which formed under conditions of elevated temperature and/or pressure. It is under such conditions that mineral grains can change shape, selectively dissolve or precipitate, and recrystallize. The net effect is the penetrative planar alignment of minerals and microstructures; that is, the presence of foliation. The presence of foliation in metamorphic rocks reflects in part a "*plasticity*" of behavior under conditions of directed stress and flow in the solid state.

Within magma, the presence of foliation may be created as a primary structure during viscous flow. The foliation may develop as the magma intrudes its way upward in the crust, flowing and shearing past the country rock. Foliation in granite may present itself as abundant, aligned disc-shaped black masses ("*enclaves*") (Figure 1.39), which were originally softball- to soccer-ball-sized globules of mafic magma within the much larger volume of granite magma. The globules changed in shape during flow.

Lineations are commonly associated with foliation and cleavage, expressed as preferred linear alignments of elements such as hornblende needles, mineral aggregates, bundles of tiny folds, or striations and grooves, visible when looking at foliation surfaces (Figure 1.40). Whether weakly or strongly developed, a lineation may reveal in its geometry and its physical character the direction of shearing or flowing of the rock. **Linear structure** is neither

Figure 1.40 Photograph of penetrative mineral lineation on layer surface in quartzite. Before weathering and erosion, the quartzite layer was in direct contact with schist, and it was within the schist that the lineation may have been best developed. Vestige of schist is seen in the mica reflections on the surface of the quartzite. [Collection of G. H. Davis. Photography and editing by Gary Mackender.vr.arizona.edu, glm@email.arizona.edu, © 2009–2010 Arizona Board of Regents.]

Figure 1.41 Linear structure in the "*extraordinary striated outcrop*" at Saqsaywaman, Peru (Feininger, 1978). To kids in Cusco this lineated rock is known as *el rodadera* (the rollercoaster). The ride is fast! To slow down, riders spit on their hands and then use the hands as brakes. Taller "kid" in background, waiting his turn, is George Davis. [Photograph by L. A. Lepry.]

delicate enough nor pervasive enough to be penetrative at the microscopic scale; instead, it forms large rod-like and groove-like structures that arise in a variety of ways, including buckling, shearing, and faulting of rocks. Under exceedingly rare circumstances some linear structure is large enough for children and structural geologists to ride (Figure 1.41).

Shear Zones

Shear zones represent a final category of basic geologic structures. Like faults, they accommodate offset, but the offset is distributed across the thickness of a tabular zone that is centimeters, meters, or even kilometers thick (Figure 1.42). Unlike ordinary fault surfaces, shear zones commonly do not display any discrete physical break. Instead, displacement is achieved without loss of cohesion and continuity, although rocks "*caught up*" in shear zones may undergo extreme changes in shape and orientation. The penetrative

Figure 1.42 Shear zone expression of rocks in Sierra Mazatan, east of Hermosillo, Sonora, Mexico. The very fine-grained dark rock (with tiny white chips of feldspar) is a rock called "*mylonite*," which forms by crystal-plastic deformation during extreme shearing. The sense of shear (right to left) is evident in the asymmetry of the folding. Geologist is "*Big Tom*" Anderson. [Photograph by G. H. Davis.]

distributed offset within shear zones can be expressed by the presence of pervasive foliation and lineation. Shear zones typically represent the deep roots of faults at levels where elevated temperature permits the crustal rocks to flow.

Virtual Collection of Geologic Structures

We wish to call to your attention a web-based virtual reference collection of geologic structures, with the hope that it will be useful to you in becoming familiar with the taxonomy and significance of geologic structures. The site is located at http://gdavis.web.arizona.edu/, "*Geologic Structures, from the Collection of George H. Davis.*" There are close to 100 pieces in the collection, and each can be rotated in space to examine 3D properties. Over the years of field work and mapping projects, I (GHD) have done almost all of my "*collecting*" with a camera, and only bring back (rarely) a rock that I graded "A+" in a number of categories: (1) fits in the palm of my hand (in part, so I can easily bring real structures to class); (2) displays beautifully one or more geologic structures; (3) lacks unsightly blemishes (e.g., rock hammer blow or miscellaneous scars and scratches); (4) contains geometry, texture, and hue that combine for a strong aesthetic appeal; and (5) reveals geologic history, including processes of formation of geologic structures.

CONCEPT OF DETAILED STRUCTURAL ANALYSIS

Experience has demonstrated that in mountain belts and other deformed regions throughout the world geologic structures and structural systems are marked by a high degree of geometric order, that the geometric order expresses the deformation, and that mechanical and tectonic principles illuminate ultimate cause and origin. Given this reality, we base our approach in this textbook on what can be described as **detailed structural analysis**. In detailed structural analysis we map the structures and systems of structures, describe the deformation, interpret (where possible) the deformation paths, and explain the origin through mechanics and tectonics. Ultimately we attempt to place the formation of the structures and systems of structures in the context of the overall geological history of the region.

We can think of the mapping, measuring, and geometric analysis required to establish the deformation as descriptive analysis. We can think of describing the deformation history, especially the strain, displacements, rotations, and deformation paths as kinematic analysis. We can think of interpreting the origin of structures in terms of force, stress, and material behavior as dynamic analysis. Incidentally, we use the term "*mapping*" in its broadest context: capturing the two-dimensional and three-dimensional properties of the deformed rock bodies, no matter how big or small, and describing the physical characteristics as carefully as possible. Indeed, "*maps are a way of organizing wonder*" (Heat-Moon, 1991, p. 1, with permission from Steinhart, 1986).

Descriptive analysis is concerned with recognizing and describing structures and measuring their locations, geometries, and orientations (see *Part III-A, Nature of Descriptive Analysis*). Geometric analysis is the core of descriptive analysis in structural geology, for it is the basis for describing deformation. This kind of geometric analysis produces the field of displacement vectors for material points in the deformed volume of rock, and thus establishes the deformation. The displacement field, once established, can be described in terms of **transformations** (equations) that can take the array of

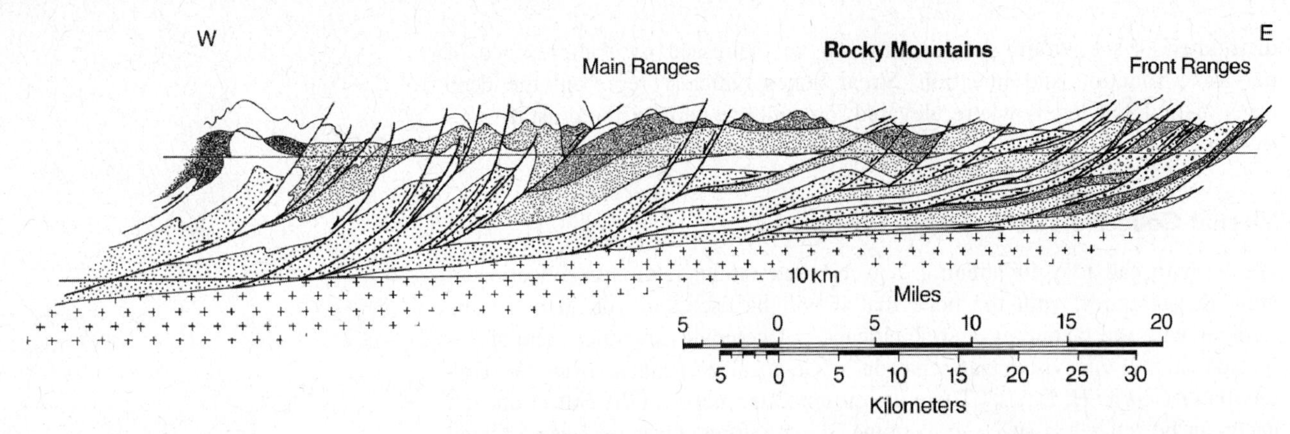

Figure 1.43 Geologic cross-section of folded and faulted strata in a part of the Canadian Rockies. [From Price and Mountjoy, 1970. Published with permission of Geological Association of Canada. © Department of Natural Resources Canada. All rights reserved.]

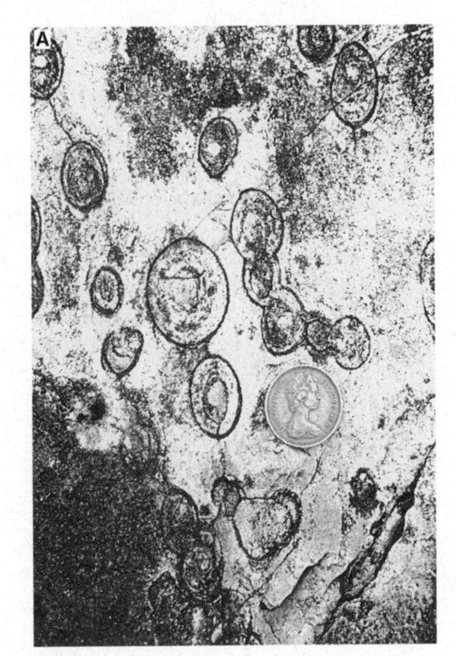

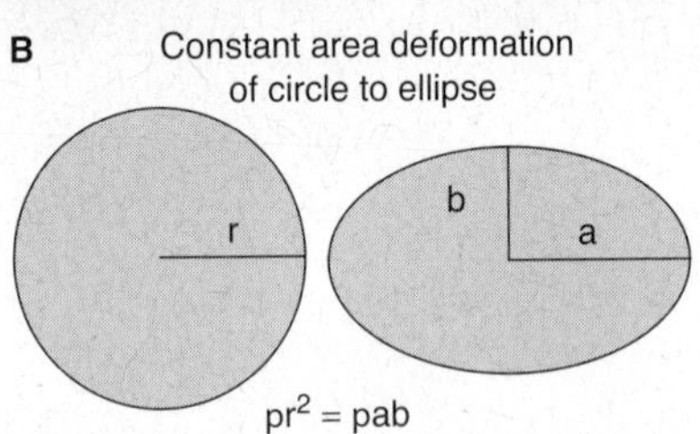

Figure 1.44 (A) Photo of originally circular, now deformed worm burrow in Cambrian sandstone along the Moine thrust near Durness, Scotland. [Photograph by G. H. Davis.] (B) Simple path to determining how the radius of the original circular worm burrow changed in length in the principal directions of strain (longest and shortest). The length of r shortened to b and stretched to a, based on the assumption that the original circular cross-sectional area of the burrow was the same as the elliptical cross-sectional area now observed.

material points in the undeformed volume and arrange them into their new array in the deformed body. The basic transformations are **translation** (change in position), **rotation** (change in orientation), **distortion** (change in shape), and **dilation** (change in size), and more generally can include all four at once.

A number of geologic studies serve as superb models for descriptive analysis. For example, in their classical structural analysis of folded and faulted rocks in the Canadian Rockies, Price and Mountjoy (1970) concisely summarized the results of an enormous amount of data. Their descriptions, plus the geologic cross-section they constructed (Figure 1.43), leave little to the imagination: They specifically called attention to types of structures (thrust faults and folds), structural orientations (southwest-dipping faults, upright folds), shapes of structures (concave-upward faults), relation of faults to bedding (faults cut up-section), and relation of the folds to the faults (thrusts die out as folds). The important descriptive statements and phrases are shown below in **boldface** type to emphasize the extent of descriptive information and the economy of word choice.

The structure of this part of the Canadian Cordillera is dominated by ***thrust faults*** *which are generally* ***southwest-dipping*** *and* ***concave upward in profile****. The* ***faults flatten with depth*** *and have the* ***upper side displaced relatively northeastward and upward****. They gradually cut* ***up through the stratigraphic layering northeastward, but commonly following the layering*** *over large areas. . . . Many of the faults* ***bifurcate upwards into numerous splays****, and the total* ***displacement along them becomes distributed among these splays****.* ***Folds are widespread*** *and have developed in conjunction with the thrusting. . . .* ***Many of the thrust faults themselves are folded*** *along with the sedimentary layering. . . .* ***The folds generally are inclined to the northeast or are upright. Many of the thrusts die out as folds****. . . .* [From Price and Mountjoy (1970), p. 10. Published with permission of Geological Association of Canada.]

Descriptive/geometric analysis is the basis for **strain analysis**: evaluating that part of the deformation that is all about distortion (changes in shape) and dilation (change in size). As a sneak preview, take a look at the deformed worm burrows shown in Figure 1.44*A*. Knowing that these elliptical outlines were circular before deformation and assuming no change in area during the

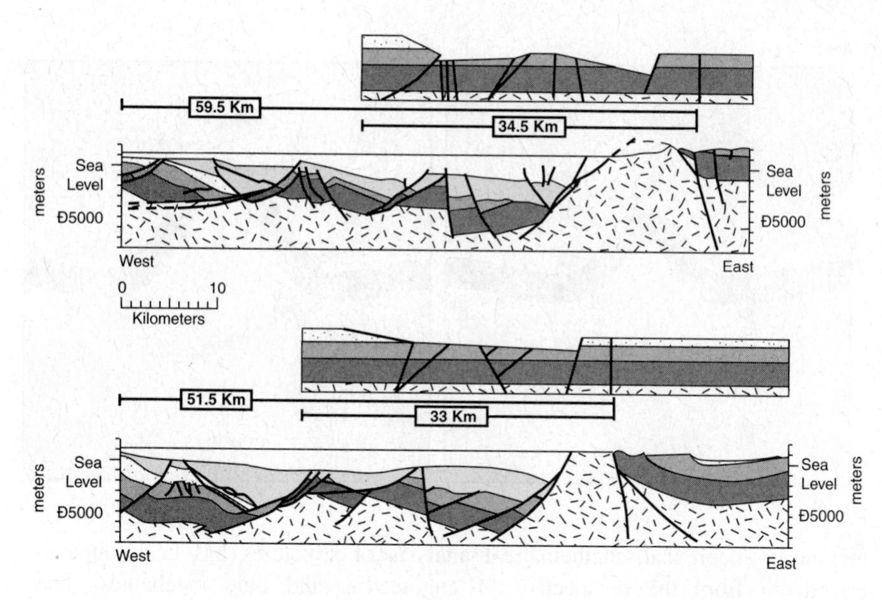

Figure 1.45 Geologic cross-sections through faulted and tilted Precambrian, Paleozoic, and Mesozoic rocks in the Virgin River depression region of Nevada and Arizona. The two longer cross-sections are present-day relationships. The two shorter cross-sections represent the way things started out. Bohannon et al. (1993) restored the section to the geometry that existed before faulting and tilting. By comparing section lengths before and after deformation, they were able to calculate 72% stretching in one example, 56% stretching in the other. [From Bohannon et al. (1993). Published with permission of the Geological Society of America.]

deformation, we can then determine the direction and amount of shortening and stretching that accompanied deformation (Figure 1.44*B*). Shortening and stretching determinations are made across regions as well. Geologic cross-sections (see *Part III-F, Preparing Geologic Cross Sections*) are used by structural geologists to measure how much shortening or stretching is accommodated by geologic structures. For example, in Figure 1.45 we see cross-sections through normal faults and associated folds in the Virgin River depression, a part of the Basin and Range province in southeastern Nevada and northwestern Arizona. Bohannon et al. (1993), who published the cross-sections, carefully "*restored*" the sections as they would have looked before faulting and tilting. By comparing the length of each cross-section of present-day relationships to each restored cross-section, they calculated regional stretching ranging from 56% to 72% (see Figure 1.45).

Kinematic analysis is the branch of mechanics concerned with motion of objects leading to the deformed state, without reference to the forces that cause the motion. Displacement vectors tell us nothing about the **deformation path** traveled by material points and/or volumes of rock, nor the **deformation rate** along each path. Donal Ragan (personal communication, 2004) described it to me this way: "*Imagine that I was in Phoenix, Arizona, two weeks ago, and Los Angeles today. A displacement vector can be used to compare my two positions. Yet there is an infinite number of possible paths I could have directly traveled from Phoenix to Los Angeles. I could have driven to Flagstaff, Arizona, and to Las Vegas (staying there for several nights), and then on to Los Angeles; or I could have flown directly from Sky Harbor Airport in Phoenix to LAX; or I could have flown to New York, London, Bombay, Singapore, San Francisco, and then to Los Angeles; or . . . Furthermore, even if you knew my travel path, exactly, there is an infinite number of possible histories along the way (including rates of travel from place to place to place).*" Kinematics thus is all about interpreting the field of displacement vectors in terms of a field of velocity vectors acting along deformation paths. *Chapter* 2 (*Displacement and Strain*) picks up on this.

Dynamic analysis interprets deformation in terms of force and stress responsible for the formation of structures, as well as evaluating the strength of the materials during deformation. Dynamic analysis is generally the most interpretive part of detailed structural analysis, but, as we shall see in *Chapter* 3 (*Force, Stress, and Strength*), it derives remarkable power from well-conceived experimental and theoretical studies drawn from principles of mechanics and

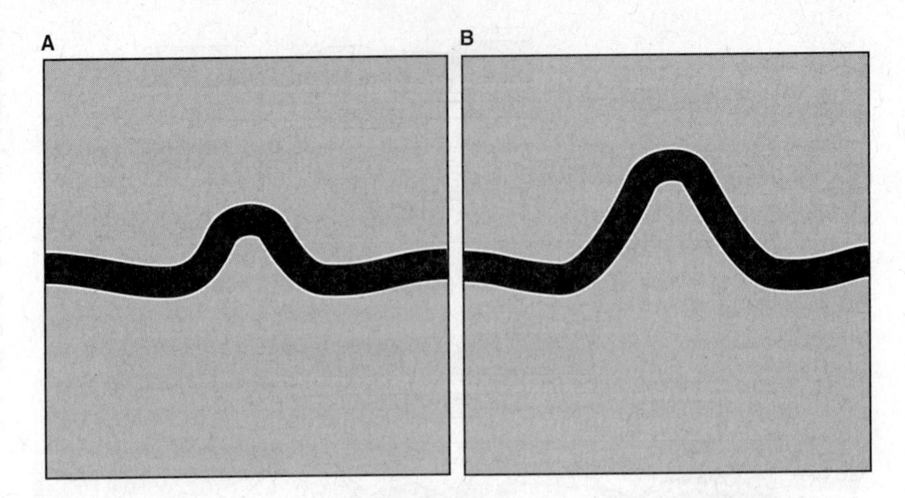

Figure 1.46 Computer simulation images of the buckling of a stiff layer (black) within a soft surrounding matrix (gray). Shortening (at 20%) is the same for each experiment, but contrast in viscosity between stiff layer and soft matrix is different. (*A*) Viscosity contrast of 5:1. (*B*) Viscosity contrast of 100 to 1. [Reprinted from *Journal of Structural Geology*, v. 15, Crystallographic preferred orientation development in a buckled single layer: a computer simulation, Zhang, Hobbs, and Jessell, p. 265–276, copyright © 1993, with permission from Elsevier.]

tectonics. Theoretical, mathematical analysis of structures has been pursued effectively from the perspective of engineering and fluid mechanics. The equations that *"picture"* the deformation are hardly digital photographs of outcrop features. They are more like abstract art.

$$L = 2\pi t \sqrt[3]{\frac{\eta}{6\eta_l}} \quad (1.1)$$

But, equations describe dynamic relationships in ways that words and photographs never could. Decoding the equations simply requires knowledge of what the variables represent. When Equation 1.1 is decoded, it reads:

The wavelength of a fold in a layer of viscosity η and thickness t is equal to the product of (1) 6.832 times the thickness and (2) the cube root of the ratio of the viscosity of the rock layer to 6 times the viscosity of the rocks in which the layer is contained.

The physical expression of this relationship is displayed in Figure 1.46, a computer simulation of a folding event in which a stiff (dark-colored) layer, embedded in a soft matrix, is shortened by 20%. The differences we see in form and size of folding (Figure 1.46*A* vs. Figure 1.46*B*) are due to significant differences in viscosity contrast between the layer and the matrix in the first versus the second experiment. The amount of shortening was the same in each experiment. The application of this piece of dynamic analysis lies in explaining *"ptygmatic"* folding in gneiss in the field (Figure 1.47).

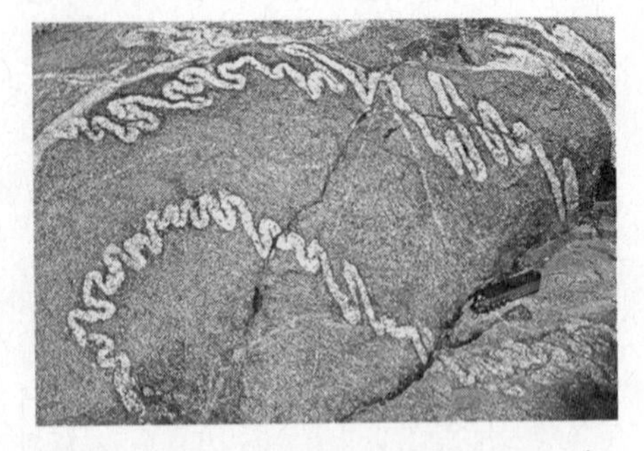

Figure 1.47 Photograph of outcrop displaying beautiful ptygmatic folding. The quartz-rich white layers were much stiffer than the granitic (magmatic) host when the layer-parallel shortening took place. As a result, the quartz-rich layers were free to fold into geometries dictated by layer thickness and ductility contrast between layer and host. Location is Monte Cristo Gulch near Mohawk Lakes, south of Breckenridge, Colorado. Swiss Army Knife for scale. [Photograph by G. H. Davis.]

THE TIME FACTOR

Detailed structural analysis feeds naturally into broader geologic synthesis, but not without a grasp of the dimension of time. Analysis of crustal movements in relation to the overall physical evolution of the Earth simply cannot be pursued outside a time reference framework. Interpreting the timing of deformational movements provides a way to recognize discrete periods of deformation. First, the relative timing of events is determined through such means as cross-cutting relationships. Then, with the help of age dates and knowledge of the ages of rock formations, the absolute timing may be established, or at least constrained. At a more sophisticated level of interpretation, it may be possible to recognize **diachronous** events. "*Diachronous*" in a structural geologic context implies that tectonic loading passed through a region and everywhere produced the same progression of "*look-alike*" structures (e.g., folds and thrust faults), but at different times. It would be wrong to conclude that because the structures look exactly the same they must have formed at the very same time. An analogy for a diachronous event is movement of a cold front from Chicago to Detroit, steadily migrating across the region through time. People in Chicago and Detroit experience the same basic changes in the environment, but the changes in Chicago and Detroit are not synchronous.

Interpreting timing permits the results of structural investigations to be integrated within a framework of broader geological processes and events. It promotes an understanding that structural deformations are just a small part of a much larger orchestration, the knowledge of which serves to clarify why certain structural relationships developed in the first place.

There is a word of caution: Ted Smiley (1964) emphasized that all dating is a matter of interpretation, whether it be a date on a structure in an archaeological site, the date of a particular rock formation, or the date of a specific geological event. He stressed that geochronological interpretations can be strengthened greatly by knowing the exact physical makeup of the sample to be dated, the complete geological history of the rock body from which the sample was removed, and the precise association between the material studied and the event whose age is being analyzed. In other words, meaningful age determinations are derived from paying close attention to the "*primary sources*." Modern radiometric isotopic determinations are yielding ages for rocks and events that are often beautifully consistent with relative age relationships determined independently on the basis of careful evaluation of the historical geology. Yet in structurally complex areas, it is not uncommon to face problems of interpreting whether a given radiometric age determination reflects an age of crystallization, an age of metamorphism, a time of uplift and cooling, or perhaps nothing geological at all.

In practice, the best way to gain initial familiarity with the timing of events and the ages of rocks within a specific region of interest is to search out and study geologic maps. The explanation of any good geologic map provides the time frame for the local or regional geological column. It is based on understanding of the established stratigraphy, knowledge of the critical contact relationships, and age determinations for igneous and metamorphic rocks. The rock column is, in effect, the timepiece to be used as a starting point in investigating the geology of the area or region of interest.

Of course the "*alphabet*" of time is the **geologic time scale**. Learning the geologic time scale places us in the proper framework for contributing, communicating, and understanding. It is first-order business. The day after graduation from high school I (GHD) entered the office of J. Von Feld, exploration geologist for Consolidation Coal Company. His first statement to me was, "*I understand you want to be a geologist.*" His second statement to me was in the form of a question: "*Do you know the geologic time scale*?"

Chapter 2 Displacement and Strain

Temple of Zeus at Olympia in the Peloponnesos, Greece. These travertine drums once belonged to one of 34 multi-drum standing columns (Doric) of the Temple of Zeus. Horizontal ground motion toppled the columns in 5th century A.D. Now the fallen drums present a beautiful analogue for kinematic analysis in the world of structural geology. The collapse was artfully achieved in a way that created for us a cross-section of domino-style faulting. It is as if we are looking at extended continental crust in the Basin and Range province of the western United States. Rotation and slip are explicit in the arrangement. It is all we can do to resist kinematic restoration. [Photograph by G. H. Davis.]

TRANSFORMATIONS

Let us begin by thinking about **transformations**, which describe how the original locations, orientations, shapes, and/or sizes of volumes of rock, or material points within volumes of rocks, become changed. To illustrate in cartoon fashion, let us start off with a familiar, mobile, deformable non-geologic object, namely, a hot air balloon (Figure 2.1). When a hot-air balloon changes position, for example during lift-off, it undergoes **translation** (Figure 2.1*A*), along with all of the material points and lines within it. When it changes orientation, slowly spinning in an eddy, it experiences **rotation** (Figure 2.1*B*), again along with all of the material points and lines within it. These two kinds of changes can occur without any change in shape of the balloon. When the balloon changes in size, due to a heating-up or a cooling-down of the air in the balloon, it undergoes **dilation** (Figure 2.1*C*), which affects the spacing of material points and the lengths, orientations, and spacing of material lines within it. When a hot air balloon changes shape, (and we never want this to happen), it experiences **distortion** (Figure 2.1*D*), and this too changes the spacing of material points and the lengths, orientations, and spacing of material lines within it. Indeed, there are four types of transformations: translation, rotation, dilation, and distortion. More than one can act on a given object at any one time.

Figure 2.2*A* provides another way to picture transformations. The original 10×10 square object lies in the center of the diagram. We might think of this as a *map view* of a volume of rock (in the deep subsurface) that over time is

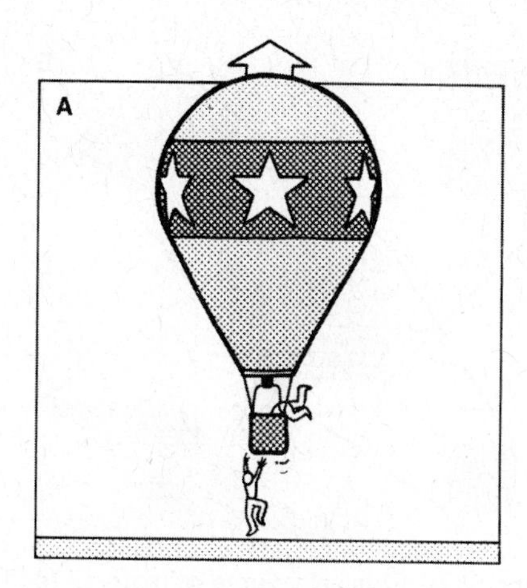

Figure 2.1 Cartoon representation of the four basic transformations: (*A*) **translation** (change in position), (*B*) **rotation** (change in orientation), (*C*) **dilation** (change in volume), and (*D*) **distortion** (change in shape). [Artwork by D. A. Fischer.]

subjected to tectonic forces. When this volume of rock, or some part of it, is forced to change its location or position, it undergoes **translation**; when forced to change its orientation, . . . **rotation;** when forced to change size, . . . **dilation**; and when forced to change shape, . . . **distortion**.

There are two different ways of thinking about the impact of transformative changes on a volume of material, and the distinction is whether the material appears to have behaved as a rigid or as a nonrigidbody in response to force and stress. This distinction between rigid and nonrigid rests on whether the material, *at the scale of observation*, moves intact without a change in shape or size (**rigid-body character**), or instead experiences a change in shape or size along the way (**nonrigid-body character**). Within a rigidbody each point maintains the same exact location *relative* to neighboring points regardless of changes in position and orientation of the volume of material (examples of translation and rotation in Figure 2.2*A*, *B*). This is not true of a nonrigidbody, where spacing of points within the volume of material changes (examples of dilation and distortion in Figure 2.2*C*, *D*).

DISPLACEMENT VECTORS AND DEFORMATION

A fundamental measure used in characterizing deformation is **displacement**, which is the shortest distance from the initial to final position of a point P in a volume of material. It is the length of "*an imaginary*" straight path. A **displacement vector** represents the length (magnitude) and direction of that

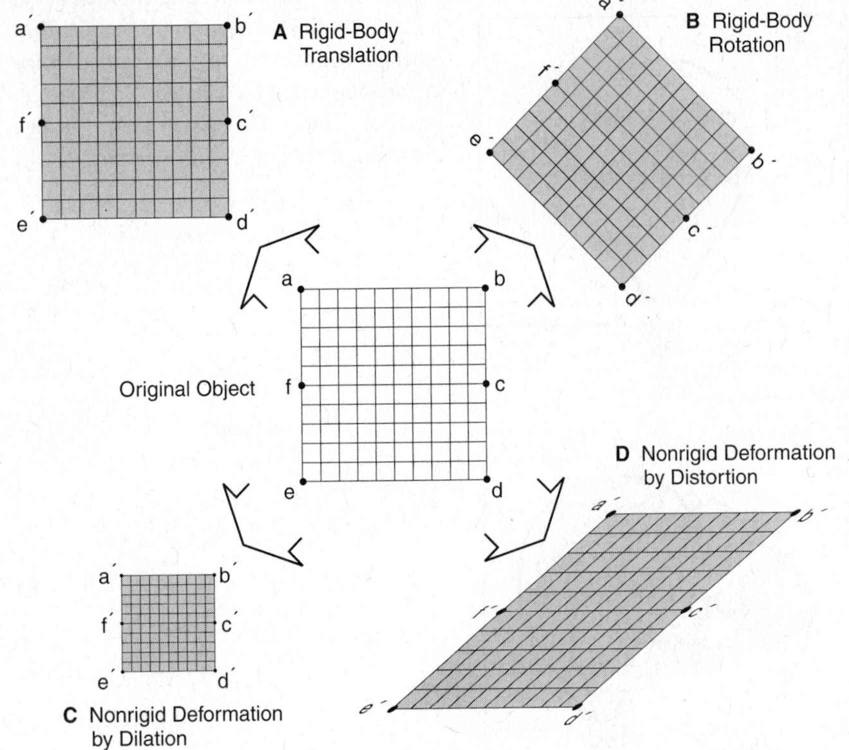

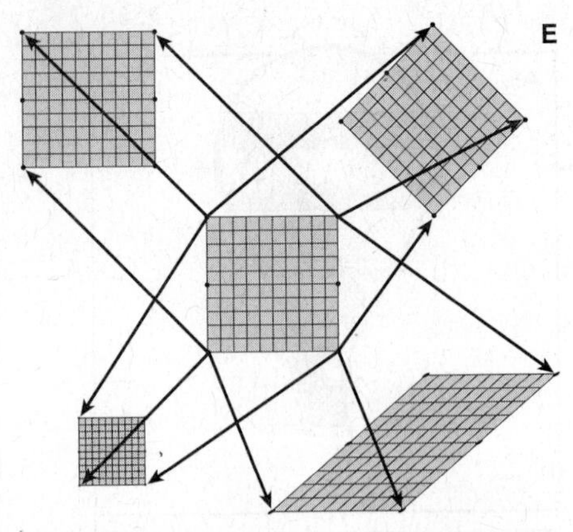

Figure 2.2 Original volume of material (*a-b-c-d-e-f*) in center of figure (i.e., square *abde*) is transformed by (*A*) translation, (*B*) rotation, (*C*) dilation, and (*D*) distortion. Final point locations in each case are *a′-b′-c′-d′-e′-f′*. Translation and/or rotation alone do not alter the relative spacing and configuration of points *within* the body. The volume of material is considered to be rigid. This is not the case for changes achieved through dilation and/or distortion, where the volume of material behaves in a nonrigid manner. (*E*) Here we show displacement vectors for the four cases: translation, rotation, dilation, and distortion.

imaginary straight line. Figure 2.2*E* displays displacement vectors for each of the four end-member transformations that were illustrated above.

The whole array of displacement vectors for a given situation is the **field of displacement vectors**, or simply the **displacement field** (Hobbs, Means, and Williams, 1976, p. 23) (Figure 2.3). **Deformation** is the name given to the displacement field. Said another way, the displacement field describes the deformation. Gaining insight regarding deformation is a descriptive art, achieved through geometric analysis; it requires no knowledge of history or causality.

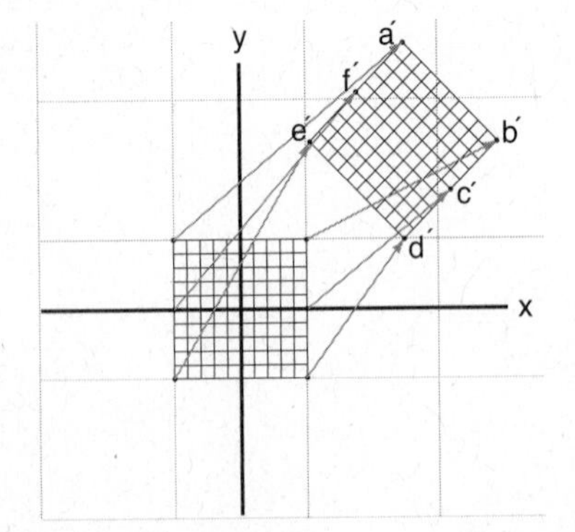

Figure 2.3 Illustration of a field of displacement vectors, also known as a displacement field. The displacement vectors each connect the starting position of a given point with the final position. The deformation, in this case, is a combination of translation and clockwise rotation.

KINEMATICS

Basics

Describing the motion responsible for deformation is the purview of **kinematics**, a branch of mechanics concerned with the aggregate movement path of a 3D volume of material and the individual movement paths of discrete points within the volume. Kinematics is carried out as descriptive analysis, *without* reference to the forces that actually caused or are causing the motion. Kinematics simply focuses on motions.

Part of kinematic analysis requires determining velocities of motion for each displacement vector, thus establishing **velocity vectors** for individual material points in a volume of material and the **field of velocity vectors** for the whole array of material points (see Figure 2.3). This is manageable in actively deforming regions where Global Position System (GPS) geodetic measurements are being carried out. (GPS is one system within the GNSS, i.e., the Global Navigation Satellite Systems). Establishing fields of velocity vectors is much more difficult when analyzing deformation achieved tens or hundreds of

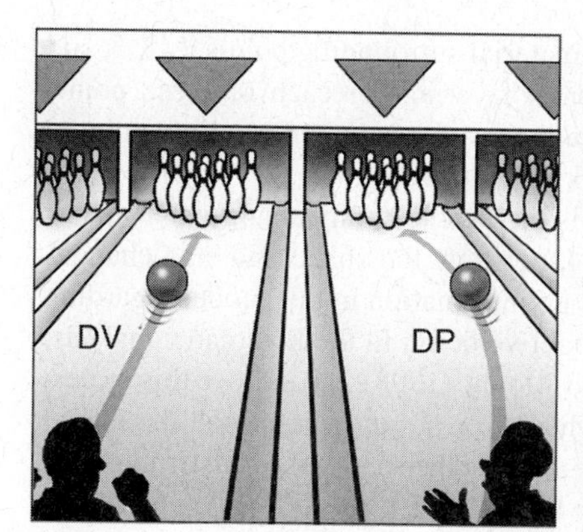

Figure 2.4 Bowlers understand the difference between displacement vectors (DV) and deformation paths (DP). [Artwork by David A. Fischer.]

millions of year ago. But, there are certain circumstances in which the ancient rock record provides the very information we need to construct velocity vectors.

There is more to kinematic analysis than velocity vectors. We want to know the actual **deformation path** followed by each material point from its starting position (in the undeformed volume) to its final position (in the deformed volume). In certain cases the displacement vector and the deformation path may coincide, but we cannot assume this to be the case, nor should we expect this to be the case as a general rule. Bowlers appreciate this difference (Figure 2.4). In fact, professional bowlers avoid at all costs any delivery technique that causes deformation paths and displacement vectors to be one and the same.

Kinematic analysis also focuses on the history of deformation rate along the deformation paths. **Distance** is the length of the path of movement of a given point from initial location to final location. Distance is measured along the deformation path. **Deformation rate** is determined along the deformation path. Again, only in rare circumstances will distance equal **displacement**, which is shortest distance between initial and final position of the material point. The definition of displacement vectors as *imaginary* straight paths is prescient, for the true paths (i.e., the deformation paths) are infinite in numbers of possibilities.

Fault/Fold Illustrations of What We are Talking About

The concept of kinematic analysis, with its focus on distance and deformation paths, can be demonstrated through a schematic of faulting and folding. Figure 2.5*A* presents a computer model of folding in sedimentary cover above a fault, which starts out in granite basement (Mitra and Mount, 1998). The block within which the folding and faulting take place is progressively shortened (by 17%) in four stages (see Figure 2.5*A*). Because Mitra and Mount placed reference points (*1–5*) within the uppermost layer within the sedimentary cover, I (GHD) found it straightforward to use their figure and compare initial and final locations of each of these points, and to construct displacement vectors for each reference point (Figure 2.5*B*). Then, I compared each displacement vector with each (actual) deformation path (see Figure 2.5*B*). Point *5* experienced no change, because the shortening had not advanced this far; the displacement vector for point *5* is nil.

Displacement vectors can be constructed for each of four of the reference points (*1–4*). Note that in each case the deformation paths are not the same as the displacement vectors, because the proportion of horizontal shortening and vertical lift changed during the *B–C* interval (see Figure 2.5*B*). Note also that as a rule the distance measured along a given deformation path is greater than the magnitude of the displacement vector. By examining the model

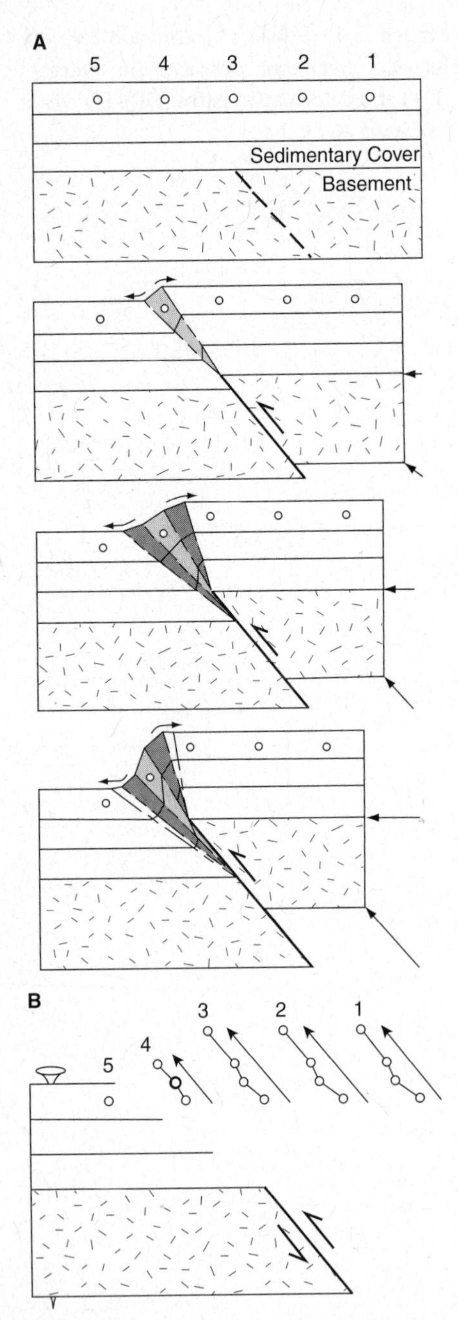

Figure 2.5 (*A*) Computer model showing a fault (dashed black line) in granite basement, followed by the "*freezing*" of three stages in the movement history of the fault. Directly above the line of faulting, in cover, is a deformation zone of folding, which progressively evolves. The reference circles (*1*–*5*) in the top diagram help us to construct deformation vectors and deformation paths. [Adapted from Mitra, S., and Mount, V. S., 1998, Foreland basement involved structures: *American Association of Petroleum Geologists Bulletin*, v. 82, Figure 5, p. 76. AAPG © 1998, reprinted by permission of the AAPG.] (*B*) The displacement vectors are the solid arrows. The deformation paths are the lines connecting the reference circles. See text for explanation.

(Figure 2.5*A*), we can conclude that the material surrounding points *1*, *2*, *3*, and *5* experienced no rotation (note horizontal layering for each of these points holds during the shortening). In contrast, the material surrounding point *4* experienced 90° degrees of counterclockwise rotation.

What is *not* revealed in Figure 2.5*B* is the extent of any change in shape or size of the material body within point *4*, nor how the shape and size changes over time. Furthermore, because there is no information in this model regarding rate(s) of motion, we can say nothing about velocity. In the kinematic analysis, it makes a difference whether Mitra and Mount (1998) carried out this experiment in 10 minutes or 10 hours, and whether or not they deformed the model continuously at the same rate of loading, and whether or not they interrupted the loading from time to time, and for how long.

A Hypothetical Illustration

Things, of course, get complicated when we try to play out the strict meaning of "*deformation*." Figure 2.6 is a cartoon of what can happen, and how under ideal circumstances we would track deformation if we had sufficient information. Portrayed in cross-sectional view are shifts in time, space, and internal configuration of an imaginary block of ground (a_0, b_0, c_0, d_0, e_0, f_0) within the crust. The inset of diagrams along the base of the figure reveals that our referenced block of ground is part of an originally horizontal sequence of sedimentary

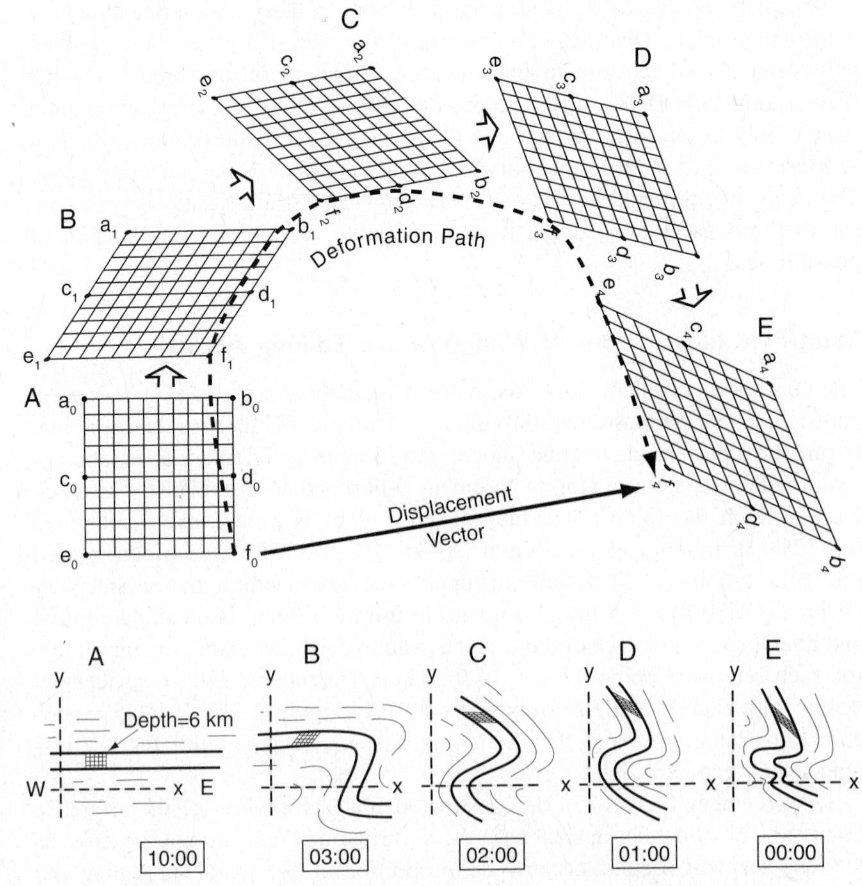

Figure 2.6 Illustration that "*tracks*" the deformation of a block of ground within an actively deforming region. Tiny digital clocks (along base of figure) keep time in millions of years before present. The original block of ground moves upward and eastward and then downward (at different rates at different times) while at the same time undergoing change(s) in size and shape. See text for explanation.

rocks, at 6 km depth, which gets caught up in folding (see Figure 2.6*A*). Throughout the folding, our block of ground remains within the flank of a fold, yet it becomes progressively deformed through time (see Figure 2.6*B–E*). As shown by the digital clock in the inset, the folding began sometime between 10 Ma and 3.0 Ma. Fast change takes place between *B* and *C*, and then things slow down somewhat between *C* and *E*. We are seeing it all happen: the block of ground moves easterly and upward (from *A* through *D*) while at the same time progressively rotates (clockwise), and changes shape. From 1.0 Ma to present the block (rock body) continues to change shape while moving downward and expanding in area (see Figure 2.6).

The net change in location of any given point (e.g., f_0) in the body of rock can be expressed as a displacement vector connecting initial and final position (e.g., $f_0 - f_4$ in Figure 2.6). A given displacement vector would rarely coincide with the actual deformation path traveled by the whole volume of rock. This fact is unambiguous in Figure 2.6*A*: the true path of point *f* is not $f_0 - f_4$.

A whole field of displacement vectors, that is, the displacement field, yields the deformation in its entirety. Go ahead and draw on Figure 2.6 the displacement vectors for each the reference points (*a* to *f*) and sketch out the deformation paths for each as well. The "*journey*" experienced by each point was roundabout, and not achieved at a steady pace!

Kinematics for the Birds

We can appreciate analysis of deformation when we watch hundreds of fast-flying birds in flight as a flock. At any second our eye picks out the 2D shape and size of the flock against the bright sky. The shape and size of the flock change continuously. However, the birds themselves do not change shape or size. They are like points. Second-by-second each bird changes position and orientation (Figure 2.7). The *geometric* challenge of this kind of bird watching is keeping track of the position and orientation of each bird as well as keeping track of the aggregate motion of the flock. Measuring this is technically very challenging; imagine the number of displacement vectors! Even more challenging is the kinematic part, that is, tracking the actual flight paths (deformation paths) and rates of travel for each bird (do they rest by a pond for a while?). Yet all of this can be achieved *without any knowledge* of what is controlling the coordination of birds in the flock.

Figure 2.7 Flock of cranes captured during their winter "*stop-over*" in southeastern Arizona. Row of standing cranes in shallow water gives a sense for original locations, shapes, sizes, and orientations of the material volumes (points) that ultimately make up the collective (flock). Imagine the challenge of following each and every bird in space and time [*x,y,z,t*] while simultaneously characterizing overall shape, size, orientation, and location of the flock as a whole. [Photograph by G. H. Davis.]

Figure 2.8 Cartoon of the paths of birds in flight. The general path of the flight for the flock as a whole is the same, but the distribution of individual bird in flight at any one point of time, and the shape of the aggregate (i.e., flock) at any one point in time, is ever changing. [Artwork by David A. Fischer.]

Kinematic analysis creates the foundation for interpreting dynamics, that is, explaining fundamental causes in terms of forces, stresses, and conditions. Craig Reynolds (1987) created a powerful computer-animation model in which he treated flocks of birds as particle systems, with the birds being the particles. To try to achieve a realistic result he built into the program certain dynamic "*laws*" of behavior, which he then tested through his program to see if it led to a "*deformation field*" and a "*system of deformation paths*" (our terms) that were consistent with the observed. Reynolds (1987) at the time emphasized three particular dynamic behaviors: each bird (1) steers a path close to average heading of the flock as a whole, (2) aims toward the average position of the flock, and (3) makes adjustments to avoid collision with flock mates. Through this work he helped explain the dichotomy of freedom of movement of individual particles (birds) yet cohesiveness of the body (flock) as a whole (Figure 2.8). We see a parallel with the aggregate kinematics of a region of rock, and the dynamic laws at work that govern motions within a system.

DEFORMATION AND KINEMATICS IN ACTIVE TECTONIC SETTINGS

Hotspots and Deformation Paths

A "*hotspots*" example can provide a sense for tracking kinematics and deformation, albeit in a limited way (translation). Absolute motions of plates were very difficult to determine following the advent of seafloor spreading and plate tectonics. It was Jason Morgan who proposed that fixed hotspots in the mantle may constitute absolute points of reference for describing plate motions, that is, translations of lithosphere over asthenosphere (Morgan, 1971, 1972). A hotspot is regarded as a thermal plume, probably of deep mantle origin, that sears the overriding lithosphere. An image that comes to mind in visualizing hotspots is one of slowly passing an old phonograph record over a blowtorch, thus steadily melting and blistering the record along the line of movement (Figure 2.9*A*).

A fascinating geological record of the passing of a plate over a hotspot is the Hawaiian Islands–Emperor Seamounts chain in the Pacific Ocean (Figure 2.9*B*). Oriented west–northwest, the Hawaiian Islands systematically increase in age from east to west. The island of Hawaii, at the eastern end of the chain, lies directly above or just northwest of the hotspot and thus still contains active volcanoes. In contrast, the Midway Islands, at the western end of the chain, are composed of 40 Ma volcanic rocks. Midway Islands formed when that part of the lithosphere was located where Hawaii is located today. Knowing that the distance between Hawaii and the Midway Islands is 3000 km, the actual relative movement of the Pacific plate with respect to the fixed hotspot in the asthenosphere can be determined. The average velocity is 3000 km/40 m.y. (7.5 cm/yr).

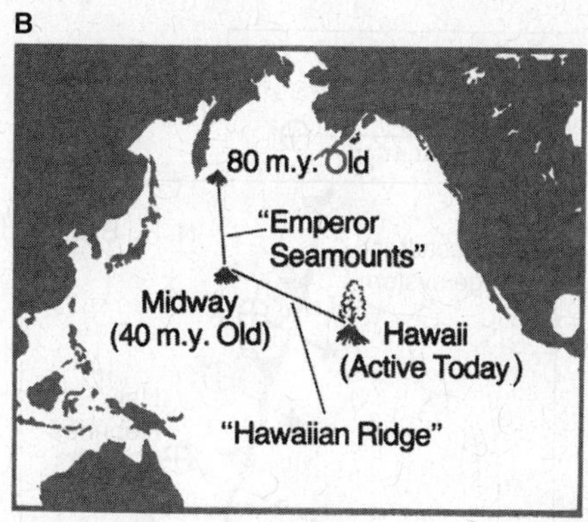

Figure 2.9 (*A*) Cartoon metaphor of translation of lithosphere over a fixed hotspot. [Artwork by D. A. Fischer.] (*B*) Map showing the Hawaiian Islands–Emperor Seamounts chain, an inferred hotspot track.

Northward from the Midway Islands, the Emperor Seamounts trend along a north–northwest line (see Figure 2.9*B*) for a distance of 3000 km, where they encounter the Aleutian Trench. Volcanic rocks of the seamounts increase in age from 40 to 80 Ma from south to north along the chain. Thus we conclude that the Pacific plate moved north–northwest with respect to a fixed hotspot in the mantle during the period 80 to 40 Ma. Combining the Emperor Seamounts and Hawaiian Islands data, Morgan postulated that 40 m.y. before present was a moment of dramatic shift in the direction of relative movement of the Pacific plate lithosphere with respect to deeper mantle beneath the plate.

This approach leads us to an exquisite example, the motion of the African plate since 100 Ma, as recorded in hotspot records. The path of movement of the African plates over hotspots is recorded in the form of linear scars of volcanoes in both the oceanic and continental lithosphere. The ages and mapped locations of the volcanic rocks yield the time-and-place data necessary to unravel displacement vectors and deformation paths (Duncan, 1981) (Figure 2.10). At the scale of mapping we see that certain hotspot tracks (A, B, I, J, K, L, and M) are short and straight, and that displacement vectors and displacement paths are identical for each. On the other hand, other hotspot tracks (C, D, E, F, G, and H) are curved, such that displacement vectors vs. displacement paths can be distinguished. For each of these, the distance along deformation paths is greater than the magnitude of the displacement vector.

What is particularly appealing about Figure 2.10 is that it is a plate-scale map of a displacement field. From it a velocity field map could be produced, and the history of deformation rates along the deformation paths could be evaluated.

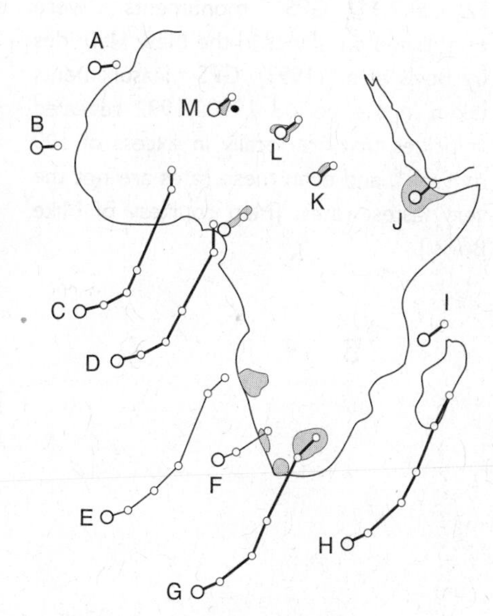

Figure 2.10 Africa hotspot motion since 100 Ma. The presently active hotspots are shown by relatively large open circles (~110 km in diameter). Smaller circles mark probable paths of translation of the African plate. Shaded areas are continental volcanic provinces. A = Canaries; B = Cape Verde; C = Ascension; D = St. Helena; E = Tristan da Cunha; F = Vema; G = Boutvet; H = Prince Edward; I = Comores; J = Afar; K = Jebel Mara; L = Tibesti; and M = Ahaggar. [Reprinted from *Tectonophysics*, v. 74, Duncan, R. A., Hotspots in the southern oceans: an absolute frame of reference for motion of the Gondwana continents, p. 29–42, copyright © 1981, with permission from Elsevier.]

Tonga Trench and Velocity Vectors

The emerging technologies that can be used to describe contemporary plate tectonic activity and deformation are extraordinary. For example, Global Positioning System (GPS) studies permit precise satellite monitoring of the four-dimensional (x,y,z,t) space–time locations of "*targets*" placed in networks on plates. Steel spikes constitute the reference targets, and they are embedded in concrete in the ground, preferably fastened to bedrock. Geoscientists then set up a tripod-supported GPS instrument above the spike. The instrument is wired to a portable computer and receives radio signals from seven or more satellites to triangulate the position of the spike in space and time. Upon revisiting the reference targets periodically over a period of months or years, changes in horizontal and vertical positions can be detected at a confidence level of millimeters.

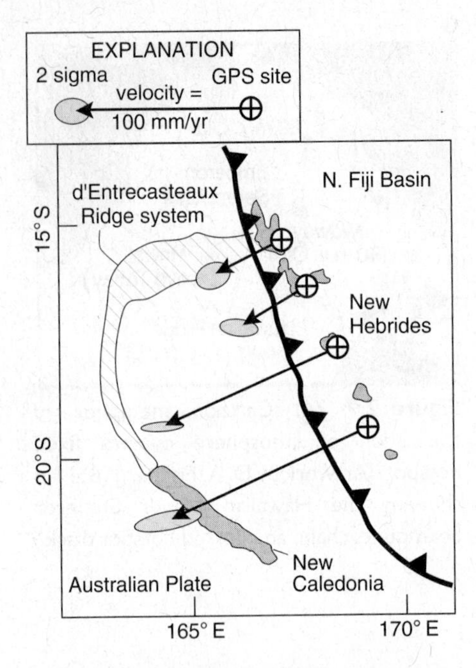

Figure 2.11 GPS monuments were established on islands in the New Hebrides by Bevis et al. (1993). GPS measurements taken in the period 1990–1992 revealed velocities that are locally in excess of 100 mm yr^{-1}, and even these rates are not the very fastest rates. [Map courtesy of Mike Bevis.]

Mike Bevis and his colleagues (1993) placed such a network of spikes on islands in the Tonga and New Hebrides island arcs in the southwest Pacific to measure the motions across a plate boundary between the Australian and Pacific plates. They in fact measured the most rapid relative plate velocity ever observed on Earth: 22 cm/yr across the Tonga Trench!! Given the combination of extraordinarily high rates and steadily improving technologies, Mike believed that he would soon be able to solve for changes in plate rates there every 3 weeks. Figure 2.11 presents the results of measurements taken (just) in the period 1990–1992 in a trench west of the New Hebrides. The velocity arrows are excellent examples of velocity vectors. The gray ellipses are error uncertainties, which are made smaller and smaller with more and more measurements.

Eastern Mediterranean Region and Active Deformation

The eastern Mediterranean region proves to be a perfect "*laboratory*" for demonstrating what is meant by deformation and kinematic analysis. For example, McClusky et al. (2000) have provided a mapping of velocity vectors based upon GPS measurement taken year after year for nearly a decade (Figure 2.12). The velocity vectors capture the westward "*tectonic escape*" of Anatolia, which is being pressed between the Arabian plate and the Eurasian plate, and which is being sucked southwestward by rollback-subduction of the African plate beneath the Aegean plate. Both the African and Arabian plates are moving generally northward (relative to Eurasia), with the Arabian plate outpacing the African plate by ~10 mm/yr. No wonder that the contact zone between them is a major (transform) fault, the Dead Sea transform (see DSTF, Figure 2.12).

Notice that the velocity vectors for stations in western Turkey are generally westward, with average velocities of approximately 20 mm yr^{-1}, whereas the velocity vectors for stations in the Aegean are southwestward and generally greater than 20 mm yr^{-1}, in places nearly 40 mm yr^{-1} (see Figure 2.12). These contrasting directions and velocities of movement are possible only if the region is internally *deforming*, in this case stretching in certain directions. The velocity vectors help us to see why the Aegean region is one such active deformation, marked by constant earthquake activity, active fault scarps (Figure 2.13), and the uplifting and submerging of shorelines (Figure 2.14). To be sure, the Aegean region is being spectacularly extended!

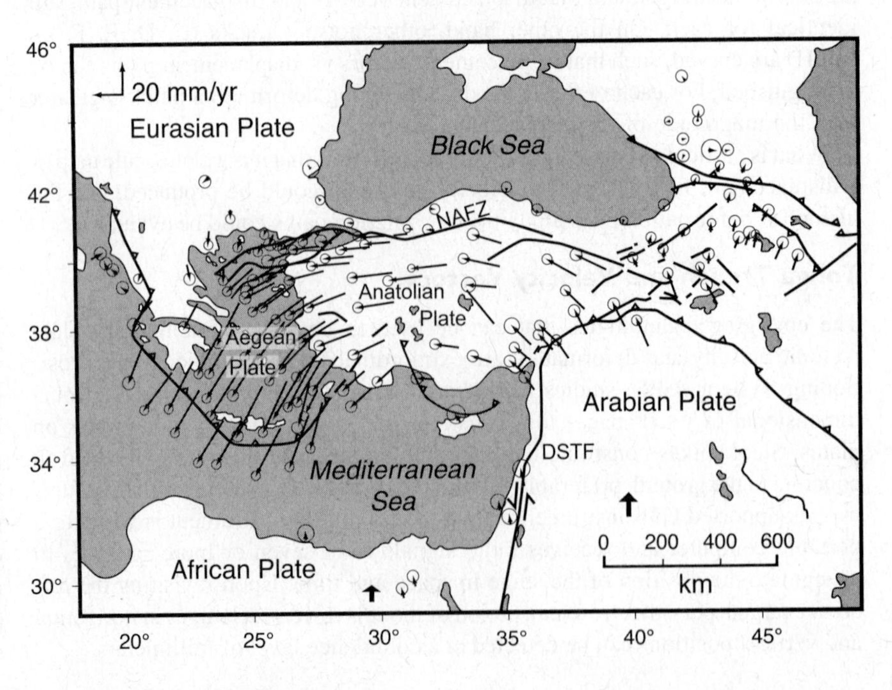

Figure 2.12 Velocity vectors in the eastern Mediterranean region based upon Global Positioning System (GPS) geodesy. Map of velocity vectors for tens of GPS monuments (stations) in the eastern Mediterranean region. See text for explanation. [From McClusky et al. Global positioning system constraints on plate kinematics and dynamics in the eastern Mediterranean and Caucasus, *Journal of Geophysical Research*, v. 105, no. B3, p. 5703, copyright © 2000 by American Geophysical Union.]

Figure 2.13 Kaparelli fault scarp in the eastern Gulf of Corinth region, Greece. The most recent movement along the fault zone took place in a 1981 earthquake sequence (greatest magnitude 6.4). Not all of the scarp shown in this photograph was produced during this earthquake. The height of the 1981 component of scarp formation was less than 1 m. Mike Davis for scale. [Photograph by G. H. Davis.]

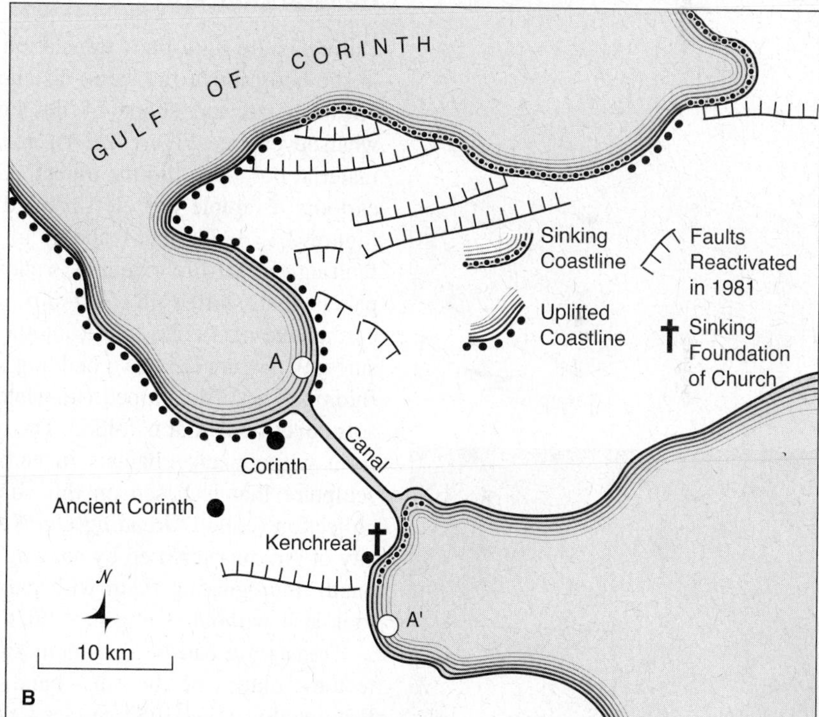

Figure 2.14 (*A*) Submerged foundation of a second-century church at Kenchreai along the southern edge of the Isthmus of Corinth. Submergence is due to the tilting of the Isthmus of Corinth by active faulting. Again, Mike Davis for scale. [Photograph by G. H. Davis.] (*B*) Map showing the Kenchreai church and pier location in relation to faults that were reactivated during earthquake activity in 1981; the sinking coastline and the uplifted coastline. (*C*) Generalized cross-section showing tilting due to faulting (and earthquakes) that is submerging Kechrie and uplifting the coast on the opposite shoreline. [Parts B and C adapted from Earth and Planetary Sciences Letters, v. 57, Jackson and others, Seismicity, normal faulting, and the geomorphological development of the Gulf of Corinth (Greece): the Corinth earthquakes of February and March, p. 377–397, copyright © 1982, with permission from Elsevier.]

Figure 2.14 *(Continued)*

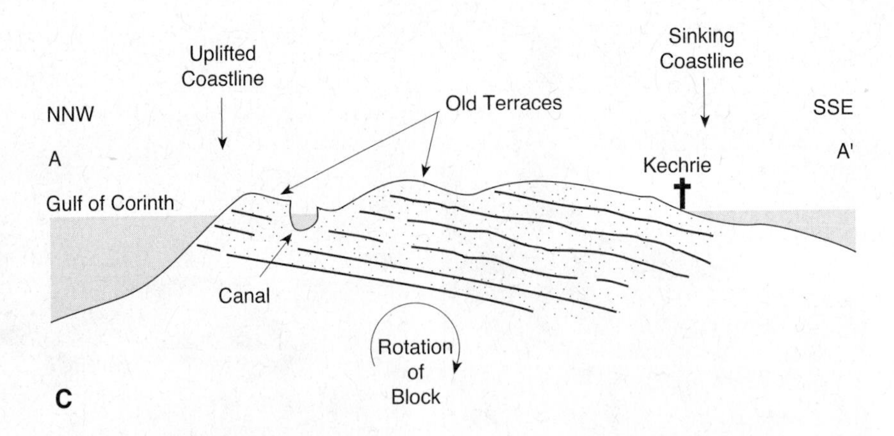

TRANSLATION

General Concept

During a pure rigid-body translation, rock is moved in such a way that all points in the body move the same distances along parallel paths. Consequently, the original size and shape of the body are maintained. There is no rotation whatsoever (see Figure 2.2*A*), nor is there a change in the configuration of material points within the object. All of the deformation paths are the same. A cartoon example of left-handed rigid-body translation is pictured in Figure 2.15*A*. Points *A* and *B* in this block were one and the same before faulting. A real-life example is shown in Figure 2.15*B*, which shows sidewalk panels that shifted in "*left-slip*" fashion during the Managua (Nicaragua) earthquake of 1972. In yet another example of rigid-body movements along surfaces (Figure 2.16 *A*), bedding surfaces were activated during flexural-slip folding. The beds slipped (translated) relative to one another, like pages in a soft-cover book that is folded. The bedding plane slippage permitted this fold to form without any changes in bed thickness. I (GHD) purchased a piece of sculpture that is based on this same concept (Figure 2.16B). It is part of a collection (called "*Reading Our Remains*") that Jessica Drenk (MFA, University of Arizona) created by soaking paperback books in water, folding them by hand, impregnating them with paraffin, and carving them modestly. I cannot look at it without seeing flexural slip.

Translation can be considered at extreme scales as well: on the one hand, tectonic plates; on the other hand, a crystal lattice. As we have seen, Global Position System (GPS) geodesy can capture plate-tectonic movements accurately and precisely. For example, Reilinger et al. (1997) reported that the Northern Anatolian fault, which is the primary tectonic boundary between Anatolia and Eurasia (see Figure 2.12), accommodates an average slip rate of 24 mm/yr, based upon a decade of measurement of the regional GPS velocity

Figure 2.15 (*A*) Schematic portrayal of horizontal left-handed slip on a vertical fault. An original reference point has been split into halves, *A* and *B*. The magnitude of translation is 5 cm. (*B*) Real-life examples of rigid-body translation in the form of sidewalk panels shifted by faulting during the 1972 earthquake in Managua, Nicaragua. Maximum translation is about 3 cm. Aggregate displacement across the 12-m long exposed zone is 28.6 cm. [Photograph by R. D. Brown, Jr. Courtesy of United States Geological Survey.]

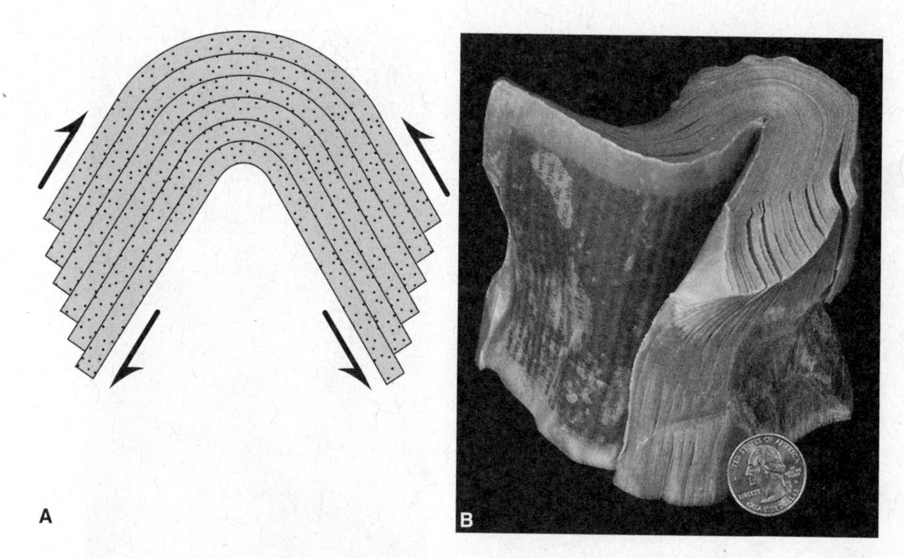

Figure 2.16 (*A*) Flexural-slip folding creates sliding along the layers as if they were pages in a book. (*B*) Flexural-slip art produced by Jessica Drenk. She soaked this used paperback book in water, shaped it, let it dry, and then went to work on it with knives, glue, and melted wax. It is from her "*Reading Our Remains*" collection, some pieces of which I (GHD) have purchased, including this one. Collection of G. H. Davis. [Photography and editing by Gary Mackender. vr.arizona.edu, glm@email.arizona.edu, © 2009–2010 Arizona Board of Regents.]

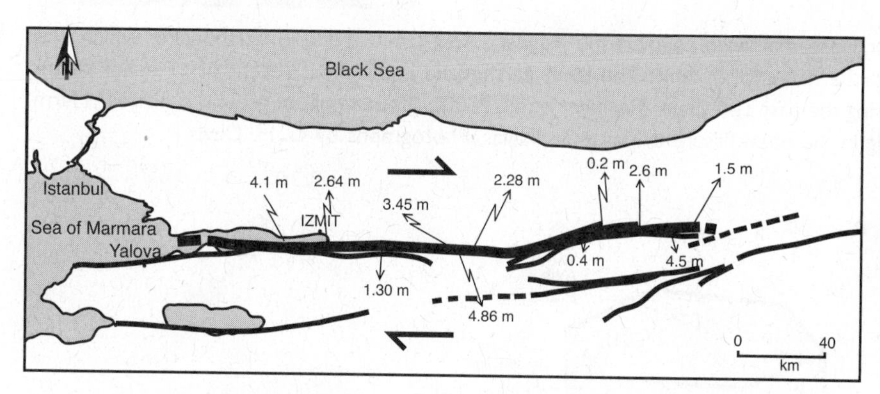

Figure 2.17 Map showing coseismic slip (right-handed) along faults and fracture zones (thick fault lines) during the August 17 and November 12, 1999 earthquakes in Turkey along the North Anatolian Fault Zone (NAFZ). Measured displacements at a number of locations are shown in meters. Faults shown as thinner lines moved during events in 1944, 1957, and 1967. [From Cemen et al., 2000, Turkish earthquakes reveal dynamics of fracturing along a major strike-slip fault zone; EOS Transactions, *American Geophysical Union*, Figure 1, p. 309.]

field. This rate of motion has held for the past 4–5 m.y. Anatolia is moving westward, relative to Eurasia, as a cohesive body along this 1000+ km long right-handed fault zone. Back in August 1999, the fault zone experienced the magnitude (M_W) 7.4 Izmit (Turkey) earthquake (Reilinger et al., 2000). The ground surface ruptured along a length of 120 km in a highly populated fertile area (Figure 2.17), with offsets up to 4.5 m. The earthquake was devastating, killing at least 30,000 people and destroying up to $6.5 billion in direct property losses. I (GHD) visited the site in October 1999, and witnessed firsthand dozens of buildings that had collapsed or had been offset during the August event (Figure 2.18). A student from Middle Eastern Technical University (META) in Ankara described to me her uncle's experience when the earthquake struck in the middle of the night. His apartment was situated directly above the source of the earthquake. He was jarred awake by the incredible shock and tried in vain to open an instantly jammed bedroom door. In order to escape his apartment he opened a window in his 4th floor bedroom and jumped out, quick to discover that even before he jumped the floor of his apartment had collapsed to ground level! Thus he "*fell through the air*" less than 1 m.

At the opposite scale of observation, individual tiny crystals, built from a lattice framework of atoms, can undergo **dislocation** when forced to do so (Figure 2.19*A*). The dislocations permit part(s) of the crystal lattice to translate relative to other parts. The directions of cleavage in a mineral disclose the crystallographic orientations in the lattice along which bonds are most easily

Figure 2.18 (*A*) Example of the building collapse that accompanied the August, 1999 Izmit (Turkey) earthquake, which was triggered by sudden movement along the Northern Anatolian fault. The Izmit earthquake is the most recent of 11 major earthquakes along the Northern Anatolian fault during the past 100 years (Reilinger et al., 2000). (*B*) Example of ground ruptures in farm fields near Golcuk, Turkey. Geologist standing in the open fissure is Hasan Sarikaya. [Photographs by G. H. Davis.]

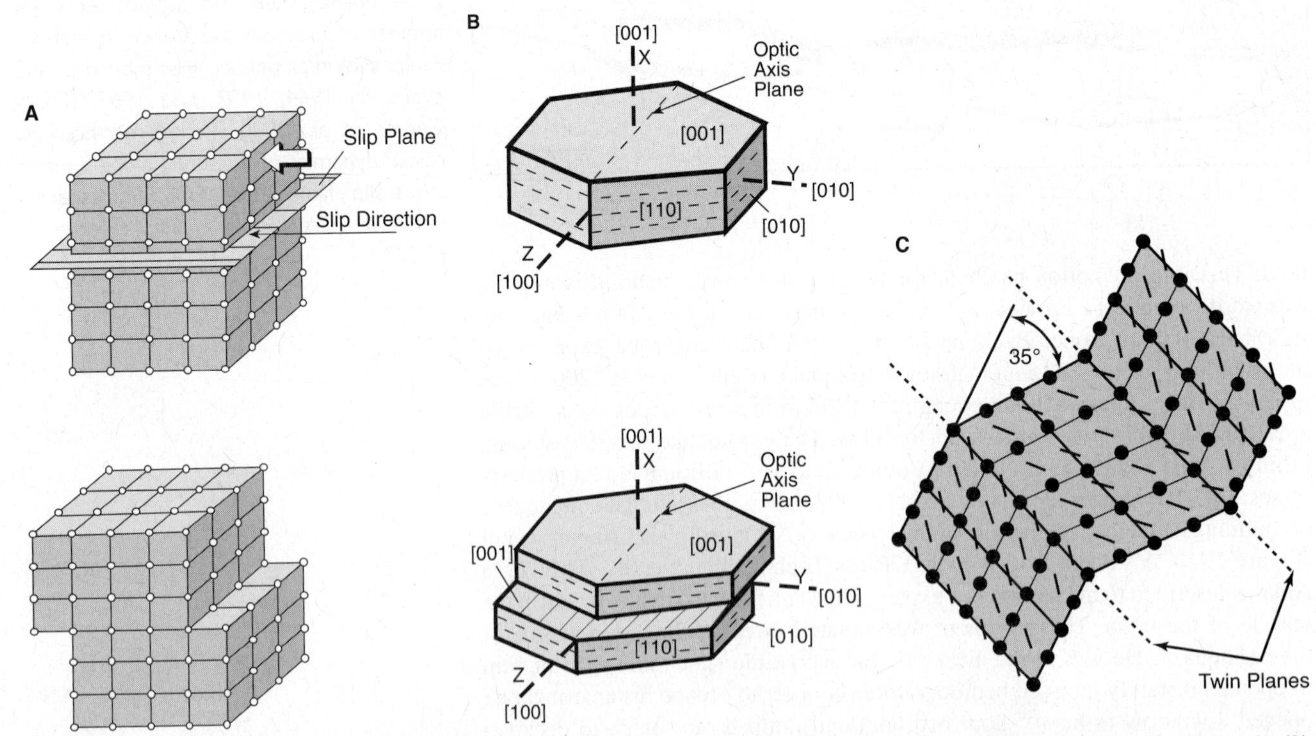

Figure 2.19 (*A*) Slip of crystal along a slip plane, both before (intact lattice showing slip plane) and after (offset lattice). (*B*) Cleavage and slip system in mica, before slip and then after. The intact crystal is marked by the familiar cleavage (001) in mica. Slip then takes place in the direction [100] on the (001) plane. [Adapted from *Crystalline Plasticity and Solid State Flow in Metamorphic Rocks: Selected Topics in Geological Sciences* by A. Nicolas and J. P. Poirier, copyright © 1976. Published with permission of John Wiley & Sons, Inc., New York.] (*C*) Mechanical twins in calcite formed by shearing of the calcite lattice parallel to a crystallographic plane. Solid dots represent Ca atoms and the short lines represent the CO_3 groups. [Adapted from Twiss and Moores, 1992, and from Hobbs, Means, and Williams, 1976.]

broken. For example, the weakest bonds in muscovite are parallel to the perfect (001) cleavage plane in mica (Figure 2.19*B*). Translations along slip planes in crystal lattices accumulate to produce true distortion of the overall crystal lattice, and this is not uncommonly expressed as a permanent change in shape of the mineral of which the lattice is a part (Figure 2.19*C*).

Displacement Vectors

As we have learned, rigid-body translations are expressed conveniently in terms of displacement vectors, which describe the initial versus final positions of points within a body. If we know the original, pre-deformation $[x,y,z]$ location of a point in the body, and also know the final $[x',y',z']$ positions, we can construct the displacement vector. The displacement vector can be described in terms of three parameters: distance of transport, direction of transport, and sense of transport. As we can imagine from the examples presented above, **distance of transport** can range from fractions of millimeters to hundreds or thousands of kilometers. **Direction of transport** is expressed as the **trend** and **plunge** (see *Part III-E, Measuring the Orientations of Structures*) of the line of movement. (For example, we might say: "*The movement was along a northeast-southwest line*"; or "*the movement was along a line plunging 10°N 40°E.*") **Sense of transport** refers to the "*polarity*" of the movement. (For example, we might say, "*the movement was from southwest to northeast*"; or "*the movement was to the southwest, along a line plunging 10° in a direction S40°W.*")

These three components of displacement vectors are illustrated in Figure 2.20, a photograph and interpretive diagram of one of the "*sliding*

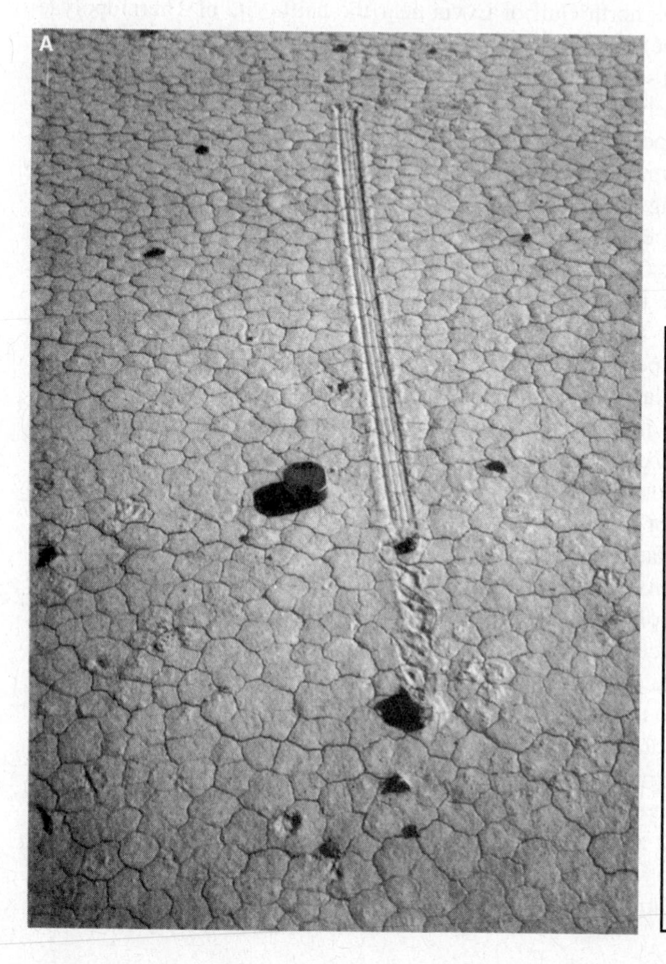

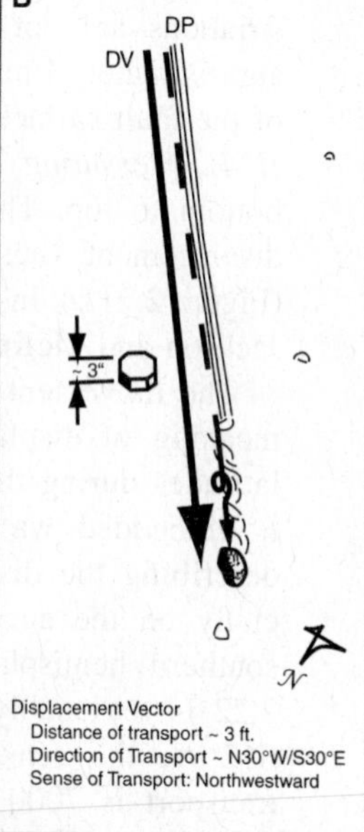

Figure 2.20 (*A*) Sliding stone on Racetrack Playa, California. [From Sharp and Carey, 1976. Published with permission of Geological Society of America.] (*B*) Displacement vector (*DV*) and deformation path (*DP*) for one of the sliding stones. Note that the displacement vector (straight!) does not coincide with the actual deformation path (slightly curved). The "*squiggle*"—quite a departure from the displacement vector—attempts to represent the flipping of the stone.

stones" on Racetrack Playa in California (Sharp and Carey, 1976). The stones that occupy the Racetrack Playa appear to have translated across slippery wet mud surfaces, propelled, apparently, by the force of the wind (see Figure 2.20*A*), though recent interpretations suggest that some of the grooves are tool marks left by rocks entrained in and projecting downward from thin sheets of ice, which were themselves blown by the wind (Reid et al., 1995).

The trail left by the stone of interest here (see Figure 2.20*B*) allows the displacement vector to be constructed. The stone appears to have slid smoothly along most its course, but then it encountered a small stone obstacle that caused it to flip over several times and come to rest. The displacement vector connects $[x,y,z]$ to $[x',y',z']$ (see Figure 2.20*B*), with a distance of transport of 3 m, a direction of transport of N30°W/S30°E, and a sense of transport from southeast to northwest.

Note again that there is a difference between a displacement vector and the actual deformation path (see Figure 2.20*B*). The deformation path along the smooth course of translation is slightly convex northeastward, and the bumpy part of the path is slightly convex northwestward. In plan view representation, we symbolized the "*flipping*" of the stone along its deformation path by a "*squiggle*." The distinction between displacement vector and deformation path is apparent in the plate tectonic and structural geological worlds as well.

A structural geologic example that highlights the distinction between "*displacement vector*" and "*deformation path*" can be seen in the work of Jackson and McKenzie (1999). They describe fresh fault striations (i.e., slickenlines), which are exceedingly well exposed on the Arkitsa fault in central Greece, along the south side of the north Gulf of Evvia near the battle site of Thermopolyae (Figure 2.21*A*). The bedrock cut by the fault is massively bedded Mesozoic limestone. The fault surface itself, which strikes N80°W and dips 60°N, had been buried by a relatively thin sheet of limestone scree (talus) coming off of the steep mountain slope. Because it is a good source of road material, scree has been quarried and removed, exposing a polished fault surface over a length of 500 to 600 m along strike. From place to place there are patches of freshly exposed fault surfaces up to 80 m in length and 40–50 m high (Figure 2.21*B*). According to Jackson and McKenzie (1999), fifty earthquake cycles were required to produce this fault surface, which is marked by corrugations in the form of lineations and grooves (Figure 2.21*C*). The orientations of fault striations are not perfectly straight up the height of the fault, but curve approximately 1 m laterally over 50 m. In other words, the slip vector at the top of the fault surface is less oblique than at the bottom, with the **rake** (*see Part III-E, Measuring the Orientations of Structures*) increasing 1° degree from bottom to top. Though this change is very subtle, it demonstrates that the displacement vector does not always coincide with the deformation path (Figure 2.21*D*). In fact, this difference is what attracted the focused attention of Jackson and McKenzie (1999).

The movement of the Indian Plate provides a plate-scale example of the meaning of displacement vectors. India was positioned in deep southerly latitudes during the early Cenozoic Era, and, later, the plate in which India is embedded was translated far to the north (Figure 2.22). The values describing the displacement vector for India's northward ride depend crucially on the accuracy of early Cenozoic reconstructions of plates in the southern hemisphere. Assuming that the reconstruction shown in Figure 2.22 is reasonably accurate, the direction of transport for India is N12°E/S12°W, the sense of transport is north–northeast, and the distance of transport is 7000 km. This is probably a close approximation of the true deformation path.

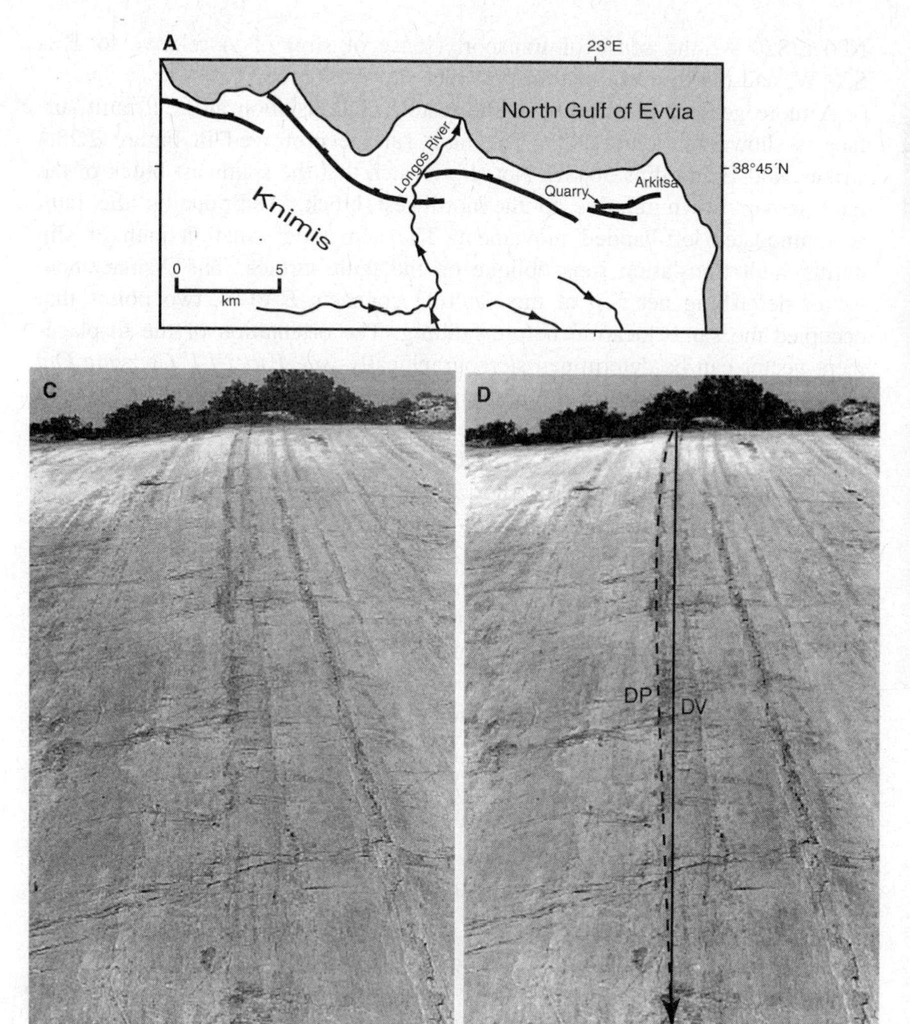

Figure 2.21 (*A*) Map of the Arkitsa fault in central Greece. The zone trends west-northwest along the south side of the north Gulf of Evvia. (*B*) South-directed photograph of fault-surface exposure in quarry. (C) Photograph of fault striations on the fault surface. (*D*) Representation of the difference between the displacement vector and the deformation path. [Parts C and D reprinted from *Journal of Structural Geology*, v. 21, Jackson, J., and McKenzie, D., A hectare of fresh striations on the Arkitsa fault, central Greece, p. 1–6, copyright © 1999, with permission from Elsevier.]

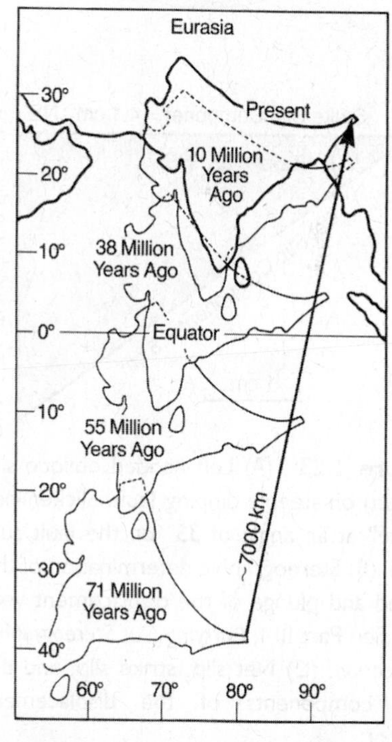

Figure 2.22 Reconstruction of the northward translation of India during the Cenozoic Era. Calculation of the displacement vector depends upon the interpretation of the starting position of India. [Reprinted from *Science*, v. 189, Molnar, P., and Tapponnier, P., Cenozoic tectonics of Asia: effect of a continental collision, p. 419–425, copyright © 1977 with permission from American Association for the Advancement of Science.]

Translation (Slip) on Faults

The concept of a displacement vector as a description of **net slip** can be applied to fault analysis. The net slip on a fault can be described by the displacement vector which "*connects*" the $[x,y,z/x',y',z']$ locations of two reference points, one on each side of a given fault, which shared a common location in the body before faulting. Net slip is *the actual relative displacement* between two points that occupied the same location before faulting. Recall Figure 2.15*A*, which shows a rigid block cut by a vertical fault that strikes N20°E. Translation on the fault was strictly horizontal and **left-handed,** which means that a displaced marker, when followed to the trace of the fault, is offset to the left. Since points *A* and *B* initially occupied the same location, they provide a means to evaluate net slip along the fault. The direction of transport (or direction of slip) is

N20°E/S20°W; the sense of transport (sense of slip) of *A* relative to *B* is S20°W; and the distance of transport (net slip) is 5 cm.

A more general illustration of the results of translation along a fault surface is shown in Figure 2.23. The fault surface portrayed in Figure 2.23*A* strikes N60°E and dips 50°SE. Net slip is such that the southeast block of the fault moves down relative to the northwest block. Additionally, the fault accommodates left-handed movement. The actual deformation path of slip during fault translation runs oblique on the fault surface. The displacement vector describing net slip of this faulting connects *B* to *A*, two points that occupied the same location before faulting. The orientation of the displacement vector can be determined stereographically (see *Part III-I, Carrying Out Stereographic Projection*) by plotting both the fault plane and the slip direction, and then interpreting trend and plunge (Figure 2.23*B*). The sense of slip of *B* relative to *A* is 26°N85°E. The magnitude of the total displacement (i.e., the **net slip**), is 5.0 cm. The horizontal component of the slip (**strike-slip** component) is 4.1 cm; and the down-the-dip component of translation (**dip-slip** component) is 2.9 cm (Figure 2.23*C*).

Reference points for determining net slip are not uncommon, particularly in regions of active faulting. For example, along active faults there are many natural and man-made reference points that permit fault translation(s) to be measured. Displacement vectors can be calculated on the basis of faulted streams (Figure 2.24*A*) and even faulted city streets (Figure 2.24*B*). The

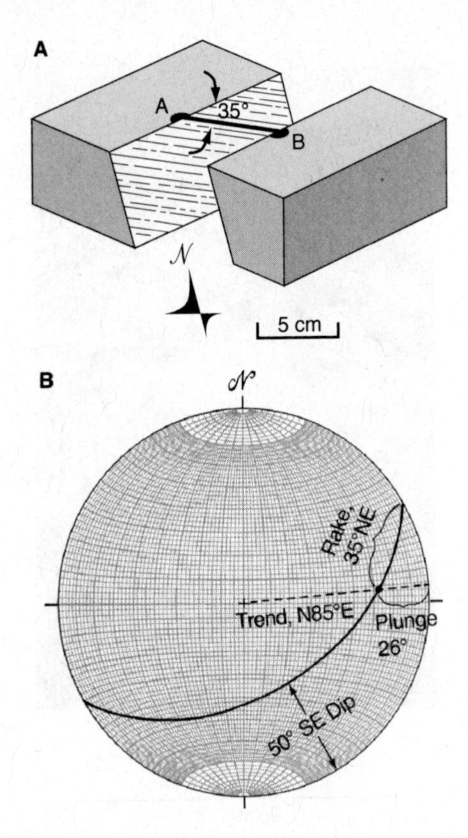

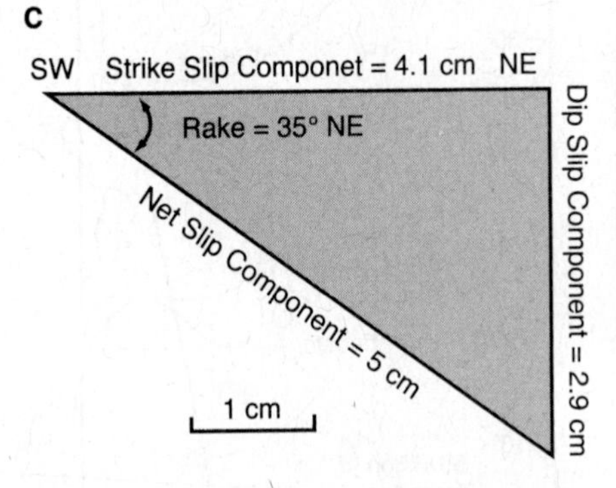

Figure 2.23 (*A*) Left-handed oblique slip (5 cm) on steeply dipping fault. Slickenlines "*rake*" at an angle of 35° on the fault surface. (*B*) Stereographic determination of the trend and plunge of the displacement vector. See Part III-I, *Carrying Out Stereographic Projection*. (*C*) Net slip, strike slip, and dip slip components of the displacement vector.

Figure 2.24 (*A*) Right-handed offset of stream due to movement(s) on the San Andreas fault as exposed in the Carrizo Plains of California. [Photograph by R. E. Wallace. Courtesy of United States Geological Survey.] (*B*) Faulting and fissuring of a street as a result of the Great San Francisco Earthquake of 1906. [Photograph of Bluxom Street, near Sixth Street by G. K. Gilbert. Courtesy of United States Geological Survey.] (*C*) Right-handed offset of Richardson Highway achieved during the Denali fault earthquake in 2002. View is from north to south, with fault trace running NW/SE. Right-handed offset is 2–2.5 m. [Photo courtesy of Professor Akihiko Ito, Utsunomiya, University of Japan.]

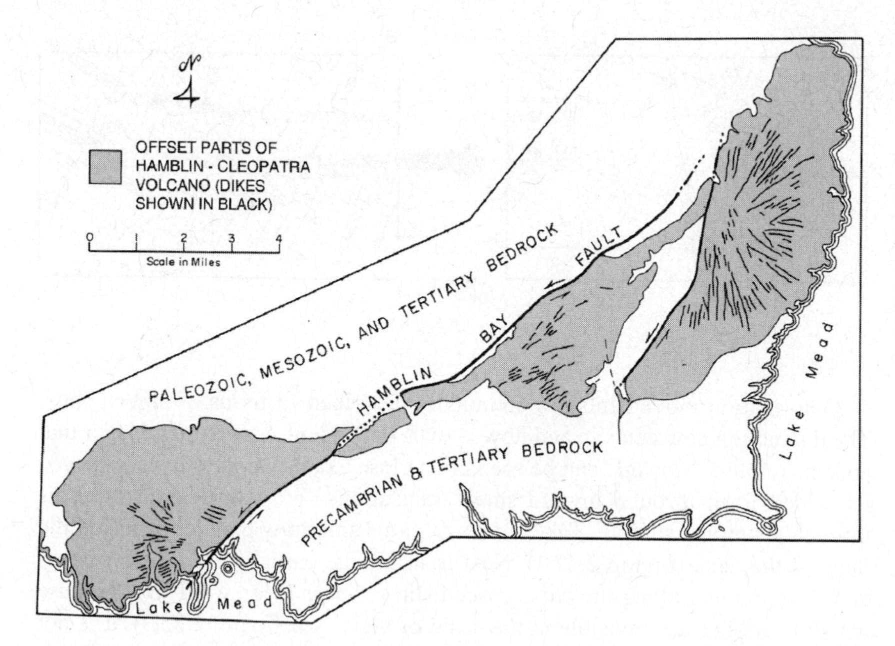

Figure 2.25 The Hamblin-Cleopatra volcano, Miocene in age, was cut in half and rearranged by faulting during the time interval 15 Ma to 10 Ma before present. The offset parts of the volcano, including its once-radial dike swarm, permit the magnitude of the displacement vector for the faulting to be calculated. Total displacement is ~19 km. [From Anderson (1973). Courtesy of United States Geological Survey.]

Richardson Highway in Alaska (Figure 2.24*C*) was cut and offset during the Denali fault earthquake (M_W 7.9), which occurred on November 3, 2002. Right-handed strike-slip offset of the highway was as great as 2.5 m.

Where faults have long been inactive, it is much more difficult to locate reference points that can be used to analyze net translation. The "*A*'s" and "*B*'s" of Figure 2.23*A* are difficult to find in nature. However, chance relationships have permitted some elegant appraisals of net slip. Anderson (1973), working in the Lake Mead region of southern Nevada, recognized a Miocene volcano, the Hamblin–Cleopatra volcano, which was cut in half between 15 and 10 MA by the Hamblin Bay fault (Figure 2.25). The northwest part of the volcano was translated approximately 19.3 km relative to the southeast half, in a left-handed sense. Discovering offset geological features like igneous plugs and volcanoes is the rare exception and not the rule. But there are ways to finesse the problem and to make the most of what we are given. In "*Descriptive Analysis*" (*Part III-K, Determining Slip on Faults through Orthographic and Stereographic Projection*) we will find opportunities to practice the analysis of net slip using methods of orthographic and stereographic projection.

The Inseparability of Rigid- and Nonrigid-body Deformation

Nonrigid-body versus rigid-body deformation is scale dependent. Fault movements are normally considered rigid-body movements, but if the faults are very closely spaced at the scale of view, the result may be viewed as a nonrigid-body deformation. This distinction is influenced by the degree to which structures are **penetrative** in the rock body under study (Turner and Weiss, 1963). For structures to be penetrative, they must be spaced so closely with respect to the size of the rock body under consideration that they appear to be everywhere. In Figure 2.26*A*, two faults, spaced kilometers apart, cut and displace a sequence of shale, limestone, and siltstone. The investigation of individual outcrops and small map areas within this geologic system would yield stratigraphic and petrologic information regarding the primary rock assemblage. The faults would be considered examples of rigid-body deformation. Seen in regional view, however, the faults would appear to be penetrative and within a system (Figure 2.26*B*), permitting the region to deform as a nonrigid-body.

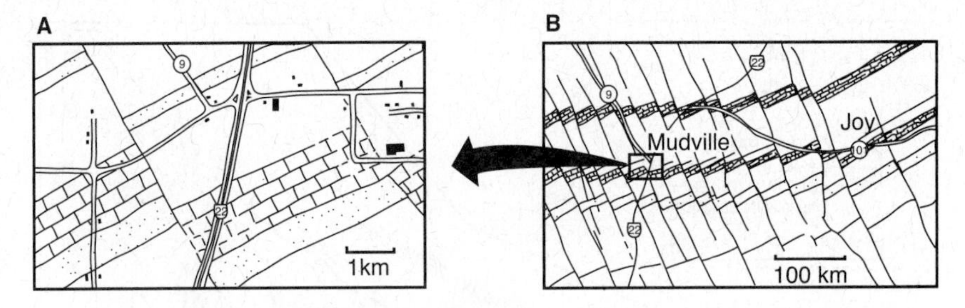

Figure 2.26 (*A*) When seen relatively close up, the faults appear to be widely spaced, and the deformation "*rigid-body*." (*B*) "*From 50,000 feet*" the faults are seen to be penetrative regionally, and the deformation is recognized as "*nonrigid-body*," changing the shape of things. [Modified from original artwork by R. W. Krantz.]

Deformation movements are commonly described in terms of **slip** or **flow**. The distinction between slip and flow is scale dependent. Sometimes a layer that appears to have "*flowed*" can be seen upon close examination to have achieved its folded form through myriad small translations. We can demonstrate this through a simple example. Take a deck of cards and draw parallel bands on the flank of the deck (Figure 2.27*A*). Next translate the cards in an irregular back-and-forth manner along the close-spaced slip surfaces (Figure 2.27*B*). Because the slip surfaces are invisible at the scale of view (see Figure 2.25*B*), it is not obvious that the "*nonrigid-body distortion*" of the original black, vertical bands was actually achieved by non-uniform pure translations. The "*flow folds*" photographed shown in Figure 2.27*C* present a quandary. They give the appearance that the rock flowed like butter and that the folds just piled up in one place. And yet we have to consider the possibility that some of the horizontal layering seen in the top of the photograph masks significant layer-parallel translation in the form of major slip or shear. The issue becomes even more significant when we map whole mountains that have this same kind of internal structure on a grand scale. This is the point that Price and Mountjoy (1970) were making (see Figure 1.43).

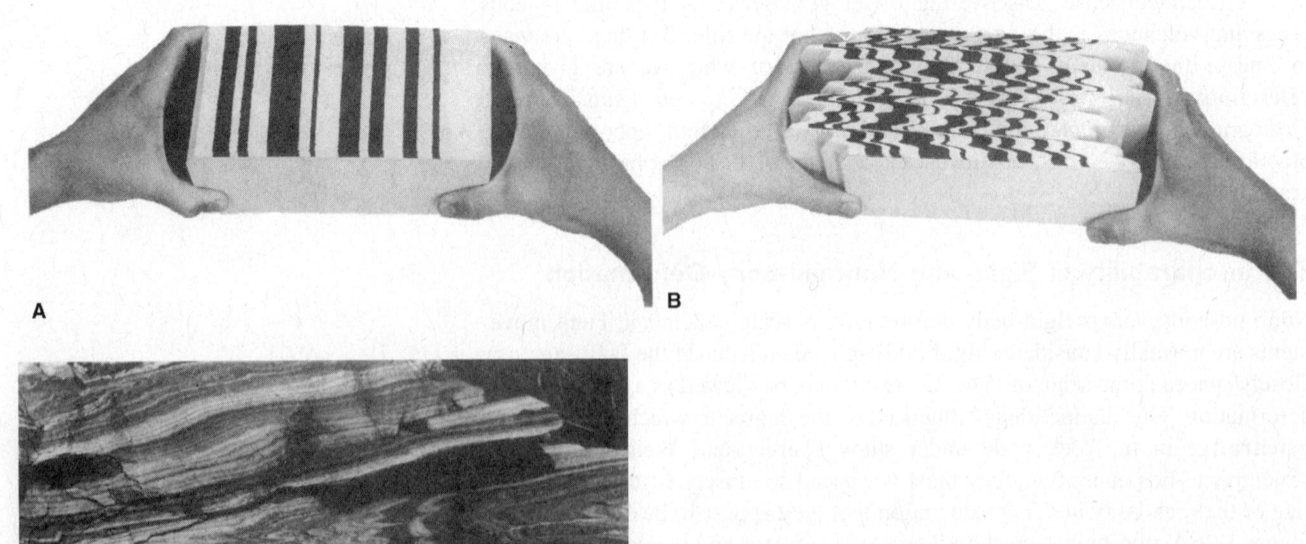

Figure 2.27 (*A*) Card deck with stripes simulating layering. (*B*) Apparent nonrigid-body distortion of the layers is achieved by simple rigid-body translations along penetrative slip surfaces. Hands are those of Paige Bausmann. [Photograph by G. Kew.] (*C*) Flow folds in metamorphic rock west of the Romback window near Bjornfeld in northern Norway. [Photograph by G. H. Davis.]

ROTATION

General Concept

During rigid-body rotation, a body changes orientation in a way that can be described as a rotation of material points about some common axis (see Figure 2.2*B*). A simple example of rigid-body rotation is expressed in a laboratory model I (GHD) made one afternoon many years ago by deforming alternating layers of modeling clay and dry, powdered kaolin clay (Figure 2.28). I "*deposited*" the clay horizontally on a "*Precambrian basement*" of pine board. But before depositing the layers, I cut a steeply dipping fault zone through the pine board with a saw, and then applied a little bit of grease along the saw cut. Next I slowly and steadily applied pressure to the butt ends of the pine board, which produced slip on the fault and shortening of the board and the overlying clay layers. Figure 2.28 reveals the degree to which the uppermost layers experienced rotation, which was counterclockwise on the left panel and clockwise within the center panel. The left and center panels experienced translation as well, for they were "*riding*" on the relatively uplifted board, which itself experienced rigid-body translation and (counterclockwise) rotation.

A field example of such deformation is seen in Figure 2.29, the Hunter's Point monocline exposed along the eastern margin of the Defiance uplift, near Window Rock, Arizona. Regionally horizontal sedimentary rocks abruptly rotate to moderately steep inclinations as a result of faulting at depth. Bedding in the middle background of this photograph is horizontal, but is rotated to a steep dip in the right background. White outcrops in foreground display near-vertical bedding. These strata in the foreground represent the along-strike projection of the steeply dipping strata in the fold in the distance.

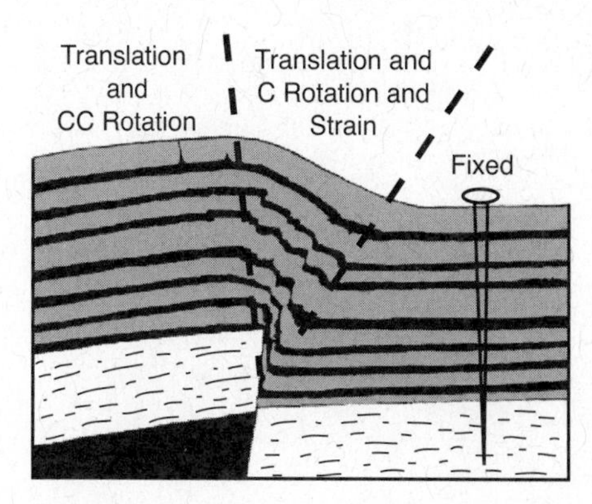

Figure 2.28 Examples of rotation. This faulting and folding was produced in the laboratory. The model consists of what were originally horizontal clay layers (alternating black and gray) resting on an originally horizontal but "*faulted*" pine board (see text for explanation). The experimental deformation caused translation and counterclockwise rigid-body rotation of the left panel as well as translation and clockwise rotation of the clay layers at the top of the wedge-shaped center panel. [Modified from Davis, G. H., 1978. Published with permission of Geological Society of America.]

Figure 2.29 North-directed view of the Hunter's Point monocline in northeastern Arizona. See text for explanation. [Photograph by G. H. Davis.]

Describing Rotations

Rotation is a rigid-body operation that changes the configuration of points in a way best described by systematic revolution about some common axis. Amusement parks thrive on rotational operations (Figure 2.30*A*). Axes of rotation may be horizontal (*Ferris Wheel*), vertical (*Merry-Go-Round*), and inclined (*The Rotor*), or all of the above (*Tilt-a-Whirl, Roll-o-Plane, The Hammer*, and other ghastly rides). The changes in locations of points are described by the orientation of the **axis of rotation** (trend and plunge), the **sense of rotation** (clockwise versus counterclockwise), and the **magnitude of rotation** (measured in degrees). Based on this set of facts, the locations of points before and after rotation can be calculated. Only those material lines that are parallel to the axis of rotation do not themselves rotate.

Designating a clockwise versus a counterclockwise sense of rotation depends partly on the direction of view. The operator behind the *Ferris Wheel* sees the cars rotating clockwise about the horizontal center axis (Figure 2.30*B*). The bystander in front, on the other hand, observes counterclockwise motion. To avoid ambiguity, a special convention is adopted in structural geology to specify sense of rotation: we describe the sense of rotation while looking *down* the axis of rotation. If the axis is strictly horizontal, we describe not only the sense of motion but also the direction in which the structure is observed. For example, the *Ferris Wheel* is rotating counterclockwise when viewed to the north along the (horizontal) axis of rotation (Figure 2.30*B*). Tracking the kinematics of rotation is sometimes best carried out through stereographic methods. *Descriptive Analysis* (*Part III-J, Evaluating Rotation Using Stereographic Projection*) offers some methodology in analyzing rotations.

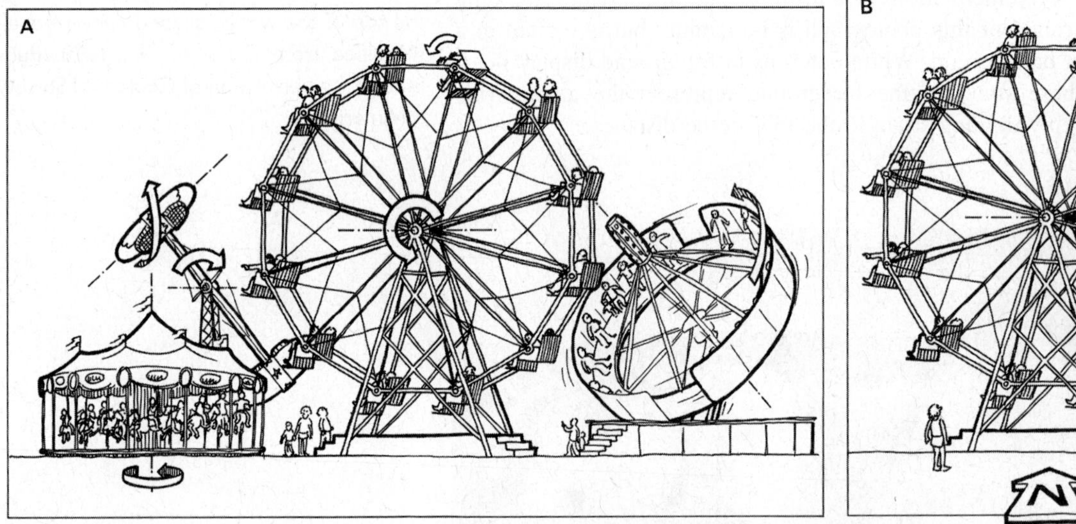

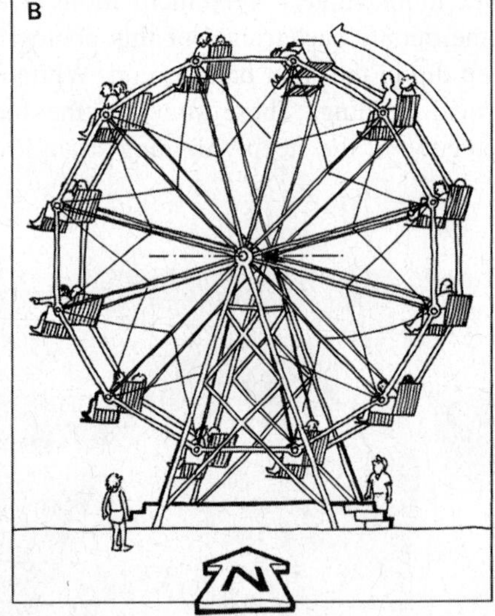

Figure 2.30 (*A*) Amusement parks thrive on rotational operations. Analysis of rotation includes describing the orientation of the axis of rotation, the sense of rotation, and the magnitude of rotation. (*B*) Sense of rotation depends on the direction of view. [Artwork by D. A. Fischer.]

Rotation does not typically occur in the absence of translation. Instead, deformation expressed by rotation is interrelated with deformation expressed by translation. With camera in hand, I (GHD) ran and caught up with a Rainbow Bread truck and captured the image shown in Figure 2.31*A*. The five slices of bread in the deformed state rotated clockwise (with respect to my direction of view) by different degrees. It would appear that the greater the rotation, the greater the net slip between adjacent slices (see Figure 2.31*B*). Displacement vectors can be interpreted for various points (before and after deformation), and once again we see that displacement vectors are commonly different from the actual deformation paths (see Figure 2.31*B*). Imagine if this deformation was taking place in real time on the back of a flatbed truck on a mountain road. Think of how this would change our calculations and impressions of displacement vectors and deformation paths (Figure 2.31*C*)!

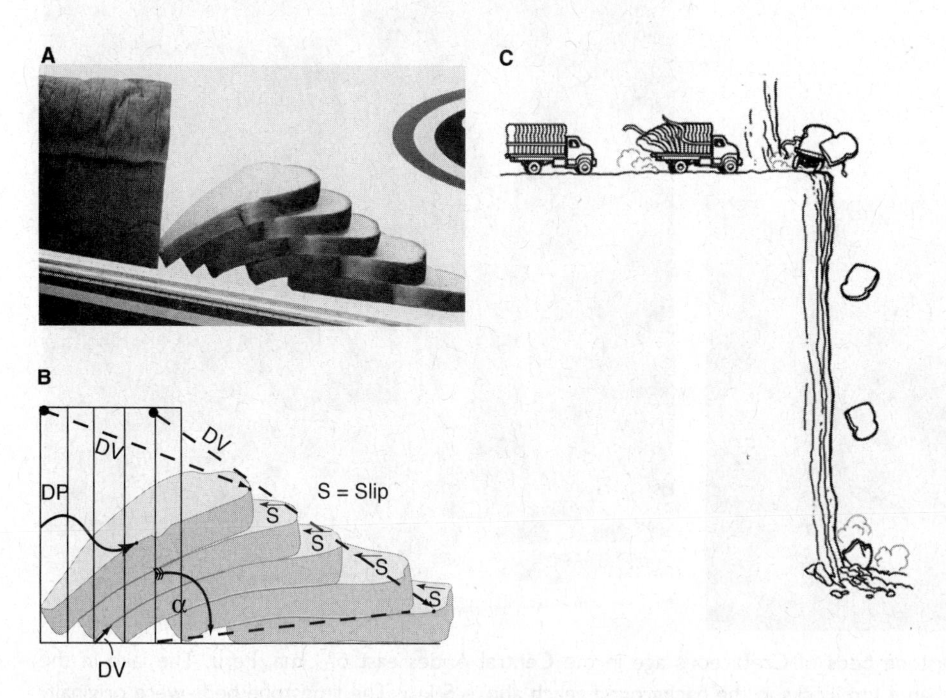

Figure 2.31 (*A*) Photograph of the side of Rainbow Bread Truck, showing bread slices, some undeformed, some deformed. [Photograph, on the move, by G. H. Davis.] (*B*) Tracing of photo that reveals *net slip* (*S*) on each fault surface, the rotation of each slice (α) is rotation of interface between 4th and 5th slice of bread, the displacement vectors for several reference points (dashed arrows mark *DV*), and a hypothetical deformation path (*DP*) for a reference point within a slice of bread (I just guessed at it; maybe someone reading this could run the experiment and trace the path!). (*C*) Imagine the potential for deformation paths as a flatbed truck with full "*bread load*" takes a corner too fast on a mountain road. [Artwork by D. A. Fischer.]

Geological Examples

A number of geological examples serve to illustrate why rotation is an important component of describing deformation. Originally horizontal layered rocks can become **inclined strata**, rotated as a result of faulting and folding processes. The strata shown in Figure 2.32 are vertical today, though in obedience to Steno's "*law of original horizontality*" they started out flat-lying. The challenge in determining the path of deformation for such strata includes figuring out whether they rotated to verticality in a clockwise or counterclockwise sense (with respect to the direction of view). Which way to "*flip*" the strata to horizontal is informed by looking for those primary sedimentary structures that can be used to determine **facing**, that is, distinguishing "*tops*" from "*bottoms*" of beds. Primary sedimentary structures and facing criteria are discussed in *Part III-D, Identifying Primary Structures.*

Another example of rotation can be seen in a type of faulting known as **listric normal faulting**. Listric normal faults curve with depth in such a way that the fault surface is concave upwards (Figure 2.33). The geometric consequence of faulting of originally flat-lying strata along such curved fault surfaces is rotation of strata to moderate or steep dips. The actual amount of rotation will depend on fault curvature and distance of displacement. Ernie Anderson (U.S.

Figure 2.32 (*A*) Steeply inclined limestone beds of Cretaceous age in the Central Andes east of Lima, Peru. The lake in the foreground is at an elevation of more than 4 km; rocks in the background reach above 5 km. The limestone beds were originally deposited below sea level! These beds were tilted to a vertical orientation in the Tertiary. [Photograph by G. H. Davis.] (*B*) Vertical beds of Miocene alluvial-fan conglomerate have been eroded to fins near the town of Tupiza in the Eastern Cordillera, southern Bolivia. The wall in left foreground is 2 m high. These beds were tilted to a vertical orientation during middle to late Miocene shortening (Horton, 1998). [Photograph by Brian Horton.]

Geological Survey) discovered and mapped such faulting in the Lake Mead region of northwestern Arizona and southeastern Nevada (Figure 2.34). There, Tertiary volcanic and sedimentary rocks unconformably overlie Precambrian basement. Prior to 20 MA, the Tertiary strata and the great unconformity were essentially horizontal. (See *Part III-C, Mapping Contact Relations*, for discussion of unconformities.) Extensional faulting beginning 12 Ma created a system of listric normal faults (see Figure 2.34), and as a result strata and the unconformity were rotated progressively to the final state shown. The axis of rotation

Figure 2.33 Illustration of listric normal faulting. as represented in cross-sections by Carney and Janecke (2005). This fault system was activated during regional detachment faulting in southeastern Idaho. The curvature along the normal faults, in combination with fault slip (translation), results in back tilting of the originally horizontal layer (black). [From Carney and Janecke, 2005. Published with permission of Geological Society of America.]

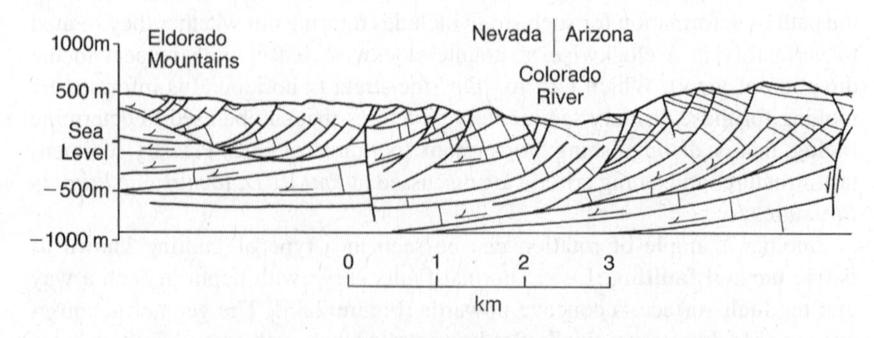

Figure 2.34 Cross-section of listric normal faulting in the Lake Mead region. Tertiary and Precambrian rocks are rotated to steep dips along curved faults. [From Anderson (1971). Courtesy of United States Geological Survey.]

is parallel to the strike of the fault, the sense of rotation is clockwise when viewed to the north, and the magnitude of rotation ranges from 10° to 90°.

Effects of rotation at the outcrop scale can be seen commonly in thin-bedded sedimentary rocks and in schists and phyllites. When compressed in ways that create layer-parallel shortening, such rocks can commonly become **kink folded**. The kink folds (Figure 2.35) are expressed as bands within which the rock is rotated out of its normal orientation. Rotations can be described as clockwise or counterclockwise with respect to the direction of view, in this case 90° counterclockwise.

Yet another expression of rotation within metamorphic rocks, this time at a microscopic scale, can be seen in garnet porphyroblasts. These garnets grew within the schist during shearing and metamorphism. Under certain conditions the shearing accompanying the metamorphism will "*roll*" the garnet and cause it to grow, like rolling a snowball. As the garnet rotates, it may entrain inclusions of surrounding material (into which it grows), thus creating spiral inclusion trails that may be seen at the microscopic scale (Figure 2.36). The sense of clockwise versus counterclockwise rotation of garnets is easy to spot where the pattern is strong, but it must be specified with respect to the direction of observation (e.g., "*the garnet rotated clockwise as viewed from south to north*"). The amount of rotation can be interpreted by measuring the degree of spiraling (see Figure 2.36).

Figure 2.35 Photograph of kink folding of thin-bedded shale and mudstone within the Uncia Formation (Silurian) in the central Andean fold-thrust belt. As seen in the direction of view, beds within the kink band rotated counterclockwise by exactly 90° relative to the orientation of the enclosing layers. Location is southwest of Cochabamba, Bolivia (McQuarrie and Davis, 2002). Field notebook for scale. [Photograph by G. H. Davis.]

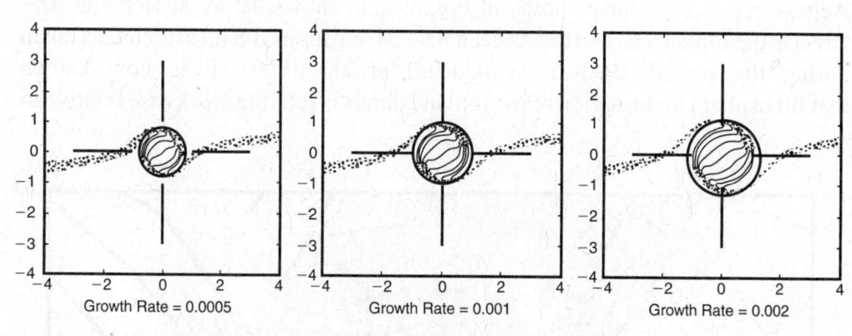

Figure 2.36 Spiral garnets can be produced by rotation, in such a way that a growing porphyroblast is simultaneously rotating in response to shear. During rotation and growth, "*trails*" of material are incorporated into the porphyroblast. These diagrams are the result of numerical modeling and show right-handed (dextral) shear and increasing growth rates from left to right. Rotation in the direction of view is clockwise. [From Stallard, Ikei, and Masuda (2002), Quicktime movies of 3D spiral inclusion trail development, in Bobyarchick, A. (ed.), Visualization teaching and learning in structural geology: *Journal of the Virtual Explorer*, v. 9, http://virtualexplorer.com.au/article/2002/63/3d-spiral-trail-development.]

There are plenty of examples of rotation at the plate and regional tectonic scales of view. A regional tectonic example is evident in the Los Angeles basin area. There, the western Transverse Ranges are oriented west-northwest at a high angle to the dominant structural grain (NW/SE). It has been known for decades that the rocks of the western Transverse Ranges started out well to the south near San Diego, and were translated northward to their present location, rotating clockwise by more than 90° before "*docking*" at their present location (Figure 2.37). The basis for this understanding is paleomagnetic analysis and geological interpretation (Crouch, 1979; Luyendyk et al., 1980; Jackson and Molnar, 1990; Nicholson et al., 1986).

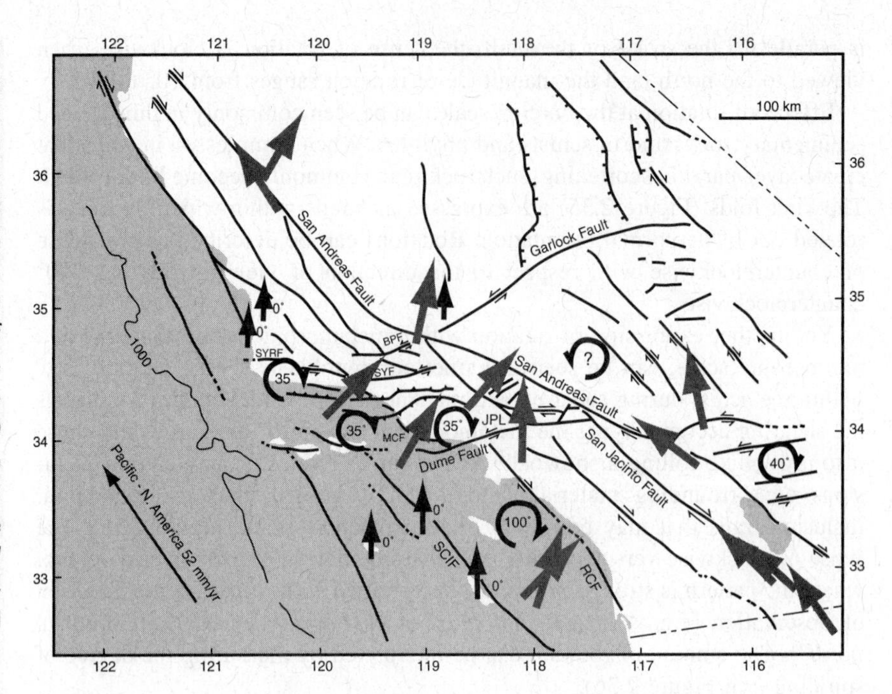

Figure 2.37 Map showing block rotations in the western United States, based on paleomagnetic analysis of rocks younger than ~10 to 15 Ma. Curved arrows give sense and magnitude of vertical-axis rotation. Paleomagnetic arrows pointing north and marked 0° give evidence for no local vertical-axis rotation in rocks at those sites. The other, larger arrows show deviation from north caused by faulting and rotation. The rocks of the western Transverse Ranges (WTR) started out at ~20 Ma near San Diego, and subsequently translated northward nearly 300 km, rotating clockwise along the way. [From Jackson and Molnar, Active faulting and block rotations in the western Transverse Ranges, California: *Journal of Geophysical Research*, v. 95, Figure 2, p. 22,075. Copyright © 1990 by American Geophysical Union.]

Another compelling active-tectonic example of rotation is displayed in the Aegean region. The map shown in Figure 2.38 shows the West Hellenic Arc, Greece, the Ionian Sea, and the Aegean Sea. Based upon GPS measurements taken during the period 1993–1998 (Cocard et al., 1999), it is now known that lithosphere in the region of the Ionian Islands is rotating clockwise relative to

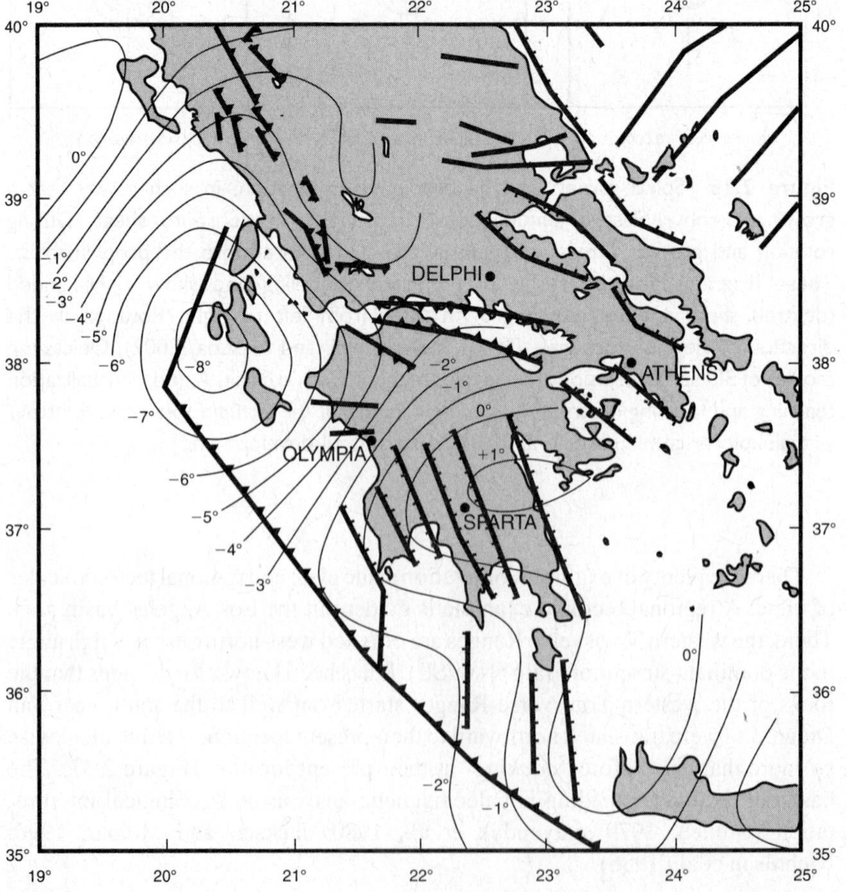

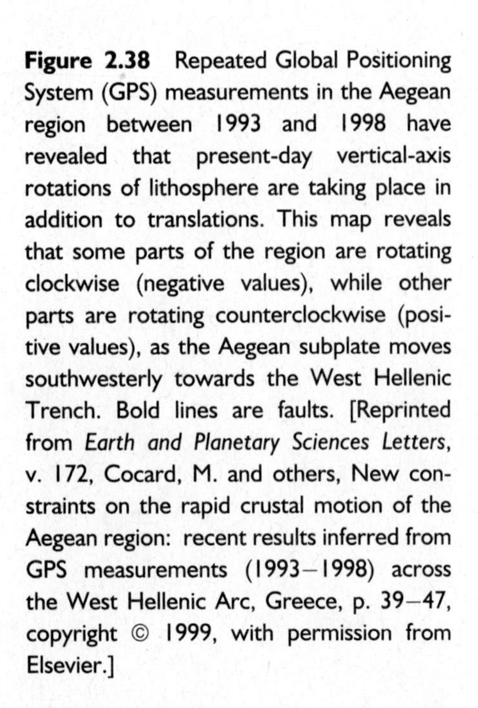

Figure 2.38 Repeated Global Positioning System (GPS) measurements in the Aegean region between 1993 and 1998 have revealed that present-day vertical-axis rotations of lithosphere are taking place in addition to translations. This map reveals that some parts of the region are rotating clockwise (negative values), while other parts are rotating counterclockwise (positive values), as the Aegean subplate moves southwesterly towards the West Hellenic Trench. Bold lines are faults. [Reprinted from *Earth and Planetary Sciences Letters*, v. 172, Cocard, M. and others, New constraints on the rapid crustal motion of the Aegean region: recent results inferred from GPS measurements (1993–1998) across the West Hellenic Arc, Greece, p. 39–47, copyright © 1999, with permission from Elsevier.]

Eurasia at a rate of 8°/MA, whereas in the southeastern Peloponessos the lithosphere is rotating counterclockwise at a rate no greater than 2°/MA. These rotations are an expression of differential movement (and distortion) of the Aegean microplate towards the West Hellenic Arc. It is remarkable how GPS technology is helping us to see translation and rotation in active tectonic environments. This provides an important part of the basis for evaluating deformation.

STRAIN

General Concept

Strain results from nonrigid-body deformation, either through a change in size (**dilation**) and/or a change in shape (**distortion**). Strain expressed as dilation or distortion results from a change in the spatial arrangement of material points within an object. The change in spatial arrangement (i.e., configuration) of points can be extraordinarily systematic.

During a sustained deformation a single object may experience both dilation and distortion. Pure dilation is a change in size without a change in shape (see Figure 2.2*C*). Dilation accompanies such nonrigid structural processes as the shrinkage of mud to produce mud cracks (see *Part III-D*, Figure D.9) and the cooling of basalt to produce columnar joints (see Figure 1.14). For a pure increase in size, the distance between all pairs of points increases by a common amount (see Figure 2.2*C*). A nice geological example of positive dilation (increase in size) is pictured in Figure 2.39, a photomicrograph of a shattered quartz crystal. It looks like an aerial photograph of sea ice during breakup; the pieces can be fit together to form a seamless smaller object.

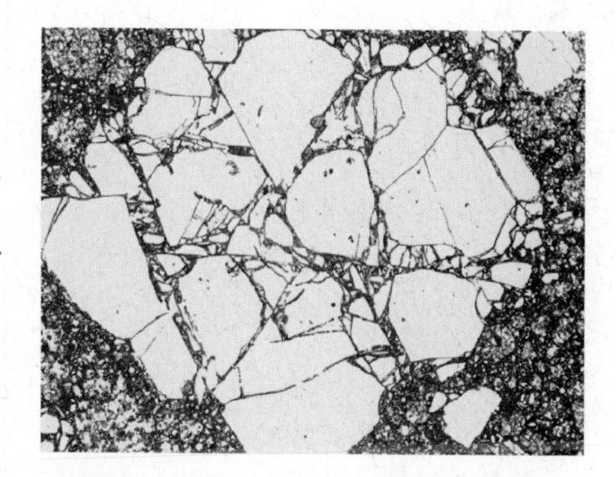

Figure 2.39 Photomicrograph of shattered quartz crystal. The fragmented crystal displays the results of nonrigid-body deformation. [Photograph by R. G. Schmidt. Courtesy of United States Geological Survey.]

Pure distortion is a change in shape without a change in size. As shown in Figure 2.2*D*, the change in shape of an object (e.g., from a square to a rhomb) takes place through systematic changes in the spacing arrangement between material points in the object. Angular relations between alignments of points change as well. A compelling outcrop example of distortion is shown in Figure 2.40, featuring distorted lapilli in tuff from the Lake District, England. Lapilli are droplets of magma blown out of a volcano during an eruption. The elliptical form of the lapilli in Figure 2.40 resulted from strain during compaction (i.e., gravitational loading) in a volcanic pile. The perfect ellipses were originally circular sections through volcanic lapilli. Strong flattening transformed the circles into ellipses through nonrigid-body distortion. Better than words, the elliptical shapes communicate the full extent of the nonrigid-body distortion

Distortion takes place at all scales during tectonic loading. For example, if we go back to the fold shown in Figure 2.28, it is easy to imagine that the clay layers in the triangular wedge immediately above the fault experienced more than simply translation and rotation. The layers became distorted, as revealed by changes in shapes of the layers.

Figure 2.40 Distorted lapilli in tuff from the Lake District of England. [From Oertel (1970). Published with permission of Geological Society of America.]

Transformation, Deformation and Strain

Insights into structural-geologic transformations of shapes of things are found in unlikely sources, for instance, a book entitled "*Mind Sights*" written by Roger Shepard, a psychologist. Within this book Shepard (1990, p. 75) presents artistic renderings of what he calls "*reversible transformations*," which are achieved through translation, rotation, dilation, and the combination of rotation and dilation (Figure 2.41). These pictures give the illusion of movements through "*figure-ground ambiguities*," which we have all seen in puzzles and cartoons: figures in the foreground suddenly disappear into the background, and what was originally background emerges in the form of figures in the

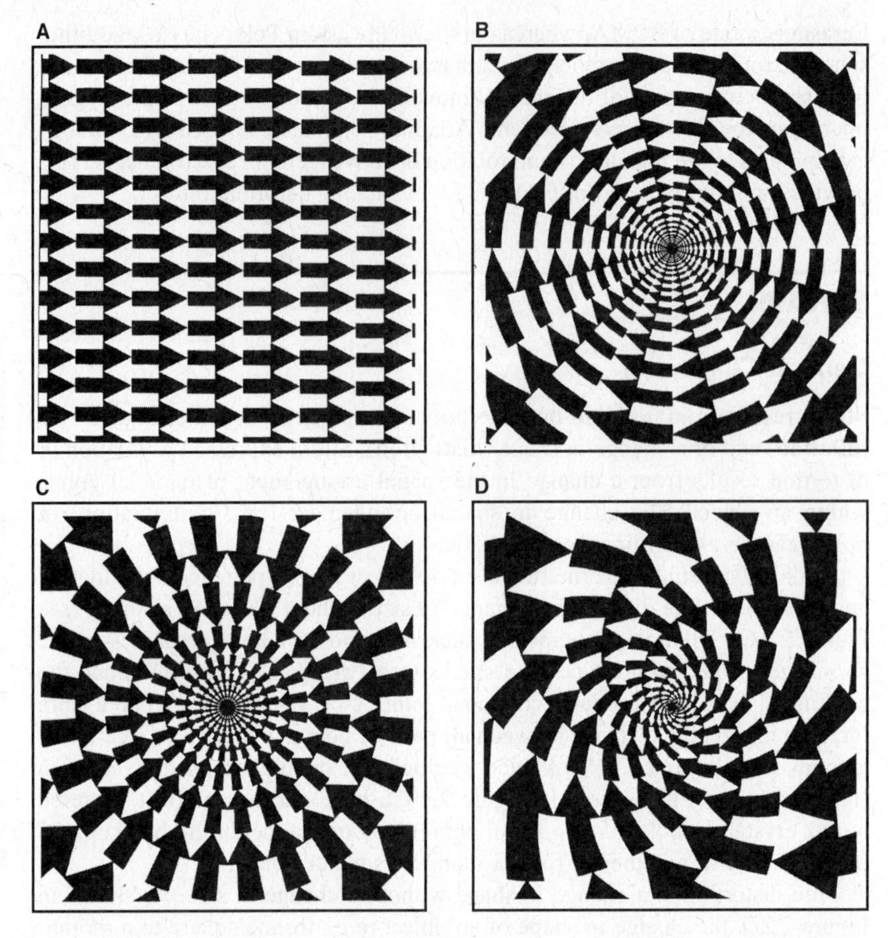

Figure 2.41 Images of reversible transformations. (*A*) Translation. (*B*) Rotation. (*C*) Dilation. (*D*) Rotation plus dilation. See text for elaboration. [From Shepard, 1990, Mind sights: original visual drawings, ambiguities, and other anomalies, Figure D7, p. 75. W. H. Freeman and Company.]

foreground (Shepard, 1990, p. 68). The figures appear to be in motion, and thus we *"see"* the motions; we *"see"* contributions to deformation! Shepard believes that these kinds of motions are rooted deeply psychologically in our evolved visual systems as human beings, giving rise to our abilities to appreciate symmetrical patterns (Shepard, 1990, pp. 138–139). Perhaps this is why the recognition and reconstruction of such patterns in matters of structural geology can be at times so richly satisfying.

Deformation describes the collective displacement vectors of points in a body. It is the three-dimensional transformation from initial to final state (van der Pluijm and Marshak, 1997), including, but usually not limited to, translation and rotation (see Figure 2.2). Part of deformation is **strain**, which includes **dilation** (a change in size) and **distortion** (a change in shape). The analysis of strain is essentially the analysis of distortion and dilation.

Translation and rotation are described relative to some external 3D reference frame: For example, during the past 71 MA, "*points*" in India moved (translated) from 80°E. Long./40°S. Lat., to 93°E. Long./27°N. Lat. (see Figure 2.22). In contrast, descriptions of strain are cast commonly with respect to an arbitrary internal reference frame, capturing changes in spacing and configuration of material points within an object without necessarily any reference to translational changes in paleolatitude or paleolongitude, or rotation of the rock body within which strained features are found. We can think about these reference frames in relation to Figure 2.8, the cartoon rendition of birds in flight. The aggregate flock may be tracked using an external reference frame. For example, if they are pigeons, locations might be expressed relative to the church tower in the center of town. Within the flock, however, the birds themselves are tracked in relation to one another, using an internal reference frame, as if the birds are 'caged' within a Cartesian coordinate system of three dimensions.

Strain is the change in the original spacing of points within the body. Once again we see that the shift of each material point from $[x,y,z]$ to $[x',y',z']$ can be described as a displacement vector, which seldom will be identical to the actual deformation path. Where pure dilation takes place without change in shape, internal points of reference spread apart or pack closer together in such a way that line lengths between points become uniformly longer or shorter (see Figure 2.2*C*). Overall shape remains the same. In contrast, during distortion the changes in spacing of points in a body are such that the overall shape of the body is altered, with or without a change in size (see Figure 2.2*D*).

As structural geologists we are indebted to John Ramsay, whose classic text, *Folding and Fracturing of Rocks* (1967), underscores the importance of strain analysis and presents the fundamentals and applications of this discipline in a thorough and enlightening way. Although mathematical in expression, strain analysis is fundamentally a geometrical challenge (Ramsay, 1967, p. 50), if not an art form (Figure 2.42). The objectives of strain analysis are formidable: we are asked to describe the changes in size and shape that have taken place in a nonrigid manner during deformation, and, in so doing, create the capacity to describe how each and every line in each and every direction in a body has changed length and relative orientation. Thankfully, there are ways to sum the effects of such changes, integrating in ways to explore the nature of changes throughout the strained object.

Figure 2.42 Some of the patterns of strain emerge, artfully, in multiply-folded metamorphic rocks. Characteristic patterns emerge through folding, then refolding of the rocks. The designs produced relate to amplitudes and wavelengths of the superposed fold sets, and their orientations relative to one another. [From *Folding and Fracturing of Rocks* by J. G. Ramsay. Published with permission of McGraw-Hill Book Company, New York, copyright @ 1967.]

The Ground Rules

Here we simplify our work in strain analysis by studying the theory in two, not three, dimensions. For more advanced applications, the three-dimensional approach to strain analysis is available in other texts, for example, Ramsay (1967), Jaeger and Cook (1976), Means (1976), Ramsay and Huber (1983, 1987), Twiss and Moores (1992), Pollard and Fletcher (2005), and Ragan (2009).

A customary simplification is to restrict strain analysis to the description of **homogeneous deformation**. Where deformation is homogeneous, lines that were straight before deformation remain straight after deformation; and lines that were parallel before deformation remain parallel after deformation. For these conditions to hold, the strain must be systematic and uniform across the body that has been deformed. A simple test for homogeneity will become obvious: homogeneous deformation transforms perfect circles into perfect ellipses, and transforms perfect spheres into perfect ellipsoids. Strained rocks found in nature typically depart from the rules of homogeneous deformation. However, we deal with this problem by subdividing regions of study into smaller subareas (**domains**) within which strain can be regarded as statistically homogeneous. The elegance and systematic rendering of Earth structure is a gift of the "*rules*" of homogeneous strain. The study and application of strain theory result in surprising discoveries of the regularity and predictability of *changes* in lengths and orientations of lines.

The Magic of Strain

A useful way to visualize the two-dimensional properties of strain is through homogeneous distortion of a circle. If a body containing a perfectly circular reference marker is subjected to perfectly homogeneous deformation, the reference circle will be transformed into a perfect ellipse. Furthermore, if the body is subjected to a second homogeneous deformation, the elliptical form of the deformed reference marker will be transformed into yet another perfect ellipse, regardless of the magnitude and orientation of the second deformation imposed on the first. There is only one exception: the special case of a second deformation that is the exact reciprocal of the first, thus returning the ellipse to its original circular form. In general, however, no matter how many times an ellipse is distorted homogeneously, it will be repeatedly transformed into a new ellipse.

Before the advent of desktop/laptop computers and advanced software design, I (GHD) found it useful to keep on hand a deck of computer 'punch-cards,' the flank of which was embossed with a perfect circle and a perfect ellipse (Figure 2.43*A*). Through shearing of the deck of cards, the outline of the

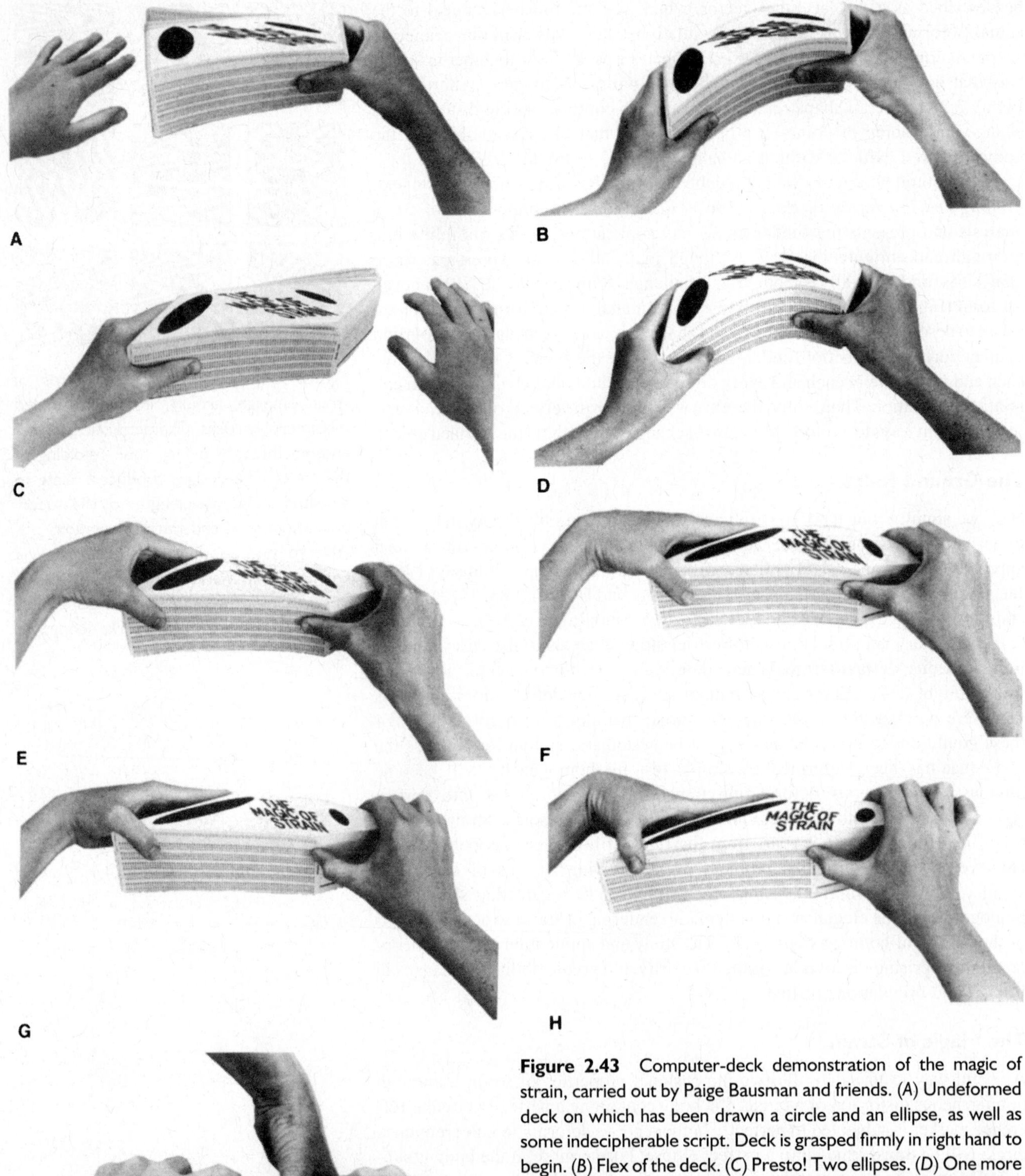

Figure 2.43 Computer-deck demonstration of the magic of strain, carried out by Paige Bausman and friends. (*A*) Undeformed deck on which has been drawn a circle and an ellipse, as well as some indecipherable script. Deck is grasped firmly in right hand to begin. (*B*) Flex of the deck. (*C*) Presto! Two ellipses. (*D*) One more flex. (*E*) Original circle is now strongly elliptical. Original ellipse is now much less elliptical. Indecipherable script begins to become decipherable. (*F, G*) Can't stop. (*H*) Original ellipse is now a circle. Original circle is as elliptical as original ellipse. (*I*) Deck becomes so thinned and stretched that it is hard to support without help. Indecipherable script is indecipherable once again, but the slant of the writing has changed. The original ellipse is an ellipse once again, but its direction of slant has also reversed. The original circle is now profoundly distorted. [Photographs by G. Kew.]

starting circle could be transformed into a perfect ellipse. This was achieved by holding tight to the right end of the card deck with the right hand (see Figure 2.43*A*), flexing the deck into the form of a fold (Figure 2.43*B*), grasping tight the left end of the deck with the left hand and then releasing the right hand (Figure 2.43*C*) (Ragan, 1969). Note that the orientations of the long and short axes of the ellipse on the cards keep changing. Note also the outline of the ellipse embossed on the flank of the deck prior to shearing (Figure 2.43*A*). It subsequently became transformed by shearing into a variety of ellipses, and in one special case into a perfect circle (Figure 2.43*H*).

By repeating the flex-and-shear drill a number of times, any ellipse can be made more and more elongate (Figure 2.43*D*–*H*). The only limit to flattening and extending an ellipse through shearing of cards is the difficulty of trying to hang onto the deck as it progressively thins and lengthens (Figure 2.43*I*). Nature has no such limit!

Note the outline of the ellipse embossed near the upper right-hand corner of the flank of the deck prior to shearing (Figure 2.43*A*). It becomes transformed into a variety of ellipses, and in one special case into a perfect circle (see Figure 2.43*H*).

The same "*magic*" holds if we subject a block, on which we have embossed the outlines of a circle and several ellipses, to a flattening (Figure 2.44*A*). As the block is progressively shortened in one direction and lengthened in another, without changing area, the circle is transformed to ellipses of greater and greater **aspect ratio** (long axis/short axis) (Figure 2.44*A*–*C*). If the progressive deformation is halted at just the right moment (Figure 2.44*B*), a given ellipse might be seen passing through the form of a circle. The orientations of the long and short axes of the strain ellipse remain constant throughout the deformation, in contrast to the steady rotation of the long and short axes in the shear deck example. More on that later.

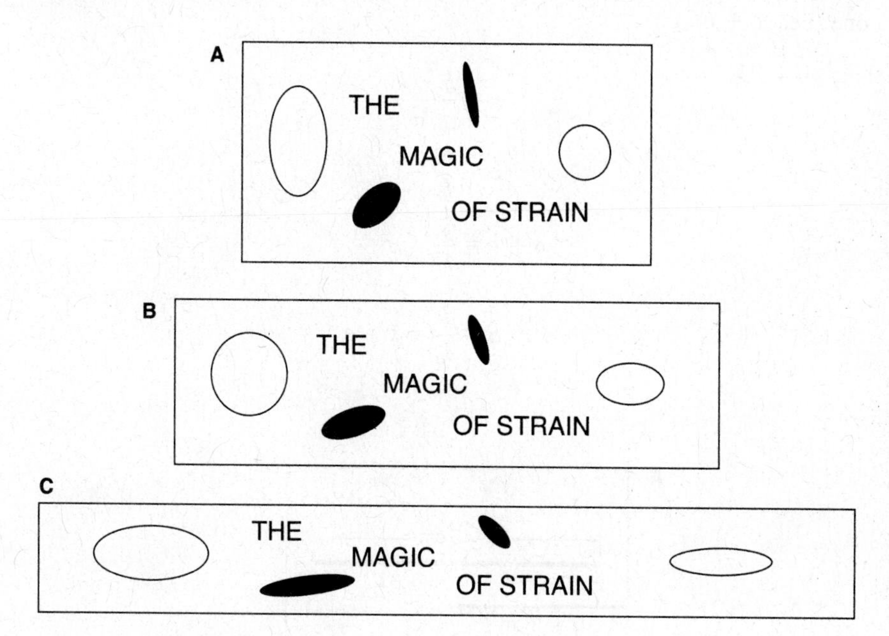

Figure 2.44 (*A*) A deformable block is embossed with a circle, a vertically oriented ellipse, two black ellipses, and words. (*B*) When flattened and extended, changes in the shapes and orientations of the reference objects on the front face of the block record the nature of the internal strain. The amount of flattening and stretching is just enough to transform the original vertical ellipse into a perfect circle. (*C*) With even more flattening and stretching, the white ellipses become tighter and tighter; the two black ellipses continuously rotate toward the direction of stretching, and the letters of the words continuously change font.

The mastery of strain analysis requires keeping track of changes in the orientations and lengths of lines as strain accrues, and keeping track of the angles between lines. If we learn to do this, we do not have to rely exclusively on the relatively uncommon occurrence of perfectly circular or spherical objects being transformed and preserved in the geologic record as perfect ellipses or ellipsoids (e.g., see Figure 2.40). Instead, we can describe the state of strain in a deformed body based on a surprisingly scant amount of information bearing on changes in lengths in lines and changes in the angles between lines. In fact, the changes in lengths of lines, and the change in angles between lines that were originally perpendicular, are sufficient to convey the magnitudes and directions of greatest shortening or stretching in the rock body as a whole!

Describing Changes in Lengths of Lines

There are two parameters that permit changes in lengths of lines to be described easily. One is **extension**, symbolized by e; the other is **stretch**, symbolized by S (Ramsay, 1967; Means, 1976). Consider line L whose **original length** (l_0) is 5 cm (Figure 2.45*A*). During deformation, the nonrigid-body in which L is contained changes shape and/or size such that the line stretches to a **final length** (l_f) of 8 cm (Figure 2.45*B*). The **change in length** (Δl) is 3 cm.

The magnitude of extension (e) in the direction of lengthening is the change in unit length of the line.

$$e = \frac{l_f - l_0}{l_0} \tag{2.1}$$

$$e = (8\text{ cm} - 5\text{ cm})/5\text{ cm} = 0.6$$

A 0.6 value for extension e corresponds to a 60% lengthening of the line. **Percent lengthening** (or **percent shortening**) is determined by multiplying e by 100%.

A second way to describe the magnitude of the change in length of line L is in terms of the **stretch**, symbolized by S. Stretch is equal to final length (l_f) divided by original length (l_0), which is also equal to the value of extension plus one (i.e., $1 + e$).

$$e = \frac{l_f - l_0}{l_0}$$

$$e = \frac{l_f}{l_0} - 1$$

$$e + 1 = \frac{l_f}{l_0}$$

$$S = \frac{l_f}{l_0} = 1 + e \tag{2.2}$$

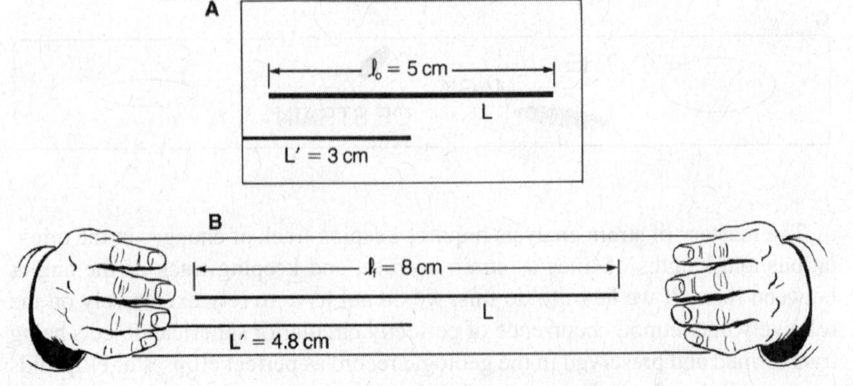

Figure 2.45 (*A*) Lines L and L' before stretching of the body within which they reside. (*B*) Lines L and L' after stretching.

Stretch tells us the final length of a line originally of unit length. A stretch of 3.0 means that a line was lengthened 3×. For the example we are considering,

$$S = 8 \text{ cm}/5 \text{ cm} = 1.6$$

If line *L* in Figure 2.45*B* lies within a body that has undergone homogeneous deformation, the values of $e = 0.6$ and $S = 1.6$ must hold for *all* the lines in the body that are parallel to *L*. Line *L′* is such a line (see Figure 2.45*A*). If the length of *L′* before deformation is 3 cm, its length after deformation can be determined by using Equation 2.1:

$$e = (l_f - l_0)/l_0$$

$$0.6 = (l_f - 3 \text{ cm})/3 \text{ cm}$$

$$l_f = (0.6)(3 \text{ cm}) + 3 \text{ cm}$$

$$l_f = 4.8 \text{ cm (see Figure 2.45}B)$$

An even simpler way to compute l_f for this line is to multiply l_0 by the stretch (*S*).

$$l_f = 1.63 \text{ cm} \times 3.0 = 4.8 \text{ cm (Figure 2.45}B)$$

Percent lengthening or shortening is determined by multiplying 100% times $(S - 1.0)$.

Change in Length of a Deformed Fossil

Extension and stretch are used in reporting the degree of stretching or shortening of geological lines. The stretched belemnite fossil featured in Figure 2.46 provides a good example. This fossil, discovered in folded rocks in the western Alps by Albert Heim in the nineteenth century (Milnes, 1979), was stretched into an array of rigid shell fragments of approximately equal size. Spaces that developed during deformation were simultaneously filled by calcite. The approximate original length (l_0) of the belemnite fossil can be determined by measuring and summing the widths of the individual shell fragments. The final length (l_f) of the belemnite is simply the total length of the fossil in its present state, including the calcite filling.

Figure 2.46 Stretched belemnite, broken into an array of separated fragments (dark) between which calcite (white) has precipitated. Lengthening is approximately equal to 130%. [From Milnes, 1979. Published with permission of Geological Society of America.]

Although the actual size of the deformed belemnite is not revealed in Figure 2.46, knowledge of the relative values of l_0 and l_f is all that is necessary to calculate extension and stretch. Using a photocopy of the belemnite illustration as originally presented in Milnes (1979), we measured an l_0 value of approximately 82 mm and an l_f value of approximately 185 mm. Based on these measurements, we calculated the values of extension and stretch for the direction along which the fossil lies.

$$e = (185 \text{ mm} - 82 \text{ mm})/82 \text{ mm} = 1.3$$

$$\% \text{ lengthening} = e \times 100 = 130\%$$

$$S = 185 \text{ mm}/82 \text{ mm} = 2.3$$

$$\% \text{ lengthening} = (S - 1) \times 100\% = 130\%$$

Not only was the belemnite stretched by 130%, but the rock in which the belemnite is encased was stretched 130% as well. This is the point!

Expressing Changes in Length due to Folding and Faulting

One of the payoffs in constructing geologic cross-sections (see *Part III-F, Preparing Geologic Cross Sections*) is being able to use them to measure how much stretch or extension is accommodated by the geologic structures. Figure 2.47 presents cross-sections through folds and thrust faults in southeastern Sicily. The sections were prepared by Butler et al. (1992). They measured final lengths (l_f) along reference lines *a–b*, *c–d*, and *e–f* (see Figure 2.47). Original lengths (l_0) were determined by measuring the folded, faulted trace lengths of a key limestone bed. Extension values (e) were found to range from −0.25 to −0.40, corresponding to percentage shortening from 25% to 40%, and stretch values (S) of 0.75 to 0.6.

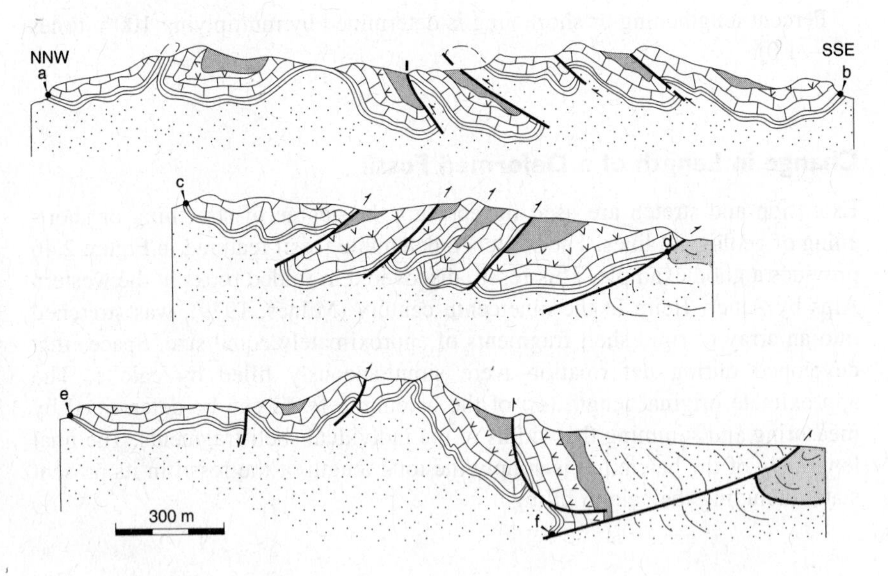

Figure 2.47 Cross-sections through folded and faulted sedimentary rocks in southeastern Sicily. After constructing the sections, Butler et al. (1992) calculated extension (e) and percent shortening values based on changes in line length. [From Butler, Grasso, and LaManna, 1992. Published with permission of The Geological Society.]

In Chapter 1, Figure 1.45, we presented cross-sections that Bohannon et al. (1993) constructed through normal faults and associated folds in the Virgin River depression in southeastern Nevada and northwestern Arizona. Bohannon et al. (1993) compared the length (l_f) of each cross-section of present-day relationships to each restored cross-section. They then determined stretch, extension, and percentage lengthening. Stretch values (S) for the two cross-sections shown were 1.56 and 1.72, respectively, corresponding to extension (e) values of 0.56 and 0.72, and lengthening of 56% and 72%, respectively.

Line Length Changes When a Circle Becomes an Ellipse

The elliptical cross-sections of the worm burrows shown in Figure 2.48*A* were approximately circular before deformation transformed their shapes. The deformed worm burrows capture the state of strain of the quartzite within which the worm burrows are preserved. In any given small outcrop containing deformed worm burrows, the local direction of lengthening (i.e., stretching) of the quartzite can be readily determined simply as the direction of the long axes of the worm-burrow ellipses, with shortening direction parallel to the short axes. Thus the deformed worm burrows, from outcrop to outcrop, can be used as strain gauges. The stretch and extension values in these two directions describe the amount of lengthening and the amount of shortening that the quartzite experienced. In this example, how do we go about determining S and e?

Stretch and extension values can be calculated if we know final length (l_f) and original length (l_0) of the long axis and the short axis of the ellipse. Determining the final-length values is easy, for we simply measure them along the two axes (Figure 2.48*B*). They are 2.6 and 2.2 cm, respectively. But how do we determine the original diameter of the circular object before deformation?

We might assume that there was no change in area during the deformation. Having made this assumption, we can say that the final cross-sectional area of the ellipse is equal to the original cross-sectional area of the circle from which the ellipse was derived (just as we did in Figure 1.44*B*):

$$\text{area of ellipse} = \text{area of circle}$$

$$\pi ab = \pi r^2$$

where a = major semiaxis of ellipse = 1.3 cm

b = minor semiaxis of ellipse = 1.1 cm

r = radius of circle

Thus

$$r = \sqrt{ab}$$

$$r^2 = ab = 1.4\ cm^2$$

$$r = \sqrt{ab} = \sqrt{1.4\ cm^2} = 1.2\ cm$$

Now we can determine stretch (S) and extension (e). Let us call the long axis line A, and set it equal to $2a$ = 2.6 cm (see Figure 2.48). Let us call the short axis line B, and set it equal to $2b$ = 2.2 cm (see Figure 2.48). Before deformation, lines A and B were the same length and equal to twice the radius of the original circle (i.e., $2r$ = 2.4). Therefore,

$$S_A = \frac{l_f}{l_0} = \frac{2a}{2r} = \frac{2.6}{2.4} = 1.1$$

$$S_B = \frac{l_f}{l_0} = \frac{2b}{2r} = \frac{2.2}{2.4} = 0.92$$

$$e_A = \frac{l_f - l_0}{l_0} = \frac{2a - 2r}{2r} = 0.083$$

$$e_B = \frac{l_f - l_0}{l_0} = \frac{2b - 2r}{2r} = -0.083$$

We conclude that lengthening parallel to the long direction of the finite strain ellipse (line A) was approximately 8.3% (i.e., $e_A \times 100\%$), and that shortening parallel to the short axis of the ellipse (line B) was approximately 8.3% (i.e., $e_B \times 100\%$).

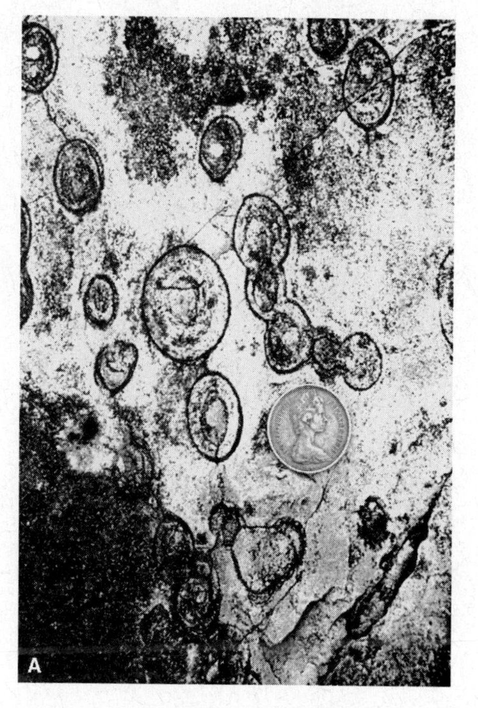

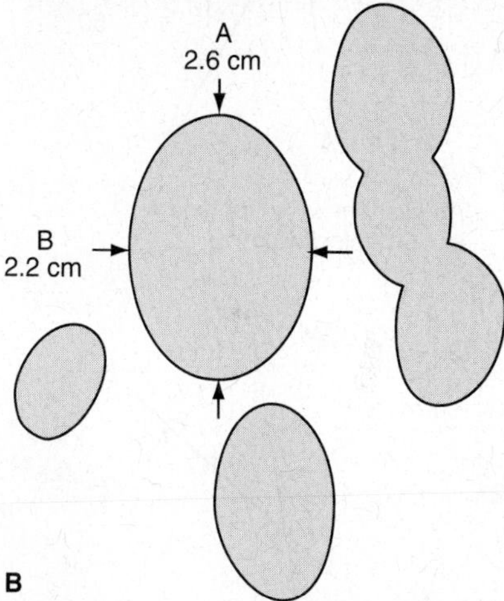

Figure 2.48 (*A*) Deformed worm burrows (genus *Skolithus*) in the Pipe Rock of Cambrian age along the Moine thrust zone of northernmost Scotland. [Photograph by G. H. Davis.] (*B*) Dimensions of one of the elliptical worm burrows. The lengths shown were measured along the long and short axes of the ellipse. Simple calculation of stretch values begins with the assumption that the original circular cross-sectional area of the worm burrow was the same as the elliptical cross-sectional area now observed in the deformed state.

Due to regional thrusting back in the early Paleozoic, the Cambrian Pipestone formation containing the deformed worm burrows is radically out of place. The Pipestone lies within the Moines thrust system in northern Scotland. The system was created during the Caledonian orogeny (500–400 Ma). Translations on individual thrust faults within the system range from 3 km to 30 km, with a total overall translation of ~50 km (McClay and Coward, 1981). The tectonic structure largely responsible for this is known as the Moine thrust. Perhaps, during thrusting, the Pipestone experienced modest rotation about a vertical axis as well. Describing the total deformation requires summing translation plus rotation plus strain. To work it out requires studying the structures at all scales, from rocks to regions!

Angular Shear: Measure of Change in Angles between Lines

Although strain parameters of extension and stretch effectively describe changes in lengths of lines in deformed bodies, they provide no information regarding changes that take place in the angles between lines. A parameter known as **angular shear**, symbolized by the Greek letter psi (ψ), comes to the rescue. To determine the angular shear along a given line, *L*, in a strained body, it is essential to identify a line that was originally perpendicular to *L*. Angular shear describes the departure of this line from its perpendicular relation with *L* (Figure 2.49). The full description requires a sign (**positive** equals **counterclockwise; negative** equals **clockwise**) and a magnitude expressed in degrees.

We illustrate the measurement of angular shear in Figure 2.50. There we see a square block (Figure 2.50*A*) about to deform into a rectangle (Figure 2.50*B*) by flattening. On the front of this block object we have drawn four reference circles (*1–4*), each containing two sets of mutually perpendicular lines (*a–b*, *c–d*, *e–f*, and *g–h*) (see Figure 2.50*A*). Because of deformation, the block becomes shortened vertically and lengthened horizontally (see Figure 2.50*B*). Furthermore, each of the lines changes length, six of the lines changing in orientation, and three sets of the lines moving out of original right-angle relationships.

It would not be difficult to calculate the extension and stretch values for each of the lines in the flattened block. We already know how to do this. Just pull out a scale, measure length before and after, and do the arithmetic. But what about determining angular shear? We can describe the angular shear (ψ) for any given line by identifying a line that was originally perpendicular to it, then measuring the angle through which the perpendicular line moved during deformation. For ellipse *ab* (see Figure 2.50*B*), there is no angular shear along line *a*, nor is there any along line *b*, for the original perpendicular relationship is preserved after deformation (see Figure 2.50*C*). For ellipse *cd* (see Figure 2.50*B*), the angular shear along *c* is +30° and the angular shear along *d* is −30° (see Figure 2.50*C*). For ellipse *ef*, (see Figure 2.50*B*), the angular shear along *e* is +38°, and the angular shear along *f* is −38° (see Figure 2.50*C*). Finally, for ellipse *gh* (see Figure 2.50*B*) the angular shear along *g* is +20°, and the angular shear along *h* is −20°.

If we keep our eyes open, we will spot expressions of angular shear. Figure 2.51*A* is a barn in Saskatchewan with an "*angularly sheared*" look, perhaps induced by the steady force of prairie wind and gravitational loading. The distorted trilobite featured in Figure 2.51*B* readily lends itself to appraisal of angular shear. Lines parallel to the original length (line *L–L′*) and to the original width (line *W–W′*) of the trilobite are assumed to have been perpendicular before distortion. Now they intersect at 60°. The angular shear along line *L–L′* is +30° (Figure 2.51*B*). This value is determined by focusing on the line *W–W′*, which was originally perpendicular to *L–L′*, and describing the sense and amount of deflection of that line. In the same fashion, the angular shear along

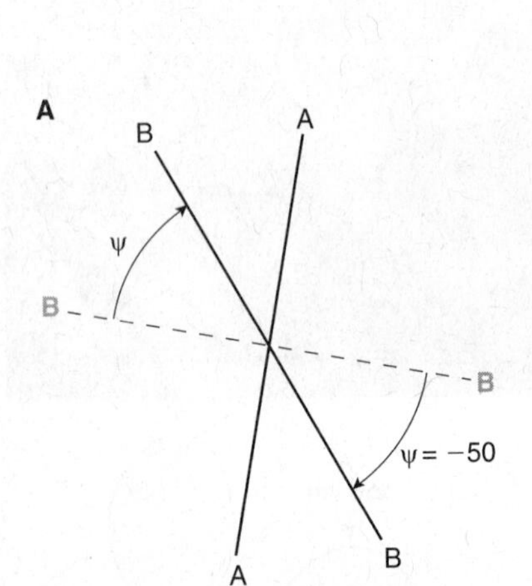

Angular shear (ψ) for Line A is −50° (clockwise!)

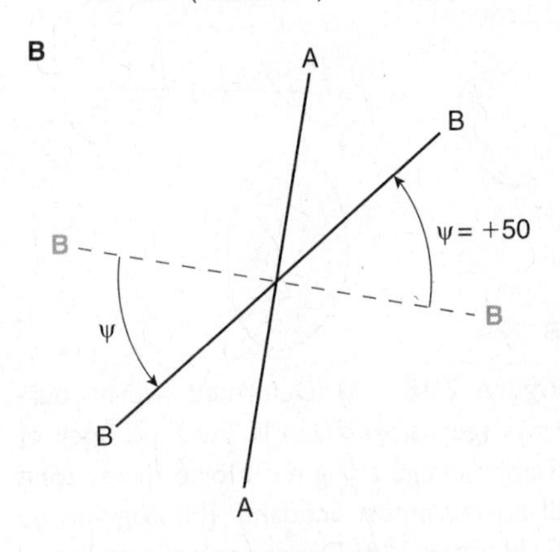

Angular shear (ψ) for Line A is +50° (counterclockwise)

Figure 2.49 Sign conventions for angular shear. (A) Determination of the angular shear of line A requires identifying a line, in this case B, which was originally perpendicular to A. The original orientation of line B relative to line A is shown by the dash line. Angular shear of line A is the shift in angle of $B_{original}$ versus B_{final}. Because the shift is clockwise, the angular shear is negative (−). (B) In this example the angular shear of line A is +50°. A counterclockwise shift is denoted by a positive (+) sign.

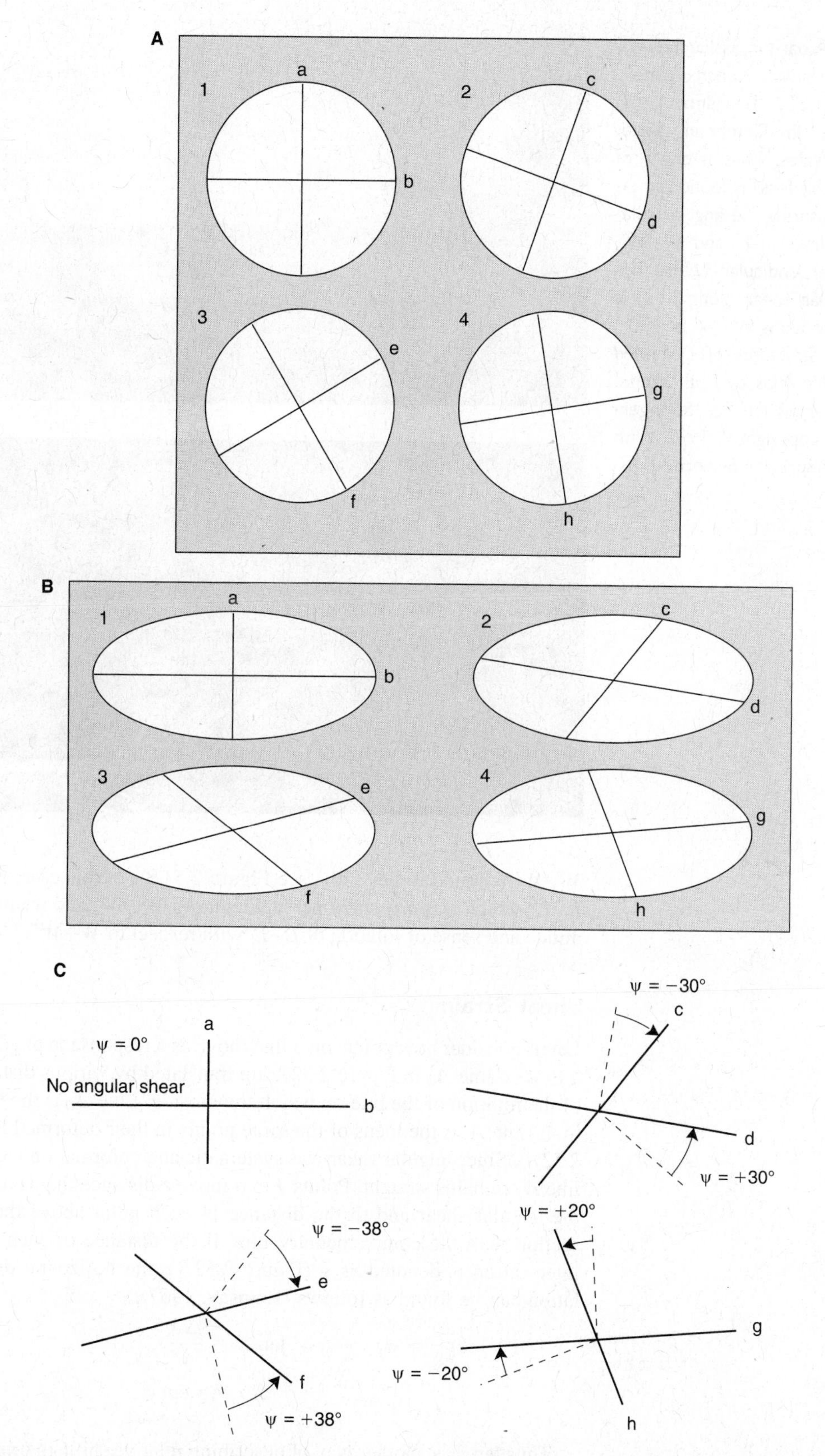

Figure 2.50 (*A*) Block containing reference circles and lines, before deformation. (*B*) Shape of the block after deformation. Original reference circles now are ellipses. The originally mutually perpendicular reference lines have all changed length, and most have changed orientation as well. (*C*) Angular shear along any line can be determined by first identifying a line originally perpendicular to it, and then measuring the angular shift. Remember, counterclockwise shifts are positive (+); clockwise shifts are negative (−). See text for explanation.

Figure 2.51 (*A*) A barn in Saskatchewan, Canada has an "*angularly sheared*" look. [Photograph by Shelby Boardman.] (*B*) Distorted trilobite in Cambrian shale, Caernarvonshire, Wales. Angular shear of rock within which this fossil is found can be determined by measuring the angular relationship between lines L–L' and W–W', lines that were perpendicular before the deformation. Angular shear along L–L' is +30°. Angular shear along W–W' is −30°. [From *The Minor Structures of Deformed Rocks: A Photographic Atlas* by L. E. Weiss. Published with permission of Springer-Verlag, New York, copyright © 1972. With kind permission of Springer Science+Business Media.]

W–W' is found to be −30° (see Figure 2.51*B*). In this case, we focus on line L–L', which was originally perpendicular to W–W', and we measure the magnitude and sense of rotation of L–L' with respect to W–W''.

Shear Strain

Let us consider how points on a line move as a response to angular shear. Points *1* to *4* on line A_0 in Figure 2.52*A* are translated by various distances as a result of the rotation of the line on which they reside. Line A_0 is the locus of points *1* to *4*. Line A_f is the locus of the same points in their deformed locations (Figure 2.52*B*). Since angular shear was systematic and deformation was homogeneous, line A_f remains straight. Points *1* to *4* move a distance that is directly related to the angular shear and to the distance of each point above the point of intersection with the complementary line. If the distance of each point above the intersection is denoted as y (Figure 2.52*B*), the horizontal distance of translation can be found as follows (Ramsay, 1967):

$$\tan \psi = \frac{\Delta x}{y} \qquad (2.3)$$

$$\Delta x = y \tan \psi$$

Thus $\tan \psi$ is another way of describing relative shifts in orientations of lines that were originally perpendicular. It is called **shear strain**, symbolized by the Greek letter gamma (γ),

$$\gamma = \tan \psi \qquad (2.4)$$

A

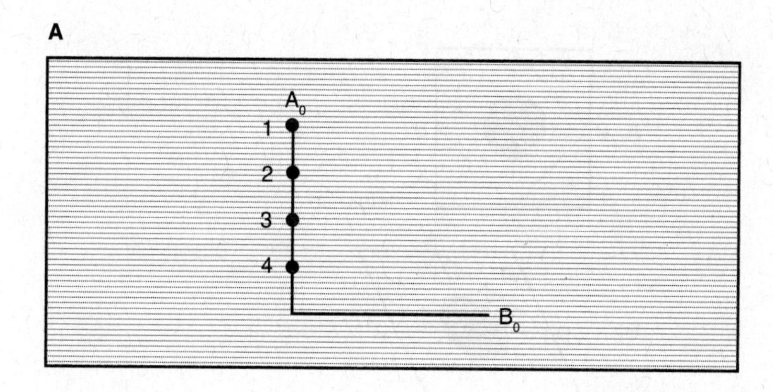

B

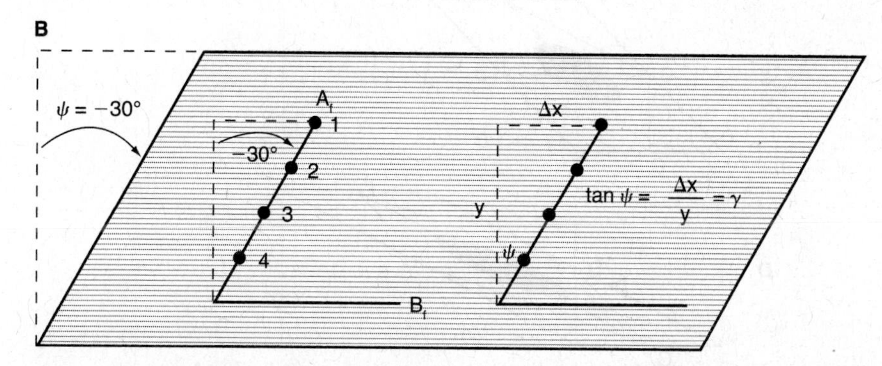

Figure 2.52 Simulation of the shearing of a computer card deck. (*A*) Deck embossed with lines A_0 and B_0 and points *1* to *4* before deformation. (*B*) Configuration of the deck, including the reference lines and points, after shearing.

Shear strain along a line (i.e., along a given direction) may be positive or negative, depending on the sense of rotation (deflection) of the line originally perpendicular to it. The range of shear strain is zero to infinity. For the example shown in Figure 2.52*B*, the shear strain of line B_f is −tan 30°, or −0.58. The shear strain of line A_f is +tan 30°, or +0.58. For the example of the distorted trilobite shown in Figure 2.51*B*, the shear strain of line L–L' is +tan 30°, or +0.58; and the shear strain of line W–W' is −tan 30°, or −0.58.

Figure 2.53 presents geometric relationships that may help draw together some of this information. The vertical faces of a square are subjected to progressive angular shear. The angular shear of the left vertical face of the undeformed block moves progressively from $\psi = 0°$ (Figure 2.53*A*) to $\psi = -71.5°$ (Figure 2.53*E*). This corresponds to shear strains ranging from $\gamma = 0.0$ to $\gamma = -3.0$ (Figure 2.53*E*). The vertical reference lines in Figure 2.53 are spaced at a distance that is equal to the length of the edge of the square before deformation, which we will consider to be one unit. Notice that when the top left-hand corner of the square moves horizontally half the distance of the original length of the side of the cube, the shear strain (γ) is −0.5 (Figure 2.53*B*). When it moves a distance of three times the distance of the original length of the side of the cube, the shear strain is −3.0. This is the tangent relationship at work.

Notice as well that the vertical edges of the original square continuously lengthen during this deformation, whereas the horizontal edges of the square do not change in length. As for the edges of the square within the square (Figure 2.53*A*), lines *DC* and *AB* first shorten, but then they lengthen during the progressive deformation (Figure 2.53*A–E*), whereas *DA* and *CB* just keep on lengthening.

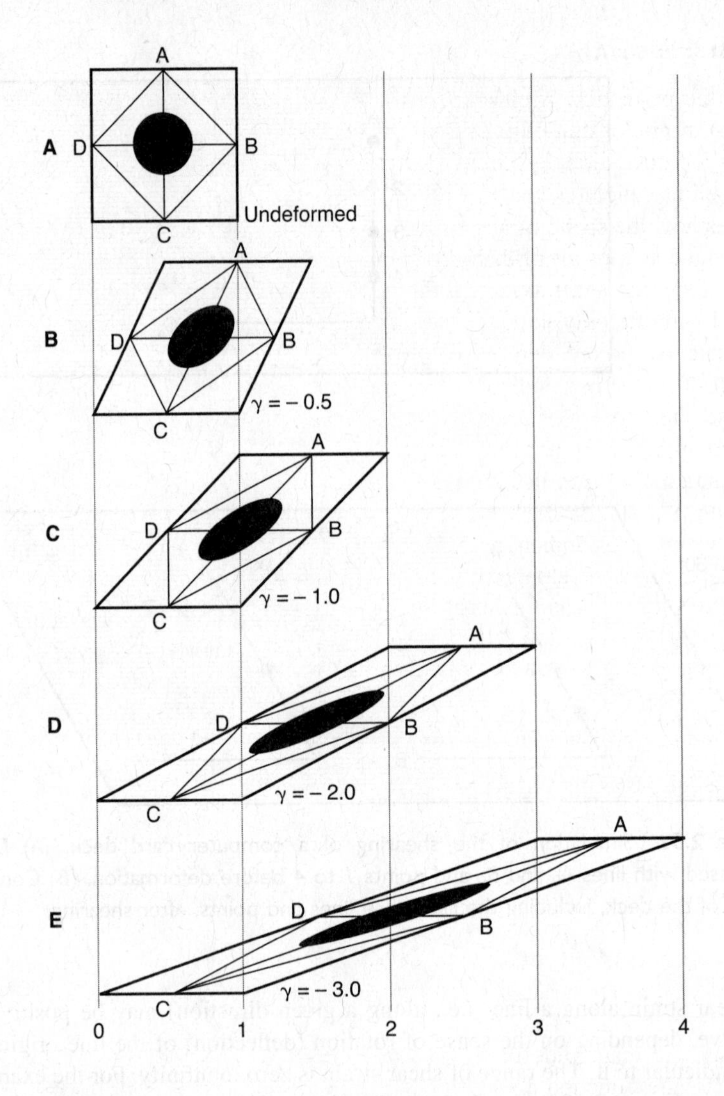

Figure 2.53 One way to get a feel for how the lengths of lines and the angles between lines change during shearing is to deform a fancy geometric object containing sets of mutually perpendicular lines of various orientations. Adding a reference circle to the ornamentation of the originally undeformed object permits changes in the shape and orientation of the strain ellipse to be tracked. Details regarding the progressive deformation from (*A*) to (*E*) are described in the text. Notice how the long and short axes of the strain ellipse, at each stage of the progressive shearing, continuously change both in length and orientation.

A flattening rendition of deformation is shown in Figure 2.54. The long and short axes of the ellipse do not rotate as the stretch (*S*) increases from 1.0 to 4.0. Although the edges of the outer square do not change orientation, they do change length. The edges of the inner square change orientation and length; they just keep on lengthening. Area is held constant.

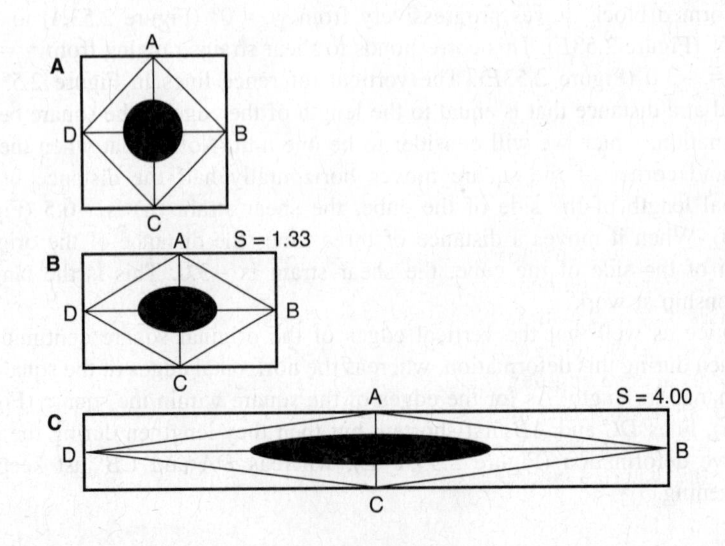

Figure 2.54 In this case, the fancy geometric object is progressively flattened, from (*A*) to (*C*). At each stage of the progressive flattening, the long and short axes of the strain ellipse change in length but not in orientation.

Strain Ellipses

Given the perfection of ellipses derived from homogeneous deformation (distortion) of circles and ellipses, it is no wonder that the strain within geologic bodies is conventionally described through the image of a **strain ellipse**. A strain ellipse pictures the distortion accommodated by a geologic body. It pictures how the shape of an imaginary or real circular reference object would be changed as a result of distortion.

The long and short axes of a strain ellipse are mutually perpendicular and aligned with the only mutually perpendicular directions in the body that were perpendicular before deformation. Imagine! It is possible to construct any number of sets of perpendicular lines within an object about to be strained. Yet, of these, the only set of such lines that started out mutually perpendicular and finished up mutually perpendicular is the set that marks the long and short axes of the strain ellipse in its final state.

When we look "*inside*" an ellipse created by homogeneous deformation of a circle, we make an important discovery. Lines parallel to the long direction of the ellipse are ones along which extension and stretch are greatest. In other words, of all the lines drawn within an originally circular body before deformation, the lines that end up parallel to the long axis of the ellipse are marked by the **greatest magnitude of extension** (e_1), and thus the **greatest stretch** (S_1). Lines that end up parallel to the short direction of the ellipse are ones along which extension and stretch are least. In other words, of all the lines drawn within an originally circular body before deformation, the lines that end up parallel to the short axis of the ellipse are marked by the **least magnitude of extension** (e_3), and thus the **least stretch** (S_3). Furthermore, lines parallel to the long or short directions of the ellipse are the only lines (i.e., the only directions) along which angular shear (ψ) and shear strain (γ) are zero.

Given the unique properties of lines parallel and perpendicular to the long and short directions of strain ellipses, these directions are given special attention in strain analysis. They are called the **principal axes of the finite strain ellipse** (Figure 2.55). The long, S_1 axis of the finite strain ellipse represents the direction and magnitude of the **maximum finite stretch**. The short, S_3 axis represents the direction and magnitude of the **minimum finite stretch**. These are the **finite stretching axes**.

The finite stretching axes are mutually perpendicular and parallel to the directions of maximum and minimum extension within the deformed body. This is illustrated in Figure 2.56. Lines A and B, drawn through the center of an undeformed reference circle (Figure 2.56*A*), are the two lines that, in this planned example, become aligned parallel to the finite stretching axes, S_1 and S_3, as a result of deformation (Figure 2.56*C*). Line A becomes line A″, changing in length from l_0 = 19.0 units to l_f = 31.5 units. Line B becomes line B″, changing in length from l_0 = 19.0 units to l_f = 11.5 units. The stretch value for line A″ is thus $S_A'' = S_1 = 1.7$. The stretch value for line B″ is $S_B'' = S_3 = 0.6$. If, patiently, we were to proceed to measure the original and deformed lengths of each and every line shown in Figure 2.56, computing the values of stretch (S) for each, we would find that no stretch value exceeds that calculated for line A″. And no stretch value is less than that computed for line B″. Thus, of all lines in the original starting circle, the one that ends up parallel to the long axis (S_1) of the strain ellipse lengthens the most. And the one that ends up parallel to the short axis (S_3) of the strain ellipse shortens the most.

The axes of the finite strain ellipse are special in yet another way. They are **directions of zero angular shear**. This means that the lines that end up parallel to finite stretching axes must have been perpendicular before deformation as well. Compare Figures 2.56*A* and 2.56*C*. A″ and B″ are the only perpendicular lines in the distorted body that were perpendicular before distortion. Along the

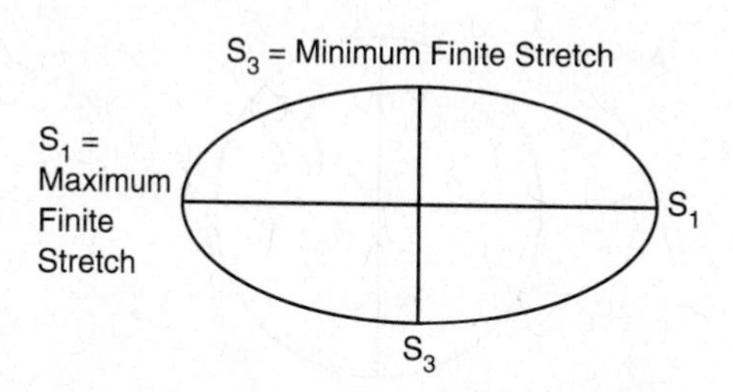

Figure 2.55 The finite strain ellipse and its principal axes. The long axis, S_1, is the direction of maximum finite stretch. The short axis, S_3, is the direction of minimum finite stretch.

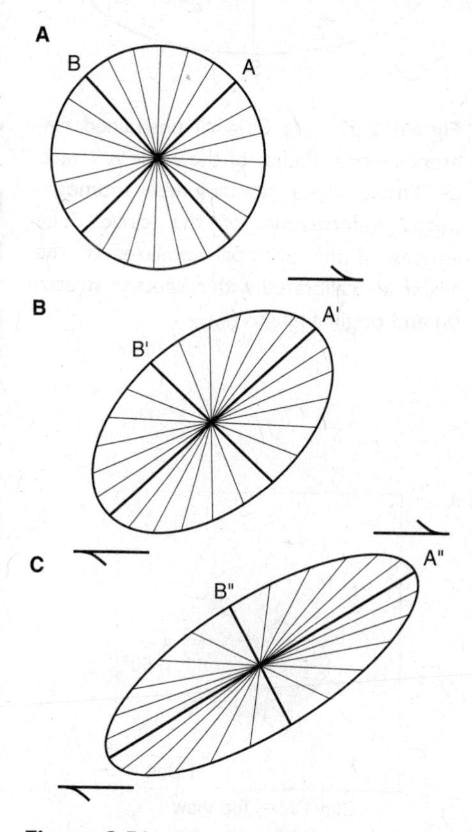

Figure 2.56 (*A*) Undeformed circular body, whose diameter is 19 units, inscribed with lines A and B of common length but different orientations. (*B*) Body is subjected to deformation by shearing. (*C*) After full deformation, almost all the lines have changed in length and orientation. Line A, as A″, ends up parallel to the direction of maximum finite stretch (S_1). Line B, as B″, ends up parallel to the direction of minimum finite stretch (S_3). Of all the lines, A lengthened the most and B shortened the most. Moreover, of all of the lines drawn in the originally undeformed circular body, only A and B were perpendicular both before and after deformation. Along the way (e.g., at stage B), lines A and B departed from being mutually perpendicular.

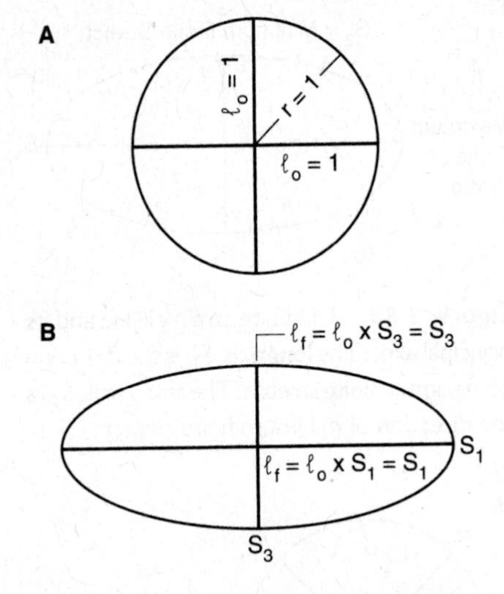

Figure 2.57 (A) Original unstrained reference circle. Radius of the circle is I unit. (B) Strain ellipse resulting from homogeneous deformation of the circle. The lengths of the principal semiaxes of the ellipse are calibrated with respect to stretch (S) and original length (l_o).

way, however, they departed from orthogonal (note the oblique relationship between A′ and B′ in Figure 2.56*B*).

Calibrating the Finite Strain Ellipse

The strain ellipse can be "*calibrated*" for purposes of evaluating changes in lengths and relative orientations of lines during nonrigid-body distortion. If the radius of an original unstrained reference circle is considered to be one unit (Figure 2.57*A*), the lengths of the finite stretching axes (S_1 and S_3) of the strain ellipse can be described in a very convenient manner (Ramsay, 1967). The final length (l_f) of a line, we have learned, is equal to its original length (l_O) multiplied by stretch (*S*). Thus the semiaxis length of S_1 (Figure 2.57*B*) is equal to the radius of the undeformed reference circle ($l_O = 1.0$) multiplied by the stretch in the S_1 direction. Since the radius of the reference circle is taken to be 1.0, the length of the semiaxis of the strain ellipse in the S_1 direction is simply S_1. Similarly, the length of the semiaxis of the ellipse measured in the S_3 direction is equal to S_3 (see Figure 2.57*B*).

Evaluating the Strain of Lines in a Body

In carrying out strain analysis, we evaluate changes in lengths and relative orientations of all lines in a geologic body, not just the special lines that end up parallel to the finite stretching axes. Figure 2.58 pictures the hypothetical experimental deformation of a clay cake. It shows how a perfectly deformable clay cake ought to distort given the rules of homogeneous strain. A reference circle with internal reference lines (Figure 2.58*A*) shows the undistorted nature of the clay cake before deformation. The deformed state of the original reference circle is shown in Figure 2.58*B*. Given this special opportunity to view both the starting materials and the finished product, we can describe the strain of lines of any orientation in the model. Let us start out by examining the change in length and the angular shear along line L (Figure 2.58*B*).

Before deformation, the length of line L was 1.0 unit; in the strained state, L is 1.11 units. The stretch and extension values for L′ in the deformed state can be determined as follows:

$$S = \frac{l_f}{l_0} = \frac{1.11}{1.0} = 1.11$$

$$e = \frac{l_f - l_0}{l_0} = \frac{1.11 - 1.00}{1.0} = 0.11$$

In its undeformed state (see Figure 2.58*A*), line L made an angle of $\theta = +50°$ with the S_1 axis. In its strained state (see Figure 2.58*B*), line L makes an angle of $\theta_d = +26.5°$ with the S_1 axis. This value is not angular shear, for the angular shear (ψ) along L is the deflection from perpendicular of a reference line that was originally normal to L.

After deformation, we can directly measure the angular shear of line M by studying its relationship with line L, which was perpendicular to M before deformation (see Figure 2.58*B*). After deformation, M and L are no longer

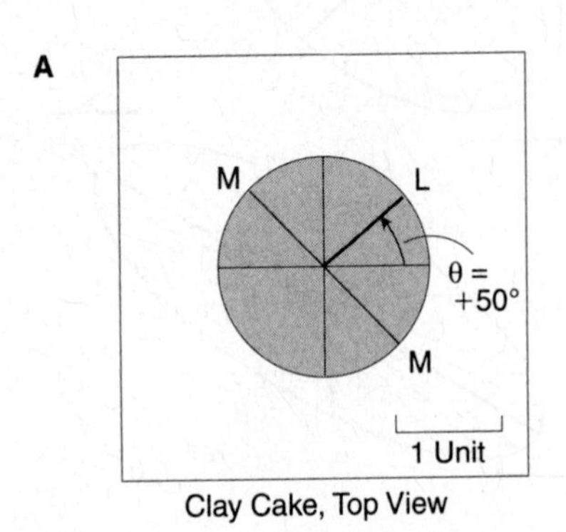

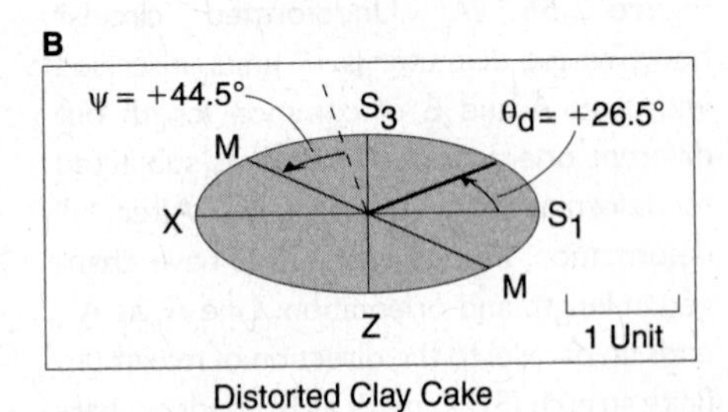

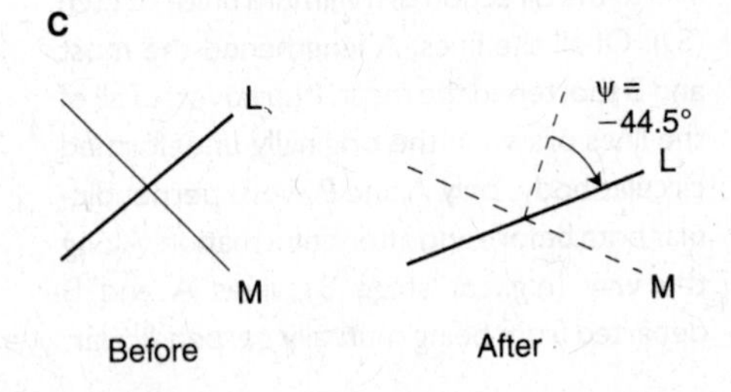

Figure 2.58 Deformation of a hypothetical clay cake, which was forced to distort in an ideally homogeneous way. Circle with lines *L* and M can be used to monitor the strain. (A) Undeformed state. (B) Deformed state. (C) Lines L and M before and after deformation. Angular shear of line M is equal to −44.5°.

perpendicular (see Figure 2.58*B*). The angular shear along M is −44.5° (Figure 2.58*C*). The shear strain, $\gamma = \tan \psi$, is −0.98.

The sign convention we use for angular shear and shear strain is **clockwise = negative, counterclockwise = positive** (Figure 2.59*A*). Similarly, when we identify the orientation (θ_d) of a line, L, with respect to S_1, we refer to the angle as **negative if measured clockwise from S_1, positive if we measure counterclockwise from S_1** (Figure 2.59*B*). These determinations of stretch and shear strain, parallel to L and M in Figure 2.58*C*, apply to the whole block along lines or directions parallel to L and M in the deformed state, assuming that the deformation of the block was homogenous. Lines L and M are simply the means to an end. It permits cautious extrapolation of strain measurements into areas for which there are no strain markers.

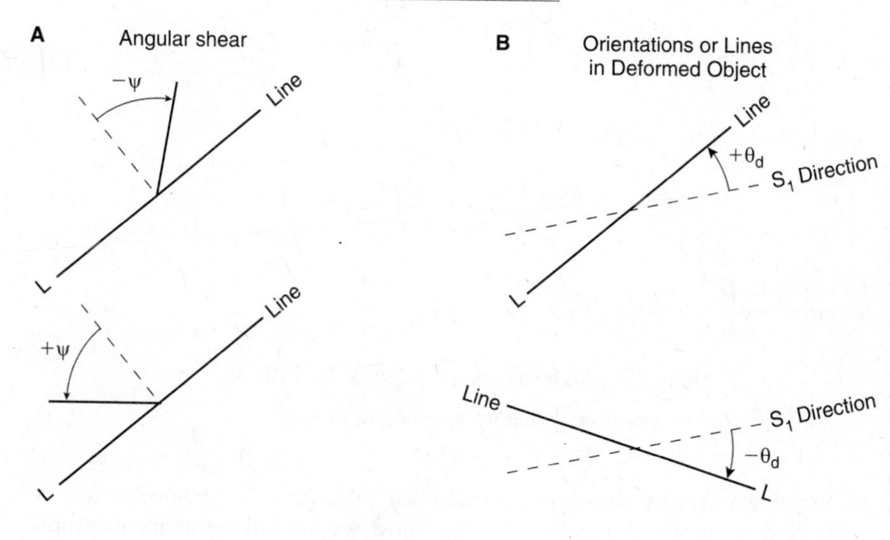

Figure 2.59 We use the convention that in all matters clockwise = negative (−), counterclockwise = positive (+). (*A*) Negative vs. positive angular shear (ψ) along line L. (*B*) Angle θ_d in strain analysis describes the angle between any given line (e.g., L), and the direction of maximum finite stretch (S_1). A counterclockwise angle from S_1 is considered to be positive (+); clockwise is negative (−).

The Fundamental Strain Equations

There are two fundamental equations that permit stretch and the shear strain to be determined along *any* direction in a strained body, not just directions for which we happen to have strain markers, provided that the stretch values of the principal axes of the finite strain ellipse (S_1 and S_3) are known, and provided that the angle (θ_d) made by a given line (e.g., L) with the direction of maximum stretch (S_1) is known. We introduce into this discussion a new parameter, lambda (λ), to set up some key equations. Lambda (λ), is called **quadratic elongation** in this context and is the square of the stretch.

$$\lambda = S^2 \tag{2.5}$$

The strain equations are built upon **reciprocal quadratic elongation** (λ'), where

$$\lambda' = \frac{1}{\lambda} = \frac{1}{S^2} \tag{2.6}$$

The **ratio of shear strain to quadratic elongation** (γ/λ) also figures importantly in the strain equations. For a given line in the deformed body, the magnitude of this ratio describes the mix of change in angle versus change in length. Lines along which the value of γ/λ approaches zero are characterized by change in length with very little shear strain. The γ/λ ratio reaches a maximum value for lines making an angle of 45° to the S_1 direction; shear strain is maximum along lines of this orientation.

The fundamental strain equations (Ramsay, 1967) look like this:

$$\lambda' = \frac{\lambda_3 + \lambda_1}{2} - \frac{\lambda_3 - \lambda_1}{2} \cos 2\theta_d$$

which also can be written

$$\lambda' = \frac{1}{2}\left(\frac{1}{\lambda_3} + \frac{1}{\lambda_1}\right) - \frac{1}{2}\left(\frac{1}{\lambda_3} - \frac{1}{\lambda_1}\right) \cos 2\theta_d \qquad (2.7)$$

and

$$\frac{\gamma}{\lambda} = \left(\frac{\lambda'_3 - \lambda'_1}{2}\right) \sin 2\theta_d$$

which also can be written

$$\frac{\gamma}{\lambda} = \frac{1}{2}\left(\frac{1}{\lambda_3} - \frac{1}{\lambda_1}\right) \sin 2\theta_d \qquad (2.8)$$

where $\lambda' = \frac{1}{\lambda}$, $\lambda'_1 = \frac{1}{\lambda_1}$, $\lambda'_3 = \frac{1}{\lambda_3}$

$\lambda = S^2$ along L, which makes an angle of θ_d with S_1

$\lambda_1 =$ greatest reciprocal quadratic elongation $= S_1^2$

When we see the subscript "*1*" attached to stretch (*S*), extension (*e*), or reciprocal quadratic elongation (λ'), we know we are talking about measurements parallel to the long axis (S_1) of the finite strain ellipse. When we see the subscript "*3*" attached to stretch, extension, or reciprocal quadratic elongation, we know we are talking about measurements parallel to the short axis (S_3) of the strain ellipse.

The strain equations can be solved readily with the aid of a calculator if values of λ_1, λ_3, and θ_d are known. To calculate λ and γ for line L in Figure 2.58*B*, the values of λ_1 and λ_3 are computed from stretch values, S_1 and S_3. Since the diameter of the original reference circle is 1.0 unit, the semiaxis lengths of the strain ellipse (see Figure 2.58*B*) are S_1 and S_3, respectively. Since the measured length of S_1 is 1.55 units,

$$\lambda_1 = S_1^2 = (1.55)^2 = 2.40$$

Since the measured length of S_3 is 0.65,

$$\lambda_3 = S_3^2 = (0.65)^2 = 0.42$$

The measured value of θ_d is +26.5°.

Using Equations 2.7 and 2.8, the reciprocal quadratic elongation and shear strain for line *L* are calculated as follows (remember, we are using reciprocals):

$$\lambda' = \frac{1}{2}\left(\frac{1}{0.42} + \frac{1}{2.4}\right) - \frac{1}{2}\left(\frac{1}{0.42} - \frac{1}{2.4}\right) \cos 53^\circ = 0.81$$

$$\lambda = \frac{1}{\lambda'} = 1.2$$

$$S = \sqrt{\lambda} = 1.1$$

Continuing with the solution of the equations:

$$\frac{\gamma}{\lambda} = \frac{1}{2}\left(\frac{1}{0.42} - \frac{1}{2.4}\right) \sin +53^\circ = +0.78$$

$$\gamma = \left(\frac{\gamma}{\lambda}\right)\lambda = (-0.78)(1.2) = +0.95$$

$$\psi = \arctan(-0.94) = +43^\circ$$

Calculating the Variations in Strain

To see the power of the strain equations (Equations 2.7 and 2.8) and the relations they depict, let us try them out in an applied, visual way. Figure 2.60*A* shows the lengths and orientations of a number of lines (lines *a–s*) that are plotted at 10° intervals in a deformed clay cake. The original orientations and lengths of lines *a–s* are shown in Figure 2.60*B*. Figure 2.60*C* extracts lines *a–s* and illustrates how each changes in length and orientation as a result of the deformation. The images serve to underscore the fundamental properties of homogeneous strain:

First, the axes of the strain ellipse are directions of maximum stretch (S_1) and minimum stretch (S_3), and zero shear strain.

Second, within any body that has undergone a plane strain, there are typically **two lines of no finite stretch** along which there has been neither lengthening nor shortening; lines oriented in these directions are characterized by stretch values of 1.0.

Third, within any homogeneously distorted body there are two directions marked by maximum shear strain.

Fourth, stretch and shear strain values increase and decrease systematically according to direction in a deformed body; specific values of reciprocal quadratic elongation and shear strain depend on the magnitudes of S_1, S_3, and θ_d.

The Mohr Strain Diagram

The fundamental strain equations (2.7 and 2.8) seem to have an unduly complicated form. But actually the form is quite elegant. Otto Mohr (1882) recognized that equations written in this way could be represented graphically as a circle. The **Mohr circle strain diagram**, the graphical construction of the strain equations, presents the systematic variations in reciprocal quadratic elongation and shear strain in a way that is both practical and versatile. We present the Mohr strain diagram, and several examples of its use, in *Part III-L, Carrying Out Strain Analysis*. If we know the principal finite strain values and directions across any outcrop or any region (e.g., across a thrust belt), we can use the fundamental strain equations and determine stretch and shear strain in any direction.

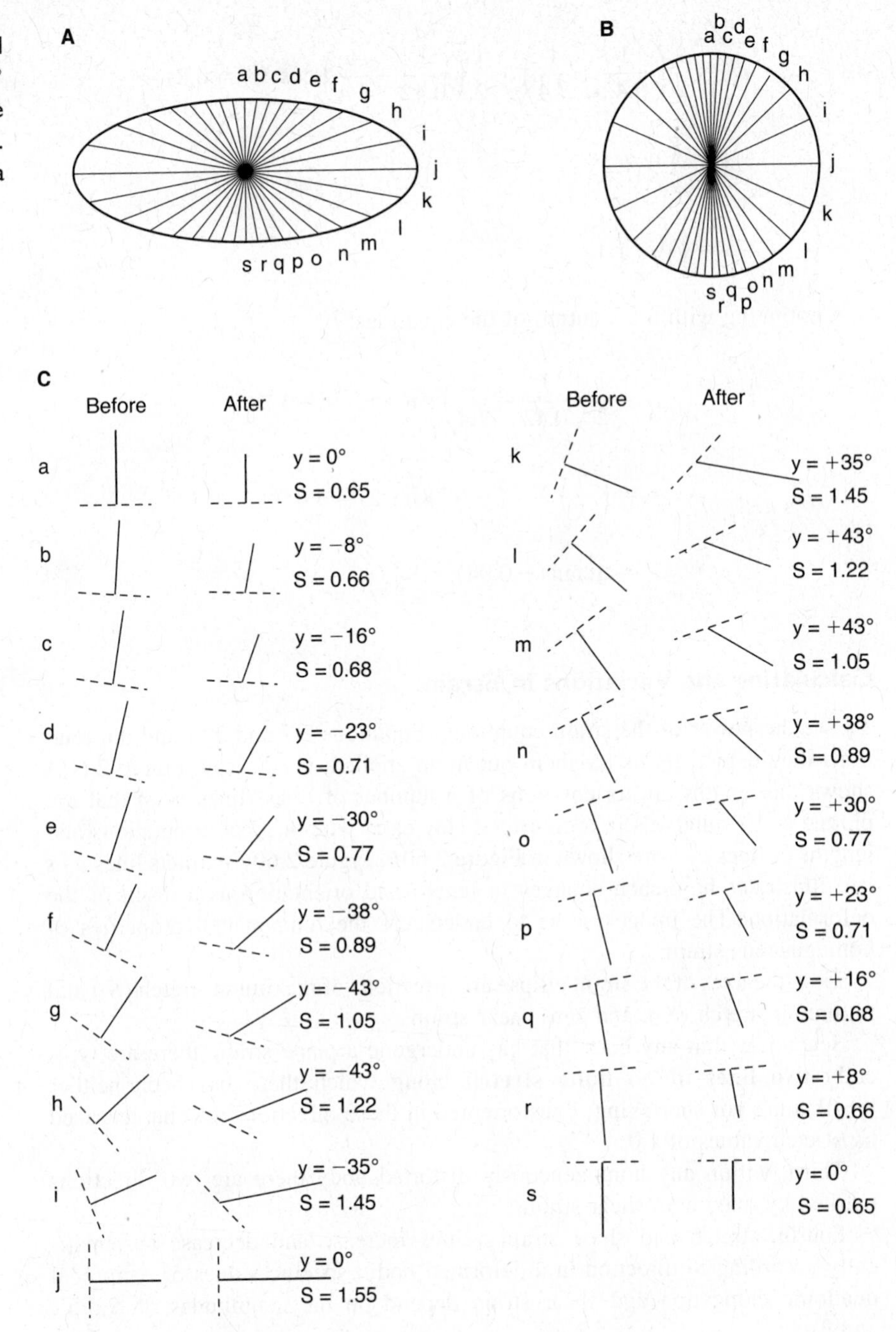

Figure 2.60 (*A*) Clay cake in deformed state, inscribed with lines *a–s* at 10° intervals. (*B*) Orientation of lines *a–s*, as the lines would look if the strain were removed. (*C*) Changes in length and orientation as a result of the distortion of lines *a–s*.

COAXIAL AND NONCOAXIAL STRAIN

Plane Strain

When geologists converge on an outcrop that displays strain, everyone starts talking about whether the deformation was "*coaxial*" or "*noncoaxial*." So that you can understand the importance of this distinction and become part of the conversation, we will explore the differences between coaxial and noncoaxial strain. But we begin by introducing the concept of "*plane strain*."

Plane strain means that the strain is essentially two-dimensional, that is, there is neither stretching nor shortening in the direction perpendicular to the plane that contains the directions of maximum and minimum finite stretch

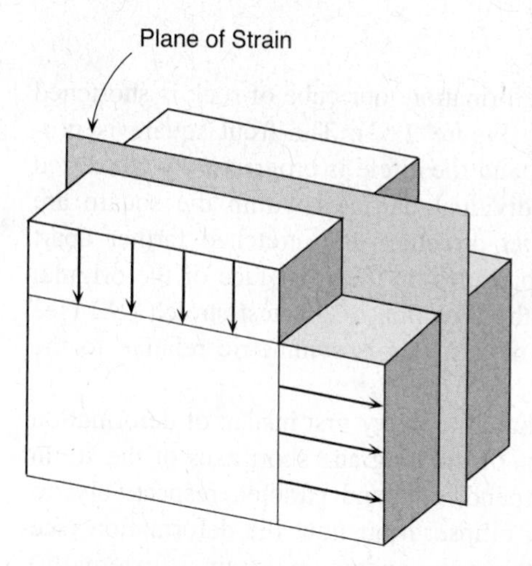

Figure 2.61 Deformation of a cube by plane strain. Vertical flattening of initial cube is accommodated by horizontal lengthening. The motion of all material particles during deformation occurs along a family of parallel planes, represented by the shaded plane of strain cutting through the material.

(S_1 and S_3). In plane strain, the motion of particles is restricted to a family of parallel planes (Figure 2.61). All strain occurs within these planes, and no strain occurs perpendicular to the planes. Maximum extension (S_1) in one direction is accommodated by maximum shortening (S_3) in a perpendicular direction within the **plane of strain**; no strain occurs in the third (S_2) dimension, normal to plane. Because no strain occurs in the third dimension, plane strain can be fully characterized using a two-dimensional strain ellipse. In 2D, the increases and decreases in the area of objects (i.e., dilations of objects) can be readily assessed on the basis of the products of the stretch values: $S_1S_3 = 1.0$ = no area change; $S_1S_3 > 1.0$ = area decrease; and $S_1S_3 < 1.0$ = area increase. Equations and concepts for analyzing three-dimensional strain and volume change are presented in more detail elsewhere (Ramsay, 1967; Ramsay and Huber, 1983, 1987; Twiss and Moores, 1992).

The undeformed object that we will employ in distinguishing coaxial vs. noncoaxial strain is shown in Figure 2.62. It is a cube of rock inscribed with a circle. A grid of points, representing material points in the deforming mass, are shown on the front of the cube so that we will be able to track the motion, first for a coaxial deformation, then for a noncoaxial deformation. In both, the deformations will be plane strain, with no motion of particles in the third dimension (i.e., into the cube).

We thus concentrate on the front face of the cube, which is a square. When we refer to a line on the square, we are really referring to a family of parallel lines that define a plane that projects along strike into the cube (see Figure 2.62). In this example, it is just for convenience that the faces of the cube are either vertical or horizontal; they could be at any angle.

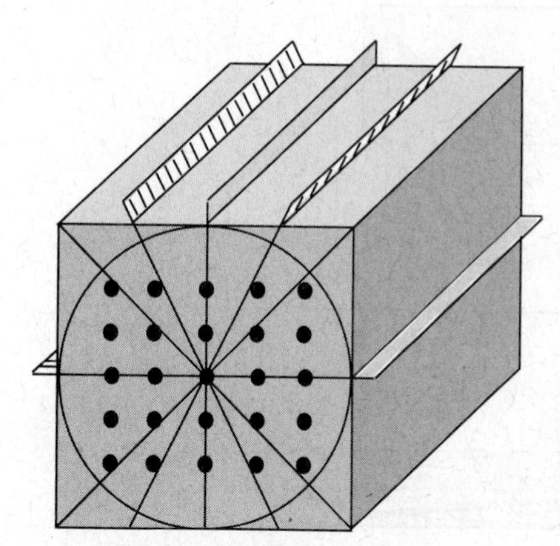

Figure 2.62 Cube before deformation. Points represent material particles and can be used to track deformation paths. Note that each line on the front, square face of the cube represents a family of lines that define a plane tracking through the volume of the cube.

Coaxial Deformation

To represent a plane-strain coaxial deformation, our cube of rock is shortened vertically and extended horizontally (Figure 2.63). The front square is progressively converted into a rectangle, and the circle is progressively converted into an ellipse. In this example, individual particles within the square are squeezed closer together in a vertical direction and stretched farther apart horizontally. Deformation paths for particles on the front face of the original cube flow inward and curve toward the direction of greatest stretch (S_1) (see Figure 2.63*E*). The movement path of particles is symmetric relative to the principal stretch directions, S_1 and S_3.

In coaxial deformation, the lines that at the very first instant of deformation were aligned parallel to the directions of the long and short axes of the strain ellipse (S_1, S_3) remain mutually perpendicular and parallel, respectively, to the long and short axes of the strain ellipse throughout the deformation (see Figure 2.63). This is true even where the object undergoing strain is (externally) rotated or translated as well.

"*Pure shear*" is commonly and at times mistakenly used in a way synonymous with coaxial strain. Pure shear is a special case of coaxial strain, one in which shortening in one direction is perfectly compensated by stretching at

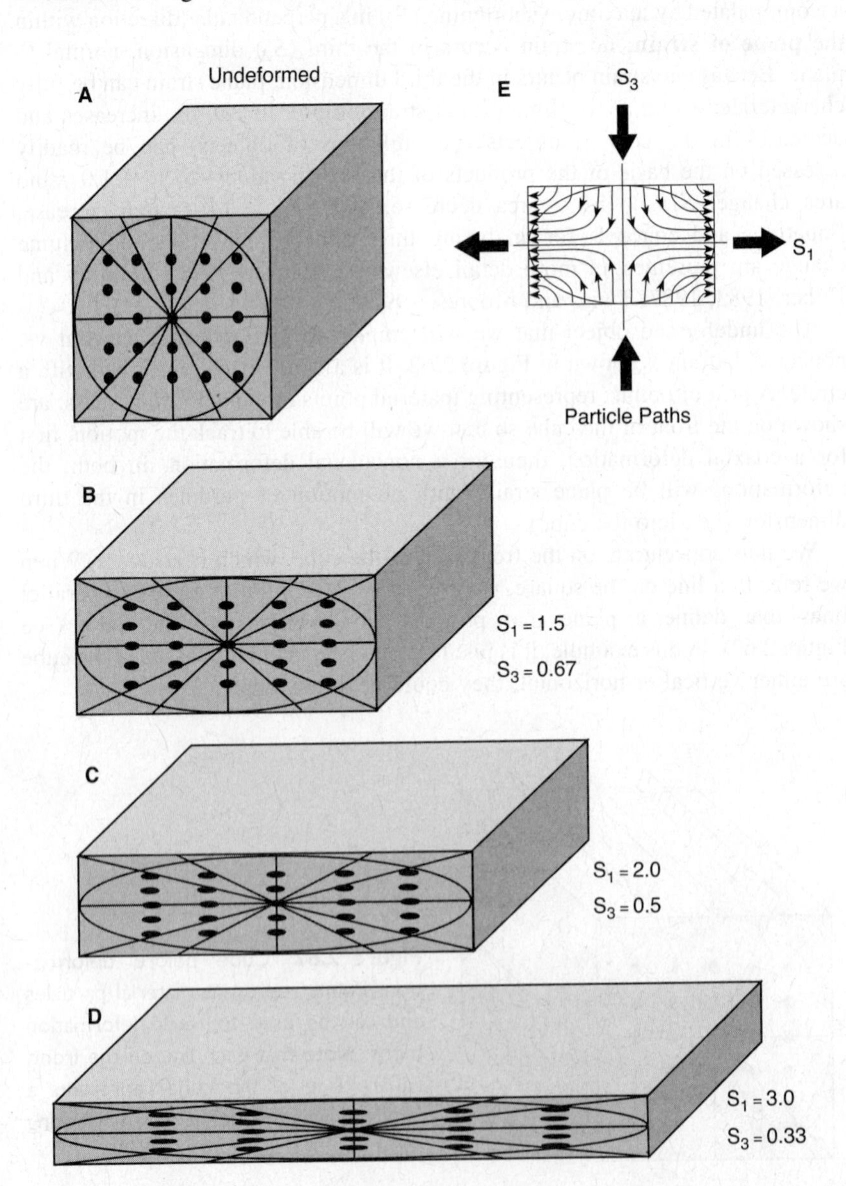

Figure 2.63 Coaxial deformation of a cube. (*A–D*) Cube is converted to progressively flatter shapes by vertical shortening and horizontal extension. The successive diagrams, from top to bottom, represent progressive increments of deformation and increasing amounts of strain. S_1 and S_3 the principal finite stretches, are shown for each stage. Note that they are reciprocals of one another, for there is no change in area. (*E*) Overall displacement paths for particles during deformation [Parts *D* and *E* adapted from Twiss and Moores, 1992, and from Hobbs, Means, and Williams, 1976.]

right angles to shortening, and with no change in volume. In 2D a square is converted to a rectangle, the original sides of the square remain parallel after deformation, and there is no change in area.

Noncoaxial Deformation

To represent a plane-strain noncoaxial deformation, the cube of rock is sheared like a deck of cards; the square is converted into a parallelogram (Figure 2.64). The top and bottom of the original cube do not change length or orientation, but both sides lengthen and change in orientation in the direction of shearing (top to the right, in this case). The height of the block does not change with deformation. All particles move parallel to the direction of shearing (horizontal in this case), but the directions of greatest stretch (S_1) and least stretch (S_3) continuously change, oblique to the direction of shearing. The relative velocity of a particle varies with its position normal to the shear plane. If the base of the square is arbitrarily fixed, the fastest relative velocities are for particles at the top (see Figure 2.64*D*).

The expression "*simple shear*" is commonly and at times mistakenly used in a way synonymous with noncoaxial strain. **Simple shear** is a special case of noncoaxial strain. In 2D a cube in transformed to a parallelogram in such a way that the top and bottom surfaces neither stretch nor shorten. Instead, they maintain their original lengths, which is the original length of the edge of the cube. In this special case there is no change in area (volume in 3D).

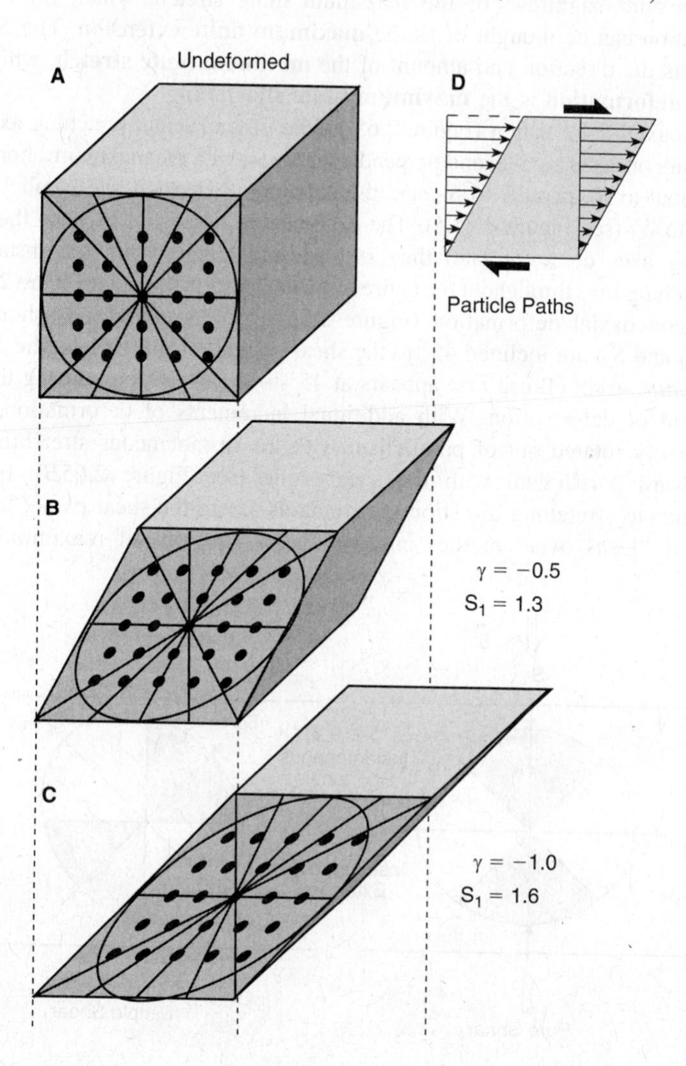

Figure 2.64 Noncoaxial deformation of a cube. (*A–C*). Cube is converted to progressively more angular, rhomb-like shapes by dextral simple shearing. The successive diagrams represent progressive increments of simple shear and increasing amounts of strain. Shear strain (γ) and S_1 are shown for each stage. (*D*) Overall deformation paths for particles during deformation [Part *D* adapted from Twiss and Moores, 1992, and from Hobbs, Means, and Williams, 1976.].

Distinguishing Instantaneous vs. Finite Strain Ellipses

As structural geologists analyzing strain in outcrops or regions, we are not often able to judge whether deformation was accomplished by coaxial or noncoaxial deformation. Without information about how the strain accrued, we can describe only the total **finite strain** as we see it. The key to distinguishing coaxial and noncoaxial strain lies in understanding how incremental strain accrues during **progressive deformation.**

It is useful to track deformation paths of material points in a cube of rock in order to understand how the very small (**infinitesimal**) increments of deformation accrue. An **incremental** or **instantaneous strain ellipse** is used to portray how a circle is affected by the smallest increments of deformation. The instantaneous strain ellipse will be nearly circular because it attempts to portray infinitesimally small amounts of strain. Thus, we must *eeexxxxaaagggeeerrraaattteee* its elliptical aspect for clarity. Because the instantaneous strain ellipse represents the strain accrued in the tiniest duration of time, it contains information about the relative *rates* of strain in different orientations. The principal axes are the **instantaneous stretching axes**, with the long axis ($\dot{S}_1$) representing the direction and magnitude of most rapid extension (lengthening) and the short axis ($\dot{S}_3$) representing the direction and magnitude of most rapid shortening.

The distinction between the *finite vs. instantaneous* strain ellipses becomes very important. The principal axes of the **finite strain ellipse** are called the **finite stretching axes**. The S_1 axis of the finite strain ellipse represents the direction and magnitude of the maximum finite stretch, which for coaxial deformation can be thought of as the **maximum finite extension**. The S_3 axis represents the direction and amount of the **minimum finite stretch**, which for **coaxial deformation** is the **maximum finite shortening**.

For coaxial deformation (Figure 2.65*A*), the instantaneous stretching axes ($\dot{S}_1$ and $\dot{S}_3$) are oriented parallel and perpendicular to S_1, with the maximum shortening rate normal to S_1, parallel to S_3, and the maximum extension rate parallel to S_1, normal to S_3 (see Figure 2.65*A*). The deformation is coaxial because the finite stretching axes do not rotate; they instead stay aligned with the instantaneous stretching axes throughout the entire history of deformation (see Figure 2.65*A*).

For noncoaxial deformation (Figure 2.65*B*), the *instantaneous* shortening axes ($\dot{S}_1$ and $\dot{S}_3$) are inclined 45° to the shear plane. For this reason, the S_1 axis of the *finite* strain ellipse first appears at 45° to the shear zone during the first increment of deformation. With additional increments of deformation, S_1 is successively rotated out of parallelism with the instantaneous stretching axis and toward parallelism with the shear zone (see Figure 2.65*B*), but the instantaneous stretching direction ($\dot{S}_1$) remains 45° to the shear plane. In other words, it "*leans over*" in the sense of shear. The axis of maximum finite

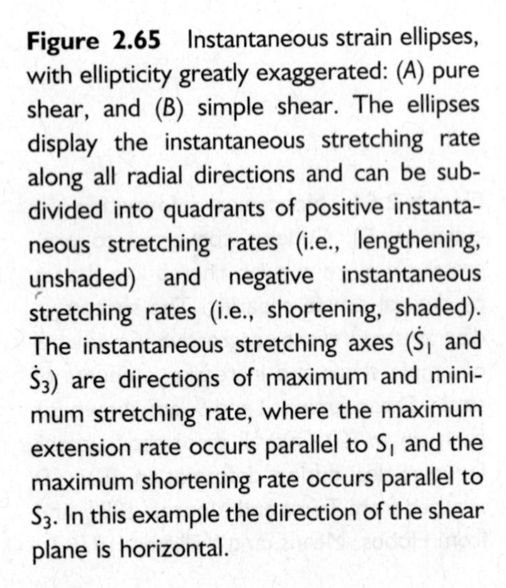

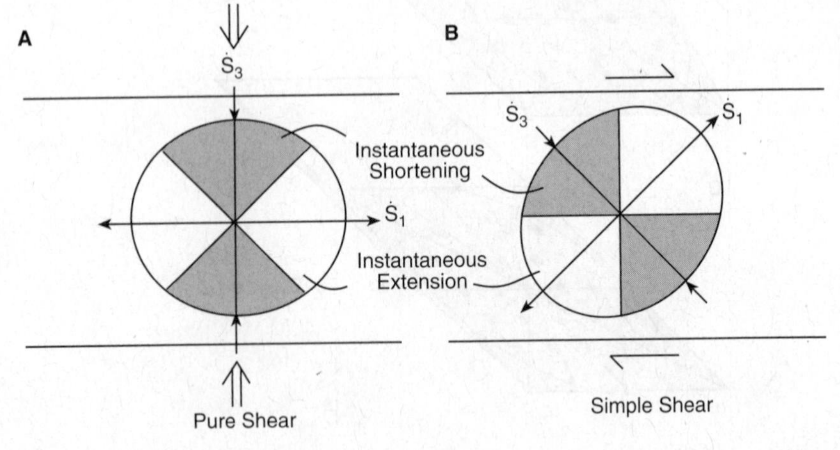

Figure 2.65 Instantaneous strain ellipses, with ellipticity greatly exaggerated: (*A*) pure shear, and (*B*) simple shear. The ellipses display the instantaneous stretching rate along all radial directions and can be subdivided into quadrants of positive instantaneous stretching rates (i.e., lengthening, unshaded) and negative instantaneous stretching rates (i.e., shortening, shaded). The instantaneous stretching axes ($\dot{S}_1$ and $\dot{S}_3$) are directions of maximum and minimum stretching rate, where the maximum extension rate occurs parallel to S_1 and the maximum shortening rate occurs parallel to S_3. In this example the direction of the shear plane is horizontal.

shortening (S_3) progressively rotates toward the normal to the shear plane. Thus, for noncoaxial strain, the finite stretching axes rotate away from parallelism with the instantaneous stretching axes as deformation proceeds, thus the expression "*noncoaxial.*" For noncoaxial deformation, the directions of maximum finite stretching and maximum finite shortening do not coincide with their instantaneous counterparts.

The distinction between instantaneous strain (incremental) versus finite strain (the total accrued strain) becomes especially important in describing and interpreting shear zones, and so we will return to and expand on this topic in *Chapter 10, Shear Zones and Progressive Deformation.*

Progressive Deformation

The ways in which incremental strains are added have a profound influence on the physical and geometric nature of structures that develop in a distorted body of rock. Whether a given line (or layer) undergoes stretching or shortening at any instant of time depends on the orientation of the instantaneous strain ellipse with respect of the line (or layer) in question. Whether a given line (or layer) undergoes total finite stretching or shortening depends on the nature of superposition of the incremental strains through time.

We can get a feel for this by observing the changes in lengths of lines L and M in Figure 2.66*A*, as the body in which they reside undergoes progressive

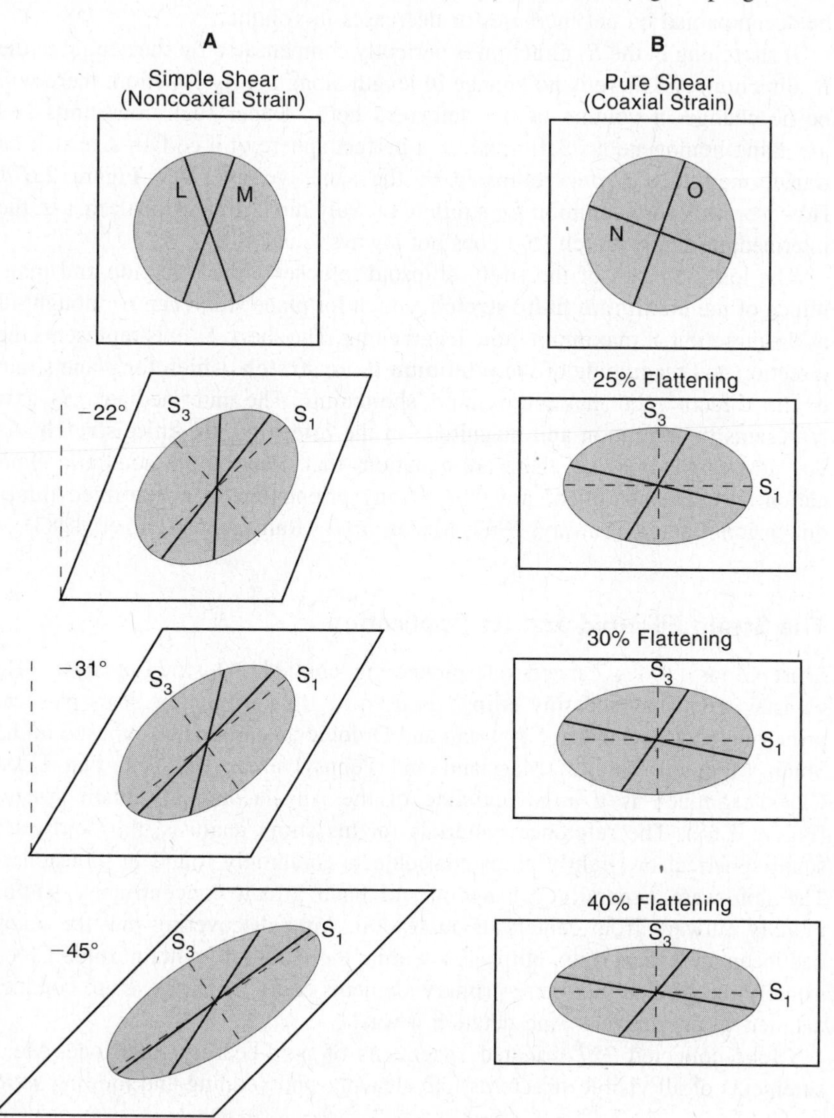

Figure 2.66 (*A*) Progressive deformation of a reference circle and two lines (*L* and *M*) by noncoaxial strain, which in this example is a simple shear. The principal strain axes rotate continuously to different orientations during the deformation. Line L at first shortens and then lengthens. Line M lengthens throughout the deformation. (*B*) Progressive deformation of a reference circle and two lines (N and O) by coaxial strain, which in this example is a pure shear. The orientations of the principal strain axes remain fixed throughout the deformation. Line N continuously lengthens. Line O continuously shortens.

noncoaxial simple shear, and lengths of lines N and O in Figure 2.66*B*, as the body in which they reside undergoes progressive coaxial deformation. L shortens at first, and then lengthens, whereas M lengthens continuously; N lengthens continuously, whereas *O* shortens continuously. We conclude that the superposition of lengthening on shortening does not necessarily mean that there have been two periods of deformation, separated in time and superimposed on one another.

THREE-DIMENSIONAL STRAIN ANALYSIS

The Finite Strain Ellipsoid

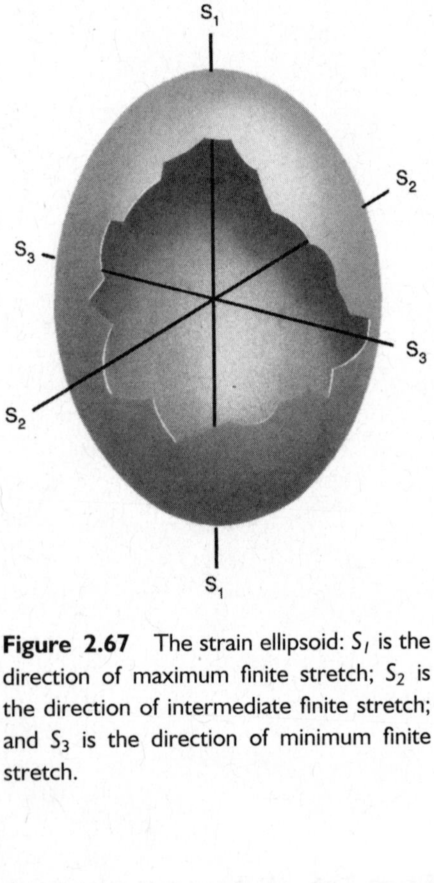

The most complete strain analyses are three-dimensional. The three-dimensional counterpart of the strain ellipse is called the **strain ellipsoid**. It pictures how the shape of an imaginary spherical reference object would be changed as a result of deformation. It is defined by three mutually perpendicular finite stretching axes (Figure 2.67). We have assumed in all examples presented so far that the strain has been a pure distortion without change in volume. This assumption has eased our entry into strain theory, but it would be misleading not to underscore the reality of changes in volume during strain. Distortion can be accompanied by net increases or decreases in volume.

If stretching in the S_1 direction is perfectly compensated by shortening in the S_3 direction, and there is no change in length along the S_2 direction, there will be no change in volume of the deformed body. Under such conditions and assuming homogeneous deformation, a perfect sphere of a certain size will be transformed to a perfect ellipsoid of the same volume (see Figure 2.67). However, it is very common for strain to be truly three-dimensional; that is, the intermediate finite stretch (S_2) does not always equal 1.0.

The long, S_1 axis of the strain ellipsoid represents the direction and magnitude of the **maximum finite stretch**, which for plane strain can be thought of as the direction of maximum finite lengthening. The short, S_3 axis represents the direction and magnitude of the **minimum finite stretch**, which for plane strain is the direction of maximum finite shortening. The intermediate, S_2 axis represents the direction and magnitude of the **intermediate finite stretch**. As you now might imagine, there are equations that describe the quadratic elongation and shear strain of any line of any orientation in a deformed three-dimensional body (Ramsay, 1967; Means, 1976; Ramsay and Huber, 1983).

Figure 2.67 The strain ellipsoid: S_1 is the direction of maximum finite stretch; S_2 is the direction of intermediate finite stretch; and S_3 is the direction of minimum finite stretch.

The Strain Ellipsoid and Its Application

Ernst Cloos (1947) carried out pioneering applied 3D strain analysis. He painstakingly analyzed tiny ellipsoids derived from originally near-spherical primary objects (ooids) in Cambrian and Ordovician carbonates exposed in the South Mountain fold in Maryland and Pennsylvania. The fold that Ernst Cloos examined is a large anticline of the Appalachian Mountain system (Figure 2.68). The reference materials for his strain analysis, the ooids, are small spherical or slightly ellipsoidal objects commonly found in limestones. The ooids are generally calcareous and have grown concentrically and/or radially outward from centers of nucleation. After discovering that the ooids had been deformed into ellipsoids within the South Mountain fold, Cloos astutely recognized that these primary elements could be harnessed to conduct detailed strain analysis. And detailed it was!

Cloos collected 227 oriented specimens of ooid-bearing limestone. Measurements of all visible structures like cleavage and bedding and jointing were

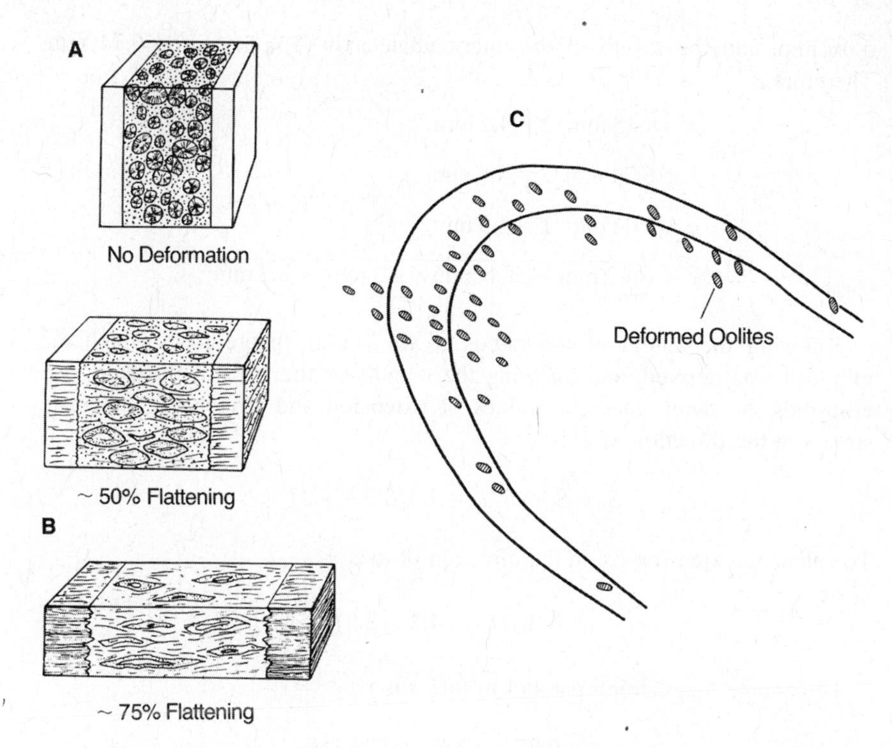

Figure 2.68 (*A*) Undeformed ooids in rocks of South Mountain, Maryland and Pennsylvania. (*B*) Ooids captured in different progressive stages of distortion. (*C*) Schematic diagram showing variation in shape and orientation of ooids as a function of position on the South Mountain fold. [From *Structural Geology of Folded Rocks* by E. T. H. Whitten, after Cloos, 1947. Originally published by Rand-McNally and Company, Skokie, Illinois, copyright © 1996. Published with permission of John Wiley & Sons, Inc., New York.]

made at each station where the specimens were collected. With the aid of a microscope that he used *in the field*, Cloos made preliminary measurements and assessments of the strain expressed by the ooids. In the laboratory he prepared 404 oriented thin sections of the ooid-bearing rocks, and these thin sections provided a view of the details of size, shape, and orientation of the strain reference features.

In calculating strain, Cloos assumed that each ooid, before distortion, had been a perfect sphere. He also assumed that distortion of the ooids was accomplished without change in volume. In effect, each perfectly spherical ooid was assumed to have had an original volume defined by

$$V_S = \frac{4}{3}\pi r^3 \tag{2.9}$$

where r is the radius.

Each sphere was transformed through distortion into an ellipsoid. The volume of an ellipsoid is defined by

$$V_E = \frac{4}{3}\pi abc$$

where a, b, and c are the long, intermediate, and short semiaxes, respectively.

Given the assumption of no change in volume, we write

$$V_S = V_E$$
$$r = \sqrt[3]{abc}$$

Thus by measuring the length of the axes of the deformed and ellipsoidal ooids, Cloos gathered data that allowed the radius of the original sphere to be determined. For example, he reported that at locality 300, the long axis of the ellipsoid ($2a$) measured 8.45 mm, the length of the short axis ($2c$) measured

5.06 mm, and the length of the intermediate axis (2*b*) measured 6.74 mm. Therefore,

$$a = 8.45 \text{ mm}/2 = 4.2 \text{ mm}$$

$$b = 6.74 \text{ mm}/2 = 3.4 \text{ mm}$$

$$c = 5.06 \text{ mm}/2 = 2.5 \text{ mm}$$

$$r = \sqrt[3]{4.2 \text{ mm} \times 3.4 \text{ mm} \times 2.5 \text{ mm}} = 3.3 \text{ mm}$$

Knowing the length of the radius of the initial sphere from which the ellipsoid was derived, and knowing the lengths of the axes of the deformed ellipsoids, he determined the values of extension and stretch. To calculate stretch in the direction of axis a,

$$S_a = S_1 = l_f / l_0 = 4.2/3.3 = 1.27$$

To calculate extension (*e*) in the direction of axis *a*,

$$e_a = e_1 = (l_f - l_0)/l_0 = (4.2 - 3.3)/3.3 = 0.27$$

Percentage lengthening parallel to line *a* is

$$0.27 \times 100\% = 27\%$$

By similar calculations, we write

$$S_b = S_2 = 1.03$$

$$e_b = e_2 = 0.03 \text{ (essentially no change)}$$

$$S_c = S_3 = 0.76$$

$$e_c = e_3 = -0.24 \text{ (24\% shortening)}$$

Some of the results of Cloos' strain findings are presented in Figure 2.68*C*, an idealized structure profile of the South Mountain fold showing the variation in orientation of the plane of flattening across the structure.

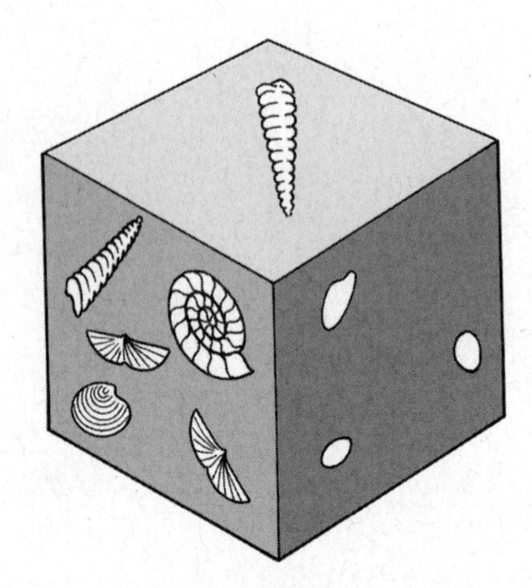

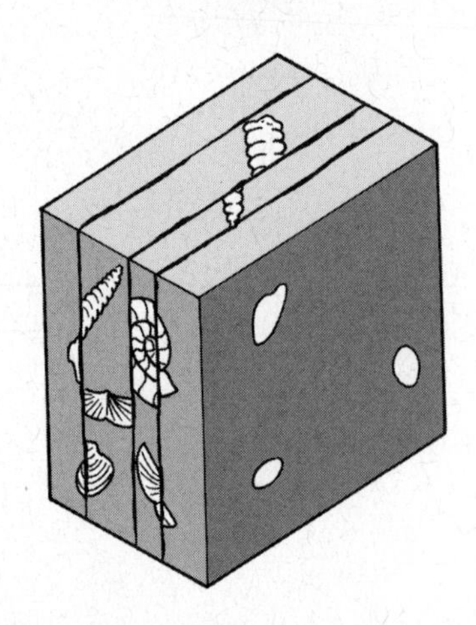

Figure 2.69 Cartoon of shortening of limestone produced by tectonic loading of a volume of rock. The directed tectonic loading increases the solubility of the calcite, which dominates the rock. "*Pressure dissolution*" removes rock volume as groundwater carries away the $CaCO_3$ in solution. Left behind are surfaces such as cleavage and stylolites. The smoking guns include fossils missing some of their parts. [Artwork by R. W. Krantz.]

Deformation Through Loss of Material

One of the stunning discoveries in structural geology is that rock can actually dissolve away in response to gravitational loading (during burial and compaction of sediments, especially carbonates) or tectonic loading (causing shortening). This process works particularly effectively in clayey limestone (i.e., marl). Solubility of calcite increases under conditions of loading, even without metamorphism. As dissolved calcite goes into solution, it is carried away as the fluids are squeezed from the sites. Insoluble material (e.g., clay; specks of organic material) remains behind and builds up (like plaque) in the sites where **pressure dissolution** takes place. The selective volume loss (negative dilation) along surfaces perpendicular to loading permits the rock to shorten!

In the case of tectonic loading (Figure 2.69), this process permits the rock to shorten even more than would be attainable through folding and faulting alone. In the case of gravitational loading during burial (Figure 2.70), this process achieves greater compaction than what would otherwise be possible. Where does the lost material end up? Commonly the dissolved material is reprecipitated as veins, oriented in favorable directions of opening of the rock. Some of the dissolved material may rise to the earth's surface and precipitate from springs to form travertine.

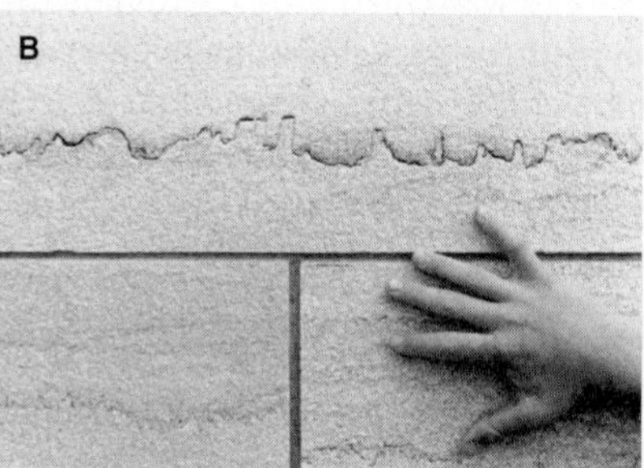

Figure 2.70 (*A*) Stylobedding in the so-called "*pancake rocks*" at Punakaiki on the west coast of the South Island of New Zealand. The accentuated bedding partings represent loci where calcite dissolved and disappeared during gravitational loading accompanying burial and compaction. [Photograph by G. H. Davis.] (*B*) Close-up photograph of a stylolite in building stone. Stylolites look like "*brain sutures.*" They are created through gravitational loading of carbonates as they are buried and compacted. The stylolite teeth are oriented in the direction of loading. The dark material highlighting the stylolites is insoluble residue. This example was photographed [by G. H. Davis] while waiting in line at the Air and Space Museum in Washington, D.C. Matt Davis' 12-year-old hand for scale.

Stylolites in well bedded limestones are normal products of simple burial, loading, and compaction (Figure 2.71). They run parallel to bedding, and the teeth/columns that make up the dominant texture of stylolites point in the direction of loading, that is, perpendicular to bedding. Deformed limestones may still preserve such "*primary*" stylolites but in addition may possess tectonic stylolites produced by directed tectonic pressure. These are most commonly oblique to bedding, with teeth aligned parallel to the tectonic shortening direction. Outcrops of limestone containing tectonic stylolites also generally full of calcite veins, representing the locations where the dissolved $CaCO_3$ was precipitated during deformation, that is, along fractures where opening could occur.

Stretch values sensitively reflect changes in area (and volume) (Figure 2.72). The formation of stylolites will tend to cause shortening in one direction (S_3),

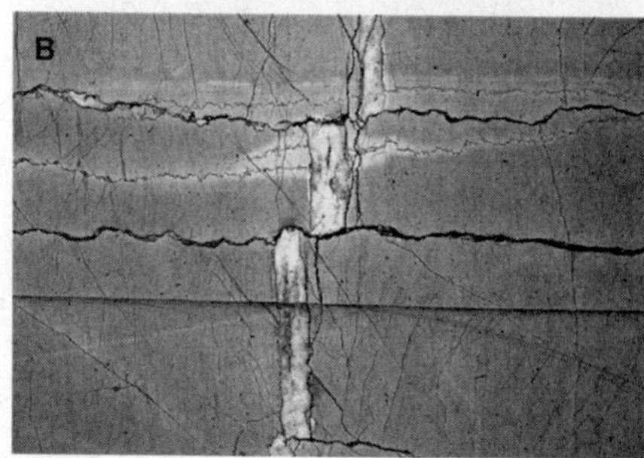

Figure 2.71 (*A*) Limestone columns (each ~1-m across) at the Temple of Apollo Epikourios at Bassi near Andretsina in Arcadia, the Peloponnesos, Greece. The wavy horizontal-spaced weathered surfaces are stylolites, evidence of gravitational loading dissolution of the limestone during primary deposition and burial. (*B*) Because the Temple of Apollo is being refurbished, there are fresh-cut replacement blocks of the limestone, such as this one. This photograph of a fresh limestone surface shows the details of stylolites, and reveals sharp truncation of a vertical calcite vein (~2-cm thick). Truncation is not by faulting, but by volume loss along a zone perpendicular to gravitational loading. [Photographs by G. H. Davis].

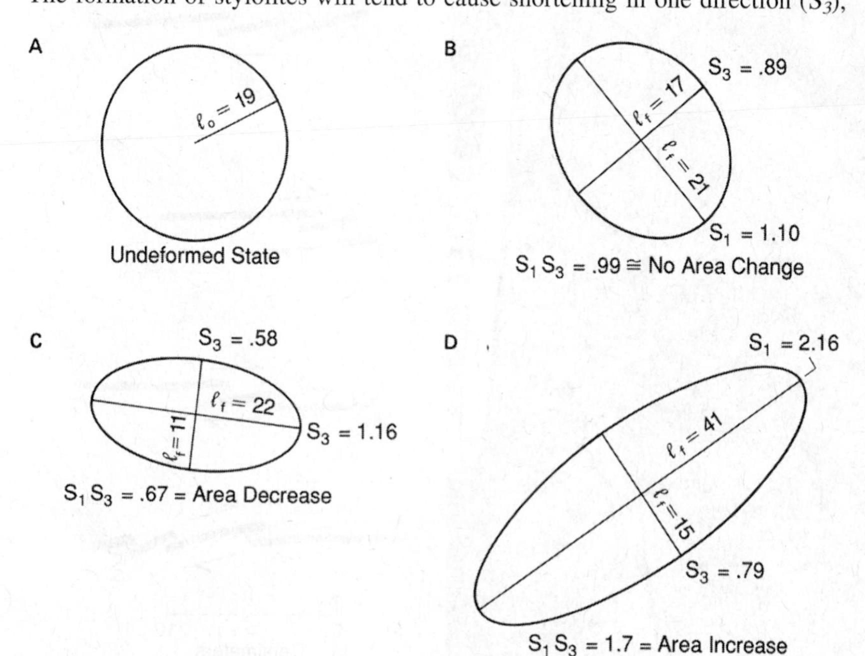

Figure 2.72 Strain ellipses properly constructed on the basis of stretch values reveal whether or not there have been changes in area (volume when considered in 3D). (*A*) No distortion, no dilation. (*B*) Distortion without dilation. (*C*) Distortion accompanied by area decrease. (*D*) Distortion accompanied by area increase.

but neither shortening nor stretching in the other principal directions (S_1, S_3). A strain ellipse depicting this state of affairs would be that in Figure 2.72*C*, where S_3 is less than 1.0 and both S_1 and S_2 are equal to 1.0. The result is an area (and volume) decrease.

Shortening and Stretching Working Together in Partnership

The interplay of shortening and stretching is wonderfully illustrated in structures mapped by Rispoli (1981) (Figure 2.73). The location of the limestone outcrop is the Languedoc region of southern France. It should be a shrine. The outcrops show left-handed strike-slip faults with small displacements. The **tip points** (i.e., end points of the fault trace) are well exposed. At one tip on one side of the fault there are stylolites projecting a short distance from the tip, and at the opposite tip point and on the same side of the fault there are calcite veins projecting the same short distance (see Figure 2.73). On the opposite side of the fault is the opposite configuration! The overall geometry is **antisymmetric**. The partnership is really quite amazing. The left-handed displacement of the limestone is absorbed by pressure dissolution along the stylolitic surface branching from the appropriate tip points and absorbing the shortening, whereas the calcite veins fill cracks that opened in the releasing end of the system as the fault block moved away. The veins branching from tip points are "*sinks*" for calcite dissolved along each stylolite. It is quite a system! But there is more. The stylolite teeth, which reflect the direction of shortening, are oriented parallel to the direction of strike-slip, and they progressively decrease in size from

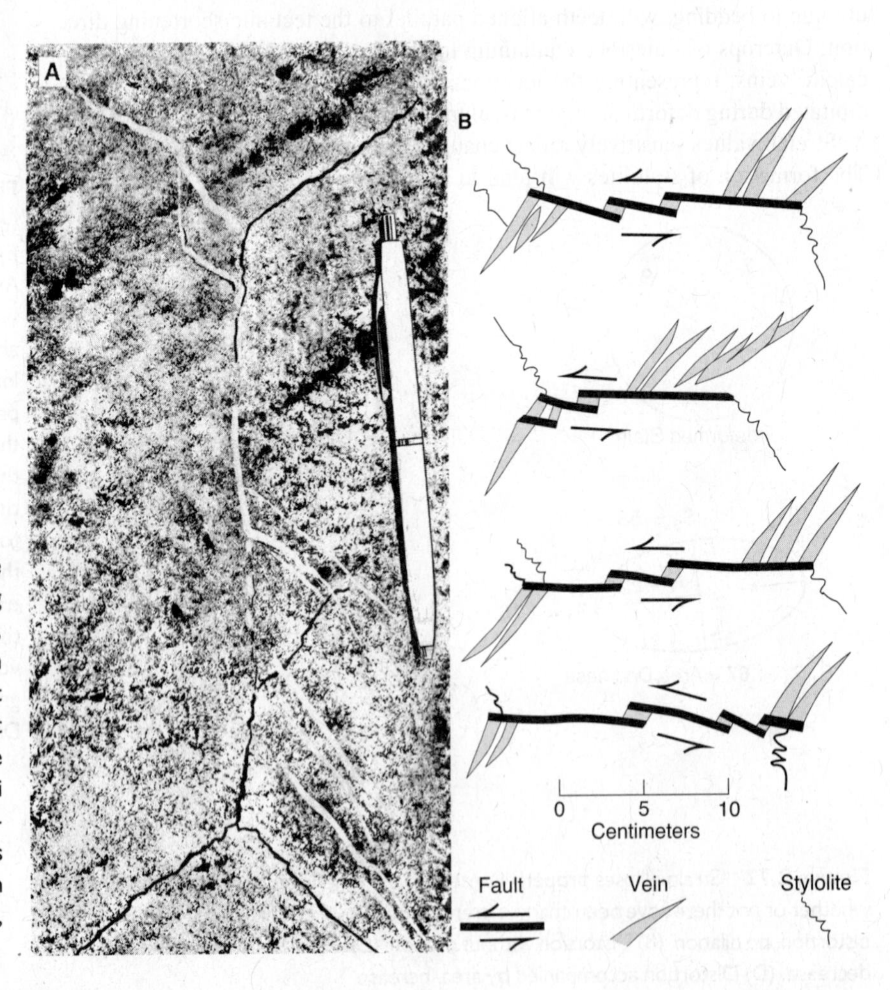

Figure 2.73 Remarkable displays of the perfect coordination that can exist among structures. In each of the four cases, microfaults (left-handed) connect an "*antisymmetric*" coupling of stylolites at one end and veins on the other. See text for explanation. These structures were discovered and described by Rene Rispoli (see Rispoli, 1981). [Reprinted from *Tectonophysics*, v. 75. Rispoli, R., The stress fields about strike-slip faults inferred from stylolites and tension gashes, p. T29–T36, © 1981, with permission from Elsevier.]

the tip point to the termination of the stylolitic surface as it tips out; the veins contain crystal fibers that are aligned with the direction of lengthening and parallel to the direction of strike-slip. The width of each vein progressively decreases from tip point to termination.

Rispoli's example is one that underscores the incredible harmony that exists within structural systems. Are these examples unusual? Only in their clarity and elegance. But most structural systems possess a cohesive internal compatibility. Everything fits. When in the face of reasonably complete information we discover that things do not fit, then we realize we are missing something important.

ON TO DYNAMICS

Translation, rotation, distortion, and dilation are responses of rocks that, under optimum conditions, can be described and interpreted in considerable detail. But what causes the displacements and strains in the first place? And how can we begin to decipher the origins of deformation, especially in ancient systems where the structures formed long ago? Suddenly we enter the arena of dynamic analysis and the interrelationships of displacement and strain to force, stress, strength, and behaviors of materials.

Chapter 3 Force, Stress, and Strength

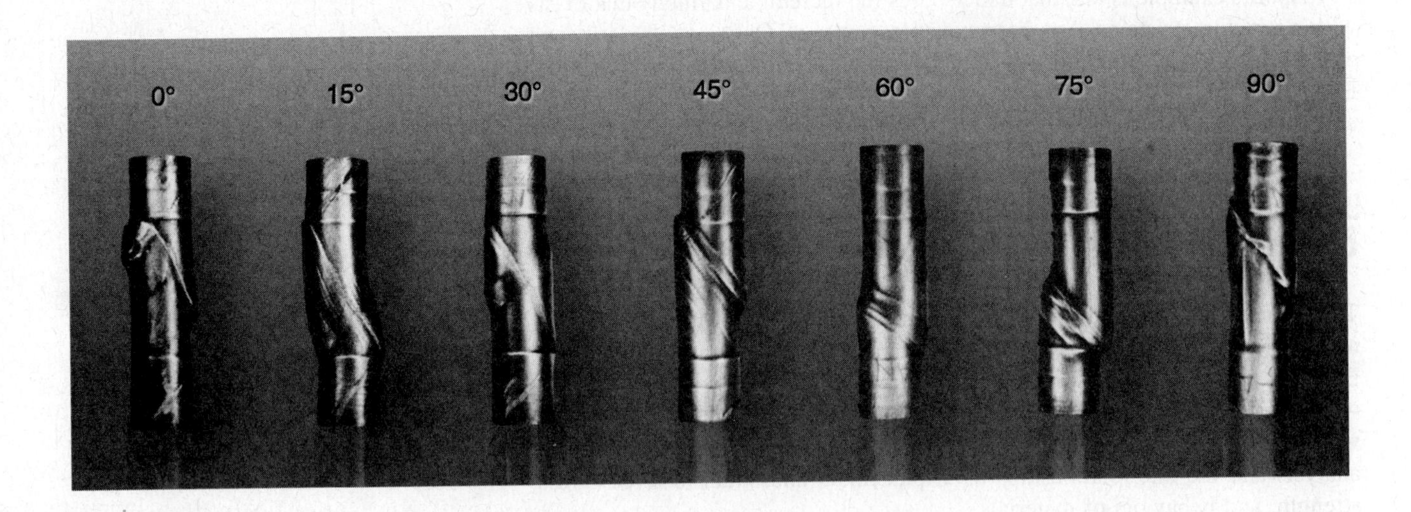

Copper-jacketed faulted cylinders of slate. When subjected to end-on loading under high confining pressure in a stainless steel deformation apparatus, the slate cylinders initially responded elastically, shortening by ~1%, a strain that is recoverable. But continued loading caused the rock to fail by faulting. The orientation and character of the faulting shows up in the jackets. The angle measurement shown above each specimen refers to the orientation of the slate's cleavage to the direction of loading. For most of these tests, it was determined that faulting followed cleavage. Only where cleavage was at a high angle to loading did the faults not utilize the preexisting weakness to accommodate shortening. There is a lot to consider in evaluating the relation of force and stress to rock deformation. [Photograph and testing by Fred Donath.]

INTRODUCTION

Strains and displacements that are "*frozen*" into bedrock and/or are forming today invite different levels of interpretation. On the one hand, the descriptions and interpretations may be broadly tectonic. Figure 3.1 shows the Alpine fault zone, which runs along the western margin of the South Island of New Zealand. It is accommodating oblique convergence between the Pacific plate to the east and the Indian-Australian plate to the west. A 'descriptive' and 'kinematic' interpretation of what is going on would look like this:

> *. . . The Alpine fault zone originated ~6.4 Ma and has accommodated 70–75% of the total oblique convergence between the Indian–Australian and Pacific plates, namely 480 km of right-handed strike-slip faulting (~36 mmyr^{-1}) and 70 km of crustal shortening (~10 mmyr^{-1}). Overall the Alpine fault strikes ~055°, dips ~50° SE, with slip trending 70°–90°. The Alpine fault zone projects to a depth of at least 25 km. Because the brittle-ductile transition is shallow (~5–8 km) and the rate of uplift high (~8mmyr^{-1}), mylonites (formed through ductile shearing) and cataclasites (formed through intensive deep brittle shearing) have been brought to the surface and are faulted atop fluvial outwash/fan gravel younger than 10,000 years. Even frictional melts (now glass) produced by earthquakes at depths of ~20 km can now be found at the surface in the form of pseudotachylite* (Figure 3.2).

But even this level of detail is insufficient. Those of us in structural geology feel a responsibility to de-construct the tectonic origin into the mechanical and dynamic analysis of the deformation caused by tectonics. This means that we not only must describe and measure the nature, geometry, strain, and displacement embodied in the geologic structures, but we must also evaluate the strength and behavior of materials in response to tectonic loading and stress, the forces that generated the stress, and the specific relationship of tectonic loading and stress to ultimate origin. Thus we enter the purview of dynamic analysis.

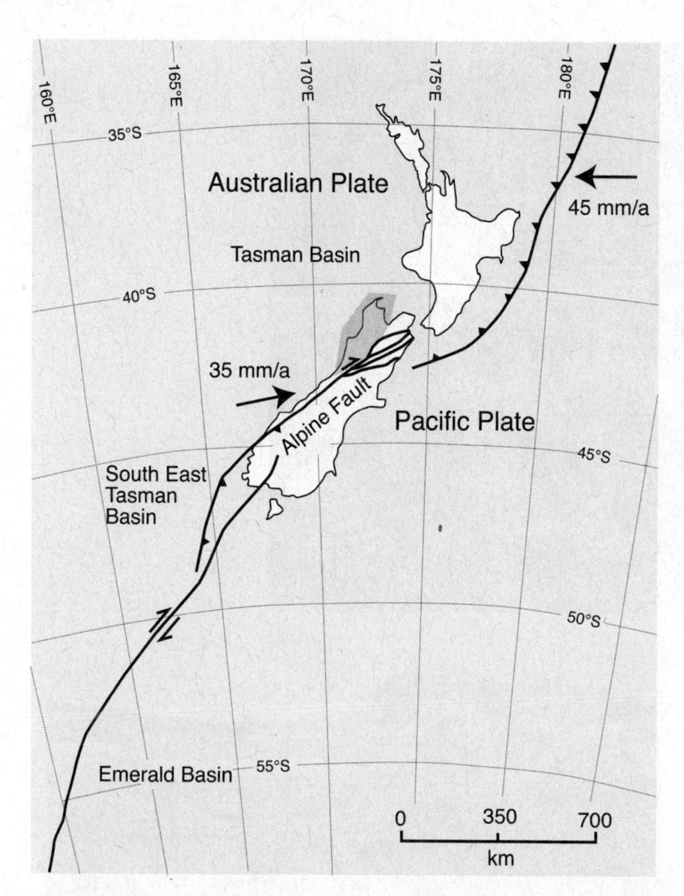

Figure 3.1 Plate tectonic setting of the Alpine fault in New Zealand. Note locations of North and South Islands of New Zealand. Relative to the Australian plate, the Pacific plate is moving west-southwest at a rate ranging from 35 to 40 mm/yr. The Alpine fault is a transform fault connecting the Puysegur trench to the south from the Hikurangi trench to the north. The overall regional setting is transpressive (i.e., strike-slip and compressive). [Reprinted from *Journal of Structural Geology*, v. 28, Ghisetti F. C. and Sibson R. H., Accommodation of compressional inversion in north-western South Island (New Zealand): Old faults versus new?, p. 1994–2010, © 2006, with permission from Elsevier.]

Figure 3.2 Photograph of specimen of pseudotachylite collected near the trace of the Alpine fault along Harold Creek on the South Island of New Zealand. [From the collection of G. H. Davis]. [Photography and editing by Gary Mackender. vr.arizona.edu, glm@email.arizona.edu, © 2009–2010 Arizona Board of Regents.]

DYNAMIC ANALYSIS

General Approach

Dynamic analysis deals with the physics of deformation. It involves interpreting the **force**, **traction**, **stress**, and **mechanics** (terms we will soon define) that give rise to structures, taking into consideration the **rheology** (strength and behavior) of the materials at the time they were deforming. For dynamic analysis to be meaningful, it must explain the physical and geometric character of the structures, the displacements and strains, and the relationships between stress and strain. The major aim of the analysis is to interpret the orientation and character of structures in the context of the orientation and magnitude of loads, forces, tractions, and stresses. This is challenging, especially when analyzing ancient geologic structures, because significant inferences must be made regarding the environment of deformation (e.g., temperature, confining pressure, fluid pressure), the strength and physical state of the materials during deformation (e.g., brittle, semi-brittle, plastic), the rate at which deformation proceeded, and the boundary conditions that framed the deformation (e.g., plate motions and plate boundaries).

Physical Models

The basis for dynamic analysis encompasses both theoretical and experimental research. Geological and engineering literature is replete with models that are

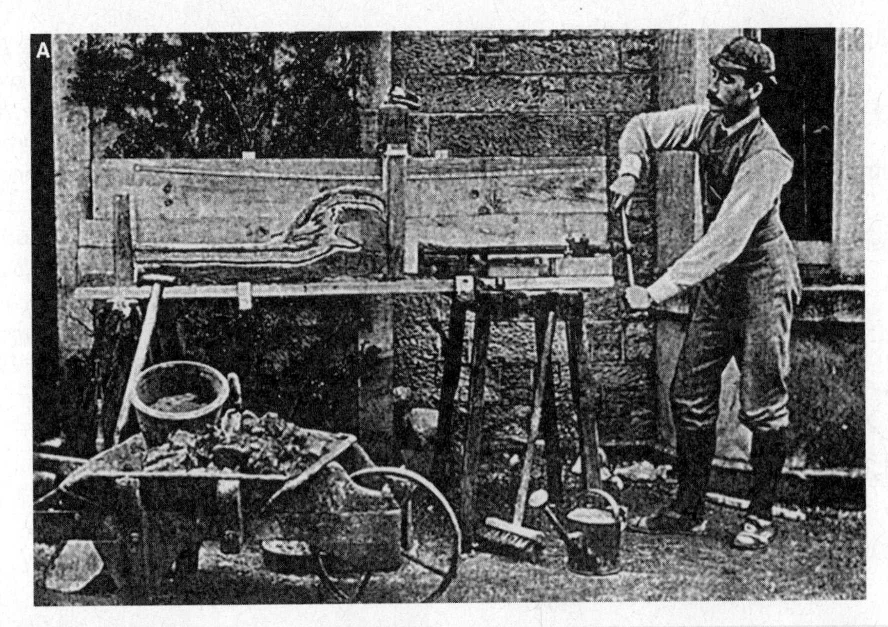

Figure 3.3 (*A*) Photograph of Hewry Cadell "*forcing*" the folding and faulting of layered materials in a squeeze box, trying to simulate the internal structure of mountains. [http//www.su.leeds.ac.uk/structure/assyntgeology/cadell/index.htm]. (*B*) Close-up photographs of progressive deformation of layers of wax, produced during a Bailey Willis experiment. Layer-parallel shortening was accommodated by folding. (Photographs by J. K. Hillers. Courtesy of United States Geological Suryvey.).

helpful in interpreting the origin of structures. The models are descriptions of the conditions under which geologic structures form.

One of the productive arenas of mechanical and dynamic modeling has been experimental deformation, ranging from relatively crude early attempts to replicate the structure of fold and thrust belts through squeezing of soft-layered materials (Figure 3.3), to sophisticated 'controlled' deformation of small cylindrical cores of rock under conditions of regulated temperature, confining pressure, rate of deformation, and fluid pressure. Empirical observations derived from experimental testing of materials have produced broad-ranging guidelines regarding the conditions under which rocks will fail by faulting, and even the orientations and characteristics of the faults thus formed.

Just to emphasize this point even further, we can describe the results of a given set of laboratory-based deformation experiments in the form of "*recipes*:"

Figure 3.4 Some of the ingredients and products of rock-deformation experiments. Left to right: a copper jacket; a 2.54 cm core of marble; a 2.54 cm core of faulted siltstone; a faulted, jacketed core of faulted siltstone; a faulted, slightly barrel-shaped core of faulted marble. [Photograph by R. W. Krantz.]

Use a core-drill device to cut a 1.3-cm diameter, 2.5-cm-long cylinder from a specimen of fine-grained sandstone (Figure 3.4). Bevel ends of the rock cylinder so that the top and bottom surfaces are planar and parallel to one another. Insert specimen in protective copper sleeve (to prevent fluids from entering the specimen) and place in stainless steel deformation apparatus. Pressurize the environment surrounding the test specimen with 1400 kg/cm^2 confining pressure, thus simulating deep burial. Load the ends of the specimen with a steadily increasing force by mechanically moving a steel piston of known elastic properties into the deformation chamber. Add the force at a rate of 400 kg/min. The specimen will fracture by faulting when the force reaches approximately 3600 kg. Expect a fault to form at an angle of approximately 30° to the direction of loading (see Figure 3.4).

In many situations, the results of theoretical studies, mechanical analysis, and numerical modeling are expressed in the form of **constitutive laws**, which are equations that describe the relations between two or more physical quantities that are specific to a given material (e.g., various kinds of rocks, various compositions of magma, etc.) and describe the specific response expected of that material when subjected to external forces under a given set of conditions. Constitutive laws are developed and tested for their validity and applicability through experiments under controlled conditions, and through comparison with what is observed in nature.

For example, a fundamental pursuit in structural geology is trying to determine the conditions under which faults will form. One of the conclusions reached is expressed in a constitutive law known as Coulomb's law of failure (Coulomb, 1773; Mohr, 1900). The law is shown below as Equation 3.1. This equation describes fundamental relationships in ways that words and photographs never could.

$$\sigma_c = \sigma_o + \tan\phi(\sigma_n) \tag{3.1}$$

where σ_c = critical shear stress required for faulting to occur

σ_o = cohesive strength of the bedrock

$\tan\phi$ = coefficient of internal friction of the bedrock

σ_n = normal stress

When Equation 3.1 is decoded, it reads as follows:

For faulting to occur in bedrock, the shear stress (force intensity that tends to cause deformation by slip or shear) must be sufficient to (1) create a fracture surface by overcoming the fundamental cohesion of the bedrock and (2) cause sliding/shear along the fracture surface by overcoming resistance to sliding friction.

Faulting is produced when the requirements of Equation 3.1 are fully met. Figure 3.5, for example, is a photograph of faults and fractures in gently dipping bedrock exposed on a wave-cut platform along Kaikoura Peninsula on the South Island of New Zealand (Figure 3.5). The long continuous fractures reveal that tectonic forces created intensities capable of overcoming the cohesion of the bedrock. Furthermore, stresses were high enough to cause faulting (i.e., actual slip along the fracture, and not just a simple opening-up along the

Figure 3.5 Photograph of fractures and faults in Oligocene bedrock exposed on a wave-cut platform on the Kaikoura Peninsula on the South Island of New Zealand. Note that traces of faults may be identified on the basis of offset of traces of the dark-weathering sandstone beds. Sense of offset is left lateral. [Photograph by G. H. Davis.]

fracture). Note in particular the modest fault offsets of the prominent dark sandstone beds, which serve as **marker beds** to track faulting.

Mathematical parameters and equations bring dynamic analysis to life. In the long run, we would like to be able to reconstruct the forces and stresses that once were active based on detailed knowledge of the structures and microstructures they produced. In achieving this, it helps to physically measure, where possible, the present-day stress in the crust of the Earth and relate measurements to actively forming geologic structures. The interpretation of the dynamic environment in which ancient structures were created is especially difficult, however, because the stresses are no longer operative, and the material behavior of the rocks at the time of deformation may not be certain. A given rock will respond to stress differently in different environments. The response may be **elastic**, where deformation is fully recovered when load is removed. The response may be **plastic**, where elasticity is lost at a certain threshold of loading and from that point on the deformation is not recoverable. Or, the response may be **viscous**, where there is no elastic or plastic strength, and thus the rock flows immediately upon loading. These differences in behavior are not always straightforward to reconstruct. Moreover, a rock that responds to stress in an elastic/plastic manner in one instant of time may shift to a viscous mode in another.

Overall, dynamic analysis is all about the interplay between the stresses that tend to deform, and the strengths that tend to resist. Gene Humphreys of Oregon State University puts it this way (Davis et al., 1996):

"The distribution of active deformation within regions expresses the interplay of force fields, stress fields, and strength fields. A balance of forces drives interconnected tectonic processes, regardless of scale. Forces are transmitted through the Earth in ways that are related to the strength of the material on which they act. Because the lithosphere is strong and mechanically decoupled from the Earth's interior by weak asthenosphere, tectonic forces can transmit stresses great distances within the lithosphere. Deformation will occur wherever and whenever material strength is insufficient to sustain the applied stress, and thus stress exceeds strength. Strength is determined by rheology, which is in turn a function of a host of internal properties, such as mineralogy, and external conditions, such as temperature and stress. Within complex active tectonic systems there is, at all scales, a ceaseless competition between stress and strength."

FORCE

Definition of Force

Translations, rotations, distortions, and dilations are responses of rocks to stresses that are generated by forces. **Force** is a push or a pull that changes, or tends to change, the state of rest or state of motion of a body. Force can cause a body in a state of rest to accelerate, and can cause a moving body to accelerate or decelerate or change direction. Forces can create **thrust** (velocity increase), **drag** (velocity decrease), and **torque** (change in rotational speed). Where forces are unevenly distributed on a body, mechanical stresses can arise, and these mechanical stresses will tend to deform the body. Pulling forces create **tension**; pushing forces create **compression**.

Force (F) is a vector quantity marked both by direction and magnitude. If several different forces act on a given body, the net effect can be determined if the direction and magnitude of each of the forces are known. The basic unit of force is the newton (N). A newton is the force required to impart an acceleration of one meter per second per second to a body of one kilogram mass. Only a force can cause something to move that had been stationary. Only a force can cause something to change its speed. Only a force can cause something to change its course of travel. A mass of 1 kg here on Earth exerts a force of ~9.8 N downward. Because on earth 1 N is the force of earth's gravity acting on a mass of ~100 g, you experience the force of ~1 N when an apple falls from a tree and hits you on the head.

Newton, through his first law of motion, described the concept of force in this way: an object at rest will remain at rest and an object in motion will continue in motion with a constant velocity unless it experiences a net force, in which case the object will accelerate or decelerate. A net force arises when forces are not balanced. In his second law of motion, Newton observed that the acceleration of an object is directly proportional to the net force on it, and inversely proportional to the mass of the object. The algebraic definition of force is, in fact, based upon **mass** (m) and **acceleration** (a):

$$\text{force} = \text{mass} \times \text{acceleration}$$
$$\mathbf{F} = ma \tag{3.2}$$

Mass and acceleration are reciprocal.

$$m_1 a_1 = m_2 a_2$$

Thus, if a given force, **F**, accelerates a 1 kg object by 3 m/s^2, it will accelerate a 2 kg object by 1.5 m/s^2, for:

$$1\,\text{kg} \times 3\,\text{m/s}^2 = 2\,\text{kg} \times 1.5\,\text{m/s}^2$$

Mass and Weight

The **mass** (m) of a body is the amount of material the body contains. The mass of a body will be expressed here in SI units. Mass can be readily calculated if volume and density are known. **Volume** (V) is the space occupied by the mass and is expressed commonly in units of cubic centimeters or cubic meters. **Density** (ρ) is the measure of the mass of a body per unit volume, and is most commonly expressed in grams per cubic centimeter or kilograms per cubic meter. The relationship among mass, volume, and density is

$$m = \rho V$$
$$\rho = \frac{m}{V} \tag{3.3}$$

Figure 3.6 The barbell creates a load, measured in kg · m/s^2, which is being supported by a force, measured in newtons. Static equilibrium! [Artwork by D. A. Fischer.]

Mass causes a body to have weight in a gravitational field, but mass and weight are not the same. The **weight** (W) of a body of a given mass is the magnitude of the force of gravity acting on the mass, and it varies according to location. The force of gravity acting on the mass of a body on the moon will be *less* than the force of gravity acting on the same body on Earth. The intensity of the Earth's gravity field under standard conditions is 9.8 N/kg.

Pumping iron in the local gym, the athlete shown in Figure 3.6 is demonstrating the relation between weight and force. The weight of the 22.7 kg mass is 22.7 kg · 9.8 m/s^2, which is 222 N. To support that weight, there is a need to exert a force of 222 N. Actually lifting the mass (i.e., giving it an upward acceleration) requires a force greater than 222 N.

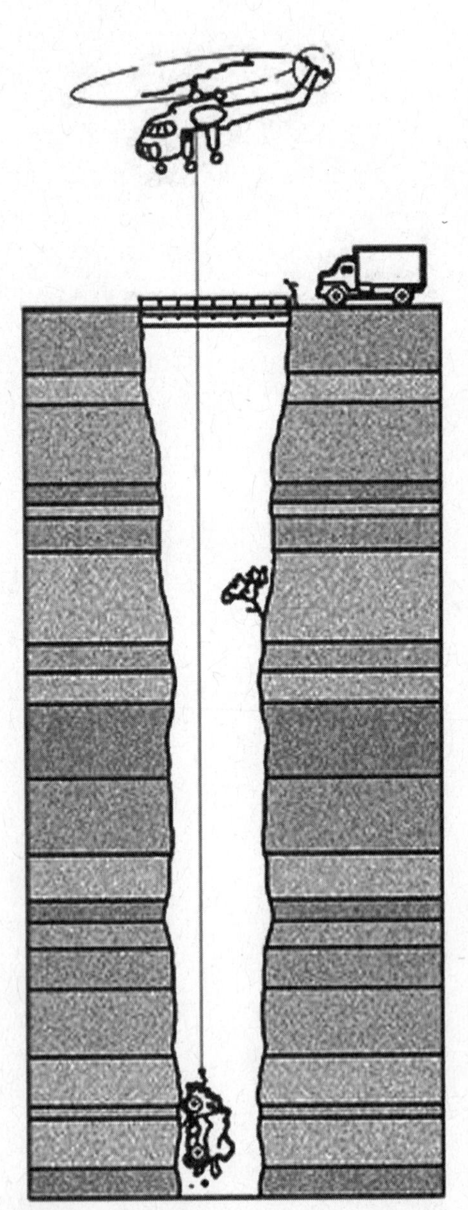

Figure 3.7 Cartoon rendering of Hell's Backbone Bridge in southern Utah, north of Escalante. [Artwork by D. A. Fischer.]

Forces as Vectors

The vector character of forces permits them to be added and subtracted using principles of vector algebra, and this in turn makes it possible to evaluate whether forces on a body are in balance. Force vectors act at point masses in bodies, such as rock bodies.

Reaction Forces

We are aware that forces can be at work without producing any visible consequence. Consider, for example, a truck about to cross the tiny, narrow bridge at Hell's Backbone north of Escalante in southern Utah. The bridge is positioned over a virtual "*Gothic*" chasm (Figure 3.7). The prudent truck driver, no matter whether driving a 4-wheeler, a 6-wheeler, or an 8-wheeler, will pause to read the signage before crossing (Figure 3.8). If the load and number of contact points (tires) are in right relationship, the driver will feel confident during the crossing because its weight (W) is balanced by the **reaction force** (**R**) due to the strength of the bridge support system. The truck therefore will not sink into the bridge or break through the bridge. We love the signage at Hell's Backbone (see Figure 3.8), because it not only introduces us to thinking of force as a **load**, for example, 19 tons, 28 tons, 36 tons, but also introduces us to the notion that the capacity to deform depends upon load per unit area, i.e., the **force intensity**, or shall we say "*load per unit* tire". If the load reaches 4.5 tons per tire, beware!

Addition and Resolution of Forces

In situations where two or more forces are exerting their effects on a body, it is possible—because forces are vectors—to resolve the two (or more) forces into a single **resultant force**. The force vectors can be added. Figure 3.9 shows two tugboats pulling a cargo ship. The resultant force is determined (magnitude and

Figure 3.8 Photograph of sign near bridge at Hell's Backbone, informing drivers of the magnitude of reaction forces that the bridge is capable of exerting. Beware of exceeding the limit! [Photograph by G. H. Davis.]

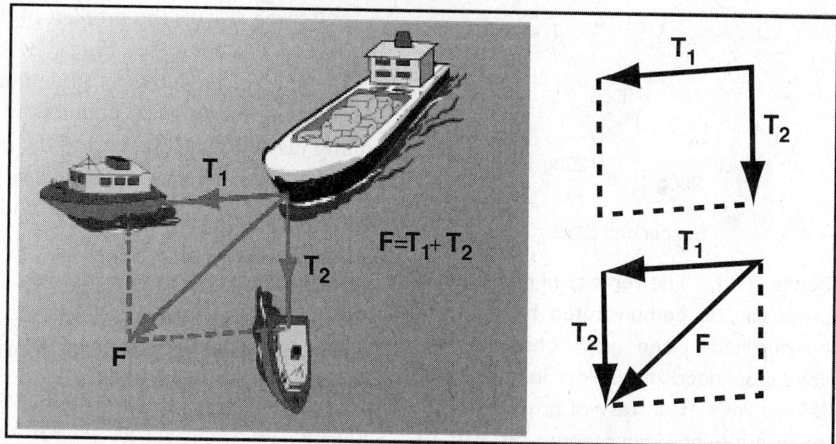

Figure 3.9 Resolution of forces in the shipping harbor. The resultant force represents the addition of the individual (tug-boat–generated) forces. The resultant force, both magnitude and direction, are determined via the "*tip-to-tail*" method. The ship's direction and the direction of the resultant force are identical. From Mechanics: Force and Motion. [http://www.hk-phy.org/contextual/mechanics/for/inert01_e.html, April 27, 2005.]

direction) using the "*tip-to-tail*" method. The direction of the resultant force is the direction in which the ship is moving at the instant of time portrayed.

There is a particularly ingenious physics laboratory demonstration that speaks to resolution of forces. The demonstration is carried out at any number of physics departments, including University of Texas at Austin, where the experiment is called "*the suspended block demonstration.*" The setup is a wooden triangular framework, the hypotenuse of which is an inclined plane (Figure 3.10*A*). The normal to the inclined plane departs from vertical by an angle of $\theta = 40°$. The lengths of the sides of the triangle are in the ratio of 3 to 4 to 5. A 1500g block is placed on the steeply inclined plane, held in place by a simple "*stop*" and connected by wires through pulleys to hooks onto which weights can be placed (see Figure 3.10*A*). This block exerts a force (**F**) of 1500g on the inclined plane. The first objective of the demonstration lab is to calculate force components generated on the inclined plane by the 1500g block, and in particular the force component perpendicular (normal) to the inclined plane (i.e., the **normal force**, $\mathbf{F_N}$) and the force component parallel to the inclined plane (i.e., the **shear force**, $\mathbf{F_S}$) (Figure 3.10*B*). The arithmetic is straightforward: the

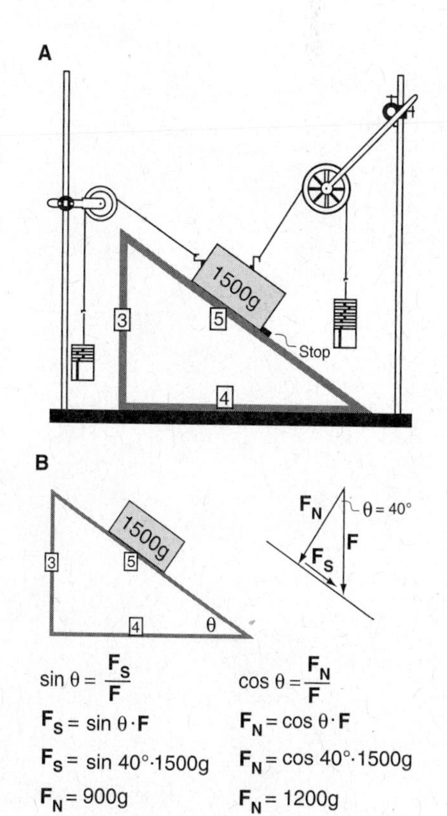

Figure 3.10 Resolution of forces: suspended block, a physics demonstration carried out at The University of Texas, Austin. See text for explanation. (*A*) The set up. (*B*) A vertical force (**F**) of 1500g is exerted upon an inclined plane. The normal to the inclined plane departs from vertical by an angle of $\theta = 40°$. **F** can be resolved into a normal force ($\mathbf{F_N}$) acting perpendicular to the inclined plane and a shear force ($\mathbf{F_S}$) acting parallel to the inclined plane.

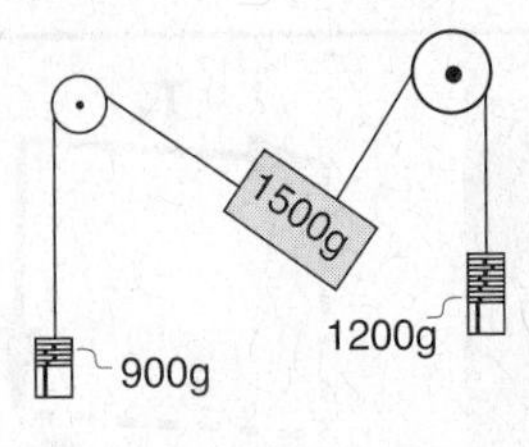

Figure 3.11 The veracity of resolution of forces can be demonstrated by removing the inclined plane and observing the block suspended motionless in static equilibrium. *Voila*. [Courtesy of http://www.ph.utexas.edu/~phy-demo/demo-txt/1j30-10.html, accessed March 27, 2011.]

normal force ($\mathbf{F_N}$) is the product of the **F** x cos θ, and the shear force ($\mathbf{F_S}$) is the product of **F** x sin θ (see Figure 3.10*B*).

The second objective is to demonstrate that the block can be held in static equilibrium by imposing countervailing normal forces and shear forces through adding just the right weights to the hooks. For suspense, this can be achieved in two steps (1) balancing the shear force and removing the "*stop*;" and then (2) balancing the normal force and removing the inclined plane. If all goes well, the block will not twitch. Instead it will be suspended motionless in space (Figure 3.11).

Forces in the Subsurface World

In the generally slow-moving tectonic world of structural geology, we largely envision forces acting on and within bodies that are at a state of rest or in very slow motion. We will most commonly be dealing with situations where rock bodies essentially are at rest, and forces appear to be balanced. When net forces produce intensities that cause deformation, they trigger accelerations that are generally of two kinds: (1) unimaginably slow acceleration of the larger geologic unit as a whole, such as a major tectonic plate that undergoes an increase in velocity from 6 cm/yr to 7 cm/yr over hundreds or thousands of years; or (2) incredibly fast, short-lived accelerations of parts of the larger body, such as fault-induced shifting of rock in the briefest period of time (seconds or fractions of seconds), achieving huge accelerations. A stunning and devastating example of the latter was the first phase of the Indian Ocean earthquake (December 26, 2004), when within 100 seconds the seafloor ruptured at a velocity of 10,000 km/h over a length of 400 km, producing on average ~15–20 m of vertical displacement. It was this magnitude of fault displacement that lifted a huge column of seawater and triggered the ravaging tsunami.

Prior to such seismic events, plate forces commonly build slowly until the strength of the crust, or some part of it, is overcome, thus triggering internal adjustments involving translation, rotation, distortion, or dilation. Some of these adjustments are recoverable when the forces are removed, namely the elastic component. Other adjustments result in a permanent damage that is non-recoverable, thus creating the possibility of a structural geologic record of the action of forces.

Body Forces and Electromagnetic Forces

There are two fundamental classes of forces that affect geologic bodies: body forces and contact forces. **Body forces** act on the mass of a body in a way that depends on the amount of material in the body but independent of the forces created by adjacent surrounding materials (Means, 1976; Twiss and Moores, 1992). **Contact forces** are pushes or pulls across real or imaginary surfaces of contact, such as a fault between adjacent parts of a rock body (Means, 1976).

From a structural geologic perspective, the most important body forces are **gravitational force** and **electromagnetic force**. The force of gravity is ultimately responsible for many geologic actions, such as the downhill flow of lava, glaciers, rock slides, rock avalanches, and debris flows; the down-basin slumping of wet sediments; the vertical ascent of low-density buoyant magmas, salt domes, and mud diapirs; the settling of heavy crystals within certain magmas, such as the Palisades Sill on the New Jersey bluffs of the Hudson River; and even the very movement of plates.

The body force called "*gravity*" can create structural deformation at a scale that is commonly large and visible. In contrast, electromagnetic forces are body forces whose structural geologic presence dwells in submicroscopic realms. Electromagnetic forces are among the significant forces that must be overcome to produce deformation (Serway, 1990). Electromagnetic forces hold individual minerals intact, and thus they hold rocks together (see *Chapter 4, Deformation Mechanisms and Microstructures*). When contact forces overwhelm electrostatic forces at some location in a mineral, the structure of the lattice can be disturbed through deformation at the submicroscopic scale. In this way, hard rocks can be forced to behave as if they were soft.

Forces as Loads

Load is a prominent word in the definition of contact force, and it presents the need to introduce yet a second way of describing the effect of a force. The first way, which we have learned, is to express the effect of a force in terms of how much acceleration the contact force will impart to the mass of a body in a given time (e.g., a hockey stick slapping a puck). The second way to describe the effect of a force is in terms of how much weight it can support (i.e., how much load it can resist) (Price and Cosgrove, 1990). In this case, the weight is called the load.

Since force equals mass times acceleration, and weight equals mass times acceleration due to gravity, it follows that there is a close relationship between force and weight. We sometimes miss that connection because the units in which load and weight are expressed are typically reported in shorthand. For example, a load or weight of 10,000 kg actually means 10,000 kg $\times$ m/s^2, which is the same as 10,000 N.

The loads that create contact forces can arise in any number of ways, some of which are secondary effects of body forces. Suppe (1985) identified three main mechanisms of loading, each of which produces contact forces: gravitational loading, thermal loading, and displacement loading. **Gravitational loading** is an omnipresent mechanism, one example of which is the effect of the weight of sediments accumulating in a sedimentary basin. Such a sedimentary column generates forces at depth that contribute to compaction and lithification. Through **thermal loading** the heating or cooling of rocks creates forces in rocks that, because of their confinement, are not able to expand or contract. Imagine the complex pattern of forces generated by heating a confined mass of tightly interlocking rocks and minerals, each of which has a different capacity for thermal expansion. **Displacement loading** generates forces through mechanical disturbance of rocks. Examples include the collision of plates, regional bending and arching, and the shouldering aside of country rock by an intruding magma or an impacting asteroid.

Plate Tectonic Forces

Advances in the understanding of the origin of plate forces have issued from the research of Forsyth and Uyeda (1975), Elasser (1968), Bott (1982), and Kearey and Vine (1990). Park (1988) nicely summarized much of this work. The principal forces and resistances that control plate movements have interesting names, such as **ridge push**, **slab pull**, and **slab suction**. To picture the whole array of forces and resistances, let us imagine a spreading center that is feeding out new oceanic lithosphere (Figure 3.12). The new oceanic lithosphere at the ridge thickens outward as it ages and cools, and the whole plate moves laterally across a vast expanse before descending into a subduction zone. Beyond the subduction zone, in this example, is continental lithosphere belonging to a different plate.

The origin of **ridge push** along spreading centers is the gravitational "*head*" created by topography, which reflects the warm buoyant character of the lithosphere along the ridge. Ridge push is applied to the separating plates, causing the plates to move (see Figure 3.12). Resistance to movement is manifest in

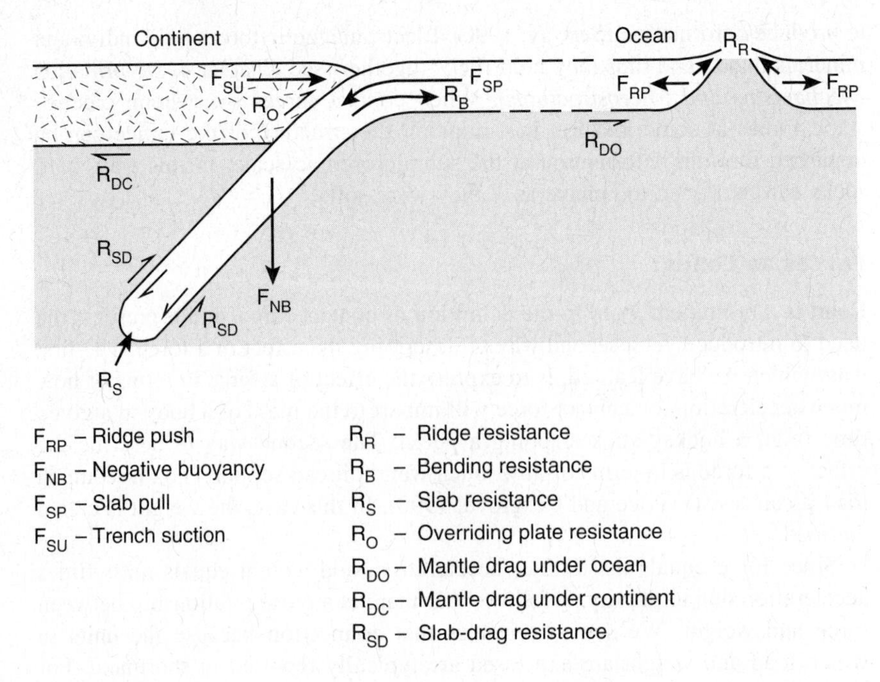

Figure 3.12 The forces acting on plates. [Adapted from Kearey and Vine, 1990. Global tectonics, Figure 5.17, p. 89. Reprinted with permission from Blackwell Science Ltd.]

shallow-focus earthquakes that originate above asthenosphere in thin brittle lithosphere. Ridge push generates compression.

A **mantle drag** resistance operates on the underside of a moving plate (see Figure 3.12), like the drag on the underside of a barge moving on a river. Kearey and Vine (1990) point out that the mantle drag beneath continental lithosphere is approximately eight times greater than the mantle drag beneath oceanic lithosphere, probably because of the deep lithospheric roots of continents (Jordan, 1988). Where the oceanic plate enters the subduction zone, it must flex to get in (see Figure 3.12). The resistance to flexing is a **bending resistance**, which slightly impedes plate motion.

Other forces arise when oceanic lithosphere is, in effect, pulled down a subduction zone. **Slab pull** is caused by negative buoyancy, created by the density contrast between the cold slab going down and the hot mantle into which the slab descends. A distinction is made between the **negative buoyancy force**, which acts vertically, and slab pull, which acts parallel to the descending slab (see Figure 3.12). As you might expect, there is resistance to the descent of a slab of oceanic lithosphere. First of all, there is a **slab resistance** force oriented perpendicular to the leading edge of the slab as it descends (see Figure 3.12). In addition, the upper and lower surfaces of the descending slab are marked by **slab drag**. Slab resistance is actually 5× to 8× larger than slab drag.

The overriding plate above the subduction zone imposes additional dynamic factors. For example, **trench suction** is a force that tends to drive the overriding plate into the trench at a faster rate (see Figure 3.12). Tension results. There are debates over the exact origin, but the effect is expressed in a number of ways, including collapse of the overriding plate into the trench and backarc spreading. In contrast, the strength and velocity of movement of the overriding plate create **overriding plate resistance** (see Figure 3.12). It is this resistance that gives rise to major earthquakes at the active plate margin.

The actual state of affairs in the descending slab depends on a number of factors (Figure 3.13). If slab pull exceeds the combination of slab resistance and overlying plate resistance, the slab is forced into a state of tension. If the opposite holds, the slab finds itself in a state of length-parallel compression. The 'proof of the pudding' lies in the earthquake record within descending

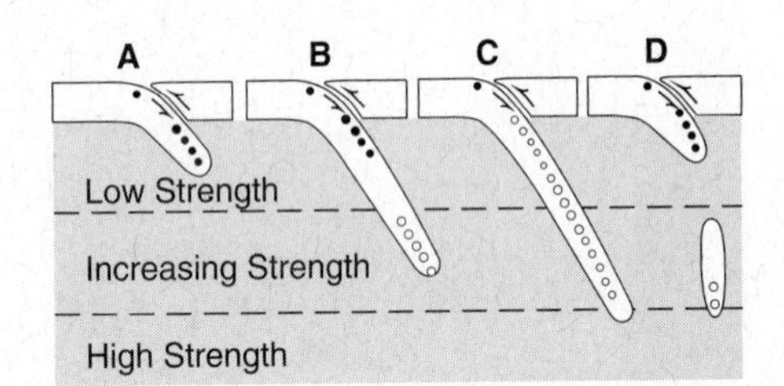

Figure 3.13 Portrayal of how forces within a descending slab of lithosphere may change with depth of projection into mantle of steadily increasing strength. (*A*) Where the slab begins to subduct relatively rapidly into weak asthenosphere, the slab experiences tensional forces. (*B*) Where the slab penetrates through the asthenosphere and feels increasing resistance to descent, the lower reaches of the slab experience compression while the upper reaches continue to experience tension. (*C*) Deep penetration into high strength regimes throws the whole slab into compression. (*D*) If the slab breaks in half, the upper part within the asthenosphere returns to an environment of tension whereas the descending part endures slab-parallel compression. [Adapted from Kearey and Vine, 1990, Global tectonics, Figure 8.15, p. 154, based on Isacks and Molnar, 1969. Reprinted with permission from Blackwell Science Ltd.]

slabs. The earthquake records can be used to monitor slab-parallel extension versus slab-parallel shortening as a function of depth (see Figure 3.13).

If the resisting forces are relatively low at the location of subduction zones, both the slab-pull forces and the suction forces will produce tension in the adjacent lithosphere. This result is in some ways counterintuitive and should make us cautious about trying to equate convergence-related subduction with collision and compression. Recall that ridge-push forces at spreading centers, induced by buoyancy forces, generate compression. There is a certain irony in the generation of compressional forces at spreading centers and tensional forces in subduction zones (Park, 1988).

TRACTIONS

Definition

Traction (**T**) is a vector quantity that considers the magnitude of a force (**F**) in relation to the area (*A*) of the surface on which it acts. (Vector quantities conventionally are shown **bold-faced**; scalar quantities are shown in *italics*). Traction is a representation of direction and magnitude of **force intensity** (Twiss and Moores, 1992, p. 130). A cartoon example of calculating traction is presented in Figure 3.14, which considers the strategy of rescuing a skater who has fallen through the ice in a frozen pond. The skater broke through thin ice because the resistance force exerted by the ice was exceeded by the force intensity of the vertical traction produced by concentration of the load of the 77 kg skater onto the thin blade of a single skate. Assuming that the 77 kg load was distributed uniformly along just one of the blades, the area of the contact between ice and blade was only about 5.1 cm^2 (Figure 3.14*A*). The magnitude of the downward vertical traction produced at the ice-blade interface must have been great enough to fracture the ice.

$$\mathbf{T}^{\text{down}} = \frac{\mathbf{F}}{A}$$

$$\mathbf{T}^{\text{down}} = \frac{\mathbf{F}}{A} = \frac{15 \text{ kg}}{\text{cm}^2} \quad \text{(see Figure 3.14}B\text{)}$$

In contrast, the 77 kg rescuer, informed in ways of dynamic analysis and tractions, moves in close to the victim while lying on a 3 × 0.45 m plank (Figure 3.14*C*). The total load of rescuer and plank is about 82 kg. The area of the plank is 13,935 cm^2 (1.39 m^2). The load of the rescuer is distributed in such

A

B

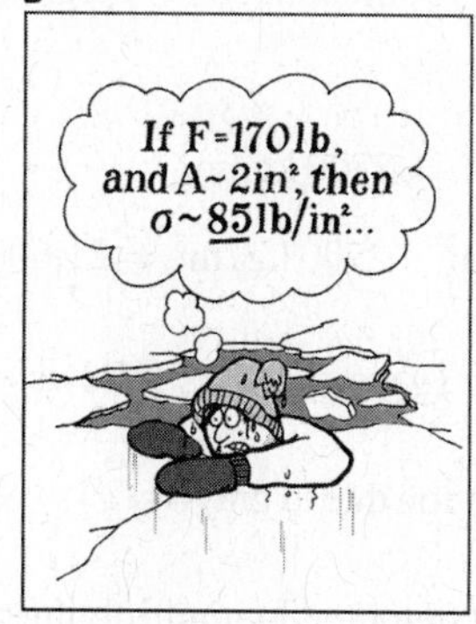

C

Figure 3.14 The "*stress*" of skating. (*A*) Gravitational loading. (*B*) Result of gravitational loading exceeding resistance forces. Traction perpendicular to ice surface is being calculated. (*C*) Intelligent "*control*" of traction during rescue. [Artwork by D. A. Fischer.]

A

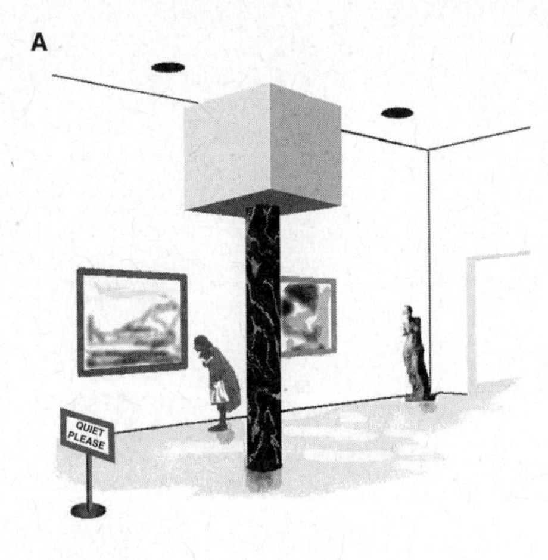

B

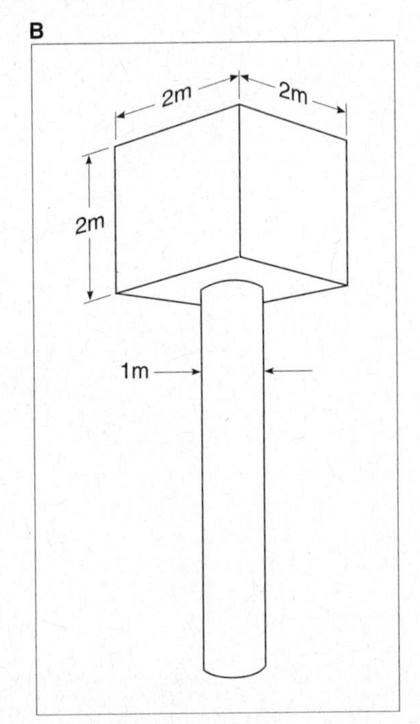

Figure 3.15 (A) This piece of sculpture of granite on marble creates an opportunity for calculating the force exerted by the granite block, and the traction generated by it on the top of the marble column. (B) Force is determined by measuring the volume of the granite block and multiplying it by its density and by the acceleration due to gravity. The traction normal to the top of the marble column is determined by dividing the force by the cross-sectional area of the marble column.

a way that the downward vertical traction (**T**^{down}) at any point under the plank is uniformly small, and well below the breaking strength of the ice.

$$\mathbf{T}^{down} = \frac{\mathbf{F}}{A}$$

$$\mathbf{T}^{down} = \frac{0.006\,\text{kg}}{\text{cm}^2}$$

Units of Traction

There are several ways to express the magnitude of traction. The preferred unit of measure is the **pascal** (Pa):

A traction of one pascal (Pa) is created by the force of one newton (N) acting on an area of one square meter (1 m^2).

Because of the small magnitude of a single pascal in comparison to the greater magnitude of tractions in the Earth, we commonly precede the term pascal with the prefix kilo-, mega-, or giga-, where:

1**kilopascal** (**kPa**) = 1000 pascals (10^3 Pa)

1 **megapascal** (**MPa**) = 1,000,000 pascals (10^6 Pa)

1**gigapascal** (GPa) = 1,000,000,000 pascals (10^9 Pa)

Though traction is normally expressed in units of N/m^2, we sometimes see traction expressed in **pounds per square inch** (**psi**) or **kilograms per square centimeter** (**kg/cm^2**).

Museum-Piece Calculation of Traction

Imagine a large block of granite resting on a marble column (Figure 3.15*A*). The weight of the granite block constitutes a gravitational load that imposes traction on the top surface of the marble column. The magnitude of this traction is found by dividing the force created by the load of the granite block by the cross-sectional area of the top of the marble column. In this example we simplify greatly by assuming that the load of the granite block is distributed *perfectly* uniformly across the upper surface of the marble column. Thus we assume that whether we calculate traction across the whole upper surface or across any small fraction of that surface, the force intensity will always be the same.

The force (**F**) created by the granite block is the product of mass (m) times acceleration (g) due to gravity. Mass (m) is volume (V) times density (ρ) (Figure 3.15*B*). In this case:

$$\text{volume } (V) = \text{width } (W) \times \text{breadth } (B) \times \text{height } (H) = 2\,\text{m} \times 2\,\text{m} \times 2\,\text{m}$$

$$V = 8\,m^3$$

$$\text{density } (\rho) = 2.7\,\text{g/cm}^3 = 2700\,\text{kg/m}^3$$

$$\text{mass } (m) = V\rho = 8\,\text{m}^3 \times 2700\,\text{kg/m}^3 = 21{,}600\,\text{kg}$$

Acceleration due to gravity (g) is given as follows:

$$\text{Acceleration due to gravity } (g) = 9.8\,\text{m/s}^2$$

Force (**F**), as just stated, is computed by multiplying mass (m) times gravity (g):

$$\text{force } (\mathbf{F}) = \text{mass } (m) \times \text{acceleration } (g) = mg$$

$$\mathbf{F} = 21{,}600\,\text{kg} \times 9.8\,\text{m/s}^2 = 211{,}680\,\text{kg} - \text{m/s}^2 = 211{,}680\,\text{N}$$

The traction ($\mathbf{T}^{\text{down}}$) created by the load of the granite block on the marble column is force divided by area:

$$\text{traction } (\mathbf{T}^{\text{down}}) = \frac{\text{force } (\mathbf{F})}{\text{area } (A)}$$

$$\text{area } (A) = \pi r^2 = 3.14 \times (0.5\,\text{m})^2 = 0.79\,\text{m}^2$$

$$\mathbf{T}^{\text{down}} = \frac{\mathbf{F}}{A} = \frac{211{,}680\,\text{N}}{0.79\,\text{m}^2} = \frac{267{,}949\,\text{N}}{\text{m}^2} = 267{,}949\,\text{Pa} = 268\,\text{kPa}$$

The traction (force intensity) produced by the granite block tends to deform the marble column. The marble column will permanently deform when its capacity to respond elastically, like a spring, is exceeded.

Calculating Traction Underground

In a manner quite similar to the museum-piece example, we can calculate the traction created by the weight of a *very* large cube of granite in the upper crust (Means, 1976, pp. 112–114). Let us picture a region of the Earth where the upper several kilometers of the crust are entirely composed of granite (Figure 3.16). Then let us calculate for a given depth level the traction ($\mathbf{T}^{\text{down}}$) created by the load of the granite upon a given level of the mine. We will choose −1000 m as the depth level of interest to us. To set up the calculation, it is helpful to visualize the −1000 m depth level as overlain by a giant cube of granite 1000 m on a side (see Figure 3.16). Our goal is to compute the magnitude of the traction at the base of the block.

The force (**F**) generated by the weight of the block is determined by multiplying the volume (V) of the block times the density (ρ) of the granite times the acceleration due to gravity (g) (see Figure 3.16):

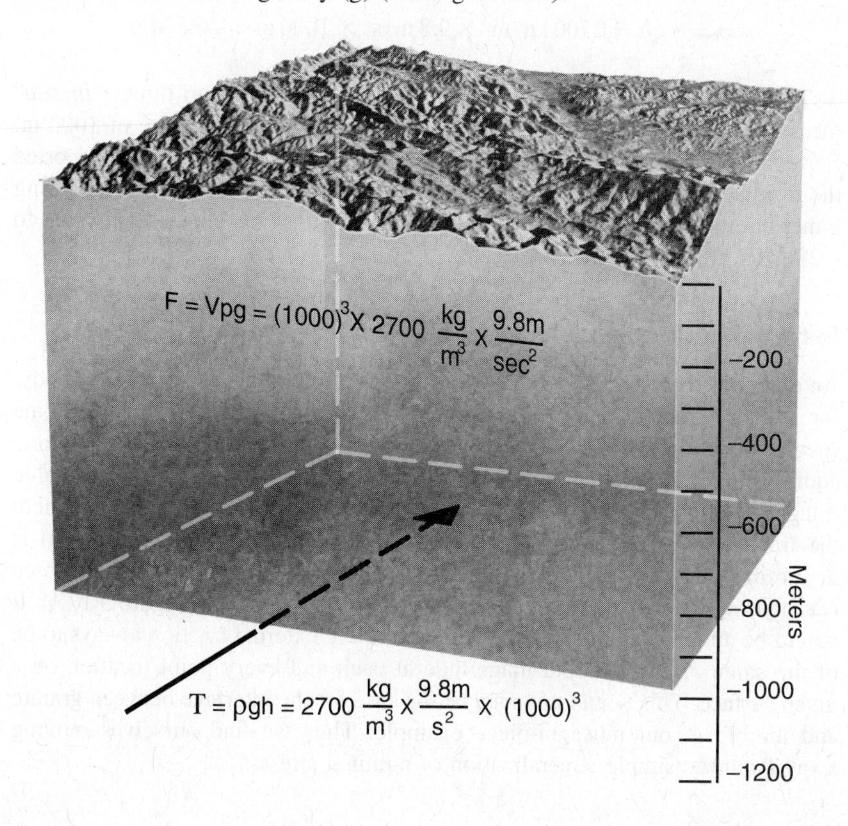

Figure 3.16 We can apply our museum-piece calculation to a huge block of granite nestled in the crust of the Earth. See text for explanation.

$$V = 1000\,\text{m} \times 1000\,\text{m} \times 1000\,\text{m}$$

$$\rho = 2700\,\text{kg/m}^3$$

$$g = 9.8\,\text{m/s}^2$$

$$\mathbf{F} = 1000\,\text{m} \times 1000\,\text{m} \times 1000\,\text{m} \times 2700\,\text{kg/m}^3 \times 9.8\,\text{m/s}^2$$

The traction ($\mathbf{T}^{\mathbf{down}}$) created by the weight of the block of granite acting on the base of the block is determined by dividing the force (**F**) by the area (A = 1000 m × 1000 m):

$$\mathbf{T}^{\mathbf{down}} = \frac{\mathbf{F}}{A} = \frac{1000\,\text{m} \times 1000\,\text{m} \times 1000\,\text{m} \times 2700\,\text{kg/m}^2 \times 9.8\,\text{m/s}^2}{1000\,\text{m} \times 1000\,\text{m}}$$

$$\mathbf{T}^{\mathbf{down}} = 1000\,\text{m} \times 2700\,\text{kg/m}^3 \times 9.8\,\text{m/s}^2 = 26{,}460{,}000\,\text{Pa} = 26.5\,\text{MPa}$$

In static equilibrium, which this example represents, the downward vertical traction ($\mathbf{T}^{\mathbf{down}}$) is resisted by an upward-vertically directed traction ($\mathbf{T}^{\mathbf{up}}$) of equal magnitude. This *pair* of equal and opposite vertical tractions created by gravitational loading is referred to in the structural geologic literature as **lithostatic stress** (σ_l), which in effect is the weight of the overlying column of rock. We need to emphasize in these examples of simple "*axial*" loading that the stress (σ) we refer to is simply a measure of the overall intensity of internal forces. It should be thought of as a scalar quantity. Such a stress is referred to by some as **engineering stress** or **nominal stress** and represents average stress (σ_{avg}) over the area, with the simplification of presumed uniform distribution. Such stresses and tractions are expressed in units of pascals. Lithostatic stress increases 26.5 MPa/km. Thus, for each 3.8 km increment downward, lithostatic stress increases by 100 MPa.

There is a shortcut to determining lithostatic stress (σ_l) at any given depth in the granite. Simply multiply density (ρ) times gravity (g) times depth (h) (see Figure 3.16):

$$\sigma_l = \rho g h = 2700\,\text{kg/m}^3 \times 9.8\,\text{m/s}^2 \times 1000\,\text{m} = 26.5\,\text{MPa}$$

The calculated stress level of 26.5 MPa is very similar to direct "*in situ*" measurements of stress in deep mines at depth levels in the range of 1000 m. For example, Suppe (1985) discussed the work of Bjorn (1970), who reported the results of direct in situ stress measurements in mines in Norway, including a measurement of ~23 MPa at a depth level of ~815 m. This is equivalent to ~28 MPa at a depth level of ~1 km.

External and Internal Tractions

An **external traction** ($\mathbf{T_e}$) on a body is force acting on the surface of the body, for instance, on the surface of a grain of sand within a sandstone at some specific location (Figure 3.17*A*). We are imagining that the body is in static equilibrium. The external traction ($\mathbf{T_e}$) is a **bound vector** because its value (magnitude and direction) depends completely on the specific location where the traction is acting. The actual magnitude of the external traction ($\mathbf{T_e}$) is determined through application of calculus, in this case, taking the limit of force ($\Delta\mathbf{F}$) per unit area (ΔA) as the area (A) approaches zero (see Figure 3.17*A*). It would be an unusual circumstance indeed for the external traction always to be of the same orientation and magnitude at each and every point location on a given surface. This would not even be the case for the interface between granite and marble in our museum-piece example. Thus we find ourselves moving away from the simple generalization of nominal stress.

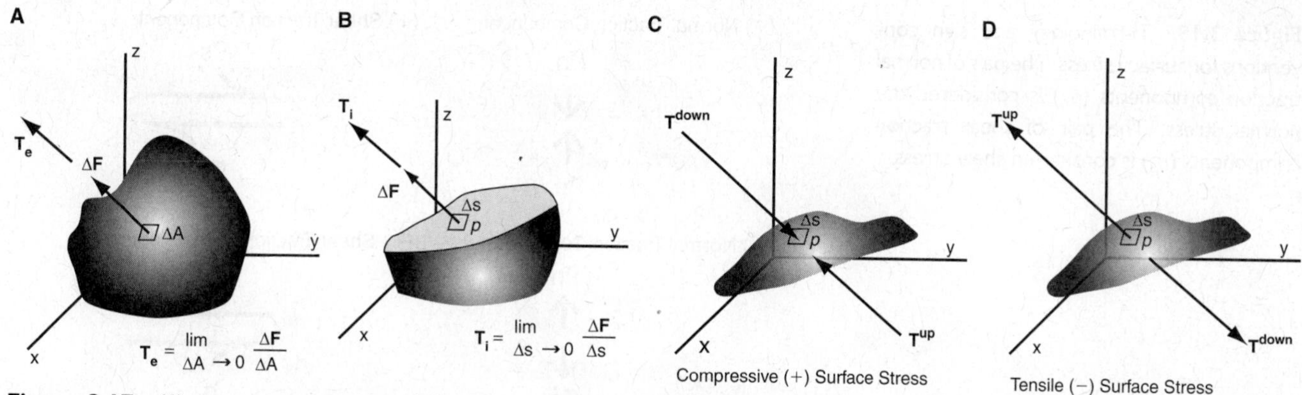

Figure 3.17 (*A*) An external traction ($\mathbf{T_e}$) acting on the surface of a body, which we imagine to be a tiny grain of sand in a sandstone formation. Its value is the magnitude of the force (**F**) divided by the area (*A*) on which the force is acting. [Figure prepared courtesy of www.efunda.com, in particular, http://www.efunda.com/formulae/solid_mechanics/mat_mechancis/stress.cfm, April 12, 2010]. (*B*) Internal traction ($\mathbf{T_i}$) acting on surface within a body. [Adapted from efunda] [Figure prepared courtesy of www.efunda.com, in particular, www.efunda.com/formulae/solid_mechanics/mat_mechancis/stress.cfm, April 12, 2010]. (*C*) The surface stress at point *p* is the pair of equal and opposite tractions. In this case the surface stress is compressive (+). (*D*) Example of tensile surface stress at point *p*, composed of pair of equal and opposite tractions.

Now, as we go inside the body (i.e., in this example the grain of sand) to an infinitesimal point *p*, we enter a space full of surfaces and discontinuities of many different orientations and sizes (Figure 3.17*B*). Acting on one such surface (Δs) at a specific point volume (*p*), we can imagine **internal tractions** ($\mathbf{T_i}$), which as vectors require both magnitude and direction for complete specification. The specific magnitude of each internal traction is determined by taking the limit of force per unit area as area approaches zero (see Figure 3.17*B*). The internal traction cannot be determined without first knowing the orientation of the datum plane, that is, the surface on which the internal traction acts.

In static equilibrium, the tractions ($\mathbf{T^{down}}$, $\mathbf{T^{up}}$) acting on a surface at a point in the body are equal in magnitude and oppositely directed. The tractions are balanced (Figure 3.17*C*, *D*). This *pair* of equal and opposite tractions is called the **surface stress** at point *p* (Twiss and Moores, 1992, p. 131). If the tractions of this surface stress act *toward* one another, the surface stress is **compressive** (see Figure 3.17*C*), which has the effect of pushing the materials together across the surface. Compression is considered to be positive (+). In contrast, if the tractions are directed *away* from one another (see Figure 3.17*D*), the surface stress is **tensile**, which has the effect of pulling material apart across the surface. Tensile is considered to be negative (−).

We see in Figure 3.18 that tractions can be resolved into **normal components** (σ_n) and **shear components** (σ_s) for compressive surface stress (Figure 3.18*A*) and tensile surface stress (Figure 3.18*B*) (Twiss and Moores, 1992, pp. 130–131). σ_n acts perpendicular to the surface on which the traction operates; σ_s acts parallel to that same surface. Again, if the normal components are directed toward one another, the surface stress is compressive (+). If, on the

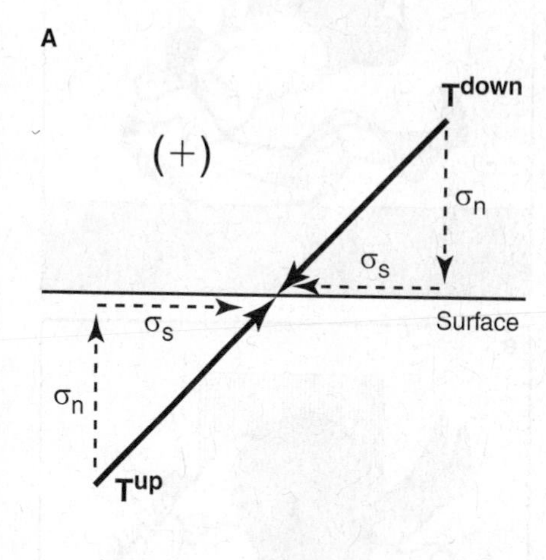

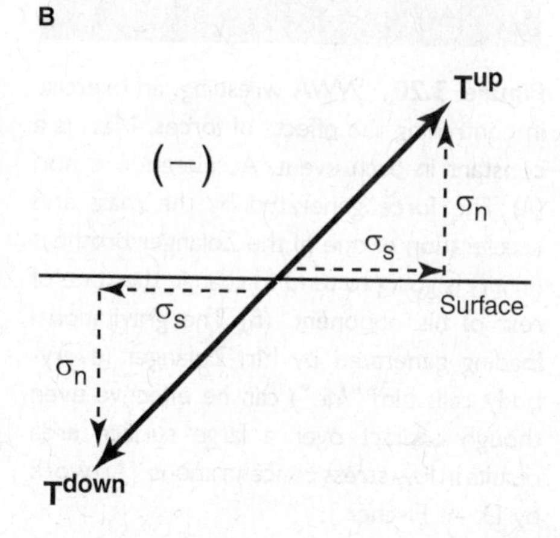

Figure 3.18 Diagrams showing tractions acting on a surface, which in this example is horizontal. (*A*) The traction operating downward from ($\mathbf{T^{above}}$) and upward from below ($\mathbf{T^{below}}$) the surface balance one another. Tractions pointing toward one another create compression. The sign convention for a compressive pair of tractions is positive (+). Each of the tractions can be resolved into normal (σ_n) and shear (σ_s) components. (*B*) As above, but in this case the tractions point away from one another, creating tensile effects. The sign convention for tensile pairs is negative (−).

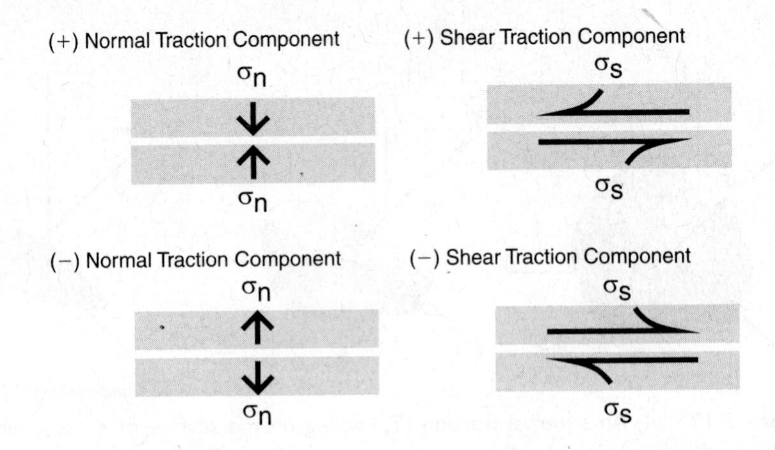

Figure 3.19 Terminology and sign conventions for surface stress. The pair of normal traction components (σ_n) is considered the normal stress. The pair of shear traction components (σ_s) is considered shear stress.

other hand, the normal components are directed away from one another, the surface stress is tensile (−) (Figure 3.19). When the shear traction components are left-handed, the surface stress is described as counterclockwise (+), but if right-handed clockwise (−) (see Figure 3.19).

STRESS

The "*Business End*" of Force

Body forces, as well contact forces created through loading, are ultimately responsible for geologic deformation. These forces work through the "*undercover agent*" known as **stress**. We can think of stress as that which tends to deform a body of material (Jaeger and Cook, 1976). Stress will deform a body if the strength of the body is exceeded.

For example, when the body of a wrestler is subjected to the force produced by the accelerating mass of a huge opponent, its state of rest or state of motion may be significantly changed (Figure 3.20*A*). The body is usually both translated and rotated. During a pin (Figure 3.20*B*), the body might even undergo slight negative dilation (decrease in volume). In rare instances, the body, or parts of the body, may endure the permanent strain of distortion, involving breaking or stretching or ripping. This does not happen often during WWA action because the wrestlers are well aware of the human body's elastic limit, and they 'pull' their punches accordingly. Whether strain is expressed in permanent deformation depends on two factors: the strength of the body and the concentration of **stress**, which is that which tends to deform a body of material.

Though traction (**T**) is a vector quantity, stress is not. Stress is a **tensor quantity** whose full three-dimensional description requires nine **normal** and **shear traction components** (reduces to six independent components), whereas tractions are vectors whose magnitudes are simply force divided by the area of the surface on which they are acting, yielding a measure of force intensity. Traction vectors act on surface elements; stress tensors act on volume elements (Babaie, 2010).

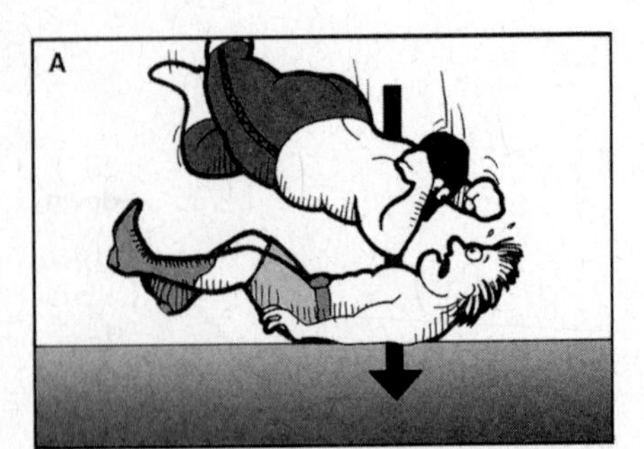

Figure 3.20 WWA wrestling: an exercise in controlling the effects of forces. Mass is a constant in each event. Acceleration is not. (*A*) The force generated by the mass and acceleration of one of the Zolanger brothers (in air) is going to tend to change the state of rest of his opponent. (*B*) The gravitational loading generated by Mr. Zolanger (everybody calls him "*Mr.*") can be effective even though contact over a large surface area results in low stress concentrations. [Artwork by D. A. Fischer.]

Going from Traction to Stress

We see in Figure 3.21*A* that any traction acting on a surface can be resolved into normal (σ_n) and shear stress (σ_s) components. This 2D resolution of tractions is simple to envision and can be achieved through construction of a scaled drawing (Figure 3.21*B*) or doing the trigonometry (Figure 3.21*C*).

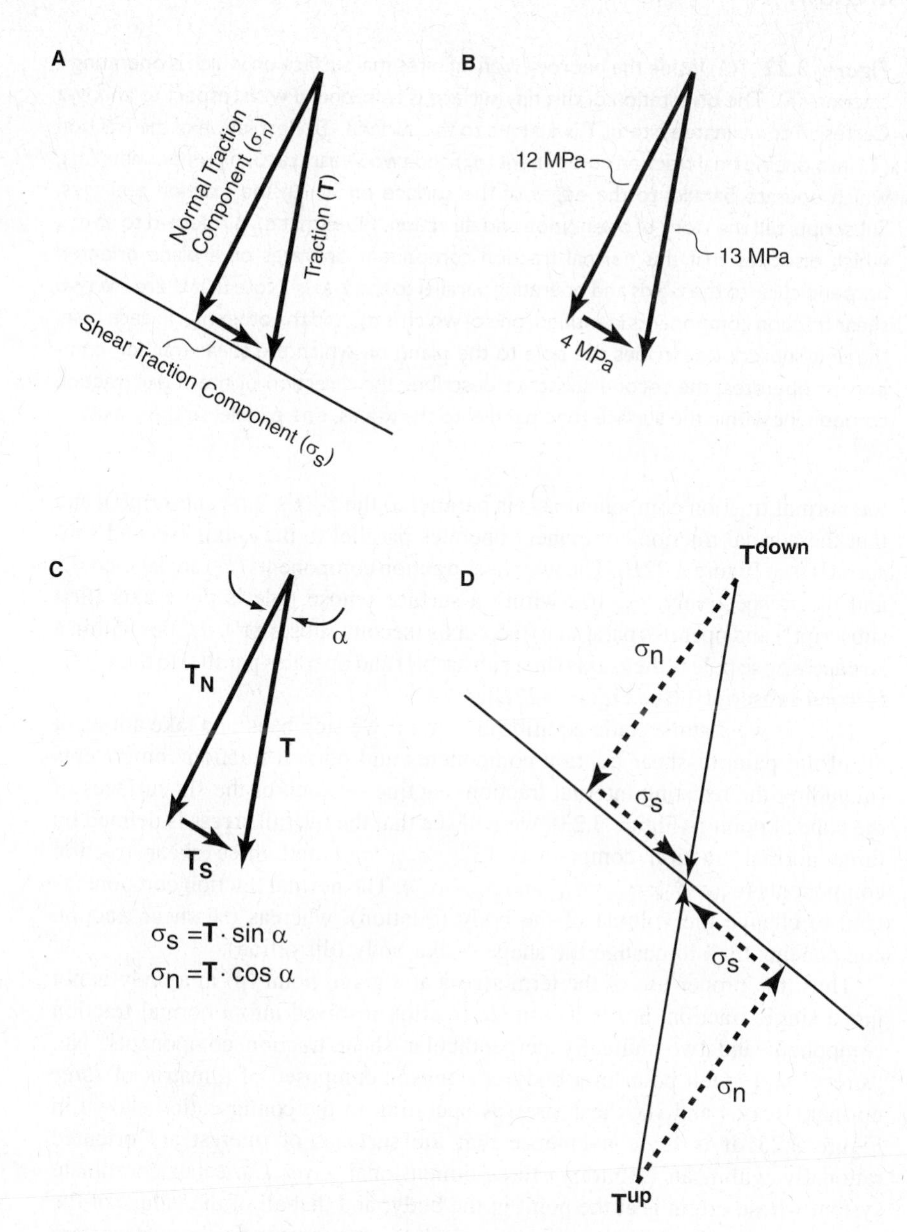

Figure 3.21 Two-dimensional approach to resolving traction (**T**) into normal traction component (σ_n) and shear traction component (σ_s). (*A*) The traction (**T**) in this example is not perpendicular to the surface on which it acts, and thus can be resolved into a normal traction component (σ_n) and shear traction component (σ_s). (*B*) Scaled-drawing solution. (*C*) Trigonometric solution. (*D*) Full surface stress showing resolution of traction.

Figure 3.21*D* reveals the full surface stress for this example, with resolution of traction components on both sides of the surface on which the tractions are acting. This 2D example helps us to distinguish traction (**T**) versus stress, for **normal stress** (σ_n) is the *pair* of equal and opposite normal traction components, and **shear stress** (σ_s) is the *pair* of equal and opposite shear traction components.

The 3D resolution of tractions works the same way, but in this case can be best visualized by a family of normal traction components and shear traction components operating on each of the surfaces of an infinitesimally small cube oriented rationally within an x-y-z Cartesian coordinate system. One such face is shown in Figure 3.22. Its orientation is parallel to the x-y plane of the coordinate system (see Figure 3.22*A*). The traction (**T**) operating on this surface is resolved into a normal traction component (σ_n), which is perpendicular to the surface and operating parallel to the z-axis of the coordinate system. In addition, the traction (**T**) is resolved into *two* shear traction components (σ_s), which run parallel to the edges of the surface on which the traction operates (Figure 3.22*B*). The normal traction component (σ_n) is described as σ_{zz}, signifying that the pole to the surface on which

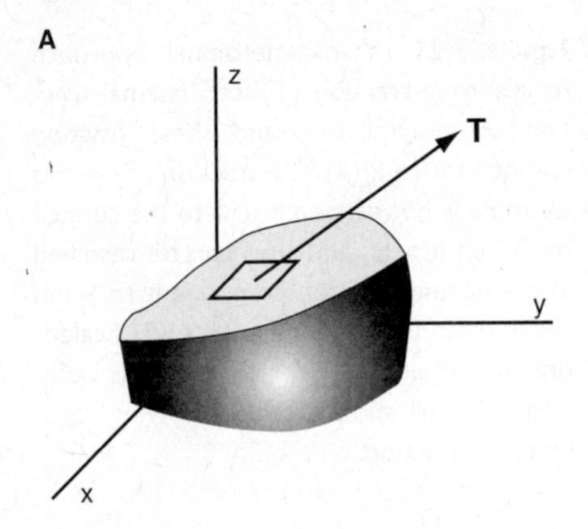

Figure 3.22 (*A*) Inside the bedrock is an infinitesimal surface on which is operating a traction (**T**). The orientation of this tiny surface is referenced with respect to an x-y-z Cartesian coordinate system. **T** is oblique to the surface. (*B*) Resolution of the traction (**T**) into one normal traction component (σ_n) and two shear traction components (σ_s), which operate parallel to the edges of the surface on which the traction operates. Subscripts tell the story of orientation and direction. Note that σ_n is referred to as σ_{zz}, which discloses that this normal traction component operates on a plane oriented perpendicular to the z-axis and operating parallel to the z-axis. Note that there are two shear traction components identified, one of which is σ_{zy} and the other σ_{zx}. In each case, the first subscript describes the pole to the plane on which the shear traction component operates; the second subscript describes the direction of the shear traction component within the surface (one parallel to the x-axis, one parallel to the y-axis).

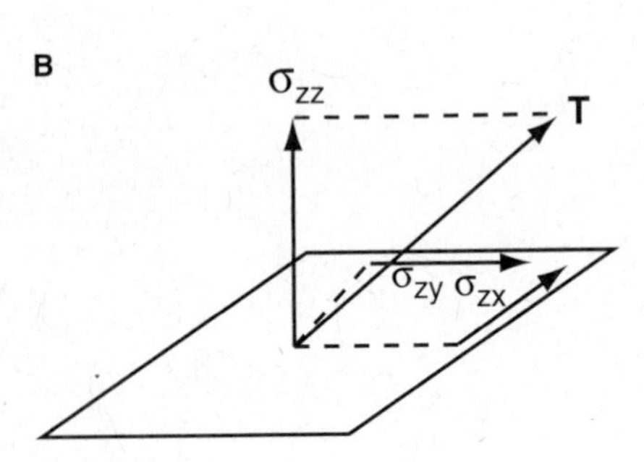

the normal traction component acts is parallel to the z-axis (first subscript!), and that the normal traction component operates parallel to the z-axis (second subscript!) (see Figure 3.22*B*). The two shear traction components (σ_s) are labeled σ_{zx} and σ_{zy}, respectively. σ_{zx} lies within a surface whose pole is the z-axis (first subscript!) and operates parallel to the x-axis (second subscript!). σ_{zy} lies within a surface whose pole is the z-axis (first subscript!) and operates parallel to the y-axis (second subscript!) (see Figure 3.22*B*).

Now, if we assume static equilibrium, and if we step back and take a look at all of the pairs of shear traction components and normal traction components (including the resisting internal tractions) acting on each of the six surfaces of the cube at point *p* (Figure 3.23), we will see that the overall **stress** is defined by three normal traction components (σ_{xx}, σ_{yy}, σ_{zz}) and three shear traction components ($\sigma_{xy}=\sigma_{yx}$; $\sigma_{xz}=\sigma_{zx}$; and $\sigma_{zy}=\sigma_{yz}$). The normal traction components tend to change the volume of the body (dilation), whereas the shear traction components tend to change the shape of the body (distortion).

Thus, the proper use of the term **stress** at a given point (*p*) in a body is not just a single traction, nor is it a single traction resolved into a normal traction component and two mutually perpendicular shear traction components. No, "*stress*" at a given point in a body is a **tensor** composed of a matrix of *three* normal stresses and *six* shear stresses operating in the configuration shown in Figure 3.23. It is for convenience that the surfaces of interest are oriented rationally within an (arbitrary) three-dimensional x-y-z Cartesian coordinate system whose origin is at the point in the body; and that all of the values of the normal traction components and shear traction components on the surfaces are determined in directions referenced to the x-y-z axes.

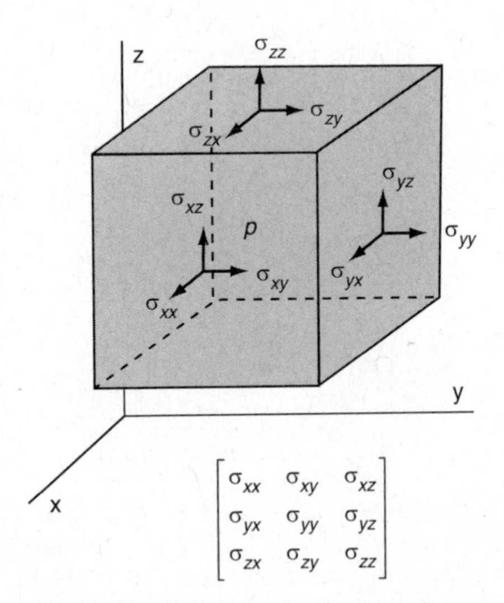

$$\begin{bmatrix} \sigma_{xx} & \sigma_{xy} & \sigma_{xz} \\ \sigma_{yx} & \sigma_{yy} & \sigma_{yz} \\ \sigma_{zx} & \sigma_{zy} & \sigma_{zz} \end{bmatrix}$$

Figure 3.23 The 3D approach to resolving traction (**T**) on the faces of a cube oriented rationally within a Cartesian x-y-z coordinate system. Stress is a tensor composed of 3 normal traction components and 6 shear traction components, the latter of which comprise 3 shear couples. See text for explanation.

Stress Analysis

We want to learn how to calculate stress at a given point (*p*) within a body of rock that is subjected to forces. We do this by calculating the normal traction components and shear traction components (magnitudes and directions) on individual surfaces of all possible orientations within the given point in the body. If we can define stress at each and every point within the body, we can then fully describe the **stress field**. When the stress field is known, it is possible to determine normal traction components and shear traction components for each and every plane of every conceivable orientation within the body. This, then, becomes the basis for predicting and/or understanding the circumstances under which rocks fail by jointing or faulting.

Setting up an Example of Stress Tensor Analysis

We now go underground again, at least in our imaginations, to a tiny point (*p*) deep in granite, located 1500 m beneath the surface (Figure 3.24). The region of crust within which point *p* resides is tectonically active and is experiencing a

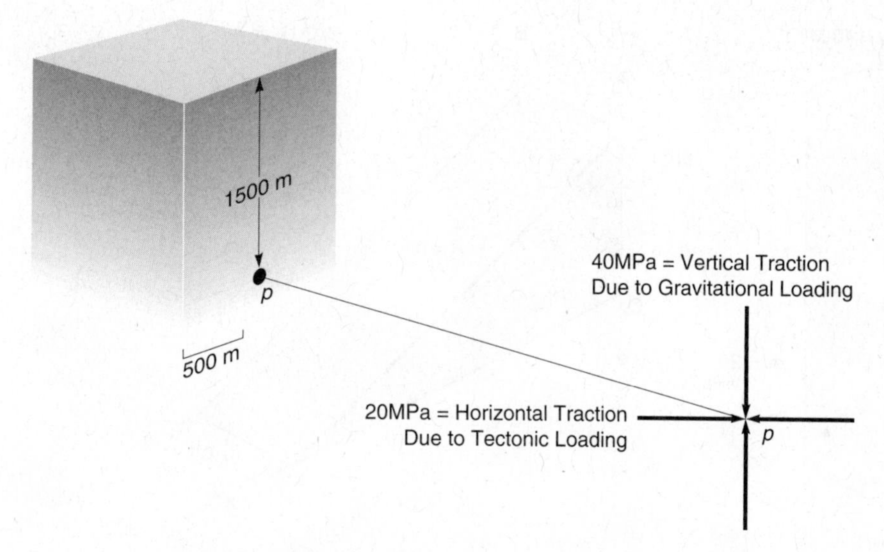

Figure 3.24 We picture a point (p) deep down in granite, which for this example we assume to be an ideal elastic body of rock. The forces acting in the subsurface in the vicinity of point p are generated by the combination of gravitational loading and displacement (tectonic) loading.

modest east-west horizontal displacement (tectonic) loading. Gravitational loading imposed by the overburden of granite impacts p, as does the smaller horizontal displacement loading of tectonic origin.

Let us picture within the interior of the granite at point p an infinite number of imaginary planes, or, if you prefer, real physical planes of all conceivable orientations. The real physical planes include such microstructures as pre-existing hairline fractures, cleavage surfaces in minerals, crystal faces, grain boundaries, and veins. Across each one of these planes "*there will exist a field of forces equivalent to the loads exerted by material on one side of the plane on the other*" (Ragan, 1985, pp. 111–112). Focusing from point to point on any one of these planes, we would discover that the forces are not uniform in either direction or magnitude. Yet, if we were to examine the magnitude and direction of force acting on a smaller and smaller surface area of one of the planes cutting through point p (i.e., if we do the calculus), we would discover that the force per unit area (i.e., the traction) would approach a fixed value, . . . fixed in both magnitude and direction.

Using Stress to Calculate Surface Tractions

Let us now calculate tractions on surfaces at point p. In our example, gravitational loading is operating vertically. At the depth of p gravitational loading has produced a lithostatic stress (σ_l) of 40 MPa. We assume that displacement loading due to tectonic conditions is operating horizontally and, at the location of point p, has a magnitude (σ_t) of 20 MPa (Figure 3.25*A*). σ_l and σ_t operate, respectively, parallel to the z- and x-axes of the Cartesian coordinate system about which point p is arbitrarily suspended.

Let us first calculate the tractions acting on the plane passing through point O and making an angle of 65° relative to the *z*-axis (see Figure 3.25*B*). We will consider the plane to be part of a small wedge of rock that is triangular in cross-section (see Figure 3.25*B*). It resembles a door-stopper. The dimensions of this wedge of rock are very, very small. The hypotenuse ($X'Z'$) of our rock triangle represents the trace of the plane for which we will calculate traction (see Figure 3.25*B*). Its length is n. The legs of the triangle represent directions (OX' and OZ'), which are parallel to the force intensities created by gravitational loading and tectonic loading (σ_l and σ_t). We will assume that the third dimension of plane $X'Z'$ is 1.0. Therefore the area of the plane is n. The area of the plane, whose trace is OZ', is $n \times \cos 65°$.

Our next step is to determine the magnitudes of the traction components acting on $X'Z'$ and operating parallel to the x- and z-axes. These are denoted

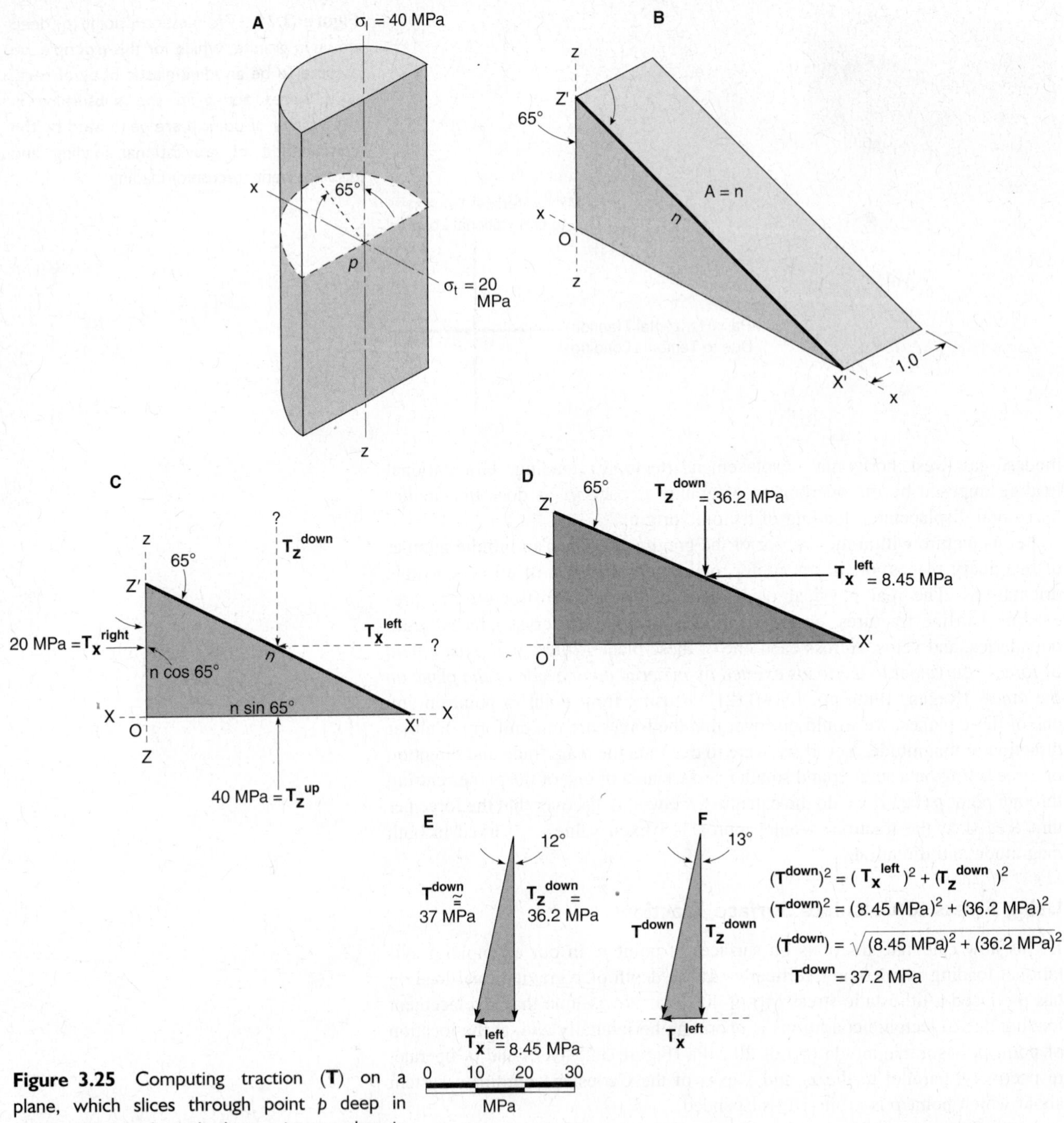

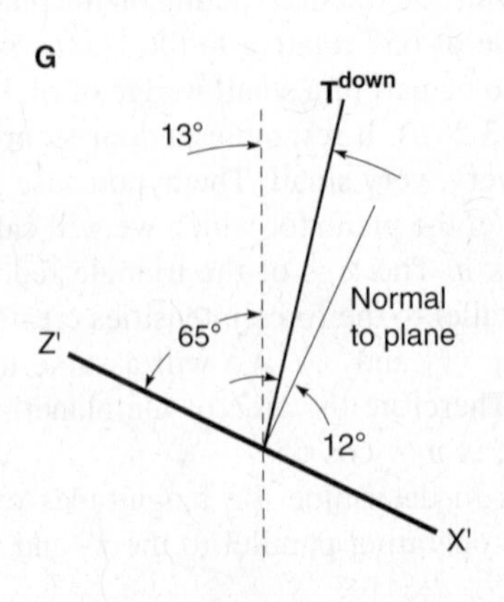

Figure 3.25 Computing traction (**T**) on a plane, which slices through point *p* deep in granite. (*A*) A look inside the granite at a plane in point *p* that is inclined 65° to the z-axis. (*B*) Block diagram view of the plane and the parcel of rock of which it is a part. (*C*) The balance of forces on the parcel of rock. (*D*) The calculated values of tractions $\mathbf{T_z}^{down}$ and $\mathbf{T_x}^{left}$ acting parallel to the z-axis (vertical) and x-axis (horizontal), respectively. (*E*) Graphical determination of the value of the traction ($\mathbf{T}^{down}$) acting on the plane. (*F*) Calculation of the traction ($\mathbf{T}^{down}$) based on application of the Pythagorean theorem. (*G*) Geometric relation of the traction ($\mathbf{T}^{down}$) to the plane.

$\mathbf{T_z}^{\mathbf{down}}$ and $\mathbf{T_x}^{\mathbf{left}}$ (Figure 3.25*C*). After we determine these intensities, we can proceed to calculate the full value of the traction ($\mathbf{T}^{\mathbf{down}}$) whose components are $\mathbf{T_z}^{\mathbf{down}}$ and $\mathbf{T_x}^{\mathbf{left}}$. Following Means (1976), we set up equations from which $\mathbf{T_z}^{\mathbf{down}}$ and $\mathbf{T_x}^{\mathbf{left}}$ can be calculated, assuming that the forces acting vertically and horizontally on this small chunk of rock are perfectly balanced, and that all the torques are balanced, for the specimen is presumed to be in static equilibrium.

The magnitude of the traction ($\mathbf{T_x}^{\mathbf{right}}$) operating on surface OZ' can be calculated by balancing forces horizontally (see Figure 3.25*C*). The force ($\mathbf{F_{Ox}}$), operating in the direction Ox, is equal to the traction ($\mathbf{T_x}^{\mathbf{right}}$) acting in the direction Ox multiplied by the area ($A_{OZ'}$) of surface OZ'.

$$\mathbf{F_{Ox}} = \mathbf{T_x}^{\mathbf{right}} \times A_{OZ'}$$

$$\mathbf{F_{Ox}} = 20\,\text{MPa} \times n \cos 65° = 20 \times n \cos 65°\,\text{N}$$

This force is balanced by the horizontal force ($\mathbf{F_{xO}}$) acting in the direction xO. The force $\mathbf{F_{xO}}$ is equal to the traction component ($\mathbf{T_x}^{\mathbf{left}}$) acting in the direction of the x-axis, multiplied by the area of the plane ($A_{X'Z'}$) whose trace is $X'Z'$.

$$\mathbf{F_{xO}} = \mathbf{T_x}^{\mathbf{left}} \times n = \mathbf{T_x}^{\mathbf{left}} \times n\,\text{N}$$

The equation that describes the horizontal balancing of forces has the following form:

$$\mathbf{F_{Ox}} = \mathbf{F_{xO}}$$

$$20\,\text{MPa} \times n \cos 65° = \mathbf{T_x}^{\mathbf{left}} \times n$$

$$20\,\text{MPa}\,(0.42) = \mathbf{T_x}^{\mathbf{left}} \times n$$

Consequently,

$$\mathbf{T_x}^{\mathbf{left}} = 8.4\,\text{MPa}$$

Similarly, the value of the traction component $\mathbf{T_z}^{\mathbf{down}}$ operating on $X'Z'$ can be calculated by balancing forces vertically (see Figure 3.25*C*). The force ($\mathbf{F_{Oz}}$), operating in the direction Oz, is equal to the traction ($\mathbf{T_z}^{\mathbf{up}}$) acting in the direction Oz multiplied by the area ($A_{OX'}$) of the plane whose trace is OX'.

$$\mathbf{F_{Oz}} = \mathbf{T_z}^{\mathbf{up}} \times A_{Ox}$$

$$\mathbf{F_{Oz}} = 40\,\text{MPa} \times n \sin 65° = 40 \times n \sin 65°\,\text{N}$$

This force is balanced by the vertical force ($\mathbf{F_{zO}}$) acting in the direction zO. The force $\mathbf{F_{zO}}$ is equal to the traction component $\mathbf{T_z}^{\mathbf{down}}$ acting in the direction of the z-axis, multiplied by the area of the plane ($A_{X'Z'}$) whose trace is $X'Z'$.

$$\mathbf{F_{zO}} = \mathbf{T_z}^{\mathbf{down}} \times n$$

Therefore, the equation that describes the vertical balancing of forces has the following form:

$$\mathbf{F_{Oz}} = \mathbf{F_{zO}}$$

$$40\,\text{MPa} \times n \sin 65° = \mathbf{T_z}^{\mathbf{down}} \times n$$

$$40\,\text{MPa}\,(0.90) = \mathbf{T_z}^{\mathbf{down}}$$

Consequently,

$$\mathbf{T_z}^{\mathbf{down}} = 36\,\text{MPa}$$

Having calculated the values of $\mathbf{T_z}^{\mathbf{left}}$ (8.4 MPa) and $\mathbf{T_z}^{\mathbf{down}}$ (36 MPa) (Figure 3.25*D*), let us determine the traction ($\mathbf{T}^{\mathbf{down}}$) for which these are components. We can do this graphically or numerically. Graphically, we map the vectors $\mathbf{T_x}^{\mathbf{left}}$ and $\mathbf{T_z}^{\mathbf{down}}$ as shown in Figure 3.25*E*. Then we construct the traction ($\mathbf{T}^{\mathbf{down}}$) from the tail of $\mathbf{T_z}^{\mathbf{down}}$ to the tip of $\mathbf{T_x}^{\mathbf{left}}$. Measuring its length according to the scale of the construction, we find that $\mathbf{T}^{\mathbf{down}}$ has a value of ~37 MPa. This traction makes an angle θ of approximately 12° with the *z*-axis. It is thus inclined at approximately 78°.

A more accurate way to determine the magnitude and orientation of $\mathbf{T}^{\mathbf{down}}$ is through the Pythagorean theorem (Figure 3.25*F*).

$$(\mathbf{T}^{\mathbf{down}})^2 = (\mathbf{T_x}^{\mathbf{left}})^2 + (\mathbf{T_z}^{\mathbf{down}})^2$$

$$(\mathbf{T}^{\mathbf{down}})^2 = (8.4\,\text{MPa})^2 + (36.2\,\text{MPa})^2$$

$$(\mathbf{T}^{\mathbf{down}})^2 = 71 + 1310 = 1381\,\text{MPa}$$

$$\mathbf{T}^{\mathbf{down}} = 37\,\text{MPa}$$

The angle that the stress makes with respect to the *z-axis* can be found trigonometrically.

$$\sin\beta = \mathbf{T_x}^{\mathbf{left}}/\mathbf{T}^{\mathbf{down}} = 8.45\,\text{MPa}/37\,\text{MPa} = 0.23$$

$$\beta = \arcsin 0.23 = 13°$$

Resolving Traction (T) into Normal and Shear Tractions

As emphasized by Means (1976), tractions are not generally perpendicular to the plane for which they have been calculated. For example, the traction ($\mathbf{T}^{\mathbf{down}}$) we just calculated is not perpendicular to surface $X'Z'$, but instead departs from 90° by 12° (Figure 3.25*G*). As we saw earlier, a given traction can be resolved into two components: one perpendicular to the surface for which the traction has been calculated, the other parallel to that surface. The normal traction component tends to inhibit sliding along a plane; the shear traction component tends to promote sliding.

Let us now resolve $\mathbf{T}^{\mathbf{down}}$ into a normal traction component (σ_n) and a shear traction component (σ_s). The graphical construction involves "*mapping*" the orientation of $\mathbf{T}^{\mathbf{down}}$ relative to the plane ($X'Z'$) on which it acts (Figure 3.26*A*). The acute angle between $\mathbf{T}^{\mathbf{down}}$ and plane $X'Z'$ is approximately 78°.

Next we show the magnitude of $\mathbf{T}^{\mathbf{down}}$ (37.2 MPa) by adding a map scale to the diagram (Figure 3.26*B*). Then, from the tail of the arrow representing the traction ($\mathbf{T}^{\mathbf{down}}$), the normal traction component (σ_n) is constructed perpendicular to the plane (see Figure 3.26*B*). Its magnitude (+36 MPa) can be read directly using the scale of the drawing. The shear traction component (σ_s) is constructed as a vector drawn from the tip of the normal traction component (σ_n) to the tip of $\mathbf{T}^{\mathbf{down}}$, and thus parallel to the plane. Its value (−8 MPa; i.e., right-handed) also can be directly read as well (see Figure 3.26*B*).

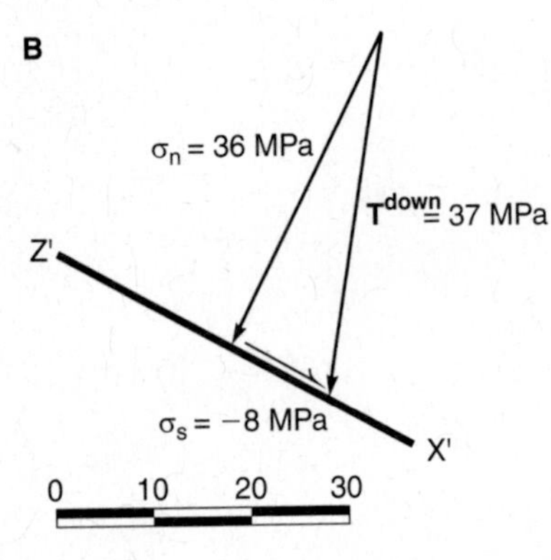

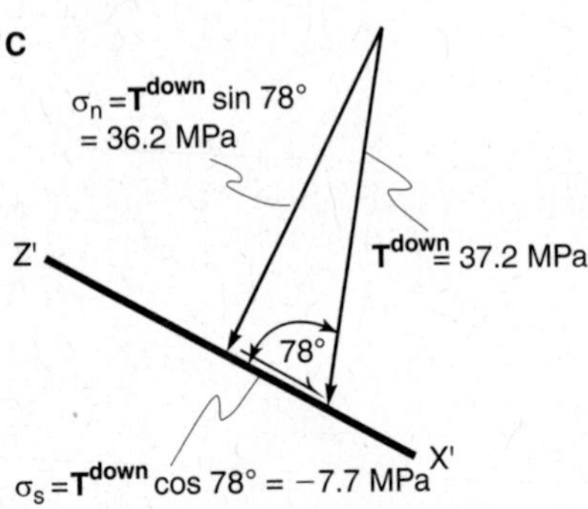

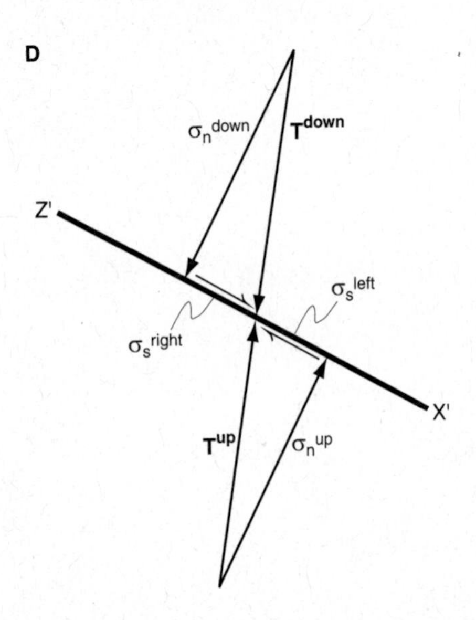

Figure 3.26 Resolving a traction ($\mathbf{T}^{\mathbf{down}}$) into normal ($\sigma_n$) and shear ($\sigma_s$) traction components. (*A*) Traction ($\mathbf{T}^{\mathbf{down}}$) is inclined 78° to plane whose trace is $X'Z'$. (*B*) Graphical determination of normal traction (σ_n) and shear traction (σ_s) components. (*C*) Trigonometric determination of normal traction (σ_n) and shear traction (σ_s). (*D*) The move from traction to stress is made in situations of static equilibrium by recognizing that normal stress (σ_N) is the pair of normal tractions and the shear stress (σ_S) is the pair of shear tractions.

The numerical solution for computing the magnitudes of the normal and shear traction components is trigonometric (Figure 3.26*C*). If 78° is the angle between the plane $X'Z'$ and the traction ($\mathbf{T}^{\mathbf{down}}$) that acts on it, the normal traction component (σ_n) can be computed as follows (Figure 3.26*C*):

$$(\sigma_n) = \mathbf{T}^{\mathbf{down}} \sin 78° = 37\,\text{MPa}\,(0.9781) = 36.2\,\text{MPa}$$

The magnitude of the shear traction (σ_s) can be determined in a comparable fashion (Figure 3.26*C*):

$$\sigma_s = \mathbf{T}^{\mathbf{down}} \cos 78° = 37\,\text{MPa}\,(\cos 78°) = -7.7\,\text{MPa}$$

By recognizing that $\mathbf{T}^{\mathbf{down}}$, σ_n, and σ_s are matched in static equilibrium by $\mathbf{T}^{\mathbf{up}}$, σ_n, and σ_s acting on the surface from beneath (Figure 3.26*D*), we can immediately shift our thinking from individual tractions and traction components to surface stresses composed of the *pair* of normal traction components operating on the surface, and the pair of shear traction components likewise acting on that same surface.

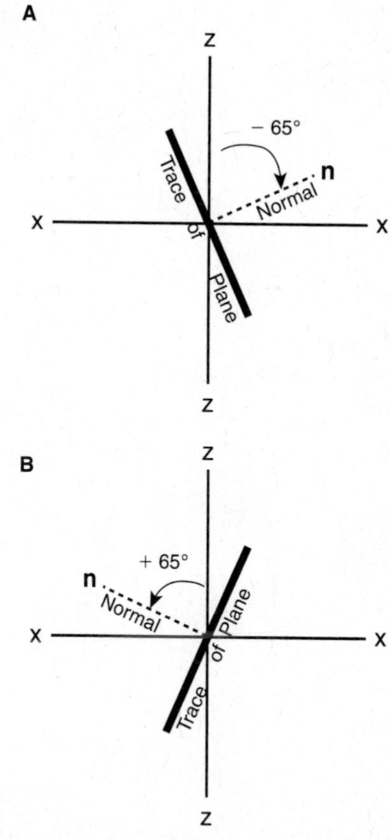

Figure 3.27 Convention for describing the orientation of a plane, or surface, for which normal and shear traction components are to be determined. (*A*) Angles measured clockwise from the z-axis to normal (n) are considered negative (−). (*B*) Angles measured counterclockwise from the z-axis to normal (n) are considered positive (+).

Determining Normal and Shear (Surface) Stresses on Any Plane

We can now imagine that pairs of normal traction components and pairs of shear traction components can be calculated for planes of any and all orientation that pass through a given point, thus yielding **normal** and **shear stresses** for any given surface. In doing this we follow the standard convention for describing the orientation of the plane for which the normal and shear traction components are to be calculated. In 2D the convention is to specify the angle made by the normal (*n*) to the trace of the plane relative to a given Cartesian coordinate axis, or relative to some other specified line or axis of reference. Counterclockwise angles are assigned positive (+) values; clockwise angles are negative (−) (Figure 3.27). We will call out this angle through the Greek letter theta (θ).

Returning to our "*underground*" example (Figure 3.28*A*), we recognize a force intensity (σ_l) of +40 MPa created by vertical lithostatic loading and a force intensity (σ_t) of +20 MPa produced by horizontal tectonic loading. We also recognize that there is a plane oriented in such a way that its normal (n) makes an angle of $\theta = -65°$ with respect to the z-axis. σ_l and σ_t resolve to tractions $\mathbf{T}^{\mathbf{down}}$ and $\mathbf{T}^{\mathbf{up}}$, which make angles of +25° with respect to the z-axis. Both $\mathbf{T}^{\mathbf{down}}$ and $\mathbf{T}^{\mathbf{up}}$ have magnitudes of 24.6 MPa (see Figure 3.28*A*). These

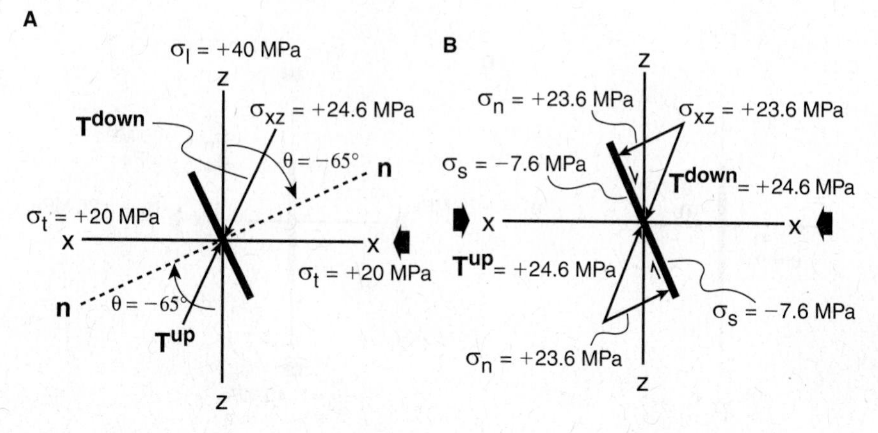

Figure 3.28 Determining normal traction component and shear traction component on plane of any given orientation. (*A*) Orientation of plane is specified by the angle (θ) its normal (n) makes with the z-axis. Magnitude and orientation of traction ($\mathbf{T}^{\mathbf{down}}$) determined by resolution of σ_l (operating vertically) and σ_t (operating horizontally). (*B*) Resolution of $\mathbf{T}^{\mathbf{down}}$ normal and shear traction components. See text for explanation.

tractions can be deconstructed into two sets of traction components, as shown in Figure 3.28*B*. The normal traction components (σ_n) prove to have magnitudes of +23.6 MPa, and the shear traction components (σ_s) have magnitudes of −7.6 MPa. (The negative sign underscores the fact that the shear traction components operate in a right-handed manner.)

The importance of determining the normal and shear surface stresses is this: by being able to determine the normal stress (σ_n) and shear stress (σ_s) components for any plane through a given point being loaded at some depth in the crust, it is possible to identify the family of planes or surfaces for which slip and displacement are most likely. High-magnitude normal stress across a surface tends to prevent slip; high-magnitude shear stress across a surface tends to favor slip. We will see that the ratio of normal stress (σ_n) to shear stress (σ_s) for a given plane or surface is one of the important factors that influences behavior.

Principal Stress Directions

For our "*underground*" example (Figure 3.29), stresses acting on vertical planes and horizontal planes are very special cases. The normal (n) to a horizontal plane makes an angle of $\theta = 0°$ with respect to the z-axis (Figure 3.29*A*), and the normal (n) to a vertical plane makes an angle of $\theta = 90°$ with respect to the x-axis (Figure 3.29*C*). The plane marked by an orientation of $\theta = 90°$ is horizontal, parallel to the x-axis (see Figure 3.29*A*). The tractions ($\mathbf{T}^{\mathbf{down}}$ and $\mathbf{T}^{\mathbf{up}}$) calculated for this plane are perpendicular to it and have the same magnitude as σ_t, namely +40 MPa. The stress on this plane is purely a normal stress and has no shear stress component (see Figure 3.29*B*). Similarly, the stress to a plane whose orientation is defined by $\theta = 0°$ (vertical plane) has the same magnitude as σ_t, namely + 20 MPa (see Figure 3.29*D*).

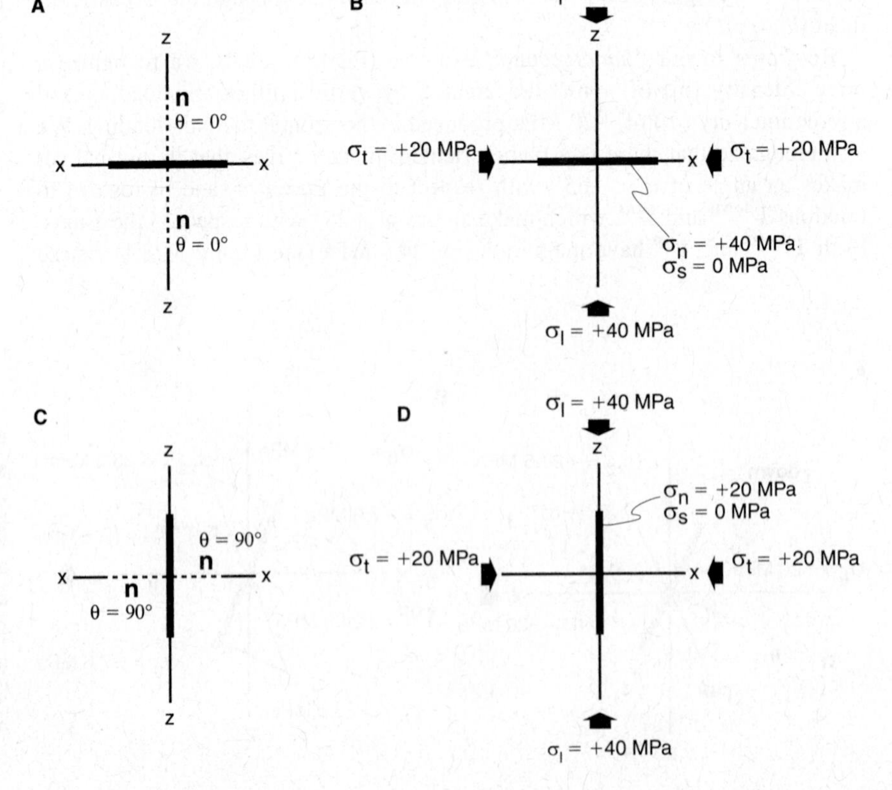

Figure 3.29 (*A*) Plane for which stresses are to be calculated is oriented at $\theta = 0°$ (i.e., normal to plane makes angle of 0° with respect to z-axis. (*B*) For such a plane, σ_n = +40 MPa and σ_s = 0 MPa. (*C*) Plane for which stresses are to be calculated is oriented at $\theta = 90°$ (i.e., normal to plane makes angle of 90° with respect to z-axis. (*D*) For such a plane, σ_n = +20 MPa and σ_s = 0 MPa.

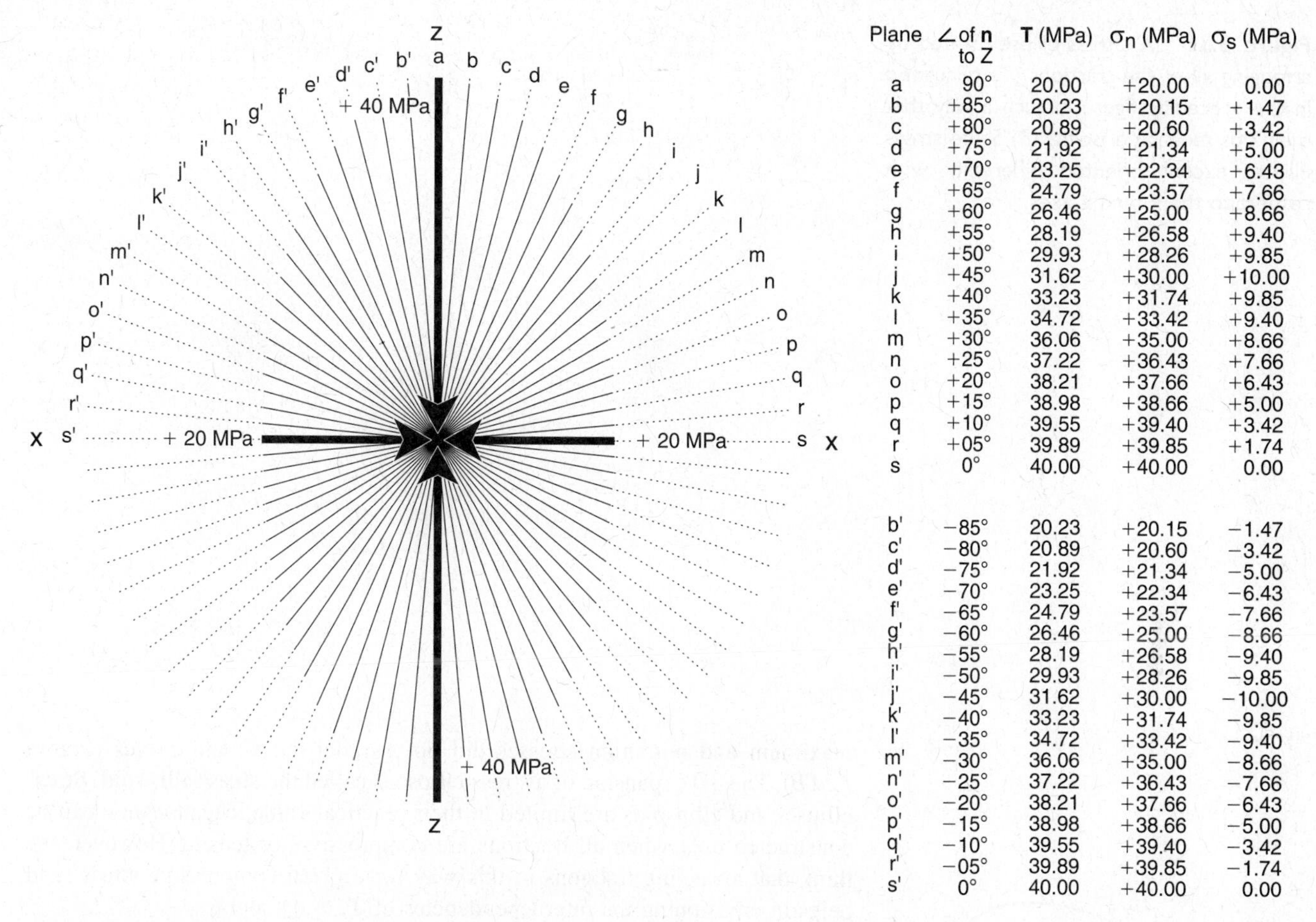

Plane	∠of **n** to Z	T (MPa)	σ_n (MPa)	σ_s (MPa)
a	90°	20.00	+20.00	0.00
b	+85°	20.23	+20.15	+1.47
c	+80°	20.89	+20.60	+3.42
d	+75°	21.92	+21.34	+5.00
e	+70°	23.25	+22.34	+6.43
f	+65°	24.79	+23.57	+7.66
g	+60°	26.46	+25.00	+8.66
h	+55°	28.19	+26.58	+9.40
i	+50°	29.93	+28.26	+9.85
j	+45°	31.62	+30.00	+10.00
k	+40°	33.23	+31.74	+9.85
l	+35°	34.72	+33.42	+9.40
m	+30°	36.06	+35.00	+8.66
n	+25°	37.22	+36.43	+7.66
o	+20°	38.21	+37.66	+6.43
p	+15°	38.98	+38.66	+5.00
q	+10°	39.55	+39.40	+3.42
r	+05°	39.89	+39.85	+1.74
s	0°	40.00	+40.00	0.00
b'	−85°	20.23	+20.15	−1.47
c'	−80°	20.89	+20.60	−3.42
d'	−75°	21.92	+21.34	−5.00
e'	−70°	23.25	+22.34	−6.43
f'	−65°	24.79	+23.57	−7.66
g'	−60°	26.46	+25.00	−8.66
h'	−55°	28.19	+26.58	−9.40
i'	−50°	29.93	+28.26	−9.85
j'	−45°	31.62	+30.00	−10.00
k'	−40°	33.23	+31.74	−9.85
l'	−35°	34.72	+33.42	−9.40
m'	−30°	36.06	+35.00	−8.66
n'	−25°	37.22	+36.43	−7.66
o'	−20°	38.21	+37.66	−6.43
p'	−15°	38.98	+38.66	−5.00
q'	−10°	39.55	+39.40	−3.42
r'	−05°	39.89	+39.85	−1.74
s'	0°	40.00	+40.00	0.00

Figure 3.30 Pictured here is a vertical force intensity (σ_z) of magnitude 40 MPa (generated by gravitational loading) and a horizontal force intensity stress (σ_x) of magnitude 20 MPa (generated by tectonic loading). The lines labeled *a–s* and *b'* to *r'* are traces of planes at 5° intervals. Traction (**T**), normal stress (σ_n), and shear stress (σ_s) have been calculated for the planes. See text for description of systematics.

Using a calculator, or better yet a computer spreadsheet that summarizes the algebra and trigonometry of these calculations, we can determine the orientations and absolute values of tractions (**T**), normal stresses (σ_n), and shear stresses (σ_s) for a host of planes inclined at 5° intervals through a point. We have done this for a condition of a vertical force intensity of 40 MPa, and a horizontal force intensity of 20 MPa (Figure 3.30). It is clear that normal stress and shear stress magnitudes vary systematically as a function of orientation: shear stress (σ_s) is zero for planes oriented parallel to the z- and x-axes ($\theta = 90°$ and 0°, respectively). Shear stress (σ_s) steadily increases from 0.0 MPa to a maximum of 10 MPa as θ changes from 90° to +45°, decreasing again to zero from $\theta = +45°$ to $\theta = 90°$. Similarly, shear stress (σ_s) steadily increases from 0.0 MPa to a maximum of −10.0 MPa as θ changes from 90° to −45°, decreasing again to zero from $\theta = -45°$ to $\theta = 0°$ (see Figure 3.30). Note how tractions (**T**) to each of these planes, when plotted accurately with respect to orientations and magnitudes (see Figure 3.30), form an ellipse.

The Stress Ellipse and the Stress Ellipsoid

The stress data posted in Figure 3.30 are incredibly systematic. This can be appreciated by plotting all the tractions (**T**) such that their tails meet at a common point, the tiny point containing the planes for which the stresses were computed. When the tractions are plotted to scale in this fashion, an elliptical form is generated, sometimes called the **stress ellipse** (Figure 3.31*A*). This ellipse is framed within the x-z coordinate system, and it just so happens in this example that stress (**T**) is maximum in the z-direction and least in the x-direction. The orientation of the coordinate system could have been such that

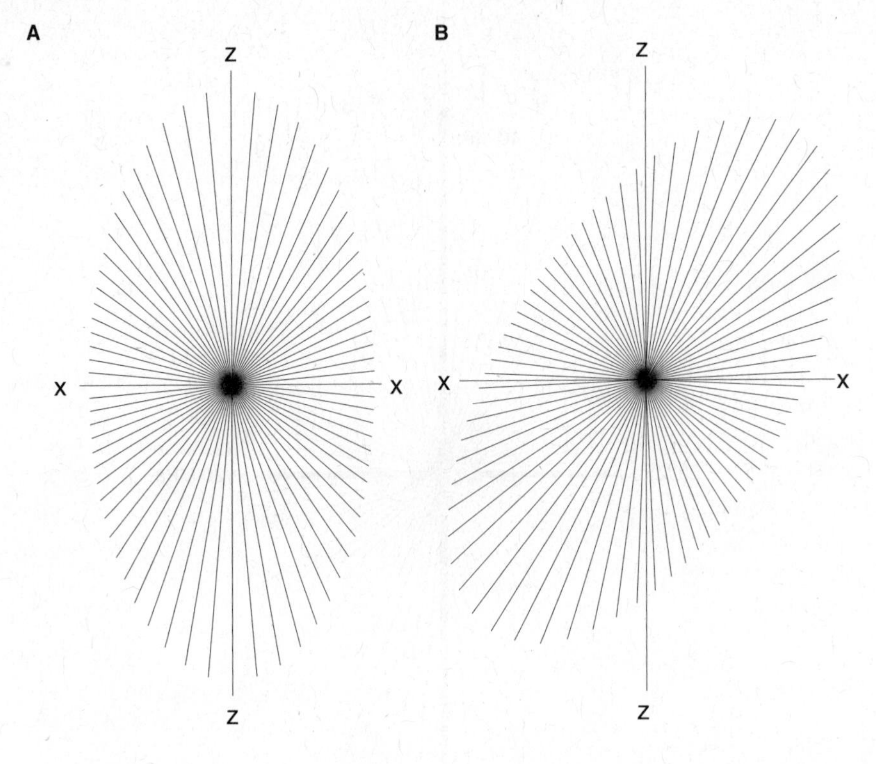

Figure 3.31 (*A*) Stress ellipse created by arranging all of the tractions (**T**) calculated in the preceding figure in such a way that their tips meet at a point. (*B*) Same stress ellipse, except oriented differently with respect to the x- and z-axes.

maximum and minimum stresses did not parallel the x- and z-axes (Figure 3.31*B*). The 3D expansion of a stress ellipse is called the **stress ellipsoid**. Stress ellipses and ellipsoids are limited in their practical value, because they can be constructed only when all tractions are compressive, or tensile. However, we think that arranging tractions in this way for a given compressive stress field helps in envisioning the interdependencies of **T**, θ, σ_n, and σ_s.

Because the orientation of the coordinate system is arbitrary, it is convenient to place axes parallel to the directions of greatest and least normal stress. Thus positioned, the axes of the stress ellipse become the **principal stress directions** (Figure 3.32). Note that they are mutually perpendicular. The long axis of the ellipse is referred to as the **axis of greatest principal stress** (σ_1); the short axis is known as the **axis of least principal stress** (σ_3). Of all the values of normal stress (σ_n) computed for stresses operating about a given point, the one marked by largest magnitude is parallel to the greatest principal stress direction (σ_1). Indeed this stress is strictly a normal stress and has no shear stress component. Similarly, the direction of least principal stress (σ_3) is marked by smallest normal stress magnitude, and by no shear stress. The specific form (i.e., the aspect ratio) and orientation of the stress ellipse depends upon the overall stress conditions affecting a tiny volume of material (see Figure 3.32). Furthermore, at a given point within a tiny volume of material, the stress conditions will change over time, in the same way that the stress conditions generally change from place to place, even within a tiny volume of material let alone a tectonic plate.

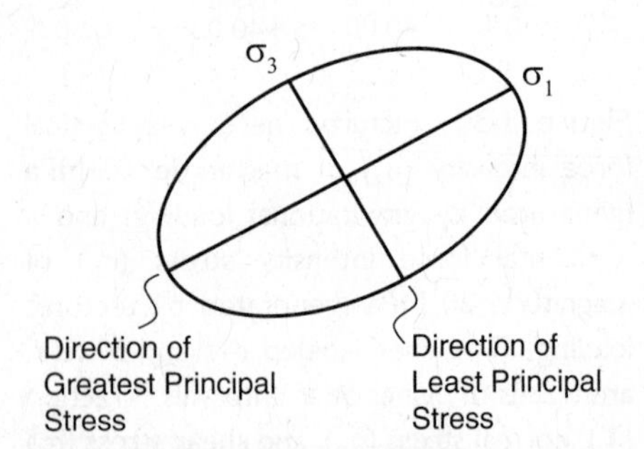

Figure 3.32 The stress ellipse, an image that shows the directions and ratios of the greatest and least principal stresses. The three-dimensional equivalent is the stress ellipsoid

In 3D dynamic analysis, it is the **stress ellipsoid** that provides the description of the stress tensor at a point. In addition to the axes of greatest (σ_1) and least principal stress (σ_3), the stress ellipsoid is characterized by an **axis of intermediate principal stress** (σ_2), which is oriented perpendicular to the plane containing σ_1 and σ_3. Like σ_1 and σ_3, axis σ_2 is a direction of zero shear stress.

The Special Case of Hydrostatic Stress

A very special stress field, called **hydrostatic**, is characterized by the absence of shear stress and the presence of identical normal stresses in *all* directions. We can experience hydrostatic stress by pausing for a few seconds at the bottom of

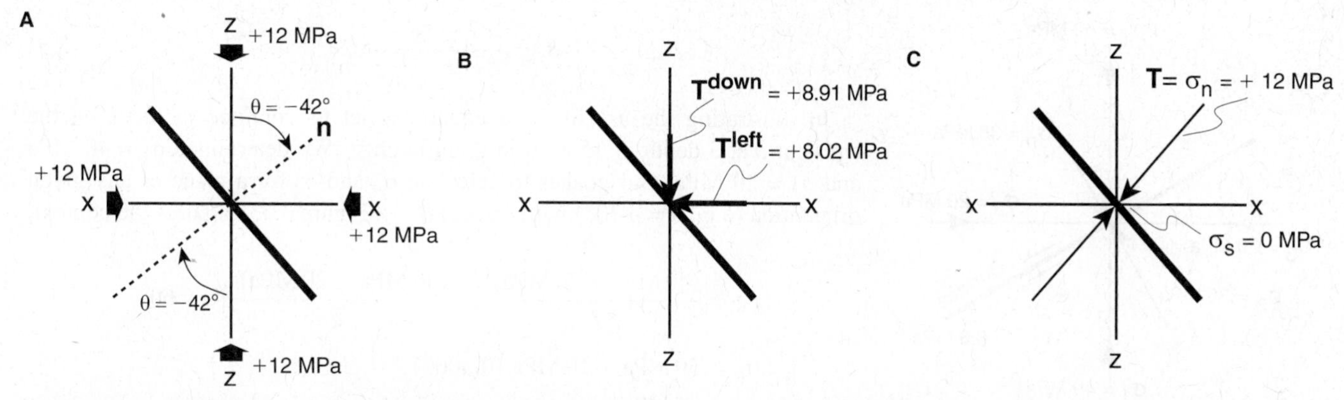

Figure 3.33 Demonstration that no shear stresses exist within a hydrostatic stress field. (*A*) The hydrostatic starting condition. Stress parallel to the *x*-axis and stress parallel to the *z*-axis are the same, 12 MPa. (*B*) Balancing of forces at point *p* results in a traction value parallel to the *z*-axis ($\mathbf{T_z}$) value of 8.91 MPa and a traction value parallel to the *x*-axis ($\mathbf{T_x}$) value of 8.02 MPa. (*C*) The single traction (**T**) for which $\mathbf{T_x}$ and $\mathbf{T_z}$ are components has a magnitude of . . . you guessed it . . . 12 MPa, and it is oriented *perfectly perpendicular* to the plane. This stress (σ) has no shear stress component.

the deep end of the swimming pool after jumping-in off the high diving board. There you will feel its pressure, a pressure that tends to dilate (negatively in this case) but not distort. Recall, in Figure 3.30, the 'drill' we carried out in calculating **T**, σ_n, and σ_s for planes covering a wide range of orientations (θ values) in a differential stress field. When we do the same within a hydrostatic stress field, we confirm that all the tractions have the same value. Moreover, we find that each traction is oriented perpendicular to the plane for which it was calculated. Thus, when a given traction is resolved into normal and shear stresses, we find that no shear stress is generated by the traction, and that normal stress magnitudes for each and every plane are all the same, in this case +12 MPa. It is no wonder that hydrostatic stresses cannot distort rocks; there are no shear stresses produced in hydrostatic stress fields.

Figure 3.33 is the proof. It pictures a location in the shallow crust where a tiny volume of rock is being loaded vertically and horizontally by the same stress intensities (12 MPa). We can choose any plane within the tiny volume of material and determine the traction acting on it (Figure 3.33*A*). We choose a plane for which $\theta = -42°$. Let us proceed immediately to determine the traction (T) on a plane inclined at $\theta = -42°$. By the balancing of forces we find that $\mathbf{T^{left}}$ = 8.02 MPa and $\mathbf{T^{down}}$ = 8.91 MPa (Figure 3.33*B*). When we compute the traction (T) for which $\mathbf{T^{left}}$ and $\mathbf{T^{down}}$ are components, we find that **T** = 12 MPa and is oriented perpendicular to the plane (Figure 3.33*C*). For this traction, the normal stress, σ_n, is 12 MPa and the shear stress, σ_s, is 0 MPa.

The Stress Equations

Thankfully, once the magnitudes and orientations of the principal stresses at a point are known, we can readily calculate the normal stress (σ_n) and shear stress (σ_s) for planes of any orientation using the **fundamental stress equations** derived in standard engineering and structural geological texts (e.g., Ramsay, 1967; Jaeger and Cook, 1976; Means, 1976). As always, we need to pay attention to the convention in specifying the orientation (θ) of the plane. θ is the angle between the direction of greatest principal stress (σ_1) and the normal (n) to the plane (Figure 3.34). The form of the stress equations is written as follows:

$$\sigma_n = \frac{(\sigma_1 + \sigma_3)}{2} + \frac{(\sigma_1 - \sigma_3)}{2} \cos 2\theta \qquad (3.4)$$

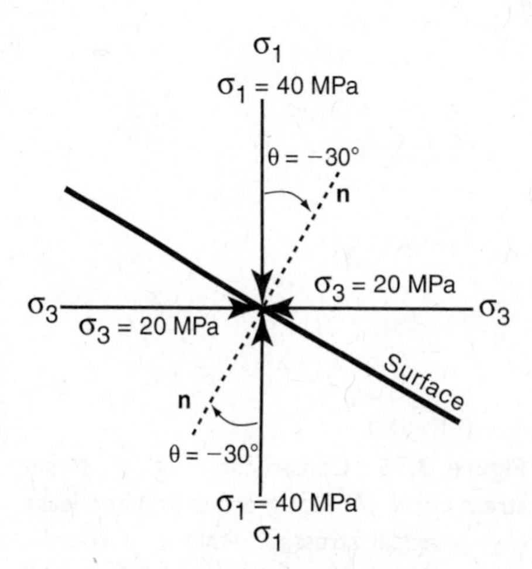

Figure 3.34 Set-up for describing the orientation (θ) of the plane for which stresses are to be determined. θ is the angle between the direction of the greatest principal stress (σ_1) and the normal (n) to the plane.

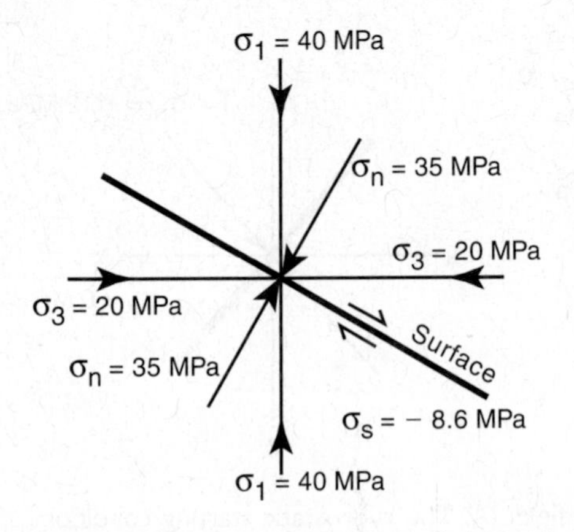

Figure 3.35 Normal stress (σ_n) and shear stress (σ_s) values determined through application of the fundamental stress equations. The negative (−) shear stress denotes right-handed (clockwise) surface shear.

$$\sigma_s = \frac{(\sigma_1 - \sigma_3)}{2} \sin 2\theta \tag{3.5}$$

In illustrating the use of these equations, let us continue to focus on the stress at p at a depth of 1500 m in granitic crust. We determined σ_1 = 40 MPa and σ_3 = 20 MPa. Our goal is to calculate σ_n and σ_s for a plane of any given orientation (e.g., $\theta = -30°$) with respect to σ_1. Figure 3.35 captures the context.

$$\sigma_n = \frac{(40\,\text{MPa} + 20\,\text{MPa})}{2} + \frac{(40\,\text{MPa} - 20\,\text{MPa})}{2}(\cos 60°)$$

$$\sigma_n = 30\,\text{MPa} + 10\,\text{MPa}\ (0.5000)$$

$$\sigma_n = 35\,\text{MPa}$$

Similarly,

$$\sigma_s = \frac{(40\,\text{MPa} - 20\,\text{MPa})}{2} \sin 60°$$

$$\sigma_s = 10\,\text{MPa}\ (0.8660)$$

$$\sigma_s = -8.6\,\text{MPa}$$

The Mohr Stress Diagram

The Mohr stress diagram gives us a very useful display of the stress (i.e., the stress tensor), for it is a clever graphical representation of the fundamental stress equations. These equations, together, describe a circular locus of paired values (σ_n,σ_s) of the normal and shear stresses that operate on planes of any and all orientations within a given body subjected to known values of σ_1 and σ_3. Using the Mohr stress diagram, we can identify a plane of any orientation (θ) relative to σ_1 and then read directly from the diagram the values of normal stress (σ_n) and shear stress (σ_s) acting on the plane.

The construction of the Mohr stress circle proceeds as follows. Principal stress values (σ_1 and σ_3) are plotted on the x-axis of the diagram (Figure 3.36*A*). A circle is drawn through these points such that ($\sigma_1 - \sigma_3$) constitutes the circle's diameter. In the example of stress at a tiny point volume where σ_1 = 40 MPa and σ_3 = 20 MPa, all the *paired* values of σ_n and σ_s as listed in Figure 3.30 exist as points on the perimeter of the circle. To define σ_n and σ_s for a specific plane (e.g., a plane oriented at $\theta = -30°$) (see Figure 3.36*A*), a radius is

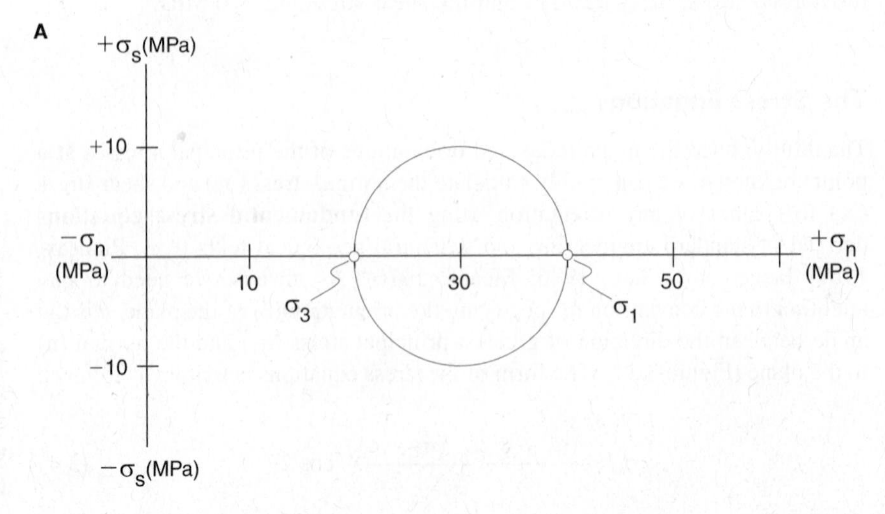

Figure 3.36 Construction of a Mohr stress circle. (*A*) Plot greatest (σ_1) and least (σ_3) principal stresses within x–y coordinate system. Normal stresses (σ_n) are plotted along x-axis; shear stresses (σ_s) are plotted along y-axis. Construct circle that passes through the principal stress values. The center of the circle lies along the x-axis.

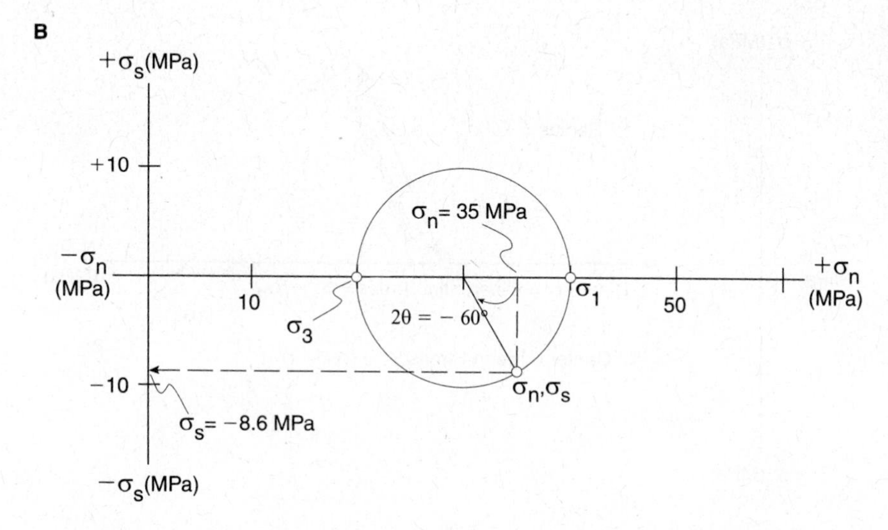

Figure 3.36 (*Continued*) (*B*) Determine the normal stress (σ_n) and shear stress (σ_s) values for a plane whose normal makes an angle of $\theta = -30°$ with respect to the direction of greatest principal stress (σ_1). Do this by constructing an angle of $-60°$ (clockwise) from the x-axis. The intersection of the radius with the circle yields a point whose (x,y) coordinates are (σ_n, σ_s).

constructed on the Mohr stress diagram at an angle of 2θ, or 60°, measured *clockwise* from the x-axis as shown in Figure 3.36*B*). Where this radius intersects the perimeter of the circle, a point is established whose (x,y) coordinates are the (σ_n,σ_s) values for the plane being interrogated. The magnitudes of these stresses are found to be $\sigma_n = +35$ MPa and $\sigma_s = -8.6$ MPa (right-handed shear) (see Figure 3.36*B*).

It is useful to work through one more example. Figure 3.37*A* shows the trace of the plane for which we will determine normal and shear stress magnitudes.

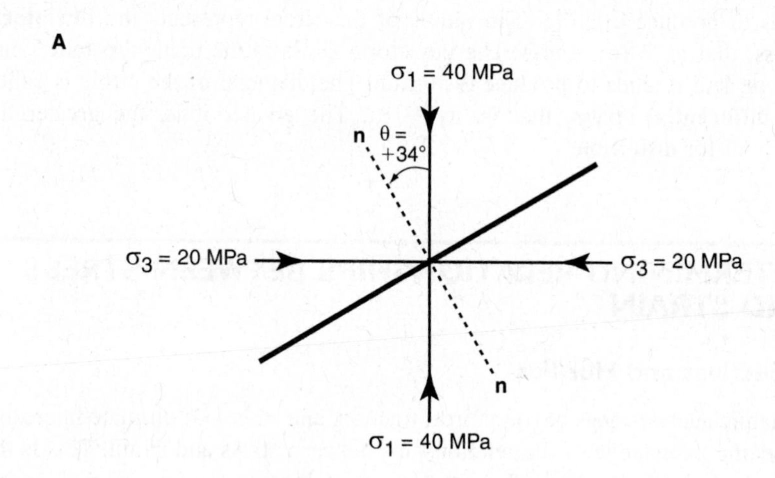

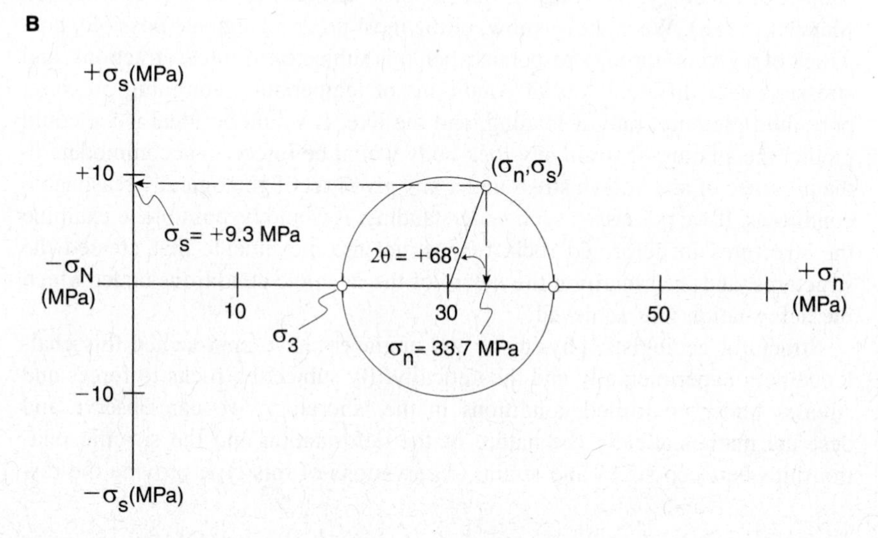

Figure 3.37 (*A*) The goal is to determine the values of normal stress and shear stress on a plane whose normal is oriented at $\theta = +34°$. (*B*) The Mohr stress circle solution. See text for explanation.

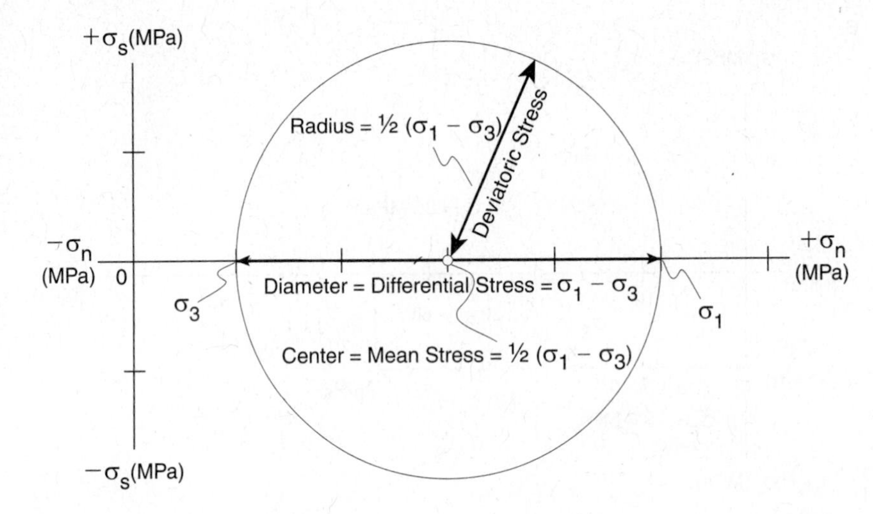

Figure 3.38 The center of the Mohr stress circle represents mean stress, which is the hydrostatic component of the stress field. Mean stress tends to produce dilation. The radius of the Mohr stress circle represents deviatoric stress, which is the nonhydrostatic component of the stress field. Deviatoric stress tends to produce distortion. The diameter of the Mohr stress circle represents differential stress. The larger it is, the greater the potential for distortion.

The angle between σ_1 and the normal (n) to the trace of this plane is $+34°$. The values of σ_n and σ_s for this plane are found by constructing on the Mohr diagram a radius of $2\theta = +68°$, this time measured counterclockwise from the x-axis (Figure 3.37*B*). The (x,y*)* coordinates of the point of intersection of this radius with the perimeter of the circle are $\sigma_n = +33.7$ MPa and $\sigma_s = +9.3$ MPa (left-handed shear).

The anatomy of the Mohr diagram is revealing (Figure 3.38). The center of the Mohr stress circle is a point that represents the **mean stress,** that is, $\frac{1}{2}(\sigma_1 + \sigma_3)$. This is the hydrostatic component of the principal stresses, and it tends to produce dilation. The radius of the circle represents the **deviatoric stress,** that is, $\frac{1}{2}(\sigma_1 - \sigma_3)$. The deviatoric stress is the nonhydrostatic component, and it tends to produce distortion. The diameter of the circle is called the **differential stress,** that is, $\sigma_1 - \sigma_3$. The greater it is, the greater the potential for distortion.

DETERMINING RELATIONSHIPS BETWEEN STRESS AND STRAIN

Objectives and Hurdles

Dynamic analysis goes beyond force, traction, and stress. Of ultimate interest is a specific knowledge of the relationships between stress and strain. This is the subject of **rheology**, the study of the response of rocks to stress (Engelder and Marshak, 1988). We want to know, in the most precise language possible, how a rock of a given lithology responds when it is subjected to forces, tractions, and stresses under different sets of conditions of temperature, confining pressure, pore fluid pressure, rate of loading, and the like. It would be ideal if we could predict the amount of strain any rock body would be forced to accommodate in the presence of any known stress under any given set of geologically reasonable conditions. If we possessed such understanding, it would be possible to examine the structures in deformed rock, reconstruct the movements that created the structures, and then interpret the nature of the dynamic conditions under which the deformation was achieved.

Structural geologists, physicists, and engineers have approached this challenge both experimentally and theoretically. By subjecting rocks to forces and stresses under controlled conditions in the laboratory, we can observe and describe mathematically the nature of the deformation and the specific relationships between stress and strain. Observations of this type provide the raw

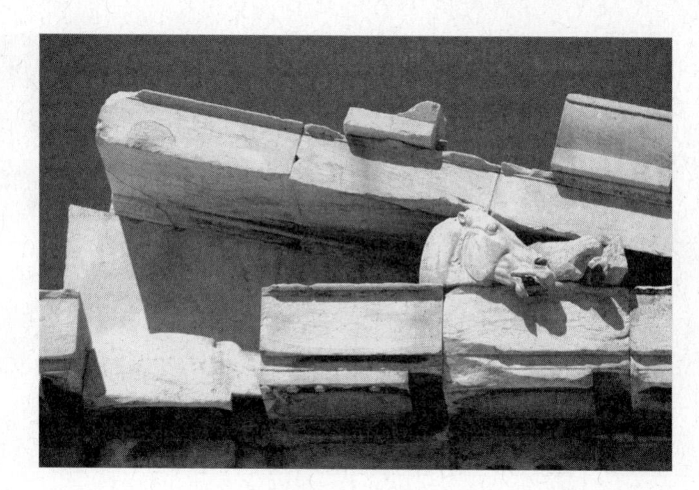

Figure 3.39 Imagining a world composed of nothing but marble (see text). This image captures part of the pediment of the Parthenon in Athens, Greece. Poseidon was the "*Earth-Shaker*" as well as the creator of horses. As a generator of earthquakes Poseidon may be captured in this art. The horses look frantic. [Photograph by G. H. Davis.]

material for assigning real numbers for the parameters in the equations that constitute the theoretical models.

If the Earth's crust were composed of merely one rock type (e.g., marble) (Figure 3.39), with just one mineral species (e.g., calcite), it would be possible to describe the relationships between stress and strain of crustal rocks through a single set of constitutive laws of behavior. This set of laws would describe exactly how the rock would respond when subjected to stress under specified conditions of temperature, confining pressure, strain rate, and the like. The relative simplicity of the system would lend itself to exhaustive, definitive analyses, at all scales. As a consequence, understanding of the mechanical behavior of such a crust would begin to match the mathematical precision with which mining engineers in the coal industry can describe the limits of strength and stability of the roof rocks above mine workings.

However, the physical and chemical character of the Earth's crust is extraordinarily heterogeneous! There are tens of different kinds of common rocks, a dozen or more common rock-forming minerals, and hundreds of different kinds of common textures and primary internal structures. Volcanic, plutonic, metamorphic, and sedimentary rock bodies are brought together into intimate, interlinked 3D contact at all scales, through extrusion, intrusion, deposition, faulting, and shearing. Then, when such heterogeneous bodies of rock are subjected to tectonic stress, each rock and each mineral in each textural configuration responds in its own way to the stress environment as a function of (1) the physical/chemical characteristics of each rock and mineral and (2) the conditions of deformation. The number of possible combinations of starting materials and deformational conditions is infinite. No single set of laws will do; only general laws that, as they become more refined, may become close approximations.

The Need for Models of Behavior

Ductile rock bodies tend to accommodate deformation without loss of cohesion; they do this by distributing the deformation throughout the body, or at the very least within broad zones (Figure 3.40). **Brittle rock bodies**, on the other hand, accommodate deformation in ways marked by pronounced loss of cohesion along through-going fractures. Instead of being distributed, the deformation is highly concentrated in narrow zones (Figure 3.41).

As useful as the terms '*brittle*' and '*ductile*' may be, they fundamentally emphasize strain. What we need instead is a set of models that helps us envision the full interplay of stress *and* strain; that helps us envision the fundamental

Figure 3.40 Expression of ductile deformation. White bedrock is marble within the Excelsior Formation (Devonian) in the Morococha District (copper-producing), Yauli Dome area, in the Central Andes, Peru. The mechanically "*soft*" character of the marble is evident in the radical changes in thickness, including places where attenuation of the marble has pinched it off completely. [Photograph by G H. Davis.]

Figure 3.41 Outcrop example of brittle deformation. This is a 1998 view of landslide-mitigation efforts along California Highway 241 east of Tustin, California, south of Los Angeles. Blading of the landscape reduced the angle of repose and created an extraordinary exposure of normal faulting. The deformation was entirely brittle, for we see no deformation-related thinning or thickening of the beds in outcrop. The exposure is now planted, and thus "*degraded*" from a structural geologic point of view [Bruce Clark, personal communication, April 21, 2010.] [Photograph by G. H. Davis.]

ways in which rocks have been found to respond to stress. Indeed, there are three basic models: **elastic**, **plastic**, and **viscous behavior**.

Elastic Behavior

Think of springs under the bed of a flatbed truck. If we load the back of a flatbed truck with solid concrete blocks, the leaf springs of the truck will shorten by an amount that is directly related to the magnitude of the load (Figure 3.42). The bed of the truck gets lower and lower as the blocks are loaded on. But when the blocks are unloaded, the leaf springs recover and so does the bed of the truck. The relationship of the load of the blocks to the displacement of the bed is an equation that describes a straight line . . . and it works perfectly no matter whether you are loading or unloading. It is the same straight-line relationship that we observe during the early, elastic stage of deformation of rock samples under controlled conditions in the laboratory.

The equation of the straight line describing the proportional relationship of stress to strain for elastic bodies is **Hooke's law** (Figure 3.43):

$$\sigma = Ee \tag{3.6}$$

where

$$E = \text{Young's modulus}$$

$$E = \frac{\sigma}{e} = \frac{\text{stress}}{\text{strain}}$$

The value of *E*, **Young's modulus**, describes the slope of a straight-line **stress/strain curve**. A stress/strain curve tracks the progressive stress levels achieved during loading of a rock specimen, and the corresponding progressive

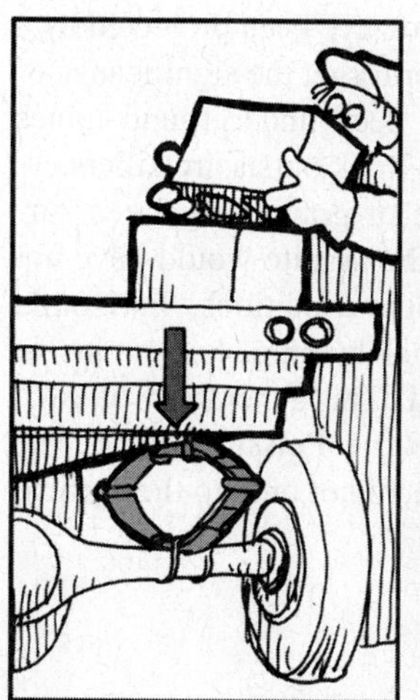

Figure 3.42 Elastic behavior is spring-like and recoverable. [Artwork by D. A. Fischer.]

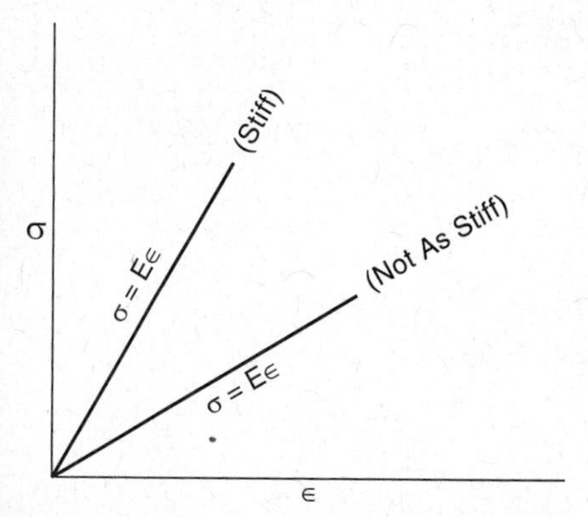

Figure 3.43 Portrayal of Hooke's law: Stress (σ) and strain (e) are directly and linearly related. The constant of proportionality (*E*) is known as Young's modulus. It is the slope of the line.

shortening (or lengthening) strain response of the specimen (see Figure 3.43). Even under the same conditions of deformation (e.g., confining pressure, temperature, strain rate), the value of *E* will vary from rock to rock, reflecting natural differences in the resistance of rock to elastic deformation. Thus the slope of a straight-line stress/strain curve is a measure of the **stiffness** of the rock. Typical values of Young's modulus are shown in Table 3.1: the higher the values, the steeper the slope, the stiffer the rock. Because extension is unit-less, Young's modulus is given in units of stress (e.g., MPa). If stress is compressive, and therefore positive (+), extension will be negative (−), and Young's modulus will be negative as well.

Young's modulus (*E*) can be thought of as an **elastic modulus** that describes how much stress is required to achieve a given amount of length-parallel elastic shortening of a rock specimen (typically in the form of a rock cylinder). A second elastic modulus, known as **Poisson's ratio** and represented by the Greek letter ν (pronounced nu), describes the degree to which a core of rock bulges as it shortens. Poisson's ratio describes the ratio of lateral strain to longitudinal strain:

$$\nu = \frac{e_{\text{lat}}}{e_{\text{long}}} \tag{3.7}$$

Poisson's ratio is unit-less, for it is a ratio of extensions. Values of Poisson's ratio are presented in Table 3.2.

Poisson's ratio meant very little to me (GHD) until one day in structural lab my students and I placed a core of granite in a deformation press and squeezed it. The core was 5 cm in diameter and about 13-cm long. We compressed it in an unconfined state (i.e., in open air) using a soil-testing ram, which features a steel piston/cylinder that very slowly and steadily presses down on the rock cylinder. The steel does not "*give*" very much (tiny elastic strain), but rocks cannot long resist shortening. We wired the core with strain gauges to monitor changes in length (longitudinal strain) and width (lateral strain). The wires were connected to a recorder to permit us to monitor magnitudes of elastic strain during the deformation. As we loaded the specimen, the rock began to shorten elastically, but to our surprise the shortening was not compensated by any increase in specimen diameter. The stress did not produce the expected lateral bulging, the "*barreling*" that we had anticipated. Volume decreased and stress somehow was being stored . . . that is, until the rock exploded with a blast like the sound of a shotgun. The largest rock fragment we found in the lab after the explosion was only

TABLE 3.1
Typical values of Young's modulus (*E*).

Rock	*E*(× 10^4 *MPa*)
Westerly granite	−5.6
Cheshire quartzite	−7.9
Karroo diabase	−8.4
Tennessee marble	−4.8
Witwatersrand shale	−6.8
Solenhofen limestone	−5.3

TABLE 3.2
Typical values of Poisson's ratio (ν)

Rock Type	
Limestone, fine grained	0.25
Aplite	0.20
Limestone, porous	0.18
Limestone, oolitic	0.18
Limestone, chalcedonic	0.18
Limestone, medium grained	0.17
Limestone, stylolitic	0.11
Granite	0.11
Shale, quartzose	0.08
Graywacke, coarse grained	0.05
Diorite	0.05
Granite, altered	0.04
Graywacke, fine grained	0.04
Shale, calcareous	0.02
Schist, biotite	0.01

pea-sized. This was brittle failure! We were grateful to have been protected by a Plexiglas shield. Based on that experience, I will never forget the significance of very low values of Poisson's ratio. Rock bursts in deep underground mines commonly represent stress releases in rocks of low ν. Rock bursts are killers.

The granite would not have exploded if the core of granite had been surrounded by a confining pressure, one so strong that the granite would have not been able to extend itself horizontally. Under such conditions, horizontal pressures would have been generated by the vertical loading, but explosive failure of the specimen would have been prevented. The generation of such horizontal stresses by vertical loading is known as the **Poisson effect**. The magnitude of such horizontal stresses ($\sigma_2 = \sigma_3$) is related not only to the vertical stress (σ_1), but also to Poisson's ratio (ν):

$$\sigma_2 = \sigma_3 = \frac{\nu}{1-\nu}\sigma_1$$

Since Poisson's ratio for common rocks is approximately 0.25, the magnitude of stress generated by the elastic phenomenon described here as the Poisson effect is approximately one-third of the greatest principal stress.

$$\sigma_3 = \frac{0.25}{1-0.25}\sigma_1 = \frac{0.25}{0.75}\sigma_1 = \frac{1}{3}\sigma_1$$

There are two other parameters that describe the elastic relationship between stress and strain (Serway, 1990). One is the **bulk modulus** (K), which is the resistance that elastic solids offer to changes in their volume. It is determined by dividing the change in hydrostatic pressure by the amount of dilation produced by the change in pressure. The other is the **shear modulus** (G), which is the resistance that elastic solids offer to the shearing of planes past each other. The shear modulus is determined by dividing shear stress (σ_s) by the shear strain (γ) it produced.

$$K = \text{bulk modulus} = \frac{\Delta_{\text{hydrostatic stress}}}{\Delta_{\text{dilation}}} \tag{3.8}$$

$$G = \text{shear modulus} = \frac{\sigma_s}{\gamma} \tag{3.9}$$

An Example of the Significance of Young's Modulus

Serway (1990) presents some wonderful examples of the use of elastic moduli. Here is one that shows the applied power of Hooke's law.

Steel railroad track has an average **coefficient of linear thermal expansion** (α) of 11 × 10^{-6}. This means that a segment of steel track 30 m long at 0°C will become 0.013 m (i.e., 1.3 cm) longer when the steel warms to a temperature of 40°C:

$$\alpha = \frac{1}{l_o} \times \frac{\Delta l}{\Delta T} = 11 \times 10^{-6}$$

$$\frac{1}{30\,\text{m}} \times \frac{\Delta l}{40°\text{C}} = 11 \times 10^{-6}, \quad \Delta l = 0.013\,\text{m}$$

The 1.3 cm change in length looks insignificant until we look up Young's modulus for steel, which is ($E = 8.67 \times 10^7$ N/m^2), and calculate first the stress, and then the force, created by the thermal expansion.

$$\sigma = Ee = \frac{\mathbf{F}}{A}$$

$$\frac{\mathbf{F}}{A} = E \times \frac{\Delta l}{l_o} = 20 \times 10^{10}\ \text{N/m}^2 \times 0.000433 = 8.67 \times 10^7\ \text{N/m}^2$$

Since

$$A = \text{area of steel rail} = 30\,\text{cm}^2$$

$$\text{force } (\mathbf{F}) = \sigma \times A = 8.67 \times 10^7\ \text{N/m}^2 \times 0.003\ \text{m}^2 = 2.6 \times 10^5\ \text{N}$$

$$\mathbf{F} = 2.6 \times 10^5\ \text{N} = 58{,}500\ \text{lb} = 29\ \text{tons!!}$$

This level of force and stress can exceed even the capacity of steel to respond elastically (Figure 3.44).

Figure 3.44 Warped trolley tracks in Asbury Park, NJ. It might at first appear as if deformation of the steel was achieved by active tectonic compression of New Jersey. However, all of the deformation took place during on an exceedingly hot July day. Deformation was due to thermal loading. [Drawing based on photograph in Serway (1990), p. 514.]

If we know Young's modulus (E), Poisson's ratio (ν), and the magnitudes of stresses in three mutually perpendicular directions (x, y, z), it is straightforward to compute the amount of elastic strain (e) that would take place in directions x, y, and z. Here are the equations. They are marvelous summary statements of the interrelationships stress and elastic strain:

$$e_z = \frac{1}{E}[\sigma_z - \nu(\sigma_x + \sigma_y)] \tag{3.10}$$

$$e_y = \frac{1}{E}[\sigma_y - \nu(\sigma_z + \sigma_x)] \tag{3.11}$$

$$e_x = \frac{1}{E}[\sigma_x - \nu(\sigma_z + \sigma_y)] \tag{3.12}$$

With these equations we can revisit Figure 3.16, whip out our calculators, assume reasonable values for E and ν, and compute in a matter of minutes the percentage of (elastic) shortening vertically, horizontally north–south, and horizontally east–west. Powerful equations!

Plastic Behavior

An unlikely metaphor for plastic deformation is the image of solid concrete blocks that must be slid from the front to the back of the bed of the flatbed truck (Figure 3.45). Let us be clear: the concrete block is not plastic. But the response of the block to our push is analogous to the response of a plastic body. Ideal

Figure 3.45 The forcing of a block to move by building stress to a critical threshold is analogous to plastic deformation. Plastic deformation is non recoverable. [Artwork by D. A. Fischer.]

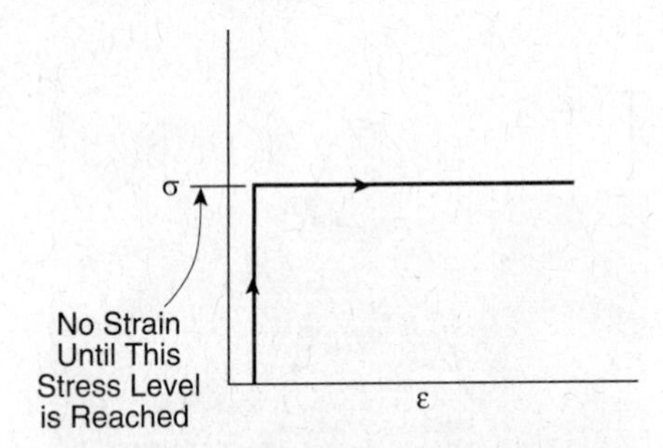

Figure 3.46 Portrayal of ideally plastic deformation. Stress (σ) is raised, but no strain (ε) accrues until a critical threshold is exceeded. From that point on, under ideal conditions, deformation continues as long as the stress level is maintained.

plastic bodies will not deform until a critical threshold of yield stress is exceeded. As we begin to push on a concrete block, nothing happens until the shear stress along the base of the block is great enough to overcome the resistance to sliding. If the floor of the bed is smooth and polished from years of this kind of activity, the resistance to sliding might be relatively low. If, however, there is leftover roofing tar smeared on the floor of the bed, the resistance to shoving might be relatively high. But once the yield stress has been overcome, and provided we are able to maintain the push at least at the level of yield stress, and assuming that the frictional resistance along the base of the bed does not change, we will be able to slide the block to the back of the truck without stopping (see Figure 3.45). Even if we managed to maintain a push at exactly the level of yield stress, the rate of movement of the block would not be uniform, for stress and strain rate are not linearly connected in the case of plastic bodies. If we stop shoving and take a rest, we will not lose ground; the block will not rebound to the front of the bed as if it were spring-loaded. The movement of the block is non-recoverable. When we decide to move the block a second time, we will need to put our backs into it to build the pressure to the yield stress and beyond.

An ideally plastic body undergoes no strain whatsoever below the yield stress, and this would be represented on a stress/strain curve by a vertical line that terminates at the value of yield stress (Figure 3.46). When the yield stress is achieved, the plastic body will strain as long as the yield stress magnitude is maintained (horizontal line).

"*Plastic*" rock bodies are not perfectly plastic. More typically, an initial elastic response is followed by steady increments of strain, each of which requires a steady, slightly increasing stress. A gentle rise of the stress/strain curve reflects this requirement, and the slope of this line is a measurement of the amount of **strain hardening** taking place within the material. It is as if the sliding friction along the floor of the bed of the truck is steadily increasing from front to back, perhaps because small clots of the roofing tar are scraped off the bed and accumulate along the leading edge of the blocks. Under some circumstances, less and less stress is required to maintain the strain: **strain softening** is taking place inside the rock. It is as if the sliding friction along the floor of the bed of the truck steadily decreased, becoming more and more slippery, from front to back.

Viscous Behavior

Think of shock absorbers underlying the bed of the flatbed truck (Figure 3.47). Viscous bodies are fluids, and the behavior of a viscous body in response to stress is analogous to the inside action of a shock absorber. Stresses parallel

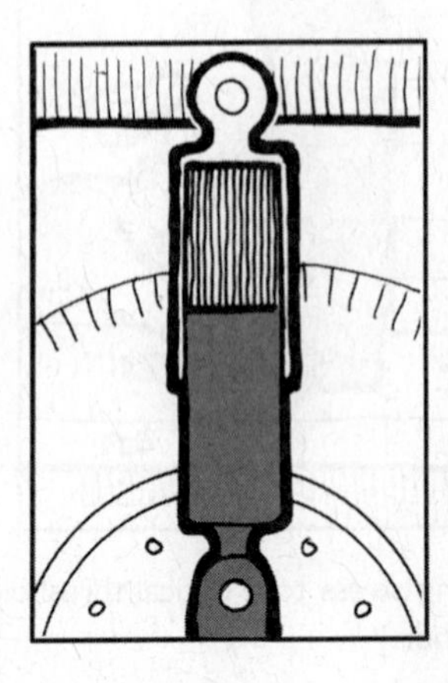
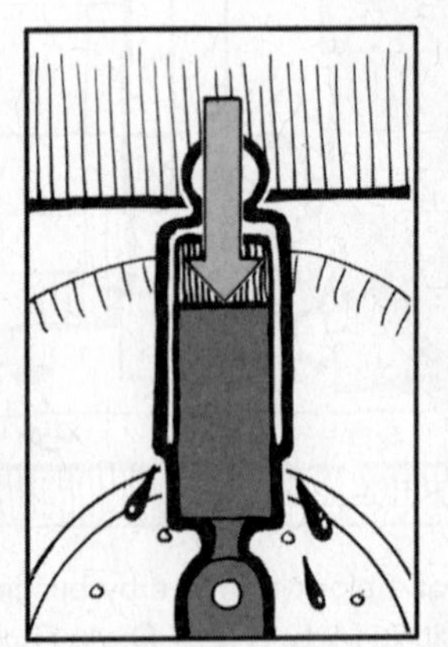

Figure 3.47 The action of shock absorbers is analogous to viscous deformation. Even the tiniest increment of stress will produce flow. Deformation is permanent, non recoverable. [Artwork by D. A. Fischer.]

to the axis of the shock absorber are generated by loading of the solid concrete blocks into the bed of the truck, sliding and jostling the blocks from the front to the back of the bed, and, once the truck is in motion, hitting bumps along the road. The loading and the bumping cause the piston within the shock to move within the cylinder, which is full of hydraulic fluid. The piston can move under the influence of stress because of opening(s) that permit the hydraulic fluid inside the shock to move from one side of the piston to the other. Even the smallest stress will displace the piston within the shock cylinder. There is absolutely no yield stress to be overcome. There is no elastic deformation. The piston simply moves at the slightest provocation. If movement of the piston instantly stops, the hydraulic fluid instantly stops deforming. The displacement is not recoverable. And there is not even an elastic recovery, for there is no elastic deformation.

The rate of movement of the piston depends upon the amount of stress and the resistance of the fluid. For an ideal **Newtonian** fluid, there is a straight-line relationship between magnitude of stress and rate of strain (Figure 3.48):

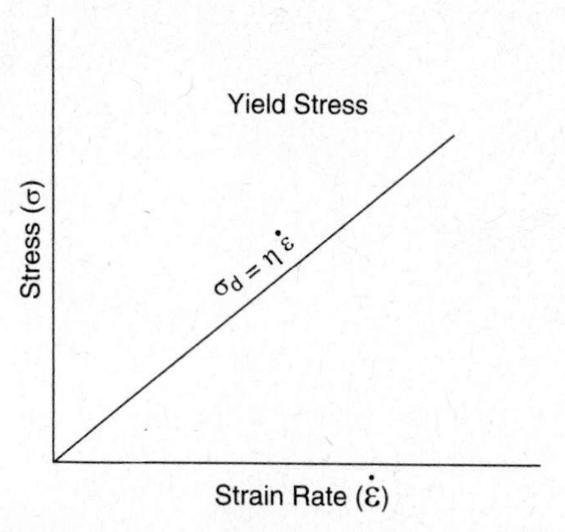

Figure 3.48 Portrayal of ideal viscous behavior on plot of stress (σ) versus strain rate ($\dot{\varepsilon}$).

$$\sigma_d = \eta\dot{\varepsilon} \qquad (3.13)$$

where σ_d = differential stress

η = viscosity

$\dot{\varepsilon}$ = strain rate

Viscosity is a measure of resistance to flow (see Figure 3.48), just as Young's modulus can be thought of as a measure of resistance to elastic deformation (see Figure 3.43). Higher and higher viscosities correlate with greater and greater internal friction. Viscosity is measured in poises: "*If a shear stress of one dyne/cm^2 acts on a liquid and gives rise to a strain rate of 1.0 sec^{-1}, the liquid has a viscosity of 1 poise*" (10 poises = 1 Pa sec) (Price and Cosgrove, 1990, p. 21). Viscosities for common fluids are presented in Table 3.3.

Years ago, I (GHD) learned from Jerry, one of the mechanics at Desert Toyota in Tucson, Arizona, that the opening between hydraulic compartments in some modern shocks is 'smart.' The opening becomes smaller when the force on the piston is greater, and expands again when the force goes back to what is considered to be a normal load. The auto engineers thus found a way to change the effective viscosity of the fluid without changing fluids. When the hole closes down, the apparent viscosity increases, for it takes more force to "*drive*" the fluid through the opening.

TABLE 3.3
Viscosities of common fluids.

Most viscous	The Earth's mantle	10^{23} poises
	Salt	10^{17}
	Rhyolite lava	10^{9}
	Roofing tar	10^{7}
	Basalt lava	10^{3}
	Corn syrup	10^{2}
	Castor oil	10^{1}
	Heavy machine oil	6
	Olive oil	.8
	Turpentine	.01
Least viscous	Water at 30°C	.008

Interrelationships of Elastic, Plastic, and Viscous Behaviors

The Earth, like the flatbed truck we loaded with concrete blocks and drove away, has some parts that are behaving elastically, some that are behaving plastically, and some that are behaving viscously. And all these behaviors are going on simultaneously (Figure 3.49). To make matters even more interesting, the parts work in tandem such that there are combined mechanical responses for every occasion: viscoelastic, elastic-plastic, viscoplastic, firmoviscous (see Ramsay, 1967; Suppe, 1985; Price and Cosgrove, 1990; Twiss and Moores, 1992). Rocks have an appropriate mechanical response for every conceivable situation.

As we carry out field work, we are struck by the contrasting mechanical responses of rocks that were side-by-side, so to speak, during tectonic loading. We see this most clearly in sedimentary sequences where stiff competent layers, such as sandstone or chert layers, are interbedded with soft, incompetent layers, such as shale or mudstone or evaporites. The more brittle layers may be faulted, and during the faulting, such layers undergo no change in original thickness, for they experience no plastic or viscous flow. On the other hand, the more ductile layers may be folded, and original stratigraphic thickness is *not* preserved because the rock experiences plastic or viscous flow. We always try to keep track of these tendencies, and based on such observations we try to build a **mechanical stratigraphy** in addition to the standard lithology-based stratigraphy. Mechanical stratigraphy underscores the likely behavior of rock layers in response to tectonic loading.

Figure 3.49 The flatbed truck, while negotiating the bumpy road, utilizes both its springs and shocks and redistributes its load of blocks in an effort to achieve a smoother ride. (Inspiration for creating this truck cartoon was Twiss and Moores, 1992, p. 367.) Rocks in any part of the crust of the Earth, in analogous fashion, utilize imaginative combinations of elastic, viscous, and plastic behaviors to relieve stresses and achieve a smoother strain. [Artwork by D. A. Fischer.]

CONDUCTING DEFORMATION EXPERIMENTS IN THE LABORATORY

Value of Laboratory Deformational Experiments

It is useful to create the very stresses whose impacts we wish to examine, and then observe the various behaviors of rocks subjected to such deformation under a variety of conditions. Some structural geologists achieve this in laboratory settings by subjecting rock specimens to controlled loading under known, prescribed conditions in a deformation apparatus. The nature and origin of structural deformation of rocks becomes clearer through images and experiences in experimental deformation. The general procedure is to fashion a

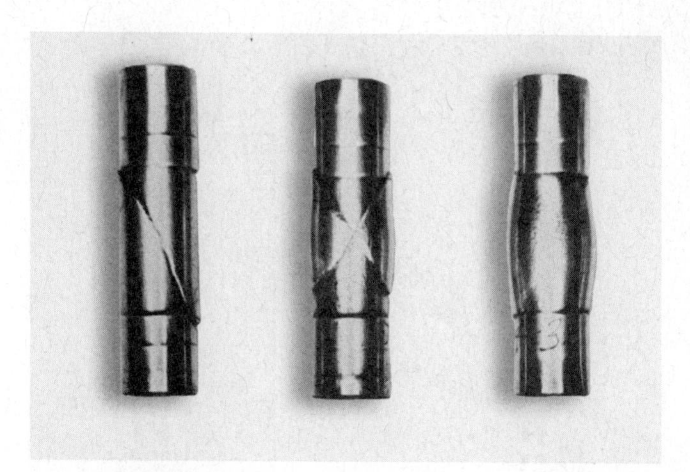

Figure 3.50 Deformation of experimentally deformed cylinders of rocks. Structure is reflected in the thin-walled copper jacket that envelops each rock specimen. *Left to right:* fault in slate, conjugate faults in sandstone, ductile flow in limestone. [F. A. Donath, 1970a. Rock Deformation Apparatus and Experiments for Dynamic Structural Geology. *Journal of Geoscience Education*, vol. 18, p. 7.]

perfectly cylindrical core of limestone with beveled ends; encase it within a thin copper jacket so that the sample does not become saturated by fluids when confined; place it inside a thick-walled steel pressure vessel; squeeze the rock cylinder by means of hydraulic loading; monitor the imposed conditions of deformation (temperature, confining pressures, strain rate, fluid pressure); and remove the now-deformed sample (Figure 3.50) and realize that it bears deformational characteristics comparable to those seen in natural outcrops. The experimental process provides a cause-and-effect glimpse of dynamics that few experiences numerical modeling can match.

Sample Preparation

A small cylinder of rock is extracted from a rock, such as limestone, through the use of a drill press and diamond-drill coring device. The ends of the cylinder of rock are beveled to smooth, planar parallel surfaces on a grinding wheel (see Figure 3.4). If the ends were not ground flat, small irregularities of rock projecting from the ends would cause high-stress concentrations, like those focused by the tips of skates on ice or the head of a nail being hammered into wood.

After the cylindrical specimen has been thus prepared, a micrometer is used to measure length (l_o) and diameter (d_o) of the specimen, preferably to three significant figures. The specimen is then placed in a sleeve or jacket of a thin-walled cylinder of copper or some other material of negligible strength. The jacket serves to seal the rock from whatever fluid (e.g., kerosene) occupies the pressure vessel. The copper jacket is essentially without strength, and thus does not in any way mechanically reinforce the rock cylinder. Once all this has been done, the jacketed specimen is fitted with an anvil at its base and an upper piston specimen holder at its top (see Figure 3.51*A*). Both of these components are made of stainless steel and equipped with O-rings to prevent entry of fluids into the jacket from above or below. Then the specimen and its trimmings are screwed into the **pressure vessel** (see Figure 3.51*B*), a steel vessel of sufficient wall thickness and inherent strength to resist fracturing under conditions of very high pressure.

During the test, the fluid surrounding the jacketed specimen in the pressure vessel is pressurized to produce a **confining pressure** to simulate burial at depth. The **temperature** of the rock core can be increased to the desired level by increasing the power to the furnace surrounding the sample in the pressure vessel. **Pore fluid pressure** within the jacketed sample can be elevated as well, and as an independent variable separate from confining pressure. Machining of the pressure vessel is so precise that fluids can be pumped at high pressure

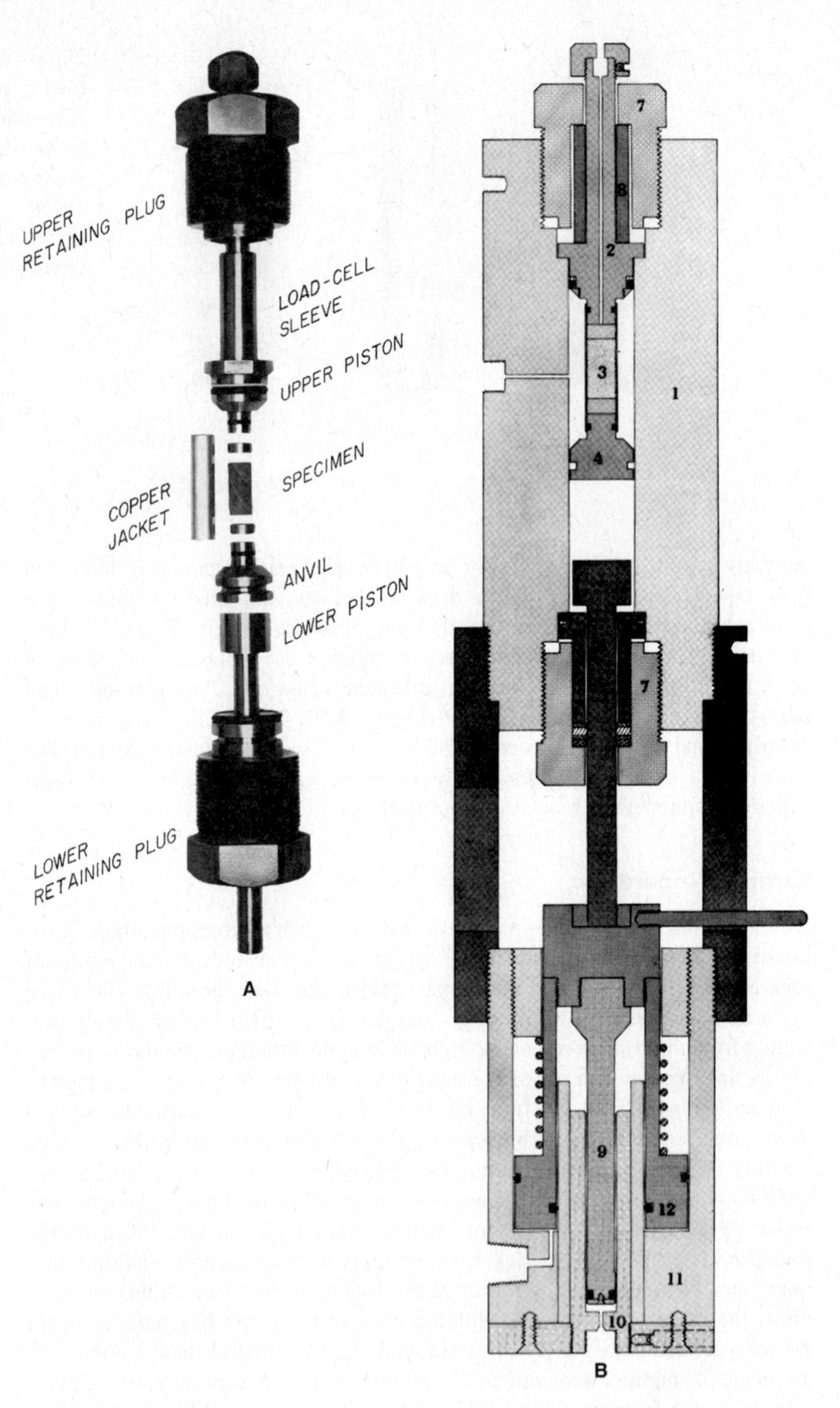

Figure 3.51 Example of the inner workings of an experimental deformation rig. (*A*) On the left, internal parts of pressure vessel, showing the relation of the cylindrical rock specimen (and copper sleeve) to pistons and anvil. [F. A. Donath, 1970. Rock Deformation Apparatus and Experiments for Dynamic Structural Geology. *Journal of Geoscience Education*, vol. 18, p, 6.] (*B*) On the right, schematic drawing of the vessel–press assembly in the Donath apparatus: 1, pressure vessel; 2, upper piston and seal; 3, specimen; 4, anvil; 5, gland and lower seal; 6, lower piston; 7, retaining plug; 8, load cell; 9, equalization piston; 10, equalization cylinder; 11, ram body; 12, ram piston; 13, collar. Practical details of the inner workings of the Donath apparatus are available in Donath (1966). [F. A. Donath, 1970. Rock Deformation Apparatus and Experiments for Dynamic Structural Geology. *Journal of Geoscience Education*, vol. 18, p. 6.]

directly into the jacketed specimen, again to whatever pressure level is reasonable for the test being conducted.

Types of Deformation Experiments

There is more than one way to squeeze a rock. In the most common procedure, the cylindrical sample is subjected to **axial compression** (Figure 3.52*A*). Vertical axial compressive stress (σ_1) parallel to the length of the core is taken to higher levels than the horizontal radial compressive **confining pressure** ($\sigma_2 = \sigma_3$) acting on the body of the rock cylinder. Because the axial stress is the greatest principal stress (σ_1) and the radial stress is the least principal stress (σ_3), the specimen undergoes length-parallel shortening.

A less common procedure is one in which the core is subjected to **axial extension** (Figure 3.52*B*). In this case, horizontal radial compressive stress, the confining pressure ($\sigma_1 = \sigma_2$) acting on the body of the rock cylinder, is greater than the vertical axial compressive stress (σ_3) parallel to the length of the sample. As a result, the specimen undergoes length-parallel extension.

Axial compression and axial extension tests carried out in this manner are referred to as **triaxial deformation experiments**, although "*triaxial*" is misleading because most equipment does not permit each of the three principal stresses to be set separately and independently.

In some tests the rock samples are not squeezed at all; rather they are pulled apart (i.e., stretched). Most commonly these tests are unconfined **tensile strength tests** (Figure 3.52*C*), which aim to determine the smallest amount of stress that will cause a rock to fail in tension. Rocks are much weaker in tension than in compression, and, if given a chance to fail in tension, will do so. As a consequence, material scientists, engineers, and structural geologists have a keen interest in the weakness of rocks in tension and have worked to calibrate it by means of a wide variety of tensile strength tests (Price and Cosgrove, 1990).

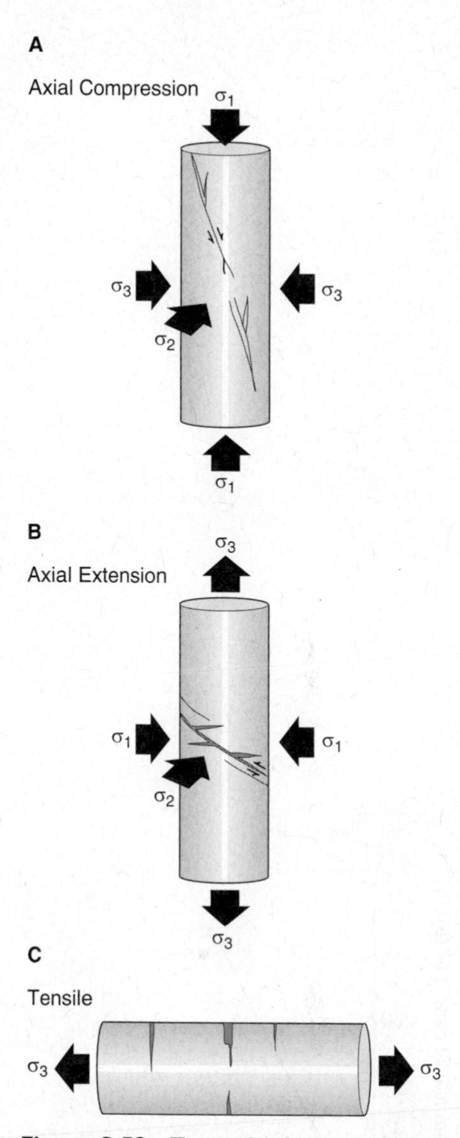

Figure 3.52 Types of deformation experiments: (*A*) Axial compression; (*B*) axial extension; (*C*) tensile.

Distinguishing Pressures, Stresses, and Loads

During experimental deformation burial pressures are imposed on test specimens by pumping fluid into the specimen chamber to produce an equal, all-sided **confining pressure** on the specimen. In nature, confining pressure at any specified level is derived from the combined effect of the weight of water occupying pore spaces in the overlying column of rock (**hydrostatic stress**) and the weight of the rock column itself (**lithostatic stress**).

The level of confining pressure can be read directly from a calibrated gauge. Raising the level of confining pressure by even small amounts results in a small reduction in volume of the specimen. The specimen quite naturally experiences negative dilation. Hydrostatic stress, even at the highest levels, cannot produce distortion. The equal all-sided compressive stress environment simply reduces the size of the specimen uniformly.

When the appropriate level of confining pressure is established, the cylinder specimen of rock is shortened by applying a vertical **axial load**. The load is applied to the specimen by manually pumping hydraulic jack fluid into the vessel-press assembly. As pressure of the hydraulic jack fluid increases, a ram forces a piston to advance slowly upward until it abuts against the anvil fixed to the base of the specimen. This proves to be a critical moment in the test, for when the piston makes contact with the anvil, a load is transmitted through the anvil into the specimen. Moreover, the load is transmitted through the specimen to the steel cap that seals the top of the thick-walled pressure vessel. The steel cap is like a very strong steel spring, which deforms elastically when subjected to stress.

By knowing the relation of stress magnitude to shortening of the steel cap, it is possible to calibrate the exact amount of load being applied by the piston at each stage of the experiment. Indeed, axial load can be tracked throughout the course of the deformation experiment. In contrast, **axial stress** is not an output during experiments. Instead, it is determined later, during data-reduction, by dividing load by the cross-sectional area of the specimen.

For example, imagine that the load at some stage of an axial compression test was 1814 kg, and assume that this load was brought to bear on a cylindrical specimen of limestone whose radius was 0.617 cm. Axial stress on the specimen is calculated as follows:

$$\sigma_a = \mathbf{T_a} = \frac{\text{load}}{\text{area}} = \frac{1814\,\text{kg}}{1.19\,\text{cm}^2} = 1524\,\text{kg/cm}^2 = 152\,\text{MPa}$$

In reality, this calculation is not as straightforward as this, due to the vast number of correction factors that are required in order to compensate for such things as the response of the various steel members that carry the load to the specimen, and to adjust for the changes that take place in the cross-sectional area of the rock cylinder.

The Measuring of Shortening

Changes in the length of the test specimen are a part of the continuous output during deformation experiments. The point of reference comes the instant the piston makes contact with the stainless steel anvil affixed to the base of the specimen. Further movement of the piston into the specimen chamber records shortening of the specimen. If rocks were perfectly unyielding, the piston could be raised no further, no matter how much load is applied to the anvil. However, rocks are deformable, and thus added increments of load always result in shortening.

When the steel piston moves into contact with the base of the specimen, the piston becomes "*seated.*" Confining pressure within the steel pressure vessel is then raised. Movement of the piston now loads the specimen itself. The initial effect is elastic deformation, revealed by a straight-line relationship between load and displacement.

The real-time information provided by the load/displacement graph becomes the basis for creating a **stress/strain curve**, which is a plot of differential stress (MPa) versus strain (%) (Figure 3.53). The length-parallel axial loading can be thought of as the greatest principal stress (σ_1). The confining pressure can be thought of as the least principal stress (σ_3). **Differential stress** is $\sigma_1 - \sigma_3$. **Strain** is extension (e) multiplied by 100%. For compression experiments, this is shortening strain. In fact, shortening of the rock cylinder at any stage of an experiment can be described in terms of its extension (e). Suppose that original specimen length was 2.3 cm and shortening was merely 0.02 cm:

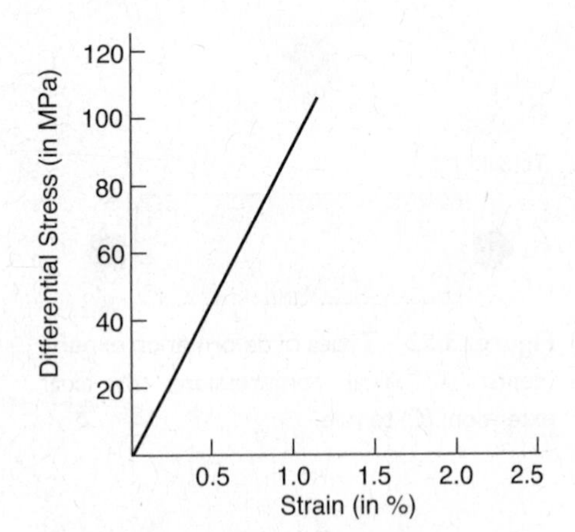

Figure 3.53 Stress/strain diagram.

$$e = \frac{\Delta l}{l_o} = \frac{-0.02\,\text{cm}}{2.3\,\text{cm}} = -0.0087$$

$$\text{shortening} = -0.0087 \times 100\% = 0.87\%$$

The Measuring of Strain Rate

The rate at which a rock is shortened or stretched has a very important bearing on how it deforms. Thus we keep track of **strain rate** ($\dot{\varepsilon}$) during deformational experiments. Strain rate ($\dot{\varepsilon}$) is determined by dividing extension (e) by time (t).

$$\text{Strain rate} = \dot{\varepsilon} = \frac{e}{t}$$

Strain rate ($\dot{\varepsilon}$) is given in units of reciprocal seconds (s^{-1}). For example, a strain rate of $10^{-5}\ s^{-1}$ means that the amount of extension per second is 0.00001. Describing strain rate in these units (s^{-1}) gives the impression, at first, that something is missing, but this is only because extension (e) is unit-less. Let us try an example. Suppose that during a one-hour experiment we shortened a cored cylinder of mudstone from its original length of 2.297 cm to a final length of 2.280 cm. What is the strain rate for this experiment?

$$\varepsilon = \frac{e}{t} = \frac{\frac{2.280 - 2.297}{2.297}}{60\,\text{min} \times 60\,\text{s/min}} = \frac{\frac{-0.017}{2.297}}{3600\,\text{s}}$$

$$\dot{\varepsilon} = \frac{-0.0074}{3600\,\text{s}} = \frac{0.0000021}{S} = -2.1 \times 10^{-6}\text{s}^{-1}$$

Let us try one more example. Suppose that we are about to subject an undeformed cored cylinder of limestone to shortening. Its length before shortening is 3.00 cm. If we were to shorten it at a constant strain rate of -10^{-5}, by how much would we need to reduce the length of the specimen in the first 5 minutes of the experiment?

$$\text{extension } (e) = \frac{\text{length change}}{\text{original length}} = \frac{-\Delta l}{l_o} = \frac{-\Delta l}{3.0\,\text{cm}}$$

$$\text{strain rate } (\dot{\varepsilon}) = \frac{\text{extension(e)}}{\text{second}} = -10^{-5}\,\text{s}^{-1}$$

$$\text{strain rate } (\dot{\varepsilon}) = \frac{-\Delta l/3\text{cm}}{300\,\text{s}} = -10^{-5}\,\text{s}^{-1}$$

$$\Delta l = -900\,\text{cm} \times 10^{-5}\,\text{s}^{-1}$$

$$\Delta l = -0.009\,\text{cm} = -0.09\,\text{mm}$$

Thus, at a strain rate of $-10^{-5}\,\text{s}^{-1}$, the limestone specimen of original length 3.00 cm will shorten by approximately one-tenth of one millimeter in the first 5 minutes of the experiment. A strain rate of 10^{-5} seems incredibly slow, but actually it is extremely fast in comparison to the strain rates responsible for the development of mountain systems, which are on the order of $10^{-14}\,\text{s}^{-1}$. For "*fun*" you might want to try to calculate how much a limestone specimen of 3.00 cm original length will shorten in the first 5 minutes of shortening at a strain rate of $10^{-14}\,\text{s}^{-1}$.

Most axial compression tests (and for that matter, axial extension tests) are **constant strain rate** experiments. The temperature, confining pressure, and pore fluid pressure are first set, and then the piston, which is seated against the specimen, is moved at a constant rate in a way that serves to shorten the specimen. Strictly speaking, the test is not carried out at a perfectly constant strain rate because each successive fraction-of-a-centimeter of shortening constitutes a comparatively larger increment of *percentage* shortening as the length of the specimen becomes progressively smaller (Twiss and Moores, 1992).

Some axial compression tests are **constant stress** experiments, but once again the name given to the procedure is somewhat misleading because it is the *load* that is kept constant while the rock sample shortens. Even while load is held constant, axial stress may steadily decrease (slightly) during the experiment as the cross-sectional area of the sample increases (slightly) as a result of flattening (Twiss and Moores, 1992). Constant stress and constant strain rate tests are driven by computers in such a way that piston displacement and amount of load are continuously adjusted to maintain the requirements for the experiment.

Example of Standard Compression Test

With this background in mind, let us carry out a compressional deformation experiment from start to finish. We will subject a specimen of limestone to a confining pressure of 28 MPa and then load it end-on. The temperature conditions will be room temperature. The purpose of the test is to observe how the limestone might respond to compression-induced shortening at very shallow crustal conditions. We will load the sample rapidly, such that the whole experiment will take only 10 or 20 minutes.

The initial length (l_o) of the specimen is 2.500 cm, and the radius (r) is 0.635 cm. When the specimen is subjected to 28 MPa confining pressure, the length of the specimen decreases by some tiny amount, approximately 0.025 mm, because of negative dilation. The decrease in radius is negligible. Thus we calculate the cross-sectional area (A) of the specimen on the basis of the initial radius.

$$A = \pi r^2 = 3.1416\ (0.635\ \text{cm})^2 = 1.267\ \text{cm}^2$$

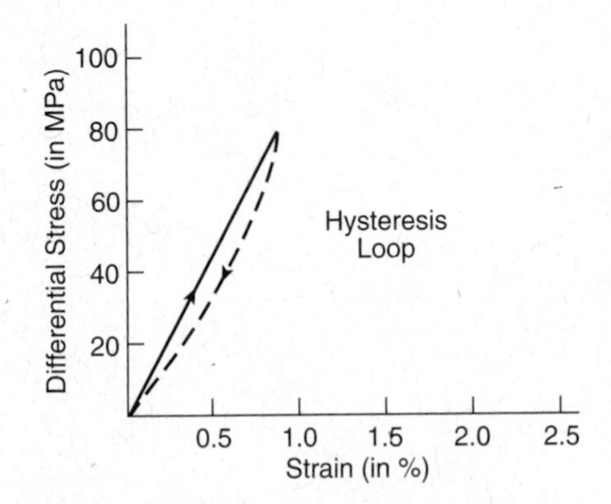

Figure 3.54 Stress–strain diagram for the start of a standard compression test under confining pressure of ~28 MPa and room temperature. The solid black line reflects shortening of the rock cylinder as axial loading takes place. The straight-line character reveals that the rock is behaving elastically. When load is removed, the rock cylinder recovers its original length (dashed line). Recovery is not instantaneous, and this accounts for the loop (i.e., hysteresis loop). See text for explanation.

In the early stage of this standard compression test (Figure 3.54, solid black line), the load–displacement graph is perfectly linear, which indicates that load and displacement are directly proportional. This indicates that the limestone specimen is behaving elastically, like a spring. The elastic behavior is derived from the rock's capacity to recover tiny nonrigid-body changes in atomic spacings in crystal lattices of its mineral components (Ramsay, 1967). During the elastic stage of deformation it is possible to remove the load and observe that the limestone cylinder almost instantly springs back to its original length (dashed line, see Figure 3.54). As load is decreased to 0.0 kg, the load–displacement diagram does not identically retrace the route of its former ascent. Instead, it loops back to its starting position, reflecting a brief time lag in the **recovery** (elimination) of all the bound-up longitudinal strain. Such a retardation is known as **hysteresis**, and the loop displayed on the graph is sometimes called a **hysteresis loop**.

If, instead of dropping the level of stress we continually raise the stress, the limestone may eventually begin to deform plastically. In essence, the range of elastic behavior is surpassed, and non-recoverable permanent strain begins to accumulate in the rock specimen. A "*hook*" develops in the load–displacement and stress/strain curve, signifying departure from a straight-line relationship between stress and strain (Figure 3.55*A*, *B*). **Plastic deformation** produces a permanent change in shape of a solid without failure by rupture.

Anyone who has tried to repair a toy metal *Slinky* that has been stretched too far knows about the non-recoverability of plastic deformation. The life expectancy of a *Slinky* is always short. (Our experience suggests that the half-life of a *Slinky* is about 4 hours). In its all-too-brief youth a *Slinky* can smoothly descend any flight of stairs. Then without warning, it becomes internally entangled. Rescue efforts typically lead to stretching segments of the metal beyond its elastic range, at which point the material becomes permanently plastically deformed. No efforts, however great, can undo the damage.

The onset of plastic deformation during the experiment occurs when the load–displacement curve (or stress/strain curve) departs from its straight-line elastic mode and begins to bend (see Figure 3.55*A*, *B*). The decrease in slope of the curve signifies that proportionally less differential stress is required to produce a given amount of shortening of the limestone. The point of departure from elastic behavior to plastic behavior is called the **elastic limit**. Its value, known as **yield strength**, is measured in load (see Figure 3.55*A*) or differential stress (see Figure 3.55*B*). Below its yield strength, a rock behaves as an elastic solid. Above the elastic limit, the rock begins to flow. If limestone behaved

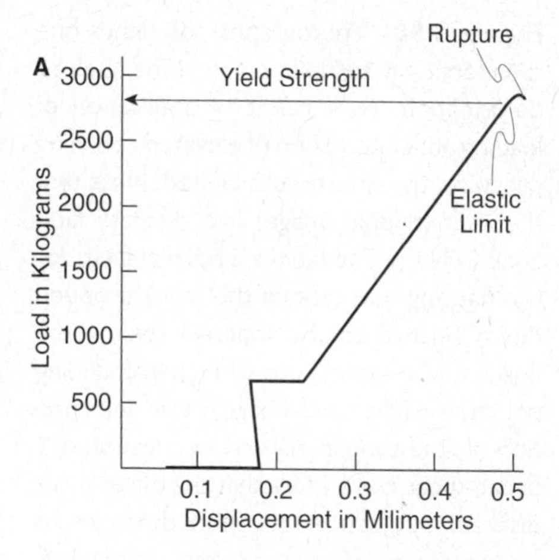

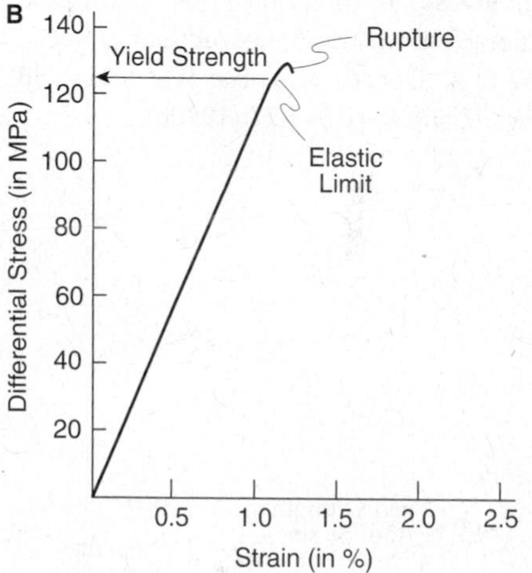

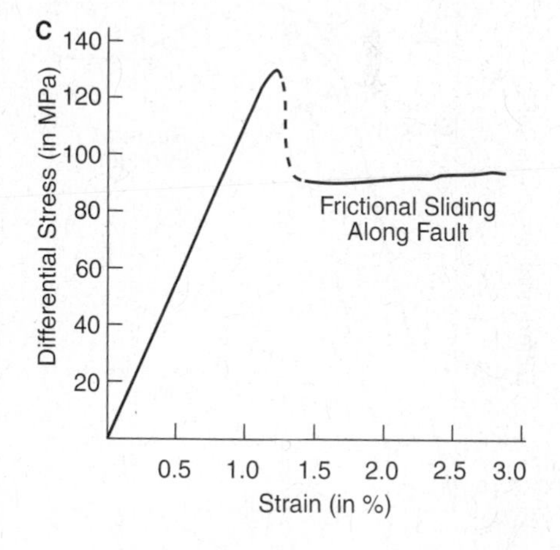

Figure 3.55 (*A*) Load–displacement diagram. Lower part of graph shows seating and reseating of the specimen. Straight line part of the graph reveals elastic deformation of the specimen during axial loading. Tiny bend at the top of the graph signifies departure from elastic behavior, followed almost immediately by rupture (i.e., faulting). The onset of the tiny bend is the elastic limit of this rock under these conditions of loading. The elastic limit is given in terms of yield strength, which in this case is ~2800 kg. (*B*) The same story, but reported in a stress/strain diagram. The rock cylinder was able to shorten elastically by ~1.2%, but then permanent deformation took over. (*C*) This stress/strain diagram reveals what happens when axial loading is continued after the rock cylinder ruptures by faulting. The dashed line shows the drop off in differential stress that accompanied faulting. The nearly flat solid line represents shortening of the faulted specimen, achieved through frictional sliding along the fault surface. The friction along the fault surface is so high that the differential stress is maintained at a relatively high level.

perfectly plastically when axially compressed under conditions of 28 MPa confining pressure, it would flow continuously without rupture as long as the axial stress were applied. However, limestone under conditions of low confining pressure is so brittle that it usually ruptures by faulting almost as soon as plastic deformation begins (see Figure 3.55*A*, *B*). Indeed, the limestone may experience a true **brittle failure** by sudden, abrupt faulting (without warning) in the straight-line range of the stress/strain curve.

Faulting of limestone under low confining pressure conditions usually is punctuated by a muffled "*pop*" within the pressure vessel, an event that is almost always marked by a sudden drop in stress level and by spectator cheering. The fault movement shortens the specimen in such a way that axial stress is at least momentarily relieved. In some experiments we conducted, the axial load dropped so fast that the soft-metal needle of the pressure gauge plummeted to 0 psi and wrapped its tip around the steel peg on which it usually gently rested. However, the drop in pressure is ordinarily modest and is marked on the load–displacement curve and stress/strain curve by a short fishhook-like bend (see Figure 3.55*A*, *B*). When axial load continues to be applied, the level of differential stress usually does not rise to its former magnitude, because even small increments of differential stress create slippage on the fault surface that now cuts the specimen (Figure 3.55*C*).

The standard compression test usually is terminated when the limestone breaks. The confining pressure is bled off and so is the axial load. Before the limestone cylinder is removed from the pressure vessel, the net change in length of the specimen is measured on the basis of the seating position of the piston, before and after deformation. For this test, the shortening was merely 0.025 cm. This is non-recoverable permanent strain. All initial elastic strain is recovered when confining pressure and axial load are removed. Total permanent strain is calculated as follows.

$$e = \frac{\Delta l}{l_o} = \frac{-0.025\text{ cm}}{2.5\text{ cm}} = -0.01, = 1\%\text{ shortening}$$

When the specimen is extracted from the pressure vessel, the imprint of the fault trace is clearly visible on the surface of the copper jacket that surrounds the specimen. The angle made by the trace of the fault with the long axis of the cylinder can be measured with a protractor. For compression tests carried out at low levels of confining pressure, this angle is commonly around 25°, corresponding to an angle of $\theta = 65°$. Upon removal of the copper jacket, the

Figure 3.56 Photographs of limestone cylinders. On far left is limestone, before deformation. As a result of application of loading under condition of elevated confining pressure, the limestone is faulted. Note that the fault (middle image) is a discrete fault zone (white). The fault is a normal fault, for the hanging wall (above the fault) dropped down relative to the footwall (below the fault). Under conditions of higher confining pressure (right-hand image), the deformation of limestone is marked by creation of a broad shear zone. Note that the offset is not discrete, but distributed. Faulted specimens of limestone. [Reprinted by permission, "Some Information Squeezed Out of Rock," by F. A. Donath, *American Scientist*, v. 58, Figs. 7 and 8, p. 54–72 (1970b).]

physical characteristics of the faulted limestone specimen can be examined (Figure 3.56).

Overall, the limestone behaved in **brittle** fashion during this test. Brittle rocks first shorten elastically and then fail by the formation of discrete fractures and faults. Failure may occur abruptly without any plastic deformation. Alternatively, failure by fracturing may follow a modest amount of plastic deformation. For rocks to be considered brittle, the amount of shortening before complete failure by fracturing must be less than 5%.

Influence of Higher Confining Pressure

Testing the response of rocks to stress in the laboratory is like eating potato chips. It is impossible to complete one test without starting on another. It is impossible to squeeze rocks without asking, *What if?* The compression test we just carried out provided information regarding the response of limestone to axial compression under conditions of 28 MPa confining pressure, room temperature, and very rapid loading. But what if we raise the confining pressure to 103 MPa? How would the limestone then respond?

Under confining pressure conditions of 103 MPa, the yield strength (elastic limit) is found to be much higher than before (curve A, Figure 3.57). Above this elevated yield strength, the limestone begins deforming plastically, as reflected in the bowing of the stress/strain curve.

Before taking the axial stress to higher levels, it is interesting to bleed off the entire axial load shortly after the limestone has begun to deform plastically (curve B, Figure 3.57). When this is done, it becomes evident that shortening due to elastic deformation is quickly recovered, but shortening due to plastic deformation is non-recoverable. Plastic deformation is bound up in the strained rock permanently. When load is reapplied to the specimen, the limestone once again behaves as an elastic solid, but this time the elastic limit is higher than the first yield strength we observed (curve C, Figure 3.57). The yield strength of the rock increases because the original fabric of the rock was modified slightly by the plastic deformation. The limestone is said to have undergone **strain hardening**, thus raising its yield strength.

But let us now deal the limestone sample its final blow. As more and more load is applied, the specimen is "*pushed*" toward failure. The stress/strain curve displays plastic deformation of the limestone in the form of a smooth, gently sloping curve (D in Figure 3.57). Unlike the test carried out at 28 MPa confining pressure, the onset of faulting does not follow immediately after a small amount of plastic deformation. Instead, the limestone seems to be able to endure a surprising amount of plastic flow. The limestone becomes so weak that the curve begins to descend to the right, steeply, signifying accelerated plastic deformation; this behavior is called **strain softening** because less stress is required to produce each new increment of strain. In real time, there is a feeling

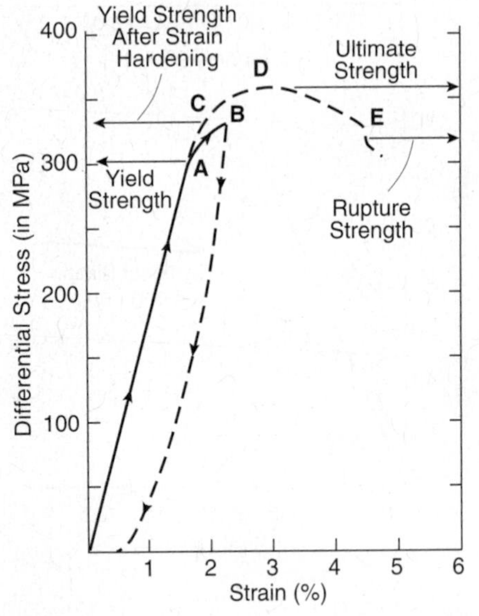

Figure 3.57 Stress/strain diagram for limestone subjected to deformation under confining pressure condition of 103 MPa. Point A marks onset of plastic deformation. Point B marks the removal of axial load. Note the non-recoverable plastic deformation (0.4%). Curve C, extending up from A, represents the now permanently strained specimen being subjected to a second loading. Curve D, plastic deformation. Point *E* marks rupture (i.e., faulting).

of impending doom by rupture. **Rupture** by faulting finally occurs, and stress drops abruptly (curve E, Figure 3.57).

The stress/strain curve accurately records the details of this deformational experience (see Figure 3.57). Above the yield strength the curve is convex upward, recording the gradual dissipation of rock strength during plastic deformation. The zenith of the curve corresponds to the **ultimate strength** of the limestone under the experimental conditions. The stress level at which faulting occurs is the **rupture strength**. Ultimate strength and rupture strength, like yield strength, are measured in terms of stress (e.g., in megapascals, MPa).

The results of our series of tests reveal that limestone becomes stronger at higher levels of confining pressure. This holds true regardless of the parameter used to describe the strength: yield strength, ultimate strength, and rupture strength are all higher for the test carried out at 103 MPa than for the test at 28 MPa. Moreover, the limestone undergoes greater plastic deformation at higher levels of confining pressure, assuming that temperature, rate of loading, and other such factors are held constant. The angle that the trace of the fault makes with the long axis of the core is seen to be greater, about 30°. (This corresponds to $\theta = 60°$). The specimen of limestone that had faulted under confining pressure conditions of 28 MPa was seen to be very fragile, somewhat powdery along the fault, and broken by a single fracture. In contrast, the specimen subjected to rupture at 103 MPa is more cohesive, slightly barrel shaped, and affected by a relatively wide zone of distributed fault surfaces, one of which accommodated the ultimate rupture. Under the higher confining pressure conditions, a greater volume of the rock was affected by the deformation. The same effect is achieved through increases in total strain under constant (low) confining pressure (Figure 3.58).

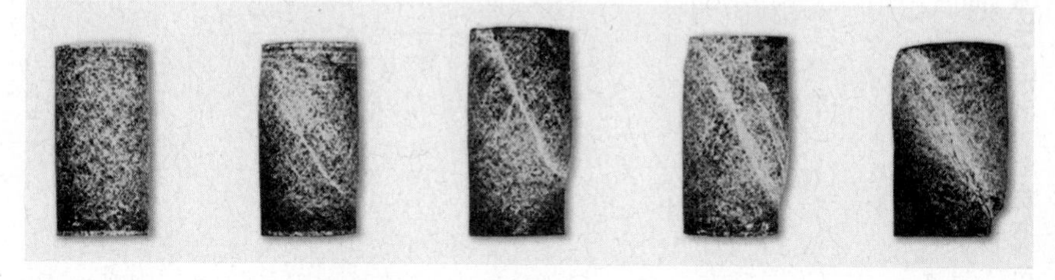

Figure 3.58 Specimens of limestone deformed to increasing total strains (2.3%, 4.5%, 6.7%, 8.6%, and 14.7%, respectively) under the same confining pressure. Increased strain caused a transition in deformational mode from brittle (toward the left) to more ductile (toward the right). [Reprinted by permission, "*Some Information Squeezed Out of Rock*," by F. A. Donath, *American Scientist*, v. 58, pp. 54–72 (1970b).]

In the above test, the rock responded in **semi-brittle** fashion. Semi-brittle rocks deform initially by elastic deformation, and they ultimately fail by the formation of discrete fractures and faults. However, the amount of plastic deformation following elastic deformation and preceding failure is much greater than that for brittle rock. For rock to be considered semi-brittle, the amount of shortening before complete failure must be between 5 and 10%.

Influence of Still-Higher Confining Pressure

When compression of the limestone is carried out at even greater confining pressure, for example 207 MPa, the limestone responds with even greater strength and ductility. There is still an initial, though limited, elastic deformation, followed by sustained plastic deformation before failure. The limestone responds in a ductile fashion. Ductile rocks have nearly continuously curved stress/strain curves. They may initially deform elastically, but if they do, it is seldom long before they begin to respond plastically. They may or may not ultimately fail by the formation of discrete fractures or faults. For rock to be considered ductile, the ultimate failure, if it does occur, takes place after the rock has shortened by at least 10%.

TABLE 3.4
Summary of compressional testing of limestone under confining pressure conditions of 27.6 MPa, 103 MPA, and 207 MPa.

	Specimen 1	*Specimen 2*	*Specimen 3*
Confining pressure	28 MPa	103 MPa	207 MPa
Differential stress at failure	124 MPa	318 MPa	552 MPa
σ_1 at failure	152 MPa	421 MPa	759 MPa
σ_3 at failure	28 MPa	103 MPa	207 MPa
Angle between fault and σ_1	25°	30°	33°

Table 3.4 presents a summary of the conditions, measurements, and results of compression tests on limestone at confining pressure levels of 28, 103, and 207 MPa. When these data are all plotted on a Mohr stress diagram (Figure 3.59), it becomes even more obvious that proportionately greater levels of differential stress ($\sigma_1 - \sigma_3$) are necessary to cause rupture when confining pressure (σ_3) is raised. In *Chapter 5, Joints*, and in *Chapter 6, Faults*, we will see that composite Mohr diagrams of this type become the basis for establishing the laws that describe the conditions under which rocks fail by fracturing and faulting.

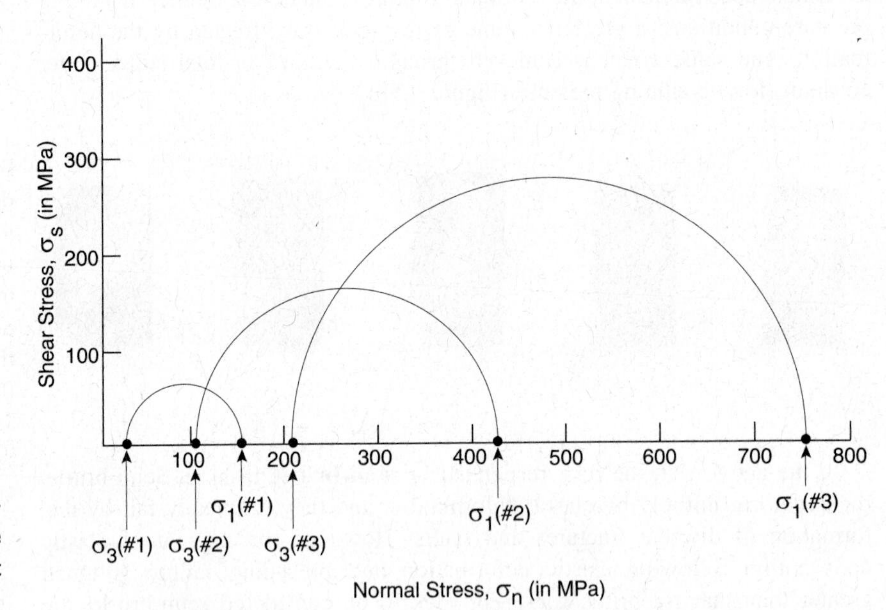

Figure 3.59 Mohr stress diagram summarizing the experimental testing of the limestone under three distinctly different confining pressure conditions.

EVALUATING MECHANICAL BEHAVIOR DURING TESTING

Assessing Strength and Ductility

The mechanical response of rocks to stress is different for different conditions. If we were to examine hundreds of stress/strain diagrams generated during laboratory testing of every rock imaginable under every conceivable set of conditions, we would see significant differences in strength and ductility from graph to graph. One of the chief conclusions that can be drawn from experimental compression tests is that measurements of rock strength and ductility are almost meaningless unless the conditions under which the deformation was achieved are also given. Furthermore, each variable such as temperature or strain rate, has a different effect on the rheology of the rock being tested.

Role of Lithology and Rock Composition

All other things being equal, it is possible to arrange the common rock lithologies in order of increasing strength for specific conditions, such as confining pressure, temperature, or rate of loading. **Stress/strain diagrams** provide the basis for the rankings. The rankings are only approximate, being strongly influenced by the composition, texture, and general condition of the rocks that happened to serve as "*representative*" test specimens for each lithology. Moreover, the nature and orientation(s) of mechanical **anisotropy** resulting from such features as fractures, layering, foliations, cleavage, and the like profoundly influence rock strength (see chapter-opening photograph).

Table 3.5 lists lithologies according to strength based on stress/strain diagrams summarizing compression tests of rocks at room temperature under low levels of confining pressure. Rocks like salt, anhydrite, shale, and mudstone are seen to be weak (and ductile). Salt, for instance, is almost entirely devoid of strength. Other rocks, such as limestone or calcite-cemented sandstones, are of intermediate strength. Quartzite, granite, and quartz-cemented sandstone are brittle and very strong by comparison.

In a sequence of different lithologies, the rocks that are likely to behave in the most ductile manner when subjected to stress are commonly referred to as **incompetent**. Rocks that are likely to deform in a brittle manner, with no obvious ductile deformation, are described as **competent**. "*Competency*" and "*incompetency*" are relative terms. The ordering of lithologies by competency may change if the conditions of deformation are changed. Changes in confining pressure, temperature, strain rate, and the presence of pore fluid may affect different rocks in different ways.

TABLE 3.5
General ranking of lithology according to strength, based on tests at room temperature and low confining pressure.

Strongest	Quartzite
	Granite
	Quartz-cemented sandstone
	Basalt
	Limestone
	Calcite-cemented sandstone
	Schist
	Marble
	Shale/mudstone
	Anhydrite
Weakest	Salt

Role of Confining Pressure and Pore Fluid Pressure

We have seen that increasing the confining pressure on a rock specimen in a compression test has the effect of increasing the strength and ductility of the rock. This was firmly documented long ago through a flurry of experimental testing of rocks (e.g., see Handin and Hager, 1957). Petroleum research divisions at that time were intensely exploring the fault and fracture properties of sedimentary rocks through triaxial testing. Much was learned, including the fact that for any given lithology, the yield strength, ultimate strength, rupture strength, and ductility attain greater and greater values with increasing confining pressure. For example, the stress/strain curves shown in Figure 3.60 reveal how the strength and ductility of the Crown Point limestone increase with increasing confining pressure.

The effect of confining pressure can be partially or completely offset by the presence of elevated **pore fluid pressure** in the rock (or test specimen) undergoing deformation. Figure 3.61 shows this nicely by means of graphs constructed by Handin et al. (1963) on the basis of a series of compressional tests on Berea sandstone. The graphs reveal that elevated fluid pressure can dramatically decrease ultimate strength, rupture strength, and ductility. In natural sedimentary basins of accumulation, water that is entrapped in sediments during deposition may be pressurized during the course of subsidence, burial, and compaction due to loading by overlying, younger sediments. As a result, a pore fluid pressure may be achieved that exceeds the hydrostatic stress level we would expect at that depth (Hubbert and Rubey, 1959).

Elevated hydrostatic pore fluid pressure conditions counteract the effects of confining pressure on strength and ductility. A measure of the net effect of

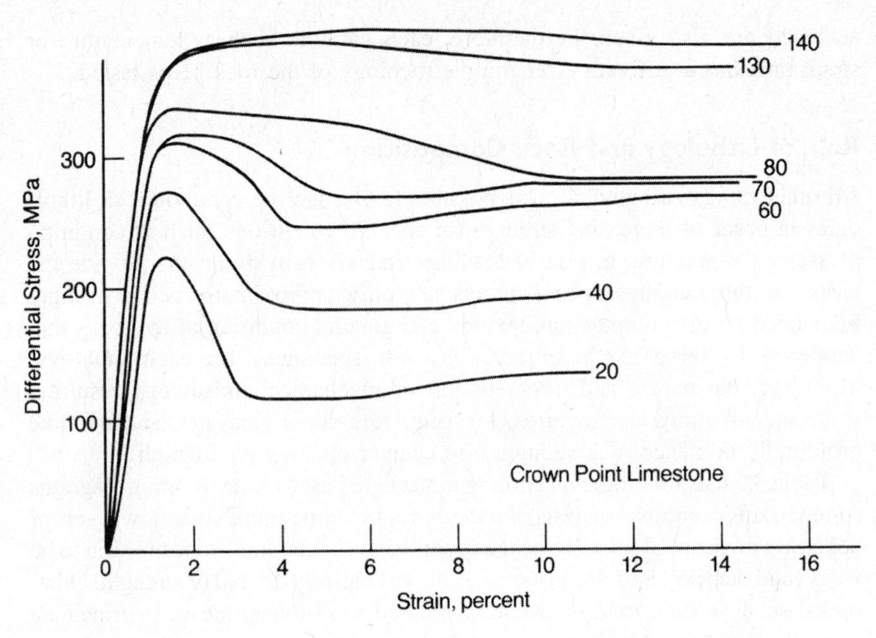

Figure 3.60 Stress/strain diagrams for Crown Point Limestone deformed at a variety of confining pressures. Tests conducted at room temperature. The magnitude of confining pressure (expressed in MPas) for each run is shown next to each curve. Both strength and plasticity increase with greater confining pressure. [Reprinted by permission, *"Some Information Squeezed Out of Rock,"* by F. A. Donath, *American Scientist*, v. 58, p. 54–72 (1970b).]

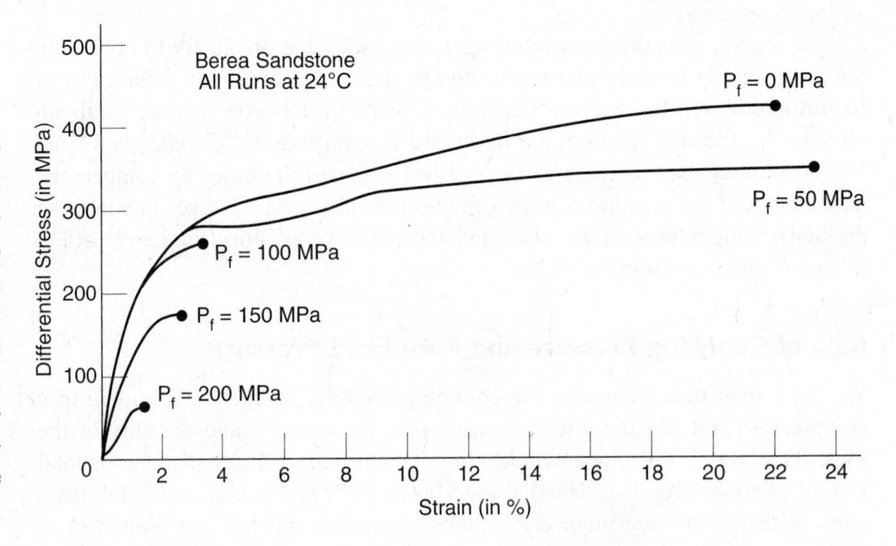

Figure 3.61 Laboratory-derived stress/strain curves showing the influence of pore fluid pressure on the strength and ductility of rocks. All tests were carried out at 200 MPa confining pressure. Fluid pressure (P_f) for each run is shown next to each curve. Based on experimental work by Handin et al. (1963). [From Marshak and Mitra, *Basic Methods of Structural Geology*, copyright © 1988, p. 201. Reprinted by permission of Prentice Hall, Upper Saddle River, New Jersey.]

confining pressure and fluid pressure is **effective stress**: effective stress equals confining pressure minus fluid pressure.

If effective stress is high, a given rock will be relatively strong and ductile; if effective stress is low, the rock will display less strength and ductility. A deeply buried sedimentary rock with anomalously high pore fluid pressure will respond to stress as if the rock were deforming in a relatively low confining pressure environment.

Role of Temperature

An increase in the temperature of a rock generally depresses yield strength, enhances ductility greatly, and lowers ultimate strength. Some rocks are more sensitive to the effects of temperature than others. Igneous rocks are less affected by modest increases in temperature than sedimentary rocks (Ramsay, 1967); they are more at home in high-temperature environments. Figure 3.62 presents a typical example of the profound influence of temperature on strength and ductility.

If heated sufficiently, rocks may deform in a plastic or viscous fashion and thus undergo very large permanent strains without ever rupturing or losing cohesion. Viscous materials, in effect, flow when subjected to *any* differential stress, no matter how weak. Unlike materials that deform plastically, truly viscous materials possess no fundamental strength threshold that otherwise would have to be overcome to induce flow. Elevated temperatures promote viscous deformation and cause rocks to flow.

Very seldom do upper crustal rocks behave in ideally viscous fashion. Even though converted to viscous fluids by greatly elevated temperature conditions, most rocks retain the capacity of behaving elastically under relatively rapid strain rates. One of the best examples of a material that possesses this dual capacity of responding both elastically and viscously is *Silly Putty* (i.e., silicone putty), which flows under conditions of very small differential stress. Yet, if subjected to rapidly applied, relatively high stresses, it will behave elastically and even break in tension.

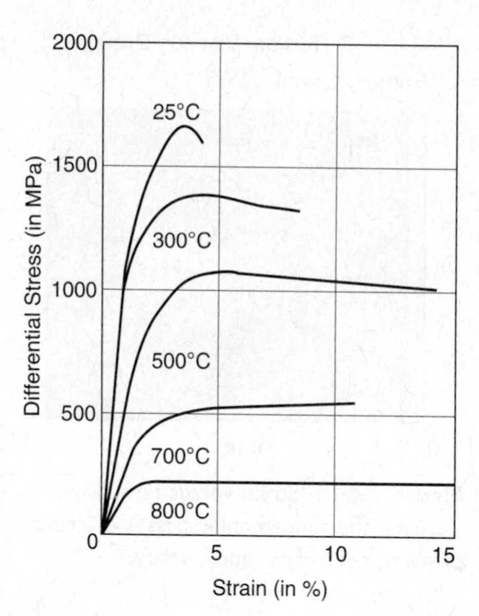

Figure 3.62 Stress/strain diagram for basalt deformed at 5-kbar confining pressure under a variety of temperature conditions. [From Griggs, Turner, and Heard, 1960, Published with permission of the Geological Society of America.]

Role of Strain Rate

Rock strength as measured in deformation experiments is partly a function of the rate at which the stress is applied. A rock can be forced to deform plastically at comparatively low levels of stress if the rate of loading is slow. Heard (1963) helped to verify this principle quantitatively through deformational experiments in which conditions were identical in every way except for the rate of deformation (Figure 3.63).

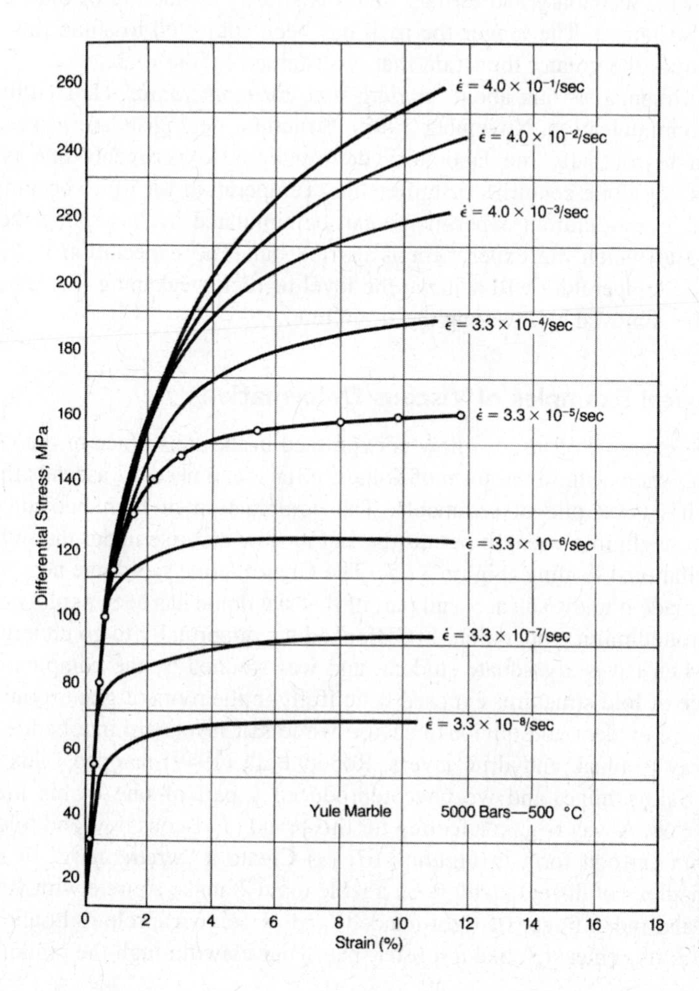

Figure 3.63 Stress/strain diagram for Yule marble deformed at different strain rate conditions. The higher the strain rate, the stronger the rock. [After Heard, 1963, Figure 18, p. 185, Copyright © 1963 by the University of Chicago. All rights reserved.]

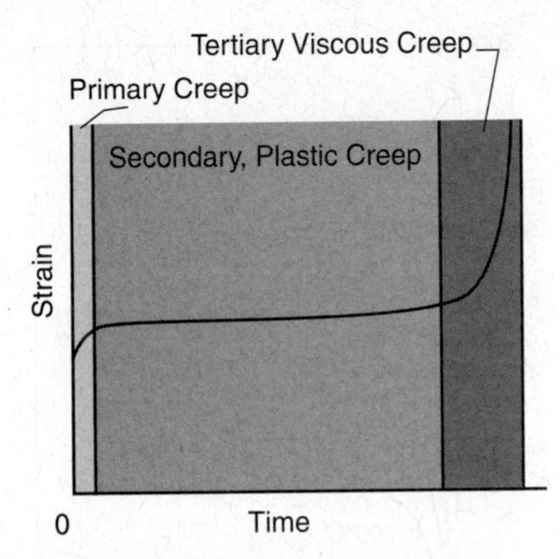

Figure 3.64 Strain versus time diagram, showing the fundamental classes of creep: primary, secondary, and tertiary.

In testing the influence of strain rate on rock strength, we can compare values of yield strength, ultimate strength, and rupture strength as a function of the rate of loading. But normal practice is to carry out a constant strain test, wherein the rate of strain is held constant. When the strain rate is relatively low, the amount of stress required to produce plastic deformation and ultimate failure is smaller than for experiments with higher strain rate.

The observable decrease in rock strength as a function of long-sustained stress is not too surprising. It can be thought of as a fatigue that sets in with time. Stress fractures in athletes are a human manifestation of the same phenomenon.

Engineers and geologists have carefully studied the time-dependent strain produced under conditions of low differential stress. The name given to this kind of strain is **creep**. Creep is the strain produced in experiments of long duration under differential stresses that are well below the rupture strength of the rock. Results of the tests produce consistent plots, like those shown in Figure 3.64. As soon as the load is applied, the rock experiences an elastic deformation. This is followed by three distinctive kinds of mechanical response, called **primary, secondary**, and **tertiary creep**.

Primary creep is a slightly delayed elastic deformation in which there is a general decrease in the amount of strain with time. This is followed by secondary, steady-state creep, a plastic deformation in which strain and time of loading are linearly related. Finally the rock, still under constant load, dramatically fatigues and discloses an accelerating rate of strain. This leads to failure by rupture. Tertiary creep approximates viscous deformation.

The initial elastic strain and primary creep are due to initial loading and are not particularly time-dependent. The amount of strain accommodated by the rock during secondary and tertiary creep is strictly a function of the time the load is sustained. The longer the rock has been subjected to some tiny stress differential, the greater the strain that is sustained by the rock.

Experimentalists talk about "*trading time for temperature*" (Jan Tullis, personal communication, November 1992). Structural geologists are just not in a position to routinely run laboratory deformational experiments that last for decades, let alone centuries or millennia. 'Temperature for time' becomes the tradeoff. Long-duration experiments can be simulated by increasing the temperature at which the experiment is carried out. The expectation is that the increased temperature will achieve the level of rock weakening that otherwise would be achieved by lowering the strain rate.

Geological Examples of Viscous Deformation

A prime example of viscous flow is expressed in the subsurface of the Gulf of Mexico, where salt, in the form of domes, pillars, and needles, ascends through the Gulf's thick pile of sediments. The combination of high buoyancy and inherent negligible strength permit the salt to flow and rise under the influence of gravitational loading (Figure 3.65). The Grand Saline salt dome near Dallas, Texas, made it to the surface, and part of this salt dome has been exploited in an underground mining operation. I (GHD) had the opportunity to go underground there when I was a graduate student, and was stunned by the complexity and elegance of fold structures exposed beautifully in the roofs of giant rooms. The folds were evident and obvious because white salt layers are interbedded with dark gray to black anhydrite layers. Robert Balk (1949) mapped folds in the Grand Saline mine, and we have reproduced a part of one of his maps as Figure 3.66. A way to gain a feeling for this brand of viscous flow and folding is shown in cartoon form in Figure 3.67: (*1*) Create a "*stratigraphy*" of 3 or 4 paper napkins of different colors on a table top; (*2*) make a circle with your left thumb and index finger (if right-handed) and place your circle right above the napkin at its center; (*3*) like a lobster, pass your claw through the circle, pinch

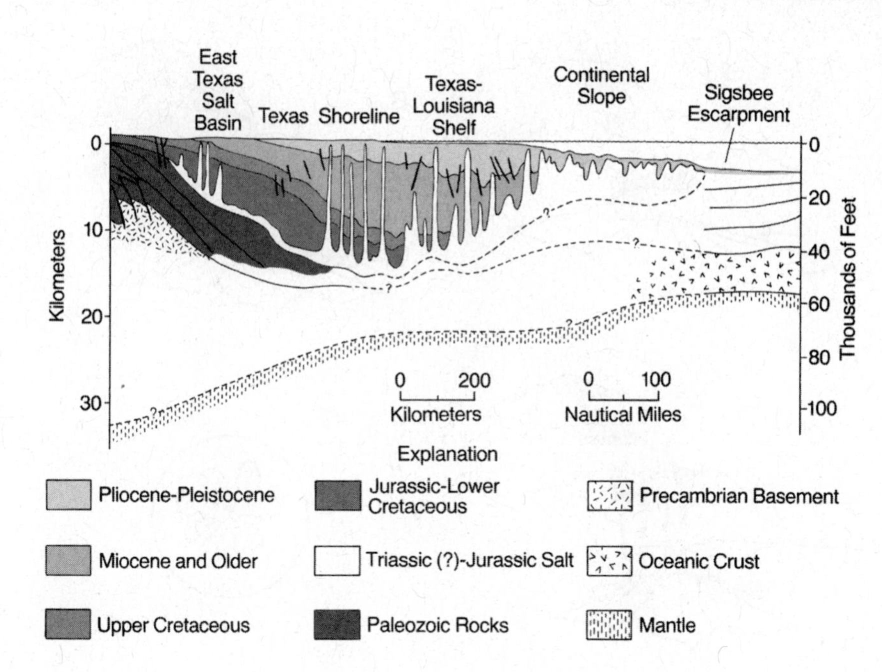

Figure 3.65 Cross-sectional profile of salt domes and salt spines in the Gulf Coast region. [From Martin (1978). Figure 4, p. 27. Published with permission of American Association of Petroleum Geologists.]

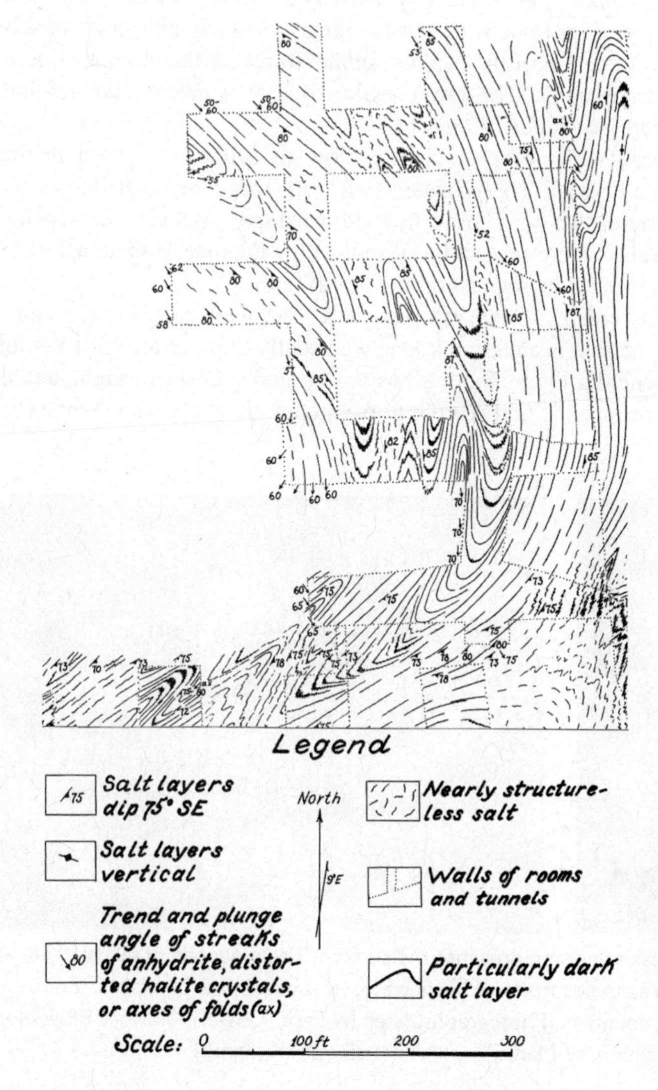

Figure 3.66 Geologic map of folds in salt in part of the subsurface of the Grand Saline salt dome, Texas. Balk, R., 1949, Structure of Grand Saline salt dome, Van Zandt County, Texas: *American Association of Petroleum Geologists Bulletin*, v. 33, p. 1791–1829. AAPG © 1949, reprinted by permission of the AAPG whose permission is required for further use.]

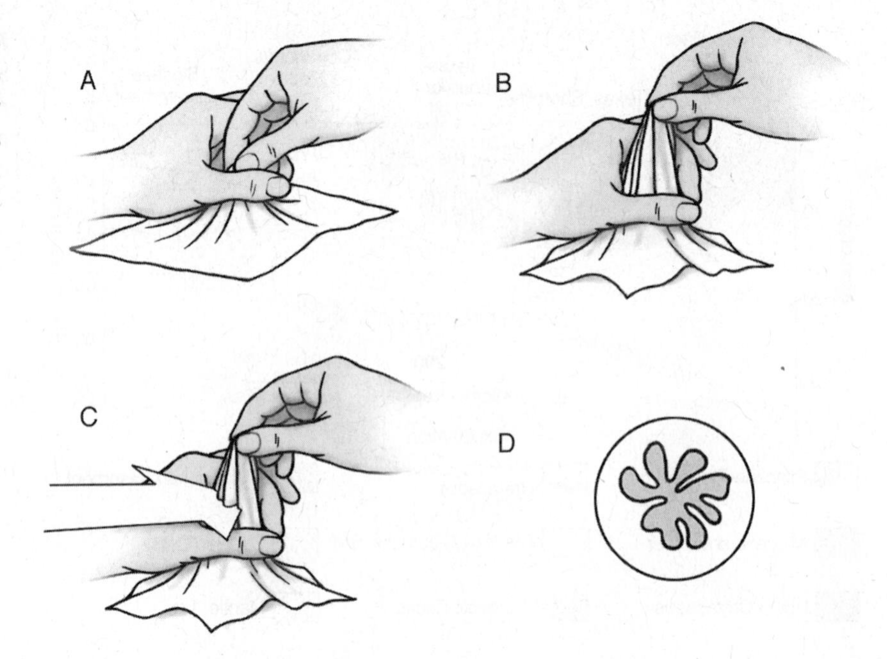

Figure 3.67 Cartoon showing how to simulate viscous flow in a salt dome, and to experience how the constrictional flow creates the distinctive internal fold patterns within salt domes. See text for explanation. [Artwork by D. A. Fischer.]

the napkin, and draw the napkin slowly but surely through the opening; (*4*) watch the napkin endure constrictional flow, accommodating the flow by creation of dozens of folds with vertical hinges; and (*5*) take a pair of scissors and cut across the grain of the napkin "*pillar*" to reveal the internal structure of the napkin stratigraphy. The cross-section that you create may remind you of Balk's map (see Figure 3.66).

A more exotic form of viscous flow of salt can be seen in the Zagros Mountains of Iran. The climate is so arid, that when salt domes breach the surface there, the salt simply flows downslope, viscously, driven by gravity. Though salt is more resistant to viscous flow than ice, it nonetheless is capable of flowing under its own weight (Figure 3.68).

Viscous creep is seen in "*real rocks*" as well, not just ice and salt. For example, marble is a real rock, one we usually think of as rigid. Yet look at the marble bench in Figure 3.69. It has bowed under its own weight, and that of its periodic occupants. Old marble grave stones show the same tendency.

A

B

Figure 3.68 (*A*) Photograph of where the Kuh-i-Namak salt dome breaches the surface. The mountain of salt is 10 km across and 1340 m high. The contact between salt and the Cretaceous rocks it penetrates is overturned in places. Location is Iran. (*B*) Photograph of the salt flowing down the flank of the mountain. [Photographs taken by Peter Llewellin (former BP geologist) from a helicopter, back in the 1970s, and brought to my attention by Mark Allen (University of Durham).]

Figure 3.69 Marble bench bent downward by its own weight and that of the occasional occupant. This marble bench is located in cemetery north of Soldiers' Home, Washington, D.C. Note that support was added underneath. We especially like the apparent puzzlement of the gentleman. And we like his 'threads.' [Photograph taken in 1925 by W. T. Lee. Courtesy of United States Geological Survey.]

Role of Preexisting Weaknesses

In preparing samples for rock deformation experiments, structural geologists are normally meticulous about using pristine, fresh rocks that have no visible cracks or other flaws. It is thought that the presence of a crack or fracture will ruin the experiment. And yet we know that outcrop-scale bodies of rock and regional-scale bodies of rock are absolutely pervaded by flaws, and at all scales: from spaced, through-going faults, fractures, and bedding surfaces; to pervasive joints and veins and stylolites; to webs of microscopic hairline cracks and voids. If we were somehow able to extract a 1 km^3 cube of granite from a very deep quarry at the surface of the Earth and place it on a solid surface of infinite strength, the cube of granite would probably simply fall apart, as if it lacked any cohesion.

The brilliant writer of creative nonfiction, Annie Dillard, tumbled onto the paradox of apparent strength in highly fractured rocks before she graduated from junior high school. She writes the following in *An American Childhood* (1987, p. 140):

> *The awesome story of earth's crust's buckling and shifting unfortunately failed to move me in the slightest. But here was an interesting find. Only a quirk of chemistry prevented the ground's being a heap of broken rubble. I hadn't thought of that. Why isn't it all a heap of broken rubble? For the bedrock fractures and cleaves, notoriously; it uplifts, crumbles, splits, shears, and folds. All this action naturally shatters the crust. But it happens that the abundant element silicon is water soluble at high temperatures. This element heals the scars. Dissolved silicon seeps everywhere underground and slips into fissures and veins; it fills in, mends, and cements the rubble, over and over, from age to age. It heals all the thick wounds on the continents' skin and under the oceans; it solidifies as it cools, uplifting, and forms pale veins of sparry quartz running through everything; it dominates the granite bedrock on which we build our cities, the granite interior of mountains, and the beds that underlie the plains.*

To be sure, quartz and other vein minerals (commonly calcite veins in limestones) strengthen rock bodies. However, still-unhealed fractures and joints abound. We find that contemporary engineering literature is full of studies of estimating the strength characteristics of large bodies of jointed and fractured rocks, having found that such macroscopic assessments of rock strength are essential in design and mining.

Without question, larger bodies of rocks are weaker than smaller bodies of rock, in part because of preexisting internal flaws and weaknesses. The influence of the presence of such weaknesses will be greater where the conditions of deformation are fundamentally brittle. Where the conditions are ductile, the whole body of rock tends to be affected by the deformation, and the influence of individual preexisting weaknesses like fractures and faults and bedding surfaces will not be as great.

As we shall see in *Chapter 6, Faults,* the ease or difficulty with which preexisting fracture surfaces can be activated depends on the combination of friction and normal stress. When normal stresses on preexisting fractures are low, the friction along the fracture surfaces is dependent upon surface roughness (Byerlee, 1967, 1978). With higher normal stress, the friction along fracture surfaces becomes independent of rock type.

Size and Scaling

Size alone has a lot to do with strength as well, as dramatically illustrated in the scaled deformation experiments carried out by structural geologists such as Vendeville et al. (1987), McClay (1989) Merle and Guillier (1989), and Brun et al. (1994). When the sizes and strengths of materials chosen to replicate regional relationships, such as all of the Himalayas or the full length of the Andes, are precisely scaled (Hubbert, 1937), the materials chosen are extraordinarily weak. If the materials chosen are too strong, they resist the deforming influence of gravity that attends all natural deformation. If the role of gravity is missing from experimental deformation, the structures that are formed will not bear exact similarity to what is found in nature. Ramberg (1967) overcame this problem by increasing the force of gravity in his models by deforming the clays while they whirled at high *G*s in a centrifuge. Ramberg ingeniously filmed the deformation live by mounting a TV camera inside the centrifuge!

Looking for a more versatile, less expensive, and less complicated solution, Peter Cobbold "*scaled down*" the strength of the materials even further, and in a way that conformed *precisely* to Hubbert's requirements. For the crust of the Earth he selected loose sand, and for the upper mantle directly below it, *Silly Putty.* For the softer mantle he chose Czechoslovakian honey, which is renowned for its perfect transparency. Then he fashioned these materials in stratified layers to form meter-scale physical models mechanically equivalent to a regional tract of the Earth's outer crust and upper mantle. The structures resulting from compressional shortening or extensional stretching of such multilayered, scaled models bore an absolutely uncanny similarity to what is found in regional systems of structures (Figure 3.70). Cobbold adds a final touch. To examine the structure of the base of the crust, he simply vacuums up the sand so that the top of the *Silly Putty* layer comes into view. To examine the structure of the base of the stiffer mantle, he looks up at the base of the *Silly Putty* layer through the transparent Czechoslovakian honey. To examine internal structure within the deformed upper crust, Cobbold adds some cement to the sand before it is stratified into layers. Then, upon completion of the experimental deformation, he adds a little water to the sand, lets it harden, and saws the deformed, scaled-down version of the upper crust along cross-sections of choice.

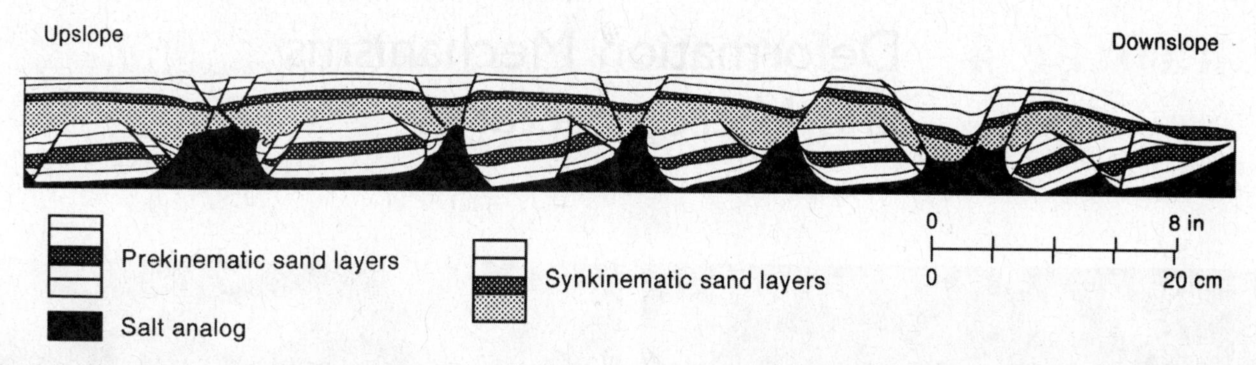

Figure 3.70 Deformation model produced through stretching of an original "*stratigraphy*" consisting of viscous silicone (black, representing salt) and sand "*deposited*" before stretching (the prekinematic sand layers). More sand was deposited while stretching proceeded (synkinematic sand layers). Stretching was achieved by spreading and gliding of material down a 2° slope. Grabens formed into which silicone diapirs rose and intruded. Total finite stretching was 75%. [Reprinted from *Marine and Petroleum Geology*, v. 9, Vendeville, B. C., and Jackson, M. P. A., "The rise of diapirs during thinned skinned extension," p. 331–353, © 1992, with permission from Elsevier.]

CONCLUSIONS

We commonly presume that the magnitude of the largest stresses that can be generated in the crust is directly related to the magnitude of the forces that create the stresses. However, the really important limitation is the strength of the rocks themselves: the weaker the rocks, the lower the stresses that can be achieved and sustained. In the upper crust, where rock temperatures are not especially high, the strength and ductility of rocks increase with depth as confining pressure is increased. The rocks behave as elastic bodies, and they deform in a brittle or semi-brittle manner. However at deeper levels, where the temperatures are higher, the rocks are weaker and more ductile, and this causes the rocks to behave as plastic or viscous bodies. The stress levels that can be achieved and sustained are quite low, especially at tectonic rates of loading.

The empirical behaviors of rocks are now well established. Stress/strain curves are becoming more and more predictable. But what is really happening? What is really happening *inside* the rock that permits it to behave elastically in one situation, semi-brittley in another, and plastically or viscously in yet another? We really must go *inside* to understand what is going on. We must move to the scale of individual crystals, and lattices of crystals, to discover how rocks can behave elastically under one set of conditions, and plastically or viscously under some different set of conditions. This is the subject of the next chapter.

Chapter 4 Deformation Mechanisms and Microstructures

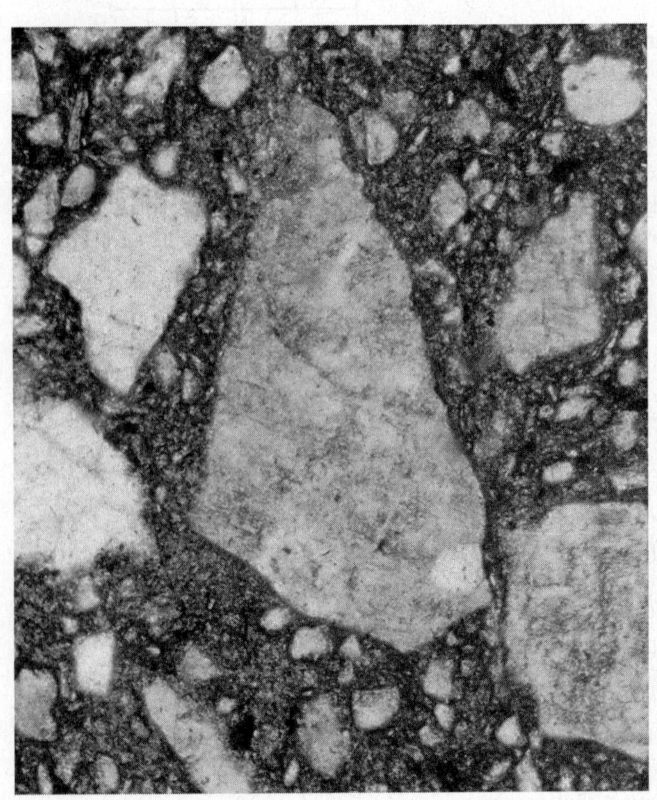
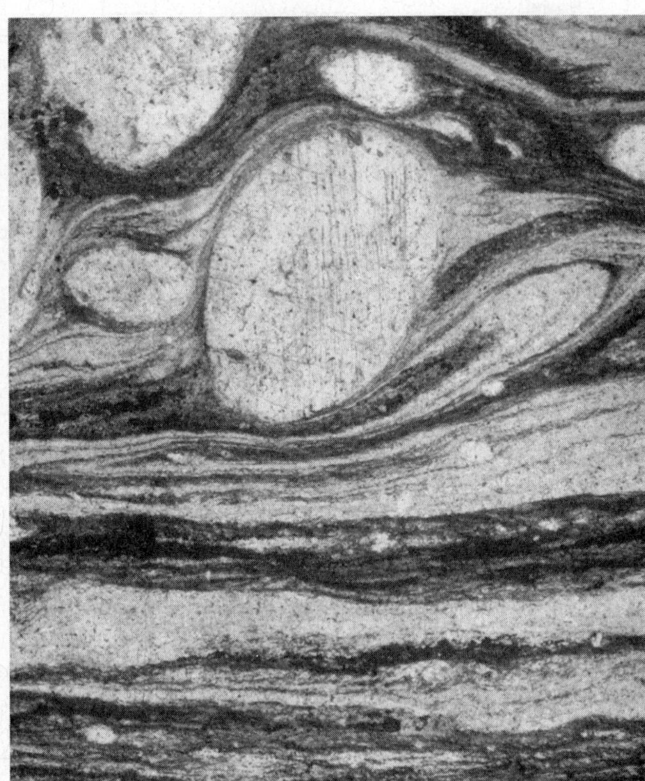

Photomicrographs of fault rock produced from deformation of granite. This double-feature shows 1-mm-tall windows into microstructures of granite deformed under contrasting sets of conditions. On the left is a photomicrograph of a rock collected from the wall of a major normal fault in the Catalina Mountains, Tucson, Arizona. It is a micro-breccia, composed of angular fragments of broken and crushed granite, dominated by the quartz and feldspar. The granite was deformed under brittle conditions of cataclasis at a depth of ~4 km. On the right is a photomicrograph of a rock collected from a shear zone in the Rincon Mountains, Tucson. It is an ultramylonite, dominated by rounded porphyroclasts of feldspar and ribbons of quartz. This granite was deformed by crystal-plastic flow at a depth of ~9 km. Temperature, pressure, strain rate, fluids, and the available deformation mechanisms make all the difference in how rocks accommodate deformation. [Photographs by G. H. Davis.]

EXPLORING AT THE FINE SCALE

The broad range of structures that we see at the scale of individual outcrops and hand specimens (Figure 4.1) draws our attention to the interior of rocks and mineral grains to see what is going on. We go "*inside*" to explore what processes operate at the microscopic and atomic scales that enable rocks, which seem so hard, to change size and shape as if they were soft. We journey inside to determine what mechanisms are available to give rocks such versatility in deforming elastically, plastically, or viscously, depending on conditions. By going inside for a while, we emerge with a better understanding of how deformation is actually achieved.

Processes that permit rocks to deform at the microscopic and atomic scales are **deformation mechanisms**, and the resulting fine-scale structures are **microstructures**. Nature has a diverse menu of deformation mechanisms to choose from, and this accounts for the wide variety of observed microstructures and textures. Some deformation mechanisms are brittle, like the splitting of minerals along microscopic cracks or cleavages; others are ductile, like the slippage on hundreds of parallel crystallographic planes or the forced-march migration of atoms from one side of a mineral to another.

We can get a rough idea of some microstructures in the field, through hands-and-knees, close-up peering at outcrop surfaces through a 10× hand lens. Even

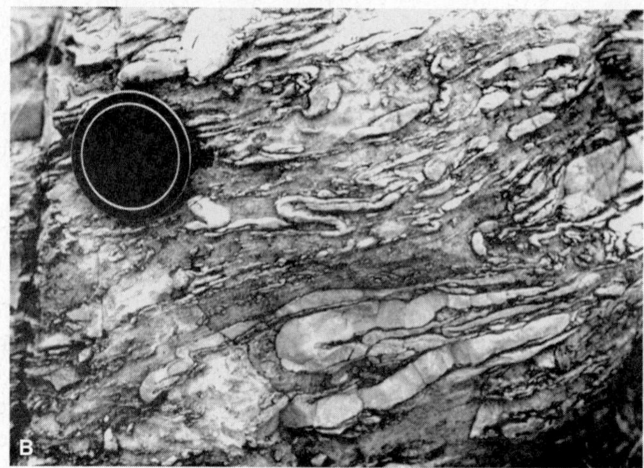

Figure 4.1 Structures in outcrop reveal the influence of different deformation mechanisms. (*A*) Breccia texture within a Tertiary landslide, Hieroglyphic Mountains, central Arizona. (*B*) Folded pebbles in a deformed carbonate-clast conglomerate, Rincon Mountains, Arizona. (*C*) Gneissic and mylonitic foliations formed by ductile mechanisms in banded gneiss, Chemiheuvi Mountains, southeastern California. [Photographs by S.J. Reynolds.]

more information comes from a thorough examination of small specimens using a petrographic microscope or an ultra–high magnification transmission or scanning electron microscope.

From the observed microstructures, we want to infer which deformation mechanisms operated during deformation. To take this interpretive step, we deform materials under controlled laboratory conditions, describe the microstructures on the basis of high magnification microscopy, compare the microstructures produced in the laboratory to those produced in nature, and apply concepts and models consistent with theoretical analyses of the strength of minerals. In some cases, we can deduce the step-by-step evolution of microscopic deformation by means of controlled, start-and-stop experiments.

We begin with a discussion of how crystalline structure and interatomic bonding influence the strength of minerals. We then present the "*players*," that is, the geologically important deformational mechanisms. Before the chapter ends, we will be better able to recognize the different types of microstructure when we see them, and to use microstructures as a basis for interpreting the deformation mechanisms and physical conditions that prevailed during deformation.

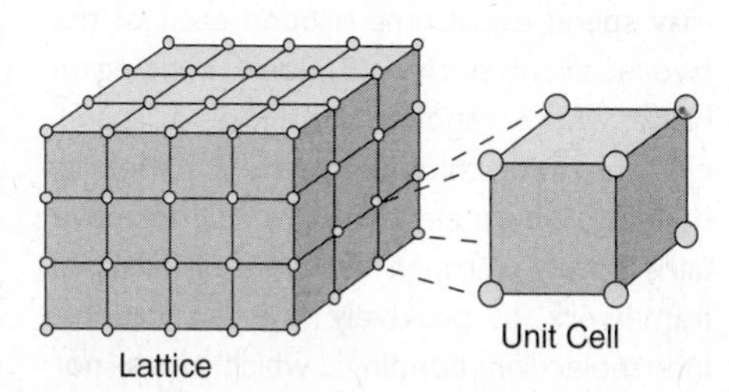

Figure 4.2 Crystalline lattice of a simple solid, showing a unit cell, the smallest known repetitive structure composing a crystal. Spheres represent atoms, ions, or molecules, and connecting lines represent bonds. [Adapted from *Crystalline plasticity and solid state flow in metamorphic rocks: selected topics in geological sciences* by A. Nicolar and J. P. Poirier, copyright © 1976. Published with the permission of John Wiley & Sons, Inc., New York.]

CRYSTALLINE STRUCTURE AND THE STRENGTH OF SOLIDS

Bonding and the Lattice of Crystals

The **lattice** of a crystal is a three-dimensional repetitive array of atoms, ions, or molecules (Figure 4.2). Both the positioning and the spacing of atoms within

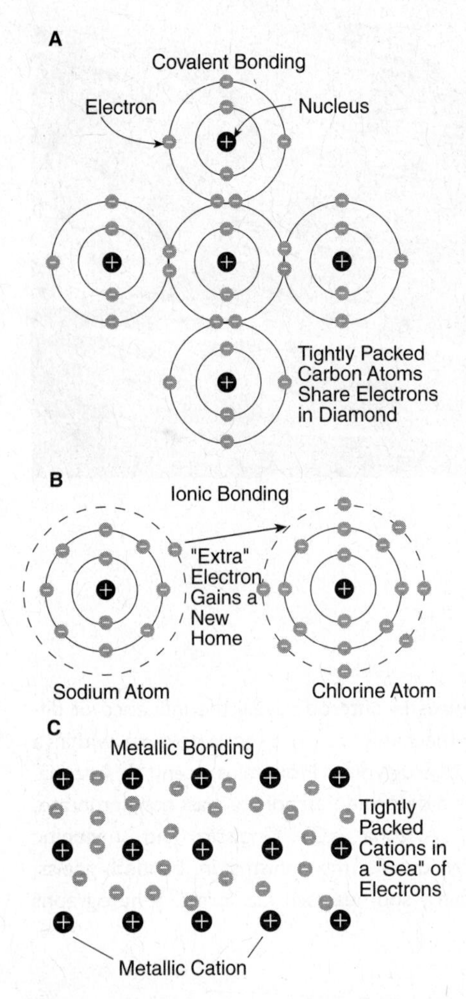

Figure 4.3 The types of bonding in minerals. (*A*) Covalent bonding in diamond, where adjacent carbon atoms share electrons. Electrons are not fixed in position, but instead may reside anywhere within the spherical electron shells. A shared electron may spend equal time around each of the two adjacent nuclei. (*B*) Ionic bonding in halite, where a sodium atom loses a "*spare*" electron to a chlorine atom. (*C*) Metallic bonding, where electrons are able to move fairly freely through a three-dimensional framework of positively charged cations. Intermolecular bonding, which does not involve loaning or sharing of electrons, is not shown

the lattice of a crystal are very systematic. The preferred configuration is one that requires minimum energy to maintain, an equilibrium configuration that reflects the delicate balancing of competing interatomic forces.

The equilibrium configuration of ions of Na^+ and Cl^- in halite is maintained by a dynamic balancing act: the positively charged nuclei of the sodium (Na) and chloride (Cl) atoms repel one another, and the Na^+ and Cl^- ions attract one another because of their opposite charges. The attracting forces tend to hold the atoms together via the process of **bonding**. If the repelling forces exceed the attracting forces, the bonds break, and the individual atoms become mobile, such as when rocks are heated up and begin to melt.

There are four fundamental types of bonding in minerals: covalent, ionic, metallic, and intermolecular. **Covalent bonds** form when two adjacent atoms share one or more electrons. In diamond, adjacent carbon atoms form covalent bonds by sharing electrons with each other (Figure 4.3*A*). **Ionic bonds** form when an atom loses one or more electrons to another atom. Halite (NaCl) is held together by ionic bonds formed when a sodium atom loses a "*spare*" electron to a nearby, electron-hungry chlorine atom (Figure 4.3*B*). **Metallic bonds** exist where electrons are able to move relatively freely through the material, rather than being strongly attached to any particular atom (Figure 4.3*C*). The mobility of the weakly attached electrons gives metals like copper their capacity to conduct electricity. **Intermolecular forces** do not involve sharing or loaning of electrons, but instead are electrical forces between a molecule and another part of the lattice. They tend to be a relatively weak bonding force, such as the forces that hold together sheets of graphite so weakly that we can smear them apart with our fingers.

Elastic Deformation of a Lattice

The normal, equilibrium spacing of atoms in a crystalline lattice is altered when the lattice is subjected to stress. When an ideal crystal (no flaws or defects!) is hydrostatically loaded (equal, all-sided stress) and forced to deform elastically, the distance between atoms is shortened by an amount related to the imposed compressive stress and to the strength of the interatomic forces (Figure 4.4*A*,*B*) (Nicolas and Poirier, 1976). When the load is removed, the stored potential energy in the compressed bonds causes the lattice to rebound to its original equilibrium configuration (see Figure 4.4*A*). Because the deformation produced is perfectly elastic, the crystal does not sustain any permanent strain, and thus there is no visible record of the deformation. Similarly, when a crystal is subjected to a small amount of tensile stress, the bonds will stretch by a small amount, permitting the lattice to expand ever so slightly (Figure 4.4*C*). As long as the stretching is less than the tensile elastic limit of the crystal, the lattice will rebound to its original configuration as soon as the stresses are removed.

Exceeding the Elastic Limit

If the tensile stress acting on a lattice exceeds the tensile elastic limit of the crystal, a row of bonds will be severed, and the crystal will break (Figure 4.4*D*,*E*). The break may irregularly crosscut the lattice, producing a **fracture**, or it may follow a crystallographic plane of weakness, producing a **mineral cleavage**. It all depends on internal structure and the relationship of the direction of stress to the internal structure. The stress required to rupture the bonds in tension is called the **tensile strength** of the crystal. Some materials, like clay minerals and calcite, are much weaker under tension and compression than tougher minerals like quartz.

Similarly, if the magnitude of shear strain exceeds the elastic capacity of the lattice to shift, either a fracture will develop, or the lattice will slip or glide. Slip occurs when atoms on one side of a crystallographic plane suddenly shear with respect to atoms on the opposite side of the plane (Figure 4.4*F*). After slip

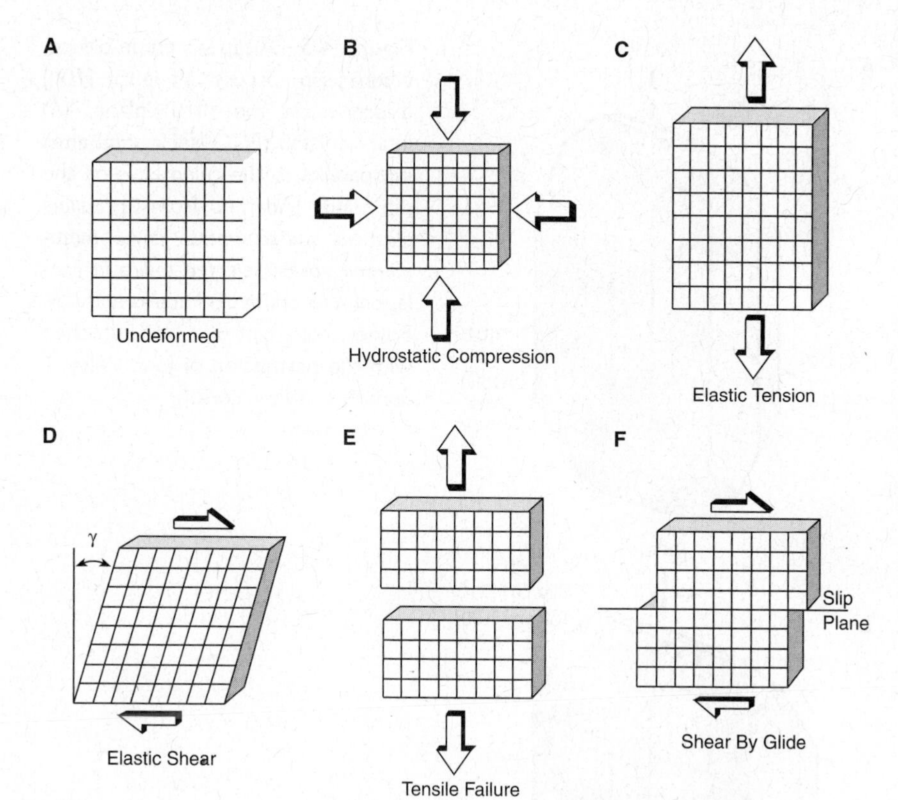

Figure 4.4 Elastic and inelastic deformation of an ideal crystal. (*A*) Starting configuration. (*B*) Elastic response to an increase in confining stress (hydrostatic compression). The bonds shorten and the crystal decreases slightly in size. The energy is stored as potential energy, as in a compressed spring. (*C*) Elastic response to an imposed tensile stress. The amount of elastic strain shown is greatly exaggerated compared to that which is possible in real geologic materials. (*D*) Elastic response to an imposed shear stress. (*E*) Rupture when tensile stresses exceed the yield strength under tension. (*F*) Shearing via glide or slippage along a crystallographic plane (slip plane). [Adapted from *Crystalline plasticity and solid state flow in metamorphic rocks: selected topics in geological sciences* by A. Nicolas and J. P. Poirier, copyright © 1976. Published with permission of John Wiley & Sons, Inc., New York.]

occurs, the lattice within the undisturbed part of the crystal remains coherent, whereas the crystal is permanently deformed where the shift occurred. The outline of the crystal itself preserves a record of deformation in the form of slight offset of its edge and slight change in shape (see Figure 4.4*F*).

Whether a crystal deforms by discrete fracture or by slip along a crystallographic plane depends on whether the stress required for fracture is greater or less than the stress required for slip. If it takes more stress to fracture than to cause slip, the crystal will deform by slip and may behave in a ductile manner. In contrast, if it takes more stress to cause slip than to produce a fracture, the crystal will behave brittlely and fracture.

Slip Systems and Crystallographic Control

Where slip along a suitably weak, suitably oriented crystallographic plane is the preferred mode of inelastic deformation, it is necessary to identify the slip plane and the direction of slip, much as in the evaluation of faulting, but at the atomic scale. The crystallographic plane that slips is the **slip plane**, which together with the line of slip within the slip plane defines the **slip system** (Figure 4.5). The kinematic character of a slip system is analogous to shearing of a microscopic set of playing cards. Each card remains undeformed, but tiny amounts of slip between each adjacent card can dramatically alter the shape of the entire deck.

The crystallographic direction along which a crystal slips is largely controlled by the strength and type of bonds, and by the geometric arrangement of atoms and bonds in the lattice (Nicolas and Poirier, 1976). Slip will occur along planes marked by the weakest bonding.

The directions of cleavage in a mineral disclose the crystallographic orientations in the lattice along which bonds are most easily broken. The relative

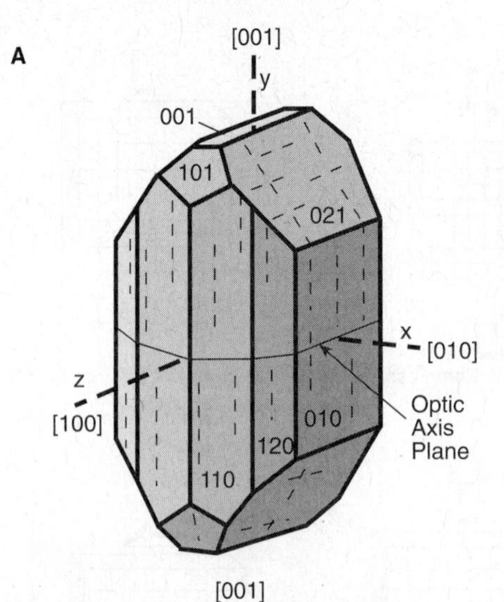

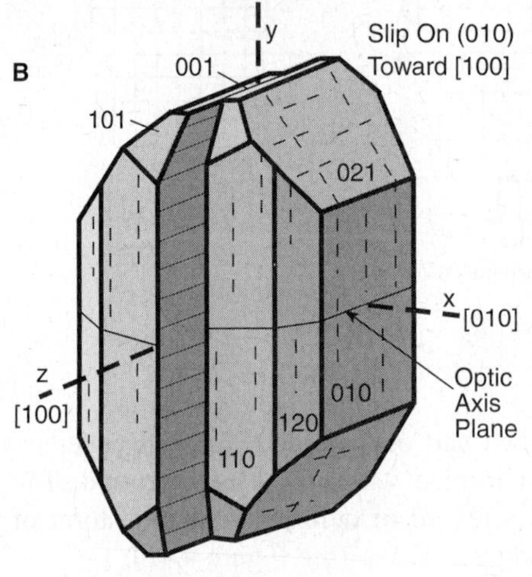

Figure 4.5 A slip system in olivine, where slip occurs along a [100] direction on the (010) plane. (*A*) Intact crystal. (*B*) Offset crystal after slip parallel to the ruled lines on the slip plane. [Adapted from *Crystalline plasticity and solid state flow in metamorphic rocks: selected topics in geological sciences* by A. Nicolas and J. P. Poirier, copyright © 1976. Published with the permission of John Wiley & Sons, Inc., New York.]

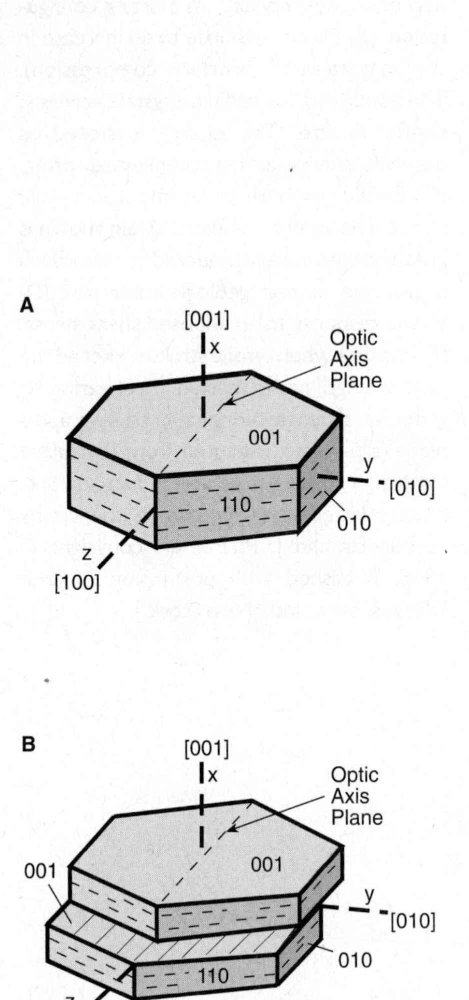

Figure 4.6 Cleavage and slip system in mica. (*A*) Intact crystal. (*B*) Offset crystal after slip along [100] on the (001) plane. [Adapted from *Crystalline plasticity and solid state flow in metamorphic rocks: selected topics in geological sciences* by A. Nicolas and J. P. Poirier, copyright © 1976. Published with the permission of John Wiley & Sons, Inc., New York.]

strength of development of cleavage in various cleavage directions reflects the relative strength of bonding along the corresponding crystallographic directions. For example, the weakest bonds in muscovite mica are parallel to the perfect (001) cleavage plane in mica (Figure 4.6).

Some minerals, including quartz, do not possess cleavage. There are no favorable pathways for fractures to cut through quartz. All crystallographic pathways through the lattice offer about the same resistance to splitting. This is why quartz crystals display conchoidal fracture, not cleavage.

SLIP SYSTEMS AND BONDING

The degree of development of cleavage and slip planes is related to the type of bonding (e.g., ionic or covalent) of minerals (Lawn and Wilshaw, 1975; Nicolas and Poirier, 1976). Take metals, for example. Metallic bonding consists of a lattice of densely packed positive ions (cations) surrounded by a "*wandering herd*" of very weakly attached electrons. Bonds are relatively weak, permitting metals to slip relatively freely. Slip is easiest in planes and directions with the closest spacing of cations. The metal slips in a direction that minimizes the distance of displacement required to move the lattice from one cation to another along the slip plane.

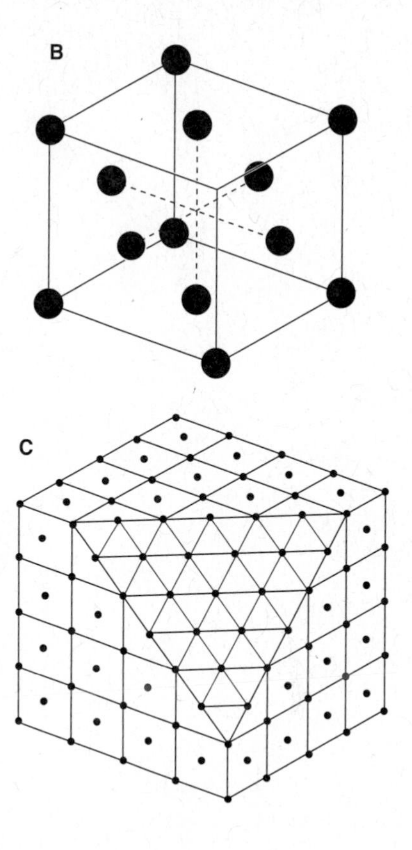

Figure 4.7 Direction of easiest slip. (*A*) Rocks in a stream bed, distributed like cations in a plane within a face-centered metal. To cross the pond with the shortest leaps, you would hop parallel to the diagonal lines. Other routes require more energy and may be impossible. [Artwork by D. A. Fisher.] (*B*) Three-dimensional arrangement of atoms in a face-centered metal. (*C*) Slip plane within a face-centered metal, oriented along the surface with the densest spacing of cations. [Reprinted from *Introduction to dislocations*, D. Hull and D. S. Bacon. Published with permission of Pergamon Press, Ltd., Oxford, © 1984.]

Imagine trying to cross a stream by jumping on large stones, which by some strange coincidence are arranged like atoms in some metals (Figure 4.7*A*). The easiest way to cross is by jumping to the nearest stone, which is along a diagonal, rather than along a row. Other directions are commonly out of reach. This arrangement of stones, like that of the atoms they represent, can be extended into three dimensions (Figure 4.7*B*). For many metals, such as copper (Cu) and aluminum (Al), the shortest *line* of slip between adjacent cations lies within the diagonal (111) plane, where cations are densely spaced (Figure 4.7*C*).

Within ionic crystals such as halite (NaCl), positive and negative ions alternate in the lattice (Figure 4.8), and easiest slip will generally be in a direction that does *not* bring ions of *like* charge into close contact (Nicolas and Poirier, 1976). In cubic crystals with ionic bonding, (e.g., a number of oxide minerals), the cubic crystal will slip within the (110) plane along the [110] diagonal.

Many minerals have complex crystalline structures, much more complex than halite or copper. Many of the common rock-forming silicates, for example, can possess covalent and ionic bonds both in the same lattice. The general rule is this: Complex silicates will fracture or slip along planes that break the weakest bonds or the fewest strong bonds. Cleavage surfaces and slip surfaces will tend to avoid breaking the bonds between silicon and oxygen atoms, which have a mixed covalent-ionic character and are especially strong (Nicolas and Poirier, 1976). Quartz, which is composed only of Si and O, is very strong because slip must break through the tough Si—O bonds. In phyllosilicates (e.g., micas), cleavage and slip are easiest along the (001) plane (see Figure 4.6), where weak ionic bonds hold a layer of cations between double sheets of strongly bonded silica tetrahedra. The relative weakness of the ionic bonds accounts for the perfect (001) cleavage of micas.

In addition to controlling the directions of fracture and slip, crystallography and bonding profoundly influence the overall **yield strength**—that is, how

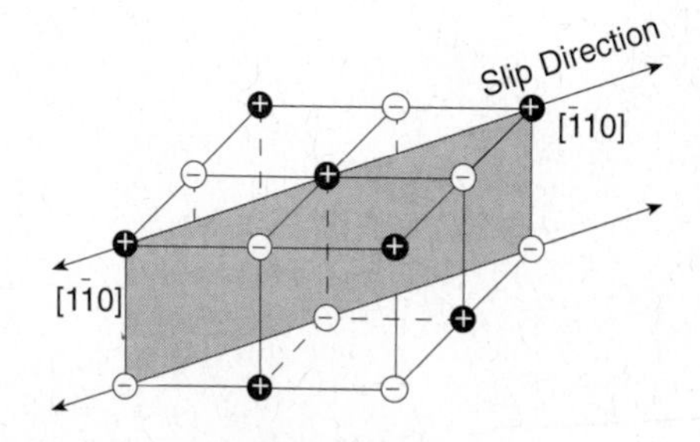

Figure 4.8 Direction of easiest slip in halite, which is along the shortest direction that does not bring ions of like charge into close proximity. [Adapted from *Crystalline plasticity and solid state flow in metamorphic rocks: selected topics in geological sciences* by A. Nicolas and J. P. Poirier, copyright © 1976. Published with the permission of John Wiley & Sons, Inc., New York.]

much elastic deformation a crystal can handle before it yields permanently by either fracture or slip. The yield strengths for quartz and diamond, which have covalent bonding, are much higher than the yield strength for halite, which has ionic bonding. The strength of halite, in turn, is higher than the yield strength for copper, which has metallic bonding.

Minerals with very low yield strengths generally can accommodate simultaneous movement on multiple slip planes, permitting ductile flow of the mineral. At room temperature, the covalent bonds in quartz and diamond are much too strong to permit ductile behavior. However, copper is a different story. The metallic bonds in copper are so weak that we can make it deform right at the workbench out in the garage, by bending a copper pipe, twisting a copper wire, or buckling a penny in a vise (Figure 4.9).

Bonding can affect strength in yet another way—by influencing the chemical properties of a mineral, especially solubility. Halite and other ionic solids are generally more soluble in hydrous fluids and more chemically reactive than covalent minerals. Soluble minerals may respond to deformation by performing a great disappearing act—by simply dissolving away under the influence of directed stress.

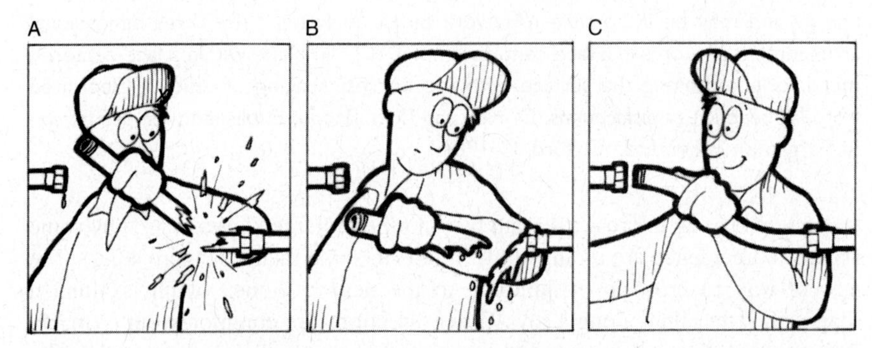

Figure 4.9 Internal atomic/crystal structure determines external material properties. If it is an optimum water connection you want, avoid pipes made of (*A*) quartz, which is too covalent and brittle, and (*B*) halite, which is too ionic and soluble. (*C*) Go for metallic-bonded copper. [Artwork by D. A. Fischer.]

Defects

Thus far we have been talking about ideal crystals and ideal crystal lattices. In reality, crystals and lattices are not perfect. They have **defects** and **flaws**, and these affect the strengths of minerals and the ways in which they are able to deform.

The observed tensile strengths of minerals, as measured through deformation experiments under conditions of room temperature, are generally significantly less than the strengths theoretically predicted for ideal crystals. The observed tensile strengths of most minerals and rocks are in fact several orders of magnitude lower than the theoretical tensile strength. What could cause minerals to be that much weaker than we would expect?

The culprits are defects and flaws—deviations from the ideal crystalline structure. **Defects** largely govern the strength of rocks and minerals by concentrating stress in ways that permit the material to fracture or slip at stress levels well below the theoretical yield strength. Rock climbers understand this principle and forever are inspecting their ropes and other equipment for possible nicks or deep scratches.

For conditions of brittle failure, microcracks are the most influential defects. Microcracks occupy individual grains and crystals at the microscopic and submicroscopic scales, but their influence is potent—the tips of the tiny cracks can invite the buildup of *very* high stress concentrations, thus permitting crack propagation and brittle failure.

There are defects other than microcracks, and their presence can influence the ease or difficulty with which a crystal or grain deforms under various conditions of temperature and pressure. Most defects tend to make it easier for a lattice to deform, but some in fact "*harden*" the lattice. At the atomic scale, there are three basic varieties of defects: **point defects, line defects**, and **planar defects** (Hobbs, Means, and Williams, 1976).

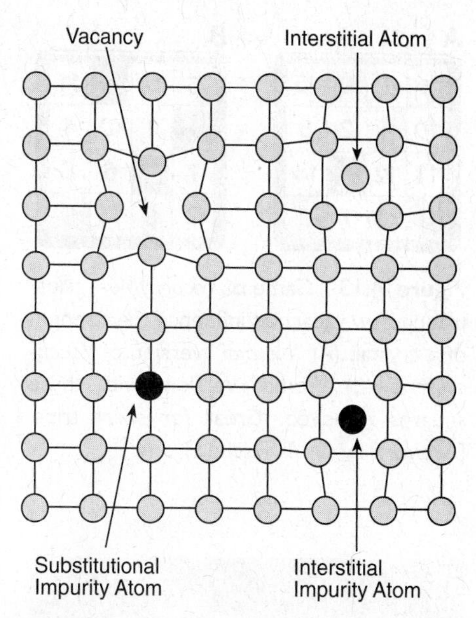

Figure 4.10 Various types of point defects within a crystal.

Point Defects

Point defects include **vacancies, interstitial atoms**, and **impurities** (Figure 4.10). A vacancy, where an atom is missing from a lattice site that is normally occupied, is like an empty parking space (Figure 4.11). Interstitial atoms are extra atoms that straddle the line between parking places that are already occupied (see Figure 4.11); they are extra atoms that rightfully belong in the lattice but are parked illegally between the normal, proper lattice positions. Impurities are unusual vehicles in parking spots normally occupied by another element (see Figure 4.11); they are foreign atoms that either are interstitial or occupy a regular space.

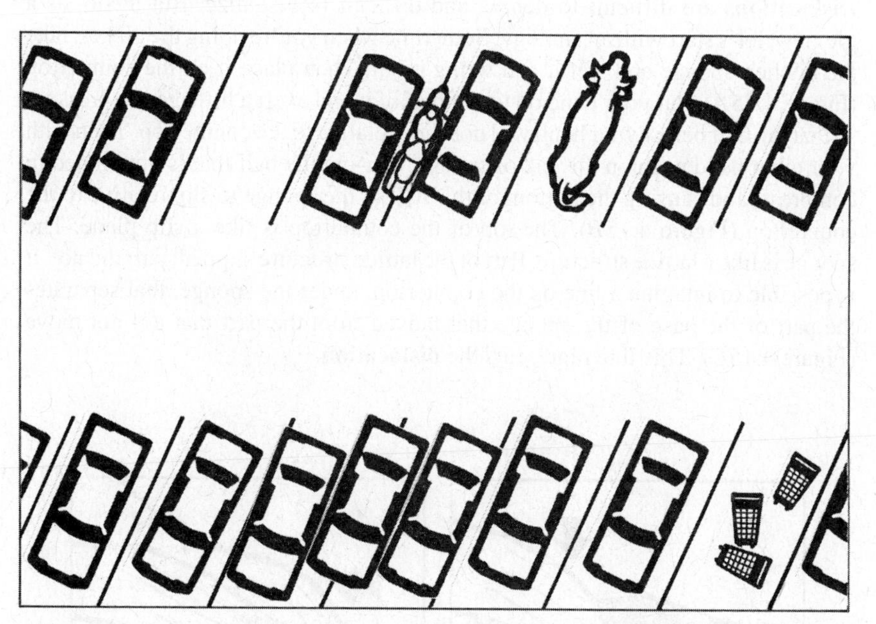

Figure 4.11 Quiz time. Can you find a vacancy, two interstitial atoms, and unexpected impurities in this parking lot analogue of point defects? [Artwork by D. A. Fischer.]

The presence of point defects affects the strength of a crystal, as can be illustrated through a preview of the deformation mechanism known as **solid-state diffusion**. Solid-state diffusion is one way in which a crystal can change its shape and volume in response to directed stress. Solid-state diffusion is achieved by the migration of atoms away from surfaces of greatest stress, toward surfaces of least stress (Figure 4.12). Vacancies make it much easier for atoms to move within a lattice during solid-state diffusion. We can envision why this is so by recalling the 15-14-13-12 . . . game that everyone plays in elementary school (Figure 4.13*A*). In Pittsburgh we used to call it "*Vacancy-Move*" because it reminded us so much of the role of point defects in crystal

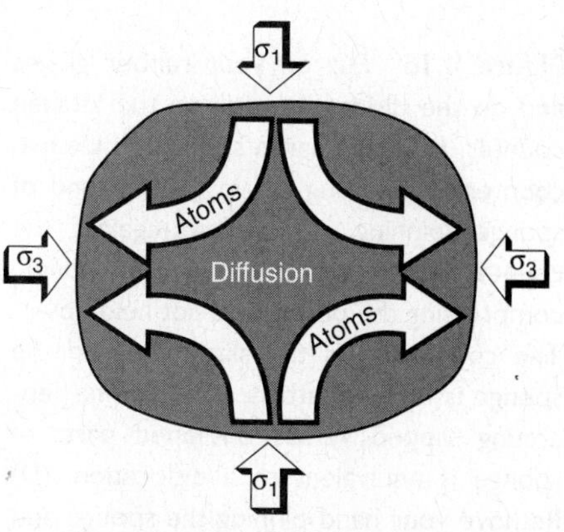

Figure 4.12 Solid-state diffusion drives atoms from surfaces of higher stress toward surfaces of lower stress.

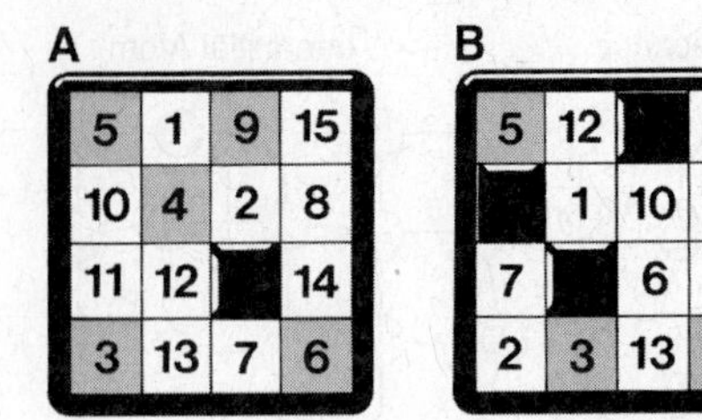

Figure 4.13 Game of *VacancyMove*, illustrating how vacancies influence the strength of a crystal. (*A*) "*Normal*" version of *VacancyMove*. (*B*) "*Turbo*" version, with three squares removed. Great for short trips. [Artwork by D. A. Fischer.]

lattices. The object of the game is to slide the small plastic squares until the numbers are in the correct reverse order. The game can be difficult for beginners because there is only one "*vacancy*." We used to pull out two of the squares and play 13-12-11-10 . . . It's a faster game! Toy manufacturers have now caught on, and they are selling the new improved "*Turbo*" version (three squares are removed) for people who don't have much time for games (Figure 4.13*B*). After playing "*Turbo*" once or twice, you will see that a crystal with abundant vacancies will be better able to rearrange itself in response to directed stress.

Interstitial atoms and impurities also affect the capacity of a mineral to deform internally. An impurity or an interstitial atom can cause a mineral to be harder to deform by hindering the movement of "*normal*" atoms. The hardening reminds me of the time back in Pittsburgh when a rival player glued down one of the squares in my *VacancyMove* game; it really slowed me down! It made the game "*harder*."

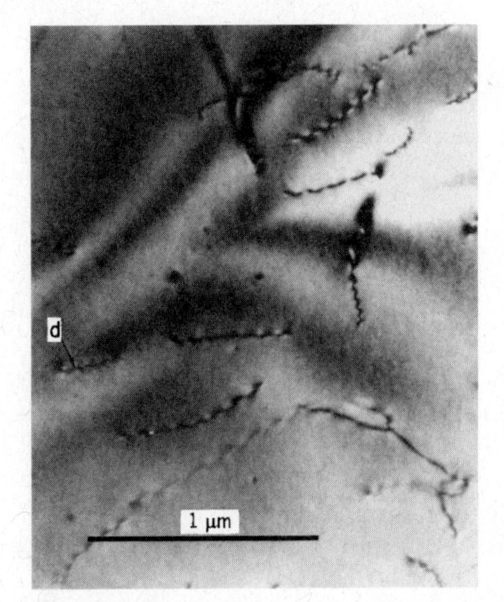

Figure 4.14 Transmitting electron microscope (TEM) image of linear defects, or dislocations. [Photograph courtesy of Andrew Kronenberg.]

Line Defects

Line defects, also known as **dislocations** (Figure 4.14), can become barriers to the otherwise smooth movement of *planes* of atoms. Dislocations are a microstructure produced by yet another deformation mechanism that we will soon examine. **Dislocations** are difficult to define, and difficult to visualize. But again, as a preview, let's start with an analogy. Sometime when you're doing the dishes, take the kitchen sponge and wet it and wring it out. Then place it on the countertop (Figure 4.15*A*). Put your hand on top of half of it and exert a little gentle pressure so that the half below your hand will not slide relative to the countertop. Then with your other hand push on the end of sponge, causing the half that is not pinned to compress, and causing the bottom of this half of the sponge to slip relative to the countertop (Figure 4.15*B*). The top of the countertop is like a slip plane. The sponge is like a lattice structure. Part of the lattice structure slipped; part did not. It is possible to imagine a line on the countertop, under the sponge, that separates the part of the base of the sponge that moved from the part that did not move (Figure 4.15*C*). This line represents the **dislocation**.

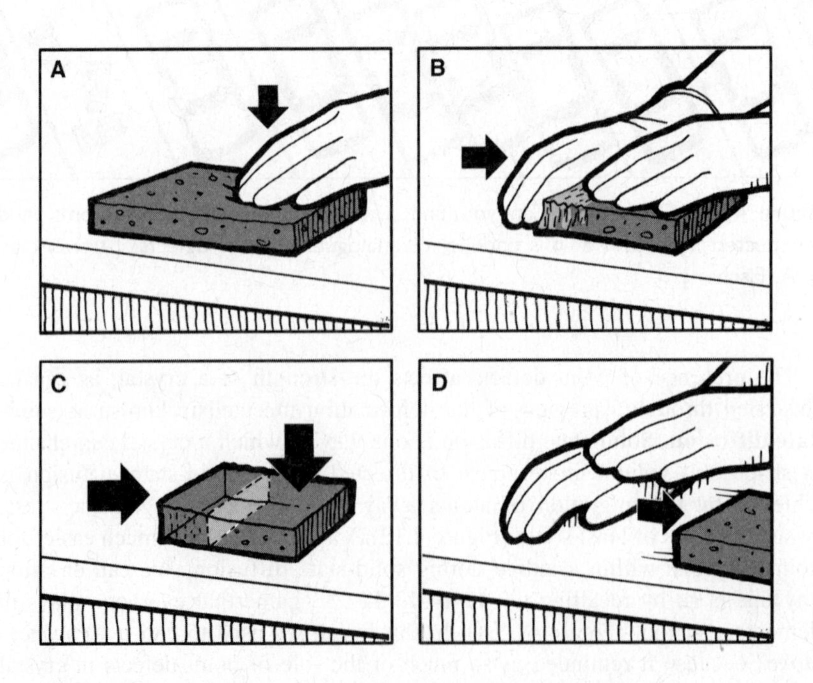

Figure 4.15 Put on your rubber gloves and do the dislocation drill on the kitchen counter. (*A*) Place damp sponge on kitchen countertop, pushing down on one end of sponge, pinning it in place against the countertop. (*B*) Push other end of sponge, compressing the part that is not held down. The countertop is the slip plane and the sponge is like the lattice. (*C*) The line separating slipped versus unslipped parts of sponge is equivalent to a dislocation. (*D*) Remove your hand pinning the sponge and observe the slip of other half of sponge. [Artwork by D. A. Fischer.]

A dislocation can be straight or curved. We have found through countless repetitions of the sponge experiment that the dislocations always seems to be a little curved and irregular. Incidentally, when you take your hand off the top of the sponge, keeping your other hand in place at the end of the sponge, the pinned part of the sponge will now slip forward as the slipped part of the sponge releases its pent-up energy (Figure 4.15*D*). As we shall see later in this chapter, this is analogous to what happens in crystals as they change their shapes and positions.

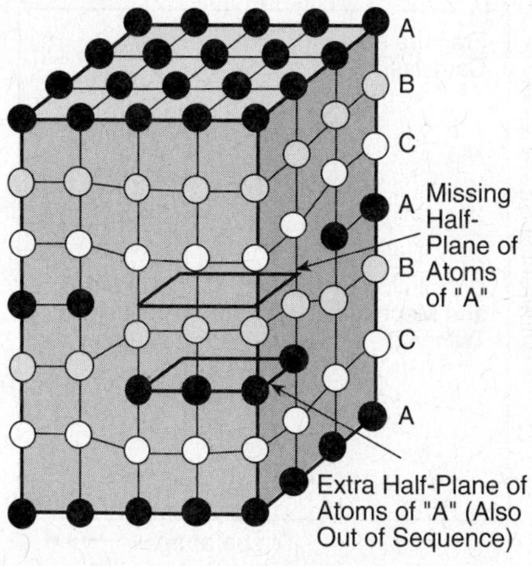

Planar Defects and Other Flaws

Planar defects can interrupt the continuity of the lattice, thereby influencing how a crystal will respond to directed stress. Planar defects of greatest interest to us are **grain boundaries, subgrains** (which are slightly misoriented regions within a crystal), **mechanical twins** (like calcite twins), and **stacking faults** (where the regular repeating pattern of a lattice is interrupted by the insertion or omission of a partial plane of atoms) (Figure 4.16). Minerals can have other flaws at the microscopic and submicroscopic scales, including fluid inclusions, veins, and especially microcracks.

A
B
C
A Missing Half-Plane of Atoms of "A"
B
C
A
Extra Half-Plane of Atoms of "A" (Also Out of Sequence)

Figure 4.16 Stacking faults, a type of planar defect in a lattice. There are two stacking faults in this example: one produced by an extra half-plane of atoms; the other by a missing half-plane.

DEFORMATION MECHANISMS

When subjected to stress, the crystalline structures and defects allow a rock to deform by one or more mechanisms. The combination of deformation mechanisms that operate within the rock at a given time and place depends on a number of factors. Of these factors, the most important are mineralogy (including grain size), temperature, confining pressure, fluid pressure, differential stress, and strain rate. For a single mineral phase (e.g., calcite) under a set of uniform conditions, one or two deformation mechanisms will typically be enough to accommodate most of the strain and will govern the type of microstructures produced. Most rocks, however, contain more than one mineral phase, and each phase may respond to deformation by a different combination of mechanisms. Furthermore, it is common for deformation to take place under continuously changing conditions. For example, temperature and confining pressure increase as rocks are buried, and both decrease as rocks are uplifted. Temperature, but not confining pressure, of rocks will increase when a magma body intrudes nearby. The magma may increase fluid pressure, causing the rocks to suddenly fracture. We can imagine countless scenarios.

Deformation can be accompanied by cyclic changes in physical conditions, especially changes related to fluid pressure and strain rate. For example, an active fault may accommodate motion via slow, steady creep for several centuries, only to instantaneously slip several meters during a single earthquake. Cycles of low-strain-rate creep would be accommodated by one set of deformation mechanisms, whereas high strain rates during an earthquake would be accommodated by a different set of deformation mechanisms, including fracture. The result could be a diverse array of microstructures and a puzzling set of crosscutting microstructural relations.

Interpreting the microstructures in deformed rocks would be nearly impossible without an understanding of deformation mechanisms. Modern research into deformation mechanisms aims to establish the range of physical/chemical conditions over which each deformation mechanism operates, and to develop detailed geometric/kinematic descriptions of the microstructures produced by each.

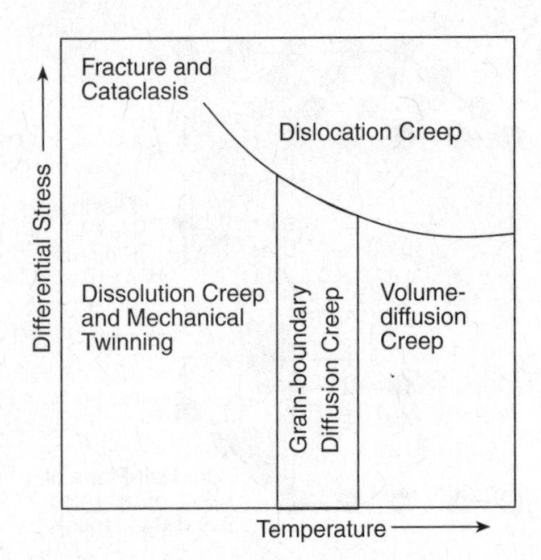

Figure 4.17 Simplified deformation map, showing the general conditions where each deformation mechanism dominates relative to one another.

The Main Mechanisms

We find it useful to arrange the main deformation mechanisms into five general categories: (1) microfracturing, cataclasis, and frictional sliding; (2) mechanical twinning and kinking; (3) diffusion creep; (4) dissolution creep; and (5) dislocation creep. These mechanisms are sometimes aided by other important processes, including recrystallization, that *lead* to less-strained lattices.

We can illustrate the physical conditions under which each mechanism prevails by use of a **deformation map** (Figure 4.17), which generally plots temperature versus differential stress or some other variable. For example, in Figure 4.17 fracture and cataclasis are shown as dominating at low temperatures and high differential stress, whereas several types of diffusion creep occur at higher temperature and lower differential stress.

Microfracturing, cataclasis, and **frictional sliding** involve the formation, lengthening, and interconnecting of microcracks; frictional sliding along microcracks and grain boundaries; and the formation and flow of pervasively fractured, brecciated, and pulverized rock and crystal fragments. **Mechanical twinning** and **kinking** are less aggressive deformational mechanisms than microfracturing and frictional sliding. Strain is achieved by bending, not breaking, of lattices.

There are several types of **creep** (see Figure 4.17), each of which uses a different strategy to change the shape and size of crystals in response to directed stress, almost as if the crystals were plastic. **Diffusion creep** changes the shape and size of crystals through the movement of vacancies and atoms within crystals and along grain boundaries. **Dissolution creep** changes the shape and size of crystals by dissolving material from one site and reprecipitating the material elsewhere, aided by fluids along grain boundaries or within pore spaces. **Dislocation creep**, the fanciest deformation mechanism, operates through intercrystalline slip of the lattice structure.

Microfracturing, Cataclasis, and Frictional Sliding

Microfracturing, cataclasis, and frictional sliding are brittle deformation mechanisms that operate at the grain and subgrain scales. As a response to stress, individual microcracks form, propagate, and link up with other microcracks to form full-blown microfractures and fractures. Individual microcracks can be opened up in tension, or they can accommodate slippage by frictional sliding. In faults and fault zones, which are characterized by the pervasive development of microcracks and fractures, the grain size of a rock may become dramatically reduced by cataclasis (crushing!), and the crushed material can move by **cataclastic flow**, as if it were a bunch of ball bearings.

Formation of Microcracks

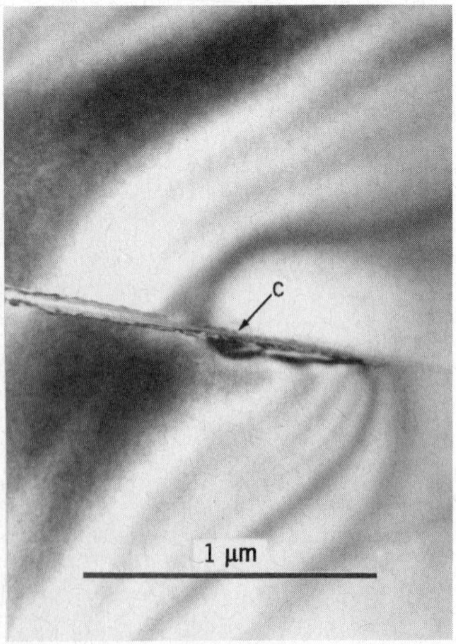

In low-temperature deformation experiments, rocks and minerals accommodate elastic strains of less than 1% before they fracture (Means, 1990). The differential stress at which they fracture is well below their predicted theoretical strength, mostly because of the role of **microcracks**, which are microscopic-to-submicroscopic cracks and surfaces (Figure 4.18). Like a magnifying glass that can focus normal sunlight to produce fire, microcracks focus ordinary low-level stresses to produce fractures. At the atomic scale, the initiation and growth of microcracks involves the severing of bonds.

Figure 4.18 TEM bright-field image of microcrack (c) in feldspar. [From A. K. Kronenberg, P. Segall, and G. H. Wolf, *Geophysical Monograph*, v. 56, Figure 15A, p. 33, copyright © 1990 by American Geophysical Union.]

Most microcracks are produced by stress that builds at points of concentration, such as grain boundaries, inclusions, pores, twins, dislocations, and earlier formed microcracks. A site where there is a large enough stress concentration or a geometric misfit between minerals may nucleate a microcrack. The stresses that cause microcracks can result from thermal, gravitational, or tectonic loading.

Microcracks can form during thermal loading (heating or cooling). As rocks are heated or cooled, neighboring minerals will expand or contract by different amounts if they have different coefficients of thermal expansion. Some minerals are anisotropic with respect to thermal expansion and contraction; that is, they expand or contract the most along one specific crystallographic direction. Quartz has an especially large and anisotropic coefficient of thermal expansion (Kranz, 1983). As quartz expands, it can indent a neighboring feldspar or other mineral that cannot get out of its way, forming microcracks. Differential thermal expansion between adjacent quartz and feldspar grains may be the main cause of microcracks in granite (Nur and Simmons, 1970).

Burial or unroofing of rocks also causes microcracks, not only because of expansion and contraction due to thermal effects, but because of gravitational loading and unloading. As rocks are buried, confining pressure increases, and grains increasingly impinge on one another because of compaction. Stress concentrations arise at grain-grain contacts, especially where grains indent or wedge into one another. Preexisting microcracks, particularly those that are low-dipping or horizontal, tend to close during burial. During uplift, microcracks will tend to open up as erosional or tectonic unroofing causes the steady reduction of confining pressure.

Microcracks also form as a result of tectonic loading, especially near preexisting microcracks and grain contacts. During semibrittle or ductile deformation, stress concentrations may develop where propagating twins and dislocations encounter obstacles such as grain boundaries, cavities, or other twins and dislocations. Microcracks form when the motion of the twin or dislocation can no longer accommodate the strain (Lloyd and Knipe, 1992).

Microcracks and Grain-Scale Fractures

Microcracks are commonly subdivided into three types: **intragranular, intergranular**, and **transgranular** (Figure 4.19) (Lawn and Wilshaw, 1975;

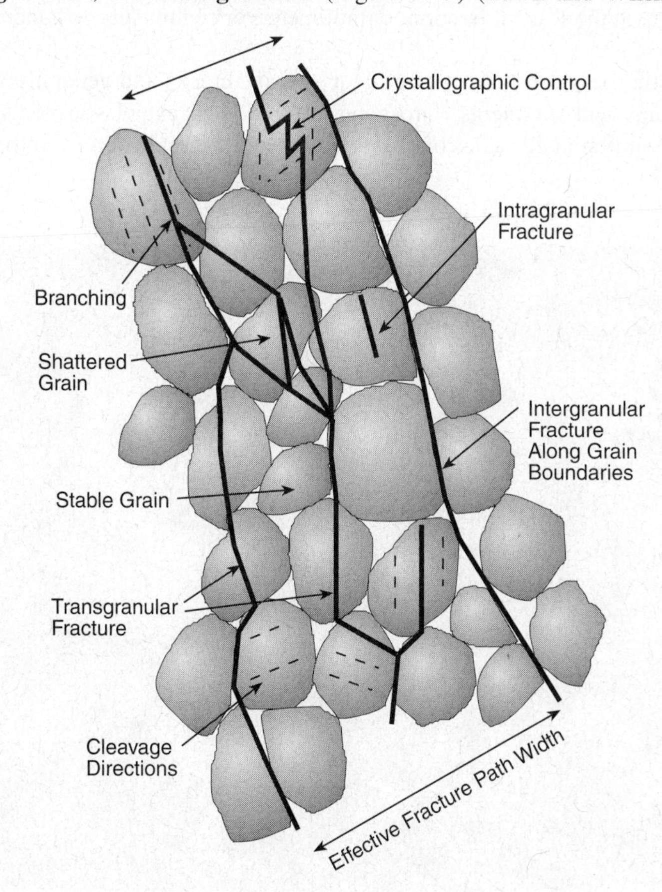

Figure 4.19 Types of microcracks, including intragranular fracture within a grain, intergranular fracture along grain boundaries, and transgranular fractures cutting across several adjacent grains [Reprinted from *Journal of Structural Geology*, Deformation mechanisms accommodating faulting of quartzite under upper crustal conditions, Lloyd, G. E., and Knipe, R. J., p. 127–143, © 1992, with permission from Elsevier.]

Kranz, 1983; Atkinson, 1987). **Intragranular microcracks** occur within a single grain, most commonly along a mineral cleavage plane, and form where the fracture strength of the grain is less than that of grain boundaries. **Intergranular microcracks** exploit grain boundaries, propagating around rather than through grains; they indicate that the grain boundaries were easier to "*crack*" than the adjacent grains. Intergranular microcracking is more difficult in coarse-grained rocks because it is less energy efficient for a propagating microcrack to "*circumnavigate*" large grains. **Transgranular microcracks** cut across adjacent grains and their mutual grain boundaries. Several conditions favor transgranular microcracking, including strong grain boundaries and similar orientations of cleavage in neighboring grains (Lawn and Wilshaw, 1975; Groshong, 1988).

Microcracks of all three types may be present in a single rock because of the diversity of minerals, crystallographic orientations, textures, and microstructures in rocks. Continuous through-going fractures, such as those observed at the scale of a hand specimen or outcrop, generally form by linkage of numerous microcracks, rather than propagation of a single fracture. Only in true tensile conditions does a single microcrack evidently propagate into a large, discrete fracture (Sammis et al., 1986, p. 69).

Cataclasis and Cataclastic Flow

Cataclasis is the pervasive brittle fracturing and granulation of rocks, generally along faults and fault zones. It produces an aggregate of highly fractured grains and rock fragments in a matrix of even smaller, crushed grains (Figures. 4.20 and 4.21). Once formed, such crushed aggregates are able to flow by repeated fracturing, frictional sliding, and rigid-body rotation of grains and fragments, a process termed **cataclastic flow**. Although cataclastic flow occurs by brittle processes, it may be homogeneous at the scale of fractions of millimeters or centimeters, to hundreds of meters.

Cataclastic rocks are characterized by pervasive cracks and generally sharp, angular grains and fragments (Groshong, 1988). Most cataclastic rocks look remarkably similar at all scales of observation, from the scale of an outcrop (see

Figure 4.20 Brecciated, shattered rock formed by fracturing and cataclasis of granite, Harcuvar Mountains, western Arizona [Photograph by S. J. Reynolds.]

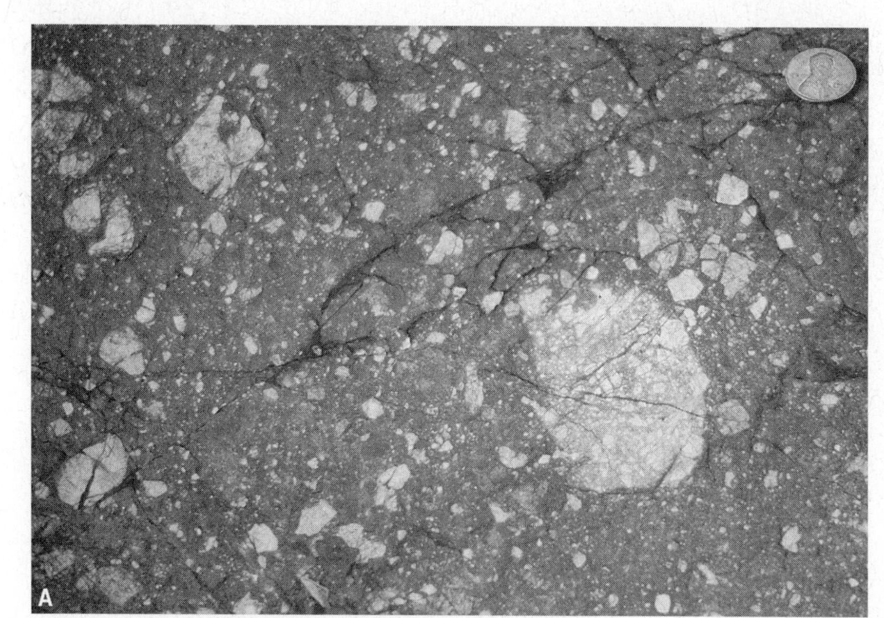

Figure 4.21 Cataclastic rocks at various scales of observation. (*A*) Outcrop of cataclastic rock along Whipple detachment fault, Whipple Mountains, southeastern California. [Photograph by S. J. Reynolds.] (*B*) Scanning electron microscope (SEM) image of fine-grained cataclastic rock. Width is 4.3 mm. (*C*) SEM close-up of part of B. Width is 0.2 mm. [Reprinted with permission from *Pure and Applied Geophysics*, v. 125, C. Sammis and others, "The Kinematics of gouge formation," 1986, Birkhauser Verlag Ag, Basel, Switzerland. Photograph courtesy of Charles Sammis.]

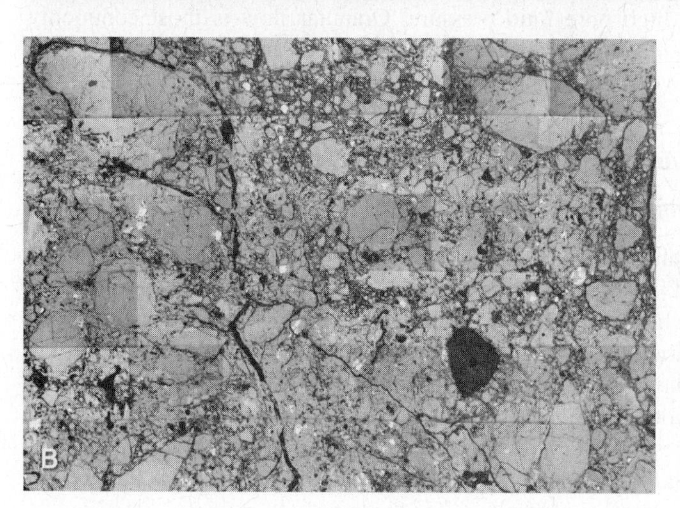

Figure 4.21*A*) down to a scale that can be resolved only through electron microscopy (see Figure 4.21*B,C*).

Cataclasis results in a progressive **decrease in grain size** as larger grains are continually broken into smaller ones. It also results in a **decrease in sorting**, as smaller and smaller fragments are produced and intermixed with larger ones. Cataclasis generally causes an *increase in volume*, a process known as **dilatancy** (Brace et al., 1966), as pore space is created between separating fragments. The dominant cause of fracturing in cataclastic rocks is stress concentration, where grains indent or impinge on one another during tectonic loading.

Cataclasis and cataclastic flow have been investigated through detailed examination of natural fault rocks and by experimental investigations of frictional sliding along precut specimens (Tullis and Tullis, 1986). The natural and experimentally produced cataclastic textures are remarkably similar (Figure 4.22), which implies that the experimentally observed behavior is likely to apply, within some limits, to natural fault zones. In laboratory experiments, cataclastic flow becomes progressively more difficult at higher confining pressures because higher confining pressures impede frictional sliding and dilatancy. As a consequence, cataclasis and

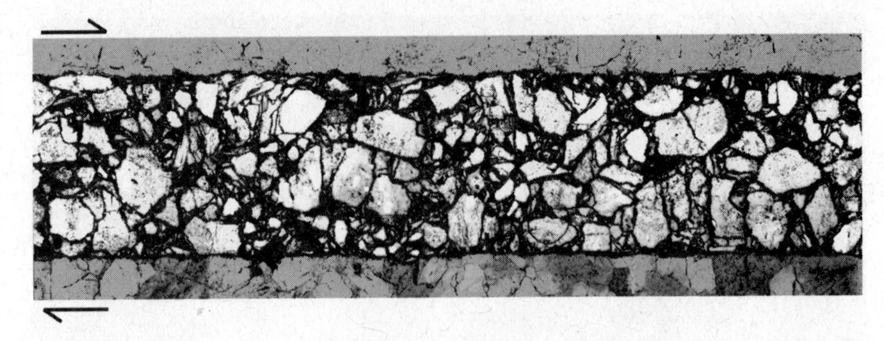

Figure 4.22 Experimentally produced cataclastic textures and fault zone. [Photograph courtesy of Charles Sammis; see Sammis et al. (1986).]

cataclastic flow are most important in the shallow parts of the crust, where they commonly occur along fault zones.

A similar mechanism is **granular flow**, which involves the frictional sliding and rolling of particles, but most grains remain intact, rather than being repeatedly fractured as in cataclastic flow. Granular flow can occur only where grain boundaries are much weaker than the individual grains and where effective confining pressure is very low, such as on or near the Earth's surface or in zones with high pore fluid pressure. Granular flow is most commonly observed in the slumping of unconsolidated sediments and other materials with little or no cohesive strength.

Mechanical Twinning and Kinking

Mechanical Twinning

Mechanical twinning is a deformation mechanism that produces a bending, rather than a breaking, of a crystalline lattice. In the simplest case, a mechanical twin is formed when the crystalline lattice is bent by shearing parallel to a favorable crystallographic plane (Figure 4.23). The crystalline lattice on one side of the twin plane is sheared by a constant angle of rotation with respect to the lattice on the other side. The amount of bending of the lattice, and thus its angle of rotation, is limited by the crystal structure of the mineral. Mechanical twinning is especially common in calcite and plagioclase feldspar, minerals that possess a suitable crystalline structure for twinning.

A different type of mechanical twin is produced when stresses cause one part of a crystal to rotate about an axis perpendicular to a favorable twin plane. In this type of twinning, the two parts of a crystal on opposite sides of a twin plane are slightly misaligned by a pivoting motion, much like opening a newly dealt hand in a card game. Compared to "*normal*" twins formed during crystal growth, mechanical twins of both types are more lenticular and more likely to end by a gradual tapering (Groshong, 1988).

Conditions Favoring Mechanical Twinning

Two conditions are necessary for twinning to occur: (1) there must be a vulnerable twin plane across which shearing or rotation can take place; and (2) the plane must be oriented such that the shear stress along the twin plane is sufficient to distort the lattice. Mechanical twinning is not particularly sensitive to the influence of confining pressure, because twinning does not involve frictional sliding or dilatancy. Moreover, susceptibility to twinning is not greatly influenced by temperature because, unlike deformation mechanisms such as diffusion, mechanical twinning does not involve a temperature-activated process. Twinning does require relatively higher differential stresses than some other deformational mechanisms, because it involves energy-intensive bending of the lattice.

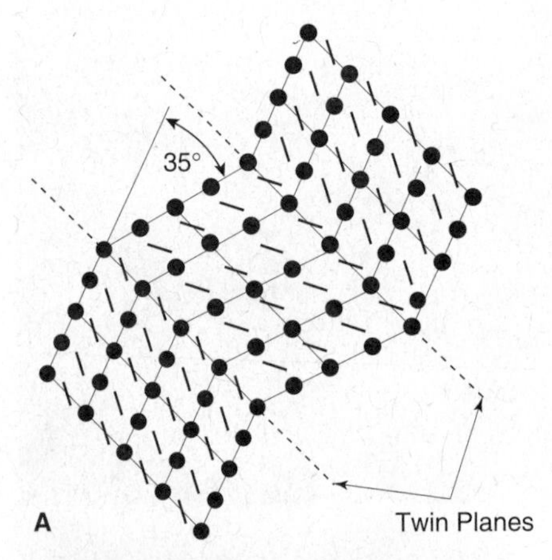

Figure 4.23 Mechanical twins in calcite. (*A*) Formation of a mechanical twin by shearing the calcite lattice parallel to a crystallographic plane. Solid dots represent Ca atoms and the short lines represent the CO_3 groups. [Adapted from Twiss and Moores (1992) and from Hobbs, Means, and Williams (1976).] (*B*) Photomicrograph of mechanical twins in grains of calcite. [Photograph by Richard H. Groshong, Jr.]

Twinning is a comparatively rapid process (Nicolas, 1987), but the amount of shear strain that can be accommodated by twinning is limited by the crystallography of the mineral. The bonds can be rotated only so far before they break. Once the lattice has been bent by the optimum and required angle to form a twin, it cannot accommodate any further strain by twinning. Any additional strain requirements must be accommodated by some other deformation mechanism, such as fracturing (Lloyd and Knipe, 1989).

We can use the geometry and appearance of calcite twins to gauge the temperature and amount of deformation (Groshong, 1988; Burkhard, 1993). Thin, straight twins represent low temperatures and low strains, whereas thick, curved twins form at higher temperatures and higher strains (Figure 4.24).

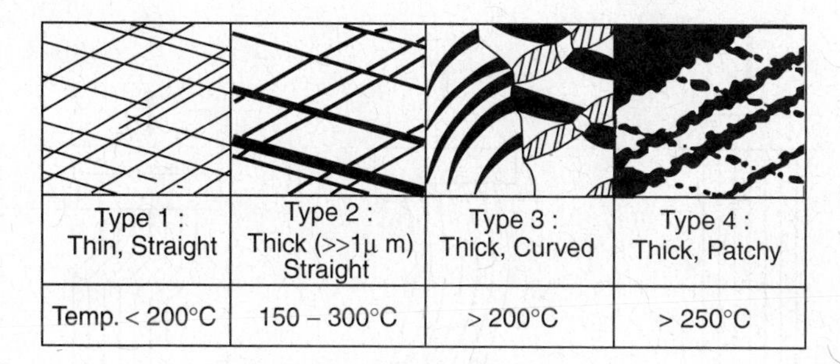

Type 1 : Thin, Straight
Type 2 : Thick (>>1μ m) Straight
Type 3 : Thick, Curved
Type 4 : Thick, Patchy
Temp. < 200°C
150 – 300°C
> 200°C
> 250°C

Figure 4.24 Types of twins as a function of temperature. [Reprinted from *Journal of Structural Geology*, v. 15, Burkhard, M., Calcite twins, their geometry, appearance and significance as stress-strain markers and indicators of tectonic regime: a review: in Casey, M., Dietrich, D., Ford, M., Watkinson, J., and Hudleston, P. J. (eds.), "The geometry of naturally deformed rocks.", p. 351–368, © 1993, with permission from Elsevier.]

Determining Strain and Stress from Mechanical Twins

The orientation of mechanical twin planes in a crystal can be used to determine the orientations of the principal directions of strain and, given a few assumptions, the orientations and magnitudes of the principal stresses (Groshong, 1988; Burkhard, 1993). We are able to reconstruct the stresses, normally a very difficult feat in structural geology, because only those potential twinning planes that are oriented favorably with respect to the imposed stresses will be activated during twinning. The ideal orientation is a twin plane at 45°, because this is the plane of maximum resolved shear stress (Figure 4.25). Various techniques have been devised, but all are applicable only to rocks with small strains (Burkhard, 1993). For calcite, the crystallographic "*c-axis*" attempts to rotate into parallelism with σ_1, the direction of greatest principal compressive stress (Figure 4.25). Therefore, the statistically-determined preferred orientation of "*c-axes*" of twin planes within the deformed rock tends to reflect the orientation of σ_1.

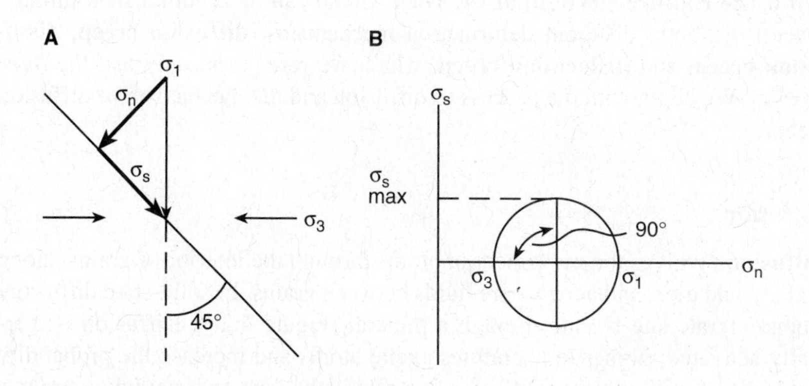

Figure 4.25 The most favorable orientation of a twin plane will be at 45° to σ_1, which is the plane of maximum resolved shear stress. (*A*) σ_1 can be resolved into a normal component perpendicular to and a shear component parallel to the twin plane. (*B*) Mohr circle showing that the maximum resolved shear stress is for surfaces at 45° to σ_1.

Kinking

Kinking, like twinning, involves the bending of a lattice, utilizing planes of weakness (Figure 4.26). Kinking commonly takes place within a discrete band through a crystal. As seen in this section, kink bands usually display a crystallographic orientation or angle of extinction that is different from the rest of the mineral. They show up microscopically as **extinction bands** (Groshong, 1988). Micas and other platy minerals are especially prone to kinking, especially when shortened in a direction parallel to cleavage. End-on compression of confined computer cards can produce similar results (Figure 4.27). Kinks can be relatively tight, because the amount of lattice rotation is *not* limited to a specified angle, as it is in the case of mechanical twinning.

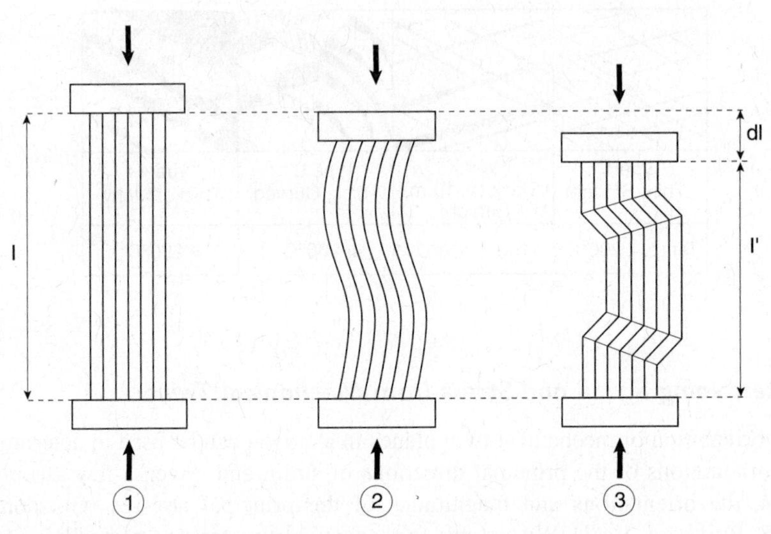

Figure 4.26 Kinking caused by end-on compression of a layered material. Initial length (*l*) is converted to *l′* with a change in length of *dl*. [From Nicholas, A., 1987, Principles of rock deformation: petrology and structural geology: D. Reidel Publishing Company, Dordrecht, Netherlands, Figure 2–22, pg. 47. With kind permission from Springer Science+Business Media B.V.]

Figure 4.27 Photograph of kink bands in a stack of computer cards experimentally deformed by applying an end-on load and a confining pressure. [Photograph and copyright by Peter Kresan.]

Diffusion Creep

Under many geologic conditions, rocks accommodate deformation by creep, rather than by fracturing, fractional sliding, or some other mechanism. **Creep** is a slow, time-dependent strain. **Creep occurs at differential stresses well below the rupture strength of the rock**. Creep can be accomplished through several distinctly different deformation mechanisms: **diffusion creep, dissolution creep**, and **dislocation creep**, which we refer to loosely as "*the three creeps.*" We begin with the process of diffusion and the mechanism of diffusion creep.

Diffusion

Diffusion involves the movement of atoms through the interior of grains, along grain boundaries, and across pore fluids between grains. In solid-state diffusion, atoms migrate site by site through a mineral (Figure 4.28). Diffusion is thermally activated; higher temperatures excite atoms and increase the probability that an atom will migrate. Diffusion is a relatively slow and inefficient process

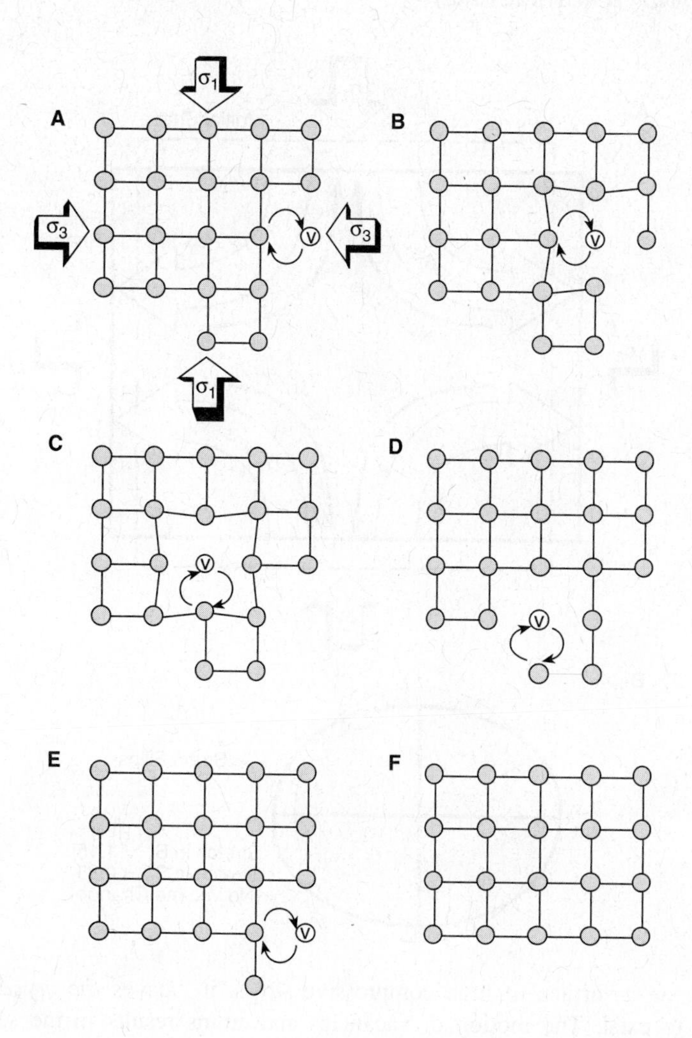

Figure 4.28 Volume-diffusion creep. Vacancy is shown by circle containing a "v". As atoms successively switch places with the vacancy, the vacancy moves through the lattice and the crystal changes shape. The atoms move from surfaces of higher stress to those of lower stress. (*A–D*) A vacancy begins at the edge of the crystal and moves one step at a time to another edge of the crystal. (*E–F*) Another vacancy moves through the crystal in two jumps (one not shown), permitting the crystal to change shape in response to the imposed stress.

in a crystalline solid, where atoms are bonded tightly together in more or less fixed positions. It is much faster and much more efficient in a fluid, where atoms can move freely.

Because diffusion through a crystalline solid relies on the presence of vacancies and other defects, diffusion is very sluggish in pure, undeformed lattices and is more rapid in lattices with abundant impurities or deformation-related imperfections. Diffusion is most effective in fine-grained rocks, where the distance that the atoms must cover is relatively short.

Volume-Diffusion Creep

At high temperatures, and in the presence of directed stress, diffusion *within* mineral grains can be fast enough to permit the grains to change their shapes. The presence of vacancies (i.e., unoccupied sites) makes this possible. Atoms systematically and sequentially swap positions with vacancies, in a way not unlike a game of checkers. The site hopping changes the shape of the mineral and allows a material to deform in response to directed stress.

When a crystal is subjected to differential stress, atoms and ions move away from sites of high compressive stress, where they are "*in the way*," to neighboring sites of lower stress (Passchier and Trouw, 2005). Vacancies display a counter-flow in the opposite direction, from low compressive stress to high compressive stress (Figure 4.29*A*). When a vacancy reaches the edge of a

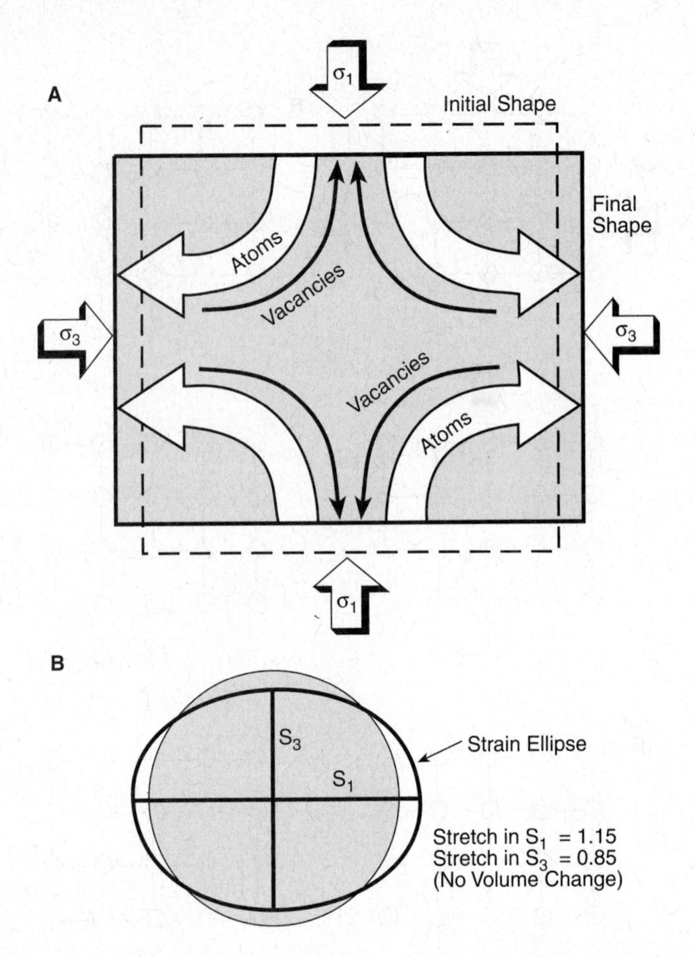

Figure 4.29 Volume-diffusion creep. (*A*) Atoms migrate from surfaces with high compressive stress and toward surfaces of lower compressive stress, whereas vacancies migrate in the opposite direction. (*B*) Finite strain ellipse produced by volume-diffusion creep in *A*. [From Nicolas, A., 1987, Principles of rock deformation: petrology and structural geology: D. Reidel Publishing Company, Dordrecht, Netherlands, Figure 14–14, pg. 40. With kind permission from Springer Science+Business Media B.V.]

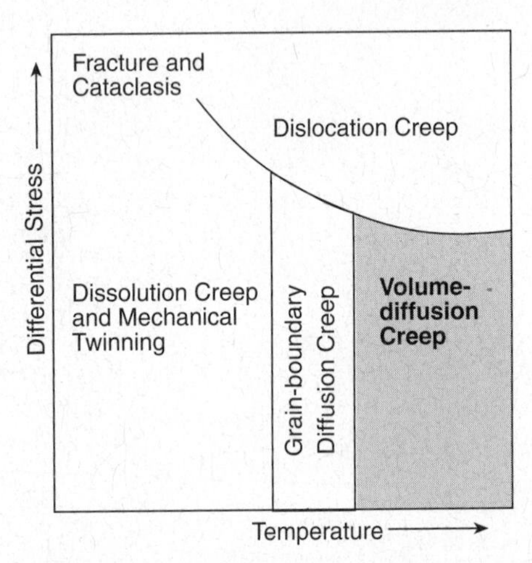

Figure 4.30 Simplified deformation map, showing conditions of volume-diffusion creep.

crystal, on a surface of high compressive stress, it "*leaves the crystal*" and ceases to exist. The motion of vacancies and atoms results in the selective removal of material from surfaces of high compressive stress and its accumulation on surfaces of low compressive stress. By this process, the crystal changes shape and accommodates strain (Figure 4.29*B*), a process known as **volume-diffusion creep**, or **Nabarro-Herring creep**. Diffusion creep is most effective in equant grains, where the diffusive path is short; it becomes progressively more difficult as grains are elongated during deformation. Volume-diffusion creep is a slow mechanism that is most important only at low differential stresses and at temperatures so high that they approach the melting temperature of the mineral (Figure 4.30).

Grain-Boundary Diffusion Creep

Minerals can also creep by diffusion of material *along grain boundaries*, a process known as **grain-boundary diffusion creep** or **Coble creep** (Figure 4.31). As in the case of volume diffusion, material migrates from surfaces of high compressive stress and accumulates on surfaces of low stress. The migration paths may be longer because the atoms and vacancies move around grains rather than through grains, but the rate of travel of vacancies and atoms is faster along grain boundaries than directly through a crystal. Therefore, grain-boundary diffusion creep tends to be more efficient than volume-diffusion creep and can occur at somewhat lower temperatures (Figure 4.32).

Superplastic Creep

Superplastic creep sounds like a dorky unpopular superhero, but actually it is another diffusion-related creep mechanism (Ashby and Verrall, 1973; Nicolas

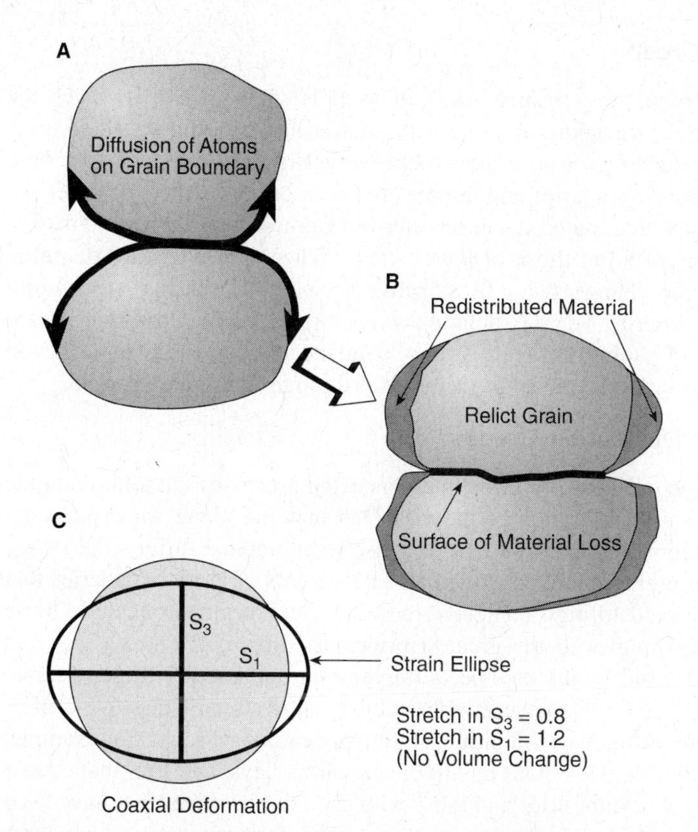

Figure 4.31 Grain-boundary diffusion creep. (*A–B*) Atoms diffuse around the grain boundary from surfaces of high compressive stress to those of low compressive stress. (*C*) Finite strain ellipse for *B*.

Figure 4.32 Simplified deformation map, showing conditions of grain-boundary diffusion creep.

and Poirier, 1976; Schmidt et al., 1981; Passchier and Trouw, 2005). Superplastic creep involves the combination of grain boundary sliding and grain-boundary diffusion (Figure 4.33). Most of the strain that accrues through this mechanism is achieved by the grain-boundary sliding. The role of the grain-boundary diffusion is to permit each of the grains to change shape as it slides so that no voids form between grains. Individual grains remain more or less equant and internally unstrained, even after significant amounts of total strain of the crystalline aggregate. Thus, almost as superheroes, the mineral aggregates can tolerate incredible levels of strain and continue to endure more punishment without becoming strain hardened.

Superplastic creep achieves higher strain rates than other types of diffusion creep, probably because grain-boundary sliding is a relatively efficient deformational process and because material needs to diffuse only a short distance along grain boundaries to accommodate the desired change in shape. The ideal conditions for superplastic creep are fine grain size (a few micrometers), to keep diffusion paths short, and relatively high temperatures (greater than half of the melting temperature), to ensure that diffusion will occur at a sufficiently high rate.

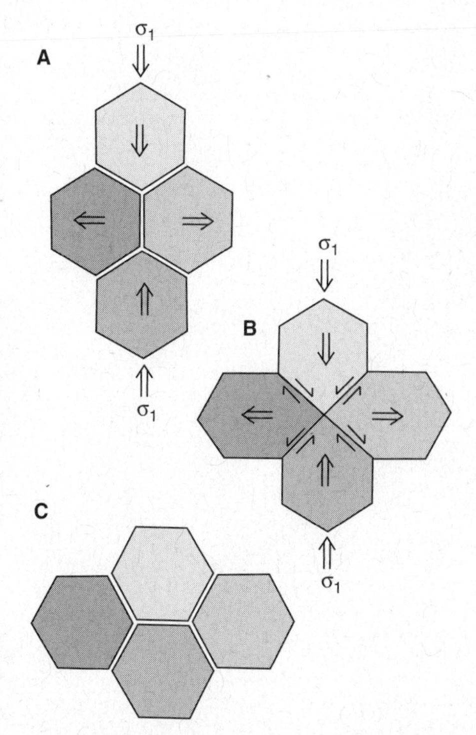

Figure 4.33 Superplastic creep. (*A*) Applied stress. (*B*) Shortening occurs via grain-boundary sliding and localized diffusion that allow the grains to change shape as they slide by one another. (*C*) Grains end up switching neighbors. [Reprinted from *Acta Metallurgia et Materiallia*, v. 21, M. F. Ashby and R. A. Verrall, Diffusion accommodated flow and superplasticity, p. 149–163, © 1973, with permission from Elsevier.]

Dissolution Creep

Dissolution creep, or **pressure solution** as it is often called, involves the selective removal, transport, and reprecipitation of material through fluid films along grain boundaries or pore fluids between grains. The presence of a fluid phase along grain boundaries and in pores between grains greatly increases the efficiency with which material can be removed from sites of high compressive stress and transported to those of lower stress. When subjected to differential stress, grains can change shape by selective dissolution, transport, and reprecipitation of material via the fluid phase (Kerrich, 1978; Groshong, 1988; Knipe, 1989). Dissolution creep differs from the various types of diffusion creep in that it involves transport of material through intergranular fluid.

Processes of Dissolution Creep

Dissolution creep depends on three interconnected processes: dissolution at the source, diffusion or migration of the dissolved material along some pathway, and reprecipitation (Figure 4.34). In response to an applied differential stress, grains become preferentially corroded along segments of grain boundaries that are being subjected to high compressive stress. Such segments include those oriented at high angles to the greatest principal compressive stress (σ_1) and those in which rigid grains or objects impinge on one another to concentrate stress (Figure 4.35). Of course, highly soluble minerals will dissolve preferentially to those with lower solubilities. In impure carbonate rocks, for example, calcite typically dissolves more readily than quartz, clays, and iron-manganese oxides. Grains with impurities and those with crystalline lattices that have been somehow damaged, perhaps by the work of other deformation mechanisms, are likewise more susceptible to dissolution than pristine grains.

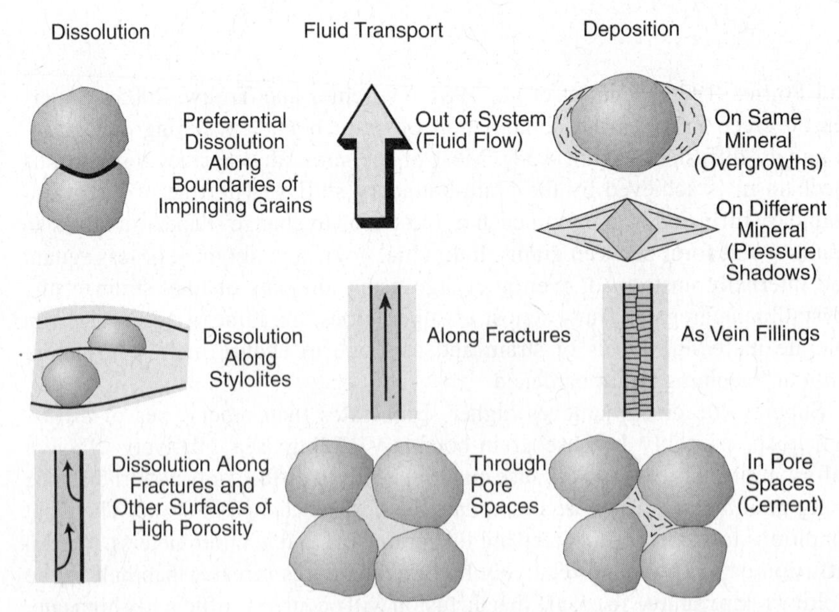

Figure 4.34 Processes of dissolution creep. [Adapted from Nicolas (1987) and Ramsay (1980b).]

As solid grains dissolve, they enrich the fluid in their constituents, especially near sites of more rapid dissolution. In contrast, little or no dissolution may occur near grain boundaries normal to the least principal compressive stress (σ_3). The differences in dissolution rates result in **chemical concentration**

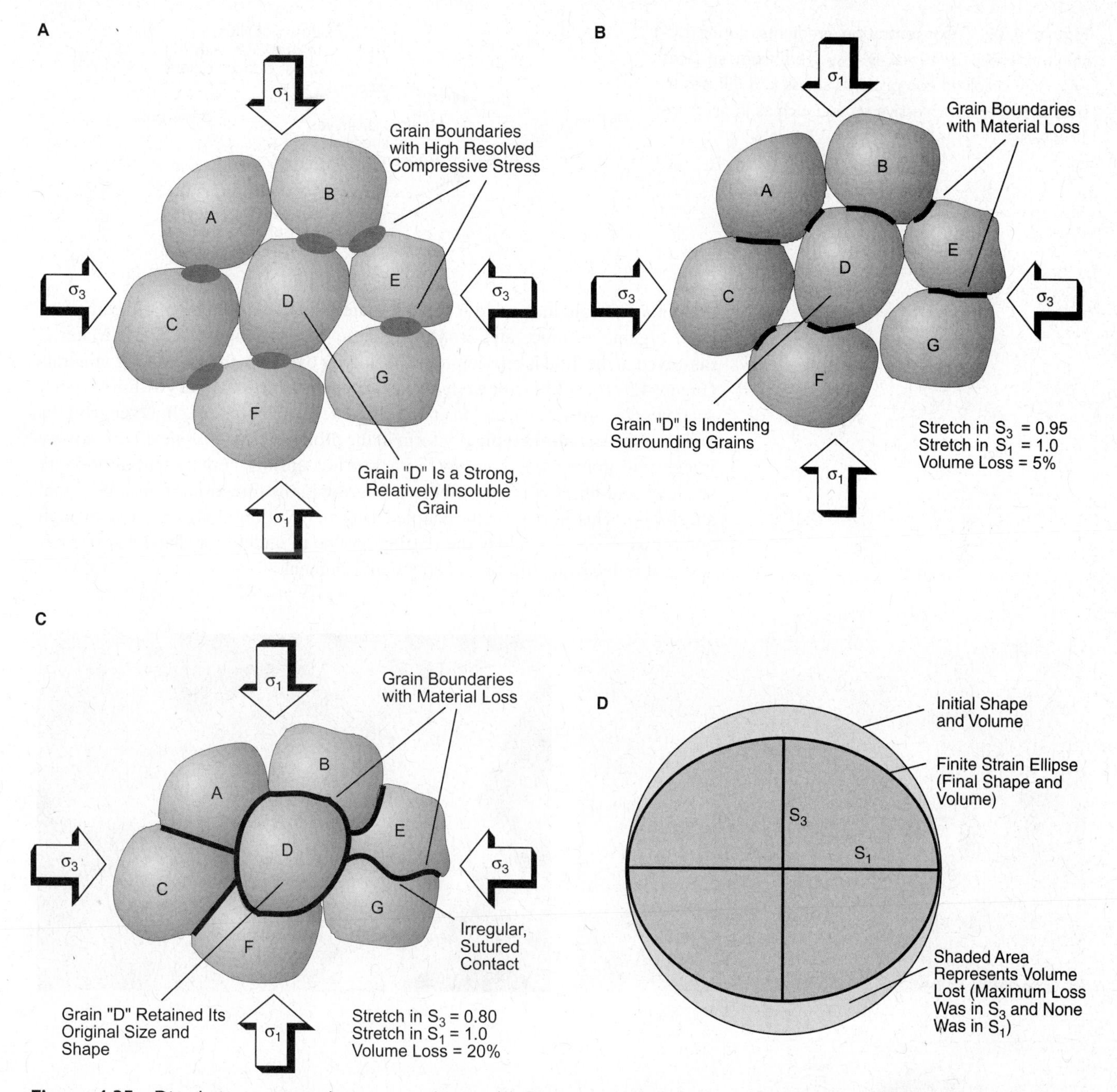

Figure 4.35 Dissolution creep and impinging objects. (*A*) Directed stress is concentrated on grain contacts oriented at a high angle to σ_1. (*B*) Grains preferentially dissolve along these contacts. Note that grain "*D*" is relatively strong and insoluble compared to surrounding grains. (*C*) Final fabric, in which rock has been shortened parallel to σ_1, with the shortening being accommodated by volume loss, rather than lengthening in a perpendicular direction. (*D*) Strain ellipse reflects shortening via volume loss.

gradients within the fluid, causing dissolved constituents to diffuse away from the dissolution sites and toward sites of lower compressive stress (Figure 4.36). Diffusion of the dissolved constituents can occur through the fluid layer adsorbed on grain boundaries or through a static pore fluid. Alternatively, the constituents can be transported by movement of their host fluid (i.e., fluid flow). Dispersal of the constituents away from dissolution sites, whether by diffusion or fluid flow, is necessary to allow further dissolution to occur.

There are telltale clues that reveal that a rock has experienced dissolution creep. Sites of continued dissolution are commonly marked by **stylolites**

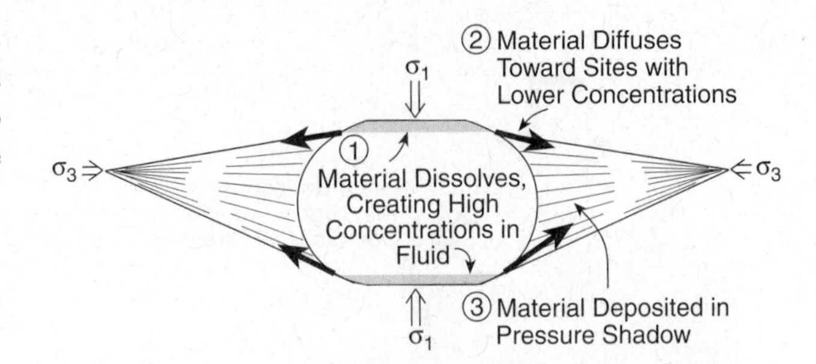
② Material Diffuses
Toward Sites with
Lower Concentrations
σ1
σ3
σ3
①
Material Dissolves,
Creating High
Concentrations in
Fluid
③ Material Deposited in
Pressure Shadow
σ1

Figure 4.36 Concentration gradients during dissolution creep. Material is dissolved from surfaces with high resolved compressive stress and diffuses to sites of lower compressive stress, such as in pressure shadows.

and the accumulations of less soluble material, such as clays, mica, carbonaceous organic residue, and iron-manganese oxides (Figure 4.37*A*). Material dissolved in the fluid is reprecipitated locally as overgrowths on existing minerals (Figure 4.37*B*), as **fibers** or wedge-shaped **beards** of crystal fibers within pressure shadows (Figure 4.37*C*), and in crystal-fiber veins (Figure 4.37*D*). **Overgrowths** and **pressure shadows** typically form in the protected lee areas next to relatively large, rigid grains, where compressive stresses are low. Fibers within pressure shadows and fibers in crystal-fiber veins grow in the direction of least principal stress (σ_3). This new growth, coupled with dissolution along surfaces of high compressive stress, enables the original grains to undergo a shape transformation that reflects the differential stress environment.

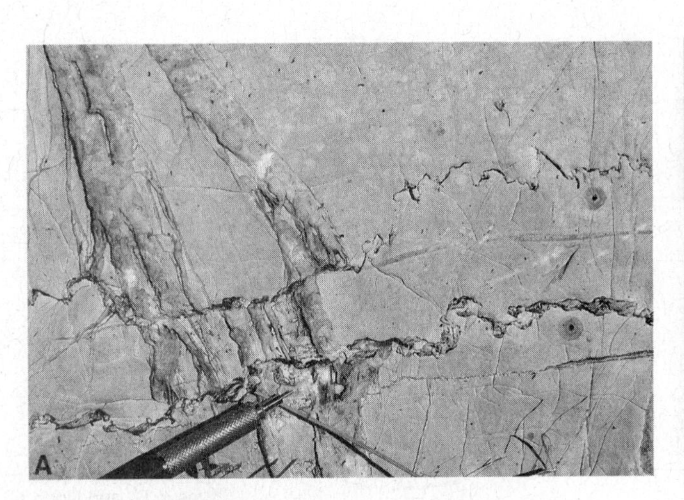
A

B

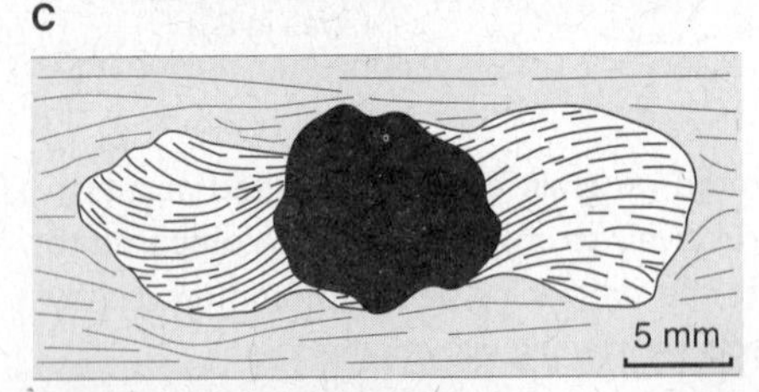
C
5 mm

D

Figure 4.37 Characteristics of dissolution creep. (*A*) Stylolites formed by dissolution of carbonate, leaving dark accumulations of more insoluble material, Papoose Flat, eastern California. [Photograph by S. J. Reynolds.] (*B*) Siliceous microveinlets and overgrowths appear dark in this Cathode Luminescence image of Silurian Tuscarora Sandstone, central Appalachian Mountains, Virginia. [Photograph by Charles M. Onasch.] (*C*) Pressure shadows with fibers around pyrite. [Reprinted from *Journal of Structural Geology*, v. 9, Etchecopar, A., and Malavielle, J., Computer Models of pressure shadows; a method for strain measurement and shear-sense determination, in Cobbold, P. R., Gapais, D., Means, W. D., and Treagus, S. H. (eds.), "Shear criteria in rocks." p. 667–677, © 1987, with permission from Elsevier.] (*D*) Crystal–fiber veins composed of gypsum in Triassic Moenkopi Formation, Lake Mead area, southern Nevada. [Photograph by S. J. Reynolds.]

Dissolution creep at a given site may ultimately become too inefficient to continue. For example, as grains change shape, the migration paths increase in length, resulting in a drop in the strain rate. Furthermore, dissolution and reprecipitation may result in cementation, compaction, and sealing of cracks and other defects, which in turn leads to strain hardening and the end of creep.

In some circumstances dissolved material may be transported by fluids great distances from the source. For example, dissolution creep is commonly accompanied by prograde metamorphism (increasing temperatures), which liberates water and other volatiles from minerals through dehydration reactions. The net increase in volume of fluid may drive fluids out of the system, carrying away the dissolved constituents. In such circumstances, the rocks that experienced diffusion creep may actually lose volume as a result of the deformation.

Conditions Favoring Dissolution Creep

Dissolution creep operates at low differential stresses (Figure 4.38) and can take place over a broad range of temperature-pressure conditions, provided an intergranular fluid is present. It can even operate efficiently at the relatively low temperatures that accompany diagenesis and low-grade metamorphism. For example, sequences of limestones are typically marked by relatively close-spaced bedding-parallel stylolitic surfaces that represent sites from which material was dissolved during the gravitational loading that accompanied burial and diagenesis. A wide range of rock types can accommodate dissolution creep, but this deformation mechanism works best in certain favored lithologies, such as shales, calcareous shales, and impure carbonate rocks. Fine-grained rocks are especially vulnerable to dissolution creep. Strain rate is inversely proportional to the cube of the grain size (Rutter, 1976); hence, grain-size reduction by dissolution leads to a quickening of the pace of dissolution, a type of strain softening.

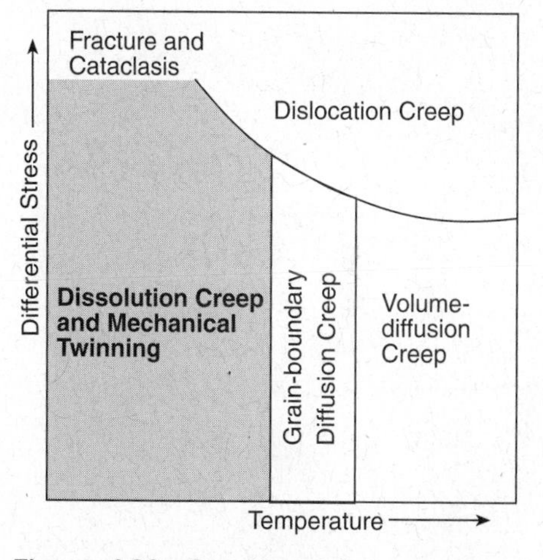

Figure 4.38 Simplified deformation map, showing conditions of dissolution creep.

Dislocation Creep

A mineral grain in a rock can also change shape by **dislocation creep**. The shape change is achieved by a combination of very localized, temporary distortion of the lattice and shearing of the lattice along a favorable crystallographic plane (the slip plane). If we eliminate for a moment the "*distortion*" part, the picture would be pretty simple.

For example, imagine a slip plane that cuts through the middle of a crystal (Figure 4.39*A*) and manages to move the entire upper half of the crystal one complete lattice spacing relative to the lower half of the crystal (Figure 4.39*B*). This mechanism is geometrically straightforward but mechanically difficult, because each increment of slip requires that *all* bonds along the slip plane be broken at the same time. Breaking all the bonds at the same time requires more shear stress than is generally available. Dislocation creep is necessary to solve the problem!

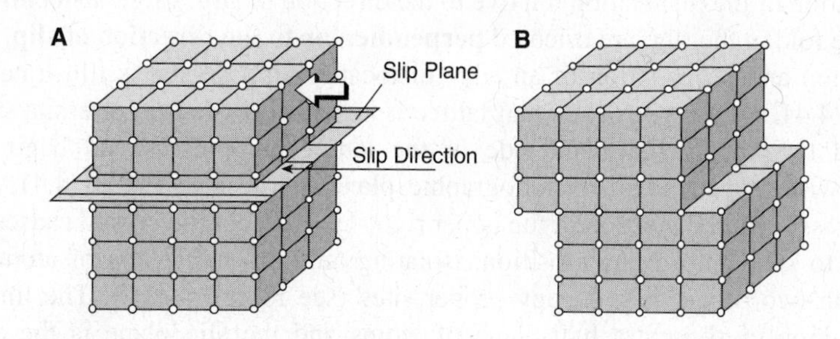

Figure 4.39 Slip of crystal along a slip plane. (*A*) Intact lattice, showing slip plane. (*B*) Offset lattice.

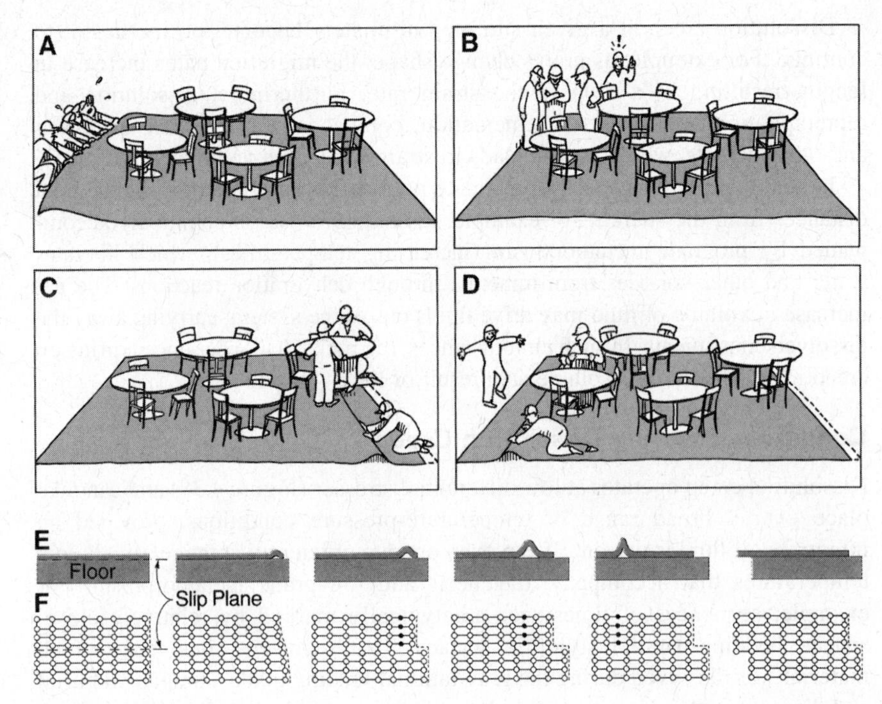

Figure 4.40 Slip systems do the impossible. (*A–D*) Moving a rug and furniture by propagating a fold across the rug. [Artwork by D. A. Fischer. Inspired by Hobbs, Means, and Williams (1976).]. (*E*) Rigid and nonrigid segments of the rug. (*F*) Rigid and nonrigid segments of a crystal during the propagation of a dislocation. [Rendering of crystal structure reprinted from *Metamorphic Textures* by A. H. Spry. Published with permission of Pergamon Press, Ltd., Oxford, © 1969.]

Nature's clever way around this problem is to activate only a small part of the slip plane at any one time, thereby breaking fewer bonds at once. Hobbs et al. (1976) compared this to the optimum way to move an enormous rug across a hardwood floor, and inspired us to create Figure 4.40. Picture a carpet so large and heavy and so laden with furniture that a line of people pulling on it from one side cannot budge it. To move (translate) the carpet without removing the furniture, it is necessary to curl the rug into a small anticlinal fold along one edge (this is the temporary, localized distortion), and systematically propagate the fold across the entire length of the rug. This way, the rug can be translated a short distance across the floor. This becomes a simple, though time-consuming job, because only a small part of the entire slip plane is active at any one time. The line that separates the part of the rug that has slipped from the part that has not slipped is a **dislocation** (see Figure 4.40*E,F*), which in the world of deformation mechanism is the boundary between slipped and unslipped parts of a crystal. (Remember the sponge cartoon, see Figure 4.15).

In a roundabout way, we have come to the definition. **Dislocation creep** is the production, motion, and destruction of dislocations through crystals and grains, accommodated by recovery and recrystallization.

Types of Dislocations

Dislocations in crystals form and propagate in several ways, depending on the orientation of the dislocation relative to the direction of slip. **Edge dislocations**, like the fold in the rug, are oriented **perpendicular to the direction of slip**. The formation and propagation of an edge dislocation in a crystal is illustrated in Figure 4.41. As the perfect crystal lattice is subjected to a shear stress, a small part of the upper right-hand side of the lattice compresses and begins to move with respect to a crystallographic plane below (see Figure 4.41A,B). Compression and distortion of the upper right-hand part of the crystal causes the lattice to snap into a new position, isolating an extra half-plane of atoms in between two others that occupy proper sites (see Figure 4.41*C*). The line of intersection of the extra half-plane of atoms and the slip plane is the edge

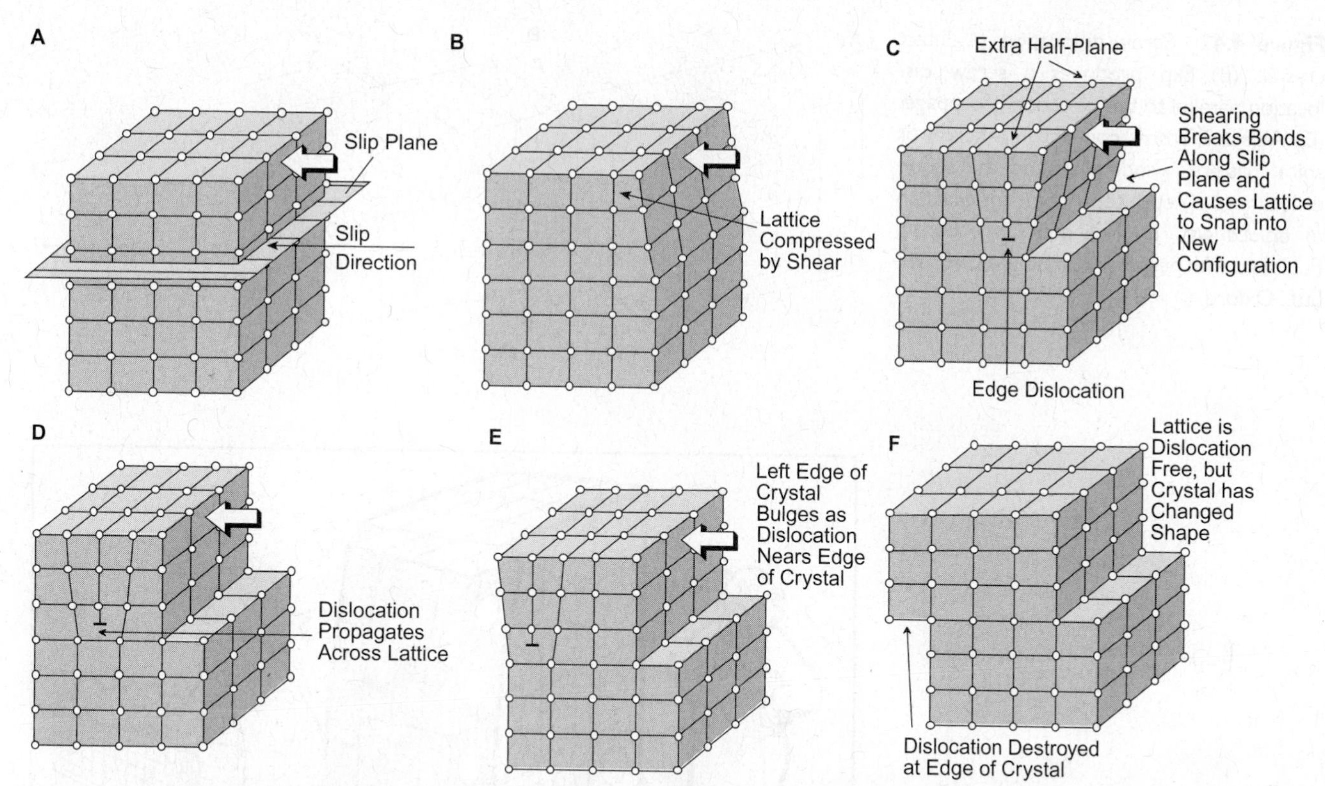

Figure 4.41 Formation and propagation of an edge dislocation. (*A*) Slip system within an undeformed crystal. White arrow indicates imposed stress on upper part of crystal. (*B*) Upper right part of crystalline lattice shortens elastically. (*C*) Stress exceeds yield strength of crystal, and lattice partially slips, forming an isolated extra half-plane of atoms above the slip plane. The bottom of the half-plane is an edge dislocation oriented perpendicular to the direction of slip. (*D* and *E*) Continued deformation causes the dislocation to propagate across the lattice, one step at a time. (*F*) Dislocation propagates to and offsets left side of crystal. The lattice is again defect free, but the crystal has changed shape. [Modified after Hobbs, Means, and Williams (1976).]

dislocation. It represents the boundary between the part of the lattice that slipped and the parts that did not. With continued stress-induced movement, the extra half-plane shifts to a proper site, forcing the previous occupant to move into an extra half-plane position (see Figure 4.41*D*). Thus the edge dislocation migrates one step at a time through the lattice.

When the extra half-plane reaches the edge of the crystal, it emerges as a small offset and the dislocation ceases to exist (see Figure 4.41*E,F*). The crystal has changed its shape ever so slightly and, for an instant, is once again defect free—until a new dislocation is produced in the same manner as the first. Through this recurring cycle of formation and propagation of edge dislocations, the crystal changes shape in a *plastic* manner. This process of propagation of a dislocation through a crystal, called **dislocation glide**, occurs on crystallographic planes along which bonds are relatively weak.

Screw dislocations are more difficult to picture. They are oriented *parallel to the direction of slip* (Figure 4.42). One way to visualize a screw dislocation is to think of a stack of cereal boxes on the shelves of the neighborhood grocery store (Figure 4.43). When originally stocked on the shelves, the boxes were neatly aligned in rows. Later, however, someone bumps the shelves, causing one column of boxes to tilt over. Boxes at the base of the tilted column are still essentially aligned with the boxes in the undisturbed column beside it, but boxes at the top of the tilted column have become aligned with the next box back in the undisturbed columns. The tilted boxes have retained their original "*lattice position*" at the base, but have been sheared backward one "*lattice*

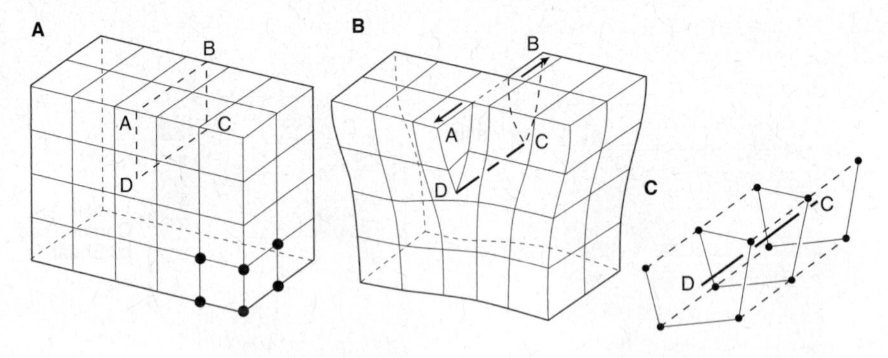

Figure 4.42 Screw dislocation. (*A*) Intact crystal. (*B*) Slip produces a screw dislocation parallel to the direction of slippage. (*C*) Adjacent atoms connect in a helical or spiral pattern, winding around the screw dislocation. [Reprinted from *Introduction to dislocations*, D. Hull and D. S. Bacon. Published with permission of Pergamon Press, Ltd., Oxford, © 1984.]

Figure 4.43 Grocery store analogy of screw dislocation. [Artwork by D. A. Fischer.]

position" at their top. This pivoting type of shear, accommodated by the vertical slip plane separating tilted and nontilted boxes to the right and left, creates a screw dislocation.

Edge and screw dislocations are generally not straight features, but instead curve via a series of offsets, either within the slip plane or from one slip plane to another. Through such offsets, an edge dislocation can curve into and become a screw dislocation, and vice versa. Dislocations can consist of both edge and screw segments and be oblique to the direction of slip. They can also form closed **dislocation loops**, which encircle the part of the crystal that has slipped. Dislocation loops may start as a point or very small loop, perhaps centered on a stress-concentrating vacancy or impurity, and progressively grow outward in multiple directions as slip continues.

Interactions with the Lattice and Strain Hardening

As a dislocation is propagated through a crystal, it may encounter and interact with other features of the lattice, such as vacancies, interstitial atoms, impurities, and even other dislocations. The motion of a dislocation may be impeded or stopped when the associated extra half-plane encounters an interstitial atom, impurity, or another dislocation. This will occur if the energy required to bypass the obstacle or incorporate it into the advancing half-plane exceeds the energy driving the dislocation through the lattice.

One dislocation can impede another, depending on the geometry of the interaction. Dislocations can overlap, becoming locked or "*pinned*," unable to move further. Dislocations moving in the same direction may act like cars and trucks driving blind in an afternoon dust storm, piling up on one another when the leading dislocation encounters an obstacle that simply cannot be overwhelmed. It takes a tremendous amount of work and effort to clear the way for traffic to resume.

Two dislocations utilizing *different slip systems* may also become pinned where they intersect. A number of dislocations can intersect and become pinned in a seemingly tangled mess, producing the atomic equivalent of a four-way stop sign (Figure 4.44). Pinning and tangling of dislocations result in a **strain hardening** of the lattice. As a result, dislocation glide becomes increasingly difficult as more and more dislocations are created within a crystal. Continued deformation, therefore, *requires* that dislocations somehow be destroyed or be able to bypass obstacles. Fortunately, nature always seems to have a solution.

Figure 4.44 TEM (transmitting electron microscope) image of a mess of tangled dislocations. [Courtesy of Andrew Kronenberg.]

Recovery and Recrystallization

Dislocation glide introduces new defects into a crystal, but the crystalline structure can be healed by eliminating as many defects as possible and by depleting leftover stored energy in the lattice. **Recovery** promotes healing through the rearrangement and destruction of dislocations. **Recrystallization** and **neomineralization** promote healing by transforming the old "*defective*" grains into brand-new grains or new configurations of grains. Recrystallization takes place within a single grain or within adjacent grains of a common mineralogy (Figure 4.45). Neomineralization, on the other hand, forms new

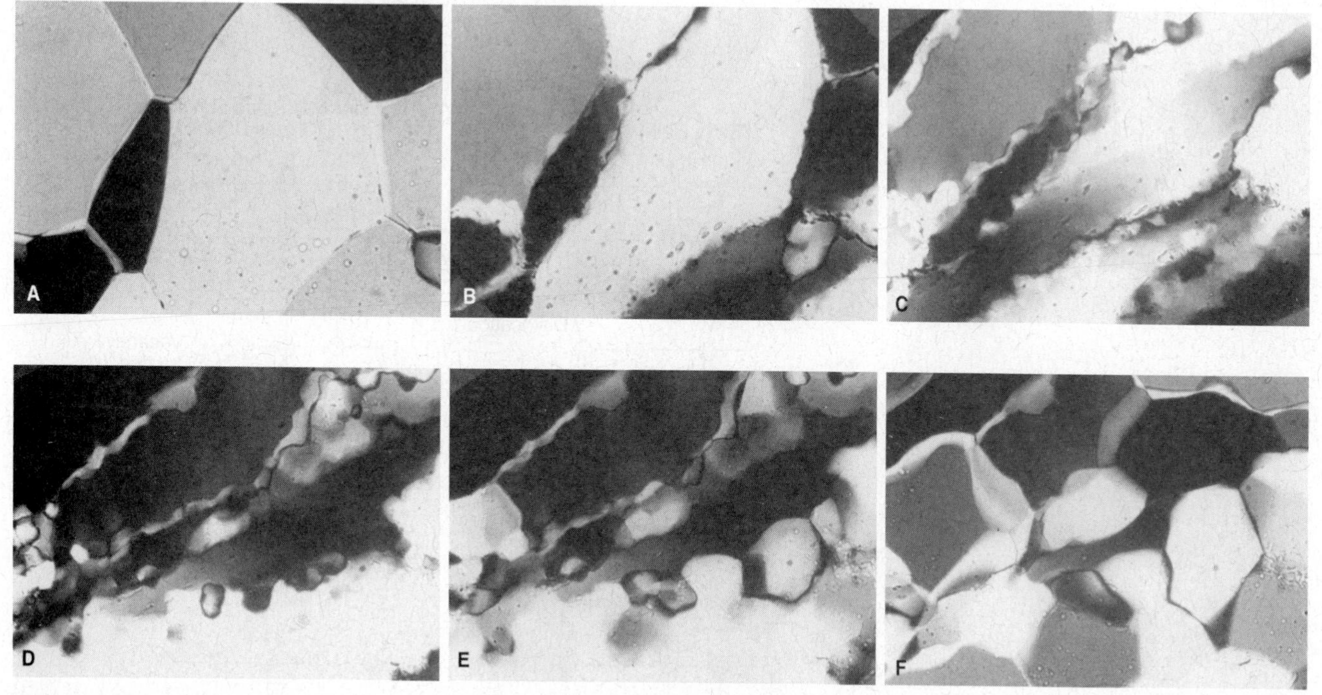

Figure 4.45 Recrystallization produced experimentally by deforming octachloropropane, a synthetic material that deforms ductilely at low temperatures. (*A*) Starting material with undeformed grains bounded by straight boundaries that meet at 120° angles. Bubbles within grains can be used to monitor the amount of strain during subsequent deformation. (*B–D*) Progressive dextral (right-handed) shear, resulting in internal deformation of grains, as recorded by the elliptical shape of bubbles (miniature strain ellipses). Dynamic recrystallization during deformation causes bulges in grain boundaries, some of which become new, small grains in *D*. (*E–F*) Recrystallization and recovery after deformation (annealing) result in growth of grains, the healing of internal defects, and the straightening of grain boundaries. Annealing leads to a fabric in *F* similar to the initial, undeformed material in *A*. [Courtesy of Win Means, SUNY, Albany.]

minerals or changes the boundary between two *different* minerals that were in contact (Urai et al., 1986).

Recovery and recrystallization *during* deformation constitute **dynamic recrystallization** (see Figure 4.45*B–D*). Recovery and recrystallization *after* deformation bring about **annealing** (see Figure 4.45*E, F*). Dynamic recrystallization counteracts strain hardening and accommodates the continued glide of dislocations, permitting the rock to sustain steady-state flow via dislocation creep. Dislocation creep is possible only where dynamic recrystallization continuously keeps pace with strain hardening, eliminating dislocations that have become pinned or tangled. Through the process of dynamic recrystallization, recovery and recrystallization essentially control the rate of dislocation creep. When dynamic recrystallization and annealing do their work, many of the ordinary telltale signatures of microstructural deformation are eliminated, creating the illusion that the minerals were never deformed.

Recovery and Rotation Recrystallization

Recovery is achieved principally through the process of **dislocation climb**, which involves the movement of a dislocation to a higher or lower slip plane when the diffusion of vacancies and atoms lengthens or shortens the associated extra half-plane of atoms (Figure 4.46). The migration of an atom toward the bottom of the extra half-plane (Figure 4.46*A, B*) causes the half-plane to lengthen and the dislocation to 'climb' downward. In contrast, the dislocation 'climbs upward' when the bottom atom in the half-plane is able to jump into a nearby vacancy (see Figure 4.46*C, D*).

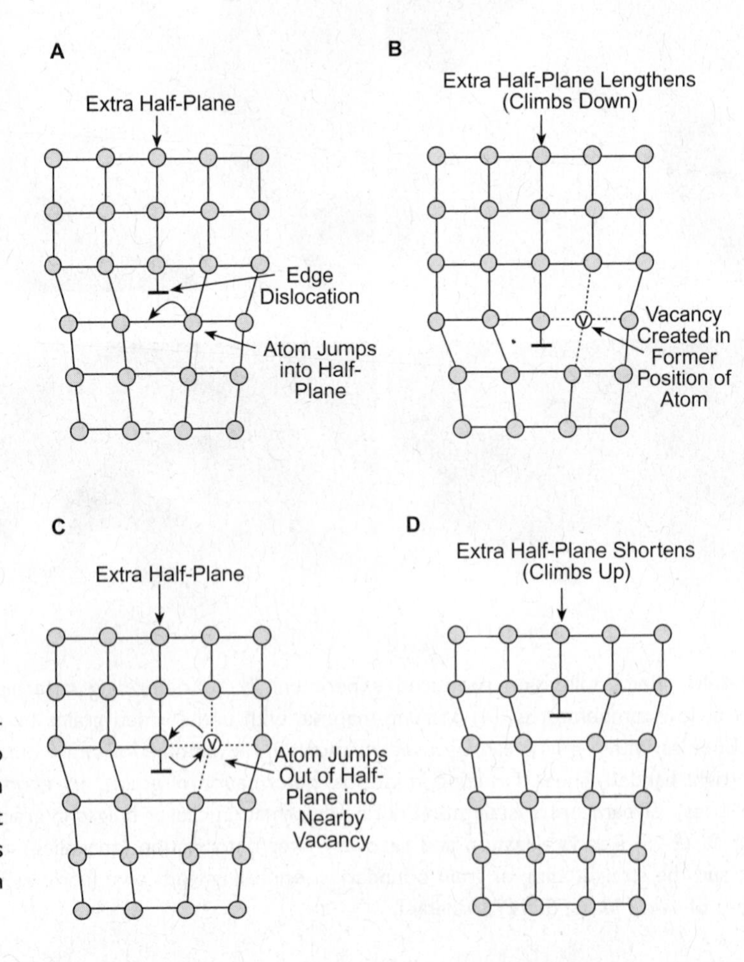

Figure 4.46 Dislocation climb. (*A–B*) Dislocation climbs down as an adjacent atom jumps into the half-plane, lengthening it. (*C–D*) Dislocation climbs up as an atom jumps into an adjacent vacancy, shortening the half-plane. Dislocation is destroyed if it can climb all the way to the grain boundary.

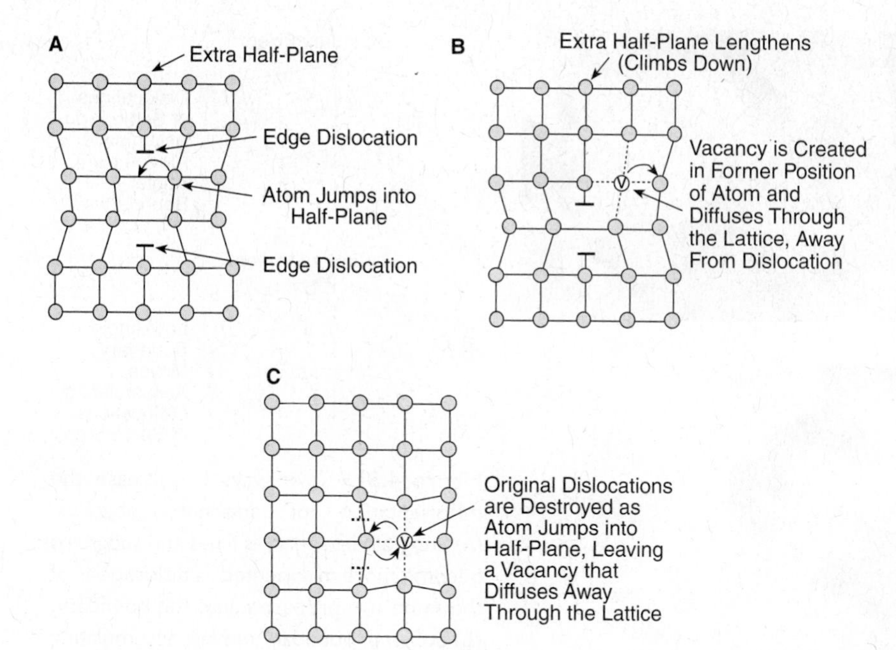
A
Extra Half-Plane
Edge Dislocation
Atom Jumps into Half-Plane
Edge Dislocation
B
Extra Half-Plane Lengthens (Climbs Down)
Vacancy is Created in Former Position of Atom and Diffuses Through the Lattice, Away From Dislocation
C
Original Dislocations are Destroyed as Atom Jumps into Half-Plane, Leaving a Vacancy that Diffuses Away Through the Lattice

Figure 4.47 Two dislocations climb into and annihilate one another.

Dislocation climb can do one of several things. Dislocation climb can permit a dislocation to bypass an obstacle and continue slipping. It can permit a dislocation to exit the grain by migration all the way to a grain boundary. Or it can permit the mutual annihilation of two dislocations migrating toward each other from opposite directions (Figure 4.47). Recovery has several other ways that dislocations can bypass an obstacle, such as by changing crystallographic planes.

Alternatively, dislocation climb may permit adjacent dislocations to arrange themselves into **walls** separating **low-angle boundaries** between slightly misoriented parts of a crystal (Figures 4.48, 4.49). These misoriented parts, termed **subgrains**, are commonly expressed microscopically in thin section as small areas of a grain with extinction angles slightly different from what is observed for the rest of the grain (Figure 4.50). Such progressive misalignment of the lattice is the cause of undulatory extinction (see Figure 4.50*B*).

The crystallographic misorientation between adjacent subgrains can increase during deformation in two main ways (Figure 4.51) (Urai et al., 1986): (1) the boundary between the subgrains can remain stationary and collect dislocations moving toward the subgrain boundary from one or both sides; or (2) the boundary can migrate through the material and accumulate dislocations as it

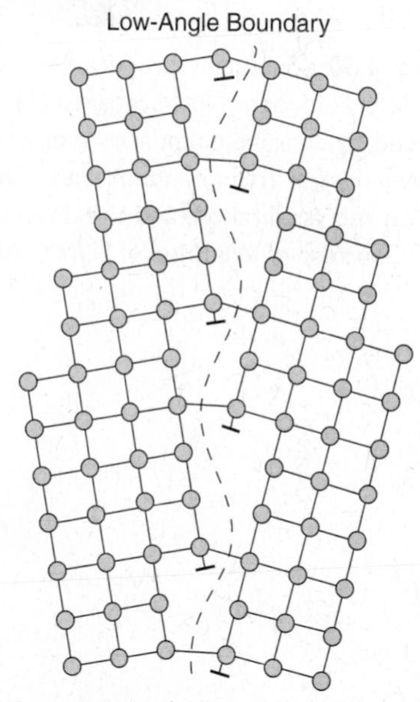
Low-Angle Boundary

Figure 4.48 Low-angle boundary, where edge dislocations permit a slight misalignment of two adjacent parts (subgrains) of a crystal. [Modified after Twiss and Moores (1992) and Hobbs, Means, and Williams (1976).]

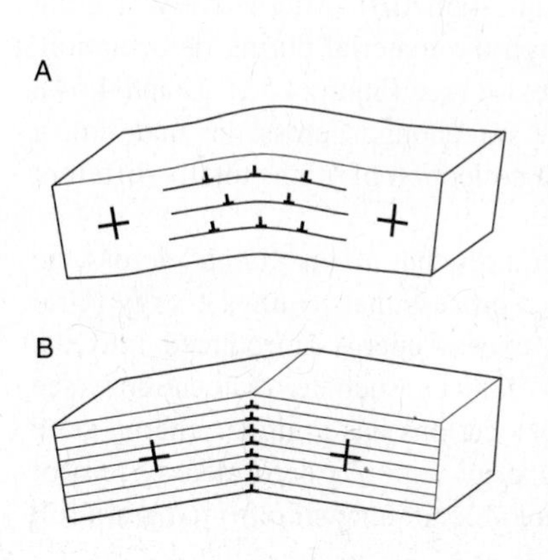
A
B

Figure 4.49 Low-angle boundary produced when dislocations become arranged into a wall. (*A*) Widely distributed dislocations of the same sign produce a gentle bending of the lattice, which would be visible in thin section as undulatory extinction. (*B*) Dislocations arranged into a wall separating two slightly misaligned subgrains. [Reprinted from *Introduction to dislocations*, D. Hull and D. S. Bacon. Published with permission of Pergamon Press, Ltd., Oxford, © 1984.]

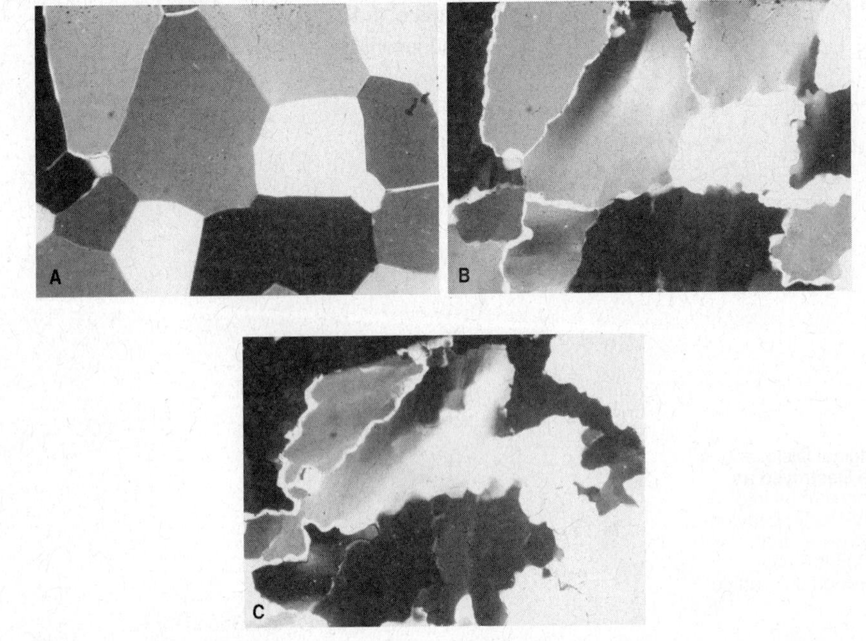

Figure 4.50 Subgrains produced by rotation recrystallization within experimentally deformed octachloropropane. (*A*) Starting material with undeformed grains bounded by straight boundaries that commonly meet at 120° angles. (*B–C*) Progressive dextral (right-handed) shear, resulting in internal deformation of grains, rotation recrystallization, and the development of subgrains and undulatory extinction. (Courtesy of Win Means, SUNY, Albany.)

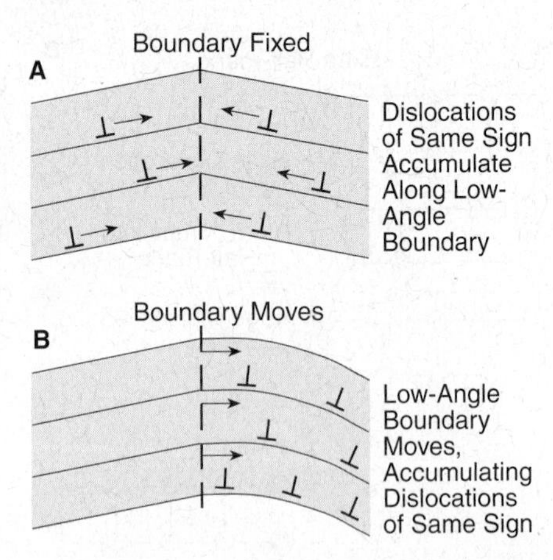

Figure 4.51 Two ways to increase the misorientation of adjacent subgrains. (*A*) Subgrain boundary is fixed and subgrains become more misoriented as dislocations of the same sign propagate into the boundary. (*B*) Subgrain boundary moves, accumulating dislocations of the same sign. [From J. L. Urai, W. D. Means, and G. S. Lister, *Geophysical Monograph*, v. 36, Figure 7, p. 167, copyright © 1986 by American Geophysical Union.]

goes. If by either process the crystallographic mismatch between the two subgrains exceeds some threshold, perhaps 10–15°, the boundary is considered a **high-angle boundary** separating distinct grains rather than subgrains.

This process of creating new grains by the formation and accentuation of low-angle boundaries and subgrains is termed **rotation recrystallization**. Evidence for the process in thin section includes **core and mantle structure** when an intact central core of a grain grades successively outward into subgrains and then into aggregates of new, recrystallized grains (Figure 4.52).

Boundary-Migration Recrystallization

Recrystallization may also occur via the migration of grain boundaries, known as **boundary-migration recrystallization** (Figure 4.53). A grain may remain stationary and grow at the expense of a neighboring grain, adding to its collection of atoms as it does so (see Figure 4.53*A*,*B*). Alternatively, a grain may be mobile and migrate laterally through the material during deformation, changing size, shape, *and* position as it does so (see Figure 4.53*C*,*D* and 4.54). If the boundaries of such a grain migrate far enough across the material, a recrystallized grain may end up enclosing a collection of atoms totally different from the atoms with which it started.

Boundary-migration recrystallization requires that atoms "*jump*" across the grain boundary from one grain to another, a process that requires energy (Urai et al., 1986). That energy source is the excess energy introduced into the grains during deformation and stored in defects, such as dislocations (see Figure 4.53*C*,*D*). A grain that is full of dislocations has a higher energy state than a less deformed grain. The overall energy within a deformed system of minerals is decreased, therefore, if atoms are able to jump ship from a grain full

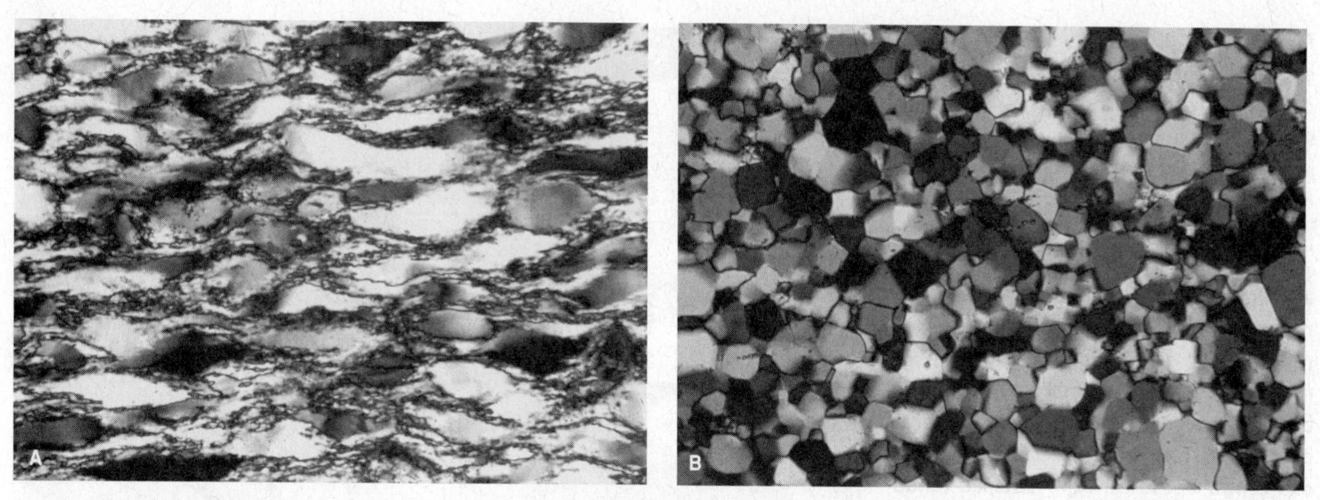
A
B

Figure 4.52 Photomicrograph of experimentally deformed Black Hills Quartzite. (*A*) Shortened 60%, producing flattened original grains with sweeping undulatory extinction, some visible subgrains, and a fringe of small recrystallized grains forming incipient core and mantle structure. Recrystallization in this sample has occurred primarily by subgrain rotation (Hirth and Tullis, 1992). (*B*) Same sample annealed by increasing the temperature after deformation had ceased. Grains are now essentially strain free, as indicated by the even extinction of individual grains and relatively straight grain-boundaries. [Photographs kindly provided by Jan Tullis, Brown University.]

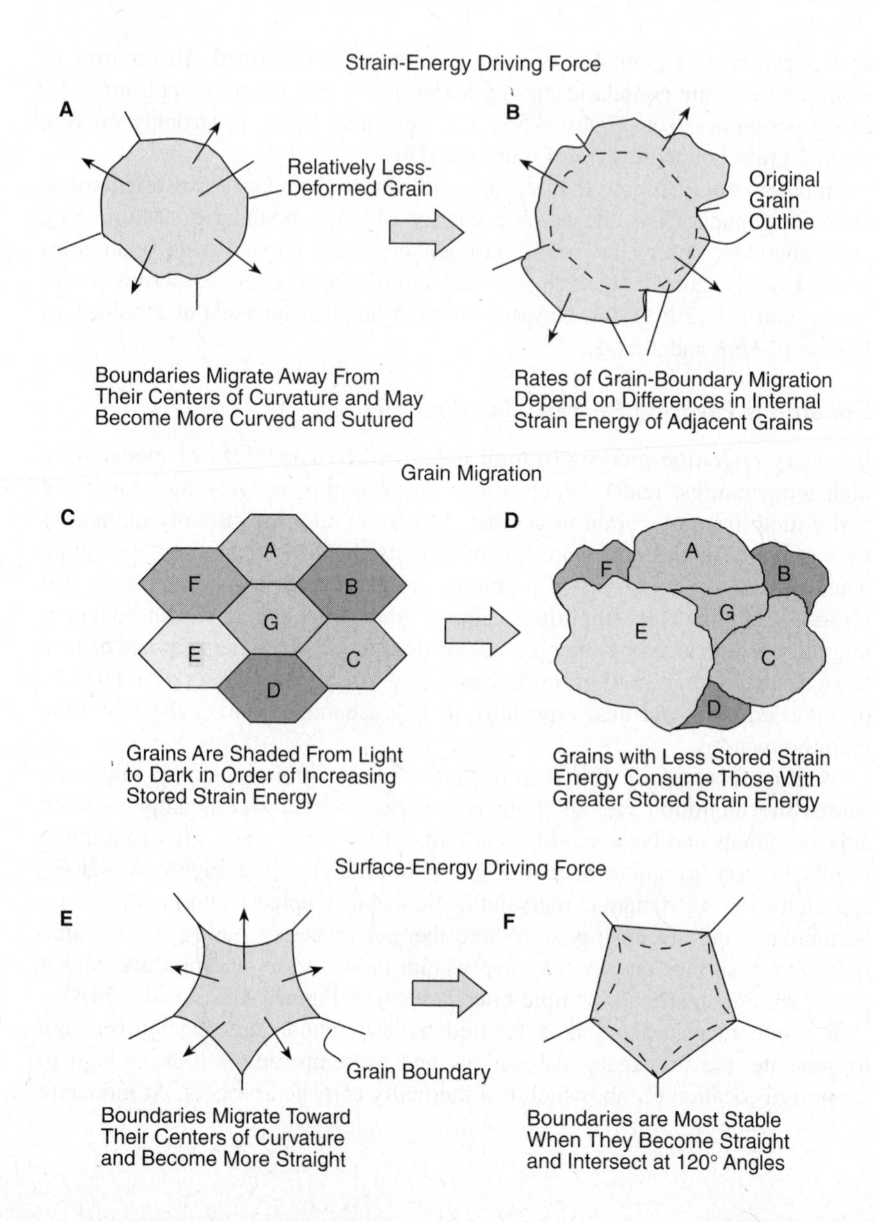
Strain-Energy Driving Force
A
B
Relatively Less-Deformed Grain
Original Grain Outline
Boundaries Migrate Away From Their Centers of Curvature and May Become More Curved and Sutured
Rates of Grain-Boundary Migration Depend on Differences in Internal Strain Energy of Adjacent Grains
Grain Migration
C
D
A
F
B
G
E
C
D
Grains Are Shaded From Light to Dark in Order of Increasing Stored Strain Energy
Grains with Less Stored Strain Energy Consume Those With Greater Stored Strain Energy
Surface-Energy Driving Force
E
F
Grain Boundary
Boundaries Migrate Toward Their Centers of Curvature and Become More Straight
Boundaries are Most Stable When They Become Straight and Intersect at 120° Angles

Figure 4.53 Processes of recrystallization via grain boundary migration. (*A–B*) Recrystallization is driven by stored strain energy, leading to more irregular, serrated grain boundaries. (*C–D*) Migration of grain boundaries as less strained grains consume more strained ones, permitting grains to migrate laterally through the rock. (*E–F*) Recrystallization is driven by surface energy, whereby grain boundaries become more straight and tend to intersect at 120° angles. [In part after Nicolas and Poirier, 1976; Urai, Means, and Lister, 1986.]

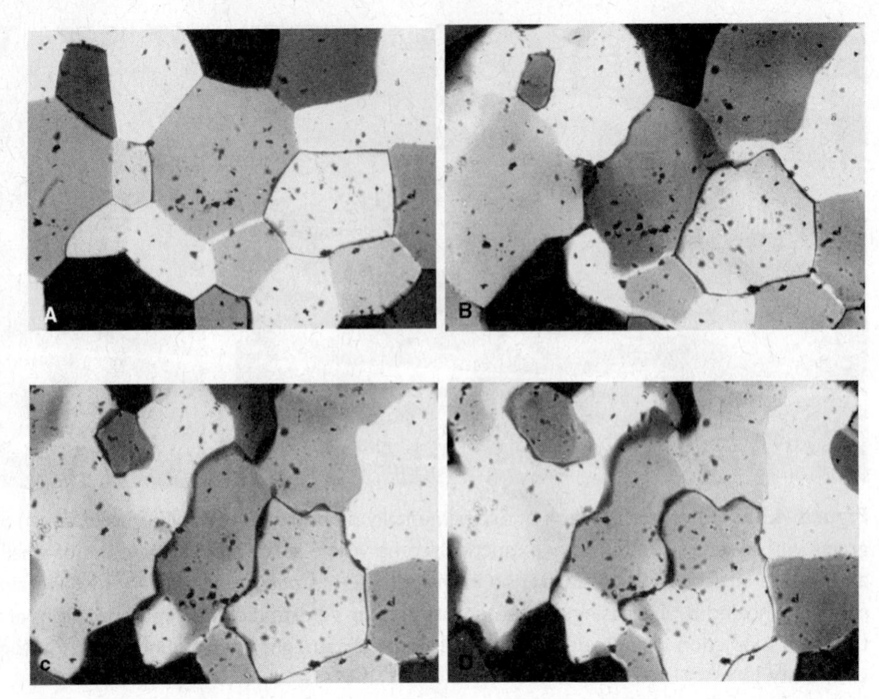

Figure 4.54 Photomicrographs of the effects of recrystallization in experimentally deformed octachloropropane. (*A*) Initial sample, containing opaque particles of grinding powder that permit us to track the motion of grain boundaries and to estimate the amount of strain. The sample displays straight boundaries with 120° grain-boundary intersections, such as is common in rocks that have been thermally annealed. (*B–D*) Progressive horizontal shortening results in dynamic recrystallization by grain-boundary migration, leaving serrated grain boundaries. Note that the dark grain at the top center becomes highly strained (undulatory extinction and subgrains) and is almost totally eaten by its less deformed neighbors. [Photographs courtesy of Win Means, SUNY, Albany.]

of dislocations to a grain that is much less internally deformed. This mutiny of atoms causes grain boundaries to migrate as defect-rich grains are consumed by defect-poor ones (see Figure 4.53*C,D*). This may result in strongly curved, sutured grain boundaries (see Figure 4.54*D*).

Another energy source that strongly influences the shapes and orientations of grain boundaries is the surface energy of grain boundaries. Minimizing grain-boundary energy favors (1) straight or gently curved grain boundaries over strongly curved boundaries, (2) large grains over numerous small grains, and (3) 120° grain-boundary intersections that intersect at a point (see Figure 4.53*E,F* and 4.54*A*).

Conditions Favoring Recrystallization

Boundary-migration recrystallization is favored by conditions of moderate to high temperatures, under which atoms are vibrating actively and can more easily jump from one grain to another. Migration rates are strongly influenced by variations in the concentration of defects, by the orientation of a grain boundary relative to the crystallographic orientation of the crystal, and by the presence of fluid or impurities along grain boundaries. Grain-boundary migration is enhanced by the presence of pore fluid, which helps atoms diffuse more easily from grain to grain. In contrast, grain-boundary migration tends to be impeded by impurities, especially if they accumulate along the migrating grain boundaries.

The interplay of all the factors that influence grain-boundary migration causes the migration rate of grain boundaries to vary significantly between adjacent grains and between different parts of the same grain. This commonly results in very irregular, **serrated grain boundaries** (see Figure 4.54*B-D*), especially during dynamic recrystallization, when defects continually form, accumulate, and are destroyed. Where thermal processes outlast deformation (annealing), surface energy may overwhelm these factors and produce equant grains with geometrically simple boundaries (see Figures 4.52*B* and 4.54*A*).

Rotation recrystallization is favored by differential stresses high enough to generate and propagate dislocations and by temperatures high enough to support dislocation climb, which is a thermally activated process. At moderate temperatures, where climb is slow, rotation recrystallization occurs, but it is outpaced by grain-boundary migration, which then dominates the generation of microstructures. Because rotation and grain-boundary recrystallization are both thermally activated, dislocation creep is most efficient at moderate-to-high temperatures (Figure 4.55).

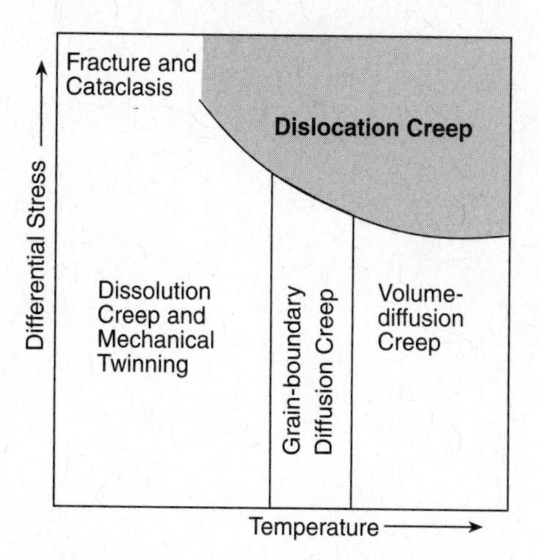

Figure 4.55 Simplified deformation map, showing the conditions of dislocation creep.

Summary

Nature has a vast array of mechanisms that permit rocks to deform. Which deformation mechanism operates depends on a long list of factors, especially mineralogy, grain size, temperature, differential stress, confining pressure, strain rate, the presence or absence of fluids, and fluid pressure. Several deformation mechanisms may be active in a deforming rock mass at the same time, and these may interact in important, but rather complex, ways. One mechanism may dominate the deformation until some strain limitation is reached, whereupon a different mechanism takes over. Dislocation glide may accommodate only a small amount of strain at low temperatures before strain hardening and stress concentration lead to brittle fracture. In other cases, two mechanisms may proceed synchronously and work together to permit more strain, as in the case of dislocation creep accompanied by dynamic recrystallization. Also, two processes may interfere with each other to such an extent that neither one can operate. For example, frictional sliding can cause microcracking and dilation, which lowers the fluid pressure, thereby inhibiting more frictional sliding and microcracking until either the shear stress or fluid pressure builds back up again.

If a rock has more than one material, such as several types of minerals, each material may deform by a different mechanism, at the same time. A strong feldspar may be fracturing while an adjacent quartz grain deforms by dislocation creep. The percentages and arrangement of different materials influences how each material deforms and controls the strength of the rock as a whole. In some rocks, strong grains are so abundant that they rest on each other, forming a **load-bearing framework** that helps shelter weaker minerals from deformation (Hardy, 1990; Tullis et al., 1990). An opposite end member is where stronger grains are completely encased in a matrix of weaker material, called an **interconnected weak layer**. Of the two end members, a load-bearing framework is inherently stronger, but over time can be unstable because the strong grains concentrate stress, causing them to fail.

DEFORMATION EXPERIMENTS

An exacting and technically challenging endeavor is the laboratory study of creep, recovery, and recrystallization. To try to generate diffusion creep, dissolution creep, and dislocation creep during short-term experiments, and to factor in the role of recovery and recrystallization, the experiments must be carried out under conditions of very high temperature and confining pressure. Moreover, it is typically necessary to use very small samples of fine-grained materials. The experimental work carried out in this way has yielded important information on the strength of minerals, the mechanisms by which minerals deform, and the conditions under which each deformation mechanism operates.

The starting materials for such experiments vary, depending on the scientific goals. Typical materials are natural or synthetic single crystals of quartz or calcite or olivine, natural quartzite, artificially sintered quartz or olivine aggregates, natural or synthetic quartz-feldspar aggregates (fine-grained granite), various types of calcareous rocks (limestone and marble), and rock salt. To evaluate the conditions favoring each deformation mechanism, the same starting material is deformed under a range of conditions of temperature, confining pressure, differential stress, and strain rate. The mechanical behavior of the material is monitored, using the relation between stress, strain, and strain rate throughout the experiment. At the end of the run, the deformed specimen is examined closely, using various microscopic techniques to characterize the microstructures and to evaluate which deformational mechanisms operated during the run under the given set of conditions.

Creep Experiments

Experiments and theories of rock deformation indicate that the three fundamental creep mechanisms (diffusion, dissolution, and dislocation) may occur in a single rock at the same time, but each mechanism will operate at a different rate, depending on temperature, pressure, differential stress, mineralogy, grain size, and availability of fluid. The physical conditions under which each creep mechanism is dominant can be illustrated on **deformation maps** (Figure 4.56), such as those that plot differential stress versus temperature *or* differential stress versus grain size (Ashby, 1972; Rutter, 1976). Strain rate is generally shown on

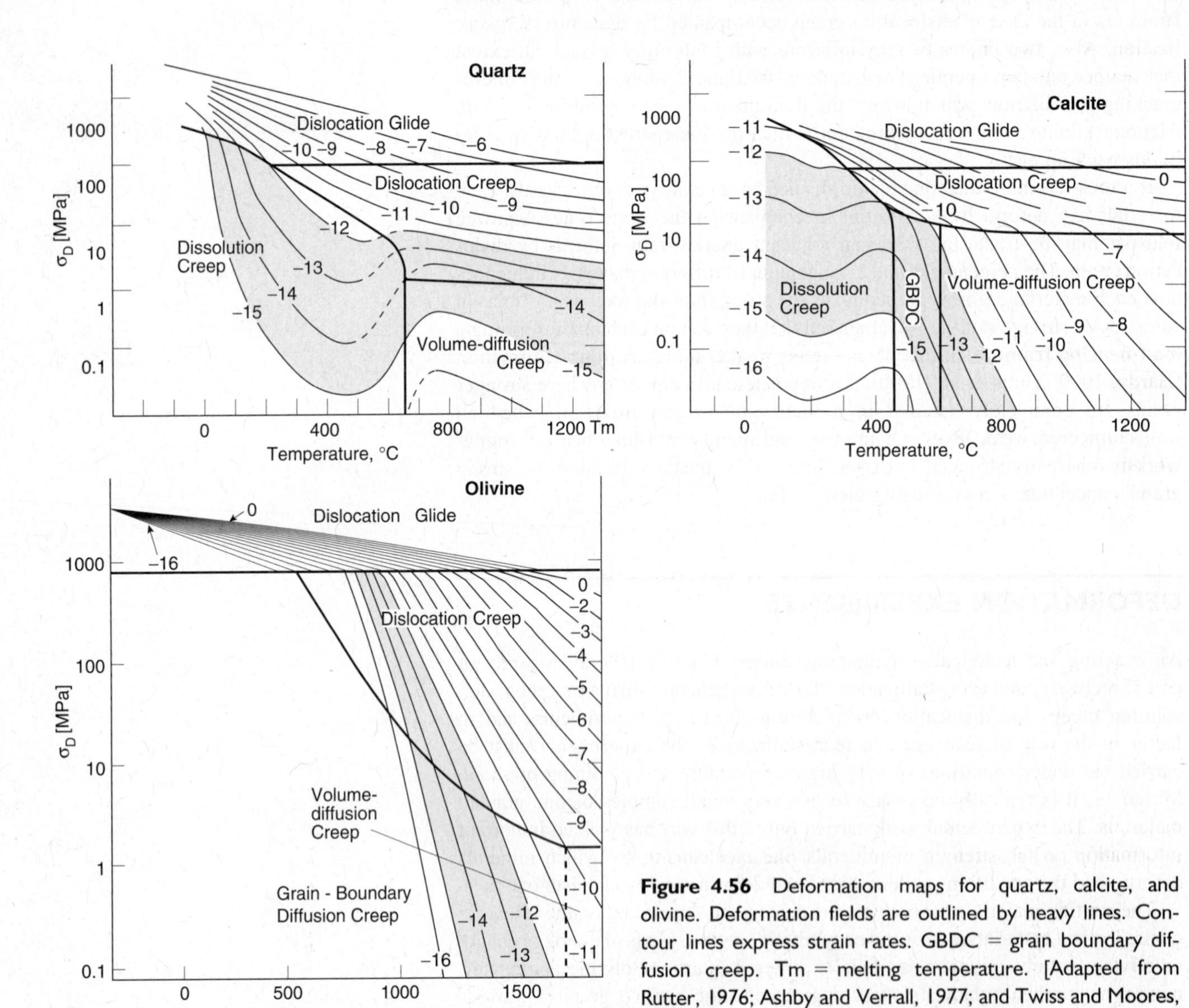

Figure 4.56 Deformation maps for quartz, calcite, and olivine. Deformation fields are outlined by heavy lines. Contour lines express strain rates. GBDC = grain boundary diffusion creep. Tm = melting temperature. [Adapted from Rutter, 1976; Ashby and Verrall, 1977; and Twiss and Moores, 1992.]

deformation maps by a series of contours that cross from one **deformation field** into another. Deformation maps are constructed partly from theories of deformation and partly from extrapolations of results from the rock deformation experiments.

Deformation maps for creep mechanisms of individual mineral species (see Figure 4.56) generally show that dissolution creep is the dominant creep mechanism at low stresses and low temperature, where "*low temperature*" is considered to be less than 40% of the melting temperature of the mineral. Higher stress favors dislocation creep, and higher temperature favors diffusion creep (see Figure 4.56).

Dislocation creep is characterized by relatively rapid strain rates compared to diffusion and dissolution creep, and it occurs over a broad range of temperatures. At low temperatures, however, dislocation creep can accommodate only a small amount of strain before dislocations become pinned and tangled by other dislocations, or grains change shape and impinge against adjacent grains in the rock. Low-temperature dislocation creep therefore leads to strain hardening and the concentration of stresses at grain boundaries and other obstacles to glide, commonly resulting in fracture (Lloyd and Knipe, 1992).

At higher temperatures or slower strain rates, dislocation creep is possible because dislocations can bypass obstacles by dislocation climb and other processes. Recovery by climb, because it involves diffusion, is thermally activated and most effective at higher temperatures. Likewise, recrystallization, which permits dislocation creep to continue by undoing the effects of strain hardening, is most effective at higher temperatures.

Theoretically derived equations for the different types of creep are listed in Table 4.1. For diffusion and dissolution creep, strain rate is directly proportional to differential stress, and thus the behavior is comparable to that of a linear viscous (Newtonian) fluid. For most models of dislocation creep (Twiss and Moores, 1992), strain rate is proportional to the differential stress raised to some power, yielding a behavior referred to as **power-law creep**. Dislocation creep is therefore capable of generating much higher strain rates for a given increment of differential stress.

For all three varieties of creep (i.e., diffusion, dissolution, and dislocation), strain rate is proportional to temperature because all are "*rate limited*" by diffusion, which is thermally activated. Diffusion creep and dissolution creep are most effective in fine-grained rocks because strain rate is inversely proportional to grain size raised to a power of 1 to 3. This means that reduction of grain size during deformation will lead to pronounced strain softening.

Recovery-Recrystallization Experiments

Experimental results have documented different regimes of deformation, which depend on the relative rates of dislocation production, dislocation climb, and grain-boundary migration (Hirth and Tullis, 1992). Each regime is characterized by a distinctive microstructure that largely reflects the operation of different mechanisms of dynamic recrystallization (Figure 4.57).

At relatively high temperatures and slower strain rates, the rate of dislocation climb is fast enough to accommodate dislocation creep via both rotation recrystallization and grain boundary migration (Tullis et al., 1990). The resulting microstructure displays abundant quartz "*ribbons*," subgrains, recrystallized grains, and irregular, sutured grain boundaries (see Figure 4.57*A*, *B*). Subgrains and recrystallized grains become larger at successively higher temperatures and begin to more closely resemble those produced by annealing (Tullis, 1990b). As a result of easy climb, dislocations have a fairly uniform distribution within grains and are curved, because segments were able to climb out of the slip plane (Figure 4.58).

TABLE 4.1
Some slip systems for common minerals

Mineral	*Slip system*
Halite	$(110)[1\bar{1}0]$
	$(001)[1\bar{1}0]$
Calcite	$(10\bar{1}1)[\bar{1}012]$
	$(\bar{2}021)[1\bar{1}02]$
Mica	$(001)[110]$
Quartz	$(0001)[11\bar{2}0]$
	$(10\bar{1}0)[0001]$
	$(10\bar{1}0)[1\bar{2}10]$
	$(10\bar{1}0)[1\bar{2}13]$
Olivine	$(100)[001]$
	$(110)[001]$
	$(110)[001]$
	$(010)[100]$

The value in parentheses is the Miller indices of the slip plane, indicating where the plane intersects the crystallographic axes for that mineral. The slip direction within that plane is indicated by the number within brackets. Some slip systems given have symmetrical equivalents. If more than one slip system is given for a mineral, the low-temperature slip systems are listed first.
Sources: Compiled from Hobbs, Means, and Williams, 1976; Nicolas and Poirier, 1976; Nicolas, 1988; and Twiss and Moores, 1992.

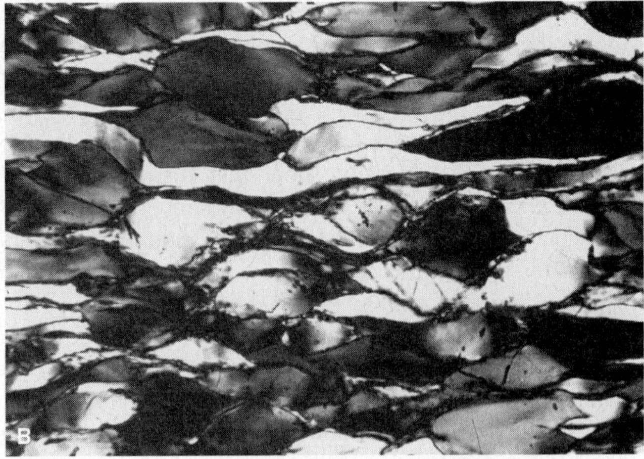

Figure 4.57 Microstructures of experimentally deformed aplite. (*A*) Starting material, consisting of fine-grained (0.15 mm) quartz and feldspar with relatively unstrained grains. (*B*) Deformation under conditions of high temperature and slower strain rate permits dislocation climb, resulting in strongly deformed ribbons and grains of quartz with subgrains. Some sutured grain boundaries also record grain-boundary migration. (*C*) Climb is limited during deformation at lower temperature and higher strain rate, resulting in a lack of subgrains. Recrystallization by grain-boundary migration produces some sutured boundaries and other evidence for grain-boundary migration (see also Figure 4.54 and Hirth and Tullis, 1992). [Photographs courtesy of Jan Tullis, Brown University].

At moderate temperatures or higher strain rates, rotation recrystallization is not important because dislocation climb, which is limited by the rate of diffusion, is too slow. Instead, dislocation creep is accommodated by grain-boundary migration (Tullis et al., 1990). The resulting microstructure lacks subgrains but contains abundant sutured grains and other evidence for grain boundary migration (see Figure 4.57*C*). Dislocations are relatively straight, and tangles are present because the dislocations could not climb out of the slip plane (see Figure 4.58*B*). Also there are large variations in dislocation density between strongly deformed "*old*" grains that contain numerous dislocations, and recrystallized "*new*" grains with relatively few dislocations. Once formed, the "*new*" recrystallized grains concentrate further deformation because they are strain-free and weaker than the strain-hardened "*old*" grains (Tullis et al., 1990).

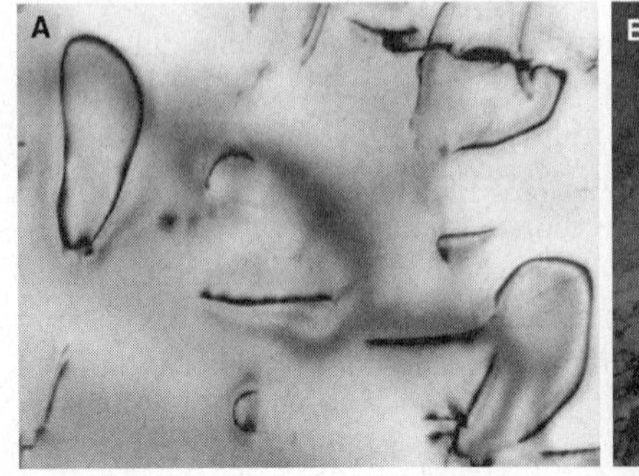

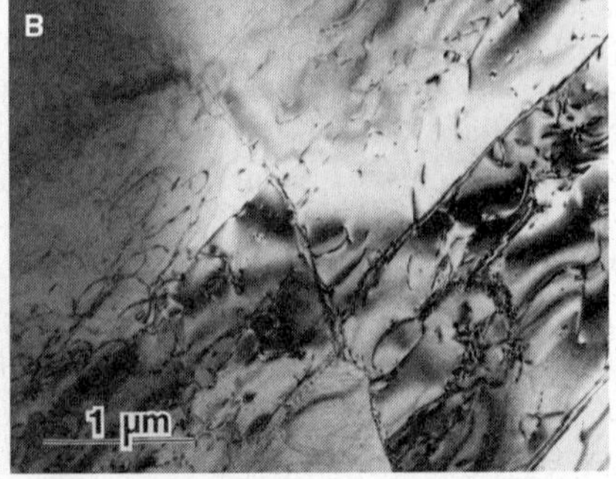

Figure 4.58 TEM (transmitting electron microscope) images of dislocations. (*A*) Curved dislocations and a fairly uniform dislocation density due to mobility of dislocations (climb was easy) due to higher temperature conditions of deformation (see Figure 4.57*B*). [Photograph by Andrew Kronenberg.] (*B*) Straight and tangled dislocations, indicative of only limited dislocation climb. These are similar to those observed in the sample in Figure 4.57*C*. [Photographs courtesy of Jan Tullis, Brown University.] [Reprinted from *Journal of Structural Geology*, Dislocation creep in quartz aggregates, G. Hirth and J. Tullis, p. 145–160, © 1992, with permission from Elsevier. Photographs kindly provided by Jan Tullis, Brown University.]

At even lower temperatures and faster strain rates, dislocation glide accommodates only a small amount of strain before strain hardening prohibits further glide. Dislocations may concentrate stress, leading to fracture. Dislocations, if present, will be straight because climb is very, very slow at low temperatures.

Experiments with Analogues and Other Easily Deformed Materials

Much of our new understanding of ductile deformation mechanisms and dynamic recrystallization comes from direct observations under the microscope of materials that deform and recrystallize at lower temperatures and higher strain rates than typical rock-forming minerals. These materials, including ice, salt, and strange mineral analogues like octachloropropane behave ductilely and recrystallize readily when deformed near room temperature. Thus, the behavior of individual grains and grain boundaries can be observed through the microscope, essentially in real time, as the material is being deformed.

Rock salt is the focus of some important deformation experiments (Urai et al., 1986), in part because it is a relatively weak, easily deformed material that flows at depth to produce salt domes and can even flow downhill on the surface in salt glaciers. Salt can be relatively impermeable, enabling it to trap oil and be a possible site for long-term storage of nuclear waste. It shows strong variations in deformation mechanism as a function of water content (Ter Heege et al., 2005), and is much stronger under dry conditions. Salt is very soluble in water, so wet conditions favor dissolution creep, where material is dissolved and reprecipitated. Water also facilitates dislocation creep by allowing fluid-assisted grain-boundary migration and thereby undoing strain hardening.

Wilson (1981, 1986) has comprehensively studied deformation mechanisms in ice as a rock analogue. Ice is an uncanny crystallographic analogue for quartz. Wilson prepared fine-grained ice samples in such a way that crystallographic *c*-axis orientations are randomly dispersed before deformation. Then he deformed the samples, step by step, by pure shear (coaxial strain) and simple shear (noncoaxial strain). Obvious deformational microstructures come and go, as ice crystals deform and then experience dynamic recrystallization. Wilson tracked the percentage of shortening at each step of the pure-shear experiments, and shear strain at each step of the simple-shear experiments, allowing us to gain a sense of the appearance of microfabrics as a function of degree of deformation. His movies of the deformation are marvelous, because in watching we suddenly realize that dislocation creep, mechanical twinning, kinking, grain-boundary migration, production of subgrains, and rotation of grains all are taking place simultaneously in different parts of the microscopic field of view. Furthermore, the movies show in real time the stereographic migration of randomly oriented *c*-axes of the ice at the beginning of a given experiment to strongly-preferred orientations of *c*-axes at the end of the given experiment. This information instills confidence that microscopic analysis of deformation fabrics can be used to distinguish whether strain has been achieved by coaxial or noncoaxial deformation.

Win Means (1980) pioneered the use of nonrock materials such as octachloropropane (think of a material like mothballs) and paradichloropropane to simulate, through deformation, the geometry and kinematics of dislocation creep and recrystallization. In some experiments he tracked the position of grain boundaries relative to some material reference frame by placing small, rigid inclusions, such as iron filings, in the materials before deformation. Figures 4.59 and 4.60 show the results of two such experiments in which the material was subjected to simple shear. In the first experiment (see Figure 4.59), octachloropropane undergoes rotation recrystallization and the formation of subgrains because climb was easy. In the second experiment (see Figure 4.60), paradichlorobenzene displays boundary-migration recrystallization: the grains

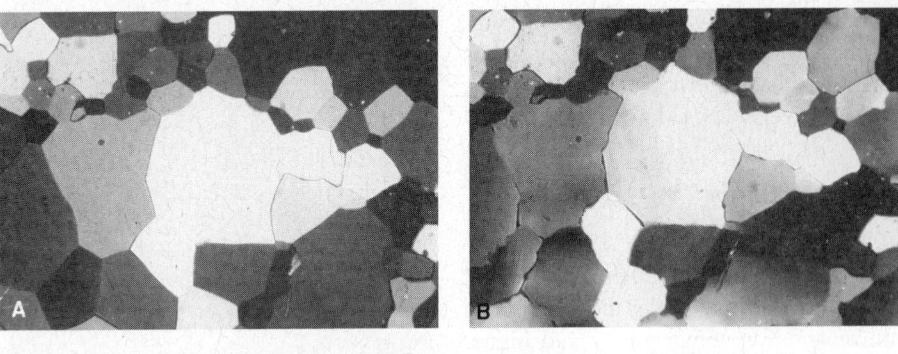

Figure 4.59 Experimental deformation of octachloropropane. (*A*) Initial specimen, showing straight boundaries and internally unstrained grains. (*B–C*) Progressive dextral shear resulting in development of undulatory extinction and subgrains via rotation recrystallization. Some grain-boundary migration is reflected by serrated grain boundaries. [Photographs kindly provided by Win Means, SUNY, Albany.]

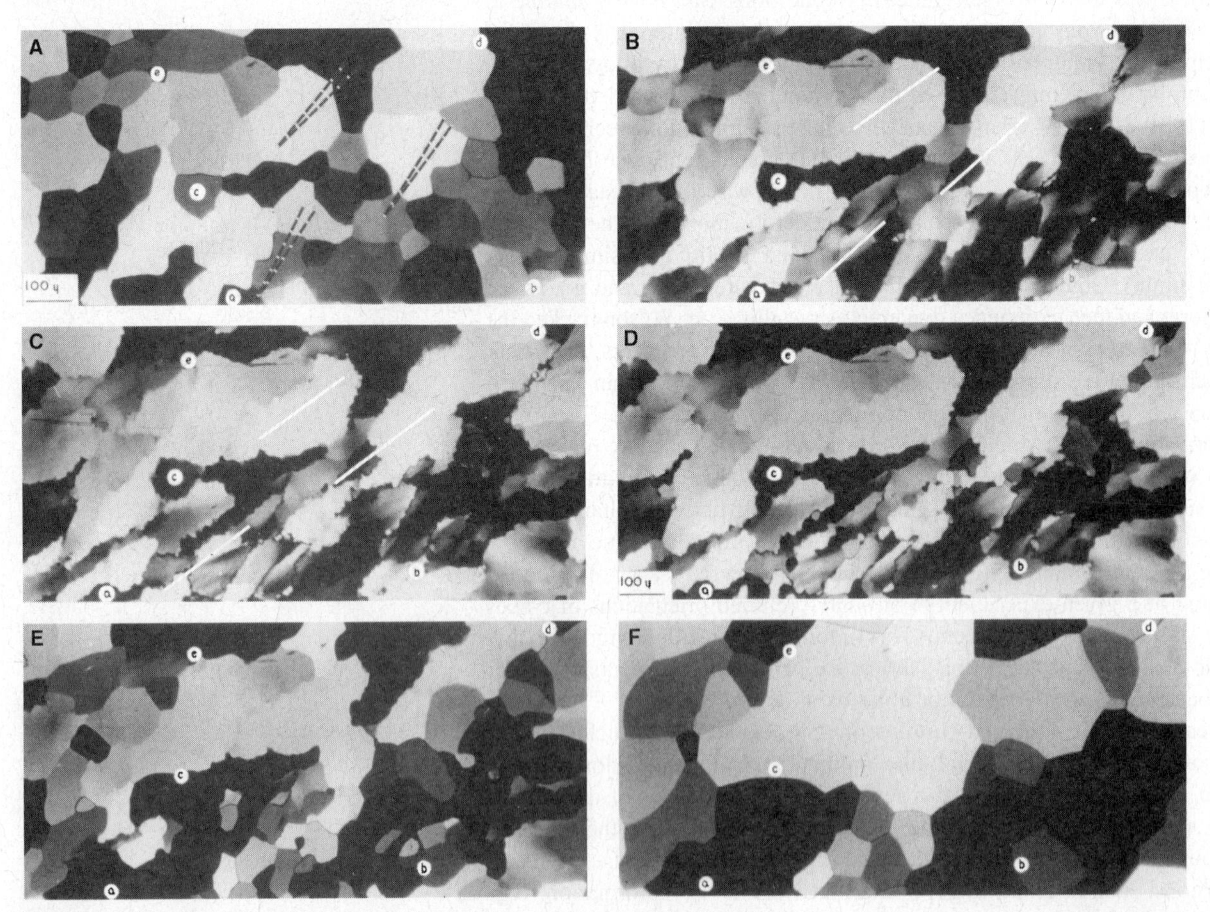

Figure 4.60 Experimental deformation and annealing of octachloropropane. (*A*) Initial specimen, showing straight boundaries and internally unstrained grains. Reference lines and selected grains are labeled. (*B–C*) Progressive dextral shear resulting in development of undulatory extinction, subgrains, and some serrated grain boundaries. (*D–F*) Progressive annealing and grain boundary migration, leading to larger undeformed grains with straight boundaries and 120° intersections. [Photographs kindly provided by Win Means, SUNY, Albany.]

change size and shape as they migrate through the material. Grains with abundant defects or excess surface energy are consumed by those with lower energy. In these experiments, the deforming material develops various microstructures, including undulatory extinction, subgrains, and sutured grain boundaries.

Computer Modeling of Deformation and the Development of Crystallographic Preferred Orientations

Increasingly, we use computer models to explore how deformation should be expressed at the microscopic level. Computer programs take the latest theoretical models of deformation and perform "*virtual deformation*" of an imaginary aggregate of grains (i.e., a "*virtual rock*"). We can observe—in real time—how the geometry and crystallographic preferred orientation of each grain changes during deformation. We can watch how differences in original crystallographic orientation or grain size cause two adjacent grains to have different behaviors during deformation and different final geometries and orientations.

Insightful computer models explored the behavior of quartz and similar minerals (Lister and Hobbs, 1980; Etchecopar and Vasseur, 1987; and Jessell, 1988a,b) (Figure 4.61). These models usually start with an aggregate of undeformed grains with random crystallographic orientations. Then the aggregate is virtually deformed, with each grain deforming via shear along the appropriately oriented slip systems. In quartz, slip commonly occurs along the basal plane (perpendicular to the *c*-axis) and parallel to each of the prismatic planes (see Figure 4.61*A*). Some models use only a few slip systems while trying to minimize the geometric misfit between adjacent grains (see Figure 4.61*B*,*C*). Other models use more slip systems to let each grain deform without becoming locked against its neighbors; five independent slip systems are required to permit this to happen. In such models, the final geometry of each grain depends on (1) which slip systems are available, (2) the orientation of these potential slip systems relative to the orientation of the imposed stresses,

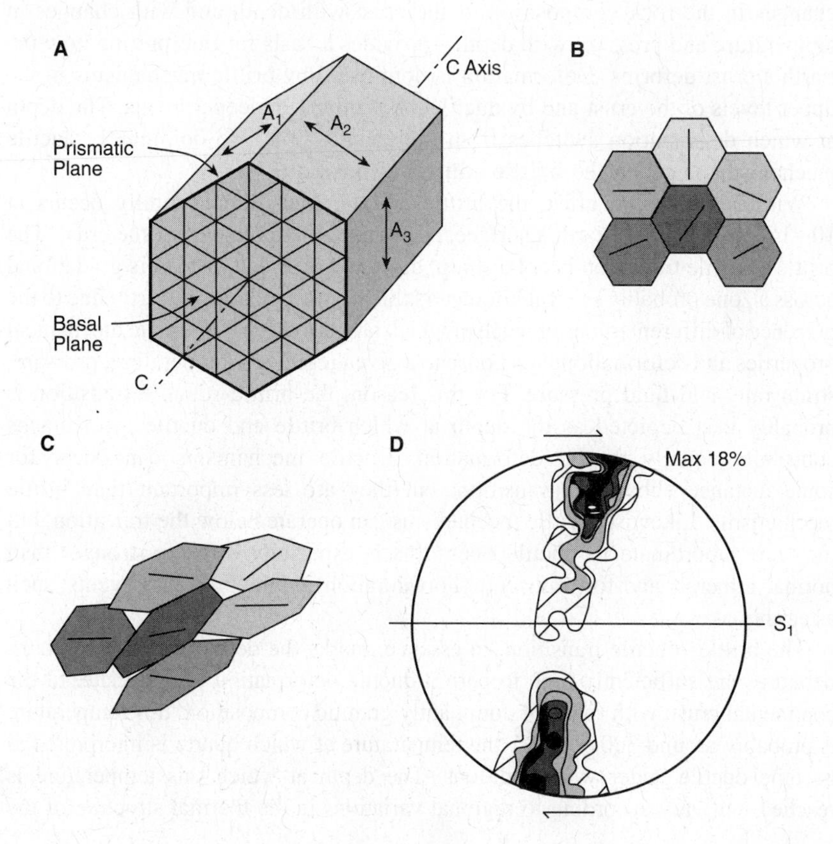

Figure 4.61 Computer simulation of the development of preferred crystallographic orientation in a mineral structure like quartz. (*A*) Basal plane and three prismatic slip planes (A_1, A_2, and A_3) in a quartz crystal. (*B*) Initial aggregate of quartz crystals, showing a single slip plane that is permitted to operate for each crystal. (*C*) Final shape of grains for dextral shear and the operation of only one slip plane for each crystal. (*D*) Stereographic distribution of *C*-axes after deformation defines a girdle that leans over in the sense of shear relative to the normal to the foliation (S_1). [Reprinted from *Journal of Structural Geology*, v. 9, A 3-D kinematic model of fabric development in polycrystalline aggregates: comparisons with experimental and natural examples, Etchecopar, A., and Vasseur, G., p. 705–717, © 1987, with permission from Elsevier.]

(3) the nature of interactions between adjacent grains, and (4) the amount of finite strain. In addition, the overall microstructure of the aggregate is controlled by the rate at which the fabric develops relative to the rate at which it is changed or destroyed by recovery and recrystallization (Jessell, 1988a,b). The most sophisticated computer models can vary the physical conditions (temperature and the like) in successive computer runs to explore how aspects such as grain-boundary migration affect the final fabric.

In addition to displaying the shape of each grain, the programs can calculate stereoplots showing the distribution of the crystallographic axes (see Figure 4.61*D*). These plots, called *c*-axis figures for short, largely reflect the amount and kind of strain imposed on the aggregate (e.g., coaxial flattening, noncoaxial simple shear, etc.). Depending on the strain and other factors, a *c*-axis figure may contain a single point maxima, a single girdle of points along a great circle (see Figure 4.61*D*), a girdle with deflections (called "*doglegs*") at the ends, or even two crossing girdles. The power of this method is to compare computer-generated *c*-axis figures for different types of strain against *c*-axis figures derived from measuring, under the microscope, hundreds or thousands of *c*-axes in naturally deformed, real rocks (Law et al., 1986). We can go further and bring in *c*-axis figures measured from some experimentally deformed rocks. Such integration demonstrates how field studies, deformation experiments in the lab, and quantitative theories of deformation as played out on a computer complement one other and provide us with much more insight than we would have gained from any single approach.

THE BRITTLE–DUCTILE TRANSITION

The wealth of information that has been generated through the study of deformation mechanisms—when integrated with basic information about changes in the rock composition of the crust with depth and with changes in temperature and pressure with depth—provides a basis for interpreting how the Earth's crust deforms. Deformation is dominated by brittle mechanisms in the upper levels of the crust and by ductile mechanisms at deeper levels. The depth at which deformation switches from dominantly brittle to dominantly ductile mechanisms is called the **brittle–ductile transition**.

Within continental crust, the brittle–ductile transition generally occurs at 10–15 km below the Earth's surface, or within the middle part of the crust. The brittle–ductile transition is not a sharp, discrete break, but instead is gradational across a zone probably several kilometers thick. This gradation is partly due to the presence of different minerals, each of which has its own characteristic mechanical properties and deformational response to a given regime of temperature, pressure, strain rate, and fluid pressure. For this reason, the brittle–ductile transition is probably best depicted as the depth at which brittle and ductile mechanisms contribute equally to the deformation. Ductile mechanisms can occur for some distance above the transition, but they are less important than brittle mechanisms. Likewise, brittle mechanisms can operate below the transition, but they are subordinate to ductile ones; this is especially true for stronger than normal minerals and for transient (short-duration), high strain rate events, such as earthquakes.

The brittle–ductile transition, in essence, marks the depth below which temperatures are sufficiently high to permit ductile deformation to dominate. In the continental crust, with rocks of dominantly granitic composition, this temperature is probably around 300–350°C, the temperature at which quartz is interpreted to become ductile under wet conditions. The depth at which this temperature is reached will vary according to regional variations in the thermal structure of the crust. In typical continental crust with a geothermal gradient of 20–25°C/km, the depth to the brittle–ductile transition might be approximately 15 km. In regions of recent tectonism, with a geothermal gradient of 50°C/km, the transition might be at 6 km. The transition may move up through the crust during an episode of regional crustal heating, for example during pluton emplacement, and down through the crust as the thermal pulse ends. In a similar manner, initially ductile rocks may be uplifted through the brittle–ductile transition and overprinted by brittle structures. Alternatively, shallow level, initially brittle rocks may become ductile if they are buried to sufficient depths and temperatures.

Using the concept of the brittle–ductile transition, we can model the strength of the entire crust by using a brittle "*frictional*" failure criterion for the upper crust and a ductile failure criterion for the lower crust (Figure 4.62). For the upper crust, we infer that the strength of the crust increases linearly with depth according to a relation called **Byerlee's law**. Byerlee used experimental testing data for a wide variety of rocks and drew attention to the linear correlation between confining pressure (depth) and rock strength. This linear relation proves to be very similar to what we would expect of a crust full of faults (i.e., surfaces of frictional sliding), each of which becomes more difficult to move as the confining pressure increases with depth.

To represent the ductile strength of the middle and lower crust, we consider the strength of granitic rocks to be controlled by their weakest abundant mineral—quartz. Accordingly, we use a curved strength envelope derived from the flow laws for wet quartz. The brittle–ductile transition is represented as the depth where the quartz flow law curve intersects the line representing Byerlee's law (Figure 4.62). At this depth, the crust has exactly the same resistance to brittle *and* ductile mechanisms. At shallower depths, the rocks fail brittlely before stresses can build up enough to cause ductile flow. At deeper levels, the rocks deform ductilely before the brittle failure criterion is reached. Note that different levels of the crust are characterized by different strain rates and by different contributions of each deformation mechanism to the total strain (Figure 4.62).

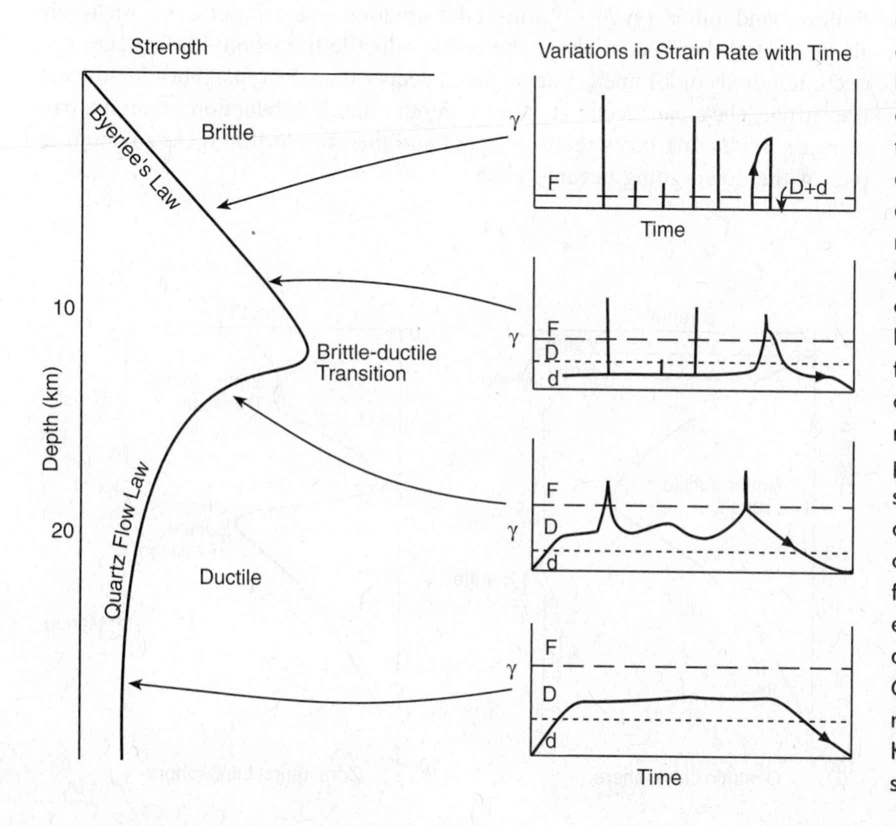

Figure 4.62 Strength envelope and the brittle–ductile transition for granitic crust. The linear part of the strength envelope represents Byerlee's law, where strength increases linearly with depth due to the increased confining pressure. The lower, curved part of the envelope is defined by the flow law for quartz. The two parts of the envelope intersect near the brittle–ductile transition, where brittle and ductile mechanisms contribute equally to deformation. To the right are plots of strain rate versus time for different levels of the crust, showing the relative contributions of diffusion and dissolution creep (d), dislocation creep (D), and faulting and other frictional processes (F). Shallow parts of the crust are characterized by short, high strain-rate, frictional events, whereas deeper parts deform via more continuous, slower strain-rate processes, like dissolution and dislocation creep. Intermediate parts of the crust may deform by continuous ductile flow, punctuated with some high strain-rate events (faulting, such as during an earthquake). [Reprinted from *Journal of Structural Geology*, v. 11, Deformation mechanisms: recognition from natural tectonites, R. J. Knipe, p. 127–146, © 1989, with permission from Elsevier.]

The concept of the brittle–ductile transition as portrayed through the **crustal strength envelope** provides a powerful way to view the entire crust. It can be used to explain why most large earthquakes in the continental crust initiate at a depth of 10–15 km: earthquakes at this depth have to break the strongest part of the crust. We can also see why the overall strength of the crust would be less in regions of higher heat flow and steeper geothermal gradients. Furthermore, we can envision how rocks may change behavior, and strength, as they are uplifted or buried through different levels of the crust.

There are numerous implications of the general restriction of ductile deformation to crustal depths near and below the brittle–ductile transition. First, when we observe ductilely deformed rocks at an outcrop on the Earth's surface, we are witnessing an uplifted exposure of the middle or lower crust. We can study the processes of the middle crust by studying these rock exposures. Second, the middle and lower crust are extremely weak and may be able to flow vertically or laterally in response to stresses, including stresses imposed by the weight of overlying rocks. Some geologists consider this deep part of the crust to be a conduit for lateral ductile flow, a process called **channel flow**. Mid-crustal channel flow may be occurring beneath Tibet in response to extremely thick crust and to erosional and tectonic removal of material from the steep and rapidly uplifting Himalaya.

The concept of strength envelopes can be extended to the entire lithosphere by characterizing the strength of the mantle using ductile flow laws for olivine, the most abundant mineral in the upper mantle. According to this model (Figure 4.63), there is an abrupt increase in strength at the Moho, where the weak, quartz-bearing lower crust is replaced downward by stronger, olivine-bearing mantle. At still greater depths, an increase in temperature and pressure results in ductile flow of olivine and a marked decrease in strength. The transition to underlying weak mantle is the lithosphere–asthenosphere boundary, with the mantle part of the lithosphere being capable of brittle deformation, whereas the asthenosphere deforms ductilely.

Deep earthquakes in subduction zones seem inconsistent with the idea that faulting and other types of brittle deformation are restricted to relatively shallow depths and occur above the brittle–ductile transition. Earthquakes can occur hundreds of kilometers deep, much deeper than the typical brittle–ductile transition. They can occur at greater depths along subduction zones in part because subduction takes relatively cool and therefore brittle rocks to depth as part of the down-going oceanic plate.

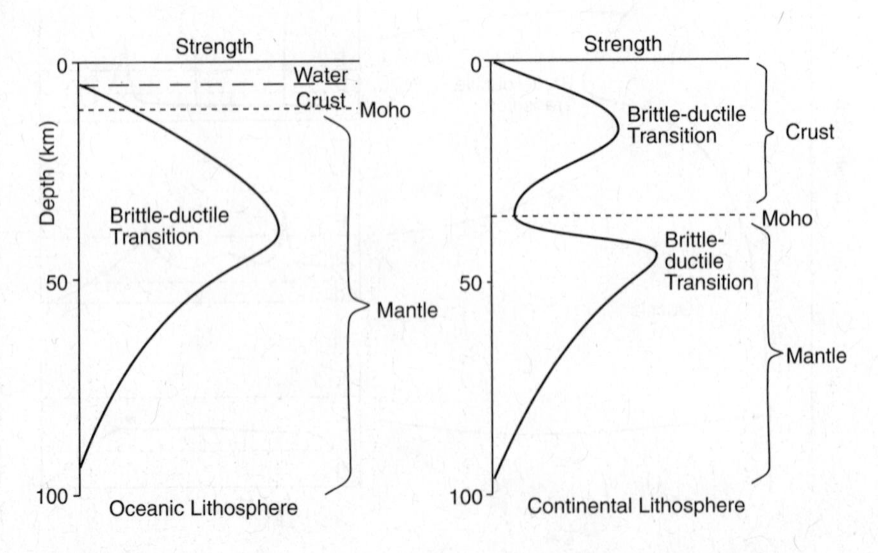

Figure 4.63 Strength envelopes and the brittle–ductile transitions for oceanic and continental lithosphere. (*A*) The strength of oceanic lithosphere increases with depth according to Byerlee's law, until the brittle–ductile transition, where the strength decreases because of the ductile flow of olivine. (*B*) The strength envelope for continental lithosphere has two maximums. Strength within the granitic crust increases downward according to Byerlee's law, until higher temperatures cause the onset of ductile flow of quartz. The strength increases at the Moho, where the quartz-rich crust gives way to the olivine-rich upper mantle. The strength of the underlying mantle part of the lithosphere decreases downward as olivine begins to flow ductilely. [Reprinted by permission from Macmillan Publishers Ltd: *Nature*, Continental tectonics in the aftermath of plate tectonics, P. Molnar, p. 131–137, © 1988, with permission from Elsevier.]

A FEW FINAL THOUGHTS

Rocks that seem so hard and rigid prove to be "*soft*" by virtue of weaknesses that are exploitable by deformation mechanisms operating at the microscopic and atomic scales. Minerals and rocks under certain conditions can flow, almost as if they had no strength, . . . almost as if they were not able to withstand even the smallest stress. William Shakespeare, uncanny as always, seemed to have understood this when, through Leonato in *Much Ado About Nothing,* he said:

> Being that I *flow* in grief
> The smallest twine may lead me.

Isn't it interesting that the study of deformation mechanisms at the microscopic and atomic scales can prove to be so essential to understanding the mechanical and rheological behavior of the entire lithosphere? In a comparable fashion, knowledge of deformation mechanisms is essential to understanding the fundamental geologic structures and structural relations, which we are now about to examine in *Part II, STRUCTURES*: *Joints* (*Chapter* 5), *Faults* (*Chapter* 6), *Folds* (*Chapter* 7), *Fault-Fold Interactions* (*Chapter* 8), *Foliation and Lineation* (*Chapter* 9), and *Shear Zones* (*Chapter* 10).

In this chapter we have examined the Earth through a microscope, looking at grain-scale deformation. After we have made a pass through the common geologic structures, we will look at the Earth through a telescope, at the plate and regional scales, focusing on *Active Tectonics* (*Chapter* 11). Throughout the journey that remains, we will keep in mind that no matter what scale of structure we are talking about, deformation mechanisms make it all possible, and place limits on just how much stress can build.

Part II

Structures

Folded and sheared metamorphic rock comprising the Laurel Mountain roof pendant in the Sierra Nevada batholith, California. Roof pendants are large, deformed metamorphic remains of country rock invaded and nearly overwhelmed by igneous intrusion. The geologic structures contained in this roof pendant have dimensions and variability that go beyond our wildest imaginations, and thus we have to be ready for anything. This is true whether our approach is one of mapping, describing, and interpreting structures in the field; creating structures in the laboratory through deformation experiments; modeling the formation of structures numerically; interpreting structures in the subsurface based on wellbore data and seismic reflection profiles; or recording active deformation through seismology and geodesy. [Photograph and copyright © by Peter Kresan.]

Chapter 5 Joints

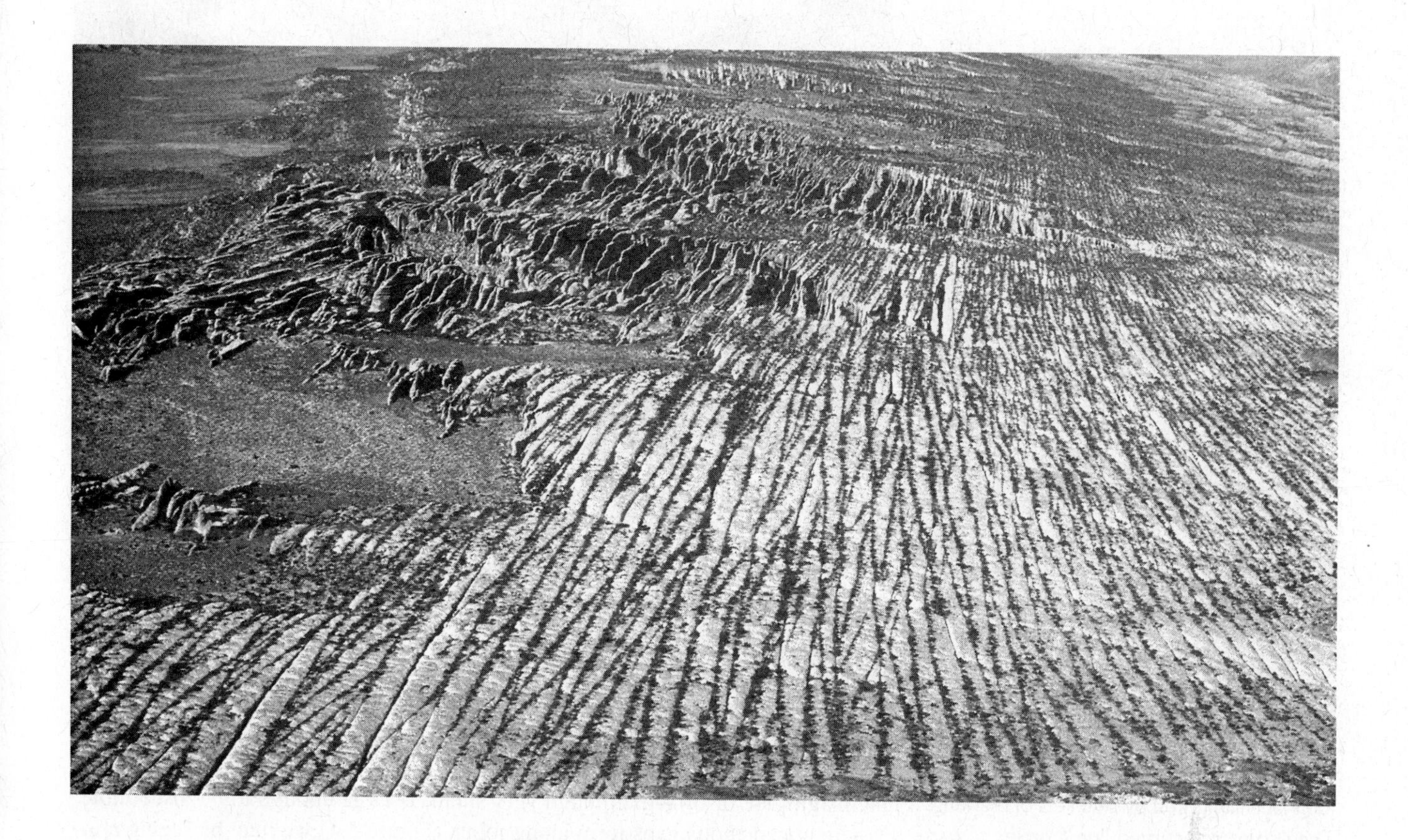

Oblique aerial view of joints and joint patterns in the Entrada Formation (Jurassic) in Arches National Park, Utah. This northwest-directed view captures among other things the prominent rock fins in the Devils Garden area. Though the fins and associated natural arches are impressive as seen on the ground, the sweeping aerial expression is mind-boggling. Joints are tension fractures along which rocks open up, very slightly. They form in orientations controlled by a combination of regional and local stress conditions. The spacing of joints is determined by the mechanical properties and thicknesses of the rock layers they occupy, as well as the amount of strain. The question of when, exactly, joints form in the 'life history' of a rock is often very difficult to answer. Joint systems are excellent plumbing systems for the flow of ground water, hydrocarbons, and mineral-bearing solutions. [Photograph and copyright © by Peter Kresan.]

DEFINITIONS AND DISTINCTIONS

General Nature of Joints

Joints are reasonably continuous and through-going planar fractures, commonly on length scales of centimeters to tens or hundreds of meters along which there has been imperceptible "*pull-apart*" movement more or less perpendicular to the fracture surface (Figure 5.1). Individual joints are planar to curviplanar surfaces that intersect the tops and flanks of outcrops as lines. Erosion and spalling of rocks along joint surfaces reveal joint faces (Figure 5.2). Joint faces are partial exposures of joint surfaces whose complete two-dimensional form in the plane of the surface is almost never seen in full. Woodworth (1896) concluded long ago, on the basis of a careful inspection of three-dimensional exposures of jointed bedrock, that joint surfaces are generally elliptical (Figure 5.3). Now it is generally known that the two-dimensional form of a joint surface is strongly influenced by the three-dimensional shape of the rock body in which the joint is contained. Joint surfaces in layered sedimentary rocks tend to be rectangular, whereas joints in massive rock bodies, such as igneous plutons, range from circular to elliptical (Pollard and Aydin, 1988).

Narr and Suppe (1991, p. 1037) have defined joints as "*planar tensile opening-mode fractures with little or no displacement parallel to the fracture plane.*" Joints are products of brittle failure, and they form when the tensile strength of stressed rock is exceeded. The tiny movement that is

Figure 5.1 Joints and joint surfaces exposed in an old rock quarry near Glen Echo, Maryland. [Photograph by G. K. Gilbert. Courtesy of United States Geological Survey.]

Figure 5.2 Joint face in the Pennsylvanian Bonaventure Formation along the shore of Chaleur Bay, New Brunswick, Canada. The joint face is marked by two sets of markings. The prominent curved line (convex toward bottom left) is a well exposed *"rib."* The fainter *"feathery"* markings are *"hackles."* As we shall learn, ribs and hackles are tell-tale indicators of mode I fracturing, which is marked by tensional pull-apart of the rock. Geologist is Wayne Nesbitt. [Photograph by G. H. Davis.]

accommodated by a joint is fundamentally an opening perpendicular to the joint surface under tension. In most cases the opening that is accommodated by jointing is nearly microscopic. Thus the amount and direction of movement represented by the existence of a joint cannot be identified in outcrop except where the joint cuts a discrete object (e.g., a pebble, fossil, or inclusion) and the magnitude of offset, although very small, is large enough to be discernible. Where conspicuous gaps form along joints (Figure 5.4) we may be seeing non-tectonic effects of expansion and contraction right at the surface of the earth, in the weathering environment. Overall, we think of joints as structures that permit minor structural adjustments to take place as the regional rock bodies

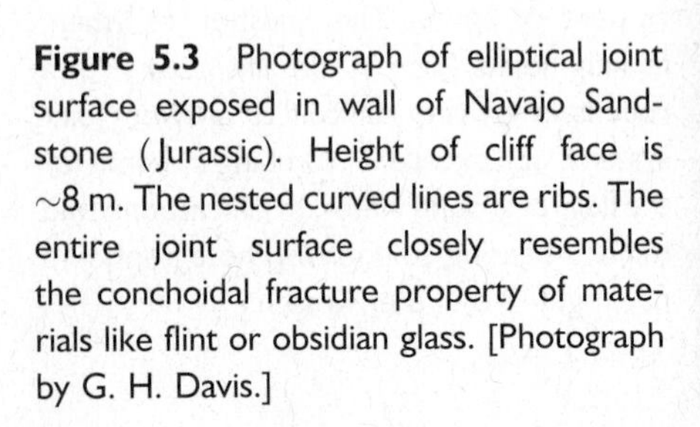

Figure 5.3 Photograph of elliptical joint surface exposed in wall of Navajo Sandstone (Jurassic). Height of cliff face is ~8 m. The nested curved lines are ribs. The entire joint surface closely resembles the conchoidal fracture property of materials like flint or obsidian glass. [Photograph by G. H. Davis.]

Figure 5.4 Photograph of rectangular slabs of sandstone. The shapes are controlled by spacing and orientation of two systematic joint sets. Weathering and erosion is causing the original horizontal sandstone to slump and tilt. Note that slumping has had the effect of creating narrow fissures along the joint traces. Carmel Formation (Jurassic), near Capitol Reef National Park, Utah. [Photograph by G. H. Davis.]

within which they are found are forced to change size or shape. The changes are a response to such actions as burial and compaction, heating and cooling, uplift and subsidence, folding and faulting, and landscape evolution.

Because joints represent very slight pull-apart of their host rock, it is not surprising that we find that some joints are marked by **veins** (Figure 5.5), for the cracks offer permeability and opportunities for fluid flow. Mineralizing solutions invade rock bodies along joints and other fractures, precipitating minerals from solution in the open space where the cracks are forced open and the chemical conditions are just right. Precipitation is triggered by favorable temperature and/or pressure conditions, and sometimes by the mixing of different fluids that happen to meet at the intersection of fracture-controlled channel ways. The quartz veins pictured in Figure 5.5 were emplaced in the granite under high

Figure 5.5 (*A*) Cross-sectional view of quartz-muscovite-pyrite-molybdenum veins occupying closely spaced fractures (joints) in a biotite-muscovite-garnet granodiorite (73 Ma) within the Tea Cup Pluton near Ray porphyry copper mine, Kearny, Arizona. This exposure represents rocks and veins formed at a depth of ~10 km, now exhumed at the surface as a result of crustal extension. Veins are ~4-cm thick. (*B*) Longitudinal view of two of the veins, one in foreground, one in background, cutting through the granodiorite. Note that the veins themselves not only occupy joints, but in turn have become jointed themselves. The orthogonal grid of fractures seen in the plane of the veins are traces of joints. Geologist photographing outcrop is Eric Seedorff. [Photographs by G. H. Davis.]

temperature and pressure conditions at a depth of ~10 km. Note that after the quartz veins were emplaced, they themselves became jointed systematically (see Figure 5.5*B*).

Joints are found in all outcrops of rock, and thus they are among the most abundant of geologic structures. Joints do not exist in isolation. Rather they are members of enormously large families of fractures with literally millions of members. Joints pervade each and every regional rock assemblage, whether a granite batholith, a plateau of sedimentary rock, a fold-and-thrust belt, a volcanic tableland, or Precambrian basement. It is impossible to try to explain the origin of every joint we see in an outcrop, let alone every joint in a regional rock assemblage. Instead, we try to explain the origin of dominant **joint sets**; that is, families of parallel, evenly spaced joints that can be identified through mapping and analysis of the orientations, spacing, and physical properties of the joints within a given system.

The most prominent joints and joint patterns are found in highly competent (stiff) rocks, such as sandstone, limestone, quartzite, and granite. Spacing of joints in a bed tends to be roughly the same as bed thickness, but we will see that the range of bed thickness to joint spacing may, in detail, vary from bed to bed, or from lithology to lithology, or from area to area. As families of fractures, joints commonly display systematic preferred orientations (Figure 5.6) and often show striking symmetry. We describe certain eye-catching joints as **systematic joints**, because they are planar, parallel, traceable for some distance, and regularly, evenly spaced at distances of centimeters, meters, tens of meters, or even hundreds of meters. A given outcrop area or region of study is typically marked by more than one set of joints (see Figure 5.6), each with its own distinctive properties such as orientation and spacing. Two or more sets constitute a **joint system.** Some joints are so irregular in form, spacing, and orientation that they cannot be readily grouped into distinctive, through-going sets. These are **nonsystematic joints**.

Joints have the effect of subdividing a rock body into myriad internal cubes and parallelepipeds, each face of which "*joins*" to a neighboring face along joints, sometimes architecturally resembling blocks set side by side (Figure 5.7). A joint itself consists of an exceedingly narrow slit, roughly rectangular, elliptical, blade-like, or circular in overall form, bounded on each side by matching planar surfaces of rock (Pollard and Aydin, 1988). The **joint surfaces** are commonly ornamented by markings that record details of

Figure 5.6 Aerial view of two sets of joints in sandstone in Canyonlands, Utah. The larger joint-bounded blocks are about 50 m on a side. Eroded aisles are zones of closely spaced joints. [Photograph by G. F. McGill; taken through a hole in the floor of a low-flying small plane.]

Figure 5.7 Sometimes human architectural products and the natural architecture of jointing bear a similarity to one another. (*A*) Built structure at the Panhellenic site of Delphi, Greece. (*B*) Natural jointing in granite on the north coast of Brittany, France. The joints are like narrow slits. [Photographs by G. H. Davis.]

joint propagation and perturbations of the local state of stress. Figure 5.8*A*, for example, shows a beautiful joint surface in a bed of radiolarian chert. The feathery features are plumes, which are the trails left by fracture propagation. The upper and lower edges of the face are the fringes of the joint surface, where energy "*pooped out*" and the nice clean surface was replaced with tens of tiny en echelon surfaces, turned at a slight angle to the main joint face. Figure 5.8*B* shows an entirely different kind of ornamentation on a joint face. In semi-arid parts of the American Southwest, rock surfaces, including long-exposed joint surfaces, can acquire thin coatings of black/brown "*desert varnish.*" The Ancient Ones (Anasazi Peoples) commonly used these as art boards, and carved through the varnish to create petroglyphs. The joint surface provided a perfect drawing surface!

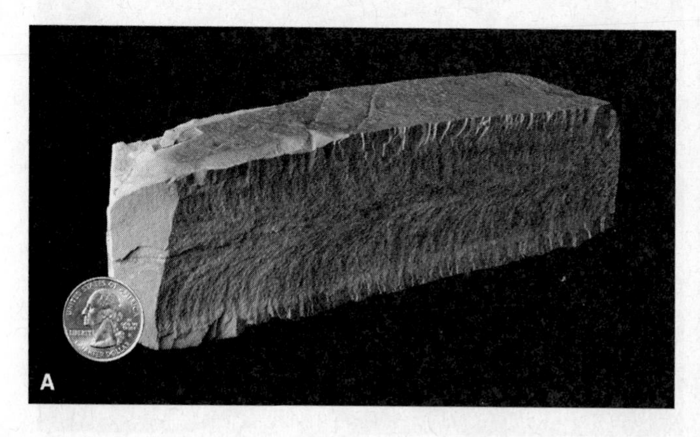

Figure 5.8 (*A*) Macro-photograph of joint surface cutting radiolarian chert (Jurassic) from the Mt. Lykaion area in the Peloponnesos, Greece. Length of specimen is 15 cm. The surface is a beauty, for feathery "*plumose*" features are preserved, which tell us something about fracture propagation; and exquisite "*fringes*" are seen near the upper and lower parts of the joint surface. Fringe areas represent places where the joint propagation was arrested, and the single planar joint surface is replaced by tiny en echelon microjoints. Collection of G. H. Davis. [Photography and editing by Gary Mackender. vr.arizona.edu, glm@email.arizona.edu, © 2009–2010 Arizona Board of Regents.] (*B*) Here is a different kind of ornamenation on a joint face. These are Anasazi petroglyphs on a joint face along the San Juan River between Bluff and Mexican Hat, Utah. Dark patina is desert varnish on the joint face, beneath which is the fresh white sandstone. Human figures are 3 m tall. [Photograph by G. H. Davis.]

Distinction between Joints and Shear Fractures

Some joint-like fractures are actually **shear fractures**, which in effect are **microfaults**. They form not by tensional opening perpendicular to a fracture face, but through shear traction parallel to the fracture surface. These fractures can be confused with joints, because the shear offset is generally not visible at the scale of outcrop or hand specimen. Some shear fractures are of the same size and scale as joints, and occur in sets of planar parallel fractures.

Shear fractures may be very difficult to distinguish from joints because of the absence of diagnostic ornamentation on what appear to be featureless fracture surfaces, or by the lack of any discernible movement or offset. Making the case for fractures being shear fractures (microfaults) may require demonstrating that they are parallel and in close spatial relationship to recognizable faults. We will learn that faulting commonly results in two sets that intersect at an angle of ~60°. As a consequence of this common observation, we tend to regard fractures that intersect in that manner to be possible shear joints (Figure 5.9). But if tiny fault-like offsets are not evident (as in Figure 1.29), or the orientations of the fracture sets do not perfectly match that of know faults, we need to '*back off*' calling these "*shear fractures*" and simply regard them generally as "*fractures*" or "*joints.*"

In trying to determine whether a given fracture *may* be the product of shearing, we also look carefully for the presence of **slickenlines** (Figure 5.10), products of shearing movement parallel to the fracture surface. The slickenlines on shear fractures are most commonly fine-scale, delicate **ridge-in-groove lineations** (Twiss and Moores, 1992) developed on the adjoining fracture surfaces, or coatings of **crystal fibers** that have grown in the direction of shear displacement. The presence of crystal fiber veins along fracture surfaces discloses that the fracture was produced by the combination of joint-like opening and fault-like shear. In Figure 5.10 we mainly see crystal-fiber lineation in the (white) quartz veins, and yet some very fine-scale ridge-and-groove lineation may be present on the dark-weathering surface of the sandstone proper.

Certain fractures do not qualify as joints, nor do they qualify as conventional shear fractures. Within fault zones, for example, rocks may be pervasively

Figure 5.9 These "*joint sets*" that intersect at an angle of ~60° actually may be shear fractures that accommodated imperceptible shearing/faulting parallel to the fracture surfaces, as opposed to typical joints that form because of opening *perpendicular* to the fracture surface. See text for words of caution. Bedrock is Lower Cretaceous sandstones of the Amole Arkose Formation, Tucson Mountains, Arizona. [Photograph by G. H. Davis.]

Figure 5.10 Crystal-fiber lineations in a quartz vein (white) that filled-in along a shear fracture cutting Jurassic siltstones in the Tucson Mountains, Arizona. The crystal-fiber lineation is a record of the direction (line) of shearing. The surface of dark-weathering sandstone also appears to be lineated ("*slickenlined*") at a very fine scale. [Photograph by G. H. Davis.]

Figure 5.11 Pervasively shattered outcrop of granite beneath a fault in southeastern California. [Photograph by Stephen J. Reynolds.]

shattered by nonsystematic, extremely closely spaced fractures (Figure 5.11). The density of the fracturing is much greater than what is reasonable for ordinary jointing and shear fracturing. Furthermore, any regularity of orientation and spacing of fracture surfaces, so typical of jointing and shear fracturing, is absent. Fractures in such shattered rocks are simply called fractures, not joints. The rocks are described as **shattered**, not jointed.

Practical Importance of Jointing

Jointing exerts profound control on weathering and erosion of bedrock, and thus on the topography of bedrock landscapes. Many scenic attractions owe their uniqueness to weathering and erosion of horizontal or gently dipping layers of rock systematically broken by joints. There are great examples of this throughout the world, such as the relationships seen in the Colorado Plateau region of the American Southwest. The landscape at Goblin Valley State Park (Utah) is a good example (Figure 5.12). The bedrock formation of

Figure 5.12 Photograph of joint-controlled 'goblins' in Goblin Valley State Park, Utah. Note the curviplanar steeply dipping joints in the sandstone bedrock at the base of the bench. The spacing of this jointing predetermines the breadth of the goblins. Matt and Jennifer Davis enjoying the site. [Photograph by Drew Skylar Davis.]

note is the Entrada Formation (Jurassic). Even though the jointing in this sandstone is not always planar and evenly spaced, the jointing nonetheless divides the rock into elements that become eroded into Halloween shapes, namely goblins. The shaping is the result of differential weathering and erosion of alternating strong and weak units.

Beyond their scenic value, joints and shear fractures constitute a structure of indisputable geologic and economic significance. Across the ages, quarry workers have taken advantage of joints in splitting and removing blocks of granite, marble, limestone and the like from bedrock (Figure 5.13). One fascinating curiosity is that the Egyptians excavated many of the tombs in the Valley of the Kings by cutting into bedrock along concentrated zones of jointing. This is a "*good news/bad news*" story. Though splitting the bedrock along the pre-existing joints eased the hard work of precise cutting of sandstone, these same joints guided surface runoff (rainwater) into the tombs, causing extensive damage over the centuries (http://www.sciencedaily.com/releases/2006/10/061023091908.htm).

Joints impart a fracture-induced permeability to bedrock, not only at and near the surface but at depth as well. Thus, assessments of infiltration of surface runoff, groundwater flow, and the unseemly and harmful migrations of pollutants and contaminants requires thorough investigations of jointing. Contamination of groundwater in countless nations now places increasing importance on the fields of hydrology and geohydrology. Computer models of water moving through sediments and rocks have become very sophisticated. The best models take into consideration the movement of fluids through fractures. The fractures, after all, contribute to the bulk porosity and permeability of the system. The more we know about the character and connectivity of fracture systems, the better the mathematical models.

Petroleum geologists evaluate the nature and degree of development of joints as one guide to the reservoir quality of sedimentary formations. In fact, to increase the yield of reservoir rocks in oil and gas fields where production is waning, it is common practice to "*frac*" the rocks artificially, either through explosives or through the high-pressure pumping of water or gels into the wells.

The natural circulation of hydrothermal fluids through joints and other fractures in hot rocks at depth constitutes a potentially significant source of energy, namely geothermal energy. For geothermal systems to be operational, thoroughly jointed rocks are as essential as heat and fluids.

Figure 5.13 Photograph of ancient limestone quarry in Greece. The natural joint orientation and spacing in this limestone predetermined optimum block shapes and sizes, and made removal of blocks more efficient and effective. This small quarry is located right along the (new) Corinth-Tripolis Highway, at a location ~22 km southwest of Ancient Corinth. The quarry was discovered during highway construction. This quarry is thought to have been worked in the Classical (6th through 4th centuries, BCE) and Hellenistic periods (4th through 1st centuries, BCE). The nearest ancient town is Kleonai. Research archaeologist is David Romano. [Photograph by G. H. Davis.]

Figure 5.14 Photograph of thick veins of galena in the Sweetwater Mine at the southern end of the Viburnum Trend District (lead belt) in the Ozark Mountains region of southeastern Missouri. The host rock is fractured and brecciated dolostone (dolomite) of the Bonneterre Formation (Cambrian). Dolomite is stiffer and stronger than ordinary calcite limestone, and thus is amenable to brittle failure, in this case by very gently dipping jointing. Age of mineralization is likely late Paleozoic. [Photograph by Ray Coveney, taken ~200 m below the surface.]

Joints and faults can serve as sites of deposition of metallic and nonmetallic minerals. In most deposits, a part of the mineralization is localized in and around fractures. The minerals are deposited either through open-space filling of joints (Figure 5.14), and/or through selective replacement of chemically favorable rocks adjacent to the fracture surfaces along which hydrothermal fluids once circulated. Even where joints and other fractures do not carry economically significant levels of mineralization, they may be marked by veins and alteration assemblages that provide clues to locations of ore deposits (Guilbert and Park, 1986).

GLIMPSE OF JOINT FORMATION IN RESPONSE TO LOADING AND STRESS

Introduction

An important goal of this chapter is interpreting the physical and geometric properties of joints in outcrop, and picturing how joints propagate. Even before presenting the physical/geometric characteristics of joints, we will benefit from a sneak preview of the way engineers, material scientists, and structural geologists think about joints forming in response to loading and the concentration of stresses. Three topics are particularly important in this regard: (1) modes of loading, (2) far-field (remote) versus local stresses, and (3) linear elastic fracture mechanics. We will touch on these topics briefly now, and expand on them as we become more familiar with the details of properties of joints in the field, and with the modeling of joint properties experimentally and numerically.

Modes of Loading

There are three fundamental end-member **loading modes (modes I, II, III)** that concentrate stresses and produce fractures, namely **opening (mode I)**, **sliding (mode II)**, and **scissoring (mode III)** (Atkinson, 1987; Engelder, 1987)

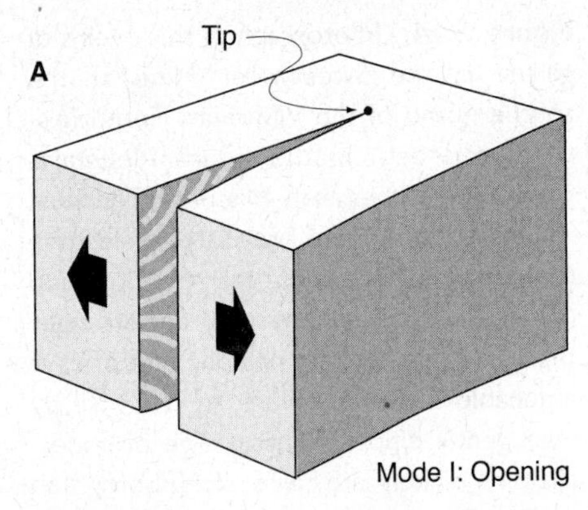

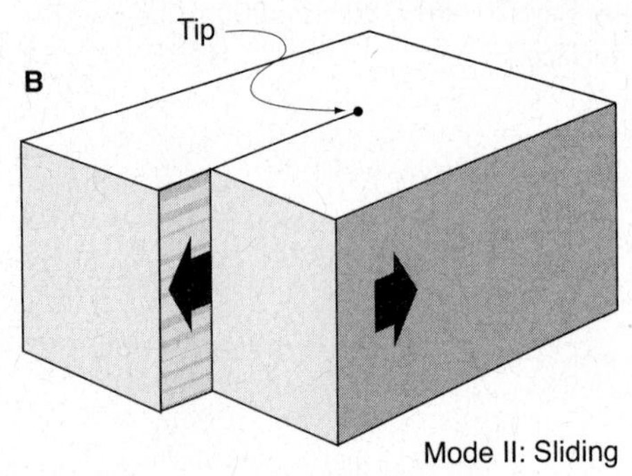

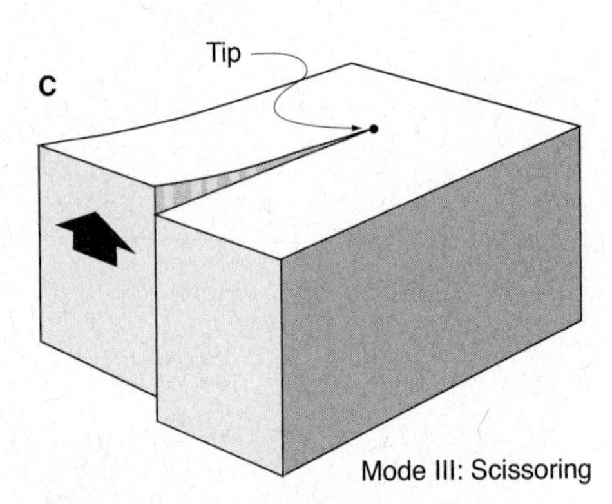

Figure 5.15 The three fundamental fracture modes: (*A*) Mode I opening perpendicular to the walls of the fracture surface. (*B*) Mode II sliding in a direction parallel to the fracture surface and perpendicular to the fracture front. (*C*) Mode III scissoring parallel to the fracture surface and parallel to the fracture front. [Modified from Atkinson (1987), Introduction to fracture mechanics and its geophysical applications.]

(Figure 5.15). These loading modes operate at the **tip line** of a fracture (see Figure 5.15), which is the name given to the leading edge of a fracture front, separating the fracture surface from yet unfractured host rock. The fracture **tip**, on the other hand, is the discrete point where the tip line intersects some 2D surface, most commonly the surface of the jointed outcrop (see Figure 5.15). The propagation of the tip of a crack can be pictured in slow-motion by ripping of a piece of paper, or the ripping apart of Velcro strips, or the splitting of firewood using a wedge and sledge hammer. In all of these examples we can actually hear the propagation of the tip line of the fracture. The sound makes palpable the significant levels of strength and resistance that must be overcome. A propagating fracture will continue to propagate as long as the tensile stress created by loading exceeds the strength of the rock along the tip line. If not, propagation of the fracture will cease; propagation will be **arrested**.

Mode I loading forms by pure extension (opening) driven by tensile stresses perpendicular to the fracture plane (Younes and Engelder, 1999). Mode I loading has no shear traction component (see Figure 5.15*A*). In contrast, mode II and mode III loading creates shear traction parallel to the plane of the fracture surface, and no opening. The distinction between mode II and mode III loading lies in the *direction* of the shear traction relative to the propagation direction of the fracture surface. Mode II loading is characterized by a sliding shear movement in the plane of the fracture surface and perpendicular to the tip line of the fracture (see Figure 5.15*B*). Mode III loading is characterized by a scissors-like shear movement in the plane of the fracture and parallel to the tip line of the fracture (Younes and Engelder, 1999) (see Figure 5.15*C*). Faults are formed through mode II and mode III loading.

Jointing is largely the outcome of mode I loading, but there are times during joint propagation where mode II and/or mode III loading plays a part. Such **mixed mode loading** occurs at different locations and/or times during the propagation history of a joint. Mixed mode loading refers to the combination of mode I and mode II loading, or mode I and mode III loading (Younes and Engelder, 1999). Some of the fine-scale physical and geometric details of joint surfaces simple cannot be understood without knowledge of mixed mode loading behaviors.

Mixed mode loading comes into play as fracture orientation adjustments take place in order to attempt to minimize shear traction on the plane of fracture. Younes and Engelder (1999, p. 221–222) put it this way: A given joint "*follows the propagation path that minimizes the shear stresses acting on the joint tip and that maximizes the tensile stress at the joint tip.*" Adjustments to achieve this ideal are commonly recorded in the form of textures and patterns (ornamentation) on fracture surfaces.

Far-Field (Remote) Versus Local Stresses

When we look for sources responsible for the loading that creates jointing, we consider both **far-field** (or **remote**) and **local stresses**. Far-field stresses have coherence across large regions, and to search for the influence of far-field stresses we need to examine the patterns and extensiveness of joint systems in relation to other regional deformational features, such as fold belts or fault systems. Joints especially capture our interest and imagination because they commonly pervade very large regions and maintain uniform or systematically changing orientations throughout (Figure 5.16). **Far-field stresses** are driven and sustained over long periods of time by large-scale processes (e.g., plate-tectonic loading) that create a regularity of stress orientation over a significant length scale. Most commonly the far-field stresses within a region are compressive, and this causes us to wonder why joints (products of tension) form in the first place.

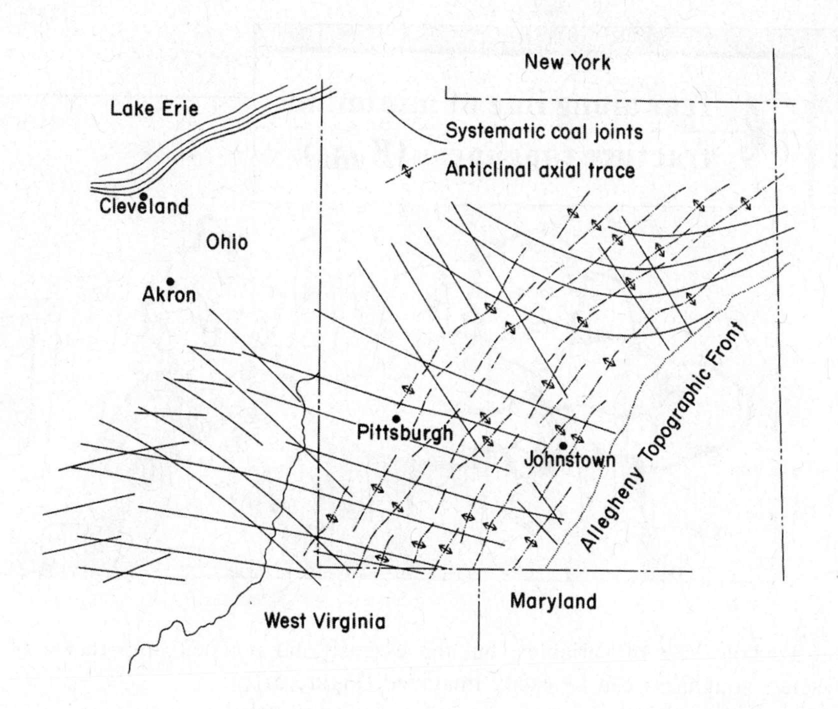

Figure 5.16 Map showing trajectories (solid lines) of major joint sets measured in coal in the Appalachian Plateau. [From Nickelsen and Hough, 1967. Published with permission of Geological Society of America.]

Enter the role of **local stresses**. Local stresses are perturbations and/or concentrations of the far-field stresses, and they may rise to magnitudes where the strength of the rock is exceeded, thus producing joints (Fischer and Polansky, 2006). Local stresses have a length scale that is much smaller compared to the expanse of the far-field stress region as a whole (Randy Richardson, personal communication, January, 2010). Local stresses can reinforce in direction and/or magnitude the far-field stress field, or they can shift the stress direction from the far-field orientation.

We will see that tiny **flaws** in bedrock are the instigators of local perturbations and concentrations of the far-field stresses. Some examples of such flaws include fossils, pebbles, sand grains, bed forms (such as the crest of a ripple mark), gas vesicles (in volcanic rock), and microcracks. Joints propagate from locations of flaws.

Linear Elastic Fracture Mechanics (LEFM)

This brand of mechanics focuses on the intensity of stress that builds up at flaws in materials that are being loaded. The context for analysis is the theory of elasticity, applied to brittle materials. The stress that builds at a flaw is a response to loading conditions, the location of the flaw, and the geometry of the flaw. Every brittle material, under a given set of conditions (e.g., temperature, confining pressure, lithology, and fluid pressure), will have certain **fracture toughness**; that is, resistance to fracturing. Fracturing (in this case, jointing) is sure to occur when the **stress intensity** exceeds fracture toughness (Atkinson and Meredith, 1987). Once the joint begins to propagate through rock, the joint propagation may actually sustain itself during certain spans of time under conditions that are **subcritical**; that is, stress intensity is less than fracture toughness. The act of jointing locally relieves the applied stress (Fischer and Polansky, 2006).

Some rocks are tougher than others when it comes to propagation of flaws to form fractures. It is a little like the difference between trying to rip a bedsheet versus trying to rip a pair of Levi's™. Toughness (K) is dependent

Tear along line of minimum fracture toughness (K_{min}).

Figure 5.17 Application of fracture toughness (K) while flying cross-country. It applies to the wings of the airplane as well. [Artwork by D. A. Fischer.]

on a whole host of variables, but the essence and practical importance of fracture toughness can be easily imagined (Figure 5.17).

DETAILED LOOK AT INDIVIDUAL JOINT SURFACES

Main Joint Face and Fringes

Now let us take a look in detail at the properties of joints. An ideal exposure of an ideal joint consists of a smooth, planar **main joint face** bordered by roughly hewn **fringes** (Hodgson, 1961) (Figure 5.18). The fringes project outward from the main

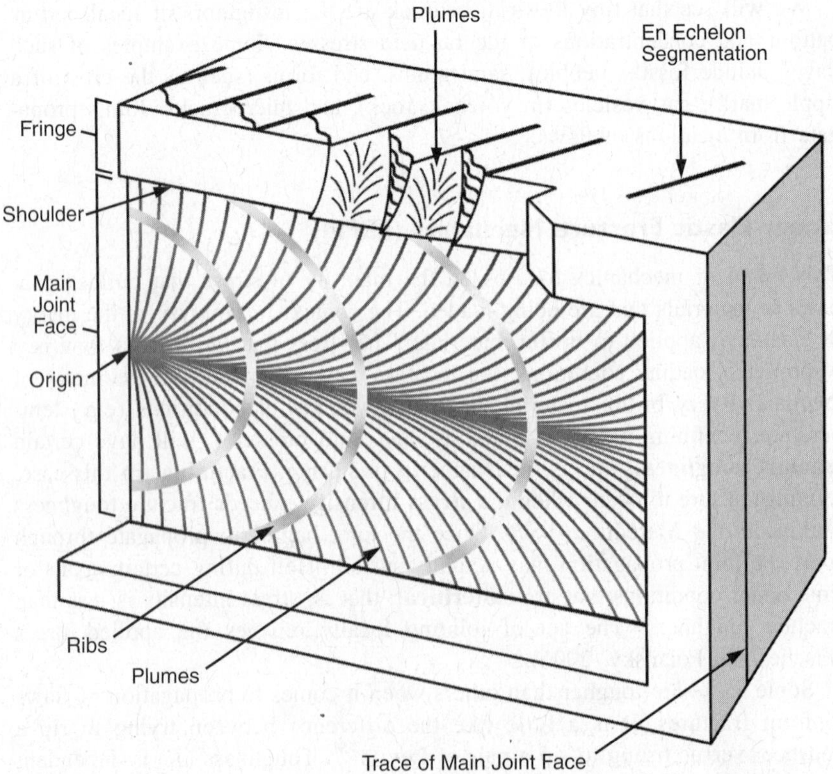

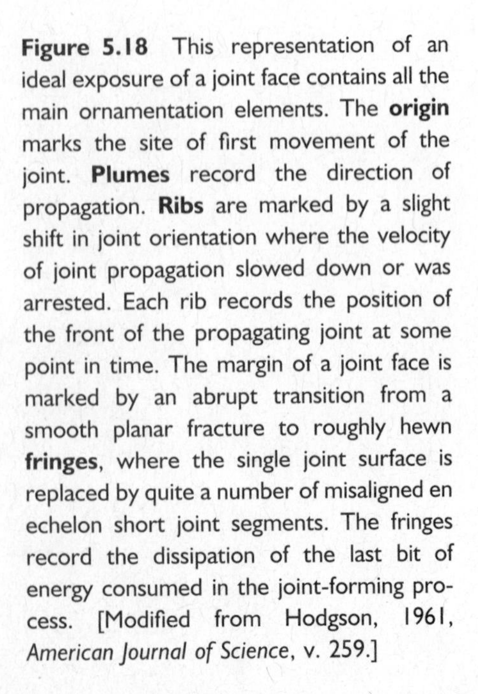

Figure 5.18 This representation of an ideal exposure of a joint face contains all the main ornamentation elements. The **origin** marks the site of first movement of the joint. **Plumes** record the direction of propagation. **Ribs** are marked by a slight shift in joint orientation where the velocity of joint propagation slowed down or was arrested. Each rib records the position of the front of the propagating joint at some point in time. The margin of a joint face is marked by an abrupt transition from a smooth planar fracture to roughly hewn **fringes**, where the single joint surface is replaced by quite a number of misaligned en echelon short joint segments. The fringes record the dissipation of the last bit of energy consumed in the joint-forming process. [Modified from Hodgson, 1961, *American Journal of Science*, v. 259.]

joint face by some small amount. Fringes of joints are the outermost margins of a given joint surface. They are terminations of the main joint face, where the energy required for propagating the joint dissipates. Well-developed and unweathered fringes display a serrated appearance produced by very closely spaced fractures that are misaligned with respect to the main face (Roberts, 1961). Figure 5.19 is a close-up photograph of a jointed rock. The very closely spaced fractures within the fringe tend to be en echelon, marked by short joints that are parallel, overlapping, and arranged in a line. En echelon joints in fringes commonly intersect the main joint face at angles of 20°–25° (see Figures 5.18 and 5.19).

Figure 5.19 Photograph of "*pie-wedge*" from thin bed of radiolarian chert, collected from the Chert Series Beds (Jurassic) near the mountain village of Lykios in the Peloponnesos, Greece. The main joint face is smooth, featureless, and is marked by a thin coating of caliche. Along the upper margin of the main joint face is the "*fringe*" of the joint face. It consists of tiny "*microjoints*" that are arranged en echelon with individual surfaces turned at a slight angle to the orientation of the main joint face. The hairline crack that runs vertically down the main joint face is yet another joint surface cutting the chert bed. Note that this joint *tips out* downward, disappearing before making it to the lower fringe. Collection of G. H. Davis. [Photography and editing by Gary Mackender. vr.arizona.edu, glm@email.arizona.edu, © 2009–2010 Arizona Board of Regents.]

The ornamentation expected in an ideal joint face is subtle but elegant (see Figure 5.18). First, there is the point of **origin** of the fracture, also referred to as the **point of initiation**. This is the very spot where fracture propagation begins. The **ribs**, also known as **undulations**, can be thought of as markings that reflect the front (tip line) of the propagating fracture at different points in time. The **plumes** are feather-like markings that reveal the actual propagation paths.

Main joint faces range in shape from circular to elliptical to blade-like to rectangular. The character of the host rock in which the joints are forming influences the shapes of individual joints. Within large homogenous rock bodies such as granite, or in very thick (e.g., 300 m) massive sandstones, the initial shapes of joints tend to be circular to elliptical. In such environments it is possible to imagine the main joint face propagating in a way that is *not* impeded by the presence of boundaries or contacts with other rock types. Bahat et al. (2003) carefully studied the shapes of main joint faces in a granite pluton (the South Bohemian Pluton) exposed in a quarry southeast of Prague in the Czech Republic. The main joint faces tended to be circular at the onset of formation, but as they propagated and grew, and as the fringe areas of the joints evolved, the composite joint surfaces became progressively more elliptical (Figure 5.20).

Joints in sedimentary layers that form in thin stiff layers sandwiched between soft beds will develop a composite shape that is rectangular or extremely

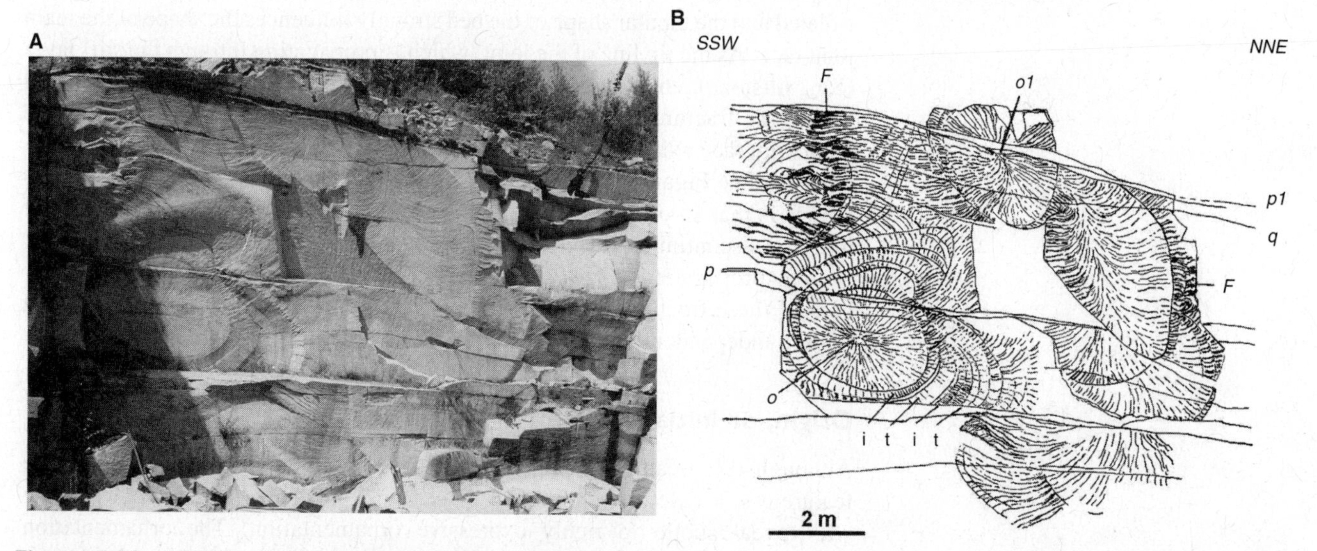

Figure 5.20 (*A*) Photograph of quarry exposure of granite of the South Bohemian Pluton. Main joint faces in the granite tend to be circular to elliptical. Here they are elliptical. (*B*) Drawing of same exposure showing details of the joints and joint properties. Plumes radiate outward from points of initiation ("*origins*"). Plume axes tend to lie along the longest dimensions of ellipses. Explanation of letters: *F* = fringe; *o* and *o1* = origins; *p* and *p1* = partings; *q* = quartz vein; *i* = inner spacing between ribs (undulations); *t* = outer spacing between ribs. Plumes rhythmically vary in their densities of development across the ribs (undulations). [From Bahat and others, 2003, Figures 9A, B, p. 154. Published with permission of Geological Society of America.]

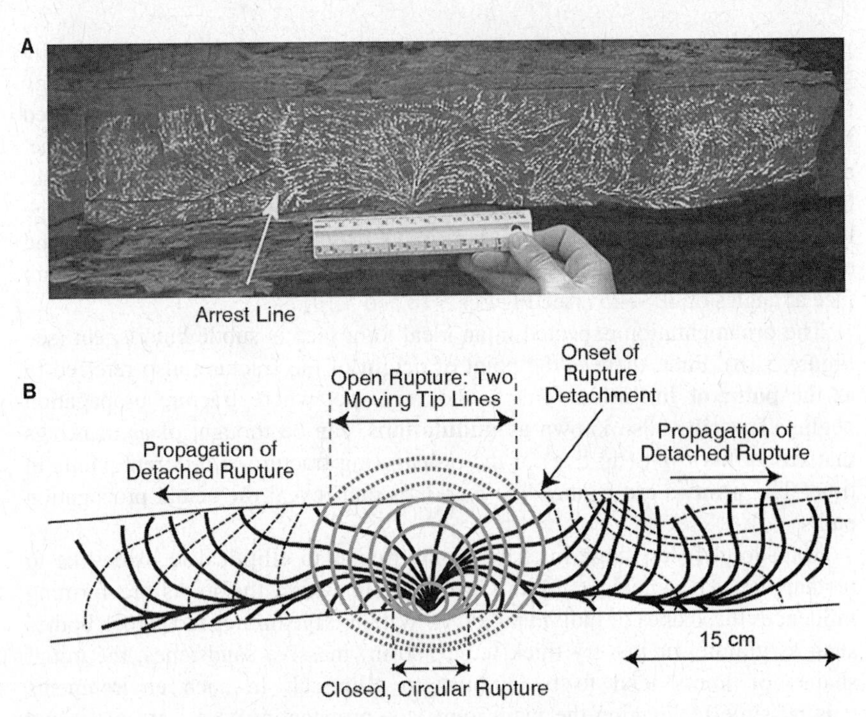

Figure 5.21 (*A*) Photograph of joint surface in Devonian Ithaca Formation, along Highway 14 between Watkins Glen and Montour Falls, New York. Chalk highlights plumes and ribs (referred to here as an arrest line). (*B*) Interpretation of propagation history and development of the shape of the composite joint face. The fracture begins as a circular rupture, initiated at the base of the bed at a point of high stress concentration. The fracture front is arrested at the top of the bed. This results in propagation of detached fracture surfaces, moving in opposite directions. [From Savali and Engelder, 2005, Figures 3A, B, p. 440. Published with permission of Geological Society of America.]

elliptical (Figure 5.21). This is because the stiff layer becomes mechanically isolated and the tabular shape of the bed strongly influences the shape of the main joint face. As the tip line of the joint, which is propagating through the stiff layer (e.g., siltstone), encounters mechanically soft beds (e.g., mudstone) above and below, the fracture will be arrested along the line of contact. Meanwhile, the main joint face will continue to propagate laterally for some distance through the siltstone (see Figure 5.21). One of the things that can happen under these conditions is that an originally circular or broadly elliptical joint surface may split into *two* discontinuous faces that expand into an overall rectangular or irregular form by propagating in *opposite* directions (Savalli and Engelder, 2005). One or both of these fracture surfaces will detach from the initial main joint face and rupture independently (see Figure 5.21) (Savalli and Engelder, 2005).

Origin, or Initiation Point

Although the greatest percentage of exposed joint faces might appear to be featureless, a careful examination of certain joint faces in favorable lighting will reveal subtle to richly expressive ornamentation. The ornamentation provides a basis for interpreting how individual fractures propagated. The most characteristic ornamentation on the main joint face consists of **origins, plumes**, and **ribs** (DeGraff and Aydin, 1987; Pollard and Aydin, 1988) (see Figure 5.18). The presence of plumes and ribs makes clear that we cannot view the formation of a single joint as if it were the product of a single smooth, continuous process. Instead, ornamentation in the form of plumes and ribs reveals a more complex and discontinuous path of fracture propagation, including multiple arrests and branching fronts (Engelder, 1987; Helgeson and Aydin, 1991).

The **origin** of a joint is the **point of initiation** of a joint surface (Figure 5.22). It is analogous to the focus of an earthquake, the site of first movement, the place where energy is first released to form the break. The origin usually coincides with a mechanical defect; that is, flaw, or irregularity in the rock (Helgeson and Aydin, 1991). Points of initiation (origins) of joints in sedimentary rocks prove to be such agents as fossil fragments, concretions, trace fossils such as worm burrows, and primary sedimentary structures. McConaughy and Engelder (2001) reported that initiation points for joints in Upper Devonian strata in the Finger Lakes area (New York) include trace fossils. More commonly, however, they recognized points of origin located on bedforms such as groove casts and flute casts (Figure 5.23). These sharp bedforms as well as the profound mechanical differences in elastic strength between siltstone (stiff) and shale (soft) at these bedform contacts created points of initiation that in turn magnified far-field stresses and triggered joint propagation.

Engelder (1987) pointed out that surrounding the point of origin of a "*joint*" in glass, or in Jell-O (Price and Cosgrove, 1990), there may be a mirror-smooth circular or elliptical area that registers the result of a slow, accelerating rupture. Outward, beyond this mirror plane the fracture formation is achieved in part by much-higher-velocity fracture propagation. Velocities can attain half the speed of sound, accompanying rapid, near-instantaneous, almost explosive snapping apart of the rock.

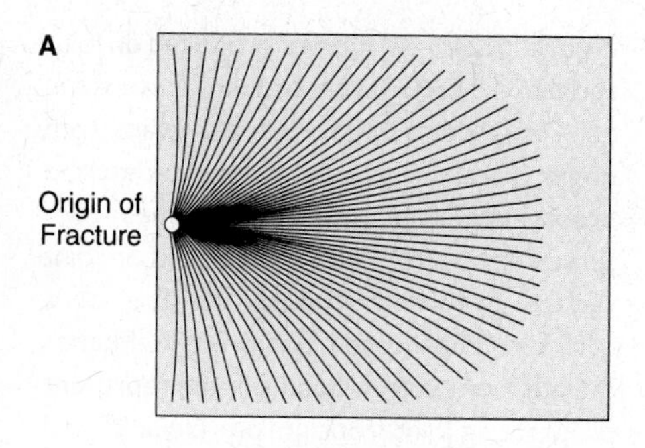

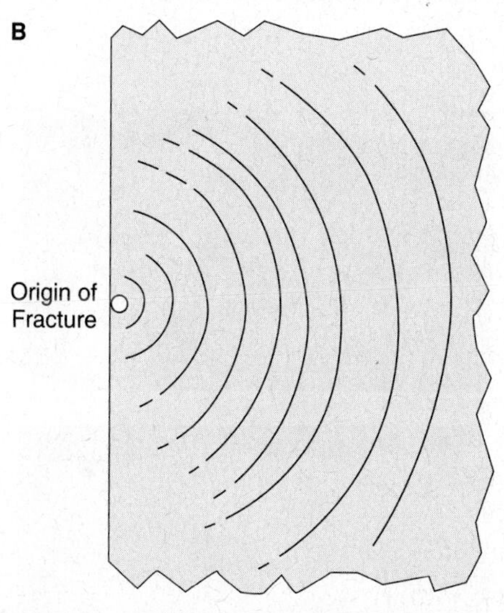

Figure 5.22 (*A*) Diagram of origin of fracture in relation to plumes. (*B*) Diagram of origin of fracture in relation to ribs. [Reprinted from Tectonophysics, v. 12, no. 5, A study of surface features relating to brittle and semi-brittle fracture, Gash, S. P. J., p. 349–391, © 1971, with permission from Elsevier.]

Figure 5.23 Photograph of joint faces marked by ornamentation, which is highlighted in chalk. Rock is siltstone (Devonian). Thickness of the ornamented siltstone layer is ~9 cm. The origin of one of the joint faces coincides with the edge of a groove cast (lowest chalked surface), where the lower boundary of the groove casts makes a turn from horizontal to vertical. [Reprinted from *Journal of Structural Geology*, v. 23, McConaughy, D. T., and Engelder T., Joint initiation in bedded clastic rocks, p. 203–221, © 2001, with permission from Elsevier.]

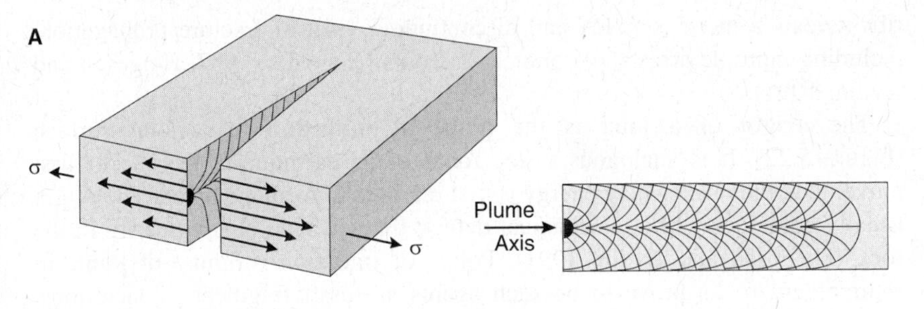

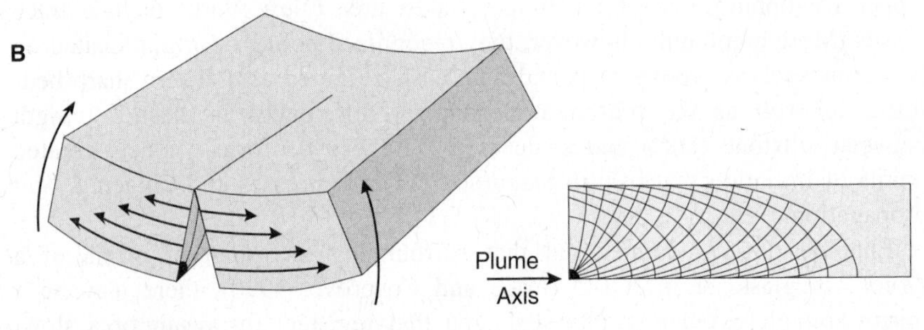

Figure 5.24 (*A*) Plumes produced on joint surface in rock layer broken in tension. Solid black dot is origin. Plumes radiate from origin. Parabolic curves in insert describe propagating joint front, sometimes preserved as ribs. (*B*) Plumes produced on joint surface produced through bending of a rock layer. Solid black dot is origin. Plumes radiate from origin. Solid curves represent propagating joint front. [From DeGraff and Aydin, 1987. Published with permission of Geological Society of America.]

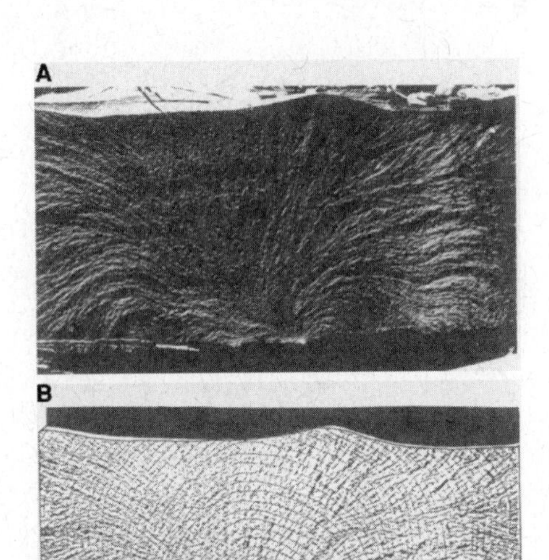

Figure 5.25 Plume patterns and the reconstruction of joint fronts from the observed hackle patterns. (*A*) Photograph of joint face in siltstone layers. Plumes converge downward to origin. [From DeGraff and Aydin, 1987. Published with permission of Geological Society of America.] (*B*) Drawing showing origin (black dot) at point of convergence of plumes. The thin, curved black lines are everywhere constructed perpendicular to plumes, and they represent collectively a map of the propagation of the joint front up and out from the origin. [Reprinted from *Journal of Structural Geology*, v. 13, Helgeson D. E., and Aydin A., Characteristics of joint propagation across layer interfaces in sedimentary rocks, p. 897–911, © 1991, with permission from Elsevier.]

Plumes and Ribs

Plumes can be seen to radiate and diverge from flaws; that is, points of origin/initiation (see Figures 5.18 and 5.23). Plumes are linear to systematically curved markings that on a joint face converge toward the origin of the joint (Figure 5.24). Plumes are physically composed of tiny ridges and troughs, a micro-topographic relief on the main joint face (Roberts, 1961). Plume patterns on adjacent rock faces that "*join*" at a common surface are complementary, for the ridges and troughs on one face nestle perfectly into the troughs and ridges on the adjacent face. At every point along a given plume, the plume line records the direction of crack propagation, which is outward from the origin in the direction of divergence of the plume lines (Engelder, 1984) (Figure 5.25). The simplest plume patterns are marked by a **straight plume axis** (Figure 5.26*A*), which lies parallel to the trace of bedding or layering (in sedimentary or volcanic rocks); the plumes radiate outward at angles of 30–35° from the central axis, gradually curving to angles of about

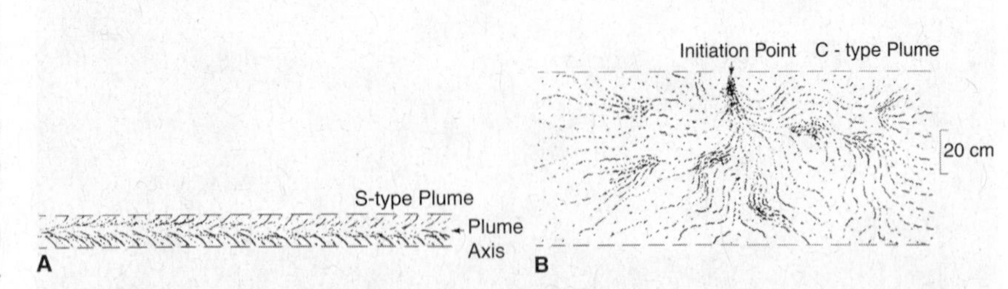

Figure 5.26 (*A*) Drawing of plumose structure with a straight plume axis. [Reprinted with permission from *the Journal of Structural Geology*, vol. 7, Engelder, T., Loading paths to joint propagation during a tectonic cycle: an example from the Appalachian Plateau (1985), Elsevier Science Ltd., Pergamon Imprint, Oxford, England.] (*B*) Drawing of curving plume axes. [Reprinted from *Journal of Structural Geology*, v. 7, Engelder, T., Loading paths to joint propagation during a tectonic cycle: an example from the Appalachian Plateau, p. 459–476, © 1985, with permission from Elsevier.]

75° near the margin of the joint surface (see Figure 5.26*A*). The straight plume axis itself coincides with the path of propagation of the tip of the fracture, typically in a direction coinciding with the longest dimension of the fracture surface (Pollard and Aydin, 1988). **Curving plume axes** are more complicated (Figure 5.26*B*). A single curved plume axis may divide into branches, and then the branches may subdivide again and again, with plumes radiating to either side of the main axis and the secondary branches.

Ribs (also known as **undulations**) are also present on many joint faces (see Figure 5.18; also see Figures 5.2 and 5.3), and they take the form of parabolically curved, spaced markings (Figure 5.27). These are the same kind of markings that on a much smaller and finer scale distinguish conchoidal fracture in obsidian or Plexiglas (Figure 5.28) (DeGraff and Aydin, 1987; Pollard and Aydin, 1988). Ribs tend to be concentric about the origin of the joint face. Furthermore, at any given point on the main joint surface, rib markings are perpendicular to plumes (Ramsay and Huber, 1987). In profile view, the change from the main joint face to a rib may be seen to be a smooth curve or a sharp kink. The steps across the ribs may be millimeters to centimeters wide, depending on several factors, including the size of the main joint face, the mechanical character of the rock, and the nature of propagation of the fracture itself.

Figure 5.27 Macrophotograph of rib display in small slab of Straight Cliffs Sandstone (Cretaceous). [Photograph and copyright © by Peter Kresan.]

Experimental work has verified that plumes and ribs on the main joint face are left as a record of the propagation of the fracture front through the host rock. Direction of propagation of the tip line of the main joint face is opposite to the direction in which the plumes on the main joint face converge. The plumes on a given joint face converge toward the origin (see Figures 5.18 and 5.24) (Secor, 1965). Ribs, on the other hand, represent positions of the joint front at some past time; they are a "*fossil record*" of a propagating joint front. In some cases they "*map*" the temporary arrest of a propagating joint front; in other cases they record the location at which the velocity of propagation of the joint front diminished.

It is thought that a main joint face tries to maintain a perpendicular orientation to the direction of the least principal stress (σ_3), which will commonly be the direction of greatest tensile stress. This orientation is commonly controlled by the orientations of far-field stresses, though the formation of any joint is the summation of far-field and local effects. A joint oriented perpendicular to least principal effective stress occupies the orientation of a principal plane (i.e., the σ_1/σ_3 plane), and thus there is no shear stress along it. Yet, as a mode I fracture propagates parallel to its own plane, it may suddenly enter a locale in which the local stress has shifted. In response the main joint face makes a "*course correction*" to maintain orientation perpendicular to the direction of least principal stress (σ_3) (Engelder, 1987).

The main joint face succeeds in making its course correction through a **tilt** or a **twist** (Figure 5.29). An **axis of twist** is in the plane of the joint surface (see Figure 5.29*A*), and parallel to the direction of propagation. During twist, the surface breaks up into discrete (i.e., "*twisted*") segments (Engelder, 1987).

Figure 5.28 Expression of ribs and plumes on a joint face produced in a block of Plexiglas. As the block of transparent Plexiglas was being subjected to differential stress, fluid was injected through a tube (at left) to the point of origin for the joint. The joint was thus produced by "*hydrofracing*" when the tensile strength of the Plexiglas was exceeded. The crack is oriented perpendicular to the direction of least principal stress, which was tensional. Height of the crack is approximately 7 cm. [From Rummel, F., 1987. Fracture mechanics approach to hydraulic fracturing stress measurements in Atkinson, B. K., Fracture mechanics of rock. Published with permission from Academic Press.]

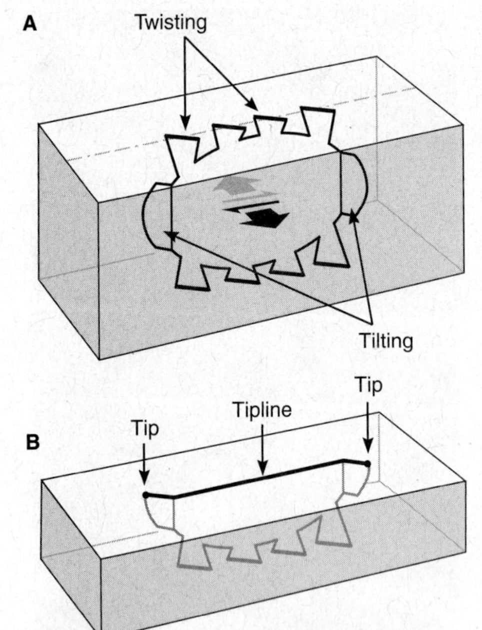

Figure 5.29 (*A*) Here is a joint (gray) in a transparent block of rock. The joint has formed as a result of the combination of mode I opening and mode II sliding (right-handed). Joints in the fringe of the main joint face are tilted and twisted, oriented slightly clockwise from the main joint face due to the right-handed shear component. The axis of tilt is perpendicular to the direction of joint propagation and lies in the plane of the main joint surface. The axis of twist is essentially parallel to the direction of joint propagation and lies in the plane of the main joint surface. The tilts are mode II. The twists are mode III. (*B*) A different level of exposure. [Reprinted from *Journal of Structural Geology*, v. 13, Cruikshank, K. M., Zhao, G., and Johnson, A. M., Analysis of minor fractures associated with joints and faulted joints, p. 865–886, © 1991, with permission from Elsevier.]

Plumes are the expression of this twisting action; they represent the line of intersection between short, discrete en echelon fractures with the main joint face (see Figure 5.18) (Pollard and Aydin, 1988). They are a product of the combination of mode I and mode III loading. In contrast, an **axis of tilt** is in the plane of the joint surface though in this case perpendicular to the direction of joint propagation (see Figure 5.29B). Ribs are an expression of tilt (Pollard and Aydin, 1988). The ribs can form as a product of bending or kinking of the leading edge of the fracture. They are a product of the combination of mode I and mode II loading.

Fringe Joints

Fringe joints by definition occupy the fringe of a joint surface, just beyond the main joint face. They are "*out of plane*" with respect to the main joint face, and yet they grew from the tip line of the main joint face (Younes and Engelder, 1999). Fringe joints are produced by mixed-mode loading, either through the combination of mode I and mode II loading, or mode I and mode III.

As we learned earlier, mixed mode I and III loading creates plumes on the main joint face through formation of very fine-scale en echelon fractures. This en echelon segmentation tends to be much better developed, and thus much more conspicuous, in the joint fringe. The fractures themselves are referred to as **en echelon fringe cracks** or as **twist hackles** (Younes and Engelder, 1999), and they too can display fine-scale delicate plumes and ribs. The twist hackles are "*twisted*" in orientation from parallelism with the main joint face, in the manner shown in Figures 5.18 and 5.29*B*. But there is more to the story. Younes and Engelder (1999) distinguish two varieties of twist hackles. **Gradual twist hackles** smoothly emerge from the main joint face, progressively rotating in orientation from main face to the top of the fringe (Figure 5.30). In contrast, **abrupt twist hackles** appear to be disconnected from the main joint face, abruptly shifted in orientation from that of the main joint face (Figure 5.31). Younes and Engelder (1999) provided clear documentation of gradual twist hackles and abrupt twist hackles in their study of fringe joints in interbedded siltstone, sandstone, and shale in the Finger Lakes region of New York. They found that the gradual twist hackles occurred in siltstone and sandstone layers, which are the same layers in which the main joint faces developed. In contrast, the abrupt twist hackles formed in shale beds situated either above or below the sandstone or siltstone layers within

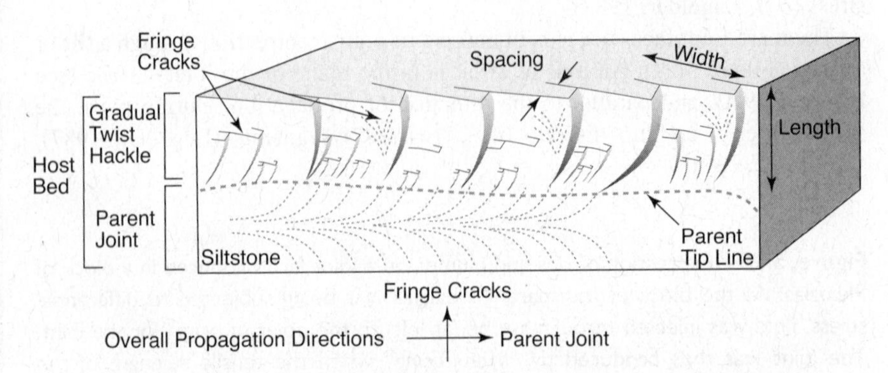

Figure 5.30 Drawing of gradual twist hackles emerging from parent joint face. Note that the number of fringe cracks decreases away from the main joint face, and thus the spacing of the cracks becomes larger progressing from the tip line of the outer edge of the fringe. [From Younes and Engelder, 1999, Figure 1, p. 220. Published with permission of Geological Society of America.]

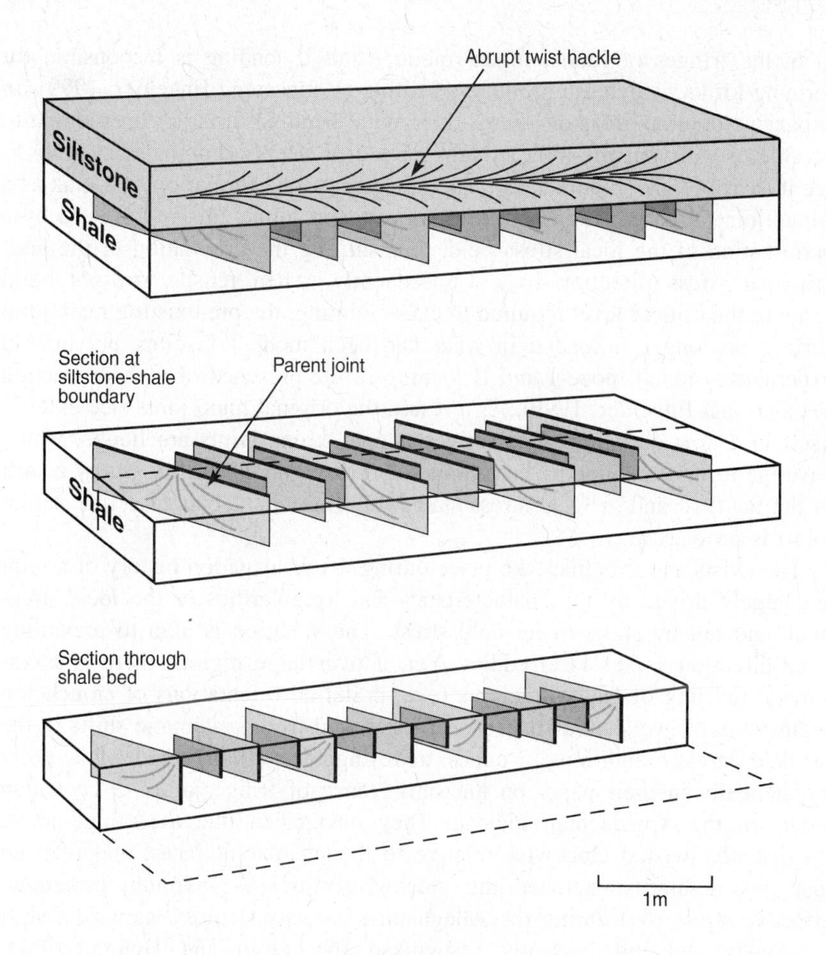

Figure 5.31 Drawing of abrupt twist hackles, which have propagated downward (in this example) from a stiff siltstone bed into a mechanically soft shale bed. [From Younes and Engelder, 1999, Figure 2, p. 220. Published with permission of Geological Society of America.]

which the main joint face formed (Figure 5.32). Thus it appears that mechanical coupling (or the lack thereof) at bedding plane boundaries in the interbedded sequence of siltstone and shale is a major factor in determining whether gradual or abrupt fringe cracks develop.

Figure 5.32 Photograph of abrupt twist hackles propagating downward from siltstone bed into a thick shale bed. These fringe joints in the upper part of the Devonian Genessee Group are exposed at Taughannock Falls State Park, New York. [Reprinted from *Journal of Structural Geology*, v. 7, Engelder, T., Loading paths to joint propagation during a tectonic cycle: an example from the Appalachian Plateau, p. 459–476, © 1985, with permission from Elsevier.]

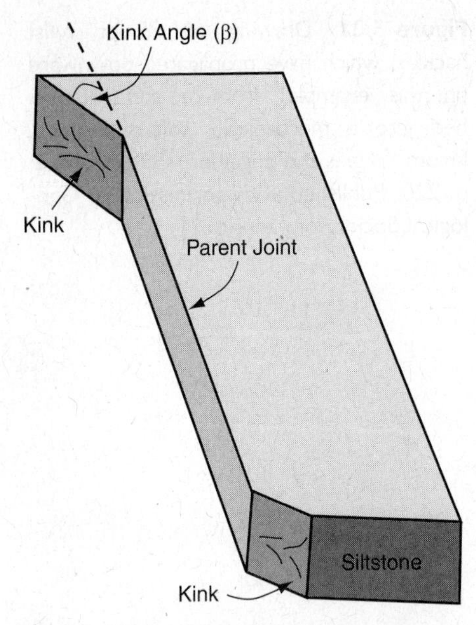

Figure 5.33 Drawing of two kinks propagating from the lateral tip lines of a parent joint. The fringe cracks are products of mixed mode loading (mode I and mode II). [From Younes and Engelder, 1999, Figure 3, p. 221. Published with permission of Geological Society of America.]

In the fringes of joints, mixed mode I and II loading is responsible for forming **kinks**, which are products of tilting (Younes and Engelder, 1999). In expansive outcrop areas of jointed, horizontal bedding, it is not uncommon to be able to walk out the long straight trace of a steeply dipping joint, only to see it sharply kink (Figure 5.33). The precursor to the formation of a kink as a fringe joint is propagation and arrest of a parent joint. This is followed by a perturbation of the local stress field, thus shifting the orientation of the least principal stress direction (σ_3). Consequently, when tensile stresses build again to the critical level required to cause jointing, the pre-existing main joint surface no longer is loaded in what had been mode I fashion, but instead experiences mixed mode I and II loading in the presence of a shear traction (Younes and Engelder, 1999). As a result, the original main joint face extends itself in a new direction. Alternatively, some fringe joints are **hooks**, which have the form of a smooth bend. Smoothly curved hooks are caused by a shift in the far-field and/or local stress state while the main joint face (the parent joint) is propagating.

The twists and tilts that take place during the propagation history of a joint are largely driven by the characteristics and irregularities of the *local* stress field, and not by shifts in far-field stress. The situation is akin to prevailing wind direction versus local eddies. Yet, if over large regions there is a systematic twisting of fringe fractures (e.g., preferred orientations of en echelon segmentation), we should suspect that it resulted from systematic shifts in the far-field stress conditions. Younes and Engelder (1999) made this point emphatically in their paper on the significance of fringe joints in Devonian strata on the Appalachian Plateau. They recognized that the fringe joints consistently twisted clockwise relative to the main joint faces, and realized that this regularity matched the clockwise shift in maximum horizontal effective stress (σ_1) during the Alleghanian orogeny (Pennsylvanian), a shift previously and independently recognized (Nickelsen and Hough, 1967). Fracture patterns may appear pretty innocuous at the outcrop scale, but they can tell a big story.

Figure 5.34 Photograph of columnar jointing, the distinctive architecture at the Giant's Causeway, along the northern coast of Northern Ireland near Bushmills. [Photograph by G. H. Davis.]

GROWTH OF JOINT SETS

Example of Columnar Jointing

We can affirm a lot of what we have learned about individual joint surfaces through close examination of columnar joints and how they form. Furthermore, we can begin to see how individual joints connect up with one another to form sets of "*coeval*" joints; that is, joints forming essentially contemporaneously and under a common set of conditions. Columnar joints result from the fracturing that accommodates contraction during congealing and shrinking of a lava flow as it cools (Figure 5.34). Columns tend to be polygonal in the same way that mud cracks resulting from shrinkage of mud are polygonal. The stress field is strictly local. Where columnar joints are well developed, the comparison of architecture and structural geology is irresistible and unavoidable (see Figure 5.34).

DeGraff and Aydin (1987) discovered kinematic relationships that prove to be invaluable to the understanding the propagation of joints in general, and the mechanical relationship of one joint surface to another. The joint faces of each column display bands that run horizontally, parallel to layering and perpendicular to the column (Figure 5.35). The bands superficially resemble stratification; however, an individual band on the flank of an individual column is in fact an individual joint face. The bands studied by DeGraff and Aydin (1987) ranged in width from 20 to 100 cm (i.e., the width of a column face)

Figure 5.35 Jim DeGraff sits in the shade of columnar joint faces in Snake River Basalt. The faces are marked by parallel, discontinuous bands oriented perpendicular to column axes. The physical expression of the bands is created by textural and geometric variations from band to band. Note how Jim and Atilla have 'chalked up' the bands in ways that show direction of propagation. [Photograph taken by Atilla Aydin along the Boise River at Lucky Peak Dam in Idaho.]

and 3–12 cm in height. Each band on each column flank is distinctive because of variations in geometry, roughness, orientation, and general texture. Most important, each band is a main joint face, possessing its own complete set of plumes, revealing that each band is a single fracture event (Figure 5.36). Thus, the flank of a column many meters high, which appears at first glance to be a single discrete joint face, is actually composed of tens to hundreds of discrete joint faces. The column grew incrementally.

Cooling of a basalt flow sets up local effective tensile stresses between the part of the flow that is contracting (because of cooling) and the hotter or more fluid part that is not (Pollard and Aydin, 1988). Since a flow cools from top down and from bottom up, so do the bands. They meet *below* the middle of the flow, because the lower half of a flow stays warmer longer. New cracks propagate at the edges of old cracks, more particularly from origins (points of initiation) located at flaws such as gas vesicles (holes) or corners of large grains or rock fragments. In this way, the cracks propagate upward or downward toward the middle of the flow. Thus: "*A column face is the net result of many discrete crack events, each of which produces a well-defined segment of the face*" (DeGraff and Aydin, 1987, p. 608).

Example of Jointing in a Siltstone/Shale Sequence

Helgeson and Aydin (1991) studied individual joint surfaces in a sequence of siltstone beds, separated by thin films of shale, in the Appalachian Plateau of New York and Pennsylvania. The shale films are so thin that the siltstone beds, above and below, are mechanically coupled to some degree. McConaughy and Engleder (2001) concluded that mechanical coupling in this situation is best achieved when shale interbeds are thinner than 1 cm. Thus, it does not take much to arrest a joint at a layer boundary. In their analysis of joint propagation through a sequence of siltstone beds, Helgeson and Aydin (1991) found that each individual siltstone bed was marked by discrete joint surfaces that did not extend beyond the upper and lower contacts of the siltstone (Figure 5.37). Because most of the joint surfaces possess their own set of plumes, ribs, and origin, it was possible to map the propagation of each joint. In doing so, Helgeson and Aydin (1991) learned that the termination

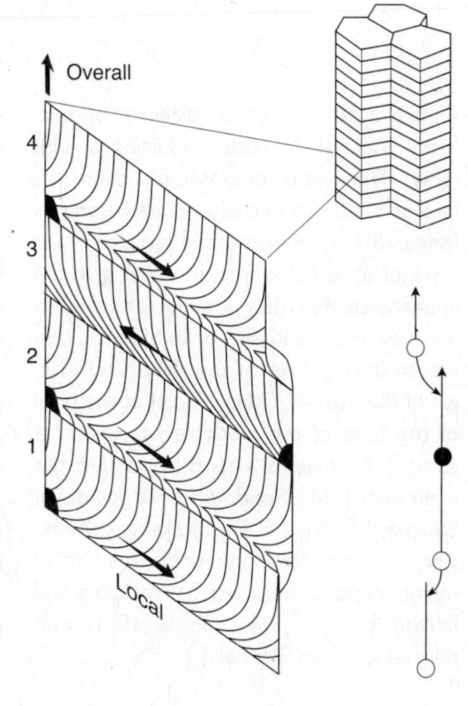

Figure 5.36 Each horizontal band of a column face comes complete with its own origin and plume pattern, indicating that each band represents an individual crack formed during a discrete fracture event. Diagrammatic representation of the progressive growth of a columnar joint face. The origins occur on the upper edges of preceding segments. The black arrows describe the horizontal propagation for each band. The numbers give the sequence of formation of the bands. [Reprinted from *Journal of Volcanology and Geothermal Research*, v. 38, DeGraff, J. M., Long, P. E., and Aydin, A., Use of joint-growth directions and rock textures to infer thermal regimes during solidification of basaltic lava flows, p. 309–324, © 1989, with permission from Elsevier.]

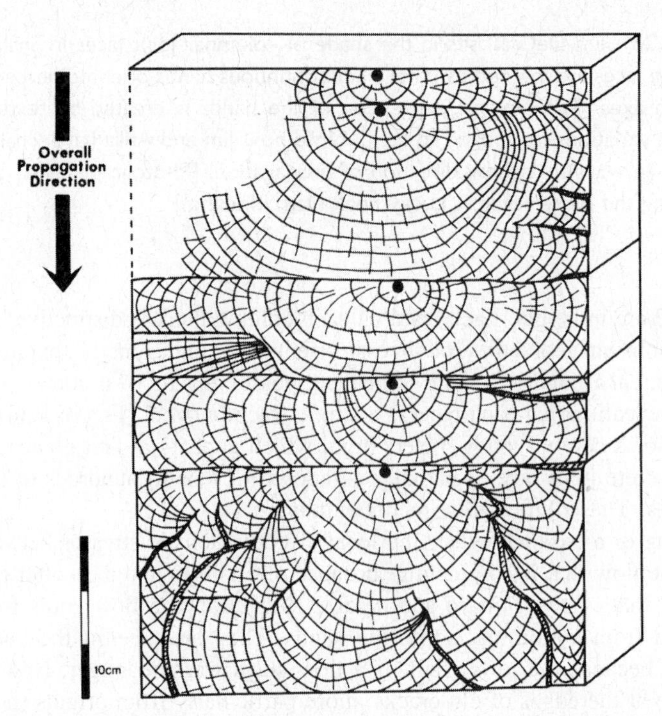

Figure 5.37 A composite joint face produced by systematic *downward* propagation through several siltstone layers separated by ever so thin shale layers. Origins are shown by solid black dots, plumes by short dash-like segments, and joint fronts by solid lines. Note that the origins are aligned vertically, one on top of another. [Reprinted from *Journal of Structural Geology*, v. 13, Helgeson, D. E., and Aydin, A., Characteristics of joint propagation across layer interfaces in sedimentary rocks, p. 897–911, © 1991, with permission from Elsevier.]

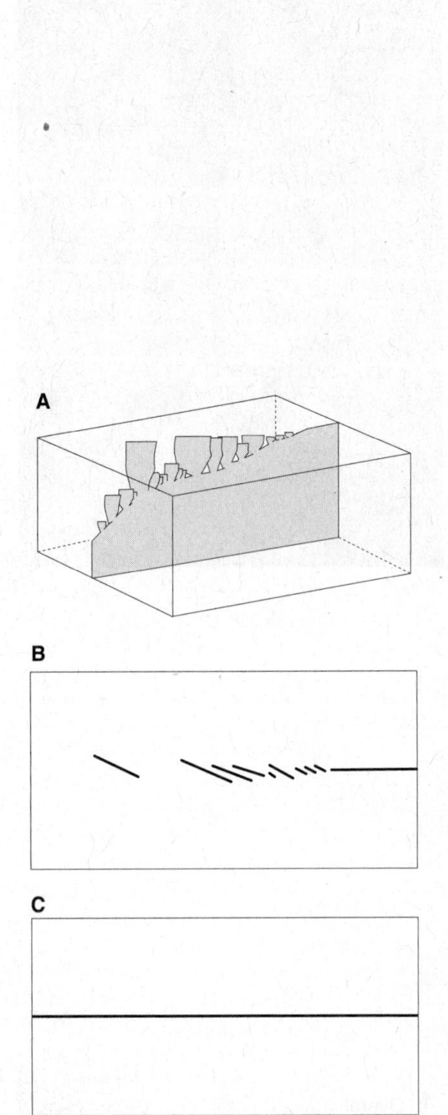

Figure 5.38 (*A*) Block diagram of perfectly transparent rock containing a joint (gray). The joint is composed of a main joint face as well as en echelon cracks along its fringe. (*B*) Map of joint at the surface reveals a set of en echelon fractures that give the appearance that they are not attached in any way to one another, or to the main face whose trace is seen at the right-hand margin of the map. (*C*) Map of joint at the level of the base of the block reveals simply a straight continuous line representing the main joint face. [Reprinted from *Journal of Structural Geology*, v. 13, Cruikshank, K. M., Zhao, G., and Johnson, A. M., Analysis of minor fractures associated with joints and faulted joints, p. 865–886, © 1991, with permission from Elsevier.]

point of a given joint at the bottom of a siltstone bed often controlled the point of origin of the next joint to form in the underlying siltstone bed. Thus Helgeson and Aydin (1991) concluded that the joints propagated incrementally *downward* through the sequence. Points of origin were almost always along the upper contacts of the siltstone layers (see Figure 5.37). The vertical alignment of origins reveals the mechanical continuity within the whole sequence.

Joint Intersections and Interpretation of Timing

Detailed observation and mapping of steeply dipping joints in outcrop reveal some common patterns in the ways in which the traces of joints intersect and terminate. First of all, the trace expression of any individual ideal joint will depend on the level of exposure relative to the geometry of the whole joint face, including its 'fringe elements' (Figure 5.38*A*). Where the level of exposure intercepts only the fringe of the joint face (Figure 5.38*B*), the trace of the joint might well be a series of en echelon lines at an angle to the main joint. Where the level of exposure intercepts the main part of the joint (Figure 5.38*C*), the trace of the joint will be a single line.

The mapping of the intersection of joints reveals some commonly repeated patterns as well. Pollard and Aydin (1988) have called attention to several intersection geometries: **Y-intersections, X-intersections**, and **T-intersections**

Figure 5.39 Joint intersection patterns. (*A*) Y-intersections of joints in a lava flow. [Reprinted with permission from *Science*, vol. 239, Aydin, A., and DeGraff, J. M., Evolution of polygonal fracture patterns in lava flows. Copyright © 1988 American Association for the Advancement of Science.] (*B*) X-intersections of joints in siltstone exposed at Kimmeridge Bay, United Kingdom. [Photography by Terry Engelder.] (*C*) T-intersection of joints in quartzite in the Salt River Canyon region, Arizona. [Photograph by G. H. Davis.]

(Figure 5.39). Y-intersections are typical of discontinuous contraction joints, like mud crack patterns or columnar joints (Figure 5.39*A*). The joints that intersect in Y-patterns meet at approximately 120° angles, which is a configuration requiring minimum energy to achieve. X-intersections are the inevitable pattern that forms when systematic continuous joints intersect at acute angles (Figure 5.39*B*). Typically one joint trace runs continuously without disruption through the intersection, whereas the other is stepped slightly to the right or left, or simply stops. In a system of two sets of vertical joints that intersect at a high angle, there is usually no consistency in which joint is stepped versus which joint continues without a break through the intersection. T-intersections are common in orthogonal joint systems, where individual joint traces meet at right angles (Figure 5.39*C*). Typically, the younger joint trace can be followed to its termination at a preexisting joint oriented at right angles.

Joints die out in trace expression in a variety of common ways. The joint trace may abruptly hook and stop (Figure 5.40*A*, *B*), or hook in such a way that forms a T-intersection with an adjacent joint trace that was coming from the other direction (Figure 5.40*C*). Truncated joints at T-intersections are younger than the joints by which they are truncated (Pollard and Aydin, 1988). The older joint surface along which a younger joint terminates is a free surface, lacking any shear stress component. Therefore, principal stresses are parallel and perpendicular to the joint surface (Suppe, 1985), "*forcing*" a younger joint to be aligned in a perpendicular arrangement. Yet another way in which a single joint trace can die out is through replacement by a series of en echelon fractures (Figure 5.40*D*). In many instances a joint trace will simply come to an end without changing trend or breaking up into segments. The manner in which joints intersect and die out can provide information on relative timing of joint set development.

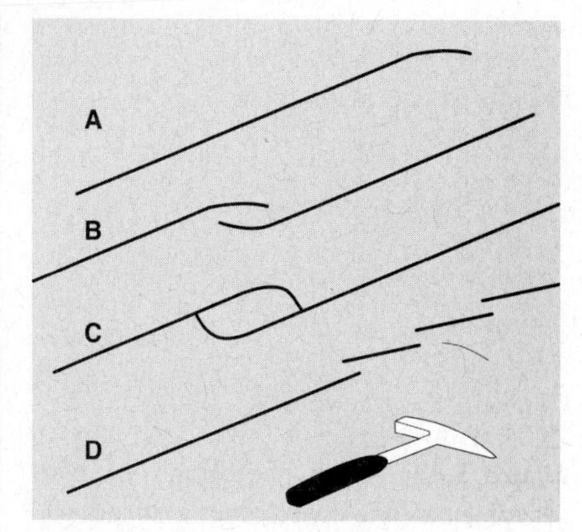

Figure 5.40 Ways in which individual joints die out. (*A*) One joint hooks and stops. (*B*) Two joints overlap, hook, and stop. (*C*) Two joints overlap, hook, and connect to form T-intersections. (*D*) Joint breaks up at its tip line into en echelon segments.

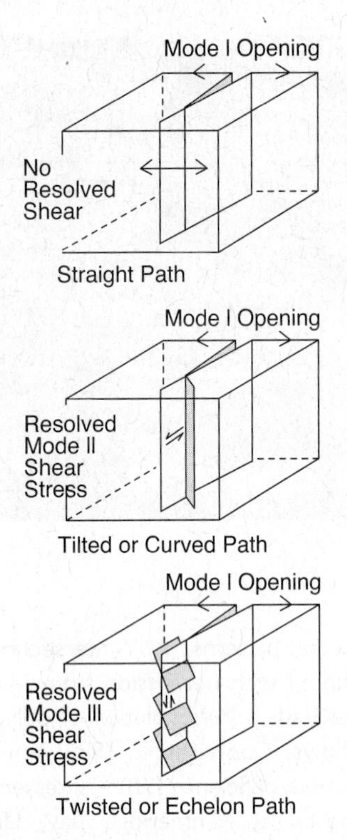

Figure 5.41 The progressive development of hook-shaped terminations and en echelon terminations. (*A*) Plain old mode I joint. (*B*) Mode I joint with mode II action at tip, creating hook-shaped termination. (*C*) Mode I joint with mode III action at tip, creating en echelon fractures at termination. [From Pollard and Aydin, 1988. Published with permission of Geological Society of America.]

The patterns of terminations of joints reveal that the local stress field around one joint can affect the growth and growth direction of a neighboring joint. Pollard and Aydin (1988), and Olson and Pollard (1988, 1989) have explained that as two joints propagate toward each other, each enhances the growth of the close neighbor by imparting an additional increment of tensile stress in the area around the joint front (Figure 5.41). However, as the tips become very close and overlap, shear stresses are induced on neighboring joints. As a consequence, each joint may tilt and curve into the other, producing the characteristic hook-shaped terminations. Or the joints may twist, creating a termination pattern of en echelon joints.

JOINT SPACING IN RELATION TO A SINGLE BED

Introduction

The goal of examining spacing of joints within a single bed relates to the search for clues regarding what controls the intensity (density) of jointing. There are any number of important variables, such as rock type, bed thickness, rock strength, loading conditions, abundance of flaws, and the interplay between far-field and local stress environments. The regularity of spacing of joints in given beds over large areas can be incredibly eye-catching (Figure 5.42). Sets of orthogonal, evenly spaced joints within rock layers have created bedrock foundations that appear to be "*engineered*" (Figure 5.43). It is nearly impossible to look at such outcrops and not ask aloud what exactly controls joint spacing and orientation. If we come to understand the factors that predetermine joint spacing, we can then apply this knowledge to predicting fracture patterns and fracture-system permeability in the subsurface. Such predictive capabilities are of great value to petroleum geologists and engineers, and to hydrogeologists.

Figure 5.42 Aerial photograph of two sets of joints (one conspicuous, one less so) that cut the Moab Member of the Entrada Sandstone (Jurassic) at Arches National Park, Utah. Spacing of joints is 25 m to 40 m. Just for fun, rotate the book 90° and look at the joint pattern again. Note how much sharper the joints appear. [Photograph by R. Dyer.]

Figure 5.43 The spacing and orientation of jointing is often so regular and systematic that it looks man-made. Compare (*A*) the jointing in limestone in northwestern New Mexico to (*B*) sidewalk-size slabs of limestone carefully placed by Romans in the ancient city of Corinth, Greece. The 'squares' in each photo are ~1m × 1m. [Photographs by G. H. Davis.]

Fracture Spacing Ratio (FSR)

The logical starting point in most studies of joint spacing is gathering joint-spacing data for a given bed or layer, and then plotting the results in ways conducive to statistical analysis. The approach begins with identifying a broad flat expanse of exposure of systematic joints cutting near-horizontal beds of measurable thickness. Next it is necessary to choose a given joint set marked by parallel planar traces cutting a particular bed. Measure the thickness of the bed, and then measure the spacing between adjacent joint traces along a line perpendicular to the overall trend of the joint set. This is carried out by laying-out a long measuring tape at right angles to the trend of the joint set, and then reading off and recording the spacings. Based on these data, a **fracture spacing ratio (FSR)** can be assigned to the data set, where FSR is bed thickness divided by median joint spacing for these data from a single bed.

A nice example of how this work is done was provided by Gross and Eyal (2007), who analyzed, near-vertical "*systematic joints*" and "*cross-joints*" along the crest of a broad anticline (the Halukim anticline) in southern Israel (Figure 5.44). Both the systematic and cross joint sets are bed-confined within carbonate units. The systematic joints are long continuous parallel evenly spaced fractures; the cross joints are short discontinuous parallel evenly spaced joints that abut against the systematic joints. During the course of their field work, Gross and Eyal (2007) measured the trends of the joints in each set, and along their scanlines measured the spacing of joints in each set, all the while keeping a record of the measured thickness of the six carbonate beds in which each spacing reading was taken. Table 5.1 is a summary of measurements of orientations and spacing data for these bed-confined joints. Gross and Eyal (2007) calculated FSR values for each data set and found that average FSR values for systematic joints were greater than the average FSR values for cross joints, namely 1.85 vs 1.15. Their fuller data presentation for these joint set measurements included rose diagrams of the trends 175 systematic joints and 284 cross joints (Figure 5.45), and histograms of number of joints versus joint spacing for all of the systematic joints and all of the cross joints (Figure 5.46).

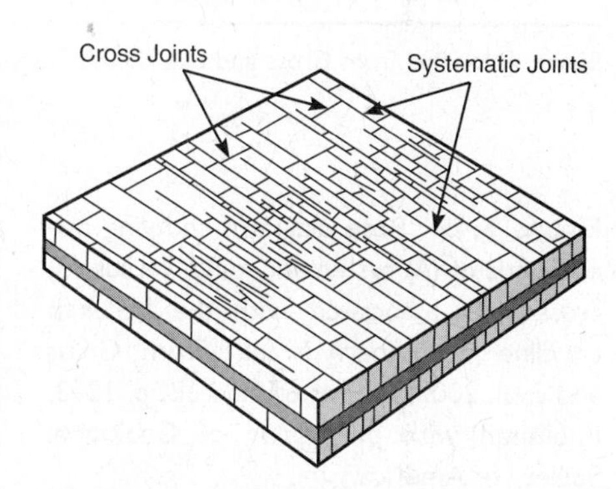

Figure 5.44 Schematic representation of systematic joints and cross-joints in carbonate beds, Halukim anticline, southern Israel. [From Gross and Eyal, 2007, Figure 13C, p. 1402. Published with permission of Geological Society of America.]

TABLE 5.1
Summary of orientation and spacing data for bed-confined systematic and cross joints measured at Halukim anticline in southern Israel. MLT-LC refers to lithology-controlled mechanical layer thickness; MLT-JC refers to joint-controlled mechanical layer thickness; FSR-LC refers to lithology-controlled fracture spacing ratio; and FSR-JC refers to joint-controlled fracture spacing ratio.

		Orientation		*Spacing*					
Scanline	*Joint set*	*Mean orientation*	*Angle between sets*	*Mean spacing (cm)*	*Median spacing (cm)*	*MLT-LC (cm)*	*MLT-JC (cm)*	*FSR-LC*	*FSR-JC*
A1	Systematic	353°/87°E	87°	34.1	31.0	42.9		1.38	
	Cross	079°/87°N		51.9	44.5	42.9	31.0	0.96	0.7
B1	Systematic	357°/88°W	79°	14.8	14.4	31.8		2.20	
	Cross	078°/85°N		34.2	27.7	31.8	14.4	1.15	0.52
B2	Systematic	353°/87°W	73°	19.8	17.0	46.3		2.72	
	Cross	088°/85°N		26.0	24.6	30.7	17.0	1.25	0.69
C1	Systematic	333°/83°E	88°	16.2	16.0	29.0		1.81	
	Cross	067°/89°N		36.9	31.7	29.0	16.0	0.92	0.51
C2	Systematic	319°/88°N	71°	19.2	19.8	34.6		1.77	
	Cross	061°/88°S		32.14	27.1	34.6	19.6	1.28	0.72
D1	Systematic	323°/89°W	62°	33.8	34.5	43.5		1.26	
	Cross	025°/88°W		42.8	32.2	43.5	34.5	1.35	1.07

Source: [Modified from Gross and Eyal, 2007, Table 1, p. 1393. Published with permission of Geological Society of America.]

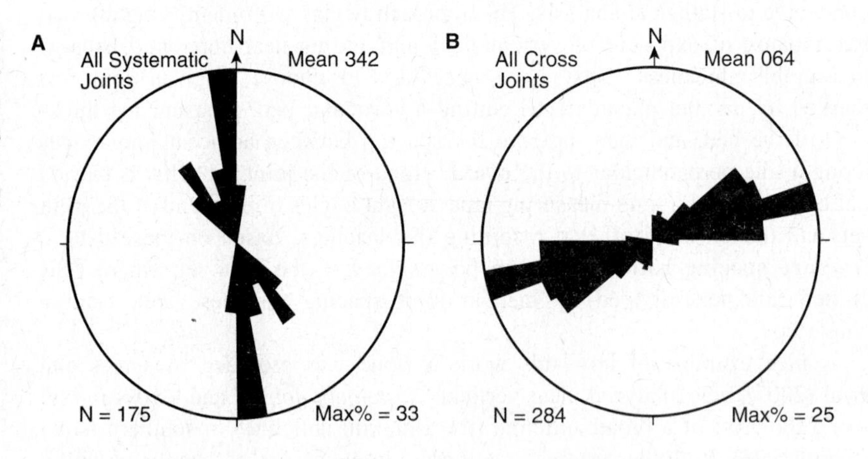

Figure 5.45 Rose diagrams showing orientations of (*A*) systematic joints versus (*B*) cross-joints associated with the Halukim anticline in southern Israel. [From Gross and Eyal, 2007, Figures 6A and 6B, p. 1393. Published with permission of Geological Society of America.]

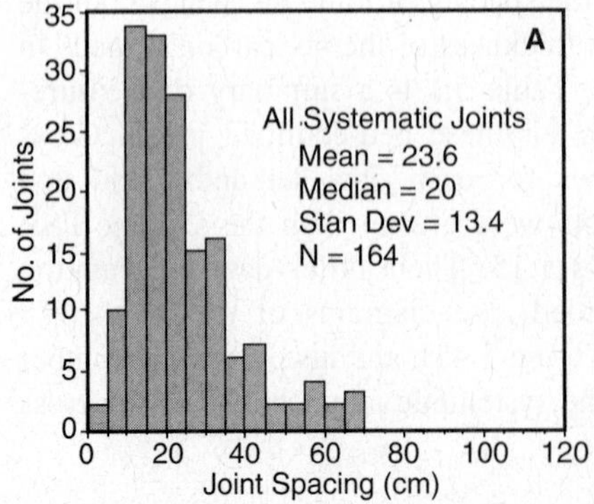

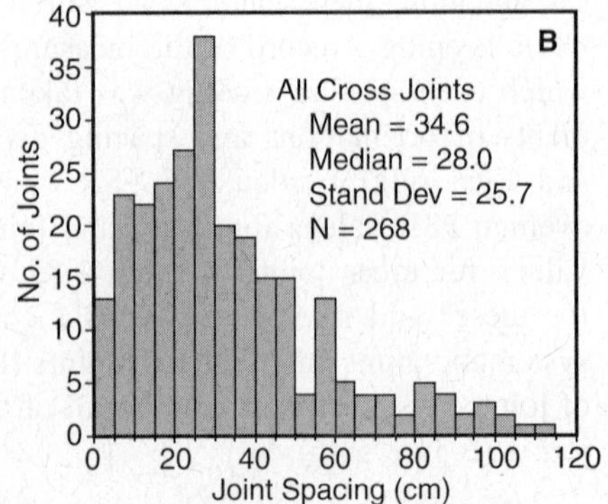

Figure 5.46 Histograms of number of joints versus spacing of joints for (*A*) all systematic joints and (*B*) all cross joints associated with the Halukim anticline. Thicknesses of beds within which these joints occur range from 29 cm to 46 cm. [From Gross and Eyal, 2007, Figures 6A and 6B, p. 1393. Published with permission of Geological Society of America.]

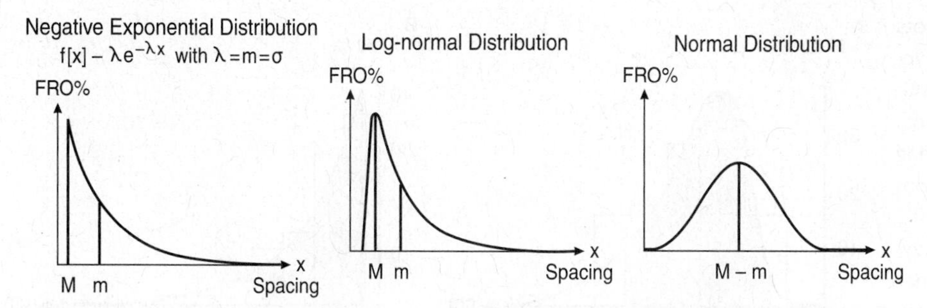

Figure 5.47 Examples of frequency distribution plots (FDPs) for jointing: (*A*) Negative exponential distribution, (*B*) log-normal distribution, and (*C*) normal distribution. M = mode; m = mean; σ = standard deviation. [Reprinted from *Journal of Structural Geology*, v. 14, Rives, T., Razack, M., Petit, J. P., and Rawnsley, K. D., Joint spacing: analogue and numerical simulations, p. 925–937, 1992, with permission from Elsevier.]

You can imagine the many ways in which FSR values can be compared and contrasted, for instance, as a function of bed thickness, where rock type remains the same; or as a function of rock type, where thickness remains the same; or as a function of location relative to major structures (faults, folds), where rock type and bed thickness remain the same. Determining Fracture Spacing Ratio (FSR) is just a starting point for characterizing joint system properties, a starting point along the path of interpreting such properties in terms of mechanics.

Frequency Distribution Plots (FDPs)

As we see in Table 5.1 and Figure 5.46, it is straightforward to construct histograms (frequency plots) of joint spacing for a given joint set within a given subarea and with respect to a given lithology and bed thickness. Frequency distribution plots for joint spacing do not all come out looking the same. For example, Figure 5.46 shows that the cross-joint histogram is quite skewed compared to the histogram for systematic joints in limestone at Halukim anticline.

Some **frequency distribution plots (FDPs)** are marked by negative exponential distributions, some by log normal distributions, some by normal distributions, and some are just irregular and annoying (Figure 5.47). Structural geologists and engineers specializing in rock mechanics have worked hard to try to correlate different frequency distributions with different fracture mechanics. The investigations by Rives et al. (1992) interpret joint frequency distributions in terms of fracture mechanics. Their approach was comprehensive and included fieldwork, physical analogue modeling, and numerical modeling. One of their several field sites was Nash Point in southernmost Wales along the Bristol Channel, where jointing in Late Triassic/Early Jurassic beds is well exposed over a broad wave-cut platform (Figure 5.48). They focused their attention on a flat-lying 1-m-thick grey limestone bed cut

Figure 5.48 Photographs of jointing in wave-cut platform at Nash Point, Wales. Strata consist of inter-bedded Liassic limestones and shales. (*A*) The wave cut platform is cut into nearly perfectly horizontal, jointed strata. (*B*) View of the very special exposures of systematic jointing at Nash Point. [Photographs courtesy of Barry Hankey, "*director at aartworld*," Newport, United Kingdom, from his collection entitled "*Wales, Glamorgan Heritage Coast, Nash Point to St. Donat's*," http://www.aartworld.com/Glamorgan.htm (29 07 10).]

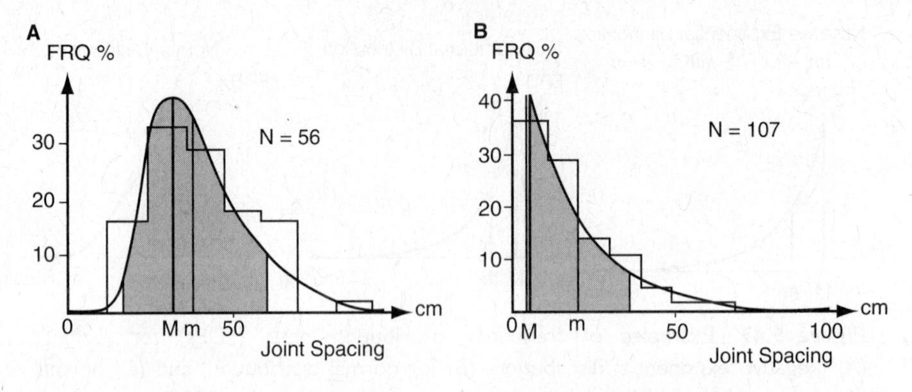

Figure 5.49 Frequency distribution plots (FDPs) for joint frequency versus joint spacing: (*A*) Joints measured at Nash Point (south Wales), showing log-normal fit, and (*B*) joints measured at Whitby (Yorkshire), showing negative exponential fit. [Reprinted from *Journal of Structural Geology*, v. 14, Rives, T., Razack, M., Petit, J. P., and Rawnsley, K. D., Joint spacing: analogue and numerical simulations, p. 925–937, © 1992, with permission from Elsevier.]

by a beautiful set of joints striking east-northeast. Their frequency plot of joint spacing for Nash Point revealed a log-normal distribution (Figure 5.49*A*). They also carried out joint analysis at Whitby, England, located on the coast of North Yorkshire along the North Sea. At Whitby the rocks are Upper Jurassic shales, cut by joints belonging to a set striking east–west. The frequency distribution plot for joint spacing at Whitby shows a negative exponential distribution (Figure 5.49*B*).

Physical Analogue Experiments

To try to explain the differences in distribution patterns, Rives et al. (1992) carried out physical analogue experiments featuring rectangular polystyrene plates (measuring 180 mm × 100 mm × 1.5 m), which they extended (stretched) under controlled conditions. Stretching was achieved by progressively bending the thin plates in ways marked by constant radius of curvature (Figure 5.50). Before bending, a thin film of alcohol was spread on the top surface of the plate; the alcohol facilitated breakdown of the polymer structure within the polystyrene. In the presence of the alcohol, traction caused by the

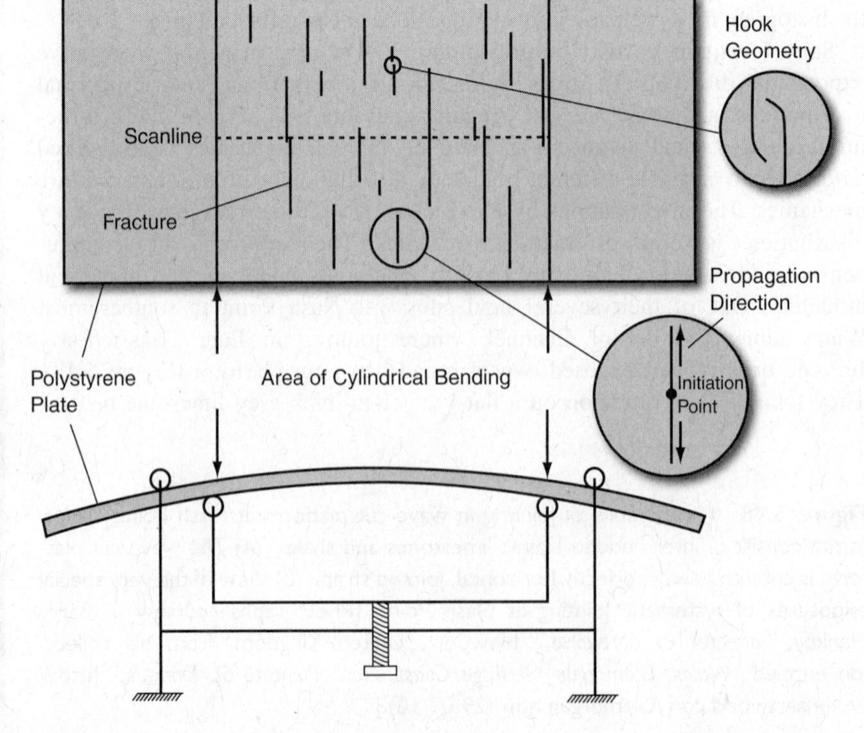

Figure 5.50 Sketch of the set-up for bending and cracking of polystyrene plates. Upper rectangle is map view of the bent plate, showing fracture traces and the location of a scanline along which crack spacing is to be measured. Note 'blow-ups' of initation point and hook geometry. Lower part shows the nature of uniform bending; that is, uniform loading. [Reprinted from *Journal of Structural Geology*, v. 14, Rives, T., Razack, M., Petit, J. P., and Rawnsley, K. D., Joint spacing: analogue and numerical simulations, p. 925–937, © 1992, with permission from Elsevier.]

bending produced a system of cracks in the polystyrene plate. Points of initiation of the cracks were tiny point defects distributed randomly throughout the polystyrene. The cracks then propagated along lines perpendicular to stretching and parallel to the axis of bending (see Figure 5.50). Individual cracks propagated in both directions from a given point defect. The crack patterns bear an uncanny similarity to jointing in bedrock (Figure 5.51).

At various stages in the stress-cracking process, Rives et al. (1992) would halt the bending and measure the crack spacing along scanlines. These stop-action evaluations provided them with the basis for evaluating frequency distribution patterns in relation to the known conditions of loading. Resumed loading would produce new cracks (splitting the distance between two existing cracks) and would cause further propagation of earlier formed cracks. Spacing between cracks became closer and closer during the course of progressive loading. Cracks would stop propagating when they got near the edges of the plates, or when they moved into very close proximity to another propagating crack. In textbook style, adjacent cracks propagating toward one another could be seen to hook into one another, to become arrested (see Figure 5.50).

Rives et al. (1992) found that in the early stage of a given experiment, the relatively few fractures presented a joint spacing marked by a **negative exponential distribution** (similar to Figure 5.49*B*). In the intermediate stage of a given experiment, a denser population of cracks presented a **log-normal distribution** (similar to Figure 5.49*A*). Where bending (and loading) progressed the longest, fracture density was highest, with fracture spacing tending toward **normal distribution**. It was noted that no crack nucleated from a point defect if the defect was located less than a certain limiting distance from any other crack. They called this limiting distance the **crack interaction zone** (Rives et al., 1992). The width of the crack interaction zone was controlled importantly by plate thickness and mechanical properties, but also appeared to be related to the loading history and total strain. During a given experiment, a thin plate would become more and more "*saturated*" with joints over time.

Figure 5.51 Tracing of photograph of analogue fracture set produced during bending and cracking of thin plate of polystyrene. [Reprinted from *Journal of Structural Geology*, v. 14, Rives, T., Razack, M., Petit, J. P., and Rawnsley, K. D., Joint spacing: analogue and numerical simulations, p. 925–937, © 1992, with permission from Elsevier.]

Fracture Spacing Index (FSI), Applied to Bedrock Units

Structural geologists have found it productive to examine the relationship of joint spacing to bed thickness across a broad range of bed thicknesses and lithologies. Narr (1991) and Narr and Suppe (1991) provided a "*best-practice*" model for this kind of study. They began by choosing a well jointed expanse of bedrock of the Monterey Formation (Miocene) in the Santa Maria basin and Santa Ynez Mountains of the Transverse Ranges of southern California (Figure 5.52). They presumed that all of the different rock types within the Monterey Formation at this general locality had experienced a common tectonic history. They then focused attention on jointing in a variety of mechanical lithologies, including discrete beds of dolomite, chert, porcelanite, and silicious shale. For each prominent steeply dipping joint set they would run a scanline perpendicular to joint trend and, for a given bed of a given thickness they would measure and record joint spacing. Their scanlines were designed to capture a variety of lithologies and a broad range of bed thicknesses.

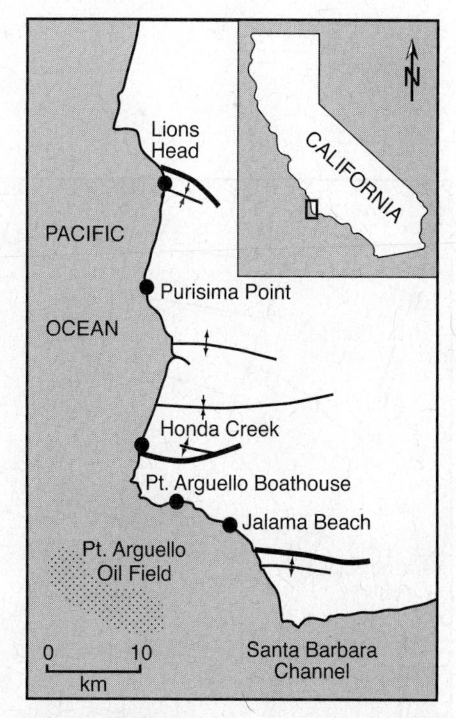

Figure 5.52 Location map showing sites (black dots) where Narr and Suppe (1991) carried out joint analysis. [Reprinted from *Journal of Structural Geology*, v. 13, Narr, W., and Suppe, J., Joint spacing in sedimentary rocks, p. 1037–1048, © 1991, with permission from Elsevier.]

Each scanline of joint-spacing data collected by Narr (1991) and Narr and Suppe (1991) represented a single data set marked by a specific rock type, bed thickness, and joint set. They then would plot the thickness/spacing data (T/S) for each data set on a simple *x-y* plot featuring bed thickness on the ordinate (y-axis) and joint spacing on the abscissa (x-axis). For cumulative plots showing multiple data sets, they plotted median thickness against median spacing (Figure 5.53). (Where such plotting produces a large dense amorphous clump of data points close to the origin, it is preferable to use *log-log plots* of bed thickness versus joint spacing instead). They produced 38 data sets incorporating up to 50

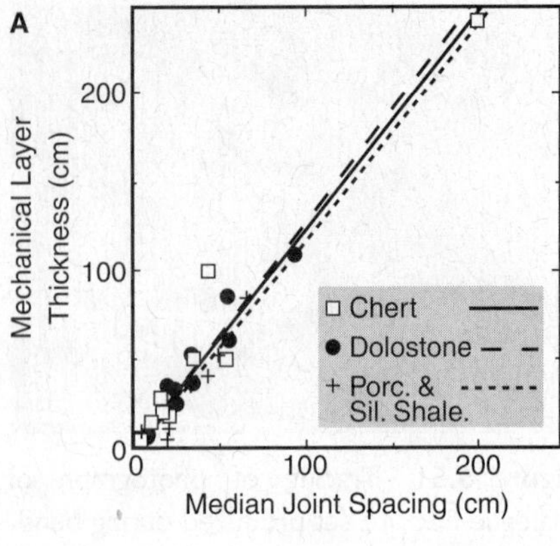

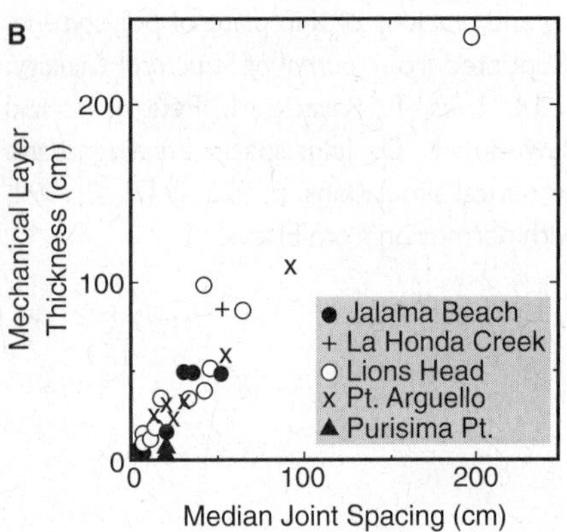

Figure 5.53 Plots of (bed thickness versus median joint spacing) in Monterrey Formation exposed. (*A*) Bed thickness versus median joint spacing at all sites, plotted by lithology. Slope of best-fit line is fracture-spacing index (FSI). (*B*) Bed thickness versus median joint spacing in all lithologies, plotted by study site. [Reprinted from *Journal of Structural Geology*, v. 13, Narr, W., and Suppe, J., Joint spacing in sedimentary rocks, p. 1037–1048, © 1991, with permission from Elsevier.]

measurements in each of 33 layers. When they plotted their data in the form of median bed thickness versus median joint spacing, each data point was identified in terms of rock type.

The final touch in the Narr (1991) and Narr and Suppe (1991) approach was to pass a best-fit (linear regression) line through data points of common rock type (see Figure 5.53*A*). They referred to slope of this line as the **fracture-spacing index (FSI)** for the given joint set in the given rock type at the given locality (or set of localities). Narr (1991) and Narr and Suppe (1991) found their results to be quite systematic across their study area (see Figure 5.53*B*), thus underscoring the important control of bed thickness on joint spacing. Whether plotting bed thickness against joint spacing, or median bed thickness versus median joint spacing, the fracture-spacing index (FSI) proved on average to be ~1.3, no matter what (mechanically stiff) lithology. Narr and Suppe (1991) cited the precedent for indexing thickness versus spacing, first introduced by Ladeira and Price (1981).

Gross (1993) applied the same methodology in the Monterey Formation, but in this case was interested in assessing the fracture spacing of non-systematic cross joints that abut at high angles against through-going systematic joints (Figure 5.54). After measuring spacing between cross joints, and the width of the joint-bounded panels within which the cross joints reside, Gross (1993) created plots of "*joint-controlled mechanical layer thickness*" versus median cross joint spacing (Figure 5.55). Fitting a regression line to these data, he determined that the fracture-spacing index (FSI) was 1.26.

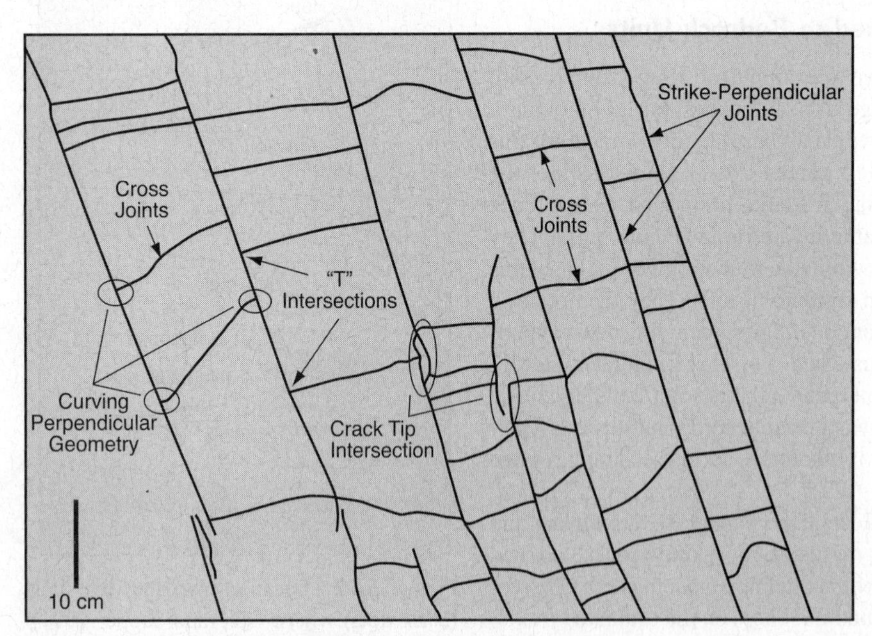

Figure 5.54 Sketch map of through-going systematic joints that partition the bed into panels occupied by non-systematic cross joints. [Reprinted from *Journal of Structural Geology*, v. 15, Gross, M. R., The origin and spacing of cross-joints: examples from the Monterey Formation, Santa Barbara coastline, California, p. 737–751, © 1993, with permission from Elsevier.]

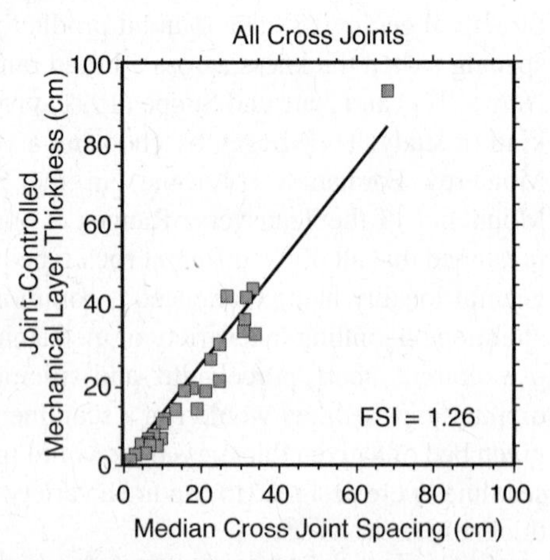

Figure 5.55 Plot of joint-controlled mechanical layer thickness versus median cross-joint spacing. Fracture spacing index proved to be 1.26. [Reprinted from *Journal of Structural Geology*, v. 15, Gross, M. R., The origin and spacing of cross-joints: examples from the Monterey Formation, Santa Barbara coastline, California, p. 737–751, © 1993, with permission from Elsevier.]

JOINT SATURATION AND JOINT IN-FILLING

Concept of "*Joint Saturation*"

What is the significance of FSI values, such as 1.29 (Narr and Suppe, 1991), or 1.26 (Gross, 1993)? Is there a value that represents a complete "*saturation*" of jointing for a given layer of a certain lithology and thickness?

Physicists define "***saturation***" as the condition in which, after a sufficient increase in causal force, no further increase in the resultant effect is possible. This applies nicely to the concept of **joint saturation**. The concept requires us to ask the following question: As a given bed of a certain lithology and thickness is being loaded by the combination of far-field and local stresses, is there a threshold above which joints stop forming?

Structural geologists have come to learn, on the basis of theory and experiments, that greater and greater saturation of jointing within a bed or layer can be achieved by new joints filling in between pre-existing joints. Recall that Rives et al. (1992) documented such infilling during the course of their tensile cracking of thin plates under controlled conditions in the laboratory (see Figures 5.50 and 5.51). The infilling led to increasing crack densities over time. These observations made Rives et al. (1992) even more aware of how difficult it is to evaluate whether the joints in a given bed in a field setting achieve some ultimate level of saturation.

Studies now demonstrate that FSI values cover a very broad range when examined from locality to locality, from region to region. For sandstones, the reported FSI values range from as low as $\leq$0.1 to as high as 10.0 (e.g., Narr and Suppe, 1991; Gross, 1993; Gross et al., 1995; Wu and Pollard, 1995). What does this mean, exactly? For a jointed sandstone bed ~1-m thick, FSI values of ~0.1 correspond to joint spacings of ~10 m, whereas FSI values of ~10.0 correspond to joint spacings of ~10 cm. Indeed, this is quite a difference! Is there a way we can proclaim that a bed is (choose one) *A*/undersaturated, *B*/ saturated, or *C*/ supersaturated with respect to jointing? If so, what is the mechanical significance of such a proclamation?

Stress-Transfer Model

One of the most important papers on the mechanical control of joint spacing in single confined beds of sedimentary rocks was published by Hobbs (1967), and later summarized in detail by Narr and Suppe (1991). Hobbss attempted to explain the roughly linear relation between joint spacing and layer thickness. The main controlling factors in his model are Young's modulus for the stiff layer that is jointed, and shear modulus for boundaries with neighboring (soft) layers. Figure 5.56*A* represents a stiff layer, such as sandstone or limestone, sandwiched between soft layers, such as shale or mudstone. The stiff layer (white) is about to become broken systematically by vertical jointing as a result of stretching and the build-up of tensile stress related to some far-field extensional strain. Far-field extensional strain is assumed to build with time (Figure 5.56*B*). Hobbs (1967) demonstrated that when the tensile strength of the bed is exceeded, and a joint forms (Figure 5.56*C*), the tensile stress within the bed is released for a certain short distance along the bed in a direction perpendicular to the joint. Otherwise, the bed continues to experience a level of stress that remains close to the critical stress required to form joints. As additional joints propagate in response to tensile stress (Figure 5.56*D*, *E*), they do so in a way marked by a characteristic spacing that is proportional to thickness and the mechanical properties of the layer. As sequential infilling takes place, a given joint set will exhibit measurably closer spacing and increasingly higher FSI values (see Figure 5.56*E*).

Thus a familiar theme we read time and again in the jointing literature is one emphasized first by Hobbs (1967): joints can form at any site along a given bed provided that the site is not too close to a pre-existing joint. The Hobbs model

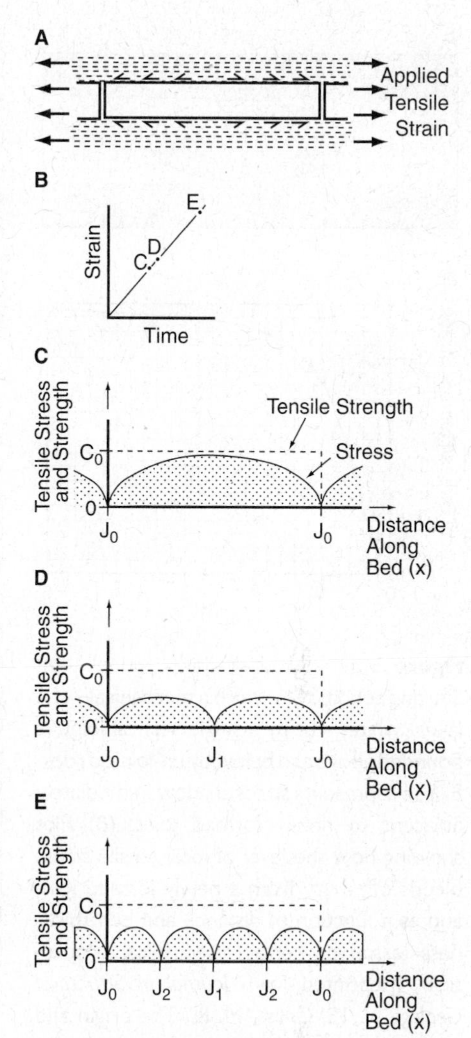

Figure 5.56 The Hobbs model of joint formation. (*A*) Stiff layer (white) sandwiched between soft layers (grey). Shear stress operates along bedding planes. Shear stress created by far-field extension. (*B*) It is assumed that far-field extensional strain increases with time. Points C, D, and E qualitatively show strain magnitudes for the examples that follow. (*C*) Plot of tensile stress magnitude (grey) and tensile strength (dashed line) just prior to formation of new joint. Points J_0 represent location of first formed joints. (*D*) New joints (J_1) have now formed, and we see the tensile stress and tensile strength magnitudes just after the formation of these joints. (*E*) At a later stage we see that still-younger joints (J_2) have formed midway between existing joints. [Reprinted from *Journal of Structural Geology*, v. 13, Narr, W., and Suppe, J., Joint spacing in sedimentary rocks, p. 1037–1048, © 1991, with permission from Elsevier.]

A

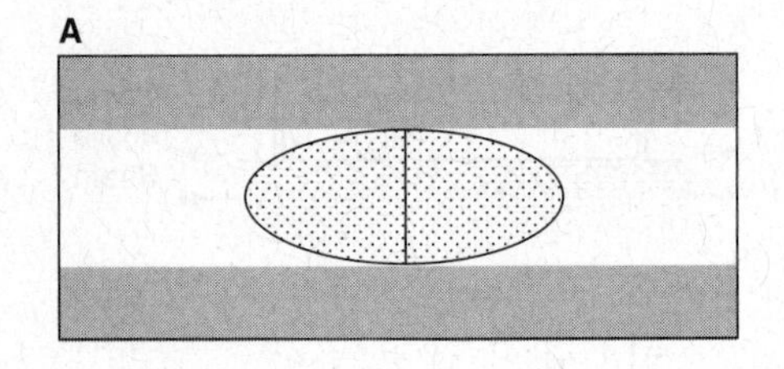

B

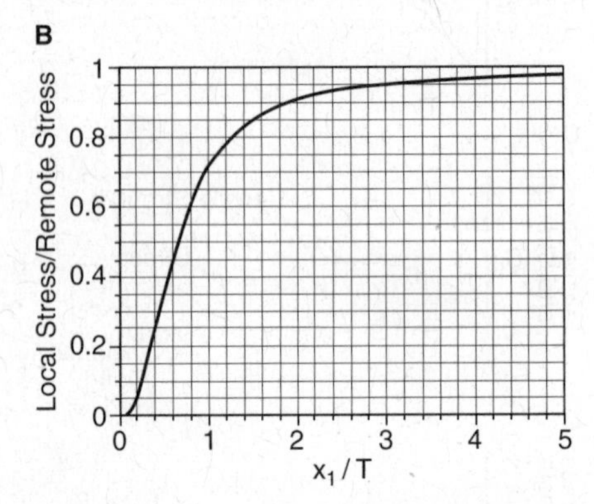

Figure 5.57 Stress-shadow model. (*A*) Starting condition in which a mechanically stiff layer (white), cut by a single vertical joint, is bounded above and below by un-jointed rock. Ellipse represents stress shadow immediately adjacent to newly formed joint. (*B*) Plot showing how the level of local tensile stress bleeds off, away from a newly formed joint and as a function of distance and bed thickness (assuming certain mechanical properties). [Reprinted from *Journal of Structural Geology*, v. 15, Gross, M. R., The origin and spacing of cross-joints: examples from the Monterey Formation, Santa Barbara coastline, California, p. 737–751, © 1993, with permission from Elsevier.]

TABLE 5.2

Joint saturation in relation to bed thickness (T) and joint spacing (S). The terms "supersaturation" and "hypersaturated" are introduced here.

Range	*Values*	*Condition*
I	$S/T > 1.2$	Undersaturated
II	$0.8 \leq S/T < 1.2$	Saturated
III	$0.3 \leq S/T < 0.8$	Supersaturated
IV	$S/T < 0.3$	Hypersaturated

Source: [Reprinted from *Journal of Structural Geology*, v. 22, Bai T., and Pollard, D. D., Fracture spacing in layered rocks: a new explanation based on the stress transition, p. 43–57, © 2000, with permission from Elsevier.]

yields the conclusion that new joints should always form exactly at the midpoint between two pre-existing joints. Yet, both Hobbs (1967) recognized and Narr and Suppe (1991) emphasized that this will not be the case generally because of the presence of flaws. Flaws, in magnifying tensile stresses locally, have an important role in influencing the locations where joints form in a bed. Because flaws are simply not randomly or uniformly distributed within and along a given bed, joints that "*in-fill*" do not generally form at the exact midpoint between earlier formed joints.

Stress-Shadow Model

A complementary way of looking at fracture spacing is through the stress-shadow model (Lachenbruch, 1961; Nur, 1982). Again, picture in cross-section a mechanically stiff flat-lying bed already broken by a vertical joint; this layer is sandwiched between beds that are un-jointed (Figure 5.57*A*) (Pollard and Segall, 1987; Gross, 1993). The very formation of the joint causes the magnitude of local tensile stress to drop to zero along the joint trace. But if you were to measure the tensile stress in the same bed away (in a perpendicular direction) from the joint trace, you would find that the local stress builds again to a magnitude that again matches the level of the far-field stress. A **stress reduction shadow** is said to exist in the region of the layer immediately adjacent to the newly formed joint (see Figure 5.57*A*). Pollard and Segall (1987) captured the concept beautifully in the graph shown in Figure 5.57B. Plotted on the ordinate is the ratio of local stress to far-field stress. Plotted on the abscissa is the ratio of distance (from the joint) to layer thickness. It is clear that with distance from the joint, local tensile stresses build to a level close to that of the far-field stresses. At the location where the magnitude of the local tensile stresses reaches some sufficient magnitude, a new joint will form, thus beginning sequential infilling. The specific manner in which the stress drops off and then builds within the shadow is a function of joint height (bed thickness) and the mechanical character of the bed.

Stress-Transition Model

Yet another model for evaluating the controls for joint spacing in relationship to layer thickness is the stress-transition model developed by Bai and Pollard (2000). Their numerical modeling revealed that when the ratio of joint spacing (S) to layer thickness (T) reaches a certain **critical value**, the normal stress acting perpendicular to joints changes (i.e., "*transitions*") from tensile to compressive, preventing jointing from occurring as long as the compressive environment is maintained. For most reasonable geological situations involving interbedded stiff and soft sedimentary rocks, the critical S/T ratio for mechanically stiff beds lies in the range of ~0.8 to ~1.2. This would correspond to a state of joint saturation.

There is a very interesting application of this stress-transition model, and it lies in explaining one of the ways in which orthogonal joint patterns in horizontal sedimentary strata can be understood. It is not uncommon to observe mutually perpendicular sets of planar parallel evenly spaced joints (Figure 5.58). It is tempting to conclude that the orthogonal pattern was produced by 90° rotations in the far-field stress field. However, the results of the work by Bai and Pollard (2000) imply that such an interpretation is not required. We can imagine a first set of systematic joints (J_1) forming perpendicular to the far-field stress direction, and through sequential joint infilling J_1 achieves joint saturation for the bedrock unit. When this happens, the local stress perpendicular to J_1 shifts from tensile to compressive, and tensile stress becomes oriented in a direction parallel to the strike of J_1. As a result, set J_2 forms as cross-joints, and this occurs without any rotation of far-field stresses.

Bai and Pollard (2000) distinguished four different degrees of joint saturation (Range I to Range IV) (Table 5.2). Range I, marked by $S/T > 1.2$,

Figure 5.58 Photograph of mutually perpendicular systematic joint sets in silty limestone beds within a sequence of Permian-Triassic sedimentary rocks in east central Utah. One set of joints (more closely spaced) strikes down the dip of the bedding. A second set of joints (more widely spaced) strikes parallel to bedding. Geologist is R. Bullerdick. [From Fisher and Polansky, *Journal of Geophysical Research*, v. 111, fig. 1, p. BV07403, © 2006 by American Geophysical Union.]

represents **undersaturated jointing** and may well be the result of very modest strain. Range II, marked by $0.8 \leq S/T < 1.2$, represents **saturated jointing**, a critical spacing that is achieved independently of applied strain. Range III, marked by $0.3 \leq S/T < 0.8$, is what I (GHD) would call **supersaturated jointing**, and Range IV, marked by $S/T < 0.3$, is what I would call **hypersaturated jointing**. Bai and Pollard (2000) suggest that Ranges III and IV may be produced where amount of strain, abundance and sizes of flaws, and levels of fluid pressure may have a strong influence on joint spacing.

Summary

Fischer and Polansky (2006) included in their paper a nice summary of how progressive jointing takes place. They remind us that field observations of regular joint spacings in sequences of sedimentary rocks cry out for mechanical explanation (Figure 5.58). Their summary of sequential parallel jointing in a stiff layer (bounded by soft layers) is quite insightful (Figure 5.59) (Fisher and Polansky, 2006). The set-up is shown in Figure 5.59*A*, featuring a stiff layer sandwiched between soft layers. All the layers are being subjected to uniform layer-parallel extension. The magnitude of the far-field (remote) stress is uniform throughout the layer. Flaws exist throughout the layer (tiny black dots and blobs) (Figure 5.59*A*, *B*), and at one of them the local stress magnitudes grow to a level that permits a joint to begin propagating. As a disk-shaped fracture, it propagates rapidly outward, and its tip line encounters the upper and lower bedding surfaces, across which there is mechanically soft material (e.g., mudstone) (Figure 5.59*C*). Joint propagation is arrested at that boundary, but that does not prevent fracture propagation from speeding length-parallel within the layer. This causes the joint surface to change shape from discoidal and elliptical to blade-like. The very propagation of the joint creates a stress reduction shadow on either side, which in turn imposes limits on joint spacing (Figure 5.59D).

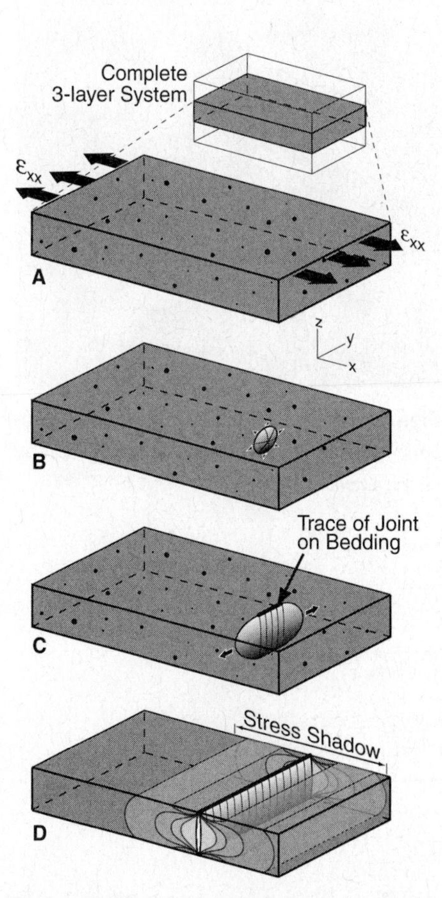

Figure 5.59 Model of sequential, parallel jointing. See text for explanation. [From Fisher and Polansky, *Journal of Geophysical Research*, v. 111, Figure 3, p. BV07403, © 2006 by American Geophysical Union.]

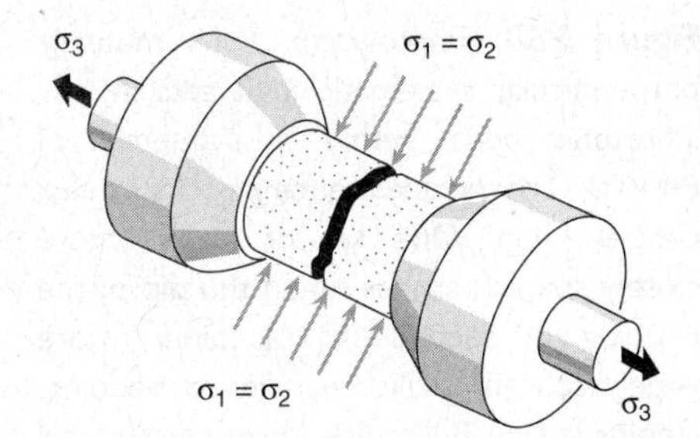

Figure 5.60 The basic setup for tensile strength tests.

Figure 5.61 Perfect samples for "*dog-bone*" tensile-strength tests. [Photograph by G. H. Davis.]

CREATION OF JOINTS IN THE LABORATORY

Tensile Strength Tests

For each experimental test, the basic procedure is to hold one principal stress constant, and then progressively increase or decrease the other to create an ever-increasing differential stress. For **tensile strength tests**, an increasing tensile stress (σ_3) is applied to the ends of the specimen while a constant confining pressure ($\sigma_1 = \sigma_2$), compressive in nature, is applied perpendicular to the flank of the cylindrical specimen (Figure 5.60). When the tensile strength of the rock is overcome, the specimen breaks in a mode I tension fracture.

A variety of tensile strength tests are carried out to evaluate the mechanical response of rocks to tension. These include, but are not limited to, straight pull tests, pull tests with clamped ends (see Figure 5.60), pull tests with cemented ends, and pull tests on "*dog-bone*"-shaped specimens (Figure 5.61). Price and Cosgrove (1990, p. 30) have emphasized that tensile strength tests are technically tricky, and unlike compressive strength tests (see Figure 5.60), the test results for any given rock are commonly inconsistent from method to method. For example, the tensile strengths measured for the Pennant Sandstone tested by Price, using a wide variety of techniques, ranged from approximately -15 MPa to approximately -65 MPa (Price and Cosgrove, 1990, p. 31). Suppe (1985) reported that the common range of tensile strength for a variety of common rocks is -5 MPa to -20 MPa. Mark Fischer emphasizes (personal communication, April 26, 2010) that test results that become summarized on Mohr circle diagrams provide insight on the basic conditions for joint formation, but should not be used to predict that exact stresses at which jointing will occur. Test results underscore the fact that joints require low differential stresses to form, and/or high fluid pressures.

The general results of tensile strength tests, when compared with the results of compressive strength tests, have made it obvious that rocks are very weak in tension. The ratio of strength in unconfined compression to strength in tension is commonly 2:1, but may exceed 30:1 (Price and Cosgrove, 1990). Everyday experience confirms this. If we want to break a Popsicle stick, we bend it, creating tension in the outer arc of the bend and compression in the inner arc. Weaker in tension, the stick snaps (i.e., fails) along the outer arc.

The sequential stages of a tensile strength experiment can be represented on a Mohr diagram. The state of stress just before tensional stresses are applied can be represented by a single point, indicating that there is no differential stress (i.e., $\sigma_1 = \sigma_3 = 0$) (Figure 5.62). A state of hydrostatic stress exists. As tensile stresses build parallel to the length of the specimen, a differential stress develops. For a small increment of tensile stress, the differential stress on the Mohr diagram is represented by a circle of very small diameter (see Figure 5.62). Stress perpendicular to the axis of the specimen essentially remains the same. It becomes by default the direction and magnitude of greatest principal stress (σ_1). During the test, stress parallel to the axis of the specimen becomes increasingly tensile. This stress is *negative* (tensile), and thus is the least principal stress (σ_3). The buildup of differential stress ($\sigma_1 - \sigma_3$) that takes place during a test is portrayed on the Mohr diagram as a family of circles of increasing diameter (see Figure 5.62). When the tensile strength of the rock is exceeded, the rock breaks perpendicular to the direction of tension (see Figures 5.60). A mode I joint is formed. The test is run again and again to pin down the average tensile strength for the lithology being examined.

The mode I fracture is parallel to σ_1 and perpendicular to σ_3; that is, perpendicular to extension. The angle (θ) between σ_1 and the normal (**n**) to the tension fracture is 90° (positive when measured counterclockwise; negative

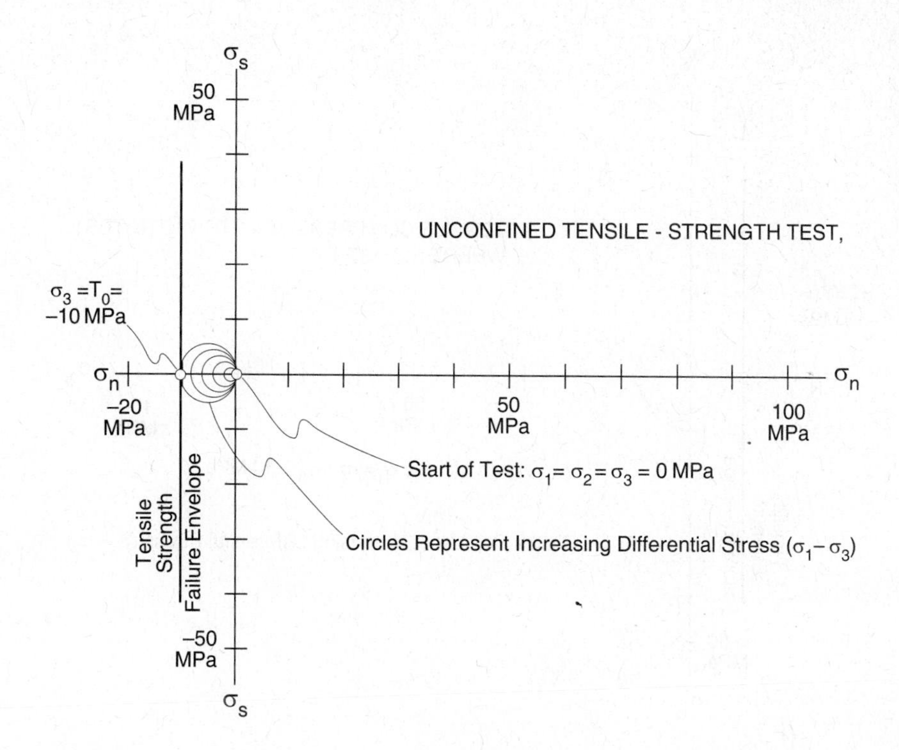

Figure 5.62 Mohr circle representation of a tensile strength test, from start to finish. There is no confinement on the specimen. Tensile stress, parallel to the length of the specimen, is gradually increased until the tensile strength of the specimen is exceeded and it breaks.

when measured clockwise). Thus, in Mohr diagram space, the radius connecting the center of the differential stress circle with the point of failure lies along the *x*-axis (normal stress axis) and makes an angle of $2\theta = \pm 180°$ with respect to the direction of greatest principal stress (σ_1).

The "*law*" formulated on the basis of the tensile strength tests is in the following form:

$$\sigma_3 = T_0 \qquad (5.1)$$

The equation simply indicates that a rock will fail by mode I fracturing if the magnitude of the least principal stress (σ_3) equals or exceeds the fundamental tensile strength T_O of the rock.

The **envelope of failure** for a given rock under conditions of unconfined uniaxial tension can be constructed on a Mohr diagram simply by drawing a vertical line through the value of tensile strength (T_O) for the rock (see Figure 5.62). The vertical line represents the value of tensile stress that must be exceeded before the rock will fail in tension. Tensile stress is negative in the convention we are using. Thus, the magnitudes of tensile stress increase from right to left along the normal stress axis to the left of origin. When the magnitude of tensile stress becomes great enough to touch the envelope, the rock fails by mode I fracturing. The extension fracture breaks perpendicular to the direction in which the specimen was pulled or stretched.

Hybrid Tensile and Compressive Strength Tests

We will now run a triaxial test in which specimens are subjected to a combination of tension and compression. We do this by applying a small compressive confining pressure to the flanks of a cylindrical specimen of sandstone, and, at the same time, imposing length-parallel tensile stress. On the Mohr diagram, a point representing the constant confining pressure (σ_1) lies along the normal stress axis to the right of the origin (Figure 5.63). Increasing levels of tensile stress are represented by points (σ_3) moving to the left of the origin along the normal stress axis. Increasing differential stress, in turn, is represented by a family of circles of larger and larger diameter, each of which passes through the point representing σ_1. Ultimately, a

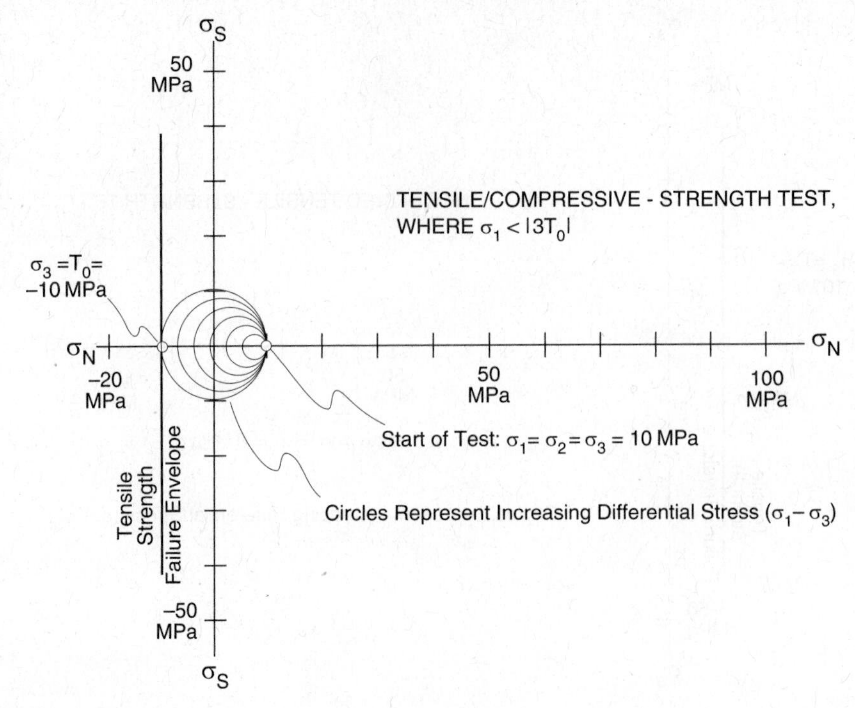

Figure 5.63 Mohr circle representation of a tensile and compressive strength test. The specimen is laterally confined by a compressive stress of 10 MPa, and then is subjected to ever-increasing tensile stress parallel to the length of the specimen. When the tensile stress, which is the magnitude of the least principal stress (σ_3), reaches the tensile strength of the rock, the rock breaks.

differential stress $(\sigma_1 - \sigma_3)$ is achieved that is sufficient to break the rock (see Figure 5.63).

According to Suppe (1985), when tensile and compressive strength tests are carried out for sandstone at confining pressures in the range $\sigma_1 = 3T_O$ to $5T_O$ (i.e., from three to five times the tensile strength of the sandstone), the collective points of failure map out a parabolic curve that rises steeply from the point on the normal stress axis representing the tensile strength (T_O) of the sandstone, then flattens slightly as it passes upward and to the right to barely cross the shear stress axis (Figure 5.64*A*).

Whenever differential stress conditions in a tensile or compressive strength test are such that the Mohr circle representing the specific state of stress touches the parabolic fracture envelope, the rock will fail by fracture. The points that define this **parabolic failure envelope** (Suppe, 1985; Twiss and Moores, 1992, p. 175) have (x,y) coordinates corresponding to the magnitudes of normal stress and shear stress on the fracture at the instant of failure (see Figure 5.64*A*).

The parabolic fracture envelope imposes sharp constraints on the types of fracture that can develop. When the confining pressure (σ_1) is less than or equal to $3T_O$, the only type of failure that is possible is mode I tensile failure. However, if the magnitude of confining pressure lies between $3T_O$ to $5T_O$, the rock will fail along fractures that are a combination of extension and shear. This response is known as **transitional tensile behavior** (Suppe, 1985). Examination of the fractured specimen will reveal that the fracture is not strictly mode I, but is a combination of extension (mode I) and shear (mode II) (Figure 5.64*B*). The extension component will be marked by dilation along the fracture. The shearing component will be evident in the form of slight offset and, perhaps, slickenlines. Unlike pure mode I fractures, the fracture that forms will not be oriented parallel to the direction of greatest principal stress (σ_1), but at an angle somewhere between 0° and 30°, depending on the magnitude of confining pressure (see Figure 5.64*B*). Thus the angle (θ) between σ_1 and the normal (n) to the transitional tensile fracture ranges from $\pm 90°$ to $\pm 60°$. Again, the angle (θ) between σ_1 and the normal (n) to the tension fracture is positive when measured counterclockwise, and negative when measured clockwise.

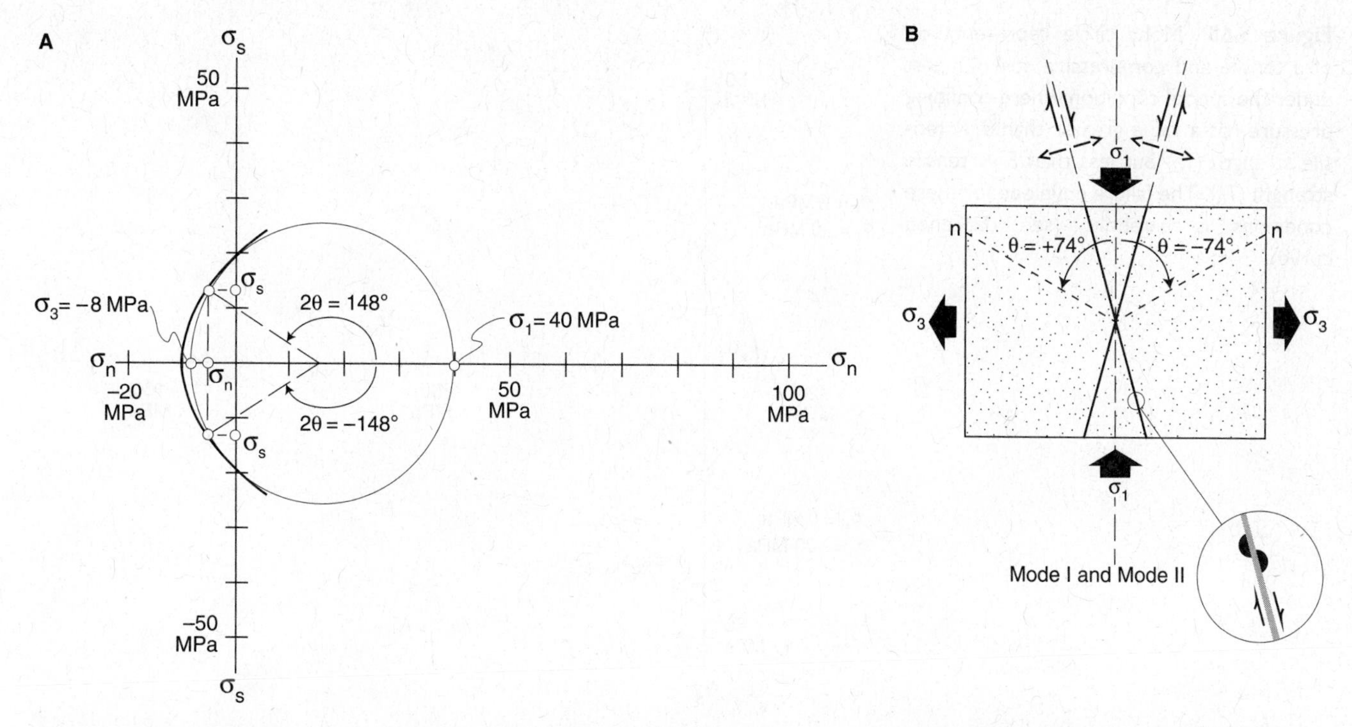

Figure 5.64 (*A*) Mohr circle representation of a tensile and compressive strength test under the special condition where confining pressure has a value greater than 3 × tensile strength (T_O) but less than 5 × tensile strength (T_O). The failure envelope for these conditions is parabolic (see darkened curve). (*B*) These conjugate fractures formed under conditions of transitional tensile behavior. Note that they are partly tensile (mode I) and partly shear. They intersect at an unusually small angle (in this case, 32°), which is bisected by the direction of greatest principal stress (σ_1).

A special condition exists when $\sigma_1 = 5T_O$ and σ_3 is just slightly less than the tensile strength of the rock. The Mohr circle representing this state of differential stress intersects the parabolic fracture envelope along the *y*-axis, at a point representing a normal stress value of 0 MPa and a shear stress value of approximately $2T_O$ (Figure 5.65). This *y*-intercept for the parabolic fracture envelope represents the fundamental **cohesive strength** (σ_0) of the rock. It represents the strength of the rock when normal stress on the potential plane of fracture equals zero.

The geometry and position of the parabolic fracture envelope can be defined by an equation, and the equation in turn can be thought of as law, the **Griffith law of failure**, which describes transitional tensile behavior (Suppe, 1985). The equation looks like this:

$$\sigma_c = \sqrt{4T_0\sigma_n - 4T_0^2} \tag{5.2}$$

where σ_c = critical shear stress required for failure
T_0 = tensile strength
σ_n = normal stress

This law of failure indicates that the critical shear stress required to break a rock in the transitional tensile field depends on tensile strength T_O and the magnitude of the normal stress (σ_n) on the fracture. As we shall see, the law is named after A. A. Griffith because he discovered the underlying reason for the behavior. More on this later.

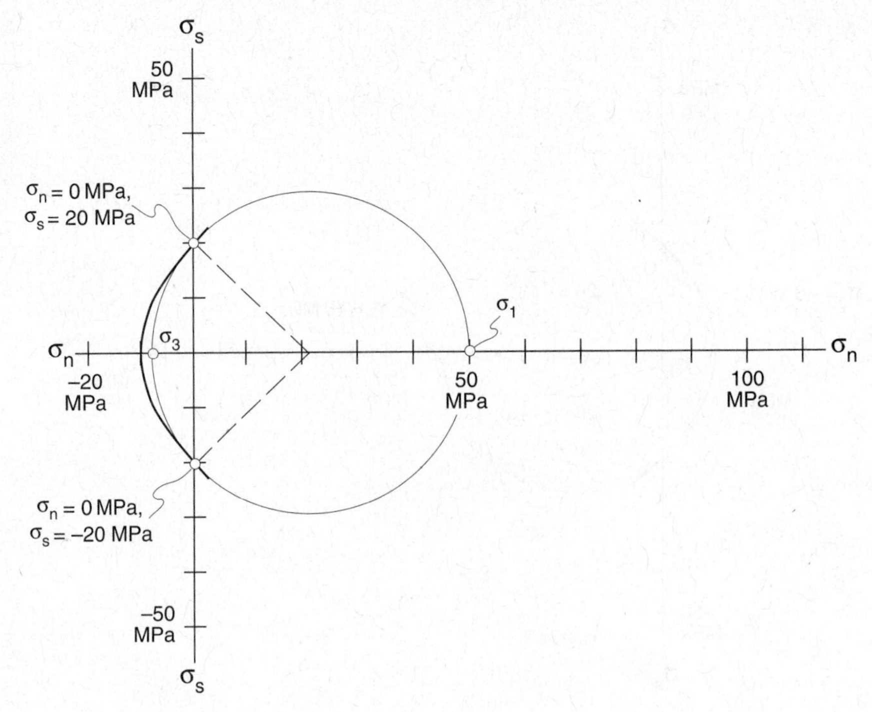

Figure 5.65 Mohr circle representation of a tensile and compressive strength test under the special condition where confining pressure has a value greater than 3 × tensile strength (T_0) but less than 5 × tensile strength (T_0). The failure envelope for these conditions is parabolic (see darkened curve).

INFLUENCE OF PORE FLUID PRESSURE ON JOINTING

The Classic Paper by Hubbert and Rubey

In a masterpiece article, Hubbert and Rubey (1959) completely transformed the way in which geologists think about joints and faults. Hubbert and Rubey were aware that petroleum companies, in drilling for oil, routinely encounter highly elevated fluid pressures in rocks within sedimentary basins, such as the Gulf of Mexico. Petroleum explorationists learned through everyday experience that sedimentary rocks in the subsurface are commonly marked by stratigraphically controlled zones of fluid pressure that are so highly elevated that they approach the value of the lithostatic stress produced by the load of the overlying sedimentary cover. In such settings, the fluid pressure essentially supports the weight of all the rock above.

There is plenty of evidence that high fluid pressures are achieved in the crust of the Earth, especially below several kilometers. Twiss and Moores (1992) point out that above 3 km, the fluid pressure is typically '*normal*'; that is, equal to **hydrostatic pressure**. Flow is unrestricted. The magnitude of hydrostatic pressure (P_h) is the product of the density of the fluid (ρ_f), gravity (g), and height of the column of groundwater (h):

$$P_h = \rho_f g h \tag{5.3}$$

Deeper than 3 km, fluid pressure generally exceeds hydrostatic pressure because of the effects of compaction. During compaction, pore fluids are forced to occupy less and less space, and fluid pressures mount. Furthermore, because of the effect of the geothermal gradient, **aquathermal pressuring** develops, causing pore fluids to acquire greater pressure as they become hotter. This occurs because water has a higher coefficient of thermal expansion than sediment. Thus pore fluid pressure can build to a much greater value than the calculated extrapolation of hydrostatic pressure based on depth. Furthermore, pore fluid pressure can rise to levels approximating, or even exceeding, the **lithostatic pressure** (i.e., the pressure derived from the weight of the entire column of overlying rock). Lithostatic pressure (P_l) is equal to the average density of the column of rock (ρ_r) multiplied by gravity (g) and depth (h):

$$P_l = \rho_r g h \tag{5.4}$$

Hubbert and Rubey introduced an expression for the **fluid pressure ratio** (λ) to describe the ratio of the pore fluid pressure (P_f) to the lithostatic pressure (P_l):

$$\lambda = \frac{P_f}{P_l} = \frac{P_f}{\rho_r g h} \tag{5.5}$$

When the fluid pressure is hydrostatic, and not elevated, λ values typically range from 0.37 to 0.47 (Suppe, 1985). Elevated fluid pressures are referred to as **abnormal fluid pressures**. Values of λ in regions of abnormal fluid pressure typically range from 0.5 to 0.9 (Suppe, 1985).

Fluid Pressure and the Development of Joints

The role of abnormal pore fluid pressure can be used to explain the paradoxical fact that tension joints can form at depths of 5 and 10 km and greater (!), within environments where the confining pressures created by overburden are strongly compressional. How can this be?

Secor (1965) found the answer by thinking carefully about the role of pore fluid pressure. His solution is most graphically illustrated on a Mohr diagram containing a **grand envelope of failure**. Shown in Figure 5.66*A* is a stress circle that displays the values of greatest and least principal stress (σ_1 and σ_3) that might exist at depth in a region of weak horizontal compressive stress. Both σ_1 and σ_3 are compressive. As we shall learn in *Chapter 6* (*Faults*), the differential stress value ($\sigma_1 - \sigma_3$) is not great enough to cause failure by faulting. If, however, the rocks at this depth possessed high pore fluid pressure, the state of stress would be radically modified. The respective values of σ_1 and σ_3 would *each* be reduced by the exact value of the fluid pressure (P_f): the **greatest principal effective stress** (σ^*_1) would equal $\sigma_1 - P_f$; and the **least principal effective stress** (σ^*_3) would equal $\sigma_1 - P_f$. When the effective stress levels of σ^*_1 and σ^*_3 are then plotted (see Figure 5.66*A*), the stress circle shifts to the left toward the field of tensional stress. If the magnitude of fluid pressure is great enough, and the differential stress is small enough, the stress circle can be "*driven*" into collision with the tensile failure envelope (see Figure 5.66*A*). As soon as σ^*_3 achieves the value of tensile strength of the rock, tensional joints form perpendicular to σ_3.

When fluid pressure builds sufficiently to trigger fracturing, the fluid pressure may be immediately dissipated, and the original magnitudes of σ_1 and σ_3 may be reestablished. This condition is represented on the Mohr diagram by a "*rebounding*" of the differential stress circle from left to right (Figure 5.66*B*). Fluids then move from pores in the rock into fractures, where the pressure builds again, and, ***POW***!! The fracture is propagated further. Bahat and Engelder (1984) have suggested that the rhythmic plume patterns, with repeated fans of hackles, reflects fracture driven by fluid pressure (Figure 5.66*C*). The "*drive*" suddenly ceases when the crack has propagated a short distance. The fluid pressure falls and tensile stress disappears . . . that is, until fluid pressure builds again for another hit.

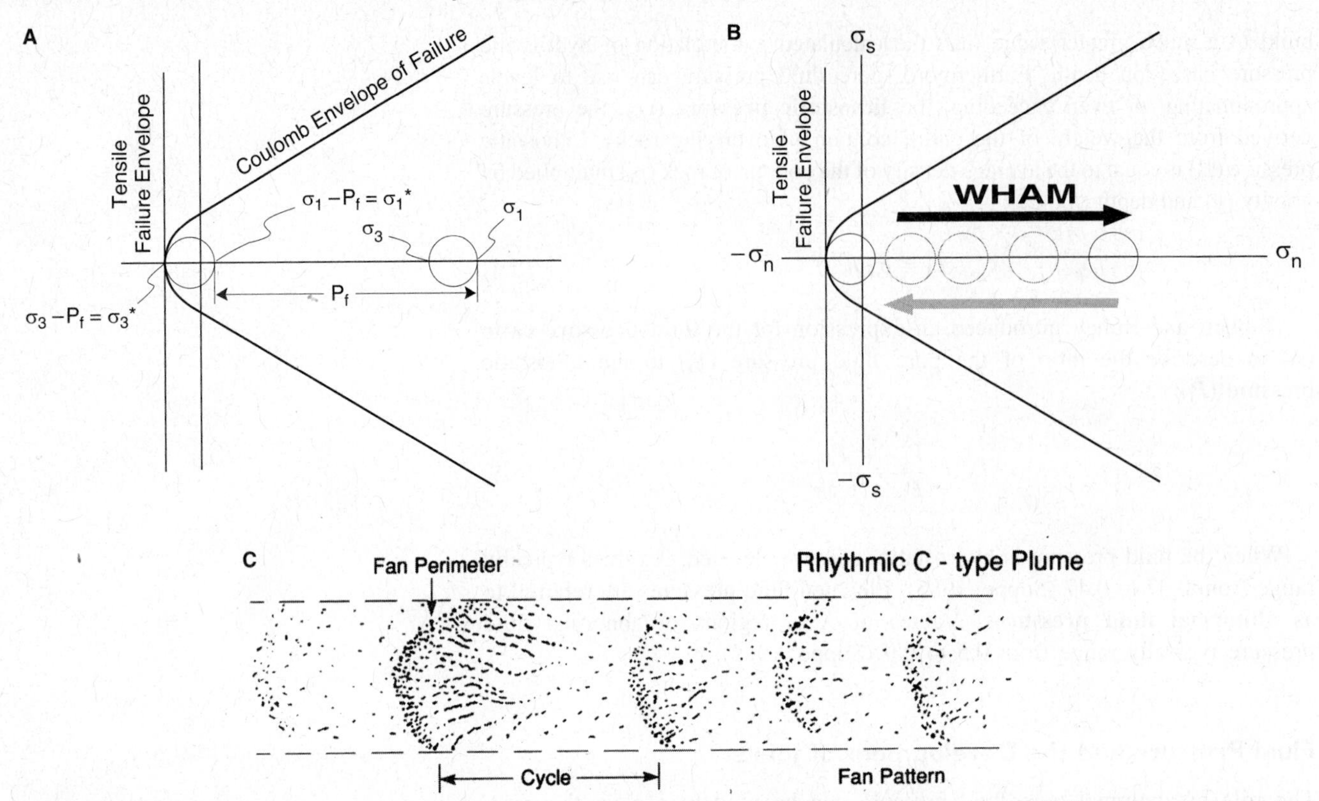

Figure 5.66 Secor's (1965) model of the role of fluid pressure in jointing. (*A*) We start with greatest (σ_1) and least (σ_3) principal stresses being compressive, and with low differential stress ($\sigma_1 - \sigma_3$). When fluid pressure (P_f) is raised, σ_1 and σ_3 are reduced by the same amount (P_f). Because differential stress was very small to begin with, the stress circle passes all the way to the tensile failure envelope, and the rock breaks by mode I failure. (*B*) Repeated tensile failure in a low differential stress field is achieved by raising the fluid pressure (circle moves left slowly), cracking the rock in tension and releasing the fluid pressure (quick move of the circle to the right!), followed again by raising the fluid pressure. (*C*) Rhythmic plumose ornamentation of joints at Watkins Glen, New York, interpreted by Bahat and Engelder (1984) to be the expression of sequential cracking responding to the rhythms of fluid-pressure buildup and fracture-induced release. [Reprinted from *Tectonophysics*, v. 104, Surface morphology on cross joints of the Appalachian Plateau, Bahat, D., and Engelder, T., p. 299–313, © 1984, with permission from Elsevier.]

Gelatin-Hydrofrac Experiment

Nick Nickelsen (personal communication, July 1993) described to me (GHD) an experiment that brings to life the role of fluid pressure in jointing. He ran the experiment at Bucknell University for years. It is an experiment originally devised by M. King Hubbert to demonstrate that **hydrofracs** (i.e., mode I joints that develop in the presence of elevated fluid pressures) form perpendicular to σ_3, the direction of least principal stress. It is an amazing experiment!

Buy some Knox Gelatin at the grocery store, mix up a batch, and pour it into a transparent plastic liter bottle (e.g., mineral water bottle). When the gelatin has set up, apply a clamp (a bent piece of metal will do) to the middle of the bottle to create a stress field (Figure 5.67*A*). The effect of the clamp will be visible distortion of the bottle from a circular to elliptical shape, as viewed in cross-section. Now stick a thin glass tube (e.g., a pipette tube), or a thin straw, vertically into the gelatin, such that the bottom of the tube (i.e., the bottom of the drill hole) is at the depth level of the clamp (Figure 5.67*B*). Next, load a syringe with a slurry of Plaster of Paris and forcefully inject the plaster of Paris down the tube into the gelatin. Immediately a disk-shaped mode I joint (i.e., a hydrofrac) will form perpendicular to the direction of least principal stress (σ_3). The location and size of the hydrofrac will be obvious for it will be filled with Plaster of Paris. Allow the plaster of Paris to harden

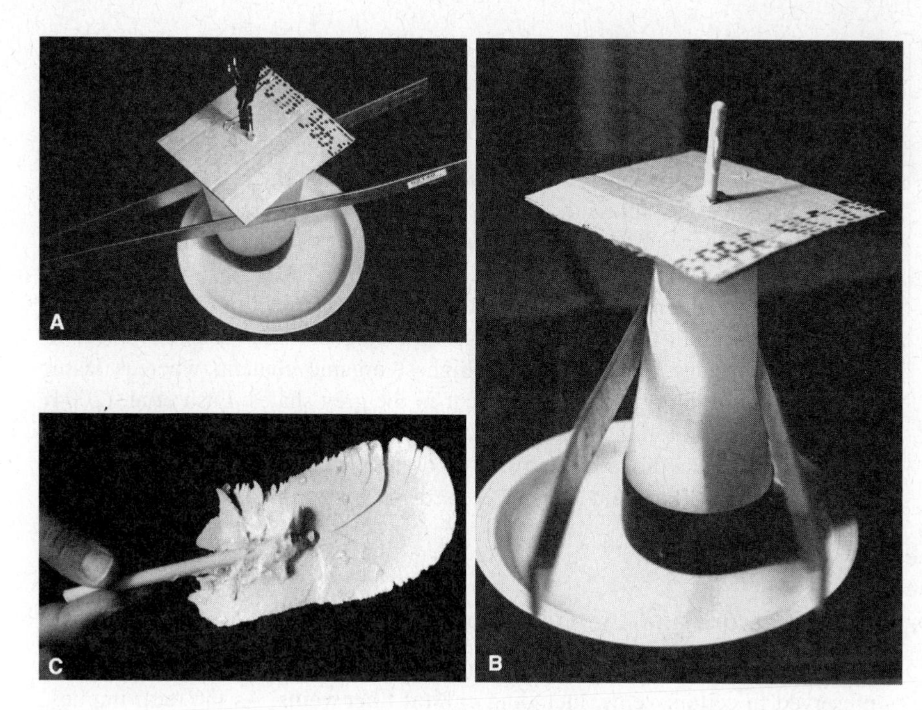

Figure 5.67 The Knox Gelatin "*hydrofrac*" experiment, as taught to us by Nick Nickelsen. (*A*) Knox Gelatin inside a clamped plastic water bottle. A glass tube, suspended by a piece of cardboard resting on top of the bottle, was inserted into the gelatin while it was solidifying. A fluid slurry of Plaster of Paris is about to be injected down the tube. (*B*) Injection of the slurry creates a mode I fracture, which fills immediately with the Plaster of Paris. (*C*) After the Plaster of Paris dries, the tube and the hardened joint filling can be extracted. [Experiment by Angela Smith. Photographs by Brian Bennon and Brook Riley.]

completely, and then cut away the bottle, reach down into the gelatin, and carefully scoop out the filled fracture.

What you will extract is a joint disk, perhaps complete with fringe joints, a perfect image of the three-dimensional geometry of a joint and its relationship to principal stress directions (Figure 5.67*C*).

The Marcellus Shale Story

A timely example of the enormous practical, economic importance of jointing in relation to overpressure is embodied in the story of the Marcellus Shale. A cogent overview is provided by The Geological Society of America in its geology.com website (Geology.com, 2010).

The Marcellus Shale is an organic-rich black shale of Devonian age that lies in the subsurface of the Appalachian Basin, primarily beneath Ohio, West Virginia, Pennsylvania, and New York (Figure 5.68). Natural gas in the Marcellus Shale is derived from the rich concentration of organic material in the black shale. Few petroleum geologists imagined, even as recently as 2002, that the amount of gas per acre could be particularly significant. But that all changed in 2008 when Terry Engelder (Pennsylvania State University) and Gary Lash (State University of New York at Fredonia) estimated that natural gas reserves in the Marcellus Shale could exceed 500 trillion cubic feet, which is sufficiently large for the Marcellus reservoir to be considered a "*super giant*" gas field. What changed?

What became recognized is that the Marcellus Shale is so thoroughly jointed that the gas contained in the rock itself and within the fractures can be drawn to well bores through the fracture system. Thus very large volumes of the shale can be drained of natural gas. *Range Resources–Appalachia, LLC* discovered that this was the case in 2003 when drilling a well into the Marcellus Shale in Washington County, Pennsylvania, just south of Pittsburgh.

Because the joints in the Marcellus Shale are vertical, it is advantageous, if not necessary, to drive the well bore horizontally through the Marcellus once the

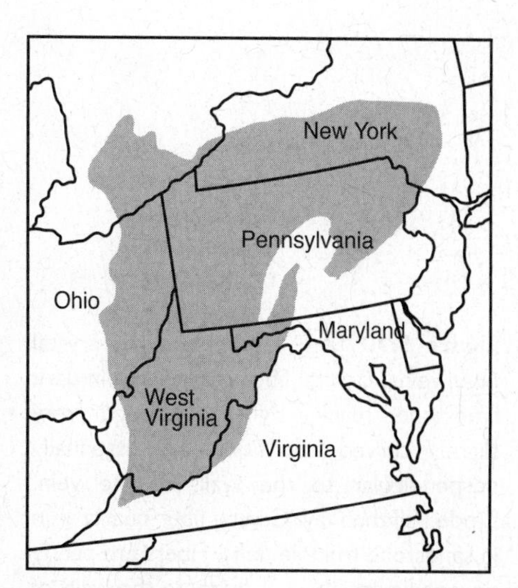

Figure 5.68 Map showing the region underlain by the Marcellus Shale (Devonian). [From http://geology.com/articles/marcellus-shale.shtml, Figure 2, p. 1.]

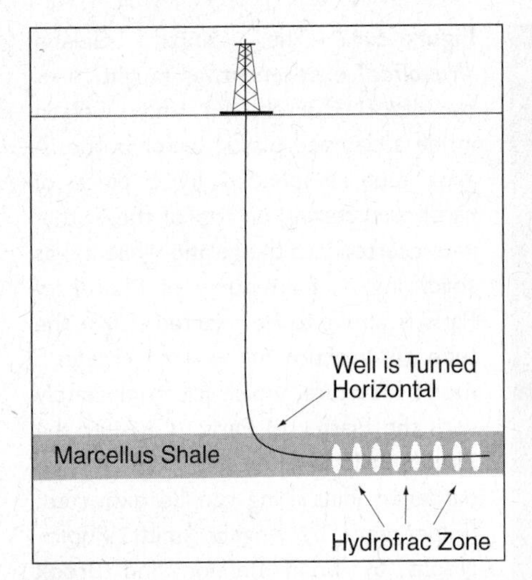

Figure 5.69 Diagram showing the manner in which natural gas within the Marcellus Shale is best recovered; that is, by drilling through the bed horizontally and at right angles to the trend of close-spaced systematic joints. [From Geological Society of America, geology.com, May 18, 2010 http://geology.com/articles/marcellus-shale.shtml, Figure 3, p. 2.]

shale is penetrated by vertical drilling (Figure 5.69). In this way a maximum number of joints is penetrated, producing maximum yield of natural gas. Yields of millions of cubic feet of gas per day have been achieved in individual wells. Furthermore, hydraulic fracturing of the shale can increase the yield further. This is accomplished by sealing off part of the well within the shale, and then injecting water and sand at high pressure into the well. The water and sand enter fractures, and the sand props open the fractures so that recovery of the natural gas can be enhanced.

There is an undeniably strong relation between production and penetrative fracture permeability with the Marcellus. The highest joint densities occur in the blackest shales, which have the highest organic content, whereas lesser joint densities and organic content occur in the grey shales. Lash et al. (2004) concluded that the jointing was enhanced by hydrofracturing during the course of generation of hydrocarbons from the organic rich shale during burial and thermal maturity. Far-field stresses caused the hydrofractures to become oriented ENE.

Fluid Pressure and the Development of Veins

The undeniably important role of fluid pressure in the opening of joints is preserved in certain veins, including **crystal fiber veins**. As the term implies, crystal fiber veins are marked by elongate fiber-like or needle-like aggregates of fine-grained minerals, commonly quartz and calcite (Figure 5.70) (Durney and Ramsay, 1973; Urai et al, 1991). The fibers may be straight or systematically curved (Figure 5.71). Where straight, they are not necessarily perpendicular to wall rock surfaces. Durney and Ramsay (1973) have shown that the formation of crystal fibers in veins is due to an opening of the walls of the fracture or joint accompanied by simultaneous vein filling, without the actual development of open space. Elevated fluid pressures contribute to the opening of the fractures.

Crystals are precipitated from the fluids during step-by-step incremental dilations of the joint. The final product, the crystal fiber vein, reflects the total movement accomplished by the dilation. In many veins the crystals

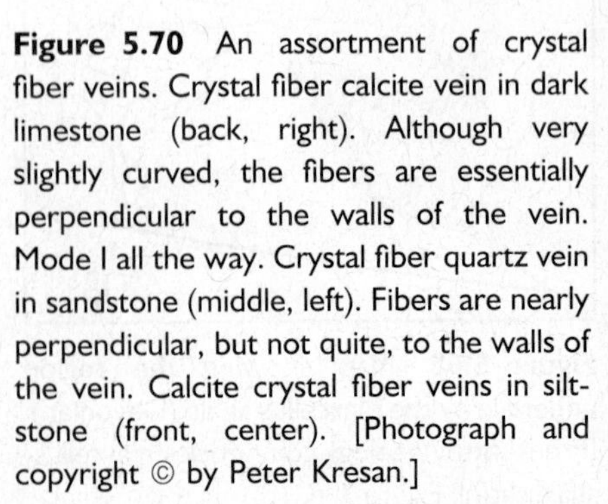

Figure 5.70 An assortment of crystal fiber veins. Crystal fiber calcite vein in dark limestone (back, right). Although very slightly curved, the fibers are essentially perpendicular to the walls of the vein. Mode I all the way. Crystal fiber quartz vein in sandstone (middle, left). Fibers are nearly perpendicular, but not quite, to the walls of the vein. Calcite crystal fiber veins in siltstone (front, center). [Photograph and copyright © by Peter Kresan.]

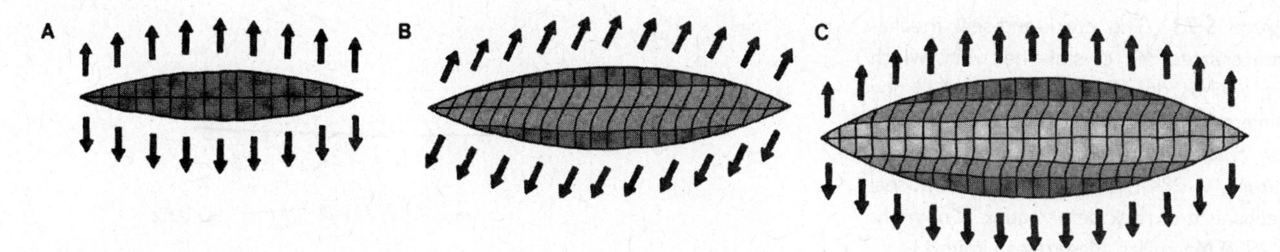

Figure 5.71 Schematic rendering of the development of curved and kinky crystal fiber veins as a response to changes in the direction of fracture opening. (*A*) First increment reflects pure mode I opening. (*B*) Second increment reflects transitional tensile behavior; that is, opening plus shear. (*C*) Final record "*maps out*" the kinematic history of the opening of the fracture within which the vein is deposited.

grow perpendicular to the walls of the joint, reflecting a pure mode I opening (Figure 5.71*A*). But where the opening is accompanied by shear components, the crystal fibers display orientations that are oblique to the vein walls (Figure 5.71*B*). The growth of the crystal fibers is ultra-sensitive to changes in the direction of progressive opening of the vein, and thus the final patterns may be strikingly curved (Figure 5.71*C*). During each incremental step, the crystal fibers grow in the direction of differential displacement, which corresponds to the direction of least principal stress. With changing directions of differential displacement, the crystal fibers take on curved forms. The curved forms "map out" the history of incremental movement along the fracture. Where the youngest crystals are added incrementally along the medial line, the center of the vein, the movement history is read from the outside in (**syntaxial veins**) (Figure 5.72*A*). Where the youngest crystals are added incrementally along the vein walls, the movement history is read from the inside out (**antitaxial veins**) (Figure 5.72*B*).

Some veins are marked by clear evidence of repeated, recurrent fracturing followed in each case by precipitation of new vein mineralization (see Figure 5.70, back right). Such veins are called **crack-and-seal veins** (Ramsay, 1980b). The presence of the fractures and the fibers may or may not be recognizable by the naked eye. For example, Ramsay (1980b) described a crystal fiber vein composed of 11 small veinlets, each approximately 45 μm wide, which form what would appear to the naked eye to be a single white vein 0.5 mm thick (Figure 5.73). Upon formation of the initial crack and initial vein, new cracks formed again and again at the vein/rock contact, and after each cracking event they were immediately sealed by a new increment of veining. Such repeated cracking typically leaves a trail of wall rock inclusions in the veins, tiny slivers of rock that were broken off at the time of cracking.

The cracks form through **hydraulic fracturing**. Fluid pressure builds, decreasing effective stress until the magnitude of the least principal stress (σ_3) matches the tensile strength (T_0) of the rock and/or vein material. After the crack has formed, and as it opens, mineral fibers are deposited incrementally from the

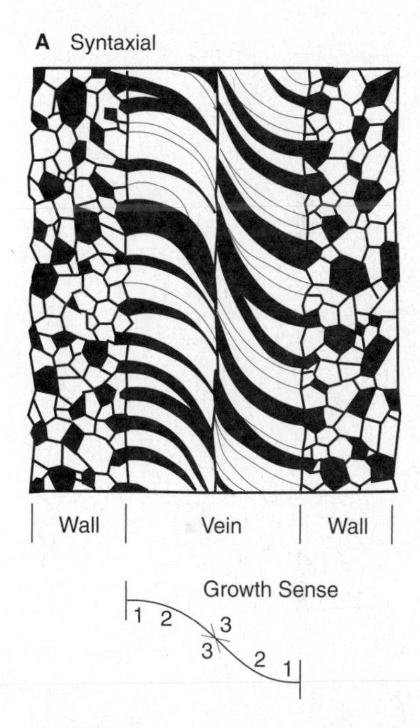

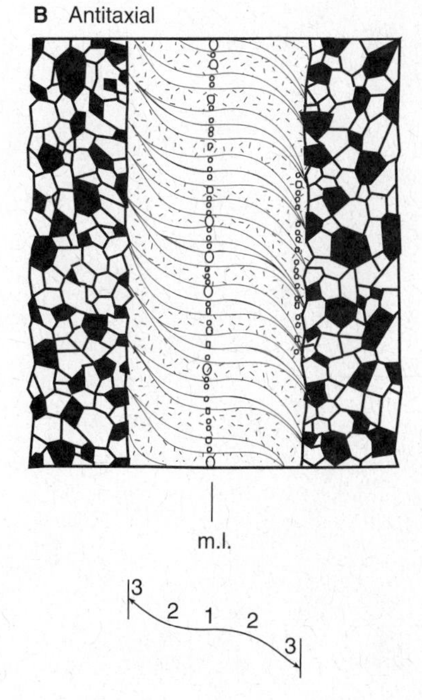

Figure 5.72 (*A*) Crystals in syntaxial crystal fiber veins grow along the medial line, and thus the fibers get younger from the walls to the middle. (*B*) Crystals in antitaxial crystal fiber grains grow along the vein walls, and so the fibers get younger from the middle to the walls. Along the middle of the vein, fibers are perpendicular to the original break line. [Reprinted from Ramsay and Huber, The techniques of modern structural geology, v. 1: Strain analysis, © 1983, with permission from Elsevier.]

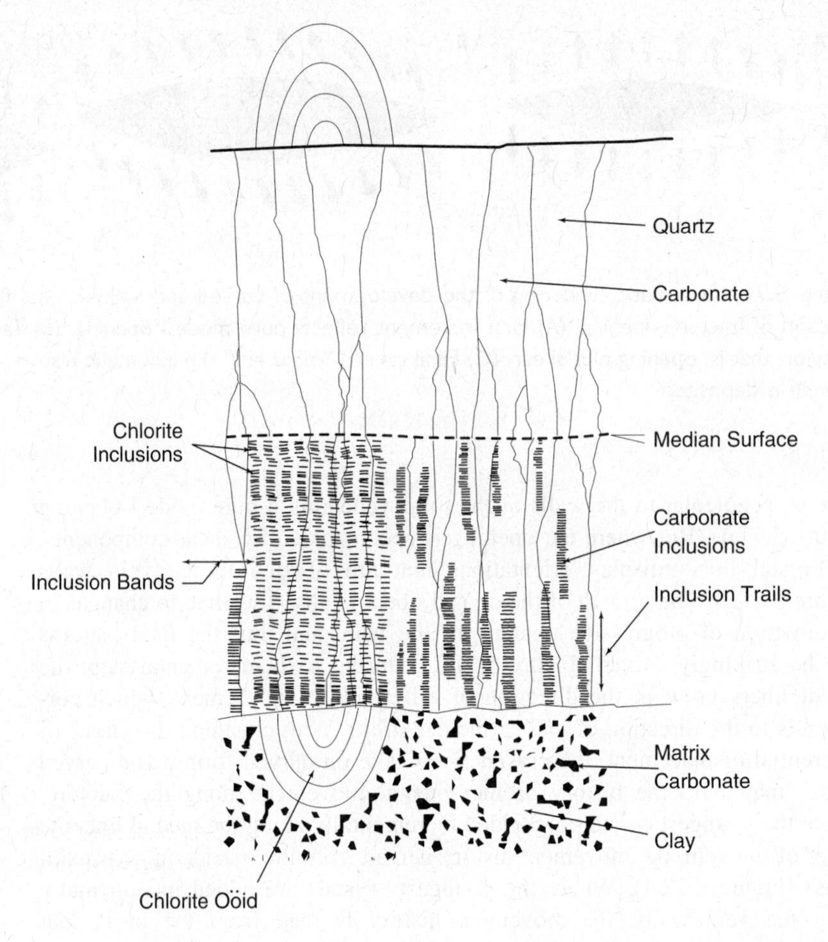

Figure 5.73 The crack-and-seal mechanism created this crystal fiber vein, which seen microscopically is composed of 11 tiny veinlets. Time and again this part of the rock was cracked and sealed. [Reprinted from Nature, v. 284, Ramsay, J., The crack-seal mechanism of rock deformation. Copyright © 1980 Macmillan Magazines Limited.]

fluids moving through the cracked rock. The crack eventually becomes completely sealed, only to be broken again when fluid pressures build to sufficient magnitude to crack the rock again . . . and again . . . and again.

A MICROSCOPIC LOOK AT THE MECHANICS OF JOINTING

The Problem

It is not at all obvious why rocks should be so weak in tension, as opposed to compression. It is also puzzling why failure values for (σ_n, σ_s) in the transitional tensile field do not fall on a straight line, but instead define a parabolic curve. Furthermore, it is disturbing that the magnitudes of stress commonly required to break a cylindrical specimen of rock in compression in the laboratory are much, much greater than the best estimates and measurements of actual in situ stress in the Earth's crust.

Although standard strength tests tell us a great deal about the conditions under which rocks will fail by fracturing, they do not in themselves reveal what is actually happening to cause rocks to fracture. The laws of failure are descriptions of what is observed, not explanations of the actual fracture mechanics. Consequently we must move to a yet finer scale to determine what may really be happening.

In particular we need to look at **stress concentrators**, imperfections that have the effect of raising the stress at a point to values that far exceed background levels. A hole would be an example, such as the axial canal of the crinoid stem shown in Figure 5.74. This opening concentrated the background stress sufficiently to cause mechanical twinning of the adjacent calcite. Yet the stress concentrators that concern us the most here are at an even finer scale.

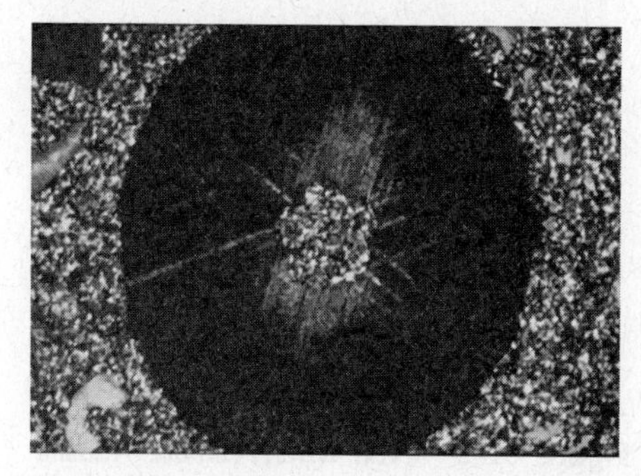

Figure 5.74 Slightly deformed crinoid stem, viewed microscopically and in cross-section. The circular opening of the crinoid columnal concentrated background stress enough to cause the mechanical twinning of the calcite. [This rare photomicrograph is courtesy of Terry Engelder.]

Microcracks

Theoretical calculations of tensile strength of rock, based on interatomic bonding, suggest that tensile strength (T_0) should be approximately one-tenth of the value of Young's modulus (E) (Price and Cosgrove, 1990). Choosing $-100{,}000$ MPa as a reasonable Young's modulus for a common rock, we would predict that the tensile strength of rock should be on the order of $-10{,}000$ MPa. Why, then, is the actual tensile strength for common rock only on the order of -10 MPa??

Griffith (1924) discovered that the mechanics of fracturing of rocks is strongly controlled by the presence of microscopic cracks. These **microcracks**, now commonly referred to as **Griffith cracks**, dramatically weaken resistance to fracturing by concentrating the stress. Stress concentrations build at the tips of favorably oriented microcracks and overcome the cohesive bonding of the rock.

Microcracks are tiny microscopic-to-submicroscopic cracks that can be visualized as ellipsoidal slits, circular to elliptical in plan, with thin apertures much smaller than the lengths of the cracks (Engelder, 1987; Pollard and Aydin, 1988). The longest dimension of a microcrack is normally between 100μm and 1000μm, whereas the aperture is normally approximately 1μm wide. Microcracks are so small that they occur wholly within individual grains as **intragranular cracks**, although sometimes they are larger and extend through a number of grains as **transgranular cracks** or along the boundaries between grains as **intergranular cracks** (Richter, 1976; Lloyd and Knipe, 1992). Microcracks do not typically display a preferred orientation; instead, they are generally randomly oriented.

The distribution of microcracks in rocks varies quite a bit. Igneous rocks are marked by a rather uniform distribution, reflecting homogeneous cooling and contraction that affected the whole body. Most rocks, whether igneous or sedimentary or metamorphic, are apt to show an increase in density of microcracking in the vicinity of major local structures, especially faults (Engelder, 1987).

Amplification of Stress at the Tips of Microcracks

Stresses become amplified by many orders of magnitude at the tips of the cracks. The amplification may create stresses of sufficient magnitude to overcome the strength of the interatomic bonding of the minerals in the rock. If this happens, the microcrack propagates. As a result of this mechanical reality, modest levels of remote stress can be amplified to create local stress magnitudes sufficient enough to create joints. The term '*remote*' is used here in reference to stresses operating from a distance, to be distinguished from '*local*' stresses that become concentrated at or near the tip of a microcrack (Pollard and Aydin, 1988).

Griffith was able to demonstrate that modest remote tensile stresses are greatly amplified to higher magnitude tensile stresses at the tips of microcracks. Furthermore, he showed that tensile stresses develop at the tips of microcracks even when the remote stress is compressive (Price and Cosgrove, 1990). Griffith's work led to the discovery that joints ultimately originate from tensional stress concentrated at flaws, including tips of microcracks.

The amplification of stresses at tips of microcracks depends on a number of factors, including orientation and the length and width of each crack (Suppe, 1985). If the remote stress (σ_3) is tensional and the microcrack is oriented at right angles to the direction of the remote tensional stress, the stress (σ_t) at the tip of a microcrack can be calculated as follows:

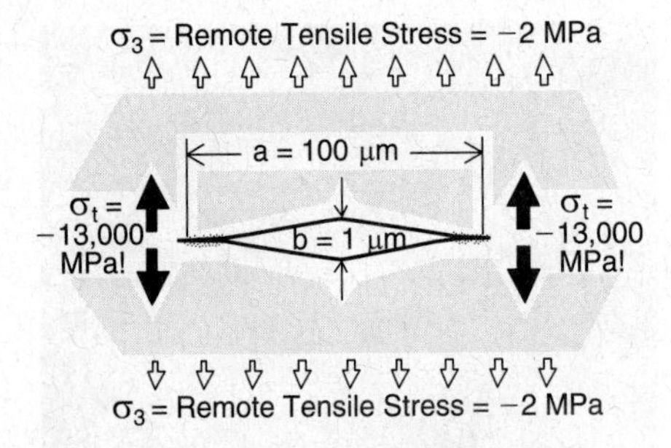

Figure 5.75 Even when "*remote*" tensile stresses (σ_3) are very, very low, huge tensional tip stresses (σ_t) can form at the tips of microcracks.

$$\sigma_t = \frac{2}{3}\,\sigma_3\left(\frac{a^2}{b}\right) \tag{5.6}$$

where a is the length of the microcrack and b is its aperture.

For example, picture a microcrack $100\mu m$ long, having an aperture of $1\,\mu m$ (Figure 5.75). Assume that it is oriented at right angles to a remote tensional stress (σ_3) whose magnitude is −2 MPa. The local stress (σ_t) that will develop at each tip of the microcrack is determined by Equation 5.6:

$$\sigma_t = \frac{2}{3}\,(-2\text{ MPa}) \times \left(\frac{100^2}{1}\right)$$

$$\sigma_t = -13{,}000\text{ MPa}$$

This value far exceeds the tensile stress required to fracture the rock beyond the tip of the crack and propagate a joint. Furthermore, if the microcrack is marked by elevated pore fluid pressure, even a lower magnitude of remote stress is sufficient to propagate a mode I joint.

Critical Orientation of Microcracks

Tensile stresses are greatest at the tips of microcracks that are most favorably oriented with respect to the principal stresses. Thus within a family of microcracks in a rock body, there will be certain favorably oriented microcracks that will nucleate through-going fractures. The angle α between the crack and σ_1 relates to the values of the principal stresses in the following way:

$$\cos 2\alpha = \frac{(\sigma_1 - \sigma_3)}{2(\sigma_1 + \sigma_3)} \tag{5.7}$$

Consistent with this relationship, microcracks oriented perpendicular to the direction of least principal stress (σ_3) are characterized by "*tip stresses*" larger than those that develop at the tips of microcracks of all other orientations. We know from standard tensile strength tests that a rock will fail by mode I fracturing when $\sigma_3 = T_0$ and $\sigma_1 = 3T_0$. If we plug these values into Equation 5.7 we find that the preferred orientation for microcracks formed in tension is parallel to σ_1 and perpendicular to σ_3:

$$\cos 2\alpha = \frac{3T_0 - (-T_0)}{2[3T_0 + (-T_0)]} = \frac{4T_0}{4} = 1$$

$$2\alpha = 0°$$

Stress/Strain Relationships at the Scale of Microcracks

At the scale of laboratory testing of cores of rocks, we conclude that jointing is purely an elastic/brittle phenomenon: when the tensile strength of an elastically deforming rock is exceeded, the rock instantaneously breaks in tension. This is not the case at the scale of the tips of microcracks (Ingraffea, 1987). Rock adjacent to the microcrack is stress free. Just beyond the tip of a microcrack, a **process zone** forms as the tip stresses build. Within the process zone, a swarm of microcracking develops as the crack comes closer and closer to propagating. Because of the dense microcracking, the rock behaves inelastically within the process zone, and the propagation of the crack to form a fracture proves to be an elastic/plastic process at this scale of observation. Beyond the process zone, the rock suffers some strain, but the behavior is strictly elastic.

If we were to build a stress/strain curve describing the behavior of rock just beyond the tip of a microcrack, it would display a period of elastic behavior

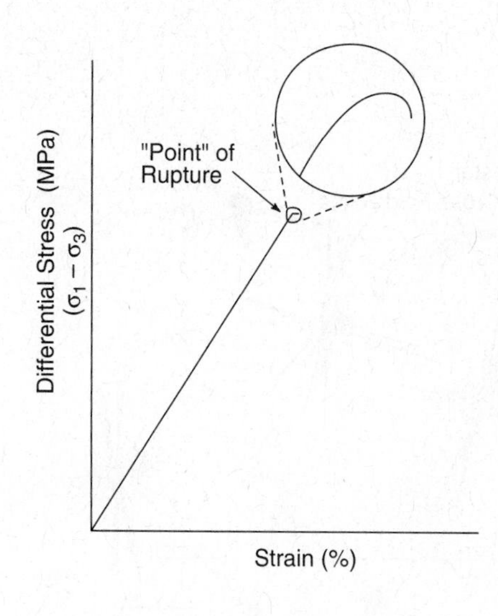

Figure 5.76 The relationship between stress and strain at the tip of a microcrack. This is what is happening at the "*point*" of rupture. Stress-strain curves can give the impression of quick snap, without plastic deformation. In reality, the failure follows inelastic deformation within the process zone.

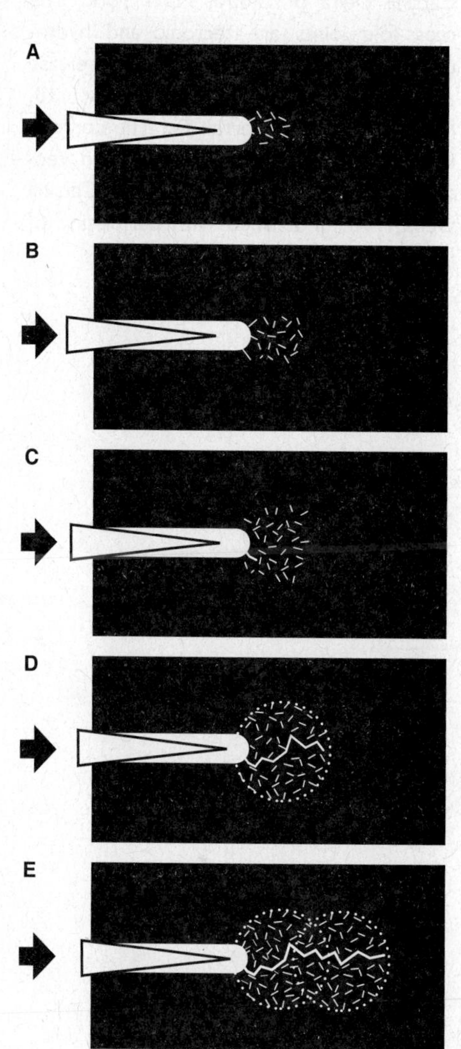

Figure 5.77 The slot experiment. (*A*) Slot is cut into rock, and, figuratively speaking, a wedge is driven into it (*B*, *C*). The outward tensional pressures amplify tensional stresses at the tip of the slot. Microcracking proliferates in the process zone (*D*, *E*). Eventually, certain favorably oriented microcracks link up in an incipient mode I fracture. [Reprinted from Atkinson, Introduction to fracture mechanics and its geophysical applications, © 1987, with permission from Elsevier.]

followed *not* by sudden instantaneous brittle failure, but by **strain softening**, in which the stress level steadily declines as the rock moves swiftly to failure by cracking (Figure 5.76) (Ingraffea, 1987). The strain softening would reflect the proliferation of microcracking as the area beyond the tip became transformed into a process zone. At failure, the crack extends from the original tip front through the process zone. Thus the elastic/brittle quick snap we observe in tensile-strength tests in the laboratory does not accurately reflect the "*micromechanisms*" by which the joint comes to life.

Relation of Microcracks to Jointing and Faulting

As failure stress is approached, the cloud of microcracks developing in the rock becomes more localized and dense; and when shear failure occurs, the surface is really the coalescence of hundreds and hundreds of microcracks (Kranz, 1983). Atkinson (1987) describes a classic experiment analogous to the ripping of a bedsheet, but it is done in rock. A slot is cut in rock, and then outward-directed pressure is placed on the walls of the slot, almost like driving a wedge into the notch, thus placing the tip of the slot in tension (Figure 5.77*A*). At first, isolated microcracks form beyond the slot, and the material behaves elastically. With further loading, a cloud of microcracks forms a process zone, and the solid behaves in a nonlinear manner (Figure 5.77*B*,*C*). Finally, a crack forms by the linking up of microcracks (Figure 5.77*D*,*E*). The crack is the incipient joint.

EXAMPLES OF INTERPRETING REGIONAL JOINT PATTERNS

Jointing in the Appalachian Plateau, New York

Engelder (1985) presented a sophisticated analysis of the development of regional jointing in the Appalachian Plateau province (Figure 5.78). He was able to distinguish the relative ages of four discrete classes of joints, and to interpret them in the context of the tectonic history of the region. **Tectonic joints** and **hydraulic joints** are "*cross-fold*" joints that formed as mode I tensile joints oriented vertically and generally parallel to the compressional direction responsible for folding (see Figure 5.78). Both sets of joints formed under conditions of elevated pore fluid pressure and directed tectonic stress.

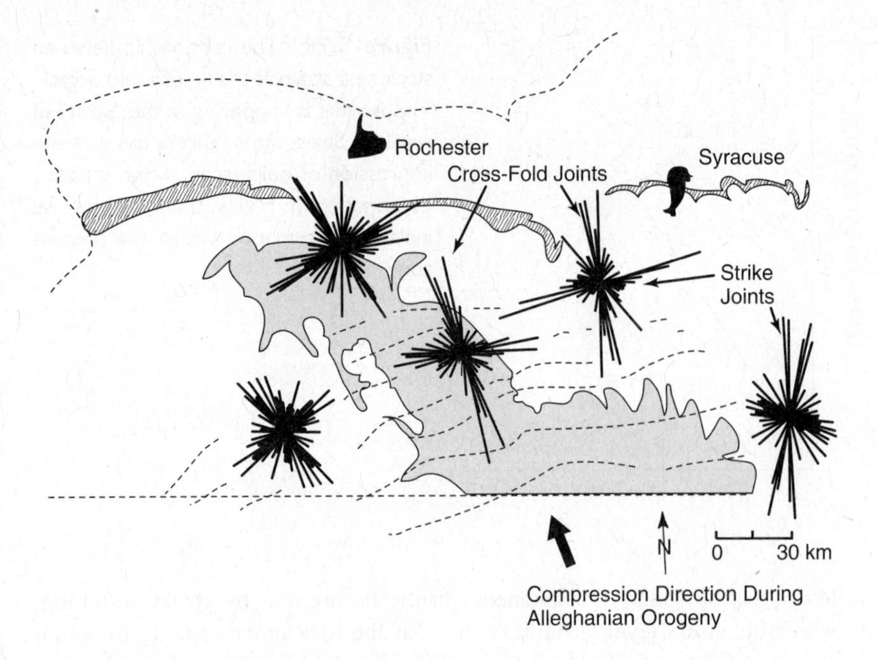

Figure 5.78 Regional pattern of cross-fold joints and strike joints within the Catskill Delta of central New York. The cross-fold joints are tectonic and hydraulic. The strike joints are due to release and unloading. [From Geology, v. 13, Engelder, T., and Oertel, G., The correlation between undercompaction and tectonic jointing within the Devonian Catskill Delta, 1985. Published with permission of the Geological Society of America.]

By sorting the cross-fold joints according to plumose ornamentations, and by noting the relative ages of each class of joints and their distributions in siltstone versus shale, Engelder concluded that the tectonic joints formed when pore fluid pressure became elevated as a result of tectonic squeezing and compression. The hydraulic joints, on the other hand, are thought to have formed when pore fluid pressure became elevated as a result of vertical gravitational loading.

In contrast, two sets of younger joints, **release joints** and **unloading joints**, formed near the surface during uplift and erosion. As the siltstones and shales were brought closer to the surface during uplift and erosion, they cooled and contracted and relaxed elastically, causing stress buildup and jointing. In effect, compressive stress was released along preexisting structural elements (such as cleavage), and the release joints formed as near-vertical mode I tension joints perpendicular to the former direction of tectonic compression. Thus they are parallel to the trends of folds (see Figure 5.78). The unloading joints are also mode I tensile joints, but they display orientations that are not systematically related to the geometry of the fold belt. Instead, they have orientations that are related to the present-day stress field within the Appalachian Plateau region.

The care with which the regional analysis of jointing on the Appalachian Plateau has been carried out is reflected in illustrations from Engelder and Geiser (1980), which reveal the map orientations of cross-fold joints (tectonic and hydraulic) and strike joints (release joints) compared with the trends of folds and with the tectonic compression direction as revealed by deformed fossils and cleavage (Figure 5.79).

Jointing in the Appalachian Plateau, Pennsylvania and West Virginia

Jointing in coal in the Allegheny Basin of the Appalachian Plateau is a good example for distinguishing far-field stresses versus local stresses. It is an example that I (GHD) grew up with, given my father's career in mining of the Pittsburgh coal just south of Pittsburgh and in northern West Virginia. Operators mining the Pittsburgh coal seam are keenly aware that two mutually perpendicular sets of

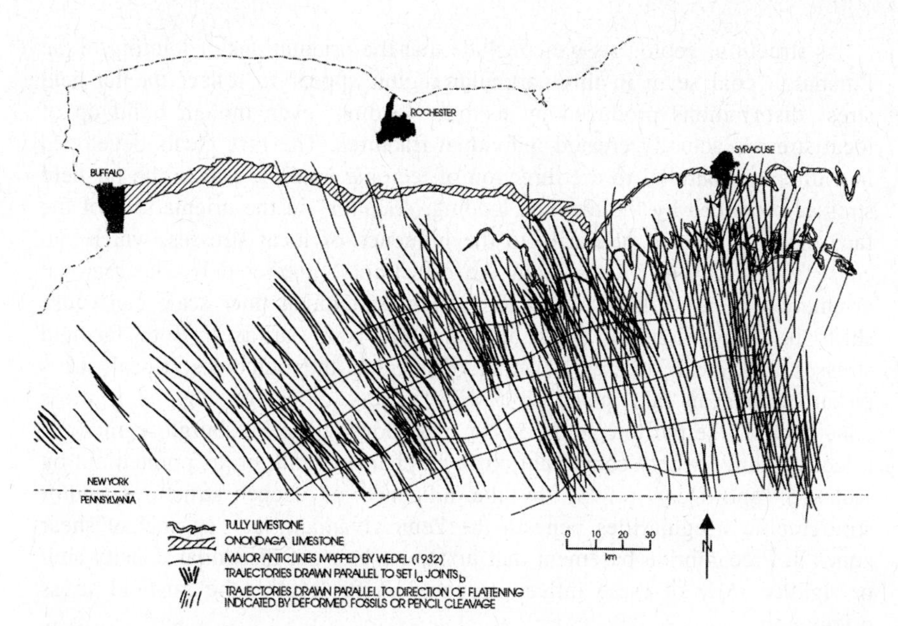

Figure 5.79 Just one example of the maps made by Engelder and Geiser (1979) to capture regional fracture patterns in relation to folding and strain in the Appalachian Plateau. [From T. Engelder and P. Geiser, *Journal of Geophysical Research*, v. 85, fig. 4b, p. 6325, © 1980 by American Geophysical Union.]

vertical fractures pervade this formation, and that these two sets are reliable in their orientation and persistence across the entire coal basin. Miners refer to these two distinctive fracture directions as **coal cleats**: a **face cleat** composed of long continuous dominant fractures (systematic joints) that extend for many meters horizontally; and a **butt cleat** composed of shorter, somewhat more poorly developed fractures that terminate at right angles against the face cleat (cross joints). Coal cleats are strategic to mining because they are the directions along which the coal breaks when mined. Main tunnels, openings, and haulage ways are driven parallel to the cleats, especially face cleats.

Diamond et al. (1975) measured the orientations of coal cleats in the Pittsburgh coal in eighteen mines over a four-county region in southwestern Pennsylvania and northern West Virginia (Figure 5.80*A*). The orientation data reveal tight groupings and mutually perpendicular orientations of the face and butt cleats (Figure 5.80*B*), with the face cleat on average trending N75°W, and the butt cleat oriented N17°E. Diamond et al. (1975) recognized a slight shift in absolute trend of these cleats in the northern versus southern parts of their study area. In particular, the face cleat orientation shifts counterclockwise from N67°W to N76°W, and the butt cleat shifts from N28°E to N17°E (see Figure 5.80*B*). This 10° counterclockwise rotation matches the identical shift in trends of folds in the region, for the face cleats were formed along a direction parallel to the shortening direction responsible for folding, a shortening direction driven by plate-scale movements that contributed to the final phases of development of the Appalachian Mountains (i.e., during the Alleghanian orogeny). As the Pittsburgh coal seam was shortened slightly in the WNW/ENE direction, it extended at right angles (NNE/SSW). This extension loaded the coal seam in ways that produced the face-cleat jointing (Nickelsen and Hough, 1967; McCulloch et al., 1974; and Diamond et al., 1975). The butt cleat, in contrast, probably formed later in response to the release of the stress that was loaded into the coal during the Alleghanian orogeny (Diamond et al., 1975). The butt cleat is clearly younger than the face cleat, for the butt cleats terminate against face cleats.

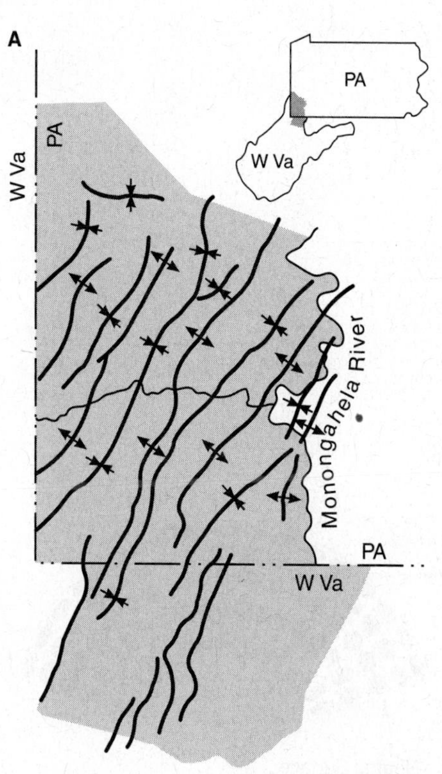

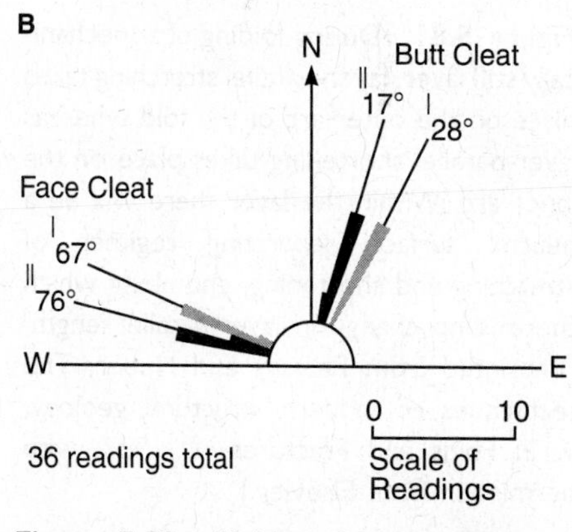

Figure 5.80 (*A*) Map of fold structures in the coal mining region of southwestern Pennsylvania and northern West Virginia. (*B*) Composite rose diagrams showing cleat orientations. Sets marked by Roman numeral I were measured in the northern half of the study area; sets marked by Roman numeral II were measured in the southern part of the study area. [From Geology, v. 3, Diamond, W. P., McCulloch, C. M., and Bench, B. M., Estimation of coal-cleat orientation using surface-joint and photolinear analysis, 1975. Published with permission of the Geological Society of America.]

As structural geologists we conclude that the orientations of jointing in the Pittsburgh coal seam in this particular region appear to reflect the far-field stress distributions produced by tectonic loading, even though build up of local stresses actually created individual fractures. The face cleats developed in a direction parallel to the direction of tectonic loading, and to the far-field stresses produced by the tectonic loading. "*Scatter*" in the orientations of the face cleats is related primarily to the influence of local stresses, which can cause shifts of joint trends from the direction(s) predicted by the regional orientation of far-field stresses, and which at a much finer scale can cause shifts in orientations of parts of individual joint surfaces. Both far-field stresses and local stresses, together, determine the outcomes. Local stress perturbations can take place at a wide variety of scales, e.g., at a stress concentrator like pyrite concretions or fossil logs in the Pittsburgh seam; or at a facies change where coal is cut out by a channel sand; or, as pointed out by Kulander and Dean (1993), by the influence of deeper structures and/or stratigraphic irregularities beneath the Pennsylvanian, such as ancient shear zones in Precambrian basement that juxtapose crust of dissimilar density and/or rigidity. Any of these influences could shift, locally, the far-field stress orientation.

Jointing Associated with Large Individual Folds

Stearns Model

When geologists imagine patterns of jointing that would develop in a thick mechanically stiff bed during the formation of a large anticline, they commonly begin by thinking about what Stearns (1968) concluded decades ago. Bergbauer and Pollard (2004) did a terrific job of summarizing Stearns's model. Folding of a stiff rock layer will be marked by stretching on the outer arc of the fold, and shortening on the inner arc, with these two regions separated by what is known as a **neutral surface** (along which there is neither shortening or stretching) (Ramsay and Huber, 1987) (Figure 5.81). Stearns (1968) underscored that the location of the neutral surface separates (above and below) distinctively different joint-forming environments. In specific, he proposed that four joint systems could be expected to form due to bending, with each consisting of two conjugate shear joint sets and one mode I joint set (Figure 5.82). The four systems correspond to four distinctive stress-field environments (see Figure 5.82). Systems 1 and 2 are marked by conjugate strike-slip shear fracture sets (and faults) and mode I joints; the systems are differently oriented above and below the neutral surface. System 3 is composed of conjugate normal shear fracture sets (and faults) and mode I joints. System 4 is composed of conjugate thrust shear fracture sets (and faults) and mode I joints.

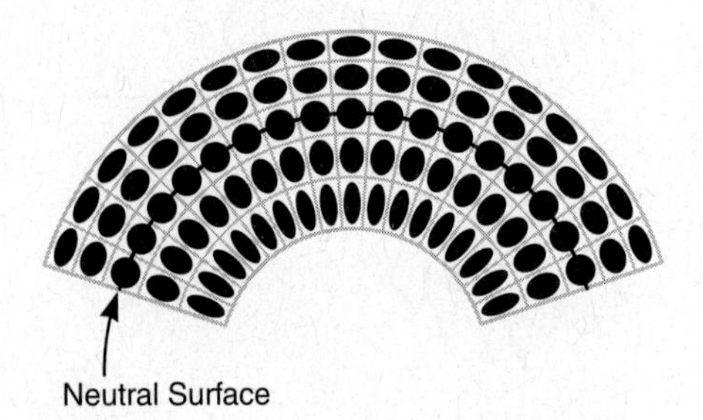

Figure 5.81 During folding of a mechanically stiff layer, layer-parallel stretching takes place on the outer arc of the fold whereas layer-parallel shortening takes place on the inner arc. Within the layer there will be a neutral surface separating regimes of stretching and shortening, and along which there is no change in layer-parallel length. [Reprinted from Ramsay and Huber, The techniques of modern structural geology, v. 2: Folds and Fractures, © 1987, with permission from Elsevier.]

Stearn's model underscores tight symmetry correspondence between the attributes of folding and the orientations of jointing and fracturing. He recognized that this is an idealized portrayal, and that in practice the joint patterns will not generally be as "*clean*," especially when considering that some joints in folded strata may have formed before the folding, and some may have formed afterwards.

Teton Anticline, Montana

Ghosh and Mitra (2009) provide a field example of the close relationship between folding and jointing. Although their example is not in strict accord with Stearn's model, there is reasonable compatibility. They studied jointing in Devonian-Mississippian carbonates (mainly dolomitic limestone) on the western

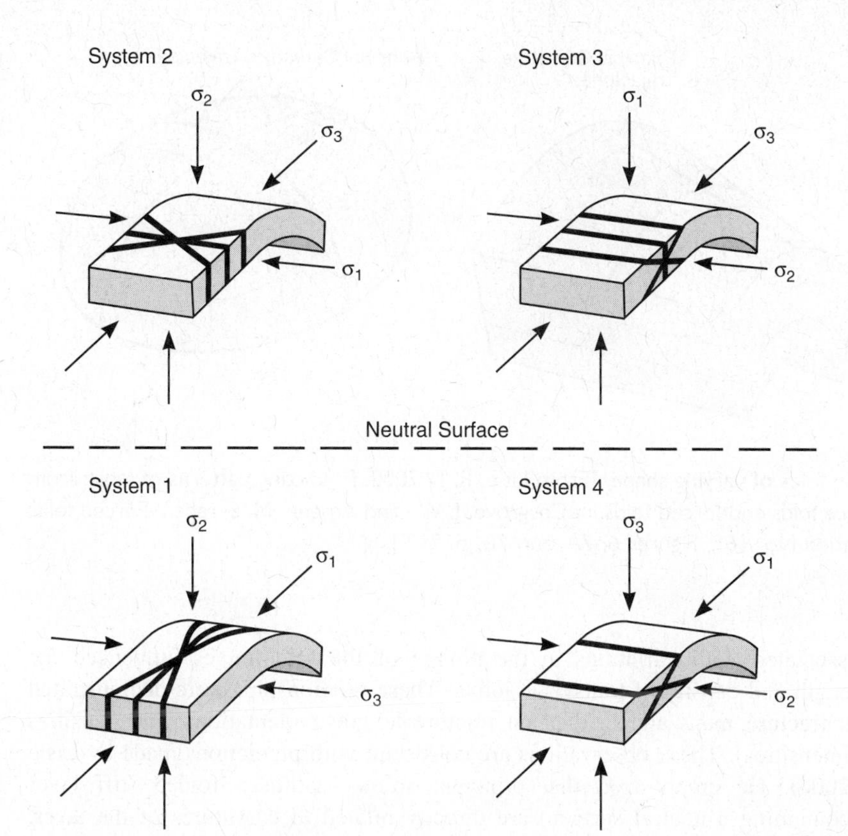

Figure 5.82 Fracture sets that, according to Stearns (1968), will form during folding of a mechanically stiff layer. Above the neutral surface there will be outer-arc stretching. Below the neutral surface there will be inner-arc shortening. Far-field stress patterns are modified by local bending characteristics. True joints are perpendicular to σ_3. The 'X' sets are shear fractures. [From Bergbauer and Pollard, 2004, Figure 1B, p. 295, after Stearns, 1968. Published with permission of Geological Society of America.]

limb of the north-trending Teton anticline located in the Sawtooth Range in Montana. This anticline formed in Late Cretaceous to early Tertiary time during east-directed thrust faulting and associated folding. The approach used by Ghosh and Mitra (2009) was to carefully map bedding orientation so that the fold properties could be pinned down. Furthermore, they mapped fracture traces, measured fracture orientations, and also determined fracture density (intensity). As a result of their work they identified two steeply dipping primary sets, a **longitudinal joint set** oriented parallel to the trend of the fold, and a **transverse joint set** oriented perpendicular to the fold axis (Figure 5.83). In addition, Ghosh and Mitra (2009) documented two **oblique joint sets**, which are not as well developed as the longitudinal and transverse joint sets (see Figure 5.83).

The amount of bending of layers around the fold hinges was found to have influenced the density (intensity) of the longitudinal jointing. The bending

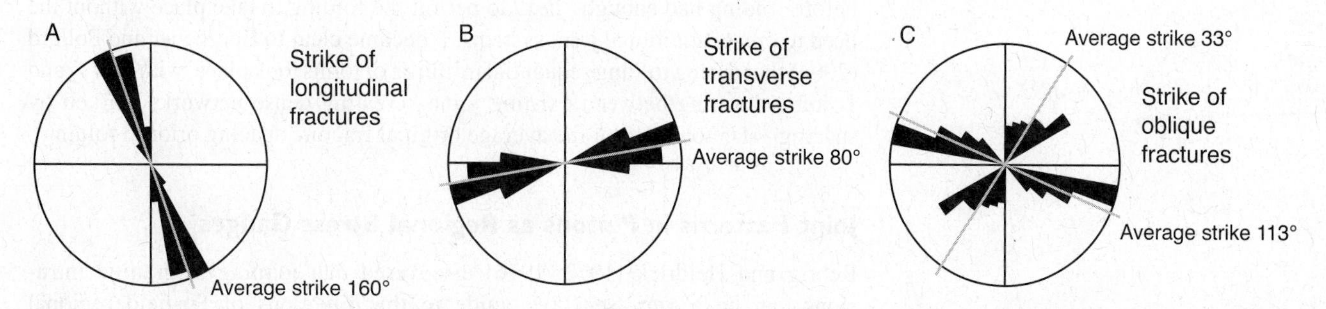

Figure 5.83 Rose diagrams showing preferred orientations of (*A*) longitudinal fractures, (*B*) transverse fractures, and (*C*) oblique fractures (shear fractures) associated with the Teton anticline. [From Ghosh and Mitra, 2009, Figures 4a, 4b, 4c. Published with permission of Geological Society of America.]

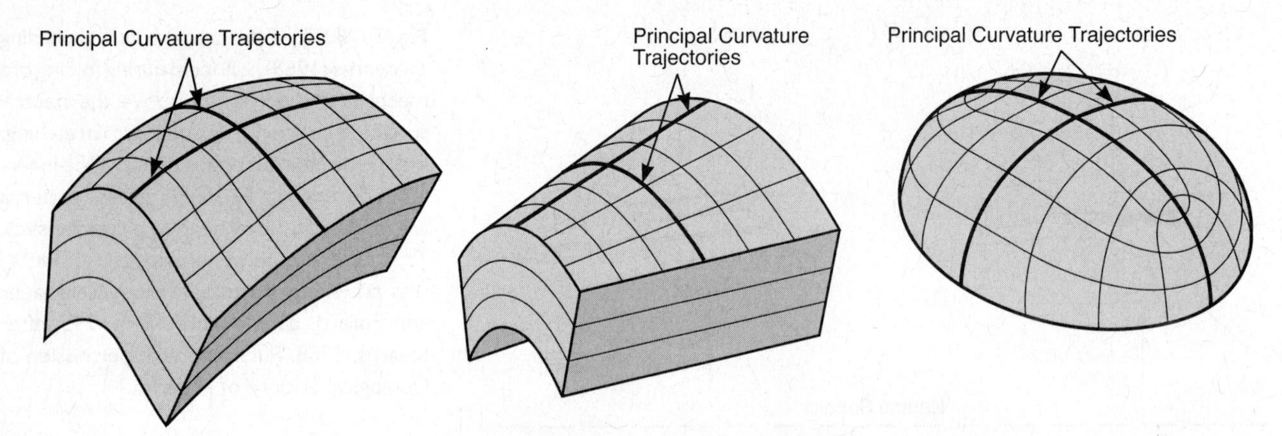

Figure 5.84 Principal curvature trajectories on folds of varying shape. [From Lisle, R. J., 2000, Predicting patterns of strain from three-dimensional fold geometries: neutral surface folds and forced folds, in Cosgrove, J. W., and Ameen, M. S. (eds.), Forced folds and fractures: Geological Society Special Publication No. 169, Figures 6, 7A, and 7B, p. 217.]

associated with variations in the plunge of the fold hinges influenced the length and density of transverse joints. These relationships were demonstrated in fracture maps and in data on fracture lengths, orientations, and densities (intensities). These observations are consistent with predictions made by Lisle (2000). He emphasized that principal strains within a folded stiff layer (containing a neutral surface) are directly related to curvatures of the layer, and by distance from the neutral surface. Strain can be predicted from curvature directions, and joint orientations can be predicted on the basis of strain. *Principal surfaces of strain are arranged parallel and perpendicular to the bedding surfaces, [and] the principal curvature directions in the bedding are principal strain directions* (Figure 5.84) (Lisle, 2000, p. 217).

Emigrant Anticline, Wyoming

Prepare yourself for surprises when assessing relationships between jointing and folding. Bergbauer and Pollard (2004) studied joint patterns in sandstone (Cretaceous Frontier Formation) exposed within the Emigrant Gap anticline near Casper, Wyoming. The anticline trends 345°. Bergbauer and Pollard (2004) identified two steeply dipping joint sets within the sandstone beds, one set (J_1) striking northwest, the other (J_2) striking northeast. The sets are mutually perpendicular, with J_2 younger and abutting against J_1. Neither set is parallel or perpendicular to the trend of the anticline, nor does the axis of the anticline bisect the trends of the joint sets. It turns out, these joint sets developed *before* folding (Figure 5.85). Apparently the joint system that was in place in the sandstones before folding had enough "*flex*" to permit the folding to take place without the need to create additional joint systems. It became clear to Bergbauer and Pollard (2004) that during folding sequential in-filling of joints took place, with new J_1 and J_2 joints forming between existing joints, creating dense networks marked by spacing ~1/5 to ~1/10 of the average original fracture spacing prior to folding.

Joint Patterns in Plutons as Regional Stress Gauges

Rehrig and Heidrick (1972, 1976) discovered that jointing in granitic intrusions can be a very sensitive guide to the directions of far-field regional tectonic stresses operative in the crust into which the plutons are emplaced. For example, they noted that Late Cretaceous to Eocene plutons contain systematic planar, parallel mode I tension fractures and parallel dikes and

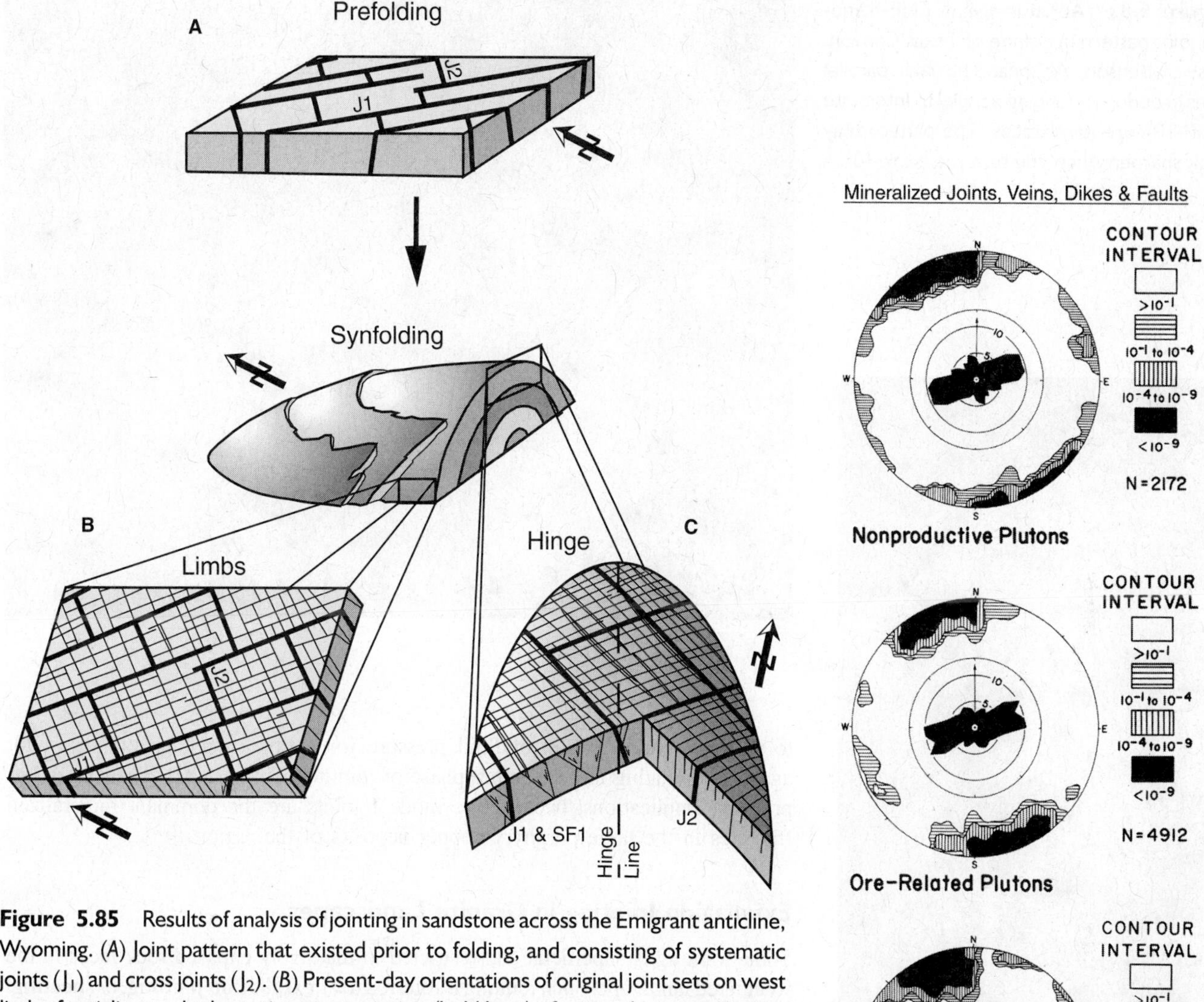

Figure 5.85 Results of analysis of jointing in sandstone across the Emigrant anticline, Wyoming. (*A*) Joint pattern that existed prior to folding, and consisting of systematic joints (J_1) and cross joints (J_2). (*B*) Present-day orientations of original joint sets on west limb of anticline, and schematic representation (bold lines) of minor shear reactivation of the original joints. (*C*) Present-day orientations of original joint sets in hinge region of fold, and schematic representation (bold lines) of shear fractures (SF_1) forming through reactivation of J_1 joints. [From Bergbauer and Pollard, 2004, Figure 13, p. 306. Published with permission of Geological Society of America.]

Figure 5.86 Pole-density diagrams and rose diagrams (in interior of each circle) show the orientations of mineralized joints, veins, dikes, and faults in Late Cretaceous to Eocene plutons in the Southwest. Look at the number (N) of structures measured and analyzed!! [From T. L. Heidrick and S. R. Titley, "Fracture and Dike Patterns in Laramide Plutons and Their Structural Tectonic Implications," Fig. 4.6 in *Advances in Geology of the Porphyry Copper Deposits: Southwestern North America*, S. R. Titley, editor. University of Arizona Press, Tucson, copyright © 1982.]

veins (Figure 5.86). At all scales it is clear that the joints occur in regular patterns over a large region.

After measuring literally thousands of joints in Late Cretaceous to Eocene plutons in southern Arizona, Rehrig and Heidrick determined that the dominant joint set trends N70°E, plus or minus 20° (see Figure 5.86). They concluded that these plutons were intruded into the upper levels of a crust and, as they cooled and congealed, cracked in directions systematically related to the regional tectonic stress field that existed at the time. Based on the patterns, they reconstructed the paleostress regime, which was marked by a horizontal north-northwest trending least principal compressive stress (σ_3). Differential stress was relatively weak. Fluid pressure buildup in the cooling pluton, manifest now in dikes and veins, had the effect of decreasing the magnitudes of the principal stresses. This produced tensional fracturing by jointing, with the dominant orientation perpendicular to the direction of least principal stress. Fracturing served

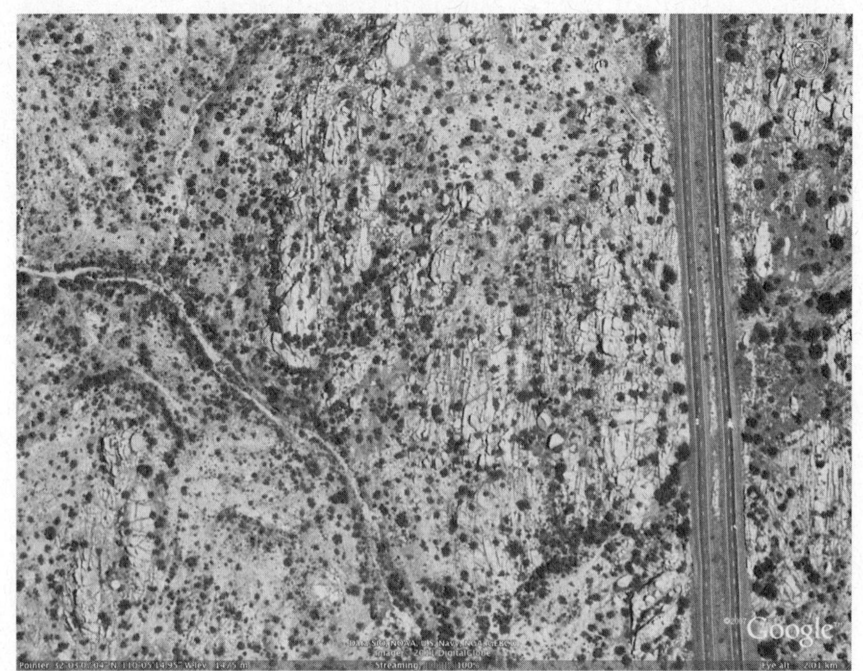

Figure 5.87 Aerial image of ENE-trending joint pattern in granite of Texas Canyon, east of Benson, Arizona. The faint parallel lines in bedrock, running parallel to Interstate 10 (I-10), are joint traces. The pattern may look sharper when you turn the book 90°.

to dissipate the elevation of fluid pressure for a time, and then pressure built again, culminating in yet another phase of jointing (Figure 5.87). Their work has practical implications, because the mode I joints are the dominant mineralized fractures in the large porphyry copper deposits of the region.

Exfoliation Jointing in Granite Landscapes

Some jointing is superficial and restricted to massively exposed rock faces in the weathered landscape. Such jointing is referred to as **exfoliation jointing**, or **sheeting**. Yosemite Valley presents one of the best displays of exfoliation jointing in the world, where the El Capitán granite (~108 Ma) is beautifully exposed. Bahat et al. (1999) applied sophisticated technologies in fractometry to the mid-region of the 1000-m-high cliffs of El Capitán (Figure 5.88*A*). Their mapping revealed that the exfoliation jointing is expressed in the form of fan-shaped fractures varying from a few meters to tens of meters in size (Figure 5.88*B*). Ribs and plumes occur in the face of the fans, and fringe areas are marked by en echelon segmentation. It is clear that mixed mode loading is at work. The fracture fans in the higher levels of the mid-region point up, and those in the lower part of the mid-region point down. Furthermore, the fans merge with one another to create composite joints (Bahat et al., 1999).

How does such exfoliation jointing develop? The greatest principal stress (σ_1) is oriented vertically, and produced by the gravitational load of the

A

B

Figure 5.88 (*A*) El Capitán at Yosemite National Park. The cliff of El Capitán Granite is marked by exfoliation jointing. (*B*) Closer view of exfoliation jointing in part of the face of El Capitán. Concentric undulations (ribs) are particularly apparent from this distance of view. [Reprinted from *Journal of Structural Geology*, v. 21, Bahat, D., Grossenbacher, K., and Karasaki, K., Mechanism of exfoliation joint formation in granitic rocks, Yosemite National Park, p. 85–96, © 1999, with permission from Elsevier.]

mountain of granite (Figure 5.89). The least principal stress (σ_3) is oriented horizontally, perpendicular to the free face. The vertical loading drives longitudinal splitting and causes outward buckling toward the free air. Thus, the main process that leads to exfoliation jointing in the mid-region of El Capitán cliffs is the coalescence of the fans, in turn driven by the overburden load. Splitting occurs upward and downward from the initiation zone, which explains the differences in fanning directions. The driving force for the actual spalling is importantly related to paleostress sealed in the granite before the granite was exhumed by erosion, but capable of being released upon exhumation and canyon cutting.

Bahat et al. (1999) explained that exfoliation jointing starts out at an initial flaw (origin), grows into a larger critical flaw, and then may propagate by unstable critical fracturing as paleostress is released. What is the magnitude of the paleostress? Bahat et al. (1999) estimate a range of between 0.01 MPa and 0.94 MPa, based upon considerations of fracture toughness, the radius of the critical flaw, radius of the mirror plane, and other factors related to calculating stress intensity. Individual joints propagate in a way that begins with slow subcritical growth, and then advances to rapid fracturing. The overall conditions and findings suggest to Bahat et al. (1999) a subcritical propagation rate of below 4×10^{-5} m/s. Thus it would take longer than 14 days to fracture a fan 50 m long (dry conditions). The propagation velocity comes out to 1.2 mm/minute, or less than 1 cm/hr.

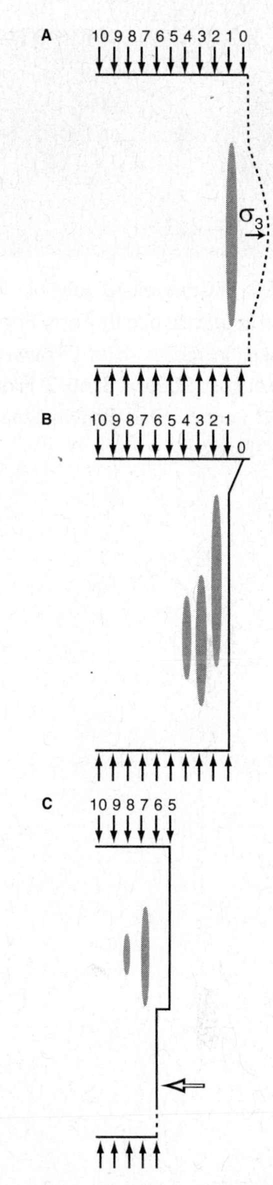

Figure 5.89 Diagram showing progressive development of exfoliation jointing by longitudinal splitting and buckling. Vertical arrows represent gravitational load created by overburden of granite. Numbers (1–10) represent positions where new exfoliation joints will progressively develop over time. Ellipses represent "*internal fractures.*" (*A*) Buckling (exaggerated) shown by dashed line. (*B*) Early slice collapse at 0, with new exfoliations developing. (*C*) Later stage yet, and we see the profile of an arch at the lower part of the cliff. [Reprinted from *Journal of Structural Geology*, v. 21, Bahat, D., Grossenbacher, K., and Karasaki, K., Mechanism of exfoliation joint formation in granitic rocks, Yosemite National Park, p. 85–96, © 1999, with permission from Elsevier.]

OPPORTUNITIES IN FRACTURE ANALYSIS

Joints readily lend themselves to many aspects of detailed structural analysis. Some additional practical methods for studying joints are presented in *Part III-N, Carrying Out Joint Analysis.* There is never a problem finding systematically fractured rocks to examine. Field and lab projects involving fracture analysis can foster firsthand experience in orientation analysis and the sorting out of structures. Furthermore, joint analysis can provide valuable experience in photogeologic interpretation, experimental deformation, kinematic analysis, and mathematical inquiry into dynamics and mechanics. Because of the difficulty encountered in establishing *exact* timing, there is always plenty of room for independent, creative, challenging inquiry.

Just to give an example of the range of applications of fracture analysis, consider the innovative (sideline) scholarship Terry Engelder (Pennsylvania State University) expressed in one of the talks he presented at colleges and universities he visited while serving as a "*Distinguished Lecturer*" for the American Association of Petroleum Geologists (AAPG). The title of this talk was "*Craquelure* [i.e., fine pattern of cracks] *in Masterpieces of the Louvre (Paris, France) as Analogue Models for Development of Joints in Fractured Reservoirs.*" The stratigraphy of the old master works, as explained by Terry, commonly begins with a base "*canvas*" of wood. This would be crystalline basement, geologically. Italian and French painters (southern Europe) used poplar as a rule, whereas Flemish and English painters (northern Europe) used oak. Poplar and oak have different "*rheologies.*" The overlying "*stratigraphy*" typically begins with ground support in the form of powdered chalk with animal glue, overlain by paint layer (binder pigment) and then varnish (wood rosin alcohol). Terry emphasized in his talk that the paint layer and varnish together constituted "stiff rock." This would be the sedimentary cover, geologically.

It takes energy to produce the "*craquelure jointing,*" and for these masterworks the source of the energy was twofold: stretching of the canvas, followed by thermoelastic contraction. The cracking serves to dissipate the

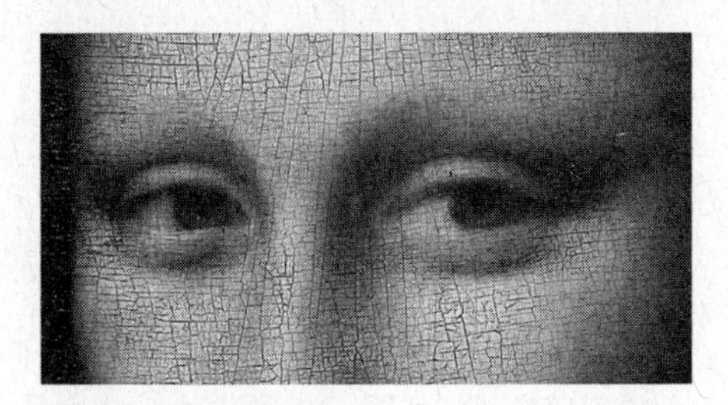

Figure 5.90 "Craquelure jointing in De Vinci's *Mona Lisa*. [Scanned by Terry Engelder. The scanned image is from Transatlantic Times, July/August, 2006, v. 3, no. 2. From the article, "Did Author Dan Brown Steal "Da Vinci Code" Plot?"]

energy. Figure 5.90 is a close-up photograph of craquelure in Leonardo da Vinci's "*Mona Lisa*."

Here is a sample from Terry's abstract:

Masterpieces in the Louvre (Paris, France) and other national galleries, like bedded sedimentary rocks, are jointed (i.e., cracked) composite materials with welded contacts between substrate and joint-bearing medium. . . . Differences in properties between the substrate and the joint-bearing medium greatly influence a number of characteristics of joint growth patterns including fractal properties (*Duccio, 1300*), propagation direction (*Duccio, 1311*), spacing vs. bed thickness (*Francesca, 1455*), orientation (*Master of St. Giles, 1500*), abutting relationships (*Clouet, 1530*), degree of systematic development (*Rembrandt, 1660*) and lack of systematic development (*Chardin, 1736*). In particular, the degree to which the joint-bearing medium is subject to "*tectonic*" stress predetermines the extent to which a number of these joint patterns develop (*Hals, 1647*). Regardless of the particulars, it is clear that the masterpieces and sedimentary rocks even share two common loading configurations: the joint-normal load (*canvas support*) and the thermoelastic load (*wood support*). When *tectonic* stresses are not present mudcracks (*Fouquet, 1455*), columnar joints, and spiral joints (*David, 1791*) propagate under thermoelastic loading.

So, before attempting to forge a master work, consider how you would ever forge the crack systems, which evolve over time in ways that are dependent upon the changing mechanical properties of canvas support and/or wood support, the rheological properties of the layered media (from base coating through oils to lacquer), and the mechanical and thermoelastic loading *over time.* Terry emphasized: with respect to old masterpieces, "*you simply can't counterfeit the crack patterns*!" Based on the substance of this chapter, we now know why.

Chapter 6 Faults

Fault scarp produced during the Gobi-Altay earthquake (M_w 8.3) of December 4, 1957. The scarp looks like it formed yesterday, for it is perfectly preserved in the cold, dry climate of this part of Mongolia. The surface ground rupture associated with this earthquake covered a trace length of ~250 km. Fault slip was oblique, with as much as 11 m of vertical offset and 3 m of left-lateral strike-slip offset. Though this was a giant quake, only 30 people were killed, for the region is very sparsely populated. The regional fault zone of which this rupture is a part is fully 40 km wide. We see that at a moment in time (i.e., no more than 60 seconds back in 1957) just a part of it was reactivated, producing at this particular location two discrete surfaces. In walking this fault, we begin to imagine how much of the critical surface record of faulting is ultimately lost to erosion, and that the formation of broad, complex fault zone is an incremental progressive process over time. [Photograph by Kurt Constenius.]

INTRODUCTION

Faults are discontinuities along which there is visible offset by shear displacement (Figure 6.1). Faults can occur as single discrete breaks along fractures (Figure 6.2), but where the rock has been repeatedly faulted, or where the rock is especially weak, no single, discrete fracture or discontinuity may be evident. What forms instead is a **fault zone** composed of countless subparallel and interconnecting closely spaced slip surfaces (Figure 6.3). Faulting is fundamentally a brittle mechanism for achieving shear displacement. At deep crustal levels, where rocks tend to deform plastically under conditions of elevated temperature and confining pressure, shear displacement is achieved by the development of **shear zones**, which is discussed in *Chapter 10, Shear Zones and Progressive Deformation.*

Faults range in length and displacement from small breaks with offsets wholly contained within individual hand specimens or small outcrops (see Figure 6.2) or larger cliff exposures, to regional crustal breaks extending hundreds to more than 1000 km and accommodating cumulative offsets of tens to hundreds of kilometers (Figure 6.4). Some shear fractures can be considered to be **microfaults**, but they generally go unrecognized as such because the tiny offset caused by shear cannot be easily resolved without aid of a microscope, or without the presence of fine-scale markers that record the displacement.

Figure 6.1 Photograph of fault exposed on freshly bladed Carmel Formation bedrock (Jurassic) in Candyland, along Cottonwood Road south of Grosvenor Arch, Utah. Puzzling over the truncation and offset are physician-scientists Setsuko Chambers (right) and Keith Joiner (left), working on a clinical diagnosis. [Photograph by G. H. Davis.]

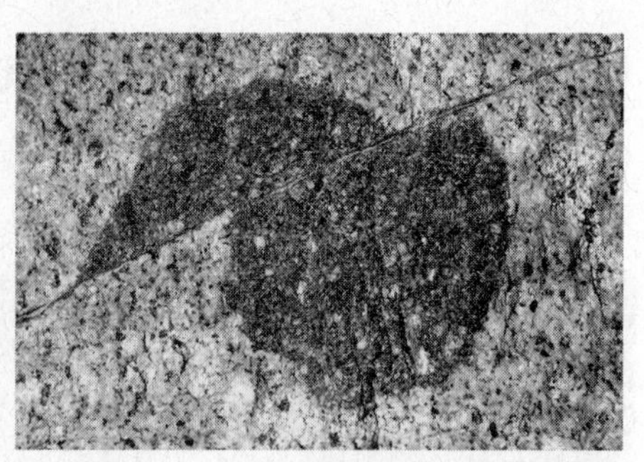

Figure 6.2 Photograph of discrete fault surface cutting and offsetting an enclave (rock inclusion) within granite. A thin quartz vein occupies part of the fault. Diameter of enclave is ~12 cm. Location is Sonora Island, British Columbia, Canada. [Photograph by G. H. Davis.]

Figure 6.3 North-directed photograph of fault zone within Carmel Formation (Jurassic) strata as exposed in the so-called Candyland locality along the Cottonwood Road, south of Kodachrome Basin State Park, southern Utah. Note that the zone is bounded both on the west and east by discrete fault surfaces, beyond which strata are intact though steeply dipping. Within the fault zone proper, bedding is really messed up, and the softest rock units appear to have been put through an egg beater. [Photograph by G. H. Davis.]

Faults should not be confused with **fissures** (Figure 6.5), which are large fractures that have accommodated conspicuous dilational opening. Fissures form at the surface of the Earth where active tectonic processes, notably earthquakes and volcanism, create local stretching and pull-apart.

Figure 6.4 Physiographic expression of a great regional fault, the San Andreas fault, as viewed northerly along the Elkhorn scarp, San Luis Obispo County, California. [Photograph by R. E. Wallace. Courtesy of United States Geological Survey.]

Figure 6.5 Fissure cutting the Hiliana Pali Road at Kalanaokuaiki Pali, Kilauea Volcano, Hawaii. The crack formed on December 25, 1965. Initially only 1 m wide, the crack soon enlarged to 2.4 m as a result of slumping along the edges. [Photograph by R. S. Fiske. Courtesy of United States Geological Survey.]

RECOGNIZING FAULTS: THE PHYSICAL CHARACTER OF FAULTS

First we need to look at ways to recognize faults, and then we will start to categorize them and analyze them more closely. Sometimes recognizing faults can be easy and sometimes not, depending on the type of fault, the size of the offset, the scale of the observations, and the quality of exposures.

Fault Scarps

Fault contacts often are easy to identify in areas where faulting is presently active. Active faulting that reaches the surface produces **offset** (i.e., displacement) of both natural and human-made objects. The offset of features such as roads, fields, streams, city streets, and beachfront property makes the locations of faults utterly conspicuous. In some cases active faulting that accompanies earthquakes may not reach the surface, but the effects of the seismic energy are nonetheless evident in the form of damage to built structures. Figure 6.6 presents expressions of the active faulting that accompanied the M_W 8.1 Izmit earthquake (Turkey) on August 17, 1999, at 3:02 a.m. Sudden slip on a strand of the North Anatolian fault zone (NAFZ), a right-handed strike-slip fault, produced the earthquake. The ground shook for 45 seconds, and the faulting produced ground rupture along a length of 133 km. The largest measured horizontal shift was 5 m, which though significant is but a tiny fraction of the ~40 km of total slip along the NAFZ. Thirty-five thousand people died as a result of this earthquake.

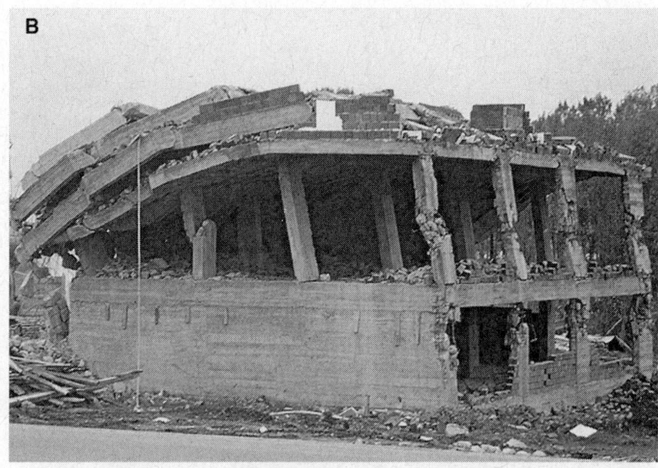

Figure 6.6 (*A*) Photograph of a small fault scarp produced in a field in Yuvacik, Turkey, as a result of the Izmit earthquake in August, 1999. (*B*) Photograph of destroyed building that collapsed during the Izmit earthquake. Location is Sakarya, Turkey, where seismic waves were amplified when they passed through the unconsolidated sediments upon which the city is built. Note the combined effects of shear and gravitational collapse. Also note the failure of the concrete, rebar-reinforced columns. [Photographs by G. H. Davis.]

Fault scarps are offsets or steps in the land surface that coincide with locations of faults (Figure 6.7*A*). They are expressions of contemporary fault movement(s). The height of a fault scarp will approximate the sum of the most recent displacements, although the effects of erosion of the raised block and depositional infilling atop the lowered block must be taken into account in order to estimate total offset. To be sure, the offset of weak erodible material indicates that the active faulting must be quite recent; otherwise the scarp would have been degraded and not easily recognized. Usually, however, scarps in the landscape that may be seen along the surface trace of a fault are not *true* fault scarps; that is, they are not a direct expression of either the dip of the fault surface or the magnitude of offset. The passing of time permits weathering and erosion to erase the original expressions of the fault offset. Fault scarps thus are gradually replaced by **fault-line scarps** (Figure 6.7*B*). Fault-line scarps are located along or near the trace of a fault and are marked by a topographic relief that simply reflects the differential resistance to erosion

Figure 6.7 (*A*) Red Canyon fault scarp in the Montana earthquake area, as it looked in August, 1959. Scarp height reflects the actual fault displacement. [Photograph by J. R. Stacy. Courtesy of United States Geological Survey.] (*B*) Fault-line scarp near Lake Mead. The topographic relief along the scarp does not reflect slip. Rather, it reflects differential erosion along the fault interface between resistant volcanic rock and nonresistant volcanic rock. The rock on which Ernie Anderson stands actually moved *upward* relative to the resistant volcanic rock on the other side of the fault. [Photograph by G. H. Davis.]

Figure 6.8 Photograph of normal faulting exposed along the Paria River in southern Utah, within the East Kaibab monocline deformation belt. In the lower part of the outcrop (Carmel Formation, Jurassic) the faulting is accommodated by a single discrete fault surface. Higher in the outcrop this surface "*branches*" into two discrete fault surfaces. The sense and magnitude of fault offset is conspicuous because of the presence of a marker layer in the form of the dark bed with a white band at its base. Geologist is Sarah Tindall. [Photograph by G. H. Davis.]

of the rocks brought into contact by faulting. Thus fault-line scarps almost always give a false impression of the magnitude of vertical displacement along a fault, and sometimes (as in Figure 6.7*B*) give a false sense of the actual relative movement.

Fault Surfaces

Where well exposed, faults are commonly expressed by the presence of a discrete fracture break or discontinuity in the rocks along a **fault surface** (Figure 6.8). The rocks on either side of a fault surface do not "*match up.*" We use the term fault *surface* instead of fault *plane* because faults are rarely perfectly planar. Some are planar; some are made up of planar segments of varied orientations; some are systematically curved; and some are highly irregular.

Individual fault surfaces do not go on forever. If we could pull discrete fault surfaces out of the ground as if they were sheets of Plexiglas and look at them, we would find that most are roughly elliptical. Some of the evidence for this is seen in the work of Barnett et al. (1987), who measured the attributes of discrete faults in the British coalfields and on seismic sections in the North Sea. They found that isolated faults have an elliptical shape with a common aspect ratio between 2:1 and 3:1 (Figure 6.9*A*). Displacements were seen to decrease to zero from the central part of each elliptical fault surface outward to the **tip line loop**, the imaginary line formed by connecting points (i.e., **tip points**) where the fault surface comes to an end (Figure 6.9*B*).

A

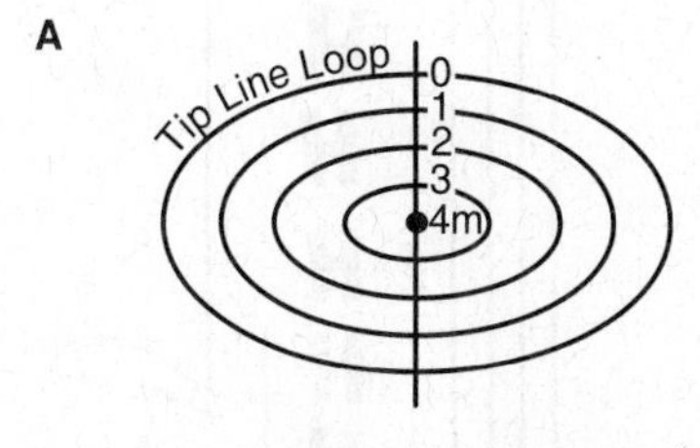

B

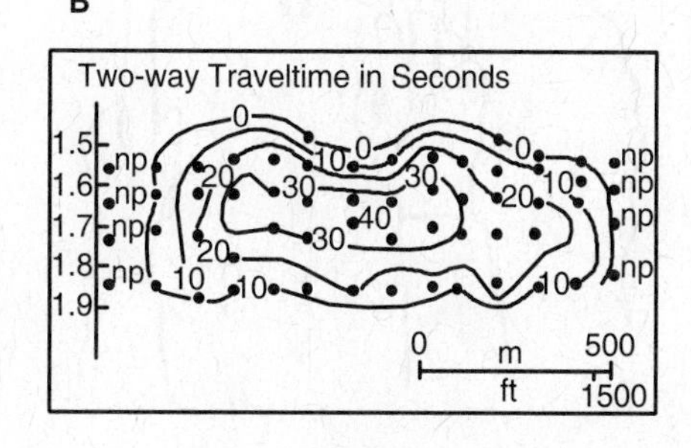

Figure 6.9 (*A*) Theoretical shape of a discrete fault surface and the gradient of displacement along it. (*B*) Displacement contour diagram of a normal fault in the North Sea. Displacement values are here not measured in distance, but in the travel time (two-way seismic wave travel times, in milliseconds) separating the marker bed on one side of the fault versus the other. The abbreviation "*np*" denotes fault not present; that is, "*np*" is noted at positions just beyond the termination tip points of the fault surface. [From Davison, I., Linked fault systems; extensional, strike-slip and contractional, in Hancock, P. L. (ed.), Continental Deformation. Copyright © Elsevier, 1994. After Barnett and others, 1987.]

Figure 6.10 Photograph of a fault surface that shows frictional polishing by faulting. This natural fault polish is referred to as "*slickensides*." Location is near La Verkin, Utah, within the Hurricane fault zone. The rock is part of the Moenkopi Formation (Triassic). Knife on top of outcrop. [Photograph by G. H. Davis.]

Figure 6.11 In bright morning light, and to the sound of clicking cameras, Edna Patricia Rodriguez points out beautifully developed slickenlines on the underside of the Rubys Inn fault surface near Bryce Canyon, Utah. The rock is Claron Formation (Eocene). [Photograph by G. H. Davis.]

Slickensides and Slickenlines

Fault surfaces are commonly finely polished, apparently the result of the fine abrasive action of frictional movement when the rock types and conditions of deformation are just right (Figure 6.10). **Slickensides** are the smooth or shiny fault surfaces themselves (Means, 1987). Some surfaces are almost mirrorlike, as if "*finished*" by Nature's equivalent of '0000' sandpaper or steel wool. Yet many fault surfaces are not slickensided at all, either because of the host rock's inability to "*take a polish*" or because of removal of an original luster by weathering and erosion. We also find that some slickensides owe their gloss not to frictional polishing, but to thin **neomineral coatings** that precipitated on the fault surface during movement.

When we examine fault surfaces in the field, it is standard procedure to inspect the surfaces closely for the presence of **slickenlines**. Slickenlines are generally straight, fine-scale, delicate skin-deep lines that occupy the fault surface itself and record the direction of slip (Figure 6.11). Slickenlines are most noticeable on slickensided surfaces, but they are by no means restricted to such glossy surfaces.

Many slickenlines we examine, such as those in Figure 6.11, prove to be **striations** produced by frictional abrasion along the fault surface during differential displacement of the wall rocks (Figure 6.12*A*,*B*). Scratches and furrows ("*tool markings*") are produced when hard **asperities**, projecting like

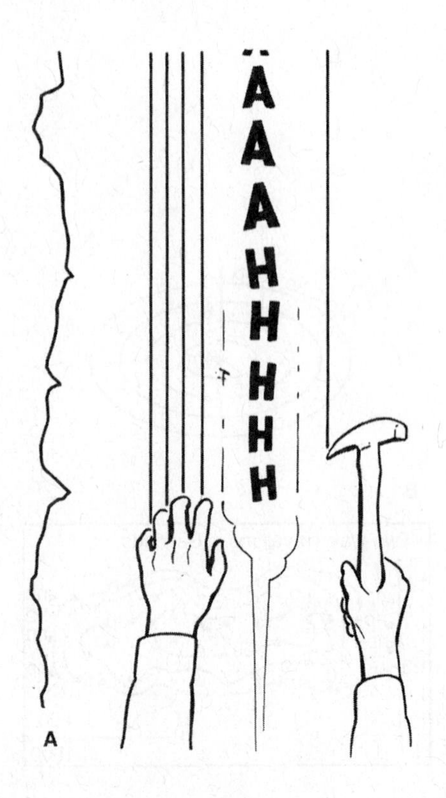

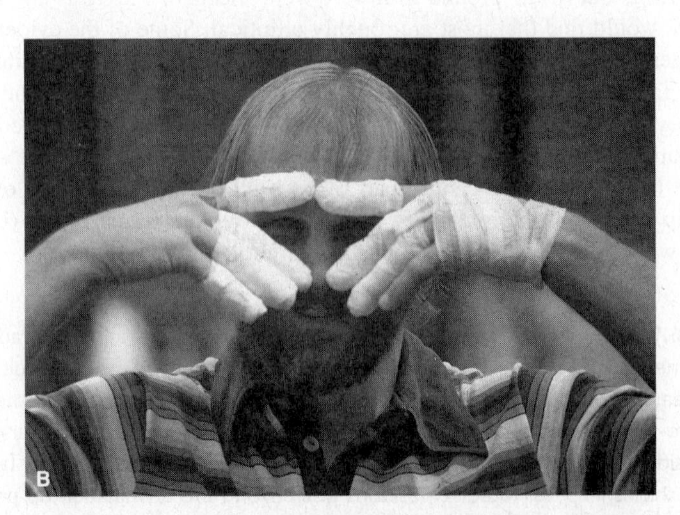

Figure 6.12 (*A*) Striae produced by differential rigid-body movement. Artwork by Wally Varner. (*B*) Geologist Steve Lingrey left such striae near the Paunsaugunt fault in southwestern Utah. [Photograph by G. H. Davis.]

A

B

Figure 6.13 (*A*) Slickensided fault surface in interstate highway cut through Paleozoic carbonates in the Virgin Mountains, northwesternmost Arizona. Polished fault grooves disclose horizontal motion. Eric Frost goes for a closer look of the slickensided surface. (*B*) Chris Menges (mustache and glasses) proudly shows off the exceptionally well exposed, mullioned fault surface that he mapped near Patagonia in southern Arizona (see Menges, 1981). Chris's hands are placed on the crest of the fault mullion. Richard Gillette's foot (*far left*) rests in trough of the adjacent concave groove. Nancy Riggs, with feet squarely planted, measures the strike azimuth of a part of the fault surface. [Photographs by G. H. Davis.]

metal cleats from the sole of a football shoe, score lines on the opposite surface against which it moves. When asperities break off, they can break up into pieces of hard debris that "*float*" along the fault surface, not unlike sand and grit underfoot. The line pattern created by scratching and furrowing is usually further enhanced by **streaks** of the fine crushed debris, which will pile up "*fore and aft*" of hard asperities, creating **tails** and **spikes** and other microridges of material (Means, 1987).

The apparent frictional development of striations can operate at a very large, coarse scale as well, producing **fault grooves** and **fault mullions** (Figure 6.13). Think of these as reflections of undulations that mark the fault surface itself, produced by the grinding and tooling of the two fault blocks moving past one another, earthquake by earthquake. The alignments of these coarse tool marks reflect the direction of slip during the faulting that produced the grooves and mullions. The surfaces resemble bedrock surfaces that have been polished and grooved by the movement of glacial advance across a bedrock land surface.

Slip-Fiber Lineations

There is yet another common expression of slickenlines, namely **slip-fiber lineations** (van der Pluijm and Marshak, 1997, 2003), produced by the preferred directional growth of minerals *during* the faulting and in the direction of movement (Figure 6.14). Slip fibers (or **crystal-fiber lineations**) on fracture surfaces are precipitated on the leeward side of tiny ridges or bumps (Figure 6.15) (Durney and Ramsay, 1973). The micro-topography in combination with shear tends to produce open space in which the crystals can grow. Narrow veins of crystal fibers grow on fracture surfaces in such a way that

Figure 6.14 Crystal fiber lineations on a thrust fault surface in the Hudson Valley fold-thrust belt in New York. Width of photo covers N/M. [From Marshak and Tabor, 1989. Published with permission of the Geological Society of America and the authors.]

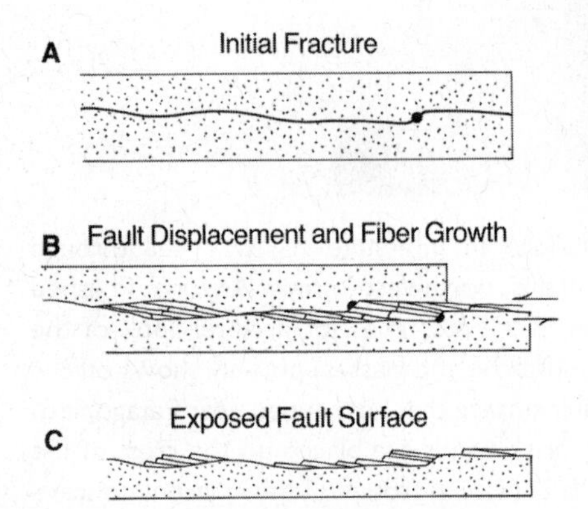

Figure 6.15 Steps in the kinematic evolution of crystal fiber lineation (a.k.a., slip-fiber lineation) on a slickenlined surface. (*A*) Formation of fault surface. (*B*) Fault displacement and simultaneous growth of crystal fibers in the direction of least stress. (*C*) Striated surface as exposed by weathering and erosion. [From Durney and Ramsay, Incremental strains measured by syntectonic crystal growths, in De Jong, K. A., and Scholten, R. (eds.), Gravity and tectonics. Published with permission of John Wiley & Sons, Inc., New York, copyright © 1973.]

individual crystal fibers trend parallel to the direction of differential displacement (see Figure 6.15); the sense of shear is disclosed by the plunge of the crystal fibers relative to the fracture surface. The length of individual fibers records the magnitude of slip accommodated by the fracture surface during the time of crystal fiber growth. You will see structural geologists checking out slip fibers surfaces in the field by rubbing their hands along them, back and forth along the direction of lineation. By seeking out the direction that feels most rough to the hand, they are trying to gage the sense of movement along the fault. Finding the smooth direction is key. The hand, like the fault block, moved in the direction that feels smoothest. (For the record, structural geologists do not gain karmic energy from this drill.)

A beautiful example of slip-fiber lineation is shown in Figure 6.16. This is a specimen I [GHD] collected along the Comb Ridge monocline (regional fold) in southern Utah. The bedrock itself is clayey limestone. The slickenlined surface is composed of slip fibers of calcite, which collectively define the "*fault*" surface. The dozens of tiny micro-cliff escarpments all face the

Figure 6.16 Photograph of slip-fiber lineation. The coating is composed of calcite. Its precipitation *during* slip resulted in the formation of slip fibers oriented in the direction of slip. Note the presence of tiny micro-cliffs, all facing in a common direction. This asymmetry produces an asymmetry of roughness-to-touch when you slide your hand along the surface in the direction of the slickenline lineation. See text for explanation. This specimen is limestone collected along the Comb Ridge monocline in southern Utah. Collection of G. H. Davis. [Photography and editing by Gary Mackender, vr.arizona.edu, glm@email.arizona.edu, © 2009–2010 Arizona Board of Regents.]

same direction. Sliding your hand against these "*micro-cliffs*" yields a rough feel; sliding your hand the opposite way, still parallel to the direction of the fibers, yields a smooth feel.

Slickolites

Slickolites are the antithesis of crystal fiber slickenlines. They form along fault surfaces where fault movement in combination with stylolitic pressure solution results in removal of material through dissolution creep. What normally might be a planar fault surface is transformed into a pock-marked surface composed of the expressions of alternating rods and cones, the stubby expression of interpenetration of the fault blocks on either side of the surface. An example is shown in Figure 6.17, a photograph of stylolitic surfaces marked by the characteristic cones and teeth of stylolites (Sutton et al., 2008). The axes of the cones and teeth are aligned in parallelism and constitute an unusual kind of lineation; that is, the **slickolite lineation**. The direction of this crude lineation records the direction of interpenetration of the fracture surface. This display is marked by especially long and needle-like stylolites developed on a fracture surface oblique to bedding in the limestone. The needle-like arrays constitute stylolitic lineation. An extraordinary exposure of the relationships of stylolite development to layer-parallel shortening and faulting is captured in Figure 6.18. Faulting, and the sense of faulting, is obvious near the top of this photograph, where a layer of limestone is folded over the relatively-uplifted block on the left. Note the rough, serrated, non-planar nature of the fault itself as it tracks vertically down the center of the photograph. The fault originally may have been planar, but during progressive deformation pressure-dissolution dissolved away much of the fault and the bordering wall rock. Moreover, in this same photograph (see Figure 6.18) we see stylolites oriented perpendicular to bedding, with teeth and cones oriented parallel to bedding. These stylolites undoubtedly formed well before faulting and helped accommodate layer-parallel shortening of the limestone.

Figure 6.17 Photograph of the stylolites in white Kasimovian limestone (Carboniferous) in the Gzhel quarry, Gzhel village, Russia. The stylolites in this exceptional example are needlelike and quite long. [Photograph courtesy of Alexander Davydov and Barry Sutton, who hold copyright for this photo. The photograph is used with permission from Barry Sutton, Alexander Davydov, and Tom Yancy.]

Figure 6.18 Photograph of combination of faulting and stylolite development in limestone. The fault runs vertically down the center of the photograph (tip of pen for scale), with offset (left side up) revealed in fold (at top) that "*drapes*" over the corner of the relatively uplifted block. The effects of pressure dissolution are evident in the rough serrated appearance of the trace of the fault zone. An additional expression of stylolites is evident in the spaced "*sutured*" zones oriented perpendicular to bedding. Teeth and cones within these stylolites are oriented parallel to bedding, revealing that the role of this stylolitization was to accommodate layer-parallel shortening. This outcrop belongs to (Cretaceous) Thick White Limestone Beds of the Pindos Group. Location is in the Three Gorges Canyon near the village of Ano Karyes, Mt. Lykaion, Peloponessos, Greece. [Photograph by G. H. Davis.]

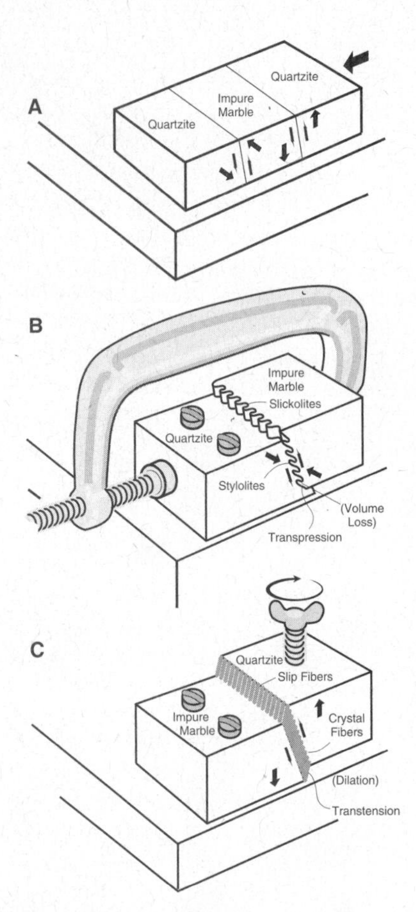

Figure 6.19 To illustrate slickolites and slip fibers, imagine a deformation machine that could create transpressive movements (slip plus shortening) and transtensile movements (slip plus extension). (*A*) We try the machine out on two blocks, one of quartzite and one of impure marble in between. (*B*) The movements plus pressure solution create stylolites along the transpressive fault surface. A view of the "*fault*" surfaces reveals slickolites. (*C*) It takes transtensile movement (slip plus extension) to create crystal slip fibers oblique to a failure surface.

Figure 6.19 is a "*thought experiment*" describing the way we might make slickolites and slip fibers in the laboratory, starting with a rock (in this case impure marble) that would lend itself to pressure dissolution (see Figure 6.19*A*). The cones and teeth of the stylolites create (in 3D) stubby lineations (i.e., the slickolites) on the fault surface and thereby record the direction of interpenetration of the fault block on either side of the fault "*surface*" (see Figure 6.19*B*). In this example the direction of interpenetration is the direction of slip, and this direction is typically not perfectly perpendicular to the fault surface. In contrast to slip fibers that are precipitated on the leeward side of tiny ridges or bumps on the fault surface (see Figure 6.19*C*), slickolites penetrate *into* the irregular fault-surface topography, reducing volume (see Figure 6.19*B*). The formation of slickolites requires lithologies that, under the conditions of deformation, can "*switch on*" pressure dissolution as a deformation mechanism.

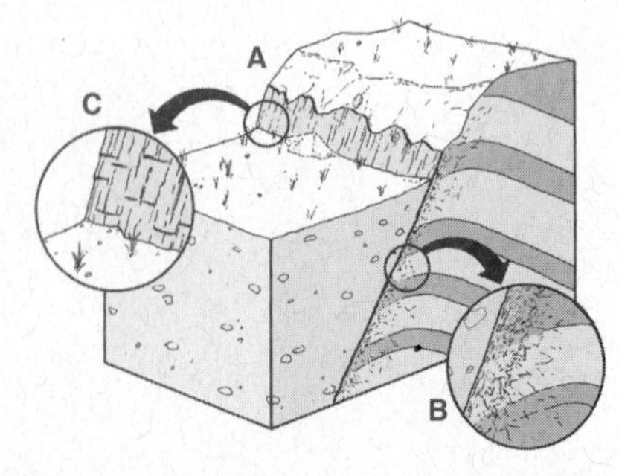

Figure 6.20 Schematic diagram showing (*A*) fault surface, (*B*) fault rocks like gouge and breccia, and (*C*) chatter marks and slickenlines.

Chatter Marks

Chatter marks commonly are formed on the surfaces of faults (Figure 6.20). These small, asymmetrical, step-like features are typically oriented perpendicular to striations. Roughness of the chatter marks is usually less than 5 mm. Sense of movement can be interpreted (very cautiously) on the basis of the geometry of chatter marks. The ideal relationship of chatter-mark geometry to differential slip is

Figure 6.21 Photograph of well developed chattermarks on fault surface cutting Cretaceous sandstone in the southern part of the Sawtooth Range, Montana. This is a loose block, not an outcrop. The chatter marks are the little micro-cliffs that face left (shadowed part) and occur in short segments. The long dimension of this block is ~50 cm. If this were an outcrop, and the fault surface was in place, we would conclude from the "*smoothness test*" (see text) that the fault block in which they are contained moved right and horizontally with respect to the fault block from which the photograph was taken. [Photograph by C. F. Kluth.]

shown in Figure 6.20: the underside of each chatter mark is in shadow, for the down-dip face of each step for this sense of fault movement was protected from mechanical grinding by the opposite block. Figure 6.21 is a photograph of chatter marks exposed on the surface of a loose block of sandstone. The exposed surface of this loose block happens to be a fault surface. Lighting conditions at the time that Chuck Kluth took this photograph nicely accentuate the "*microcliff-like*" expression of chattermarks.

The asymmetry of chatter marks is not a completely reliable indicator of displacement sense. However, the presence of chatter marks is helpful in confirming that a particular surface is actually a fault. In the absence of any other sense-of-movement criteria, study of chatter mark orientations and asymmetries may be used to yield the statistically most probable direction and sense of slip.

Fault Zones

Fault zones consist of numerous closely spaced fault surfaces, commonly separating masses of broken rock (Figure 6.22). Zones of intensely fractured and crushed rocks associated with faults vary in thickness from less than several centimeters to a kilometer or more. In general, the thicker the fault zone, the larger the amount of displacement on the zone (Schulz, 1990). A single fault zone, however, may thicken or thin along strike as individual fault strands merge with or split away from the zone and as the fault encounters rocks of varying mechanical properties. Changes in the width and character of fault zones are especially common near bends or jogs. The wall rocks outside

Figure 6.22 The Lincoln Ranch fault zone in western Arizona carries strongly deformed Precambrian rocks over Miocene sandstone. Note how the sedimentary layers are bent, kinked, and broken as they approach the zone of deformation. Ivo Lucchitta of the U.S. Geological Survey is admiring the result of this faulting. [Photograph by S. J. Reynolds.]

Figure 6.23 (*A*) Southeast-directed photograph of drag folding produced by the faulting obvious in the immediate foreground. The sense of curvature reveals that the block that Keith Joiner (immunologist) occupies moved left (east) relative to the block from where the photo was taken. Rock is Carmel Formation (Jurassic). Location is Cottonwood Road in Candyland, south of Kodochrome Basin State Park. [Photograph by G. H. Davis.] (*B*) Drag folding of hanging-wall beds records normal movement on this Tertiary low-angle fault, Crete. [Photograph by S. J. Reynolds.]

of a brittle shear zone may be largely unaffected by the faulting or may show a zone of **drag folding** flanking the zone (Figure 6.23). Whether faulting produces a discrete fault surface or a fault zone depends on many factors, none of which can be precisely known for a given fault. Rock strength, strain rate, the physical environment of deformation, and duration of faulting are all important factors.

BRITTLE FAULT ROCKS

Comminution and Cataclasis

During frictional movement along a fault or within a fault zone, rocks may be transformed into a variety of **brittle fault rocks**, notably **breccias**. **Comminution** is a word that we use to describe rocks being broken and ground down into finer and finer size during faulting. The deformation mechanisms at work are **frictional sliding** and **cataclasis** (see *Chapter 4, Deformation Mechanisms and Microstructures*). The resulting breccias are marked by unsorted angular clasts in a finer-grained crushed matrix. As in the case of all breccias, there is a general absence of preferred orientations of the rock fragments (regardless of average sizes). Instead, breccias possess random fabrics that appear surprisingly similar when viewed at widely different scales of observation, from submicroscopic to the scale of outcrops.

The specific nature of fault rocks will depend upon starting materials, conditions, and deformation mechanisms. Our focus here is on fault rocks formed under conditions that caused the rocks to be damaged in brittle fashion. In *Chapter 10, Shear Zones and Progressive Deformation*, we cover fault rocks formed under conditions of ductile deformation. Fault rocks (whether produced under brittle or ductile conditions) can be thought of as "*fault-related rocks*" (Snoke et al., 1998a); that is, rocks transformed by conditions of faulting and shearing.

The basic varieties of brittle fault rocks are shown in Table 6.1, which is adapted from the classification developed by Rick Sibson (1977). Sibson

TABLE 6.1
Brittle fault rocks

Breccia Series

These brittle fault rocks are marked by angular clasts in a finer matrix. Generally there is no preferred orientation of clasts, and thus the fabric is "*random.*" Breccia series rocks are noncohesive to compacted, except where silicified or mineralized. The four main varieties are based on size of clasts:

Megabreccia	clast size > 0.5 m
Breccia	clast size > 1 mm < 0.5 m
Microbreccia	clast size > 0.1 mm < 1 mm
Gouge	clast size < 0.1 mm

Cataclasite Series

These brittle fault rocks are marked by angular clasts in a finer matrix, and generally there is no preferred orientation of clasts. Cataclasite series rocks are cohesive and strongly indurated. The two main varieties are based on size of clasts:

Cataclasite	clast size > 0.1 mm < 10 mm
Ultracataclasite	clast size < 0.1 mm

Pseudotachylite

Unlike breccias and cataclasites, this brittle fault rock is created by frictional melting, not by grinding and fracturing. It is glassy to cryptocrystalline and generally brown, gray, or black in color. There may be tiny crystals (less than 1 μm) in an isotropic glassy groundmass. Pseudotachylite occurs in veinlike arrays, some of which may be spiderlike in form.

Source: Adapted from Sibson (1977).

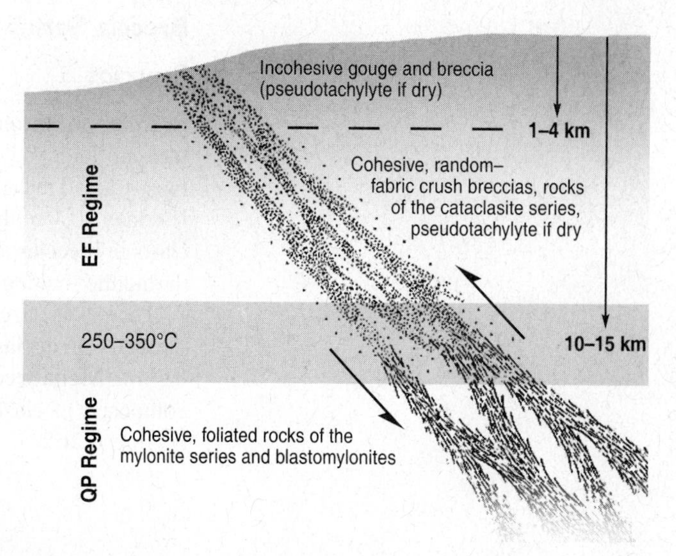

Figure 6.24 Schematic crustal-scale cross-section showing the brittle fault rock fabrics in the context of depth and temperature. "EF" = elastico-frictional processes; "QP" = quasi-plastic processes. [Adapted from Sibson, 1977. Reprinted with permission from The Geological Society Publishing House, Bath, England.]

distinguishes **brittle fault rocks** that are incohesive (can be made to fall apart in your hands) from those that are cohesive (rock solid). Incohesive brittle fault rocks (the "***breccia series***") form in the uppermost crust (shallower than ~4 km) and include breccia and gouge (see Table 6.1). Cohesive brittle fault rocks (the "***cataclasite series***") include cataclasites, ultracataclasites, and pseudotachylite. These fault rocks form at greater depths than the breccia-series fault rocks, in fact as deep as ~15 km if factors such as temperature and strain rate are just right. At depths greater than ~10 to 12 km, crystal plastic deformation takes over, and the fault rocks produced are mylonites and ultramylonites (which are presented in *Chapter 10, Shear Zones and Progressive Deformation*). Figure 6.24 captures Rick Sibson's thinking, underscoring elastico-frictional (EF) processes at work at upper levels, and quasi-plastic (QP) processes at work at deeper levels.

Snoke and Tullis (1998b) do a nice job in summarizing Sibson's (1977) contributions on fault rocks and the environments within which they form. They underscore the effectiveness of Sibson's model (see Figure 6.24) with its two basic regimes of fault-rock production, an upper **frictional regime** where frictional strength prevails and rock strength increases linearly with depth. Frictional behavior progressively gives way to a **crystal plastic regime** across a depth transition where temperatures increase from ~250° to ~350°C. Mylonites and ultramylonites form through crystal-plastic processes in an environment where rock strength decreases with depth (see Figure 6.24).

Breccia Series Fault Rocks

Breccias

Breccias are incohesive within the breccia series (see Table 6.1 and Figure 6.24); they are fault rocks composed of angular fragments (clasts) of wall rock set in a finer-grained matrix of crushed wall-rock material (Figure 6.25). Matrix, comprising less than 30% of the breccia, is subordinate to clasts (Figure 6.26*A*). Ordinarily the clasts in breccias appear to have translated and rotated with respect to one another during the faulting and shearing.

There are size cutoffs for the classification of breccias (see Table 6.1). Ordinary **breccias** contain clasts that are larger than 1 mm and smaller than 0.5 m. **Megabreccias** contain clasts larger than 0.5 m. **Microbreccias** are composed of clasts that are smaller than 1 mm but greater than about 0.1 mm (Figure 6.26).

Figure 6.25 Brecciated granite. Note the slickensided fault surface on the underside of the ledge above the brecciated granite. [Photograph by S. J. Reynolds.]

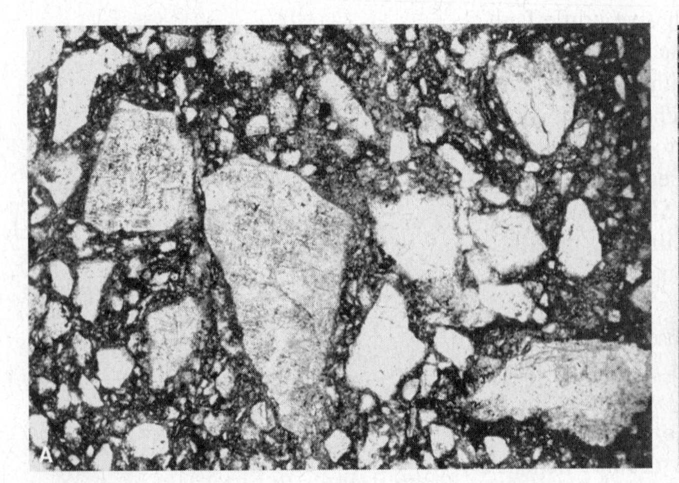

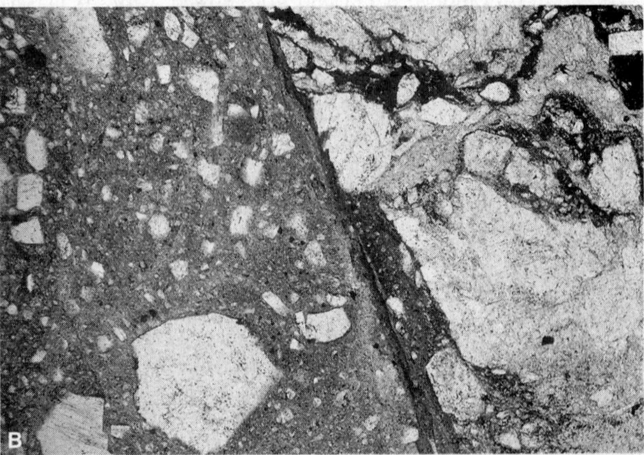

Figure 6.26 (*A*) Photomicrograph of microbreccia, a fault rock also produced by comminution of the Catalina Granite by the Pirate fault. Width of field of view is ~4 mm. Angular clasts are granitic. Dark matrix is mixture largely of silica and hematite. (*B*) Another photomicrograph of microbreccia derived from the same granite as above. Note the presence of a discrete slip surface. Width of field of view is ~4 mm. [Photomicrographs by G. H. Davis.]

Dilation and volume increase are characteristic of brecciation, which implies that the formation of breccias is favored in environments of low confining pressure and/or high fluid pressure. Voids representing pull-apart and dilation may not be completely filled by the finer-grained crushed matrix at the time of faulting. However, the void space may later be partially or completely filled by precipitation of minerals through circulation of groundwater and/or hydrothermal solutions. This can make the breccias interesting from a mining geology standpoint.

When pore fluid pressures are elevated during the faulting, brecciated material may be partly flushed into tensional openings, forming **breccia dikes** (Figure 6.27). Such dikes do not ordinarily show the telltale faulting signatures, such as slickensides or slickenlines, nor are they igneous intrusions. Instead their presence calls attention to proximity to fault-induced brecciation and conditions of high fluid pressures.

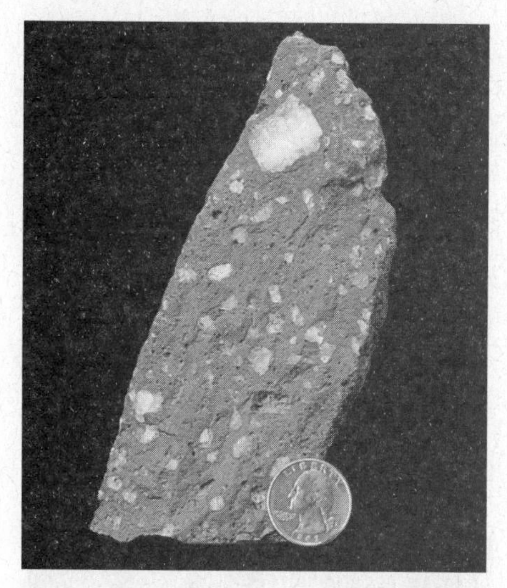

Figure 6.27 Photograph of specimen of breccia dike material, collected from breccia dikes invading Tertiary granite in Sierra de Hualfín, Argentina (see García and Davis, 2004). Longest dimension of this piece is 20 cm. The dark matrix is dominantly silica and hematite. The angular clasts are composed of quartz-feldspar granitic material, presumably crushed during faulting. Collection of G. H. Davis. [Photography and editing by Gary Mackender. vr.arizona.edu, glm@email.arizona.edu, © 2009–2010 Arizona Board of Regents.]

Gouge

Gouge is yet another fault-rock of the breccia series. Gouge is a light-colored *very* fine-grained clayey fault rock that is commonly found along fault surfaces and within fault zones (Figure 6.28). Grain size generally is less than 0.1 mm. In its dry state gouge feels like a loose to moderately compacted talcum powder, although admittedly a poor grade of talcum because it retains bits and pieces of grains that did not completely succumb to the crushing processes. At locations where wet gouge crops out as sticky clay, globs of the gouge will stick on your rock hammer as you try to dig into the stuff; we call it **sticky gouge**. Zones of gouge may be wispy and thin, or they may be meters wide. The weak, friable character and high clay content of gouge indicates that it forms under relatively low-temperature and low-pressure conditions.

Arrangements of Breccia-Series Rocks

The distribution of breccia-series fault rocks within a fault zone will vary considerably from one fault zone to another, but it is important to map the distributions, not just for structural geologic insight but also for insight regarding how to model the flow of fluids (groundwater, or hydrothermal solutions) along the plumbing system afforded by fault zones. This was the emphasis of Storti et al. (2003) in examining fault zones cutting carbonate rocks in the Apennines (Italy). They built a conceptual model that portrays the structural architecture of fault zones in their study area (Figure 6.29). Their conceptual model features a **fault zone core** within which most of the fault displacement is achieved. Within the fault zone core there are breccias of varying clast size (from fine to coarse) as well as gouge within shear bands (see Figure 6.29). **Boundary zones** separate the fault zone core from **damage zones** outside of the fault zone (see Figure 6.29). The damage zones lack fault rocks; instead the bedrock is intensely fractured. The physical reality of this conceptual model is, in part, captured in Figure 6.30, a photograph of a fault core between damage zones in a fault zone in limestone.

Figure 6.28 Fault gouge (white material beneath fracture surface) formed along low-dipping, nearly horizontal detachment fault in the Rincon Mountains, Arizona. The gouge was derived from crushing and grinding of Precambrian granite. Clipboard and compass in foreground. [Photograph by G. H. Davis.]

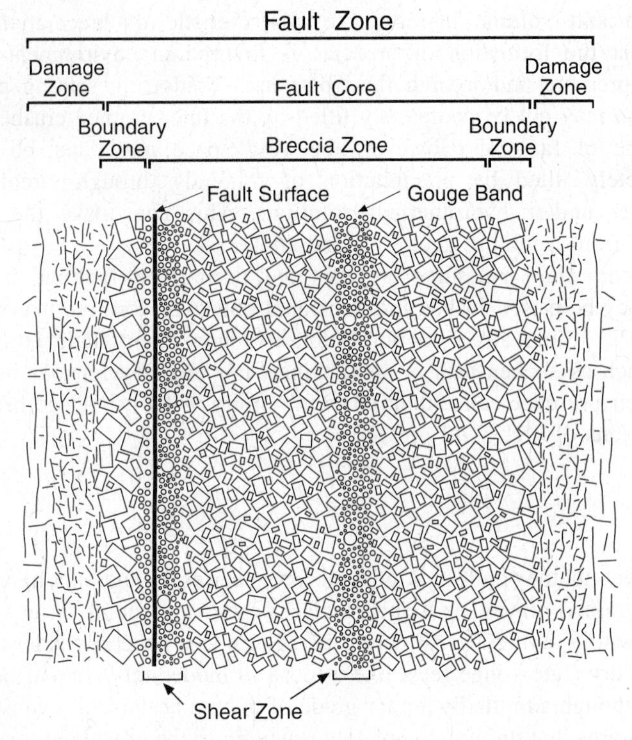

Figure 6.29 Conceptual model of the structural architecture of a fault zone, with emphasis on brittle fault-rock development. This is based on detailed examination of fault zones in carbonate rock in the Apennines (Italy), carried out by Storti et al. (2003). [Reprinted from *Earth and Planetary Science Letters*, v. 206, Storti, F., Billi, A., and Salvini, F., Particle size distributions in natural carbonate fault rocks: insights for non-self-similar cataclasis, p. 173–186, © 2003, with permission from Elsevier.]

Figure 6.30 Contact between damage zone and fault zone core. Fault zone in Pindos Group rocks, near Neda, Greece. [Photograph by G. H. Davis.]

Categories of Brecciation

Breccias form by processes involving **attrition**, **crushing**, and **implosion** (Sibson, 1986). The relationship of these classes of breccias to fault-zone architecture is shown in Figure 6.31. **Attrition brecciation** is a grinding of the host rock by progressive frictional wear along fault surfaces (see Figure 6.31). The chips and clasts of brecciated host rock show shapes of having been rolled. Individual clasts are marked by high internal deformation, notably microfracturing. Where impediments and jogs in fault zones cause the fault to lock up, the host rock on either side of the fault will jam into one another, producing distributed **crush brecciation** (Sibson, 1986) (see Figure 6.31). As in the case of attrition brecciation, clasts will be

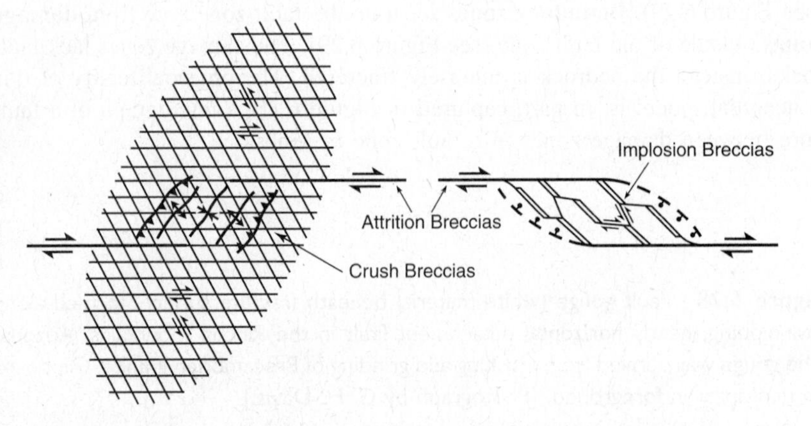

Figure 6.31 Drawing of a fault system (map view) marked by straight runs, jogs that restrain (jam), and jogs that release (open). Attrition brecciation is favored along the straight runs. Crush brecciation is favored along the restraining jogs. Implosion brecciation takes place along releasing jogs. [Reprinted with permission from *Pure and Applied Geophysics*, v. 124, Sibson, R. H., "Brecciation processes in fault zones: Inferences from earthquake rupturing," 1986, Birkhauser Verlag Ag, Basel, Switzerland.]

marked by pervasive microfracturing. In places along faults there may be jogs that, when fault slip occurs producing an earthquake, a volume of space will open for a split second. Because "*nature abhors a vacuum*," the walls of the opening will implode, causing **implosion brecciation** (Sibson, 1986) (Figure 6.31). Jigsaw and fitted fabric textures characterize such breccias that implode into the open space. There is no grinding or rolling of the fragments, and the fragments lack internal deformation. The pieces can be fit together, like a puzzle. The openings surrounding the implosion breccia fragments tend to be filled in by minerals deposited from hydrothermal solutions, for the sudden dilatancy sucks fluids into the volume, and the pressure drop causes minerals to precipitate (Sibson, 1986).

Cataclasite Series Fault Rocks

Cataclasite

Under conditions of the higher confining pressure and temperature that reside at depths of ~4 km to ~8 km (see Figure 6.24), faulting produces **cataclasite**. Cataclasites are strongly indurated; that is, hard and cohesive. They display a random fabric (absence of parallel, planar alignments of the broken fragments) (Figure 6.32). Cataclasites are commonly so indurated that they display pronounced resistance to erosion, as can be seen in outcrops in the Whipple Mountains of California (Figure 6.33).

Cataclasites typically range in clast size from 0.1 mm to 10 mm, but the finer-grained equivalent, **ultracataclasite**, possesses an average clast size less than 0.1 mm, which is equivalent to that of gouge (see Table 6.1). Grain size of cataclasite is like that of gouge, typically less than 0.1 mm, although it can contain up to 50% visible though still fine-grained clasts. Ultracataclasites are ultra-fine-grained and not uncommonly are nearly glassy in appearance (Figure 6.34*A*). Yet, when examined microscopically under very high magnification, ultracataclasite is seen to be marked by angular fragments and tiny fault surfaces (Figure 6.34*B*).

The indurated cohesive nature of cataclasite reflects comminution under somewhat elevated temperature and pressure conditions. The crushing also increases the total surface area of the minerals in the rock and makes it more susceptible to chemical reactions and cementation. The tiny particles of crushed materials, are cemented by hydrothermal or metamorphic minerals.

Figure 6.32 Photograph of cataclasite derived through pervasive crushing and grinding along a low-angle detachment fault. Width of field of view is approximately 11 cm. This sample was collected along the Catalina detachment fault in the Rincon Mountains (Arizona). Composition is quartzo-feldspathic, for the rock (protolith) from which it was derived is granitic. [Photograph by G. H. Davis.]

Figure 6.33 Ledge of cataclasite in the Whipple Mountains, southeastern California. The top surface of this ledge marks the position of the Whipple Mountains detachment fault. [Photograph by S. J. Reynolds.]

A

B

Figure 6.34 (A) Photograph of black ultracataclasite, from a ledge of ultracataclasite exposed on Tanque Verde Mountain east of Tucson, Arizona. The host rock from which this dark, extraordinarily fine-grained ultracataclasite was derived was granitic mylonitic granite, which seems almost unimaginable given its present character. Collection of G. H. Davis. [Photography and editing by Gary Mackender. vr.arizona.edu, glm@email.arizona.edu, © 2009–2010 Arizona Board of Regents.] (B) Photomicrograph of this same ultracataclasite, revealing angular clasts and tiny fault surfaces. Width of field of view is ~4 mm.

Fractal Character

Experiments and field studies reveal that cataclastic fault rocks tend to have a **self-similar character** (Sammis et al., 1986). Cataclastic rocks are fractal and thus look the same at a wide range of magnifications, from outcrop scale to microscopic. At each scale of observation cataclastic fault fabrics have about the same overall appearance, including the same proportion of clasts to matrix. Under progressively higher magnification, what appears at lower magnification to be the fine-grained matrix is found to be clasts in a yet finer-grained matrix. This pattern repeats itself down to extremely small, microscopic scales.

Pseudotachylite

Pseudotachylite is a fault rock all to its own (see Table 6.1). The term **pseudotachylite** is reserved for a rarely exposed but telling component of certain major fault zones, particularly those that have faulted the deep reaches of the Earth's crust. **Tachylite** is an old name for basaltic volcanic glass, but pseudotachylite is *not* a volcanic product. It is fault melt that freezes to glass. Pseudotachylite occurrences are in essence fossil records of deep earthquakes (Sibson, 1980). To form pseudotachylite there must be a powerful shock under very high-pressure conditions and an instantaneously imposed strain. The frictional heating of faulting causes the already hot rocks to melt. Thus pseudotachylite along fault zones is thought to represent a small volume of melt produced by frictional heating during the seismic event (Sibson and Toy, 2006). During the earthquake the fault melt invades dilational cracks, and there it freezes to glass (Figure 6.35*A*). Sibson (1977) allows that pseudotachylite can form at depths as shallow as 4 km, but mostly pseudotachylite forms at the depth-level of earthquakes, typically deeper than 4 km.

Figure 6.35 (A) Photograph of cobble Otaga Schist (medium gray) invaded by psedudotachylite (black). This "*fossil earthquake*" formed in the Alpine fault zone, South Island, New Zealand. The Otaga Schist was deep, buried, and hot when a sudden earthquake locally added frictional heat, causing melting. The melt then squirted into fractures, freezing to glass. Location is the Harold Creek in Westland, a locale recommended to me (GHD) by Rick Sibson. Collection of G. H. Davis. [Photography and editing by Gary Mackender. vr.arizona.edu, glm@email.arizona.edu, © 2009–10 Arizona Board of Regents.]. (B) Photomicrograph of fault glass, pseudotachylite. Glass (black) contains microbreccia fragments. The banded laminae are ultramylonite. [Reprinted from *Journal of Structural Geology*, v. 2, R. H. Sibson, Transient Discontinuities in Ductile Shear Zones, p. 165–171, © 1980, with permission from Elsevier.]

Pseudotachylite is a dark, very fine-grained, originally glassy rock, and can contain scattered microbreccia fragments derived from crushing and grinding of the host rock (Figure 6.35*B*). Pseudotachylite is sometimes seen microscopically to be completely glassy, but in most cases the original glass has partially to totally **devitrified** to a fine-grained crystalline matte. Pseudotachylite generally forms thin veins or selvages along discrete fault surfaces, which are the **generation surfaces**. Pseudotachylite may be restricted to the generation surface, although small injection veins typically branch out from the main generation surface into the wall rocks.

MAP AND SUBSURFACE EXPRESSIONS OF FAULTS

Geologic Map Expressions

Systematic geologic mapping has proved to be an extremely effective method for locating faults (see *Part III-B, Geologic Mapping*), particularly where the faults cut a geologic column of sedimentary or volcanic rocks whose stratigraphy is well known. Geologic maps reveal the plan-view expression(s) of fault patterns, and books have been written to explore the relationships (e.g., Bolton, 1989; Spencer, 1993). Faults are identified and tracked on the basis of mapped patterns that disclose **truncation** and **offset** of one or more bedrock units (Figure 6.36).

Truncation and offset can take many forms, depending on the orientations of the faults, the movements on them, and the orientations of the rock layers that are cut and displaced. Truncation and offset usually result in an apparent horizontal (i.e., map-view) shifting of mapped bedrock units.

Where faults trend parallel to the strike of bedding, translation does not produce the simple but conspicuous horizontal shifting of layers. Instead, the presence of such faults must be recognized on the basis of **inconsistent stratigraphic patterns**. The rock formation shown in Figure 6.37 is composed of 10 distinctive units (numbered in the figure) and is cut by two faults that strike parallel to bedding. The two faults dip such that they cut across the bedding at steep angles. The presence of the two faults might not be

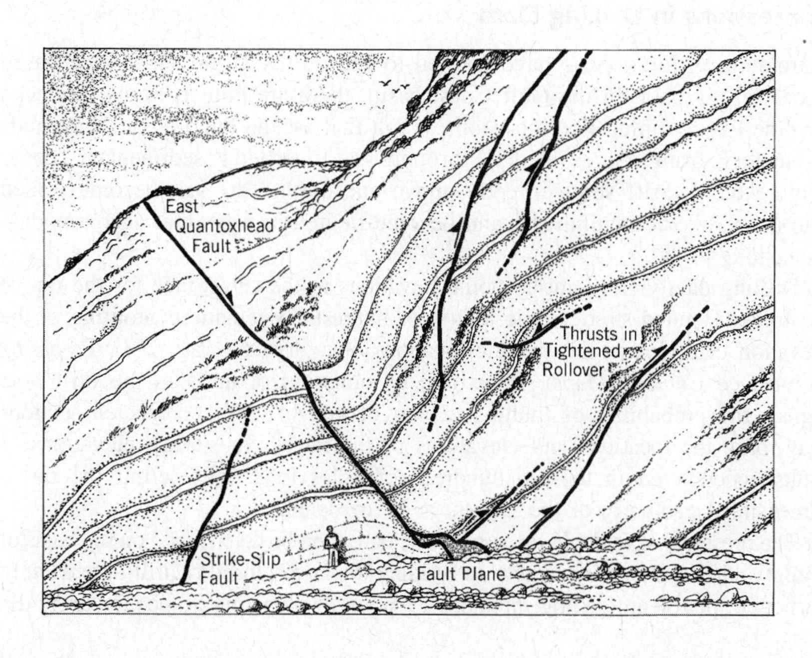

Figure 6.36 Faults can be discovered and mapped on the basis of truncation and offset of bedrock units. Here the East Quantoxhead fault and associated faults reveal themselves in the disturbed stratigraphy. [From Peacock and Sanderson, 1992, Figure 7A. Reprinted with permission from The Geological Society Publishing House, Bath, England.]

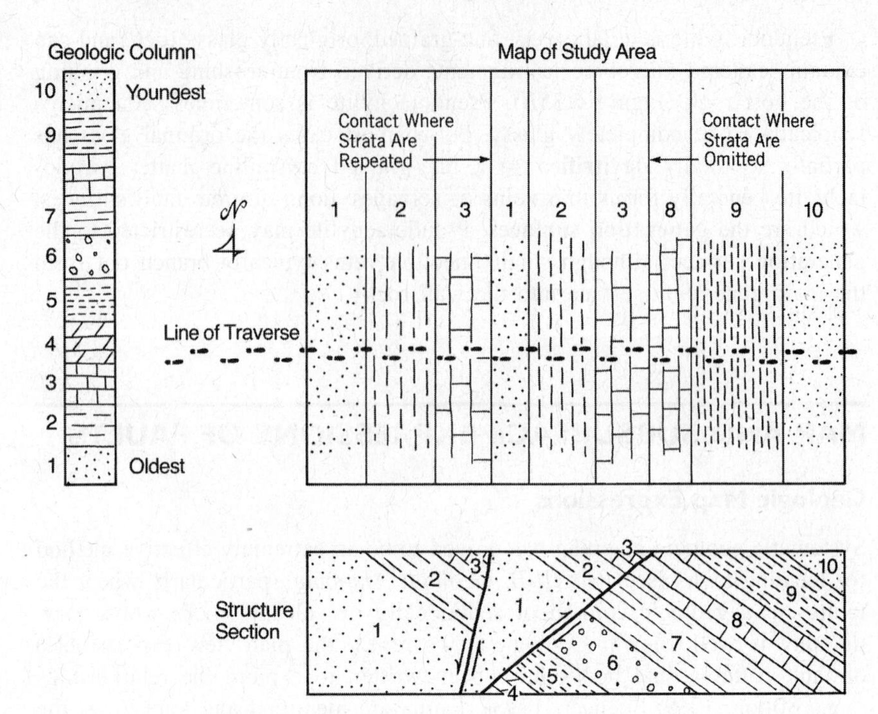

Figure 6.37 Identification of locations of faults on the basis of inconsistent stratigraphic patterns. One fault is marked by repetition of strata, the other by omission of strata.

immediately apparent in the geometry of the map pattern, but knowledge of the stratigraphy of the formation permits the faults to be recognized by **repetition** and **omission of strata**. If, during a traverse across the map area (see Figure 6.37), we move up-section from unit 1 to unit 3, only to find ourselves crossing back unexpectedly into a *repeated* section of the same units, we would recognize that we had crossed a fault. And if during our continued traverse we were to cross directly from unit 3 into unit 8, the omission of units 4 through 7 would alert us to the fact that we had crossed a second fault. Unconformities can also account for omission of strata, so be careful! Cross-sections through faulted bedrock reveal repetitions and omissions of strata (see *Part III-F, Preparing Geologic Cross-Sections*).

Expressions in Drilling Data

Petroleum explorationists have learned to spot faults in the subsurface based upon drilling data. In the Gulf Coast basin, there are fault patterns produced by pipe-like and plug-like intrusions of salt that ascend hundreds to thousands of meters from source bed(s) through the overlying sedimentary cover (Figure 6.38). Stiff strata directly above and around the salt become folded into dome-like configurations, and become faulted in ways that accommodate extension.

Drilling data yield point-by-point elevations in the subsurface for the top of the folded, faulted stratigraphic unit(s) of interest. Systematic contouring of the elevation data reveals the specific locations of faults (see *Part III-G, Preparing Subsurface Contour Maps*). Repetition or omission of units in the drill hole signals the probability of faulting. Zones of gouge between repeated sections may mark the locations and elevations of the faults. Subsurface elevations of faults encountered in tens to hundreds of holes in a field permit the three-dimensional geometry of the fault(s) to be modeled.

The presence of faults in the subsurface can often be determined on the basis of abrupt and/or systematic changes in bedding dip as disclosed in drill-hole data. In particular, petroleum geologists make it a point of preparing graphs featuring dip

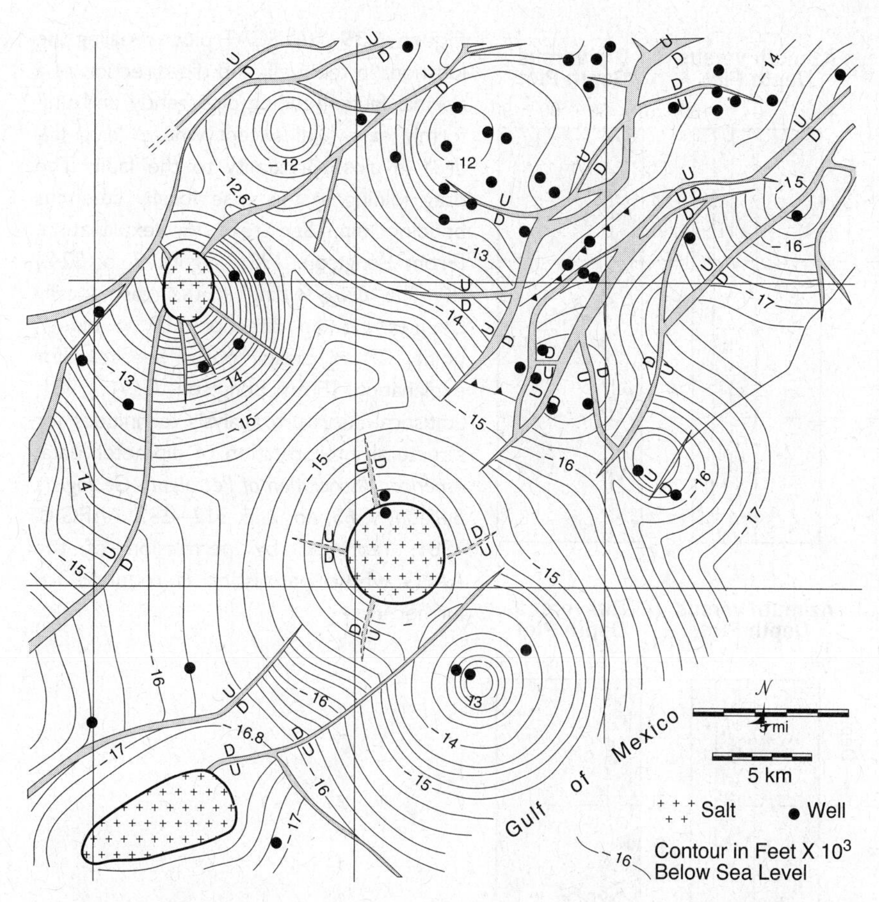

Figure 6.38 Wells provide some of the subsurface control in defining the pattern of doming and faulting of the Tertiary Frio Formation in the Gulf Coast region. Contours are the top of the so-called T5 horizon within the Frio Formation. The numbers (e.g., −15) refer to thousands of feet below sea level. The mounds, domes, and ridges are expressions of salt intrusion. The faults are shown as the light gray branching forms. Note how contour lines are offset across the faults. [From Capuano, R. M., 1993, Evidence of fluid flow in microfractures in geopressured shales: *American Association of Petroleum Geologists Bulletin*, v. 77, no. 8, p. 1303–1314. AAPG © 1993, reprinted by permission of the AAPG whose permission is required for further use.]

inclinations and directions (measured by dip meters during drilling) in the subsurface (Bengston, 1981). Graphs of such data are known as **SCAT plots.** (Bengston, 1981). SCAT does not refer to what bears leave on trails, but instead are "*statistical curvature analysis technique plots.*" They are used to interpret fault and fold geometry in the subsurface. Abrupt changes in dip inclination and/or direction may mark the presence of a discrete fault. Intervals of chaotic dip inclination and/or direction may represent fault zones and/or brecciated bedrock. Intervals of no discernable dip data whatsoever may be locations where originally bedded sedimentary rock has been reduced by faulting to fault gouge. Gradual and continuous changes of dip on both sides of a suspected fault zone may indicate the presence of folding, including drag folding, associated with the faulting.

We present here two examples of SCAT plots for recognition of faulting in the subsurface, based upon the dipmeter record of drag folding. Figure 6.39*A* shows an east-dipping normal fault in the subsurface, cutting through strata that dip evenly to the east at ~10°–20°. Note that the faulting produced drag folding, and that the drag folding has steepened the dip of beds (to ~45°) in close proximity to the fault. The SCAT plot to the right of the cross-section (see Figure 6.39*A*) was produced simply by plotting dip data (magnitude and direction of dip) as a function of depth, in this case as projected onto a cross-section perpendicular (transverse) to the strike of bedding and the fault. Because the drag folding associated with faulting had the effect of steeping bed dip on either side of the fault, the SCAT plot shows an east-pointing cusp, the tip of which points in the direction of fault dip and marks the depth where the well penetrates the fault (see Figure 6.39*A*). The two dashed horizontal lines crossing the SCAT plot demarcate the full interval of the drag folding, which can be thought of as a damage zone.

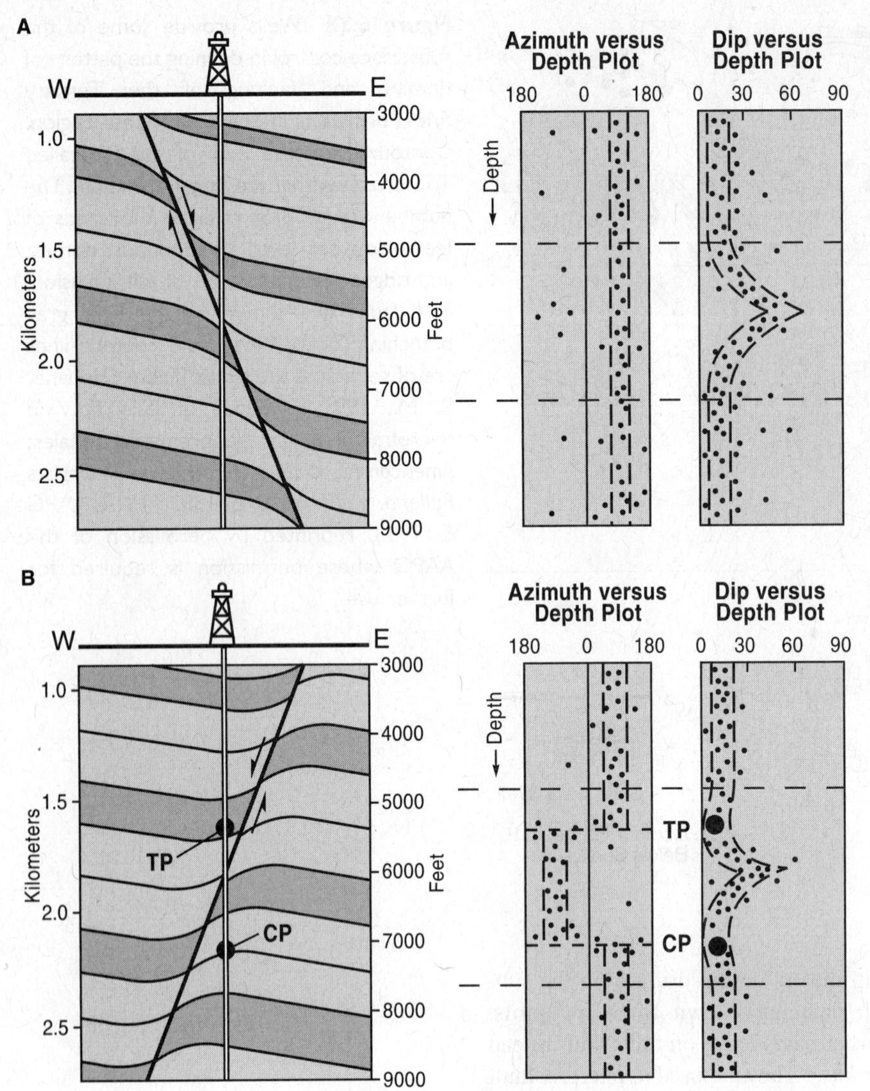

Figure 6.39 (*A*) SCAT plot revealing the location (in the well) and dip-direction of a normal fault. Bedding dips gently and uniformly eastward, except where "*drag folded*" in close proximity to the fault. The drag folding in this case locally steepens bedding dip. See text for explanation. [From Bengtson, 1981, Figure 9, p. 324.] (*B*) The drag folding in this case locally changes the dip direction. TP = trough point. CP = crestal point. See text for explanation. [From Bengtson, C. A., 1981, Statistical curvature analysis techniques for structural interpretation of dipmeter data: *American Association of Petroleum Geologists Bulletin*, v. 65, no. 2, p. 312–332. AAPG © 1981, reprinted by permission of the AAPG whose permission is required for further use.]

Figure 6.39*B* presents another situation where a normal fault cuts and displaces strata that dip evenly to the east at ~10°–20°. However, in this case the fault dips westward, and the effect of drag folding is to flatten the beds in close proximity to the fault. The SCAT plot constructed (see Figure 6.39*B*) is marked by a cusp positioned at the depth where the well penetrates the fault; the cusp points opposite fault dip. Moreover, the interval of drag folding shows up clearly in the SCAT plot. Within the interval of drag folding trough points (TP) and crestal points (CP) can be identified on folded beds. The SCAT plot approach can work wonders during "*ordinary*" field mapping, as demonstrated in *Part III-Q, Deciphering Structure in Boreholes.*

Seismic Expression

The "*bread-and-butter*" of geophysical methodology used by petroleum exploration companies is **seismic data**. In carrying out seismic-reflection profiling, sound waves are propagated into the subsurface by some means of producing a concussion, such as setting off charges of dynamite or air guns under water or by shaking the ground surface with large truck-mounted vibrators. The waves speed to depth, where they eventually are reflected by

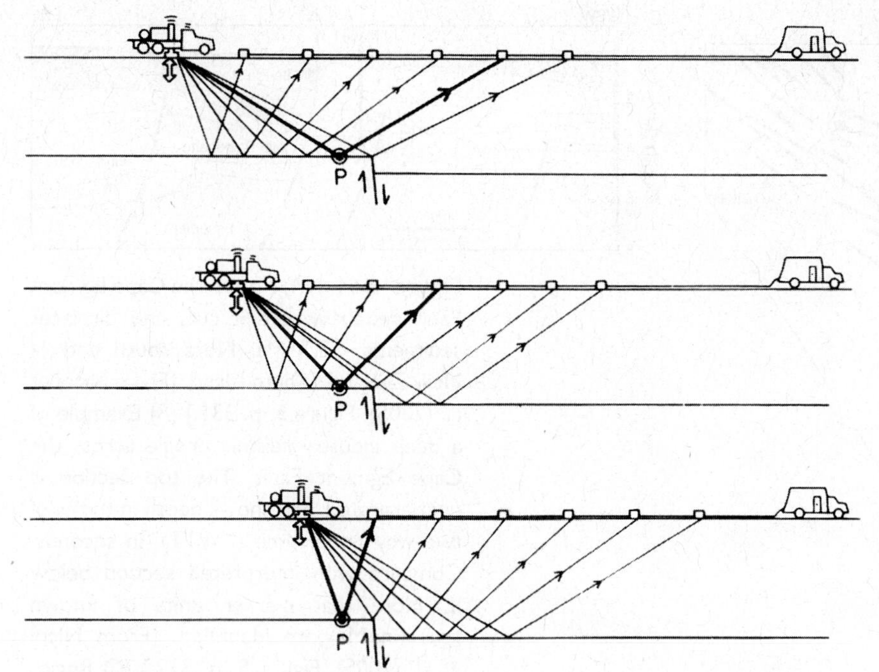

Figure 6.40 Schematic illustration of the process of seismic-reflection profiling. [From "The Southern Appalachians and the Growth of Continents" by F. A. Cook, L. D. Brown, and J. E. Oliver, copyright © 1980 by Scientific American, Inc. All rights reserved. Published with permission of George Kelvin.]

discontinuities marking sharp contrasts in density and rigidity, hence the velocity of sound between rocks above and below (Figure 6.40). The physics of this is analogous to what happens on a still night, namely when you can easily pick up conversations taking place on the far side of the lake. The difference in the sonic velocity between the water and air causes the sound waves to bounce off of the interface. Interfaces in the earth's subsurface include not only bedding or other lithologic boundaries within the geologic column, but also faults, which bring rocks of different physical properties into direct contact. Thus faults are among the discontinuities that are capable of causing sound waves to be reflected. The sound energy radiates back to the surface, where its character and time of arrival are collected and measured through vibration sensors known as geophones. The challenge becomes one of deciphering which incoming signals "*bounced*" off the same reflector, and to determine the depth of each reflector at every point along the seismic line.

There are countless examples of how reflection seismology can disclose the characteristics of fault systems in the subsurface, in ways analogous to CAT-Scanning the earth. For example, not far off the southwest "*corner*" of the South Island of New Zealand, there is a tectonically active basin that is marked by earthquakes and active "*stretching*" accommodated by the Cape Egmont fault (system) (Nicol et al., 2005). Exploring the attributes of this offshore, subsurface fault system required seismic reflection surveying along a relatively closely spaced (2 km) gridwork centered on the trace of the fault system, ~70 km in length (Figure 6.41*A*). Wells and logging (e.g., Maui 1 and 2, see Figure 6.41*A*) established the underlying stratigraphy. The traces of fault segments within the Cape Egmont fault system were mapped on the basis of interpretation of the seismic reflection profiles (see Figure 6.41*A*). The quality and clarity of individual seismic lines can be appreciated by comparing, in Figure 6.41*B*, the uninterpreted versus interpreted sections. Integrating seismics and drilling, Thrasher et al. (1995) and Nicol et al. (2005) were able to construct meaningful geological cross-sections (Figure 6.41*C*). They concluded that the Cape Egmont fault dips ~65° E and has a total maximum offset of ~2.3 km (which required hundreds of large-magnitude earthquakes). Moreover, they concluded that the fault experienced

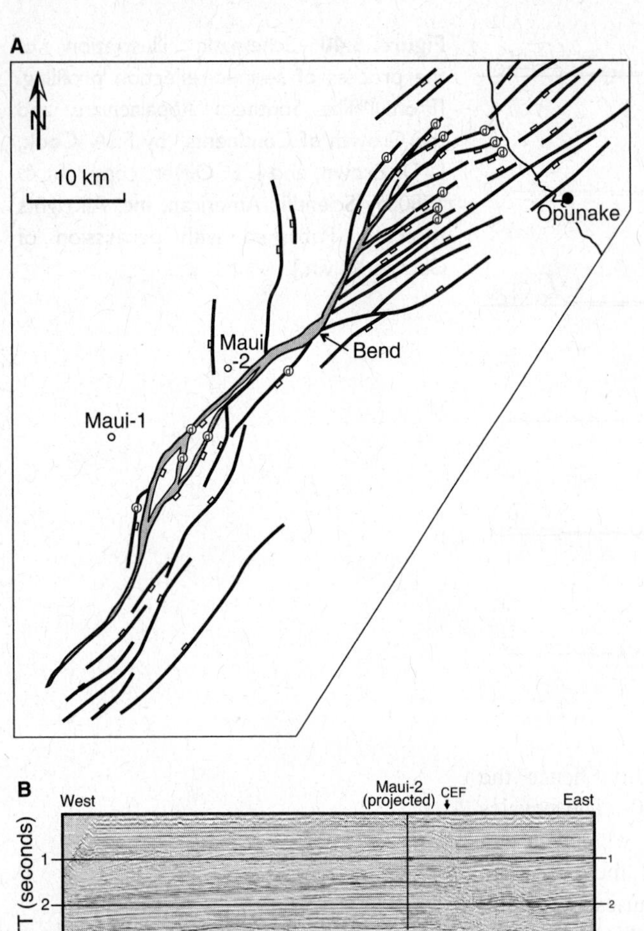

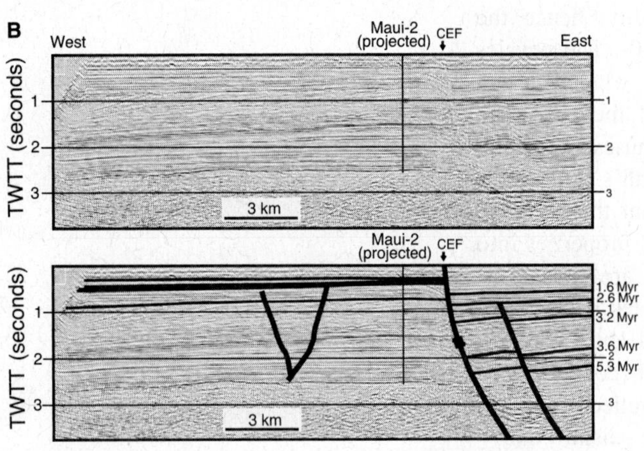

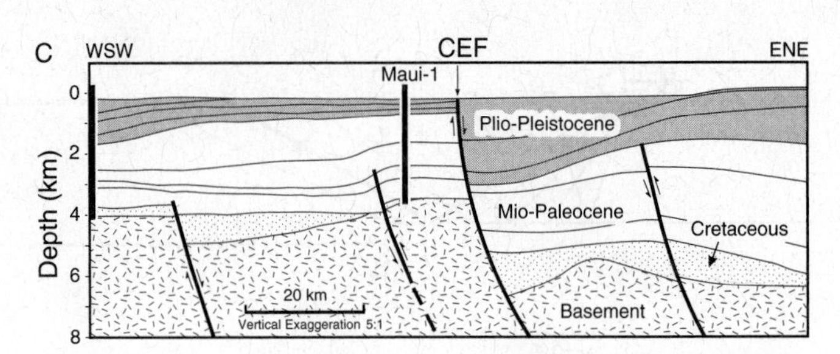

Figure 6.41 (*A*) Map of the Cape Egmont Fault system where it cuts and displaces sediments ~3.7 Ma. Note south end of New Zealand's South Island. [From Nicol et al. (2005), Figure 5, p. 331.] (*B*) Example of a deep industry seismic profile across the Cape Egmont Fault. The top section is uninterpreted, and shows depth in terms of two-way travel time (TWTT), in seconds. Compare with interpreted section below it. Note that marker units of known ages (in Ma) are identified. [From Nicol et al. (2005), Figure 5, p. 332.] (*C*) Regional cross-section across the Cape Egmont Fault; depth shown in kilometers. [Reprinted from *Journal of Structural Geology*, v. 27, Nicol, A., Walsh, J., Berryman, K., and Nodder, S., Growth of a normal fault by the accumulation of slip over millions of years, p. 327–342, © 2005, with permission from Elsevier.]

down-to-the-east displacement during Late Cretaceous and Pliocene-Pleistocene, and down-to-the-west displacement in late Miocene (Nicol et al., 2005).

THE NAMING AND CLASSIFICATION OF FAULTS

Slip and Separation

Most faults that we see in the field, or in the subsurface, are named and described on the basis of the dip of the fault surface and the direction and sense of offset. The ideal goal is to establish the magnitude of displacement, direction of displacement, and sense of displacement. These three components constitute the **slip**, which is the *actual* relative displacement between the hanging wall and the footwall. Where the basis for a slip determination is unavailable, we settle for **separation**, which is the *apparent* relative displacement. Distinguishing slip from separation is one of the most important steps in fault analysis (Crowell, 1959; Hill, 1959).

Slip Classification

If we can establish the three components of slip for a given fault, we can use special terms to "*name*" the fault. **Strike-slip faults**, which may be left-handed or right-handed (Figure 6.42*A*), are marked by horizontal movement. Movement is essentially parallel to the strike of the fault. Ideal exposures of strike-slip faults are marked by horizontal or very low-raking slickenlines.

Dip-slip faults (Figure 6.42*B*) are marked by movement directly up or down the dip of the fault surface. Movement on a dip-slip fault is described with reference to the relative movement of **hanging wall** and **footwall** (see cartoon in Figure 6.42*B*). "*Hanging wall*" and "*footwall*" terminology only applies to inclined faults; that is, non-vertical or non-horizontal. The hanging wall is the fault block toward which the fault dips. The footwall is the fault block on the underside of the fault. The terms "*hanging wall*" and "*footwall*" are derived from old mining jargon: a prospector working in a drift that he has tunneled along an inclined fault

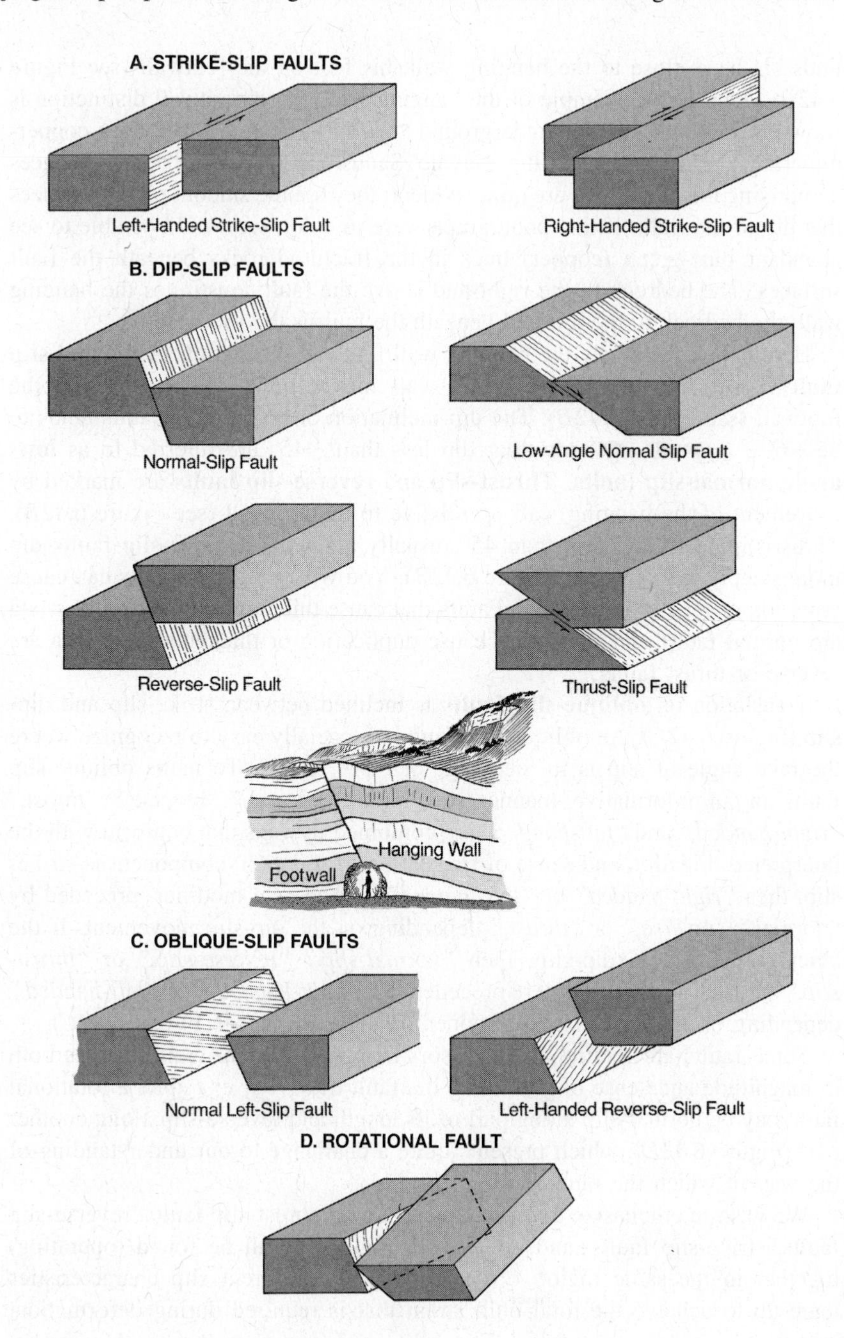

Figure 6.42 Slip classification of faulting. (*A*) Block diagrams showing left-handed and right-handed strike-slip faulting. (*B*) Block diagrams showing dip-slip faults, including normal-slip, reverse-slip, low-angle normal-slip, and thrust-slip faulting. Also shown is convention for designating "*hanging wall*" versus "*footwall*." (*C*) Examples of oblique-slip faults, including a normal left-slip fault and a left-handed reverse-slip fault. (*D*) Schematic block diagram of a rotational fault.

Figure 6.43 Northwest-directed photograph in the Hattie Green mine. This tunnel was "*driven*" along a mineralized fault zone. The smooth rock surface in the right foreground is a fault surface that dips to the right (i.e., northeastward). The straight-line trace of a second, parallel fault is quite evident right above the tunnel. Hanging wall rocks lie above the fault surface. Footwall rocks lie beneath the fault surface. Sarah Tindall showed me this mine, located in southern Utah along the East Kaibab monocline. The mine lies within the Fivemile Valley 7.5-minute topographic quadrangle. [Photograph by G. H. Davis.]

finds his head close to the hanging wall, his feet on the footwall (see Figure 6.42*B*). A beautiful example of the hanging wall versus footwall distinction is shown in Figure 6.43. The underground "*drift*" (i.e., tunnel) follows a copper-mineralized fault zone cutting Navajo Sandstone (Jurassic). Fault surfaces comprising this fault zone are quite evident; they are the smooth planar surfaces that dip to the right. If this photograph were in color, we would be able to see abundant blue-green (copper) hues in the fractured rocks beneath the fault surfaces. The bedrock to the right and above the fault constitutes the hanging wall; the bedrock to the left and beneath the fault is the footwall.

Here is how we apply the hanging-wall/footwall distinction: A **normal-slip fault** is one in which the hanging wall moves down with respect to the footwall (see Figure 6.42*B*). The dip-inclination of normal-slip faults tends to be ~60°. Normal-slip faults that dip less than ~45° are referred to as **low-angle normal-slip faults**. **Thrust-slip** and **reverse-slip faults** are marked by movement of the hanging wall *up* relative to the footwall (see Figure 6.42*B*). Thrust-slip faults dip less than 45°, usually at ~30°. Reverse-slip faults dip more steeply than 45° (see Figure 6.42B). You will recall that faults can cause omission or repetition of strata. Faults that cause thinning or omission of strata are normal faults and those that cause duplication or thickening of strata are reverse or thrust faults.

Translation on **oblique-slip faults** is inclined between strike-slip and dip-slip (Figure 6.42*C*). An oblique-slip fault is especially easy to recognize where the rake angle of slip is in the range of ~30° to 60°. To name oblique-slip faults in an informative manner, the terms "*normal*," "*reverse*," "*thrust*," "*right-handed*," and "*left-handed*" are combined in ways that conform with the interpreted direction and sense of translation. If the main component is strike-slip, then "*right-handed*" or "*left-handed*" is used as a modifier, preceded by "*normal*," "*reverse*," or "*thrust*," depending on the dip-slip movement. If the chief component is dip-slip, then "*normal-slip*," "*reverse-slip*," or "*thrust-slip*," is used as the modifier, preceded by "*right-handed*" or "*left-handed*," depending on the strike-slip component of movement (see Figure 6.42*C*).

Some faults are rotational or scissors-like. A rotational fault changes both in magnitude and sense of slip along the fault trace. For example, a rotational fault may be normal-slip along part of its length and reverse-slip along another part (Figure 6.42*D*), which presents quite a challenge to our understanding of the way in which the fault formed.

We wish to emphasize that normal-slip faults, thrust-slip faults, reverse-slip faults, strike-slip faults, and oblique-slip faults can all be found (operating) together in the same region of rock. Faults of different slip characteristics team up to achieve the total bulk strain that is required during deformation.

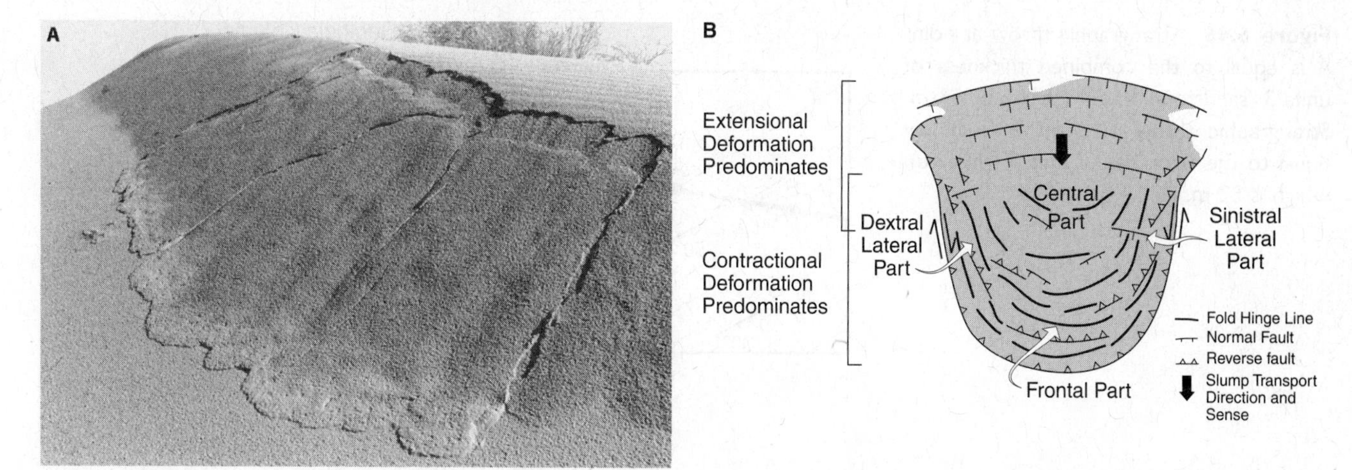

Figure 6.44 (*A*) Landslide on the face of sand dune in Sonora, Mexico. Note the breakaway normal faulting at the top, strike-slip faulting along the margins, and thrust faulting at the toe. [Photograph and copyright © by Peter Kresan.] (*B*) Map of the structural geology of a slump. Normal faults comprise the breakaway zone along the top, thrust faults along the toe, and strike-slip faults along the left and right margins. [Reprinted from *Journal of Structural Geology*, v. 31, Debacker, T. N., Dumon, M., and Matthys, A., Interpreting fold and fault geometries from within the lateral to oblique parts of slumps: A case study from the Anglo-Brabant Deformation Belt, p. 1525–1539, © 2009, with permission from Elsevier.]

Thus we should not suffer under the impression that normal faults will be exclusively in one place, and thrust faults in another. We can illustrate this concept in a rather unique way, using an extraordinary set of relationships photographed by Peter Kresan (Figure 6.44*A*). The photo shows a small landslide on the face of a sand dune in Sonara, Mexico. At the top of the "*sandslide*" is a scarp representing the break-away. At the base is the toe of the slide. Along the side is a lateral shear. We can think of these three elements as normal-slip faults, thrust-slip faults, and strike-slip faults, respectively. Figure 6.44*B* makes this clear. It is a diagrammatic map of faults and folds in a slump sheet (Debacker et al., 2009). Their sketch map is based upon their studies of slump sheets in Ordovician strata in Belgium, as well as detailed studies carried out by Farrell and Eaton (1987) in the Pyrenees and by Bradley and Hanson (1998) in Maine. The map reveals that slumping was accommodated by strike-slip tearing on either side of the slump sheet, normal-slip faulting in the breakaway zone, thrust-slip faulting at the toe of the slump, and folding in the down-slope body of the slump sheet (Debacker et al., 2009). The several different parts of this system provide reference markers for determining the overall displacement vector for the slump mass.

Slip is fully determined when we know for individual faults, and for the whole faulted body, the **magnitude of displacement**, the **direction of displacement**, and the **sense of displacement**.

The Separation Problem

Interpreting the direction and sense of slip on faults is often complicated by deceptive patterns created by the interaction of structure and topography and by the absence of minor structures that, if present, would have helped to define the slip path. **Separation** is the offset observed along faults in 2D views of outcrops, maps, or geologic cross-sections. Separation refers only to the *apparent* sense and magnitude of offset along faults; it is the *apparent* relative movement. The faulted basalt layer shown in Figure 6.45*A* (a map view) displays 200 m of right-lateral *separation*. Yet in cross-sectional view (Figure 6.45*B*) we see that the fault is a normal fault, marked by ~6 m of

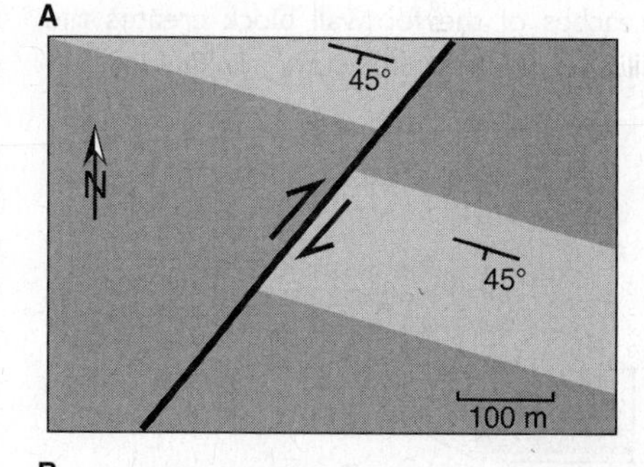

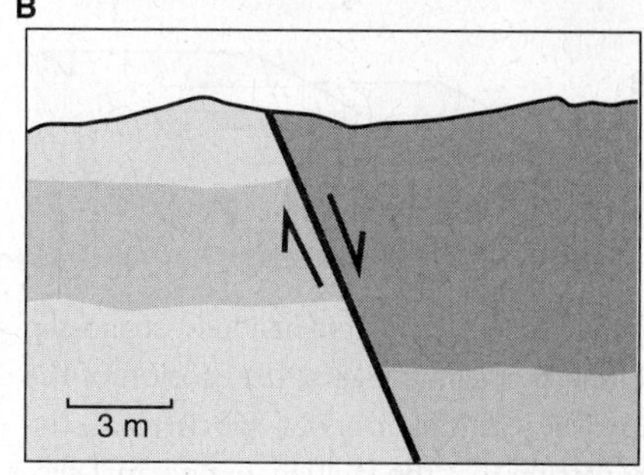

Figure 6.45 (*A*) Map view of right-lateral separation of a basalt layer. (*B*) Cross-sectional view of normal separation of the basalt layer.

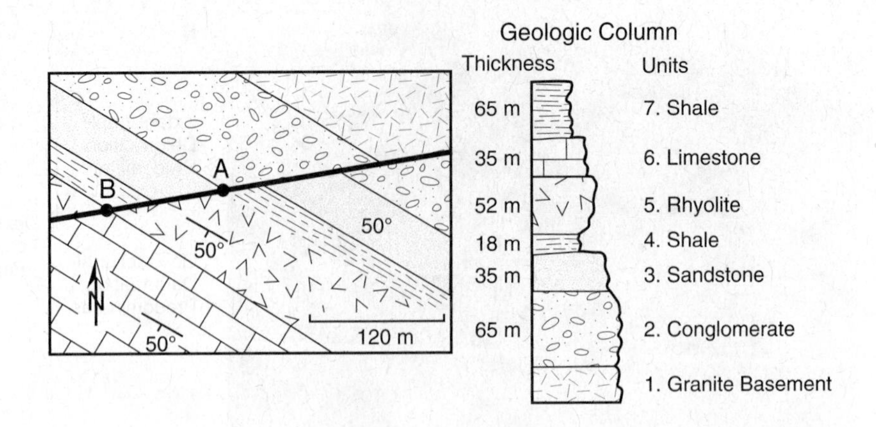

Figure 6.46 Stratigraphic throw at point *A* is equal to the combined thickness of units 3 (sandstone) and 4 (shale), or 53 m. Stratigraphic throw at point *B* is simply equal to the thickness of unit 5 (rhyolite), which is 52 m.

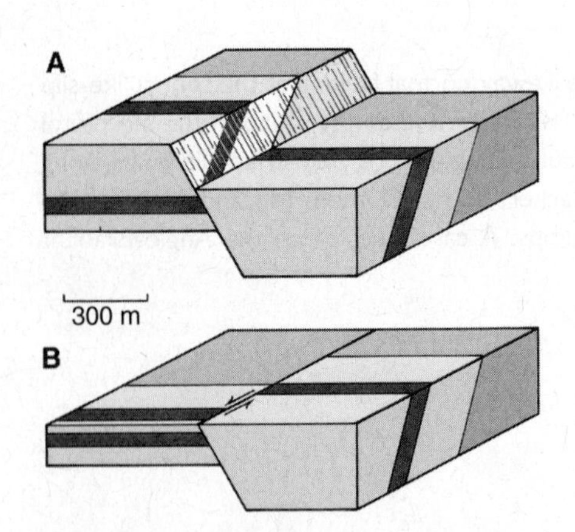

Figure 6.47 Block diagrams that underscore the difference between slip and separation. (*A*) Inclined layers are displaced along a normal-slip fault. (*B*) Erosion of the upper reaches of the footwall block creates the illusion of left-handed strike-slip faulting.

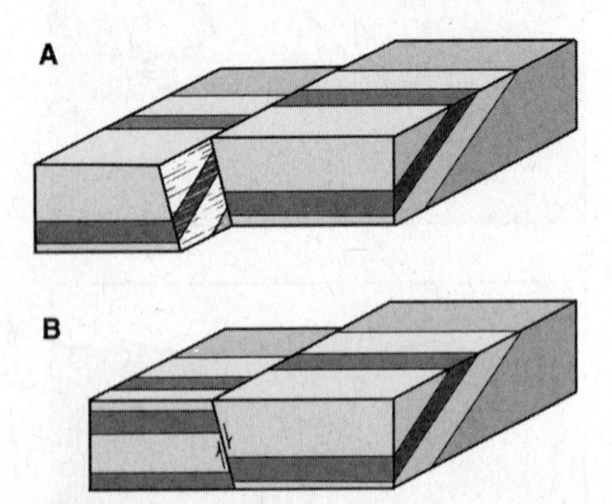

Figure 6.48 (*A*) Left-handed strike-slip faulting of inclined layers. (*B*) Erosion of the front end of the footwall block gives the impression that the faulting was normal-slip.

normal *separation.* We cannot conclude based on this cross-section and this cross-section alone that the faulting here was normal-slip faulting.

Stratigraphic throw is a special measurement of separation, one that is commonly used in exploration geology as a convenient measure of the magnitude of faulting. Stratigraphic throw is the thickness of the stratigraphic interval *between* two beds that are brought into contact by faulting (Figure 6.46). At point *A* in Figure 6.46, faulting has brought the base of unit 3 into contact with the top of unit 4. Stratigraphic throw equals the combined thickness of units 3 and 4, namely 53 m. At point *B,* the base of unit 5 is faulted into contact with the base of unit 6. In this case stratigraphic throw equals the thickness of unit 5, namely 52 m (see Figure 6.46).

Apparent relative movement seldom corresponds with *actual* movement between the fault blocks. The separation that we view on geologic maps and in cliff faces is a product of many influences: orientation of layering; strike and dip of the fault surface; slip on the fault, including direction, sense, and magnitude of displacement; and the orientation of the exposure in which separation is viewed. Consider some simple examples. The tilted sedimentary rocks shown in Figure 6.47*A* are cut by a normal-slip fault whose magnitude of net slip is 182 m. The fault strikes at right angles to the strike of bedding. Following erosional beveling of the faulted terrain to a common level (Figure 6.47*B*), the outcrop relationships convey the false impression that left-handed strike-slip faulting had taken place. Conversely, left-handed strike-slip faulting of a tilted sequence of rocks (Figure 6.48*A*) can produce structural relationships that in cross-sectional view appear to be the result of normal-slip faulting (Figure 6.48*B*). The possibilities are, unfortunately, limitless! So it is important to keep the difference between separation and slip in mind as you interpret faults.

Separation Classification

Separations viewed in cross-sectional exposures are described simply as "*normal*," "*thrust*," and "*reverse*" (Figure 6.49*A*). Normal separation is marked by apparent offset of the hanging wall downward relative to the footwall. If fault dip is less than 45°, this separation would be described as resulting from low-angle normal faulting. Thrust and reverse faults, in the separation sense, are characterized by offset of the hanging-wall rocks upward relative to the footwall. Thrust faults dip less than 45°, and reverse faults dip more steeply than 45°. Separations viewed in plan view are described as "*left lateral*" or "*right lateral*" (Figure 6.49*B*). The suffix "*-handed*" is reserved for proclaiming slip.

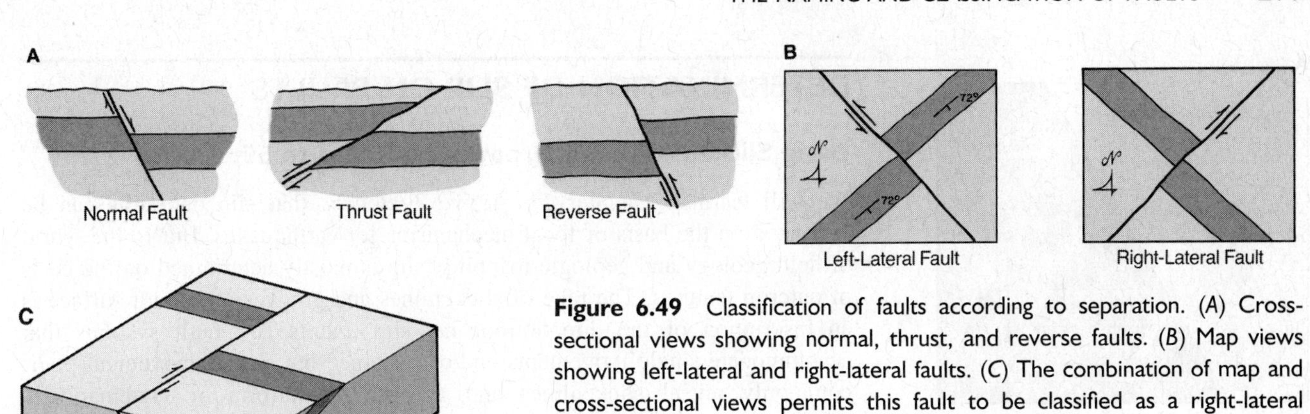

Figure 6.49 Classification of faults according to separation. (*A*) Cross-sectional views showing normal, thrust, and reverse faults. (*B*) Map views showing left-lateral and right-lateral faults. (*C*) The combination of map and cross-sectional views permits this fault to be classified as a right-lateral reverse fault.

Complete descriptions of separation are based both on plan-view and cross-sectional observations of offset (Figure 6.49*C*). This fault would be described as a reverse right-lateral fault. The map view (see Figure 6.49*C*) reveals right-lateral faulting. The front-face cross-section shows reverse faulting. Because the right-lateral separation is greater than the reverse separation, the name given to the fault is reverse right lateral.

To sum up: If the main fault separation is lateral, the name given to the fault is "*right-lateral*" or "*left-lateral*", preceded by the modifier "*normal*", "*reverse*", or "*thrust*", depending on the separation seen in cross-section. If the main separation is normal, reverse, or thrust, then one of these terms is used as name of the fault, preceded by the modifier "*right-lateral*" or "*left-lateral*", depending on the separation observed in plan. Figure 6.50 presents our separation classification and suggestions regarding map symbols for distinguishing separation versus slip.

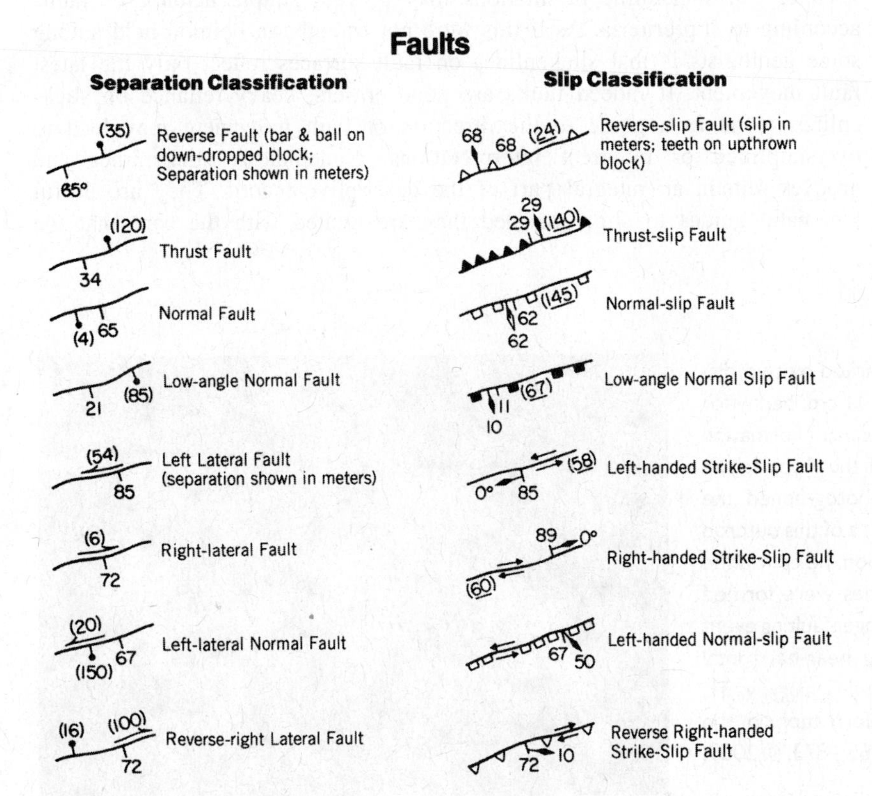

Figure 6.50 Our suggested map symbols for distinguishing between separation and slip.

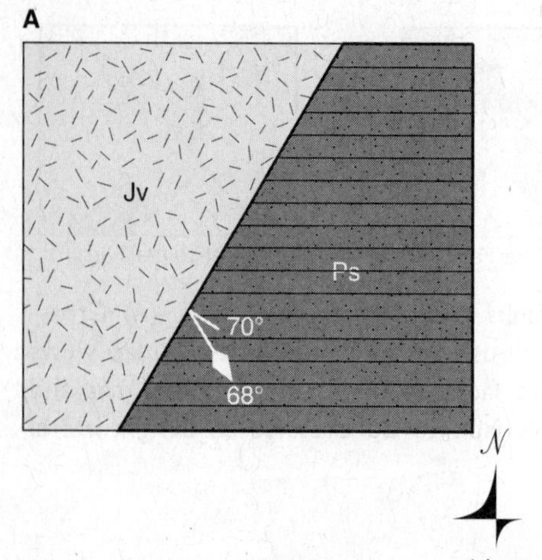

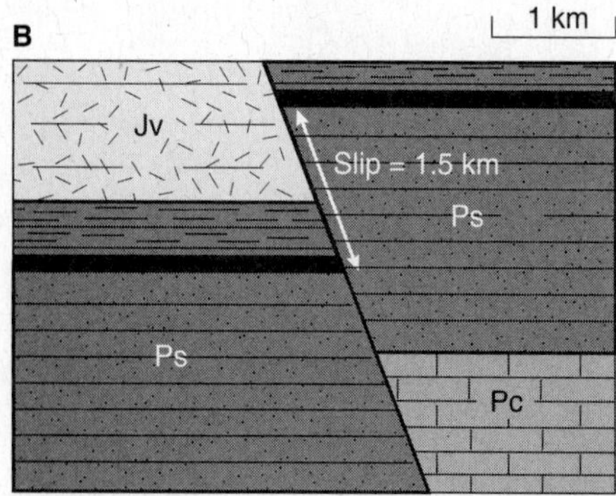

Figure 6.51 (*A*) Map of fault contact between Jurassic and Permian strata. Note that through symbols a distinction is made between the dip of the fault and the trend and plunge of slickenlines on the fault surface. The fault is a left-handed reverse-slip fault. (*B*) The actual slip along the fault can be measured along a vertical cross-section drawn parallel to the trend of the slickenlines, and then matching up a faulted layer.

DETERMINATION OF SLIP ON FAULTS

Using Slickensides and Grooves as Guides to Slip

We will learn in *Chapter 11, Active Tectonics*, that slip on faults can be evaluated on the basis of focal mechanisms for earthquakes. But in the world of field geology and geologic mapping, slip is mostly determined on the basis of outcrop features. The rake of slickenlines and grooves on a fault surface is an inscription of the direction of net slip. Faults and fault systems that accommodate single movements and/or sustained but simple movement plans commonly reveal slickenlines and grooves of uniform or systematically changing rake.

For example, consider the mapped fault relationship shown in Figure 6.51*A*. The fault strikes N30°E and dips 70°SE. It separates volcanic rocks of Jurassic age on the footwall from Permian sedimentary rocks on the hanging wall. Permian strata were faulted upward with respect to the Jurassic volcanic rocks, but along what line? Slickenlines on the fault surface reveal the direction of slip. They plunge 68°S 30°E, which corresponds to a rake of slickenlines on the fault surface of 80°SW. This overall geometry and separation indicates oblique-slip faulting, featuring a major dip-slip component of translation. Given the orientation of the slickenlines and the age relationships of the rocks on hanging wall and footwall, the fault is classified as a "*left-handed reverse-slip*" fault. The actual magnitude of slip is determined by constructing a vertical cross-section parallel to the trend of the slip direction, and measuring the distance between the **footwall cutoff** and the **hanging-wall cutoff** of a specific geologic contact (Figure 6.51*B*).

There may be difficulties in using striations and other slickenlines as a guide to slip. Many faults undergo multiple, complex histories of movement. Although not particularly common, some slickensided surfaces can be marked by sets of slickenlines of two (or more) different orientations (Figure 6.52). "*Scatter*" in slickenline orientations may prevent simple naming of faults according to slip criteria. As if this were not enough, an opinion held among some geologists is that slickenlines on fault surfaces reflect only the latest fault movement. If indeed faults are good erasers, heavy reliance on slickenlines alone as a guide to the direction of fault movement may lead to oversimplified or incorrect interpretations. Nonetheless, slickenlines and grooves remain an integral part of the descriptive record. They are useful kinematic guides to slip, provided they are treated with the care that the

Figure 6.52 Photograph of fault surface marked by two distinct sets of slickenlines. Width of the photo is 11 cm; penny for scale. The quartzite bedrock is part of the Tuscarora Formation (Silurian) in the Valley and Ridge Province of the Appalachian Mountains. Nickelsen (2009) spotted and photographed the "*double-slickenlining*" of this surface. On the basis of this outcrop as well as analysis of the fault history of the region, he concluded that the steeply raking near-vertical slickenlines were formed earliest. These were later overprinted by a younger faulting event reflected, in this outcrop, by the low-raking near-horizontal slickenlines. [Reprinted from *Journal of Structural Geology*, v. 31, Nickelsen, R. P., Overprinted strike-slip deformation in the southern Valley and Ridge in Pennsylvania, p. 865–873, © 2009, with permission from Elsevier.]

foregoing words of caution suggest. Serious slickenline aficionados go after large statistical samples to work with. For example, Eric Erslev and his students (Erslev and Larson, 2006; Erslev and Koenig, 2009) measured more than 10,000 slip indicators on Late Cretaceous and Paleogene faults in the Rocky Mountain region of the western United States. On the basis of these data and fault surfaces they interpreted regional paleo-stress directions.

Using Drag Folds as Guides to Slip

Drag folds can be used to determine the direction and sense of slip during faulting. Drag folding is a distortion of bedding (or other rock layering) resulting from shearing of rock bodies past one another. Consider how fault movements can distort a layered sequence of sedimentary rocks (Figure 6.53*A*). Strata close to the fault surface are deformed by frictional drag into folds that are convex in the direction of relative slip. The truncated ends of dragged layers point away from the sense of actual relative movement. Hanging-wall strata of a thrust-slip or reverse-slip fault may be dragged into an anticline, whereas the footwall strata may be dragged into a syncline (see Figure 6.53*A*). Similarly, under ideal circumstances, hanging-wall strata of a normal-slip fault may be dragged into a syncline, whereas the footwall strata may be dragged into an anticline (Figure 6.53*B*). We are already familiar with recognition of drag fold patterns through the examples we presented on SCAT analysis of dipmeter data for strata cut and displaced by normal faulting (see Figure 6.39).

Drag folds resulting from strike-slip faulting can be spectacular in map view, especially where steeply dipping layers are radically folded. Right-handed and left-handed patterns of drag are distinctly different (Figure 6.53*C*). Within broad fault zones marked by spaced strike-slip faults, individual layers can become completely fault bounded. And if each end of a fault-bounded layer is curled by drag, **sigmoidal drag folds** result (Figure 6.53*D*). The layers are doubly curved. Sigmoidal drag folds formed by right-handed simple shear are gently *S*-shaped. Those formed by left-handed simple shear resemble backward *S*'s, the kind that kids paint on signs. Folds shaped like backward *S*'s are referred to as *Z*-shaped. A wonderful example of sigmoidal drag folding is shown in Figure 6.54, which shows the detailed structure geology of part of the Stockton Pass fault, southern Arizona, as mapped by Monte Swan (1976). The "*dragged*" features are foliation and lithologic layering.

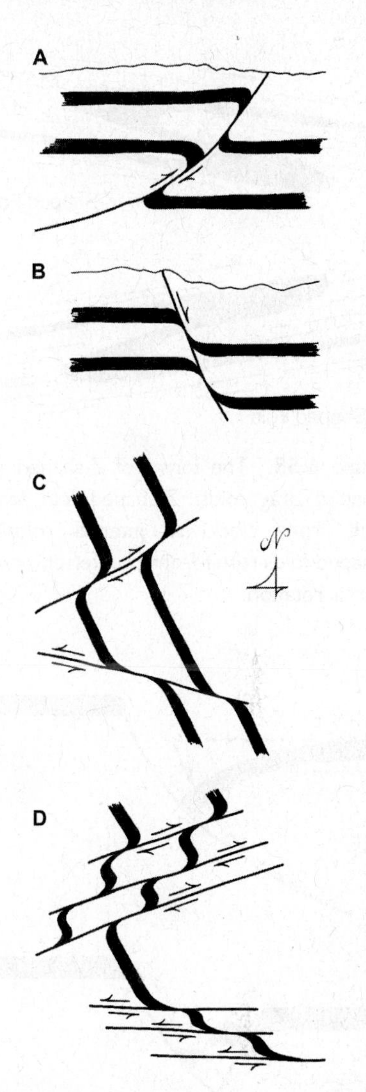

Figure 6.53 Examples of drag folds. (*A*) Strata on the hanging wall of thrust-slip and reverse-slip faults are dragged into an anticline, whereas strata on the footwall are dragged into a syncline. (*B*) Drag fold relationships along a normal-slip fault. (*C*) Patterns of drag folding along right-handed and left-handed strike-slip faults. (*D*) Sigmoidal drag folding along closed-spaced, distributed right-handed and left-handed strike-slip faults.

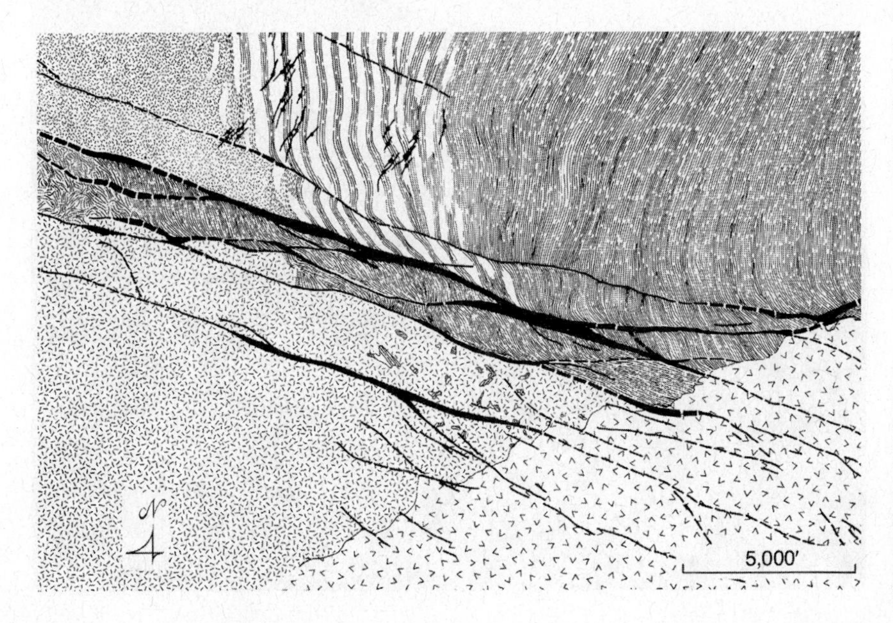

Figure 6.54 Drag folding, including sigmoidal drag folding, of foliation in Precambrian gneiss, Stockton Pass, Arizona. Drag folding was caused by left-handed strike-slip faulting along the Stockton Pass fault zone. [From Swan, 1976. Published with permission from the University of Arizona Press.]

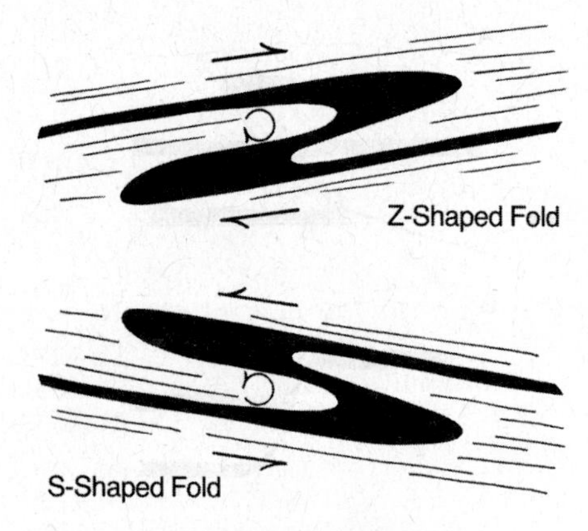

Figure 6.55 The forms of *Z*-shaped and *S*-shaped drag folds: *Z*-shaped fold forms result from clockwise internal rotation; *S*-shaped folds result from counter-clockwise internal rotation.

Note that where dip-slip faults (i.e., normal-slip, reverse-slip, thrust-slip) cut and drag horizontal strata, the axes of drag folds will be oriented approximately horizontal, perpendicular to the direction of slip (see Figure 6.53*A*,*B*). Where pure strike-slip faulting operates along vertical surfaces cutting steeply dipping layers, the axes of resulting drag folds will be steep, roughly perpendicular to the direction of movement (see Figure 6.53*C*,*D*). The orientations of slickenlines and drag folds tend to be mutually perpendicular in this circumstance.

Tight asymmetrical drag folds commonly occur in the interior of fault zones; these are very useful in evaluating the sense of simple shear movement during faulting. The asymmetrical folds, when viewed down plunge, can be described as *S*-shaped or *Z*-shaped (Figure 6.55). The clockwise rotation that typifies asymmetric, *Z*-shaped drag folds reflects right-handed shear. The counter-clockwise rotation that characterizes asymmetric *S*-shaped drag folds reflects left-handed shear. Learning to interpret asymmetric drag folds is an important aid in naming faults according to slip.

Bold generalizations are dangerous when it comes to drag folding, because rocks do not always cooperate in the ideal ways just described. The most notorious exception to the rule is **reverse drag** on normal-slip faults. Originally horizontal bedding in the hanging wall of a normal-slip fault can become folded in such a way that it actually dips toward the fault surface, thus forming a **rollover anticline** (Figure 6.56*A*). Hamblin (1965) first recognized this phenomenon in faulted horizontal strata on the Colorado Plateau in Utah and Arizona. He reasoned that reverse drag may be unique to listric fault geometries. **Listric faults** are curved faults that flatten, or decrease in dip, with depth (see Figure 6.56). Sustained movement on a listric normal-slip fault favors a pulling away of the hanging-wall block from the footwall. Such a gap never actually develops because of such folding (see Figure 6.56*A*), or **antithetic faulting** into the zone of potential separation (Figure 6.56*B*). A nice example of reverse drag, in the form of a rollover anticline, is shown in Figure 6.57.

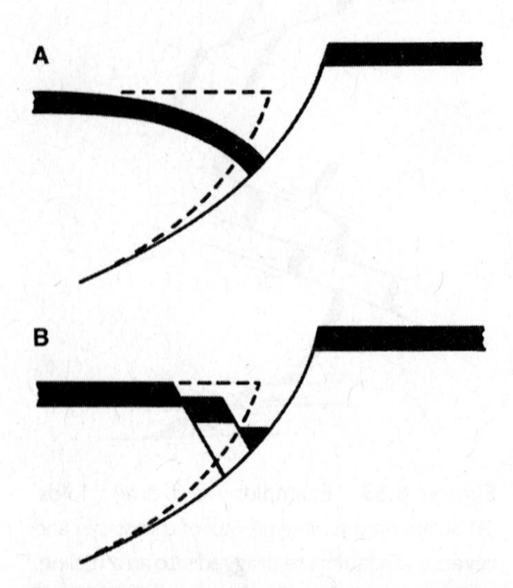

Figure 6.56 Formation of reverse drag through (*A*) down-bending of strata into the zone of potential separation and/or (*B*) antithetic faulting. [From Hamblin, 1965. Published with permission of Geological Society of America.]

Figure 6.57 Photograph of Cambrian dolostones in the Muddy Mountains (east of Las Vegas, Nevada) cut and offset by a large, left-dipping normal fault. Note that the hanging wall strata (above the fault) are curved into the form of a rollover anticline, presumably because of a shallowing of dip of the normal fault at depth. [Photograph by G. H. Davis.]

Using Gash Fractures and Tight Drag Folds as Guides to Slip

Direction and sense of slip due to faulting can be evaluated on the basis of two particular minor structures, gash fractures and tight folds. Think of fault zones as brittle zones of *noncoaxial* shear that distort rocks. During shear, at each instant of time, distortion stretches the rock at a 45° angle to the fault zone, and shortens it at right angles to the direction of **maximum instantaneous stretch** ($\dot{S}_1$). If you could sit and watch the shearing for a time, you would see that the directions of instantaneous stretching and instantaneous shortening do not shift relative to the fault zone. You might also see that **en echelon tension fractures**, which are mode I features created in noncoaxial shear, form at right angles to the direction of maximum instantaneous stretching ($\dot{S}_1$) (Figure 6.58*A*). These en echelon tension fractures will rotate during progressive shear, causing them to open, inviting infiltration of mineral-bearing fluids that produce **en echelon veins**. The dilation of tension fractures may be so great relative to fracture length that the fractures are best referred to as **gash fractures**, or—if filled—**gash veins** (see Figure 1.28).

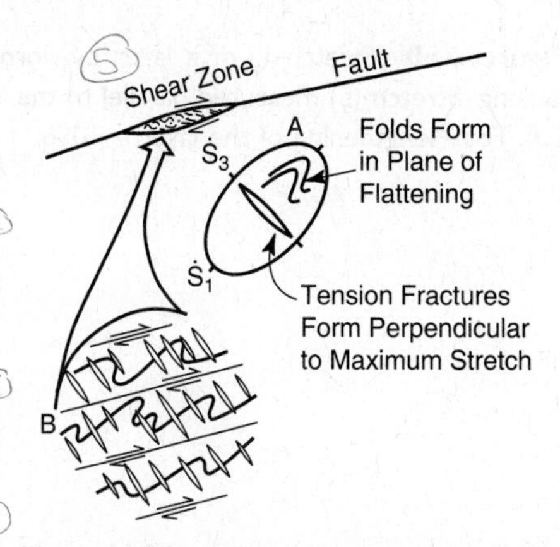

Figure 6.58 Simple shear origin of tension fractures and tight folds within a shear zone. (*A*) Instantaneous strain. (*B*) Finite strain. See text for details.

In addition, progressive noncoaxial shear associated with faulting can produce **asymmetric folds**, which lean in a direction consistent with the sense of shear (see Figure 6.58*A*). The folds initially orient themselves at right angles to the direction of **minimum instantaneous stretch** ($\dot{S}_3$).

Interestingly, as shearing continues, the earlier formed gash fractures rotate in a way that is sympathetic with the sense of shear within the fault zone (Figure 6.58*B*), while brand new gash fractures form perpendicular, as always, to the direction of maximum instantaneous stretch ($\dot{S}_1$). Meanwhile, the asymmetric folds rotate and become tighter and tighter, maintaining an orientation perpendicular to the direction of minimum *finite* stretch ($\dot{S}_3$). Given the geometry of this progressive deformation, the combination of gash fractures and tight folds virtually establishes the sense of slip that took place during faulting.

Analysis of minor structures, such as gash fractures, in fault zones has important practical application in mining geology. Suppose a rich ore vein is abruptly intercepted by a fault and translated safely out of sight. How can its offset position be found using gash fractures as a guide? Consider the steeply dipping N15°E-striking fault zone shown in map view in Figure 6.59*A*. Within the zone are vertical gash fractures filled with quartz. They strike approximately N70°E. Horizontal slickenlines on the fault surfaces indicate that the fault zone accommodated strike-slip movement. However it is not known whether translation was right-handed or left-handed. To solve for sense of slip, assume that the gash fractures formed at right angles to the direction of maximum instantaneous stretching ($\dot{S}_1$) of the rock during faulting (Figure 6.59*B*). The attitude of the gash fractures in this example is compatible with right-handed strike-slip faulting.

Part III-K (*Determining Slip on Faults through Orthographic and Stereographic Projection*) provides some additional material regarding determining the slip on faults.

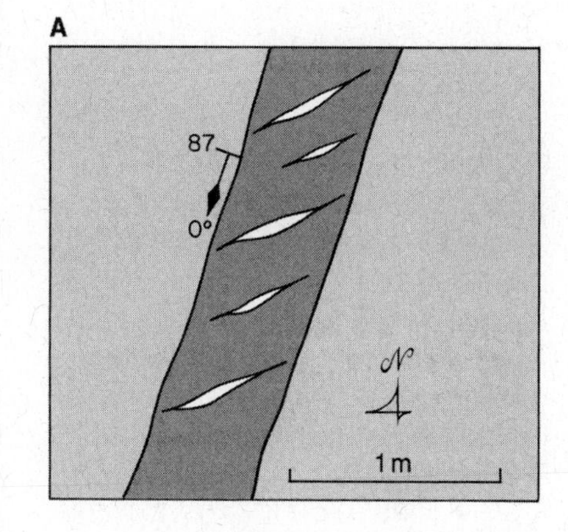

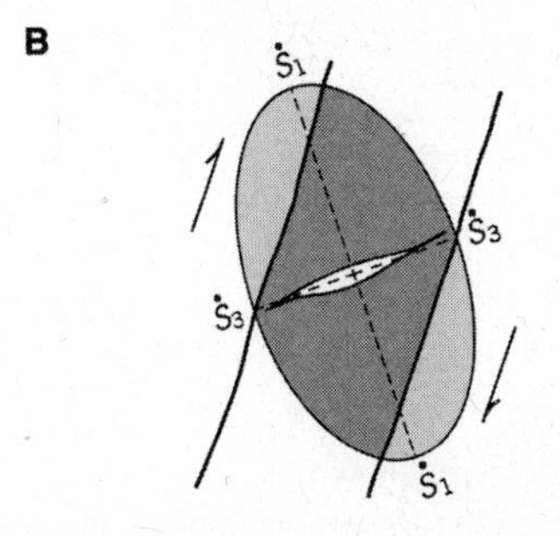

Figure 6.59 Use of gash fractures to determine direction and sense of translation on a fault. (*A*) Map showing steeply dipping fault zone and vertical gash fractures. (*B*) Geometric relationship between the orientation of gash fractures and the direction of greatest stretching allows the fault to be recognized as a right-handed strike-slip fault.

STRAIN SIGNIFICANCE OF FAULTS

The strain-related outcome of faulting is to shorten the crust in one direction and to stretch it in another. The directions of shortening and stretching are mutually perpendicular. In examining how this is accomplished, we generally assume plane strain. In other words, we assume that volume is conserved during faulting and that there is neither stretching nor shortening in the direction of intermediate finite stretch (S_2).

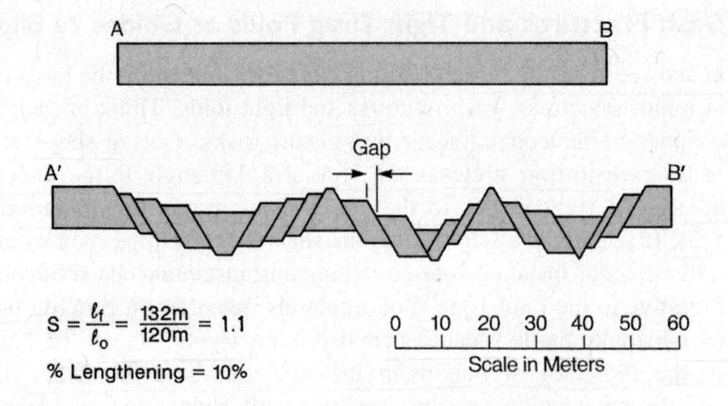

Figure 6.60 Stretching of a layer by normal-slip faulting. Stretch (S) measured parallel to the layer is 1.1. Total lengthening of the layer is 10%.

Strain Significance of Normal-Slip Faults

Normal-slip faults accommodate *horizontal* stretching accompanied by thinning (which is a kind of shortening or flattening). Consider the flat-lying sedimentary layer (*A-B*) 120 m long and 10 m thick shown in Figure 6.60. When this layer is cut and displaced by twenty 60°-dipping normal-slip faults, the end points of the faulted layer (*A′-B′*) are shifted to positions 132 m apart (see Figure 6.60). The layer is said to be stretched, even though we all know that it was not stretched like a rubber band or a rubber sheet. The 12 m of stretching is the sum of the **gaps** between offset layers (see Figure 6.60). Stretch (*S*) measured parallel to the layer is easily computed:

$$S = \frac{l_f}{l_0} = \frac{132 \text{ m}}{120 \text{ m}} = 1.1$$

$$\% \text{ stretching} = (S - 1) \times 100\% = (1.1 - 1.0) \times 100\% = 10\%$$

If the layer had been uniformly and penetratively stretched to its 132-m final length, its thickness would have been reduced from 10 m to 9 m. Final average thickness, $t_f = 9$ m, can be calculated by assuming that the cross-sectional area (A) of the layer remained the same before and after deformation.

$$A_o = A_f$$

$$(l_o)(t_o) = (l_f)(t_f)$$

$$(120 \text{ m})(10 \text{ m}) = (132 \text{ m})(t_f)$$

$$t_f = \frac{1200 \text{ m}^2}{132 \text{ m}} \approx 9 \text{ m}$$

There is beauty in moving from schematics to reality when it comes to normal faulting. For example, Figure 6.61 captures superbly the final condition of stretching and thinning of a rock layer through normal faulting. This is an outcrop example of distributed normal faulting. Janos Urai took this photograph at a quarry near Kinidaros on the Greek island of Naxos (Urai et al., 2008, p. 1201). The faulted layer is pegmatite, encased in marble. The faulted surface of the block (just 1.5 m across) is a miniature example of the block-faulted mountains that contribute to the rugged overall topography of Greece. As is evident, the normal faulting was accompanied by block rotation as extension was accommodated.

As improbable as it may sound, it appears that normal faults "*communicate*" with neighboring faults so that the contributions of slip and displacement by each fault result in an overall uniform strain. As an example, Nicol et al.

Figure 6.61 Photograph of an outcrop-scale system of normal faulting as exposed in a quarry near Kinidaros, Naxos, Greece. The faulted layer is pegmatite, which is encased in marble. This is a beautiful example of stretching achieved by normal faulting. It is a miniature example of what can happen to regions of rocks. Note the slickenlines on the fault faces! [Reprinted from *Journal of Structural Geology*, v. 30, Urai, J. L., Schenk, O., van der Zee, W., and Blumenthal, M., Photograph of the month, p. 1201, © 2008, with permission from Elsevier. Photography courtesy of Janos Urai.]

(2010) examined the history of normal faulting and earthquake activity for an active fault system within the Taupo Rift, New Zealand. Some 30 paleoseismology trenches expose faulted beds ranging in age from 0 to 26 ka. Furthermore, they studied the history of surface rupturing of the general landscape, using marker horizons dated in the age range of ~18 to 340 ka. Nicol et al. (2010) were able to show that from fault to fault, and from earthquake to earthquake, there are complementary changes in displacements, displacement rates, and earthquake histories among the players. The faults interact with one another, with each fault doing what it has to do in order to accommodate uniform stretching of the region of rock in which the faults are found. Figure 6.62 captures the heart of the matter. It portrays a fault system composed of a number of faults, displacements along which have created uniform extension for the system as a whole. Filled circles in the "*Earthquake Timing*" panel on the left-hand side of this block diagram identify the timing of earthquakes on individual faults. In achieving uniform extensional strain by normal faulting, "*recurrence intervals of earthquakes on individual faults are variable and complementary*" (Nicol et al., 2010, p. 1102).

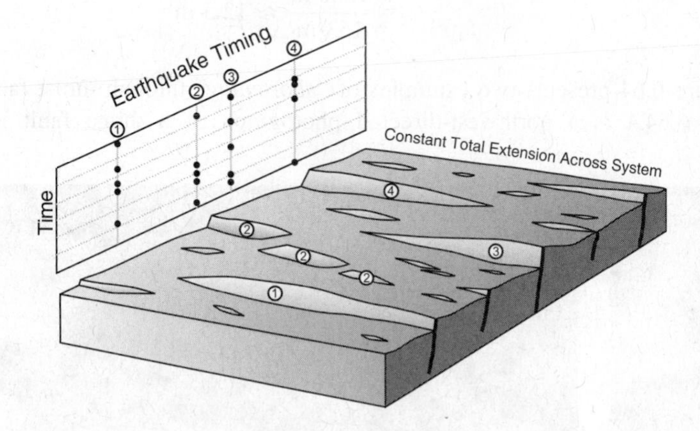

Figure 6.62 Block diagram showing normal fault system. There is constant total extension across the system, achieved through variable though complementary displacements along the individual faults. Filled circles in the graph to the left of the block diagram indicate the timing of earthquakes on faults 1–4, which are the highest displacement structures in the system. [Reprinted from *Journal of Structural Geology*, v. 32, Nicol, A., Walsh, J. J., Villamor, P., Seebeck, H., and Berryman, K. R., Normal fault interactions, paleoearthquakes and growth in an active rift, p. 1101–1113, © 2010, with permission from Elsevier.]

Strain Significance of Thrust- and Reverse-Slip Faults

Thrust-slip and reverse-slip faults perform the opposite function of normal-slip faults. Thrust-slip and reverse-slip faults are called upon to contract (shorten) a layer of rock horizontally and thicken it vertically. Figure 6.63 shows how this is achieved. A layer 188 m long and 10 m thick is cut and displaced by a combination of thrust-slip and reverse-slip faults. Translation on each of the

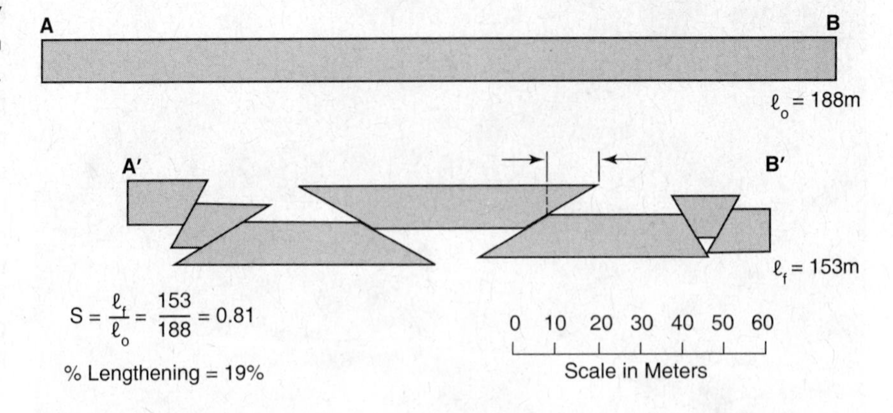

Figure 6.63 Shortening of a layer by thrust-slip and reverse-slip faulting. Stretch (S) measured parallel to the layer is 0.81. Shortening of the layer is 19%.

faults creates an **overlap** between offset layers that effectively shortens the initial layer. The sum of the overlaps in this example is 35 m. Final length (l_f) of the layer is 153 m (see Figure 6.63). Stretch (*S*) measured parallel to the layer can be calculated as follows:

$$S = \frac{l_f}{l_o} = \frac{153 \text{ m}}{188 \text{ m}} = 0.81$$

$$\% \text{ shortening} = (0.81 - 1.00) \times 100\% = 19\%$$

If the layer had been uniformly and penetratively shortened to its 153-m final length, its thickness would have been increased from 10 m to ~12 m. This result can be calculated by equating the cross-sectional areas of the layer, before and after deformation.

$$A_o = A_f$$

$$(l_o)(t_o) = (l_f)(t_f)$$

$$(188 \text{ m})(10 \text{ m}) = (153 \text{ m})(t_f)$$

$$t_f = \frac{1880 \text{ m}}{153 \text{ m}} \approx 12.3 \text{ m}$$

Figure 6.64 presents two examples of "*shortening*" through thrust faulting. Figure 6.64*A* is a northwest-directed photograph of a thrust fault in the

Figure 6.64 (*A*) Photograph of thrust faulting of well exposed, cliff-forming Bolsa Quartzite (Cambrian) in the Huachuca Mountains, southeastern Arizona. (Photograph by G. H. Davis.) (*B*) Thrust duplication of a single iron-rich limestone layer within the Pennsylvanian Stellarton Formation. Location is Stellarton, Nova Scotia. [Photograph by John Waldron, and published with his permission.]

Huachuca Mountains in southeastern Arizona. The well exposed cliff-forming resistant sandstones belong to the Bolsa Quartzite (Cambrian). Note that the thrust faulting causes the Bolsa Quartzite to override itself, thus achieving contraction and thickening. In Figure 6.64*B* we see thrust faulting of a single layer (siderite, i.e., iron-rich limestone) within the Stellarton Formation (Pennsylvanian) in an exposure near Stellarton, Nova Scotia. Note how the siderite layer has been shortened through thrust faulting. The thrust fault itself is essentially invisible, "*lost*" in the weak shaley beds that "*sandwich*" the siderite. Quite literally we see a doubling of bed thickness in parts of this outcrop.

Strain Significance of Strike-slip Faults

If we view the cross-sections shown in Figs. 6.60 and 6.63 as if they are plan-view maps of fault patterns, we see the effects of two-dimensional strain produced by strike-slip faulting. Strike-slip faulting results in stretching and shortening horizontally. If bedding is horizontal before pure strike-slip faulting, the thickness of bedding will not be altered by the faulting. Although the directions of extension and shortening are mutually perpendicular, the absolute orientations of these directions depend on the relative degree of development of right-handed versus left-handed strike-slip faults. If translations along right-handed and left-handed strike-slip faults are balanced, strain is coaxial. On the other hand, if one sense of strike-slip faulting predominates, the strain is non-coaxial. Figure 6.65 is block diagram summarizing the nature of strike-slip faulting cutting Navajo Sandstone (Jurassic) in the Sheets Gulch area, Capitol Reef National Park, Utah. Principal fault zones are shaded. One is left-handed, the other right-handed, and each zone is made of discrete segments that overlap in transfer zones. The overall "*purpose*" of this strike-slip faulting was to contract the Navajo Sandstone horizontally E-W, and to stretch it horizontally N-S (see Figure 6.65). The front face of the block diagram reveals that the strike-slip faulting was not pure; rather there was a small component of vertical displacement.

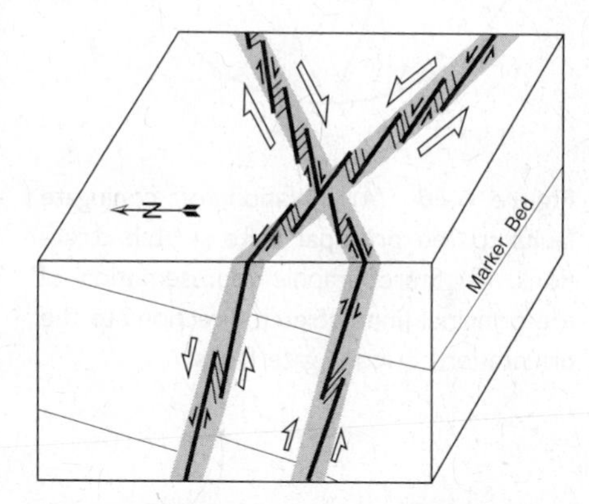

Figure 6.65 Block diagram summarizing the orientation, geometry, and sense-of-slip along strike-slip faults mapped in the Sheets Gulch area of Capitol Reef National Park. The host rock is Navajo Sandstone (Jurassic). Right-handed and left-handed strike-slip fault sets are each well developed. The faulting permitted the Navajo Sandstone at this location to shorten horizontally E–W and to stretch horizontally N–S. [From Davis, 1999. Published with permission of the Geological Society of America.]

Strain Function Held in Common by All Classes of Faults

The main classes of faults are distinctive because the absolute directions of stretching and shortening are different for each. Normal-slip faulting accommodates horizontal stretching and thinning, whereas thrust-slip and reverse-slip faulting accommodate horizontal shortening and thickening. With regard to strike-slip faulting, stretching and shortening *both* take place in the horizontal plane. Taken together, the main classes of faults are a versatile array of structures: *faults can distort the crust in any way required by plate tectonics and/or local stresses.* The potential for such versatility enlarges when we consider the fact that oblique-slip faulting is commonplace.

Because the main fault classes all have the same strain function, to achieve simultaneous stretching and shortening, we might guess that each brand of faulting would produce fault systems of *identical* physical properties, irrespective of actual orientations. However, such is not the case. Each class of faults has its own special and distinctive properties. The differences result largely from the way each of the major classes of faults interacts with preexisting structures, especially bedding and preexisting faults. We will pursue the unique geometric properties of thrust versus normal versus strike-slip faults in the final section of this chapter and in *Chapter 8* (*Fault-Fold Interactions*).

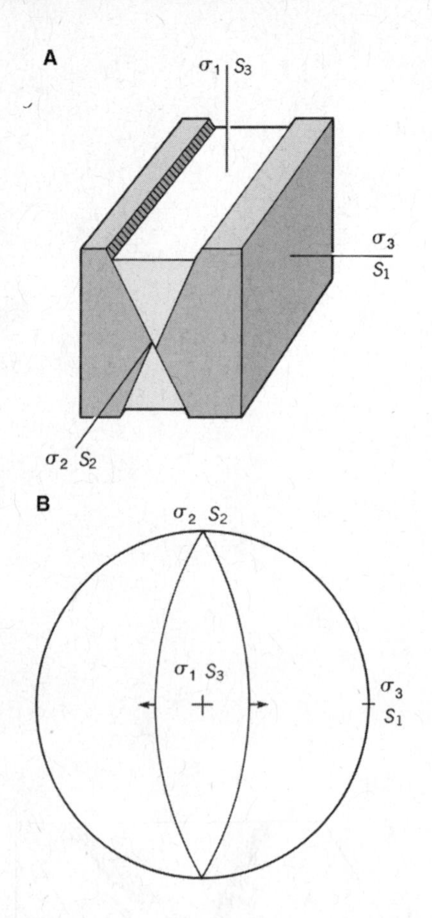

Figure 6.66 (*A*) Relation of conjugate faults to the principal finite stretch directions. (*B*) Stereographic representation of the principal finite stretch directions to the orientations of conjugate faults.

Relation of Faults to Principal Finite Strain Directions

It is not uncommon for faults to occur in **conjugate** sets (Figure 6.66*A*) that intersect in an acute angle, commonly ~60°, except where modified by flattening brought about by internal rotations. Whether they are thrust-slip, normal-slip, or strike-slip, the conjugate faults generally intersect in a line that is parallel to the direction of intermediate finite stretch (S_2) (Figure 6.66*A*). The directions of maximum and minimum finite stretch (S_1 and S_3) occupy a plane that is perpendicular to the direction of intermediate finite stretch (S_2). The direction of minimum finite stretch (S_3) bisects the conjugate angle between faults. Slickenlines are parallel to the line of intersection of each fault with the S_1–S_3 plane (Figure 6.66*A*). Figure 6.66*B* presents these relationships stereographically.

MECHANICS OF FAULTING

Dynamic Analysis

Dynamic analysis of faulting allows us to understand why faults can be so conveniently separated into normal-slip, thrust-slip, and strike-slip categories. We will learn that there is a dynamic basis for the fact that, on average, strike-slip faults dip vertically, normal-slip faults dip at 60°, and thrust-slip faults dip at 30°. Dynamic analysis of faulting is concerned both with stress conditions under which rocks break and with the orientations of faults relative to stress patterns. Results of laboratory testing of rock strength and behavior laid the groundwork for these understandings.

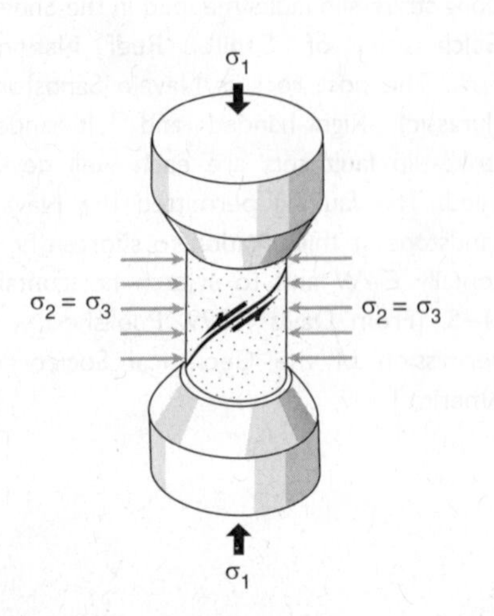

Figure 6.67 The basic setup for compressive strength tests.

Compressive Strength Tests

For **compressive strength tests**, increasing compressive stress (σ_1) is applied to the ends of the test specimen while a constant compressive confining pressure ($\sigma_3 = \sigma_2$) is applied perpendicular to the flank of the cylindrical specimen (Figure 6.67). Under conditions of low temperature and low-to-moderate confining pressure, the specimen will fail by faulting when the compressive strength of the rock is overcome. In such testing, the magnitudes of σ_1 and σ_3 at the point of failure are calculated and recorded. The specimen is then removed from the pressure vessel, whereupon the orientation of the fault can be measured relative to the direction of loading. Based on this information, we can then determine the values of normal stress (σ_n) and shear stress (σ_s) on the fault at the instant of failure.

As we learned in Chapter 3, the Mohr circle diagram provides a graphical means to describe the values of normal stress (σ_n) and shear stress (σ_n) on any plane within a body subjected to known values of greatest (σ_1) and least (σ_3) principal stress. We use the Mohr diagram to "*map*" the values of normal stress (σ_n) and shear stress (σ_s) at the instant of faulting for each deformation experiment, as well as the conditions of greatest (σ_1) and least (σ_3) principal stress that prevailed during faulting (Figure 6.68*A*). After a series of tests has been carried out on the same lithology under different conditions of confining pressure, a whole family of failure values (or **points of failure**) can be identified. Collectively these define an **envelope of failure** (Figure 6.68*B*), which separates the differential stress conditions under which the rock will remain unfaulted and stable, versus the differential stress conditions under which the rock will fail by faulting.

As an example, let us subject cylindrical cores of sandstone to compressive loading under confining pressure conditions of 10, 20, and 30 MPa

(Figure 6.69). We shall calculate axial stress at the instant of faulting for each of the tests. After we remove each faulted specimen from the pressure vessel, we shall measure and record the angle (θ) between the direction of greatest principal stress (σ_1) (i.e., direction of loading) and the normal (n) to the fault trace (see Figure 6.69). In compressive strength tests the angle θ typically lies in the range of $\pm 55°$ to $\pm 65°$, corresponding to faults at angles in the range of 35° to 25° with respect to the direction of greatest principal stress (σ_1). The average is $\sim 30°$ and the consistency is remarkable. The fractures are mode II, marked by fault offset and by slickenlines in the direction of slip along the surface(s). There is no opening (mode I) component on these faults.

To proceed from testing to representing the results on a Mohr diagram, we plot the values of σ_1 and σ_3 at failure for each test and draw circles through each set of points (see Figure 6.69). We then construct a radius that makes an angle 2θ with the σ_n axis of the diagram. Recall that the 2θ angle is measured counterclockwise if positive, clockwise if negative. The values of normal stress (σ_n) and shear stress (σ_s) on each fault plane at the instant of failure are reflected by the coordinates of the failure points marking the intersection of each circle with its radius (see Figure 6.69).

By connecting the failure points for each of the stress circles, we are able to construct the failure envelope for sandstone in the compressive regime (see Figure 6.69). In this case the envelope is a straight line, rising from lower left to upper right. It intersects the shear stress axis (*y*-axis) of the Mohr diagram at small positive values, with actual positioning a function of rock type. The slope and the straightness of the envelope reveal that the compressive strength of the sandstone increases linearly with increasing confining pressure. The actual angle of slope (ϕ) (see Figure 6.69) is called the **angle of internal friction**. The envelope itself is called the **Coulomb envelope**, in reference to the **Coulomb fracture criterion**, a law that captures the specific conditions under which a rock will fail by faulting under compressive stress conditions.

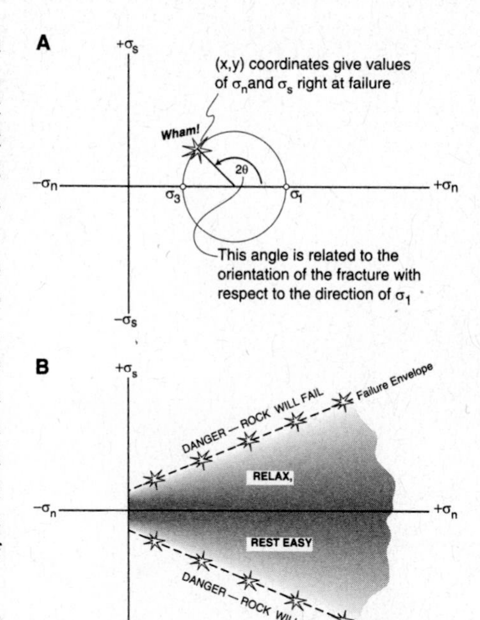

Figure 6.68 (*A*) The Mohr stress diagram can be used to plot the stress conditions of a test, in this case a compressive strength test where both the greatest principal stress (σ_1) and the least principal stress (σ_3) are positive. In this case the differential stress ($\sigma_1 - \sigma_3$) was great enough to cause failure by fracturing. The splattered star represents the values of normal stress (σ_n) and shear stress (σ_s) on the fracture at the instant of failure. (*B*) By carrying out a number of tests under a variety of differential stress conditions, it is possible to define "*failure envelopes*" within which the rock will not fail, beyond which it will.

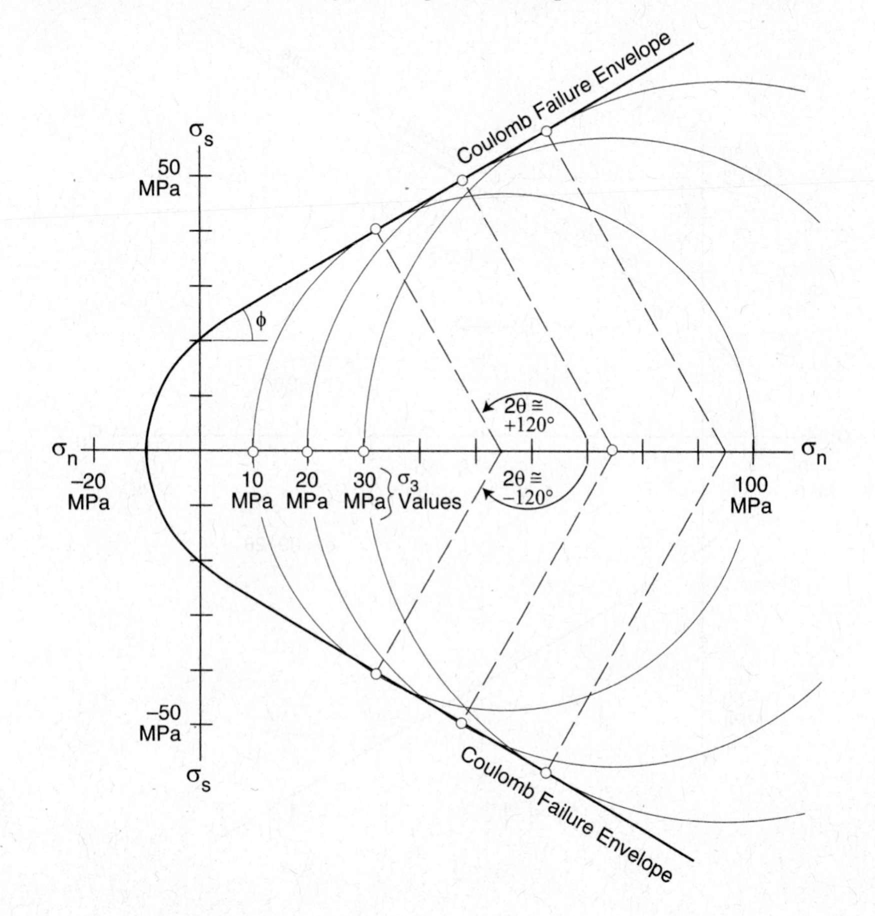

Figure 6.69 Mohr circle representation of a series of compressive strength tests carried out under different conditions of confining pressure (10, 20, 30 MPa). Points of failure lie on a straight line, the Coulomb envelope of failure.

The Coulomb Law of Failure

The Law Itself

Coulomb's law of failure is based on dynamic/mechanical models developed by Coulomb (1773) and Mohr (1900). The law is an equation that describes the height and slope of the linear envelope of failure for rocks in compression (Figure 6.70):

$$\sigma_c = \sigma_0 + \tan\phi\,(\sigma_n) \tag{6.1}$$

where σ_c = critical shear stress required for faulting
σ_0 = cohesive strength
$\tan\phi$ = coefficient of internal friction
σ_n = normal stress

The Coulomb law of failure takes on meaning when we see its geometric expression in the Mohr diagram. Figure 6.70 shows, once again, the envelope of failure for sandstone. A point of failure on the Coulomb envelope has [*x*,*y*] coordinates that reveal the magnitudes of normal stress (σ_n = 43 MPa) and shear stress (σ_s = 47 MPa) on the fault at the time of failure. In terms of the Coulomb law of failure, the shear stress value of 47 MPa is the critical shear stress (σ_c) necessary for faulting to occur. Part of its magnitude is cohesive strength (σ_0), expressed in units of stress (see Figure 6.70). The value of σ_0 can be read directly from the Mohr diagram as the *y*-intercept of the envelope of failure (see Figure 6.70). The rest of σ_c is the stress required to overcome internal frictional resistance to triggering movement on the fracture. This component is labeled $\tan\phi\,(\sigma_n)$ in Figure 6.70, a value expressed in terms of the normal stress (σ_n) acting on the fault plane and the angle of the internal

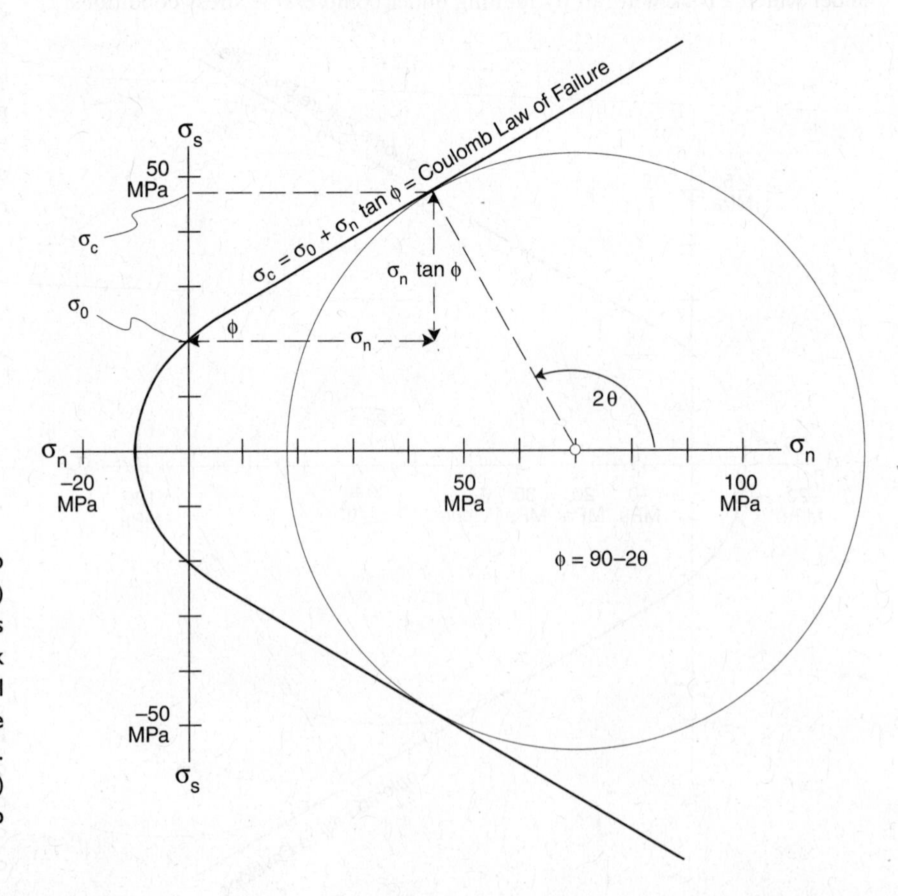

Figure 6.70 According to the Coulomb law of failure, the critical shear stress (σ_c) required to break a rock by shear failure is equal to the cohesive strength of the rock (σ_0) plus another increment of shear equal to the product of normal stress (σ_n) and the coefficient of friction of the rock ($\tan\phi$). Graphically, the angle of internal friction (ϕ) is simply the slope angle of the Coulomb envelope.

friction (ϕ), which is the slope of the envelope of failure. Thus, as indicated in Figure 6.70,

$$\sigma_c = \sigma_0 + \tan\ \phi\ (\sigma_n)$$

We learn from this that the stress level at which faults will form in compression is strongly influenced by ϕ, the angle of internal friction. The angle of internal friction for most rocks lies between 25° and 35°. Consequently, the coefficient of internal friction, tan ϕ, commonly ranges from 0.466 to 0.700. If we can assume that cohesive strength is a very small part of the critical shear stress required for faulting, most faults form when shear stress on the plane of failure reaches a level that is slightly more than 50% of the normal stress acting on the surface. The exact optimum ratio of shear stress to normal stress will vary from rock to rock.

The angle of internal friction (ϕ) determines (θ), the angle between the direction of greatest principal stress (σ_1) and the normal (n) to the fault surface. Since most rocks in nature possess an angle of internal friction (ϕ) of about 30°, the value of θ for most faults is ~60°. This means that the angle between the fault surface and the direction of greatest principal stress (σ_1) is ~30°. Twiss and Moores (1992) refer to this angle as α, **the angle of failure**. Where a pair of faults is formed during a compression test, they are referred to as **conjugate faults**. Conjugate faults are bisected by the direction of greatest principal stress (σ_1), intersect along a line parallel to the direction of intermediate stress (σ_2), and have an acute angle of intersection of 2α = ~60°. A beautiful field example of conjugate faulting is shown in Figure 6.71. In this case the faults are thrust faults, and the acute bisector (direction of σ_1) is horizontal (parallel to bedding) and responsible for bed-parallel shortening.

Figure 6.71 Photograph of "*conjugate*" thrust faulting in Straight Cliffs Sandstone (Cretaceous) in the Hillsdale Canyon area near Bryce Canyon, Utah. The acute angle between the conjugate thrusts is bisected by the direction of greatest principal stress (σ_1). The direction of greatest principal stress (σ_1) in this case is perfectly parallel to bedding. The thrusting accommodated bed-parallel shortening. The thickness of this faulted bed is ~2 m. [Photograph by G. H. Davis.]

Applications of the Coulomb Criterion

Having established the failure envelope for this rock, suppose we wanted to know the magnitude of the greatest principal stress that would be required to fracture sandstone under confining pressure conditions of 18 MPa (Figure 6.72). Would 60 MPa be high enough? Not a chance! A stress circle drawn through points on the σ_n axis of the Mohr diagram corresponding σ_1 = 60 MPa and σ_3 = 18 MPa is *not* large enough to intersect the envelope of failure (see Figure 6.72). The stress circle resides entirely within the field of stability. For the differential stress ($\sigma_1 - \sigma_3$) to be great enough to cause faulting, σ_1 must be raised much higher than 60 MPa. Indeed, if σ_1 is raised to the level of 123 MPa, while confining pressure is held constant, the stress circle becomes tangential to the envelope of failure at σ_n = 43 MPa, and σ_s = 47 MPa (see Figure 6.72). On this basis we would surmise that σ_1, when raised to 123 MPa, would cause the specimen to fault. Moreover, the angle (θ) between σ_1 the normal (n) to the fault surface could be predicted to be ~60° and α ~30° (see Figure 6.72).

Predicting the point of a rock is some of the fun of rock deformation experiments. Great shouts, screams, and moans, not to mention the clinking of nickels and dimes, once issued forth from our rock deformation lab. Guessing the value of greatest principal stress (σ_1) required to break the very first specimen in a series of tests is almost mindless speculation. But after stress data and θ-values for two or three tests have been posted on a Mohr diagram, the predicted stress values for failure begin to cluster within a *very* narrow range. Beads of sweat begin to form as data are plotted and failure envelopes are constructed.

Deformation experiments thus confirm that faults are oriented systematically with respect to stress directions (Figure 6.73). Vertical compression of rock cylinders under controlled laboratory conditions usually produces a single

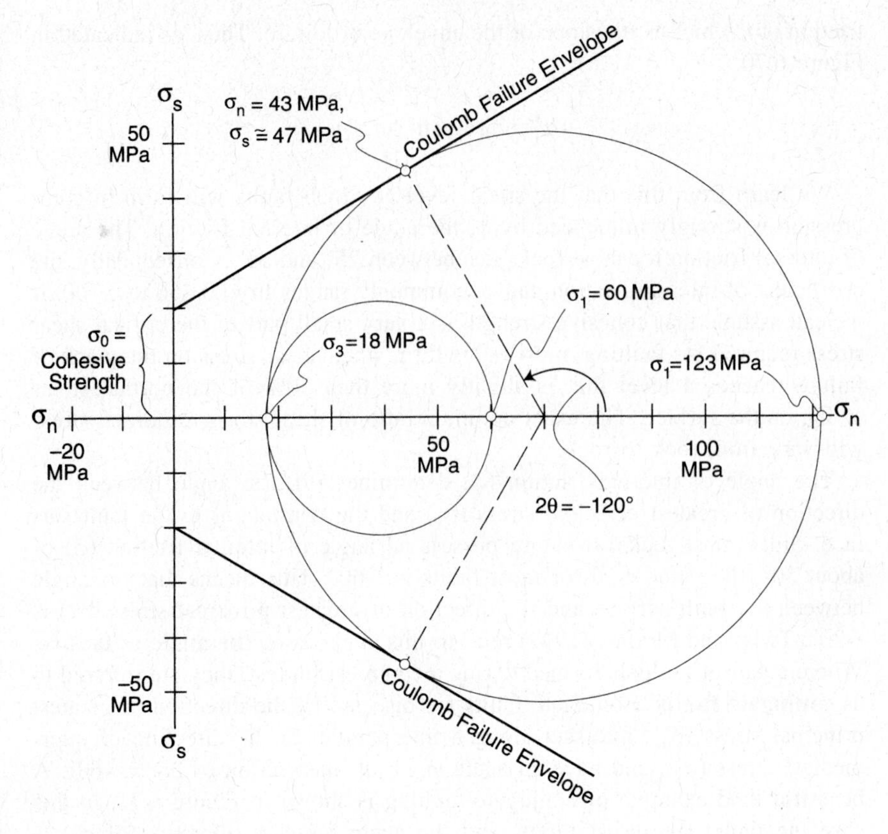

Figure 6.72 If confining pressure is set at 18 MPa, will an increase of greatest principal stress (σ_1) to 60 MPa be enough to cause shear failure? Not a chance! On the other hand, 123 MPa would be just the right amount.

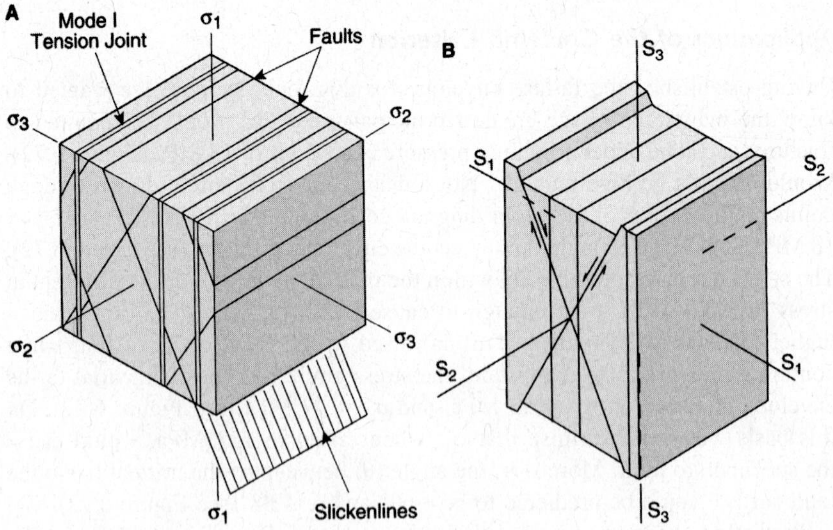

Figure 6.73 Drawings of conjugate faults as well as mode I tension fractures in a block of rock that had been subjected to length-parallel shortening. Note the orientations of the principal stress directions with respect to the orientations of faults, slickenlines, and mode I joints. Blocks are framed with respect to (*A*) principal stress directions and (*B*) principal strain directions.

discrete fault dipping approximately 60°, or—less commonly—conjugate faults, each of which dips approximately 60°. When conjugate faults form, the **conjugate angle** between the faults is bisected by the direction of greatest principal stress (σ_1), which is the direction that the piston bears down on the rock cylinder. Slickenlines on the fault surfaces are visible when the specimen is pulled apart. Slickenline orientations on a given fault are defined by the intersection of the fault surface with the σ_1/σ_3 plane. If mode I tension fractures open up, they form parallel to the direction of greatest principal stress (σ_1) and perpendicular to the direction of least principal stress (σ_3) (see Figure 6.72).

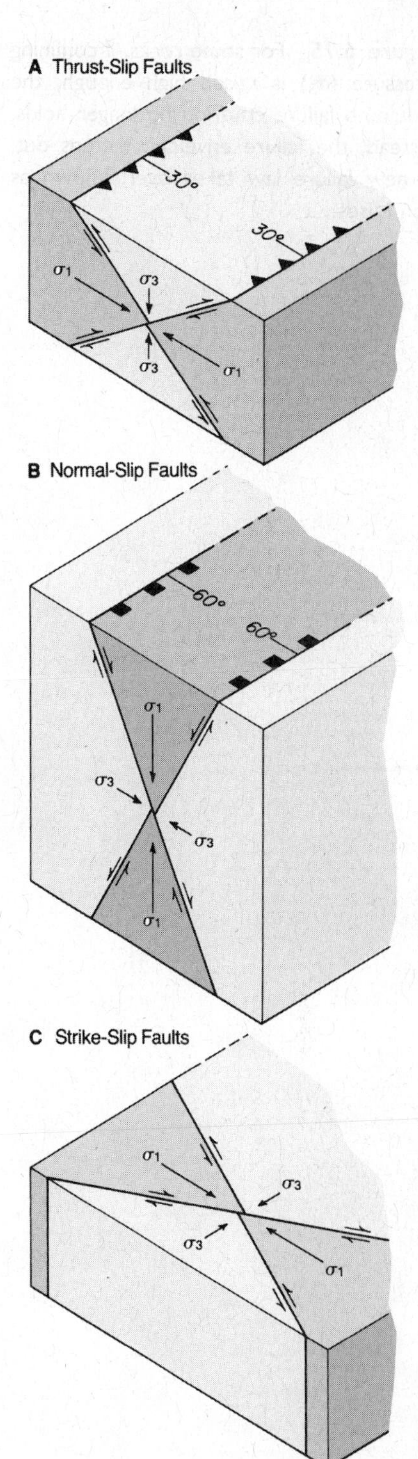

Figure 6.74 Schematic representation of (*A*) thrust faults, (*B*) normal faults, and (*C*) strike-slip faults at or near the surface of the earth. These are the likely orientations since each of the three principal stress directions at or near the surface of the Earth is either horizontal or vertical, and since the angle of internal friction for rocks is almost always ~30°.

The geometric systematics that prevail among conjugate faults, tension fractures, and principal stress directions provide a basis for interpreting paleostress directions in rocks deformed millions of years ago. *Part III-K, Determining Slip on Faults Through Orthograhic and Stereographic Projection,* focuses on how to interpret principal stress directions on the basis of fault attributes, and how to predict fault attributes on the basis of principal stress directions. This work involves stereographic projection as an aid to dealing with the three-dimensional geometries.

Andersonian Application of Coulomb Criterion

Anderson (1951) concluded that if principal stress directions are vertical or horizontal at or near the surface of the Earth, and if the angle of internal friction for most rocks is about 30°, only normal-slip, strike-slip, and thrust-slip faults should be able to form at or near the Earth's surface (Figure 6.74). Thrust-slip faults form when σ_3 is vertical; normal-slip faults form when σ_1 is vertical; and strike-slip faults form when σ_2 is vertical. *Determining the Relationship of Faults to Principal Stress Directions is* presented in *Part III-K.*

Departure From Coulomb Behavior

Were we to subject specimens of sandstone to compressional deformation at high confining pressures, we would discover that pure Coulomb's criterion for faulting no longer holds. With increasing confining pressure, rocks behave in a less brittle, more ductile fashion. We learned this in *Chapter 3, Force, Stress, and Strength,* particularly as expressed in stress/strain curves departing from a strict elastic linear relationship. In a somewhat analogous manner, we learn through compressive strength tests that the linear (Coulomb) relationship between fracture strength and confining pressure only goes so far, and beyond a certain point, the rock begins to weaken. The expression of this in the world of Mohr diagrams is departure from the straight-line Coulomb envelope to a concave inward envelope of lesser slope (Figure 6.75).

The flattening of the envelope expresses two things: (1) with each increment of increased confining pressure proportionally less shear stress is required to fracture rock in the brittle/ductile transition, and (2) when rocks in the brittle/ductile transition do fracture, they fail by faults that make an angle of failure (α) greater than 30° with respect to the direction of greatest principal stress (σ_1). In fact, α approaches 45° as the rock becomes increasingly ductile.

The law that describes the deformational behavior above the brittle/ductile transition is called the **von Mises criterion** (Ramsay, 1967). When the critical yield stress (which varies from rock to rock) is surpassed, the rock will fail by ductile faults along "*planes*" of maximum shear stress, oriented at 45° with respect to the direction of greatest principal stress (σ_1) (see Figure 6.75).

Failure Envelopes, All Together

Figure 6.75 is a summary diagram that combines envelopes for tensile behavior, transitional tensile behavior, Coulomb behavior, and von Mises behavior into a single "*grand envelope*" for failure (Suppe, 1985; Twiss and Moores, 1992). The form of the grand envelope is common for almost all

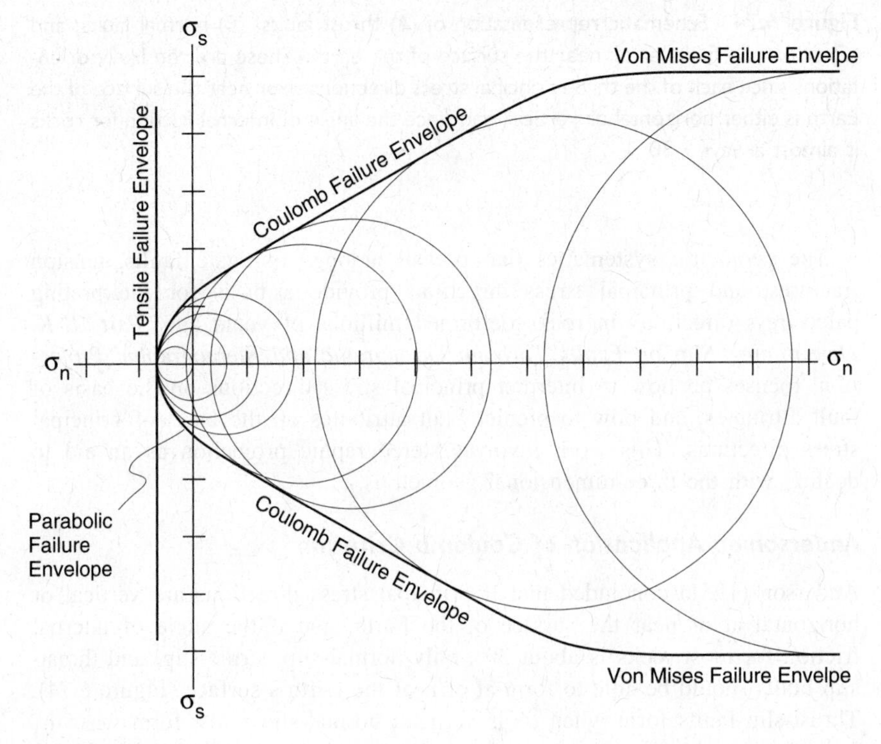

Figure 6.75 For some rocks, if confining pressure (σ_3) is raised high enough, the Coulomb failure criterion no longer holds. Instead, the failure envelope flattens out. A new failure law takes over, known as von Mises.

rocks, although the positioning and shape of the envelope will vary according to tensile strength (T_0), cohesive strength (σ_0), and internal friction (ϕ). The values for tensile strength, cohesive strength, and internal friction for common rock types are shown in Figure 6.76.

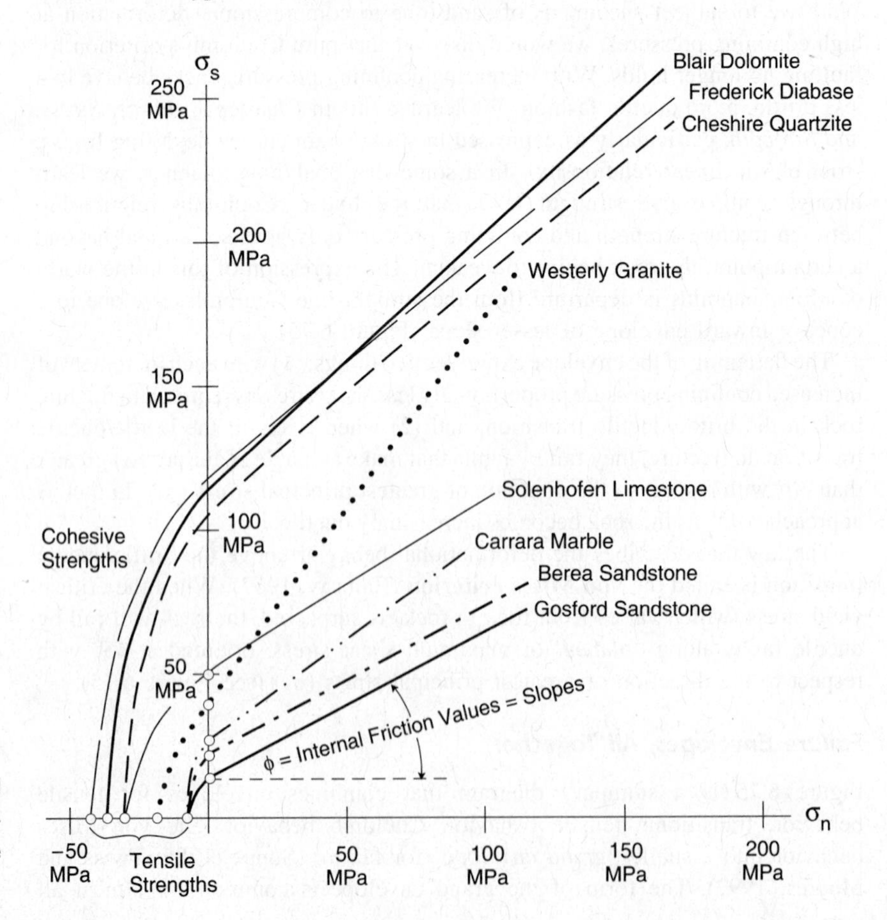

Figure 6.76 Mohr diagram of actual measured values of tensile strength, cohesive strength, and internal friction for a wide range of rock types.

Adding Fluid Pressure to the Mix

Just as fluid pressure influences the conditions under which jointing occurs (see *Chapter 5, Joints*), so too does fluid pressure influence faulting. Hubbert and Rubey were able to show that elevated fluid pressures in rocks tend to offset the magnitude of lithostatically-produced normal stress (σ_n) acting on fault and fracture surfaces. They introduced the concept of **effective stress** (σ^*), which is the difference between normal stress and fluid pressure ($\sigma_n - P_f$). They then modified the Coulomb law of failure in a way that makes it clear that reducing the normal stress by fluid pressure also lowers the value of critical shear stress required to produce faulting.

The Coulomb law of failure looks like this:

$$\sigma_c = \sigma_0 + \tan\phi\,(\sigma_n) \tag{6.1}$$

When modified by Hubbert and Rubey (1959), it becomes this:

$$\sigma_c = \sigma_0 + \tan\phi\,(\sigma_n - P_f) \tag{6.2}$$

The "*simple*" change was replacing normal stress with effective stress. But what a difference it makes! It reduces significantly the stress magnitude required for failure. In fact, when effective stress becomes zero, which it does in certain circumstances, one entire term of the Mohr-Coulomb law of failure is completely eliminated:

$$\tan\phi\,(\sigma_n - P_f) = \tan\phi\,(0) = 0$$

What is left is the expression:

$$\sigma_c = \sigma_0$$

When this state exists, a fault can form simply by overcoming the cohesion of the rock. And if the rock is already broken along a preexisting fracture of favorable orientation, the job is even easier. Exhaustive experimental testing has substantiated the importance of effective stress. Time and again it has been demonstrated that rock samples fail by faulting (and jointing) under significantly smaller loads when pore fluid pressure is added to the sample (Paterson, 1978; Tullis and Tullis, 1986).

In stress fields of high differential stress, the gradual elevation of fluid pressure will "*drive*" the stress circle into collision with the Coulomb failure envelope in the compressional field (Figure 6.77). In this situation, faults form, and they form at lower stress levels than would normally be expected.

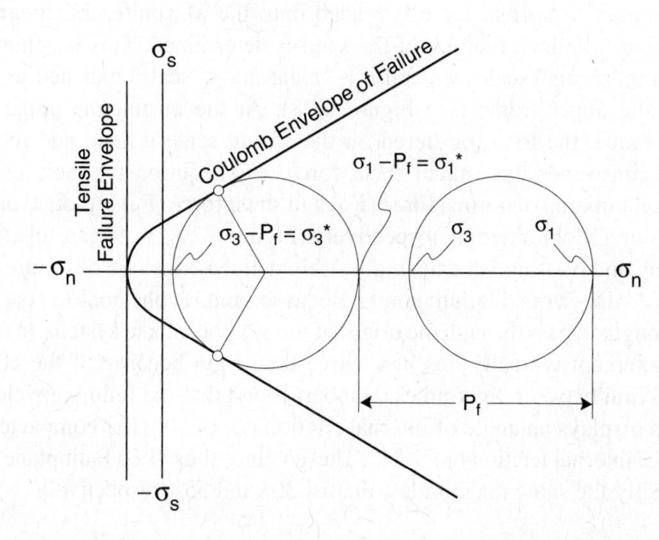

Figure 6.77 Elevated fluid pressure can cause Coulomb-style faulting to take place even where differential stress normally would not be up to the task. The effect of increasing fluid pressure is to shift the differential stress circle laterally into "*collision*" with the failure envelope, thereby producing faulting.

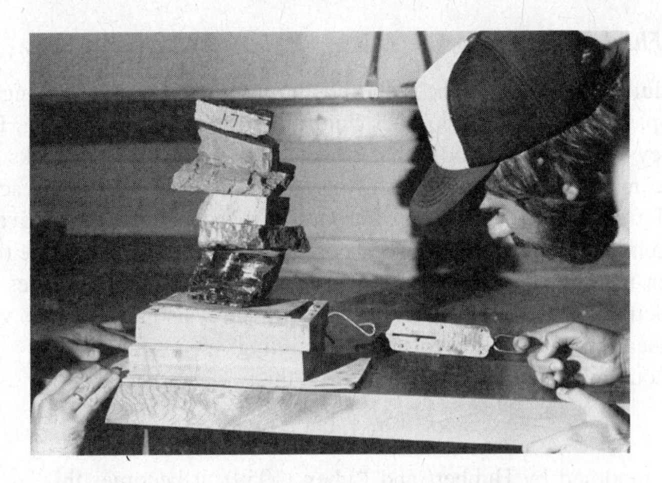

Figure 6.78 Construction of the failure envelope for sand. Experimental setup consists of two wooden frames, sand, a piece of Masonite, rocks of known weight, a good spring scale, and a baseball cap. The experiment itself involves measuring the amount of shear force required to move the upper frame for a number of given conditions of normal force. Andrew Arnold demonstrates the technique. [Photograph by R. W. Krantz.]

Hubbert's Sandbox Illustration of Coulomb Theory

Appreciation of Anderson's theory of faulting and the Coulomb law of failure can be gained in a simple but elegant set of experiments described by Hubbert (1951). The experiments allow us to develop a working knowledge of constructing failure envelopes and applying the results to fault analysis. Sand is substituted for rock, but otherwise all conditions hold.

Constructing an Envelope for Sand

The first step is to construct an envelope of failure for sand. The "*no frills approach*" simply requires two small, lightweight wooden or metal frames stacked one on top of the other (Figure 6.78). The surfaces of the frames should be very smooth so that the frames can effortlessly slide past each other. The interior of the double-frame box is filled with fine white sand (what sedimentologist R. L. Folk used to call "*spittoon sand*"). A piece of Masonite is cut to fit snugly into the interior of the upper frame, in order to distribute the load evenly. When positioned, the Masonite sheet rests entirely on sand.

To construct an envelope of failure for sand, we determine the level of shear force that is required to cause differential movement of the upper and lower frames under a variety of conditions of normal force. For the frames to move with respect to each other, the sand must actually "*fault*".

Normal force ($\mathbf{F_N}$) is applied to the Masonite by placing rock samples of known weight on top of the Masonite (see Figure 6.78). The rocks are weighed as the experiment proceeds, and the values (in ounces or grams, pounds or kilograms) are written directly on them using a marking pen. Once a given amount of normal force is loaded onto the Masonite, the **shear force ($\mathbf{F_S}$)** required to induce faulting of the sand is determined. This is achieved by pulling on a "*spring*" scale (we call it a "*meat-hook*" scale) attached to an eye screw on the upper frame (see Figure 6.78). At the instant the upper frame begins to move, the load registered on the spring scale is read and recorded. This load constitutes the critical shear force ($\mathbf{F_S}$) required for faulting.

Each combination of normal force ($\mathbf{F_N}$) and shear force ($\mathbf{F_S}$) is plotted on the *x*- and *y*-axes of a Mohr diagram, respectively (Figure 6.79). With careful attention to weighing and plotting, an amazingly straight-lined envelope of failure can be fit to the half-dozen or so failure points. Because sand is cohesionless, the failure envelope ought to pass through the origin of the *x*-*y* coordinate system. In point of fact, the intercept typically lies just above the origin because of the effect of sliding friction between the frames. Hubbert found that the failure envelope for loose sand displays an angle of internal friction (ϕ) of 30°. For compacted sand the angle of internal friction (ϕ) is 35°. These values thus yield fault plane angles (α) of exactly the same magnitudes; that is, 30° and 35°, respectively.

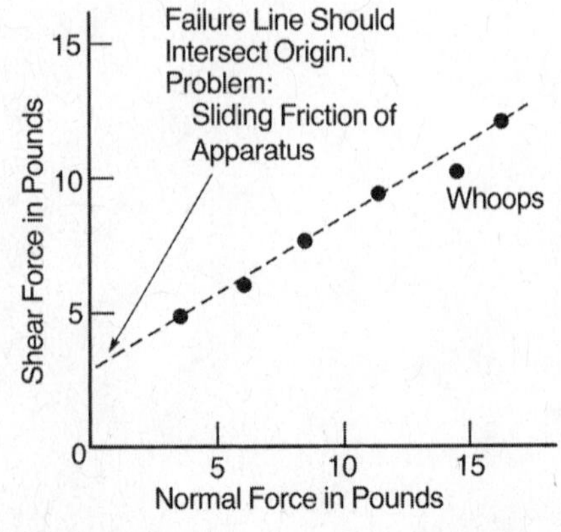

Figure 6.79 Mohr diagram showing the envelope of failure for sand based on paired measurements of shear force ($\mathbf{F_S}$) and normal force ($\mathbf{F_N}$). We ran this experiment "*back in the day*," when we were measuring weights in pounds.

The Sandbox Experiment Itself

The second part of Hubbert's experiment involves generating normal-slip and thrust-slip faults in sand. A sandbox of the type shown in Figure 6.80 is filled with layers of fine white sand separated by marker horizons of white dry powdered clay. A vertical wooden or metal partition serves to separate two compartments, a smaller one on the left in which normal-slip faults will form, and a larger one on the right in which thrust-slip faults will form. Deformation of the sand simply requires moving the partition to the right by means of a manually driven worm-screw arrangement. The instant that the partition moves, a normal-slip fault develops in the left-hand compartment (see Figure 6.80). Its dip is typically ~60°, that is, the complement of the angle of internal friction for sand. As the partition is forced to move further and further to the right, the first-formed normal-slip fault increases in displacement. Other normal-slip faults form as well. As a result of the normal faulting, the upper surface of the sand in the left-hand compartment develops a fault-scarp topography.

Compression-induced shortening of sand in the right-hand compartment eventually forces the development of a thrust-slip fault. The sand first arches slightly and then is cut by a thrust-slip fault, which slices up-section at an angle of ~30°. After a certain amount of translation has been accommodated

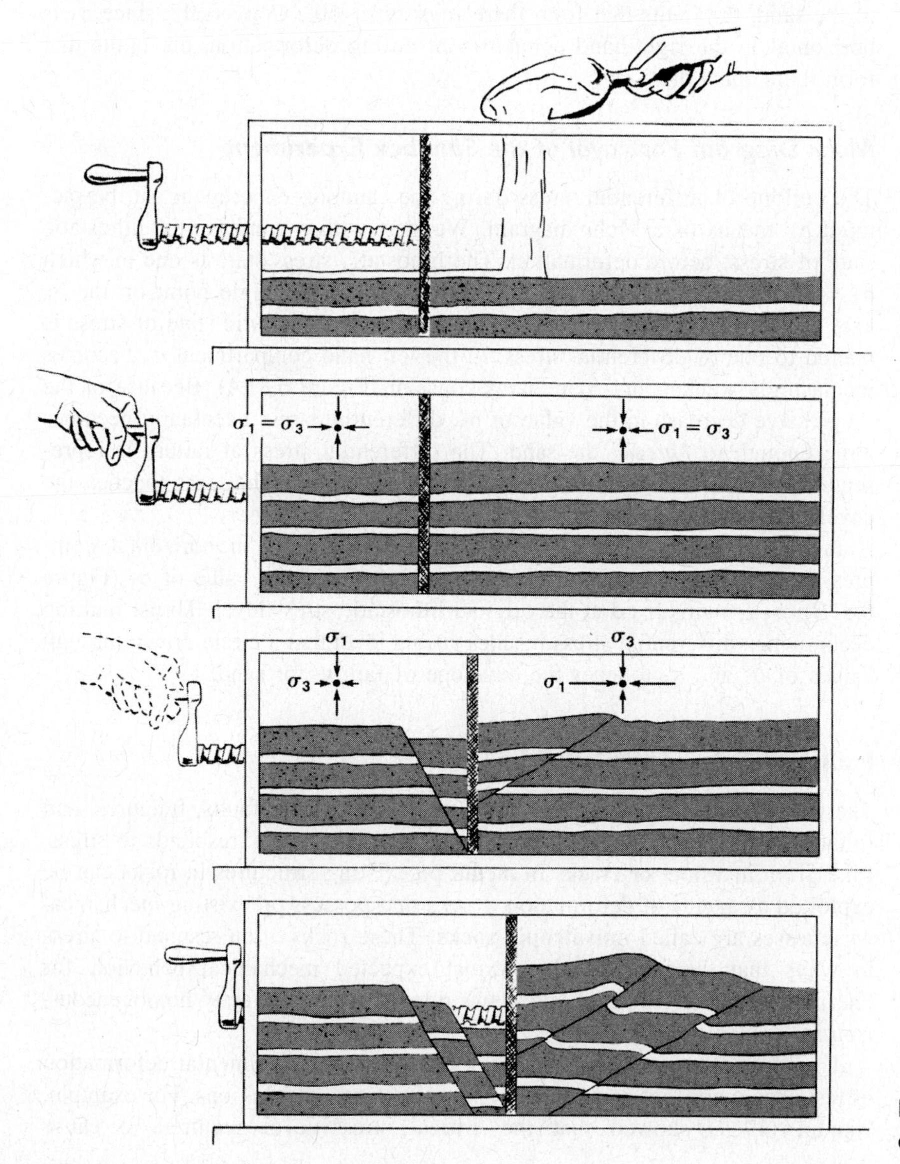

Figure 6.80 The famous sandbox experiment. See text for the narrative.

by the thrusting, a second thrust-slip fault develops; this fault typically forms beneath the first, cutting up to the surface beyond it in the direction of tectonic transport. Translation on the first thrust ceases the *instant* that translation is initiated on the second. Steady translation on the second thrust results in folding of the first. Sometimes **back-limb, antithetic** thrusts develop that dip oppositely to the main thrust-slip faults.

The structural relationships produced in the sandbox experiment conform to what we have learned about dynamic analysis of faulting. Before movement of the partition in the sandbox, both compartments of sand are marked by a state of lithostatic stress in which stress is equal and all-sided (see Figure 6.80). Movement of the partition to the right relieves horizontal stress in the left compartment. At the same time, horizontal stress intensifies in the right-hand compartment. Thus, very early in the experiment, different states of stress evolve in the two compartments. In the left compartment, σ_1 is vertical, and σ_3 is horizontal. In the right compartment, σ_1 is horizontal, and σ_3 is vertical. Given these principal stress directions and our knowledge of the angle of internal friction (ϕ) for sand, it becomes possible for us to predict the orientations of the faults that must develop in the sandbox experiment. Faults form in loose sand at angles of $\alpha = \sim 30°$ to the direction of greatest principal stress. Since σ_1 is vertical in the left compartment during ongoing deformation of the sand, the faults that form there *must* dip ~60°. Conversely, since σ_1 is horizontal in the right-hand compartment during deformation, the faults that form there *must* dip ~30°.

Mohr Diagram Portrayal of the Sandbox Experiment

The buildup of differential stress during the sandbox experiment can be pictured by means of a Mohr diagram. We start with a condition of lithostatic state of stress, before deformation. The lithostatic stress state is one in which $\sigma_1 = \sigma_3$; it is represented on the Mohr diagram by a single point on the σ_n axis. As soon as the partition begins to move, the lithostatic state of stress is altered to one of differential stress. In the left-hand compartment σ_3 becomes increasingly weaker, but σ_1 remains constant (Figure 6.81*A*). Because of the progressive decrease in the value of σ_3, differential stress eventually becomes large enough to "*break*" the sand. The differential stress at failure is represented by values of σ_1 and σ_3, which define a circle that just touches the envelope of failure for sand.

In the right-hand compartment, the steady increase in horizontal compressive stress is represented by a steady increase in the value of σ_1 (Figure 6.81*B*); σ_3 remains fixed at the original lithostatic stress level. Thrust faulting occurs when differential stress reaches such a level that a circle drawn through values of σ_1 and σ_3 touches the envelope of failure for sand.

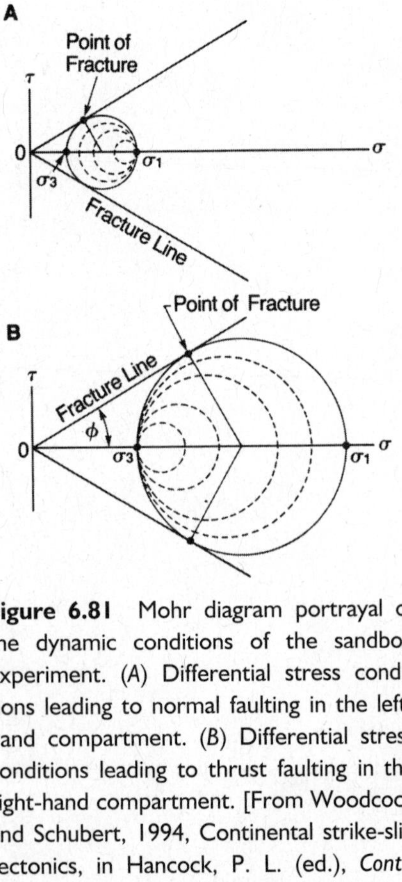

Figure 6.81 Mohr diagram portrayal of the dynamic conditions of the sandbox experiment. (*A*) Differential stress conditions leading to normal faulting in the left-hand compartment. (*B*) Differential stress conditions leading to thrust faulting in the right-hand compartment. [From Woodcock and Schubert, 1994, Continental strike-slip tectonics, in Hancock, P. L. (ed.), *Continental deformation*: Pergamon Press, New York, p. 251–263. Copyright Elsevier, 1994.]

Faulting of Anisotropic Rocks

The presence of preexisting structures, such as bedding, faults, fractures, and foliations, can substantially alter the way in which a rock responds to stress. Like grain in wood, or cracks in metal, preexisting structures in rocks can be exploited as agents of deformation. Rocks that possess preexisting mechanical weaknesses are called **anisotropic rocks**. These rocks often respond to stress in ways that depart from the normal expected mechanical behavior, for "*normal*" behavior is based upon laws developed for perfectly homogeneous, isotropic rocks.

The clearest examples of this kind are based on experimental deformation of test specimens that contain preexisting fractures or foliations. For example, Handin (1969) showed that the critical stress level required to cause

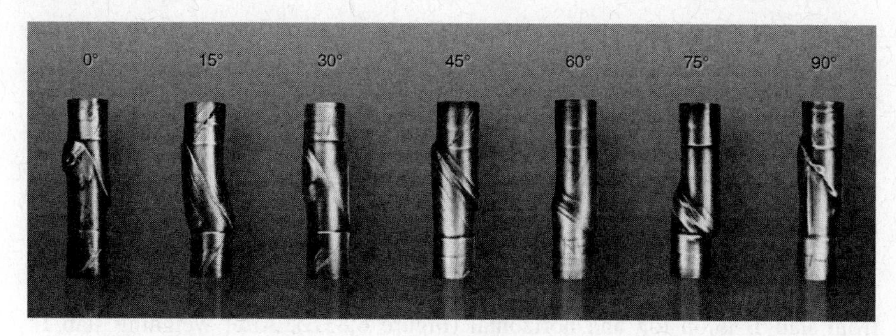

Figure 6.82 Specimens of anisotropic rock compressed at various angles to foliation. The orientation of faulting, in each case, is influenced by the angular relationship between the orientation of foliation and the direction of greatest principal stress (σ_1). [From Donath, 1961. Published with permission of the Geological Society of America.]

reactivation along preexisting fracture surfaces is *less* than that required to break an unfractured specimen of the same lithology. Furthermore, Handin (1969) was able to show that fractures oriented at angles as high as $\alpha = 65°$ to the direction of greatest principal stress (σ_1) can be *reactivated* as faults. $\alpha = 65°$ is a long way from $\alpha = 30°$, which is the expected angle of failure for isotropic rocks in a perfect world. Thus if we were to disregard the possible influence of anisotropy and assume that all faults form close to $\alpha = 30°$, we would get into deep trouble in our attempts to reconstruct the paleo-orientations of greatest principal stress directions.

As yet another example, Donath (1961) demonstrated the degree to which the orientation of foliation (e.g., schistosity or slaty cleavage) influences the orientation of faults that develop in test specimens (Figure 6.82). When foliation is oriented at a very high angle to the greatest principal stress direction (σ_1), the specimen will fault, as usual, at an angle such that the fault plane angle (α)—that is, the angle between σ_1 and the fault—is approximately equal to the angle of internal friction (ϕ) for the faulted rock. But if foliation is at a closer angle to σ_1, the orientation of the fault will differ from what we would expect for a homogeneous specimen. For example, if foliation is parallel to σ_1, the angle α will be very small, perhaps 10° or 20°. If foliation is inclined between 25° and 45° to σ_1, the fault surface will commonly develop right along the foliation, thereby overriding the influence of internal friction. If foliation is inclined more than 45° to σ_1, the fault surface will form at an angle (α) that obeys Coulomb's law of failure, almost as if the rock were homogeneous.

Faulting of Pre-Fractured Rocks

When an already fractured rock is tested, it is found to possess neither tensile strength nor cohesive strength. As a result, failure of the specimen simply requires movement on one or more of the preexisting fractures. Since there is no cohesive strength to overcome, the critical level of shear stress required to produce movement on a preexisting fracture is simply related to the frictional resistance to sliding.

Determining whether or not preexisting structures such as joints, faults, and foliations will reactivate by sliding to form faults requires focusing on frictional behavior. **Friction** is the measure of resistance to sliding along a surface. Friction increases as normal force on the surface increases, but not as a function of the area of the surface (Suppe, 1985). Normal force, as we know from *Chapter 3, Force, Stress, and Strength*, varies with the orientation of the surface relative to principal stress directions. As a consequence of these factors, the level of stress required to activate a preexisting fracture surface depends mainly upon the combination of two factors: the **friction** along the surface and the **orientation** of the surface. Of course friction is related to the character of the surface and to rock type. Common everyday experience tells us that the magnitude of frictional resistance is less where surfaces are smooth and planar and greater where the surfaces are rough and textured. To be sure, frictional resistance on fractures increases where there are **asperities** (i.e., tiny protuberances, bumps, and other irregularities) (Suppe, 1985).

Resistance to sliding on a given fracture in a given rock can be quantitatively measured. Tests are carried out in a technically sophisticated way, but we can picture the process by carrying out a "*thought experiment.*" We start by collecting slabs (*A, B, C, D, E*) from a mountain of rock that is split by a set of natural joints (Figure 6.83*A*). Hauling these out on a flatbed truck, we lay slab *A* on its side, such that the joint surface that originally separated it from slab *B* is on top and horizontal (Figure 6.83*B*). After weighing slab *B* and resting it on slab *A*, we measure how much shear force is required to initiate movement along the joint surface. The values of normal force (weight of slab *B*) and shear force (spring-scale reading) are recorded. We then weigh and add slab *C* to the stack, and once again determine the level of shear force required to trigger movement along the joint surface between slabs *A* and *B*.

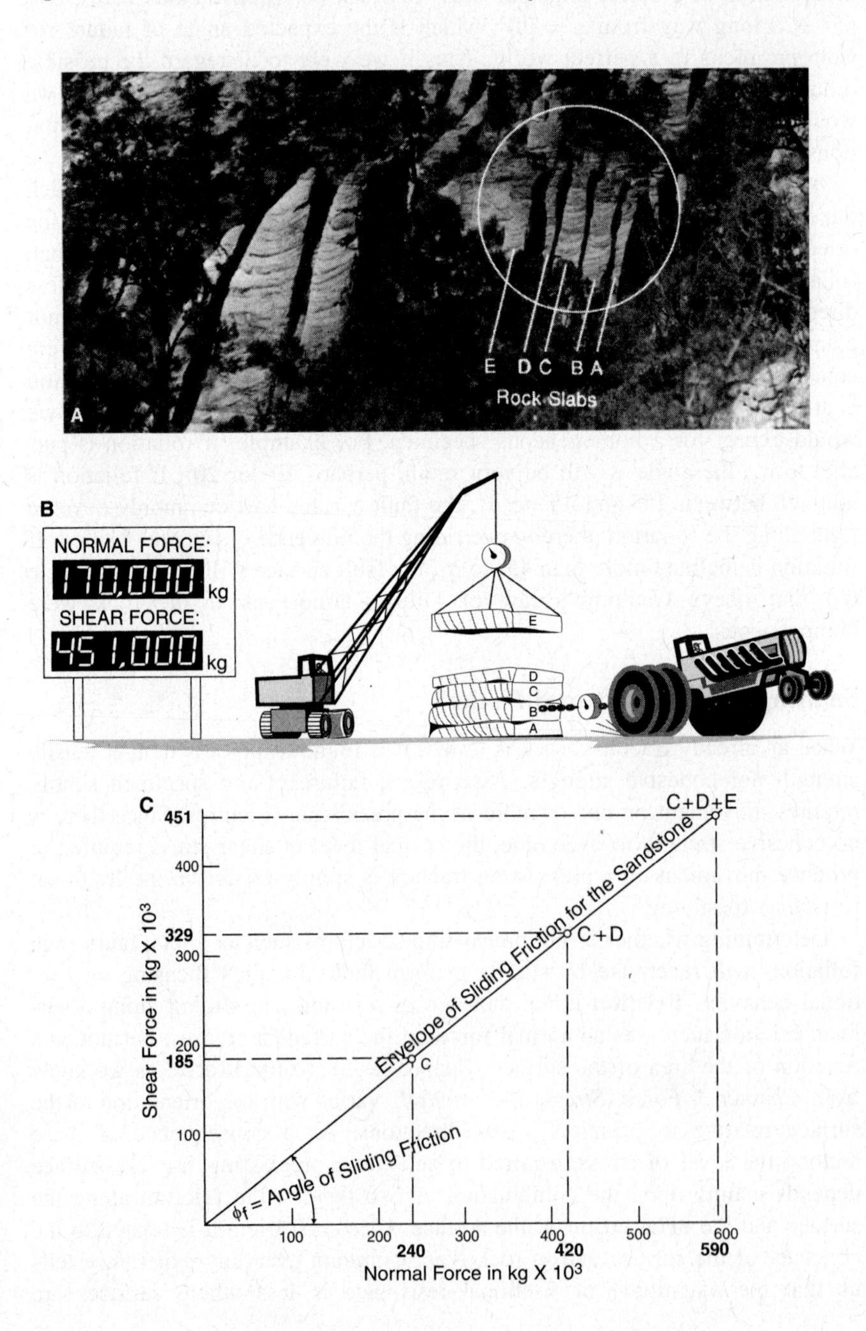

Figure 6.83 (*A*) Close-spaced jointing in Zion National Park has broken the Navajo Sandstone (Jurassic) into slabs. In virtual reality we collect slabs *A-E*. [Photograph by G. H. Davis.] (*B*) Tractor-pull extravaganza features measuring the relationship between shear force and normal force. [Artwork by Wally Varner.] (*C*) Mohr-diagram plot shows the results: a straight-line relationship!

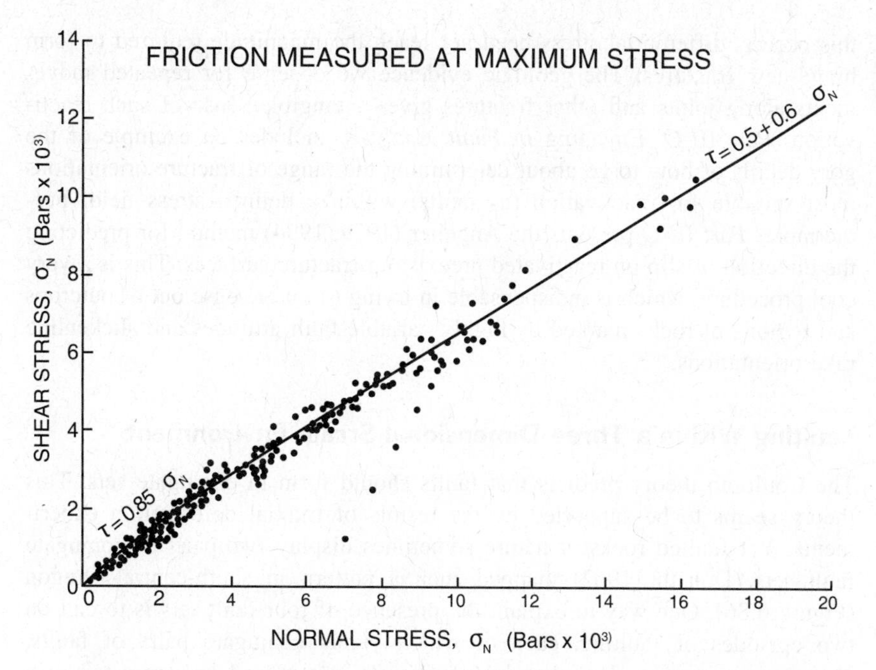

Figure 6.84 Original plot of Byerlee's law of sliding friction, which is based on hundreds of sliding friction experiments on a wide variety of rock types. [Reprinted with permission from *Pure and Applied Geophysics*, v. 116, Byerlee, J. D., "Friction of rocks," 1978, Birkhauser Verlag Ag, Basel, Switzerland.]

Then slab *D* is added, and the shear force required to cause movement is measured again. By plotting all three sets of paired values of normal force and shear force, it is possible to construct a **failure envelope for frictional sliding** (Figure 6.83*C*). The slope of the straight-lined envelope is called the **angle of sliding friction** (ϕ_f) (see Figure 6.83*C*). The tangent of the angle of sliding friction is known as the **coefficient of sliding friction**:

$$\mu_f = \tan \phi_f \qquad (6.3)$$

The coefficient of sliding friction expresses the ratio of shear stress to normal stress on the fracture surface at the instant the that sliding initiates (see Figure 6.83*C*).

The law that describes the conditions under which an existing fracture will move is a modification of the Coulomb law of failure. Since a fractured rock has no cohesion, the term describing cohesive strength is removed from the Coulomb equation. What is left is a simple expression:

$$\sigma_c = \tan \phi (\sigma_n) \qquad (6.4)$$

This equation, called **Byerlee's law**, says that the critical shear stress necessary to cause reactivation of an existing fracture is equal to the coefficient of sliding friction of the rock multiplied by the normal stress acting on the fracture surface. To the surprise of most, Byerlee (1978) showed that the maximum coefficient of sliding friction ($\mu_f = \tan \phi_f$) is the same for almost all rocks under conditions of moderate to high normal stress (Figure 6.84). For relatively low magnitudes of confining pressure, the average value for the coefficient of sliding friction is ~0.85, corresponding to an angle of sliding friction of ~40° (see Figure 6.84). For moderate to higher levels of confining pressure, the average angle of sliding friction is ~35° (see Figure 6.84). The coefficient of sliding friction becomes smaller as frictional sliding pulverizes mineral grains into clayey gouge.

The envelope for frictional sliding lies beneath the Coulomb envelope (Figure 6.85). Rock will preferentially fail by movement on preexisting fractures, provided that the preexisting fractures are suitably oriented. When

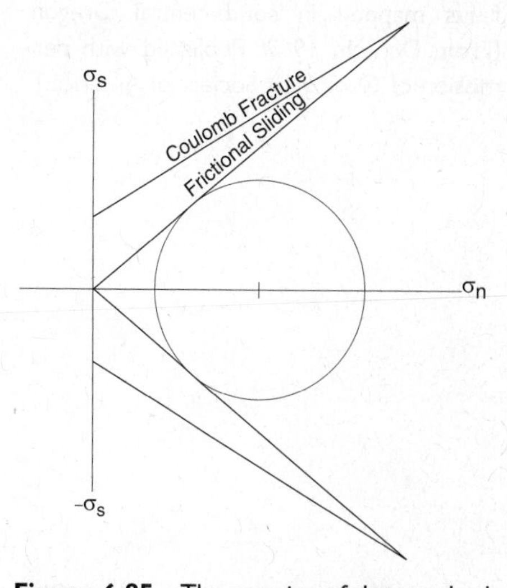

Figure 6.85 The merging of the standard envelope of failure with the frictional law of failure on a Mohr diagram. In this example, greatest principal stress (σ_1) has been raised to the point where the differential stress circle touches the frictional sliding envelope. No need to form brand new faults. Faulting instead will take place along preexisting fractures, if any are suitably oriented. Differential stress may never grow to reach the Coulomb envelope. [From Suppe, J., Principles of Structural Geology, © 1985, p. 163. Reprinted by permission of Prentice-Hall, Upper Saddle River, New Jersey.]

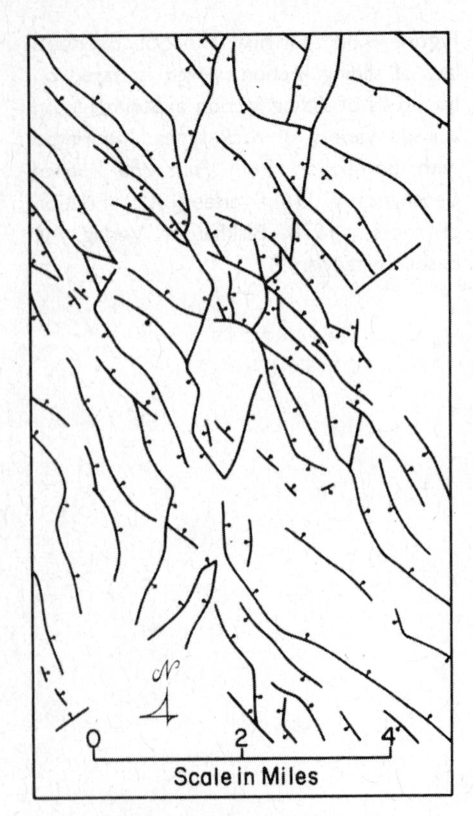

Figure 6.86 Geologic map pattern of faults mapped in south-central Oregon. [From Donath, 1962. Published with permission of Geological Society of America.]

this occurs, differential stress need not reach the magnitude required to form brand-new fractures. The geologic evidence we observe for repeated movements along joints and other fractures gives a tangible sense of such reactivation. *Part III-O, Engaging in Fault Analysis,* includes an example of the gory details of how to go about determining the range of fracture orientations most suitable for reactivation (as faults) within a defined stress field. Furthermore, *Part III-O* presents the Angelier (1979, 1994) method for predicting the direction of slip on reactivated preexisting fracture surfaces. This is a very cool procedure, which is indispensable in trying to make sense out of outcrops and regions of rocks marked by highly variable fault attitudes and slickenline rake orientations.

Faulting Within a Three-Dimensional Strain Environment

The Coulomb theory predicts that faults should form in conjugate sets. This theory seems to be supported by the results of triaxial deformation experiments. Yet faulted rocks in nature sometimes display *two* pairs of conjugate fault sets. Donath (1962) mapped such a pattern in south-central Oregon (Figure 6.86). One way to explain the presence of four fault sets is to call on two episodes of faulting, each of which yields conjugate pairs of faults. However, it is now clear that four fault sets can indeed be generated in a single event. How is this possible?

Reches (1978a) emphasized that the presence of three or four fault sets in a given region of study is the natural result of faulting within a three-dimensional strain field. The patterns seem peculiar only because rock deformation experiments traditionally have been carried out under conditions of two-dimensional coaxial stress and strain. Where specimens are allowed to shorten (or stretch) by different amounts in *three* mutually perpendicular directions, the characteristic fault pattern that emerges is one of three or more sets arranged in orthorhombic symmetry (Figure 6.87). Such a fault pattern was produced by Oertel (1965) in a clay cake subjected to stretching in a three-dimensional strain field. Also, Reches and Dieterich (1983) produced such patterns in cubes of sandstone, granite, and limestone that were subjected to compression in a three-dimensional strain field. Reches (1983) further showed that the angular relationships among fault sets formed in this way relate not only to the angle of internal friction (ϕ) of the host rock, but also to the ratio of strain along the principal finite stretch directions (S_1, S_2, and S_3). The work by Reches and Dieterich (1983) puts us on notice that rocks are highly sensitive to the difference between plane strain and three-dimensional strain.

Bob Krantz (1988) carried out an elegant analysis of faulting in a three-dimensional strain field. His field area was within the San Rafael Swell, a

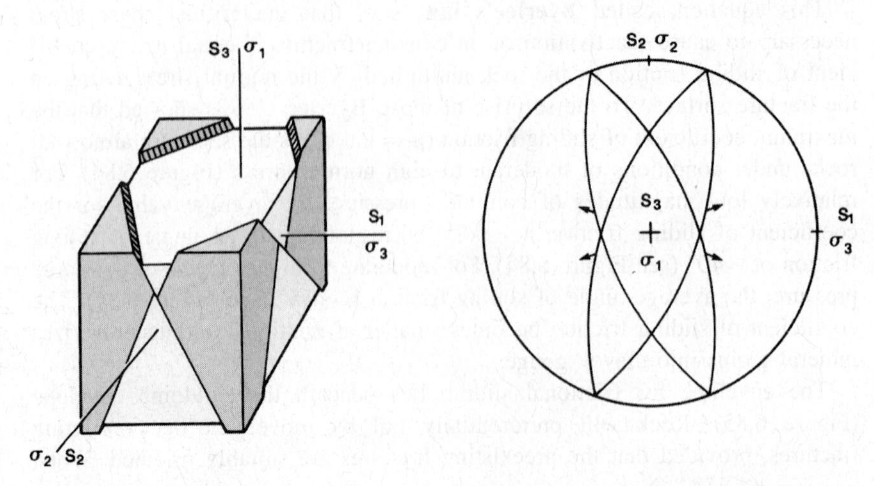

Figure 6.87 Fault sets produced in three-dimensional strain field. [Reprinted from *Tectonophysics*, v. 95, Reches, Z., Faulting of rocks in three-dimensional strain fields: II. Theoretical analysis, p. 133–156, © 1983, with permission from Elsevier.]

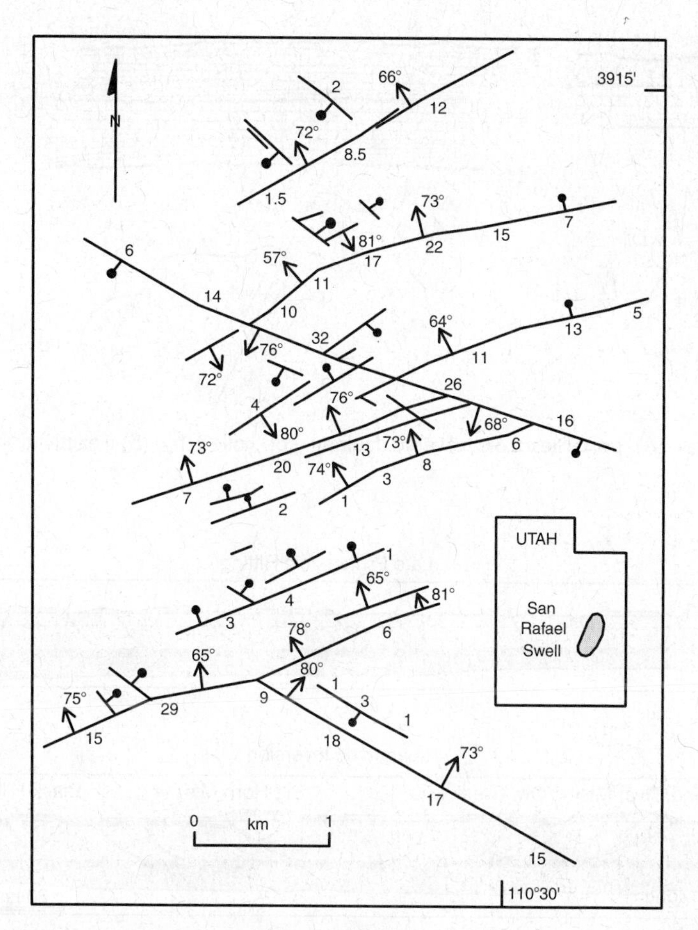

Figure 6.88 Bob Krantz prepared an extraordinary map of normal faults in the Chimney Rock area of the San Rafael Swell, Utah. The faults cut and displace the contact between Jurassic Navajo Sandstone and overlying Carmel Formation, and thus it is easy to determine amount of offset. The numbers along the fault traces denote offset, in meters. [From Krantz, 1986. Published with permission from the University of Arizona Press.]

major Colorado Plateau uplift. Exposures there are exceptional, and he was able to map out the full lengths and displacement gradients along the faults within the system (Figure 6.88). He measured large numbers of slickenlines on slickensided surfaces in order to determine the slip directions and their variations along each of the principle fault surfaces. On the basis of the fundamental theory developed by Reches, as well as a clever stereographic methodology he developed himself, Krantz determined both the directions and magnitudes of the principal strains responsible for the faulting. Testing these results with those calculated from line-length measurements in detailed cross-sections, he found there was a very close correspondence.

The Problem of Reverse Faults

Where does reverse faulting fit into the Anderson model of faulting and Coulomb theory? Reverse-slip faults typically dip 60° or more and accommodate crustal shortening. They are found in orogenic belts around the world, yet they are not featured in Anderson's fundamental classes of faults.

One obvious explanation for the origin of some reverse-slip faults is that they occupy former sites of normal faulting. Where reverse faulting is a result of **fault reactivation**, the dip of the reverse-slip fault is largely inherited from a previous event (Figure 6.89).

A regional example of this at the system scale is seen across the Wyoming Province and Colorado Plateau of the western United States. Late Proterozoic rifting produced by extension caused the formation of a system of normal faults (Figure 6.90*A*). During Late Cretaceous and early Tertiary time, regional (Laramide) contraction caused reactivation of many of these normal faults as reverse faults (Figure 6.90*B*). Transfer of reverse faulting of the

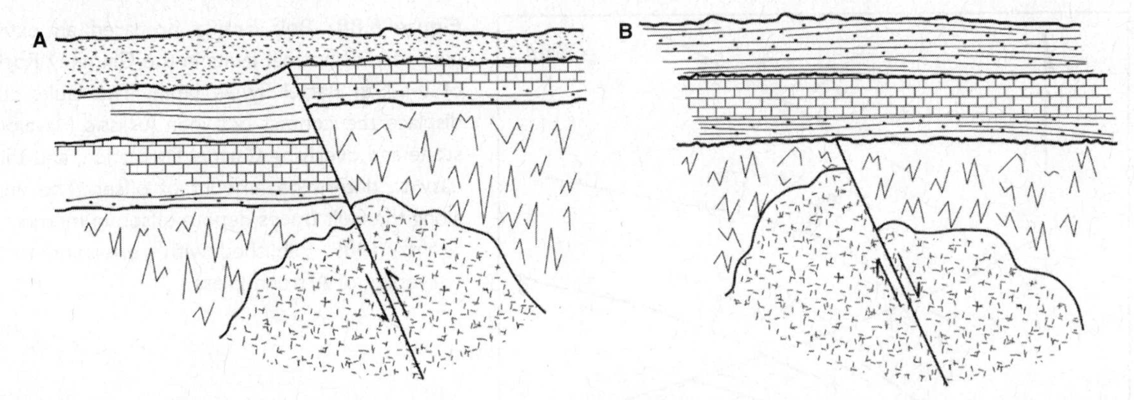

Figure 6.89 These geological cross-sections portray (*A*) post-Mesozoic reverse faulting controlled by (*B*) reactivation of a Precambrian normal fault.

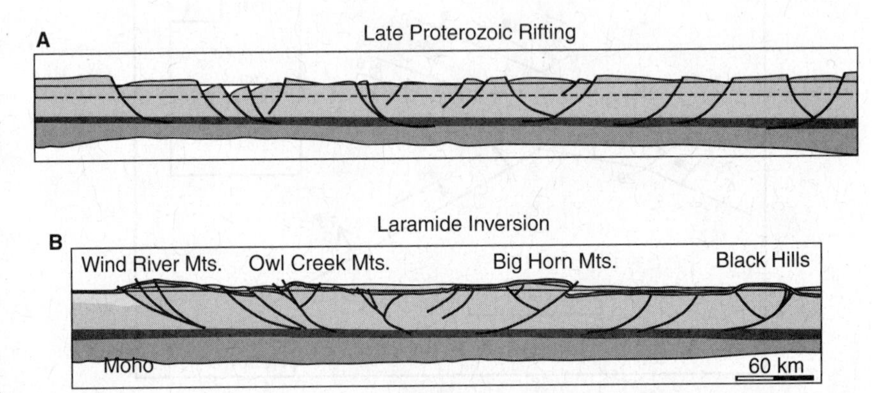

Figure 6.90 Reverse faults can form as a result of reactivation of normal faults. We see a system of such reactivation within the Wyoming Province and Colorado Plateau of the western United States. (*A*) Normal faults accommodated crustal extension in the Late Proterozoic. (*B*) Laramide contraction in Late Cretaceous and the early Tertiary caused stress-induced reactivation of the normal faults, converting them to reverse faults. Adapted from Marshak et al. (2000), Figs. 3A and 3B, p. 737. [Artwork from Davis and Bump (2009), Figure 15, p. 113. Published with permission of the Geological Society of America.]

basement into the overlying sedimentary cover is what produces the magnificent monoclines within the Colorado Plateau (Figure 6.91).

Another explanation of reverse faulting is that principal stress directions, though horizontal and vertical at the earth's surface, may twist and become inclined at depth. **Stress trajectories** become inclined and/or curved as the result of changes in the state of stress both laterally and vertically. Hafner (1951) demonstrated this theoretically. He showed, for example, that the

Figure 6.91 North-directed photograph of the East Kaibab monocline, southern Utah. This is a segment of the "*type*" monocline first described and defined by John Wesley Powell. In the foreground are Cretaceous shales (dark gray) and sandstones (light), east dipping. In the left (west) background are nearly flat lying Jurassic sandstones (Navajo Sandstone). [Photograph by G. H. Davis.]

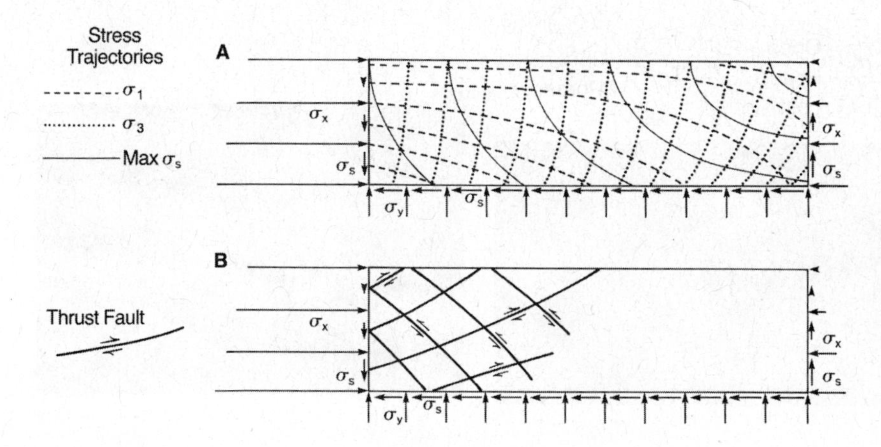

Figure 6.92 (*A*) Pattern of stress trajectories that would be produced in a block subjected to strong horizontal compressive stresses that die out laterally. (*B*) The pattern of curved faults that would emerge within the field of curved stress trajectories. [From Hafner, 1951. Published with permission of the Geological Society of America.]

dissipation of horizontal compressive stresses can result in a strain field marked by curved stress trajectories (Figure 6.92*A*). Curved compressive stress trajectories give rise to continuously curved faults, parts of which are thrust-slip faults, and parts of which are reverse-slip faults (Figure 6.92*B*).

Hafner's work (1951) showed that compression-induced thrusts at depth may steepen upward into reverse-slip faults. In contrast, Sanford (1959), Stearns (1978), and Friedman et al. (1976) have carried out experiments indicating that upper-level thrusts can steepen downward into reverse-slip faults (Figure 6.93). Interpreting the dynamics of formation of reverse faulting thrust requires full three-dimensional views of faults and fault systems.

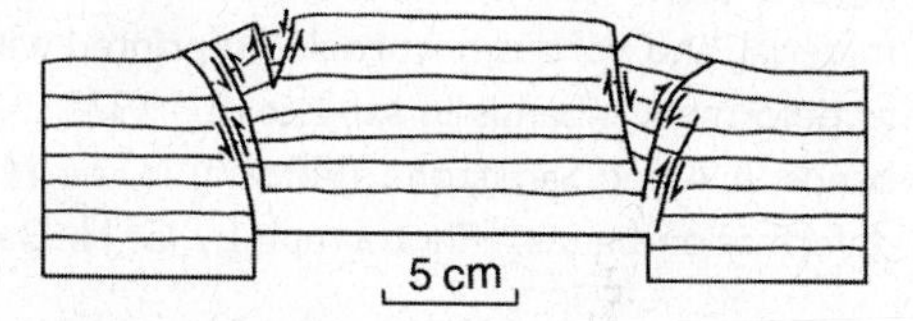

Figure 6.93 Curved reverse-slip faults produced by differential vertical uplift in deformation experiment. [From Sanford, 1959. Published with permission of the Geological Society of America.]

Deformation Bands: A Peculiar Brand of Faulting

Coulomb theory assumes that faulting proceeds without any significant dilation of the volume of rock that is deformed within a given fault zone. Yet there are certain rocks, notably porous pure-quartz sandstones, which fault in a special way that requires reduction of rock volume brought about by collapse of pore space. The physical and geometric characteristics of such faulting are quite different from what we are accustomed to seeing.

Credit Atilla Aydin and Arvid Johnson for bringing this kind of faulting to the attention of the structural geology community. **Deformation bands** (Aydin, 1978; Aydin and Johnson, 1978) occur preferentially in highly porous sandstones (e.g., aeolian sandstones with ~20% porosity) that have sheared under conditions that would produce in normal rocks slickenlined shear surfaces and faults with conspicuous offset. Deformation bands are brittle shear zones formed in ways accompanied by loss of porosity and volume (Figure 6.94). Deformation bands tend to be strongly resistant to erosion and thus stand out like veins, ribs, or fins in outcrop (Davis, 1998) (Figure 6.95). Commonly they occur in concentrated *zones* of deformation bands. These are not ordinary faults, for they form through a combination of shear and volume reduction.

Aydin's (1978) classic portrayal of the internal anatomy of deformation bands is pictured in Figure 6.96. The band itself is the dark core, composed of tiny angularly crushed quartz and nearly complete loss of porosity, in stark contrast to the (optimum) aeolian sandstone host rock (light gray, Figure 6.96) with its rounded sand grains and abundant (~20%) porosity. The formation of individual deformation bands begins with a stress-induced collapsing of pore spaces. Grains move closer and closer together, and eventually come into contact, eliminating porosity (see Figure 6.96). Internal friction will have increased to the point that further deformation must involve grain-scale microfracturing and cataclasis within the inner zone of the deformation band. A thin shear band thus develops (Figure 6.96*A*), marked by tiny offset. This

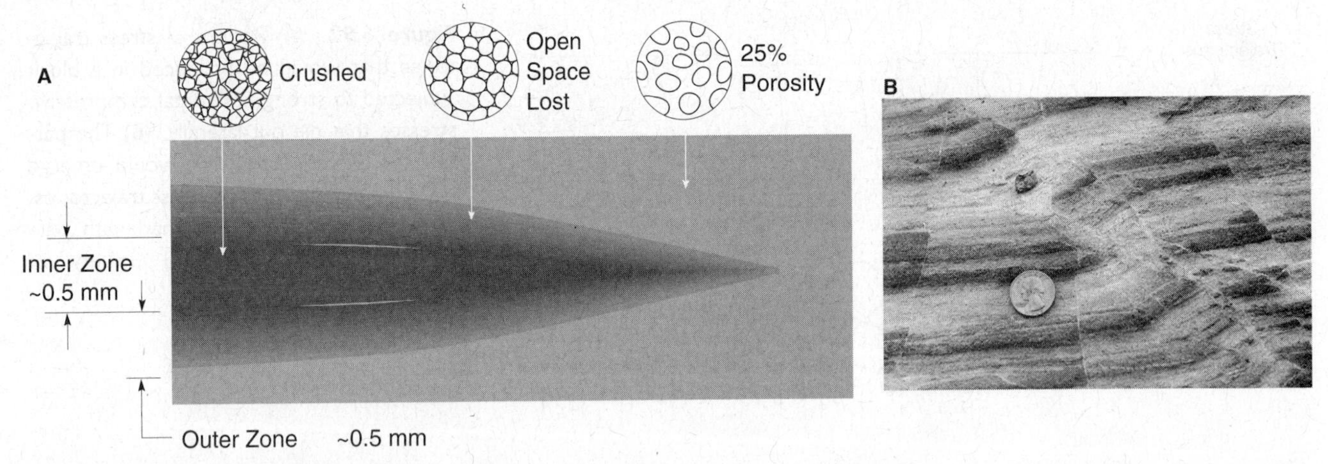

Figure 6.94 (*A*) Idealized anatomy of a deformation band, showing inner zone of crushed material, outer zone of compressed material, and porous host rock. [Reprinted with permission from *Pure and Applied Geophysics*, v. 116, Aydin, A., "Small faults formed as deformation bands in sandstone," 1978. Birhauser Verlag Ag, Basel, Switzerland.] (*B*) Photograph of thin, delicate deformation bands in Aztec Sandstone (Jurassic) in the Muddy Mountains, east of Las Vegas, Nevada. Note tiny fault offsets associated with deformation bands. [Photograph by G. H. Davis.]

Figure 6.95 (*A*) Photograph of an unusually good example of resistance-to-erosion of deformation band shear zones. Notice the several "*fault fins*" that project upward from Navajo Sandstone (Jurassic) along the East Kaibab monocline near Cottonwood Road south of Kodachrome Basin State Park in southern Utah. Pilar García is geologist. These deformation bands accommodated normal faulting. [From Davis, 1998, Fault-fin landscape, in *Geological Magazine*, v. 135, no. 2, p. 283–286. Published with permission of Cambridge University Press.] [Photograph by G. H. Davis.] (*B*) Photograph of outcrop-scale system of deformation bands (white). Camera case for scale. Differential resistance to wind erosion creates an outcrop appearance that Randy Tufts dubbed "*radiator rock*." This deformation band system is made up of conjugate faults. Location along East Kaibab monocline near Cottonwood Road. [Photograph by G. H. Davis.]

deformation band "*strain hardens*" to the point that if continued porosity-reduction and grain-fracturing are to continue, they will take place by initiating a new band alongside (Figure 6.96*B*). A zone of deformation bands will evolve in this way (Figure 6.96*C,D*) (Aydin 1978; Aydin and Johnson 1978, 1983). Oddly enough, the formation of a zone of deformation band transforms

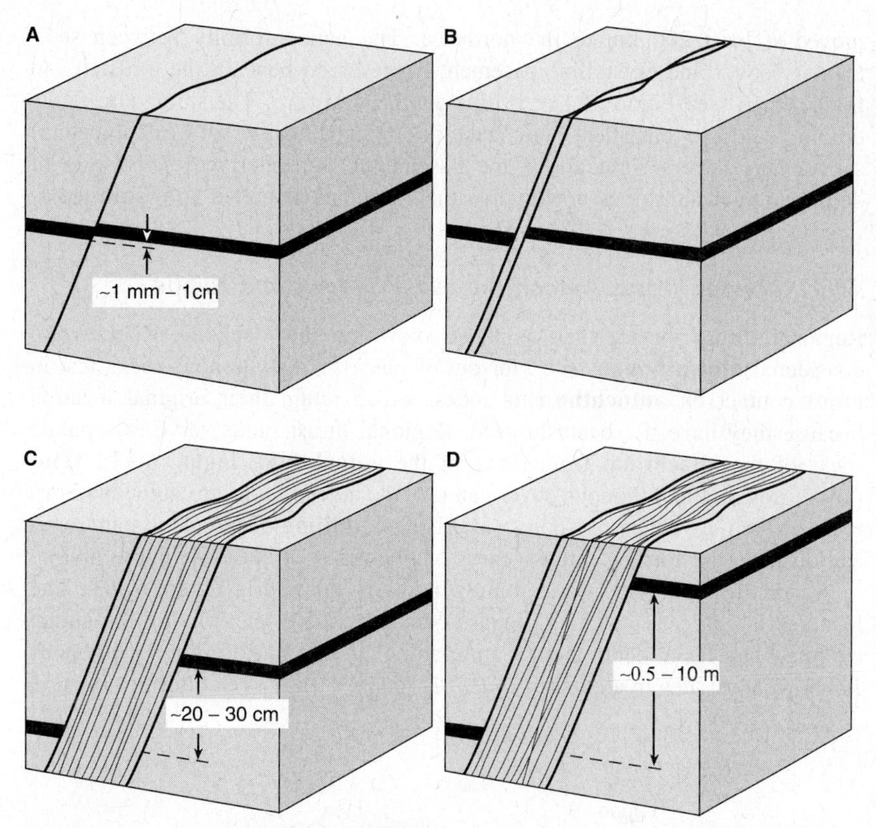

Figure 6.96 Progressive development of a zone of deformation bands, starting with a single band (*A*) and ending with full zone (*D*). [Reprinted with permission from *Pure and Applied Geophysics*, v. 116, Aydin, A., and Johnson, A. M., "Development of faults as zones of deformation bands and as slip surfaces in sandstone," 1978. Birhauser Verlag Ag, Basel, Switzerland.]

Figure 6.97 This fault surface with horizontal (strike-slip) slickenlines marks the outer surface of a deformation band shear zone within a system exposed in the Sheets Gulch area along the Waterpocket fold in Capital Reef National Park, southern Utah. Progressive porosity reduction as deformation bands formed ultimately produced a rock of standard mechanical strength, capable of faulting in a conventional manner. [Photograph by G. H. Davis.] [From Davis, 1999, Figure 132, p. 122. Published with permission of the Geological Society of America.]

porous rock to a much more rigid counterpart, which thus prepared is able to fault in a conventional manner (Figure 6.97).

THRUST FAULT SYSTEMS

Now that we know about the basic physical and geometric characteristics of faults at outcrop scale, and fault mechanics based upon laboratory experiments, it is useful to "*crank up*" the scale and consider the behavior of thrust faults, normal faults, and strike-slip faults as parts of larger fault systems. As we carry out this exploration, we inevitably get involved in an interplay of geometry, kinematics, and dynamics. We start here with thrust faults!

Regional Overthrusting

Canadian Rockies

The structural characteristics of thrust faults and the systems within which they arise are beautifully portrayed in regional cross-sections, such as those constructed for the Canadian Rockies of Alberta and British Columbia see Figure 1.43. Great expanses of strata have been deformed by thrusting in this region as a result of tectonic processes related to plate tectonic convergence. Price and Mountjoy (1970) estimated that the original sedimentary basin of the Canadian fold-and-thrust belt was about 500 km wide, tapering eastward from a thickness of 12 km to 2 km. The westernmost strata were moved at least 200 km to the northeast. The nonconformity between sedimentary cover and crystalline basement is preserved beneath the thrusted and folded mass (see Figure 1.43); it dips gently westward. The thrust-slip faults do not cut into basement, but feed into mechanically soft (incompetent) sedimentary layers right above the basement/cover interface. This style of deformation, which does not involve the basement, is called **thin skinned**.

Allochthonous Versus Autochthonous, Windows and Klippe

Regional thrust sheets, such as those shown in the Canadian Rockies are considered **allochthonous** (i.e., far out of place). Allochthonous rocks rest in thrust contact on **autochthonous** rocks, which retain their original location because they have not been thrusted. Regional thrust faults serve to separate rocks of the **allochthon** from rocks of the **autochthon** (Figure 6.98). **Windows** through allochthonous cover can provide a deep look into autochthonous rocks, which are otherwise concealed. Isolated **klippe** of allochthonous rocks can disclose the former extensiveness of overthrust strata (see Figure 6.98).

A beautiful example of a tectonic window is pictured in Figure 6.99*A*. The location is just east of Las Vegas, Nevada, where the Muddy Mountain overthrust places allochthonous Cambrian dolostones on top of autochthonous Jurassic sandstones in a classic "*older on younger*" thrust relationship. Erosion

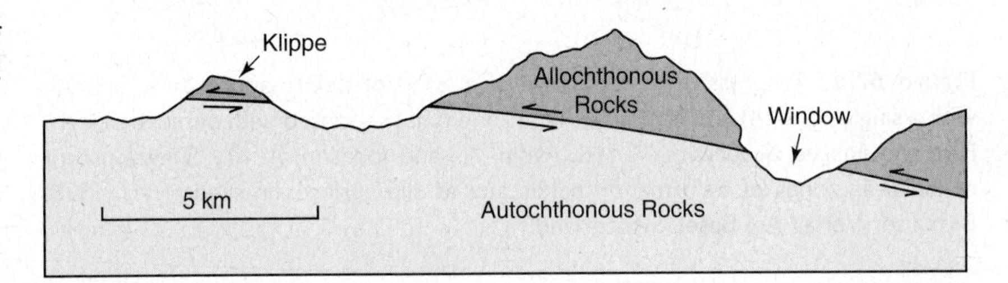

Figure 6.98 Cross-sectional view of allochthonous and autochthonous rocks, as well as klippe and window.

Figure 6.99 (*A*) Photograph of part of the Buffington (tectonic) window, which provides a view *through* the Muddy Mountains overthrust. The dark-colored rocks at higher elevations are Cambrian dolostones comprising here the upper plate of the Muddy Mountains overthrust. The light-colored rocks at lower elevations are Jurassic sandstones (Aztec Sandstone) comprising the lower plate of the thrust. Trace of thrust is marked. (*B*) An extraordinary cross-sectional exposure of the Muddy Mountains thrust. Again, the thrust places Cambrian rocks (dark rocks) on Jurassic rocks (light rocks); that is, older on younger, the tell-tale signature of thrust faulting. Trace of thrust does not need to be marked. These photographs were taken in the Buffington window, east of Las Vegas, Nevada. [Photographs by G. H. Davis.]

of the hanging wall (upper plate) has created a proverbial "*window*," permitting us to look down into the footwall (lower plate) of the thrust. This window is the "*Buffington window*." The dark-colored rocks shown in Figure 6.99*A* are the Cambrian dolostones of the upper plate; the light-colored rocks are the Jurassic sandstones of the lower plate (see Figure 6.99*B*). Without the presence of this window, it would be difficult to imagine that the Muddy Mountain thrust is so flat and, moreover, created kilometers of overlap! Figure 6.99*B* is another shot of the Muddy Mountain thrust, this time viewed in a more conventional cross-sectional view.

A nice example of a tectonic klippe is shown in Figure 6.100*A*. The location is the Sanctuary of Zeus, Mt. Lykaion in Arcadia, within the Peloponessos of Greece. The mountain peak itself comprises the St. Elijah klippe. The base of the klippe is the low-angle Lykaion thrust fault. In map view this klippe resembles an island (Figure 6.100*B*), with the thrust trace circumnavigating the mountain and everywhere marking the base of the klippe. This klippe relationship is made more complicated because of active normal faulting that cuts and down-drops segments of the thrust fault. As a result, the thrust trace is not a continuous "*loop*" around the base of the mountain, and in some places the thrust trace is covered by large landslide deposits coming off of the active faults. Concealed contacts are dotted (see Figure 6.100*B*).

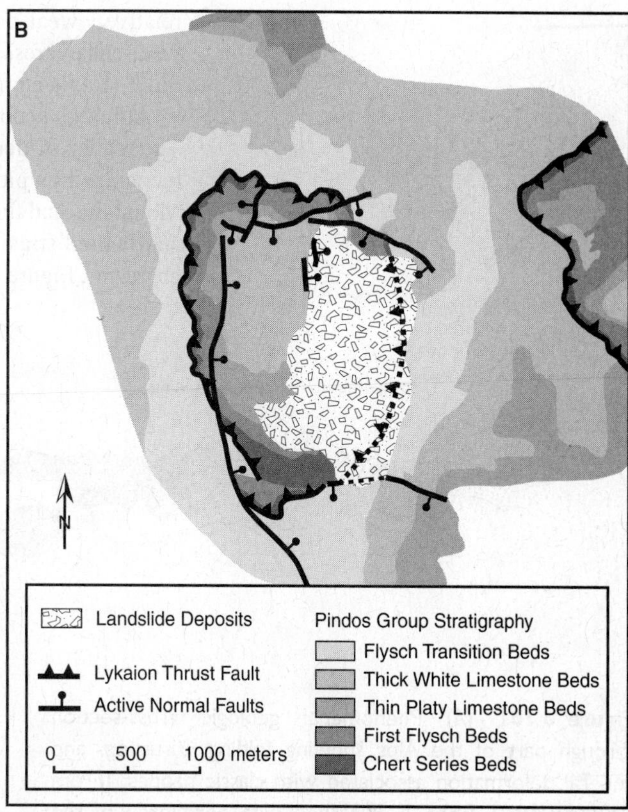

Figure 6.100 (*A*) South-directed photograph of Mt. Lyakion. The part of the mountain above the Lykaion thrust (marked) comprises the klippe, which is composed of folded and faulted Jurassic to Paleocene strata. The thrust itself dips merely 10° westward. It so happens that the Lykaion thrust separates the upper and lower sanctuaries at the Sanctuary of Zeus, Mt. Lykaion, near Megalopolis, Greece. The lower sanctuary includes a hippodrome (chariot racing track), baths, and a processional way. The upper sanctuary includes a mountaintop (bone) ash altar. Note that Thick White Limestone Beds in the foreground are below the thrust, and then repeated by thrusting in such a way they reappear in the upper left at much higher elevation. [Photograph and interpretation by G. H. Davis.] (*B*) Map of the Lykaion klippe. Note that the thrust fault trace encircles the mountain. The trace is not entirely continuous, because in places it is cut and offset by active normal faults. Mapping by G. H. Davis.

Thin-Skinned Deformation and Decollement

The contradiction of overthrusting is revealed in the structure profile of the Canadian Rockies (see Figure 1.43): strongly folded, faulted sequences of "*miogeoclinal*" strata rest atop basement that clearly has not been involved in the distortion. Dahlstrom, in reference to the Canadian Rockies belt, wrote:

> ***. . . there is never a balance between the length of Mesozoic and Paleozoic beds and the length of basement. The "cover" beds are always too long, which is explained in cross-section by a sole fault along which the upper beds have been moved (shoved? glided?) into the cross-section from the west.*** (From "Balanced Cross-sections" by D. C. A. Dahlstrom, 1969, p. 746. Reproduced by permission of National Research Council of Canada from *Canadian Journal of Earth Sciences.*)

"*Sole fault,*" as used above by Dahlstrom, has other names as well. Some call it a **basal shearing plane** (DeSitter, 1964). The commonly used term is **decollement**, although as DeSitter points out the term refers not to a structure but to a process: "*the detachment of the upper cover from its substratum.*" The position of a decollement is typically along an interface marked by major ductility contrast, permitting cover to become distorted independently of what is underneath. The actual basal decollement itself is commonly a weak horizon, such as clay or shale or salt (Davis and Engelder, 1985). At depths of 10 km or less, clay and shale are relatively weak compared to most rocks, and salt is *vastly* weaker. Salt will flow when shear stress is 1 MPa or less and can be thought of (structurally) as a sort of fluid! (Davis and Engelder, 1985; Carter and Hansen, 1983).

Although decollement surfaces are easy to describe, they are tough to explain. The reality of **decollement faulting** and **thin-skinned overthrusting** was first recognized by pioneer geologists of the nineteenth century mapping in the Jura Mountains and the Alps. They mapped enormous, far-traveled **nappes** of folded and faulted strata resting on crystalline rocks of an entirely different structural character. Figure 6.101*A* gives a sense of the huge scale of nappes, as well as the

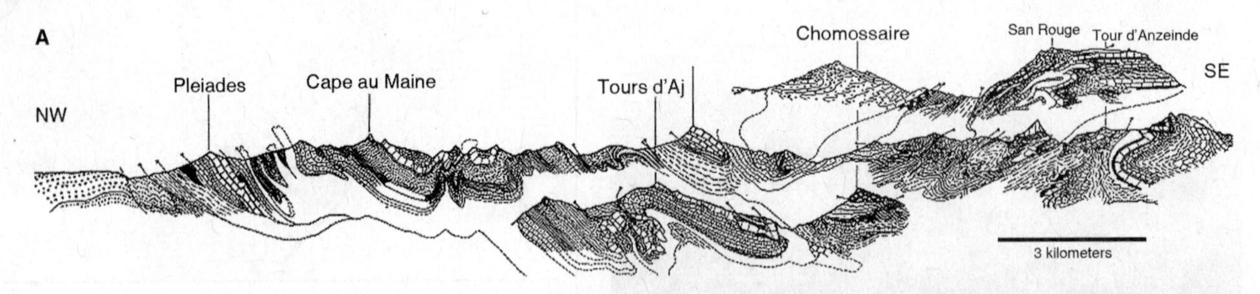

Figure 6.101 (*A*) Phenomenal geologic cross-section through part of the Alps showing folding, thrusting, and general deformation associated with classic nappes. [From Trûmpy, 1960, Plate 2. Published with permission of the Geological Society of America.] (*B*) Miniature example of nappe structure, marked by recumbent folds (i.e., folds lying on their sides) and a decollement sparating folded rocks above from unfolded rocks below. From Caledonian nappe terrain in northern Norway. Collection of G. H. Davis. [Photography and editing by Gary Mackender. vr.arizona.edu, glm@email.arizona.edu, ©2009–10 Arizona Board of Regents.]

profound internal deformation of the strata that have been thrusted for such distances. The folding within nappes is spectacular at the scale of mountains, and can be just as spectacular in outcrops and hand samples (Figure 6.101*B*). Nappe structures give reality to the expression, "*soft as a rock*."

The image of thin-skinned overthrusting and the formation of nappes haunted geologists for decades and decades, especially in terms of questions regarding the dynamics of overthrusting: If the thrusts formed by compressional shortening, why was the basement not affected? Or was it? If the deformed cover moved as a whole, what magnitude of force was required to translate it by more than 150 km? This conundrum is referred to as the **paradox of overthrusting**, for the force required to move the thrust sheets horizontally such a long distance is much greater than the crushing strength of the rock.

The Mechanical Paradox of Overthrusting

Show-Stopping "Breakthrough" by Hubbert and Rubey

One of the most provocative papers on the mechanics of thrust faulting was written by Hubbert and Rubey (1959), in a paper entitled "*The Role of Fluid Pressure in Mechanics of Overthrust Faulting.*" Hubbert and Rubey underscored the dominant, if not essential, role of elevated fluid pressure during low-angle tectonic transport of great overthrust sheets. They calculated the approximate amount of force that would be required to translate allochthonous thrust terranes of the size known to exist in the Canadian Rockies, the Western Cordillera of the United States, and the Appalachians. They concluded that the calculated force, if applied to the rear of the mass of crust, would far exceed the crushing strength of granite. In fact they concluded that the longest thrust sheet that could be pushed across a horizontal surface would be limited to ~10 km or so. Considering the possibility that thrust sheets are not pushed, but rather slide "*downhill*" under the influence of gravity, Hubbert and Rubey calculated as well the dynamic requirements of gravitational models. The results were the same: large thrust sheets should not exist, even if gravity is the propelling mechanism.

Their solution to the paradox of overthrusting involved modifying the Coulomb law of failure, in the way that we examined earlier in this chapter. They concluded that the critical shear stress (σ_c) required to cause slip on a thrust fault, which has no cohesive strength (σ_0), must be equal simply to the product of the coefficient of internal friction (tan ϕ) and the effective normal stress ($\sigma_n{}^*$), which is normal stress (σ_n) minus pore fluid pressure (P_f):

$$\sigma_c = \sigma_0 + \tan\ \phi\,(\sigma^*{}_n)$$

$$\sigma_c = \sigma_0 + \tan\ \phi\ (\sigma_n - P_f)$$

Modern sedimentary basins, such as the Gulf of Mexico, are marked by bedding-parallel zones of fluid pressure that are so highly elevated that they approach the value of the lithostatic stress of the overlying sedimentary cover. Under such circumstances, *the fluid pressure essentially supports the weight of all the rock above!* When effective normal stress approaches zero, resistance to thrusting becomes almost inconsequential. All that is required to move a thrust sheet is to overcome sliding friction. A visual picture of this effect is an air hockey table. Before the air is turned on and comes up through the holes in the table to "*float*" the hockey puck, force is required to slide the puck across the surface of the table. In contrast, when compressed air is turned on and forced up through the many closely spaced pin holes in the table, a mere flick of a finger will send the puck careening and bouncing across the surface. The air decreases the effective stress and makes sliding much easier.

The Famous Beer Can Experiment

Hubbert and Rubey (1959) demonstrated the fundamentals of the fluid pressure model in their now-famous beer can experiment. Sample preparation consists of drinking two beers, preferably out of non-aluminum cans (Figure 6.102*A*). Place one of the empties in the freezer (Figure 6.102*B*), and remove a window from your house or apartment or lab (Figure 6.102*C*). Clean the glass with detergent, rinse, and leave it wet with a thin film of water. Place the can that is not in the freezer, top down, on the pane of glass. Now lift one end of the glass to form an inclined plane, and, with protractor in hand, measure the angle at which the beer can commences movement down the plane (Figure 6.102*D*). Hubbert and Rubey (1959) report typical angles of $\sim 17°$, corresponding to a coefficient of sliding friction of metal on wet glass of ~ 0.3. After the can in the freezer has been chilled, quickly pull it out and perform the same exercise (Figure 6.102*E*). This time the beer can begin to move down the inclined plane at negligible angles of slope (somewhere around 1°). It moves easily because fluid pressure derived from expansion of the warming air inside the can offsets the normal stress exerted by the can on the glass. Hours can be spent enjoying experiments on the role of fluid pressure in overthrusting (Figure 6.102*F*).

The beer can experiment is compelling in more ways than one. However, almost as soon as Hubbert and Ruby (1959) suggested that high fluid pressure could resolve the paradox of overthrusting, many geologists worried about how high fluid pressures in thrust belts can be maintained over the long-lived geologic time spans during which the thrusting takes place. This dilemma was solved to some extent by imagining that an active thrust fault operating at depth creates a compressional stress environment out in front of its propagating tip line. When this compression front impacts a yet unfaulted flat-lying sedimentary sequence, the interstitial pore fluids within these yet-unfaulted sedimentary rocks experience a build-up in fluid pressure. As the fluid pressure builds, effective stress decreases, making it possible for the thrust front to

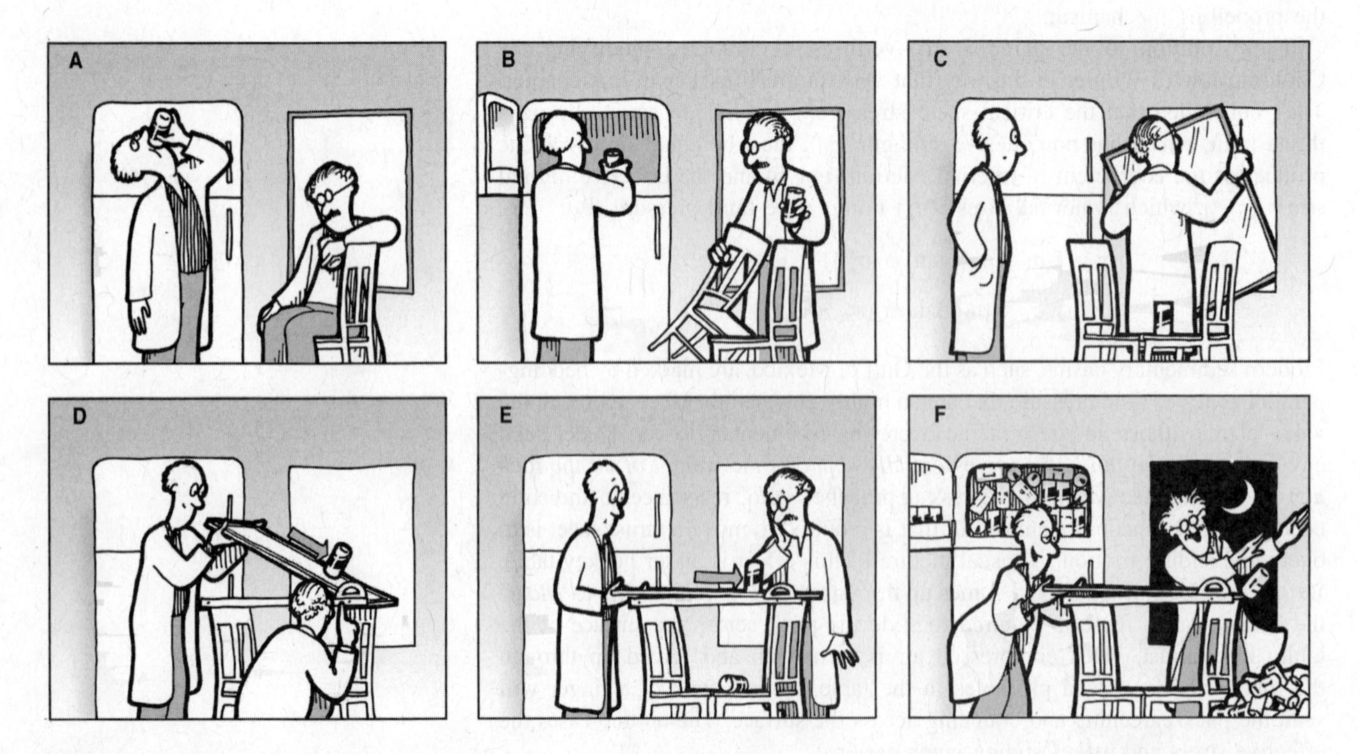

Figure 6.102 Picturing the beer can experiment in cartoon fashion. See text for explanation. [Artwork by D. A. Fischer.]

move forward, buoyed on the pressurized footwall. Although relatively rare, there are places along thrust faults where the signs of such **transient high fluid pressures** are obvious, e.g., in the form of sandstone dikes.

The Aha Moment: Thrust Sheets Don't Move as One Big Mass

It turns out that study of active thrust faulting, especially the lessons learned from earthquake seismology, has provided much-needed insight regarding how to solve the mechanical paradox of overthrusting. Ray Price (1988) concluded that a regional thrust sheet does not move forward all at once as a single entity. Instead, a given thrust earthquake is a response to the release of strain energy brought about by movement of only some very tiny part of the regional overthrust system. The movement of the entire thrust sheet proceeds a little like the movement of an inch worm (*aka* 2.54 cm worm). Slip on one part of the fault surface will release strain energy at one location and yet "*load up*" other parts of the fault surface with strain energy, thus setting the stage for the next increment of slip.

The amount of stress required to activate any tiny part of the overall thrust fault is, of course, orders of magnitude less than that which would be required to advance the entire thrust sheet all at once. In retrospect, we "*all should have known*" that the classic assumption ("*it all moves at once*!") was a bad assumption. Elevated fluid pressure probably plays an important role in thrusting, but its role and expression normally may be quite transient and local, building up to some critical level, triggering slip along a fraction of the entire thrust fault, and then dissipating as slip occurs. Again and again the fluid pressure builds to levels sufficient for thrusting to occur, establishing a cyclic process called **hydraulic pumping**. Such a mechanism is consistent with what earthquake seismologists "*see*" during an earthquake/slip event.

The Wedge Model of Overthrusting

Chappel's Insight into Wedge-Shaped Geometry of Thrust Belts

Yet another major insight ultimately resolved completely the mechanical paradox of overthrusting, though it took some time for the implications to sink in. In 1978 an extraordinary structural geologist by the name of Bill Chappel pointed out that thin-skinned fold and thrust belts are in fact **wedge shaped**, not sheet-like (Figure 6.103), thicker at the back end from which the thrusts come. Furthermore, Chappel emphasized that such regional wedges build up on top of an underlying weak decollement, and that the wedges are strongly internally deformed (Chappel, 1978, p. 1189). Chappel incorporated these fundamental characteristics into modeling of the mechanics of thin-skinned thrusting. No more blocks sliding on tabletops, nor beer cans moving down imperceptible gradients on glass table tops.

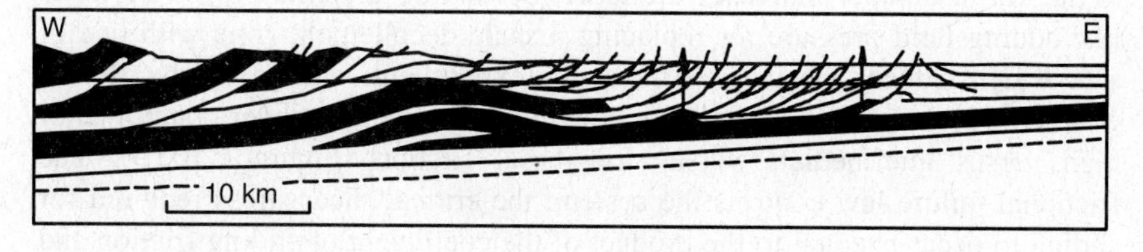

Figure 6.103 Wedge-shaped nature of thrust belts, as illustrated by the Canadian Rockies. [From D. M. Davis, J. Suppe, and F. A. Dahlen, *Journal of Geophysical Research*, v. 88, fig. 1a, p. 1154, copyright © 1983 by American Geophysical Union.]

Critical Taper

Davis et al. (1983) and Dahlen et al. (1984) constructed wedge-shaped geometries into their numerical and experimental models of thin-skinned folding and thrusting. As a result of their modeling, Davis et al. (1983) came

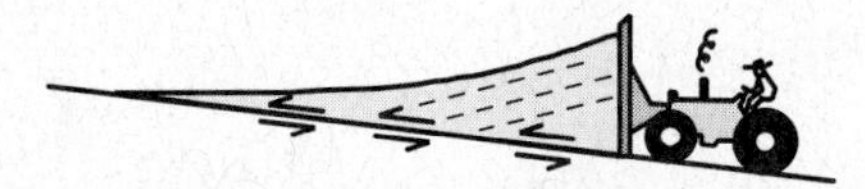

Figure 6.104 The mechanics of development of a thrust belt is analogous to the movement of a mass of snow, sand, or dirt as it piles up and eventually slides *en masse* in front of the blade of a tractor. This is a far different image from pushing a large rectangular block of rock in front of the blade. [From F. A. Dahlen, J. Suppe, and D. M. Davis, *Journal of Geophysical Research*, v. 89, fig. 3, p. 10,088, copyright © 1984 by American Geophysical Union.]

up with the concept of **critical taper** for thrust belts. They assumed that wedge-shaped thrust belts are populated by "*Coulomb materials*," which by definition fail in strict accord to Coulomb failure criteria. They envisioned that the wedge-shaped geometric form of a thrust belt is analogous to the wedge-shaped volume of sand or gravel or snow that builds up in front of the blade of a bulldozer (Figure 6.104). Next time you are standing around with three or four others watching a bulldozer in action, you will see that the blanket of cohesionless material (e.g., sand) will not slide as a mass until a wedge shape is first attained (see Figure 6.104). Then, and only then, the whole mass moves.

If you look at the action even more closely, and squint at the wedge through a transparent plastic protractor, you will see that not just any wedge shape will do. Instead, before moving as a mass, the wedge must attain a certain critical taper whose angle, measured between the pavement and the upper surface of the wedge, relates to the internal friction of the material being shoved (sand versus gravel versus snow) and the external sliding friction of the surface over which the material is piled (concrete aggregate versus blacktop versus cobblestone). The critical taper builds through internal deformation of the material within the wedge.

Definitive Critical-Taper Experiment

Davis et al. (1983, p. 1153) summarized critical taper theory in this way:

> ***Material deforms until a critical taper is attained, after which it slides stably, continuing to grow at constant taper as additional material is encountered at the toe. The critical taper is the shape for which the wedge is on the verge of failure under horizontal compression everywhere, including the basal decollement.***

Davis et al. (1983) built a deformation apparatus to create wedge models of thrusting in the laboratory. The results of their experiments were so compelling that other workers followed their lead right away. For example, Huiqui et al. (1992) built the apparatus shown in Figure 6.105*A*. The original Davis-Dahlen-Suppe apparatus design permits changes in the dip of the decollement (Figure 6.105*B*). Sand is stratified into color-coded layers atop a Mylar sheet, which serves as a decollement. The Mylar is wound around a spool in such a way that the Mylar can be "*reeled in*," thereby transporting the overlying sand against the back wall of the wooden frame. A little graphite on the sidewall reduces friction. The sand thickens initially at the wooden buttress, and then the locus of thickening moves forward until a uniform smooth taper is achieved (Figure 6.105*C*). Thickening is achieved through thrusting and folding.

Ultimately, when just the right taper is achieved, internal deformation of the sand ceases and the Mylar slips freely beneath the load of sand. Swapping Mylar for sandpaper increases the taper. Sprinkling graphite on the Mylar is like adding fluid pressure, or replacing a shale decollement zone with one of salt, which reduces the taper. Among the experiments carried out by Huiqui et al. (1992) were ones to illustrate how the taper changes under conditions of high versus intermediate versus low basal friction (Figure 6.105*D*). The frictional failure law controls the system: the critical shear stress required for sliding to occur is equal to the product of the coefficient of sliding friction and the effective normal stress!

A beautiful context for the wedge experiments is the literature on regional thrusting in Taiwan contributed by John Suppe. He has carefully worked out the geometry of the regional thrusting (Figure 6.106). Furthermore, as one of the principal developers of the wedge model, he has confirmed the

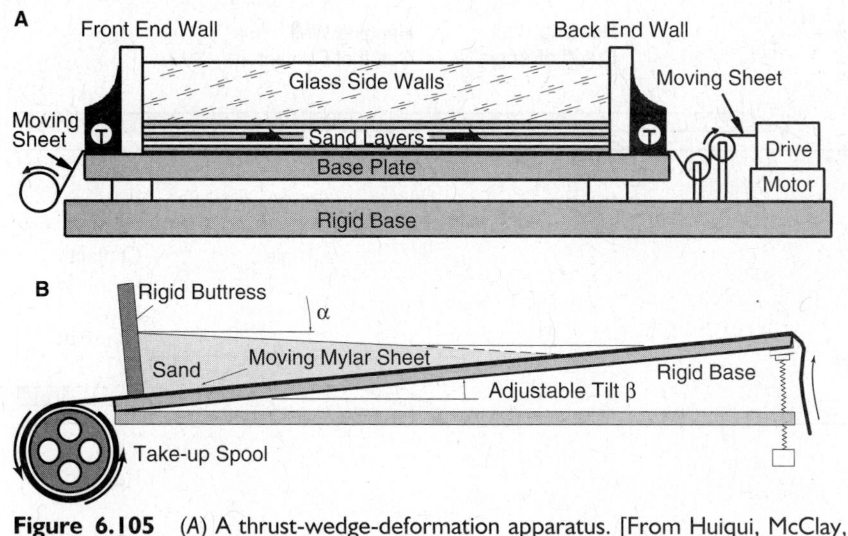

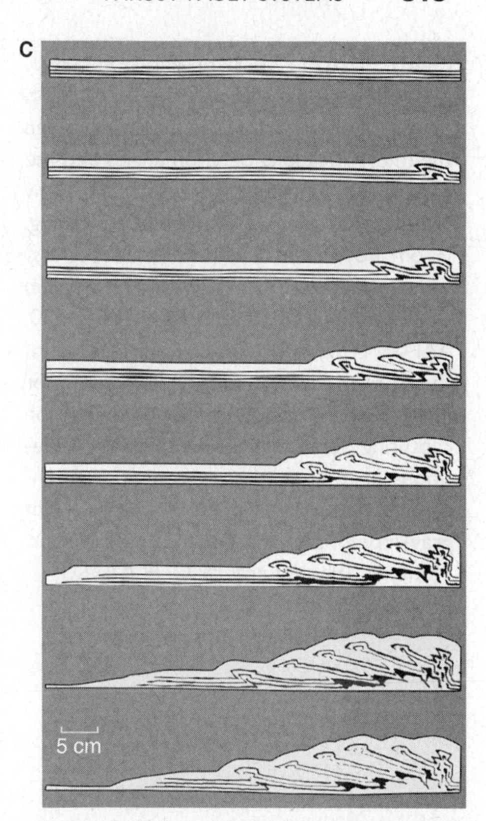

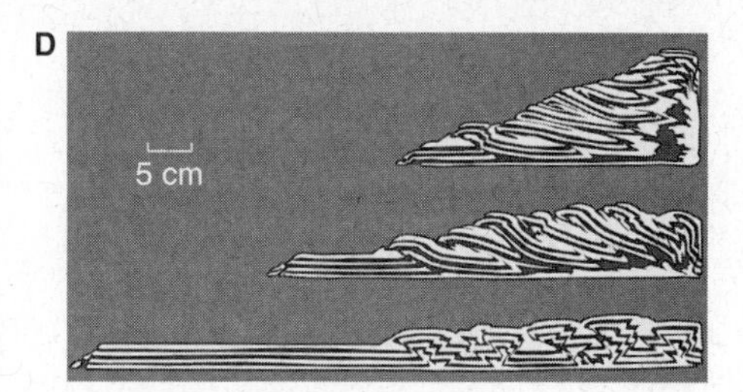

Figure 6.105 (*A*) A thrust-wedge-deformation apparatus. [From Huiqui, McClay, and Powell, 1992, Figure 1, p. 72, reprinted with permission from Chapman and Hall.] (*B*) The original design of the thrust-wedge-deformation apparatus creates the opportunity to vary the dip of the decollement. [From D. M. Davis, J. Suppe, and F. A. Dahlen, *Journal of Geophysical Research*, v. 88, fig. 2, p. 1155, copyright © 1983 by American Geophysical Union.] (*C*) Steady motion of the Mylar sheet beneath the sand draws the stratigraphic sequence into the rigid buttress, creating a stable wedge via thrusting and folding of the sand layers. The progressive shortening values in this sequence are 0%, 3%, 9%, 18%, 27%, 39%, 48%, and 49%. [From Huiqui, McClay, and Powell, 1992, Figure 4, p. 74, reprinted with permission from Chapman and Hall.] (*D*) Marvelous portrayal of the influence of basal friction on wedge geometry. From top to bottom, high to intermediate to low basal friction. [From Huiqui, McClay, and Powell, 1992, Figure 7, p. 78, reprinted with permission from Chapman and Hall.]

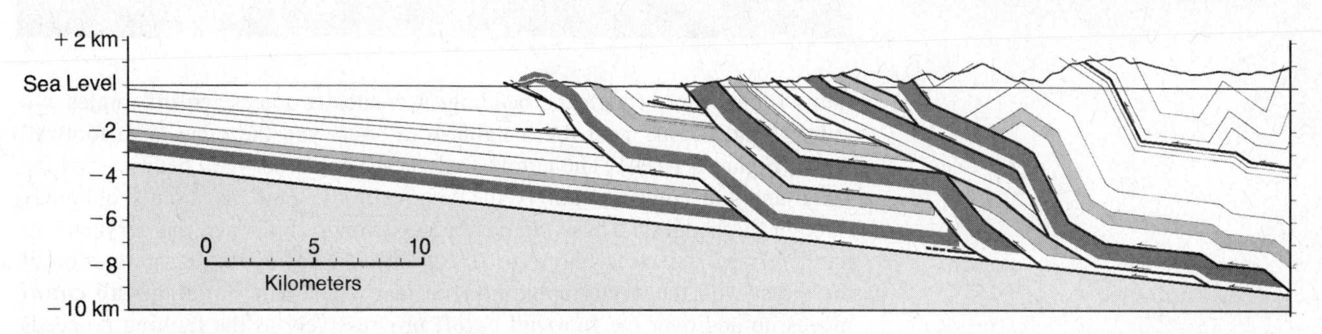

Figure 6.106 Regional structure section showing an array of stacked thrusts in Taiwan. [From Suppe, 1980a. Published with permission of Geological Society of China.]

applicability of the wedge hypothesis to the earthquake activity taking place today in Taiwan (Suppe, 1980; Carena, et al., 2002; and Yue et al., 2005).

Ramp-Flat Geometry and Kinematics

Major thrust faults typically have a stair-step, or **ramp-flat** geometry (Figure 6.107). Individual fault surfaces step from layer-parallel segments (flats) within soft, incompetent layering and cut obliquely across stiff, competent beds (ramps) en route to the next favorable incompetent unit. Where a thrust

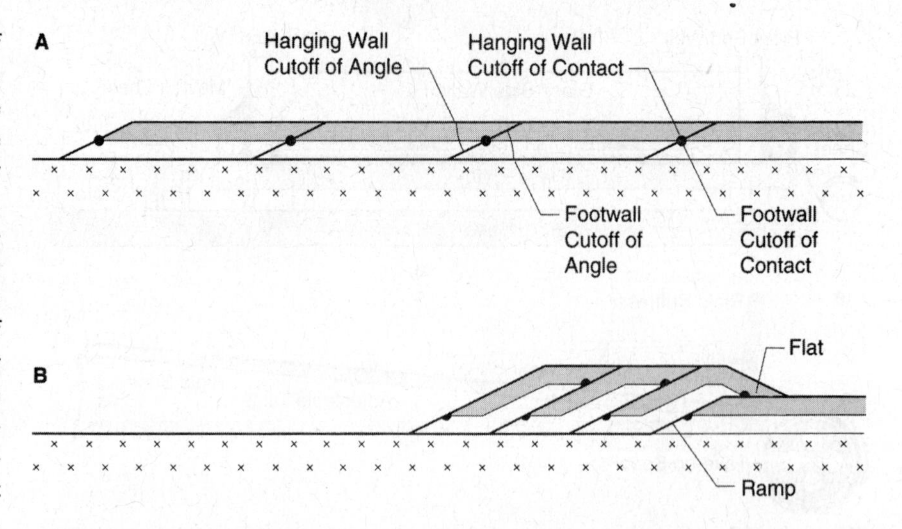

Figure 6.107 The ramp-flat geometry of a typical regional thrust (Boyer and Elliot, 1982). (*A*) Before thrusting starts, we see the future locations of cutoff angles and the hanging wall and footwall cutoffs. (*B*) Now we see the end result of the thrusting. [From Boyer and Elliot, 1982, Thrust systems: *American Association of Petroleum Geologists Bulletin*, v. 66, p. 1196–1230. AAPG©1982, reprinted by permission of the AAPG whose permission is required for further use.] (*C*) South-directed photograph capturing aerial expression of thrust repetitions, in this case repetition-upon-repetition of Mississippian limestone (light gray) comprising the Sawtooth Mountains, Montana. [Photograph by G. H. Davis.]

fault "*ramps*" up through the bedding, it creates two acute **cutoff angles** and **cutoff lengths**, one with the hanging-wall strata and the other with footwall strata (Figure 6.107*A*). The cutoff angle is the angle between bedding and the fault, and the cutoff length is the length of the fault as it cuts obliquely through the bedding. These characteristics strongly influence the geometry of the deformed state (see Figure 6.107*B*). **Cutoff lines** mark the intersection of the thrust with the stratigraphic horizon that is cut. The **hanging-wall cutoff** moves up and over the **footwall cutoff** progressively as the faulting proceeds (Figure 6.107*B*) (Boyer and Elliott, 1982). The net result is repetition-upon-repetition of older-on-younger relations (see Figure 6.107*B*). Such repetitions are well displayed in the Sawtooth Mountain, Montana (Figure 6.107*C*).

We can picture the progressive development of the thrust fault as the leading edge (the **tip line**) of the fault moves its way along horizontal bedding (flat), then acutely across the bedding (ramp), and then back into bedding (flat). The "*ramp-flat*" terminology helps us to describe three possible fault-bedding arrangements based on the bedding orientation relative to fault plane orientation in the hanging wall versus footwall (Figure 6.108). These relationships include (1) hanging-wall flat on footwall flat (called "*flat on flat*"); (2) hanging-wall flat on footwall ramp ("*flat on ramp*"); and (3) hanging-wall ramp on footwall flat ("*ramp on flat*"). Woodward et al. (1985) emphasized

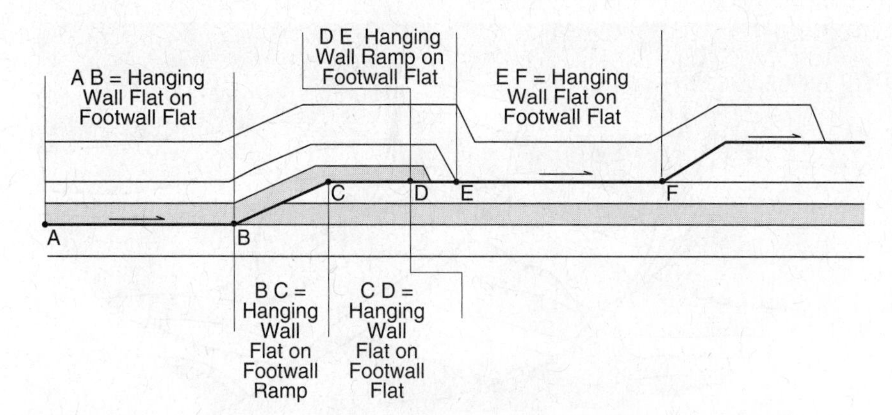

Figure 6.108 Representation of the three types of fault-contact relationships associated with ramp-flat thrust systems. [From Marshak/Mitra, Basic methods of structural geology, © 1988, p. 305. Reprinted by permission of Prentice Hall, Upper Saddle River, New Jersey.]

that movement along a **flat** (i.e., a flat-fault segment parallel to bedding) is not reflected in stratigraphic separation until the **hanging-wall flat** moves onto another unit at the **footwall ramp**.

Lateral Ramps, Tear Faults, and Compartmental Faults

Some ramps are **lateral ramps** (Figure 6.109). These are places where a thrust flat abruptly cuts up-section laterally along the strike of the thrust sheet. It

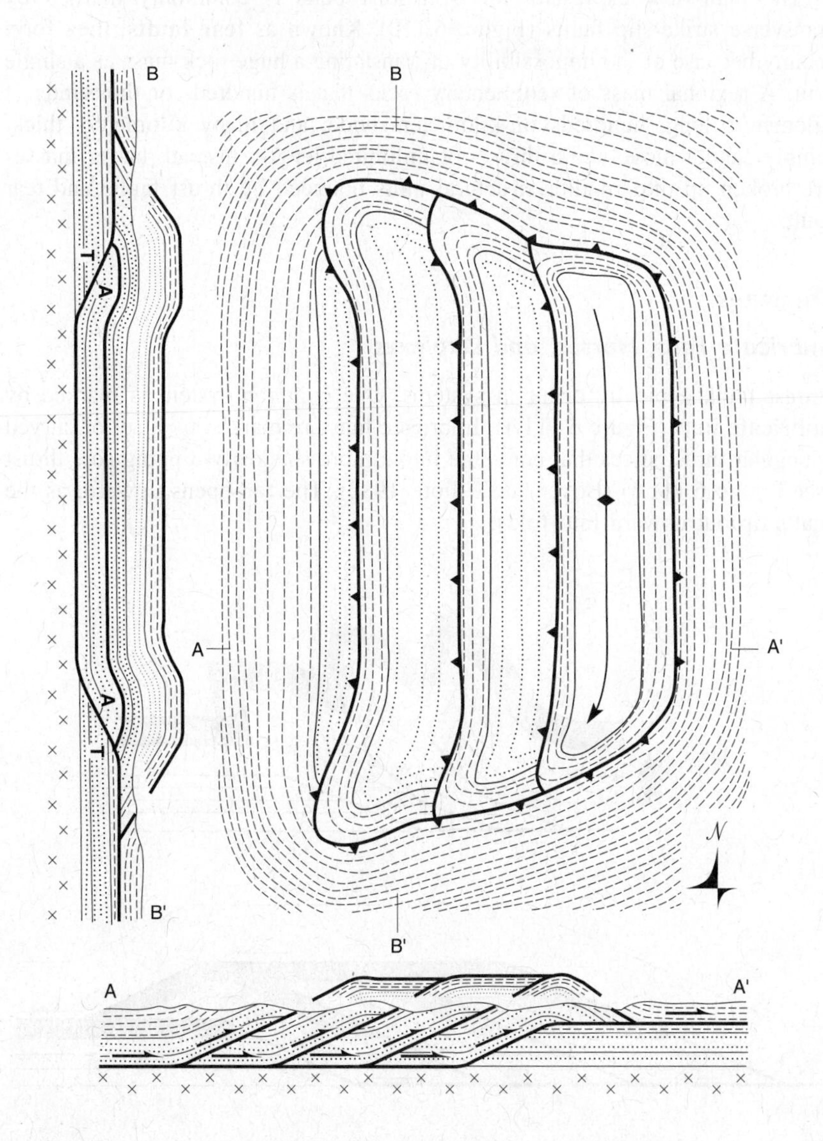

Figure 6.109 Cross-section B-B' is a longitudinal cross-section through a thrust system. The direction of thrusting is from west to east, and the view of section B-B' is in the direction of tectonic transport. Hanging wall of the lowest thrust is moving away ("*A*") from the viewer. Relative movement of footwall is toward ("*T*") the viewer. Note that hanging-wall material has not only moved laterally in strike-slip fashion, but it has also ramped up and over the footwall on the flanks. [From Boyer and Elliot, 1982, Thrust systems: *American Association of Petroleum Geologists Bulletin*, v. 66, p. 1196–1230. AAPG©1982, reprinted by permission of the AAPG whose permission is required for further use.]

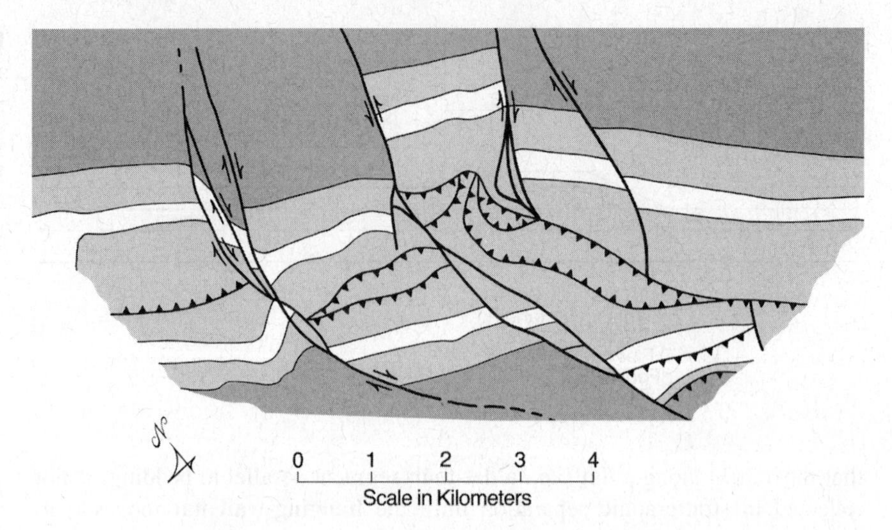

Figure 6.110 Map showing tear faults within a regional system of folds and thrust faults. [Adapted from Price, Mountjoy, and Cook (1978).]

ramps upward in a way that is discordant to bedding until it reaches a higher glide horizon, where once again it becomes a flat. Fault movement along a lateral ramp includes a significant component of strike-slip, for the trend of the ramp is usually subparallel to the overall direction of thrusting.

The plan-view expression of overthrust belts is commonly marked by transverse strike-slip faults (Figure 6.110). Known as **tear faults**, they form mainly because of the impossibility of translating a huge rock mass as a single unit. A regional mass of sedimentary rocks that is hundreds or thousands of kilometers long, hundreds of kilometers wide, and many kilometers thick, simply cannot move along thrusts of unlimited length. Instead, larger masses are broken up into smaller structural units bounded by thrust faults and tear faults.

Duplexes

Imbricate Fans, Horses, and Duplexes

Thrust faults typically occur in systems. One common system is marked by **imbricate fans** (Figure 6.111*A*). Imbricate fans are marked by a set of curved triangular thrust slices that converge into a more shallowly dipping sole thrust (see Figure 6.111*A*) (Boyer and Elliott, 1982). The fan opens upward, as the faults tip out upward into folds.

A

B

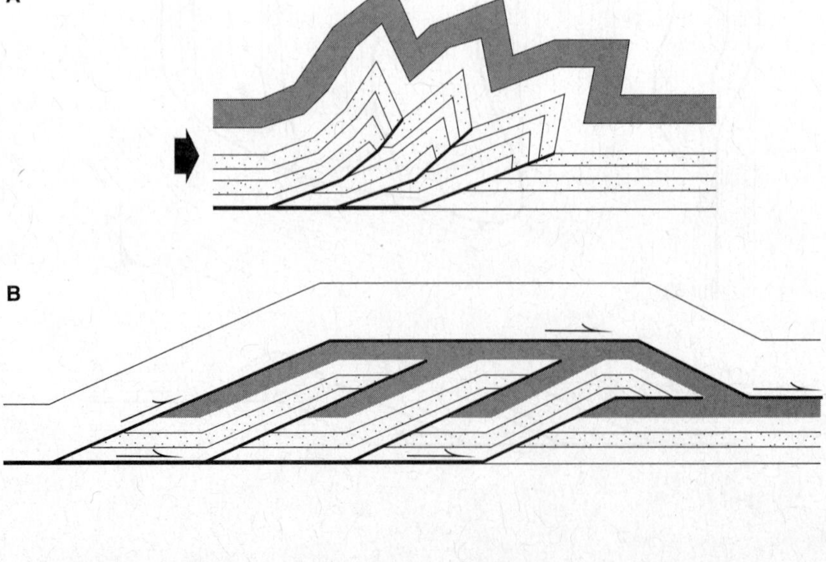

Figure 6.111 Two types of thrust systems. (*A*) Imbricate fan. (*B*) Duplex. [Reprinted from *Journal of Structural Geology*, v. 31, Nickelsen, R. P., Overprinted strike-slip deformation in the southern Valley and Ridge in Pennsylvania, p. 865–873, © 2009, with permission from Elsevier.]

A

B

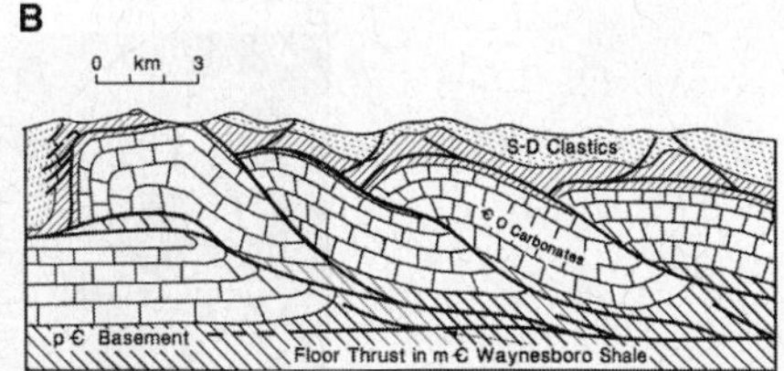

Figure 6.112 Images of duplexes, including the folding of horses within duplexes. (*A*) Several horses within outcrop-scale duplex structure associated with the Moine thrust. Knife for scale. [Photograph by G. H. Davis.] (*B*) Duplex structure underlying folds in the central Appalachian Valley and Ridge province. [From Boyer and Elliott, 1982, Thrust systems: *American Association of Petroleum Geologists Bulletin*, v. 66, p. 1196–1230. AAPG©1982, reprinted by permission of the AAPG whose permission is required for further use.]

An equally common system is marked by **duplexes** (Figure 6.111*B*). **Duplexes** cannot be described in any simple way, but the term itself reminds us again of the close connection between structural geology and architecture. McClay (1992) does a nice job of sorting out the terminology related to thrust faulting and the ways that thrusts link, geometrically, kinematically, and mechanically. To picture a duplex, visualize a set of thrust faults in which the flats and ramps all tie together, thus completely isolating individual fault-bound blocks (see Figure 6.111*B*). Such fault-bounded blocks are called **horses**. "*Horse*" in this geologic sense is an old mining term meaning a body of rock completely enveloped by faults that branch and merge around them, yet all dipping in the same general direction (Bailey, 1938, p. 607). The amount of movement along individual thrusts may be modest, but collectively there is an impressive sum of shortening (i.e., "*telescoping*") of the section (Bailey, 1938).

Horses within a duplex may resemble imbricate tiles resting on top of one another (see Figure 6.111*B*), each of which is separated not only by ramp thrusts but also, below and above, by thrust flats. The thrust flat below is called the **floor thrust**, and the one above the **roof thrust**. Within each horse there may be an anticline/syncline combination (i.e., a **fold pair**) (see Figure 6.111*B*) depending on the size of the horses and the amount of displacement on the thrusts that flank the horses (Boyer and Elliott, 1982). Horses become folded as a natural part of duplex development, and the folding takes place irrespective of the scale of the duplex system (Figure 6.112).

The net effect of duplex development is not unlike the piggyback arrangement of truck cabs on flatcars (Figure 6.113*A*), an efficient arrangement that, compared to an end-to-end lineup, shortens the length of the line of truck cabs. Outcrop examples of duplexes can be stunning, such as those photographed by Stuart Thomson in Crete (Figure 6.113*B*) (Thomson et al., 1998).

A

Figure 6.113 (*A*) Duplex image of pickup trucks. These arrangements save room, advantageous both in business and geology.

Figure 6.113 (*Continued*) (*B*) Thrust duplex composed of tens of horses. The thrust-repeated layer is part of a thin-bedded deepwater turbidite sequence in the Arvi unit of the Pindos Group stratigraphy, exposed just east of Lendas, southern Crete. The turbidites range in age from Cretaceous to Eocene. The duplex was formed during the Oligocene, during accretion of the host formation along the Hellenic subduction zone. Note that the duplex is "*sandwiched*" between layers that look as if they spent their life in Kansas. [Photograph by Stuart Thomson. This photograph was the cover photograph of Geology, v. 26, no. 3, March, 1998. Published with permission of the Geological Society of America.]

The magnitude of **slip** on a thrust fault in comparison with the **length** of the next-lower thrust slice will determine the actual shape and geometry of a duplex (Figure 6.114). Where slice length is greater than fault slip, a **normal duplex** forms in which the horses dip gently backward; that is, opposite the direction of overall thrust transport (see Figure 6.114*A*). Where slice length and fault slip are about the same, an **antiformal duplex** develops in which the horses are actually arched (see Figure 6.114*B*). Where fault slip exceeds slice length, a **forward-dipping duplex** develops (Figure 6.114*C*) (Boyer and Elliott, 1982; Mitra, 1986).

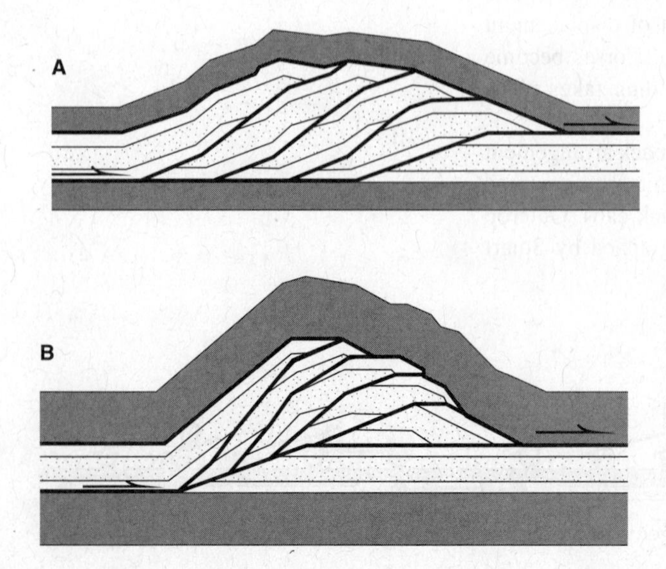

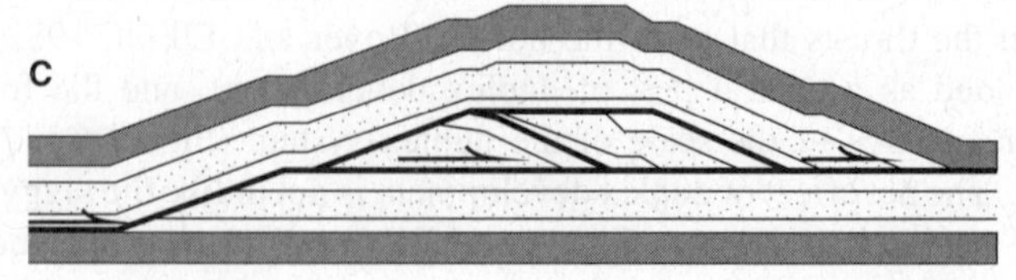

Figure 6.114 Form of duplexes depends on spacing of ramps and the amount of slip. (*A*) A normal duplex develops where slice length exceeds fault slip. (*B*) An antiformal duplex develops where slice length and fault slip are essentially equal. (*C*) A forward-dipping duplex develops where fault slip is greater than slice length. Terminology from Boyer and Elliot (1982). [After McClay, 1992a, Figures 24–26, p. 426, in McClay, K. R. (ed.), Thrust tectonics: Chapman and Hall, London, p. 419–433. With kind permission from Springer Science+Business Media B.V.]

The Sequence of Thrusting in Duplexes

Early ideas of the evolution of ramp-flat thrusts assumed that major faults cut up-section, one at a time, in the direction of tectonic transport. Overall, the major thrusts split off from one another at **branch points**, and the resulting pattern is marked typically by **in-sequence** faulting in which successively younger faults form in the direction of tectonic transport (Dahlstrom, 1969; Royse et al., 1975).

The concept of critical taper of an orogenic wedge helps explain some of the out-of-sequence faults that we observe. For example, when a thrust belt moves forward as a mass, it serves to lengthen its wedge shape, and this tends to lower the belt below its critical tape angle. In order for the wedge to move forward again, it must thicken in order to achieve critical taper for its new length. This thickening sometimes is accomplished by out-of-sequence faulting within the wedge itself. Once the wedge is thickened to its critical taper angle again, the whole process can repeat itself.

Active Versus Passive-Roof Duplexes

In true duplexes, the thrust faults that laterally confine each horse are parallel to the very frontal ramp of the duplex, as in Figure 6.114*A*. This would be a type of **active roof duplex** breaking toward the flat-lying undeformed strata in the foreland. There is another class of duplexes, called **passive roof duplexes**. In this case the so-called roof thrust "*passively*" evolves as the duplex underthrusts strata in the foreland (Banks and Warburton, 1986; McClay, 1992). In this situation, the sequence of strata above the roof thrust has not been transported toward the foreland, but instead the strata are underthrusted and lifted up over the duplex proper (Figure 6.115) (Couzens-Schultz et al., 2003).

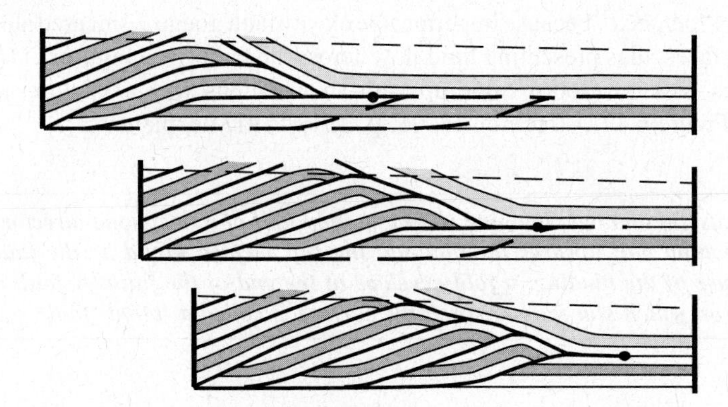

Figure 6.115 Passive-roof duplex, in which the flat-lying foreland strata are "*simply*" underthrust and lifted up by the advancing tip lines of successive in-sequence thrusts. [Reprinted from *Journal of Structural Geology*, v. 25, Couzens-Schultz, B. A., Vendeville, B. C., and Wiltschko, D. V., Duplex style and triangle zone formation: insights from physical modeling, p. 1623–1644, © 2003, with permission of Elsevier.]

Couzens-Schultz et al. (2003) concluded that passive-roof duplexes form in cases where the decollement is strong, thus providing strong coupling between the decollement and cover, favoring underthrusting. In this situation, underthrusting takes place. Where the decollement is weak, horse blocks are transported long distances, and active-floor duplexes predominate. The blocks within active-roof duplexes are less internally deformed than blocks within passive-roof duplexes (Couzens-Schultz et al., 2003).

Blind Thrusts

Just as a reminder, some thrusts do not reside in duplex systems, and do not "*start*" and "*stop*" in floor and roof thrusts. Rather, they occur as discrete entities, slashing up from depth towards the Earth's surface and commonly transferring their fault

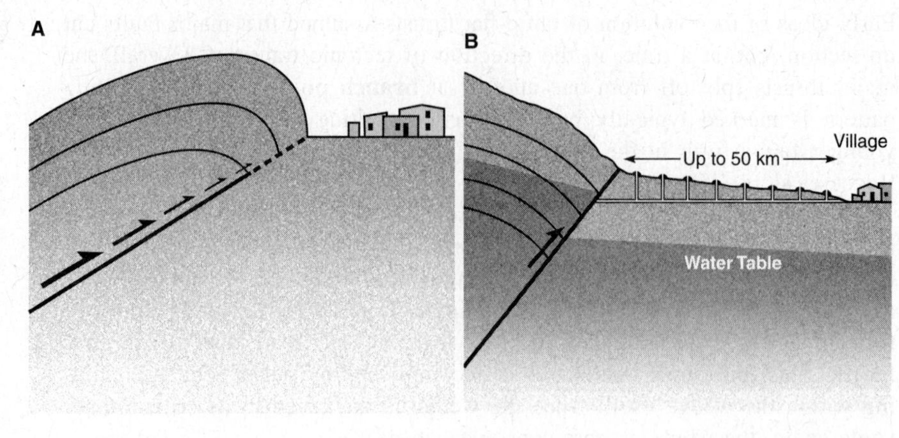

Figure 6.116 (*A*) Schematic cross-section of "*blind*" thrust (or reverse) fault adjacent to village. Slip on fault is denoted by black arrows. Note that fault slip dies out toward the surface, which in turn deforms by folding. (*B*) Schematic cross-section of blind thrust. The combination of incremental reverse fault slip and clay smear (gouge) along the fault creates the perched groundwater table on which villagers depend. Access to water is achieved through digging irrigation tunnels (qanats), a relatively straightforward task through footwall alluvium. Jackson (2006) notes that the qanats can be up to 100 m deep at the range front. [From Jackson, J., Fatal attraction: living with earthquakes, the growth of villages into megacities, and earthquake vulnerability in the modern world: *Philosophical Transactions of the Royal Society* A, v. 364, Figures 2 and 3, pp. 1913 and 1916, 2006.]

slip into folds (Figure 6.116*A*). Such a "*hidden*" thrust is called a **blind thrust**. Strictly speaking, many blind thrusts are actually reverse faults, which dip 45° or greater. As we shall see in *Chapter 11, Active Tectonics*, blind thrusts lurk beneath Quaternary folds in the Los Angeles basin, . . . hidden time bombs waiting to release strain energy and wreak havoc on the mega-city above. James Jackson (2006) emphasized that earthquake hazards associated with blind thrusts have been keenly experienced in Iran, where village settlements developed astride blind thrusts (Figure 6.116*B*). This fault/village association persists even though earthquakes from time to time level the settlements and kill large percentages of local populations, . . . for the basis of the association is water! The water table becomes elevated on the hanging walls of these active blind faults. The water table becomes "*perched*" because impermeable clayey fault gouge is smeared along the fault surfaces, thus presenting fluid flow across the faults (see Figure 6.116*B*).

James Jackson (2006) came up with an ingenious way to picture a blind thrust (Figure 6.117). Jackson (2006, p. 1912–1913) explains:

> ***As a useful analogy, imagine sliding the top half of a telephone directory over the bottom half towards the binding; the slip surface would be the fault but, because of the binding, a fold develops at the end of the fault. A fault of this type, on which slip fails to reach the surface is called a 'blind' fault.***

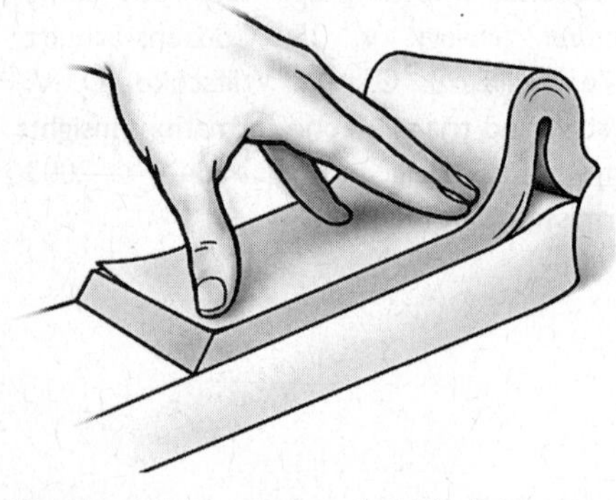

Figure 6.117 Cartoon of blind thrusting, inspired by James Jackson (2006, pp. 1912–1913). All you need is a telephone book, or some other thick, soft-covered, bound document. Note how faulting "*gives way*" upward to folding. [From Jackson, J., Fatal attraction: living with earthquakes, the growth of villages into megacities, and earthquake vulnerability in the modern world: *Philosophical Transactions of the Royal Society* A, v. 364, Figures 1 and 2, p. 1912–1913, 2006.]

The Moine Thrust, Northern Scotland

The Moine thrust (Scotland) is a classic structural locality. It was "*in the Moines*" that Peach and Horne (e.g., Peach et al., 1907) carried out phenomenal structural mapping of duplexes and imbricates; where mylonites were described for the first time (Lapworth, 1885); and where outstanding British geologists have picked apart the fine detail of the systems of structures (e.g., McClay and Coward, 1981).

The structures are incredible, including, for example, clean outcrop exposures of flat segments of the Moine thrust separating Precambrian rock *above* from Cambrian rock *below*. Imagine hiking up the long ridge shown in the distance in Figure 6.118. You would start out in Precambrian basement and after a while cross the great unconformity into Cambrian. Then as you hike up the ridge still farther, you would find yourself back in Precambrian, having crossed the very gently dipping Moine thrust.

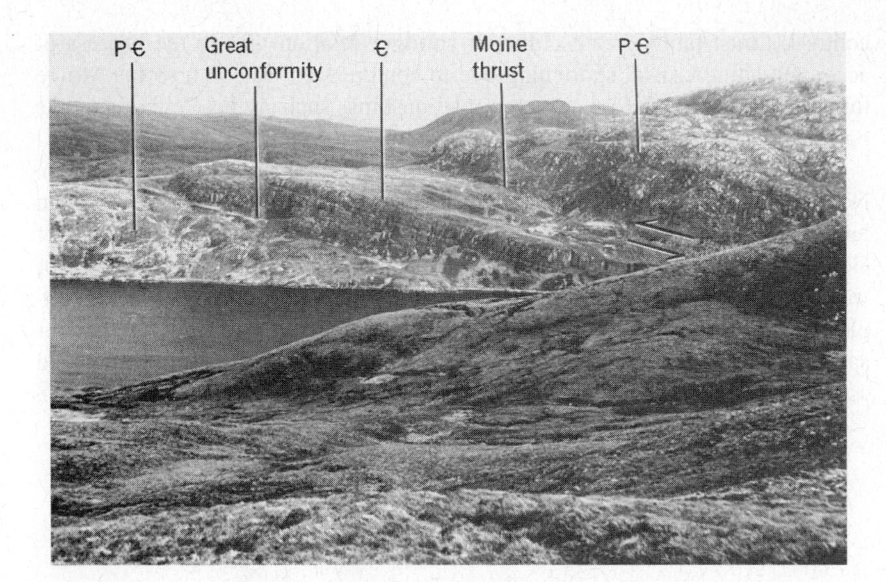

Figure 6.118 Distant view of the Moine thrust, showing thrust repetition of the Great Unconformity. Unreal. See text for explanation. [Photograph by G. H. Davis.]

NORMAL FAULTING

Younger-On-Older Relationships

Normal faulting accommodates extension of the Earth's crust, and it does so by movement of hanging-wall rock downward with respect to footwall (Figure 6.119*A*). As a result, relatively high-level, commonly younger strata in the hanging wall are brought downward into contact with relatively deep-level, commonly older strata in the footwall. We call this a **younger-on-older** relationship, created as a result of crustal stretching due to extension. In

Figure 6.119 (*A*) Normal fault cutting Jurassic sandstone and mudstone in a roadcut west of Moab, Utah. [Photograph by S. J. Reynolds.] (*B*) Close-up of a left-dipping normal fault. The "*triplet*" of marker beds is offset in a way that the hanging wall (above the fault surface) drops relative to the footwall (underside of fault). The polished rock slab is one of George Davis' belt buckles, given to him as a present from economic geologist John Guilbert. Among structural geologists, the buckle gives rise to severe cases of "*belt-buckle envy*."

contrast, thrust faults create **older-on-younger** relationships in the process of accommodating crustal shortening due to compression. Recall how the Moine thrusting created older-on-younger relationships, such as Precambrian rocks on top of Paleozoic rocks (see Figure 6.118).

Indeed, normal fault systems form where crustal rocks undergo extension. Normal faulting is the way individual rock layers, or the crust itself, can lengthen and stretch in brittle or semi-brittle fashion. The Rhine graben of Germany, the East African rift valleys, and the Basin and Range province of western North America emerged as among the foremost early classical examples of regions extended by normal faulting. With the discovery of seafloor spreading and plate tectonics in the 1960s, normal faulting became recognized as the dominant fault class along mid-oceanic spreading centers (Figure 6.120).

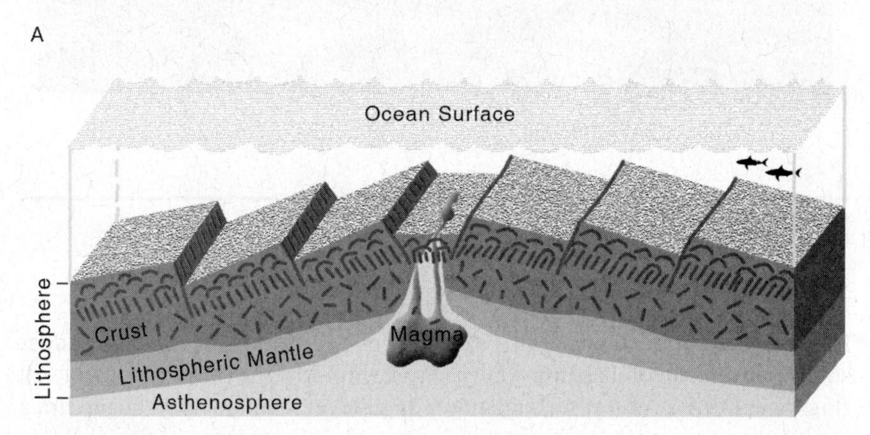

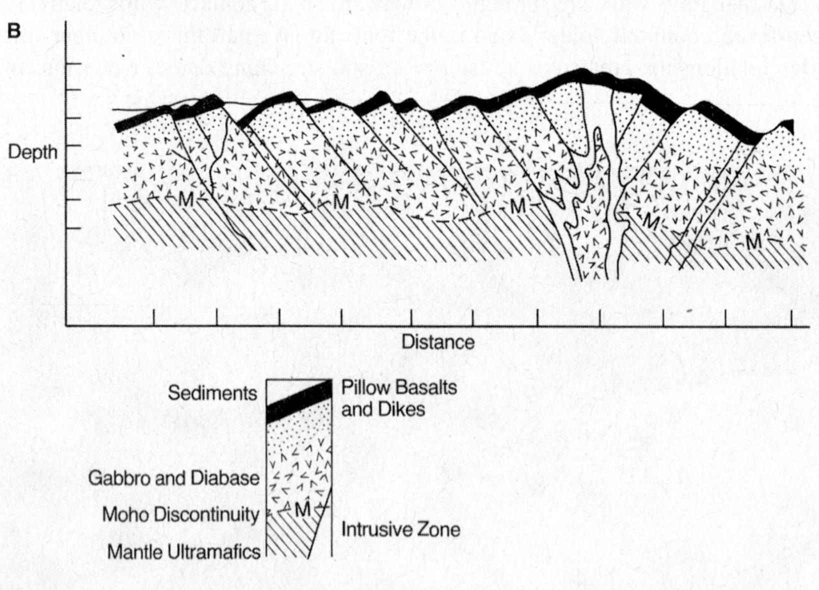

Figure 6.120 (*A*) Mid-oceanic spreading centers are marked by crustal stretching, which is accommodated by normal faulting. [Adapted from Redfern, 1986. Published with permission of the American Geological Institute.] (*B*) Vertically exaggerated schematic rendering of extensional processes at work in an ocean ridge system. Note that the normal faults cut all the way through the oceanic crust. [From Anderson and Noltimier, 1973, v. 34, fig. 5, p. 144. Published with permission of Geophysical Journal of the Royal Astronomical Society.]

Modeling Normal Faulting in the Laboratory

Early Modeling by Hans and Ernst Cloos

The characteristics of normal faulting were elegantly modeled in simple clay deformation experiments by Hans Cloos (1936), and later by Ernst Cloos (1955). The delicate, intricate fault patterns that can emerge in deformed clay that is properly prepared and deformed are astonishingly similar to natural patterns.

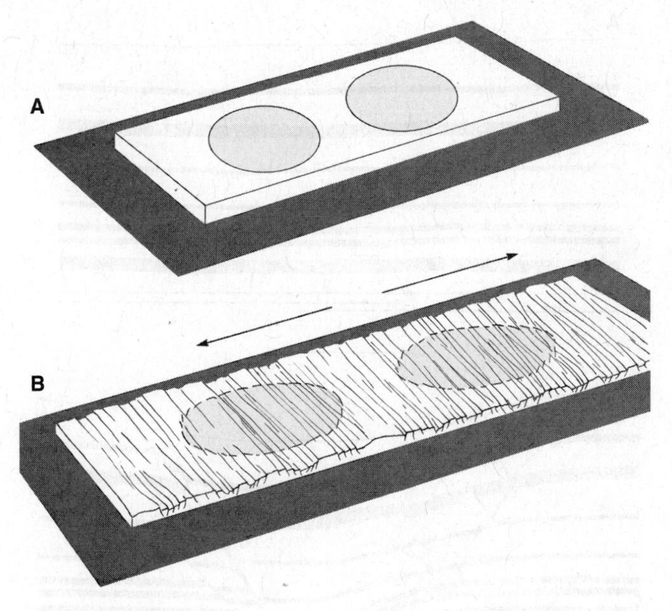

Figure 6.121 Clay cake simulation of normal faulting. (*A*) Clay cake on a rubber sheet, before stretching. Note reference circles. (*B*) Clay cake after stretching. Note fault pattern. Reference circles transformed to ellipses. [From Cloos, 1955. Published with permission of the Geological Society of America.]

The clay cake to be deformed is prepared by mixing dry kaolin with water until a soft buttery consistency is achieved. The clay is spread as a layer ± 4-cm thick, approximately 15 × 30 cm in size, and smoothed with a putty knife. Reference circles may be gently impressed onto the surface of the clay cake to serve as a strain marker.

Hans and Ernst Cloos recommended two different methods for deforming the clay. In the first experiment the clay cake is built on top of an elastic rubber sheet (Figure 6.121*A*). Then the rubber sheet is stretched slowly and uniformly. Since the base of the clay cake is "*coupled*" (i.e., "*sticks*") to the rubber sheet, the entire clay layer experiences the effects of the stretching. As stretching commences, normal-slip faults along with tear faults emerge to accommodate the extension (Figure 6.121*B*). Gradually the faults become uniformly distributed throughout the clay cake. The top of a clay cake stretched in this way reveals numerous closely spaced faults and fault scarps, which typically are oriented at a high angle to the direction of stretching. Some fault traces are straight, whereas others are curved. Fault surfaces exposed along the fault scarps are marked by dip-slip striations. When viewed from a distance, the final deformed state of the clay cake shows lengthening in the direction of stretching, and thinning of the layer as a whole. We emphasize this by showing the contrast between starting reference circles (see Figure 6.121*A*) and their conversion into ellipses in the deformed state (see Figure 6.121*B*). Extension and stretch values for the deformed clay cake can be determined by comparing original versus final thickness, and original versus final lengths.

The cross-sectional view of a clay cake stretched on a rubber sheet reveals a system of fault-bounded blocks (Figure 6.121*B*). The bounding faults are all normal-slip faults, dipping on the average of 60°. Sets of oppositely dipping normal-slip faults are about equally developed. Relatively uplifted blocks bounded by the normal-slip faults are **horsts**; down-dropped blocks between horsts are called **graben** (Figure 6.122*A*). Look closely as well for listric normal faulting, in which the fault surfaces are curved in such a way that they dip less and less with depth (Figure 6.122*B*). Faulting along curved fault surfaces causes the hanging wall block to rotate in a way marked by backtilting into the fault (reverse drag) and/or deform by antithetic faulting.

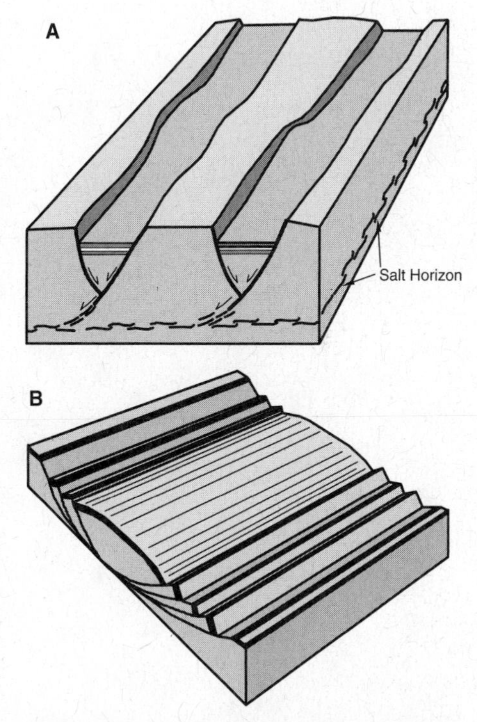

Figure 6.122 (*A*) Horsts and graben are classical terms describing fault-bounded uplifted and down-dropped blocks, respectively, in extended regions. This drawing is based on an example in the Canyonlands graben system, Utah. [After Trudgill and Cartwright, 1994. Published with permission of the Geological Society of America.] (*B*) Normal faults that curve with depth are known as listric normal faults. Movement along these causes rotation of the hanging-wall blocks.

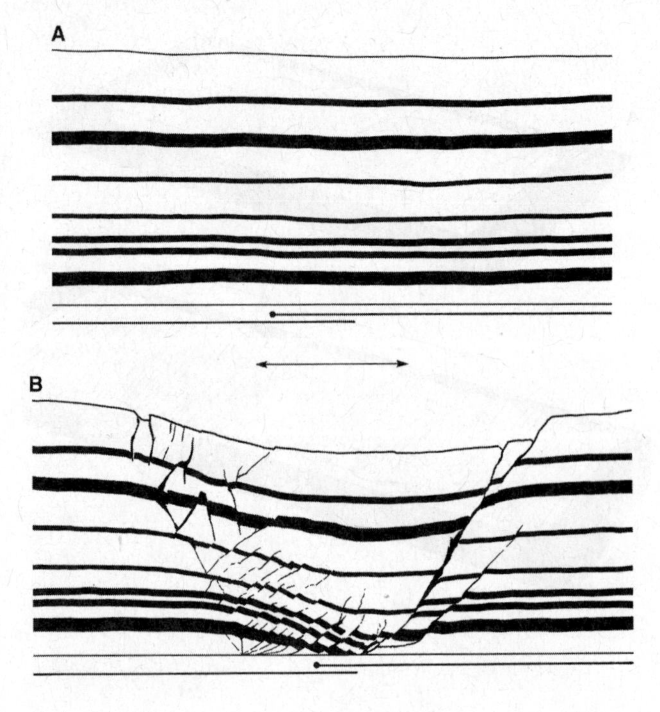

Figure 6.123 A second clay cake simulation of normal faulting. (*A*) Deformation of a clay cake on top of overlapping panels of sheet metal. Stretching and normal faulting are achieved by pulling apart the sheet metal panels. (*B*) Cross-sectional view of the deformed clay cake shows a series of tilted fault blocks. [From Cloos, 1968, Experimental analysis of Gulf Coast fracture patterns: *American Association of Petroleum Geologists Bulletin*, v. 52, p. 420–444. AAPG©1968, reprinted by permission of the AAPG whose permission is required for further use.]

In a second experiment designed by Hans Cloos and Ernst Cloos, the clay cake is placed on top of overlapping sheets of sheet metal. To achieve stretching of the clay, the ends of the sheets of metal are pulled slowly and uniformly away from each other (Figure 6.123*A*). This movement causes stretching of the clay in the region directly above the join of the metal sheets. The area affected by the extension increases in breadth as the sheets are pulled farther and farther apart. This produces a single complex graben (Figure 6.123*B*). The surface of the clay layer bows downward at the site of the graben, and the whole width of the graben is pervaded by normal-slip faults. The major boundary faults of graben formed in this way dip inward, at ~60°. In fact, almost all the faults in the clay cake tend to dip inward toward the center of the graben. There are, however, exceptions. Antithetic faults form close to the major boundary faults but dip in the opposite direction (see Figure 6.123*B*).

Modern Scaled Analogue Modeling of Normal Faulting

Building on the experimental approach originally developed by Peter Cobbold and Jean Pierre Brun, Ken McClay and his colleagues carried out world-class **scaled analogue modeling** of normal faulting (McClay et al., 1991a,b). McClay used very fine-grained sand to simulate brittle upper crust, and after dyeing the sand he would deposit a distinctive stratigraphy of alternating colored and white sands. In order to create listric faulting, he built a rigid footwall block with a concave curved upper surface (Figure 6.124*A*). Then he draped a movable plastic sheet over the footwall block, and attached the sheet to a moving wall. Movement and strain rate were computer automated. McClay's models of simple listric faults show the development of rollover in the hanging-wall strata, as well as the crestal collapse of the rollover to produce graben structures (Figure 6.124*A–C*).

The models that McClay created are especially realistic and informative because new sand was continuously deposited *during* faulting. As a result, patterns of **growth fault sedimentation** develop in which the most depressed topographic areas receive the greatest sand. Whereas each layer deposited

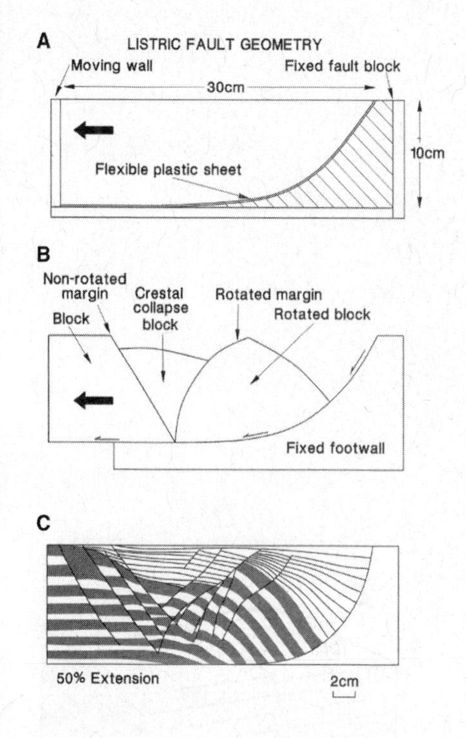

Figure 6.124 (*A*) Experimental setup for producing simple listric fault geometries. [From Ellis and McClay, 1988, Figure 2c, p. 57. Reprinted with permission of Blackwell Science, Ltd.] (*B*) Structural elements produced during simple listric normal faulting. Two of the major features are the rollover anticline and the crestal collapse graben. [From Ellis and McClay, 1988, *Basin Research*, Fig. 3, p. 59. Reprinted with permission of Blackwell Science, Ltd.] (*C*) Tracing of photograph of model produced through simple listric faulting, showing prefaulting sediments in black and white bands, with synfaulting sediments stippled. [From Ellis and McClay, 1988, *Basin Research*, v. 1 Fig. 4b, p. 59. Reprinted with permission of Blackwell Science Ltd.]

before faulting (i.e., **pre-fault layers**) shows uniformity of thickness (even though offset from place to place), the layers deposited during faulting (i.e., **syn-fault layers**) show dramatic changes in thickness. In particular, if we were to follow any given syn-fault layer in the hanging wall, it would be seen to thicken significantly in the direction of the fault (Figure 6.124*C*). Layers deposited after faulting ceased (i.e., **post-fault layers**) locally are in sharp angular unconformity with underlying syn-fault layers. The models show interesting detail. For example, McClay observed that the crestal collapse graben typically are bounded by a planar **antithetic fault** on the basinward side and a **synthetic fault**, which is listric in the syn-fault sequence but is **anti-listric** (i.e., convex upward), in the pre-fault sequence.

McClay added another wrinkle by showing what happens when scaled analogue modeling incorporates normal faulting along a surface that has flats and ramps (Figure 6.125) (Ellis and McClay, 1988). McClay figured out that by sprinkling mica along the top of a sand layer and then burying it with more sand, the coefficient of frictional sliding is reduced and flats can operate. Notice that such faulting can produce folds and reverse faults (see Figure 6.125*C*), giving the false impression that the deformation was achieved through shortening!

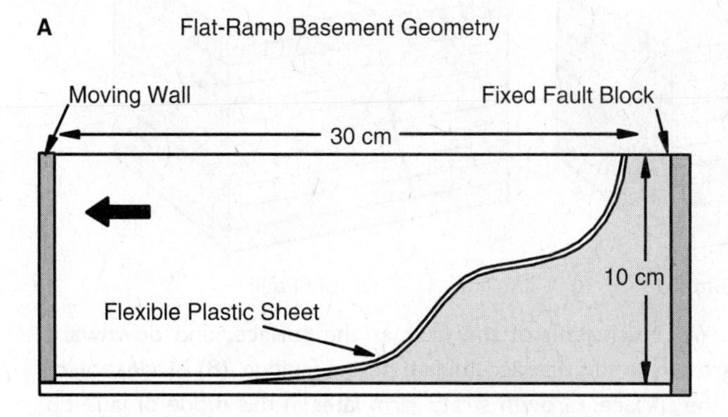

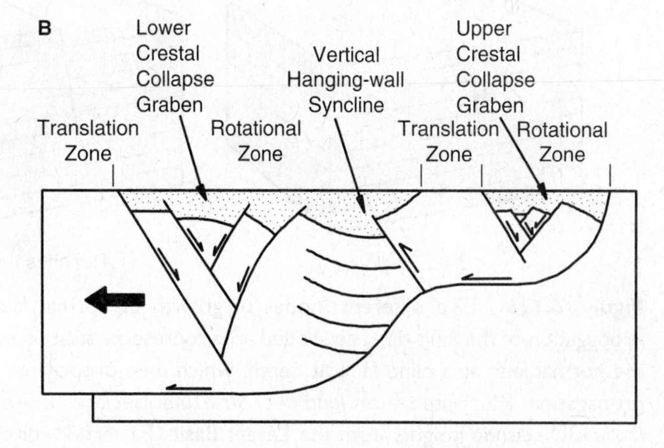

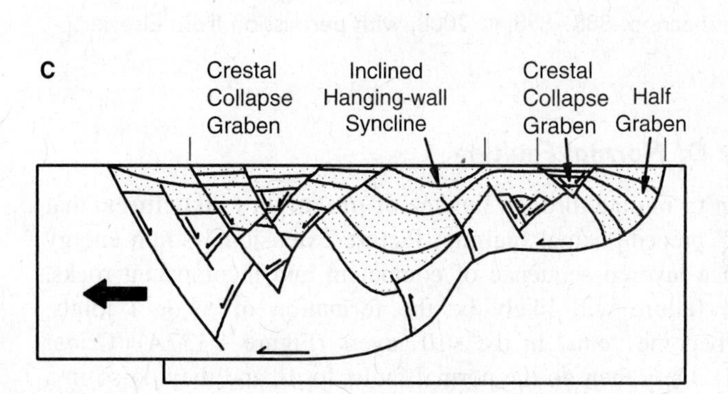

Figure 6.125 (*A*) Experimental setup for normal faulting accommodated by a ramp-flat basement geometry. (*B*) Ramp-flat listric fault geometry showing hanging-wall deformation features in the prefaulting sediments. Low extension. (*C*) Ramp-flat listric fault geometry showing deformational features in the syn-rift sediments (stippled). High extension. [From Ellis and McClay, 1988, *Basin Research*, v. 1, Figure 2b, p. 57, Fig. 22a, p. 69, and Figure 22b, p. 69. Reprinted with permission of Blackwell Science Ltd.]

Behavior of Normal Faults Within Systems

Propagation of Individual Normal Faults

Baudon and Cartwright (2008) provide us with perspective regarding nucleation and propagation of normal faults. Their work was based on studying seismic profiles of systems of normal faults in the Eastern Mediterranean region. They considered the two end-member cases. On the one hand, a normal fault may initiate at the surface, and as time passes the tip of the fault cuts deeper and deeper into the crust, earthquake by earthquake, as the fault grows (Figure 6.126*A*). On the other hand, the fault may form at depth as a blind normal fault, and from there grow in size (surface area) and length, eventually reaching the surface and cutting and displacing the free face (Figure 6.126*B*). When the surface becomes cut and displaced, the fault operates as a **growth fault**, which means that as the normal faulting takes place so too does sedimentation within the fault-controlled basin that progressively develops.

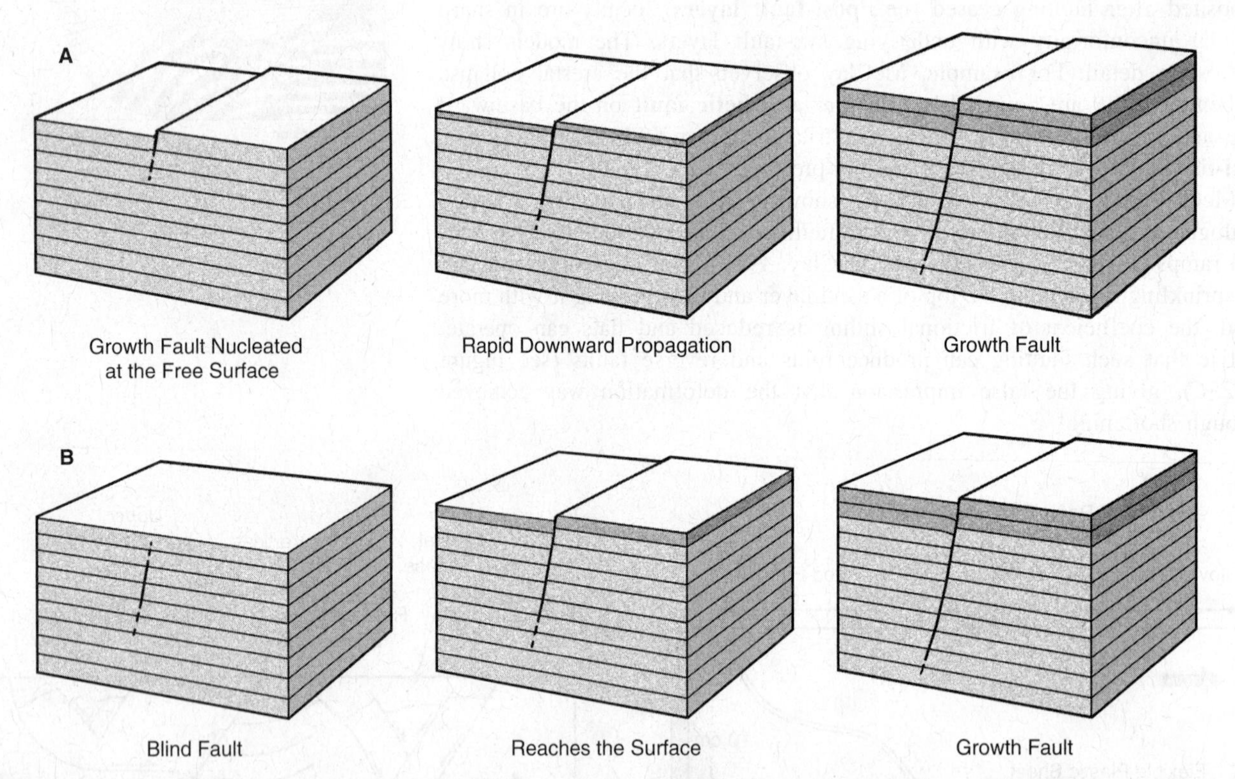

Figure 6.126 Two different modes of growth of normal faults. (*A*) Nucleation of the fault at the surface, and downward propagation of the fault tip. The shaded area represents strata (i.e., growth strata) that accumulate during faulting. (*B*) Nucleation of the normal fault as a blind fault at depth, which then propagates to the surface. Growth strata form later in this mode of fault-tip propagation. [Reprinted from *Journal of Structural Geology*, v. 30, Baudon, C., and Cartwright, J., Early stage evolution of growth faults: 3D seismic insights from the Levant Basin, Eastern Mediterranean, p. 888–898, © 2008, with permission from Elsevier.]

Down-Dip Onset Of Normal Faulting

Normal faults seem to initiate through linkage of **precursory structures**; that is, ones that "*just*" precede actual faulting. Picture extensional strain energy building up within a layered sequence of competent and incompetent rocks. The first onset of failure will likely be the formation of mode I joints, which commonly become veins, in the stiff layers (Figure 6.127*A*) (Crider and Peacock, 2004). Only then do the normal faults form, and they do so in a

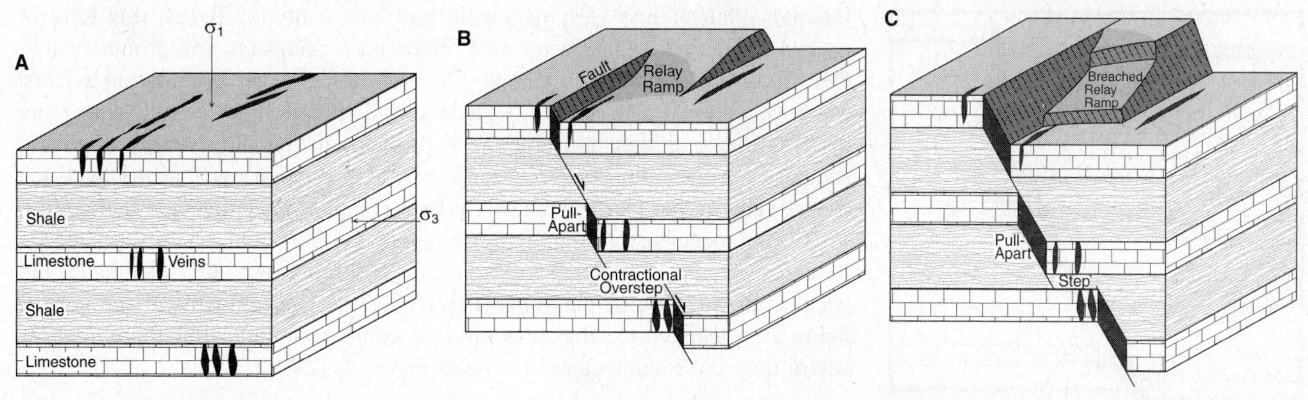

Figure 6.127 Development of normal faults in an interbedded limestone-mudstone sequence. (*A*) As a result of layer-parallel extension, mode I joints and veins initiate as precursory structures in the stiff limestone beds. (*B*) Continued extension dilates the mode I fractures and veins. Normal faults develop in the mudstones, and tip out in the mode I joints and veins. As seen in map view, relay ramps develop where adjacent faults overstep one another. (*C*) The normal fault segments may eventually link up completely and breach the relay ramps. [Reprinted from *Journal of Structural Geology*, v. 26, Crider, J. G., and Peacock, D. C. P., Initiation of brittle faults in the upper crust: a review of field observations, p. 691–707, © 2004, with permission from Elsevier.]

zig-zag geometry marked by an inclined fault segment running through less competent beds and near-vertical segments that reactivate the mode I joints in the competent beds (Figure 6.127*B*, *C*). Because of the zig-zag geometry of such normal faults, slip along the inclined segment causes pull-apart opening of the mode I precursory joints.

Yet another way in which normal faults initiate is through precursory formation of transitional-tensile fractures (as opposed to mode I joints) in the strongest beds (Ferrill and Morris, 2003). Consider a perfectly flat-lying sequence of interbedded stiff and weak units (Figure 6.128). The segments of transitional-tensile (hybrid) normal faults that cut the stiffest beds would be marked by very steeply dipping surfaces, but not vertical surfaces. In contrast, the weak beds would fail by standard Coulomb failure, producing dip angles close to 60°. Because the faulting occurs along zig-zagging failure surfaces, dilational openings are produced in the competent beds (see Figure 6.128). This type of faulting is favored by deformation under conditions of very low confining pressure.

Ferrill and Morris (2003) have emphasized the practical consequences of **dilational normal faulting** along such zig-zag faults. For example, on the

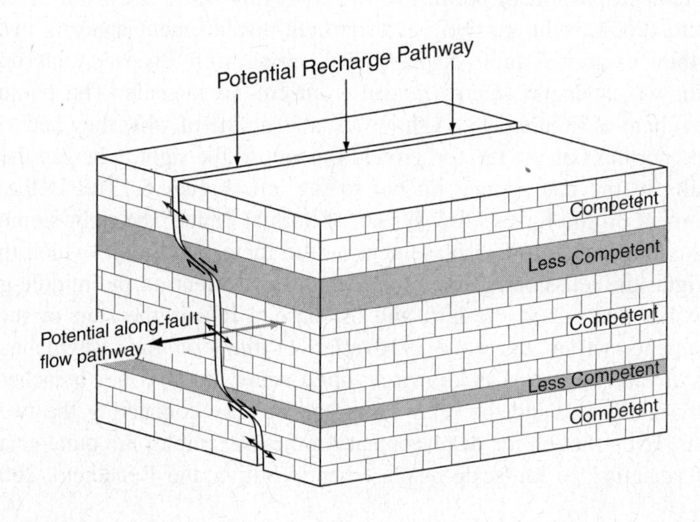

Figure 6.128 Dilational normal faults created by hybrid transitional-tensile failure across strongest beds within this sequence of Cretaceous limestones on the Edwards Plateau, Texas. Note that the fault architecture creates pathways for flow of groundwater, and for recharge. [Reprinted from *Journal of Structural Geology*, v. 25, Ferrill, D. A., and Morris, A. P., Dilational normal faults, p. 183–196, © 2003, with permission from Elsevier.]

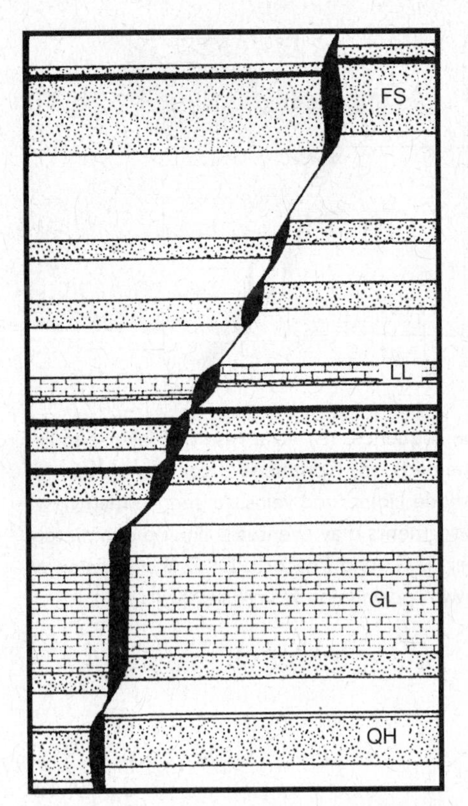

Figure 6.129 Illustration from Wallace (1861) showing lead-rich fissure veins (black) in the North Pennine Orefield in the UK. Strata are Carboniferous, consisting of alternating competent (limestone and sandstone) and incompetent (shale and coal) layers. Fault offsets are in the range of ~1 to ~10 m. "FS," and "QL," and "GH" are simply references to bedrock units. [Reprinted from *Journal of Structural Geology*, v. 25, Ferrill, D. A., and Morris, A. P., Dilational normal faults, p. 183–196, © 2003, with permission from Elsevier.]

Edwards Plateau just west of Austin and San Antonio, Texas, this kind of normal faulting has produced easy circulation pathways for ground water. Over time, the dilational openings are made even larger by solution effects. Moreover, Ferrill and Morris (2003) called attention to certain (rare) ore deposits that owe their existence to hydrothermal fluid flow within dilational staircase normal fault zones. In particular, they pointed out that the lead-rich fissure veins in the North Pennine Orefield (England) occupy such fault zones. The veins occur exclusively within mode I joints and transitional-tensile fractures in the competent beds, thus accounting for the systematic but peculiar distribution of the veins (Figure 6.129). The segments of the normal faults are steep where they cut through competent units, but more shallow where they cut incompetent units (see Figure 6.129).

Along-Strike Displacement on Normal Faults

Individual faults, including normal faults, "*tip out*" along strike. You see "*start-and-stop*" traces of discrete fault surfaces in maps of systems of normal faulting. Individual discrete fault traces are known as **fault segments**. **Segmentation** is a property of fault systems.

In the examples we will consider here, the tips of adjacent normal faults overlap one another, in effect creating a transfer of offset from one fault segment to another. We can examine the gradient of transfer of offset from one fault segment to another through what are called **fault-displacement profiles**. Take, for example, the simple isolated fault segment shown in Figure 6.130*A*. Note that the map view of this fault is shown in Figure 6.130*A*, with both tips exposed and identified. The fault-displacement profile for this normal fault (see Figure 6.130*A*) is constructed parallel to the strike of the fault. Fault displacement (on the *y*-axis) is plotted against distance along the fault trace (on the *x*-axis). In this example, the maximum displacement occurs at the midpoint of the map-view trace of the fault.

In contrast, Figure 6.130*B* shows a map view of two overlapping fault segments, with a **relay ramp** in the zone of overlap. Note that the fault-displacement profile (see Figure 6.130*B*) shows two profiles, one for each of the faults. These two displacement profiles overlap. For any give point (on the *x*-axis) along the fault trace, the amount of offset is the *sum* of the displacement achieved by both faults (Solvia and Benedicto, 2004). Figure 6.130*C* is yet a third example, one that places emphasis on the **relay ramp** in relation to the two fault segments. This displacement profile shows how overall fault displacement is preserved as one fault dies out and the other fault picks up, for normal faulting is a shared enterprise in accommodating stretching.

The fault-displacement profiles shown in Figure 6.130 are from Soliva and Benedicto (2004), who carried out their fault-displacement analysis in Spain, where they examined fault displacements along well exposed outcrop-scale faults for which displacements ranged from cm- to m-scale. The photograph presented here as Figure 6.131*A* shows us an example of what they had to work with: the normal fault in the foreground tips out to the right, whereas the other two faults in the background tip out to the left. Figure 6.131*B* is the fault-displacement profile for this tiny system of normal faults. The right-sloping line represents the gradient of displacement on the foreground fault, which tips out to the right; the left-sloping line represents displacement on the middle-ground fault, which tips out to the left; and the dashed line is the sum of the fault displacements within the zone of overlap of the two fault segments. This overlap of fault segments is known an **open relay**, for it is not breached by a fault physically crossing the relay ramp and thereby connecting the two fault segments. Even though the displacements in these examples are quite small, the results "*scale-up*" to km-scale displacements (Soliva and Benedicto, 2004).

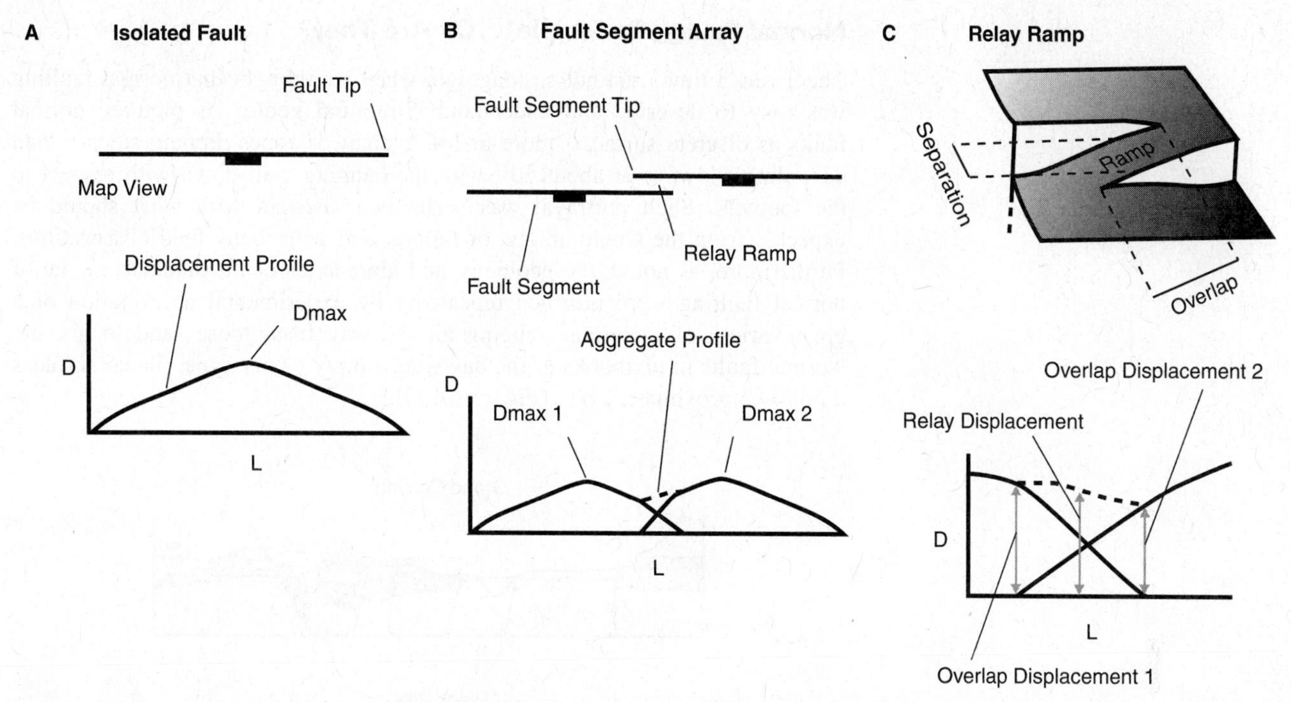

Figure 6.130 (*A*) Map view of simple isolated fault segment, with accompanying displacement profile showing the relationship of displacement (D) to distance along the trace of the fault segment, from tip to tip. (*B*) Map view of two parallel normal fault segments, arranged so that tips overlap. Between the overlapping tips is a relay ramp. Accompanying displacement profile shows aggregated displacement (D) along the traces of the two faults. The low spot in the aggregate profile in the region of overlap is due to NOT taking into consideration the effect of dipping strata within the relay ramp. (*C*) Here we see the geometry of the relay ramp in 3D view. The displacement profile demonstrates that when both fault displacement and bed displacement are taken into consideration, the overall displacement is transferred evenly from one fault to the other. [Reprinted from *Journal of Structural Geology*, v. 26, Soliva, R., and Benedicto, A., A linkage criterion for segmented normal faults, p. 2251–2267, © 2004, with permission from Elsevier.]

Peacock (2003) studied relay ramps in faults exposed in Lower Jurassic limestone and shale along the Somerset coast of England. Figure 6.132 is one of Peacock's wonderfully clear drawings of a relay ramp; that is, a **fault displacement transfer zone**. The drawing features two overstepping normal faults, both dipping in the same direction. Note how the direction and magnitude of the dip of bedding along the relay ramp is precisely that required to accommodate smooth transfer of displacement between the two fault segments (see Figure 6.132). It is quite extraordinary how faults "*team up*" and "*communicate*" with one other in order to neatly and efficiently take care of business. We see the same arrangement in Figure 6.127*B*, though in the late stages of such normal faulting the relay ramp may become breached and thus hard linked (see also Figure 6.127*C*).

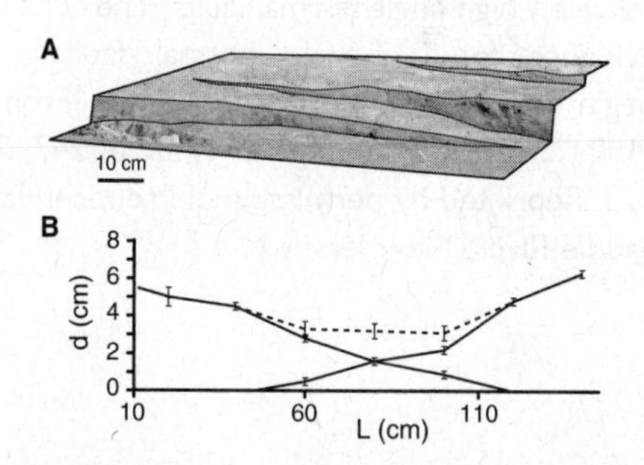

Figure 6.131 (*A*) Outcrop photograph of relay ramp between two high-angle normal faults whose tips overlap. (*B*) Displacement profiles for each of the two faults as well as the displacement (dash line) accommodated by dipping layering within the relay ramp. Error bars are shown on displacement profiles. [Reprinted from *Journal of Structural Geology*, v. 26, Soliva, R., and Benedicto, A., A linkage criterion for segmented normal faults, p. 2251–2267, © 2004, with permission from Elsevier.]

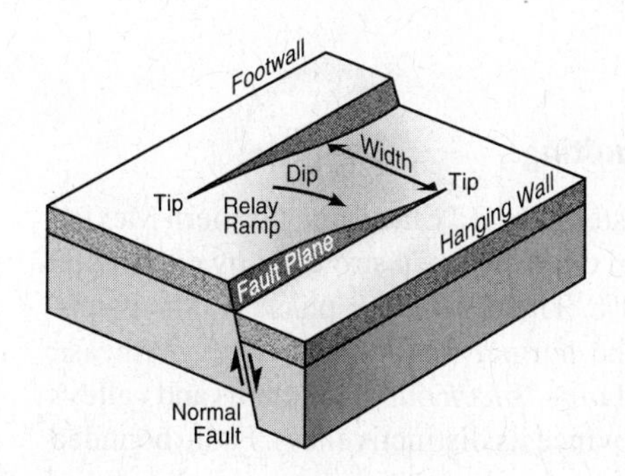

Figure 6.132 Block diagram of a relay ramp between the overstepped segments of two normal faults. [Reprinted from *Journal of Structural Geology*, v. 25, Peacock, D. C. P., Scaling of transfer zones in the British Isles, p. 1561–1567, © 2003, with permission from Elsevier.]

Normal Faults, So Simple... Or Are They?

There was a time, and not so long ago, when we thought that normal faulting was easy to describe and understand. Structural geologists pictured normal faults as discrete surfaces, more or less planar, or zones dipping steeper than 45°, almost always at about 60°, with the hanging wall down with respect to the footwall. Such portrayal was perfectly consistent with what should be expected from the Coulomb law of failure, and with many field observations. Furthermore, as noted, the geometry and kinematics of the standard model of normal faulting were affirmed repeatedly by experimental deformation of a great variety of materials, ranging all the way from loose sand to granite. Normal faults in textbooks of the day were simply of one type: planar surfaces dipping approximately 60° (Figure 6.133).

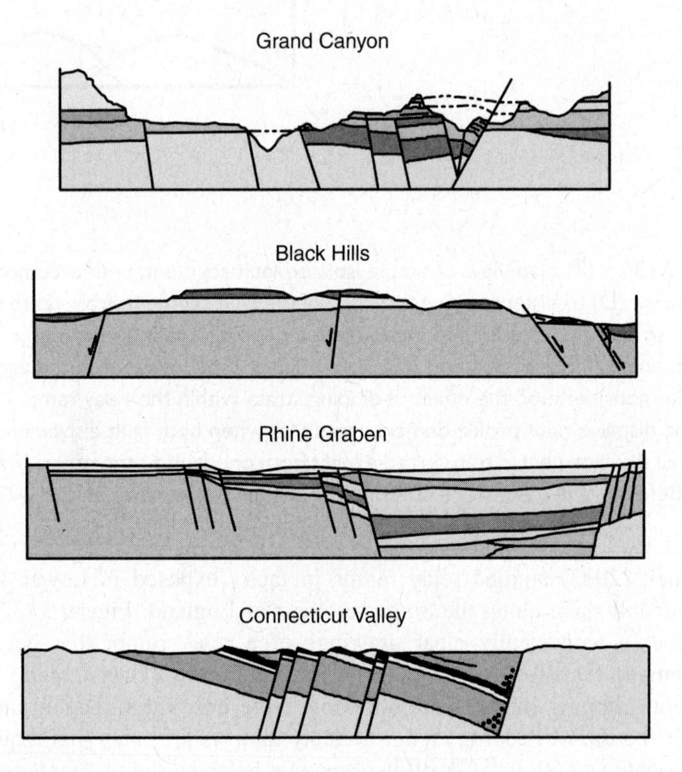

Figure 6.133 These cross-sections convey the classical view of extended regions; that is, marked by relatively high-angle normal faults. The concept of and evidence for low-angle normal faulting did not begin to emerge until the late 1960s. [From Billings, M. P., Structural geology, © 1972, p. 247, 249, 251, 523. Reprinted by permission of Prentice-Hall, Upper Saddle River, New Jersey.]

The possibility that some shallow-dipping faults could be normal faults (as opposed to thrust faults) was not entertained: at face value it violated Coulomb's law. This in itself created a subliminal bias such that geologists tended to believe that any fault dipping 30° or less must be a thrust fault produced by compression. This false bias was one of the factors that delayed recognition of the full complement of geometric details of normal faulting.

Low-Angle Normal Faulting

Discovery of Low-Angle Normal Faulting

The Basin and Range province of the western United States and northern Mexico occupies a huge region where continental crust has been stretched by distributed normal faulting. Extension exceeds 100%. The still-active phase of this extensional faulting began ~12 m.y. ago. The normal faults trend north-northeast, north-south, and north-northwest. These faults "*block out*" mountains and valleys in ways that give the Basin and Range province its distinctiveness. Fault-bounded basins and ranges are generally 25 to 70 km broad and 50 to 300 km long. The map-view expression of the ranges in the southern part of the Basin and Range was once described by United States Geological Survey geologist Clarence Dutton as "*an army of caterpillars marching northward out of Mexico*" (King, 1959, p. 152).

It was thought that the Basin and Range province had been fashioned exclusively by high-angle normal faulting, which produced horsts and graben. However, it is now clear that much of the extension was accomplished by low-angle normal faulting. Regional low-angle normal faults are now known as **detachment faults**. Low-angle normal faults were co-discovered in the late 1960s by Ernie Anderson (1971), who mapped them in the Lake Mead region of southeastern Nevada; Dick Armstrong (1972), who recognized them within western Utah and eastern Nevada; and John Proffett (1977), who mapped them in the Yerington district of Nevada. During the 1960s and early 1970s John Proffett, working as research geologist for Anaconda, had been evaluating the structural geologic setting of a porphyry copper deposit southeast of Reno in west-central Nevada. East-dipping concave-upward fault surfaces accommodated normal faulting and rotation of hanging-wall strata (Figure 6.134). During the faulting, ignimbrite volcanics of Oligocene age were back-tilted from horizontal to steep westward dips. Some rocks rotated by more than 90°. Relatively old normal-slip faults were rotated to shallow dips and became inactive, later to be cut and rotated to even shallower dips by relatively young faults. Indeed, the youngest faults in the Yerington district are the steepest, whereas the oldest faults typically dip shallowly (see Figure 6.134).

Proffett constructed detailed structural profiles of the fault relationships on the basis of geological mapping and subsurface drilling. From these he computed slip magnitude for the major faults, determining that the largest offset was ~4 km. The strain implications of the fault system in the Yerington district proved to be astonishing: more than 100% stretching! Based on structure profiles, Proffett estimated the original east–west width of the study site to have been 7.3 km. The extended width, after normal faulting, is 17.3 km. On this basis, stretch and percent stretching could be calculated:

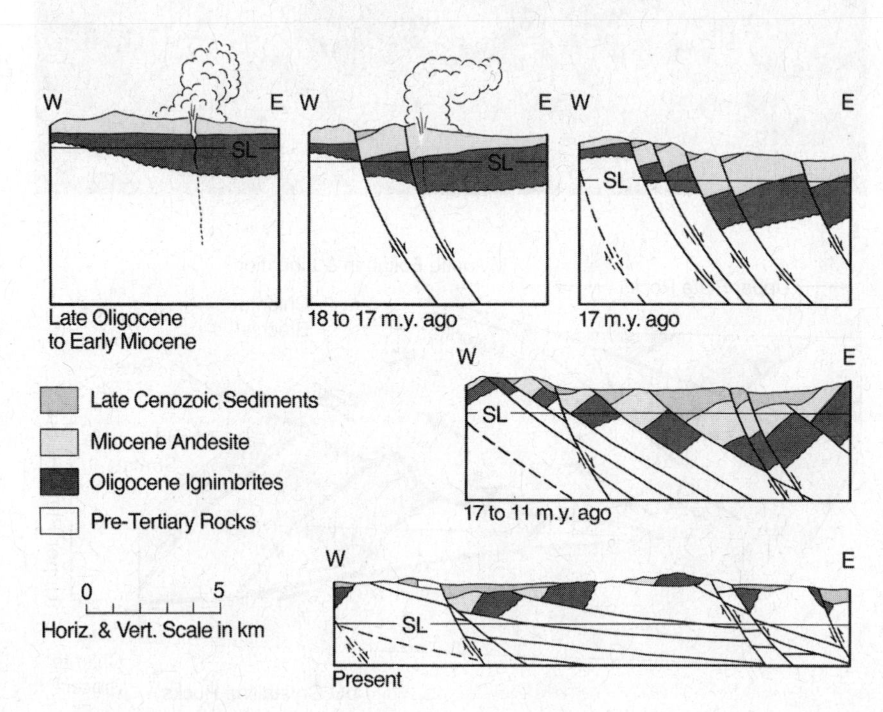

Figure 6.134 Kinematic evolution of normal faulting in the Yerington district, west-central Nevada. See text for explanation. [From Proffett, 1977. Published with permission of Geological Society of America.]

$$S = \frac{17.3 \text{ km}}{7.3 \text{ km}} = 2.37$$

$$\% \text{ lengthening} = (2.37 - 1.00) \times 100 = 137\%$$

Detachment Faults

Detachment faults are regional subhorizontal to very low-angle normal faults that place upper-level relatively young rocks on deep-level relatively old rocks. Detachment faults are at such a low angle that we choose to describe offset and movement in relation to the **upper plate** and the **lower plate** of the fault, not "*hanging wall*" and "*footwall*." Detachment faulting can be thought of as a **tectonic denudation** of upper crustal cover rock to expose progressively deeper and deeper levels, a fact that becomes quite obvious when we find 20 Ma volcanic rocks in the upper plate resting directly in low-angle fault contact on 1400 Ma crystalline rocks in the lower plate.

The Whipple detachment fault in southeastern California (G. A. Davis et al., 1980) is a perfect example of a detachment fault. Exposures there are unencumbered by vegetation. The Whipple detachment surface itself dips less than 10°, with a knife-sharp contact between the upper plate and lower plate (Figure 6.135*A*). The

A

B

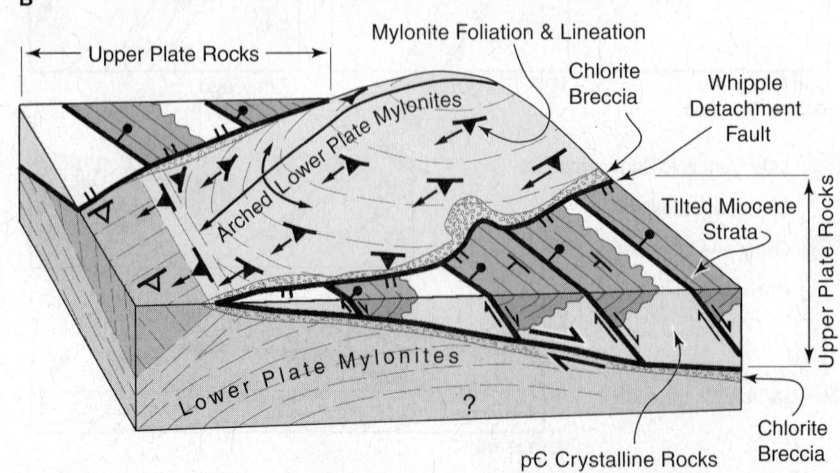

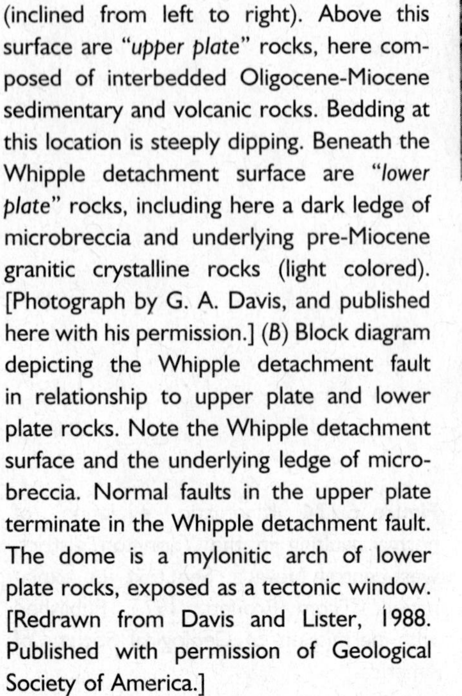

Figure 6.135 (*A*) Northwest-directed photograph of the trace of the Whipple detachment fault, southwestern Whipple Mountains, California. The detachment fault is the prominent gently dipping surface (inclined from left to right). Above this surface are "*upper plate*" rocks, here composed of interbedded Oligocene-Miocene sedimentary and volcanic rocks. Bedding at this location is steeply dipping. Beneath the Whipple detachment surface are "*lower plate*" rocks, including here a dark ledge of microbreccia and underlying pre-Miocene granitic crystalline rocks (light colored). [Photograph by G. A. Davis, and published here with his permission.] (*B*) Block diagram depicting the Whipple detachment fault in relationship to upper plate and lower plate rocks. Note the Whipple detachment surface and the underlying ledge of microbreccia. Normal faults in the upper plate terminate in the Whipple detachment fault. The dome is a mylonitic arch of lower plate rocks, exposed as a tectonic window. [Redrawn from Davis and Lister, 1988. Published with permission of Geological Society of America.]

lower plate rocks beneath the Whipple detachment fault contain Precambrian, Mesozoic, and Cenozoic igneous and metamorphic rocks (Figure 6.135*B*). The lower plate rocks commonly are converted into beautiful mylonites and cataclastites, testaments to the fact that these lower-plate rocks were brought up to the surface from depths of 10 to 15 km (G. A. Davis et al., 1986). The faulted upper plate rocks include Miocene volcanic and sedimentary rocks, which through detachment faulting were brought into *younger-on-older* fault contact with the lower plate rocks (see Figure 6.135). Fault displacement was several tens of kilometers, down-to-the east along a line of ~N50°E (G. A. Davis et al., 1986). Directly beneath the Whipple detachment fault, on the very top of the lower plate, the rocks are typically transformed over several meters or several tens of meters into microbreccia. Formation of these microbreccias was the natural result of the frictional effects of kilometers of slip as the footwall was "*raised*" from depth toward the surface.

Some low-angle normal faults (such as those mapped by Proffett at Yerington) started out at a higher angle but were rotated progressively to shallower dips as the crust extended. On the other hand, detachment faults such as that mapped by Greg Davis in the Whipple Mountains actually originated as shallow-dipping faults (Reynolds et al., 1988). Thus we now know that "*younger-on-older*" normal faulting accommodating extensional strain can be accomplished in any number of ways (Wernicke and Burchfiel, 1982). These include **high-angle normal faulting** along planar faults without any rotation of the faults or the strata; **domino-style normal faulting** in which planar normal faults rotate to progressively shallower dips and the originally flat-lying strata, now fault-bounded, rotate to progressively steeper dips (Figure 6.136*A*); **listric normal faulting** accompanied by reverse drag accomplished by collapse with or without antithetic faulting (Figure 6.136*B*); **imbricate listric normal faulting,** again with rollover (Figure 6.136*C*); and listric normal faulting bounding a family of planar normal faults (Figure 6.136*D*).

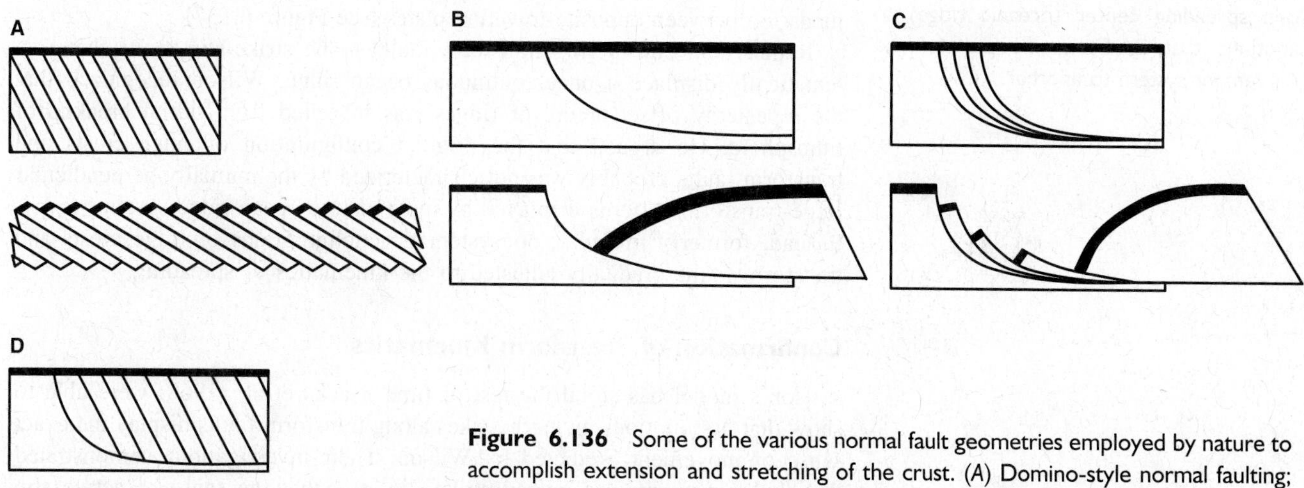

Figure 6.136 Some of the various normal fault geometries employed by nature to accomplish extension and stretching of the crust. (*A*) Domino-style normal faulting; (*B*) listric normal faulting with reverse drag; (*C*) imbricate listric normal faulting; and (*D*) listric normal faulting bounding a family of planar normal faults. [Reprinted from *Journal of Structural Geology*, v. 4, B. Wernicke and B. C. Burchfiel, Modes of extension tectonics, p. 105–115, © 1982, with permission from Elsevier.]

STRIKE-SLIP FAULTING

Plate Tectonic Significance of Major Strike-slip Faults

Strike-slip faults are steeply dipping faults along which horizontal slip has occurred. They comprise a fundamental class of faults whose geometric and kinematic properties are quite distinctive. *Major* strike-slip faults in continental settings are counterparts to the mid-oceanic **transform faults**. Like oceanic transform faults, most of the truly large strike-slip faults in continental terranes are fundamental plate boundaries. The San Andreas fault in California, the Motagua fault in Guatemala, the Alpine fault in New Zealand, and the North Anatolian fault in Turkey are examples. These faults, and some others like them, are thousands of kilometers long and have accommodated hundreds of kilometers of strike-slip displacement. Although strike-slip faults occur as compartmental faults and tear faults within systems of thrust, reverse, and normal faulting, these cannot compare in size, slip, and overall importance with those that currently are, or formerly were, major plate boundaries.

Discovery of Transform Faults

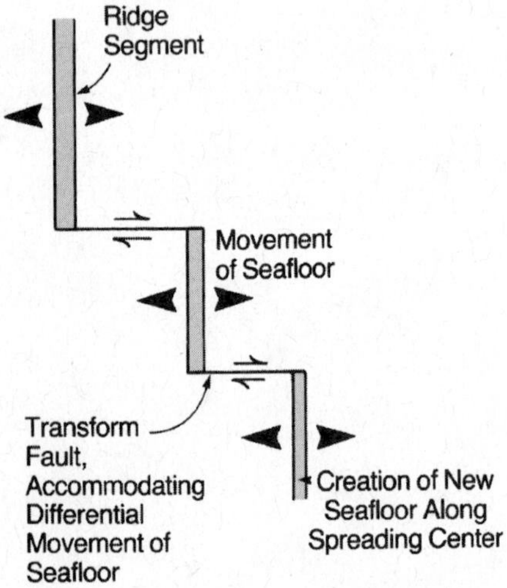

Transform faults, discordant to the trend of ocean ridge segments, are an integral part of the ocean ridge system. The topographic expression of these faults is boldly revealed in the bathymetry of the oceans (Figure 6.137). The faults appear to offset ridge segments within the otherwise continuous ocean ridge system. Shallow-focus earthquakes emanate from parts of these faults. The seismic activity expresses faulting and frictional sliding of thin lithosphere.

It was J. Tuzo Wilson who unlocked the geometric and kinematic significance of what had been considered "*conventional*" strike-slip faults, albeit occurring in oceanic settings. The stepping-stone for Wilson's break-through interpretation was the model of seafloor spreading as set forth by Dietz (1961) and Hess (1962). Wilson (1965) reasoned that the function of "*transcurrent intraoceanic faults*" was to connect the end of one ridge segment (spreading center) to another, thus enabling the entire movement system along the spreading axis to be integrated (see Figure 6.137). Because the faults "*transform*" the motion from one ridge to another, Wilson named these faults transform faults. Transform faults are *bona fide* plate boundaries that serve as zones of strike-slip accommodation between opposite-traveling plates (see Figure 6.137).

Figure 6.137 Transform fault linkage of one spreading center (oceanic ridge) to another; that is, from one part of the movement system to another.

Rather than considering transform faults to be strike-slip faults that systematically displace a once-continuous ocean ridge, Wilson recognized that the repeatedly offset nature of ridges was inherited from initial breakup of lithosphere. He argued that the original configuration of ridge crests and transform faults probably was not characterized by the mutually perpendicular ridge-transform patterns displayed by most mature ocean ridge systems today. Instead, formerly irregular, nonsystematic configurations of ridge crests and transform faults gradually adjusted to the kinematics of spreading.

Confirmation of Transform Kinematics

Wilson's model has stood the test of time. Isacks et al. (1968) were able to show that first motions on earthquakes along transform faults display the exact sense of movement predicted by Wilson. Their investigations demonstrated that the movements are indeed strike-slip, and that the sense of active slip along a given transform fault conforms to the relative motions of seafloor spreading along ridges joined by the transforms. The displacement vector for a given transform fault is *opposite* to the sense of separation that would be required to explain the offset of the ridge crest (Isacks et al., 1968). For the

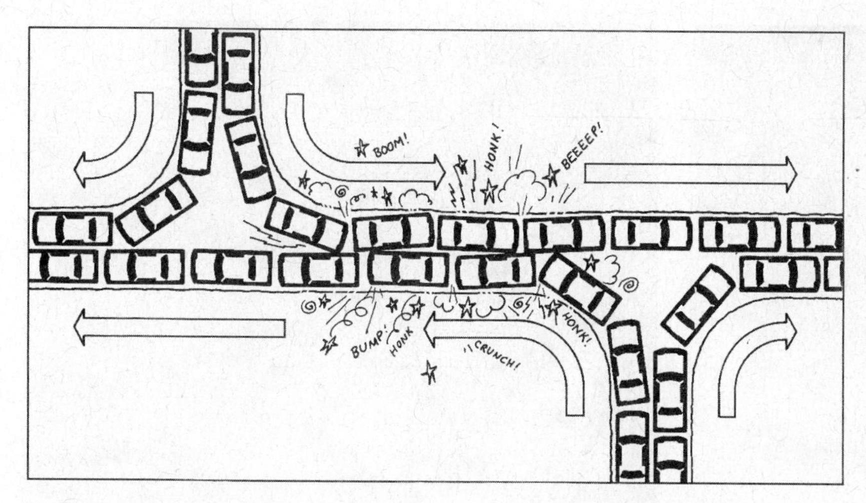

Figure 6.138 The noisy part of a transform fault is restricted to the interior segment, which lies between the ends of spreading centers connected by the fault. [Artwork by D. A. Fischer.]

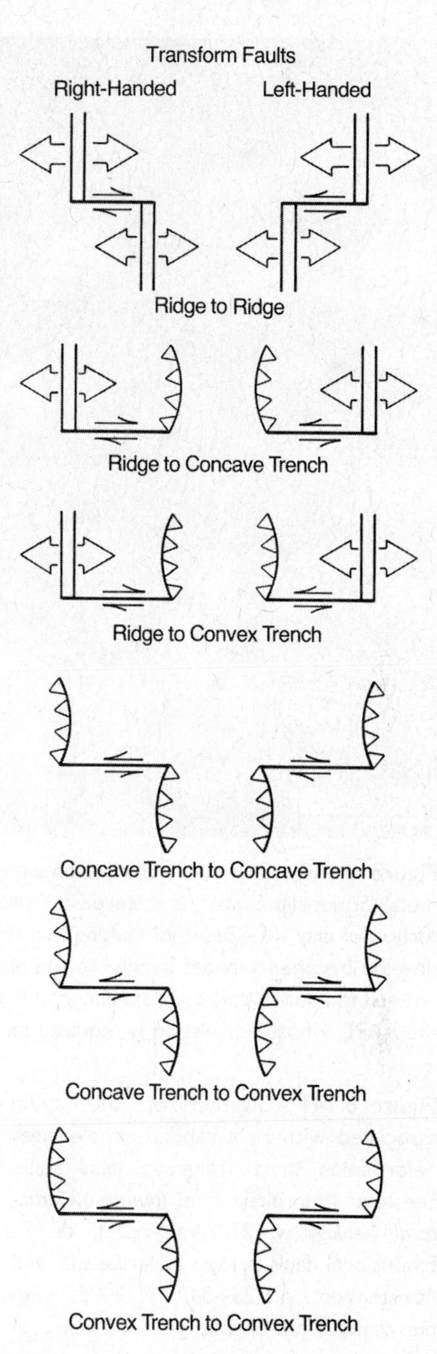

Figure 6.139 The classes of transform faults. Teeth are on overriding plate. [Reprinted by permission from Macmillan Publishers Ltd: *Nature*, A new class of faults and their bearing on continental drift, Wilson, J. T., p. 343–347, © 1965, with permission from Elsevier.]

transform faults shown in Figure 6.137, slip is right-handed whereas the sense of ridge offset is left-lateral. Lynn Sykes (1967) determined that the only seismically active part of a **ridge-to-ridge transform fault** is just where you would expect it to be: along the interior segment between the ridge crests connected by the transform (see Figure 6.137). If one were to figuratively stand in the active part of a transform fault zone, it would be like standing on the narrow median strip of a Los Angeles highway at 5:25 p.m. watching traffic slowly go by on each side in opposite directions (Figure 6.138). The brushing of the flanks of oppositely moving cars, as well as shouts of irate drivers, would create the "*seismic*" noise.

Varieties of Intraoceanic Transform Faults

Transform faults that connect offset ridge segments are the most abundant and perhaps the easiest to understand kinematically. However, two other fundamental types of transform fault exist as well, and these too were recognized by Wilson (1965). One links the tip of a ridge with the tip of a trench, which is the surface expression of a subduction zone. The second connects the tips of two subduction zones. Like **ridge-ridge** transform faults, **ridge-trench** and **trench-trench** transform faults exist because they serve to accommodate differential strike-slip movement of adjacent plates. Figure 6.139 presents the full spectrum of transform faults and the movements they accommodate. One of the great transform faults is the San Andreas fault, which we shall visit in *Chapter 11, Active Tectonics.*

Modeling Strike-Slip Faulting in the Laboratory

Wilcox et al. (1973) designed some clay cake experiments to evaluate the structural patterns that develop as a result of strike-slip faulting (Figure 6.140). Clay cakes are prepared on panels of sheet metal that can be moved oppositely and uniformly past one another during the course of an experiment. The line of contact between the metal panels predetermines the location and orientation of the linear zone of strike-slip faulting that develops in the overlying clay. Reference circles are impressed on the surface of the clay cakes before deformation so that strain, translation, and rotation can be evaluated (Figure 6.140*A*). Initial strike-slip movement of the metal panels results in distortion of the clay in such a way that the reference circles are

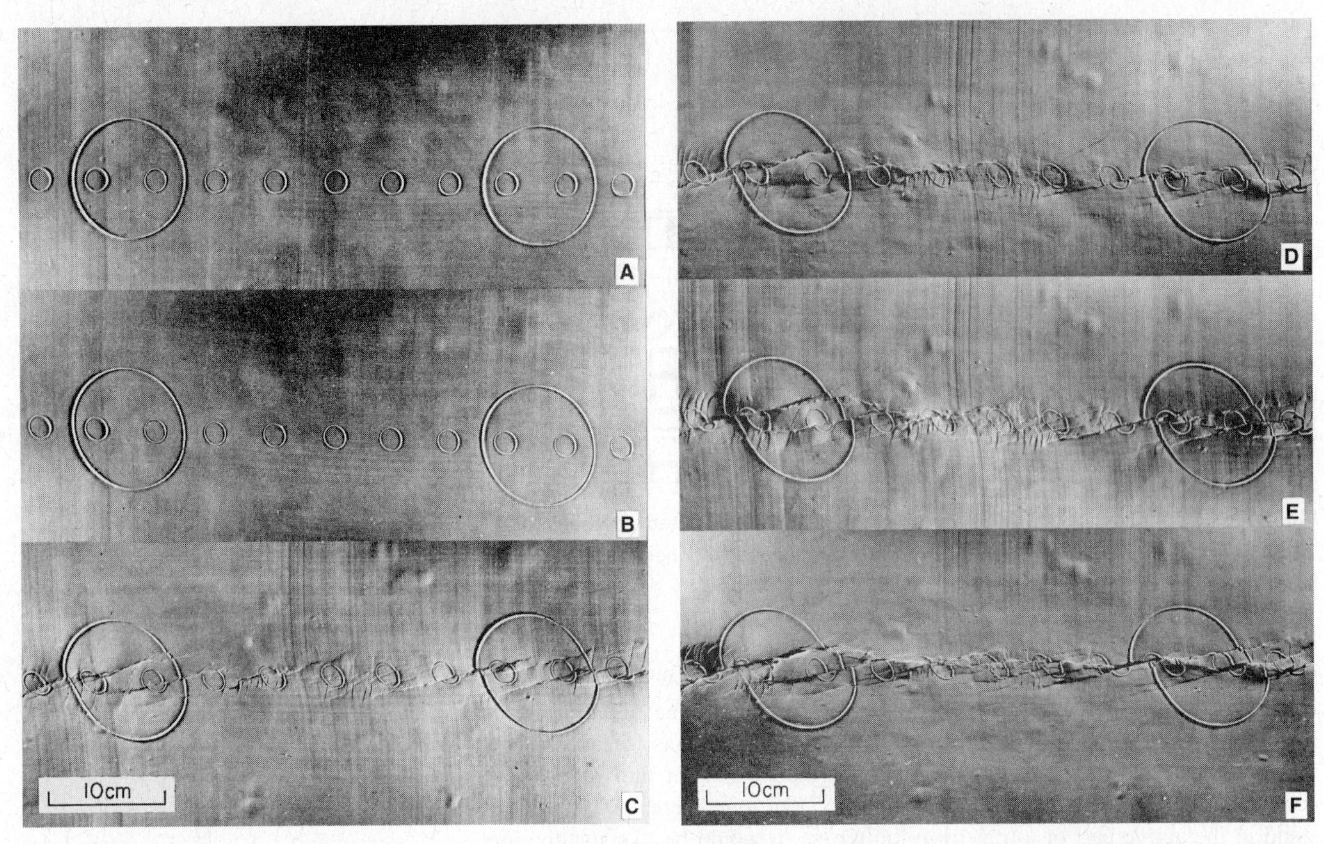

Figure 6.140 Clay cake deformation experiments simulating strike-slip faulting. Clay cake is placed on adjoining panels of sheet metal. Strike-slip faulting is achieved by shifting the panels horizontally past one another. (*A*) Starting configuration. (*B*) Initial distortion of clay. (*C*) Onset of faulting and the formation of synthetic and antithetic faults. (*D*)–(*F*) Continued faulting. Folds that develop become oriented parallel to the direction of maximum finite stretch (S_1). [From Wilcox, Harding, and Seely, 1973, Basic wrench tectonics: *American Association of Petroleum Geologists Bulletin*, v. 57, p. 74–96. AAPG © 1973, reprinted by permission of the AAPG whose permission is required for further use.]

transformed to ellipses (Figure 6.140*B*). Eventually the clay begins to fault within a zone parallel to the underlying metal panels (Figure 6.140*C-F*).

Waldron (2005) does a nice job in summarizing the details of faulting and folding in the interior of such strike-slip fault zones formed in clay. His example is that of a right-handed strike-slip fault zone (Figure 6.141). As the fault zone initiates, a discrete system of structures forms that consists of

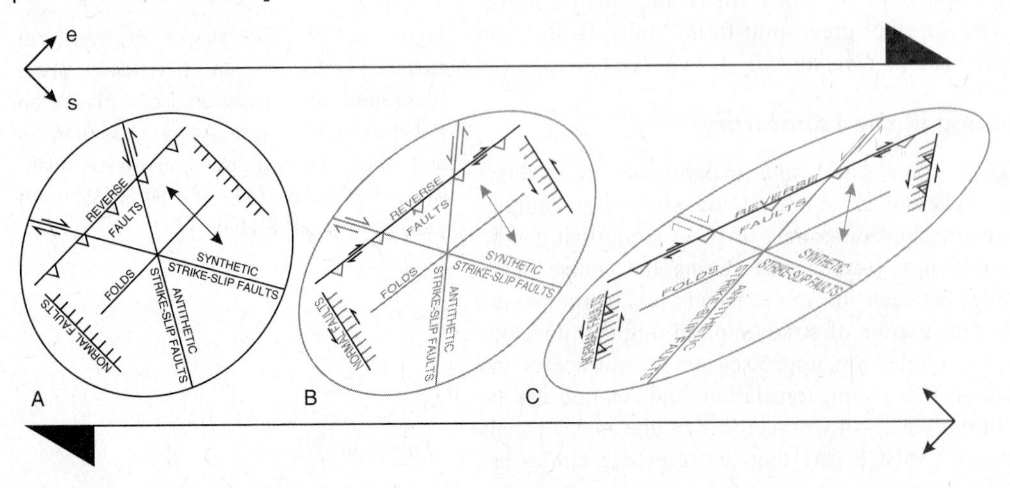

Figure 6.141 Diagram of the strain associated with right-handed simple-shear deformation in a strike-slip fault zone. See text. [Reprinted from *Journal of Structural Geology*, v. 27, Waldron, J. W. F., Extensional fault arrays in strike-slip and transtension, p. 23–34, © 2005, with permission from Elsevier.]

(1) conjugate strike-slip faults, (2) folds, (3) reverse faults, and (4) normal faults (Figure 6.141*A*). The strike-slip fault sets are marked by an initial conjugate angle of intersection of ~60° (see Figure 6.141*A*). Of the two conjugate fault sets, the one whose sense of slip is the same as that of the main zone of faulting is called **synthetic**; in this example, the right-handed strike-slip faults are the synthetic ones. The set whose sense of slip is opposite to that of the main zone is called **antithetic**; in this example, the left-handed strike-slip faults are the antithetic ones (see Figure 6.141*A*). The synthetic faults are typically oriented at a small acute angle to the trace of the main fault zone; the antithetic faults are oriented at a very high angle to the main zone (see Figure 6.141*A*). Initially formed folds and reverse faults orient themselves perpendicular to the direction of greatest instantaneous shortening, and thus trend/strike parallel to the direction of greatest instantaneous stretch (see Figure 6.141*A*). In contrast, normal faults orient themselves perpendicular to the direction of greatest instantaneous stretch, and thus they strike in a direction parallel to the direction of greatest instantaneous shortening.

Waldron (2005) goes on to emphasize that continued strike-slip shearing results in rotation of the elements in the system, just as you would expect during non-coaxial shear. As these old elements rotate to new orientations (see Figure 6.141*B*,*C*), some arrive at orientations that invite fault reactivation. For example, in an intermediate stage of rotation (see Figure 6.141*B*) the earlier formed normal faults accommodate left-handed strike-slip motion, and the reverse faults accommodate right-handed strike-slip motion. At a more extreme stage of rotation (see Figure 6.141*C*), the normal faults may accommodate oblique slip, expressed as left-handed reverse-slip faulting. Folds formed in the interiors of strike-slip fault zones will arrange themselves in en echelon patterns, with axes of individual folds aligned parallel to the direction of maximum finite stretch (S_1) (Figure 6.142) (Wilcox et al. 1973). Such folds are described as **right-handed** or **left-handed**, depending on the sense of shear along the main fault zone (see Figure 6.142). "*Right-handed folding*" means that if we were to walk the length of one of the en echelon folds, to the point where it disappears, we would turn to the right to search and discover the next fold in line (see Figure 6.142*A*).

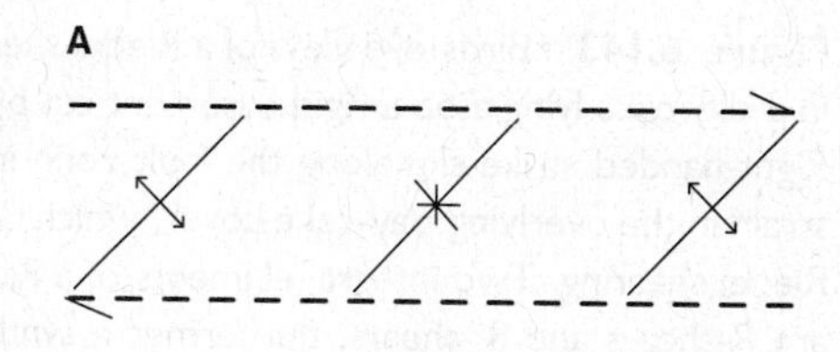

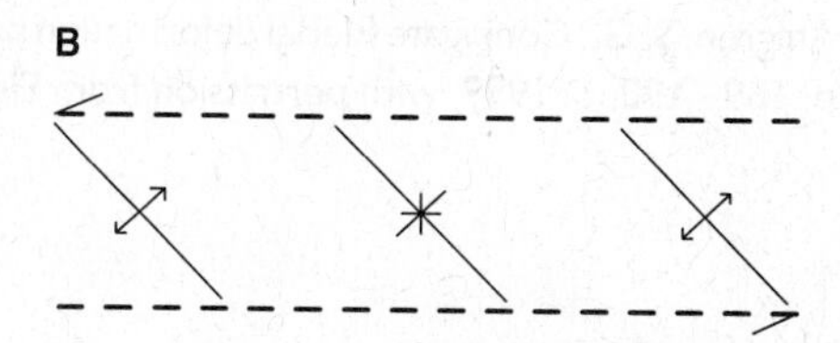

Figure 6.142 (*A*) Right-handed en echelon folds created within right-handed strike-slip fault zone. (*B*) Left-handed en echelon folds created within left-handed strike-slip fault zone.

Riedel Shears

The faulting produced in the experiments carried out by Wilcox et al. (1973) beautifully exemplifies the geometry and kinematics of **Riedel shearing**, which is characteristic of, but not limited to, strike-slip faulting (Davis et al., 1999). What we now know and love as Riedel shears were first observed and documented by Cloos (1928) and Riedel (1929) in clay-cake deformation experiments featuring an underlying "*basement block*" cut longitudinally by a vertical "*fault*" in the basement. When strike-slip motion is imposed on the fault in basement, "*Riedel shear patterns*" form in the overlying (moist) clay cake (Figure 6.143). The synthetic strike-slip faults are **Riedel shears** (**R-shears**), and they form at an acute angle of about 15° to the main line of faulting (see Figure 6.143). Their arrangement is **en echelon**, which means that they are parallel to one another and arranged along a common line of bearing. The antithetic strike-slip faults are **conjugate Riedel shears** (**R′-shears**), and they form at a high angle of about 75° to the main line of faulting (see Figure 6.143). The direction of greatest principal stress (σ_1) bisects the angle between R and R′. Figure 6.144 is a map-view drawing of Riedel shearing within a left-handed strike-slip fault zone. The drawing is a composite of ~16 photographs. The R-shears are oriented at ~15° to the trace of the fault zone; the R′-shears are mainly developed where the R-shears overlap one another.

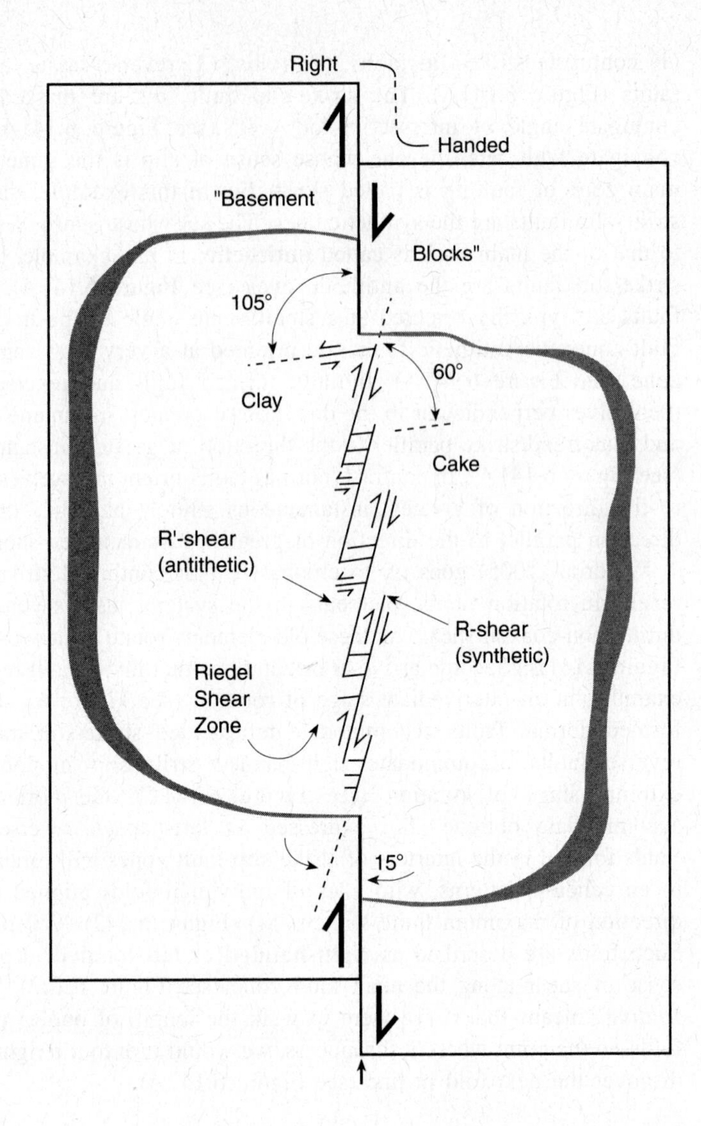

Figure 6.143 Birds-eye view of a Riedel shear zone formed in a clay cake lying atop a rigid basement cut by a "*fault zone.*" Right-handed strike-slip along the fault zone imposes a shear strain in the overlying clay-cake cover, which deforms through Riedel shearing. Two integral elements of a Riedel shear zone are R-shears and R′-shears, the former a synthetic shear, the latter an antithetic shear. [Reprinted from *Journal of Structural Geology*, v. 22, Davis, G. H., Bump, A. P., García, P. E., and Ahlgren, S. G., Conjugate Riedel deformation band shear zones, p. 169–190, © 1999, with permission from Elsevier.]

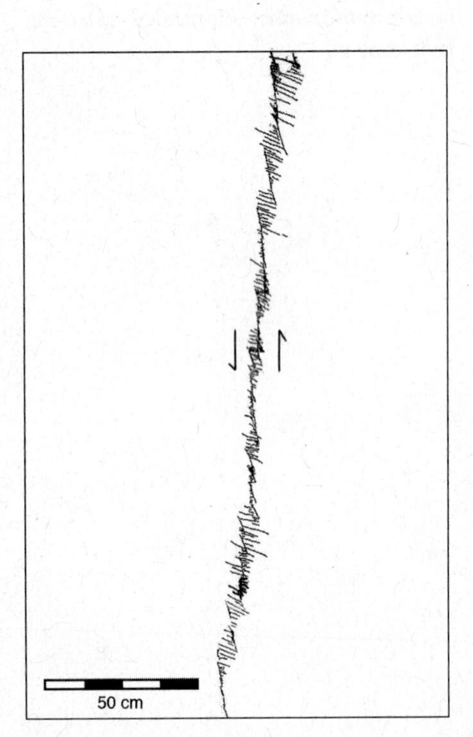

Figure 6.144 Drawing (based on ~16 close-range photographs) of Riedel shears within a left-handed strike-slip fault zone in Navajo Sandstone (Jurassic) in the Sheets Gulch area, Capitol Reef National Park, Utah. The R-shears (synthetic) deviate by ~15° from the trace of the fault; the R′-shears are at a high angle to the trace of the fault, and are particularly well developed in areas where the R-shears overlap one another. [Reprinted from *Journal of Structural Geology*, v. 22, Davis, G. H., Bump, A. P., García, P. E., and Ahlgren, S. G., Conjugate Riedel deformation band shear zones, p. 169–190, © 1999, with permission from Elsevier.]

As strike-slip faulting along the main zone proceeds, R-shears achieve a closer angle with the main line of faulting, and R′-shears may be rotated to a higher angle. Furthermore, a new set of synthetic shears known as **P-shears** may develop, and these form at a small acute angle (~10°) to the main line of faulting. Figure 6.145 shows the relationships between R-shears, R′-shears, and P-shears for a left-handed strike-slip fault. The arrangement shown in Figure 6.145 works for thrust, reverse, and normal faults is well. All you have to do is rotate the diagram such that both the fault trace and sense-of-slip are oriented properly with respect to the fault zone that you are investigating.

Structures Developed Along Bends of Strike-Slip Faults

Crowell (1974) emphasized that bends in strike-slip faults invite high concentrations of strain. The strain is distributed within distinctive suites of structures. Movement along perfectly planar strike-slip faults results in coherent structural patterns. Wall rocks slide past each other without much interference (Figure 6.146*A*). Branching and braiding of faults is minimal. In contrast, if a strike-slip fault is marked by an abrupt bend, or even a gradual bend, complications arise (Figure 6.146*B*). Country rock adjacent to the fault zones are required to adjust to stress buildups by stretching or shortening.

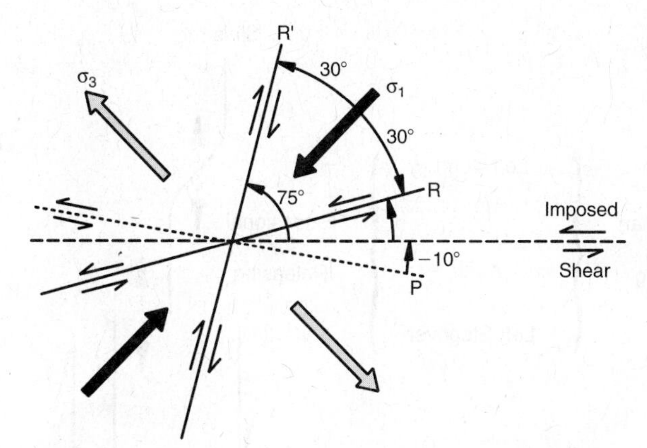

Figure 6.145 Summary diagram of Riedel shearing. Context is left-handed strike-slip faulting. R and R′ are conjugate Riedel shears; R is synthetic, whereas R′ is antithetic to the main movement. R and R′ form first. P shears "*come in*" a little later. The direction of greatest principal stress (σ_1) bisects the conjugate Riedel shears and makes an angle of ~45° to the main line of faulting. [After Woodcock and Schubert, 1994, *Continental deformation*, Figure 12.8, p. 257, in Hancock, P. L. (ed.), Continental deformation: Pergamon Press, New York, p. 251–263.]

Fault curvature can be described in terms of **bends** or **jogs**. Movement of wall rock along a fault with a bend leads to convergence or divergence, depending on the sense of motion and the sense of curvature (see Figure 6.146*B*). **Releasing bends** tend to create open space, whereas **restraining bends** are sites of crowding. In either case, the bends in the strike-slip faults are very efficient in changing horizontal movement into vertical movement.

Deformation at releasing bends is marked by extensional deformation, especially normal faulting. Fault-bounded grabens formed at releasing bends can form rapidly as the strain changes from horizontal movement to vertical movement. Therefore these structures are capable of receiving thousands of meters of clastic basin deposits in a very short time interval. In mineral deposit settings, releasing bends are probable sites of open-space filling and ore deposition. In contrast, restraining bends are marked by shortening, which is achieved by thrusting and folding. A common response to such crustal shortening is vertical uplift of the thickened block.

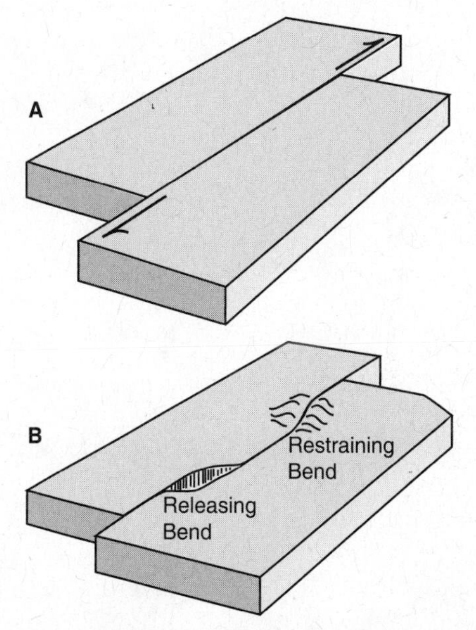

Figure 6.146 (*A*) Strike-slip movement along perfectly planar faults produces neither gaps nor overlaps. (*B*) Strike-slip movement along irregularly curved faults produces gaps at releasing bends and crowding at restraining bends. [From Crowell, 1974. Published with permission of Society of Economic Paleontologists and Mineralogists.]

Fault-bounded **wedges** of rock may form during the evolution of a bend (Figure 6.147). Resistance at a restraining bend may become so large that a newly formed fault cuts around the restraint, thus isolating a wedge of country rock. Wedges can be large or small; some associated with the San Andreas system are 100 km long. A wedge, during continued fault movement, can rotate such that one tip subsides to form a basin while the other tip rises to produce an uplift. The uplift becomes a source area for clastic debris, which inevitably is deposited in the nearest accommodation space.

Restraint and release along bends in strike-slip faults clearly cause bedrock to become uplifted or down-dropped. Thus, during the course of mapping of a regional high-angle fault suspected of being a strike-slip fault, it is not always straightforward to be certain that the faulting is indeed primarily strike-slip in nature. Yet, when we see that adjacent fault blocks seem to be "*porpoising*"; that is, rising and then falling along the fault trace, we become more certain that we are dealing with a strike-slip fault.

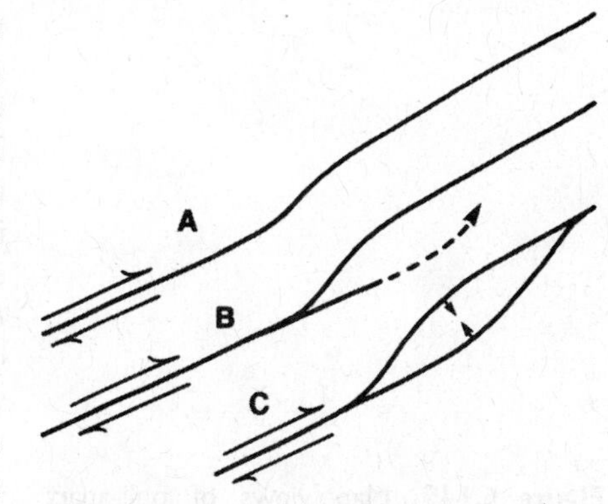

Figure 6.147 (*A*) to (*C*) The progressive development of a fault-bounded structural wedge at a restraining bend of a strike-slip fault. [From Crowell, 1974. Published with permission of Society of Economic Paleontologists and Mineralogists.]

Transfer of Strike-Slip Displacement at Stepovers

We learned in studying normal faults that slip and displacement can be transferred from one fault segment to the other through transfer zones in the form of relay ramps (e.g., see Figs. 6.130 and 6.131). Within strike-slip fault systems, such transfer of slip and displacement takes place at **stepovers**. At stepovers, one strike-slip fault segment ends, and another one, of essentially the same trend, begins (Figure 6.148) (Mann et al., 1983). If the fault segments do not actually interconnect at the stepover, the transfer of displacement is said to be **soft linked**. The actual transfer of slip within a soft-linked system

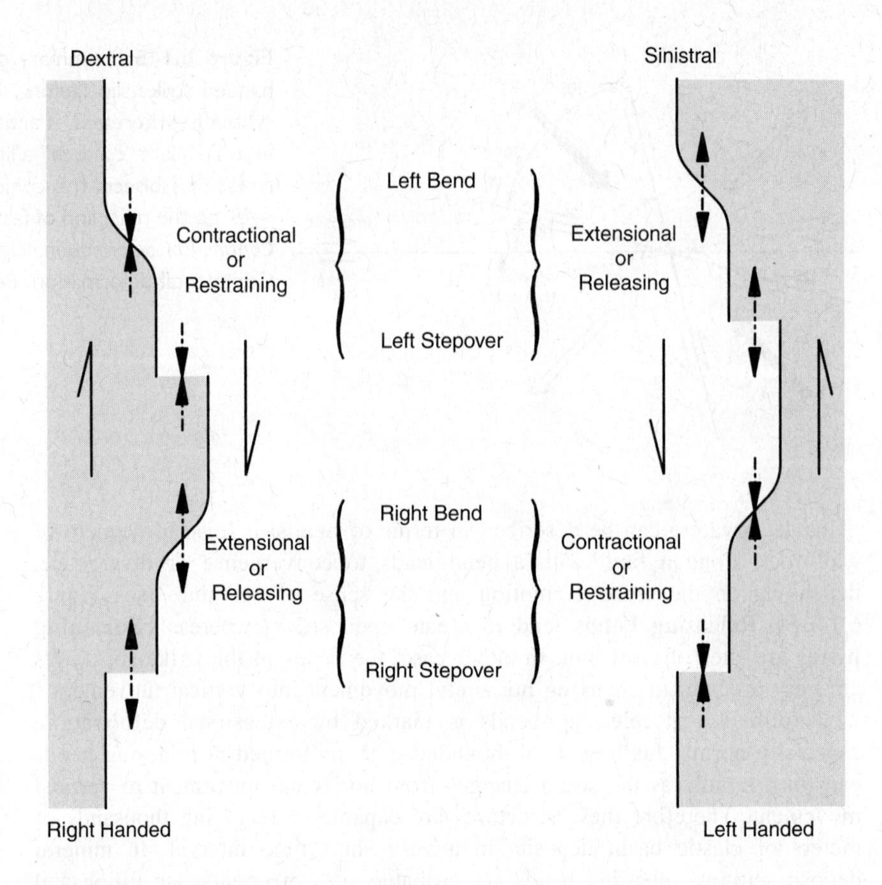

Figure 6.148 Portrayal of the difference between bends and stopovers, both in right-handed and left-handed strike-slip systems. [From Twiss and Moores, *Structural Geology*, Fig. 7.5, p. 116. Copyright © 1992, W. H. Freeman and Company, with permission.]

takes place through strain within the intervening crust between the strike-slip fault segments (see Figure 6.148).

There are both **right stepovers** and **left stepovers**. The distinction is based on whether we turn to the right or left to find the next fault segment as we walk beyond the tip of a given strike-slip fault segment (see Figure 6.148). It is satisfying, we believe, to think through whether a given stepover will be marked by extension/release or contraction/restraint. In doing so, we must factor in whether the stepover is right or left, and whether the fault slip is right-handed or left-handed. The various combinations are shown in Figure 6.148. At extension/release locations we see pull-apart basins, normal faulting, and other such evidence for crustal stretching (Figure 6.149*A*). At contraction/restraint locations we see pop-ups, thrusting (and/or reverse faulting), folding, and other such evidence for crustal shortening (Figure 6.149*B*). The term **pull-apart basin** was introduced by Burchfiel and Stewart (1966) in interpreting the tectonic origin of Death Valley. A more terrifying name, introduced earlier by S. Warren Carey (1958), is **rhombochasm**, a very deep depression bounded by master strike-slip faults that are stepped. Because "*the bottom falls out*" of pull-apart basins, they become sites of thick accumulation of sediments. In contrast, **pop-ups** are topographic and structural highs from which sediments are shed. The amount of extension or shortening in the releasing or restraining bends and/or stopovers is related to the amount of strike-slip displacement on the bounding faults. By measuring that extension or

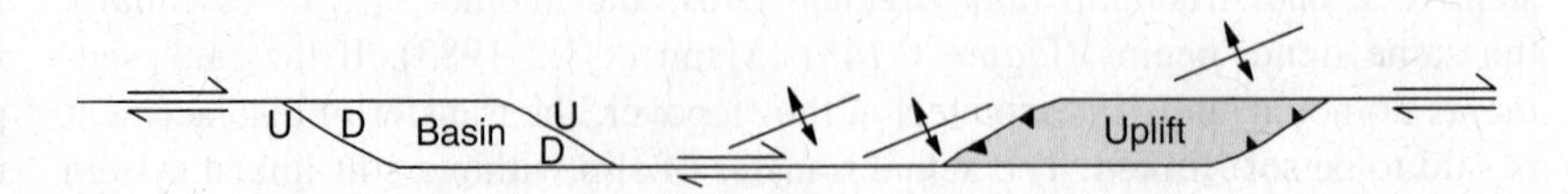

Figure 6.149 Map views of pull-apart basins and pop-ups and their relations to bends in right-handed vs. left-handed strike-slip systems. [From Suppe, J., Principles of Structural Geology, © 1985, p. 280. Reprinted by permission of Prentice-Hall, Upper Saddle River, New Jersey.]

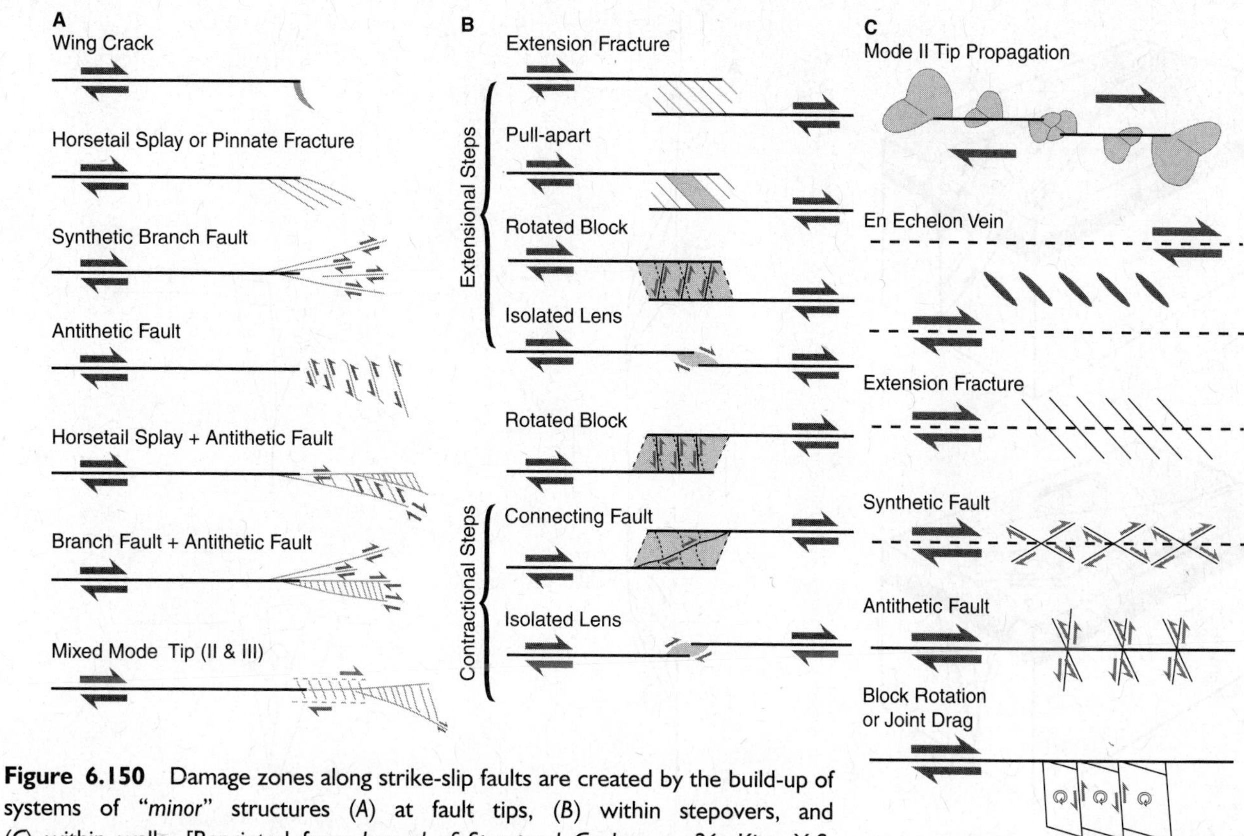

Figure 6.150 Damage zones along strike-slip faults are created by the build-up of systems of "*minor*" structures (*A*) at fault tips, (*B*) within stepovers, and (*C*) within walls. [Reprinted from *Journal of Structural Geology*, v. 26, Kim, Y-S, Peacock, D. C. P., and Sanderson, D. J., Fault damage zones, p. 503–517, © 2004, with permission from Elsevier.]

shortening, we can derive an estimate of the amount of strike-slip displacement on the fault system, when otherwise no reliable markers are present.

Stepover structures associated with strike-slip faulting can be observed at the outcrop scale as well. In fact, the deformation required at stopovers may be the primary contributor to the **damage** so commonly seen in fault zones. Kim et al. (2004) think about "*fault damage zones*" as the volume of rock around a fault surface that becomes deformed as a result of the slip history of the faulting, from precursory structures that develop, to structures and fabrics that result from the history of slip along the fault surface. Most of the damage builds up at the tips of the faults, and in the stepovers between faults segments. The features include wing cracks (mode I fractures), pull-apart zones, conjugate faults, rotated blocks, connecting faults, and antithetic faults. Kim et al. (2004) subdivide the damage into three domains: at tips (Figure 6.150*A*), in stepovers (Figure 6.150*B*), and along the walls (Figure 6.150*C*).

Strike-Slip Duplexes

Bend and stepover geometries invite the development of **strike-slip duplexes**. Crust within a bend, or within the interior of a stepover, becomes progressively faulted. Movement along one master fault is transferred to another along a connecting fault (the equivalent of a ramp in thrust-fault duplexes). As the duplex evolves, connecting faults die out one by one, and new ones form (Figure 6.151). Each fault connects with the master fault at depth, thus assuring a linkage and kinematic organization. The geometric evolution of strike-slip duplexes is akin to what we would see if we turned an evolving

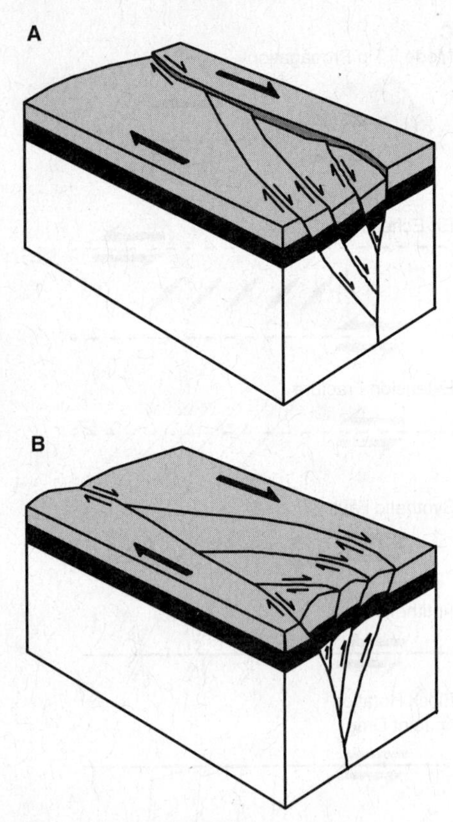

Figure 6.151 Strike-slip duplexes may form at bends along the major strike-slip fault. (*A*) Extensional duplexes form at releasing bends. (*B*) Compressional duplexes form at restraining bends. [From Twiss and Moores, *Structural Geology*, Figs. 7.6 and 7.7, p. 118 and 119. Copyright © 1992, W. H. Freeman and Company, with permission.]

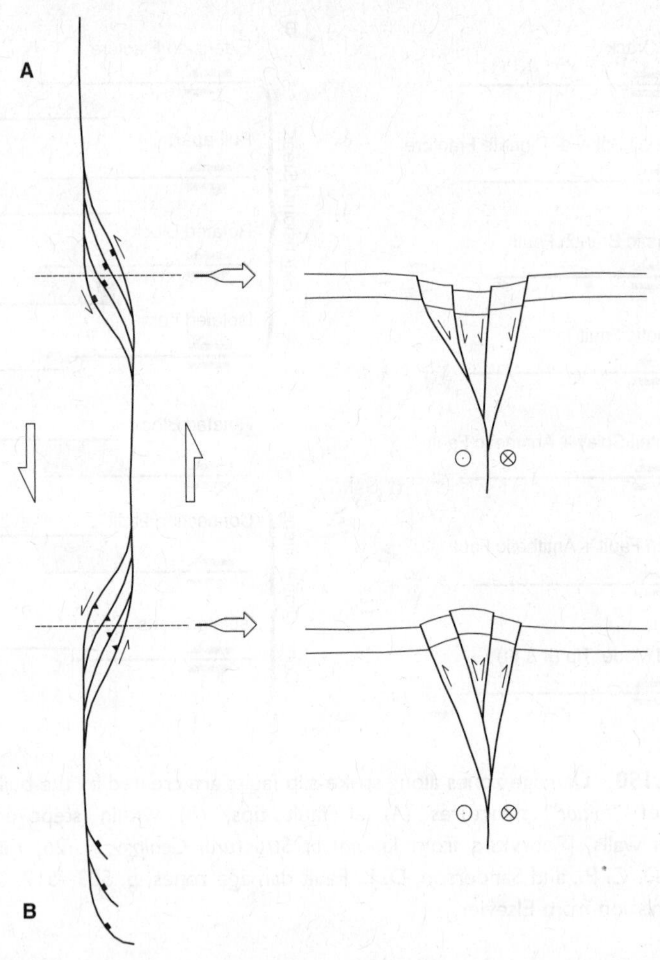

Figure 6.152 (*A*) Map and cross-section of extensional strike-slip duplex at a releasing bend, and the formation of a negative flower structure. (*B*) Map and cross-section of contractional strike-slip duplex at a restraining bend and the formation of a positive flower structure. [Reprinted from *Journal of Structural Geology*, v. 25, Woodcock, N. H., and Rickards, B., Transpressive duplex and flower structure: Dent Fault System, NW England, p. 1981–1992, © 2003, with permission from Elsevier.]

thrust-fault duplex or normal-fault duplex on end, so that it could be seen in a horizontal view. Depending upon whether there is shortening or stretching across the zone, the strike-slip duplex will be compressional or extensional. The fault patterns evident in cross-sections of strike-slip duplexes have been described as **flower structures**: bouquets of steeply dipping fault splays that diverge upward. An extensional strike-slip duplex is often referred to as a **negative flower structure**, because the rocks are dropped downward (Figure 6.152*A*). A compressional strike-slip duplex is often referred to as a **positive flower structure**, because material is pushed upward (Figure 6.152*B*). In either restraining bend strike-slip duplexes or releasing bend strike-slip duplexes, a key to recognizing them in cross-section is the upward divergence of the faults. They converge downward into a steep zone of the main strike-slip fault at depth.

A nice example of duplex formation in a strike-slip setting was worked out by Woodcock and Rickards (2003). They studied the Dent fault on the eastern margin of the (beautiful) Lake District, England. Faulting and associated deformation took place in late Carboniferous. The Dent fault zone emerged through reactivation of an ancient fault in the basement. High-angle reverse

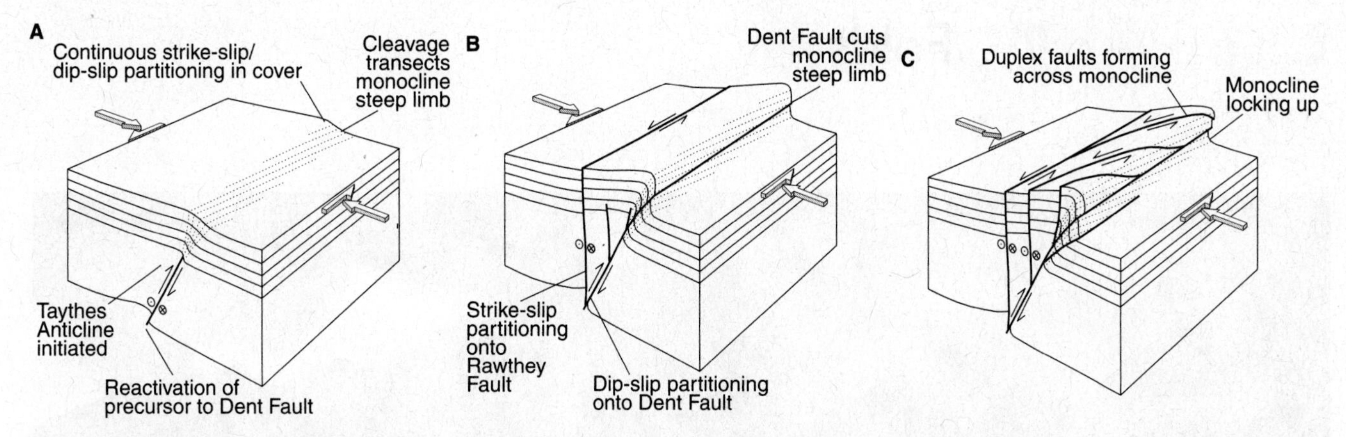

Figure 6.153 Block diagrams showing the formation of a positive flower structure and a contractional duplex, brought about by transpression across the Dent fault, eastern Lake District, England. (*A*) Reactivation of basement fault and early development of Taythes anticline. (*B*) Emergence of Dent fault, cutting steep limb of monocline. (*C*) Duplex faults forming across monocline. [Reprinted from *Journal of Structural Geology*, v. 25, Woodcock, N. H., and Rickards, B., Transpressive duplex and flower structure: Dent Fault System, NW England, p. 1981–1992, © 2003, with permission from Elsevier.]

faulting produced the Taythes anticline and drove left-handed strike-slip faulting (Figure 6.153*A*). Next, a second major fault, the Rawthey fault, formed (Figure 6.153*B*). The two faults quickly "*reached an understanding*:" the Dent fault would manage all of the reverse faulting, and the Rawthey fault would handle all of the strike-slip displacement. As deformation proceeded in this way, a duplex formed, marked by a positive flower structure built through folding and reverse faulting within the transpressive fault zone (Figure 6.153*C*). Faults of all categories work together!

CONCLUDING REMARKS

The study of faulting encompasses geometric, kinematic, and dynamic analysis in the broadest possible way. If we are to understand faulting, we need to integrate field, experimental, and theoretical research. The rewards for such activities are significant. Faulting is one of the most important mechanisms of rock deformation; it produces regional distortion; it traps petroleum and controls ore deposition; it channels ground water; it provides clues to rearrangements of the Earth's architecture through time. Furthermore, faulting and the associated earthquakes present one of the most significant natural hazards to society. Experimental deformation and theoretical studies help us immensely in understanding the properties of faults that we find in nature.

Chapter 7 Folds

Giant syncline in Cretaceous flysch, west of Shigatse, Tibet, China. This is just one expression of the tectonic shortening that worked its way inboard from the collision of the Indian–Australian plate and the Eurasian plate. Commonly it is the anticlines that get all of the attention, but this syncline — with village for scale — is of such prodigeous scale that it cannot be ignored. Folding of the solid rock was achieved largely through flexural slip along bedding. Original bed thicknesses were reasonably maintained during the folding, but as a result the fold was forced to tighten upward toward its core, which is exposed on the ridge line. Normally when we see well-exposed large-scale structures in the field, we immediately begin to project geometry downward into subsurface cross-sectional views. In this case we clearly see the need to project the structural geometry skyward as well. [Photograph and copyright © by Peter Kresan.]

INCENTIVES FOR STUDY

Visual Impact

Folds are visually the most spectacular of Earth's structures. They are extraordinary displays of strain, conspicuous natural images of how the original shapes of rock bodies can be changed during deformation. The physical forms and orientations of folds seem limitless. Some are upright (Figure 7.1); some lie on their sides (Figure 7.2); some are inclined (Figure 7.3). Some show neatly arranged, uniformly thick layers; others are sloppy. Fold size varies too, from anticlines that fit into the palm of a hand to regional folds best seen through the eyes of a satellite. Mapping the forms of folds is pure pleasure, unless of course the folds turn into geometric nightmares. Constructing cross-sections of folded terrains becomes a fundamental tool for structural geologic insight (*Part III-F*).

Mechanical Contradiction of Folding

It is almost impossible to view "*waves*" of solid rock without wondering how materials we regard as strong can be folded in a manner that makes them seem so weak. Bailey Willis (1894), in trying to model the structures of the Appalachian fold belt, found it necessary to represent sedimentary strata with

Figure 7.1 Upright anticlines and synclines in Cretaceous limestones within the Sanctuary of Zeus, Mt. Lykaion, in the Peloponessos. Geologist is Randy Goosen. [Photograph by G. H. Davis.]

Figure 7.2 Folds in Oligocene-Miocene turbidites in Kii Peninsula, southwest Japan. [Photograph by W. R. Dickinson.]

Figure 7.3 Inclined fold structure exposed in a Utah Highway 276 road cut through Carmel Formation (Jurassic). Geological location is on the west side of Mt. Holmes in the Henry Mountains, Utah. Height of this fold exposure is ~5 m. [Photograph by G. H. Davis.]

layers of very soft materials like clay, putty, cheese, and wax (see Figure 3.3). One the one hand, it is easy to imagine that slow, incremental steady-state creep and flow will produce folds under metamorphic conditions of highly elevated temperatures and pressures. When the mechanical influence of bedding or other layering is subdued by elevated temperature and confining pressure, and the layers become "*soft*," folding is achieved in ways that may be marked by profound changes in layer thickness across the fold (Figure 7.4*A*), and/or the development of cleavage oriented systematically with respect to the fold properties (Figure 7.4*B*). Cleavage marks locations where pressure dissolution caused by the influence of directed tectonic stress resulted in loss of material (yet another way of accommodating shortening) (see *Chapter 9, Cleavage, Foliation, and Lineation*). The cleavages cut across the lithological layering of folded beds *as if the beds had no mechanical influence whatsoever, . . . as if the bed did not exist.*

It is difficult to picture how massively thick, stiff, strong rock layers within regional rock assemblages can fold, seemingly effortlessly, in response to

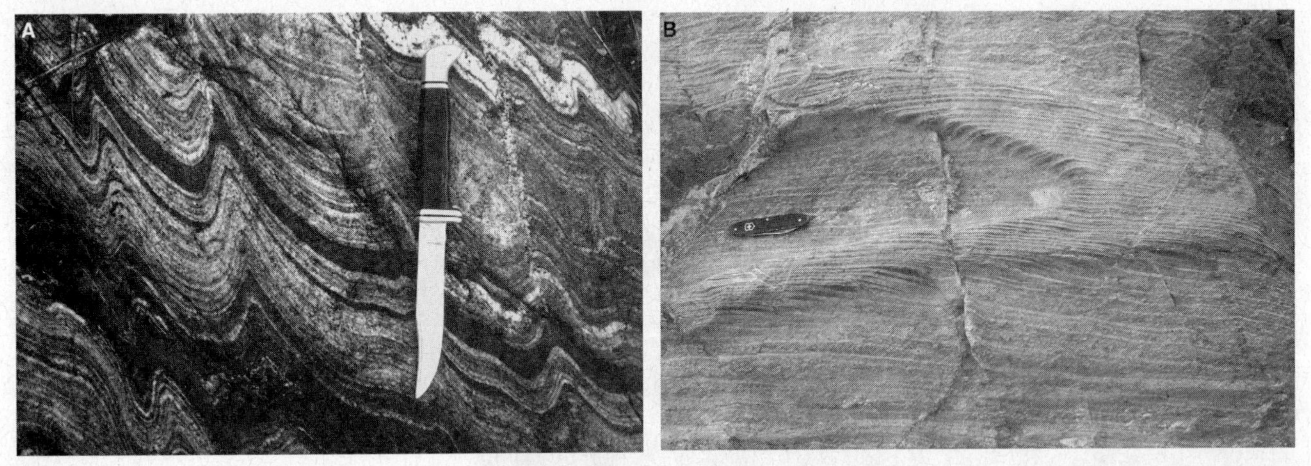

Figure 7.4 (*A*) Photograph of fold in banded gneiss in north Norway, within the Ballangen region south of Narvik. Folding and shearing took place in early Paleozoic, and under deep-seated metamorphic conditions that caused the rock to flow during folding. In parts of the region the basement rocks experienced depths of ~25 km. Note the thickening and thinning of individual layers. [Photograph by G. H. Davis.] (*B*) Photograph of passive fold in fine-grained quartz-muscovite schist of Precambrian age. Location is Hemlo Mine area, Ontario, Canada. The lamination crossing the image horizontally is not bedding, but rather cleavage. Bedding is expressed in the form of the tightly folded layer, subtly expressed through lighter hue and differential resistance to weathering and erosion. [Photograph by S. J. Reynolds.]

Figure 7.5 Photograph of broad syncline in the thick, massive Navajo Sandstone (Jurassic), as exposed near Bluff, Utah along Comb Ridge. Geologist in foreground is Steve Kidder. [Photograph by G. H. Davis.]

tectonic loading (Figure 7.5). Oftentimes in the apparent absence of any major faulting, mechanically stiff, strong beds bend by buckling, with flexure accommodated by slip along bedding. Interbeds of weak shale and mudstone expedite such flexural folding, for they are zones of relatively easy glide (Figure 7.6). We learn that where strata are very well bedded and relatively stiff, folding is usually achieved by a process called **flexural slip**. The individual layers buckle and fold (the "*flexural*" part), and as they do they slide past each other by layer-parallel slippage (the "*slip*" part) (Donath and Parker, 1964). Like the pages of a slick magazine, beds can easily slip past one another along bedding surfaces during folding (see Figure 2.16).

Part of the mechanical contradiction of folding is resolved when we recognize the intimacy of relationships between folding and faulting (see *Chapter 8,* Fold-Fault Interactions). Sometimes the faulting that "*drives*" the folding is out of sight. This is most certainly the case of blind thrusting that "*feeds slip*" into folding (see Figure 6.116). Yet, when, after an earthquake, we witness the results of catastrophic failure by faulting, it is hard to picture how earthquake events and folding are compatible.

To be sure, folding seems to be full of mechanical contradictions. But regardless of mechanism (either earthquake by earthquake; continuous incremental flow and creep; everything in between), folding of rock layers to very steep dips is a process that requires time, and lots of it! Pondering how solid rock can fold at all is a first-order wonder in earth science, and presents a pathway to begin to appreciate the vastness of deep time.

Figure 7.6 Kink-like fold in limestone at Durdle Door in the Purbeck Country, England. Note that limestone (light) forms stiff, resistant layers that alternate with weaker shale beds. Within the fold proper, bedding-plane slip takes place in a way made easier by the strength contrast between limestone and shale. [Photograph by G. H. Davis.]

Geometric Pleasures

Analysis of the geometry of folds invites us to probe more deeply into stereographic analysis of structural data (see *Part III-I, J, N*). This, in turn, leads to more expansive awareness of geometrical concepts and methods. Simple stereographic procedures permit us to calculate and describe the orientations of folds, to measure fold tightness, to unfold folds and reconstruct original orientations of primary structures, and to describe the total integrated geometry of fold systems. Turner and Weiss (1963), through their classic text, *Structural Analysis of Metamorphic Tectonites*, inspired the full use of stereographic analysis in modern geometric analysis of folds and fold-related structures.

Informative Minor Structures

Folding results in the formation of a delightful and curious array of small-scale structures, typically referred to as **minor structures**. The adjective "*minor*" belies the usefulness of this class of structures in evaluating the kinematics of folding. Within the family of minor structures are small folds and faults; slickenlines and crystal fiber lineations; a variety of cleavages and penetrative lineations; joints, shear fractures, veins, and stylolites; and some other structures as well, such as mullion. Different kinds of minor structures form in different parts of folds, because different parts of folds are marked by different local strain environments. Some parts of folded beds are stretched; others are shortened; some lose volume; some gain volume; still other parts suffer no distortion or dilation whatsoever.

Most students of geology enjoy the challenge of trying to interpret the orientations and structural locations of minor structures in light of the overall geometry and kinematics of folding. Learning to do so is fundamental to mapping and unraveling regional assemblages of folded rocks. For example, when we see asymmetric folds in an outcrop, we name them *Z-shaped* (Figure 7.7*A*) or *S-shaped* (Figure 7.7*B*), based on their form. During mapping of what appears to be a straightforward dipping panel of beds, we might spot a change from *Z*-shaped minor fold (Figure 7.7*C*, left) to *S*-shaped (Figure 7.7*C*, middle), and then back to Z-shaped (Figure 7.7*C*, right). This would alert us to the presence of otherwise hidden fold structures (Figure 7.7*D*).

Interpreting folds, including the formation of minor structures, is a fruitful exercise in applying principles and methods in kinematic and dynamic analysis. Kinematic analysis of folding requires integration of all kinematic movements: translation, rotation, dilation, and distortion. Dynamic analysis requires identifying the physical and mechanical variables that influence the nature of folding. Dynamic analysis is pursued both theoretically and experimentally. Fold analysis provides an unusual opportunity for integrating and synthesizing principles

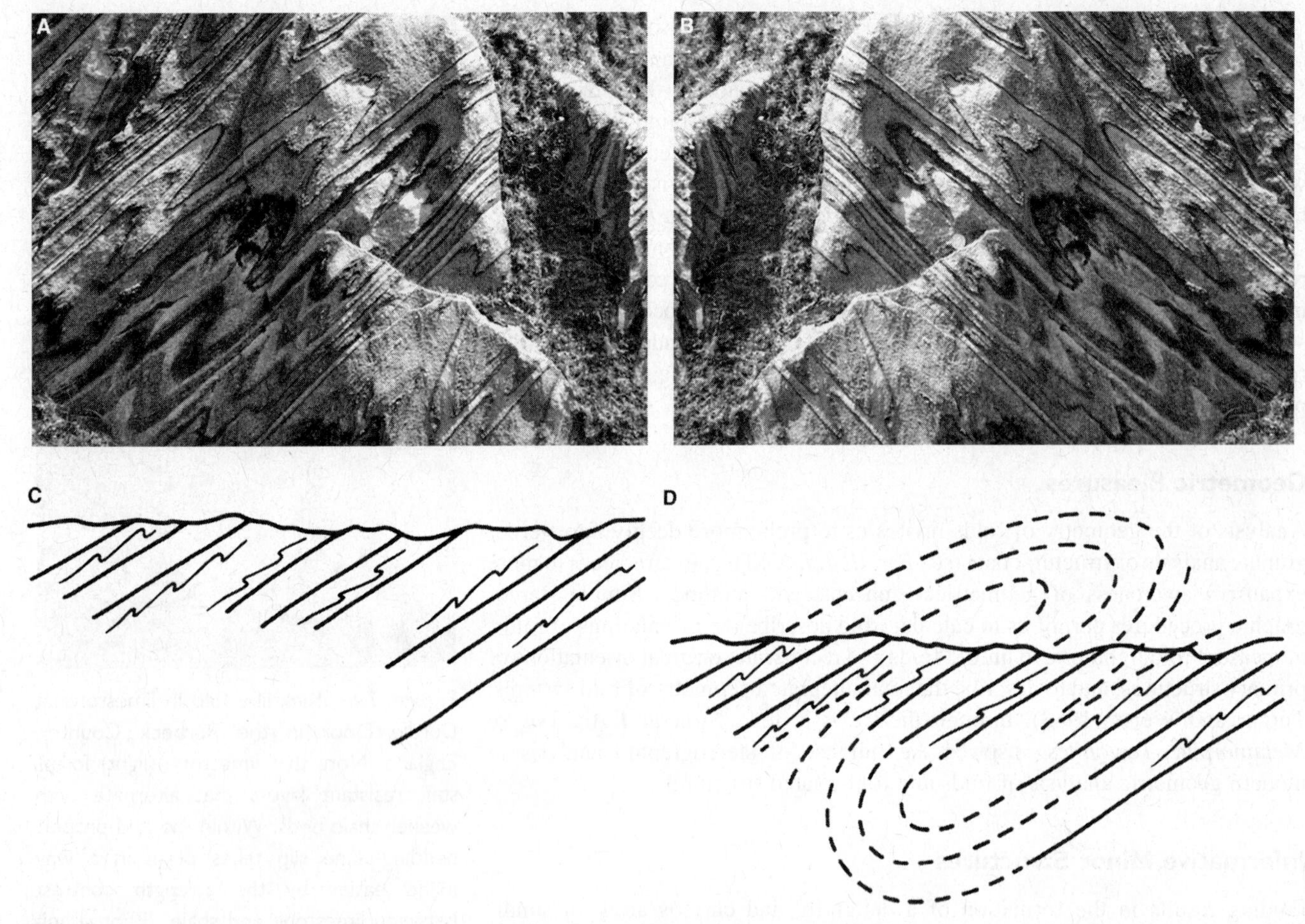

Figure 7.7 The asymmetry of folds can be used to disclose hidden structures. (*A*) Z-shaped minor fold in left-dipping bed of marble. (*B*) S-shaped minor fold in left-dipping bed of marble. (*C*) Z-shaped and S-shaped minor folds in a dipping panel of beds may denote (*D*) hidden folding.

of structural geology. Friends of folds have John Ramsay (1967) to thank for *Folding and Fracturing of Rocks*, which presents the foundation for modern quantitative fold analysis (see also Ramsay and Huber, 1987).

Guides to Exploration and Production of Hydrocarbons and Minerals

There are practical incentives to come to know folds and their properties. Foremost, there is a legacy of discovery of oil and gas in **structural traps** created by folding (Figure 7.8). Folds commonly serve as collection sites for oil and gas that migrate up the dip of strata from hydrocarbon source beds below. The migration of oil and gas takes place within a permeable reservoir rock, such as porous sandstone or vuggy limestone. If the reservoir bed is overlain depositionally by a tight, impermeable layer, such as shale, the top of the trap is effectively sealed to prevent escape of the oil and gas during up-dip migration. Where a sealed reservoir is folded into a dome or anticline, the **closure** of the fold prevents further upward migration of the oil and gas. The fluids may collect there to form a rich pool. "*Pool*," however, is a misnomer. The fluids themselves reside in the pore spaces between the grains in the rock, and not in a continuous "*pool*" like we picture with a swimming pool. The reason that we can extract fluids such as water or hydrocarbons out of rock is that the pore spaces within the reservoir rock are connected.

Figure 7.8 Folds commonly trap petroleum accumulations. This fold occurs in thin-bedded Monterey formation exposed along Soto Street, north of Alhambra Avenue, Los Angeles. [Photograph by M. N. Bramlette. Courtesy of United States Geological Survey.]

Where ore deposits are localized within folded rocks, prospects for exploration and mining offer clear incentives to understand the three-dimensional geometry of folds. Of special importance is learning methods for projecting the forms and trends of folds to depth as an aid in estimating reserves and in planning underground mining. **Saddle reef deposits** represent perhaps the most intimate of the associations between folding and mineralization. Saddle reefs are **lodes** of quartz and precious metals that occupy the cores of folds, in openings where bedding and/or foliation has been separated by fold-forming movements (Figure 7.9*A*). Although saddle-shaped in cross-sectional view, they are pipe-like in three dimensions, trending parallel to the axis of folding. The most renowned saddle reef deposits are in the Bendigo goldfields of Victoria, Australia (McKinstry, 1961; Guilbert and Park, 1986; Park and MacDiarmid, 1964), and in the Salmon River gold district of Nova Scotia (Malcolm, 1912).

Strata-bound ore deposits are commonly associated with folded, metamorphic rocks. The lead–zinc–silver deposits of Broken Hill, Australia, constitute a fine example (Gustafson et al., 1950) (Figure 7.9*B*). Shaped like ordinary beds of volcanic or sedimentary rocks, these deposits are abundantly rich in sulfides and/or precious metals. Although traditionally viewed as replacements of chemically favorable host rocks, most strata-bound sulfide deposits are now interpreted to have been submarine chemical sediments and/or volcanic-exhalative precipitates. In the course of the normal geologic history of active tectonic margins, these unusual metalliferous rock layers become folded, faulted, and metamorphosed, just like their ordinary sedimentary and volcanic counterparts. For my (GHD) dissertation research, I mapped a strata-bound sulfide deposit in eastern Canada, and discovered through detailed structural analysis that the body had been folded *three* times, creating wonderful geometries (Figure 7.10) (Davis, 1972).

Folded ore bodies, and ore bodies in folds, are challenging to exploration and mining geologists. The step-by-step underground geologic mapping of ore bodies in folded rocks, carried out during the normal course of mining, has yielded illuminating documentaries on the three-dimensional anatomy of folds. A good example is the geological treatise on the Homestake gold mine, South Dakota, written by Caddey and others (1991).

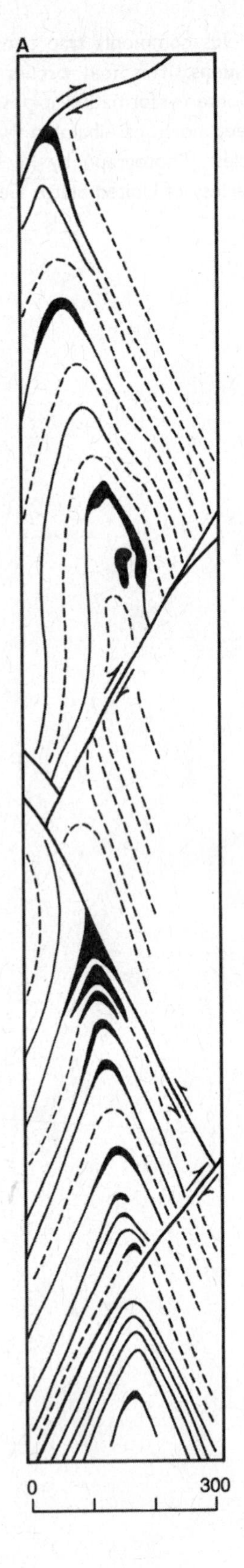

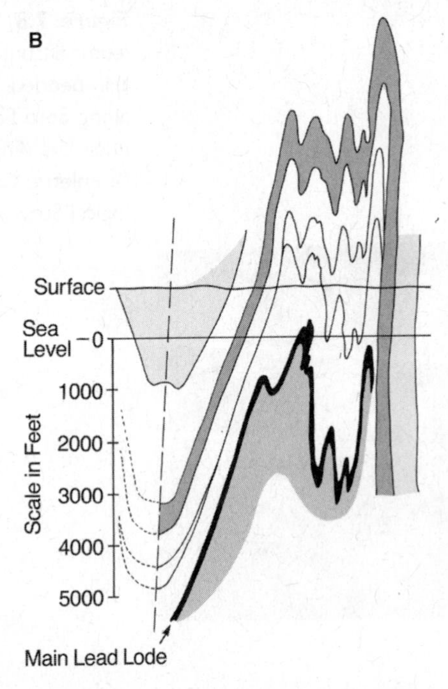

Figure 7.9 (*A*) Saddle reef deposit in the Bendigo goldfield, Bendigo, Victoria, Australia. Scale in feet. [From McKinstry, H. E., Mining Geology, © 1961. Published with permission of Prentice-Hall, Inc., Englewood Cliffs, New Jersey.] (*B*) Structure section through part of the folded, strata-bound Broken Hill ore deposit. [From Gustafson, Burrell, and Garretty, 1950. Published with permission of Geological Society of America.]

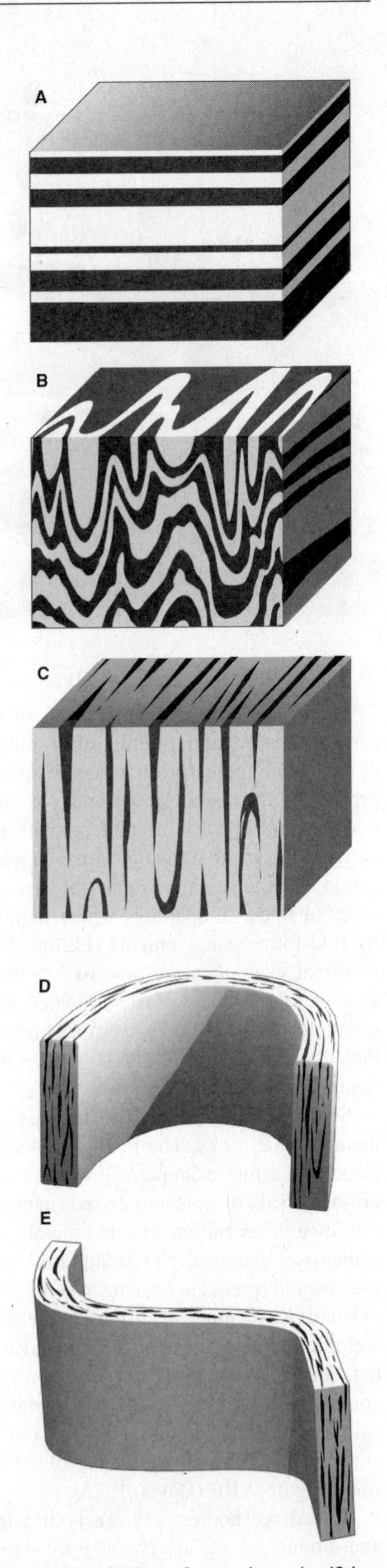

Figure 7.10 Schematic rendering of progressive and superposed folding of strata bound sulfides within a sequence of Ordovician volcanics and sediments, northern New Brunswick, Canada. (*A*) Horizontal bedding, before folding. (*B*) Moderate to tight folding. (*C*) Tight folding, flattening, and internal shearing of the sequence. (*D*) Refolding of the original folds. (*E*) Yet another episode of refolding. [From Davis, 1972. Published with permission of Society of Economic Geologists.]

ANTICLINES AND SYNCLINES

Basic Definitions

"*Anticline*" and "*syncline*" are part of the ABCs for all who have been introduced to geology. Ordinarily we think of the complete definition of "*anticline*" as an upright, convex-upward fold, such as those shown in Figure 7.11*A*, *B*; and a "*syncline*" as an upright, concave-upward fold (Figure 7.11*C*, *D*). But there is more to it. An **anticline** is a fold that is convex in the

Figure 7.11 (*A*) North-directed photograph of anticline expressed in folded Cretaceous limestone beds in the Three Gorges Ravine near the village of Ano Karyes, west of Megalopolis, Greece. In the lower part of the exposure this anticline is upright and symmetrical. However, the form of the anticline abruptly changes upward and becomes asymmetrical, leaning leftward (to the west). [Photograph by G. H. Davis.] (*B*) Another exposure of this same anticline. View is to the north–northwest. Three distinct sedimentary rock formations are exposed. The youngest formation is the white limestone in the right foreground, which wraps around the anticline (note west-limb expression of this limestone in upper left). The oldest formation comprises the core of the anticline. It is sandstone. Trees here prefer to grow on sandstone, not limestone. The arc of dense black woods is in fact the expression of the folded sandstone. (These woods burned during the wildfires in the summer of 2007.) Sandwiched between the resistant limestones (on top) and the folded sandstone (below) is a thin-bedded limestone unit, expressed in the landscape as the grassy slope in between, again wrapping around the anticline. [Photograph by G. H. Davis.] (*C*) View of upright syncline at Split Rock, Cheverie, Nova Scotia. The folded sandstone is part of the Horton Bluff Formation (Early Mississippian). [Photograph by John Waldron, and published here with his permission. The Structure Pages, © 2011 John Waldron, http://www.ualberta.ca/~jwaldron/structurespage.html] (*D*) Photograph of syncline in Cambrian quartzite in north Norway near Bjornfeld. Beneath the quartzite is Precambrian granite and mylonitic granite. Steve Naruk stands like "*Rocky*" for scale. [Photograph by G. H. Davis.]

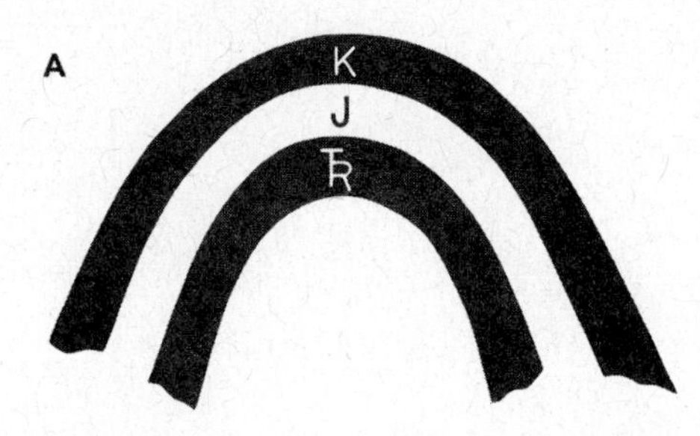

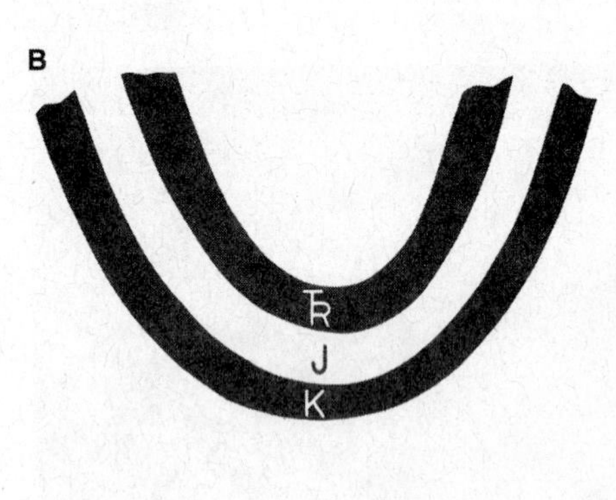

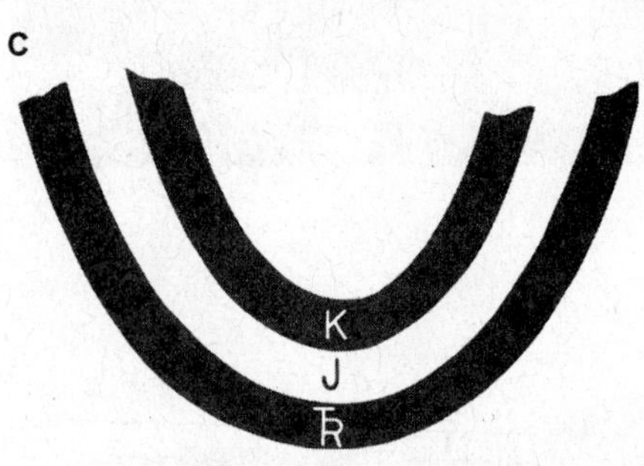

Figure 7.12 Anticlines and synclines, the cornerstones of fold terminology. Oldest layer is Triassic (TR); youngest layer is Cretaceous (K); in-between is Jurassic (J). (*A*) Anticline. (*B*) Synformal anticline. (*C*) Syncline. (*D*) Antiformal syncline.

direction of the youngest beds in the folded sequence (Figure 7.12*A*). But even the "*upside-down*" anticline shown in Figure 7.12*B* is called an anticline, albeit a special kind known as a **synformal anticline**. The adjective "*synformal*" means that the fold is concave upward. A **syncline** is a fold that is convex in the direction of the oldest beds in the folded sequence (Figure 7.12C). These too can be oriented in any way, from perfectly upright synclinal folds to convex-upward **antiformal synclines** (Figure 7.12*D*).

Overturned Folds

It is very rare for anticlines and synclines to be completely upside-down, but it is not at all uncommon for them to be **overturned**. The distinction between upside-down and overturned is this: upside-down folds are totally inverted, like the antiformal syncline shown in Figure 7.12*B*. In contrast, a fold is "*overturned*" if at least one of its **limbs** (i.e., flanks) is overturned. Saying that the limb of a fold is overturned does not mean that the fold is completely upside-down. It simply means that one limb has been rotated beyond vertical such that the **facing direction** (i.e., the direction that was originally up; see *Part III-D*) of the limb points downward at some angle. A schematic rendering of an overturned anticline that joins two overturned synclines is shown in Figure 7.13*A*. Each fold is marked by a right-side-up limb dipping at about 30°W and an overturned limb dipping at about 80°W. An example of the real thing is shown in Figure 7.13*B*. Conventional usage identifies the overturned limb as the **front limb**. The more gently dipping right-side-up limb is considered the **back limb**. More generally, overturned folds are typically **asymmetric folds**, and the dip of the front limb is steeper than that of the back limb. Where asymmetric folds occur within shear zones or fault zones, the asymmetry may accurately reveal the direction of differential shear or (fault) slip (**Figure** 7.14). However, caution should be exercised, for reasons that will become more apparent later in this chapter when we talk about kink folds.

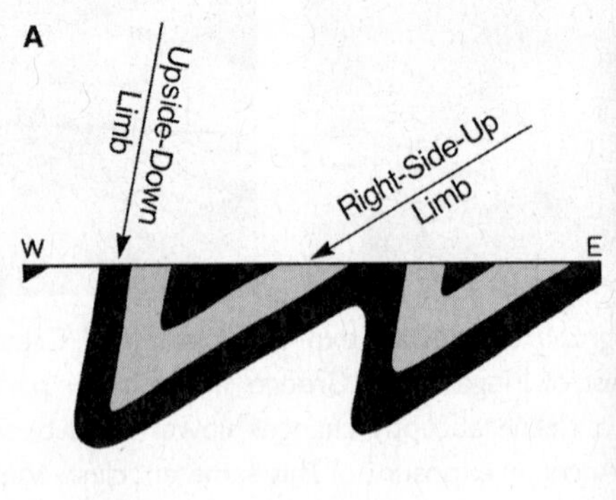

Figure 7.13 (*A*) Schematic rendition of overturned anticline and syncline. (*B*) The real thing! Overturned anticline and syncline in sedimentary rocks in the Funeral Mountains, Death Valley, California. Geologist barely visible in core of main anticline is Stan Ballard. [Photograph by S. J. Reynolds.]

Figure 7.14 Drawing of folding of banded gneisses within a shear zone, exposed near Cochie Springs, Tortolita Mountains, north of Tucson, Arizona. The folds are asymmetrical. We name them Z-folds, for the combination of long limb/short limb/long limb spells out a Z. Based on this asymmetry, we infer the sense of simple shear shown by the arrows. Saguaro cactus in left foreground is ~2.5 m tall. [Field sketch by G. H. Davis.]

Antiforms and Synforms

Use of the term "*anticline*" or "*syncline*" implies that stratigraphic succession within the folded sequence has been worked out on the basis of the established geological column or the determination of facing. If facing and stratigraphic order cannot be established, these terms must be scrapped in favor of **antiform** and **synform**. An antiform is, very simply, a fold that is convex upward. A synform is a fold that is concave upward.

"*Antiform*" and "*synform*" are normally used in reference to folds in sedimentary and/or volcanic sequences within which facing and/or stratigraphic order is either unknown or uncertain. But these terms are also appropriately used in describing folds in metamorphic rocks such as gneisses and schists, and in igneous rocks where, for example, folds may have formed in stiff magma along the margin of a dike as the magma, during emplacement, moved past wall rock. If layering and/or foliation in a rock body is not related to normal depositional process, nor to conventional stratigraphic succession, it is meaningless to attempt to describe facing and stratigraphic order (Figure 7.15).

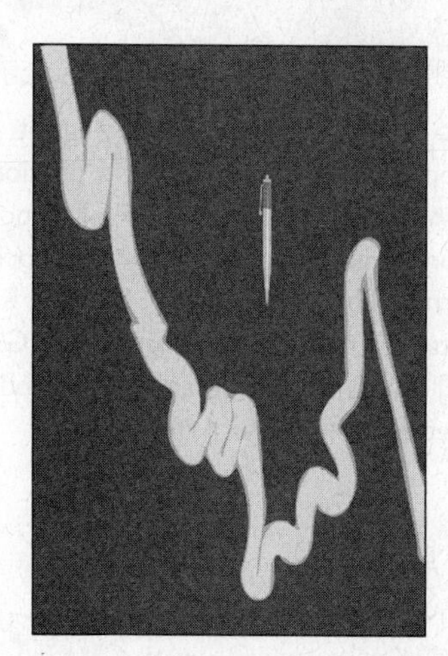

Figure 7.15 The aplite dike shown here has been folded back and forth on itself like an intestine. We would refer to these folds as antiforms and synforms.

Furthermore, the terms "*anticline*" and "*syncline*" may have only limited application in describing **superposed folds** in a sedimentary and/or volcanic sequence. By way of example, the early-formed fold shown in structure profile view in Figure 7.16*A* may be described as a **recumbent**, **isoclinal** anticline. "*Recumbent*" means that the fold lies on its side. "*Isoclinal*" means that the limbs of the fold are equally inclined. Figure 7.16*B* shows the early-formed isoclinal fold after being modified by a second folding. The anticline is no longer perfectly recumbent. Rather, it is deflected into convex-upward and convex-downward superposed folds. These late folds cannot be described as anticlines and synclines because there is no overall internal coherence of stratigraphic order and facing within the layers that define the folds. The second-stage folds are appropriately described as antiforms and synforms. An outcrop example that presents the same dilemma is shown in Figure 7.17. Note the isoclinal fold in the lower/center part of the photo is itself folded about an upright antiform.

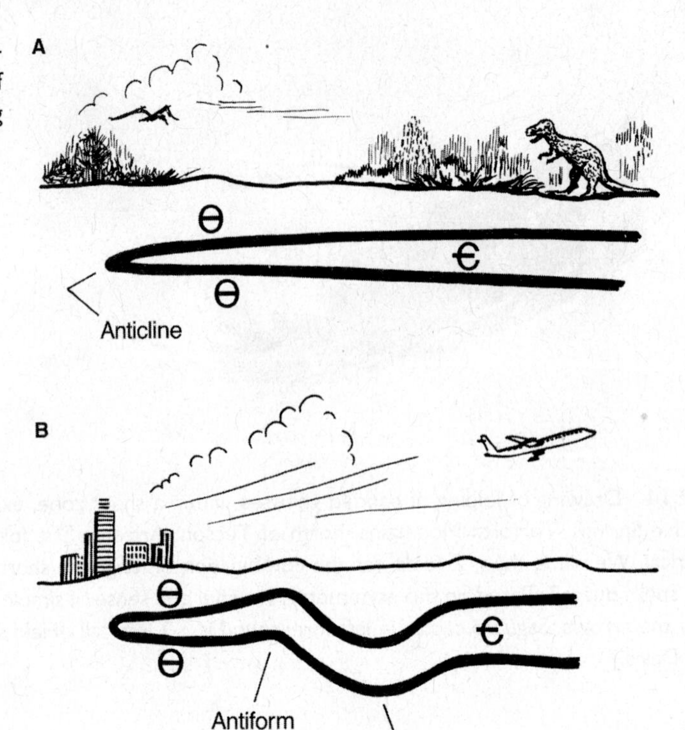

Figure 7.16 An example of regional-scale superposed folding. (*A*) Recumbent, isoclinal anticline, of pre-Jurassic age. (*B*) Antiformal and synformal folding of the recumbent, isoclinal anticline.

Figure 7.17 Superposed folding at the outcrop scale. In the right/center portion of this photograph there is a well-exposed overturned antiform that refolds an isoclinal antiform. This outcrop is in the Ruby Mountains, Nevada. [Photograph by G. H. Davis.] Original photograph is rotated so that second folding is antiformal.

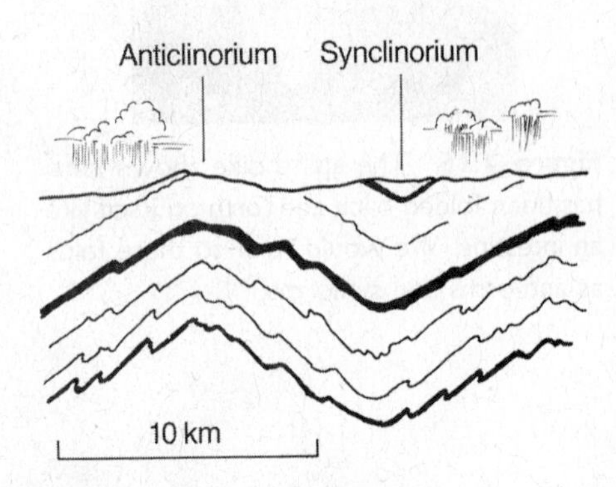

Figure 7.18 Schematic rendering of an anticlinorium and a synclinorium.

Anticlinoria and Synclinoria

Regional fold belts contain very large anticlines and synclines, kilometers across, that are themselves marked by the presence of smaller anticlines and synclines (Figure 7.18). We call such a regional anticline an **anticlinorium**, and such a syncline a **synclinorium**. Anticlinoria and synclinoria will be marked by the presence of reasonably systematically spaced second-order and third-order anticlines and synclines (see Figure 7.18). But keep in mind that folds in nature do not look like systematic wave trains, such as the ones we see in physics class or in analytical geometry. Instead, second-order and third-order folds, and even the anticlinoria and synclinoria themselves, can change abruptly in size and form depending on rock type, strain environment, and the geometry of regional faults with which they are associated.

Large anticlines and synclines, as well as even larger anticlinoria and synclinoria, are found in fold belts thoughout the world. The very best examples in the United States are in the Appalachian fold belt, especially as expressed in the Valley and Ridge sector of central Pennsylvania. Aerial and satellite images capture the form and elegance of this fold belt (Figure 7.19), which was produced through plate tectonic convergence between the African plate and the North American plate during the Alleghany orogeny, ~350 to 300 Ma (Carboniferous).

Monoclines

Monoclines constitute yet another category of regional-scale folds. The type locality for monoclines is the Colorado Plateau of the American Southwest. In fact it was John Wesley Powell who first recognized and named this category

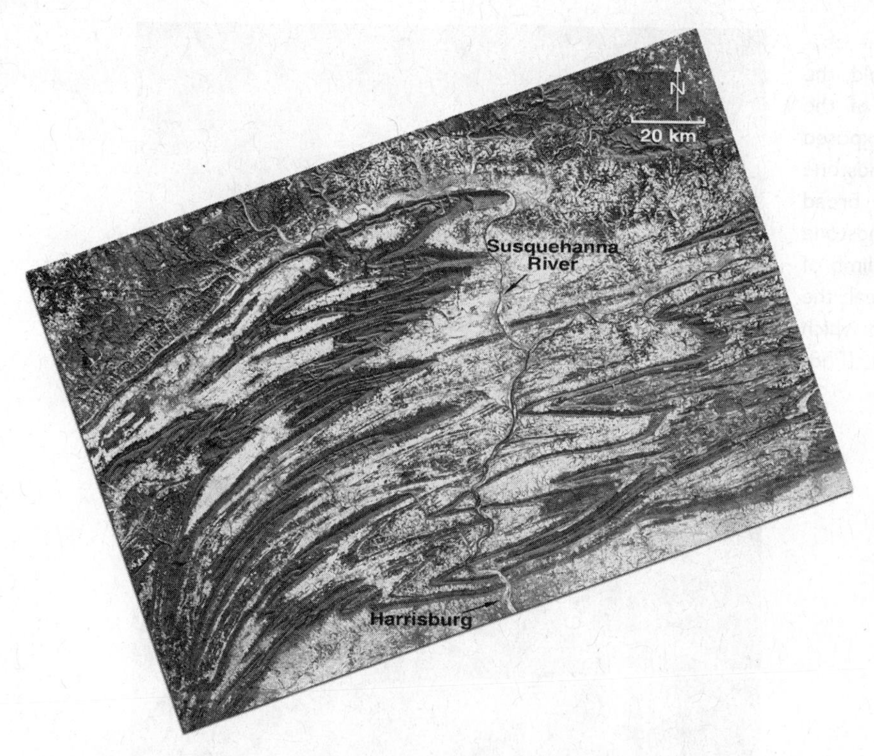

Figure 7.19 Astronaut photo over Harrisburg, Pennsylvania, capturing an ~100 km-wide swath of the Valley and Ridge province of the Appalachian Mountains. The landscape expression of folding is quite spectacular. This is a winter scene, with snow on the ground. The center point of the image is 40.5°N. Latitude and 77°W Longitude. The Susquehanna River crosses the image. [Image courtesy of the Image Science and Analysis Laboratory, NASA Johnson Space Center. Display record (Mission/Roll/Frame): ISS026/E/19757. Access through http://eol.jsc.nasa.gov.]

of folds. His "*discovery*" monocline was the East Kaibab monocline, which borders the Kaibab uplift in Utah and Arizona (Figure 7.20*A*). These folds are huge regional structures that commonly trend for more than 150 km (Figure 7.20*B*). They are broad step-like folds that cause otherwise horizontal or very shallowly dipping strata to bend abruptly to steeper inclinations within very narrow zones (Figure 7.20*B*). Asymmetric in profile form, monoclines are marked by two hinges (one anticlinal, one synclinal) connected by a middle limb (see Figure 7.20*A*). The middle-limb strata are generally smoothly curved and continuous, but sometimes they are broken by faults. The most spectacular monoclines display middle-limb inclinations of 90°

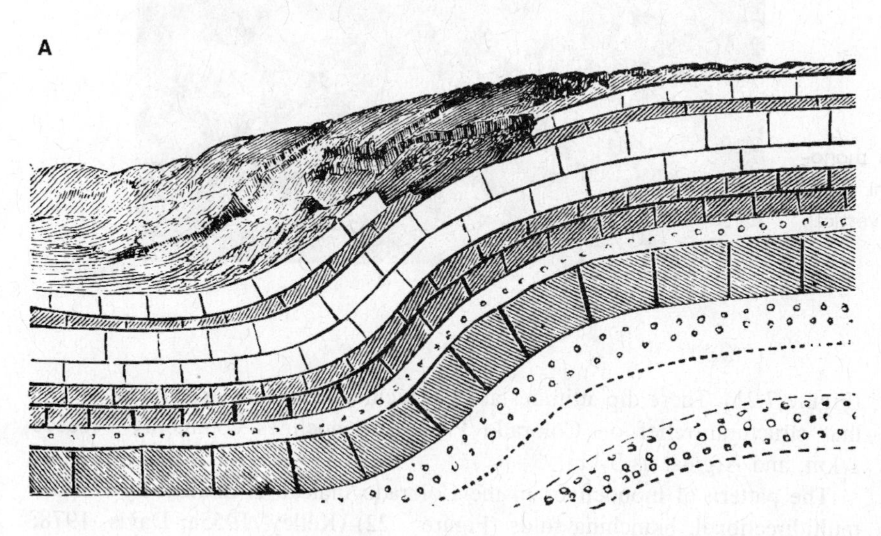

Figure 7.20 (*A*) First illustration of a monocline, by John Wesley Powell (1873). View to the south. [From Kelley, 1955a. Published with permission of the Geological Society of America.]

Figure 7.20 *(Continued)* (*B*) North-directed oblique aerial photograph of the Waterpocket fold, the monocline that forms the eastern margin of the Circle Cliffs uplift, Utah. The jointed, boldly exposed bedrock in the left foreground is Wingate Sandstone (Triassic). This formation is arched into a broad anticline. The white beds are Navajo Sandstone (Jurassic). They occupy the exposed middle limb of the monocline. This image begins to reveal the magnitude of monoclines and the uplifts with which they are associated in the Colorado Plateau. [Photograph by K. Constenius.]

Figure 7.21 An aerial view of the Nutria monocline along the western margin of the Zuni uplift, New Mexico. Strata in middle limb are vertical. [Photograph courtesy of Vince Kelley.]

(Figure 7.21). These dip attitudes are spectacular, especially when considering that structural relief on Colorado Plateau monoclines commonly exceeds 1 km, and is great as 3 km!

The pattern of monoclines in the Colorado Plateau is marked by sinuous multidirectional, branching folds (Figure 7.22) (Kelley, 1955a; Davis, 1978;

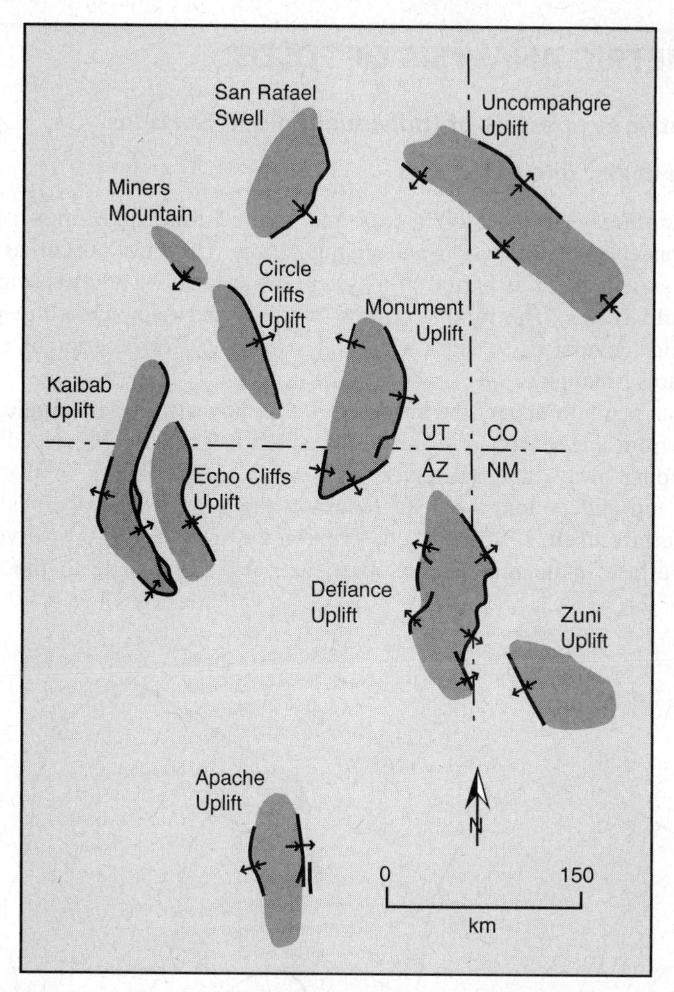

Figure 7.22 Map of Colorado Plateau uplifts. [From Davis and Bump, 2009, Figure 2A, p. 103. Published with permission of the Geological Society of America. Adapted from Kelly, 1955b, and Davis and others, 1981.]

Davis and Bump, 2009). The major monoclines serve to mark the boundaries between the great uplifts and basins of the Colorado Plateau (see Figure 7.22).

Monoclines appear to be associated with ancient, reactivated steeply dipping fault zones. This association is especially clearly revealed in the Grand Canyon, where deep erosion has exposed the "*roots*" of the East and West Kaibab monoclines (Figure 7.23). Separation relationships disclose the presence of ancient, reactivated Precambrian faults (Huntoon and Sears, 1975).

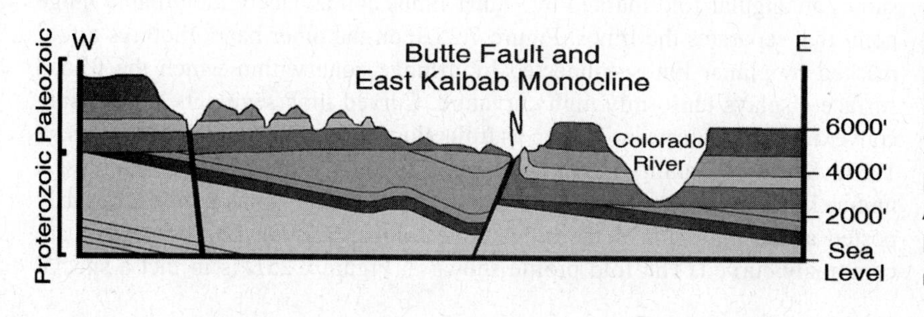

Figure 7.23 The East Kaibab monocline "*roots into*" the Butte fault, as represented here in a structure section cross part of the Grand Canyon. During the Neoproterozoic, movement on the Butte fault created normal displacement. During the latest Cretaceous and early Tertiary, reactivation of the Butte fault produced reverse displacement, tipping out upward into monoclinal folding. The magnitude of reverse throw was less than Neoproterozoic normal throw, and thus net offset of Mesoproterozoic basement remains a normal displacement. [From Davis and Bump, 2009, Figure 5, p. 105. Published with permission of the Geological Society of America. Adapted from Tindall, 2000a, Figure A5, p. 54, and used here with permission of Sarah Tindall.]

GEOMETRIC ANALYSIS OF FOLDS

Geometric Properties of Individual Folded Surfaces

Limbs, Hinges, and Inflections

Folded surfaces vary greatly in size and form. Furthermore, they come in every conceivable orientation and configuration. They may occur as single isolated structures, or as part(s) of a system of repeated waveforms sometimes called **fold trains**. The infinite variety of folds and their three-dimensional complexity compel us to learn a special vocabulary for descriptive analysis and geologic mapping.

We can begin to dissect the anatomy of a fold by extracting a **single folded surface** from the deformed sequence of which it is a part (Figure 7.24*A*,*B*). By stripping away the rock layers that serve as boundaries to the folded surface, top and bottom, we can focus on the geometric properties of the folded surface itself. If it is a 2D description we are looking for, we view the folded surface in **normal profile view**, i.e., at a right angle to the folding (Figure 7.24*C*).

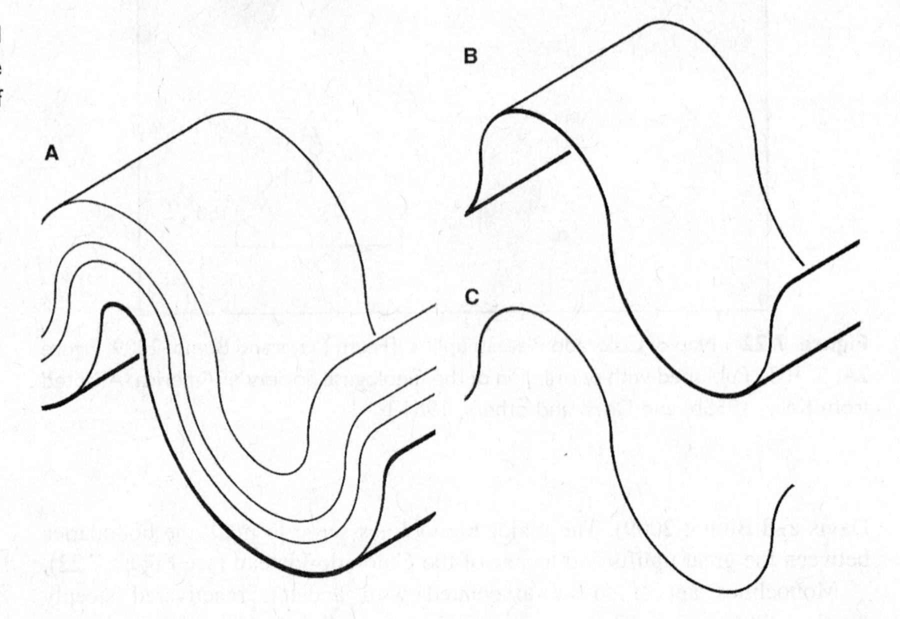

Figure 7.24 (*A*) Sequence of folded layers. (*B*) Single folded surface within the folded sequence. (*C*) Normal profile view of the single folded surface.

Considered two-dimensionally in normal profile view, folded surfaces can be subdivided into **limbs** and **hinge points** (Figure 7.25). Limbs are the flanks of folds, and these are joined at the hinge point. The hinge of a sharp, angular folded surface is sometimes a single point, called the **hinge point**. More commonly, the hinge is a zone, **the hinge zone**, distinguished by the maximum curvature achieved along the folded surface (Ramsay, 1967). Figure 7.25*A* shows an angular fold marked by planar limbs and an easily identifiable hinge point that separates the limbs. Figure 7.25*B*, on the other hand, pictures a fold marked by planar limbs connected by a hinge zone within which the folded surface displays uniformly high curvature. Curved limb segments of opposing convexity join at locations known as **inflection points** (Figure 7.25*C*) (Ramsay, 1967). Strictly speaking, the hinge zone of a folded surface does not possess a unique hinge point. But for descriptive purposes a hinge point can be arbitrarily posted at the midpoint of the hinge zone (Figure 7.25*D*). Limbs of folds are commonly curved. The fold profile shown in Figure 7.25*D* is in fact a special case where the surface has been folded into a series of perfectly circular arcs. Folded surfaces of this type lack fold limbs, *per se*. Instead, each discrete circular arc may be thought of as a hinge zone marked by uniform curvature. The hinge point of each fold is taken as the midpoint of each circular arc. The inflection points of the folded surface separate circular arcs of opposing convexity (see Figure 7.25*D*).

The vocabulary used in describing folded surfaces in normal profile view is summarized in Figure 7.25*E*, which shows a fold marked by slightly curved to planar limbs connected by a narrow hinge zone within which the surface displays a pronounced and high degree of curvature. The hinge zone is the zone of maximum curvature, the midpoint of which is the hinge point. Inflection points are taken to be the midpoints of the planar limb segments, but technically, planar limbs do not actually have an inflection point.

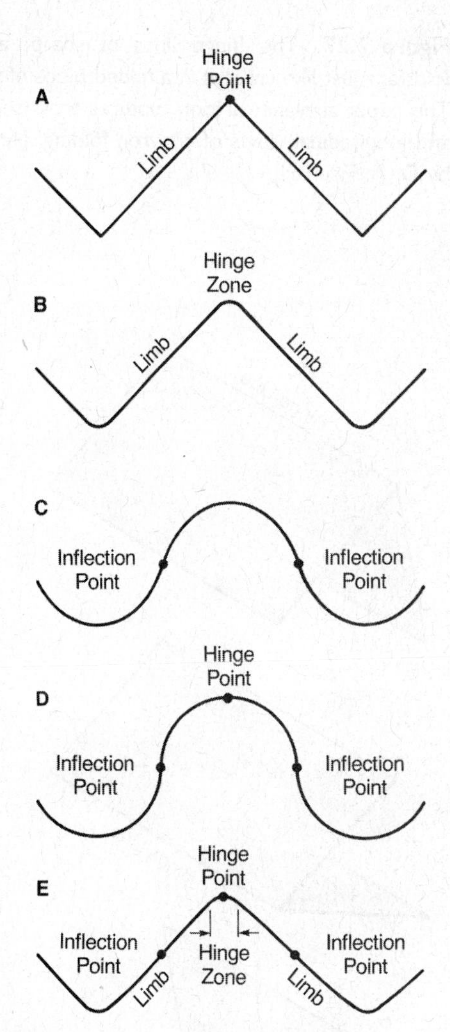

Figure 7.25 Geometric and physical elements of single folded surfaces, seen in normal profile view. See text for discussion.

Hinge Lines, Axial Surfaces, and Axial Traces

The three-dimensional geometric characteristics of folded surfaces invite a yet fuller nomenclature (Figure 7.26). The hinge line is a line that is defined by the points on the surface of maximum curvature (or minimum radius of curvature). In some cases the hinge line of a folded surface is perfectly straight (Figure 7.26*A*), just like creases in a paper airplane (Figure 7.27). More commonly hinge lines are systematically curved (see Figure 7.26*B*) or, more rarely, terribly irregular (see Figure 7.26*C*) (Turner and Weiss, 1963).

The orientation of a folded surface can in part be specified by the orientation of its hinge line. The orientation of a hinge line is described conventionally in terms of trend and plunge. A single measurement of trend and plunge is adequate for hinge lines that are perfectly straight (see Figure 7.26*A*). Where folded surfaces are marked by hinge lines that are not straight, it is necessary to document the variations in hinge line orientation through a number of measurements of representative segments of the hinge. In practice this is achieved by subdividing the folded surface into 3D domains within which the hinge line approaches a straight line (Figure 7.26*D*).

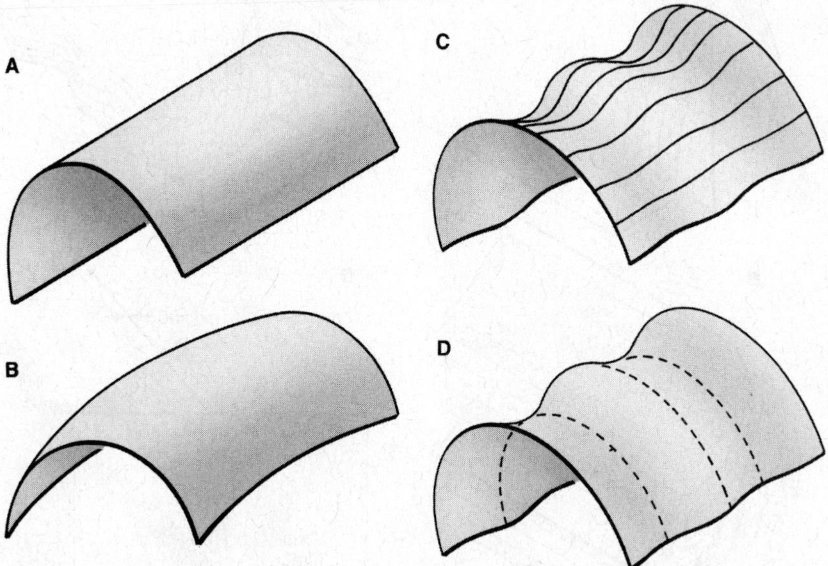

Figure 7.26 (*A*) Fold with straight hinge line. (*B*) Fold with systematically curved hinge line. (*C*) Fold with irregularly curved hinge line. (*D*) Subdivision of fold with irregularly curved hinge line into domains marked by nearly straight hinge lines.

Figure 7.27 The hinge lines of sharp, angular folds are just like creases in a folded piece of paper. This paper airplane cartoon captures cross-sectional and longitudinal views of chevron folding. [Artwork by D. A. Fischer.]

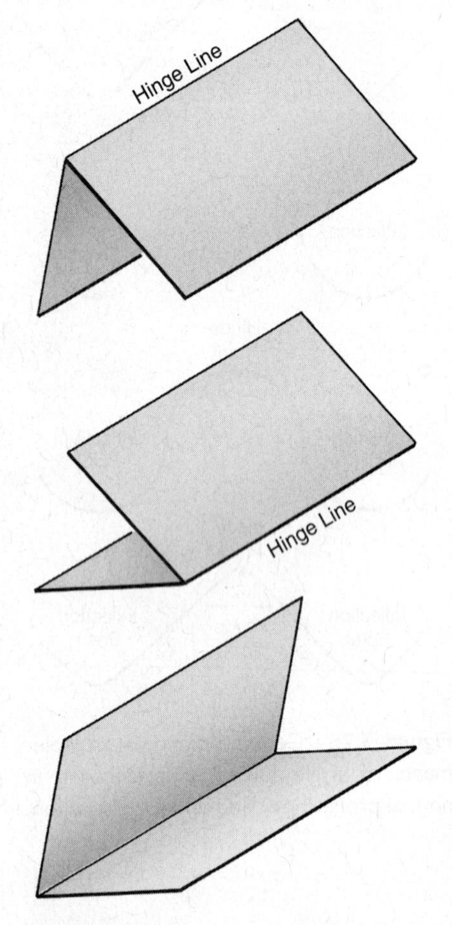

Figure 7.28 The trend and plunge of the hinge line of a fold does not uniquely define the overall orientation of the fold. See for yourself.

Knowing the orientation of the hinge line of a fold does not uniquely establish the orientation, or **attitude**, of the fold. Folds having the same hinge line orientation can have strikingly different configurations (Figure 7.28). To describe unambiguously the attitude of a fold, it is necessary to measure yet another structural element, a geometric element known as an **axial surface**. The axial surface of a fold passes through successive hinge lines in a stacking of folded surfaces (Figure 7.29) (Ramsay, 1967; Dennis, 1972). By definition,

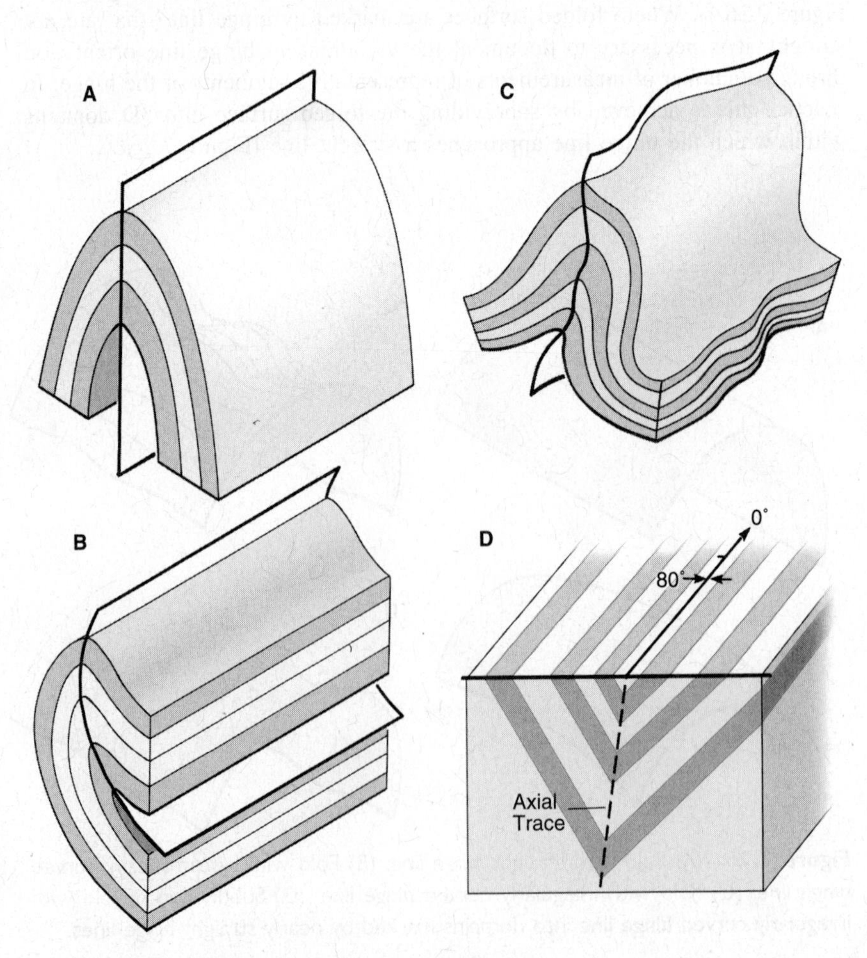

Figure 7.29 (*A*) Fold with a planar axial surface. (*B*) Fold with a systematically curviplanar axial surface. (*C*) Fold with an irregularly curviplanar axial surface. (*D*) Axial trace of a fold, as seen in cross-section and in map view.

the axial surface passes through all of the points of maximum curvature in all of the folded layers. The axial surface of a fold may be planar (Figure 7.29*A*), in which case it is called an **axial plane**. More commonly the axial surface of a fold is either systematically curved (Figure 7.29*B*) or nonsystematically irregular (Figure 7.29*C*), in which case **axial surface** is the appropriate term (Turner and Weiss, 1963).

In normal profile view, the trace of the axial surface of a fold can be seen to pass through successive hinge points in the stacking of folded surfaces (Figure 7.29*D*). This line is called the **axial trace** of the fold in profile view. In a more general sense, **axial trace** refers to the line of intersection of the axial surface with *any* other surface, the cut of an open-pit mine, the steep flank of a mountain, the faces of a block diagram of folded layers, or, in the field, just the ordinary ground surface (see Figure 7.29*D*).

Because axial surfaces of folds are planar or curviplanar surfaces, their orientations are described in terms of strike and dip. A single strike-and-dip measurement is all that is necessary to describe the orientation of the axial plane of a fold (e.g., the vertical axial plane for the fold shown in Figure 7.29*A*). For a nonplanar axial surface, however, a number of strike-and-dip measurements are required to document the full spectrum of orientations of the axial surface (Figure 7.29*B*). Imagine, for example, the number of measurements that would be required to "*capture*" the orientation and form of the axial surface shown in Figure 7.29*C*).

Knowing the orientation of the axial surface does not fix uniquely the attitude of the fold. Folds having a common axial surface orientation can have radically different configurations (Figure 7.30). Only when the orientations of both the hinge line *and* the axial surface of a fold are known can the configuration of the fold be firmly established.

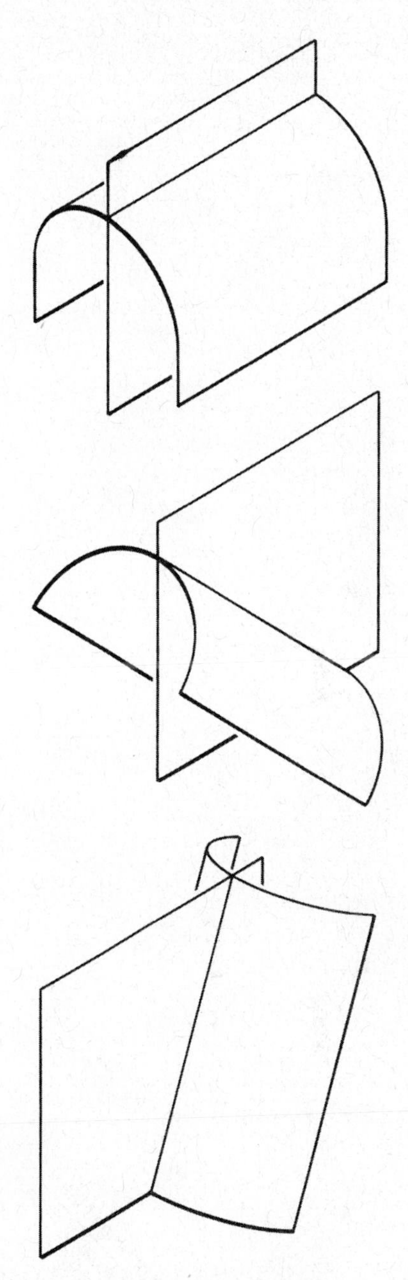

Figure 7.30 The strike and dip of the axial surface of a fold does not uniquely define the overall orientation of the fold.

Crestal Surfaces Versus Axial Surfaces

Especially in the petroleum industry, it is useful to pay attention to locations of crests of folds, for these locations in just the right stratigraphic circumstances may coincide with structural traps for oil and gas. Consequently, we want to be clear on the locations and forms of axial surfaces, crestal surfaces, and inflection surfaces. Figure 7.31 makes these distinctions (in cross-sectional view) for a large asymmetrical fold in the subsurface. We see the trace of the

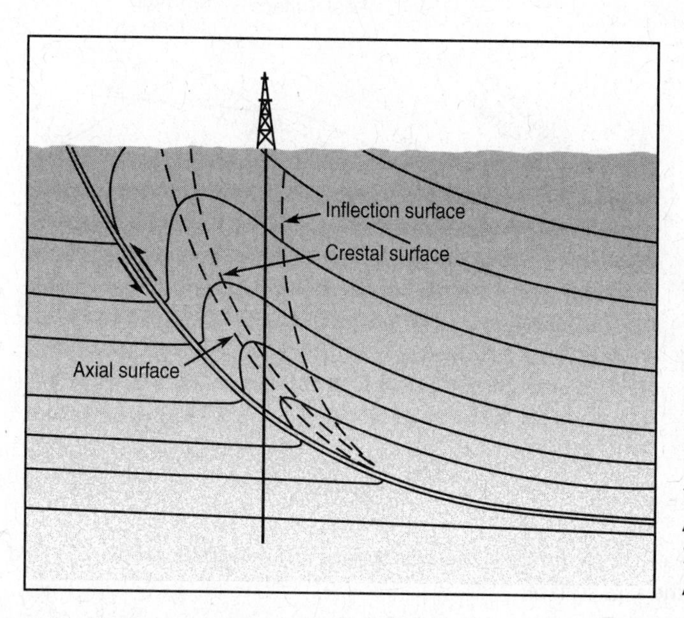

Figure 7.31 Cross-section of large asymmetrical fold in the hanging wall of a thrust fault. The shape of the fold was determined through seismics and drilling. Here we show the distinction between the trace of the crestal surface, the trace of the axial surface, and the trace of the inflection surface. [From Bengtson, C. A., 1981, Statistical curvature analysis techniques for structural interpretation of dipmeter data: *American Association of Petroleum Geologists Bulletin*, v. 65, no. 2, p. 312–332. AAPG © 1981, reprinted by permission of the AAPG whose permission is required for further use.]

crestal surface, which passes through points of highest elevation for each bed within the folded sequence. Only in perfectly upright folds will the axial surface and the crestal surface be one and the same. For the fold shown in Figure 7.31, we see that the trace of the axial surface, which passes through the hinge points for each bed, does not coincide with the trace of the crestal surface. Note also in Figure 7.31 the location of the trace of the inflection surface, passing through inflection points for each bed within the folded sequence.

If, in the field, you are having lunch right on the very crest of a non-plunging fold, you will recognize that strata there dip 0°. If, however, on the very crest of a fold you observe that the strata are dipping, you may conclude that the fold is a plunging fold, and that the dip direction and magnitude reveal trend and plunge of the fold.

Geometric Coordination of Hinge Lines and Axial Surfaces

Because of the manner in which "*hinge line*" and "*axial surface*" are defined, the hinge line of a fold must lie within the fold's axial surface. This constraint notwithstanding, hinge lines and axial surfaces can be combined in many more ways than we might at first imagine. It is relatively easy to picture a hinge line that is parallel to the strike of the axial surface (Figure 7.32*A*, *B*). Although this arrangement is common, it is a very special case: for in general the trend of the hinge line of a fold may be parallel *or* oblique *or even* perpendicular to the strike of the axial surface and still remain within the axial surface (Figure 7.32*C*). Said another way, the rake of the hinge line in an axial surface can range from 0° to 90°.

The breadth of the geometrically permissive relative orientations of hinge lines and axial surfaces, combined with the even greater range of absolute orientations that hinge lines and axial surfaces can assume, present us with limitless possible fold configurations. To deal with such broad-ranging geometries, Fleuty (1964) created a useful classification scheme, one that is based on both the relative orientations and the absolute inclinations of hinge lines and axial surfaces. The classification scheme, which he presented in the

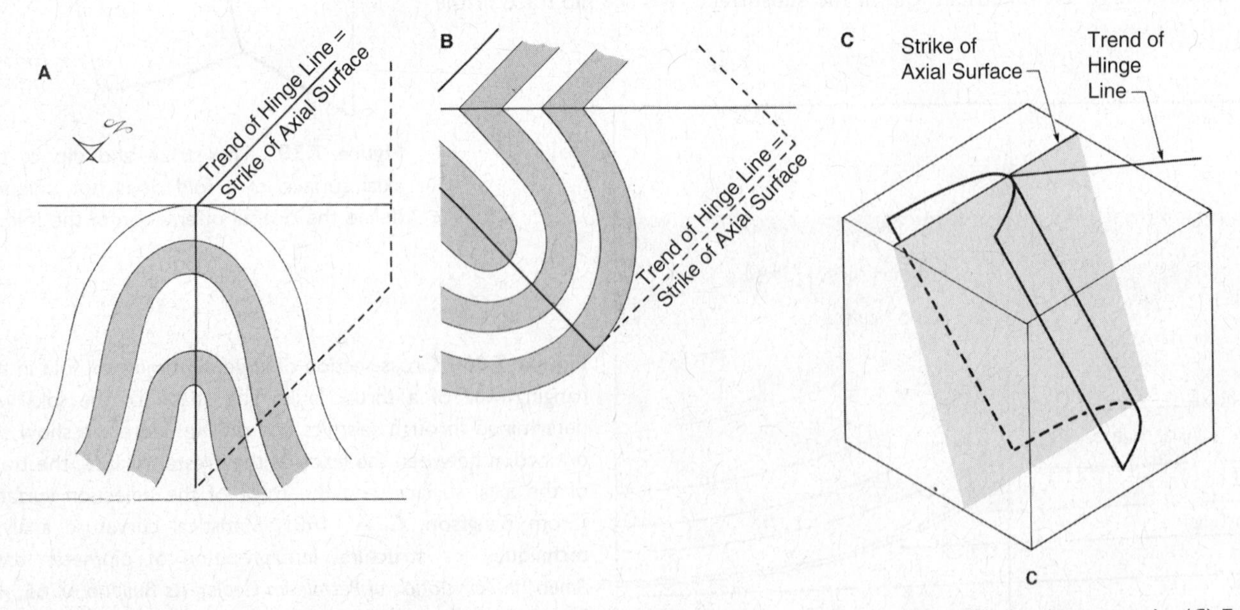

Figure 7.32 (*A*) Fold marked by hinge line whose trend is parallel to the strike of the axial surface. (*B*) Another example. (*C*) Fold marked by hinge line whose trend is discordant to the strike of the axial surface.

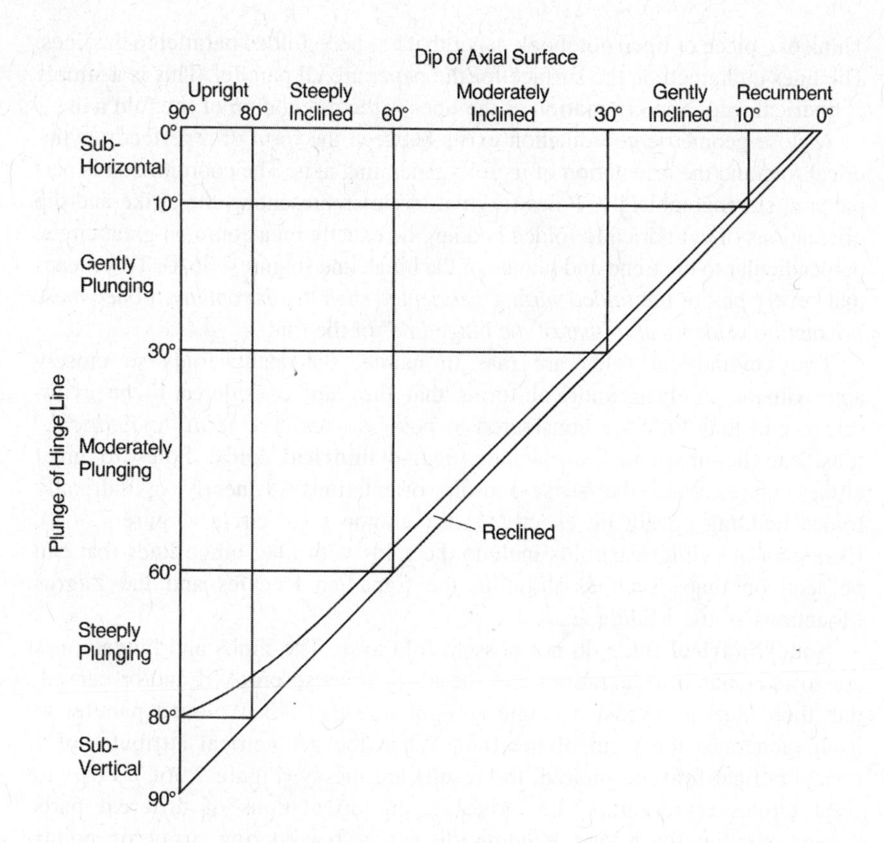

Figure 7.33 Fleuty diagram used for describing folds on the basis of the geometric interrelationship of axial surface and hinge line. [Reprinted from *Geological Association Proceedings*, v. 75, Fleuty, M. J., The description of folds, p. 461–492, © 1964, with permission from Elsevier.]

form of a diagram (Figure 7.33), permits folds to be named according to fold configuration. Along the *x*-axis of Fleuty's diagram is plotted the **dip** of the axial surface; along the *y*-axis is plotted the **plunge** of the hinge line. On the basis of cutoffs of 0°, 10°, 30°, 60°, 80°, and 90°, sixteen categories of folds are identified. A fold whose hinge line plunges 5°N38°E and whose axial surface orientation is N40°E, 82°NW can be described as a **subhorizontal upright fold** (see Figure 7.33). A fold whose hinge line plunges 50°S72°W and whose axial surface orientation is N80°W, 70°SW is classified as a **moderately plunging**, **steeply inclined fold**. A fold with hinge line and axial surface orientations of 20°N65°E and N30°W, 20°NE, respectively, is called a **gently plunging**, **gently inclined**, **reclined fold**. A **reclined** fold is one whose hinge line plunges directly down the dip of the fold's axial surface.

Cylindrical versus Noncylindrical Folds

It is easy to fall into the trap of using the terms **hinge line** and **fold axis** interchangeably. But beware. Strictly speaking, a "*fold axis*" is a geometric (thus imaginary) linear structural element that does not possess a fixed location. It is the closest approximation to a straight line that when moved parallel to itself generates the form of the fold (Figure 7.34) (Donath and Parker, 1964; Ramsay, 1967).

Folds that possess axes are **cylindrical folds**. The expression "*cylindrical folds*" should bring to mind parts of cans and pipes. A cylindrical fold can have a form best described as *part* of a single, perfect cylinder (Figure 7.35*A*). But more commonly a cylindrical fold has the form of a co-linear arrangement of parts of cylinders of different diameters (Figure 7.35*B*). The distinctive geometric characteristic of cylindrical folds is that every part of the folded surface is oriented such that it contains a line whose orientation is identical to that of the hinge line.

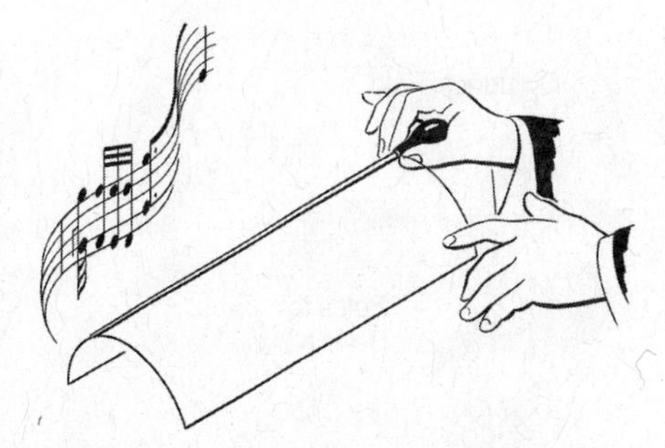

Figure 7.34 A fold axis is the closest approximation to an imaginary straight line, which when moved parallel to itself, defines the form of the fold.

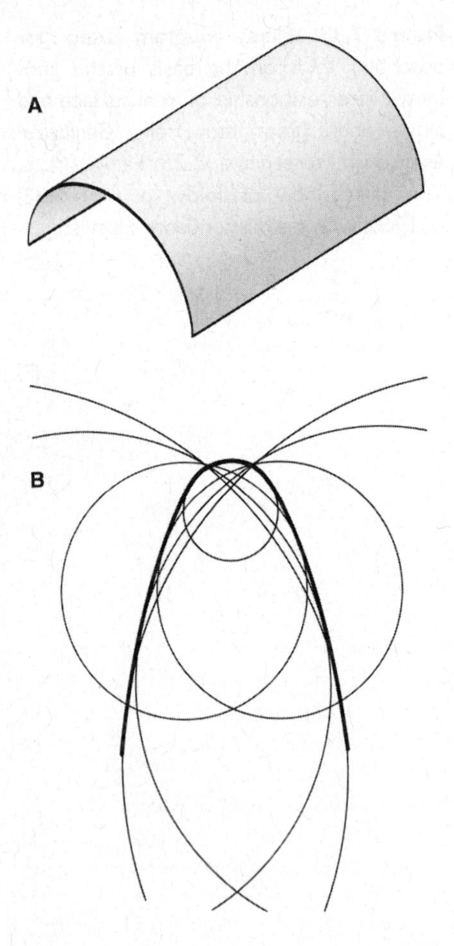

Figure 7.35 Here are two perfectly acceptable cylindrically folded surfaces. (*A*) One conforms to the outline of a single, perfect cylinder. (*B*) The other (bold dark normal profile of a folded surface) is composed of a co-linear arrangement of parts of cylindrical surfaces (parts of circles in this view) of different curvatures.

Think of a piece of lined notebook paper that has been folded parallel to the lines. The lines contained on the surface for the paper are all parallel. This is a simple cylindrical fold. The orientation of the lines is the orientation of the fold axis.

A close geometric coordination exists between the form of a perfectly cylindrical fold and the orientation of its fold-generating axis. The coordination is best pictured stereographically. Poles to great circles representing the strike-and-dip orientations of cylindrically folded bedding lie exactly on a common great circle, perpendicular to the trend and plunge of the hinge line (Figure 7.36*A*). This means that "*every part of the folded surface is oriented such that it contains a line whose orientation is identical to that of the hinge line*" of the fold.

Truly cylindrical folds are rare in nature, but many folds so closely approximate purely cylindrical forms that they are considered to be cylindrical, and thus they are considered to possess axes. The term "*cylindrical,*" thus, can be broadened to include **near-cylindrical folds**. Poles to great circles representing the strike-and-dip orientations of nearly cylindrically folded bedding do not lie exactly on a common great circle (Figure 7.36B). Examples of cylindrical folds include the folds with long hinge lines that can be seen on maps such as those in the Canadian Rockies and the Zagros Mountains in the Middle East.

Noncylindrical folds do not possess fold axes. The limbs and hinge zones are so irregular in orientation, and the hinge lines so crooked and/or curved, that there does not exist a single straight line that when moved parallel to itself, generates the form of the fold. When the geometrical attributes of a noncylindrical fold are plotted, the results are messy (Figure 7.36*C*). Poles to great circles representing the strike-and-dip orientations of different parts of noncylindrically folded bedding display a bewildering array of points that cannot be fit to a common great circle. To penetrate the chaos of noncylindrical folding in the course of detailed structural analysis, it is necessary to subdivide noncylindrical folds into domains of cylindrical or near-cylindrical folds, each of which is marked by a relatively short, relatively straight hinge line. When this is done, and the strike-and-dip orientations of the folded surfaces are plotted domain by domain, the stereographic patterns clean up immeasurably. Strain analysis of noncylindrical folds is also quite challenging, requiring full 3D analysis as opposed to relying on "*simple*" cross-section representations of the fold. 3D analysis is required because noncylindrical folding involves "*flow*" of material in and out of the cross-section (Shackelton and Cooke, 2007).

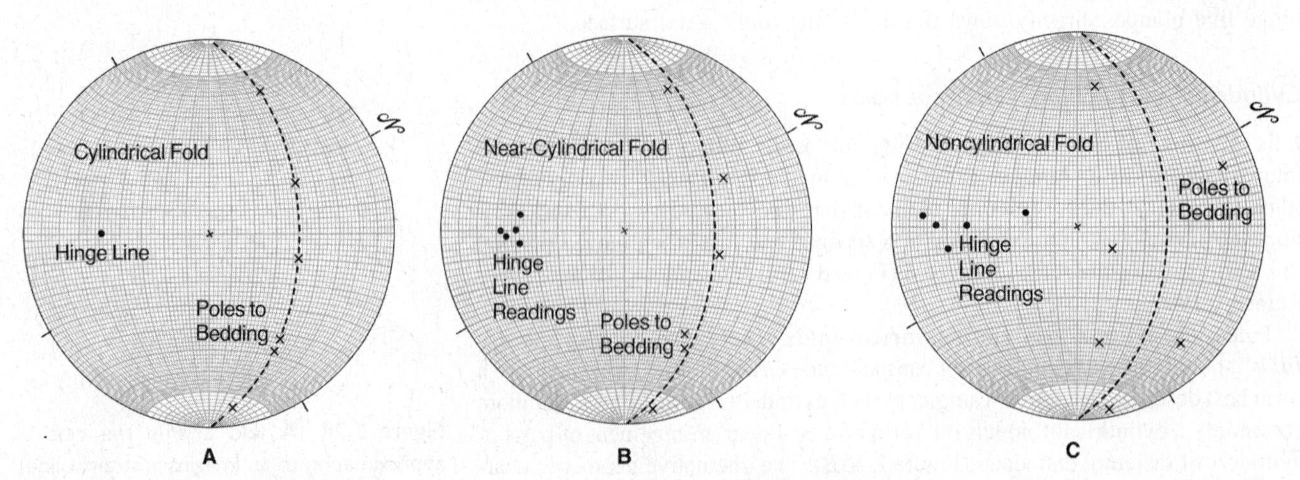

Figure 7.36 Stereographic geometry of (*A*) a perfectly cylindrical fold, (*B*) a nearly cylindrical fold, and (*C*) a noncylindrical fold. See text for explanation.

Conical Folds

Purely cylindrical folds have the unique property that they cannot die out, though they may be cut off and displaced by a fault, such as a tear fault. Yet, one manner in which cylindrical folds die out is through shape change into a **conical** form. An ice cream cone is a useful visual image for the nature of a conical fold (Figure 7.37*A*). Part of the geometry of a conical fold is defined by the trend and plunge of the **center-line axis**, which runs right down the center of cone and passes through the **vertex**, which lies in the direction of the tip of the cone (Figure 7.37*B*). Every part of the surface of a conical fold makes the same angle with a fixed (center-line) axis The angle between the axis and any part of the conically folded surface is known as the **apical angle**. Classifying a fold as "*conical*" requires demonstrating that poles to bedding attitudes taken on a folded bed or layer plot stereographically along a small circle. The axis about which the small circle "*revolves*" is the axis of the fold. There is no independent way of determining trend and plunge of the cone axis. For conical folds, there is no physical counterpart of a tangible "*hinge line*" whose trend and plunge can be measured directly in the field. Visualizing conical folds is not always straightforward. It is important to bear in mind the position of the vertex. If the vertex lies down-plunge, as is the case in Figure 7.37, the conical fold is considered to be a type II conical fold; if, however, the position of the vertex lies up-plunge, the conical folds is considered to be a type I conical fold (Bengtson, 1981).

Structural geologists did not worry much about conical folds in the past. While working in 2D, they assumed that their cross-sections of folded rocks were capturing the properties of cylindrical folds, and thus geologists projected data *into* the lines of cross-sections as if the folds were cylindrical. Modern techniques of fold analysis, including robust analytical and visualization software, make it both possible and important to distinguish between cylindrical and conical folds. In 3D it really does matter whether or not you establish the precise geometry of the folds that you are studying. Even in the 2D world of cross-sections, if you do not project data into the line of the cross-section accurately, you have no chance of interpreting the structure correctly.

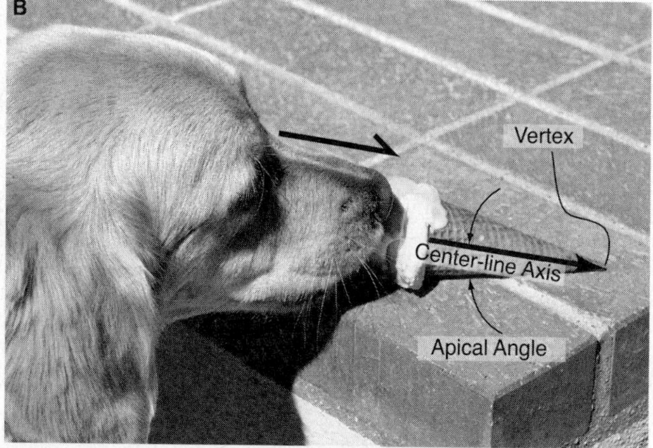

Figure 7.37 (*A*) An ice cream cone is a useful image for a conical fold. Imagine measuring the strike-and-dip of each and every tiny square inscribed by the manufacturing process on the surface of the cone, and then plotting the poles to these planes stereographically. The result would depict the constancy of angle between the cone axis and any and all parts of the cone surface. (*B*) The top of Lillie's snout is just about parallel to the trace of the center-line axis of the cone. In an ideal conical fold, the apical angle between the center-line axis and any part of the conically folded surface is constant. [Photographs by G. H. Davis.]

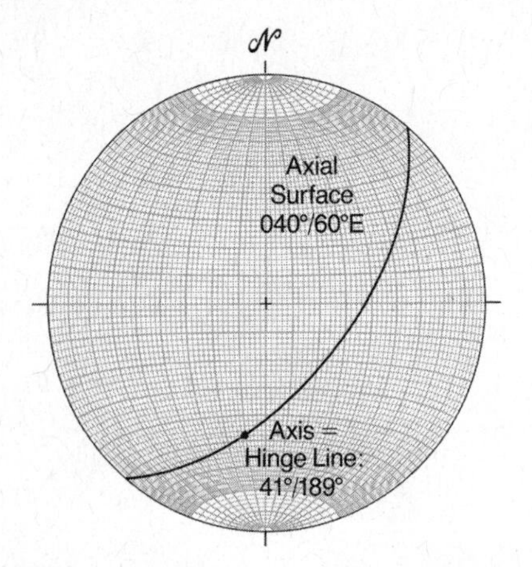

Figure 7.38 Geometric coordination of the axis and axial surface of a fold, as portrayed stereographically.

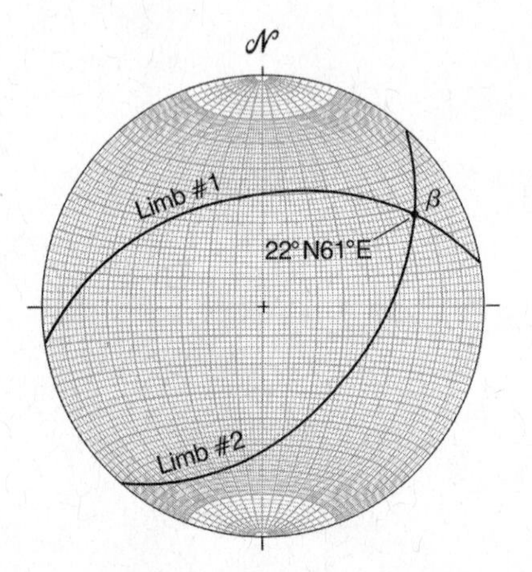

Figure 7.39 Simple β diagram. The orientation of the axis of the fold (β) is the line of intersection of the fold limbs.

Stereographic Determination of Fold Orientations

It is usually impossible to measure directly, in an accurate way, the axis and axial surface orientations of a large fold that spills beyond the expanse of a single outcrop. Instead, the trend and plunge of the fold axis and the strike and dip of the axial surface are normally calculated stereographically on the basis of representative strike-and-dip measurements of the folded surface. A stereographic plot of the axis and axial surface orientations measured for a single fold at a single outcrop displays the geometric coordination of these elements. An axial surface, by definition, passes through the hinge lines of successive folded surfaces within a given fold. Expressed stereographically, the point representing the trend and plunge of the hinge line (or fold axis) lies on the great circle that describes the orientation of the axial surface (Figure 7.38).

In the simplest case, the trend and plunge of the axis of a fold can be determined stereographically on the basis of merely two strike-and-dip measurements, one for each limb of the folded surface. The orientations of the two limbs are plotted stereographically as great circles (Figure 7.39). The intersection of the great circles, labeled β, represents a close approximation to the trend and plunge of the hinge line of the folded surface. The hinge line orientation, in turn, is taken to be a close approximation to the fold axis orientation. This specific stereographic construction is called a **beta (β) diagram**, where β refers to the trend and plunge of a fold axis deduced stereographically in the manner described.

Another way to calculate stereographically the orientation of a fold is through the construction of a **pi (π) diagram**. This requires plotting the limbs of the folded surface as poles, not as great circles (Figure 7.40*A*). Once plotted, the poles to each limb are fitted to a common great circle, known as a **π circle** (Figure 7.40*B*). The special geometric property of a π circle is that it represents the strike and dip of a plane that is perfectly perpendicular to the hinge line of the fold. The pole to this great circle, known as the **π-axis** (Figure 7.40*B,C*), expresses stereographically the orientation of the fold axis; "*π-axis*" refers to the trend and plunge of a fold axis as deduced stereographically in this manner.

The key to comfort in understanding π-diagram construction is visualizing that poles to a cylindrically folded surface indeed lie geometrically in a plane oriented at a right angle to the hinge line of the fold. Figure 7.41, an extraordinary view along the mined-out hinge of an anticline, can help us

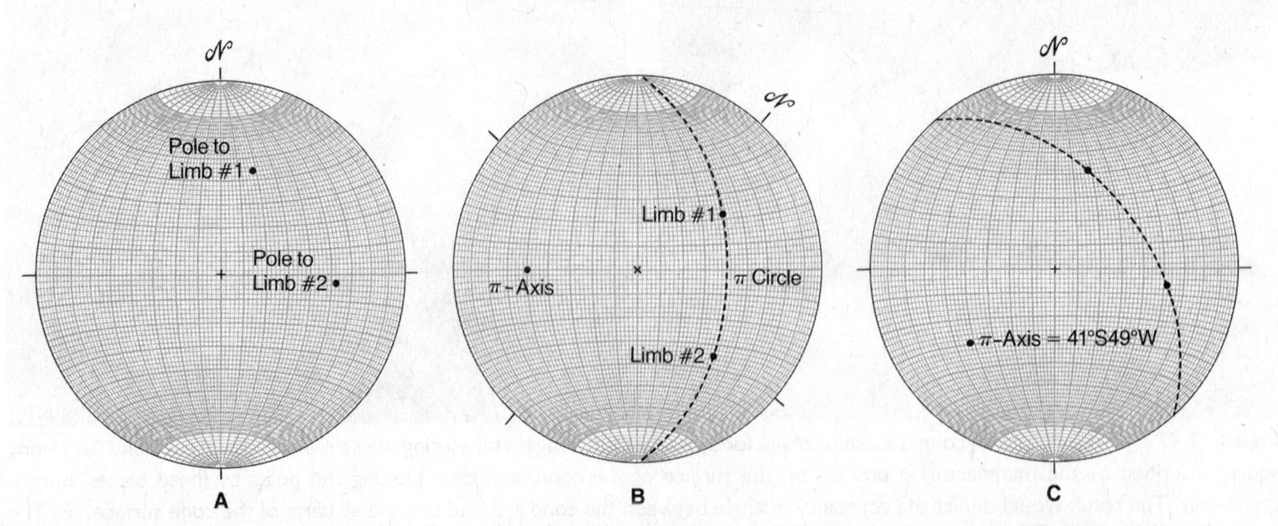

Figure 7.40 Steps in the construction of a π diagram. (*A*) Stereographic portrayal of poles to limbs. (*B*) The fitting of the poles to a common great circle (the π circle). Identification of the pole to the π circle (the π axis). (*C*) Trend and plunge of the π axis.

Figure 7.41 Visual image of the geometry of a π-diagram. Note that the roof supports (timber) are everywhere perpendicular to bedding, which is folded. The roof supports are like poles to bedding in the stereographic world. The trend-and-plunge readings for the timber supports, virtually, lie within a common plane, whose orientation is at right angles to the fold axis. This tunnel, in fact, is oriented parallel to the fold axis of this deformed bedding. [Photograph by C. D. Walcott. Courtesy of United States Geological Survey.]

picture this relationship. Each of the timbers that supports the mine roof is oriented as a pole to the bedding it supports. The orientations of the timbers, taken together, define a plane at right angles to the hinge line. The miner with hands on hips in the deep recesses of the tunnel is smiling because he recognizes how closely his timber support system captures the inherent stereographic geometry of π-diagrams.

Axial surfaces of small folds can normally be measured directly in outcrop, by holding a field notebook parallel to the envisioned orientation of the axial surface, and then taking a strike-and-dip reading on the clipboard or notebook. However, stereographic procedures are required to determine the axial surface orientations of large folds and to come up with consistently accurate orientations for axial surfaces of outcrop-scale folds. The simplifying premise in the stereographic calculation is that the **bisecting surface** of a fold is a close approximation to the axial surface. For a given folded surface, the bisecting surface passes through the hinge line and splits the angle (the **interlimb angle**) between the limbs (Figure 7.42*A*). When the bisecting surface of a single folded surface is compared to the axial surface of the fold as a whole (Figure 7.42*B*), minor differences in orientation may sometimes be evident. But generally the differences are so slight that the strike and dip of the bisecting surface can be taken as the strike and dip of the axial surface.

The stereographic procedure in computing the orientation of the bisecting surface of a folded surface is reasonably straightforward. Although shortcuts can be taken, the full flavor of the method emerges by combining a β-diagram and a π-diagram on a common projection. First, plot the attitudes of the fold limbs as great circles to form a simple β-diagram, and then identify the intersection of the great circles as β (Figure 7.43*A*). Next a π-diagram is added, by plotting the two poles to the two fold limbs and fitting these poles to a common great circle (the π circle) (Figure 7.43*B*). The π-axis of this great circle is coincident with β. The bisecting surface of the fold is the great circle that passes through the hinge line (through β) and perfectly bisects the angle between the two poles as measured along the π circle (Figure 7.43*C*). In the example we are considering here, the strike and dip of the bisecting surface proves to be N60°W, 66°SW (Figure 7.43*D*).

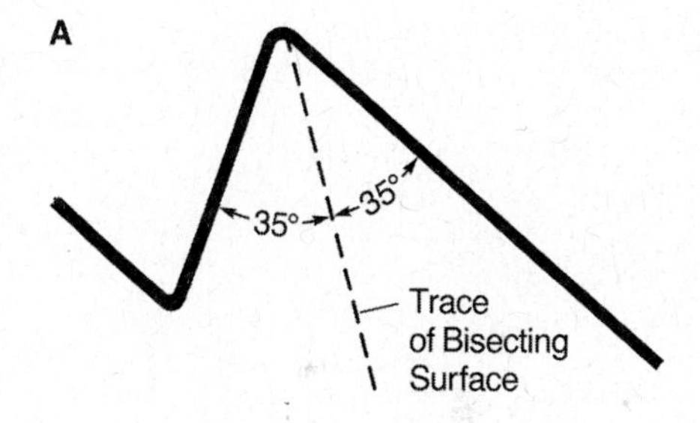

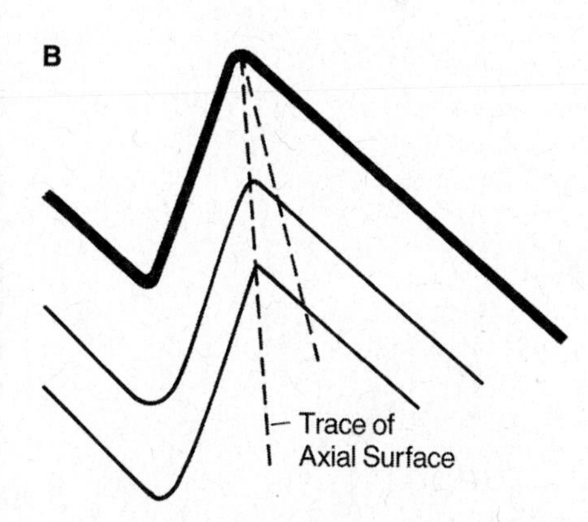

Figure 7.42 (*A*) The bisecting surface of a fold splits the angle between the limbs. (*B*) The axial surface of a fold passes through the hinge points of successive fold surfaces. The bisecting surface of a given fold is not necessarily the same as the axial surface. [From *Structural Analysis of Metamorphic Tectonites* by F. J. Turner and L. E. Weiss. Published with permission of McGraw-Hill Book Company, New York, copyright © 1963.]

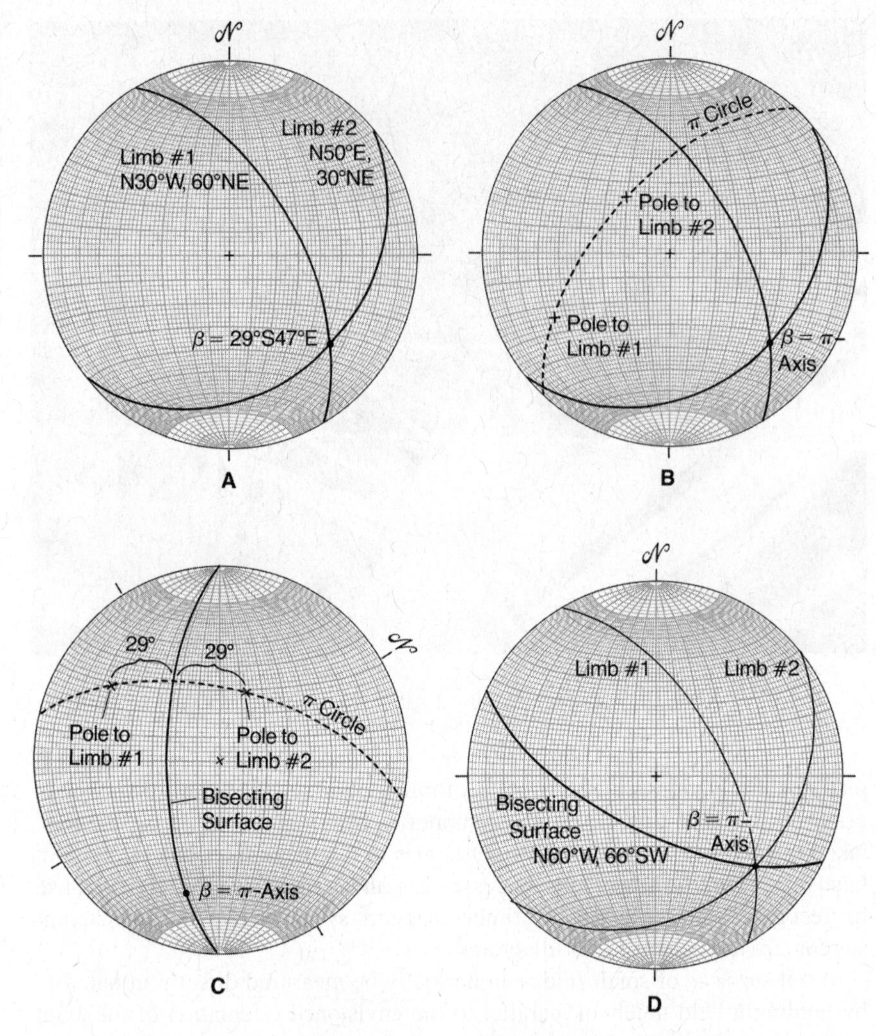

Figure 7.43 Steps in stereographically determining the orientation of the bisecting surface of a fold. (*A*) Plot fold limbs as great circles and identify β. (*B*) Plot poles to limbs. (*C*) Measure the angle between the poles to the limbs. Fit a great circle to the bisector of this angle and to β. (*D*) Stereographic configuration of the bisecting surface, in proper orientation.

There is always some ambiguity in stereographically computing the orientation of a bisecting surface. There are actually two different points on the π circle that serve as bisectors to the limbs (see Figure 7.43*C*): one bisects the acute angle between the bedding traces; the other bisects the obtuse angle. One of these bisectors, but not both, must be used along with β as a reference point for constructing the great circle representing the orientation of the bisecting surface. To select the appropriate bisector, it is necessary to keep in mind the fold form whose orientation is being sought. If the fold is upright, the proper bisector is one that yields a relatively steeply inclined bisecting surface. If the fold is overturned or recumbent, the proper bisector is one that yields a relatively low-dipping bisecting surface.

Polar Tangent Determinations of Fold Orientations

Basics

The **polar tangent plot** is yet another device for determining the trend and plunge of a fold on the basis of bedding orientations. Its practical usefulness was advocated and demonstrated by Andy Bengtson (1980, 1981). Like the stereographic projection, the polar tangent plot is a circular template (Figure 7.44*A*). On the perimeter of this template are azimuths, plotted in 2°-increments clockwise from 0° to 360°. Dip inclinations are plotted as a set of concentric circles, representing 5°-intervals from the center to the periphery

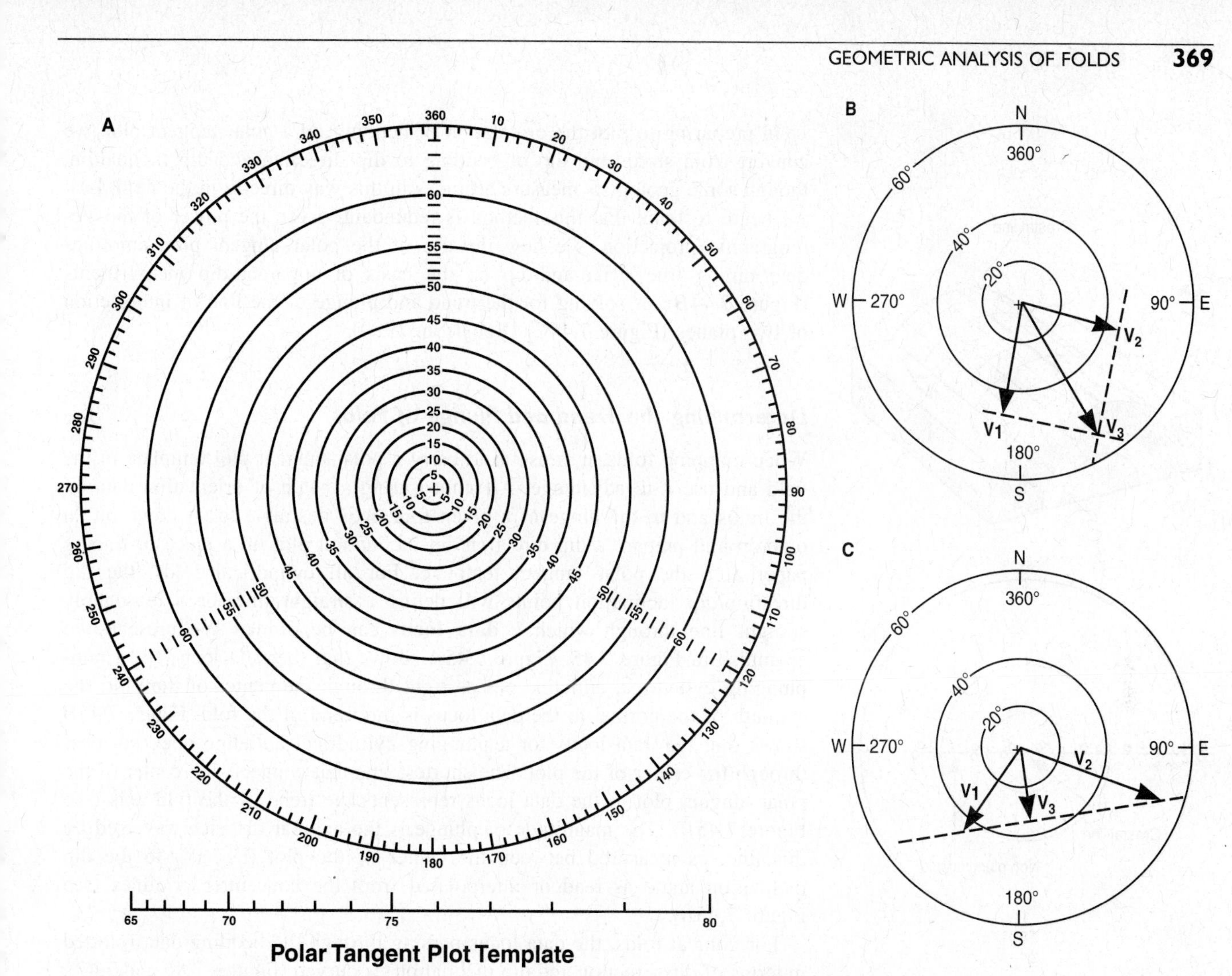

Figure 7.44 (*A*) The polar tangent diagram is a polar coordinate graph in which the direction or azimuth of dip is shown around the perimeter of the circle, and the dip inclination is shown by concentric circles of increasing dip away from zero dip at the center. The radius of each dip-inclination circle is proportional to the tangent of the dip angle. Within the diagram, dips up to 65° can be plotted. Note below the diagram there is an auxillary scale for treating dip angles in the range of 65° to 90°. [From Bengtson, 1980, Figure 1, p. 599, Geology, v. 8, p. 599–602. Used with permission of the Geological Society of America.] (*B*) This drawing shows the ease of determining true dip (**V₃**) on the basis of two apparent dips (**V₁** and **V₂**). First, plot the trend and plunge of each of the two apparent dip measurements. Second, draw perpendicular lines through the end points of the vectors. Third, simply read the true dip and dip direction from the intersection of the perpendicular construction lines. [From Bengtson, 1980, Figure 5A, B, p. 601, Geology, v. 8, p. 599–602. Used with permission of the Geological Society of America.] (*C*) Determining the trend and plunge of the line of intersection of two planes is a "*piece of cake*," and can be applied to determining the orientation of the hinge line of a fold. First, plot the true dip vectors (**V₁** and **V₂**) for the two planes. Second, connect the end points of these two vectors. Third, construct a perpendicular (**V₃**) from the origin to the construction line. **V₃** represents the trend and plunge of the line of intersection of the two planes. [From Bengtson, 1980, Figure 3B, p. 600, Geology, v. 8, p. 599–602. Used with permission of the Geological Society of America.]

(i.e., from 0° to 90°) (see Figure 7.44*A*). Note that the concentric circles that represent dip inclinations are *not* equally spaced, but vary by the tangent function of the dip angle (though the tangent function of 90° is not defined). Thus the distance from the 40°-circle to the 50°-circle is greater than distance separating the 10°- and 20°-circles. Note that periphery of this polar tangent plot template represents inclinations of 65°, requiring that greater inclinations be plotted through the use of the auxillary scale shown beneath the plot.

In preparing to plot the orientation of bedding on a polar tangent plot, we convert from strike and dip of bedding to dip direction and dip inclination, though some geologists measure attitudes in this way directly in the field. Lest we begin to think that this method is redundant, given the power of the stereographic projection, see how the use of the polar tangent plot simplifies determining true strike-and-dip on the basis of apparent dip measurments (Figure 7.44*B*), or solving for the trend and plunge of the line of intersection of two planes (Figure 7.44*C*) (Bengtson, 1980).

Determining the Trend and Plunge of Folds

When mapping folds, it is useful to carry a polar tangent plot template in the field and use it to advantage. After measuring a batch of orientation data on the limbs and in the hinge of a given fold, it is desirable to sit down on an outcrop and prepare a dip direction/dip inclination plot on a piece of tracing paper atop the polar tangent template. For all cylindrical folds, the dip direction/dip inclination points will define a straight line, or a reasonably straight line through which a **data locus** can be drawn. We present two examples in Figure 7.45. Figure 7.45*A* shows that the data locus for a nonplunging cylindrical anticline passes right through the center of the plot; the azimuth of the normal to the data locus is the trend of the fold. Figure 7.45*B* shows that the data locus for a plunging cylindrical anticline does *not* pass through the center of the plot. The shortest line that connects the center of the polar tangent plot to the data locus represents the trend of the fold axis (see Figure 7.45*B*). The magnitude of plunge is the angular distance traversed by this line, as measured between the center of the plot (i.e., 0°) to the dip inclination angle as read or interpolated from the concentric gridlines (see Figure 7.45*B*).

For conical folds, the data locus passing through the bedding data (plotted in terms of dip direction and dip inclination) is curved (Figures 7.46 and 7.47). The amount of curvature of the data locus reflects the magnitude of the apical angle of the conical fold. As the apical angle increases, the amount of

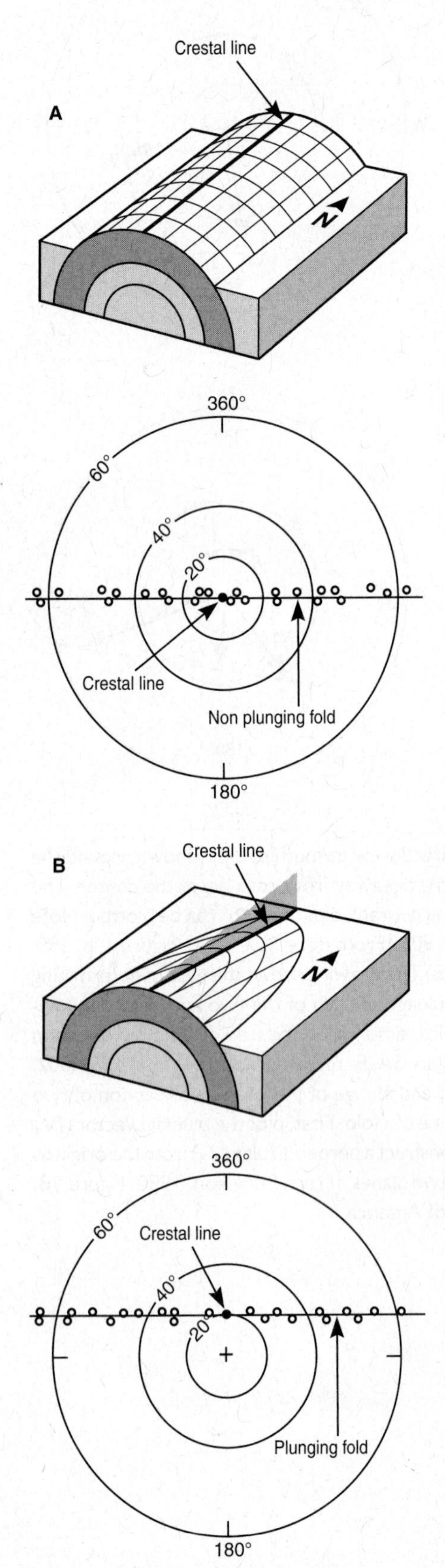

Figure 7.45 (*A*) Top drawing shows a north-trending nonplunging upright anticline. Below it is a polar tangent plot showing a straight-line distribution of bedding orientation data (points) plotted in terms of dip inclination and dip direction. The data locus passing through these points is a straight line, for the fold is cylindrical. Because the data centerline passes through the center of the plot, we know that the fold does not plunge. The normal to the data locus trends north, which is the trend of the fold. [From Bengtson, 1980, Figure 4A, B, p. 600, Geology, v. 8, p. 599–602. Used with permission of the Geological Society of America.] (*B*) Top drawing shows a north-plunging upright anticline. Below it is a polar tangent plot showing a straight-line distribution of bedding orientation data (points) plotted in terms of dip inclination and dip direction. The data locus passing through these points is a straight line, for the fold is cylindrical. Because the data locus *does not* pass through the center of the plot, we know that the fold plunges. Note that the trend of the fold corresponds to the azimuth (measured along the periphery of the plot) of the shortest line connecting the center of the template to the data centerline. The plunge of the fold is the angular distance (in degrees) between the center of the plot and the closest point on the data locus, using the concentric gridlines as a guide. [From Bengtson, 1980, Figure 5A, B, p. 601, Geology, v. 8, p. 599–602. Used with permission of the Geological Society of America.]

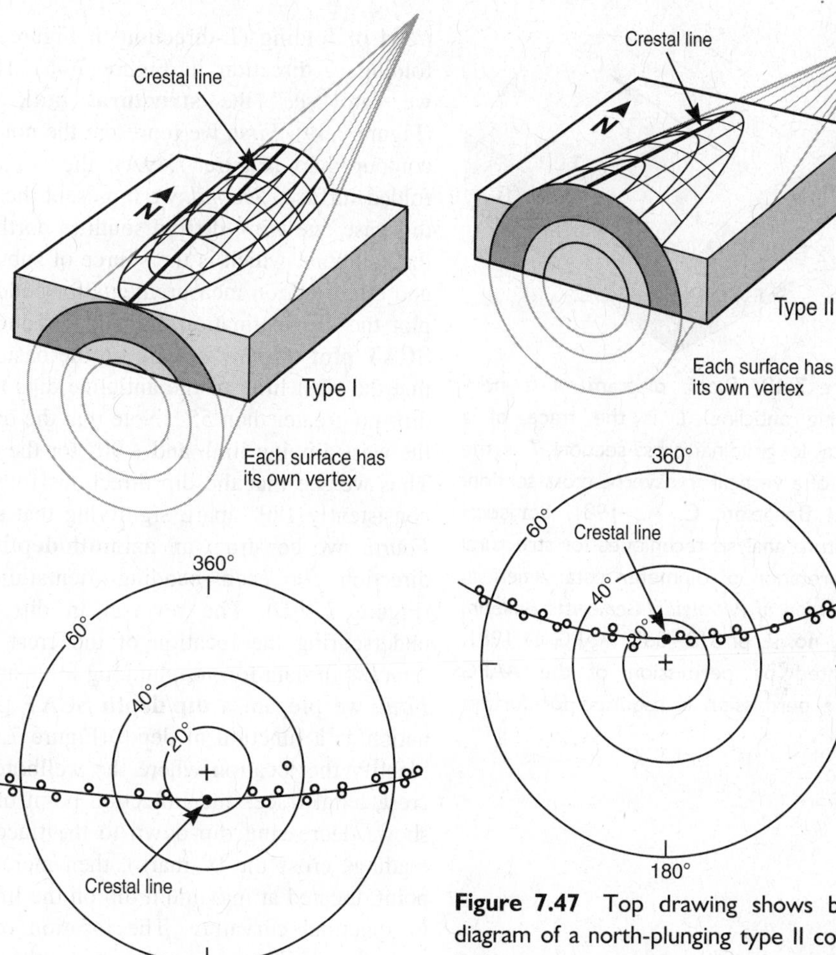

Figure 7.46 Top drawing shows block diagram of a south-plunging type I conical fold, i.e., one in which the fold's vertex lies up-plunge. Beneath is a polar tangent plot of dip direction/dip inclination bedding orientation data for this fold. The data locus is curved, for the fold is not cylindrical. Furthermore, the data locus passes south of the center of the plot and is convex in the direction of plunge. Note that the trend of the fold corresponds to the azimuth (measured along the periphery of the plot) of the shortest line connecting the center of the template to the data locus. The plunge of the fold is the angular distance (in degrees) between the center of the plot and the closest point on the data locus, using the concentric gridlines as a guide. [From Bengtson, 1980, Figure 6A, B, p. 601, Geology, v. 8, p. 599–602. Used with permission of the Geological Society of America.]

Figure 7.47 Top drawing shows block diagram of a north-plunging type II conical fold, i.e., one in which the fold's vertex lies down-plunge. Beneath is a polar tangent plot of dip direction/dip inclination bedding orientation data for this fold. The data locus is curved, for the fold is not cylindrical. Furthermore, the data locus passes north of the center of the plot and is concave in the direction of plunge. Note that the trend of the fold corresponds to the azimuth (measured along the periphery of the plot) of the shortest line connecting the center of the template to the data locus. The plunge of the fold is the angular distance (in degrees) between the center of the plot and the closest point on the data locus, using the concentric gridlines as a guide. [From Bengtson, 1980, Figure 7A, B, p. 601, Geology, v. 8, p. 599–602. Used with permission of the Geological Society of America.]

curvature of the data becomes greater and greater. The trend and plunge of the fold can be determined quickly, in the same way as we determine trend and plunge for cylindrical folds (see Figures 7.46 and 7.47).

SCAT Plots Determination of Fold Properties

In *Chapter 6, Faults*, we introduced SCAT plots as a methodology for identifying faults in the subsurface, and for evaluating the shapes and orientations of folds associated with faulting. SCAT plots conventionally are used in the oil/gas industry for interpreting subsurface structure on the basis of dip direction and dip inclination as a function of depth in a borehole (see Figure 6.39). "*SCAT*" stands for "*Statistical Curvature Analysis Technique*," the utility of which was recognized by Andy Bengtson (1980, 1981). Although originally designed for dip data from oil wells, SCAT plots can be constructed from field data as well, with dip direction and dip azimuth data collected as a function of distance along a given traverse. Using the SCAT approach in the field, we find that a little data goes a long way.

As an example, let us use the SCAT approach to determine the signature of a nonplunging anticline in the subsurface (Figure 7.48). For nonplunging folds, strike remains constant throughout the structure. Typically we evaluate the form of folds by constructing cross-sections oriented parallel to the

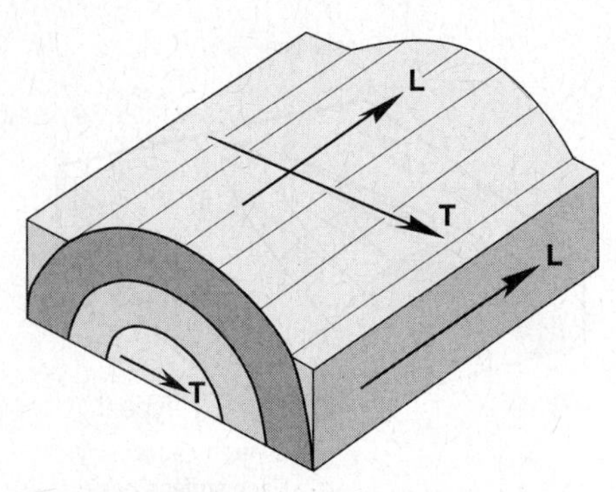

Figure 7.48 Block diagram of a nonplunging anticline. *L* is the trace of a vertical longitudinal cross-section. *T* is the trace of a vertical transverse cross-section. [From Bengtson, C. A., 1981, Statistical curvature analysis techniques for structural interpretation of dipmeter data: *American Association of Petroleum Geologists Bulletin*, v. 65, no. 2, p. 312–332. AAPG © 1981, reprinted by permission of the AAPG whose permission is required for further use.]

trend of folding (*L*-direction in Figure 7.48) and transverse to the trend of folding (*T*-direction in Figure 7.48). However, using the SCAT approach, we can "*see*" the **structural bulk curvature** in much greater detail (Figure 7.49). *First,* we represent the nonplunging anticline through a structure contour map (Figure 7.49*A*); the contours show subsurface depths to the folded surface. *Second,* we represent the fold in a **transverse cross-section**, in this case one viewed from south to north (Figure 7.49*B*). Note the location of the wellbore, which is the source of subsurface information for dip inclination and dip direction measurements for bedding as a function of depth. *Third,* we plot the dip inclination and dip direction measurements on an **azimuth/dip SCAT plot** (Figure 7.49*C*), on the basis of which we immediately recognize that the west limb of the anticline dips no greater than 25°, and the east limb dips no greater than 55°. Note that the dip direction holds steady at ~270° for the west-dipping limb and ~90° for the east-dipping limb (see Figure 7.49*C*). Thus we see that the dip directions for west limb and east limb bedding are consistently 180° apart, signifying that strike is constant throughout the fold. *Fourth,* we construct an **azimuth/depth SCAT plot** that represents the dip direction for each bedding-orientation reading as a function of depth (Figure 7.49*D*). The reversal in dip direction "*jumps out*" in this plot, underscoring the location of the crest of the fold. The distinctive parallel "*tracks*" of data for nonplunging folds are called "*railroad tracks*" in the trade. *Fifth,* we prepare a **dip/depth SCAT plot**, which shows apparent dip inclination as a function of depth (Figure 7.49*E*). Using this plot it is possible to identify the location where the wellbore happens or traverse to intersect the crest, axial trace, and inflection point of folded layers. In particular this plot shows decreasing dip down to the trace of the crestal plane (dip-inclination readings cross the 0° mark), then increasing dip downward to the inflection point, located at maximum dip on the limb where anticlinal curvature changes to synclinal curvature. The location of the intersection of the axial trace with the wellbore corresponds to the greatest rate of change of the dip-inclination data (see Figure 7.49*E*). *Sixth,* we create a **transverse dip/depth SCAT plot**, which integrates apparent dip inclination, dip direction, and depth in ways that complement the dip/depth SCAT plot (Figure 7.49F). This knowledge of specific changes in gradient of dip inclination and dip direction is critically important to explorationists as they seek to identify structural traps in the subsurface. *Seventh,* we end with the **longitudinal dip/depth SCAT plot** (Figure 7.49*G*), which in this example looks quite boring, for *all* apparent dips hover around 0°, which is just what we would expect for a nonplunging fold.

It takes a while to adjust our eyes to SCAT plots, but once mastered they become invaluable aids to interpretation. For example, what do SCAT plots look like for a plunging fold, such as that shown schematically as Figure 7.50? Our plotting approach is the same as for nonplunging folds, but the resulting "*signatures*" plot by plot are much different (Figure 7.51). Comparing and contrasting the equivalent diagrams, we can appreciate how SCAT diagrams sensitively distinguish folds of different forms and/or orientations. Data patterns for plunging folds are special in that there is no 0°-dip anywhere on the fold. The lowest dip observed when mapping a plunging fold corresponds to the plunge-value itself. For plunging folds, the azimuth/dip SCAT plots show a "*horseshoe*" pattern (Figure 7.51*C*). Just as when we map around the nose of a fold, the lowest dip is the angle and direction of plunge. The lowest point on the horseshoe corresponds to the trend of the fold axis, and the dip amount that corresponds to the lowest point on the horseshoe is the plunge amount. The longitudinal dip/depth SCAT plot (Figure 7.51*G*) shows an unchanging, gentle north dip, which is the plunge of the fold.

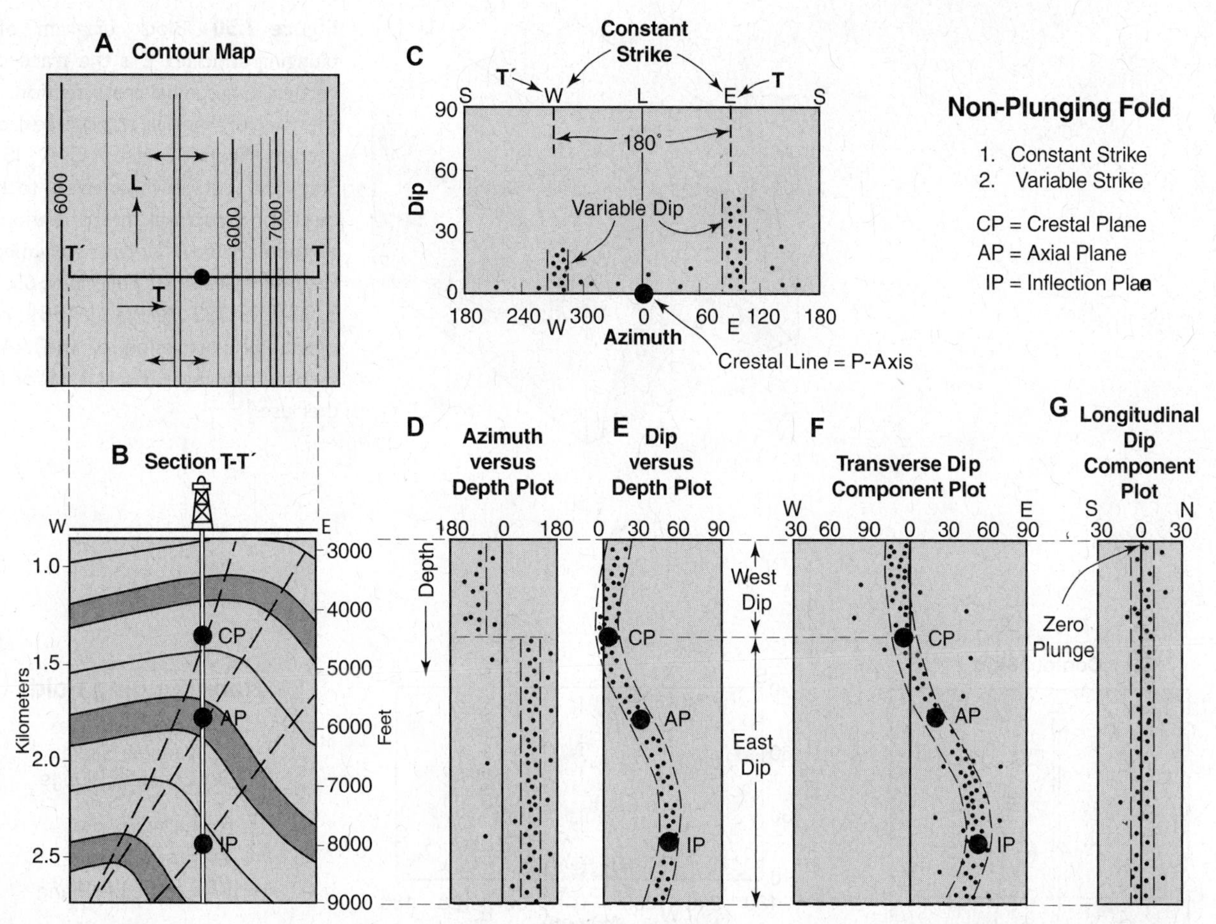

Figure 7.49 (*A*) Structure-contour map of non-plunging anticline. This anticline trends north-south. Line T′–T is a "*transverse*" line of cross-section, at right angles to the trend of the axis of the fold. The contour lines (2000-ft interval) represent depth to a folded datum surface. (*B*) Cross-section T′–T of the anticline, as viewed from south to north. Note the location of the wellbore. CP is intersection of wellbore and the trace of the crestal plane of the anticline. AP is the intersection of the wellbore and the trace of the axial plane of the anticline. IP is the intersection of the wellbore and the trace of inflection plane of the east limb of the anticline. (*C*) Azimuth/dip SCAT plot. All bedding data are plotted according to dip inclination (*y*-axis) and dip direction (*x*-axis). Points representing westerly dips fall to the left of the vertical centerline; points representing easterly dips fall to the right of the centerline. Note that lowest dips fall closest to the base of the plot. The "*railroad-track*" clusters of points are straight, parallel, and separated by 180°, all consistent with expectations for a nonplunging fold. (*D*) Azimuth/depth SCAT plot. Dip direction is plotted in relation to depth. Points lying left of the vertical centerline correspond to westerly dips, and points lying to the right of the centerline correspond to easterly dips. The crossover in dip direction takes place at a depth of ~4500 ft. The two bands of data are parallel to one another, signifying that the fold is nonplunging. (*E*) Dip/depth SCAT plot. Data points above the horizontal dashed line are west dipping; those below are east dipping. CP is point at the crest of the fold as penetrated by the wellbore. AP is a point on the axial trace of the fold. IP is an inflection point on the east limb of the fold. (*F*) Transverse dip/depth SCAT plot. It is marked by a vertical centerline separating west-dipping apparent dip inclinations (on the left) from east-dipping ones (on the right). Again, points CP, AP, and IP are noted. (*G*) Longitudinal dip/depth SCAT plot. Apparent dips are plotted in the direction of the trend of the fold. All hover at very low angles, consistent with the fact that this fold has no plunge. [From Bengtson, C. A., 1981, Statistical curvature analysis techniques for structural interpretation of dipmeter data: *American Association of Petroleum Geologists Bulletin*, v. 65, no. 2, p. 312–332. AAPG © 1981, reprinted by permission of the AAPG whose permission is required for further use.]

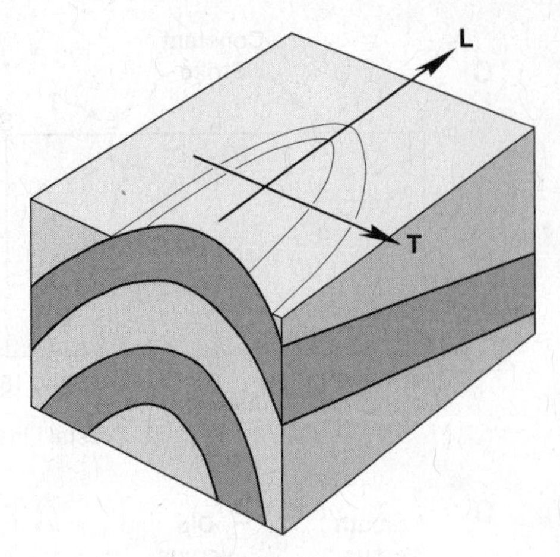

Figure 7.50 Block diagram of a plunging anticline. *L* is the trace of a vertical longitudinal cross-section. *T* is the trace of a vertical transverse cross-section. [From Bengtson, C. A., 1981, Statistical curvature analysis techniques for structural interpretation of dipmeter data: *American Association of Petroleum Geologists Bulletin*, v. 65, no. 2, p. 312–332. AAPG © 1981, reprinted by permission of the AAPG whose permission is required for further use.]

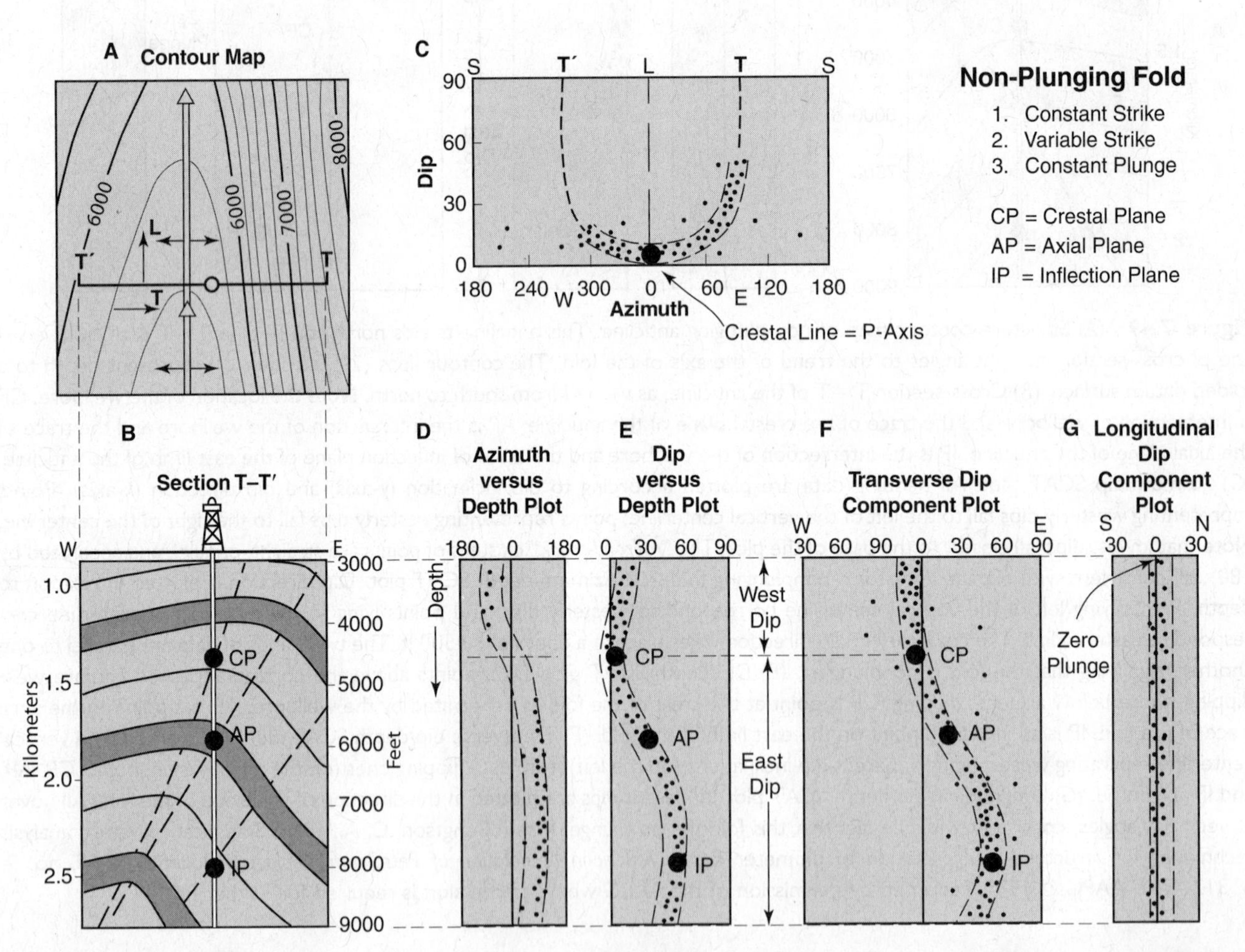

Figure 7.51 The SCAT approach applied to data for a plunging anticline. (*A*) Structure contour map of north-plunging anticline. (*B*) Transverse cross-section of anticline, viewed from south to north. (*C*) Azimuth/dip SCAT plot. (*D*) Azimuth/depth SCAT plot. (*E*) Dip/depth SCAT plot. (*F*) Transverse dip/depth SCAT plot. (*G*) Longitudinal dip/depth SCAT plot. [From Bengtson, C. A., 1981, Statistical curvature analysis techniques for structural interpretation of dipmeter data: *American Association of Petroleum Geologists Bulletin*, v. 65, no. 2, p. 312–332. AAPG © 1981, reprinted by permission of the AAPG whose permission is required for further use.]

Describing the Shape and Size of a Folded Surface

Common Fold Shape

As part of the overall description of a folded surface, it is useful to convey a sense of the shape of the fold, including its tightness. Fold shape is described in normal profile view. Normal profile views of folded surfaces can be afforded by outcrop exposures, photographs, geologic cross-sections, rock slabs, and thin sections *of proper orientation.*

All the conventional terms for describing the profile shape of a folded surface attempt to convey a picture of the form and the configuration of limbs and hinge (Figure 7.52). A **chevron fold**, for example, is marked by planar limbs that meet at a discrete hinge point or at a very restricted subangular hinge zone (Figure 7.52*A*). A **cuspate fold** exhibits curved limbs that are opposite in sense of curvature to those of most ordinary folds (Figure 7.52*B*). An upright, cuspate anticline displays limbs that are concave upward; an upright, cuspate syncline has limbs that are concave downward. Oddly enough, there is no conventionally used term to describe a folded surface whose profile form is wholly part of a circular arc (Figure 7.52*C*), nor is there a term to describe a folded surface whose profile form is part of an ellipse (Figure 7.52*D*).

Some folded surfaces have two hinges. **Box folds** (or conjugate folds) are composed of three planar limbs connected by hinge points or narrow, restricted subangular hinge zones (Figure 7.52*E*). Upright box folds are characterized by flat crests. **Teardrop folds** are continuously curved folded surfaces shaped, of course, like teardrops. They are involuted and curve back on themselves (Figure 7.52*F*).

Fold shapes in nature are endlessly variable, products of any number of combinations of layer rheology, fold mechanisms, and magnitude of tectonic strain. In Figure 7.53 we see folding of lower Paleozoic dolostones exposed in the Muddy Mountains (Nevada) just beneath a major thrust sheet. The form of the fold is cuspate (see similarity to Figure 7.52*B*). Note how the form of the fold changes from bottom to top.

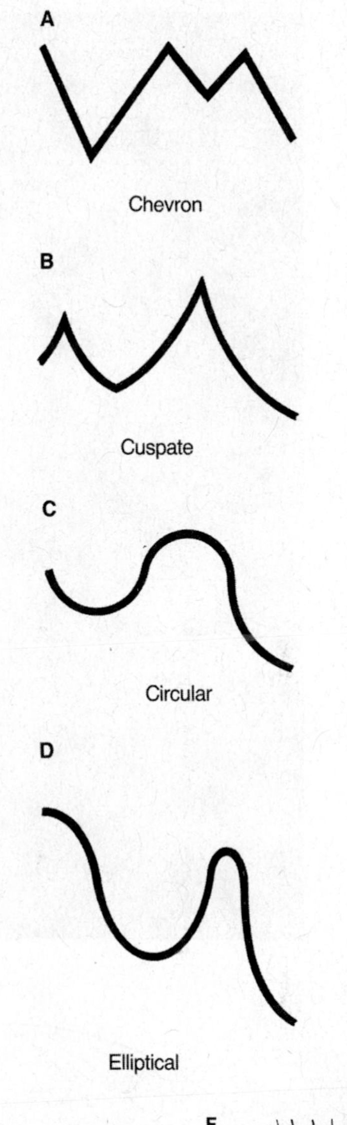

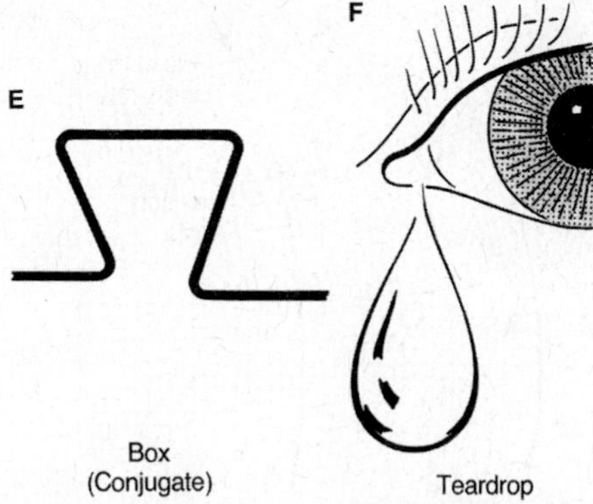

Figure 7.52 Some common fold shapes.

Fold Tightness

Fold tightness is described in terms of **interlimb angle** (Ramsay, 1967), the internal angle between the limbs of the folded surface. Although the interlimb angle of a folded surface can be measured with a protractor on the surface of a profile exposure of a small fold, or from a profile-view photograph of a large fold, profile views of folds are the exception, not the rule. Consequently it is usually necessary to calculate interlimb angles stereographically (Figure 7.54). This is achieved by taking the strike and dip of the folded surface at each inflection point, plotting the orientations stereographically as poles, fitting the poles to a common great circle, and measuring the angle between the poles along the common great circle (see Figure 7.54). To know whether the acute or obtuse angle between the poles is the appropriate interlimb angle, it is necessary to keep clearly in mind the general form of the fold. The interlimb angle of a very tight fold is acute. The interlimb angle of a very open fold is obtuse.

The measured value of interlimb angle provides a basis for choosing an adjective that describes fold tightness. Figure 7.55 shows a classification scheme adapted but slightly modified from the nomenclature proposed by Fleuty (1964). **Gentle folds** are marked by interlimb angles ranging from 170° to 180°. Nice examples are seen in highway exposures on the eastern edge of

Figure 7.53 Photograph of large anticline in lower Paleozoic dolostones in the Muddy Mountains, Nevada. The fold is cuspate in form, except for the tight elliptical forms at the midpoint of the exposure, and the gentle rounded forms near the base. [Photograph by G. H. Davis.]

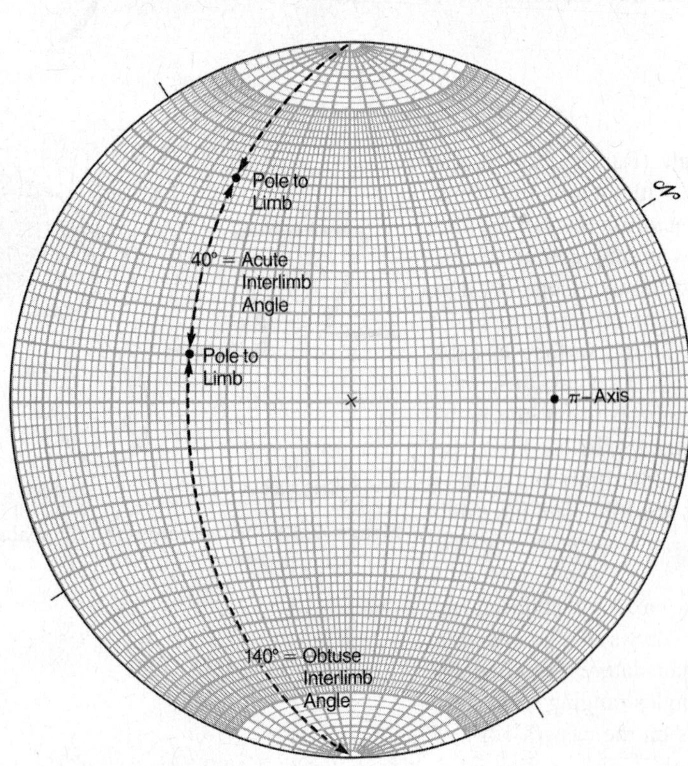

Figure 7.54 Stereographic determination of the interlimb angle of a fold.

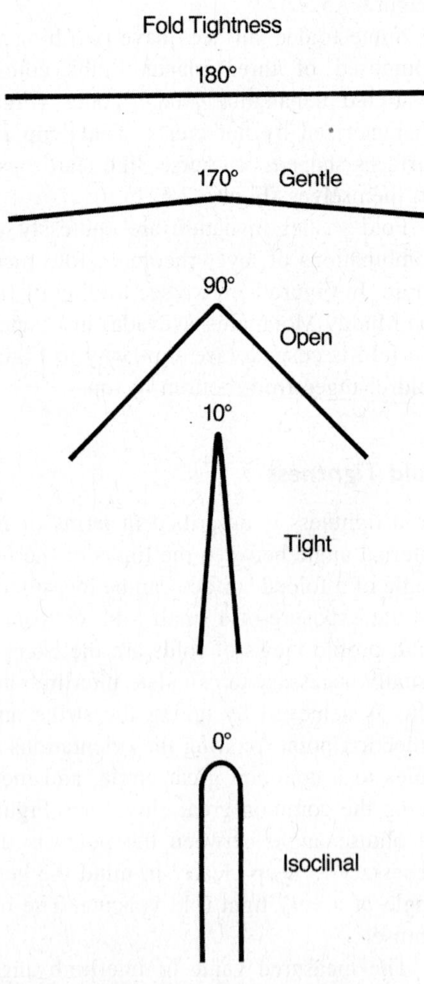

Figure 7.55 Classification of folds according to tightness, based on the size of the interlimb angle. [Reprinted from *Geological Association Proceedings*, v. 75, Fleuty, M. J., The description of folds, p. 461–492, © 1964, with permission from Elsevier.]

Figure 7.56 Folds across the spectrum of tightness. (*A*) Gently folded Pennsylvanian sedimentary rocks exposed in a highway cut just east of Morgantown, West Virginia. Full-blown anticlines and synclines of the Valley and Ridge province of the Appalachians begin just a few kilometers to the east. [Photograph by G. H. Davis.] (*B*) Tight folds in the Mazatzal Formation (late Paleoproterozoic) in Barnhardt Canyon, Mazatzal Mountains, central Arizona. [Photograph by S. J. Reynolds.] (*C*) Mountain-sized recumbent isoclinal fold nappe in metamorphosed Mesozoic rocks in the Patagonian Andes, near the famous granitic spires of Mt. Fitzroy and the border between Chile and Argentina. [Photograph by S. J. Reynolds.]

the Appalachian Plateau, in northern West Virginia (Figure 7.56*A*) and western Pennsylvania. **Open folds** have interlimb angles ranging from 90° to 170° (see Figure 7.55). Open folds are commonplace in mountain belts. Folds are considered to be **tight** if they display interlimb angles in the range of 10° to 90° (Figure 7.56*B*). **Isoclinal folds** are marked by interlimb angles in the range of 0° to 10° (Figure 7.56*C*). Cutoffs for isoclinal, tight, open, and gentle are easy to remember: 10°, 90°, and 170°. We find the Fleuty (1964) classification useful when summarizing changes, if any, in fold tightness across a region or area of study, or between folds or fold systems of different tectonic generations.

Fold Size

We wish that fold size were as easy to measure and describe as fold tightness. The standard measures of **wavelength** and **amplitude** can seldom be employed because so many folds occur as solitary, isolated, "*decoupled*" structures, and not as obvious parts of continuous, repeated, sinusoidal waveforms. Many folds encountered in the field are not linked structurally to other folds: they are **rootless,** cut off on either side by faults and/or shear zones. Some folds that appear to be rootless may in fact be continuous with other folds, but the connection cannot be demonstrated because of the misfortunes of erosion and/or the poor quality of exposure. Even if a fold can be

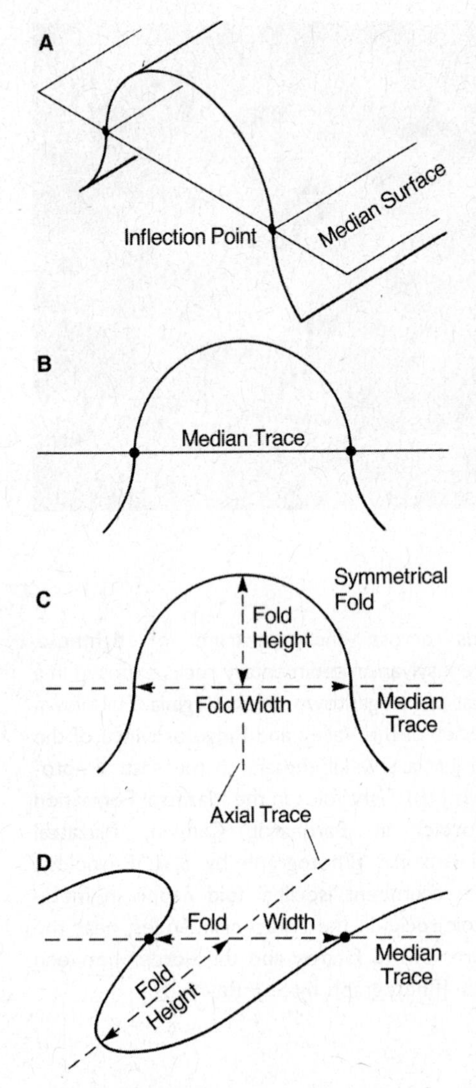

Figure 7.57 Geometric nature of (*A*) the median surface and (*B*) the median trace of a fold. Our convention for measuring fold height and fold width of (*C*) symmetrical and (*D*) asymmetrical folds.

shown to occur within an interconnected system of folds, the shapes and sizes of individual folds within a wave train may vary tremendously, not at all like an ideal wave. Given these problems and limitations, we find it practical to describe the size of a folded surface in terms of two measures: **fold height** and **fold width**, as measured in profile view.

To describe exactly what is meant by fold height and fold width, it is necessary to introduce the idea of a **median surface**. A median surface of a fold is an imaginary, geometric surface that passes through all the inflection points of a given folded surface (Figure 7.57*A*) (Ramsay, 1967). Both fold height and fold width are measured with respect to the **median trace** of the folded surface, that is, the trace of the median surface as seen in a profile view of the fold (Figure 7.57*B*). We find it practical to describe **fold height** as the distance between the median trace and the hinge point of the folded surface as measured along the axial trace of the fold, and **fold width** as the distance between inflection points on a folded surface as measured along the median trace (Figure 7.57*C, D*).

Fold Symmetry

The overall **symmetry** of a fold can be described in terms of the angular relationship of its median trace and axial trace. **Symmetrical folds** are characterized by a median trace and an axial trace that are mutually perpendicular; thus fold height and fold width are measured along mutually perpendicular lines (see Figure 7.57*C*). **Asymmetrical folds**, on the other hand, are marked by limbs of different lengths; thus the median trace and the axial trace of an asymmetric fold intersect at some oblique angle (see Figure 7.57*D*).

Overall Form

The overall form of a folded surface owes its character to a combination of factors, including shape, tightness, symmetry, and ratio of height to width. Thus, strings of adjectives are normally required to describe adequately the profile form of a given folded surface: for example, "*the fold is best described as a tight symmetrical cuspate anticline.*" If we add adjectives describing the geometric configuration of the folded surface, the string of adjectives becomes even longer: "*the fold is best described as a gently plunging, moderately inclined, tight attenuated, symmetrical cuspate anticline.*"

With the goal of trying to convey more detail about fold form in fewer words, Hudleston (1973) devised a visual classification scheme that aids in categorizing the forms of folded surfaces. Using the Hudleston classification, the shape of a folded surface, from hinge point to inflection point, is compared to 30 idealized fold forms arranged systematically by number (*1* to *5*) and letter (*A* to *F*) (Figure 7.58*A*). In using Hudleston's scheme it is simplest to reproduce the 30 basic fold forms on a plastic or Mylar template, then compare the forms to the folded surface in question by peering through the template toward the outcrop, photograph, or geologic cross-section portraying the fold. The payoff in using this technique comes in discovering that specific rock types and/or structural domains may be characterized by specific fold shapes.

More recently, Srivastava and Lisle (2004) introduced a "*rapid analysis*" approach that is applied to the digital image of fold trace curvature for a single limb between the hinge point and the inflection point. The profile form of such a curved trace is compared first with reference curves generated by the Bezier drawing tool (found in many software products). The standard reference curves range from L = 0 (straight-limbed chevron fold) to L = 1 (rounded fold with uniform curvature) (Figure 7.58*B*). The second step is evaluation of the ratio (R) of fold amplitude to wavelength (see Figure 7.58*B*). Fold shape groups emerge from graphing L against R (see Figure 7.58*B*). Moreover, Srivastava and Lisle show how their fold shape analysis can be nicely wedded with Fleuty's (1964) classification of fold tightness (Figure 7.58*C*).

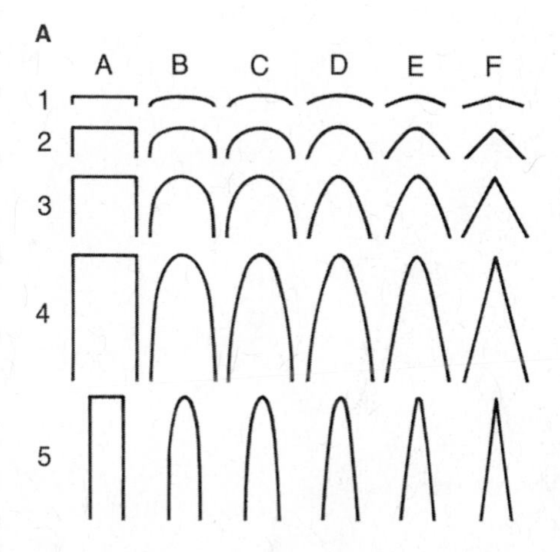

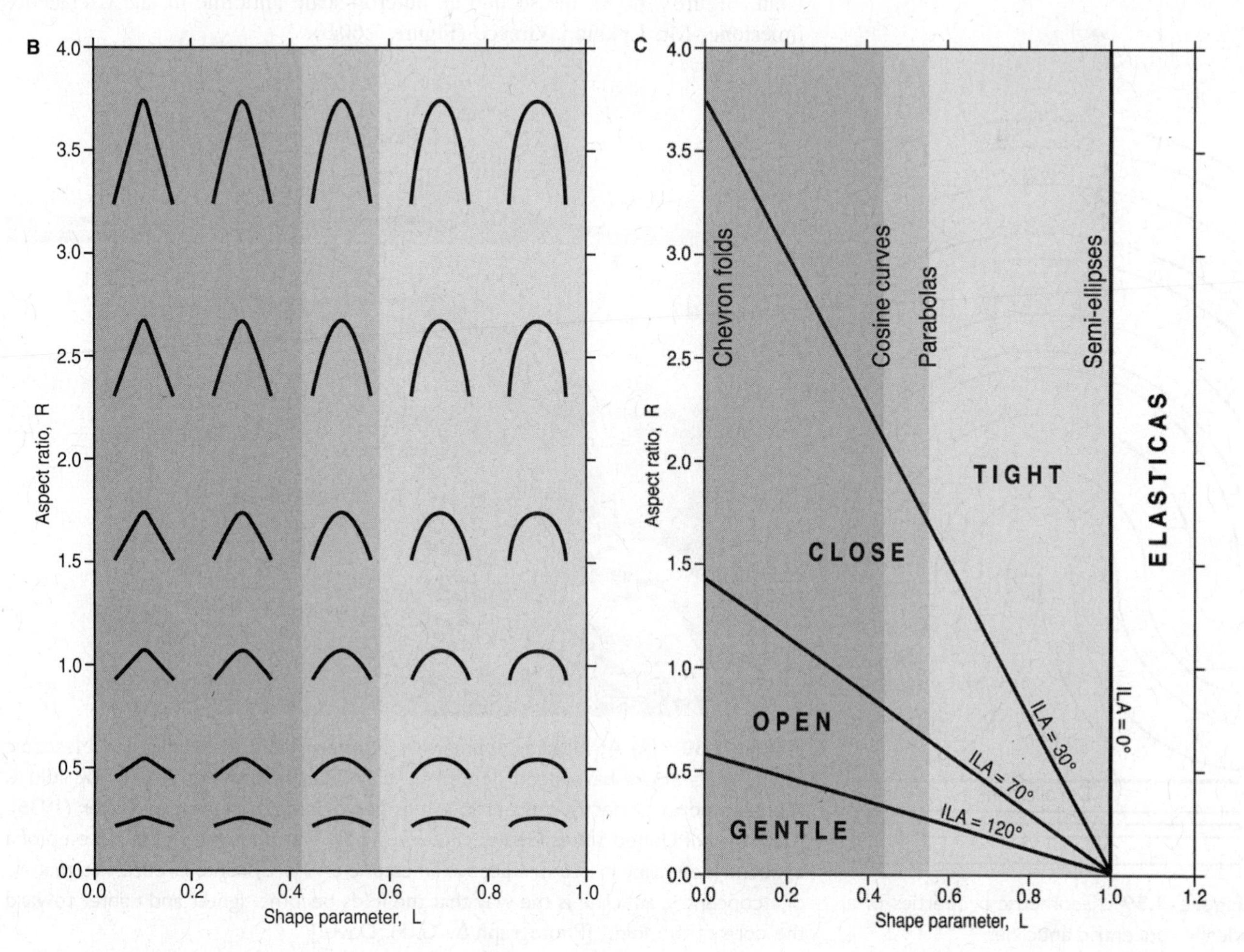

Figure 7.58 (*A*) Visual classification of the shape(s) of individual folded surfaces. [Reprinted from *Tectonophysics*, v. 16, Hudleston, P. J., Fold morphology and some geometrical implications of theories of fold development, p. 1–46, © 1973, with permission from Elsevier.] (*B*) Fold shape classification of Srivastava and Lisle (2004) is based on the confluence of aspect ratio (R), which is the ratio of fold amplitude/fold wavelength; and a shape parameter (L), which varies from L = 0 (chevron) to L = 1 (uniformly rounded). (*C*) Srivastava and Lisle (2004) also integrated their fold shape classification with Fleuty's (1964) fold tightness classification. "*ILA*" = interlimb angle. [Reprinted from *Journal of Structural Geology*, v. 26, Srivastava, D. C., and Lisle, R. J., Rapid analysis of fold shape using Bezler curves, p. 1553–1559, © 2004, with permission from Elsevier.]

Classifying Folds on the Basis of Changes in Layer Thickness

Thickness Changes, Reflections of Distortion

Some **folded layers** maintain uniform thickness across the full profile view of the fold. Other layers show striking, systematic variations in layer thickness. Whether the thickness of a rock layer is modified during folding depends on the internal stresses it is forced to bear and the rock's strength to resist. Remarkably, the degree of distortion from one layer to the next is nearly always somehow perfectly regulated to assure a perfectly compatible fit among layers within the folded sequence. Sometimes noticeable gaps and overlaps testify to the difficulty in achieving a perfect fit. The achieving of **strain compatibility** from folded layer to folded layer is yet more of the magic of strain.

Concentric, Parallel Folds

Individual folded layers that are marked by uniform thickness were originally known as **concentric folds** (Figure 7.59) (Van Hise, 1896). The profile forms commonly are circular or elliptical. Surfaces that separate individual folded layers in an ideal concentric fold are perfectly parallel, like the rails of a curved train track at a bend in the line. Because of this distinctive geometric characteristic, and because the individual beds do not really have the same center of curvature, concentric folds are also known as **parallel folds**. Figure 7.60 presents concentric, parallel folds of strikingly different magnitudes, the first a broad anticlinal flexure at Navajo Mountain, southern Utah (Figure 7.60*A*), the second an outcrop-scale anticline in late Cretaceous limestone, Mt. Lykaion, Greece (Figure 7.60*B*).

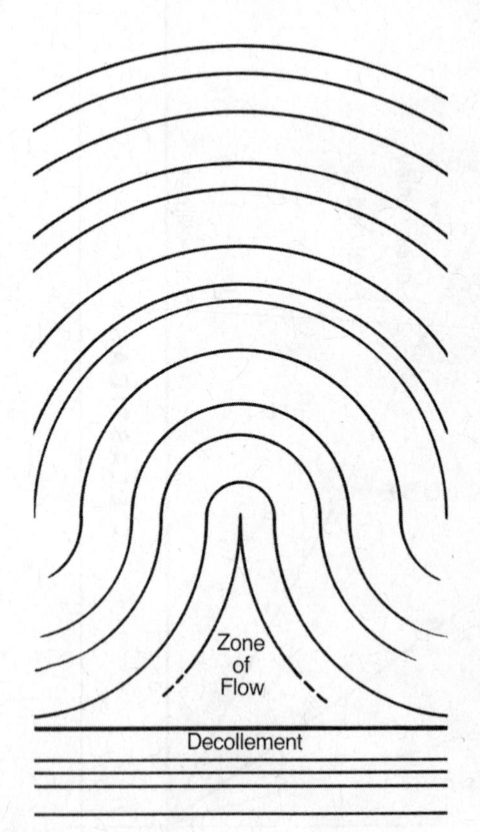

Figure 7.59 Geometric properties of an ideally concentric anticline.

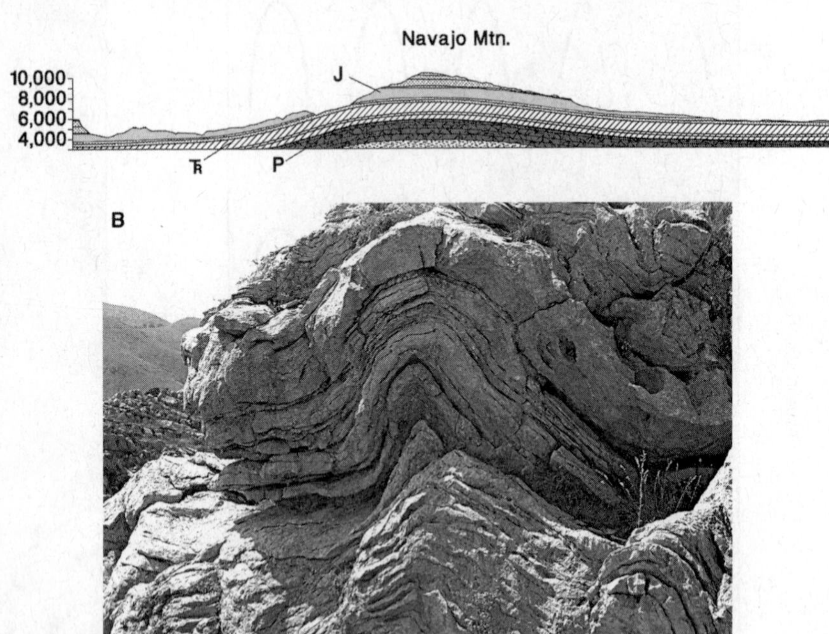

Figure 7.60 (*A*) At Navajo Mountain in southern Utah, Paleozoic and Mesozoic sedimentary rocks have been domed by a middle Tertiary intrusion. The fold is represented as perfectly concentric in this geologic cross-section by Baker (1936). [Courtesy of United States Geological Survey.] (*B*) This photograph is a close-up of a concentric anticline in late Cretaceous limestone on Mt. Lykaion, Greece. A signature of a concentric anticline is the way that the folds become tighter and tighter toward the core of the fold. [Photograph by G. H. Davis.]

An unexpected but very important, geometric peculiarity arises from parallel folding: the profile form of folded layers must continuously change upward and downward within the folded sequence, until the folds disappear altogether. That is, the folds tighten into their cores and finally detach from those above or below. As is evident in Figure 7.59, an upright anticline becomes progressively tighter downward within a concentrically folded sequence, ultimately transforming into a narrow, pinched, cuspate anticline before being cut off by a detachment fault and completely dying out. Upward, the concentric anticline progressively flattens into a very gentle arc before vanishing. Synclines behave in the opposite manner. Simply look at Figure 7.59 upside-down to see that an upright concentric syncline will pinch-out upwards in very tight cuspate folds. Downward it will gradually become a gentle dish-like fold before subtly merging with deeper, unfolded layers.

If we look carefully at a diagram of an upright anticline/syncline pair of parallel folds "*embedded*" in a uniformly shortened block (Figure 7.61), we can appreciate that each bed shortens by the same amount, and that at any given stratigraphic level the fold forms of the anticline and adjacent syncline are complementary. This is achieved in a way that the lowermost beds in the syncline take up more of the shortening than do the lowermost beds of the anticline, whereas the uppermost beds in the anticline take up more of the shortening than the uppermost beds of the syncline.

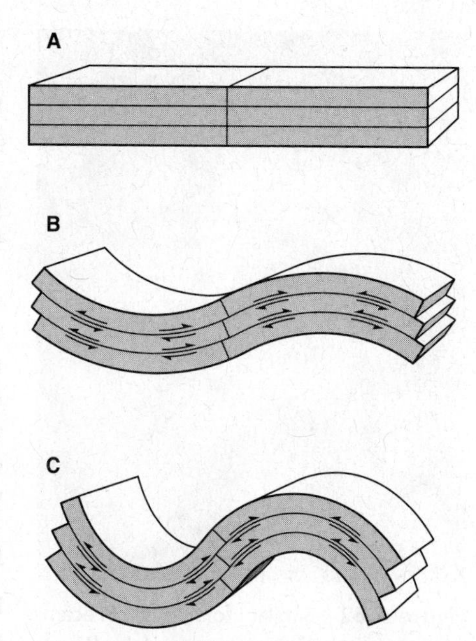

Figure 7.61 Nature of bedding-plane slip in concentric folding. Progressive concentric folding involves flexing of the beds and relative movement of the beds along bedding planes. (*A*) Original attitude of beds. Note the vertical reference line. (*B*) Flexing to create gentle to open folds. The original vertical reference line is now cut and displaced by bedding-plane slip. Sense of slip is such that beds move out of the synclinal hinge toward the anticlinal hinge. (*C*) Tight folding has now been achieved. We see that relative movement is zero at the anticline and synclinal hinge points and greatest at the inflection points. The thickness of the beds, measured at right angles to bedding, does not change as the result of the folding. At each stratigraphic level the bedding slip for the anticline and syncline are complementary.

The geometric idiosyncrasies of parallel folding can be more fully appreciated by graphically constructing a structure profile view of an upright, circular concentric anticline. Concentric circular arcs representing folded surfaces are drawn with a drafting compass (as in Figure 7.59). The circular arcs serve to define the boundaries of individual folded layers, which maintain uniform thickness *as measured perpendicular to layering*. As the arcs are drawn, one by one, it becomes more and more difficult to propagate the form of the anticline to depth. A space problem develops, and it becomes impossible to fit a decent circular arc into the available space. The space problem is satisfied by replacing the folded layers above with unfolded flat-lying layers below. In essence, the folded layers become **detached** from their underlying foundation. Nature achieves detachment through formation of a surface of "*unsticking*," a **decollement zone** of layer-parallel slippage and rock flowage. The last remaining vestige of the concentric anticline that can be constructed is a tiny cuspate fold (see Figures 7.59 and 7.60*B*). Between the cuspate anticline and the flat-lying strata below, a small amount of open space is created. In natural systems this open space would be filled by soft incompetent rock, capable of distortional flow during folding. Detachments are relatively easy to recognize because they produce *different structural geometries at different stratigraphic levels.* We have seen that before, e.g., Figure 7.53.

Detachment takes place in the cores of concentric folds because there is a space problem in the core of the fold and further shortening of the rock layers cannot be accommodated. As a result, the fold becomes detached from underlying strata (in the case of an upright anticline) and overlying strata (in the case of an upright syncline). The shortening accommodated by the folding is, on the other side of the detachment, transferred to some other location.

Similar Folds

Individual folded layers that display thickening in the hinge and thinning on the limbs, and nearly the same geometry from one layer to the next one are known, generally, as **similar folds** (Figure 7.62). Perfectly similar folds are

Figure 7.62 Similar folding in Precambrian banded gneisses exposed in Gjeroy Island, Nordland, Norway. [Reprinted from *Journal of Structural Geology*, v. 14, Lisle, R. J., Strain estimation from flattened buckle folds, p. 369–371, © 1992, with permission from Elsevier.]

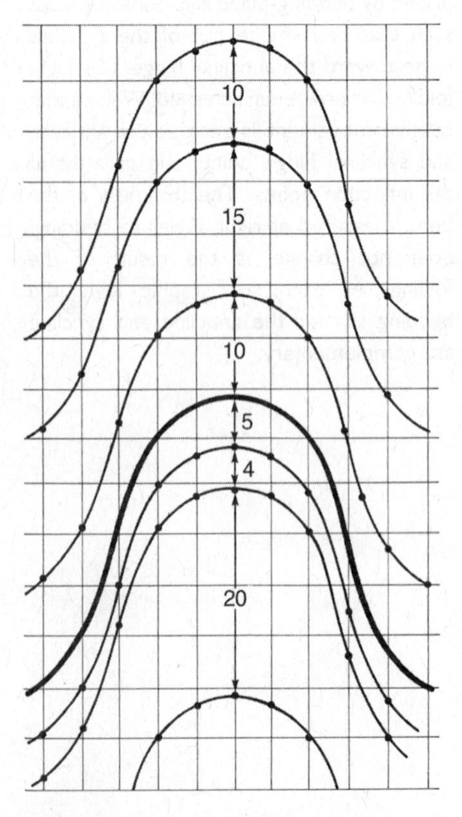

Figure 7.63 Geometric construction of an ideally similar anticline. See text for details.

marked by layers whose upper and lower surfaces are virtually identical in shape (Van Hise, 1896). Because this is so, the form of an ideally similar fold can be propagated upward and downward for any distance without change and requires no detachments. This sets them apart from parallel folds. The secret to the geometry of an ideal similar fold is that layer thickness, as *measured parallel to the axial trace of the fold,* remains constant for each layer or bed.

The geometric intrigue of similar folding can be appreciated through a graphical construction (Figure 7.63). The first step is to draw the profile form of a single folded surface in the middle of a long sheet of paper. Any upright fold form will do. The next step is to construct folded layers above and below this folded surface. Carefully build each folded layer by measuring and maintaining a constant thickness parallel to the axial trace. When this construction is carried out with care and precision, there is never a departure of individual folded surfaces from the starting profile form.

Full Range of Shape of Folded Layers

Parallel folds and similar folds are simply two special cases within a broad range of possible shapes of folded layers. Ramsay (1967, pp. 359–372) was able to demonstrate that fundamental classes of folded layers can be distinguished on the basis of **relative thickness** of the folded layer in the hinge versus the limbs. He showed that the **relative curvature** of the upper and lower bounding surfaces of an individual folded layer is also a sensitive index to systematic variations in layer thickness.

John Ramsay (1967) distinguished three main classes of folds by comparing the curvature of the **outer arc** of a given folded layer to the curvature of the **inner arc** of the same layer in the same fold (Figure 7.64). **Class 1 folds** are marked by a curvature of the inner arc that is greater than that of the outer arc. **Class 2 folds** are ideal similar folds, distinguished by identical curvatures of the inner and outer arcs. **Class 3 folds** are marked by curvature of the outer arc that is greater than that of the inner arc.

Ramsay further subdivided class 1 folds into three types on the basis of thickness variations (see Figure 7.64). **Class 1A folds** are marked by a layer thickness in the hinge that is less than layer thickness on the limbs. **Class 1B folds** are ideal concentric folds, distinguished by uniform layer thickness across the whole fold profile. **Class 1C** are intermediate between ideal concentric folds (class 1B) and ideal similar folds (class 2). They show a modest thickening in the hinge, and a modest thinning on the limbs.

Distinguishing Fundamental Fold Classes

The assignment of a given folded layer to one of the five fundamental fold classes can be carried out in a qualitative way on the basis of an eyeball estimate of relative curvature and relative thickness. But the power of Ramsay's approach is best appreciated through the actual measurement of relative curvature and relative thickness. Normal profile views of the folded layers are used as the database for carrying out the necessary constructions and measurements.

By convention, the fold profile under study is first rotated into the orientation of a perfectly upright antiform (see Figure 7.64). Next, **dip isogons** connecting points of equal inclination on the outer and inner bounding surfaces of the folded layer are constructed graphically. Once constructed, the dip isogon pattern sensitively reveals differences in outer arc and inner arc curvature, thus providing a basis for assigning the folded layer to class 1, 2, or 3.

As can be seen in Figure 7.64, Class 1 folds are distinguished by dip isogons that converge downward, signifying that the curvature of the outer arc

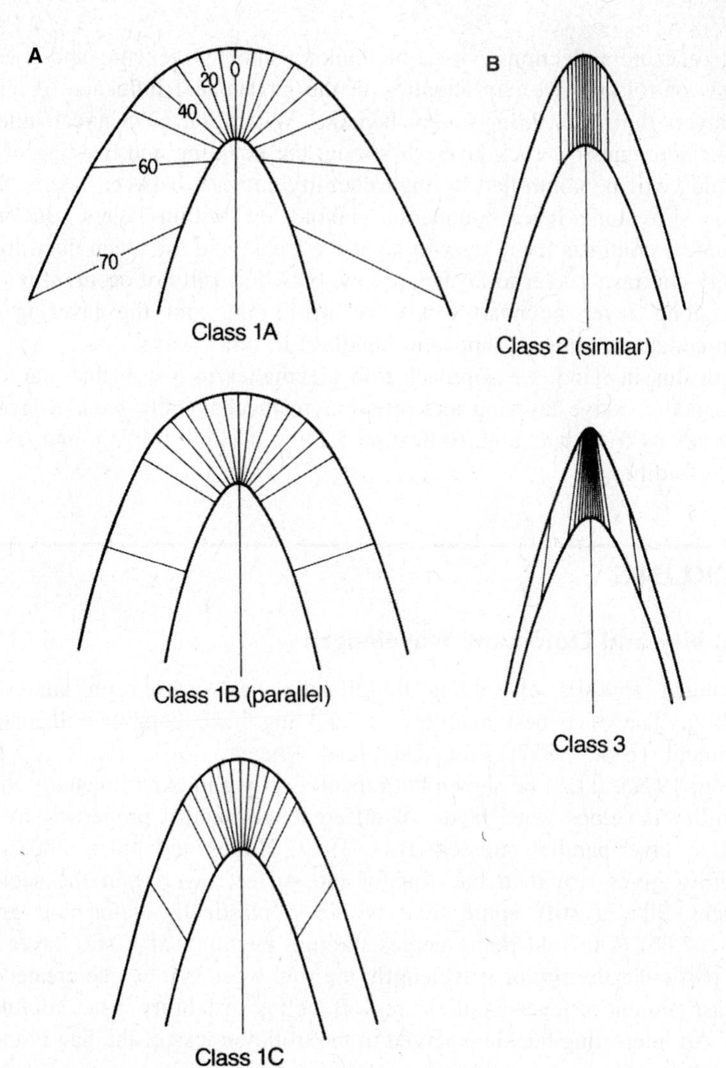

Figure 7.64 The fundamental classes of shapes of folded layers. See text for explanation. [From Folding and Fracturing of Rocks by J. G. Ramsay. Published with permission of McGraw-Hill Book Company, New York, copyright © 1967.]

is less than that of the inner arc. Dip isogons drawn for class 2 folds are strictly parallel, revealing that curvature of the outer arc matches exactly the curvature of the inner arc of the fold. Class 3 folds are marked by dip isogons that diverge downward, because outer arc curvature exceeds inner arc curvature.

Dip isogon patterns are especially revealing when they are drawn for a series of folded layers of different shapes (Figure 7.65). The divergence, convergence, and parallelism of dip isogons, as they cut through a folded sequence of layers, draws attention to the variety of classes of folded layers that can be represented in a single structure. Dip isogon diagrams of folds call attention to layer shape distortion as a function of rock type. An approach to using fold layer shape as a guide to strain is presented in *Part III-P, Carrying Out Fold Analysis.*

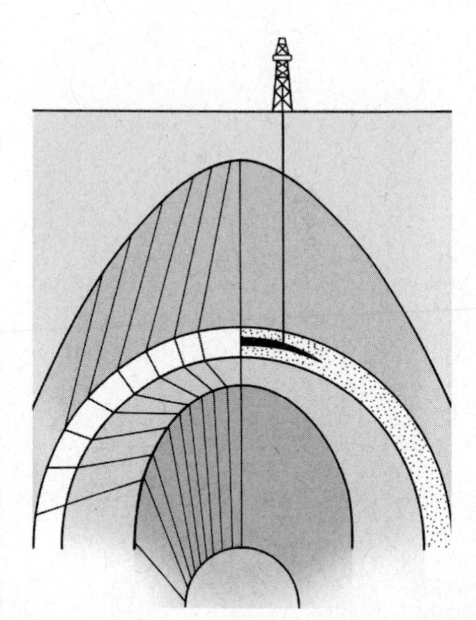

Figure 7.65 Schematic diagram showing how changes in inclination of dip isogons reflect changes in shape(s) of folded layers.

TRANSITION FROM GEOMETRY TO DYNAMICS

We have gained an appreciation of how to go about determining the orientations and shapes of folds based upon orientation data (bedding, layering) and the construction of maps and cross-sections. Furthermore, we have learned how to distinguish folds according to differences in layer shape (concentric, similar) based upon evaluation of thickness changes observed/measured from hinge zones to inflections. Now, in thinking about kinematic and dynamic analysis of folding, we pay attention to the mechanical influence, if any, of rock layers during buckling and/or bending. Where the mechanical influence and distinctiveness of rock layers is strong, the buckling and bending of beds into folds will be controlled by high ductility contrast between layers. There will be slip along layer boundaries and/or flow within layers. But where mechanical contrasts from layer to layer are weak, and the mean ductility and ductility contrasts (layer to layer) are low, buckling will not occur, slip and/or shear along layer boundaries will be negligible, and the layering must accommodate bending (or apparent bending) in other ways.

With this in mind, we approach fold kinematics in a way that starts with mechanically active layering and proceeds to mechanically passive layering. This takes us from buckling, to flexural folding, to kink folding, and lastly to passive folding.

BUCKLING

Instability and Dominant Wavelength

Mechanical analysis of folding traditionally has focused on analysis of **buckling.** The very best analyses of buckling have combined theory and experiment (Biot, 1957; Biot, Ode, and Roever, 1961; Ramberg, 1967; Johnson, 1977). It can be shown both theoretically and experimentally that an **instability** develops when layers of different mechanical properties are subjected to layer-parallel stresses (Biot, 1957; Hobbs and other, 2008). The instability gives rise to a buckling of the stiffest layer(s) in the sequence of rocks, like a stiff aplite dike within a plastically deforming granite (Figure 7.66). The fold that emerges through buckling of a stiff layer is of some particular **dominant wavelength**, the fold wave that can be created with the least amount of layer-parallel stress. Buckling instability is not confined to rocks. An interesting buckle emerged in the trolley tracks of the San Francisco streets during the great earthquake of 1906 (Figure 7.67).

Knowledge gained from mechanical analysis makes it possible to predict the dominant wavelength that will emerge when a single folded layer, or a multilayer sequence, is shortened. Predictions are based on hard-earned

Figure 7.66 Buckled, intestine-like, ptygmatically folded aplite dike, within granite. The relatively stiff aplite dike (white) was free to buckle through shortening within the rheologically much softer (plastic) granite during folding. The bedrock is Precambrian basement. The location lies within the Colorado Rockies, south of Breckenridge, Colorado, ~2.5 km east of the Mt. Bross summit. [Photograph by G. H. Davis.]

mathematical descriptions that relate dominant wavelength to the strength and thickness properties of the layers to be deformed.

A word of caution: the sites of specific folds may not relate so much to predictable dominant wavelengths as to unpredictable sites of flaws in the multilayer sequence. Willis (1894) recognized through experimental modeling that 1°–2° changes in the initial dip of sedimentary layers can predetermine the sites at which fold hinges will emerge (see Johnson, 1970). Such observations underscore one of the great contradictions that emerge from the mechanical analysis of folds: *buckling cannot occur in perfectly planar multilayers that are shortened by stresses that are perfectly layer parallel* (Biot, 1959; Jeng and others, 2002). Fortunately for fold enthusiasts, the smallest imperfections in the primary geometry of layering can trigger fold-forming processes.

Figure 7.67 Buckling of rails by compression on Howard Street (South Van Ness Avenue) near 17th Street, San Francisco. The buckling was caused by movements related to the earthquake of 1906. [Photograph by T. L. Youd. Courtesy of United States Geological Survey.]

Simple Buckling of a Single Layer, in Theory

The dependence of wavelength on layer thickness and strength is most simply expressed in equations that describe the buckling of a stiff layer embedded in a softer medium. Dominant wavelength depends not only on the thickness and strength of the stiff layer, but also on the strength of the weak, confining medium. Thickness is easy to deal with both mathematically and experimentally. But how is "*stiffness*" of a rock modeled quantitatively? How stiff, exactly, are those rocks shown in Figure 7.68?

As it turns out, the mathematical description of layer strength depends on whether the mechanical properties of the single stiff layer and its confining medium are viewed as elastic or viscous. If an elastic model of deformation is applied, the strengths of layers are described in terms of the fundamental elastic moduli: **Young's modulus** (E) and **Poisson's ratio** (ν) (Figure 7.69*A*). However, if a viscous model of deformation is used, the strengths of the layer and its confining medium are expressed in terms of **viscosity coefficients** (Figure 7.69*B*). The ratio of viscosities is thought to control the geometry of folds formed in volcanic rocks where all the layers are not yet cold enough to be solidified. In hybrid models where a stiff elastic layer is considered to be embedded in a soft, nonelastic confining medium, the strength of the stiff layer

Figure 7.68 The spheres are granite cannonballs, stacked next to a cannon that once guarded the entrance of Le Mont St. Michelle. Personally, if I had a choice, I would rather get hit by a Nerf cannonball having a much lower Young's modulus than that of granite. [Photograph by G. H. Davis.]

A

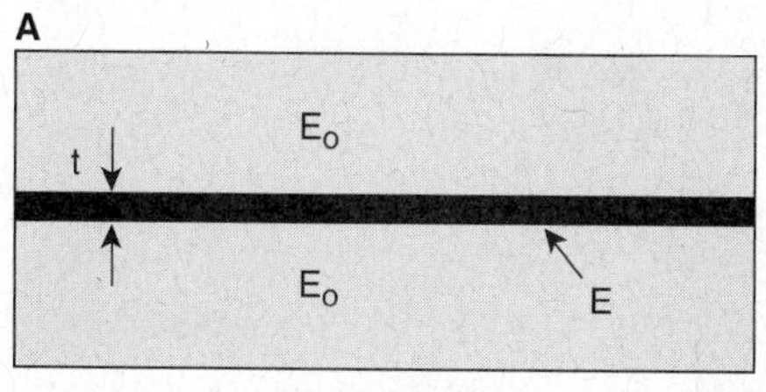

Elastic Model

B

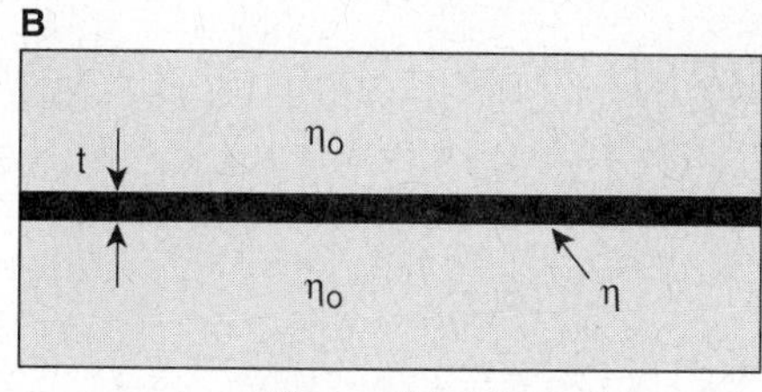

Viscous Model

Figure 7.69 The modeling of the mechanical properties of layers about to be deformed by layer-parallel shortening and buckling. (*A*) Elastic model. (*B*) Viscous model.

is described in terms of elastic moduli, and the strength of the confining medium is specified by a coefficient of viscosity. These are referred to by some as elasto-viscous layer-matrix models, i.e., a stiff elastic layer embedded within a softer, viscous matrix (Jeng et al., 2002).

Bijlaard (1946) modeled the mechanics of folding of a single layer in terms of a stiff elastic plate in a soft elastic medium. What he discovered was a surprisingly straightforward relationship:

$$L = 2\pi t\left(\frac{B}{6B_0}\right)^{1/3} \tag{7.1}$$

where, L = dominant wavelength
t = thickness of stiff layer
B = elastic modulus of stiff layer
B_0 = elastic modulus of confining medium

Elastic moduli B and B_0 are not mystery-variables pulled from the sky. Rather they express strength in terms of Young's modulus and Poisson's ratio:

$$B = \frac{E}{1-\nu_2} \tag{7.2}$$

where E is Young's modulus and ν is Poisson's ratio.

Currie, Patnode, and Trump (1962) reexamined the mechanics of folding of a stiff elastic layer in a soft, elastic confining medium. In doing so they chose to eliminate Poisson's ratio as a variable, arguing that the influence of this modulus on folding is small, especially considering the uncertainty in trying to describe precisely the value of ν for real rock layers at the time of folding. The resulting equation has the form of Equation 7.3.

$$L = 2\pi t\left(\frac{E}{6E_0}\right)^{1/3} \tag{7.3}$$

where, Young's moduli of the stiff layer and the confining medium are E and E_0, respectively.

Biot (1959) and Ramberg (1959) treated the folding of a single folded layer in the perspective of viscous deformation. Their independently derived mathematical analyses uncovered the same kind of relationship reported for the case of elastic deformation:

$$L = 2\pi t\left(\frac{\eta}{6\eta_0}\right)^{1/3} \tag{7.4}$$

where, L = dominant wavelength
t = thickness of stiff layer
η = coefficient of viscosity of stiff layer
η_0 = coefficient of viscosity of confining medium

Coefficients of viscosity are expressed in **poises**, the standard measure of resistance to flow of a viscous material.

Adding Layer-Parallel Internal Strain, in Theory

Just because the various buckling equations have the same form does not mean that they are the last word on the folding of single layers. Folds and folding continue to keep us humble. Sherwin and Chapple (1968), for example, demonstrated that the dominant fold wavelength that arises during layer-parallel shortening is responsive to the amount of layer-parallel strain absorbed by the layer before buckling. Thus in addition to strength and thickness, **layer-parallel strain** emerges as an important variable that must be taken into account.

Layer-parallel strain is specified in terms of a parameter that is familiar to us: stretch (S). Sherwin and Chapple found it necessary to describe stretch in two directions within the plane of layering, both parallel and perpendicular to the direction of layer-parallel shortening. Hudleston (1973) rewrote the Sherwin–Chapple equation in a form that can be directly compared with the Biot (1959) and Ramberg (1962) equations,

$$L = 2\pi t \left[\frac{\eta(s-1)}{6\eta_0(2s^2)} \right]^{1/3} \tag{7.5}$$

where, t = thickness of stiff layer
η = coefficient of viscosity of stiff layer
η_0 = coefficient of viscosity of confining medium
$s^2 = S^2{}_1/S^2{}_2$

As in all matters of science, closer and closer scrutiny of the Earth at work always seems to lead to a greater appreciation of the delicacy of dynamic process. Even for "*simple*" systems of geologic structures, dynamics of formation are found to be influenced by a much broader range of variables than we may originally perceive.

Buckle Folding of a Single Layer, in Practice

Part of the fun of mechanical analysis is testing equations to see if they really work. Biot did not wait for others to test the equation he derived for the folding of a single layer, viewed viscously (Equation 7.4). Instead, he himself teamed up with two colleagues to test his findings (Biot et al., 1961). Together the investigators set up a series of experiments that included the layer-parallel shortening of single layers of stiff pitch (i.e., tar), which they deformed in a confining medium of corn syrup (Figure 7.70). Layers of pitch of different thicknesses were fabricated in molds of different depths. Viscosities of both the pitch and the syrup were carefully measured before the start of the experiments, because, of course, they were interested in the effects of the contrast of viscosity on folding. Therefore, the folding of pitch in corn syrup could be used as an analogue to folding in rocks if the ratio of the viscosities is appropriate, because it would take too long to fold actual rocks.

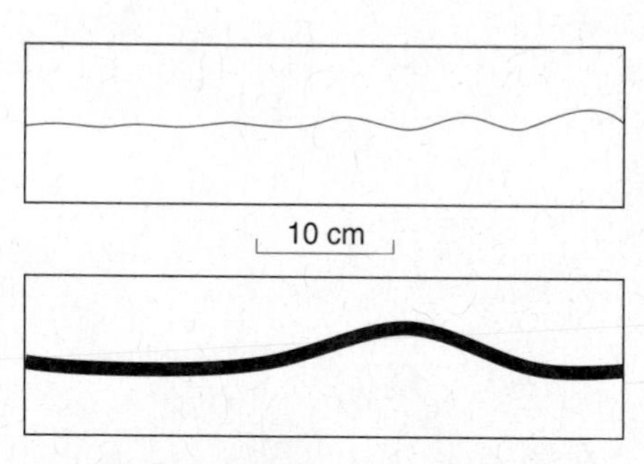

Figure 7.70 Layer-parallel shortening of pitch layers of different thicknesses in a medium of syrup. [From Biot, Ode, and Roever, 1961. Published with permission of Geological Society of America.]

On the basis of strength and thickness data (Table 7.1), Biot, Ode, and Roever calculated the dominant wavelengths predicted by Biot's equation.

TABLE 7.1
The testing of Biot's equation, $L = 2\pi t \sqrt[3]{\mu/6\mu_0}$

Thickness (tl) of Pitch Layer	*Viscosity (μ) of Pitch Layer*	*Viscosity (μ_0) of Corn Syrup*	*Predicted Fold Wavelength (L_p)*	*Observed Fold Wavelength (L_o)*
0.35 cm	3×10^7 poise	1.35×10^4 poise	15.78 cm	12.4–18.0 cm
0.87 cm	3×10^7 poise	1.35×10^4 poise	39.24 cm	34.0–41.0 cm
1.08 cm	3×10^7 poise	1.35×10^4 poise	48.71 cm	38.0–52.0 cm

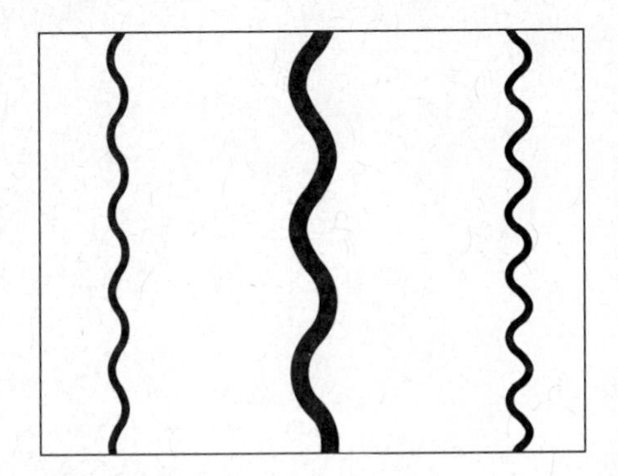

Figure 7.71 Layer-parallel shortening of gum rubber strips in a medium of gelatin. The outside strips are 4 mm thick. The middle strip is 8 mm thick. [From Currie, Patnode, and Trump, 1962. Published with permission of Geological Society of America.]

Then they subjected each of the three pitch layers to layer-parallel shortening and measured the range of wavelengths of folds that emerged in each buckled layer. Experimental results were found to be quite consistent with the predictions of theory!

As part of their research, Currie, Patnode, and Trump (1962) experimentally tested their equation for folding of a single elastic plate in an elastic medium (Equation 7.3). They found it practical to deform thin gum rubber strips of known thickness within a medium of gelatin (Figure 7.71). The gum rubber used for their experiments yielded, on testing, a Young's modulus E of −69 kPa. The gelatin in which the gum rubber layers were embedded was mixed from scratch, in such a way that the Young's modulus E_0 for each gelatin specimen could be predetermined, within a range of −6.9 kPa to −69 kPa. I (CFK) have always been impressed with the ingenuity of this study: embedding rubber strips into clear gelatin, squeezing the mass, and passing polarized light through the model to evaluate stress/strain, . . . ultimately leading to insights about deformation in rocks. Yet today the combination of elegantly scaled physical analogue modeling and numerical modeling delivers astonishingly effective results, and quite efficiently. For example, Figure 7.72 shows the power of **finite element modeling** techniques in addressing folding by buckling. Cruikshank and Johnson (1993) applied fundamental buckling equations to virtual materials with stiffness contrasts prescribed in terms of viscosity contrast. Figure 7.72*A* holds viscosity contrast constant (at 50:1), thus showing the influence of increasing degrees of layer-parallel shortening on fold form. Figure 7.72*B* holds the degree of layer-parallel shortening constant (40%), thus showing the influence of viscosity contrast on fold form.

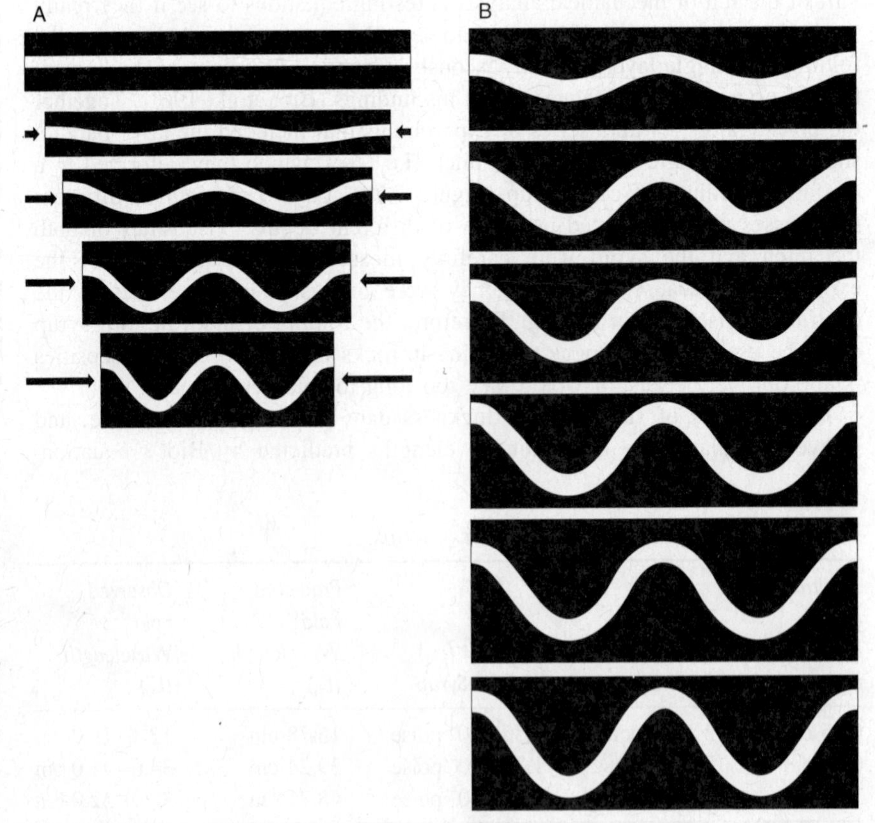

Figure 7.72 Computer simulations of buckling carried out by Cruikshank and Johnson (1993). (*A*) A stiff layer (white), 50 times stiffer than the surrounding medium (black), is progressively shortened. There is an equal amount of shortening between each step. (*B*) In each of these six experiments, the stiff layer (white) is shortened by 40% ($S = 0.6$). Differences in profile form are due to the viscosity ratio between the stiff layer (white) and its surrounding medium. The viscosity ratios, from the top experiment to the bottom, were 10, 20, 30, 40, 50, and 100. [Reprinted *Journal of Structural Geology*, v. 15, Cruikshank, K. M. and Johnson, A. M., High amplitude folding of linear-viscous multilayers, p. 79–94, © 1993, with permission from Elsevier.]

Influence of Competency Contrast on Fold Form

Ramsay and Huber (1987) did a beautiful job summarizing the influence of **competency contrast** (between a stiff layer sandwiched between soft layers) on fold form (Figure 7.73). Where the competent layer is much stiffer than the surrounding materials, the **amplification rate** of buckling is very high, and the competent layer deflects robustly into the material above and below. Large wavelength, rounded forms are produced, the best example of which is the **ptygmatic fold** (see uppermost folded layer, Figure 7.73). On the other hand, where competency contrast is low, the amplification rate of buckling is very small. As a result, the folds that are created are of short wavelength, and the typical forms are **cuspate–lobate folds** (see lowermost folded layer, Figure 7.73). The deflection of less competent (i.e., softer) rock into more competent (i.e., stiffer) rock produces pointed, cuspate fold forms. The cusps point into the stiffer rock. Thus in outcrops dominated by cuspate–lobate fold forms, it is possible to know at a glance whether at the time of folding a given layer was relatively stiff or relative soft, compared to beds on either side (see lowermost folded layer, Figure 7.73).

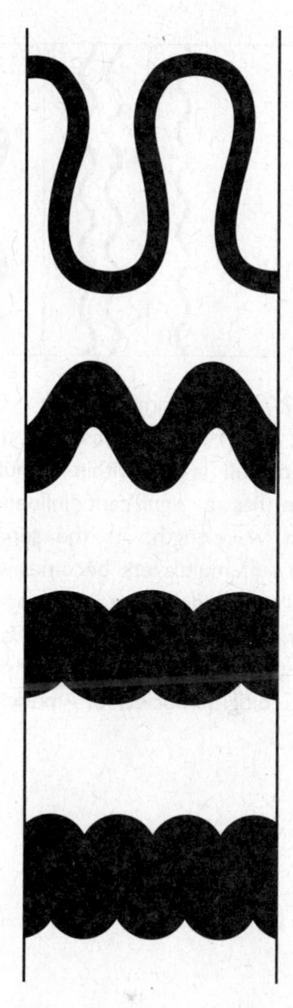

Figure 7.73 The shapes of buckle folds reflect the competency (stiffness) contrast between the stiffer layer and the less competent medium that it occupies. High contrast leads to ptygmatic folds. Low contrast leads to cuspate–lobate folds. Medium contrast creates fold forms that are intermediate between ptygmatic and cuspate–lobate. [Reprinted from Ramsay and Huber, The techniques of modern structural geology, v. 2: Folds and Fractures, © 1987, with permission from Elsevier.]

Influence of Strain Rate on Fold Form

Strain rate makes a difference in the progressive evolution of buckle folds. This was demonstrated by Jeng and others (2002) who applied numerical modeling to a hypothetical sequence of strata marked by elasto-viscous layer-matrix properties (Figure 7.74). They determined that under conditions of fast strain rate (10^{-11}sec^{-1}), the response of a stiff elastic layer embedded within viscous materials will respond elastically (see Figure 7.74, top left and top right). When they increased the rate of loading to 10^{-10}sec^{-1}, buckling continued and the deformation was dominated by a typical elastic response (see Figure 7.74, middle left and lower left), of the type we have described. However, when they slowed the strain rate to (10^{-14}sec^{-1}), the matrix viscosity reduced the input energy as well as the magnitude of elastic energy stored in the stiff layer. As a result, buckling was dampened, and the earlier formed folds simply amplified (see Figure 7.74, middle right and lower right). What is fast and what is slow in the real world? Jeng and others (2002) note that the mean crustal strain rate in plate tectonic zones of convergence is on the order of 10^{-14}sec^{-1}. This is what they refer to as slow. An interim shift to 10^{-10}sec^{-1} would be a profound shift to a fast rate.

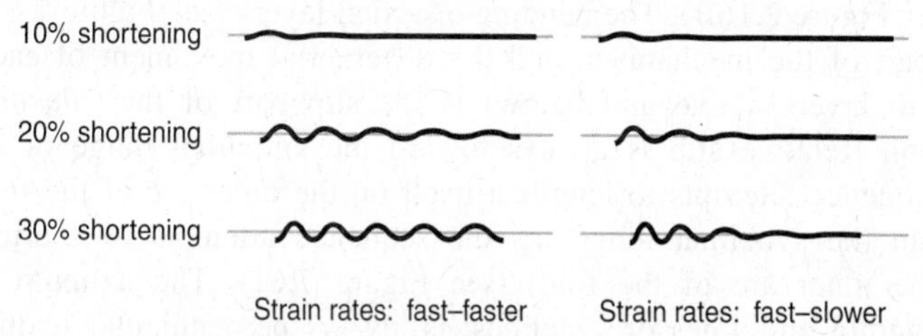

Figure 7.74 Line drawing showing the effect of strain rate on buckling of an elasto-viscous sandwich. The black layer is a stiff, competent, elastic layer, which is embedded in a viscous matrix. The gray layer is the original undeformed length and orientation. Numerical modeling shows three stages of shortening, namely 10%, 20%, and 30%. The left-hand set of diagrams shows the nature of folding when strain rate begins fast (10^{-11}sec^{-1}), but then at 15% shortening gets faster (10^{-10}sec^{-1}). The right-hand set of diagrams shows the nature of folding when strain rate begins fast (10^{-11}sec^{-1}), but then at 15% shortening slows down (10^{-14}sec^{-1}). See text for explanation. [Reprinted from *Journal of Structural Geology*, v. 24, Jeng, F. S., Lin, M. L., Lai, Y. C., and Teng, M. H., Influence of strain rate on buckle folding of an elasto-viscous single layer, p. 501–516, © 2002, with permission from Elsevier.]

Figure 7.75 Experiments by Currie, Patnode, and Trump demonstrated that the spacing of stiff layers within a multilayer sequence has a significant influence on dominant wavelength. As the separation between stiff multilayers becomes smaller and smaller, the dominant wavelength gets bigger and bigger. [From Currie, Patnode, and Trump, 1962. Published with permission of Geological Society of America.]

Buckle Folding of Multilayers

Complicated mathematical expressions are required to describe the behavior of **multilayer sequences** containing layers of widely different strength and thickness. The mathematical expressions must include variables above and beyond those already mentioned, notably the spacing of stiff layers within the sequence and the degree of cohesive strength between layers within the sequence.

The ratio of dominant wavelength to thickness of a folded stiff layer is greatly reduced when the layer belongs to and is analyzed as part of a multilayer sequence (Bijlaard, 1946; Johnson, 1977). Gum rubber and gelatin experiments carried out by Currie et al. (1962) reveal this quite clearly (Figure 7.75). Widely separated gum rubber strips display short-wavelength fold waves, but the dominant wavelength steadily increases as the gum rubber strips are brought into closer and closer contact.

One of the mechanical idiosyncrasies of layer-parallel shortening of multilayers is that thinner layers in the sequence may buckle into short-wavelength folds before the folding of the entire sequence (Ramberg, 1963). Such "*minor*" folding is an expression of the layer-parallel strain that constitutes the preliminary step in the formation of most drag folds.

FLEXURAL FOLDING

Flexural flow versus Flexural Slip

Donath and Parker (1964) recognized two fundamental mechanisms of folding: flexural folding and passive folding. **Flexural folding** takes place when the mechanical influence of layering in a rock is very strong. The layers actively participate in the folding by bending and flexing. Flexural folding can take place by flexural slip, by flexural flow, or by a combination of these. Depending on the mechanical properties of the layered sequence, one or both of these mechanisms are initiated when layer-parallel resistance to shortening is overcome and the layers of rock begin to actively buckle. **Flexural-slip folding** accommodates the buckling by layer-parallel slip along contacts between layers (Figure 7.76*A*). The layers slip like pages in a slick magazine when the magazine is rolled up (Figure 7.76*B*), or when a paperback book is flexed (see Figure 2.16B). The bending of actual layers (see Figure 7.61) is the flexural part of the mechanism, and the differential movement of each layer (relative to layers above and below) is the slip part of the "*flexural slip*" mechanism. Relative slip is always toward the anticlinal hinge of the fold (as the sequence attempts to lengthen itself on the outer arc of the fold), and away from the synclinal hinge (as the sequence attempts to shorten itself around the inner arc of the fold) (see Figure 7.61). The azimuth of slip, whether within anticlines or synclines, is always perpendicular to the hinge line of the folding that is generating these movements.

Flexural-flow folding accommodates the bending by layer-parallel flow or shear within mechanically soft units sandwiched between stiff units (Figure 7.77*A*). The flow of the rock is a little like the flow of ice cream when squeezed between the stiff, competent cookie layers on top and bottom of an ice-cream sandwich (Figure 7.77*B*). This analogy is especially apt if the ice cream rheology is soft. Such ice cream, under conditions of opposable finger/thumb-loading, will flow in a way that is "*channeled*" by the confining cookie layers, providing that you do not press so hard that the cookie layers fracture or fault, in which case there will be intrusion and/or eruption.

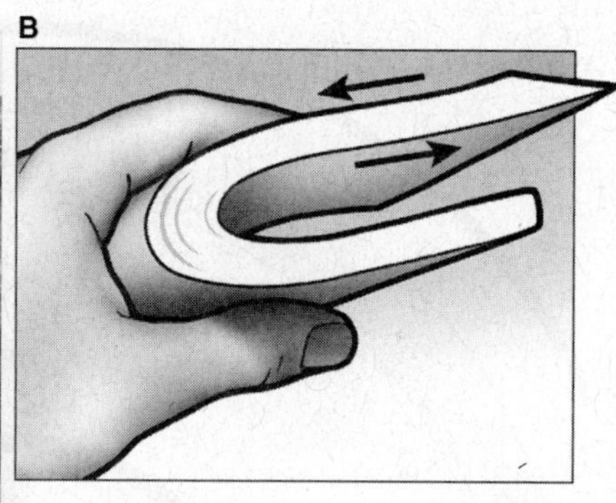

Figure 7.76 (*A*) Flexural-slip folding manifest in bedding in the Mazatzal Quartzite (late Paleoproterozoic), with its interbeds of quartzite and schist. Quartzite bed thicknesses have *not* been altered by the flexural slip. Location is Barnhardt Canyon, Mazatzal Mountains, Arizona. [Photograph by S. J. Reynolds.] (*B*) Differential slip of the pages of a slick magazine as the magazine is rolled up. [Artwork by D. A. Fischer.]

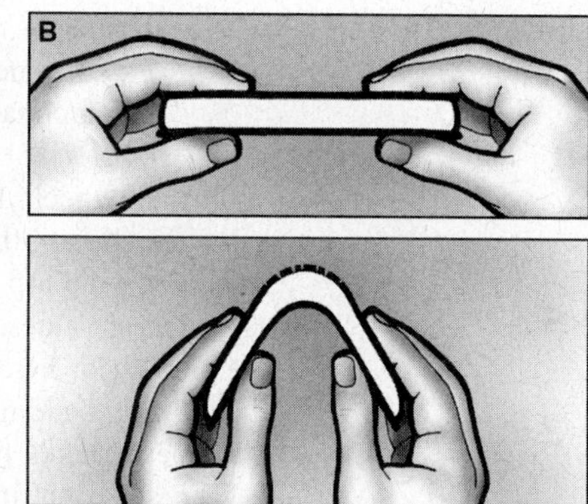

Figure 7.77 (*A*) Flexural-flow fold in alternating marble (gray) and calc-silicate layers (white) in metamorphosed Pennsylvanian rock in Happy Valley, southeastern Arizona. Each calc silicate layer tends to retain uniform thickness from limb to hinge, in stark contrast to the marble beds that, almost like ice cream, thicken radically in the hinge zones and pinch to nearly nothing on the limbs. India-ink tracing of G. H. Davis photograph by David O'Day. (*B*) Cartoon of flexural flow of an ice cream sandwich. [Artwork by D. A. Fischer.]

Flexural-Slip Kinematics

Flexural-slip displacements between layers (or pages) are tiny when viewed individually, but the sum of the displacements is always enough to accommodate a true bending of a rock body (or book). The actual amount of slippage along the top of any layer is easy to calculate (Ramsay, 1967, pp. 392–393). As in the analysis of layer shape, the fold form to be analyzed is rotated into the orientation of a perfectly upright antiform (Figure 7.78). Then the locations where slip is to be calculated are specified by the inclination values (α) of the top of the folded layer at the chosen sites (see Figure 7.78). Slip is then determined using the following formula:

$$s = t\alpha \quad (7.6)$$

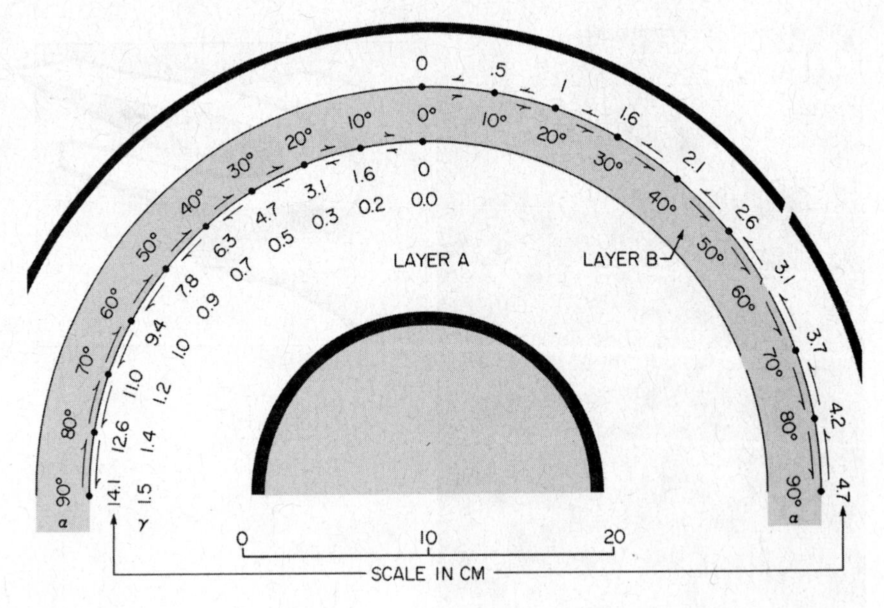

Figure 7.78 The amount of slip between layers of a flexural-slip fold depends on layer thickness and limb inclination. Shear strain depends on limb inclination alone. See text for explanation.

where, s = slip
t = thickness of the folded layer
α = inclination in radians ($1° = 0.0175$ radian)

For the mock fold shown in Figure 7.78, we can use Equation 7.6 to calculate slip at 10 sites on the top of layer *A* (thickness = 9 cm) and 10 more sites along the top of layer *B* (thickness = 3 cm). The sites are located at 10° dip interval values, from $\alpha = 0°$ to $\alpha = 90°$. The calculations demonstrate that the amount of layer-parallel slip increases both with layer thickness and with dip of the bedding. In fact, the calculations show that no interlayer slip whatsoever takes place at the actual hinge point of a folded layer.

The shear strain (γ) due to interbed slip can be calculated too (Ramsay, 1967, p. 393),

$$\gamma = \alpha \tag{7.7}$$

where, α is the inclination in radians.

Figure 7.78 shows the results of calculations of shear strain for layer *A*. The distribution of values of shear strain reveal that shear strain due to flexural-slip folding is greatest at the inflection point of a fold but is negligible at the hinge.

Donath and Parker (1964) emphasized that layered sequences that readily fold by flexural slip are marked by strong, stiff layers, the contacts of which are marked by **low cohesive strength**. Thin- to medium-bedded sandstone, siltstone, and limestone sequences are especially susceptible to flexural slip. Individual layers that are folded by the flexural-slip mechanism tend to retain their primary, original thicknesses, in the same way that the pages of a book or magazine neither thicken nor thin when flexed. Thus layer shape of flexural-slip folds tends to be class 1B; that is, parallel folds. As flexural slip takes place in the real world, we find that there is initiation of new (bedding-parallel) movement horizons as well as continued slip on existing bedding surfaces and discontinuities; and that such slip is aided commonly by fluid pressure (Horne and Culshaw, 2001).

Layer Strain Associated with Flexural-Slip Folding

Even though individual layers tend to retain their original thickness during flexural-slip folding, they nonetheless generally enjoy some internal distortion. The distortion takes place mainly in the hinge zone of the folded layer, where curvature is greatest. When an individual layer is actively buckled, rock on the outer arc of the hinge undergoes **layer-parallel stretching**, and rock on the inner arc of the hinge experiences **layer-parallel shortening** (Kuenen and DeSitter, 1938) (Figure 7.79). Layer-parallel strain decreases toward the middle of each folded layer, toward the **neutral surface** of no strain (see Figure 7.79). The neutral surface separates an outer arc domain of layer-parallel stretching from an inner arc domain of layer-parallel shortening. The location of the neutral surface depends on the physical properties of the rocks and the ability of the beds to slip past one another. When the units can slip past one another easily, the beds slip from the syncline towards the anticline easily, and thus the ability of the sequence to lengthen around the outer arc of the fold is enhanced. In this case, the neutral surface is positioned relatively high in a given competent layer in the hinge of the fold because less internal strain, in the form of extension, is required to accomplish the bending. Just the opposite is the case where the bedding plane slip is not very efficient. In such cases beds cannot slip as easily out of the syncline toward the anticline, and thus more internal deformation, in the form of extensional faulting and fracturing, is required. When thinning of outer arc rocks by layer-parallel stretching is perfectly compensated by thickening of inner arc rocks by layer-parallel shortening, the folded layer retains a class 1B form.

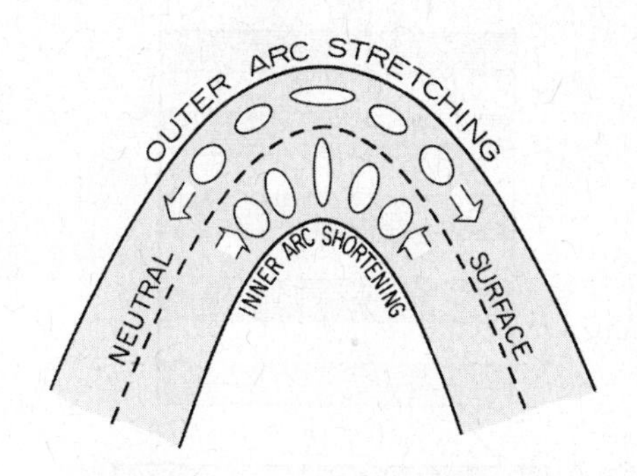

Figure 7.79 Layer-parallel stretching and layer-parallel shortening associated with folding. The layer-parallel stretching occurs within the outer arc of the fold, and the layer-parallel shortening occurs within the inner arc.

Layer-parallel stretching on the outer arc of a folded layer can be accommodated in a number of ways, depending on the strength of the layer (Figure 7.80). Stiff layers respond to the stretching by the formation of tension fractures and normal-slip faults. Tension fractures, including veins, form perpendicular to the direction of layer-parallel stretching. Conjugate normal-slip faults form in such a way that their line of intersection is parallel to the axis of folding. Such inward-dipping conjugate normal faults may define the boundaries of a graben, which are classic expressions of stretching on the outer arc of a folded layer (see Figure 7.80). We sometimes refer to well developed grabens as **keystone grabens**, for they closely resemble the wedge-shaped stone pieces that stone masons place at the apex of an arch (Figure 7.81). The "*keystone*" is the final piece to be put in place, and permits a mortarless arch to bear weight.

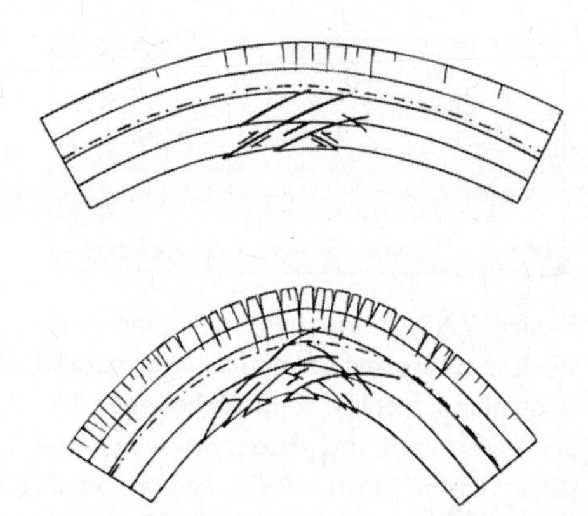

Figure 7.80 Minor structures associated with layer-parallel stretching in the outer arc of a folded layer, and layer-parallel shortening in the inner arc. [From Folding and Fracturing of Rocks by J. G. Ramsay. Published with permission of McGraw-Hill Book Company, New York, copyright © 1967.]

Figure 7.81 The contacts between the blocks comprising this mortarless arch resemble the normal faulting and tension fracturing that results from bending on the outer arc of an anticline. The uppermost block on the apex of this arch is the keystone, which from a dynamic standpoint holds the arch in place. This is a remnant of the ancient tunnel through which athletes entered the stadium in the ancient Olympic games. Spectator is Merrily Davis. [Photograph by G. H. Davis.]

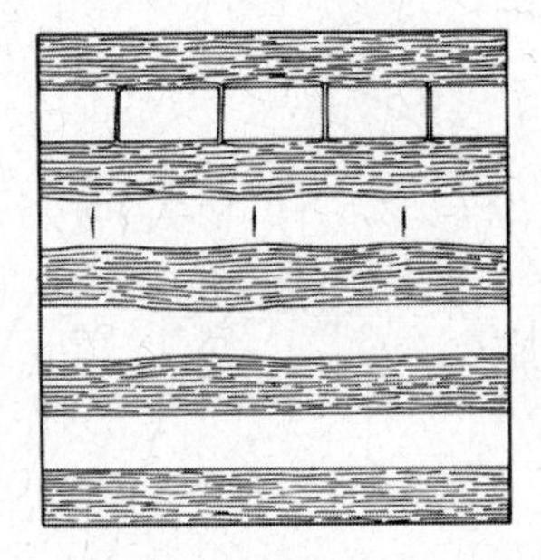

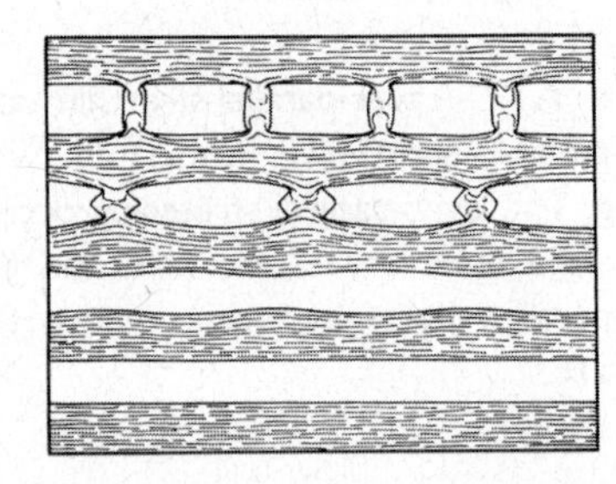

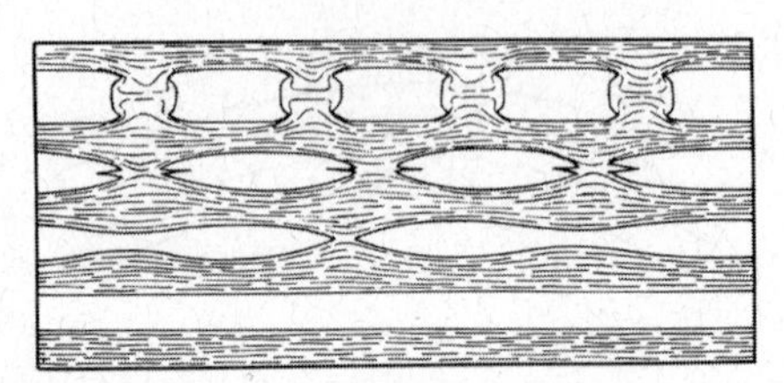

Figure 7.82 Formation of pinch-and-swell structure and boudins by layer-parallel stretching. Ductility contrast between layers determines the extent to which the stiffer layers pinch, neck, and/or break. [From Folding and Fracturing of Rocks by J. G. Ramsay. Published with permission of McGraw-Hill Book Company, New York, copyright © 1967.]

Figure 7.83 Boudins in a stretched calc-silicate layer (white) covered top and bottom by marble (gray). The boudins are bounded by tiny faults. Movement on the faults has permitted the layer to stretch its length. Happy Valley area, southeastern Arizona. The host rock was originally Pennsylvanian-Permian sedimentary rock of the Earp Formation. Happy Valley area, southeastern Arizona. [Photograph by E. G. Frost.]

If the layering in the outer arc of a fold is an interbedded composite of soft and stiff layers, stretching is commonly achieved by boudinage and pinch-and-swell structure (Figure 7.82). This strain mechanism is particularly common in very tightly compressed flexural-flow folds. **Boudins** (**Fr**, sausages) form in sequences of alternating soft and stiff layers that have been subjected to flattening and extension. Stiffer layers tend to break or neck, and the softer layers tend to flow and fill in, wherever required (Figure 7.83). The isolated tight recumbent fold shown in Figure 7.84 is a dramatic example of boudins that formed preferentially in a stiff layer on the outer arc of a fold. (We will learn more about boudinage in *Chapter 9, Foliation and Lineation*).

Layer-parallel shortening on the inner arc of a folded layer gives rise to symmetrical minor folds, conjugate thrust faults, pressure dissolution (including spaced cleavage), and more intense cleavage development (e.g., slaty cleavage)

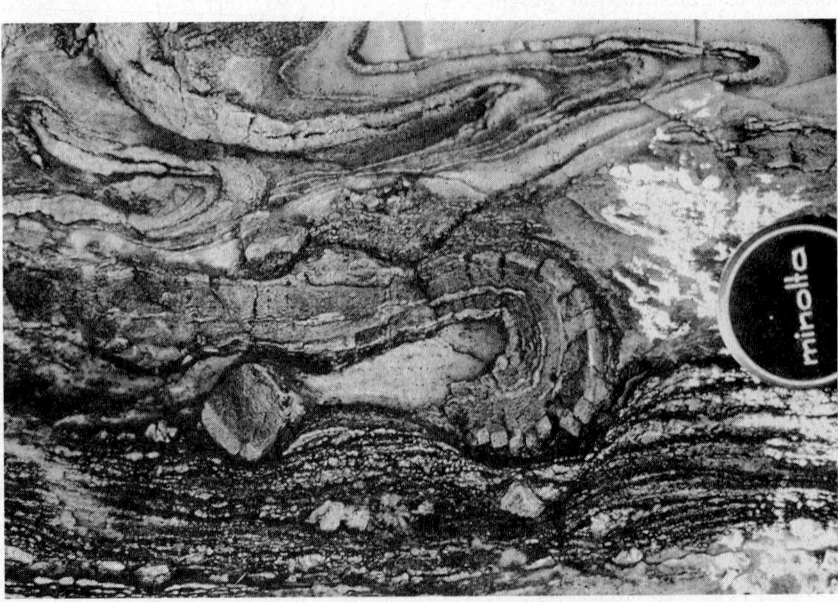

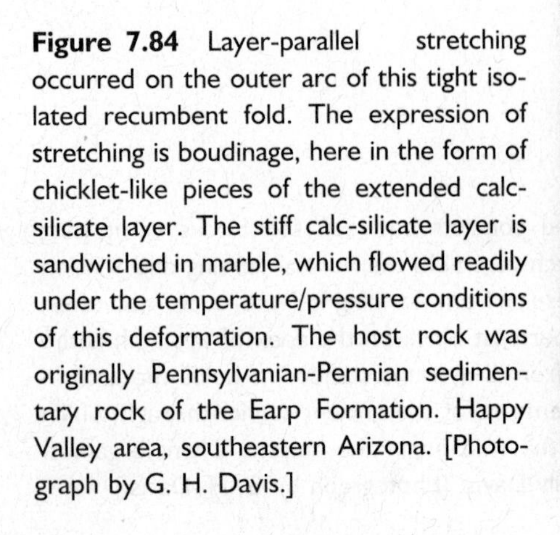

Figure 7.84 Layer-parallel stretching occurred on the outer arc of this tight isolated recumbent fold. The expression of stretching is boudinage, here in the form of chicklet-like pieces of the extended calc-silicate layer. The stiff calc-silicate layer is sandwiched in marble, which flowed readily under the temperature/pressure conditions of this deformation. The host rock was originally Pennsylvanian-Permian sedimentary rock of the Earp Formation. Happy Valley area, southeastern Arizona. [Photograph by G. H. Davis.]

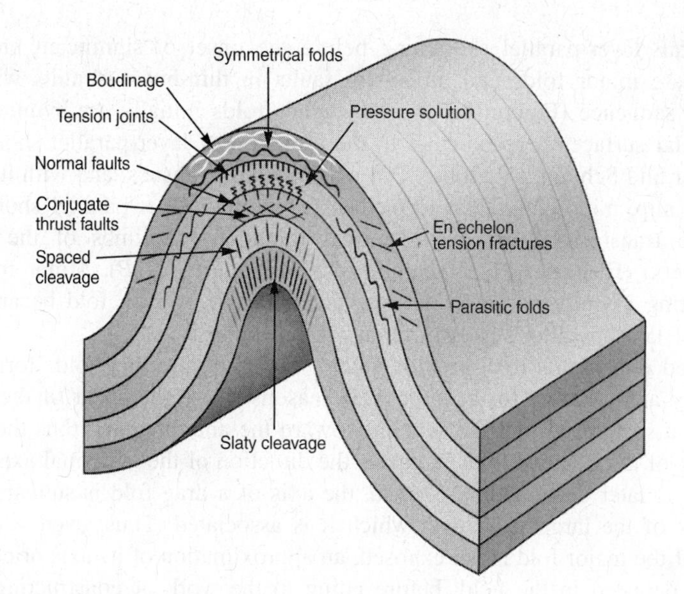

Figure 7.85 Schematic profile view of folded layers, with portrayal of the kinds of minor structures that develop on the outer arc, inner arc, and limbs of folds. The layers depicted here vary in the rheological properties, and thus see a variety of fold profiles, ranging from parallel to similar.

(Figure 7.85). These structures work together to accommodate the room problem created when the inner arc of a layer closes in on itself. The minor symmetrical folds are coaxial with the axis of the major fold. Conjugate thrust-slip faults intersect in a line parallel to the axis of folding. The cleavage that forms as a response to layer-parallel shortening strikes parallel to the axial surface of the major fold, with dip distributions generally symmetrically disposed about the axial surface (see Figure 7.85). **Layer-parallel stretching** on the outer arc of a folded layer gives rise to tension fracturing, boudinage, and normal faulting (see Figure 7.85). **Tangential longitudinal strain** on the limbs of folds gives rise to parasitic folds and en echelon tension fractures (see Figure 7.85).

Taken together, the minor structures that occur within folded layers create a marvelous addition to the architecture of deformed layered rocks (Figure 7.86).

Figure 7.86 We believe that the architects and masons who designed and fashioned this elegant entry to the Durham Cathedral (Durham, England) would have been very interested in learning about minor structures in folded layers. [Photograph by G. H. Davis.]

Additional Minor Structures Created during Flexural-Slip Folding

As we are learning, flexural-slip folding creates an informative array of minor structures. The minor structures reflect a combination of four complementary mechanisms of deformation: overall layer-parallel shortening, layer-parallel slip on the fold limbs, layer-parallel stretching on the outer arc of the hinge zone, and layer-parallel shortening on the inner arc of the hinge.

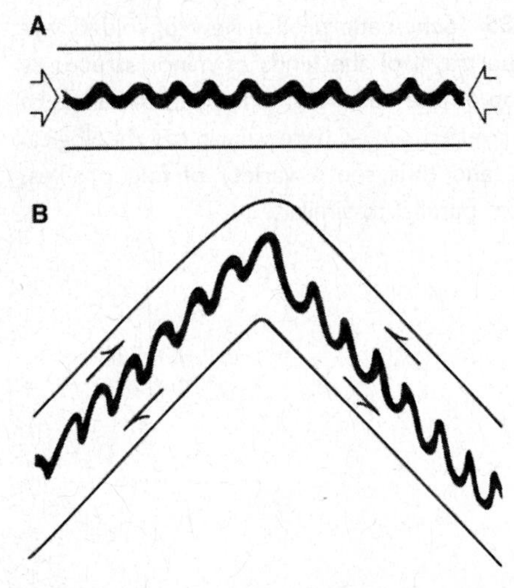

Figure 7.87 The generation of drag folds. (*A*) Layer-parallel shortening before buckling creates an array of upright, symmetrical anticlines and synclines. (*B*) Buckling and the onset of flexural-slip folding transform the symmetrical folds into asymmetrical folds.

Overall layer-parallel shortening before the onset of significant buckling can create minor folds and thrust-slip faults in thin-but-stiff units within a layered sequence (Figure 7.87*A*). The minor folds initially are symmetrical, with axial surfaces perpendicular to the direction of layer-parallel shortening (Frehner and Schmalholz, 2006). But when buckling ensues, and with it layer-parallel slip, minor folds that formed during overall layer-parallel shortening may be transformed into asymmetrical folds on the limbs of the major structure(s) (Frehner and Schmalholz, 2006) (Figure 7.87*B*). Minor folds of contrasting asymmetry mark the opposite limbs of a given fold because the sense of layer-parallel slip is different on opposing limbs.

Called **drag folds** (or **parasitic folds**), asymmetric minor folds formed in this way are valuable for at least three reasons. *First,* the shear of the upper beds in a section is from the syncline toward the anticline, and thus the sense of shear of the *S* and *Z* folds indicates the direction of the anticlinal axis, even if there is later deformation. *Second,* the axis of a drag fold is subparallel to the axis of the larger fold with which it is associated. Thus, even when the hinge of the major fold is not exposed, an approximation of its axis orientation can be deduced in the field, before going to the work of constructing a π-, β-diagram, or polar tangent plot. *Third,* in terranes characterized by isoclinal and poorly exposed folds, the location of the axial trace of a major fold can be identified on the basis of the shift in minor-fold asymmetry from *Z* to *S,* or *S* to *Z* (see Figure 7.87).

The use of drag folds to interpret fold patterns can be pictured more easily than described in words. Figure 7.88*A* shows the predicament. The axial trace of a major isoclinal fold is discovered and mapped on the basis of asymmetry of drag folds. What remains uncertain is whether the fold is an antiform or a synform. When an antiformal fold form is fitted to the configuration of the axial trace and the fold limbs (Figure 7.88*B*), the expected sense of layer-parallel slip on each limb of the antiform is contradicted by the sense of asymmetry of the drag folds. On the other hand, when a synformal fold form is fitted to the axial trace–fold limb configuration, the observed drag fold pattern is wholly consistent with the expected sense of layer-parallel slip (Figure 7.88*C*).

Hinge-Parallel Extension During Folding

We may recall from *Chapter 5, Joints,* that jointing commonly appears to be closely coordinated with folding and/or the far-field tectonic loading that drove the folding. We see this at the outcrop scale in Figure 7.89, a photograph of an exposure of Ordovician metasedimentary rocks along the Bay of Fundy near St. John, New Brunswick, Canada. From a distance (Figure 7.89*A*) we see that steeply dipping systematic joints display evenness of spacing and perfect parallel alignment. Up close (Figure 7.89*B*) we recognize that the rocks are penetratively folded, and that the joint faces are oriented perfectly perpendicular to the fold hinges. These joints would be considered to be **cross joints**. They formed during folding as a result of fold-hinge-parallel extension. Apparently the strain environment was truly three-dimensional, with most of the shortening accommodated by layer-parallel shortening

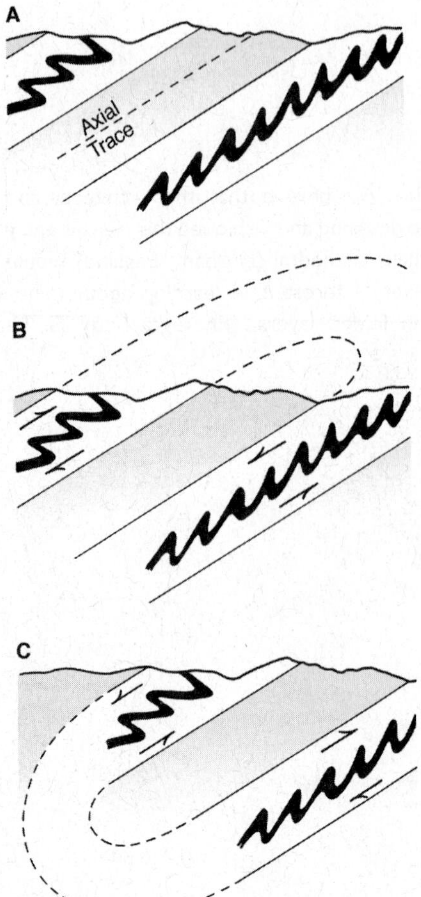

Figure 7.88 (*A*) Is the fold an overturned synform or an overturned antiform? (*B*) The fold is *not* an antiform, for the asymmetry of drag folds contradicts the sense of bedding-plane movements that would characterize the limbs of an antiform. (*C*) The drag fold pattern conforms perfectly to flexural slip on the limbs of an overturned synform.

Figure 7.89 (*A*) Cross joints in folded metamorphic rocks along the Bay of Fundy, near St. John, New Brunswick, Canada. Note the uniformity of spacing and orientation. (*B*) Close-up view of one of the joint faces shows that these cross joints are indeed perfectly perpendicular to the hinges of folds in the Ordovician metasedimentary rocks. After taking this shot, I (GHD) was overly eager to remove one of the beautiful folds. The resounding blow of the 3-pound hammer was cushioned by my finger, splitting my fingernail and producing a resounding cry. [Photographs by G. H. Davis.]

perpendicular to the fold hinges, yet with minor (extensional) escape of material parallel to the fold hinges

KINK FOLDING

Importance of Preexisting Foliation and Loading Direction

Strongly foliated rocks like schists and phyllites, and some thin-bedded sedimentary or volcanic rocks, commonly display **kink folds** (Figure 7.90). Kink folds are distinguished by sharp hinges, straight limbs, and an asymmetry expressed by a short limb connecting two longer limbs. Superficially, kink folds resemble buckle folds or flexural-slip folds, but they are really a distinct class.

Z-shaped kink folds are called **dextral**, whereas *S*-shaped kink folds are called **sinistral** (Figure 7.91). Axial surfaces of kink folds are referred to as **kink planes** (see Figure 7.91). The narrow zones where foliation is kinked are called **kink bands** (see Figure 7.91), and such bands show up quite distinctively in outcrop views (see Figure 7.90*A*). These folds are not truly flexural-slip folds, because the slip between layers is highly localized; that is, just within the kink bands. Schistosity or thin bedding or laminae outside of the kink bands proper do not experience layer-parallel (flexural) slip.

Paterson and Weiss (1966) eliminated some of the mystery of kink folding by successfully reproducing the progressive evolution of kink folding in highly foliated, real rock specimens, which they subjected to layer-parallel shortening under confining pressure (Figure 7.92). They demonstrated that there is a close relationship among the geometry of kink fold systems, the orientation of the strongly developed **planar anisotropy** (i.e., strong foliation), and the direction of loading. Layer-parallel loading first produced conjugate kink bands (Figure 7.92*A*, *B*). As shortening progressed, the widths of the kink bands increased to the point that they were replaced by a pervasive array of symmetrical chevron folds (Figure 7.92*C*, *D*, *E*). This is a progression from very localized slip (within kink bands only) to pervasive flexural slip throughout.

Figure 7.90 (*A*) Kink bands in Devonian phyllite exposed near Morthoe, Devonshire, England. [From *The Minor Structures of Deformed Rocks: A Photographic Atlas* by L. E. Weiss. Published with permission of Springer-Verlag, New York, copyright © 1972.] (*B*) Photomicrograph of kink folds in schist. [Photograph by A. L. Albee. Courtesy of United States Geological Survey.] (*C*) Close-up view of kink fold collected from Ordovician quartz sericite schist in northern New Brunswick, eastern Canada. [Photograph by G. H. Davis.]

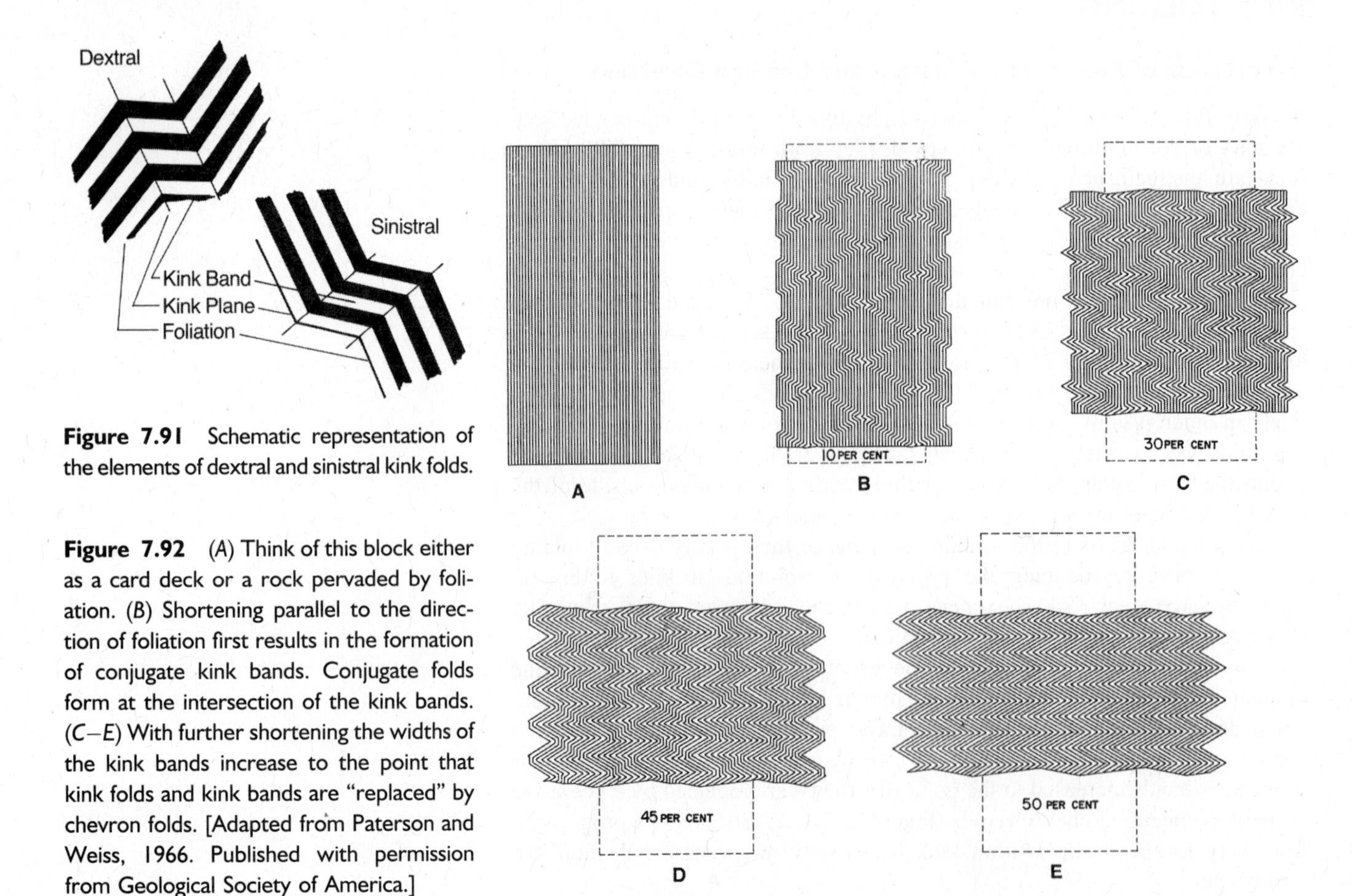

Figure 7.91 Schematic representation of the elements of dextral and sinistral kink folds.

Figure 7.92 (*A*) Think of this block either as a card deck or a rock pervaded by foliation. (*B*) Shortening parallel to the direction of foliation first results in the formation of conjugate kink bands. Conjugate folds form at the intersection of the kink bands. (*C–E*) With further shortening the widths of the kink bands increase to the point that kink folds and kink bands are "replaced" by chevron folds. [Adapted from Paterson and Weiss, 1966. Published with permission from Geological Society of America.]

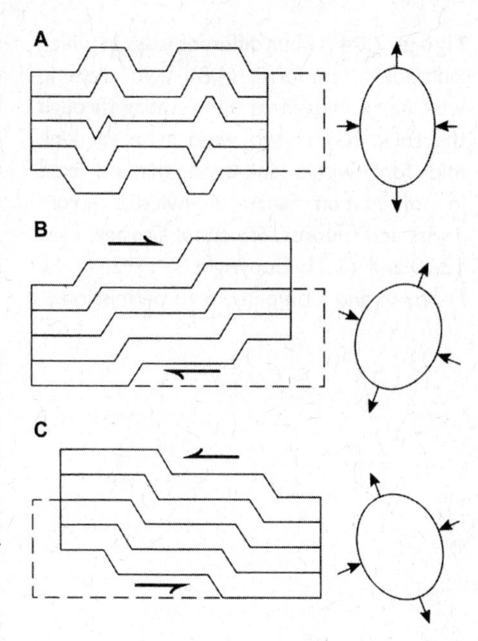

Figure 7.93 Relationship of total finite strain to the formation of kink folds. (*A*) Symmetrical sets of equally developed dextral and sinistral kink folds form when the direction of least stretch is oriented parallel to the direction of foliation. (*B*) Sinistral kink folds or (*C*) dextral kink folds develop when shortening is inclined to the layering. Note that sinistral kink folds form when there is a right-handed shear component parallel to foliation, and that dextral kink folds form when there is a left-handed shear component parallel to foliation. [Reprinted from Ramsay and Huber, The techniques of modern structural geology, v. 2: Folds and Fractures, © 1987, with permission from Elsevier.]

Ramsay and Huber (1987) emphasized the relationship of kink-fold geometry to strain. They pictured how the finite strain ellipse is oriented with respect to the overall orientation of foliation for three different situations: equally developed dextral and sinistral kink folds (Figure 7.93*A*); sinistral kink folds only (Figure 7.93*B*); and dextral kink folds only (Figure 7.93*C*). Where layer-parallel shortening is strictly layer parallel, no shear stress is generated parallel to layering; the result is roughly equal development of dextral and sinistral kink folds. More commonly, where layer-parallel shortening is applied at an angle not strictly parallel to layering, there *will* be a shear stress component in the plane of layer. Ironically, a right-handed component of shear creates sinistral kink folds (see Figure 7.93*B*), whereas a left-handed component of shear generates dextral kink folds (see Figure 7.93*C*). This is opposite to what we learned for asymmetric "*drag*" folds and reflects the different mechanism of folding for kink versus drag folds.

Importance of Cohesive Bonding between Layers

Kink folding requires strong contacts between layers (Reches and Johnson, 1976). It is this cohesion that must be overcome, locally, for slip between layers to occur locally, thus forming kink bands. Ghosh (1968) had recognized the degree to which the formation of kink folds depends on the tightness of cohesive bonding between layers. Once again there is irony. Ghosh (1968) investigated the influence of layer-boundary cohesion on fold form through a series of experiments. In each test, he subjected layers of modeling clay to end-on loading. In each run, strength, thickness, and spacing of layers was always the same. However, he varied the layer-boundary cohesion by adding or reducing the amount of lubrication (grease) he applied between layers of the modeling clay as he built his multilayer models for the end-on loading. Ghosh (1968) was able to create a striking array of fold forms without changing in any way the strength, thickness, or spacing of layers, or the direction of loading. He found that when he liberally greased the layer boundaries, layer-parallel shortening created smooth, rounded, sinusoidal folds. At the other extreme, Ghosh (1968) produced kink folds when he placed the layers of modeling clay in direct frictional contact with one another without any lubricant in between. The kink folds he produced in the laboratory were identical to those found so abundantly in nature.

Modes of Kink Folding

Twiss and Moores (1992) have nicely summarized different ways in which kink folds evolve. In some cases the kink bands migrate through the material (Figure 7.94*A*, *B*). For example, a kink band can nucleate perpendicular to foliation and then progressively rotate and expand in width to accommodate layer-parallel shortening (Figure 7.94*A*). Or the kink band may form at an oblique angle to foliation and simply expand in width (without changing orientation) to accomplish layer-parallel shortening (Stewart and Alvarez, 1991) (Figure 7.94*B*). In other modes, the kink bands do not migrate through the

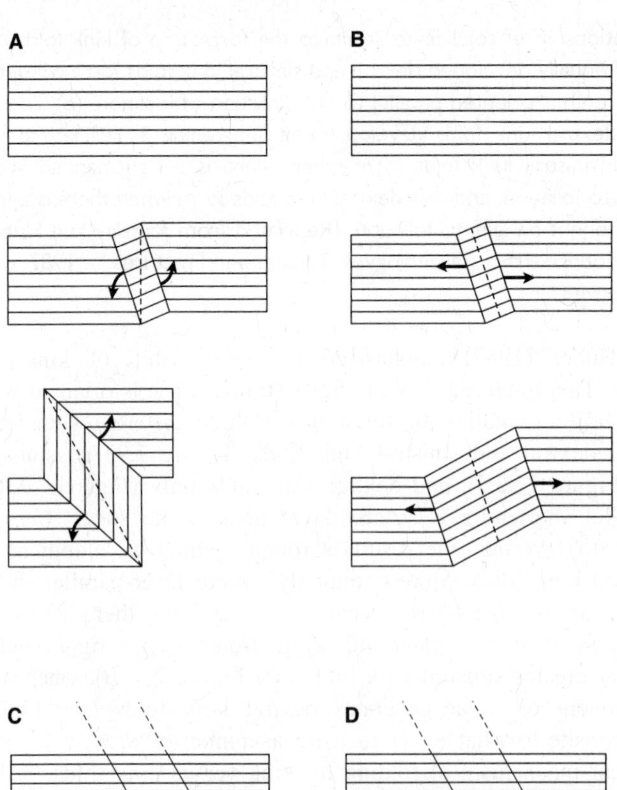

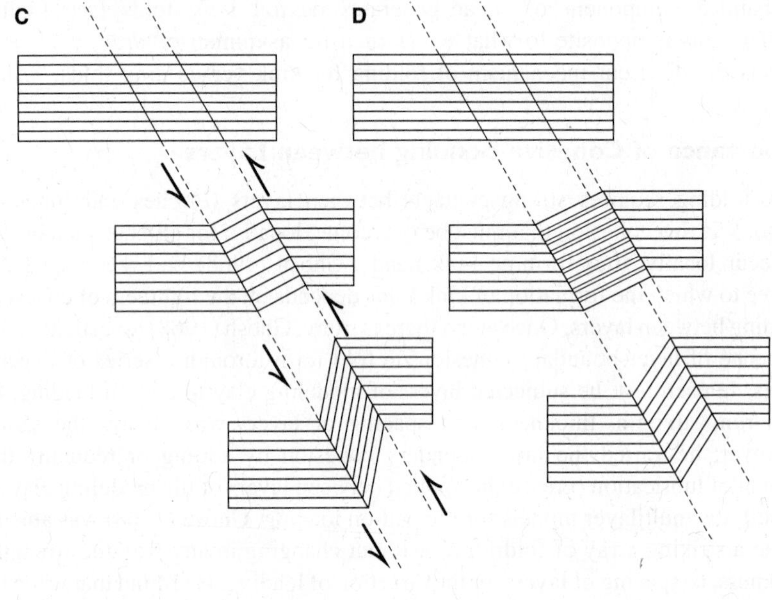

Figure 7.94 Four different ways in which kink folds can form. (*A,B*) Two ways in which kink folds form by migrating through the rock. (*C,D*) Two ways in which kink folds form within kink bands that are fixed in orientation and in width. [From Twiss and Moores, *Structural Geology*, Figs. 12.20 and 12.21. Copyright © 1992, W. H. Freeman and Company, with permission.]

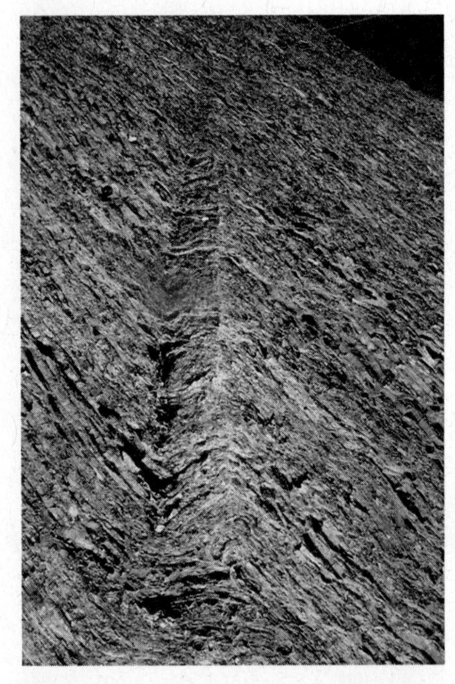

Figure 7.95 Photograph of a kink band within interbedded siltstone and shale of the Uncia Formation (Silurian). Width of band at base of outcrop is ~35 cm. Tectonic location is the Central Andean backthrust belt, Boliva, east of Lake Poopo (see map, McQuarrie and Davis, 2002, Figure 2). Note that the kink band tips out upward. The interlimb angle of the kink band decreases from the tip downward toward the maximum kink-band width exposed. [Photograph by G. H. Davis.]

material. Boundaries remain fixed in orientation and spacing throughout the shortening event. Layer-parallel shortening is accomplished by simple shear parallel to the fixed boundaries (Figure 7.94*C*) or by rigid rotation of the laminations within the kink band (Figure 7.94*D*). Tell-tale geometries can be identified in outcrops to distinguish among the possible origins.

I (GHD) encountered an exposure of a kink band in Bolivia that helped me visualize the progressive development of a kink band of the type that grows in width (McQuarrie and Davis, 2002). The host rock is interbedded siltstone and sandstone, with laminated bedding. It is shown as Figure 7.95. The tipping out (upward) of the kink band is completely exposed. By sequentially following (by eye) the changes in kink band structure from the tip point downward to the band at its maximum exposed width, we can treat the structure as a movie of how the band developed through time; that is, widening and progressively achieving a smaller and smaller interlimb angle.

PASSIVE FOLDING

Characteristics of Passive Folds

Passive folds characteristically display profile forms that are class 1C, 2, or 3, typified by some degree of apparent layer thickening in the hinge and thinning on the limbs (Figure 7.96). Passive folds typically occur in metamorphic rocks, and are commonly marked by axial plane cleavage, a nice example of which is shown in Figure 7.97. We will learn in *Chapter 9, Foliation and Lineation,* that cleavage develops at the micro- and hand-specimen scale, expressed as a set of penetrative, parallel, planar to slightly wavy structures. In 2D cleavage resembles a wood grain running through the rock. The function of cleavage is to accommodate tectonic shortening, . . . a degree of shortening that cannot be accommodated by folding alone! We will learn in *Chapter 9* (*Foliation and Lineation*) that deformation mechanisms responsible for

Figure 7.96 (*A*) Passive fold in metasedimentary rocks (late Paleoproterozic) in the Salt River Canyon region, Arizona. Height of fold is approximately 1.5 m. Rock [Photograph by F. W. Cropp.] [From Davis and others, 1981. Published with permission of Arizona Geological Survey.] (*B*) Passive fold in polished slab of pyritic ore from the Caribou strata-bound sulfide deposit in the Bathurst mining district of New Brunswick, Canada. Cleaved black layers represent original bedding. Cleavage is axial planar to the folded layering. [Photograph by G. Kew.] (*C*) Recumbent passive folds in marble derived from Pennsylvanian-Permian limestone in Happy Valley, southeastern Arizona. [Photograph by G. H. Davis.]

Figure 7.97 In the surface of this tabletop (width of view is 50 cm) we see passive folding of metamorphic rock. The core of this complex antiform is composed of marble (mottled white) and the overlying rock is schist (gray). The cleavage runs roughly parallel to the axial surface of the antiform. The thin gray seams along the cleavage in the marble represents insoluble residue, left behind during the pressure dissolution accompanying cleavage development. The white layer within the schist is a calcite vein, probably derived from pressure dissolution of marble. Note that within the schist the trace of the cleavage bends upward to the right, maintaining reasonable parallelism with the axial trace of the fold. [Photograph by S. J. Reynolds.]

cleavage formation importantly include pressure dissolution. The intimate relationship between passive folding and cleavage is obvious in the geometric layout of the two: cleavage will be symmetrically disposed across the fold, tending to align itself subparallel to the axial surface of the fold, not commonly "*fanning*" about the axial trace. It is clear that the tectonic loading that produced folding also produced the cleavage, and that together tectonic shortening was accomplished.

Conditions Favoring Passive Folding

In contrast to flexural folding, **passive folding** is the favored mechanism when the mechanical influence of layering in a sequence of rocks is very weak (Donath and Parker, 1964). Sequences that are especially vulnerable to passive folding are distinguished by uniformly soft, weak layers. Donath and Parker (1964) describe this mechanical condition for passive folding as one of **high mean ductility** and **low ductility contrast**. The dominant geologic circumstances that leads to sedimentary or volcanic rocks acquiring high mean ductility and low ductility contrast is regional metamorphism. Rocks that are perfectly stiff and strong under relatively low temperature and pressure conditions may lose most of their strength under metamorphic conditions of elevated temperature and confining pressure. The heat and pressure blot out the mechanical influence layering might have had under ordinary, nonmetamorphic conditions of deformation. When such rocks are compressed, passive folding may be one of the expressions of deformation.

Mechanisms of Passive Folding

During passive folding, layers take on a folded form without really having been bent as much as they appear. The layering is passive. It is not active, but is acted upon.

In most cases the onset of cleavage development and passive folding probably takes place after some amount of buckling, bending, and flexural folding has already been achieved. The onset of pressure solution permits a rock to shorten to a degree not possible by further rotation of the limbs of a fold. In effect, a point is reached in the folding process at which material needs to be force out of the inner arc region of the fold. **Flattening** is required to eliminate the room problem. This requirement is addressed by the development of cleavage, which in turn requires pressure-dissolution loss of material. We will see in *Chapter 9* (*Foliation and Lineation*) that as pressure dissolution takes place and material is removed, shortening proceeds in a direction parallel to S_3, the direction of minimum finite stretch. Fold limbs naturally steepen to greater inclinations. Fold profiles change from concentric to similar.

Figure 7.98 The stunning aesthetics of big folds. (*A*) The Sheep Mountain anticline in the Big Horn basin, Wyoming. (*B*) Folding at Sheep Mountain, Wyoming. [Photographs by K. Constenius.]

REGIONAL TECTONIC FOLDING

Tectonic loading at the regional scale has produced spectacular folds in mountain belts the world over. Aesthetically, they are the favorite structure of most geologists (Figure 7.98). They can form in a variety of ways. Interpretations must be based on a solid descriptive foundation, as much subsurface control as available, and effective cross-sections.

Particularly as we have focused so much of this chapter on buckling and flexural folding, it is tempting to interpret individual regional folds and regional systems of folds as products of **free folding**. The concept of "*free folding*," asserts that the properties of folds depend entirely on the physical–mechanical properties of the layers that are shortened. The origin of folds would be considered exclusively on the basis of buckling, flexural-slip, ductility contrast, and degree of cohesion along layer boundaries. Interpretation of the evolution of free folds affecting sequences of beds hundreds of meters in thickness might proceed as follows: (1) Layer-parallel stresses are initially accommodated by layer-parallel elastic and inelastic shortening. (2) A buckling instability develops, and the character of the fold (e.g., dominant wavelength) will be related to the mechanical character and locations of the thickest, strongest layers within the sequence. (3) Depending on ductility contrast between the most competent layers and the less competent rocks on either side, the fold is amplified at a relatively high or a relatively low rate. (4) As the thickest, stiffest layers buckle as single units, the multilayer sequence as a whole will undergo flexural-slip folding. (5) Tightening of the fold and continued flexural folding favors tangential longitudinal strain within the limbs (Bastida et al., 2007; Toimil and Fernández, 2007) as well as outer-arc stretching and inner-arc shortening. (6) When flexural folding gives out, the fold may continue to tighten by flattening (Srivastava and Shah, 2006; and Toimil and Fernández, 2007).

In *Chapter 8*, (*Fault-Fold Interactions*) we will discover that detachment folding is the purest form of free folding. Detachment folding is favored in flat-lying sequences of sedimentary rocks with thick beds of evaporates, which are mechanically weak. "*Fold trains*" of anticlines and synclines will give the impression that they formed quite independently of faulting (Figure 7.99). Yet geological mapping combined with subsurface data will reveal that the detachment folding commonly occupies a transfer zone between thrust faults.

In contrast to free folding, **forced folding** is a general category of folding in which the geometric characteristics and overall form of the folds are "*forced upon*" the layers by virtue of the orientation and form of faults with

Figure 7.99 (*A*) North-directed photograph of detachment folding in evaporite beds of the Carmel Formation (Middle Jurassic) in the Reed Wash area on the west flank of the San Rafael Swell, Utah (Royse, 2003). (*B*) Closer view of the detachment fold forms in this gypsiferous stratigraphic sequence. [Photographs by G. H. Davis.]

which the folding is associated, and by which the folds are driven (Figure 7.100). The beds are not "*free*" to fold, nor are they transmitting through the layers much in the way of layer-parallel stresses. Instead, they just go along for the ride, and some of the beds happen to find themselves in awkward places and are required to stretch or bend from one step to another. In *Chapter 8, Fault-Fold Interactions,* we will discover that **fault-bend folding** and **fault-propagation folding** are the main categories and expressions of forced folding.

CONCLUSIONS

Though the mechanical properties of a sequence of rocks strongly influence the geometry and kinematics of folding, we will see in *Chapter 8, Fold-Fault Interactions,* that the origin and character of folds and fold systems are also intimately related to faulting. It is literally impossible to understand and appreciate folding without studying the details of how fault slip and fold shortening work together in partnership to accommodate deformation, regionally and locally, big and small. That is why we will devote a full chapter to this topic.

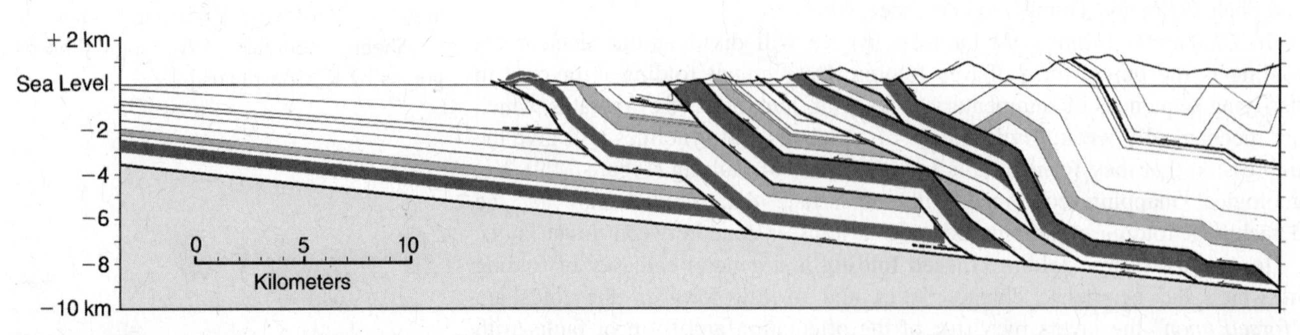

Figure 7.100 Fault–bend fold geometry "*forced*" upon beds as they are moved along ramp–flat fault geometries. Taiwan. Adapted from Suppe (1980a). Published with permission of Geological Society of China.

Chapter 8 Fault-Fold Interactions

Bedding in Paleozoic formations in the Sheep Mountain Range north of Las Vegas, Nevada. The stark exposures of bedding create ledgy expression due to the bench-forming resistance of carbonate units. Bedding is folded, but not smoothly folded. We see eye-catching kink-like geometries, marked by panels within which bedding attitude is uniform, yet distinctive from that of adjacent panels. The orientations of the kink boundaries, and the shifts in bedding orientation from one panel to the next, are a record of the shape of the fault that produced the folding. As bedding is faulted along a deep 'flat,' and then up a 'ramp,' and then along an upper 'flat,' it will be bent and folded in ways that record such a zig-zag fault shape in the subsurface. If we are to reconstruct the fault geometry accurately, we must map and measure the changes in strike and dip of the bedding very precisely. By doing this we can exploit what we can see in order to construct what is hidden. [Photograph and copyright © by Peter Kresan.]

INCENTIVES FOR EVEN FURTHER STUDY

We have seen that faults and folds are interesting to study in their own right, and that commonly the formation of folds and the formation of faults are intertwined and interdependent. Geological cross-sections constructed across major fold-thrust belts provide at a glance the intimate association of folding and faulting. Figure 8.1*A*, for example, is one of the products of the mapping and analysis by Pete DeCelles and Jim Coogan (2006) in the Utah-Nevada part of the Sevier thrust belt in the western United States. This section is rich in fold-fault details. Keep in mind that the Canyon Range thrust fault alone accommodated 120 km of tectonic transport! When we examine such a geological cross-section closely, we gain an appreciation for the challenges in interpreting the interplay of thrust faulting and folding. The Pavant thrust faults are bedding parallel. In contrast, the Canyon Range thrust "*runs*" parallel to bedding and then suddenly "*ramps*" up *across* bedding (see Figure 8.1*A*). All of the thrusts appear to be folded. Furthermore, lurking near the skyline are normal faults, which in this reconstruction have not yet formed. Yet, DeCelles and Coogan (2006) found it necessary to "*back off*" the fault displacements accomplished by normal faulting in order to generate their interpretation of the fundamental geometry of folding and thrusting.

One of the things we will learn is that the shape of a fault, combined with the movement of the hanging wall along it, has a direct influence on the shape

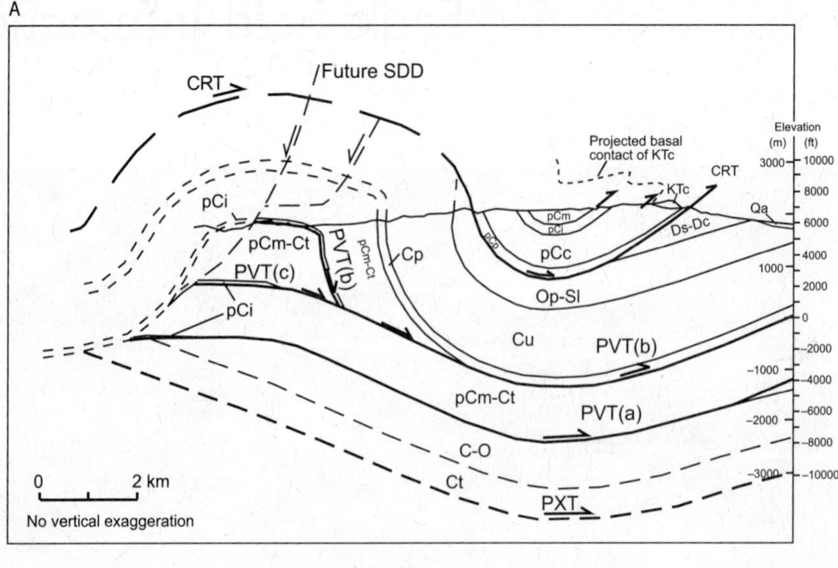

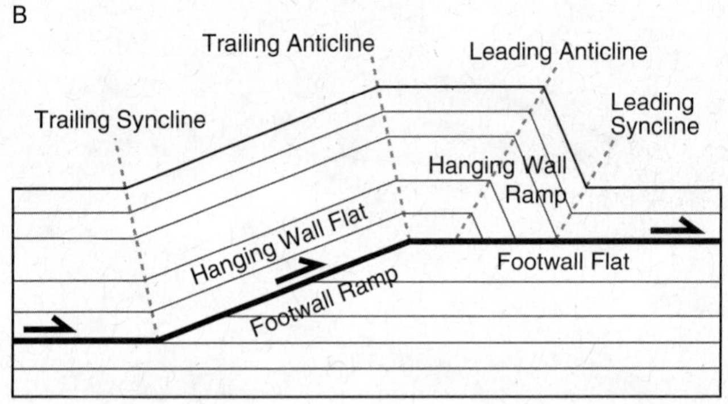

Figure 8.1 (*A*) Interplay of faulting and folding as captured in a geological cross-section through part of the Sevier thrust belt. The location is the Canyon Range, Utah. Thrust faults from bottom to top: PXT = Paxton thrust; PVT(a) = main Pavant thrust; PVT(b) and PVT(c) = hanging wall "*imbricate*" faults of the Pavant thrust; CRT = Canyon Range thrust. SDD = Sevier Desert detachment (post-thrusting extension). Ages of tectonic assemblages from oldest to youngest: pC = Precambrian (Neoproterozoic); C-O = Cambrian-Ordovician in footwall of Pavant thrust; D = Devonian; O-S = Ordovician and Silurian; K-T = Cretaceous/ early Tertiary; Q = Quaternary. [From DeCelles and Coogan, 2006, Figure 5, p. 847. Published with permission of the Geological Society of America.] (*B*) Diagrammatic cross-section showing the properties of a "*fault-bend fold*." The dominant thrust fault is bedding-parallel in its deepest reaches, but ramps upward from left to right, only to become bedding-parallel at its highest level. As the hanging wall is translated (earthquake by earthquake) by thrust faulting, the strata within it acquire a folded form, marked by a geometry exactingly related to the shape of the fault. [From http://commons.wikimedia.org/wiki/File:Thrust_-with_fault_bend_fold.svg.]

of the rocks "*riding*" along the fault. One example of this is the "*fault-bend fold*" relationship shown in Figure 8.1*B* (*http://commons.wikimedia.org/wiki/File:Thrust_with_fault_bend_fold.svg*). We see a thrust fault that is oriented parallel to bedding at depth, ramps upward across bedding, and then re-enters a bedding-parallel position, but at a higher level than before. The hanging wall moves upward relative to the footwall, earthquake by earthquake. The shape of the footwall imposes a very particular structural geometry on the hanging wall. The leading syncline, leading anticline, trailing anticline, and trailing syncline (see Figure 8.1*B*) all owe their shapes and locations to the deformation path experienced by the hanging wall.

To be sure, the geometry and orientation of a fault, and the earthquake-by-earthquake displacements along it, may exert a strong control on the geometry, orientation, and shape of bedding and folding within the hanging wall. This is true both of thrusting faulting and normal faulting; and the same principles can be applied to strike-slip faulting and associated folding.

Knowing this provides even greater incentive for mapping bedding, folds, and faults as meticulously as possible, for we can use these data to construct the locations, orientations, and shapes of major faults in the subsurface. Alternatively, if our "*map base*" is not surface mapping but bedding, fold, and fault data derived from seismic reflection profiles and well bore data, we can use this information to construct the locations, orientations, and shapes of first-order faults in the deeper subsurface.

GENERAL MODEL OF FAULT-RELATED FOLDING

Early workers recognized that there was a relationship between the shape of the fault and the geometry of structures in the hanging wall (Rich, 1934). However, there were insufficient observations and models to describe those relationships quantitatively. When John Suppe began mapping in the thrust belt in Taiwan, he encountered there a stratigraphic section that is thinly bedded and therefore had abundant planes that could "*take up*" interbed shear (Suppe, 1983). The rock column lent itself to the formation of chevron folds in the form of giant kink folds, marked by straight planar limbs and abrupt, angular hinges. Suppe realized that quantitative description of folding of this particular geometry held promise for geometric and kinematic analysis, and his work on same led to what is now called **dip domain fold theory** (Figure 8.2*A*).

Dip domain fold theory became a basis for constructing geological cross-sections for fold-thrust systems marked by constant bed thickness and parallel folds. One of the rules in such constructions is that bed length must be conserved. Thus the length of any given bed shown in such a cross-section will remain unchanged, before versus after deformation. The folds themselves are constructed in ways that they are comprised of homoclinal panels (see Figure 8.2*B*), each with a characteristic, constant dip. Part of the value of fault-related fold models is that any number of possible geometries can be built and tested efficiently. Furthermore, the fault-related fold structures are self-similar, and thus common geometric relationships prevail regardless of size of structures. Dip domain fold theory produces predictable geometrical relationships, but not necessarily perfectly accurate renditions. In some cases

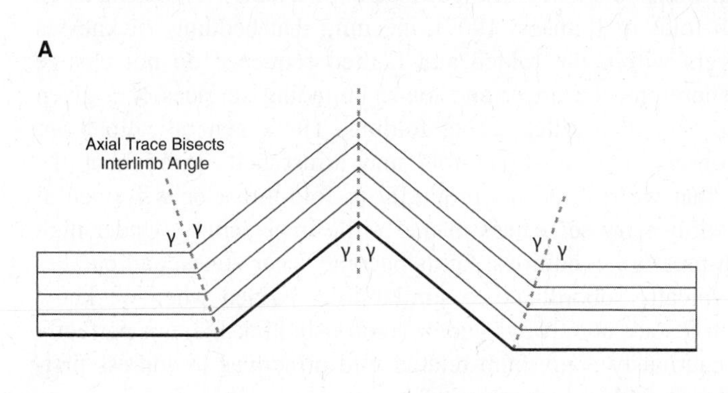

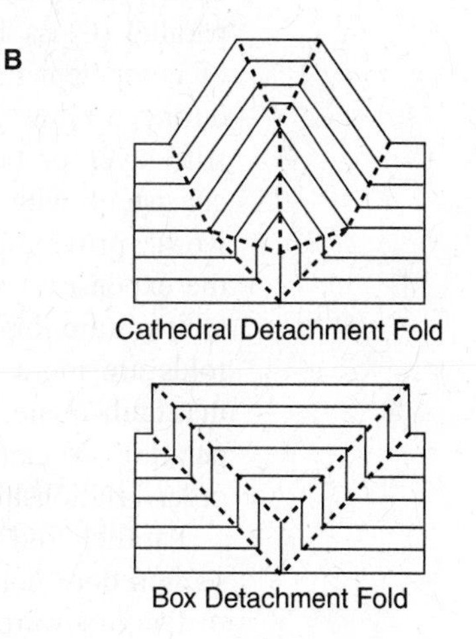

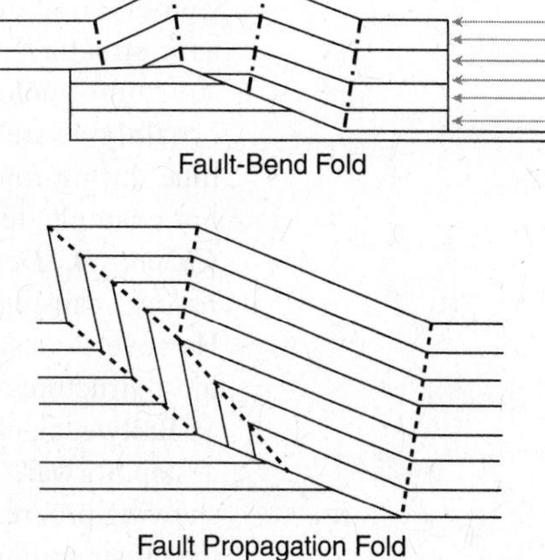

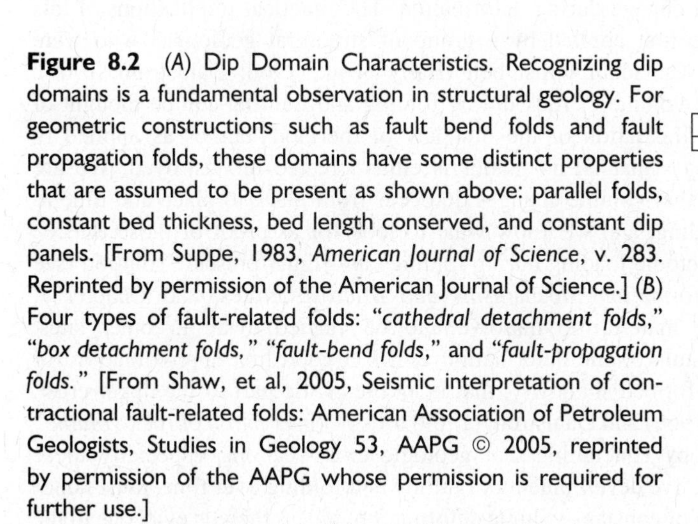

Figure 8.2 (A) Dip Domain Characteristics. Recognizing dip domains is a fundamental observation in structural geology. For geometric constructions such as fault bend folds and fault propagation folds, these domains have some distinct properties that are assumed to be present as shown above: parallel folds, constant bed thickness, bed length conserved, and constant dip panels. [From Suppe, 1983, *American Journal of Science*, v. 283. Reprinted by permission of the American Journal of Science.] (*B*) Four types of fault-related folds: "*cathedral detachment folds,*" "*box detachment folds,*" "*fault-bend folds,*" and "*fault-propagation folds.*" [From Shaw, et al, 2005, Seismic interpretation of contractional fault-related folds: American Association of Petroleum Geologists, Studies in Geology 53, AAPG © 2005, reprinted by permission of the AAPG whose permission is required for further use.]

the geology is not as "*cooperative*" as the assumptions would lead us to believe. The precise mechanisms by which a given fold forms may not conform to basic fault-related fold models. For example, the fold mechanism in nature may have produced bed thickening in the fold hinge and bed thinning on the limbs, and thus the fold is not strictly speaking a parallel fold.

Dip-domain fold analysis obeys sets of trigonometric equations that describe the angular interrelationships that evolve during the course of folding. The geometric requirements are applied to bedding orientation data (facts) measured at the earth's surface, or interpreted from drilling information and/or seismic reflection profiles. Products are cross-sections, and sequential cross-sections showing progressive development of fold-fault interactions through time. The fundamental dip domain fold-theory equations have been incorporated into computer software so that the hard work of generating subsurface interpretations of geologic structure is made easier. Computer animations of dip-domain folding are especially useful in visualizing the interaction of faulting and folding over time.

ASSUMPTIONS ABOUT DIP DOMAINS AND FAULT-RELATED FOLDS

Bed Length, Bed Thickness, and Bed Volume

Regardless of size, fault-related folds tend to be kink-like in shape (Faill 1973), with planar limbs and narrow hinges (see Figure 8.2*B*). They tend to be parallel (Class 1b folds of Ramsay, 1967), meaning that bedding thicknesses of competent layers within the folded and faulted sequence do not change during folding. Therefore, the upper and lower bounding surfaces of a given stiff layer or bed remain parallel during folding. These generalizations are consistent with observations made in fold and thrust belts throughout the world, providing that we restrict this primarily to fold-thrust belts formed at the expense of sedimentary sequences in the brittle upper crust. (Under high temperature, high-pressure conditions, faults become shear zones, and parallel folds are more typically subordinate to similar folds.) Even when we know that fault-related folds are not always and/or rigorously kink-like nor perfectly parallel, we can cautiously apply fault-related fold principles to address first-order relationships on fault-fold interactions.

Parallel fold geometries imply that the volume of the folded stratigraphic section does not change during deformation. The practical implications of this realization were first applied by a group of structural geologists who were working in the Canadian thrust belt (Bally et al., 1966; Dahlstrom, 1969). Conservation of volume of rock during folding and faulting can be thought of as a structural illustration of the first law of thermodynamics as applied to structural geology; that is, no matter is either created or destroyed. We are certainly aware that volume changes do occur from place to place and time to time during folding, and we know what to look for as proof of those effects. For example, tectonic loading during folding can trigger pressure solution (see *Chapter 4, Deformation Mechanisms and Microstructures,* and *Chapter 7, Folds*), causing material to dissolve and be carried away to other sites. However, constant-volume deformation is still a good first approximation for most structures formed shallowly; that is, in the brittle part of the upper crust.

Bally et al. (1966) and Dahlstrom (1969) took "*conservation of rock volume*" a step forward by concluding that geologic cross-sections, especially those showing progressive development of faulting and folding over time, must honor the presumption of constant-volume deformation, unless there is evidence to the contrary. Thus in preparing cross-sections (then without the aid of computing technologies), they made certain that the bed length of each competent bed showed neither lengthening shortening at any stage of the progressive deformation (see *Part III-F, Preparing Geologic Cross-sections*). If thickness does not change, and bed length does not change, then the area of the folded/faulted section cannot change. They then tested the validity of their cross-section by restoring the structure sections to the undeformed state; that is, "*unfolding*" the folds and "*unfaulting*" the faults, making certain that the stiff beds (horizontal and undeformed after restoration) all had the same lengths. In practical terms, we can measure the line length of the beds in the competent units in the deformed state and assume that those beds will be the same length(s) in the restored state. This provides a powerful constraint on cross-section interpretations, if our assumption of constant line length is valid. With this approach was born the concept of a **balanced cross-section**, which we will discuss later in the chapter (also see *Part III-F, Preparing Geologic Cross-sections*).

Bed Shear

Another implication of the assumption of parallel folding is that the amount of bed-parallel shear on the flank of a fold (e.g., an anticline) is matched by the amount of shear on the opposite limb (though with the opposite sense of shear). Recall that the interbed shear in a parallel fold is out of the synclinal hinge and into the anticlinal hinge (see Figure 7.61). The net shear across a given fold is assumed to be zero. This implies negligible interbed shearing outside of the folds, where strata are flat lying.

Axial Surfaces

Another consideration in thinking about fault-related parallel folds focuses on the axial surfaces of the folds. If (in 2D analysis) the axial trace of a fold truly bisects the angle between the limbs of the fold, the cross-sectional area of the beds is preserved (see "*Box Fold*" in Figure 8.2*B*). Restoration of the cross-section can test this assumption. However, if the axial trace is not the bisecting plane of the fold limbs, the axial region of the fold will be faulted (see lower reaches of "*Fault Propagation Fold*" in Figure 8.2*B*). Otherwise, volume (or area in two dimensions) will not be preserved.

Implications

These assumptions about bed thickness, bed length, bed shear, and axial surfaces commonly hold quite well and are born out by observations, especially in sedimentary sequences deformed in the brittle parts of the upper crust. We breathe a sigh of relief where these conditions are met, for we know we will be able to leverage fold geometry as a guide to interpreting fault geometry and structural evolution. This is the 'bread and butter' for exploration geologists in the oil and gas industry, and for 'academics' in the field of structure-tectonics, for the geometry of any part of a major fold of this nature may be quantitatively related to the geometry of the rest of the structural system.

FAULT-BEND FOLDS

General Character

One of the major categories of folding described by the fault-related fold model is **fault-bend folding**. Fault-bend folds are formed when the hanging wall rock changes shape as it moves over a change in dip; that is, a bend, in the fault surface. The fault-bend fold model is a relatively simple geometric model that attributes changes in bedding orientations in the hanging way to changes in the orientation of the fault surface (Figure 8.3).

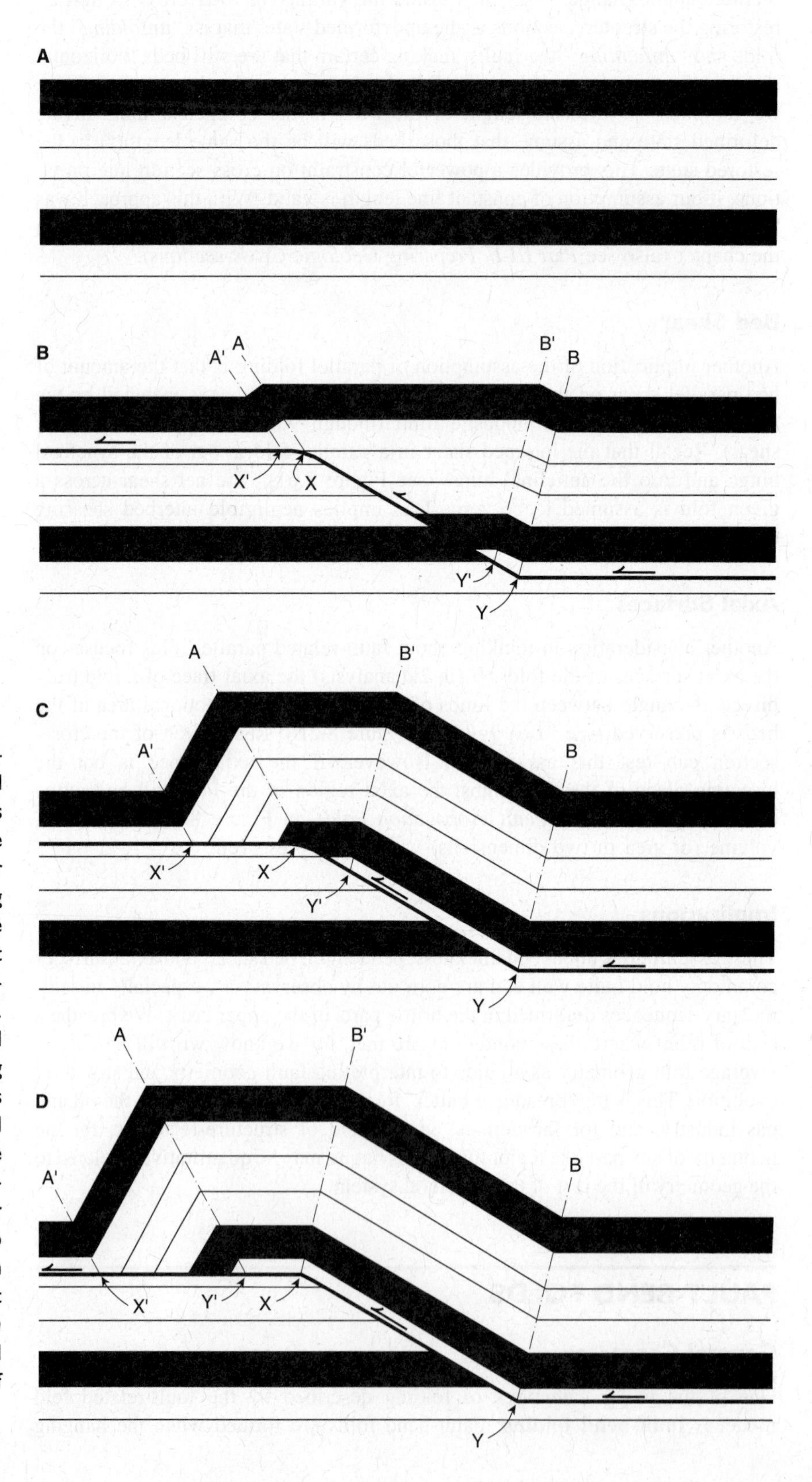

Figure 8.3 Progressive evolution of fault-bend fold geometries. Fault-bend fold geometry and kinematics. Fault-bend folds are common in areas of thrusting, where the thrusts form a 'ramp' from a lower stratigraphic unit to a higher one. Modeling gives the impression that all parts of the thrust surface are active when displacement takes place, but this is not true. (*A*) Flat-lying sequence of sedimentary rocks. (*B*) Thrusting begins. X and Y are footwall cutoffs; X′ and Y′ are the respective hanging wall cutoffs. Beds flex into kink-like folds as they move from flat to ramp positions, and from ramp to flat positions. (*C*) Anticline grows in height and width. Width of kink-like fold gets larger and larger. (D) After hanging-wall cutoff Y′ makes it to the top of the ramp, the fold ceases to grow in height; instead it grows in width. See text for details. [From Suppe, 1983, *American Journal of Science*, v. 283. Reprinted by permission of the American Journal of Science.]

It is important to first place fault-bend folding in the context of tectonic shortening. Picture a thrust fault cutting up through a horizontal stratigraphic section from a lower-level bedding-parallel fault horizon (i.e., flat) to a higher-level bedding-parallel fault horizon (Figure 8.3*A, B*). Hanging-wall beds and footwall beds are **cut off** sharply and obliquely along the **ramp** portion of the thrust fault. Following Suppe (1983), it is useful to call special attention to the **footwall cutoffs** (X and Y, see Figure 8.3*B*) at the top and bottom of the ramp; and the **hanging-wall cutoffs** (X′ and Y′, see Figure 8.3*B*) also at the top and bottom of the ramp. The deformation path for this kind of thrusting and fault-bend folding is relatively easy to track for any material point in the hanging wall of the fault. Attributing Rob Knipe, Groshong (1988) pointed out that "*rocks do not suffer deformation; rather, they enjoy it.*" Thus we can follow exactingly how the rocks "*enjoy*" deformation at each stage along the way.

When movement along the thrust fault begins (see Figure 8.3*B*), hanging-wall strata begin moving up the ramp. As this happens, the orientation of the ramp is forced upon the orientation of the beds that move up the ramp. As beds in the hanging wall move from the flat to the ramp, they become creased into the form of a chevron syncline, which has the form of a giant kink fold (axial trace B–Y in Figure 8.3*B*). Depending on magnitude of thrusting, some length of the hanging-wall section that was originally horizontal and above the lower flat will be "*fed*" up the ramp and be creased by folding at the bend in the fault plane (point Y, see Figure 8.3*B*). Thus, at the earliest stages of movement of hanging wall strata up the ramp, a narrow kink band forms and extends from the bottom of the ramp to the top of the section. One boundary of the kink band is B–Y, a fixed axial trace through which bedding becomes bent. The other boundary of the kink band is B′–Y′, which is an axial trace that simply passively moves as part of the hanging wall (see Figures 8.3*B, C*).

At the very onset of thrusting (see Figure 8.3*B*), an anticline forms at the top of the ramp, where the thrust becomes an upper flat. The axial trace of the anticline is A–X, and it is pinned to the footwall cutoff (X). A second kink band develops at this position as thrusting continues. One boundary of the kink band is in fact A–X, a fixed axial trace through which bedding becomes bent. The other boundary of the kink band is A′–X′, which is an axial trace that passively moves as part of the hanging wall (see Figures 8.3*B, C*).

As thrusting progresses, and as the hanging-wall cutoff (Y′) moves further and further up the ramp (see Figure 8.3*C*), the kink band B–Y/B′–Y′ continuously grows in breadth. In the same fashion, as the hanging-wall cutoff (X′) at the top of the ramp progressively moves outward along the upper flat, the kink band A–X/A′–Y′ grows in breadth (see Figure 8.3*C*). The fold reaches a maximum amplitude when Y′, the hanging-wall cutoff formed at the base of the ramp, reaches X, the footwall cutoff formed at the top of the ramp (see Figure 8.3*C*). From that point on, the fold grows in breadth but not in height as thrusting continues (Figure 8.3*D*).

Axial Traces

We see that as the hanging-wall strata move along the "*flat-ramp-flat*" fault surface, they encounter two angular bends, one concave up, the other concave down (see Figure 8.3). The distinguishing fault-bend folding geometry is marked by four axial traces (as viewed in cross-section). Two of the axial traces (B–Y and A–X) are **active axial traces** and the two other (B′–Y′ and A′–X′) are **inactive axial traces** (see Figure 8.3). This distinction is based on whether or not rocks are actively moving through the axial trace (axial surface in 3D) and being folded. Active axial traces are labeled with plain letters (e.g., B–Y, A–X); inactive axial traces are labeled with primed letters (e.g., B′–Y′, A′–X′).

The two active axial traces are located at the very bottom and very top of the fault ramp. Beds in the hanging wall move through these axial traces conveyor-belt-like, becoming "*kinked*" as faulting proceeds (see Figures 8.3*B*, *C*). No bed passing through these active axial traces escapes folding. Axial trace B–Y is "*pinned*" geometrically to the base of the fault ramp, where the fault changes orientation from bedding-parallel to ramp-parallel. Axial trace A–X is "*pinned*" geometrically to the top of the fault ramp, where the fault changes orientation from ramp parallel to upper-flat parallel. The hanging-wall beds above the lower flat are bent upward as they move up the fault ramp (see Figure 8.3*B*, *C*). When these same beds pass through the very top of the ramp, they are bent downward as they move from the thrust ramp to the upper flat (see Figure 8.3*C*, *D*).

At the very instant of first faulting, the active and inactive axial traces coincide in position and orientation. But as fault slip proceeds, the inactive axial traces (B′–Y′ and A′–X′) ride along the fault surface, separating themselves spatially from the active axial traces. Axial trace B′–Y′ starts out at the base of the ramp, whereas axial trace A′–X′ starts out at the top of the ramp. When inactive axial trace A′–X′ passively translates to the top of the thrust ramp, it becomes pinned there and changes to an active axial trace because the beds that were cut off by the ramp have now all moved up and over the upper bend in the fault surface. From that point onward, the formerly inactive trace (A′–X′) separates beds that started out above the lower flat from those that originated above the ramp. The old active surface (A–X) is now rendered inactive, separating rocks that started out on the ramp from those that originated above the upper flat. Thereafter, axial trace A–X joins axial trace A′–X′ in passively moving horizontally along the upper flat as shortening continues (see Figure 8.3), content to ride horizontally along on the upper flat as shortening continues.

Folding as a Guide to Faulting

The fold that was created by our example of fault-bend folding is a giant anticline, with a relatively short steep forelimb and a relative long gentle backlimb (see Figure 8.3*D*). What can we learn about faulting by looking at the folding? First, we see that the dip of the back limb of the anticline matches the dip of the fault ramp below (see Figure 8.3*D*). Thus, by identifying and measuring the dip of the back limb of a fault-bend fold, we can estimate the dip of the fault ramp. Second, we observe that axial traces are either pinned to bends in the fault surface, or are formed passively separating domains of different attitude and thickness. The inactive axial traces occur in the hanging wall at places located originally at fault bends. Thus, by recognizing and mapping the passive axial traces (axial surfaces in three dimensions) we can estimate the length of the fault ramp. Moreover, if we have some knowledge of stratigraphic thicknesses, we can estimate the height of the ramp. In these practical ways we can leverage fold geometries in the hanging wall to infer the overall structural geometry of the system.

All of this assumes that the *entire* fault, including both flats and the ramp connecting the flats, was active right from the start of shortening. However, the whole fault may not have been active all once, particularly early in the development of the structure. An alternative model, proposed by Peter Verrall and illustrated by Clint Dahlstrom (1970), features the possibility that two separate bedding-parallel faults form initially at two different structural levels (Figure 8.4). These fault segments later become linked by a single through-going fault that connects the two detachment levels. Simply based upon the final geometry of the fault-fold system, it is difficult to differentiate between these two models. Frankly, the practical geometric utility of fault-bend folding holds in either case.

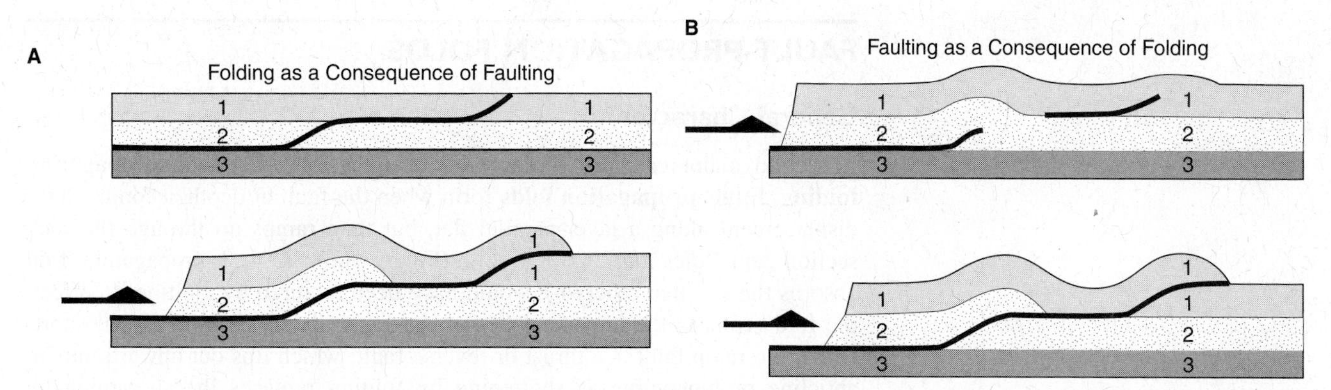

Figure 8.4 (*A*) The classic fault-bend fold model is one in which the entire flat-ramp-flat structure is active from the very beginning of the fault movement. The hanging wall moves over the bends in the fault surface and becomes distorted into a fault bend fold. (*B*) Beware of the possibility that originally there were two discrete flat thrusts, one at depth, and one at higher stratigraphic level. Layer-parallel shortening results in detachment folds forming above the flat detachments. Over time these detachments may link together, producing a flat-ramp-flat geometry. The final geometry produced is similar to classic fault-bend folding, but fold genesis and timing in this second case is quite different. [From Dahlstrom, 1970, Bulletin Canadian Petroleum Geologists, CSPG© 1970, reprinted by permission of CSPG whose permission is required for further use.]

Variations on the Theme

Important variables that influence the shape of fault bend folds include the size of the cut-off angles as the fault changes dip angle, the amount of slip, the shape of the fault, the internal shearing (if any) within incompetent beds, and the depth to the lower flat. Structural geologists have investigated the contributions of such variables, and results have delivered greater precision in interpreting fault geometry in the subsurface. For example, modest changes in fault shape can deliver major differences in fold form (Suppe, 1983). Figure 8.5*A* reveals the differences in fold form resulting from a shift from planar to convex fault shape, whereas Figure 8.5*B* compares differences resulting from a shift from planar to concave fault shape.

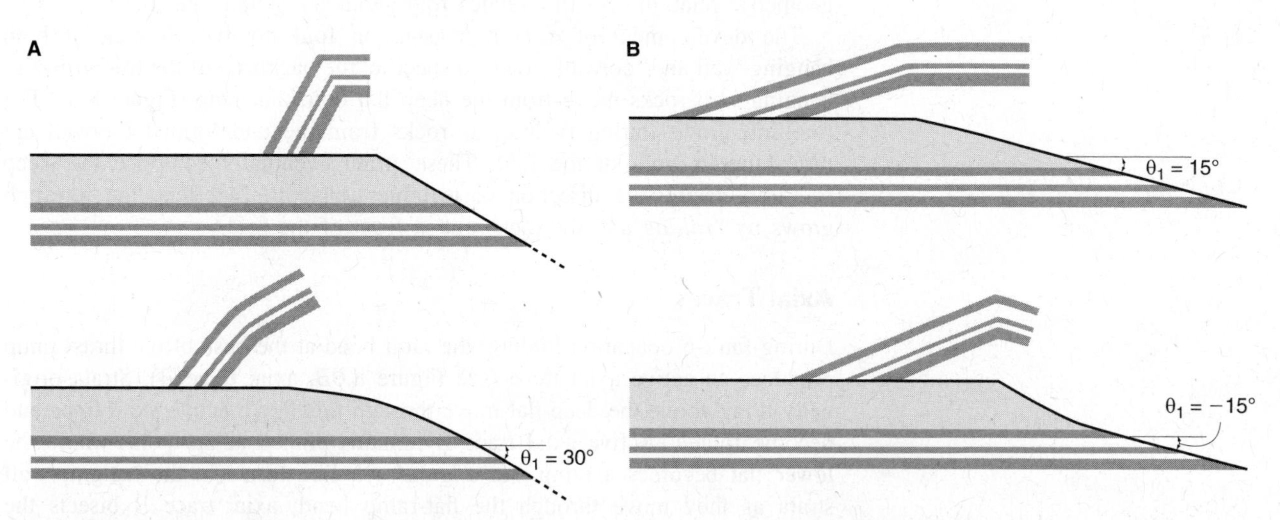

Figure 8.5 Fault-bend folds may have strikingly different fold forms even in cases where differences in fault geometry are pretty subtle. (*A*) Fold shape resulting from a planar ramp versus one that is slightly convex. (*B*) Fold shape resulting from a planar ramp versus one that is slightly concave. [From Suppe, 1983, *American Journal of Science*, v. 283, Figure 13, p. 702. Reprinted by permission of the American Journal of Science.]

FAULT-PROPAGATION FOLDS

General Character

A second major category of fault-related folds is called **fault-propagation folding**. **Fault-propagation folds** form when the fault at depth accommodates displacement along a layer-parallel flat, but then ramps up through the rock section and "*dies out*" into a fold (Figure 8.6). A fault-propagation fold absorbs the slip that otherwise would have been required by faulting. In effect, the fold replaces the upper reaches of the fault. In the case of crustal shortening, the ramp fault is a thrust or reverse fault, which tips out upward into an anticline or monocline as shortening by folding replaces the shortening by faulting. A fault-propagation fold is like a "*process zone*" at the advancing tip of a fault. The expressions "*dies out*" and "*is replaced by*" are actually misnomers, because what is really happening is that beds are being folded as the tip of the fault propagates upwards. But these expressions are firmly embedded in our vocabulary, and they will stick.

In the western United States some of the best-exposed fault-propagation folds are Laramide style, formed in late Cretaceous/early Tertiary, and are found in the Wyoming province and Colorado Plateau (see Figures 6.91, 7.20, and 7.98) (Chester et al., 1991, Mitra, 1990b, Erslev and Rogers, 1993, Schmidt et al., 1993). Today there are impressive fault-propagation folds forming in the Sierras Pampeanas, Argentina, within the easternmost part of the Andes (Jordan and Allmendinger, 1986; Allmendinger, 1986; García and Davis, 2004).

At each stage of development of fault-propagation folds, hanging-wall strata move along a lower flat and up a ramp, but the ramp does not tie into an upper flat. Instead, the ramp thrust is replaced upward by an anticline that is asymmetrical in the direction of transport (see Figure 8.2*B*). As long as the thrusting proceeds, an unbroken fault-propagation fold just beyond the tip of the fault will be transitory, and will eventually be overtaken, sliced, and offset by the fault as it advances. Thus faulting, exclusively, will mark the lowest reaches of a fault-propagation fold system. A faulted fold will be seen in the middle reaches of a fault-propagation-fold system. The faulted fold will transition upward to an unfaulted fold. Suppe (1985) worked out the precise geometric relationships that relates fold geometry to fault geometry.

The development of a fault-propagation fold involves folding of both hanging-wall and footwall strata. In specific, the backlimb of the fold grows as hanging-wall rocks move from the deep flat onto the ramp (Figure 8.6). The forelimb grows during faulting as rocks from the undeformed footwall are turned up in front of the fold. These strata eventually comprise the steep forelimb, facing the direction of tectonic transport. In effect, the forelimb grows by "*rolling up*" the rocks out in front of the fold.

Axial Traces

During fault-propagation folding, the fault bend at the base of the thrust ramp produces an active axial trace (see Figure 8.6*B*, axial trace B). Strata originally lying above the deep flat move through this fixed, active axial trace and become folded. Active axial trace B remains pinned at the point where the lower flat becomes a ramp. An angular syncline forms in the hanging-wall strata as they move through the flat-ramp bend; axial trace B bisects the interlimb angle of this syncline (see Figures 8.6*B*,*C*). The beginnings of a backlimb are evident almost as soon as the fault changes dip and thrusting begins (see Figure 8.6*B*). The backlimb grows in breadth, but remains

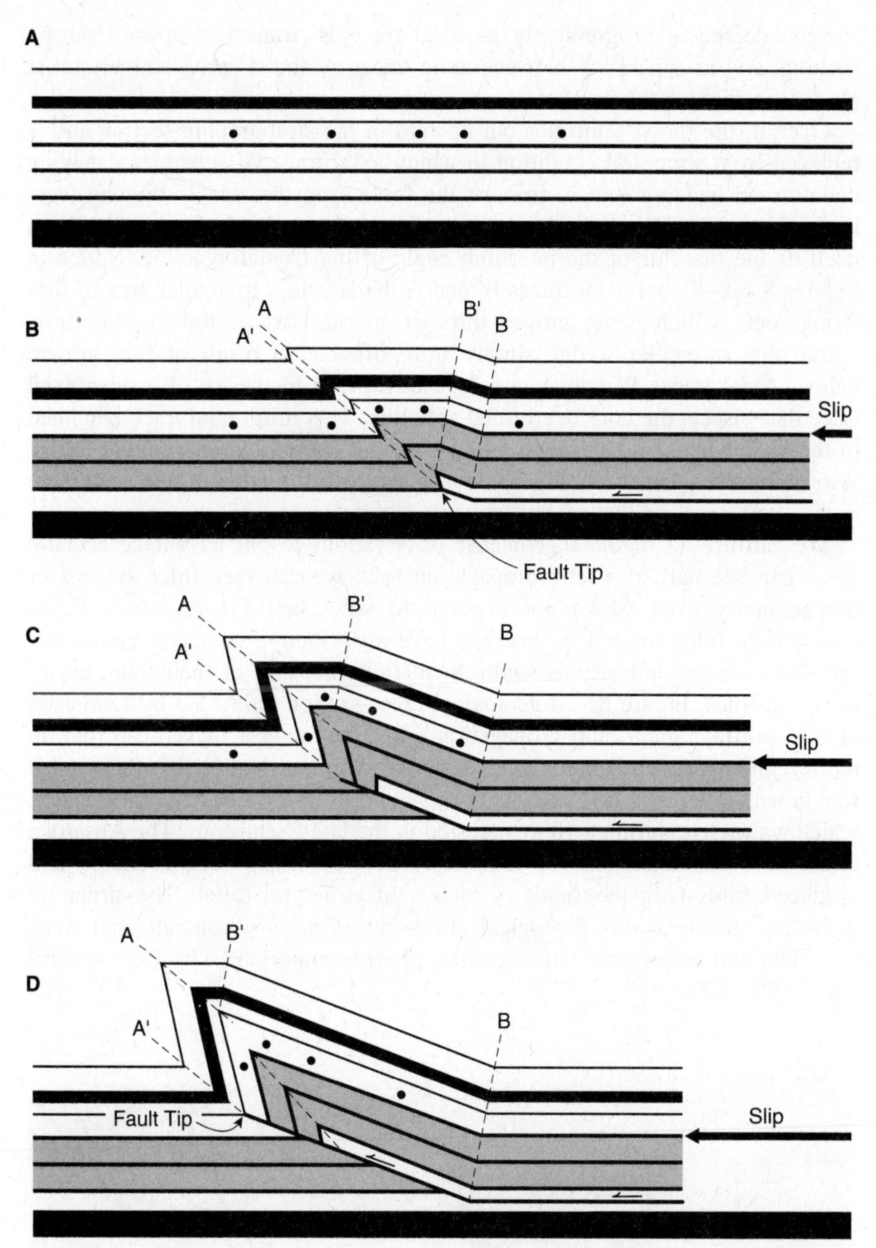

Figure 8.6 Fault propagation folds form when faults die out (tip out) upwards into folds. Progressive evolution of fault-propagation fold geometries. (*A*) Flat-lying sequence of sedimentary rocks. (*B,C,D*) As thrusting takes place, the fault tip migrates steadily upward. Beyond the tip the fault is "*replaced*" by an overturned syncline. Above the fault tip in the hanging wall there is a complementary fault-propagation fold: an overturned anticline whose axial surface dips more steeply than the thrust. See text for details. [From Suppe, J., Principles of Structural Geology, © 1985, p. 163. Reprinted by permission of Prentice-Hall, Upper Saddle River, New Jersey.]

constant in dip (and parallel to the fault ramp) as the inactive axial trace B′ separates itself from B and progressively moves up the ramp (see Figures 8.6*B,C*).

The trickiest fold geometry lies out front of the tip region of the thrust fault. In that region a syncline/anticline pair develops as the propagation of the fault proceeds through the section (see Figures 8.6*B,C*). The deepest reach of the axial trace of the syncline (see A′, Figure 8.6*B*) connects with the tip point of the thrust fault, no matter where the fault tip resides. This axial trace (A′) moves through the rocks as the fault tip propagates upward and laterally through the stratigraphic section. Parallel to A′ is the axial trace (A) of the emerging and growing anticline (see Figures 8.6*A,B*). Throughout fault-propagation folding, axial traces A and B′ converge downward to meet at a point of intersection, thus defining a triangular wedge of bedding that is itself not folded (see Figures 8.6*A–C*). The size of this triangular wedge at any horizon decreases progressively as axial trace B′ translates upward during faulting, and as separation between axial traces A and A′ progressively takes place (see Figures 8.6*B–D*).

Overall, the thrust fault dies out upward in the stratigraphic section and is replaced by a structural condition in which axial trace A′ separates flat-lying undeformed bedding (out in front of the fold) from the steeply dipping forelimb of the fold. This frontal axial trace is pinned to the fault tip and orients itself as the bisector of the interlimb angle of the frontal syncline. Notice in Figures 8.6*A–C* that axial traces B′ and A demarcate a triangular area of flat-flying rock, which never moved through an axial trace. Rather, the strata within this triangular wedge simply were lifted as a result of fold growth below. Axial traces B′ and A converge downward to the tip of a third axial trace that bisects the core overturned anticline. This third axial trace is pinned to the fault where fold shortening replaces fault shortening (see Figure 8.6*D*). We can observe that the stratigraphic level of the bifurcation of this axial trace is the same level as that of the fault tip.

We can use all of these geometric observations to our advantage because *if* we can see part of a fault-propagation fold, we can then infer the rest of the geometry even where not exposed to view (see Figure 8.6*C*). Fault-propagation folds formed in this way have a distinctive geometry commonly referred to in the industry as **snake head folds** because of their tight, asymmetric profiles. Figure 8.7, a geological cross-section published by Lamerson (1982), portrays giant fault-propagation folds, including a snake head fold, in the Ryckman Creek oil field, southwestern Wyoming. Notice that these folds root downward into splay faults that ramped up from the Absaroka thrust fault, which was active during a 10 m.y. period in the late Cretaceous. The Absaroka thrust accommodated tens of kilometers of west-to-east displacement, and produced folds with thousands of meters of structural relief. The structural geometry shown in this geological cross-section is very complex, but well-bore data and seismic reflection profiles provide important subsurface control (see Figure 8.7).

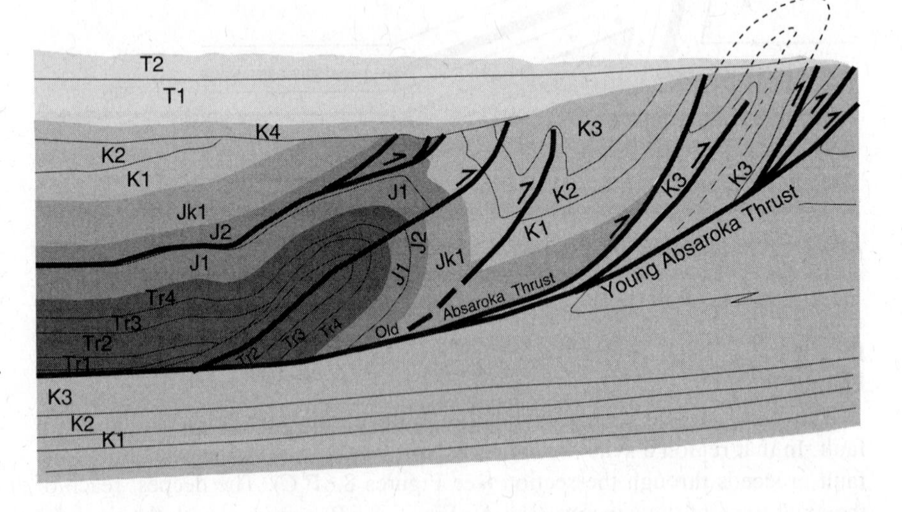

Figure 8.7 Geologic cross-section through part of the Ryckman Creek oil field, southwestern Wyoming. Splay faults ramp upward from the Absaroka thrust. "*Snake head folds*" formed as fault-propagation folds beyond the tips of the splay faults. A number of wells have been drilled into this geological system. Dip meter readings (in addition to seismic data) provide some subsurface control for this set of interpretations. [From Lamerson, 1982, Plate 5. Published with permission of the Rocky Mountain Association of Geologists.]

Caveats

Fault-related folding models assume constant bedding thickness during the progressive deformation. Furthermore, it is generally assumed that the gentle back limb of the major fold is always parallel to the dip of the fault surface at depth. Field observations indicate, however, that bedding thickness does not always remain constant during such folding (especially in the front limb of a fault-propagation fold), and that the back limb is not always parallel to the fault. Consequently, some structural geologists have modified the fault-propagation modeling equations to accommodate thickness changes on the front limb and mismatches between orientations of backlimbs versus faults (Jaimison 1987, Suppe et al., 2004). Such modifications to the original geometric relationships permit analysis of a wider range of observations, and they have a certain predictive value. Ultimately there should be a tight relationship between observable geometries in the real world and the modeled geometries in the computer world. If, in a given project, we fail to achieve such correspondence, then "*we better regroup.*"

Trishear Fold-Fault Models

Another way to explain the geometries of fault-propagation folds is through **trishear** (Figure 8.8*A*). Eric Erslev (1991) developed the **trishear model**, calling attention to the triangular zone of folded strata that converges downward to the underlying fault tip (see Figure 8.8*A*). In particular Erslev focused upon constructing velocity vectors and deformation paths for material points throughout the triangle zone. This continuum-mechanics approach to modeling fault-fold interactions was advanced further by Hardy and Ford (1997), Allemendinger (1998), and Hardy and Finch (2006).

In trishear fault-fold models we see the fold develop in advance of the upward-propagating fault tip, with material points of rock moving in rays radiating outward and upward from the underlying fault tip (Figure 8.8*B*). The material points pinned to each ray move at a common velocity, but each ray has its own unique velocity during folding. Along a line drawn perpendicular to the projection of the fault from its tip, the particle velocity for each point decreases in a linear fashion to zero (see Figure 8.8*B*) (Allmendinger, 1998). The coordination of ray velocities and point velocities is such that a zone forms progressively above the fault tip as the fault tip propagates through the rock column. The geometry of deformation modeled by trishear is similar to what is observed in the field (Figure 8.9).

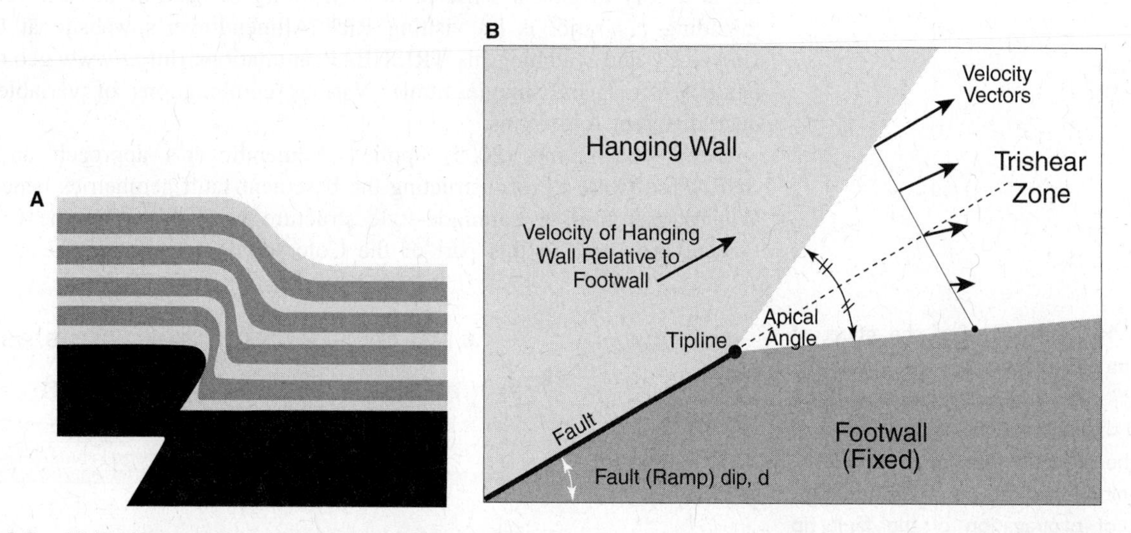

Figure 8.8 (*A*) Trishear model of fault-fold interaction. Note how the reverse fault affects basement (black) and lowest part of the cover. The fault slip is accommodated upward by folding and related deformation in a triangular wedge in front of the advancing fault tip. [From Erslev, 1991, *Geology*, v. 19, Figure 4, p. 619. Published with permission of the Geological Society of America.] (*B*) Within the triangular deformation zone, the velocity of particles decreases along a line that is perpendicular to the fault trace. The differential motion leads to deformation in the zone that matches what we observe. [From R. W. Allmendinger, *Tectonics*, v. 17, no. 4, Figure 1b, p. 641, copyright © 1998 by American Geophysical Union.]

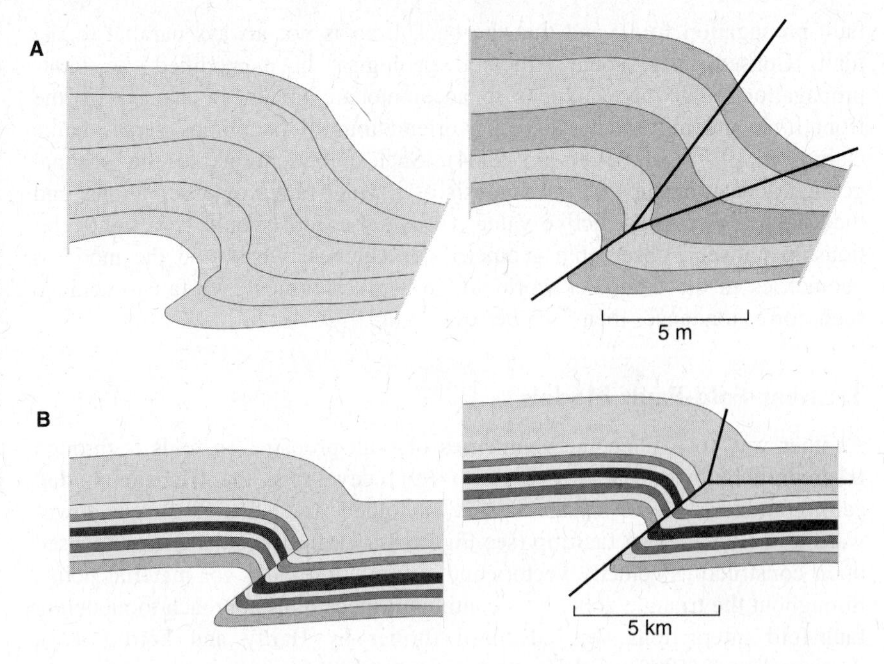

Figure 8.9 (*A*) Structural profile of meters-scale fault-propagation fold (left) from the Hudson Valley fold and thrust belt, and superimposed trishear elements (right). The structural profile is a simplified tracing of a photograph from Mitra (1990b, Figure 16). [From R. W. Allmendinger, *Tectonics*, v. 17, no. 4, Figure 11, p. 651, copyright © 1998 by American Geophysical Union.] (*B*) Structural profile of the kilometers-scale Rangely anticline (northwestern Colorado) (left) constructed by Mitra and Mount (1998), and superimposed trishear elements (right). [From R. W. Allmendinger, *Tectonics*, v. 17, no. 4, Figure 13, p. 653, copyright © 1998 by American Geophysical Union.]

Allmendinger (1998, 2004) developed a mechanical basis for trishear modeling, including a set of programs that can be used to efficiently carry out both forward and inverse numerical modeling of fault-propagation folds. His *TRISHEAR* (2D) program incorporates the five specific variables that Erslev (1991) emphasized as being critical to the final fault-fold geometries: initial location of the fault tip, fault dip, total fault slip, trishear angle, and propagation-to-slip (p/s) ratio. "*Propagation*" refers to rate of advance of the fault tip upward through the section. "*Slip*" refers to the rate of slip on the fault. A p/s ratio of 0 fixes the trishear to the footwall of the fault, whereas a p/s ratio of 1 fixes the trishear area to the hanging wall. Figure 8.10 shows how variations in p/s ratio play out in trishear modeling of fault-fold interactions. But the best way to gain a sense of the sensitivity of each of these 5 variables, including p/s ratio, is by visiting Rick Allmendinger's website at Cornell University and watching his TRISHEAR animations (http://www.geo.cornell.edu/RWA/trishear/TSmovies.html). Various combinations of variables produce different fold forms.

Davis and Bump (2009) applied Allmendinger's approach and *TRISHEAR* software to reconstructing the basement fault geometries beneath the Waterpocket fold, a Laramide-style structure in Utah within the Colorado Plateau province. In this part of the Colorado Plateau there are no "grand

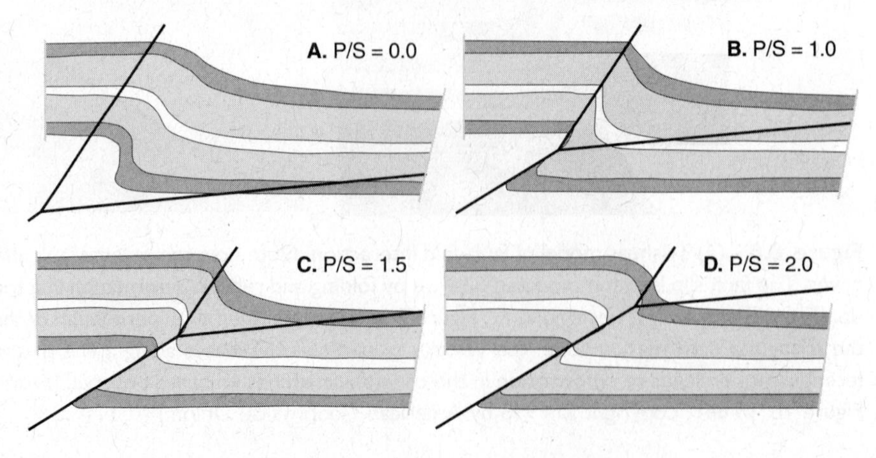

Figure 8.10 Illustration of the effect of varying the P/S ratio in trishear modeling, where "P" is the rate of propagation of the tip line of the fault and "S" is the amount of slip on the fault. In this set of examples, the magnitude of slip is held constant, but rate of propagation of the fault tip increases from (*A*) to (*D*). Note the wide variation of structural geometries that can result from the variation P/S. [From R. W. Allmendinger, *Tectonics*, v. 17, no. 4, Figure 2, p. 642, copyright © 1998 by American Geophysical Union.]

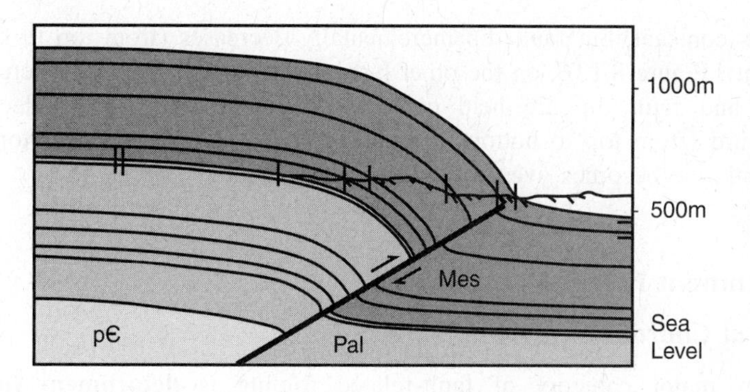

Figure 8.11 Best-fit forward model for the Waterpocket monocline on the east margin of the Circle Cliffs uplift, Utah. The present-day topographic profile is shown, along with bedding-dip attitudes. pC = pre-Cambrian; Pal = Paleozoic; Mes = Mesozoic. [From Davis and Bump, 2009, Figure 16, p. 115. Published with permission of the Geological Society of America.]

canyons' available to hike down to the deeper reaches of the big monoclines. The data we "*fed*" into *TRISHEAR* included (1) meticulously measured bedding dips across the fold structures, and (2) well-log based thicknesses for each stratigraphic formation. Based on running 200 *TRISHEAR* forward and inverse models that employed different combinations of the principal variables, Davis and Bump (2009) concluded that the Waterpocket fold "*roots*" into rather gently dipping thrust faults. The best fit for the Waterpocket fold proved to be one with a fault ramp angle of 30°, a trishear angle of 100°, and an initial fault tip 2.3 km below the basement-cover contact (Figure 8.11). The very best solutions revealed a p/s ratio of 6.0 within basement, reducing to 2.1 in cover for the remainder of the total 3.5 km of fault displacement.

Variations on the Theme

As in the case of fault-bend folding, we become aware that fault-propagation fold form is highly sensitive to differences in ramp height, fault dip, presence or absence of bed shear, and other factors. Structural geologists have found it productive to invest major effort in exploring the details of fault-propagation folding through numerical modeling, physical analogue modeling, and integrated field/subsurface investigations. As an example, Figure 8.12*A* shows four scenarios of fault-propagation folding in a situation where ramp height

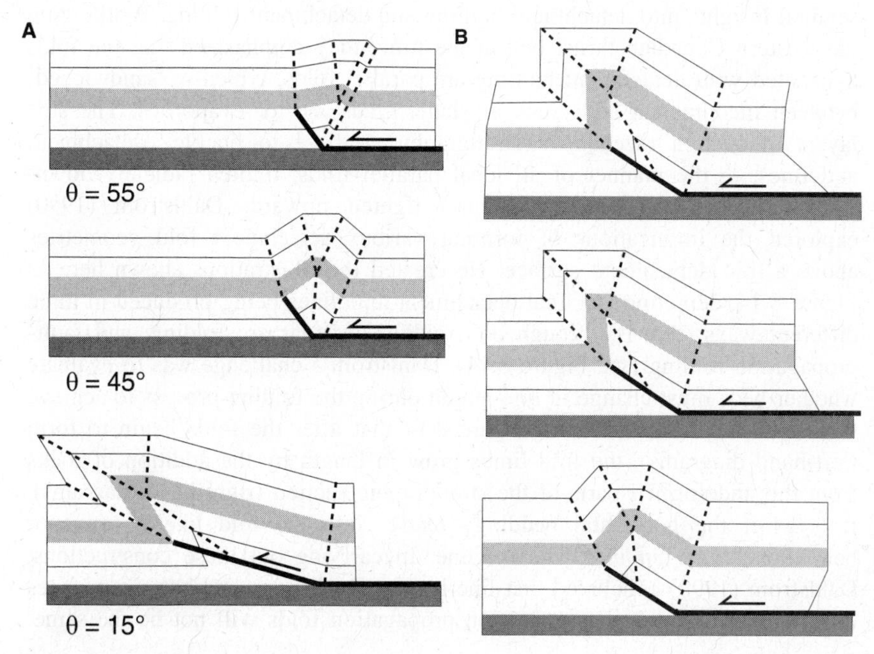

Figure 8.12 Fault-propagation fold form will vary with different conditions. (*A*) In these examples ramp height is held constant, but fault angle decreases from top to bottom diagram. (*B*) In these examples ramp height and fault angle are held constant, but bedding-parallel shear decreases from top to bottom. [From Mosar and Suppe, 1992, Figure 6, p. 126, and Figure 7, p. 128, in McClay, K., (ed.), Thrust tectonics: Chapman & Hall, London, p. 123–132. With kind permission from Springer Science+Business Media B.V.]

remains constant, but fault dip incrementally decreases (from top to bottom diagram). Figure 8.12*B*, on the other hand, shows four scenarios where ramp height and fault dip are held constant, but bed-parallel shear decreases downward (from top to bottom diagram). Where shear is greatest (top), the core anticline becomes overturned (Mosar and Suppe, 1992).

Detachment Folds

General Character

Another major category of fault-related folding is **detachment folding**. **Detachment folds** form in restricted stratigraphic intervals above **detachment faults**, not to be confused with "*detachment faults*" associated with metamorphic core complexes (see *Chapter 6*, Faults, and *Chapter 10, Shear Zones and Progressive Deformation*). Detachment faults 'run' parallel or subparallel to horizontal bedding, occurring directly beneath a fold or wave train of folds (Figure 8.13). Unlike faults associated with fault-bend folding and fault-propagation folding, detachment faults do not turn upward or downward into ramps. Fault slip "*detaches*" the rocks above the fault surface from those below, and for reasons that will become clear, fault shortening is replaced by fold shortening. This replacement occurs in a way that the structural characteristic of rocks above the detachment fault will be markedly different than those below (see Figure 8.13). Indeed, one of the ways we identify detachment faults "*hidden*" within incompetent bedding (e.g., shale, evaporates) is by looking for different structural geometries at different stratigraphic levels.

Figure 8.13 Detachment folding in the Jura Mountains of the Swiss Alps. The original is by Buxtorf (1916), then published by Shankar Mitra (2003). [Reprinted from *Journal of Structural Geology*, v. 25, Mitra, S., A unified kinematic model for the evolution of detachment folds, p. 1659–1673, © 2003, with permission from Elsevier.]

Detachment folds have been recognized for a long time (Willis, 1894). Dahlstrom (1960, 1969) and Bally et al. (1966) in the oil industry, provided seminal insights into detachment faulting and detachment folding. Working on the western Canadian thrust belt at the time, they emphasized that the folds associated with detachment faulting are parallel folds, typically 'sandwiched' between incompetent beds such as shale, mudstone, or evaporites. The stiff layers in such a mechanical stratigraphy are free to buckle. Detachment anticlines, in the manner of all ideal parallel folds, tighten radically downward, whereas detachment synclines tighten upward. Dahlstrom (1990) explored the implications of forming various detachment fold geometries above a thin detachment surface. He created the illustrations shown here as Figure 8.14, depicting isoclinal detachment anticlines being produced in three different ways; that is, through box folding, circular arc folding, and fault-propagation folding (see Figure 8.14). Dahlstrom's challenge was to evaluate whether beds must change in line-length during the folding process to achieve these extreme forms. Note in Figure 8.14 that after the folds begin to form (left-hand diagrams), the fold limbs grow in length by the addition of rocks from the undeformed parts of the stratigraphic section (right-hand diagrams). It is as if the horizontal bedding "*feeds*" into the fold like a conveyor belt, then "*turns up into*" the isocline. In carrying out these constructions, Dahlstrom (1990) concluded that line-lengths for the top and bottom surfaces of circular-arc and fault-propagation propagation folds will not be the same.

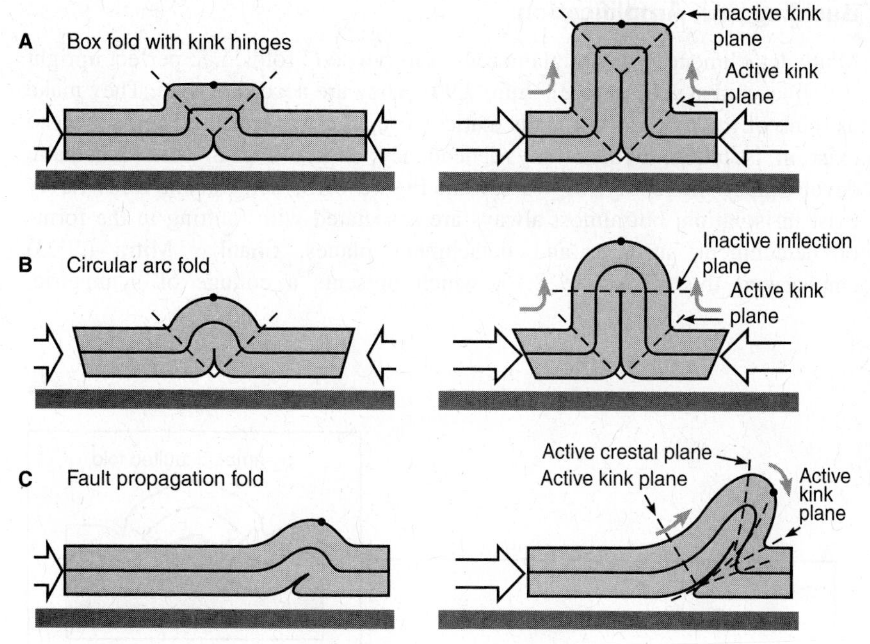

Figure 8.14 The evolution of three types of isoclinal detachment anticlines: (*A*) box fold; (*B*) circular-arc fold; and (*C*) fault-propagation fold. Note that following the initial formation of the anticlines (left), the folds grow through limb lengthening (right). See text for details. [From Dahlstrom, 1990, Geometric constraints derived from the law of conservation of volume and applied to evolutionary models for detachment folding : *American Association of Petroleum Geologists Bulletin*, v. 74, p. 336–344. AAPG © 1990, reprinted by permission of the AAPG whose permission is required for further use.].

This mismatch requires that other deformational mechanisms (e.g., faulting) must be at work to maintain the parallel fold forms.

Detachment Planes and Detachment Zones

One of Dahlstrom's most masterful contributions to understanding detachment folding is seen in Figure 8.15. Featured is a beautiful concentric fold panel, yet the "*beauty*" collapses in the pinched isoclinal synclines and anticlines. This overall fold geometry managed to evolve because the stiff concentric fold panel is overlain and underlain by incompetent, ductile, "*mobile*" rocks that were able to be squeezed around. These volumes of distorted incompetent rocks are **detachment zones**, and each of these (lower and upper) is overlain by unfolded strata (see Figure 8.15). Moreover, the base of the lower detachment zone is a **detachment plane**, as is the top of the upper detachment zone (see Figure 8.15).

The elegance and subtleties of such detachment-fold architecture were grasped thoroughly by Clint Dahlstrom (1960, 1969, 1990). Note, for example, how he points out (see Figure 8.15) that the lower detachment plane dies out to the right, whereas the upper detachment plane dies out to the left. He concluded that the concentric fold panel provides the linkage to transferring slip from the lower detachment plane to the upper detachment plane. This is another example of the way nature works wonders.

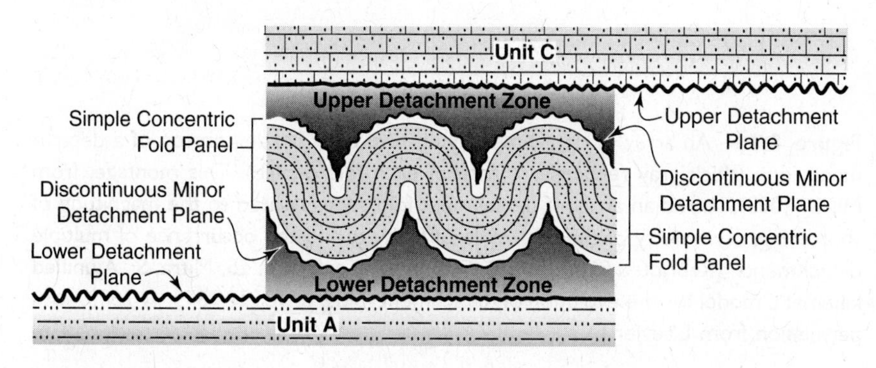

Figure 8.15 The classic Dahlstron detachment-fold analysis. Parallel folds tighten into their cores, both downward in the anticline and upward in the synclines. This tightening results in a volume problem in the fold cores. There is just too much rock to fit into the increasingly small space. Thus the folds detach both below the 'fold train' and above. The concentric fold panel provides the linkage to transfer the slip from the lower detachment plane to the upper detachment plane. [From Dahlstrom, 1990, Geometric constraints derived from the law of conservation of volume and applied to evolutionary models for detachment folding : *American Association of Petroleum Geologists Bulletin*, v. 74, p. 336–344. AAPG © 1990, reprinted by permission of the AAPG whose permission is required for further use.]

Buckling and Amplification

Many detachment folds are glamorous "*Hollywood*" folds, i.e., perfect upright anticlines and synclines (see Figure 7.99). They are accordion-like. They make us think of the physics and mathematics of regular waveforms. They appear to exist in isolation, without any dependency on faulting in their kinematic development. In reality, as we saw in Figure 8.15, detachment folds rarely exist in isolation, but almost always are associated with faulting in the forms of detachment surfaces and detachment planes. Shankar Mitra (2003) emphasized this in Figure 8.16, which presents a collage of symmetric,

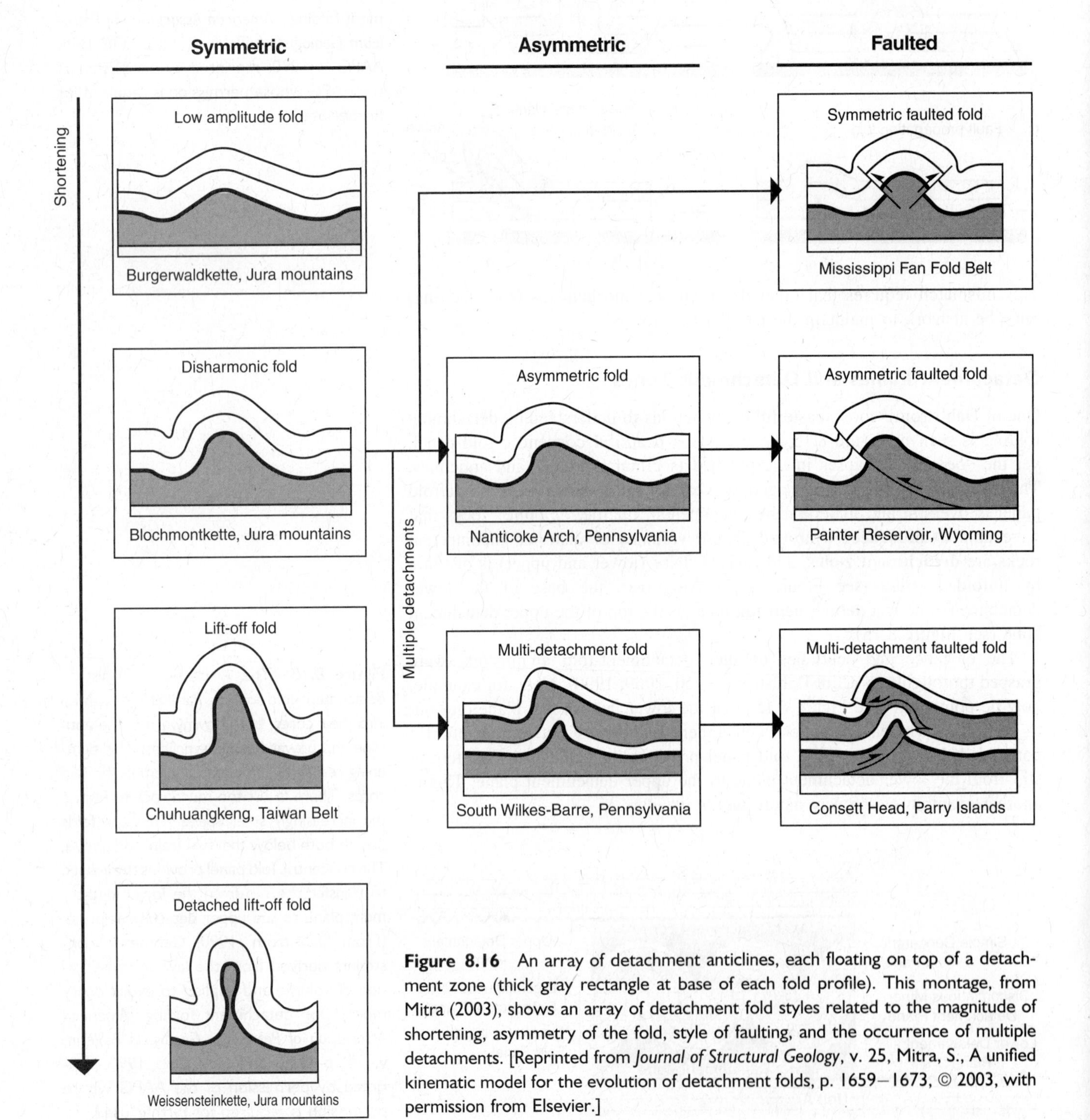

Figure 8.16 An array of detachment anticlines, each floating on top of a detachment zone (thick gray rectangle at base of each fold profile). This montage, from Mitra (2003), shows an array of detachment fold styles related to the magnitude of shortening, asymmetry of the fold, style of faulting, and the occurrence of multiple detachments. [Reprinted from *Journal of Structural Geology*, v. 25, Mitra, S., A unified kinematic model for the evolution of detachment folds, p. 1659–1673, © 2003, with permission from Elsevier.]

asymmetric, and faulted detachment folds, each of which is seen to "*float*" on detachment zones and detachment planes. The variations relate to magnitude of shortening, asymmetry, style of faulting, and influence of multiple detachments (Mitra, 2003).

We conclude that detachment anticlines are initiated in competent (stiff) units where they are underlain by incompetent (soft) strata, such as shale, mudstone, and evaporite (Figure 8.17). The specific loci of fold formation may be at a point where there already exists a hinge or crest of a very subtle anticline, produced simply by shifts in the primary initial dips of layers or some other stratigraphic/structural change that acts as a "*seed*" for the fold growth. Such a trigger may determine the exact site where buckling is initiated and where amplification then takes place. The competent beds buckle under these conditions and form parallel folds by flexural slip. These folds then become true detachment folds as the softer lithologies (e.g, shale, mudstone, evaporites) are fed into the core of the fold by displacement along the sub-horizontal detachment fault (see Figure 8.17). If the detachment layer is thin (Figure 8.17*A*), the fold will not amplify much because there is not enough mobile material to flow into the core of the fold. If, however, the detachment layer is thick, significant fold amplification can be accommodated (Figure 8.17*B*) (Stewart, 1996). The growth of the detachment anticline continues as long as buckling of the stiff layer is supported (underneath) by the flow of incompetent rock into the core of the anticline. The detachment surface itself is the fault interface between incompetent layers above it and the undeformed competent layer below the fault in the footwall.

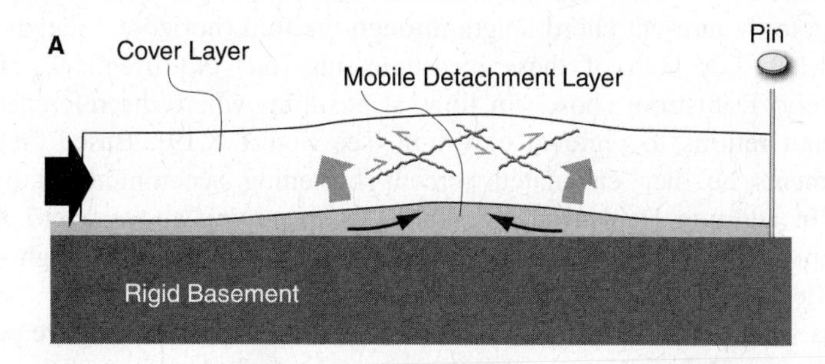

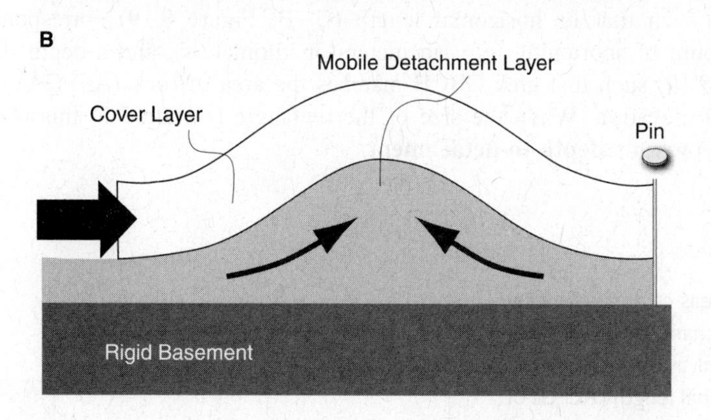

Figure 8.17 Thickness and mobility of detachment layer will influence whether deformation is by pure folding or by a combination of folding and faulting. (*A*) In this case the mobile detachment layer is very thin, and consequently the fold cannot be amplified very much. Continued shortening requires thrust faulting. (*B*) In this second case the mobile detachment layer is thick, and there is plenty of material to support the fold as it grows. [Reprinted from *Journal of Structural Geology*, v. 18, Stewart, S., Influence of detachment layer thickness on style of thin-skinned shortening., p. 1271–1274, © 1996, with permission from Elsevier.]

Determining Depth to Detachment

Imagine you have just carried out geologic mapping of detachment folding in northern Wyoming and have prepared a geological cross-section (Figure 8.18) that is consistent with map relations and all subsurface data available. Then someone walks up to you and insists that you predict the depth to the

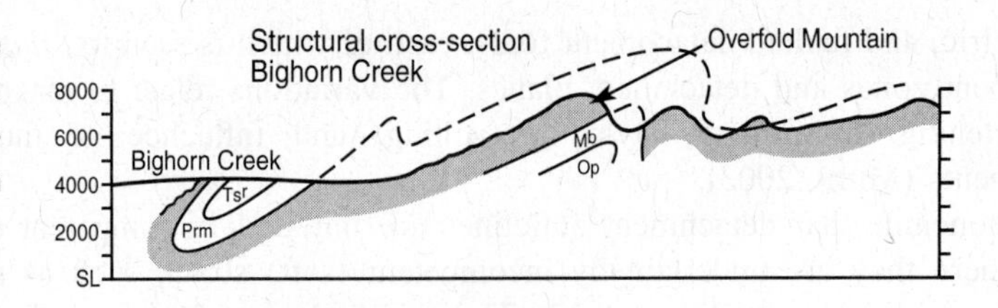

Figure 8.18 Geologic cross-section showing the structural geology of the Bighorn anticline, Wyoming. [From Dahlstrom, 1969a. Reproduced with permission of the National Research Council of Canada from *Canadian Journal of Earth Sciences*.]

detachment surface based upon your geological cross-section. How on earth do you proceed?

Area balancing (Chamberlain, 1910, Mitra and Namson, 1989) is the means to calculating depth of detachment. The method for doing this relies on the premise that the volume of folded/faulted rock remains constant, before and after deformation (i.e., area is conserved). In the case of crustal shortening, the depth-to-detachment calculation must balance the area of folded rock raised vertically above regional elevation against the area of rock that enters into the section laterally through shortening. **Regional elevation** for any bed is the elevation at which it would reside if the bed were undeformed.

Dahlstrom (1969b) presented a beautiful example of the method used for determining depth to detachment (Figure 8.19). He started with a cross-section showing the geometry of detachment folding as mapped in the field (see Figure 8.18). Based upon this cross-sectional geometry, he calculated the amount of layer-parallel shortening achieved by the folding. He accomplished this by measuring folded bed length (trace A-E-F-G, Figure 8.19), and compared this to the present chord length through the fold (horizontal length A-G, Figure 8.19). For both of these measurements (our old friends l_o and l_f, respectively) Dahlstrom chose **pin lines** at positions where the reference bed departs and returns to regional elevation (see Figure 8.19). Based on these measurements he then calculated percent-shortening accommodated by the folding. In addition Dahlstrom measured the area lying above chord A–G, which represents the amount of material in the fold that was, through shortening, lifted above regional elevation (see Figure 8.19).

This is where Dahlstrom's approach becomes ingenious. To achieve perfect area balancing, Dahstrom constructed a rectangle at the butt end of the cross-section, such that its horizontal length (G–B, Figure 8.19) corresponds to the amount of shortening (e.g., in meters or kilometers), and a depth (G–H, Figure 8.19) such that area GBCH matches the area of rock (AEFGA) above regional elevation. When the size of the rectangle is crafted in this way, its depth represents **depth-to-detachment**.

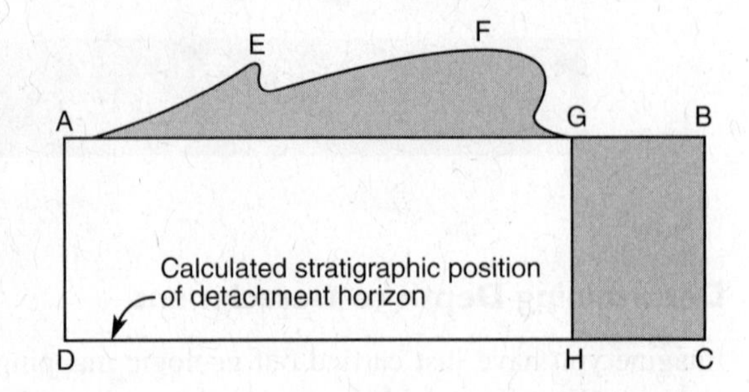

Figure 8.19 Schematic sections showing the basic ideas of area balancing and depth of detachment calculation. Deformation has changed the shape of area ABCD to AEFGBHD. Shortening can be measured by subtracting the current length (AG) of the deformed bed from the original length (AEFG) of the undeformed bed. The area (AEFG) of the rocks below the marker bed that are *above* their regional elevation is equal to the shortened area (GBCH). Once we know the shortening (GB) and the area (AEFG) of the rocks moved above their regional, we can divide area by shortening to determine the depth (BC) to the detachment surface. The secret to all of this is knowing that area GBCH = area AEFG. [From Dahlstrom, 1969b, Figure 12. Reproduced with permission of the National Research Council of Canada from the *Bulletin of Canadian Petroleum Geology*.]

Other Examples of Balance and Depth to Detachment

R.T. Chamberlain (1910, 1919) first used the area-balance method on whole thrust belts, and yet it works for individual folds as well (Mitra, 2002). A classy display of the modern methodology is pictured in Figure 8.20, representing the work of Epard and Groshong (1993). This is an example for which everything is known, allowing us to grasp fully the interrelationships. There are two key beds (#1 and #2), which started out horizontal and undeformed. The lower bed (#1) was originally positioned at a height h_1 above the detachment surface, and the higher bed (#2) sat at a height h_2 above the detachment. Both beds were then shortened by a known amount, D (Figure 8.20*A*). Excess area was measured for each of the two beds, and these areas are S_1 and S_2, respectively. **Excess area** is that part of the deformed stratigraphic sequence lifted above regional elevation (see Figure 8.20*A*). All of these values and relationships are then plotted graphically in an **excess-area diagram** (Figure 8.20*B*) constructed by plotting excess area for S_1 and S_2 on the *y*-axis and height-above-detachment for S_1 and S_2 on the *x*-axis (see Figure 8.20*B*). The slope of the line passing through (h_1,S_1) and (h_2,S_2) is the magnitude of displacement on the detachment (see Figure 8.20*B*).

But what if we do not know the depth to detachment? We might infer it through knowledge of the mechanical properties of the stratigraphy in the subsurface, but that is normally not good enough. Epard and Groshong (1993) provide a clever solution. Because depth-to-detachment is not known for this second example, an **arbitrary reference level** must be chosen, both in the cross-section (Figure 8.21*A*) and in the excess-area diagram (Figure 8.21*B*). The line that links the plotted points in the excess-area diagram intersects this reference level and the horizontal axis of the graph (see Figure 8.21*B*). The intersection point along the *x*-axis has a value that represents true depth to detachment. Moreover, the slope of the line through the data points is the magnitude of slip on the detachment surface.

The "*sky is the limit*" in evaluating depth-to-detachment and calculating total slip on a detachment. Figure 8.22, again from Epard and Groshong

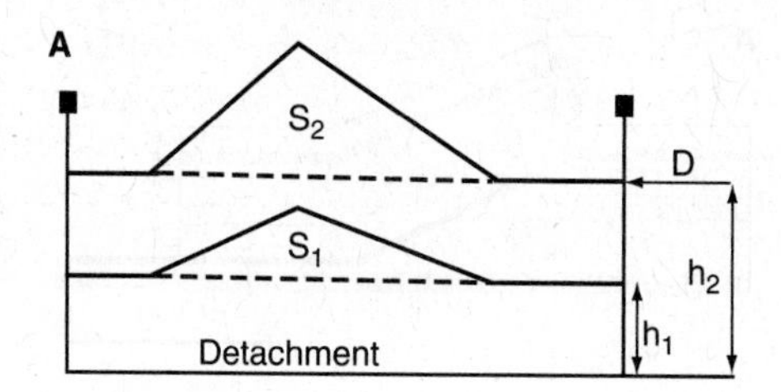

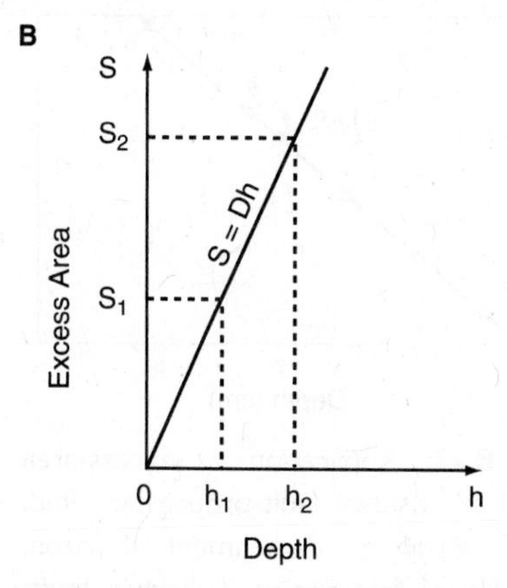

Figure 8.20 Area balancing and depth to detachment surface. (*A*) Cross-section showing what is known at the start of this example: amount of shortening (*D*), position of detachment, heights of each of two layers above the detachment surface (h_1 and h_2), and excess area of folded beds above regional (S_1 and S_2). (*B*) Graph of these data on an "*excess area plot.*" Excess area is plotted against height above detachment for each layer. Slope of the best-fit line corresponds to amount of slip on the detachment surface. See text for details. [From Epard and Groshong, 1993, Excess area and depth to detachment: *American Association of Petroleum Geologists Bulletin*, v. 77, no. 8, p. 1291–1302. AAPG © 1993, reprinted by permission of the AAPG whose permission is required for further use.]

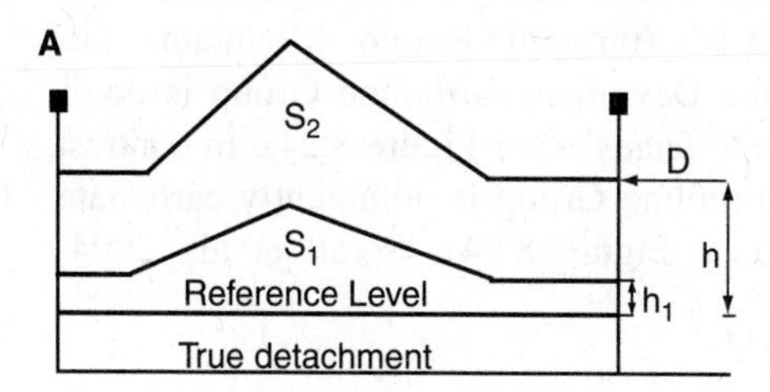

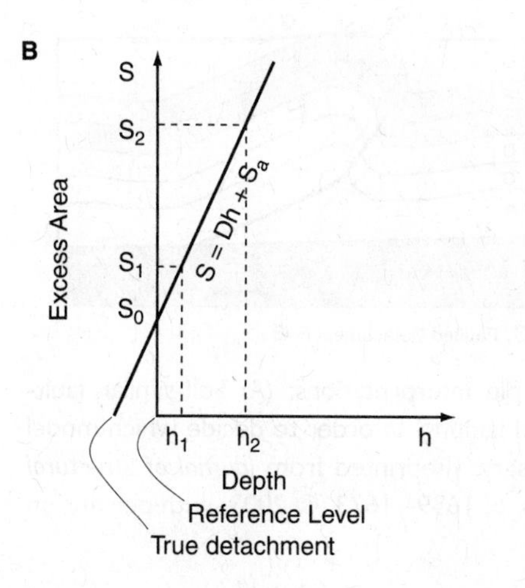

Figure 8.21 Method for determining depth to detachment. (*A*) A "*reference level*" is introduced, with respect to which heights of key beds are measured. (*B*) The data are plotted as "excess area" against heights above reference level. The best-fit line (whose slope magnitude is displacement, D) intersects the *x*-axis at a value corresponding to the depth of the detachment surface beneath the reference level. See text for details. [From Epard and Groshong, 1993, Excess area and depth to detachment: *American Association of Petroleum Geologists Bulletin*, v. 77, no. 8, p. 1291–1302. AAPG © 1993, reprinted by permission of the AAPG whose permission is required for further use.]

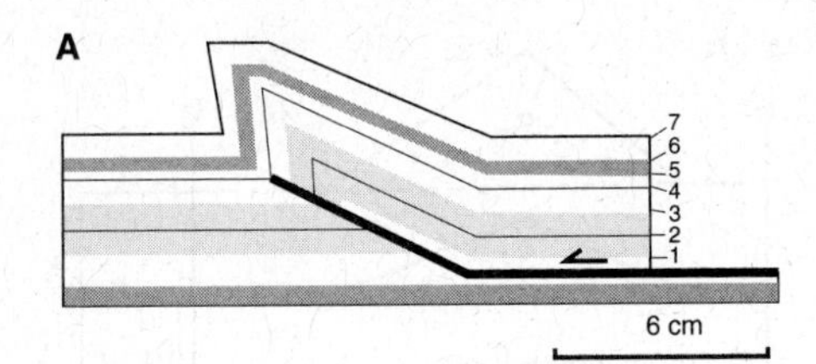

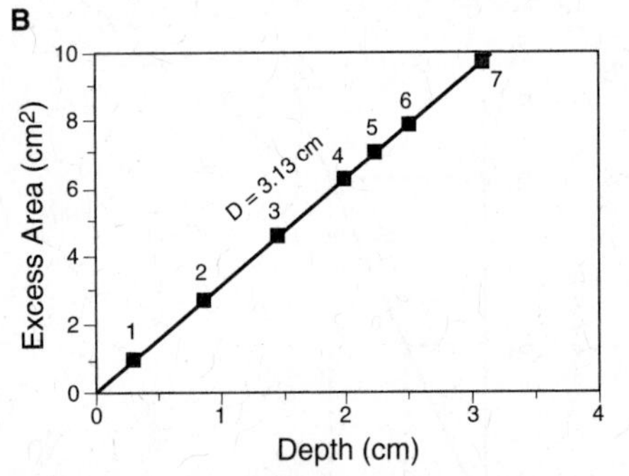

Figure 8.22 Application of excess-area method to an ideal fault-propagation fold, formed atop a detachment horizon. (*A*) Profile of fold shows 7 marker beds. (*B*) Excess area was determined for each of 7 reference beds. Slope of line is the displacement on the detachment surface. [From Epard and Groshong, 1993, Excess area and depth to detachment: *American Association of Petroleum Geologists Bulletin*, v. 77, no. 8, p. 1291–1302. AAPG © 1993, reprinted by permission of the AAPG whose permission is required for further use.]

(1993), makes this clear. Excess area for each of 7 stratigraphic horizons were measured (see Figure 8.22*A*) and then plotted (see Figure 8.22*B*) in relation to height above a known detachment horizon. A best-fit line is drawn through the data points, underscoring a tight solution and the relative ease in evaluating total slip (D) along a detachment.

Differentiating Fault-Propagation Folds From Detachment Folds

You may have noticed that the uppermost geometries of fault-propagation folds and asymmetrical detachment folds look very similar to one another. Think how challenging it is to interpret subsurface structure. We can fit different structural interpretations to the same data base of bedding and faulting mapped in the field. Without subsurface control, it is frankly very difficult to come up with a unique solution. This point is illustrated in Figure 8.23, from Mitra (2003), where surface data are fit to three different fault/fold models: self-similar fault-propagation folding, tri-shear fault-propagation folding, and faulted detachment folding.

Structural Style and Mechanical Stratigraphy

We are seeing that large differences in mechanical stratigraphy can predetermine whether deformation is achieved by fault-bend folding, or fault-propagation folding, or detachment folding. Up until now, our examples have been ones of gross differences in mechanical stratigraphy. However, even slight differences in mechanical stratigraphy can make a big difference in predetermining fault/fold mechanisms and final structural style.

A really good example of this was reported by Spratt et al. (2004), based on their work in the Canadian Rockies. In particular, they compared and contrasted the fold/fault styles at two localities only ~20 km apart, and along a common thrust (the Bare thrust) in the Canadian fold-thrust belt. Figure 8.24 is a structure/tectonic map that shows the relevant structural geology and stratigraphy, including the traces of major west-dipping thrust faults. Note in particular the trace of the Bare thrust, and lines of cross-sections A-A′ (through Mount Mitchener) and B-B′ (through Rough Mountain) (see Figure 8.24). At Mount Mitchener, the Devonian Fairholme Group is dominantly shale; that is, the so-called "*shale facies*" (see Figure 8.24). In contrast, at Rough Mountain, the Devonian Fairholme Group is dominantly carbonate; that is, the so-called "*reef facies*" (see Figure 8.24) (Spratt et al., 2004).

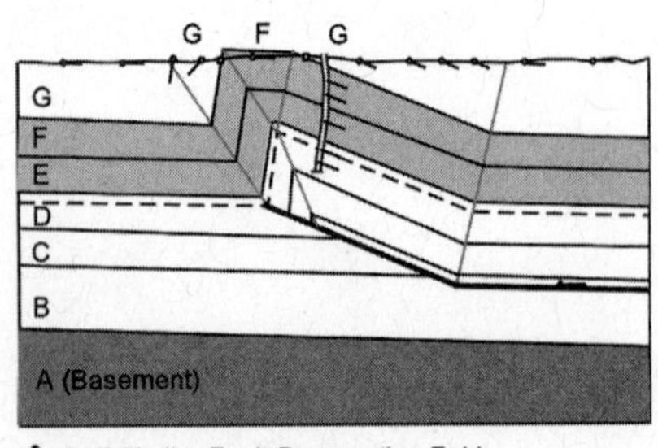

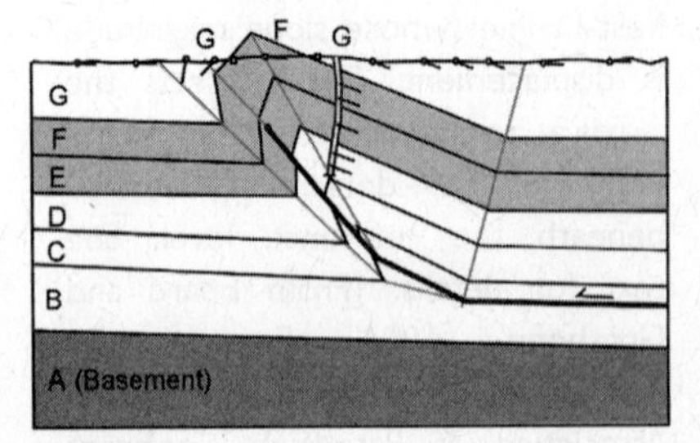

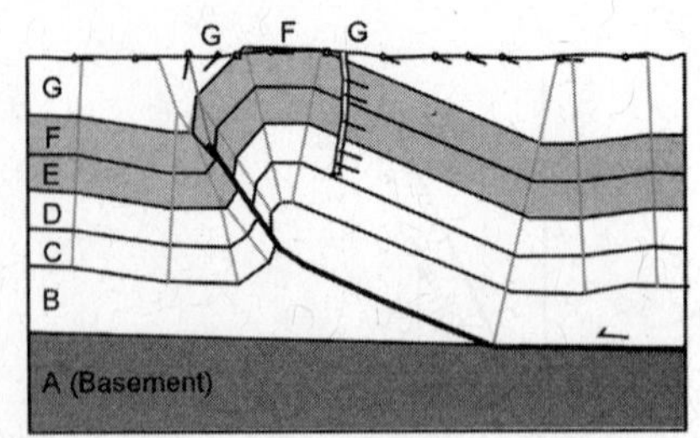

Figure 8.23 Fundamental map data and subsurface control can lend itself to multiple interpretations: (*A*) Self-similar fault-propagation folding; (*B*) trishear fault-propagation folding; and (*C*) detachment folding and faulting. In order to decide which model to use, additional subsurface data and knowledge of mechanical stratigraphy are necessary. [Reprinted from *Journal of Structural Geology*, v. 25, Mitra, S., A unified kinematic model for the evolution of detachment folds, p. 1659–1673, © 2003, with permission from Elsevier.]

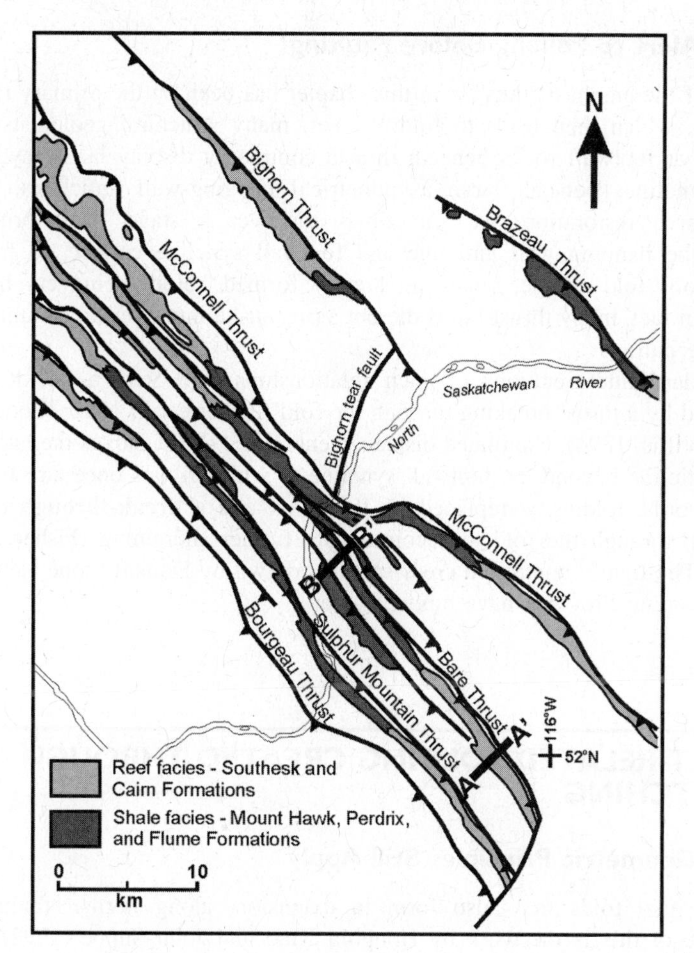

Figure 8.24 Geologic map showing thrust faults and shale and reef facies of Devonian Fairholme Group in the Canadian Rockies near the North Saskatchewan River. A–A′ is location of cross-section shown in Figure 8.25*A*. B–B′ is the cross-section location shown in Figure 8.25*B*. [From Spratt et al., 2004, Changes in structural style controlled by lithofacies contrast across transverse carbonate bank margins—Canadian Rocky—Mountains and scaled physical models, in K. R. McClay (ed.), Thrust tectonics and hydrocarbon systems: *American Association of Petroleum Geologists Memoir*, v. 82, p. 259–275. AAPG © 2004, reprinted by permission of the AAPG whose permission is required for further use.]

The facies differences exerted profound influence on style of deformation! Tight similar folding is the preferred fault/fold style in the Devonian-Mississippian strata that overlies the shale facies at Mount Mitchener (Figure 8.25*A*). In contrast, parallel folds and brittle faulting characterizes Devonian-Mississippian strata that overly the reef facies at Rough Mountain (Figure 8.25*B*). Only the shale facies supports detachment folding.

This sensitivity of structural style to mechanical stratigraphy is really quite extraordinary. Between the detachment folds in Mount Mitchner to the imbricate thrusting in Rough Mountain there is just a small shift (i.e., 10%) in the amount of (soft) shale versus (stiff) carbonate in the stratigraphic section. Yet this is sufficient to predetermine fold/fault style.

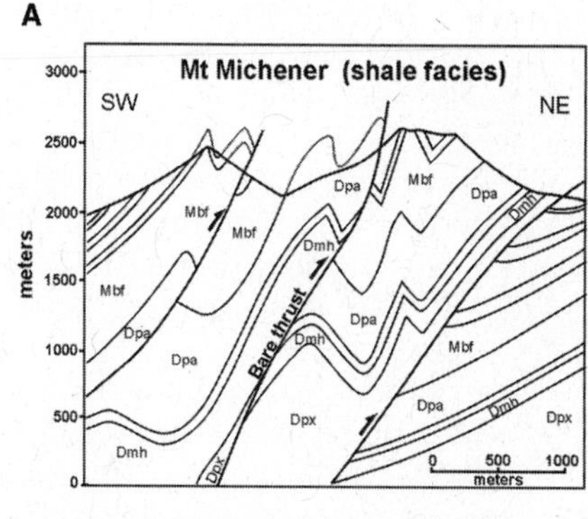

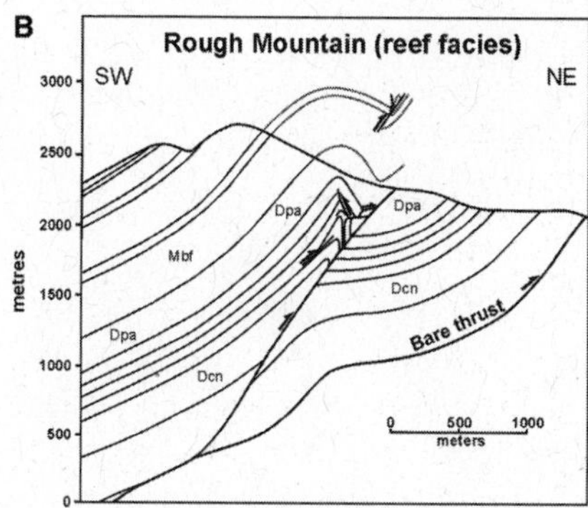

Figure 8.25 Cross-sections showing the change in structural style of Devonian and Mississippian rocks overlying the facies boundary in the Fairholme Group. Both sections are at the same scale, and without vertical exaggeration. (*A*) At Mt. Michener tight similar folds occur above the shale facies. (*B*) At Rough Mountain, parallel folds and brittle faulting occurs above the reef facies. [From Spratt et al., 2004, Changes in structural style controlled by lithofacies contrast across transverse carbonate bank margins—Canadian Rocky Mountains and scaled physical models, in K. R. McClay (ed.), Thrust tectonics and hydrocarbon systems: *American Association of Petroleum Geologists Memoir*, v. 82, p. 259–275. AAPG © 2004, reprinted by permission of the AAPG whose permission is required for further use.]

Keep Alert to Folding Before Faulting

Much of the emphasis thus far in this chapter has been on the primary role of faulting, which then leads to folding. Yet, many structural geologists have noted that footwall rocks beneath thrusts commonly display large asymmetrical synclines beneath large asymmetrical hanging-wall anticlines. Upon progressive restoration, we sometimes discover a stage in deformation where the hanging wall anticline and footwall syncline connect to form a continuous fold couple, which apparently formed before being cut by the thrust. In fact, many thrust faults die out along strike into overturned anticline/syncline pairs.

Standard interpretations of such relationships start with a buckle fold, followed by a thrust breaking through the fold (Fisher et al., 1992; Woodward 1997; Willis, 1894). Continued displacement on the thrust moves the hanging-wall anticline beyond the footwall syncline to which it was once attached. In other words, folding is replaced by thrusting, and a **break-through thrust** steps up through the fold to accommodate further shortening (Fisher et al., 1992). The tightly controlled cross-sections drawn by Donald Stone (1993) in the Wyoming Province have this character.

FAULT-RELATED FOLDING CREATED THROUGH STRETCHING

Basic Geometric Principles Still Apply

Fault-related folds may also form in extension, along normal faults. An example of this is the work by Hongbin Xiao and John Suppe (1992), who reconstructed the geometry of normal faults in the subsurface of the Gulf of Mexico based on fold geometries in the hanging walls of the faults. The example shown in Figure 8.26 is one in which the normal fault is by no means planar but instead is marked by a steeply dipping upper segment and a more shallowly dipping lower segment. We can track offset in this example by being attentive to the deformation path of a common point *x*, which was located right at the bend before faulting. Upon initiation of faulting, reference point *x* is split in half, and the hanging-wall part of it moves progressively down the fault plane. Figure 8.26*A* is deliberately fanciful, showing what would happen during normal faulting if the hanging wall rock was unbendable. In this situation, normal-slip along the fault bend produces a wide gap. But in the real world, **hanging wall collapse** accompanies the normal faulting, and thus a fissure-like gap never forms (Figure 8.26*B*). There are different forms of hanging-wall collapse, but in this case the hanging wall "*collapses*" toward the fault by kink-like folding, with axial surfaces dipping antithetically (i.e., oppositely) to the dip of the fault. One of the axial surfaces is fixed and active, with its trace rising from the footwall cutoff. The other axial surface is inactive and moves along with the hanging-wall cutoff (see Figure 8.26*B*).

We see via Figure 8.26 that if a normal fault changes dip and becomes progressively less steeply inclined with depth, the hanging wall strata must change in dip and/or shape as they move through the bend. The width of the kink band progressively increases as faulting takes place; that is, as more and more hanging-wall strata move through the active axial surface and become '*bent out of shape*' (see Figure 8.26*B*). Such movement of hanging-wall strata through the bend in a normal fault (during extensional stretching) can be analyzed in a way comparable analyzing folds related to shortening.

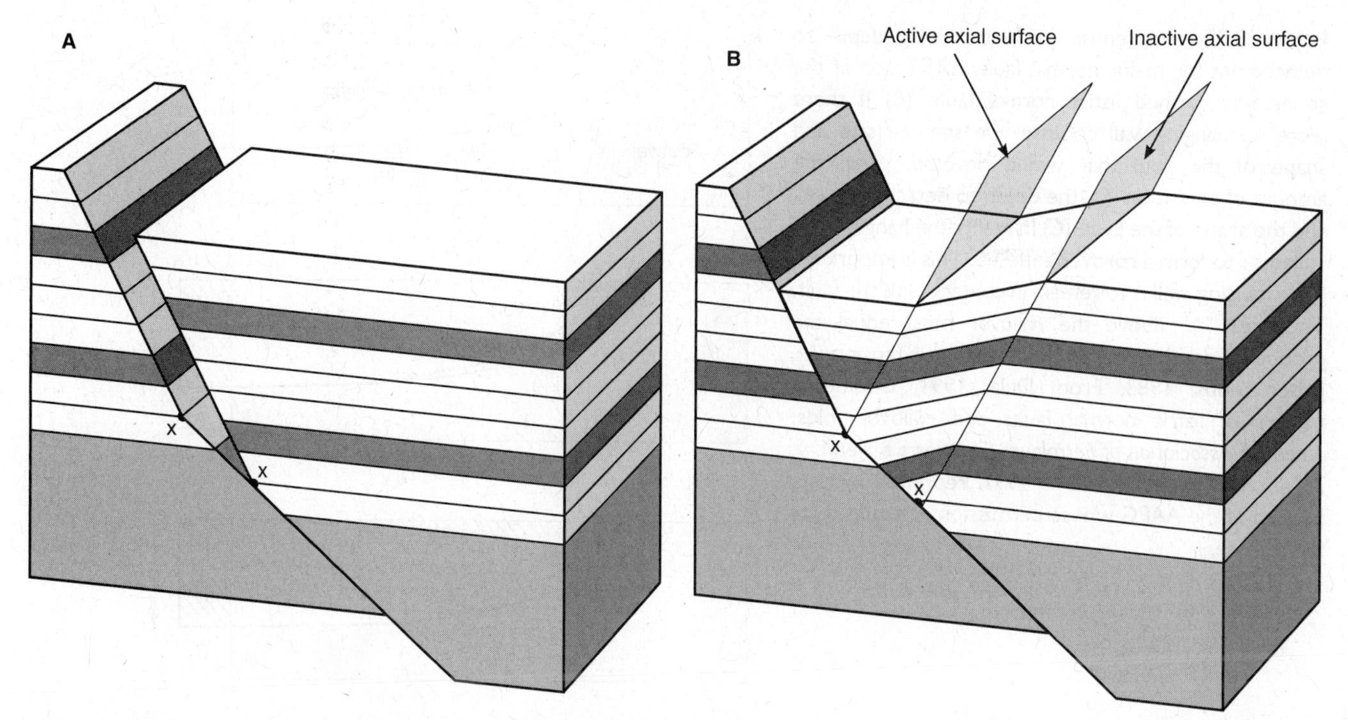

Figure 8.26 Block diagram showing an example of extensional fault-bend folding. (*A*) This is what will not happen, i.e., the development of a wide fissure as displacement on a bent normal fault takes place. (*B*) Instead, as the hanging wall incrementally moves away and downward from the footwall of a fold, the hanging wall collapses (incrementally) back against the footwall. Point x is a point that was right at the bend in the fault at the initiation of faulting. When slip begins, this reference point is 'split' in half as hanging-wall and footwall cutoffs. During faulting, x in the hanging wall is the point to which an inactive axial surface is pinned. In contrast, x in the footwall remains at the bend in the fault plane and is the point to which an active axial trace is pinned. It is at this point where hanging-wall rocks moving down the fault change attitude as they move through the bend in the fault surface. The band of dipping rocks widens as the fault displacement increases. [From Xiao and Suppe, 1992, The origin of rollover: *American Association of Petroleum Geologists Bulletin*, v. 76, p. 509–529. AAPG © 1992, reprinted by permission of the AAPG whose permission is required for further use.]

Hanging-Wall Growth Strata

As the hanging-wall strata drops relative to original regional elevation during normal faulting, **accommodation space** develops within which **growth strata** are deposited and preserved. In the early stages of faulting (Figure 8.27*A*), we see a listric normal fault curving into parallelism with bedding in the pre-fault stratigraphy. The early growth strata (white) "*roll over*" toward the trace of the normal fault, as do the overlying growth strata (light gray). The growth strata shown in gray in Figure 8.27*A* reveal the wedge-like form of this stratigraphic unit as well as the profound increase in its thickness closer and closer to the fault. In the late stages of faulting (Figure 8.27*B*) the same geometry prevails, yet the scale of the thickness of these growth strata is profound! Such growth sequences are termed "*synkinematic*," for they are deposited *during* faulting.

In areas such as deltas on a passive plate margin (e.g., the U.S. Gulf Coast or the Niger Delta in West Africa), the growth strata have a folded form, and the fold geometry can be related exactingly to the curvature of the normal fault with which the hanging wall is associated.

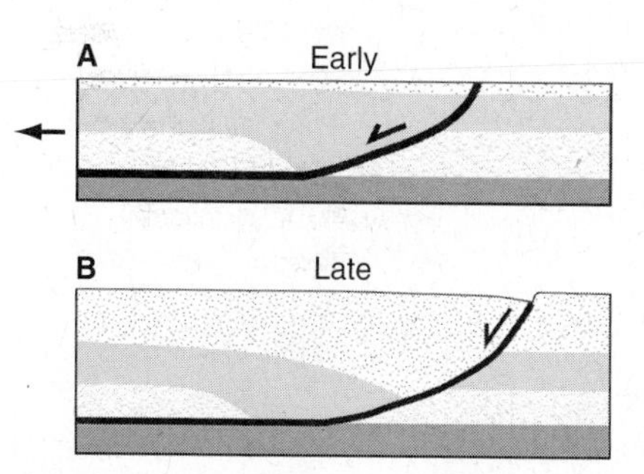

Figure 8.27 Simple schematic cross-sections of the deposition of growth strata during normal faulting. Note how the normal fault is listric, tucking into parallelism with horizontal strata at depth. The curved upper part of the fault, in combination with fault displacement, causes progressive 'rollover' of strata as they are deposited and buried. Synkinematic growth sequences give away their identity, because they thicken toward the normal fault, and they 'roll over' toward the master fault. (*A*) Early stage. (*B*) Late stage.

Depth-To-Detachment and Area Balancing

In considering normal faulting, the meaning of depth-to-detachment is slightly different from our usage in discussing detachment folding. Here,

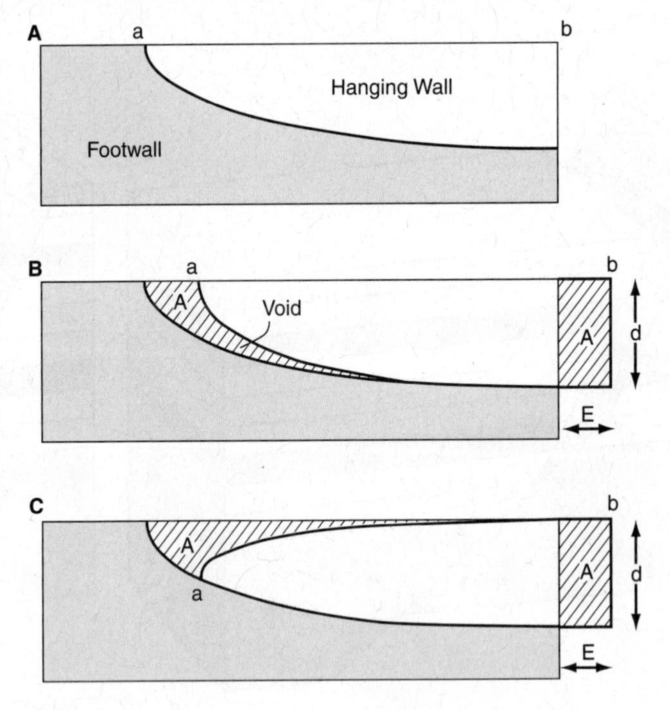

Figure 8.28 A premise for determining depth to detachment for major normal faults. (*A*) Trace of the soon-to-be formed listric normal fault. (*B*) If there were no hanging-wall collapse, we see the size and shape of the void that would develop, given the amount of extension (E), the depth to detachment (d), and the shape of the fault. (*C*) In reality the hanging wall collapses to form a rollover anticline. This is imperative if the hanging wall is to remain in contact with the fault. The area "A" above the rollover must equal the product of depth to detachment (d) and extension (E). [After Gibbs, 1983. From Dula, 1991, Geometric models of listric normal faults and rollover folds: *American Association of Petroleum Geologists Bulletin*, v. 75, p. 1609–1625. AAPG © 1991, reprinted by permission of the AAPG whose permission is required for further use.]

depth-to-detachment is the depth at which a given normal fault becomes parallel with bedding (or with the earths surface) (Figure 8.28). The shape of the hanging wall fold will reflect the shape of the normal fault surface, because the hanging wall arrived at its present position by moving (translating) over the geometry of the underlying fault. This is a requirement, because the hanging wall remains in direct contact with the fault throughout movement. Thus it is not surprising that fault shape, hanging wall geometry, depth to detachment, and amount of extension are all related (see Figure 8.28).

Geologists working in the oil industry were the first to develop a technique for determining the fault shape and depth to detachment from hanging wall geometry related to normal faulting. These geologists had powerful incentives to come up with reliable methodologies, for by knowing the depth-to-detachment of normal faults they could identify the locations of probable hydrocarbon trap in hanging wall settings. The method that they devised is yet another application of **area balancing** that we described above. Alan Gibbs (1983) showed the way, emphasizing that (in 2D) the area occupied by growth strata in the hanging wall of a major listric normal fault must be equal to the product of amount of extension and depth to detachment (see Figure 8.28).

Constructing Normal Fault Geometries on the Basis of Hanging Wall-Geometry

It is one thing to "*see*" the complete profile of a listric normal fault and associated growth strata in a seismic reflection profile. It is quite another to infer the complete profile based just upon knowledge of the location and orientation of the upper reaches of the fault, the cross-sectional form of growth strata in the hanging wall, and the position of regional elevation. But this is what exploration geologists are now able to do, for they have created methods to address the challenge of projecting faults and the hanging wall strata deep into the subsurface.

The solution brings us once again to **displacement vectors**, in this case constructing a rational family of displacement vectors that connect control points on a horizontal reference bed at regional elevation *before* faulting to final positions *after* faulting. We go to all of this trouble because seismic data do not typically image the deeper reaches of the normal faults, and thus we are forced to base interpretations on what the seismic sections *do* reveal, i.e., the internal structure of the hanging wall.

The starting point is drawing a cross-sectional portrayal of (1) the location and form of at least one stratigraphic horizon (**reference bed**) in the hanging wall of the fault; (2) the location and geometry of the upper part of the normal fault, including the depth at which it cuts the reference bed; and (3) the original regional elevation of the reference bed (Figure 8.29*A*). With this information in hand, we begin to make assumptions about how points on the reference horizon moved during faulting. In this first example, we assume that each point on the reference horizon moved downward in a way that was marked by constant **heave**, which is the horizontal distance between the hanging wall cutoff and the footwall cutoff of the reference bed (Figure 8.29*B*). The construction requires drawing a series of vertical lines on the cross-section with spacing equivalent to heave (see Figure 8.29*B*). We then assume that each given point on the reference bed in the hanging wall moved from its original pre-faulting location on the regional elevation to its present position along a path parallel to the trace of the fault. Thus the solid black arrows in Figure 8.29*C* are displacement vectors for each of the reference points. By sequentially taking each of these displacement vectors and "*hooking*" them nose to tail along the fault trace, we are able to fill out the full cross-sectional geometry of the fault at depth (Figure 8.29*D*). In this way, we succeed in capturing the geometry of a fault that is hidden from view.

Though this method had been used for some time by explorationists in the oil industry, it was Peter Verrall that brought this method to the attention of a much broader audience. It became known as the "*Verrall Method*," even though Verrall himself made no claim to having devised the technique. He sent out word to colleagues that it should be called something different, and consequently the next set of such papers referred to this technique as the "*Chevron Method*," after the corporation for whom Peter worked. Now it is commonly referred to as the **vertical shear** method, because Peter referred to the displacement-vector construction lines (see Figure 8.29*D*) as constituting a "*vertical angle of shear*."

Refinements of the construction employ different "*angles of shear*" for different situations. Some reconstructions use 60° **antithetic shear,** because Coulomb failure criteria predicts that an ideal normal fault dips 60° (Figure 8.30). Results of "*antithetic-shear*" reconstructions conform quite nicely to the internal geometry of normal faulting derived through physical analogue modeling.

The more we learn about how deformation in the hanging wall actually behaves, point by point, the more successful we will be in reconstructing fault geometries at depth. The range of possibilities is daunting, even starting with the same fault/bedding data at the surface (Dula, 1991). Take, for example, Figure 8.31*A*. This is a constant heave construction, which preserves area balance. The construction is one with which we are now familiar, and derived from knowledge of the dip of the uppermost part of the normal fault and the shape of the hanging-wall reference bed at the base of the growth strata. Yet Dula (1991) provides us with four others, each of which carries an assumption different from "*constant heave*." Figure 8.31*B* is based on an antithetic shear construction, where points move on parallel inclined planes dipping oppositely to the master fault. Figure 8.31*C*, in contrast, assumes that all reference points have moved parallel to the fault surface in the dip direction of the master fault. Then there is the reconstruction shown as Figure 8.31*D*,

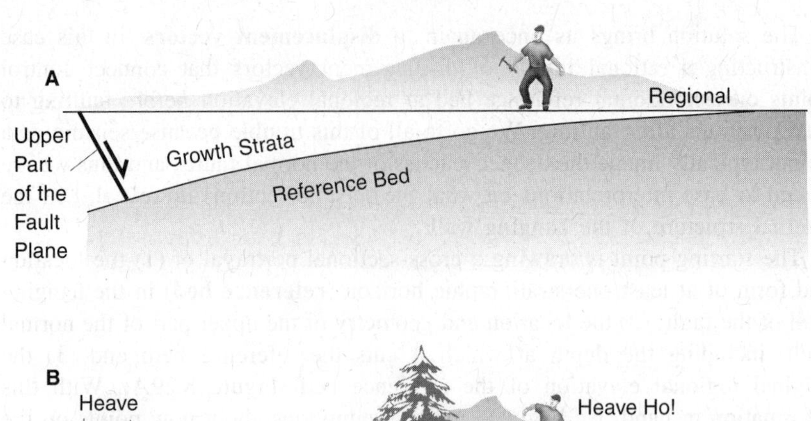

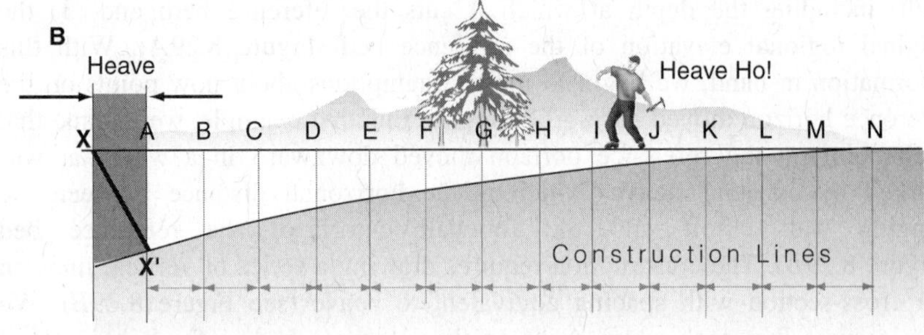

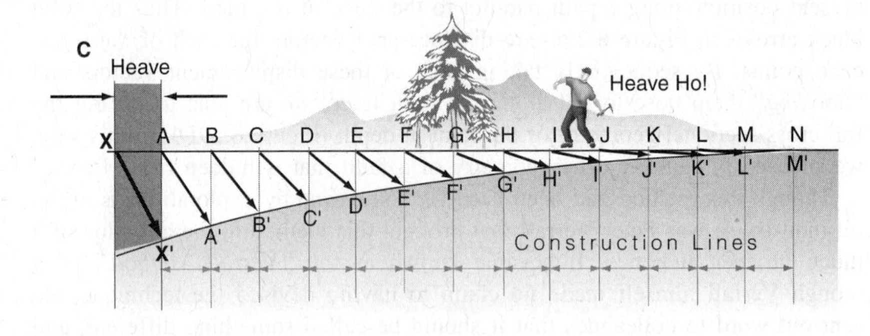

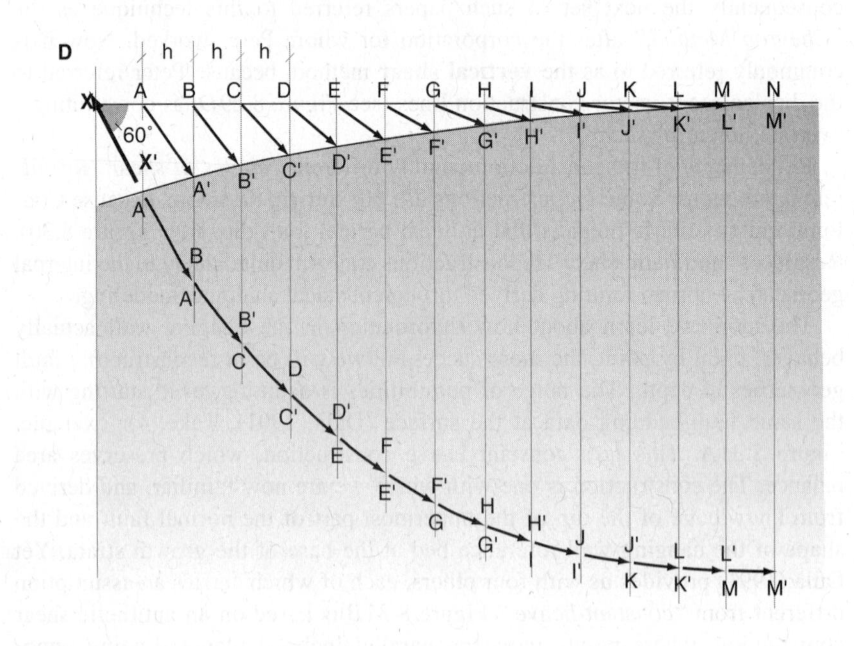

Figure 8.29 Approach used to determine depth to detachment for a listric normal growth fault. (*A*) The method starts with drawing a cross-section of the upper part of the system. This cross-section shows the upper part of the normal fault, the profile of the top surface of the growth strata (or if necessary a surface within the growth strata sequence), and regional elevation. (*B*) Next we measure the heave, which is the horizontal component of the fault displacement. Heave is measured from the footwall cutoff of the bed to the hanging wall cutoff of the same bed. Then we mark off vertical construction lines spaced one 'heave distance' apart. The intersections of each vertical construction line with the upper surface of growth strata set us up for the next step. (*C*) For we now construct a family of displacement vectors, each of which connects a given (starting) reference point on the original horizontal unfaulted/unfolded reference bed to its post-fault/fold position in the subsurface. (*D*) Finally, and this is the really clever part, we construct the subsurface profile of the normal fault by connecting each displacement vector nose to tail, from top to bottom. The 'nose to tail' construction of the displacement vectors reveals the depth and geometry of the fault plane that produced that hanging wall form, providing the assumption (in this case) of constant heave is valid.

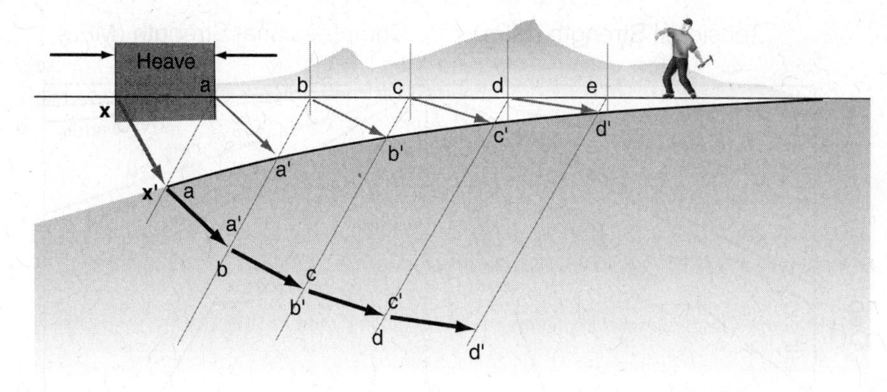

Figure 8.30 The assumption of 'antithetic shear' in the hanging wall can also be used in the 'depth-to-detachment' construction. The first construction line is drawn through the hanging wall cutoff at an angle that is antithetic to the dip of the fault plane. The remainder of the construction is the same as the 'constant heave method,' except the construction lines are all oriented at the presumed antithetic angle. Note that this construction results in a fault profile that reaches the detachment level at a shallower depth than that of the previous example.

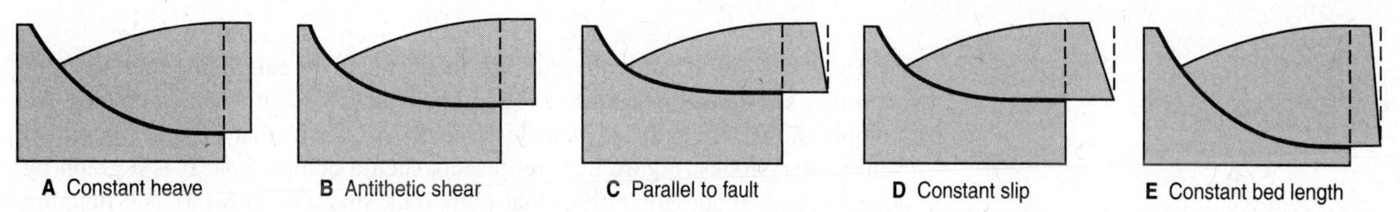

Figure 8.31 Here are 5 examples of depth-to-detachment constructions, which all start with the same facts regarding dip of uppermost part of the master normal fault, the rollover geometry of the top of the growth sequence, and regional elevation. Each construction carries different assumptions [After Gibbs, 1983. From Dula, 1991, Geometric models of listric normal faults and rollover folds: *American Association of Petroleum Geologists Bulletin*, v. 75, p. 1609–1625. AAPG © 1991, reprinted by permission of the AAPG whose permission is required for further use.]

which assumes constant displacement along the fault, and Figure 8.31*E*, which assumes constant bed length. Each solution carries different implications regarding the shape of the fault, the shape of the hanging-wall block, and depth-to-detachment. This is not the consistency you want when identifying petroleum targets in the subsurface and preparing to drill, but at least it is a start.

SALT-RELATED STRUCTURES

Strength of Evaporites

We have been emphasizing here the ways in which mechanical stratigraphy influences progressive deformation and fault/fold interactions. There is no 'bigger player' in the game of mechanical stratigraphy than salt. Astonishingly weak and viscous compared to all other sedimentary rocks, salt, especially where present in thick layers, profoundly influences structural geology and structural associations. Thus we will invest some time here on the topic of **halokinesis**, which refers to the mobilization and flow of subsurface salt, its emplacement, and the effects on structure and stratigraphy.

Evaporites form via evaporation of seawater, and the sedimentary layers produced are dominated by such salts as gypsum, anhydrite, halite, and sylvite. Notably, the initial phases of plate-tectonic rifting of a continental margin will create the ideal circumstances for evaporites to accumulate in astounding thicknesses (thousands of meters). Because the minerals comprising evaporates are weak, evaporite-dominated sections of sedimentary rocks are extremely incompetent and often act as detachments for overlying structures. But in the process of deformation and flow they produce a unique set of structures that are often associated with faults.

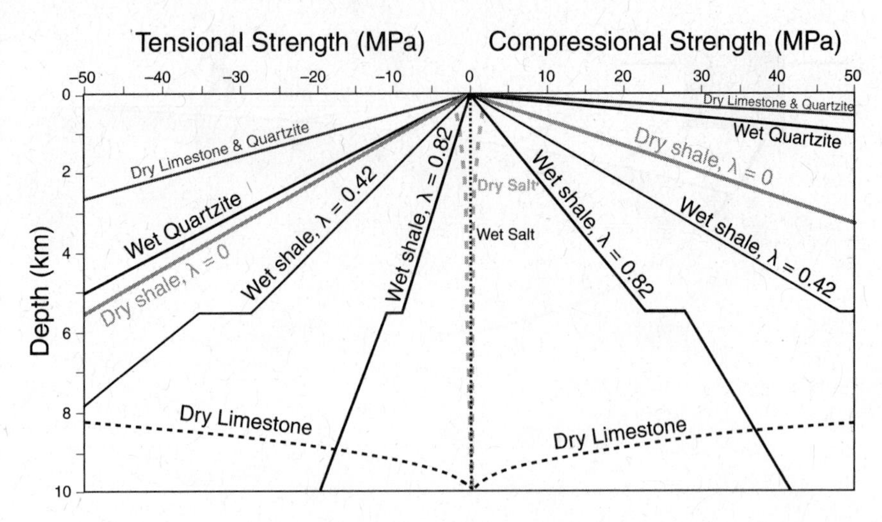

Figure 8.32 Comparison of creep and frictional strengths with dry and wet sedimentary rocks. Rock strength (in MPa) is plotted as a function of depth (kms). To the right of center are compressional strengths, and to the left of center are tensile strengths. Note that salt has essentially no strength under most crustal conditions. See text for details. [From Jackson and Vendeville, 1994, Figure 6, p. 65. Published with permission of the Geological Society of America.]

Though gypsum and anhydrite are mechanically weak, halite and some of the other salts are almost without strength. This is evident when comparing the compressional strength and tensile strength of salt to that of other common sedimentary rocks. Figure 8.32 represents such a comparison. It is a graph by Jackson and Vendeville (1994) that plots rock strength (in MPa) as a function of depth (in kms). To the right of center are compressional strengths, while to the left of center are tensile strengths. Note that Jackson and Vendeville (1994) were not content simply to show the strength of rocks when dry, but rather include strength data for "*wet*" conditions, i.e., where the rock contains water, which is a weakening agent. Moreover, strength for a variety of fluid pressure conditions is also represented. Note in Figure 8.32 the parameter λ, which is the pore-pressure coefficient, and recall that it is the ratio of fluid pressure to lithostatic pressure (see Figure 8.32). Wet salt has no strength, whether close to the surface or down at 10 km. The resistance of wet salt to deformation and flow is 0.0 MPa. Dry salt has *very* low finite strength at burial depths less than 2.5 km, and *no* strength below that depth (see Figure 8.32). Even shale is strong relative to salt!

Not only is salt weak compared to even the most poorly lithified rocks, its viscosity is also very low. Recall that viscosity is a measurement of the ratio of applied stress to strain rate. Viscosities for salt have been determined through experimental deformation, results of which take the form shown as Figure 8.33, where after a nearly insignificant response by elastic deformation and primary creep, the salt behaves perfectly viscously. Thus salt-dominated

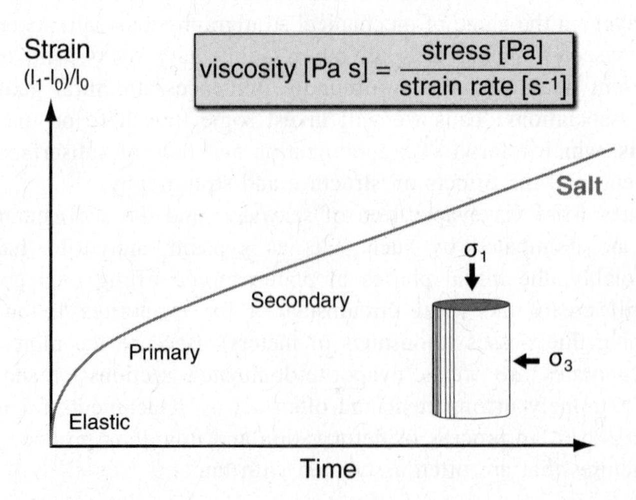

Figure 8.33 The viscous behavior of salt can be captured through experimental deformation. In this case a constant stress is applied to a cylinder of salt. The response of the salt is plotted as a function of strain versus time. Note that after miniscule elastic deformation and some primary creep, the salt behaves as a perfectly viscous fluid. [From Kukla et al., 2010. Reprinted with permission from The Geological Society Publishing House, Bath, England.]

rock layers behave as if they were viscous fluids, and they will literally flow from locations of high overburden pressure concentration to low pressure areas. The driving force is **hydrostatic head**, which is gravity at work. The hydrostatic head causes salt bodies to rise. To picture this, take a board, drill some big holes through it, fill your bathtub with water, and toss the board into it. Water rises up into the holes. Salt would do the same thing if you filled a football stadium with a cohesive layer of salt (up to the base of the "*nose-bleed*" section) and then placed on top of the salt layer a 25-m-thick steel lid with large-diameter holes drilled through it. After a few weeks or months, salt would rise up through the holes in the steel.

There are historic artifacts of the effects of the flow of salt, one of which is quite macabre:

> *. . . in salt mines in the Austrian Salzburg district, old tunnels were discovered in which bodies of ancient Celtic miners were found with their mining tools, embedded in salt.* (From "Structure of Grand Saline Salt Dome, Van Zandt County, Texas" by R. Balk, 1949, p. 1823. Published with permission of American Association of Petroleum Geologists.)

The viscosity of salt for geologically relevant diapiric strain rates is *very* low (10^{17} to 10^{18} Pa · s), less than 1/100th of "normal" sedimentary rocks. The density of salt is also quite low: 2270 kg/m^3 compared to densities of ~2700 kg/m^3 for common sedimentary rocks. Because of this combination of low viscosity and low density, salt-rich layers tend to be buoyant. Therefore, given the opportunity, salt will rise compared to the denser sediments that overlie or surrounding it. We wish to emphasize that the density contrast between salt and normal sedimentary rock is not as great as the viscosity contrast, so do not expect that the buoyancy of salt is great enough to pierce overlying sedimentary layers, even where these layers are poorly lithified. There simply is not enough buoyancy pressure to cause such piercing unless the overlying sedimentary rocks are thin.

Behavior of Salt During Normal Faulting

One of the most efficient detachment materials is salt. Faults and fault-related folds are often associated with flow of salt in the subsurface. This has been born out by intensive exploration for oil and gas in sedimentary basins such as the Gulf of Mexico (Jackson et al., 1994). Seismic reflection profiles and extensive deep-water drilling have revealed a close association between oil/gas targets and salt tectonics.

Because of its low viscosity, salt will *immediately* begin to flow when normal faulting creates nearby loci of markedly decreased overburden pressure. During extension and normal faulting, lithostatic pressure decreases as the hanging wall of the fault moves down and away from the footwall. Because it is unlawful in nature for a great hole to form in the deep subsurface, the potential opening must be simultaneously filled by antithetic faulting or reverse drag or, if salt layers are present and sufficiently thick, by the flow of salt. During normal faulting, newly deposited growth sediments will not yet be lithified or compacted, and thus these new additions on the hanging-wall side will not be as dense as the rock column in the footwall. The growth sediments will weigh less than adjacent lithified sedimentary counterparts and therefore will place less overburden pressure on the underlying salt. As a result, the salt will flow toward the area of low pressure along the fault, and from there the salt will ascend along the fault surface.

An outstanding example of the interrelationship of extensional faulting and salt diapirism is seen in the geological record across the Whale and Horseshoe salt basins in the southern Grand Banks off Newfoundland (Jackson and

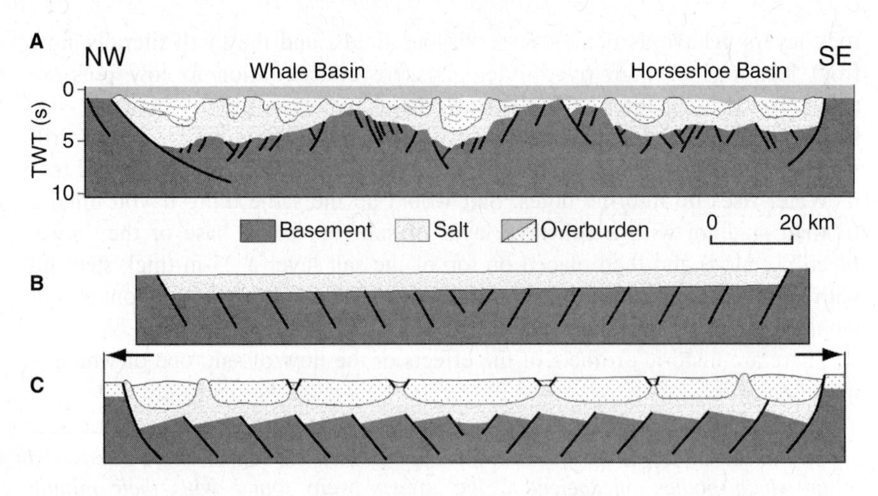

Figure 8.34 Interaction of extension and salt flowage. (*A*) Subsurface interpretation of regional seismic profile across the southern Grand Banks of Newfoundland. The salt (white with speckles) is Triassic-Jurassic in age and has risen as diapiric walls. (*B*) Portrayal of pre-extension layer-cake geology of salt over basement. (*C*) Extension produced obvious normal displacement on basement faults and stretching of the (older) overburden directly above the salt. As extension proceeded, the diapiric walls widened. Post-extension stratigraphy is represented by uppermost cover, middle Cretaceous in age (see part A). [From Jackson and Vendeville, 1994, Figure 4, p. 62. Published with permission of the Geological Society of America.]

Vendeville, 1994). The salt is Triassic-Jurassic in age, having formed there during the initial stages of the opening of the North Atlantic via sea-floor spreading. Seismic reflection images capture huge masses of salt that ascended diapirically to their present stratigraphic positions as the overburden was extended by faulting (Figure 8.34*A*) (Jackson and Vendeville, 1994). The diapiric walls widened as extension accrued. Deposition of Cretaceous strata (Aptian, ~125–112 Ma) post-dated the diapiric ascent of the salt walls (see Figure 8.34*A*). Jackson and Vendeville (1994) created drawings picturing salt deposition (Triassic/Jurassic) atop faulted basement (Figure 8.34*B*), followed by extension of the cover (Figure 8.34*C*).

The viscous diapiric ascent of weak salt produces **salt diapers** and **salt walls**, such as those forming today in the northern Red Sea region (Figure 8.35). The salt walls are intimately associated with grabens and half grabens produced by ongoing regional extension of thin overburden above an upper Miocene salt layer (~7–5 Ma). As fault slip occurs, bit-by-bit, more and more salt flows to fill the void produced by separation of the hanging wall and the footwall.

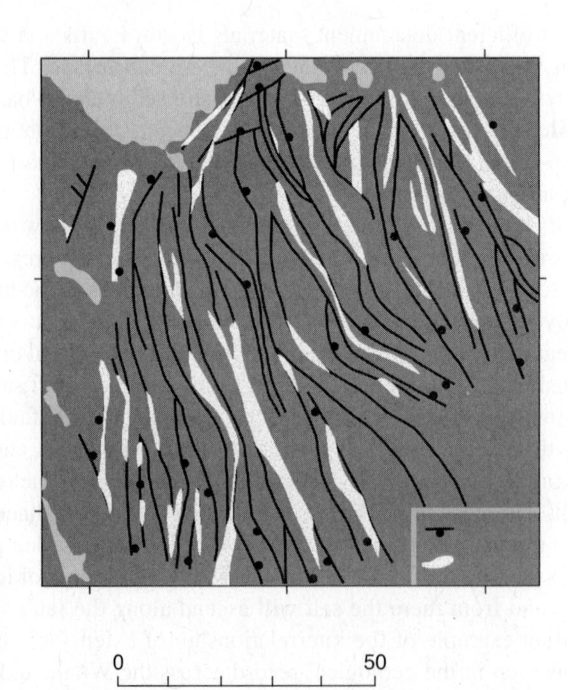

Figure 8.35 Map of northern Red Sea region, with distribution of diapiric salt walls (white) parallel to and closely associated with normal faults (black; ball on downthrown block). The faults border grabens and half-grabens produced during regional extension. [After Mart and Ross, 1987, later published by Jackson and Vendeville, 1994, Figure 2, p. 60. Published with permission of the Geological Society of America.]

The relationship of extensional faulting to the formation of salt diapers and salt walls has been investigated through physical analogue modeling as well (e.g., see Figure 3.69). Loose sand is used to represent the scaled mechanical properties of overburden, whereas silicone (polydimethslsiloxane, a long-chain tongue-twisting viscous silicone polymer, code name SGM-36) is used to represent the scaled behavior of salt. Figure 8.36 is the product of a 6.5-hour run (Jackson and Vendeville, 1994). The starting configuration reveals a simply layer-cake stratigraphy of sand on silicone (Figure 8.36*A*). When normal faulting was initiated, salt began to diapirically insert itself into the deep footwall reaches of normal fault #2 (Figure 8.36*B*). When normal fault #3 formed (Figure 8.36*C*), it too triggered a diapiric rise of salt in its deep footwall. As extension continued (Figures 8.36*D*, *E*), a single salt wall grew to dominate. In the end the roof became pierced by the salt (see Figure 8.36*E*) (Jackson and Vandeville, 1994).

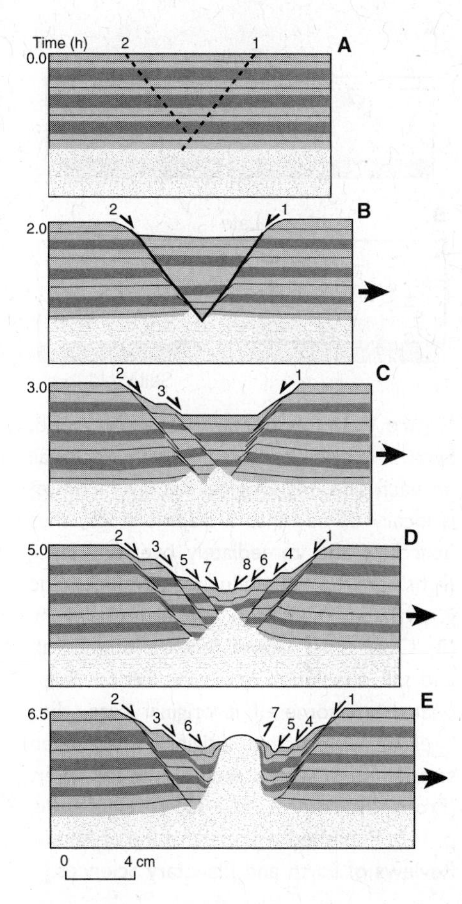

Figure 8.36 These are tracings of cross-sections through stages in the physical analogue modeling of salt diapirism. Salt is represented by the (white) substratum of polydimethylsiloxane, a viscous silicone polymer. Overlying strata are represented by dry sand. These drawings capture extensional faulting and diapirism after 2 cm, 3 cm, 5 cm, and 6.5 cm of total extension. [From Vendeville and Jackson, 1994, Figure 10, p. 68. Published with permission of the Geological Society of America.]

Vertical Loading of Salt

Vertical loading of the upper surface of a thick salt layer can trigger flow. Jackson et al. (1994) have described the process in a way that gives us a starting point. Imagine a small circular depression forming on the seafloor bottom on the surface of a salt layer or salt wall (Figure 8.37*A*). Such a depression will fill with sediments (Figure 8.37*B*). The thickness of the sedimentary columns at such a location will be slightly greater than normal. The weight of accumulation of a thicker and thicker sedimentary column creates greater overburden pressure, which in turn causes the immediately underlying salt to flow into neighboring lower-pressure areas (see Figure 8.37*B*). As this takes place, a feedback loop is created that eventually produces over time a minibasin (relatively depleted in salt) surrounded by domes of salt. The tops of adjacent circular domes of salt will remain near the seafloor, with elevation being maintained by the underlying inflow of salt from juxtaposed minibasins (Figure 8.37*C*). This process will continue until the supply of salt is exhausted, at which time the tops of the circular domes can no longer be maintained at the level of the seafloor.

Salt Welds and Fault Welds

During sedimentary loading, as the supply of salt flows from higher-pressure to lower-pressure sites, the original thickness of a given salt layer will

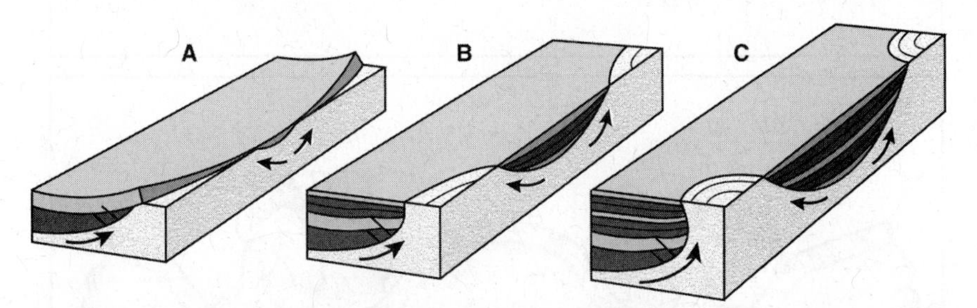

Figure 8.37 The weight of sediments can cause flow of underlying salt. (*A*) The starting condition features a salt wall that has emerged at the seafloor bottom. (*B*) and (*C*) Progressive sedimentation into a minibasin creates a gravitational load that causes the salt to flow (see arrows) from underneath the area of accumulating sediment, which in turn increases subsidence and sedimentation. Adjacent stocks of salt grow and become more rounded. [From Jackson et al., 1994, Figure 2, p. 100. Published with permission of Annual Reviews of Earth and Planetary Sciences.]

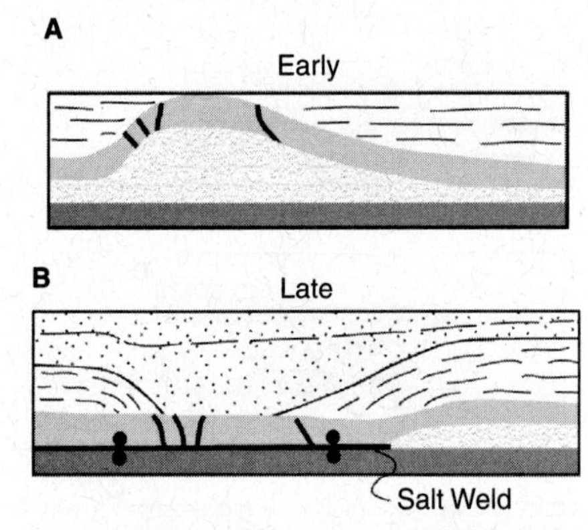

Figure 8.38 Formation of a salt weld. Symbol is black contact line with black ball on each side. (*A*) Original salt layer (white) is locally domed into a diapiric stock, thus uparching the immediately overlying layer (light gray) and influence synkinematic growth sedimentation (uppermost layer). (*B*) Over time, lateral flowage of the salt and salt dissolution cause the salt to disappear from some of it original sites, thus "*welding*" units that originally had been stratigraphically separated by the salt layer. [From Jackson et al., 1994, Figure 8c, p. 110. Published with permission of Annual Reviews of Earth and Planetary Sciences.]

become steadily reduced because of the viscous outflow. Moreover, part of the thickness reduction of a salt layer may be related to dissolution of the salt, i.e., the dissolving of the salt through the introduction of water undersaturated in salts. We can well imagine one of the effects: the sedimentary beds *above* the salt are slowly let down onto the sedimentary beds directly *below* the salt. Figure 8.38, based on the work of Jackson et al. (1994) helps us to picture this process. We start with a framework where a salt diapir (white) has domed a prekinematic overlying sedimentary formation (light gray) as well as the seafloor bottom (Figure 8.38*A*). Maximum sedimentary accumulation (uppermost layer, Figure 8.38*A*) thereafter takes place on the flanks of the dome. Over time (Figure 8.38*B*), as the salt becomes progressively buried, viscous outflow of the salt and salt dissolution cause the original dome of salt to disappear altogether. A **salt weld** forms, a contact where the sedimentary formation originally immediately *above* the salt layer is pressed down into contact with the sedimentary formation that originally resided immediately *below* the salt layer (see Figure 8.38*B*). We can picture the development of a salt weld by imagining what would happen over the course of an hour if we all decided to hang out on a long bench made of polydimethslsiloxane (Figure 8.39). We would become "*welded*" to the grass.

Some welds are known as **fault welds**, where salt has been eliminated by the combination of outflow, dissolution, and shearing. The net result, again, is one in which strata once overlying the salt is brought into contact with strata that originally resided beneath the salt. Figure 8.40*A* shows the starting configuration, namely a listric normal fault that decreases in dip with depth, becomes parallel to bedding, inserts itself into a subhorizontal salt layer (white) whose immediate footwall is a salt diapir. Shearing of the salt is producing an incipient fault weld. At a later stage (Figure 8.40*B*), a sizable fault weld has formed. A salt weld developed as well. Generally speaking, if normal faulting continues after the supply of salt is exhausted, a weld will

Figure 8.39 Cartoon analogy of the formation of a 'welded' contact. On a summer afternoon we all decide to hang out on a bench made of polydimethylsiloxane, a viscous silicone polymer. Before long, aseismically, we are welded to the grass, not even noticing what is happening. [Artwork by David A. Fischer.]

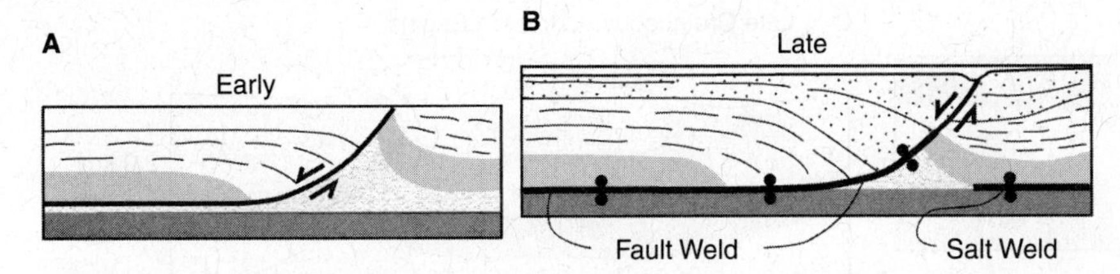

Figure 8.40 Formation of a fault weld. Symbol is black contact line with black ball on each side. (*A*) Early condition shows a listric normal fault, the footwall of which is a salt diapir (white). The listric normal fault soles downward into a horizontal salt layer. Shearing of the salt has produced an incipient fault weld. (*B*) At a later stage the salt diapir is diminished in size. Faulting has created a significant fault weld. A salt weld has formed at the lower right, due to salt flowage and dissolution. [From Jackson et al., 1994, Figure 8b, p. 110. Published with permission of Annual Reviews of Earth and Planetary Sciences.]

subside deeper into the subsurface. If, on the other hand, normal faulting ceases, the weld will hold its position in the subsurface. The weld will effectively separate the accommodation space (close to the fault) from the remaining entrapped salt layer beyond the weld.

Growth of Salt Diapirs

Rates of extension and volume of salt supply are limiting factors for the growth of salt diapirs and salt walls (Jackson and Vendeville, 1994). Initially, normal faulting produced by extension permits salt to well up to fill accommodation space during the so-called **reactive phase** of salt diapir development (Figure 8.41*A*). The top of the diapir is like an arrowhead, inserting itself as extension makes room for it. As the salt diapir ascends closer and closer to the elevation of the seafloor, it acquires sufficient buoyancy to force the small thickness of overburden aside (Figure 8.41*B*). This is called the **active phase** of salt diapir development. The diapir forces it way up, making room for itself. It thins the overburden through shearing, and creates shortening accommodated by monoclinal folding and/or thrusting. When the salt actually reaches the seafloor, the salt diapir can rise continuously on its own as long as there is source material to feed the outflow. There is no room problem whatsoever in this the **passive phase** of salt diapir development (Figure 8.41*C*).

Where a salt diapir breaches the surface, what happens along the edge of the salt will depend upon salt supply and the relative rates of sedimentation and flow. If the rate of salt flow is *greater* than the rate of sediment supply, the salt will flow out onto the seafloor. In contrast, if the rate of salt flow is *less* than the rate of sediment supply, the salt will be buried in sediment. It is as if a competition is being waged, and every round of the competition is recorded in the **halokinetic sequence**, which is the name for the growth stratigraphic record that evolves along the edges of a salt diapir (Giles and Lawton, 2002).

Tectonic shortening may cause a salt diapir to flow to the surface to form a **salt sheet extrusion** (Figure 8.42*A*–*C*). Salt extrusion takes place at the expense of the 'mother salt' layer at depth. Salt welds form at locations where this salt layer becomes fully depleted by flow and dissolution (see Figure 8.42*C*). Extrusion continues until the subsurface feeder for the salt is pinched off by reverse faulting and shearing (Figure 8.42*D*). Fault welds form where the salt feeder is pinched off (see Figure 8.42*D*). By the time shortening ceases, the extruded salt sheet lies buried (Figure 8.42*E*).

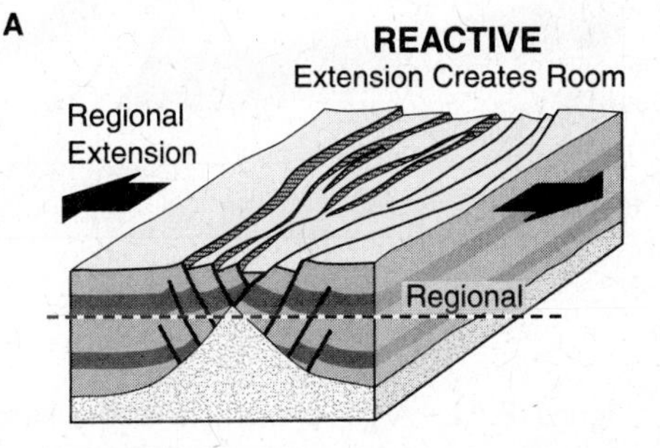

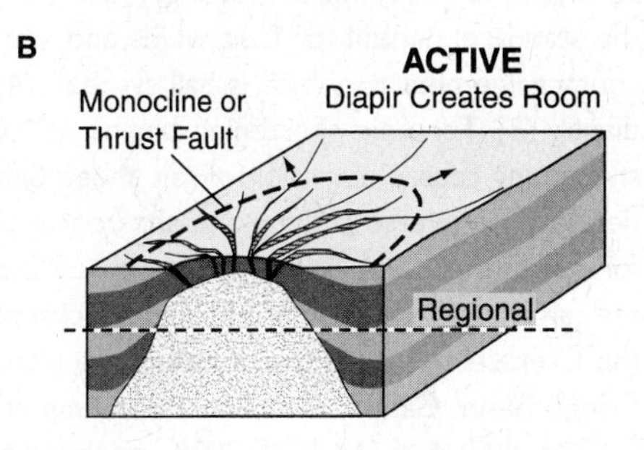

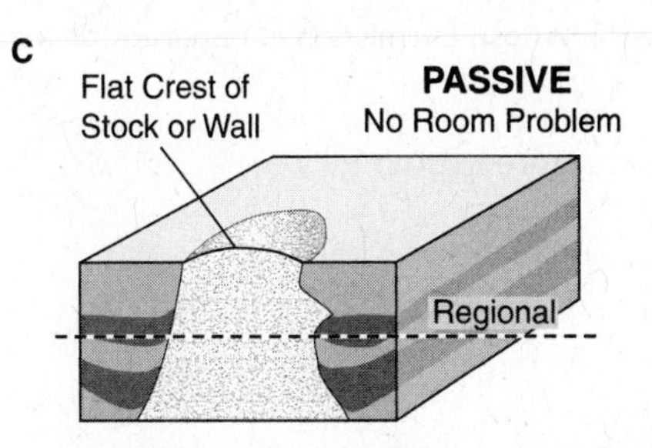

Figure 8.41 Stages of diapiric salt growth. (*A*) Reactive phase: Overlying rocks are being stretched by normal faulting. The salt diapir moves into the region of decreased overburden pressure. (*B*) Active phase: The salt diapir has sufficient bouyancy to penetrate the thin sedimentary cover. (*C*) Passive diapirism: Salt continuously flows onto the surface, supplied by an ample source of salt in the subsurface. [From Jackson et al., 1994, Figure 1, p. 100. Published with permission of Annual Reviews of Earth and Planetary Sciences.]

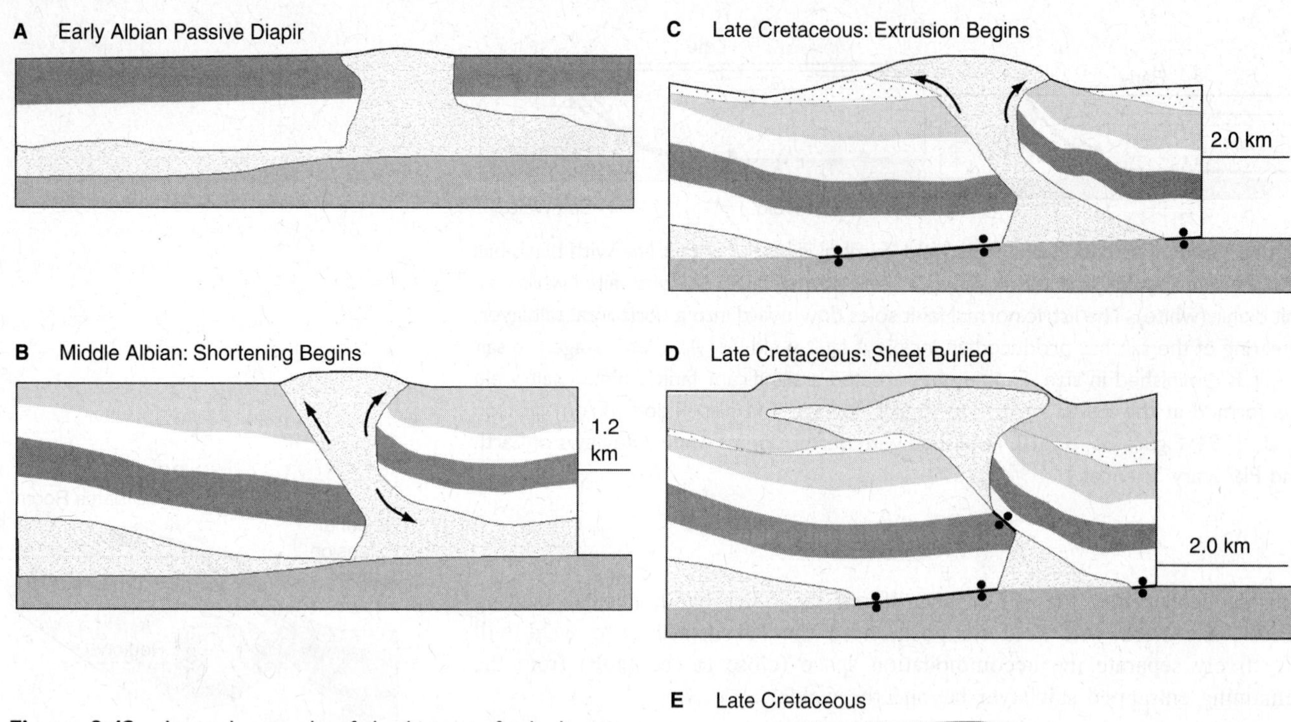

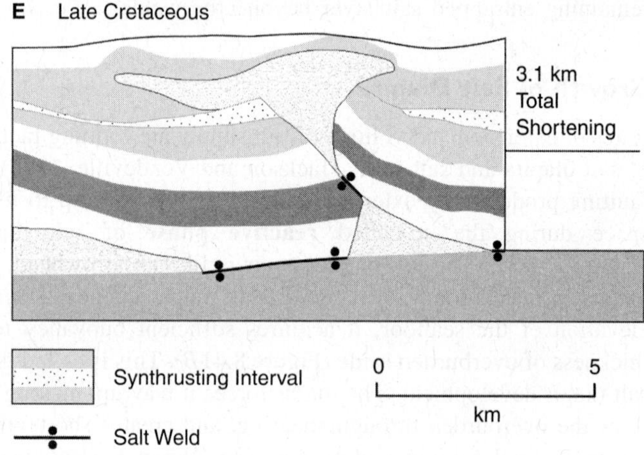

Figure 8.42 A good example of the history of salt diapirism accompanying tectonic shortening. Note the scale. Also note that the standard symbol for fault welds and salt welds is used, i.e., contact line next to which is ball symbol. (*A*) The initial passive diapir. (*B*) Tectonic shortening begins. (*C*) Continued tectonic shortening causes extrusion of salt sheet. Salt weld produced at depth. (*D*) Reverse faulting and shortening pinch off the feeder for salt extrusion. A fault weld forms. (*E*) Shortening ceases and the salt sheet becomes buried. [From Jackson, 2008, Evolution of the Cretaceous Astrid thrust belt in the ultradeep-water Lower Congo Basin, Gabon: *American Association of Petroleum Geology*, v. 92, p. 487–511. AAPG © 2008, reprinted by permission of the AAPG whose permission is required for further use.]

Salt Glaciers and Salt Canopies

In areas where salt supply markedly exceeds sediment supply, **salt glaciers** form. In *Chapter 3, Force, Stress, and Strength*, we showed photographs of the Kuh-i-Namak salt glacier (Figure 3.68) in Iran. Figure 8.43 captures this salt glacier again, as well as two of its siblings. The satellite images underscore the size and shape of classic salt glaciers. Imagine them being buried in sediment. If they had extruded in a submarine environment as part of a halokinetic sequence, the toe regions of these glaciers might have become salt flanges.

Salt glaciers may coalesce to form a continuous cover known as a **salt canopy**. A giant example occupies the northern part of the sea bottom of the Gulf of Mexico (Figure 8.44). After this image was published by Diegel et al. (1995) the geology of the Gulf of Mexico has never looked quite the same. The Luann Salt (Jurassic) moved to the surface diapirically, and from there salt glaciers began to flow, coalescing into a widespread canopy above the sedimentary section. The dimples on the surface (see Figure 8.44) are active

A

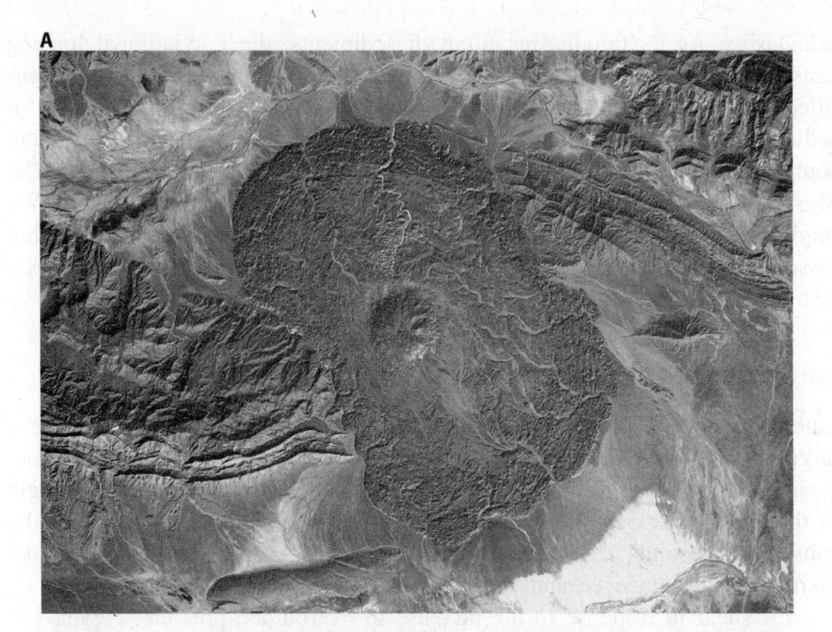

B

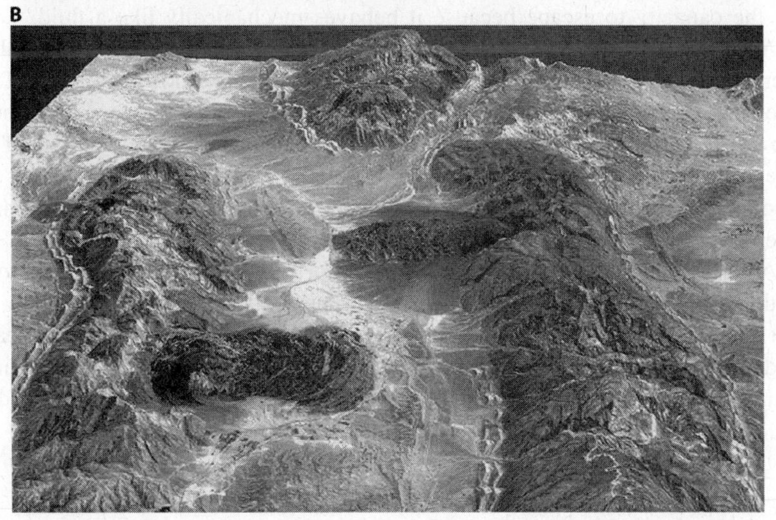

Figure 8.43 (*A*) LANDSAT TM7 Space satellite image of the Kuh-i-Namak salt glacier in Iran. [Processed and enhanced by Jim Ellis. Original imagery courtesy of U. S. Geological Survey.] (*B*) Oblique perspective view of two glaciers (black) emerging from breached salt anticlines and flowing into valley near Karmastaj, Iran. The salt glacier closest to viewer is ~7 km long.

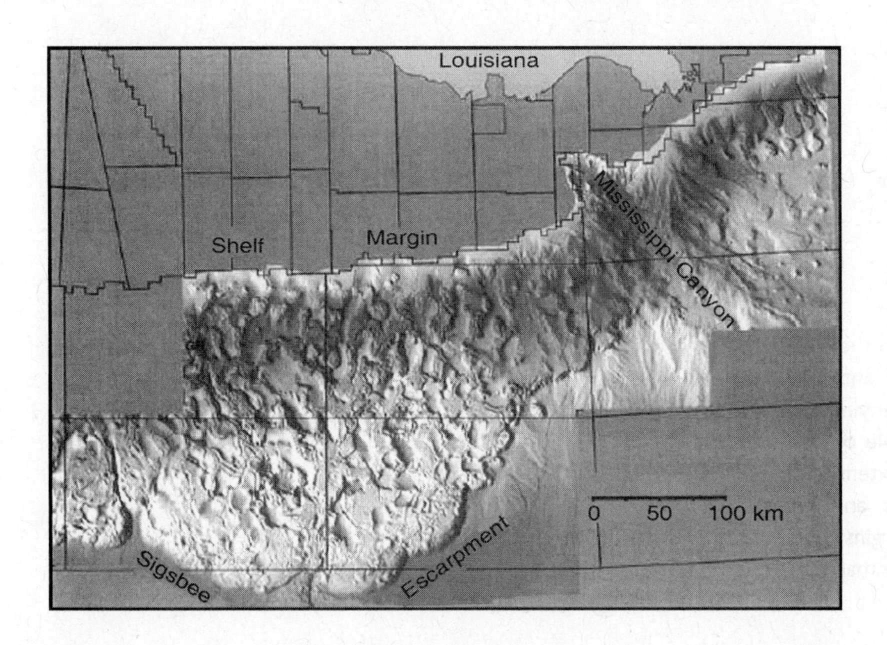

Figure 8.44 When salt glaciers flow out onto the seafloor, they may coalesce to form a continuous masses called a salt canopy. This example is from the Gulf of Mexico. Note small round depressions (minibasins) formed as sediment accumulates and, through gravitational loading, forces the salt to flow laterally away from areas of higher overburden pressure. [From Diegel et al., 1995, Cenozoic structural evolution and tectonostratigrapphic framework of the northern Gulf coast continental margin, in Jackson, M. P. A., Roberts, D. G., and Snelson, S. (eds.), Salt tectonics: a global perspective: *American Association of Petroleum Geologists Memoir*, 65, p. 109–151. AAPG © 1995, reprinted by permission of the AAPG whose permission is required for further use.]

minibasins. As the minibasins fill with sediments, the gravitational loading caused the salt in the underlying canopy to flow basinward. The minibasins filled sequentially in a progression, from locations closer to the source of sediments to locations farther down slope and further from the sediment source. This progression is sometimes described as "*fill and spill.*" The Sigsbee Escarpment is at the downslope end of the advancing salt canopy. We might imagine that submarine salt canopies will be dissolved by seawater, but dissolution is generally resisted because continuous sedimentation produces a thin protective armor on the surface of the canopy.

Lateral Shear-Loading of Salt

Lateral loading is yet another trigger for causing flowage of salt. Picture a large river system that dumps sediments at its mouth and over time builds out a giant deltaic deposit. This is the situation that exists on the northern margin of the Gulf of Mexico and the western margin of Africa. As salt layers in the subsurface become gravitationally loaded in this fashion, the effect is like taking a giant squeegee and shearing the salt basinward. The salt flows by viscous shear in response to the increase in overburden pressure. Again, salt has the capacity to escape because it behaves mechanically like a fluid.

As the salt flows downslope toward the continental margin, the salt plus overlying deltaic sediment column behaves like a mega-landslide system. The extension at the up-dip end is accommodated by normal faulting. The total slip at the up-dip breakaway must be accommodated somehow. Most of it is expressed by thrusting at the toe of the moving mass. Some of it is accommodated by the upward movement of salt diapers *into* the moving and extending mass.

An example of this is evident in Mississippi Canyon in the seafloor bottom of the Gulf of Mexico (Figure 8.45). Sediments there are deformed in ways that reflect gliding of the underlying salt sheets, driven by lateral shear-loading. The main lobe is ~25 km long and as wide as ~15 km. It is sliding southeastward. A break-away system of normal faults is evident at its up-dip

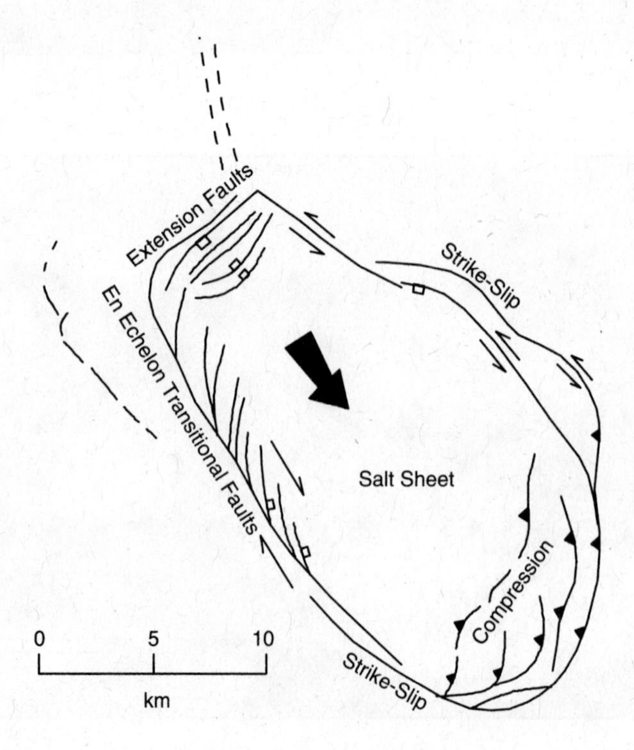

Figure 8.45 Here at Mississippi Canyon in the sea-floor bottom of the Gulf of Mexico, sediments are deformed in ways that reflect gliding of underlying salt sheets. This salt sheet is a beautiful example of system-wide structural compatibility, with extensional faults at the breakaway, strike-slip faults and en echeon transtensional faults along the margins, and thrust faults at the toe. [Courtesy of Kerry Inman and the British Petroleum Company.]

end, and a toe of compression-induced shortening, achieve through thrust faulting, marks the down-dip end (see Figure 8.45). Strike-slip faults along the lateral margins are required to permit this mega-landslide to move.

A Word on Overpressured Shales

The behavior of shale accompanying faulting and/or differential loading in sedimentary basins superficially resembles the behavior of salt. This is the case where shale is **overpressured**; that is, where pore-fluid pressure is higher than lithostatic (load) pressure. Fluid in overpressured shales forces the clay particles apart and lubricates them in ways that the behavior of the shale body acquires a very low viscosity.

Shale diapers and related structures do indeed resemble salt structures, but commonly they seem underdeveloped when compared to salt structures The explanation is simple: Maintaining an overpressured state requires that the shale be "*sealed*," preventing escape of the fluids. Yet faulting can readily break through overpressured shale bodies, thus creating the plumbing along which the fluids can be expelled. When fluid is expelled from overpressured shale, mud volcanoes may form on the seafloor bottom or on the land. We see this, for example, in Azerbaijan (Davies and Stewart, 2005), where muddy water erupts at the surface.

As shale moves from an overpressured state to a normal state, the mechanical character of the shale body abruptly changes, and the capacity of the shale to deform viscously becomes sharply compromised (Morley and Guerin, 1996). In contrast, the mobile character of salt is not dependent on high pore fluid pressure; it habitually responds as a viscous fluid.

STRUCTURAL INVERSION

Fault Reactivation And Inversion Structures

Millions of years or even hundreds of millions of years after a fault zone ceases to be active, it can be reactivated when the loading conditions are sufficient in magnitude and where the principal stress directions are properly oriented. This is because faults, once formed, commonly become zones of weakness in the crust.

Fault reactivation rarely duplicates the exact sense of slip and direction of slip that first prevailed. In fact, it is not uncommon for the sense of slip on the reactivated fault to be quite the opposite to that of the original displacement. When faults are reactivated in opposition to that of the original sense of displacement on the fault, the structures that form are known as **inversion structures** (Williams et al., 1989).

There are two major categories of inversion structures (Figure 8.46). Inversion structures are classified as **positive inversion structures** if earlier normal faults are reactivated as reverse faults (Figure 8.46*A*). They are classified as **negative inversion structures** if earlier reverse faults are reactivated as normal faults (Figure 8.46*B*).

It is interesting to explore the basin-stratigraphy story of positive inversion (Figure 8.47*A*). Note the care with which Yamada and McClay (2004) distinguished the structural-geologic significance of each stratigraphic unit within a sequence that experienced positive inversion. At the base of the column are **prerift rocks** deposited *before* normal faulting. The growth strata deposited during normal faulting are the **synrift sediments** (see Figure 8.47*A*). Note that the synrift growth strata show rollover (before inversion),

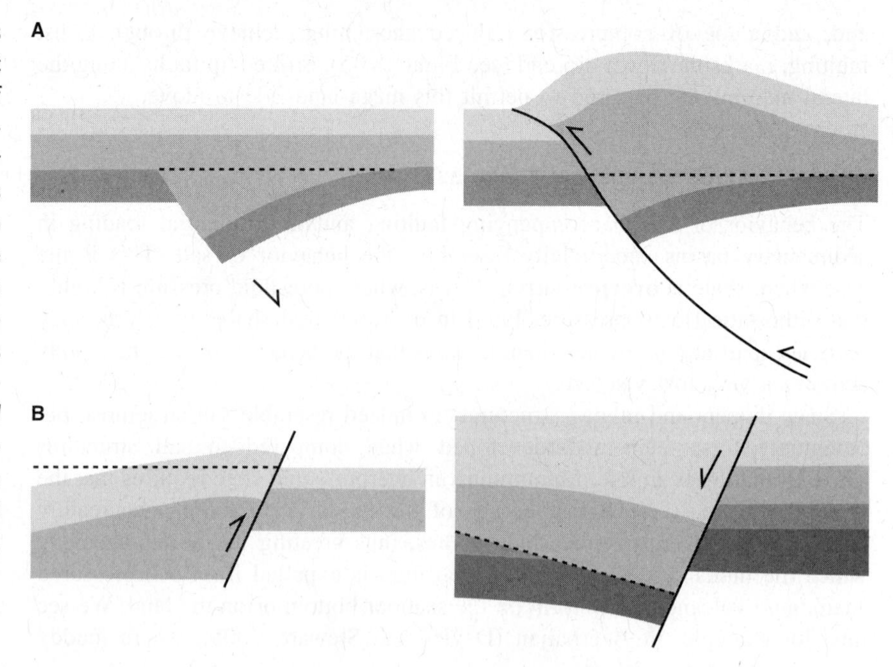

Figure 8.46 (*A*) Positive inversion begins with normal faulting in an environment of extension followed by reactivation of normal faults as reverse faults in an environment of tectonic shortening. [After Williams et al., 1989, Inversion tectonics Figure 1, p. 3. Reprinted with permission from Blackwell Science Ltd.] (*B*) Negative inversion begins with reverse faulting in an environment of tectonic shortening, followed by reactivation of reverse faults as normal faults in an environment of extensional tectonics. [Reprinted from *Marine and Petroleum Geology*, v. 26, Su et al., Geometry styles and quantification of inversion structures in the Jiyang depression, Bohai Bay Basin, eastern China, p. 25–38, © 2009, with permission from Elsevier.]

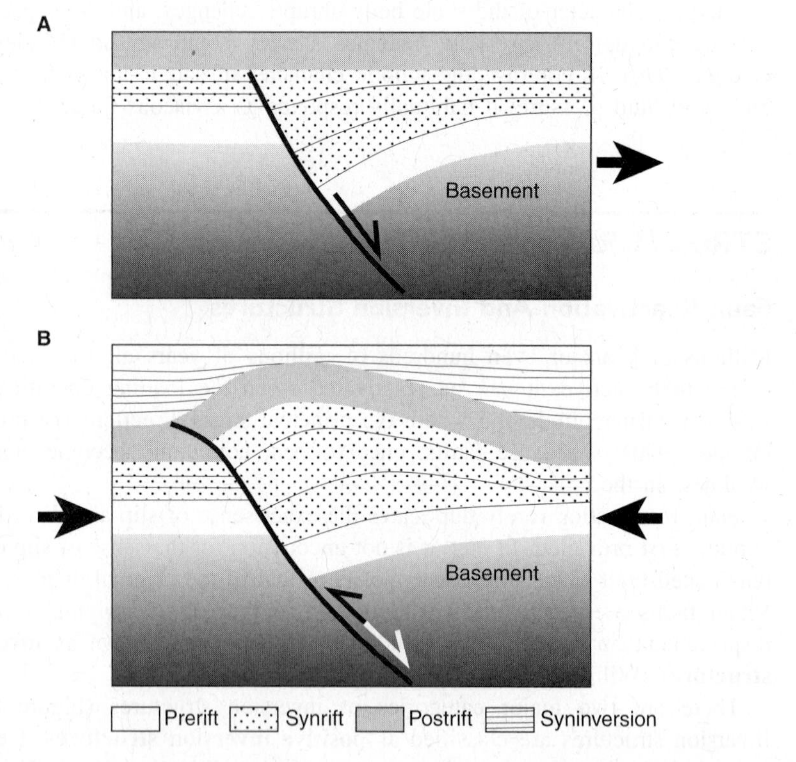

Figure 8.47 Basin stratigraphy records positive inversion. (*A*) Synrift sediments accumulate to greatest thicknesses in the hanging wall of the normal fault. Simultaneous normal faulting and sedimentation result in rollover. Postrift sediments lie unconformably on synrift units, denoting the timing of cessation of the normal faulting. (*B*) Reverse-fault reactivation of the original normal fault uplifts, folds, and displaces postrift units and is accompanied by syninversion sedimentation. Note the thickness variations of syninversion stratigraphy in relation to the uplift. [From Yamada and McClay, 2004, Analog modeling of inversion thrust structures, in McClay, K. R. (ed.), Thrust tectonics and hydrocarbon systems: *American Association of Petroleum Geologists Memoir*, 82, p. 276–301. AAPG © 2004, reprinted by permission of the AAPG whose permission is required for further use.]

and that the total sediment thickness in the hanging wall is much, much greater than on the footwall. The normal faulting phase is topped off by **postrift sediments** (see Figure 8.47*A*). During positive inversion (Figure 8.47*B*), the structural geometry of the prerift, synrift, and postrift formations changes markedly. Notice how the postrift stratigraphy, in this example, becomes deformed by folding and thrust faulting. As positive inversion takes place, the form of the rising structure imposes a marked influence on the thickness and distribution of syninversion sediments (see Figure 8.47*B*). Syninversion sediments are thinnest on top of the uplift. Finally, notice that

Yamada and McClay (2004) chose to represent the active faulting through the use of a black arrow, reserving a white arrow for the first interval of faulting. In this example of positive inversion, the magnitude of reverse slip is quite modest (see offset of postrift layer), and thus the *net* offset of the prerift layer is still one of normal displacement (see Figure 8.47*B*).

Positive inversion by reverse faulting will commonly produce a fault-propagation fold (see Figure 8.47*B*). The "*smoking-gun*" that clinches positive inversion is the presence of a thicker stratigraphy on the hanging wall compared with that of the footwall (see Figure 8.46*A*). Normally the hanging wall of a reverse fault will display a thinner stratigraphy (compared to the footwall) because (1) hanging-wall erosion accompanies uplift and folding, and (2) less accommodation space is available above the structure. However, prior to positive inversion, the early normal faulting would have accommodated deposition of growth strata on the hanging-wall side (see Figure 8.46*A*). The net result, even after reverse-fault inversion, is that the hanging-wall side of the fault will have the thicker sedimentary section. In contrast, and based on similar arguments, it follows that negative inversion structures have anomalously thin stratigraphic thicknesses on the hanging wall of the fault (see Figure 8.46*B*, especially the deepest part of the lower panel).

Null Points

It is helpful to imagine the deformation path of fault cutoffs of stratigraphic horizons during positive inversion (Williams et al., 1989). First, during the normal faulting phase (Figure 8.48*A*), prerift beds are cut off and move with the hanging wall as it drops relative to bed **cutoffs** left behind on the footwall. A growth sequence builds incrementally during the full history of normal

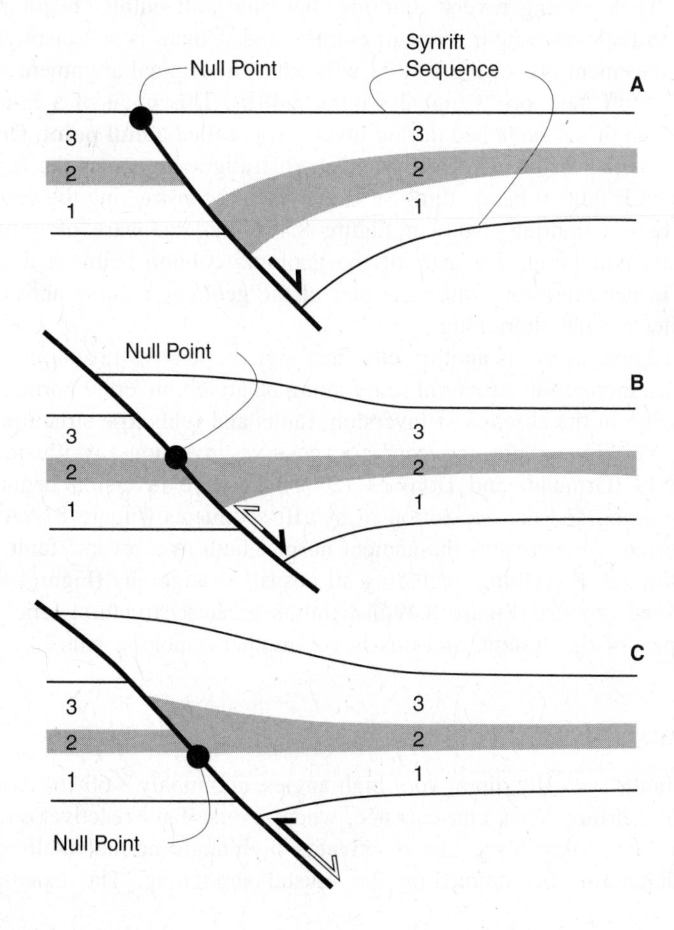

Figure 8.48 To understand the meaning of "*null point*," let us track unit #2 through positive inversion. (*A*) Offset of unit #2 by normal faulting. (*B*) During reactivation of the normal fault as a reverse fault, we capture here a stage where the hanging-wall and footwall parts of the top of unit #2 are reunited at their regional elevation. The null point is shown, i.e., the point where net throw is nil. (*C*) Note that the null point moves *down* the fault plane as progressively older beds are uplifted during inversion and brought back to regional elevation. Rocks above the null point reveal net shortening, while those below reveal net extension. [After Williams et al., 1989, Inversion tectonics, Figure 8.53. Reprinted with permission from Blackwell Science Ltd.]

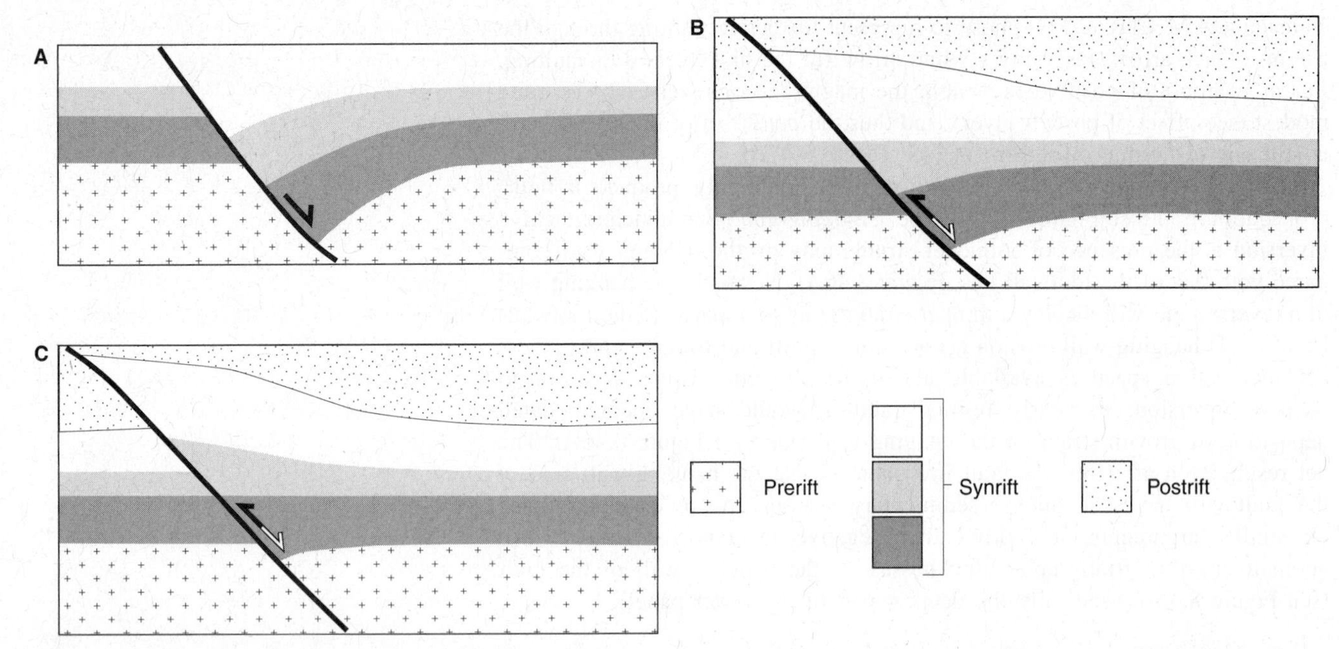

Figure 8.49 One of the characteristics of positive inversion is upward increase in structural relief. (*A*) Initial normal faulting and growth sedimentation. (*B*) Onset of positive inversion. (*C*) Final product of positive inversion. Note that stratigraphic throw near top of uppermost synrift unit is much greater than that of the immediately underlying synrift unit. [From Grimaldi and Dorobek, 2011, Fault framework and kinematic evolution of inversion structures: Natural examples from the Neuquén Basin, Argentina: *American Association of Petroleum Geologists Bulletin*, v. 95, no. 1, p. 27–60. AAPG © 2011, reprinted by permission of the AAPG whose permission is required for further use.]

faulting. Then, during reverse faulting, hanging wall cutoffs begin moving back up to their respective footwall cutoffs, and if there is sufficient reverse-fault displacement one of the cutoffs will return to original alignment with its footwall cutoff (see top of unit 2, Figure 8.48*B*). This point of a re-attained juxtaposition of the same bed during inversion is called a **null point**. Only at a null point is it possible to directly correlate stratigraphy across the fault surface, which is quite a handy thing to know when reconstructing the geological column before faulting. Note in Figure 8.48*C* that the base of unit 2 has reached its null point. The part of the geologic column below a null point remains in net extension, while the part of the geologic column above a null point reflects a net shortening.

We become aware of another clue that signals positive inversion, namely the upward increase in structural relief along positively inverted normal faults. Customarily, in the absence of inversion, faults and folds lose structural relief upward. Yet this is not the case for positive inversion, as illustrated in Figure 8.49 (Grimaldi and Dorobek, 2001). Positive inversion begins with normal faulting and the deposition of synrift sediments (Figure 8.49*A*). Tectonic shortening reactivates the ancient normal fault as a reverse fault, which cuts up through everything, including all postrift stratigraphy (Figures 8.49*B*). The finished product (Figure 8.49*C*) exhibits greatest structural relief in the highest part of the system, and this is yet another 'smoking gun.'

Shortcuts

Normal faults typically dip at very high angles, commonly ~60° or more near the earth's surface. As a consequence, when reverse-fault reactivation begins and shortening takes place, the reactivated high-angle normal faults are not very efficient in accommodating the crustal shortening. The hanging-wall

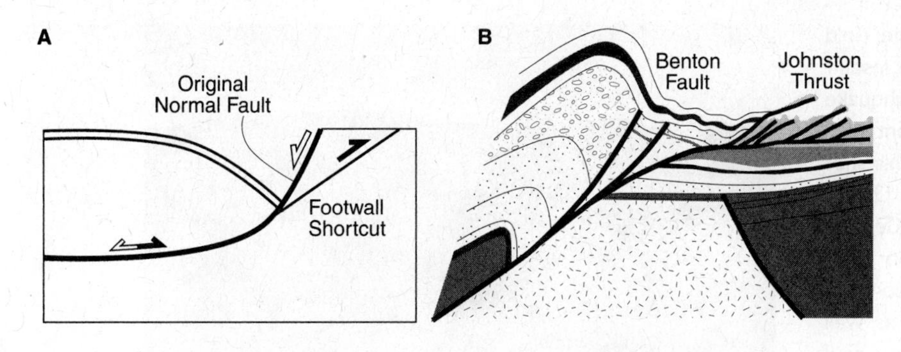

Figure 8.50 Footwall shortcut. (*A*) A listric normal fault forms during the first phase of positive inversion. During inversion, the deeper part of the listric normal fault is reactivated as a footwall shortcut. [From Williams et al., 1989, Inversion tectonics. Reprinted with permission from Blackwell Science Ltd.] (*B*) Positive inversion and the formation of a footwall shortcut. Location is southwest Wales. During the Devonian, the Benton fault originated as a normal fault. In Carboniferous, positive inversion took place, and the Benton fault was reactivated as a reverse fault and footwall shortcut. [From Powell, 1989, p. 445. Reprinted with permission from The Geological Society Publishing House, Bath, England.]

block must be uplifted disproportionately to accommodate even modest amounts of shortening.

Nature can always be counted upon to find a solution. In this case, positive inversion may activate a brand new new low-angle thrust, which is called a **footwall shortcut** (Figure 8.50*A*). This new thrust branches from the reactivated normal fault at a level where the master fault has a relatively low dip. Faulting associated with the new thrust will cut off a corner of the former footwall. The hanging wall then incorporates this piece of original footwall into the hanging wall of the new thrust (see Figure 8.50*A*).

A nice example of a footwall cutoff is shown in Figure 8.50*B*, which portrays in cross-section a fault/fold system in southwest Wales (Powell, 1989). The Benton fault began life as a normal fault during crustal extension during the Devonian. In late Carboniferous, tectonic shortening triggered positive inversion of the Benton fault, creating not only folding of growth strata in the hanging wall but also a new low-angle thrust in the footwall. This low-angle fault, known as the Johnston thrust is a perfect example of a footwall shortcut associated with inversion (see Figure 8.50*B*).

FOLDS ASSOCIATED WITH STRIKE-SLIP FAULTING

Strike-slip fault systems produce folds, and thus once again there are opportunities to evaluate the interactions of faulting and folding. In some cases, folds form between convergent strike-slip fault segments. We can think of such an environment as one of "*crowding*," which produces fault-perpendicular folding. At times these take place on a tiny scale, for example during an earthquake, in which case they are referred to as **pressure ridges** (McCalpin, 2009). For example, a small pressure ridge formed during the strike-slip Hector Mine earthquake ($M_W = 7.1$) on October 16, 1999, near Twentynine Palms, California. We have Paul "*Kip*" Otis-Diehl of the United States Marine Corps to thank for photographing this pressure ridge while it was still fresh and unweathered (Figure 8.51).

But folds are associated with strike-slip faults at the tectonic scale as well. In fact, en echelon folding is one of the singular characteristics of folding associated with strike-slip faulting. Systems of en echelon folds are forming today in the California Continental Borderland region off the west coast of southern California (Figure 8.52*A*). This active tectonic setting is perfect for producing structures associated with strike-slip faulting, for it is the Pacific/North America plate boundary region and the San Andreas Fault is located just inboard (see Figure 8.52*A*). Through application of high-resolution multi-channel seismic profiling, details of the active deformation can be imaged on the seafloor bottom. Figure 8.52*B* is one example, described by Legg et al. (2007). The San Diego Trough fault is a right-handed strike-slip

Figure 8.51 Photograph of pressure ridge (fold structure) produced in the landscape at the instant of the Hector Mine strike-slip earthquake (M_W = 7.1) on October 16, 1999, near Twentynine Palms, California. U.S. Marines for scale. [This shot was taken by Paul "Kip" Otis-Diehl of the United States Marine Corps. Courtesy of the U.S. Geological Survey. See Andrew Alden's website on "*Tectonic Landforms*," http://geology.about.com/od/structureslandforms/ig/tectoniclandforms/ as well as the USGS site, http://pasadena.wr.usgs.gov/hector/photos.html.]

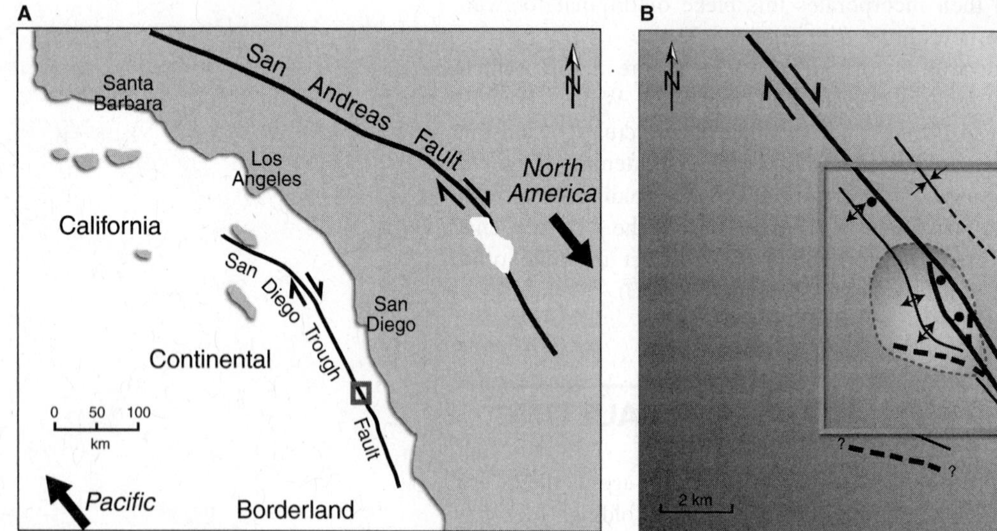

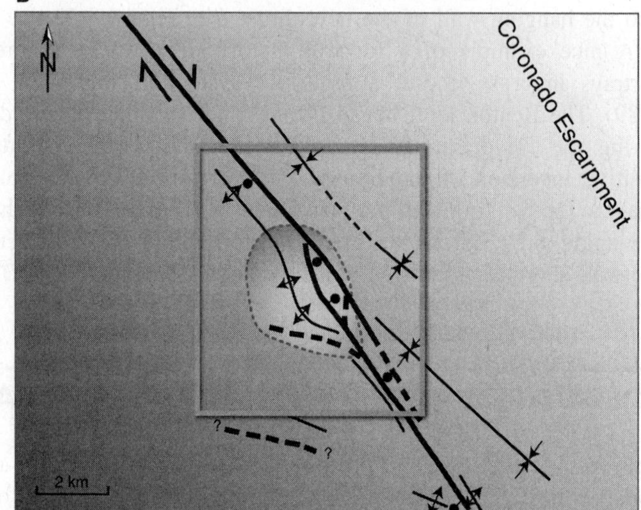

Figure 8.52 (*A*) Regional map showing the California Continental Borderland in relation to southern California (gray). Note trace of San Andreas fault. The small box identifies an area where en echelon folds have developed. (*B*) Map showing trace of the San Diego Trough fault. Slight bend in the fault, combined with on-going right-handed strike-slip faulting, creates a restraining bend and a pop up. Another expression of the restraining bend is the set of en echelon folds. [From Legg et al., 2007, Figure 2, p. 145, and Figure 17, p. 162. Reprinted with permission from The Geological Society Publishing House, Bath, England.]

fault cutting the seafloor bottom. A slight bend (just 12°, clockwise) in the trend of the fault has created a restraining bend (see Figure 8.52*B*), and at that bend there is a pop-up measuring ~2.3 km × ~3.8 km, and displaying 700 m of structural/topographic relief (Legg et al., 2007). Furthermore, as can be seen in Figure 8.52B, a system of en echelon folds is actively forming

Fault-bend folds, fault-propagation folds, and detachment folds are all 'fair game' in strike-slip settings, and the kind of fault/fold interaction that takes place is a function of a variety of factors, including mechanical stratigraphy and the role (if any) of pre-existing faults. But for folding to take place in strike-slip settings, there must be environments that create layer-parallel shortening or layer-parallel stretching. Such environments, as we learned in *Chapter 6, Faults*, are associated with **restraining bends**, **releasing bends**, and **stepovers** (see Figures 6.146, 6.148, and 6.150). At restraining bends there is shortening and uplift; at releasing bends there is stretching and subsidence. At step-overs *between* discrete fault segments, the sense of strike-slip displacement (right-handed versus left-handed) in combination with the type of step-over (right-stepping versus left-stepping) determines whether the step-over region will be marked by shortening and uplift, or stretching and subsidence (see Figure 6.150). Not surprisingly, the most dramatic folds within major strike-slip systems will form in the regions of shortening and uplift.

In *Chapter 11, Active Tectonics*, we will take a close look at folds formed during strike-slip faulting, in particular the folds associated with major thrust faults in the Los Angeles basin. This active deformation is the consequence of shortening at the "*Big Bend*" in the San Andreas fault.

ROLE OF STRUCTURAL DEVELOPMENT IN SEDIMENTATION

Introduction

When the combination of faulting and folding deforms the earth's surface, the changes in topography and/or bathymetry that that result will influence where sediments accumulate as the structures continue to grow. The topography/bathymetry produced by active deformation influences sediment thickness as well as the distribution of sedimentary facies. If the ages of the sediments in sedimentary growth sections are known, then the rate(s) of growth fault- and fold-induced uplifts can be determined. The age of the oldest unit within the growth sequence provides the approximate age of the onset of faulting and/or folding. The age of the uppermost unit within the sedimentary section provides the approximate time of when faulting and/or folding came to an end.

Sedimentary Growth Sequences

As we have seen, the sediments that a tectonic basin receives accumulate to form a **sedimentary growth sequence**, controlled in large part by the nature of the faulting and folding that produced the accommodation space (Suppe et al., 1997; Suppe, 2004; Namson and Davis, 1988; Lawton et al., 1999; Bernal and Hardy, 2002). In some tectonic environments, bed-form geometry of growth sequences can be explained through scenarios of progressive deformation involving fault-bend folding or fault-propagation folding. Let us take the simplest case, where tectonic shortening is producing folding and faulting, but under circumstances where sedimentation rates are so high that there is never any topographic or bathymetric expression of the uplift. Suppe et al. (1992) describe this for fault-bend folding (Figure 8.53*A*) as well as fault-propagation folding (Figure 8.54*A*). Though the geometry is elegantly frightful, this is a good thing, for details of the bending of strata in the growth sequence can be used to interpret fault geometry in the subsurface (Suppe et al., 1992).

Preservation of growth sequences on the very top of uplifts is by no means the rule, especially when the rates of uplift exceed sedimentation rates (see Figures 8.53*C* and 8.54*C*). In contractional settings, when the hanging wall is lifted above regional elevation, critical parts of the structural geometry will become lost to erosion. Yet there *are* situations in contractional settings where the rate of sedimentation is so high that no topographic or bathymetric

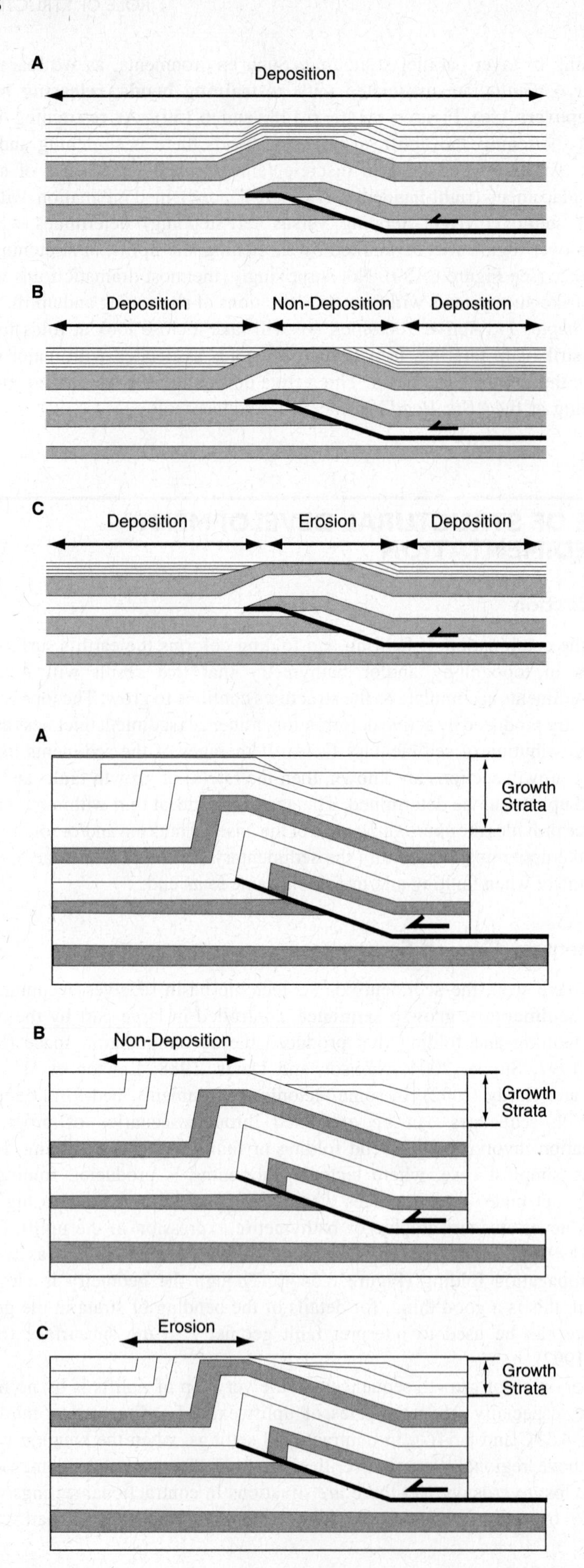

Figure 8.53 Growth-sequence stratigraphy deposited during fault-bend folding (*A*) Rate of sedimentation greater than the rate of uplift; crest of the uplift becomes buried. (*B*) Rate of uplift is equal to rate of sedimentation; neither erosion nor deposition on crest of uplift. (*C*) Rate of uplift greater than rate of sedimentation; crest of uplift is subjected to erosion. [From Suppe et al., 1992, Figure 13, p. 112, in McClay, K. R., (ed.), Thrust tectonics: Chapman and Hall, London, p. 105–121. With kind permission from Springer Science+Business Media B.V.]

Figure 8.54 Growth-sequence stratigraphy deposited during fault-propagation folding. (*A*) Rate of sedimentation greater than rate of uplift; crest of the uplift becomes buried. (*B*) Rate of uplift equal to rate of sedimentation; neither erosion nor deposition on the crest of the uplift. (*C*) Rate of uplift greater than rate of sedimentation; crest of the uplift is subjected to erosion. [From Suppe et al., 1992, Figure 20, p. 115, in McClay, K. R., (ed.), Thrust tectonics: Chapman and Hall, London, p. 105–121. With kind permission from Springer Science+Business Media B.V.]

expression develops, and the subsurface record of growth stratigraphy is fully preserved (Suppe et al., 1992) (see Figures 8.53*A*, 8.54*A*). In general, however, growth sequences are more likely to be more fully preserved in extensional settings.

Forward Modeling

Balanced forward modeling of the development of fault/fold relations, including the structural geology of the growth sequences, is an essential component of oil/gas exploration. Yes, the final structural product is important, but even more important is how it came to be. An important emphasis is on the interplay of sedimentation, erosion, and progressive deformation over time.

In order to carry out balanced forward modeling, it is essential to determine where each volume of rock was deposited with respect to the growing structure. Once this is achieved, it is then necessary to reconstruct the deformation path for each volume of rock during faulting and/or folding. Figure 8.55 is a model that assumes that deformation took place through a growing, migrating kink fold, linked to a fault (not shown) at depth. The starting configuration (Figure 8.55*A*) shows reference points #1–#9 that we can "*track*" through time. Also shown is the swath (i.e., "*active hinge*") within which the kink band will initiate. Reference points to the left of the kink band will just 'sit there' minding their own business, until they are overtaken by the left edge of the migrating kink band. Points to the right of the kink band (in this case point #9) will remain encased in horizonal strata at the top of the uplift, though they will travel up and to the left relative to pinned lower-left corner of the model. All the reference points, including #8 and #9,

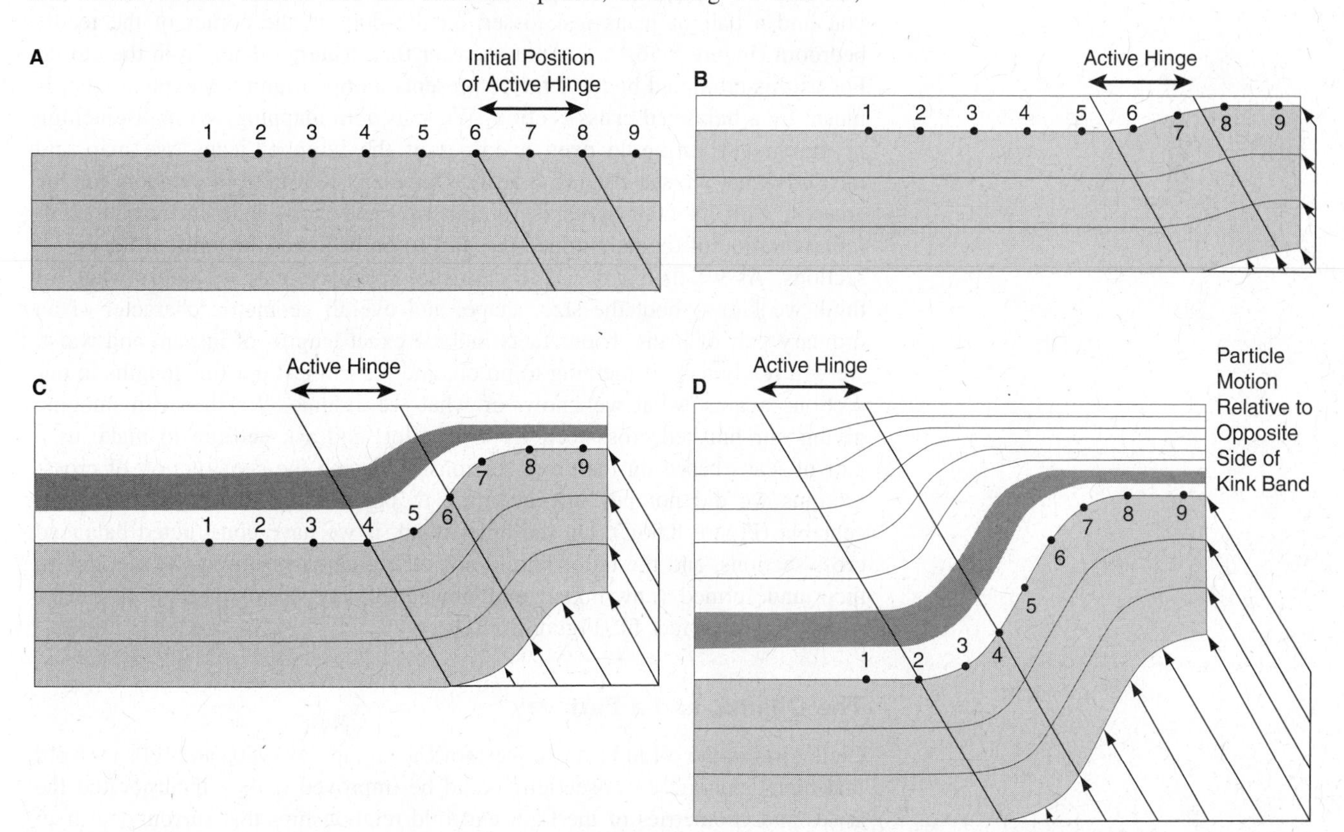

Figure 8.55 Details of forward modeling the sequential development of folding and growth sedimentation. The level of attention required is such that we need to 'track' the migration of individual points, i.e., the deformation paths. See text for details. [Reprinted from *Journal of Structural Geology*, v. 19, Suppe, J., et al., Bed-by-bed fold growth by kink-band migration: Sant Llorenc de Morunys, eastern Pyrenees, p. 443–461, © 1997, with permission from Elsevier.]

become buried by growth strata. When contraction, uplift, and folding begin (Figure 8.55*B*), strata become bent within the active hinge, and immediately the positions of the upper and lower hinge zones of the fold become apparent. The sedimentation rate (relative to uplift rate) is such that a growth sequence begins to form. With progressive contraction, uplift, and folding (Figure 8.55*C, D*), the active hinge sweeps through strata and reference points. Reference points #1 and #2 remain embedded in horizontal, unstrained strata, . . . but are they in for a surprise!

STRUCTURAL BALANCE

Introducing the Concept

When we think of the art and science of making geologic cross-sections of faulted and folded rocks, the first thing that comes to mind is creating a 'picture' that is completely consistent with all the facts we have in hand, notably the mapped geology and/or subsurface information derived from drilling and seismics. These facts, if abundant and comprehensive, enable us to represent accurately the shapes and sizes of the deformed geological formations. Based on these facts we construct geologic cross-sections in such a way that when they are restored, the shapes and sizes of the geological formations look reasonable in their undeformed states, and have lengths and thicknesses that conform to all that we know about them. This general concept is known as **structural balance**.

We introduce here the concept of structural balance through an unconventional example. As most parents of teenagers know, there are times when you find a pair of jeans just tossed on the floor in the corner of the teen's bedroom (Figure 8.56*A*), . . . much easier than hanging them up in the closet. For us this crumpled body of fabric presents an opportunity to explain what is meant by a **balanced cross-section**. We start with mapping. Without touching or moving or lifting-up even a corner of the lump of jeans, we map and measure what we see (Figure 8.56*B*). Our map, with its fold symbols for the creases, with its fault symbols for the rips and tears, with its normal fault representation of the unbuttoned fly, and so on becomes the **control** for cross-sections. As we then proceed to construct cross-sections, we apply what we think we know about the size, shape, and overall geometric character of an ordinary pair of jeans. If our facts include exact lengths of inseam and waist, we will really have something to go on, and we can test our line lengths in our sections against what we know, or what we assume. We then can imagine giving our finished cross-sections to a tailor, and ask her/him to make us a pair of jeans based on the cross-sections. If during the construction of cross-sections we did not pay any attention to balance, the result will be unacceptable (Figure 8.56*C*). On the other hand, if we have constructed balanced cross-sections, and the tailor knows something about restoring the sections to their undeformed (unwrinkled and unwadded up) shape, there is a greater chance for a perfect fit (Figure 8.56*D*).

The Dilema, and a Pathway

Geologists in the oil industry in western Canada in the 1950s and '60s realized that their geologic cross-sections could be improved if they incorporated the styles and geometries of the fault and fold relationships that surrounded them in the spectacular exposures of the Canadian Rockies (Figure 8.57). Furthermore, they developed some practical kinematic rules that could be used to test and improve their interpretations of the subsurface geology in the fold-thrust belt of the Canadian Rockies.

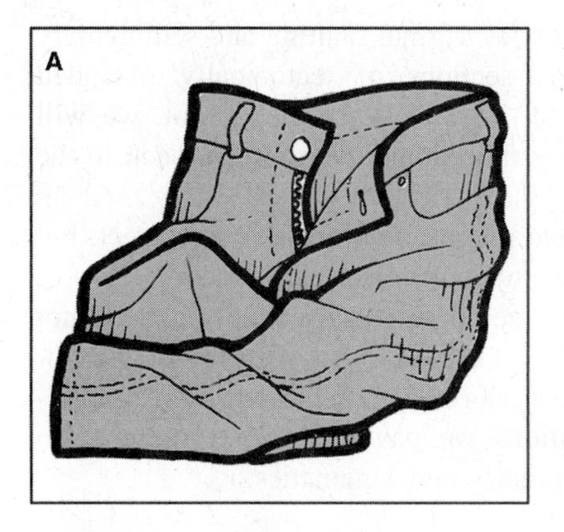

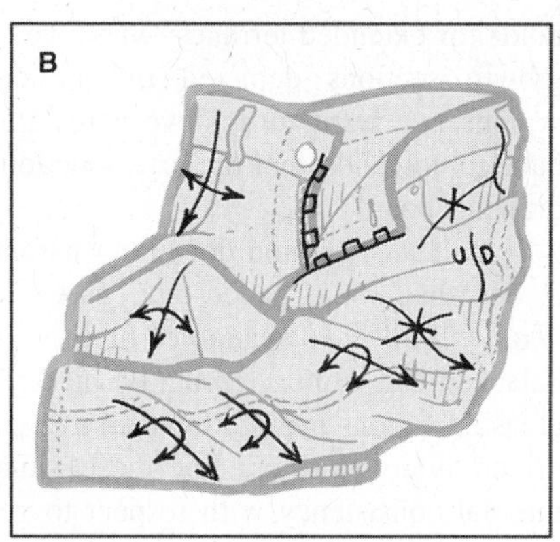

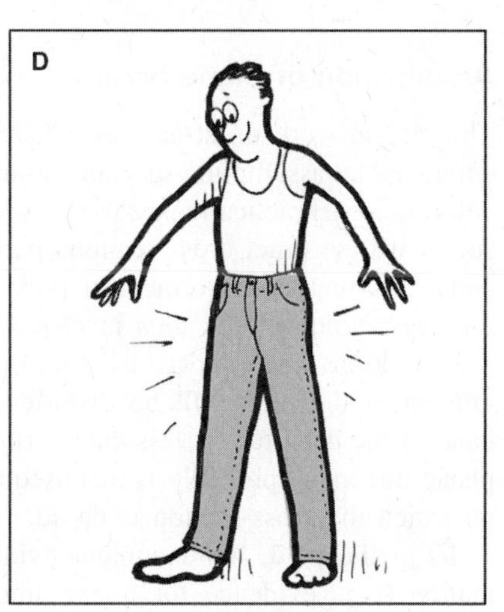

Figure 8.56 Cartoon example of the importance of achieving structural balance in geologic cross-sections. (*A*) The structural geologic system, about to be mapped. (*B*) The map. (*C*) Reconstruction based on ineffective, unbalanced cross-sections. (*D*) Reconstructions based on effective, unbalanced cross-sections. See text for details. [Artwork by D. A. Fischer.]

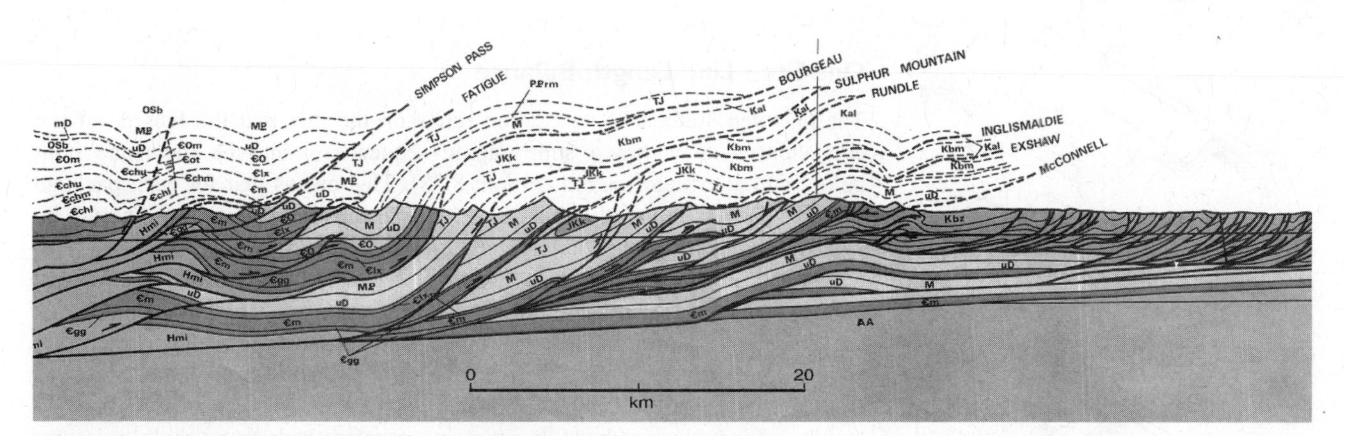

Figure 8.57 Balanced cross-section through a part of the Canadian Rockies. [From Price and Fermor, 1985. © Department of Natural Resources Canada. All rights reserved.]

Clint Dahlstrom (1969) advocated structural balance, and was among those who underscored the utility of constructing balanced cross-sections. **A balanced cross-section must be restorable to its undeformed state**. In other words, when a cross-section of thrust faulted and folded strata is stretched back to its pre-faulted, pre-folded state, we want the restoration of the geological column not only to look reasonable, but to conform with *all* that we know about the original size, shape, and thickness of bedrock units. The same holds for extended terranes: when we "*undo*" normal faulting and sedimentary growth sections depicted in our cross-sections of tectonically extended regions, we want to achieve a restoration that makes sense. If not, we will have to conclude that the cross-section is flawed, and we must go back to the drawing board.

A balanced section does not guarantee that an interpretation is correct, but is one that *could* be correct, given its internal consistency and the way it "*honors*" all data available. In contrast, geologic cross-sections that are not balanced are almost certainly incorrect. Balanced cross-sections drawn for areas or regions marked by sparse or poor data must rely more fully on model-driven interpretations. For these situations we pay very close attention to internal consistency with respect to geometry and kinematics.

Assumption of Plane Strain

The classic work on structural balance carried out in the Canadian Rockies embraced an assumption of plane strain. Plane strain implies constant-volume deformation. Practically speaking, when we assume constant-volume plane strain, we construct cross-sections parallel to the tectonic transport direction, and we assume no movement of rock in or out of the plane of the section. In this way, volume (i.e., *area* in cross-sectional view) is preserved. In reality, cross-sections that depart by more than ~10° or 15° from the tectonic movement direction will be distorted and not truly balanced. On the other hand, if the departure is less than ~10–15°, the distortion imparted by out-of-plane movement probably is inconsequential, given the uncertainty of the data on which the cross-section is based.

If, in the field, we document evidence for loss of volume during deformation (e.g., evidence for major amounts of pressure dissolution), we will need to make appropriate adjustments in our balanced cross-sections. This is yet another example of taking full advantage of all that we learn from studying the rocks and structures exposed in available outcrops.

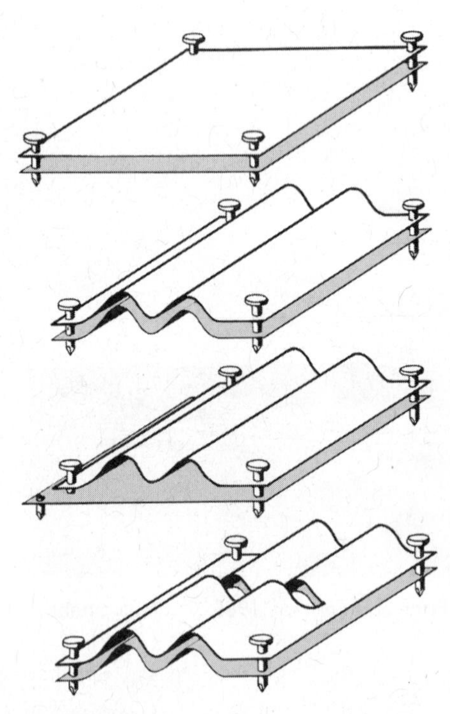

Figure 8.58 Balanced cross-sections are constructed in a way to preserve the original cross-sectional areas of beds that are involved in the deformation. [From Dahlstrom, 1969a. Reproduced with permission of the National Research Council of Canada from *Canadian Journal of Earth Sciences*.]

One-Step Line Length Balance

The one-step method of restoration relies on the fact that the lengths of the competent beds within a stratigraphic sequence should be the same both before and after deformation (Figure 8.58). This means that for each package of stratigraphic units, the line lengths of each competent bed should be identical both before and after tectonic shortening. If not attentive to line-length balance, we find ourselves drawing folds and faults in ways that cannot be restored. For example, we might discover to our dismay that the trace of the upper surface of a given massive bed is much shorter (or longer) than that of the lower surface. This is permissible only if there are geological facts to explain why these line lengths should be different. 'All bets are off' for line lengths of incompetent beds, because incompetent beds (e.g., shale, gypsum, and salt) are so ductile that their bed lengths, areas, and volumes can become distorted during deformation. The "*balancing*" of incompetent beds is achieved through **area balancing**. Area balance requires that the original thickness of the unit be determined. This may be possible for shales in some cases, but may often be quite difficult for salt because there may be no place where the original formation thickness is preserved.

Mitra and Namson (1989) presented an excellent example of one-step balancing, using the combination of line-length balancing and area balancing (Figure 8.59). Their starting point was a cross-section through the southern Appalachian Valley and Ridge province published by Roeder et al. (1978). It

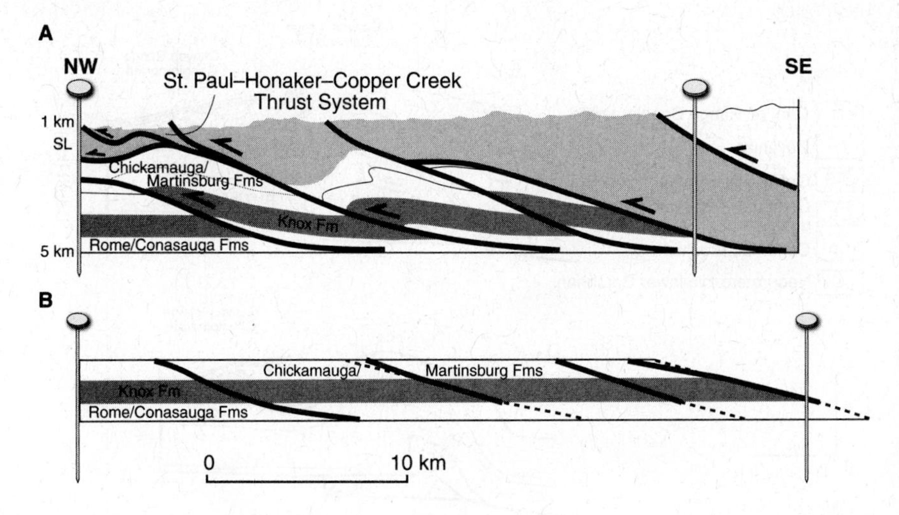

Figure 8.59 Mitra and Namson (1985) applied one-step balancing to a geological cross-section through the southern Appalachians, published by Roeder et al. (1978). They restored all faults and folds to their presumed pre-deformation geometry in a single move. The lengths of the unit boundaries of competent rocks, such as the Knox Formation, in their deformed states were assumed to be the same as they were *before* deformation. Restoration of the incompetent units, such as the Rome Formation, was achieved through area balancing because it is known that line-lengths of incompetent beds may change during deformation. [From Mitra and Namson, 1989, *American Journal of Science*, v. 289, Figure 2, p. 568. Reprinted by permission of the American Journal of Science.]

was this section that they balanced. Mitra and Namson (1989) applied line-length balancing to the competent Knox Formation (Cambro-Ordovician), and area balancing to the incompetent interval of Rome Formation through the Conasauga Formation (Cambrian). The Rome/Conasagua sequence required area balancing because these rocks were distorted in such a way during folding and faulting that thickening and thinning occurred from place to place.

Measuring line lengths requires **pin lines**, the end points for the measurements. We see in Figure 8.59 the pin lines chosen by Mitra and Namson (1989) in their line-length balancing of the Knox Formation. In general, pin lines are positioned at right angles to bedding where the beds return to their regional elevation. If there is no opportunity to position pin lines beyond the system of folding and faulting, it is best to draw the pin lines parallel to the axial traces of the folds, for at these locations the pin lines will be perpendicular to bedding.

"Last-In, First-Out" Line Length Balance

One-step line-length balance works pretty well, but better results are achieved if the structures themselves are restored incrementally, and in an order opposite to their formation. We call this the "*last-in, first-out*" rule of restoration. Using this approach, faults and fault-related folds within the cross-section are restored one at a time, with the youngest structures restored first. This method generally assumes that an individual fault forms and dies out before the next fault initiates, which is not always the case.

The "*last-in, first-out*" approach can be helpful in sorting out the sequencing of faulting and the question of whether some or all of the faults overlapped temporally. Sections that resist this mode of restoration may be ones in which several faults were active at the same time. This incremental approach quickly illuminates geometries that simply are not geologically reasonable. Seismic reflection profiles and drilling data can place tight constraints on incremental restoration.

DeCelles and Coogan (2006) published a superb example of balanced, incremental retrodeformation of their cross-section through the Canyon Range, Utah. In fact we led off this chapter with one of their geologic cross-section (see Figure 8.1*A*). Their sequential cross-sections, presented here in Figure 8.60, were based upon very detailed geological mapping, a sophisticated grasp of thrust tectonics, and preparation of balanced cross-sections. In Figure 8.60*A* we see the initial state of affairs, following thrust-displacement

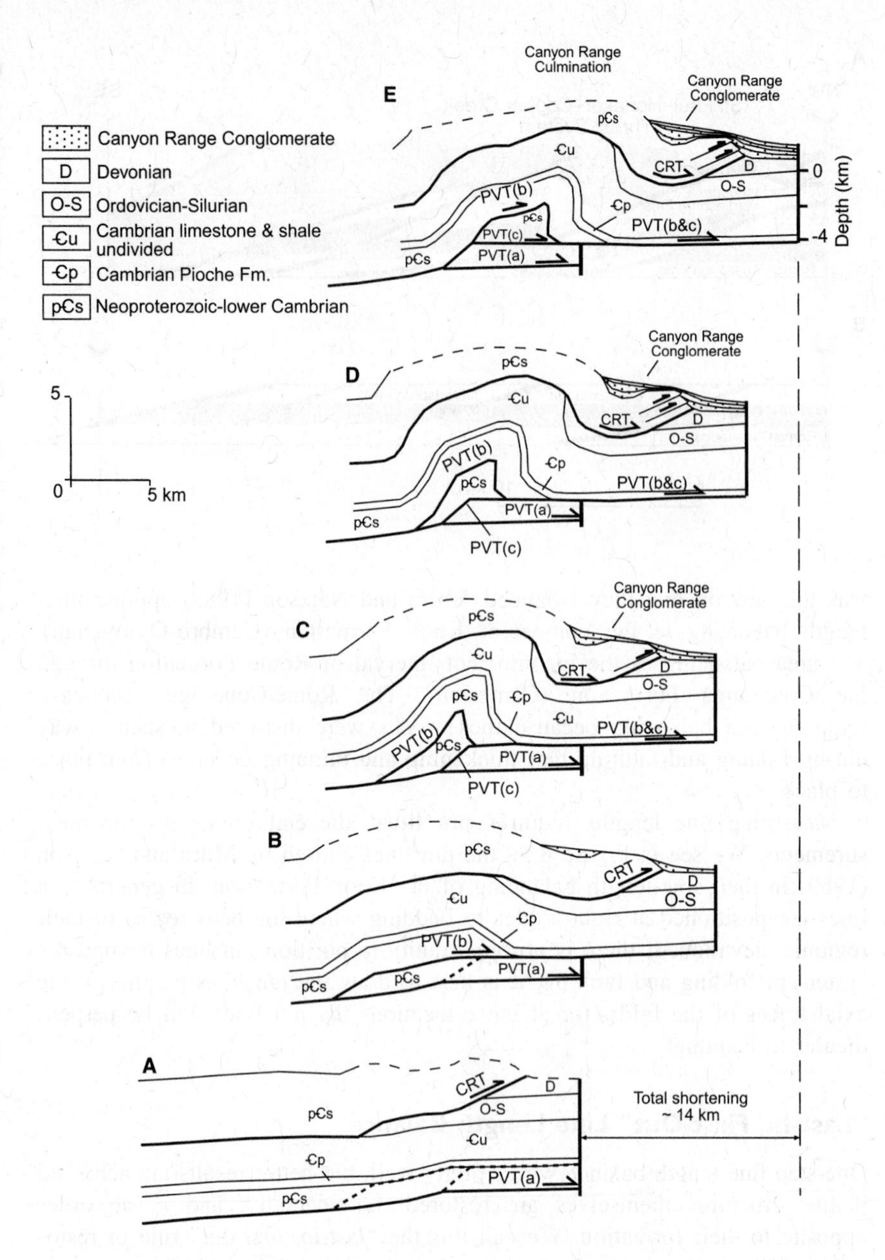

Figure 8.60 This set of cross-sections is a balanced, incremental restoration of thrust faulting and folding in the Canyon Range, Utah, based on the work of DeCelles and Coogan (2006). See text for details. [From DeCelles and Coogan, 2006, Figure 9, p. 855. Published with permission of the Geological Society of America.]

on the Pavant thrust (PVTa). Minimum slip along the main Pavant thrust was ~40 km. In Figures 8.60*B–E* the DeCelles-Coogan interpretation shows the building of the Canyon Range duplex. This involved slip along PVT(b) and PVT(c), which are hanging-wall imbricate thrust splays ramping upward from the main Pavant thrust. Fault slip on these imbricates contributed an additional 42 km of slip to the Pavant thrust system. Duplex formation simultanteously created the huge anticline/syncline pair of the Canyon Range (see Figure 8.60*C–E*). All of this deformation took place between ~110-86 Ma (DeCelles and Coogan, 2006).

Estimating Strain from Balanced Sections

Balanced cross-sections of regional thrust relationships can provide a strong foundation for estimating regional strain. Price and Mountjoy (1970) estimated regional strain produced by crustal shortening in the Canadian Rockies (see

Figure 1.43). There, thrust faulting and fault-related folding occurred between late Jurassic and Eocene. The westernmost strata were moved at least 200 km to the northeast. The nonconformity between sedimentary cover and crystalline basement is preserved underneath the thrusted and folded mass (see Figure 1.43); it dips gently westward. The thrust-slip faults do not cut into basement, but feed into ductile sedimentary layers right above the basement/cover interface. Based on restoration of their balanced cross-sections, Price and Mountjoy (1970) determined that the original miogeoclinal sediment package occupied an east-west distance of ~475 km wide, which was then reduced to ~240 km by folding and thrusting.

$$S = \frac{240 \text{ km}}{475 \text{ km}} = 0.51$$

$$\% \text{ shortening} = (0.51 - 1.00) \times 100 = 49\%$$

For both of these regions, the geological cross-sections are superb, marked by balance and internal consistency.

Consistency of Displacement

A region of bedrock that has enjoyed deformation will display a vast array of large-scale geologic fold/fault structures, which, taken together, accommodate the total strain and displacement. Individual structures seldom cross an entire region, and thus strain and displacement must be transferred, or linked, from one major structure to another. Dahlstrom (1969) recognized this clearly and as a result advocated the importance of preparing maps and geologic cross-sections marked by **consistency of displacement**. As applied to faults, "*consistency of displacement*" means that offset measured along any given fault should be uniform in sense and magnitude (if the rocks are perfectly rigid) or should change systematically (Figure 8.61). If data show that offset varies from place to place along a fault, this circumstance must be explained by interpretations that involve the **interchange** of different degrees of folding and faulting to accommodate a common shortening or the **replacement** of one fault by a series of **imbricate splay faults.**

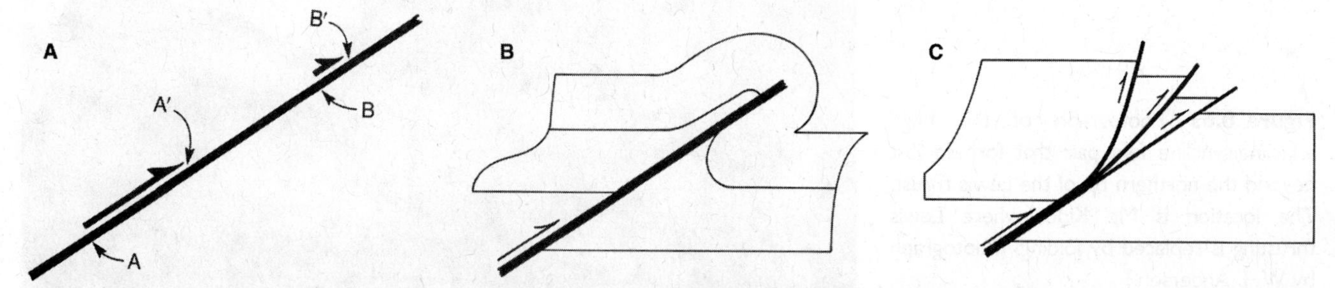

Figure 8.61 (*A*) Fault showing an inconsistency of displacement. The slip along the lower reaches of the fault is much greater than that toward the surface. The inconsistency can be accommodated by (*B*) the interchange of faulting and folding, or (*C*) the replacement of the single fault by several splay faults. [From Dahlstrom, 1969a. Reproduced with permission of the National Research Council of *Canada from Canadian Journal of Earth Sciences*.]

A wonderful example of interchange and replacement is seen just south of Banff, Alberta, in the Canadian Rockies. Figure 8.62*A* is a map that shows decreasing displacement along the Lewis thrust as it dies out northward. This steady decrease in displacement is evident in the steady northward reduction

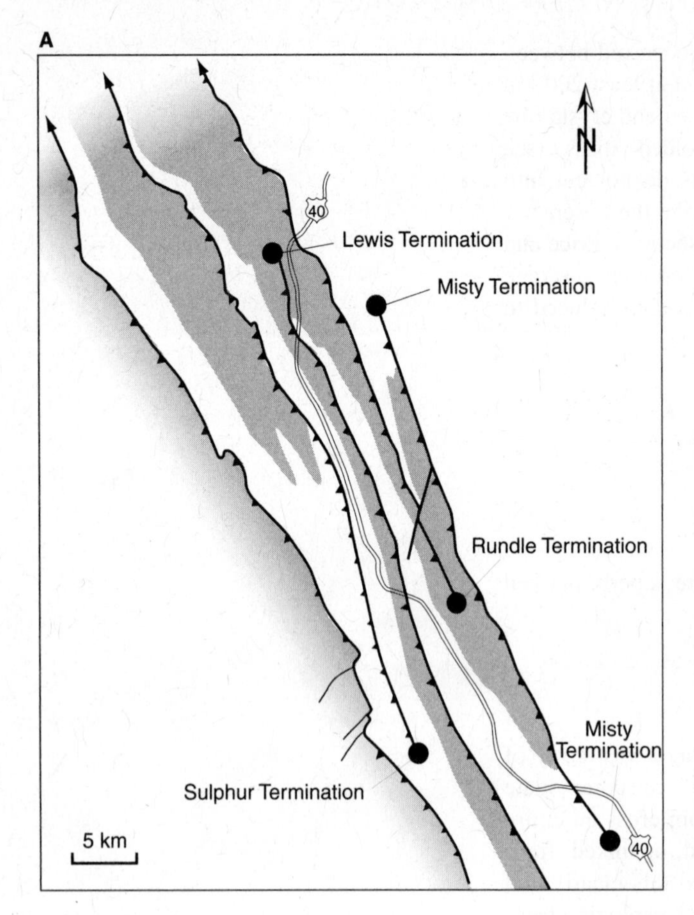

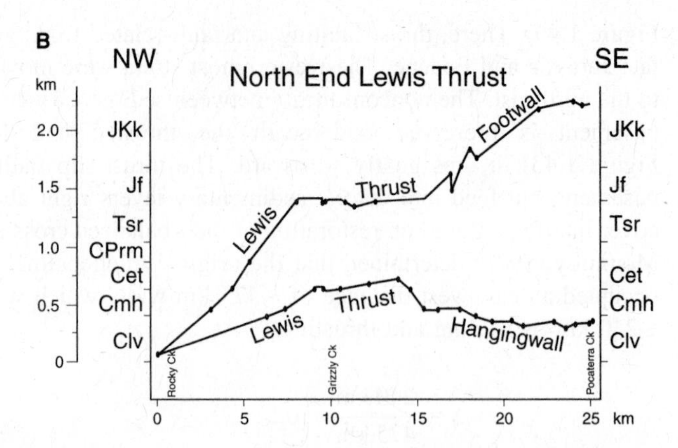

Figure 8.62 (*A*) Map showing Sulfur, Lewis, Rundle, and Misty thrust faults within the Canadian Rockies in the region south of Banff, Alberta. Black circles are tip points where thrusts die out along strike. The map captures the full trace length of the Misty thrust and captures the southern termination points for the Rundle and Sulfur thrusts and the northern tip for the Lewis thrust. (*B*) Fault displacement diagram for the Lewis thrust, showing stratigraphic formations in immediate hanging wall versus immediate footwall as a function of location along the fault trace (*x*-axis). Note how the overall throw decreases northward. Scale along *y*-axis shows stratigraphic thicknesses. Clv = Carboniferous Livingston Member of Rundle Group; Cmh = Carboniferous Mountain Head Member of Rundle Group; Cet = Carboniferous Etherington Member of Rundle Group; CPrm = Carboniferous/Permian Rocky Mountain Formation; Tsr = Triassic Spray River Formation; Jf = Jurassic Fernie Shale; and JKk = Jurassic/Cretaceous Kootnai Formation. [From McMechan, 2005.]

Figure 8.63 Photograph of the huge anticline/syncline fold pair that formed just beyond the northern tip of the Lewis thrust. The location is Mt. Kidd, where Lewis thrusting is replaced by folding. [Photograph by W. J. Anderson.]

of stratigraphic throw along the thrust (Figure 8.62*B*). Just south of Mt. Kidd the Lewis thrust dies out completely, replaced to the north by the huge anticline/syncline pair that dominates Mt. Kidd (Figure 8.63).

Thus we see that displacement on a single fault may decrease along strike to zero, even though regional shortening along the belt as a whole remains strong and robust. This apparent contradiction can be understood by recognizing that displacement (slip) can be transferred from one thrust fault to another (Figure 8.64). **Transfer zones** lie between the overlapped ends of faults in which

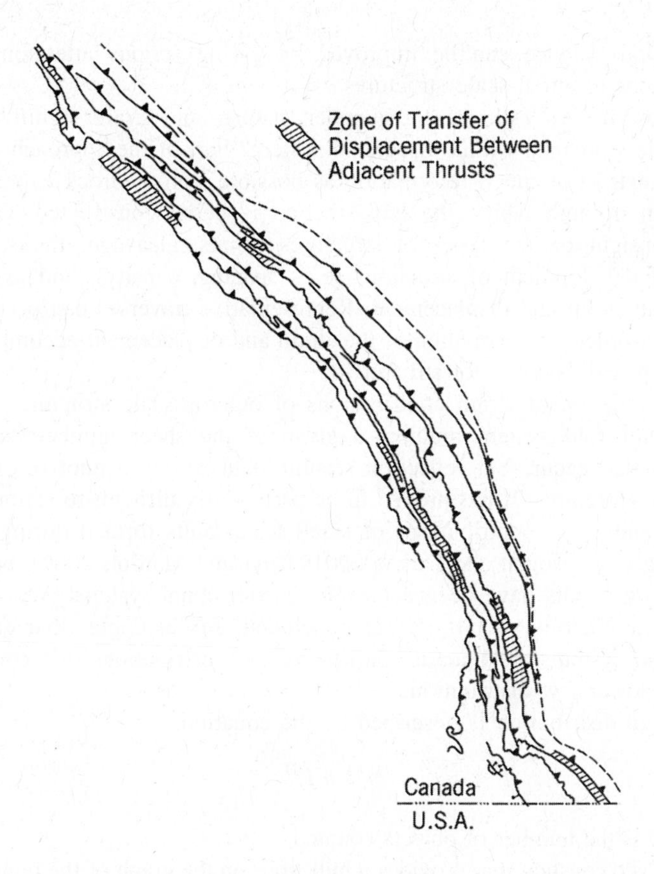

Figure 8.64 Map of regional thrust faults in the Canadian Rockies. Cross-hatched symbols identify transfer zones where decreasing tranlation on one 'dying' fault is compensated by increasing translation on an 'emerging' fault. [From Dahlstrom, 1969a. Reproduced with permission of the National Research Council of Canada from *Canadian Journal of Earth Sciences*.]

decreasing slip on one fault is compensated by increasing slip on the other (Dahlstrom, 1969). This type of linkage is possible only if (1) the linked faults join a common decollement in the subsurface, and (2) the individual faults in the linked system all were active at approximately the same time.

The idea of consistency of displacement and transfer zones is of equal importance in extended terranes. The direction of movement on individual faults need not be the same from one fault to the next in the array. As is the case in tectonically shortened terranes, the transfer of displacement from one fault to another can take place by overlap of the fault tips in map view, or by interleaved faults that transfer the displacement from one to another within the array. In this way, the **linked system of structures** can maintain uniform strain across a region even though individual structures within the system come and go. This concept is one of **regional structural balance**. A corollary to regional structural balance is that overall strain does not abruptly change from place to place, unless, for example, there is some additional strain component observed in a part of the region created by an earlier tectonic event or some later superposed event. The interesting paradox is that regional structural balance is generally maintained even though individual structures can in fact abruptly come and go.

SMALL-SCALE STRUCTURES, AND SCALING

The early pioneers who developed balanced cross-sections focused on the major structures exclusively, and essentially ignored in their balancing the strain and displacement contributed by outcrop-scale structures (e.g., folds, faults, joints, cleavage, boudinage). Now we emphasize that the 'big picture' of structural balance can be improved by giving serious attention to the contributions of small-scale structures.

But how do we estimate the number, nature, and strain significance of small-scale structures within a regional system? Part of the approach is, once again, to extract as much information as possible from bedrock exposures in the region through which the cross-section is being constructed. Outcrops provide insight on the types of structures (joints, cleavage, faults, folds), degree of development of structures (e.g., spacing, density), and associated small-scale strain and displacement. Representative traverses can be made to gain a first-order approximation of the strain and displacement accommodated by outcrop-scale systems of structures.

The evaluation of strain contributions of outcrop-scale structures in subsurface fault-fold system requires a grasp of the sheer numbers of small structures that occur. Yet reflection seismic data simply cannot resolve outcrop-scale structures. It has proved to be particularly difficult to estimate raw numbers and strain contributions of small-scale faults formed during crustal shortening (e.g., Hogan and Dunne 2001; Koyi and Maillot, 2007), but more encouraging results have been achieved in extensional systems. Walsh et al. (1991) and Yielding et al. (1992) concluded, for example, that faults in extensional systems are **fractal**, marked by size distributions that commonly follow a power law distribution.

A fractal distribution is described by the equation:

$$N = a\,L^{D} \tag{8.1}$$

where, N is the number of objects counted,
a is a constant that provides a bulk shift on the graph of the relationship depending on the sample size,
L is the actual size such as throw or displacement of a fault, and
D is the "*fractal dimension.*"

Where fractal dimension is large, small objects are disproportionately abundant. For example, a larger fractal dimension implies that there are proportionately more small displacement faults for every large displacement fault in the data set. If we know the fractal dimension as well as the number of large faults (e.g., within a given size range that we observe), we can predict the number of faults that are too small to be seen (Figure 8.65). Thus this relationship gives us a way to estimate the size distribution of the faults that we cannot observe in reflection seismic data.

How do we carry this off in practice? Because $(\text{Log N}) = (\text{Log a}) - \text{D}\,(\text{Log L})$, a log-log plot of this relationship will be a straight line with D as its slope (Figure 8.66). By identifying this slope, we can project the best-fit line and predict the number of small objects (faults in the example) that are present in the size range that we cannot see.

Recognition of a fractal size distribution of faults in a region of rock does not provide any insight regarding where the small faults are located. However, studies reveal that small-scale faults tend to cluster around the bigger faults, and may be an integral part of the damage zone. This is probably because the small faults were formed early in the history of the propagation of the fault tip through the rock. We know that as a fault tip advances, it is led by a process zone marked by small-scale fracturing and faulting. Ultimately some of these small-scale faults in the process zone coalesce and link to form larger faults. When a single through-going fault is established by this process, many of the earlier-formed small-scale faults find themselves 'abandoned,' external to the fault zone. Assuming this is the case, it is not surprising that there are relatively large numbers of small-scale faults in the vicinity of major fault zones.

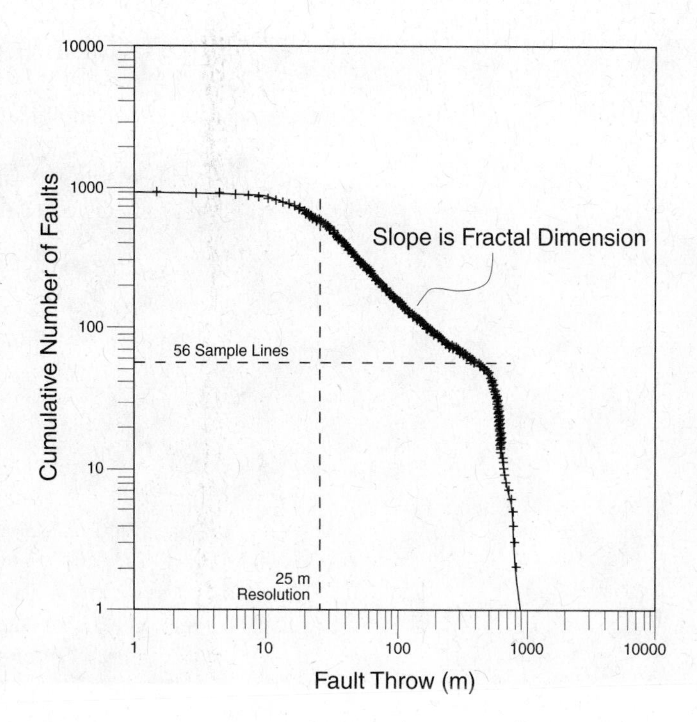

Figure 8.65 Log-log plot of cumulative number of faults versus fault throw (m). Measurements taken from seismic lines, where resolution on throw is ~25 m. As shown here, the size distribution of faults generally follows a power-law distribution. The curve tails off upward and to the left because of the difficulty in resolving very small throws. The curve also tails off down and to the right because faults marked by very large throw tend to be very widely spaced, and thus might not be captured in the sample area. The slope of the straight middle part of the plot is the "*fractal dimension.*" [From Yielding et al., 1992, First Break, 10, p. 449–460.]

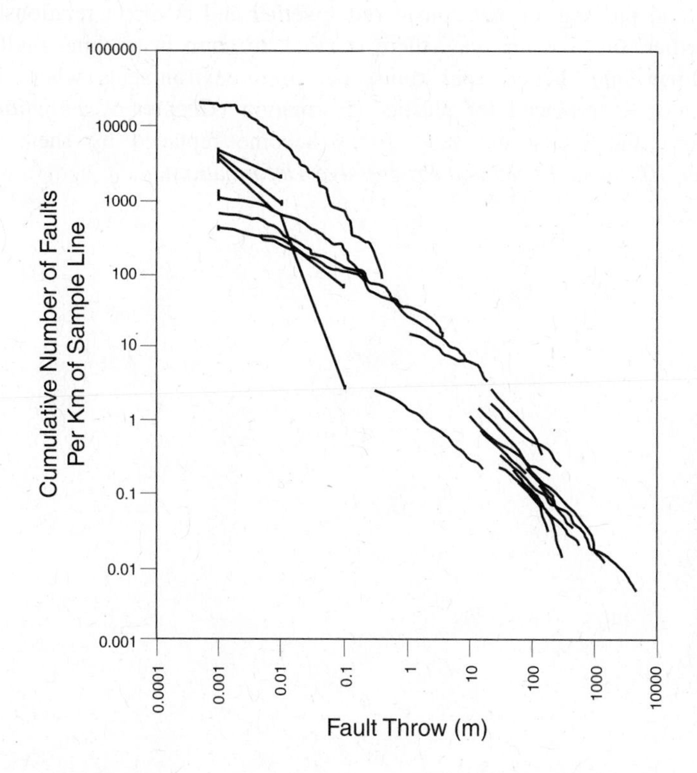

Figure 8.66 Log-log plot of cumulative number of faults intercepted per kilometer of sample line versus fault throw. The distribution is fractal. Not all the fractal dimensions are the same due to other factors such as strain rates and mechanical stratigraphy. Yet, knowledge of the fractal dimension for faults in a given subregion permit intelligent estimations of the full range and frequency of faults and their offsets. [Yielding et al., 1992, First Break, 10, p. 449–460.]

CONCLUDING THOUGHTS

For decades it was conventional to learn about faults and folds as separate distinct entities, and folding and faulting as separate distinct processes. We hope that this chapter demonstrates that faulting and folding operate together (Figure 8.67). The opportunities for robust interpretations are diminished if

Figure 8.67 Photograph of Rattlesnake Mountain, Wyoming. The steep forelimb faces SW. The backlimb dips gently and consistently NE. Precambrian basement rocks are exposed in the core. Folded and tilted cover rocks are Paleozoic. One look at this magnificent exposure draws us into subsurface thinking about fault-fold interactions. [Photograph by Eric Erslev and published with his kind permission.]

folding and faulting are not considered *together* and in direct relationship to one another. In the same way, there is much to learn in tracking faults and folds down into their deeper roots, i.e., into environments where brittle deformation is replaced by plastic deformation (*Chapter 9, Foliation and Lineation*) and faults and fault zones become replaced by shear zones (*Chapter 10, Shear Zones and Progressive Deformation*).

Chapter 9 Foliation and Lineation

Photograph of spectacularly deformed and metamorphosed rocks from the Silvretta region of the central Alps, near the border between Austria and Switzerland. Most of the rocks are banded gneiss and schist that show well-developed foliation and compositional layering that is sheared and locally folded. The darker veins are psuedotachylyte, a glass produced by frictional melting associated with fault movement during seismic events (earthquakes). In the upper right corner and left edge of the outcrop are pockets of psuedotachylyte containing angular fragments of gneiss in a glassy matrix. These pockets probably represent implosion breccias, where fault movement opened up a small, depressurized cavity, into which the wall rock imploded during faulting. The various fabrics are associated with the regional emplacement of the well-known Silvretta Nappe, which formed as slices of Africa and a microcontinent were thrust over Europe during formation of the Alps. [Photograph by S. J. Reynolds.]

NATURE OF FOLIATION AND LINEATION

Expression in Rocks

Rocks that have been internally strained commonly contain planar and linear deformational fabrics. The general term for *planar* deformational fabrics in rocks is **foliation**, which encompasses cleavage, schistosity, gneissic banding, and several other types of planar features (Figure 9.1*A*,*B*,*C*). Foliation typically forms by tectonic strain, but can also form during the flow of

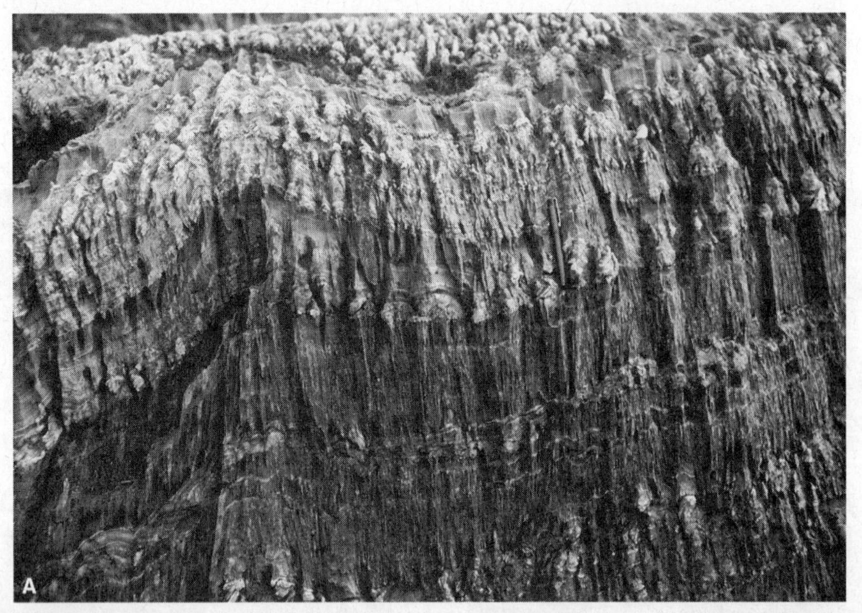

Figure 9.1 Photographs of foliation and lineation. (*A*) Steeply dipping cleavage cutting across gently dipping but folded bedding in slate and quartzite of the Ordovician Meguma Group, Nova Scotia. Note that cleavage is parallel to the axial surfaces of the folds. (*B*) Foliation and relics of compositional layering in schist and gneiss at Pemaquid Point, Maine. (*C*) Gneissic banding and foliation of Cretaceous age, Harcuvar metamorphic core complex, western Arizona. (*D*) Lineation defined by smeared mica and other mineral grains, Chemehuevi Mountains, southeastern California. [Photographs by S. J. Reynolds.]

magma, the gravity-loading compaction (welding) of pyroclastic igneous rocks, and some other processes we do not typically think of as "*deformation*." The term for linear deformation fabrics in rocks is **lineation**, which refers to linear fabrics (Figure 9.1*D*) defined by the alignment of elongated features in the rock, including aligned prismatic minerals in a metamorphic rock or the long axes of stretched pebbles in a strained conglomerate. Strained rocks typically contain both a foliation and a lineation or only foliation. Some strained rocks have lineation but without an accompanying foliation, but this is relatively uncommon.

Foliation and lineation can develop in a variety of settings and conditions, but they record some type of strain, shearing, or flow. In this chapter we discuss the character, strain significance, and formative processes of foliation and lineation, beginning with cleavage, a type of foliation typically associated with folded rocks that are weakly to moderately metamorphosed, though it is not uncommon to recognize cleavage in strained but unmetamorphosed sedimentary rocks.

NATURE OF CLEAVAGE

General Outcrop Appearance

Folded sedimentary and low-grade to medium-grade metamorphic rocks commonly display a fundamental internal grain known as **cleavage**. The term "*cleavage*" is difficult to define: it broadly refers to closely spaced, aligned, planar to curviplanar surfaces that tend to be associated with folds and to be oriented parallel to subparallel to the axial surfaces of folds (Figure 9.2). The presence of cleavage in a rock permits the rock to be split into thin plates and slabs. When a rock possessing cleavage is smacked with a hammer, the rock will typically break along the cleavage. Similarly, when rocks possessing cleavage are subjected to scores of centuries of persistent weathering, the worn-down rock that survives in outcrop is commonly marked by sharp-edged, finlike projections that express the presence and general orientation of its internal fabric (Figure 9.3). The slabby, platy nature of cleaved outcrops

Figure 9.2 Steep, penetrative slaty cleavage cutting across folded bedding, Great Smoky National Park, Tennessee. [Photograph by S. J. Reynolds.]

Figure 9.3 Weathered expression of cleaved Paleozoic carbonate rocks, Inyo Mountains, California. [Photograph by S. J. Reynolds.]

sometimes misleads us into thinking that cleavage is akin to fracturing. In truth, cleavage forms *without apparent loss of cohesion*, and in this respect alone cleavage surfaces are much different from fracture surfaces. Cleavage typically cuts bedding discordantly, without much regard to the orientation of bedding.

Domainal Character of Cleaved Rocks

When any cleaved rock is examined closely, cleavage is determined to be an expression of systematic variations in mineralogy and fabric, where fabric refers to the total sum of the shapes, sizes, orientations, and configuration of grains in a rock (Sander, 1930, 1970). The systematic variations in mineralogy and fabric that give rise to cleavage are not primary features related to the formation of the rock. Rather, they are expressions of the changes in mineralogy and fabric that were required to accommodate distortion of the rock body.

Figure 9.4 Penetrative cleavage (vertical) as seen in photomicrograph of quartz-sericite schist from the Caribou mine area, New Brunswick, Canada. Folded black layer is composed of fine-grained pyrite. [Photograph by G. H. Davis.]

Cleavage can appear to be penetrative at both the outcrop and microscopic scales (Figure 9.4), but closer examination reveals a systematic variation in mineralogy and fabric that defines different **domains** in the rock. This **domainal structure** produces a kind of structural layering or lamination composed of domains of strongly developed cleavage that alternate with domains with less-developed cleavage (Figure 9.5). The strongly cleaved domains, or **cleavage domains,** are thin, anastomosing to subparallel, mica-rich laminae within which the fabric of the original host rock has been strongly rearranged and/or partially removed. Minerals and mineral aggregates within cleavage domains show a strongly preferred dimensional and/or crystallographic orientation. The less cleaved domains, called **microlithon domains** or simply **microlithons**, are narrow lensoidal to trapezoidal slices of rock within which the mineralogy and fabric of the original host rock remain essentially preserved. Unless the microlithons are composed of rock that contains a preexisting cleavage, minerals and mineral aggregates in microlithons tend to be equigranular, lacking a conspicuous preferred orientation. Microlithons are sharply or gradationally bounded on either side by cleavage domains. Although the domainal structure of some cleaved rocks is apparent in outcrop and/or thin section, in many cases it is visible only when the cleaved rock is scrutinized microscopically at very high levels of magnification (Figure 9.6).

Figure 9.5 Excellent example of domainal structure in a quartz-mica schist exposed near Loch Leven, Inverness-shire Scotland. The cleavage domains are the dark, fine-grained micaceous zones. The microlithon domains are the light-colored, coarser grained zones of crenulated laminae of quartz and mica. [With kind permission from Springer Science+Business Media: The Minor Structures of Deformed Rocks, 1972, L. E. Weiss.]

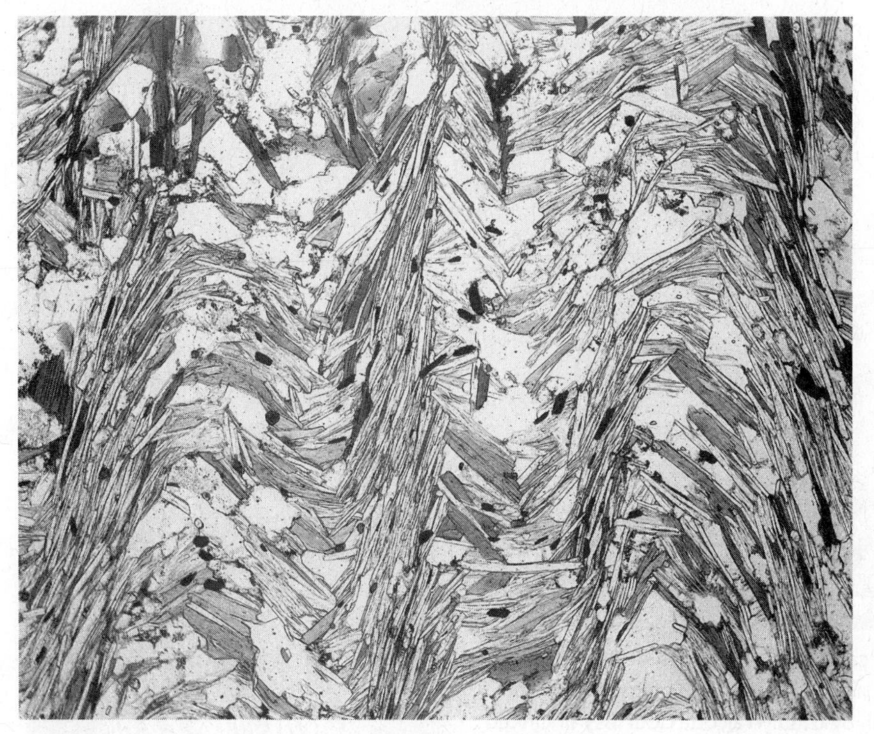

Figure 9.6 Photomicrograph showing domainal structure in mica schist. Oriented micas comprise cleavage domains. The cleavage domains separate microlithon domains of quartz, feldspar, and mica. [Photograph by D. M. Sheridan. Courtesy of United States Geological Survey.]

Types of Cleavage

There are many ways to name and classify cleavage. Most classifications are anchored in the insightful work of Dennis (1972) and Powell (1979). Both Dennis and Powell recognized that it is practical to subdivide cleavage into two classes on the basis of the *scale* at which the domainal character of cleavage can be recognized. Where the distinction between cleavage domains and microlithons is too fine to be resolved without the aid of a petrographic or an electron microscope, the cleavage is described as **continuous cleavage**. Where domainal structure can be seen with the unaided eye, the cleavage can be described as **disjunctive cleavage**, meaning interrupted.

Continuous Cleavage

The main types of continuous cleavage are **slaty cleavage** and **phyllitic structure**. They are generally associated with strongly strained, generally folded, metasedimentary and metavolcanic rocks. Slaty cleavage and phyllitic structure are reasonably similar in outcrop expression, but they differ in grain size and in the scale of development of domainal structure.

Slaty cleavage is typically associated with very fine-grained (<0.5 mm) pelitic (shaley) rocks metamorphosed to low grade. Where slaty cleavage is well developed, it imparts to rocks an exquisite splitting property. Indeed, the presence of slaty cleavage allows a rock to be cleaved into perfectly tabular, thin plates or sheets (Figure 9.7). Roofing slates and old-fashioned slate blackboards owe their existence and usefulness to slaty cleavage.

Phyllitic structure is similar to slaty cleavage but is slightly coarser in grain size, causing phyllites to display a soft, pearly, satiny luster (Figure 9.8). They glisten in the sun, but lack the distinct individual mica grains that characterize schists and other coarser grained metamorphic rock. Phyllites exhibit the capacity to split neatly but not perfectly, in part because phyllitic cleavage commonly is warped by kinks, crenulations (see below), and other folds that cause it to be no longer be planar.

Figure 9.7 (*A*) The phenomenal splitting capacity of slates, as displayed in the "*middle quarry*" of the Penrhyn Slate Company, Washington County, New York. Note quarryman at bottom right. [Photograph by C. D. Walcott. Courtesy of United States Geological Survey.] (*B*) Specimen of Martinsburg Slate from the Delaware Water Gap region, New Jersey. The planar surface on the left is a natural break controlled by the cleavage. The texture of the cross-section resembles wood fiber; it is the trace of cleavage. Note that a band can be traced across the flank of the specimen. This is relict bedding, oriented discordantly to the slaty cleavage. [Photograph and copyright © by Peter Kresan.]

Figure 9.8 Steep phyllitic structure bent by near-horizontal kink folds, Santa Rosa Peak, southern Arizona. Note the pearly sheen produced by the phyllitic structure. [Photograph by S.J. Reynolds.]

Disjunctive Cleavage

There are two main types of disjunctive cleavage: spaced cleavage and crenulation cleavage. **Spaced cleavage** consists of an array of parallel to anastomosing, irregular to smooth, fracture-like partings that typically are occupied by clayey and carbonaceous matter (Nickelsen, 1972). Spaced cleavage is most common in folded but unmetamorphosed sedimentary rocks, especially impure limestone and marl (Figure 9.9*A*,*B*), and in some impure sandstones as well. Spacing of the partings (i.e., the cleavage domains) typically ranges from 1 to 10 cm, and thus the microlithons are quite thick compared with all other cleavages. Thickness of the partings commonly is on the order of 0.02–1 mm, although they may be as thick as 1 cm or more.

A fundamental characteristic of spaced cleavage is the offset (i.e., separation) of bedding markers along the cleavage. Offset of bedding along spaced cleavage is commonly seen in outcrop (Figure 9.9*C*). Cleavage surfaces are *not* faults, even though the offsets along them might give that impression. Cleavage domains associated with bedding offsets are not marked by the slickenlines or polish that characterize faults. Also, truncation of fossils at the boundaries of microlithons cannot be completely restored by fault-slip motions, for material has been lost (Groshong, 1975a).

Crenulation cleavage, the second type of disjunctive cleavage, is very distinctive in that it cuts a host rock that possesses a preexisting continuous cleavage, especially phyllitic structure or schistosity (Figure 9.10). In rocks that contain crenulation cleavage, a preexisting continuous cleavage is typically "*crenulated*" into microfolds. Two kinds of crenulation cleavage are recognized. One is discrete; the other is zonal (Gray, 1977a). **Discrete crenulation cleavage** is a disjunctive cleavage in which very narrow cleavage domains sharply truncate the continuous cleavage of the microlithons, almost like tiny faults

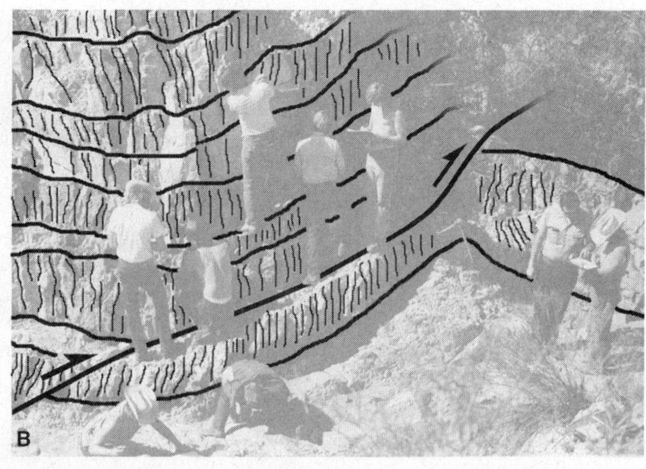

Figure 9.9 (*A*) Photograph of outcrop expression of spaced cleavage in folded Permian limestone in the Agua Verde area near Vail, Arizona. Outcrop is being assaulted by a structure class from the early '70s. (*B*) Drawing of the spaced cleavage. (*C*) Even closer view showing truncation of bedding laminations along cleavage surfaces. [Photographs by G. H. Davis.]

(Figure 9.11*A*). **Zonal crenulation cleavage**, on the other hand, is marked by wider cleavage domains that coincide with tight, appressed limbs of microfolds that contort preexisting continuous cleavage in the **microlithons** (Figure 9.11*B*). Whether discrete or zonal, cleavage domains in rocks possessing crenulation cleavage are closely spaced, generally between 0.1 mm and 1 cm. Discrete crenulation cleavage tends to form in slate. Zonal crenulation cleavage tends to form in phyllite and coarser grained metamorphic rocks, like schist. Whether crenulation cleavage is discrete or zonal is determined by the degree of mechanical anisotropy of the preexisting fabric and by the orientation of that fabric versus the newly imposed stress (Naus-Thijssen, et al., 2010).

Schistosity

In terms of metamorphic character, schistosity is the next step up from phyllitic structure. Rocks with **schistosity** are typically medium-grained

Figure 9.10 Spaced crenulation cleavage cutting steeply across folded fabric and compositional layering, Port Elizabeth Formation, Willard Beach, Maine. [Photograph by S. J. Reynolds.]

Figure 9.11 (*A*) Discrete crenulation cleavage in metamorphosed tuffaceous rocks (Mesozoic), Granite Wash Mountains, western Arizona. Thin white bands are calcite veins, formed during a previous deformation event. The cleavage developed during thrusting. (*B*) Zonal crenulation cleavage in Mesozoic metasedimentary rocks, Granite Wash Mountains, western Arizona. [Photographs by S. J. Reynolds.]

(1–10 mm) or even coarse grained, with flakes of mica and other minerals that are visible in hand specimen (Figure 9.12*A*). The grain size of schist, which is larger than that of slates and phyllites, reflects greater recrystallization accompanying metamorphism. The most obvious outcrop characteristic of schistosity is the parallel, planar alignment of micas, including muscovite, biotite, chlorite, and sericite. Schistosity can also be defined by deformed, disc-shaped minerals and by prismatic minerals whose long axes are oriented within the schistosity. Schists typically have quartz and feldspar, in addition to mica, and can contain a variety of minerals, including calcite, sillimanite, and garnet. Most schists have some compositional variation expressed as layers parallel to the schistosity (Figure 9.12*B*). Such layers can be inherited from the original rock (e.g., beds that existed before metamorphism) or can be formed or accentuated by metamorphism and deformation. There is a complete gradation from slaty cleavage to phyllite and then schist, largely reflecting an increase in grain size, which is in turn largely due to an increase in metamorphic temperatures during deformation. Slate forms at low metamorphic grades, and schist forms at higher ones. Schistosity is best developed in pelitic metasedimentary rocks and certain volcanic rocks metamorphosed to

Figure 9.12 (*A*) Schist with visible grains of mica and other minerals, Black Hills, South Dakota. (*B*) Schistosity parallel to compositional layers, as photographed underground in a gold mine at Hemlo, Ontario, Canada. [Photographs by S. J. Reynolds.]

Figure 9.13 Outcrop and landscape expression of schistosity near the CalaSerena ravine, Cap de Creus, Spain. Geologist is Elena Druguet, preparing to take a group photo and responding to questions about shear zone geometrics. [Photograph by G. H. Davis]

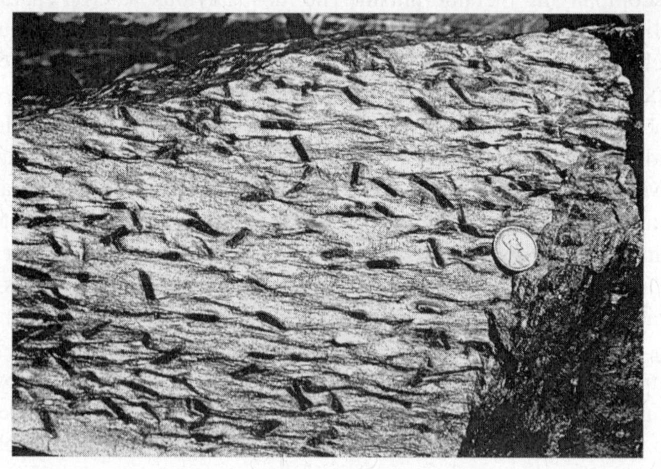

Figure 9.14 Mica schist with porphyroblasts of andalusite, Pioneer Mountains, Idaho. [Photograph by S. J. Reynolds.]

medium or high grade. It is locally present in some granitic rocks within which shearing under fluid-rich conditions caused feldspar to be converted into abundant white mica.

The splitting capacity of schist is not nearly as elegant as that of slates, but it is nonetheless very pronounced (Figure 9.13). Schists seldom split cleanly and evenly when struck with a hammer. Instead, chunks break off in the form of discoidal to crudely tabular hand specimens or slabs. The tendency for schist to break apart and weather easily causes most schists to have relatively subdued outcrop characteristics. In spite of this fact, my (GHD) brother the dentist shouts "*SCHISTOSITY*!" indiscriminately whenever we drive by any bold outcrop of any rock anywhere. Many hand specimens and outcrops of schist have a lovely sheen (Figure 9.14).

MICROSCOPIC PROPERTIES OF CLEAVAGE AND SCHISTOSITY

Slaty Cleavage

Microscopic, high-magnification examination of rocks possessing slaty cleavage reveals a fabric marked by discoidal to lenticular aggregates of

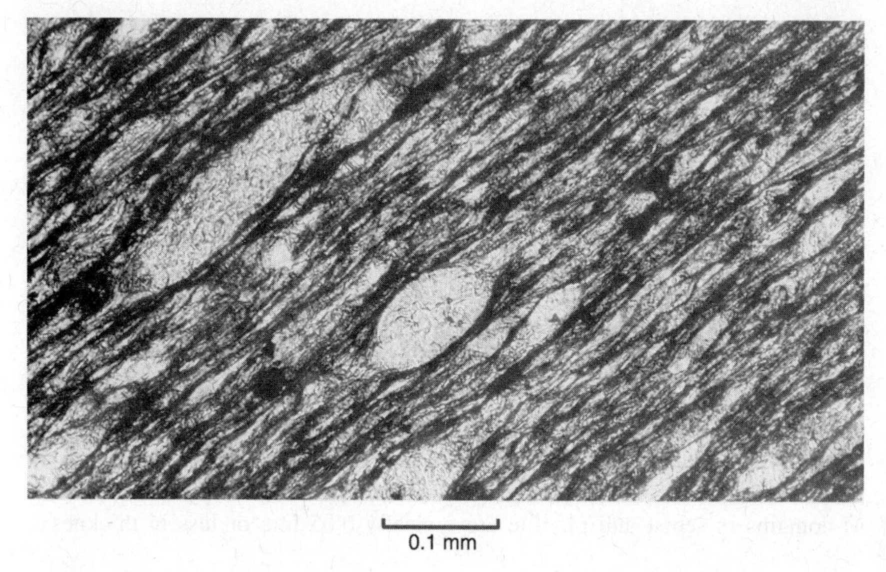

Figure 9.15 Domainal microfabric in slaty cleavage from the Ribagorzana Valley area, Spanish Pyrenees. Mica-rich domains (M-domains) are the black laminae that 'anastomose' around large quartz grains and aggregates of the QF-domains. [Photograph by W. C. Laurijssen. From *An Outline of Structural Geology* by B. E. Hobbs, W. D. Means, and P. F. Williams. Published with permission of B. E. Hobbs.]

quartz, feldspar, and minor mica enveloped by anastomosing, discontinuous mica-rich laminae (Figure 9.15). The micaceous laminae known as **M-domains**—that is, **mica-rich domains**—constitute cleavage domains. The discoidal, lenticular quartz–feldspar aggregates known as **QF-domains**—that is, **quartz–feldspar domains**—comprise microlithons. The scale of development of domainal structure in slaty cleavage is mighty small. Thickness of the QF-domains typically ranges from 1 mm to less than 10 μm. The M-domains are typically only 5 μm thick (Roy, 1978).

The QF-domains in rocks possessing slaty cleavage provide a glimpse of the nature of the original host rock. Except for micas, individual grains and mineral aggregates tend to be equigranular, lacking a conspicuous preferred orientation. In sharp contrast to the QF-domains, the M-domains are zones within which the original fabric of the rock is almost completely reconstituted, transformed into intergrowths of mica, quartz, and feldspar crystals that have a strong planar alignment. Micas show the most conspicuous preferred orientation, but hidden among the micas are flat to lensoidal quartz and feldspar grains, whose shapes are oriented parallel to the overall planar alignment of micas and the M-domains.

The lenslike, "*flattened*" nature of individual grains in slaty cleavage is further accentuated by **overgrowths** of chlorite, quartz, sericite, and other minerals. The overgrowths are like **beards**, growing from the "*chins*" of relatively large grains of quartz, feldspar, and pyrite (Figure 9.16*A*) (Roy, 1978). Crystal fiber beards grow in the "*plane*" of cleavage, as a response to the influence of directed stress. In Figure 9.16*B* we see a hand-specimen-scale example of quartz-sericite fibers growing as a "*pressure shadow*" from the sheltered ends of a clast of quartzite.

One of the surprising revelations of high magnification examination of slaty cleavage is that the M-domains are somewhat curved and anastomosing. This irregularity at the microscopic scale seems to be inconsistent with the capacity of slates to split along "*perfectly*" planar and parallel surfaces. This apparent inconsistency reminds us once again of the influence of scale on our geologic observations.

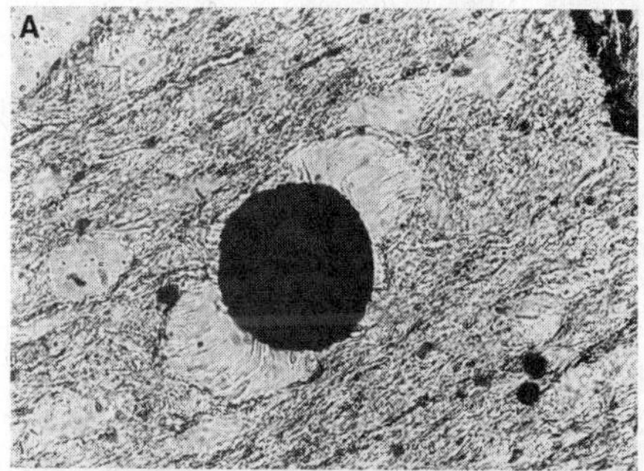

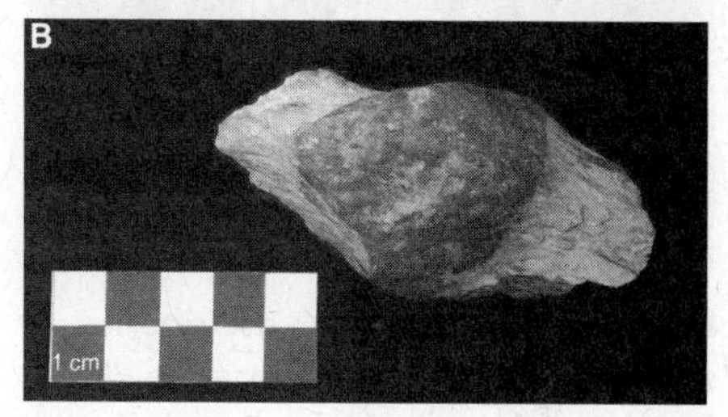

Figure 9.16 (*A*) Photomicrograph of overgrowths of chlorite and quartz on pyrite (black). The "*beards*" are oriented subparallel to slaty cleavage. They grew under the protection of the strong pyrite grain that refused to flatten parallel to cleavage. Trace of slaty cleavage is from lower left to upper right. Martinsburg Slate, Delaware Water Gap, New Jersey. [Photograph by E. C. Beutner.] (*B*) Hand-specimen-scale example of overgrowths of quartz fibers, forming "*beards*." The dark rock is from Late Paleoproterozoic meta-quartzite and metaconglomerate in the Cottonwood Cliffs area, northwestern Arizona. Pull-apart of the rock was accompanied by synkinematic precipitation of fibers of quartz and sericite. The direction of greatest finite stretching is reflected by the orientation of the fibers. [Collection of G. H. Davis. Photography and editing by Gary Mackender. vr.arizona.edu, glm@email.arizona.edu, © 2009–2010 Arizona Board of Regents.]

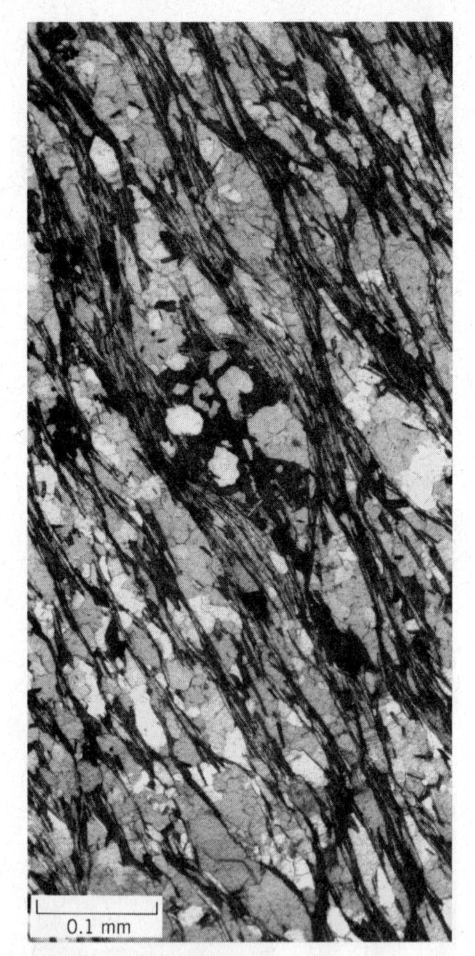

Figure 9.17 Domainal microfabric in schist from Ducktown, Tennessee. Micas form films that envelope aggregates composed principally of quartz. [Photograph by W. C. Laurijssen. From *An Outline of Structural Geology* by B. E. Hobbs, W. D. Means, and P. F. Williams. Published with permission of B. E. Hobbs.]

Phyllitic Structure and Schistosity

Like the fabric of slaty cleavage, the microscopic fabric that gives expression to phyllitic structure and schistosity is composed of anastomosing M-domains (cleavage domains) and lenticular QF-domains (microlithons). The parallelism of micas within the M-domains imparts to phyllites and schists their fundamental splitting capacities (Figure 9.17). These mica-rich cleavage domains contain lens-like and disk-like quartz and feldspar grains, which are commonly overgrown by chlorite, quartz, and other minerals at their tips. Quartz and feldspar grains within the QF-domains may contain a record of the fabric of the original host rock, albeit one that was slightly reconstituted by the effects of recrystallization.

The fundamental distinction between phyllitic cleavage and schistosity is simply one of grain size. Phyllites tend to be fine-grained, with average grain diameter less than 1 mm. Schists tend to be medium grained or coarser, with average mica diameter ranging from 1 to 10 mm. Wispy anastomosing M-domains in schist and phyllite are typically 0.05 mm or less in thickness.

Crenulation Cleavage

The microscopic fabric of crenulation cleavage is quite distinctive. The cleavage domains are M-domains packed with aligned, interlocking micas surrounding lensoidal quartz and feldspar grains as well as opaque minerals and clots of carbonaceous material (Gray, 1979) (see Figure 9.15). The microlithons are QF-domains composed of a preexisting continuous cleavage, like slaty cleavage, phyllitic cleavage, or schistosity. They are relatively rich in quartz and feldspar but relatively poor in micas (Marlow and Etheridge, 1977; Gray, 1979). The continuous cleavage that makes up the microlithons of crenulation cleavage is typically crenulated into unbroken waveforms of tiny folds. Axial surfaces of the folds are subparallel to crenulation cleavage.

The physical and geometric relation of M-domains to QF-domains depends on whether the crenulation cleavage is discrete or zonal. M-domains associated with discrete crenulation cleavage are relatively narrow mica-rich laminae along which the continuous cleavage of adjacent microlithons is abruptly and sharply truncated. They are abrupt (Figure 9.18*A*). In contrast, zonal

Figure 9.18 (*A*) Example of discrete crenulation cleavage. Folded schistosity is abruptly truncated along crenulation cleavage. Vishnu Schist in the Grand Canyon. [Photograph by S. J. Reynolds.] (*B*) Photomicrograph of zonal crenulation cleavage (vertical) coincident with the steep limbs of asymmetric folds in schistosity. The zonal cleavage domains are carbonaceous and micaceous. They have a distinctively lower proportion of quartz than that of the initial fabric. [From Gray, 1979, *American Journal of Science*, v. 279. Reprinted by permission of the American Journal of Science.]

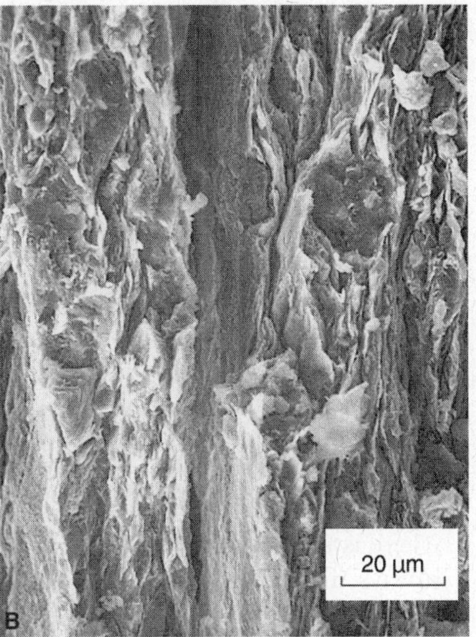

Figure 9.19 (*A*) Photomicrograph of anastomosing, dark, undulating spaced cleavage seams. (*B*) Scanning electron micrograph of the spaced cleavage seams. Composed of densely packed clays, these seams are markedly straight in their trace expression (vertical). Clays in intervening microlithons are more loosely packed, not as preferentially oriented. [Reprinted from *Tectonophysics*, v. 78, Gray, D. R., Compound tectonic fabrics in singly folded rocks from southwest Virginia, U.S.A., p. 229–248, © 1981, with permission from Elsevier.]

crenulation cleavage (Figure 9.18*B*) is marked by M-domains that are relatively wide and serve as unbroken fold limbs connecting microfolds in the continuous cleavage of adjacent microlithons. Whether associated with discrete or zonal crenulation cleavage, micas within M-domains are oriented within 5° of the orientation of the cleavage domains as a whole.

Spaced Cleavage

Microscopically, cleavage domains in rocks cut by spaced cleavage reveal that they are sharp discontinuities lined with seams or films of clayey and/or carbonaceous material (Figure 9.19). The clayey material in the seams and films tends to be aligned parallel to the cleavage itself. The spaced cleavage may be stylolitic, but more commonly it is anastomosing, or smooth.

Microlithons between cleavage domains in rocks cut by spaced cleavage typically lack a preexisting continuous cleavage. Rather, the host rock for most spaced cleavage is simply unmetamorphosed impure sedimentary rock, especially limestone, marl, and clayey sandstones.

STRAIN SIGNIFICANCE OF CLEAVAGE

The Issues

The origins and significance of cleavage were debated vigorously for more than a century. There are many things that require explanation! These include the relationship between folds and cleavage, the mechanical role of cleavage in the folding process, the kinematic development of preferentially oriented mineral grains and cleavage domains, the division of micas and quartz–feldspar aggregates into domainal structure, the presence of clay-filled partings in rocks possessing spaced cleavages, the development of oriented beardlike overgrowths of chlorite and quartz, the whereabouts of the missing parts of truncated fossils, and the overall strain significance of cleavage.

The debate on the origin of cleavage once focused on the relative contributions of **rigid-body rotation** of individual platy minerals within the original host rock; preferred **directional recrystallization** of individual

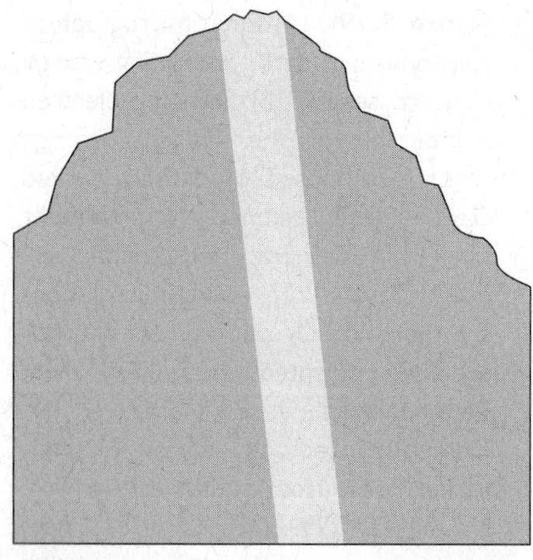

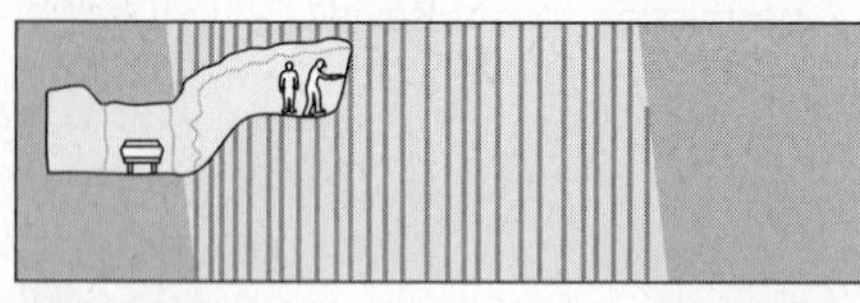

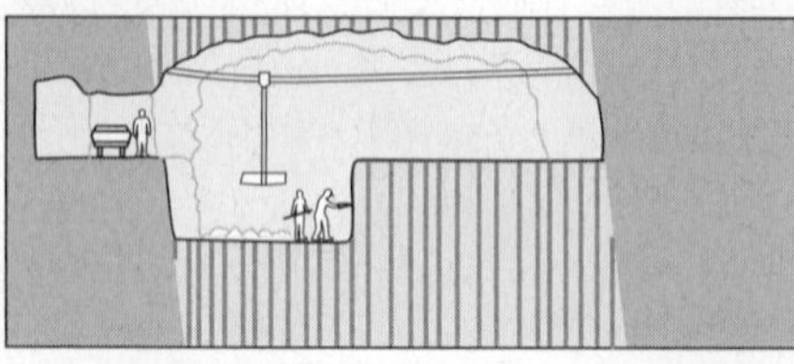

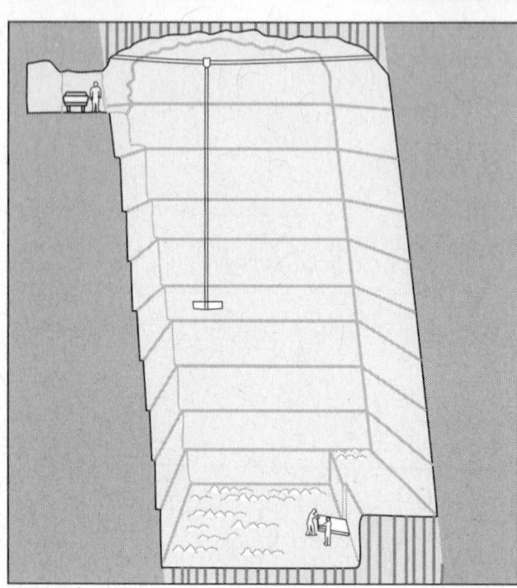

Figure 9.20 *"Diagramau yn dangos y whthien o'r lechen a'r dull o'l chloddio."* (*"Diagrams showing the vein of slate and how it is extracted."*) This illustration, based on a display in the Visitor's Center in Machynlleth, Wales, shows beautifully the difference between bedding and cleavage. The slate occurs in what the miners call "*veins*." Veins are the beds themselves (lighter-toned steeply inclined band in top diagram). Cleavage is discordant to the bedding and is vertical in this example. Thus when the miners split out the slate, they do so along a direction that is oblique to bedding. [Re-drawn from photograph by G. H. Davis.]

minerals within the original host rock; and fault-like **simple-shear translation** along extremely closely spaced fracture-like surfaces. Hardly a thought was given to whether the deformation might have been accompanied by volume loss or volume gain. Thus, constant volume deformation was an underlying but unspoken assumption in the debates. Although each of the three above-mentioned mechanisms contributes to cleavage development, it now seems clear that the dominant mechanism for cleavage formation is **pressure dissolution** removal of original host rock. Wholesale removal of rock by pressure solution is perhaps the supreme strain accommodation to directed stress. In other words, when all else fails, get out of the way!

Strain Significance of Slaty Cleavage

Expression of Shortening

There is unmistakable evidence that the orientation of slaty cleavage is almost always discordant to bedding (Figure 9.20). As a structural fabric generally associated with folded rocks, slaty cleavage typically is parallel to, or symmetrically "*fanned*" about, the axial surfaces of associated folds (Figure 9.21). The intimate coordination of the geometry of folding to the orientation(s) of cleavage leaves little doubt that folding and slaty cleavage development are, at least in part, synchronous. Thus, if folds are considered to be products of shortening, slaty cleavage must be considered a product of shortening as well. In fact, fossils, reduction spots, and other primary structures that are found preserved in slates typically are flattened in the "*plane*" of cleavage. Flattened fossils and reduction spots thus provide a dramatic statement that slaty cleavage is indeed an expression of severe shortening.

The deformation of fossils in slates was recognized and appreciated more than a century ago in the slate quarries of Wales. Phillips (1844) was the first to point out the close relation between fossil distortion and slaty cleavage. Sharpe (1847) went a step further than Phillips, pointing out that the most highly distorted fossils are associated with the most highly cleaved slates. Noting that the fossils are flattened in the cleavage surfaces and that the cleavage surfaces tend to be parallel and/or symmetrically disposed about the axial surfaces of associated folds, Sharpe concluded that *slaty cleavage forms perpendicular to the direction of greatest shortening.*

To estimate the amount of shortening accommodated by the formation of slaty cleavage, Sorby (1853, 1856) cleverly used reduction spots as a guide to distortion. He concluded that the presence of slaty cleavage can signal levels of distortion as great as 75%. Always ahead of his time, Sorby was able to demonstrate through strain analysis that the plane of cleavage in slates is statistically perpendicular to the direction of greatest shortening.

Since the time of the classic work by Phillips, Sharpe, and Sorby, many other geologists have addressed the strain significance of cleavage. The results are the same, time and time again. Oertel (1970), for example, analyzed slaty cleavage in a volcanic tuff unit in the Lake District of England, using ellipsoidal objects within the tuff as guides to strain. Oertel assumed that the objects were initially spherical, but were transformed into ellipsoids during folding and the development of slaty cleavage. He proceeded to show that cleavage surfaces in the tuff developed perpendicular to the direction of greatest shortening (Figure 9.22). Tullis and Wood (1975), like Sorby long before, used reduction spots to evaluate the state of strain in Cambrian slates in northern Wales. They concluded that the direction of greatest shortening responsible for the formation of the slaty cleavage was oriented precisely perpendicular to the cleavage. Shortening averaged about 65% in the rocks they examined.

Figure 9.21 (*A*) Syncline cut by strong penetrative axial plane cleavage. The bedding/cleavage relationship is best seen in the right half of this photo, where the bedding dips gently to the left in contrast to the slaty cleavage, which dips steeply to the left. These slates are located at "*old quarry 2*" at Slatington, Lehigh County, Pennsylvania. [Photograph by E. B. Hardin. Courtesy of United States Geological Survey.] (*B*) Example of fanning cleavage across a syncline of folded sandstones and shales. Geologist is C. W. Hayes, who is sitting on bedding that dips steeply to the right. Note how the cleavage crosses bedding throughout the outcrop. Location is 5 km west of Hancock, Maryland. [Photograph by C. D. Walcott. Courtesy of United States Geological Survey.]

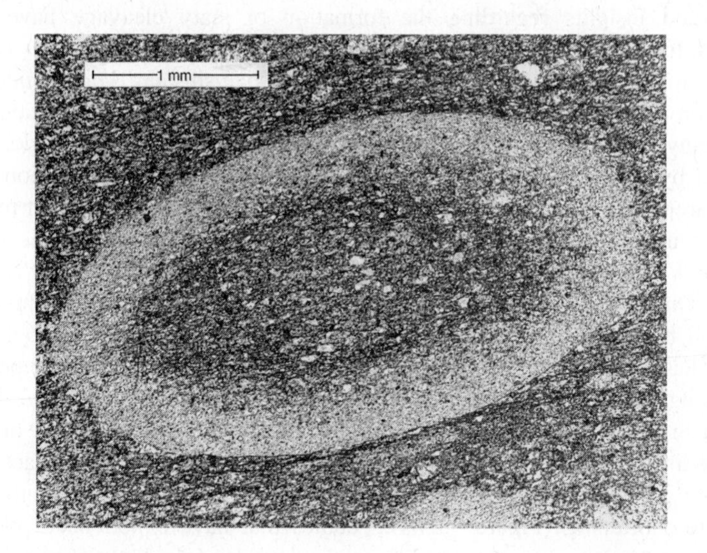

Figure 9.22 Distorted lapillus in slate derived from tuff. Lake district, England. [From Oertel, 1970. Published with permission of the Geological Society of America.]

Role of Recrystallization

The role of **directional crystallization** in the formation of slaty cleavage includes altering the shapes of grains and enhancing the flattened, elongated appearance of minerals and mineral aggregates. In essence, new mineral growth takes place in the plane of cleavage—for example, within **pressure shadows** next to relatively large rigid mineral grains that can provide shelter from the harsh directed stresses that would otherwise inhibit the growth of new minerals. Pressure shadows of chlorite and fiber quartz are very common in M-domains in slates (Figure 9.23*A*). The shadows grow as microscopic beards from the tips of pyrite, feldspar, and quartz grains. The direction of crystal growth is the direction of incremental extension, regardless of the attitude of bedding cut by the cleavage (Figure 9.23B).

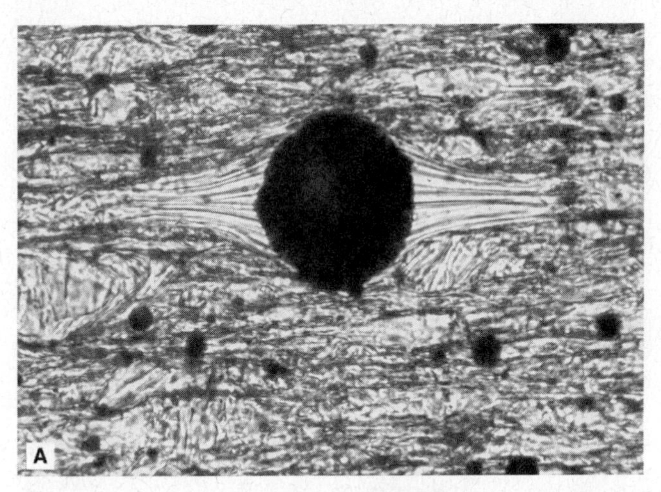

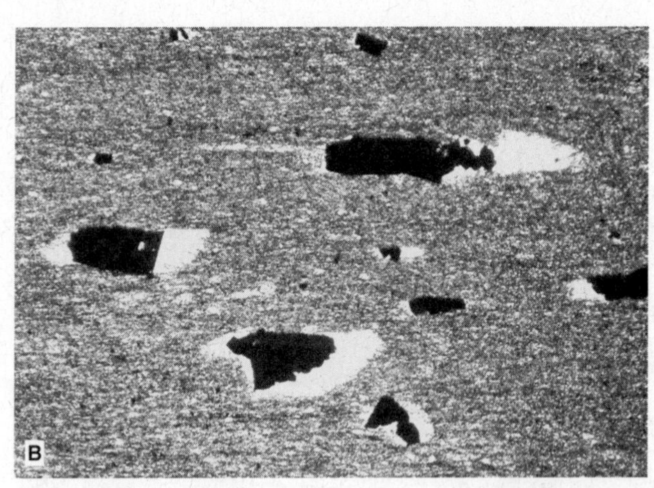

Figure 9.23 (A) Photomicrograph of pressure shadows containing fibrous quartz and chlorite. The pressure shadows are "*attached*" to a spherical pyrite aggregate. Diameter of pyrite is 36 μm. From fold in Martinsburg Slate, Delaware Water Gap, New Jersey. [Photograph by E. C. Beutner.] (*B*) Photomicrograph of feathery pressure shadows (crystal fiber beards) of quartz at the ends of pyrite crystals and calcareous slate. Note the faint horizontal trace of bedding, dipping 50 degrees to the right. The pyrite occurs mostly along the bedding, but the pressure shadows have formed parallel to cleavage (horizontal). [Photograph by L. Pavlides. Courtesy of United States Geological Survey.]

Pressure Solution Origin of Slaty Cleavage

Significant insights regarding the formation of slaty cleavage have been derived from fuller appreciation of the role of pressure solution in crustal deformation (recall *Chapter 4, Deformation Mechanisms and Microstructures*). Based on an explosion of research since the early 1970s, it is now recognized that many of the attributes of slaty cleavage seem to be mineralogical and textural by-products of a shortening achieved through pressure solution.

Awareness that minerals or parts of minerals can be removed from rock in solution undermines the commonly held premise of constant-volume deformation. Volume losses of 50% and more are not unusual in highly cleaved rocks! One example of clear insight into the role of pressure solution is afforded by the deformational characteristics of quartzite pebble conglomerates. It has been known for some time that when neighboring quartzite pebbles in a conglomerate are forced into contact during strong, penetrative deformation in a low grade metamorphic environment, the quartz in one or both pebbles is capable of dissolving (or diffusing) at the site of contact. The quartzite pebbles thus indent one another as a means of accommodating the requisite shortening (Figure 9.24*A*). Pebbles that have experienced pressure dissolution creep, when extracted from the bedrock, display concave indentations not unlike chin-dimples. The concavities testify to removal of material. Rims of the dimples may be marked by **stylolitic halos**, where dark insoluble residue accumulated (Figure 9.24*B*). It is the presence of a stress-induced chemical potential gradient that drives quartz away from sites of high stress concentration. The quartz reprecipitates into "*sheltered*" areas of low stress concentration, where the quartz can recrystallize as pressure shadows, veins, and/or beardlike crystal fiber overgrowths.

Microscopic characteristics of slaty cleavage show all the signs of pressure solution. Lens-like and trapezoidal grains of quartz and feldspar are corroded relics of what were once larger, more equant grains. The cleavage-parallel flanks of the lensoidal grains are facets marking the extent of advancement of pressure solution. So are the cleavage-parallel flanks of the lensoidal mineral aggregates that comprise QF-domains. The formation of the ultra-thin quartz and feldspar grains in the M-domains reminds me of the LifeSaver-candy game my (GHD) brothers and I would play as kids while riding for hours in

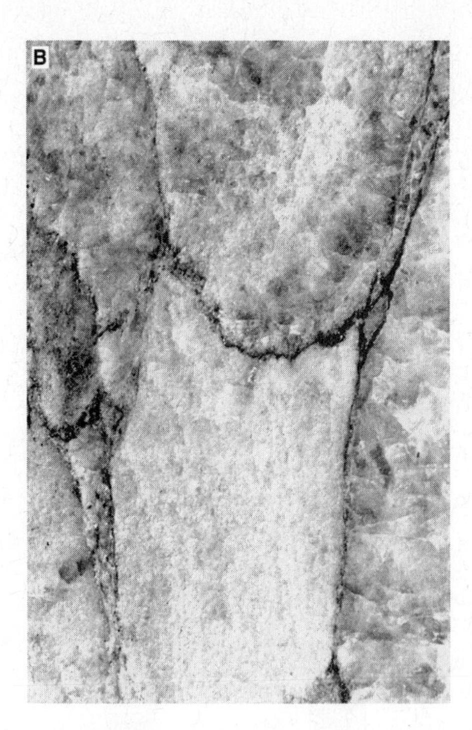

Figure 9.24 (*A*) Mesozoic conglomerate containing flattened quartzite pebbles that experienced pressure dissolution and other ductile deformation during deformation. Pebbles were forced into contact with one another, causing some to indent into others. Little Harquahala Mountains, western Arizona. [Photograph by S. J. Reynolds.] (*B*) This photograph shows the contact area between quartzite pebbles of the Barnes Conglomerate (late Mesoproterozoic) that were forced tectonically into contact with one another. The quartz experienced pressure dissolution creep during the deformation. The upper pebble seems to have had greatest resistance to pressure dissolution, and it penetrated into the pebble below it. Insoluble residue (black) can be seen at the contact. [Photograph and copyright © Peter Kresan.]

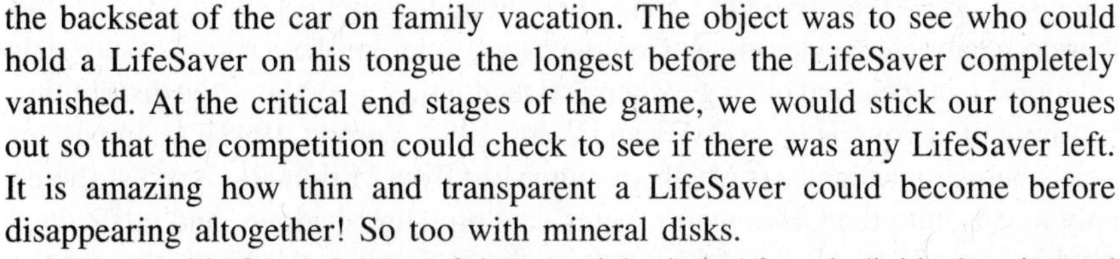

the backseat of the car on family vacation. The object was to see who could hold a LifeSaver on his tongue the longest before the LifeSaver completely vanished. At the critical end stages of the game, we would stick our tongues out so that the competition could check to see if there was any LifeSaver left. It is amazing how thin and transparent a LifeSaver could become before disappearing altogether! So too with mineral disks.

In the world of rocks, some of the material missing from individual grains and grain aggregates may be accounted for in the presence of overgrowths and pressure shadows. But much of the dissolved rock must have passed completely out of the system, perhaps along the M-domains and along fractures. The densely packed concentrations of micas and carbonaceous matter in **cleavage seams** represent the accumulations of less soluble to insoluble residue of the original host rock. The strong preferred orientation of the micas in the M-domains, in some cases, may be the natural result of strain-induced progressive rotation of micas in response to the stress-induced removal of the surrounding rock matrix.

Alignment of Mica without Grain Rotation

Ed Beutner (1978) demonstrated that the preferred orientation of micas in slaty cleavage can take place without grain rotation. Beutner analyzed the structural geology and structural petrology of the Martinsburg Slate, a favorite target of structural geologists working in the eastern part of the central Appalachians.

Beutner's work focused on the chlorite grains, which he interpreted to be part of the mineral assemblage of the original pelite from which the slate was derived. Orientation analysis of the chlorite grains led him to conclude that the grain alignment of the chlorite was a by-product of the progressive systematic, selective corrosion of each original chlorite grain. The general mechanism is pictured in Figure 9.25*A*. Beutner originally suggested that randomly oriented chlorite grains can become dimensionally aligned by virtue of preferential grain size reduction at right angles to the direction of greatest shortening. Depending on the original

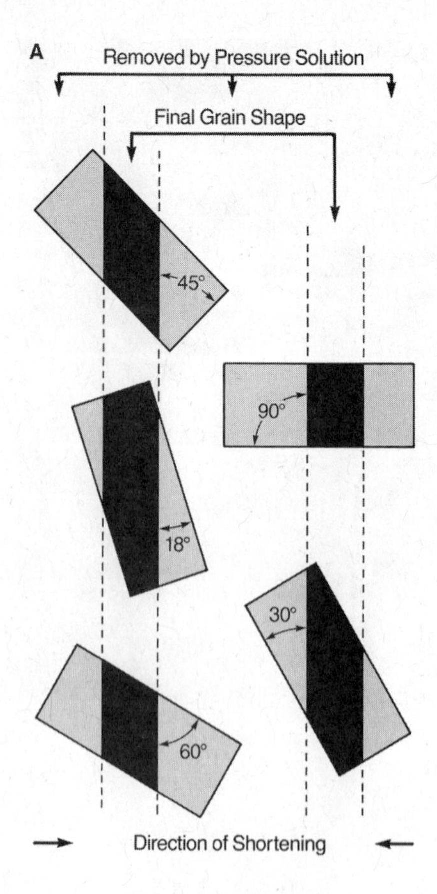

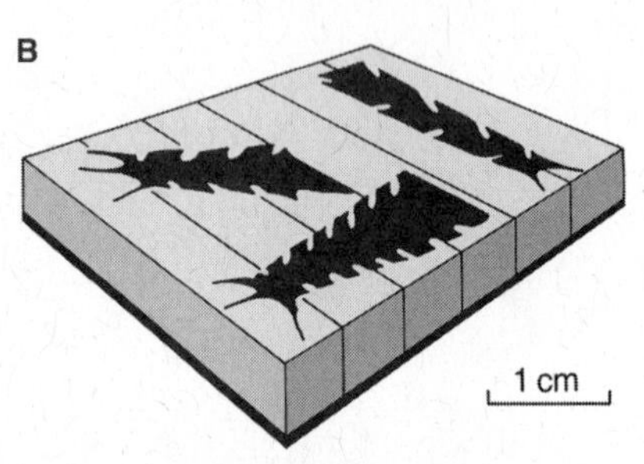

Figure 9.25 (A) The formation of a preferred dimensional alignment of chlorite through progressive pressure solution. [From Beutner, 1978, *American Journal of Science*, v. 278. Reprinted by permission of the American Journal of Science.] (B) Telltale signs of pressure solution of graptolite-bearing slate. The observed size and shape of each distorted graptolite are systematically related to the orientation of each graptolite with respect to the direction of shortening. [From Wright and Platt, 1982, *American Journal of Science*, v. 282. Reprinted by permission of the American Journal of Science.]

orientation of an individual chlorite grain with respect to the direction of greatest shortening, final grain shape due to corrosion can be a parallelogram, a rectangle, or a rhombus. The resulting fabric is marked by preferred dimensional orientation of grains, but not by the preferred crystallographic orientation of grains that would be expected if grain rotation was dominant.

Estimates of Volume Loss

The amount of shortening and volume loss accommodated by the pressure solution removal of host rock can be evaluated quantitatively in rocks containing abundant primary objects of known original shape and size. Wright and Platt (1982) analyzed shales in the Martinsburg Formation, using the geometric and dimensional properties of fossil graptolites in the slate as a guide to distortion. The deformed graptolites proved to be magnificent strain indicators. When Wright and Platt compared the sizes and shapes of the deformed graptolites with those of the undeformed counterparts, they found that graptolites oriented parallel to the trace of cleavage in the plane of bedding were narrower than normal (Figure 9.25*B*). Graptolites oriented perpendicular to cleavage in the plane of bedding were shorter than normal. (We would be too if acted on by the stresses that created the slates of the eastern Appalachians.)

Examining deformed graptolites oriented at *all* angles to the direction of cleavage, Wright and Platt were able to show that the Martinsburg Formation shales were shortened by an average of 50%, perpendicular to the orientation of the slaty cleavage. Because shortening was accommodated by volume loss, not by constant volume deformation, the rock was not required to stretch out in any direction as a compensation for the shortening. Consequently, the shortening by an average of 50% reflects an average volume loss of 50%!

Studies of slaty cleavage have taken structural geologists and structural petrologists into the twilight zone of submicroscopic investigations and thermodynamic considerations. Geologists use microprobes, mass-balance calculations, isotopes, and other geochemical techniques to evaluate the distribution and amount of volume loss (Erslev and Ward, 1994; Vernon, 1998). An important aspect being investigated is whether volume lost from M-domains is redistributed only locally into the QF-domains, veins, and pressure shadows, and other sites, resulting in no overall loss of volume by the rock mass. Alternatively, are the chemical components dissolved from the M-domains transported completely out of the deforming rock mass by fluids, causing an overall loss in rock volume?

Strain Significance of Crenulation Cleavage

The evaluation of the strain significance of crenulation cleavage serves to further underscore the importance of pressure solution in the formation of cleavage. Pressure solution nicely explains the conspicuous domainal fabric of crenulation cleavage, both discrete and zonal. Furthermore, it provides a means of understanding why micas are concentrated in thin bands and laminae in the M-domains, whereas QF-domains contain more quartz and feldspar.

Gray and Durney (1979) emphasized that the development of crenulation cleavage involves a physical/chemical redistribution of minerals as a function of relative solubilities and chemical mobilities. They pictured the cleavage domains as sites of the removal, by pressure solution, of substantial amounts of host rock, leaving behind insoluble residues of clayey and carbonaceous material. According to Gray and Durney, the pressure solution takes place on grain and/or layer boundary discontinuities oriented perpendicular to the direction of minimum finite stretch (the S_3-direction of the strain ellipsoid). Movement of dissolved material follows paths controlled by chemical potential gradients that relate in magnitude and direction to the local stress environment. Cleavage domains emerge along the limbs, or along the *former* positions of limbs, of microfolds in the continuous cleavage of the host rock

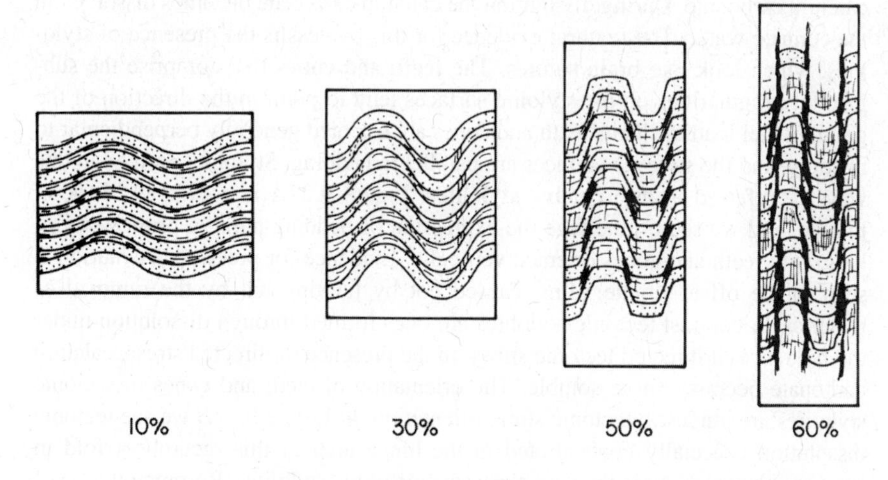

Figure 9.26 Accommodation of shortening through the development of crenulation cleavage. [From Gray, 1979, *American Journal of Science*, v. 279. Reprinted by permission of the American Journal of Science.]

(Figure 9.26). Spacing of the cleavage is related to the dominant wavelength of the microfolds and to the amount of solution-induced shortening across the limbs of the microfolds (Gray, 1977b, 1979).

Shortening accommodated by the formation of crenulation cleavage is achieved through a kind of progressive deformation through time, as illustrated in Figure 9.26. Continuous microfold waveforms are buckled into existence by layer-parallel and/or layer-inclined shortening of the preexisting continuous cleavage (Marlow and Etheridge, 1977; Gray, 1979). The fold forms that emerge reflect the influence of thickness and mechanical character (including ductility contrast) of continuous cleavage laminae, degree of cohesion between the multilayers, and magnitude of shortening. If the shortening required by the strain environment surpasses what can be achieved through folding alone, the rock will begin to shorten by pressure solution loss of material. Dissolution takes place along the loci of fold limbs, and soon cleavage domains emerge within which quartz and feldspar become relatively depleted compared to adjacent microlithon (QF) domains. When soluble mineral phases are dissolved along the fold limbs, they are transported in solution along chemical potential paths to fold hinges, where new minerals are deposited in the form of overgrowths and/or thin laminae. Some of the dissolved material exits the system altogether.

Cleavage domains (M-domains) may initially form at a variety of angles with respect to the direction of overall shortening, but progressive strain eventually brings the cleavage domains into subparallel alignment, perpendicular to the direction of greatest shortening. Substantial pressure solution can lead to the complete removal of fold limbs, leaving fault-like truncations of continuous cleavage within the microlithon (QF) domains.

Strain Significance of Spaced Cleavage

Stylolitic Surfaces

Spaced cleavage appears to be yet another product of pressure solution. Indeed, a revolution of thought emerged from the discovery of the significant role of pressure solution in the formation of spaced cleavage in unmetamorphosed strata, especially impure limestones and marls. It is not surprising to find spaced cleavages associated with tightly folded strata. It *is* surprising, however, to discover that spaced cleavage can develop in essentially flat-lying strata, lacking conspicuous folds.

One of the clearest indications of pressure solution is the presence of **stylolites.** The most commonly occurring stylolites form as a natural result of the formation of limestone during initial burial and compaction under a "*weighty*" load of sediments. The gravitational load exerted on limey beds at depth causes dissolution of calcium carbonate. During dissolution the calcium carbonate becomes dissolved in the connate water. The textural evidence for this process is the presence of stylolites, which look like brain sutures. The **teeth** and **cones** that comprise the subparallel irregularities on the stylolite surfaces tend to point in the direction of the gravitational loading. Thus teeth and cones are oriented generally perpendicular to bedding, and the stylolite surfaces are parallel to bedding. Stylolites formed in this way are referred to as **primary stylolites**. Figure 9.27*A* is a slabbed block of marble, and we see on the face the expression of bedding-parallel primary stylolites, with teeth and cones oriented vertically. Evidence for pressure dissolution is seen in the offset calcite veins, caused not by faulting but by the removal of material. In contrast tectonic stylolites are ones formed through dissolution under the influence of directed tectonic stress. In the presence of directed stress, calcium carbonate becomes more soluble. The orientation of teeth and cones of tectonic stylolites are guides to tectonic stress orientation. In Figure 9.27*B* we see tectonic dissolution especially concentrated in the hinge area of this recumbent fold in limestone bedrock. Note the very close spacing of the stylolites. Re-precipitation of the dissolved calcium carbonate is expressed in the abundance of calcite veins.

An excellent description of tectonic stylolites is provided by Dean et al. (1988). They documented a perfect geometric and kinematic relation between tectonic stylolites and Appalachian folds and thrusts in southeastern West Virginia. They found abundant tectonic stylolites in the Greenbrier Limestone (Mississippian) (Figure 9.28*A*) and measured the strike and dip of the stylolitic surfaces and the trend of plunge of the teeth. When plotted on a structural geologic map (Figure 9.28*B*), these data were seen to be systematically arranged with respect to folds (e.g., the overturned Glen Lyn syncline) and thrust faults (e.g., the St. Clair fault). Outcrop expressions of the stylolites are quite impressive (Figure 9.28*C*).

Stylolitic surfaces produced by tectonic stress (as opposed to burial and compaction) typically are axial planar with respect to associated folds, a

Figure 9.27 (*A*) Photograph of freshly slabbed face of a block of marble. Primary stylolitic surfaces are oriented horizontally, parallel to bedding, with teeth and cones oriented vertically, i.e., perpendicular to bedding. These stylolites formed during burial and compaction of the limestone. This cut block was being prepared as a part of the conservation work at the Temple of Apollo Epicurius at Bassai in the Peloponessos, Greece. Note that pressure dissolution produced offset of calcite vein. (*B*) Here we see the same formation as above, but in this case in bedrock marked by tectonic stylolites. Stylolites are especially densely developed in the fold. The abundant calcite veins are derived from the pressure dissolution. Location near Ano Karyes, Peloponessos. [Photographs by G. H. Davis.]

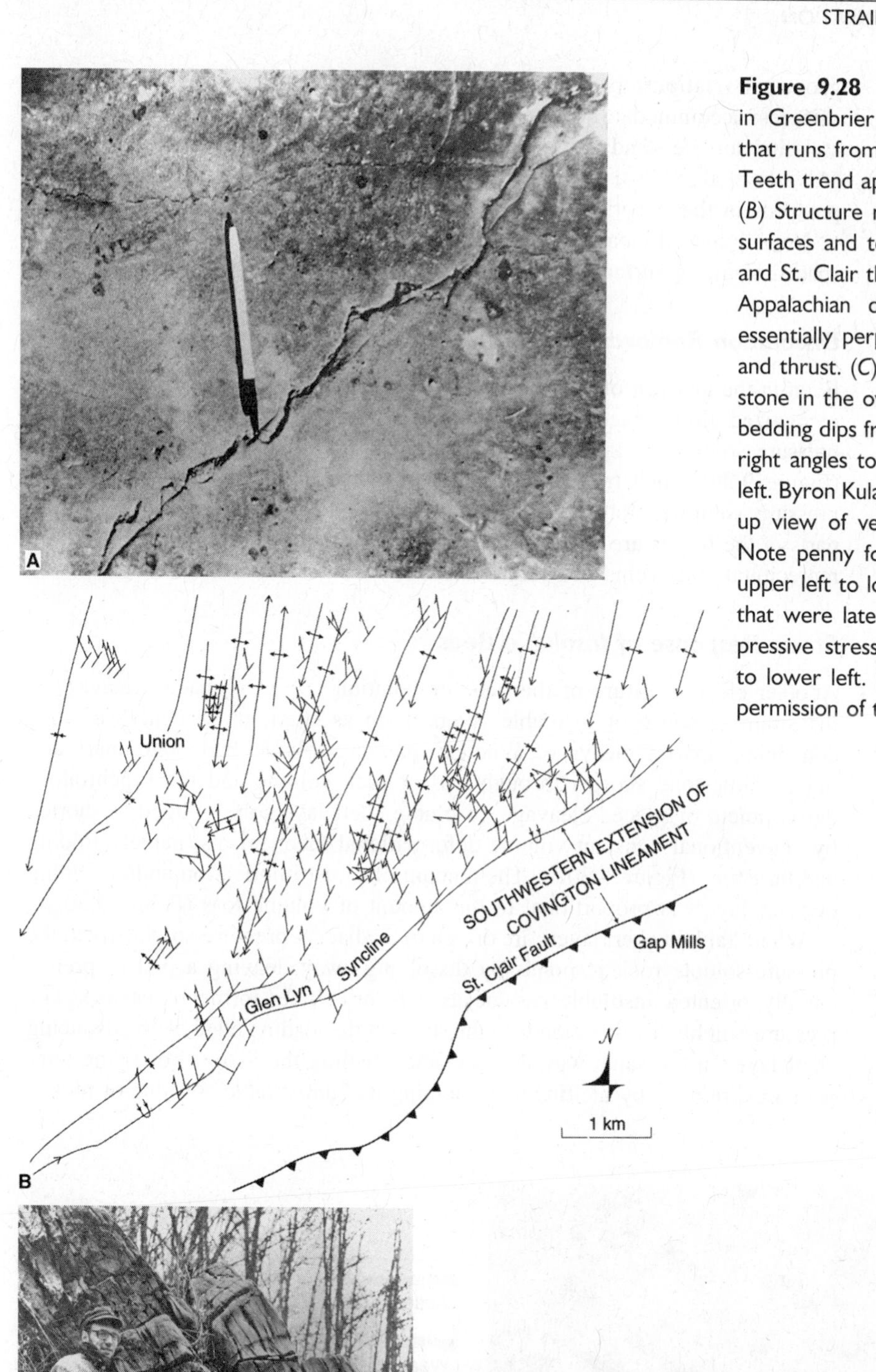

C

D

Figure 9.28 (*A*) Bedding plane view of stylolites in Greenbrier limestone. The major tectonic stylolite that runs from upper right to lower left trends N30°E. Teeth trend approximately N10°W (orientation of pen). (*B*) Structure map showing the orientations of stylolitic surfaces and teeth in relation to the Glen Lyn syncline and St. Clair thrust, both of which were formed during Appalachian deformation. Note that the teeth are essentially perpendicular to the trace of the major fold and thrust. (*C*) Outcrop expression of Greenbrier limestone in the overturned limb of a syncline. Overturned bedding dips from upper left to lower right. Stylolites, at right angles to bedding, dip from upper right to lower left. Byron Kulander, dressed for the weather. (*D*) Close-up view of vertical face of the Greenbrier limestone. Note penny for scale. Bedding (overturned) dips from upper left to lower right. Early formed extension joints, that were later modified into stylolites by lateral compressive stresses before folding, dip from upper right to lower left. [From Dean et al., 1988. Published with permission of the Geological Society of America.]

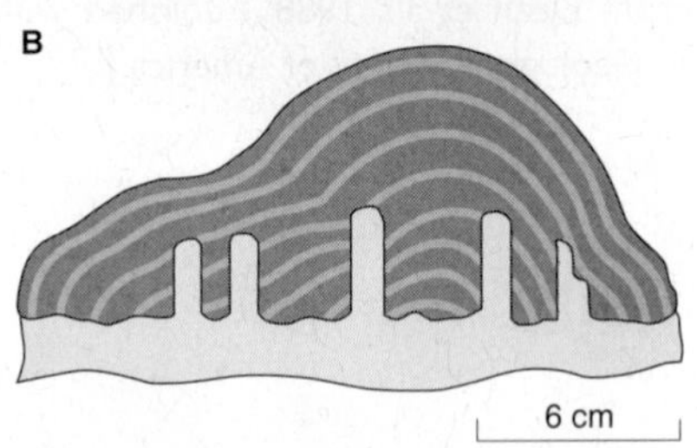

Figure 9.29 (*A*) Photomicrograph of incomplete fusulinid fossil truncated along dark stylolitic seam of insoluble residue. [Photograph by A. Bykerk-Kauffman.] (*B*) Diagram of stromatoporoid fossil into which several stylolitic columns have penetrated. [Redrawn from Stockdale (1943), Figure 2, p. 10.]

geometric relationship that is quite compatible with the notion that stylolitic surfaces accommodate shortening (Nickelsen, 1972). Teeth and cones of stylolitic surfaces tend to be oriented perpendicular to axial surfaces of folding (Alvarez et al., 1976), an observation that supports the premise that dissolution proceeds in the direction of greatest principal stress (σ_1) (see Figure 9.27*B*).

Not all spaced cleavage surfaces are stylolitic. In fact it would seem that spaced cleavage surfaces become smoother as dissolution proceeds.

Dissolution Removal of Fossils

Even in the absence of stylolitic surfaces, the role of pressure solution can be recognized along spaced cleavage surfaces. For example, a clear signature of pressure solution is the abrupt truncation of fossils along cleavage surfaces (Figure 9.29). Such truncated fossils record a stealing away of material by pressure solution, not offset by faulting or extensional fracturing. Missing parts of the fossils are never to be found; they go into solution and largely are reprecipitated as veins or overgrowths.

Strain Response of Insoluble Beds

Another clear signature of the pressure solution origin of spaced cleavage is the strain response of insoluble layers, such as chert, in sequences of rock containing spaced cleavage. Whereas pressure-soluble beds like marl and impure limestone shorten through loss of rock volume and the synchronous development of spaced cleavage, insoluble chert layers are obliged to shorten by conventional constant volume deformational mechanisms, namely folding and thrusting (Figure 9.30*A*). The amount of thrust imbrication and/or folding of chert layers is proportional to the amount of volume loss (Figure 9.30*B*).

Where faulted chert layers are driven into adjacent pressure-soluble rock, the pressure-soluble rock responds by dissolving away, leaving a pod of preferentially oriented insoluble residue as a record of its former existence. The pressure-soluble layer responds to the stress at the leading edge of an advancing chert layer in the same way that a glacier, feeling the "*stress*" of rising temperatures, retreats by melting and dumping its "*unmeltable*" residue of rocks.

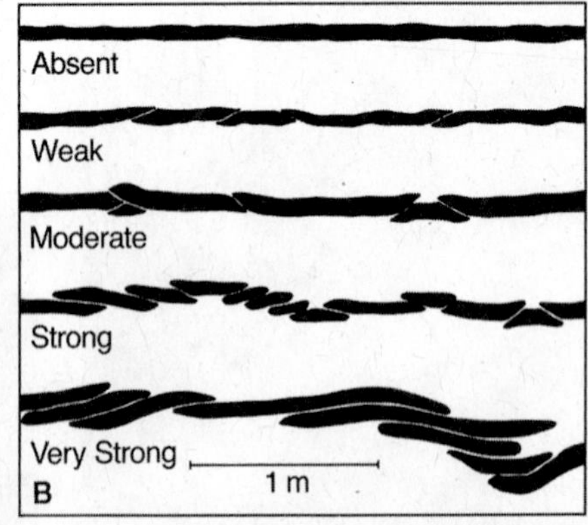

Figure 9.30 (*A*) Spaced cleavage in a strongly cleaved impure limestone. Arrows point out thrust-fault imbrication of insoluble black chert layer that was incapable of shortening by pressure solution. [From Alvarez, Engelder, and Lowrie (1976). Published with permission of Geological Society of America and the authors.] (*B*) Degree of fault imbrication of insoluble chert layers (black) corresponds to the intensity of development of cleavage. [From Alvarez, Engelder, and Lowrie, 1976. Published with permission of the Geological Society of America.]

Kinematic Significance of Offset Bedding

Yet another signature of pressure solution in rocks with spaced cleavage is the offset of bedding or other laminae along the spaced cleavage surfaces (Figure 9.31). The offset is due to removal of material along a spaced cleavage surface. The magnitude and sense of offset mostly depend on a number of factors, including the orientation of the spaced cleavage surface relative to the orientation of the marker bed, the orientation of the spaced cleavage surface relative to the direction of minimum finite stretch (S_3), and the amount of dissolution. Only in two conditions will pressure solution along a cleavage surface fail to cause offset of the marker bed that is cut by the cleavage: when cleavage and bedding are (1) mutually perpendicular or (2) strictly parallel, assuming that the direction of shortening in each case is perpendicular to the cleavage surface. Otherwise, anything goes!

An example of bedding offset due to pressure solution is portrayed in Figure 9.32. Figure 9.32*A* is a geologic map of a nonplunging, overturned anticline cut by an axial planar spaced cleavage. At station 36 on the right-side-up western limb of the fold, there is a revealing exposure of the interrelationship of bedding and cleavage (Figure 9.32*B*). Both bedding and cleavage strike 350°, but they dip westerly by different degrees. The bedding dips 20°W, and the cleavage dips 50°W. Bedding is repeatedly offset along the cleavage by very small amounts. Separation is in all cases normal, averaging 0.5 cm.

Assuming that the direction of shortening was perpendicular to the orientation of cleavage, it is possible to graphically determine the amount of dissolution required to account for the offset along any given cleavage surface.

Figure 9.31 Macro photograph of cleavage (wavy black lines slanting down-to-the-right. Note that white marker shows apparent fault offsets where cut by cleavage. Rock is marl (clayey limestone). Pindos Group, Mt. Lykaion, Greece. [Photograph by G. H. Davis.]

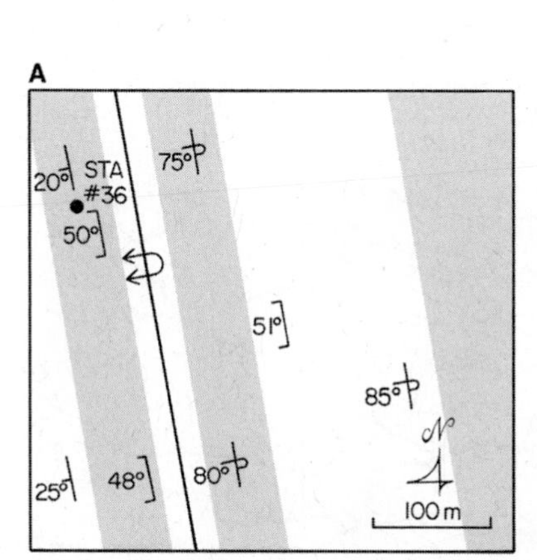

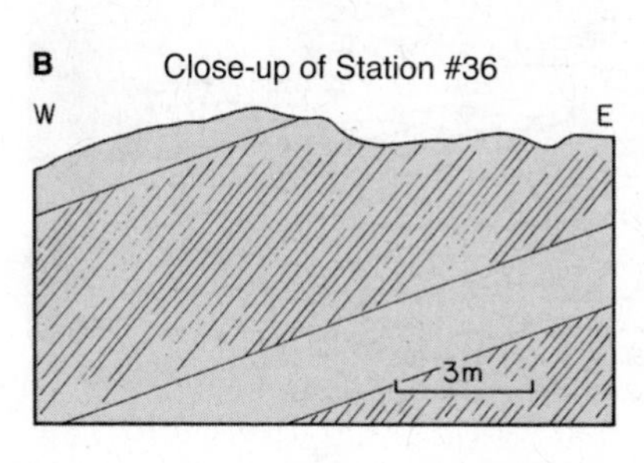

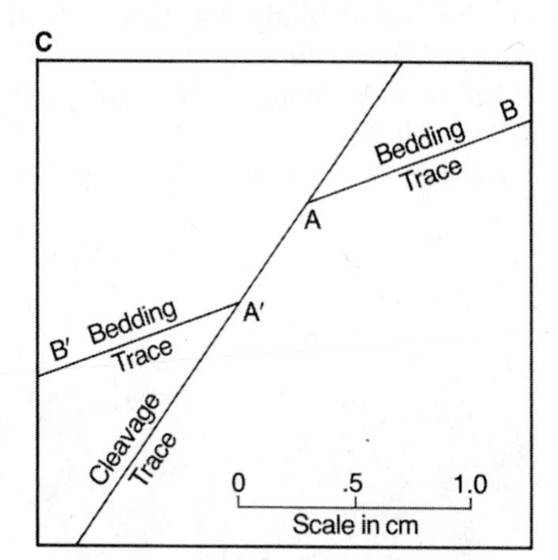

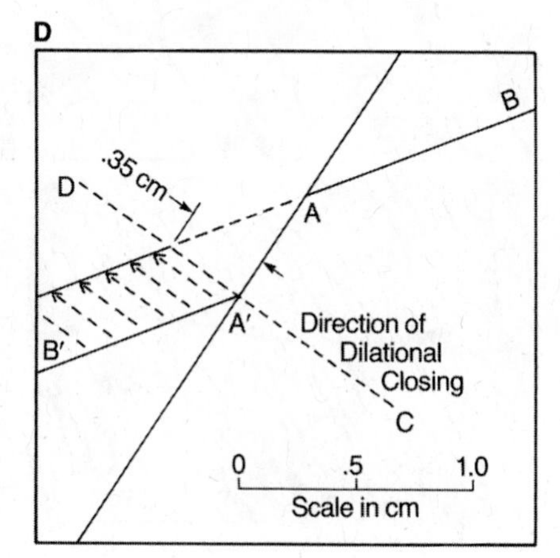

Figure 9.32 Kinematic analysis of bedding offset due to shortening by pressure solution. (*A*) Map showing overturned fold cut by axial plane cleavage. (*B*) Bedding/cleavage relationship exposed at station 36. (*C, D*) Graphical determination of the magnitude of "*dilational closing*" required to explain the offset of bedding along the cleavage surface.

Figure 9.32*C* shows the relationships that need to be explained. Lines *AB* and *A′B′* are bedding traces that formerly were in alignment. Distance *AA′* is the magnitude of normal separation, namely 0.5 cm.

To determine the magnitude of dilational closing, simply "*back off*" bedding trace *A′B′* along the direction of dilational closing (path *DC*, Figure 9.32*D*), maintaining the 20° dip of *A′B′*. When aligned with the projection of bedding trace *AB*, line *A′B′* is situated in its restored position, *relative to AB*, before deformation. The magnitude of dilational closing is measured along the direction of dilational closing (*DC*) between the restored location of *A′* and the trace of cleavage. It measures 0.35 cm. Calculations of this type, when averaged over an array of cleavage surfaces in a fold, provide the basis for estimating total volume loss due to dissolution (Crespe, 1982).

Classification of Spaced Cleavage

Alvarez et al. (1978) designed a classification of spaced cleavage, one that provides the means to estimate shortening within a rock layer on the basis of the nature and spacing of cleavage surfaces. Their approach was to systematically correlate the properties of spaced cleavage with quantitative estimates of shortening deduced from bedding offsets, truncated fossils, and the degree of folding and thrusting of insoluble chert layers.

The categories of spaced cleavage that Alvarez, Engelder, and Geiser recognized are described as weak, moderate, strong, and very strong. These correspond to shortening percentages of 0–4, 4–25, 25–35%, and greater than 35%, respectively. Cross-sectional and bedding-plane views of two of the intensities of cleavage (moderate and strong) are shown in Figure 9.33.

The strongest cleavages are the most closely spaced. The weakest cleavages are the most stylolitic. Moderate and strong spaced cleavages tend to display clear-cut intersecting sets that are symmetrically disposed about the direction of greatest shortening. Very strongly developed cleavage is marked by sigmoidal cleavage and abundant calcite veining. Veining is yet another expression of pressure-induced mobilization of rock constituents, in this case, calcite.

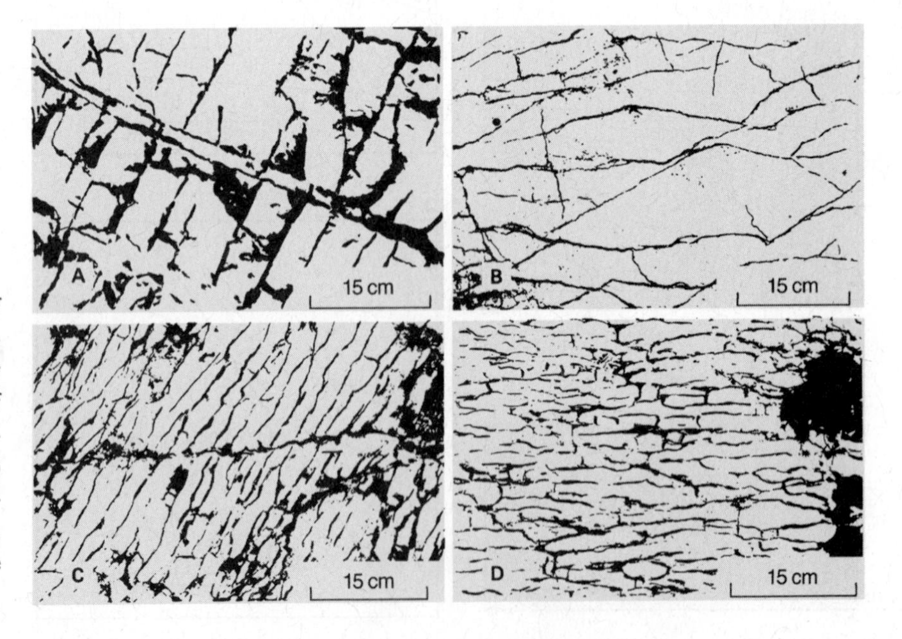

Figure 9.33 Geometry and spacing of moderately and strongly developed spaced cleavage. (*A*) Cross-sectional view of "*moderate*" cleavage; (*B*) expression of cleavage on the bedding surface. (*C*) Cross-sectional view of "*strong*" cleavage; (*D*) expression of cleavage on the bedding surface. [From Alvarez, Engelder, and Geiser, 1978. Published with permission of the Geological Society of America.]

RELATIONSHIP OF CLEAVAGE TO OTHER STRUCTURES

Geometric Relationship of Cleavage to Folding

Cleaved rocks are generally folded rocks, but folded beds are not always cleaved. A close geometric coordination exists between the orientation(s) of cleavage surfaces and the configuration of folded bedding. Ordinarily, cleavage surfaces either are parallel to the axial surface of folding, or they are disposed symmetrically about the axial surface in a **fan** of orientations (Figure 9.34). In either case, the cleavage surfaces comprise an **axial plane cleavage**. Folded bedding in an upright fold is cut by cleavage surfaces that everywhere are steeper than the inclination of bedding. In overturned folds, cleavage can dip less steeply than bedding. *An axial plane cleavage that dips in the same direction as bedding, but less steeply than bedding, is a warning signal that bedding may be overturned.*

Figure 9.34 Incipient development of spaced cleavage in limestone. The expression of the cleavage is the set of curviplanar surfaces that have a "*fan-like*" orientation distribution across the hinge zone of the fold. These limestones belong to the Flysch Transition Beds (Paleocene) within the Pindos Group. Location is Mt. Lykaion, Greece. [Photograph by G. H. Davis.]

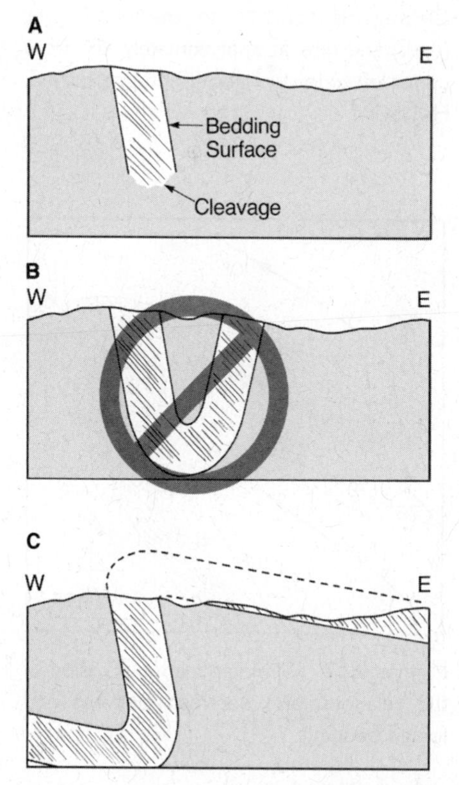

Figure 9.35 Use of the orientations of bedding and cleavage to construct the form of the fold with which the bedding and cleavage are associated. (*A*) The outcrop relationships. (*B*) Misfit between the cleavage orientation and the interpreted fold form. (*C*) A good fit!

The relationship of axial plane cleavage to bedding in folded rocks can often be used to evaluate the likely **facing** of a bed, and to construct the configuration of folds in profile view. A useful application of the fundamental geometric relationship between bedding and axial plane cleavage is illustrated in Figure 9.35. In the outcrop shown in Figure 9.35*A*, bedding is cut by cleavage surfaces that are known to be "*axial planar.*" Bedding dips 80°E, but the cleavage surfaces dip just 45°E. Knowing that the cleavage is an axial plane cleavage, we can determine the fold configuration of which the bed is a part. This is achieved by drawing a folded surface that maintains an axial planar relationship to the cleavage surfaces. If a fold profile is drawn in such a way that the east-dipping bedding represents the west limb of an upright syncline (Figure 9.35*B*), the form of the syncline cannot be fit in an axial planar manner to the cleavage surfaces. If, on the other hand, the east-dipping bedding is considered to be part of the overturned west limb of an overturned

Figure 9.36 Photograph of part of Van Hise Rock, an official historic landmark in Wisconsin. Bedding is nearly vertical. Cleavage is confined to the shale layer (dark) and dips at approximately 40° from upper left to lower right. [Photograph by G. H. Davis.]

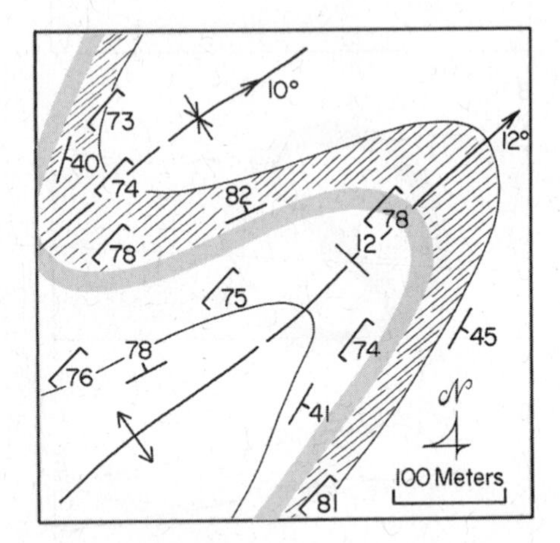

Figure 9.37 Geologic map expression of the relationship(s) between cleavage and folded bedding.

anticline (Figure 9.35*C*), the form of the fold is perfectly compatible with a 45°E-dipping axial plane cleavage.

A reference location for recognition of this relationship is Van Hise Rock in Wisconsin (Figure 9.36), where the difference in orientation between bedding and cleavage is striking. The outcrop is like a vertical sandwich, with cross-bedded sandstone layers on the outsides and dark shale in the middle. Cleavage in the shale dips from upper left to lower right, in the same manner pictured in Figure 9.36.

Bedding and cleavage surfaces are carefully distinguished on geologic maps. The common map symbol for cleavage is shown in Figure 9.37, a simplified geologic map of a plunging anticline/syncline pair. Cleavage symbols in combination with bedding symbols serve to highlight the interrelationships of bedding and cleavage across folds. Where cleavage surfaces cut through the hinge of a fold, there is a maximum discordance between bedding and cleavage. At the hinge point proper, the discordance is 90° (i.e., bedding is perpendicular to cleavage). At each point on the limb of a fold, cleavage surfaces generally cut bedding at some small, acute angle. The angle of intersection steadily decreases from the hinge to the inflections of a fold. Isoclinal folds present the special case in which cleavage surfaces and bedding on the fold limbs are perfectly parallel to each other.

One way to keep track of bedding-cleavage relationships in the field is to view the cleavage as being somewhat fixed in orientation, and noting whether bedding is rotated clockwise or counterclockwise relative to cleavage. In this reference frame, bedding will be clockwise on one limb and counterclockwise on the other; it is at 90° in the hinge. Try it out with Figure 9.35*C*!

The orientation of cleavage surfaces in the hinge of a fold is a close approximation to the axial surface of the fold. However, the orientation of cleavage surfaces at any one point on the limb of a fold usually does not reflect the orientation of the axial surface of folding, simply because cleavage surfaces generally display a fanning of orientations across the folded surfaces. Also, cleavage can bend, or **refract**, slightly as it passes between two rocks with differing mechanical properties, such from shale to sandstone and back to shale.

The clear geometric harmony between cleavage and folding leads to the conclusion that cleavage forms as a response to shortening and flattening. The surfaces of cleavage lie in the S_1S_2 plane of the strain ellipsoid perpendicular to the direction of minimum finite stretch (S_3). More on this later.

Palinspastic Reconstruction of Folds and Layers

Palinspastic reconstruction (the stretching back) of folds and fold belts marked by spaced cleavage requires not only the unfolding of folded layers, it also requires an accounting of losses due to pressure solution. The folds must be stretched out like an accordion, to account for pressure solution losses, before we rotate bedding back to horizontal, thereby restoring deformation due to flexural-slip folding and/or buckling. Where the development of spaced cleavage preceded folding, the palinspastic reconstruction steps must be reversed.

A classic example of palinspastic reconstruction of folded layers cut by spaced cleavage was presented by Groshong (1975). His analysis focused on a single buckle fold in impure limestone. Spacing of cleavage in the fold was found to be 0.2–0.5 cm, on average (Figure 9.38*A, B*). Bedding in the fold is repeatedly offset along the cleavage surfaces, especially in the inner arc, core region of the fold. Small fossils in the limestone were truncated and corroded along the cleavage surfaces. After evaluating the average displacements and pressure solution losses along the cleavage surfaces, Groshong palinspastically

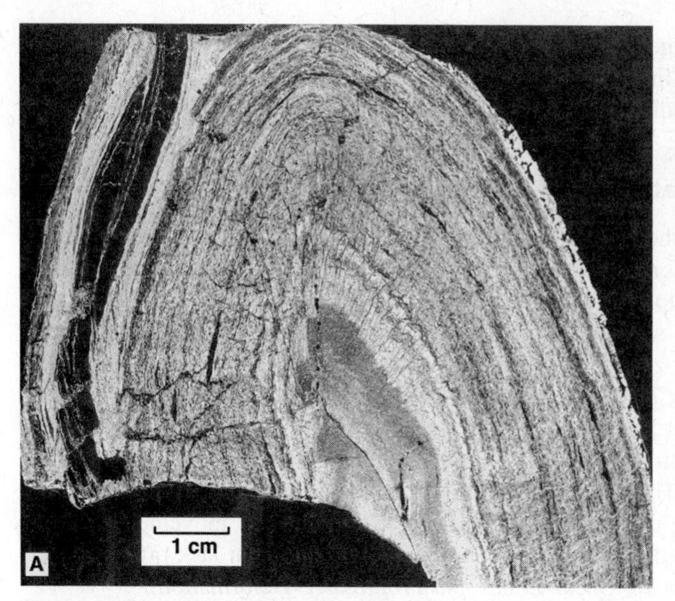

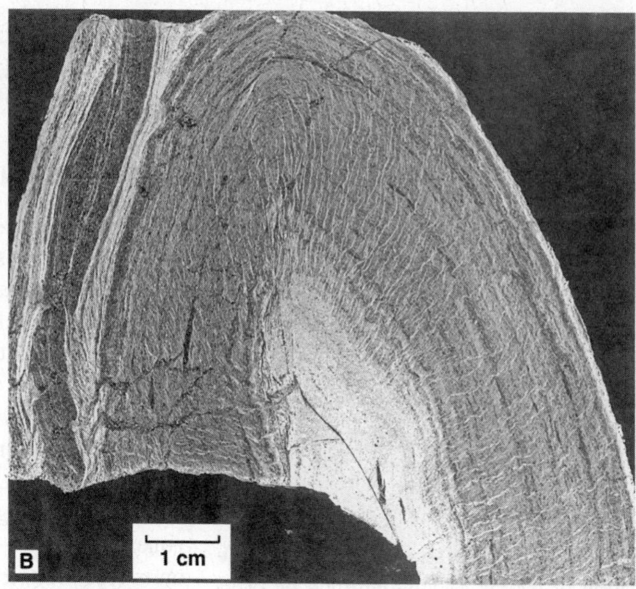

Figure 9.38 Small fold in impure limestone. (*A*) Photograph taken in normal lighting emphasizes nature of bedding. (*B*) Photograph taken in polarized light emphasizes nature of cleavage. [From Groshong, 1975a. Published with permission of the Geological Society of America.]

restored the fold to a more open configuration by separating the microlithons. Minimum volume loss was estimated to be 18%.

Volume losses well above 18% are recorded in highly deformed folded terranes. Magnitudes of 40–50% dissolution are not at all uncommon. Significant volume losses are also recorded in certain foreland terranes; for instance, the Appalachian foreland. Bedding in the Appalachian foreland region of Pennsylvania and New York is essentially flat lying, folded about very gentle folds with limb dips less than 3–4°. The innocent-looking, flat-lying nature of these foreland strata is sharply contradicted by the measured strain state of these rocks. Engelder and Engelder (1977) demonstrated a layer-parallel shortening of 10–15%, most of which was accommodated by the formation of spaced cleavage.

Any attempts to prepare balanced cross-sections of folded and thrusted terranes must take into account the realities of pressure solution loss of material. Line–length balancing alone will not suffice. We cannot simply unfold folds if analysis of spaced cleavage in the cores of some folds reveals a 40% loss of volume. The shortening component of the loss of volume *must* be taken into consideration.

In most cases, the onset of cleavage development probably takes place after some amount of buckling and flexural folding has already been achieved. The onset of pressure solution permits a rock to shorten to a degree not possible by further rotation of the limbs of a fold. In effect, a point is reached in the folding process at which material needs to be forced out of the inner arc region of the fold. **Flattening** is required to eliminate the room problem.

Flattening can be accomplished by pressure-induced removal of material. Class 1B folds can be transformed into class 1C, 2, or 3 folds by removal of narrow, parallelogram-shaped sections of rock along directions parallel to the axial surface of the folds with which the cleavages are associated (Figure 9.39*A*). As pressure solution takes place and material is removed, shortening keeps pace to prevent the creation of open space. Adjacent microlithons bordering on a common cleavage domain move toward one another and interpenetrate one another along a direction that is parallel to S_3, the direction of minimum finite stretch. Fold limbs naturally steepen to greater inclinations, and

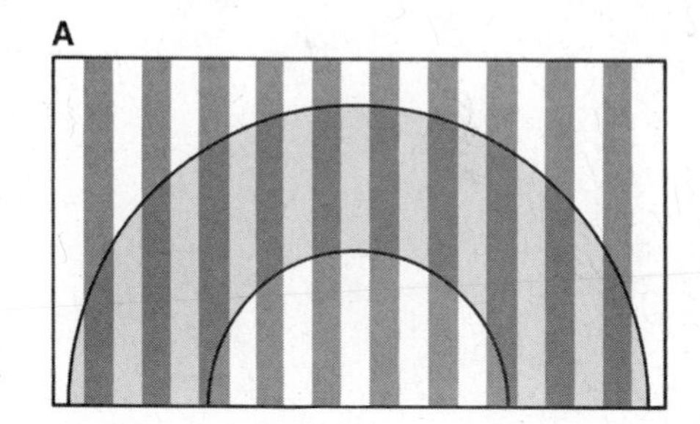

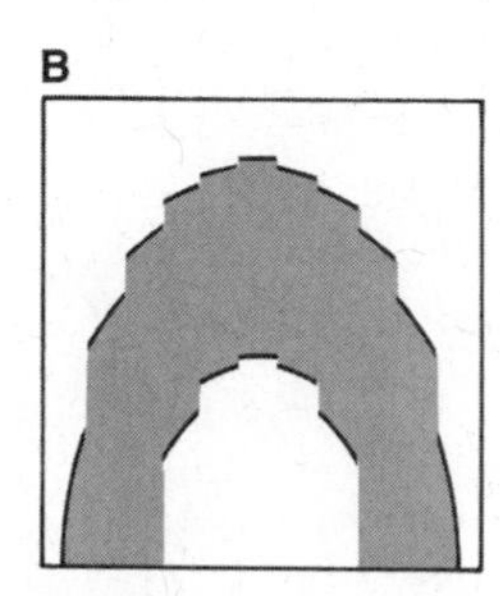

Figure 9.39 Schematic rendering of the transformation of (*A*) a class 1B fold to (*B*) a class 1C fold by pressure solution. Such a transformation can be simulated easily with a deck of cards, not by displacing the cards in simple shear fashion, but by removing domains of material at spaced intervals within the deck.

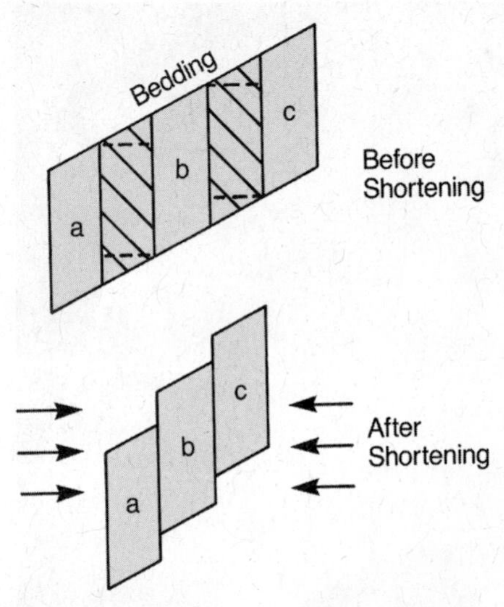

Figure 9.40 The steepening of the inclination of overall bedding attitude by pressure-solution loss of material. [From Alvarez, Engelder, and Lowrie, 1976. Published with permission of the Geological Society of America.]

fold profiles change from concentric to similar (Figure 9.39*B*). The effect of pressure solution imposed on an early fold form is the creation of shear-like separations of bedding along cleavage surfaces (Figure 9.40).

Geometric Relationship of Cleavage to Shearing

Fault zones and shear zones may contain cleaved rocks, even when no folds are present. We sometimes see subtle, delicate, penetrative cleavage surfaces in gouge zones along faults (Figure 9.41*A*). When cleavage is found in this kind of structural setting, its orientation is typically aligned at a small acute angle to the fault zone itself. Brittle–ductile shear zones can show the same relationship (see *Chapter 10, Shear Zones and Progressive Deformation*).

As we shall see, cleavage in fault zones and shear zones occupies an orientation of flattening, corresponding to the S_1S_2 plane of the strain ellipsoid (Figure 9.41*B*). This plane generally "*leans over*" in the sense of shear, thereby providing us with a way to interpret the sense of movement of fault zones and brittle–ductile shear zones, even in outcrops that lack offset marker units (see *Chapter 10, Shear Zones and Progressive Deformation*).

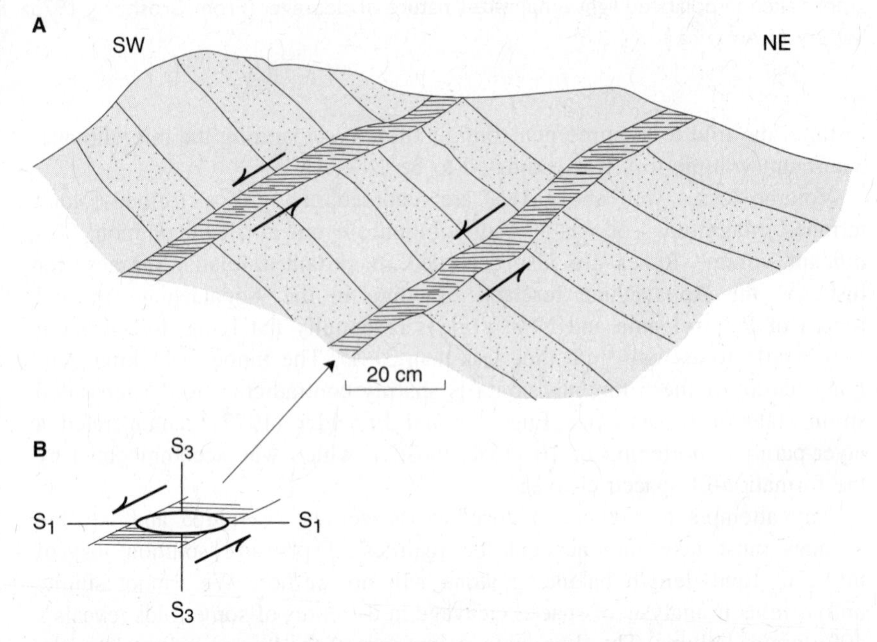

Figure 9.41 (A) Sketch based on outcrop relations in the San Manuel Formation (Miocene) near Tucson, Arizona. Sandstone and siltstone of the San Manuel Formation is cut by normal faults. The faults contain clayey gouge, and some of the gouge zones display a delicate penetrative cleavage. (B) The orientation of the cleavage in the gouge discloses the sense of movement, aligning itself with the direction of maximum finite stretch (S_1) and perpendicular to the direction of minimum finite stretch (S_3).

Transposition

Strongly cleaved sequences of metasedimentary and metavolcanic rocks in fold belts characteristically display a parallelism of foliation and lithologic layering. The lithologic layers, distinguished on the basis of color, texture, mineralogy, and general outcrop appearance, tend to pinch out along strike, giving way to other combinations of lithologic layers (Figure 9.42). The mapped relationships give the impression that foliation developed parallel to bedding in the original host rock, a host rock that was marked by abundant abrupt facies changes.

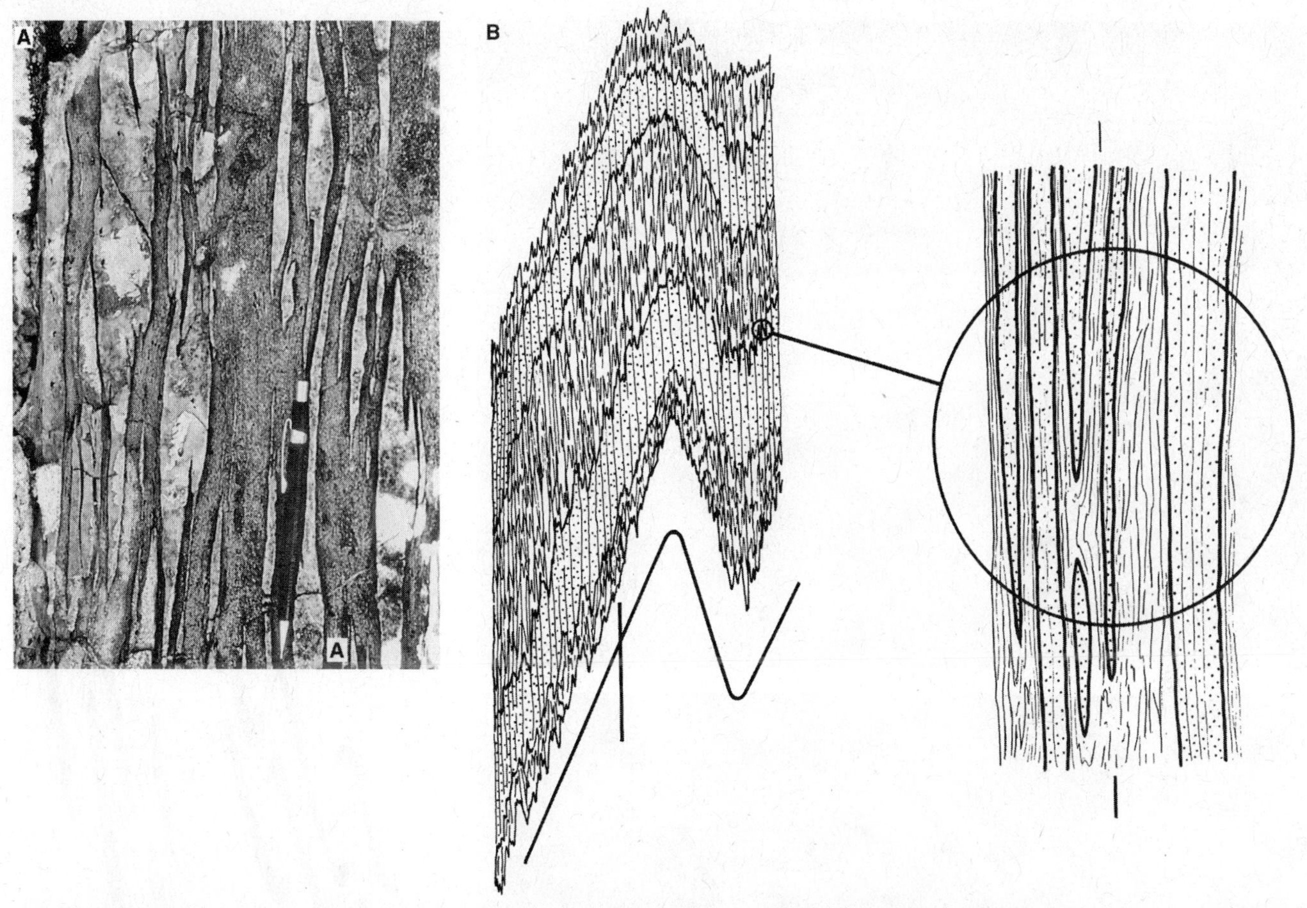

Figure 9.42 (*A*) Pseudostratigraphy in metasedimentary rocks in the Happy Valley region of the Rincon Mountains, near Tucson, Arizona. [Photograph by S. H. Lingrey.] (*B*) Pseudostratigraphy within folded rock. [From *Structural Analysis of Metamorphic Tectonites* by F. J. Turner and L. E. Weiss. Published with permission of McGraw-Hill Book Company, New York, copyright © 1963.]

Structural relationships are not always what they seem. Careful inspection of terranes containing "*bedding-plane foliation*" usually uncovers rare exposures of **tight to isoclinal intrafolial folds** that are cut by the dominant foliation in axial planar fashion (Figure 9.43). The folds typically are passive, class 1C to 3 in form, and are sandwiched by through-going foliation and lithologic layering. According to Turner and Weiss (1963) and Whitten (1966), intrafolial folds in strongly deformed metamorphic rocks commonly reflect **bedding transposition** in which tight folding of the original beds is accompanied by shear parallel to the axial planes of developing flexures (Figure 9.44*A,B*). Individual fold limbs are attenuated by progressive shear and pressure solution. Ultimately they are separated from their hinge zones (Figure 9.44*C,D*). The planes of shear and dissolution are in fact the foliation, occupying the S_1S_2 plane of the finite strain ellipsoid.

Transposition creates a **pseudostratigraphy** containing disrupted and rotated segments of once-continuous beds. The entire sequence is pervaded by structural discontinuities, one expression of which is apparent bedding-plane cleavage. Transposed sequences show no internal consistency of facing: right-side-up and upside-down beds are stacked together (Whitten, 1966). The main clue that all this has happened is the presence of intrafolial folds. The concept of bedding transposition teaches us to be very cautious in trying to identify original bedding. Unless it can be firmly demonstrated that the layering represents an internal, undisrupted, coherent, original stratigraphy, lithologic layering in strongly deformed metamorphic rock is simply described as "*layering*" rather than bedding.

Figure 9.43 Some preserved isoclinal folds in transposed strata cut by axial plane foliation. (*A*) Isoclinal fold in metamorphic rocks of the Bucksport Formation, Pemaquid point, Maine. [Photograph by S. J. Reynolds.] (*B*) Folds and schistosity in upper part of outcrop become sheared out and transposed in lower part, along the Cala Prona shear zone, Cap de Creus, Spain. [Photograph by G. H. Davis.]

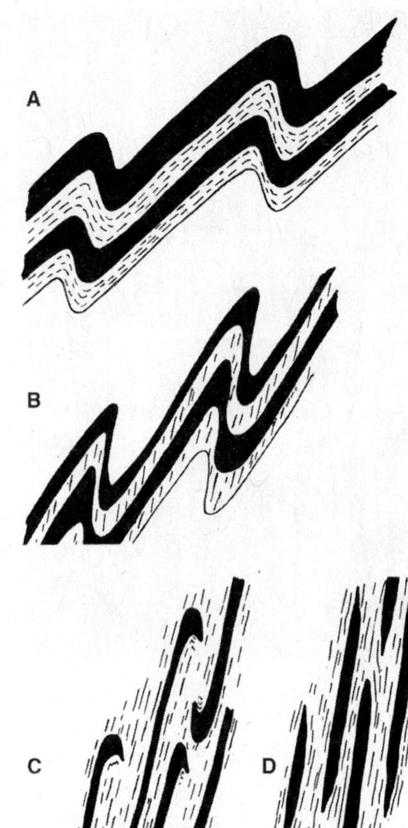

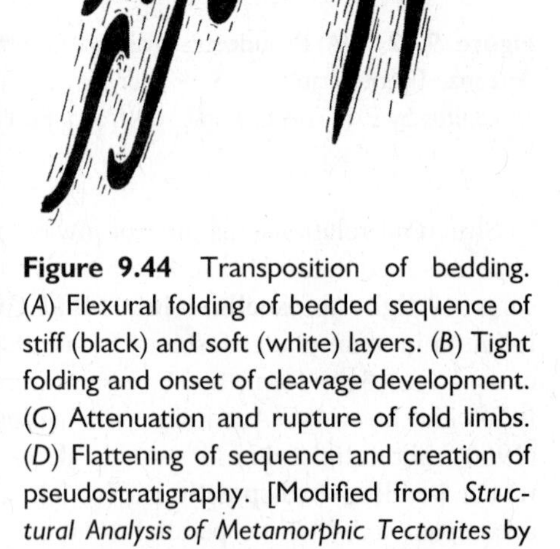

Figure 9.44 Transposition of bedding. (*A*) Flexural folding of bedded sequence of stiff (black) and soft (white) layers. (*B*) Tight folding and onset of cleavage development. (*C*) Attenuation and rupture of fold limbs. (*D*) Flattening of sequence and creation of pseudostratigraphy. [Modified from *Structural Analysis of Metamorphic Tectonites* by F. J. Turner and L. E. Weiss. Published with permission of McGraw-Hill Book Company, New York, copyright © 1963.]

FOLIATION

Definition of Foliation

Cleavage is just one brand of foliation. **Foliation** is any mesoscopically penetrative parallel alignment of planar fabric elements in a rock, usually in metamorphic rocks (Figure 9.45). The definition of foliation is full of code words.

Figure 9.45 Banded gneiss with compositional layering defined by alternating light-colored layers of granitic composition and dark layers that are more biotite rich. Location is Alamo Dam, western Arizona. [Photograph by S. J. Reynolds.]

"*Planar fabric element*" can refer to any number of features, such as domainal structure in slaty cleavage, flattened pebbles in a metaconglomerate, or compositional banding in a banded gneiss. "*Parallel alignment*" means roughly parallel, with a lot of latitude. It can refer to perfectly parallel slaty cleavage, approximately parallel pebbles in a flattened pebble metaconglomerate, or converging/diverging domainal structure in a spaced cleavage. "*Planar*" can mean perfectly planar, like a slaty cleavage surface in outcrop; curviplanar, such as the face of a flattened, discoidal pebble in metaconglomerate; or irregular, like the margins of pegmatitic lenses in granitic gneiss.

The code words in the definition of foliation attempt to set limits and to be encompassing, both at the same time. The definition of "*foliation*" must be broad enough to include the extraordinarily wide variety of foliation fabrics that exist.

Primary Versus Secondary Foliation

Some foliations are **primary foliations**, and these develop during the flow of magma, the compaction of pyroclastic rocks, or the deposition and compaction of sediment. One type of primary foliation is **flow banding**, which forms when magma flows and shears in a molten state (Figure 9.46*A*). Flow banding is defined by distinct-to-subtle bands and lenses marked by variations in color, abundance of vesicles, degree of devitrification, or some other aspect. Tabular and prismatic crystals, as well as flattened vesicles, in flow-banded rocks can also have a planar alignment, or **flow foliation**, parallel to the flow bands. Flow banding is most common in lava flows of felsic (e.g., rhyolitic) and intermediate (e.g., andesitic) composition, but flow foliations are present in volcanic rocks of all compositions. Flow banding and flow foliations in felsic lava flows can be beautifully folded and contorted, with isoclinal folds, refolded folds, and folds that have been sheared apart. Patterns of flow foliation are typically complex at the scale of an outcrop and the scale of a map (Figure 9.46*C*). Flow foliation, and in some cases flow banding, can also form in igneous rocks from viscous flow of magma against wall rocks along the

Figure 9.47 (*A*) Basalt dike cutting Tertiary conglomerate in northwestern Arizona. Peter Coney is seen checking out a hand specimen of the dike rock. (*B*) Close-up of margin of the dike. Note the flow foliation expressed by banding and lamination in the dike. The foliation is oriented parallel to the contact with wall rock. [Photographs by G. H. Davis.]

Figure 9.46 (*A*) Flow banding in a rhyolitic lava flow, Wickenburg, Arizona. [Photograph by S. J. Reynolds.] (*B*) Map showing the orientations and configuration of flow foliation defined by alignment of feldspars, mica, and inclusions in granitic rocks in the Temple Mountain area of the Sierra Nevada. [Mapping by E. B. Mayo.]

margins of dikes, sills, and plugs (Figure 9.47). In this case, the foliation is typically parallel to the margin of the dike, sill or plug.

Granitic rocks, especially those with larger phenocrysts, may retain a record of flow of magma in the form of subtle or not-so-subtle flow foliations defined by aligned phenocrysts and other crystals (Figure 9.48*A*) and by flattened inclusions. In granitic rocks with such fabrics, there is always a question if the fabric developed while the material was still molten, in which case it is a **primary foliation** and is commonly called a **magmatic fabric**. If the fabric developed after the magma had solidified, it is a **secondary foliation** and is called a **solid-state fabric**. Intrusive rocks can also display **schlieren**, which are wispy dark and light bands representing the segregation of different minerals within the magma (Figure 9.48*B*). Schlieren are more dramatic if the compositional variations are accompanied by variations in grain size (Figure 9.48*C*).

Volcanic ash-flow tuffs can display **eutaxitic structure**, a foliation created through flattening of pumice fragments and gas bubbles as the volcanic unit

Figure 9.48 (*A*) Flow foliation defined by a preferred orientation of feldspar phenocrysts in a granite, Date Creek Mountains, central Arizona. (*B*) Schlieren defined by wispy layers and lenses of lighter-colored granodiorite and darker diorite, Granite Wash Mountains, western Arizona. (*C*) Schlieren defined by variations in composition and grain size in a dike of granite and pegmatite, at the locally famous "*garnet schlieren locality*," Summerhaven, Arizona. The dark crystals in the thin bands are all red garnet—what an outcrop! [Photographs by S. J. Reynolds.]

flattens under its own weight during cooling and compaction (Figure 9.49). Such fabrics are evidence that the tuff is **welded**, and only occur if the ash- and pumice-rich material accumulates while it is still hot, as in a pyroclastic flow, and in a sufficient thickness to cause compaction of its lower parts.

Sediment can also develop true primary foliations, such as weakly developed **axial plane cleavage** associated with slump folds that form in wet, barely consolidated sediments. Shales exhibit a tendency to break apart into thin flakes, lenses, and sheets, a characteristic called **fissility** (Figure 9.50) that is in some ways similar to cleavage. Fissility originates partly during the settling and stacking up of aligned clay particles during deposition in quiet water and partly from compaction as a rock is buried, compacted, and lithified. It is a type of primary foliation.

With further burial and compaction, sediment can develop a weak foliation that is parallel to bedding and best expressed by a preferred orientation of detrital mica grains in shales and other fine-grained rocks. The mica grains attained this preferred orientation largely via physical rotation of the micas as the rock was compacted and pore spaces between the micas were reduced by the pressure. Such bedding-parallel foliation, sometimes called a **diagenetic**

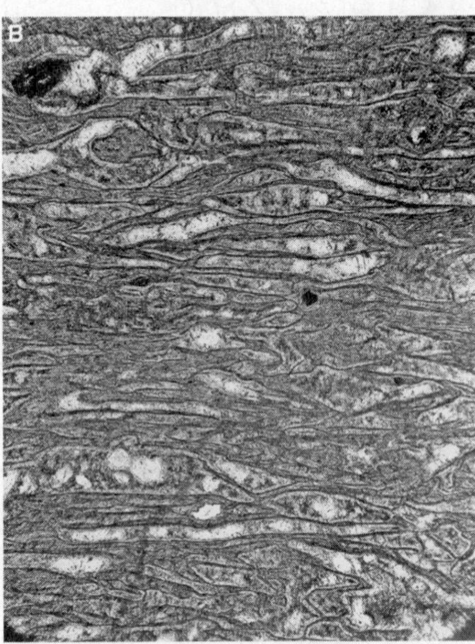

Figure 9.49 (*A*) Eutaxitic structure within a Tertiary ash flow sheet in Nevada. Lower right half of outcrop is composed of densely welded tuff. Partings in the tuff follow concentrations of pumice lapilli, which have been preferentially weathered out. [Photograph by P. W. Lipman. Courtesy of United States Geological Survey.] (*B*) Photomicrograph of primary igneous foliation in rhyolite from the Creede caldera, San Juan Mountains, Colorado. [Photograph by J. C. Ratte. Courtesy of United States Geological Survey.]

Figure 9.50 Fissility in Devonian shale, weathering to small chips near Elkins, West Virginia. [Photograph by S. J. Reynolds.]

foliation (Passchier and Trouw, 2005), is not strictly a primary foliation, since it forms partly after the rock is lithified. It is also not a classic secondary cleavage because it is not related to folds and other types of tectonic deformation. However it is classified, diagenetic bedding-parallel foliation, because it represents a mechanical anisotropy, can greatly influence subsequent tectonic deformation of the rock (Passchier and Trouw, 2005).

In contrast to primary foliations, **secondary foliations** form as a result of the deformation and distortion of sedimentary, volcanic, or intrusive igneous rocks, almost always under metamorphic conditions of elevated temperature and pressure. Throughout the remainder of this chapter, we will continue to concentrate on secondary foliations that typify classic regionally deformed and metamorphosed rocks, such as slates, phyllites, schists, and gneisses. We will also examine **mylonitic foliations**, which are mainly associated with shear zones formed by highly concentrated strain.

Foliations in Typical Metamorphic Rocks

Foliation is a fundamental characteristic of regionally metamorphosed rocks. We have already discussed in some detail the foliations known as cleavage, phyllitic structure, and schistosity. **Slaty cleavage** develops as fine-grained sedimentary and volcanic rocks are transformed into slate during low-grade metamorphism. **Phyllitic structure** and **schistosity**, on the other hand, form when fine- to medium-grained sedimentary and volcanic rocks are deformed during more intense metamorphism, at low grades for phyllites and medium to high grades for schist.

Gneissic structure, another major category of foliation, typically develops as a result of recrystallization of igneous or sedimentary rocks during medium- to high-grade metamorphism. It is generally composed of medium- to coarse-grained minerals and can be defined by various features. Gneissic structure is characterized by a preferred planar orientation of platy, tabular, or prismatic minerals, and by subparallel lenticular mineral grains and grain aggregates (Figure 9.51*A*). In many cases, the aggregates have a planar orientation, but individual grains within those aggregates have more equant shapes; this texture indicates that recrystallization was able to keep pace with deformation. In other cases, individual grains exhibit deformed, strained shapes. Such textures indicate that crystals were being deformed fast enough that they could not totally recrystallize into more stable, equant shapes. With these two end-members nature is showing us the competition between strain rates and crystallization rates. At high temperatures and slower strain rates, crystallization wins.

Gneissic structure occurs in metamorphic rocks of almost every composition and original rock type (protolith). It can develop in granitic rocks that have been deformed and metamorphosed at medium to high grades, or in sedimentary and volcanic rocks that have been metamorphosed at relatively high grades. It is easier to make gneiss out of granite than sedimentary or volcanic rocks because granitic rocks already have the medium-to-coarse grain size needed for us to call a rock "*gneiss*."

Most, but not all gneisses, also have some type of compositional banding, which may occur at any scale, from thick, continuous bands or layers that can be mapped across an entire field area, to discontinuous lenses and laminae that pinch out within individual outcrops, hand specimens, or thin sections (Figure 9.51*B*). Such compositional banding may be the expression of relict original bedding or volcanic layering. It can also express metamorphic segregation and redistribution of dark and light-colored components of the rock, producing bands in a rock that began with no layering. Gniessic banding can be accentuated by lenticular quartz veins formed via the redistribution and precipitation of silica by hot fluids. In many gneisses, the light-colored layers are not produced by metamorphic segregation but instead are thin magmatic intrusions (sills and dikes) of granite and pegmatite. In high-grade rocks, the associated thin layers and pods of granite and pegmatite can become so abundant that we call the rocks **migmatite**, signifying a swirled and mixed rock composed of metamorphic and igneous material (Figure 9.51*C*). The complex appearance of many gneisses and migmatites is due to the fact that the rock has compositional banding from all these origins (original layering, metamorphic segregation, hydrothermal precipitation, and igneous injections).

Foliation in typical metamorphic rocks is also expressed through preferentially oriented "*odds and ends*." For example, strongly flattened and metamorphosed conglomerates may contain pancake-shaped pebbles all in parallel, planar alignment, thus defining a conspicuous **flattened-pebble foliation**

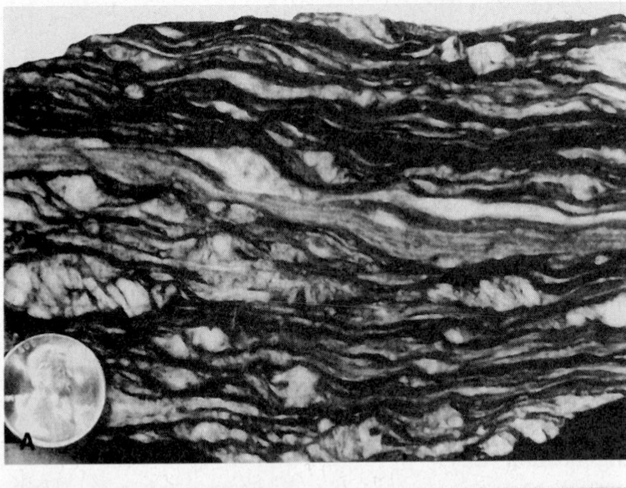

Figure 9.51 (*A*) Mylonitic granitic gneiss, with gneissic foliation defined by feldspar porphyroclasts (white) and "*ribbons*" of quartz with mica. [Photograph by Carol Simpson.] (*B*) Precambrian banded granitic gneiss of very high metamorphic grade (granulite), Chuckwalla Mountains, California. (*C*) Migmatite zone near the southwestern boundary of the Cala Serena shear zone, Cap de Creus, Spain. Pegmatite (white) formed through partial melting during deformation, and thus it is sheared like the surrounding schists. [Photograph by G. H. Davis.]

(Figure 9.52). Strongly flattened and metamorphosed volcanic rocks may contain flattened lapilli inclusions, all in parallel, planar alignment, thus expressing foliation (see Figure 9.22). The flattening that accompanies the formation of most foliation can cause stiff compositional layers surrounded by softer layers to neck and pull apart into **boudins**, sausage-shaped structures whose interesting forms can accentuate the appearance of the foliation (Figure 9.53).

Figure 9.52 Flattened quartzite pebbles define a foliation in highly deformed Barnes Conglomerate (Precambrian), Tortolita Mountains, Arizona. [Photograph by G. H. Davis.]

Figure 9.53 Boudins in layered gneiss, Pemaquid point, Maine. [Photograph by S. J. Reynolds.]

Foliations in Mylonitic Rocks

A structurally important family of foliated rocks are the strongly deformed rocks of the **mylonite series** (Table 9.1). Mylonitic rocks are characterized by a grain size has been reduced as a result of intense shearing (Figure 9.54). The grain-size reduction is a result of ductile or mixed brittle–ductile deformation mechanisms, especially dislocation creep, dynamic recrystallization, and fracturing of brittle grains. The formation of mylonitic rocks, called **mylonitization**, commonly occurs within the high strain environment of ductile shear zones (*Chapter 10, Shear Zones and Progressive Deformation*).

Most mylonitic rocks have a distinctive, somewhat platy appearance because of a well-developed foliation. The foliation, which has been likened to the flattening fabric seen in welded ash flow tuffs, is defined by a planar-parallel arrangement of flattened grains, mineral aggregates, broken mineral grains, and small shear surfaces. Mylonitic foliation commonly contains very lenticular, individual crystals, termed **ribbons** (see Figure 9.51A). Foliation can be fairly planar in monomineralic rocks, like mylonitic quartzite, but more commonly has a distinctly lensoidal or anastomosing aspect, especially in rock containing both quartz and feldspar. In part, the lensoidal foliation reflects how different minerals responded to mylonitization. If a rock contains more than one mineral, ductile strain may be accommodated preferentially by the mineral that is weakest. This can result in lenses or ribbons of the weaker mineral wrapped across less deformed crystals of the stronger one. Such an appearance is common in mylonitized granitic rocks, where strongly flattened quartz ribbons drape less deformed feldspars. In such rocks, quartz typically looks as if it has flowed like butter, whereas the feldspars have behaved brittlely, breaking into chips, which may become oriented within the foliation

TABLE 9.1
Mylonite series rocks

Rock Name	*Matrix Grain Size*	*Percentage Matrix*
Protomylonite	>50 μm	<50%
Mylonite	<50 μm	50%–90%
Ultramylonite	<10 μm	>90%

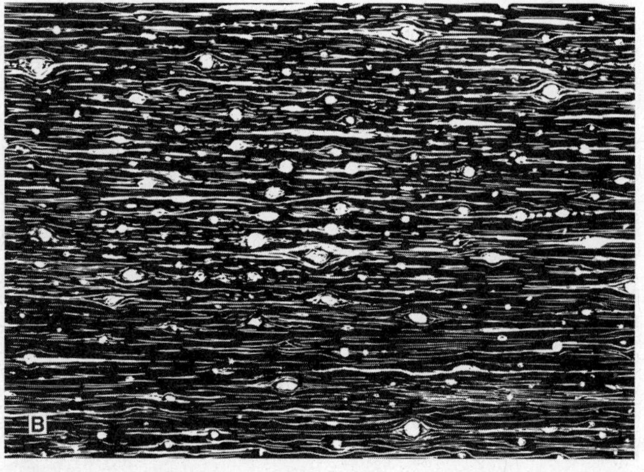

Figure 9.54 (A) The microscopic texture of mylonite. The white represents fragmented grains (porphyroclasts) and breccia streaks. The black represents plastically deformed quartz. (B) The microscopic texture of ultramylonite. [From Higgins (1971). Courtesy of United States Geological Survey.]

(see Figure 9.51*A*). Through the process of mylonitization, a coarse-grained protolith, such as granite, can be transformed into an ultra-fine-grained, laminated rock.

The nomenclature of mylonitic rocks is based mainly on the grain size and the proportion of fine-grained matrix to larger relict grains (see Table 9.1). In essence, the nomenclature scheme establishes arbitrary boundaries within the progression from undeformed protolith to an extremely deformed and uniformly fine-grained mylonitic rock. The initial stages of mylonitization produce a weakly to moderately mylonitic rock, called a **protomylonite**. Protomylonite contains less than 50 percent fine-grained matrix (i.e., is at least 50 percent relict grains), and in thin section commonly displays **mortar texture**, where small recrystallized grains surround larger relict crystals. Foliation ranges from very subtle to well defined and is defined by flattened, inequant grains and locally by very thin, nonpenetrative zones of high shear strain. With further deformation and grain-size reduction, a protomylonite is converted into a **mylonite**, which contains 50 to 90 percent matrix. Most mylonites are very strongly foliated and lineated, and contain porphyroclasts scattered throughout the fine-grained, laminated matrix. Mylonitization culminates in the formation of an **ultramylonite**, a more thoroughly deformed and fine-grained rock containing more than 90 percent matrix and less than 10 percent relict grains. **Ultramylonites** are mylonites taken to the edge of recognition.

Under certain conditions, shearing results in rocks that are intermediate in character between mylonitic rocks and more-common metamorphic rocks, such as gneiss and schist. A rock intermediate in character between a mylonite and a gneiss is a **mylonitic gneiss**. A rock that is transitional between a mylonite and a schist is a **mylonitic schist** or, if finer grained, a **phyllonite**.

Foliation in mylonitic rocks is also commonly defined by lenticular grains or clasts called **augen** or **porphyroclasts** (Figure 9.55). Augen are commonly relics from the protolith, being relatively strong minerals that resisted mylonitization while surrounding grains were progressively converted into the fine-grained matrix. Many mylonitized granites, for example, contain lenticular feldspar augen floating in a fine-grained matrix of ductilely deformed or recrystallized quartz and mica. With continued mylonitization, even the resistant augen may be destroyed, leaving a fine-grained rock with few clues about its heritage.

Some foliation in mylonitic rocks is actually the expression of thin shear zones. Within these zones, the normal foliation of the mylonitic rock is smeared

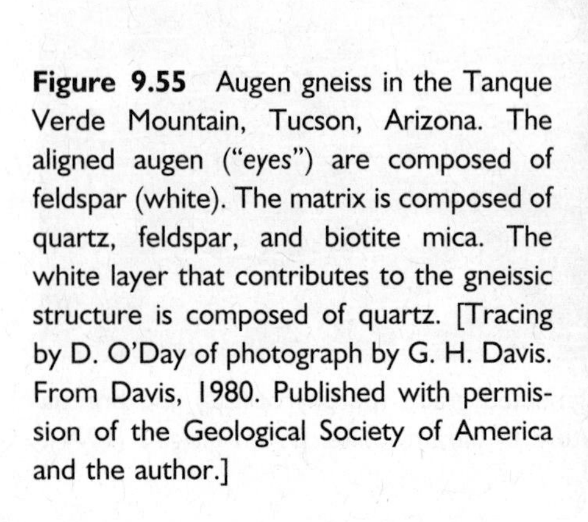

Figure 9.55 Augen gneiss in the Tanque Verde Mountain, Tucson, Arizona. The aligned augen ("eyes") are composed of feldspar (white). The matrix is composed of quartz, feldspar, and biotite mica. The white layer that contributes to the gneissic structure is composed of quartz. [Tracing by D. O'Day of photograph by G. H. Davis. From Davis, 1980. Published with permission of the Geological Society of America and the author.]

out into ultrathin laminations. These zones may be at an acute angle to the main foliation, imparting a lensoidal aspect to the overall rock (see *Chapter 10, Shear Zones and Progressive Deformation*).

The wide range of appearances of foliations in typical metamorphic rocks and in mylonitic rocks reflects differences in the composition, mineralogy, and texture of the original protolith from which the metamorphic rock was derived; differences in the temperature–pressure environments of metamorphism and deformation; and differences in strain. Each of these factors influences the appearance of deformed metamorphic rocks, and such knowledge helps us to better interpret the protolith and history of metamorphic rocks we encounter in the field.

LINEATION

Definition and Expression

The definition of lineation, like the definition of foliation, is loaded with code words. **Lineation** is the subparallel to parallel alignment of elongate, linear fabric elements in a rock body, commonly penetrative at the outcrop and/or hand specimen scales of observation (Figure 9.56), and commonly at the microscopic scale as well. Some lineations are so penetrative that the lineated rock looks like driftwood with a pronounced etched grain. Other lineations are expressed in the form of such large aligned parallel elements that it is best to refer to the lineation as **linear structure**. Just like foliation, lineation has innumerable physical expressions and occurs in metamorphic, igneous, and sedimentary rocks (Cloos, 1946; Turner and Weiss, 1963; Weiss, 1972).

Some lineation and linear structure is **primary**, produced in flowing lava and other magma or in sediments *prior to lithification*. **Ropy lava** (**pahoehoe**) is an example of primary linear structure (Figure 9.57*A*). It forms during the flow of relatively low viscosity basalt when local lava currents drag the plastic skin of the flow, contorting it into a series of nested arcs, which tend to be convex in the sense of current flow. Just as volcanic and plutonic rocks can have flow foliation, they can have flow lineation defined by aligned phenocrysts, stretched vesicles,

Figure 9.56 (*A*) India-ink rendering of lineation on foliation surface in mylonitic gneiss in the Santa Catalina Mountains, Tucson, Arizona. [Tracing by D. O'Day of photograph by G. H. Davis. From Davis, 1980. Published with permission of the Geological Society of America and the author.] (*B*) Penetrative mineral lineation in strongly deformed quartzite from the Coyote Mountains, southeastern Arizona. Collection of G. H. Davis. [Photography and editing by Gary Mackender. vr.arizona.edu, glm@email.arizona.edu, © 2009–10 Arizona Board of Regents.]

Figure 9.57 Two examples of *primary* lineation in igneous rocks. (*A*) Ropy lava (pahoehoe) in a basalt flow on the island of Hawaii. Width of field of view is ~2m. [Photograph by S. J. Reynolds.] (*B*) This thing that looks like a tree trunk is actually igneous rock containing penetrative mineral lineation. The lineation formed in stiff magma ascending as if through a pipe toward the surface. Big Bend country of West Texas. Height of "*tree*" is ~2.5m. [Photograph by G. H. Davis.]

stretched pumice, and other linear objects. In the Big Bend country of West Texas, there is an outcrop (Figure 9.57*B*) that looks just like a petrified tree! It is a tiny analogue of the "*neck*" of a volcano. The grain of the "*tree*" reflects penetrative lineation that formed when magma flowed upward through a tight constricted neck toward the surface.

Sediments can have primary lineation too. **Parting lineation** is a subtle primary structure that commonly occurs in siltstone and sandstone. Expressed as a faint linear grain on bedding surfaces, it records the current direction at the time of deposition of the sand and silt (Figure 9.58). The physical expression of parting lineation is due to the subparallel alignment of the longest dimensions of silt or sand grains within bedding laminae. The grains are aligned in the direction of paleocurrent (Conybeare and Crook, 1968).

Primary lineation and linear structure can provide a sense of paleocurrent in an ancient depositional basin, as well as flow directions in ancient volcanic fields and

Figure 9.58 Parting lineation in the plane of cross-stratification in the Navajo Sandstone, Zion National Park. Knife is aligned parallel to the lineation. The parting lineation records the wind direction during an instant of time in the Jurassic when the Navajo sand was being formed in a great desert dune field. [Photograph by G. H. Davis.]

magma ascent patterns in ancient igneous intrusions. Secondary lineations give us different information: the nature of strain within rocks forced to change size or shape during deformation. We will focus on the **secondary** lineation and linear structure that forms when sedimentary, igneous, or metamorphic rocks are deformed during metamorphism and/or shearing.

Telling the Difference Between Lineation and Foliation

Once when I (GHD) was doing field work in northern Norway, Arild Andresen showed me an unusually interesting outcrop of white medium-grained granite containing distinct feldspar crystals. In one part of the outcrop the feldspar was distributed and arranged in a way that had resulted in foliation. In another part of the same outcrop the feldspar was arranged as a lineation.

I collected two samples, one of foliated granite, the other of lineated granite. At first glance the samples looked identical. They both appeared to be foliated. However, when I turned each specimen through my fingers, I saw an important difference. In the first sample, the feldspars were arranged in crude planes, not lines, thus defining foliation (Figure 9.59*A*). (I was able to measure the strike and dip of this foliation in outcrop.) In the second sample, the feldspars are arranged in lines, not planes, thus defining lineation (Figure 9.59*B*), even though the flanks of this specimen look *just like* the trace of foliation. Thus, it was only when I looked at the ends of the specimens that I recognized that the feldspar crystals are aligned like lines, intersecting the top and bottom surfaces of the sample as black equant objects. (I was able to measure the trend and plunge of this lineation in outcrop.)

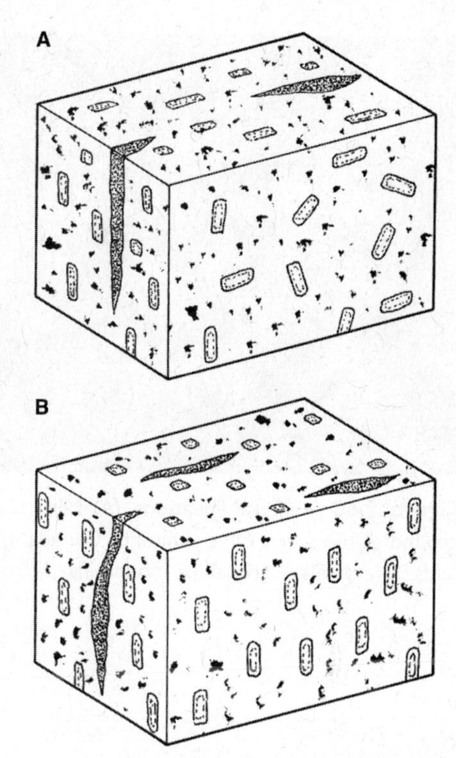

Figure 9.59 These feldspar drawings by Evans B. Mayo illustrate the fabrics I saw in a granite outcrop in Norway. Both rocks appear to be foliated, yet one is and one is not. (*A*) This drawing shows a granite that contains foliation defined by feldspars, but no lineation. (*B*) This drawing shows a granite that contains lineation defined by feldspars, but no foliation.

Types of Lineation

The dominant types of lineation are mineral lineation, intersection lineation, and crenulation lineation. Each type represents a different attribute in a deformed rock, and each has a different structural significance.

Mineral Lineation

A **mineral lineation** is the expression of aligned minerals or mineral aggregates in a rock, generally occurring within the foliation plane in a metamorphic rock. Mineral lineation can be expressed in a number of ways. The classic example of mineral lineation is one where the lineation is defined by *aligned inequant mineral grains,* commonly of prismatic minerals such as hornblende and aluminosilicate minerals (e.g., andalusite, kyanite, and sillimanite). This type of mineral lineation results when grains become aligned in a metamorphic rock by growing or recrystallizing in a preferred direction or by rotating into a preferred orientation during deformation (Figure 9.60*A*). Minerals also can become elongated by pressure dissolution in one or more directions accompanied by precipitation in another direction (see *Chapter 4, Deformation Mechanisms and Microstructures*). Mineral grains in metamorphic rocks can be flanked by beardlike pressure shadows with associated crystal fibers growing in a preferred orientation from the tips of the sheltering mineral grain.

Another type of mineral lineation (Figure 9.60*B*) is defined by *elongated mineral aggregates,* like an elongated collection of quartz crystals that together define a long, stretched pebble of quartz. Another example would be elongated quartz-mica aggregates that, prior to metamorphism and deformation, were pumice fragments in a pyroclastic rock. Lineation defined by deformed preexisting objects is commonly called a **stretching lineation**, especially if the long axis of the elongated objects is interpreted to be parallel to the maximum stretch (S_1). Some geologists avoid the term "*stretching lineation*" because of its interpretive nature.

Figure 9.60 Mineral lineations. (*A*) Mineral lineation defined by aligned light-colored crystals of the mineral sillimanite, underground at the Geco Mine, Ontario, Canada. (*B*) Mineral lineation defined by deformed mineral aggregates, in this case, stretched feldspar crystals, Harquahala Mountains, Arizona. [Photographs by S. J. Reynolds.]

Another type of mineral lineation, called **mylonitic lineation**, is inherent to mylonitic rocks and ductile to brittle-ductile shear zones. Mylonitic lineation has a streaky appearance (Figure 9.61), with relatively long (more than several cm), aligned streaks and lenses of minerals and mineral aggregates. These streaks and lenses can reflect an alignment of some combination of the following: (1) elongated grains, whether they represent newly grown or recrystallized crystals (i.e., a typical mineral lineation); (2) deformed crystals, pebbles, and other objects; (3) deformed remnants of original features in the rock, like gneissic layering; (4) mineral fibers that crystallized or precipitated parallel to lineation; and (5) bits of broken grains arranged in lines parallel to mineral lineation. These broken bits formed when more rigid grains were fractured and disaggregated (comminuted) during deformation, and then smeared out by continued shearing. Mylonitic lineation is commonly a hybrid of these different types of linear features, and the dominant expression of the lineation can change with time as rocks become more sheared (Sengupta and Ghosh, 2007).

The relative roles of ductile strain of preexisting objects, crystal growth, and deformation-caused reduction in grain size (whether by brittle comminution or ductile recrystallization) are a function of the mineralogy of the rock, the size and arrangement of grains, and the temperature, pressure, strain rate, and fluid conditions during deformation. As a result of this competition between recrystallization and comminution, mylonitic lineation commonly resembles slickenlines (see Figure 9.61). But unlike slickenlines, the lineation is not restricted to a single surface or a thin zone of faulting or shearing. Instead it pervades a substantial part of the body of the mylonitic rock. Where initially equant objects, such as pebbles or inclusions, were present in the rocks prior to mylonitization, they are elongated parallel to lineation and flattened into parallelism with foliation. In other words, lineation in most mylonitic rocks is a stretching lineation parallel to S_1.

Figure 9.61 Streaky mineral lineation on the plane of foliation in mylonite gneiss in Tanque Verde Mountain, Tucson, Arizona. [Photograph by G. H. Davis.]

Intersection Lineation

A common lineation in slates and more highly metamorphosed rocks is **intersection lineation**, which consists of geometric lines created by the intersection of two sets of planes (Figure 9.62). Intersection lineation can be

Figure 9.62 Intersection of bedding and cleavage produce here an 'intersection lineation' on the flat cleavage plane in slate, Selkirk Mountains, British Columbia, Canada. [Photograph by S. J. Reynolds.]

the intersection of foliation with bedding or any other compositional layering, or the intersection of two (or more) foliations. As we will soon learn, multiple foliations are a characteristic of strongly deformed metamorphic rocks. The intersection of two closely spaced, penetrative, parallel foliations can result in well-developed lineation, a linear grain if you will, marked in outcrop by a myriad of closely spaced subparallel-to-parallel lines. Intersection lineations can be exceptionally long, many meters in large outcrops. In practice, finding a very long lineation in the field should always raise the suspicion that the feature is an intersection lineation. A similarly long and linear appearance results from the intersection of foliation or compositional layering with the outcrop surface, but this is not a true lineation, nor does it have much structural significance (Passchier and Trouw, 2005).

When intersection lineation is identified as such, it leads quite naturally to the recognition of two or more intersecting planar fabrics. The orientation of intersection lineation is the trend and plunge of the intersection of the two planes; this can be confirmed stereographically using the strike and dip of each plane. Intersection lineation, unlike some other types of lineation, may not reflect the axes of the strain ellipse since it is simply the intersection of two planes, one of which existed before some of the deformation. A lineation marking the intersection of bedding and cleavage, however, is commonly parallel to local fold hinges, because cleavage is parallel to axial surfaces, and a fold hinge marks the intersection of the axial surface with bedding (Figure 9.63). Tough to describe, easier to visualize!

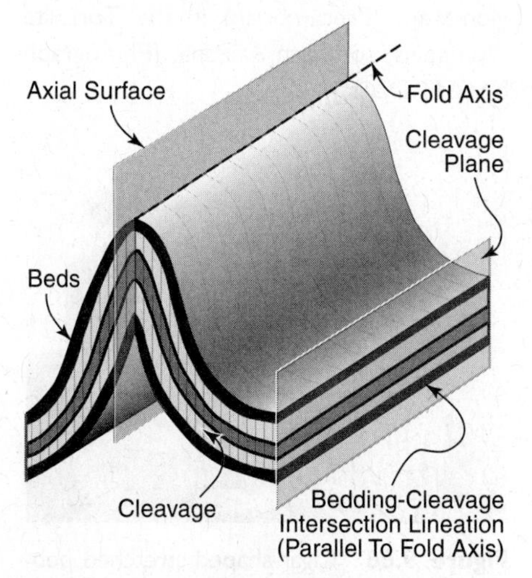

Figure 9.63 Intersection lineation, defined by bedding-cleavage intersections parallel to local fold axes.

Crenulation Lineation

Crenulation lineation is a lineation expressed in the form of bundles of tiny, closely spaced fold hinges (Figure 9.64). Crenulation lineation is especially well developed in phyllites and schists that have been repeatedly deformed by folding. Viewed in the plane of cleavage, crenulation lineation is an array of straight to slightly curved, discontinuous **crests** and **troughs** of folds, more folds than anyone could ever hope to measure, even for "*fold junkies*" like us.

Crenulation lineation is especially pronounced in mica schists, where the conspicuousness of the structure is enhanced by contrasts in light reflected from the variously oriented micaceous surfaces. It is not surprising that phyllites and schists so characteristically display crenulation lineation. Phyllitic structure and schistosity impart to rocks an **anisotropy** that makes

Figure 9.64 Crenulation lineation, defined by crests and troughs of minor folds in quartz-sericite schist. [Photograph by G. Kew.]

the rock highly vulnerable to kink folding in the presence of stresses that are approximately layer parallel. Mica-rich phyllites and schists readily crinkle, almost like paper.

Types of Linear Structure

Linear structures are relatively large compared to mineral lineations or have attributes that otherwise distinguish them from typical lineations. The most common examples of linear structure are stretched-pebble conglomerate, rodding, mullion, pencil structure, and boudins.

Figure 9.65 Elongate flattened pebbles in outcrop of strongly deformed Barnes Conglomerate (Precambrian) in the Tortolita Mountains, southern Arizona. [Photograph by G. H. Davis.]

Stretched-Pebble Conglomerate

Stretched-pebble conglomerate is composed of closely packed elongate clasts (Figure 9.65), a linear structure fashioned through the distortion of cobbles and pebbles, and in some cases boulders. Many different stretched-pebble shapes are possible: cigar-shaped pebbles (Figure 9.66) and pebbles shaped like tongue depressors (*ahhhh*, what pebbles!) are especially fun to collect. The expression of stretched-pebble linear structure is commonly enhanced by the presence of mineral lineation in the rock matrix and by the development of crystal fiber beards emanating from the tips of the stretched clasts. The longest stretched pebble I [GHD] ever extracted from an outcrop measured 20 in. (50 cm). Longer ones are common, but they are difficult to remove from outcrop in one piece.

Rodding is a linear outcrop-scale structure that is defined by a penetrative array of parallel, highly elongate bodies of rock (Figure 9.67) (Wilson, 1961). Some are made simply of milky or icy quartz. Rods typically vary in size from flashlights to the length of an arm, although some resemble long tree trunks. Viewed end-on, they are circular, oblate, or lensoidal.

Figure 9.66 Cigar-shaped stretched pebble, enjoyed by Stan Keith in the Tortolita Mountains, southern Arizona. [Photograph by G. H. Davis.]

Quartz rods and stretched quartz-pebble conglomerate clasts can resemble each other to a remarkable degree. Quartz rods can be formed in a number of ways. Some may be the boudined, necked expressions of once-continuous layers of quartz. Some may be thought of as the linear equivalent of veins, products of open-space filling and/or replacement, not along fractures but rather along the hinge zones of penetrative folds. Quartz in such rods may in part represent the reprecipitation of quartz made available by local pressure solution of the same host rock in which the rods were found.

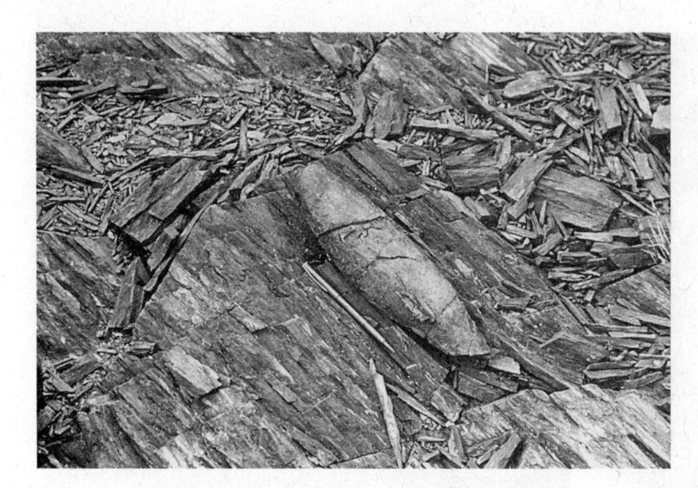

Figure 9.67 This rod occurs in a strongly deformed outcrop of Devonian Excellsior Phyllite in the central Andes of Peru. The landscape within this general outcrop area was strewn with rods, many of them streamlined. [Photograph by G. H. Davis.]

Mullion

The term **mullion** reminds us again of the kinship of structural geology and architecture. Architecturally, mullions are the long, vertical stone members that separate adjacent window openings of Gothic churches (Figure 9.68*A*) (Hobbs et al., 1976). The mullion face that projects into open air is convex outward. If the mullions were laid out and aligned on the ground in the manner shown in Figure 9.68*B*, the array of stone would look exactly like mullion structure in rock.

Viewed in outcrop (Figure 9.69*A*), mullion structure displays regular, repeated, fold-like forms, ranging in wavelength from centimeters to meters. The cuspate–lobate forms are very distinctive, consisting of linked circular or elliptical arcs. They bear a likeness to oscillation ripple marks but are much more linear and systematic, and they are usually larger. Mullions are not composed of newly introduced minerals like quartz, but rather are always fashioned from the host rock itself.

Perhaps the most important descriptive relationship bearing on mullion structure is its occurrence along the **interface** between a mechanically soft and a mechanically stiff layer—slate and quartzite, respectively, for example. The fold-like mullion forms are convex in the direction of the mechanically soft layer, with the pinched, cuspate foldlike forms pointing toward the mechanically stiff layer (Figure 9.69*B*). Mullion structures arise from **buckling instability** produced by layer-parallel shortening of a contact separating two rock layers of contrasting mechanical strength. Verification of this can be achieved in the laboratory by subjecting two-layer models to strong layer-parallel shortening (Ramsay, 1967). Experiments of this type demonstrate that **dominant wavelength** of mullion structure is determined by the viscosity ratio of the stiff and soft layers.

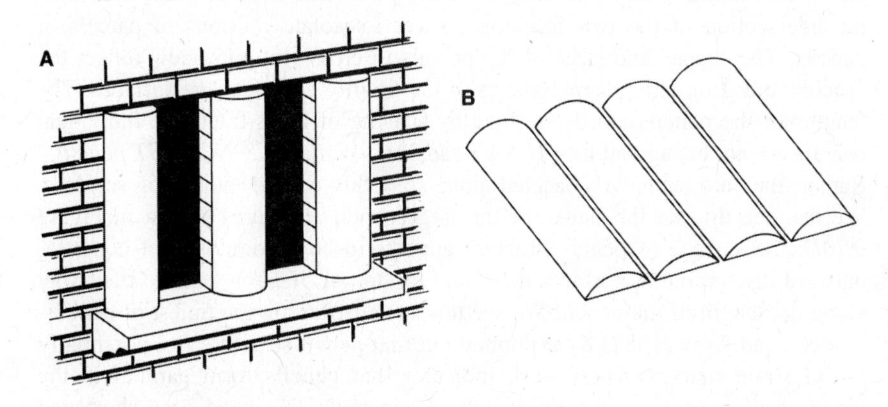

Figure 9.68 Architectural mullions (*A*) adorning a Gothic church and (*B*) lined up on the ground in a way resembles geologic mullions.

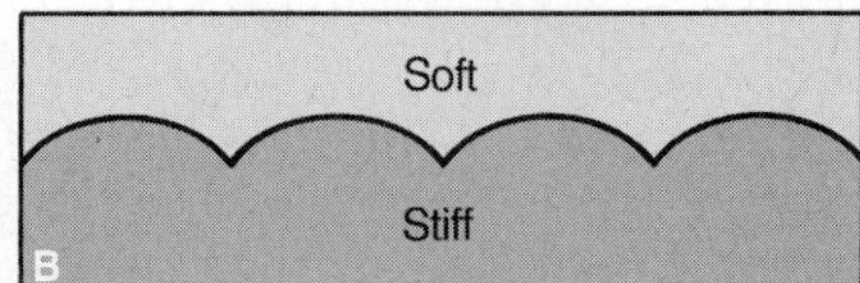

Figure 9.69 (*A*) Mullion structure, formed by cuspate–lobate folds along the interface between quartzite and slate, Barnhardt Canyon, Mazatzal Mountains, Arizona. [Photograph by S. J. Reynolds.] (*B*) Mullions form preferentially at the interface between mechanically soft vs. mechanically stiff rocks. Buckling instability due to layer parallel compression produces the cuspate–lobate pattern. (C) Example of cuspate-lobate pattern at contacts between pegmatite (white) and schist (dark). Pegmatite is stiffer. [Photograph by G. H. Davis.]

Pencil Structure

Pencil structure is a very distinctive linear structure associated with folded and cleaved mudstones and siltstones. Outcrops pervaded by pencil structure are strewn with unnatural-looking "*pencils*" of rock (Figure 9.70*A*). Tiny ones are more like short toothpicks. Larger ones are like magic wands (Figure 9.70*B*). After publication of the first edition of this book, Larry Rogers sent me a picture of himself striking an heroic pose while standing on cleaved rock in Spitsbergen, Norway, supporting part of his weight on a "*walking stick*" pencil structure 1.5 m long (Figure 9.70*C*). He claimed the new record, but it did not last. Shortly after the publication of our second edition, another geologist (whose email with name I have lost!) sent me (GHD) photographs of pencil structure in Tibet used for roof support (Figure 9.70*D*). These are not likely to be surpassed in length and quality, but we never know.

Pencil structure generally occurs in folded strata. In fact, the orientation of pencil structure is dependably subparallel with the axes of associated folds. The actual physical expression of most pencil structure is formed by the intersection of **bedding fissility** and cleavage (Reks and Gray, 1982). In folded sedimentary layers, bedding-parallel fissility can be from original sedimentation, compaction, layer-parallel shear, or most likely some combination of these. Where layers containing bedding fissility are cut by penetrative axial plane cleavage, the intersection of the two foliations serves to isolate millions of parcels of pencils. The shapes and sizes of the pencils in cross-sectional view reflect the spacing and geometric characteristics of the fissility and cleavage surfaces. The lengths of the pencils are determined by spacing of cross-fractures. Individual pencils are not hexagonal like "*USA Readibond Wallace CONQUEST pencils.*" Rather, they are irregularly faceted along smoothly curved interfering surfaces. We can see this on the flanks of the large pencil structures shown in Figure 9.70*D*: the surface of pencil structure appears to be a composite of concave-outward discontinuous surfaces that seem to "*torque*" and/or "*spiral*" their way along the length of the pencil. We see this same pattern in the tiniest of pencils.

Reks and Gray (1982) have pointed out that pencil structure is a potentially useful strain marker. Their work indicates that pencils form parallel to the direction of intermediate finite stretch (S_2) in rocks that have been shortened

Figure 9.70 Pencil structure. (*A*) Outcrop of pencil structure in fine-grained calcareous siltstone at Agua Verde near Tucson, Arizona. (*B*) A former record-breaking pencil structure, proudly displayed by Ralph Rogers, the former record holder. The pencil was extracted from a wonderful display in Devonian phyllite in central Peru. [Photographs by G. H. Davis.] (*C*) Pencil structure being used as a walking stick by its discoverer, Larry Rogers (no relation to Ralph), who claimed the new record for the 2nd edition of this text. Location is on the slopes of Spitsbergen. This outcrop of pencil structure occurs in a graded mud/sandstone of Proterozoic age from the Hecla Hock succession. [Photograph by friend of Larry Rogers.] (*D*) The new record in the form of pencils that are used as beams in housing construction projects in Tibet. This is the new record. Would the geologist who sent me (GHD) this photograph please write me again so that we can properly acknowledge you in the 2nd printing of this 3rd edition.

by 9–26%. It is within this strain range that bedding fissility and axial plane cleavage are equally well developed.

Some pencil structures are the result of two *cleavages* that intersect at fairly high angles. Specifically, a second cleavage, typically a crenulation cleavage, cross-cuts an earlier cleavage. If you encounter pencil structures in the field, it is a good idea to investigate whether there is more than one cleavage.

Boudin [Fr., "Sausage"]

A final linear structure of significant importance is one we have already encountered—the **boudin** (Figure 9.71). Boudins form as a response to layer-parallel extension (and/or layer-perpendicular flattening) of stiff layers enveloped top and bottom by mechanically soft layers. The way in which the stiff layer stretches depends mainly on the ductility contrast of the participating layers and on the magnitude of the stretches (S_1, S_2, S_3) describing the level of strain parallel and perpendicular to the layering (Ramsay, 1967). Boudinage is most common in highly deformed sequences with interlayered rock types of different strengths (Figure 9.72).

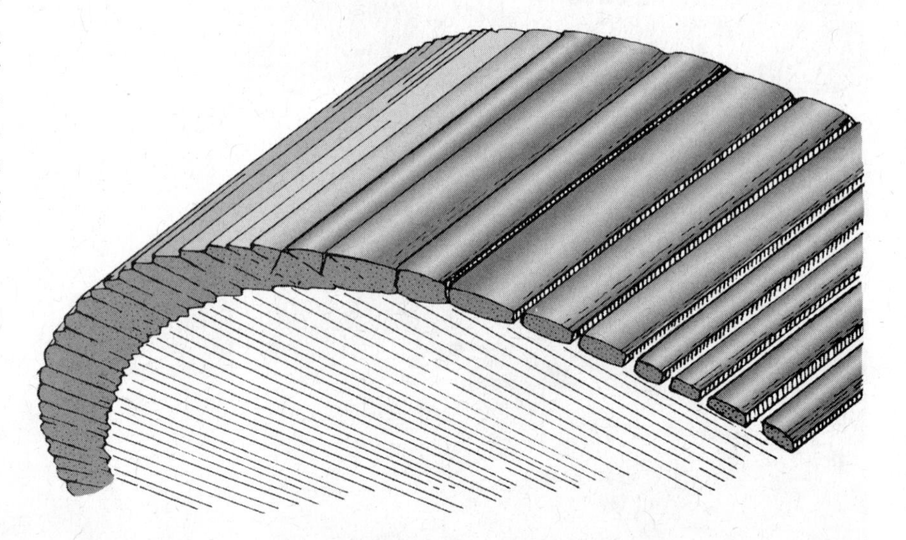

Figure 9.71 Boudins on the flank of a fold. Note how boudinage produces linear structure. [From *Introduction to Small-Scale Geological Structures* by G. Wilson. Published with permission of George Allen & Unwin (Publishers) Ltd., London, copyright © 1982.]

Figure 9.72 (*A*) The relatively stiff, dark mafic igneous rock in northern Norway was flattened and stretched to the point that it "*necked*," creating boudins. Surrounding rock is a mechanically softer, highly foliated metasedimentary/metavolcanic sequence. (*B*) Asymmetric boudins bounded by normal faults. Clot de Vellosa locale, Cap de Creus, Spain. [Photographs by G. H. Davis.] (*C*) Geologists love boudinage, in this case in pegmatite (white). Cap de Creus, Spain. [Photograph by G. H. Davis.]

Boudinage of the strong layer may occur via tension fracturing, shear fracturing, ductile necking (pinch-and-swell structures), or some combination, depending on the ductility contrast between the boudinaged layer and its matrix. Once formed, individual boudins may behave rigidly or may deform somewhat along with the matrix, and this behavior influences the final shape of the boudin (Ghosh, 1993).

Rectangular and **rhombic** boudins (Figure 9.73) form where there is a large ductility contrast between a rigid boudinaged layer and the adjacent rocks. Rectangular boudins separate via tension fractures, whereas rhombic boudins form by shear fracturing or by tension fracturing followed by rotation and deformation of individual boudins. Openings between boudins of both types are filled by plastic flow of the enveloping soft layers or by infillings of "*vein*" material, such as calcite and quartz. Rectangular and rhombic boudins, because they develop from layers that are very strong compared to the flowing matrix, generally remain relatively rigid and unstrained after separation.

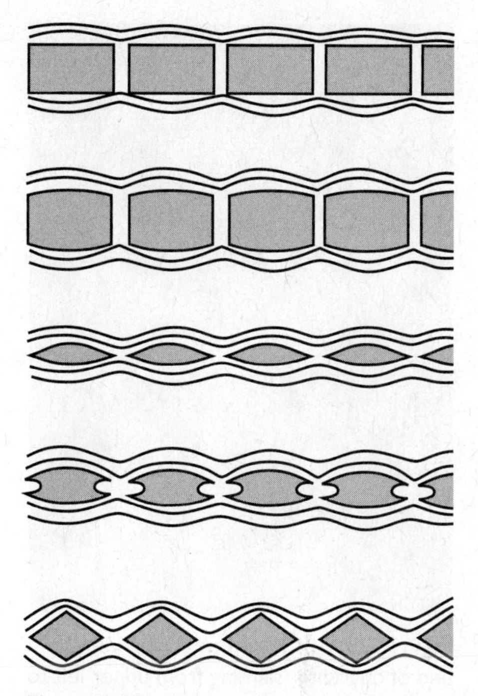

Figure 9.73 Some shapes of boudins. [Reprinted from *Structural Geology: Fundamentals and Modern Developments*, S. K. Ghosh. Published with permission of Pergamon Press, Ltd., Oxford, © 1993.]

Where the strength contrast is moderate, the stiff layer will tend to deform into boudins whose cross-sectional forms are stubby lenses or barrels. **Barrel-shaped** boudins reflect some amount of ductile necking of the stiff layer prior to tension fracture, and they tend to deform somewhat after separation (see Figure 9.73). **Fish-head** boudins begin as rectangular or barrel-shaped boudins, but are strongly deformed after separation, causing the lateral walls of the boudin to collapse inward. Some material that flows into the opening between boudins is trapped in the mouth of the fish (see Figure 9.73). Where strength contrast is very small, boudinage occurs via complete ductile necking of the relatively stiff layer, without the involvement of tension fractures. This forms simple **pinch-and-swell** structure and **lenticular** boudins (see Figure 9.73). Once formed, lenticular boudins may deform similarly to the matrix material.

Although their shape is quite variable, boudins tend to have consistent aspect ratios, being commonly one to four times longer than they are wide when viewed on surfaces cut parallel to lineation and perpendicular to foliation. Boudins longer than this tend to be broken again by the shear stresses generated as the surrounding matrix flows past the top and bottom of the boudin. Once the boudin has become shorter than about twice its thickness, the shear stresses are insufficient to break it again (Ghosh, 1993).

Irrespective of the profile forms of boudins, the process of boudinage creates in rock a linear structure in the plane of layering (see Figure 9.71). The linear structure produced by pinch and swell is an array of parallel-to-subparallel furrows. The linear structure produced by true boudins, in which stiff layers are repeatedly disrupted, consists of a subparallel array of stripes that reflect the presence and structural configurations of the stiff and soft layers. Where the stiff layer is extended in two directions parallel to the layering, the boudins are more equant or blocky, as viewed on the foliation surface. This is called **chocolate tablet** boudinage and commonly looks like rows of lumps, like a colony of sleeping fur seals.

DESCRIPTIVE/GEOMETRIC ANALYSIS OF FOLIATION AND LINEATION

The Problem

Foliated, lineated rocks present some special challenges in geologic mapping and detailed structural analysis. A wide variety of physical and geometric forms can give rise to foliation(s) and lineation(s), and we must be prepared to identify and describe them and to measure their orientations. Furthermore, some rocks possess multiple, **superposed** foliations and lineations. Unscrambling the interrelationships of these is sometimes geometrically very difficult. Some rocks display a half-dozen foliation and lineation elements in a single exposure!

Figure 9.74 Close-up outcrop view of a bed of quartzite (slanting from upper left to lower right) cut by near-vertical cleavage. Bedding is encoded as S_0; cleavage as S_1. From a road cut in folded and cleaved Precambrian rocks, Negaunee, Michigan. [Photograph by S. J. Reynolds.]

The Coding System

Foliations

There are established methods for sorting and classifying foliations and lineations in the course of mapping and analyzing metamorphic rocks (Turner and Weiss, 1963). Foliations are coded with the letter S, meaning "*planar surface,*" like cleavage or schistosity. Each S-surface is subscripted (e.g., S_1, S_2 S_3) according to the apparent relative order of formation within the rock, with numbers increasing from earlier fabrics to later fabrics. For example, close examination of an outcrop reveals that bedding is cut by a cleavage (Figure 9.74). Bedding is entered into the field notebook as S_0 (Figure 9.75). Cleavage is entered as S_1, not to be confused with the direction of greatest finite stretch (S_1). (This distinction is always obvious in context.) If the cleavage is then cut by crenulation cleavage (Figure 9.76), the

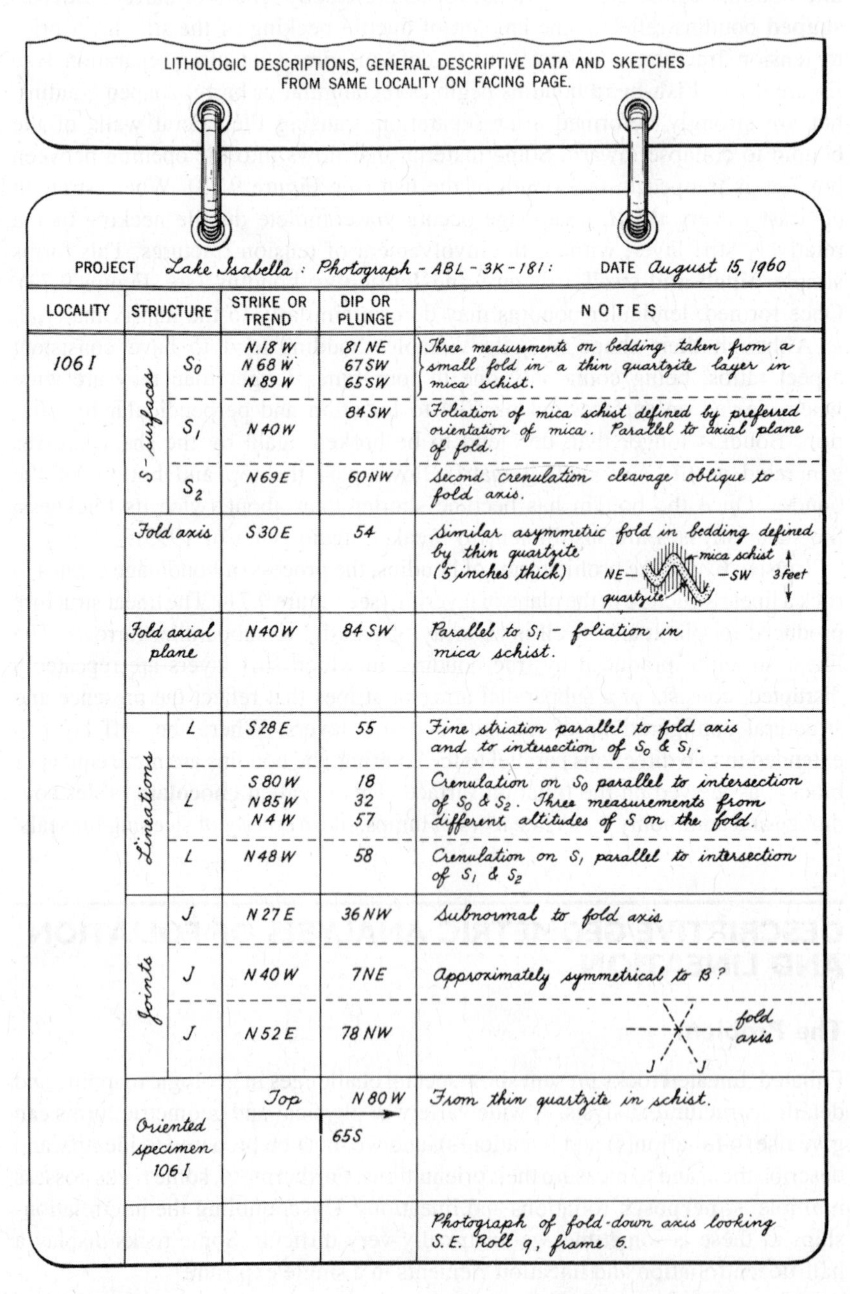

LITHOLOGIC DESCRIPTIONS, GENERAL DESCRIPTIVE DATA AND SKETCHES FROM SAME LOCALITY ON FACING PAGE.

PROJECT *Lake Isabella : Photograph - ABL - 3K - 181:* DATE *August 15, 1960*

LOCALITY	STRUCTURE		STRIKE OR TREND	DIP OR PLUNGE	NOTES
106 I	S-surfaces	S_0	N.18 W N 68 W N 89 W	81 NE 67 SW 65 SW	Three measurements on bedding taken from a small fold in a thin quartzite layer in mica schist.
		S_1	N 40 W	84 SW	Foliation of mica schist defined by preferred orientation of mica. Parallel to axial plane of fold.
		S_2	N 69 E	60 NW	Second crenulation cleavage oblique to fold axis.
	Fold axis		S 30 E	54	Similar asymmetric fold in bedding defined by thin quartzite (5 inches thick) NE– mica schist quartzite –SW 3 feet
	Fold axial plane		N 40 W	84 SW	Parallel to S_1 – foliation in mica schist.
	Lineations	L	S 28 E	55	Fine striation parallel to fold axis and to intersection of S_0 & S_1.
		L	S 80 W N 85 W N 4 W	18 32 57	Crenulation on S_0 parallel to intersection of S_0 & S_2. Three measurements from different altitudes of S on the fold.
		L	N 48 W	58	Crenulation on S_1 parallel to intersection of S_1 & S_2
	Joints	J	N 27 E	36 NW	Subnormal to fold axis
		J	N 40 W	7 NE	Approximately symmetrical to B?
		J	N 52 E	78 NW	fold axis J J
	Oriented specimen 106 I		Top	N 80 W 65 S	From thin quartzite in schist.
					Photograph of fold-down axis looking S.E. Roll 9, frame 6.

Figure 9.75 Field notebook entries showing the record keeping of foliation, lineation, and folding. [Modified from *Structural Analysis of Metamorphic Tectonites* by F. J. Turner and L. E. Weiss. Published with permission of McGraw-Hill Book Company, New York, copyright © 1963.]

crenulation cleavage is entered as S_2 (see Figure 9.75). The physical properties and orientations of each foliation are described and posted before moving on to the next outcrop. If, at the next outcrop, the crenulation cleavage (S_2) is cut by yet another foliation, perhaps a spaced cleavage, the crenulation cleavage retains the status S_2 and the spaced cleavage is awarded the symbol S_3. Any type of foliation can have any number, but S_0 is reserved for original bedding.

From time to time in the course of a field investigation, new foliations are discovered that must be inserted between already "*established*" subscripted S-surfaces. Just when we feel confident that there are three, and only three, foliations within the metamorphic rock under examination, we roll back the moss on an outcrop and discover another crenulation cleavage, one that postdates S_2 but predates S_3. When this happens, the subscripts in the entire notebook must be edited according to the change. Alternatively, the newly discovered crenulation cleavage is temporarily assigned the notation S_{2a} with the understanding that eventually all the foliations must be reordered. If we determine through further field work that this new crenulation is very limited in the degree and extent of its development, we may elect to keep referring to it as S_{2a}, retaining the integers (S_1, S_2, S_3, . . .) for the other more widespread and penetrative fabrics. We might also retain the S_{2a} numbering if we conclude that the S_{2a} fabric is related to the original S_2 fabric, perhaps being a late phase of the same deformation event.

Figure 9.76 An early cleavage (S_1) is essentially horizontal and is cut by a nearly vertical crenulation cleavage (S_2). Devonian phyllites in the central Andes, Peru. Part of lens cover for scale. [Photograph by G. H. Davis.]

Lineations

Lineations are coded with the letter L, meaning lineation or linear structure. As in the case of foliations, each lineation is subscripted according to the apparent relative order of development within the rock being studied. The symbol L_0 is reserved for primary lineations or primary linear structure within the rock (i.e., the protolith) from which the metamorphic rock was derived.

Some lineations are simply intersection lineations, produced as the passive geometric product of the intersection of bedding and foliation or of two (or more) foliations. It is important to recognize intersection lineations as such, and not to confuse them with *bona fide* physical lineations, like a mineral lineation. Again, the presence of a conspicuous intersection lineation forces us to recognize the presence of the foliations whose intersection they reflect, foliations that in some cases might otherwise be subtle and hard to spot. Intersection lineations are seldom entered as part of the hierarchy of subscripted L's. The L-symbols are reserved for real, physical lineations.

Folds

Penetrative sets of folds are coded F. Since most bodies of highly deformed metamorphic rocks are marked by superposed folds, it is necessary to assign subscripts to identify the relative order of the development of folds. As in the case of the "*S's and L's*," F_0 is reserved for any folds of primary origin, such as folds produced by slumping of sediment prior to lithification. Folds produced after the original rock was formed are coded F_1, F_2, F_3, . . . , in the order in which they formed.

Within fold belts in regionally metamorphosed terranes, it is the rule, not the exception, to find at least two clearly defined fold sets in phyllite and schist. The earliest fold set (F_1) is typically composed of tight-to-isoclinal passive folds that are axial planar to the earliest formed foliation (S_1). These F_1 folds, along with S_1, are typically refolded by a set of younger folds (F_2) that is composed of tight to open flexural-slip folds, especially kink folds. Key outcrops show the interrelationship and relative timing between two (or more) fold sets in multiply deformed areas. In the field, it is essential to determine what exactly is being folded. Is bedding being folded, or is cleavage being folded? If cleavage is being folded, this requires a cleavage-forming event that preceded folding.

Fold structures entered into the field notebook as F_1, F_2, and F_3 are described carefully according to physical and geometric properties. Furthermore, the orientations of representative folds are recorded. Axial surface orientations are presented in terms of strike and dip; axis orientations are recorded in terms of trend and plunge. A critical component of the observations and measurements are sketches, clearly labeled with orientations, and showing the fold shape and the relationship between the fold, bedding, foliations, and lineations.

The Grouping of Structural Elements

Foliations, lineations, and folds that are interpreted to have formed at the same time are given the same subscript rank. If a field notebook bears entries that discuss and describe S_1, L_1, and F_1, it is understood that all these structural elements formed contemporaneously in the deformed rock. A typical scenario is this: an S_1 schistosity formed as an axial plane cleavage to F_1, a passive fold; and L_1, a mineral lineation, formed parallel to the axis of F_1 folding. Structural elements of a given discrete event, or phase of a progressive deformation, are gathered under the heading D and assigned to D_1, D_2, D_3, and so on.

Lineations (or foliations or folds) that are quite dissimilar in physical expression may be assigned the same subscript rank on the basis of compatibility of geometric orientations, provided that they are constrained to being in the same position in the deformational sequence (e.g., D_2) and that there is a logical explanation for differences in the physical expressions. For example, a given lineation may have a physical expression that varies from lithology to lithology because of differences in the texture and mineralogy of the host rocks from which the metamorphic rock was derived. The clue that the physically dissimilar lineations formed as a set is expressed in the compatibilities of geometric orientations and relative age in the deformational sequence. The geometric compatibility may be disclosed as a constancy of preferred orientation, or as a systematically changing array of orientations. The final check on whether the lineations should be grouped together is based on interpreting whether all the lineations could have formed in the same temperature–pressure environment by a common structural process. Even though slickenlines and mineral lineation in high-grade gneiss may have a common orientation, they are not compatible physical elements. Slickenlines are products of brittle faulting; mineral lineation in high-grade gneiss is a product of deformation and/or recrystallization under metamorphic conditions of elevated temperature and pressure (Figure 9.77).

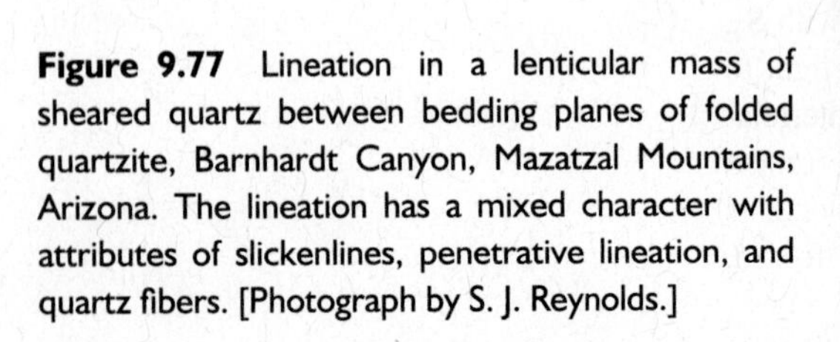

Figure 9.77 Lineation in a lenticular mass of sheared quartz between bedding planes of folded quartzite, Barnhardt Canyon, Mazatzal Mountains, Arizona. The lineation has a mixed character with attributes of slickenlines, penetrative lineation, and quartz fibers. [Photograph by S. J. Reynolds.]

Yet another sorting process is required to decide which foliations, lineations, and folds may have formed together as a system in response to a common event. Decisions for this brand of sorting are based on the geometric interrelationships among the elements and, once again, the compatibility of physical forms. Correctly interpreting the structural compatibility of foliations, lineations, and folds is a critical step in beginning to evaluate the strain significance of particular fabrics.

Correlation of Fabric and Fold Elements in Time and Space

The subscripted rankings of foliation, lineation, and folds give the impression that structures of different rankings formed during different structural events, perhaps even different tectonic events. This is not necessarily the case. A given set of penetrative folds (F_1) might be systematically refolded by a second set of folds (F_2) during a continuum of **progressive deformation**. In similar fashion, a given cleavage (S_1) that develops in the early stages of progressive simple shear may be systematically rotated to near-parallelism with the plane of shear and be cut by a newly formed cleavage (S_2) that is genetically related to the same continuum of simple-shear deformation. Determining whether structural elements are products of a continuum of deformation or are products of discrete events widely separated in time is based, once again, on physical, geometric, and temporal compatibility. In many cases geochronology and metamorphic petrology (e.g., P-T conditions) are needed to firmly resolve the problem.

Subscripted rankings may be misleading in yet another way. It is tempting to assume that an S_2 structure defined by a geologist working in one part of a deformed belt may correlate with an S_2 structure mapped and defined by some other geologist working in the same deformed belt, but in a different area. Keep in mind that one person's S_2 may be another person's S_3! Correlation of structural elements from place to place within a region requires careful appraisal of descriptive and geometric properties of the elements.

A final interpretive pitfall is linked to the assumption that a foliation–lineation–fold suite in one area formed synchronously with a physically identical foliation–lineation–fold suite in another area within the same orogen. This is not necessarily the case. The development of fabric elements can act in a **diachronous** fashion. A cold front is a good example of a diachronous event. A cold front delivers the same orderly sequence of events at every point along its path across a region—sudden drop in temperature, snow flurries, 3 feet of snow, . . . first in Chicago, then in Cleveland, then in Pittsburgh, then in Philadelphia. The compaction foliation fabric at the base of the 3 feet of snow in Chicago is not the same age as that at the base of the new snow in Cleveland, even though it looks identical physically and geometrically.

STRAIN, SHEARING, AND FABRIC DEVELOPMENT

Tectonites

Rocks that are *pervaded* by foliation and/or lineation are known as **tectonites** (Figure 9.78). Tectonites are rocks that have flowed in the solid state in such a way that few parts of the rock body escaped the distortional influence of flow, at least when observed at the scale of a single hand specimen and/or outcrop. Tectonites have a well-developed, penetrative, and strongly aligned foliation and/or lineation, an expression of the **state of strain**. Although tectonites, by definition, are rocks that have been able to flow in the solid state (Turner and

Figure 9.78 Rocks that are penetratively foliated and/or lineated are known as tectonite. This tectonite, replete with isoclinal recumbent folds, occurs near Bjornfeld in north Norway. [Photograph by G. H. Davis.]

Weiss, 1963), we now are fully aware that what we perceive as flow is scale dependent. Microscopically, the flow of tectonites is accommodated by a combination of slip, crystallization, or dissolution along exceedingly closely spaced discontinuities.

Most tectonites, and thus most foliations and lineations, form in environments of elevated temperature and confining pressure. Metamorphic and igneous environments are ideal. However, tectonites also form through distortion of soft sediments, excessively weak lithologies like salt or gypsum, or rock types that are especially vulnerable to stress-induced dissolution.

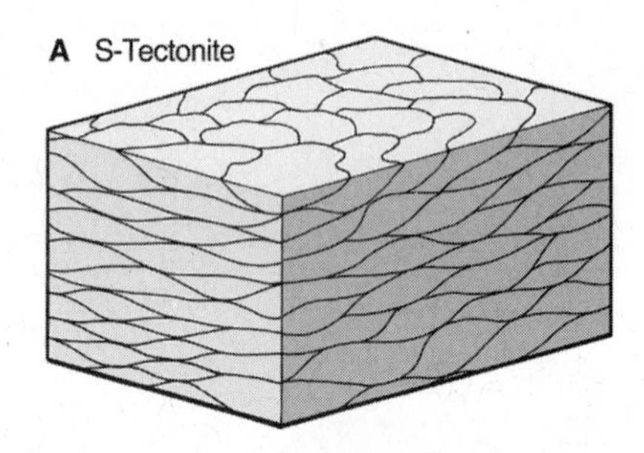

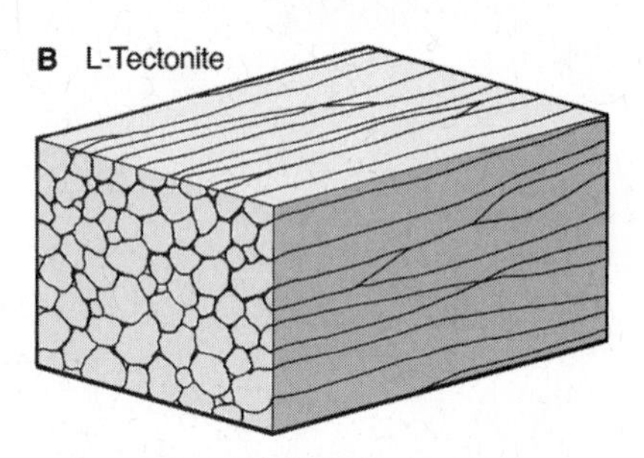

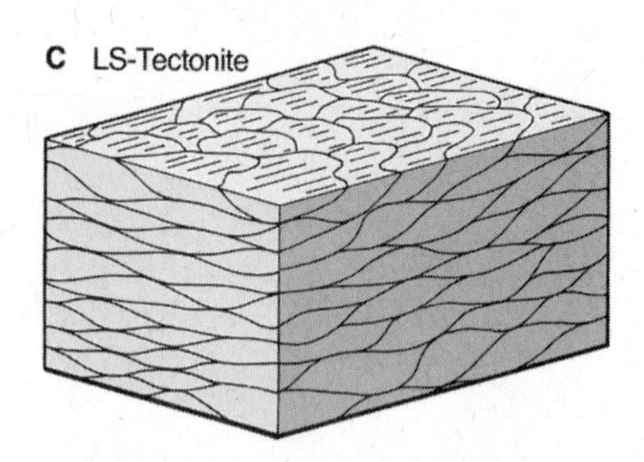

Figure 9.79 Schematic portrayal of S-, L-, and LS-tectonites. (*A*) S-tectonites are marked by a single, penetrative foliation. (*B*) L-tectonites are marked by pervasive lineation, but no foliation. (*C*) LS-tectonites are marked both by foliation and lineation. The lineation in LS-tectonite lies in the plane of foliation.

Types of Tectonites

Several classes of tectonites can be distinguished on the basis of whether the tectonite contains foliation, lineation, or both. These three types are known as **S-tectonites, L-tectonites**, and **LS-tectonites** (Figure 9.79), where "*S*" and "*L*" refer to foliation and lineation, respectively. S-tectonites are tectonites marked by foliation but not lineation. L-tectonites are tectonites marked by lineation but not foliation. LS-tectonites contain both. Of all the tectonites, LS-tectonites are the most common. These are the tectonites marked both by foliation and lineation. Lineation in LS-tectonites lies in the plane of foliation.

Strain Significance of Tectonites

Overall Objective

The presence of foliation or lineation in a rock is a signal that the rock has undergone significant distortion, with or without dilation. One of the goals of detailed structural analysis is to interpret the strain significance of tectonites; that is, to try to interpret the magnitudes and directions of distortion and/or dilation that were accommodated by the development of foliation and lineation. This, of course, is not a simple task. Success in interpreting the strain significance of tectonites depends to a large extent on discovering and analyzing distorted primary objects, of known original shape and/or size, as guides to the extent of dilation and/or distortion. Finding such treasures is more the exception than the rule.

Flattening, Constriction, and Plane Strain

The evaluation of distorted primary objects in tectonites has revealed that **S-tectonites** tend to be products of **flattening**, a state of finite strain in which the magnitudes of finite stretch are such that $S_1 = S_2 > S_3$. Flattening is the kind of distortion that transforms an original sphere into an **oblate strain ellipsoid** (Figure 9.80*A*). This type of strain, where shortening in the S_3 direction is accommodated by extension equally along S_1 and S_2, is most easily explained by volume loss due to pressure solution. Beware, however, because the lack of lineation in some rocks may be due to a predominantly micaceous character or other lithologic influence that either obscures the lineation or prevents it from forming even though the strain conditions are appropriate.

L-tectonites tend to be products of unidirectional stretching or **constriction**, a state of strain in which $S_1 > S_2 = S_3$. Constriction transforms an original sphere into a **prolate strain ellipsoid** (Figure 9.80*B*). L-tectonites, although not exceedingly common, are present in some shear zones, probably because extension parallel to S_1 was accommodated by semiequal shortening along S_2 and S_3.

LS-tectonites, probably the most common type of tectonite, typically represent something close to "*plane strain*", where stretching in one direction is compensated by flattening at right angles to the direction of stretching, with neither stretching nor shortening in the intermediate direction. Plane strain commonly results from "*noncoaxial deformation*", such as simple shear. Simple shear can transform an original sphere into a **triaxial ellipsoid** in which the state of strain is marked by $S_2 = 1$, and $S_1 > S_2 > S_3$ (Figure 9.80*C*). Flattening, constriction, and plane strain are distortional strains that may or may not be accompanied by gains or losses in volume of the rock during its conversion to tectonite.

We can create a kitchen image that distinguishes between L-tectonites, S-tectonites, and LS-tectonites. Imagine mixing up a thick pancake batter and pouring it from a pitcher onto the griddle (Figure 9.81). As the batter hangs in midair between pitcher and griddle, it stretches itself out into a narrow column of L-tectonite (see Figure 9.81). Air bubbles in the batter are prolate ellipsoids: one long axis parallel to the direction of stretching, and two smaller axes equal in length. When the batter hits the griddle, flattens, and spreads out equally in all directions, a thin, planar, S-tectonite is created (see Figure 9.81). Air bubbles in the flattened batter are oblate spheroids: one short axis perpendicular to the direction of flattening, and two long axes equal in length. Now imagine the batter flowing off the griddle, downslope onto the stove (see Figure 9.81). As the batter flows downslope, a stretching lineation will be evident in the plane of the batter; the lineation will be oriented down-dip. The air bubbles will evolve into true triaxial ellipsoids: three axes each of different length; the long axis becomes progressively longer, the short axis becomes progressively shorter, and the intermediate axis stays the same length throughout the downslope journey. Now if only we had some fresh blueberries!

It is not difficult to imagine structural environments in nature in which tectonites can be formed through flattening, constriction, and plane strain. Some slaty cleavage is undoubtedly a product of distortion by flattening. An axial planar slaty cleavage could form through significant stress-induced shortening perpendicular to the "*plane of cleavage*," accommodated by pressure solution rather than appreciable stretching or shortening within the plane of cleavage. Examples of the development of an L-tectonite through constriction could include the emplacement of a salt diapir, and the magmatic intrusion of a rhyolitic plug. Plane strain is a state of strain that typifies many shear zones (see *Chapter 10, Shear Zones and Progressive Deformation*). Progressive simple shear can create LS-tectonites whose physical and geometric characteristics express the nature of the simple shear process itself. Foliation would occupy the direction of flattening within the shear zone. Lineation would form in the plane of flattening (S_1S_2), oriented parallel to S_1.

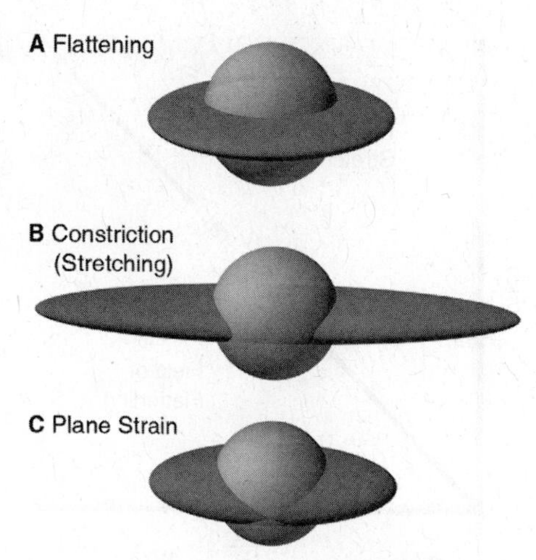

Figure 9.80 Shapes of three-dimensional strain ellipsoids provide images for visualizing the idealized strain significance of (*A*) S-tectonites (an accommodation to flattening), (*B*) L-tectonites (an accommodation to unidirectional stretching), and (*C*) LS-tectonites (commonly an expression of plane strain). There is no volume change for any of the examples shown here.

Figure 9.81 A thick viscous pancake batter, plus the right equipment, can result in early-morning experimental production of stretching fabrics (L-tectonites), flattening fabrics (S-tectonites), and combo fabrics (LS-tectonites). We prefer the S variety.

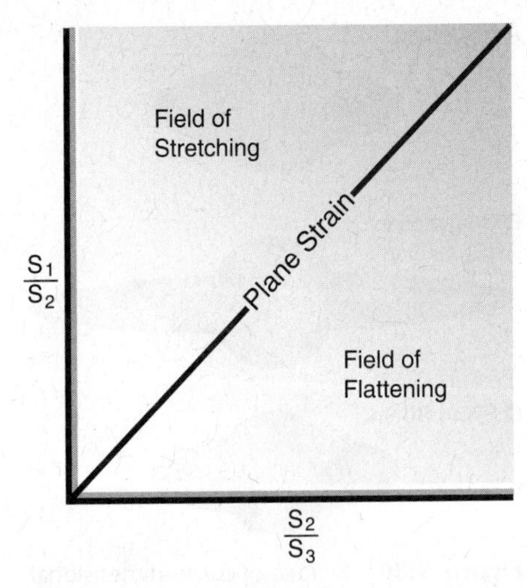

Figure 9.82 The Flinn diagram, a device for portraying the state of strain of deformed rock.

Representing Strain on Flinn Diagrams

The full range of three-dimensional strains that may be reflected in the physical and geometric properties of tectonite goes well beyond the special cases of pure flattening, pure constriction, and plane strain. Each of these strain states can be achieved with or without volume changes. Furthermore, stretching and flattening may team up in a spectrum of combinations that are limitless.

A convenient way to visualize the possibilities of three-dimensional strain, with or without volume change, is through a device known as a **Flinn diagram**. Introduced by Zingg (1935) and expanded by Flinn (1962), the Flinn diagram is a simple *x-y* graph that pictures ellipsoids that result from distortion and/or dilation of an original reference sphere. Along the *y*-axis of a Flinn diagram is plotted the strain ratio S_1/S_2 (Figure 9.82). Along the *x*-axis of a Flinn diagram is plotted the strain ratio S_2/S_3. Values of S_1, S_2, and S_3 are derived from strain analysis of distorted primary objects in the tectonite under study, objects like pebbles, fossils, and reduction spots.

If a tectonite forms through pure flattening, the coordinates (S_2/S_3, S_1/S_2) that describe the state of strain of primary objects in the tectonite will plot along the *x*-axis of a Flinn diagram. Conversely, if a tectonite forms through pure constriction, the coordinates (S_2/S_3, S_1/S_2) will plot along the *y*-axis. If a tectonite forms by plane strain, the coordinates (S_2/S_3, S_1/S_2) will plot along a 45°-sloping line that intersects the origin of the plot. The lines of pure flattening, pure constriction, and constant volume plane strain are simply three in number. There is plenty of room in a Flinn diagram to plot the limitless combinations of flattening and constriction that one might encounter in nature.

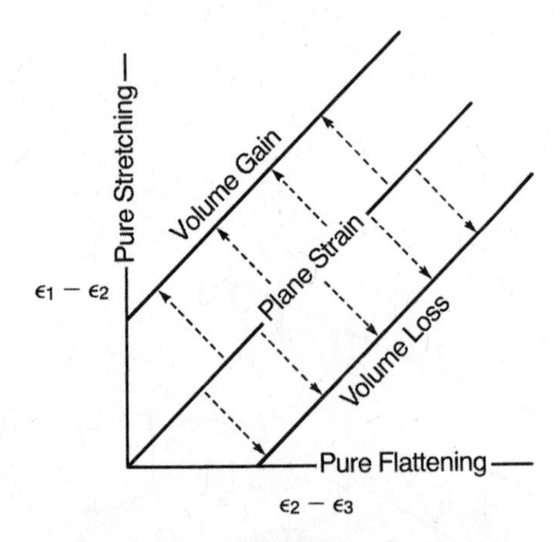

Figure 9.83 The logarithmic Flinn diagram is even more useful than the Flinn diagram because it can be used to evaluate changes in volume.

Representing Strain on Logarithmic Flinn Diagram

An even more useful way to present the state of strain of tectonite is through the use of a logarithmic Flinn diagram (Figure 9.83). This modification of the Flinn diagram was introduced by Ramsay (1967), and it can be used to keep track of changes in volume that might accompany distortion.

Logarithmic strain (ε), otherwise known as **natural strain** or **true strain**, is equal to the logarithm of the finite stretch *S* (Ramsay, 1967):

$$\varepsilon = \log_e(S) \tag{8.1}$$

where ε is logarithmic strain, and *S* is stretch.

Logarithmic strain (ε) can be instantly calculated if values of stretch are known. To represent on a logarithmic Flinn diagram the state of strain of a given tectonite, values of S_1, S_2, and S_3 must first be transformed into values of ε_1, ε_2, and ε_3. Once these conversions have been made, $\varepsilon_1 - \varepsilon_2$ is plotted along the *y*-axis of the logarithmic strain Flinn diagram, and $\varepsilon_2 - \varepsilon_3$ is plotted along the *x*-axis (see Figure 9.83). As in the ordinary Flinn diagram, pure flattening is described by values of ($\varepsilon_2 - \varepsilon_3$, $\varepsilon_1 - \varepsilon_2$) that plot along the *x*-axis, and pure constriction is represented by values of ($\varepsilon_2 - \varepsilon_3$, $\varepsilon_1 - \varepsilon_2$) that plot along the *y*-axis of the diagram. Points representing the condition of constant-volume plane strain fall along a 45°-sloping line that intersects the origin of the plot (see Figure 9.83). If a tectonite forms through plane strain accompanied by dilation, the coordinates $\varepsilon_2 - \varepsilon_3$ and $\varepsilon_1 - \varepsilon_2$ describing the state of strain of primary objects in the tectonite will plot along 45°-sloping lines that do not intersect the origin of the plot. Instead they intersect the *x*- or *y*-axis, depending on whether the dilation was accompanied by a volume decrease or a volume increase.

Deformation Paths

Flinn diagrams and logarithmic Flinn diagrams take on value and meaning when we put them to use. Let us consider a simple example of plane strain that we can model with a card deck. Figure 9.84 shows progressive simple shear of a deck embossed with a reference circle. Five stages of incremental, progressive deformation are shown—*A* through *E*. At each stage of the deformation, the value of S_2, measured perpendicular to our view of the simple shear, is 1.0. In other words, there is neither shortening nor stretching in the third dimension. If we know the size of the original reference circle, we can calculate the values of S_1 and S_3 at each stage of the deformation. These values in turn can be converted to logarithmic strain ε (see Figure 9.84).

With these strain data in hand, it becomes a simple matter to plot on Flinn and logarithmic Flinn diagrams the state of strain representing deformational stages *A* through *E* (Figure 9.85). Portrayal of these states of strain on both diagrams illustrates what we know to be true: that the deformation was achieved by constant volume plane strain. Points *A* through *E* approximate a 45°-sloping line that intersects the origin of the graph. These points, taken together, represent the **deformation path** that was followed during the progressive deformation.

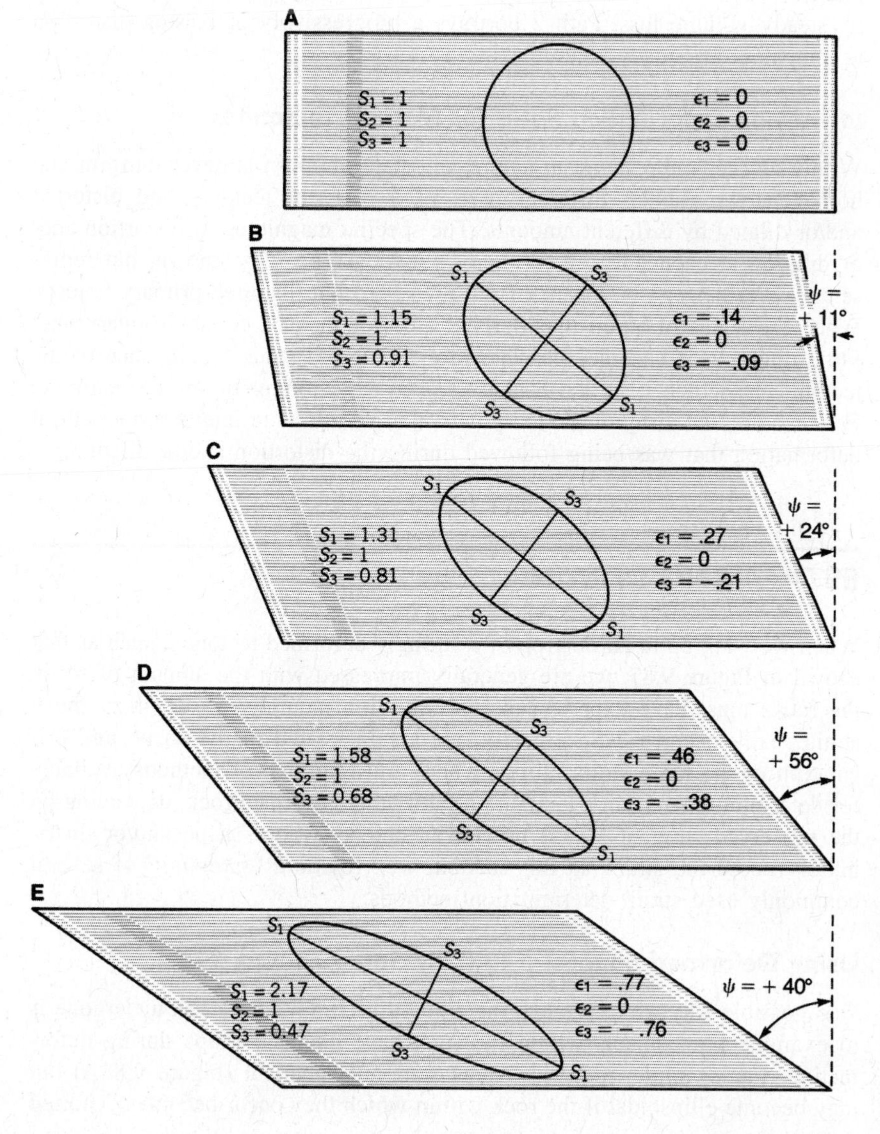

Figure 9.84 Card deck modeling of simple shear provides a database for preparing Flinn diagrams and logarithmic Flinn diagrams, to see how they work. Stretch (S) values and logarithmic strain (ε) are computed for five stages of progressive simple shear—*A*–*E*.

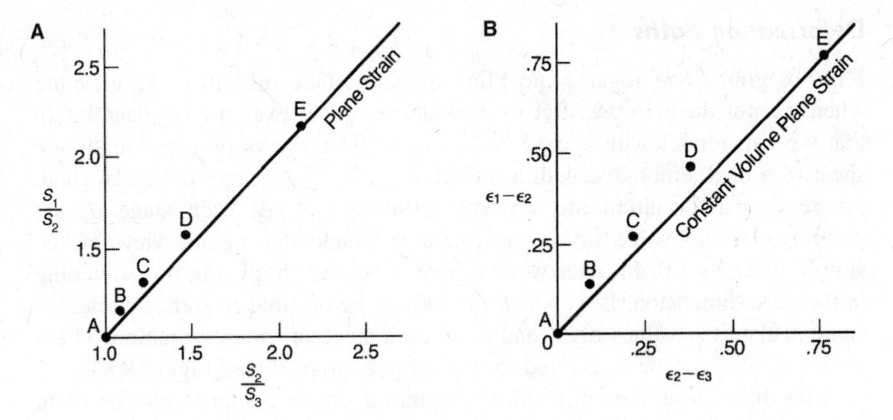

Figure 9.85 (*A*) Flinn and (*B*) logarithmic Flinn diagrams showing the deformation path of the progressive simple shear illustrated in Figure 9.84. Departures from the straight line are simply due to measurement errors.

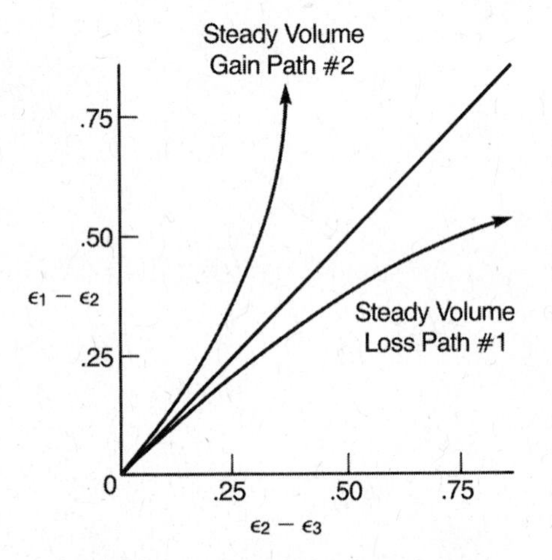

Figure 9.86 Logarithmic Flinn diagram picturing progressive simple shear accompanied by progressive volume loss (path 1), and progressive simple shear accompanied by progressive volume gain (path 2).

Progressive simple shear accompanied by volume changes would follow a deformational path different from that represented by path *A–E* in Figure 9.85. By way of example, the logarithmic Flinn diagram pictured in Figure 9.86 shows two deformation paths, each representing a different combination of distortion and dilation. Path 1 describes a progressively increasing distortion accompanied by steady volume loss. Path 2 portrays a progressively increasing distortion accompanied by steady volume gain.

Identifying Deformation Paths for Natural Tectonites

Within any tectonite found in nature, the state of strain is never uniform and homogeneous. Rather, different parts of a body of tectonite are distorted and/or dilated by different amounts. The specific magnitude of distortion and/or dilation at each point in the body will never be fully known, but representative values can be gleaned from preserved but distorted primary objects. When the states of strain thus derived are plotted on a common logarithmic Flinn diagram, it becomes possible to evaluate the strain significance of the tectonite, whether it be an S-tectonite, an L-tectonite, or an LS-tectonite. Furthermore, the array of plotted points may permit us to interpret the path of deformation that was being followed during the distortion and/or dilation.

ESTIMATING STRAIN

When we walk up to an outcrop of a strongly deformed tectonite, such as that shown in Figure 9.87, we are generally impressed with the amount of strain the rocks must have experienced. The question naturally arises—how much strain really occurred? Specifically, we are interested in the shape and orientation of the finite strain ellipse. There are a number of methods available for determining strain in a tectonite or any other deformed rock, depending on the preserved suite of **strain markers**—objects whose shape and/or distribution record the strain. In this section, we provide a "*sampling*" of several commonly used strain determination methods.

Using Deformed Shapes of Initially Spherical Objects

One possible way we can gauge the amount of strain a rock has undergone is to examine how objects within the rock have changed shape during deformation. For example, reduction spots start out spherical (Figure 9.88*A*) but may become ellipsoidal if the rock within which they occur becomes distorted

Figure 9.87 (*A*) Outcrop of highly cleaved sandstone (light) and argillite (dark) of the Meguma Group (Ordovician) of Nova Scotia. (*B*) Close-up view shows the vertical cleavage along which pressure-solution removal of material has taken place. Moreover, the close-up view reveals distorted cross-bedding and ripples in the sandstone. [Photograph by S. J. Reynolds.]

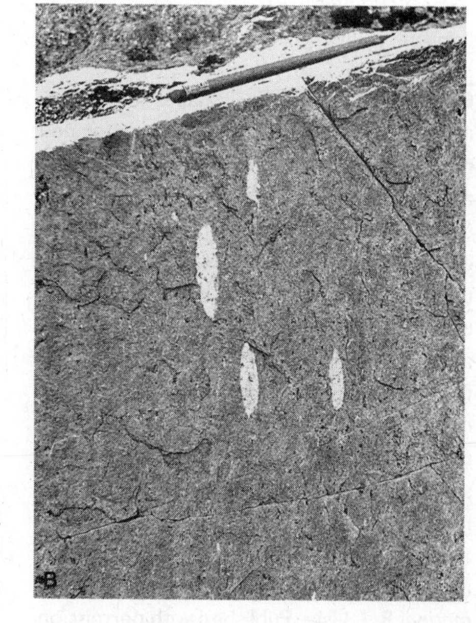

Figure 9.88 (*A*) Undeformed reduction spots (white) in red sandstone of Proterozoic age in the Grand Canyon. Note that some of the spots have been eroded in a way to reveal their exact centers, i.e., black specks of reducing agents, such as pyrite or organic fragments. [Photograph by S. J. Reynolds.] (*B*) Deformed reduction spots. The distortion reflects a secondary strain. [Photograph by O. T. Tobisch.]

(Figure 9.88*B*). When this happens we can measure the **aspect ratios** (e.g., length to height) of such objects on three different surfaces, ideally those cut parallel to the principal axes of the finite strain ellipsoid (see Figure 9.89). The ideal surfaces include one parallel to foliation, one parallel to lineation and perpendicular to foliation, and one perpendicular to both lineation and foliation. For each surface, we determine the ratio of two principal stretches, and then combine results for the three surfaces to define the shape, but not the absolute dimensions, of the three-dimensional finite strain ellipsoid (see Figure 9.89). We can discover the true dimensions (actual values of the principal finite stretches) only if we know the average initial diameter of the objects, which is unlikely. If we assume constant volume deformation, however, we can use the volume of the ellipsoid to determine the diameter of a sphere with the same volume. Each principal stretch is then determined by dividing its length by the diameter of the sphere.

Figure 9.89 This slab of "*rock*" containing deformed reduction spots has been cut in a way to reveal the "aspect ratios" of ellipses on each of the three principal planes. By combining the ratios derived from each plane, we can construct the shape of the strain ellipsoid for the outcrop from which the specimen was collected. Clearly there is a lot to chew on when it comes to 3D fabrics.

Using the R_f/ϕ Method

We need to use a slightly more involved approach if the objects were originally elliptical, rather than spherical. The so-called R_f/ϕ method (Ramsay, 1967) is commonly used to measure the strain in deformed conglomerates, most of which start with nonspherical clasts. The basis of the method is recognizing that final shapes and orientations of ellipsoidal pebbles in the deformed rock (Figure 9.90*B*) are the product of the original shapes and orientations in the undeformed rock (Figure 9.90*A*) and the shape and orientation of the finite strain ellipsoid to which the undeformed rock was subjected (see Figure 9.90*B*).

For a number of deformed objects exposed on a single planar surface, we measure present-day ellipticity and long-axis orientation, which are a function of (1) the original ellipticity and orientation of the objects and (2) the orientation and ellipticity of the finite strain ellipse. **Ellipticity** is defined as the ratio of the long axis of the ellipse to the short axis, as measured on the surface we are observing. Using these measurements we can derive, either graphically or mathematically, the orientation and shape of the finite strain ellipse on the exposed surface (Figure 9.90*C*). To more fully characterize the strain, we repeat the technique on several surfaces, especially those three special directions: parallel to foliation, parallel to lineation and perpendicular to foliation, and perpendicular to both foliation and lineation.

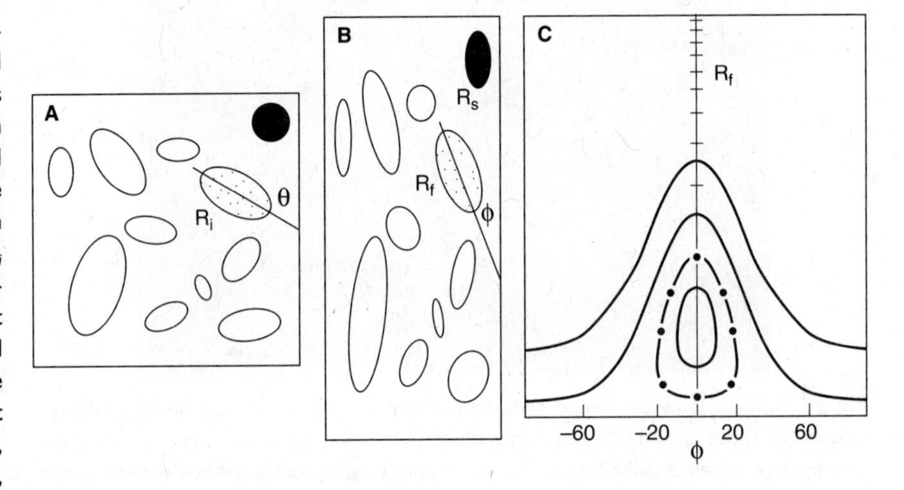

Figure 9.90 The final shapes and orientations of ellipsoidal objects in deformed rock is a product of (*A*) the original shapes and orientations of the ellipsoidal objects in the undeformed rock and (*B*) the shape and orientation of the finite strain to which the undeformed rock was subjected. Thus an original elliptical object of initial axial ratio R_i and initial orientation θ becomes transformed by a finite strain of R_s (oriented at $\phi = 0°$) to an elliptical object of final axial ratio R_f and final orientation ϕ. (*C*) The shapes of these onion curves of the R_f/ϕ plot reveal the interrelationships between R_f, R_i, and R_s. [Reprinted from Palaeostrain analysis, in Hancock, P. L. (ed.), *Continental deformation*, R. J. Lisle. Published with permission of Pergamon Press, Ltd., Oxford, © 1994.]

Using Thinning of Layers of Known Original Thickness

In other cases we can get an idea of the amount of strain by seeing how deformation has thinned or thickened stratigraphic units or other layers whose initial thickness we know. A classic example of this is in the Big Maria Mountains of southeastern California (Figure 9.91*A*), where Warren Hamilton (1964, 1987) recognized that an upright Paleozoic section 1 km thick could be followed around the hinges of huge synmetamorphic folds and into the overturned limbs, where it was attenuated to 30 m or less! The units are thinned to 3% of their normal thickness, representing a stretch of 0.03 ($S = 30$ m/1000 m). Incredibly, each of the distinctive formations is still present and can be recognized, even in the most extremely attenuated sections. These formations are equivalent, in both name and original thickness, to the well-known stratigraphic section of the Grand Canyon. Imagine the entire Grand Canyon section

Figure 9.91 In the Big Maria Mountains, southeastern California, the familiar Paleozoic sequence of the Grand Canyon (from the Cambrian Tapeats to Permian Kaibab) has been reduced in places to a tiny percentage of its original thickness. (*A*) View looking east showing overturned fold. Thicknesses are close to normal (1000 m) in the lower upright limb (right-hand side). The overturned upper limb (layering on left in photo) is as thin as 10 m, i.e., 1% of its original thickness! (*B*) In the nearby Granite Wash Mountains, the entire thickness of the middle part of the Paleozoic section can be measured with an on-the-outcrop photo scale. All that is left of the mighty Redwall Limestone is the thin white band to the right of the scale. We have Warren Hamilton to thank for recognizing this profound thinning, and then convincing the world. You haven't lived until you have taken three or four giant steps across the entire Grand Canyon Paleozoic. [Photographs by S. J. Reynolds.]

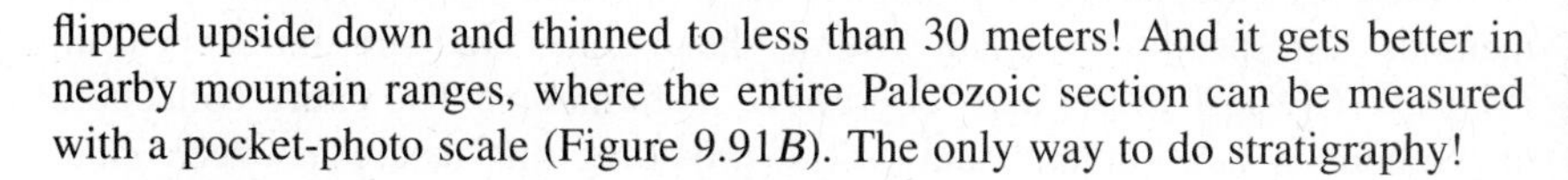

flipped upside down and thinned to less than 30 meters! And it gets better in nearby mountain ranges, where the entire Paleozoic section can be measured with a pocket-photo scale (Figure 9.91*B*). The only way to do stratigraphy!

Using the Center-to-Center Technique

The **center-to-center technique**, also called the **Fry method** after its inventor, Norman Fry (1979), is used to calculate the strain in rocks that before deformation contained randomly scattered objects or particles, such as phenocrysts in a deformed granitic rock or oolites in a limestone. During deformation, the centers of adjacent objects become closer in directions in which the rock is being shortened and farther apart in directions in which the rock is being lengthened. The average distance between adjacent particles, called **nearest neighbors**, should be least in directions parallel to S_3 and greatest in those parallel to S_1 (Figure 9.92*A*). To perform the analysis, we commonly use as a data base a photograph of an exposed surface or thin section. We overlay the photo with a piece of tracing paper, the center of which is marked by a tiny reference cross. The process is simple. Place the reference cross on a given object and mark, on the overlay, points representing the locations of the neighboring objects. Then slide the tracing paper such that the reference cross lies on a second object, and again mark the locations of nearest neighbors. Repeat and repeat and repeat, being careful *never* to rotate the tracing paper (Figure 9.92*B*). The shape of the finite strain ellipse magically appears! The entire process can be done using computers, although we miss out on the fun of watching the strain ellipse progressively appear as we hand plot the data. The method works as long as the objects we are studying were initially randomly distributed and were not clustered in any way.

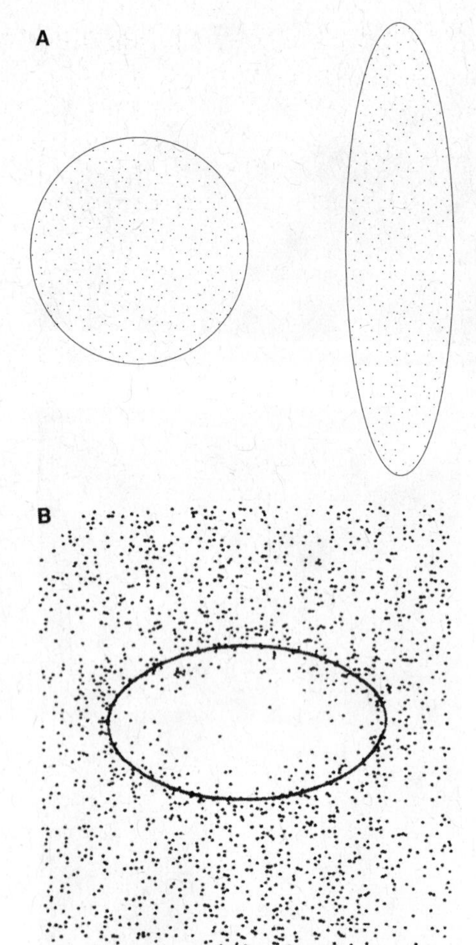

Figure 9.92 *(A)* The success of the Fry method depends on the initial presence of an "*anticlustered*" distribution of points within the undeformed body. When the body is deformed (as represented by the change in the circle to an ellipse), the spacing between points systematically changes. [From Simpson, in Marshak and Mitra, Basic Methods of Structural Geology, copyright © 1988. Reprinted by permission of Prentice Hall, Upper Saddle River, New Jersey.] *(B)* Example of plot produced by the Fry method as applied to an oolitic limestone. [Reprinted from Palaeostrain analysis, in Hancock, P. L. (ed.), *Continental deformation*, R. J. Lisle. Published with permission of Pergamon Press, Ltd., Oxford, © 1994.]

Using Asymmetric Magnetic Susceptibility

Another method for strain analysis uses a specific magnetic character of rocks called **anisotropy of magnetic susceptibility** or, thankfully, AMS for short (Borradaile and Jackson, 2004). In the field we collect oriented samples of strained rocks and, back in the lab, subject them to a low-strength magnetic field in order to determine differences, if any, in the magnetic susceptibility of the sample in different directions. Many rocks possess a measurable anisotropy of magnetic susceptibility. The magnetic susceptibility of a rock is influenced by many factors, including the types, sizes, and shapes of magnetic minerals. If all the factors are favorable, we can use the collected magnetic data to infer the *orientation* of the strain ellipse. Even though the AMS method does not permit strain magnitudes to be determined, it nonetheless is a vital tool in permitting the investigation of structural problems that were previously intractable. For example, Morgan et al. (2008) used AMS to determine the emplacement directions in magma sheets as they formed the now-famous laccoliths of the Henry Mountains of Utah (Figure 9.93). AMS has many other applications, including rapid characterization of the orientation of the strain ellipse over an entire area, checking whether apparently undeformed lake sediments have experienced any deformation, and examining the directions of magma emplacement and sea-floor spreading in mid-ocean ridges.

Other Methods

There are many other methods for determining strain in special circumstances (see Ramsay, 1967; Ramsay and Huber, 1983.) One involves evaluating changes in the angles between initially perpendicular lines, such as for deformed brachiopods (see *Chapter 2, Displacement and Strain*). Other methods are specific to strain within shear zones and are introduced in *Chapter 10, Shear Zones and Progressive Deformation.* The entire collection of methods is quite diverse and useful, providing that nature has been kind enough to preserve the appropriate strain makers.

To sum up

Terranes of tectonites force us to apply principles of strain on a vast scale. They require us to think of flattening, constriction, and plane-strain simple shear at scales that are quite dramatic, almost unimaginable. Furthermore, they require us to think of the progressive, step-by-step development of the structures in foliated rocks (Figures 9.94 and 9.95).

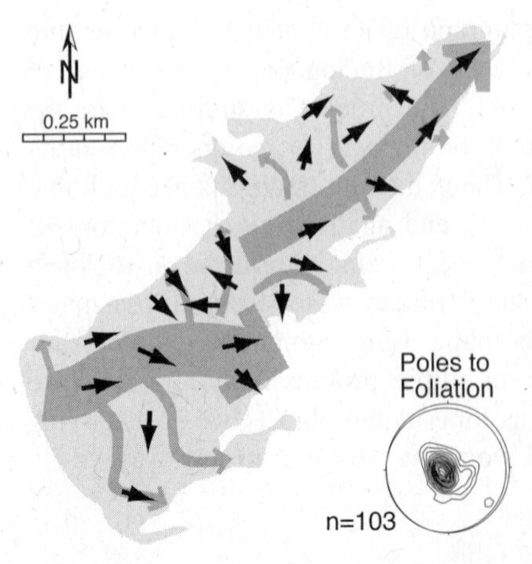

Figure 9.93 Map showing directions of magmatic flow as recorded by anisotropy of magnetic susceptibility (AMS). Short black arrows are individual AMS data points, which help define overall flow patterns (longer gray arrows) in a magmatic sill within a laccolith, Henry Mountains, Utah. [Reprinted from *Journal of Structural Geology*, v. 30, Morgan and others, Emplacement of multiple magma sheets and wall rock deformation: Trachyte Mesa intrusion, Henry Mountains, Utah, p. 491–512, © 2008, with permission from Elsevier.]

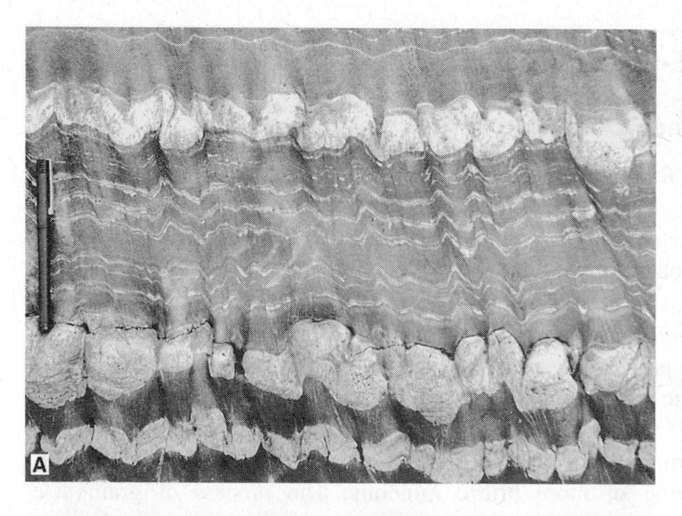

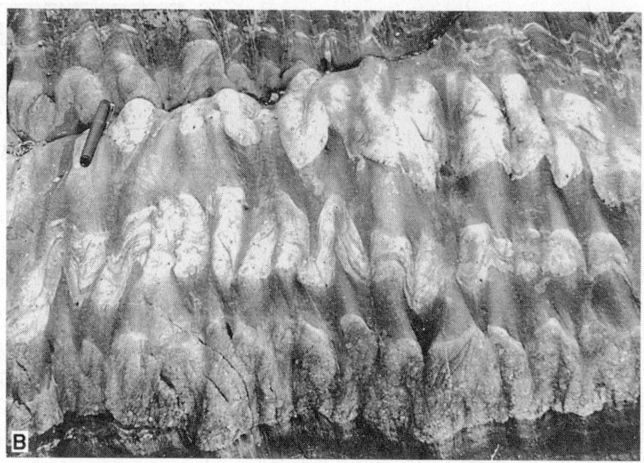

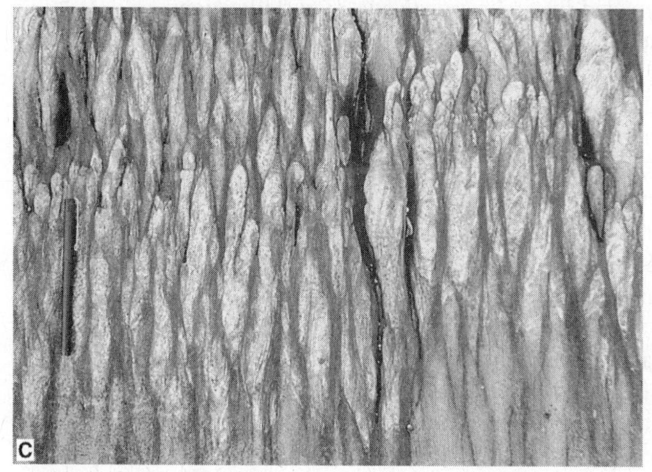

Figure 9.94 Progressive development of axial plane cleavage, and transposition, within sequence of interbedded argillites and sandstones of Meguma Group (Ordovician), eastern coast of Nova Scotia. (*A*) Open folding of sandstone layers (light) and slaty cleavage development in argillite (dark). (*B*) Tight folding of sandstone layers, strong slaty cleavage development in argillite, and incipient spaced cleavage development in sandstone. (*C*) Pervasive cleavage development in both argillite and sandstone. Few fold hinges recognizable. [Photographs by S. J. Reynolds.]

Figure 9.95 These photos of Vishnu Schist in Vishnu Canyon, Grand Canyon, are marvelous displays of progressive transposition. (*A*) The original schistosity of the Vishnu is preserved in the microlithon domains (white). Note that the schistocity has become crenulated with axial surfaces parallel to the newly developed cleavage. (*B*) Progressive deformation builds cleavage domains (dark bands) at the expense of microlithons. (*C*) Where transportation is nearly complete, you need to look long and hard for vestiges of the original schistocity, for most of the rock is composed of dark cleavage domains. [Photographs by S. J. Reynolds.]

"*TECTONITE-FORMING*" GEOLOGIC SETTINGS

Relation Between Deformation and Metamorphism

Deformation that forms tectonites occurs within the context of metamorphism, and so we are clearly interested in the relationship between structural processes and metamorphic ones. As in any deforming rock mass, several processes are all occurring at once, often in competition with one another. The processes involved in the formation of tectonites are the same processes that form more "*normal*" metamorphic rocks, such as gneiss and schist. In tectonites, though, the processes operate in different relative amounts or are carried out to an extreme degree. The deformation mechanisms that dominate the formation of most tectonites, especially mylonites, are the reduction of grain size during deformation via ductile dislocation creep, dynamic recrystallization, and fracturing of more brittle minerals. The process of grain size reduction is counteracted by grain growth via recrystallization.

Grain size reduction dominates at low temperatures or high strain rates, where the introduction of strain into crystals outpaces recrystallization, producing fine-grained rocks. As a result, tectonites deformed under greenschist and lower amphibolite facies conditions are commonly platy mylonitic rocks.

At higher temperatures or lower strain rates, recrystallization of existing minerals, the growth of new minerals, and other forms of annealing become dominant, leading to typical high-grade metamorphic rocks, such as schist and gneiss. Such rocks represent middle amphibolite to granulite–facies conditions, where metamorphism may also be accompanied by the formation of granitic layers and pods via metamorphic differentiation, in situ melting, and the injection of granitic magma parallel to foliation.

Between these extremes are conditions that are just right to form rocks intermediate in character between true mylonitic rocks and "*normal*" metamorphic rocks. These rocks are usually medium grained and include mylonitic gneiss, mylonitic schist, and phyllonite.

The overall fabric and appearance of a rock strongly depends on its *subsequent* metamorphic and structural history. During **prograde metamorphism**, for example, rocks become progressively deeper and hotter as a result of tectonic burial, perhaps related to the emplacement of an overlying thrust sheet. In this case, tectonite fabrics that formed early in the deformation history may be overprinted by higher grade metamorphic conditions; or they may become annealed during the higher temperatures and overgrown by new, higher grade metamorphic minerals.

Some tectonites are instead formed during **retrograde metamorphism**, such as when rocks are uplifted in the footwall of major normal faults (Davis, 1983; Wernicke, 1985; Reynolds and Lister, 1987). In this case early ductile fabrics may be overprinted by lower-grade mineral assemblages, but they should not be extensively annealed. Many retrograde mineral reactions are hydration reactions and cannot proceed in the absence of water. Since high-grade metamorphic rocks have generally lost most of their water during dehydration reactions that accompanied the *prograde* part of their pressure–temperature path, many retrograde reactions cannot occur unless water is somehow reintroduced into the metamorphic rock. Often this does not happen, which is why we find relatively pristine high-grade metamorphic rocks and structures exposed at the surface. If retrograde reactions were more successful, metamorphic petrologists and structural geologists would have a much harder time studying processes that occur at depth.

The predictable differences in the character of the total fabric of tectonites for different **metamorphic paths** provide the means to distinguish fabrics

Figure 9.96 Metasomatism is not uncommonly associated with formation of replacement ore deposits, when hydrothermal solutions from a granitic body at depth move up and through a sequence of sedimentary rocks, notably limestones and dolomites. (*A*) Here we see the mined-out Ludwig Lode, a copper-bearing replacement lode mined in the early 19th century in the Yerington District, Nevada. Covey of exploration geologist for scale. (*B*) Original argillaceous limestone was locally transformed by metasomatism into garnet-pyroxene skarn. [Photographs by G. H. Davis.]

formed in different tectonic settings. This approach has been used to distinguish tectonites formed along retrograde paths during crustal extension from those formed along prograde paths during thrusting (Reynolds et al., 1988).

The differences in the availability of fluid between prograde and retrograde metamorphic paths influence the formation of tectonites in yet one other way. Dehydration reactions during prograde metamorphism typically liberate large amounts of fluid during deformation. These fluids may promote deformation and/or may deposit veins of quartz, calcite, and other minerals. Multiple generations of veins in various states of deformation are common in tectonites, attesting to the important interplay between deformation and fluids. The passing fluids, most of which are composed of variable proportions of water, CO_2, methane, and other volatiles, may leach some elements and deposit others in a process known as **metasomatism**. Metasomatism may obscure or obliterate previously formed fabrics and change the way the rock deforms as old minerals are dissolved and new minerals are formed (Figure 9.96).

Relation Between Deformation and Plutonism

When studying tectonites and other metamorphic rocks in the field, we inevitably encounter plutons whose relation to deformation turns out to be an important, if not critical, part of the story (Figure 9.97). Plutons are incredibly informative because they generally provide us with the clearest, most unambiguous way to determine the age of deformation. Also, because they are emplaced in a relatively short time period compared to deformation, plutons may help us determine whether any time intervened between two phases of deformation (e.g., D_1 versus D_2). Plutons may also represent an important potential source of heat responsible for the metamorphism.

We are primarily concerned with determining whether a given intrusion (e.g., pluton, dike, sill) was emplaced before, during, or after deformation. An intrusion emplaced before deformation and metamorphism is called **pre-kinematic** and should reveal evidence of having experienced deformation *after* it had become totally solidified. Such an intrusion might contain foliation

Figure 9.97 Light-colored granite intruding across darker colored metamorphic rocks, all of Cretaceous age, in the Harcuvar Mountains of west-central Arizona. Small roof pendants of dark metamorphic rocks occur in the central granite intrusion, which is affectionately called "*The Claw*" by local geologists. [Photograph by S. J. Reynolds.]

and lineation parallel to that of the country rock it intrudes, indicating that the intrusive rock and the wall rock experienced a common deformational history. After it forms, a prekinematic intrusion might be transformed into an elliptical shape much different from its original shape. Some prekinematic intrusive rocks are deformed only along their margins, remaining undeformed in their centers. The degree to which the center of a prekinematic intrusive rock is deformed will depend on such factors as ductility contrast, metamorphic conditions, and intensity of the local or regional strain.

An intrusion emplaced *during* deformation is **synkinematic** and should display evidence that it was partially molten during deformation (Figure 9.98). Criteria for recognizing synkinematic intrusion include (1) evidence for injection of melt into synchromously developing structures in the country rock; (2) recognition that early phases of the pluton are deformed whereas later phases cross-cut the fabrics; and (3) geometric and kinematic coordination between primary magmatic fabric and the deformation of the country rock (Passchier et al., 1990; Paterson and Tobish, 1992; Karlstrom, 1989; Ghosh, 1993; Passchier and Trouw, 2005). A synkinematic intrusion have magmatic fabrics, not solid-state ones.

An intrusion emplaced *after* deformation is **postkinematic** and should lack the fabric observed in the surrounding deformed country rock. Postkinematic

Figure 9.98 Granite emplaced during deformation, Harcuvar Mountains, western Arizona. The granitic sills have a foliation parallel to their margins and to that in the wall rocks, and they show evidence of being partially molten when deformed. [Photograph by S. J. Reynolds.]

Figure 9.99 Margins of a granitic dike that crosscuts fabric in the wall rocks and contains inclusion of foliated wall rocks, Marshall Point, Maine. [Photograph by S. J. Reynolds.]

intrusions commonly cross-cut the fabric of the wall rock and may contain inclusions of previously foliated wall rocks (Figure 9.99). Contact metamorphic aureoles associated with a postkinematic intrusion should be superimposed on deformation-related fabric and minerals in the wall rocks.

As always, Nature has some playful little tricks. Postkinematic intrusions may inject along foliation, giving the impression that the igneous rock and the metamorphic country rock became foliated at the same time. Some prekinematic intrusions of the "*right*" petrology, such as coarse-grained pegmatites, do not easily "*pick up*" fabric, even though they were around for the entire deformation history. Also, as we will see in the next chapter, a prekinematic dike in one orientation relative to the imposed strain may appear essentially undeformed, whereas a nearby prekinematic dike of the same age but a different orientation will be strongly folded.

ON TO SHEAR ZONES

Some of the most beautiful foliated and lineated rocks in the world lie concentrated along ductile shear zones. Ductile shear zones are "*hot faults*" along which displacement is achieved through distributed shear and distortion across a zone of deformation. Within shear zones, the components of the array of structures and fault rocks lie oriented sympathetically to the progressive strain history the rocks have experienced. Thus, they provide valuable information that can be used to decipher structural and tectonic history.

Equipped with some knowledge about deformation mechanisms, strain, cleavage, foliation, and lineation, we will now enter the realm of shear zones and put this knowledge to further use. We will discover in *Chapter 10, Shear Zones and Progressive Deformation,* that not all shear zones are "*hot*" and ductile; some are "*cold*" and brittle. And we will be back to looking even deeper into the nature of fault zones. The study of shear zones provides a wonderful opportunity for synthesis.

Chapter 10 Shear Zones and Progressive Deformation

Outcrop exposure of the spectacular Sa Ferradura shear zone that crops out in Cap de Creus, Spain. The shear zone, formed during the Variscan (Carboniferous), strikes north-south and cuts across large pegmatite bodies and mylonitic schists. In this photograph we are looking westward and can see a pegmatite body (white) becoming dragged and thinned as it enters the heart of the shear zone. The sense of curvature of the strained pegmatite reveals that sense-of-shear is dextral (right-handed). Total horizontal displacement accommodated by this shear zone is ~75 m. Universitat Autonoma de Barcelona undergraduate geology students Emma Soldevila Moldes (left) and Ingrid Terricabres Polo stand firmly on opposite (west) block of shear zone. [Photograph by G. H. Davis.]

THE NATURE OF SHEAR ZONES

At scales ranging from a thin section to an outcrop to a region, deformation is commonly not homogeneous, but is instead concentrated into highly deformed zones within less deformed rocks. The general term for such zones is **high-strain zones**, but most high-strain zones have a noncoaxial shear component to the deformation and so are simply called **shear zones** (Passchier and Trouw, 2005). Shear zones are tabular to sheetlike, planar or curviplanar zones in which rocks are more highly strained than rocks adjacent to the zone (Figure 10.1). The intensity with which rocks can be deformed in shear zones is astonishing. Granites can be so strongly and thoroughly sheared that they resemble, and have been mistakenly mapped as, metarhyolite or metasedimentary schist (Figure 10.2*A*). Conglomerates can be smeared out to such a degree that individual clasts resemble thin sedimentary layers (Figure 10.2*B*).

Shear zones have certain characteristics that permit us to recognize them in the field, in thin sections, and on geologic maps and cross-sections. The distinguishing characteristics vary, depending on whether the shear zone formed under brittle, ductile, or intermediate conditions. A fault zone is a shear zone formed under brittle conditions. Displacement is taken up on

Figure 10.1 Shear zones in the field: (*A*) Steeply inclined brittle shear zone cutting Proterozoic quartzite, 75 Mile Canyon, Grand Canyon, Arizona; note marking pen in center for scale. [Photograph by S. J. Reynolds.] (*B*) Steeply inclined ductile shear zone cutting Scourie pyroxenite, Ballcall Bay, north Scotland. [Photograph by Carol Simpson.]

a network of closely spaced faults (see Figure 10.1*A*). When shear zones form under ductile conditions, deformation is accompanied by metamorphism and produces rocks with foliation, lineation, folds, and related features (see Figure 10.1*B*). Some shear zones develop under conditions that are intermediate between strictly brittle and strictly ductile deformation. These may consist of zones that are partly faults and partly ductile shear zones

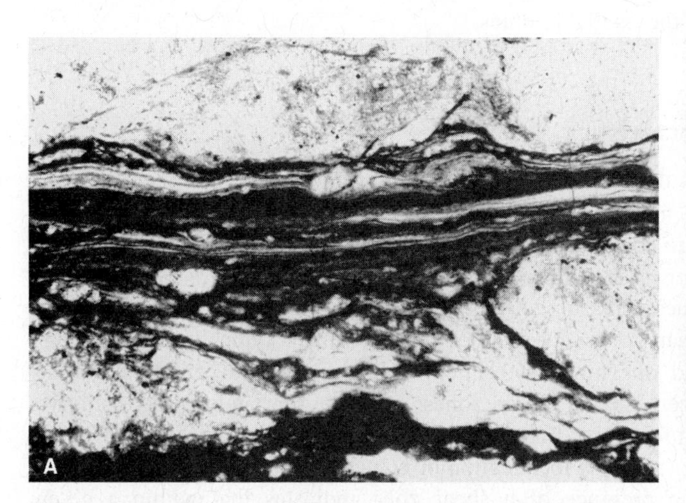

Figure 10.2 Rocks deformed in shear zones. (*A*) Thin ductile (mylonitic) shear zone cutting Tertiary granodiorite, South Mountains, central Arizona. Width of dark shear zone is 1 mm. (*B*) Strongly deformed Mesozoic conglomerate within a thrust zone in the Granite Wash Mountains, western Arizona. White 'lenses' are flattened pebbles. [Photographs by S. J. Reynolds.]

Figure 10.3 (*A*) Brittle–ductile shear zone cutting metasedimentary rocks, Mosaic Canyon, Death Valley, California. Width of area shown is approximately 1 m. (*B*) Brittle–ductile shear zone produced by interlayered ductilely deformed marble and brittlely fractured quartzofeldspathic gneiss, Plomosa Mountains, western Arizona. [Photographs by S. J. Reynolds.]

(Figure 10.3*A*), and may have formed in interlayered rocks with contrasting strengths (Figure 10.3*B*).

For many geologists, including us, shear zones are some of the most fun and fascinating structures to work with. Shear zones contain the most intensely deformed rocks known on Earth! The geometry and interplay among different structural elements within shear zones can be truly elegant, providing a startlingly detailed record of the history of deformation. Most shear zones contain features that permit us to determine the *sense of displacement* along the zone. In favorable situations, we may be able to reconstruct the *amount of displacement.* Deformed objects within shear zones can be used to quantify the *amount of strain* imposed on the rocks during shearing. Finally, a shear zone provides us with a wonderful "*mystery story*," whose storyline we try to reconstruct from the available clues.

General Characteristics

Most shear zones that we encounter are much longer and wider than they are thick, commonly having relative dimensions somewhere between a sheet of paper and an audio compact disc (of *soft*-rock music, of course). They exist at all scales. The largest are hundreds of kilometers long and tens of kilometers thick, with displacements of tens to hundreds of kilometers (Figure 10.4*A*). The smallest shear zones observable in outcrop are typically several centimeters long and one millimeter thick, and may have a centimeter or so displacement. Even smaller shear zones, with appropriately scaled relative dimensions, are informative only when viewed in thin section (Figure 10.4*B*).

All shear zones reflect a localization or concentration of deformation into a narrow zone. The presence of a shear zone indicates that within a given deforming rock mass, the distribution of strain was heterogeneous rather than homogeneous. As a result, shear zones are characterized by spatial gradients in the amount of strain. The amount of strain is generally highest near the center of a shear zone, decreasing outward into the wall rocks adjacent to the zone. If

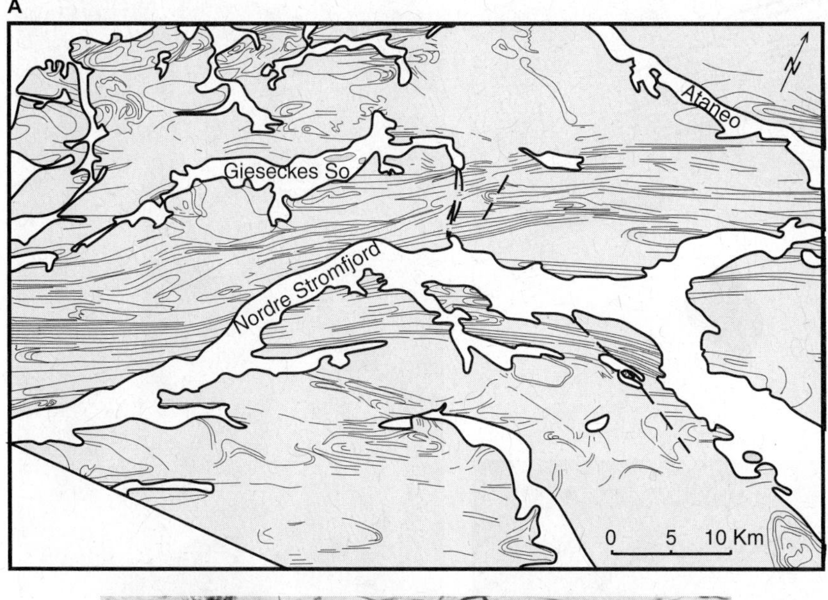

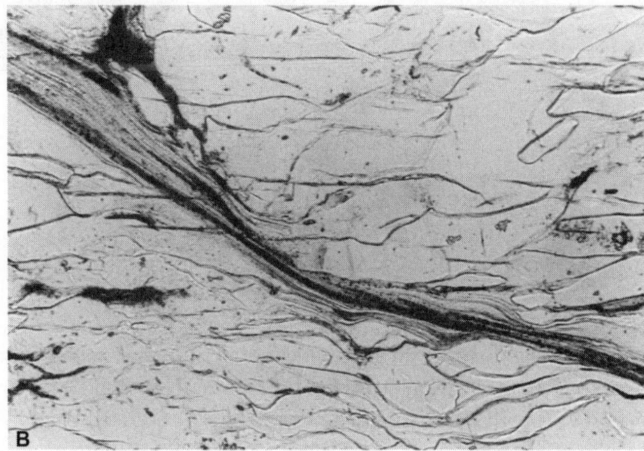

Figure 10.4 The big and little of shear zones: (*A*) Map of lithologic boundaries within and near the Nordre Stromfjord shear zone, Greenland. [From Sorensen, K., *Journal of Geophysical Research*, v. 88, no. B4, copyright © 1983 by American Geophysical Union.] (*B*) Shear zone in a thin section of experimentally deformed aplite. Shear zone is approximately 0.1 mm wide. [Photomicrograph courtesy of Jan Tullis.]

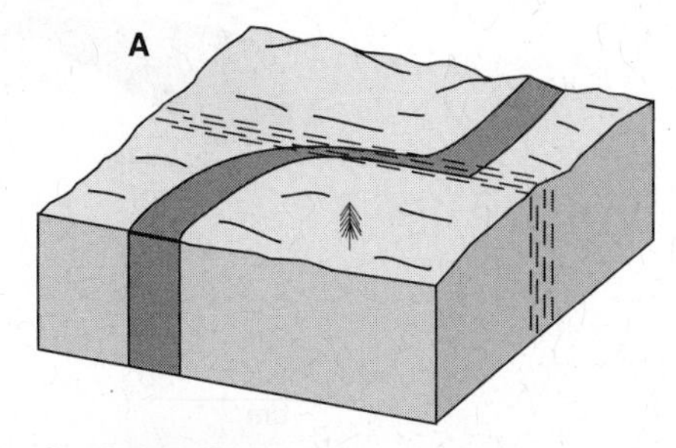

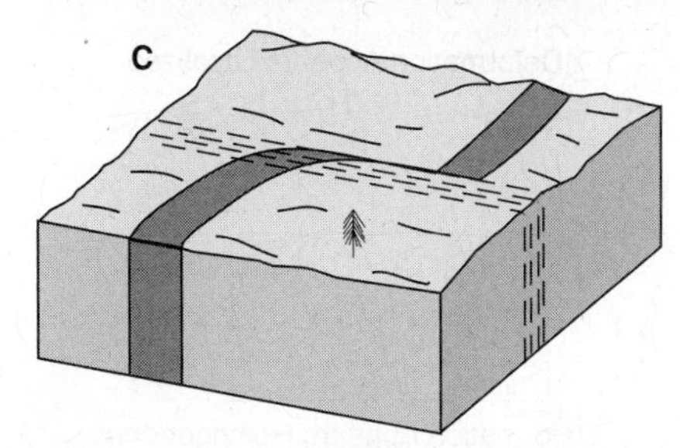

Figure 10.5 Continuous and discontinuous shear zones: (*A*) Continuous shear zone deflecting a marker that passes uninterrupted through the shear zone. (*B*) Outcrop example of pegmatite entering a shear zone. Cap de Creus peninsula, Spain. [Photograph by G. H. Davis.] (*C*) Discontinuous shear zone that truncates a marker.

the decrease in strain away from the zone is gradual without any distinct physical break, the shear zone is considered to be **continuous** (Figures 10.5*A*,*B* and 10.6*A*,*B*). Continuous shear zones most commonly form under ductile conditions, where the rocks flow in the solid state without loss of cohesion. If the decrease is more abrupt, with clear discontinuities, the zone is considered to be **discontinuous** (Figures 10.5*C* and 10.6*C*). In most discontinuous shear zones, strongly deformed rocks within the zone are juxtaposed against much less deformed rocks along a sharp physical break or a very thin band along one or both margins of the shear zone. This occurs, for example, when a discrete fault surface forms that accommodates all further shearing. Whether a strain gradient appears abrupt or gradual, and whether deformation *appears* homogeneous or localized, depends on the scale at which we observe the structures (Figure 10.7). A shear zone that *appears* discontinuous in outcrop may show, in thin section, a continuous gradient from weakly to strongly deformed rock over several millimeters.

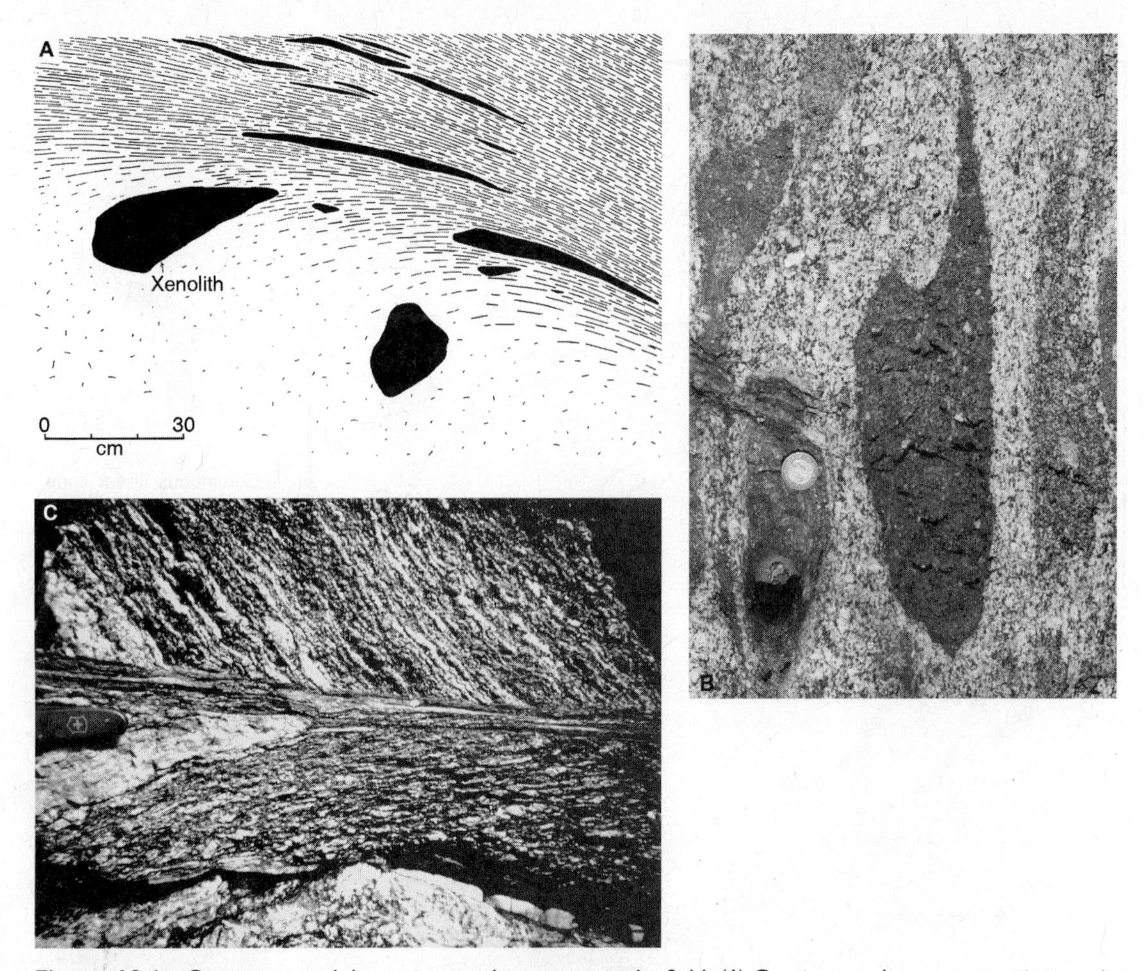

Figure 10.6 Continuous and discontinuous shear zones in the field: (*A*) Continuous shear zone cutting granite and inclusions, Lepontine Alps, Switzerland. [Reprinted from Ramsay and Huber, The techniques of modern structural geology, v. 1: Strain analysis, © 1983, with permission from Elsevier.] (*B*) Sheared enclave in Roses Granodiorite, Cap de Creus peninsula, Spain. [Photograph by G. H. Davis.] (*C*) Thin discontinuous, ductile shear zone cutting foliation in mylonitic granite, Harcuvar Mountains, western Arizona. [Photograph by S. J. Reynolds.]

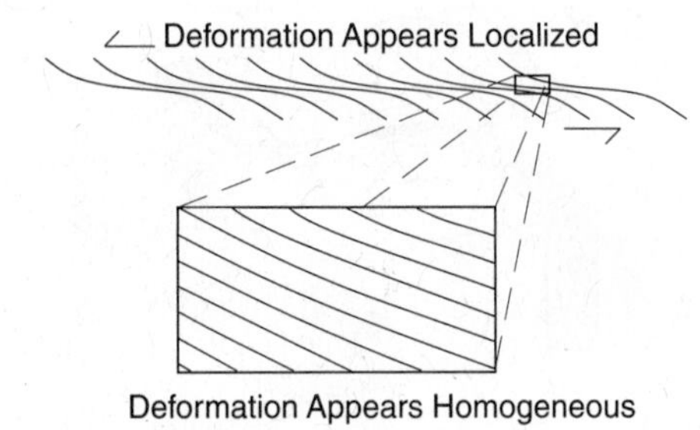

Figure 10.7 Whether deformation appears homogeneous or localized depends on the scale of observation.

Geometries

Shear zones are typically planar to gently curved, but some can have complex geometries. Most shear zones have subparallel margins and retain a fairly consistent thickness over much of their length (Figure 10.8*A*). Where the margins diverge, the shear zone becomes wider (Figure 10.8*B*). Widening is most common near the ends of a shear zone, where strongly deformed rocks within the zone grade into a wider zone of less deformed rocks. A shear zone may also thin or taper as the margins converge, such as where a shear zone passes near or between rigid objects (Figure 10.8*C*). This thinning or tapering is generally accompanied by an increase in the degree of deformation as strain becomes concentrated in a progressively thinner zone.

Shear zones are commonly arranged in networks or sets composed of a number of individual shear zones. They may occur in subparallel sets, may deflect toward one another and link up in an anastomosing pattern, or may crosscut and displace one another (Figure 10.9).

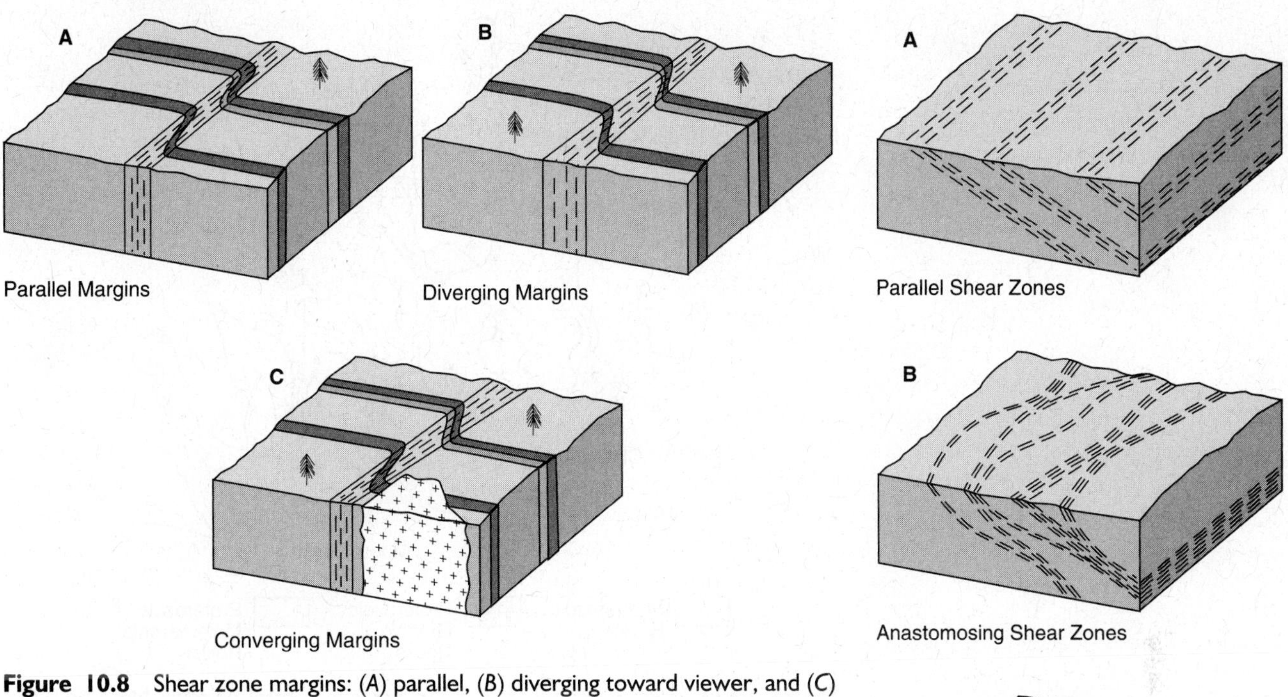

Figure 10.8 Shear zone margins: (*A*) parallel, (*B*) diverging toward viewer, and (*C*) converging near a rigid pluton.

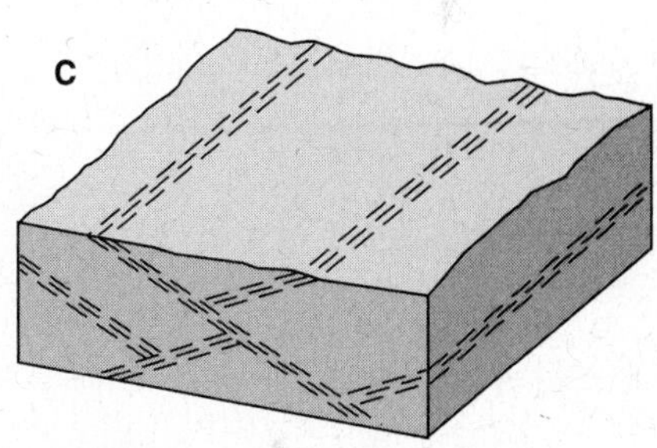

Figure 10.9 Shear zone sets: (*A*) parallel, (*B*) anastomosing, and (*C*) conjugate.

Some shear zones have a curviplanar or folded geometry. Such a geometry may indicate that an originally planar shear zone (Figure 10.10*A*) was folded or warped by subsequent deformation (Figure 10.10*B*). Alternatively, many shear zones form with an original curviplanar geometry, encompassing and wrapping around more rigid, less deformed objects (Figure 10.10*C*). On a regional scale, such objects may include a relatively rigid pluton or volcanic pile surrounded by less competent shale, schist, and marble (Figure 10.11). Examples of smaller rigid objects include inclusions or pebbles of a strong rock in a weaker matrix, and strong crystals of one mineral in a matrix of other, weaker minerals.

Offset and Deflection of Markers

When a shear zone cuts across a preexisting feature, such as a dike or compositional layer (Figure 10.12), it can truncate, offset, or deflect the feature, or do all three. In a *continuous* shear zone, a preexisting feature will generally

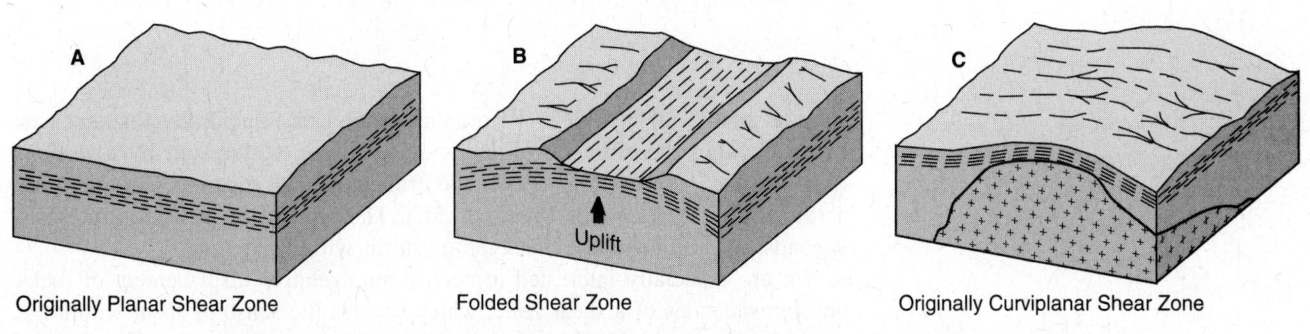

Figure 10.10 Curviplanar shear zones: (*A*) originally planar shear zone, (*B*) folding and erosion of same zone to expose a curved shear zone, and (*C*) shear zone formed with an originally curviplanar geometry around a rigid object, in this case a pluton.

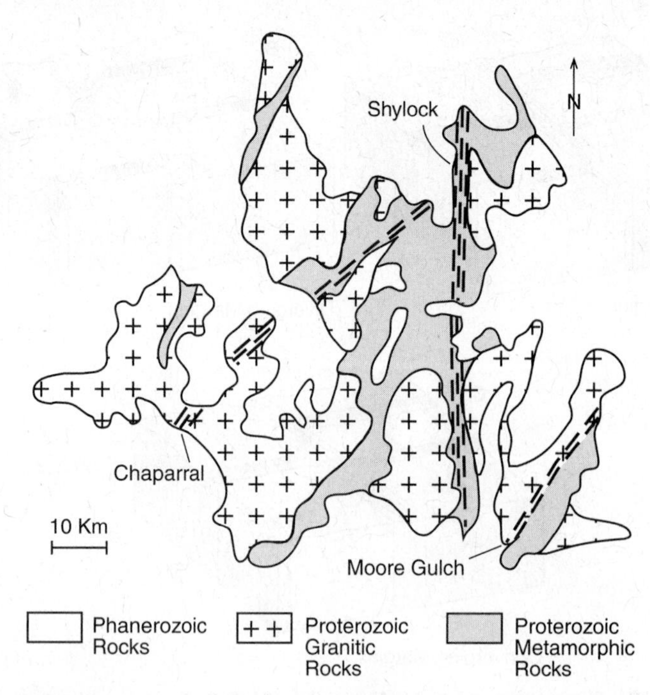

Figure 10.11 Map of Proterozoic shear zones in central Arizona. Note how the Shylock shear zone occurs between and on the edges of more rigid plutons.

Figure 10.12 Brittle shear zone cutting Tertiary sandstone, Waterman Hills, Mojave Desert, California. [Photograph by S. J. Reynolds.]

show a gradual deflection in its orientation and be structurally thickened or thinned within the zone (see Figures 10.5*A,B*). A feature cut by a *discontinuous* shear zone will generally be truncated and show an abrupt offset across the shear zone (see Figure 10.5*C*). The amount of deflection or offset depends on the magnitude and type of strain within the zone.

We are especially interested in determining relative displacement of rocks on *opposite* sides of a shear zone, which reveals the *sense of shear* within the zone. In keeping with fault terminology (*Chapter 6*), shear zones can be strike-slip, normal, reverse, and oblique-slip. **Strike-slip shear zones** may be right-handed (**dextral**) or left-handed (**sinistral**) (Figure 10.13*A,B*). **Normal-slip shear zones** are marked by hanging wall displacement *downward* relative

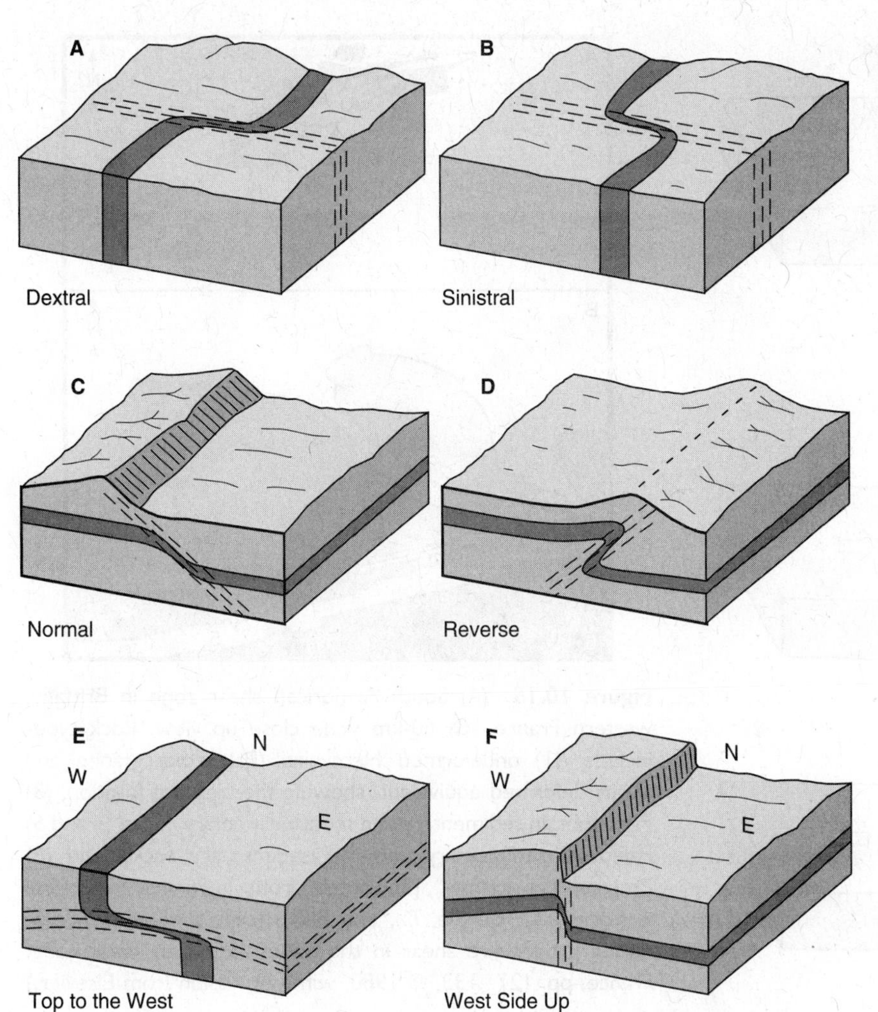

Figure 10.13 Deflection and offset across shear zones: (*A*) right-handed or dextral, (*B*) left-handed or sinistral, (*C*) normal, (*D*) reverse, (*E*) top to the west, and (*F*) west side up.

to the footwall (Figure 10.13*C*). **Reverse-** and **thrust-slip shear zones** are marked by hanging wall displacement *upward* relative to the footwall (Figure 10.13*D*). **Oblique shear zones** have components of both strike-slip and dip-slip.

Describing the sense of shear on subhorizontal or variably dipping shear zones is done by specifying which way the hanging wall moved, such as "*top to the west*" (Figure 10.13*E*). For vertical shear zones with a dip-slip component of motion, we use phrases such as "*west-side up*" (Figure 10.13*F*) or "*northeast-side down*" to convey the sense of shear.

Tectonic Settings

Shear zones form in a wide variety of tectonic settings (Figure 10.14), including plate boundaries of all types. They are undoubtedly forming at depth today in any region with abundant earthquake activity or other manifestations of active deformation. For example, shear zones are present along seismically active strike-slip zones, such as the San Andreas fault of California, the Alpine fault of New Zealand, and the numerous strike-slip faults that dissect China and Tibet north of the India–Asia continental collision. Shear zones also mark the sites of past strike-slip zones, such as the South Armorican shear zone of western France (Figure 10.15) and the Great Glen fault of Scotland.

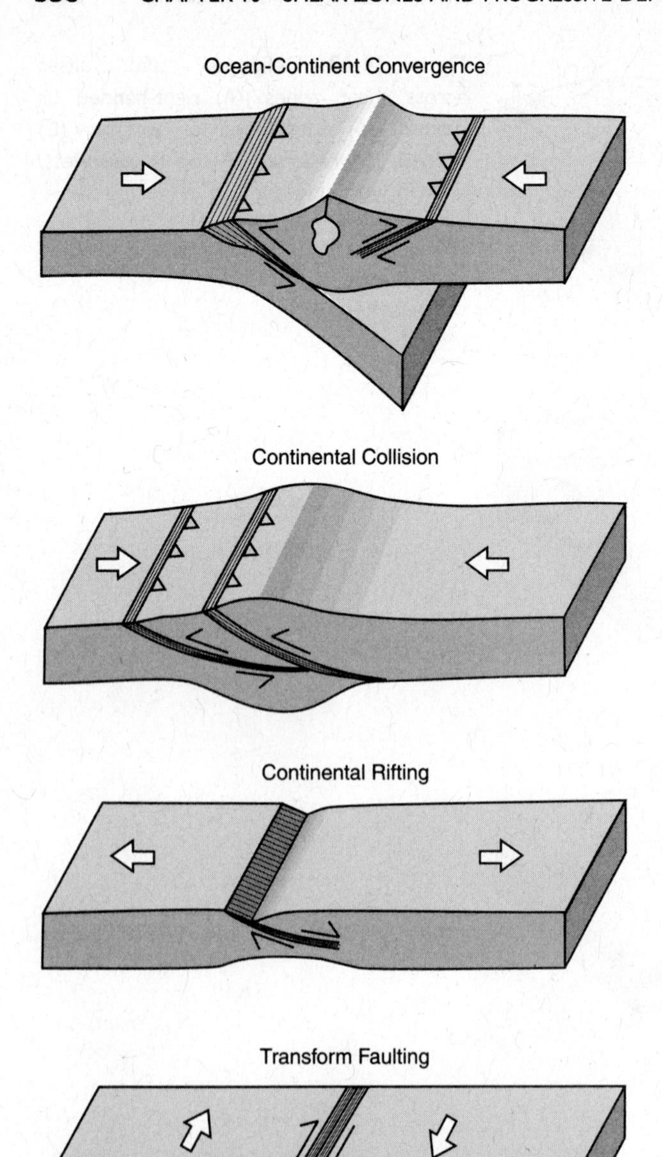

Figure 10.14 Plate tectonic settings of some shear zones.

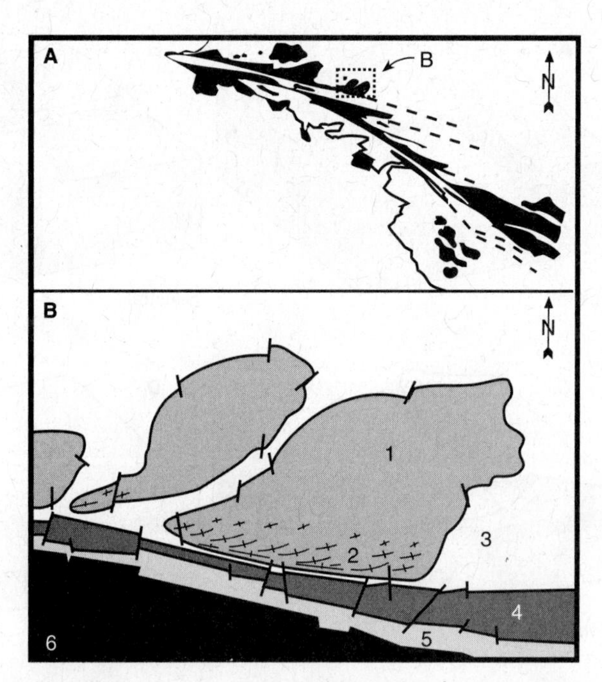

Figure 10.15 (*A*) South Armorican shear zone in Brittany, western France. (*B*) 60-km wide close up view. Rock types include: (1) undeformed Hercynian (Paleozoic) granite and (2) its deformed equivalent, showing the trend of foliation, (3) Precambrian sedimentary and metasedimentary rocks, (4 and 5) Paleozoic sedimentary and metasedimentary rocks, and (6) Precambrian granite. [Reprinted from *Journal of Structural Geology*, v. 2, Berthé, D., and Brun, J. P., Evolution of folds during progressive shear in the South Amorican shear zone, France, pp. 127–133, © 1980, with permission from Elsevier.]

Shear zones that form during plate convergence and crustal shortening commonly have thrust displacements that typically bring older, deeper rock up against younger, higher level rock (Figure 10.16*A*). Huge, impressive thrust shear zones occupy nearly the entire length of the Alpine–Zagros–Himalayan belt, which is associated with the collision of Africa and India with the southern flank of Europe and Asia. Some ductile thrust zones represent the deep-level equivalents of thin-skinned fold and thrust belts, such as those in the Canadian Rockies, the Appalachian and Ouachita Mountains of the eastern United States, the "*Outback*" of central Australia, the Cape Fold belt of South Africa, and the Moine thrust of Scotland. Others form beneath basement-cored uplifts, such as the Wind River Mountains in the Rockies of Wyoming.

Shear zones that accommodate crustal extension place high-level rock in the hanging wall down against deeper rocks of the footwall (Figure 10.16*B*),

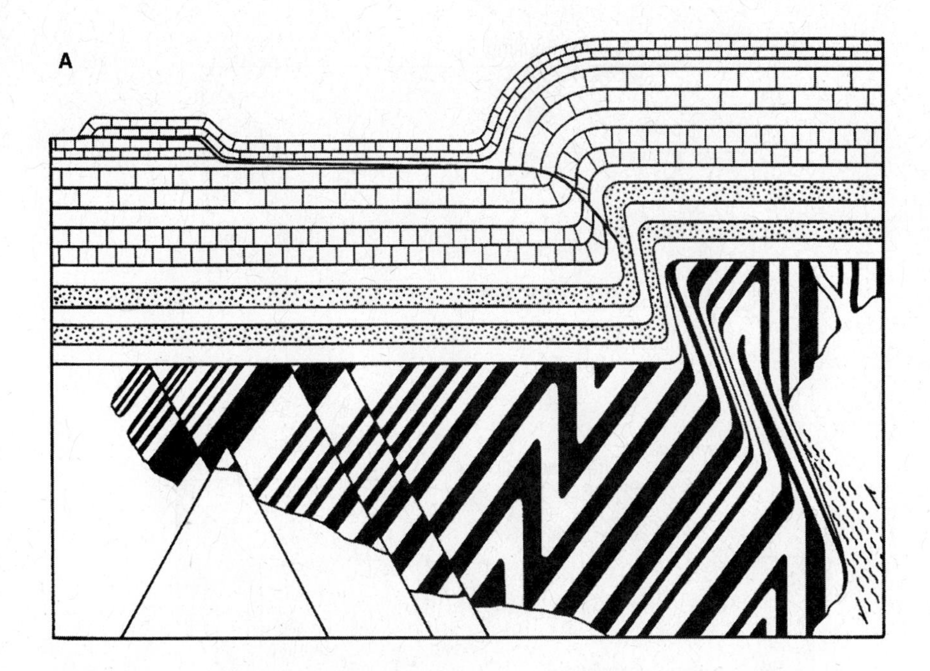

Figure 10.16 Shear zones that continue up dip into faults: (*A*) Reverse displacement during crustal shortening, (*B*) Normal displacement during crustal extension. [Reprinted from *Journal of Structural Geology*, v. 2, Ramsay, J. G., Shear zone geometry: a review, pp. 83–99, © 1980, with permission from Elsevier.]

commonly in a "*younger-on-older*" relationship (younger rocks above, older rocks below). Extensional shear zones are interpreted as forming at depth in regions of active continental rifting, such as the African rift, Greece, and the Basin and Range province of western North America. Sites of past extreme extension are represented by shear zones in metamorphic core complexes of western North America, the Aegean, and many other mountain belts (Figure 10.17). Extensional shear zones can even accompany continental collisions, as in Tibet, where extensional shear along the north-dipping, brittle-ductile South Tibetan detachment has aided in unroofing high-grade metamorphic rocks in the Himalaya (Burchfiel et al., 1992).

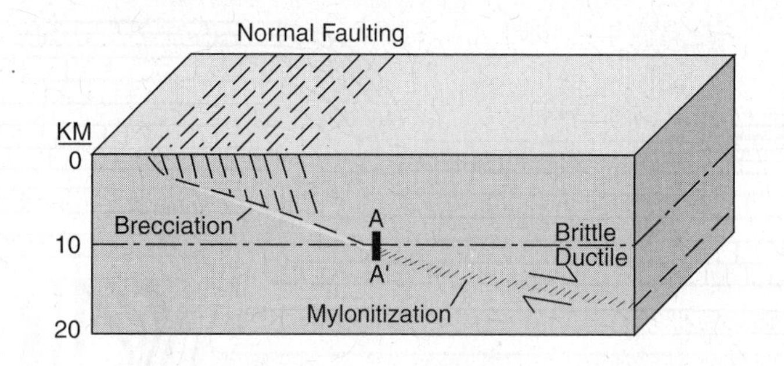

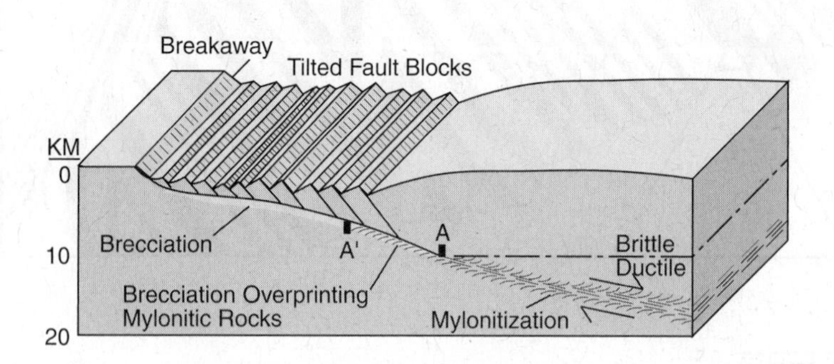

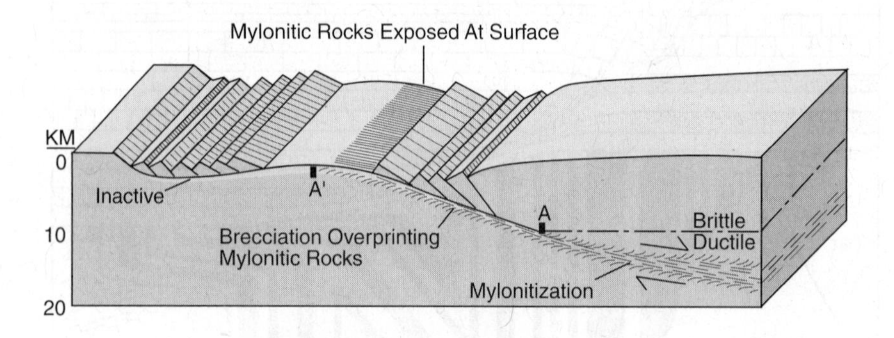

Figure 10.17 Schematic evolution of a low-angle shear zone with both ductile and brittle segments. Normal displacement along the shear zone progressively unroofs footwall rocks, causing early ductile fabrics to be overprinted by brittle ones as the rocks are isostatically uplifted and cooled. This is one model for the origin of metamorphic core complexes. [Modified from Reynolds et al., in Ernst, W. G. (ed.), Metamorphism and Crustal Evolution of the Western United States, © 1988, p. 493. Reprinted by permission of Prentice Hall, Upper Saddle River, New Jersey.]

TYPES OF SHEAR ZONES

When we encounter a shear zone in the field, we generally are impressed with the variety of structures and the many ways in which the rocks deformed. Such variety exists because shear zones form under many different conditions and from a variety of preexisting rock types. The deformation mechanisms that operate within a shear zone depend on the mineralogy and grain size of the affected rock and on the physical conditions that prevailed during deformation. Brittle deformation mechanisms tend to dominate when temperatures are cooler, pressures are lower, strain rate is faster, and fluid pressure is higher. Ductile deformation mechanisms become dominant when temperatures and pressures are higher, strain rate is slower, and fluid pressure is lower. Of course, the threshold from brittle to ductile depends on rock type: rock units composed of halite and gypsum will deform ductilely under conditions in which quartz and feldspar are decidedly brittle.

We subdivide shear zones into four general types, based on the characteristic type of deformation. A **brittle shear zone** contains fractures and other features formed by brittle deformation mechanisms. A **ductile shear zone**

Figure 10.18 Brittle shear zone cutting Miocene volcanic rocks, River Mountains, Nevada. Slickenlined fault surface forms an overhang. [Photograph by S. J. Reynolds.]

displays structures, such as foliation and lineation, that have a metamorphic aspect and record shearing by ductile flow. **Semibrittle shear zones** include en echelon veins and stylolites, and involve mechanisms such as pressure solution and cataclastic flow. **Brittle–ductile shear zones**, which show evidence for both brittle and ductile deformation, form where conditions during shearing were intermediate between brittle and ductile, or where conditions changed from ductile to brittle or from brittle to ductile.

Brittle Shear Zones

Brittle shear zones form in the shallow parts of the crust, generally within 5–10 km of the Earth's surface, where deformation is dominated by brittle mechanisms, such as fracturing and faulting. Brittle deformation is also favored by the relatively rapid strain rates that occur during seismic events (many earthquakes occur within the upper 10–15 km of the crust). Accordingly, shear zones formed in this environment are characterized by closely spaced faults, numerous joints and shear fractures, and brecciation (Figure 10.18). **Brittle shear zones** are in effect **fault zones**, and they are marked by fault gouge and other rocks of the *breccia series* (see *Chapter 6, Faults*).

The dominance of faulting and fracturing in brittle shear zones results in abrupt, discontinuous margins that truncate and offset markers. Closely spaced faults define brittle shear zones composed of numerous discrete fault surfaces (Figure 10.19) and a chaotic assemblage of strongly fractured and brittlely disrupted rocks. Zones of intensely fractured and crushed rocks associated with faults vary in thickness from less than a millimeter to a kilometer or more. In general, the thickness of a brittle shear zone increases with the amount of displacement accommodated by the zone. A single brittle shear zone, however, may thicken or thin along strike as individual fault strands merge with or bifurcate away from the zone, such as when the zone encounters rocks of varying mechanical properties. Changes in the width and

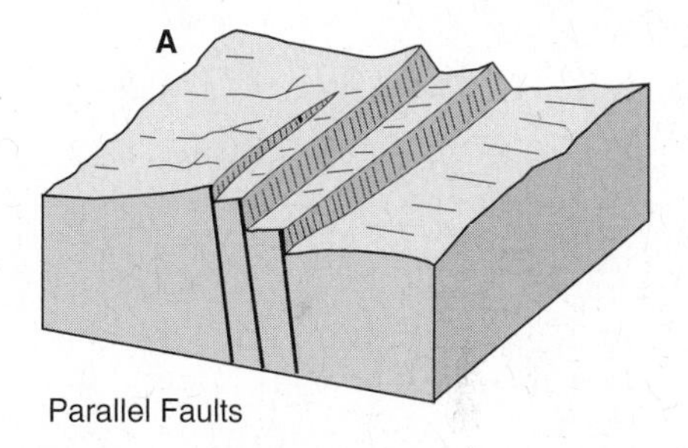

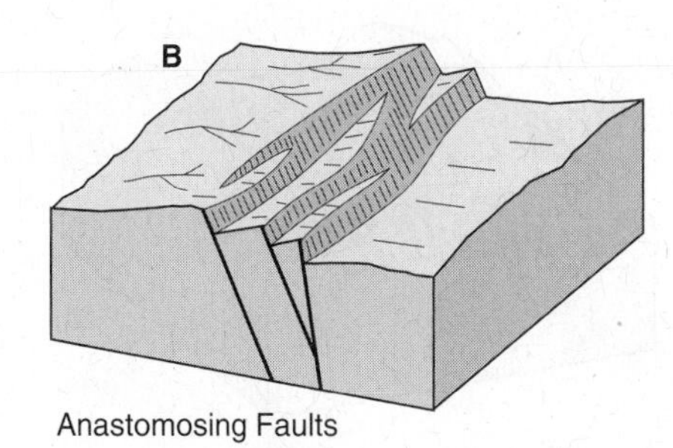

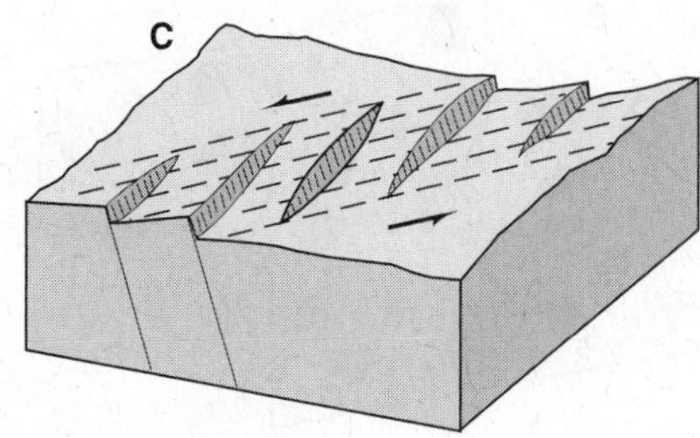

Figure 10.19 Sets of brittle shear zones (faults and fault zones): (*A*) parallel, (*B*) anastomosing, and (*C*) en echelon.

Figure 10.20 Ductile shear zones in the field: (*A*) Thin shear zones cutting Proterozoic gneiss, South Mountains, central Arizona. [Photograph by S. J. Reynolds.] (*B*) Shear zone cutting folded metasedimentary rocks, Cap de Creus Peninsula, axial zone of the Pyrenees. [Photograph by G. H. Davis.]

character of fault zones are especially common near bends or jogs. The wall rocks outside a brittle shear zone may be largely unaffected by the faulting or, alternatively, may show a zone of drag folding flanking the zone.

Ductile Shear Zones

Ductile shear zones are formed by shearing under ductile conditions, generally in the middle to lower crust and in the asthenosphere. For the most common crustal rocks (e.g., granite), brittle deformation at shallow crustal levels gives way downward into ductile deformation at the **brittle–ductile transition** (see *Chapter 4, Deformation Mechanisms and Microstructures*). A similar brittle-to-ductile transition is present within the mantle, probably near the lithosphere–asthenosphere boundary. In both cases, rocks below the brittle–ductile transition are at temperatures and pressures so high that they respond to imposed stresses by ductile flow rather than by faulting and brittle fracture. Accordingly, we generally see ductile shear zones developed at the expense of rocks we would expect to find in the middle crust and deeper—gneiss, schist, marble, amphibolite, granulite, migmatite, large intrusions, pegmatite, and deep-level mafic and ultramafic rocks. A few exceptionally weak rock types, including salt, gypsum, and clay, are able to deform ductilely at relatively low temperatures and at very shallow crustal levels (Ter Heege et al., 2005).

Most ductile shear zones form under *metamorphic* conditions, and the resulting sheared rocks are metamorphic in character, typically possessing foliation and metamorphic minerals (Figure 10.20). Rocks within a ductile shear zone may be so changed by the intense shear, by metamorphism, and by fluids passing through the shear zone that it becomes very difficult, if not impossible, to decipher the original rock—the protolith. Some rock types that form in ductile shear zones are sufficiently distinctive in appearance to merit a terminology separate from that for "*normal*" metamorphic rocks. Such rocks are sometimes simply called **tectonites**, or are assigned to the important family of **mylonitic rocks** (see *Chapter 9, Foliation and Lineation*).

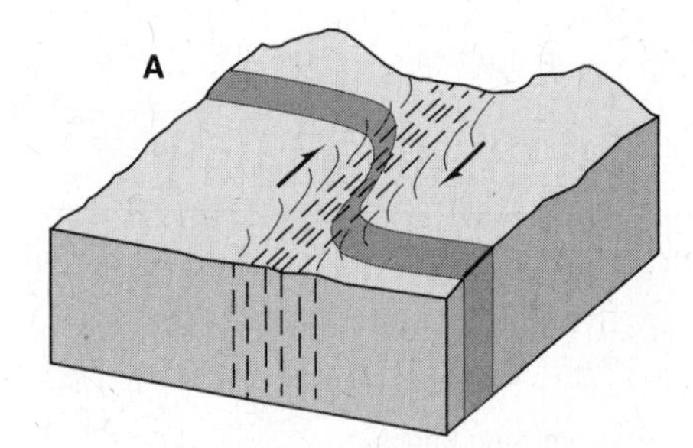

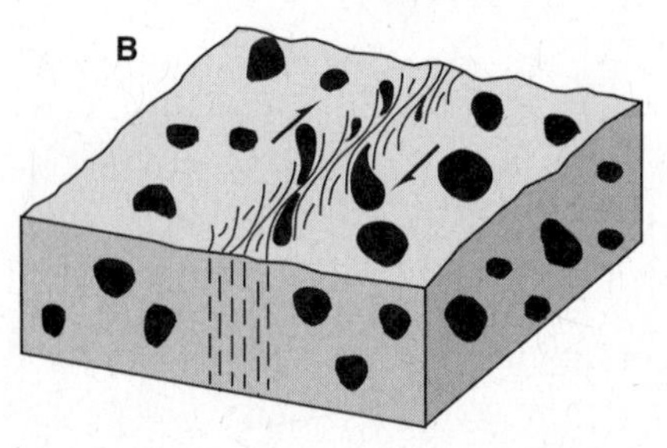

Figure 10.21 Ductile shear zones: (*A*) marker offset by continuous, dextral shear zone, (*B*) shear zone cutting plutonic rocks with inclusions and isotropic initial fabric.

Ductile shear zones, unlike typical fault surfaces or brittle shear zones, commonly do not display any discrete physical break (Figure 10.21). Instead, differential translation of rock bodies, separated by the shear zone, is achieved entirely by ductile flow. Markers pass through ductile shear zones without necessarily losing their continuity, but the effects of the shearing are recorded by distortion of the markers and by the development of foliation, lineation, and other shear-related fabrics.

Between Brittle and Ductile: The Middle Ground

Many shear zones have characteristics intermediate between those of brittle and ductile shear zones (Ramsay, 1980a). These shear zones display some distinctly brittle aspects, like fractures, in combination with more ductile aspects. Brittle versus ductile character may change along a shear zone as it encounters rocks of contrasting mechanical properties. Some rock types affected by the shear zone may respond brittlely, whereas others respond by ductile flow. In other cases, a shear zone may operate under progressively changing physical conditions, from ductile to brittle (e.g., progressive uplift accompanied by a drop in temperature and pressure). Alternatively, a shear zone may be reactivated under physical conditions totally different from those under which it first formed. All these situations can produce a shear zone that is neither strictly brittle nor strictly ductile. We subdivide such shear zones into two general types: **semibrittle** and **brittle–ductile**.

Semibrittle Shear Zones

Although dominated by brittle deformation mechanisms like fracturing and cataclastic flow, **semibrittle shear zones** contain some ductile aspects as well (Figures 10.22, 10.23). A common example of a semibrittle shear zone is a zone of **en echelon veins** or **en echelon joints** (Figures 10.22, 10.23*A*). Deformation along the zone is accommodated by brittle mode I fractures, now filled by veins, and by distributed deformation between the veins. Another common example is a zone of **en echelon stylolites**, formed by pressure

Figure 10.22 Semibrittle shear zones in the field defined by en echelon quartz veins. (*A*) A single set of veins defining a shear zone crossing diagonally across the photograph. [Photograph by G. H. Davis.] (*B*) Conjugate sets of en echelon calcite veins crossing in left center of photograph, as well as subvertical tension veins bisecting the acute angle between the shear zones. Jasper Park, Canada. [Photograph by Shelby J. Boardman.]

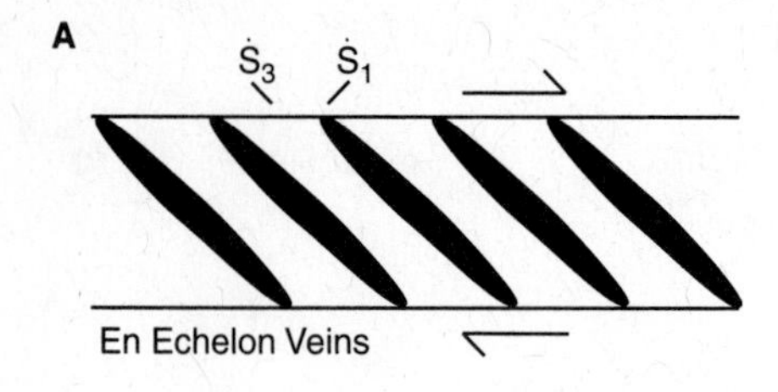

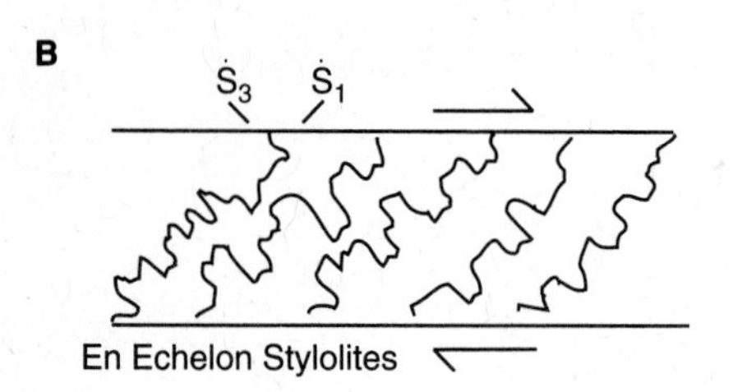

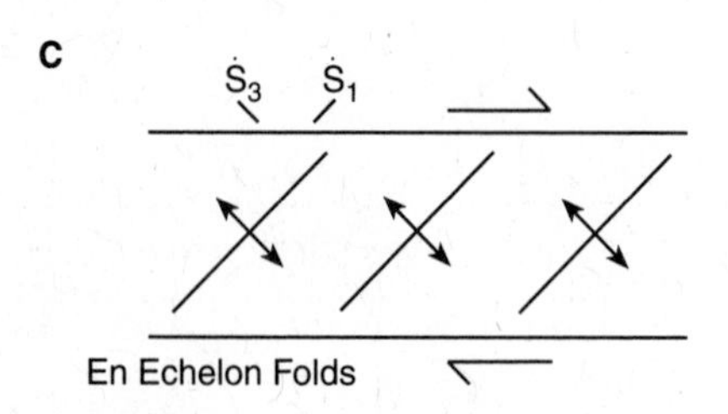

Figure 10.23 Semibrittle shear zones: (*A*) en echelon extension veins, (*B*) en echelon stylolites, and (*C*) en echelon folds. $\dot{S}_l$ is the axis of maximum instantaneous stretching, and $\dot{S}_3$ is the axis of minimum instantaneous shortening.

solution (Figure 10.23*B*). Some shear zones contain both veins and stylolites, so arranged that the shortening direction for the stylolites is approximately perpendicular to the extension direction indicated by the veins.

Shear zones defined by **en echelon folds** (Figure 10.23*C*) can be either semibrittle or ductile, depending on the conditions under which they form and on the character of associated structures. Many zones of en echelon folds are associated with faults and are probably best classified as semibrittle shear zones. The faults are brittle features, but the folding may occur by ductile mechanisms, such as pressure solution, without loss of cohesion of the rocks. Alternatively, folding may be accommodated by brittle or semibrittle mechanisms, including layer-parallel slip and pervasive small-scale faulting, jointing, and cataclasis. The folds appear ductile from a distance, but up close the rocks look as if they have deformed in brittle fashion. Some zones of en echelon folds are formed at greater depths, where truly ductile mechanisms prevail.

Brittle–Ductile Shear Zones

Brittle–ductile shear zones contain evidence of deformation by both brittle and ductile mechanisms and come in many flavors. Nearly all contain some type of tectonite fabric (Figure 10.24), such as mylonitic foliation and lineation, but this fabric may be well developed only in some mineralogic phases of the rock or in the more easily deformed rock types in a lithologically diverse sequence. Many brittle–ductile shear zones contain boudins, rock fragments, and porphyroclasts of the more brittle minerals and rock types, all floating in a tectonite matrix of more easily deformed minerals and rocks. In some brittle–ductile shear zones, tectonite fabric may resemble a mylonitic fabric in outcrop, but have many brittle aspects in thin section, including microfaults, grain-scale fractures, and zones of microbreccia and cataclasite. The margins of many brittle–ductile shear zones are sharply defined and faultlike, or are zones of concentrated fracturing and brecciation.

Brittle-ductile shear zones form when (1) the physical conditions permit brittle and ductile deformation to occur at the same time, (2) different parts of a rock have different mechanical properties, (3) a shear zone "*strain hardens*," (4) a short-term change in physical conditions, such as in strain rate, causes the rock to switch from ductile to brittle mechanisms or vice versa, (5) physical conditions change systematically during deformation, or (6) a shear

Figure 10.24 Brittle–ductile shear zones cutting Archean metavolcanic rocks along the Cadillac Break, Malartic Hygrade Mine, Quebec, Canada. [Photograph by S. J. Reynolds.]

zone is reactivated under physical conditions different from those in which the shear zone originally formed. We discuss each of these conditions below.

Many brittle–ductile shear zones form under physical conditions that permit brittle and ductile deformation mechanisms to be active at the same time (Figure 10.25*A*). This occurs because different deformation mechanisms overlap appreciably in the physical conditions under which they operate. Even in a monomineralic rock, like quartzite or marble, it is possible to have adjacent grains deforming by different mechanisms. Some calcite grains in a marble may deform by twin gliding while other nearby grains deform by microcracking. Such different responses could be controlled by variations in **crystallographic orientation** of the grains; the slip planes may be favorably oriented for slip in one grain but in a "*hard*" orientation in an adjacent grain. Different-sized grains of the same mineral may also accommodate deformation by different mechanisms. Even within a single grain, deformation may proceed by one mechanism until the grain experiences strain hardening and begins deforming by fracturing or some other mechanism.

A mixed brittle–ductile style of deformation is also expected if different parts of a rock have different mechanical properties (Figure 10.25*B*). Most rocks contain more than one mineral, and each mineral may deform by a different mechanism, even under the same physical conditions (Holyoke and Tullis, 2006). In a marble containing both calcite and dolomite, the calcite begins behaving ductilely at temperatures at which dolomite is still brittle. Deformation at this temperature therefore yields a rock with brittlely fractured dolomite grains floating in a matrix of ductilely flowed calcite. At higher temperatures, dolomite is more ductile than calcite (Piane et al., 2008). A similar discrepancy in style of deformation may occur where a shear zone cuts through a lithologically, and therefore mechanically, heterogeneous sequence of rocks. Shear zone fabric may be well developed where the zone cuts a relatively weak rock type, such as a calcite marble, but poorly developed or of a different structural style where the zone cuts a stronger or more brittle rock unit, such as quartzite or pegmatite. A shear zone cutting different rock types may contain boudins of relatively stiff rocks floating in a matrix of more easily deformed ones (Figure 10.25*B*).

Some shear zones may switch from ductile to brittle simply as a result of strain hardening. For example, some shear zones are fundamentally ductile in character but have abrupt margins and faultlike attributes. Such shear zones may begin as continuous ductile shear zones but develop into sharp, faultlike features when the shear strain is too great or too fast to be accommodated by strictly ductile means.

A mixed brittle–ductile character to a shear zone may also indicate that physical conditions fluctuated during deformation. The conditions most likely to experience short-term changes are strain rate and fluid pressure. A short-term increase in strain rate, such as during a seismic event, can cause a rock to switch from ductile to brittle behavior. Likewise, an increase in fluid pressure acts to reduce the effective confining stress and can cause a ductilely deforming rock to fracture. Ductile deformation may resume after this short-term, brittle event.

In many cases, the brittle–ductile character of a shear zone indicates either that the physical conditions systematically changed during deformation or that the shear zone formed under one set of conditions and was later reactivated under much different conditions. When conditions change from *ductile to brittle*, brittle structures, such as fractures, will overprint an earlier ductile fabric in the shear zone (Figure 10.26*A*). It is more difficult recognizing a shear zone formed during a change from *brittle to ductile* conditions, because early, brittle structures may be totally overprinted and "*healed*" by later ductile fabric and metamorphic minerals (Figure 10.26*B*).

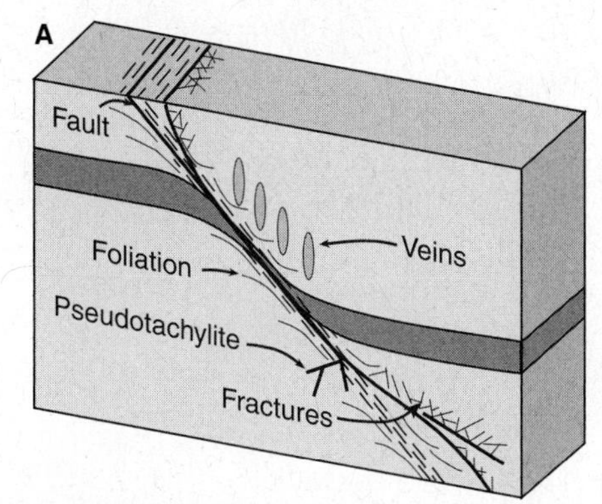

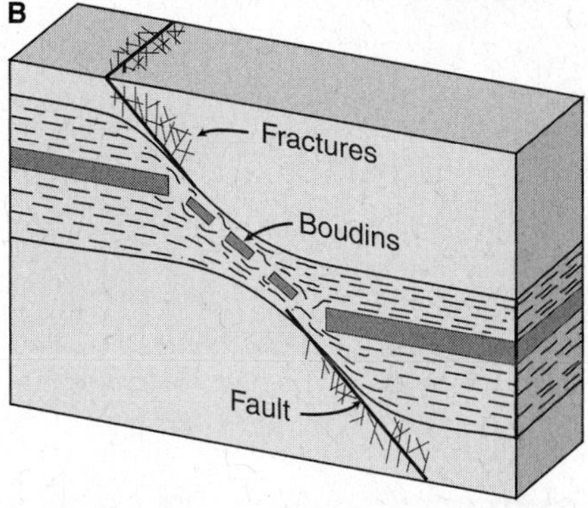

Figure 10.25 Brittle–ductile shear zones. (*A*) Formed by intermediate (brittle–ductile) conditions, where brittle, ductile, and semi-brittle deformation occurred during the same event, in part due to variations in strain rate and fluid pressure. (*B*) Formed in interlayered rocks with differing rheologies and responses to deformation. The thin, middle gray layer was relatively rigid and formed boudins because it is enveloped in a less competent and more foliated unit that deformed ductilely. The top and bottom layers were more competent and deformed by faulting and fracturing.

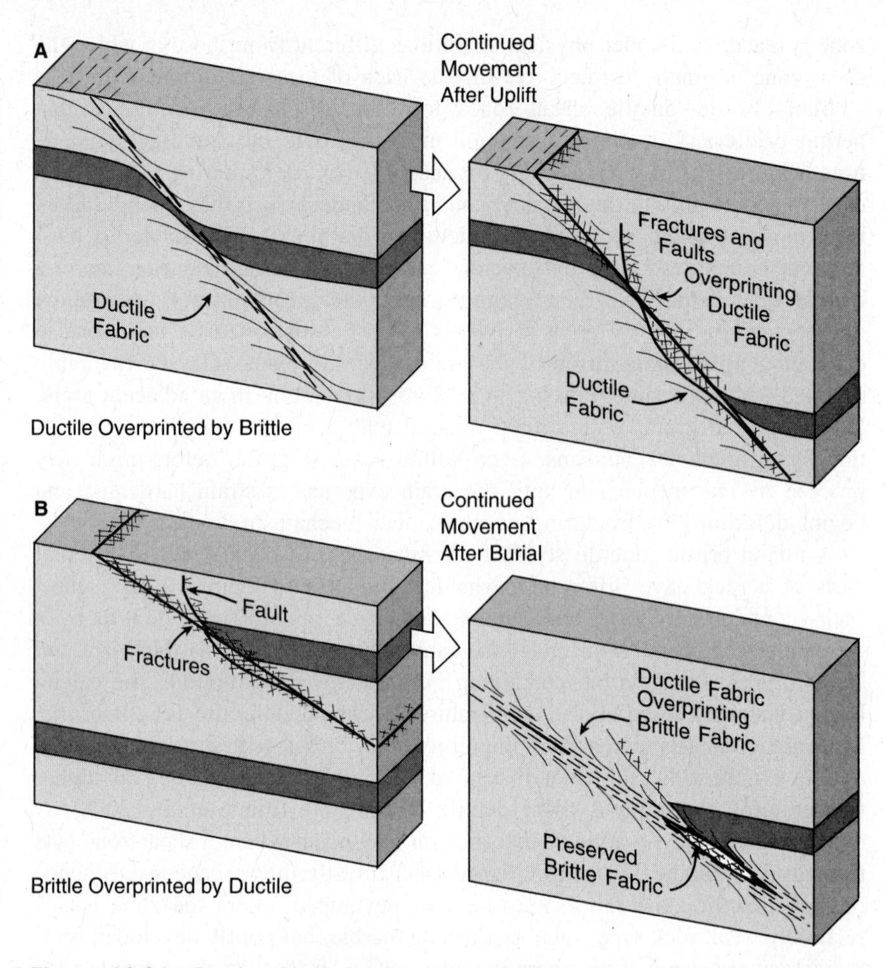

Figure 10.26 Brittle–ductile shear zones formed by a change in physical conditions during shearing or by reactivation of the shear zone under conditions different from those in which it first formed. (*A*) Ductile structures overprinted by brittle ones. A similar shear zone is shown in Figure 10.17. (*B*) Brittle structures overprinted and largely obliterated by a ductile overprint as rocks are buried.

WHY SHEAR ZONES FORM, THIN, AND THICKEN

The formation of a shear zone requires the concentration of deformation into a relatively thin zone, rather than a dispersion throughout the entire rock mass. As a shear zone develops and accommodates displacement over a long period of time, the *actively* deforming part of the zone may become thinner with time or may become wider by encroaching outward into less sheared protolith. What accounts for these behaviors? The explanation resides in how the different deformation mechanisms proceed and interact.

Softening Processes in Ductile Shear Zones

The localization of deformation into shear zones indicates that it is easier to continue deforming rocks *within the zone* than it is to broaden the zone by deforming the wall rocks. For this to occur, some processes must have caused rocks within the zone to be weakened or "*softened*" relative to rocks outside the zone (White et al., 1980). Four main types of **strain softening** operate in ductile shear zones: softening due to grain-size reduction, geometric softening, reaction softening, and fluid-related softening. Shear zones also may become softer if (1) metamorphism results in compositional bands, some of which are dominated by relatively weak minerals (Handy, Wissing et al., 1999; Park et al., 2006); (2) the zone becomes hotter than surrounding rocks because of shear heating (White et al., 1980; Pavlis, 1986; Molnar and England, 1990); or (3) weak, magma-rich zones develop as a result of partial melting or injection of magma (Druguet and Carreras, 2006).

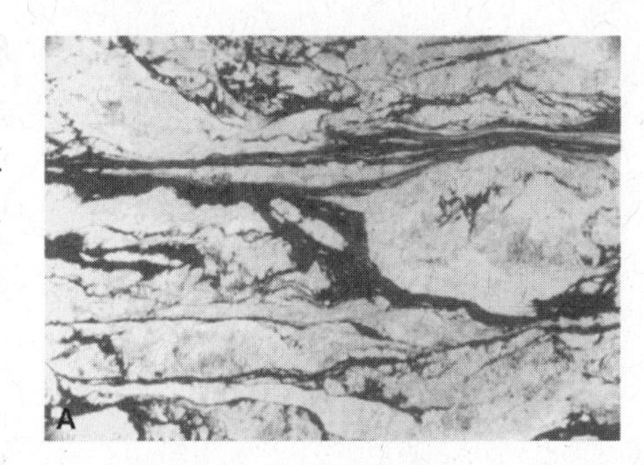

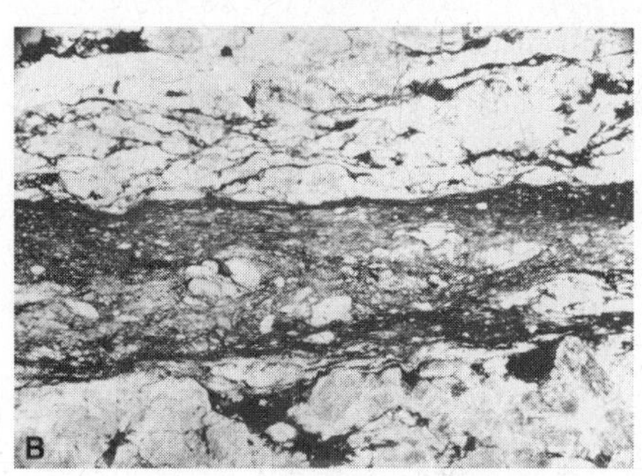

Figure 10.27 Localization of strain by fine-grained zones within Tertiary granodiorite, South Mountains, central Arizona. Dark shear zones in both (*A*) and (*B*) represent glass (pseudotachylite) formed by frictional melting during movement along small faults. The broad spectrum of deformational style, from elongated, ductilely deformed grains to angular, fractured, brittlely deformed ones, reflects the differing mechanical responses of feldspar (brittle) versus quartz and glass (both ductile), as well as likely variations in strain rate. Both photographs display an area 1.1 cm wide. [Photographs by S. J. Reynolds.]

Softening due to Grain-Size Reduction

In most ductile shear zones, there is a reduction of grain size of the affected rocks during deformation. Since many deformation mechanisms are most efficient in fine–grained rocks, a reduction in grain size can lead to strain softening and continued localization of shear in the most deformed, finest grained parts of the shear zone (Figure 10.27). A reduction in grain size may also permit a change in the dominant deformation mechanism. For example, a deforming rock may switch from dislocation creep to diffusion-accommodated grain-boundary sliding once the grain size has reached an acceptably small size. After the switch, deformation can continue through very high shear strains because it leaves the grains relatively defect-free, and thus there is no strain hardening. The abundance and distribution of grain sizes, along with the average grain size, controls the strength of a rock and how it softens during deformation (Herwegh et al., 2005).

Geometric Softening

In **geometric softening**, grains rotate during deformation until their slip systems are more favorably oriented for slip. Favorably oriented grains can easily continue slipping, whereas grains in "*hard*" orientations may rotate by rigid-body rotation or may shatter, dissolve, or accommodate deformation in some other way. Eventually, most grains end up with their slip systems aligned favorably for slip, and this leads to a **crystallographic preferred orientation** and an overall weakening of the rock.

The proportion and arrangement of weak versus strong minerals greatly controls the softening behavior, overall rock strength, and resulting fabric (Holyoke and Tullis, 2006). A key factor is whether weaker minerals form an **interconnected weak phase** that takes up most of the deformation, thereby softening the rock, or whether stronger minerals form a **load-bearing framework** that bears most of the stress, strengthening the rock by sheltering the weaker minerals (Handy, 1990).

Reaction Softening

Reaction softening occurs when deformation is accompanied by the formation of new minerals that deform more easily than the minerals from which they were derived (White and Knipe, 1978; Beach, 1980; White et al., 1986; Jefferies et al., 2006). Serpentine, an exceptionally weak mineral, commonly forms in shear zones cutting otherwise strong, ultramafic rocks. In rocks of felsic to intermediate composition, mylonitization under greenschist–facies conditions can cause feldspar to break down into white mica and cause amphibole to break down into biotite or chlorite, provided there is sufficient water for these **hydration reactions** reactions to proceed. The newly formed micas, which slip easily parallel to their well-developed basal cleavage, are generally easier to deform than either feldspar or amphibole. Therefore, as a rock becomes more micaceous during mylonitization, it becomes weaker and will tend to concentrate additional strain. This process may make the deeper parts of fault zones weaker than we would predict from Byerlee's Law (O'Hara, 2007).

Fluid-Related Softening

A ductilely deforming rock may also become weakened by other types of interaction with fluids. For example, fluids may dissolve and remove grains that otherwise resist ductile deformation. The dissolved material may not form new minerals, as in reaction softening, but may instead be reprecipitated as the same mineral in pressure shadows or veins, or else be entirely lost from the rock as the fluid flows out of the shear zone. Alternatively, fluids entering a shear zone may deposit new minerals, such as calcite, that are weaker than the preexisting minerals. Such fluids are responsible for many synkinematic veins of calcite and quartz that form parallel to foliation in ductile shear zones. The veins may concentrate further strain because they are (1) composed of weaker minerals than the wall rocks, (2) finer grained than the wall rocks, and (3) initially less deformed and therefore easier to deform than strain-hardened grains in the host rocks.

The presence of fluids may also cause a change in the dominant deformation mechanism. It is well known that fluids cause **hydrolytic weakening** of quartz, thereby permitting quartz to deform by dislocation creep. An increase in the *abundance* of fluids may cause dissolution creep or grain boundary diffusion creep to become relatively more important (Holyoke and Tullis, 2006). An increase in fluid pressure can cause strong grains (i.e., those that are relatively resistant to ductile flow) to rupture by hydrofracture. This in turn may cause the entire rock to become weaker as the strong grains accommodate strain by fracture and are dispersed as fragments into the ductilely flowing matrix, which then acts as an interconnected weak phase.

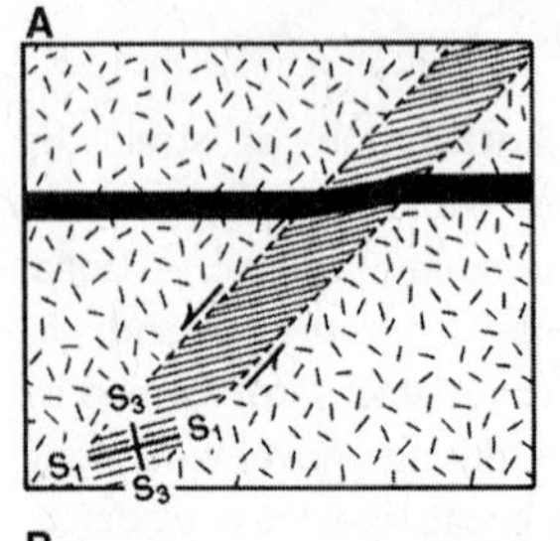

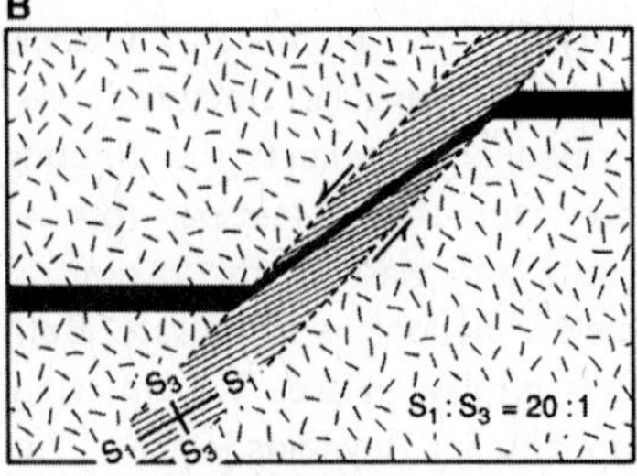

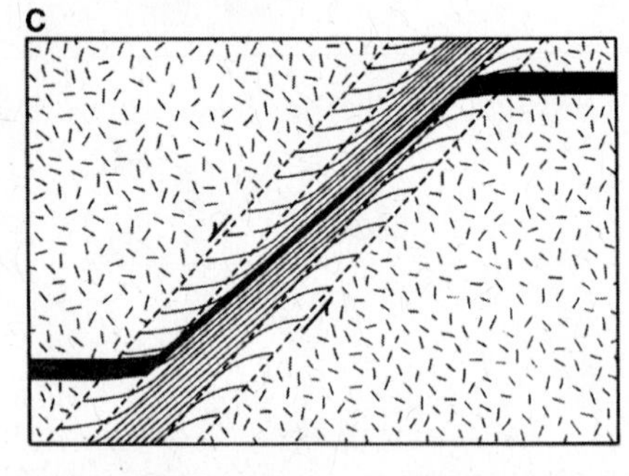

Figure 10.28 Kinematic development of a zone of simple shear. (*A*) Shearing results in stretching and thinning of the marker and initial development of foliation at 45° to the shear zone. (*B*) With progressive deformation, foliation leans over in the direction of shear and toward parallelism with the shear zone. (*C*) The shear zone widens, probably due to strain hardening, and foliation attains a sigmoidal form.

Strain Hardening in Ductile Shear Zones

If left unabated, the various types of strain softening should cause deformation to be continually concentrated into the same, very narrow zones, and we should see only thin shear zones. However, we actually observe some shear zones that are tens of kilometers wide. Also, we observe that shear zones with large displacements are generally wider than those with less displacement. This implies that ductile shear zones, like fault zones, widen as they accommodate increasing amounts of displacement (Figure 10.28). For a shear zone to widen, it must be *more difficult* to continue deforming rocks in the shear zone than it is to deform the adjacent wall rocks. In other words, the shear zone eventually undergoes **strain hardening** relative to the wall rocks. Strain hardening mostly occurs when highly deformed grains attain a high density of dislocations that become pinned or tangled and thereby impede further dislocation glide (see *Chapter 4, Deformation Mechanisms and Microstructures*). It also can result from mismatches along grain boundaries as adjacent grains attempt to deform against one another. The balance between strain softening and strain hardening is controlled by factors such as strain rate and temperature, and it determines whether a shear zone develops as a thin zone of extremely strained rocks or as a wider zone of less strained rocks.

STRAIN IN SHEAR ZONES

Deformation within shear zones can be accommodated by **displacement on discrete breaks**, by **straining of rocks** within or adjacent to the shear zone, and by **rotation** of relatively rigid objects entrained in the shear zone. Displacement on discrete breaks is easily visualized and can be evaluated via the offset of a marker, such as a dike, across the break (Figure 10.28). Strain within shear zones is typically heterogeneous and may have components of **distortion** (change in shape) and **dilation** (volume change). It can involve components of **coaxial deformation**, like pure shear, and **noncoaxial deformation**, such as simple shear. Objects entrained within a shear zone can exhibit complex, and perhaps unexpected, rotations.

Following the pioneering work of John Ramsay and others, we explore the geometry and distribution of strain in shear zones and their wall rocks, showing that strain can vary in some remarkably systematic and useful ways. These systematic variations contain important clues about the history of a shear zone, including its sense of shear and amount of displacement, and the kinds of strain (simple shear, dilation, etc.) that were involved.

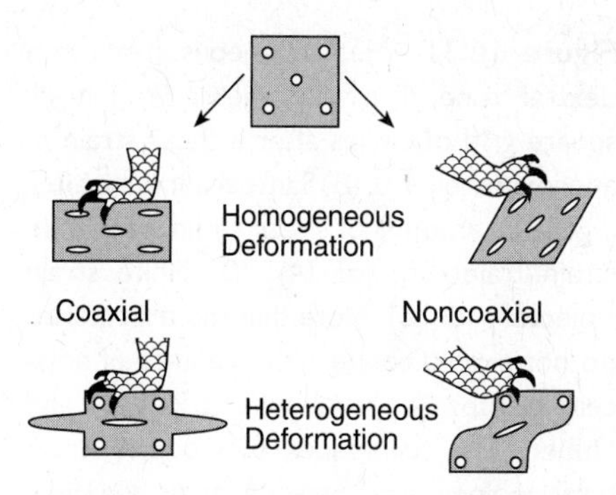

Figure 10.29 Homogeneous versus heterogeneous, and coaxial versus noncoaxial strain. [With kind permission from Springer Science+Business Media: Field geology of high-grade gneiss terrains, 1990, Passchier et al.]

Heterogeneous Strain

The mere presence of a shear zone indicates that the distribution of strain was heterogeneous rather than homogeneous—strain is clearly localized preferentially in the shear zone relative to the wall rocks. But how about strain *within* the shear zone? Is it also heterogeneous, and if so are the variations in strain systematic? How exactly does strain decrease from the shear zone into the wall rocks?

For deformation to be **homogeneous**, strain must be uniform in orientation and magnitude throughout the entire volume of material being considered (Figure 10.29). Lines that were straight before deformation are straight after deformation, and any two lines that were parallel before deformation remain so during the entire deformation history. Circles are transformed into ellipses, and spheres become ellipsoids.

In **heterogeneous** deformation, strain is not uniform (Figure 10.29). Some parts of the rock mass are more strongly deformed than others. Lateral variation in strain causes some lines that were originally straight to become curved, and may rotate two lines out of parallelism. A circle may be deformed into the outline of a teardrop or a boomerang, and a sphere may be converted into a shape that would be right at home in the produce section of your neighborhood grocery store.

In trying to understand strain in shear zones, we seek to compare the size, shape, and orientation of a deformed rock mass with its *initial* condition. Since size, shape, and orientation are three dimensional aspects, we ideally are interested in a three-dimensional representation of strain — the strain ellipsoid. For many shear zones, however, it is convenient to consider the special two-dimensional type of strain called **plane strain** (Figure 10.30; see *Chapter 2, Displacement and Strain*). In plane strain, shortening in one direction (S_3) is accommodated by lengthening in a perpendicular direction (S_1), with no strain in the S_2 direction. The two-dimensional character of plane strain allows it to be fully characterized with a two-dimensional strain ellipse. We begin with one variety of plane strain: simple shear.

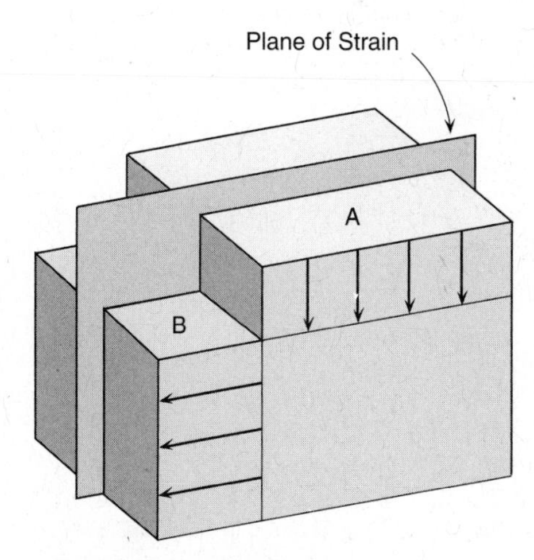

Figure 10.30 Deformation of a cube by plane strain, in this case, pure shear. Vertical shortening of initial cube "A" is accommodated by horizontal lengthening, resulting in the boxlike shape of "B." The motion of all material particles during deformation occurs along a family of parallel planes, represented by the shaded plane of strain cutting through the material.

The Geometry of Heterogeneous Simple Shear

The distribution of strain within a zone of heterogeneous simple shear after two large increments of strain is shown in Figure 10.31*A,B*. Because strain in the zone is **heterogeneous**, we have subdivided the rock mass into a number of small **domains** that are small enough to allow us to consider the deformation within each domain to be more or less **homogeneous**. We represent the state of strain within each domain with a finite strain ellipse, the shape and orientation of which varies between different domains.

The shear zone is defined by a zone of concentrated deformation cutting horizontally through the center of the material. The wall rocks are undeformed

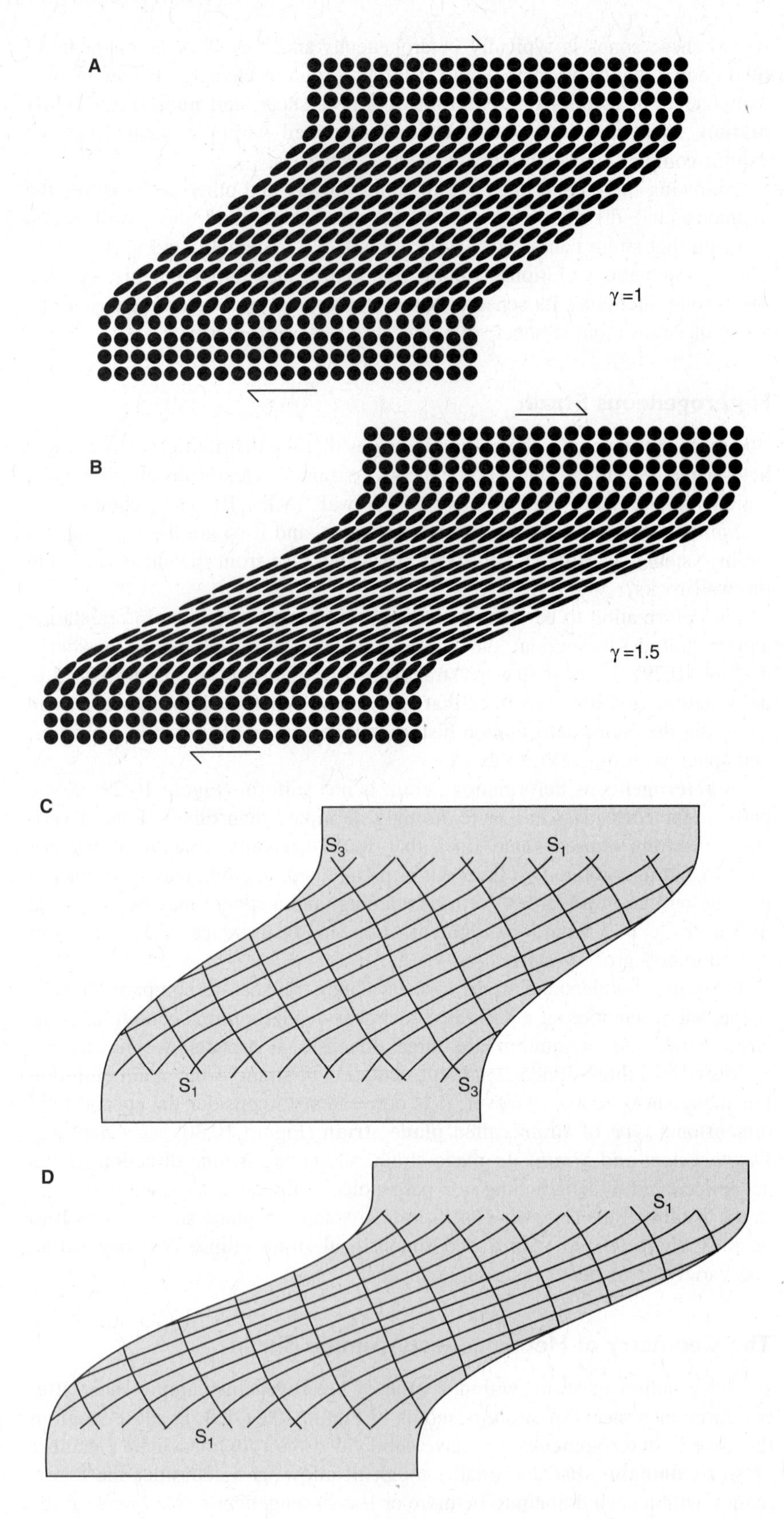

Figure 10.31 Heterogeneous strain in a dextral zone of simple shear. (*A*) Initially square grid of circles after a shear strain of approximately 1.0. (*B*) Same shear zone after a greater amount of shear strain. (*C*) Finite strain trajectory for (*A*). (*D*) Finite strain trajectory for (*B*). Note that the trajectories do not simply connect the centers of adjacent ellipses. [Reprinted from Ramsay and Huber, The techniques of modern structural geology, v. 1: Strain analysis, © 1983, with permission from Elsevier.]

away from the shear zone, but strain increases systematically into the zone. The finite strain ellipses become progressively more elliptical toward the center of the zone, without major discontinuities, recording a continuous strain gradient *perpendicular* to the shear zone. There is, however, no strain gradient *parallel* to the shear zone; strain ellipses are the same for all points that are the same distance from the center of the shear zone (i.e., in the same row in this figure). The concept of a **strain gradient** is illustrated in Figure 10.32.

If we were to overlay the shear zones in Figure 10.31*A* and *B* with tracing paper, we could mark the orientation of the long S_1 axis of the finite strain ellipse for each domain (Ramsay and Huber, 1983). We could then interpolate between the domains and draw smooth curves, called **finite strain trajectories**, showing how the orientation of S_1, the maximum finite stretch, varies across the shear zone (Figure 10.31*C*). Repeating this process for the short axis of the ellipse yields a second set of curves representing the trajectories of S_3. The trajectories representing S_1 are at 45° to the shear zone at the margins of the zone, where the strain is lowest, and curve toward lower angles with the shear plane as they approach the more highly strained center of the zone. This increase in strain toward the center of the shear zone causes the trajectories to be **sigmoidal**, especially after a greater amount of shear strain on the entire zone (Figure 10.31*D*).

The S_3 trajectories will be perpendicular to the S_1 trajectories at any point because the two strain axes are by definition always perpendicular. Accordingly, the S_3 trajectories first form at 45° to the shear zone, but with the opposite inclination of S_1. They rotate toward the normal to the shear zone with increasing strain. Because the S_1 and S_3 directions of the finite strain ellipse have the special properties of being originally perpendicular before deformation, the S_1 and S_3 trajectories were two sets of straight, perpendicular lines prior to deformation (Ramsay and Huber, 1987). When we look at finite strain trajectories, we are therefore seeing what a grid of originally perpendicular lines would look like after deformation.

In addition to being sigmoidal, the S_1 trajectories converge and diverge, being most closely spaced in the center of the zone. The convergence of S_1 trajectories reflects the increase in strain within the zone and becomes more pronounced at higher overall shear strains (Figure 10.31*D*). S_1 trajectories, and therefore foliation patterns, will always converge as they pass from regions of low strain to those of higher strain, except for some special geometries of dilation (Ramsay and Huber, 1983, 1987).

Finite strain trajectories are an incredibly useful way to study shear zones. For example, any feature in the rock that develops parallel to S_1 should display the same trajectory as S_1 across the shear zone. The trajectories reveal the underlying explanation for the curved geometry of foliation across a shear zone, as displayed in Figure 10.20*A*. Later, we will show how such trajectories can be used to determine the *amount* of shear strain across the entire zone (Ramsay and Huber, 1983).

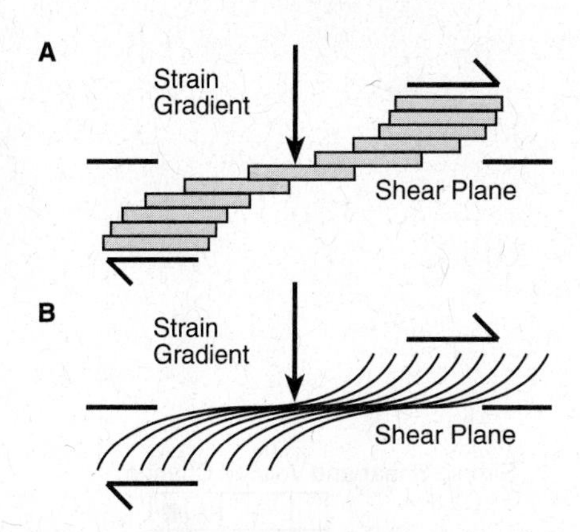

Figure 10.32 Strain gradient normal to a (*A*) discontinuous or (*B*) continuous shear zone. [Modified from Hanmer and Passchier, 1991. Reproduced with the permission of Natural Resources Canada 2011, courtesy of the Geological Survey of Canada, Paper 90–17.]

Components of Strain in Shear Zones

Ramsay and Graham (1970) demonstrated that only three main components of strain are possible in planar, parallel-sided shear zones. As illustrated in Figure 10.33, they are (1) heterogeneous simple shear, with the shear plane oriented parallel to the shear zone walls; (2) heterogeneous dilation, with the volume change occurring via displacement perpendicular to the shear zone walls; and (3) homogeneous strain of any type that affects the shear zone and its wall rocks. Although all three components of strain are possible, only the first two are **heterogeneous**, hence actually able to form a shear zone.

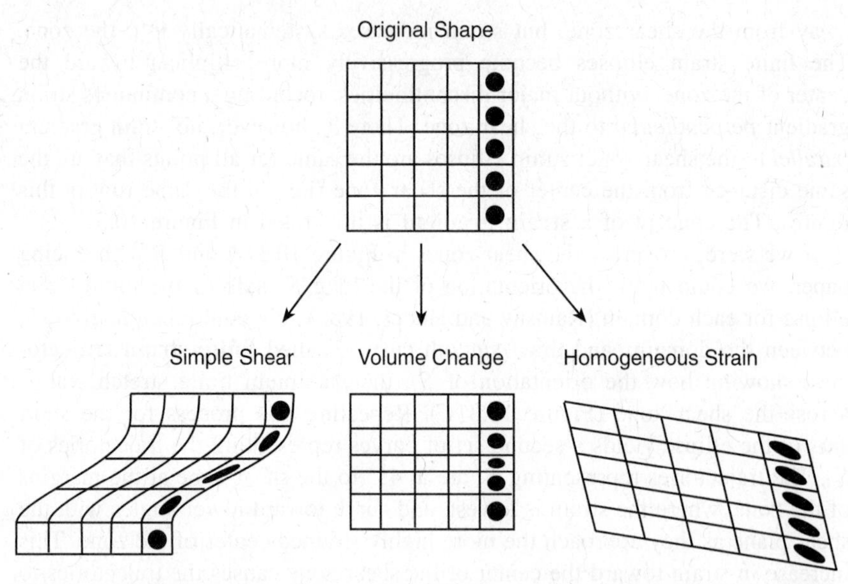

Figure 10.33 Three components of strain are possible in a planar, parallel-sided shear zone with no discontinuities along its margins. Components are simple shear, volume change perpendicular to the shear zone, and homogeneous strain affecting the zone and its wall rocks uniformly. [Reprinted from Ramsay and Huber, The techniques of modern structural geology, v. 2: Folds and Fractures, © 1987, with permission from Elsevier.]

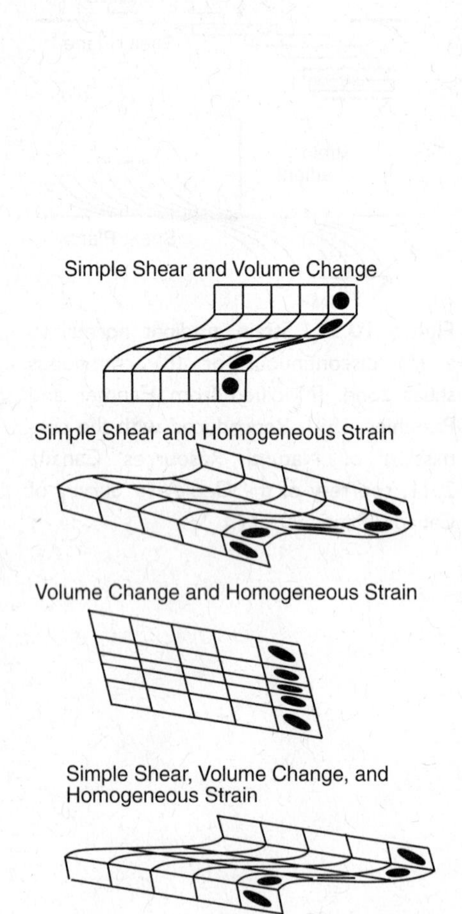

Figure 10.34 Various combinations of the three components of strain in Figure 10.33. [Reprinted from Ramsay and Huber, The techniques of modern structural geology, v. 1: Strain analysis, © 1983, with permission from Elsevier.]

These three types of strain can be combined in any proportions (Figure 10.34). We can also think of these three types of strain as the three components of strain possible within a parallel-sided shear zone. By treating each component separately, a concept known as **strain factorization** (Ramsay and Huber, 1987), we can mathematically isolate and predict the strain effects for each component. Equations describing the displacement caused by each component can then be mathematically combined, most easily through the use of matrix multiplication, to define the total state of strain and other aspects of shear-zone evolution (Fossen and Tikoff, 1993; Tikoff and Fossen, 1999).

In thinking about different components of strain within shear zones, it is helpful to consider shear zones that represent end members. One end member (Figure 10.35*A*) is zone of noncoaxial, simple shear, where the two sides of the zone shear past each other but there is no strain perpendicular to the zone. The opposite end member (Figure 10.35*B*) is a high-strain zone formed by strictly coaxial deformation, with shortening perpendicular to the zone but no shearing of one side of the zone relative to the other. A shear zone could have both components: **noncoaxial shearing** parallel to the zone, combined with **coaxial shortening** perpendicular to the zone, a situation called **transpression** (Figure 10.35*C*). If **noncoaxial shearing** is accompanied by **coaxial stretching** perpendicular to the zone, it is **transtension** (Figure 10.35*D*). Other variations are possible, such as where the coaxial strain is imposed at some angle to the zone, rather than being strictly perpendicular. Numerical models of various combinations of coaxial and noncoaxial strain (Fossen and Tikoff, 1993; Tikoff and Fossen, 1999) have provided important insight, such as the surprising fact that lineations in such zones can be perpendicular to the noncoaxial shearing direction if the amount of coaxial deformation overwhelms the contribution from shearing. As discussed in *Chapter 11, Active Tectonics*, transpression and transtension are an important aspect of many active and ancient shear zones.

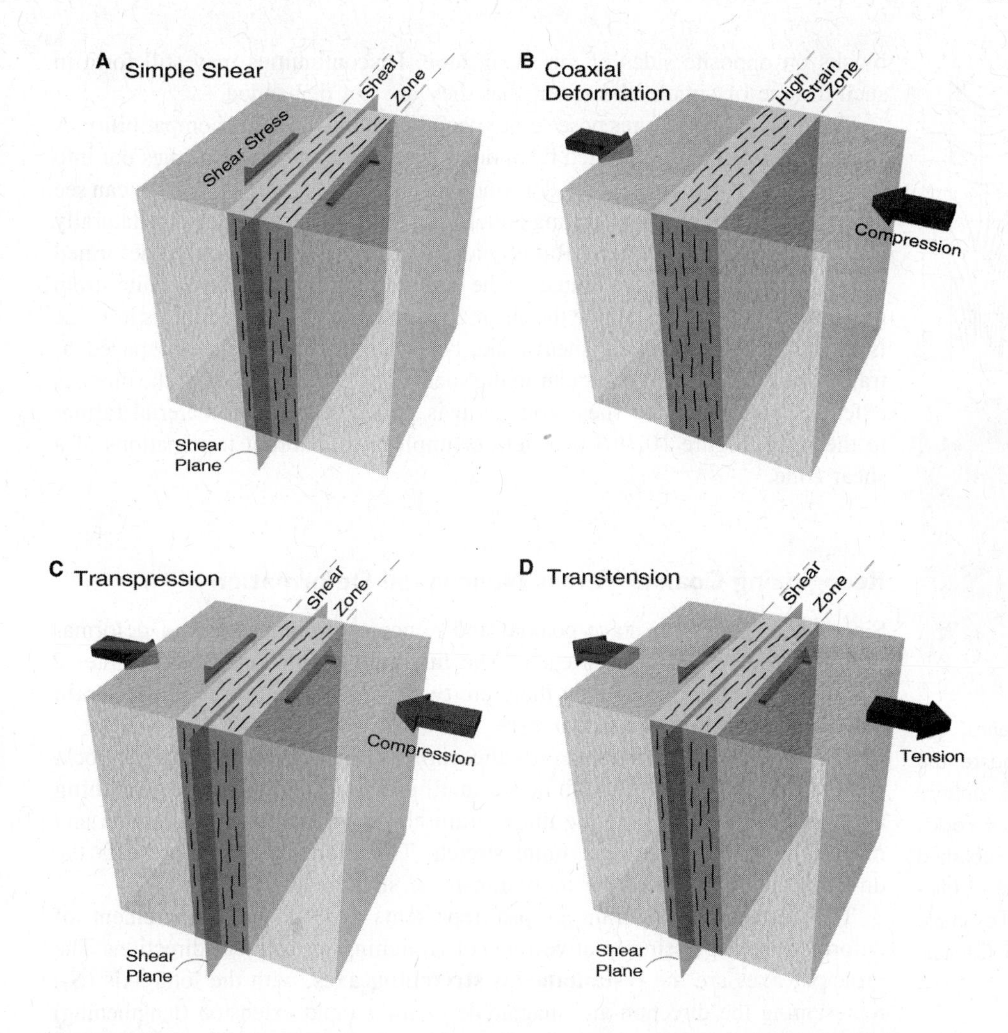

Figure 10.35 End members and combinations of coaxial and noncoaxial strain. (*A*) Zone of noncoaxial deformation, in this case simple shear; (*B*) Zone of coaxial deformation, with shortening perpendicular to the zone; (*C*) Transpression, the combination of noncoaxial shearing and coaxial shortening perpendicular to the zone; and (*D*) Transtension, the combination of noncoaxial shearing and coaxial stretching perpendicular to the zone.

Strain Compatibility in Shear Zones

An important aspect to consider in shear zones is **strain compatibility**, which specifies how strain can vary within a heterogeneously deformed material without causing structural discontinuities, holes between domains, or abrupt changes in the type of strain (Ramsay and Huber, 1983). Strain in two adjacent domains must be compatible; otherwise, room problems, such as gaps and overlaps, will develop along the interface. If material in one domain stretches or rotates more than the adjacent material, an intervening structural discontinuity must form to alleviate the resulting room problems and mismatches. The high-strain zone in Figure 10.35*B* has major strain compatibility issues, unless strain is occurring by volume loss or else the deforming material within the zone can be squeezed laterally or vertically out of the way.

A planar zone of simple shear (see Figures 10.31 and 10.35*A*) does not have such strain compatibility problems because (1) the noncoaxial strain in each domain is compatible with that in the adjacent domains, and (2) strain within the shear zone can be accommodated by the relative translation of

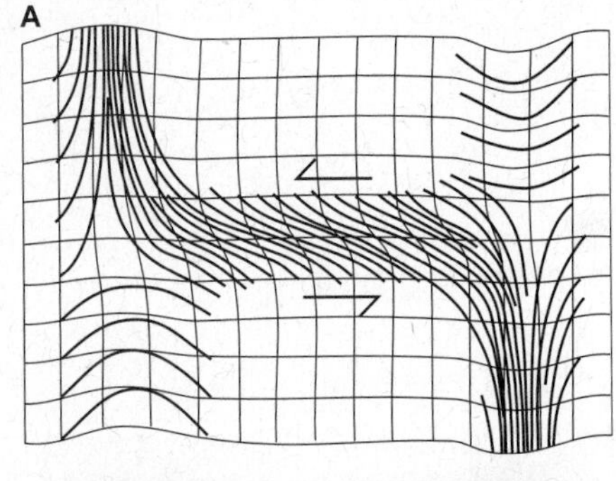

Figure 10.36 End of shear zones. (*A*) Strain pattern that can accommodate the termination of a planar zone of sinistral simple shear cutting undeformed wall rocks. [After Ramsay and Allison, 1979. Published with permission of the Swiss Society of Mineralogy and Petrology.] (*B*) A natural example in a granitic rock, Sudbury, Ontario, Canada. [Photograph by S. J. Reynolds.]

blocks on opposite sides of the shear zone. Discontinuities may still form in such a zone of simple shear, but that they are not demanded.

The ends of shear zones pose some special problems in strain compatibility. A shear zone is either terminated by another shear zone or somehow dies out into undeformed or homogeneously deformed material. In Figure 10.36*A*, we can see how a zone of simple shear cutting undeformed wall rocks might change laterally into a complex strain pattern (Ramsay and Allison, 1979). We see the deformed shapes of originally square domains; the darker lines represent the S_1 finite strain trajectory. As the block above the shear zone is translated to the left, its left side is shortened parallel to the shear zone, becoming a zone of closely spaced S_1 trajectories that are perpendicular to the shear zone. The right side of the block is extended parallel to the shear zone as it is pulled away from material farther to the right. Figure 10.36*B* is a field example of the abrupt terminations of a shear zone.

Recognizing Coaxial Versus Noncoaxial Deformation

Since shear zones can have coaxial and noncoaxial components of deformation, how do we distinguish each? The first step is to recall from *Chapter 2 (Displacement and Strain)* that there are two types of strain axes: **finite strain axes** and **instantaneous strain axes**.

The finite strain ellipse records the *entire* strain experienced by the rock, and the principal axes of the finite strain ellipse are called the **finite stretching axes**. The long, S_1 axis of the finite strain ellipse represents the direction and magnitude of the maximum finite stretch. The short, S_3 axis represents the direction and amount of the minimum finite stretch.

The instantaneous strain ellipse represents only a small increment of deformation, expressing relative *rates* of stretching in different directions The principal axes are the **instantaneous stretching axes**, with the long axis ($\dot{S}_1$) representing the direction and magnitude of most rapid extension (lengthening) and the short axis ($\dot{S}_3$) representing the direction and magnitude of most rapid shortening.

For coaxial deformation, such as pure shear, the instantaneous stretching axes are parallel and perpendicular to the shear plane (shear zone). The maximum shortening rate ($\dot{S}_3$) is normal to the shear plane, and the maximum extension rate ($\dot{S}_1$) is parallel to the shear plane (Figure 10.37*A*). The long axis of the *finite* strain ellipse (S_1) starts out parallel to the shear plane from the onset of deformation and remains so during the entire deformation. The finite stretching axes do not rotate, so the deformation is coaxial.

For simple shear, the *instantaneous* shortening axes are inclined 45° to the shear plane (Figure 10.33*B*). For this reason, the S_1 axis of the finite strain ellipse also starts at 45° to the shear zone, parallel to the maximum extension rate ($\dot{S}_1$). As deformation proceeds, however, S_1 successively rotates out of parallelism with the instantaneous stretching axis and toward parallelism with the shear zone, leaning over in the sense of shear. The finite stretching axes rotate away from parallelism with the instantaneous stretching axes, so the deformation is noncoaxial.

Combinations of coaxial and noncoaxial deformation produce intermediate results. For transpression, the component of coaxial shortening perpendicular to the walls of the shear zone causes the instantaneous strain ellipse to be at a lower angle to the shear zone than it is for simple shear; that is, the ellipse leans farther over in the direction of shear (Figure 10.37*C*). This orientation lies between the simple-shear and pure-shear end members, the result of combining the coaxial and noncoaxial components. This type of deformation, with a simple-shear component and a coaxial component applied

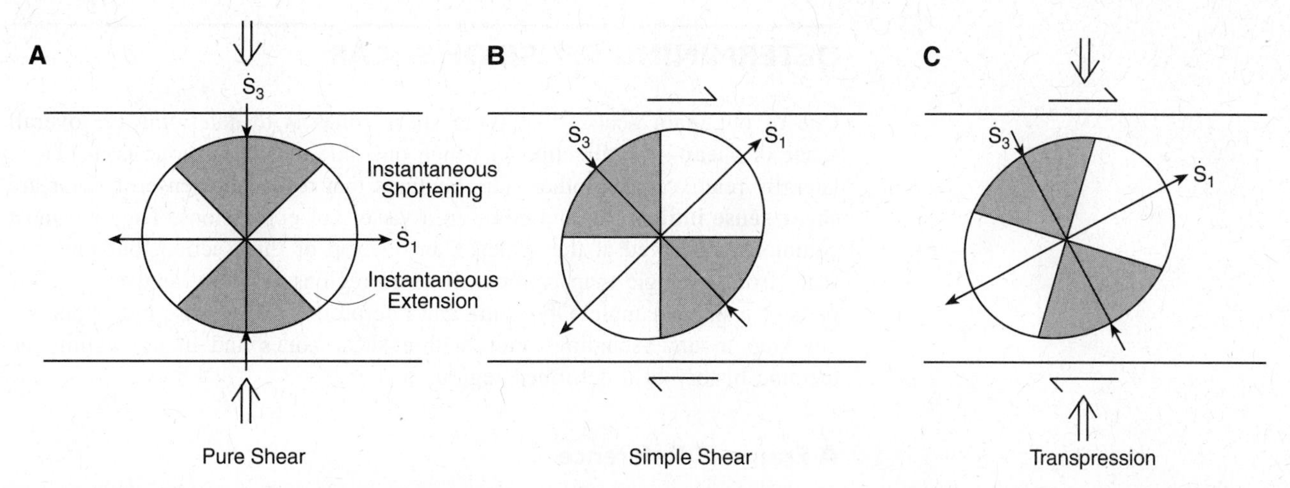

Figure 10.37 Instantaneous strain ellipses: (*A*) pure shear, (*B*) simple shear, and (*C*) transpression. The ellipses display the instantaneous stretching rate in all directions and can be subdivided into quadrants of positive instantaneous stretching rates (i.e., extension or lengthening) and negative instantaneous stretching rates (i.e., shortening). The instantaneous stretching axes are directions of maximum and minimum stretching rate, where the maximum extension rate is along $\dot{S}_1$ and the maximum shortening rate is along $\dot{S}_3$.

perpendicular to the shear zone, is **subsimple shear** (Simpson and De Poar, 1993). Subsimple shear is analogous to shearing a deck of cards, except the entire deck is composed of modeling clay and so can be stretched (thinned) during shearing. The noncoaxial component causes shearing of the cards, and the coaxial component causes the thinning. Subsimple shear is itself a type of **general shear** (Simpson and De Poar, 1993), which can involve the coaxial component being applied along any direction, including one not perpendicular to the zone. In general shear, the rock can deform in all three directions (not plane stain), resulting in potentially complex structural fabrics (Tikoff and Fossen, 1999; Iacopini, et al., 2007). General shear is analogous to shearing the deck of clay cards, while squeezing or pulling it from some random direction.

A measure of whether and by how much the finite stretching axes rotate is **vorticity**, which quantifies the amount of rotation (Means et al., 1980; Lister and Williams, 1983). If the axes do not rotate, as in coaxial deformation, the vorticity is zero. If the axes do rotate, as in any noncoaxial deformation, the vorticity is not zero. To compare different types of deformation, vorticity is normalized for strain rate, producing a dimensionless value called the **kinematic vorticity number** (W_k). Kinematic vorticity number is a measure of the relative contributions of the coaxial and noncoaxial components (Hanmer and Passchier, 1991; Simpson and De Paor, 1993; Weijermars, 1993). W_k is zero for coaxial deformation, 1.0 for simple shear, and in between (e.g., 0.6) for subsimple shear where there is shortening perpendicular to the zone (e.g., see Figure 10.37*C*). The closer the deformation is to being purely coaxial, the closer W_k is to zero.

At this point, we suspect that we should be able to distinguish coaxial and noncoaxial deformation (i.e., whether vorticity is zero) based on whether the associated small-scale structures are inclined to the shear zone. We also suspect that we should be able to tell the sense of shear on a noncoaxial shear zone by determining which way S_1 is inclined with respect to the zone. Specifically, S_1 will lean over to the right for dextral sense of shear and to the left for sinistral shear. As we shall see, we can indeed often do this by carefully observing the small-scale structures.

DETERMINING SENSE OF SHEAR

One of our main goals in studying shear zones is to determine the overall **sense of shear**—the direction in which one side of a shear zone is displaced laterally relative to the other side. Features that reveal the sense of shear are **shear-sense indicators**, and come in a variety of expressions. They are most commonly observed at the scale of an outcrop or thin section, but range in scale from geologic map patterns of large regions to X-ray analyses of tiny parts of a single sample. Observing and interpreting shear-sense indicators is a key step in understanding strain within shear zones and in unraveling the tectonic history of a deformed region.

A Frame of Reference

To determine the sense of shear for a shear zone, we need to have a convenient frame of reference, and we need to know in which direction we should look (Figure 10.38). This is true whether we are examining a natural outcrop or a thin section. For most ductile shear zones, the shear zone itself is the most convenient reference frame because generally it is a tangible feature that can be seen in the field and is parallel to the shear plane of the deformation. If we cannot see enough of the shear zone to determine its orientation—for example, when neither margin is exposed—we can use the foliation as a reference frame, keeping in mind that foliation will generally be inclined to the shear zone.

Once we have determined the orientation of the shear zone or foliation, we need to find the **line of transport**—the direction within the shear zone along which relative displacement occurred. If the rock has a stretching lineation

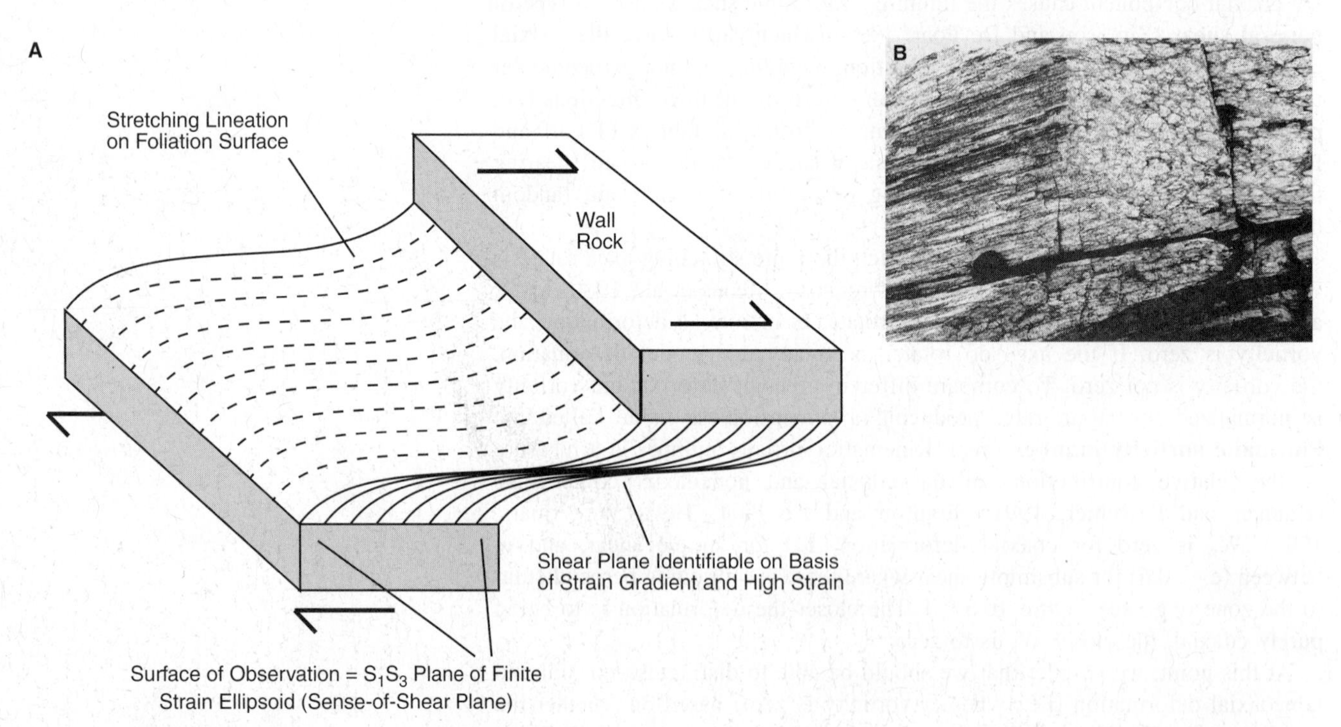

Figure 10.38 Relation of the sense-of-shear plane to structures in the rock. (*A*) Choose a rock face cut parallel to lineation and perpendicular to foliation to determine the sense of shear. This face will be parallel to the S_1S_3 plane of the finite strain ellipsoid. [Modified from Hanmer and Passchier, 1991. Reproduced with the permission of Natural Resources Canada 2011, courtesy of the Geological Survey of Canada, Paper 90–17.] (*B*) Mylonitic quartz-pebble conglomerate, Tortolita Mountains, southern Arizona. Sense-of-shear plane is face on left, which is parallel to lineation and perpendicular to foliation. Face on right, oriented perpendicular to lineation, reveals the conglomeratic nature of the protolith. [Photograph by S. J. Reynolds.]

(i.e., lineation parallel to S_1), this is a good place to start since displacement is generally parallel to stretching lineations for simple shear. Therefore, for a rock deformed by simple shear, a plane cut *parallel* to the stretching lineation and *perpendicular* to the shear zone or foliation (see Figure 10.38) will contain the line of transport. This is the plane in which all displacements occur in plane strain (S_1S_3), and is the plane we should inspect for shear-sense indicators. For convenience, we will call this plane the **sense-of-shear plane**, or simply the "*SOS*" plane — shear-sense indicators 'save our souls.'

Below, we briefly describe the most common shear-sense indicators. We restrict the discussion to shear-sense indicators found in ductile, brittle–ductile, and semibrittle shear zones, since criteria for sense of slip on faults (i.e., brittle shear zones) were discussed in *Chapter 6, Faults.* We will revisit some key shear-sense indicators later in this chapter to explore how they form and how they can be used to estimate the *amount* of strain in a shear zone. Even more detail on shear-sense indicators can be found elsewhere (Ramsay and Huber, 1983, 1987; Simpson and Schmid, 1983; Lister and Snoke, 1984; Simpson, 1986; Cobbold et al., 1987; Hanmer and Passchier, 1991; and Passchier and Trouw, 2005).

Offset Markers

Perhaps the most unambiguous shear-sense indicator is an offset marker, such as a distinctive dike, lithologic layer, or other rock unit (Figure 10.39*A*). In this case, we can determine both the amount and sense of displacement, as long as we are certain that the similar-appearing features on opposite sides of the shear zone are indeed equivalent and were originally continuous across the zone. In observing offsets, we should always question whether our perspective of the offset is correctly oriented (i.e., observing the SOS plane) to observe the total amount of offset. Also, the amount of apparent offset (separation) across a shear zone, like that along a fault, depends on the geometric relationships

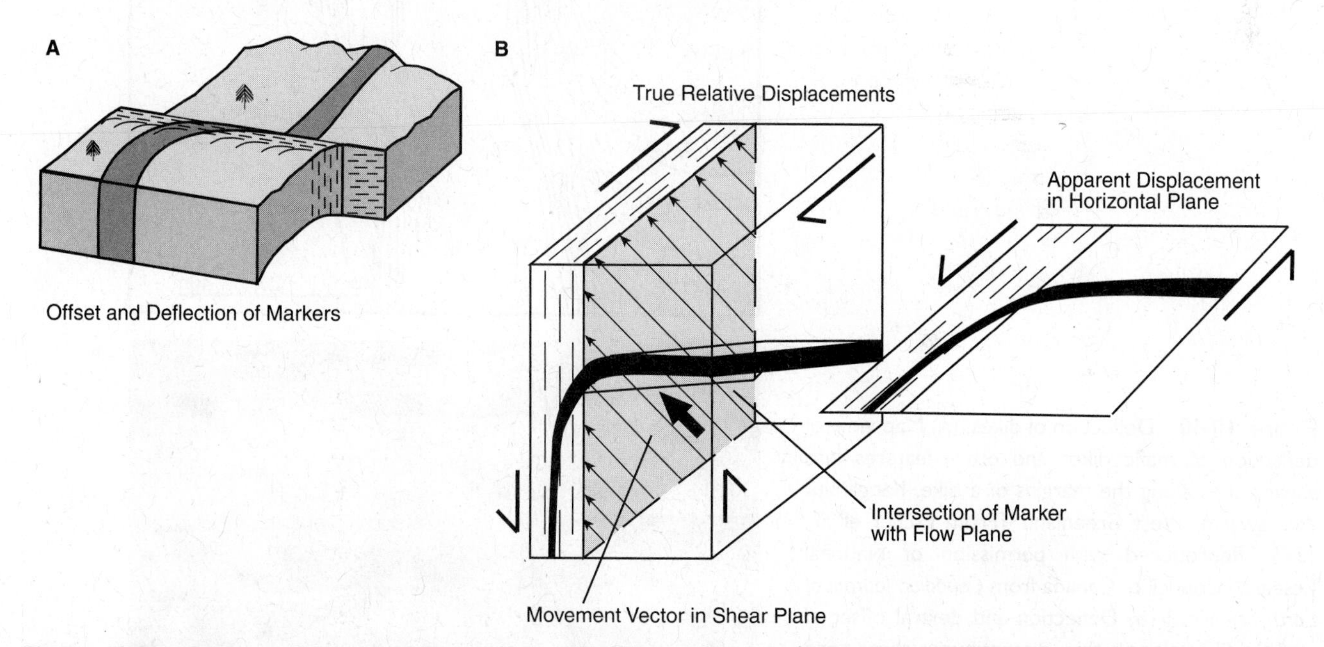

Figure 10.39 Offset and deflection of markers in ductile shear zones. (*A*) Discontinuous shear zone with dextral offset. Note deflection of marker into shear zone prior to its truncation. (*B*) Apparent relative displacement on a single improperly oriented surface (horizontal plane on right) may not reflect true displacement (arrows within plane of shear zone on left). [Modified from Hanmer and Passchier, 1991. Reproduced with the permission of Natural Resources Canada 2011, courtesy of the Geological Survey of Canada, Paper 90–17.]

between orientation of the offset feature, the orientation of the shear zone, and the actual displacement vector.

Deflection of Markers

Although sometimes we cannot find the actual offset of a marker, it is possible to observe how a marker is deflected as it goes from the wall rocks into the shear zone (see Figure 10.39*A*). The sense of the deflection reflects the sense of shear, as long as we are looking at the SOS plane. It is very easy to be fooled by an apparent sense of deflection on the wrong outcrop face (see Figure 10.39*B*).

Deflections of markers are not restricted to outcrop-scale features but may be visible on a larger scale, such as on a local or regional geologic map (Figure 10.40). A dike swarm or any other linear feature that trends into the shear zone at a high angle is especially diagnostic, provided the shear zone actually cuts the dikes. Post-shearing dikes may bend into a shear zone, even though they postdate all displacement, because they were preferentially intruded along the platy shear zone fabric, rather than breaking a new path across the zone. One of nature's playful tricks.

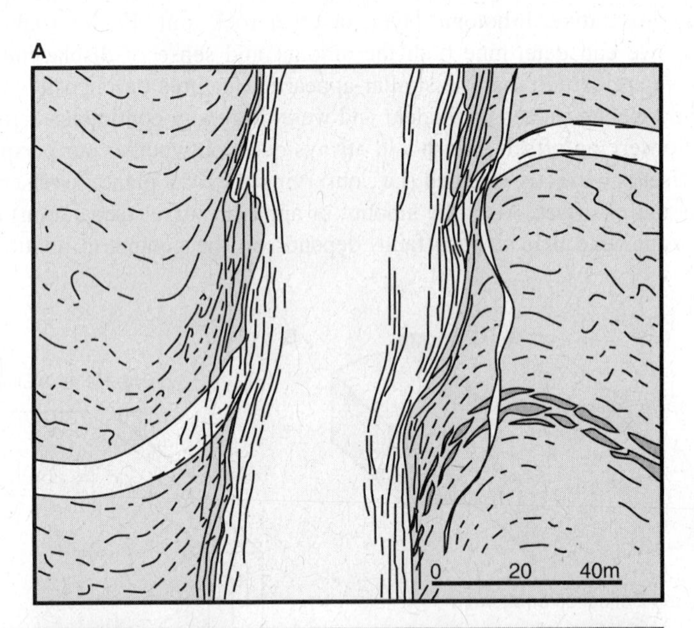

Figure 10.40 Deflection of dikes: (*A*) Map view of deflection of mafic dikes and other features into shear zones along the margins of a dike, Kangâmiut dike swarm, West Greenland. [From Escher et al., 1975. Reproduced with permission of National Research Council of Canada from *Canadian Journal of Earth Sciences*.] (*B*) Deflection and dextral offset of granitic dikes along a thin, discontinuous shear zone cutting Proterozoic granite, Harquahala Mountains, western Arizona. Note the shear zone fabric developed in rocks directly above the zone. [Photograph by S. J. Reynolds.]

Foliation Patterns

The systematic variations in the orientation of foliation across a shear zone, as expressed by foliation trajectories, provide one of the most useful shear-sense indicators. Foliation, because it generally reflects the S_1S_2 plane of the finite strain ellipsoid, is inclined to the shear zone for any noncoaxial deformation, and leans over in the sense of shear (Figure 10.41*A*). It commonly has a curved or a sigmoidal shape *across* a shear zone, with a sense of deflection consistent with the overall sense of shear (Figure 10.41*B*). Such foliation trajectories mimic variations in the orientation of the finite strain ellipse. Foliation will be rotated toward parallelism with the shear zone where the strain is higher, such as in the center of a shear zone (Figure 10.41*C*). Such foliation patterns are especially clear in thin shear zones, where the entire shear zone is exposed in a single outcrop (e.g., see Figure 10.20A). If only one margin of the shear zone is exposed (e.g., see Figure 10.20B), we should expect to see the foliation curve into the shear zone, reflecting the increase in shear strain toward the center of the zone.

Diagnostic foliation patterns may also exist across an entire field area or within a thick, regional shear zone (Figure 10.42). They appear in cross-sections of some dipping shear zones, where they provide an important shear-sense indication (Figure 10.43). Documenting foliation patterns in regional maps and cross-sections, especially in areas of poor exposure, may require

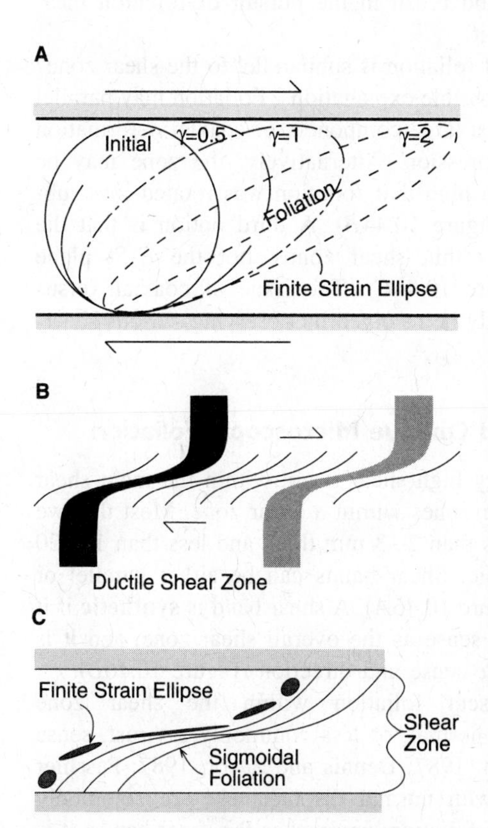

Figure 10.41 Foliation patterns in shear zones. (*A*) Finite strain ellipse and associated foliation lean over in the sense of shear and rotate toward the shear zone during progressive deformation. (*B*) Sigmoidal foliation patterns in a shear zone. (*C*) Variations in finite strain ellipse based on foliation patterns.

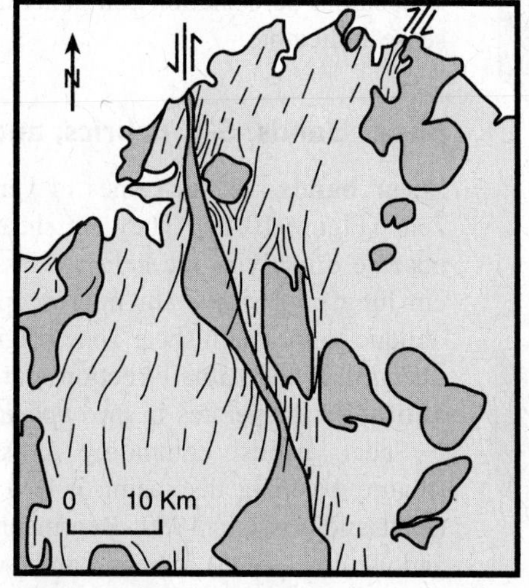

Figure 10.42 Foliation patterns in regional shear zones in the Iberian Arc, Galicia, Spain. [Reprinted from *Journal of Structural Geology*, v. 2, Ponce de Leon, M. I. and Choukroune, P., Shear zones in the Iberian Arc, pp. 63–68, © 1980, with permission from Elsevier.]

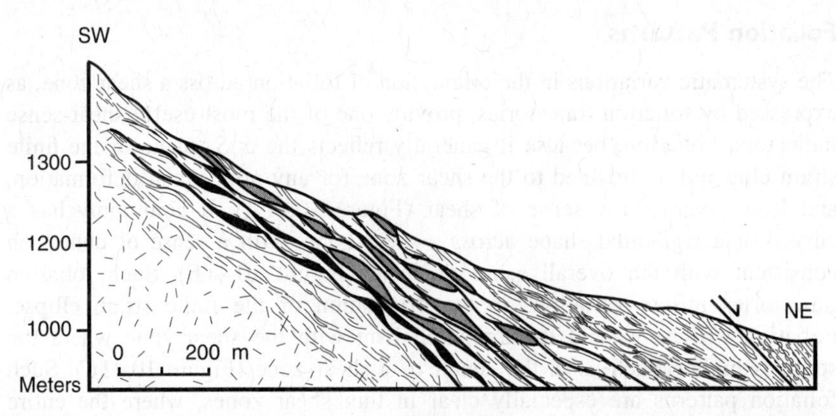

Figure 10.43 Foliation patterns in cross-section, Coyote Mountains, southern Arizona. [Reprinted from *Journal of Structural Geology*, v. 9, Davis et al., Shear zone origin of quartzite mylonite and mylonitic pegmatite in the Coyote Mountains, Arizona, U.S.A., pp. 289–298, © 1987, with permission from Elsevier.]

many traverses and very carefully taking numerous foliation measurements. But when a pattern emerges that documents the kinematics of the entire zone, the close encounters with cactus and brush in the pursuit of foliation measurements become suddenly worth it.

So what do we do if we find that foliation is subparallel to the shear zone? We should entertain at least three possible explanations. Foliation may parallel the shear zone because it expresses a large component of coaxial deformation (Figure 10.44*A*), such as in transpression. Alternatively, the zone may be noncoaxial, but shear strains are so high that foliation was rotated into subparallelism with the shear zone (Figure 10.44*B*). A third option is that the observed foliation itself represents thin shear zones, not the S_1S_2 plane of the finite strain ellipsoid (Figure 10.44*C*). The issue of coaxial versus noncoaxial deformation can generally be resolved by observing various shear-sense indicators.

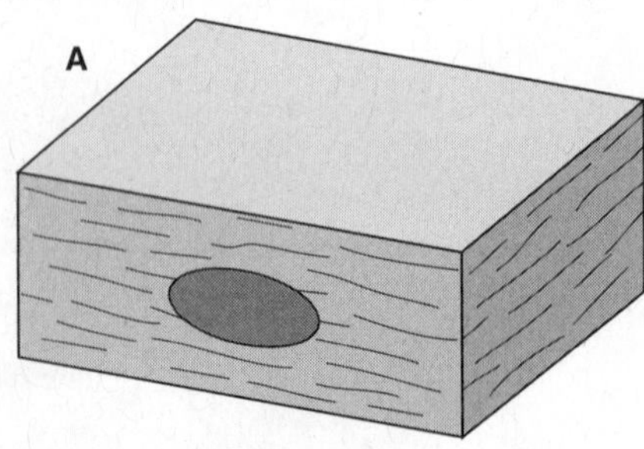

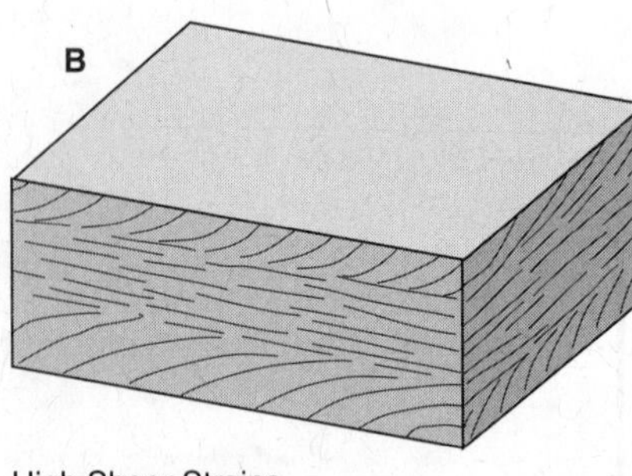

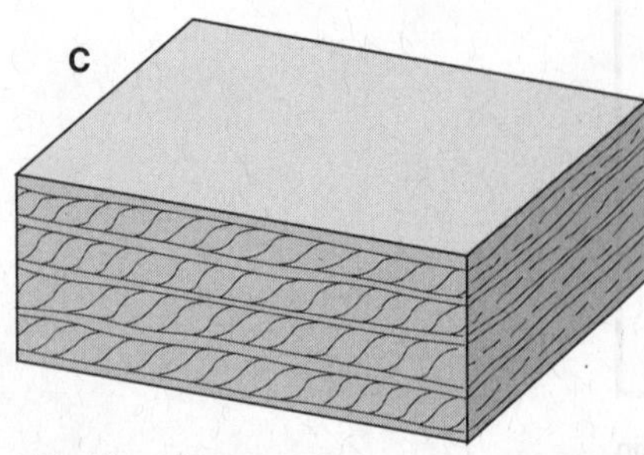

Figure 10.44 Foliation oriented parallel to a shear zone may be the result of (*A*) coaxial deformation, (*B*) high shear strain, or (*C*) the presence of thin discrete shear zones.

Shear Bands, S-C Fabrics, and Oblique Microscopic Foliation

Shear bands are thin zones of very high shear strain within the main shear zone (Figure 10.45). They are shear zones within a shear zone. Most that we observe directly in the field are less than 2–3 mm thick and less than 10–20 cm long, and many are microscopic. Shear bands can be either parallel or oblique to the main shear zone (Figure 10.46*A*). A shear band is **synthetic** if it operates in the *same* direction and sense as the overall shear zone, and it is **antithetic** if operates in an *opposite* sense and direction (Figure 10.46*B*).

Shear bands commonly crosscut foliation within the shear zone (Figure 10.46*B*), displacing it in a normal, or less commonly a thrust, sense (Platt and Vissers, 1980; Behrmann, 1987; Dennis and Secor, 1987; Passhier and Trouw, 2005). Shear bands with normal displacement are commonly called **extensional shear bands**. As foliation approaches the shear bands, it is deflected toward parallelism with the bands because of the associated increase in shear strain. The foliation patterns associated with shear bands therefore mimic those observed in a typical shear zone, and the same sense-of-shear principles apply.

S-C fabrics are among the most useful sense-of-shear indicators in ductile shear zones (Berthé et al., 1979; Simpson and Schmid, 1983; Lister and

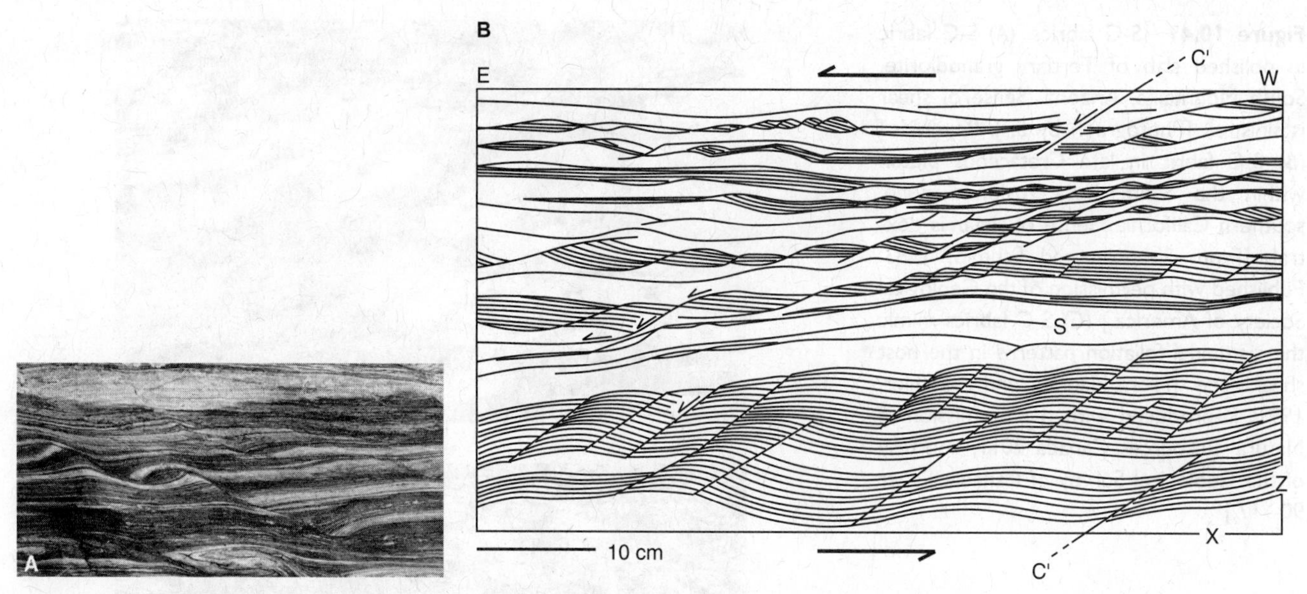

Figure 10.45 Shear bands: (*A*) Cuttting Rabassers quartzite near Cadaques, Spain. [Photograph by G. H. Davis.] (*B*) Cutting mica schist and quartzite, Raft River Mountains, Utah. Shear bands are labeled C′. [Reprinted from *Journal of Structural Geology*, v. 9, Malavieille, J., Kinematics of compressional and extensional ductile shearing deformation in a metamorphic core complex of the northeastern Basin and Range, pp. 541–554, © 1987, with permission from Elsevier.]

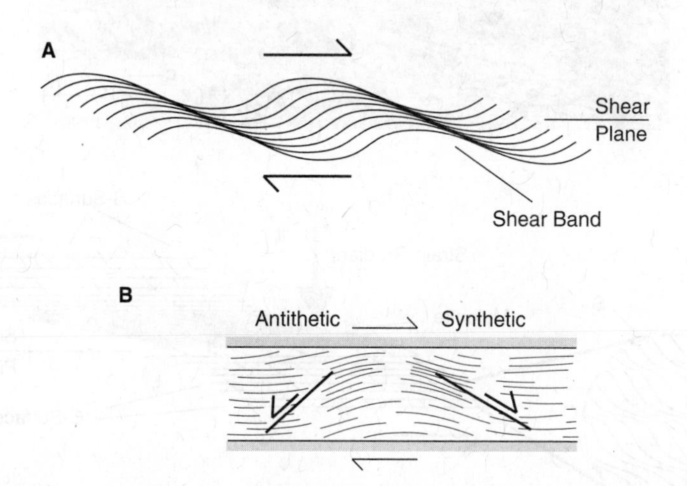

Figure 10.46 Geometries of extensional shear bands: (*A*) Shear bands commonly cut across the shear plane at approximately 30°. [Modified from Hanmer and Passchier, 1991. Reproduced with the permission of Natural Resources Canada 2011, courtesy of the Geological Survey of Canada, Paper 90–17.]

Snoke, 1984; Simpson, 1986; Hanmer and Passchier, 1991; Passchier and Trouw, 2005). They consist of two sets of planes or surfaces: foliation and shear bands (Figure 10.47*A*–*C*). The foliation planes are called **S-surfaces** (from the French term for "*schistosity*"), whereas the shear bands are denoted **C-surfaces** because they are zones of high shear strain ("*cisaillement*" is French for "*shear*"). C-surfaces typically are discrete zones of high shear strain 2–20 cm long and less than 1 mm thick. Most are aligned parallel to the shear zone and crosscut foliation. S-surfaces (i.e., foliation) deflect toward parallelism with a C-surface as they approach it and have a distinctive

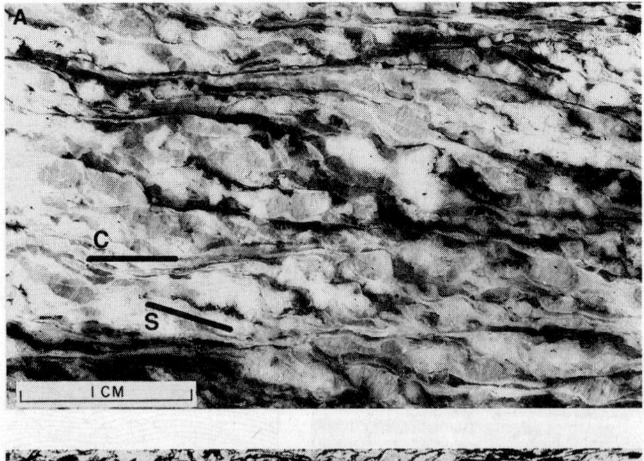

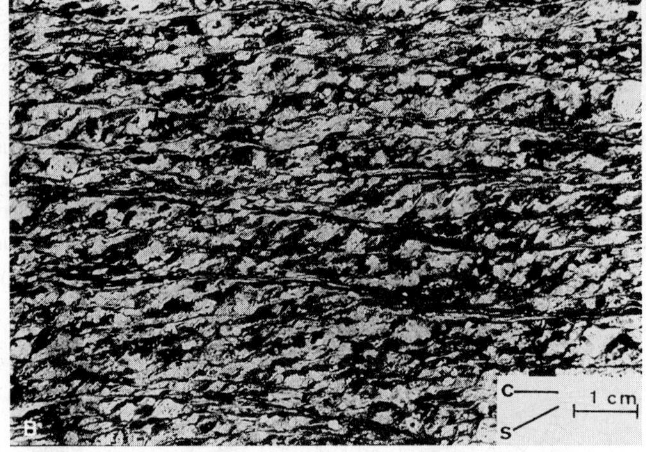

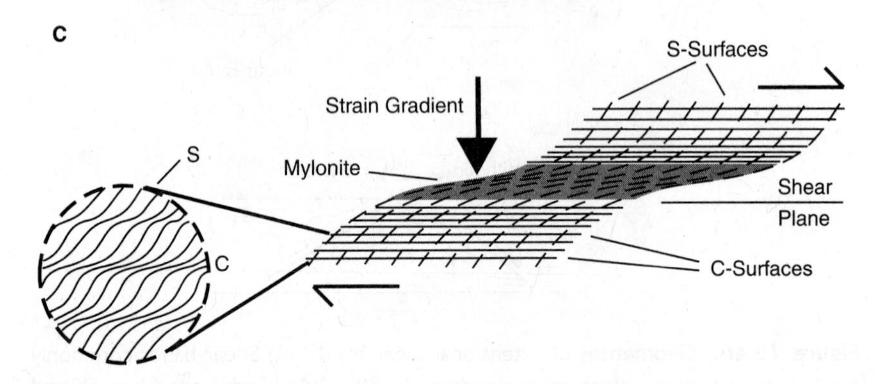

Figure 10.47 S-C fabrics. (*A*) S-C fabric in polished slab of Tertiary granodiorite, South Mountains, Arizona. Sense of shear is sinistral. [Photograph by S. J. Reynolds.] (*B*) S-C fabric in late Cretaceous pluton within the Santa Rosa mylonite zone, southern California. Sense of shear is dextral. [From Simpson and Schmid, 1983. Published with permission of the Geological Society of America.] (*C*) S-C fabrics mimic the sigmoidal foliation patterns in the host shear zone. [From Hanmer and Passchier, 1991. Reproduced with the permission of Natural Resources Canada 2011, courtesy of the Geological Survey of Canada, Paper 90–17.]

sigmoidal shape between adjacent C-surfaces. S-surfaces lean over in the direction of shear, relative to the C-surfaces. S-C fabrics are commonly visible in outcrop, once we have trained our eyes to recognize them. The clearest examples are in mylonitic granitic and gneissic rocks that contain coarse porphyroclasts or augen of feldspar. In some rocks, S-C fabrics are not obvious in outcrop but are well developed in thin section. Such thin sections should be cut parallel to the SOS plane (parallel to lineation, perpendicular to foliation), and from samples oriented in the field, so that the sense of shear interpreted from thin section can be related back to the field area.

Thin sections of mylonitic rocks commonly contain another shear-sense indicator, expressed as a microscopic foliation oblique to the main mylonitic foliation. The **oblique microscopic foliation** is defined by aligned subgrains

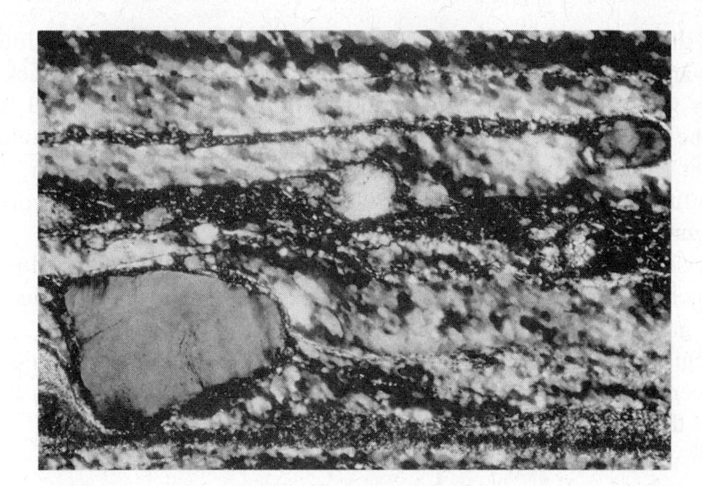

Figure 10.48 Photomicrograph of oblique foliation (diagonal, upper left to lower right) in quartz ribbons in mylonite. Sense of shear is sinistral. [Photograph by Carol Simpson.]

oblique to the long axis of larger individual grains and ribbons (Figure 10.48). The oblique foliation, like the S surfaces in S-C fabrics, leans over in the direction of shear relative to the main foliation defined by the larger grains. Later in the chapter, we explore how these fabrics form and whether they record all of the finite strain.

Mica Fish

Many mylonitic rocks contain lenticular porphyroclasts of muscovite and biotite, which have been termed **mica fish** (Lister and Snoke, 1984). The name refers to their troutlike shape (Figure 10.49). Mica fish, which are commonly asymmetric with respect to the mylonitic foliation or to shear bands, make excellent shear-sense indicators. Asymmetric mica fish are generally observed in thin section but can also be visible in hand specimen. Commonly, they are oriented approximately parallel to S-surfaces in S-C fabrics, and thus lean over in the sense of shear.

In many cases, the asymmetry of mica fish can be determined by observing their reflections in the sunlight (Reynolds and Lister, 1987; Simpson, 1986), an approach we enjoy referring to as the **fish-flash** method. Simply examine the outcrop for a potential hand specimen with a visible stretching lineation on foliation planes or on a shear band. Mark a north arrow on the top, so that later you can relate your shear-sense results back to the outcrop. With your

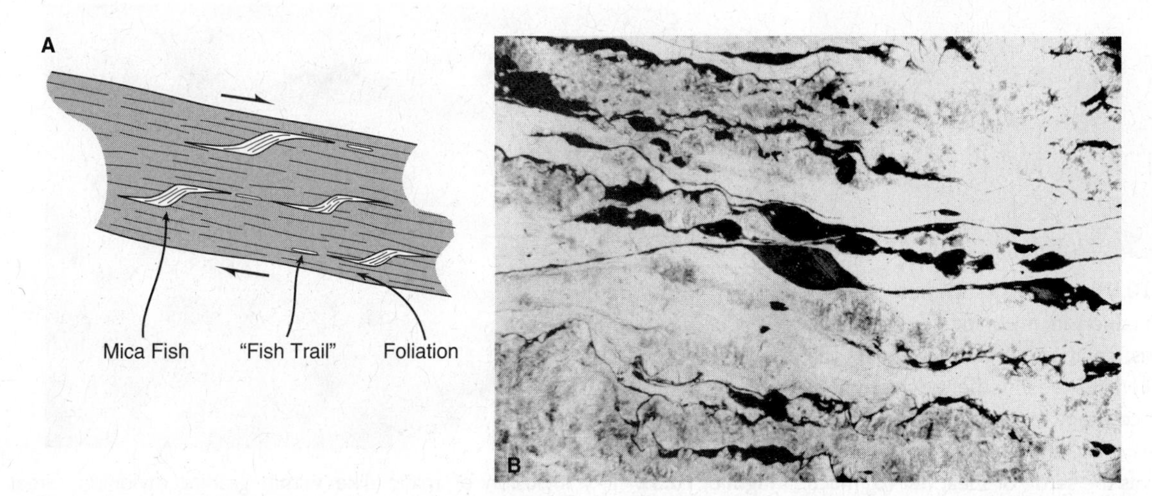

Figure 10.49 Mica fish. (*A*) Mica fish lean over in the sense of the dextral shear, and their tails are commonly parallel to C surfaces and to the shear zone. (*B*) Mylonitic Tertiary granodiorite, South Mountains, Arizona. Thin, dark tails streaming off the dark, 1 mm long, fish-shaped biotite crystal in the center of the photograph reflect sinistral shear. [Photograph by S. J. Reynolds.]

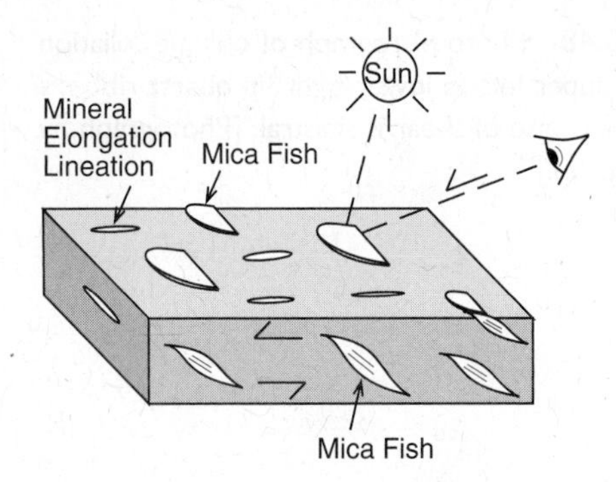

Figure 10.50 Fish-flash. Most mica fish in a rock will lean over in the same direction, resulting in a maximum reflectivity when looking down the lineation, in the same direction as the sense of shear. [C. Simpson, 1986. Determination of movement sense in mylonites. *Journal of Geoscience Education*, vol. 34.]

back to the sun and the sample held in front of you, look down the foliation surface and along the lineation, tilting the sample forward and backward, and note whether the sample looks dull or "*flashy*" (Figures 10.50 and 10.51). From the position of the sun and the rock, it is easy to determine the dominant inclination of the mica fish and thus the sense of shear. When you see the flash, you are looking in the same direction as the sense of shear (see Figure 10.51*A*).

The fish-flash technique works best on deformed rocks with several percent mica, including mica-bearing quartzites, dikes, and many fine- to medium-grained granitic rocks. It is less useful in rocks with abundant mica, such as mica schists and phyllonites, where the flash is commonly scattered or is nearly perpendicular to the foliation. In some rock types, the fish-flash technique is the only one available. "*Fishing*" permits are not required!

Inclusions

Outcrops of shear zones commonly contain various **inclusions**—discrete objects that are lithologically or mechanically distinct in some way from the main mass, or **matrix**, of the shear zone (Figure 10.52). Inclusions in shear zones are mostly centimeters to meters in diameter, but range from single grains less than a centimeter in diameter to rock fragments that are hundreds

Figure 10.51 Fish-flash viewed parallel to the lineation in mylonitic schist, Rincon Mountains, southern Arizona. Keeping the lighting conditions and the orientation of foliation constant, the sample shows (*A*) strong flash when viewed in the same direction as the sense of shear but (*B*) much less flash when viewed in the opposite (180°) direction. [Photographs by S. J. Reynolds.]

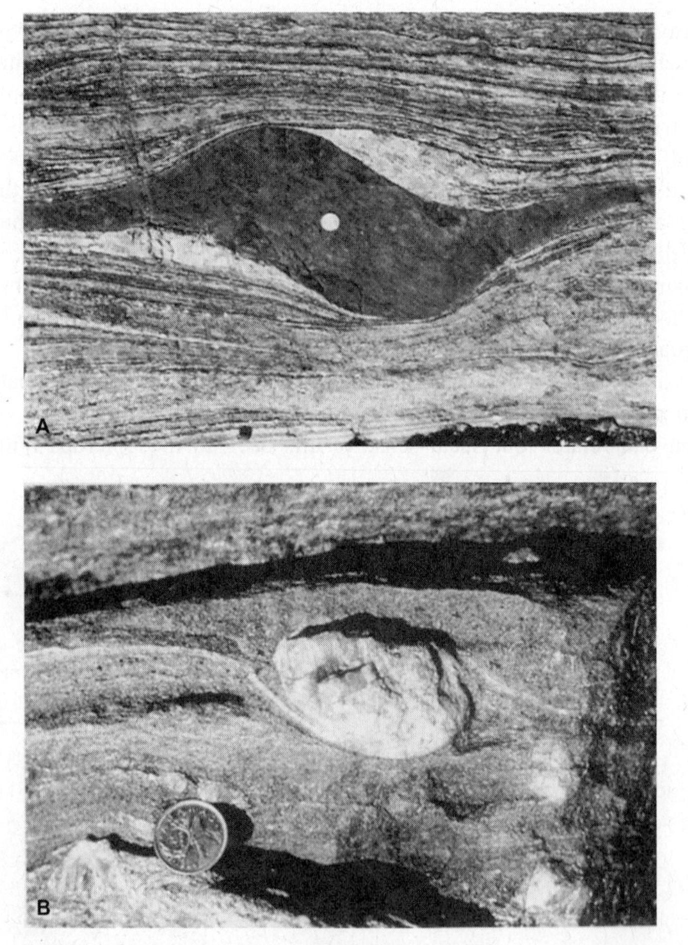

Figure 10.52 (*A*) Inclusion of mafic dike within granite mylonite, Great Slave Lake shear zone, Canada. Pressure shadows of melt flank the inclusion, reflecting dextral shear. [Photograph by Simon Hanmer.] (*B*) Inclusion with coiled wings, Sinistral shear. [Courtesy of Carol Simpson.]

of meters long. The can be nearly equant or be strongly elliptical, lenticular, or sigmoidal. They can have sharp or rounded corners. Most large inclusions are fragments of one rock type in another, such as original igneous inclusions or boudins derived from fracturing of a relatively rigid, competent layer. Crystal-sized inclusions may be fragments of older, large crystals, or they may be crystals that grew during metamorphism.

Many inclusions are more rigid than the surrounding, ductilely deformed matrix and resist distortion, whereas others deform similarly to the matrix. The term **inclusion**, as used in the context of shear zones, generally refers to an object that responded to deformation in a *more rigid* manner than the surrounding, ductilely deformed matrix (Hanmer and Passchier, 1991). There is a complete spectrum, however, from rigid inclusions to those that deform as if they were matrix. The shape and behavior of an inclusion help indicate whether it was rigid or deformable during shearing (Passchier and Trouw, 2005).

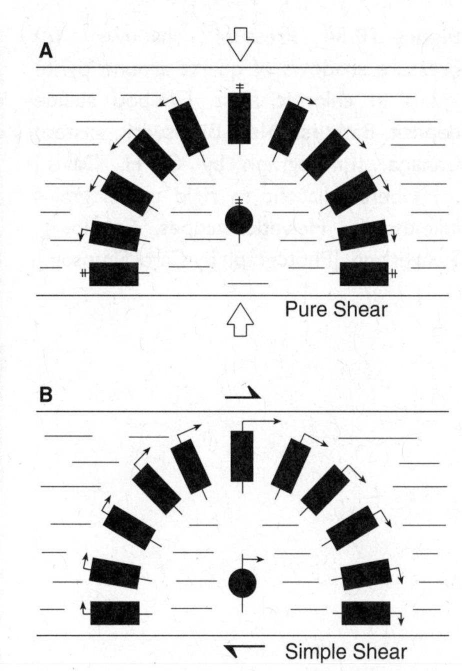

Figure 10.53 Rotation rates of inclusions as a function of orientation. The sense of rotation (clockwise versus counterclockwise) for each inclusion is shown by the direction of the attached arrow, with the relative rotation rates being proportional to the lengths of the arrows. Lines with double ticks do not rotate. (*A*) In a zone of coaxial strain, in this case pure shear, inclusions oriented parallel to either instantaneous stretching axis (parallel to and perpendicular to the shear zone) do not rotate. An equant inclusion (center of figure) does not rotate. Rotation rates are fastest for inclusions oriented at 45° to the shear zone. Note that some inclusions rotate clockwise, whereas others rotate counterclockwise. Rotation rates shown are for rigid inclusions with a 2:1 aspect ratio; longer inclusions may rotate faster. (*B*) In a zone of noncoaxial simple shear, all rigid inclusions, including equant ones, rotate in the same direction as the sense of shear. The slowest rate is for inclusions oriented nearly parallel to the shear zone, and the fastest rate is for those oriented perpendicular to the zone. Nonrigid, passive markers (not shown) do not rotate if they are strictly parallel to the shear zone. Equations are presented in Ghosh (1993).

Rotation of Inclusions

Inclusions provide sense-of-shear information by the way they rotate, deform, recrystallize, and interact with their matrix. The specifics of how an inclusion rotates relative to the matrix are more complex than we might suspect. Whether an inclusion rotates depends on its shape, rigidity, and orientation relative to the imposed strain, and on the type of strain (e.g., coaxial versus noncoaxial).

For coaxial deformation, a **rigid, equant inclusion** does not rotate, but an **inequant or deformable inclusion** may rotate, depending on its orientation (Figure 10.53*A*). For coaxial deformation and an initially random distribution of orientations of inequant or deformable inclusions, equal numbers should rotate with clockwise and counterclockwise senses. An individual inclusion, therefore, may have an asymmetry reflecting its sense of rotation, but the sense of rotation for a number of inclusions should be inconsistent and statistically symmetric with respect to the shear zone.

The story is different for a noncoaxial shear zone. During simple shear, inclusions will rotate with a uniform sense but at different rates, depending on their shape, orientation, and rigidity compared to the matrix (Figure 10.53*B*). Shear-sense indicators recording the rotation of a single inclusion and of the entire population of inclusions should be asymmetric and should reflect the overall shear sense of the zone.

Pressure Shadows

When inclusions are strong compared to the matrix, they help shield the matrix on the flanks of the inclusion from strain. These shielded areas, or **pressure shadows**, are wedge-shaped areas composed of less deformed matrix or of minerals that grew or recrystallized during deformation (Figure 10.54). Most pressure shadows are microscopic, and those visible in outcrop are typically less than 1 cm long. They are thickest adjacent to the associated inclusion and taper outward, commonly becoming aligned parallel to or at a low angle to foliation and lineation in the surrounding matrix.

Pressure shadows tend to be the locus of crystallization of quartz, calcite, chlorite, and other materials that are relatively mobile during deformation. These materials are deposited either as void fillings or as metamorphic or metasomatic replacements. Many pressure shadows, especially those of quartz, contain aligned fibrous or platy minerals connecting the inclusion and matrix (Ramsay and Huber, 1987).

Growth of minerals in pressure shadows largely occurs within zones of extension and decoupling between the rigid inclusion and the flowing matrix.

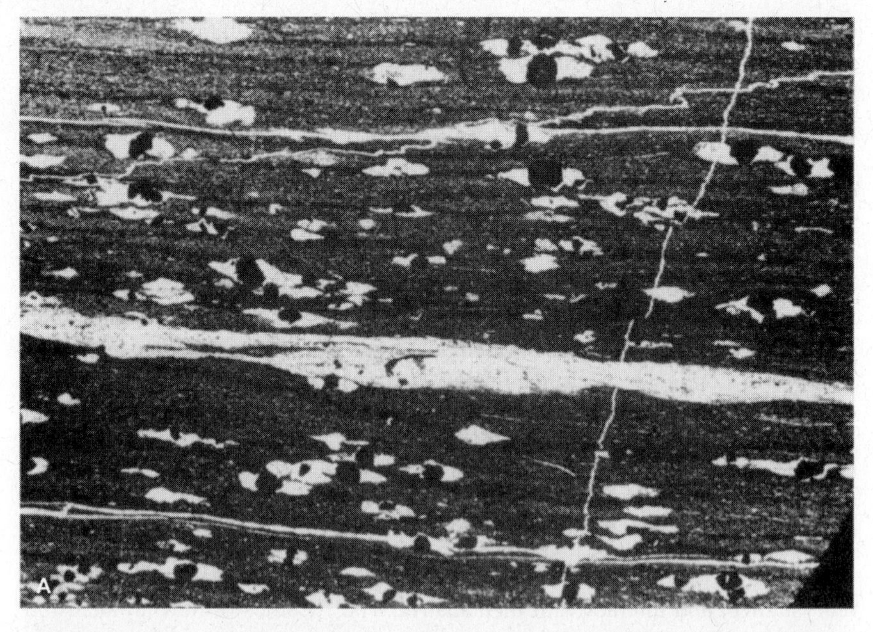

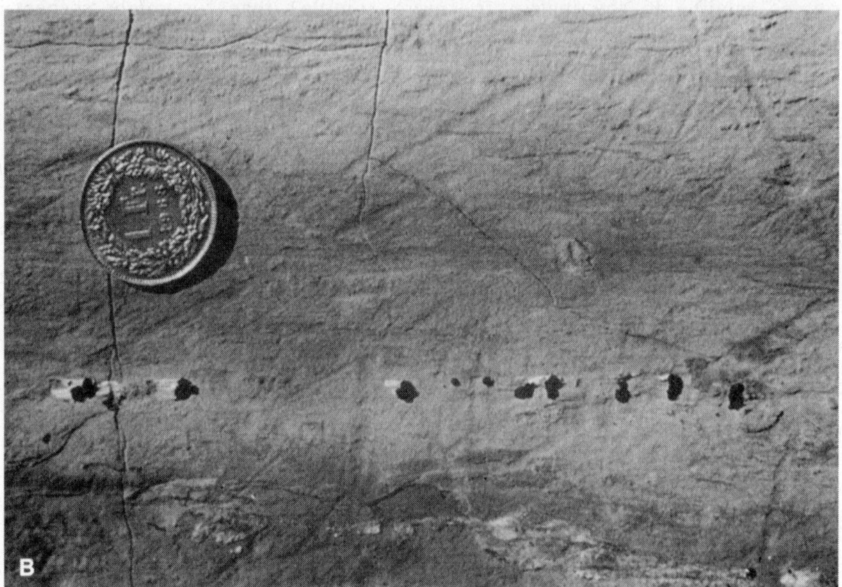

Figure 10.54 Pressure shadows. (*A*) Pressure shadows of quartz around pyrite grains in chloritic shist, Caribou sulfide deposit, Bathurst, New Brunswick, eastern Canada. [Photograph by G. H. Davis.] (*B*) Fibers adjacent to rigid pyrite grains inlimestone, Helvetic nappes, Engelberg, Switzerland. [Photograph by Carol Simpson.]

Mineral growth occurs gradually, accompanying each small increment of extension. For kinetic reasons, it may be easier to continue growing an existing crystalline grain than it is to form a new grain. As deformation proceeds, therefore, existing grains grow into long fibers or plates (Figure 10.54). Fibers can grow outward from grains in the inclusion or inward toward the inclusion from grains in the matrix (Durney and Ramsay, 1973; Ramsay and Huber, 1983). Alternatively, they grow as new mineral grains unrelated to those in either the inclusion or the matrix; this generally requires an external derivation of the constituents of the fibers — for example, redistribution of material from one rock type to another as a result of fluid flow during deformation.

Geometry of Fibers

The geometry of fiber growth in pressure shadows is controlled either by the incremental displacement direction between the inclusion and its matrix, or by the orientation of the faces of the inclusion. Some fibers within pressure shadows grow parallel to the local displacement direction between the inclusion and its matrix (Figure 10.55*A*), and are called **displacement-controlled fibers**. They record the displacement history between the inclusion and its matrix, thereby providing information about the progressive strain history of the rock. The shape of the inclusion controls the distribution, but not the orientation, of the fibers.

In other cases, fiber growth is controlled by the orientation of the faces of the inclusion, rather than local displacement directions (Figure 10.55*B*). Such **face-controlled fibers** generally grow perpendicular to the faces of the inclusion. Fibers growing from two adjacent faces meet at a **suture line** that extends outward from the corners of the inclusion (see Figure 10.55*B*). The geometry of individual fibers does not provide a direct record of progressive displacement, but the suture line tracks the displacement path of the associated corner of the inclusion as it was progressively pulled away from the matrix. Short fiber segments truncated by the suture line represent fibers that stopped growing when they were no longer in contact with the inclusion.

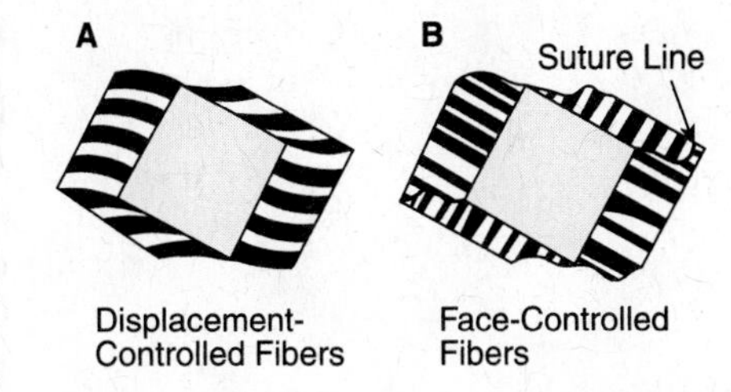

Figure 10.55 Geometry of fibers within pressure shadows. (*A*) Displacement-controlled fibers grow incrementally, parallel to the displacement direction between inclusion and matrix. (*B*) The geometry of face-controlled fibers is governed by the faces of the inclusion. The suture line records the progressive displacement of the corner of the crystal away from the matrix. [Reprinted from Ramsay and Huber, The techniques of modern structural geology, v. 1: Strain analysis, © 1983, with permission from Elsevier.]

Determining Sense of Shear from Pressure Shadows and Fibers

Pressure shadows form incrementally, generally approximately parallel to the axis of maximum *instantaneous* extension ($\dot{S}_1$). They should be parallel to the high-strain zone for coaxial deformation, but they will grow obliquely to the shear zone for noncoaxial simple shear (Figure 10.56), leaning over in the sense of shear.

The geometry of fibers within the shadows is also a shear-sense indicator (Figure 10.56). Displacement-controlled fibers grow incrementally, parallel to the direction of maximum instantaneous extension, and so lean over in the sense of shear. Face-controlled fibers can provide a sense of which way an inclusion is rotating during deformation, and therefore also reflect the sense of shear.

Beautiful, but geometrically complex pressure shadows and fibers can arise when the inclusion rotates during deformation (Figure 10.57). Etchecopar and Malavieille (1987) have successfully used computer models to explain how such ornate, batlike creatures grow. Other geometries can develop around inclusions that rotate, and these require some care when interpreting sense of shear (Hanmer and Passchier, 1991; Passchier and Trouw, 2005).

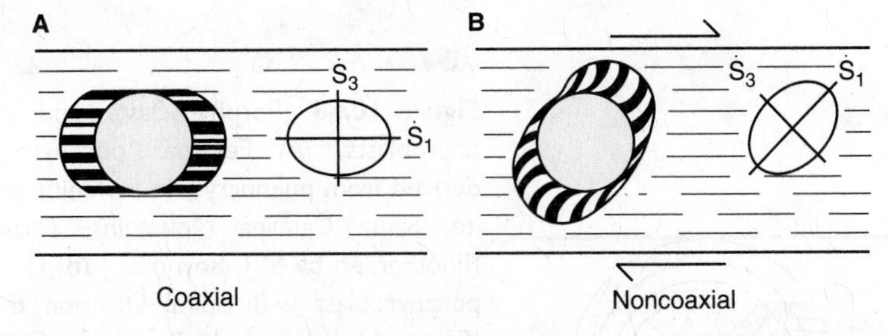

Figure 10.56 Pressure shadows, fibers, and instantaneous strain ellipses for coaxial and noncoaxial deformation. (*A*) Pressure shadows and fibers grow parallel to the shear zone during coaxial deformation, but (*B*) are inclined over in the direction of shear for noncoaxial deformation. In both cases, the shadows and fibers grow parallel to $\dot{S}_1$, the long axis of the instantaneous strain ellipse. [Reprinted from Ramsay and Huber, The techniques of modern structural geology, v. 1: Strain analysis, © 1983, with permission from Elsevier.]

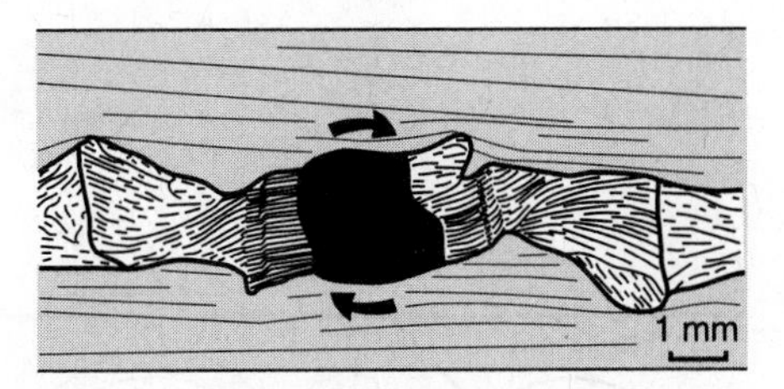

Figure 10.57 Bat-winged pressure shadow around pyrite. [Reprinted from *Journal of Structural Geology*, v. 9, Etchecopar, A. and Malavieille, J., Computer models of pressure shadows; a method for strain measurement and shear sense determination, pp. 667–677, © 1987, with permission from Elsevier.]

Porphyroclasts and Porphyroblasts

Ductile shear zones commonly contain relatively rigid grains of one mineral within a more strongly deformed, fine-grained matrix having a different mineralogy. Such rigid grains may represent relics from the original protolith, or they may be new metamorphic grains that grew during or after deformation. Grains interpreted to be brittlely reduced relics from the protolith are called **porphyroclasts** because they commonly represent fragments or **clasts** of original phenocrysts or detrital grains (Figure 10.58*A*). Many ductilely deformed granites, for example, contain large feldspar porphyroclasts, which represent original feldspar phenocrysts, now scattered in a more deformed, fine-grained matrix of quartz, feldspar, and mica. Likewise, metavolcanic and metasedimentary rocks may contain deformed original phenocrysts or detrital clasts of feldspar, mafic minerals, or quartz, which "*float*" in a strongly deformed, fine-grained, micaceous matrix. Under the proper conditions, porphyroclasts may flow ductilely or recrystallize during deformation.

If the large grains instead are new metamorphic grains that grew during or after deformation, they are called **porphyroblasts** (Figure 10.58*B*). Porphyroblasts that grow during deformation are **synkinematic** and those that grow after deformation are **postkinematic**. Porphyroblasts, once formed, may be very rigid compared to the matrix, as is the case for garnet. Other porphyroblasts, including andalusite and muscovite, are more deformable, and their response to deformation may be similar to that of the matrix. The distinction between a deformable inclusion and a rigid inclusion is important, because the two types may be associated with different suites of small-scale structures. Deformable porphyroblasts are less likely to be associated with pressure shadows and may be less instructive as shear-sense indicators.

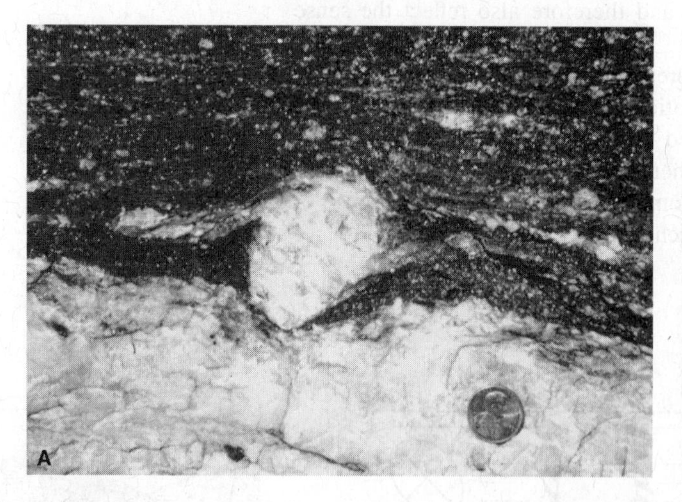

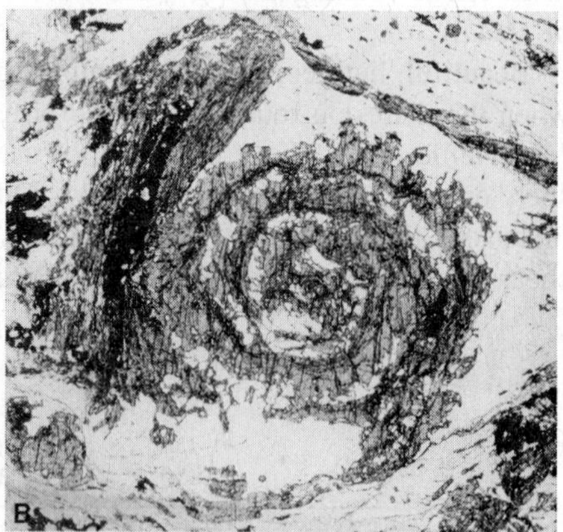

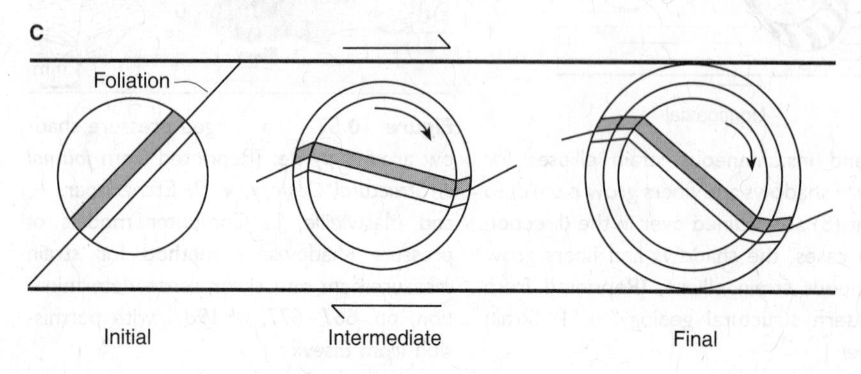

Figure 10.58 Porphyroclasts and porphyroblasts. (*A*) Feldspar porphyroclast derived from phenocryst in mylonitic granite, Santa Catalina Mountains, Arizona. [Photograph by S. J. Reynolds.] (*B*) Garnet porphyroblast with spiral inclusion trails. [From Hanmer and Passchier (1991). Courtesy of the Geological Survey of Canada and the authors.] (*C*) Sequential rotation and growth of a porphyroblast, from left to right. [From Hanmer and Passchier, 1992. Published with permission of Geological Association of Canada.]

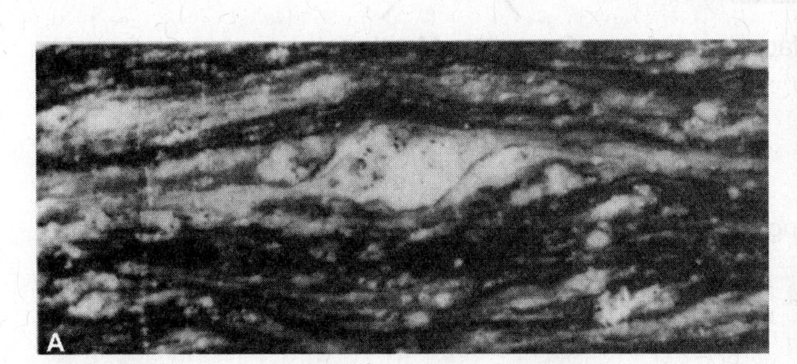

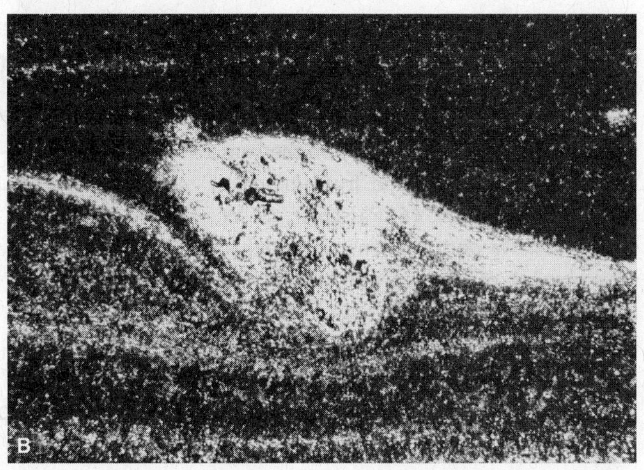

Figure 10.59 Sigma- and delta-type porphyroclasts. (*A*) σ-type porphyroclast in polished slab of mylonitic granite, Santa Catalina Mountains, Arizona. Dextral shear. [Photograph by S. J. Reynolds.] (*B*) δ-type porphyroclast in mylonitic granite. Sinistral shear. [Photograph by Carol Simpson.]

Some rigid porphyroblasts contain sigmoidal or spiral **inclusion trails** of another mineral (Figure 10.58*B*,*C*). These trails represent a succession of small crystals that were entrapped as the porphyroblast grew and rotated during deformation. The shape of the inclusion trail reveals which way the inclusion rotated relative to the matrix, and documents the sense of shear. Alternatively, some porphyroblasts may grow around and entrap fabric formed during an earlier deformation event or successive increments of one event, possibly providing the only record of fabrics that have been largely or totally obliterated by continued deformation in the matrix (Bell, 1985).

As porphyroclasts, porphyroblasts, and other inclusions interact with the shear zone matrix, their outer edges may preferentially accumulate crystalline strain and recrystallize (Passchier and Simpson, 1986; Passchier, 1987, 1994). The recrystallized **mantle** is composed of numerous small crystals and is weaker than the monocrystalline core. As a result, the mantle tends to be sheared out away from the porphyroclast, forming "*wings*" or "*tails*" that extend out into the matrix (Figure 10.59). The centerlines of wings on opposite sides of the porphyroclast may be straight and aligned, or they may be curved and markedly asymmetric with respect to the porphyroclast.

The Greek Squadron of Winged Porphyroclasts

Five main geometries of winged porphyroclasts have been identified based on whether the centerlines of the wings are straight or curved, and symmetrical or asymmetrical with respect to an imaginary reference line through the center of the clast (Figure 10.60). Four of these have been named after Greek letters whose shape they resemble (Simpson and Schmid, 1983; Passchier, 1994). **Theta** (θ) objects have round to elliptical mantles, but no real wings. Wings on **phi** (ϕ) type porphyroclasts have fairly straight centerlines and are largely symmetrical with respect to the porphyroclast; they resemble the Greek letter *phi* turned onto its side. **Sigma** (σ) type have wings with gently curved centerlines and are asymmetric; the wing extends off the top of one side and off the bottom of the opposite side, a pattern referred to as "*stair stepping*" (Figures 10.59*A*, 10.60). **Delta** (δ) type have strongly curved wings that are asymmetric with respect to the porphyroclast (Figure 10.59*B*). The final type consists of **complex porphyroclasts** with several sets of wings. The general shear-sense implication of each type is illustrated in Figure 10.60. In three dimensions, wings are lenticular to rodlike (Figure 10.61).

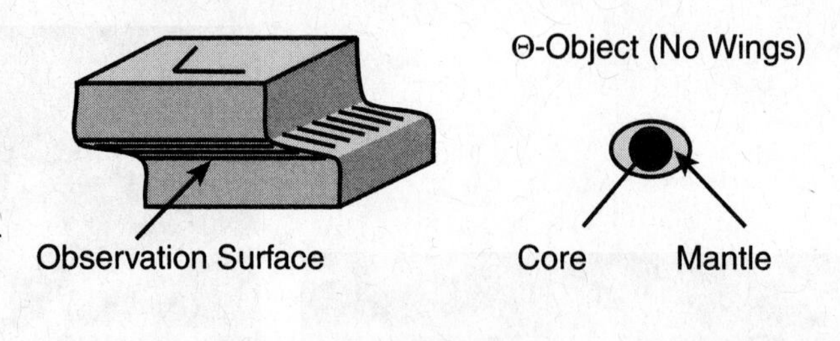

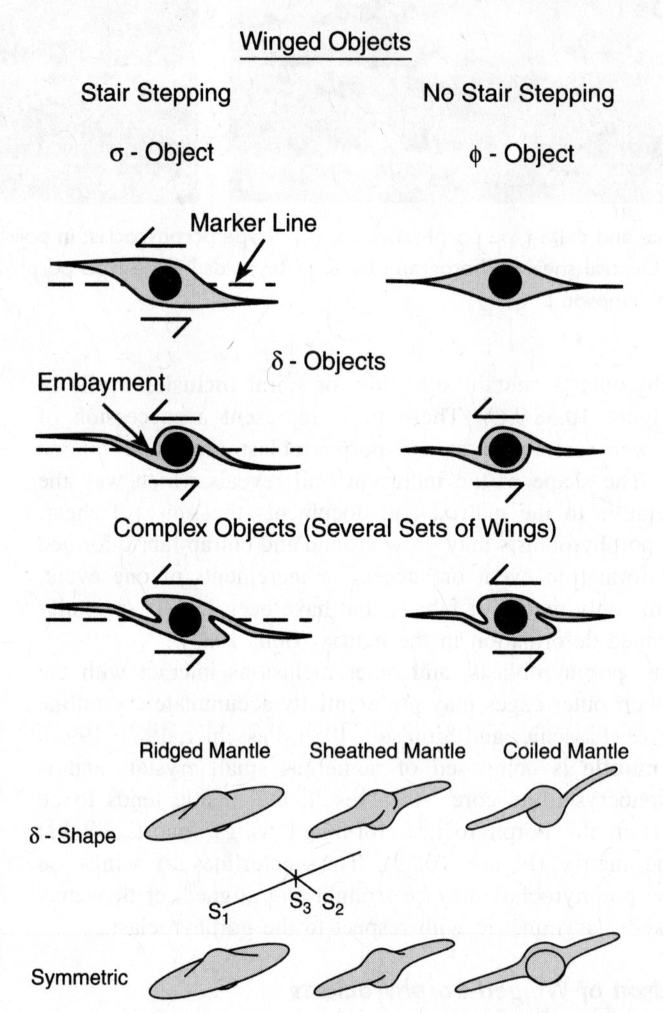

Figure 10.60 Greek squadron of winged porphyroclasts. The darkly shaded center is the rigid core and the more deformable recrystallized mantle is lightly shaded. Views are cross-sections as observed on the sense of shear plane. [Reprinted from *Journal of Structural Geology*, v. 16, Passchier, C. W., Mixing in flow perturbations: a model for development of mantled porphyroclasts in mylonites, pp. 733–741, © 1994, with permission from Elsevier.]

Figure 10.61 Three-dimensional view of winged porphyroclasts. [Reprinted from *Journal of Structural Geology*, v. 15, Passchier, C. W., and Sokoutis, D., Experimental modelling of mantled porphyroclasts, pp. 895–909, © 1993, with permission from Elsevier.]

The formation of the different wing types has been studied in natural rocks and in experimental simulations using analogue materials that can be deformed at low temperature and high strain rates (Passchier, 1987, 1994; van den Driessche and Brun, 1987; Passchier and Sokoutis, 1993). From the ingenious experiments of Passchier (1994), we learn that all wings initially develop parallel to the maximum *instantaneous* extension direction (S_1') and subsequently become drawn out in the direction of maximum *finite* extension (S_1) (Figure 10.62). Several distinct geometries of wings evolve with additional strain, depending on the amount of strain and the rate of recrystallization relative to the strain rate. The rate of recrystallization is key because it controls the relative proportions of mantle material to rigid core. If the rigid core is large compared to its mantle, it deflects flow in the shearing material, protecting its mantle from being sheared away. In contrast, a relatively smaller core does not deflect the flow enough to protect its mantle. In either case, both the rigid core and the mantle rotate, reflecting the sense of shear of the zone.

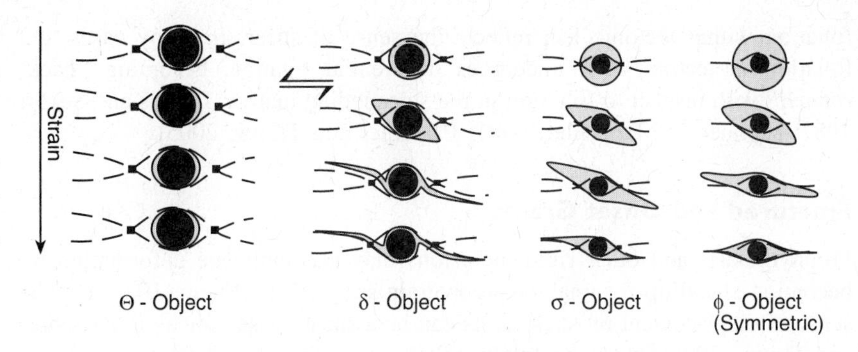

Figure 10.62 Cross-sections of experimentally produced mantled porphyroclasts. The darkly shaded center is the rigid core, and the more deformable recrystallized mantle is lightly shaded. The shear plane is horizontal. The dashed lines represent a flow-regime boundary, separating more laminar flow on the outside and more circular flow paths on the inside. The recrystallized mantle remains symmetrical if it remains inside the flow-regime boundary, but becomes asymmetric if it extends outside the boundary, as would occur if the flow perturbation caused by the rigid core shrank with time as the core recrystallizes (e.g., the progression from left to right). See text. [Reprinted from *Journal of Structural Geology*, v. 16, Passchier, C. W., Mixing in flow perturbations: a model for development of mantled porphyroclasts in mylonites, pp. 733–741, © 1994, with permission from Elsevier.]

According to the experiments, a porphyroclast that recrystallizes slowly or not at all becomes a θ object (Figure 10.62). In this case, the rigid core protects the relatively small mantle. At rates of recrystallization that are higher, but still low relative to the strain rate, the mantle is smeared out into wings, which become rolled along with the porphyroclast to form δ-type wings. Where recrystallization is fast compared to the strain rate, the supply of recrystallized mantle material is fast enough to permit each wing to develop into a continuous wedge-shaped mass that is smeared out from the porphyroclast, either symmetrically (ϕ-type) or asymmetrically (σ-type).

The shape of the inclusion also influences the type of wings that form, in part because it affects the rate at which the inclusion rotates compared to the matrix. For example, δ-type wings are commonly associated with equant inclusions, which can continue rotating during shearing (Passchier and Trouw, 2005). Other types of wings are generally found around more elliptical inclusions. The experimental results of Passchier and others are exciting because they help us appreciate why so many different kinds of winged porphyroclasts may "*fly*" within a single shear zone.

Foliation Fish

A different kind of inclusion forms when part of the shear zone becomes structurally isolated between discrete shear bands, fractures, or some other type of structural discontinuity. Once formed, such discontinuity-bounded material may behave mechanically differently from the surrounding matrix, being deformed into lenticular or sigmoidal masses called **tectonic fish** or **foliation fish** (Figure 10.63). As you might suspect, the sigmoidal shape of

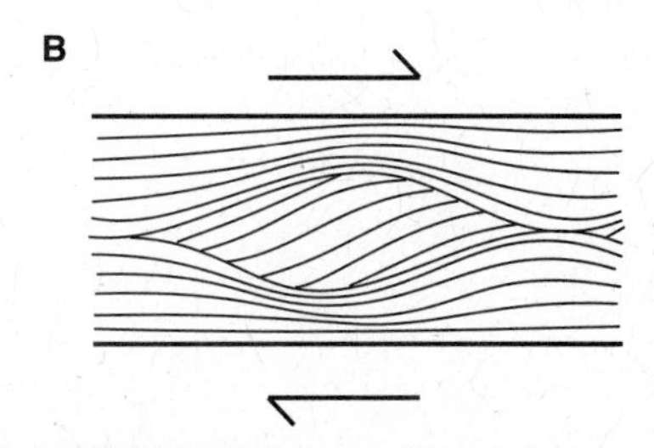

Figure 10.63 Foliation fish: (*A*) With back-rotated foliation, within mylonitic gneiss, Chemihuevi Mountains, southeastern California. Sinistral shear. [Photograph by S. J. Reynolds.] (*B*) Back-rotated foliation within fish. [Reprinted from *Journal of Structural Geology*, v. 14, Stock, P., A strain model for antithetic fabric rotation in shear band structure, pp. 1267–1275, © 1992, with permission from Elsevier.]

foliation within tectonic fish reflects the sense of shear. In many cases, the foliation in tectonic fish undergoes differential rotation, becoming "*back-rotated*" with respect to foliation in the surrounding matrix (Dennis and Secor, 1987; Hanmer and Passchier, 1991; Passchier and Trouw, 2005).

Fractured and Offset Grains

Porphyroclasts and other rigid inclusions may accommodate deformation by becoming sliced up by small-scale or grain-scale faults (Figure 10.64*A*). The sense of displacement on such faults can be a shear-sense indicator (Passchier and Trouw, 2005), but not a totally reliable one. When such faults are at low angle to the shear zone, they generally have a sense of shear that is **synthetic** (i.e., the same) as the shear zone (Figure 10.64*B*). Like shear bands, synthetic faults can displace the inclusion with either a normal or a thrust sense when viewed in the SOS plane. When the faults are at a high angle to the shear zone, they commonly have a sense of shear that is **antithetic**, or opposite, to the shear zone (see Figure 10.64*B*). This antithetic slip helps accommodate synthetic rotation of the inclusion, much as the slip surfaces between books on a shelf permit the books to topple over when the bookend is removed, or knocked over by the cat (Figure 10.64*C*).

The kinematic significance of such features can be ambiguous, hence should not be the sole basis for interpreting the sense of shear (Hanmer and Passchier,

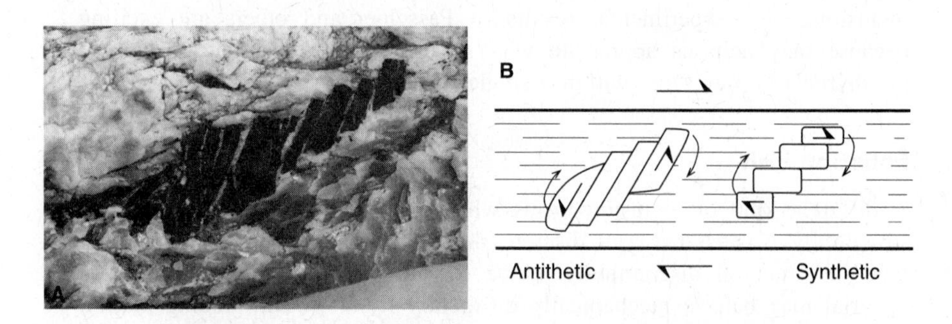

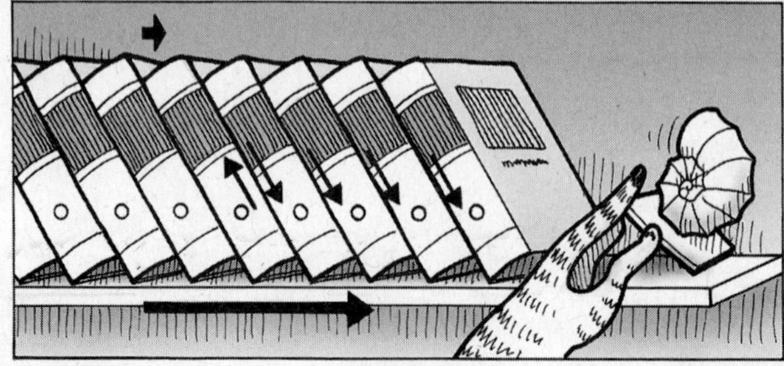

Figure 10.64 Fractured and offset grains: (*A*) Antithetic shear fractures cutting tourmaline in protomylonitic pegmatite. Northern slopes of Puig Alt Petit, Cap de Creus, Spain. [Photograph by G. H. Davis.] (*B*) Synthetic and antithetic offset of grains. Curved arrows adjacent to grains indicate direction of rotation of long axis of grain. (*C*) Antithetic rotation of books due to "cat tectonics." [Artwork by D. A. Fisher.]

1991). For example, a structure that can be mistaken for an offset inclusion arises where adjacent inclusions rotate into one another, piling up in a process called **tiling** (Figure 10.65). We might get the wrong sense of shear if we misinterpret tiled grains as a single offset grain. Also, computer models indicate that we need to observe many instances of tiling before being confident to use tiling as our main indicator of sense of shear (Mulchrone et al., 2005).

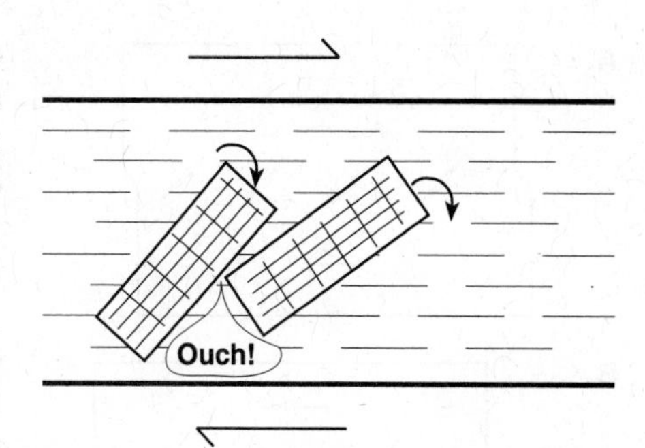

Figure 10.65 "Tiling, . . . when one grain rotates into another one.

Veins

As we map and analyze shear zones, we are generally impressed with the abundance of veins (Figure 10.66). Most shear zone–related veins contain quartz and calcite, but feldspar, mica, gypsum, and iron-oxide minerals are also common in certain settings and rock types. These minerals are deposited from the fluids that helped "*prop*" open the fracture while it was filled by the vein material.

Veins can be excellent and reliable shear-sense indicators because their initial orientations are commonly controlled by the instantaneous stretching axes. Most veins form perpendicular to the axis of maximum *instantaneous* extension ($\dot{S}_1$), because this is the direction in which tension fractures form. Veins should be oriented *perpendicular* to foliation and lineation for coaxial deformation, but not for noncoaxial deformation (Figure 10.67). For simple shear, veins will form at 45° to the shear zone—that is opposed to the direction of foliation and to the inclination of most other shear-sense indicators we have examined so far. Once formed, the veins may be shortened and partially rotated over, toward the direction of shear (Figure 10.68). New veins, however, will continue to form in the same original orientation, with their tips aligned in accordance with the instantaneous stretching axes, *irrespective of changes in the orientation of the finite strain ellipse with time.* Multiple generations of earlier, deformed veins and younger, less deformed ones are

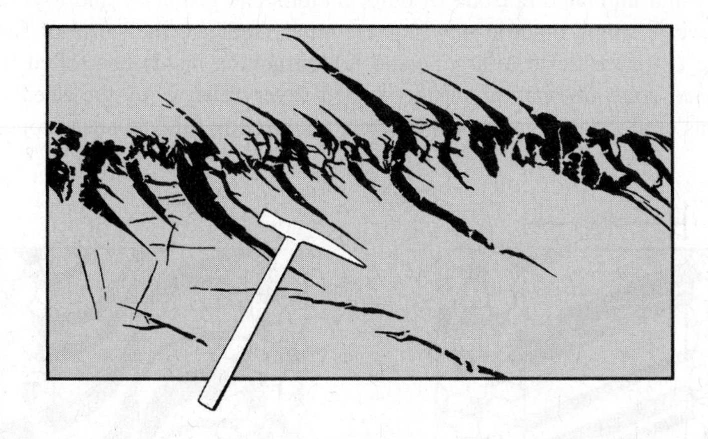

Figure 10.66 En echelon, sigmoidal quartz veins defining a dextral shear zone. [Reprinted from *Journal of Structural Geology*, v. 2, Ramsay, J. G., Shear zone geometry: a review, pp. 83–99, © 1980, with permission from Elsevier.]

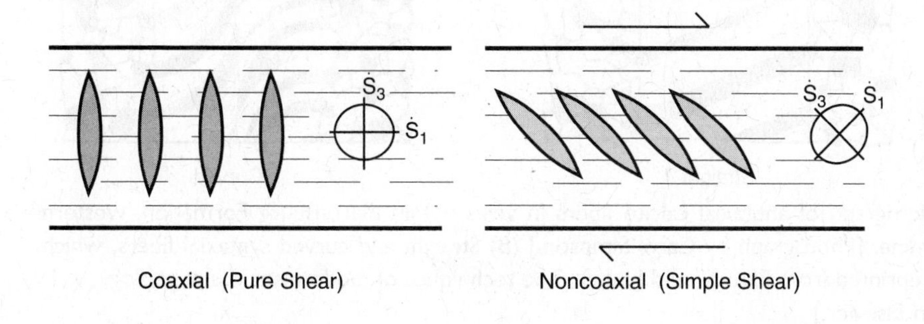

Figure 10.67 Orientation of veins compared to the instantaneous stretching axes for coaxial and noncoaxial deformation. Veins form parallel to the maximum shortening rate ($\dot{S}_3$) and perpendicular to the maximum stretching rate ($\dot{S}_1$).

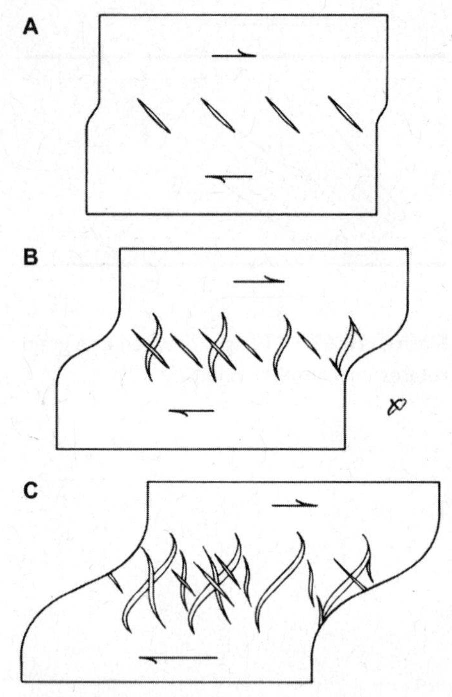

Figure 10.68 Progressive formation and folding of en echelon veins within a zone of simple shear. Tension fractures are initially oriented at 45° to the shear zone and progressively rotate in the direction of the sense of shear during subsequent deformation. Later veins form in the original orientation. [From Durney and Ramsay, Incremental strains measured by syntectonic crystal growths, in De Jong, K. A., and Scholten, R. (eds.), Gravity and tectonics. Published with permission of John Wiley & Sons, Inc., New York, copyright © 1973.]

commonly present in a single outcrop, and the deflection of the older veins and how they are buckled are kinematic indicators (see Figures 10.66, 10.68). A curved vein geometry can also locally arise because of complex stress fields created by interactions between the tips of adjacent, closely spaced veins.

Veins, like pressure shadows, may contain crystal fibers that grow one increment at a time during opening of the vein (Figure 10.69; see *Chapter 5, Joints*). The fibers grow via a **crack–seal mechanism**: a small increment of extension fracturing and opening of the vein is followed by growth of fibers from existing crystals along the walls of the fracture (Ramsay, 1980b). The new fiber increment will be in optical continuity with the existing crystal, but the boundary may be marked by a thin train of fluid inclusions trapped as the fiber grew. The fibers grow parallel to the direction of maximum instantaneous extension ($\dot{S}_1$) and therefore lean over in the direction of shear. The fibers may be straight or may exhibit curved shapes that record shearing parallel to the walls or changes in the orientation of the vein with respect to the instantaneous stretching axes (Figure 10.69*B*).

Folds

Shear zones contain several types of folds (Carerras et al., 2005), some of which can be used as shear-sense indicators (Figure 10.70). The vergence of asymmetric **intrafolial folds** commonly reflects the overall sense of shear (Figure 10.70*A*), especially for folds formed late in the shear-zone evolution. Intrafolial folds may be folding lithologic layering that existed before deformation (e.g., bedding, metamorphic banding) or foliation that formed during the earlier phases of shear zone evolution. With progressive deformation, either type of fold may be rotated into subparallelism with lineation, at which point it becomes nondiagnostic in terms of sense of shear. Crenulations and other folds that cut *across* the shear zone fabric should not be used to determine shear sense of the shear zone because they may be caused by a younger and unrelated episode of deformation. The geometry and asymmetry of folds in a single outcrop may not accurately indicate the sense of shear if they are (1) parasitic to a larger-scale fold structure, or (2) controlled by the *pre–shear zone orientation* of the folded layer relative to the shear zone (Carerras et al., 2005). In our experience, interpreting folds in shear zones can

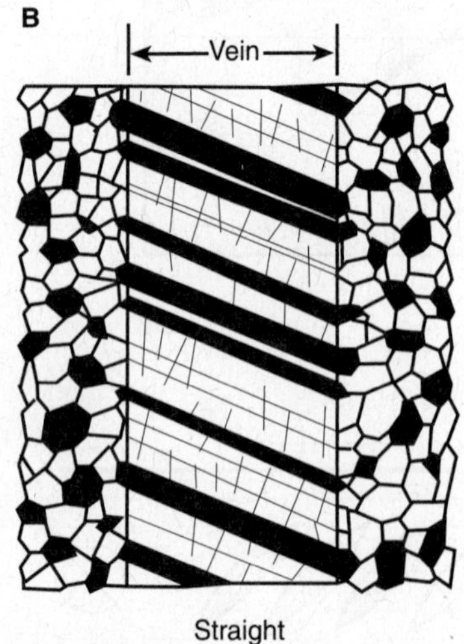

Figure 10.69 Fibers in veins: (*A*) Photomicrograph of antitaxial calcite fibers in veins within Martinsburg Formation, western Maryland. Fibers grow outward from center-line. [Photograph by Carol Simpson.] (*B*) Straight and curved syntaxial fibers, which grow inward from grains in the wall rocks. [Reprinted from Ramsay and Huber, The techniques of modern structural geology, v. 1: Strain analysis, © 1983, with permission from Elsevier.]

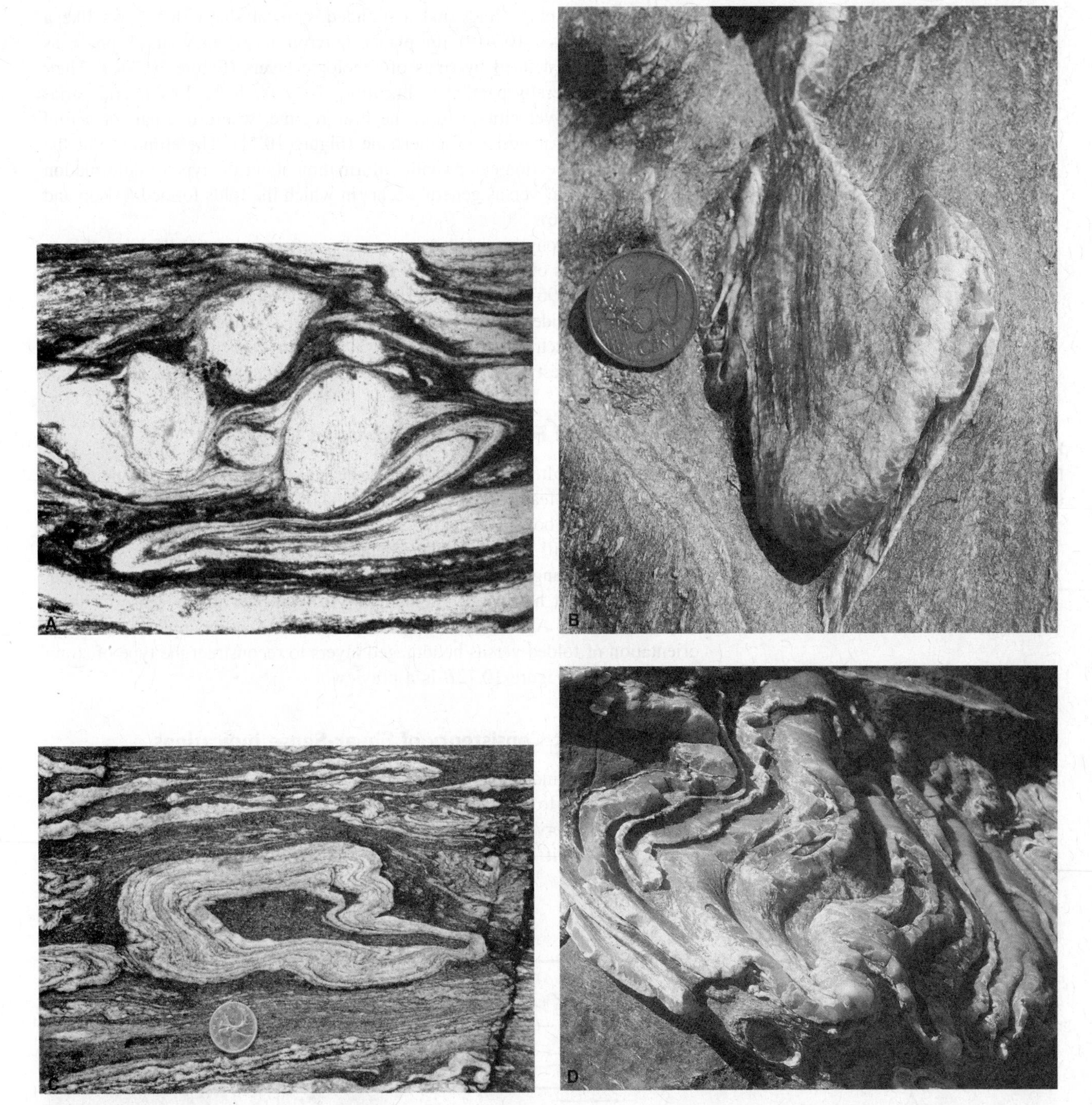

Figure 10.70 Folds in shear zones. (*A*) Asymmetric interfolial fold in mylonitic granite, Santa Catalina Mountains, southern Arizona. Dextral sense of shear indicated by fold is supported by the asymmetrical wings of the overlying, round porphyroclast of feldspar. [Photograph by G. H. Davis.] (*B*) Perfect sheath folding of mylonitic quartz vein. Cala Prona locale, Cap de Creus, Spain. [Photograph by G. H. Davis.] (*C*) Eye-shaped fold, reflecting a curved fold axis, within banded gneiss. [Photograph by Simon Hanmer.] (D) Extreme isoclinal and sheath folding of mylonitic quartz veins. Note eye-shaped fold, ~3 cm in diameter Cala Prona locale, Cap de Creus, Spain. [Photograph by G. H. Davis.]

be a little tricky unless the folds show a consistent asymmetry over many outcrops.

Sheath folds are unusual, noncylindrical folds formed in some strongly deformed rocks (Carreras et al., 1977; Quinquis et al., 1978; Cobbold and Quinquis, 1980; Ramsay, 1980a; Alsop and Holdsworth, 2006). They have

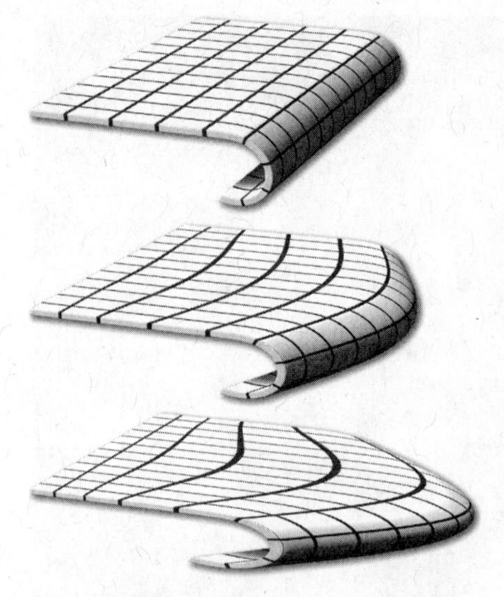

Figure 10.71 Formation of sheath folds. Differential flow causes the middle part of this fold to progressively bow out, forming a sheath fold.

strongly curved hinge lines and a rounded, conical shape that looks like a wind sock (Figure 10.70*B*). In planar outcrop faces, they may appear as elliptical "*eyes*" defined by rings of lithologic layers (Figure 10.70*C*). Their long axis is generally parallel to lineation. They are formed by lateral variations in particle velocities within the flow regime, where one part of a fold flows faster than material to either side (Figure 10.71). The ellipticity of the "*eyes*" in cross-section can provide information about the type of deformation (e.g., simple shear versus general shear) in which the folds formed (Alsop and Holdsworth, 2006).

Folds with strongly curved axes can also reflect superposed folding due to multiple episodes of deformation or to progressive deformation during a single event (Ramsay, 1967; Ramsay and Huber, 1987). Overprinting fabrics, if present, will usually be evident after careful observations in a number of outcrops. As with any aspect of structural geology, it is always best to see as many outcrops, representing the broadest geographic spread possible, before jumping to conclusions.

Orientation of Folded and Boudinaged Layers

Finally, we can often infer the sense of shear by comparing the orientations of layers or other features that have been shortened versus those that were lengthened and boudinaged (Figure 10.72*A*). Any feature that is aligned parallel to S_1 will have been lengthened, perhaps by boudinage, whereas a feature at right angles to this will have been folded or otherwise shortened. The features that have been lengthened will lean over in the sense of shear (Figure 10.72*A*). At the end of this chapter, we see an example of using the orientation of folded versus boudinaged layers to reconstruct the type of strain in shear zones. Figure 10.72*B* is a preview.

A Word about Consistency of Shear-Sense Indications

The shear-sense indicators described above represent a powerful collection of tools with which to analyze the kinematics of shear zones. Their diversity may seem a little overwhelming, but this same diversity enables us to use them confidently, albeit carefully. When we study a shear zone, we can use the

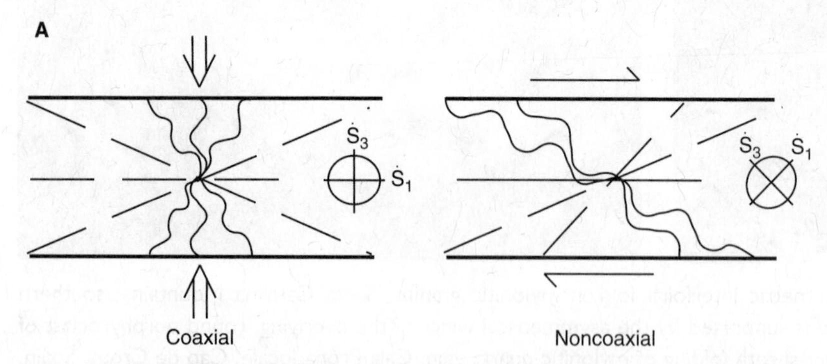

Figure 10.72 Folded and boudinaged layers: (*A*) Orientations of folded versus boudinaged lines for coaxial and noncoaxial deformation. The distribution of boudinaged versus folded lines is symmetrical with respect to the shear zone and is normal for coaxial deformation, but not for noncoaxial deformation. The instantaneous strain ellipses are shown for each type of deformation. (*B*) Pegmatite dikes oriented at a high angle to the gneissic foliation are folded, whereas a mafic layer and pegmatite dikes parallel to the foliation are boudinaged. Proterozoic Vishnu Schist in Clear Creek, Grand Canyon, northern Arizona. [Photograph by S. J. Reynolds.]

different shear-sense indicators as mutual checks. We can also evaluate our results with the general geologic context of the shear zone. If our shear zone is placing older, deep-level metamorphic rocks over younger, shallow-crustal rocks, then we might suspect a thrust history. But there are other ways to produce this geometry, and we should not discount the *observed* sense of shear if it appears inconsistent with the geometry or with some conceptual model we have in mind. The observed sense of shear may be telling us that the shear zone started out as a thrust but has been reactivated with a normal sense of displacement. The recognition of such a series of geologic events may have major implications for our understanding of the tectonic evolution of the region or for modeling the distribution of petroleum or other natural resources.

Thankfully, most shear zones that have been studied display a fairly consistent sense of shear, reflecting a clear noncoaxial component of deformation. Other shear zones have contradictory shear-sense indicators or lack obvious indicators, possibly because they record coaxial deformation, an usual strain regime, or overprinting episodes of deformation with different senses of shear. But even these should be decipherable with careful work and a trained eye. The complexities only add to the "*shear joy*" we feel in analyzing a shear zone and finally figuring it out.

FABRIC DEVELOPMENT AND ITS RELATION TO THE AMOUNT OF STRAIN IN SHEAR ZONES

Strain in shear zones results in the formation of fabrics defined by the shapes of deformed objects. The deformed objects are typically individual grains or grain aggregates, but range from tiny subgrains visible only in thin section to slivers or boudins hundreds of meters long. The fabrics in shear zones can be subdivided into three general types based on how well the fabrics reflect the *total* strain in the shear zone. Some fabrics, referred to as **strain-sensitive fabrics**, fully record the finite strain and so can be used to determine the amount of strain in a shear zone. In contrast, **strain-insensitive fabrics** do not record the total finite strain because some competing process prohibits a fully strain-sensitive fabric from developing. **Composite shape fabrics** may or may not record the finite strain; typically, they develop when discrete, subsidiary shear bands form parallel to or at an angle to the main shear plane.

In this section, we explore how each type of fabric forms, and how we can use such fabrics to estimate the amount of strain within a shear zone. We present only methods of determining strain that are applicable to ductile and brittle–ductile shear zones. Additional methods suitable for determining strain in metamorphic rocks in general are covered in *Chapter 9, Foliation and Cleavage.*

Strain-Sensitive Shape Fabrics

Strain-sensitive shape fabrics are parallel to the finite strain ellipse and therefore record the total finite strain of the rock. They include typical foliation and lineation, such as the fabric in a deformed conglomerate that contained initially spherical clasts (Figure 10.73). As long as the clasts and their matrix responded identically to deformation, the shape and orientation of each clast would reflect the finite strain ellipse. For any plane strain, the clasts would be flattened in the S_3 direction and correspondingly lengthened parallel to S_1. The clasts would define a foliation parallel to the S_1S_2 plane and perpendicular to S_3. A stretching lineation, if present, is parallel to S_1. The amount of strain in such a rock can be evaluated by the R_f/ϕ, center to center,

Figure 10.73 Shape fabric defined by the preferred orientation of clasts in a strongly deformed Proterozoic metaconglomerate, Phoenix Mountains, central Arizona. Note less deformed, granitic cobble, which does not (black discontinuous bands) reflect the finite strain of the entire rock. [Photograph by S. J. Reynolds.]

and other conventional strain methods discussed separately (see *Chapter 9, Foliation and Lineation*) (see also Ramsay and Huber, 1983).

Within shear zones, we can also use the orientations of foliation relative to the shear zone to determine the amount of strain. Recall that in noncoaxial deformation, foliation starts forming at an angle to the shear zone and rotates into lower angles with the shear zone at higher shear strains. It turns out that for simple shear, the angle between foliation and the shear zone, as measured parallel to the shear direction on the SOS plane, is directly related to the shear strain via the extremely useful little equation (Figure 10.74) (Ramsay and Graham, 1970):

$$\gamma = \frac{2}{\tan 2\theta} \tag{10.1}$$

where γ is the shear strain and θ is the angle between the foliation and shear zone.

We can use Equation 10.1 to calculate the shear strain at any one site within a shear zone.

We can also calculate the principal stretches directly from θ (Ramsay and Graham, 1970):

$$S_1 = \cot \theta$$

$$S_3 = \tan \theta$$

Alternatively, if we can determine the principal stretches by some other method, we can use them to calculate θ, which then gives us the shear strain. Note that these equations are valid only for simple shear. The angle between foliation and shear zone will change if there is any non–simple shear component, such as a coaxial component imposed on the shear zone. If these other components cause the shear zone to be *thinned* during deformation, θ will

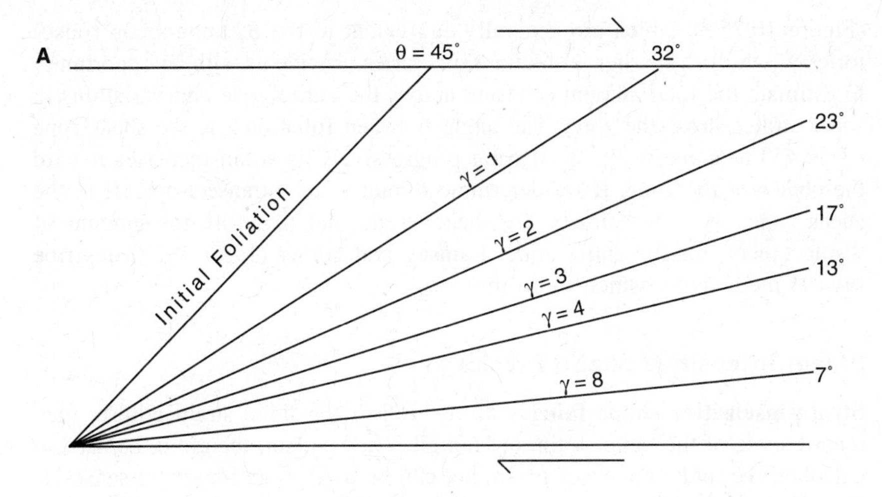

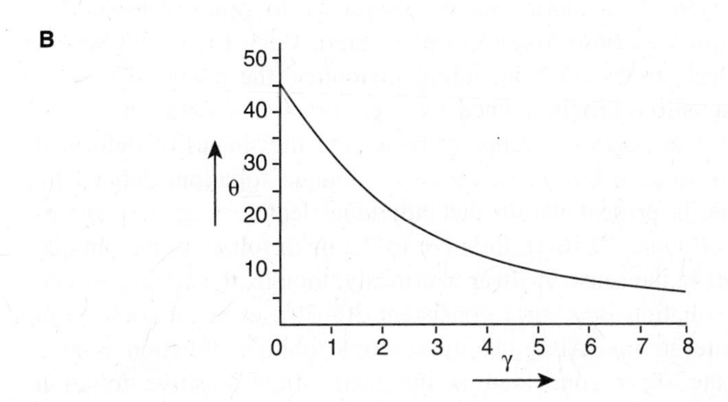

Figure 10.74 The angle between foliation and shear zone (θ) as a function of shear strain (γ): (*A*) a zone of simple shear, (*B*) plot of θ versus γ for simple shear. [Part *B* reprinted from *Structural geology: fundamentals and modern developments*, S. K. Ghosh. Published with permission of Pergamon Press, Ltd., Oxford, © 1993.]

decrease compared to its expected value for simple shear, and the equations will overestimate the shear strain. If extension normal to the shear zone causes the shear zone to be *thickened*, θ will be greater than its expected value for simple shear, and the equations will underestimate the shear strain.

An ideal ductile shear zone shows a gradual decrease in strain away from its center, as is reflected by the sigmoidal shape of foliation, lineation, and other strain-sensitive fabrics across the zone (Figure 10.75*A*). The pattern of foliation we observe in the field can be portrayed as **foliation trajectories**

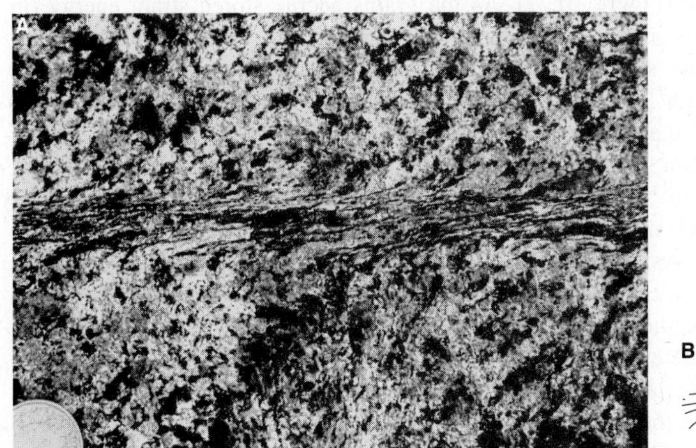

B

Figure 10.75 Dextral shear zone: (*A*) Cutting granite, Laghetti, Ticino, Switzerland. [Photograph by Carol Simpson.] (*B*) Foliation trajectories for shear zone in (*A*).

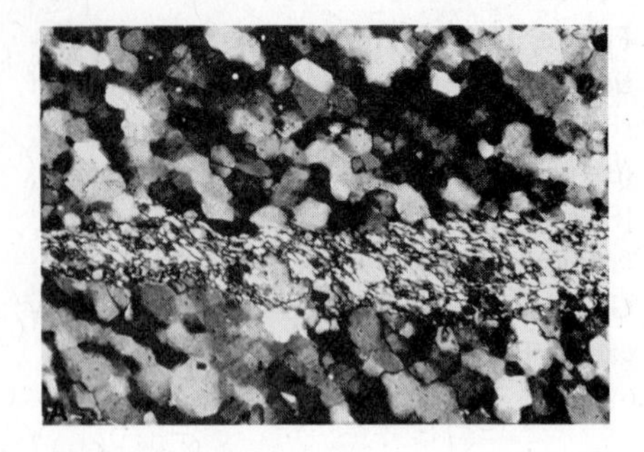

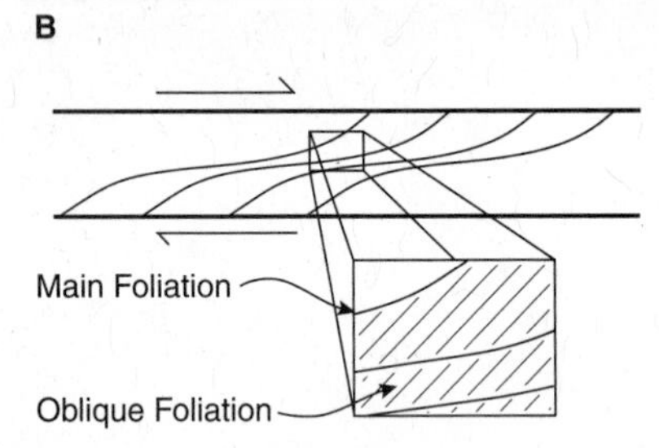

Figure 10.76 Strain-insensitive oblique fabric in mylonitic rocks. (*A*) Photomicrograph of deformed pegmatite, Borrego Springs mylonite zone, southern California. Main foliation is subhorizontal, parallel to the strongly deformed feldspar grain that defines a band across the center of photograph. The oblique foliation is expressed by the alignment of quartz and feldspar subgrains from upper left to lower right (sinistral sense of shear). The quartz and feldspar subgrains have a slightly different orientation because the quartz subgrains recrystallize more easily than do the feldspar grains and so are more nearly parallel to the direction of maximum instantaneous extension. (*B*) Geometric relations between strain-sensitive main foliation, strain-insensitive oblique foliation, and the shear zone.

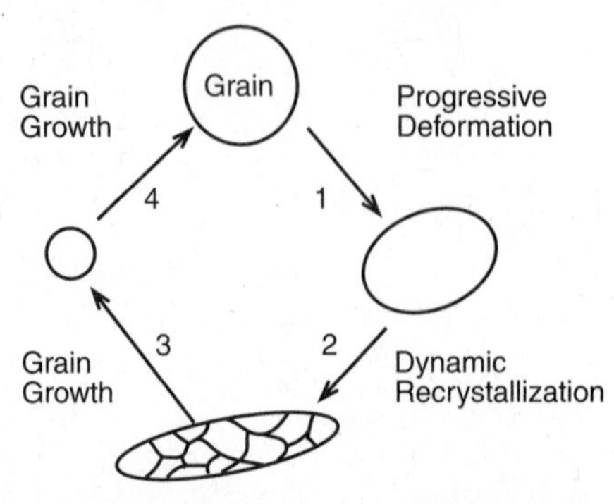

Figure 10.77 Cycle of progressive deformation leading to dynamic recrystallization, leading to grain growth and grain boundary migration, resulting in a "*new*" less-strained grain. [From Hanmer and Passchier, 1991. Reproduced with the permission of Natural Resources Canada 2011, courtesy of the Geological Survey of Canada, Paper 90–17.]

(Figure 10.75*B*), which are generally equivalent to the S_1 finite-strain trajectories we explored earlier. Foliation trajectories provide us with an opportunity to estimate the total amount of strain across the entire zone and variations in shear strain across the zone. The angle between foliation and the shear zone (θ) is 45° near the walls, decreasing progressively as strain increases toward the center of the zone. If we determine θ values for a transect normal to the shear zone, we can estimate the shear strain, and therefore the amount of displacement, for the entire zone. Ramsay and Huber (1983, 1987) describe several methods by which to do this.

Strain-Insensitive Shape Fabrics

Strain-insensitive shape fabrics do not reflect the finite strain ellipse; they record some of the finite strain, but not all. They cannot, therefore, be used to calculate the total amount of strain, but can be used as an important sense-of-shear indicator.

The most common strain fabric that is, insensitive to strain magnitude is **microscopic oblique foliation** (Simpson and Schmid, 1983; Lister and Snoke, 1984). In most shear zone rocks, including mylonites, the main foliation is largely a **strain-sensitive fabric** defined by a preferential orientation of lenticular grains and aggregates of grains, ribbons, and the shapes of deformed inclusions. In thin section, however, a second, oblique foliation, defined by **aligned subgrains**, is present within the individual lenticular grains, aggregates, and ribbons (Figure 10.76*A*). Relative to the main foliation, the oblique foliation leans over in the sense of shear. Curiously, though, the angle between the two types of foliation is a fairly consistent 20–30°, even for rocks with very different finite strains. Also, the microscopic oblique foliation is at a higher angle to the shear zone than is the main strain-sensitive foliation (Figure 10.76*B*); it cannot record the same amount of finite strain as the main foliation. Therefore, the angle between the oblique foliation and either the shear zone or the main foliation cannot be used to estimate the shear strain. Instead, the angle remains approximately constant, irrespective of the finite strain.

Strain-insensitive fabrics require some process that can form a foliation that is oblique to the finite strain ellipse, but that does not rotate with increased finite strain. Win Means (1980) proposed that for a shape fabric to avoid rotation during noncoaxial deformation, processes that form the fabric must be balanced by those that try to obliterate it. As grains begin to deform and rotate in accordance with the finite strain, dynamic recrystallization modifies the resulting fabric (Figure 10.77). As the grains accrue stored strain energy (in the form of dislocations), they may recrystallize into "*new*" grains or subgrains with less stored strain energy, with more equant or more irregular shapes, and with less of a preferred orientation. Such recrystallization effectively "*resets*" the finite strain clock for each recrystallized grain, or at least causes the grain to have an imperfect memory of the total finite strain. Recrystallization therefore decreases the overall preferred orientation of the grains and weakens the appearance of the fabric. This process is aided by the tendency of recrystallization to preferentially affect the most deformed, strain-hardened grains, the very ones that will have the most preferred orientation and best define the fabric.

However, there is more to the story. The newly recrystallized grain, once formed, will again experience strain as deformation continues. In fact, such grains, because they are relatively strain-free, will deform preferentially relative to adjacent grains that are still strain hardened (Passchier and Trouw, 2005; Holyoke and Tullis, 2006). As the newly recrystallized grain begins to deform, it changes shape incrementally, in accordance with the instantaneous stretching axes, *not* the finite strain axes. In simple shear, the first increment of strain will form a foliation that will be at 45° to the shear plane and will be oblique to the finite-strain-sensitive foliation in the surrounding rock.

The entire process represents a cycle in which grains are continually being strained, recrystallized, and strained again. At any given time, different grains will be in different phases of the cycle, and the overall fabric will represent a composite or hybrid state of strain. The oblique foliation will be intermediate in position between the finite and instantaneous strain ellipses at any given time, and its angle to the shear plane may partially reflect the relative success of the fabric-forming and fabric-weakening processes. The process has been modeled using computer simulations (Jessel, 1988a,b).

The temperature during deformation influences the degree of development of such strain-insensitive fabrics, since it controls the rate of recrystallization. At higher temperatures, recrystallization is relatively fast and should be most efficient at "*resetting the finite strain clock.*" It will be less able to keep up with the imposed strain at lower temperatures or higher strain rates, where the resulting fabrics should more accurately reflect the finite strain.

Composite Shape Fabrics

Composite shape fabrics contain shear bands in addition to either a strain-sensitive or strain-insensitive fabric (usually foliation and lineation). The shear bands are subsidiary to the main ductile shear zone and can be either parallel or oblique to it. The two main types of composite fabric that contain shear bands are **S-C fabrics** and **extensional shear bands**.

S-C Fabrics

S-C fabrics consist of foliation (S-surfaces) and shear bands (C-surfaces). The S-surfaces and C-surfaces are at an angle to one another (Figure 10.78), indicating the sense of shear (Berthé et al., 1979). Not all S-C fabrics are alike, with differences due mostly to the character of the S-surfaces (Lister and Snoke, 1984).

In most S-C fabrics, S-surfaces are a typical foliation defined by flattened and stretched, lenticular grains and aggregates of grains. These S-surfaces are interpreted to be a strain-sensitive foliation parallel to the S_1S_2 plane of the finite strain ellipsoid. C-surfaces in these rocks are typical, crosscutting shear bands (Figure 10.78*A*). This is the dominant type of S-C fabric in most mylonitic quartzofeldspathic rocks.

In other S-C fabrics, S-surfaces are defined by the oblique microscopic foliation described in the preceding section, and by mica fish. The S-surfaces are partly or largely a strain-insensitive fabric. C-surfaces are thin shear bands marked by trails of deformed mica sheared ("*scaled*") off the fish (Figure 10.78*B*). This type of S-C fabrics is most common in micaceous quartzite and some strongly mylonitic rocks.

In both types of S-C fabric, the C-surfaces are commonly aligned parallel to the shear zone and are associated with the same strain gradients as any continuous shear zone (Figure 10.78). These gradients cause the sigmoidal shape of the accompanying S-surfaces as they approach and rotate toward parallelism with the C-surfaces. Some C-surfaces display a more discontinuous strain gradient, whereby they offset individual grains along a very abrupt contact. Such sharp C-surfaces are most common adjacent to a grain that behaved as a rigid object during deformation.

We can use the angle between S- and C-surfaces in the first type of S-C fabrics to estimate the amount of strain, using the same equation we used for the angle between foliation and the main shear zone. For simple shear, S

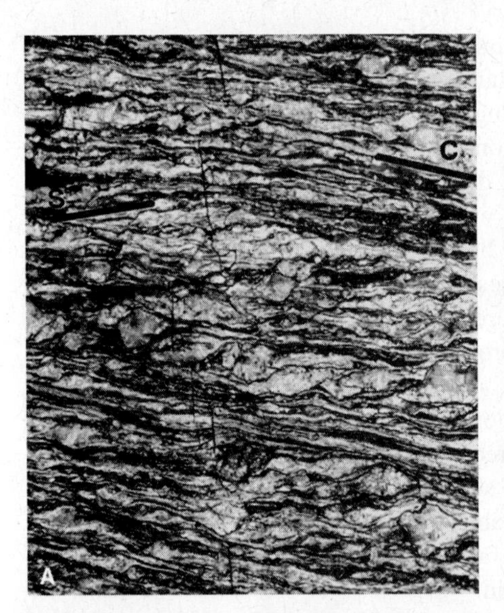

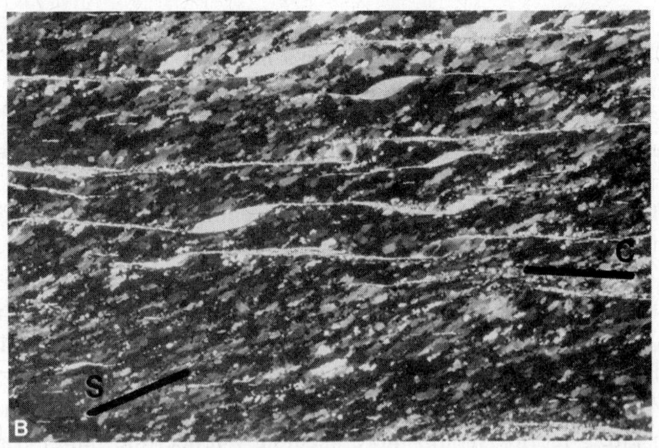

Figure 10.78 S-C mylonites: (*A*) S-C mylonite, Santa Catalina Mountains, Arizona. C-surfaces are defined by the thin zones of very high shear strain, whereas S-surfaces are defined by the obliquely inclined feldspar porphyroclasts. Dextral shear sense. [Photograph by S. J. Reynolds.] (*B*) Photomicrograph of S-C mylonite, Coyote Mountains, southern Arizona. S surfaces are defined by mica fish and oblique subgrains in quartz, whereas C surfaces are defined by the long, subhorizontal tails (bright in photograph) extending off the tips of mica fish. Dextral shear sense. [Reprinted from *Journal of Structural Geology*, v. 9, Davis et al., Shear zone origin of quartzite mylonite and mylonitic pegmatite in the Coyote Mountains, Arizona, U.S.A., pp. 289–298, © 1987, with permission from Elsevier.]

surfaces, like any strain-sensitive foliation, start to form at 45° to the shear zone, but at this stage they are too weakly developed to be seen. As the strain increases, they become better developed and rotate toward the shear zone along with S_1. They first become visible in the field when they are approximately 35° to the shear zone, equivalent to a shear strain (γ) of 0.7. With further strain of the entire rock or as they approach a C-surface, the S-surfaces rotate toward parallelism with and become asymptotic to the C-surfaces.

The presence of both S- and C-surfaces accentuates the already-lenticular appearance of most mylonitic rocks. Both fabrics can contribute to the overall foliated appearance of a rock, and the *main* foliation that we tend to see and measure in the field can be either one, or an average of both, depending on how each fabric is expressed and how closely we look at the rocks. In our experience, we tend to notice and measure S-surfaces in weakly deformed rocks and C-surfaces in strongly deformed ones. In moderately deformed rocks, we may see and measure either one, or, if we are not careful, some average of the two that has with no real kinematic significance.

Ideally, we would like to measure the strike and dip of both S-surfaces and C-surfaces in the field. We would then know something about the orientation of the main shear zone, even if its margins are not exposed, because C-surfaces generally are subparallel to the overall orientation(s) of the shear zone boundaries. We could also use a stereonet to compute the angle between S and C in the line of transport, and thereby estimate the associated shear strain. In many cases, however, we cannot accurately measure the strike and dip of both types of surfaces because we see them exposed on only a single rock face, and we cannot tell how the planes continue into the rock. In such a case, we can still assess the angular relation between S and C along the line of transport. One approach (Reynolds and Lister, 1990) is to measure the

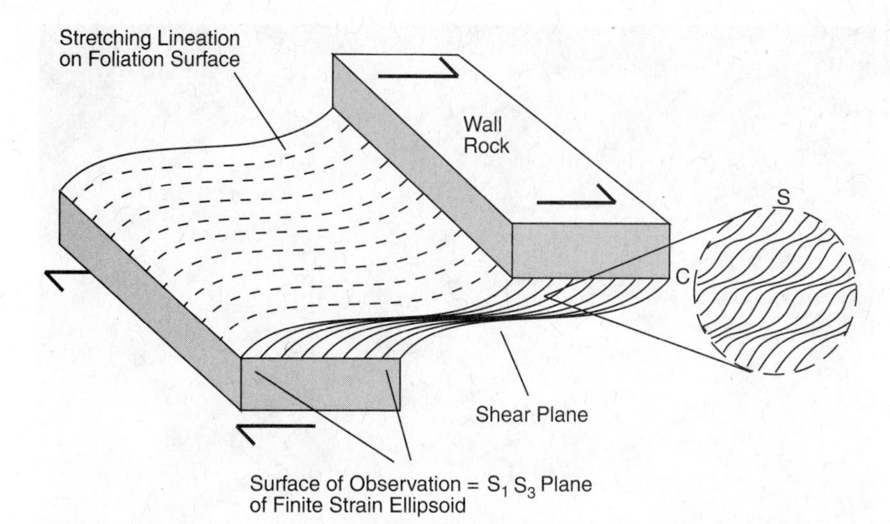

Figure 10.79 Measuring the angle between S-surfaces and C-surfaces on the sense of shear plane. [From Hanmer and Passchier, 1991. Reproduced with the permission of Natural Resources Canada 2011, courtesy of the Geological Survey of Canada, Paper 90–17.]

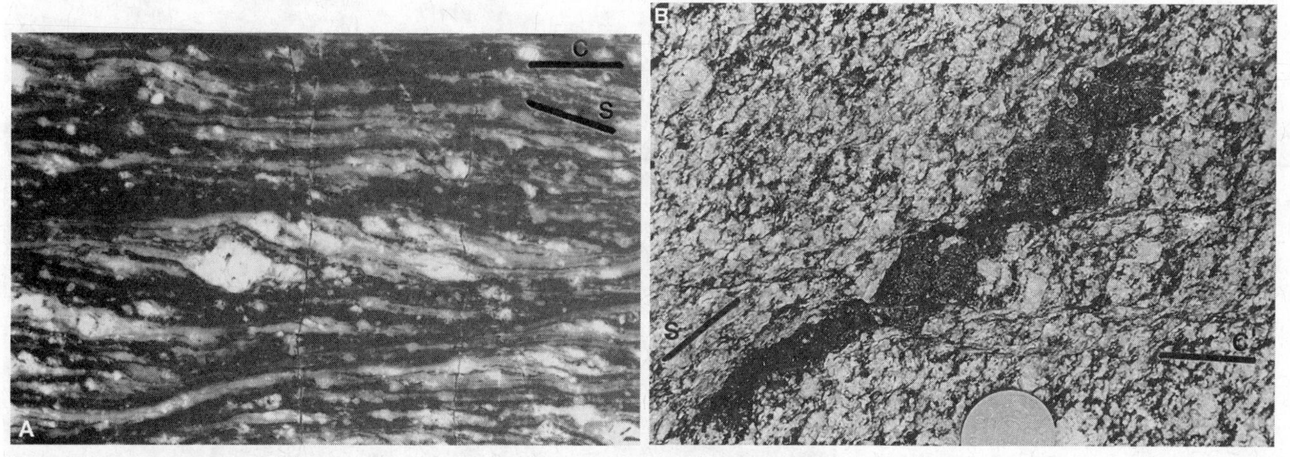

Figure 10.80 Try measuring the angle between S-surfaces and C-surfaces on the sense of shear plane in actual rocks. (*A*) Polished slab of mylonitic granodiorite, South Mountains, Arizona. C-surfaces are expressed as the dark, continuous, strongly deformed bands, whereas S-surfaces are defined by the individual elongated, light-colored, feldspar grains. The relatively low angle (15–20°) between S and C is consistent with the highly strained appearance of the rock. [Photograph by S. J. Reynolds.] (*B*) Outcrop of mylonitic late Cretaceous tonalite containing a mafic inclusion, which is parallel to S. The high angle (45°) between S and C is inconsistent with magnitude of shear strain. This may indicate that the C surfaces formed after much foliation was already developed. [Photograph by Carol Simpson.]

apparent dips of S-surfaces and C-surfaces on a rock face cut parallel to the SOS plane (Figs. 10.79 and 10.80). In this way, we directly measure the angle between S and C in the field and immediately calculate the shear strain, assuming simple shear (Equation 10.1). We are able to evaluate the variations in shear strains as we hike along, as questions about the geometry and kinematics of the shear zone arise. It is in the field where the key observations and answers related to shear zones surely lie.

The Origin of S-C Fabrics

The origin of S-C fabrics has been primarily evaluated by tracing undeformed protoliths into shear zones and documenting how the S- and C-surfaces develop (Berthé et al., 1979; Lister and Snoke, 1984). These studies reveal that the S- and C-surfaces are coordinated fabrics, not not unrelated, superimposed fabrics. As a rock is progressively deformed, the S-surfaces generally

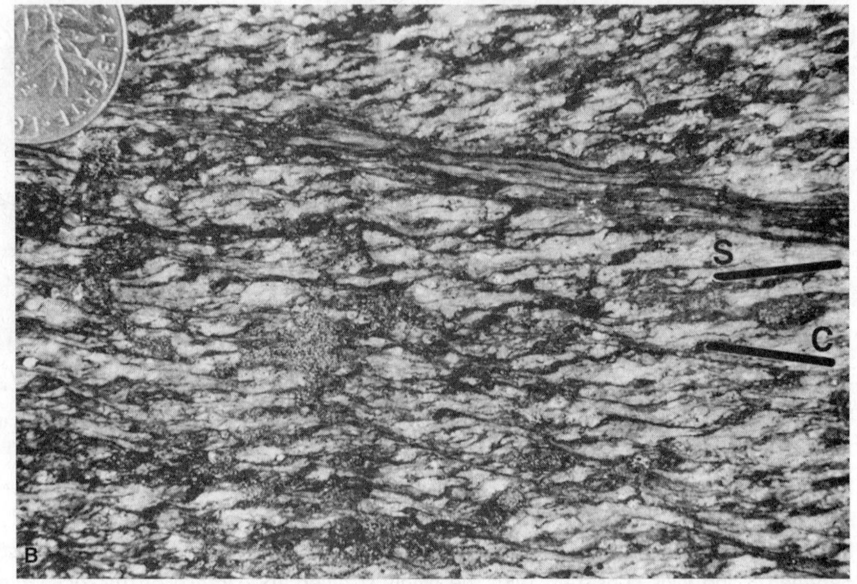

Figure 10.81 Formation of S-C fabrics in granite, South Armorican shear zone near Lescastel, Brittany, France. (*A*) Initial formation of S-C fabric, with weakly developed S-foliation, thin C-surfaces, and a large angle between S and C. Sinistral sense of shear. (*B*) Strongly developed S-C fabric in which S and C are nearly parallel. The thin, well-developed shear zone is a late-stage shear band (C′). Dextral shear sense. [Photographs by Carol Simpson.]

form first and in weakly deformed rocks may be the only type of surface present. With increasing shear strains, C-surfaces start to form, and then S and C continue to develop together (Figure 10.81*A*). The S-surfaces progressively rotate toward C, and the C-surfaces may become more closely spaced and better developed (Figure 10.81*B*). At very high shear strains, S becomes subparallel to and indistinguishable from C.

The formation of S-C fabrics may be triggered by rigid grains that cause the strain distribution to become heterogeneous on a small scale (Hanmer and Passchier, 1991; Passchier and Trouw, 2005; Holyoke and Tullis, 2006). These grains act as stress raisers to concentrate stress, perhaps causing a C-surface to nucleate where the boundaries of adjacent rigid grains are aligned parallel to the shear plane. When strain starts to become concentrated into a zone adjacent to the rigid grains, strain softening takes over, causing additional localization of strain, and a C-surface is born. This sequence of events may explain why S-C fabrics are generally found in rocks that contain some rigid grains interspersed with ductile ones; such rocks include quartzofeldspathic rocks (e.g., granite, gneiss, metarhyolite) deformed at moderate temperatures, where quartz flows ductilely but feldspar is more rigid. S-C fabrics are less common in monomineralic rocks and in quartzofeldspathic rocks (granites and gneisses) deformed at higher temperatures, where both quartz and feldspar deform by ductile mechanisms and dynamic recrystallization.

Extensional Shear Bands

Extensional shear bands are oblique to the main shear zone (White et al., 1980; Hanmer and Passchier, 1991; Passchier and Trouw, 2005). Shear bands of this type crosscut the main foliation, displacing it in a normal sense when viewed on the SOS plane (Figure 10.82). Although sporadically developed in some outcrops, they may be the most obvious shear bands in others, imparting a second foliation to the rocks. As the main foliation approaches the shear bands, it is rotated into parallelism with the bands. This results in an apparent folding of the main foliation (Platt and Vissers, 1980). They are designated as

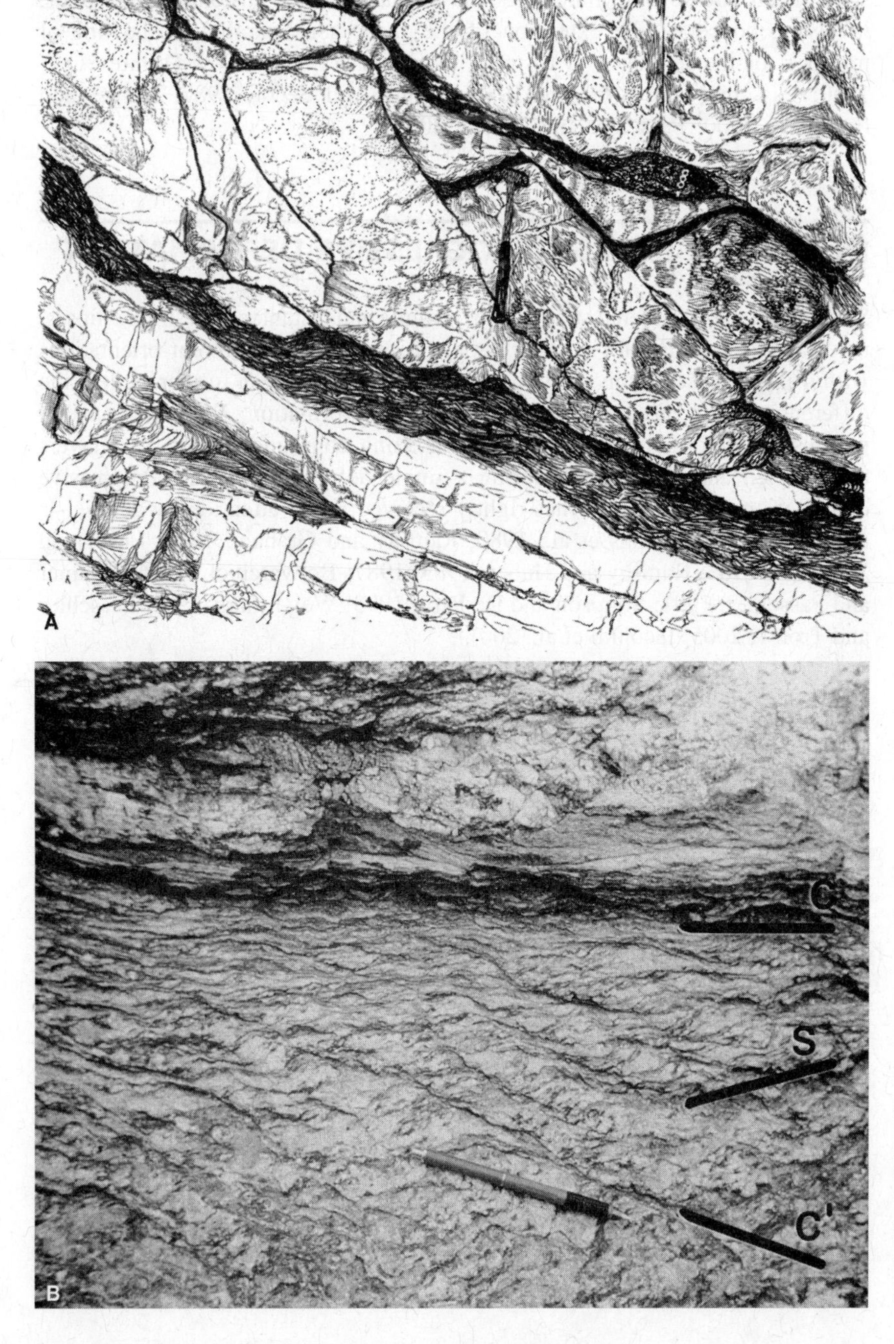

Figure 10.82 Extensional shear bands: (*A*) Cutting pegmatitic layer in banded gneiss. [David O'Day tracing of photograph by G. H. Davis. From Davis, 1980. Published with permission of Geological Society of America and the author.] (*B*) Cutting mylonitic granite, Harquahala Mountains, western Arizona. [Photograph by S. J. Reynolds.]

C′ ("***C prime***") to differentiate them from normal C-surfaces and to emphasize that they generally form after most of the main foliation has been established.

Extensional shear bands locally form conjugate sets, but more typically they occur in one dominant set that is **synthetic** to the overall shear sense of the main shear zone. The sense of shear on this set of shear bands is generally the same as the overall sense of shear for the zone. Thus, they are asymmetrical and may be a sense-of-shear indicator for the main zone. In certain cases, however, extensional shear bands or other late-stage shear bands have the sense of shear opposite to that of the main shear zone (Behrmann, 1980; Reynolds and Lister, 1990). We therefore suggest that extensional shear bands be used somewhat cautiously, especially if they appear from cross-cutting relations to be very late-stage features that formed after most of the main shear zone fabric. Such late-stage shear bands may have metamorphic mineral assemblages different from those of the main shear zone fabric or may have other distinctive characteristics, such associated veins or a brittle–ductile style. There are several models for the formation of extensional shear bands, involving concepts like extension of the shear zone parallel to the line of transport or the influence of a strong fabric on continued flow (White et al., 1980; Passchier and Simpson, 1986; Dennis and Secor, 1987; Hafner and Passchier, 2000; Passchier and Trouw, 2005).

INSIDE THE ELLIPSE: PROGRESSIVE DEFORMATION

So far, we have considered strain using the finite and instantaneous strain ellipses, not really accounting for what is happening to lines of different orientations *within* each ellipse. In this section, we go inside the ellipse to further explore the differences between coaxial and noncoaxial deformation. We also consider progressive deformation and its rather amazing consequences, such as layers that shorten and then stretch during a single deformation event. Progressive deformation in shear zones has rightly received much attention by geologists (Ramberg, 1975; Means et al., 1980; Ramsay and Graham, 1980; Lister and Williams, 1983; Ramsay and Huber, 1983, 1987; Bobyarchick, 1986; Hanmer and Passchier, 1991; Simpson and De Paor, 1993; Weijermars, 1993; Passchier and Trouw, 2005; Iacopini et al., 2007).

The Instantaneous Strain Ellipse

We have used the instantaneous strain ellipse to represent the instantaneous stretching rates along every radial line within the plane of the ellipse (Figure 10.83*A*). If we superimpose a circle representing the unstrained state on top of this ellipse (Figure 10.83*B*), positive stretching rates (extension) occur along radii where the ellipse is *outside* the initial circle and negative stretching rates (shortening) occur where the ellipse is *inside* the circle. Two radial lines passing through the intersections of the ellipse and circle represent orientations with *zero* rates of instantaneous stretching and are called **lines of zero stretching rate**. As usual, we are limiting this discussion to two-dimensional plane strain.

The principal axes of the instantaneous strain ellipse, or the **instantaneous stretching axes**, are $\dot{S}_1$ and $\dot{S}_3$, with the dot over the letter indicating that we are referring to stretching rates and the **instantaneous strain ellipse**, rather than the **finite strain ellipse**. The long, $\dot{S}_1$ axis of the ellipse represents the direction and magnitude of most rapid extension. The short, $\dot{S}_3$ axis of the ellipse is the direction with the maximum instantaneous shortening rate. The stretching rate progressively decreases for lines at a higher angle to

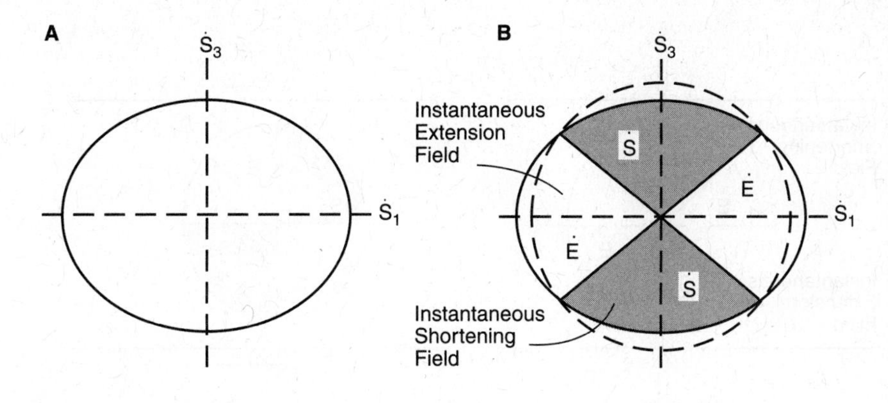

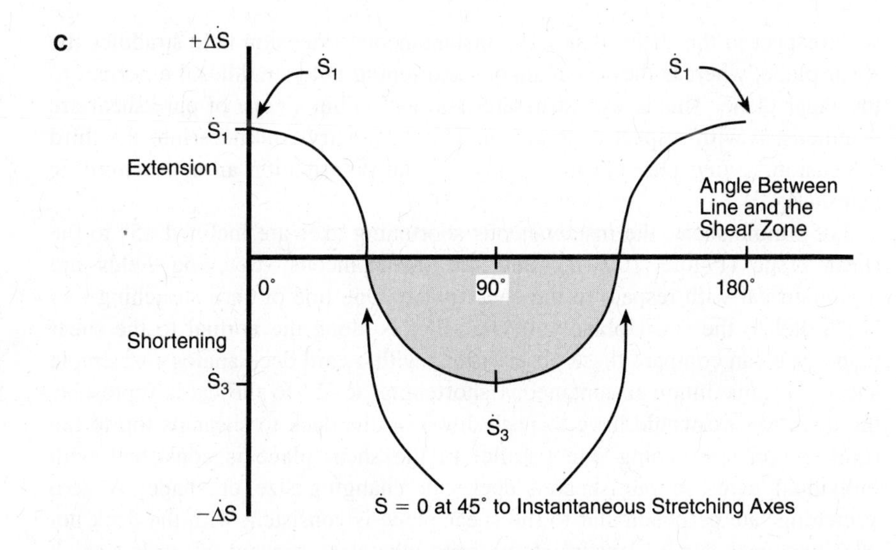

Figure 10.83 Instantaneous strain ellipse: (*A*) ellipse showing instantaneous stretching axes, (*B*) superimposed circle (dashed line) on ellipse, showing instantaneous stretching fields, and (*C*) stretching rates of lines as a function of orientation relative to the stretching axes. The maximum extension rate is along $\dot{S}_1$ and the maximum shortening rate is along $\dot{S}_3$. [From Hanmer and Passchier, 1991. Reproduced with the permission of Natural Resources Canada 2011, courtesy of the Geological Survey of Canada, Paper 90–17.]

the instantaneous stretching axes, reaching zero along the lines of zero stretching rate at 45° to the instantaneous stretching axes (Figure 10.83*C*).

The lines of zero stretching rate subdivide the instantaneous strain ellipse into four quadrants (Figure 10.83*B*), which we will refer to as the **instantaneous stretching fields**. The shaded quadrants in Figure 10.83*B* encompass all radial lines that are undergoing instantaneous shortening and comprise the **instantaneous shortening field**. The unshaded quadrants include all lines that are undergoing instantaneous extension and comprise the **instantaneous extension field**. The instantaneous stretching fields are symmetrical to and bisected by the instantaneous stretching axes. Figure 10.83*B* is the foundation for what now follows in this chapter.

Instantaneous Strain Ellipses for Pure and Simple Shear

An important distinction between pure and simple shear, or between any coaxial and noncoaxial deformation, is the inclination of the instantaneous stretching axes to the shear plane. The instantaneous strain ellipses for pure and simple shear are shown in Figure 10.84.

For pure shear, the maximum extension rate ($\dot{S}_1$) is parallel to the shear plane, and the maximum shortening rate ($\dot{S}_3$) is normal to the shear plane (Figure 10.84*A*). The lines of zero stretching rate are at 45° to the shear plane, and both the instantaneous shortening and extension fields are symmetrical

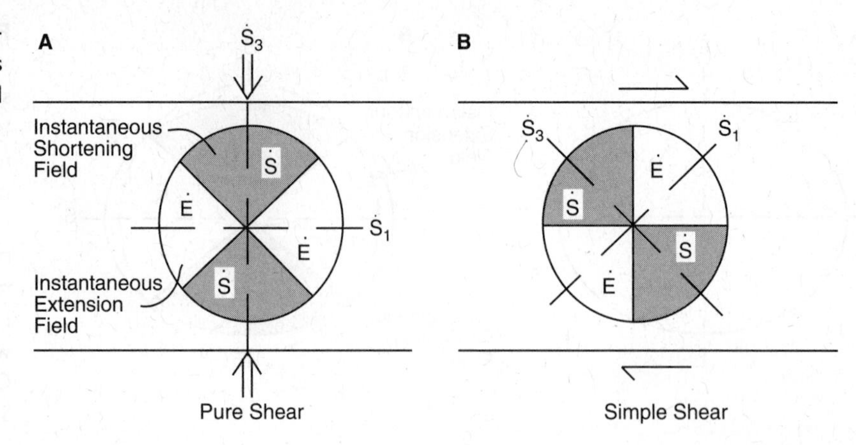

Figure 10.84 Instantaneous strain ellipses, showing instantaneous stretching axes and stretching fields, for (*A*) pure shear and (*B*) simple shear.

with respect to the shear plane. The instantaneous extension field straddles the shear plane, whereas the instantaneous shortening field straddles the normal to the shear plane. This is why structures formed within a zone of pure shear are symmetrical with respect to the zone. This symmetry continues into the third dimension, giving pure shear and any coaxial deformation an **orthorhombic symmetry**.

For simple shear, the instantaneous shortening axes are inclined 45° to the shear plane (Figure 10.84*B*), and the instantaneous stretching fields are asymmetrical with respect to the shear plane. One line of zero stretching rate is parallel to the shear plane, and the other is along the normal to the shear plane. We can compare these observations with a card deck analogy of simple shear. The maximum instantaneous shortening at 45° to the cards represents the direction we would have to push down on the deck to shear its top to the right. A zero stretching rate parallel to the shear plane is consistent with individual cards in our sheared deck not changing size or shape. A zero stretching rate perpendicular to the shear plane is consistent with the deck not changing in thickness, except at the ends where we run out of cards (which would be unlikely in real rocks). The only plane of symmetry for simple shear is parallel to the shear plane, so simple shear has a **monoclinic symmetry**. As you might expect, the symmetry of the deformation is reflected in the symmetry of the resulting fabrics (Turner and Weiss, 1963; Tikoff and Fossen, 1999).

Rotation Rates of Lines

We can also evaluate the **instantaneous rotation rates** (angular velocities) of lines, rather than their stretching rates. With rotations, we need to choose a convenient frame of reference that we can consider to be fixed in position. Possible reference frames include the shear zone, the instantaneous stretching axes, the axes of the finite strain ellipse, or some *absolute* geographic reference frame. In the field, the shear zone commonly is the most practical reference frame.

Like the stretching rates, instantaneous rotation rates of lines vary systematically with respect to orientation—for both pure and simple shear. In pure shear (Figure 10.85*A*), lines parallel to the instantaneous stretching axes do not rotate. The rotation rate is very slow for lines at a low angle to the instantaneous stretching axes; such lines are nearly parallel to or perpendicular to the shear plane. The rotation rate progressively increases for lines at higher angles to the instantaneous stretching axes, reaching a maximum for lines at 45° to the axes. These lines of maximum rotation rate are also the lines of zero

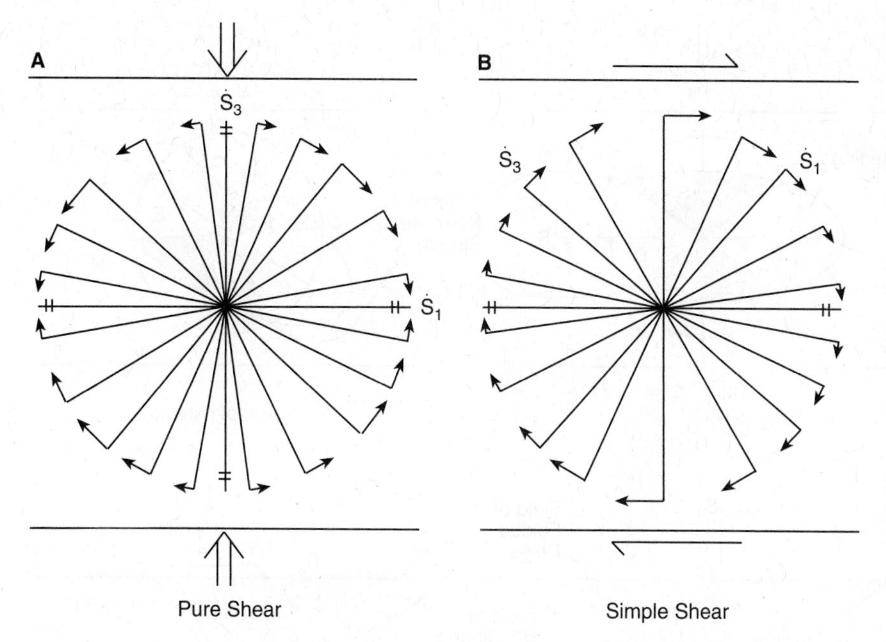

Figure 10.85 Rotation rates of different orientations of lines (passive markers). The relative rate of rotation of each line is represented by the length of the arrow on the end of that line. Lines that do not rotate are shown with double ticks. (*A*) In coaxial strain, in this case pure shear, lines parallel to either instantaneous stretching axis (parallel to and perpendicular to the shear zone) do not rotate. Rotation rates are fastest for lines oriented at 45° to the shear zone, with some lines rotating clockwise and others rotating counterclockwise. (*B*) In noncoaxial simple shear, all lines that rotate do so in the same direction as the sense of shear. The fastest rate is for lines perpendicular to the zone. Lines parallel to the shear zone do not rotate.

stretching rate. In other words, the lines that rotate the fastest are those that stretch the slowest, and vice versa.

In simple shear, all lines rotate except those parallel to the shear plane (Figure 10.85*B*). The rotation rates are very slow for lines at low angles to the shear plane, but they progressively increase for lines at higher angles to the shear plane, reaching a maximum for lines normal to the shear plane. In simple shear, all lines that rotate do so in the same direction.

The Finite Strain Ellipse

As small increments of stretching and rotation add up, they result in significant strain that can be represented with the **finite strain ellipse**, whose principal axes S_1 and S_3 are the **finite stretching axes** (Figure 10.86). If we superimpose the original circle on the finite strain ellipse, there are two radial lines that pass through points where circle and ellipse intersect (Figure 10.86*A* and *B*). These lines are the same length as they were at the start and are called **lines of no finite stretch**. They separate the finite strain ellipse into four sectors, called the **finite stretching fields**. The shaded sectors represent families of radial lines that have been shortened, whereas the unshaded sectors represent lines that have been extended. The sectors of shortening on the finite strain ellipse are called the **finite shortening field,** and the sectors of extension are called the **finite extension field**.

We can use the positions of the finite stretching fields to predict the structures we would see from deformation of a rock containing dikes of various initial orientations (Figure 10.86*C* and *D*). Dikes with orientations in the finite shortening field are represented as being folded, whereas dikes oriented within the finite extension field are shown as boudinage. The geometric disposition of folded versus boudinaged dikes can be used in some circumstances to determine, in the field, the approximate orientations of the finite shortening and extension fields, and the sense of shear. What a concept!

We can explore the finite strain differences between pure and simple shear by progressively deforming identical circles by each mechanism and comparing the end results (Figures 10.87 and 10.88). The undeformed circle contains four radial lines spaced 45° apart, which we can watch to see how they rotate. In each case, we will deform the circle in such a way that the

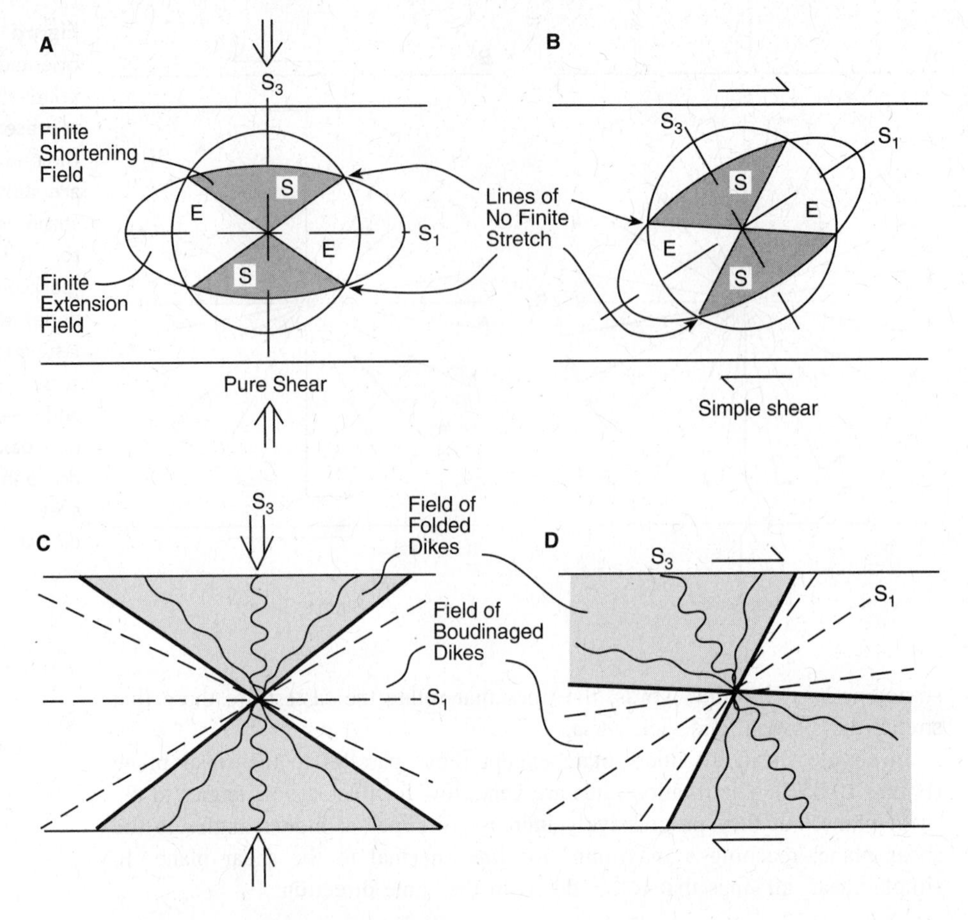

Figure 10.86 Finite strain ellipses and expected orientations of structures for pure and simple shear. Finite strain ellipses are shown for (*A*) pure shear and (*B*) simple shear, along with finite stretching axes (S_1 and S_3), lines of no finite stretch, and finite stretching fields. Expected orientations of folded and boudinaged dikes for a zone of (*C*) pure shear differ from those for (*D*) simple shear.

resulting shear plane (shear zone) will be horizontal, parallel to line *P* on each figure. Line *N* begins normal to the shear plane, and the other two lines, labeled *L* and *R*, begin at 45° to the shear plane. The shaded sectors between lines *L* and *R* on the initial circle represent the finite shortening field that will exist after the first infinitesimal increment of deformation, which is too small to show with an ellipse.

The Finite Strain Ellipse during Pure Shear

For pure shear, deformation consists of shortening perpendicular to the shear plane (parallel to line *N*) and associated extension parallel to the shear plane (parallel to line *P*). At the start of deformation (Figures 10.87*A* and *B*), the instantaneous stretching axes are parallel to lines *P* and *N*, and the lines of zero stretching rate are parallel to *L* and *R*. After some strain, lines *L* and *R* have lengthened only slightly, but have rotated toward the shear plane (Figure 10.87*C*). In contrast, lines *N* and *P* have substantially changed length but have not rotated at all. With continued deformation (Figure 10.87*D* and *E*), lines *L* and *R* continue rotating toward the shear plane and lengthening with time. At very high strains, lines with diverse original orientations end up at a low angle to the shear plane.

The lines of no finite stretch, marked by the intersections of the finite strain ellipse and the initial circle, also change orientation as the finite strain ellipse becomes more elliptical. As they rotate, the finite shortening field widens in angle, and the finite extension field narrows. The lines of no finite stretch are not fixed to any material lines during deformation. They start out parallel to material lines *L* and *R*, but rotate more slowly toward the shear plane. In contrast, the finite stretching axes (the axes of the ellipse) do not rotate, but instead stay aligned with the instantaneous stretching axes throughout the entire history of deformation. This is characteristic of *coaxial* deformation,

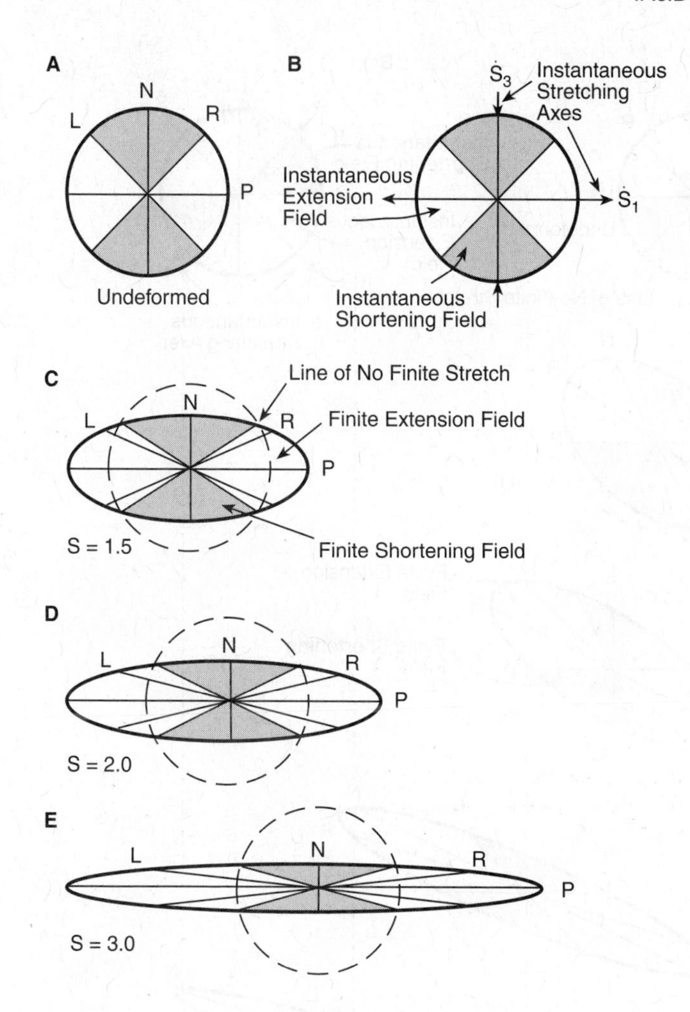

Figure 10.87 Evolution of finite strain ellipses for pure shear. (*A*) An initial circle will be deformed in such a way that the resulting shear plane (shear zone) is horizontal, parallel to line *P* on each figure. Line *N* begins normal to the shear plane, and the other two lines, labeled *L* and *R*, begin at 45° to the shear plane. (*B*) The shaded sectors between lines *L* and *R* on the initial circle represent the finite shortening field that will exist after the first infinitesimal increment of deformation, which is too small to show with an ellipse. The instantaneous stretching axes are along $\dot{S}_1$ and $\dot{S}_3$. (*C–E*) Progressive deformation causes lines *L* and *R* to rotate into the finite extension fields, whereby they are lengthened. The intersection of the finite strain ellipse and the initial circle (dashed line) defines lines of no finite stretch, which have the same length with which they started. The finite shortening field is shaded and grows with time as the lines of no finite stretch rotate.

where the material lines that at first were stretched the fastest ended up being stretched the most.

Let us examine what happens to lines *L* and *R* as they rotate (Figure 10.87). At the start of deformation, both lines were parallel to the lines of zero instantaneous stretching rate, which are also lines of maximum rotation rate. Accordingly, lines *L* and *R* rotate relatively fast, but change length slowly. As deformation proceeds, their new orientations progressively cause them to rotate more slowly and to lengthen faster. Their rate of rotation steadily decreases as they approach the shear plane. At very high strains, such lines will be greatly lengthened and will be at a very low angle to the shear plane.

The Finite Strain Ellipse during Simple Shear

In simple shear, S_1 first appears at 45° to the shear plane, parallel to $\dot{S}_1$ (Figures 10.88*A* and *B*). With additional increments of deformation, S_1 is successively rotated over, out of parallelism with $\dot{S}_1$ and toward parallelism with the shear plane (Figure 10.88*C*, *D*, and *E*). S_3 becomes increasingly shortened and progressively rotates toward the normal to the shear plane and away from $\dot{S}_3$. Both finite stretching axes therefore rotate away from parallelism with the instantaneous stretching axes as deformation proceeds, a characteristic of a *noncoaxial* deformation. An implication of noncoaxial deformation is that different material lines pass through both the *instantaneous* stretching axes and the *finite* stretching axes during deformation. S_1 starts out subparallel to line *R* at the onset of deformation but ends up nearly aligned with line *N* after a shear strain of 3 (Figure 10.88*F*).

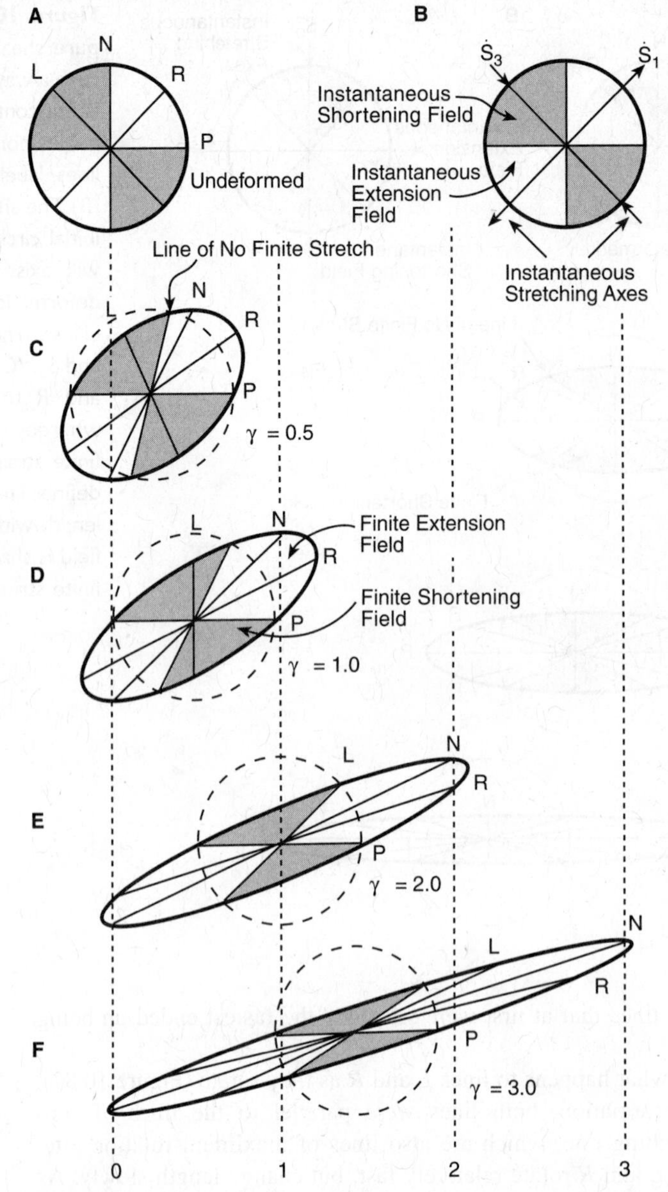

Figure 10.88 Evolution of finite strain ellipses for simple shear. Symbols and lines are the same as described in Figure 10.93. Note that line *N*, normal to the shear zone, rotates, but line *P*, parallel to the shear zone, does not. All other lines rotate in the same direction as the sense of shear. Also, the long axis of the finite strain ellipse progressively leans over in the direction of the sense of shear.

There are two lines of no finite stretch that pass through the intersection of the finite strain ellipse and initial circle. One is fixed parallel to the shear plane and does not rotate or stretch during deformation (line P in Figure 10.88). The other line of no finite stretch starts out perpendicular to the shear plane at the onset of deformation but then rotates in the same direction as the finite strain ellipse. It rotates more slowly than material lines; it started out parallel to line *N* but becomes increasingly discordant to *N* with progressive deformation. As this line of no finite stretch rotates, the finite shortening field expands in angular width at the expense of the finite stretching field. The other boundary of the finite shortening field, parallel to line *P*, remains parallel to the shear plane and does not rotate.

Note that lines *L*, *N*, and *R* all rotate in the same direction as the sense of shear. Line *N* starts out normal to the shear plane and rotates the fastest during the initial stages of deformation. Line *R* rotates progressively more slowly as it rotates into a lower angle with the shear plane. Line *L* starts out rotating at the same rate as *R*, but accelerates until it has reached the normal to the shear plane, after which it rotates more slowly. As with pure shear, lines rotate very slowly as they approach the shear plane.

Progressive Deformation

The behavior of lines during the two types of shear illustrates a fundamental concept in structural geology—**progressive deformation**. Strain is built into rocks one small increment at a time, and the finite strain of the rock is the sum of all these small increments. A material line, because it may rotate during deformation, may not be affected in the same way by each increment of deformation. A line may start out in the instantaneous shortening field but progressively rotate into the instantaneous extension field; in other words, it may shorten and then lengthen during the course of a single kinematically consistent deformation.

To explore the implications of progressive deformation, we superimpose the instantaneous stretching fields onto the finite strain ellipse (Figure 10.89). After any amount of shear, one or both lines of no finite stretch will have rotated out of parallelism with its originally corresponding line of zero stretching rate. The finite stretching fields, therefore, will not coincide with the instantaneous stretching fields even after the first increment of strain.

The lack of correspondence of the finite and instantaneous stretching fields results in three different sectors of the finite strain ellipse that contain lines with similar stretching histories (Figure 10.89*D*). The sectors are labeled with two letters: the first indicates whether radial lines that fall within that sector have been shortened or lengthened overall. Lines with orientations that fall within the **finite shortening field** have been shortened overall and are designated *S*. Lines that fall into the **finite extension field** have been lengthened (extended) overall and are designated *E*.

The second letter denotes whether a line is being instantaneously shortened or lengthened. Lines that fall in the **instantaneous shortening field** are in the process of being shortened and are designated with an $\dot{S}$. Lines that fall in the **instantaneous extension field** are in the process of being lengthened and are designated with an $\dot{E}$.

The $S\dot{S}$ sector is inside both the *instantaneous* and finite shortening fields, and it represents radial lines that have been shortened overall and are still being shortened (Figure 10.89*D*). The sector retains the same 90° angular width during progressive deformation because the instantaneous shortening field does not widen with time. In the field, dikes, layers, or other long

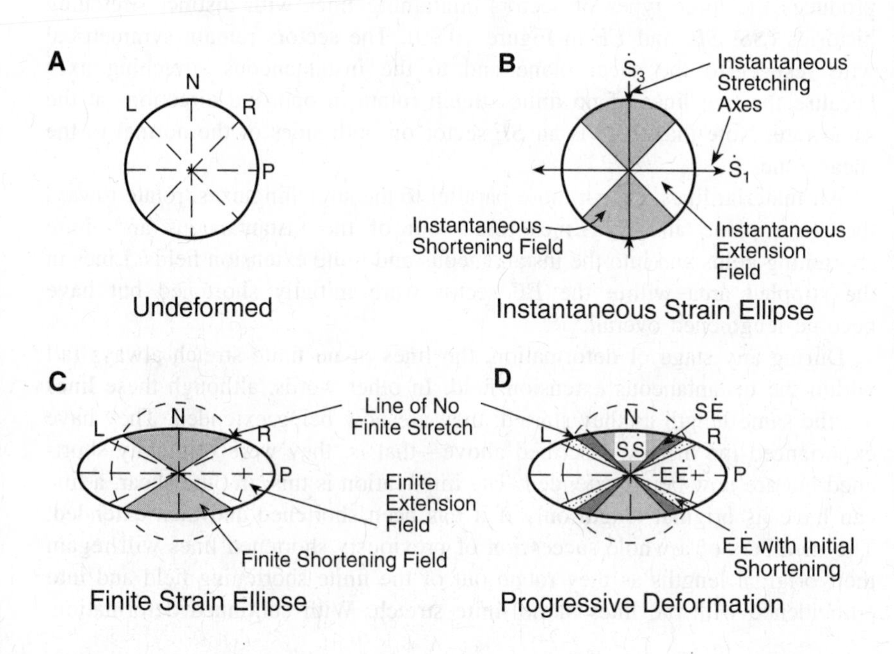

Figure 10.89 Strain ellipses for progressive deformation. (*A*) Undeformed circle containing lines *P* (parallel to shear zone), *N* (normal to shear zone), *L*, and *R*. (*B*) Instantaneous strain ellipse, showing the instantaneous stretching axes and stretching fields. (*C*) Finite strain ellipse with finite stretching fields bounded by the lines of no finite stretch. (*D*) Ellipse for progressive deformation, derived by superimposing the instantaneous and finite strain ellipses. The lightly shaded $S\dot{S}$ sectors indicate where the finite and instantaneous extension fields overlap. The darkly shaded $S\dot{E}$ sectors indicate where lines are shortened overall (inside the lines of no finite stretch) but are now being extended. The $E\dot{E}$ sectors indicate the orientation of lines that are extended overall and are still being extended. Some such lines may have had an initial history of shortening ($E\dot{E}$ sectors with initial shortening).

features with orientations within this sector would be folded or would display some other evidence of continual shortening.

The unshaded to lightly stippled $E\dot{E}$ sector is within the instantaneous and finite extension fields and contains lines that have been lengthened overall and are still being lengthened. It decreases in angular width during deformation as the lines of no finite stretch rotate.

The intervening $S\dot{E}$. sector is within the finite shortening field but also within the instantaneous extension field. It encompasses lines that have been shortened overall but are being instantaneously lengthened. It widens with progressive deformation as the lines of no finite stretch rotate toward $\dot{S}_1$. Material lines that fall in this sector have moved from the instantaneous shortening field into the instantaneous extension field. They were originally shortened but later rotated into orientations where they could be lengthened. They are still shorter than they were at the start because they are in the finite shortening field. This sector might be expressed in the field as folds that have been partially unfolded or dissected into incipient boudinage.

A more surprising stretching history is represented by the lightly stippled area of the $E\dot{E}$ sector. This area is within both the instantaneous and finite extension fields, and so lines within it have been lengthened overall and are still being lengthened. They began, however, in the instantaneous shortening field, but were rotated out of this field and into the $S\dot{E}$ sector, where they began to become lengthened. With enough strain, they were rotated into the $E\dot{E}$. sector, where they were lengthened to more than their original length!

Note that no lines rotate from the instantaneous extension field into the instantaneous shortening field. As a result, we do not expect to see boudins that have been folded. If such features occur, they may indicate that the rocks experienced more than one deformation event.

Progressive Pure Shear

To examine how progressive deformation works for pure shear, we superimpose the instantaneous stretching fields on the finite strain ellipses (Figure 10.90). As expected, the finite stretching axes (S_1 and S_3) do not rotate out of parallelism with the instantaneous stretching axes. In contrast, both lines of no finite stretch rotate toward the shear plane, resulting in a widening of the finite shortening field compared to the instantaneous shortening field. This produces the three types of sectors containing lines with distinct stretching histories ($S\dot{S}$, $S\dot{E}$, and $E\dot{E}$ in Figure 10.90). The sectors remain symmetrical with respect to the shear plane and to the instantaneous stretching axes because the two lines of no finite stretch rotate in opposite directions at the same rate. Note that there is an $S\dot{E}$ sector on both sides of the normal to the shear zone.

All material lines, except those parallel to the stretching axes, rotate toward the shear plane and $\dot{S}_1$. Lines rotate out of the instantaneous and finite shortening fields and into the instantaneous and finite extension fields. Lines in the stippled area within the $E\dot{E}$ sector were initially shortened but have become lengthened overall.

During any stage of deformation, the lines of no finite stretch always fall within the instantaneous extension field. In other words, although these lines are the same length as they started, they are now being extended. They have experienced the history described above—that is, they were originally shortened but are now being extended. The implication is this: in pure shear, a line can have its original length only if it has been shortened and then extended. The members of a whole succession of previously shortened lines will regain their original lengths as they rotate out of the finite shortening field and into coincidence with the lines of no finite stretch. With continued deformation,

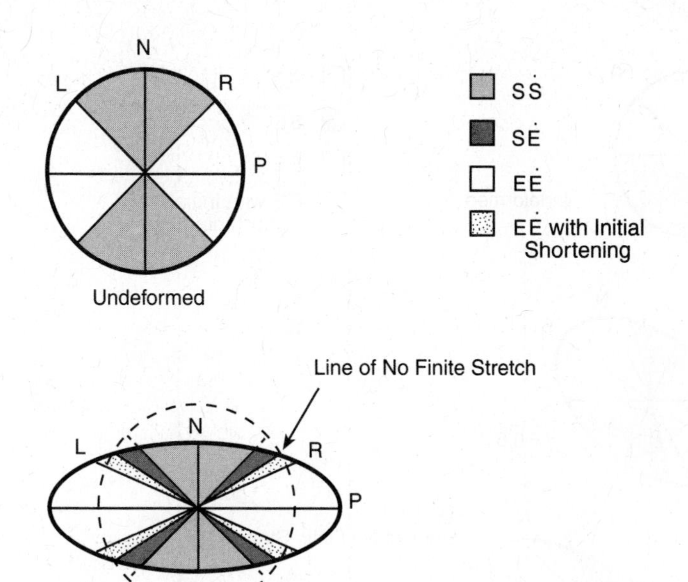

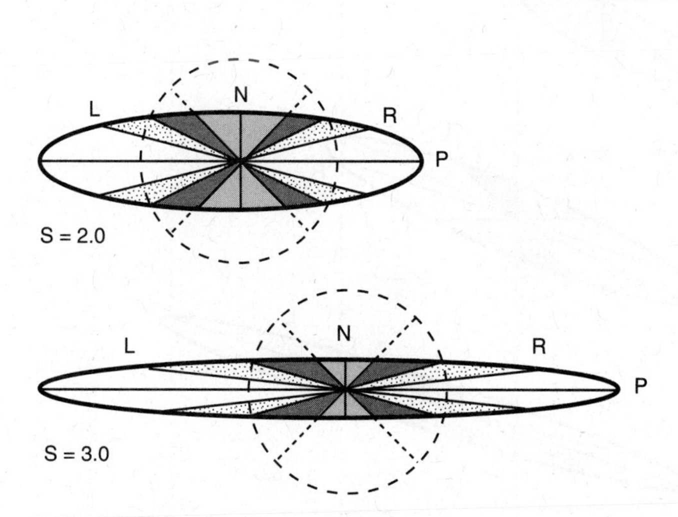

Figure 10.90 Progressive deformation during pure shear. Same initial conditions and sequence of deformation as in Figure 10.89. The different sectors ($S\dot{S}$, $S\dot{E}$, and $E\dot{E}$) are described in Figure 10.89.

however, they are rotated away from the lines of no finite stretch and into the instantaneous extension field, where they will be lengthened further.

Progressive Simple Shear

During progressive simple shear, the instantaneous and finite stretching fields again rotate out of coincidence (Figure 10.91). The finite shortening field becomes wider than the instantaneous shortening field as the lines of no finite stretch rotate. There are the same three sectors containing lines with similar stretching histories, but the evolution of the boundaries of the sectors is much different during progressive simple shear than in progressive pure shear. This is because in simple shear, the finite stretching fields are asymmetric with respect to the shear plane.

The instantaneous and finite stretching fields share a common boundary parallel to the shear plane, where one line of no finite stretch coincides with a line of zero stretching rate (Figure 10.91). This boundary stays fixed during deformation because, in simple shear, lines parallel to the shear plane never stretch and never rotate. The $S\dot{E}$ sector widens as the other line of no finite stretch rotates in the same direction as the overall sense of shear. From the first increment of deformation, it is asymmetric with respect to the

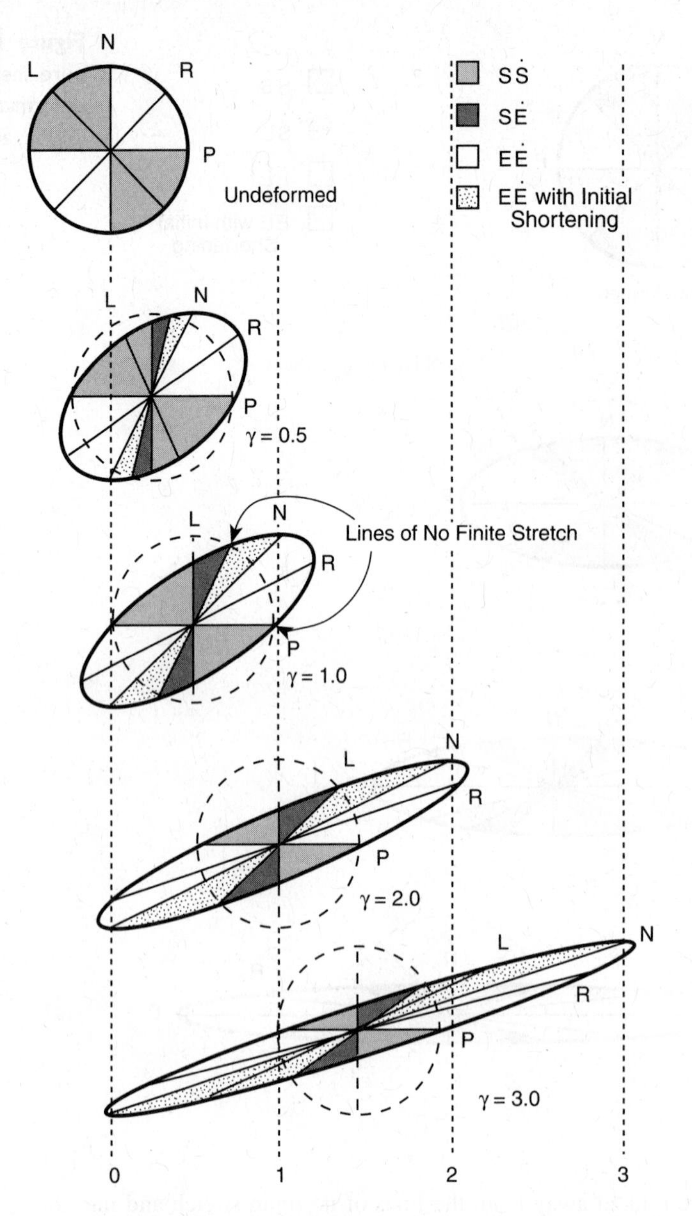

Figure 10.91 Progressive deformation during simple shear. Same initial conditions and sequence of deformation as in Figure 10.88. Symbols are the same as described for Figure 10.89. Note the difference in symmetry between this figure and the one for pure shear (Figure 10.90).

instantaneous stretching axes and to the shear plane. The adjacent $E\dot{E}$ sector narrows with time and is similarly asymmetric. The asymmetry reflects the sense of shear and the rotation of all material lines in the same direction, except lines parallel to the shear zone.

Applying the Principles of Progressive Deformation to the Field

As an example of how to apply the principles of progressive deformation in the field, Figure 10.92*A* portrays a composite of several exquisite outcrops of mylonitically foliated gneiss that I (SJR) discovered while rafting down the Colorado River through the depths of the Grand Canyon. At first glance, the outcrop seemed to be a chaotic mess, with a variety of folded, boudinaged, and apparently undeformed pegmatite dikes. In addition to the dikes, a lone thin shear zone and a few S-C fabrics cut across the foliation, also at a low angle. Luckily, lineation within the thin shear zone and in the mylonitic host rock was roughly parallel to the outcrop surface (i.e., I was observing the SOS plane).

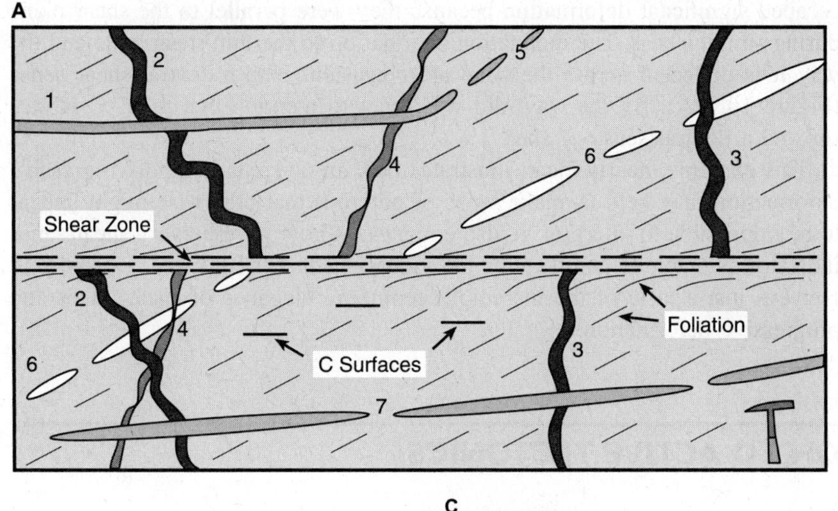

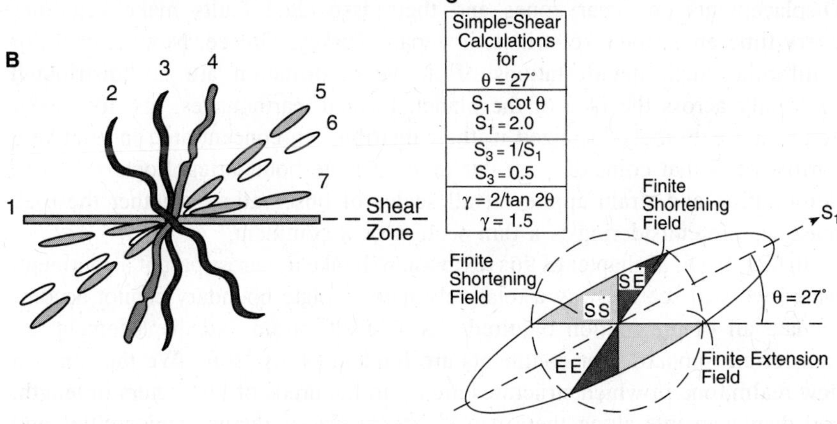

Figure 10.92 Progressive deformation near a shear zone in the Grand Canyon, Arizona. (*A*) Sketch of relations observed in several adjacent outcrops, showing numbered dikes of differing orientations and their response to deformation. (*B*) Plot of style of deformation (folded, boudinaged, or undeformed) versus orientation of dike. (*C*) Orientation of the finite stretching fields derived from (*B*). The asymmetric distribution of the stretching fields with respect to the shear zone indicates noncoaxial deformation and a dextral sense of shear. If we conclude that deformation was by simple shear and that S_l bisects the finite extension field, we can calculate the shape and orientation of the finite strain ellipse from the angle between S_l (foliation) and the shear zone. The undeformed character of dike 1 implies that it coincides with a line of no finite stretch, which for simple shear lies parallel to the shear plane. Symbols same as in Figure 10.89.

So what can we tell about the strain that affected this outcrop? Where do we start? After describing and sketching the outcrop in detail (Figure 10.92*A*), we might decide to make a plot of how each dike was strained versus its apparent trend on the outcrop (Figure 10.92*B*). This plot shows that the folded dikes all trend from upper left to lower right across the outcrop, whereas dikes that had been boudinaged have a more restricted range of trends in the opposite orientation. A few dikes, with a narrow range of orientations, are boudinaged folds, and some dikes that are largely undeformed are parallel to the lone thin shear zone. We can see that the plot has different sectors, each confining orientations of dikes with a specific structural style (Figure 10.98*C*). The folded dikes are assigned to the finite shortening field, whereas the boudinaged dikes must be in the finite extension field. Dikes containing boudinaged folds fall in an intervening sector where they have been shortened and then lengthened.

This figure looks strikingly similar to ones we just examined for progressive deformation. But does the distribution of our sectors look more like coaxial deformation (pure shear) or noncoaxial deformation (simple shear)? And what is the sense of shear, if any? The sectors on this figure most closely resemble simple shear because they have a clear asymmetry, displaying only one sector of boudinaged folds, rather than the two we expect for pure shear (Figure 10.90). The pattern suggests a dextral sense of shear, along a shear plane at a low angle to foliation. This calculated shear plane is approximately parallel to the thin shear zone and to the C-surfaces, which is reassuring. It is also approximately parallel to the least deformed dikes, which must have escaped significant deformation because they were parallel to the shear plane during simple shear. The inclination of foliation to the thin shear zone and the way it is deflected across the zone are consistent with a dextral shear sense (Figure 10.92A). By the way, did we forget to mention that the S-C fabrics showed a dextral sense of shear?

This example, nearly ideal, illustrates how an understanding of progressive deformation may help us make sense of outcrops that otherwise might appear to be impossible to interpret. It also emphasizes how we use every shear-sense indicator available to compare the individual results for consistency. Finally, it conveys just a taste of the incredible geometric elegance of shear zones and progressive deformation.

ON TO ACTIVE TECTONICS

Displacements on shear zones and their associated faults make headlines every time an earthquake strikes, in Japan, Turkey, Greece, New Zealand, or California. Such manifestations of *active* deformation are not distributed uniformly across the face of the planet. Instead, earthquakes, just like shear zones, are extremely localized in their distribution, concentrated in relatively narrow belts that coincide with our modern plate boundaries. Localization of deformation and strain applies at all scales of observation, whether the rock mass we are considering is a thin section or a continent.

In fact, the final chapter of this textbook will take us across part of a continent, where we will see the way a relatively narrow plate boundary cannot accommodate all of the motion required. As a result, strain and displacement are transferred 'inboard,' for continents are fundamentally 'soft.' We thus enter a new realm, one in which structures are up to hundreds of kilometers in length, and displacements along them can — overnight — disrupt major cities and impact millions of lives. We now enter the realm of **active tectonics**!

Chapter 11 Active Tectonics

North-directed photograph of the Laguna Salada fault, bordering here the western flank of Sierra los Cucapa and the eastern margin of the Laguna Salada salt flat, Baja Norte, Mexico. The fault, which strikes NNW and dips westward, straddles the U.S./Mexico border. Earthquakes and displacements along the Laguna Salada fault are accommodating some of the North America/Pacific plate motion, which is ~52 mm/yr right lateral shear. Slip rate along the Laguna Salada fault is ~2–3 mm/yr. Offset along it is a combination of right-lateral and normal faulting. We know this because the last earthquake along the Laguna Salada fault to cause surface rupture produced 4 m of right-lateral slip and 3.5 m of normal slip; this happened on February 23, 1892. The Laguna Salada fault cuts and displaces alluvial surfaces that range in age from late Pleistocene to historic. Paleoseismological studies reveal that big earthquake events along the Laguna Salada fault take place on average every 1000 to 2000 years. [Photograph and copyright © by Peter Kresan.]

STRUCTURAL GEOLOGY AND ACTIVE TECTONICS

Regions of present-day tectonic activity can be thought of as natural laboratories within which deformation of the Earth's crust can be studied. The field of structural geology is benefiting enormously from study of actively deforming regions, by means of integration of detailed structure-tectonic mapping, Global Position System (GPS) geodesy, earthquake seismology, and paleoseismology. Until very recently, and with many notable exceptions, there seemed to have been a "*property boundary*" between the seismology community, which addressed the study of earthquakes and active faults, and the structural geology community, which studied no-longer-active, ancient geological systems. Thankfully, structural geology, earthquake seismology, and geodesy are now more closely integrated than ever.

Actively deforming regions are distinguished by unique opportunities to measure and monitor the physical environments in which deformation takes place. We can "*see*" how structures are forming in direct relationship to plate motions and rates of movements; magnitudes and directions of strains and displacements; loading conditions and stresses; and earthquake phenomena. "*Seeing*" deformation in action gives insight into how structures form (Figure 11.1).

Geologic structures commingle. This is obvious when we closely examine actively deforming regions. Rocks utilize the whole range of deformation mechanisms available to them to accommodate deformation. If a certain amount of regional shortening needs to be accommodated during regional

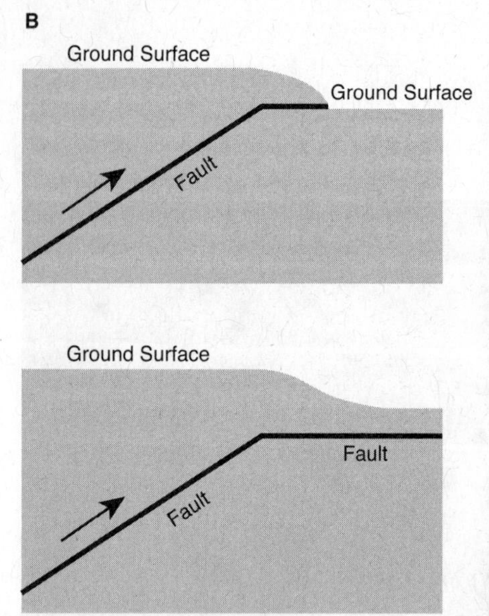

Figure 11.1 (*A*) Photograph of deformation zone produced in Taiwan during the 1999 Chi-Chi earthquake. Terrace deposits are thrust faulted over Pleistocene sediments. Folding of the Pleistocene sediments created a fold scarp, up to 2 m high. On average the thrust fault dips 24°NE. The deformation zone was exposed by trenching (the Chushan trench) carried out by the Taiwan Central Geological Survey. (*B*) The well exposed fold-scarp profile helps to constrain the faulting mechanism. Whether a fold of this type has a parabolic profile (top) versus a monocline profile (bottom) gives insight regarding the underlying fault geometry. [Courtesy of Owen Huang and Arvid Johnson.]

compression, the strain can be "*partitioned*" through folding with some thrust faulting; thrust faulting with some folding; folding and thrust faulting developed about equally; reverse faulting and shearing; and so on. The displacements that in one subregion are accommodated by faulting can be transferred to any number of other structures or combinations of structures within an adjacent subregion. There is no way we know how to predict, *a priori*, what combination of structures will be employed by nature to accommodate a given regional strain. Yet, by studying active tectonic systems, we gain a sense for the range of possibilities.

PLAN OF ACTION FOR THIS CHAPTER

The purpose of this chapter is to interrogate actively deforming regions to gain insight regarding how geologic structures and systems of geologic structures form. We intend this chapter to be a way of drawing together how major structures organize themselves in response to the requirements of plate tectonics and plate-scale deformation.

The actively deforming "*laboratory*" we will be examining is the western United States, specifically the region extending eastward from the San Andreas fault (SAF) system to the western edge of the Colorado Plateau (Figure 11.2). Though it was tempting to construct this chapter in ways that draw upon examples from around the globe, we found it advantageous to take a comprehensive look of just one (large) region, in order to emphasize the relationships of the parts to the whole. Moreover, the western United States is attractive because of the enormous level of detail known about this region through studies of plate motions, GPS motions, earthquake seismology, faulting and folding, paleoseismology and subsurface exploration. For example, Figure 11.2*A* displays velocity vectors for GPS monuments scattered throughout our study region. We see the effect of P/NA plate interactions, which cause western North America to "*flow*" westward, northwestward, even

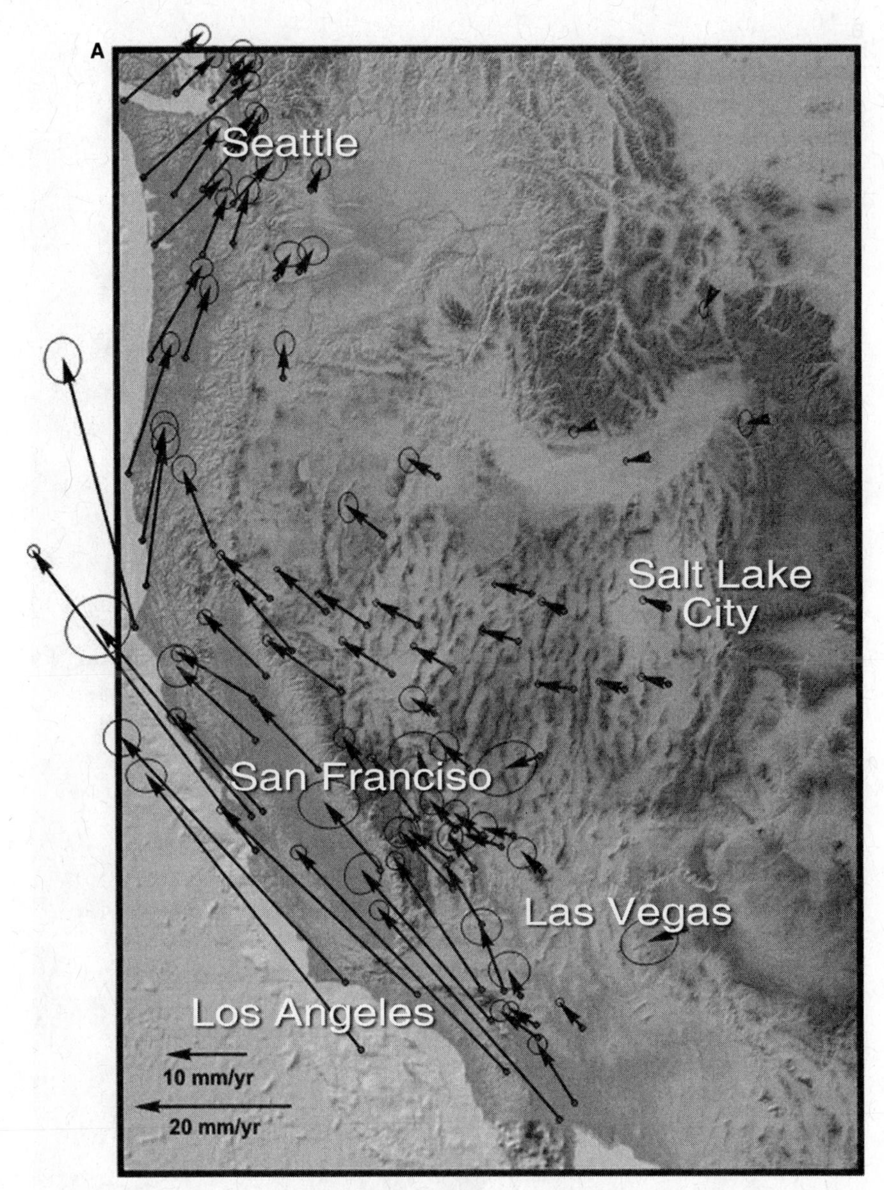

Figure 11.2 (*A*) Map of velocity vectors of GPS reference monuments within the region of study. Velocities vectors are scaled in mm/yr and referenced with respect to stable North America. [Map courtesy of John Oldow.]

northward with respect to an (arbitrarily) fixed Colorado Plateau. This picture captures the fact that the western part of the North American plate is *not* rigid, and that the plate interactions cause structures to form far inboard from the plate boundary.

There is a broad range of regional strain environments in western North America (Figure 11.2*B*), assuring us that we will be examining a full and diverse range of regional geologic structures. Along the boundary between the Pacific plate and the Sierran microplate, shortening takes place along a NNE/SSW direction (Figure 11.2*B*), and there we will have the chance to examine closely the San Andreas fault system and the structure of the Los Angeles basin, including its record of big earthquakes and inversion tectonics. Along the boundary between the Sierran microplate and the Western Great Basin (see Figure 11.2*B*), east-west extension is producing the combination of normal faulting and right-handed strike-slip faulting within the Eastern California Shear Zone and the Walker Lane. Walker Lane is a brand new (continental) transform fault, feeding fault-slip northward into Cascadia (see Figure 11.2*B*), contributing to the N-S shortening in the vicinity of Seattle in the form of

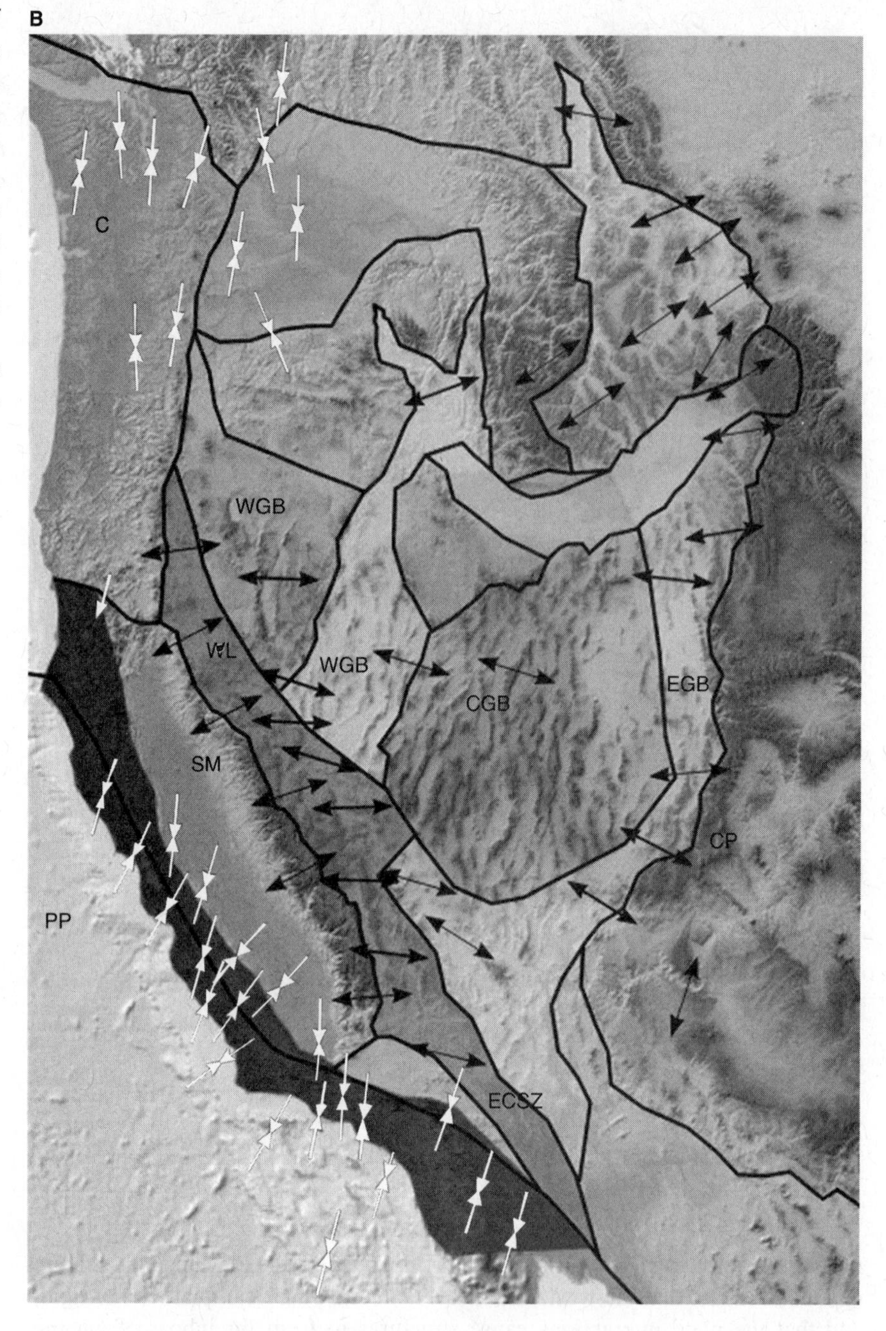

Figure 11.2 (*Continued*) (*B*) Map of active regional strain domains within the western United States, based on GPS, earthquake focal mechanisms, and inversion of fault-slip data. Bold dark arrows show extension; white arrows show shortening. Tectonic provinces referred to in text are as follows: PP = Pacific Plate; SM = Sierran Microplate; C = Cascadia; ECSZ = Eastern California Shear Zone; WL = Walker Lane; WGB = Western Great Basin; CGB = Central Great Basin; EGB = Eastern Great Basin; CP = Colorado Plateau. [Map courtesy of John Oldow.]

thrust faults and fault-related folds. Crossing the Central Great Basin (Figure 11.2*B*), we will see that east-west crustal stretching is taking place, and yet at different rates in different parts. Through examining the boundary between the (stable) Colorado Plateau and the Eastern Great Basin to the west (Figure 11.2*B*), we will treat ourselves to the properties of great normal faults (such as the Wasatch fault zone) that are responding to east-west extension. Were we to continue on this trip to the Rio Grande rift, we would see more in the way of extension, and yet note the contrast between the giant size of the rift and the slow pace of the present rates of stretching.

Some of these subregions are dominated by strike-slip fault systems, others by thrust or normal fault systems, and still others by hybrid combinations of all the major classes of faulting. We will discover that some subregions are marked by 3D nonplane strain in the form of transpression (producing shortening and strike-slip) or transtension (producing extension and strike-slip), and others by the ordinary plane strain (both coaxial and noncoaxial). Because we will be able to examine the development of major structures in real time as well as during the recent past, we have the chance to report rates of deformation (including rates of regional strain), slip on individual major faults, and growth folds. We will also witness the remarkable ways in which deformation (e.g., expressed as earthquakes and ground rupture) 'jumps around' from place to place over time.

Throughout this chapter we will call attention to the "***Lesson Learned***" from studying the active tectonic environments in a structural geologic context. These lessons bridge active and ancient tectonic environments, where "*ancient*" in this context is shorthand for structural geological systems that are no longer active.

Part of the insight into how regions are actively deforming is captured in contemporary earthquake activity. The radiation that emanates from an earthquake hypocenter where a fault slip occurs can be evaluated in terms of fault plane orientation and slip vector orientation. The radiation pattern discloses whether the faulting event was one of normal faulting, thrust faulting, reverse faulting, right-handed strike-slip faulting, left-handed strike-slip faulting, or some combination of these. *Part III-S, Determining Focal Mechanisms for Earthquakes*, describes how focal mechanisms are determined, and presents the standard symbols (**beach ball diagrams**) for picturing the faulting associated with a given earthquake. Determining a focal mechanism on an earthquake is equivalent to taking the strike-and-dip of an active fault, trend-and-plunge of slickenlines, and sense of offset.

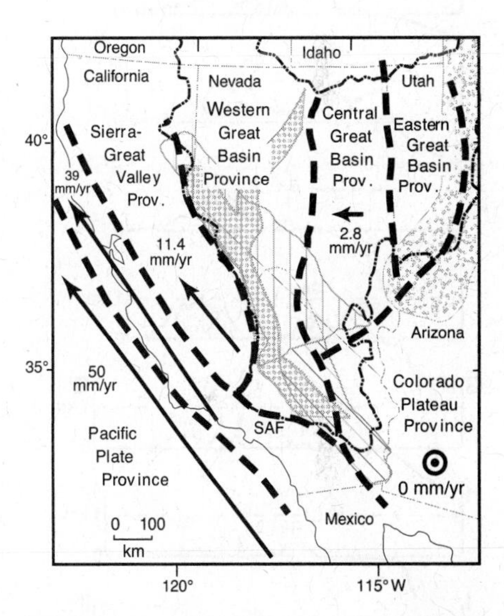

Figure 11.3 Shown here are geodetically determined strain provinces of the western United States. Note that the Pacific plate moves ~50 mm/yr relative to the Colorado Plateau, the arbitrarily fixed point of reference (i.e., stable North American plate). Movement of the Central Great Basin (2.3 mm/yr) is accommodated in part by diffuse deformation across the eastern Great Basin. Movement of the Sierra Nevada / Great Valley region (11.9 mm/yr) is accommodated in part by diffuse deformation along the Western Great Basin. The San Andreas fault is accommodating ~39 mm/yr of Pacific plate movement relative to the Sierra Nevada/Great Valley region [From Bennett et al. 2003, Figure 10, p. 3–20.].

WESTERN UNITED STATES

The Big Picture, As Sharpened Through GPS

Although geoscientists might quibble over a millimeter or two, the motion between the Pacific (P) and North American (NA) plates is ~51 mm/yr, which is ~51 kilometers in a million years (see Figure 11.3) (Kreemer and Hammond, 2007). Most people generally think of the San Andreas fault (SAF) as the P/NA plate boundary, but it is not. We now know that only about three-quarters of the P/NA plate motion is being accommodated by the San Andreas fault. The slip rate is ~39 mm/yr (see Figure 11.3). Tanya Atwater (1970) figured this out well before the emergence of the technologies we have available today. This caused her to conclude that an important part of the total P/NA plate motion, which we now know to be ~12 mm/yr, is taken up via a broad transform fault zone that occupies western North America (Atwater, 1970).

It *is* accurate to emphasize that the SAF forms the boundary between the Pacific (P) plate and the Sierran (S) microplate, which encompasses the Sierra Nevada and the Great Valley (see Figure 11.3). The Sierran microplate is moving ~12 mm/yr northwestward relative to the North American plate. It does so as a perfectly rigid block without internally deforming. The Sierran microplate may be thought of as "*tectonically escaping*" northwestward toward the Cascadian subduction zone (see Figure 11.2*B*) (Kreemer and Hammond, 2007).

Plate motions are commonly presented in Mercator projection, maps constructed in such a way that the plate boundary proper is a great circle about the pole of rotation (relative motion) between the two plates represented. One of the advantages of plotting plate-boundary fault systems on a Mercator projection is that we can pick out those individual fault segments that are perfectly parallel to plate motion, and those that are not. When we apply this approach to P/NA motion, the result is quite illuminating (Figure 11.4*A*). Comparing the precise direction of P/NA plate motion to the trend of the SAF, we discover that there are segments of the SAF that deviate from ideal strike slip. Where a particular fault segment trends more northwesterly from P/NA, there is a slight divergence along it, expressed as a component of extension. On the other hand, where a particular fault segment trends more northerly than the SAF, there is a slight convergence along it, expressed as a component of shortening (see Figure 11.4*A*).

The overall effect of imperfect alignments between specific fault segments within the SAF zone and the exact trajectory of P/NA motion is ~3.3 mm/yr of shortening perpendicular to the average trace of the SAF (Argus and Gordon, 2001). The fault segments that are most misaligned in a clockwise sense are along mountain blocks that are higher and wider. This is because fault-normal convergence produces thrust faulting, reverse faulting, buckling, and overall crustal thickening. Elevations are higher in such regions. Indeed, there is a very close relationship between the present rates of fault-normal convergence from segment to segment within the SAF and the sizes of individual mountains along the California Coast Ranges (Figure 11.4*B*). Larger, broader mountains occur where convergence rates are high, and lower narrower mountains are found where convergence rates are low. Argus and Gordon (2001) graphed their data in a

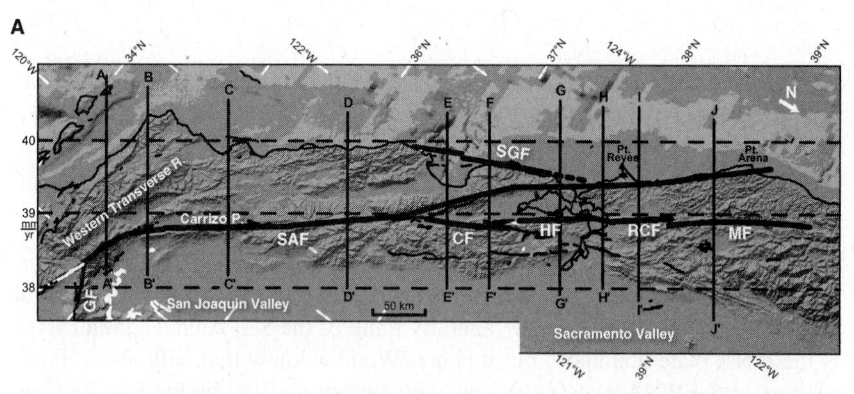

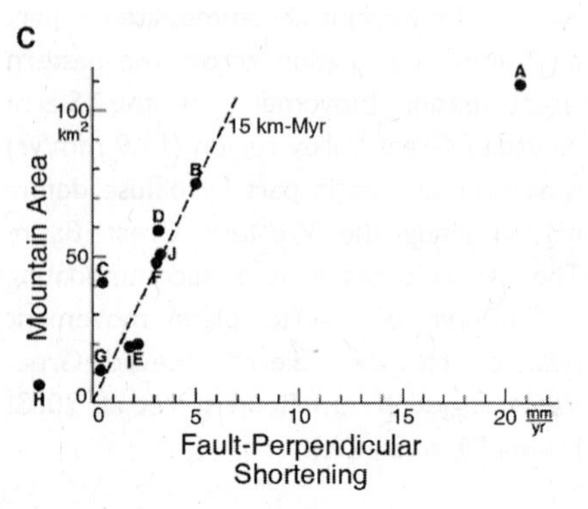

Figure 11.4 (*A*) Mercator projection showing the direction of Pacific-Sierran plate motion compared with strikes of main faults in San Andreas fault system. Faults are shown in black. Plate velocities shown in horizontal dashed lines labeled 40 mm/yr, 39 mmyr^{-1}, and 38 mmyr^{-1}. Background is digital topography. Lines A-A′ to J-J′ are lines along which topographic profiles are constructed. Fault abbreviations: GF = Garlock; SAF = San Andreas; HF = Hayward; RCF = Rodgers Creek; MF = Maacama; CF = Calaveras; SGF = San Gregario. [From Argus and Gordon, 2001, Figure 3, p. 1583.] (*B*) Topographic profiles along the ten lines of section (A-A′ to J-J′). Each line of section is perpendicular to the direction of Pacific-Sierran plate motion. The area above sea level along each profile is given. Vertical exaggeration is 19.1. [From Argus and Gordon, 2001, Figure 5A, p. 1587.] (*C*) Graph of area above sea level (for each topographic profile) versus rate of fault-perpendicular shortening shows a direct relation between shortening and mountain building. Overall rate of fault-perpendicular shortening is ~14 mm/yr. [From Argus and Gordon, 2001, Figure 5B, p. 1587.]

unique way; that is, size of a given mountain (cross-sectional area in km^2) versus fault-perpendicular shortening (expressed in mm/yr), and by doing so they discovered a surprisingly systematic relationship (Figure 11.4*C*). The rate of growth of mountains along this part of the plate boundary amounts to ~15 km/my.

Lesson Learned #1 Precise mapping of transform plate-boundary motions and associated active strike-slip faulting reveals that even the slightest misalignments of specific faults with respect to the trend of the boundary itself require either a component of fault-perpendicular shortening (contraction) or fault-perpendicular stretching (divergence). Knowing this, when we analyze the kinematics of major ancient strike-slip shear zones, we need to factor-in the possibility of overall shortening/ thickening, or overall stretching/thinning of the deformation zone as a whole, and not just assume that everything is the result of a simple "*sliding past*."

Testing of the Big Picture Through Geology

The rates and directions of plate motions between the Pacific (P), Sierran (S), and North American (NA) plates are based on sophisticated geodetic technologies (e.g., very long baseline interferometry, satellite laser ranging, the Global Positioning System); standard plate-tectonic approaches utilizing transform-fault azimuths; earthquake slip vectors; and spreading rates from marine magnetic anomalies. But in addition, detailed field-based fault analysis contributes plate motion insight. Our example is the work of two outstanding geoscientists from the University of Oregon, Gene Humphreys and Ray Weldon, who decided to evaluate whether the P/NA plate rates and directions could be corroborated on the basis of the rates and directions of fault slip along active fault zones in the western United States. They were intent on locating the ~12 mm of North American/Pacific plate motion declared "*missing*" along the SAF system. They imagined that the total motion (~51 mm/yr) represents the entire 'budget' of relative motion between the North American and Pacific plates, and they believed it should be possible to locate where each bit of missing motion is being accommodated in the interior of the western United States. They summed up the rates and directions of fault slip along three traverses across the southwesternmost part of the North American plate (Figure 11.5*A*). It was as if they were looking for loose change.

Humphreys and Weldon (1994) extracted from the literature all published slip rates for major active faults. They used fault slip rates at scores of locations in the western United States, calculating the total amount of motion being contributed fault-by-fault and subregion-by-subregion. The fault-slip rates are largely based upon careful mapping of offsets of marker beds of precisely known ages; GPS determinations capturing displacements during earthquakes (the coseismic strain); and GPS determinations capturing displacements between earthquake events (the interseismic strain).

The budget analysis turned out well. They found most of the loose change! Based on fault slip rates and directions, they determined an overall P/NA relative plate motion of 48 ± 3 mm/yr, which compares closely with the ~51 mm/yr velocity determined independently through geodetic methods. Similarly, they computed the motion of the Pacific plate relative to North America, and found that it deviated only 5° to 9° (counterclockwise) from the N45°W $\pm 2°$ direction of movement determined on the basis of geophysical methods (Figure 11.5*B*). Just as Tanya Atwater (1970) predicted, the fault displacement along the San Andreas is in arrears, and the missing slip is distributed broadly across the western United States.

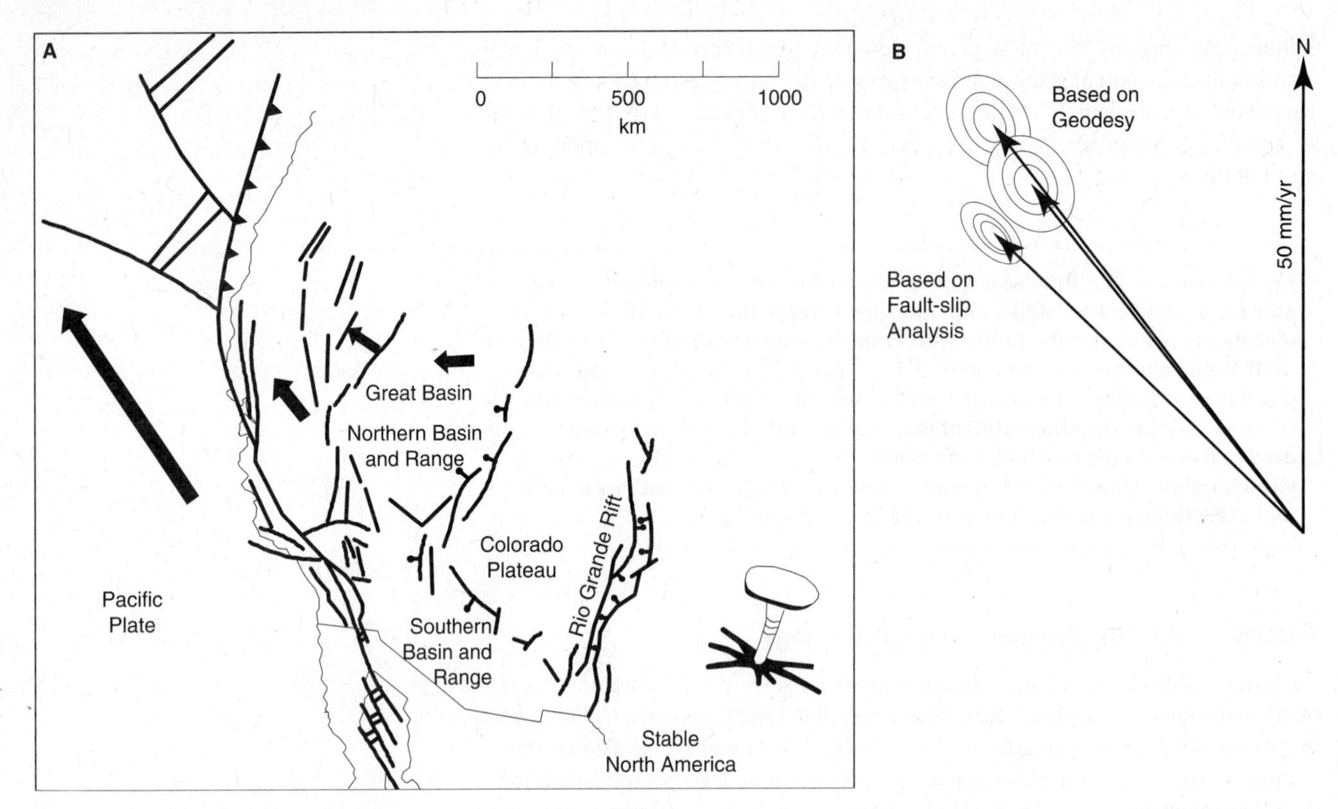

Figure 11.5 (*A*) The region across which Humphreys and Weldon (1994) added up fault-slip rates in order to see if the summation matched the plate velocity vector describing the relative motion of the Pacific plate (P) relative to the North American plate (NA). [Modified from Humphreys and Weldon, *Journal of Geophysical Research*, Figure 1, p. 19,976, copyright © 1994 by American Geophysical Union.] (*B*) Plot of velocity vectors describing movement of Pacific plate relative to fixed North America. Two of the vectors are geodesy based. The other is based on the fault-slip direction/rate determinations made by Humphreys and Weldon. Note the beautiful correspondence! Ellipses encompass range of uncertainty in results. [From Humphreys and Weldon, 1994, Figure 8, p. 19,986.]

Lesson Learned #2: When we do the fieldwork and the math, we find there is a remarkable correspondence between active plate-tectonic *inputs* (movement rates and directions over time) and active continental deformation *outputs* (distributed strain and displacements over time). Knowing this we should keep a ledger as we sum-up strains and displacements in ancient tectonic systems, comparing the results to what the plate history appears to demand. Of course, all of this underscores the fact that continental parts of plates are not rigid. For continental crust, plate forces drive deformation well inboard from plate boundaries.

THE SAN ANDREAS FAULT

Its Three Segments

The San Andreas fault is ~1300 km in length, but we should resist any notion of thinking of it as a perfectly straight break marked by uniform behavior along its entire length. In fact, there are three discrete segments (Figure 11.6): a northern and a southern segment both marked by seismic stick-slip fault behavior, and in between a central segment marked by aseismic creep. The northern segment of the SAF is ~310 km-long, the central creeping segment ~180-km long, and the southern segment ~800 km-long (see Figure 11.6).

The seismic **stick-slip faulting** behavior that takes place within both the northern and southern segments of the SAF occurs in cycles marked by long silent interseismic build-up of strain energy that can lull you to sleep. Then, abruptly, there will be a sudden coseismic release of pent-up strain energy, producing the earthquake itself, accompanied by fault movement (slip) and rebound (elastic strain recovery). Strain energy builds because faults in the upper crust become "*locked*" along geometric irregularities (**asperities**) while in the lower crust below there is continuous, viscous, aseismic shearing taking place at a steady rate.

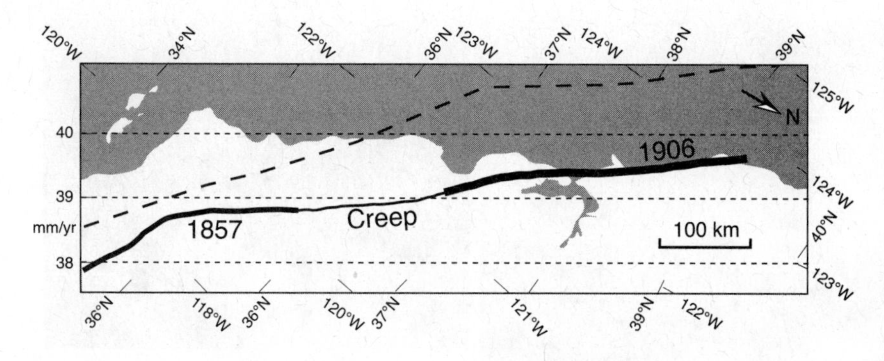

Figure 11.6 Mercator projection. Dark is the Pacific Ocean; white is (mostly) California. The San Andreas fault is composed of three discrete segments: northern, central, and southern. The northern segment is highlighted by the rupture zone of the 1906 earthquake. The southern segment, which runs from Parkfield, California, to the East Pacific rise in the Gulf of California, is highlighted by the rupture zone of the 1857 earthquake. Between these two segments is the central, creeping segment. The three fault segments are compared with the direction of Pacific-Sierran plate motions. Plate velocities shown in horizontal dashed lines labeled 40 mm/yr, 39 mmyr^{-1}, and 38 mm/yr. [From Argus and Gordon, 2001, Figure 7.]

When and where the SAF is locked, slip is prevented, and the fault-slip budget goes into the red. Coseismic events (i.e., slip during earthquakes) represent "*catch-up*." ***Pow***! The catch-up is like a new roll of the dice, creating changed conditions in the state of stress in the upper crust. At some locations, strain energy is relieved by the fault slip. At other locations, the fault slip creates a stress transfer that brings rocks closer to their rupture point, and thereby sets the stage for the next location of the next earthquake.

In contrast to stick-slip behavior along the northern and southern segments of the SAF, slip along the central segment of the SAF takes place by continuous, steady movement, absent major earthquakes. In the central segment the fault zone apparently is quite weak and cannot withstand any appreciable build-up of strain energy. The creeping central segment of the SAF is managing a (time averaged) steady slip of ~3 cm/yr.

Northern Segment of the San Andreas Fault

We all have heard about the 1906 Great San Francisco earthquake. It took place on April 18, 1906 (at 5:12 A.M. local time) and ruptured the northern segment of the SAF (Figure 11.7). The rupture length was fully 477 km, with

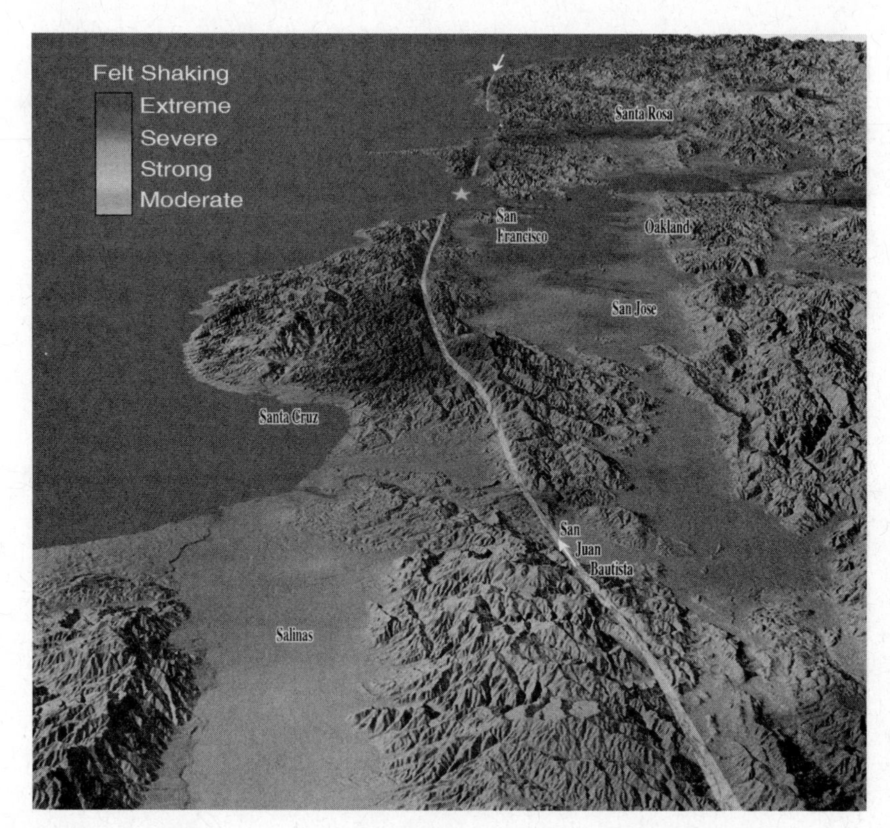

Figure 11.7 Virtual oblique aerial image of part of the rupture trace of the 1906 San Francisco earthquake. Note epicenter (star). [This image is a black-and-white version of the cover image of the 1906 San Francisco earthquake centennial field guides, published as Geological Society of America Field Guide 7. The image was created by Michael Rymer, and the ground motions are from Boatwright and Bundock (2005).] [From Prentice and Moores (2006).]

Figure 11.8 Historic photographs taken in the aftermath of the San Francisco earthquake of 1906. (*A*) Offset of fence located ∼1 km northwest of Woodville, California. View is northeast. Fence is offset in right-handed fashion by a distance of 2.6 m. [Photograph taken by G. K. Gilbert. ID. Gilbert, G.K. 2845 ggk02845. Courtesy of the U. S. Geological Survey.] (*B*) Offset of road and fence, with horse and buggy for scale. Road located between Upper and Lower Crystal Springs Reservoirs, currently Highway 92. [Photograph courtesy of Bancroft Library, University of California, Berkeley.] (*C*) Train thrown down by the earthquake at Point Reyes Station. This locomotive was standing on a siding when the April 18th earthquake pounded the region with seismic shock waves. [Photograph taken by G. K. Gilbert. ID. Gilbert, G. K. 3400 ggk03400. Courtesy of U. S. Geological Survey.]

right-handed strike-slip offset as great as 6 m. The magnitude (M_W) of the 1906 earthquake is estimated to have been somewhere between 7.7 to 7.9. Historic photographs leave nothing to the imagination in regard to magnitude of offset and ground motion (Figure 11.8). Nor does the landscape expression of San Andreas fault leave anything to imagination (Figure 11.9).

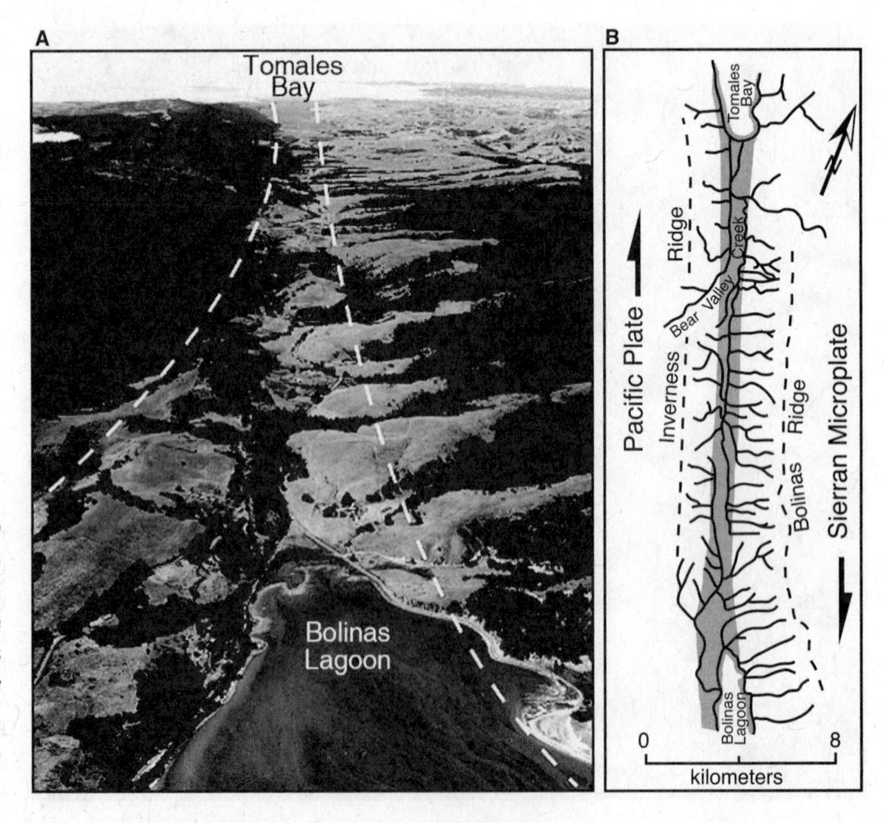

Figure 11.9 (*A*) Oblique aerial photo view of trace of San Andreas fault from Bolinas Lagoon to Tomales Bay, California. The dashed lines show the approximate limits of the active zone of the San Andreas fault. (*B*) Inset map showing plate boundary and drainage pattern. Photograph by R. E. Wallace. [From Hall and Niemi, 2008, Figure 2.]

One of the finest geologists ever to live, G. K. Gilbert, made first-hand observation of SAF offset and damage in the Point Reyes area (Figure 11.10) just one week after the earthquake occurred (Niemi et al., 2006). G. K. Gilbert's observations were the ones that confirmed an instantaneous right-handed strike-slip on the fault. This was big news, because at the time most geologists only thought of normal, thrust, and reverse as the acceptable behavior for faults.

G. K. Gilbert easily tracked the rupture zone, because ground damage and disruption produced furrows and "*moletracks*" (Figure 11.11*A*), still recognizable at Point Reyes. Moreover, at Skinner Ranch, Gilbert documented 5 m of right-handed offset, using as a basis the offset of a row of raspberry bushes (4.4 m); a path to the house (4.6 m); a fence (4.7 m); and the southeast corner of the Skinner Ranch cow barn (4.9 m) (Figure 11.11*B*) (Niemi et al., 2006, p. 168). The part of the fault zone that accommodated these displacements during the 1906 earthquake proved to be just a single fault surface within a zone no wider than 4.5 m (Niemi et al., 2006). Paleoseimology investigations have revealed a minimum of ten earthquake events during the past 2500 years in the Point Reyes area. Niemi et al., (2006) estimated that the average slip rate during the Holocene has been ~24 mm/yr, which is comparable to the average slip rate over the past 200–400 k.y., but only ~60% of the expected total slip rate for the SAF.

We should keep in mind that the SAF system is not just the San Andreas fault proper, but includes other faults as well. For example, the Hayward fault

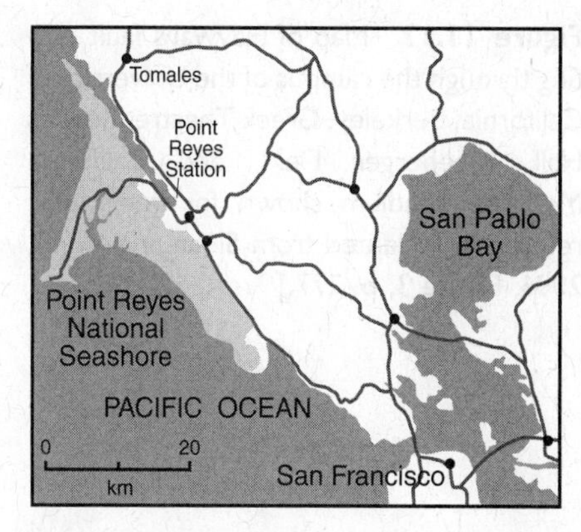

Figure 11.10 Location map of Point Reyes National Seashore relative to San Francisco. [Adapted from visitor brochure handed out at Point Reyes National Seashore.]

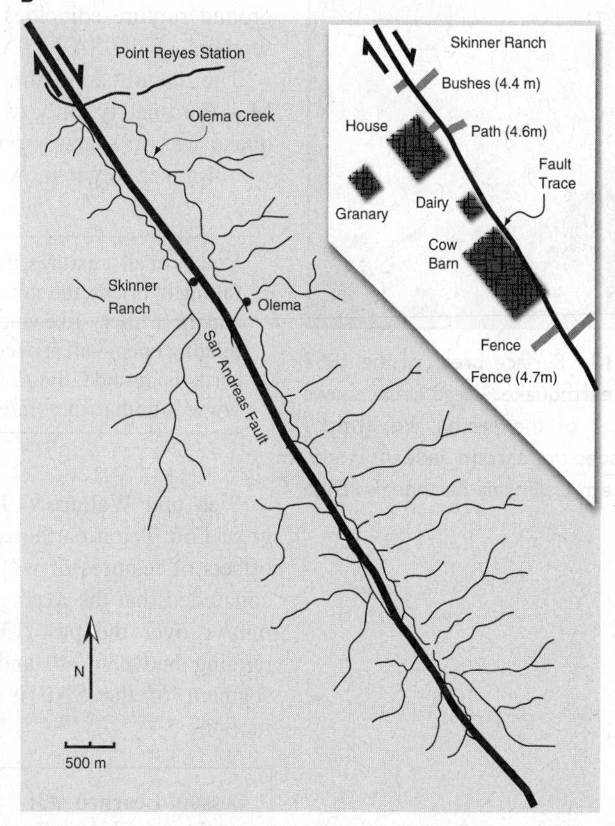

Figure 11.11 (*A*) Northwest directed view toward Skinner Ranch. The trend of the San Andreas fault is evident in the scarp-like topography. "*Moletracks*" are evident, and this represents ground rupture and displacement that accompanied the 1906 earthquake. [Photo by G. K. Gilbert (photo number 2924). Courtesy of the U. S. Geological Survey.] (B) Map of drainage system in Olema Valley, which occupies the trace of the San Andreas fault. Note location of Point Reyes Station. Inset shows sketch map of displaced cultural features as measured by G.K. Gilbert at Skinner Ranch. Broken lines show positions of features before earthquake in relation to same objects to west of fault. [From Niemie and others, 2006, Figure 8, p. 168.]

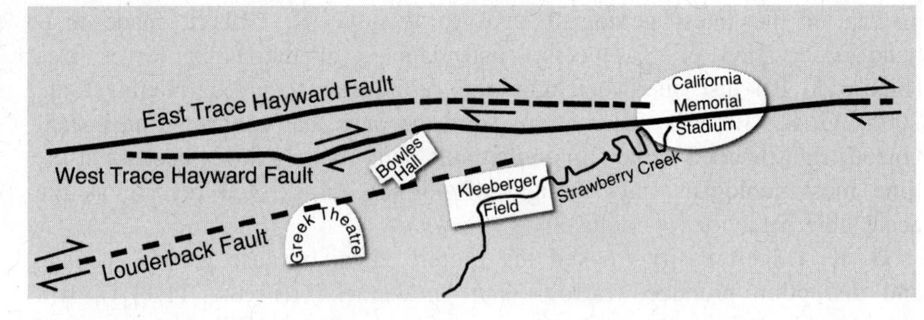

Figure 11.12 Map of Hayward fault cutting through the campus of the University of California, Berkeley. Greek Theatre, Bowles Hall, Kleeberger Field, and California Memorial Stadium shown for geographic reference. [Adapted from Sloan and Wells, 2006, Figure 3, p. 277.]

is an active fault that parallels the northern segment of the SAF and lies east of the San Francisco Bay. In fact, it passes right through the campus of the University of California at Berkeley. (Figure 11.12) (Sloan and Wells, 2006). To be sure, there is no shortage of active faults in the region separating the Pacific plate from the Sierran microplate!

Southern Segment of the San Andreas Fault

The stick-slip behavior of the southern segment of the San Andreas fault revealed itself profoundly during the Fort Tejon earthquake, which took place on January 9, 1857, 8:20 A.M. local time. It was one of the greatest earthquakes ever to have occurred in the United States. Approximately 360 km of the southern segment suddenly ripped and slipped in right-handed fashion, with a maximum offset of ~9.5 m (Figure 11.13). The northernmost extent of ground rupture coincides with the southernmost part of the central creeping segment of the SAF.

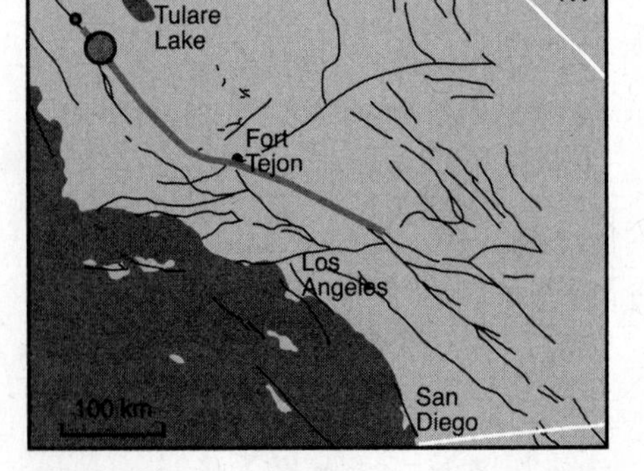

Figure 11.13 Surface break of the 1857 Fort Tejon earthquake. Large circle shows the epicenter of the earthquake. [http://www.data.scec.org/chrono_index/forttejo.html. Southern California Earthquake Data Center.]

The magnitude of the Fort Tejon earthquake is estimated to have been 7.9 M_W. Fortunately, only two people were killed, for the epicentral region of the quake was so sparsely populated. Yet those who experienced this earthquake would never forget it. Augusta J. Crocheron (1885) recalled:

> Voices of all creatures, the rattling of household articles, the crackling of boards, the falling of bricks, the splashing of water in wells, the falling of rocks in the mountains and the artillery-like voice of the earthquake, and even that awful sound of the earth rending open—all at once, all within a few seconds, with the skies darkened and the earth rising and falling beneath the feet—were the work of an earthquake" (http://www.johnmartin.com/earthquakes/eqpapers/00000039.htm).

Sieh and Wallace (1987) recognized a 10-m offset produced during the great Fort Tejon earthquake of 1857. On the basis of measuring distances of offsets of features of wide-ranging but known ages, Sieh and Wallace (1987) concluded that the average slip rate along this part of the SAF has been ~34 mm/yr over the past 13,000 years. They estimated the repeat interval as ranging between 240 and 450 years. It thus would appear that the southern segment of the SAF is reasonably keeping pace with the requisite plate motions.

Lesson Learned #3: Active faults are almost always the products of earthquake-by-earthquake stick-slip incremental movements. Each slip event requires build-up of strain energy followed by sudden energy release. Knowing this, as we study the details of ancient faults with brittle expressions, we should place emphasis on the fact that these faults are really shear zones representing the cumulative nested product of individual slip-displacements that took place over long periods of time.

Central Segment of the San Andreas Fault

The central segment of the SAF lies between the south end of the 1906 surface rupture and the north end of the 1857 surface rupture (see Figure 11.6). Within the creeping segment of the San Andreas fault there are three distinct sections (northwest, central, and southeast). The distinctiveness of these sections became apparent as the result of very precise surveying of 25 control lines laid out and studied by the U.S. Geological Survey during the 1960s and 1970s, before the advent of GPS (Figure 11.14). The concept used was very simple: set up straight control lines crossing and perpendicular to the SAF; give the SAF adequate time to creep; and come back later (perhaps several times) to measure the offset and/or deflection of each line.

The survey monuments for each control line were placed at ~30 m to ~200 m, and the spacing of the lines themselves was ~7 km (see Figure 11.14). The U.S. Geological Survey reoccupied and resurveyed these lines over a five-year period revealed, and this provided the basis for evaluating the nature and rates of fault movement.

Data for the Eade Ranch control line (site #13) is a good example of what Burford and Harsh (1980) reported (Figure 11.15). Open circles represent the monuments in their original straight-line alignment; the filled circles are the displaced positions of the monuments. The main zone of slip (creep) was found to be ~19 m wide. Creep rates across it increased from 25.9 to 30.8 mm/yr during the course of the study; no one would have expected that the creep rate would be absolutely constant. The average creep rate proved to be 31.3 mm/yr, which matches that determined for the creeping section as a whole (Burford and Harsh, 1980). This rate is ~80% of that for the SAF overall, which is pretty amazing considering the short time interval over which

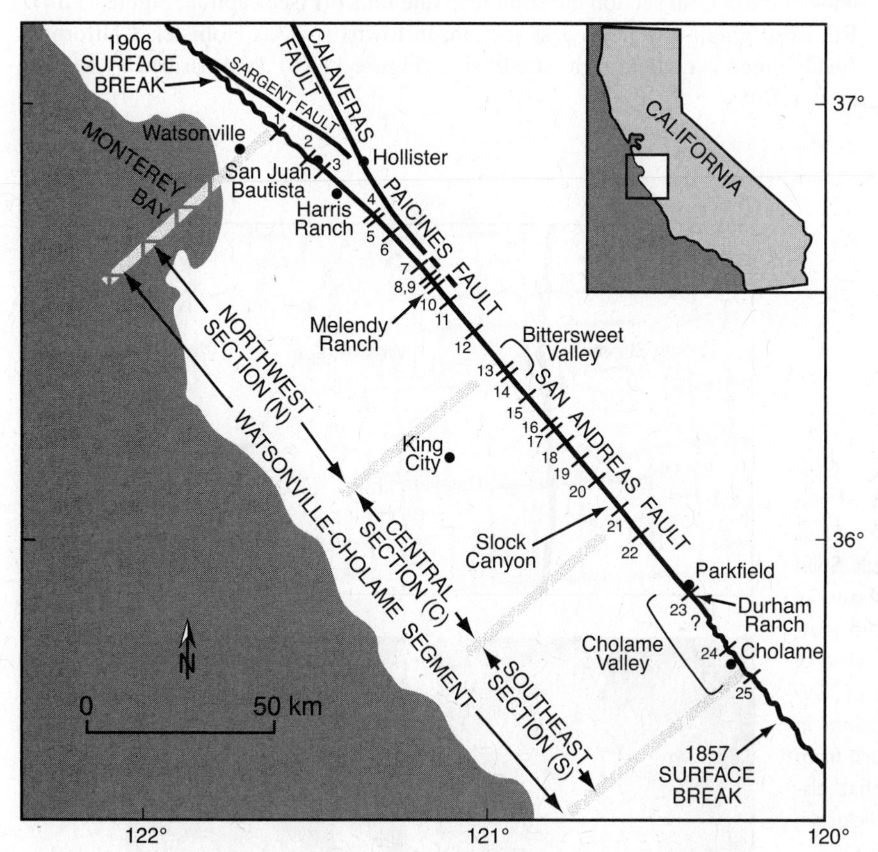

Figure 11.14 Map of the central (creeping segment) of the San Andreas fault. The central segment is subdivided into three sections (northwest, central, southeast) based on creep rates. From stations #1 to #12 in the Northwest Section, creep rates increase from ~1 mmyr^{-1} to ~22 mm/yr. From stations #13 to #21 in the Central Section, creep rates range up to ~30 mm/yr. From stations #22 to #25 in the Southeast Section, creep rates diminish from ~24 mm/yr to ~1 mm/yr. [Modified from Burford and Harsh, 1980, Figure 1, p. 1234.]

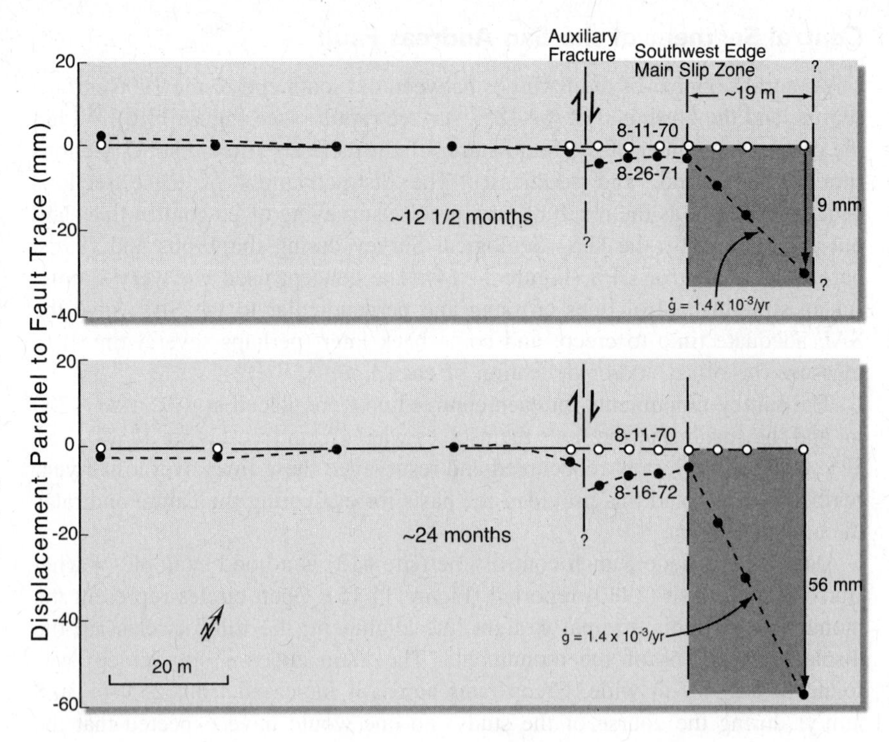

Figure 11.15 Mapped basis for the 31.3 mm/yr creep-rate determination at the Eade Ranch site (station #13, Figure 11.14). Open circles are original monument positions, which were tightly surveyed along a straight line oriented perpendicular to the trace of the creeping section of the San Andreas fault. Displacements of these monuments were monitored by revisiting and resurveying the Eade Ranch array four times between August 1970, and September 1975. Filled circles are the "*slipped*" monument positions. Dark shaded zone is main slip zone, approximately 19 m wide. During the period in which monitoring was carried out, the slip rate increased fro 25.9 to 30.8 mm/yr. [From Burford and Harsh, 1980, Figure 5, p. 1241.]

data were gathered. This close correspondence suggests that very little strain energy is being stored in the central, creeping section of the SAF.

The central creeping section was found to be ~55 km-long, with the steady rate of ~32 mm/yr. Yet it is clear that the northwest and southeast sections of the creeping segment are storing some strain energy, for both at the north and south ends of the central section the slip/creep rate falls off (see caption, Figure 11.14). But even within the northwest section, in towns such as Hollister, California, steady creep is evident right at curbside (Figure 11.16), as described by Rymer et al. (2006).

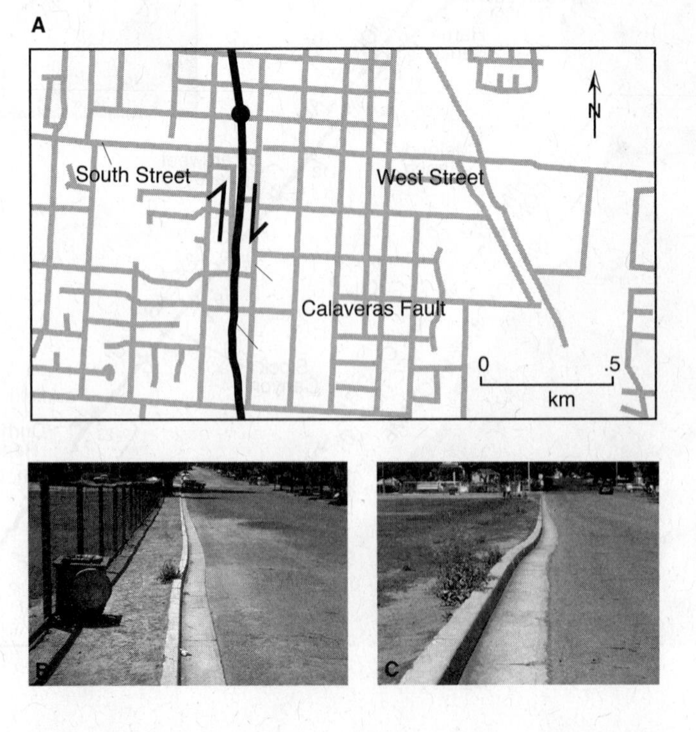

Figure 11.16 (*A*) Street map of Hollister, California. Hollister lies at the very north end of the creeping segment of the SAF (see Figure 11.14). Fault shown is the active strand of the Calaveras fault. Solid dot at Seventh Street is location of photos shown in Figure 11.16*B*, *C*. [From Rymer, Hickman, and Stoffer, 2006, Figure 5, p. 242.] (*B*) West-directed view of right-handed offset of curb along south side of Seventh Street in Hollister, CA. Photograph taken in June, 1989. (*C*) Photograph taken of same curb from the same perspective in June 2005; note that displacement has increased. [From Rymer, Hickman, and Stoffer, 2006, Figures 9A and 9B, p. 245.]

Lesson Learned #4: Most sections of active faults are marked by stick-slip behavior, but there are some sections that creep continuously, and some that are transitional between creep and stick-slip behavior. This makes us want to know more about the physical character of the different active fault sections at depth, with the idea of then comparing what we find with the physical character of exposed ancient fault zones and shear zones.

A Deeper Look at the Central Creeping Section of the San Andreas Fault

The San Andreas Fault Observatory at Depth (SAFOD) was established in 2002, with the purpose of examining the behavior of the San Andreas fault. SAFOD is funded by the National Science Foundation and operated in partnership with the United States Geological Survey. Dr. Mark Zoback of the U.S. Geological Survey leads the project, and he and his colleagues have been reporting some very interesting observations.

SAFOD thinks big, and the best example of this was its success in drilling into and through a part of the creeping segment of the San Andreas fault. Geoscientists on the project have sampled just about everything you would want to sample along an active fault zone, such as fluids and fault rocks along an active fault. SAFOD deliberately targeted a location along the creeping segment of the SAF, where displacement is being achieved through the combination of aseismic creep and repeated microearthquake activity. The pilot hole was drilled at a location 1.8 km southwest of Parkfield, California (see Figure 11.14) (Zoback et al. 2009), where seismological data show a repeated history of $M_W \sim 2$ earthquakes (Hickman et al. 2004). There the SAF creeps at ~2 cm/yr within a zone of active deformation no wider than 10 m. During the course of the drilling, rock samples were collected and microseismicity was recorded continuously. It appears that the microearthquakes are being triggered by slip on fault surfaces that are no larger than 100 m^2.

By collaring the drill hole west of the San Andreas fault, it was possible to drill through the entire fault zone, which proved to be only ~10 m thick (Figure 11.17). This drilling project represents a superb technological achievement, given the requirement to go as deep as 1.6 km and then to bend the drilling path from vertical to a 40° inclination so that the drill stem would pass at a high angle to the fault zone (Hickman et al. 2010).

Not long after successfully drilling through the fault zone, it was determined that the drill casing itself was being bent by creep along the SAF. The deformation was measured by a 40-finger caliper tool. Mark Zoback explained (personal communication, May 27, 2010), which we paraphrase here:

Deformation of the casing at 3302 m was apparent in October, 2005, only 6 weeks after the casing was cemented in place. At 3192 m, the casing deformation was not apparent until a subsequent caliper survey was carried out in June, 2007, at which time appreciably more deformation of the casing at 3302 m was observed. At the present time, it is very difficult to reconstruct how the hole is shearing, and the rate at which it is shearing. The hurdles include achieving a common orientation of the tool in successive runs; dealing with the geometry of a curved hole trajectory; and maintaining exact positioning of the tool with respect to the wellbore axis as a function of depth and from run to run.

Among the specific goals of SAFOD was to determine why parts of the fault are so weak, and in particular, why the northern and southern segments of the San Andreas fault "*stick and slip*," whereas the central section "*creeps?*" There is an extensive literature suggesting that the San Andreas

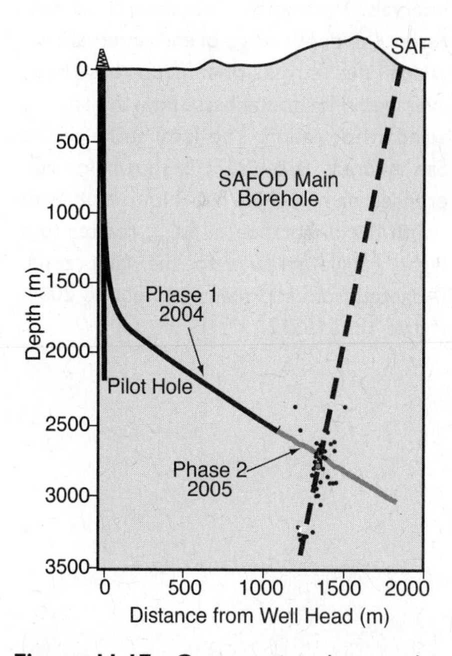

Figure 11.17 Cross section showing the line of drilling of the SAFOD hole relative to the orientation of the San Andreas fault (SAF). The drilling was accomplished in three steps, starting with a pilot hole in 2002 and proceeding with phase 1 drilling in 2004 and phase 2 drilling in 2005. Black dots show microseismicity. Vertical exaggeration is 0.83X. [From Zoback, Hickman, and Ellsworth, 2009, Figure 2, p. 30. Courtesy of EarthScope, http://www.earthscope.org/data/safod.]

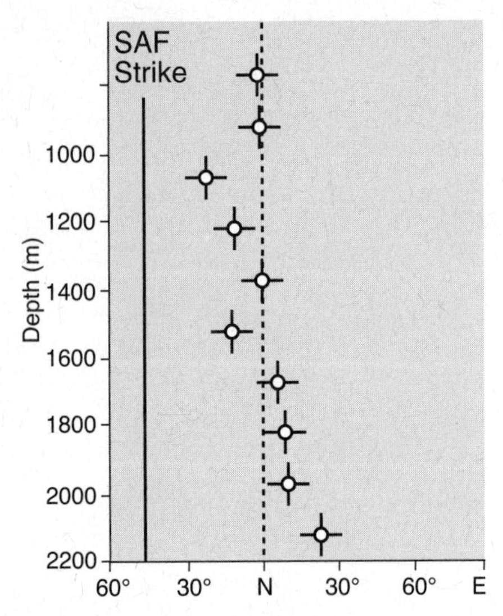

Figure 11.18 Graph showing the orientation of the direction of maximum horizontal stress (S_{Hmax}) as a function of depth in the SAFOD pilot hole. The mean orientation of S_{Hmax} (shown as crosses) is presented for each of 10 discrete 150-m depth intervals. Vertical bars on each cross represent the depth range of each interval over which the stress orientation has been averaged. Horizontal bars show the angular standard deviation. The local strike of the San Andreas fault (SAF) is shown for reference, namely N46°W. Note that with depth the orientation of S_{Hmax} rotates to a higher angle relative to the fault zone. [Adapted from Hickman and Zoback, 2004, Figure 1b, L15S12.]

fault is weak at depths down to 15 to 20 km, where "*weak*" in this case is a shear strength estimated to be less than 20 MPa, and "*strong*" between 50 and 100 MPa (Hickman and Zoback, 2004). In fact, the San Andreas fault is able to move under circumstances of *very* low shear stress. Two observations have supported this view: (1) there is no detectable heat flow related to friction; and (2) the direction of greatest principal stress (σ_1) is at a high angle to the fault, indicating that the fault orientation is close to that of a principal plane, and therefore shear stress (σ_S) on the fault must be very low.

By monitoring the orientation of greatest principal stress with depth, SAFOD confirmed that the SAF is weak. The guide to the orientation of greatest principal stress (σ_1) is the orientation of hydrofracs and borehole breakouts. For the SAFOD hole, there is with depth an increase in the angle of failure between σ_1 and the strike of the fault, such that the angle is 25° from 100 to 1500 m depth, increasing to 69° at 2050–2200 m depth (Hickman and Zoback, 2004) (Figure 11.18).

Chips of talc were recovered in drilling mud coming up from the creeping section of the fault. This created a rush of excitement among investigators, and the immediate publishing of the idea that weakness of this part of the SAF may be due to the presence of talc, which is a rock seriously lacking in strength (Wibberley, 2007; Moore and Rymer, 2007). Talc is a very weak hydrous magnesium silicate derived from hydrothermal alteration of the original country rocks (in this case serpentinite). Hot (hydrothermal) solutions are flowing through and along the fault zone, with water temperatures in the range 100° to 400° C, as measured in the SAFOD drill hole.

Lesson Learned #5: Some segments of active faults are extremely weak and achieve displacement through nearly continuous creep. This knowledge gives us another reason to pay close attention to the nature of fault rocks along major shear zones. Excessive amounts of clayey rock and/or abundant evidence for elevated fluid pressure during faulting might signal the possibility that fault displacement was achieved by creep mechanisms. This knowledge provides an incentive to reconstruct paleo-principal stress orientations along ancient faults, using as a basis outcrop-scale structures such as joints, faults, crystal-fiber veins, and stylolites. The results might inform us as to whether a given fault was weak or strong.

Folding, Faulting, and Earthquake Activity in the Kettleman Hills

The Kettleman Hills, which lie ~35 km east-northeast of Parkfield, California, are an expression of shortening perpendicular to the trace of the San Andreas fault (Figure 11.19). The structural geology of the Kettleman Hills is marked by the combination of a blind thrust fault beneath a perfect anticline, which is exposed at the surface (see Figure 11.19). So often we tend to think of anticlines forming through buckling achieved through "*continuous shortening*" in a vice-like tectonic grip. More and more, however, we realize that anticlines are formed earthquake-by-earthquake, even along strike-slip shear zones.

The Kettleman North Dome anticline is marked by ~1250 m of structural relief (Figure 11.20). In 1985 there was a series of earthquakes (as powerful as M_W 6.1) that occurred beneath the northwest-trending Kettleman Hills North Dome anticline. Ekstrom et al. (1992) carefully studied the three-dimensional relations between the earthquakes and geologic structure, concluding that the main earthquake was associated with sudden rupture on a shallow-dipping thrust fault buried at ~10 km depth. Aftershocks extended all along the 20-km trace length of the anticline, specifically in advance of the fault tip. Focal mechanisms for the main shock and the aftershocks are thrust mechanisms (Figure 11.21).

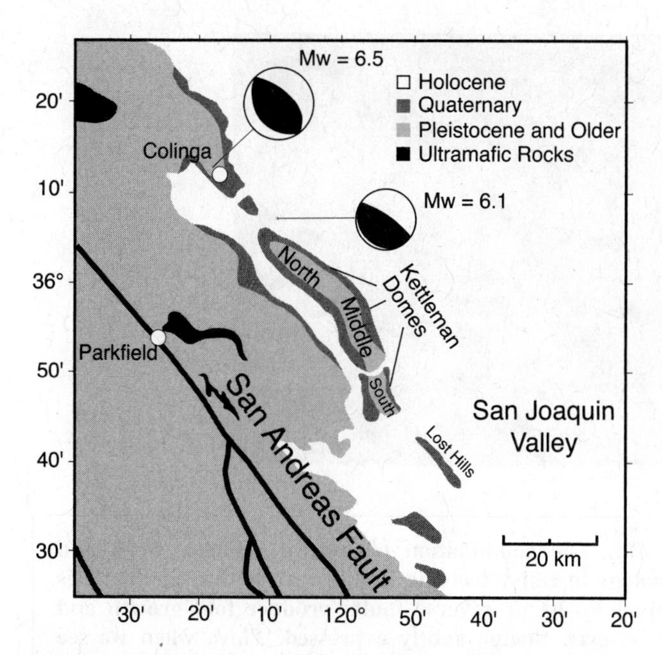

Figure 11.19 Location of the Kettleman Hills anticline relative to Colinga, Parkfield, and the San Andreas fault. Focal mechanisms are shown for the $M_W = 6.5$ Colinaga earthquake (1983) and M_W 6.1 Kettleman Hills earthquake (1985). [From Ekstrom et al. 1992, Figure 1, p. 4844.]

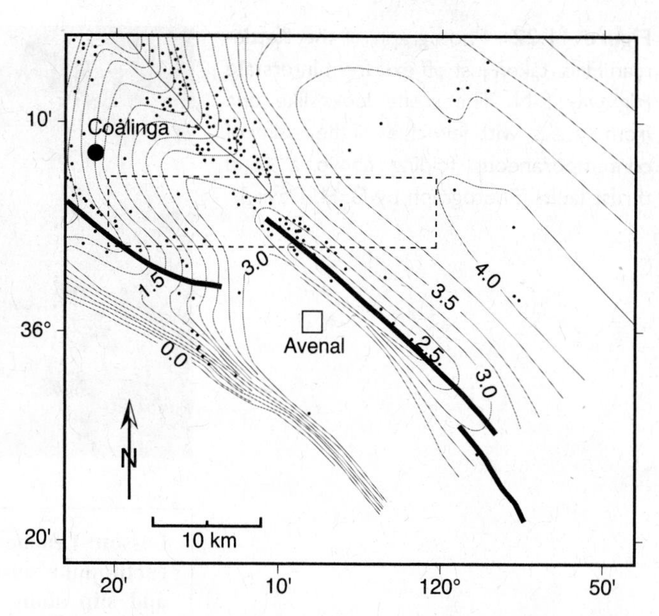

Figure 11.20 Structure contour map of the Kettleman Hills. The contours (in kilometers) represent depth to the top of the Middle Eocene Kreyenhagen Formation. Dots show the locations of oil wells, which reach the top of this formation. The box contains the location of a seismic reflection line. [From Ekstrom et al. 1992, Figure 2B, p. 4845.]

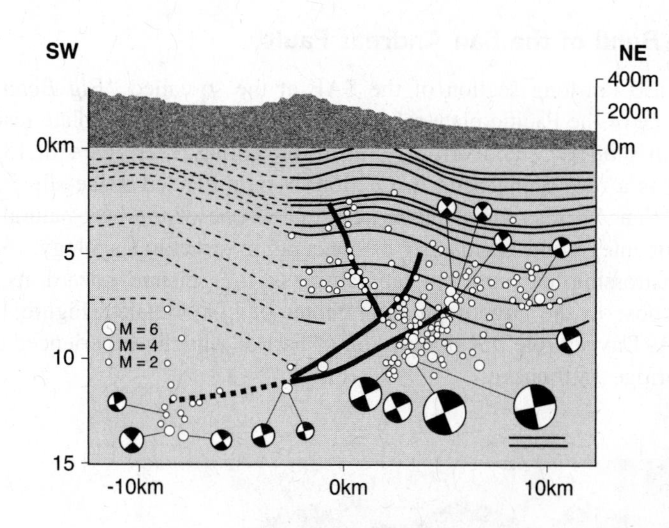

Figure 11.21 Cross-section showing subsurface structure of Kettleman Hills, as projected to the plane of the seismic reflection line shown in Figure 11.20. First motion focal mechanisms are shown in back-hemisphere projections. The double-underlined beach ball focal mechanism symbol is the preferred solution for the main shock. [From Ekstrom et al. 1992, Figure 3B, p. 4846.]

Ekstrom et al. (1992) compared the elevations of benchmarks along the Kettleman Hills, before and after the earthquake. They recognized that the earthquake caused 7 cm of coseismic uplift, produced by $\sim$0.5 m of fault slip. Based on geologic cross-sections of deformation and stratigraphy, it became apparent to Ekstrom et al. (1992) that the fold could be no older than 2.5 Ma. All things considered, Ekstrom et al. (1992) concluded that the Kettleman Hills North Dome anticline is growing vertically at a rate of $\sim$0.5 mm/yr, driven by a fault-slip rate of $\sim$2.9 mm/yr. None of this is obvious driving along Interstate 5-N from Los Angeles to San Francisco (Figure 11.22).

Figure 11.22 Photograph of the Kettleman Hills, taken just off exit from Interstate Highway 5-N. The scene looks like one from Kansas, with few clues of the reality of contemporaneous folding above a blind thrust fault. [Photograph by G. H. Davis.]

Lesson Learned #6: The combination of careful geodetic work and earthquake seismology in active tectonic setting reveals that earthquakes and slip along blind thrust or reverse faults produce fold growth and amplification above, even though subtly expressed. Thus, when we see evidence in outcrop for flexural-slip deformation along folded bedding in ancient settings, we can imagine that such flexural slip may have accompanied a major earthquake event, or accompanied aftershocks reflecting flexural-slip adjustments beyond the tip line of the master fault.

THE LOS ANGELES BASIN

The Big Bend of the San Andreas Fault

Along a 150 km-long section of the SAF at the so-called "*Big Bend*", the convergence of the Pacific plate relative to the North American plate results in north-south shortening and vertical crustal thickening (see Figure 11.13). The Big Bend is a restraining bend in relation to right-handed strike-slip faulting along the San Andreas fault, and it has created one of the best natural laboratories for integrating all manner of concepts in structural geology.

The distressing part of this laboratory is the seismic hazard its inner workings pose to the huge population center of Los Angeles (Figure 11.23). Gregory A. Davis wrote this description of part of what he experienced during the Northridge earthquake:

Figure 11.23 Damage to Ara's Pastry Shop on Hollywood Boulevard. Damage was sustained during the Northridge earthquake in 1994. [Photograph by Greg Davis.]

> In total city-wide darkness, my three-story building shook and flexed and rolled with a ferocity so unimaginable to me that I knew it would fail. A horrendously loud and ugly cacophony—the dissonant, overprinted sounds of furniture toppling, glass breaking, wood flexing, screams coming from an adjacent condo, and car alarms blaring—accompanied the intense lurching and vibration of the building around me.

Astonishingly and fortunately Greg's building did not fail. And he was able to see a bit of the lighter side of how "*we*" deal with disasters:

> Sign on the front of a condemned San Fernando Valley apartment house: THE FAT LADY HAS SUNG. Quote from a radio traffic announcer after the earthquake: "The traffic is stopped, but the freeways are moving." A pre-earthquake FOR SALE sign on a damaged home in Northridge is joined by a hand-lettered notice: SOME ASSEMBLY REQUIRED. And to the south in heavily battered Sherman Oaks, a barber shop announces it's open for business with this notice: SHAKE AND A HAIRCUT, TWO BRICKS.

Tectonic Setting of the LA Basin

It is useful, first, to step back and look at the big-picture relationship of the Los Angeles basin to the San Andreas fault. Figure 11.24 helps us do this. It is a 3D block diagram prepared by Fuis et al. (2001), based upon the results of the deep seismic refraction survey they ran from the Pacific coast to the Mojave Desert (see Figure 11.24). They determined that the Los Angeles basin is ~8–9 km deep along the surveyed line of the basin; and that well below this depth in the middle crust, there appears to be a low dipping thrust-fault detachment, upward from which some of the major thrusts of the Los Angeles basin branch (see Figure 11.24). The San Andreas fault appears to dip ~83°NE, extending downward to at least the crust/mantle boundary (i.e., to

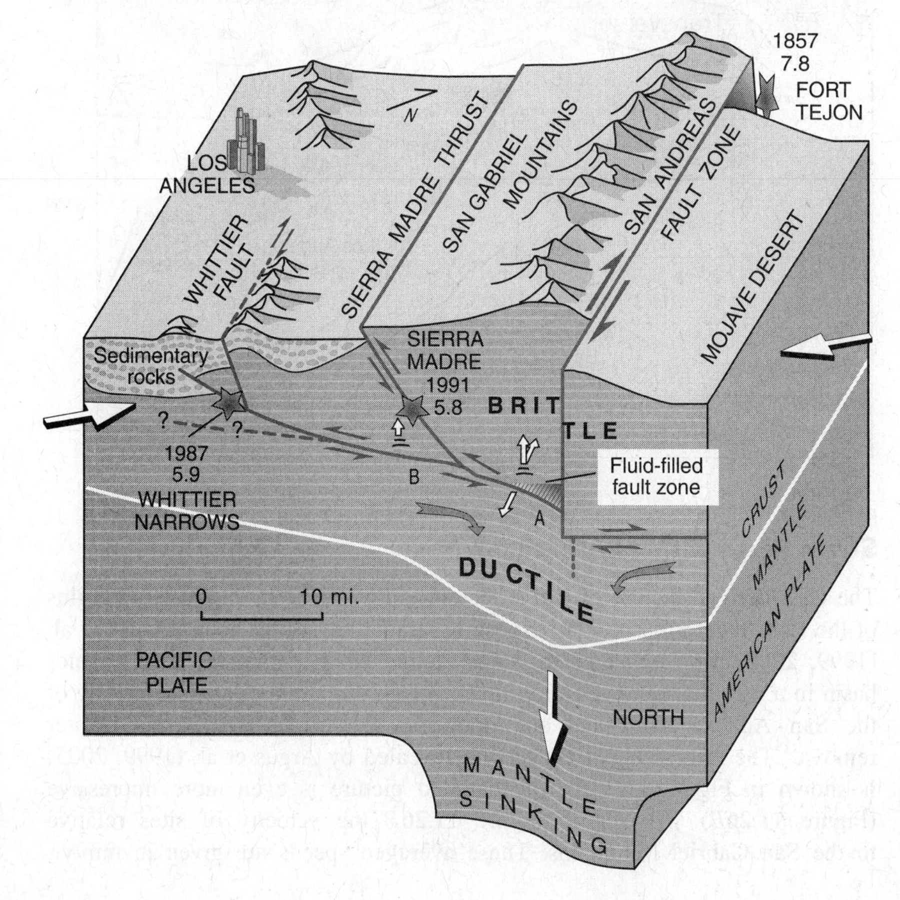

Figure 11.24 3D block diagram of part of southern California, based on results of a deep seismic refraction survey running from south of Los Angeles to the Mojave Desert. Large white arrows show direction of convergence between the North American (NA) plate and the Pacific (P) plate. [From Fuis et al. 2001, Figure 3, p. 17.]

the Moho) (see Figure 11.24). The plate boundary (SAF) between the North American plate and the Pacific plate is distinct in the upper, brittle crust, but is more diffuse in the ductile lower crust and upper mantle. Overall, the Big Bend restraining bend causes the SAF to lock (at depths of ~13 to ~20 km), to accumulate strain energy, and to undergo crustal shortening through active thrust and reverse faulting accompanying earthquakes. Among these earthquakes is the Fort Tejon earthquake, which occurred near the northern end of the bend in the SAF (see Figures 11.13 and 11.24).

The Los Angeles basin developed in the Miocene as a deep extensional basin bounded by major normal faults. In the thickest part of the LA basin, there are more than 10 km of Miocene to Quaternary strata (Tsutsumi et al. 2001). Beginning in the late Miocene with the onset of the SAF system, the basin experienced inversion tectonics brought about by contraction and shortening at the Big Bend (Tsutsumi et al. 2001). As an expression of inversion tectonics, some of the original normal faults are reactivated during earthquake events as reverse and thrust faults. These faults commonly tip out upward into active fault-propagation folds.

The crustal shortening at the Big Bend explains the high mountain ranges bordering the Los Angeles region as well as the big earthquakes that have wracked the city: such as the San Fernando earthquake (1971), the Coalinga earthquake (1983), the Whittier Narrows earthquake (1987), and the Northridge earthquake (1994) (Figure 11.25). Paleoseismic investigations in the Los Angeles basin reveal that during the past 12,000 years there has been a history of episodic strain release marked by brief bursts of energy followed by lulls (Dolan et al. 2007).

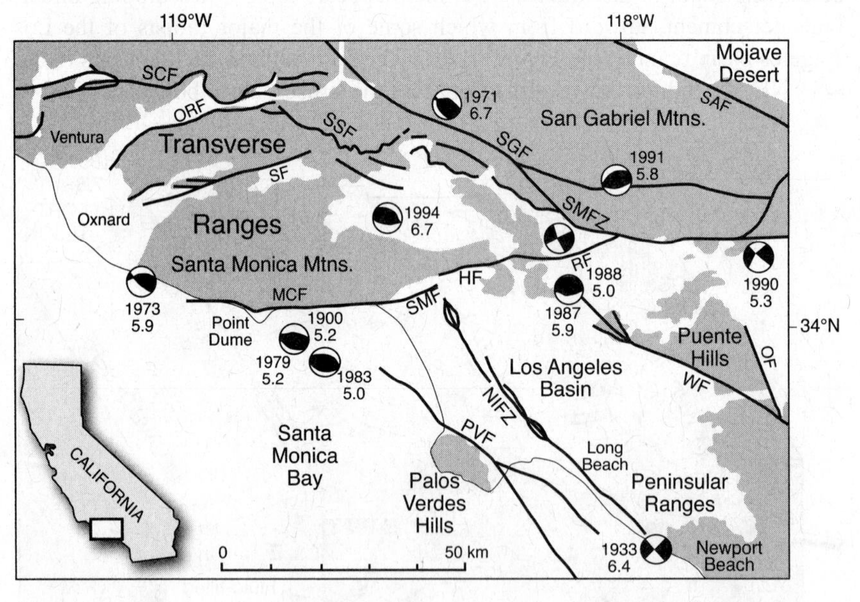

Figure 11.25 Map of part of the western Transverse Ranges, Los Angeles basin, and the northern end of the Peninsular Ranges. Black lines are major late Quaternary faults. Abbreviations for active faults: CF = Chino fault; HF = Hollywood fault; MCF = Malibu Coast fault; NIFZ = Newport-Inglewood fault zone; ORF = Oak Ridge fault; PVF = Palos Verdes fault; RF = Raymond fault; SAF = San Andreas fault; SCF = San Cayetano fault; SF = Simi fault; SGF = San Gabriel fault; SMF = Santa Monica fault; SMFZ = Sierra Madre fault zone; SSF = Santa Susana fault; WF = Whittier fault. Focal mechanism of earthquakes with magnitudes 4.9 or greater since 1970 include the 1971 San Fernando, 1973 Point Mugu, 1979 Malibu, 1987 Whittier Narrows, 1988 Pasadena, 1989 Malibu, 1990 Upland, 1991 Sierra Madre, and 1994 Northridge earthquakes. [From Tsutsumi, Yeats, and Hutile, 2001, Figure 1, p. 455.]

Strain Rates within LA Basin

The GPS-derived picture of active tectonic movements in and at the margins of the Los Angeles basin is an expose of strain and displacement. Argus et al. (1999, 2005) have managed to isolate active strain within the Los Angeles basin in a way that removes all influence of elastic strain related to locking of the San Andreas fault and anthropogenic shifts related to ground-water removal. The large-scale strain picture revealed by Argus et al. (1999, 2005) is shown in Figure 11.26*A*. The detailed picture is even more impressive (Figure 11.26*B*). We see in Figure 11.26*B* the velocity of sites relative to the San Gabriel Mountains. These averaged speeds are given in mm/yr.

A

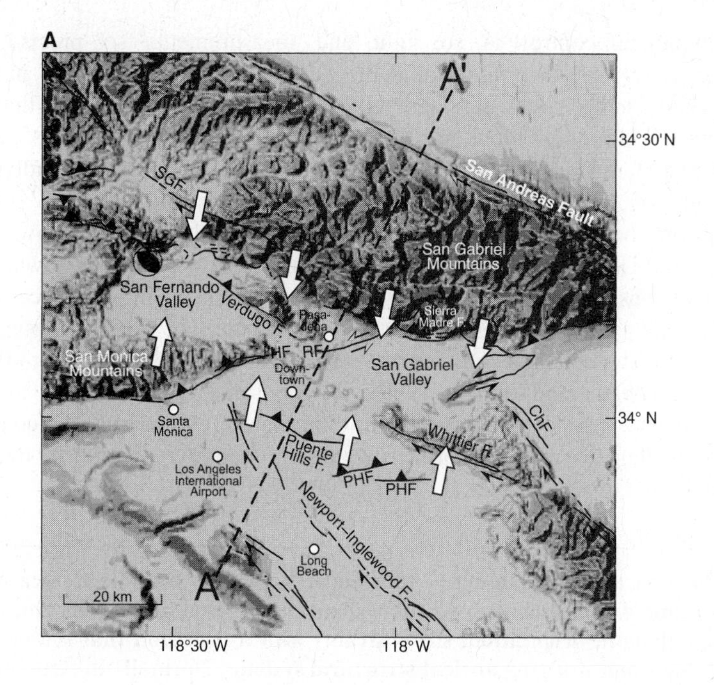

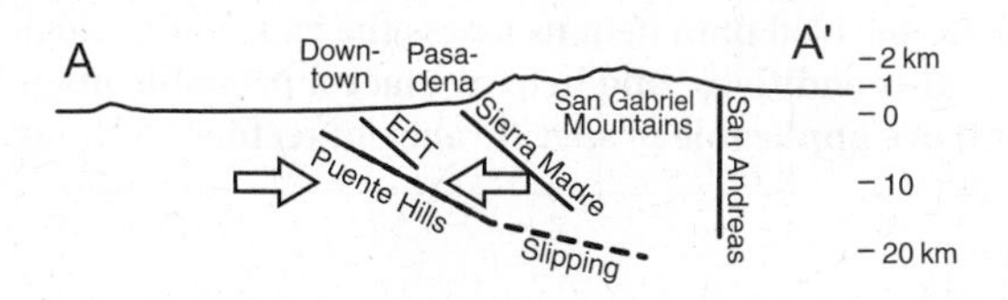

B

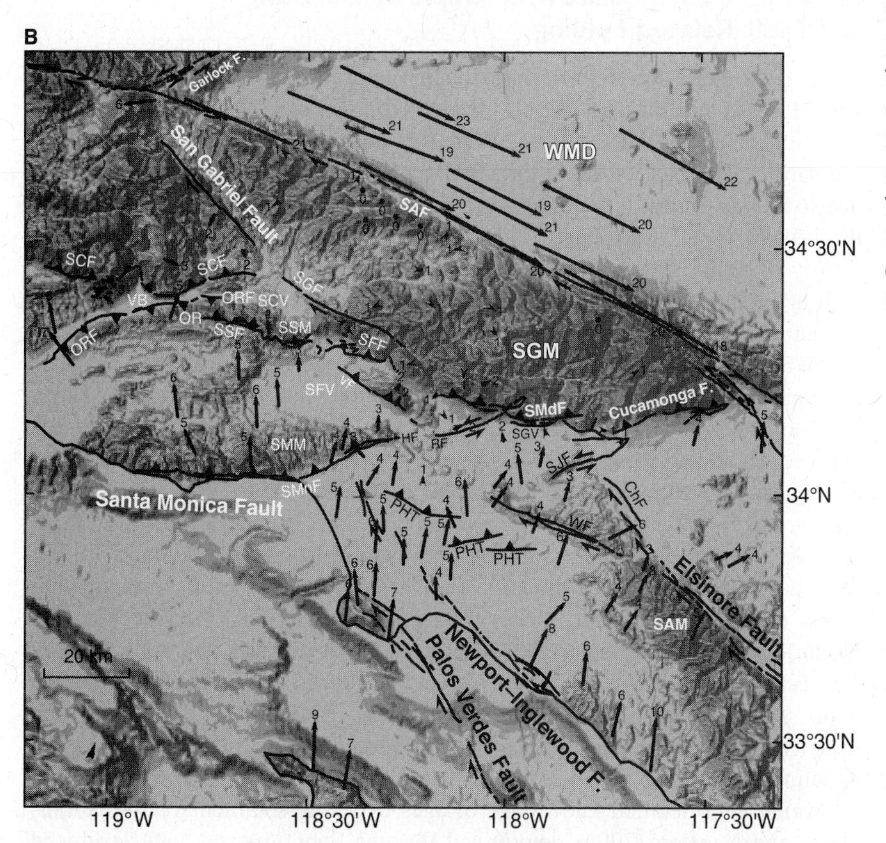

Figure 11.26 (*A*) Map of southern California showing crustal shortening based upon GPS-based tectonic analysis. SGF = San Gabriel fault; HF = Hollywood fault; RF = Raymond fault; PHF = Puento Hills fault; ChF = Chino fault. Focal mechanism is for San Fernando earthquake. Line A-A' is line of cross-section, below which are the Puente Hills, Elysian Park thrust, and Sierra Madre faults in relation to the San Andreas fault. [Courtesy of Donald F. Argus, Jet Propulsion Laboratory.] (*B*) Map of site velocities relative to the San Gabriel Mountains. These are residual velocities (see text). Site locations are noted (data collected per site ranges from 3 to 17 years). Speeds are expressed in mm/yr. SGM, San Gabriel Mountains; WMD, west Mojave desert. Faults are ChF, Chino fault; HF, Hollywood fault; ORF, Oak Ridge fault; RF, Raymond fault; SAF, San Andreas fault; SCF, San Cayetano fault; SFF, San Fernando fault; SGF, San Gabriel fault; SJF, San Jose fault; SMdF, Sierra Madre fault; SMNF, Santa Monica fault; SSF, Santa Susana fault; VF, Verdugo fault; WF, Whittier fault. Mountains and Valleys are OR, Oak Ridge; SAM, Santa Ana Mountains; SCV, Santa Clarita valley; SGV, San Gabriel valley; SFV, San Fernando valley; SMM, Santa Monica Mountains; SSM, Santa Susana Mountains; VB, Ventura Basin. Site abbreviations: CC = Claremont College; CI = Catalina Island; CIT = California Institute of Technology; MW = Mount Wilson; PV = Palos Verdes; UCLA = University of California Los Angeles; USC = University of Southern California; WC = Whittier College. [From Argus et al. 2005, Figure 1.]

GPS-monument control is so tight and measurements so precise that Argus et al. (1999) determined that the San Gabriel Mountains are moving south relative to the Channel Islands at a rate of ~8 mm/yr. Not only that, the San Gabriel Mountain block is rotating clockwise relative to the Channel Islands at a rate of ~2° per m.y. The Ventura basin is contracting rapidly, at a strain rate of $1.2 \times 10^{-6}\text{sec}^{-1}$, accommodated by thrusting and reverse faulting on the San Cayetano and Oak Ridge faults. In downtown Los Angeles, the University of Southern California (USC) is moving toward the Jet Propulsions Laboratory at a velocity of ~6 mm/yr, in part because of thrusting along the Hollywood fault. Relative to the San Gabriel Mountains, the present distance between these two sites is ~30 km. Strain rate is $\sim 0.2 \times 10^{-6}\text{sec}^{-1}$ (Argus and others, 1999). At the same time, USC is moving northward at ~6 mm/yr relative to the San Gabriel Mountains. In contrast, Whittier College is moving northeastward at ~6 mm/yr relative to the San Gabriel Mountains (Argus et al. 2005).

Lesson Learned #7: In active tectonic settings, it is possible to measure strains and displacements over subregions and regions, and report rates of change (length, orientation, shear strain) with a precision that seldom is available when studying ancient structural systems. Normally in structural geology classes we mention strain rates only in the context of deformation experiments, we mention rotations only in the context of "*paleomag*" research, and we mention strains in the context of deformed (small) objects. The new geodetic-derived data help us assess the range of geologically reasonable strain-rate conditions, and help us place a probable magnitude for strain rates that are applicable to specific ancient tectonic settings.

Northridge Earthquake as Example of Thrusting and Fault-Related Folding

Since 1970, there has been a flurry of big earthquakes in the western Transverse Ranges and the Los Angeles basin (see Figures 11.25 and 11.26). What is insidious about the earthquakes in the Los Angeles region is that they ordinarily take place on blind thrusts. Because of this the mitigation of losses due to seismic hazards requires learning and applying all we can about how the geometry of fault-related folding relates to the geometry and movement history of faults to which they connect at depth.

It is one thing to determine the epicenter, hypocenter, and focal mechanism for an earthquake, such as the Northridge earthquake (1994). It is quite another thing to interpret the location and focal mechanism of a major earthquake in terms of the local structural geology and basin tectonics. Davis and Namson (1994) provide an illustration of how to do this. They constructed detailed balanced cross-sections through the hypocenter of the Northridge earthquake (~18 km beneath the surface) and concluded that the hypocenter was on the Pico fault (near its intersection with the Elysian Park thrust), which lies beneath and along an enormous fault-bend fold complex (Figure 11.27). During the Northridge earthquake, thrust-fault-slip was transferred upward along the Pico fault into fault-propagation folding. To be specific, the Northridge earthquake delivered ~2.5 m of thrust displacement. The folding was accommodated by the Santa Susana Mountains anticlinorium and the Santa Clara Valley synclinorium (see Figure 11.27). The vast common limb these two folds share dips ~60° to 70° northeast. Total structural relief on the anticlinorium is ~6 km!

Armed with detailed knowledge of ages of young sedimentary formations, Davis and Namson (1994) determined that the Santa Susana fold developed

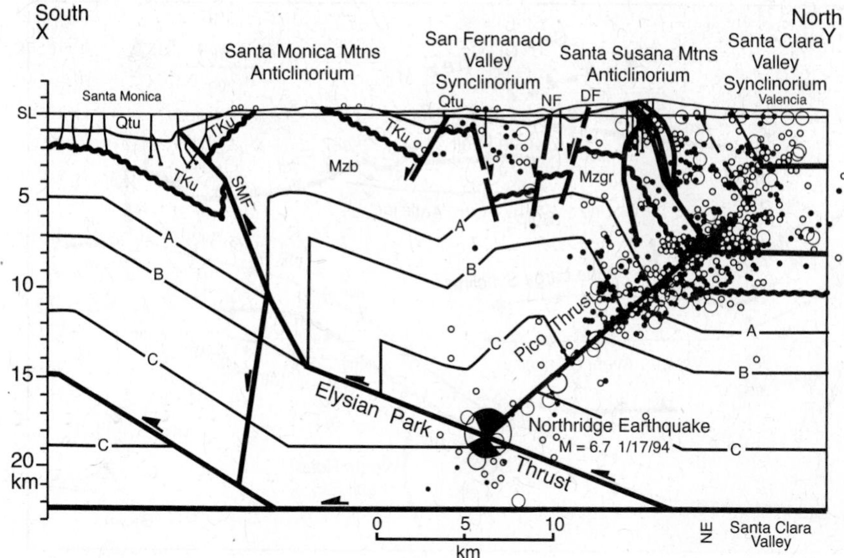

Figure 11.27 Balanced north-south-trending cross-section across the Transverse Range fold-and-thrust belt. The section passes through the hypocenter of the Northridge 1994 earthquake (see focal mechanism). Aftershocks are plotted as well. Note that the hypocenter for the main shock occurs on the Pico thrust at its intersection with the Elysian Park thrust. The shallow batch of aftershocks reflects fold growth (through flexural slip) through 'feeding' thrust slip up and into the Santa Susana Mountains anticlinorium and the Santa Clara Valley synclinorium. [From Davis and Namson, 1994, Figure 2A, p. 168.]

over a period of several million years. They concluded that the Northridge earthquake proved to be a snapshot of "*how folds fold,*" for the earthquake produced active movement along bedding-plane faults, which could be mapped right at the surface. Such bedding-parallel movement is of course quite consistent with flexure-slip formation of the anticlinorium (Davis and Namson, 1994). Based on structure and stratigraphy as well as the magnitude of the Northridge slip event, Davis and Namson (1994) concluded that the repeat time for the Pico fault ranges from ~1500 to ~1800 years. If the Northridge slip event can be considered typical, it would take ~1300 such earthquake events to make the Santa Susana Mountains as we see them today!

Lesson Learned #8: Balanced cross-sections through areas of active faulting and folding can become the critical map projection upon which seismicity is plotted, in order to better evaluate (retrospectively) the structural context of a past earthquake or (prospectively) the seismic hazards picture of future earthquakes. Knowing this, when we study ancient fault/fold systems, we will surely not separate folds versus faults in our analysis, but carry out analysis that fully integrates the total system of movements and connections. Most large folds in the brittle crust are the products of earthquake-by-earthquake incremental movements. Thus when we are mapping the attributes of folds and capturing fold growth over time, we should picture amplification and tightening of the fold being achieved during myriad earthquake-by-earthquake movements.

The Ventura Avenue Fold, Another Product of Plate-Scale Transpression

One of the best-studied active fault-propagation folds is the Ventura Avenue anticline, which is located just north of Los Angeles in the Ventura basin (Figure 11.28). A photograph of its western extension, known as the San Miguelito anticline, is shown as Figure 11.29*A*. The San Miguelito anticline is especially impressive because of its beautifully exposed form and its staggering size. Moreover, it is an actively producing structural trap within the San Miguelito oil field (see pumpjack in Figure 11.29*A*). Art Sylvester (from Santa Barbara University) took my students and me (GHD) to see the San Miguelito anticline. We had to watch where we stepped, for in places oil seeps right out of the ground.

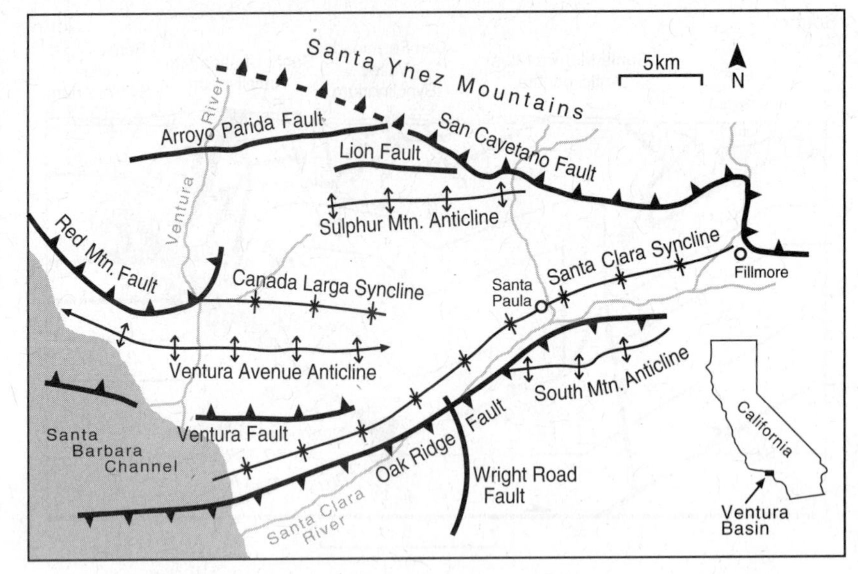

Figure 11.28 Map showing the location of the Ventura Avenue anticline within the overall context of major faults and folds in the Ventura basin, southern California. The Red Mountain fault is linked to the Ventura Avenue fold. [From Azor et al. 2002, Figure 1, p. 746.]

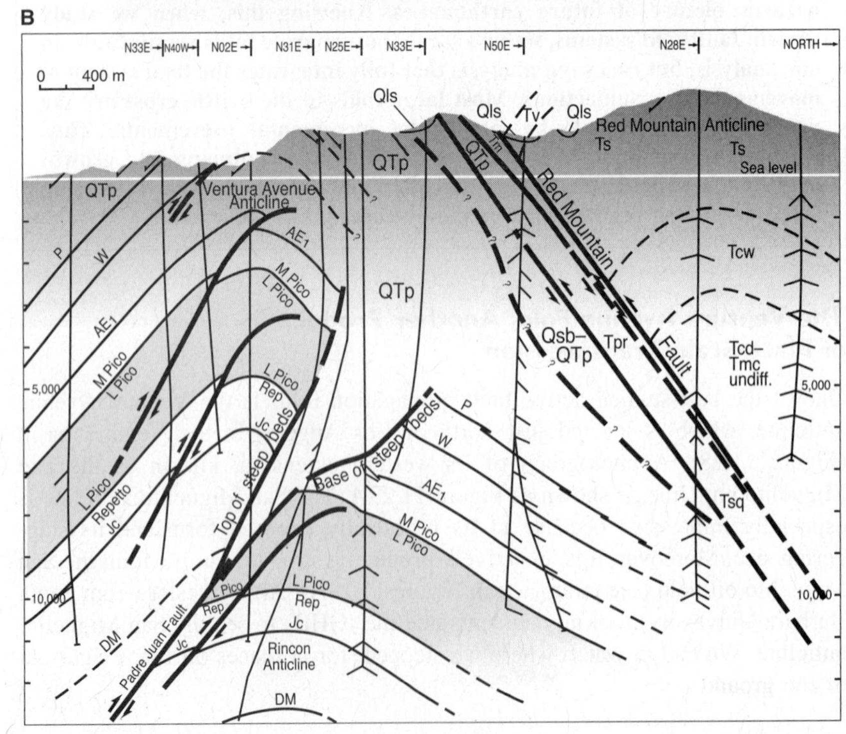

Figure 11.29 (*A*) Photograph of the San Miguelito anticline, which is the westward extension of the Ventura Avenue anticline. Note the pumpjack (a.k.a., nodding donkey) on the ridge (upper left) lifting oil from the subsurface. [Photograph courtesy of Art Sylvester.] (*B*) North-south geologic cross-section through the Ventura Avenue anticline. Note the tight subsurface control, including dip and stratigraphic information derived from drilling. The Pico Formation (QTp), which crops out at the surface at the very south end of this profile, is of Pliocene-Pleistocene age. The oldest strata within this cross-section are of Eocene age, exposed in the subsurface on the hanging wall of the Red Mountain fault. [From Yeats and Grigsby, 1987, Figure 5, p. 222.]

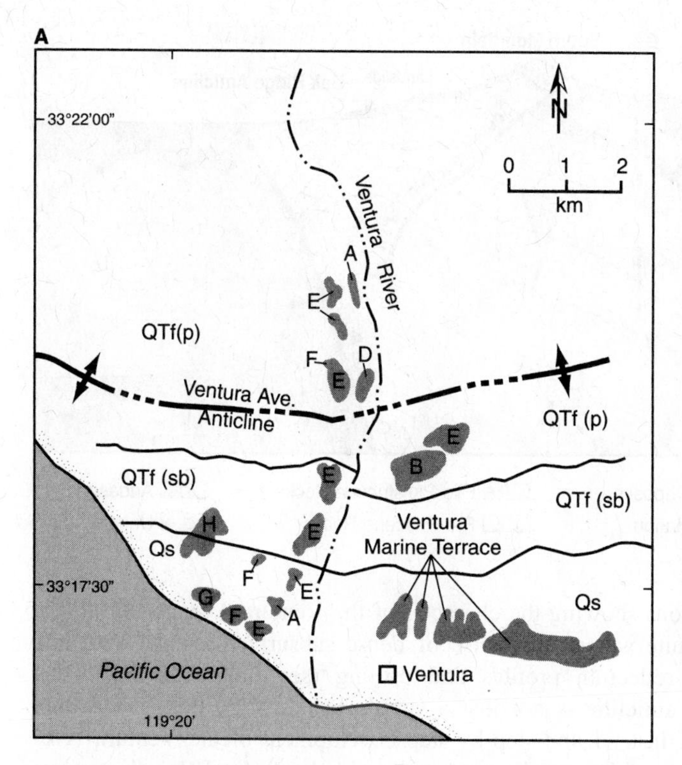

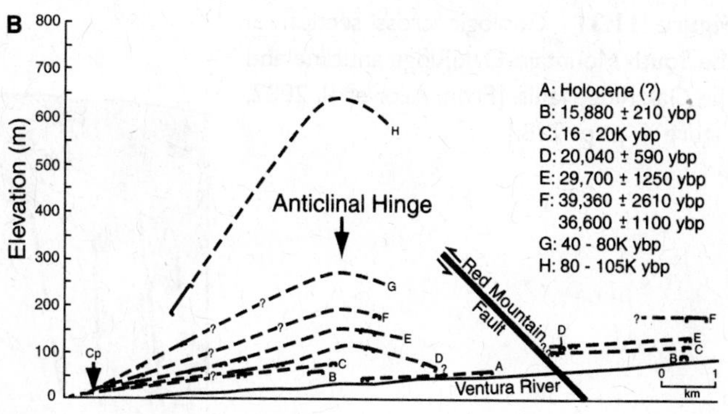

Figure 11.30 (*A*) Map showing the distribution of patches of late Quaternary marine terraces and Ventura River fluvial terraces. Letters correspond to curves showing deflection of river terraces (part *B* of figure). [From Rockwell et al. 1988, Figure 3, p. 853.] (*B*) Ventura River terrace profiles over the Ventura Avenue anticline and Red Mountain fault. Ages of terraces shown in upper right. [From Rockwell et al. 1988, Figure 4, p. 853.]

Folding began ~200,000 years ago (Yeats and Grigsby, 1987), and today the Ventura Avenue fold is rising vertically at an averaged rate of ~4 mm/yr, in response to an averaged shortening rate of the Ventura basin of ~20 mm/yr. As the fold has grown, the Ventura River has cut across it, with the result that even the river terraces are folded! Upper Quaternary marine terraces and Ventura River fluvial terraces cross the anticline (Figure 11.30*A*) (Rockwell et al., 1988). The oldest terraces are uplifted and folded the most, and the youngest terraces are uplifted and folded the least (Figure 11.30*B*).

Rates of uplift and folding are available in this active tectonic setting. The data are so robust that it is possible to declare that the rates of uplift and tilting have been decreasing since the fold began to develop 200 Ka (Rockwell et al., 1988). The minimum averaged rate of uplift along the crest of the anticline has decreased from ~14 mm/yr to ~ 2 mm/yr. The rate of tilting has decreased from ~5.8 μrad/yr to ~1.2 μrad/yr. Let's have some fun with these peculiar numbers and units. One radian is equal to 57.295°. One microradian (μrad) is equal to .057°, which in the military amounts to a rifle-shot deflection of 1 m over the distance of 1 km. One μrad/yr of tilt is equal to a change in limb dip of a mere 0.0000572°/yr. This does not sound like much, but such a rate will result in 57.2° of limb dip in a million years. Rockwell et al. (1988) concluded that the average shortening rate responsible for the development of the Ventura Avenue fold has been 9 mm/yr, which is about half that of what Yeats and Grigsby (1987) estimated.

Lesson Learned #9: When we are asked about rates of folding, we might suggest that 60° of limb dip can be achieved in a million years, with an average tilt increment of only 0.0000572° per year. If we indeed respond in this way, we will surely be asked, "How on earth did you come up with those numbers?" The more such examples like this we can find, the more we can begin to frame a range of plausible rates, and then tie those rates to prevailing plate-tectonic boundary conditions. As a result of this, we can look at ancient fold systems and be able to better estimate the time frame required for fashioning such a system.

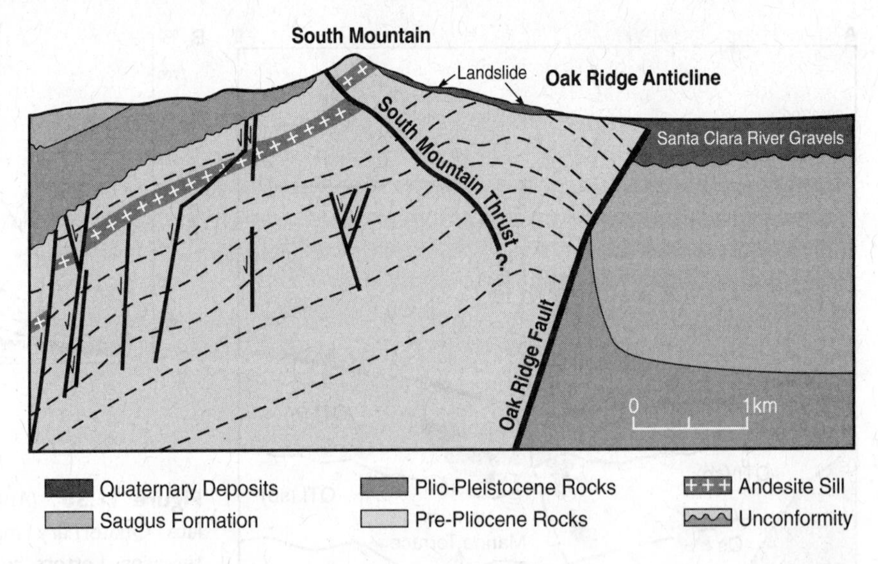

Figure 11.31 Geologic cross section of the South Mountain-Oak Ridge anticline and the Oak Ridge fault. [From Azor et al. 2002, Figure 4, p. p. 748.]

Structure sections showing the character of the Ventura Avenue fold in the subsurface are quite accurate because of *dense* subsurface control available through seismic reflection profiles and drilling (see Figure 11.29*B*). The Ventura Avenue anticline is not just a simple buckle fold, nor is it a pure flexural-slip fold. Instead, the step-by-step development of the Ventura Avenue anticline has been intimately connected to, and coordinated with, reverse faulting. Active fault slip has in places broken right through the steel casing of oil wells. Some reverse faults, such as the Red Mountain fault (see Figure 11.29*B*), formed during the very early stages of the folding. As a consequence, some of these faults became folded over the crest of the Ventura Avenue anticline (Yeats and Grigsby, 1987). We always think of bedding being folded; we less often think of faults being folded!

An even larger fold in the Ventura basin is the South Mountain–Oak Ridge anticline, an east-west trending fault-propagation fold occupying the hanging wall of the (blind) Oak Ridge reverse fault (see Figure 11.31). The folded stratigraphy includes a Pliocene-Pleistocene sedimentary sequence up to several kilometers thick (Figure 11.31) (Azor et al. 2002). Fold growth in the past 0.5 Ma has been taking place in response to a shortening rate in the range of ~7–10 mm/yr. The master fault was originally a normal fault. Now the master fault is an oblique-slip fault, marked by the combination of reverse- and left-handed slip.

The South Mountain–Oak Ridge anticline is growing as we speak. In the past 0.5 Ma it has extended itself westward along strike a distance of 15 km as the fold progressively grew in amplitude. Azor et al. (2002) interpreted this westward growth as the result of decrease in vertical slip from east to west along the fault. Indeed, the nature of the slip changes systematically from east to west, from dominantly reverse, to reverse-oblique, to left-handed strike slip. The change in vertical slip rate dies westward from ~5 m/ky to ~1–2 m/ky The backlimb of the fold decreases in dip from east to west from ~35° to ~20° (Azor et al. 2002).

Lesson Learned #10: In active tectonic settings we can observe that the amplitude of a major fold may die out along trend, and we can demonstrate that this reflects a systematic decrease in magnitude of slip on the master fault associated with the folding. Furthermore, we can determine that even the direction of slip may change as the fault tips out. Given this knowledge, when we are mapping an ancient reverse fault, we must hold open the possibility that the rake of slip might change from that of pure reverse slip to oblique slip and to strike-slip along the trace to its tip. It is therefore not sufficient to "*document*" slip (based on slickenlines) at just a

few locations on the fault and declare it to be 'simply' a reverse fault. Rather, exposures permitting, it is necessary, 'point by point along the fault trace,' to compile data that show both slip direction and slip magnitude, comprehensively. All of this forces us as well to imagine that a given portion of a fault surface was once a tip region but then (as the tip passed beyond) became simply a part of the 'main' fault surface.

Northern Los Angeles Fault System

The rubber meets the road along the northern edge of the Los Angeles basin, where the northern Los Angeles fault system is lifting and shifting the Santa Monica Mountains. To the north of this fault system lie the Transverse Ranges. Along the very northern edge of the Los Angeles basin is a complex east-northeast striking fault zone that links the Santa Monica fault, the Hollywood fault, and the Raymond fault (Figure 11.32), all active faults. These are steep north-dipping faults along which slip is a combination of reverse-left-handed movement (Tsutsumi et al. 2001). Big impressive fault-propagation folds are associated with each of these faults. Hidden beneath basin sediments are blind thrust faults, comprising a system within which the Santa Monica, Hollywood, and Raymond faults are a part. The subsurface character of the Los Angeles basin is shown in Figure 11.33, and includes glamorous names such as the Hollywood fault, the East Beverly Hills anticline, and the East Beverly Hills–Las Cienegas monocline. Though the Santa Monica and Hollywood faults have average slip rates less than 1 mm/yr, they are still capable of generating moderate-sized earthquakes, which could have disproportionately intense consequences given the population density (Tsutsumi et al. 2001).

Fold scarps and fault scarps can be identified even in the urban center of Los Angeles because . . . "*the builders [of Los Angeles] simply draped the city across the existing landscape with minimal cutting and filling*" (Dolan et al. 2000, p. 1562). The faults and folds are mainly in alluvium. Most of us are not accustomed to mapping fault and fold scarps in urban settings, but this type of work is essential in trying to get a handle on seismic risk in Los Angeles. Based on paleoseismic investigations, Dolan et al. (2000) were able to conclude that the oblique, left-lateral-reverse motion along the Santa Monica fault is distributed near the surface into wide zones of closely spaced

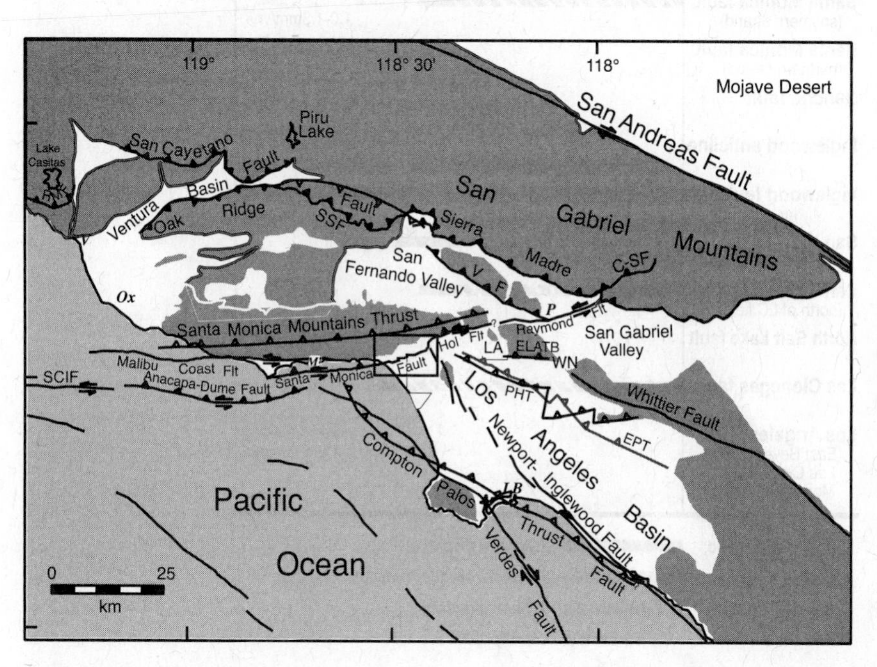

Figure 11.32 Regional neotectonic map for metropolitan southern California showing major active faults. The Santa Monica fault is a 40-km-long active fault within the 215-km –long Transverse Ranges Southern Boundary fault system, which encompasses the Raymond, Hollywood, Santa Monica, Malibu Coast, Anacapa-Dume, Santa Cruz Island, and Santa Rosa Island faults, as well as blind thrust faults. Abbreviations: C-SF—Clamshell-Sawpit fault; ELATB—East Los Angeles blind thrust system; EPT—Elysian Park blind thrust fault; Hol Flt—Hollywood fault; PHT—Puente Hills blind thrust fault; RMF—Red Mountain fault; SCIF—Santa Cruz Island fault; SSF—Santa Susana fault; SJcF—San Jacinto fault; SJF—San Jose fault; VF—Verdugo fault; LA—Los Angeles; LB—Long Beach; NB—Newport Beach; Ox-Oxnard; P—Pasadena; V—Ventura; WN—Whittier Narrows. Downtown Hollywood is centered between Hol and Flt in figure. [From Dolan et al. 2000, Figure 1, p. 1560.]

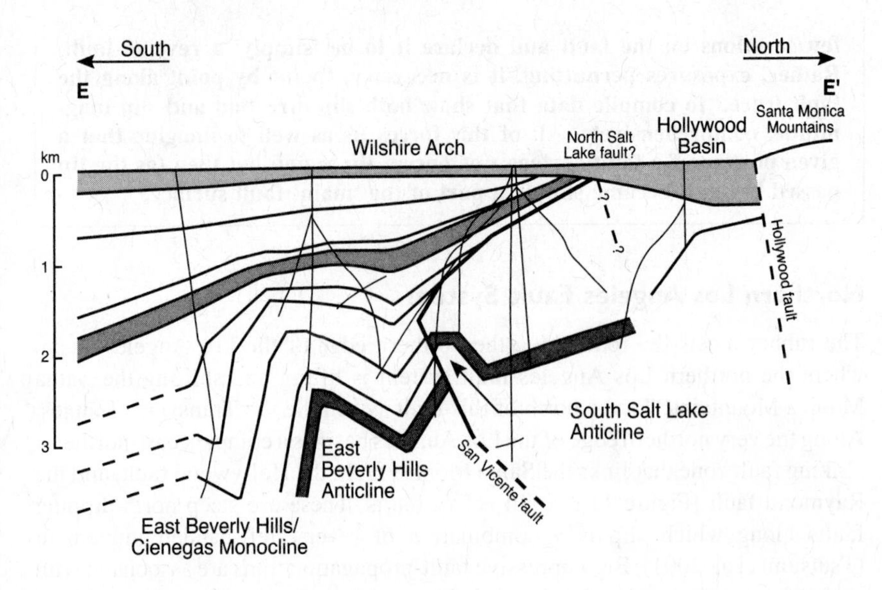

Figure 11.33 North-south cross-section through the northern part of the Los Angeles basin, showing Santa Monica Mountains, the Hollywood basin, the Wilshire arch, and the underlying geologic structure, including the Hollywood fault, the San Vicente fault, the East Beverly Hills anticline, and the East Beverly Hills–Las Cienegas monocline. Middle to late Quaternary strata are shown in light gray. Youngest bedrock beneath Quaternary strata belongs to the Pico Formation, which is Late Pliocene and Pleistocene in age. At a deeper level, the unit highlighted in medium gray is Pliocene in age. Highlighted unit near base of section (dark gray) is Middle to Late Miocene in age. Note drill holes, which provided control for formation boundaries as well as bed dips. [From Tsutsumi et al. 2001, Figure 4E, p. 458.]

strike-slip faults, with thrust-faulting components. They were able to underscore to officials that the fault is active and capable of producing damaging earthquakes in the urban setting of northwestern Los Angeles (Dolan et al. 2000).

Tsutsumi et al. (2001) found a nice way to summarize the results of paleoseismic investigations in the northern part of the LA Basin, in a way that emphasizes the order in which faults and formed develop, uncertainties in interpretation, kinematics, absolute timing, and extent of synchronicity (Figure 11.34). Fault and fold structures are listed in the far left column. The

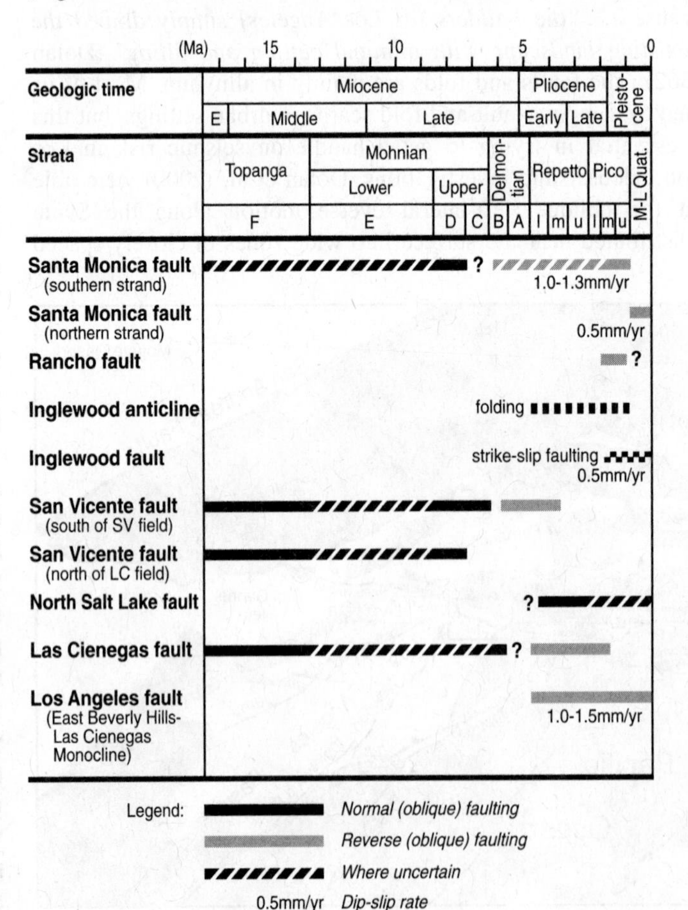

Figure 11.34 Diagram showing the timing of faulting and folding along structures in the northern part of the Los Angeles basin. Note that present-day reverse and strike-slip faulting are superimposed upon a Miocene history of normal faulting, underscoring the role of inversion tectonics in the Los Angeles basin. See text for explanation. [From Tsutsumi et al. 2001, Figure 5, p. 462.]

relevant part of the geological time scale is posted along the uppermost row. The body of the chart reveals which faults and folds were active when, and the nature of the faulting (e.g., normal faulting, reverse faulting, strike-slip faulting).

Lesson Learned #11: The flurry of earthquakes that have taken place in the Los Angeles Basin region from 1970 to the present have brought to light the important role of blind faults in relation to structural geology generally and seismic-hazards mitigation specifically. Based on this knowledge, when we study inactive reverse or thrust fault cutting through layered stratigraphy we need to imagine that beyond the tip (or tip line) of this fault there likely could have been transfer of displacement to a fault-propagation fold, or a fault-bend fold, that itself developed and grew earthquake-by-earthquake through slip movements on the master fault. Again, we must pay close attention to the relationship between the faulting and any major folds that might be present.

THE EASTERN CALIFORNIA SHEAR ZONE

General Character

Now, with the San Andreas fault to our back, we head east from the Los Angeles basin to the Eastern California Shear zone (ECSZ), in the Mojave Desert region. Though lacking the broad name recognition of the San Andreas fault, the ECSZ is quite an extraordinary feature, and its partnership with the San Andreas fault is quite extraordinary as well.

The Eastern California Shear Zone (ECSZ) is a 50 to 100 km wide zone of north-northwest-trending, primarily right-handed strike-slip faults, complete with an important normal slip component (Figure 11.35). Its northern continuation is the Walker Lane (WL) (Figure 11.34), which is a broader and more diffuse zone of faulting that also is accommodating strike-slip faulting and extension (Wesnousky, 2005a). The ECSZ and WL together separate the Sierra Nevada and Basin and Range physiographic provinces. In plate-tectonic terms they separate the Sierran (S) microplate from the North America (NA) plate (see Figure 11.34). The ECSZ and WL presently are presently accommodating ~25% of P/NA relative plate motion, and over time have accounted for ~9% to 23% of the total relative plate motion between the North America and Pacific plates.

Dokka and Travis (1990) named the "*Eastern California Shear Zone*" in reference to this major regional zone of faulting, which is marked by oblique-slip deformation (Figure 11.36). The faulting represents part of the accommodation to Pacific-North America transform motion. Strain is transtensional, that is, a combination of stretching (normal faulting) and horizontal shear (strike-slip faulting) (see also Le and others, 2007).

The Eastern California shear zone appears to "*communicate*" with the San Andreas fault zone. When faulting is most active and intense in the Los Angeles basin, things are very quiet within the Eastern California shear zone, and vice versa. Dolan et al. (2007) worked this out based on study of the paleoseismic records of fault activity in the past 12,000 years. There is an "*anticorrelation*" between times of earthquakes within the San Andreas fault system versus the East California shear zone system (Figure 11.37). Dolan et al. (2007) believe that the Big Bend of the San Andreas fault may have a lot to do with this: fault slip along the Big Bend may tectonically load the ECSZ in a direction perpendicular to the ECSZ, thus inhibiting fault movement.

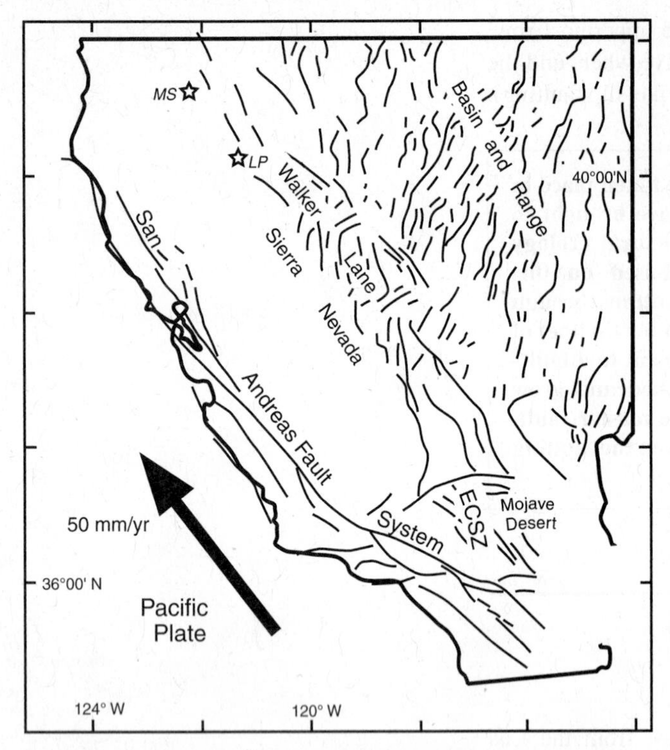

Figure 11.35 Generalized fault map of the western United States showing the locations of the Eastern California shear zone and the Walker Lane in relationship to the Basin and Range province (to the east) and the San Andreas fault system (to the west). Bold arrow represents P/NA plate motion at ~50 mm/yr at N50°W. [From Wesnousky, 2005a, Figure 1, p. 1506.]

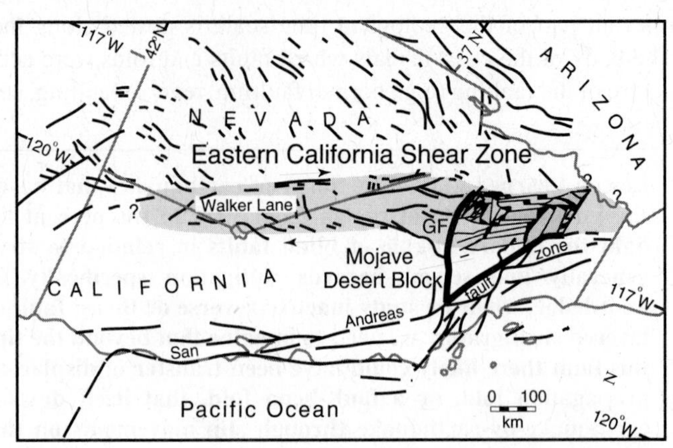

Figure 11.36 Location of the Eastern California shear zone (ECSZ) in relation to the Pacific–North America transform boundary in the western USA. [From Dokka and Macaluson, 2001, Figure 1, p. 30,626. Published with permission of Journal of Geophysical Research, Figure 1.]

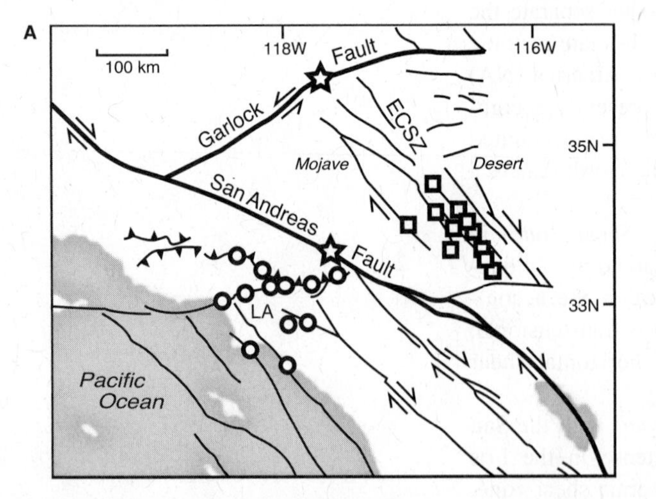

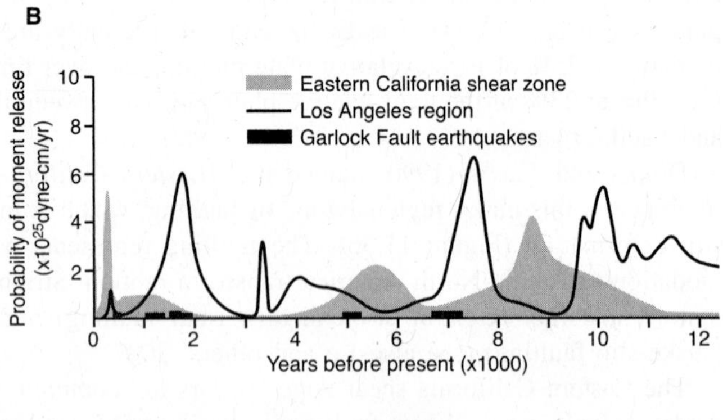

Figure 11.37 (A) Map of major active faults of Southern California showing locations of paleoseismological trench sites used to construct Figure 11.37B. Open circles denote Los Angeles region sites. Open squares show trenches in the Mojave section of the Eastern California shear zone (ECSZ). Stars denote Wrightwood trench site on Mojave section of the San Andreas fault and the Garlock trench site on the Garlock fault. [From Dolan, Bowman, and Sammis, 2007, Figure 1, p. 856.] (B) Comparison of cumulative seismic moment released through time for fault networks in Los Angeles region (dark gray) versus Mojave Desert part of Eastern California shear zone (light gray). There is an apparent temporal clustering of seismic moment release in both regions, and peaks and lulls in moment release for the two regions that appear to be anticorrelated. [From Dolan, Bowman, and Sammis, 2007, Figure 2B, p. 856.]

Landers Earthquake, Real-Time Example of Oblique Fault Slip in the ECSZ

The Landers earthquake occurred within the Eastern California shear zone (ECSZ) at 4:57 A.M. local time on June 28, 1992 (Figure 11.38). It was big. Initially the estimated magnitude was 7.8, later downgraded to 7.3. The ground shook for 2 to 3 minutes. There were more than 40,000 foreshocks and aftershocks! At the time it occurred, the Landers earthquake was the largest earthquake to have occurred in the contiguous United States in 40 years.

The Landers earthquake produced an 80-km-long system of surface ruptures in the form of dominantly right-handed belts of shear zones and fractures zones, striking approximately N10°W, but curving slight clockwise southward. Average maximum displacements were ~3 m. Arvid Johnson and his colleagues mapped the belts of surface ruptures in excruciating detail, discovering *broad*

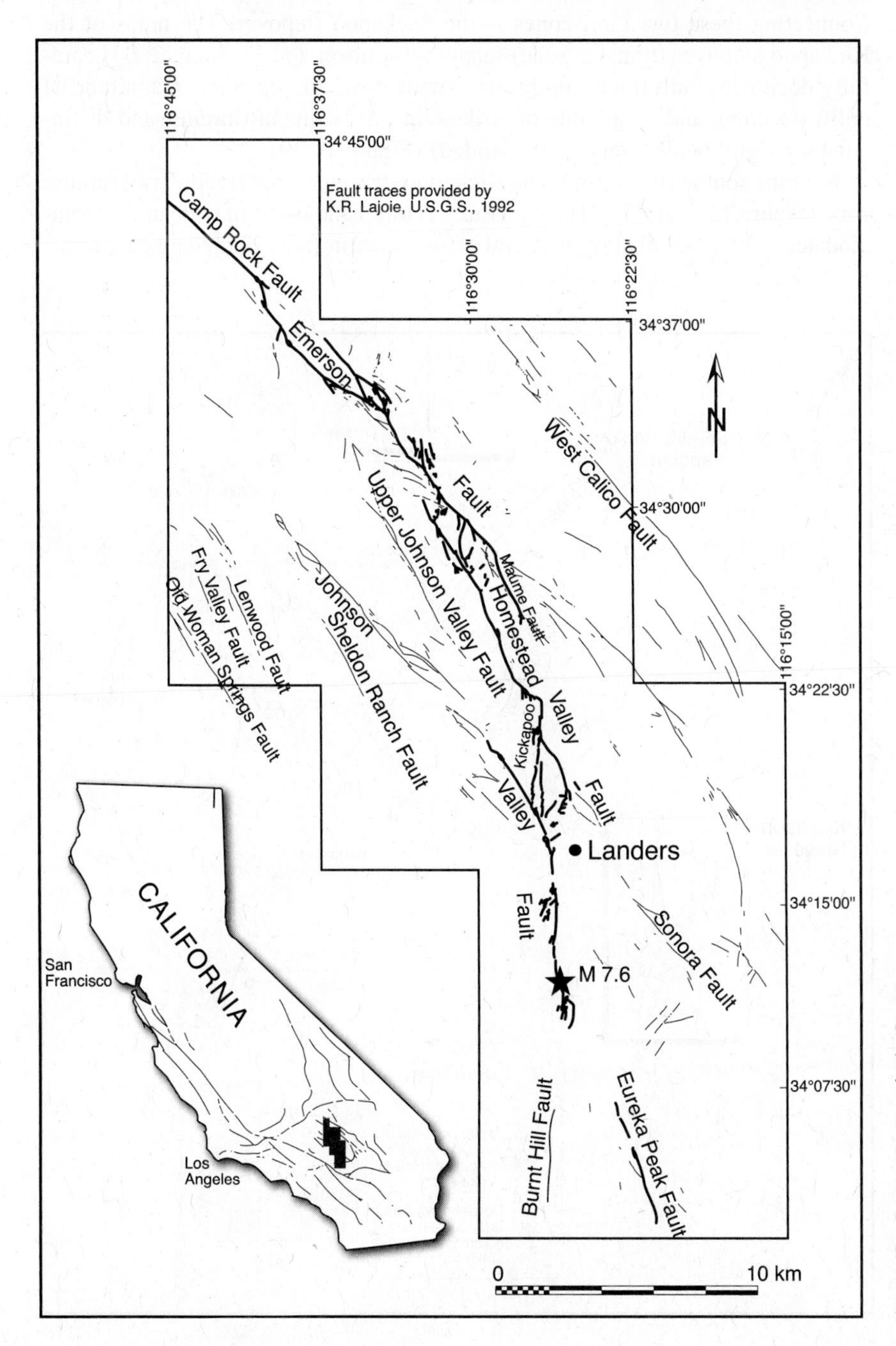

Figure 11.38 Map of the four major en echelon fault zones along which rupture took place during the Landers earthquake. These four zones are West Calico, Camp Rock, Emerson, and Johnson Valley. Inset map shows regional location of the Landers rupture zone within California. Solid star denotes the epicenter of the Landers earthquake. Fault traces provided by K. R. Lajie, U. S. Geological Survey, 1992. [Courtesy of Arvid Johnson.]

fracture zones of mild shearing (generally 50 to 200 m wide), containing narrow fracture zones of more intense shearing (5 to 20 m wide), containing faults or fracture zones of very intense shearing (0.5 to 2 m wide)" (Johnson and Fleming, 1994).

The main features comprise a system of at least four broadly en echelon fault zones, each ~10 km long, and marked by branch connections to one another in the form of right-lateral faults and shear zones within right-stepping transfers. The overall system of surface rupturing produced during the Landers earthquake is shown in Figure 11.38. The main fault zones (from northeast to southwest) are the West Calico, Camp Rock, Emerson, Homestead Valley, and Johnson Valley. Johnson and Fleming (2004) carefully documented details of the surface rupturing. For example, the Johnson Valley rupture belt is ~100–200 m wide, with right-lateral offsets up to 2.6 m. The Homestead rupture belt is up to ~500 m wide and accommodated up to 4.5 m of right-lateral offset (see Figure 11.38). Connecting these two fault zones is the Kickapoo stepover. The maps of the Kickapoo stepover (transfer zone) made by Johnson and Fleming, 2004) carefully document fault traces, upthrown versus downthrown sides, magnitude of uplift (in cms), and magnitude of strike-slip displacement (in cms, and distinguishing right-handed versus left-handed) (Figure 11.39).

Near the southeast margin of the Kickapoo stepover is the Happy Trail rupture zone (Figure 11.40*A*). The Happy Trail fracture zone is ~4 m wide and accommodated ~30 to 40 cm of right-lateral offset. Shearing was marked by pure right-

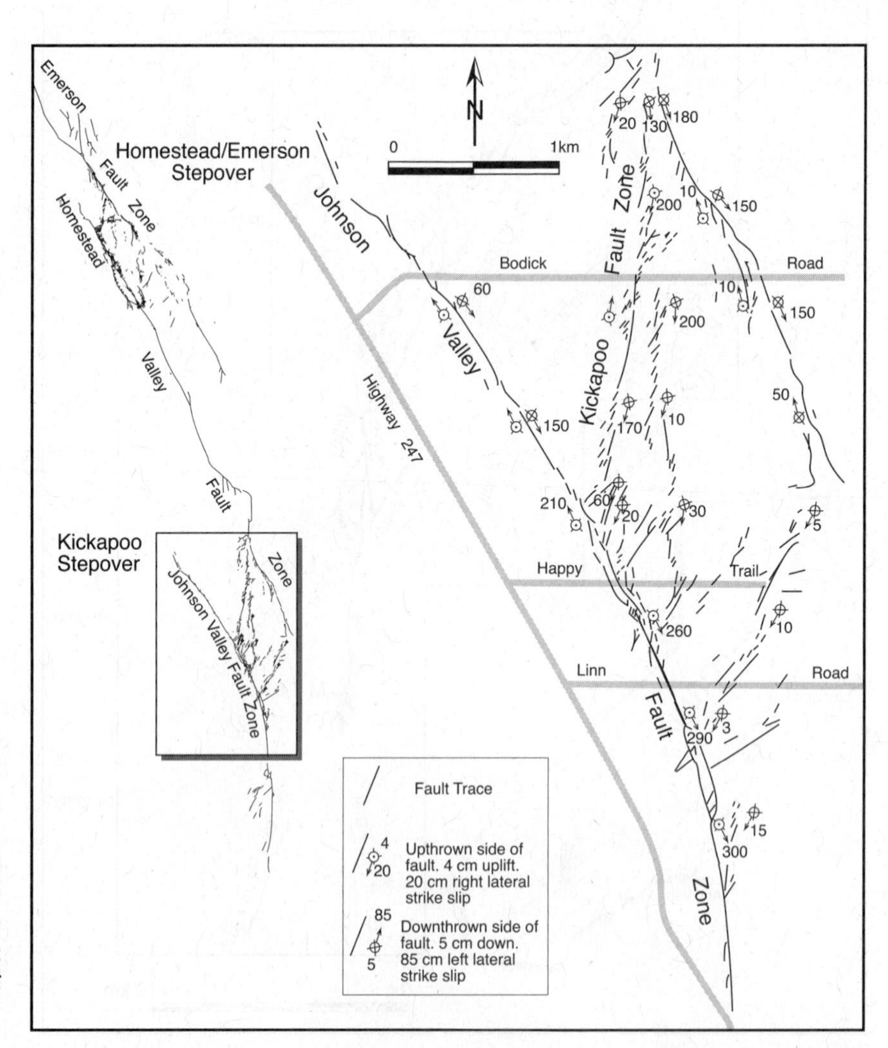

Figure 11.39 Detailed structural geologic map of faults and fractures comprising the Kickapoo rupture belt, which transfers displacement from the Homestead Valley fault to the Johnson Valley fault. [Courtesy of Arvid Johnson.]

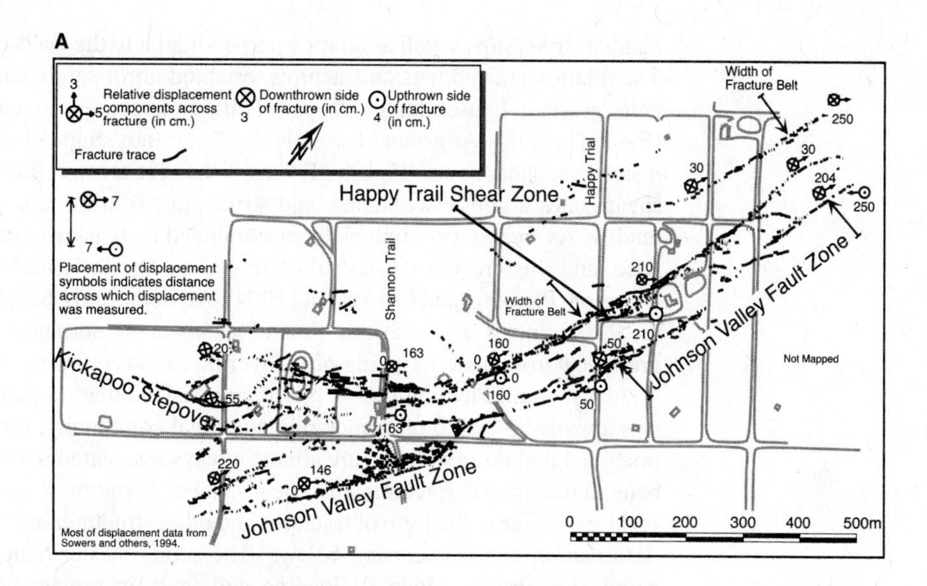

Figure 11.40 (*A*) Happy Trail fracture zone in the area of the junction of the Johnson Valley rupture belt and the Kickapoo Stepover. Happy Trail is an E–W road. The Happy Trail fracture zone lies a long a N–S road indicated by the arrow. The Johnson Valley rupture belt is approximately 200 m at this location, and accommodated ~280 cm of right-lateral offset. The Happy Trail rupture zone is approximately 120 m wide and accommodated ~210 cm of offset. [From Johnson and Fleming, 21 October 2004, Figure 40, p. 62.] (*B*) Aerial photograph of part of the Happy Trail fracture zone. This rupture zone is on the east side of the belt. The roads are generally E–W or N–S. Northwest is to the left in this figure. In the words of Johnson and Fleming (2004, p. 63), the Happy Valley fracture zone "*makes a perfect hit on a house.*" [From Johnson and Fleming, 21 October 2004, Figure 41A, p. 63.] (*C*) Photograph of part of the Happy Trail shear zone at ground level. View is toward N15°W, diagonally from the SW side to the NE side of the shear zone. Width of shear zone is ~4 m. En echelon cracks are oriented ~20° clockwise from the direction of view, and these are generally reactivated as left-lateral strike-slip faults. Barbs are in hanging wall of small thrust fault. [From Johnson and Fleming, 21 October 2004, Figure 43B, p. 65.]

handed strike-slip as well as dilation perpendicular to the walls of the shear zone. The dilation created tension fractures, oriented north-south. One of the fracture belts grazes a house along "*Happy*" Trail, as is evident in aerial photography (Figure 11.40*B*). At ground level, the faults are not single discrete surfaces but instead are shear zones (Figure 11.40*C*). When the tension fractures opened, the shear zones abruptly weakened, and strike-slip shearing took place even more readily. As shearing continued, blocks bounded by tension cracks rotated clockwise, and this created left-lateral offsets along the (original) tension fractures (Johnson, Fleming, and Cruikshank, 1994). All of this must have happened during the 2–3 minutes of co-seismic ground shaking; yet normally we think of progressive deformation as taking place over thousands or millions of years.

Johnson and Fleming (2004, p. 95) emphasized that the Landers earthquake was a wake-up call to the structural geological community, for most of us have imagined that discrete faults are almost always associated with earthquakes, not belts and zones of fractures and shearing. Furthermore, they emphasized that surface damage in the form of fractures and other structures are "*merely guides to deformation that is occurring below.*" The actual master fault or shear zone is below the surface. Mode III loading and fault-tip propagation produced the damage seen at Landers. Even though motion along the master fault was strike-slip, the tip line of the fault propagated vertically toward the surface. In advance of the tip line a process zone absorbed the near-field effects in the form of zones and belts of right-lateral dilational fracturing and shearing (Johnson et al. 2001).

Lesson Learned #12: We normally think of shear-zone deformation as occurring beneath the surface, "*safely out of sight*" at depth, where overall conditions favor a more semi-brittle or semi-ductile deformation environment. The near-perfect exposures within active tectonic settings such as Landers reveal that in an "*instant*" of fault slip an entire outcrop system of discrete fractures, faults, tension fractures, and shear zones can be produced. In mapping ancient faults and discrete shear zones, we should pay attention to the character of fracture and shear systems just beyond the tip line(s), for this domain of observation and mapping might well be marked by an observable transition in structural style.

Northern Part of the ECSZ

The northern part of the Eastern California shear zone (ECSZ) contains some big-name fault zones, including the Owens Valley-Lone Pine fault zone; the White Mountains fault zone; and the Furnace Creek–Death Valley fault zone (Figure 11.41). These are faults you will want to visit sometime!

Based on GPS, right-handed strike-slip motion appears to average as follows: 2.8 mm/yr of slip along the Furnace Creek-Death Valley fault zone; 2.5 mm/yr along the White Mountains fault zone; and 5.3 mm/yr along the Owens Valley–Lone Pine fault zone. Today the average slip rate for all of the ECSZ faults, taken together, corresponds nicely with the slip rate determined through geodetic measurements over the past 40 years, namely ~6.7 mm/yr ± 1.3mm (Dokka and Travis, 1990; Le and others, 2007). To put things into perspective, this is ~10% of the total P/NA plate motion.

Lesson Learned #13: When someone asks us (when we least expect it) about rates of slip on faults, we can now venture estimates based upon what we learn in various active tectonic settings. We are already seeing (for western U.S. examples) that reasonable responses might be in the range of fractions of millimeters to several millimeters of slip per year.

Several millimeters of slip per year corresponds to several kilometers of slip per million years. We can carry this knowledge with us when we study ancient, large-displacement fault zones, and gain a more tangible sense for how much time was required for total net slip to accrue.

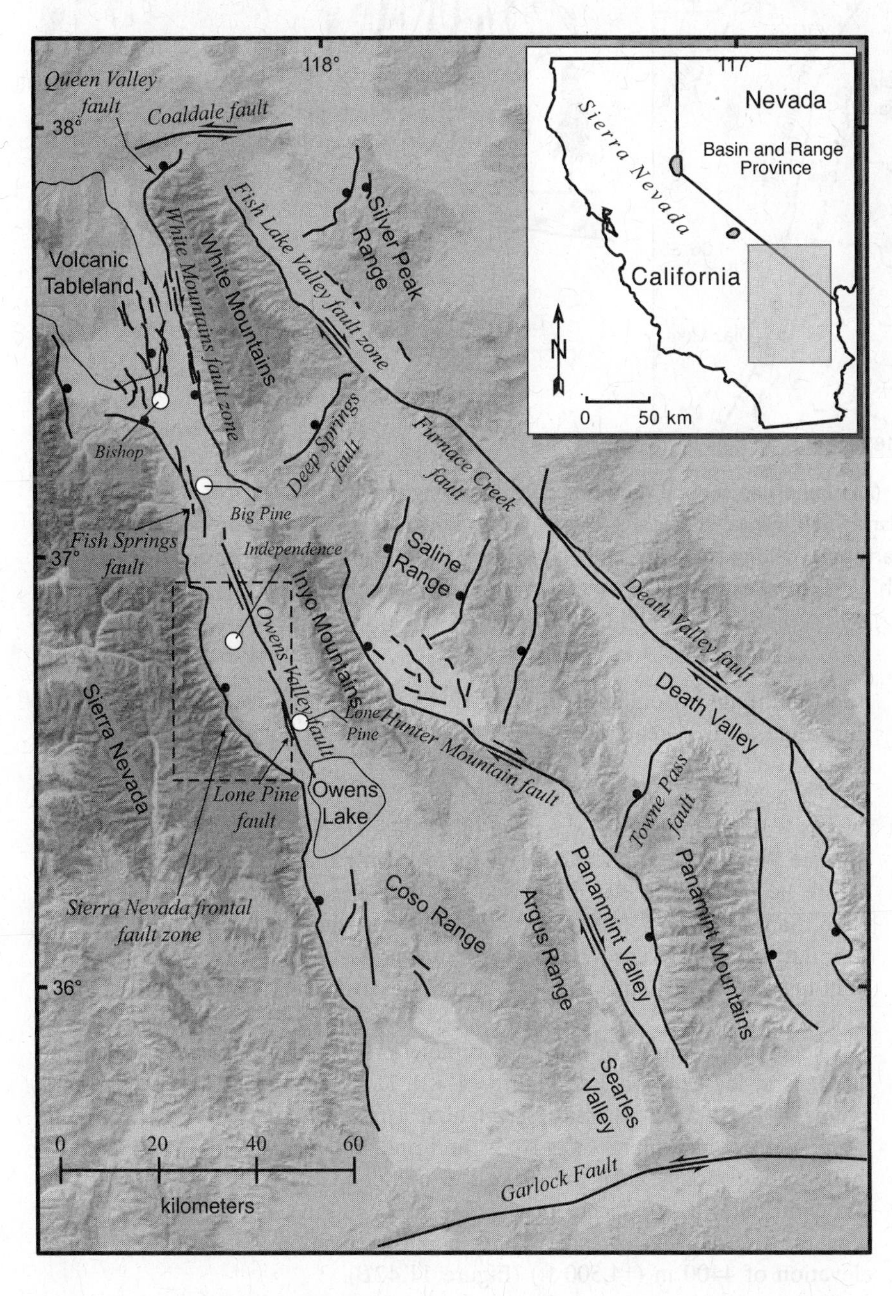

Figure 11.41 Map of northern part of the ECSZ, with the 3 main fault zones: the Fish Springs–Owens Valley–Lone Pine fault zone; the White Mountains–Hunter Mountain–Panamint Valley fault zone; and the Fish Lake Valley–Furnace Creek-Death Valley fault zone. Also shown is the Sierra Nevada frontal fault zone and the Volcanic Tableland. Inset provides regional geographic context. [From Le, Lee, Owen, and Finkel, 2007, Figure 1, p. 241.]

Owens Valley-Lone Pine Fault Zone

The Great 1872 Owens Valley Earthquake

The 1872 Owens Valley earthquake ($M_W = 7.5+$) ranks right up there as one of the three largest historic earthquakes in California (Lubetkin and Clark, 1987; Bacon and Pezzopane, 2007). It occurred at 2:30 A.M. on March 26, 1872. The tiny towns of Lone Pine and Independence experienced violent shaking, causing almost all of the adobe buildings and some of the

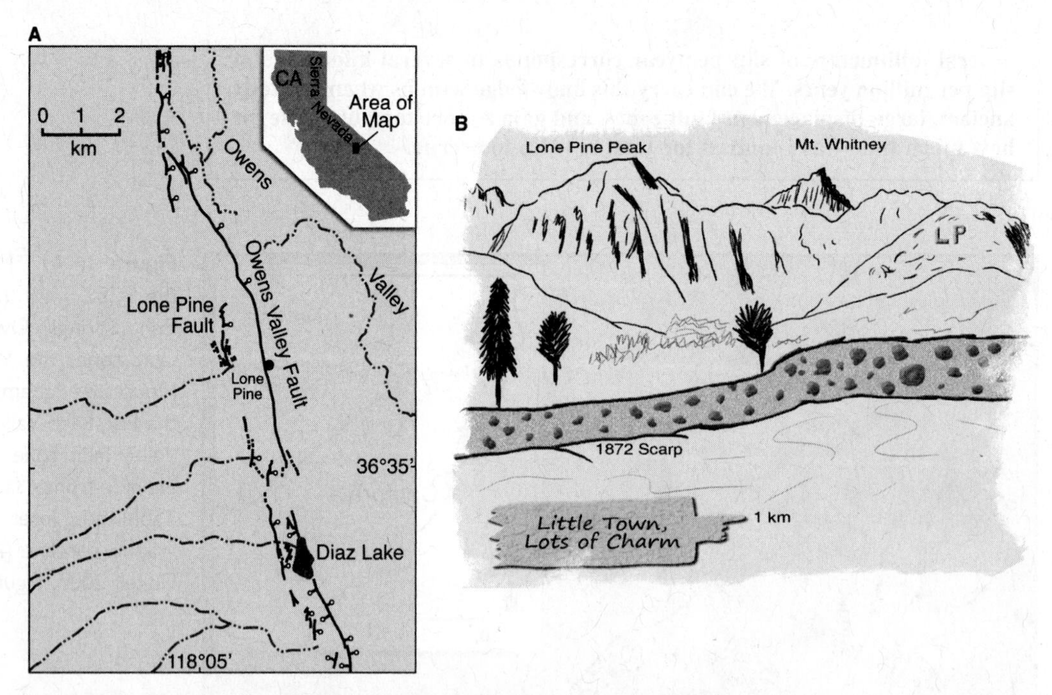

Figure 11.42 (*A*) Map of the surface rupturing created during the Owens Valley earthquake. [From Lubetkin and Clark, 1987, Figure 2, p. 151.] (*B*) My (GHD) sketch of a fault scarp produced during the Owens Valley earthquake. Lone Pine Peak and Mt. Whitney (~4400 m) in the background. Sketch was drawn on March 26th, 1999, exactly 127 years following the great earthquake.

brick buildings to collapse (Figure 12.42*A*). Twenty-seven people in Lone Pine died, which represented 10% of the local population (Beanland and Clark, 1982).

The geologist that first described the effects of faulting during the great Owens Valley earthquake was none other than G. K. Gilbert. He made a visit to Lone Pine in August, 1883, 11 years following the earthquake and 23 years before he traveled to San Francisco to study the ground expressions of the 1906 San Francisco earthquake. Gilbert was the first to note that fault slip during the Owens Valley earthquake was oblique, marked both by right-handed strike slip and down-to-the-east normal slip (Beanland and Clark, 1982).

The mapped length of the fault rupture produced during the great earthquake is at least 90 km (see Figure 11.42). Down-to-the-east normal faulting averaged 1 $\pm$0.5 m, whereas right-lateral offset was much larger, averaging ~6m, with a maximum of ~10 m at Lone Pine. The normal component, though more modest than the strike-slip component, is still quite substantial, and contributes to the spectacular setting of Lone Pine. Directly west of Lone Pine the Sierra Nevada achieves its highest elevation: Mt. Whitney at the elevation of 4400 m (14,500 ft) (Figure 11.42*B*).

The Lone Pine fault lies 1.4 km west of the town of Lone Pine and is expressed as a prominent east-facing scarp that cuts a Holocene alluvial fan, full of grossly unsorted boulders, pebbles, and cobbles (Figure 11.43). One of the more spectacular expressions of the Owens Valley faulting is located south of Lone Pine at the mouth of McGee Creek along the east flank of the Sierras. An ancient valley glacier carved the great valley that McGee Creek now occupies. Remnants of glaciation include ribbon-shaped lateral moraines, which at McGee Creek are oriented at a high angle to the trace of the Owens Valley fault. The lateral moraines are now cut and offset by the Owens Valley

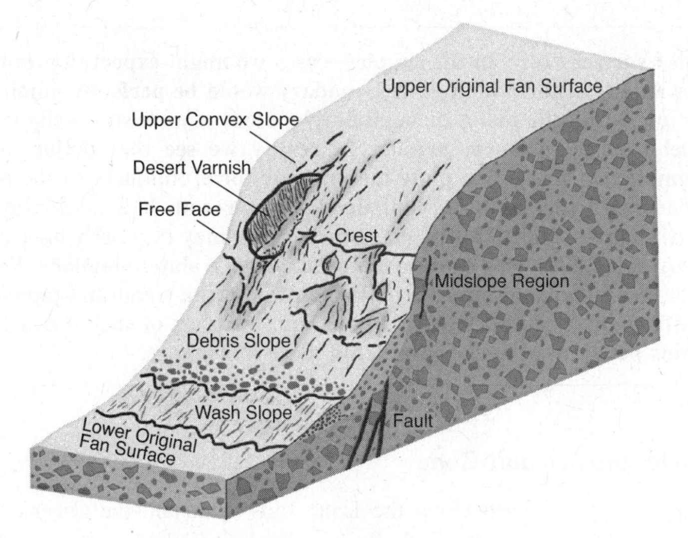

Figure 11.43 Idealized profile of a fault scarp, which fits the picture of scarps along the Owens Valley fault. [Modified from Wallace by Lubetkin and Clark, 1987, Figure 4, p. 152.]

fault (Figure 11.44). The crest line of the lateral moraine can be used as a structural geology "*piercing line*" to determine offset. We are not surprised to conclude that the offset is consistent with normal right-handed slip (see Figure 11.44).

The oblique-slip fault displacement during the Owens Valley earthquake proves to be quite consistent with the overall net slip along the Owens Valley fault, which is marked by ~2500 m of east-side-down normal displacement and between ~10 to ~20 km of right-handed strike slip! The oblique fault slip is consistent with the fact that overall orientation of the Owens Valley–Lone Pine fault zone is not perfectly parallel to the direction of Sierran microplate/North American plate motion. Rather, the trend of this zone is oriented ~20° to ~30° clockwise from the pure strike-slip direction, thus creating a "transtensional" geometry (Bacon and Pezzopane, 2007).

Figure 11.44 Northwest-directed photograph of the mouth of McGee Creek on the east flank of the Sierras. In the foreground is the fault scarp, down on the east. It is snow covered south of the path. In the middle ground, there is a ridge slanting down gently to the right (east). This is a lateral moraine. Where the crest of the moraine hits the fault, it shifts southward. This is a good locale to see the combination of normal and right-handed slip along the Owens Valley fault system. [Photograph by G. H. Davis.]

Lesson Learned #14: In the simplest cases we might expect that motion related to a transform-fault plate boundary would be perfectly mimicked in the interior of the plates through development of pure strike-slip faults parallel to the transform motion. In reality we see that major faults accommodating transform plate motion may form obliquely to the plate boundary, and accommodate oblique slip as a result. It is fascinating to imagine that for ancient faults the nature of slip may originally have been *perfectly compatible* with motion along the closest plate boundary. When we measure strike-and-dip of a major fault, and the trend-and-plunge of slickenlines on it, we normally do not think in terms of such *direct* relationships between fault kinematics and plate-kinematics.

White Mountains Fault Zone

If we were to travel north along the Lone Pine fault and the Owens Valley fault, we would come to the Fish Springs fault, which dies out at Bishop, California (see Figure 11.41). Just to the north of the town of Bishop lies the Volcanic Tableland. East of the Volcanic Tableland, across the valley, is the range-bounding White Mountains fault zone. It links southward with the Hunter Mountain fault and the Panamint Valley fault (see Figure 11.41).

The White Mountains fault zone, like the Owens Valley fault zone, is marked by the combination of right-handed strike-slip faulting and normal faulting, with the right-handed component being most dominant. These faults are both normal-right-handed even though the White Mountains fault zone dips west and the Owens Valley fault zone dips east! The White Mountains fault zone and the Owens Valley fault zone connect with one another via a right-handed step-over (see Figure 11.41) (Evans and Bradbury, 2004). The Volcanic Tableland is located within the step-over.

The onset of oblique slip along the White Mountains fault zone appears to have begun ~3 Ma. Total displacement from Pliocene time until today may be as large as 2.5 km, which would correspond to an average slip rate of ~1 mm/yr (Kirby et al. 2006).

Volcanic Tableland

The dominantly flat bedrock that supports the Volcanic Tableland is the Bishop Tuff, an ash deposit blown out of Long Valley caldera ~760,000 years ago (Bateman, 1965). The ash cloud collapsed, flowed, and settled to form an ignimbrite sheet with a perfectly smooth, planar upper surface. The Volcanic Tableland is cut by more than 200 NNW-trending active normal faults (Figure 11.45). Some of the faults are west dipping, others are east dipping, and thus the system as a whole is a conjugate normal fault system. There is more than 300 km of trace length of surface rupture (Sheridan, 1975; Pinter, 1995; Evans and Bradbury, 2003). The faulting has accommodated ENE/SWS stretching of the Bishop Tuff. Geologic cross sections drawn through the Volcanic Tablelands reveal that the normal faulting has achieved ~2% to ~5% of Quaternary extension of the Bishop Tuff (Pinter, 1995).

The fault scarps on the Volcanic Tableland are the most beautiful you would ever want to see (Figure 11.46). Exposures are so crisp that the scarps are revealed from tip to tip. When viewed face on, the top of a given scarp will have a longitudinal profile resembling a gentle smooth arc. Pinter (1995) has us view the scarps as "*lens-shaped rips*" in the surface of the Bishop Tuff! The arc profiles of the scarps are plainly seen because the top of the Bishop Tuff is so smooth, and because changes in topographic relief mimic exactly the gradient of slip along each faults (Figure 11.46*A*, *B*). Commonly it is evident in the landscape that as one

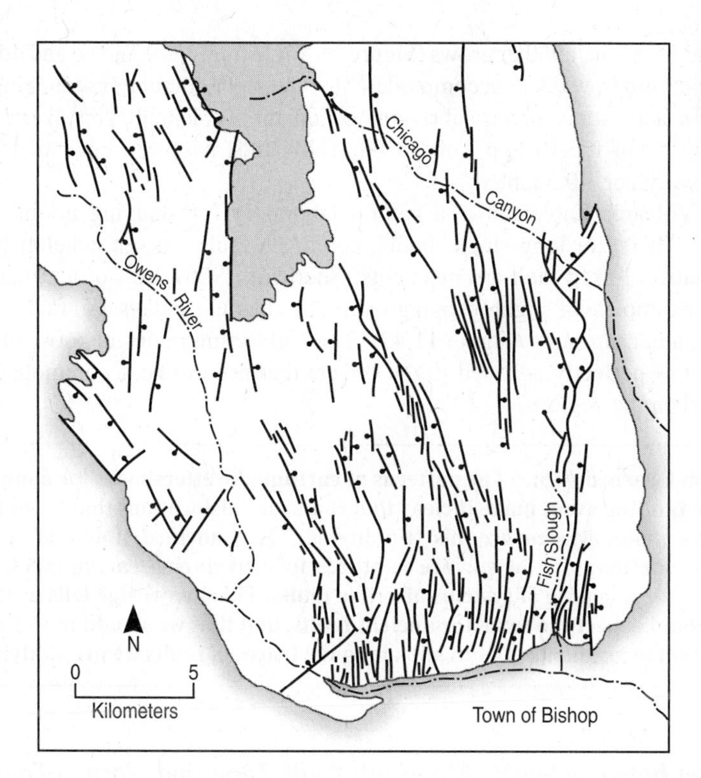

Figure 11.45 Map of normal faulting within the southern lobe of the Bishop Tuff, Volcanic Tablelands. [From Sheridan, 1975, Figure 12, p. 108.]

A

B

C

Figure 11.46 (*A*) Photograph of a number of fault scarps, produced by normal faulting of the Bishop Tuff on the Volcanic Tableland. These scarps are associated with normal faults that dip toward the viewer. (*B*) Photograph of a single fault scarp (normal faulting; fault dips toward viewer) on the Volcanic Tableland. Note that the conspicuous scarp steadily decreases in height to the left as the fault slip systematically "*tips out.*" (C) Photograph of two main fault scarps. The fault in the foreground tips out to the left. The fault in the far background tips out to the right. In the middle ground we can see yet another scarp, which tips out to the right. [Photographs by David Ferrill.]

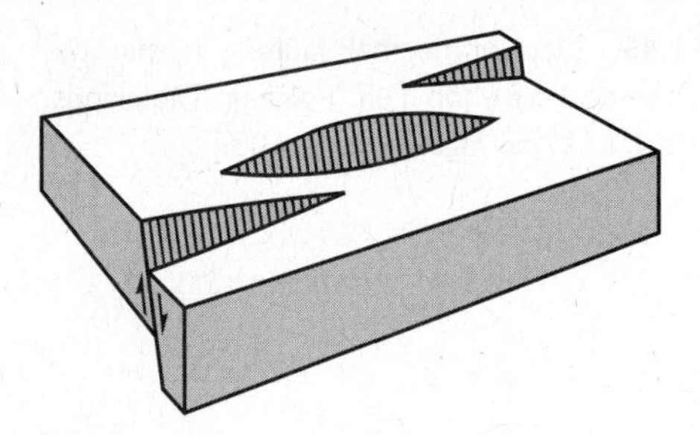

Figure 11.47 Schematic illustration of left-stepping set of fault offsets on Volcanic Tableland. Note the tipping out of the faults, and the "*lens-shaped*" form of individual complete fault scarps. Lines on fault surface simulate slip direction, but they are not particularly well displayed in this field setting. [From Pinter, 1995, Figure 3, p. 75.]

fault scarp dies out, another grows (Figure 11.46*C*): they come and go and double-up or triple-up in ways to accommodate just the right amount of stretching and extension across any given subregion within the Tableland. The overall rate of extension of the Bishop Tuff in the Tablelands has been estimated to be ~300 m/m.y., or ~0.3 mm/yr.

The Volcanic Tableland is a natural laboratory for studying normal fault systems. There are long single faults, conjugate fault sets, en echelon faults, and beautiful linked fault arrangements (Sheridan, 1975). Best of all, there are terrific examples of right-stepping and left-stepping relays of fault offset between adjacent faults (Figure 11.47). The connecting ramps in between have orientations perfectly adjusted to the offsets that need to be accommodated in the overlapping regions.

Lesson Learned #15: The patterns of earthquake aftershocks for a major active faulting event make it clear that faults are dislocations that tip out in all directions. Moreover, this 'tipping-out' is confirmed time-and-again during the mapping of surface ruptures formed during earthquakes, or during the mapping of systems of active faults. This knowledge tells us that we should expect the same for ancient faults, and that we should pay closer attention to gradients of fault slip along the traces of faults we are studying.

Relation between White Mountain Fault Zone and Normal Faulting in the Tablelands

The White Mountains lie directly east of the Volcanic Tableland, just across the northernmost stretch of Owens Valley (see Figure 11.41). It turns out that the normal faulting of Bishop Tuff in the Volcanic Tableland is related to oblique-slip faulting along the White Mountains fault zone. We should not think of the White Mountains fault zone as perfectly planar, for as it dips westward into the subsurface it curves into the form of a listric normal fault (Figure 11.48). Bishop Tuff in the Volcanic Tableland lies in the hanging wall of this listric fault, and in response to the normal component of faulting along the White Mountains fault zone, the Bishop Tuff warps into a rollover anticline. Outer arc extension along the crest of the broad anticline is creating the system of normal faulting. With each and every major earthquake along the White Mountain fault, the hanging wall slips downward, flexing the Bishop Tuff and progressively amplifying the rollover anticline. Evidence for this interpretation 'suddenly' showed up one day in the 1986, when the Chalfant earthquake sequence occurred (Smith and Priestly, 2000). This earthquake produced a normal oblique rupture along the White Mountain fault, and also produced a discontinuous series of ruptures through the Volcanic Tableland (David Ferrill, personal communication, October 20, 2009). David interpreted the ruptures in the Tableland as accommodations to outer arc extension above the listric White Mountain (Lienkaemper et al. 1987; Ferrill and Morris, 1999).

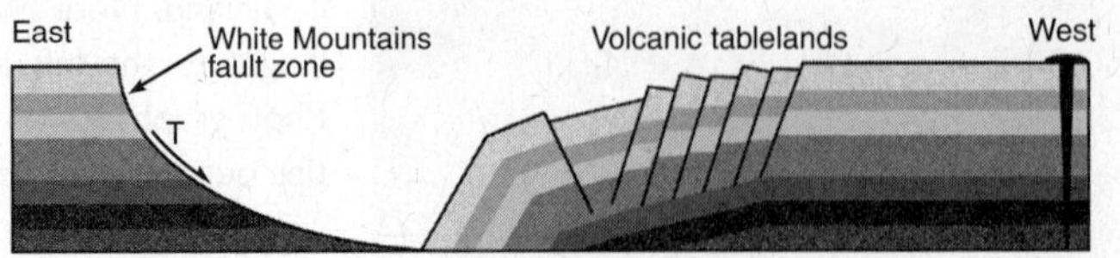

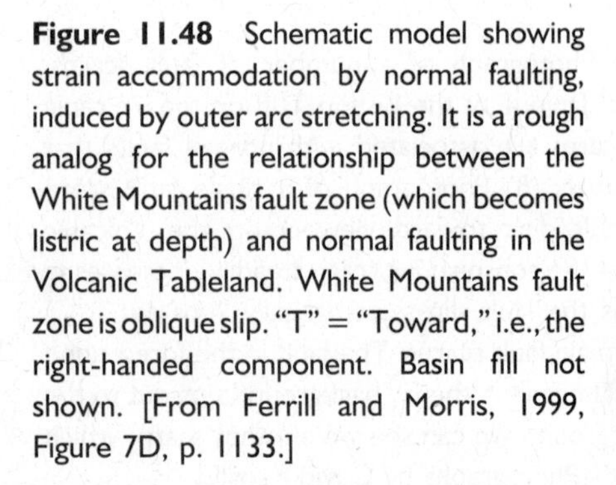

Figure 11.48 Schematic model showing strain accommodation by normal faulting, induced by outer arc stretching. It is a rough analog for the relationship between the White Mountains fault zone (which becomes listric at depth) and normal faulting in the Volcanic Tableland. White Mountains fault zone is oblique slip. "T" = "Toward," i.e., the right-handed component. Basin fill not shown. [From Ferrill and Morris, 1999, Figure 7D, p. 1133.]

Lesson Learned #16: When we interpret rollover anticlines forming on normal faults in ancient records, we seldom consider rollover anticlines developing in association with dominantly strike-slip faults with oblique normal-slip movement. When we think we are seeing outer-arc extension on big fold structures formed long ago, we tend to think of the folding itself as a product of compression and shortening, and not the combination of extension and strike-slip. Active tectonic settings provide a reality check on the limitless combinations of structural associations, and tend to move us beyond unfounded biases we have developed.

Death Valley-Furnace Creek Fault Zone

Throughout the ECSZ, as we have seen, individual major faults generally 'do it all'; they carry both the strike-slip and normal-slip components of oblique fault movement, which is a behavior consistent with the regional need to accommodate transtension between the Pacific and North American plates. In contrast, there are places within the ESCS where the transtensional strain is '***partitioned***,' with one fault or fault zone accommodating most of the strike-slip component, and a different fault or fault zone accommodating most of the normal slip. We see this at work in the Death Valley–Furnace Creek fault zone (see Figure 11.41). This fault system accommodates ~9.3 mm/yr of right-handed displacement between the Sierran (S) microplate and the Colorado Plateau, along a direction of N37° ±3° W, which is almost indistinguishable from the direction of P/NA movement (Bennett et al. 2003).

There are actually *two* Death Valley fault zones: the Northern Death Valley–Furnace Creek fault zone, which tips out southeastward; and the South Death Valley fault zone, which tips out northwestward (Figure 11.49). These faults are parallel, strike northwesterly, and overstep one another by ~20 km. Between the two lies a high mountain, known as the Black Mountains block

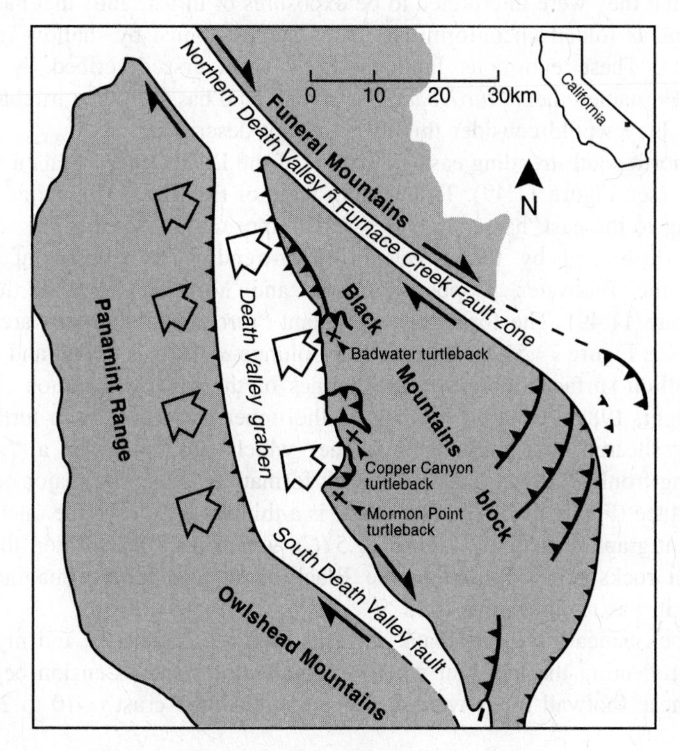

Figure 11.49 Structural geologic map of the Death Valley region. Hachured lines mark positions of major normal faults. Full arrows show inferred direction of crustal extension. Half arrows show relative displacement on strike-slip fault zones. Note the locations of the Black Mountains and the Panamint Range. Also shown are the locations of the three *"turtlebacks,"* which from north to south are the Badwater, Coppper Canyon, and Mormon Point turtlebacks. [From Troxel and Wright, 1987, Figure 7, p. 129.]

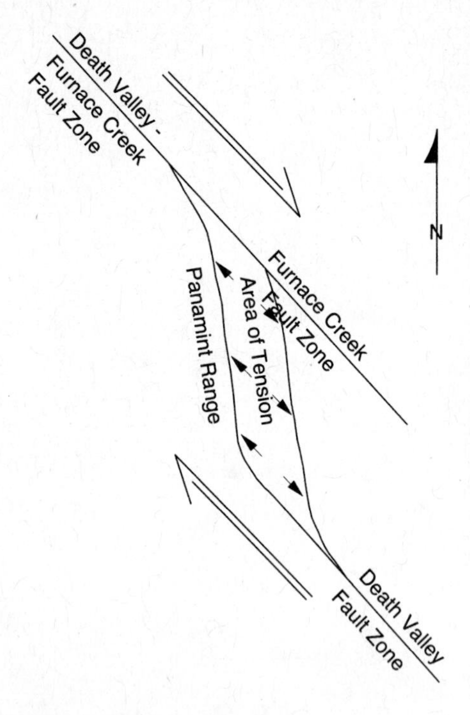

Figure 11.50 Diagrammatic map showing interpretation of strike-slip movement along right-stepping faults creating Death Valley as a pull-apart basin. [From Burchfiel and Stewart, 1966, Figure 2, p. 440.]

(elevation 1934 m), and Death Valley, which is the lowest land in North America (85 m below sea level). This overall system was envisioned early on as a giant "*pull apart*" of the Mojave Desert region (Figure 11.50) (Burchfiel and Stewart, 1966). The pull apart basin contains at least 3000 m of basin fill.

Thus the location in which Death Valley resides is a huge releasing bend (see Figure 11.50), the very antithesis of the setting of Los Angeles, which lies in a restraining bend. The releasing bend occurs at the right-stepping transfer zone between the southern and northern strands of the Death Valley fault zone. The Death Valley graben, which formed at this releasing bend, is ~80 km long and oriented nearly north south. Its orientation is distinctly different from the northwest trends of the (dominantly) strike-slip fault zones that become linked by this structural depression. The pull-apart basin opened along a regional extension direction of N60°W. Troxel and Wright (1987) estimated a maximum of 11 km of right-handed strike-slip faulting near the junction of the Furnace Creek fault zone and the northern strand of the Death Valley fault zone. Death Valley has experienced little strike-slip faulting, precisely because it is a pull-apart between the ends of two overstepping faults.

The Turtlebacks of Death Valley

Recall that faults are marked in places by fault mullions, which display smooth, sometimes polished, curviplanar surfaces, and whose linear orientations record the direction of fault movement. Normally we think of fault grooves and mullions of such a size and scale that they can be recognized entirely within single small outcrops of competent bedrock, though at times we see larger examples having wavelengths and amplitudes on the scale of meters or a few tens of meters. However, the fault mullions that occur along the western flank of the Black Mountains next to Death Valley are "*off-scale*;" they are enormous! (Figure 11.51*A*). They are so large that they were not even recognized as fault mullions until 1974 (Wright et al. 1974; Otton, 1974). Before that they were interpreted to be exposures of thrust faults that had been folded, or as folded unconformities, or as arches domed by shallow igneous intrusions. These enormous fault mullions were first described by Curry (1938). He named them "*turtlebacks.*" Anyone who has seen the turtlebacks of Death Valley would consider this to be an apt description.

The north-south-trending eastern margin of the Death Valley graben is ~60 km long (see Figure 11.49). The western edge of the Black Mountains block, bordering to the east, has a slightly serrated appearance. Notably the serrated edge is fashioned by the three northwest-trending turtlebacks of Death Valley: the Badwater, Copper Canyon, and Mormon Point turtlebacks (see Figure 11.49). The forms of these giant "*turtleback*" mullions are spectacular (see Figure 11.51*A*). The mullions plunge ~10° to 25°NW, and on the fault-mullion surfaces there are slickenlines of the same orientation (Troxel and Wright, 1987). In large patches on the upper surface of each turtleback there is sheared and faulted limestone, which proves to be a "*smear*" remaining from the down-faulting of rock formations along the major normal-fault surface (Figure 11.51*B*). The smear is a thinned residue of the once-thick thick stratigraphic sections (Figure 11.51*C*), which once rested atop the Precambrian rocks now exposed in the Black Mountains block. Imagine how difficult it was to figure this out!

Bedrock beneath the turtleback surfaces includes cataclasite and mylonite fabrics, reflecting the fact that when normal faulting and extension began to occur, these footwall rocks were at a deep level in the crust (~10 to 20 km)

Figure 11.51 (*A*) Northeasterly directed photograph of the Badwater turtleback, a huge fault mullion bordering Death Valley. The plunge angle of the nose of the turtleback is ~25° NW, which is an expression of the low-angle faulting represented by the fault surface. (*B*) Southwest-directed photograph of the Copper Canyon turtleback, showing the fault contact between Precambrian crystalline rock in the footwall and Tertiary sedimentary rocks in the hanging wall, i.e., the upper plate. (*C*) Bennie Troxel examining the contact between the Precambrian crystalline rock and sheared Tertiary cover along the Copper Canyon turtleback. [Photographs by G. H. Davis].

(Figure 11.52). Earthquake-by-earthquake they made their way to the surface. In the early going, under conditions of high temperature and pressure, some of this rock was transformed into mylonite. Later, at shallower levels (e.g., 3 to 6 km), the mylonites were overprinted by cataclastic deformation, and microbreccias were formed. At shallowest levels extension was achieved through conventional but low-angle brittle normal faulting, producing slickenlines and fault breccias (see Figure 11.52).

Taken together, the pieces of this system are beautifully coordinated. The turtleback orientations are perfectly parallel to the trends of the southern and northern strands of the Death Valley and Death Valley-Furnace Creek fault zones (see Figure 11.49). Death Valley is a half-graben that opened northwesterly, guided laterally by the trends of the southern and northern fault strands, and along a slip direction perfectly parallel to the turtlebacks. Some of the Precambrian and Paleozoic strata that once occupied the Black Mountain block are now resting northwest of Death Valley in the Panamint Mountains. The net fault transport was ~80 km (Troxel and Wright, 1987). This is big-time tectonics, ... reflecting the kind of tectonic action necessary to create a situation where the lowest elevation point in the continental U.S. (i.e., Death Valley) is just a 'stone's throw away' (120 km) from the highest point (i.e., Mt. Whitney).

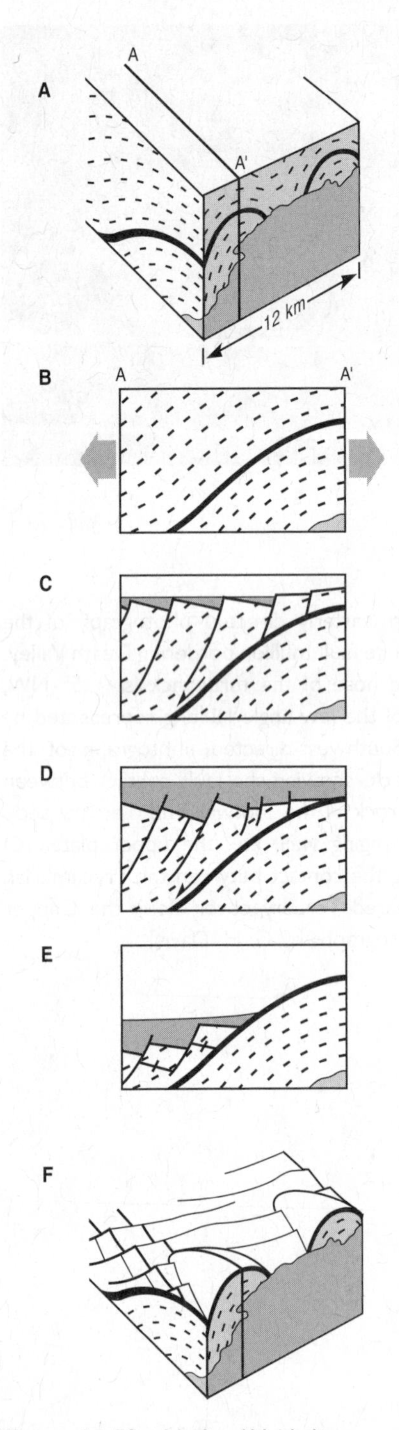

Figure 11.52 Idealized block diagrams and cross-sections, illustrating pull-apart concept of turtleback formation. Abbreviations: c = carbonate layers; ms = mixed metasedimentary rock; Qs = Quaternary sediments; tf = turtleback fault; Ts = Tertiary sedimentary rocks; vf = valley floor. [From Troxel and Wright, 1987, Figure 8, p. 130.]

Lesson Learned #17: Active tectonic settings reveal clear evidence for two end-member ways in which regional transtension can be accommodated; that is, oblique normal strike-slip on individual faults, *or* partitioning of pure strike-slip movement on some faults and pure normal-slip movement on others. There is comparable versatility in managing transpression, through oblique thrust strike-slip motions, *or* by partitioning between pure strike-slip and pure thrust slip motions. What holds up regardless of the details is a certain systematic kinematic behavior, and whether in the active or ancient record, we can track it with greater understanding and predictability.

WALKER LANE

Response to Transtension

Walker Lane is the northern projection of the Eastern California Shear Zone, though it is generally broader and more diffuse (see Figures 11.35 and 11.36). Wesnousky (2005a) compiled a beautifully detailed map of active faults and historical earthquake surface ruptures within the Walker Lane (Figure 11.53). Wesnousky's map reveals that the Walker Lane is full of faults, and is a source region for significant earthquakes! In studying the map it becomes apparent that there is no single through-going fault zone in the Walker Lane (see Figure 11.53). This situation is distinctively different from the San Andreas fault zone (Wesnousky, 2005a) and the Eastern California shear zone, where there are major through-going faults.

Walker Lane is accommodating right-lateral shear of the Sierran microplate relative to the central Great Basin (Faulds et al. 2005). Based on GPS-derived velocity maps, the Sierran block is moving N56°W at a velocity of ~12 mm/yr with respect to the Colorado Plateau and stable North America (Dixon et al. 2000; McClusky et al. 2001; Unruh et al. 2003). Furthermore, it is rotating very slowly counterclockwise (0.28° per m.y.). The rate of right-handed strike-slip motion parallel to the trend of the Walker Lane is ~9 mm/yr. The rate of east-west extension across the Walker Lane is ~1 mm/yr (Bennett et al. 2003).

What helps in visualizing the significance of the faulting within the Walker Lane are GPS-derived facts derived from deconstructing the plate-motion vector for the Sierran block into two components (Figure 11.54): a right-handed shear of ~9 mm/yr oriented 330°, thus parallel to the overall trend of the Walker Lane; and an extensional component of ~1 mm/yr oriented 240°, thus perpendicular to the overall trend of the Walker Lane (Bennett et al. 2003; Wesnousky, 2005a). The residual extensional component is what creates a transtensile environment within the Walker Lane. In contrast (see Figure 11.54), the San Andreas fault system is marked by a residual contractional component of ~3 mm/yr, creating within it an overall transpressional environment.

Lesson Learned #18: If each of the two largest active strike-slip fault systems in western North America are not pure strike slip (one transpressive, the other transtensile), it follows that pure strike-slip faulting is probably the exception and not the rule, . . . even for ancient strike-slip systems. We first learn and then tend to think of the major classes of faulting (normal, thrust, reverse, strike-slip) in their purest forms, when in fact we always need to be prepared for hybrid combinations.

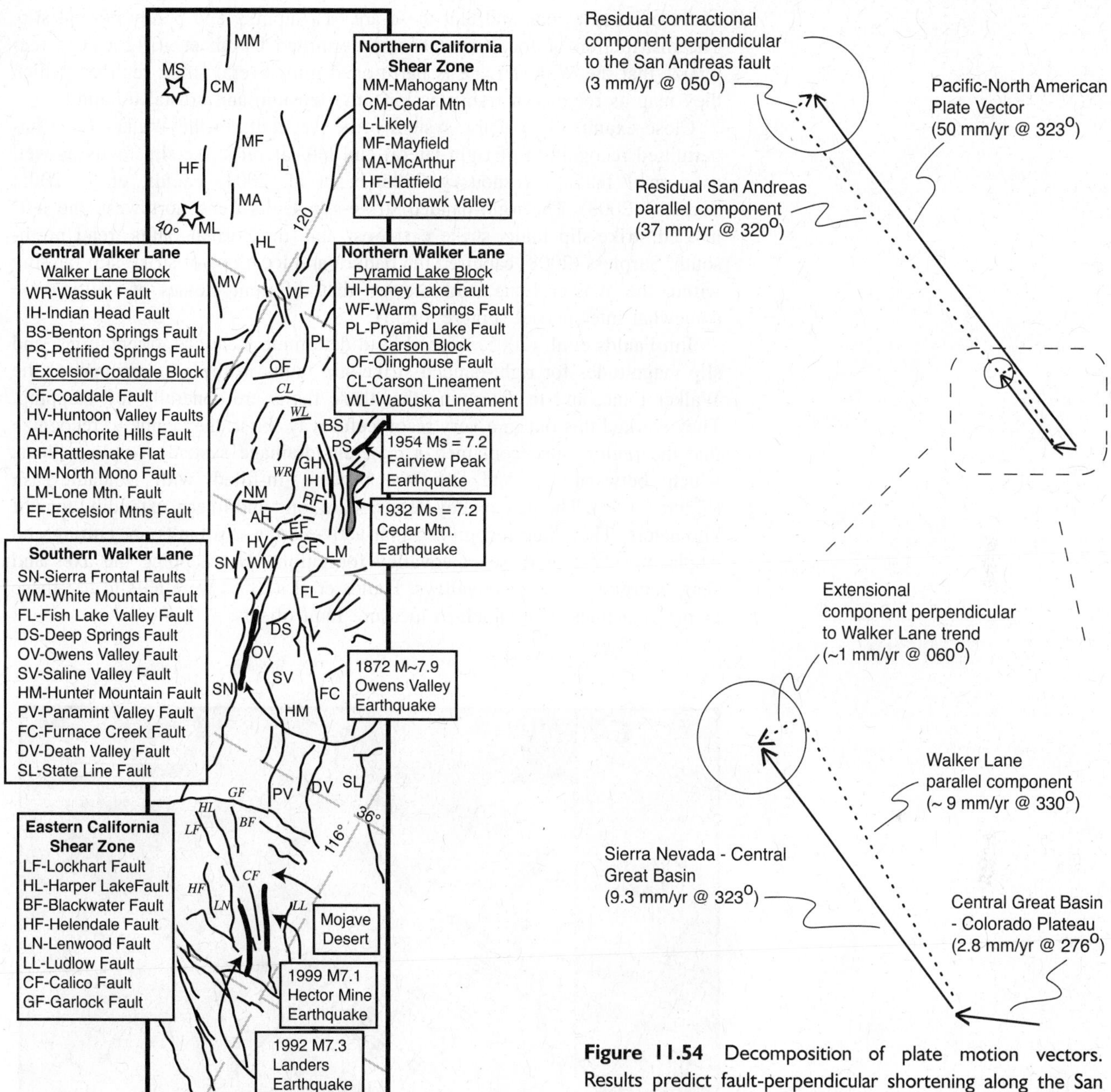

Figure 11.53 Active faults of the Walker Lane. Also shown (in bold lines) are historical earthquake ruptures. [From Wesnousky, 2005a, Figure 2, p. 1507.]

Figure 11.54 Decomposition of plate motion vectors. Results predict fault-perpendicular shortening along the San Andreas fault, and fault-perpendicular extension along the Walker Lane. [From Wesnousky, 2005, Figure 3, p. 1508, originally modified from Bennett et al. (2003).

Basic Fault Pattern

The active faulting in the Walker Lane is well adapted to accommodating plate-scale transtension (Unruh et al. 2003; Wesnousky, 2005a; Surpless, 2008). Focal mechanisms based on earthquake data reveal that the region is being simultaneously sheared (in a right-handed manner) and stretched (east-west) Unruh et al., 2003. The focal mechanisms show that northwest-trending faults are either pure right-handed strike-slip or oblique right-handed normal slip. Focal mechanisms also reveal north-trending faults along the eastern edge of the Sierra Nevada, and that these are east dipping and purely normal slip. The distribution of focal mechanisms permitted Unruh et al. (2003) to recognize that the Walker Lane is partitioned into several strain regimes, which they map as regions of transpression, transtension, and crustal thinning.

Close examination of the system of active faults in the Walker Lane has permitted recognition of right-lateral and left-lateral strike-slip faults as well as normal faults (Wesnousky 2005a,b; et al. 2003; Faulds, et al. 2005; Surpless, 2008). The right-handed strike-slip faults trend northwest, the left-handed strike-slip faults strike east-west, and the normal faults trend north-south. Surpless (2008) captured the pattern and locations of strike-slip faulting within the Walker Lane, and showed that different classes of faulting are somewhat interspersed (Figure 11.55).

Jim Faulds et al. (2005) were able to determine locations, orientations, and slip magnitudes for right-handed strike-slip faults in the northern part of the Walker Lane, and to determine that these faults are generally left stepping. They worked this out in a very resourceful way. First, they came to recognize that the region once contained a coherent drainage system of paleovalleys, which between 31 and 28 Ma became in-filled with volcanic ash (Figure 11.56). The paleovalley deposits are long and linear, extending tens of kilometers. They then recognized that active strike-slip faults are cutting and displacing some of these paleovalley formations. By mapping the axes and sharp margins of the paleovalleys, Faulds et al. (2005) were able to use these as piercing lines with which to measure fault offsets.

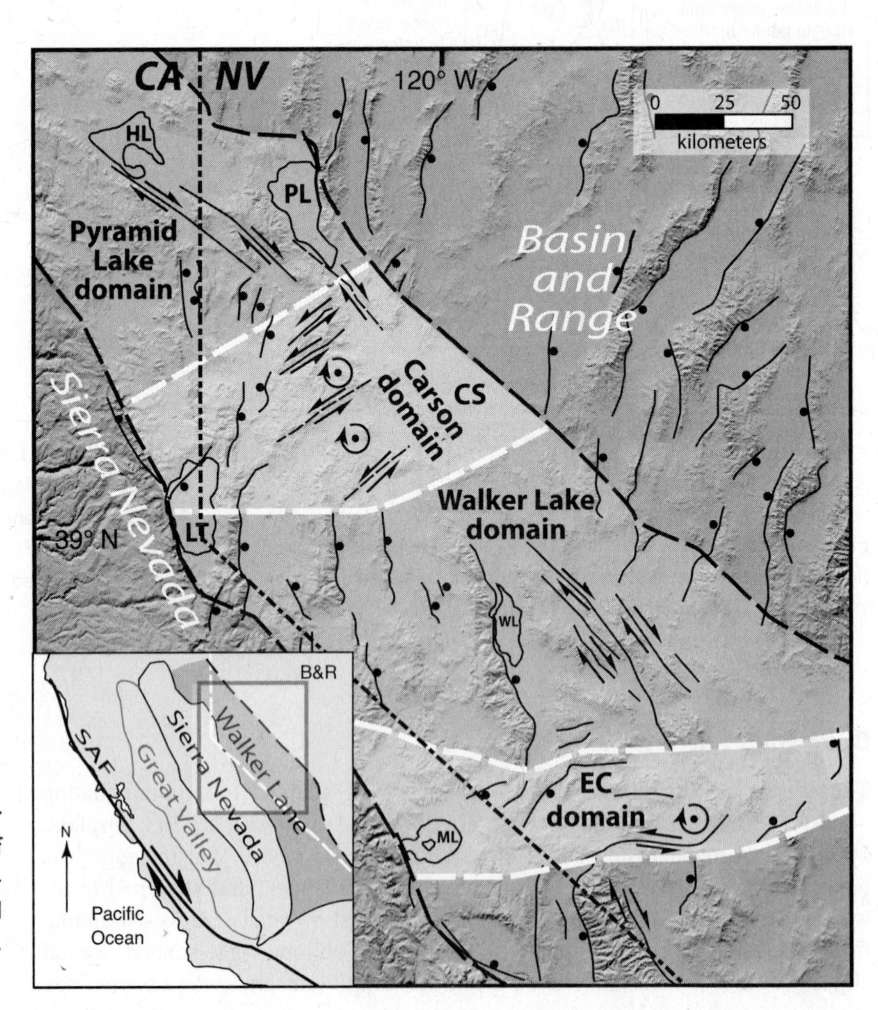

Figure 11.55 Map of faults in Walker Lane. Background is digital shaded relief map. Right-handed strike-slip faulting, left-handed strike-slip faulting, and normal faulting are interspersed within this region. [From Surpless, 2008, Figure 1, p. 240.]

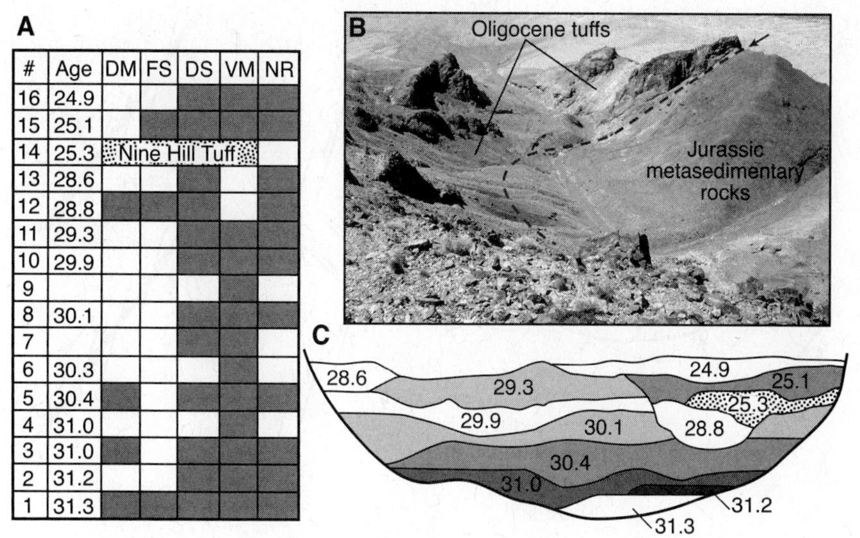

A
Age DM FS DS VM NR
16 24.9
15 25.1
14 25.3 Nine Hill Tuff
13 28.6
12 28.8
11 29.3
10 29.9
9
8 30.1
7
6 30.3
5 30.4
4 31.0
3 31.0
2 31.2
1 31.3
B
Oligocene tuffs
Jurassic metasedimentary rocks
C
28.6
29.3
24.9
25.1
25.3
29.9
30.1
28.8
30.4
31.0
31.2
31.3

Figure 11.56 (*A*) Shaded boxes indicate presence of tuff in paleovalley segments. Ages are in millions of years. (*B*) Northern margin (dashed line) of paleovalley, Nightingale Range (looking west). (*C*) Schematic drawing of paleovalley fill at Dogskin Mountain; ages correspond to individual tuff deposits. [From Faulds, Henry, and Hinz, 2005, Figure 2, p. 506.]

A good example of the analysis by Faulds et al. (2005) is strike-slip faulting along the Pyramid Lake fault (Figure 11.57). On the east side of this fault, within the Nightingale Range, they mapped a 10-km-wide paleovalley marked by an abrupt northern margin. The paleovalley is filled with a 700+ m sequence of tuff (subdivided into 11 map units!), which pinches out abruptly against the northern paleovalley wall. On the west side of the Pyramid Lake fault, in the Virginia Mountains, they mapped the offset counterpart: a 600-m-thick stack of 15 ash units, with strong similarity to what was found in the Nightingale Range. On the basis of this match, they determined that right-handed strike-slip offset along the Pyramid Lake fault is at least 5 km and perhaps as great as 10 km (see Figure 11.57).

Both Faulds (2005) and Wesnousky (2005b) concluded that the pattern of faulting in the northern Walker Lane corresponds with the properties of Riedel shearing (see Chapter 6, *Faults*), on a grand scale. The northwest-trending,

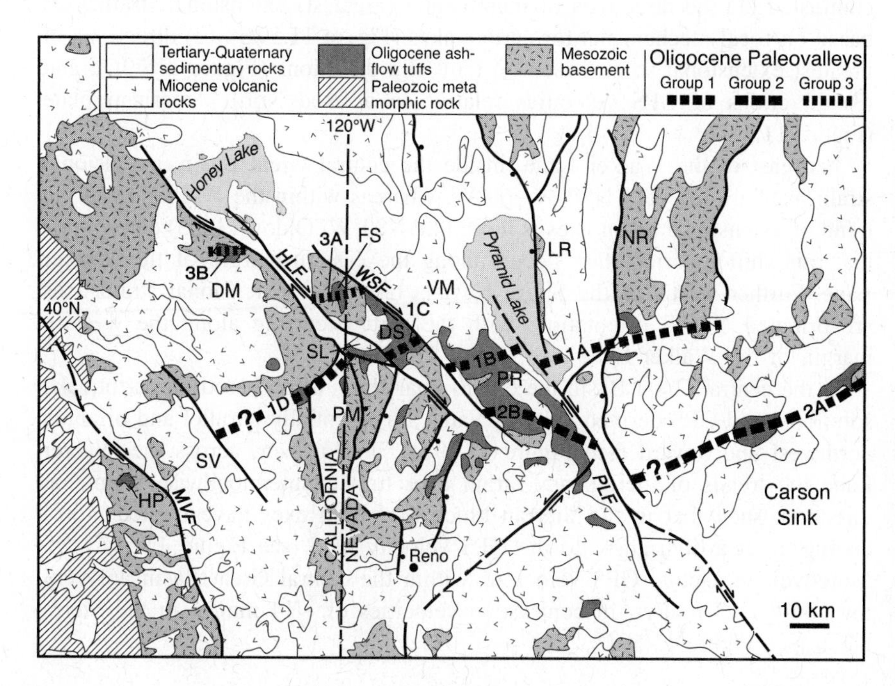

Figure 11.57 Northern Walker Lane showing left-stepping NW-striking dextral faults and offset paleovalley axes. Paleovalleys are grouped into three sets. Individual segments in each set are inferred to represent originally continuous paleovalley. Abbreviations: DM = Diamond Mountains; DS = Dogskin Mountain; FS = Fort Sage Mountains; HLF = Honey Lake fault; HP = Haskell Peak; LR = Lake Range; MVF = Mohawk Valley fault; NR = Nightingale Range; PLF = Pyramid Lake fault; PM = Peterson Mountain; PR = Pah Rah Range; SL = Seven Lakes Mountain; SV = Sierra Valley; VM = Virginia Mountains; WSF = Warm Springs Valley fault. [From Faulds, Henry, and Hinz, 2005, Figure 3, p. 506.]

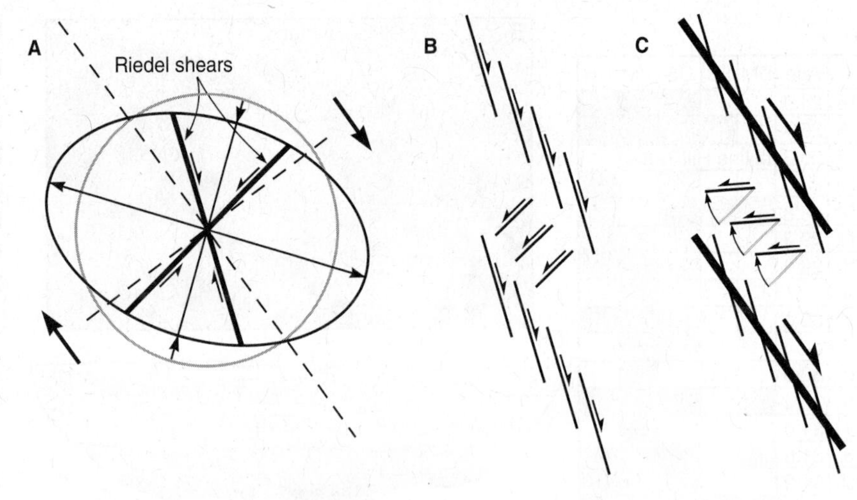

Figure 11.58 Riedel interpretation of Walker Lane fault pattern. (*A*) Image showing strain (see ellipse) produced at a point in time by right-handed regional shear couple. Two Riedel shear sets are produced: R, which is right-handed (and synthetic with respect to regional shear), and R′, which is left-handed (and antithetic with respect to regional shear). (*B*) Left-handed R′ shears forming within the stepover region of zones of R shears. (*C*) Rotation of the R′ shears and coalescence of R shears along main shear zones. [From Wesnousky, S. G., 2005b, Figure 35.]

left-stepping right-handed strike-slip faults are the R shears; the east-northeast-trending, right-stepping left-handed strike-slip faults are the R′ shears (Figure 11.58). Wesnousky (2005b) points out that more-and-more slip along the R-shears results in clockwise rotation of the R′ shears (see Figure 11.58). We will see that Riedel shearing may be one of the ways in which strike-slip fault patterns get their start.

Strain and Displacement Analysis

The fault pattern in Walker Lane makes perfect sense when it is recognized that it is accommodating transtension and transpression. We recall that Unruh et al. (2003) recognized this, identifying regions marked by transpression, transtension, or crustal thinning. John Oldow (2003) reached similar conclusions. The region he framed for analysis is shown in Figure 11.59*A*, which includes the Eastern California shear zone (ECSZ), the Walker Lane (WL), the Central Nevada seismic belt (CNSB), and the central Great Basin (CGB). The density of earthquake activity of this region underscores its active-tectonic nature (Figure 11.59*A*,*B*). Within this region Oldow (2003) compared and contrasted (1) the directions of incremental (greatest) extensional strain axes based on focal mechanisms for earthquakes (Figure 11.59*B*), (2) directions of greatest extensional strain based on fault slip directions (Figure 11.59*C*), and (3) directions of GPS velocities relative to a fixed North American plate (Figure 11.59*D*).

Just east of the Walker Lane, inside the central Great Basin, extensional strain axes are oriented N50°W–60°W, whereas within the Walker Lane the trend of extensional strain axes is more like N80°W. Oldow (2003) determined that this shift in orientation occurs along the eastern margin of the Walker Lane. Further west into the Sierra Nevada block, the extensional strain axes are oriented N75°E orientation, with the shift occurring along the western margin of the Walker Lane (see Figure 11.59*C*).

In the central Great Basin the extensional strain axes (based on earthquake solutions) and GPS-derived velocity paths are essentially parallel and oriented northwest and toward (see Figure 11.59*C*, *D*). However, across the Walker Lane the trends of incremental strain axes turn counterclockwise from that direction, such that in the Sierran block the strain axes have rotated ~50°, ending up at a high angle to the GPS velocity field (see Figure 11.59*C*, *D*). Moreover, the actual GPS velocities within the central Great Basin are very low, only 2–3 mm/yr, in contrast to velocities of ~14 mm/yr in the Sierra block (see Figure 11.59*D*).

Lesson Learned #19: Just as strain analysis of the ancient record discriminates between instantaneous versus finite strain, so too can this be done in active settings, for the instantaneous records are provided by GPS and earthquake mechanisms, and the finite strain record is provided by the full 3D attributes of the fault system.

Oldow (2003) put all of this together and concluded that the Eastern California shear zone and the Walker Lane, taken together, are a transtensile boundary zone between the Sierra Nevada and the Great Basin. Within this boundary the crust is not deforming by plane strain, but is deforming in a 3D nonplane strain manner marked by constriction. This active constriction is being achieved through extension-dominated transtension in the western part of the Walker Lane, and wrench-dominated transtension in the eastern part (Figure 11.59*E*). The direction of movement of the Sierran block is not quite parallel to the eastern margin of the block, but rather it points a bit counterclockwise from this margin. This difference accounts for what is producing transtension. If we go way back to Figure 11.2*A*, we see that it beautifully complements the conclusions reached regarding active tectonics of the Walker Lane. It shows "*the tectonic flow*" of the ESCZ/WL in a GPS velocity-vector perspective that is broad enough to include the Basin and Range on the east, the Sierran microplate and the SAF system to the west, and the northwest 'corner' of the continental United States.

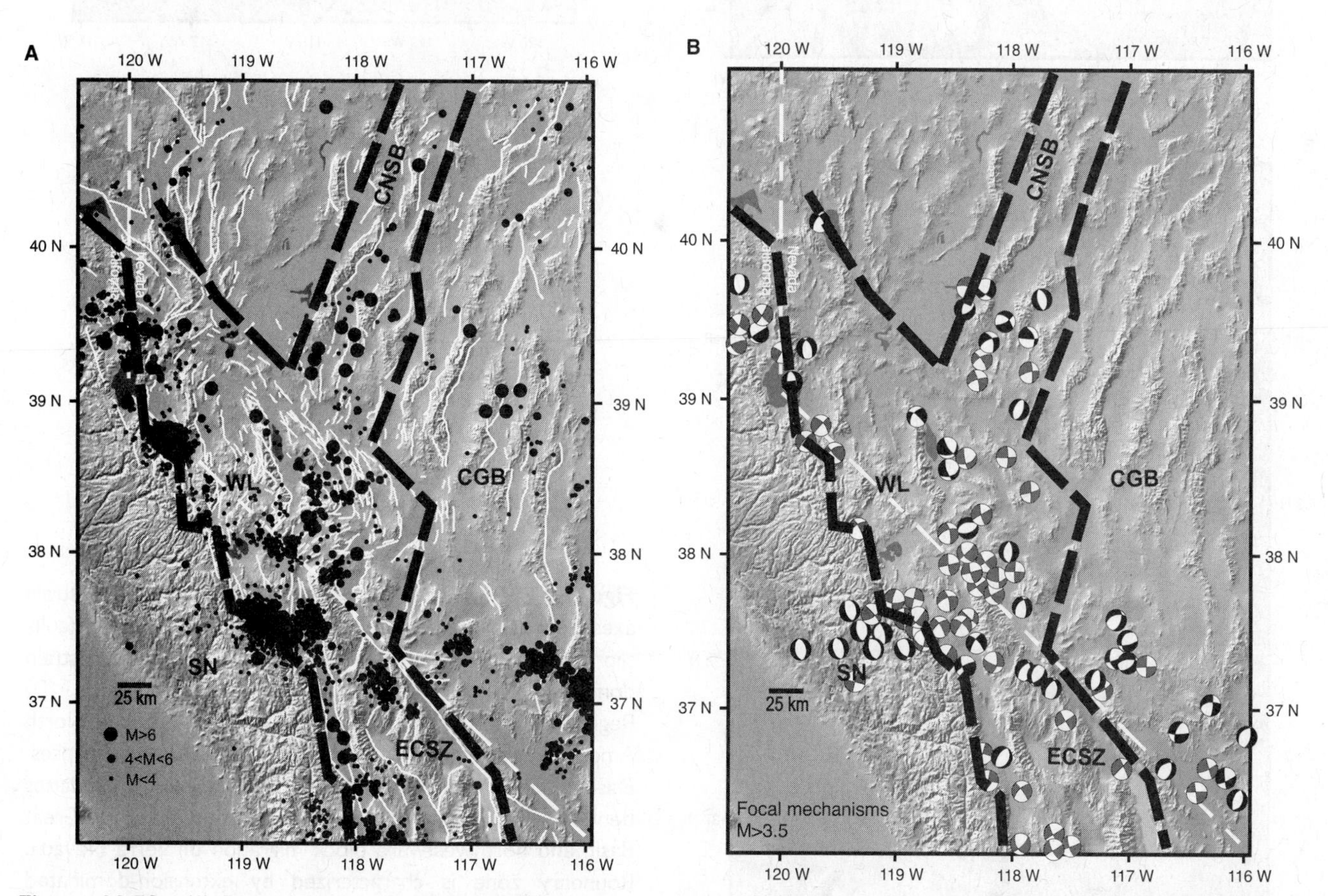

Figure 11.59 (*A*) Digital topography of Sierra Nevada and western Great Basin showing tectonic domains (dashed lines) and seismicity, 1976–2000 for M < 6.0 and 1850–2000 for M > 6.0. CNSB = central Nevada seismic belt; CGB = central Great Basin; WL = Walker Lane; SN = Sierra Nevada; ECSZ = eastern California shear zone. Late Cretaceous faults shown in white. (*B*) Focal mechanisms for M > 3.5 earthquakes.

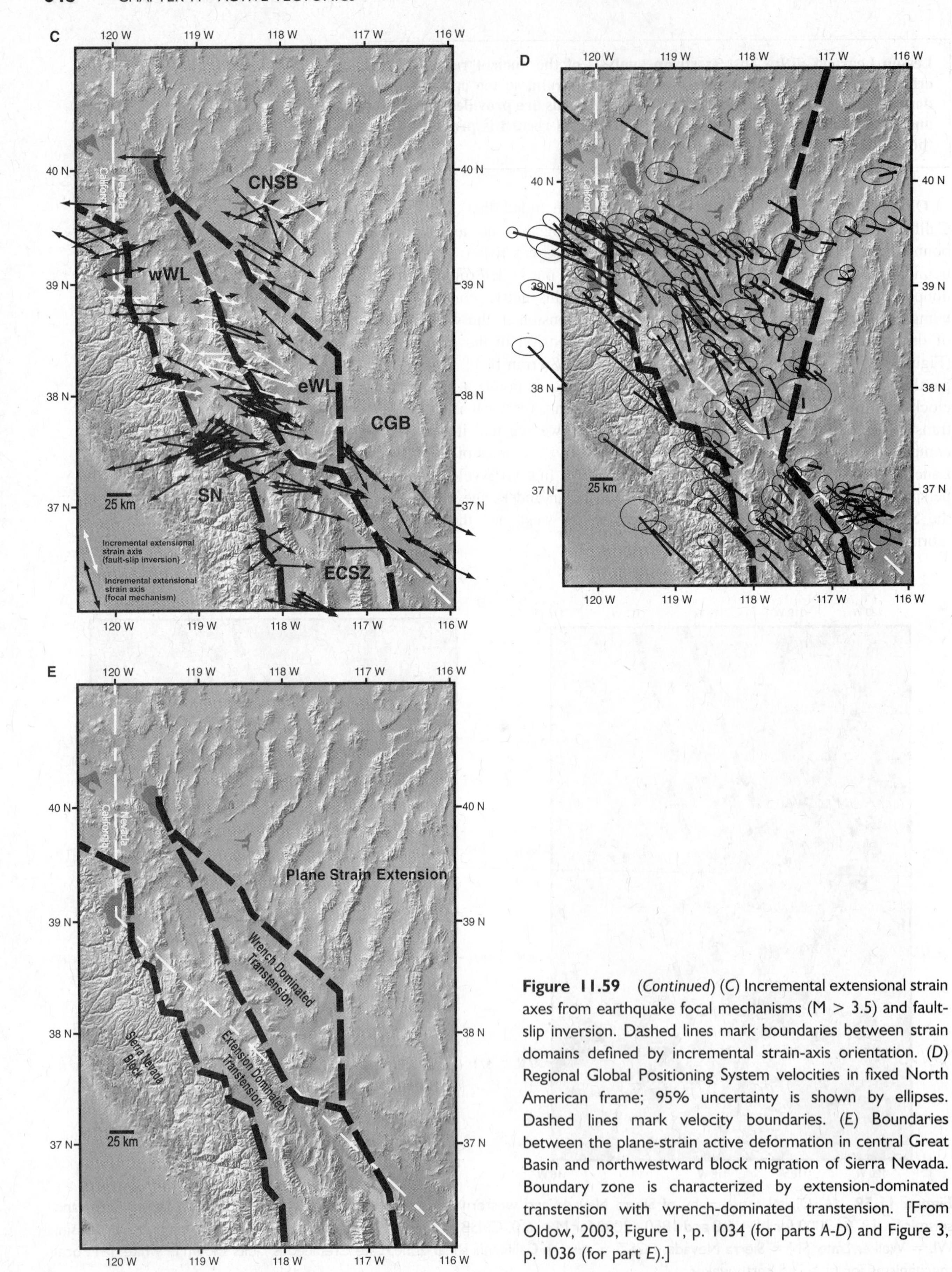

Figure 11.59 *(Continued)* (*C*) Incremental extensional strain axes from earthquake focal mechanisms (M > 3.5) and fault-slip inversion. Dashed lines mark boundaries between strain domains defined by incremental strain-axis orientation. (*D*) Regional Global Positioning System velocities in fixed North American frame; 95% uncertainty is shown by ellipses. Dashed lines mark velocity boundaries. (*E*) Boundaries between the plane-strain active deformation in central Great Basin and northwestward block migration of Sierra Nevada. Boundary zone is characterized by extension-dominated transtension with wrench-dominated transtension. [From Oldow, 2003, Figure 1, p. 1034 (for parts *A-D*) and Figure 3, p. 1036 (for part *E*).]

Lesson Learned #20: Transpression and transtension both represent nonplane, non-coaxial strain. Examples in the structural geological literature of nonplane, non-coaxial strain are typically discussed with respect to outcrop-scale structures. Through studying active deformation we become aware that there are huge regions marked by nonplane, non-coaxial strain, and some of these are transition zones between regions marked by "*conventional*" coaxial plane strain. This knowledge motivates us to map regional tracts of the ancient record in a way that more precisely underscores domains of plane versus nonplane strain. Furthermore, it motivates us to elevate fault-system data into the integrated regional strain picture.

Special Significance of the Walker Lane

Decades ago Tanya Atwater (1970) demonstrated that the San Andreas transform fault has been propagating northward since its inception nearly 30 Ma (Atwater and Stock, 1998). The "*inboard*" counterpart of this is the northward propagation of the Eastern California shear zone together with the Walker Lane. Within the ECSZ/WL system, there are greater amounts of offset toward the south, and no offset at its northwest terminus. Of the ~48–60 km of dextral offset accomplished by the ECSZ/WL system during the last 27 my, most has been achieved since 10–15 Ma (Cashman and Fontaine, 2000; Surpless, 2008). This picture highlights the fact that the Walker Lane is an especially youthful, immature transform fault system. This is the reason that the fault pattern within the Walker Lane is so diffuse and lacking single giant faults of the type we see within the San Andreas fault system, or even within the Eastern California shear zone.

To be sure, Faulds et al. (2005) emphasized that the Walker Lane is the least developed and youngest part of the transform boundary between the Pacific plate and the North American plate. They regard the Walker Lane to be the perfect laboratory within which to study the birth of transform faults within continental crust.

Lesson Learned #21: For the longest time (in pre-plate tectonic murk) the expression "*Walker Lane*" was tossed around in vague reference to a regionally extensive right-handed strike-slip fault zone in Nevada. During a time in which some geologists were writing elaborate regional tectonic interpretations of regional alignments and lineaments (think about the word "*Lane*"), there was a lot of skepticism regarding whether a Walker Lane strike-slip fault zone even existed!! Who would have guessed that such a zone would prove to be absolutely necessary to explain part of the budget deficit between NA/P plate motion and the slip along the San Andreas fault? Who would have imagined that such a diffuse zone of start-and-stop faulting (left-handed, right-handed, and normal faults) would be in any way effective in creating tectonic shifts? When complexities such as this become so evident in 'well exposed' active tectonic settings, we enter the ancient tectonic settings quite humbled indeed, but hopeful.

RELATION TO CASCADIA, AMERICAN NORTHWEST

Side Trip to Puget Lowland, Washington

Though tempted immediately to charge eastward across the Great Basin to the Wasatch Front, we will instead travel "*north by northwest*" to Cascadia, which is located in the northwesternmost corner of the continental United States. During this excursion we will discover that when it comes to active deformation of large regions, *all* motions are interrelated! Furthermore, we will add thrust faulting and thrust-related folding to the varieties of structural systems we have been 'collecting' since our journey began along the northern segment of the San Andreas fault.

Plate Setting of Cascadia

The first-order plate tectonic setting of the American Northwest is not a transform boundary. Instead, the North American plate and the Juan de Fuca plate (to the west) meet along a convergent plate-tectonic boundary (Figure 11.60). With respect to a stable NA plate, the Juan de Fuca plate is converging N60°E toward NA at a velocity of 36 mmyr^{-1} (Sherrod et al. 2004). Based on the plate tectonic configuration of Cascadia, we might expect that any and all faulting and folding within the Cascadia forearc would be aligned parallel to the Cascadia trench (i.e., N10°W), or perhaps along a N30°W trend perpendicular to the direction of convergence. However, there is more to the story.

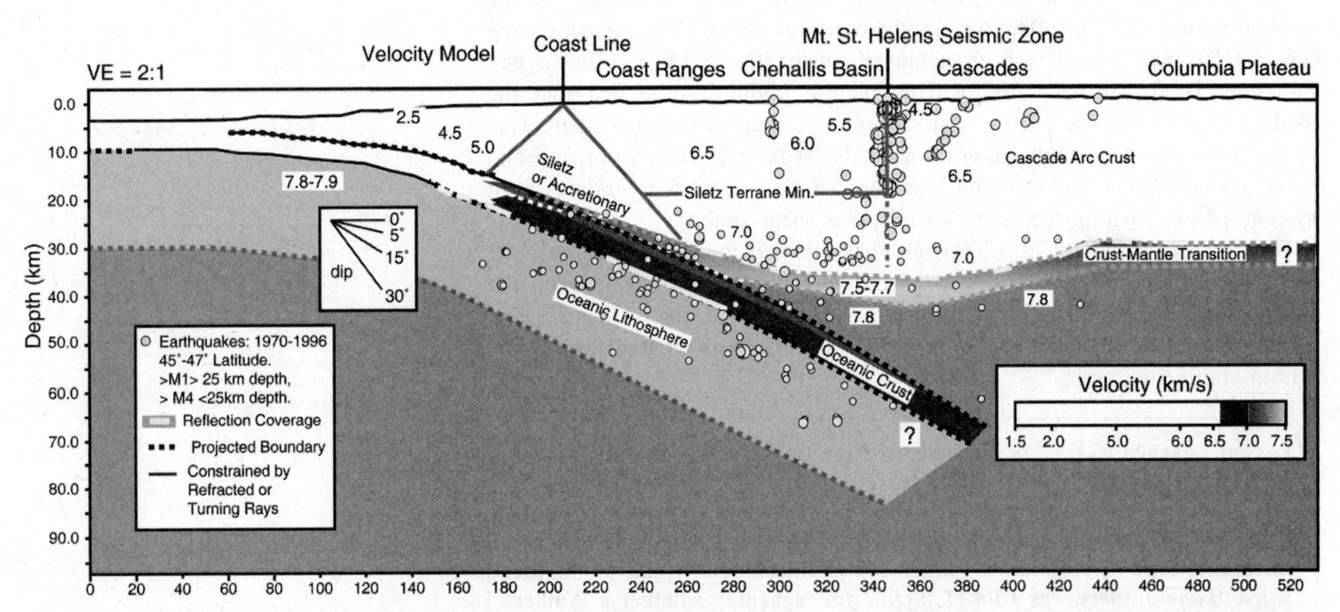

Figure 11.60 Cross-section of Cascadia, based on seismic transect by Parsons et al. (1998). The "*stratigraphy*" of the lithosphere is based on the seismic velocities determined during this experiment; velocities are given in terms of km/sec. Note the handy "*dip inclination*" scale to track the descent of the down-going Juan de Fuca slab. See text for explanation. [From Parsons et al. 1998, Figure 2B, p. 200.]

As we have seen, the Sierran (S) microplate is moving ~11 mm/yr northwesterly (Figure 11.61) (Wells et al. 1998; Sherrod et al. 2004). In so doing, it applies end-on pressure to the Cascadia forearc, including the Oregon Coast Ranges. Just north of the Oregon Coast Ranges is the Puget Lowland region (Washington), which contains the urban centers of Seattle and Tacoma. To the north of the Puget Lowland lies the Canadian Coast Ranges (British Columbia). Because the Canadian Coast Ranges form a resistant backstop, the Puget Lowland is caught in a 'vice' closing from the south (see Figure 11.61) (Wells et al. 1998; ten Brink et al. 2002).

Lesson Learned #22: Even though the plate tectonic setting of Cascadia appears to be perfectly "*Andean*," marked by ocean/continent convergence/subduction, we see that the actual kinematics are being strongly influenced by forces generated along an entirely different plate boundary (the San Andreas transform), and then coming around the 'back door.' In reconstructing the ancient plate tectonic settings that existed at the time of formation of ancient systems of regional structures, we need to resist paying attention only to the nearest, largest plate boundary, and do our best to reconstruct the network of all potentially relevant plate boundaries in a full 4D manner.

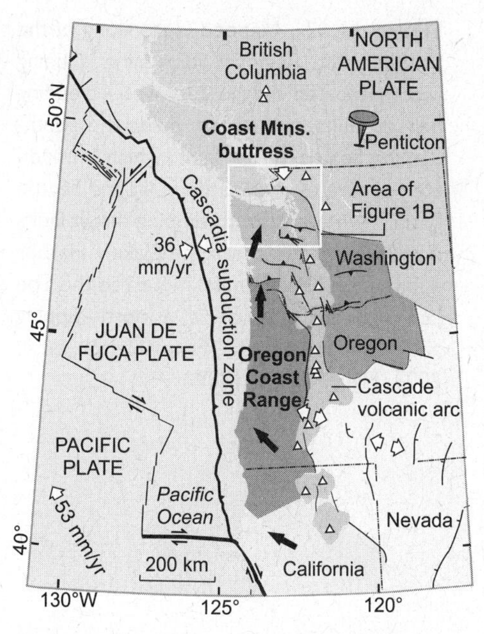

Figure 11.61 Tectonic setting of Cascadia. Juan de Fuca plate is converging N60°E at a velocity of ~36 mm/yr. Western Washington region is experiencing transpression. Puget Lowland, including Seattle and Tacoma, is experiencing north-south shortening (manifest in folds and reverse faults), caught between the northerly movement of the Sierra plate and the Coast Mountains 'buttress' in British Columbia. Bold arrows indicate motions of tectonic blocks inferred from geologic and geodetic data. [From Sherrod et al. 2004, Figure 1, p. 9.]

Major Active Structures in the Puget Lowland

We thus see that the 'prime mover' of crustal deformation in the Puget Lowland is not the Juan de Fuca plate but rather the Sierran microplate. The Puget Sound Lowland is experiencing north-south shortening at a rate of ~4 to ~7 mm/yr (Figure 11.62) (Johnson et al. 1999; Nelson et al. 2003; Unruh et al. 2003). The dominant structures are active east-west-trending thrust faults and compression-generated uplifts and basins. The uplifts and basins are roughly 30–50 km wide (Blakely et al. 2002; Nelson et al. 2003), and on the order of 70 km in length.

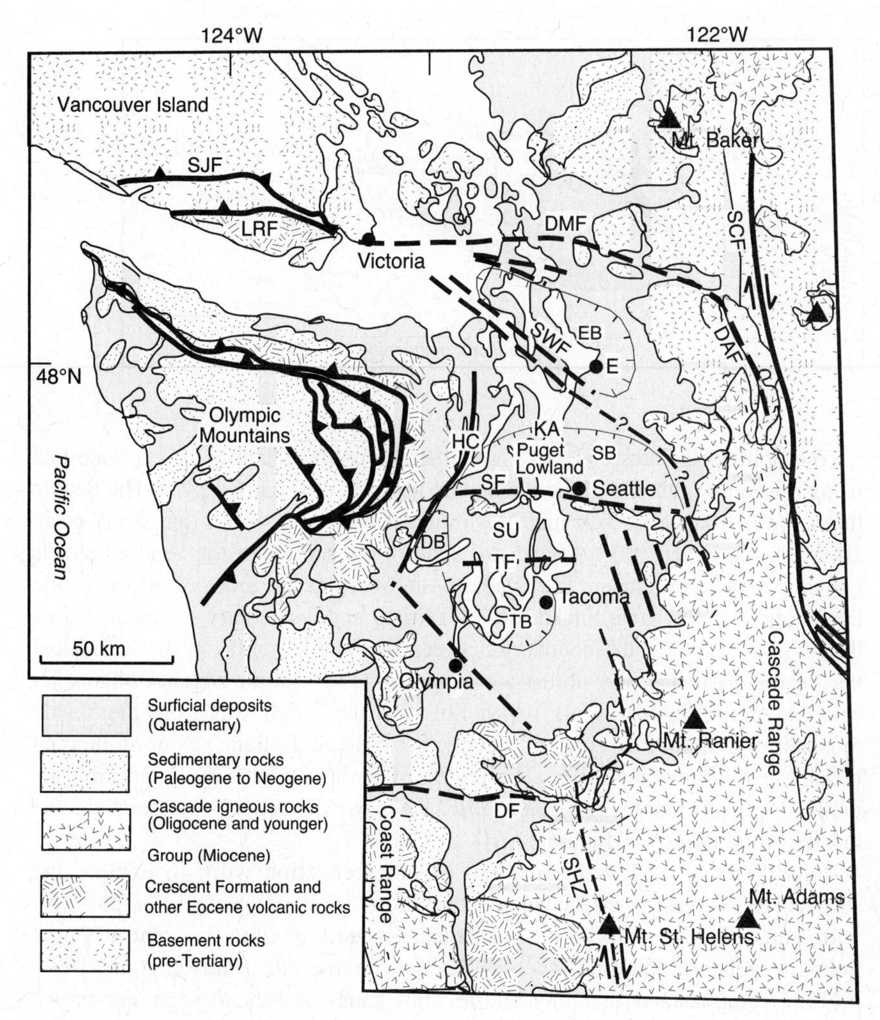

Figure 11.62 Details of structural deformation in the Puget Sounds Lowland. Abbreviations for faults: CBF = Coast Range Boundary fault; DAF = Darrington fault; DF = Doty fault; DMF = Devils Mountain fault; LRF = Leech River fault; SB = Seattle basin; SCF = Straight Creek fault; SF = Seattle fault; TF = Toe Jam fault; SJF = San Juan fault; SWF = southern Whidbey Island fault. [From Johnson et al. 1999, Figure 1, p. 1043.]

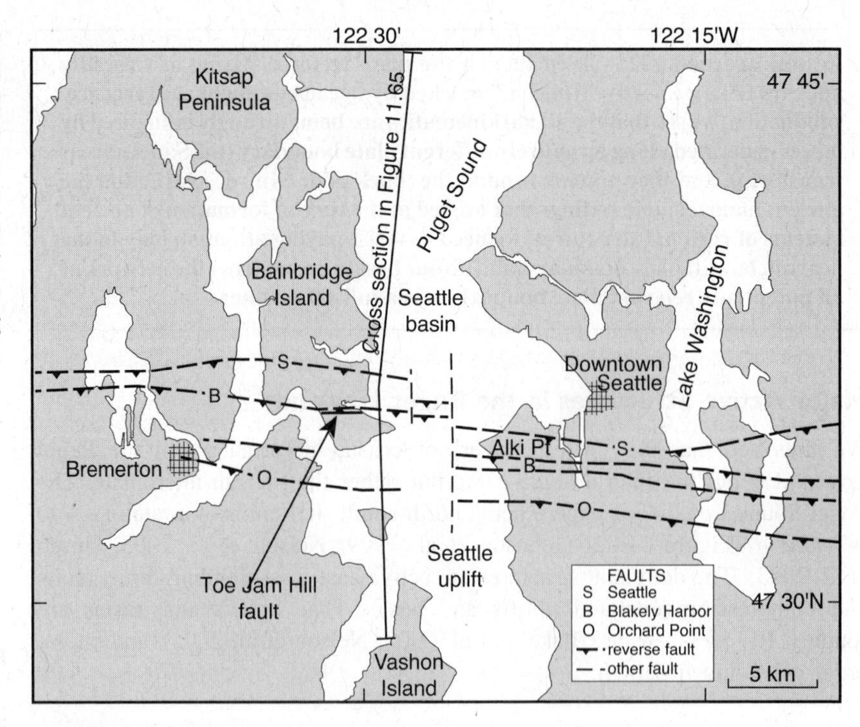

Figure 11.63 Mapped faults along of the Seattle uplift. Regional shortening is being accommodated by the east-west–trending belt of faults, the largest being the south-dipping Seattle thrust fault, which immediately separates the Seattle uplift and Seattle basin. Two other south-dipping thrust faults on the Seattle uplift are the Blakely Harbor and Orchard Point faults. Note also the Toe Jam Hill fault, which is a north-dipping antithetic thrust fault. [From Nelson et al. 2003, Figure 1B, p. 1389.]

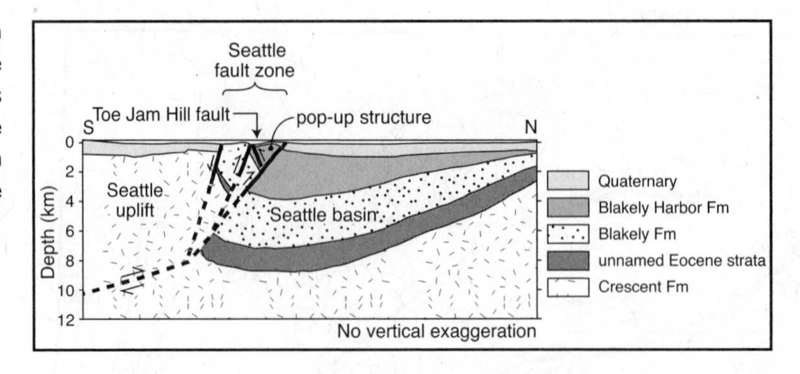

Figure 11.64 North-south cross section showing relation of the Seattle uplift to the Seattle basin. The Seattle fault zone is marked by a north-dipping monocline, the Seattle thrust, and the antithetic Toe Jam Hill fault. [From Nelson et al. 2003, Figure 1C, p. 1389. Figure 12.63.]

The uplift of greatest celebrity is the Seattle uplift, which is 40-km wide and being raised by fault slip along the Seattle thrust fault (Figure 11.63). The Seattle thrust fault strikes east-west and dips roughly 40° to 60° south (ten Brink et al. 2002; Nelson et al. 2003). At and just beneath the surface of the Seattle uplift is Eocene volcanic basement. The Seattle basin occupies the footwall of the Seattle thrust, and contains great thicknesses of Tertiary and Quaternary sediments (more than 5 km!). Right at the boundary between the Seattle uplift (on the south) and the Seattle basin (on the north) are steep (60°–80°) north-dipping Oligocene-Miocene strata (Figure 11.64), which Kelsey et al. (2008) refer to as the Seattle monocline. Within this zone of steeply north-dipping strata are north-dipping bedding-parallel reverse faults. The largest of these antithetic faults goes by the rather disgusting name "*Toe Jam Hill*" *fault*, which appears to "*root*" in the Seattle thrust fault (see Figure 11.64).

Johnson et al. (1999) carried out seismic reflection work to explore the subsurface expression of the Seattle fault zone in as much detail as possible, with particular interest in the earthquake-hazard picture for central Puget Sound in general, and cities such as Seattle in particular. Johnson et al. (1999, p. 1042) emphasized that the Seattle fault zone "*strikes through downtown Seattle in the densely populated Puget Lowland of western Washington.*" They

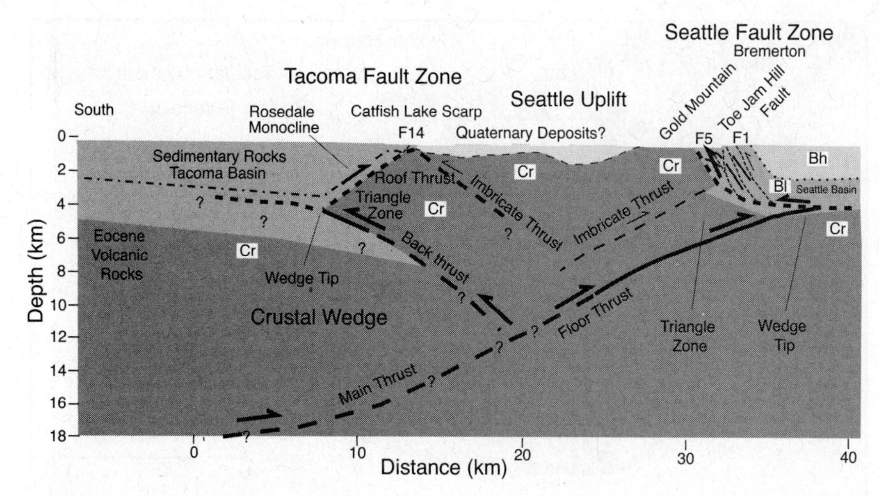

Figure 11.65 Schematic structural cross-section along the Puget Sound, showing the relationship of the Seattle uplift, the Seattle basin, and the Tacoma basin to the south-dipping Seattle thrust, the north-dipping back thrust, the Toe Jam Hill fault, and bedding-plane reverse faults. [From Brocher et al. (2004), Figure 10C, p. 1396.]

concluded that the fault zone is 4–6 km wide, with Quaternary sediments folded and faulted along 3 or 4 major faults. They interpreted the rate of slip along the Seattle thrust to be ~0.7–1.1 mm/yr.

The Seattle uplift is not small, as can be seen in subsurface interpretation by Brocher et al. (2004) (Figure 11.65). The total structural relief separating the top of the Seattle uplift to the floor of the Seattle basin is ~5 km (Kelsey et al. 2008). Blakey et al. (2002) concluded that the overall displacement along the Seattle thrust at least 7 km, and could be as much as 10 km. In their view the Seattle thrust dips moderately steeply (40° to 70°) in the upper 6 km of the crust but then flattens to an angle as shallow as 20°. This flattening takes place above a major low-angle thrust detachment, which resides at a depth range of 14 to 20 km.

Active Nature of the Major Structures

For the longest time it was not obvious to geoscientists (not to mention residents and city leaders of Seattle and Tacoma) that the Seattle thrust and associated structures in the Puget Lowland are tectonically active. Part of the reason it took so long for this to be recognized is that glacial outwash, forests, lakes, and urban infrastructure cover most of the bedrock. Moreover, major earthquakes have simply *not* been part of the history of the Puget Lowlands during the period since white settlers moved into the region. The last major earthquake to have occurred in this region was a M_W ~7 earthquake, *circa* A. D. 900–930 (Blakely et al. 2002). This was more than 100 years before the Battle of Hastings (1066 A.D.)!

Advanced technologies paved the way for productive active tectonic studies in the Puget Lowland. The technology of choice proved to be airborne laser mapping using LiDAR ("*light detection and ranging*"), which can detect fault scarp heights as low as several centimeters. Moreover, LiDAR technology is capable of seeing through the forests to the topographic relief of the ground surface, and at the finest scales. Figure 11.66 is an example of the application of airborne laser swath mapping (ALSM). The pronounced linear pattern in this image reflects the fact that the land surface in this region is marked by pervasive glacial grooves and furrows, which were carved into bedrock when glaciers advanced over this region. Doing our best to ignore the grooving and looking closely into this image we see that LiDAR picked up traces of the fault scarp associated with the Toe Jam Hill fault (see Figures 11.63 and 11.65) (Nelson et al. 2003).

The Toe Jam Hill scarp crosses the southern part of Bainbridge Island within the Puget Lowland. Knowing exactly where the scarp resides beneath

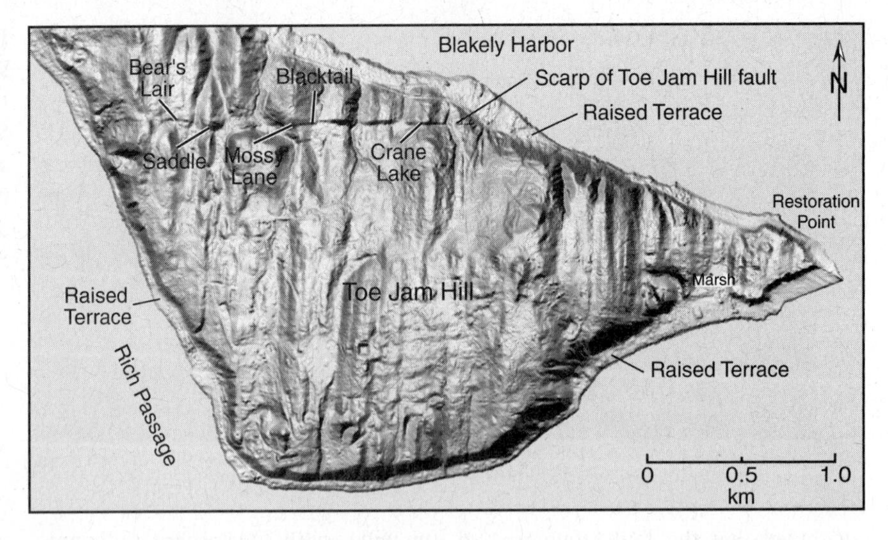

Figure 11.66 Airborne laser swath mapping (ALSM) digital elevation model of the southernmost part of Bainbridge Island, Washington. North-south lineations are glacial striations. The south-facing scarp of the Toe Jam Hill fault trends east-west across the island. The clarity of the trace of this fault permitted informed placement of trenches for paleoseismic investigations. The five trench sites are shown: Bear's Lair, Saddle, Mossy Lane, Blacktail, and Crane Lake. [From Nelson et al. 2003, Figure 2, p. 1390.]

forest canopy, it was possible for Nelson et al. (2003) to make intelligent choices regarding exactly where to carry out trenching for paleoseismological investigations. Then, based on trenching and dating, Nelson et al. (2003) discovered paleo-earthquakes, each responsible for ~1–2 m of surface folding and faulting. These earthquakes may well have been M_W ~7, corresponding with surface ruptures greater than 36 km in length.

Sherrod et al. (2004) also harnessed airborne laser mapping to identify active fault scarps (Figure 11.67). They were interested in knowing more about the Tacoma fault, an active east-west–striking, north-dipping reverse fault that passes through Tacoma, Washington. The Tacoma fault marks the southern boundary of the Seattle uplift (Figure 11.62). The airborne laser mapping permitted Sherrod et al. (2004) to see right through the forest cover in the Puget Lowland and to pick out scarps with topographic relief as tiny as 15 cm (Figure 11.67*A*); this aided site selection of paleoseismology trenches (Figure 11.67*B*). On the basis of all of this work, Sherrod et al. (2004) determined that the Tacoma fault zone contains three south-facing scarps that cut across the striated glacial landscape. The scarps could be traced to places

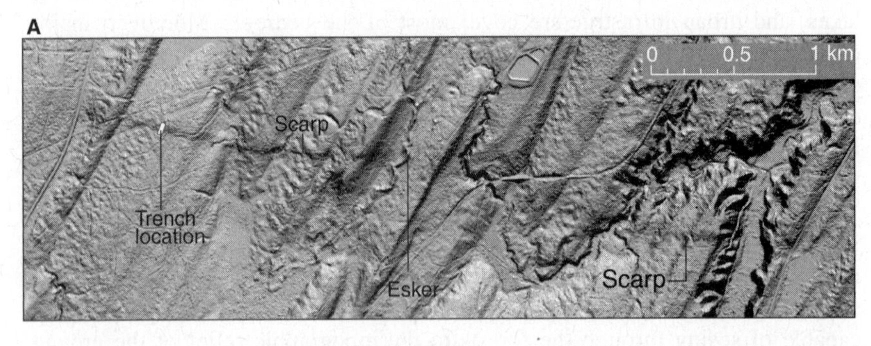

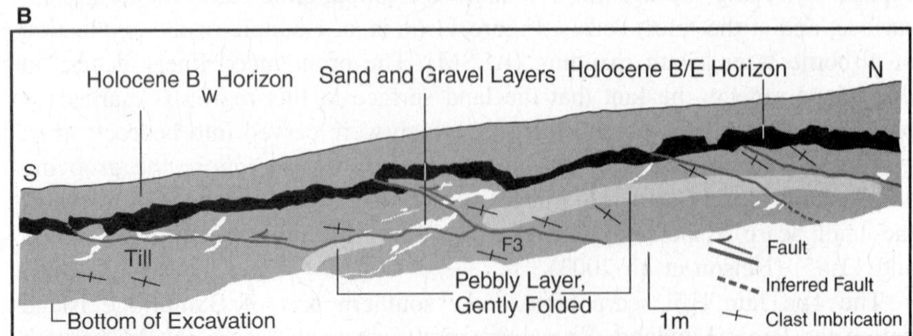

Figure 11.67 (*A*) Hill-shaded digital-elevation model of LiDAR data from area along the Tacoma fault. Lineations are northeast-trending glacial striation. Note locations of fault scarps and fault-uplifted tidal flats. (*B*) Diagram of trench wall, showing deformation by faulting and folding. [From Sherrod et al. 2004, Figure 3, p. 11.]

where deformed shorelines straddle the faults. They documented coseismic uplift of a tidal flat by at least 1.5 m. Similarly Kelsey et al. (2008) mapped a raised shore platform, which was lifted ~5–7 m during the great earthquake ~1100 years ago. In addition they recognized peculiar late Quaternary fault scarps along the Seattle monocline. These scarps were produced by reverse faulting along faults parallel to the bedding.

Lesson Learned #23: Technologies are making all the difference in discovery of the full three-dimensional characteristics of major faults in active tectonic settings. We continue to rely on earthquake focal mechanisms, seismic reflection profiling, and resourceful approaches in paleoseismology, but to these are added new and emerging advances in GPS-geodesy and remote sensing. In the ancient record various geophysical approaches are essential for *"seeing"* the geometric nature of the finished product. The active and ancient go together nicely. The ancient settings reveal the finished products. The active settings teach us about the progressive steps that lead to finished products.

What Is Really Going On in the Puget Lowland?

A really interesting picture has emerged that helps explain how the Seattle uplift, the Seattle basin, the Seattle thrust, the Tacoma thrust, the Toe Jam Hill fault, the Seattle monocline, and the bedding-parallel reverse faulting along the monocline all fit together. As is almost always the case, the picture emerged from a series of interpretations, one building on the other (ten Brink et al. 2002; Brocher et al. 2004; and Kelsey et al. 2008). The basic model is one of fault-bend folding, but with emphasis on a **triangle zone** and a **wedge tip** that form at the leading edge of the master thrust (Jones, 1996) (Figure 11.68).

The master fault has four parts: a lower flat (floor thrust), a ramp, an upper flat, and a fault tip. Fault bend folding takes place above the master thrust, and the geometry of folding conforms to the geometry of the master fault (see Figure 11.68). A key dimension of the folding is the formation of a broad angular anticline above the ramp and upper flat. Forward of the anticline is a large angular syncline (see Figure 11.68). The connecting piece between the anticline and syncline is a monocline. The axial trace of the syncline can be followed downward, where it dies out precisely at the tip of the master thrust. With each (coseismic) increment of faulting along the master thrust, bedding in the fault-bend syncline move upward out of the hinge by bedding-parallel reverse faulting up the monocline. At the top of the monocline, the antithetic reverse fault flattens into a roof thrust (see Figure 11.68).

The upper flat thrust fault and the antithetic reverse fault together define a **wedge tip** (**triangle zone**) that drives forward, earthquake-by-earthquake (see Figure 11.68). As the wedge drives forward, splay faults, antithetic to the master thrust, peal off and upward from the roof thrust.

The cross-section of the Puget Lowland region prepared by Brocher et al. (2004) and presented earlier as Figure 11.65 is based on the wedge thrust/fault-bend folding concept. They recognized that the leading edge of the Seattle thrust may well be a wedge-tip triangle zone (see Figure 11.65); and that the wedge tip itself is bounded by thrust faults of opposite movement directions. Brocher et al. (2004) interpret the presence of triangle zones at both the southern (Tacoma thrust) and northern (Seattle thrust) margins of the Seattle uplift (see Figure 11.65).

Kelsey et al. (2008) subscribe to this basic model as well (Figure 11.69), developed it further, and spelled out implications in relation to earthquake phenomena. They portray the Seattle thrust ramping upward from a depth

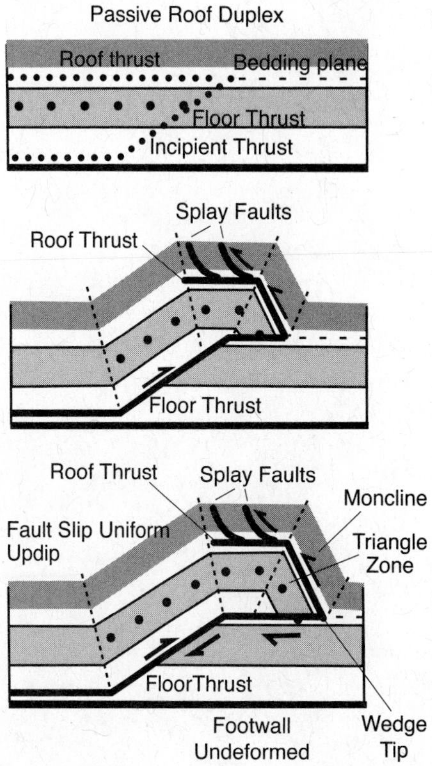

Figure 11.68 Illustration of development of passive-roof duplex. [From Brocher et al. 2004, Figure 3C, p. 1383, based principles outlined by Jones (1996).]

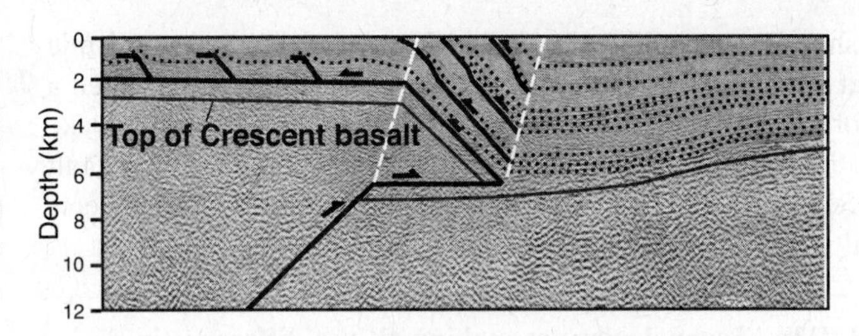

Figure 11.69 Wedge thrust model for the Seattle fault zone. Note that folds and thrusts are doubly vergent. Bedding-plane reverse faults root in the synclinal hinge. Note top of the Eocene Crescent Formation basalt. Steep dash lines mark axial traces of folds. [From Kelsey et al. (2008), Figure 9B, p. 1591.]

greater than 12 km, and then flattening at a depth of ~7 km into an upper flat. They show the Seattle thrust fault tipping out northward at the base of the Seattle monocline. The antithetic reverse fault that climbs up out of the monocline flattens as a roof thrust at a depth of 2 km into a roof thrust (see Figure 11.69). Fault-bend folding propels this reverse faulting.

Large-magnitude ***"folding earthquakes,"*** as large as M_W 6, can occur as a result of this tectonic wedging (Kelsey et al. 2008). It works this way. The coseismic slip on the master thrust results in growth of the fault-bend fold at the tip of the wedge. As the folding takes place, slip occurs on flexural-slip bedding plane surfaces. In some cases the coseismic activity that accompanies this slip occurs simultaneously with the coseismic activity that triggered it on the master thrust. But, according to Kelsey et al. (2008), the folding earthquake may be delayed, a matter of days, or years, or thousands of years. Thus folding earthquakes constitute yet another seismic hazard, in the form of $M_W\sim$ 6 events generated at shallow depth (less than 6 km).

> **Lesson Learned #24: Active fault-bend folding in the Puget Lowland region has proceeded to a special stage where we can witness the peculiar geometries and kinematics of a wedge tip driving forward, earthquake-by-earthquake, to accomplish shortening. This action, which we see in ancient systems, is the crowding of strata in the hinge of a syncline, forcing strata there to *"flow"* by bedding-parallel slip out of the syncline and move toward the hinge of the adjacent anticline. We learn in Cascadia that this action produces 'folding' earthquakes from time to time. Insight into the interaction of faulting, folding, and stick-slip earthquake behavior is being uncovered by the urgent need to mitigate loss of life and property due to earthquake hazards in Cascadia. What is learned will be immensely beneficial in recognizing comparable systems, both large and small, in the ancient record.**

THE BASIN AND RANGE

Returning now to Walker Lane, we will head eastward across the Great Basin in order to examine the ways in which active faulting is being expressed.

Geographic and Tectonic Divisions

The Basin and Range province can be subdivided into three provinces based on active strain: the Western Great Basin province (200- to 500-km wide), the Central Great Basin province (500-km wide), and the Eastern Great Basin province (100-km wide) (Figure 11.70). As we have learned, the Western Great Basin is bordered on the west by the combination of the Walker Lane and Eastern

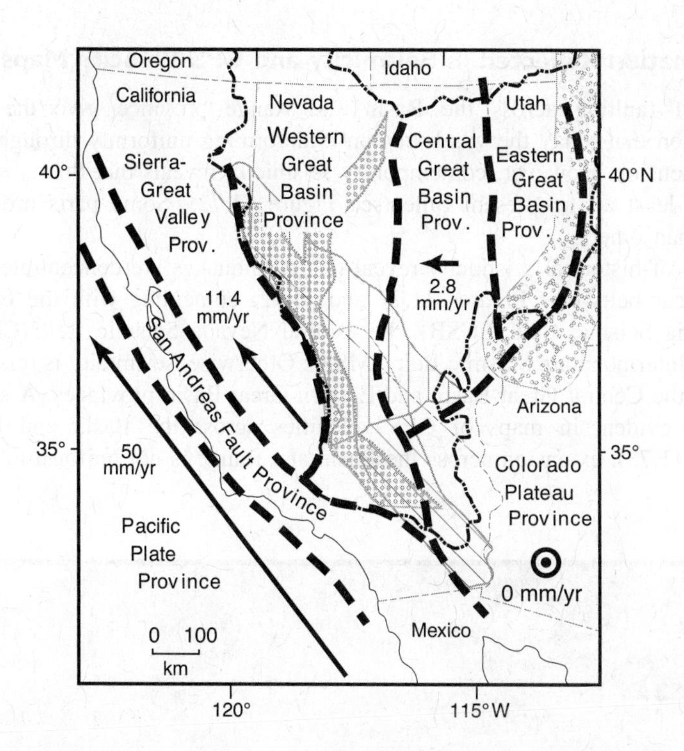

Figure 11.70 Geodetic provinces of the Southwestern United States. Note in particular the geography of the Basin and Range Province relative to the Sierra–Great Valley Province and the San Andreas Fault Province. The Basin and Range Province is itself composed of three subprovinces: the Western Great Basin, the Central Great Basin, and the Eastern Great Basin. [Adapted from Bennett et al. 2003, Figure 11.]

California shear zone. The Eastern Great Basin borders the Colorado Plateau (see Figure 11.70). The great breadth of distributed normal faulting within the Basin and Range province is really quite extraordinary, making this province a classic laboratory for studying extensional tectonics. Machette et al. (1996–2004) report that there are ~750 faults in the Great Basin that have been active in the Quaternary; and that approximately 320 of these faults have been active within the past 130,000 years. Yet in the past 150 years only 14 earthquakes in the Basin and Range province have produced surface ruptures (Figure 11.71).

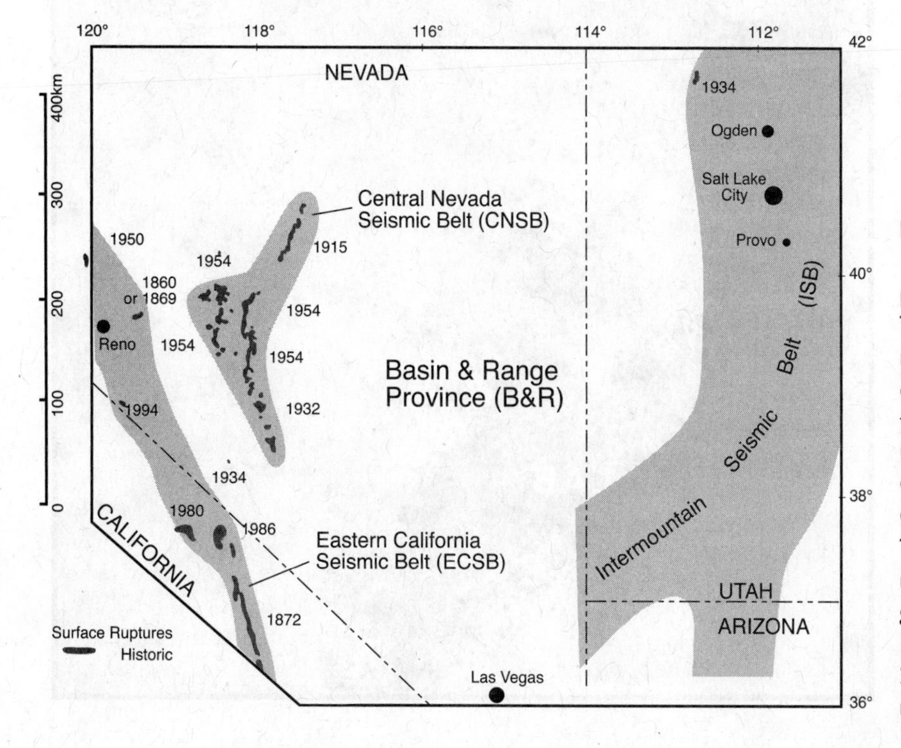

Figure 11.71 In the past 150 years, only 14 earthquakes in the Basin and Range province have produced ground rupture. This map shows the traces of the surface rupture, and the years in which surface-damaging earthquakes occurred. Note that these historic ground-rupturing earthquakes have occurred exclusively within the Eastern California Seismic Belt (ECSB) and the Central Nevada Seismic Belt (CNSB). The Intermountain Seismic Belt (IMSB) is nearly devoid any earthquake-induced ground rupture in the past 150 years. [Adapted from Machette et al. 1996–2004, Figure 3, http://earthquake.usgs.gov/regional/imw/imw_bnr_faults/.]

Deformation Reflected in Seismicity and GPS Velocity Maps

Maps of faulting across the Basin and Range province give the false impression that today the whole region is deforming uniformly through east-west extension. However, contemporary seismicity reveals that this is *not* the case, at least at the present time (see Figure 11.72). Some parts are more active than others.

Maps of historical seismicity reveal that earthquakes are concentrated into long linear belts (see Figures 11.71 and 11.72) coinciding with the Eastern California Seismic Belt (ECSB), the Central Nevada Seismic Belt (CNSB), and the Intermountain Seismic Belt (IMSB). Otherwise, seismicity is relatively light in the Central Great Basin and Eastern Great Basin provinces. A similar story is evident in maps of GPS velocities across the Basin and Range (Figure 11.73): extension across the Basin and Range is not uniform in either

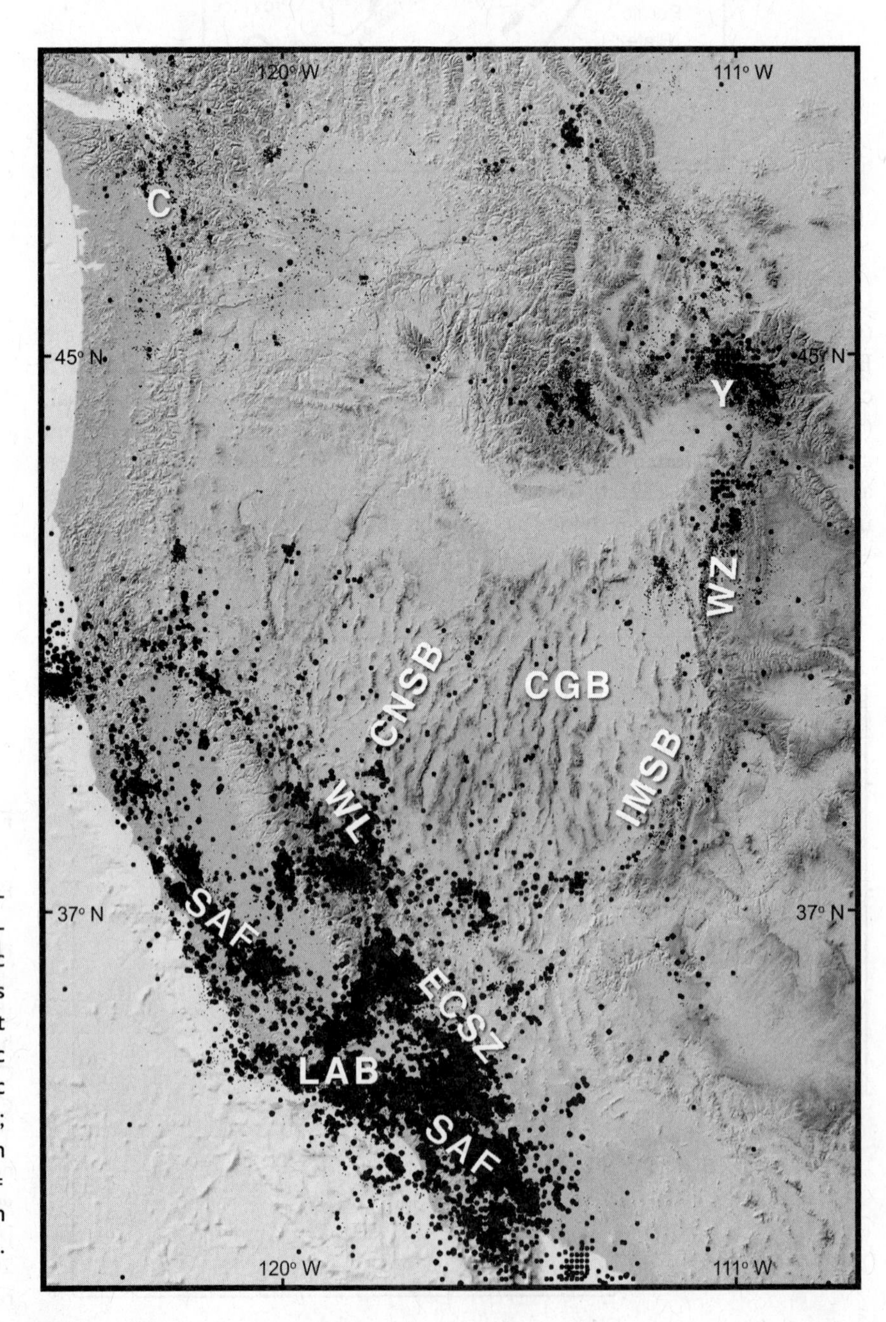

Figure 11.72 Physiography of the western U.S. Cordillera with earthquake epicenters M>4 [Advanced National Seismic System, ANSS.] Earthquake epicenters shown as black dots. CGB = Central Great Basin; CNSB = Central Nevada Seismic Belt; ECSZ = Eastern California Seismic Zone; IMSB = Intermountain Seismic Belt; LAB = Los Angeles Basin SAF = San Andreas Fault; WL = Walker Lane; WZ = Wasatch Zone; Y = Yellowstone. [From Oldow and Singleton, 2008, Figure 1, p. 538.]

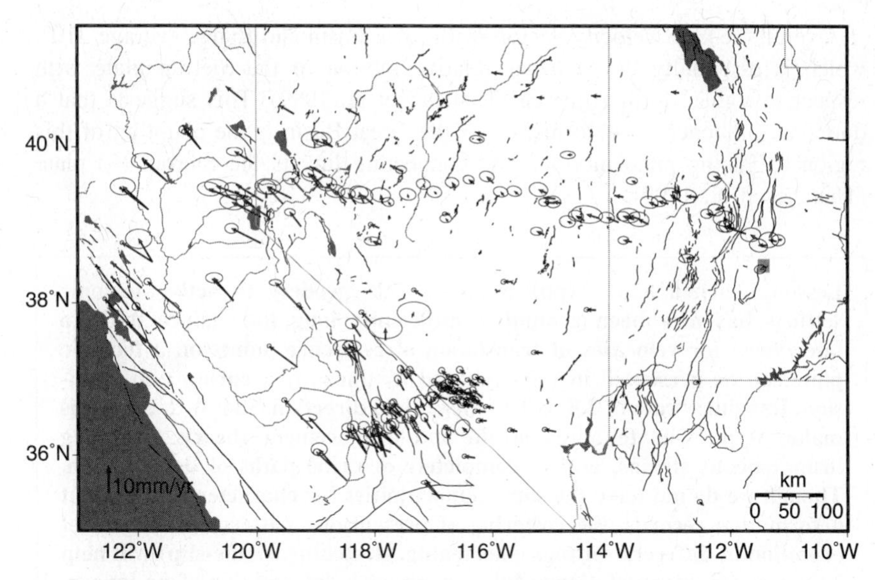

Figure 11.73 Map of GPS velocities of sites in the Sierra Nevada-Great Valley microplate, northern Basin and Range, and Colorado Plateau regions with respect to North America. Error ellipses represent the 95% confidence level. Deformation is evident as far east as the Wasatch fault zone. Also shown are selected Quaternary faults. [From Bennett et al. 2003, Figure 2.]

magnitude or direction (Thatcher et al. 1999; Bennett et al. 2003). Based on GPS data, extension within the Central Great Basin region is negligible (see Figure 11.73), revealing that this subregion is moving as a single coherent province relative to the Colorado Plateau. The path of movement of the Central Great Basin (relative to the Colorado Plateau) is N84°W $\pm 5°$, with a rate is 2.8 ± 0.2 mm/yr (see Figure 11.3) (Bennett et al. 2003).

Lesson Learned #25: The Basin and Range province appears to have deformed homogeneously and uniformly. But the record of the distribution and orientation of faults today within this region says nothing about the tempo and distribution of fault formation through time. In ancient structure systems we may have discovered this to be true for fold-thrust belts, which seem to move their way across regions, almost as a wave. We are not generally accustomed to thinking about faulting starting-and-stopping from place to place and from time and time within a broad tectonic province, creating a chaotic time/place pattern marked by disorder. But at the heart of understanding earthquakes and active or ancient faulting, we must think in these terms.

Velocity Vectors, Great Basin

The velocity and direction of westward movement of the Central Great Basin are entirely consistent with the active west-northwest extension that is documented for the Eastern Great Basin and the Wasatch fault zone (Thatcher et al. 1999). The east-west extension in the Eastern Great Basin is reflected by north-northeast-trending normal faults. This can all be seen in the combination of GPS velocity vectors and traces of normal faults mapped along the boundary zone between the Colorado Plateau and Eastern Great Basin (see Figure 11.73).

Based on our examination of the Walker Lane, we can anticipate that active deformation in the Western Great Basin is quite different from what is observed in the Eastern Great Basin (see Figure 11.73). First of all, the direction of extension swings more northwestward, and the fault traces swing more northeastward (Bennett et al. 2003). Velocities pick up as well, with the highest velocities ($\sim$3.7 mm/yr) being measured across the Central Nevada Seismic Belt (Thatcher et al. 1999) (see Figures 11.71 and 11.73).

Overall the GPS velocity vectors within the Basin and Range average 310°, which is essentially the same as relative motion of the Sierran plate with respect to stable North America (Thatcher et al. 1999). This suggests that a dominant influence on deformation of the Great Basin is the coupling of this region to Pacific plate motion. The connecting link is the Sierran (S) plate (Thatcher et al. 1999).

Lesson Learned #26: Application of GPS geodesy to active tectonic settings has now been around a sufficiently long time (since mid- to late-1990s) for velocities of translation of reference points on a tectonic plate can be presented in ways marked by uncanny accuracy and precision. Imagine: "rate of 2.8 ± 0.2 mm/yr, in a direction N84°W ± 5°." This makes it possible to apply strain and displacement theory, including instantaneous strains, and to completely describe paths of deformation. Though we do not have the same opportunities for characterizing ancient deformation records, the rostering of present-day patterns of structural evolution in different settings (shortening, stretching, strike-slip) will help improve our interpretations of the progressive deformation of no-longer-active tectonic provinces.

Central Nevada Seismic Belt

The Central Nevada Seismic Belt (CNSB), which is distinguished by a diffuse zone of earthquakes fully 300 km long, crosses northeastward from the Western Great Basin to the Eastern Great Basin (Figure 11.71). Seven major earthquakes took place within it between 1915 and 1954. We see four of these in Figure 11.74, a map showing (from north to south) the epicenters of the Pleasant Valley earthquake (1915, M_W 7.6), the Dixie Valley earthquake (1954, M_W 6.8), the Fairview Peak earthquake (1954, M_W 7.2), and the Cedar Mountain earthquake (1932, M_W 7.2). The focal mechanisms for these earthquakes underscore the differences between the active deformation operating in each of these provinces (Gourmelen and Amelung, 2005). Toward the north and east, the earthquakes conform to pure normal faulting and east-west extension. Toward the south and west, the earthquakes are strike-slip. In between, they tend to be hybrid, partly normal, partly strike-slip (see Figure 11.74).

The Pleasant Valley earthquake was a purely normal fault earthquake and produced scarps up to 4 m in height. The Fairview Peak earthquake was marked by oblique slip, combining right-handed and normal slip. The vertical component of offset was 3.8 m, and the dextral component was 2.9 m. The Cedar Mountain earthquake was pure strike-slip, with 6 to 12 m of right-handed strike slip offset (see Figure 11.74).

The Central Nevada Seismic Belt (CNSB) in general, and the focal mechanisms for these three earthquakes, provides living proof of the shift in strain and displacement westward from the Great Basin into the Walker Lane.

Lesson Learned #27: We observe in regions of active faulting today that there can be progressive, substantial shifts in the type of faulting (i.e., normal vs. oblique vs. strike slip) along a belt of deformation. Thus once again we must resist assuming that if we capture the fault geometry and slip direction in one part of the belt we can then conclude that this will hold for the belt as a whole. This kind of assumption does not necessarily hold up where we have earthquake records, and it will not necessarily hold up in ancient belts of deformation.

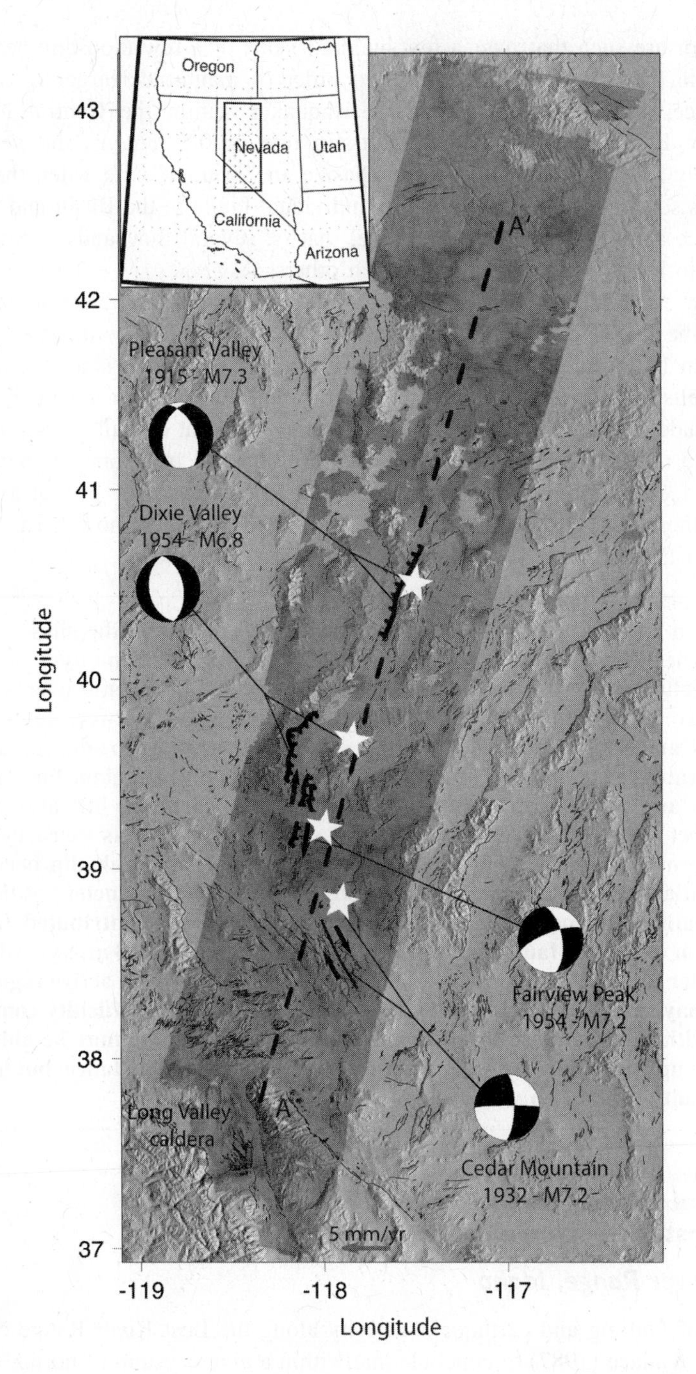

Figure 11.74 Map of the epicenters and magnitudes of the Pleasant Valley, Fairview Peak, and Cedar Mountain earthquakes. Focal mechanisms are shown in the standard manner, i.e., as "*beach balls*," which are similar to lower-hemisphere stereographic projections of fault orientations. The focal mechanism for the Pleasant Valley and Dixie Valley earthquakes are marked by normal faulting. The focal mechanism for the Farview Peak earthquake is hybrid (strike-slip and normal). The focal mechanism for the Cedar Mountain earthquake is strike-slip. Line A-A′ is the general orientation of the Central Nevada Seismic Belt. [From Gourmelen and Amelung, 2005, Figure 1, p. 1474.]

Slip Rates on the Normal Faults in the Central Great Basin

The big, active normal faults in the Basin and Range province tend to be high angle (60° or more), and they appear to be locked to depths of 10 to 20 km, which causes the strain energy to build over time (Thatcher et al. 1999). Slip rates are relatively low, and thus repeat intervals are relatively long (Machette et al. 1996–2004). The GPS velocity maps, province by province, might lull us into thinking that slip rates on these active normal faults are uniform and predictable. But life is never that simple.

Structural geologic mapping combined with paleoseismic analysis of normal faults in the Basin and Range has revealed that that there are significant variations in fault-slip rates. Slip rates along major normal faults have been found to average anywhere from 0.01 to 1.0 mm/yr, and yet variations in behavior are such that over a few hundred years or a few thousand years the slip rate for a given fault may be an order of magnitude larger or smaller (Wallace, 1987). The 1915–1954 earthquakes within the Central Nevada Seismic Belt reflect an overall fault-slip rate of 0.5 mm/yr, and yet it is probably not meaningful to overemphasize an "*average*" rate when the variation is so large. Repeat intervals on individual faults in the Basin and Range province also are also highly variable; that is, several thousands of years on some, to longer than 100,000 years on others.

Slow fault-slip rates and the long recurrence intervals make us wonder about the flurry of big-earthquake activity that took place within the CNSB between 1915 and 1954. Wallace (1987) concluded that this is an example of how belts of activity migrate regionally, in pulses. This can result in large-magnitude earthquakes occurring suddenly in belts that overall are marked by low slip rates. Not only do earthquakes jump around from one segment of a master fault to another, they appear to jump around from one part of a region to another. This is the kind of thing that underscores just how hard it is to forecast the sites and general timing of future earthquakes.

Lesson Learned #28: Paleoseismic investigations at specific sites (often trenches) yield rates of slip on active faults (e.g., 0.08 mm/yr), which typically emerge from an averaging of repeated slip events over some relatively long duration. We always need to distinguish between fault slip rates, and strain rates. To determine strain rates we need to address fault-slip rates for a number of active faults across a given region, for strain rates are not only determined with respect to duration, but also with respect to the length (or size) of the area being deformed. As we carry out this work, accurate measurements and/or estimates of fault dip become essential, just as they are in our investigations of faults in ancient settings. For any given period of time, the extensional strain contributed by a system of normal faults marked by slip rates averaging 0.08 mm/yr will be greater if the faults average 45° versus 85° in inclination. In active regions, we may 'nail' fault-slip rate in a given trench, but have difficulty coming up with average fault dip at depth. In ancient regions we may be able to come up with information on fault dip over a significant distance, but have difficulty coming up with rate(s) of fault slip.

Normal Faulting in the Eastern Great Basin along the Lost River Range, Idaho

Lost River Range, Idaho

Study of faulting and earthquake activity along the Lost River Range caused Robert Wallace (1987) to conclude that within a given system of normal faults or fault segments in the Great Basin, the surface faulting skips around from one fault to another, and from one part of a given fault to another. Some segments of a given fault zone may show repeated slip events, whereas other segments within the same fault zone may have slept (our word) the whole time. But then the dormant fault segment(s) may wake up, and the once-active segment may again become dormant. Wallace (1987) referred to this as the **grouping of fault displacements**, which is analogous to **clustering of earthquakes**. Wallace gained these insights in part by observations he made in the Eastern Great Basin, especially while working on the Lost River fault.

Lost River Fault

The Lost River fault defines the western margin of the north-northwest-trending Lost River Range (Figure 11.75) (Wallace, 1987; Susong et al. 1990).

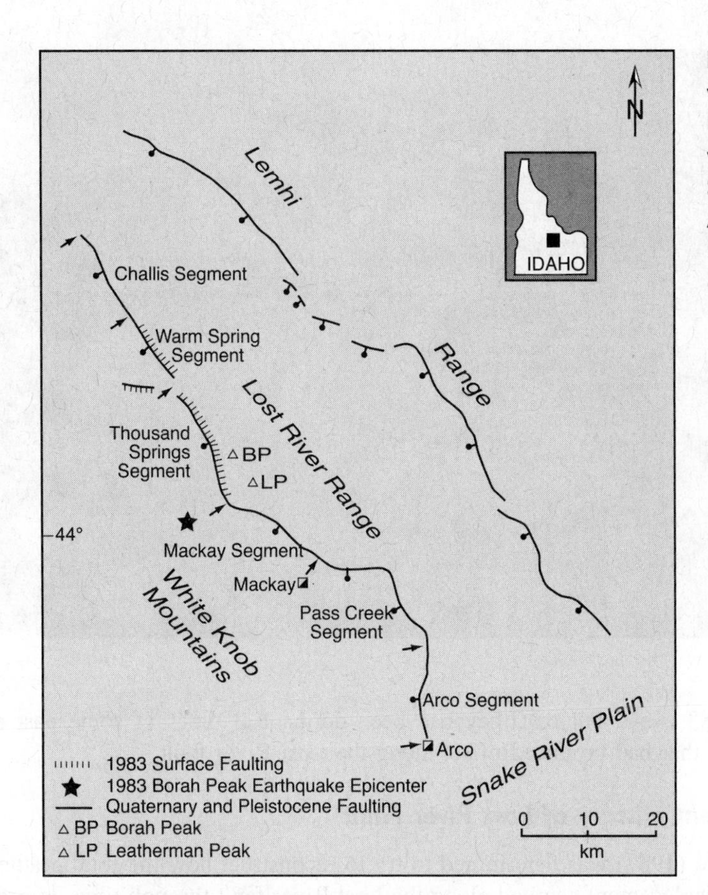

Figure 11.75 Location map for the Lost River fault, which borders the west flank of the Lost River Range, Idaho. The Lost River fault is composed of segments, which from north to south, are the Challis, Warm Spring, Thousand Springs, Mackay, Pass Creek, and Arco segments. The epicenter of the 1983 Borah Peak earthquake is shown as well. [From Susong and others, 1990, Figure 1, p. 58.]

The Lost River fault has a trace length of 160 km. The maximum amount of cumulative normal offset along the Lost River fault appears to be on the order of 6 km (Crone et al. 1987). This fault is not simply one continuous break that, in each and every earthquake, accommodates movement along its entire length. The energy required to do so would be too great; something would 'give' before the requisite energy level would be achieved. Instead, the Lost River fault is segmented, and one or more of these segments is completely or partially activated earthquake-by-earthquake. Six segments comprise the Lost River fault. From north to south these are the Challis, Warm Spring, Thousand Springs, Mackay, Pass Creek, and Arco segments (see Figure 11.75).

Borah Peak Earthquake

The Lost River fault was the site of the M_W 7.3 Borah Beak earthquake, which occurred 8:06 A.M. local time on October 28, 1983. This was a remarkable surface-rupture event, in fact the largest earthquake ever recorded in Idaho in historic time. This earthquake produced surface displacement along ~36 km of the Lost River fault (Figure 11.76) (Crone et al. 1987). Maximum net throw was up to 2.7 m; width of ground damage was as great as 140 m (Crone et al. 1987). Those who came to visit sites of surface rupture were not disappointed, for the earthquake caused dramatic changes to the landscape. In places the faulting was achieved along multiple breaks, creating families of fault scarps.

Fault slip was concentrated mainly along two segments: the Thousand Springs and Warm Springs segments (see Figure 11.75). The Thousand Springs segment strikes N10°W and projects downward along a dip of ~49° WSW to a depth of ~15 km (Susong et al. 1990). Fault slip was mainly normal during the Borah Peak event, but experienced modest left-handed strike-slip displacement as well (Crone et al. 1987; Suson et al. 1990). Wallace (1987) observed that the Thousand Springs and Mackay segments were the ones that were active during

Figure 11.76 Photograph of surface rupture accompanying the Borah Peak earthquake, which took place on October 28, 1983. This photograph was taken by on November 8th, a few days after the earthquake, at a site approximately 50 m south of the Doublesprings Pass road. The stadia rod is 3-m long, and the person holding it is Susan Goter. The rounded slope immediately above the fresh free-face of the 1983 scarp is the upper part of the pre-1983 fault scarp on the Lost River fault. This pre-1983 scarp had been trenched (and then filled back up) a few years earlier before the earthquake, and then the 1983 rupture cut and exposed (in the new rupture face) the filled trench. Quite a coincidence! Photograph by A. J. Crone, 8 November 1983. [From Crone et al. 1987, Figure 14, p. 756.]

the 1983 event, but that they had been dormant at times of some past earthquakes that had produced offset along the Lost River fault.

Segment History of Lost River Fault

Wallace (1987) was determined to try to reconstruct how the total gradient of fault displacement accrued along the Lost River fault through time. In order to do this, he examined the paleoseismology findings of Scott et al. (1985), which provided a picture of specific times in the past when faulting had taken place along the Lost River fault, and the locations of each fault-rupture event. He then plotted these data on a graph that identified the locations of segments and segment boundaries (Figure 11.77).

What became clear to Wallace is that fault-slip during successive earthquakes will cause slip on one segment, then on another, and then on yet another, in a manner that may appear to be quite disordered. However, over time, the net displacement along the fault zone as a whole evolves in a surprisingly systematic, cohesive way. He demonstrated this by constructing a longitudinal topographic profile of the crest of the Lost River Range, and pointing out that there is a systematic tapering-off of elevation at each end (see Figure 11.77). He concluded that this reflects the overall net slip along bounding fault zones over a long period of time. The tip-to-tip tapered fault-scarp profile is not unlike what we see on a much smaller scale in the Volcanic Tableland.

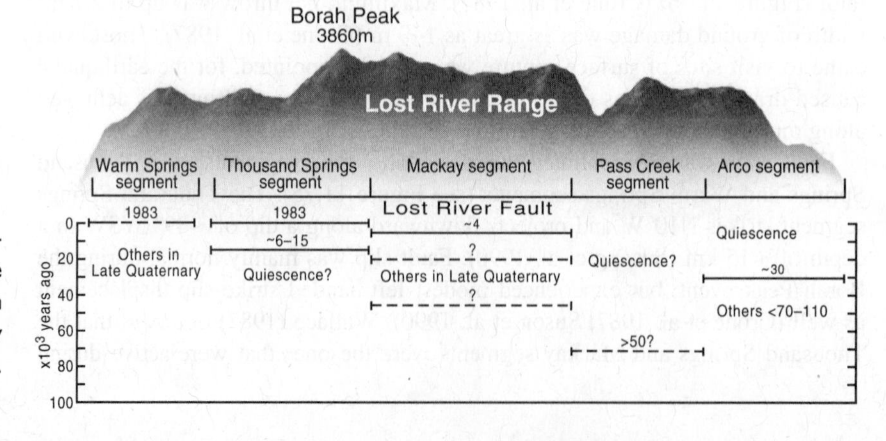

Figure 11.77 Image showing the distribution of displacement events along the Lost River fault, Lost River Range, Idaho, in 1983 and previously during late Quaternary time. [From Wallace, 1987, Figure 1, p. 869. Mountain artwork by David A. Fischer.]

Lesson Learned #29: Regional-scale active fault zones are segmented, no matter whether they are strike-slip, normal, reverse, or thrust. We see this in the earthquake record, in maps of surface ruptures following earthquakes, and in the physiography and geomorphology of mountain fronts. Individual segments typically link-up through branching patterns of faulting, including transfer zones between fault segments. This knowledge makes us more attentive to recognizing segmentation of ancient faults we happen to be studying.

One implication of the grouping concept of Wallace (1987) is that slip rates along a given active segment may be much greater than the average long-term slip rate, and *vice versa.* Wallace imagines that uplift of a range is achieved *"by first one segment of a range-front fault then another having a series of displacement events, each series being followed by a period of quiescence"* (Wallace, 1987, p. 869).

Lesson Learned #30: The history of offset within certain active faults reveals that earthquakes and fault-displacements jump around from one segment of the fault to another, or from several segments of the fault to several other segments, over a time scale of hundreds of thousands to millions of years. Somehow the net effect tends to be a smooth gradient of displacement over the length of the fault. This knowledge should help us search hard for an overall order in the net cumulative displacements along major fault zones in the ancient record.

Eye Witness Accounts of Active Faulting accompanying the Borah Peak Earthquake

Eyewitnesses to surface rupturing were on the ground at the time of the earthquake, and they later were interviewed by Robert Wallace and shared with him the drama of what they experienced (Wallace, 1984). In particular, Mrs. Lawana Knox was hunting elk with her husband, and in the moments just before the earthquake was keeping a watchful eye on a broad hill slope to her north, where she anticipated her husband would be driving elk toward her. Suddenly ground shaking rocked the ground on which she was standing. In an instant a 1.0–1.5 m scarp formed on the hill slope right in front of her. According to Mrs. Knox, the scarp reached its full height instantaneously; it just locked into place, not adjusting up or down, or oscillating back and forth (Wallace, 1984, p. 1091). Mrs. Knox reported that she actually saw surface rupture tear the ground over a distance of several kilometers and within just a few seconds. She said the fault scarp ran across the land . . . *"just as though one took a paint brush and painted a line along the hill"* (Wallace, 1984, p. 1092). Both Mrs. Knox and her husband reported that the earthquake began with a noise. The ground shaking that Mrs. Knox experienced snapped her head back and forth, giving her both a headache and neck ache. Mr. Knox *"said he felt sick at the beginning of the shaking and instantly went down on one knee. As the strong shaking increased, he lay on his stomach and was rolled back and forth by the motion of the ground"* (Wallace, 1984, p. 1093). He had the presence of mind, however, to realize that the ground motion oscillated in a north-south direction.

Balancing Extension and Shortening at a Plate Scale

Given all of the tectonic stretching taking place in the Basin and Range province, we might logically ask: Is the continent getting larger? This is the question that Kreemer and Hammond (2007) asked, and then answered!! They first determined that ongoing tectonic stretching of the crust in the Basin and Range province is causing aerial growth of ~5200 m^2/yr. They then surmised that expansion is offset completely by comparable contraction, resulting in essentially *no change* in area of the western United States as a whole.

Kreemer and Hammond (2007) figured this all out by closely examining a data set of 1477 GPS velocities! The net change they determined was a mere ~0.3%. The data reveal that the increase in area being experienced by the Basin and Range is compensated fully by contraction in northern California in the region of the California Coast Ranges and the Klamath Mountains (Kreemer and Hammond, 2007); that is, by the same contraction that is forcing thrust faulting and folding in Cascadia. This is all quite extraordinary!

Lesson Learned #31: The new technologies permit asking questions that heretofore would never have been considered. Imagine, in real time, attempting to determine whether the increase in area of the entire Basin and Range province is being compensated anywhere in the western United States by an equivalent amount of contraction? We should attempt to find ways to make such comparisons in the ancient regional tectonic record, comparing and contrasting the overall strain of multiple provinces that were actively deforming at the same time(s) in geological history.

THE WASATCH FRONT

Its Nature and Location

The Wasatch Front marks the boundary between the Eastern Great Basin (on the west) and the Colorado Plateau (on the east) (see Figure 11.70). It is generally marked by a bold, composite physiographic escarpment, which indeed marks the eastern boundary of the Basin and Range province and the westernmost extent of the Colorado Plateau (Figure 11.78). The Wasatch front

Figure 11.78 Eastward oblique aerial photo view of the Wasatch front near Provo, Utah. Note triangular facets along the front of the Wasatch mountain block, as well as active fault scarps. [Photograph by C. Glass, G. Brogan, and L. Cluff. Courtesy of Woodward-Clyde.]

is the product of the history of cumulative offset along Wasatch fault, which is ~370 km long, extending from southern Idaho to central Utah. The Wasatch fault is composed of individual, linked segments ranging from 30 to 60 km in length (Figure 11.79): from north to south the Brigham City (BC), Weber (WB), Salt Lake City (SLC), Provo (PV), Nephi (NP), and Levan (LV) segments (Bruhn et al. 2005). We are no longer surprised to learn that a major fault zone, such as this, is segmented!

The Wasatch fault is truly one of the great active faults of North America, and it passes just east of Salt Lake City and Provo, Utah. In fact, 80% of Utah's 2 million people live along the Wasatch Front (Chang and Smith, 2002). The Wasatch fault lies within the Intermountain Seismic Belt (IMSB) (Smith and Sbar, 1974) (see Figures 11.71 and 11.72). Unlike the regions we refer to as the Central Nevada Seismic Belt (CNSB) and the Eastern California Seismic Belt (ECSB), the Wasatch Front region does not get much national public attention because there has not (yet) been a M_W 7.0 earthquake along the Wasatch fault in historic times. Furthermore, paleoseismological studies

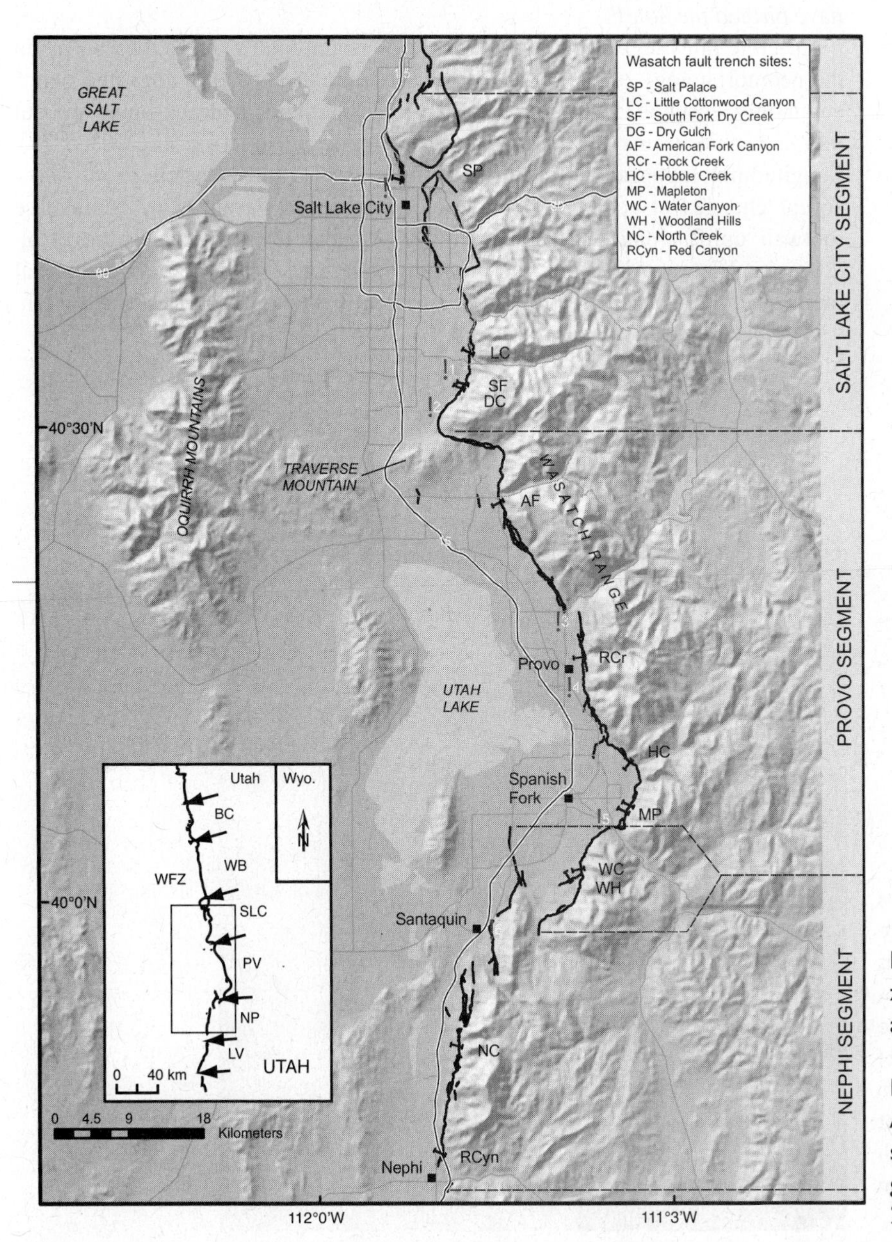

Figure 11.79 Map of the Wasatch fault zone (bold) showing (in the inset) six active segments: BC = Brigham City; WB = Weber; SLC = Salt Lake City; PV = Provo; NP = Nephi; and LV = Levan. Also shown are locations of paleoseismology trenches, which are also useful geographic references. [From Bruhn et al. 2005, Figure 1, p. 3.]

show that the repeat intervals for faulting along segments of the Wasatch fault are very, very long. Thus it is not surprising that the Wasatch Front has been, until recently, "*out of sight, out of mind*" with respect to seismic risk. However this is no longer the case.

Segmentation of the Wasatch Fault

The Wasatch fault dips westward, as do its segments. Fault slip is purely normal-slip. Some exposures of the fault surfaces are quite impressive, and display the telltale signatures of fault slip, such as slickenlines, grooves, fault mullions, and fault gouge (Figure 11.80). There are also good examples of fault scarps, including a Holocene fault scarp produced by slip along the Wasatch fault near American Fork Canyon (Figure 11.81*A*). The fault scarps and benches near American Fork Canyon are now incorporated into the landscape of a golf course sited there (Figure 11.81*B*). When golfers think of hazards (e.g., sand traps, water, deep rough), they seldom think of experiencing an earthquake just as they are lining up a putt. (We can just hear the excuses: "*If only that P-wave hadn't arrived, I would have birdied the hole!*")

Discrete fault segments comprising the Wasatch fault express themselves in the geomorphology of the Wasatch Front (Bruhn et al. 2005). Jogs and bends commonly occur at segment boundaries. Furthermore, bedrock spurs jut out into adjacent basins at boundaries between segments (see Figure 11.79). Longitudinal topographic profiles of the crest of the Wasatch Front show abrupt changes at segment boundaries, reflecting differences in cumulative footwall uplift along the Wasatch fault, segment by segment. Bruhn et al. (2005, p. 2) emphasized that the Wasatch fault zone is a collection of normal faults, each of which nucleates independently and grows in strike length,

Figure 11.80 Photograph of outcrop of a part of the Seven Peaks fault surface along the Wasatch fault. The slip surface is polished carbonate-clast breccia with dip-parallel slickenlines. Material in the distance consists of fault gouge (dark gray), blocks of limestone (light gray), and displaced Lake Bonneville deposits (bedded gray units at base). Undergraduate field trip for Utah Geology class led by Ron Bruhn. [From Bruhn and others, 2005, Figure 8, p. 10.]

A

B

Figure 11.81 (*A*) Oblique aerial photograph of Holocene fault scarp on the Provo segment of the Wasatch fault near American Fork Canyon. [The original photograph was taken by Michael Machette in the early 1980s when, as a United States Geological Survey geologist he was mapping and studying the Wasatch fault zone.] The scarp here is the dark escarpment just beyond the highway. (*B*) This Holocene fault scarp is now incorporated into the design of the Alpine Country Club golf course. Furthermore, a housing development is encroaching onto the scarp from the south. (Anthony Crone, personal communication, 2009.) [Image captured by Anthony Crone on Google Earth.]

linking up with neighboring faults and producing a "*continuous normal fault zone.*" Sounds just like the Lost River Range, and the Lost River fault.

Nephi Segment of the Wasatch Fault

We can gain a deeper appreciation for fault segmentation along the Wasatch fault zone by taking a look at the Nephi segment, which is 42 km in length and the connecting link between the Provo segment to the north and the Levan segment to the south (see inset map on Figure 11.79) (Bruhn and others, 2005). As can be seen in Figure 11.82, there is a right-stepping overlap between the north end of the Nephi segment and the south end of the Provo segment. In contrast, a 15-km gap devoid of faulting separates the south end of the Nephi segment and the north end of the Levan segment (see Figures 11.79 and 11.82). The Nephi segment proper consists of two fault strands, the Santaquin strand (17 km long) and the Nephi strand (25 km long) (see Figure 11.82).

Fault offsets along the Nephi segment are particularly well exposed at Red Canyon (Rcyn, Figure 11.79). Seen from the air, the Red Canyon area reveals a conspicuous west-facing fault scarp (Figure 11.83). Along it there is a 20m vertical displacement of a late Pleistocene fan remnant, and a 6m vertical displacement of a Holocene alluvial fan surface. Trenching of the Holocene fan surface revealed that the 6m scarp was produced by three earthquake events in the past 5–6 ka (Bruhn et al. 20005). Trenching of the late Pleistocene fan remnant revealed that the 20-m scarp was produced by at least 9 earthquake events, and that vertical displacement averaged ~2 m per event. Paleoseismological investigations such as these have been carried out all along the Wasatch fault, and locations of a number of these paleoseismological sites are shown in Figure 11.79.

Speaking of paleoseismological investigations, there is one particular figure from Bruhn et al. (2005) that underscores the fastidious research that is being carried out in order to understand the detailed history of individual fault

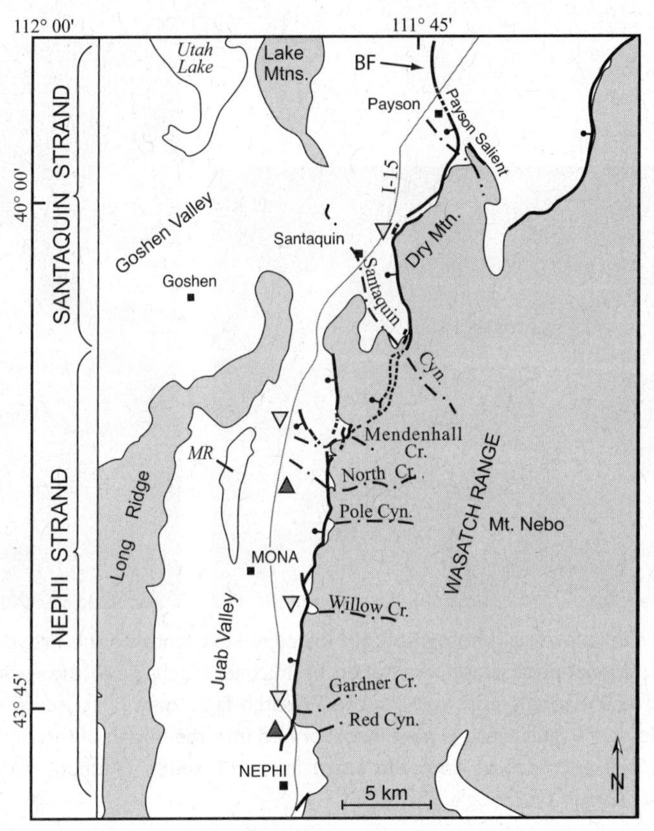

Figure 11.82 Map showing the character of the Nephi segment of the Wasatch fault. The Nephi segment is composed of two strands, namely the Nephi and Santaquin strands. Note right-stepping relationship between the Nephi and Provo segments. Paleoseismolgy trenches are marked with triangle symbols, and those along the Nephi segment include [From Bruhn et al. 2005, Figure 3, p. 5.]

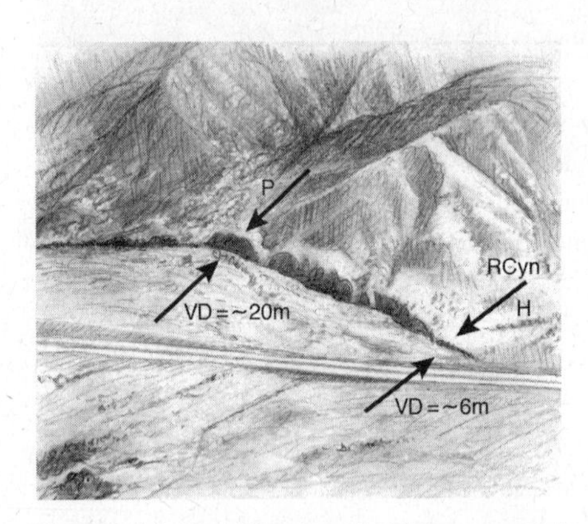

Figure 11.83 Sketch of oblique aerial photograph of the Red Canyon fault scarp locality near the southern end of the Nephi segment of the Wasatch fault. View is toward the east, with Interstate Highway I-15 in the middle ground. Note the increase in vertical displacement (VD) of the scarp from ~6 m on the Holocene (H) alluvial surface to ~20 m across the late Pleistocene (P) fan remnant. RCyn=Red Canyon fault trench site. [Based upon photograph from Bruhn et al. 2005, Figure 2, p. 4. Artwork by David A. Fischer.]

segments (Figure 11.84). The diagram efficiently assembles the critical data. The main part of the figure is a graph that shows vertical fault displacement (in meters) plotted as a function of location along the fault segment (distance from the north tip-out point of the fault, in kilometers). The curves within the graph are drawn for each of 5 dated alluvial surfaces, and plotted in such a way to represent cumulative net vertical slip at each point along the Nephi segment (see Figures 11.79 and 11.82). Juxtaposed on the top of the graph is a map of the Nephi segment, complete with the Santaquin and Nephi strands, flipped on its side. Thus this figure beautifully shows that total net vertical displacement is greatest along the central 25 km of the Nephi segment, dying out to zero at its southern and northern tips.

Lesson Learned #32: Plotting vertical fault displacement as a function of location along a fault trace for beds of different ages is a 'natural' for studying active normal faulting. The equivalent approach for ancient normal faults is plotting stratigraphic throw in longitudinal profile along fault traces. It really does not matter whether displacement magnitudes are simply plotted directly on structural geological maps or longitudinal profiles. What matters is finding some way to keep track of variations of displacement along fault traces, so that gradients of displacements can be seen as well as abrupt terminations of gradients. These data thereby provide a basis for recognizing individual fault segments and envisioning progressive deformation by faulting. Practical applications for recognition of segment boundaries are many. In active settings, knowledge of the locations of segment boundaries will influence assessments of seismic risks in relation to earthquakes. In some ancient settings, fault-controlled ore 'lodes' may thin down and disappear at segment boundaries, and knowledge of this will influence projections of reserves.

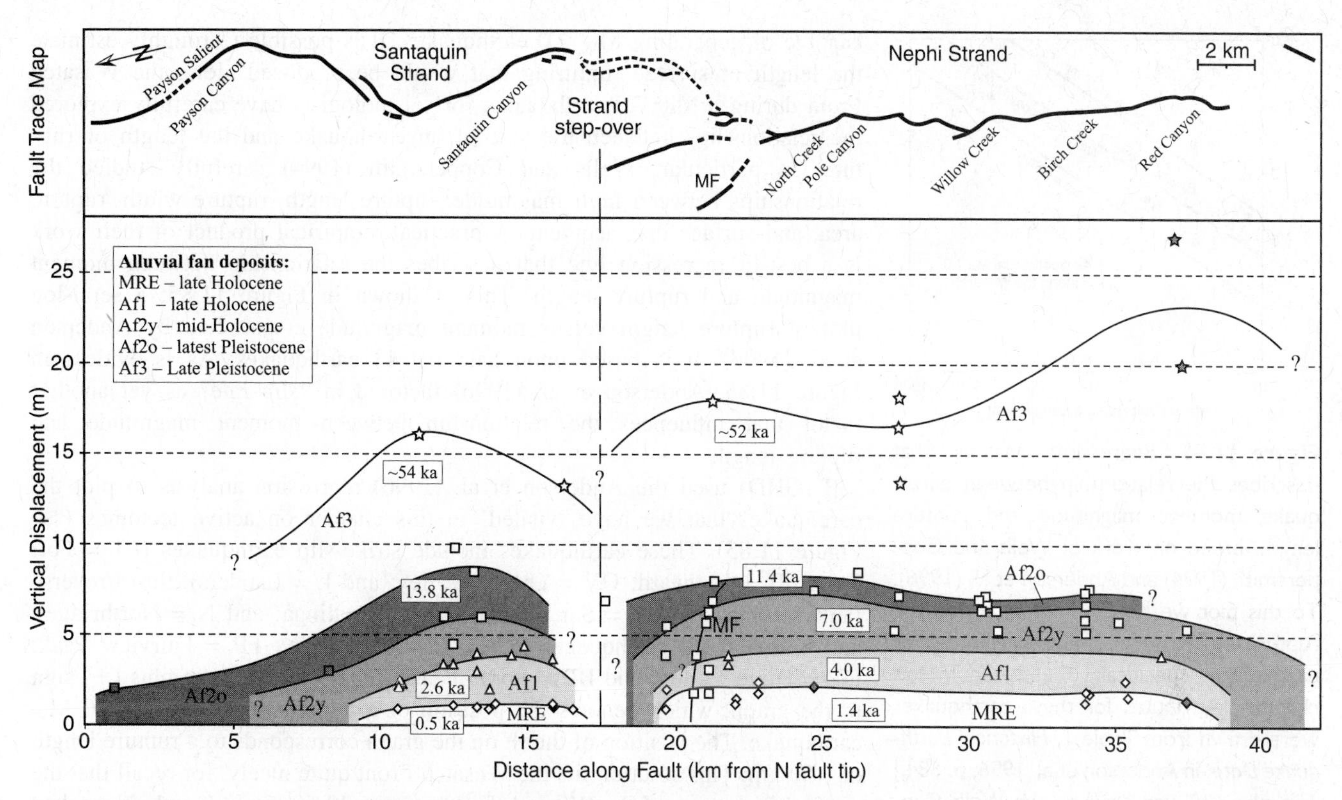

Figure 11.84 An amazing plot! Along the top is a map showing the geometry and dip direction of the Santaquin and Nephi fault segments (strands). Distance scale for the map relations runs along the very bottom of the plot. The curved lines summarize results of paleoseismological investigations at sites along the Santaquin and Nephi segments where alluvial fan surfaces of known age(s) have been displaced. Each curved line represents a dated earthquake event (see ages in small boxes), and the height variation for each curve is a measure of vertical fault displacements (see scale on left margin of plot). The cumulative heights of the curves give a picture of the fault-displacement gradient along the length of the Santaquin and Nephi segments. Diamonds—MRE (most recent event); Af—datum points on alluvial fans. [From Bruhn et al. 2005, Figure 5, p. 8.]

Age and Activity of the Wasatch Fault

Recurrence intervals for major earthquakes along the Wasatch fault range from 1200 yrs to 2400 yrs for each of the five central segments of the Wasatch fault (see Figure 11.79) (Weber, Salt Lake City, Provo, Nephi, and Levan segments) (Bruhn et al. 2005). Paleoseismology-based findings lead us to conclude that there should be one major fault event roughly every 500 years within this grouping of five segments (Schwartz and Coppersmith, 1984). Five hundred years is a long time in a young country, and thus the tectonic setting of this part of Utah has seemed peaceful and quiet; and who would *not* want to build and live immediately adjacent to the spectacular backdrop of the Wasatch Range?

The longer, geological view, informed by paleoseismology, tells us that the Wasatch fault got its initial start nearly 18 Ma, and was 'humming' by 10–12 Ma (Bruhn and others, 2005). When we average the vertical slip components along the Wasatch fault over the past 10 Ma, the footwall of the Wasatch fault has been rising at a rate of 0.4 to 0.7 mm/yr (Bruhn et al. 2005). This is equivalent to uplift of ~0.5 km in a million years: not too shabby, and great for skiing. All in all, the Wasatch fault zone accommodates ~2–3 mm/yr of extension (Kreemer et al. 2010).

Estimating Rupture Lengths for Faults along the Wasatch Front

Lest we run the risk of underestimating the seismic hazard posed by the Wasatch fault, keep in mind that each segment of the Wasatch fault is quite

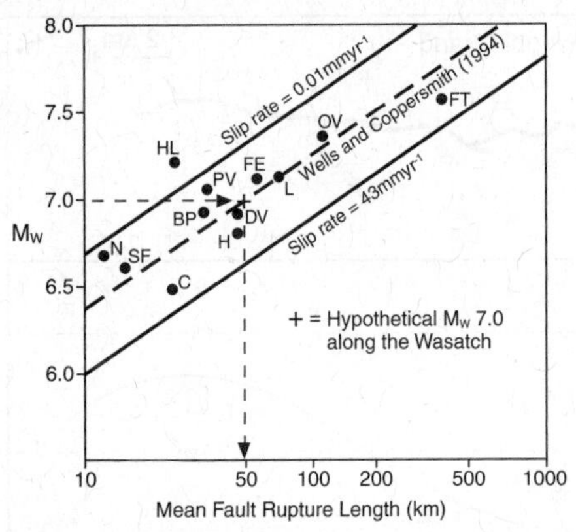

Figure 11.85 Regression analysis that describes the relationship between earthquake moment magnitude and rupture length, based on work of Wells and Coppersmith (1994) and Anderson et al. (1996). To this plot we have ourselves added the rupture lengths and moment magnitudes for earthquakes specifically mentioned in this chapter. [Attributes for these earthquakes were drawn from Table 1, *Historical Earthquake Data*, in Anderson et al. 1996, p. 684.] The plus (+) sign lying on the Wells-Coppersmith regression line corresponds to a M_W 7.0 earthquakes, which then predicts a fault rupture length of ~50 km. Abbreviations are as follows. Strike-slip fault earthquakes: FT = Fort Tejon; H = Hayward; OV = Owens Valley; and L = Landers. Thrust/reverse fault earthquakes: SF = San Fernando; C = Coalinga; and N = Northridge. Normal fault earthquakes: PV = Pleasant Valley; FP = Fairview Peak; DV = Dixie Valley; and BP = Borah Peak.

capable of generating M_W 7.0 earthquakes. It is possible to roughly estimate the length of surface rupturing that would be produced along the Wasatch Front during a M_W 7.0 earthquake, for seismologists have carefully explored the relationships between the size of an earthquake and the length of rupture. In particular, Wells and Coppersmith (1994) carefully studied the relationships between fault magnitude, rupture length, rupture width, rupture area, and surface displacement. A practical, empirical product of their work is a best-fit regression line that describes the relationship between moment magnitude and rupture length. This is shown in Figure 11.85, a semi-log plot of rupture length versus moment magnitude constructed by Anderson et al. (1996). It is based upon data for 43 earthquakes. As is evident in Figure 11.85, Anderson et al. (1996) factored in "*slip rate*" as yet another factor that influences the relationship between moment magnitude and rupture length.

I (GHD) used the Anderson et al. (1996) regression analysis to plot the earthquakes that we have 'visited' in this chapter on active tectonics (see Figure 11.85). These earthquakes include strike-slip earthquakes (FT = Fort Tejon; H = Hayward; OV = Owens Valley; and L = Landers), thrust/reverse fault earthquakes (SF = San Fernando; C = Coalinga; and N = Northridge), and normal fault earthquakes (PV = Pleasant Valley: FP = Fairview Peak; DV = Dixie Valley; and BP = Borah Peak). Finally I added the plus (+) sign to the graph, which denotes the position of a hypothetical Wasatch 7.0 M_W earthquake. The position of the + on the graph corresponds to a rupture length of ~50 km. This solution fits the Wasatch Front quite nicely, for recall that the six fault segments of the Wasatch fault average 30-60 km in length. Not a bad correspondence!

Lesson Learned #33: For large-magnitude earthquakes, we should get into the habit of comparing magnitude with rupture length. In this way we can develop a 'feel' for the seismic significance of ancient faults that we are studying. We can better imagine the minimum magnitudes required to cause slip along fault sections or segments of known length.

Stress and the Wasatch Fault

Recall that GPS-derived velocity vectors for the Eastern Great Basin and the Central Great Basin relative to stable North America are oriented approximately east-west. However, these velocity vectors "*twist*" as they approach the Wasatch Front. The present-day velocity vectors mapped along the Basin-and-Range/Colorado Plateau boundary are essentially perpendicular to the Wasatch Front, thus perpendicular to the Wasatch fault zone. Chang and Smith (2006) pointed this out, based on GPS vector-velocity data collected along the Wasatch Front during the period 1992 to 2003 (Figure 11.86). This twist suggests that stresses causing faulting along the Wasatch Front are in part generated by the elevated topography of the Wasatch Range relative to the low desert to the west (Thatcher et al. 1999).

Indeed, the orthogonal relation of velocity-vector azimuths to the trend of the Wasatch Front is consistent with an extensional stress field produced by the 1-km increase in elevation and the 15 km increase in crustal thickness from west to east across the Wasatch Front. Chang and Smith (2006) determined that the horizontal displacement rate across the Wasatch Front is 1.6 ±0.4 mm/yr. This may not seem like much, but it amounts to fully 50% of the active crustal deformation within the 200-km-wide Eastern Great Basin province.

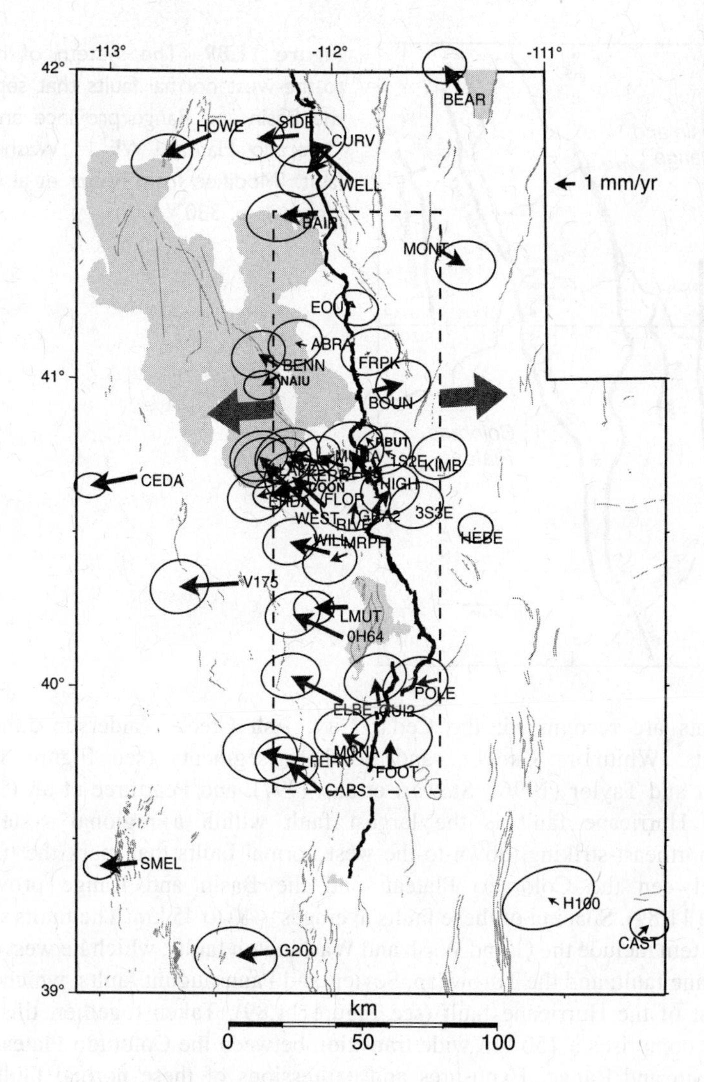

Figure 11.86 Horizontal velocity vectors for reference monuments (abbreviations) along the Wasatch Front. Reference frame is with respect to a stable North America plate. Data derived through continuous campaign observations from 1992 to 2003. Ellipses represent 95% confidence intervals for measurements at each site. Gray lines are Quaternary faults, and black lines highlight the Wasatch fault. The gray arrows represent the direction of principal extension for the Wasatch fault. [From Chang and Smith, 2006, Figure 5.]

Lesson Learned #34: The magnitudes and orientations of stresses in actively deforming regions are strongly influenced by topography and crustal thickness. When we think about the directions of stress that caused deformation observed in the ancient record, we think it is profitable to imagine what a velocity-vector map might have looked like (e.g., back in the Cretaceous), and bring into consideration regional directions of shortening and/or stretching and/or strike-slip shear, as well as the stress effects of topography and/or thickened crust.

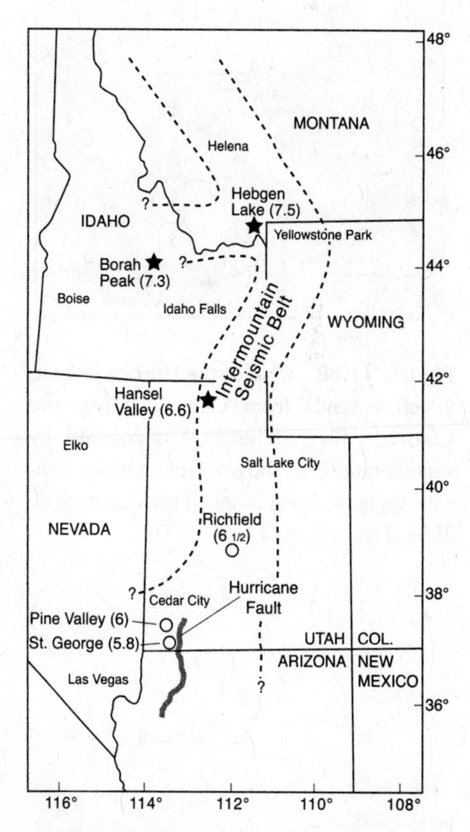

Figure 11.87 Location of the Hurricane fault with respect to the Intermountain Seismic Belt and the Wasatch fault. Note sites of major historical earthquakes, including the Borah Peak earthquake. Earthquake magnitudes are given in parentheses; stars indicate earthquakes produced surface rupture. [From Lund et al. 2006, Figure 4, p. 5.]

THE HURRICANE FAULT

Location and Character

If we keep heading south along the Wasatch Front, we will come upon the Hurricane fault (pronounced "*Hurr ih kun*"). The Hurricane fault lies within the southern part of the Intermountain Seismic Belt, picking up where the Wasatch fault zone gives out (Figure 11.87). The Hurricane fault is yet another major, active, high-angle normal fault zone. It has a trace length of 250 km, with the main expression extending from Cedar City, Utah, to the Colorado River, Arizona (where the river exits the Grand Canyon) (Figure 11.88). Like the Wasatch fault, the Hurricane fault is segmented. Six

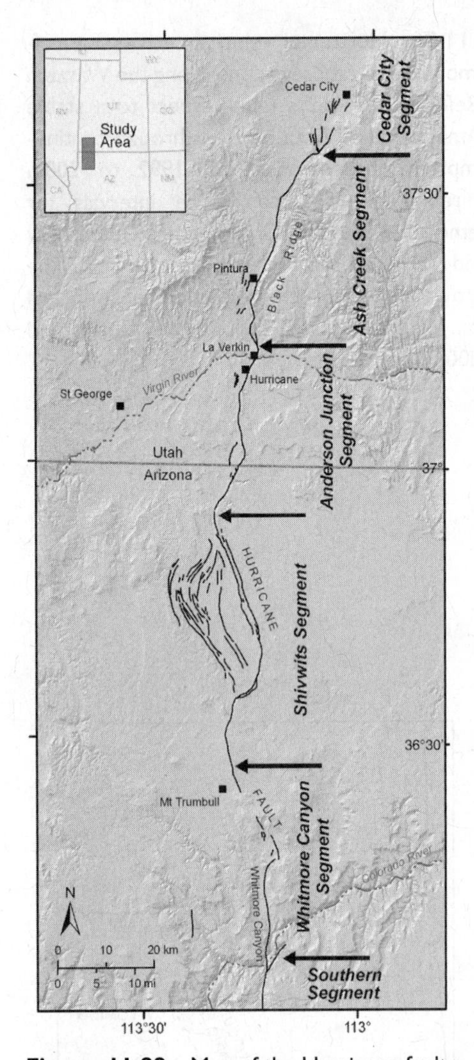

Figure 11.88 Map of the Hurricane fault, which extends from Cedar City to the Colorado River. The fault is marked by segmentation, as shown. Bold arrows indicate segment boundaries. [From Lund et al. 2006, Figure 2, p. 3.]

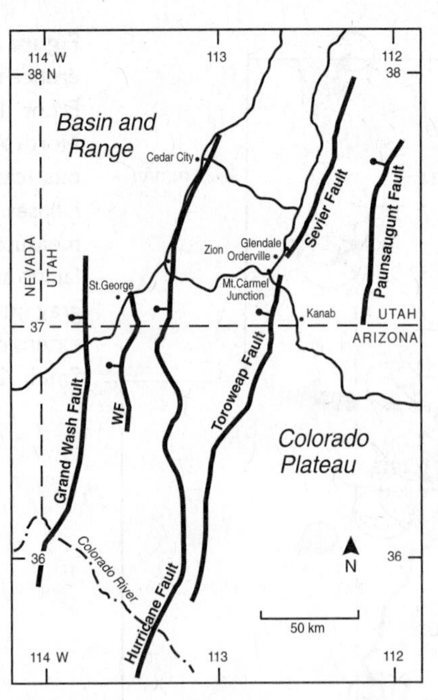

Figure 11.89 The system of down-to-the-west normal faults that separate the Basin and Range province and the Colorado Plateau. WF = Washington fault. [Modified from Reber et al. 2001, Figure 1, p. 380.]

segments are recognized: the Cedar City, Ash Creek, Anderson Junction, Shivwits, Whitmore Canyon, and Southern segments (see Figure 11.90) Stewart and Taylor (1996), Stewart et al. (1997), and Pearthree et al. (1998).

The Hurricane fault is the largest fault within a regional system of north-northeast-striking, down-to-the-west normal faults that mark the transition between the Colorado Plateau and the Basin and Range provinces (Figure 11.89). Spacing of these faults averages ~40 to 45 km. The faults within this system include the Grand Wash and Washington faults, which lie west of the Hurricane fault; and the Toroweap, Sevier, and Paunsaugunt faults, which lie to the east of the Hurricane fault (see Figure 11.89). Taken together, this fault system comprises a 150-km wide transition between the Colorado Plateau and the Basin and Range. Exposures and expressions of these normal faults are especially fine within the Colorado Plateau, where flat-lying layer-cake stratigraphy makes it easy to recognize and measure offsets (Figure 11.90).

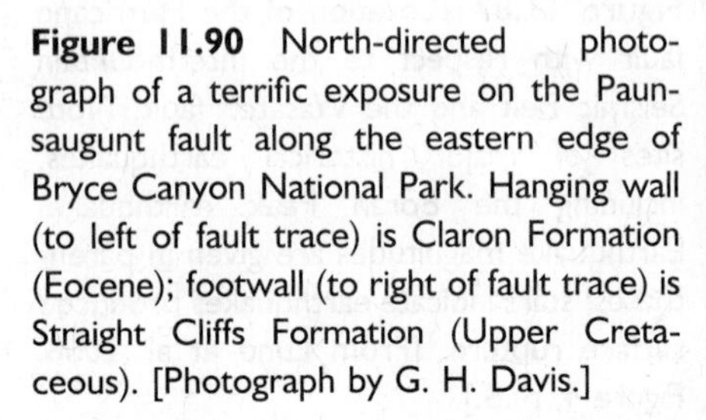

Figure 11.90 North-directed photograph of a terrific exposure on the Paunsaugunt fault along the eastern edge of Bryce Canyon National Park. Hanging wall (to left of fault trace) is Claron Formation (Eocene); footwall (to right of fault trace) is Straight Cliffs Formation (Upper Cretaceous). [Photograph by G. H. Davis.]

Figure 11.91 South-directed view of normal fault along a part of the Hurricane Cliffs, near La Verkin, Utah. Note the smooth curved changes in fault orientation (mullions!) and the polished nature of the surface. [Photograph by G. H. Davis.]

The Hurricane fault zone is typically narrower than 1 km, and marked by high-angle fault surfaces. Where the faults cut and displace just the right rock types, the fault surfaces are beautifully embossed with polish (slickensides), fault lines (slickenlines), and fault grooves (mullions) (Figure 11.91). Where the Moenkopi Formation (Triassic) is caught up in the faulting, the offsets are conspicuous and readily measured (in meters), for the 7-member stratigraphy of the Moenkopi is quite distinctive in its expression.

Lesson Learned #35: Active environments of normal faulting are commonly marked by imposing topographic relief on the footwall side. In examining the Wasatch Front and the Hurricane Cliffs, it is not difficult to imagine how much of the footwall structural story will be lost to erosion, and never show up in the geologic record. To be sure, the unraveling of structural geological history of normal faulting is particularly dependent upon the geology of hanging wall basin regions. In this regard, well exposed footwall domains in active tectonic settings, such as the footwall regions of the Wasatch and Hurricane faults, are invaluable structural geological laboratories.

Segmentation along the Hurricane Fault

The Hurricane fault zone is segmented, in a manner quite similar to what we see along the Wasatch fault to the north (Reber et al. 2001) (see Figure 11.88). Taylor et al. (2001) emphasized that the reality of segmentation of the Hurricane fault is not particularly surprising if a basic rule of thumb is applied, namely, "*no normal fault whose along-strike length exceeds 70 km can slip along its whole length during a single event.*"

The segment *boundaries*, when viewed in detail, are located at distinctive geometric bends along the mapped trace of the Hurricane fault (Figure 11.92). The bends prove to be quite informative in disclosing how normal faults grow along strike. Consider an early period in the development of the Hurricane fault, when individual discrete normal faults are forming and growing through lengthening along strike (Figure 11.93). At some stage two such faults move into positions, relative to one another, where they are underlapping (Figure 11.93*A*) or overlapping (Figure 11.93*B*)). At such stages of fault system development, there is a high probability that the stress fields near the tip of each fault will

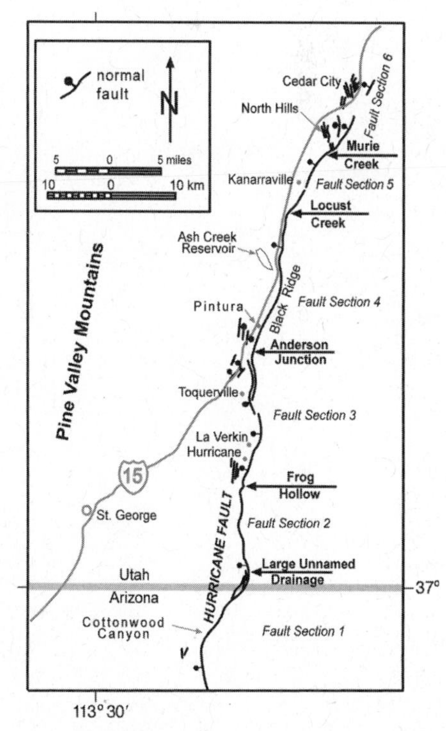

Figure 11.92 Here it is the segment *boundaries* that are highlighted on a map of the Hurricane fault zone. [From Lund et al. 2006, Figure 6, p. 8.]

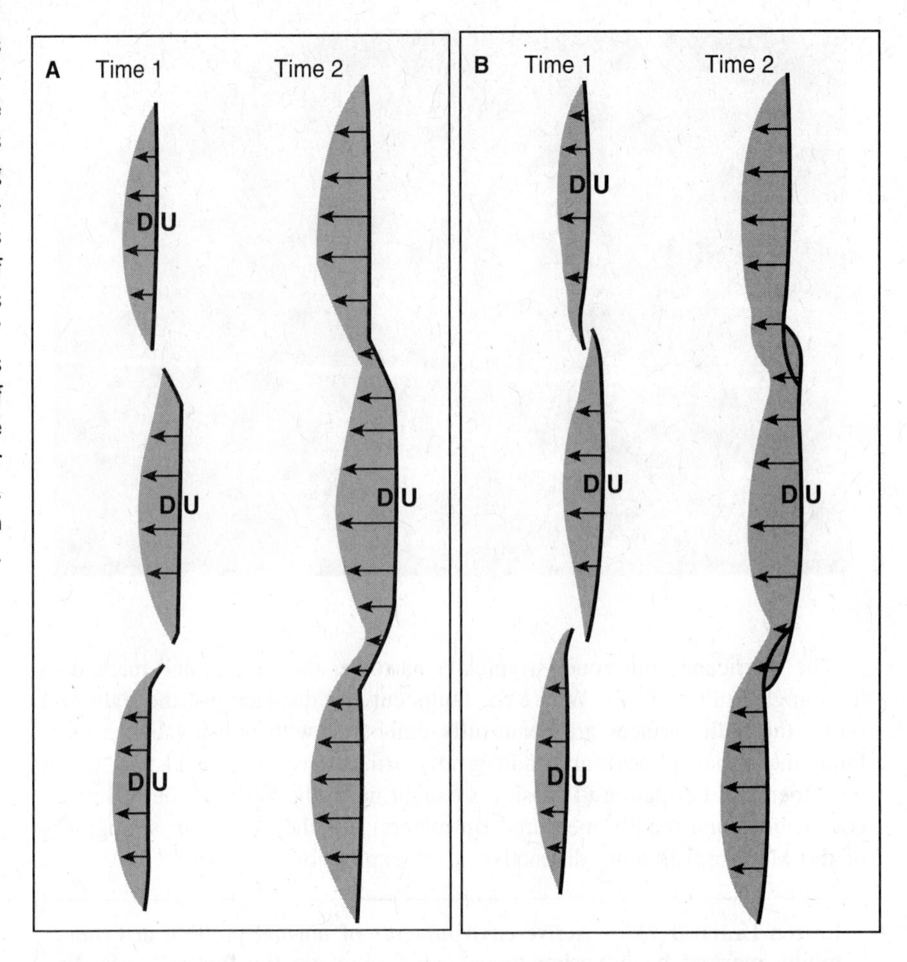

Figure 11.93 Conceptual diagrams showing the map view propagation of normal faults as well as segment linkage and the formation of geometric segment boundaries and segments. (*A*) Linkage of underlapping fault segments. "*Time 1*" shows early discrete underlapping faults; "*Time 2*" shows merging of these faults. (*B*) Linkage of overlapping fault segments. "*Time 1*" shows early discrete overlapping faults; "*Time 2*" shows merging of these faults. The arrows show direction and relative magnitude of fault slip projected into plan view. Notable changes in fault slip magnitude occur at or near fault tips and sites of segment linkage. D = down-dropped block. U = up thrown block. [From Reber et al. 2001, Figure 2, p. 381.]

interfere and cause the faults to curve towards one another prior to actually linking up (see Figure 11.93) (Reber et al. 2001; Taylor et al. 2001).

Lesson Learned #36: Mapping of fault segments associated with active normal faulting alerts us to the fact that fault strike may abruptly change, and that this change may also be associated with a decrease in fault offset. Such shifts may be indicative of reaching the termination of a fault segment. We should keep our eyes open for the same 'signatures' in ancient systems, thereby being on the lookout for segment boundaries.

Segment boundaries prove to be important locations in considering seismic hazards, for the complexities at such boundaries can control the exact locations where earthquakes are triggered. Conversely, segment boundaries might have the effect of retarding further propagation of a rupture into the next adjacent segment. Reber et al. (2001) emphasized that very sharp fault-slip gradients may occur near sites of segment linkages.

Slip Characteristics of the Hurricane Fault

Total net offset due to Hurricane faulting appears to range between 600 and 850 m (Anderson and Mehnert, 1979; Lund et al. 2006). Based on paleoseismological studies incorporating trenching and dating, Bill Lund et al. (2006) came up with long-term and short-term slip rates and recurrence intervals. Average long-term slip rates are in the range of 0.21 to 0.57 mm/yr, increasing from south to north along the Hurricane fault. The repeat

Figure 11.94 Northerly directed photograph of construction pad at very base of the Hurricane Cliffs. Excavated face is fault surface. For scale is a party of geologists, awed by what they are seeing and discussing how very challenging it is to communicate knowledge of geoscience hazards in ways that will positively affect public policy and practice. [Photograph taken by G. H. Davis in November, 2007.]

times based upon long-term history range from 2600 to 4100 years. Short-term slip rates (based on Late Quaternary faulting) range from 0.12 to 0.4 mm/yr, with repeat intervals ranging from 3800 to 15,000 years. Kreemer et al. (2010) emphasized that the combined Quaternary slip rate for the Hurricane-Toroweap-Sevier fault system is <0.4 mm/yr, which is about one-fourth of the equivalent across the Wasatch fault zone to the north.

Estimates of earthquake magnitude for events along the Hurricane fault are impressive, from M_W 6.4 to 7 (Lund et al. 2006). The known severity of M_W 7.0 earthquakes gives rise to the urgent, practical incentive for this work on the Hurricane fault: mitigation of seismic hazard is at stake. The towns of Hurricane (pop. 8250) and La Verkin (pop. 3392) are built immediately adjacent to the Hurricane fault, and the long-term stability of the cliffs makes the area appear sound and stable to residents. New homes are going up along the fault as we speak. The Hurricane Cliffs provide the immediate backdrop for these new residential communities. Homes have been built right on fault strands. Gouge can be collected at *cul de sacs* in neighborhoods (Figure 11.94).

FINISHING UP

Brief Word on the Rio Grande Rift

The plate-boundary motions between the Pacific and North American plates are felt all the way eastward from the west coast to the very western margin of the Great Plains. The final bits of the 52 mm/yr slip budget are found along the Rio Grande rift, which extends for a distance of more than 1000 km from near Leadville, Colorado, in the north to northern Mexico in the south (Figure 11.95). Because the Rio Grande rift is so remote from the actual plate boundary, and because the slip components are so small, the expressions of this active deformation are somewhat underwhelming in comparison to what we have already examined. The long term average for extension accommodated by the Rio Grande rift is $\sim$0.2 mm/yr (Humphreys and Weldon, 1994).

GPS velocities mapped across the Rio Grande rift are less than 1 mm/yr, perhaps related to a very slight clockwise rotation ($\sim$0.1°/m.y.) of the Colorado Plateau relative to a fixed North America (Bennett et al. 2003). The challenging part has been actually recognizing and demonstrating that some of the deformation along the Rio Grande rift is attributable to P/NA plate motion, as advocated by Humphreys and Weldon 1994), and by others as well.

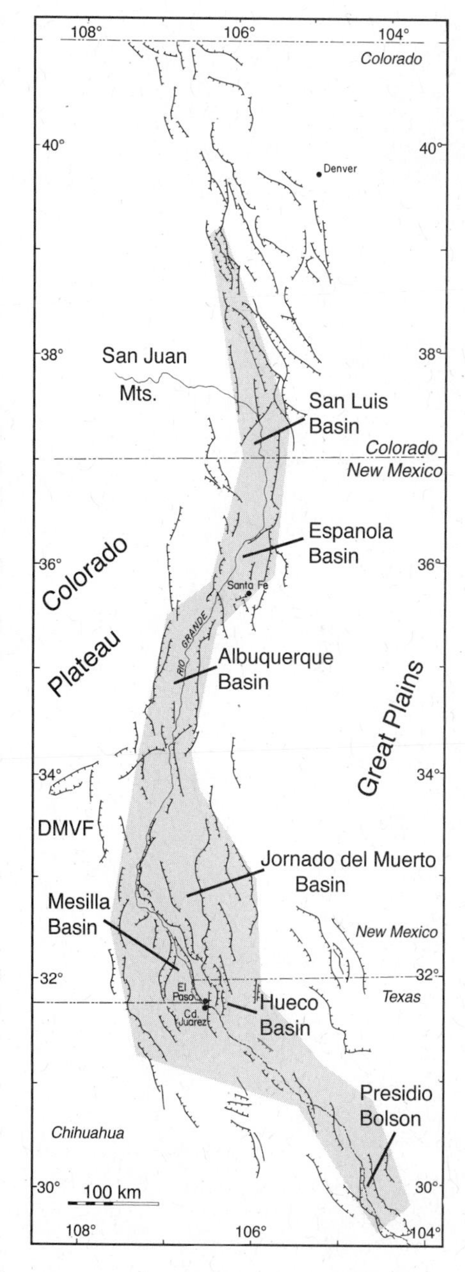

Figure 11.95 Regional location map for the Rio Grande rift. Major late Cenozoic normal faults are shown as dark lines. Note the major basins. [From Keller et al. 1999, Figure 1, p. 122.]

The initiation of Pacific/North America plate interaction caused a shift in the direction of regional extension within the region now known as the Rio Grande rift. Before the Pacific and North American plates came into transform-fault contact, the direction of greatest extension in the Rio Grande rift area had been WSW/ENE. P/NA interaction caused a clockwise rotation of the direction of greatest extension to WNW/ESE (Golombek et al. 1983). As a result, the grain of active normal faulting is north-northeast/south-southwest, and this shift in fault orientation is part of the criteria for identifying those faults created through P/NA plate interaction.

The fault zone that has received the most attention in the Rio Grande rift is the Pajarito fault, which is ~50 km in length and, as described above, crops out in places along the western margin of the Espanola basin (Figure 11.95). The Pajarito fault lends itself to study because it is exposed as a steep fault scarp 100 m high, cutting 1.1 m.y. Bandelier Tuff, a beautiful ash flow tuff that erupted from the Valles caldera ~1.1 Ma (Golombek, 1981). Bandelier Tuff is offset ~60–100m along the Pajarito fault. Where Golombek (1981) studied it, the Pajarito fault has a near-vertical dip at the very surface, but is thought to dip ~60° at greater depth (Golombek, 1981). The near-vertical dips at the surface are an expression of the tensional nature of the extension, and the transtensional nature of normal faulting under conditions where confining pressure is 'zilch' (Golombek, 1981).

In reviewing the literature on studies of individual faults in the Rio Grande rift, we found it puzzling that the references that popped up were almost always about the Pajarito fault, starting with the work of Golombek (1981), followed by many others. It turns out that this north- to northeast-tending *active* fault zone runs right through the town of Los Alamos, New Mexico, home of the *Los Alamos National Laboratory*!

Lavine et al. (2003a, b) and Lewis et al. (2010) carried out high-precision geological mapping of the Pajarito fault zone in Los Alamos, using an electronic total station (Figure 11.96*A*). Their incentive was to assess the potential for seismic hazards generally, and in particular the potential for surface rupture. Mapping was advantaged by the presence of a reliable rock horizon, namely the Bandelier Tuff (age, 1.25 Ma), with plenty of internal stratigraphic units to gage fault throws. They applied high-resolution digital elevation data to track precisely the variations in throw along the traces of the faults (Lewis et al. 2010), permitting them to map faults with offsets as small as 30 cm. As a result of the mapping by Lavine et al. (2003) and Lewis et al. (2010), we now recognize that the Pajarito fault is not expressed by merely a single strand or trace, but rather is marked by multiple strands and segments within a zone 100–200 m wide (see Figure 11.96*B*). The fault segments typically range from ~8 km to ~14 km in length. In some cases the throw is maximum near the center of the trace length, but not uncommonly maximum throw is shifted toward one end (Lewis et al. 2010). The structural geologic expressions along the fault zone are one of three types: single large normal fault escarpments; faulted monoclines; or zones of distributed normal faulting, . . . down to the east. Fault growth appears to have taken place through linkage of overlapping segments.

The Final Bits of Slip

If studies of the Pajarito fault can be generalized, the amount of extension that is taking place along the Rio Grande rift as a result of P/NA interaction is very small. Golombek (1981) was the first to stick his neck out, concluding that the rate of extension across the Pajarito fault zone in the past 1.1 m.y is ~0.05 mm/yr. Going further, Golombek et al. (1983) reached the conclusion that the rate of extension across the entire Espanola basin was ~0.5 mm/yr from 10 to

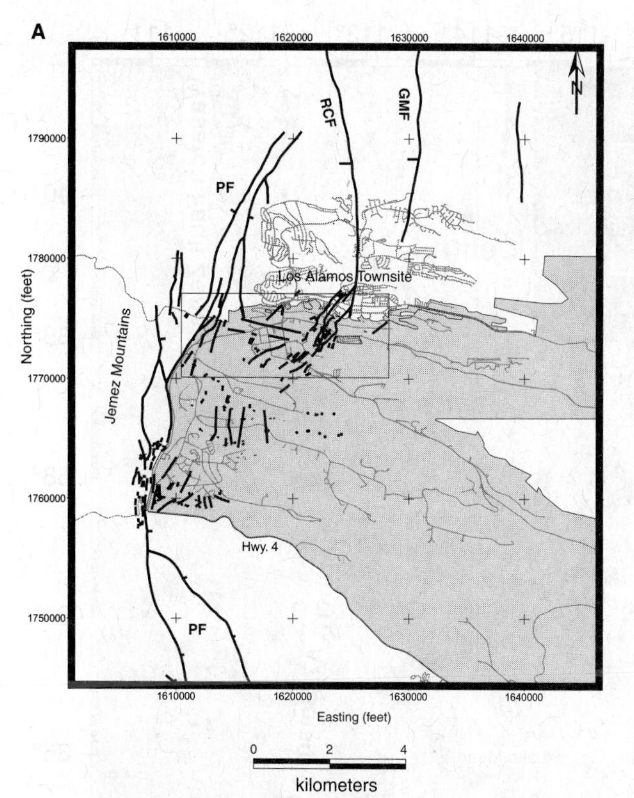

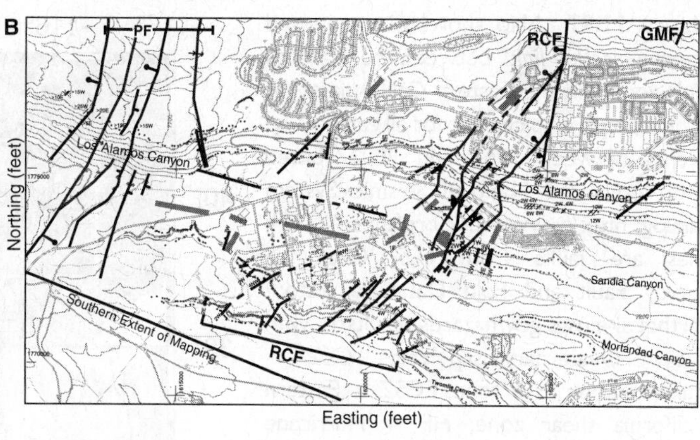

Figure 11.96 (*A*) Map showing the Pajarito fault system in the vicinity of Los Alamos National Laboratory (shaded gray). PF= Pajarito fault, RCF= Rendija Canyon fault, GMF= Guaje Mountain fault. Roads are shown in light grey. Many areas where no faults are shown have not yet been mapped. Grid is in the State Plane Coordinate System (in feet), New Mexico Central Zone, NAD83. [From Lavine et al. 2003, Figure 2.] (*B*) Map of surveyed points and structures from the northwestern portion of Los Alamos National Laboratory (see box, Figure 11.104*A*). Black lines represent faults and folds. PF= Pajarito fault, RCF= Rendija Canyon fault. Balls and bars on downthrown block of faults; arrows on folds show direction of downward flexure. Roads and buildings are shown in gray. Coordinates are in the State Plane Coordinate System (in feet), New Mexico Central zone, NAD83. [From Lavine et al. 2003a, Figure 7; modified from Gardner et al., 1999.]

5 Ma, and ~0.14 mm/yr from 5 Ma to present. The extension rate accommodated by faulting within the Rio Grande rift is much smaller than that for the Eastern Great Basin. But, the Rio Grande contributes what it needs to add the final bits of slip to the overall P/NA slip budget.

Kreemer et al. (2010) have thought hard about the drop in rate of extension as we proceed southward along the Wasatch fault zone to the Hurricane-Toroweap-Sevier zone, and the absence of much active extension in the southern Basin and Range and in the Rio Grande. They believe that the answer partly lies in a 'communication link' between the Wasatch fault zone and the Eastern California shear zone. Through their study of faulting and seismicity and through continuous global positioning system geodesy, they have recognized the presence of a NE-SW-trending shear zone that broadens southwestward from 100 m to 400 km and links the southern part of the Wasatch with the Eastern California shear zone (Figure 11.97). They call this transfer zone the Pahranagat shear zone, which lies along the Southern Nevada Transverse (seismic) Zone (see Figure 11.97). It is left-handed, and accommodates different strain rates to north and south. We conclude that this discovery affirms, once again, that everything is connected.

Lesson Learned #37: All in all, the study of actively deforming regions results in greater understanding of the geometry, kinematics, mechanics, rates, and overall strain significance of geologic structures and systems of geologic structures. Ancient structures and systems of structures provide records of total finite strain and displacement, including elegant and complete three-dimensional geometries. Actively deforming structures give us insights into progressive deformation, including deformation paths and rates. We really cannot understand one without the other! And, for those pursuing structural geology, active tectonics provides additional pathways for career contributions, including assessing seismic risk and helping to mitigate loss of life and property related to earthquake phenomena.

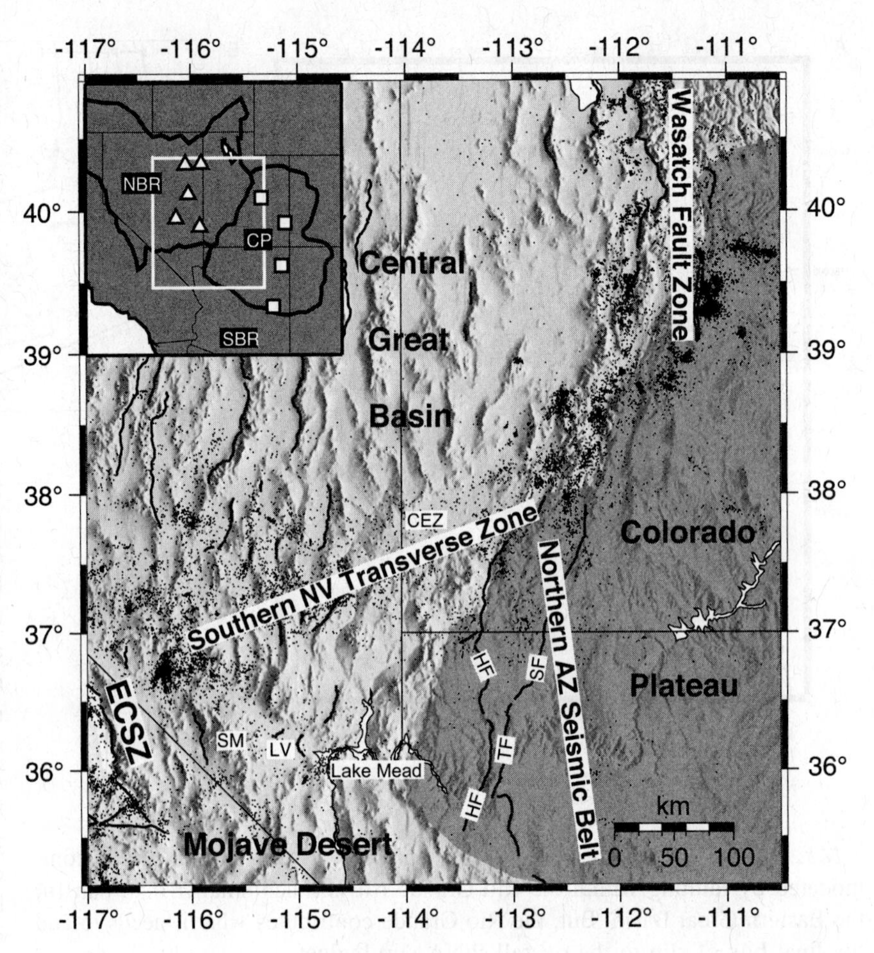

Figure 11.97 Map showing a proposed "*communication link*" between the Wasatch fault zone and the Eastern California shear zone. Dots are epicenters for all earthquake events between 1964 and 2003 (based upon Advanced National Seismic System catalog; http://www.ncedc.org/cness/). Dark lines are faults from U.S. Geological Survey Fault and Fold Database (http://earthquake.usgs.gov/hazards/qfaults/) with known Quaternary slip rates. CEZ = Caliente-Enterprise zone; ECSZ = Eastern California shear zone; HF = Hurricane fault; LV = Las Vegas; SF = Sevier fault; SM = Spring Mountains; TF = Toroweap fault. Inset shows location of study area with outlines of the Colorado Plateau (CP), the northern Basin and Range (NBR), and the southern Basin and Range (SBR). Triangles and squares are locations of global positioning system (GPS) stations used to determine the reference frames. [From Kreemer et al., 2010, Figure 1A, p. 476.]

SUMMING UP

It seems fitting to close this chapter with images that provide an overview of the whole region we have been exploring. With indebtedness to Rick Bennett and Joshua Spinler (Department of Geosciences, University of Arizona) we add two figures specially prepared for this moment. Figure 11.98*A* presents a mapping of GPS-derived velocity vectors. Figure 11.98*B* presents at the same scale a mapping of all faults known to have accommodated some active slip since 1.6 Ma. From a tectonic point of view, this combination of maps underscores the way in which Pacific/North America plate motion is accommodated across a vast region of the continent. From a structural geologic point of view, this combination of maps underscores the fact that flow versus faulting is in the "*eye of the beholder*," and depends importantly on scale of observation. It depends upon our point of view, from "*ROCKS to REGIONS*."

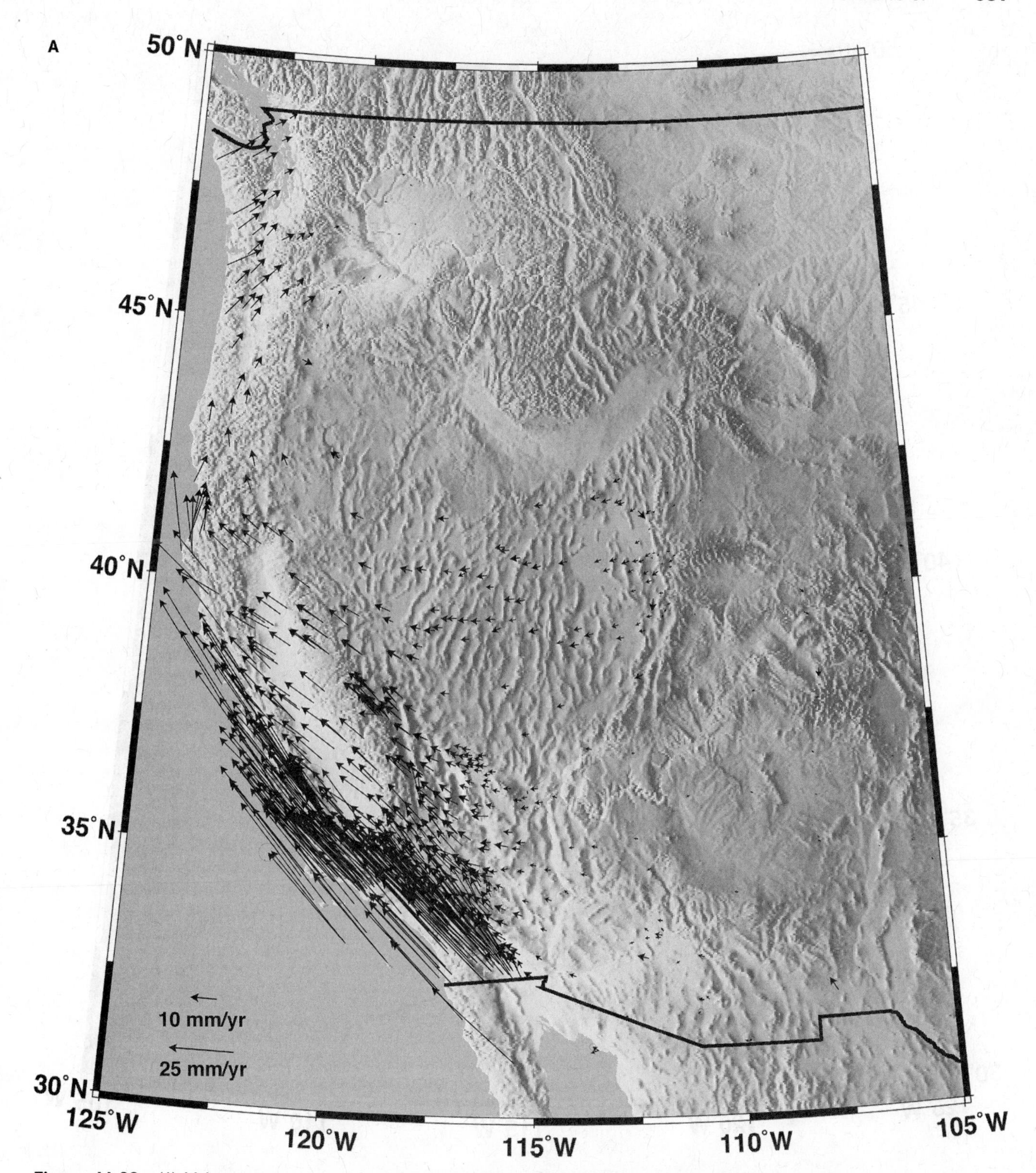

Figure 11.98 (*A*) Velocity vector map of western U.S. (*B*) Active faults map western U.S. [Previously unpublished. Prepared specifically for this chapter by Rick Bennett and Joshua Spinler, Department of Geosciences, The University of Arizona.]

B

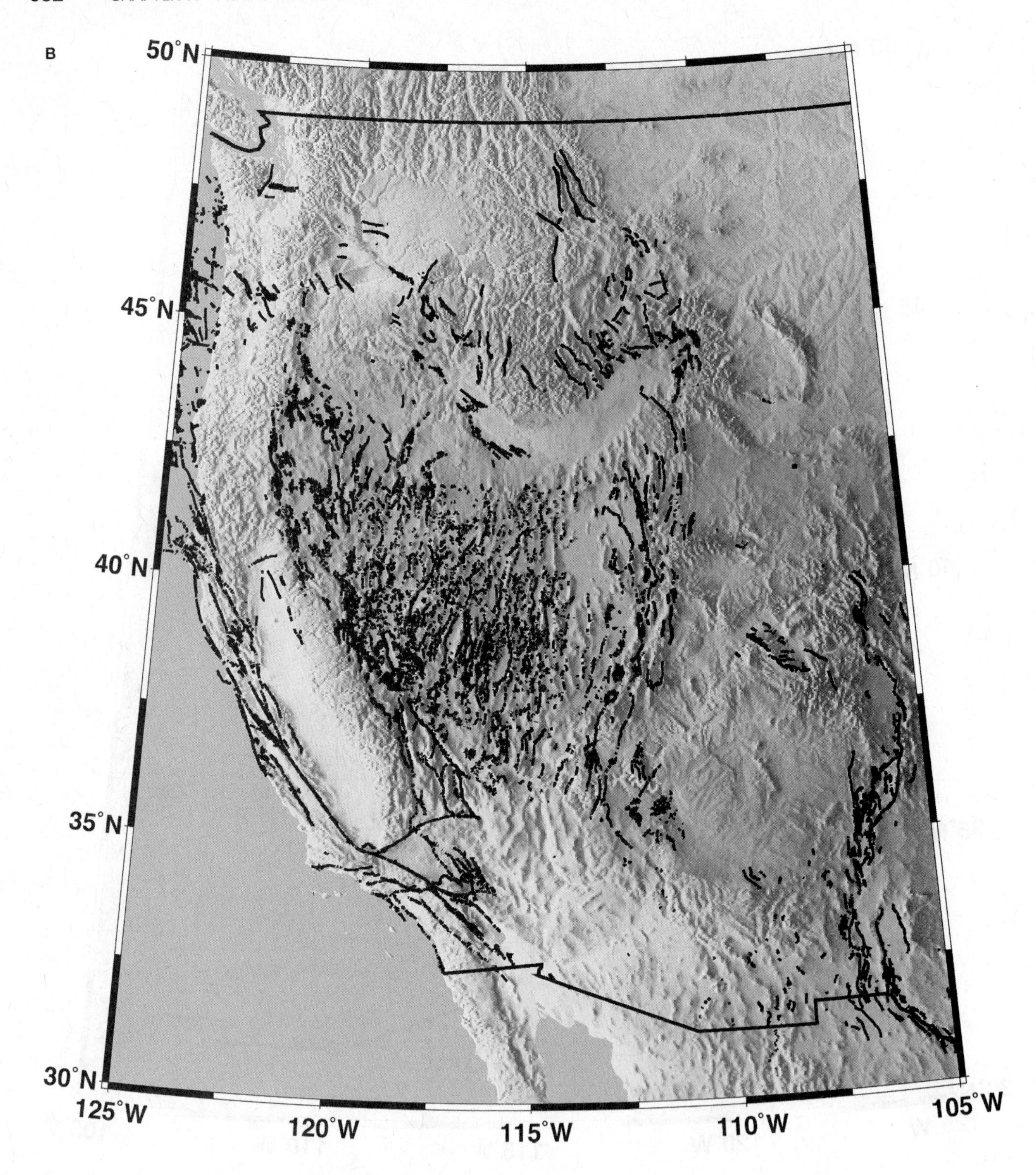

Part III

Descriptive Analysis

Southeast view of the Raplee anticline breached by the San Juan River near Mexican Hat, Utah, along the Comb moncline. It is quite a nice Colorado Plateau experience to raft through the Raplee anticline and check out stratigraphy and structure. The San Juan River meanders through the anticline, and thus the internal anatomy of this fold can be viewed in any number of directions. Jointing is splendidly developed in mechanically stiff units, with a spacing that relates to bed thickness. Folding is achieved in a smooth continuous fashion, with very little faulting. Though this photograph of the Raplee is appealing, the red-rock reality is even more inspiring. Those geologists on the raft are having a hands-on ground-truth experience, one that begins with descriptive documentation of physical and geometric properties of the deformed beds. [Photograph and copyright © by Peter Kresan.]

How to function in the field, and how to reduce the data

A. NATURE OF DESCRIPTIVE ANALYSIS

The Basis

In descriptive analysis we recognize structures, measure their orientations, and describe, literally inside and out, their physical and geometric components (Figure A.1). Descriptive analysis results in facts regarding the physical properties, orientations, and internal configuration of the structures. The basis for descriptive analysis is broad-ranging: direct observation of field relationships, examination of rocks deformed experimentally in the laboratory; drilling into the subsurface; geophysical monitoring and probing of the subsurface; exploring the structures of the ocean floor; and studying the stratigraphy of the rocks in which the structures occur.

The foundation for solid descriptive analysis in structural geology is geologic mapping. Geologic mapping reveals the nature of contacts between rock bodies, which in turn discloses the sequential history of major events. Moreover, geologic mapping results in a three-dimensional portrayal of the geometric architecture of the rock systems under investigation, which in turn becomes a basis for interpreting the structural and tectonic history.

Geologic mapping of contacts, in combination with compass measurements of contacts and structures, yields a database that can be used to construct geologic cross-sections. Geologic cross-sections are as important or more important than the maps themselves, for "*structure sections*" represent the "*best*" interpretations of the geology as it projects into the subsurface or up into the sky.

Petroleum exploration companies, mining exploration companies, and firms and agencies specializing in hydrogeology commonly need more than standard geologic maps and structure sections to carry out their work. Instead, they set up drilling programs and directly sample the subsurface geology, not only the rocks themselves, but also the depths at which specific rock formations are penetrated. On this basis, a variety of subsurface structural geologic maps can be generated, including structure contour maps and isopach maps.

Figure A.1 Photograph of geologist Paul Stockwell taking a bedding reading on the underside of a sandstone layer within the Moenkopi Formation (Triassic) near La Verkin, Utah: an example of going "*inside*" bedrock in order to describe fully the structural geology. [Photograph by G. H. Davis.]

Structural Elements

What we must remember above all is that shifts in scale of observation are absolutely necessary to make a complete inventory of the rocks and structures within any deformed body of rock. Learning to recognize and describe the main types of structure is not quite enough. Many different varieties of each type of structure exist in nature. Furthermore, each structure is composed of **structural elements** that in turn must be identified and described, to permit us to carry out a complete descriptive analysis. The variety and complexity of structures and structural forms is astonishing (Figure A.2).

Structural elements are the physical and geometric components of structures. The **physical elements** are real and tangible (Figure A.3), and they have measurable geometry and orientation. The **geometric elements** are imaginary lines and surfaces, invisible but identifiable in the field; they too have measurable geometries and orientations. As an example, the fold shown in Figure A.4, consists of folded layers, the bedding-plane discontinuities

Figure A.2 Photograph of folded limestone near the village of Ano Karyes in Arcadia, the Peloponessos, Greece. This is an example of the wonderful, sometimes astonishing structures that we encounter in carrying out structural geology. The bedding in this Cretaceous limestone may not be true bedding, for compaction during burial of the original limey sediments caused dissolution of calcium carbonate, producing stylolites and zones of stylolites along the original bed contacts. [Photograph by G. H. Davis.]

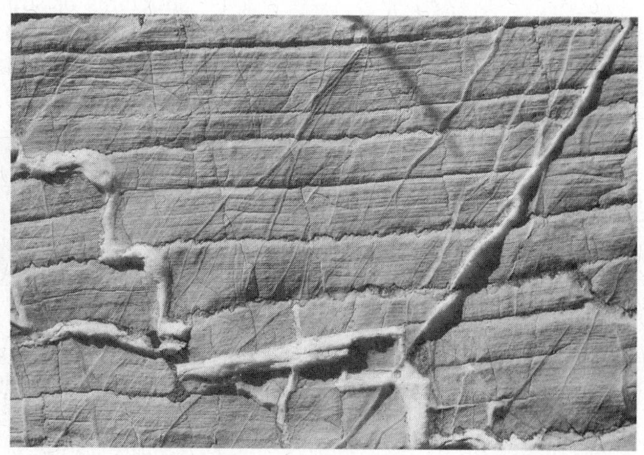

Figure A.3 Close-up photograph of real, tangible physical structural elements in the same Cretaceous formation as shown in the previous figure. The two most prominent elements are stylolites (brain-suture-like in appearance and "*running*" along bedding) and calcite veins (white). Edges of some of the veins are marked by stylolites, revealing that some parts of some veins experienced pressure dissolution in the face of tectonic stress. There is yet a third set of structural elements in the form of what appear to be hair-like cracks (throughout the rock). These too are calcite veins, but on the tiniest of scales. [Photograph by G. H. Davis.]

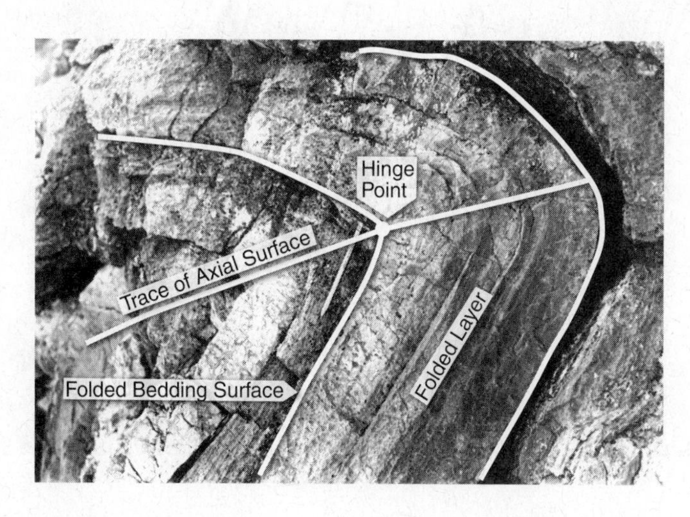

Figure A.4 Physical and geometric elements of a chevron fold. [Photograph by G. H. Davis.]

between layers (bedding surfaces), the hinge point (point of maximum curvature of top or bottom of folded layer), and axial surface (passes through hinge lines, from one layer to the next to the next). The folded layers are physical and real, composed of the rocks that have been folded. The hinge is also real, fixed in position and contained in real rock. The axial surface and bedding-plane discontinuities are geometric and imaginary. The bedding-plane discontinuities separate each of the folded layers. The axial surface is a convenience for helping to define the orientation and form of the fold.

The degree of geometric order in a deformed rock body is evaluated by measuring the orientations of large numbers of structural elements. The orientations are plotted graphically and evaluated statistically to discern the quality of preferred orientation. Through this process, **sets** of structures or structural elements can be defined. Sets are composed of elements sharing common geometric and/or physical appearance and parallel orientation. For example, the joints shown in Figure A.5 may be subdivided into three sets on the basis of orientation, continuity, and spacing. Two or more sets of like structures or structural elements constitute a **system**. All such systems, taken together, plus all structures that do not conveniently arrange themselves into sets, comprise the **total structural system**.

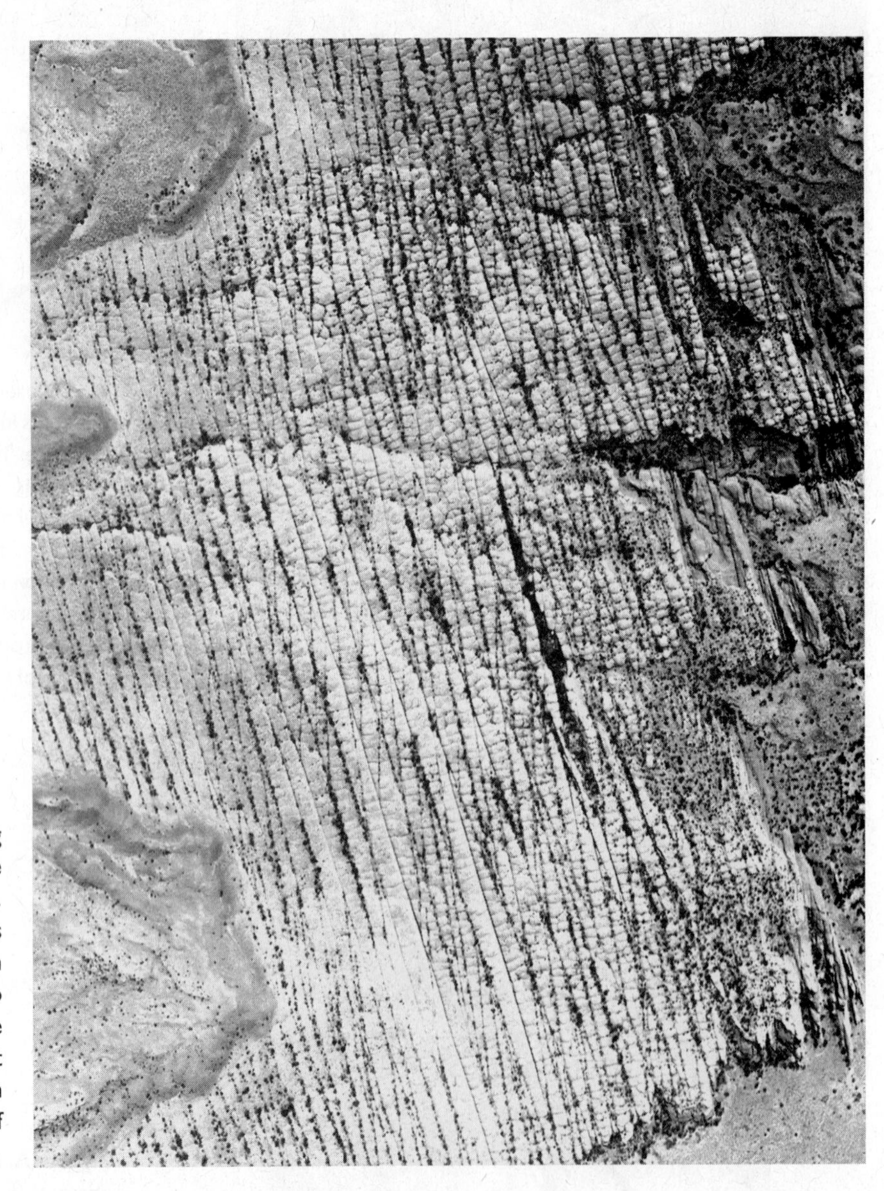

Figure A.5 Aerial photograph of jointing in the Entrada Sandstone (Jurassic) near the campground at Arches National Park, Utah. Two prominent sets of long continuous joints come into clear, crisp resolution when the photo is viewed at a low angle to the plane of the page and parallel to the trend(s) of the joints. Moreover, a third set of short discontinuous joints can be seen "*running*" at a high angle to the trace(s) of the other two. [Photograph by R. Dyer.]

B. GEOLOGIC MAPPING

Philosophy and Mind Set

Geologic mapping is the heart of descriptive analysis in structural geology. The geologic map provides an image of the distribution of rock formations within an area (Figure B.1). At the same time, it discloses the form and structural geometry of rock bodies.

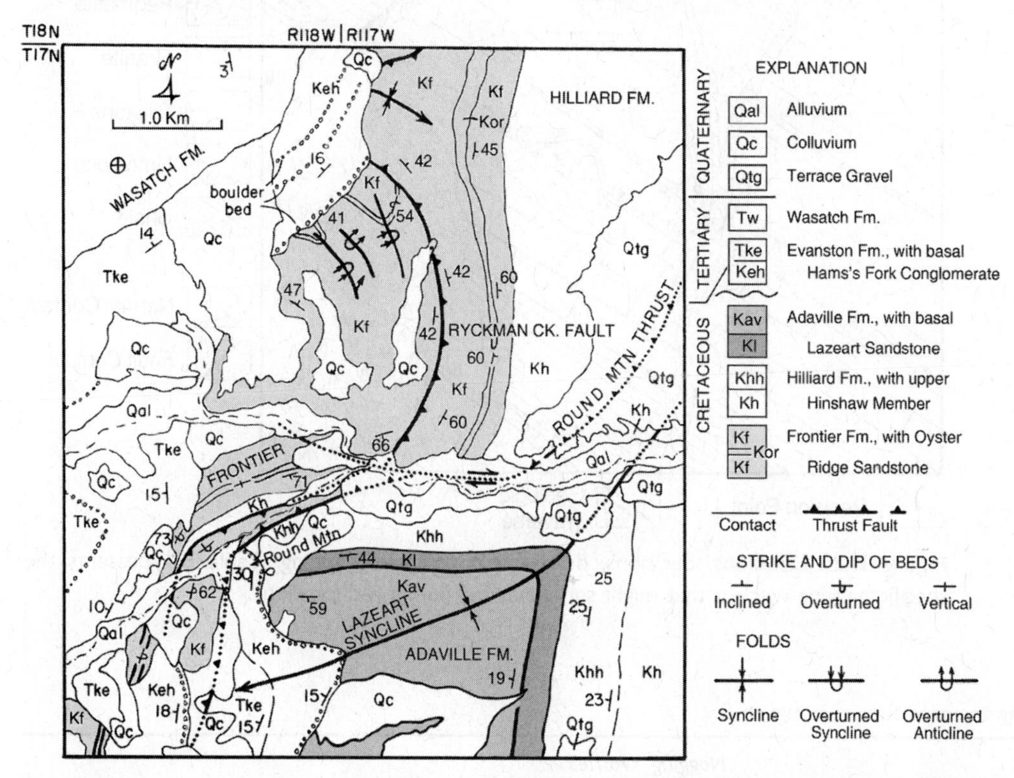

Figure B.1 Example of informative geologic map, Round Mountain area, Wyoming. [From Worrall, 1977. Published with permission of Wyoming Geological Association.]

Geologic mapping is more than an activity. It is a powerful method for systematic structural analysis and scientific discovery. A proper geologic map cannot have loose ends. If a geologist decides that one of the essential formations to be mapped is the Nugget Sandstone, then the top and bottom of the Nugget Sandstone must be tracked across the entire map area. Its upper and lower contacts with adjacent formations must be followed (figuratively speaking) step by step. Where the contacts are covered with alluvium, this must be shown. Where the contacts are offset by faults, or intruded by dikes, the offsets or intrusions also must be shown. If the formation "*suddenly disappears,*" the geologist must cope until the formation is "*found*" or until a satisfactory explanation is discovered that can be portrayed through the use of standard geological symbols.

Hundreds of decisions have to be made in the course of a single day of geologic mapping. Consider the map area shown in Figure B.2. The geology consists of northeast-trending sedimentary rocks that first were faulted, and later intruded by granite and pegmatite. As we begin our mapping through this hypothetical geology, we might as well be walking through a corridor with walls so high that we are cut off from all geologic relationships except those exposed directly underfoot. Problems and nagging worries can accumulate exponentially in even the simplest geologic circumstances (see Table B.1). We need to remember to keep a sense of humor and a sense of purpose during the initial stages of mapping, and resist being overwhelmed by the number of loose ends in the early going.

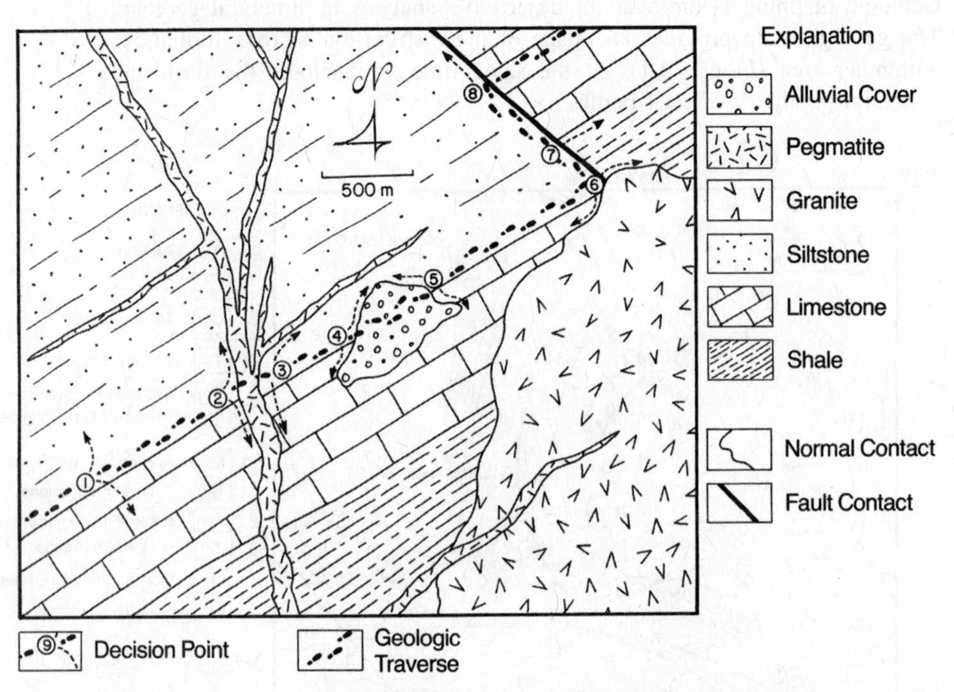

Figure B.2 Decisions, decisions, decisions in geologic mapping. Table B.1 presents the specific nagging worries that might surface at the numbered locations.

TABLE B.1
Nagging worries while mapping the area shown in Figure B.2

Point	*Decision*	*Nagging Worries*
1	To map the contact between the limestone and siltstone units, following it to the northeast.	When am I going to map the contact to the southwest? When am I going to map northwestward across the siltstone? When am I going to traverse southeastward across the limestone?
2	To search for the contact between limestone and siltstone on the east side of the pegmatite dike.	When am 1 going to follow the west edge of the dike to the north? When am I going to follow the western contact of the dike to the south?
3	To continue to follow the contact between the limestone and siltstone to the northeast.	When am I going to map the east edge of the dike to the north? When am I going to map the east edge of the dike to the south? Where does the dike branch?
4	To cross the alluvial cover and rediscover the contact between the limestone and the siltstone.	When am I going to map-out the contact between alluvium and the siltstone unit? When am I going to map the contact between alluvium and the limestone unit?
5	To continue to follow the contact between limestone and siltstone to the northeast.	When am I going to traverse northwestward across the siltstone? When am I going to traverse southeastward across the limestone?
6	To map the fault northwestward and to search for the offset limestone unit.	When am I going to map the contact between limestone and granite to the southwest? When am I going to map the contact between shale and granite to the southeast? Where is the offset limestone bed? What if I don't find it??
7	To continue to map the fault contact and to search for the siltstone/limestone contact.	When am I going to map the contact between limestone and shale to the east?
8	To continue to follow the contact between limestone and siltstone.	When am I going to map the fault northward? When is lunch??

Base Maps

Ideal control for geologic mapping is the combination of topographic maps and aerial photographs. Topographic maps show details of physiography and culture, which permit accurate positioning of contacts and structures. For those mapping in the United States, United States Geological Survey (USGS) publishes both 15- and 7.5-minute maps, scaled at 1:62,500 and 1:24,000, respectively. (These maps cover areas corresponding to 15′ and 7.5′ of latitude and longitude). For regional mapping and compilation, 30′ × 60′ topographic sheets (scaled at 1:100,000), or 1° × 2° topographic sheets (scaled at 1:250,000) are extremely valuable. A geologic map of an entire state or country may be at a scale of 1:500,000, 1:1,000,000, or 1:5,000,000. It is advantageous to map on a copy of the topographic map that is enlarged compared to the intended presentation of the final map.

In some instances the most valuable base for geologic mapping is a set of **aerial photographs**, preferably with enough overlap to afford stereoscopic coverage. Aerial photographs of some areas reveal the bedrock distribution and structure so explicitly that the mapping of contacts can be done effortlessly (Figure B.3). The total display of rock, vegetation, and physiographic and cultural features permits the locations of control points and contacts to be posted easily and accurately.

Figure B.3 Steep east-dipping Jurassic beds along east margin of San Rafael Swell, just southwest of Interstate 50, near Green River, Utah. White rocks are Navajo Sandstone, overlain by Carmel Formation. [From Google Earth.]

Let us imagine we are about to undertake a detailed structural geologic mapping of a relatively small area (less than 25 km^2) somewhere in the United States. An effective way to proceed is to first retrieve existing geologic maps that contain the area of interest. This is readily done by going to the USGS National Geologic Map Database at http://ngmdb.usgs.gov/ngmdb/ngm_catalog.ora.html. Within any luck at all you will be able to secure a quadrangle map, in color, with descriptions and distributions of all of the geologic formations, and representations of the fundamental structural geology. Such maps provide the big picture. Next we would want to download a USGS topographic map, by going to the USGS Topographic Map website, http://topomaps.usgs.gov/. Additionally, we would go to *Google Earth* and download images of a variety of scales, including images that present the map area in fly-over modes. Having *Google-Earth images* allows us to take full advantage of the aerial expressions of rocks and structures, expressed in living color. It is particularly useful to carry a GPS instrument to plot lat/long or AMS coordinates for sites where we take readings or enter notebook descriptions.

Components of a Geologic Map

Learning how to make a geologic map is aided by learning the fundamental components of such a map. Ridgeway (1920) long ago described the components (Figure B.4). The map shows the distribution of rock formations by means of color, patterns, and letter symbols. Each color, pattern, and symbol on the map should coincide exactly with the distribution of the corresponding rock formation. The meaning of the colors, patterns, and symbols is provided in the **Explanation**: the series of boxes, suitably colored, patterned, and labeled, that provides the identification of formation name and usually a brief lithologic description.

If geologic maps are printed in black and white, line and symbol patterns must be used in lieu of color to show the distribution of rock formations. Line symbols are difficult to choose in structural studies because most patterns interfere with and/or mask the fundamental **line work** of structural information and symbology. Therefore, it is useful to represent the various mappable units with **lithologic symbols** that convey an impression of the nature of the

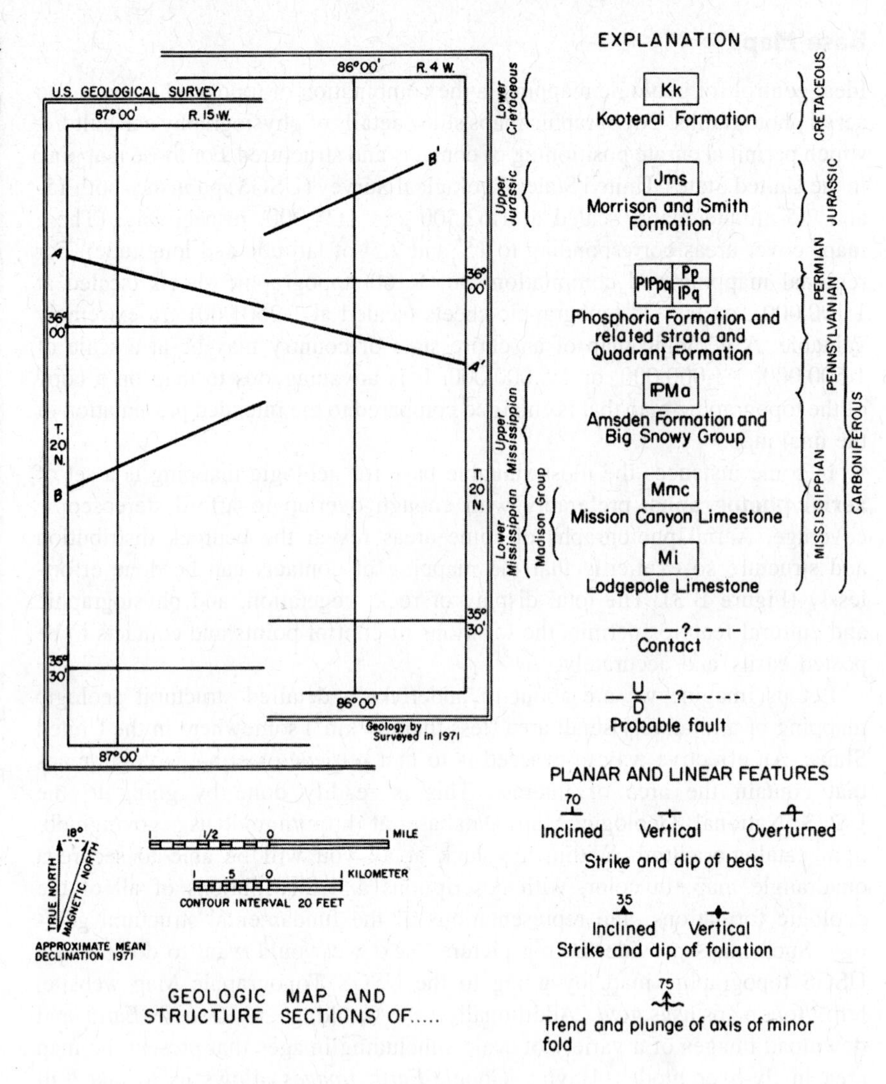

Figure B.4 Elements of the layout of a geologic map. [From Ridgeway (1920). Courtesy of United States Geological Survey.]

dominant bedrock units. An example of the use of lithologic symbols is shown in Figure B.5.

The formation symbols and ages are arranged vertically in the Explanation, in the order of relative age (see Figure B.4). Astride the vertical array of boxes are entered the ages of the rock formations, as known or as inferred. Each of the formations shown on a geologic map must be enclosed by a contact, except of course where the formations extend outside the map area. There they are intercepted ("*cut off*") by the border of the base map. Depositional and intrusive contacts are represented by thin lines; fault contacts are represented by thicker lines (see Figure B.1).

Showing the distribution and contacts of rock formations is not enough. To these must be added symbols that disclose the major structures and the internal geometry of the rock system. The most informative structural geologic maps display abundant **structural symbology** (see Figure B.1), which conveys the geometric and physical nature of the structures present, such as contacts, bedding, joints, shear fractures, veins, faults, cleavage, foliation, lineation, and shear zones (Table B.2). Examples of the use of symbols are presented in Figure B.6.

In order to designate the size and orientations of features shown on a geologic map, all maps include **scales** and **north arrows** (see Figure B.4). Both bar and ratio scales are normally presented. Bar scales should be labeled

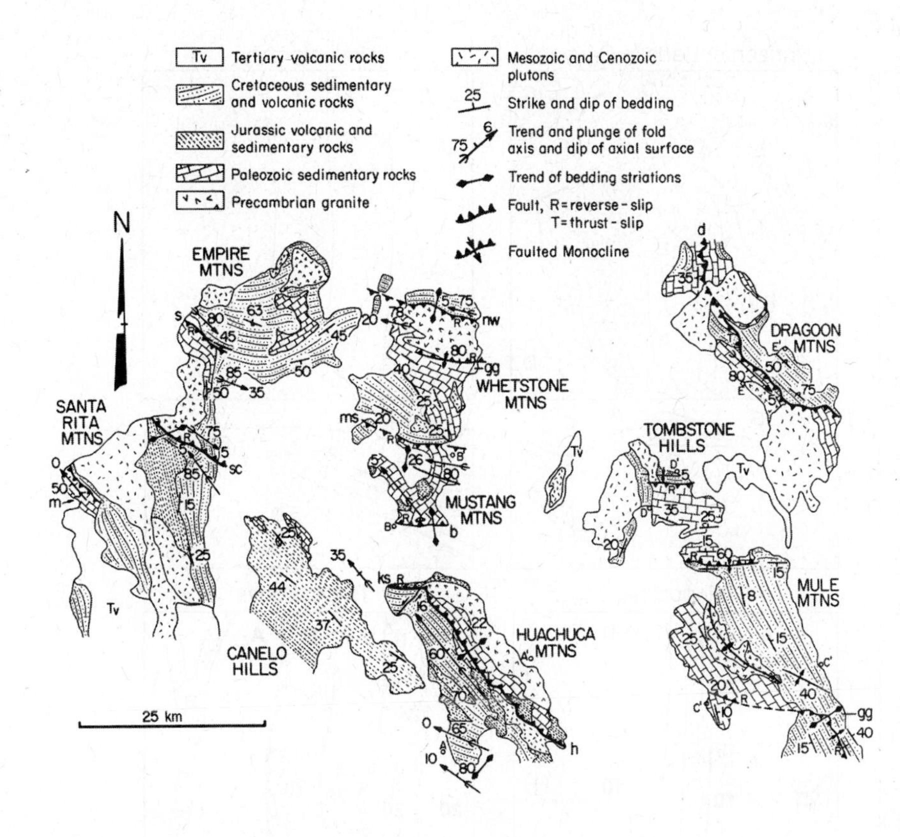

Figure B.5 Structural geologic map of southeastern Arizona showing the use of lithologic symbols to distinguish among the mapped units. [From Davis, 1979, *American Journal of Science*, vol. 279. Reprinted by permission of American Journal of Science.]

TABLE B.2
Map symbols for bedding, foliation and cleavage, and lineation

Symbol	Meaning
Bedding	
20	Strike and dip of bedding
60	Strike and dip of overturned bedding
	Strike and dip of vertical bedding
⊕	Horizontal bedding
Foliation and Cleavage	
30	Strike and dip of foliation
	Strike of vertical foliation
80	Strike and dip of cleavage
	Strike of vertical cleavage
Lineations	
30	Trend and plunge of lineation
25 40	Strike and dip of foliation, and trend and plunge of lineation in the plane of foliation.
65	Strike and dip of bedding, and trend and plunge of lineation in the plane of bedding.

both in metric (km, m) and U.S. customary (mi, ft) units. Two north arrows are shown on a geologic map, one pointing to **true north**, and the other to **magnetic north**. **Magnetic declination** is specified in degrees east or west of true north. If the base map is a topographic map, the contour interval is also shown.

Borders and **title** are the final components of a geologic map (see Figure B.4). Most geologic maps do not need a border; this feature is used only for maps that are very large or irregular. The title of the map is generally placed along the lower margin, and it conveys the name and general location of the map area. The name of the geologist who prepared this map appears either beneath the title or in the lower right-hand corner of the map. The source of the base map should appear in the lower left-hand corner.

Some Basic Procedures

The geologic mapping process is aided by keen observational skills, breadth of geological background, a curious and questioning disposition, attention to geometric details, accuracy, neatness, and patience. Some maps are better than others because of the attention paid to detail, the quality of the line work, and the depth of thought during the mapping process. Really good maps are often reflections of really good thinking, mechanics and techniques aside. The geologic map is a description. The more complete and accurate the description, the greater the impact of the map.

The basic tools for geologic mapping include covered clipboard, hardback field notebook, protractor scale, pencil(s) (hardness No. 3-4), colored pencils, drafting pen, hand lens, rock hammer, 2-m tape, and compass. To this list we would add a handheld GPS unit. All these items can be carried handily on a belt and in a day pack, together with other gear that might be useful (water bottles, camera, chisel, binoculars, first-aid kit, pocket altimeter, 50-ft tape,

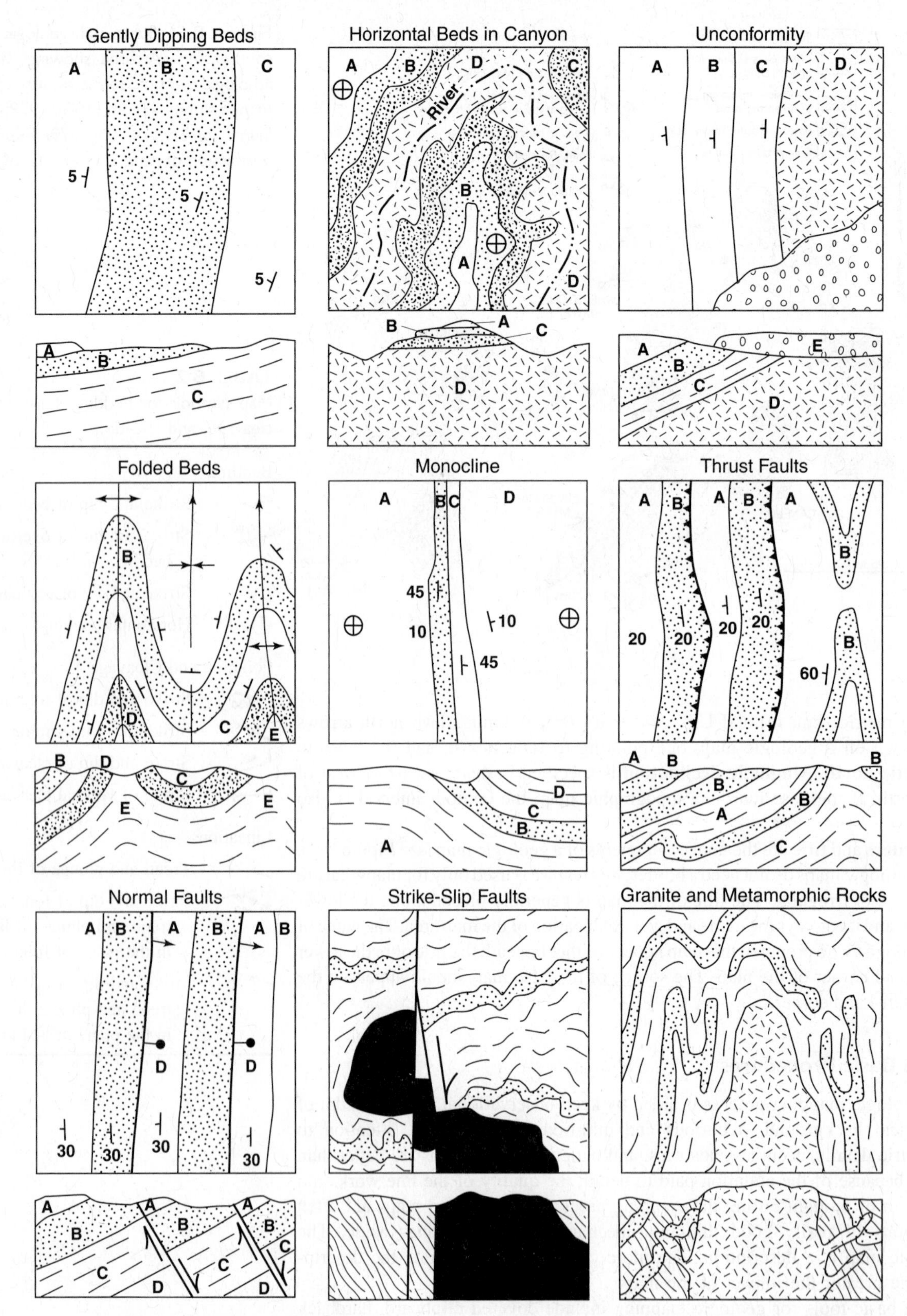

Figure B.6 Examples of symbols used in presenting structures in map view and cross section. Note how different structural associations have different and distinctive map patterns.

sunscreen, raincoat, and tools required by local circumstances e.g., tweezers to remove cactus spines if mapping in desert environments).

The actual process of geologic mapping begins with establishing **map units** appropriate to the project at hand. In mapping a 7.5′ or 15′ quadrangle (scales: 1:24,000, and 1:62,500, respectively), it is customary to map out the formally established stratigraphic formations. In mapping small areas at a large scale (1:12,000 or larger), it is customary to "*break out*" distinctive units *within* single formations. In mapping very large areas at a small scale (1:125,000 or smaller), it is customary to combine groups of established regional formations into tectonic assemblages formed under different tectonic conditions or circumstances (e.g., arc magmatic belt, miogeoclinal prism, accretionary wedge, etc.).

The actual selecting of map units involves both reconnaissance fieldwork and the reading of pertinent literature. Through this preparation, we can become reasonably familiar with the lithologic characteristics and the contact relationships within the geologic system of interest. Beyond that, we can develop a specific and detailed knowledge of the characteristics of each formation or marker unit or assemblage by visiting **type localities** and by **measuring sections** of the rock in the area of study.

Geologic mapping may proceed in a number of ways, but significant time and effort will always be invested in tracing out contacts of each of the map units. It is useful early in the mapping to traverse across the grain of the rocks and structures, to become familiar with each of the rock formations, their contacts, and their internal structures. It is also important to *walk out* specific contacts, in order to follow their course, their geometry, and their continuity (or lack of) to the limits of the area.

Rock and structural data are collected along each traverse. We assign **station numbers** to localities where we collect data and where we make observations and interpretations that we believe might be included in the final map or final report. These numbers are posted on the base map at the exact locations where the measurements and observations were made. Data are entered in the field notebook under the appropriate station number.

For purposes of detailed structural analysis, it is often helpful to identify and map **marker units** as an aid in establishing the structural style of deformation. Markers are chosen on the basis of distinctiveness, resistance to erosion, and continuity. Figure B.7 displays a number of marker beds in

Figure B.7 This is a photograph of part of the fold array that is captured on the cover of this book. All of the strata within this frame belong to the Horquilla Formation (Pennsylvanian), an interbedded sequence of shale (slope former) and limestone (ledge former). The limestone beds are useful "*marker beds*" in mapping internal deformation within the Horquilla Formation. Location is near Colossal Cave, east of Tucson, Arizona. [Photograph by G. H. Davis.]

Figure B.8 Detailed structural geologic map of the marker-bed strata shown in Figure B.7. Note symbology used in presenting structures in map view and cross section. [From Davis et al., 1974, *Journal of Geological Education*, v. 22, pp. 204–208. Published with permission of National Association of Geology Teachers.]

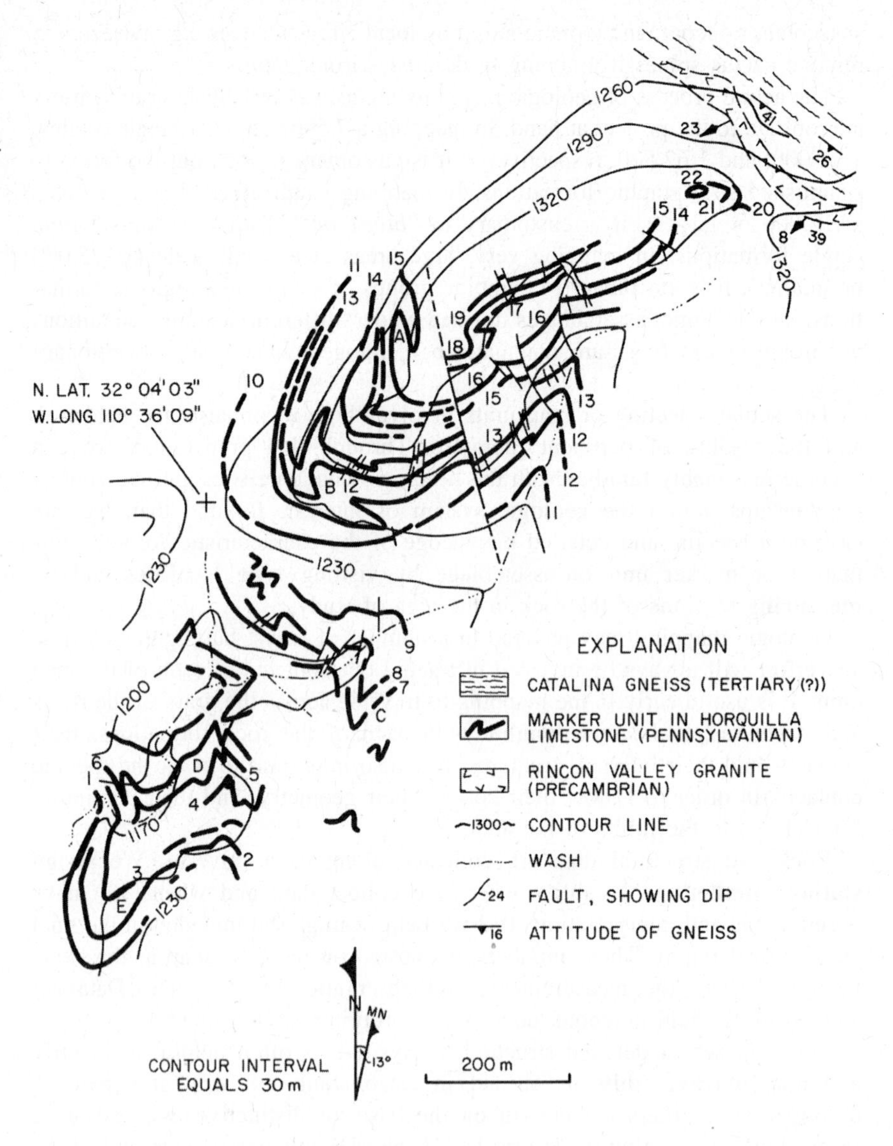

complexly folded and faulted Paleozoic strata in the foothills of the Rincon Mountains, east of Tucson. Such continuously exposed resistant beds are a godsend in complexly folded and faulted sequences of rocks, for they can be traced out. The limestone marker units captured in Figure B.7 are interbedded with shale beds, which are poorly exposed, occupying the covered zones between the bold white limestone outcroppings. My (GHD) students and I made good use of these marker beds in creating the structural geologic map shown in Figure B.8. We presented the marker units in a distinctive line-style, and in the Explanation made clear that all of these marker beds are part of the Horquilla Formation (Pennsylvanian). Mapping the upper and lower contacts of the Horquilla Formation would not have been sufficient in determining internal structural geometry.

Figure B.9 Veteran prospector-geologist mapping at close range in the field. [G. H. Davis encountered and photographed this person in the Whetstone Mountains, southeastern Arizona.]

Geologic maps are made in the field (Figure B.9), not constructed in the office at the end of the day on the basis of notebook data and memory. Contact lines are drawn on the base map as geologic mapping proceeds, and map units are colored progressively as their distributions become established. Orientations of bedding, foliation, folds, cleavage, fractures, and other structures are entered in the field notebook, and representative readings are posted directly on the base map as well. It is easier but inadvisable to record orientation readings in the notebook only, waiting to plot them on the map back at the office. Plotting readings immediately on the map in the field is essential in order to visually track the evolving patterns of orientations and geometries and thereby recognize inconsistencies and departures that will require explanation. Careful study of the map as it develops will provide direction in determining where to go next for geologic insight.

Record Keeping in Field Notebooks

There is no one way to arrange your notebook. Structural geologists use different approaches based largely on personal preference. Turner and Weiss (1963) recommended a notebook layout that is practical when large numbers of orientation and physical measurements are to be taken (Figure B.10). One side of the notebook is reserved for the measurements. The data are recorded

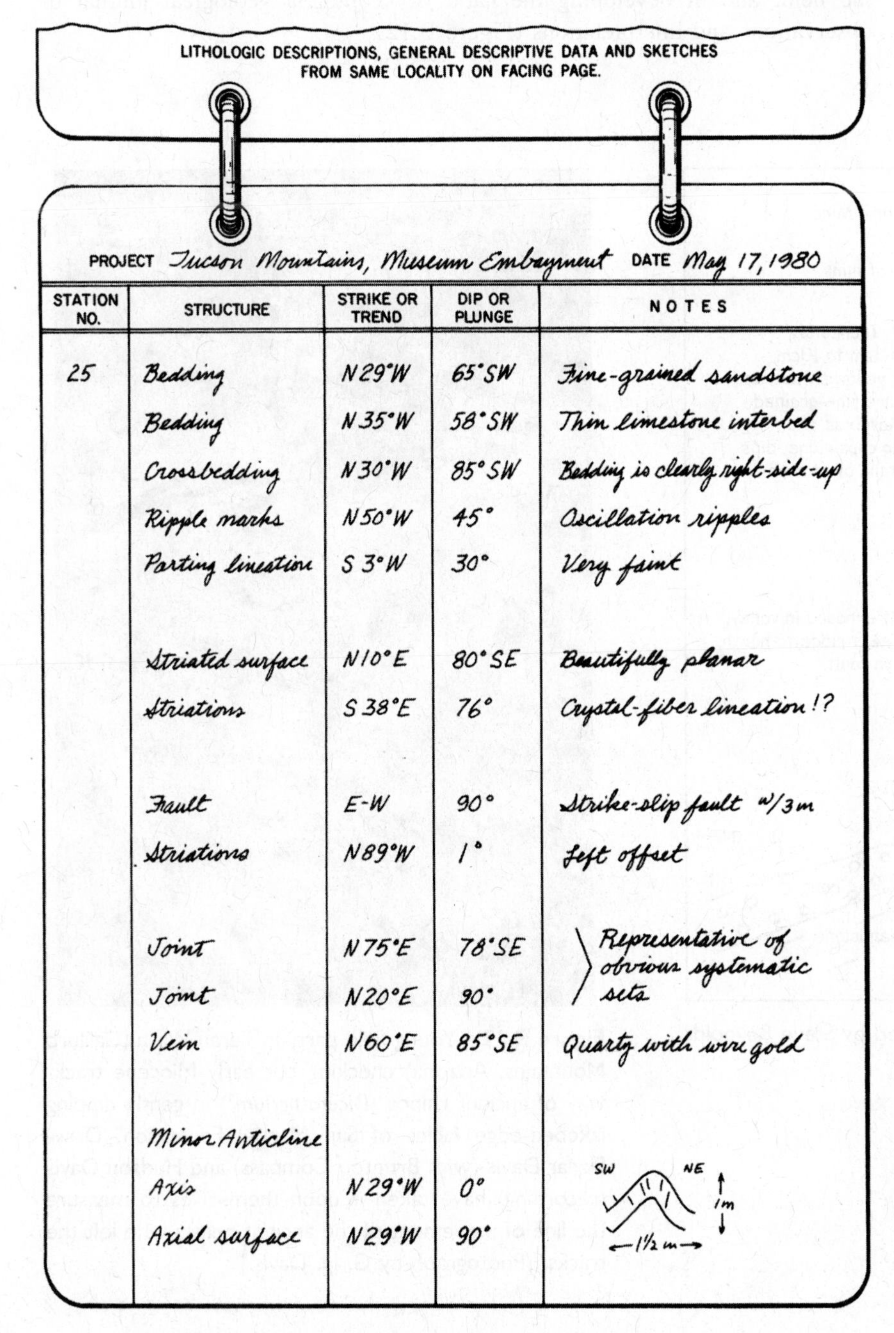

LITHOLOGIC DESCRIPTIONS, GENERAL DESCRIPTIVE DATA AND SKETCHES FROM SAME LOCALITY ON FACING PAGE.

PROJECT *Tucson Mountains, Museum Embayment* DATE *May 17, 1980*

STATION NO.	STRUCTURE	STRIKE OR TREND	DIP OR PLUNGE	NOTES
25	Bedding	N29°W	65°SW	Fine-grained sandstone
	Bedding	N35°W	58°SW	Thin limestone interbed
	Crossbedding	N30°W	85°SW	Bedding is clearly right-side-up
	Ripple marks	N50°W	45°	Oscillation ripples
	Parting lineation	S3°W	30°	Very faint
	Striated surface	N10°E	80°SE	Beautifully planar
	Striations	S38°E	76°	Crystal-fiber lineation!?
	Fault	E-W	90°	Strike-slip fault w/3m
	Striations	N89°W	1°	Left offset
	Joint	N75°E	78°SE	Representative of obvious systematic sets
	Joint	N20°E	90°	
	Vein	N60°E	85°SE	Quartz with wire gold
	Minor Anticline			
	Axis	N29°W	0°	
	Axial surface	N29°W	90°	

Figure B.10 Notebook arrangement recommended by Turner and Weiss. [From Structural Analysis of Metamorphic Tectonites by F. J. Turner and L. E. Weiss. Published with permission of McGraw-Hill Book Company, New York, copyright © 1963.]

in an orderly way in columns (not scattered within paragraphs) to enable efficient recovery when it is time to plot data on maps or enter data into a computer. In the approach used by Turner and Weiss, rock descriptions are placed on facing pages, along with sketches or cross-sections of important relationships, code numbers for specimens collected, photograph numbers, and most important, ideas that come to mind during fieldwork.

My approach (SJR) is to take notes on normal lined or gridded notebook paper carried beneath the map in a covered clipboard. I sketch, take notes, and enter measurements on the sheets in an organized way so that I can quickly extract what I need at a later time (Figure B.11). I have found this method to be efficient, because my map and note sheets are all in one place, and I never have to fish for my notebook. I prefer not having to carry *all* of my notes with me *all of the time* in the field.

You cannot start out too early in life in mastering effective note-taking in the field, and in developing the habit of keeping a geological journal of observations and interpretations (Figure B.12).

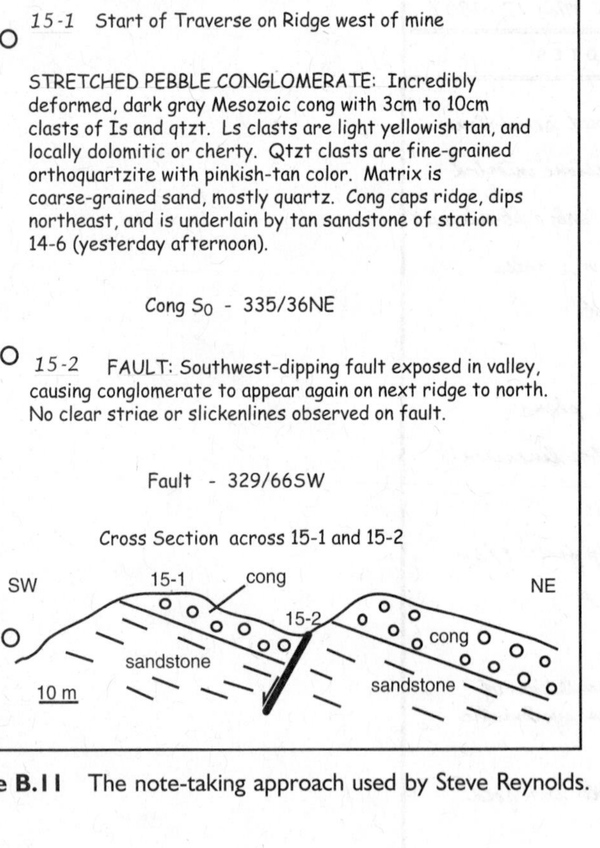

April 15, 1995 Granite Wash Mountains, Yuma Mine

15-1 Start of Traverse on Ridge west of mine

STRETCHED PEBBLE CONGLOMERATE: Incredibly deformed, dark gray Mesozoic cong with 3cm to 10cm clasts of ls and qtzt. Ls clasts are light yellowish tan, and locally dolomitic or cherty. Qtzt clasts are fine-grained orthoquartzite with pinkish-tan color. Matrix is coarse-grained sand, mostly quartz. Cong caps ridge, dips northeast, and is underlain by tan sandstone of station 14-6 (yesterday afternoon).

Cong S_0 - 335/36NE

15-2 FAULT: Southwest-dipping fault exposed in valley, causing conglomerate to appear again on next ridge to north. No clear striae or slickenlines observed on fault.

Fault - 329/66SW

Cross Section across 15-1 and 15-2

SW 15-1 cong NE
15-2
cong
sandstone
sandstone
10 m

Figure B.11 The note-taking approach used by Steve Reynolds.

Figure B.12 Young field party in Teran Wash (Galiluro Mountains, Arizona) checking out early Miocene track-way of ancient rhinos (*Diceratherium*?) in gently dipping lakebed-edge facies of San Manuel Formation. Drew Skylar Davis (with Brunton Compass) and Hudson Davis (recording) have taken it upon themselves to measure the line of movement of the ancient beasts who left the tracks. [Photograph by G. H. Davis.]

C. MAPPING CONTACT RELATIONSHIPS

Among the most important structural features to identify, interpret, map, and describe in the course of fieldwork are the **geologic contacts** between different rock formations. Rock bodies are the building blocks of the crust of the Earth. They come in all sizes, shapes, and strengths. They are modules, of sorts, that make up the whole. Through processes of deposition, intrusion, faulting, and/or shearing, they generally fit together perfectly along tight contacts. A wonderful nongeologic image of the ultimate in perfect fit is seen in the mortarless contacts between limestone blocks in walls of the Inca architectural marvel known as Saqsaywaman, Cuzco, Peru (Figure C.1). Geologic symbols for lithologic contacts (contacts of deposition or intrusion) and fault contacts are presented in Table C.1. Contacts are simple to distinguish in theory. But where structures are complex, deformations are multiple, and exposures are poor, the job of recognizing the nature of contacts can be very difficult and challenging.

Figure C.1 Limestone blocks in Inca wall fit together perfectly along mortar-less contacts. Window-like opening is ~1 m high. Like many geologic contacts, these man-made contacts are difficult to interpret: What processes were required to bring the limestone blocks into perfect contact? [Photograph by G. H. Davis.]

Conformable Depositional Contacts Within Formations

Within any given geologic formation, such as the Madison Limestone or the Berea Formation, it is always possible to identify beds that are distinctive because of properties such as color, composition, lithology, texture, and thickness. Such beds *within* a formation of sedimentary or volcanic rocks may be nearly the same age, and are essentially parallel to one another. Such beds comprise a conformable sequence within the formation. The beds accumulated through an essentially continuous deposition, or intermittently in a way that the age difference(s) between adjacent beds are negligible, at least geologically speaking. We refer to such contacts formed between layers and within such sequences as **conformable depositional contacts** (Figure C.2). Conformable depositional contacts are usually planar to slightly irregular in form. We routinely measure the orientations of the beds and the contacts between the beds as a part of geologic mapping (Figure C.3), but we only map such **intraformational conformable depositional contacts** if they occur on either side of a distinctive marker bed whose structure we wish to represent.

TABLE C.1
Geological symbols for lithologic contacts and fault contacts

Symbol	Meaning
Lithologic Contacts	
(line with tick, 60)	Contact, showing dip
- - - - -	Contact, approximate location
. . . .	Contact, concealed
Fault Contacts	
(heavy line with tick and ball)	Fault, showing dip
▬ ▬ ▬ ▬ ▬ ▬	Fault, approximate location
. . . .	Fault, concealed
▬ ? ▬ ? ▬	Fault, existence uncertain
Fault Symbology	
(line with bar and ball)	Fault, bar and ball on downthrown block
(line with teeth)	Thrust fault, teeth on hanging wall
(line with hachures)	Normal fault, hachures on hanging wall
(paired half-arrows)	Strike-slip fault

Unconformities

Rock sequences are made more challenging to interpret because of the presence of contacts that represent time-gaps within the geologic column, where conformable deposition was interrupted, or where erosion during long intervals removed a part of the rock record. A drop in sea level will cause a shallow sea to recede from the continental interior it once occupied, resulting in the cessation of deposition of marine limestones, sandstones, and shales, and the onset of erosion. Long afterward, when the sea advances again and renews marine deposition in the continental interior, younger sediments will be deposited on the eroded top of sedimentary rocks of the earlier period. The contact between the base of the younger sequence and the eroded top of the older sequence will be marked by an **unconformity,** which might later be preserved in the geologic record.

Alternatively during uplift of mountains in a region occupied by the sea the marine sediments and the older rocks beneath will be elevated high above sea level. Tens of millions of years of erosion may eventually reduce the mountain to a coastal plain, and when the sea again advances, it will deposit much younger sediments unconformably on the upturned ends of older sedimentary rocks, and perhaps on the deeply eroded igneous and metamorphic rocks that once made up

Figure C.2 (*A*) Normal depositional contacts within rhyolite ignimbrite in the barranca country of Sinaloa, Mexico. Contacts are beautifully exposed in this region, the most deeply dissected part of the Sierra Madre Occidental. The volcanic rocks are part of a magmatic arc, approximately 400 × 1500 km in surface area. The volcanics of the arc erupted almost entirely during Miocene time. [Photograph by D. J. Lynch.] (*B*) Normal depositional contacts within sequence of sedimentary rocks exposed along the Goosenecks of the San Juan River, Utah. [Photograph by G. H. Davis.]

Figure C.3 Good example of a planar intraformational conformable depositional contact within Cretaceous sandstones just outside of Keystone, Colorado. Vivian Chang and Drew Davis each straddle the contact. [Photograph by G. H. Davis.]

the deep core of the mountain belt. The unconformity in places will separate young horizontal sedimentary rocks from older, tilted, deformed sedimentary rocks. In other places the young horizontal sedimentary rocks will rest directly on an erosionally carved surface on old granite or schist. Again, the unconformity marks a gap in the rock record. Time is missing.

In short, an **unconformity** is a depositional contact between two rocks of measurably different ages. Unconformities are divided into three major classes: **nonconformities**, **angular unconformities,** and **disconformities** (Figure C.4), and their characteristics permit their identification during geologic mapping.

Nonconformities are depositional surfaces separating distinctly younger sedimentary or volcanic rocks above from distinctly older igneous or metamorphic rocks beneath (see Figure C.4). Nonconformities include the **great unconformities** that separate Precambrian crystalline basement rocks from overlying Paleozoic, Mesozoic, or Cenozoic sedimentary or volcanic strata (Figure C.5*A*). But they also include the unconformities separating post-Precambrian crystalline rocks from overlying, younger, sedimentary and/or volcanic strata (Figure C.5*B*) Standing on a nonconformity is something special, particularly if the time gap between crystalline and cover rock represents a billion years (Figure C.6).

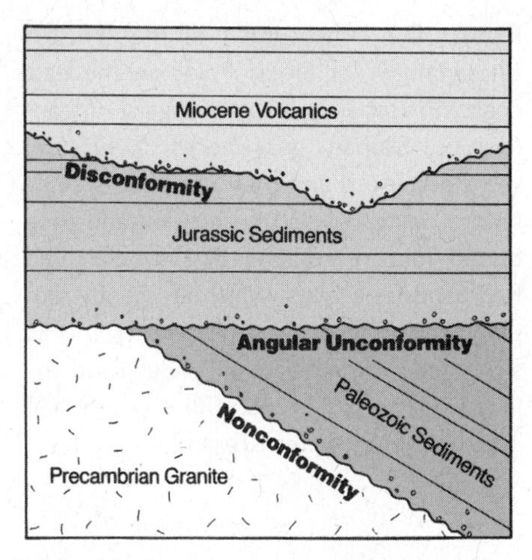

Figure C.4 Schematic portrayal of the three kinds of unconformity: nonconformity, angular unconformity, disconformity.

Figure C.5 (*A*) Nonconformity between flat-lying Cambrian Tapeats Sandstone and underlying Precambrian crystalline rocks, as exposed in Granite Gorge of the Grand Canyon. [Photograph by N. W. Carkhuft. Courtesy of United States Geological Survey.] (*B*) Nonconformity between mid-Tertiary andesitic volcanic rocks (black) and underlying early Tertiary Gunnery Range Granite, southwestern Arizona. (Not all contacts are as plain as black and white.) [Photograph by D. J. Lynch.]

Figure C.6 George Noble Davis has his back to Precambrian granite and his hand on the steep contact between granite and overlying Pennsylvanian sandstone. He recognizes that the contact is a nonconformity. Locale is Red Rocks Amphitheater in the Front Range of the Rocky Mountains just west of Denver, Colorado. [Photograph by George Davis.]

Figure C.7 Photograph of the "*Hutton Unconformity*" at Siccar Point on the east coast of Scotland. It is an angular unconformity. Silurian greywackes (~425 Ma) were deposited and then upturned steeply before being beveled by erosion and then buried (unconformably) by Devonian Old Red Sandstone beds, which are gently dipping. This outcrop was discovered and its signficance (immediately) understood by James Hutton in 1788. [Photograph and copyright © by Peter Kresan.]

An **angular unconformity** is an unconformity that separates layers above and below that are not parallel. Classical angular unconformities are horizontal depositional surfaces separating relatively young horizontal strata above from older steeply dipping strata below (see Figure C.4). Spectacular angular unconformities are sometimes displayed in single outcrops or cliff exposures, most notably the discovery outcrop known as the "*Hutton Unconformity*" (Figure C.7). In contrast, subtle angular unconformities marked by very slight angular discordance can be "seen" only through careful regional mapping.

A **disconformity** is an unconformity separating strata that are parallel to each other (see Figure C.4). The physical presence of a disconformity may be hard to detect. Recognition may require complete knowledge of the ages of beds within the sequence of strata that contains the disconformity. In the Paleozoic geologic column of the Grand Canyon, Ordovician and Silurian rocks are completely missing, and Devonian sedimentary rocks are spotty. Thus there exists a disconformity separating the flat-lying Mississippian Redwall Formation above from the flat-lying Cambrian Muav Limestone below.

For all three types of unconformity, the surface marking the unconformity itself is parallel to the bedding or layering of the rocks above the unconformity. The bed directly above an unconformity commonly contains a **basal conglomerate**, normally composed of clasts of the rock directly beneath the unconformity. Basal conglomerates advertise erosional intervals. The basal conglomerate may range in coarseness from a thin fine granule conglomerate to a thick coarse boulder conglomerate. Surfaces of unconformity may locally possess **topographic relief** that can be recognized as the product of ancient erosion, perhaps even including the preservation of the cross-section of an old stream channel. One such channel can be seen in the sidewall of Blacktail Canyon in the Grand Canyon (Figure C.8). Under ideal conditions, fossil soil profiles, called **paleosols**, are preserved in rocks directly beneath the old erosion surface. These may be baked where overlain by lava flows.

Intrusive Contacts: Magmatic

The mapping of **intrusive contacts** requires identifying locations where magma has solidified against the country rocks through which it once flowed. The term **country rock** refers to the rock assemblage, whatever its nature, which hosts the intruder. Country rock may be sedimentary, igneous, or metamorphic. The intrusive contact proper is the interface between country rock and the intrusive body.

Figure C.8 Deep down in the Grand Canyon, a hike up Blacktail Canyon takes you to this amazing exposure of the Great Unconformity between Cambrian sandstone above, and Proterozoic schist beneath. Right at the contact is a Cambrian channel, full of cobbles derived from quartz veins in the underlying schist. This is relative dating at its best. [Fish-eye view photograph by S. J. Reynolds.]

Figure C.9 (*A*) Photograph of tonguelike apophyses branching from Kuna Crest granodiorite dikes (~95 Ma). Location near Marie Lakes, northwestern Ritter Range, Inyo National Forest, California. Parts of the dikes are pegmatitic. [Photograph kindly provided by Vali Memeti.] (*B*) Dark host rock (granite gneiss) is here intruded by older set of pegmatite dikes (white, running upper right to lower left), which are in turn intruded by a younger set. Maggi Davis ascending this Paleoproterozoic host. Location is Frisco, Colorado. [Photograph by G H. Davis.]

Igneous intrusive contacts may be recognized on the basis of a number of features. Outcrops of the country rock close to an igneous intrusive contact may be invaded by **apophyses** of irregular tongue-like injections into country rock (Figure C.9). Pieces of fractured country rock, known as **xenoliths** (Ancient Greek, "*foreign rock*"), may become detached from the wall of contact during intrusion to become incorporated within the main igneous intrusion (Figure C.10).

The form of igneous intrusions varies with depth, the composition of magma, and the presence or absence of structural control of the site of intrusion. As a consequence, the description of the contact relationships between igneous intrusions and country rock can be quite challenging to map and visualize, though we show some of the basics in a composite diagram created by Bob Krantz (Figure C.11). Aspects of the vocabulary are stereotypical of geology and its propensity for colorful taxonomy: **batholith**, **laccolith**, **lopolith**, and **phacolith**, not to mention **harpolith** and **bysmalith**.

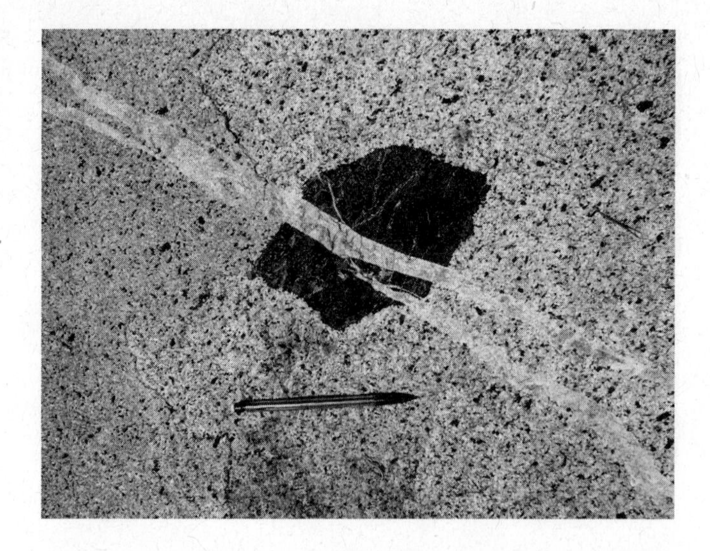

Figure C.10 Amphibolite xenolith in ~90 Ma equigranular Half Dome granodiorite (Memeti et al., 2010) crosscut by late-stage aplitic dikelets near contact to Cretaceous Alaskite of Grace Meadow (Wahrhaftig, 2000) southeast of Tilden Lake, northern Half Dome lobe, Tuolumne batholith. [Photographs kindly provided by Vali Memeti.]

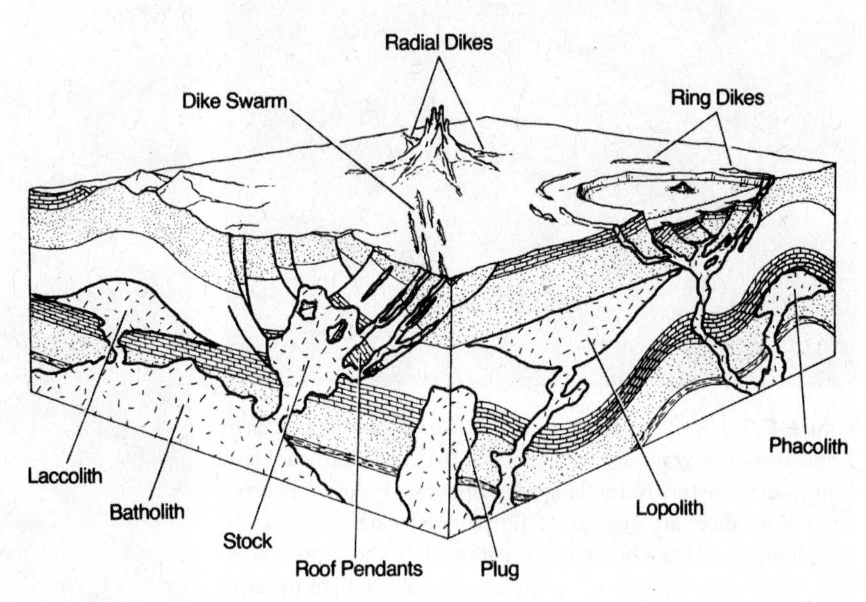

Figure C.11 Composite block diagram showing some of the major kinds of igneous intrusions. [Artwork by R. W. Krantz.]

Large granitic bodies, especially, may have internal flow structure that reveals aspects of the emplacement of the magma, and even the 3D shapes of the body. The primary magmatic flow structures take the form of foliations. Not uncommonly magmatic flow structure is most intensely developed near and parallel to contacts with country rock. One of the most conspicuous foliation expressions is created by alignment of **enclaves** Figure C.12), where the term *enclave* refers to dark elongate microgranitoid enclaves, representing magma globules (Patterson et al., 1989). Pure magmatic flow may be seen as well in preferred orientations of primary igneous minerals, such as feldspars, and the marked absence of plastic deformation or recrystallization of these minerals (Patterson and others, 1989). During primary magmatic flow, crystals (such as feldspars) in the melt are able to rotate without collisions with nearby minerals. Furthermore biotite and hornblende, as independent euhedral grains, may be aligned.

A good example of the relationships between the macroscopic geometry of lobes of granitic intrusion, and the internal foliations within the body, are evident in the mapping carried out by Zak and colleagues (2007) in the Late Cretaceous Tuolumne batholith, central Sierra Nevada, California. Their

Figure C.12 Photograph of enclaves in the Catalina Granite (~25 Ma), exposed near Cargadero Canyon, Catalina Mountains, north of Tucson, Arizona. Note the preferred orientations of disk-shaped black microgranitoid enclaves. [Photograph by G. H. Davis.]

map of the Kuna Crest Granodiorite (a part of the batholith) is presented as Figure C.13. Note how the steeply dipping foliations (four varieties) define a pattern that is consistent with the shapes of the various mappable compositional units within the body (I, II, III, IV, V). Within the plane of some of the foliations there is lineation, representing the (ascending) flow direction.

Country rocks invaded by magma respond to the heat by recrystallization and metamorphism. Sedimentary rocks, unaccustomed to hot environments, are particularly vulnerable to thermal alteration. The intrusive magma itself may quickly cool upon contact with the wall rock, forming a **chill zone** of very fine-grained igneous rock at the border of the igneous body. Igneous intrusions that penetrate country rocks at high, relatively cool levels in the crust may impart to the wall rocks a local **contact metamorphism** that produces a restricted halo or **aureole** of metamorphism in wall rocks along the contact (Figure C.14). For example, one day you might find yourself mapping a limestone unit, only to see it progressively transform into marble, or a **skarn** (limestone altered to a calcium silicate rock). As you walk farther, and if exposures permit, you will come into contact with the igneous intrusive rock that "*baked*" the limestone.

Fault and Shear Zone Contacts

Fault contacts can be recognized at the outcrop scale by virtue of an array of characteristic physical properties, not all of which will be assembled at any one place. Commonly we should expect to find a discrete fracture break or discontinuity, a **fault surface** (Figure C.15) whose orientation and position satisfactorily describe the contact between rocks representing two different formations. Alternatively, a **fault zone** may be present instead of a discrete fault surface. Fault zones consist of numerous closely spaced fault surfaces, commonly separating masses of broken rock, including **gouge** and **breccia**, and perhaps some minor folds. Fault surfaces may be finely polished to **slickensided surfaces** as a result of differential movement, and the slickensided surfaces are almost always marked by **striations** or **grooves** that reflect fault movement.

Shear zones are fault-like in that they accommodate displacements. Unlike ordinary fault surfaces, however, shear zones often do not display any physical break. Instead, the movement is achieved by penetrative deformation within a zone that may be centimeters to kilometers across (Figure C.16). One day you may be mapping a 3-m-thick dike (see Figure C.16), and suddenly find that it bends abruptly and tapers quickly to 6 cm, or to nothing at all, and then just as abruptly reappears to bend in the opposite direction and return to its original

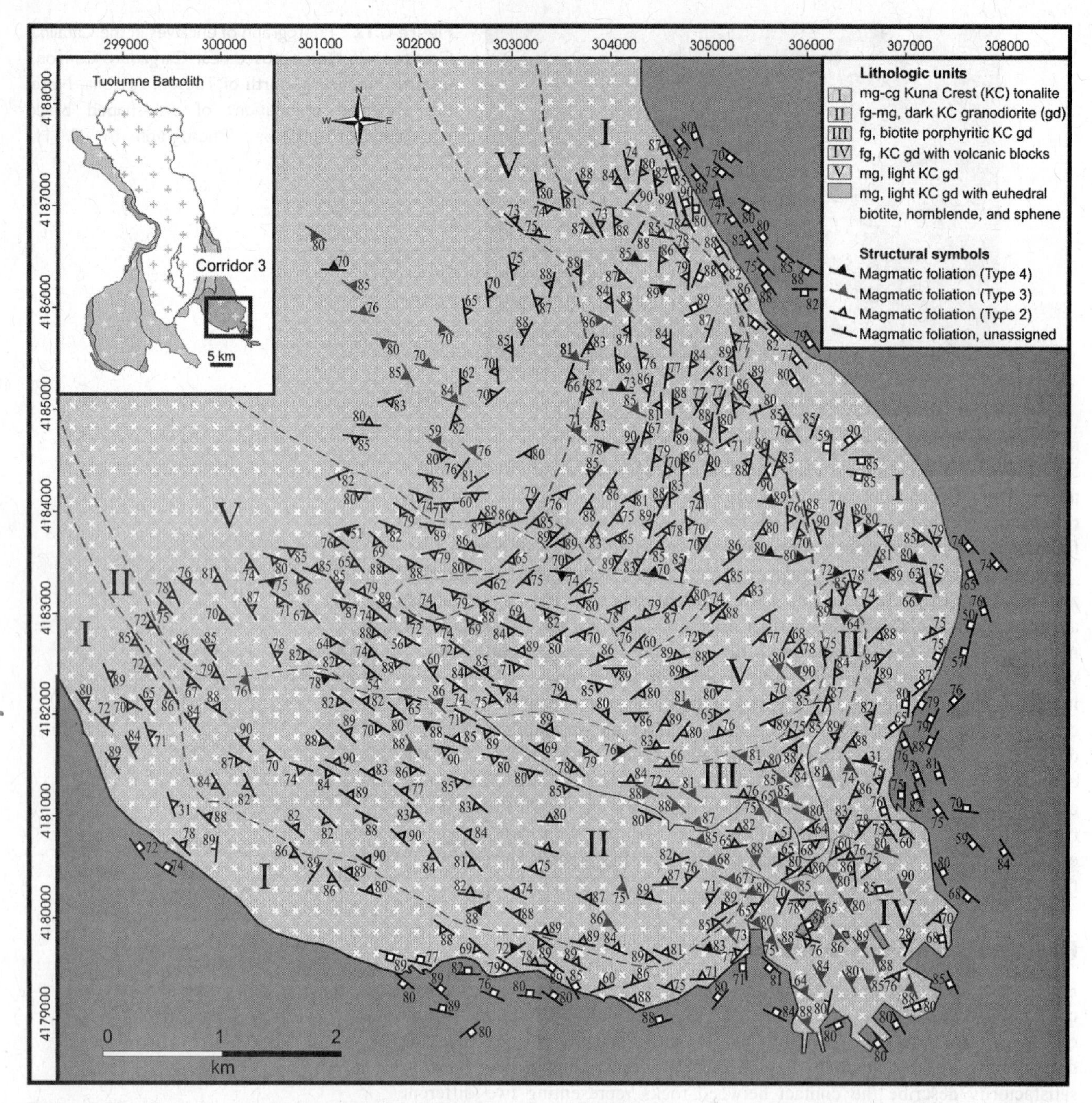

Figure C.13 Geologic map of the southern lobe of the Kuna Crest Granodiorite, part of the Late Cretaceous Tuolumne Meadow batholith, central Sierra Nevada, California. Steeply dipping magmatic foliations are parallel to the outer margins of the lobe, forming an arcuate pattern. Four different types of magmatic foliations are shown. Type 1 was produced by highly localized magma flow. The other three record strain along the boundary of the batholith as well as superimposed regional tectonic strain. [From Zák et al., 2007, Figure 5, p. 191. Published with permission of the Geological Society of America.]

thickness (see Figure C.16). Other layers do the same. Lines representing shear zone contacts are drawn to show the margins of the shear zone. They can be dashed if all of the sheared layers are present, though thinned. But if a layer shears down to nothing, then the contact should be shown solid, denoting that units are actually missing, and that the formations now in contact are normally not in contact where unsheared. The formations are welded together, much in the same manner as salt welds and fault welds in geological environments where salt is sheared to zero thickness.

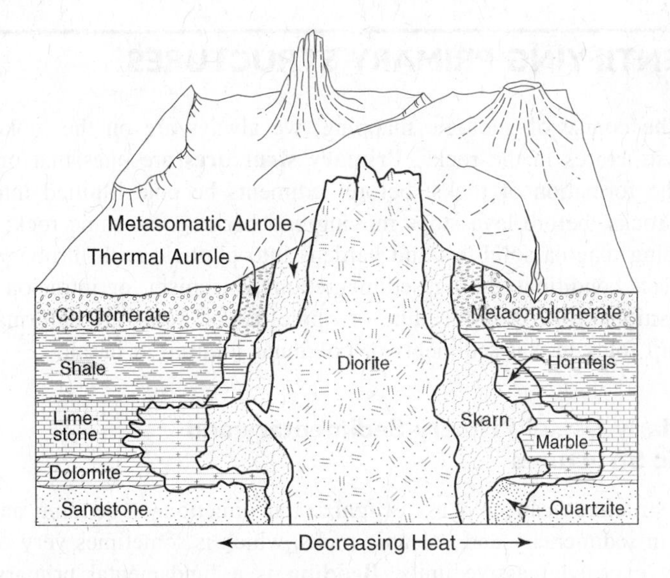

Figure C.14 Block diagram showing contact metamorphism surrounding an igneous intrusion of diorite. The intrusion has invaded a sequence of sedimentary rocks, and converted each to its contact-metamorphic equivalent (e.g., limestone to marble) as shown. Two "*aureoles*" are distinguished: a metasomatic aureole, dominated by chemical alteration of rock by convective circulation of hydrothermal fluids; and a thermal aureole, impacted dominantly by heat conduction in the absence of significant fluid flow. [Diagram courtesy of Lynn S. Ficthter, Department of Geology and Environmental Science, James Madison University, http://csmres.jmu.edu/geollab/fichter/Fichter/Fichterls.html.]

Figure C.15 North-directed photograph of west-dipping reverse-fault contact between (older) Navajo Sandstone (white) on the left and (younger) Carmel Formation (dark) on the right. Location is southern Utah, specifically where the Cottonwood Road crosses the Paria River. Geologist is Sarah Tindall. [Photograph by G. H. Davis.]

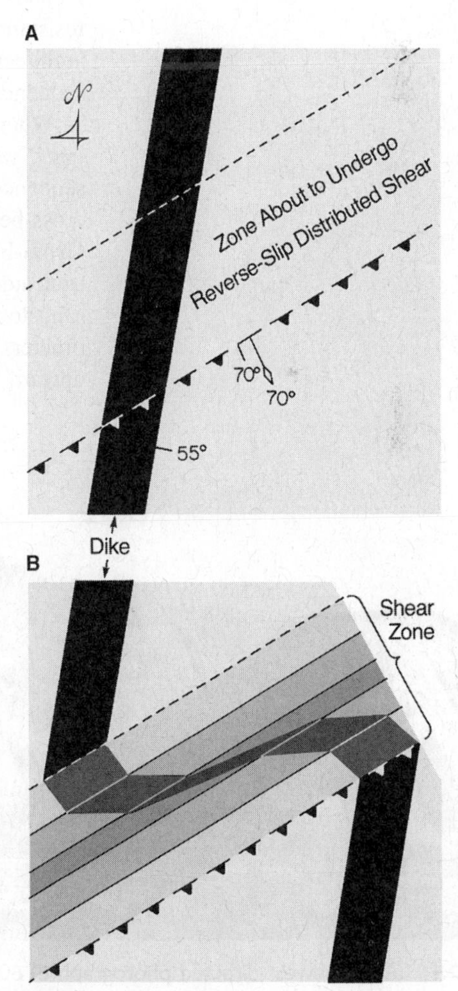

Figure C.16 (*A*) Map view of dike about to be deformed by ductile shearing. (*B*) Distortion of dike achieved by ductile shearing. Contacts of shear zones separate distorted rock from undistorted rock.

D. IDENTIFYING PRIMARY STRUCTURES

During the course of geologic mapping, we always are on the lookout for primary structures in the rocks. **Primary structures** are ones that originate during the formation of rocks: before sediments become lithified into sedimentary rocks; before lava stops moving and becomes volcanic rock; before an intruding magma solidifies and hardens into plutonic rock. Primary structures reflect conditions during sedimentation, volcanism, or intrusion. Some are depositional, like cross-bedding or ripple marks. Some are deformational, like slump folds in wet sediments or columnar jointing in basalt.

Particularly Useful Primary Sedimentary and Volcanic Structures

Primary structures can help us recognize the direction of bedding and flow layering in sedimentary and volcanic rocks, which is sometimes very difficult in poorly exposed massive units. **Bedding** is a fundamental primary sedimentary structure. Distinctive because of color, texture, composition, and resistance to erosion, bedding imparts to sedimentary rocks their fundamental architecture (Figure D.1). **Flow layering** is the counterpart in volcanic sequences.

When we are mapping within a sequence of sedimentary rocks that are "*on end*," we look for primary structures to determine the "*way up*" within the sequence (Figure D.2). The most commonly used **facing indicators** are cross-bedding, graded bedding, oscillation ripple marks, and mud cracks. **Cross-bedding** sweeps into parallelism with the base of a bed and is sharply truncated along the top (Figure D.3). **Graded bedding**, which occurs in sandstones and conglomerates, is marked by intervals on the order of centimeters or meters within which grain or clast size decreases systematically upward toward the top of the bed (Figure D.4).

A

Figure D.1 (*A*) Southwest-directed photograph of 60°-dipping bedding in resistant arkosic sandstone (fluvial, nonmarine) of the Fountain Formation (Pennsylvanian). Location is the Roxborough Park south of Denver. Golf course in foreground is *Arrowhead Golf Course*: "*Our kind of golf course*." Hills in the background are underlain by Precambrian basement (~1400 Ma). The arkoses rest unconformably upon the Precambrian rocks, and were derived from erosion of the Ancestral Rocky Mountains. (The Kluths' home sits on the Great Unconformity). Tilting was late Cretaceous/early Tertiary. (*B*) Northwest-directed photograph showing Fountain Formation bedding at close range. Location is Reds Rocks Amphitheatre near Morrison, Colorado. Mike and Thomassen Davis in foreground. [Photographs by G. H. Davis.]

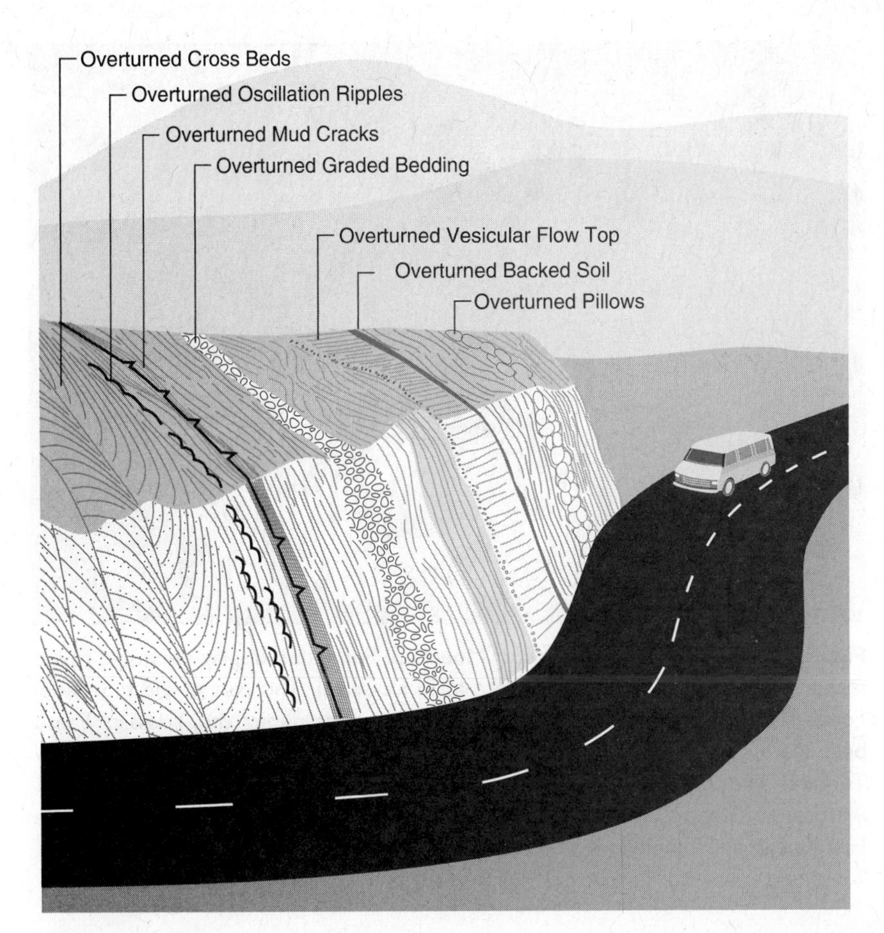

Figure D.2 Diagrammatic representation of an array of facing indicators in a roadcut of interlayered sedimentary and volcanic rocks. Facing indicators reveal the way up within the sequence.

Figure D.3 Cross-bedding in Navajo Sandstone near Checkerboard Mesa, Zion National Park. Geologist is Drew Davis. Truncation of cross-strata along upper bedding surface where Drew is sitting reveals that the Navajo Sandstone is right-side up. [Photograph by Drew's dad.]

Figure D.4 This photograph shows graded bedding within flat-lying Todos Santos Formation (Jurassic) in northwestern Guatemala. Note how the base of the bed is marked by pebble conglomerate. Upward there is a steady decrease in clast and grain size. Then, abruptly, there is a return to coarse material, which represents the base of the overlying (younger) depositional unit. [Photograph by G. H. Davis.]

Ripple marks can be a useful facing indicator, providing you are looking at the right kind, namely **oscillation ripple marks**. Oscillation ripple marks display a symmetrical, concave-upward form, with cusps that point toward the top of the bed (Figure D.5*A*). On the other hand, current ripple marks (Figure D.5*B*)

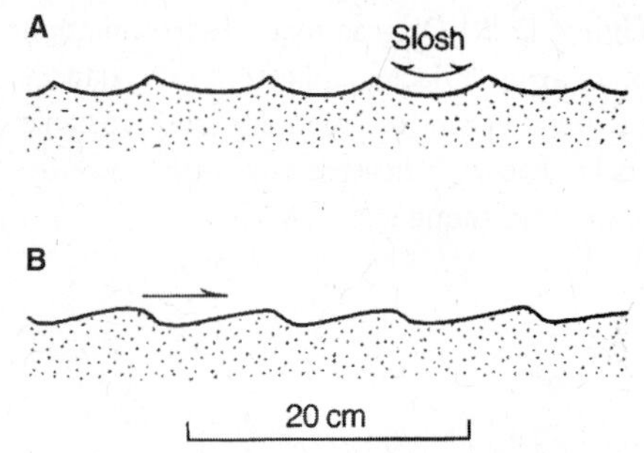

Figure D.5 (*A*) Oscillation ripple marks are distinguished by their cuspate, symmetrical concave-upward forms. The sharp tips point up in the direction of the youngest beds in the sequence. (*B*) Current ripple marks, on the other hand, are asymmetrical and look identical whether upside down or right-side up. Though they disclose nothing about facing, they record current direction.

Figure D.6 Nearly vertical beds of Moenkopi Formation display pervasive ripple markings in the plane of bedding (note knife for scale). This part of the Moenkopi Formation was deposited in a tidal flat environment during the Triassic. These beds were subsequently "cranked up to vertical" during the late Cretaceous/early Tertiary. The location of this outcrop is along the old highway from Cameron, Arizona, to Grand Canyon Village. This steeply dipping bedding occurs where the East Kaibab and Grandview monoclines join. [Photograph by G. H. Davis.]

look exactly the same whether viewed right-side up or upside down. What they lack as a facing indicator they make up for as a current-direction indicator (see Figure D.5*B*). Note that current ripple marks are marked by an asymmetrical profile, with the steepest face inclined in the direction in which the current was moving at the time in which these ripple marks formed. When we discover current ripple marks in tilted rocks (Figure D.6), our task becomes one of properly rotating bedding to horizontal, and then interpreting paleocurrent direction based upon the orientation and asymmetry of the ripple marks.

We are all quite familiar with the phenomenon of mudcracks and their formation, for they are so commonly seen along the edges of rivers or lakebeds, where once-wet muds have dried out. The polygonal patterns are characteristic of desiccation, brought about through uniform shrinkage (Figure D.7). Because mudcracks commonly taper downward toward the base of the bed in which they form, they hold potential for determining facing direction. A nice example of this is shown in Figure D.8. The host layer is a red mud layer, now shale. Shortly after the cracks formed, wind-blown sand filled them. Thus, as can be seen in cross-sectional view (see Figure D.8), the mudcracks are preserved as downward-tapering sandstone wedges.

Figure D.7 Mud cracks in the Rio Grande, west Texas. [Photograph by G. H. Davis.]

Volcanic rocks also carry primary structures that are useful in determining facing. Of these, **pillow structure** is perhaps the most useful. Where basaltic lava is extruded underwater, or where lava pours into the sea, pillow lavas instantly form (Figure D.9*A*). Pillows are generally flattened, with rounded tops. Where a pile of pillows has been formed, the bottoms of the pillows conform in shape to the rounded tops of those beneath them, so that in vertical cross-section they show lobes projecting downward between the intersecting curved tops of the underlying pillows (Figure D.9*B*). Given their cross-sectional forms, pillows are valuable as facing indicators, even within metamorphosed basalts (i.e., greenstones) of Precambrian age.

Figure D.8 These mudcracks in the Snowslip Formation are preserved in the form of the steeply inclined, downward-tapering sandstone casts in red mudstone (dark). The cracks first formed in mud, and then the cracks were filled with wind-blown sand (white). The fact that the cracks tend to "lean" a little reflects differential compaction of the sedimentary sequence. The sand casts (of the mudcracks) behaved like relatively stiff struts compared to the highly compactable mud. The sand casts could not compact as much as the mud, and thus they accommodated compaction, in part, by rotating. Location is along the Going-to-the-Sun Highway, Glacier National Park. [Photograph by G. R. McGimsey, National Park Service, Glacier National Park. Courtesy of the U.S. Geological Survey.]

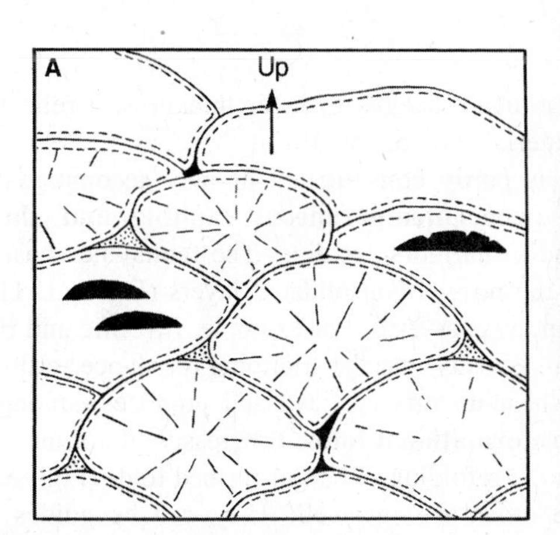

Figure D.9 (*A*) Sketch of pillow lava showing the characteristic right-side-up form of pillows. [From Macdonald, 1967, Forms and structures of extrusive basaltic rocks *in* Hess, H. H., and Poldervaart, A. (eds.), Basalts - the Poldervaart treatise on rocks of basalt composition. Reprinted by permission of John Wiley and Sons, Inc., New York, copyright ©1967.] (*B*) Outcrop photograph of pillows in steeply dipping basalt. Top of flow is to the left; bottom is to the right. Location is Janex massive sulfide prospect, Ontario, Canada. [Photograph by S. J. Reynolds.]

Primary Deformational Structures

We have to be careful not to confuse **soft sediment deformational structures** with conventional geologic structures that formed much later when the rocks were fully consolidated. **Penecontemporaneous folds** can form during deformation of unconsolidated water-rich sediments. Such folds are typically **intraformational**; that is, they are restricted to a given bed or a given sequence of beds. Beds and/or laminae above and below are not deformed, and this is the "give-away." The restriction of penecontemporaneous folds to narrow stratigraphic intervals is consistent with folding on the floor of a depositional basin or at very shallow depths of burial. Figure D.10 is such an example of soft-sediment folding. If we were to see such a fold in ancient rocks, we might incorrectly assume that the deformation was tectonic in origin! Thankfully, most penecontemporaneous folds are obvious products of soft-sediment deformation, for they are highly dispersed in orientations; they do not propagate upward or downward in any systematic, predictable manner;

Figure D.10 Intraformational soft-sediment deformation in Jurassic beds. [Photograph by S. J. Reynolds.]

Figure D.11 Soft-sediment faulting in Tertiary beds, near Bullard, Arizona. [Photograph by S. J. Reynolds.]

and they commonly show radical changes in layer thickness, a reflection of mobility of the mass of material thus distorted.

If sedimentary material is partly consolidated, it may accommodate soft sediment deformation by **penecontemporaneous faulting and shearing**, rather than by slumping and disharmonic folding. The faults and shear zones serve to extend or shorten the not-yet-consolidated layers (Figure D.11).

Primary folding occurs in lavas as well. For example, rhyolitic and rhyodacitic flows are of such high viscosity that the frictional resistance to flow produces internal shear. The shear distorts the lava and produces, among other structures, **asymmetric intraformational folds**. Progressive deformation with continued flow can result in the refolding of earlier formed folds (Figure D.12).

Primary structures have other uses as well. They can be guides to the flow direction of rivers, ocean currents (Figure D.16), wind currents, and lava (Figure D.13). Furthermore, when primary structures *of* known original shape and size become distorted as a result of **secondary deformation** (i.e., after lithification), they can provide a basis for quantitative strain analysis (Figure D.14).

Figure D.12 Fold within a fold in rhyodacite specimen collected in the Chiricahua Mountains, Arizona. The tight nearly hidden folds formed "early on" when the rhyolitic flow was "*on the move*." Then, at some stage, this part of the flow became refolded about what here resembles an upright anticline. [Photography and editing by Gary Mackender. vr.arizona.edu, glm@email.arizona.edu, © 2009–2010 Arizona Board of Regents.]

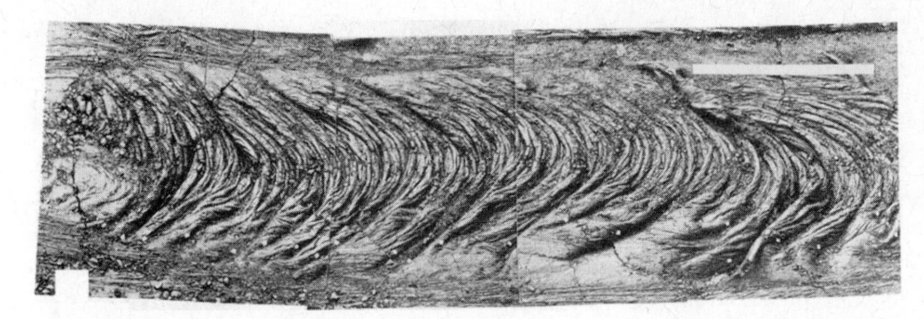

Figure D.13 Ropy pahoehoe structure, Kilauea Volcano, Hawaii. Bar scale represents 2 m. [From Fink, 1980, *Geology*, v. 8. Published with permission of the Geological Society of America.]

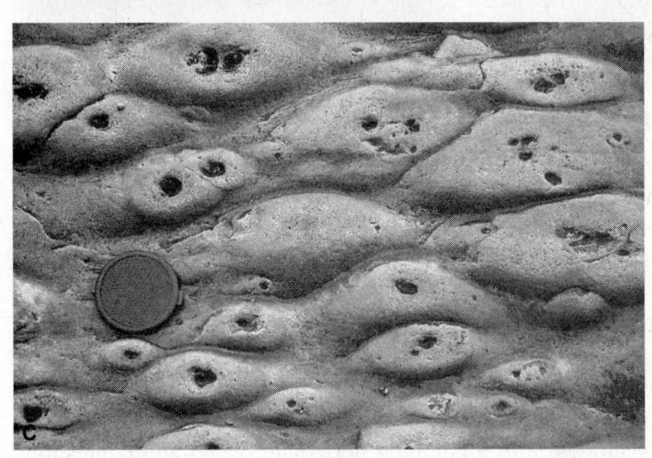

Figure D.14 (*A*) Distorted trilobite in Cambrian shale, Maentwrog, Wales. The width of the fossil is 3 cm. [With kind permission from Springer Science+Business Media: The Minor Structures of Deformed Rocks, 1972, L. E. Weiss.] (*B*) Very slightly distorted crinoid stems in limestone, Appalachian Plateau sector of New York State. [From Engelder and Engelder, 1977, *Geology*, v. 5. Published with permission of the Geological Society of America.] (*C*) Distorted sand volcanoes in low-grade metasedimentary rocks, Meguma Group, Nova Scotia, Canada. [Photograph by S. J. Reynolds.]

E. MEASURING THE ORIENTATIONS OF STRUCTURES

There are countless oriented structures in rocks, each with its own peculiar geometry. This is why a compass is *the* indispensable tool in geologic mapping. Measurements of the three-dimensional orientations of geological features provide the backbone of structural analysis.

The Brunton compass is the standard compass used by geologists in the United States (Figure E.1*A*), but the Silva compass (Ranger) has become increasingly popular because of price and compactness (Figure E.1*B*). Each instrument is equipped with the means to set **magnetic declination**, the angle between true north and magnetic north for a specific locality. When we take compass readings, we make sure that magnets, rock hammers, and metal clips are a safe distance away.

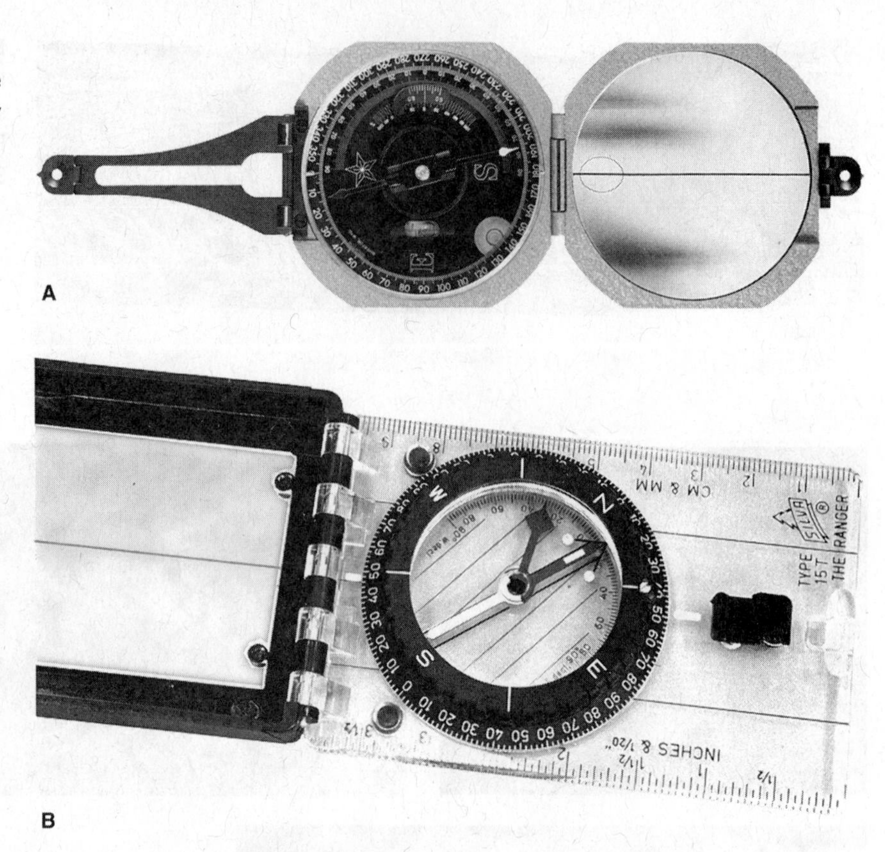

Figure E.1 (*A*) The Brunton compass. [Courtesy of the Brunton Company.] (*B*) The Silva compass (Ranger). [Photograph by G. Kew.]

Trend and Plunge

Compasses are used in structural analysis to measure trend and inclination. **Trend** refers to the azimuth or bearing of a line. **Azimuth** is measured in degrees clockwise from north (e.g., 120°; 267°). **Bearing** can be measured in degrees east or west from north or south (e.g., N60°E; S21°W) or in straight azimuth (60°; 201°). **Inclination** is the angle, measured in degrees, between an inclined line and horizontal. Its value may range from 0° to 90°.

Figure E.2 depicts the trend and inclination of a rather unusual line, the tallest human ladder ever constructed on a steeply inclined fault surface. The trend is determined by projecting both the foot of the ladder and the head of the ladder vertically upward into a common horizontal plane, and connecting these points of projection. The azimuth of this line is measured with a compass. The inclination of the human ladder is the angle between the line of bodies and horizontal, *as measured in a vertical plane.*

In practice, the orientation of a line in space is expressed in terms of **trend and plunge**, where plunge is a measure of inclination. Geologic lines, like grooves on a fault surface, are called **linear elements**. The trend of a linear element is measured by holding the compass level while aligning its edge parallel to the direction of the line (Figure E.3*A*). The compass is pointed parallel to the **vertical projection** of the line onto an imaginary horizontal plane. Trend is read directly from the Brunton compass after the compass needle has come to rest (Figure E.3*B*). To measure the azimuth of trend using the Silva compass, the calibrated outer ring on the face of the compass must be rotated until the rotatable outline of the compass arrow coincides with the actual free-spinning magnetic needle (Figure E.3*C*).

Plunge is measured by turning the compass on its side and aligning its edge along the linear element, or parallel to it (Figure E.3*D*). If the Silva compass is used, a plumb-bob–like inclination needle points automatically to the value of

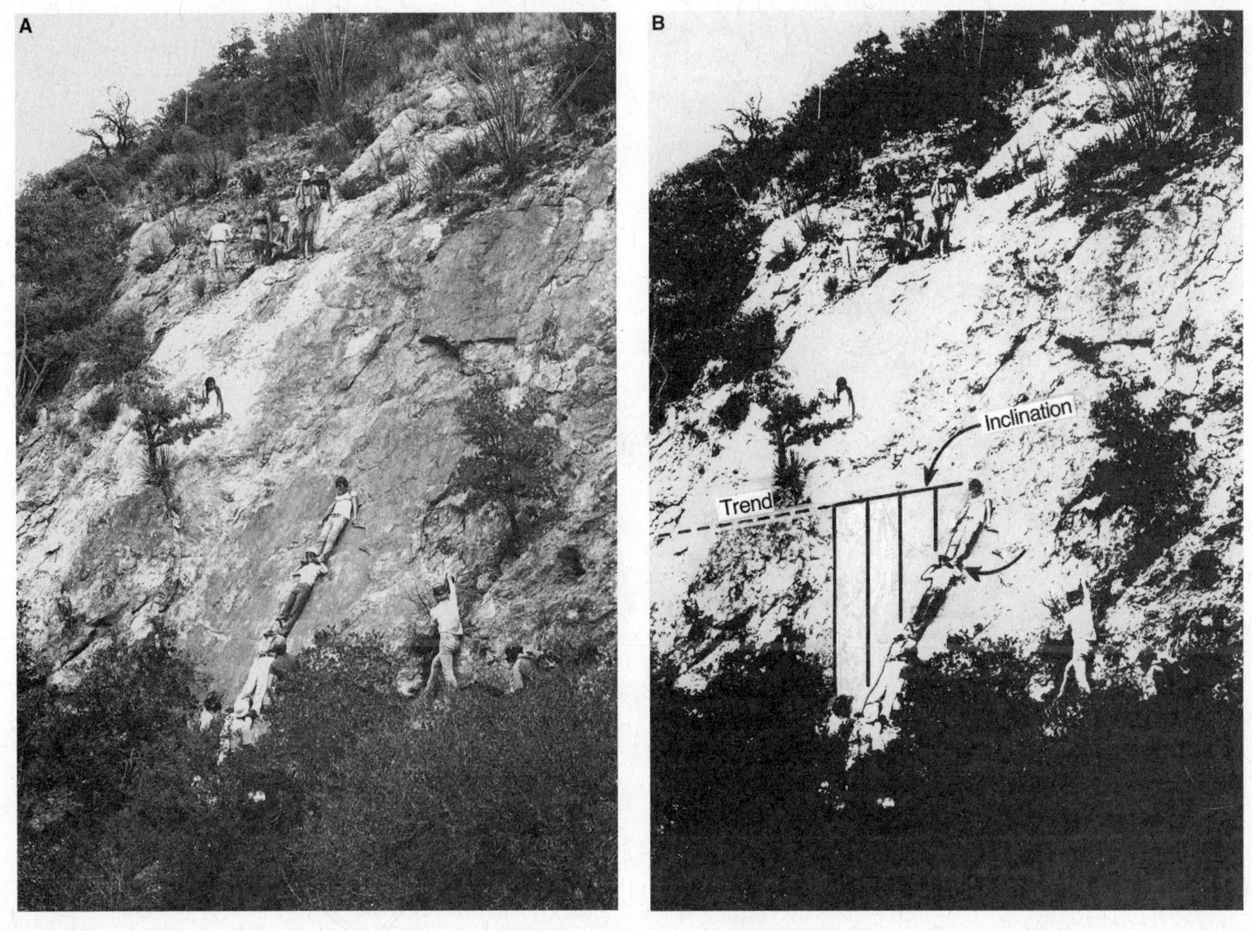

Figure E.2 (*A*) Human ladder constructed on face of a steeply inclined fault, near Patagonia, Arizona. [Photograph by G. H. Davis.] (*B*) Azimuth and inclination of human ladder.

the plunge when the compass is thus oriented (Figure E.3*E*). (The outer dial of the Silva has to be set on 90°; or 270°; so that the inclinometer reads correctly.) The Brunton compass does not have a free-swinging inclination needle. Instead, a calibrated scale known as a **clinometer** and located inside the compass is used to measure plunge. The clinometer, attached to a small carpenter's level, can be moved back and forth by means of a lever on the outside base of the compass (Figure E.3*F*). Holding the Brunton such that its edge is positioned parallel to the line being measured, the clinometer lever is moved until the bubble in the carpenter's level becomes centered (see Figure E.3*F*). The value of plunge is read directly using the inner scale embossed on the inside base of the compass.

The full description of the trend and plunge of a line in space can be recorded in two different ways. For example, 20° N60°E refers to the orientation of a line that plunges 20° along an azimuth 60° east of north; N60°E is the **sense of direction** of the down-plunge end of the line. Similarly, 45° S20°E describes the orientation of a line plunging 45° in a direction that is 20° east of south. For a compass calibrated in azimuth from 0° to 360°, these measurements would be recorded as 20°/060°; and 45°/160°, respectively.

Strike and Dip

Measuring the orientation of a planar feature is handled differently. If the orientations of two lines are known, the orientation of the plane that contains

Figure E.3 Steps in measuring trend and plunge. See text for details.

Figure E.4 (*A*) Photograph of part of the Kaparelli fault surface (Greece) which was reactivated in a 1981 earthquake sequence. The white rectangular corridor runs exactly down the dip direction of the fault. The white color is fresh limestone, exposed during careful sampling of the weathered surface of the fault for cosmogenic dating by Benedetti et al. (2003). The geochemical investigations help establish times of movement of this fault. Here we use the photograph to underscore dip direction of a surface. Geologist (actually lawyer) is Peter Collins. [Photograph by G. H. Davis.] (*B*) Schematic view showing orientations of lines of strike and dip.

these lines is also known. Two lines determine a plane. The measuring of orientations of bedding planes, fault planes, dikes (Figure E.4*A*), and other geological **planar elements** is based on this relationship. If the orientations of two lines that lie in a plane can be established, the orientation of the plane itself is established as well. Any two lines will do, as long as they are not parallel or close to being parallel. For convenience, the two lines in a plane that are chosen are a horizontal line and the line of *steepest* inclination (Figure E.4*B*). These two lines are at right angles to each other. The first is called the **line of strike**; the second is the **line of dip**. For the special case of a strictly horizontal plane, all lines are strike lines.

Strike and dip are the measurements required to define the orientation of a plane. **Strike** is the trend of the line of strike, that is, the trend of a horizontal line in a plane. Because its inclination is by definition 0°, the value of strike is recorded simply in terms of degrees of azimuth or bearing. For compasses with trend calibrated by quadrant, strike is always expressed in terms of north: for instance, N72°E, or N68°W, or N1°W. For compasses calibrated in azimuth from 0° to 360°, azimuth is presented in accord with the right-hand rule, i.e., view azimuth in direction such that plane dips to the right. The **dip** of a plane is the inclination of the line of dip, that is, the line of steepest inclination in a plane. It is recorded in terms of inclination angle and the dip of the plane (SW, NW, NE, SE, N, S, E, W). The specific azimuth of dip direction is not directly measured because its value can be determined from the strike of the plane and/or knowledge of the right-hand rule. For example, the dip direction of a N35°W-striking plane is either N55°E or S55°W. Only the general value of dip direction is recorded in the notebook; that is, SW or NE for this example. This distinction permits the two possible dip directions to be distinguished. On the other hand, there are many geologists who prefer NOT to measure strike, and instead define the orientation of a plane through measuring the azimuth and inclination of the 'line of dip' of the plane. There are compasses 'built' to take these readings quickly and accurately.

The procedure for measuring the strike and dip direction of a plane using a Brunton compass is as follows (Figure E.5). To find the line of strike, first set the clinometer to 0° so that the compass may be used as a carpenter's level (Figure E.5*A*). Place a side or edge of the compass flush against the plane (Figure E.5*B*), or against a field notebook or nonmagnetic clipboard held parallel to the plane; rotate the compass until the carpenter's level bubble is centered. When centered, the edge of the compass held against the plane is horizontal and oriented parallel to the line of strike. To determine azimuth of the strike line, rotate the compass downward (Figure E.5*C*), still keeping the lower-side edge of the compass fixed against the surface until the bull's-eye bubble is centered. The compass needle now swings freely. Dampen the compass needle to stop it from swinging and read the azimuth (Figure E.5*D*).

To measure the strike of a plane using a Silva compass, it is useful to carry an auxillary level, or to attach a small level to the compass itself. The level can be used to quickly identify the line of strike. Once found, its orientation can be measured by aligning the edge of the Silva parallel to the line of strike and rotating the calibrated outer ring until the reference compass needle is aligned with the actual magnetic needle (Figure E.5*E*).

To measure dip with a Brunton, place the compass on a side face on the inclined plane such that the compass is aligned in the direction of the line of dip (Figure E.5*F*). Then measure the inclination of the line by rotating the clinometer until the carpenter's level bubble is centered. The Silva compass has an inclination needle for measuring dip directly, as long as the outer ring is set at E or W to properly align the inclinometer and the edge of the compass (Figure E.5*G*).

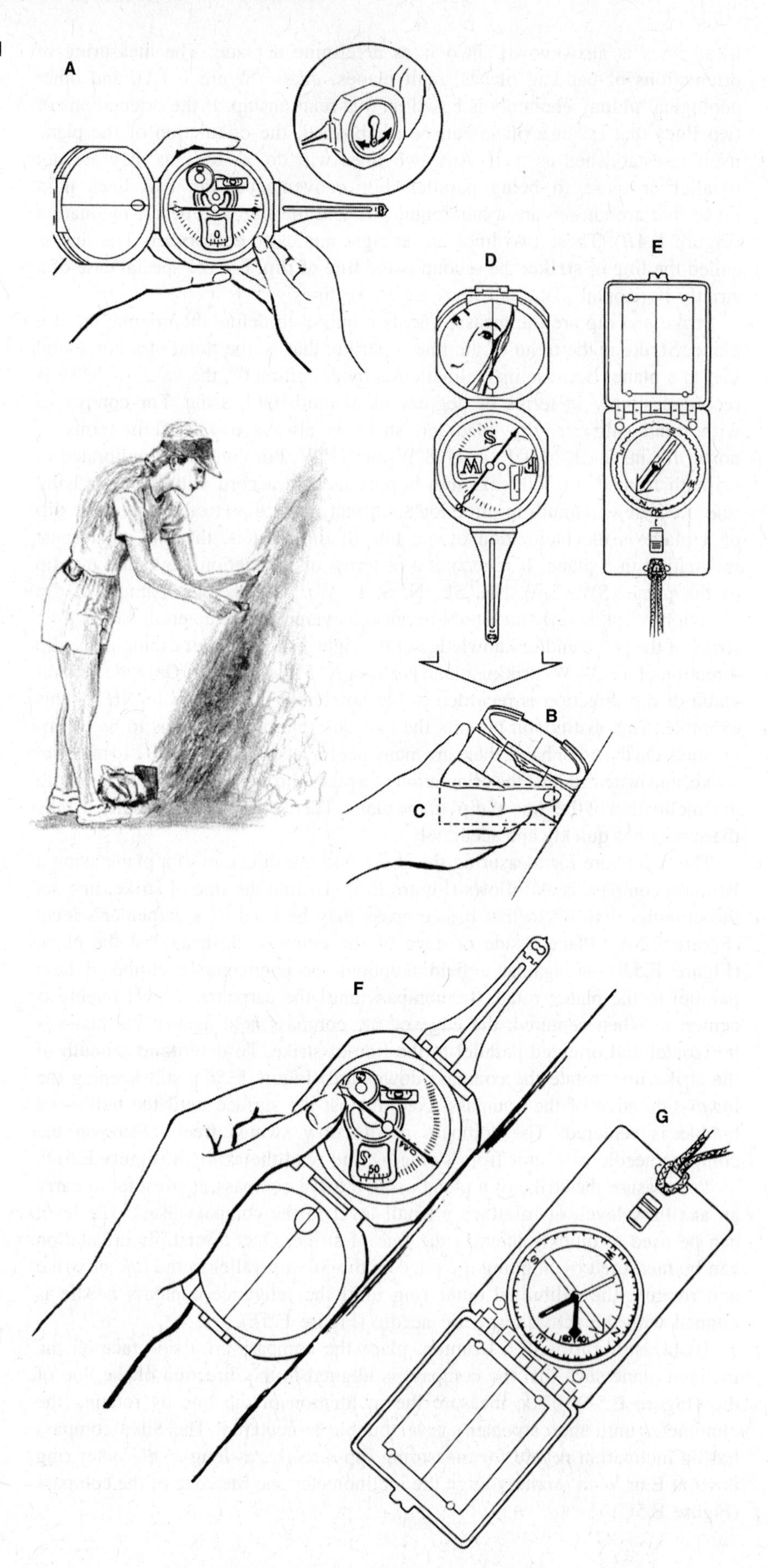

Figure E.5 Steps in measuring strike and dip. See text for details.

Figure E.6 The do's and don'ts of the sighting method for measuring strike and dip.

Strike-and-dip readings can be taken by the **sighting method** as well. This method is especially useful when beds or layers do not crop out as convenient resistant planes for direct measurement, and/or when attempting to measure the average strike and dip of an area of rock that is larger than outcrop size. The method is illustrated in Figure E.6. The trick is to position yourself in the proper location so that your line of sight is a strike line in the plane of the dipping layer whose orientation is being determined. When viewed in this way, the dipping layer appears in strict cross-sectional view, with no expression of the surface of the layer. The azimuth of the line of sight constitutes the strike of the dipping layer. The inclination of the layer as seen from this unique line of sight is true dip. Watch where you stand!

Recording the Measurements

Strike-and-dip and trend-and-plunge orientations measured in this way are recorded in the field notebook. Representative readings are placed on the geologic map as well. If a Brunton compass is used, a protractor or protractor scale is used to accurately plot the strike-and-dip readings. The base map is scribed lightly with penciled N−S guidelines so that strike or trend can be measured and plotted with relative efficiency. The Silva compass has the added advantage of being functional as a protractor for plotting strike or trend on the base map. Without disturbing the compass setting for trend or strike, the compass is placed on the map in such a way that the red lines on the inner base of the compass coincide with the N−S guidelines scribed on the map. When this is accomplished, the straight edge of the compass matches the azimuth of strike or trend.

Measurements collected and plotted in this way give geometric life to the geologic map (Figure E.7). There emerges from the map a physical and geometric expression of the form and internal structure of the rock formations. Furthermore, the orientation measurements stored in the field notebook become the basis for subsequent analysis and interpretation. Finally, the distribution of plotted measurements on the maps gives the map reader an idea of which areas have exposures and were "*covered*" by the geologist who made the map.

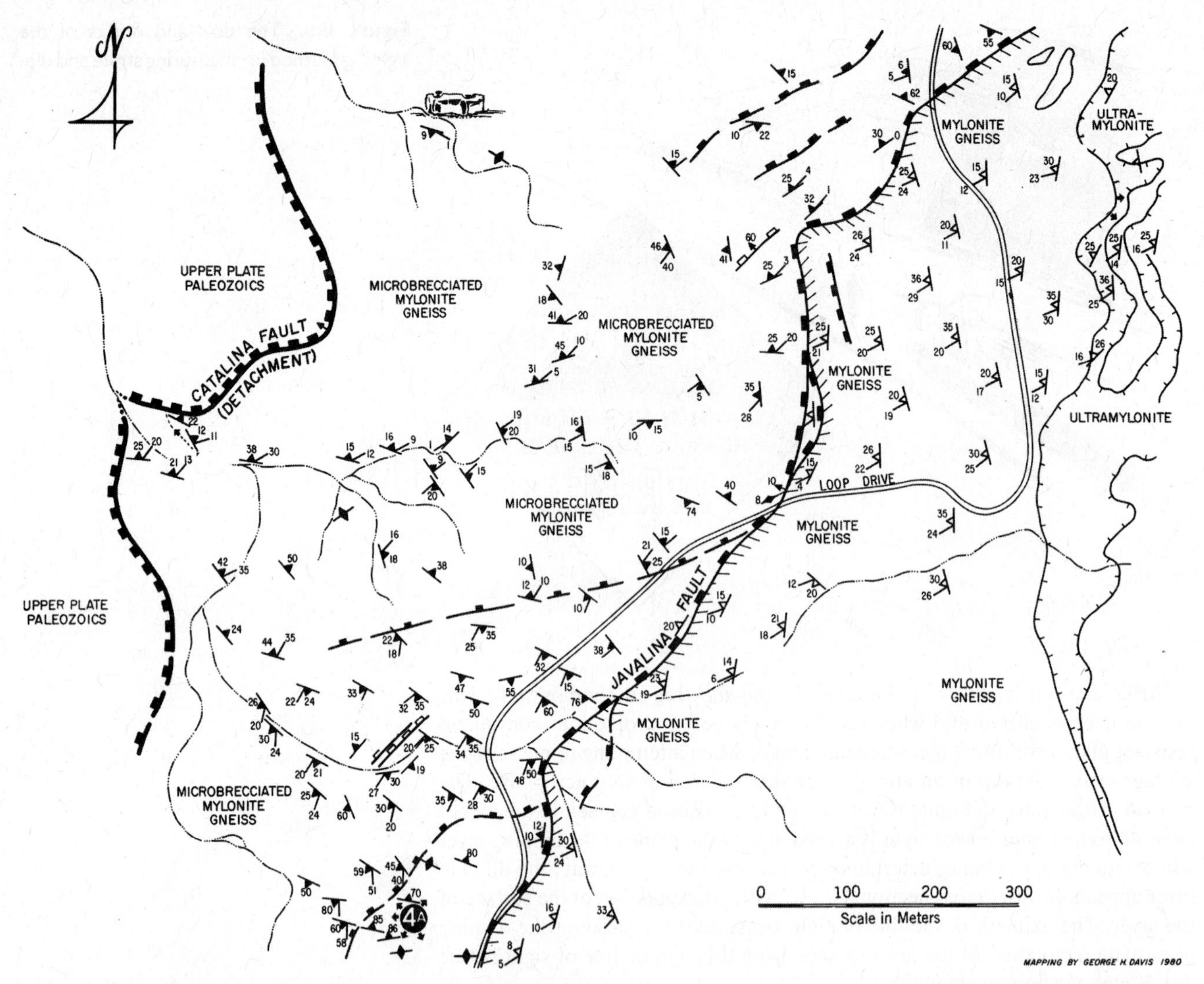

Figure E.7 A plethora of strike-and-dip readings brings geometric life to maps. This map, rendered by G. H. Davis, shows internal structure within mylonitic and cataclastic rocks in the Saguaro National Park, Rincon Mountain District, Tucson, Arizona. [From Davis, 1987. Published with permission of the Geological Society of America.]

F. PREPARING GEOLOGIC CROSS-SECTIONS

Geologic maps are used as a basis for constructing **geologic cross-sections**, which can be thought of as vertical slices through a map area showing a **profile view** of the subsurface structure (Figure F.1). Geological cross-sections reveal interpretations of what lies below the surface. This is where "*the rubber meets the road*," because geologic cross-sections get most of the attention in spotting drill holes in the exploration for oil and gas or geothermal energy; in locating the best drill sites or exploration shafts in the search for metals; in interpreting the subsurface expression of active fault zones; or in communicating the fundamental structural style within mountain belts, like the Alps.

Drawing realistic geologic cross-sections is very difficult, and the construction steps have become quite sophisticated. The difficulty is that the geologic map, on which the cross-section is based, only describes the relationships right at the surface. Therefore, the locations and compass-measured orientations of faults, intrusive contacts, unconformities, bedding, and other

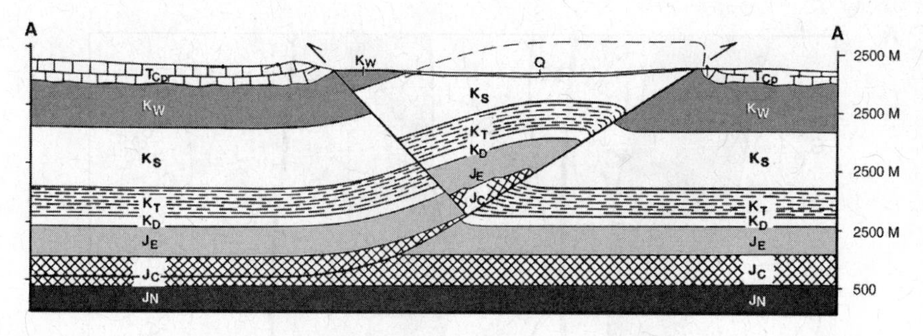

Figure F.1 Geologic cross-sections constructed by Eric Lundin (1989) across thrust-faulted strata in the Bryce Canyon region, Utah. Drilling and seismic information reveal that the major thrust "*soles*" into weak salt horizons within the Jurassic Carmel Formation. [From Lundin, 1989. Published with permission of the Geological Society of America.]

structures hold only for the surface of the area, and it cannot be presumed that they hold for any depth below the surface. As a consequence, the subsurface geology somehow must be portrayed in a way that is everywhere consistent with the **geologic control** at the surface and the style of deformation seen in the map area.

Steps in Constructing Cross-Sections

Let us take a look at the basic steps in preparing a geologic cross-section for an area marked by straightforward geologic relationships. The geologic map (Figure F.2*A*) shows that the area of interest contains three formations of sedimentary rock of Cambrian age, resting nonconformably on Precambrian granite. The Cambrian rocks and the nonconformity are inclined approximately 30°E. Two north–south-trending faults, each inclined 60°W, cut and displace the rock units, west-side-down (see Figure F.2*A*). We are interested in showing the geologic relationships as we believe they might exist in the subsurface.

First, select a **line of section** (line A–A′ in Figure F.2*A*) that is strategically placed along a line that trends east–west, perpendicular to the trend of the faults, and perpendicular to the trend of the Cambrian rock formations and the major unconformity. Let the east–west–trending line of section pass through a part of the area where the rock exposures are good and the geologic control at the surface is solid. Next, construct a **topographic profile** along the line of section, using the same vertical scale as the map scale, that is, *no vertical exaggeration.* This is accomplished by laying a work sheet along the line of section, transferring the end points (A and A′ on Figure F.2*A*), and transferring and labeling each of the topographic contour lines that crosses the line of section as well (Figure F.2*B*). This information is used to construct the topographic profile (Figure F.2*B*). The final preliminary step is to lay the work sheet once again along the line of section, and this time transfer and label the **control points** where the contacts for each of the Cambrian formations crosses the line of section, and the control point where the faults cross the line of section. Using a protractor, plot tiny **control lines** from each of the respective control points, showing the 60°W inclination of the faults and the 30°E inclinations for each of the Cambrian formations and the unconformity (Figure F.2*B*). With this step, the topographic profile starts to become a geologic cross-section.

To fill in the geology at depth, extend each of the rock formations along the line of section down into the subsurface along the 30°E trajectory (Figure F.2*C*). Unless you have reason to do otherwise, do not change the thickness of any of the formations as you project them into the subsurface. Next you might extend the faults into the subsurface along the 60°W trajectory, using a heavier hand, and thus a heavier line, than that for the formation contacts (see Figure F.2*C*). Again, unless you have reason to do otherwise, do not change the basic angle of inclination of the faults as you project them into

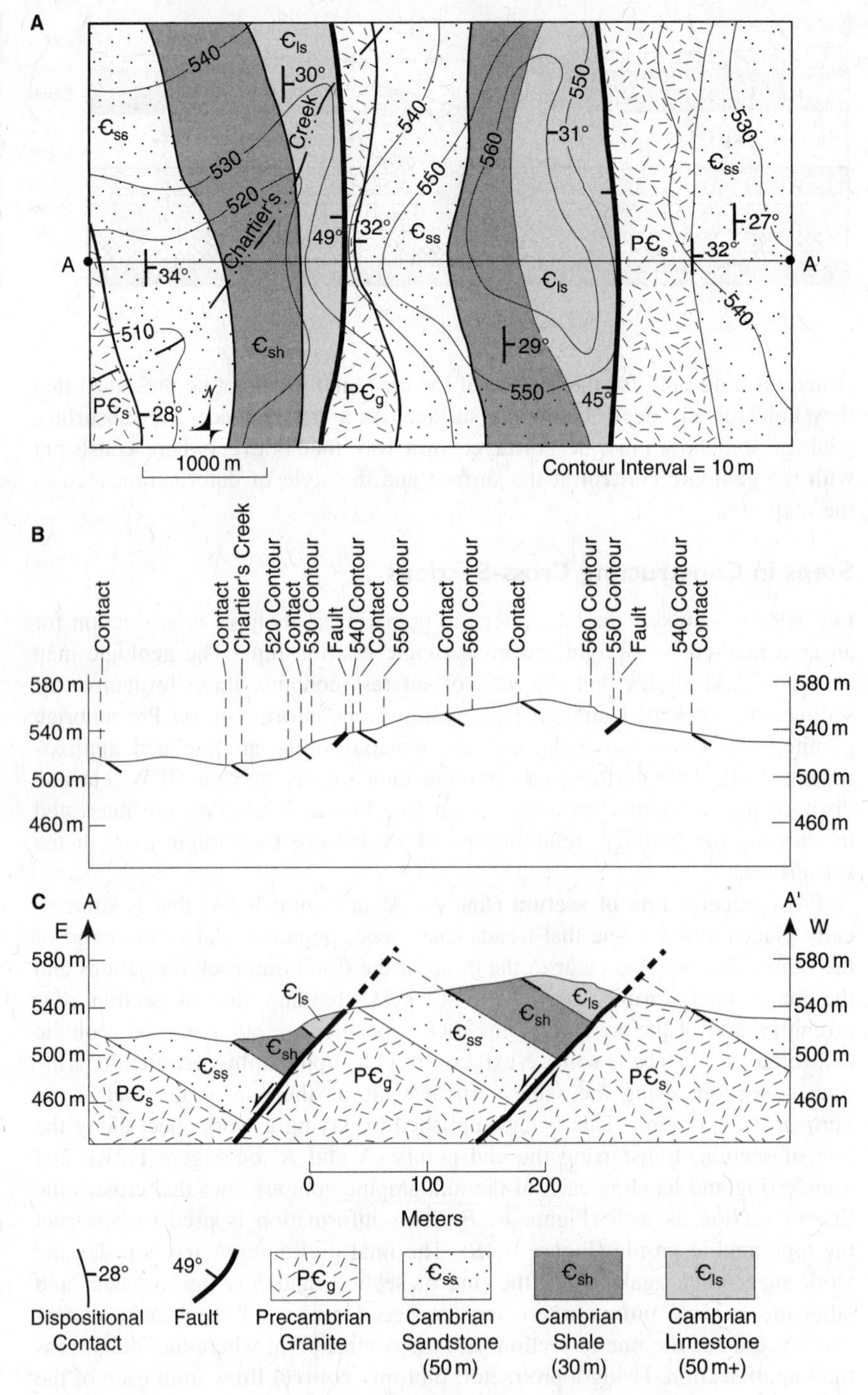

Figure F.2 Basic steps in constructing a simple geologic cross section. (*A*) Geologic map and trace of line (A-A′ along which geologic cross-section will be constructed). (*B*) Topographic profile along trace A-A′, and control lines denoting dip inclinations of bedding and faults. (*C*) Geologic cross section and explanation of units and symbols. See text for explanation.

the subsurface. Finally, extend each of the rock formations in the area down into the subsurface along the 30°E trajectory, terminating the formations at the faults (see Figure F.2*C*). Add color or line symbols and your cross-section is complete.

If asked, "*How would you determine how much separation there is on the faults*?" simply project (using dashed lines) the Cambrian rock formations up into the air to the west, terminating them at the skyward projection of the faults. The amount of stratigraphic separation can then be directly measured on the cross-section, using as a reference any one of the formations, or the unconformity (Figure F.2).

Correcting for Apparent Dip

Most lines of section will be oblique to some of your strike readings. Thus, it is necessary to make a correction from the true dip measured in the field to the apparent dip that should show up on your geologic cross-section. There are two practical ways to do this. One is to do it stereographically (see *Part III-I*). The other is to use the chart shown in Figure F.3. The point is this: Only sections that are perpendicular to strike will reveal true dip. All other sections will reveal apparent dip, which will be *less* than true dip. To use the correction chart presented in Figure F.3, go to the map itself and measure (with a protractor) the angle between the line of section and the line of strike. Now go to the chart (Figure F.3) and find the value of this angle in the right-hand column. Lay a ruler from this value in such a way that it passes through the magnitude of true dip in the left-hand column. The intersection of the ruler edge and the middle column marks the value of the apparent dip.

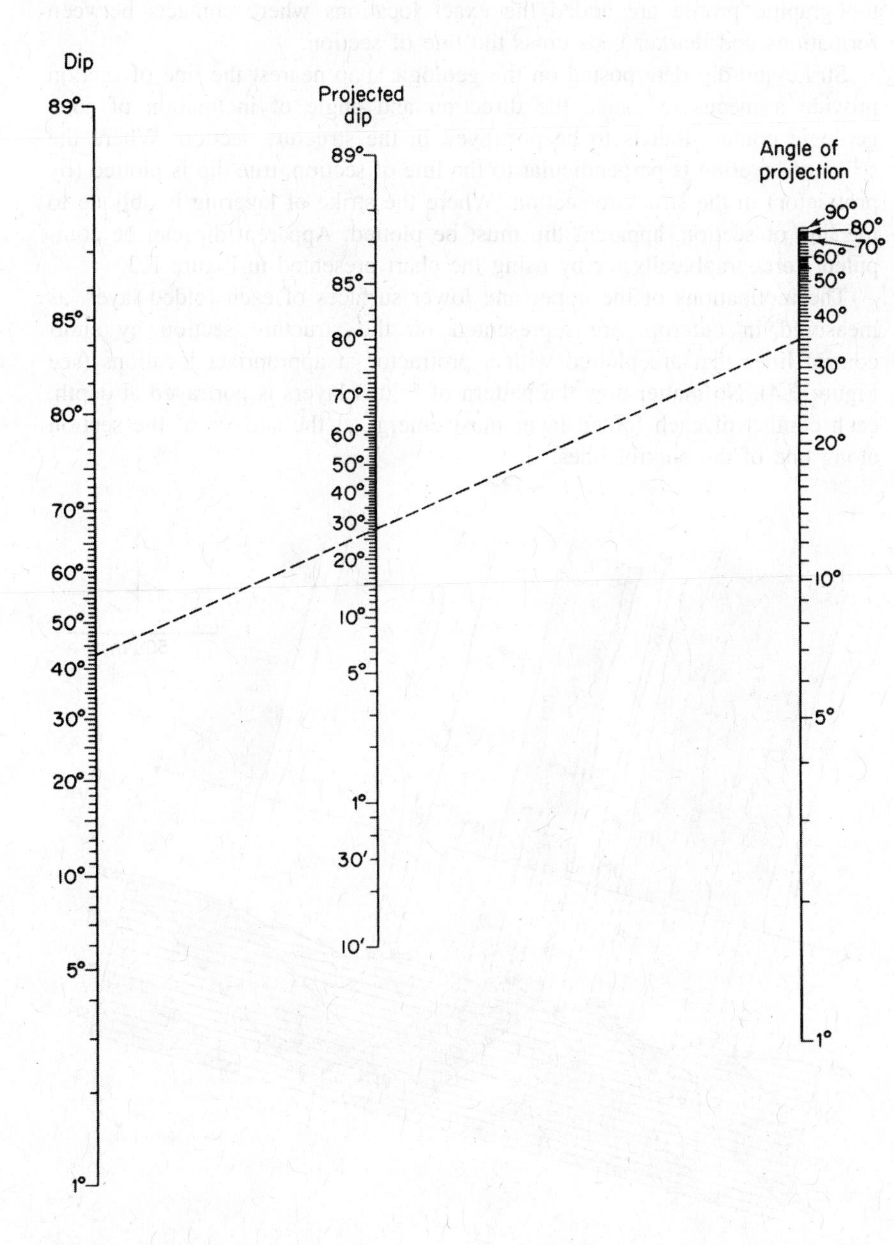

Figure F.3 Chart used to convert from true dip to apparent dip when constructing geologic cross-sections. If the true dip is 43°, the apparent dip on a vertical section making a 35° angle with the strike of the bedding would be 28°. [From Billings, M. P., Structural Geology, © 1972, p. 523. Reprinted by permission of Prentice Hall, Upper Saddle River, New Jersey.]

Constructing Normal Profiles of Folds in Vertical Sections

Geometric analysis of the shapes of individual folded surfaces and folded layers requires views of folds in **normal profile**. In areas or regions where folds are nonplunging, normal profile views are obtained by constructing vertical structure sections. The base of control is a geologic map (Figure F.4). The line of section along which the structure section is to be fashioned is laid out perpendicular to the average trend of fold axes, an average determined stereographically (see *Part III-I*) if necessary. End points of the line of section are marked on the map (A, A′, Figure F.4). The line of section is positioned where strike-and-dip data for the folded layering are reasonably abundant and where the fold patterns are especially interesting and informative.

Once the line of section has been chosen and positioned on the geologic map, a topographic profile is constructed using elevation control afforded by topographic contour lines (see Figure F.4). The topographic profile, like the structure section itself, is drawn without vertical exaggeration in order to avoid introducing distortion of the true form of the folded layers. To the topographic profile are added the exact locations where contacts between formations and marker beds cross the line of section.

Strike-and-dip data posted on the geologic map nearest the line of section provide a means to gauge the direction and angle of inclination of each geologic contact that is to be portrayed in the structure section. Where the strike of layering is perpendicular to the line of section, true dip is plotted (by protractor) in the structure section. Where the strike of layering is oblique to the line of section, apparent dip must be plotted. Apparent dip can be computed stereographically, or by using the chart presented in Figure F.3.

The inclinations of the upper and lower surfaces of each folded layer, as measured in outcrop, are represented on the structure section by short control lines that are plotted with a protractor at appropriate locations (see Figure F.4). No matter how the pattern of folded layers is portrayed at depth, each contact of each folded layer must emerge at the surface of the section along one of the control lines.

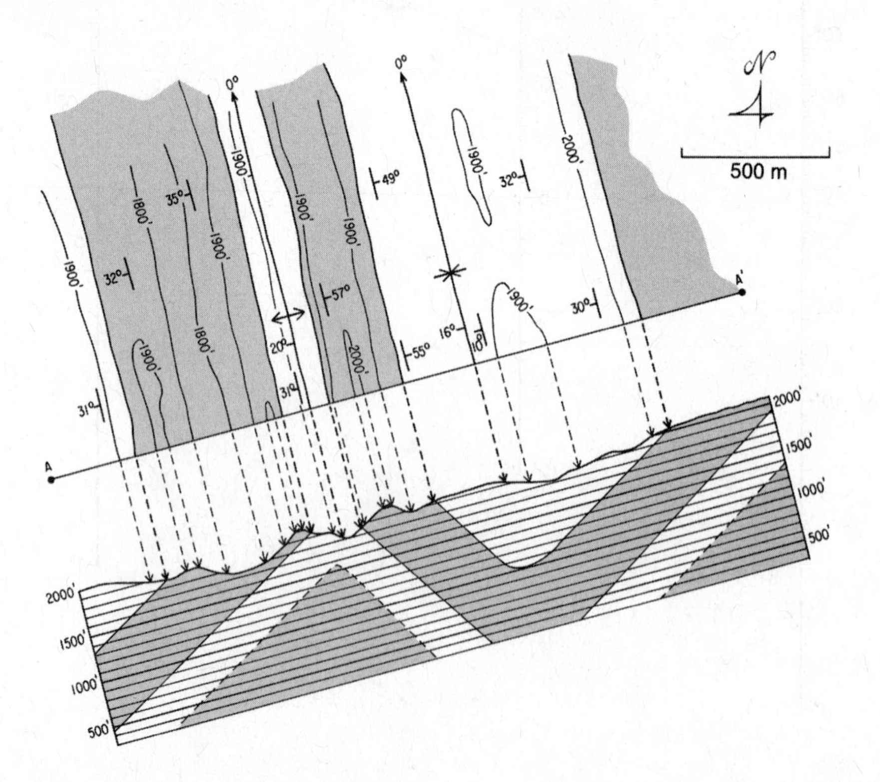

Figure F.4 Preparation of a vertical structure section on the basis of geologic map relationships. See text for details.

The manner in which the folded layers are portrayed in the subsurface (e.g., amount of thickening and thinning) is guided by field observations regarding the response of each folded layer to the distortional influence of folding. Similarly, the 'kinkiness' versus the roundness of fold hinges is influenced in part by the styles of folding directly observed. As a final check on the internal consistency of the subsurface interpretation, the trace lengths of the midlines of each of the folded layers are measured to determine whether the section is **balanced**. An example of the constructing balanced cross-sections is presented later in this section.

Down-Structure Method of Viewing Folds

Geologic maps of plunging fold structures can become cross sections, if viewed from the proper angle. This is achieved by applying J. Hoover Mackin's "*down-structure method*" of viewing geologic maps' (Mackin, 1950). Normal profile views of plunging folds can be seen at a glance by viewing the geologic map patterns in the direction of plunge, at an angle of inclination of view corresponding to the amount of plunge. Try it on the map pattern shown in Figure F.5*A*, comparing what is seen to the graphically constructed geologic cross-section (Figure F.5*B*). Then try out the method on the hieroglyphics shown in Figure F.5*C*). On some streets in most cities in the United States we see this concept practiced by highway departments. Be on the lookout for STOP AHEAD painted on the street in exaggerated down-structure form, perfectly "*angled*" so that the driver perceives normal, undistorted letters.

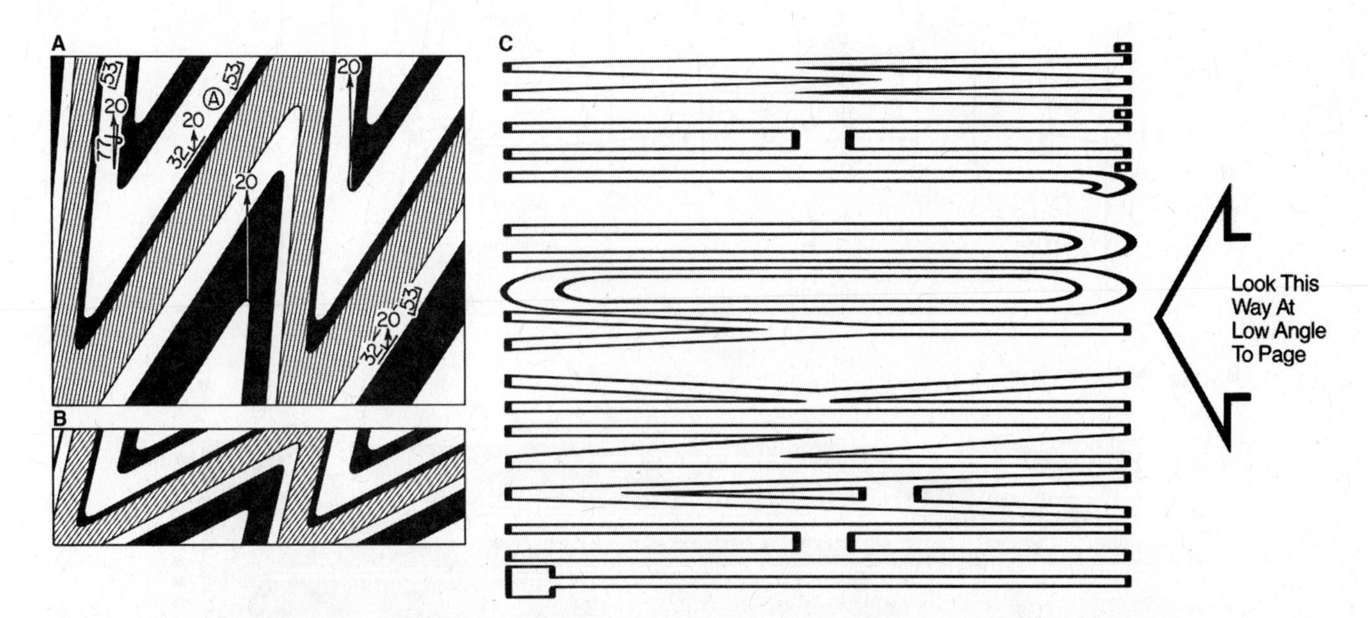

Figure F.5 (*A*) Map pattern of folds and (*B*) down-structure view of same. [From Whitten, 1966, Structural geology of folded rocks. Reprinted by permission of John Wiley and Sons, Inc., New York, copyright ©1966.] (*C*) Glance at this pattern in the down-structure view.

Preparing Balanced Cross-Sections

The concept of "*balanced*" cross-sections is presented in *Chapter 8, Fault-Fold Interactions.* The methods were developed for folded and thrusted regions, but they can be applied to extended regions as well. Perhaps the best way to visualize the preparation of balanced sections is to see examples of some good ones, such as those prepared by Shankar Mitra (1988) for the Pine Mountain thrust region in the southern Appalachians. The locations of his cross-sections with respect to the major regional structures are shown in Figure F.6*A*, and three representative sections are presented in Figures F.6*B*,*C*, and *D*. Woodward, Boyer, et al. (1985), and Marshak and Woodward et al. (1988), have presented guidelines for preparing such sections. To begin, choose a line of section that lies parallel to the direction of thrusting and avoids major lateral structures, such as tear faults and lateral ramps. Establish the thickness (and changes in thickness) of each stratigraphic unit. Estimate the depth to the floor thrust, sole thrust, or decollement based upon seismic information, regional geologic information, or drilling information (see *Chapter 8, Fault-Fold Interactions*). Plot along the surface profile of the cross-section the locations of fault contacts and formation contacts as well as dip information for faults and bedding.

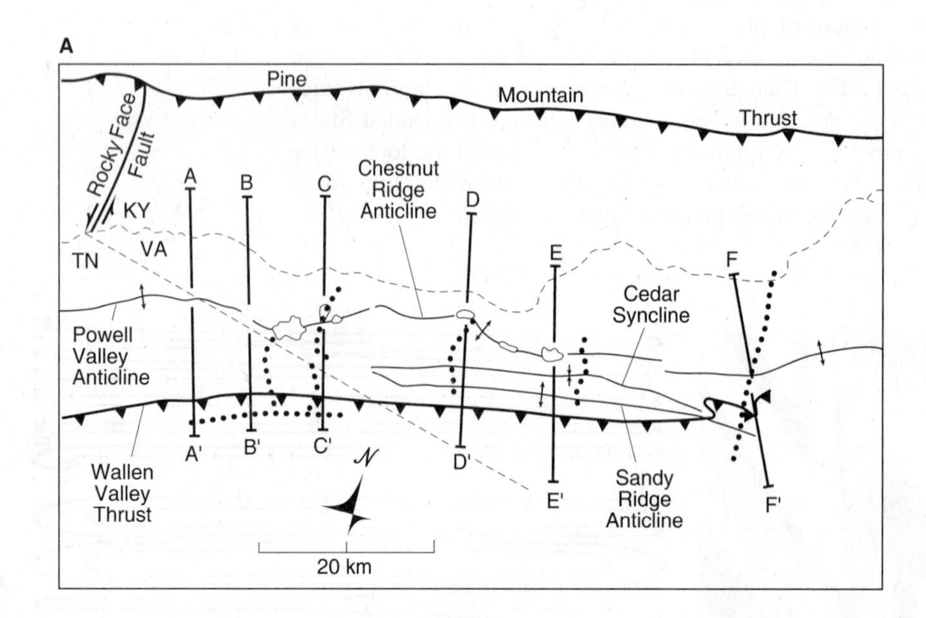

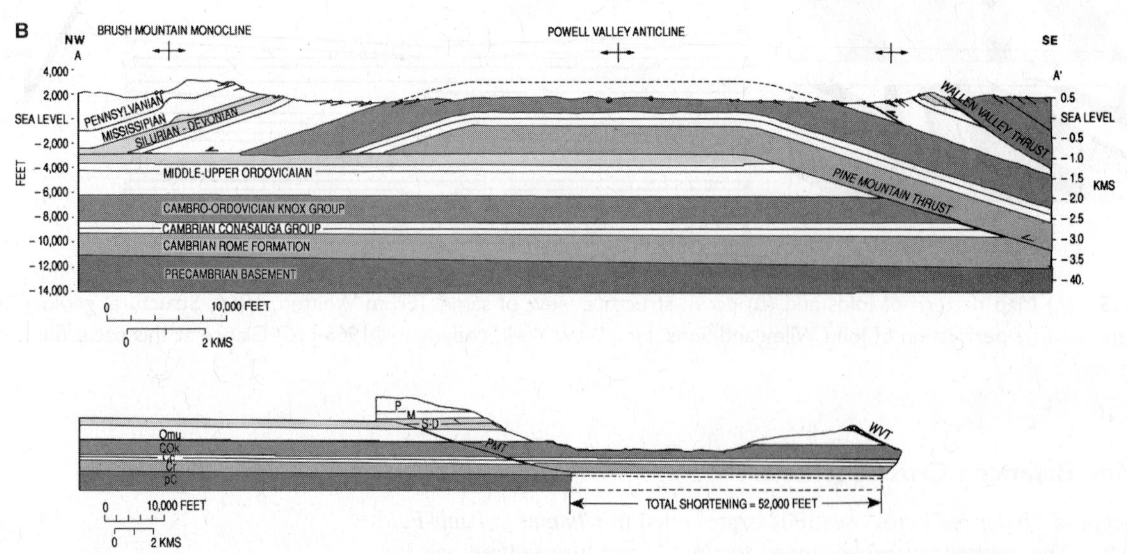

Figure F.6 (*A*) Map showing locations of the cross-sections prepared by Mitra (1988). Dotted lines are seismic lines. Also shown are location of wells. (*B*) Balanced cross-section and restored counterpart through the Wheeler area (A–A′).

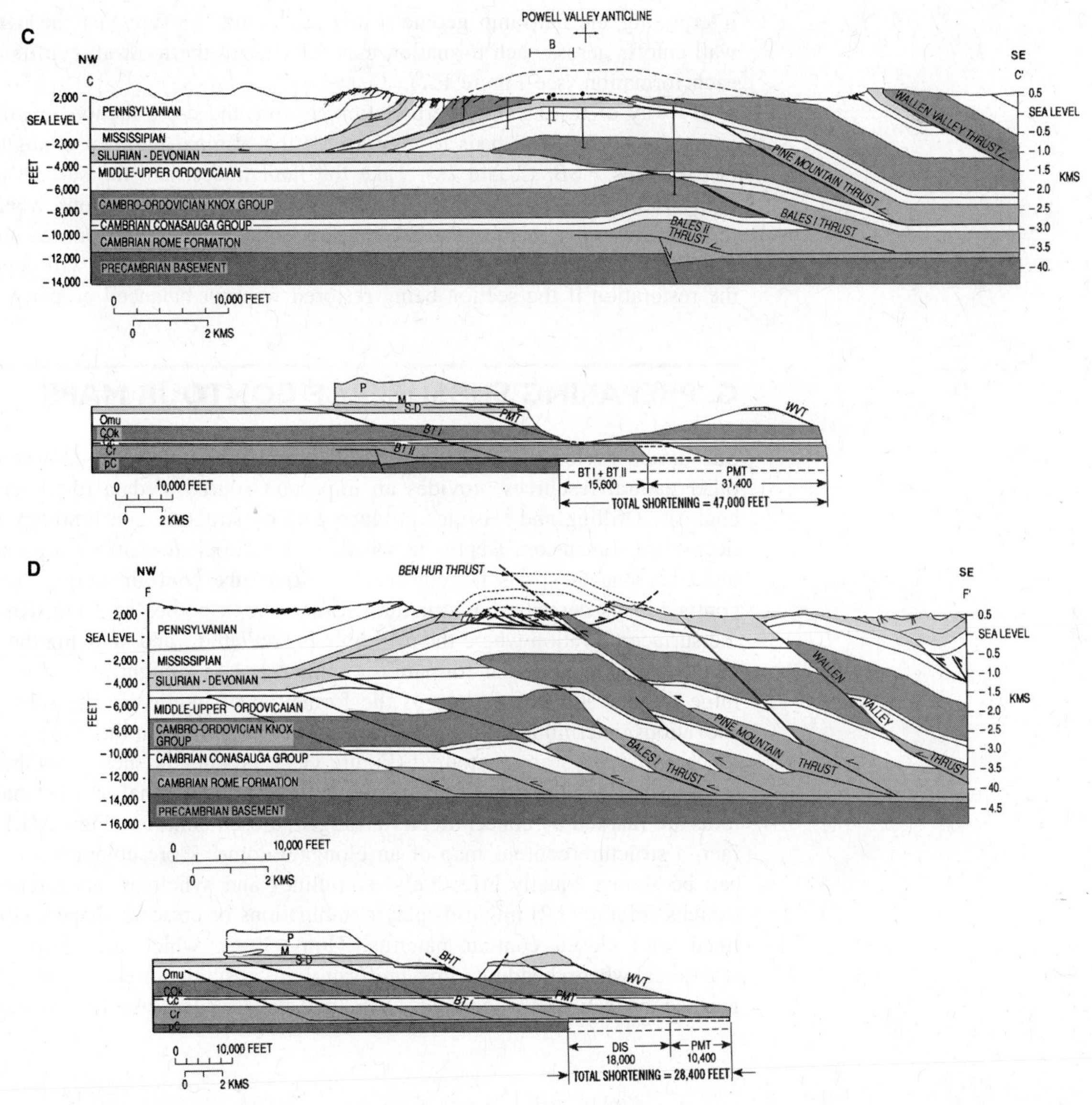

Figure F.6 *(Continued)* (*C*) Balanced cross-section and restored counterpart through the Martin Creek window (C–C′). (*D*) Balanced cross-section and restoration through the Big Stone Gap area (F–F′). [From Mitra, 1988. Published with permission of the Geological Society of America.]

By way of execution, identify the footwall ramps and flats for each fault at depth by using the geometry of the hanging-wall structure as a kind of template (Figure F.7). The spacing between dip panels in the hanging-wall rock will often correspond to the lengths of flats and ramps on the footwall. Upon

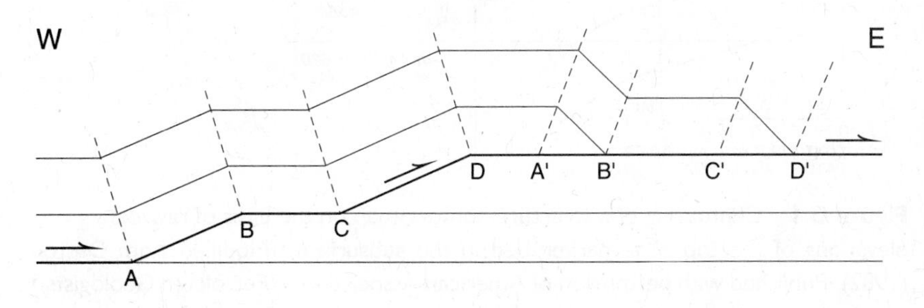

Figure F.7 Cross-section drafted in a way that the hanging-wall ramps and flats can be restored such that they match the footwall ramps and flats. [After Marshak and Woodward, *in* Marhsak and Mitra, Basic Methods of Structural Geology, © 1988, p. 313. Reprinted by permission of Prentice Hall, Upper Saddle River, New Jersey.]

interpreting the flat/ramp geometry for each fault, be sure that the hanging-wall cutoffs across each formation match in length the footwall cutoffs across each formation (see Figure F.7).

By way of testing the interpretation, restore the cross-section by straightening out each of the beds to conform to the original regional configuration (see Figures F.6*B*, *C*, and *D*.). Take the fault traces, and the surface topography too, along for the ride. The original geographic locations where the thrusts ramp up-section toward the surface will become apparent, as will the degree of regional shortening. Unsightly gaps and mismatches will appear in the restoration if the section being restored was not balanced properly.

G. PREPARING SUBSURFACE CONTOUR MAPS

Subsurface exploration in the search for petroleum, metals, groundwater, and other natural resources provides an important source of data for descriptive analysis. Drilling and seismics produce data on structure and lithology in the deep third dimension. Depths at which specific rock formations are encountered become the basis for constructing **structure contour maps**. Structure contour maps describe the structural form of rock bodies at depth. Knowing the surface elevation where the drill hole is "*collared*," and knowing the depth to the top of a particular formation of interest, it is straightforward to determine the elevation of the top of the formation of interest in the subsurface. Elevations determined in this way can then be contoured in the same way that topographic maps are contoured (Figure G.1). Frequently such maps are used to describe the structure of sedimentary formations. Domal and basinal patterns are marked by concentrically arranged, closed contours. Figure G.1 is, in fact, a structure contour map of an elongate dome. More complex structures can be shown equally effectively. Anticlines and synclines, and arches and troughs (Figure G.2) might display combinations of crescent-shaped, straight-lined, and closed contour patterns. **Homoclines**, which are simple tilted structures where bedding dips uniformly in a single direction, are distinguished by subparallel contour lines that steadily decrease (or increase) in

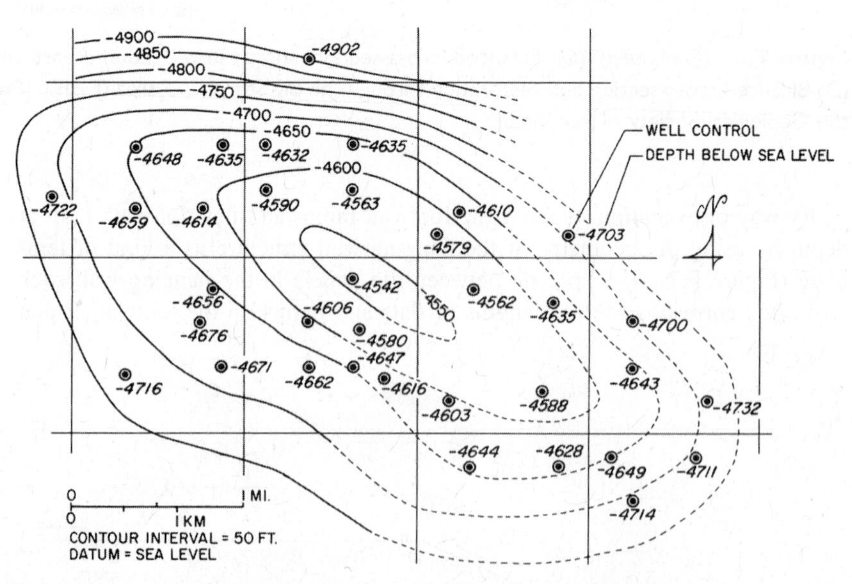

Figure G.1 Contouring of a structure contour map on the basis of raw data giving elevations of the top of a marker bed in the subsurface. [Modified from Dutton (1982). Published with permission of American Association of Petroleum Geologists.]

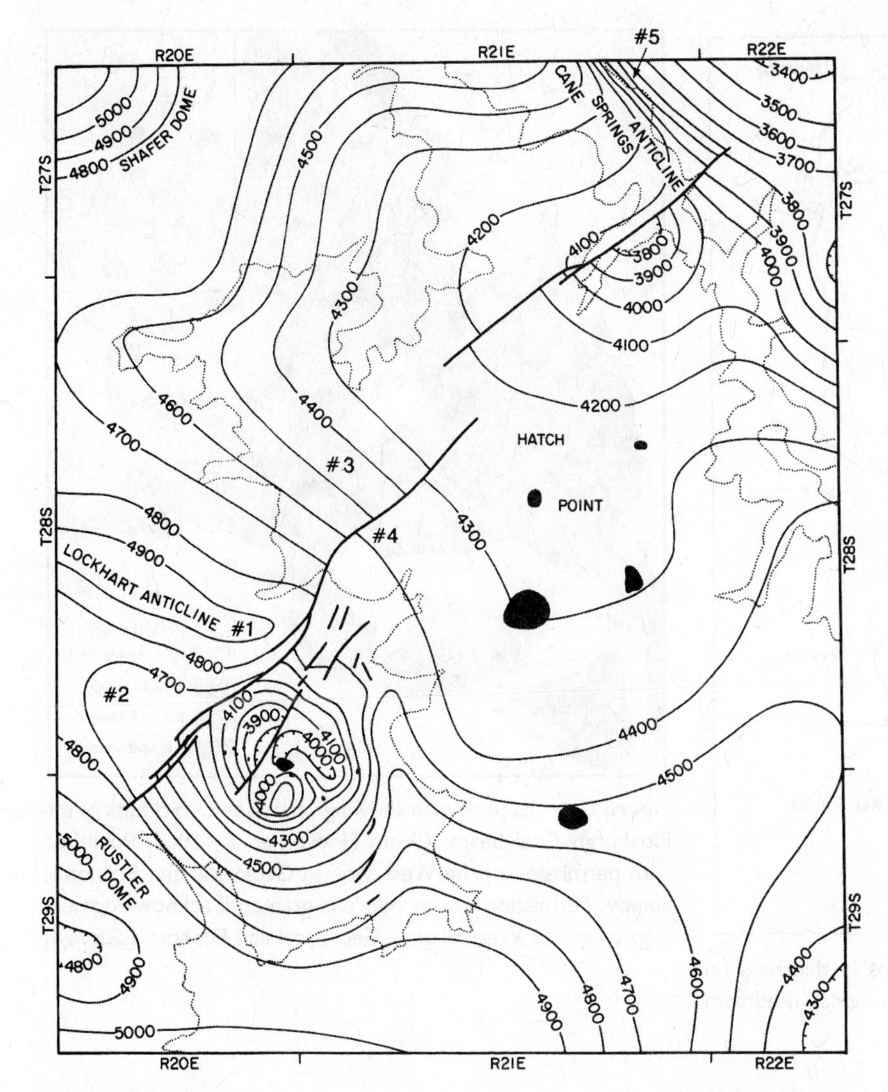

Figure G.2 Structure contour map of a part of the Canyonlands region, Utah, depicting examples of representations of (1) an anticline, (2) a syncline, (3) a homocline, (4) a fault, and (5) steeply dipping strata. [From Huntoon and Richter, 1979. Published with permission of the Four Corners Geological Society.]

elevation value across the map (see Figure G.2). Faults are marked by offset contour lines (see Figure G.2). Where a rock formation gradually steepens, the contour lines come closer and closer together. The contour lines actually merge into one where the top of the rock formation dips vertically.

Isopach maps are contour maps that describe formation thicknesses. Thicknesses are compiled on the basis of geologic mapping, underground mapping and mining, and subsurface drilling. Values of thicknesses are posted on a base map and contoured. The resulting patterns describe the variations in thickness of a particular formation and/or series of formations. The isopach map shown in Figure G.3 shows the variations in thickness of windblown Jurassic sandstone (Nugget and Navajo) in the Rocky Mountain region. The thickness variations indicate the shape and scope of the ancient dune field. I (GHD) find the isopach map shown in Figure G.4 to be especially interesting, having grown up in Pittsburgh, the son of a father who, as a mining engineer, worked the Pittsburgh Coal Seam. The map provides a picture of the thickest accumulations of organic mass in the ancient Pennsylvanian swamps.

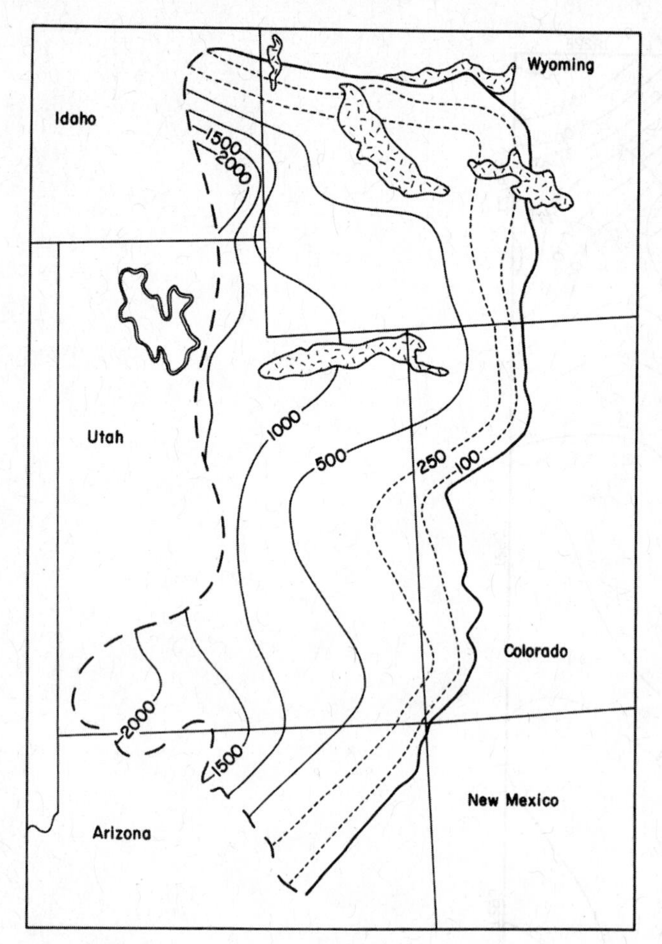

Figure G.3 Isopach map showing variations in thickness (in feet) of the Navajo and Nugget Sandstones (Jurassic). [Redrawn from Jordan (1965).]

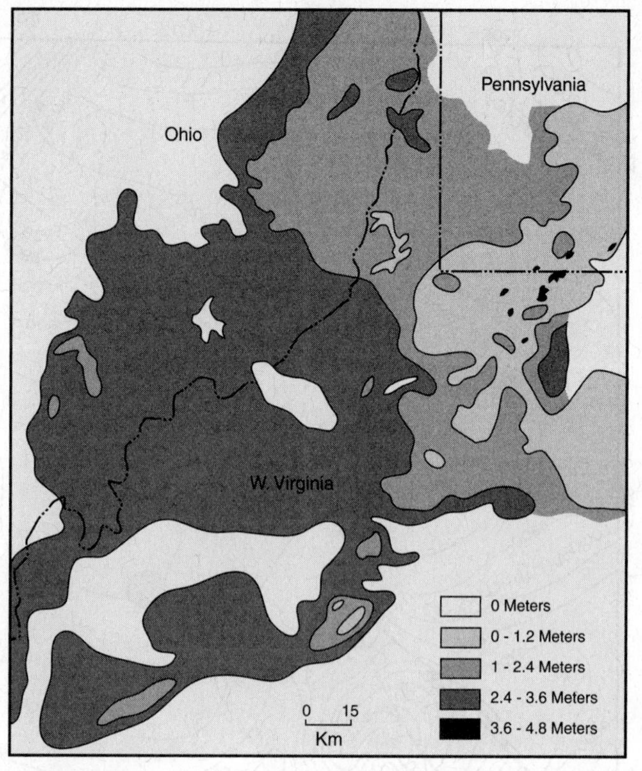

Figure G.4 Isopach map showing variations in thickness of the Pittsburgh Coal Seam. [From Hoover et al., 1969. Published with permission of the West Virginia Geological and Economic Survey. Permission to reproduce is granted if acknowledgment is given to the West Virginia Geological and Economic Survey.]

H. USING ORTHOGRAPHIC PROJECTION

A traditionally useful method for solving geometric problems is a kind of descriptive geometry known as **orthographic projection**. In essence, line-drawing constructions are prepared as a means of determining angular and spatial relationships in three dimensions. The constructions are difficult to visualize at first because the drawings convert map relationships into mixtures of maps and cross-sections. Fundamental to the procedure is constructing structure profiles and structure contour lines.

Constructing Simple Structure Profiles

Cross-sectional profiles, or **structure profiles**, are generally drawn at right angles to the trend or strike of structural features. These profiles show the **traces** of plunging or dipping structures, as they would appear in a vertical "*cut*" through the uppermost part of the earth. Consider a limestone bed that crops out in a perfectly flat area at the location shown in Figure H.1*A*. The bed strikes N40°E and dips 60°SE. A structure profile view of the bed is constructed in a vertical cut

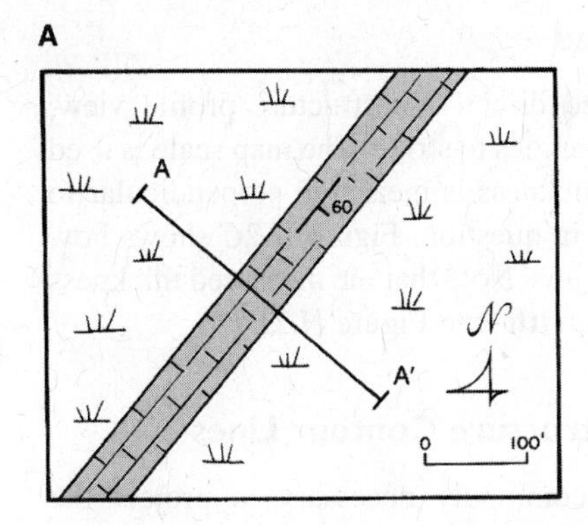

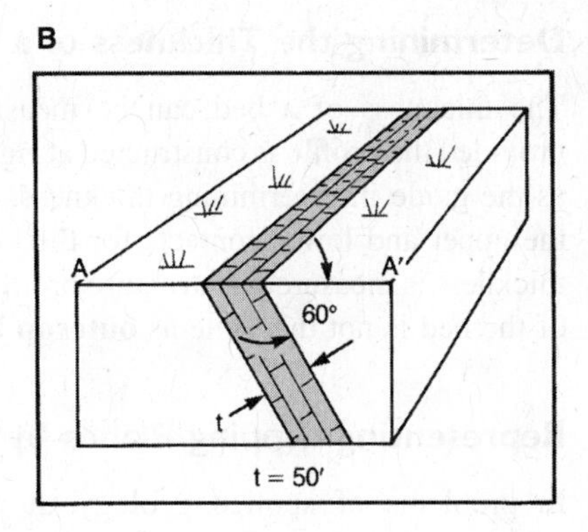

Figure H.1 (*A*) Plan view of a limestone bed that strikes northeast and dips southeast. (*B*) Structure profile view of the dipping bed. The profile view is the front face of the block diagram. It is oriented at right angles to the line of strike.

along A−A′ at right angles to the line of strike (Figure H.1*B*). The true dip of the bed is exposed to full view, as is the true thickness of the bed.

Using orthographic projection, let us construct step by step a structure profile for the limestone layer shown in Figure H.1*A*. First choose the orientation and location of the **profile line** along which the structure section is to be constructed (Figure H.2*A*). Points A and B are identified along the profile line such that A is on the lower contact of the limestone and B is on the upper contact at a location directly along the dip direction from A. Project points A and B from the interior of the map toward the edge of the map (or off the map onto another sheet of paper), where there is more available working space (Figure H.2*B*). Project each reference point by the same distance and in a direction strictly parallel to the strike of the limestone bed.

Draw a line through the **projected points** A′ and B′ (see Figure H.2*B*). This line represents the topographic **surface profile** for the location of the profile line where the limestone crops out. In this special case the surface profile is perfectly horizontal. In the general case, the surface profile would be marked by some **topographic relief**. Showing such relief in profile would be part of the construction process. Topographic control would be afforded by topographic contour lines on the base map.

Using a protractor, construct the angle of dip of the limestone bed in the subsurface, beneath the surface profile (Figure H.2*C*). For this example the dip is 60°SE. From A′ and B′ draw the 60°-dipping lines that correspond to the lower and upper contacts of the limestone bed. This completes the structure profile (see Figure H.2*C*). It represents an approximation of the structural form and the **attitude** (i.e., orientation) of the limestone bed at depth. It is a "*projection*" of the form and the dip of the limestone unit based on surface exposures.

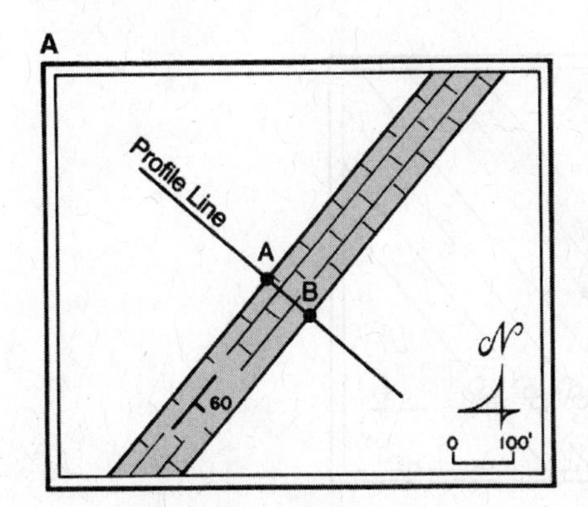

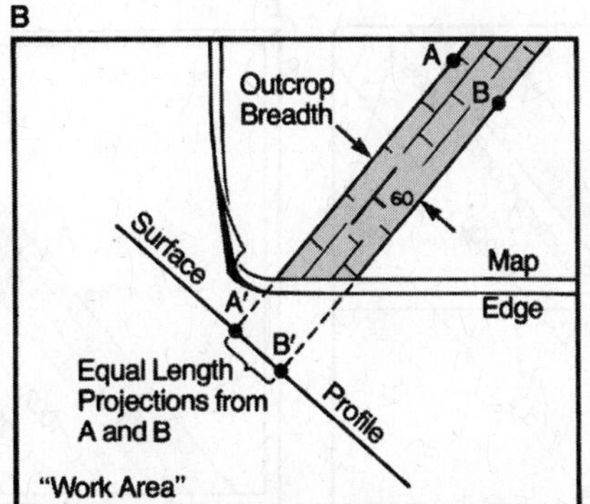

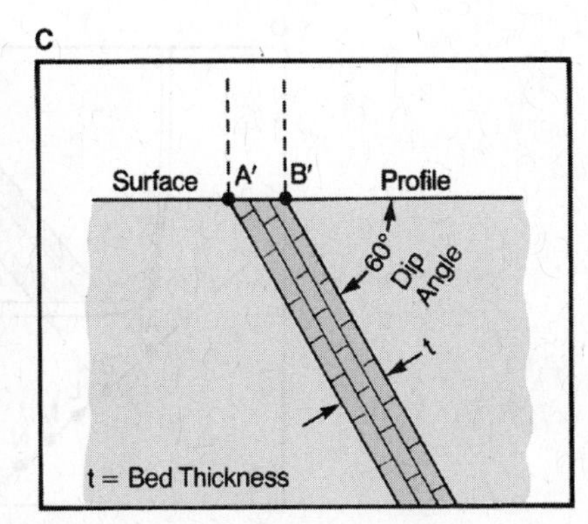

Figure H.2 Steps in constructing a structure profile. (*A*) Draw the profile line (AB) for which the structure profile is to be constructed. (*B*) Step the profile line to the edge of the map, or off the map sheet onto another piece of paper, where there is working space. (*C*) Plot the true dip of the bed and draw the subsurface expression of the top and bottom of the bed.

Determining the Thickness of a Bed

The thickness of a bed can be measured directly in structure profile view, provided the profile is constructed at right angles to strike. The map scale is used as the guide in determining thickness. Thickness is measured perpendicular to the upper and lower contacts for the bed in question. Figure H.2*C* shows how thickness is measured in structure profile view. Note that the measured thickness of the bed is not the same as **outcrop breadth** (see Figure H.2*B*).

Representing Dipping Planes by Structure Contour Lines

In problems of applied geology, it is commonly necessary to project the location of rock layers, contacts, and structures at depth. For example, if the surface outcrop of the limestone bed in the preceding example was found to be mineralized, the location of the limestone in the subsurface would be of economic interest. A first-order approximation of the limestone bed in the subsurface would be achieved through preparing a structure contour map of the limestone bed, using the upper contact of the limestone bed as a reference datum. Structure contour lines would connect points of equal elevation on the upper surface of the limestone. Because contour lines, by definition, connect points of equal elevation, they are lines of strike. Each contour line represents a strike line on the limestone bed at some specified elevation.

If the designated contour interval for the structure contour map is, for example, 100 ft (30 m), the pattern of the corresponding contour lines can be found by a series of orthographic construction steps. First, construct a structure profile for the limestone bed, and add horizontal lines to the structure profile below the surface profile such that the lines are spaced vertically at 100-ft (30-m) intervals (Figure H.3*A*). The map scale is used as the reference

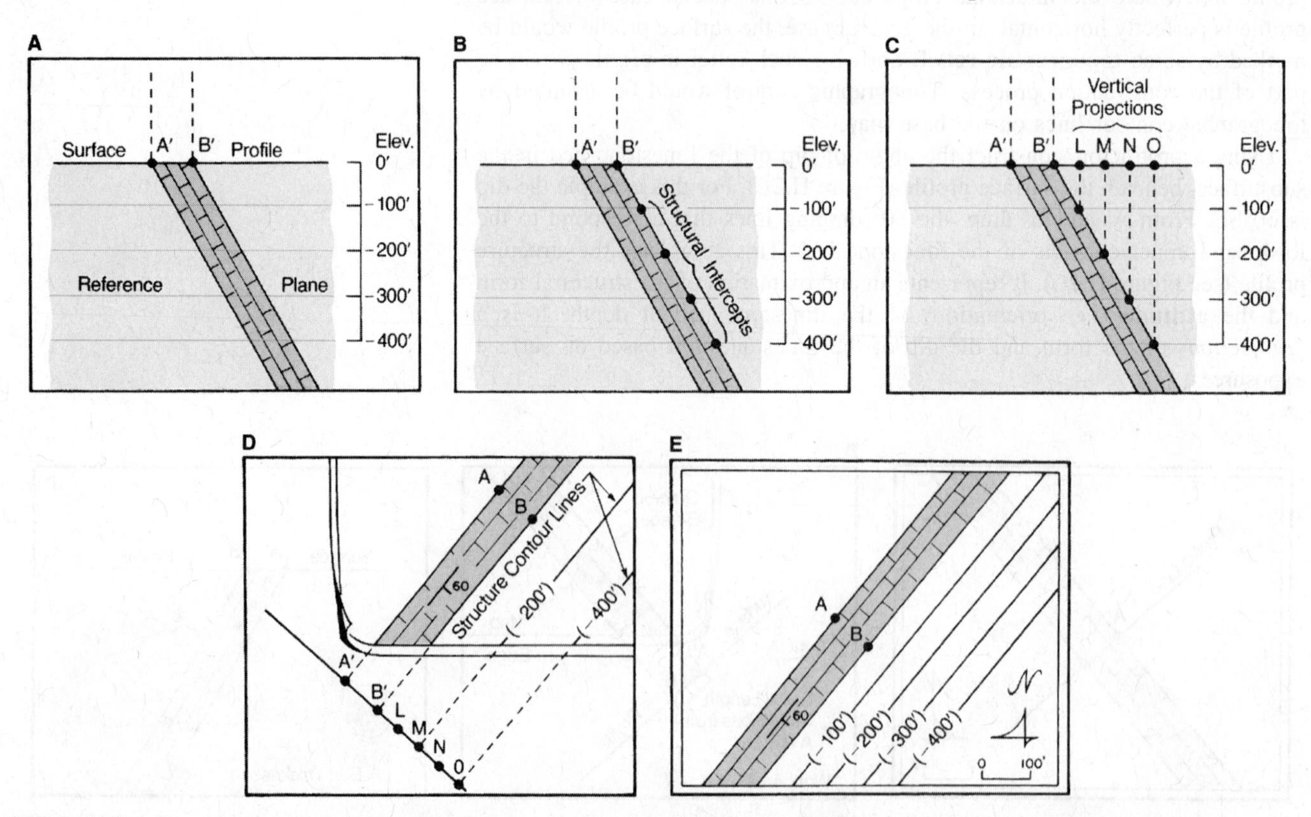

Figure H.3 Construction of structure contour lines representing the upper surface of a dipping bed. (*A*) Starting information: the structure profile view of a dipping bed. (*B*) Using the structure profile view, identify the structural intercepts of the top of the bed with the elevation reference planes shown in the subsurface. (*C*) Plot the vertical projections of each of the structural intercepts. (*D*) Project the vertical projections parallel to the line of strike. (*E*) The finished structure contour map.

for positioning the lines. Each of the lines represents the intersection of the plane of the structure profile with a horizontal plane of some given elevation. The horizontal planes are called topographic reference planes in the subsurface, or simply **reference planes**.

The next step is to identify the points of intersection of the upper surface of the limestone bed and the trace of each of the reference planes (Figure H.3*B*). These **structural intercepts** become the basis for positioning the structure contour lines.

Project each of the structural intercepts vertically to the surface profile (Figure H.3*C*). These projected points are the **vertical projections** of the structural intercepts of the upper contact of the limestone bed with each of the horizontal reference planes. Vertical projections are fundamental to orthographic projection. Consider vertical projection points M and O as examples. Point M lies directly above the point where the top of the limestone bed is exactly 200 ft (61 m) below the surface. Point O lies directly above the point where the upper contact of the limestone lies exactly 400 ft (122 m) below the surface.

Finally, we shift our construction from structure profile to map view (Figure H.3*D*), and we project **strike lines** from M and O into the interior of the map. These lines are **structure contour lines**, one representing −200 ft (i.e., 200 ft below the surface), the other −400 ft (−122 m). Since point M is the vertical projection of a point on the top of the limestone bed at elevation −200 ft (−61 m), every point on the strike line through M must also lie 200 ft (61 m) below the surface. The structure contour map is completed by drawing strike lines through all the vertical projections (Figure H.3*E*).

The ability to construct structure profiles and structure contour maps on the basis of surface or mine map patterns provides the means to solve a number of practical structural geologic problems.

Measuring Apparent Dip

Apparent dip is the inclination of the trace of a plane in a direction other than the true dip direction. Using the structure contour map displayed in Figure H.3*E*, we can solve for apparent dip of the limestone bed in any direction, for example, along a north−south line. First draw a north-trending profile line from one structure-contour line to another (Figure H.4*A*). The profile line in Figure H.4*A* connects a point on the −400-ft (−122-m) contour line with a point on the −100 ft (−30 m) contour line. Project the end points of this profile line to the edge of the map, or off it, and draw the surface profile (Figure H.4*B*). The end points of the surface profile are vertical projections

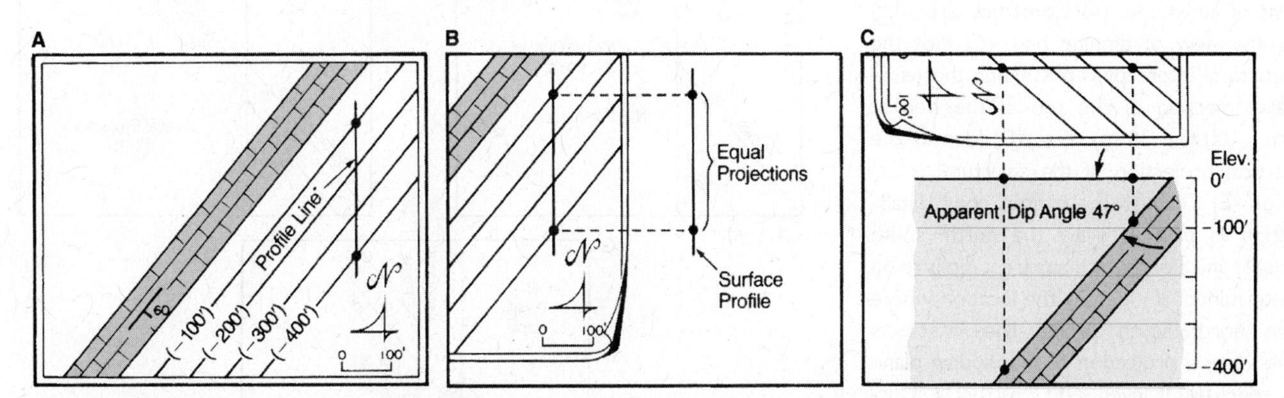

Figure H.4 Orthographic construction method for determining the apparent dip of a bed. (*A*) Designate the location and trend of the profile line along which apparent dip is to be determined. (*B*) Step the profile line to a location where working space is available. (*C*) Construct horizontal reference planes for the −100-ft (−30 m) and −400 ft (−122 m) levels. Project reference points on profile line to corresponding reference planes. Connect these structural intercepts to display apparent dip.

from the upper surface of the limestone bed at elevations corresponding to the values of the structure contour lines.

Begin to fashion the structure profile by constructing the horizontal reference planes that correspond to the contour lines on which the reference points of the profile line rest (Figure H.4*C*). Then project lines vertically down from the reference points on the surface profile to the corresponding structural intercepts of the limestone bed and the horizontal reference planes. A line connecting the structural intercepts at the −100-level and −400 ft level represents the upper contact of the limestone bed. Its angle of inclination, as measured from the horizontal, is the apparent dip. Its value, as measured with a protractor, is 47° (see Figure H.4*C*). This compares to true dip of 60°.

The standard orthographic solution to the apparent dip problem is a shortcut to the orthographic method just described. Here is how to do it. On a sheet of paper designate a **control point** that lies on the upper (or lower) contact of the limestone bed (Figure H.5*A*). Through it draw a strike line

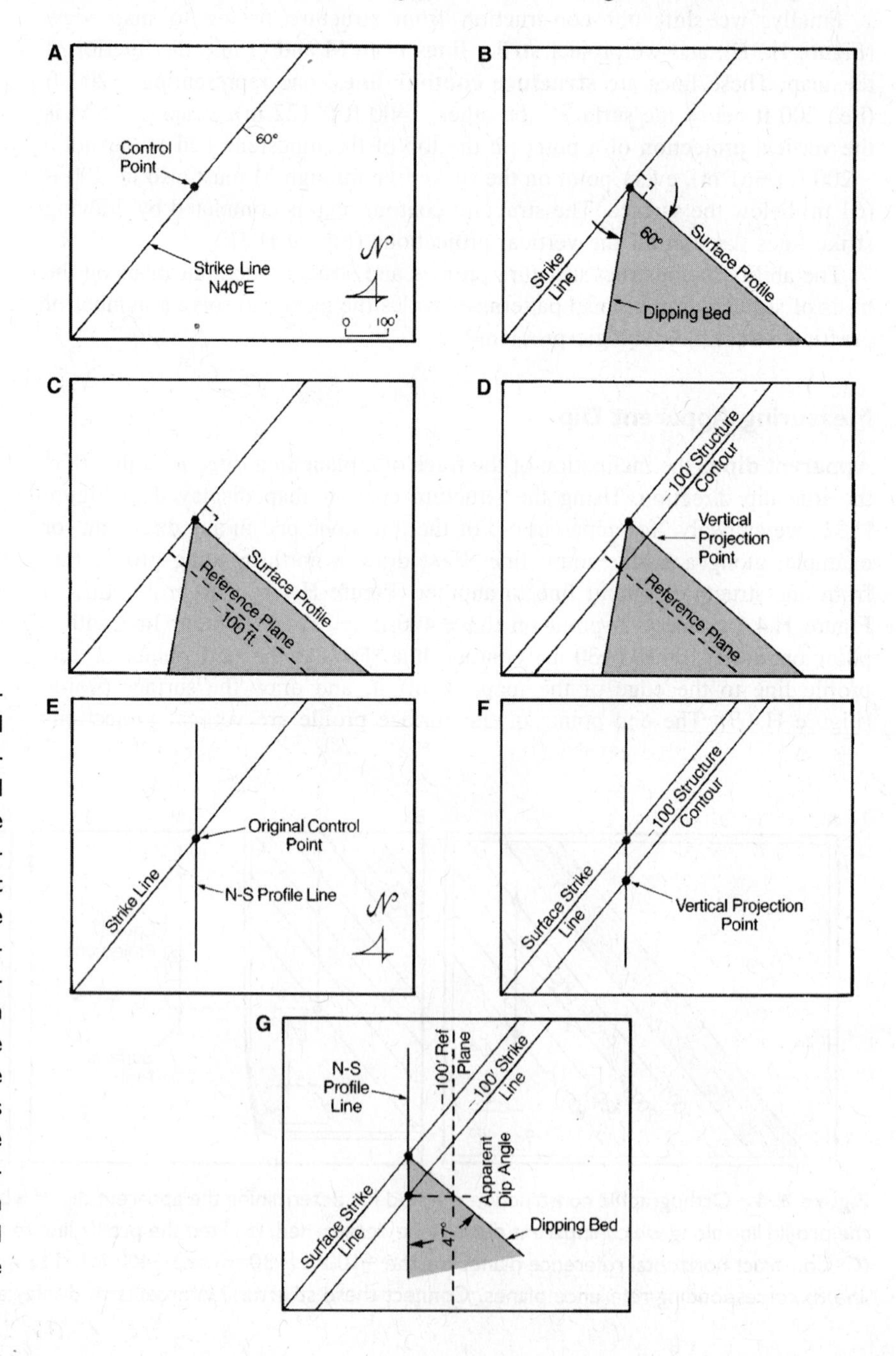

Figure H.5 "*Shortcut*" method for determining apparent dip. (*A*) Designate control point at convenient location on lower contact of limestone. (*B*) Construct structural profile view of dipping bed. (*C*) Plot the structural intercept representing the vertical projection of the top of the bed at the −100 ft (−30 m) level. (*D*) Identify the vertical projection of the structural intercept. (*E*) Through the control point established in part *A*, draw the north–south profile line along which apparent dip is to be determined. (*F*) Identify the location where the north–south profile line intersects the vertical projection of the dipping plane at the −100 ft level. (*G*) Construct a structure profile on the basis of the known elevation of the surface control point A and the −100 ft structural intercept. Measure the apparent dip with a protractor.

(N40°E) representing the strike attitude of the bed. At right angles to the strike line, construct a surface profile and draw the structural profile view of the dipping bed (Figure H.5*B*). Add to this profile view a horizontal reference plane that is positioned some arbitrary but known distance beneath the surface, for example, −100 ft (−30 m) (Figure H.5*C*)). Plot the vertical projection of the structural intercept of the reference plane and the dipping bed (Figure H.5*D*), and project a structure contour line across the map from the vertical projection point.

To find apparent dip as viewed in a north-trending vertical exposure, draw a north−south-trending profile line northward from the original control point (Figure H.5*E*). Where this profile line crosses the strike line of value −100 ft, there lies the vertical projection of the structural intercept of the upper contact of the limestone bed and the −100 ft structure contour (Figure H.5*F*). The apparent dip of the limestone is found by constructing a structure profile along the north-trending surface profile (Figure H.5*G*). The value of the apparent dip is 47°S. Apparent-dip constructions of this type make it clear that true dip of bedding or any other planar structure can be viewed only in profiles oriented *perpendicular* to strike. Exposures of structure profiles oriented *parallel* to strike reveal 0° apparent dip. Sections oriented obliquely to strike disclose intermediate values of dip between 0° and the true dip.

Constructing the Line of Intersection of Two Planes

Determining the trend and plunge of the line of intersection of two planes is fundamental to a number of geometric and geologic problems. Let us solve for the trend and plunge of the line of intersection of a dike and a limestone bed (Figure H.6*A*). Map relationships show that the dike strikes N68°E and dips 45°NW; and the limestone bed strikes N39°W and dips 35°NE. We assume that the dike and the limestone bed are perfectly planar and that the land surface is perfectly flat. For simplicity, no shifting of the limestone bed due to dike emplacement is shown in Figure H.6*A*.

Let the intersection of the northwest margin of the dike with the upper contact of the limestone bed be a control point for the constructions that follow (Figure H.6*B*). The control point is one of the two intersection points that we need to define the trend and plunge of the line of intersection of the two planes. The second control point will be found in the subsurface, where the top of the limestone bed and the northwest margin of the dike intersect at some known depth.

To find the second control point, first construct structure profile views of both the limestone bed and the dike (Figure H.6*C*). For each structure profile, construct a horizontal reference plane at some specified elevation, such as −1000 ft (−300 m) below each surface profile (see Figure H.6*C*).

Identify the structural intercept of the upper surface of the limestone bed and the horizontal reference plane, and then define its vertical projection to the surface profile (Figure H.6*D*). In the same manner, plot the vertical projection of the structural intercept of the dike and the horizontal reference plane.

Construct a strike line through the vertical projection of the limestone/reference plane structural intercept. This line is a map view of the vertical projection of the intersection of the top of the limestone bed with the −1000 ft reference plane (Figure H.6*E*). It is a structure contour on the limestone bed at elevation −1000 ft. In the same fashion, establish a −1000 ft contour line for the dike.

The intersection of the −1000 ft (−300 m) structure contour lines for the dike and the top of the limestone bed, respectively, is the vertical projection of the intersection of the dike and the limestone bed at elevation −1000 ft (Figure H.6*F*). Connect this intersection point with the original control point

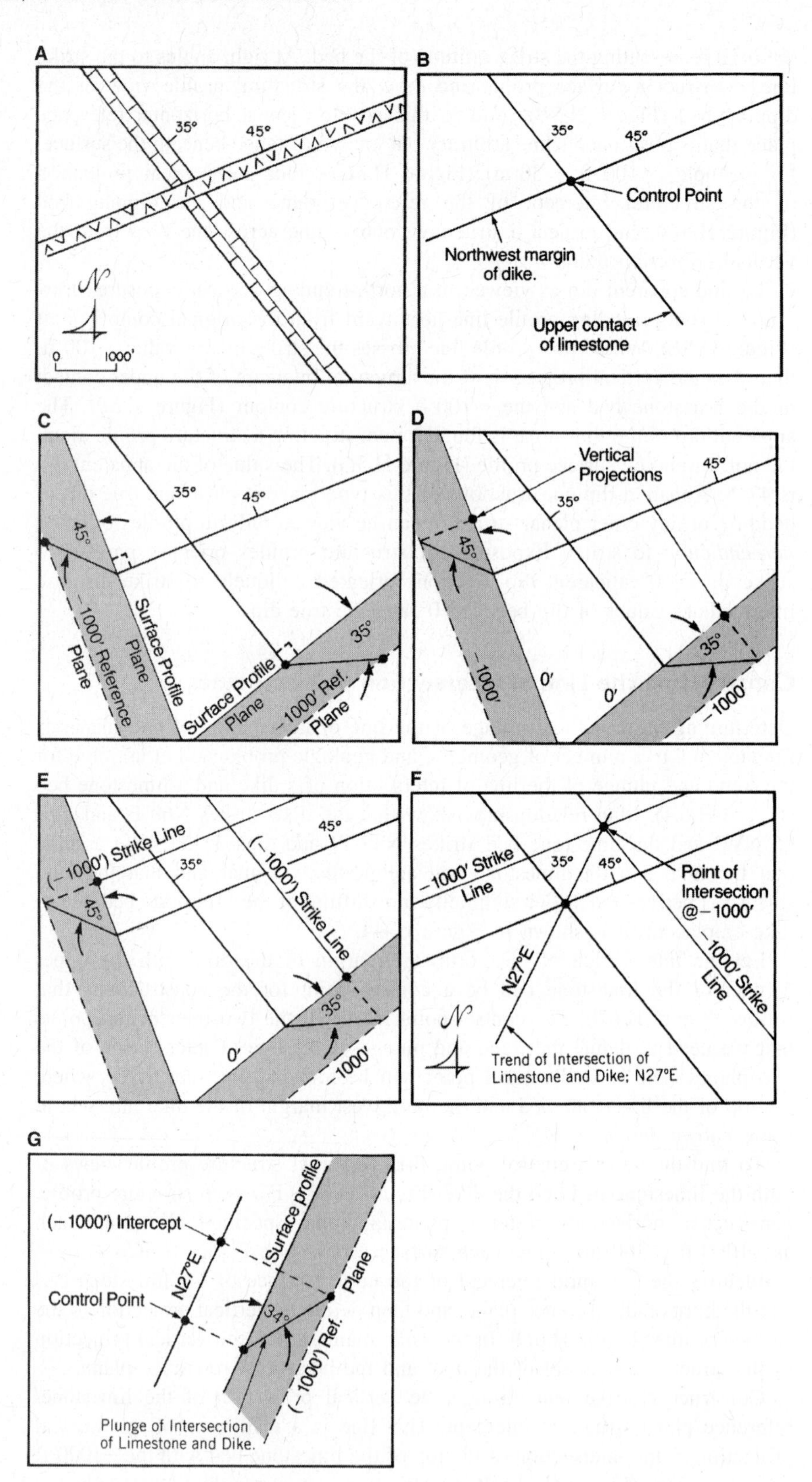

Figure H.6 Orthographic construction for determining the trend and plunge of the intersection of two planes. (*A*) Map of limestone bed that is intruded by a dike. (*B*) Simplified map of the same limestone bed, showing only the elements that are required to solve the problem at hand. (*C*) Draw two structure profiles—one for the dike, another for the limestone bed. Construct each profile, as always, at right angles to the line of strike. (*D*) Identify the vertical projection of the dike with the −1000 ft (−300 m) elevation reference plane. Also identify the vertical projection of the top of the limestone bed with the −1000 ft (−300 m) elevation reference plane. (*E*) Identify the vertical projection of the intersection of the top of the limestone bed with the northwestern margin of the dike at the −1000 ft (−300 m) level. (*F*) Connect the point of intersection of the northwest margin of the dike and the top of the limestone bed at the surface with the vertical projection of the same intersection at the −1000 ft (−300 m) elevation level. This line is the trend of the line of intersection of the dike and the limestone bed. (*G*) Construct a structure profile parallel to the trend of the line of intersection of the dike and the limestone bed. The plunge of the line of intersection is determined by drawing a line in profile view that connects the point of intersection of the dike and limestone bed at the surface level with the point of intersection of the dike and limestone bed at the −1000 ft (−300 m) level.

to define the trend of the intersection of the two planes. Its value is N27°E. To determine the plunge, construct a structure profile of the line of intersection of the dike and the limestone bed (Figure H.6*G*). The plunge measures 34°NE.

The Three-Point Problem

In some studies, it may be necessary to assume that a specific structural feature, such as an unconformity, is planar within a given area. If such an assumption is reasonable, the average strike and dip of the unconformity over a relatively large area may be determined through an orthographic construction known as a **three-point problem** (Figure H.7). Solving a three-point problem requires elevation control for at least three points *that lie in a common plane* in this case the surface of unconformity. The elevation control may be derived from topographic maps used in conjunction with geologic maps and/or subsurface drilling information. The three control points used in three-point constructions should define a triangular array of relatively widely spaced points, which mark different elevations.

Consider points A, B, and C, at elevations of 3400, 2700, and 2400 ft (1030, 818, and 727 m), respectively (Figure H.7*A*). Each point lies on the angular unconformity whose attitude is sought. Point B is intermediate in elevation between points A and C. The trace of the unconformity as seen in a cross-section passing through points A and C is constructed in vertical profile (Figure H.7*B*). It is inclined at an apparent dip of 28° toward C. Somewhere along its trace is a point whose elevation is the same as B. The location of this point (D) is found by constructing a horizontal reference plane whose elevation is the same as that of B (Figure H.7*C*). The intersection of the trace of the unconformity with the reference plane is projected vertically to D′, which represents the vertical projection of the point on the unconformity whose elevation is 2700 ft. By connecting B and D′ (i.e., points of equal elevation on the angular unconformity), the line of strike is defined (Figure H.7*D*). Its trend is N44°E.

To determine the dip of the unconformity, it is necessary to construct a structural profile at right angles to the line of strike, and to project the trace of the unconformity into this plane. Such a profile is shown in Figure H.7*E*, with elevations projected into it from control points A, B, and C. The inclination of the unconformity in this special profile is a measure of true dip, namely 33°SE.

Three-point constructions are not restricted to determining the strike and dip of unconformities. The three-point method can be used effectively to establish the strike and dip of any extensive planar structure, such as a bed, formation, fault, or intrusive contact.

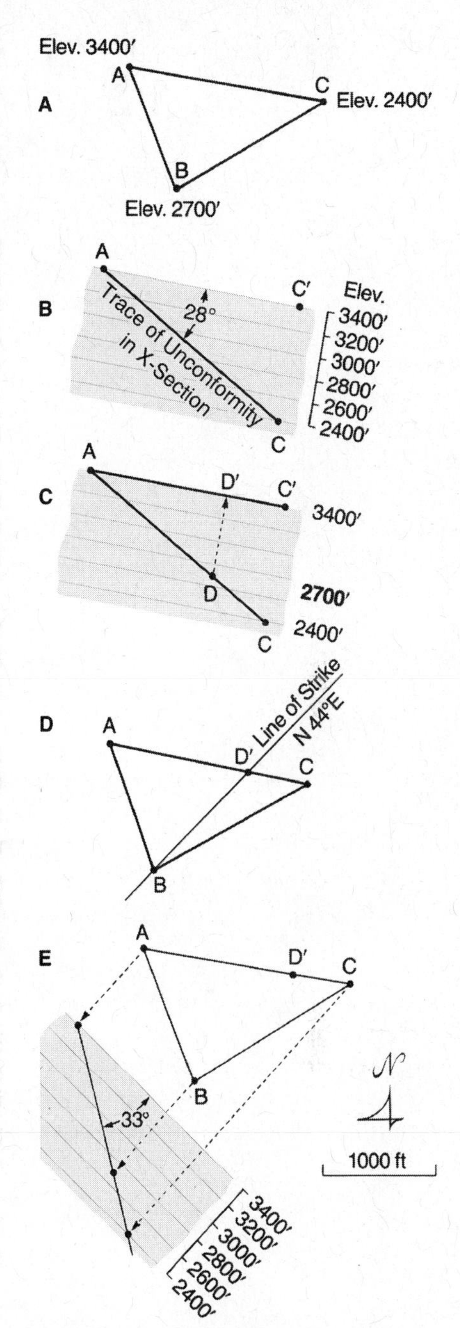

Figure H.7 Three-point problem. (*A*) Starting data: locations and elevations of three points on a common plane, in this case a planar unconformity. (*B*) Trace of unconformity as seen in cross-section through points A and C. (*C*) Identification of point D on the trace of unconformity where elevation is the same as point B. Identification of vertical projection of point D to the line of section AC. (*D*) Line of strike connects points B and D′. (*E*) Determination of true dip of unconformity.

I. CARRYING OUT STEREOGRAPHIC PROJECTION

Stereographic projection is a powerful method for solving geometric problems in structural geology (Bucher, 1944; Phillips, 1971). Stereographic projection differs from orthographic projection in a fundamental way: orthographic projection preserves spatial relations among structures, but stereographic projection displays geometries and orientations of lines and planes without regard to spatial relations.

The use of stereographic projection is preferable to orthographic projection in solving many geometric problems, simply because of ease of operations. Solving for apparent dip, the trend and plunge of the intersection of two planes, and angles between lines and planes in space can be carried out rapidly and accurately using stereographic projection. Orthographic projection, in contrast, requires the slow, careful construction of line drawings. But, orthographic projection remains the only effective way to solve geometric problems when topographic relief, map relationships, and depth to structures in the subsurface are integral to the solution of structural problems. In practice, we combine orthographic and stereographic projection techniques in ways that are practical, efficient, and complementary.

Geometry of Projection

Think of stereographic projection as a procedure comparable to using a three-dimensional protractor. A two-dimensional protractor is simple to use. Using a protractor we can plot trends of lines, measure angles between lines, construct **normals** (i.e., perpendiculars) to lines, and rotate lines by specified angles. Stereographic projection permits the same kinds of operation, but in three-dimensional space. Moreover, both lines and planes can be plotted and analyzed. Equipped with a three-dimensional protractor, we can do the following: plot orientations of lines; plot orientations of planes; determine the orientation of the intersection of two planes; determine the angle between two lines; determine the angle between two planes; measure the angle between a line and a plane; and rotate lines and planes in space about vertical, horizontal, or inclined axes.

All the preceding operations would be simple if it were possible to assemble real lines and planes in space, like Tinkertoys, and measure their geometric properties directly. Using stereographic projection techniques, we figuratively assemble lines and planes within a reference sphere.

The line or plane to be stereographically represented can be thought of as passing through the center of a reference sphere and intersecting its lower hemisphere (Figure I.1*A*). Planes intersect the lower hemisphere in the form of **great circles**; lines intersect the lower hemisphere in **points**. Stereographic projection of lines and planes to points and great circles constitutes a systematic reduction of three-dimensional geometry to two dimensions. The "*flattening*" to two dimensions is achieved by projecting the lower hemisphere intersections to an **equatorial plane** of reference that passes through the center of the sphere (Figure I.1*B*). This is the plane of stereographic projection. The lower hemisphere intersections are projected as rays *upward* through the horizontal reference plane to the **zenith** of the sphere. Where the rays of projection pass through the horizontal reference plane, point or great-circle intersections are produced, and these are **stereograms** or **stereographic projections** of lines and planes. Details of the projection geometry are presented in Phillips (1971).

Steep-plunging lines stereographically project to locations close to the center of the horizontal plane of projection; shallow-plunging lines project to locations near the perimeter of the plane of projection (Figure I.2*A*). Steeply

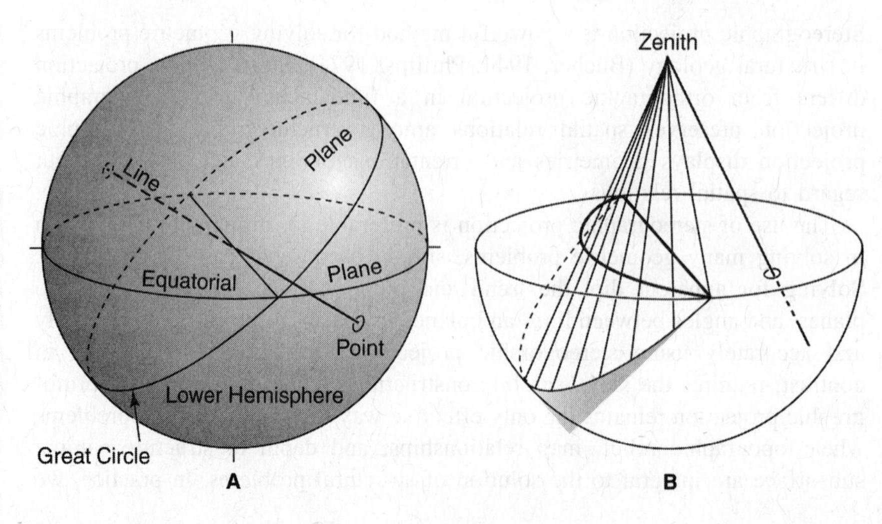

Figure I.1 The inherent three-dimensional geometry of stereographic projection. (*A*) Projection of a plane and a line through the center of a reference sphere. The plane intersects the lower hemisphere of the reference sphere as a great circle. The line intersects the lower hemisphere as a single point. (*B*) Projection of intersection points from the lower hemisphere of the reference sphere to the zenith of the projection.

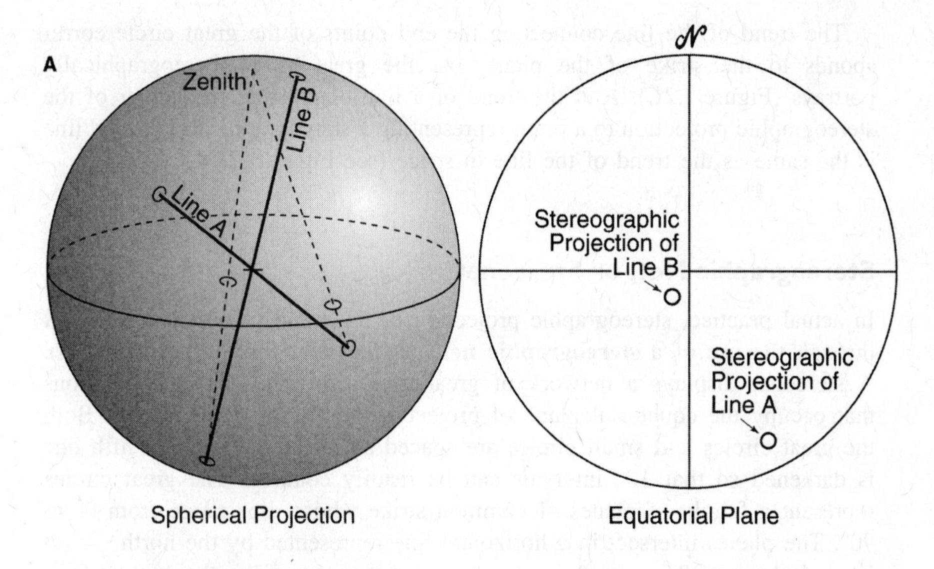

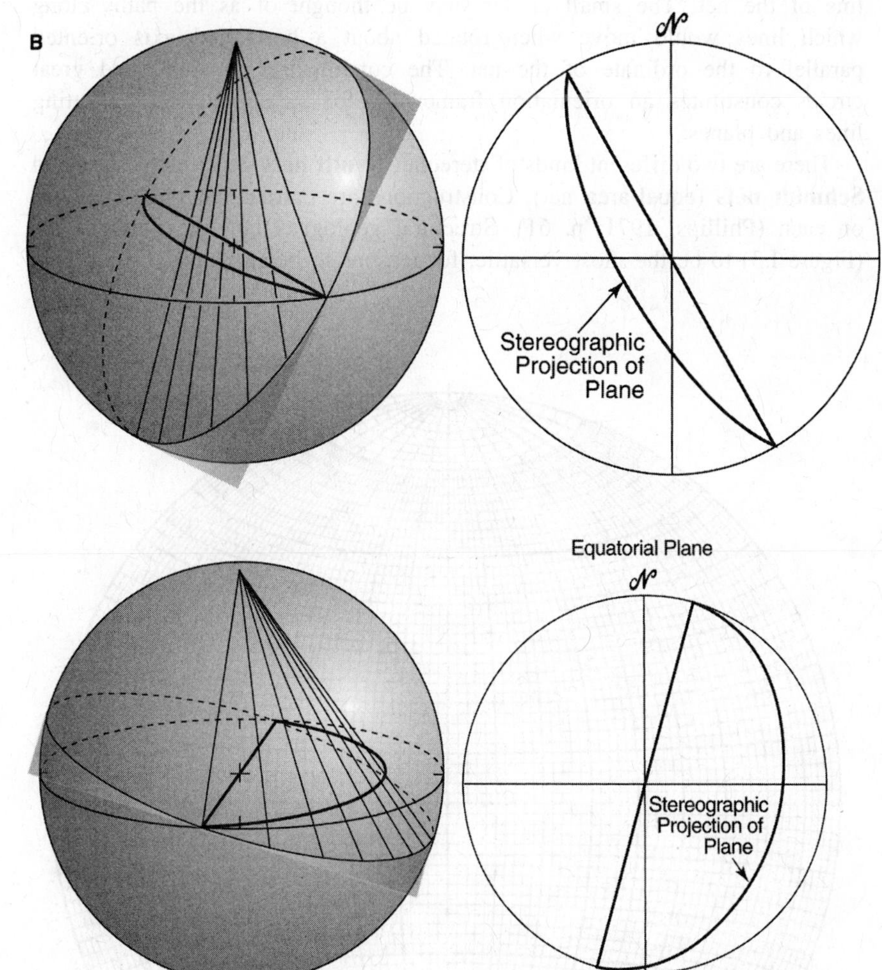

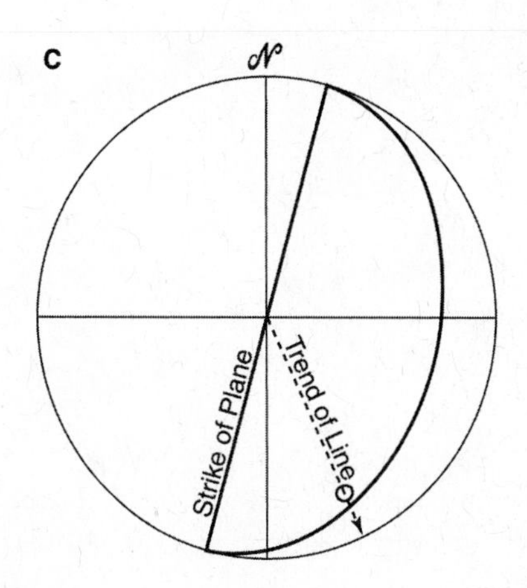

Figure I.2 The distance that a great circle or point lies from the center of the equatorial plane of projection is a measure of the inclination of a plane or line. (*A*) Shallow plunging lines project close to the perimeter of the equatorial plane; steeply plunging lines project close to the center. (*B*) Great circles that represent the orientation of steeply dipping planes pass close to the center of projection; shallow dipping planes are represented by great circles that pass close to the perimeter of the equatorial plane. (*C*) Stereographic representation of the strike of a plane and the trend of a line.

dipping planes stereographically project as great circles that pass near the center of the plane of projection; gently dipping planes project as great circles passing close to the perimeter of the horizontal plane of projection (Figure I.2*B*). The distance that a great circle or point departs from the center of the plane of projection is a measure of the degree of inclination of the plane or line that has been stereographically plotted.

The trend of the line connecting the end points of the great circle corresponds to the strike of the plane that the great circle stereographically portrays (Figure I.2*C*). And the trend of a line connecting the center of the stereographic projection to a point representing a stereographically plotted line is the same as the trend of the line in space (see Figure I.2*C*).

Stereographic Net, or Stereonet

In actual practice, stereographic projection of lines and planes is carried out through the use of a **stereographic net**, or **stereonet** for short (Figure I.3). A stereonet displays a network of great-circle and small-circle projections that occupy the equatorial plane of projection of the reference sphere. Both the great circles and small circles are spaced at 2° intervals; every fifth one is darkened so that 10° intervals can be readily counted. The great circles represent a family of planes of common strike whose dips range from 0° to 90°. The planes intersect in a horizontal line represented by the north–south line of the net. The small circles may be thought of as the paths along which lines would move when rotated about a horizontal axis oriented parallel to the ordinate of the net. The combination of small and great circles constitutes an orientation framework for stereographically plotting lines and planes.

There are two different kinds of stereonet: **Wulff nets** (equal-angle net) and **Schmidt nets** (equal-area net). Constructions are carried out the same way on each (Phillips, 1971, p. 61). Structural geologists find the Schmidt net (Figure I.3) to be the most versatile, for reasons to be explained later.

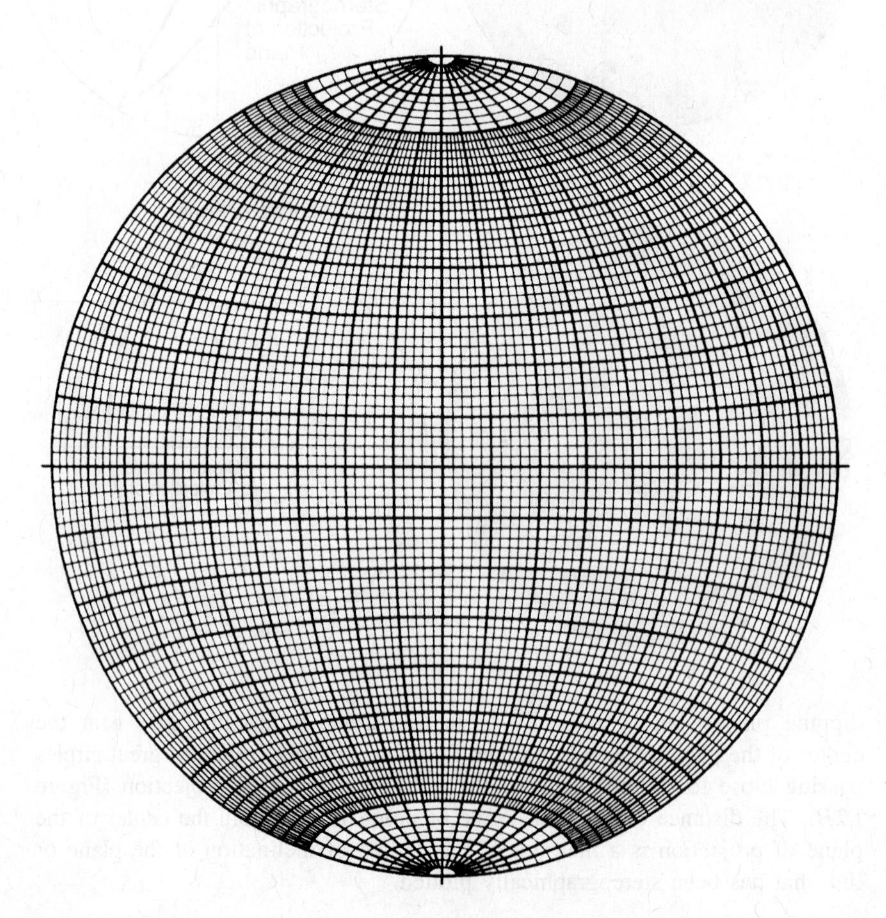

Figure I.3 The Schmidt net, an equal-area net, is the stereographic template that structural geologists typically use.

To prepare the Schmidt net for use, tape or glue it to a heavy backing, such as cardboard or Masonite. Insert a thumbtack through the backing and through the exact center of the net, taping the tack to the underside of the net so it cannot fall free (Figure I.4). A sheet of tracing paper is placed on the net so that the paper, punctured by the thumbtack, can rotate about the tack. A small square of clear tape applied to the back of the tracing paper, covering the area where the thumb tack will go through, prevents the paper from ripping during plotting. All construction work is carried out on the tracing paper, which is oriented with respect to north by marking a **north index** at a point corresponding to the top of the north–south line. This geographically orients the overlay for the constructions to be carried out.

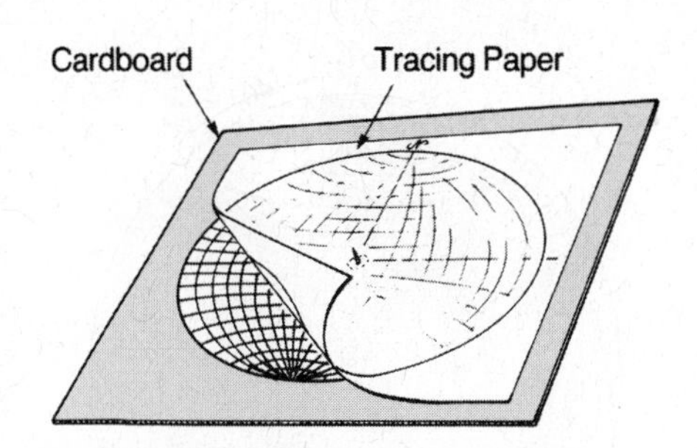

Figure I.4 Stereographic net ready for action. [From Whitten, 1966, Structural geology of folded rocks. Reprinted by permission of John Wiley and Sons, Inc., New York, copyright ©1966.]

Plotting the Trend and Plunge of a Line

Before any problems can be solved stereographically, it is necessary to learn how to represent the orientations of lines and planes on a stereonet. Lines are easiest to plot. Consider a line that plunges 26° N40°E. To represent the trend and plunge of this line stereographically, first plot the trend, in degrees, on the outer perimeter of the stereographic net. To do this, measure east from north by 40° (Figure I.5*A*). This can be accomplished simply by using the stereographic net as you would a protractor, counting clockwise from the north index on the tracing paper along the periphery of the net to 40°. The 40° azimuth corresponds to the 40° small-circle intercept on the perimeter of the net. Mark the point at 40° with a **trend-index mark** (t) (Figure I.5*A*).

To plot the 26° plunge, first rotate the overlay clockwise until "t" comes to rest on the right end of the east–west line of the net (Figure I.5*B*). This is one of two lines (the N–S axis being the other) where inclination can be directly measured and plotted. The plunge is measured by counting 26° from the perimeter of the net along the east–west line toward the center of the net. Point L represents the 26° plunging line (see Figure I.5*B*). By rotating the tracing paper counterclockwise such that the north index again becomes aligned with the top of the north–south line ("*home position*"), point L can be viewed in its proper orientation framework (Figure I.5*C*). As a general check, it can be seen that point L lies in the northeast quadrant, corresponding to a northeast trend. Furthermore, it falls relatively close to the perimeter, reflecting a rather shallow plunge.

One shortcut is available. Plotting the N40°E trend can be achieved simply by rotating the tracing paper counterclockwise such that the north index comes to rest on 40° W (Figure I.5*D*). This automatically orients the N40°E trend line along the north–south line of the net, along which the 26°plunge can be directly measured.

As a second example, let us plot stereographically the orientation of a line plunging 75° S65° W. First define the trend by measuring 65° west of south (Figure I.6*A*). Point "t" represents the trend of this line. Next rotate "t" until it coincides with the left end of the east–west line (Figure I.6*B*). Measure the value of the plunging line by counting inward 75° from the perimeter of the net (L marks the 75°-plunging line). Rotate the tracing paper back to home position and view L in its proper orientation (Figure I.6*C*). Note that L plots close to the center of the net because of its steep plunge.

Plotting the Strike and Dip of a Plane

Let us now stereographically plot the orientation of a plane. Consider a plane striking N40°W and dipping 30° SW. The strike line of this plane is found by rotating the tracing paper clockwise until the north index comes to rest on 40°E (Figures I.7*A*). When the strike line of the plane is rotated into

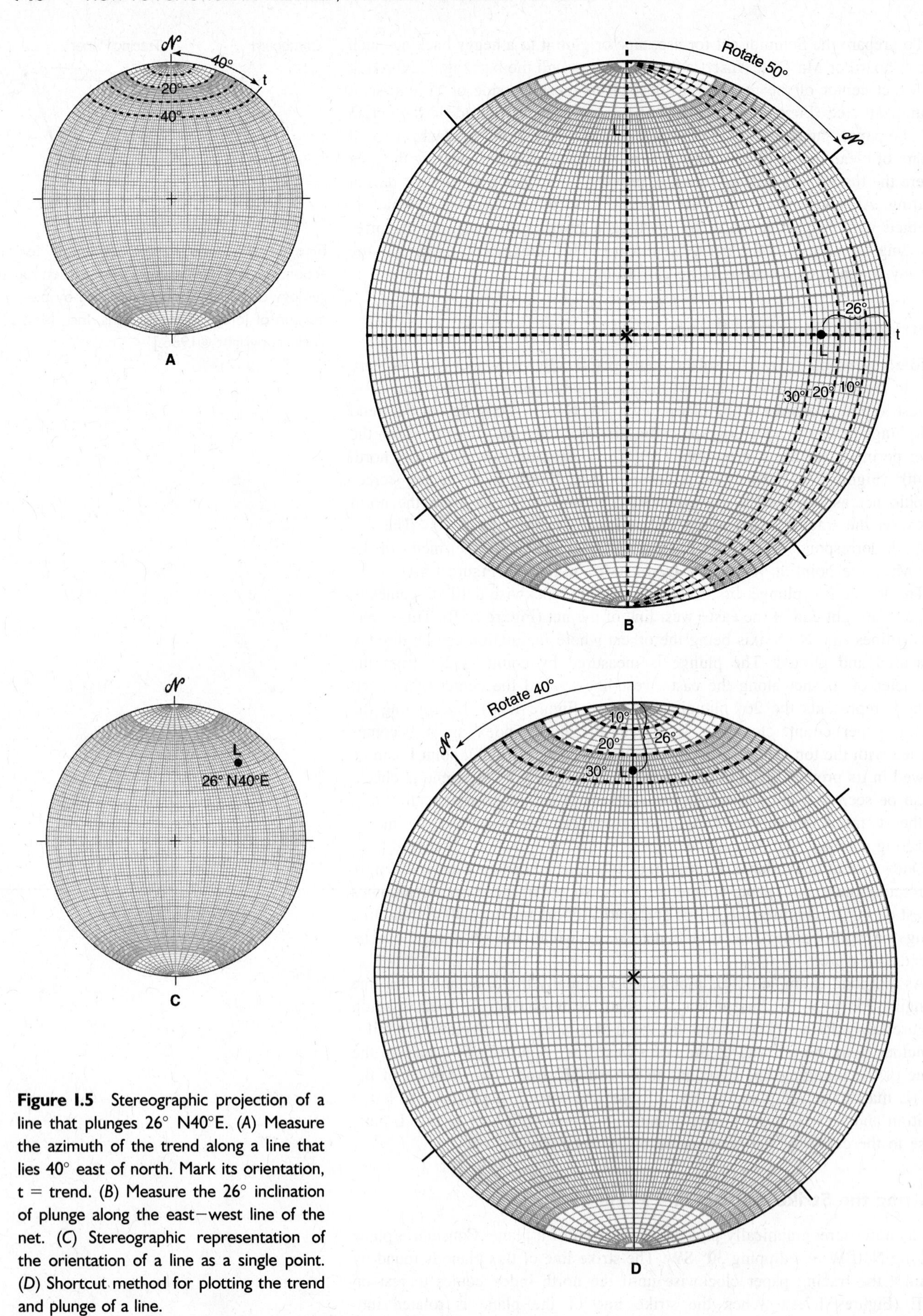

Figure I.5 Stereographic projection of a line that plunges 26° N40°E. (*A*) Measure the azimuth of the trend along a line that lies 40° east of north. Mark its orientation, t = trend. (*B*) Measure the 26° inclination of plunge along the east–west line of the net. (*C*) Stereographic representation of the orientation of a line as a single point. (*D*) Shortcut method for plotting the trend and plunge of a line.

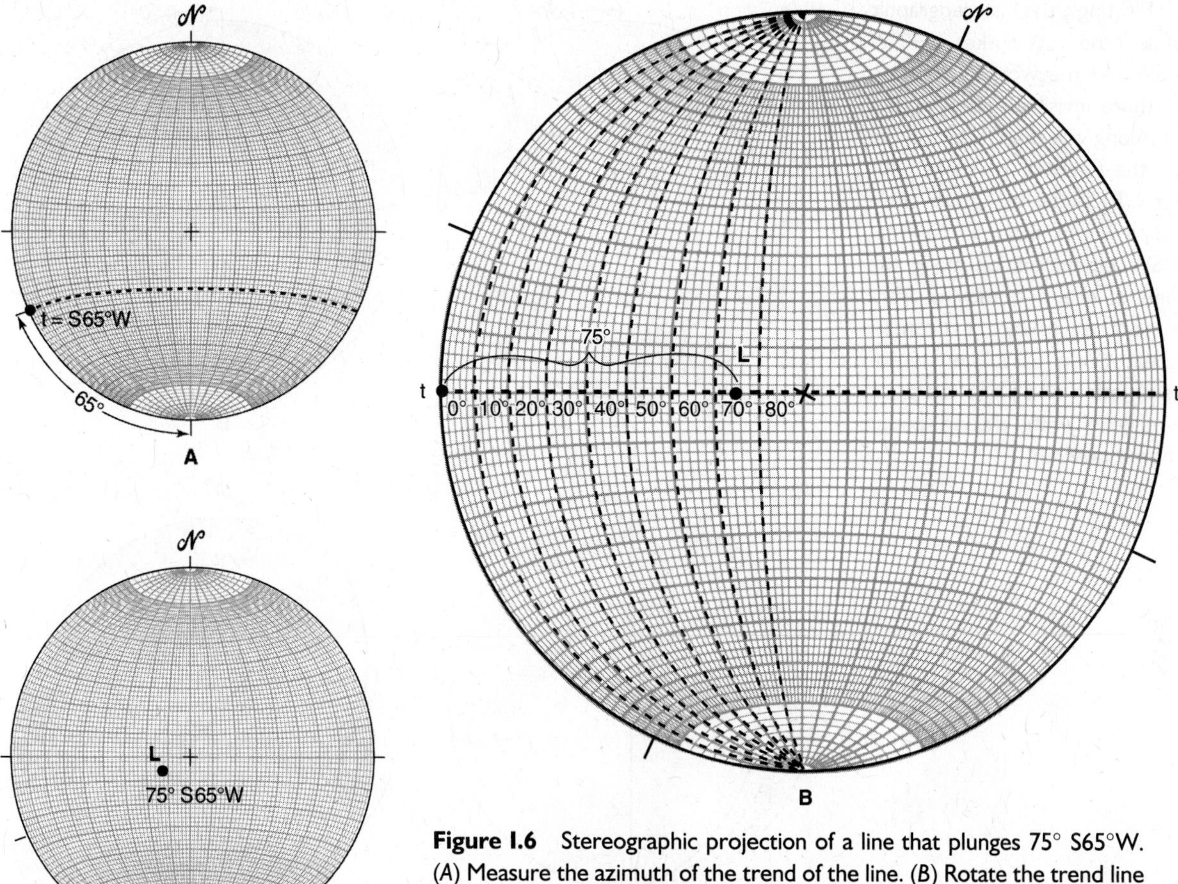

Figure I.6 Stereographic projection of a line that plunges 75° S65°W. (*A*) Measure the azimuth of the trend of the line. (*B*) Rotate the trend line to the east–west line of the net. Measure the plunge inclination along the east–west line, moving from the edge of the net toward the center. (*C*) Stereographic result as seen when overlay is in home position.

coincidence with the north–south line of the stereonet, the dip of the plane can be plotted (Figure I.7*B*). Count inward 30° from the perimeter of the net along the east–west line, which is the line of dip when the strike line of the plane coincides with the north–south line of the net. Then trace the great circle that coincides with the 30°SW dip (see Figure I.7*C*). By rotating the north index back to home position (Figure I.7*D*), the great circle becomes aligned in an orientation that corresponds stereographically to a plane striking N40°W and dipping 30°SW.

When large numbers of planes must be plotted stereographically on a single projection, it is far more practical to plot each plane as a pole, not as great circles. The resulting diagram is cleaner and more conducive to interpretation of preferred orientations. The orientation of any plane in space can be described uniquely by the orientation of a line perpendicular to the plane. If the trend and plunge of a normal (i.e., **pole**) to a plane is known, the orientation of the plane itself is also established. The pole to a vertical plane is horizontal, and it stereographically plots as a point on the perimeter of the stereonet. The pole to a horizontal plane is vertical, and it plots stereographically as a point at the very center of the stereonet. The pole to an inclined plane plots as a point somewhere in the interior of the net, but not at its center. To visualize this, place a pencil between your fingers, pointing it down and away from your palm. The pencil represents the pole to the plane of your hand. When your hand is horizontal, the pencil points straight down. When your hand is held vertically, the pencil points horizontally.

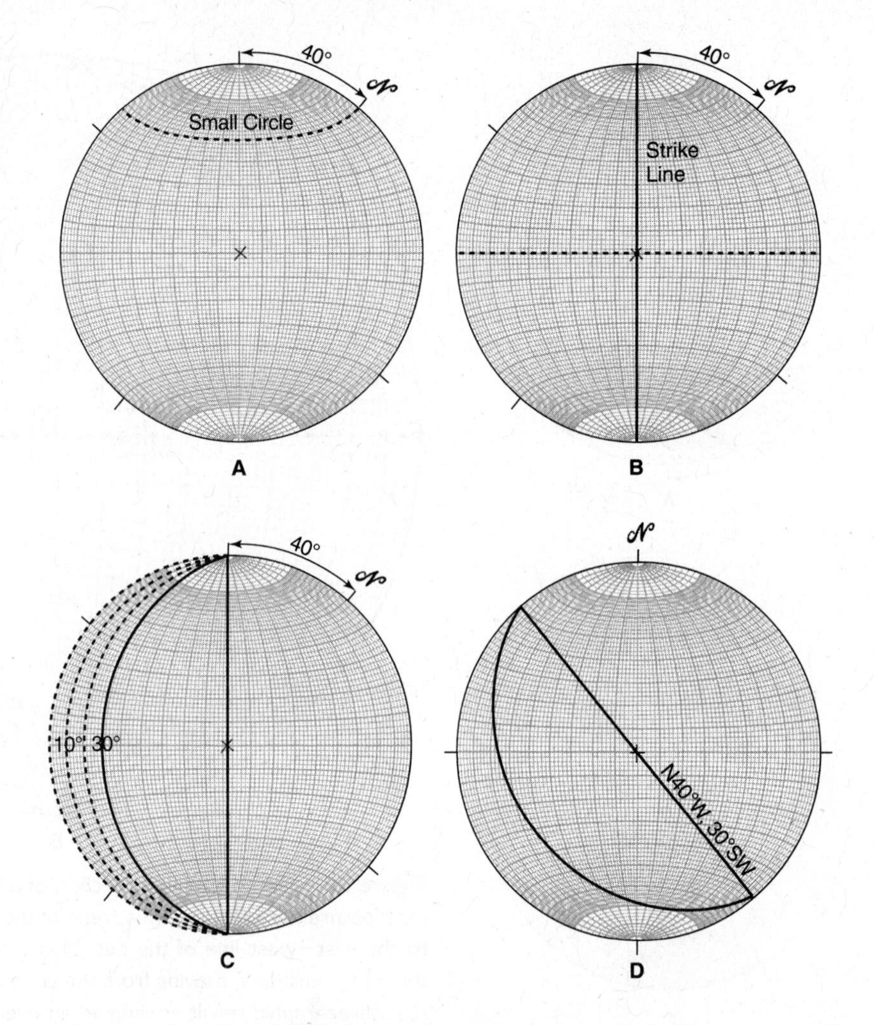

Figure I.7 Plotting the stereographic projection of a plane that strikes N40°W and dips 30°SW. All the work is done on tracing paper (here invisible) overlying the template. (*A*) Along the periphery of the net, measure the azimuth of strike, 40° west of north. (*B*) Note the great circle along which dips will be plotted. (*C*) Measure the 30°SW dip inclination along the east–west line of the net, moving from the edge of the net toward its center. Trace the great circle. (*D*) The stereographic representation of a plane is a great circle, shown here with tracing paper restored to home position.

The procedure for plotting poles to planes stereographically is reasonably straightforward. Consider the pole to a plane that strikes N80°E and dips 20°SE. Figure I.8*A* shows the orientation of the plane plotted stereographically as a great circle. By definition, the pole to this plane is oriented 90° to the plane, measured in a vertical plane perpendicular to strike. To plot this pole stereographically, rotate the strike azimuth so that it coincides with the north–south line of the stereonet (Figure I.8*B*). In this orientation, the line of true dip in the plane lies on the east–west line of the net. From the point representing the line of true dip of the plane, measure 90° along the east–west line (Figure I.8*C*). The position of the pole is 20° beyond the center point of the net and is plotted as point P. Rotating the north index back to home position results in placement of the pole in its proper orientation (Figure I.8*D*).

In practice, plotting a pole to a plane need not include plotting the plane as a great circle. Rather, the strike line of the plane is rotated so that it coincides with the north–south line of the stereographic net (as in Figure I.8*C*); then the pole to the plane is found by measuring along the east–west line *outward* from the center of the net, into the quadrant opposite the dip direction of the plane.

Plotting the Orientation of a Line in a Plane

Many structural relationships involve the presence of a line in a plane, like slickenlines on a fault surface. Stereographically, the point representing the trend and plunge of a line in a plane must lie on the great circle representing the strike and dip of the plane. Consider the geometry of a fault that strikes N10°E and dips 44°NW, containing slickenlines that plunge 40° N50°W. If

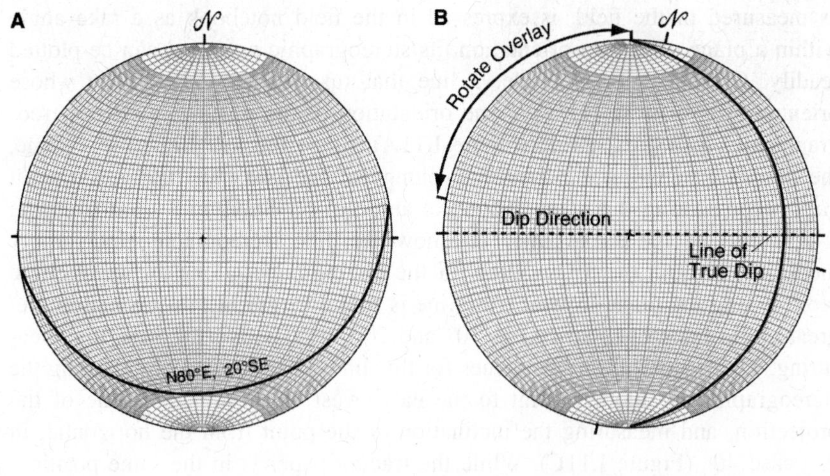

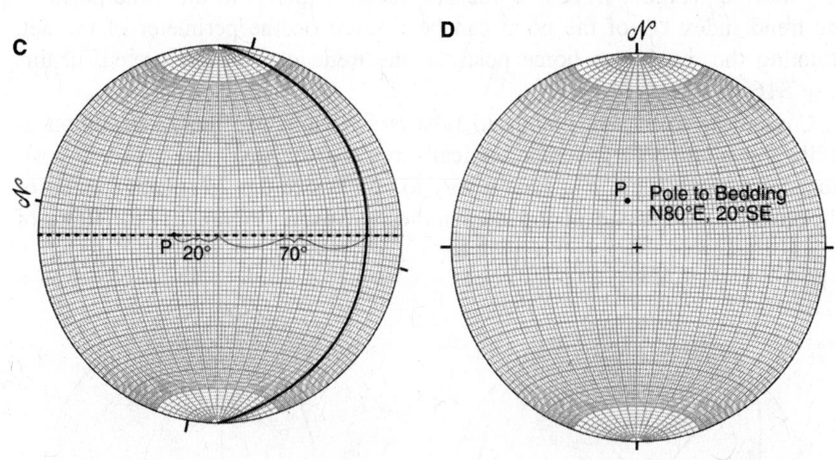

Figure I.8 Stereographic projection of the pole to a plane. (*A*) Great-circle stereographic representation of a plane that strikes N80°E and dips 20°SE. (*B*) Rotate strike line into parallelism with the north–south line of the net. (*C*) Identify the pole to the plane by measuring 90° along the east–west line of the net from the point that represents the inclination of true dip. (*D*) Final portrayal of the stereographic representation of the plane as a pole.

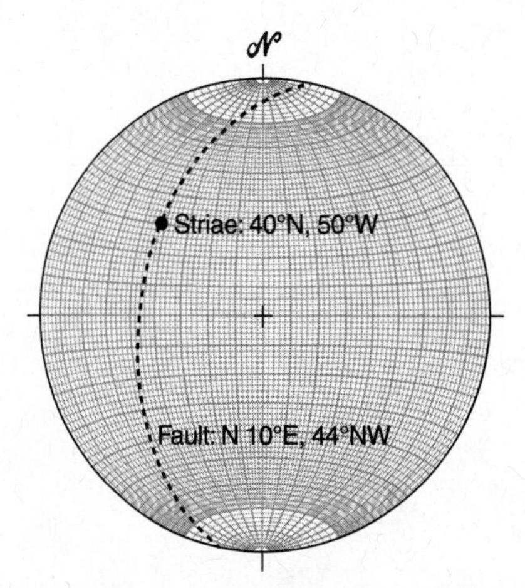

Figure I.9 If a line lies in a plane, the stereographic projection of the line as a point must fall on the great circle that stereographically represents the orientation of the plane.

the stereographic orientations of these two elements are plotted independently, it is found that the trend and plunge of the slickenlines are represented by a point that falls on the great circle corresponding to the strike and dip of the fault (Figure I.9).

Another way to describe the orientation of a line in a plane is to measure the **rake** (or **pitch**) of the line. Rake is the angle between a line and the strike line of the plane in which it is found (Figure I.10). If the orientation of a line,

Figure I.10 Barnyard conversation about tools and recreation. [Artwork by D. A. Fisher.]

as measured in the field, is expressed in the field notebook as a rake angle within a plane of known orientation, its stereographic portrayal can be plotted readily. Consider, for example, a line that rakes 65°SE in a plane whose orientation is N40°W, 45°SW. The orientation of the plane is plotted stereographically as a great circle (Figure I.11*A*). Since the line lies in this plane, the point representing the trend and plunge of the line must lie on the great circle representing the plane. The rake angle of 65°SE is measured from the S40°E end of the strike line. To show the line stereographically, simply measure 65° from the SE quadrant of the tracing paper along the great circle representing the plane in which the line is found (Figure I.11*B*). Small-circle/great-circle intercepts, spaced at 10° and 2° intervals, are the basis for measuring. The trend-and-plunge values for this line are interpreted by rotating the stereographically plotted point to the east–west or north–south line of the projection, and measuring the inclination of the point from the horizontal, in this case 40° (Figure I.11*C*). While the tracing paper is in the same position, the trend index "t" of the point can be marked on the perimeter of the net. Rotating the overlay to home position, the trend can be interpreted, in this case S16°W (Figure I.11*D*).

Converting trend and plunge to rake is a reasonably smooth operation as well. The fault surface stereographically represented in Figure I.12*A* is positioned stereographically in such a way to emphasize the 23° **plunge** angle of slickenlines. The **trend** of the slickenlines is S44°E (Figure I.12*B*). **Rake** of

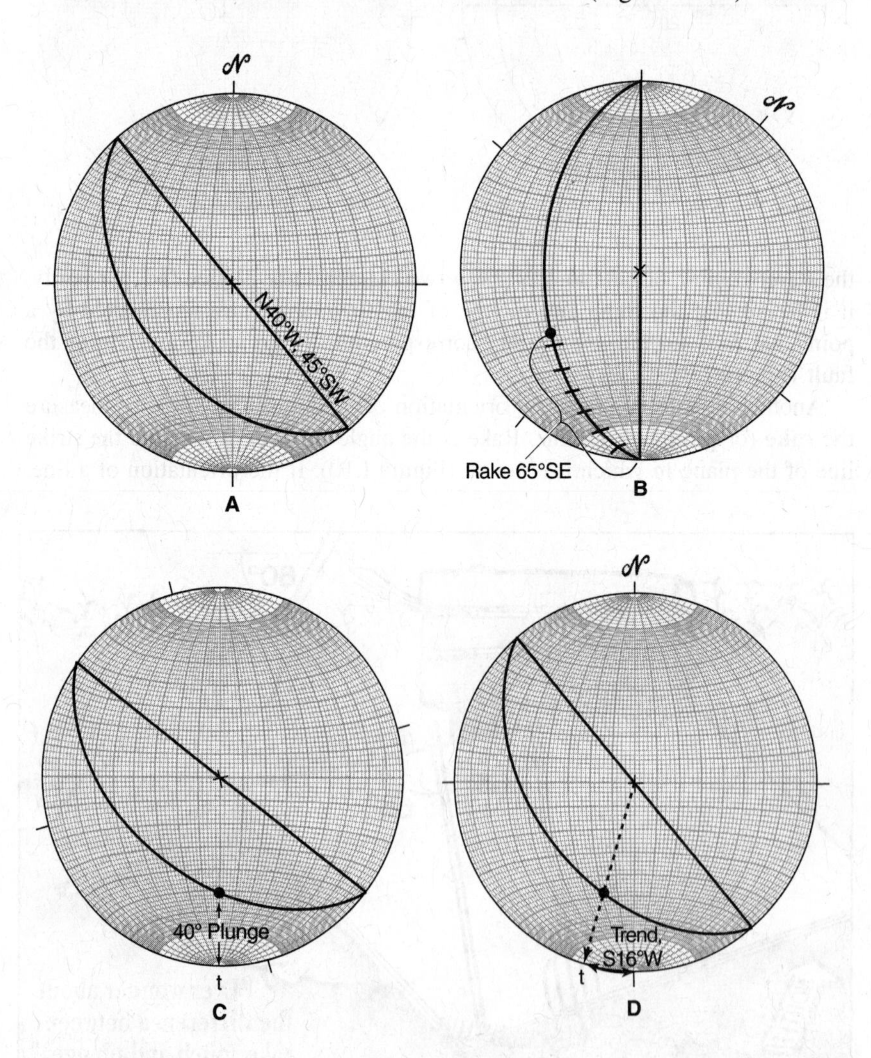

Figure I.11 Stereographic meaning of rake. (*A*) Great-circle stereographic representation of a plane that strikes N40°W and dips 45°SW. (*B*) Counting the 65° rake angle (along the great circle) that describes the orientation of a line in the plane. (*C*) Measurement of the plunge of the line contained in the dipping plane. (*D*) Measurement of the trend of the line contained in the dipping plane.

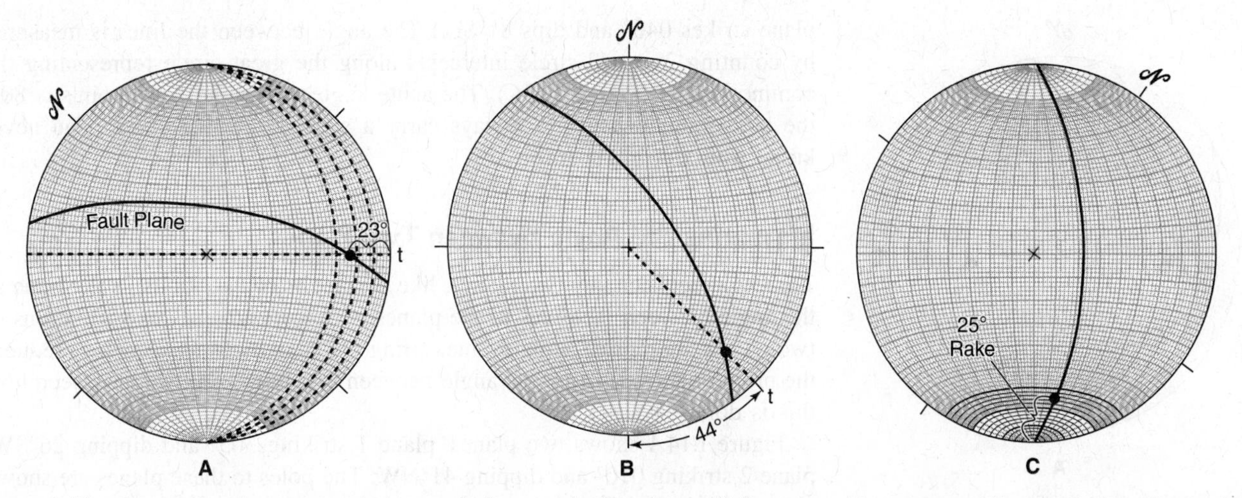

Figure I.12 Converting the trend and plunge of a line to rake in a plane. (*A*) Stereographic representation of a fault surface and the slickenlines on the fault. The great circle portraying the orientation of the fault plane is oriented such that the plunge of the slickenlines can be measured directly. (*B*) With the north index mark in home position, the trend of slickenlines can be measured as S44°E. (*C*) The rake of slickenlines (25°S) can be measured directly when the great circle that represents the fault orientation is rotated to coincide with an appropriately oriented great circle on the stereographic net.

the slickenlines is found by first aligning the great circle representing the fault with the corresponding great circle on the stereographic net (Figure I.12*C*). The rake of 25°SE is measured along the great circle, inward from the perimeter to the point representing the trend and plunge of slickenlines.

Measuring the Angle Between Two Lines

If someone walked up to you on the street and asked you to compute the angle between two lines in space, one plunging 16° N42°E and the other plunging 80° S16°E, how would you do it? The stereographic solution is based on knowledge that two lines define a plane, and that the angle between the two lines is measured in the plane common to both. The orientations of the lines are given as 16°/042° (line 1) and 80°164° (line 2). We can plot these two lines stereographically, as shown in Figure I.13*A*. The plane defined by these two lines is found by rotating the tracing paper overlay until the stereographic points representing the lines lie on a common great circle (Figure I.13*B*). (The

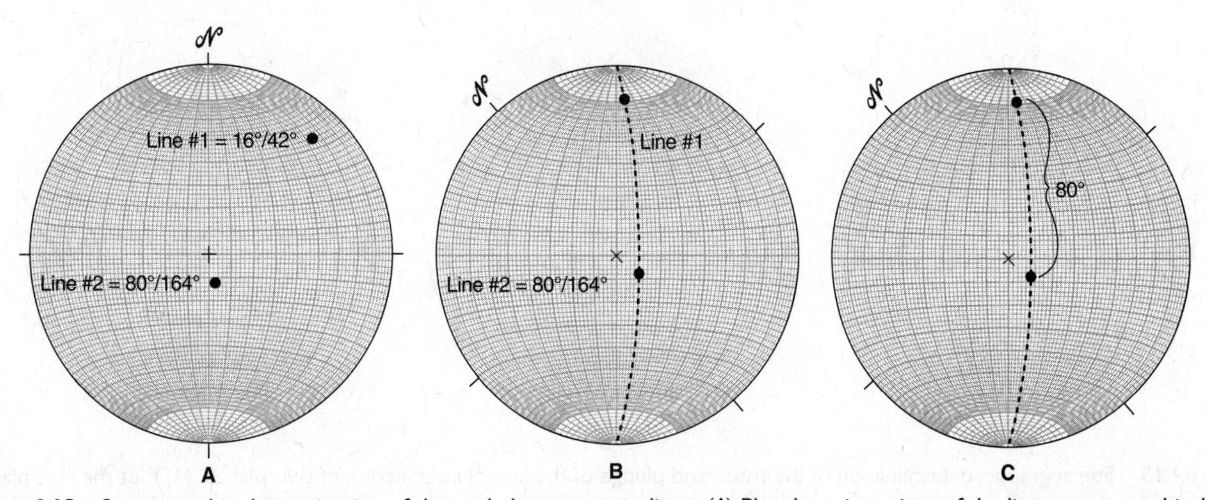

Figure I.13 Stereographic determination of the angle between two lines. (*A*) Plot the orientations of the lines stereographically as points. (*B*) Fit the lines to a common great circle (i.e., to a common plane). (*C*) Measure the acute angle between the two lines by counting along the common great circle that connects the points.

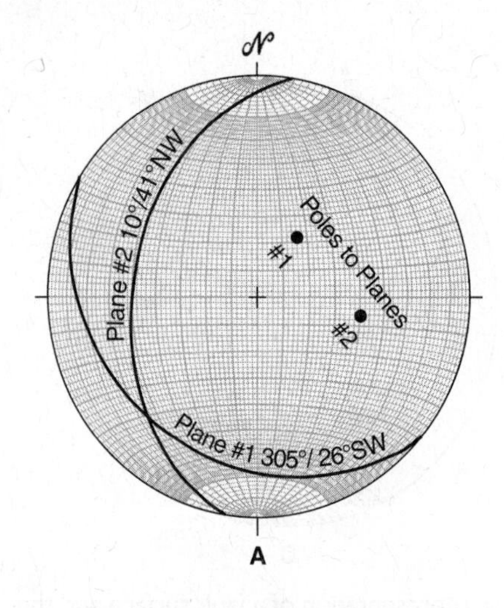

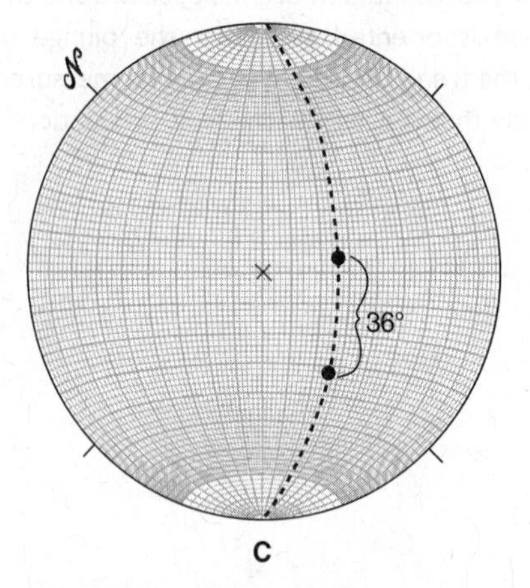

Figure I.14 Stereographic determination of the angle between two planes. (*A*) Plot the orientations of the planes both as great circles and as poles. (*B*) Fit the poles of the planes to a common great circle, and then measure the acute angle between the poles.

plane strikes 040° and dips 81°SE.) The angle between the lines is measured by counting 2° small-circle intercepts along the great circle representing the common plane (Figure I.13*C*). The acute angle separating these points is 80°; the obtuse angle is 100°. Always carry a stereonet in the street. You never know who you might meet.

Measuring the Angle Between Two Planes

The angle between any two planes, like two faults or two joints, is the same as the angle between the poles to the planes. Consequently, if the orientations of two planes are plotted as poles, measuring the angle between the poles reduces the problem to measuring the angle between two lines. And we have seen how this is done.

Figure I.14*A* shows two planes, plane 1 striking 305° and dipping 26°SW, plane 2 striking 010° and dipping 41°NW. The poles to these planes are shown stereographically. To measure the angle between the planes, simply align the two poles on the same great circle (Figure I.14*B*) and measure the acute angle between the poles by counting 2° small-circle intercepts along the great circle. For this example, the angle is 36°.

Determining the Orientation of the Intersection of Two Planes

The payoff for learning the principles of stereographic projection is derived from the ease with which certain geometric problems can be solved. One of the best examples of the effectiveness of stereographic projection is determining the trend and plunge of the intersections of two planes. Consider two planes, one striking N49°E and dipping 42°SE, the other striking N10°W and dipping 65°NE (Figure I.15*A*). The intersection of the great circles is a point "i" whose orientation is that of the line of the intersection of the two planes. The trend of the line "i" can be determined by drawing (projecting) a straight line from the center of the projection through line "i" to the perimeter, and measuring the orientation of this line with respect to north or south. For this example, the trend is S34°E (Figure I.15*B*). The plunge of the line is found by rotating line "i" to the east–west or north–south lines and measuring its inclination from the perimeter (Figure I.15*C*). The plunge as measured in this example is 42°.

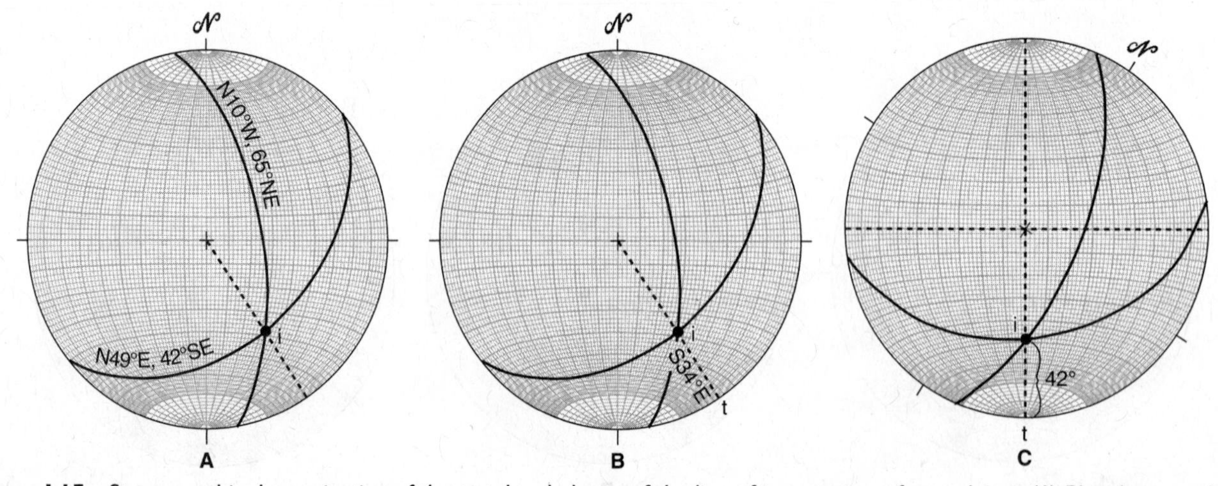

Figure I.15 Stereographic determination of the trend and plunge of the line of intersection of two planes. (*A*) Plot the two planes stereographically as great circles. Identify the intersection (*A*) of the two great circles, recognizing that its orientation reflects the trend and plunge of the intersection of the two planes. (*B*) Interpret the trend of the line of intersection. (*C*) Measure the plunge of the line of intersection along either the east–west line or the north–south line of the net.

Figure I.16 Application of stereographic methods for the determination of apparent dip. (*A*) A vertical cliff exposure along the coast trends N40°E. Inclined strata in the cliff strike N24°E and dip 79°SE. (*B*) Apparent dip of strata in the cliff exposure is determined by identifying the intersection of the orientation of the cliff (N40°E, 90°) and the orientation of the bedding (N24°E, 79°SE).

Apparent dip problems are a special case of determining the line of intersection of two planes. Figure I.16*A* shows bedrock in a seacoast exposure. Bedding strikes N24°E and dips 79°SE. What would be the apparent dip for these beds, as observed in a vertical cliff face that trends N40°E?

To solve, plot the orientation of the bed and the orientation of the vertical face stereographically (Figure I.16*B*). The intersection of the two planes is a point that represents the trend and plunge of the line of intersection of the bed and the vertical cliff face. The plunge value is in fact the apparent dip of the plane, namely 54°NE. Examples like this begin to scratch the surface of the power of the three-dimensional protractor known as the stereographic net.

Stereographic Projection as a Statistical Tool

We are often required in structural analysis to determine the average orientation of a certain structural element, or to determine whether the range of orientations of a particular structure is in some way systematic. One way to identify **preferred orientations** of structures is to plot lines or poles to planes stereographically on a Schmidt net (see Figure I.3) and to evaluate the extent to which the plotted points tend to cluster or to spread in systematic ways. The geometry of projection of the Schmidt net (see Figure I.3) is such that 2° areas bounded by great and small circles are the same size across the net (Phillips, 1971). Since 2° great-circle/small-circle areas are the same size across the entire face of a Schmidt net, randomly distributed orientations will appear random, whereas nonrandom concentrations of stereographically plotted points reflect preferred orientations. Contouring the values of the **density distribution** of plotted points provides a measure of the degree of preferred orientation.

Evaluating Preferred Orientations

A stereographic projection of 65 poles to bedding in Cretaceous strata in the Mule Mountains near Bisbee, Arizona, is shown in Figure I.17*A*. The general concentration of points is near the center of the projection, signifying that the beds whose orientations are plotted are rather gently dipping. But what is the specific orientation of the bedding, as expressed in strike and dip? And

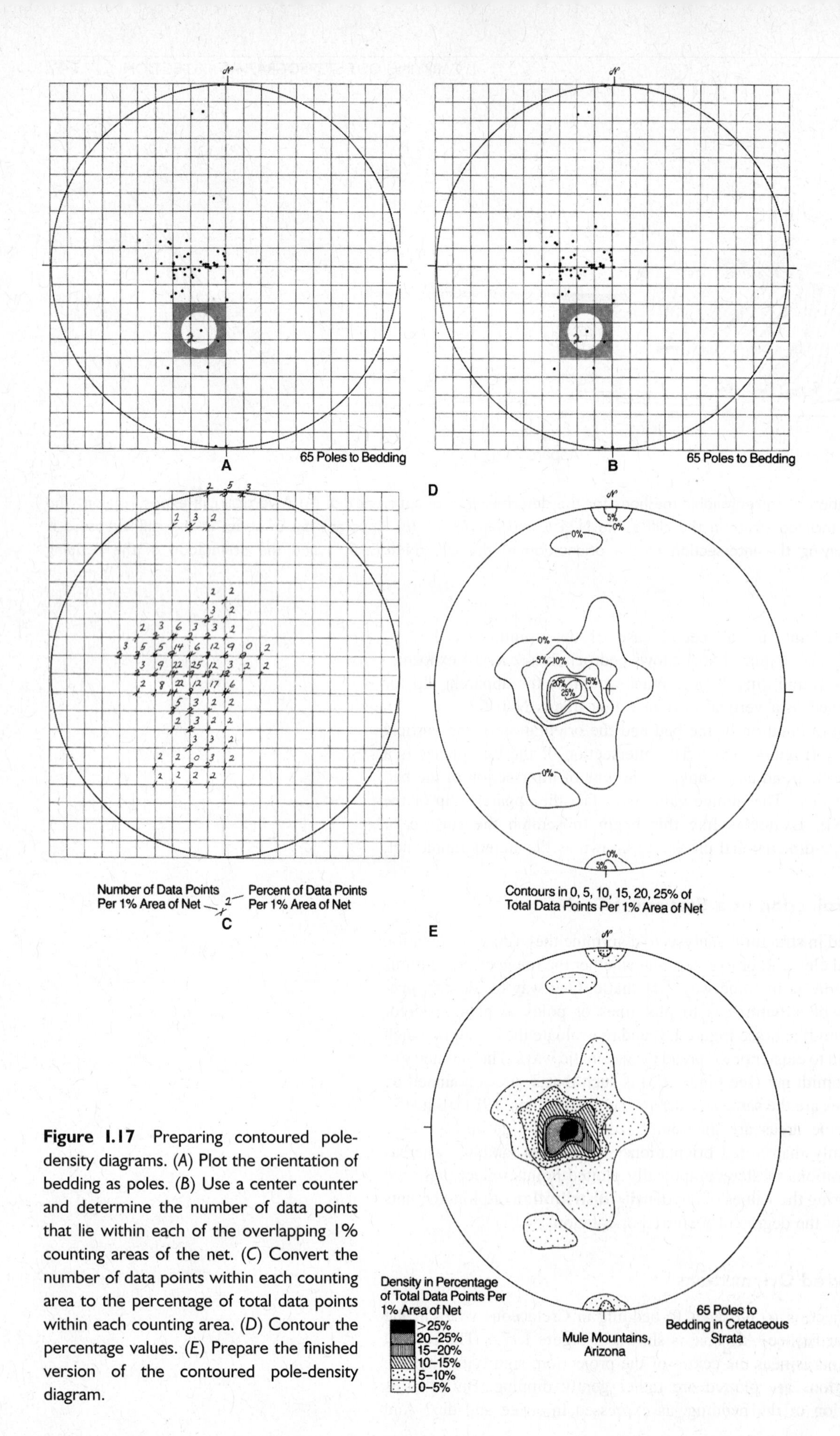

Figure I.17 Preparing contoured pole-density diagrams. (*A*) Plot the orientation of bedding as poles. (*B*) Use a center counter and determine the number of data points that lie within each of the overlapping 1% counting areas of the net. (*C*) Convert the number of data points within each counting area to the percentage of total data points within each counting area. (*D*) Contour the percentage values. (*E*) Prepare the finished version of the contoured pole-density diagram.

what is the strength of the preferred orientation? Answering these questions requires contouring the density distribution of the plotted points on the face of the stereonet.

To evaluate density distribution, the equal-area net is subdivided into a gridwork of many overlapping circular areas, each of which corresponds to 1% of the area of the stereographic projection. Density is described in terms of percentage of total data points falling within a given 1% area of the stereographic projection.

Pole-density distribution is thus calculated as follows:

$$\text{Density }(\%) = \frac{\text{number of points within 1\% area of net}}{\text{Total number of data points}} \times 100$$

To subdivide a 20-cm-diameter stereogram into overlapping 1% circular areas, a square grid is constructed such that the spacing of grid intersections is 1 cm. The square gridwork is overlain by the tracing paper on which the data points were originally stereographically plotted (Figure I.17*B*). The grid intersection points are used as control points for systematically moving a **center counter** whose area is 1% that of the stereogram (Whitten, 1966, pp. 20–26). In this manner the number of data points that lie within each of the overlapping 1% areas is counted. The counts are posted on yet another overlay (Figure I.17*C*).

A calculator is used to convert the numbers of data points in each of the 1% areas to percentage of total data points (see Figure I.17*C*). It is the **percentage values** of density distribution that are contoured (Figure I.17*D*). Since the 1% areas are overlapping, and because most data points are counted more than once, the density distribution values represent a smoothed portrayal of the raw orientations.

Ideally, the finished contour diagram (Figure I.17*E*) should be marked by some constant contour interval, with the number of contour-line values not exceeding five or six. To emphasize the density distribution, it is common practice to shade the diagrams in such a way that the zones of highest density are darkest and most pronounced. For the example of bedding orientations in the Mule Mountains, the completed diagram discloses a bull's-eye, **unimodal distribution** of points. The eye or center of the contoured pattern corresponds to the preferred orientation of the bedding.

The counting and contouring process is relatively straightforward, except when considering density distribution values close to or on the perimeter of the projection. Consider the point plot of 85 measurements of mineral lineation in gneiss in the Rincon Mountains of southern Arizona (Figure I.18*A*). The lineations are low plunging, as can be judged by their closeness to the perimeter of the projection. The number of data points falling within each 1% area in the interior of the stereographic projection can be counted using a center counter. But in this example, individual 1% area counting circles sprawl beyond the perimeter of the projection, and thus the points near or at the periphery occupy areas less than 1% of the area of the projection. To count these points, a **peripheral counter** is used (Figure I.18*B*). The orientations of structural elements represented by points lying in peripheral counting areas correspond closely to those that fall within the peripheral counting area diametrically (180°) opposite. In fact, each point on the perimeter of the net corresponds exactly in orientation to the point that lies on the perimeter 180° away. Recognizing this equivalence in orientation, and given that each peripheral counting area is less than 1% of the area of the net, an extra step is required to convert from *number of data points* to *percentage of data points.* It is necessary to add the number of data points that fall within *each pair* of supplementary partial circles on the perimeter of the net and to assign this sum

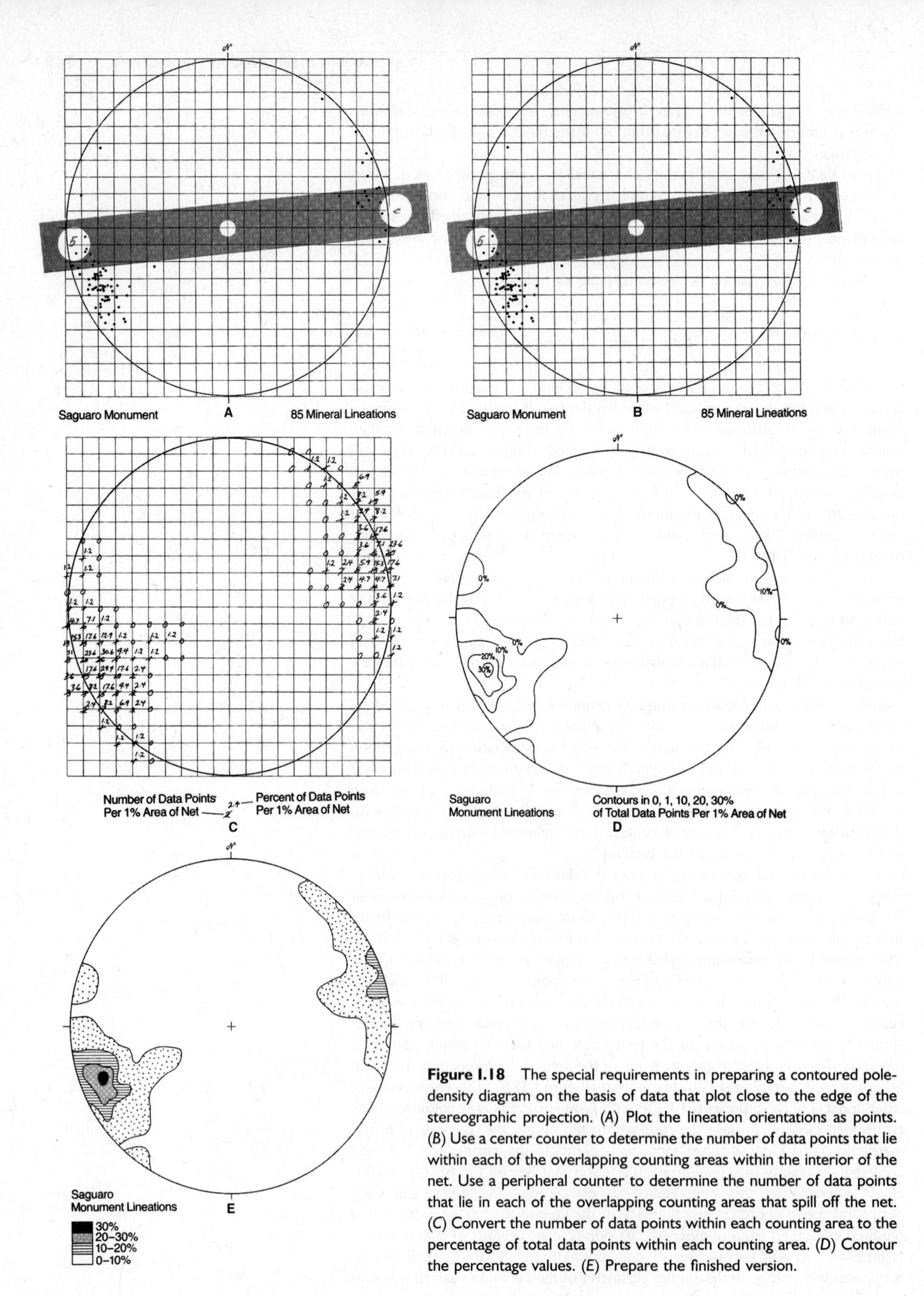

Figure I.18 The special requirements in preparing a contoured pole-density diagram on the basis of data that plot close to the edge of the stereographic projection. (*A*) Plot the lineation orientations as points. (*B*) Use a center counter to determine the number of data points that lie within each of the overlapping counting areas within the interior of the net. Use a peripheral counter to determine the number of data points that lie in each of the overlapping counting areas that spill off the net. (*C*) Convert the number of data points within each counting area to the percentage of total data points within each counting area. (*D*) Contour the percentage values. (*E*) Prepare the finished version.

to *each* of the partial circles. These values are converted to percentages, along with the values assigned to center counting areas in the interior of the net (Figure I.18*C*). When the values are contoured (Figure I.18*D*), special care must be taken to assure that each point of intersection of a contour line with the perimeter of the net is matched by a corresponding intersection point on the perimeter 180° away. The finished product is shown in Figure I.18*E*.

Contour diagrams prepared in this way are called **pole-density diagrams**, where "*pole*" refers loosely to a stereographically plotted point, regardless of whether it represents the trend and plunge of a linear element or the trend and plunge of a pole to a plane. Pole-density diagrams describe the range in distribution and the preferred orientation(s), if any, of the measured structures. They are useful in summarizing large quantities of geometric data. Each diagram should be clearly labeled according to area of study (e.g., where the data were collected), structural element, number of measurements, and contour-line values (see Figures I.17*E* and I.18*E*).

Commonly an array of such diagrams is necessary to describe the full orientation range for a system of structures. Where many diagrams are presented, individual diagrams should be prepared so that contour lines are of a common interval, in percentage. When there are fewer than 50 orientation measurements, pole-density diagrams are not particularly meaningful statistically, and thus point diagrams suffice.

Pole-density diagrams should be viewed critically in regard to number of data points, values of pole density, and patterns. Statistical tests are available to evaluate the significance of pole-density values in light of the number of data points (Kamb, 1959). Many types of pattern are possible. The nature and the symmetry of pole-density diagrams have important implications for the geometry and kinematics of structural systems.

Software for Stereographic Projection

Stereographic projection, including the production of contour diagrams, is made fast and easy through computer software, such as that generously made available as freeware for noncommercial use by Rick Allmendinger, Cornell University (*http://www.structuralgeology.org/2009/02/stereonet.html*). Students and faculty around world are using such programs to plot, contour, and analyze their data much faster than by hand. Yet learning by hand is the very best way to begin.

J. EVALUATING ROTATION USING STEREOGRAPHIC PROJECTION

Certain stereographic techniques are indispensable in the kinematic analysis of rotational operations. Stereographic techniques can be used to picture the rotation of lines and planes in geometric space. A stereonet is used to plot the path or **locus** of points representing lines and planes at various stages of rotation. The plotting procedure is based on facts regarding the orientation of the axis of rotation, sense of rotation, and the magnitude of rotation in degrees. Let us consider some examples.

Example of Rotational Faulting

Figure J.1*A* shows strata in various stages of rotation during listric normal faulting. Orientation of strata at each stage is portrayed stereographically in Figure J.1*B*, both through the use of great circles representing strike and dip of

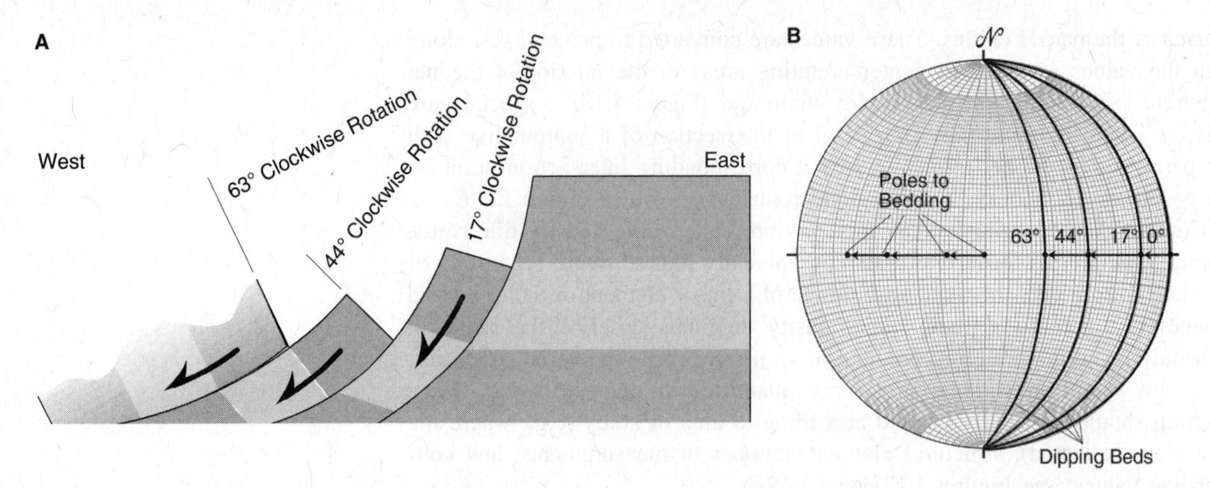

Figure J.1 (*A*) Rotation of strata by listric faulting, portrayed at different stages. (*B*) Stereographic representation of the rotation of strata.

bedding and through poles to bedding. Poles provide the most convenient representation. The track of poles, taken together, is the locus of points representing progressive rotation of the strata by faulting. Note that the sequence of poles to bedding is aligned along a small circle perpendicular to the axis of rotation.

Example of Rotation of Layers by Folding

Progressive folding of a layer of rock is portrayed in Figure J.2*A*. In Figure J.2*B*, the strike and dip of bedding on the limbs of the fold are shown stereographically at each stage of deformation. The poles to bedding, taken together, represent the locus of points describing the rotation of bedding during folding. The locus is in the form of a great circle that lies 90° from the axis of rotation, just like the great-circle distribution of poles to bedding in the example of rotation during listric faulting (see Figure J.1*B*).

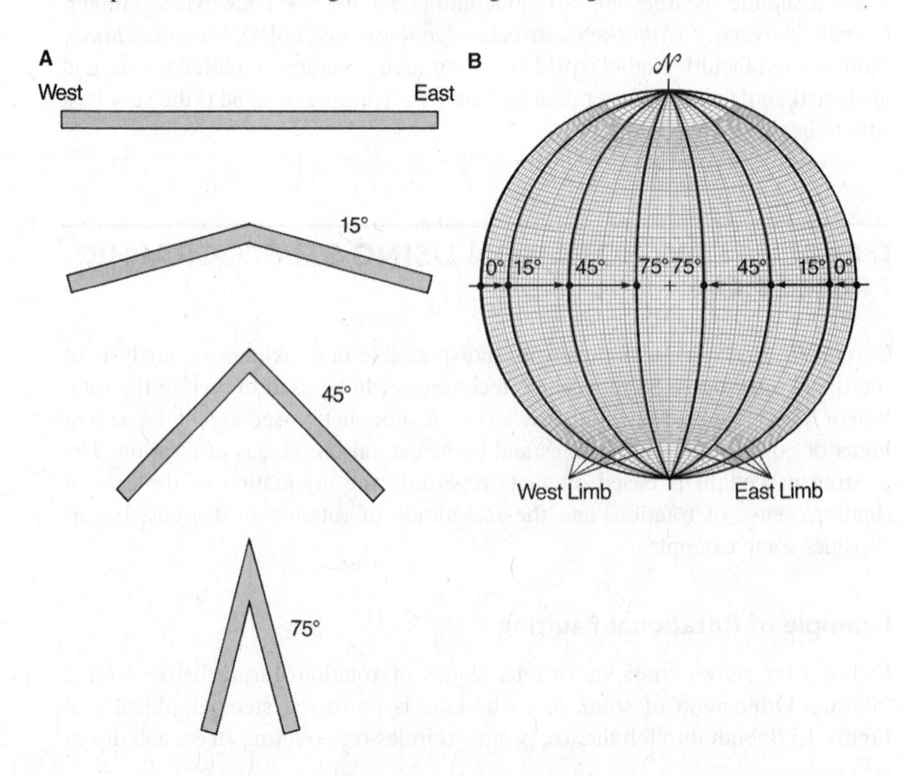

Figure J.2 (*A*) Rotation of strata during progressive folding. (*B*) Stereographic representation of the rotation of bedding during the folding.

Stereographic Unfolding of Layers with Ripple Marks

The general case of rotation of strata by folding is shown in Figure J.3, where layered strata are flexed about an inclined axis. The axis of rotation plunges 30° S45°E. The locus of poles to folded bedding describes a great circle lying 90° from the axis.

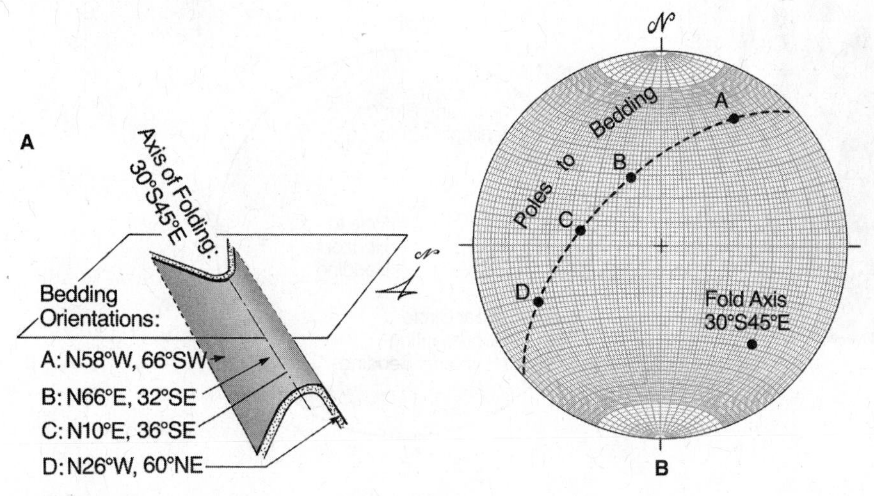

Figure J.3 (*A*) Folding of bed about an inclined axis. (*B*) Stereographic portrayal of poles to the rotated, folded bedding, measured at locations A–D. Note that the poles lie on a great circle whose pole is the fold axis.

The ability to stereographically rotate lines in space is essential in structural studies. The geological lines we focus on now are the crests and troughs of ripple marks (Figure J.4*A*). Consider a horizontal layer of sandstone containing parallel-aligned current ripple marks (Figure J.4*B*). The ripple marks lie in the plane of bedding, and thus a point that stereographically portrays the trend and plunge of the ripple marks must fall on the great circle representing the orientation of bedding (Figure J.4*C*). If the horizontal sandstone layer is subjected to 48° of southeastward tilting about a horizontal, N70°E-trending axis (Figure J.4*D*), how will the orientation of the ripple marks change?

When the bed is tilted 48° about a horizontal axis of rotation, poles to bedding at the various stages of tilting trace out part of a great circle, which lies at right angles to the axis of rotation (Figure J.4*E*). In contrast, points describing the orientation of ripple marks during tilting trace out a small circle on the stereonet. At each stage of tilting, the point describing the orientation of the ripple marks remains on the great circle representing the strike and dip of bedding. These stereographic relations can be seen more clearly by rotating the fold axis to the north–south line of the net and observing that poles to bedding all lie on the *x*- axis of the net, and noting the correspondence between the movement path of ripple mark orientations and one of the small-circle traces embossed on the underlying net (Figure J.4*F*). The angle between the ripple mark orientation and the axis of rotation, at each stage of tilting, remains constant, matching the angle between the trend of the ripple marks and the trend of the axis of rotation before the tilting commenced (see Figure J.4*B*).

Suppose as field geologists we encounter an outcrop of tilted strata revealing both bedding and ripple marks (Figure J.5*A*). How can we determine the original orientation of the ripple marks? We begin by measuring the strike and dip of bedding (N80°W, 80°SW) and the trend and plunge of ripple marks (39° N88°W). Alternatively, we measure the strike and dip of bedding (N80°W, 80°SW) and the rake of the ripple marks in the plane of bedding (39°W). When these data are plotted stereographically (Figure J.5*B*), the point representing the trend and plunge of the ripple marks is seen to lie on the great circle that describes the strike and dip of bedding. The original orientation of

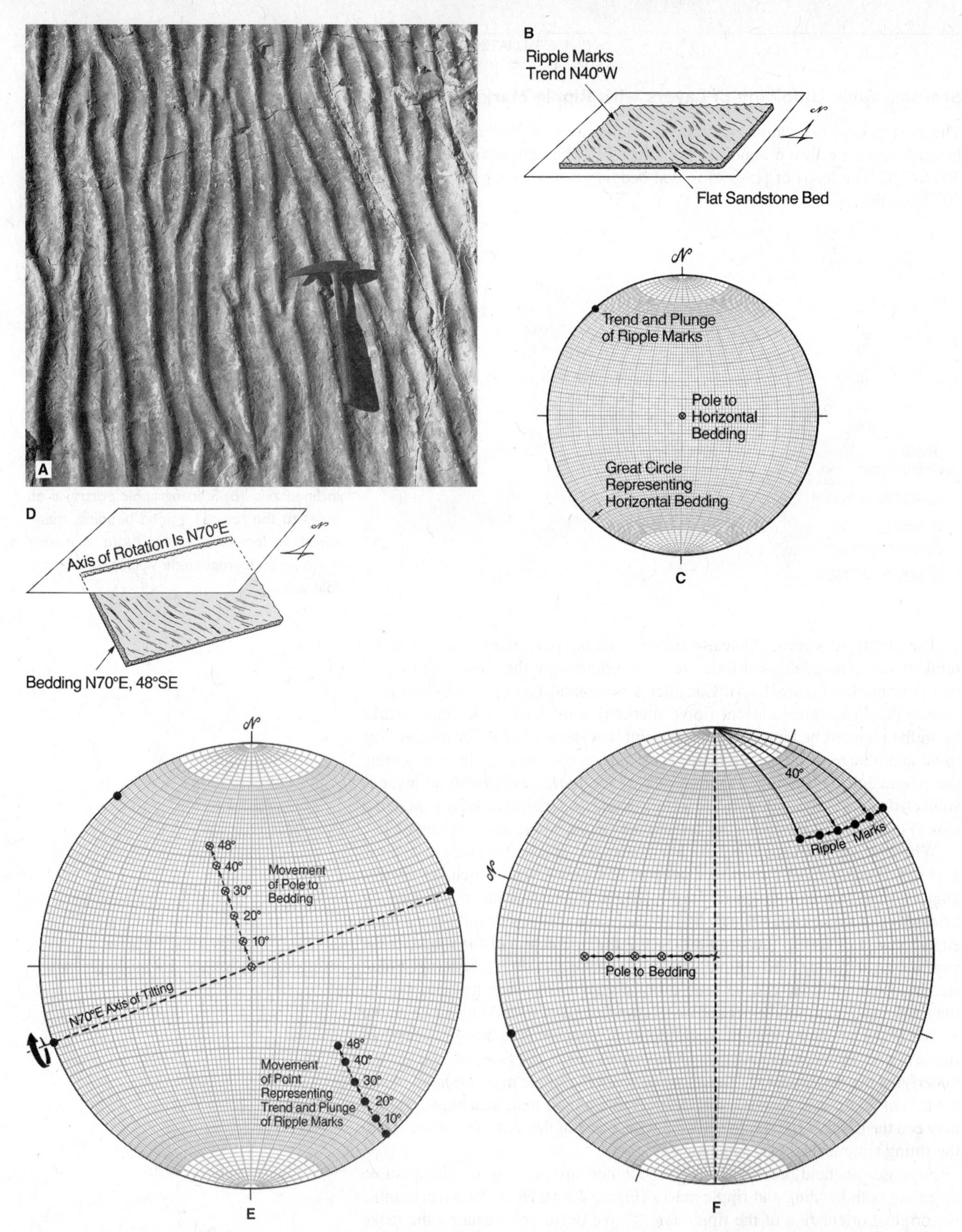

Figure J.4 (*A*) Geologic lines in a plane, namely, crests and troughs of current ripple marks in the Dakota Sandstone. [Photograph by J. R. Stacy. Courtesy of United States Geological Survey.] (*B*) Ripple marks trending N40°W in horizontal sandstone bed. (*C*) Stereographic portrayal of the orientations of the sandstone bed and the ripple marks before tilting. (*D*) Bedding and ripple marks after tilting. (*E*) Stereographic portrayal of the tilting of the sandstone bed and the consequent change in orientation of the ripple marks. (*F*) Stereographic view showing how the movement path of the pole to bedding follows the great-circle trace along the east–west line of the net, whereas the movement path of the point representing the orientation of the ripple marks follows a small-circle path.

the ripple marks is found by rotating bedding to its inferred original horizontal orientation. To do this an axis of rotation must be chosen.

For this problem, the line of strike of bedding is the right choice. Its orientation is N80°W. To rotate stereographically about any horizontal axis, the point(s) representing the selected axis of rotation must be brought into alignment with north (or south) on the perimeter of the stereonet. This is achieved simply by rotating the tracing overlay clockwise by 80° (Figure J.5*C*). Thus positioned, the axis of rotation lies as the central axis to the small-circle paths of rotation. By rotating the pole to bedding and the point representing trend and plunge of ripple marks about this axis, and in the proper sense, the bed containing the ripple marks is, in effect, lifted from its steeply inclined orientation to horizontal (see Figure J.5*C*). The points traverse small circles from the interior of the net to the perimeter. Once accomplished, the original trend of the ripple marks can be measured: namely, S61°W.

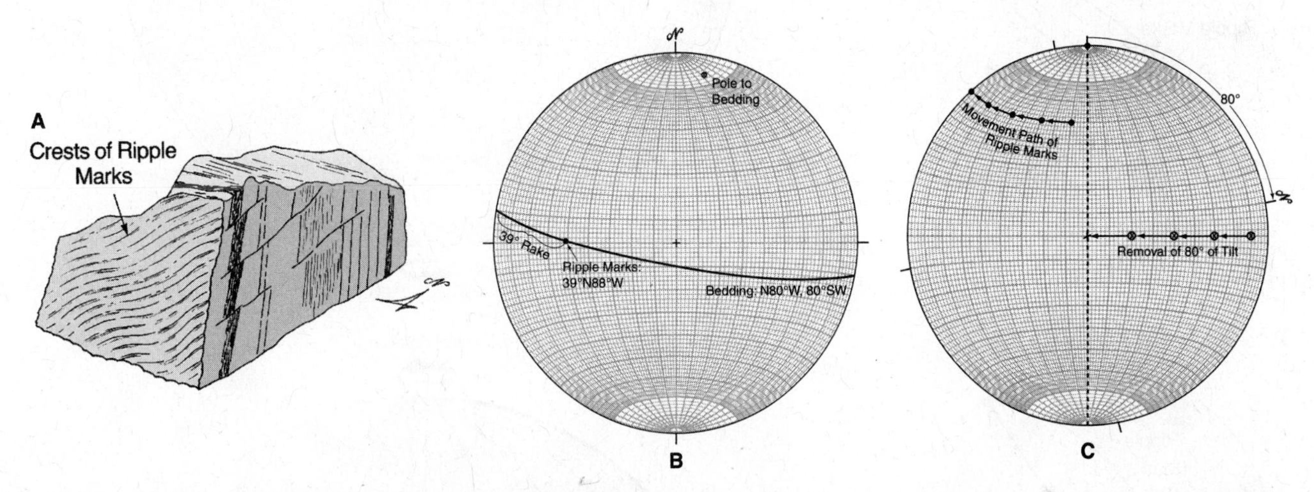

Figure J.5 (*A*) Outcrop of steeply dipping faulted sedimentary strata. Front face of outcrop is the plane of bedding, revealing ripple mark lineation. (*B*) Stereographic representation of the orientations of bedding and ripple marks, as measured in outcrop. (*C*) Stereographic restoration of bedding and ripple marks to their orientations before tilting.

Multiple Rotations

A final example of rotational kinematics is required to present the general operation(s) that can be adapted to all rotational problems, regardless of complexity. The goal is to be able to carry out successive rotational operations about axes of different orientations. The order in which rotational operations are performed vitally influences the final orientation of the rotated body. Multiple rotations are simply a series of single rotations applied in sequence. If we can do one, we should be able to do more than one.

Suppose we are interested in determining the original orientation of ripple marks that lie in a steeply inclined limb of a plunging fold (Figure J.6*A*). The fold axis plunges 30° N36°E. Bedding on the west limb of this fold strikes N50°E and dips 65°NW. The ripple marks rake 80°SW. These geometric data are plotted stereographically in Figure J.6*B*. The first step in the restoration is to rotate the fold axis to horizontal, and at the same time to rotate the pole to bedding and the ripple marks by the same amount. The rotation axis chosen to achieve this plunges 0° N54°W, at right angles to the trend of the fold axis. The rotation axis is brought to the north position on the perimeter of the net (Figure J.6*C*) and is lifted to horizontal, traversing 30° along the east–west line of the net to the perimeter. The pole to bedding and the point defining the trend and plunge of ripple marks move 30° in the same sense but along small-circle paths.

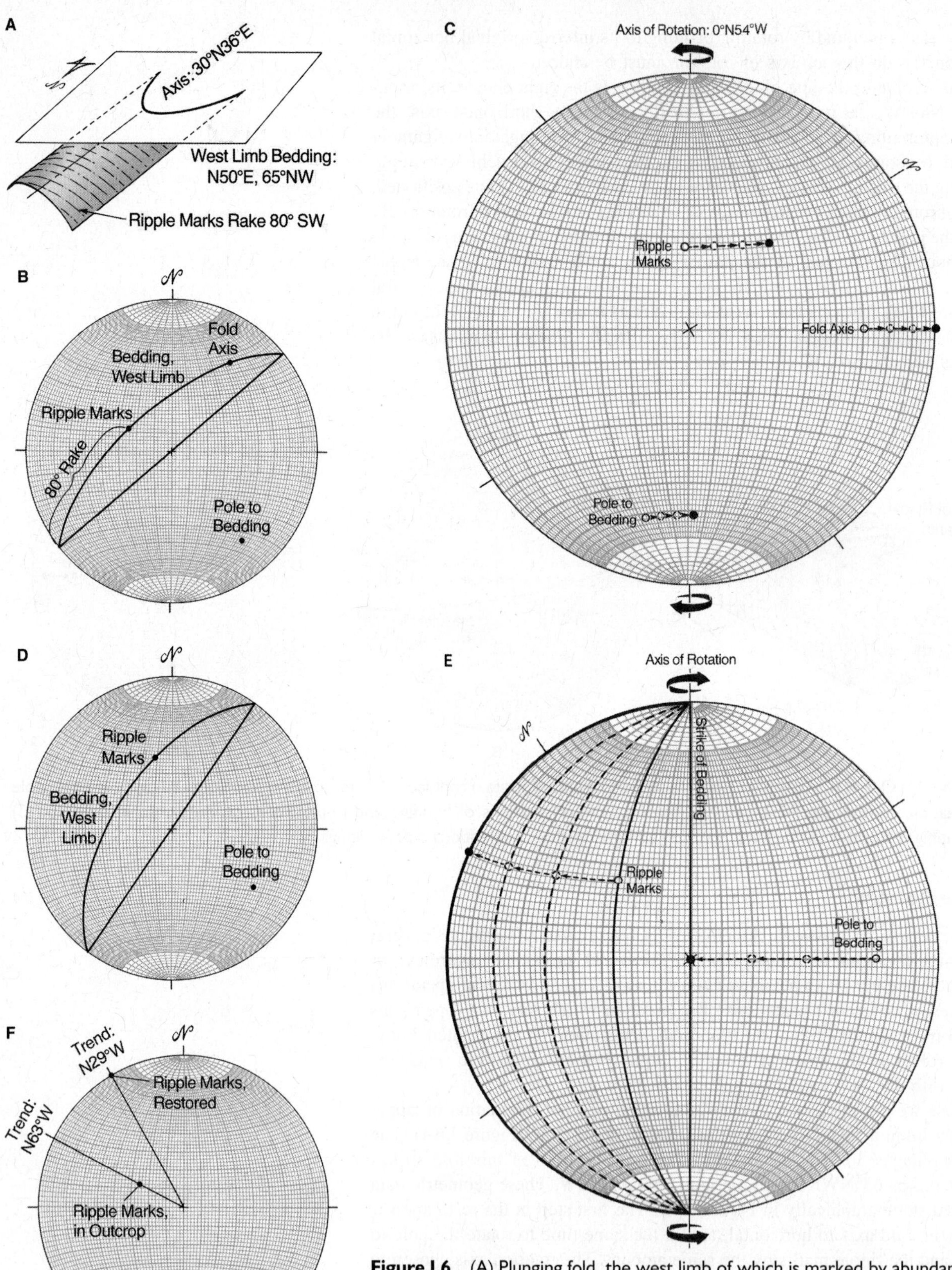

Figure J.6 (A) Plunging fold, the west limb of which is marked by abundant ripple marks. (*B*) Stereographic representation of bedding, ripple marks, and the axis of folding. (*C*) Rotation of fold axis to horizontal, and rotation of the pole to bedding and the ripple marks by a like amount. (*D*) Orientation of west-limb bedding and the ripple marks following rotation of the fold axis to horizontal. (*E*) Rotation of bedding to horizontal about an axis parallel to the strike of bedding. (*F*) Measurement of the trend of ripple marks in the restored configuration.

Once this initial step has been accomplished, the great circle perpendicular to the pole to bedding is plotted (Figure J.6*D*). It passes through the rotated position of the ripple mark axes. The strike line of the bedding is then aligned along the north–south line of the net (Figure J.6*E*), serving as the axis of rotation about which the final operation is completed. Bedding and the ripple marks are lifted to horizontal, traversing small-circle paths to the perimeter of the net (see Figure J.6*E*). Returning the tracing paper to home position (Figure J.6*F*), the original trend of the ripple marks can be directly measured. Its value is N29°W, substantially different from N63°W, the trend of the ripple marks as they appear in outcrop.

K. DETERMINING SLIP ON A FAULT THROUGH ORTHOGRAPHIC AND STEREOGRAPHIC PROJECTION

The Challenge

Sometimes it is necessary to determine the direction, magnitude, and sense of slip from geologic map relationships, using the combination of orthographic and stereographic projection. There are a number of ways to do it, but all require the identification of a real or geometric line that has been offset by faulting. A classic method in evaluating slip on faults is one made famous in *Structural Geology* by Billings (1954, 1972). The "*Billings*" fault had been a rite of passage in undergraduate structural geology labs.

There are few opportunities to determine the net-slip for a large fault on the basis of the offset of a tangible physical geological linear object, because such objects are rare. Exceptions do exist, however, such as the offset paleovalley deposits mapped and analyzed by Faulds and colleagues (2005) in the northern Walker Lane subprovince of the Basin and Range (see Figures 11.56 and 11.57). On the other hand, geometric lines that can be used as references for net slip do in fact exist in deformed geological systems, and we simply must be watchful and recognize them. Billings (1954, 1972) drew attention to the kind of geometric line that is formed by the intersection of two geologic planes, for instance, a dike and a bed. If such a line is cut and displaced by a fault, a basis for evaluating translation is readily available. The concept is relatively simple (Figure K.1). The actual solution requires 3D visualization.

Consider a fault that strikes N70°E and dips 75°SE (Figure K.2*A*). Assume that it is exposed in a perfectly flat area whose elevation is 2000 ft (609 m). A dike and a distinctive limestone bed crop out on both the north and south sides of the fault. The dike strikes Nl0°E and dips 45°SE. The limestone bed strikes N60°W and dips 50°SW. What is the displacement vector describing the movement of the south block of the fault relative to the north block?

The intersection of the dike and the limestone bed provides the geometric line of reference for establishing slip on this fault. The trend and plunge of the reference line is determined stereographically by plotting the intersection of the great circles representing the dike and the limestone bed, respectively (Figure K.2*B*). The orientation of the line is 32° S29°E. The vertical projection of the line of intersection of the dike and the limestone bed on the north block of the fault is plotted as line AA′ (see Figure K.2*A*) This line passes through the surface intersection of the trace of the dike and the trace of the limestone bed as seen on the surface in outcrop. The counterpart of this line on the south block of the fault is BB′.

To find the displacement vector for fault translation, it is necessary to determine where the two geometric lines of intersection, one on each side of

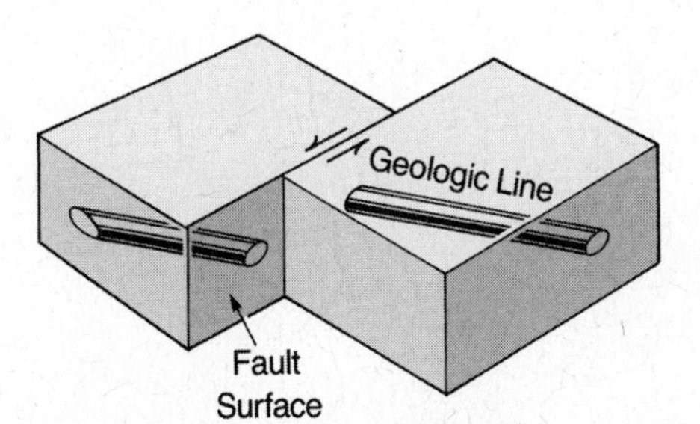

Figure K.1 Schematic portrayal of the offset of a once-continuous line by faulting. Reconstruction of the line permits the direction, sense, and magnitude of the translation vector due to faulting to be calculated.

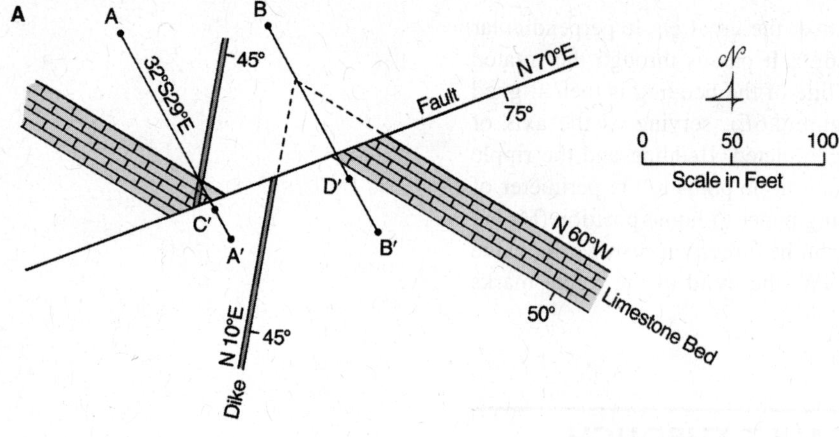

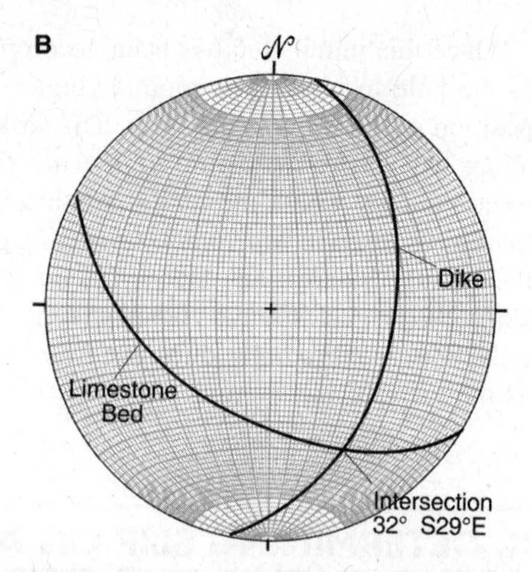

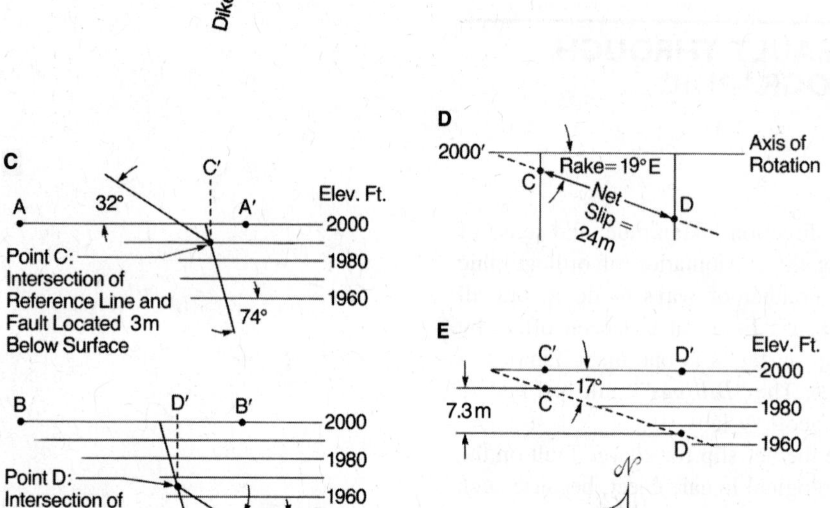

Figure K.2 Determination of the displacement vector for a fault by reconstructing the faulted line of intersection of a dike and a limestone bed. (*A*) Faulted limestone bed and faulted dike. Lines AA′ and BB′ are the vertical projections of the limestone–dike intersections on the north and south sides of the fault, respectively. (*B*) Stereographic determination of the trend and plunge of the line of intersection of the limestone bed and the dike. (*C*) Cross-sections showing the relation of the fault trace to the line of intersection of the limestone bed and dike. (*D*) View in the plane of the fault showing net slip and rake of net slip. (*E*) Cross-section showing plunge of the displacement vector.

the fault, pierce the fault surface. This step is ordinarily the most difficult to visualize and to construct graphically. To see the relationship, construct vertical cross-sections that pass through the vertical projections (AA′ and BB′) of the intersection of the dike and the limestone bed (Figure K.2*C*) The 32° plunge of the line of intersection of the dike with the limestone bed is plotted on each of the cross-sections. The *apparent* dip of the fault is also plotted, and its value is determined stereographically. In this example, the apparent dip of 74° happens to be merely 1° less than true dip. Where the line of intersection of the dike and the limestone bed meets the inclined trace of the fault, *there* lies one of the key reference points needed to solve for net slip. For the north block of the fault, point C marks this reference point, and point C′ marks its vertical projection (see Figure K.2*C*). For the south block of the fault, the intersection point (reference point on the fault) is point D, and its vertical projection is D′ (see Figure K.2*C*). Points C and D originally occupied the same position before faulting. They are equivalent points that can be connected by a displacement vector. Elevations of these reference points are evident in the cross-sections shown in Figure K.2*C*: the elevation of point C is 1990 ft (606 m), and the elevation of point D is 1966 ft (599 m). Now we are ready to finish it off!

To determine the net slip, points C and D must be viewed in the plane of the fault, for slip is measured in the plane of movement. To achieve this view, the fault surface is figuratively rotated to the surface about its strike line (Figure K.2*D*). The fault plane is thus laid out in the plane of the paper so that the **net slip** can be directly measured. The axis of rotation, the strike line, is the intersection of the fault with the horizontal surface of the terrane. Its elevation is 2000 ft. The trace of this axis of rotation is shown in Figure K.2*D*. Reference points C and D are plotted along lines perpendicular to the axis of rotation, and at down-dip distances corresponding to the elevations of C and D. Once C and D have been plotted, the net slip can be directly measured. The magnitude of the net slip is 80 ft (24 m). The rake of the net slip on the fault plane is 19°E. Net slip can be subdivided into two components, a strike–slip component of 75 ft (23 m) and a dip–slip component of 25 ft (7.6 m) (see Figure K.2*D*).

The trend of the displacement vector is the azimuth of the line C′D′ (see Figure K.2*A*) connecting the vertical projections of reference points C and D. The azimuth measures N78°E. The plunge of the displacement vector is found by constructing a structure profile along C′D′ (Figure K.2*E*), showing the relative positions of C and D at depth. The plunge measures 17°. Thus, the southeast block moved down and to the east with respect to the northwest block, along a displacement vector of magnitude 80 ft, along a line of 17° N 78°E.

Using the Faulted Hinge of a Fold to Determine Slip

The faulted hinge line of a fold also can be used as a basis for determining fault slip. The hinge of a folded layer is like the crease in a folded piece of paper (see Figure 7.27). In the mapped relationship shown in Figure K.3*A*, the hinge of a syncline plunges 22° S10°E. The fold hinge is offset by a fault that strikes N75°W and dips 65°SW. (This example is based on mapping in folded, faulted Cretaceous rocks in Cabin Wash in the Tucson Mountains, Arizona). Displaced parts of the faulted fold hinge crop out on both sides of the fault (locations A and B, see Figure K.3*A*). Before faulting, these hinges were part of a straight, continuous hinge line in a sandstone bed. Slip on the fault is deduced by projecting each of the offset hinge segments into the plane of the fault. Once accomplished, the distance and direction between these projected lines can be measured.

The orthographic solution involves constructing vertical structural profiles oriented parallel to the trend of the fold axis (N10°W) (Figure K.3*B*). On the footwall side, the hinge line projects downward at 22° toward the projected position of the fault in the subsurface. The projection of the hinge line pierces the fault at point C at elevation 2610 ft (791 m). On the hanging-wall side, the hinge line rises 22° skyward from its outcrop elevation to point D (elevation 2645 ft; 801 m) where it pierces the upward projection of the fault plane.

As in the previous example, the net slip on the fault is oblique, neither parallel to the strike of the fault or down the dip of the fault. The net slip can be divided into two components: strike–slip and dip–slip (Figure K.3*C*). The actual strike–slip and dip–slip components of translation are measured in the plane of faulting (see Figure K.3*C*). To accomplish this, points C and D are plotted in the plane of the fault at their respective elevations and locations. The strike–slip, dip–slip, and net–slip components of translation are 86 ft, 35 ft, and 91 ft (26 m, 11 m, and 28 m), respectively (see Figure K.3*C*). The rake of the displacement vector on the fault plane is 23°W. This corresponds to a net-slip orientation of 21° N86°W, as computed stereographically.

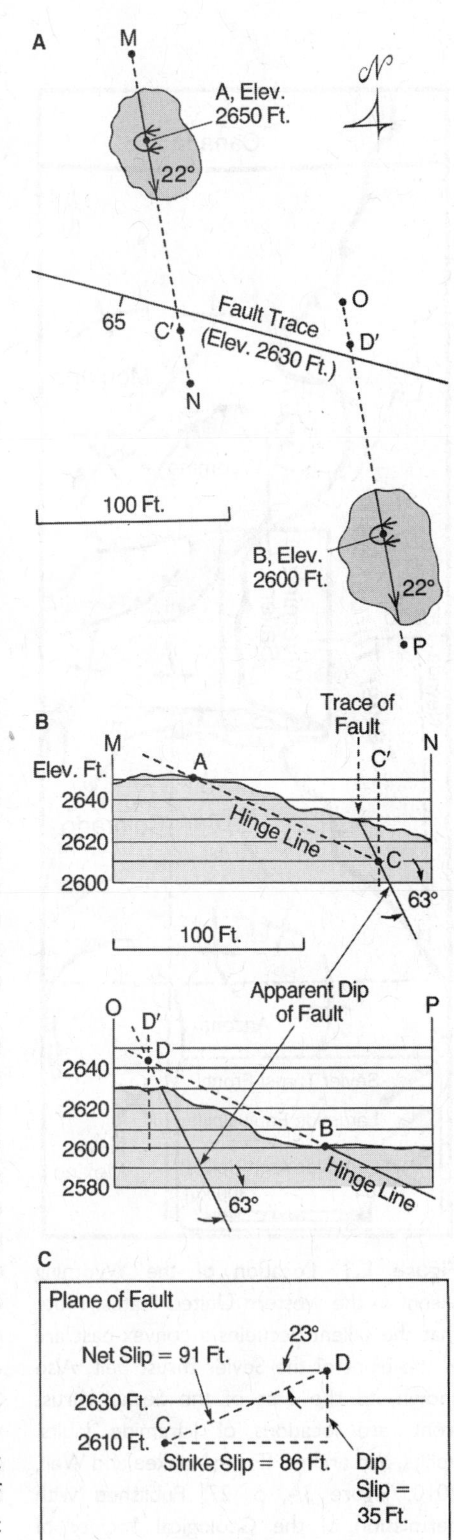

Figure K.3 Determination of the translation along a fault on the basis of the offset of the hinge line of a fold. (*A*) Map relationships showing the fault offset of the trace of an overturned syncline. (*B*) Structure profiles showing where hinge lines intersect fault surface. (*C*) Slip components as seen in the fault plane.

L. CARRYING OUT STRAIN ANALYSIS

Background

There are superb textbooks and manuals that describe methodologies for carrying out detailed strain analysis of outcrop-scale deformation. These include publications by Ramsay (1967), Ramsay and Huber (1983), and Ragan (1985, 2009). Furthermore, Declan De Paor over the years has published prodigiously on methodologies in strain analysis such as De Paor (1988) and Simpson and De Paor (1997). He has advanced the techniques, innovated new ones, and written software that permits computer technologies to carry the heavy analytical load required in strain analysis, a load that otherwise had to be done by hand through difficult, repetitive, time-consuming iterations (e.g., De Paor, 2001). Part of the reason that Declan's contributions have advanced the field is that he has made the new software broadly available to the structural geology community. See, for example, the part of his web site that features *Structural Analysis: An Interactive Course for Earth Science Students* (*http://www.lions.odu.edu/~ddepaor/Structural_Analysis/index.swf*).

In comparison to strain analysis of outcrop-scale deformation, regional analysis has received less attention. Ramsay (1969) laid down part of the framework for doing same in his important paper on the measurement of strain and displacement in orogenic belts. Furthermore, productive and revealing strain analysis has been fundamental to structural-geological analysis of certain select areas, including the Moine thrust belt in northern Scotland (see, for example, Wilkinson et al., 1975; Butler, 1982; Coward, 1984; and Law et al., 1984).

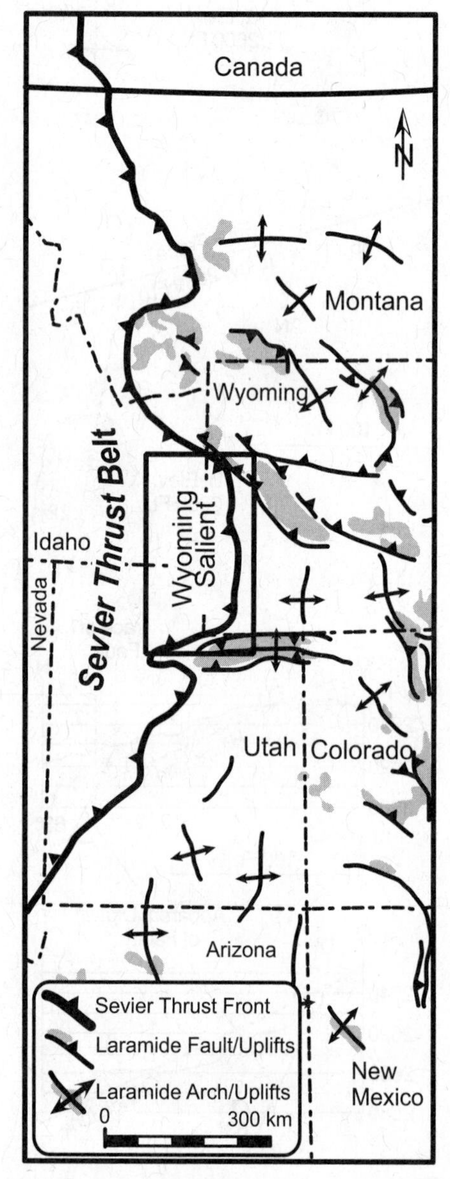

Figure L.1 Location of the Wyoming salient in the western United States. Note that the salient occupies a convex-east arc in the front of the Sevier thrust belt. Also shown, to the east of the Sevier thrust front, are locations of Laramide faults, uplifts, and arches. [From Yonkee and Weil, 2010, Figure 3A, p. 27. Published with permission of the Geological Society of America.]

Field-Based Strain Analysis, Cross Scales Of Observation

Here we call attention to an exceptionally well conceived and executed regional strain analysis that was carried out by Adolph Yonkee and Arlo Weil. Yonkee and Weil (2010) took on the task of describing the total deformation of the Wyoming salient of the Sevier thrust belt (largely Cretaceous deformation). In thrust belt context, a salient is a large (tens of km's long) distinctive convex/concave arc (map view) in a fold-thrust belt. The Wyoming salient is pictured in Figure L.1. The salient is convex in the direction of thrusting, in this case eastward toward the direction of the foreland.

Yonkee and Weil (2010) wanted to figure out the origin of the salient. Was the Wyoming salient originally straight, and later deformed into its arcuate shape? Did the Wyoming salient start out with a primary arcuate shape and then later increase in length and curvature during the course of thrusting? Yonkee and Weil believed that the analysis of strain features at outcrop study sites throughout the belt could provide the means for constraining the origin of the salient. On the basis of strain analysis they reached the conclusion that at the very beginning of thrusting there was a primary salient, with a small overall range of trend of just 6° (Figure L.2). In this early stage the shortening direction, all along the salient, was due east. But soon after the initiation of thrusting, slip directions within the system became radial, and the length and curvature of the salient increased, such that today the range of trend is 34° (see Figure L.2). What proved critical was determining (1) the internal strain within the thrust sheets, (2) comparing layer-parallel shortening (LPS) to arc-parallel tangential extension (TE), and (3) determining the direction(s) of thrusting over time.

What strain recorders did they have to work with? At 154 sites in red beds of the Ankareh Formation (Triassic), they determined local 3D strain based on

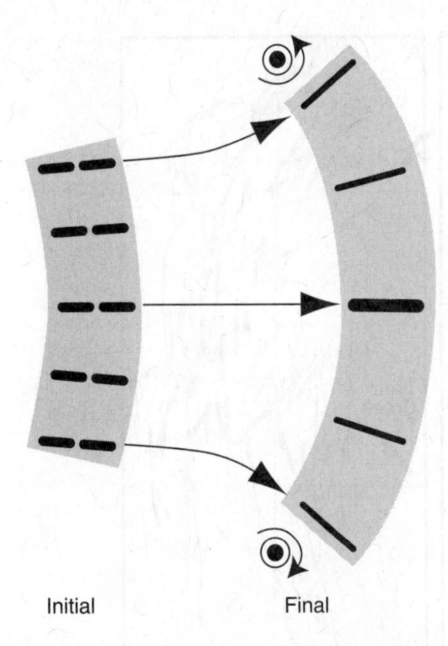

Figure L.2 Yonkee and Weil (2010) concluded that only a small part of the arc-like nature of the salient was due to its primary shape. Initially the trend of the eastern outer arc of the salient varied only 6 degrees. The full arcuate shape developed as a result of east-west shortening along thrust slip directions that diverged eastward, tending toward a radial pattern. Thus the initial curvature of the eastern outer arc was increased, such that arcuate variation became 36°. The model recognizes a difference between early tectonic direction of layer parallel shortening, basically east west, to the more radial pattern. Overall the northern part of the salient rotated counter-clockwise, and the southern clockwise. This pattern is verifiable through paleomagnetic investigation. Dark dash lines on left show initial orientation of trend of layer parallel shortening. Solid dark lines on right show final layer-parallel shortening directions. [From Yonkee and Weil, 2010, Figure 21, Case 3, p. 45. Published with permission of the Geological Society of America.]

Figure L.3 Photograph of reduction spots in the Ankareh Formation. They are deformed into ellipsoidal shapes, with flattening direction subparallel to cleavage (S1). Inset photograph shows reduction spots where undeformed. [From Yonkee and Weil, 2010, Figure 14A, p. 37. Published with permission of Geological Society of America.]

deformed reduction spots (Figure L.3), and they determined shortening directions and magnitudes through study of spaced cleavage and tectonic stylolites. At 144 sites in the limestones of the Twin Creek Formation (Jurassic), they determined amounts and directions of extension based on orientations of cross-strike joints, and again determined directions and magnitudes of shortening through study of cleavages and tectonic stylolites. Interpretation of the strain data required sequential restoration of the thrust belt through time. This was a big job!

All of this can be summed up in a few of the diagrams they produced. Figures L.4 and L.5 show the *restored* directions of layer-parallel shortening and tangential extensional strain. Figure L.6 shows bed-parallel strain for the Ankareh Formation, with a notable increase in strain ratio westward and

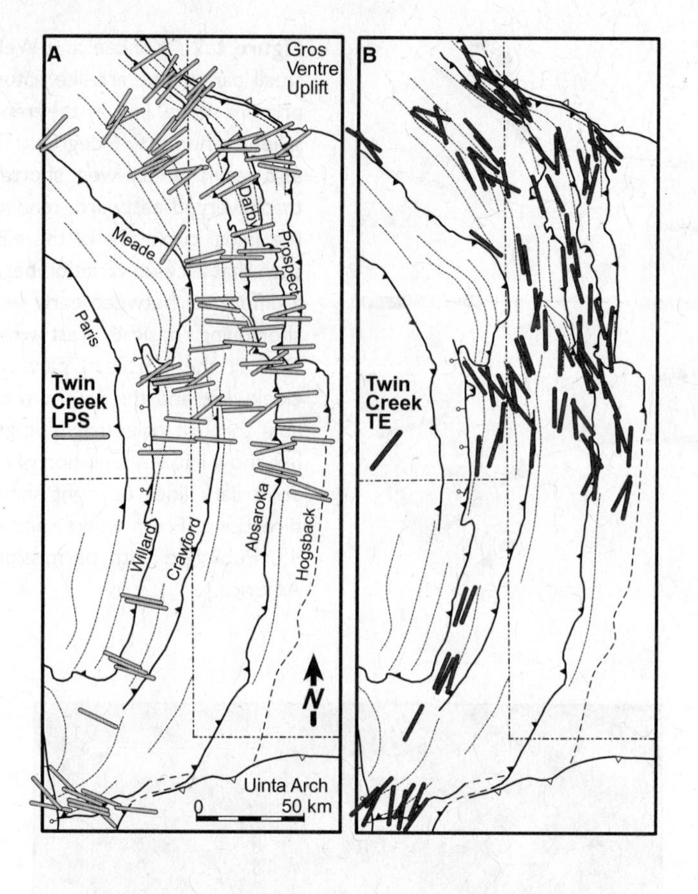

Figure L.4 Map of restored deformation fabric elements in Twin Creek Formation (Jurassic) reveals (*A*) a radial distribution of layer-parallel shortening (LPS) directions (gray rectangles) estimated from orientations of cleavage and tectonic stylolites, and (*B*) a pattern of tangential extension (TE) (black rectangles) estimated from cross-strike veins. [From Yonkee and Weil, 2010, Figure 10, p. 35. Published with permission of the Geological Society of America.]

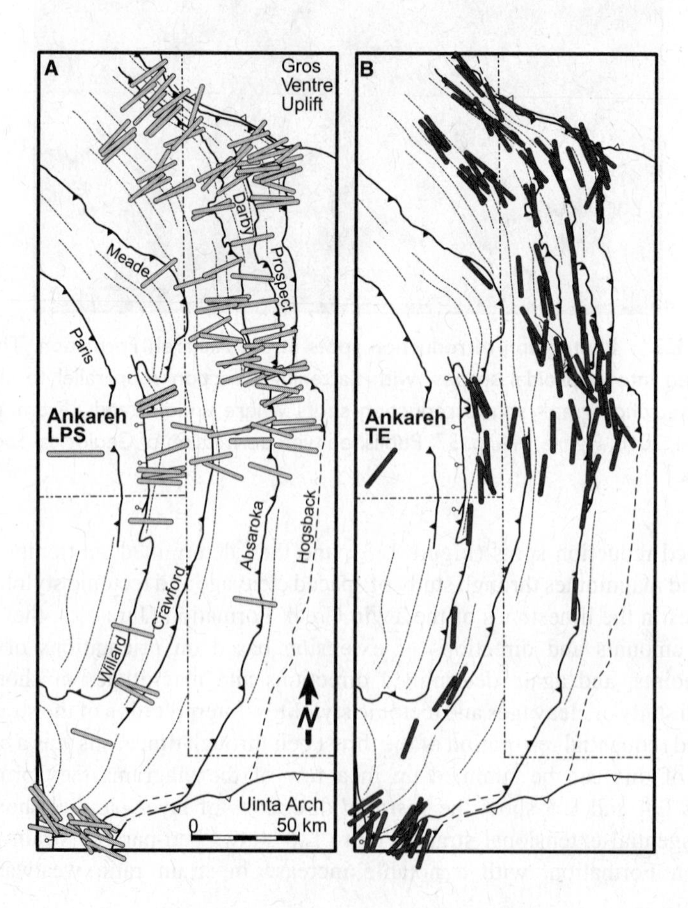

Figure L.5 Map of restored deformation fabric elements in Ankareh Formation (Triassic) reveals (*A*) a radial distribution of layer-parallel shortening (LPS) directions (gray rectangles) estimated from orientations of cleavage and high-angle fractures, and (*B*) a pattern of tangential extension (TE) (black rectangles) estimated from cross-strike fractures and veins. [From Yonkee and Weil, 2010, Figure 11, p. 36. Published with permission of the Geological Society of America.]

toward the northern and southern tips of the salient. Maps such as these show how beautifully outcrop-scale strain analysis and regional strain analysis can be integrated in addressing even larger-scale tectonic questions.

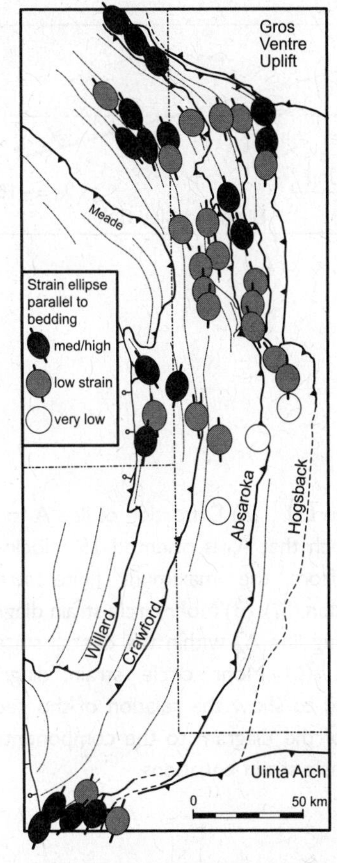

Figure L.6 Map of bed-parallel strain for the Ankareh Formation (Triassic). The strain ellipses reflect the ratio of layer-parallel shortening (LPS) to tangential extension (TE). The shortening directions show a radial pattern. The strain ratios increase westward and toward each of the tips of the salient. [From Yonkee and Weil, 2010, Figure 17, p. 39. Published with permission of the Geological Society of America.]

Mohr Strain Analysis

The Mohr Strain Diagram

We introduced the fundamental strain equations in *Chapter 2, Displacement and Strain.* These are equations that permit shear strain and stretch to be determined along any direction within a strain field, whose principal stretch directions and values are known. The key parameters in these equations (see below) are reciprocal quadratic elongation (λ'), shear strain (γ), and the reference angle (θ). Recall that λ is the square of the stretch (S), and λ' is the reciprocal of λ. The reference angle, θ_d, is the angle between the direction of greatest principal stretch (S_1) and the line for which stretch (S) and shear strain (γ) are to be determined. Positive values of θ_d are counterclockwise measurements from the direction of principal stretch; negative values of θ_d are clockwise measurements.

Here are the fundamental equations:

$$\lambda' = \frac{\lambda_3' + \lambda_1'}{2} - \frac{\lambda_3' - \lambda_1'}{2} \cos 2\theta_d \qquad \text{(L.1)}$$

where

$$\lambda' = \frac{1}{\lambda}, \ \lambda_1' = \frac{1}{\lambda_1}, \ \lambda_3' = \frac{1}{\lambda_3}$$

and

$$\frac{\gamma}{\lambda} = \frac{\lambda_3' - \lambda_1'}{2} \sin 2\theta_d \qquad \text{(L.2)}$$

which also can be written

$$\frac{\gamma}{\lambda} = \frac{1}{2}\left(\frac{1}{\lambda_3} - \frac{1}{\lambda_1}\right) \sin \theta_d$$

Let us construct a Mohr diagram to represent the state of strain in the hypothetically deformed clay cake shown here as Figure L.7*A*. The single reference line is line A_d. Line A_d makes an angle of $\theta_d = -15°$ (clockwise acute angle!) with the S_1 axis of the finite strain ellipse. The principle strains for this overall deformation are $\lambda_1 = 2.40$ ($S_1 = 1.55$); $\lambda_3 = 0.42$ ($S_3 = 0.65$).

Our goals are to construct a Mohr circle strain diagram on the basis of this information and to determine the values of quadratic elongation (λ) and stretch (S), as well as shear strain (γ) and angular shear (ψ) for line (direction) A_d. The Mohr circle strain diagram is plotted in $x-y$ space such that values of λ', λ_1', and λ_3' are plotted and measured on the *x*-axis (Figure L.7*B*). The λ_1' and λ_3' values are the reciprocals of the maximum and minimum quadratic elongations. A circle centered on the *x*-axis and drawn through the values of λ_1' and λ_3' is the Mohr circle proper. Its circumference can be thought of as the locus of hundreds of points whose $x-y$ coordinates are paired values of λ' and γ/λ. These values permit quadratic elongation and shear strain to be calculated for lines of *every* orientation in the deformed body shown in L.7*A*.

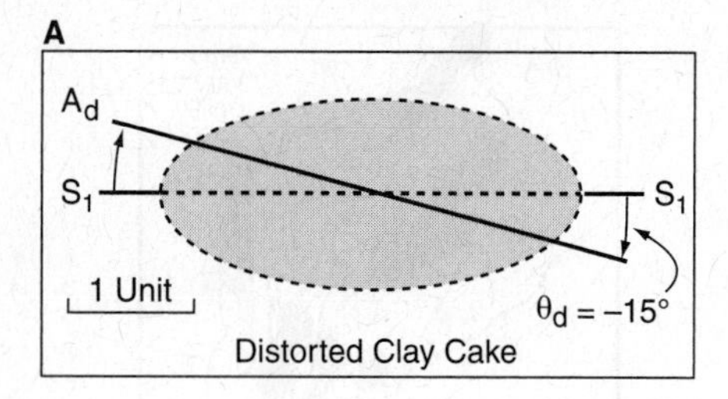

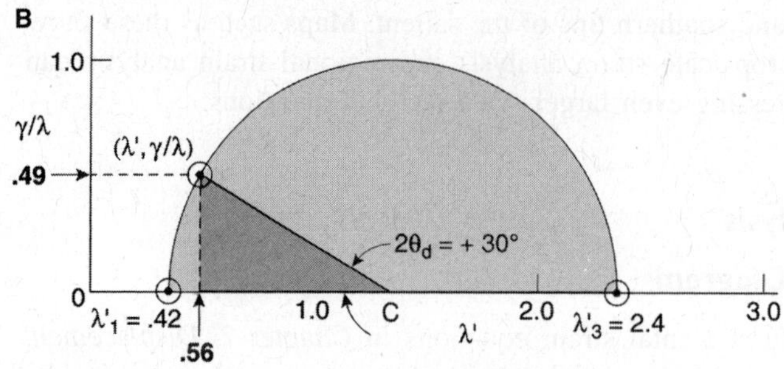

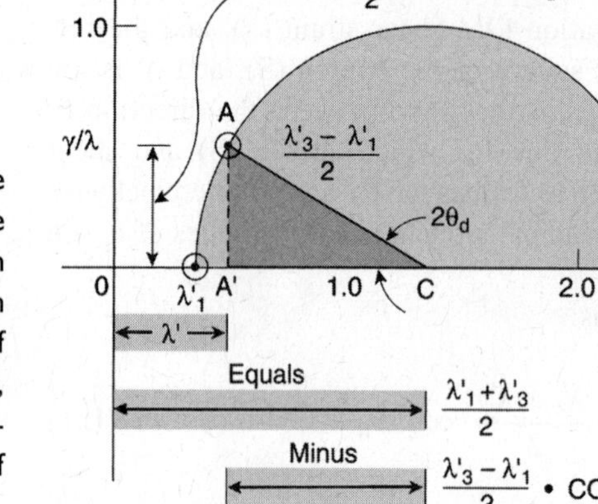

Figure L.7 (*A*) Distortion of line A to line A_d, such that A_d is oriented 15° clockwise (−) from the maximum finite stretch direction (S_1). (*B*) Mohr circle strain diagram showing line A_d within the overall state of strain. (*C*) Mohr circle strain diagram, labeled to show the relation of the geometry of the diagram to the components of the basic strain equations.

For example, λ_1' has (*x*, *y*) coordinates of (0.42, 0.0), corresponding to the reciprocal of quadratic elongation for lines parallel to the maximum finite stretch direction (S_1). Lines thus oriented are marked by 0.0 shear strain. Similarly, λ_3' has (*x*, *y*) coordinates of (2.4, 0.0), the *x* value being the reciprocal of quadratic elongation for lines parallel to the minimum finite stretch direction (S_3). The *y* value of zero again reflects 0.0 shear strain for lines thus oriented.

The paired values of (λ', γ/λ) for line A_d lie at a point somewhere on the perimeter of the Mohr circle. But where? Line A_d lies 15° clockwise from the maximum finite stretch direction (S_1) (see Figure L.7*A*). The location of the point on the Mohr circle representing strain values for this line can be found by plotting a radius *clockwise* from λ_1' by an angle of $2\theta_d = -30°$ (Figure L.7*B*). The point where the radius intersects the perimeter of the circle has (*x*, *y*) values that correspond to γ/λ. For line A_d,

$$\frac{\gamma}{\lambda} = 0.49$$

$$\lambda' = 0.56$$

$$\lambda = \frac{1}{\lambda'} = \frac{1}{0.56} = 1.88$$

$$\gamma = \left(\frac{\gamma}{\lambda}\right)\lambda = (0.49)(1.8) = +0.88 \text{ (positive = counterclockwise)}$$

Since

$$\gamma = \tan\psi$$
$$\psi = \arctan\gamma = 41°$$

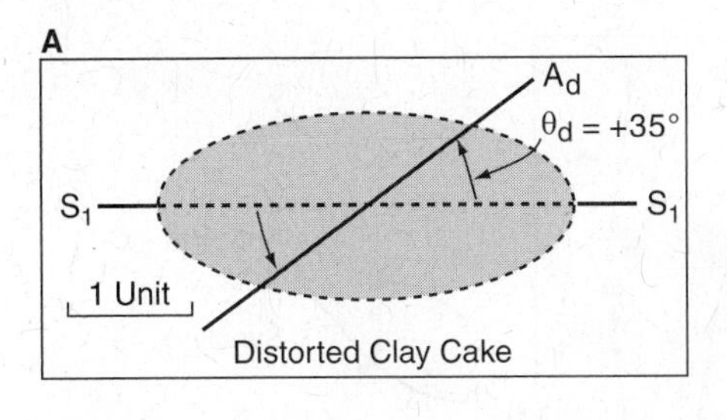

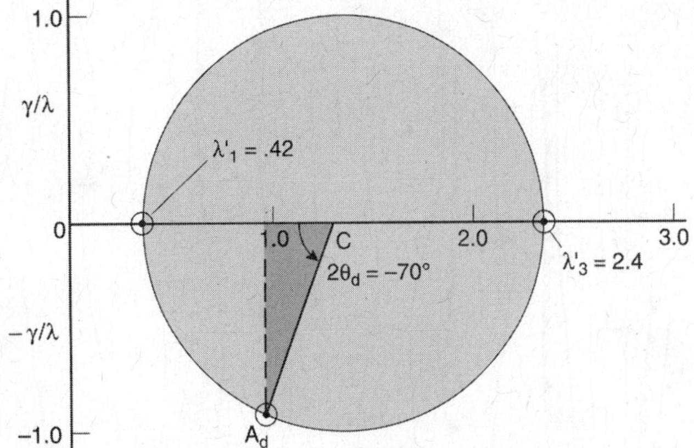

Figure L.8 (*A*) Distortion of line A to line A_d, such that A_d is oriented 35° counterclockwise (+) from the maximum finite stretch direction (S_1). (*B*) Mohr circle strain diagram showing line A_d within the overall state of strain.

This is an elegant way to go about things, for the Mohr circle diagram devolves nicely into the fundamental strain equations (Figure L.7*C*).

What if line A_d were to make a counterclockwise acute angle with respect to the direction of maximum finite strain (S_1)? What would this look like on the Mohr circle strain diagram? Line A_d in Figure L.8*A* makes an angle of $\theta_d = +35°$ with respect to S_1. The state of strain of this line may be found in the Mohr diagram (Figure L.8*B*) by constructing a radius at an angle of +70° counterclockwise from the x−axis.

Retrodeforming Regional Strain

Suppose that mapping of an ancient shoreline deposit within a large region of strongly deformed sedimentary rocks revealed the trend line of an ancient beach. We suspect that the present mapped trend of N60°E (Figure L.9*A*) is a distortion of the original line of beach. We want to determine the original trend of the shoreline deposit. Fortunately we have more information, which we have drawn from our geologic cross-sections. We know that S_3, the magnitude of stretch in the direction of greatest horizontal shortening (east−west), is equal to 0.6; and S_2, the magnitude of stretch in the direction of intermediate horizontal lengthening (north−south), is 1.1 (Figure L.9*B*). In this example, S_1 is vertical.

A very important equation presented by Ramsay (1967) is the basis for establishing the rotation that a line endures during distortion (compare the trend of the shoreline deposits in part *A* versus *B* in Figure L.9). The equation reads as follows:

$$\tan \theta_d = \tan \theta \left(\frac{S_3}{S_2}\right) \qquad \text{(L.3)}$$

where

θ = angle between the line of interest (L) and the direction of intermediate finite stretch in the undeformed state

and

θ_d = angle between the line of interest (L) and the direction of intermediate) finite stretch in the *deformed* state

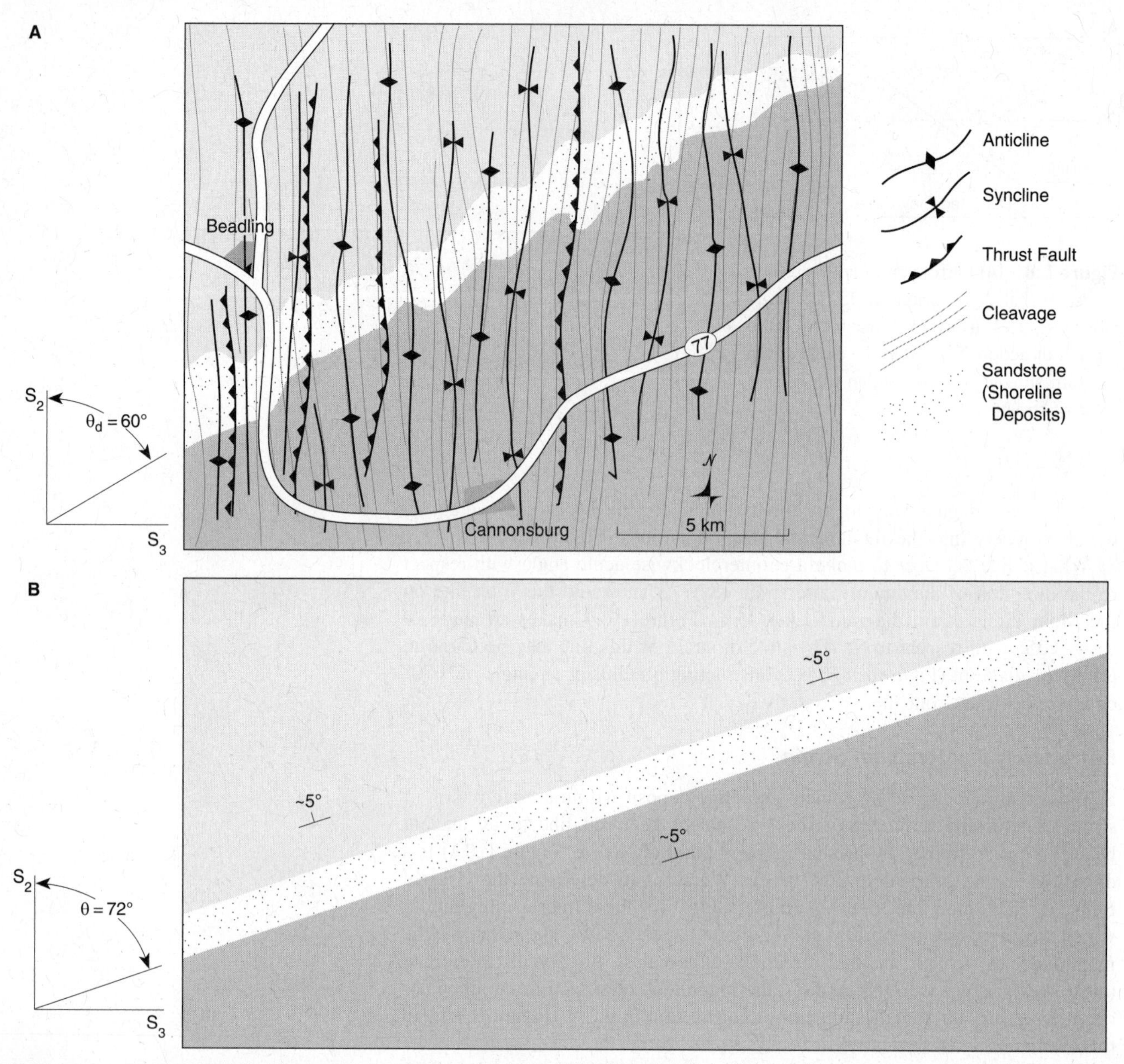

Figure L.9 (*A*) Map of folded and thrusted sedimentary rocks. The white stippled pattern denotes a sedimentary facies representing an ancient beach separating nonmarine deposits (dark gray) to the south and shallow marine limestones (light gray) to the north. The direction of minimum finite stretch (S_3) is east–west; the direction of intermediate stretch (S_2) is north–south; and the direction of maximum finite stretch (S_1) is vertical. (*B*) When the folded and thrusted region is "*retrodeformed*" to its original state, the beach line is seen to run N72°E, different from its N60°E trend in the deformed state.

If the plane of view contained S_1 and S_3, Equation L.3 would read:

$$\tan \theta_d = \tan \theta \left(\frac{S_3}{S_1}\right)$$

In our example (Figure L.9), the rotation depends on only two factors: the initial orientation made by the line in question with the direction of intermediate finite stretch (S_2), and the ratio of the principal stretch values, S_3/S_2.

We can use Equation L.3 to solve for the *original* trend of the shoreline deposit. From Figures L.9*A* we know that $\theta_d = -60°$. We already know that

$S_2 = 1.1$ and $S_3 = 0.6$. Substituting these values into Equation L.3,

$$\tan 60^\circ = \tan \theta\,(0.6/1.1)$$

$$1.7 = \tan \theta\,(0.55)$$

$$\tan \theta = \frac{1.7}{0.5} = 3.1$$

$$\theta = \arctan 3.1756 = 72^\circ$$

We conclude (Figure L.9*B*) that before deformation, the shoreline deposit was originally along a trend 72° east of north.

M. DETERMINING THE RELATIONSHIP OF FAULTS TO PRINCIPAL STRESS DIRECTIONS

Interpreting Fault Orientations from Stress Directions

The likely orientations of faults that should form in a given stress field can be evaluated stereographically, provided the angle of internal friction is known. In interpreting likely fault orientations, we assume that rocks are homogeneous and isotropic, and that the Coulomb law of failure holds.

Let us consider what kinds of faults should develop in a stress field where $\sigma_1 = 0°$ N20°E, $\sigma_2 = 0°$ N70°W, and σ_3 is vertical. Recall that normally the pole to a fault plane makes an angle of $\theta = 60°$ to the direction of greatest principal stress (σ_1). Let us assume in this example that indeed $\theta = 60°$. We know that thrust-slip faults should form simply on the basis of the vertical orientation of σ_3. The probable orientations of the thrusts can be predicted by applying what we have learned in *Chapter 6, Faults*, about the formation of conjugate faults. Conjugate faults intersect in σ_2. The trace of each fault, when viewed in the σ_1/σ_3 plane, is oriented at an angle of 30° with respect to σ_1.

To give these relationships geometric reality, we stereographically plot points that portray the orientations of the principal stress directions (Figure M.1*A*). We then define the σ_1/σ_3 principal plane by fitting σ_1 and σ_3 to a common great circle (Figure M.1*B*). Counting 30° along the great circle from σ_1, we locate two reference points (1 and 2) that represent the intersection of the fault planes with the σ_1/σ_3 principal plane. By fitting σ_1 and reference point 1 to a common great circle, one of the thrust-slip faults is defined (Figure M.1*C*). And by fitting reference point 2 and σ_2 to a common great circle, the second thrust-slip fault of the conjugate set is established. The actual orientations of the faults can then be determined by normal stereographic procedures (Figure M.1*D*). Comparable solutions for defining the orientations of strike–slip and normal–slip faults within appropriate stress fields are shown in Figures M.2 and M.3, respectively.

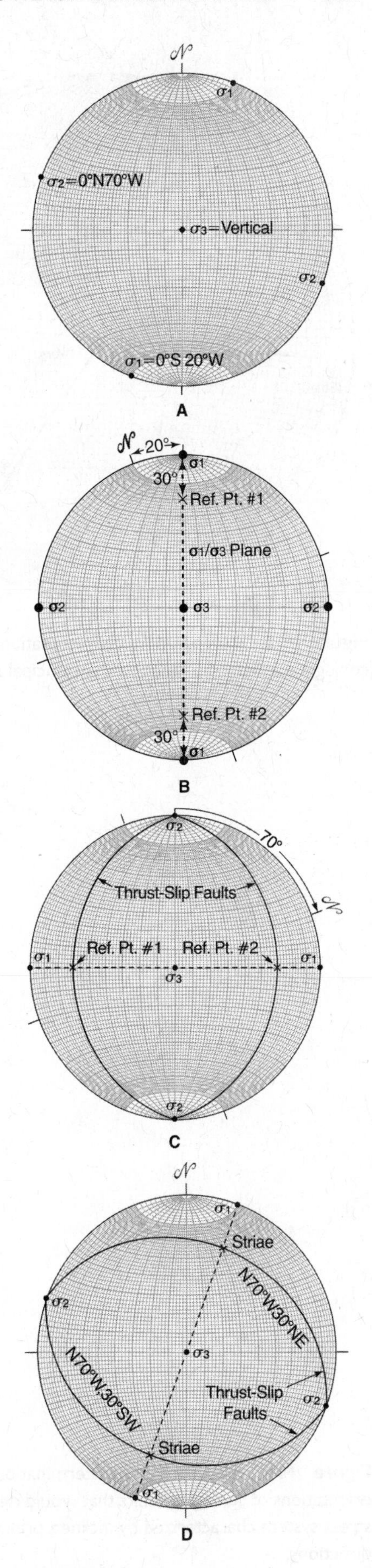

Figure M.1 Stereographic representation of the relation of faults to the principal stress directions. (*A*) Stereographic portrayal of principal stress directions. (*B*) Identification of the σ_1/σ_3 plane, and the plotting of reference points (1 and 2) at 30° angles from σ_1 on the σ_1/σ_3 great circle. (*C*) The orientation of one of the thrust faults is represented by a great circle that passes through reference point 1 and σ_2. The orientation of the second thrust fault is defined by the great circle that passes through reference point 2 and σ_2. (*D*) Portrayal of actual orientations of the faults and the slickenline striae that occur along them.

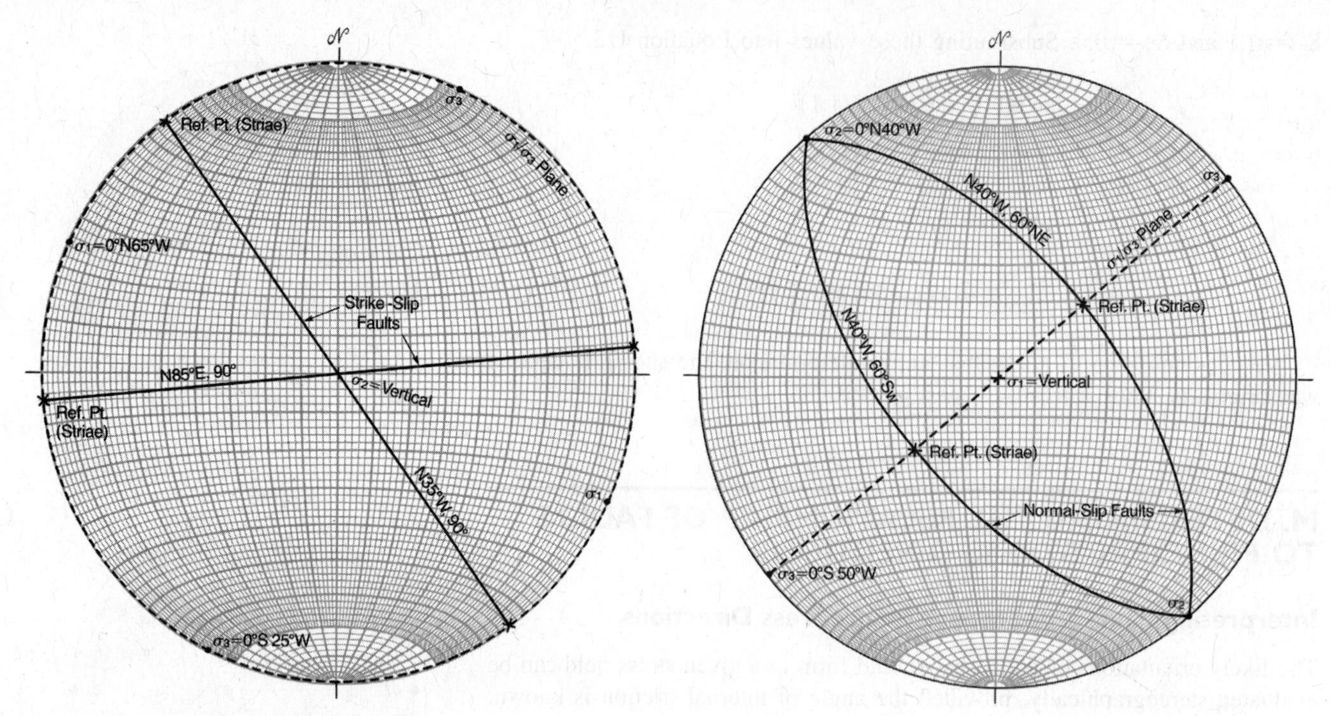

Figure M.2 Stereographic representation of the relation of conjugate strike–slip faults to the principal stress directions.

Figure M.3 Stereographic representation of the relation of conjugate normal–slip faults to the principal stress directions.

If one or more of the principal stress directions is inclined, the solution is more difficult to visualize, but the stereographic operations are the same. Consider the general situation wherein none of the principal stress direction are vertical or horizontal; instead all are inclined (Figure M.4). As before, we define the σ_1/σ_3 principal plane by fitting σ_1 and σ_3 to a common great circle, and then we set off reference points from σ_1 along the great circle at distances corresponding to 30°. The orientations of the fault planes are each defined by one of the reference points and σ_2 (see Figure M.4).

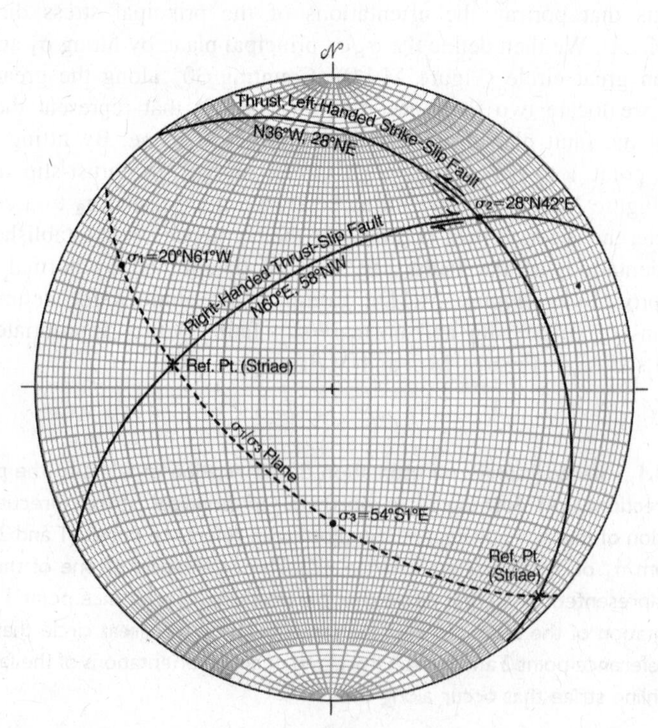

Figure M.4 Stereographic determination of the orientations of conjugate faults that would develop in a stress system characterized by inclined principal stress directions.

Using Faults to Interpret Stress

In the course of field-oriented structural analysis, it is common to work through this problem backward, as an **inverse problem**. Suppose we want to interpret principal stress directions on the basis of the orientations of two conjugate faults. The faults are first plotted stereographically as great circles, as in Figure M.5*A*. The intersection of the two great circles is taken to be the orientation of σ_2; in this example, σ_2 plunges 20° N50°W. We know that the σ_1/σ_3 principal plane is perpendicular to σ_2. Given the trend and plunge of σ_2, the σ_1/σ_3 plane must strike N40°E and dip 70°SE (Figure M.5*B*). σ_1 lies somewhere along the great circle representing the σ_1/σ_3 planes, but where?

We know from experiments that σ_1 bisects the acute angle between conjugate faults. Stereographically this means that we can locate σ_1 by bisecting the angle between the fault traces that lie in the σ_1/σ_3 plane. To do this, we rotate the overlay such that the strike of the σ_1/σ_3 plane becomes aligned along the north–south line of the net (Figure M.5*C*). Then measure the acute angle between the reference points. Since the acute angle measures 64°, the bisector σ_1 is located 32°from each reference point along the σ_1/σ_3 plane. σ_3 is perpendicular to σ_1. The orientations of the principal stress directions are ultimately defined in terms of trend and plunge (Figure M.5*D*).

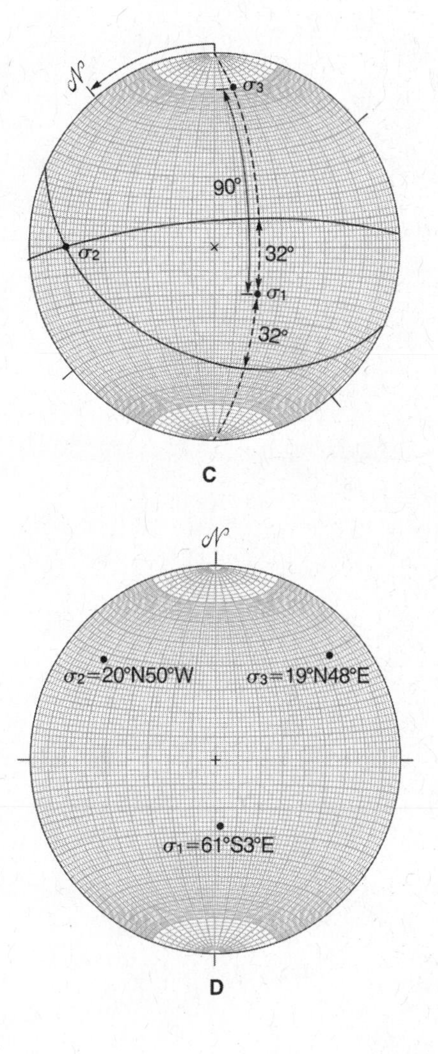

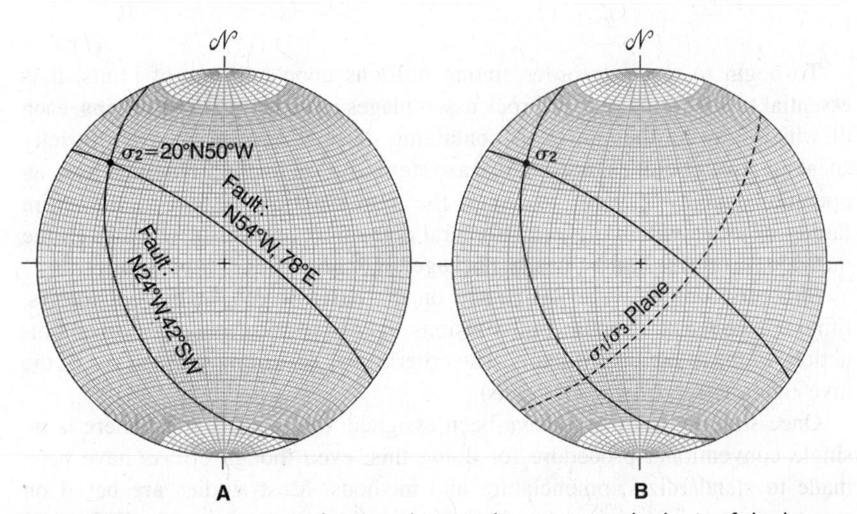

Figure M.5 Interpretation of principal stress directions on the basis of the known orientations of conjugate faults. (*A*) Stereographic representation of the faults as great circles. The faults intersect in σ_2. (*B*) Identification of the σ_1/σ_3 plane, the great circle whose pole is σ_2. (*C*) Determination of the stereographic locations of σ_1 and σ_3. (*D*) Portrayal of the orientations of σ_1, σ_2, and σ_3, as solved.

N. CARRYING OUT JOINT ANALYSIS

Establishing Structural Domains

Joints do not exist in isolation. Rather they are members of enormously large families of fractures with literally millions of members. Every regional rock assemblage, such as a granitic batholith or a plateau of sedimentary rock, is pervaded by jointing. It is impossible to try to explain the origin of every joint in an outcrop, let alone every joint within a regional rock assemblage. Instead, we try to explain the origin of dominant **sets** of joints that can be identified through statistical analysis of the orientations and physical properties of the joints within a given system (Figure N.1).

Figure N.1 Beautiful exposure of two sets of joints (mutually perpendicular) exposed on tessellated pavement at Eaglehawk Neck, Tasmania. Light-colored strips along each joint trace is alteration due to fluid flow along the joints. Rock is sandstone. [Photograph by S. J. Reynolds.]

To begin to discover order among millions upon millions of joints, it is essential to subdivide regional rock assemblages into **structural domains**, each of which may be thought of as containing its own fractures system. Strictly speaking, two or more **sets** comprise a **system**. But we find it preferable, from an operational point of view, to regard the term **fracture system** as the entire family of joints within a given structural domain, regardless of whether all the individual surfaces can be neatly packaged into sets of like orientation.

Structural domains are designated on the basis of geographic boundaries, lithologic contacts, structural subdivisions, ages of rock formations, and combinations of these and other factors. The criteria vary according to the scope of the investigation (see *Chapter 5, Joints*).

Once structural domains have been assigned, the work begins. There is no single conventional procedure for doing this, even though efforts have been made to standardize nomenclature and methods. Most studies are based on structural analysis of joints at selected **stations** within each structural domain. Less commonly, and usually for very specific applied purposes, **fracture pattern maps** are generated.

The Mapping of Joints

In regions where joints are distinctly expressed in the weathered landscape, the mapping of the fractures is best achieved photogeologically. Aerial images often display remarkable portrayals of fracture patterns. With patience and care, the intricacies of joint patterns can be accurately reproduced in inked tracings, such as those rendered by Cruikshank and others (1991) for the Arches National Park area, Utah (Figure N.2). Conventional large-scale aerial photography (either black and white or color) serves nicely as a base for most studies. But for regional analysis of fracture systems, it is advantageous to use some of the extraordinary space-satellite imagery that is available, including that available on Google Earth. Advances in remote-sensing techniques that have accompanied the proliferation of regional imagery make photogeologic analysis of jointing even more attractive.

There are shortcomings associated with using the photogeologic approach, exclusively, in the mapping of joints. Low-dipping fractures are automatically

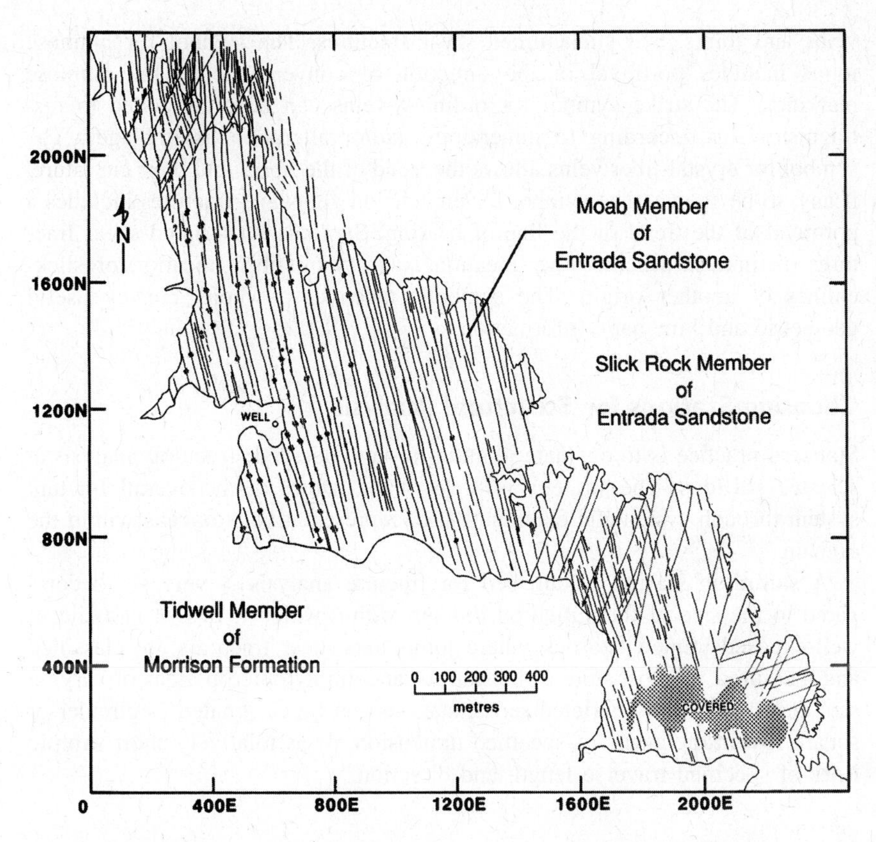

Figure N.2 Ink tracing of jointing in the Garden area, Arches National Park region, Utah. The traces of systematic fractures include joints, faulted joints, and jointed faults. There are three sets of joints in the area, trending about N10°W, N30°E, and N60°E. [Reprinted with permission from *Journal of Structural Geology*, v. 13, Cruikshank, K. M., Zhao, G., and Johnson, A. M., Analysis of minor fractures associated with joints and faulted rocks, 1991, Elsevier Science Ltd., Pergamon Imprint, Oxford, England.]

screened from view, thus biasing the data toward moderate to steep-dipping fractures. Furthermore, it is impossible to measure from photographs the dip and dip direction of joints and to gather information regarding the types of fractures that exist in each structural domain.

Where jointing is not well expressed in the landscape, or is too subtle or fine to be resolved on aerial images, the mapping of fractures begins with systematically measuring the orientations of representative joints in the field. Generally, little attempt is made to show the actual physical traces of the fractures, for they are generally far too numerous and much too short to portray at reasonable map scales. Instead, the orientations of the dominant systematic sets of joints are portrayed through standard joint symbology, plotted as close as possible to the locations where the fractures are measured. The symbols are drafted in dense overlapping clusters that literally fill the map to overflowing. Under ideal circumstances such fracture-pattern maps show at a glance the chief types of joint and their respective orientations. Under less favorable circumstances, the maps are a collage of confusion, doing little to clarify the basic elements of the fracture pattern.

It is common practice to measure eye-catching joints as a part of the normal geologic mapping process. Representative structures are plotted according to type and orientation. The few and scattered data that are typically posted on geologic maps do not constitute a basis for structural analysis. Instead, they simply provide a preliminary indication of the dominant joint sets that might exist in the area of investigation.

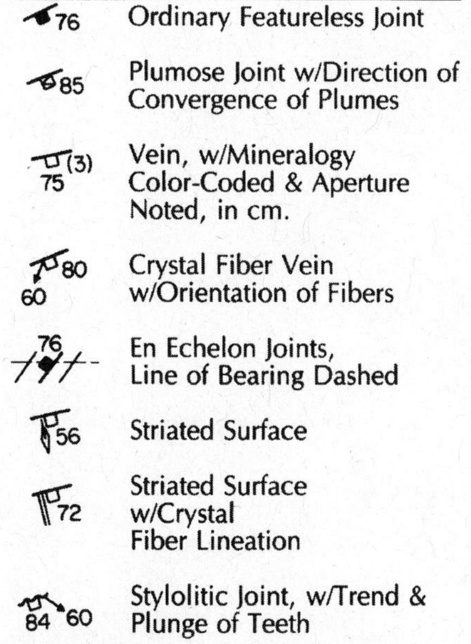

Figure N.3 Map symbology for joints, veins, shear fractures, and stylolites.

Map Symbology

Map symbology for representing joint data on fracture-pattern maps and normal geologic maps is presented in Figure N.3. Symbols distinguish ordinary featureless joints, plumose joints, veins, crystal fiber veins, en echelon veins and joints, and slickenlined shear fractures. The symbol for plumose joints includes portrayal of the direction of convergence of the plumose markings. The strike symbol for ordinary veins can be color-coded to distinguish veins according to mineralogy and/or alteration assemblages. The symbol for crystal-fiber veins allows the trend of the fibers and their curvature, if any, to be shown. The symbol for en echelon veins and fractures includes a portrayal of the trend of the line of bearing. Symbols for striated shear fractures distinguish whether the lineation is a crystal fiber lineation or slickenlines of another origin. The symbols, taken as a whole, convey useful geometric and kinematic information.

Choosing Stations for Structural Analysis

Standard practice is to evaluate jointing through detailed structural analysis at selected **stations**. The strategy is to learn the nature of the overall fracture system through systematic examination of representative **subareas** within the domain.

A **sampling station** established for fracture analysis is very small compared to the size of the structural domain within which it lies. It is a site of well-exposed jointed bedrock where joints and shear fractures are classified and measured. In most studies the stations are simply outcrop areas of varying size and shape. In a restricted sense, stations can be designated as circular or square **inventory areas** of specified dimension, or as relatively short **sample lines** of specified traverse length and direction.

Measuring Orientations of Joints

In practice, there are two basic approaches that are used in collecting orientation data at sample stations. One, which we might call the **selection method**, involves selecting only certain joints for measurement and study. The second approach, which we might call the **inventory method**, requires measuring and classifying every single joint at a station site.

The basis of the selection method is to restrict analysis to joints and shear fractures that are continuous and through-going, and are conspicuously associated with other fractures of similar appearance and orientation. It is neither very easy nor very objective to decide which fracture surface should be measured at each station, and which should be left alone. But in spite of this difficulty, many workers have found the selection method to be practical and useful. Parker (1942) restricted his joint-orientation measurements in the Appalachian Plateau region of New York to those fractures that appeared straight and continuous. Hodgson (1961), in analyzing the fracture pattern of part of the Colorado Plateau, measured only smooth planar continuous joints, especially those that displayed plumose markings. Rehrig and Heidrick (1972, 1976) measured joints and veins that occurred in sets of three or more, ignoring all others.

In contrast to the selection method, the inventory method requires measuring all the joints and shear fractures that occupy a sampling station. Measurements are taken within an inventory area or along a sample line. In sampling along a line it is common practice to stretch out a tape measure to some prescribed length (normally less than 20 m), and to classify and measure the orientation of every joint that is intercepted by the tape (see *Chapter 5, Joints*).

A novel, effective inventory approach is one we refer to as the **circle-inventory method** (Titley, 1982). First a circle of some known and predetermined diameter (normally less than 3 m) is traced out on a bedrock surface of perfectly exposed fractured rock (Figure N.4). The circle is drawn with a piece of (carpenter's) chalk attached to a string of suitable length. Then the orientation and trace length of each fracture within the circle are

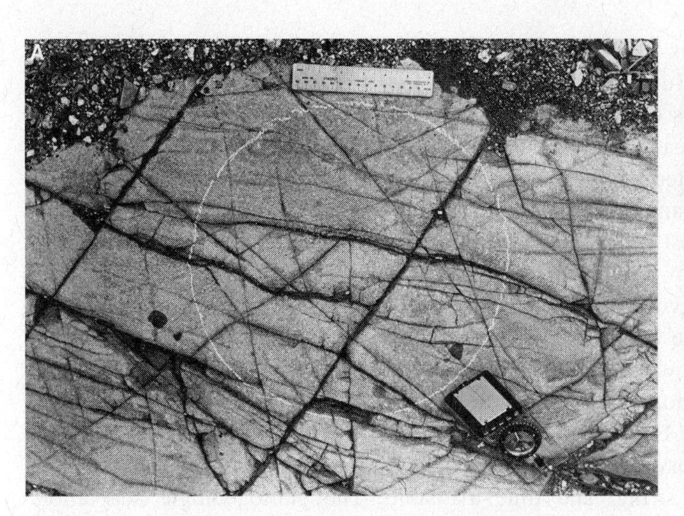

Figure N.4 Circle-inventory method for sampling the orientations of joints and joint-related structures. (*A*) The circle, drawn on bedrock with chalk. (*B*) With joints measured and traced, Matt Davis and Katie let their handiwork be admired. Katie still has her chalk in her paw. [Photographs by G. H. Davis.]

measured. Data are recorded in the manner shown in Table N.1. To avoid measuring the same fracture surface twice, it is helpful to trace out the full length of each joint or shear fracture with chalk after it is measured. Outcrops are generally available in which the three-dimensional expression of each

TABLE N.1
Circle-inventory method for evaluating fracture density

Station #34, Tucson Mountains
Lower Cretaceous Sandstone
Radius of Inventory Circle = 19 cm; Area = 1134 cm²

Trend of Fracture	*Length of Fracture (cm)*
N79E	35
N46W	36
N04W	30
N32E	25
NS	22
EW	20
N16E	20
N42W	39
N41W	13
N44W	17
N43W	11
N30W	28
N41E	18
N82W	13
N64W	13
NS	6
N12W	20
N23E	14
N11E	7
	387 cm = Cumulative Fracture Length

$$\text{Fracture Density} = \frac{387\ \text{cm}}{1134\ \text{cm}^2} = \boxed{.34\ \text{cm}^{-1}}$$

fracture surface is clear, allowing orientations to be measured in terms of strike and dip. But where outcrops are so smooth and flat that only the straight-line traces of the fractures are evident in the bedrock surface, trends alone can be measured.

The time and tedium of the circle-inventory operation depends largely on the size of the sampling circle and the abundance of the joints and shear fractures that must be measured. Some initial trial-and-error planning may permit selection of an optimum circle diameter. When analyzing fractures in sedimentary and volcanic rocks, it is useful to select a circle radius whose magnitude is some function of layer thickness. In fact, setting the radius equal to layer thickness works out reasonably well in thin- to medium-bedded rocks. However, if the host rock for the fractures is very thick-bedded or massive, for example 15 to 150 m (50 to 500 ft) or more, it becomes necessary to adapt the circle-inventory method to a photogeologic approach, unless there is plenty of chalk, string, and time available. The photogeologic adaptation requires drawing an inventory circle of appropriate diameter onto an aerial photograph, or a transparent overlay of the photograph, then measuring the trends and trace lengths of fractures directly from the photograph. North arrow and scale provide the control. Subsequent field investigations can focus on measuring dip magnitudes of fracture sets and classifying the fractures according to their physical and kinematic characteristics.

Measuring Density of Joints and Shear Fractures

The abundance of joints and shear fractures at a given station is described through the evaluation of **fracture density**. Fracture density can be measured and described in a number of ways: average spacing of fractures; number of fractures in a given area; total cumulative length of fractures in a specified area; surface area of all fractures within a given volume of rock.

The measure of fracture density used in conjunction with the circle-inventory method is the summed length of all fractures within an inventory circle, divided by the area of the circle:

$$\rho_f = \frac{L}{\pi r^2}$$

where

ρ_f = fracture density

L = cumulative length of all fracture traces

R = radius of inventory circle

An example of the computation is shown in Table N.1. The fracture density is expressed in units of length/area (e.g., ft/ft^2, cm/cm^2, m/m^2, km/km^2). In practice, the values of fracture density are converted to the reciprocal form (ft^{-1}, cm^{-1}, m^{-1}, km^{-1}). Thus a fracture density of 0.57 ft/ft^2 would be expressed as 0.57 ft^{-1}. Comparative analysis of fracture density, however expressed, is especially useful in applied studies. Petroleum geologists Harris, Taylor, and Walper (1960) compared joint density to degree of curvature of the flanks of two oil-producing domes in Wyoming. They were interested in evaluating the relationship between fracture-induced permeability and degree of folding. As part of their study they found it necessary to normalize the natural variations in joint density that are due to differences in rock type and bedding thickness.

Haynes and Titley (1980) compared quantitative differences in fracture density in veined, mineralized rocks of the Sierrita porphyry copper deposit south of Tucson. Their purpose was to explore for centers of intrusion and/or mineralization, using joint-density variations as a guide. Wheeler and Dickson (1980) sought to evaluate whether systematic changes in joint density can disclose the locations of known or hidden faults. They prepared contour maps showing variations in joint density in a part of the Central Appalachians, and compared these maps with the known fault distribution as revealed on geologic maps of the same region.

We saw in *Chapter 5, Joints*, that the most productive ways to evaluate density of jointing is to focus on joint spacing. Based on these data, a **fracture spacing ratio (FSR)** is assigned to each data set, where FSR is bed thickness divided by median joint spacing for these data from a single bed (Narr, 1991; Narr and Suppe, 1993; Gross and Eyal, 1999; Eyal and others, 1999). It is straightforward to construct histograms (frequency plots) of joint spacing for a given joint set within a given subarea and with respect to a given lithology and bed thickness (Rives et al., 1992). **Frequency distribution plots** for joint spacing do not all come out looking the same, and this gives insights into mechanics of origin. These approaches permit **fracture-spacing indices (FSI)** to be assigned to a given joint set in a given rock type at a given locality (or set of localities). There are plenty of examples of fracture-spacing methodologies in *Chapter 5, Joints*.

Recording the Data

It is important to keep a systematic record of the descriptive characteristics of joints that are examined at each sampling station. Notebook entries should include site location; rock type; orientation of bedding or foliation (if present); bed or layer thickness (if applicable); offset (if detectable) along joint or; geometry and orientation of plumose and rib structure; nature and locations of origins; orientation of joints and shear fractures; geometry of joint intersections and terminations; and measurements pertinent to computing fracture density. Block diagrams that schematically portray the array of joints and joint-related structures are also a good idea (Figure N.5).

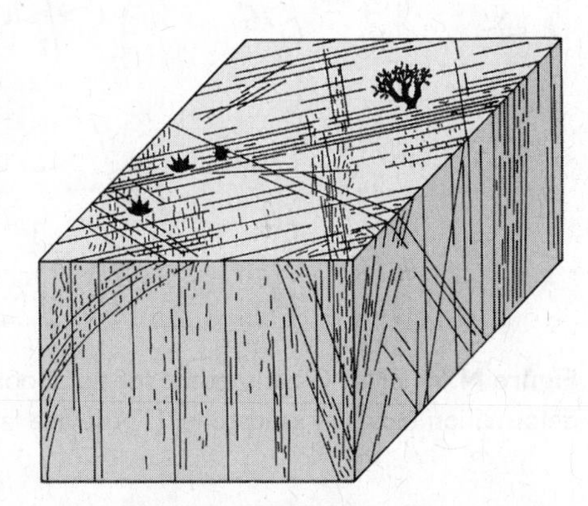

Figure N.5 Block diagram portrayal of the nature and orientation of dikes, joints, and joint-related structures in Laramide granitic plutons in southern Arizona. [From T. L. Heidrick and S. R. Titley, 1982. Published with permission from the University of Arizona Press.]

Nick Nickelson uses a method for recording fracture and fault data at a given station during the course of mapping, and during reconnaissance evaluation of jointing. He learned the approach from Ernst Cloos. Rather than recording the orientation data in columns, Nick fills a page of his notebook with strike-and-dip symbology, based on his measurements (Figure N.6). No attempt is made to portray the relative locations within the station outcrop where each reading was taken. However, the bearing of the strike of each feature, or the trend of slickenlines, is constructed accurately using a protractor. In this way, the drawing, at a glance, provides a visual portrayal of preferred orientations and systematic relations, if indeed they exist. The method takes just a little longer than lining up readings neatly in columns, but the results are *much* more conducive to clear thinking about the meaning of the structures. This method of course works for all minor structures in outcrops. As Nick points out: "*You will never record a structure improperly because when you complete your data collection, you orient your notebook north and compare your symbols with the outcrop, seeing if you have shown proper strike, dip, rake, or plunge directions*" (Nick Nickelson, personal communication, 1993).

Systematically recorded fracture data provide a basis for answering the typical questions that are raised in the structural analysis of joints and shear fractures. Are certain fractures associated with specific rock types? How does fracture density vary according to lithology and layer thickness? Does fracture

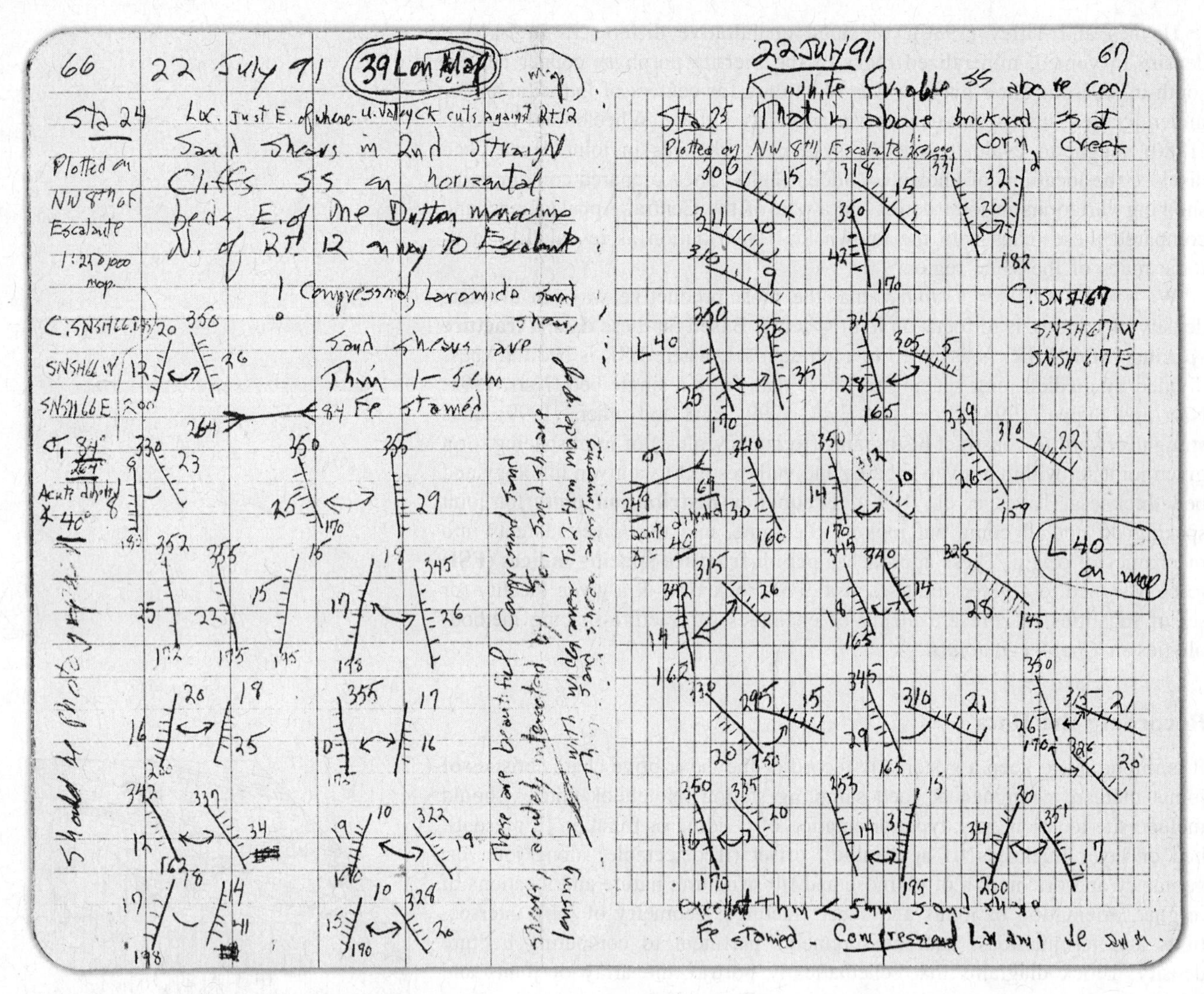

Figure N.6 Ernst Closs method for notebook-recording of structural data, including joint data. These data reflect orientations of deformation bands in sandstone. [These are facing pages in one of Nick Nickelson's field notebooks, and kindly provided.]

density and/or orientation change according to structural location? Are rocks of different ages characterized by different fracture patterns? Can favored directions of vein mineralization be recognized? The data that are collected and recorded, station by station, provide a playground for comparative analysis of a statistical nature.

Preparing Fracture-Orientation Diagrams

Orientation data collected during the course of fracture analysis may be summarized in **pole diagrams** and **pole-density diagrams**. Pole diagrams (and contoured pole-density diagrams) are three-dimensional stereographic displays of strike-and-dip data (Figure N.7*A*). Preferred orientations of fractures emerge from pole diagrams as dense clusterings of poles to joints. The orientation of the center of distribution represents the average orientation of the set.

Where three-dimensional control on the attitude of joints is not attainable, either due to the nature of the bedrock surfaces or because the fracture orientations are gathered from aerial photographs, it is appropriate to present the orientation data on **rose diagrams** or **strike histograms**, that is, on two-dimensional plots (Figure N.7*B*, *C*). In preparing rose diagrams and strike histograms, the trend and/or strike data are first organized into **class intervals**

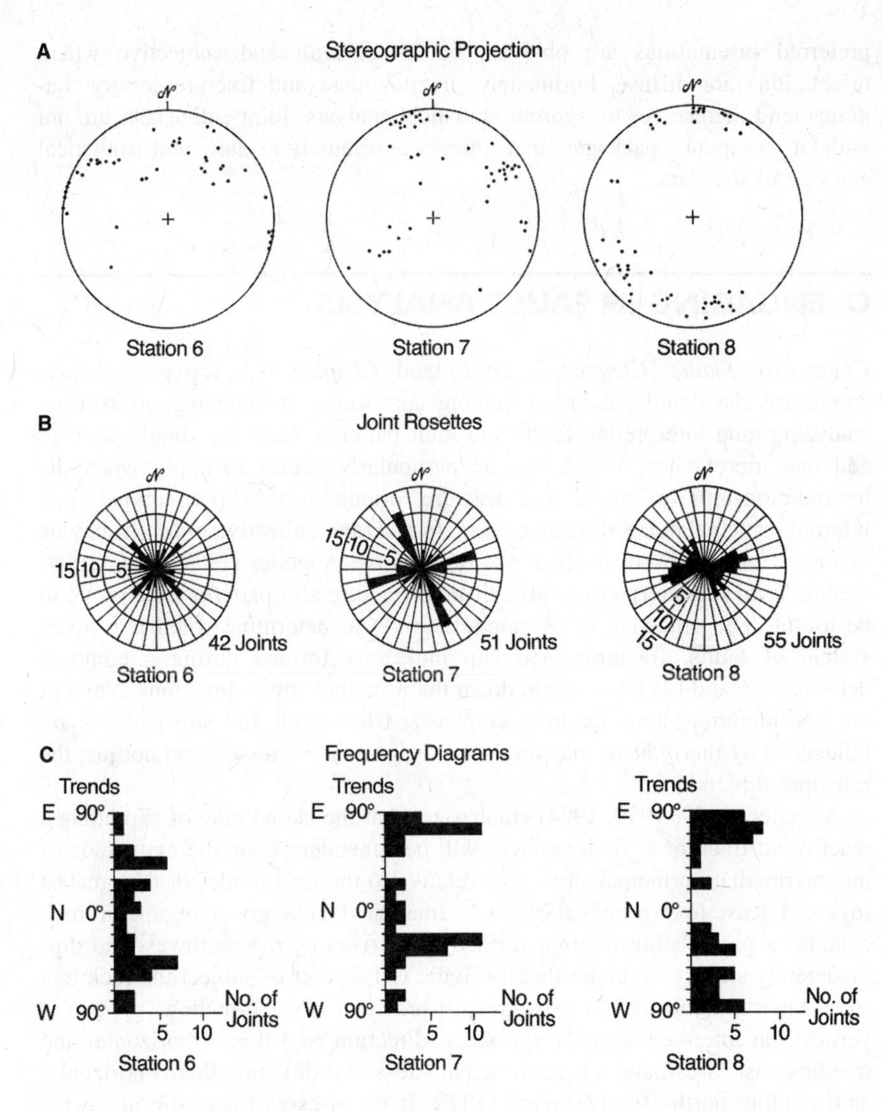

Figure N.7 Three kinds of plots for displaying the orientations of joints: the data reflect orientations of joints in Cretaceous sandstones in the Tucson Mountains, Arizona. (*A*) Stereographic pole diagrams. (*B*) Joint rosettes (rose diagrams). (*C*) Frequency diagrams. [Data collected and plotted by R. Chavez.]

of 5° or 10°, encompassing the orientation range from west through north to east. The number and percentage of readings that fall within each class interval are then tallied. Data thus arranged are plotted in one of two ways. For rose diagrams, class intervals are distinguished by rays subtending arcs of 5° or 10° that extend outward from a common point (Figure N.8). A family of concentric circles provides scaled control for the number (or percentage) of fracture-orientation readings that occupy each class interval. For strike histograms, class intervals are plotted along the y−axis of an $x-y$ plot; numbers (or percentages) of readings are plotted along the x−axis (Figure N.8). Trends of dominant fracture sets coincide with high-frequency peaks.

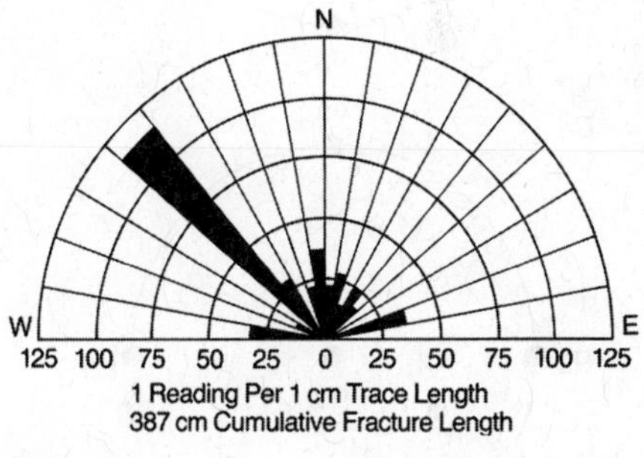

Figure N.8 Rose diagram of joint orientation data.

Rehrig and Heidrick (1972) effectively used rose diagrams in presenting the orientations of steeply dipping Late Cretaceous to Eocene dikes, veins, and elongate plutons in the southern Basin and Range of Arizona and New Mexico. Their pole-density diagrams display the orientations of 12,412 mineralized joints, veins, dikes, and faults in Late Cretaceous-Eocene plutons in the American Southwest. The data were subdivided according to three domains: productive plutons, ore-related plutons, and nonproductive plutons. The data in each diagram were contoured according to statistical probability of random distribution.

Preferred orientations of sets of joints or shear fractures are commonly estimated by eye from fracture-orientation diagrams. This is easy to do where preferred orientations are obvious, but is difficult and subjective where orientations are diffuse. Fortunately, fracture data and fracture-density diagrams lend themselves to rigorous statistical analysis. Joint enthusiasts are not without computer packages that integrate orientation data and statistical analysis of the data.

O. ENGAGING IN FAULT ANALYSIS

Chapter 6, Faults, Chapter 7, Folds, and *Chapter 11, Active Tectonics,* emphasize the details of broad ranging approaches to mapping, measuring, analyzing, and interpreting faults and fault patterns. Here we simply want to add one more piece, which can be particularly useful to apply when the interrelationships of fault and fracture orientations, slip-directions, and inferred principal stress directions seem to lack any collective order. It may be that all efforts to discover order will fail, but the Angelier (1979, 1990, 1994) method used for interpreting difficult fault/fracture slip patterns may prove to be useful. The objective of the method is (1) to determine whether a given system of faults, fractures, and slip indicators formed during a common deformation, and (2) if so, to pin down the principal stress directions. None of our considerations thus far have emphasized that fault and slip patterns are influenced by the *relative magnitudes* of the principal stresses, and not just the principal directions.

Angelier (1979, 1990, 1994) emphasizes that the actual rake of slip along a reactivated fracture or fault surfaces will be dependent upon the magnitude of the intermediate principal stress (σ_2) relative to the magnitudes of the greatest (σ_1) and least (σ_3) principal stresses. Imagine that a given region of rock contains a pre-existing fracture surface that strikes north–northwest and dips moderately steeply to the northeast (Figure O.1*A*). Let us subject the rock to a stress environment marked by a greatest principal stress direction (σ_1) that is vertical, an intermediate principal stress direction (σ_2) that is horizontal and trending east–west, and a least principal stress (σ_3) direction that is horizontal and trending north–south (Figure O.1*B*). If the stresses cause slip along the pre-existing fracture surface, the actual slip direction will depend upon the magnitude of the intermediate stress (σ_2) relative to the magnitudes of σ_1 or σ_3. Angelier (1979) demonstrated that the slip direction will be parallel to the direction of greatest shear stress resolved along the pre-existing fracture surface.

How do we predict the direction of slip? Following Angelier, if the magnitude of the intermediate principal stress (σ_2) is the same as the magnitude of

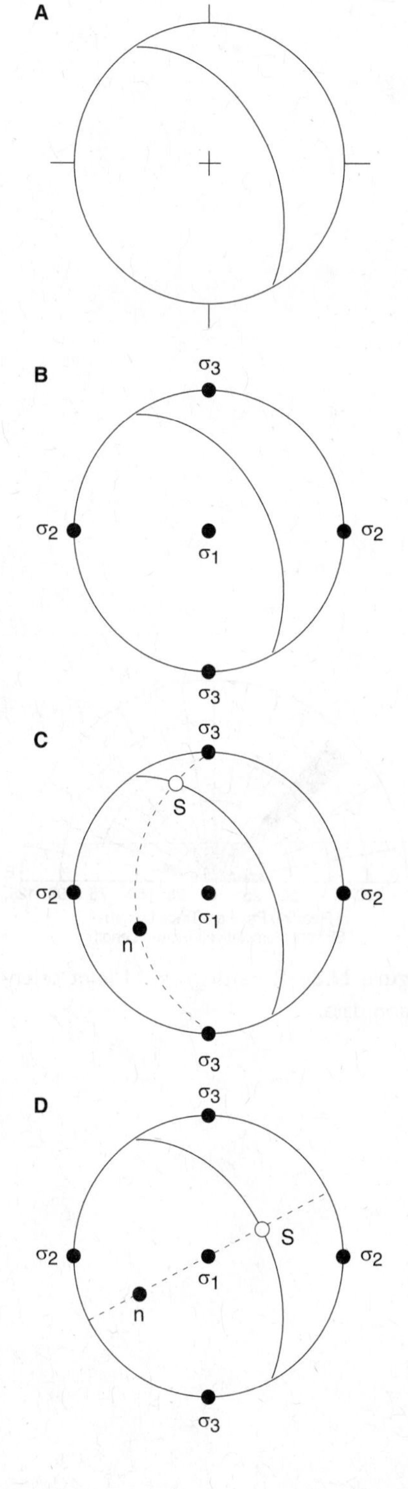

Figure O.1 Lower-hemisphere stereographic projections that illustrate the heart of the Angelier method. (*A*) A fracture surface strikes north–northwest and dips moderately steeply to the northeast. (*B*) The rock body within which the preexisting fracture surface is found is subjected to stress such that greatest principal stress (σ_1) is vertical; intermediate principal stress (σ_2) is horizontal, east–west, and least principal stress (σ_3) is horizontal, north–south. (*C*) Slip direction (S) on preexisting fracture surface, assuming that the magnitudes of σ_2 and σ_1 have the same values. Note that n is the normal to fracture surface). (*D*) Slip direction (S) on preexisting fracture surface, assuming that the magnitudes of σ_2 and σ_2 are equal to one another. [Reprinted from Angelier, J., Fault slip analysis and palaeostress reconstruction, *in* Hancock, P. L. (ed.), Continental deformation, Figure 4.29, p. 67. Published with permission of Pergamon Press, Ltd., Oxford, © 1994.]

the greatest principal stress (σ_1), then the slip direction is determined stereographically in two steps: (1) by finding the great circle that contains (σ_3) and the pole (n) to the fracture surface, and (2) by finding the intersection of this great circle with the great circle representing the orientation of the fracture surface (Figure O.1*C*). This intersection will coincide with the direction of slip on the reactivated fracture surface. If, on the other hand, the magnitude of the intermediate principal stress (σ_2) is the same as the magnitude of the least principal stress (σ_3), then the slip direction is determined stereographically in a slightly different way: (1) by finding the great circle that contains σ_1 and the pole (n) to the fracture surface, and (2) by finding the intersection of this great circle with that representing the orientation of the fracture surface (Figure O.1*D*)

Jacques Angelier "*inverted*" this approach to produce what is now known as the Angelier method. This is a method of reconstructing stress directions and stress ratios on the basis of slip directions disclosed by slickenlines on preexisting fracture surfaces in deformed rocks. Angelier has worked in regions throughout the world testing the validity of his method and interpreting paleostresses (e.g., Angelier et al., 1985; Angelier, 1990, 1994).

P. CARRYING OUT FOLD ANALYSIS

Direct Measurement of Axes and Axial Surfaces

It might be useful to review from start to finish the steps involved in measuring the orientations of axes and axial surfaces of folds (Figure P.1). Where folds are so small that they are completely exposed in single outcrops, the orientations of fold axes and axial surfaces are usually easy to visualize and can be measured directly with a compass. Conditions of measurement are ideal when the hinge of one or more of the folded layers has weathered out as a pencil-like or rodlike linear form. The hinge is taken to be a line whose orientation is parallel to the axis of the fold. Its orientation is measured and described in terms of trend and plunge and is entered in the field notebook under "*fold axis*" orientation. To help visualize the hinge line when it is not weathered out in full three-dimensional relief, it is useful to align a pencil parallel to the orientation of the hinge line. The pencil provides a tangible, physical guide in shooting the trend and plunge.

Measuring the orientation of the axial surface of a fold in outcrop requires a degree of physical dexterity. Since axial surfaces are geometric elements,

Figure P.1 Roadcut expression of minor fold associated with the Virgin anticline, Utah. Edna Patricia Rodriquez stands along the vertical western limb of the asymmetrical syncline. Axial trace inclined from upper right to lower left. [Photograph by G. H. Davis.]

there is no real physical surface on which to directly measure the axial surface orientation unless, by chance or by nature's design, there are cleavage surfaces or joint surfaces that are exactly parallel to the axial surface. In the absence of a natural physical surface on which to take a direct measurement, the strike-and-dip orientation is measured on a clipboard or field notebook held in one hand and oriented parallel to the axial surface. To be certain that the clipboard or notebook is aligned properly, the trace of the axial surface must be visible on at least two surfaces of the outcrop. Using as reference two (or more) axial traces common to a single axial surface, the clipboard (or notebook) can be aligned in the unique attitude that satisfies the trend of each. Axial surface orientation is measured and then posted in the field notebook in terms of strike and dip.

Map Determination of Axial Surface Orientation

There is a useful way to determine the orientation of the axial surface of a large fold, using a topographically controlled geologic map as a guide as well as some stereographic projection. This method does not depend on measuring the orientation of the bisecting surface as an approximation to the axial surface. The procedure is to plot stereographically the trend and plunge of two or more axial traces of the fold as points, and to fit these points (representing the orientations of lines) to a common great circle. Two lines define a plane. The great circle, constructed in this manner, describes the orientation of the plane.

An example of the use of this method is shown in Figure P.2. The data base is a simplified geologic map of an overturned anticline. As revealed by the

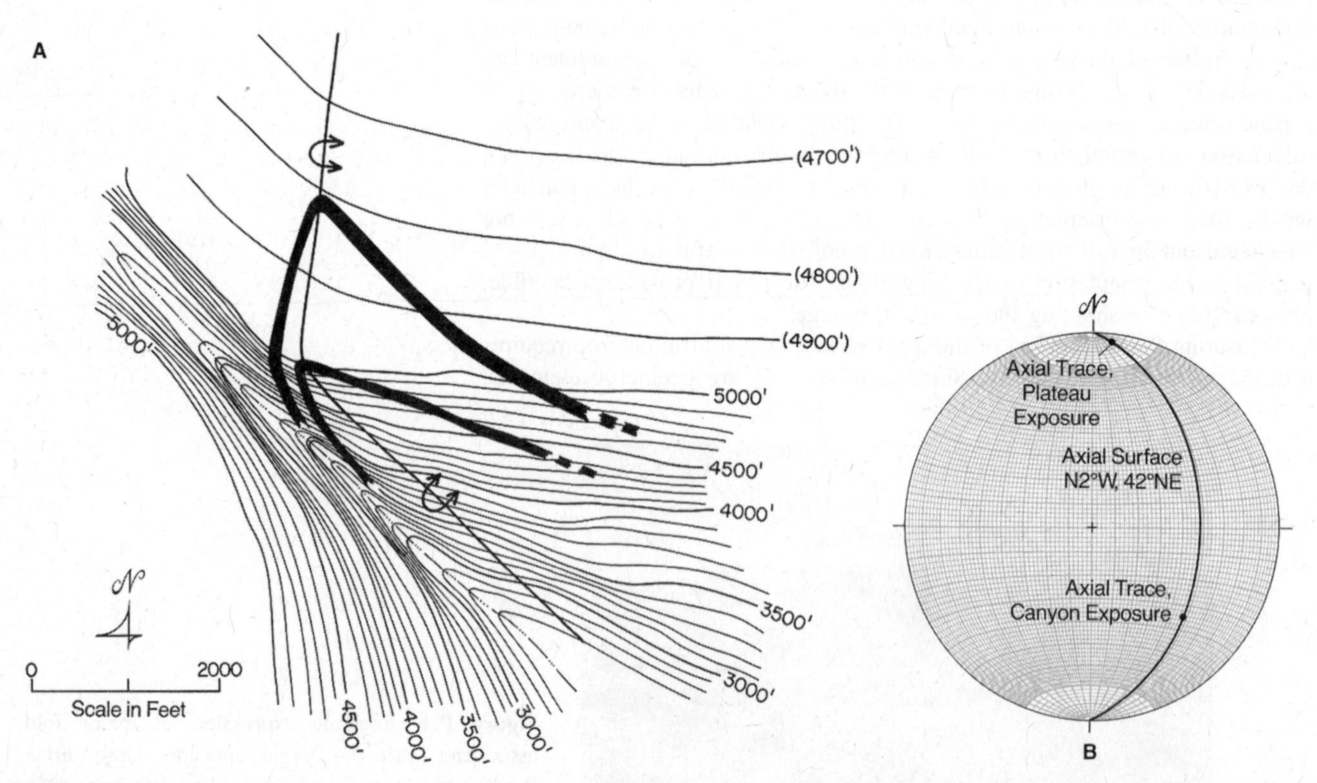

Figure P.2 Determination of the orientation of the axial surface of a fold on the basis of axial trace orientations. (*A*) Geologic map expression of folded beds (thick black bands). Two nearly straight axial traces can be identified, one crossing the high, flat tableland, the other exposed along the canyon wall. Orientations of these axial traces can be measured from the map relationships. (*B*) The axial surface orientation is determined stereographically by fitting the two axial trace orientations to a common great circle.

topographic contour lines, the fold is exposed in a funky plateau-like terrain cut by a steep-walled canyon (Figure P.2*A*). The topography affords excellent exposures of the axial trace of the anticline, both in map and cross-sectional views. On the plateau surface the axial trace of the fold is seen to trend N5°E along a reasonably straight line that passes, as it should, through hinge points of successively folded layers. The plunge of the axial trace, as calculated trigonometrically (or graphically) from the elevation control available, is 6°NE; that is, very shallow. The axial trace of the fold exposed to view in the northeast wall of the canyon trends S47°E. Its plunge is 32°SE (see Figure P.2*A*). The two axial trace orientations are lines, one plunging 6° N5°E, and the other plunging 32° S47°E. When these two lines are plotted stereographically and fitted to a common great circle, the overall attitude of the axial surface (axial plane) is found to be N2°W, 42°NE (Figure P.2*B*).

Plotting Fold-Orientation Data on Geologic Maps

Fold-orientation data, whether measured directly in the field or computed stereographically from representative strike-and-dip orientations, should become part of the geologic map record. How many of the orientation data are plotted depends largely on the size and scale of the final, rendered map. Orientation data that do not become part of the final geologic map are not lost. Rather, they constitute an integral part of stereographic displays of the preferred orientations of bedding, axes, and axial surfaces, displays that support the map relationships in a complementary way.

Formation contacts and marker beds that are plotted in the normal course of geologic mapping serve to outline the general form of map-scale folds. Strike-and-dip data representing the orientations of bedding and/or foliation help to disclose the fold configurations. The main fold-orientation information, however, is represented by symbols for axial trace, axis, and axial surface attitude. The axial trace of each fold is drawn not as a ruled straight line but a smooth, straight to curved line that passes through the hinge points of the folded formations and marker beds (Figure P.3). The trend and plunge of the fold axis and the strike and dip of the axial surface are plotted at one or more locations along each axial trace. Plotting axis and axial surface data at multiple locations along a given axial trace is necessary to describe the changes in fold orientation and configuration across the area of investigation.

The conventional symbols used in representing map-scale folds on geologic maps are listed in Table P.1 The symbols are used to identify whether a given fold is an anticline or syncline, or an antiform or synform. Overturned folds receive special attention. Like the standard symbols for anticlines and synclines, the symbols for overturned anticlines and synclines represent the dip directions of the limbs with small arrows. For overturned folds, these arrows point in a common direction, like the dip of the limbs of the overturned folds. The arrowheads point toward the U-shaped closure on the symbol of an overturned syncline. They point away from the U-shaped closure on the symbol for an overturned syncline.

Representative minor folds are plotted on geologic maps using special symbols (Table P.1). Each symbol portrays the strike and dip of the axial surface, the trend and plunge of the fold axis, and the generalized form of the folded surface as it is viewed down plunge. Symmetrical folds are shown as M-shaped forms. Asymmetrical folds are shown as Z-shaped and S-shaped forms. It is visually useful to orient the M's, S's, and Z's on the map in such a way that the axial trace of the minor fold bisects the form of the fold. If the scale of mapping permits, strike-and-dip orientations of the limbs of the minor folds should be posted as well.

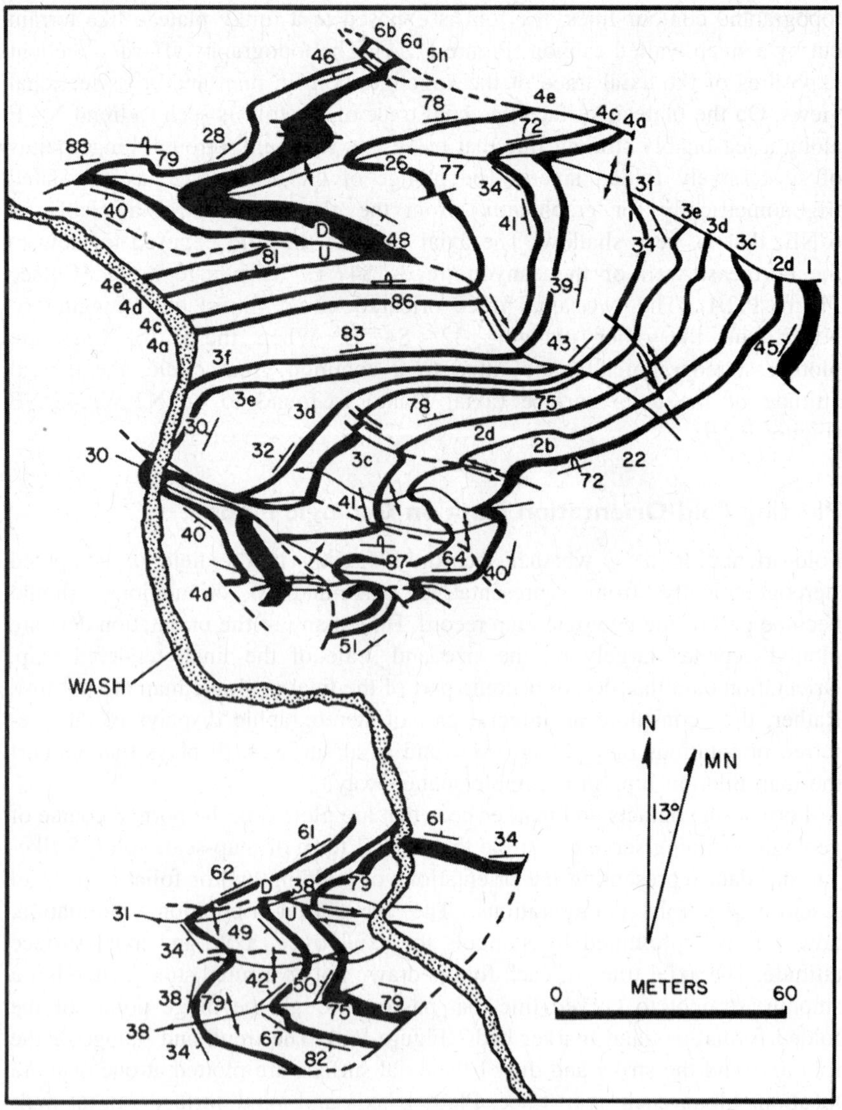

Figure P.3 Structural geologic map of folded strata at the Agua Verde locality near Tucson, Arizona. Alphanumeric symbols like "3f" serve to identify marker beds. [Map based on mapping by undergraduate student, D. Gossett.]

TABLE P.1
Geologic map symbology for folds and fold elements

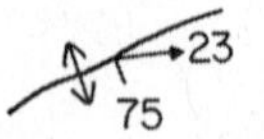 Axial Trace of Anticline, Showing Strike & Dip of Axial Surface, & Trend & Plunge of Axis
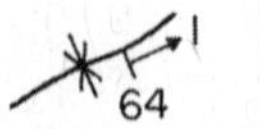 Axial Trace of Syncline, Showing Strike & Dip of Axial Surface, & Trend & Plunge of Axis
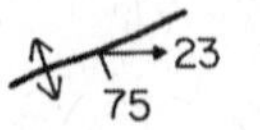 Major Anticline, Showing Strike & Dip of Axial Surface, & Trend & Plunge of Axis
 Major Syncline, Showing Strike & Dip of Axial Surface, & Trend & Plunge of Axis
 Overturned Anticline, Showing Strike & Dip of Axial Surface, & Trend & Plunge of Axis
 Overturned Syncline, Showing Strike & Dip of Axial Surface, & Trend & Plunge of Axis
 Dextral Minor Fold, Showing Strike & Dip of Axial Surface & Trend & Plunge of Axis
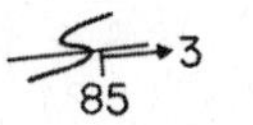 Sinistral Minor Fold, Showing Strike & Dip of Axial Surface & Trend & Plunge of Axis
 Symmetrical Minor Fold, Showing Strike & Dip of Axial Surface & Trend & Plunge of Axis

Suggestions for Describing Folds in Outcrop

Analysis of the orientations of shapes of folded surfaces demands care in systematically collecting and recording the pertinent data. Suppose we encounter a well-exposed fold in profile view in outcrop, of the type shown in Figure P.4*A*. What should we measure and describe before leaving the outcrop? Here are our suggestions. First, make a sketch of the fold in profile view, showing as clearly as possible the shape of one or two or more of the individual surfaces (such as bedding surfaces) that separate the folded layers (Figure P.4*B*). Label each end of the profile sketch with a direction indicator, such as NW (northwest), and (unlike the photo taken for Figure P.4*A*) place a simple bar scale below the sketch for approximate size reference. If fold analysis is a serious part of the ongoing study, photograph the fold, in normal profile view if possible, making sure that there is some reference scale in the picture. The photograph will capture details of fold shape, fold size, and fold tightness, which can be studied later in the laboratory.

Next measure representative strike-and-dip attitudes of the folded layers. For a house-size fold, measure two or three readings on each limb, and three

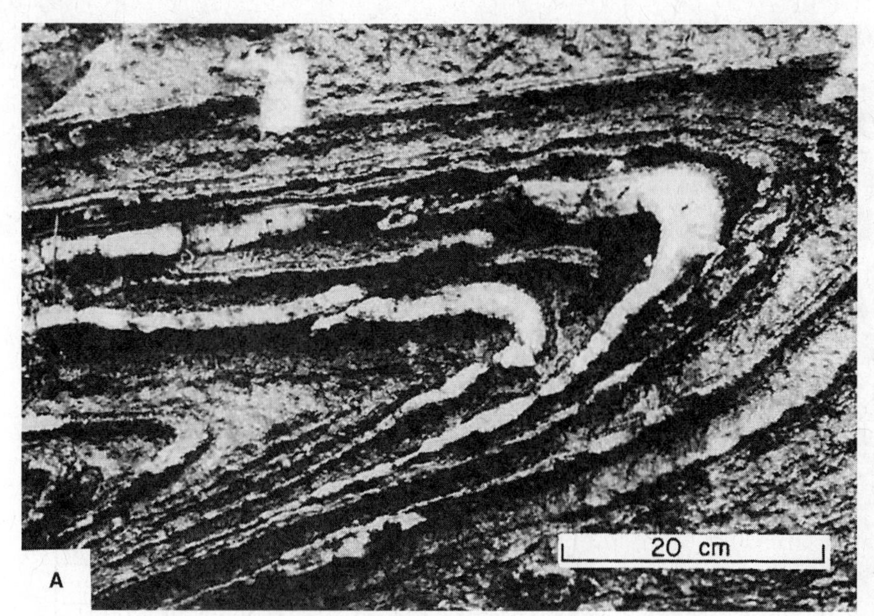

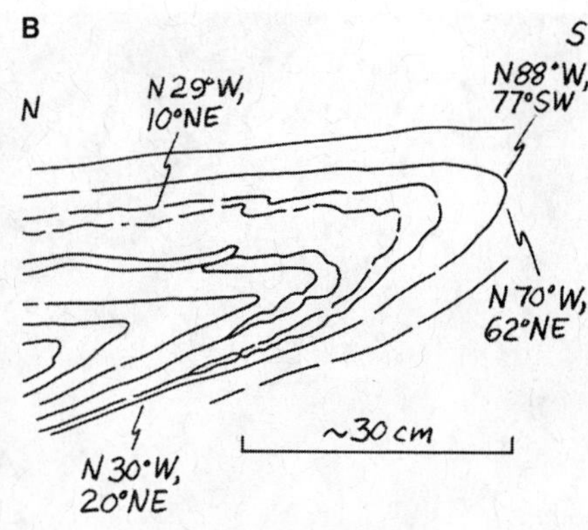

Figure P.4 (*A*) Profile view of fold in highly deformed sequence of marble and fine-grained quartzite. [Photograph by G. H. Davis.] (*B*) Notebook sketch of the fold, including orientation measurements.

or four more within the hinge zone. For desk-size folds, one reading on each limb and two readings in the hinge zone, if possible, will suffice. Record on the profile sketch of the fold the locations of each reading (see Figure P.4*B*). Finally measure the orientations of the axial surface and the hinge line of the fold, recording these in the field notebook along with the rest of the data.

With this information in hand, the fundamental field facts pertaining to shape, size, and orientation analysis are completed. The only remaining step before heading to the next outcrop is posting the orientation data on the geologic map.

Classifying Folds According to Layer Shape

Ramsay (1967) presented a sensitive approach to sorting out the shapes of folded layers according to classes. The basis for analysis is a geologic cross-section or a profile-view photograph (Figure P.5*A*). The method requires comparing layer thickness measured at a number of locations on the limbs of a fold with layer thickness measured in the hinge. The thickness that is measured is **true thickness**. Limb thickness (t_α) is identified according to **limb inclination** (α) as measured when the fold is positioned as a perfectly upright antiform (Figure P.5*B*). Limb thickness is taken to be the spacing between tangents drawn through control points of equal inclination value on the upper and lower surfaces of the folded layer (see Figure P.5*B*). The thicknesses thus measured are recorded and become the basis for constructing a graph allowing fold class to be precisely assigned.

The graph that reveals layer shape is a plot of **relative thickness** (t') versus limb inclination (α) (Figure P.5*C*) Relative thickness (t') is the ratio of thickness (t_α) measured at a location on the fold limb to thickness (t_o) measured in the hinge of the fold:

$$t' = \frac{t_o}{t_o},$$

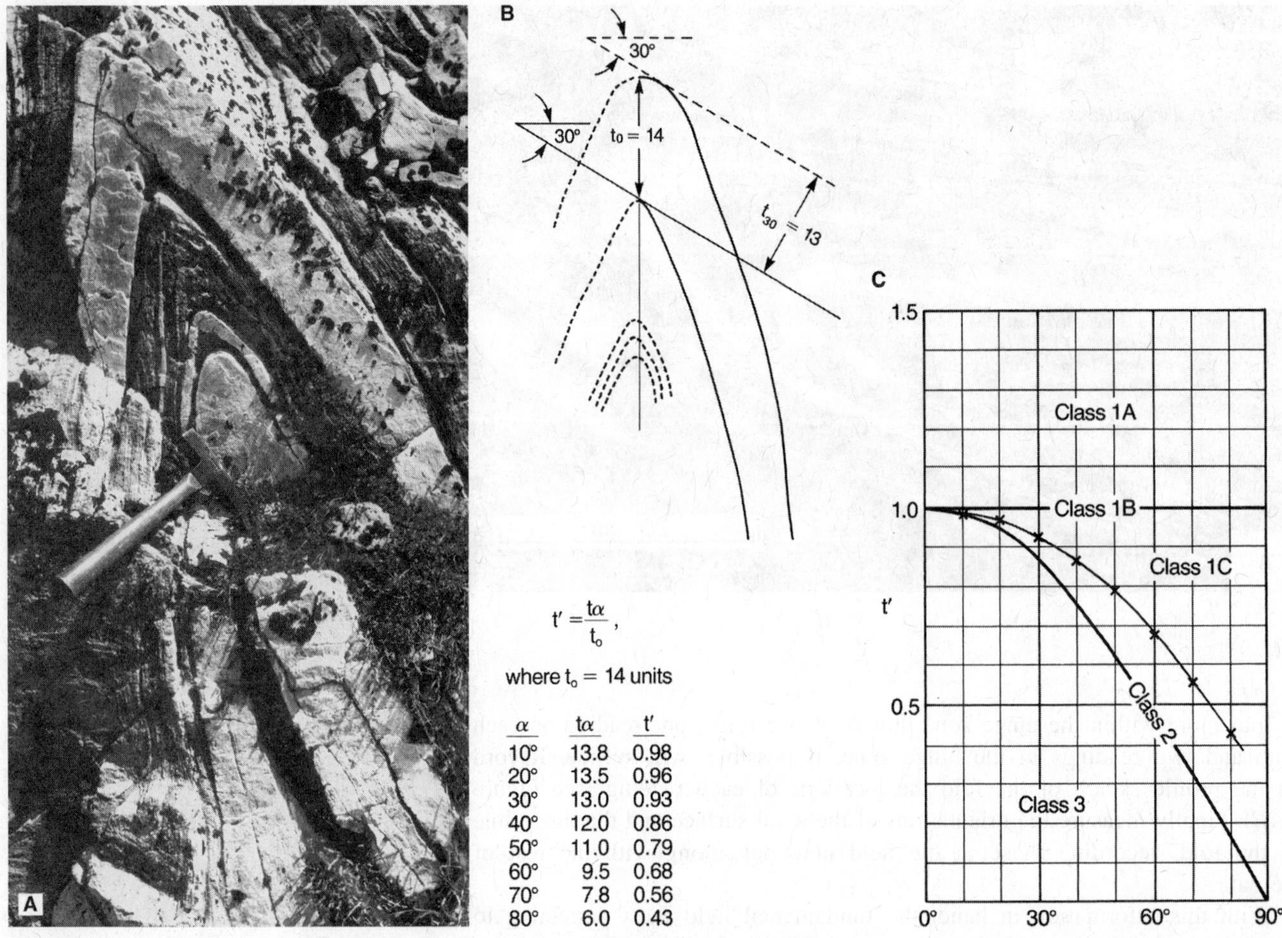

Figure P.5 (*A*) Normal profile view of a tightly folded sequence of layers. (*B*) Construction steps used in determining layer thickness at layer inclinations of 0° and 30°. (*C*) The standard graphical expression of the variations in layer thickness that characterize each of the major classes of folds. Note that the fold shown in *A* plots as a class 1C fold. [From Folding and Fracturing of Rocks by J. G. Ramsay. Published with permission of McGraw-Hill Book Company, New York, copyright © 1967.]

where t' = relative thickness

t_α = limb thickness

α = limb inclination

t_o = layer thickness in hinge

$\alpha = 0°$

Using their measured values of t_α and t_o, t' is calculated and recorded in the manner shown in Figure P.5*B*. Then each set of t' values is plotted to form the graphic construction shown in Figure P.5*C*. The plotted points for a given fold layer are connected by a smooth curve. The locus of the curve within the graph can then be compared with the subareas and lines occupied by the fundamental classes of folds. Graphs of this type show that the full range of layer shapes of folds is very expansive and that concentric folds (class IB) and similar folds (class 2) are indeed very special cases.

A Special Touch

As an elegant follow-up to this layer shape analysis, Richard Lisle (1992) shows us how to estimate strain in flattened buckle folds. His simple, direct method

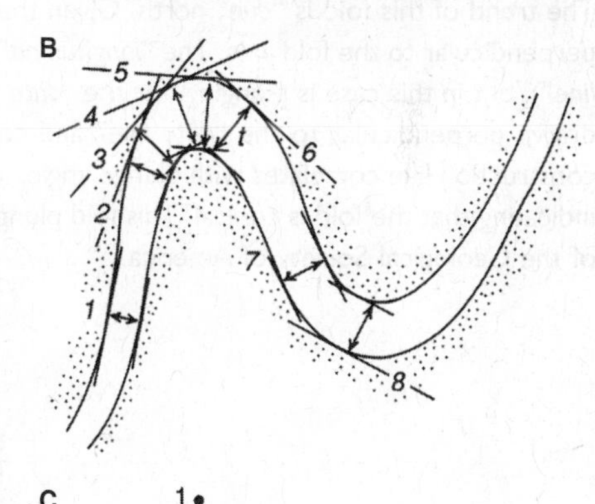

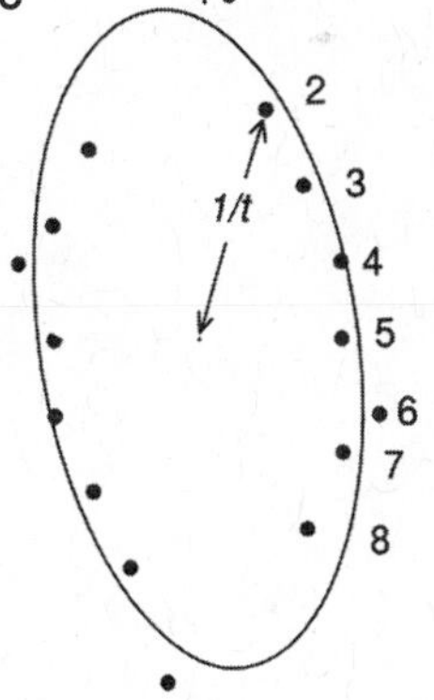

Figure P.6 (*A*) Photograph of folds in Precambrian banded gneiss. [Reprinted from *Journal of Structural Geology*, v. 14, Lisle, R. J., Strain estimation from flattened buckle folds, pp. 369–371, © 1992, with permission from Elsevier.] (*B*) Measurement of orthogonal thicknesses of folded layer. Note that measurements are made perpendicular to tangents 1–8. (*C*) Construction of strain ellipse reflecting flattening. Simply plot inverse thicknesses from a common point in the direction of the appropriate line of tangent. [Reprinted from *Journal of Structural Geology*, v. 13, Lisle, R. J., Strain estimation from flattened buckle folds, pp. 369–371, © 1992, with permission from Elsevier.]

assumes that a given class IC, II, or III fold had a class IB (parallel) shape before flattening. To determine the amount of flattening in a fold such as shown in Figure P.6*A*, measure the orthogonal thicknesses (t) perpendicular to tangents drawn to folded layering (Figure P.6*B*), and then plot inverse thicknesses (1/t) from a common point, each in the direction of the line of tangency (P.6*C*). A strain ellipse emerges that discloses strain due to flattening!

Q. DECIPHERING STRUCTURE IN BOREHOLES

Dipmeter Data and SCAT Plots

Structural geologists working in the petroleum industry have devised methods for extracting more information than might be ordinarily "*squeezed out* of" dip data collected (dipmeters and image logs) in well bores. We emphasized this point both in *Chapter 6, Faults*, and *Chapter 7, Folds.* We re-emphasize these methods here in the hopes they may become part of lab/field exercises, and in order to point out that these same methods can be used to interpret bedding-dip data collected along dip-direction traverses across *field areas*!

Dipmeter data can be displayed on wellbore columns as "*tadpoles*" (Figure Q.1). The head of each "*tadpole*" is located on the plot at the appropriate depth level and in a way that matches the appropriate dip magnitude. The "*tail*" of the tadpole points in the dip direction (see Figure Q.1). Just a glance at a tadpole plot gives us an immediate sense for the structural

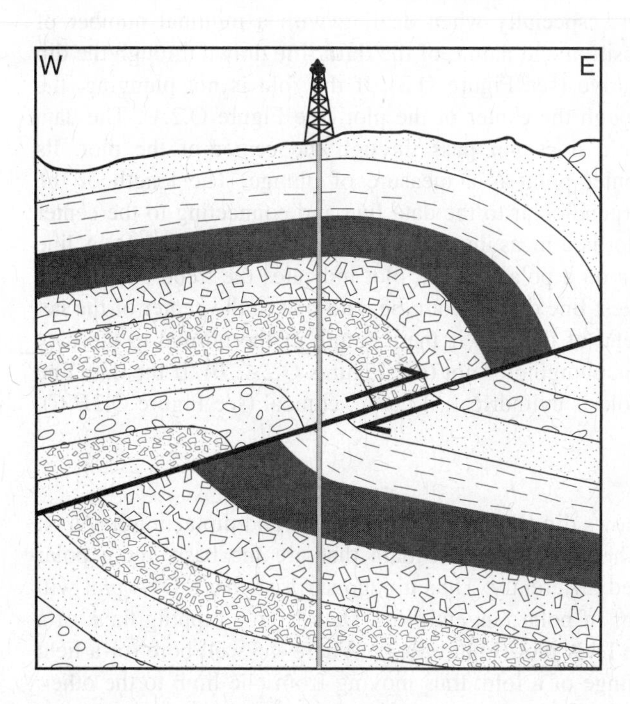

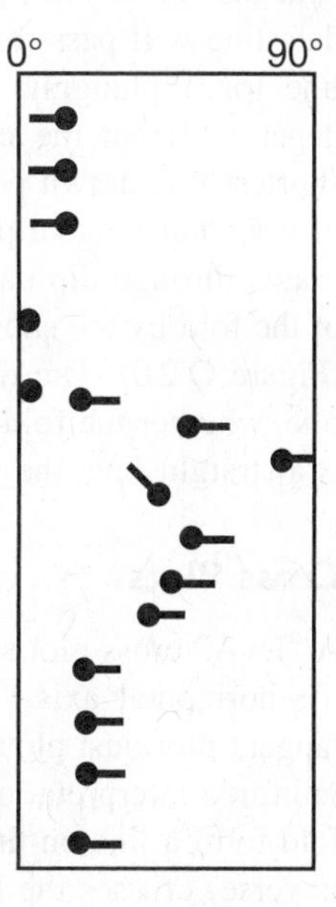

Figure Q.1 Schematic cross section across a thrust fault. The right hand plot shows the "*tadpoles*" for dip readings down the well. The "*head*" of the tadpole indicates the amount of dip of the beds, whereas the "*tail*" of the tadpole indicates the direction of dip. [Modified from Schlumberger (1970).]

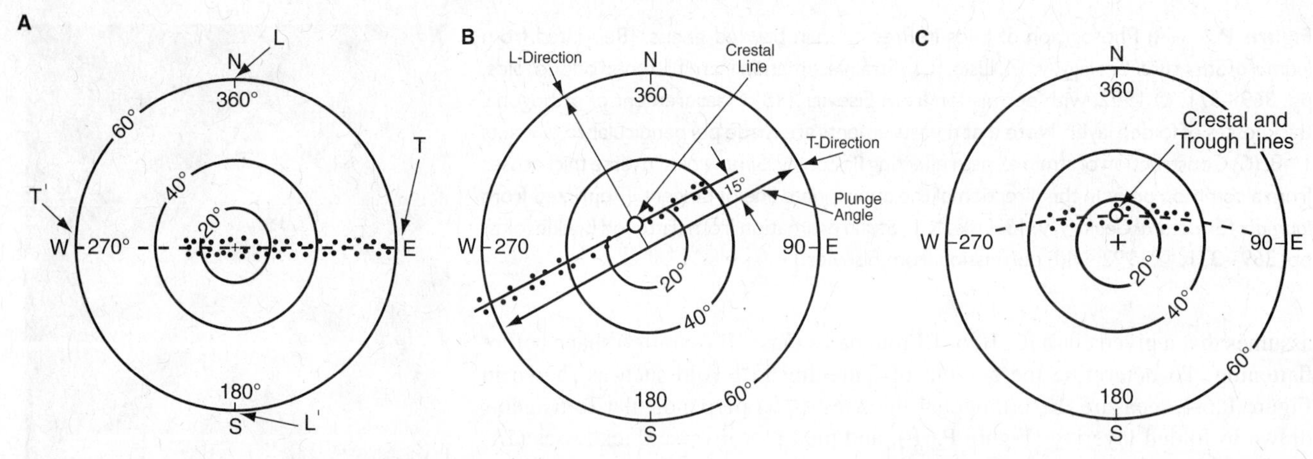

Figure Q.2 Polar tangent plot of dip data from three folds. Note that the "*data line*" (or "*mid line*") through the measurements serves as a guide for "eye-balling" the data, and interpreting. (*A*) The data points are aligned in a straight line, indicating that the fold is cylindrical. The "data line" passes through the center of the plot, indicating that the fold hinge is horizontal, i.e., a nonplunging fold. The trend of this fold is "*due*" north. Given the north trend, the "*transverse*" direction of this polar tangent plot is east–west, i.e., perpendicular to the fold axis. The "*longitudinal*" direction is north–south. (*B*) The fold plotted here is also cylindrical (straight "*data line*"), but in this case is plunging, for the "*data line*" is offset from the center of the net. Recall that the azimuth of the shortest line drawn perpendicular to the "*data line*" and connecting to the center of the plot is the trend of the fold axis. The length of this construction line correlates with plunge angle. This fold thus plunges 15° towards 330°. (*C*) In this example the "*data line*" is curved, indicating that the fold is conical. This fold plunges 15° north. [From Bengtson, C. A., 1980, *Geology*, v. 8. Published with permission of the Geological Society of America.]

geometry penetrated by the well. Yet more information and insight emerges from plotting a series of SCAT plots of the dip data, where "*SCAT*" refers to statistical curvature analysis techniques (Bengtson, 1981).

The first step in dip analysis using SCAT is to determine the trend and plunge of any folds crossed by the wellbore or the traverse. We do this using polar tangent plots of dip magnitude and dip direction azimuth (Figure Q.2). Polar tangent plots are even more efficient and effective than stereonets in defining fold attributes, especially when dealing with a minimal number of data points. It is the position and nature of the **data line** drawn through the dip data that is so informative (see Figure Q.2). If the fold is not plunging, the data line will pass through the center of the plot (see Figure Q.2*A*). The data line for a plunging fold does not pass through the center of the plot. Its departure from the center point is a measure of plunge. The length of the shortest line drawn perpendicular to the data line and connecting to the center of the polar tangent plot is a measure of the plunge (see Figure Q.2*A*). A line drawn through dip data on a polar tangent plot indicates the trend and plunge of the fold by the shortest line drawn from the data line to the center of the net (Figure Q.2*B*). The trend of the fold is the azimuth of the normal to the data line, whether the fold is plunging or not (see Figures Q.2*A*, *B*). If the data line is a straight line, the fold is cylindrical; if curve, conical (see Figure Q.2*B*,*C*).

Cross Plots

A "DvA" cross plot shows dip (D) on the vertical axis and dip azimuth (A) on the horizontal axis. These are the same data that are the basis for a polar tangent plot, just plotted differently. Thus it is no surprise that a DvA plot will reinforce interpretations of polar tangent plots. Data from a dipping flank of a fold form a line on the DvA plot (Figure Q.3). Where the well bore (or a field traverse) crosses the hinge of a fold, thus moving from one limb to the other,

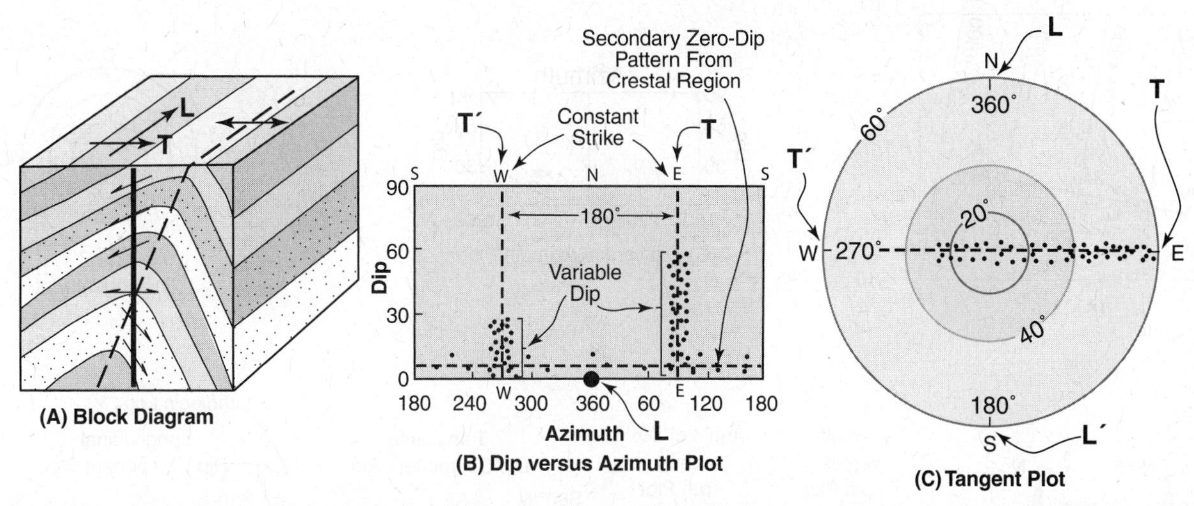

Figure Q.3 Cross plot data. This example is for a nonplunging cylindrical fold. (A) Block diagram portrayal of nonplunging cylindrical fold. A well bore is shown on the front face of the block. Dip arrows show the inclination of the beds encountered in the well during drilling. (B) The bedding dips produce straight "*data lines*" when plotted on a DvA plot. Dip inclinations on the west flank decrease downward toward the crest of the fold, to the point where the well crosses over to the east flank. From there the beds increase in dip from the crest downward into the core of the fold. The overall pattern is that of a parallel fold. (C) The overall "*data line*" is straight and passes through the center of the plot, thus revealing that the fold is cylindrical and nonplunging. [From Bengtson, C. A., 1981, Statistical curvature analysis techniques for structural interpretation of dipmeter data: *American Association of Petroleum Geologists* Bulletin, v. 65, no. 2, pp. 312–332. AAPG ©1981, reprinted by permission of the AAPG whose permission is required for further use.]

the DvA pattern becomes horseshoe-shaped (Figure Q.4). The trend and plunge of the axis of the fold will be the coordinates of the lowest point on the horseshoe, just like the lowest dip reading taken along of a traverse across a fold marks the trend and plunge of the hinge of the fold. The plunge is read on the vertical axis, the trend is read on the horizontal axis, again at the lowest

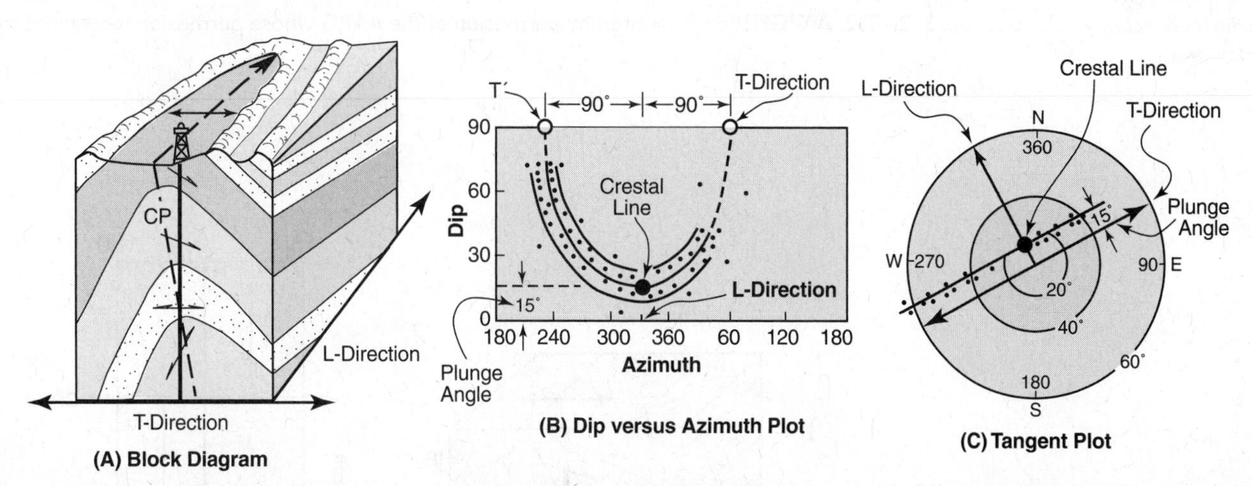

Figure Q.4 This is much like the fold presented in Figure Q.3, but in this case the fold plunges. (A) Block diagram representation of this plunging fold. Note dip arrows. (B) DvA plot showing shows "*horseshoe*" pattern, representing a continuous line of dip trends from the western limb of the fold to the eastern limb. The "*horseshoe*" is raised above the horizontal axis because in cases of plunging fold, there is no location where bedding is horizontal. The lowest point of the horseshoe corresponds to the trend and plunge of the fold axis, in this case 15° 330°. (C) In this polar tangent plot we see that the "*data line*" is straight (parallel fold) but offset from the center of the net. The line that connects the center with the data line indicates that the fold axis plunges 15° 330°. The DvA plot and the polar tangent plot thus agree! [From Bengtson, C. A., 1981, Statistical curvature analysis techniques for structural interpretation of dipmeter data: *American Association of Petroleum Geologists* Bulletin, v. 65, no. 2, pp. 312–332. AAPG © 1981, reprinted by permission of the AAPG whose permission is required for further use.]

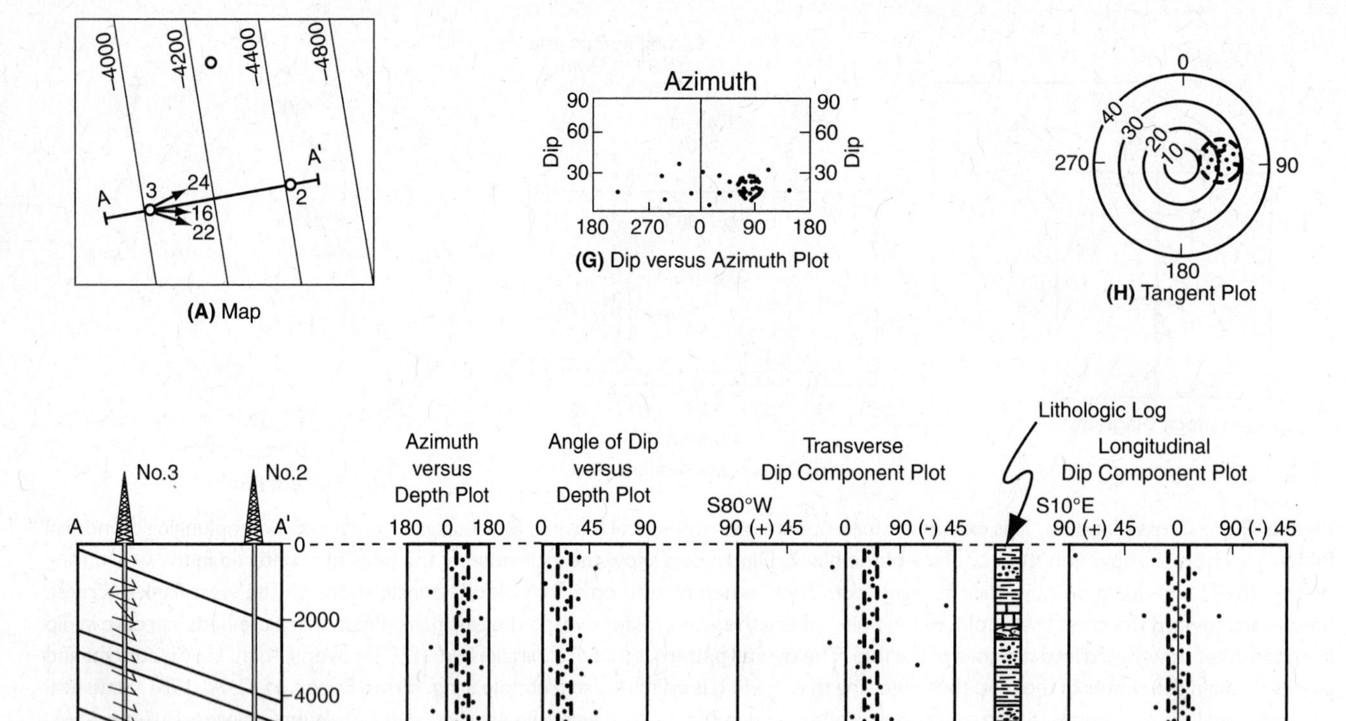

Figure Q.5 (*A*) Map and (*B*) section of homoclinal dipping beds showing wells and dip arrows. Following determination of trend and plunge of the fold, we examine the details of structural geometry using the following plots: (*C*) Azimuth vs. Depth, (*D*) Dip vs. Depth, (*E*) Transverse Dip vs. Depth, (*F*) Longitudinal Dip vs. Depth, (*G*) Dip vs. Azimuth, and (*H*) Tangent. [From Bengtson, C. A., 1981, Statistical curvature analysis techniques for structural interpretation of dipmeter data: *American Association of Petroleum Geologists Bulletin*, v. 65, no. 2, pp. 312–332. AAPG©1981, reprinted by permission of the AAPG whose permission is required for further use.]

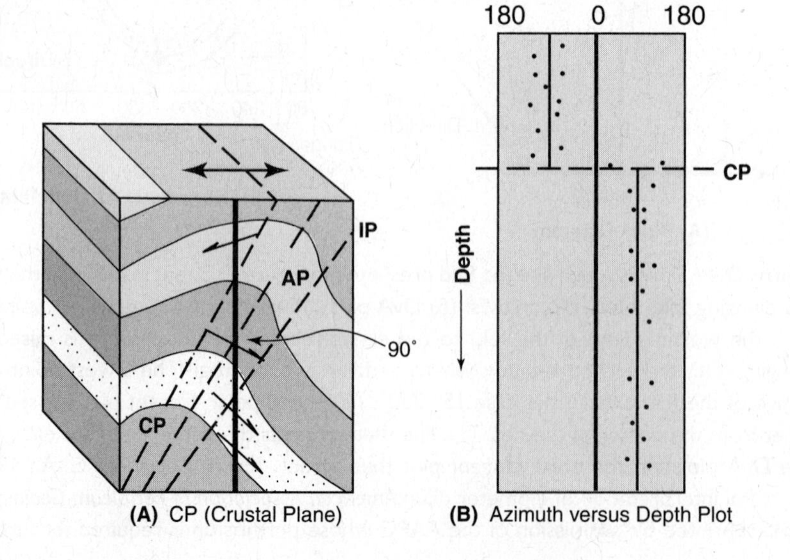

Figure Q.6 (*A*) Block diagram of a non-plunging fold. Well is shown on the front face of the block along with the axial plane (AP), crestal plane (CP), inflection plane (IP), and dip arrows in the well bore. (*B*) Azimuth vs. Depth plot for the nonplunging fold. Note on this plot that the azimuth of dip changes abruptly, indicating the well has passed from the west limb to the east limb.

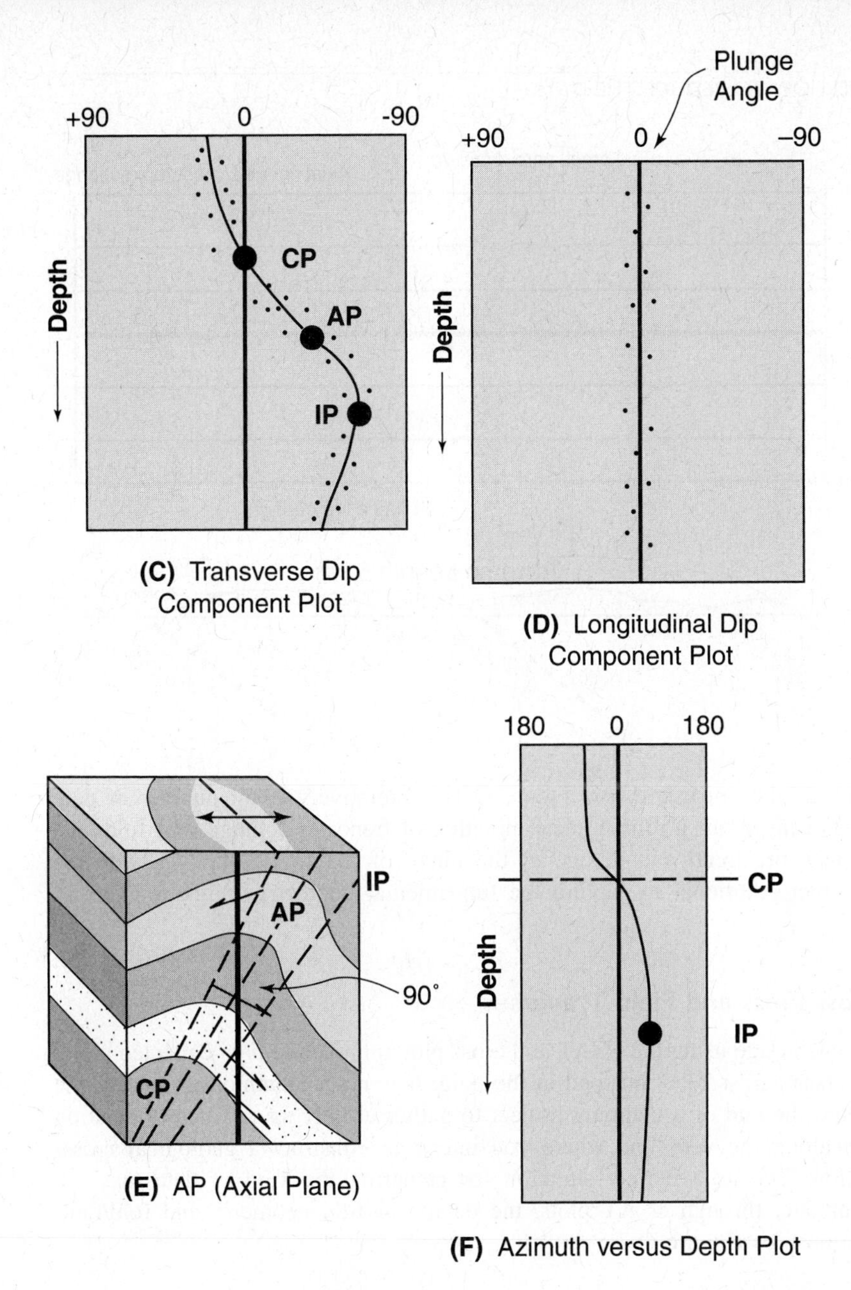

Figure Q.6 (Continued) (*C*) These are Transverse Dip Component vs. Depth Plot. Bedding dip gradually decreases to the zero crossing, which is the crestal plane (CP). The location of maximum rate of change of the transverse dip component marks the axial plane (AP). The maxima of the dip component shows where anticlinal curvature changes to synclinal curvature at the inflection plane (IP). (*D*) Longitudinal Dip Component vs. Depth Plot. The fold reveals "zero" dip because the plot displays dips in the plunge direction. (*E*) Block diagram for a plunging fold. (*F*) Azimuth vs. Depth Plot for the plunging fold. (*G*) Longitudinal Dip vs. Depth for the plunging fold. (*H*) Longitudinal Dip vs. Depth for the plunging fold. [From Bengtson, C. A., 1981, Statistical curvature analysis techniques for structural interpretation of dipmeter data: *American Association of Petroleum Geologists Bulletin*, v. 65, no. 2, pp. 312–332. AAPG © 1981, reprinted by permission of the AAPG whose permission is required for further use.]

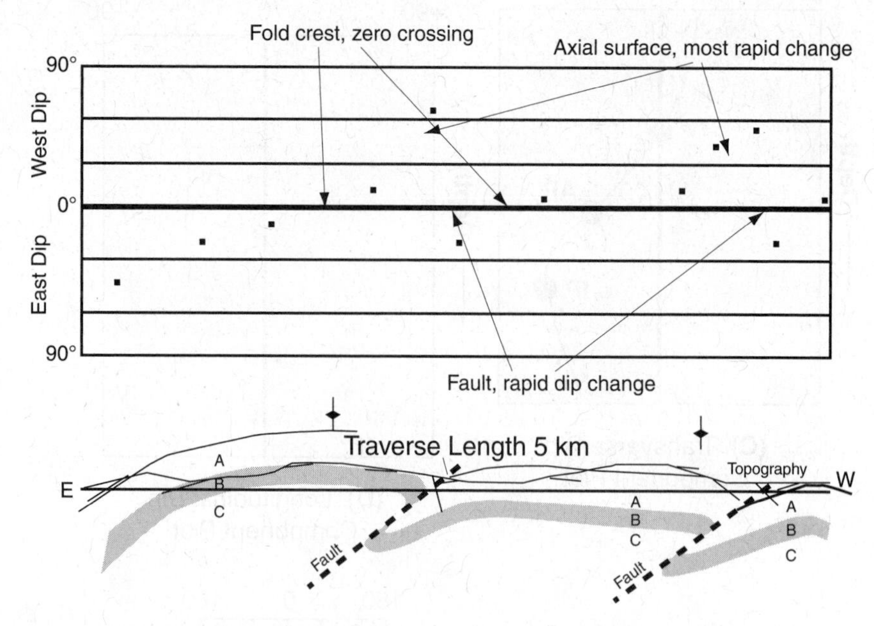

Figure Q.7 Plot and interpretation of Dip Traverse, and Apparent Dip Component vs. Traverse Length, illustrating usefulness of SCAT plots in interpreting field data. The analytical procedure is the same as illustrated for well data. The traverse data can be thought of as a well laying on its side.

point on the horseshoe (see Figure Q.4). A relatively small number of data points can permit a reliable determination of trend and plunge of a fold.

Once the trend and plunge of the hinge of the fold is known, we may construct additional to capture the fundamental geometry. (Figures Q.5 and Q.6).

Cross Plots and Field Traverses

We encourage using the SCAT and cross plot approach to data collected along traverses across folds mapped in the field. It is not uncommon to invest time toward the end of a mapping project to gather closely spaced bed-orientation data along the very line where you intend to construct a geological cross-section. This is a perfect situation for gathering the kind of data that can illuminate, through SCAT plots, the details of fold geometry and fold/fault relationships (Figure Q.7).

R. STUDYING SHEAR ZONES IN THE FIELD

When we encounter a shear zone in the field, there are numerous things to describe, measure, and think about. First we need to study and map the big picture relationships, and in the example we present here, the big picture is seen in a cliff face at Lower Tanque Verde Falls, Tucson, Arizona (Figure R.1). The view is to the southeast. Evident in the cliff face is a large shear band (see *Chapter 10, Shear Zones and Progressive Deformation*), in the form of a large brittle–ductile normal shear zone. Drag along the shear band (see Figure R.1) discloses the fact of extensional shearing, top to the southwest. Granitic rock has been transformed into mylonite (dark). Pegmatite and aplite layers (white) are involved in the deformation and display (at outcrop scale) mylonitic fabrics. It is at this scale of observation that we begin to describe the character of the shear zone, including its length, width, enclosed rock types, style of deformation, and amount of separation on any offset markers. This stage of work ultimately will require mapping for kilometers along strike. The big picture observations permit us to intelligently address the outcrop scale

Figure R.1 Southeast-directed view of cliff face at Lower Tanque Verde Falls, Tucson, Arizona. Banded mylonites derived from granite (dark) and pegmatite/aplite (white) are well exposed. The major structure is a brittle-ductile shear zone that is moderately inclined to the southwest west (right). Note the footwall drag of mylonite. [Photograph by G. H. Davis.]

fabrics. Realistically, fieldwork of this type is conducted at dual scales all along the way.

At the base of Lower Tanque Verde Falls we see that shear band structures flatten into parallelism with gently dipping mylonitic layers (Figure R.2), and that in the latest stages of deformation the shearing was brittle, in the form of low-angle detachment faulting (see Figure R.2). Positioned at this location, we ask the question whether the mylonites in the lower plate of the detachment fault show a sense-of-shear that is the same as that of the shear band; that is, top to the southwest. The mylonites are well exposed in Tanque Verde Wash (Figure R.3). To evaluate sense-of-shear within the mylonites we work at very close range, where, for example, the direction of shear is recorded in the

Figure R.2 Southwest–west directed photograph of low-angle detachment fault exposed near base of Lower Tanque Verde Falls. Undoubtedly there was progressive deformation in which ductile and semiductile shearing preceded the brittle phase. Slickenlines on underside of detachment fault surface, combined with drag, reveal S60W-directed slip, i.e., top to the southwest. Geologist is Bob Krantz, sitting on ultramylonite (black). [Photograph by G. H. Davis.]

Figure R.3 Outcrop expanse of mylonite. View is to NE. White porphyroclasts of feldspar are evident even from a distance. [Photograph by G. H. Davis.]

Figure R.4 Macrophotograph of mineral lineation in mylonite derived from granite protolith. Alignment of streaky lineation is ~N60°E/S60°W.NE is toward top of image. [Photograph by G. H. Davis.]

"*plane*" of mylonitic foliation (Figure R.4). It is at this outcrop scale where we measure and record the orientations of foliation and lineation. If foliation and lineation vary in orientation across the zone, multiple measurements (strikes and dips) will help us evaluate whether the variation is systematic across the shear zone. (In Figure R.1 we gain a glimpse of the variations in attitudes of foliations). We need to carefully describe what exactly is defining the foliation and lineation (quartz ribbons, compositional banding), so that we can evaluate their strain significance.

If we look at an appropriately oriented rock face, we may recognize features indicative of the sense of shear (*Chapter 9, Foliation and Lineation*; *Chapter 10, Shear Zones and Progress Deformation*). The proper rock face to examine for sense-of-shear indicators is a face cut parallel to the stretching lineation and perpendicular to the shear zone or foliation (Figure R.5). Such indicators can be very subtle and may require us to really put our nose to the outcrops and look carefully. Here we recognize S-C fabrics, which convey a top-to-the-southwest sense of shear (see Figure R.5). When we recognize sense-of-shear indicators, we describe them, recording sense of shear (dextral, or top to the right; sinistral, or top to the left; top to the west; northeast side up, etc.). As we draw sketches and take photographs, we note the direction in which we are viewing these features (e.g., from southeast to northwest), for a

Figure R.5 Northwest-directed photograph of SOS plane within mylonite derived from granite protolith. The white porphyroclasts are feldspars. The gray streaks are dominantly quartz. Note the spaced "*fracture-like*" surfaces that dip very gently to the right (northeast). These are the expressions of C-surfaces. Note also the tendency for the feldspar porphyroclasts to be slightly oblique to the C-surfaces, dipping more steeply to the right (northeast). Sense of shear is top-to-the-left (southwest). [Photograph by G. H. Davis.]

shear zone that is dextral when viewed one way will be sinistral when viewed from the opposite direction. We are particularly careful in properly measuring and recording angular relations between the feature, foliation, and other fabrics. These angular relationships will have strain significance.

What do we avoid? An easy trap to fall into is to walk up to an outcrop, see a single sense of shear indicator, record the sense of shear, and walk away without really looking at the entire outcrop. Instead, we take the time to see how many sense of shear indicators are present and whether they are consistent with one another. If shear indicators are lacking in a host rock that should ordinarily record strain, we record this observation. Also, when we see evidence for conflicting sense of shear, we don't just walk away, but record these observations as well.

S. DETERMINING FOCAL MECHANISMS FOR EARTHQUAKES

What Earthquake Seismologists and Structural Geologists Have in Common

Following major earthquake events the public does not hear very much about "*focal mechanisms.*" The media emphasizes earthquake magnitude. Focal mechanisms, which we will soon learn to be proxies for fault type, do not rise to media attention, unless a big thrust event in an oceanic setting triggers a tsunami; or a blind thrust rocks Los Angeles, in which case *Today Show* commentators have been known to use the expression "*blind thrust.*" Yet, earthquake seismologists and structural geologists are on the same page in regard to the importance of paying attention to earthquake magnitude.

Consider this: moment magnitude (e.g., $M_w = 7.1$) is based on multiplying (1) the area of the fault surface that accommodated slip during the earthquake; (2) the length of the maximum displacement vector; and (3) the rigidity modulus for crust broken by faulting. Moment magnitude is a measure of energy released. A one step rise in moment magnitude (e.g., from $M_w = 7.1$ to $M_w = 8.1$) represents a 32× increase in amount of energy released! Field-oriented structural geologists, in their own way, evaluate virtual moment magnitude as they (1) map the full trace length of a fault no longer active, (2) infer the approximate fault surface area of which faulting took place, (3) determine the maximum amount of slip along it, and (4) plug in a reasonable rigidity modulus. This is not truly a moment magnitude, for the large ancient faults we map over great distances are products of thousands of earthquake events. Perhaps we should call it cumulative moment magnitude, as we compare the sizes of faults within a system, and speculate on amounts of energy released. Regardless, the moment magnitude of earthquake seismologist is something structural geologists can relate to.

The emphasis here is on focal mechanism as a description of fault type. In deciphering fault kinematics for structures no longer tectonically active, structural geologists use facts regarding strike-and-dip of the fault surface, trend-and-plunge of the slip indicators, and sense-of-slip (sense of offset) in order to determine the type of fault: thrust-slip fault, reverse-slip fault, normal-slip fault, low-angle normal-slip fault, right-handed strike-slip fault, left-handed strike-slip fault, or oblique-slip (e.g., right-handed reverse-slip fault movement). This pursuit is exactly the same for earthquake seismologists, who interrogate the motions produced at the earthquake hypocenter (i.e., the earthquake source) in order to determine the **focal mechanism** for the earthquake. Seismologists refer to the orientation of the fault surface and the direction of slip in an earthquake as the focal mechanism. Establishing the focal mechanism thus requires determining the strike-and-dip of the fault surface, the trend-and-plunge of the slip direction, and sense-of-slip for the fault/earthquake event. Based on this information, earthquake seismologist can then determine fault type. The methods used by structural geologists and earthquake seismologist are different, but they pursue the same outcome. Both geological communities recognize that determining fault type is prerequisite to interpreting the stress field responsible for the faulting and the relationship of the stress field to tectonic movements.

First Motions, in 2D

First motion produced by an earthquake is the direction of ground motion as the P wave arrives at a seismic station. P-waves are sometimes informally called "push-pull" waves, because P waves shake the ground back and forth along the same line along which the wave is traveling. At any given seismic station, the first motion received will either be a "*push*" or a "*pull.*" A "*push*" moves the seismic station away from the earthquake source, whereas a "*pull*" draws the station toward the earthquake source. This is counter-intuitive, for we might think that a sudden release of energy at an earthquake source would produce a radial wave pattern such that all stations across the globe would be "*pushed.*" It turns out that earthquakes create a seismic radiation pattern that is described as quadrantal (Scholtz, 1990). "*Quantrantal*" signifies four lobes of seismic energy, two of which are compressional (C) (creating the "*push*"), two of which are tensional (T) (creating the "*pull*"). Figure S.1*A* portrays these lobes simply in two-dimensions, even though the full portrayal must be three-dimensional. The four lobes may be neatly divided by the traces of two **nodal planes** (np), one of which is the **fault plane** (F) itself, the other the **auxillary plane** (A), which is oriented perpendicular to the fault plane. The auxillary plane has no geological significance or expression. Directions of compression and tension are referred to as the P-axis and the T-axis, respectively; these bisect the compressional (C) and tensional (T) lobes (Figure S.1*B*).

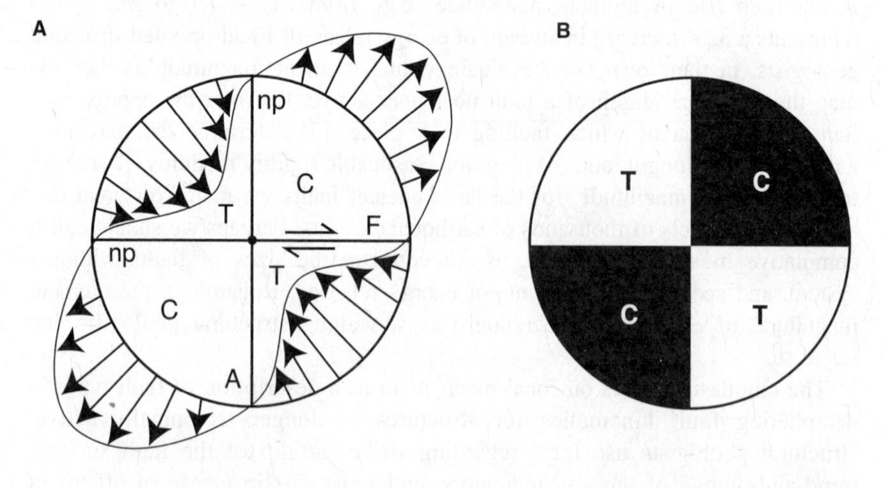

Figure S.1 (*A*) An earthquake radiates seismic energy in a way that is "*quadrantal,*" with an outward "*push*" received in compressional lobes (C) and inward "*pull*" in tensional lobes (T). The "*push*" and "*pull*" movements in the left-hand diagram are exaggerated for effect. Boundaries between the compressional and tensional lobes are the nodal planes (np). One of the nodal planes is the fault plane (F), and the other is the auxillary plane (A). (*B*) Simple stereographic portrayal of tensional (T) and compressional (C) lobes. [Source is http://en.wikipedia.org/wiki/File:Focal_mechanism_01.jpg.]

First Motions, in 3D

Let us now step back and imagine seismic stations distributed all across the globe. A large earthquake releases energy at its source. Seismic waves and first motions ("*push*" versus "*pull*") are received at each of countless stations. When the "*push*" and "*pull*" signatures are plotted globally, the locations of the nodal planes become obvious because the "*push*" and "*pull*" signatures

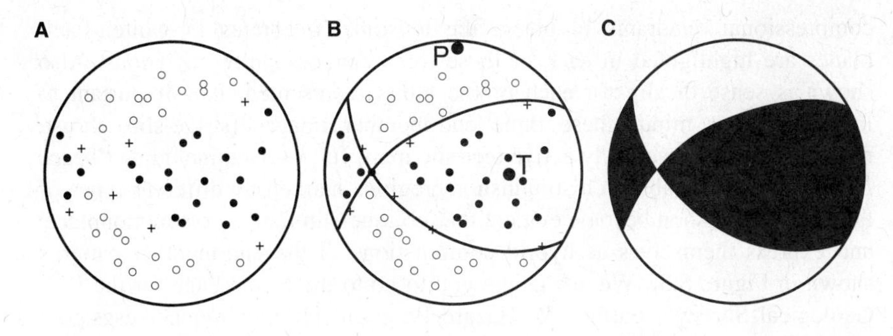

Figure S.2 Stereographic projections representing first motions and focal mechanisms. (*A*) Global (stereographic) projection showing first motions (black = push, open circles = pull) received at broadly distributed seismic stations. The distribution of first motions is "*quantrantal*." (*B*) The boundaries between quadrants are the nodal planes. The P-axis bisects the compressional lobe; the T-axis bisects the tensional lobe. (*C*) Simple stereographic portrayal of focal mechanism. Without additional information, it is impossible to judge which of the nodal planes represents the actual fault, and which nodal plane is "*auxillary*." In either case we know that the faulting will be reverse oblique slip: note the orientation of the P-axis in relation to the nodal planes. [Source is http://en.wikipedia.org/wiki/File:Focal_mechanism_02.jpg.]

organize themselves into quadrants (Figure S.2*A*). The 3D convention is to plot stereographically the nodal planes (Figure S.2*B*), which separate the compressional versus tensional quadrants, bisectors of which are the P- and T-axes, respectively. Conventionally we represent the compressional (push) lobes in black, and the dilational (pull) lobes in white (see Figure S.2*C*)

One of the nodal planes becomes identified as the fault plane (see Figure S.1*A*). How so? If there is ground rupture, it becomes straightforward to distinguish the fault plane orientation from the auxillary plane orientation, because the orientation of the fault "*plane*" can be measured on the ground. Otherwise, distinguishing the fault plane from the auxillary plane is achieved by comparing each nodal plane orientations with local mapped fault patterns. Furthermore, 3D patterns of earthquake aftershocks will normally concentrate along the fault plane, and not the auxillary plane.

Once the nodal planes have been determined, it is tempting to consider the P- and T- axes to correspond to the greatest and least compressive stress directions. But as Scholz (1990) warns, the P- and T-axes are oriented *45°* to the fault plane, which is not at all consistent with Coulomb failure. Keep in mind that earthquakes are typically triggered along pre-existing faults, and thus the stress conditions/directions that produced the original Coulomb fault break may be quite different from the stress conditions/directions that caused slip reactivation.

Focal Mechanisms

This brings us to representation of focal mechanisms as "**beach ball**" **diagrams**, and grasping the relationships between the beach balls and fault types (Figure S.3). The stereographic projections (lower hemisphere) again portray

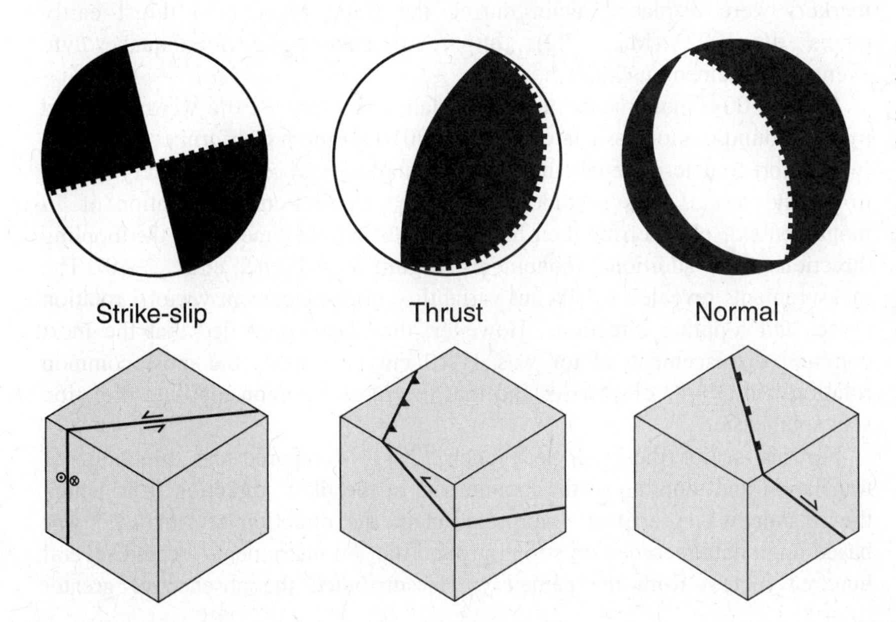

Figure S.3 These "beach balls" represent the three end-member types of faulting: strike-slip, thrust, and normal. Note that fault planes here are highlighted (dotted margin) in the "*beach ball*" focal mechanism diagrams. The focal mechanisms are stereographic projections, and thus we can "see" the correspondence between fault orientations shown stereographically and presented in the block diagrams. [Source is http://en.wikipedia.org/wiki/File:Focal_mechanism_03.jpg.]

compressional quadrants in black, the tensional quadrants in white. Fault planes are highlighted in each of these three stereographic projections. Also shown is sense of slip for each of the faults represented. It is important to implant in our minds these three end-member images (strike-slip, thrust, normal) so that as we study active tectonic maps full of data, including "*beach balls*," we will be able to distinguish subregions marked by different types of faulting. It will soon become evident that oblique-slip faults are commonplace, and express themselves as hybrid combinations of the end-member patterns shown in Figure S.3. We encourage you to go to the Home Page for the U.S. Geological Survey's Earthquake Hazard Program (http://earthquake.usgs.gov/earthquakes/), where focal mechanisms are an essential part of the subsection on "*Scientific and Technical Information.*"

Exploring Ground Evidence for First Motions

We are all *too* aware of the damage produced by seismic energy. Built structures collapse; lives are lost. There are widespread examples where the toppling and displacement of built structures is "*directional.*" An ancient example is seen today in the remains of the Temple of Zeus at Olympia (see chapter-opening photograph, *Chapter 2, Displacement and Strain*). Big earthquakes destroyed the site in 522 and 528 C.E. It is obvious that a number of temple columns at Olympia fell in the same exact direction, which presumably relates to the direction and nature of the seismic energy that first arrived.

Sue Cashman and her undergraduate structural geology students at Humboldt State University (Arcata, California) designed a structural field trip intended to explore directionality of earthquake-induced collapse of built structures. The impetus for their investigation was the M_w 6.5 offshore northern California earthquake of January 10, 2010. The earthquake was triggered by left-lateral strike-slip faulting on a northeast-trending fault within the subducting Gorda Plate (*http://earthquake.usgs.gov/earthquakes/eqinthenews/2010/nc71338066/*). Sue learned that the earthquake damaged grave markers and monuments in the Ferndale Cemetery, an active yet historic cemetery in Ferndale, California. Seismic damage was not new to the stewards of the Ferndale Cemetery, for all of the grave-site monuments were knocked down in the 1906 earthquake, as described by Ellin Beltz (*http://ebeltz.net/fieldtrips/1906quake-fdale.html*). Furthermore, many monuments and grave markers were displaced again during the Cape Mendocino thrust earthquake of 1992 ($M_w = 7.2$) (http://earthquake.usgs.gov/earthquakes/dyfi/events/nc/capemend/us/index.html).

Of the 300+ monuments in the Ferndale Cemetery, ~30% were displaced by the ground motion associated with the 2010 northern California earthquake. In an effort to determine whether the deformation field of the cemetery related to ground motion, Sue's students measured translation and rotation of 79 monuments displaced from their bases. In addition, they measured the toppling directions of 18 additional monuments (Figure S.4) (French et al., 2010). The measurements revealed significant variability in displacement vectors, rotation sense, and toppling directions. However, the class concluded that the most common displacement vector was 1–10 cm northwest; the most common rotation was 3°–8° clockwise; and that the most common toppling direction was southeast.

For his senior thesis, Kyle French (2011) compared the directions of translation and toppling of the monuments to the likely direction from which the seismic waves arrived. Interpretation of the direction of "*arrivals*" was based upon data recorded on strong ground motion instruments located several hundred meters from the cemetery. He attributed the absence of greater

Figure S.4 Photographs taken during the Sue-Cashman-led field trip to Ferndale Cemetery, California. (*A*) Amanda Admire and Lindsay Drosos measure the trend of fallen monument. (*B*) Brian Gerber and Janetta Kelley measure the trend of the displacement vector, which describes the shift of this translated, rotated monument. (*C*) Brian Gerber and Janetta Kelley take measurement related to the magnitude of the displacement vector describing the above-mentioned shift.

consistency in the deformation field to a certain amount of liquifaction, and the stabilizing effects of certain built features such as retaining walls and concrete pads. French noted that the toppling direction was more consistent than displacement vectors for displaced monuments that did not topple. He attributed the relative consistency of toppling to an instant of toppling experienced upon arrival of the first seismic waves. In contrast, French attributed the relative inconsistency in displacement vectors to reflect slippage back and forth and sideways during the full interval of ground motion (French, 2011).

We think that this example very underscores creative design of undergraduate structural geology class experiences, and the attractiveness of integrating structural geology, earthquake seismology, and active tectonics in ways that tap basic principles of displacement and strain analysis.

References

Allmendinger, R. W., 1986, Tectonic development, southeastern border of the Puna Plateau, northwest Argentine Andes: Geological Society of America Bulletin, v. 97, p. 1070–1082.

Allmendinger, R. W., 1998, Inverse and forward numerical modeling of trishear fault-propagation folds: Tectonics, v. 17, no. 4, p. 640–656.

Allmendinger, R. W., Zapata, T. R., Manceda, R., and Dzelalija, F., 2004, Trishear kinematic modeling of structures, with examples from the Neuquén Basin, Argentina, in McClay, K., ed., Thrust tectonics and hydrocarbon Systems: Tulsa, American Association of Petroleum Geologists, 356 p.

Alsop, G. I., and Holdsworth, R. E., 2006, Sheath folds as discriminators of bulk strain type: Journal of Structural Geology, v. 28, p. 1588–1606. Alvarez, W., Engelder, T., and Geiser, P. A., 1978, Classification of solution cleavage in pelagic limestones: Geology, v. 6, p. 263–266.

Alvarez, W., Engelder, T., and Lowrie, W., 1976, Formation of spaced cleavage and folds in brittle limestone by dissolution: Geology, v. 4, p. 698–701.

Anderson, E. M., 1951, The dynamics of faulting and dyke formation with applications to Britain: Oliver & Boyd, Edinburgh, 206p.

Anderson and Mehnert, 1979, Reinterpretation of the history of the Hurricane fault in Utah: Basin and Range Symposium, Rocky Mountain Association of Geologists, p. 124–165.

Anderson, J. G., Wesnousky, S. G., and Stirling, M. W., 1996, Earthquake size as a function of fault slip rate: Bulletin of the Seismological Society of America, v. 86, p. 683–690.

Anderson, R. E., 1971, Thin-skin distension in Tertiary rocks of southeastern Nevada: Geological Society of America Bulletin, v. 82, p. 43–58.

Anderson, R. E., 1973, Large magnitude Late Tertiary strike slip faults, north of Lake Mead, Nevada: United States Geological Survey Professional Paper 794, 18 p.

Anderson, R. N., and Noltimier, H. C., 1973, A model for the horst and graben structure of midocean ridge crests based upon spreading velocity and basalt delivery to the oceanic crust: Geophysical Journal of the Royal Astronomical Society, v. 34, p. 137–147.

Angelier, J., 1979, Determination of the mean principal directions of stresses for a given fault population: Tectonophysics, v. 56, p. T17–T26.

Angelier, J., 1994, Fault slip analysis and palaeostress reconstruction, *in* Hancock, P. L. (ed.), Continental deformation: Pergamon Press, Oxford, p. 53–100.

Argus, D. F., and Gordon, R. G., 2001, Present tectonic motion across the Coast Ranges and San Andreas fault system in central California: Geological Society of America Bulletin, v. 113, no. 12, p. 1580–1592.

Argus, D. F., Heflin, M. B., Donnellan, A., Webb, F. H., Dong, D., Hurst, K. J., Jefferson, D. C., Lyzenga, G. A., Watkins, M. M., and Zumberge, J. F., 1999, Shortening and thickening of metropolitan Los Angeles measured and inferred by using geodesy: Geology, v. 27, no. 8, p. 703–706.

Argus, D. F., Heflin, M. B., Peltzer, G., Crampe, F., and Webb, F. H., 2005, Interseismic strain accumulations and anthropogenic motion in metropolitan Los Angeles: Journal of Geophysical Research, v. 110, B04401, doi:10.1029/20033JB002934.

Armstrong, R. L., 1972, Low-angle (denudation) faults, hinterland of the Sevier orogenic belt, eastern Nevada and western Utah: Geological Society of America Bulletin, v. 83, p. 1729–1754.

Ashby, M. F., 1972, A first report on deformation–mechanism maps: Acta Metallurgica et Materialia, v. 20, p. 887–897.

Ashby, M. F., and Verrall, R. A., 1973, Diffusion accommodated flow and superplasticity: Acta Metallurgia et Materialia, v. 21, p. 149–163.

Atkinson, B. K., 1987, Introduction to fracture mechanics and its geophysical applications, *in* Atkinson, B. K. (ed.), Fracture Mechanics of Rock: Academic Press, London, p. 1–26.

Atkinson, B. K., and Meredith, P. G., 1987, The theory of subcritical crack growth with applications to minerals and rocks, *in* Fracture mechanics of rock, B. K. Atkinson (ed.): Academic Press Geology Series, London, p. 111–166.

Atwater, T. M., 1970, Implications of plate tectonics for the Cenozoic tectonic evolution of western North America: Geological Society of America Bulletin, v. 81, p. 3513–3536.

Atwater, T., and Stock, J., 1998, Pacific-North American plate tectonics of the Neogene southwestern United States: International Geology Review, v. 40, no. 5, p. 375–402.

Aydin, A., 1978, Small faults formed as deformation bands in sandstone: Pure and Applied Geophysics, v. 116, p. 913–930.

Aydin, A., and DeGraff, J. M., 1988, Evolution of polygonal fracture patterns in lava flows: Science, v. 239, p. 471–476.

Aydin, A., and Johnson, A. M., 1978, Development of faults as zones of deformation bands and as slip surfaces in sandstone: Pure and Applied Geophysics, v. 116, p. 931–942.

Aydin, A., and Johnson, A. M., 1983, Analysis of faulting in porous sandstone: Journal of Structural Geology, v. 5, p. 19–31.

Azor, A., Keller, E. A., and Yeats, R. S., 2002, Geomorphic indicators of active fold growth: South Mountain–Oak Ridge anticline, Ventura basin, southern California: Geological Society of America Bulletin, v. 114, no. 6, p. 745–753.

Babaie, H., April 14, 2010, Force, traction, and stress: *http://www.gsa.edu/~geohab/pages/geo/4013/lectures.htm.*

Bacon, S. N., and Pezzopane, S. K., 2007, A 25,000-year record of earthquakes on the Owens Valley fault near Lone Pine, California: implications for recurrence intervals, slip rates, and segmentation models: Geological Society of America Bulletin, v. 119, no. 7/8, p. 823–847.

Bahat, D., 1999, Single-layer burial joints vs. single-layer uplift joints in Eocene chalk from the Beer Sheva syncline in Israel: Journal of Structural Geology, v. 21, no. 3, pp. 293–303.

Bahat, D., and Engelder, T., 1984, Surface morphology on cross-fold joints of the Appalachian Plateau, New York and Pennsylvania: Tectonophysics, v. 104, p. 299–313.

Bahat, D., Bankwitz, P., and Bankwitz, E., 2003, Preuplift joints in granite: evidence for subcritical and postcritical fracture growth: Geological Society of America Bulletin, v. 115, no. 2, p. 148–165.

Bahat, D., Grossenbacher, K., and Karasaki, K., 1999, Mechanism of exfoliation joint formation in granitic rocks, Yosemite National Park: Journal of Structural Geology, v. 21, no. 1, pp. 85–96.

Bai, T., and Pollard, D. D., 2000, Fracture spacing in layered rocks: a new explanation based on the stress transition: Journal of Structural Geology, v. 22, p. 43–57.

Bailey, E. B., 1938, Eddies in mountain structure: Geological Society of London Quarterly Journal, v. 94, p. 607–625.

Balk, R., 1949, Structure of Grand Saline salt dome, Van Zandt County, Texas: American Association of Petroleum Geologists Bulletin, v. 33, p. 1791–1829.

Bally, A. W., Gordy, P. L., and Stewart, G. A., 1966, Structure, seismic data, and orogenic evolution of the southern Canadian Rocky Mountains: Bulletin of Canadian Petroleum Geology, v. 14, p. 337–381.

Banks, C. J., and Warburton, J., 1986, "Passive-roof" duplex geometry in the frontal structures of the Kirthar and Suaiman mountain belts, Pakistan: Journal of Structural Geology, v. 66, p. 1196–1230.

Barnett, J. A. M., Mortimer, J., Rippon, J. H., Walsh, J. L., and Watterson, J., 1987, Displacement geometry in the volume containing a single normal fault: American Association of Petroleum Geologists Bulletin, v. 71, p. 925–938.

Bastida, F., Aller, J., Toimil, N. C., Lisle, R. J., and Bobillo-Ares, N. C., 2007, Some considerations on the kinematics of chevron folds: Journal of Structural Geology, v. 29, no. 7, p. 1185–1200.

Baudon, C., and Cartwright, J., 2008, Early stage evolution of growth faults: 3D seismic insights from the Levant Basin, Eastern Mediterranean: Journal of Structural Geology, v. 30, p. 888–898.

Beach, A., 1980, Retrogressive metamorphic processes in shear zones with special reference to the Lewisian complex, *in* Carreras, J., Cobbold, P. R., Ramsay, J. G., and White, S. H. (eds.), Shear zones in rocks: Journal of Structural Geology, v. 2, p. 257–263.

Beanland, S., and Clark, M. M., 1982, The Owens Valley fault zone, eastern California, and surface faulting associated with the 1872 earthquake: U. S. Geological Survey Bulletin 1982, 29p.

Bebout, D. G., Loucks, R. G., and Gregory, A. R., 1978, Frio sandstone reservoirs in the deep subsurface along the Texas Gulf Coast: University of Texas Bureau of Economic Geology Report of Investigation 91, 93 p.

Behrmann, J. H., 1987, A precautionary note on shear bands as kinematic indicators: Journal of Structural Geology, v. 9, no. 5–6, p. 659–666.

Bell, T. H., 1985, Deformation partitioning and porphyroblast rotation in metamorphic rocks: a radical reinterpretation: Journal of metamorphic geology, v. 3, no. 2, p. 109–118.

Benedetti, L., Finkel, R., King, G., Armijo, R., Papanastassiou, D., Ryerson, F. J., Flerit, F., Farber, D., and Stavrakakis, G., 2003, Motion on the Kaparelli fault (Greece) prior to the 1981 earthquake sequence determined from 36Cl cosmogenic dating: Terra Nova, v. 15, p. 118–124.

Bengtson, C. A., 1980, Structural use of tangent diagrams: Geology, v. 8, p. 599–602.

Bengtson, C. A., 1981, Statistical curvature analysis techniques for structural interpretation of dipmeter data: American Association of Petroleum Geologists Bulletin, v. 65, no. 2, p. 312–332.

Bennett, R. A., Wernicke, B. P., Niemi, N. A, Friedrich, A. M., and Davis, J. L., 2003, Tectonics, v. 22, no. 2, doi:10.129/2001TC001255.

Bergbauer, S., and Pollard, D. D., 2004, A new conceptual fold-fracture model including prefolding joints, based on the Emigrant Gap anticline, Wyoming: Geological Society of America Bulletin, v. 116, no. 3/4, pp. 294–307.

Bernal, A., and Hardy, S., 2002, Syn-tectonic sedimentation associated with three-dimensional fault-bend fold structures: a numerical approach: Journal of Structural Geology, v. 24, p. 609–635.

Berthé, D., and Brun, J. P., 1980, Evolution of folds during progressive shear in the South Armorican shear zone, France: Journal of Structural Geology, v. 2, p. 127–133.

Berthé, D., Choukroune, P., and Jegouza, P., 1979, Orthogneiss, mylonite and noncoaxial deformation of granites: the example of the South Armorican shear zone: Journal of Structural Geology, v. 1, p. 31–42.

Beutner, E. C., 1978, Slaty cleavage and related strain in Martinsburg Slate, Delaware Water Gap, New Jersey: American Journal of Science, v. 278, p. 1–23.

Bevis, M., Schutz, B., Taylor, F. W., and Recy, J., 1993, Direct observations of convergence and back-arc spreading at the Tonga Arc (1988–1992): Geological Society of America Abstracts with Programs, v. 25, p. 242.

Bijlaard, P. P., 1946, On the elastic stability of thin plates supported by a continuous medium: Royal Dutch Academy of Science Proceedings, v. 49, p. 1189–1199.

Billings, M. P., 1972, Structural geology (3rd edition): Prentice-Hall, Englewood Cliffs, New Jersey, 606 p.

Biot, M. A., 1957, Folding instability of a layered viscoelastic medium under compression: Royal Society of London Proceedings, Series A, v. 242, p. 211–228.

Biot, M. A., 1959, On the instability of folding deformation of a layered viscoelastic medium under compression: Journal of Applied Mechanics, v. 26, p. 393–400.

Biot, M. A., Ode, H., and Roever, W. L., 1961, Experimental verification of the folding of stratified viscoelastic media: Geological Society of America Bulletin, v. 72, p. 1621–1630.

Bishop, R. S., 1977, Shale diapir emplacement in south Texas: Laward and Sherriff examples: Transactions of the Gulf Coast Association of Geological Societies, v. 27, p. 20–31.

Bjorn, L. I., 1970, Natural stress-values obtained in different parts of the Fennoscandian rock masses: Proceedings of the Second Congress of the International Society for Rock Mechanics, "Jaroslav Cerni" Institute for Development of Water Resources, Belgrade, Yugoslavia, p. 209–212.

Blackwood, J. R. (ed.), 1969, College talks: Oxford University Press, New York, 177 p.

Blakely, R. J., Wells, R. E., Weaver, C. S., Johnson, S. Y., 2002, Location, structure, and seismicity of the Seattle fault zone, Washington: Evidence from aeromagnetic anomalies, geologic mapping, and seismic-reflection data: Geological Society of America Bulletin, v. 114, no. 2, p. 169–177.

Boatwright, J., and Bundock, H., 2005, Modified Mercalli intensity maps for the 1906 San Francisco earthquake plotted in ShakeMap format: Open-file Report, U.S. Geological Survey (2005), OF 2005–1135.

Bobyarchick, A. R., 1986, The eigenvalues of steady flow in Mohr space: Tectonophysics, v. 122, p. 35–51.

Bohannon, R. G., Grow, J. A., Miller, J. J., and Blank, R. H., 1993, Seismic stratigraphy and tectonic development of Virgin River depression and associated basins, southeastern Nevada and northwestern Arizona: Geological Society of America Bulletin, v. 105, p. 501–520.

Bolton, T., 1989, Geologic maps; their solution and interpretation: Cambridge University Press, Cambridge, United Kingdom, 144 p.

Borradaile, G. J., and Jackson, M., 2004, Structural geology, petrofabrics and magnetic fabrics (AMS, AARM, AIRM): Journal of Structural Geology, v. 32, no. 10, p. 1519–1551.

Bott, M. H. P., 1982, The mechanism of continental splitting, *in* Hales, A. L. (ed.), Geodynamics final symposium: Tectonophysics, v. 81, p. 301–309.

Boyer, S. E., and Elliott, D., 1982, Thrust systems: American Association of Petroleum Geologists Bulletin, v. 66, p. 1196–1230.

Brace, W. F., Paulding, B. W., Jr., and Scholz, C., 1966, Dilatancy in the fracture of crystalline rocks, Journal of Geophysical Research, v. 71, p. 3939–3953.

Bradley, D., and Hanson, L., 1998, Paleoslope analysis of slump folds in the Devonian Flysch of Maine: Journal of Geology, v. 106, p. 305–318.

Brocher, T. M., Blakely, R. J., and Wells, R. E., 2004, Interpretation of the Seattle uplift, Washington, as a passive-roof duplex: Bulletin of the Seismological Society of America, v. 94, p. 1379–1401.

Bronowski, J., 1973, The ascent of man: Little, Brown, & Company, Boston, 448 p.

Bruhn, R. L., DuRoss, C. B., Harris, R. A., and Lund, W. R., 2005, Neotectonics and paleoseismology of the Wasatch fault, *in* Pederson, J., and Dehler, C. M., (eds.), Interior Western United States: Geological Society of America Field Guide 6, 20 p., doi: 10.1130/2005.fld006(11).

Brun, J.-P., Soukoutis, D., and Van Den Driessche, J., 1994, Analogue modeling of detachment fault systems and core complexes: Geology, v. 22, p. 319–322.

Bucher, W. H., 1944, The stereographic projection, a handy tool for the practical geologist: Journal of Geology, v. 52, p. 191–212.

Burchfiel, B. C., and Stewart, J. H., 1966, "Pull-apart" origin of the central segment of Death Valley, California: Geological Society of America Bulletin, v. 77, p. 439–441.

Burford, R. O., and Harsh, P. W., 1980, Slip on the San Andreas fault in central California from alinement array surveys: Bulletin of the Seismological Society of America, v. 70, p. 1233–1261.

Burkhard, M., 1993, Calcite twins, their geometry, appearance and significance as stress-strain markers and indicators of tectonic regime: a review, *in* Casey, M., Dietrich, D., Ford, M., Watkinson, J., and Hudleston, P. J. (eds.), The geometry of naturally deformed rocks: Journal of Structural Geology, v. 15, p. 351–368.

Butler, R. W. H., Grasso, M., and LaManna, F., 1992, Origin and deformation of the Neogene–Recent Maghrebian foredeep at the Gela Nappe, Southeast Sicily: Journal of the Geological Society, v. 149, p. 547–556.

Byerlee, J. D., 1967, Frictional characteristics of granite under high confining pressure: Journal of Geophysical Research, v. 72, p. 3639–3648.

Byerlee, J. D., 1978, Friction of rocks: Pure and Applied Geophysics, v. 116, p. 615–626. Caddey, S. W., Bachman, R. L., Campbell, T. J., Reid, R. R., and Otto, R. P., 1991, The Homestake gold mine, an Early Proterozoic iron-formation–hosted gold deposit, Lawrence County, South Dakota: U. S. Geological Survey Professional Paper 1857-J, 67p.

Capuano, R. M., 1993, Evidence of fluid flow in microfractures in geopressured shales: American Association of Petroleum Geologists Bulletin, v. 77, no. 8, p. 1303–1314.

Carena, S., Suppe, J., and Kao, H., 2002, Active detachment of Taiwan illuminated by small earthquakes and its control of first-order topography: Geology, v. 300, no. 10, p. 935–938.

Carey, S. W., 1958, The tectonic approach to continental drift, *in* Carey, S. W. (eds.), Contiental drift: Hobart, Tasmania, University of Tasmania Geology Department Symposium 2, p. 177–355.

Carey, S. W., 1962, Folding: Journal of Alberta Association of Petroleum: Geologists, v. 10, p. 95–144.

Carney S. M., and Janecke, S. U., 2005, Excision and the original low dip of the Miocene-Pliocene Bannock detachment system, SE Idaho: Northern cousin of the Sevier Desert detachment?: Geological Society of America Bulletin, v. 117, p. 334–353.

Carreras, J., Estrada, A., and White, S., 1977, The effects of folding on the c-axis fabrics of a quartz mylonite: Tectonophysics, v. 39, p. 3–24.

Carter, N. L., and Hansen, F. D., 1983, Creep of rock salt: Tectonophysics, v. 92, p. 275–333.

Cashman, P. H., and Fontaine, S. A., 2000, Strain partitioning in the northern Walker Lane, western Nevada and northeastern California: Tectonophysics, v. 326, p. 111–130.

Cemen, I., Gokten, E., Varol, B., Kilic, R., Ozaksoy, V., Emre, C., and Pinar, A., 2000, Turkish earthquakes reveal dynamics of fracturing along a major strike-slip fault zone; EOS Transactions, American Geophysical Union, v. 81, no., 28, p.309 and 313.

Chamberlain, R. T., 1910, The Appalachian folds of Central Pennsylvania: Journal of Geology, v.18, p. 228–251.

Chamberlain, R. T., 1919, The building of the Colorado Rockies: Journal of Geology, v. 27, p. 225–251.

Chang, W. L., and Smith, R. B., 2002, Integrated seismic-hazard analysis of the Wasatch Front, Utah: Bulletin of the Seismological Society of America, v. 92, no. 5, p. 1904–1922.

Chang, W. L., and Smith, R. B., 2006, Contemporary deformation of the Wasatch Fault, Utah, from GPS measurements with implications for inter-seismic fault behavior and earthquake hazard: observations and kinematic analysis: Journal of Geophysical Research, v. 111, B11405, doi:10.1029/2006JB004326.

Chapman, D. S., and Pollack, H. N., 1977, Regional geotherms and lithospheric thickness: Geology, v. 5, p. 265–268.

Chappel, W. M., 1978, Mechanics of thin-skinned fold-and-thrust belts: Geological Society of America, v. 89, p. 1189–1198.

Chase, C., 1978, Plate kinematics: the Americas, east Africa, and the rest of the world: Earth and Planetary Sciences Letters, v. 37, p. 355–368.

Chester, J. S., Logan, J. M., and Spang, J. H., 1991, Influence of layering and boundary conditions on fault-bend and fault-propagation folding: Geological Society of America Bulletin, v. 103, p. 1059–1072.

Cloos, E., 1946, Lineation: A critical review and annotated bibliography: Geological Society of America Memoir 18, 122 p.

Cloos, E., 1947, Oolite deformation in the South Mountain fold, Maryland: Geological Society of America Bulletin, v. 58, p. 843–918.

Cloos, E., 1968, Experimental analysis of Gulf Coast fracture patterns: American Association of Petroleum Geologists Bulletin, v. 52, p. 420–444.

Cloos, H., 1936, Einführung in die Geologie, ein Lehrbuch der inneren Dynamik: Borntraeger, Berlin, 503 p.

Cobbold, P. R., and Quinquis, H., 1980, Development of sheath folds in shear regimes: Journal of Structural Geology, v. 2, p. 119–126.

Cobbold, P. R., Gapais, D., Means, W. D., and Treagus, S. H. (eds.), 1987, Shear criteria in rocks: Journal of Structural Geology, v. 9, 778 p.

Cocard, M., Kahle, H.-G., Peter, Y., Geiger, A., Veis, G., Felekis, S., Paradissis, D., and Billiris, H., 1999, New constraints on the rapid crustal motion of the Aegean region: recent results inferred from GPS measurements (1993–1998) across the West Hellenic Arc, Greece: Earth and Planetary Science Letters, v. 172, p. 39–47.

Conybeare, C. E. B., and Crook, K. A. W., 1968, Manual of sedimentary structures: Commonwealth of Australia, Department of National Development, Bureau of Mineral Resources, Geology and Geophysics Bulletin, v. 102, 327 p.

Cook, F. A., Brown, L. D., and Oliver, J. E., 1980, The southern Appalachians and the growth of continents: Scientific American, v. 243, p. 156–168.

Coulomb, C. A., 1773, Sur une application des regles de maximis et minimis à quelques problemes de statique relatifs à l'architecture: Academie Royale des Sciences, Memoires de Mathematique et de Physique par divers Savants, v. 7, p. 343–382.

Couzens-Schultz, B. A., Vendeville, B. C., and Wiltschko, D. V., 2003, Duplex style and triangle zone formation: insights from physical modeling: Journal of Structural Geology, v. 25, p. 1623–1644.

Cox, A., and Hart, R., 1986, Plate tectonics: how it works: Blackwell Scientific Publications, London, 392 p.

Crider, J. G., and Peacock, D. C. P., 2004, Initiation of brittle faults in the upper crust: a review of field observations: Journal of Structural Geology, v. 26, p. 691–707.

Crone, A. J., Machette, M. N., Bonilla, M. G., Lienkaemper, J. J., Pierce, K. L., Scott, W. E., and Bucknam, R. C., 1987, Surface faulting accompanying the Borah Peak earthquake and segmentation of the Lost River fault, Central Idaho: Bulletin of the Seismological Society of America, v. 77, no. 3, p. 739–770.

Crouch, J. K., 1979, Neogene tectonic evolution of the western Transverse Ranges and the California Continental Borderland: Geological Society of America Bulletin, v. 90, p. 338–345.

Crowell, J. C., 1959, Problems of fault nomenclature: American Association of Petroleum Geologists Bulletin, v. 43, p. 2653–2674.

Crowell, J. C., 1974, Origin of late Cenozoic basins in southern California, *in* Dickinson, W. R. (ed.), Tectonics and sedimentation: Society of Economic Paleontologists and Mineralogists Special Publication 22, p. 190–204.

Cruikshank, K. M., and Johnson, A. M., 1993, High-amplitude folding of linear-viscous multilayers: Journal of Structural Geology, v. 15, p. 79–94.

Cruikshank, K. M., Zhao, G., and Johnson, A. M., 1991, Analysis of minor fractures associated with joints and faulted joints: Journal of Structural Geology, v. 13, p. 865–886.

Currie, J. B., Patnode, A. W., and Trump, R. P., 1962, Development of folds in sedimentary strata: Geological Society of America Bulletin, v. 73, p. 655–674.

Curry, H. D., 1938, "Turtleback" fault surfaces in Death Valley, California [abs]: Geological Society of America Bulletin, v. 49, p. 1875.

Cutler, A., 2003, The seashell on the mountain; a story of science, sainthood, and the humble genius who discovered a new history of the earth: Dutton, New York, 228 p.

Dahlen, F. A., Suppe, J., and Davis, D. M., 1984, Mechanics of fold-and-thrust belts and accretionary wedges (continued): Cohesive Coulomb theory: Journal of Geophysical Research, v. 89, p. 10087–10101.

Dahlstrom, D. C. A., 1960, The upper detachment in concentric folding: Bulletin of Canadian Petroleum Geology, v. 17, p. 326–346.

Dahlstrom, D. C. A., 1969a, Balanced cross sections: Canadian Journal of Earth Sciences, v. 6, p. 743–757.

Dahlstrom, D. C. A., 1969b, The upper detachment in concentric folding: Bulletin of Canadian Petroleum Geology, v. 17, no. 3, p. 323–346.

Dahlstrom, D. C. A., 1970, Structural geology in the eastern margin of the Canadian Rocky Mountains: Bulletin Canadian Petroleum Geologists, v. 17, p. 326–346.

Dahlstrom, D. C. A., 1990, Geometric constraints derived from the law of conservation of volume and applied to evolutionary models for detachment folding: American Association of Petroleum Geologists Bulletin, v. 74, no. 3, p. 336–344.

Davies, R. J., and Stewart, S. A., 2005, Emplacement of giant mud volcanoes in the south Caspian Basin: 3D seismic reflection imaging of their route zones: Journal of the Geological Society (London), v. 162, p. 1–4.

Davis, D. M., and Engelder, T., 1985, The role of salt in fold-and-thrust belts: Tectonophysics, v. 119, p. 67–88.

Davis, D. M., and Engelder, T., 1985, The role of salt in fold-and-thrust belts, *in* Carter, N. L. and Uycda (eds.), Collision tectonics: Deformation of continental lithosphere: Tectonophysics, v. 119, p. 67–88.

Davis, D. M., Suppe, J., and Dahlen, F. A., 1983, Mechanics of fold-and-thrust belts and accretionary wedges: Journal of Geophysical Research, v. 88, p. 1153–1172.

Davis, G. A., and Lister, G. S., 1988, Detachment faulting in continental extension: Perspectives from the Southwestern U.S. Cordillera: Geological Society of America Spec. Paper 218 (John Rodgers Symposium Volume), p. 133–159.

Davis, G. A., Anderson, J. L., Frost, E. G., and Shackelford, T. J., 1980, Mylonitization and detachment faulting in the Whipple-Buckskin-Rawhide Mountains terrane, southeastern California and western Arizona, *in* Crittenden, M. D., Jr., Coney, P. J., and Davis, G. H. (eds.), Cordilleran metamorphic core complexes: Geological Society of America Memoir 153, p. 79–129.

Davis, G. A., Lister, G. S., and Reynolds, S. J., 1986, Structural evolution of the Whipple and South mountains shear zones, southwestern United States: Geology, v. 14, no. 1, p. 7–10.

Davis, G. H., 1972, Deformational history of the Caribou strata-bound sulfide deposit, Bathurst, New Brunswick, Canada: Economic Geology, v. 67, p. 634–655.

Davis, G. H., 1978, The monocline fold pattern of the Colorado Plateau, *in* Matthews, V. (ed.), Laramide folding associated with basement block faulting in the western United States: Geological Society of America Memoir 151, p. 215–233.

Davis, G. H., 1979, Laramide folding and faulting in southeastern Arizona: American Journal of Science, v. 279, p. 543–569.

Davis, G. H., 1980, Structural characteristics of metamorphic core complexes, *in* Crittenden, M. D., Jr., Coney, P. J., and Davis, G. H. (eds.), Cordilleran metamorphic core complexes: Geological Society of America Memoir 153, p. 35–77.

Davis, G. H., 1983, Shear-zone model for the origin of metamorphic core complexes: Geology, v. 11, p. 348–351.

Davis, G. H., 1987, Saguaro National Monument, Arizona: outstanding display of structural characteristics of metamorphic core complexes, *in* Hill, M. L. (ed.), Centennial Field Guide Volume 1, Cordilleran Section of the Geological Society of America: Geological Society of America, Boulder, p. 35–40.

Davis, G. H., 1998, Fault-fin landscape: Geological Magazine, v. 135, no. 2, p. 283–286.

Davis, G. H., and Reynolds, S. J., 1996, Structural geology of rocks and regions: John Wiley and Sons, New York, 776p.

Davis, G. H., Bevis, M., Blom, R. G., Brandon, M., Bull, W., Cowan, D., Doka, R. K., Fink, J., Humphreys, E., Johnson, A., Mayer, L., Lay, T., McNutt, M., Minster, B., Sieh, K., Smith, R. B., Sutter, J., Tullis, J., Yeats, R. S., Zoback, M. L., 1996, Active tectonics and society: a plan for integrative science: National Science Foundation, Washington, D. C., 56p.

Davis, G. H., Bump, A. P., García, P. E., and Ahlgren, S. G., 1999, Conjugate Riedel deformation band shear zones: Journal of Structural Geology, v. 22, p. 169–190.

Davis, G. H., Eliopulos, G. J., Frost, E. G., Goodmundson, R. C., Knapp, R. B., Liming, R. B., Swan, M. M., and Wynn, J. C., 1974, Recumbent folds—focus of an investigative workshop in tectonics: Journal of Geological Education, v. 22, p. 204–208.

Davis, G. H., Gardulski, A. F., and Lister, G. S., 1987, Shear zone origin of quartzite mylonite and mylonitic pegmatite in the Coyote Mountains: Journal of Structural Geology, v. 9, p. 289–298.

Davis, G. H., Showalter, S. R., Benson, G. S., McCalmont, L. S., and Cropp, F. W., 1981, Guide to the geology of the Salt River Canyon region, Arizona: Arizona Geological Society Digest, v. 13, p. 48–97.

Davis, G. H., 1999, Structural geology of the Colorado Plateau region of southern Utah, with special emphasis on deformation band shear zones: Geological Society of America Special Paper 342, 157 p.

Davis, G. H., and Bump, A. P., 2009, Structural geologic evolution of the Colorado Plateau, *in* Kay, S. M., Ramos,

V. A., and Dickinson, W. R., (eds.), Backbone of the Americas: Shallow subduction, plateau uplift, and ridge and terrane collision: Geological Society of America Memoir 204, p. 99–124.

Davis, T., and Namson, J., 1994, A balanced cross section of the 1994 Northridge earthquake, southern California: Nature, v. 372, p. 167–169.

Davison, I., 1994, Linked fault systems; extensional, strike-slip and contractional, *in* Hancock, P. L. (ed.), Continental Deformation: Pergamon Press, Oxford, p. 121–142.

De Paor, D. G., 1988, R_f/ϕ strain analysis using an orientation net: Journal of Structural Geology, v. 10, p. 323–333. DeSitter, L. U., 1964, Structural geology (2nd edition): McGraw-Hill Book Company, New York, 551 p.

De Paor, D. G., 2001, New interactive animations for analysis of stress, strain, and flow: Journal of the Virtual Explorer, v. 3, ISSN 1441–8126.

De Sitter, L. U., 1964, Structural geology: McGraw-Hill, New York, 551p.

Dean, S. L., Kulander, B. R., and Skinner, J. M., 1988, Structural chronology of the Alleghenian orogeny in southeastern West Virginia: Geological Society of America Bulletin, v. 100, p. 299–310.

Debacker, T. N., Dumon, M., and Matthys, A., 2009, Interpreting fold and fault geometries from within the lateral to oblique parts of slumps: a case study from the Anglo-Brabant Deformation Belt: Journal of Structural Geology, v. 31, p. 1525–1539.

DeCelles, P. G., and Coogan, J. C., 2006, Regional structure and kinematic history of the Sevier fold-thrust belt, central Utah: Geological Society of America, v. 118, p. 841–864.

DeGraff, J. M., and Aydin, A., 1987, Surface morphology of columnar joints and its significance to mechanics and direction of joint growth: Geological Society of America, v. 99, p. 605–617.

DeGraff, J. M., Long, P. E., and Aydin, A., 1989, Use of joint-growth directions and rock textures to infer thermal regimes during solidification of basaltic lava flows: Journal of Volcanology and Geothermal Research, v. 38, p. 309–324.

Dennis, A. J., and Secor, D. T., 1987, A model for the development of crenulations in shear zones with applications from the Southern Appalachian Piedmont: Journal of Structural Geology, v. 9, p. 809–817.

Dennis, J. G., 1972, Structural geology: Ronald Press, New York, 532 p.

Dewey, J. F., 1972, Plate tectonics: Scientific American, v. 226, p. 56–72.

Dewey, J. F., and Bird, J. M., 1970, Mountain belts and the new global tectonics: Journal of Geophysical Research, v. 75, p. 2625–2647.

Diamond, W. P., McCulloch, C. M., and Bench, B. M., 1975, Estimation of coal-cleat orientation using surface-joint and photolinear analysis: Geology, v. 3, no. 12, p. 687–690.

Diegel, F. A., Karlo, J. P., Schuster, D. C., Shoup, R. C., and Tauvers, P. R., 1995, Cenozoic structural evolution and tectono-stratigrapphic framework of the northern Gulf coast continental margin, *in* Jackson, M. P. A., Roberts, D. G., and Snelson, S. (eds.), Salt tectonics: a global perspective: American Association of Petroleum Geologists Memoir 65, p. 109–151.

Dietz, R. S., 1961, Continent and ocean basin evolution by spreading of sea floor: Nature, v. 190, p. 854–857.

Dietz, R. S., 1968, Shatter cones in cryptoexplosion structures, *in* French, B. M., and Short, N. M. (eds.), Shock metamorphism of natural materials: Mono Book Corporation, Baltimore, p. 267–285.

Dillard, A., 1987, An American childhood: Harper & Row, New York, 257p.

Dixon, T. H., Miller, M., Farina, F., Wang, H., and Johnson, D., 2000, Present-day motion of the Sierra Nevada block and some tectonic implications for the Basin and Range Province, North American Cordillera: Tectonics, v. 19, p. 1–24.

Dokka, R. K., and Travis, C. J., 1990, Role of the Eastern California shear zone in accommodating Pacific-North American plate motion: Geophysical Research Letters, v. 19, p. 1323–1326.

Dolan, J. F., Sieh, K., and Rockwell, T. K., 2000, Late Quaternary activity and seismic potential of the Santa Monica fault system, Los Angeles, California: Geological Society of America Bulletin, v. 112, no. 10, p. 1559–1581.

Dolan, J. L., Bowman, D. D., and Sammis, C. G., 2007, Long-range and long-term fault interactions in southern California: Geology, v. 35, p. 855–858.

Donath, F. A., 1961, Experimental study of shear failure in anisotropic rocks: Geological Society of America Bulletin, v. 72, p. 985–989.

Donath, F. A., 1962, Analysis of Basin-Range structure, south-central Oregon: Geological Society of America Bulletin, v. 73, p. 1–16.

Donath, F. A., 1966, A triaxial pressure apparatus for testing of consolidated or unconsolidated materials subjected to pore pressure, *in* Johnson, A. I, and Wallace, G. B. (eds.), Testing techniques for rock mechanics: American Society for Testing and Materials, STP 402, p. 41–51.

Donath, F. A., 1970a, Rock deformation apparatus and experiments for dynamic structural geology: Journal of Geological Education, v. 18, p. 1–12.

Donath, F. A., 1970b, Some information squeezed out of rock: American Scientist, v. 58, p. 54–72.

Donath, F. A., and Parker, R. B., 1964, Folds and folding: Geological Society of America Bulletin, v. 75, p. 45–62.

Druguet, E., and Carreras, J., 2006, Analogue modeling of synthetic leucosomes in magmatic schists: Journal of Structural Geology, v. 28, p. 1734–1747.

Dula, W. F., 1991, Geometric models of listric normal faults and rollover folds: American Association of Petroleum Geologists Bulletin, v. 75, p. 1609–1625.

Duncan, R. A., 1981, Hotspots in the southern oceans: an absolute frame of reference for motion of the Gondwana continents: Tectonophysics, v. 74, p. 29–42.

Dunne, L. A., Manoogian, P. R., and Pierini, D. F., 1990, Structural style and domains of the Northern Oman Mountains (Oman and United Arab Emirates), *in* Robertson, et al. (eds.), The geology and tectonics of the Oman region: Geological Society of London Special Publication 49, p. 375–386.

Durney, D. W., and Ramsay, J. G., 1973, Incremental strains measured by syntectonic crystal growths, *in* De Jong, K. A., and Scholten, R. (eds.), Gravity and tectonics: John Wiley & Sons, New York, p. 67–96.

Dutton, S. P., 1982, Pennsylvanian fan-delta and carbonate deposition, Mobeetie Field, Texas Panhandle: American Association of Petroleum Geologists Bulletin, v. 66, p. 389–407.

Ekstrom, G., Stein, R. S., Eaton, J. P., and Eberhart-Phillips, D., 1992, Seismicity and geometry of a 110-km-long blind thrust fault 1. The 1985 Kettleman Hills, California, earthquake: Journal of Geophysical Research, v. 97, no. B4, p. 4843–4864.

Elasser, W. M., 1968, The mechanics of continental drift in Gondwanaland revisited: new evidence for continental drift: Proceedings of the American Philosophical Society, v. 112, p. 344–353.

Elliott, D., and Johnson, M. R. W., 1980, Structural evolution in the northern part of the Moine thrust belt, northwest Scotland: Royal Society of Edinburgh Transactions, v. 71, p. 69–96.

Ellis, P. G., and McClay, K. R., 1988, Listric extensional fault systems: results of analogue model experiments: Basin Research, v. 1, p. 55–70.

Engelder, T., 1984, Loading paths to joint propagation during a tectonic cycle: an example from the Appalachian Plateau, U.S.A.: Eos, American Geophysical Union Transactions, v. 65, p. 1118.

Engelder, T., 1985, Loading paths to joint propagation during a tectonic cycle: an example from the Appalachian Plateau, U.S.A.: Journal of Structural Geology, v. 7, p. 459–476.

Engelder, T., 1987, Joints and shear fractures in rock, *in* Atkinson, B. K. (ed.), Fracture mechanics of rock: Academic Press, London, p. 27–69.

Engelder, T., 1999, Transitional-tensile fracture propagation: a status report: Journal of Structural Geology, v. 21, p. 1049–1055.

Engelder, T., and Engelder, R., 1977, Fossil distortion and decollement tectonics of the Appalachian Plateau: Geology, v. 5, p. 457–460.

Engelder, T., and Geiser, P., 1979, The relationship between pencil cleavage and lateral shortening within the Devonian section of the Appalachian Plateau, New York: Geology, v. 7, p. 460–464.

Engelder, T., and Geiser, P., 1980, On the use of regional joints sets as trajectories of paleostress fields during the development of the Appalachian Plateau, New York: Journal of Geophysical Research, Series B, v. 85, p. 6319–6341.

Engelder, T., and Marshak, S., 1988, Analysis of data from rock-deformation experiments, *in* Marshak, S., and Mitra, G. (eds.), Basic methods of structural geology: Prentice-Hall, Englewood Cliffs, New Jersey, p. 193–212.

Engelder, T., and Oertel, G., 1985, The correlation between undercompaction and tectonic jointing within the Devonian Catskill Delta: Geology, v. 13, p. 863–866.

Epard, J.-L., and Groshong, R. H., Jr., 1993, Excess area and depth to detachment: American Association of Petroleum Geologists Bulletin, v. 77, no. 8, p. 1291–1302.

Ernst, W. G., 1988, Metamorphism and crustal evolution of the western United States: Prentice Hall, Upper Saddle River, New Jersey.

Erslev, E. A., 1991, Trishear fault propagation folding: Geology, v. 19, p. 617–620.

Erslev, E. A., and Koenig, N. B. 2009, 3D kinematics of Laramide, basement-involved Rocky Mountain deformation, U.S.A.: insights from minor faults and GIS-enhanced structure maps, *in* Kay, S., Ramos, V., and Dickinson, W. R. (eds.), Backbone of the Americas: Shallow subduction, plateau uplift and ridge and terrane collision: Geological Society of America Bulletin, v. 120, p. 877–892.

Erslev, E. A., and Larson, S. M., 2006, Testing Larmide hypotheses for the Colorado Front Range arch using minor faults, *in* Raynolds, R., and Sterne, E. (eds.), Mountain Geologist Special Issue on the Colorado Front Range, v. 43, p. 45–64.

Erslev, E. A., and Rogers, J. L., 1993, Basement-cover geometry of Laramide fault-propagation folds, *in* Schmidt, C. J., Chase, R. B., and Erslev, E. A., (eds.), Laramide basement deformation in the Rocky Mountain foreland of the western United States: Geological Society of America Special Paper 280, p. 125–146.

Erslev, E. A., and Ward, D. J., 1994, Element and volume flux in coalesced slaty cleavage: Journal of Structural Geology v. 16, p. 531–554.

Escher, A., Escher, J. C., and Watterson, J., 1975, The reorientation of the Kangamiut dike swarm, West Greenland: Canadian Journal of Earth Sciences, v. 12, no. 2, p. 158–173.

Etchecopar, A., and Malavieille, J., 1987, Computer models of pressure shadows: a method for strain measurement and shear-sense determination, *in* Cobbold, P. R., Gapais, D., Means, W. D., and Treagus, S. H. (eds.), Shear criteria in rocks: Journal of Structural Geology, v. 9, p. 667–677.

Etchecopar, A., and Vasseur, G., 1987, A 3-D kinematic model of fabric development in polycrystalline aggregates: comparisons with experimental and natural samples: Journal of Structural Geology, v. 9, p. 705–717.

Evans, J. P., and Bradbury, K. K., 2004, Faulting and fracturing of nonwelded Bishop Tuff, Eastern California: deformation mechanisms in very porous materials in the vadose zone: Vocose Zone Journal, v. 3, p. 602–623.

Eyal, Y., Gross, M. R., Engelder, T., and Becker, A., 2001, Joint development during fluctuation of the regional stress in southern Israel: Journal of Structural Geology, v. 23, no. 2/3, pp. 279–296.

Faill, R. T., 1973, Kind-band folding, Valley and Ridge Province, Pennsylvania: Geological Society of America Bulletin, v. 84, p. 1289–1314.

Farrell, S. G., and Eaton, S., 1987, Slump strain in the tertiary of Cyprus and the Spanish Pyrenees. Definition of palaeoslopes and models of soft-sediment deformation, *in*, Jones, M. E., and Preston, R. M. F. (eds.) Deformation of sediments and sedimentary rocks, Geological Society, London, Special Publications v. 29, p. 181–196.

Faulds, J. E., Henry, C. D., and Hinz, N. H., 2005, Kinematics of the northern Walker Lane: an incipient transform along the Pacific-North American plate boundary: Geology, v. 33, no. 6, p. 505–508.

Feininger, T., 1978, The extraordinary striated outcrop at Saqsaywaman, Peru: Geological Society of America Bulletin, v. 89, p. 494–503.

Ferrill, D. A., and Morris, A. P., 1999, Geometric considerations of deformation above curved normal faults and salt evacuation surfaces: The Leading Edge, v. 16, p. 1129–1133.

Ferrill, D. A., and Morris, A. P., 2003, Dilational normal faults: Journal of Structural Geology, v. 25, no. 2, pp. 183–196.

Fink, J., 1980, Surface folding and viscosity of rhyolite flows: Geology, v. 8, p. 250–254.

Fischer, M. P., and Polansky, A., 2006, Influence of flaws on joint spacing and saturation: results of one-dimensional mechanical modeling: Journal of Geophysical Research, v. 111, B07403, doi:10.1029/2005JB0041152006.

Fisher, M. P., Woodward, N., and Mitchell, M., 1992, The kinematics of break-thrust folds: Journal of Structural Geology, v. 14, p. 451–460.

Fleuty, M. J., 1964, The description of folds: Geological Association Proceedings, v. 75, p. 461–492.

Flinn, D. 1962. On folding during three-dimensional progressive deformation: Quarterly Journal of the Geological Society of London, v. 118, p. 385–433.

Forsyth, D. W., and Uyeda, S., 1975, On the relative importance of the driving forces of plate motion: Geophysical Journal of the Royal Astronomical Society, v. 43, p. 163–200.

Fossen, H., and Tikoff, B., 1993, The deformation matrix for simultaneous simple shearing, pure shearing, and volume change, and its implications to transpression/transtension tectonics: Journal of Structural Geology, v. 15, p. 413–422.

Frehner, M., and Schmalholz, S. M., 2006, Numerical simulations of parasitic folding in multilayers: Journal of Structural Geology, v. 28, no. 9, p. 1569–1724.

French, K. S., 2011, Displacement patterns of cemetery monuments in Ferndale, CA during the Mw 6.5 offshore northern California earthquake of January 10th, 2010: Senior Thesis, Department of Geology, Humboldt State University.

French, K. S., Cashman, S. M., and Structural Geology Class, Spring 2010, 2010, Displacement patterns of cemetery monuments in Ferndale, CA, during the Mw 6.5 offshore California earthquake of January 10, 2010: American Geophysical Union, Fall Meeting, 2010, Abstract No. S51B-1930.

Friedman, M., Handin, J., Logan, J. M., Min, K. D., and Stearns, D. W., 1976, Experimental folding of rocks under confining pressure. Part III. Faulted drape folds in multilithologic layered specimens: Geological Society of America Bulletin, v. 87, p. 1049–1066.

Fry, N., 1979, Random point distributions and strain measurement in rocks: Tectonophysics, v. 60, p. 69–105.

Fuis, G. S., Ryberg, T., Godfrey, N. J., Okaya, D. A., and Murphy, J. M., 2001, Crustal structure and tectonics from the Los Angeles basin to the Mojave Desert, southern California: Geology, v. 29, no. 1, p. 15–18.

García, P. E. and Davis, G. H., 2004, Evidence and mechanism for folding of granite, in Sierra de Haulfín basement-cored uplift, northwest Argentina: American Association of Petroleum Geologists, v. 88, no. 9, p. 1255–1276.

Gardner, J. N., Reneau, S. L., Lewis, C. J.,Lavine, A., Krier, D., Wolde Gabriel, G., and Guthrie, G. 2001. Geology of the Pajarito fault zone in the vicinity of S-Site (TA-16), Los Alamos National Laboratory, Rio Grande rift, New Mexico. Los Alamos National Laboratory Report LA-13831-MS, 86 pp.

Gash, S. P. J., 1971, A study of surface features relating to brittle and semi-brittle fracture: Tectonophysics, v. 12, no. 5, p. 349–391.

Ghisetti, F. C., and Sibson, R. H., 2006, Accommodation of compressional inversion in northwestern South Island (New Zealand): Old faults versus new? Journal of Structural Geology, v. 28, p. 1994–2010.

Ghosh, K., and Mitra, S., 2009, Structural controls of fracture orientations, intensity, and connectivity, Teton anticline, Sawtooth Range, Montana: American Association of Petroleum Geologists Bulletin, v. 93, no. 8, pp. 995–1014.

Ghosh, S. K., 1968, Experiments of buckling of multilayers which permit interlayer gliding: Tectonophysics, v. 6, p. 207–249.

Ghosh, S. K., 1993, Structural geology: fundamentals and modern developments: Pergamon Press: Oxford, 598 p.

Gibbs, A. D., 1983, Balanced cross-section construction from seismic sections in areas of extensional tectonics: Journal of Structural Geology, v. 141, p. 121–129.

Giles, K. A., and Lawton, T. F., 2002, Halokinetic sequence stratigraphy adjacent to El Papalote diapir, La Popa basin, northeastern Mexico: American Association of Petroleum Geologists Bulletin, v. 86, p. 823–841.

Glazner, A. F., Bartley, J. M., Coleman, D. S., Gray, W., and Taylor, R. Z., 2004, Are plutons assembled over millions of years by amalgamation from small magma chambers?: GSA Today, v. 14, no. 4/5, doi: 10.1130/1052-5153(2004)014<0004:APAOMO>2.0.CO;2.

Golombek, M. P., 1981, Geometry and rate of extension across the Pajarito fault zone, Española basin, Rio Grande rift, northern New Mexico: Geology, v. 9, p. 21–24.

Golombek, M. P., McGill, G. E., and Brown, L., 1983, Tectonic and geologic evolution of the Española basin, Rio Grande rift: structure, rate of extension, and relation to the state of stress in the western United States: Tectonophysics, v. 94, p. 483–507.

Gourmelen, N., and Amelung, F., 2005, Postseismic mantle relaxation in the Central Nevada Seismic Belt: Science, v. 310, no. 5753, p. 1473–1476.

Gray, D. R., 1977a, Morphologic classification of crenulation cleavages: Journal of Geology, v. 85, p. 229–235.

Gray, D. R., 1977b, Some parameters which affect the morphology of crenulation cleavages: Journal of Geology, v. 85, p. 763–780.

Gray, D. R., 1979, Microstructure of crenulation cleavages: An indication of cleavage origin: American Journal of Science, v. 279, p. 97–128.

Gray, D. R., 1981, Compound tectonic fabrics in singly folded rocks from southwest Virginia, U.S.A.: Tectonophysics, v. 78, p. 229–248.

Gray, D. R., and Durney, D. W., 1979, Investigations on the mechanical significance of crenulation cleavage: Tectonophysics, v. 58, p. 35–79.

Grieve, R. A. F., and Pesonen, L. J., 1992, The terrestrial impact cratering record: Tectonophysics, v. 216, p. 1–30.

Griffith, A. A., 1924, Theory of rupture: Proceedings of the First International Congress on Applied Mechanics, Delft, the Netherlands, p. 55–63.

Griggs, D. T., Turner, F. J., and Heard, H. C., 1960, Deformation of rocks at 500° to 800°C, *in* Griggs, D. T., and Handin, J. (eds.), Rock deformation: Geological Society of America Memoir 79, p. 39–104.

Grimaldi, G. O., and Dorobek, S. L., 2011, Fault framework and kinematic evolution of inversion structures: Natural examples from the Neuquén Basin, Argentina: American Association of Petroleum Geologists Bulletin, v. 95, no. 1, p. 27–60.

Groshong, R. H., Jr., 1975, "Slip" cleavage caused by pressure solution in a buckle fold: Geology, v. 3, p. 411–413.

Groshong, R. H., Jr., 1988, Low-temperature deformation mechanisms and their interpretation: Geological Society of America Bulletin, v. 100, p. 1329–1360.

Gross, M. R., 1993, The origin and spacing of cross-joints: examples from the Monterey Formation, Santa Barbara coastline, California: Journal of Structural Geology, v. 15, pp. 737–751.

Gross, M. R., and Eyal, Y., 2007, Throughgoing fractures in layered carbonate rocks: Geological Society of America Bulletin, v. 119, no. 11/12, pp. 1387–1404.

Gross, M. R., Fischer, M. P., Engelder, T., and Greenfield, R. J., 1995, Factors controlling joint spacing in interbedded sedimentary rocks: integrating numerical models with field observations from the Monterey Formation, U.S.A., *in* Ameen, M. S. (ed.), Fractography: Fracture topography as a tool in fracture mechanics and stress analysis: Geological Society [London] Special Publication 92, pp. 215–233.

Guilbert, J. M., and Park, C. F., Jr., 1986, The geology of ore deposits: W. H. Freeman, New York, 985 p.

Gustafson, J. K., Burrell, H. C., and Garretty, M. D., 1950, Geology of the Broken Hill ore deposit, New South Wales, Australia: Geological Society of America Bulletin, v. 61, p. 1369–1437.

Hafner, M., and Passchier, C. W., 2000, Development of S-C' type cleavage in paraffin wax using a circular shear rig: Journal Virtual Explorer, v. 2, p. 16–17.

Hafner, W., 1951, Stress distribution and faulting: Geological Society of America Bulletin, v. 62, p. 373–398.

Hall, N. T., and Niemi, T. M., 2008, The 1906 Earthquake fault rupture and paleoseismic investigations of the northern San Andreas fault at the Dogtown site, Marlin County, California: Bulletin of the Seismological Society of America, v. 98, no. 5, p. 2191–2208.

Hamblin, W. K., 1965, Origin of "reverse drag" on the downthrown side of normal faults: Geological Society of America Bulletin, v. 76, p. 1145–1164.

Hamilton, W., 1964, Geologic map of the Big Maria Mountains NE Quadrangle, Riverside County, California, and Yuma County, Arizona: U.S. Geological Survey Geologic Quadrangle Map GQ–350.

Hamilton, W., 1987, Mesozoic geology and tectonics of the Big Maria Mountains region, southeastern California, *in* Dickinson, W. R., and Klute, M. A. (eds.), Mesozoic rocks of southern Arizona and adjacent areas: Arizona Geological Society Digest, v. 18, p. 33–47.

Handin, J., 1969, On the Coulomb-Mohr failure criterion: Journal of Geophysical Research, v. 74, p. 5343–5348.

Handin, J., and Hager, R. V., 1957, Experimental deformation of sedimentary rocks under confining pressure: tests at room temperature on dry samples: American Association of Petroleum Geologists Bulletin, v. 41, p. 1–50.

Handin, J., Hager, R. V., Jr., Friedman, M., and Feather, J. N., 1963, Experimental deformation of sedimentary rocks under confining pressure: pore pressure tests: American Association of Petroleum Geologists Bulletin, v. 47, p. 717–755.

Handy, M. R., 1990, The solid-state flow of polymineralic rocks: Journal of Geophysical Research, v. 95, p. 21, 613–21, 634.

Handy, M. R., Wissing, S. B., and Streit, L. E., 1999, Frictional-viscous flow in mylonite with varied bimineralic composition and its effect on lithospheric strength: Tectonophysics, v. 303, p. 175–191.

Hanmer, S., and Passchier, C., 1991, Shear-sense indicators: a review: Geological Survey of Canada Paper 90–17, p. 72.

Hardy, S., and Finch, E., 2006, Discrete element modeling of the influence of cover strength on basement-involved fault-propagation folding: Tectonophysics, v. 415 (1–4), p. 225–238.

Hardy, S., and Ford, M., 1997, Numerical modeling of trishear fault propagation folding: Tectonics, v. 16, p. 841–854.

Haynes, F. M., and Titley, S. R., 1980, The evolution of fracture-related permeability within the Ruby Star granodiorite, Sierrita porphyry copper deposit, Pima County, Arizona: Economic Geology, v. 75, p. 673–683.

Heard, H. C., 1963, Effect of large changes in strain rate in the experimental deformation of Yule marble: Journal of Geology, v. 71, p. 162–195.

Heat-Moon, W. L., 1991, PrairyErth: Houghton Mifflin, New York, 622p.

Heidrick, T. L., and Titley, S. R., 1982, Fracture and dike patterns in Laramide plutons and their structural and tectonic implications: American Southwest, *in* Titley, S. R. (ed.), Advances in geology of the porphyry copper deposits: Southwest North America: University of Arizona Press, Tucson, Arizona, p. 73–91.

Helgeson, D. E., and Aydin, A., 1991, Characteristics of joint propagation across layer interfaces in sedimentary rocks: Journal of Structural Geology, v. 13, p. 897–911.

Hendry, H. E., and Stauffer, M. R., 1977, Penecontemporaneous folds in cross-bedding: inversion of facing criteria and mimicry of tectonic folds: Geological Society of America Bulletin, v. 88, p. 809–812.

Herwegh, M., De Bresser, J. H. P., and Ter Heege, J. H., 2005, Combining natural microstructures with composite flow laws: an improved approach for the extrapolation of lab data to nature: Journal of Structural Geology, v. 27, p. 503–521.

Hess, H. H., 1962, History of ocean basins, *in* Engel, A. E. J., James, H. L., and Leonard, B. F. (eds.), Petrologic studies: a volume in honor of A. F. Buddington: Geological Society of America, Boulder, Colorado, p. 599–620.

Hickman, S., and Zoback, M. D., 2004, Stress measurements in the SAFOD pilot hole: implications for the frictional strength of the San Andreas fault: Geophysical Research Letters, v. 31, L15S12.

Hickman, S., and Zoback, M., 2004, Stress orientations and magnitudes in the SAFOD pilot hole: Geophysical Research Letters, v. 31, L15S12, doi:10.1029/2004GL020043.

Hickman, S., Zoback, M., and Ellsworth, W., 2004, Introduction to special section: Preparing for the San Andreas Fault Observatory at Depth: Geophysical Research Letters, v. 31, L12S01.

Higgins, M. W., 1971, Cataclastic rocks: United States Geological Survey Professional Paper 687, 97 p.

Hill, M. L., 1959, Dual classification of faults: Geological Society of America Bulletin, v. 43, p. 217–221.

Hirth, G., and Tullis, J., 1992, Dislocation creep regimes in quartz aggregates: Journal of Structural Geology, v. 14, p. 145–160.

Hobbs, B. E., Means, W. D., and Williams, P. F., 1976, An outline of structural geology: John Wiley & Sons, New York, 571 p.

Hobbs, B., Regenauer-Lieb, K., and Ord, A., 2008, Folding with thermal-mechanical feedback: Journal of Structural Geology, v. 30, no. 12, p. 1572–1592.

Hobbs, D. W., 1967, The formation of tension joints in sedimentary rocks: an explanation: Geological Magazine, v. 104, pp. 550–556.

Hodgson, R. A., 1961, Classification of structures on joint surfaces: American Journal of Science, v. 259, p. 493–502.

Hogan, J. P., and Dunne, W. M., 2001, Calculation of shortening due to outcrop-scale deformation and its relation to regional deformation patterns: Journal of Structural Geology, v. 23, p. 1508–1529, doi: 10.1016/S0191-8141(01)00016-5.

Holyoke, C. W., III, and Tullis, J., 2006, Mechanisms of weak phase interconnection and the effects of phase strength contrast on fabric development: Journal of Structural Geology, v. 28, p. 621–640.

Hoover, J. R., Malone, R., Eddy, G., and Donaldson, A., 1969, Regional position, trend, and geometry of coals and sandstones of the Monongahela Group and Waynesburg formation in the Central Appalachians, *in* Donaldson, A. C. (ed.), Some Appalachian coals and carbonates: models of ancient shallow-water deposition: West Virginia Geological and Economic Survey, Morgantown, West Virginia, p. 157–192.

Horne, R., and Culshaw, N., 2001, Flexural-slip folding in the Meguma Group, Nova Scotia: Journal of Structural Geology, v. 23, no. 10, p. 1631–1652.

Horton, B. K., 1998, Sediment accumulation on top of the Andean orogenic wedge: Oligocene to late Miocene basins of the Eastern Cordillera, southern Bolivia: Geological Society of America Bulletin, v. 110, p. 1174–1192.

Hubbert, M. K, 1937, Theory of scale models as applied to the study of geologic structures: Geological Society of America Bulletin, v. 48, p. 1459–1519.

Hubbert, M. K., 1951, Mechanical basis for certain familiar geologic structures: Geological Society of America Bulletin, v. 62, p. 355–372.

Hubbert, M. K., and Rubey, W. W., 1959, Role of fluid pressure in mechanics of overthrust faulting. Part 1: Geological Society of America Bulletin, v. 70, p. 115–166.

Hudleston, P. J., 1973, Fold morphology and some geometrical implications of theories of fold development: Tectonophysics, v. 16, p. 1–46.

Huiqi, L., McClay, K. R., and Powell, C. M., 1992, Physical models of thrust wedges, *in* McClay, K. R. (ed.), Thrust tectonics: Chapman and Hall, New York, p. 71–81.

Hull, D., and Bacon, D. S., 1984, Introduction to dislocations: Pergamon Press, New York, 251 p.

Humphreys, E. D., and Weldon, R. J., 1994, Deformation across the western United States: a local estimate of Pacific-North America transform deformation: Journal of Geophysical Research, v. 99, p. 19,975–20,010.

Hunt, C. B., Averitt, P., and Miller, R. L., 1953, Geology and geography of the Henry Mountain region, Utah: United States: Geological Survey Professional Paper 228, 234 p.

Huntoon, P. W., and Richter, H. R., 1979, Breccia pipes in the vicinity of Lockhart Basin, Canyonlands area, Utah, *in* Baars, D. L. (ed.), Permianland: Four Corners Geological Society Guidebook, 9th Field Conference, p. 47–53.

Huntoon, P. W., and Sears, J. W., 1975, Bright Angel and Eminence faults, eastern Grand Canyon, Arizona: Geological Society of America Bulletin, v. 86, p. 465–472.

Iacopini, D., Passhier, C. W., Koehn, D., and Carosi, R., 2007, Fabric attractors in general triclinic flow systems and their application to high-strain zones: a dynamical systems approach: Journal of Structural Geology, v. 29, p. 298–317.

Ingraffea, A. R., 1987, Theory of crack initiation and propagation in rock, *in* Atkinson, B. K. (ed.), Fracture mechanics of rock: Academic Press, London, p. 71–110.

Isacks, B., and Molnar, P., 1969, Mantle earthquake mechanisms and the sinking of the lithosphere: Nature, v. 223, p. 1121–1124.

Isacks, B., Oliver, J., and Sykes, L. R., 1968, Seismology and the new global plate tectonics: Journal of Geophysical Research, v. 73, p. 5855–5899.

Jackson, J. A., Gagnepain, J., Houseman, G., King, G. C. P., Papadimitriou, P., Soufleris, C., and Virieux, J., 1982, Seismicity, normal faulting, and the geomorphological development of the Gulf of Corinth (Greece): the Corinth earthquakes of February and March, 1981: Earth and Planetary Science Letters, v. 57, p. 377–397.

Jackson, J., 2006, Fatal attraction: living with earthquakes, the growth of villages into megacities, and earthquake vulnerability in the modern world: Philosophical Transactions of the Royal Society A, v. 364, p. 1911–1925.

Jackson, J., and McKenzie, D., 1999, A hectare of fresh striations on the Arkitsa fault, central Greece: Journal of Structural Geology, v. 21, p. 1–6.

Jackson, J., and Molnar, P., 1990, Active faulting and block rotations in the western Transverse Ranges, California: Journal of Geophysical Research, v. 95, p. 22,073–22,089.

Jackson, M. P. A., 2008, Evolution of the Cretaceous Astrid thrust belt in the ultradeep-water Lower Congo Basin, Gabon: American Association of Petroleum Geology, v. 92, p. 487–511.

Jackson, M. P. A., and Vendeville, B. C., 1994, Regional extension as a geologic trigger for diapirism: Geological Society of America Bulletin, v. 106, no. 1, p. 57–73.

Jackson, M. P. A., Vendeville, B. C., and Schultz-Ela, D. D., 1994, Structural dynamics of salt systems: Annual Reviews of Earth and Planetary Sciences, v. 22, p. 93–117.

Jaeger, J. C., and Cook, N. G. W., 1976, Fundamentals of rock mechanics: Halsted Press, New York, 585 p.

Jefferies, S. P., Holdsworth, R. E., Wibberley, C. A. J., Shimamoto, T., Spiers, C. J., Niemeijer, A. R., and Lloyd, G. E., 2006, The nature and importance of phyllonite development in crustal-scale fault cores : an example from the Median Tectonic Line, Japan: Journal of Structural Geology, v. 28, p. 333–352.

Jeng, F. S., Lin, M. L., Lai, Y. C., and Teng, M. H., 2002, Influence of strain rate on buckle folding of an elastoviscous single layer: Journal of Structural Geology, v. 24, no. 3, p. 501–516.

Jessell, M. W., 1988a, Simulation of fabric development in recrystallizing aggregates; I, Description of the model: Journal of Structural Geology, v. 10, p. 771–778.

Jessell, M. W., 1988b, Simulation of fabric development in recrystallizing aggregates; II, Example model runs: Journal of Structural Geology, v. 10, p. 779–793.

Johnson, A. M., 1970, Physical processes in geology: Freeman, Cooper, and Company, San Francisco, 577 p.

Johnson, A. M., 1977, Styles of folding: mechanics and mechanisms of folding of natural elastic materials: Elsevier Scientific Publishing Company, Amsterdam, 406 p.

Johnson, A. M., and Fleming, R. W., 2004, Ground distortion and lurching—causes of near-fault damage to engineering structures, Part II, Fracture belts: p. 54–124.

Johnson, A. M., Fleming, R. W., and Cruikshank, K. M., 1994, Broad belts of right-lateral surface rupture along simple segments of fault zones that slipped during the 28 June 1992 Landers, California, earthquake: Bulletin of the Seismological Society of America, v. 84, p. 499–510.

Johnson, A. M., Johnson, K. M., Durdella, J., Sozen, M., and Gur, T., 2001, Main rupture and adjacent belt of right-lateral distortion detected by viaduct at Kaynasli, Turkey, 12 November 1999, Düzce earthquake: Earthquake Engineering Group School of Civil Engineering and Harry Fielding Reid Earthquake Rupture Laboratory, Department of Earth and Atmospheric Sciences, Purdue University, April 3, 2001, 33 p.

Johnson, S. Y., Dadisman, S. V., Childs, J. R., and Stanley, W. D., 1999, Active tectonics of the Seattle fault and central Puget Sound, Washington—Implications for earthquake hazards: Geological Society of America Bulletin, v. 111, no. 7, p. 1042–1053.

Jones, P. B., 1996, Triangle zone geometry, terminology and kinematics: Canadian Petroleum Geology Bulletin, v. 44, p. 139–152.

Jordan, T. E., and Allmendinger, R. W., 1986, The Sierras Pampeanas of Argentina: A modern analogue of Rock Mountain foreland deformation: American Journal of Science, v. 286, p. 737–764.

Jordan, T. H., 1988, Structure and formation of the continental tectosphere, in Menzies, M., and others (eds.), Oceanic and continental lithosphere; similarities and differences: Journal of Petrology, Special Lithosphere Issue, p. 11–37.

Jordan, W. M., 1965, Regional environmental study of the Early Mesozoic Nugget and Navajo Sandstones: Ph.D. dissertation, University of Wisconsin, Madison, 206 p.

Karato, S. I., and Wu, P., 1993, Rheology of the upper mantle: a synthesis: Science, v. 260, p. 771–778.

Kearey, P., and Vine, F. J., 1990, Global tectonics: Blackwell Scientific Publications, Boston, 302 p.

Kelley, V. C., 1955a, Monoclines of the Colorado Plateau: Geological Society of America Bulletin, v. 66, p. 789–804.

Kelley, V. C., 1955b, Regional tectonics of the Colorado Plateau and relationship to the origin and distribution of uranium: New Mexico University Publications in Geology, no. 5, 120 p.

Kelsey, H. M., Sherrod, B. L., Nelson, A. R., Brocher, T. M., 2008, Earthquakes generated from bedding plane-parallel reverse faults above an active wedge thrust, Seatlle fault zone: Geological Society of America Bulletin, v. 120, no. 11/12, p. 1581–1597.

Kerrich, R., 1978, An historical review and synthesis of research on pressure solution: Zentralblatt für Geologie, Mineralogie und Palaeontologie, v. 1, p. 512–550.

Kim, Y.-S., Peacock, D. C. P., and Sanderson, D. J., 2004, Fault damage zones: Journal of Structural Geology, v. 26, p. 503–517.

Knipe, R. J., 1989, Deformation mechanisms: recognition from natural tectonites: Journal of Structural Geology, v. 11, p. 127–146.

Koyi, H. A., and Maillot, B., 2007, Quantifying the effect of ramp dip and friction on thickness change of hanging wall units: Journal of Structural Geology, v. 29, 924–932.

Krantz, R. W., 1986, The odd-axis model: orthorhombic fault patterns and three-dimensional strain fields: Ph.D. dissertation, University of Arizona, Tucson, 97 p.

Krantz, R. W., 1988, Multiple fault sets and three-dimensional strain: theory and application: Journal of Structural Geology, v. 10, p. 225–237.

Kranz, R. L., 1983, Microcracks in rocks; a review, *in* Friedman, M., and Toksoez, M. N. (eds.), Continental tectonics: structure, kinematics and dynamics: Tectonophysics, v. 100, p. 449–480.

Kreemer, C., and Hammond, W. C., 2007, Geodetic constrains on areal changes in the Pacific-North America plate boundary zone: what controls Basin and Range extension?: Geology, v. 35, no. 10, p. 943–947.

Kreemer, C., Blewitt, G., and Hammond, W. C., 2010, Evidence for an active shear zone in southern Nevada linking the Wasatch fault to the Eastern California shear zone: Geology, v. 38, no. 5, p. 475–478.

Kronenberg, A. K., Segall, P., and Wolf, G. H., 1990, Hydrolytic weakening and penetrative deformation within a natural shear zone, *in* Duba, A. G., Durham, W. B., Handin, J. W., and Wang, H. F. (eds.), The brittle-ductile transition in rocks: Geophysical Monograph, v. 56, p. 21–36.

Kuenen, P. H., and DeSitter, L. U., 1938, Experimental investigation into the mechanism of folding: Leidse Geological Medediingen, v. 9, p. 217–239.

Kukla, P. A., Urai, J. L., and Mohr, M. 2008, Dynamics of salt structures, *in* Litke, R., Bayer, U., Gajewski, D., and Nelskamp, S., (eds.), Dynamics of complex intracontinental basins: The Central European Basin System: Springer-Verlag, Berlin Heidelberg, p. 291–306.

Kukla, P., Urai, J. Warren, J., Reuning, L., Becker, S., Schoenherr, J., Morh, M., van Gent, H., Abe, S., Li, S., Desbois, G., Scheder, Z., and de Keijzer, M., 2010, An integrated, multi-scale approach to salt dynamics and internal dynamics of salt structures—Salt tectonics, sediments and prospectivity: The Geological Society, Burlington House, London: reproduced in American Association of Petroleum Geologists Search and Discovery.

Kulander, B. R., and Dean, S. L., 1993, Coal-cleat domains and domain boundaries in the Allegheny Plateau of West Virginia: American Association of Petroleum Geologists Bulletin, v. 77, no. 8, p. 1374–1388.

Lachenbruch, A. H., 1961, Depth and spacing of tension cracks: Journal of Geophysical Research, v. 66, p. 4273–4292.

Ladeira, F. L., and Price, N. J., 1981, Relationship between fracture spacing and bed thickness: Journal of Structural Geology, v. 3, p. 179–183.

Lamerson, P. R., 1982, The Fossil Basin and its relationship to the Absaroka system, Wyoming and Utah: *in* Powers, R. B. (ed.), Geologic studies of the Cordilleran thrust belt: Rocky Mountain Association of Geologists, v. 1, p. 279–340.

Lapworth, C., 1885, The highland controversy in British geology: its causes, course and consequences: Nature, v. 32, p. 558–559.

Lash, G. G., and Engelder, T., 2009, Tracking the burial and tectonic history of Devonian shale of the Appalachian Basin by analysis of joint intersection style: Geological Society of America Bulletin, v. 121, p. 265–277.

Lash, G., Loewy, S., and Engelder, T., 2004, Preferential jointing of Upper Devonian black shale, Appalachian Plateau, U.S.A.: evidence supporting hydrocarbon generation as a joint-driving mechanism, *in* Rogers, C. M., and Engelder, T. (eds.), The initiation, propagation, and arrest of joints and other fractures: Geological Society Special Publication 231, London, pp. 129–151.

Lavine, A., Gardner, J. N., and Reneau, S. L., 2003a, Total station geologic mapping: an innovative approach to analyzing surface-faulting hazards: Engineering Geology, v. 70, no. 1–2, p. 71–91.

Lavine, A., Lewis, C. J., Katcher, D. K., Gardner, J. N., and Wilson, J. E., 2003b, Geology of the north-central to northeastern portion of the Los Alamos National Laboratory, New Mexico: Los Alamos National Laboratory Report: LA-14043-MS, 44 p.

Law, R. D., Casey, M., and Knipe, R. J., 1986, Kinematic and tectonic significance of microstructures and crystallographic fabrics within quartz mylonites from the Assynt and Eriboll regions of the Moine thrust zone, NW Scotland: Royal Society of Edinburgh Transactions, v. 77, p. 99–125.

Law, R. D., Knipe, R. J., and Dayan, H., 1984, Strain path partitioning within thrust sheets: microstructural and petrofabric evidence from the Moine Thrust zone at Loch Eriboll, northwest Scotland: Journal of Structural Geology, v. 6, no. 5, p. 477–497.

Lawn, B. R., and and Wilshaw, T. R., 1975, Fracture of brittle solids: Cambridge University Press, Cambridge, 204 p.

Lawton, T. F., Roca, E., Guimera, J., 1999. Kinematic-stratigraphic evolution of a growth syncline and its implications for tectonic development of the proximal foreland basin, southeastern Ebro basin, Catalunya, Spain. Geological Society of America Bulletin 111 (3), 412–431.

Le, K., Lee, J., Owen, L. A., and Finkel, R., 2007, Late Quaternary slip rates along the Sierra Nevada frontal fault zone, California: slip partitioning across the western margin of the Eastern California shear zone-Basin and Range province: Geological Society of America Bulletin, v. 119, no. 1–2, p. 240–256.

Legg, M. R., Goldfinger, C., Kamerling, M. J., Chaytor, J. D., and Einstein, D. E., 2007, Morphology structure, and evolution of California Continental Borderland restraining bends: Geological Society of London Special Publications, v. 290, p. 143–168.

Lewis, C. J., Gardner, J. N., Schultz-Fellenz, E. S., Lavine, A., Reneau, S., and Olig, S., 2010, Fault interaction and along-strike variation in throw in the Pajarito fault system, Rio Grande rift, New Mexico: Geosp;here, v. 5, p. 252–269.

Lienkaemper, J. J., Pezzopane, S. K., Clark, M. M., and Rymer, M. J., 1987, Fault fractures formed in association with the 1986 Chalfant Valley, California, earthquake sequence: preliminary report: Bulletin of the Seismological Society of America, v. 7, p. 297–305.

Lisle, R. J., 1992, Strain estimation from flattened buckle folds: Journal of Structural Geology, v. 14, p. 369–371.

Lisle, R. J., 1994, Palaeostrain analysis, *in* Hancock, P. L. (ed.), Continental deformation: Pergamon Press, New York, p. 28–42.

Lisle, R. J., 2000, Predicting patterns of strain from three-dimensional fold geometries: neutral surface folds and forced folds, *in* Cosgrove, J. W., and Ameen, M. S. (eds.), Forced folds and fractures: Geological Society Special Publication No. 169, p. 213–222.

Lister, G. S., and Hobbs, B. E., 1980, The simulation of fabric development during plastic deformation and its application to quartzite: the influence of deformation history: Journal of Structural Geology, v. 2, p. 355–370.

Lister, G. S., and Snoke, A. W., 1984, S-C mylonites: Journal of Structural Geology, v. 6, p. 617–638.

Lister, G. S., and Williams, P. F., 1983, The partitioning of deformation in flowing rock masses: Etheridge, M. A., and Cox, S. F. (eds.), Deformation processes in tectonics: Tectonophysics, v. 92, p. 1–33.

Lloyd, G. E., and Knipe, R. J., 1992, Deformation mechanisms accommodating faulting of quartzite under upper crustal conditions: Journal of Structural Geology, v. 14, p. 127–143.

Lorenz, J. C., Warpinski, N. R., and Teufel, L. W., 1996, Natural fracture characteristics and effects: The Leading Edge, v. 15, no. 8, p. 909–911.

Lubetkin, L. K. C., and Clark, M. M., 1987, Late Quaternary fault scarp at Lone Pine, California: location of oblique slip during the great 1872 earthquake and earlier earthquakes, *in* Hill, M. L. (ed.), Centennial Field Guide Volume 1: Cordilleran Section of the Geological Society of America: The Decade of North American Geology Project Series, Geological Society of America, Boulder, CO, p. 151–156.

Lund, W. R., Hozik, M. J., and Hatfield, S. C., 2006, Paleoseismic investigation and long-term slip history of the Hurricane Fault in Southwestern Utah: Utah Geologic Survey Special Study 119, 52p.

Lund, W. R., Taylor, W. J., Pearthree, P. A., Stenner, H. D., Amoroso, L., and Hurlow, H. A., 2006, Structural development and paleoseismology of the Hurricane fault, southwestern Utah and northwestern Arizona, *in* Lund, W. R. (ed.), Field Guide to geologic excursions in southwestern Utah and adjacent areas of Arizona and Nevada: Utah Geological Survey Open File Report 02–172, p. 1–84.

Lundin, E. R., 1989, Thrusting of the Claron Formation, the Bryce Canyon region, Utah: Geological Society of America Bulletin, v. 101, p. 1038–1050.

Luyendyk, B. P., Kamerling, M. J., and Terres, R. R., 1980, Geometric model of Neogene crustal rotations in southern California: Geological Society of America Bulletin, v. 91, p. 211–217.

Macdonald, G. A., 1967, Forms and structures of extrusive basaltic rocks in Hess, H. H., and Poldervaart, A. (eds.), Basalts: the Poldervaart treatise on rocks of basalt composition: Wiley-Interscience Publishers, New York, v. 1, p. 1–61.

Machette, M., Haller, K., Dart, R., and Rhea, S., 1996–2004, Summary of the Late Quaternary tectonics of the Basin and Range province in Nevada, Eastern California, and Utah: Derived from USGS database on Quaternary faults and folds in the United States: http://earthquake.usgs.gov/regional/imw/imw_bnr_faults/

Mackin, J. H., 1950, The down-structure method of viewing geologic maps: Journal of Geology, v. 58, p. 55–72.

Malavieille, J., 1987, Kinematics of compressional and extensional ductile shearing deformation in a metamorphic core complex of the northeastern Basin and Range: Journal of Structural Geology, v. 9, p. 541–554.

Malcolm, W., 1912, Gold fields of Nova Scotia: Canadian Geological Survey Memoir 20-E, 331 p.

Mann, P., Hempton, M. R., Bradley, D. C., and Burke, K., 1983, Development of pull-apart basins: Journal of Geology, v. 91, p. 529–554.

Manton, O. O., 1965, The orientation and origin of shatter-cones in the Vredefort Ring: New York Academy of Science Annals, v. 123, p. 1017–1049.

Marlow, P. C., and Etheridge, M. A., 1977, Development of a layered crenulation cleavage in mica schists of the Kanmantoo Group near Macclesfield, South Australia: Geological Society of America Bulletin, v. 88, p. 873–882.

Marshak and Mitra, 1988, Basic methods of structural geology: Prentice-Hall, Englewood Cliffs, New Jersey, 446p.

Marshak, S., and Tabor, J. R., 1989, Structure of the Kingston orocline in the Appalachian fold-thrust belt, New York: Geological Society of America Bulletin, v. 101, no. 5, p. 683–701.

Marshak, S., and Woodward, N., 1988, Introduction to cross-section balancing, *in* Marshak, S., and Mitra, G. (eds.), Basic methods of structural geology: Prentice-Hall, Englewood Cliffs, New Jersey, 303–326.

Marshak, S., Karlstrom, K., and Timmons, J. M., 2000, Inversion of Proterozoic extensional faults: An explanation for the pattern of Laramide and Ancestral Rockies intracratonic deformation: Geology, v. 28, p. 735–738.

Martin, R. G., 1978, Northern and eastern Gulf of Mexico continental margin, stratigraphic and structural framework, *in* Bouma, A. H., Moore, G. T., and Coleman, J. M. (eds.), Framework, facies, and oil-trapping characteristics of the upper continental margin: American Association of Petroleum Geologists Studies in Geology 7, p. 21–42.

McClay, K. R., (ed.), 1992, Thrust tectonics: Chapman and Hall, London, 447 p.

McClay, K. R., 1989, Analogue models of inversion tectonics, *in* Cooper, M. A., and Williams, G. D. (eds.), Inversion tectonics: Geological Society of London Special Publication No. 44, Blackwell Scientific Publications, London, p. 41–59.

McClay, K. R., and Coward, M. P., 1981, The Moine thrust zone: an overview, *in* McClay, K. R., and Price, N. J.

(eds.), Thrust and nappe tectonics: Geological Society of London Special Publication 9, Blackwell Scientific Publications, p. 241–260.

McClay, K. R., and Scott, A. D., 1991a, Experimental models of hangingwall deformation in ramp-flat listric extensional fault systems, *in* Cobbold, P. R. (ed.), Experimental and numerical modelling of continental deformation: Tectonophysics, v. 188, p. 85–96.

McClay, K. R., Waltham, D. A., Scott, A. D., and Abousetta, A., 1991b, Physical and seismic modelling of listric normal fault geometries, *in* Roberts, A. M., Yielding, G., and Freeman, B. (eds.), 1991, The geometry of normal faults: Geological Society of London Special Publication No. 56, p. 231–239.

McClusky, S. C., Bjornstad, S. C., Hager, B. H., King, R. W., Meade, B. J., Miller, M. M., Monastero, F. C., and Souter, B. J., 2001, Present day kinematics of the Eastern California shear zone from a geodetically constrained block model: Geophysical Research Letters, v. 28, no. 17, p. 3369–3372.

McClusky, S., Balassanian, S., Barka, A., Demir, C., Ergintav, S., Georgiev, I., Gurkan, O., Hamburger, M., Hurst, K., Kahle, H., Kastens, K., Kekelidze, G., King, R., Kotzev, V., Lenk, O., Mahmoud, S., Mishin, A., Nadariya, M., Ouzounis, A., Paradissis, D., Peter, Y., Prilepin, M., Reilinger, R., Sanli, I., Seeger, H., Tealeb, A., Toksoz, M. N., and Veis, G., 2000, Global positioning system constraints on plate kinematics and dynamics in the eastern Mediterranean and Caucasus: Journal of Geophysical Research, v. 105, no. B3, p. 5695–5719.

McConaughy, D. T., and Engelder, T., 2001, Joint initiation in bedded clastic rocks: Journal of Structural Geology, v. 2/3, pp. 203–221.

McCulloch, C. M., Deul, M., and Jeran, P. W., 1974, Cleat in bituminous coalbeds: U.S. Bureau of Mines Report of Investigation, No. 7910, 25 p.

McKinstry, H. E., 1961, Mining geology: Prentice-Hall, Englewood Cliffs, New Jersey, 680 p.

McQuarrie, N., and Davis, G. H., 2002, Crossing the several scales of strain-accomplishing mechanisms in the hinterland of the central Andean fold-thrust belt, Bolivia: Journal of Structural Geology, v. 24, no. 10, p 1587–1602.

Means, W. D., 1976, Stress and strain: Springer-Verlag, New York, 339 p.

Means, W. D., 1980, High temperature simple shearing fabrics; a new experimental approach: Journal of Structural Geology, v. 2, p. 197–202.

Means, W. D., 1987, A newly recognized type of slickenside striation: Journal of Structural Geology, v. 9, p. 585–590.

Means, W. D., 1990, Review paper: Kinematics, stress, deformation, and material behavior: Journal of Structural Geology, v. 12, no. 8, p. 953–971.

Means, W. D., Hobbs, B. E., Lister, G. S., and Williams, P. F., 1980, Vorticity and non-coaxiality in progressive deformations: Journal of Structural Geology, v. 2, no. 3, p. 371–378.

Memeti, V., Paterson, S., Matzel, J., Mundil, R., and Okaya, D., 2010, Magmatic lobes as "snapshots" of magma chamber growth and evolution in large, composite batholiths: an example from the Tuolumne intrusion, Sierra Nevada, California: Geological Society of America Bulletin, v. 122, no. 11–12, p. 1912–1931.

Menges, C. M., 1981, The Sonoita Creek basin: implications for late Cenozoic evolution of basins and ranges in southeastern Arizona: M.S. thesis, University of Arizona, Tucson, 239 p.

Merle, O., and Guillier, B., 1989, The building of the Central Swiss Alps: an experimental approach: Tectonophysics, v. 165, p. 41–56.

Milnes, A. G., 1979, Albert Heim's general theory of natural deformation (1878): Geology, v. 7, p. 99–103.

Mishra, R. K., 2009, How do scissors work?: http://www.laparoscopyhospital.com/PRO3.HTM

Mitra, S., 1986, Duplex structures and imbricate thrust systems: geometry, structural position and hydrocarbon potential: American Association of Petroleum Geologists Bulletin, v. 70, p. 1087–1112.

Mitra, S., 1988, Three-dimensional geometry and kinematic evolution of the Pine Mountain thrust system, southern Appalachians: Geological Society of America Bulletin, v. 100, p. 72–95.

Mitra, S., 1990, Fault-propagation folds: Geometry, kinematic evolution, and hydrocarbon traps: American Association of Petroleum Geologists Bulletin, v. 74, p. 921–945.

Mitra, S., 1990, Geometry and kinematic evolution of inversion structures: American Association of Petroleum Geologists Bulletin, v. 77, 1159–1191.

Mitra, S., 2003, A unified kinematic model for the evolution of detachment folds: Journal of Structural Geology, v. 25, p. 1659–1673.

Mitra, S., and J. S. Namson, 1989, Equal-area balancing: American Journal of Science, v. 289, p. 563–599.

Mitra, S., and Mount, V. S., 1998, Foreland basement involved structures: American Association of Petroleum Geologists Bulletin, v. 82, p. 70–109.

Mohr, O., 1882, Ueber die Darstellung des Spannungszustandes und des Deformationszustandes eines Körperelementes und über die Anwendung derselben in der Festigkeitslehre, Civilingenieur 28, pp. 113–115.

Mohr, O. C., 1900, Welche Ümstande bedingen die Elastizitätsgrenze und den Bruch eines Materials: Zeitschrift der Vereines Deutscher Ingenieure, v. 44, p. 1524–1530 and 1572–1577.

Molnar, P., 1988, Continental tectonics in the aftermath of plate tectonics: Nature, v. 335, p. 131–137.

Molnar, P., and England, P., 1990, Temperatures, heat flux and frictional stress near major thrust faults: Journal of Geophysical Research, Series B: Solid Earth and Planets, v. 95, p. 4833–4856.

Molnar, P., and Tapponier, P., 1975, Cenozoic tectonics of Asia: effect of a continental collision: Science, v. 189, p. 419–425.

Moore, D. E., and Rymer, M. J., 2007, Talc-bearing serpentinites and the creeping section of the San Andreas fault: Nature, v. 448, p. 795–797.

Morgan, S., A. Stanik, E. Horsman, B. Tikoff, M. de Saint Blanquat, and G. Habert, 2008, Emplacement of multiple magma sheets and wall rock deformation: Trachyte Mesa intrusion, Henry Mountains, Utah, J. Struct. Geol., v. 30, p. 491–512, doi:10.1016/j.jsg.2008.01.005.

Morgan, W. J., 1968, Rises, trenches, great faults, and crustal blocks: Journal of Geophysical Research, v. 73, p. 1959–1982.

Morgan, W. J., 1971, Convection plumes in the lower mantle: Nature, v. 230, p. 42–43.

Morgan, W. J., 1972, Convection plumes and plate motions: American Association of Petroleum Geologists Bulletin, v. 56, p. 203–213.

Morley, C. K., and Guerin, G., 1996, Comparison of gravity-driven deformation styles and behavior associated with mobile shales and salts: Journal of the Geological Society of London, v. 155, p. 475–490.

Mosar, J., and Suppe, J., 1992, Role of shear in fault-propagation folds, *in* McClay, K., (ed.), Thrust tectonics: Chapman & Hall, London, p. 123–132.

Mulchrone, K. F., Grogan, S., and De, P., 2005, The relationship between magmatic tiling, fluid flow and crystal fraction: Journal of Structural Geology, v. 27, p. 179–197.

Namson, J. S., and Davis, T. L., 1988, Seismically active fold and thrust belt in the San Joaquin Valley, central California: Geological Society of America Bulletin, v. 100, p. 257–273.

Narr, W., 1991, Fracture density in the deep subsurface: techniques with application to Point Arguello oil field: American Association of Petroleum Geologists Bulletin, v. 66, p 1231–1247.

Narr, W., and Suppe, J., 1991, Joint spacing in sedimentary rocks: Journal of Structural Geology, v. 13, p. 1037–1048.

Naus-Thijssen, F. M. J., Johnson, S. E., and Koons, P. O., 2010, Modeling crenulation cleavage: a polymineralic approach: Journal of Structural Geology, v. 32, p. 330–341.

Nelson, A. R., Johnson, S. Y., Kelsey, H. M., Wells, R. E., Sherrod, B. L., Pezzopane, S. K., Bradley, L.-A., Koehler III, R. D., and Bucknam, R. C., 2003, Late Holocene earthquakes on the Toe Jam Hill fault, Seattle fault zone, Bainbridge Island, Washington: Geological Society of America Bulletin, v. 115, no. 11, p. 1388–1403.

Nicholson, C. L., Seeber, P., Williams, P., and Sykes, L. R., 1986, Seismicity and fault kinematics through the eastern Transverse Ranges, California: Block rotations, strike-slip faulting, and low-angle thrusts: Journal of Geophysical Research, v. 91, p. 4891–4908.

Nickelsen, R. P., 1972, Attributes of rock cleavage in some mudstones and limestones of the Valley and Ridge province, Pennsylvania: Pennsylvania Academy of Science Proceedings, v. 46, p. 107–112.

Nickelsen, R. P., 2009, Overprinted strike-slip deformation in the southern Valley and Ridge in Pennsylvania: Journal of Structural Geology, v. 31, p. 865–873.

Nickelsen, R. P., and Hough, V. D., 1967, Jointing in the Appalachian Plateau of Pennsylvania: Geological Society of America Bulletin, v. 78, p. 609–630.

Nicol, A., Walsh, J. J., Villamor, P., Seebeck, H., and Berryman, K. R., 2010, Normal fault interactions, paleoearthquakes and growth in an active rift: p. 1101–1113.

Nicol, A., Walsh, J., Berryman, K., and Nodder, S., 2005, Growth of a normal fault by the accumulation of slip over millions of years: Journal of Structural Geology, v. 27, p.327–342.

Nicolas, A., 1987, Principles of rock deformation: petrology and structural geology: D. Reidel Publishing Company, Dordrecht, Netherlands, 208 p.

Nicolas, A., and Poirier, J. P., 1976, Crystalline plasticity and solid state flow in metamorphic rocks: selected topics in geological sciences: John Wiley & Sons, London, 444 p.

Niemi, T. M., Hall, N. T., and Dahne, A., 2006, The 1906 earthquake rupture trace of the San Andreas fault north of San Franscisco, with stops at points of geotechnical interest, *in* Prentice, C. S., Scotchmoor, J. G., Moores, E. M., and Kiland, J. P. (eds.), 1906 San Francisco Earthquake Centennial Field Guides: Field trips associated with the 100th Anniversary Conference, 18–23 April 2006, San Francisco, California: Geological Society of America Field Guide 7, p. 157–176.

Nur, A., 1982, The origin of tensile fracture lineaments: Journal of Structural Geology, v. 4, p. 31–40.

Nur, A., and Simmons, G., 1970, The origin of small cracks in igneous rocks: International Journal of Rock Mechanics and Mining Science, v. 7, p. 307–312.

O'Hara, K., 2007, Reaction weakening and emplacement of crystalline thrusts: diffusion control on reaction rate and strain rate: Journal of Structural Geology, v. 29, p. 1301–1314.

Oertel, G., 1965, The mechanism of faulting in clay experiments: Tectonophysics, v. 2, p. 343–393.

Oertel, G., 1970, Deformation of a slaty, lapillar tuff in the Lake District, England: Geological Society of America Bulletin, v. 81, p. 1173–1188.

Oldow, J. S., 2003, Active transtensional boundary zone between the western Great Basin and Sierra Nevada block, western U. S. Cordillera: Geology, v. 31, no. 12, p. 1033–1036.

Oldow, J. S., and Singleton, E. S., 2008, Application of Terrestrial Laser Scanning in determining the pattern of late Pleistocene and Holocene fault displacement from the offset of pluvial lake shorelines in the Alvord extensional basin, northern Great Basin, Utah: Geosphere, v. 4, no. 3, p. 536–563.

Olson, J., and Pollard, D. D., 1988, Inferring stress states from detailed joint geometry, *in* Cundall, P. A., Sterling, R. L., and Starfield, A. M. (eds.), Key questions in rock mechanics: Proceedings of the 29th U. S. Symposium on Rock Mechanics, A. A. Balkoma, Rotterdam, p. 159–167.

Olson, J., and Pollard, D. D., 1989, Inferring paleostresses from natural fracture patterns: a new method: Geology, v. 17, p. 345–348.

Otton, J. K., 1974, Geologic features of the central Black Mountains, Death Valley, California, *in* Guidebook, Death Valley region, California and Nevada: Shoshone, Californaia, Death Valley Publishing Company, p. 65–72.

Park, C. F., and MacDiarmid, R. A., 1964, Ore deposits: W. H. Freeman, San Francisco, 475 p.

Park, R. G., 1988, Geological structures and moving plates: Blackie, Glasgow, 337 p.

Park, Y., Yoo, S.-H., and Ree, J.-H., 2006, Weakening of deforming granitic rocks with layer development at middle crust: Journal of Structural Geology, v. 2006, p. 919–928.

Parsons, T., Trehu, A. M., Luetgert, J. H., Miller, K. C., Kilbride, F., Wells, R. E., Fisher, M. A., Flueh, E. R., ten Brink, U. S., and Christensen, N. I., 1998, A new view into the Cascadia subduction zone and volcanic arc; implications for earthquake hazards along the Washington margin: Geology, v. 26, no. 3, p. 199–202.

Passchier, C. W., 1987, Stable positions of rigid objects in non-coaxial flow: a study in vorticity analysis, *in* Cobbold, P. R., Gapais, D., Means, W. D., and Treagus, S. H. (eds.), Shear criteria in rocks: Journal of Structural Geology, v. 9, p. 679–690.

Passchier, C. W., 1994, Mixing in flow perturbations: a model for development of mantled porphyroclasts in mylonites: Journal of Structural Geology, v. 16, p. 733–741.

Passchier, C. W., and Simpson, C., 1986, Porphyroclast systems as kinematic indicators: Journal of Structural Geology, v. 8, p. 831–843.

Passchier, C. W., and Sokoutis, D., 1993, Experimental modeling of mantle porphyroclasts: Journal of Structural Geology, v. 15, p. 895–909.

Passchier, C. W., and Trouw, R. A. J., 2005, Microtectonics: Springer, Berlin, 366p.

Passchier, C. W., Myers, J. S., and Kroner, A., 1990, Field geology of high-grade gneiss terrains: Springer-Verlag, New York, 150 p.

Paterson, M. S., and Weiss, L. E., 1966, Experimental deformation and folding of phyllite: Geological Society of America Bulletin, v. 77, p. 343–374.

Paterson, M. S., Fowler, T. K., and Miller, R. B., 1996, Pluton emplacement in arcs: a crustal-scale exchange process: Transactions of the Royal Society of Edinburgh: Earth Sciences, v. 87, p. 115–123.

Paterson, W. A., 1978, A technique for approximating values for the formation factor parameters of "m" and "a" and formation water resistivities of shaly formations: The Log Analyst, v. 19, p. 12–22.

Patterson, S. R., Vernon, R. H., and Tobisch, O. T., 1989, A review of criteria for the identification of magmatic and tectonic foliations in granitoids: Journal of Structural Geology, v. 11, no. 3, p. 349–363.

Pavlis, T. L., 1986, The role of strain heating in the evolution of megathrusts: Journal of Geophysical Research, v. 91, p. 6522–6534.

Peach, B. N., Horne, J., Gunn, W., Clough, C. T., Hinxman, L. W., and Teall, J. J. H., 1907, The geological structure of the Northwest Highlands of Scotland: Memoir of the Geological Survey of the United Kingdom.

Peacock, D. C. P., 2003, Scaling of transfer zones in the British Isles: Journal of Structural Geology, v. 25, p. 1561–1567.

Peacock, D. C. P., and Sanderson, D. J., 1992, Effects of layering and anisotropy on fault geometry: Journal of the Geological Society of London, v. 149, Part 5, p. 793–802.

Pearthree, P. A., Lund, W. R., Stenner, H. D., and Everitt, B. L., 1998, Paleoseismic investigation of the Hurricane fault in southwestern Utah and northwestern Arizona—Final project report: Arizona Geological Survey (Tucson) and Utah Geological Survey (Salt Lake City), Final Technical Report to the U.S. Geological Survey National Earthquake Hazard Reduction Program, award no. 1434-HQ-97-GR-03047, 131 p.

Phillips, F. C., 1971, The use of stereographic projection in structural geology: Edward Arnold, London, 90 p.

Phillips, J., 1844, Orientation movements in the parts of stratified rocks: British Association for the Advancement of Science Report 1843, p. 60–61.

Piane, C. D., Burlini, L., Kunze, K., Brack, P., and Burg. J. P., 2008, Rheology of dolomite: large strain torsion experiments and natural examples: Journal of Structural Geology, v. 30, p. 767–776.

Pinter, N., 1995, Faulting on the Volcanic Tableland, Owens Valley, California: Journal of Geology, v. 103, p. 73–83.

Platt, J. P., and Vissers, R. L. M., 1980, Extensional structures in anisotropic rocks: Journal of Structural Geology, v. 2, p. 397–410.

Pollard, D. D., and Aydin, A., 1988, Progress in understanding jointing over the past century: Geological Society of America Bulletin, v. 100, p. 1181–1204.

Pollard, D. D., and Fletcher, R. C., 2005, Fundamentals of structural geology: Cambridge University Press, Cambridge, 500p.

Pollard, D. D., and Segall, P., 1987, Theoretical displacements and stresses near fractures in rock: with applications to faults, joints, veins, dikes, and solution surfaces, *in* Atkinson, B. K. (ed.), Fracture mechanics of rock: Academic Press, London, p. 277–349.

Ponce De Leon, M. I., and Choukroune, P., 1980, Shear zones in the Iberian arc: Journal of structural geology, v. 2, p. 63–68.

Powell, C. McA., 1979, A morphological classification of rock cleavage, *in* Bell, T. H., and Vernon, R. H. (eds.), Microstructural processes during deformation and metamorphism: Tectonophysics, v. 58, p. 21–34.

Powell, C., 1989, Structural controls on Palaeozoic basin evolution and inversion in southwest Wales: Journal of the Geological Society (London), v. 146, p. 439–446.

Prentice, C. S., and Moores, E. M., 2006, Introduction, *in* Prentice, C. S., Scotchmoor, J. G., Moores, E. M., and Kiland, J. P. (eds.), 1906 San Francisco earthquake centennial field guides: The Geological Society of America Field Guide 7, p. vii–xi.

Price, N. J., and Cosgrove, J. W., 1990, Analysis of geological structures: Cambridge University Press, Cambridge, England, 502 p.

Price, R. A., and Mountjoy, E. W., 1970, Geologic structure of the Canadian Rocky Mountains between Bow and Athabasca rivers progress report, *in* Wheeler, J. O. (ed.): Geological Association of Canada Special Paper 6, p. 7–25.

Price, R. A., Mountjoy, E. W., and Cook, G. G., 1978, Geologic map of Mount Goodsir (west half), British Columbia: Geological Survey of Canada, map 1477A, 1:50,000.

Price, R. A., Beaumont, C., Nguyen, M., and Lee, B., 1988, Mechanics of thin-skinned fold-and-thurst belts: insights from numerical models, in Sears, J. W., Harms, T. A., and Evenchick, C. A., (eds.), Whence the mountains? Inquiries into the evolution of orogenic systems: a volume in honor of Raymond A. Price: Geological Society of America Special Paper 433, p. 63–98.

Proffett, J. M., Jr., 1977, Cenozoic geology of the Yerington district, Nevada, and its implications for the nature and origin of Basin and Range faulting: Geological Society of America Bulletin, v. 88, p. 247–266.

Quinquis, H., Audren, C., Brun, J. P., and Cobbold, P. R., 1978, Intense progressive shear in Île de Groix blueschists and compatibility with subduction or obduction: Nature, v. 273, p. 43–45.

Ragan, D. M., 1969, Introduction to concepts of two-dimensional strain and their application with the use of card-deck models: Journal of Geological Education, v. 17, p. 135–141.

Ragan, D. M., 1973, Structural geology, an introduction to geometrical techniques (2nd edition): John Wiley and Sons, New York, 208 p.

Ragan, D. M., 1985, Structural geology, an introduction to geometrical techniques (3rd edition): John Wiley and Sons, New York, 416p.

Ragan, D. M., 2009, Structural geology, an introduction to geometrical techniques (4th edition): Cambridge University Press, Cambridge, 624 p.

Ramberg, H., 1959, Evolution of ptygmatic folding: Norsk Geologisk Tidsskrift, v. 39, p. 99–151.

Ramberg, H., 1962, Contact strain and folding instability of a multilayered body under compression: Geologische Rundschau, v. 51, p. 405–439.

Ramberg, H., 1963, Evolution of drag folds: Geological Magazine, v. 100, p. 97–106.

Ramberg, H., 1967, Gravity, deformation and the Earth's crust as studied by centrifuged models: Academic Press, New York, 214 p.

Ramberg, H., 1973, Model studies in gravity-controlled tectonics by the centrifuge technique, *in* De Jong, K. A., and Scholten, R. (eds.), Gravity and tectonics: John Wiley & Sons, New York, p. 49–66.

Ramberg, H., 1975, Particle paths, displacement and progressive strain applicable to rocks: Tectonophysics, v. 28, p. 1–37.

Ramos, V. A., 2009, Darwin at Puente del Inca: Observations on the formation of the Inca's Bridge and mountain building: Revisita de la Associatión Geológica Argentina, v. 64, no. 1, p. 170–179.

Ramsay, J. G., 1967, Folding and fracturing of rocks: McGraw-Hill Book Company, New York, 560 p.

Ramsay, J. G., 1969, The measurement of strain and displacement in orogenic belts, *in* Kent, P. E., Satterthwaite, G. E., and Spencer, A. M. (eds.), Time and place in orogeny: Geological Society of London Special Publication 3, p. 43–79.

Ramsay, J. G., 1980a, Shear zone geometry: a review: Journal of Structural Geology, v. 2, p. 83–99.

Ramsay, J. G., 1980b, The crack-seal mechanism of rock deformation: Nature, v. 284, p. 135–139.

Ramsay, J. G., and Graham, R. H., 1970, Strain variation in shear belts: Canadian Journal of Earth Sciences, v. 7, p. 786–813.

Ramsay, J. G., and Allison, I., 1979, Structural analysis of shear zones in Alpinised Hercynian granite, Maggia Nappen, Pennine Zone, Central Alps: Schweiz. Miner. Petronr., v. 59, p. 251–279.

Ramsay, J. G., and Huber, M. I., 1983, The techniques of modern structural geology, v. 1: Strain analysis: Academic Press, London, 307 p.

Ramsay, J. G., and Huber, M. I., 1987, The techniques of modern structural geology, v. 2: Folds and fractures: Academic Press, London, 381 p.

Reber, S., Taylor, W. J., Stewart, M., and Schiefelbein, I. M., 2001, Linkage and reactivation along the northern Hurricane and Sevier faults, Southwestern Utah, in Erskine, M. C., Faulds, J. E., Bartley, J. M., and Rowley, P. D. (eds.), The geologic transition, High Plateaus to Great Basin—a symposium and field guide (The Mackin Volume): Utah Geological Association Publication 30, p. 379–400.

Reches, Z., 1978a, Development of monoclines: Part 1. Structure of the Palisades Creek branch of the East Kaibab monocline, Grand Canyon, Arizona, *in* Mathews III, Vincent, (ed.), Laramide folding associated with basement block faulting in the western United States: Geological Society of America Memoir 151, p. 235–272.

Reches, Z., 1978b, Analysis of faulting in three-dimensional strain field: Tectonophysics, v. 47, p. 109–129.

Reches, Z., 1983, Faulting of rocks in three-dimensional strain fields: II. Theoretical analysis: Tectonophysics, v. 95, p. 133–156.

Reches, Z., and Dieterich, J. H., 1983, Faulting of rocks in three-dimensional strain fields: 1. Failure of rocks in polyaxial, servo-control experiments: Tectonophysics, v. 95, p. 111–132.

Reches, Z., and Johnson, A. M., 1978, Development of monoclines. Part II. Theoretical analysis of monoclines, *in* Matthews, V. (ed.), Laramide folding associated with basement block faulting in the western United States: Geological Society of America Memoir 151, p. 273–311.

Redfern, R., 1986, The making of a continent: American Geological Institute, Royal Smeets, 242 p.

Rehrig, W. A., and Heidrick, T. L., 1972, Regional fracturing in Laramide stocks of Arizona and its relationship to porphyry copper mineralization: Economic Geology, v. 67, p. 198–213.

Rehrig, W. A., and Heidrick, T. L., 1976, Regional tectonic stress during the Laramide and late Tertiary intrusive periods, Basin and Range province, Arizona: Arizona Geological Digest, v. 10, p. 205–228.

Reid, J. B., Bucklin, E. P., Copenagle, L., Kidder, J., Pack, S. M., Polissar, P., Pratigya, J., and Williams, M. L., 1995, Sliding rocks at the Racetrack, Death Valley: What makes them move?: Geology, v. 23, p. 819–822.

Reilinger, R. E., Ergintav, S., Bürgmann, R., McClusky, S., Lenk, O., Barka, A., Gurkan, O., Hearn, L., Feigl, K. L., Cakmak, R., Aktug, B., Ozener, H., Töksoz, M. N., 2000, Coseismic and postseismic fault slip for the 17 August 1999, M=7.5, Izmit, Turkey Earthquake: Science, v. 289, p. 1519–1524.

Reks, I. J., and Gray, D. R., 1982, Pencil structure and strain in weakly deformed mudstone and siltstone: Journal of Structural Geology, v. 4, p. 161–176.

Reynolds, C. W., 1987, Flocks, herds, and schools: A distributed behavioral model, in Computer Graphics, v. 21, no. 4, SIGGRAPH '87 Conference Proceedings, p. 25–34.

Reynolds, S. J., and Lister, G. S., 1987, Structural aspects of fluid-rock interactions in detachment zones: Geology, v. 15, p. 362–366.

Reynolds, S. J., and Lister, G. S., 1990, Folding of mylonitic zones in Cordilleran metamorphic core complexes: evidence from near the mylonitic front: Geology, v. 18, p. 216–219.

Reynolds, S. J., Richard, S. M., Haxel, G. B., Tosdal, R. M., and Laubach, S. E., 1988, Geologic setting of Mesozoic and Cenozoic metamorphism in Arizona, *in* Ernst, W. G. (ed.), Metamorphism and crustal evolution of the western United States (Rubey Volume VII): Prentice Hall, Englewood Cliffs, New Jersey, p. 466–501.

Rich, J. L., 1934, Mechanics of low-angle overthrust faulting as illustrated by Cumberland thrust block, Virginia, Kentucky, and Tennessee: American Association of Petroleum Geologists Bulletin, v. 18, p. 1584–1596.

Richter, D., 1976, Allgemeine Geologie: Walter de Gruyter, Berlin, 366 p.

Ridgeway, J., 1920, Preparation of illustrations for the reports of the United States Geological Survey, with brief descriptions of processes of reproduction: United States Geological Survey, Washington, D. C., 101 p

Riedel, W., 1929, Zur Mechanik geologischer Brucherscheinungen: ein Beitrag zum Problem der "Fiederspälten": Centralblatt für Mineralogie, Geologie, und Paleontologie, Part B, p. 354–368.

Rispoli, R., 1981, The stress fields about strike-slip faults inferred from stylolites and tension gashes: Tectonophysics, v. 75, p. 29–36.

Rives, T., Razack, M., Petit, J. P., and Rawnsley, K. D., 1992, Joint spacing: analogue and numerical simulations: Journal of Structural Geology, v. 14, pp. 925–937.

Roberts, J. C., 1961, Feather fractures and the mechanics of rock jointing: American Journal of Science, v. 259, p. 481–492.

Rockwell, T. K., Keller, E. A., and Dembroff, G. R., 1988, Quaternary rate of folding of the Ventura Avenue anticline, western Transverse Ranges, southern California: Geological Society of America Bulletin, v. 100, p. 850–858.

Roeder, D., Gilbert, O. E., and Witherspoon, W. D., 1978, Evolution and macroscopic structure of Valley and Ridge thrust belt, Tennessee and Virginia: Studies in Geology, Department of Geological Sciences, University of Tennessee, Knoxville, 25p.

Roy, A. B., 1978, Evolution of slaty cleavage in relation to diagenesis and metamorphism: a study from the Hunsrückschiefer: Geological Society of America Bulletin, v. 89, p. 1775–1785.

Royse, F., Jr., 2003, Detachment fold train, Reed Wash area, west flank San Rafael swell, Utah: An example of a limb lengthening, roll-through folding process on the eastern margin of the Sevier thrust belt: The Mountain Geologist, v. 33, p. 45–64.

Royse, F., Jr., Warner, M. A., and Reese, D. L., 1975, Thrust belt structural geometry and related stratigraphic problems, Wyoming-Idaho-northern Utah, *in* Bolyard, D. W. (ed.), Deep drilling frontiers of the central Rocky Mountains Symposium: Rocky Mountain Association of Geologists, p. 41–54. Rummel, F., 1987, Fracture mechanics approach to hydraulic fracturing stress measurements, *in* Atkinson, B. K., Fracture mechanics of rock: Academic Press, London, p. 217–239.

Rutter, E. H., 1976, The kinetics of rock deformation by pressure solution: Philosophical Transactions of the Royal Society of London, Series A, v. 283, p. 203–219.

Rymer, M. J., Hickman, S. H., and Stoffer, P. W., 2006, A field guide to the central, creeping section of the San Andreas fault and the San Andreas Fault Observatory at Depth, *in* Prentice, C. S., Scotchmoor, J. G., Moores, E. M., and Kiland, J. P. (eds.), 1906 San Francisco earthquake centennial field guides: The Geological Society of America Field Guide 7, p. 237–272.

Sammis, C. G., Biegel, R., and King, G., 1986, A self-similar model for the kinematics of gouge deformation: American Geophysical Union Transactions, v. 67, p. 1187.

Sammis, C. G., King, G., and Biegel, R., 1987, The kinematics of gouge formation: Pure and Applied Geophysics, v. 125, p. 777–812.

Sander, B., 1930, Gefügekunde der Gesteine: Springer-Verlag, Vienna, 352 p.

Sander, B., 1970, The study of fabrics of geological bodies: Pergamon Press, New York, 641 p.

Sanford, A. R., 1959, Analytical and experimental study of simple geologic structures: Geological Society of America Bulletin, v. 70, p. 19–52.

Savalli, L., and Engelder, T., 2005, Mechanisms controlling rupture shape during subcritical growth of joints in layered rocks: Geological Society of America Bulletin, v. 177, no. 3/4, p. 436–449.

Schmidt, C. J., James, C., and Shearer, J. N., 1981, Estimate of displacement in major zone of tear-faulting in fold and thrust belt, southwest Montana: American Association of Petroleum Geologists Bulletin, v. 65, p. 986–987.

Schmidt, R. B., Chase, R. B., and Erslev, E. A., 1993, Laramide basement deformation in the Rocky Mountain foreland of the western United States: Geological Society of America Special Paper 280, 365p.

Scholz, D. H., 1990, The mechanics of earthquakes and faulting: Cambridge University Press, New York, 439p.

Schreiber, J. F., Jr., 1974, Field descriptions of sedimentary rocks, *in* Davis, G. H. (ed.), Geology field camp manual: University of Arizona, Tucson, Arizona, p. 97–110.

Schwartz, D. P., and Coppersmith, K. J., 1984, Fault behavior and characteristic earthquakes: examples from the Wasatch and San Andreas fault zones: Journal of Geophysical Research, v. 89, no. B7, p. 5681–5698.

Scott, W. E., Pierce, K. L., and Hait, M. H. Ur, 1985, Quaternary tectonic setting of the 1983 Borah Peak earthquake, central Idaho: Bulletin of the Seismological Society of America, v. 75, no. 4, p. 1053–1066.

Secor, D. T., Jr., 1965, Role of fluid pressure in jointing: American Journal of Science, v. 263, p. 633–646.

Sengupta, S., and Ghosh, S. K. 2007, Origin of striping lineation and transposition of linear structures in shear zones: Journal of Structural Geology, v. 29, no. 2, p. 273–287.

Serway, R. A., 1990, Physics for Scientists and Engineers (3rd edition): Saunders, Philadelphia: 336 p.

Shackelton, J. R., and Cooke, M. L., 2007, Is plane strain a valid assumption in non-cylindrical fault-cored folds?: Journal of Structural Geology, v. 29, p. 1229–1240.

Sharp, R. P., and Carey, D. L., 1976, Sliding stones, Racetrack Playa, California: Geological Society of America Bulletin, v. 87, p. 1704–1717.

Sharpe, D., 1847, On slaty cleavage: Geological Society of London Quarterly Journal, v. 3, p. 74–105.

Shaw, J. H., Conners, C., and Suppe, J., 2005, Seismic interpretation of contractional fault-related folds: American Association of Petroleum Geologists, Studies in Geology 53, 156p.

Shepard, R. N., 1990, Mind sights: original visual illusions, ambiguities, and other anomalies: W. H. Freeman and Company, New York, 213 pp.

Sheridan, M. F., 1975, Tectonic displacement of the Bishop Tuff: California Geology, May, 1975, p. 107–110.

Sherrod, B. L., Brocher, T. M., Weaver, C. S., Buckham, R. C., Blakey, R. J., Kelsey, H. M., Nelson, A. R., and Haugerud, R., 2004, Holocene fault scarps near Tacoma, Washington, USA: Geological Society of America Bulletin, v. 32, no. 1, p. 9–12.

Sherwin, J. A., and Chapple, W. M., 1968, Wavelengths of single layer folds: a comparison between theory and observation: American Journal of Science, v. 266, p. 167–179.

Sibson, R. H., 1977, Fault rocks and fault mechanisms: Journal of the Geological Society (London), v. 133, p. 191–213.

Sibson, R. H., 1980, Transient discontinuities in ductile shear zones: Journal of Structural Geology, v. 1, p. 165–171.

Sibson, R. H., 1986, Brecciation processes in fault zones: inferences from earthquake rupturing: Pure and Applied Geophysics, v. 124, p. 159–175, doi: 10.1007/BF00875724

Sibson, R. H., and Toy, V., 2006, The habitat of fault-generated pseudotachylyte: presence vs. absence of friction-melt, *in* Earthquakes: radiated energy and the physics of faulting: Geophysical Monograph Series 170, American Geophysical Union, p. 153–166.

Sieh, K., and Wallace, R. E., 1987, The San Andreas fault at Wallace Creek, San Luis Obispo County, California, *in* Hill, M. L. (ed.), Centennial Field Guide Volume 1: Cordilleran Section of the Geological Society of America: The Decade of North American Geology Project Series, Geological Society of America, Boulder, CO, p. 233–238.

Simpson, C., 1986, Determination of movement sense in mylonites: Journal of Geological Education, v. 34, p. 246–261.

Simpson, C., 1988, Analysis of two-dimensional finite strain, *in* Marshak, S., and Mitra, G. (eds.), Basic methods of structural geology: Prentice-Hall, Englewood Cliffs, New Jersey, p. 333–359.

Simpson, C., and De Paor, D. G., 1993, Strain and kinematic analysis in general shear zones: Journal of Structural Geology, v. 15, p. 1–20.

Simpson, C., and Schmid, S. M., 1983, An evaluation of criteria to deduce the sense of movement in sheared rocks: Geological Society of America Bulletin, v. 94, p. 1281–1288.

Sloan, D., and Wells, D., 2006, The Hayward fault, *in* Prentice, C. S., Scotcmoor, J. G., Moores, E. M., and Kiland, J. P., (eds.), 1906 San Francisco earthquake centennial field guides: Field trips associated with the 100th anniversary conference, 18–23 April 2006, San Francisco, California: Geological Society of America Field Guide 7, p. 273–332.

Smith, K. D., and Priestly, K. F., 2000, Faulting in the 1986 Chalfant, California, sequence: local tectonics and earthquake source parameters: Bulletin of the Seismological Society of America, v. 90, no. 4, p. 813–831.

Smith, R. B., and Sbar, M. L., 1974, Contemporary tectonics and seismicity of the western United States with emphasis on the Intermountain Seismic Belt: Geological Society of America Bulletin, v. 85, p. 1205–1218.

Snoke, A. W., and Tullis, J., 1998b, An overview of fault rocks, in Snoke, A. W., Tullis, J., and Todd, V. R. (eds.), Fault-related rocks: a photographic atlas: Princeton University Press, Princeton, N. J., p. 3–18.

Snoke, A. W., Tullis, J. A., and Todd, V. R. (eds.), 1998a, Fault-related rocks: a photographic atlas: Princeton University Press, Princeton, N.J., 617 p.

Soliva, R., and Benedicto, A., 2004, A linkage criterion for segmented normal faults: Journal of Structural Geology, v. 26, p. 2251–2267.

Sorby, H. C., 1853, On the origin of slaty cleavage: Edinburgh New Philosophical Journal, v. 55, p. 137–148.

Sorensen, K., 1983, Growth and dynamics of the Nordre Stromfjord shear zone: Journal of Geophysical Research, v. 88, no. B4, p. 3419–3437.

Spencer, E. W., 1993, Geologic maps; a practical guide to the interpretation and preparation of geologic maps for geologists, geographers, engineers and planners: Macmillan Publishing Company, New York, 147 p.

Spratt, D. A., Dixon, J. M., and Beattie, E. T., 2004, Changes in structural style controlled by lithofacies contrast across transverse carbonate bank margins—Canadian Rocky Mountains and scaled physical models, *in* McClay, K. R. (ed.), Thrust tectonics and hydrocarbon systems: American Association of Petroleum Geologists Memoir, v. 82, p. 259–275.

Spry, A. H., 1969, Metamorphic textures: Pergamon Press, Oxford, 350 p.

Srivastava, D. C., and Lisle, R. J., 2004, Rapid analysis of fold shape using Bezler curves: Journal of Structural Geology, v. 26, no. 9, p. 1553–1559.

Srivastava, D., and Shah, J., 2006, A rapid method for strain estimation from flattened parallel folds: Journal of Structural Geology, v. 28, no. 1, p. 1–8.

Stallard, A., Ikei, H., and Masuda, T., 2002, Quicktime movies of 3D spiral inclusion trail development, *in* Bobyarchick, A. (ed.), Visualization, teaching and learning in structural geology: Journal of the Virtual Explorer, v. 9, http://virtualexplorer.com.au/article/2002/63/3d-spiral-trail-development.

Stearns, D. W., 1968, Certain aspects of fractures in naturally deformed rock, *in* Riecker, R. E. (ed.), NSF Advanced Science Seminar in Rock Mechanics,Air Force Cambridge Research Laboratories.

Stearns, D. W., 1978, Faulting and forced folding in the Rocky Mountain foreland, *in* Matthews, V. (ed.), Laramide folding associated with basement block faulting in the western United States: Geological Society of America Memoir 151, p. 1–37.

Stewart, K. J., and Alvarez, W., 1991, Mobile-hinge kinking in layered rocks and models: Journal of Structural Geology, v. 13, p. 243–259.

Stewart, M. E., and Taylor, W. J., 1996, Structural analysis and fault segment boundary identification along the Hurricane fault in southwestern Utah: Journal of Structural Geology, v. 18, p. 1017–1029.

Stewart, M. E., Taylor, W. J., Pearthree, P. A., Solomon, B. J., and Hurlow, H. A., 1997, Neotectonics, fault segmentation, and seismic hazards along the Hurricane fault in Utah and Arizona—an overview of environmental factors in an actively extending region: Brigham, Young University Geologic Studies, v. 42, part II, p. 235–277.

Stock, P., 1992, A strain model for antithetic fabric rotation in shear band structure: Journal of Structural Geology, v. 14, p. 1267–1275.

Stockdale, P. B., 1922, Stylolites: their nature and origin: Indiana University Studies, Bloomington, v. 9, 97 p.

Stone, D. S., 1993, Basement-involved thrust-generated folds as seismically imaged in the subsurface of the central Rocky Mountain foreland, *in* Schmidt, C. J., Chase, R. B., and Erslev, E. A., (eds.), Laramide deformation in the Rocky Mountain foreland of the western United States: Geological Society of America Special Publication 280, p. 271–318, with 3 plates.

Storti, F., Billi, A., and Salvini, F., 2003, Particle size distributions in natural carbonate fault rocks: insights for non-self-similar cataclasis: Earth and Planetary Science Letters, v. 206, p. 173–186.

Su, J., Zhu, W., Lu, H., Xu, M., Yang, W., and Zhang, Z., 2009, Geometry styles and quantification of inversion structures in the Jiyang depression, Bohai Bay Basin, eastern China: Marine and Petroleum Geology, v. 26, p. 25–38.

Suppe, J., 1980, A retrodeformable cross section of northern Taiwan: Geological Society of China Proceedings, no. 23, p. 46–55.

Suppe, J., 1983, Geometry and kinematics of fault-bend folding: American Journal of Science, v. 283, p. 684–721.

Suppe, J., 1985, Principles of structural geology: Prentice-Hall, Englewood Cliffs, New Jersey, 537 p.

Suppe, J., Chou, G. T., and Hook, S. C., 1992, Rates of folding and faulting determined from growth strata, *in* McClay, K. R., (ed.), Thrust tectonics: Chapman and Hall, London, p. 105–121.

Suppe, J., Connors, C. D., and Zhang, Y., 2004, Shear fault-bend folding, *in* McClay, K. R. (ed.), Thrust tectonics and hydrocarbon systems: American Association of Petroleum Geologists Memoir, 82, p. 303–323.

Suppe, J., Sabat, F., Munoz, J. A., Poblet, J., Roca, E., and Verges, J., 1997, Bed-by-bed fold growth by kink-band migration: Sant Llorenc de Morunys, eastern Pyrenees: Journal of Structural Geology, v. 19, no. 3–4, p. 443–461.

Surpless, B., 2008, Modern strain localization in the central Walker Lane, western United States: Implications for the evolution of intraplate deformation in transtensional settings: Tectonophysics, v. 457, p. 239–243.

Susong, D. D., Janecke, S. U., and Bruhn, R., 1990, Structure of a fault segment boundary in the Lost River fault zone, Idaho, and possible effect on the 1983 Borah Peak earthquake rupture: Bulletin of the Seismological Society of America, v. 80, no. 1, p. 57–68.

Sutton, B., Davydov, A., and Yancey, T., 2008, Diagenesis and alteration in Kasimovian limestones, Gzhel quarry, Gzhel village, Russia; Carboniferous fossils of Russia, Stratigraphy and sediments, www.lakeneosho.org/Russia/Stratig3.html

Swan, M. M., 1976, The Stockton Pass fault: an element of the Texas lineament: M.S. thesis, University of Arizona, Tucson, 119 p.

Sykes, L. R., 1967, Mechanism of earthquakes and nature of faulting on the mid-oceanic ridges: Journal of Geophysical Research, v. 72, p. 2131–2153.

Tapponnier, P., Peltzer, G., Le Dain, A. Y., Armijo, R., and Cobbold, P., 1982, Propagating extrusion tectonics in Asia; new insights from simple experiments with plasticine: Geology, v. 10, p. 611–616.

Taylor, W. J., Stewart, M. E., and Orndorff, R. L., 2001, Fault segmentation and linkage: examples from the Hurricane fault, Southwestern U.S.A., *in* Erskine, M. C., Faulds, J. E., Bartley, J. M., and Rowley, P. D. (eds.), The geologic transition, High Plateaus to Great Basin—A symposium and field guide (The Mackin Volume): Utah Geological Association Publication 30, p. 113–126.

ten Brink, U. S., Molzer, P. C., Fisher, M. A., Blakely, R. J., Bucknam, R. C., Parson, T., Crosson, R. S., and Creager,

K. C., 2002, Subsurface geometry and evolution of the Seattle fault zone and the Seattle basin, Washington: Bulletin of the Seismological Society of America, V. 92, no. 5, p. 1737–1753.

Ter Heege, J. H., De Bresser, J. H. P., and Spiers, C. J., 2005, Rheological behaviour of synthetic rocksalt: the interplay between water, dynamic recrystallization, and deformation mechanisms: Journal of Structural Geology, v. 27, p. 948–963.

Thatcher, W., Foulger, G. R., Julian, B. R., Svarc, J., Quilty, E., and Bawden, G. W., 1999, Present-day deformation across the Basin and Range province, western United States: Science, v. 283, no. 5408, p. 1714–1719.

Thomson, S. N., Stöckhert, B., and Brix, M. R., 1998, Thermochronology of the high-pressure metamorphic rocks of Crete, Greece: implications for the speed of tectonic processes: Geology, v. 26, no. 3, p. 259–262.

Thrasher, G. P., King, P. R., and Cook, R. A., 1995, Taranaki Basin Petroleum atlas. 50 maps plus booklet. Institute of Geological and Nuclear Sciences Ltd, Lower Hutt.

Tikoff, B., and Fossen, H., 1999, Three-dimensional reference deformations and strain facies: Journal of Structural Geology, v. 21, p. 1497–1512.

Tindall, S., 2000, Development of oblique-slip basement-cored uplifts: insights from the Kaibab Uplift and from physical models: Ph.D. dissertation, The University of Arizona, 261p.

Titley, S. R. (ed.), 1982, Advances in geology of the porphyry copper deposits: Southwest North America: University of Arizona Press, Tucson, 560 p.

Toimil, N. C., and Fernández, F. J., 2007, Kinematic analysis of symmetrical natural folds developed in competent layers: Journal of Structural Geology, v. 29, p. 467–480.

Troxel, B. W., and Wright, L. A., 1987, Tertiary extensional features, Death Valley region, eastern California, *in* Hill, M. L. (ed.), Centennial Field Guide Volume 1: Cordilleran Section of the Geological Society of America: The Geological Society of America, Boulder, Colorado, p. 121–132.

Trudgill, B., and Cartwright, J., 1994, Relay-ramp forms and normal-fault linkages: Geological Society of America Bulletin, v. 106, p. 1143–1157.

Trümpy, R., 1960, Paleotectonic evolution of the central and western Alps: Geological Society of America Bulletin, v. 71, p. 843–908.

Tsutsumi, H., Yeats, R. S., and Huftile, G. J., 2001, Late Cenozoic tectonics of the northern Los Angeles fault system, California: Geological Society of America Bulletin, v. 113, no. 4, p. 454–468.

Tullis, J., 1990, Experimental studies of deformation mechanisms and microstructure in quartzo-feldspathic rocks, *in* Barber, D. J., and Meredith, P. G. (eds.), Deformation processes in minerals, ceramics and rocks: Unwin Hyman, London, p. 190–227.

Tullis, J., Dell'Angelo, L., and Yund, R. A., 1990, Ductile shear zones from brittle precursors in feldspathic rocks; the role of dynamic recrystallization, *in* Duba, A. G., Durham, W. B., Handin, J. W., and Wang, H. F. (eds.), The brittle-ductile transition in rocks: Geophysical Monograph, v. 56, p. 67–82.

Tullis, T. E., and Tullis, J., 1986, Experimental rock deformation techniques, *in* Hobbs, B. E., and Heard, H. C. (eds.), Mineral and rock deformation; laboratory studies; the Paterson volume: Geophysical Monograph, Monash University, Clayton, Victoria, Australia, v. 36, p. 297–324.

Tullis, T. E., and Wood, D. S., 1975, Correlation of finite strain from both reduction bodies and preferred orientation of mica in slate from Wales: Geological Society of America Bulletin, v. 86, p. 632–638.

Turner, F. J., and Weiss, L. E., 1963, Structural analysis of metamorphic tectonites: McGraw-Hill Book Company, New York, 560 p.

Twiss, R. J., and Moores, E. M., 1992, Structural geology: W. H. Freeman & Company, New York, 532 p.

United States Geological Survey, 1988, Crater Lake National Park and vicinity, Oregon: 1:62,500-scale topographic map with text and block diagrams, U.S. Geological Survey, Denver.

Unruh, J. Humphrey, J., and Barron, A., 2003, Transtensional model for the Sierra Nevada frontal fault system, eastern California: Geology, v. 31, no. 4, p. 327–330.

Urai, J. L., Means, W. D., and Lister, G. S., 1986, Dynamic recrystallization of minerals, in Hobbs, B. E., and Heard, H. C. (eds.), Mineral and rock deformation: laboratory studies; the Paterson volume: Geophysical Monograph, v. 36, p. 161–199.

Urai, J. L., Schenk, O., van der Zee, W., and Blumenthal, M., 2008, Photograph of the month: Journal of Structural Geology, v. 30, p. 1201.

Urai, J. L., Williams, P. F., and van Roermund, H. L. M., 1991, Kinematics of crystal growth in syntectonic fibrous veins: Journal of Structural Geology, v. 13, p. 823–836.

Van den Driessche, J., and Brun, J.-P., 1987, Rolling structures at large shear strain, *in* Cobbold, P. R., Gapais, D., Means, W. D., and Treagus, S. H. (eds.), Shear criteria in rocks: Journal of Structural Geology, v. 9, p. 691–704.

van der Pluijm, B. A., and Marshak, S., 1997, Earth Structure: an introduction to structural geology and tectonics (1st edition): WCB/McGraw Hill, New York, 493p.

van der Pluijm, B. A., and Marshak, S., 2003, Earth Structure: an introduction to structural geology and tectonics (2nd edition): W.W. Norton, New York, 674p.

Van Hise, R., 1896, Principles of North American pre-Cambrian geology: United States Geological Survey 16th Annual Report, Part 1, p. 581–844.

Vendeville, B. C., and Jackson, M. P. A., 1994, The rise of diapirs during thin-skinned extension: Marine and Petroleum Geology, v. 9, p. 331–353.

Vendeville, B., Cobbold, P. R., Davy, P., Brun, J. P., and Choukroune, P., 1987, Physical models of extensional tectonics at various scales, *in* Coward, M. P., Dewey, J. F., and Hancock, P. L. (eds.), Continental extensional tectonics: Geological Society of London Special Publication, v. 28: Blackwell Scientific Publications, London, p. 95–107.

Vernon, R. H., 1998. Chemical and volume changes during deformation and prograde metamorphism of sediments, *in* Treloar, P. J., and O'Brien, P. J. (eds.), What drives metamorphic reactions?: Geological Society, London, Special Publications, v. 138, p. 215–246.

Waldron, J. W. F., 2005, Extensional fault arrays in strike-slip and transtension: Journal of Structural Geology, v. 27, p. 23–34.

Wallace, R. E., 1984, Eyewitness account of surface faulting during the earthquake of 28 October 1983, Borah Peak, Idaho: Bulletin of the Seismological Society of America, v. 74, no. 3, p. 1091–1094.

Wallace, R. E., 1987, Grouping and migration of surface faulting and variations in slip rates on faults in the Great Basin province: Bulletin of the Seismological Society of America, v. 77, p. 868–876.

Walsh, J. J., Watterson, J., and Yielding, G., 1991, The importance of small-scale faulting in regional extension: Nature, v. 351, p. 391–393.

Weijermars, R., 1993, Pulsating strains: Tectonophysics, v. 220, p. 51–67.

Weiss, L. E., 1972, The minor structures of deformed rocks: a photographic atlas: Springer-Verlag, New York, 431 p.

Wells, D. L., and Coppersmith, K. J., 1994, New empirical relationships among magnitude, rupture length, rupture width, rupture area, and surface displacement: Bulletin of the Seismological Society of America, v. 84, no. 4, p. 974–1002.

Wells, R. E., Weaver, C. S., Blakely, R. J., 1998, Fore-arc migration in Cascadia and its neotectonic signficance: Geology, v. 26, no. 8, p. 759–762.

Wernicke, B. P., 1985, Uniform-sense normal simple shear of the continental lithosphere: Canadian Journal of Earth Sciences, v. 22, p. 108–125.

Wernicke, B., 1981, Low-angle normal faults in the Basin and Range province: nappe tectonics in an extending orogen: Nature, v. 291, p. 645–648.

Wernicke, B., and Burchfiel, B. C., 1982, Modes of extension tectonics: Journal of Structural Geology, v. 4, p. 105–115.

Wesnousky, S. G., 2005a, The San Andreas and Walker Lane fault systems, western North America: transpression, transtension, cumulative slip and the structural evolution of a major transform plate boundary: Journal of Structural Geology, v. 27, p. 1505–1512.

Wesnousky, S. G., 2005b, Active faulting in the Walker Lane: Tectonics, v. 24, TC3009, doi:10.1029/2004TC001645.

White, S. H., and Knipe, R. J., 1978, Transformation and reaction-enhanced ductility in rocks: Geological Society of London Journal, v. 135, Part 5, p. 513–516.

White, S. H., Bretan, P. G., and Rutter, E. H., 1986, Fault-zone reactivation; kinematics and mechanisms: Royal Society of London Philosophical Transactions, Series A: Mathematical and Physical Sciences, v. 317, p. 81–92.

White, S. H., Burrows, S. E., Carreras, J., Shaw, N. D., and Humphreys, F. J., 1980, On mylonites in ductile shear zones, *in* Carreras, J., Cobbold, P. R., Ramsay, J. G., and White, S. H. (eds.), Shear zones in rocks: Journal of Structural Geology, v. 2, p. 175–187.

Whitten, E. T. H., 1966, Structural geology of folded rocks: Rand-McNally, Skokie, Illinois, 663 p.

Wibberley, C., 2007, Talc at fault: Nature, v. 448, no. 16, p. 756–757.

Wilcox, R. E., Harding, T. P., and Seely, D. R., 1973, Basic wrench tectonics: American Association of Petroleum Geologists Bulletin, v. 57, p. 74–96.

Wilkinson, P., Soper, N. J., and Bell, A. M., 1975, Skolithos pipes as strain markers in mylonites: Tectonophysics, v. 28, no. 3, p. 143–157.

Williams, G. D., Powell, C. M., and Cooper, M. A., 1989, Geometry and kinematics of inversion tectonics, *in* Cooper, M. A., and Williams, G. D. (eds.), Inversion tectonics: Geological Society of London Special Publication No. 44, Blackwell Scientific Publications, London, England, p. 3–15.

Willis, B., 1894, The mechanics of Appalachian structure: United States Geological Survey 13th Annual Report, Part 2, p. 213–281.

Wilson, C. J. L., 1981, Experimental folding and fabric development in multilayered ice: Tectonophysics, v. 78, p. 139–159.

Wilson, C. J. L., 1986, Deformation induced recrystallization of ice: the application of in situ experiments: Geophysical monograph 36, p. 213–232.

Wilson, G., 1961, The tectonic significance of small-scale structures and their importance to the geologist in the field: Annales de la Société Géologique de Belgique, v. 84, p. 424–548.

Wilson, G., 1982, Introduction to small-scale geologic structures: George Allen & Unwin Ltd., London, 128p.

Wilson, J. T., 1965, A new class of faults and their bearing on continental drift: Nature, v. 207, p. 343–347.

Witkind, I. J., and Stickney, M. C., 1987, The Hebgen Lake earthquake area, Montana and Wyoming, in Beus, S. S. (ed.), Centennial Field Guide Volume 2: Rocky Mountain Section of the Geological Society of America, p. 89–94.

Woodcock, N. H., and Rickards, B., 2003, Transpressive duplex and flower structure: Dent Fault System, NW England: Journal of Structural Geology, v. 25, p. 1981–1992.

Woodward, N. B., 1997, Low-amplitude evolution of break-thrust folding: Journal of Structural Geology, v. 19, p. 293–301.

Woodward, N. B., Boyer, S. E., and Suppe, J., 1985, An outline of balanced cross sections: Studies in Geology (Knoxville), v. 11, 170 p.

Woodworth, J. B., 1896, On the fracture system of joints, with remarks on certain great fractures: Boston Society of Natural History Proceedings, v. 27, p. 163–184.

Worrall, D. M., 1977, Structural development of Round Mountain area, Uinta County, Wyoming, *in* Heisey, E. L. (ed.), Rocky Mountain thrust belt, geology and resources: Wyoming Geological Association Guidebook 29, p. 537–541.

Wright, L. A., Otton, J. K., and Troxel, B. W., 1974, Turtleback surfaces of Death Valley viewed as phenomena of extensional tectonics: Geology, v. 2, p. 53–54.

Wright, T. O., and Platt, L. B., 1982, Pressure dissolution and cleavage in the Martinsburg Shale: American Journal of Science, v. 282, p. 122–135.

Wu, H., and Pollard, D. D., 1995, An experimental study of the relationships between joint spacing and layer thickness: Journal of Structural Geology, v. 17, pp. 887–905.

Xiao, Hongbin, and Suppe, J., 1992, The origin of rollover: American Association of Petroleum Geologists Bulletin, v. 76, p. 509–529.

Yamada, Y., and McClay, K., 2004, Analog modeling of inversion thrust structures, in McClay, K. R. (ed.), Thrust tectonics and hydrocarbon systems: American Association of Petroleum Geologists Memoir 82, p. 276–301.

Yeats, R. S., and Grigsby, F. B., 1987, Ventura Avenue anticline: amphitheater locality, California, *in* Hill, M. L. (ed.), Centennial Field Guide Volume 1: Cordilleran Section of the Geological Society of America: The Decade of North American Geology Project Series, Geological Society of America, Boulder, CO, p. 219–223.

Yielding, G., Walsh, J. J. and Watterson, J. 1992, The prediction of small-scale faulting in reservoirs: First Break, v. 10, p. 449–460.

Yonkee, A., and Weil, A. B., 2010, Reconstructing the kinematic evolution of curved mountain belts: Internal strain patterns in the Wyoming salient, Sevier thrust belt, U.S.A.: Geological Society of America Bulletin, v. 122, no 1/2, p. 24–49.

Younes, A. I., and Engelder, T., 1999, Fringe cracks: key structures for the interpretation of the progressive Alleghanian deformation of the Appalachian plateau: Geological Society of America Bulletin, v. 111, no. 2, pp. 219–239.

Yue, L-F., Suppe, J., and Hung, J.-H., 2005, Structural geology of a classic thrust belt earthquake: the 1999 Chi-Chi earthquake Taiwan (Mw = 7.6): Journal of Structural Geology, v. 27, no. 11, p. 2058–2083.

Zák, J., Paterson, S. R., and Memeti, V., 2007, Four magmatic fabrics in the Tuolumne batholith, central Sierra Nevada, California (USA): Implications for interpreting fabric patterns in plutons and evolution of magma chambers in the upper crust: Geological Society of America Bulletin, v. 119, no. 1/2, p. 184–201.

Zhang, Y., Hobbs, B. E., and Jessell, M. W., 1993, Crystallographic preferred orientation development in a buckled single layer: a computer simulation: Journal of Structural Geology, v. 15, p. 265–276.

Zingg, T., 1935, Beitrag zur Schotteranalyze: Schweizer Mineralogische und Petrographische Mitteilungen, v. 15, p. 39–140.

Zoback, M., Hickman, S., and Ellsworth, W., 2010, Scientific drilling into the San Andreas fault zone: Eos, v. x91, no. 22, June 2010, p. 197–199.

Author Index

Subject Index